Karl K. Klett, Jr

Dept. of Physics & Astronomy

Univ. of Wyoming

Laramie, WYO. 82071

9 May, 1986

HANDBOOK OF MATHEMATICAL FUNCTIONS

WITH FORMULAS, GRAPHS, AND MATHEMATICAL TABLES

Edited by

Milton Abramowitz and Irene A. Stegun

DOVER PUBLICATIONS, INC., NEW YORK

The text relating to physical constants and conversion factors (page 6) has been modified to take into account the newly adopted Système International d'Unites (SI).

ERRATA NOTICE

The original printing of this Handbook (June 1964) contained errors that have been corrected in the reprinted editions. These corrections are marked with an asterisk (*) for identification. The errors occurred on the following pages: 2–3, 6–8, 10, 15, 19–20, 25, 76, 85, 91, 102, 187, 189–197, 218, 223, 225, 233, 250, 255, 260–263, 268, 271–273, 292, 302, 328, 332, 333–337, 362, 365, 415, 423, 438–440, 443, 445, 447, 449, 451, 484, 498, 505–506, 509–510, 543, 556, 558, 562, 571, 595, 599, 600, 722–723, 739, 742, 744, 746, 752, 756, 760–765, 774, 777–785, 790, 797, 801, 822–823, 832, 835, 844, 886–889, 897, 914, 915, 920, 930–931, 936, 940–941, 944–950, 953, 960, 963, 989–990, 1010, 1026.

Published in Canada by General Publishing Company, Ltd., 30 Lesmill Road, Don Mills, Toronto, Ontario.

Published in the United Kingdom by Constable and Company, Ltd., 10 Orange Street, London WC 2.

This Dover edition, first published in 1965, is an unabridged and unaltered republication of the work originally published by the National Bureau of Standards in 1964.

This ninth Dover printing conforms to the tenth (December 1972) printing by the Government Printing Office, except that additional corrections have been made on pages 18, 79, 80, 82, 408, 450, 786, 825 and 934.

Standard Book Number: 486-61272-4
Library of Congress Catalog Card Number: 65-12253

Manufactured in the United States of America
Dover Publications, Inc.
180 Varick Street
New York, N.Y. 10014

Preface

The present volume is an outgrowth of a Conference on Mathematical Tables held at Cambridge, Mass., on September 15–16, 1954, under the auspices of the National Science Foundation and the Massachusetts Institute of Technology. The purpose of the meeting was to evaluate the need for mathematical tables in the light of the availability of large scale computing machines. It was the consensus of opinion that in spite of the increasing use of the new machines the basic need for tables would continue to exist.

Numerical tables of mathematical functions are in continual demand by scientists and engineers. A greater variety of functions and higher accuracy of tabulation are now required as a result of scientific advances and, especially, of the increasing use of automatic computers. In the latter connection, the tables serve mainly for preliminary surveys of problems before programming for machine operation. For those without easy access to machines, such tables are, of course, indispensable.

Consequently, the Conference recognized that there was a pressing need for a modernized version of the classical tables of functions of Jahnke-Emde. To implement the project, the National Science Foundation requested the National Bureau of Standards to prepare such a volume and established an Ad Hoc Advisory Committee, with Professor Philip M. Morse of the Massachusetts Institute of Technology as chairman, to advise the staff of the National Bureau of Standards during the course of its preparation. In addition to the Chairman, the Committee consisted of A. Erdélyi, M. C. Gray, N. Metropolis, J. B. Rosser, H. C. Thacher, Jr., John Todd, C. B. Tompkins, and J. W. Tukey.

The primary aim has been to include a maximum of useful information within the limits of a moderately large volume, with particular attention to the needs of scientists in all fields. An attempt has been made to cover the entire field of special functions. To carry out the goal set forth by the Ad Hoc Committee, it has been necessary to supplement the tables by including the mathematical properties that are important in computation work, as well as by providing numerical methods which demonstrate the use and extension of the tables.

The Handbook was prepared under the direction of the late Milton Abramowitz, and Irene A. Stegun. Its success has depended greatly upon the cooperation of many mathematicians. Their efforts together with the cooperation of the Ad Hoc Committee are greatly appreciated. The particular contributions of these and other individuals are acknowledged at appropriate places in the text. The sponsorship of the National Science Foundation for the preparation of the material is gratefully recognized.

It is hoped that this volume will not only meet the needs of all table users but will in many cases acquaint its users with new functions.

ALLEN V. ASTIN, *Director*.

Washington, D.C.

Preface to the Ninth Printing

The enthusiastic reception accorded the "Handbook of Mathematical Functions" is little short of unprecedented in the long history of mathematical tables that began when John Napier published his tables of logarithms in 1614. Only four and one-half years after the first copy came from the press in 1964, Myron Tribus, the Assistant Secretary of Commerce for Science and Technology, presented the 100,000th copy of the Handbook to Lee A. DuBridge, then Science Advisor to the President. Today, total distribution is approaching the 150,000 mark at a scarcely diminished rate.

The success of the Handbook has not ended our interest in the subject. On the contrary, we continue our close watch over the growing and changing world of computation and to discuss with outside experts and among ourselves the various proposals for possible extension or supplementation of the formulas, methods and tables that make up the Handbook.

In keeping with previous policy, a number of errors discovered since the last printing have been corrected. Aside from this, the mathematical tables and accompanying text are unaltered. However, some noteworthy changes have been made in Chapter 2: Physical Constants and Conversion Factors, pp. 6-8. The table on page 7 has been revised to give the values of physical constants obtained in a recent reevaluation; and pages 6 and 8 have been modified to reflect changes in definition and nomenclature of physical units and in the values adopted for the acceleration due to gravity in the revised Potsdam system.

The record of continuing acceptance of the Handbook, the praise that has come from all quarters, and the fact that it is one of the most-quoted scientific publications in recent years are evidence that the hope expressed by Dr. Astin in his Preface is being amply fulfilled.

LEWIS M. BRANSCOMB, *Director*
National Bureau of Standards

November 1970

Foreword

This volume is the result of the cooperative effort of many persons and a number of organizations. The National Bureau of Standards has long been turning out mathematical tables and has had under consideration, for at least 10 years, the production of a compendium like the present one. During a Conference on Tables, called by the NBS Applied Mathematics Division on May 15, 1952, Dr. Abramowitz of that Division mentioned preliminary plans for such an undertaking, but indicated the need for technical advice and financial support.

The Mathematics Division of the National Research Council has also had an active interest in tables; since 1943 it has published the quarterly journal, "Mathematical Tables and Aids to Computation" (MTAC), editorial supervision being exercised by a Committee of the Division.

Subsequent to the NBS Conference on Tables in 1952 the attention of the National Science Foundation was drawn to the desirability of financing activity in table production. With its support a 2-day Conference on Tables was called at the Massachusetts Institute of Technology on September 15–16, 1954, to discuss the needs for tables of various kinds. Twenty-eight persons attended, representing scientists and engineers using tables as well as table producers. This conference reached consensus on several conclusions and recommendations, which were set forth in the published Report of the Conference. There was general agreement, for example, "that the advent of high-speed computing equipment changed the task of table making but definitely did not remove the need for tables". It was also agreed that "an outstanding need is for a Handbook of Tables for the Occasional Computer, with tables of usually encountered functions and a set of formulas and tables for interpolation and other techniques useful to the occasional computer". The Report suggested that the NBS undertake the production of such a Handbook and that the NSF contribute financial assistance. The Conference elected, from its participants, the following Committee: P. M. Morse (Chairman), M. Abramowitz, J. H. Curtiss, R. W. Hamming, D. H. Lehmer, C. B. Tompkins, J. W. Tukey, to help implement these and other recommendations.

The Bureau of Standards undertook to produce the recommended tables and the National Science Foundation made funds available. To provide technical guidance to the Mathematics Division of the Bureau, which carried out the work, and to provide the NSF with independent judgments on grants for the work, the Conference Committee was reconstituted as the Committee on Revision of Mathematical Tables of the Mathematics Division of the National Research Council. This, after some changes of membership, became the Committee which is signing this Foreword. The present volume is evidence that Conferences can sometimes reach conclusions and that their recommendations sometimes get acted on.

v

Active work was started at the Bureau in 1956. The overall plan, the selection of authors for the various chapters, and the enthusiasm required to begin the task were contributions of Dr. Abramowitz. Since his untimely death, the effort has continued under the general direction of Irene A. Stegun. The workers at the Bureau and the members of the Committee have had many discussions about content, style and layout. Though many details have had to be argued out as they came up, the basic specifications of the volume have remained the same as were outlined by the Massachusetts Institute of Technology Conference of 1954.

The Committee wishes here to register its commendation of the magnitude and quality of the task carried out by the staff of the NBS Computing Section and their expert collaborators in planning, collecting and editing these Tables, and its appreciation of the willingness with which its various suggestions were incorporated into the plans. We hope this resulting volume will be judged by its users to be a worthy memorial to the vision and industry of its chief architect, Milton Abramowitz. We regret he did not live to see its publication.

P. M. Morse, *Chairman*.
A. Erdélyi
M. C. Gray
N. C. Metropolis
J. B. Rosser
H. C. Thacher, Jr.
John Todd
C. B. Tompkins
J. W. Tukey.

Contents

Handbook of Mathematical Functions

with

Formulas, Graphs, and Mathematical Tables

Edited by Milton Abramowitz and Irene A. Stegun

1. Introduction

The present Handbook has been designed to provide scientific investigators with a comprehensive and self-contained summary of the mathematical functions that arise in physical and engineering problems. The well-known Tables of Functions by E. Jahnke and F. Emde has been invaluable to workers in these fields in its many editions[1] during the past half-century. The present volume extends the work of these authors by giving more extensive and more accurate numerical tables, and by giving larger collections of mathematical properties of the tabulated functions. The number of functions covered has also been increased.

The classification of functions and organization of the chapters in this Handbook is similar to that of An Index of Mathematical Tables by A. Fletcher, J. C. P. Miller, and L. Rosenhead.[2] In general, the chapters contain numerical tables, graphs, polynomial or rational approximations for automatic computers, and statements of the principal mathematical properties of the tabulated functions, particularly those of computa-tional importance. Many numerical examples are given to illustrate the use of the tables and also the computation of function values which lie outside their range. At the end of the text in each chapter there is a short bibliography giving books and papers in which proofs of the mathematical properties stated in the chapter may be found. Also listed in the bibliographies are the more important numerical tables. Comprehensive lists of tables are given in the Index mentioned above, and current information on new tables is to be found in the National Research Council quarterly Mathematics of Computation (formerly Mathematical Tables and Other Aids to Computation).

The mathematical notations used in this Handbook are those commonly adopted in standard texts, particularly Higher Transcendental Functions, Volumes 1–3, by A. Erdélyi, W. Magnus, F. Oberhettinger and F. G. Tricomi (McGraw-Hill, 1953–55). Some alternative notations have also been listed. The introduction of new symbols has been kept to a minimum, and an effort has been made to avoid the use of conflicting notation.

2. Accuracy of the Tables

The number of significant figures given in each table has depended to some extent on the number available in existing tabulations. There has been no attempt to make it uniform throughout the Handbook, which would have been a costly and laborious undertaking. In most tables at least five significant figures have been provided, and the tabular intervals have generally been chosen to ensure that linear interpolation will yield four- or five-figure accuracy, which suffices in most physical applications. Users requiring higher precision in their interpolates may obtain them by use of higher-order interpolation procedures, described below.

In certain tables many-figured function values are given at irregular intervals in the argument. An example is provided by Table 9.4. The purpose of these tables is to furnish "key values" for the checking of programs for automatic computers; no question of interpolation arises.

The maximum end-figure error, or "tolerance" in the tables in this Handbook is $\frac{6}{10}$ of 1 unit everywhere in the case of the elementary functions, and 1 unit in the case of the higher functions except in a few cases where it has been permitted to rise to 2 units.

---

[1] The most recent, the sixth, with F. Loesch added as co-author, was published in 1960 by McGraw-Hill, U.S.A., and Teubner, Germany.
[2] The second edition, with L. J. Comrie added as co-author, was published in two volumes in 1962 by Addison-Wesley, U.S.A., and Scientific Computing Service Ltd., Great Britain.

3.   Auxiliary Functions and Arguments

One of the objects of this Handbook is to provide tables or computing methods which enable the user to evaluate the tabulated functions over complete ranges of real values of their parameters. In order to achieve this object, frequent use has been made of auxiliary functions to remove the infinite part of the original functions at their singularities, and auxiliary arguments to cope with infinite ranges. An example will make the procedure clear.

The exponential integral of positive argument is given by

$$\mathrm{Ei}(x) = \int_{-\infty}^{x} \frac{e^u}{u} du$$

$$= \gamma + \ln x + \frac{x}{1 \cdot 1!} + \frac{x^2}{2 \cdot 2!} + \frac{x^3}{3 \cdot 3!} + \ \cdots$$

$$\sim \frac{e^x}{x} \left[1 + \frac{1!}{x} + \frac{2!}{x^2} + \frac{3!}{x^3} + \ \cdots \right] (x \to \infty)$$

The logarithmic singularity precludes direct interpolation near $x=0$. The functions $\mathrm{Ei}(x) - \ln x$ and $x^{-1}[\mathrm{Ei}(x) - \ln x - \gamma]$, however, are well-behaved and readily interpolable in this region. Either will do as an auxiliary function; the latter was in fact selected as it yields slightly higher accuracy when $\mathrm{Ei}(x)$ is recovered. The function $x^{-1}[\mathrm{Ei}(x) - \ln x - \gamma]$ has been tabulated to nine decimals for the range $0 \leq x \leq \frac{1}{2}$. For $\frac{1}{2} \leq x \leq 2$, $\mathrm{Ei}(x)$ is sufficiently well-behaved to admit direct tabulation, but for larger values of x, its exponential character predominates. A smoother and more readily interpolable function for large x is $xe^{-x}\mathrm{Ei}(x)$; this has been tabulated for $2 \leq x \leq 10$. Finally, the range $10 \leq x \leq \infty$ is covered by use of the inverse argument x^{-1}. Twenty-one entries of $xe^{-x}\mathrm{Ei}(x)$, corresponding to $x^{-1} = .1(-.005)0$, suffice to produce an interpolable table.

4.   Interpolation

The tables in this Handbook are not provided with differences or other aids to interpolation, because it was felt that the space they require could be better employed by the tabulation of additional functions. Admittedly aids could have been given without consuming extra space by increasing the intervals of tabulation, but this would have conflicted with the requirement that linear interpolation is accurate to four or five figures.

For applications in which linear interpolation is insufficiently accurate it is intended that Lagrange's formula or Aitken's method of iterative linear interpolation[3] be used. To help the user, there is a statement at the foot of most tables of the maximum error in a linear interpolate, and the number of function values needed in Lagrange's formula or Aitken's method to interpolate to full tabular accuracy.

As an example, consider the following extract from Table 5.1.

x	$xe^x E_1(x)$	x	$xe^x E_1(x)$
7.5	.89268 7854	8.0	.89823 7113
7.6	.89384 6312	8.1	.89927 7888
7.7	.89497 9666	8.2	.90029 7306
7.8	.89608 8737	8.3	.90129 60°3
7.9	.89717 4302	8.4	.90227 4695

$$\begin{bmatrix} (-6)3 \\ 5 \end{bmatrix}$$

The numbers in the square brackets mean that the maximum error in a linear interpolate is 3×10^{-6}, and that to interpolate to the full tabular accuracy five points must be used in Lagrange's and Aitken's methods.

Let us suppose that we wish to compute the value of $xe^x E_1(x)$ for $x=7.9527$ from this table. We describe in turn the application of the methods of linear interpolation, Lagrange and Aitken, and of alternative methods based on differences and Taylor's series.

(1) Linear interpolation. The formula for this process is given by

$$f_p = (1-p)f_0 + pf_1$$

where f_0, f_1 are consecutive tabular values of the function, corresponding to arguments x_0, x_1, respectively; p is the given fraction of the argument interval

$$p = (x - x_0)/(x_1 - x_0)$$

and f_p the required interpolate. In the present instance, we have

$$f_0 = .89717\ 4302 \qquad f_1 = .89823\ 7113 \qquad p = .527$$

The most convenient way to evaluate the formula on a desk calculating machine is to set f_0 and f_1 in turn on the keyboard, and carry out the multiplications by $1-p$ and p cumulatively; a partial check is then provided by the multiplier dial reading unity. We obtain

$$f_{.527} = (1 - .527)(.89717\ 4302) + .527(.89823\ 7113)$$
$$= .89773\ 4403.$$

Since it is known that there is a possible error of 3×10^{-6} in the linear formula, we round off this result to .89773. The maximum possible error in this answer is composed of the error committed

[3] A. C. Aitken, On interpolation by iteration of proportional parts, without the use of differences, Proc. Edinburgh Math. Soc. **3**, 56–76 (1932).

by the last rounding, that is, $.4403 \times 10^{-5}$, plus 3×10^{-6}, and so certainly cannot exceed $.8 \times 10^{-5}$.

(2) Lagrange's formula. In this example, the relevant formula is the 5-point one, given by

$$f = A_{-2}(p)f_{-2} + A_{-1}(p)f_{-1} + A_0(p)f_0 + A_1(p)f_1 + A_2(p)f_2$$

Tables of the coefficients $A_k(p)$ are given in chapter 25 for the range $p=0(.01)1$. We evaluate the formula for $p=.52$, $.53$ and $.54$ in turn. Again, in each evaluation we accumulate the $A_k(p)$ in the multiplier register since their sum is unity. We now have the following subtable.

x	$xe^4E_1(x)$		
7.952	.89772 9757		
		10622	
7.953	.89774 0379		-2
		10620	
7.954	.89775 0999		

The numbers in the third and fourth columns are the first and second differences of the values of $xe^xE_1(x)$ (see below); the smallness of the second difference provides a check on the three interpolations. The required value is now obtained by linear interpolation:

$$f_p = .3(.89772\ 9757) + .7(.89774\ 0379)$$

$$= .89773\ 7192.$$

In cases where the correct order of the Lagrange polynomial is not known, one of the preliminary interpolations may have to be performed with polynomials of two or more different orders as a check on their adequacy.

(3) Aitken's method of iterative linear interpolation. The scheme for carrying out this process in the present example is as follows:

n	x_n	$y_n = xe^xE_1(x)$	$y_{0,n}$	$y_{0,1,n}$	$y_{0,1,2,n}$	$y_{0,1,2,3,n}$	$x_n - x$
0	8.0	.89823 7113					.0473
1	7.9	.89717 4302	.89773 44034				$-.0527$
2	8.1	.89927 7888	.89774 48264	.89773 71499			.1473
3	7.8	.89608 8737	2 90220	2394	.89773 71938		$-.1527$
4	8.2	.90029 7306	4 98773	1216	16	89773 71930	.2473
5	7.7	.89497 9666	2 35221	2706	43	30	$-.2527$

Here

$$y_{0,n} = \frac{1}{x_n - x_0} \begin{vmatrix} y_0 & x_0 - x \\ y_n & x_n - x \end{vmatrix}$$

$$y_{0,1,n} = \frac{1}{x_n - x_1} \begin{vmatrix} y_{0,1} & x_1 - x \\ y_{0,n} & x_n - x \end{vmatrix}$$

$$y_{0,1,\ldots,m-1,m,n} = \frac{1}{x_n - x_m} \begin{vmatrix} y_{0,1,\ldots,m-1,m} & x_m - x \\ y_{0,1,\ldots,m-1,n} & x_n - x \end{vmatrix}$$

If the quantities $x_n - x$ and $x_m - x$ are used as multipliers when forming the cross-product on a desk machine, their accumulation $(x_n - x) - (x_m - x)$ in the multiplier register is the divisor to be used at that stage. An extra decimal place is usually carried in the intermediate interpolates to safeguard against accumulation of rounding errors.

The order in which the tabular values are used is immaterial to some extent, but to achieve the maximum rate of convergence and at the same time minimize accumulation of rounding errors, we begin, as in this example, with the tabular argument nearest to the given argument, then take the nearest of the remaining tabular arguments, and so on.

The number of tabular values required to achieve a given precision emerges naturally in the course of the iterations. Thus in the present example six values were used, even though it was known in advance that five would suffice. The extra row confirms the convergence and provides a valuable check.

(4) Difference formulas. We use the central difference notation (chapter 25),

x_0	f_0				
		$\delta f_{1/2}$			
x_1	f_1		$\delta^2 f_1$		
		$\delta f_{3/2}$		$\delta^3 f_{3/2}$	
x_2	f_2		$\delta^2 f_2$		$\delta^4 f_2$
		$\delta f_{5/2}$		$\delta^3 f_{5/2}$	
x_3	f_3		$\delta^2 f_3$		
		$\delta f_{7/2}$			
x_4	f_4				

Here

$$\delta f_{1/2} = f_1 - f_0, \ \delta f_{3/2} = f_2 - f_1, \ldots,$$
$$\delta^2 f_1 = \delta f_{3/2} - \delta f_{1/2} = f_2 - 2f_1 + f_0$$
$$\delta^3 f_{3/2} = \delta^2 f_2 - \delta^2 f_1 = f_3 - 3f_2 + 3f_1 - f_0$$
$$\delta^4 f_2 = \delta^3 f_{5/2} - \delta^3 f_{3/2} = f_4 - 4f_3 + 6f_2 - 4f_1 + f_0$$

and so on.

In the present example the relevant part of the difference table is as follows, the differences being written in units of the last decimal place of the function, as is customary. The smallness of the high differences provides a check on the function values

x	$xe^xE_1(x)$	$\delta^2 f$	$\delta^4 f$
7.9	.89717 4302	$-2\ 2754$	-34
8.0	.89823 7113	$-2\ 2036$	-39

Applying, for example, Everett's interpolation formula

$$f_p = (1-p)f_0 + E_2(p)\delta^2 f_0 + E_4(p)\delta^4 f_0 + \ldots + pf_1 + F_2(p)\delta^2 f_1 + F_4(p)\delta^4 f_1 + \ldots$$

and taking the numerical values of the interpolation coefficients $E_2(p)$, $E_4(p)$, $F_2(p)$ and $F_4(p)$ from Table 25.1, we find that

$10^9 f_{.527} = .473(89717\ 4302) + .061196(2\ 2754) - .012(34)$
$\qquad + .527(89823\ 7113) + .063439(2\ 2036) - .012(39)$
$\qquad = 89773\ 7193.$

We may notice in passing that Everett's formula shows that the error in a linear interpolate is approximately

$$E_2(p)\delta^2 f_0 + F_2(p)\delta^2 f_1 \approx \tfrac{1}{2}[E_2(p) + F_2(p)][\delta^2 f_0 + \delta^2 f_1]$$

Since the maximum value of $|E_2(p) + F_2(p)|$ in the range $0 < p < 1$ is $\tfrac{1}{8}$, the maximum error in a linear interpolate is approximately

$$\frac{1}{16}\,|\delta^2 f_0 + \delta^2 f_1|, \text{ that is, } \frac{1}{16}\,|f_2 - f_1 - f_0 + f_{-1}|.$$

(5) Taylor's series. In cases where the successive derivatives of the tabulated function can be computed fairly easily, Taylor's expansion

$$f(x) = f(x_0) + (x - x_0)\frac{f'(x_0)}{1!} + (x - x_0)^2\frac{f''(x_0)}{2!}$$
$$+ (x - x_0)^3\frac{f'''(x_0)}{3!} + \ldots$$

can be used. We first compute as many of the derivatives $f^{(n)}(x_0)$ as are significant, and then evaluate the series for the given value of x. An advisable check on the computed values of the derivatives is to reproduce the adjacent tabular values by evaluating the series for $x = x_{-1}$ and x_1.

In the present example, we have

$$f(x) = xe^x E_1(x)$$
$$f'(x) = (1 + x^{-1})f(x) - 1$$
$$f''(x) = (1 + x^{-1})f'(x) - x^{-2}f(x)$$
$$f'''(x) = (1 + x^{-1})f''(x) - 2x^{-2}f'(x) + 2x^{-3}f(x).$$

With $x_0 = 7.9$ and $x - x_0 = .0527$ our computations are as follows; an extra decimal has been retained in the values of the terms in the series to safeguard against accumulation of rounding errors.

k	$f^{(k)}(x_0)/k!$	$(x - x_0)^k f^{(k)}(x_0)/k!$
0	.89717 4302	.89717 4302
1	.01074 0669	.00056 6033 3
2	−.00113 7621	−.00000 3159 5
3	.00012 1987	.00000 0017 9
		.89773 7194

5.   Inverse Interpolation

With linear interpolation there is no difference in principle between direct and inverse interpolation. In cases where the linear formula provides an insufficiently accurate answer, two methods are available. We may interpolate directly, for example, by Lagrange's formula to prepare a new table at a fine interval in the neighborhood of the approximate value, and then apply accurate inverse linear interpolation to the subtabulated values. Alternatively, we may use Aitken's method or even possibly the Taylor's series method, with the roles of function and argument interchanged.

It is important to realize that the accuracy of an inverse interpolate may be very different from that of a direct interpolate. This is particularly true in regions where the function is slowly varying, for example, near a maximum or minimum. The maximum precision attainable in an inverse interpolate can be estimated with the aid of the formula

$$\Delta x \approx \Delta f / \frac{df}{dx}$$

in which Δf is the maximum possible error in the function values.

Example. Given $xe^x E_1(x) = .9$, find x from the table on page X.

(i) Inverse linear interpolation. The formula for p is

$$p = (f_p - f_0)/(f_1 - f_0).$$

In the present example, we have

$$p = \frac{.9 - .89927\ 7888}{.90029\ 7306 - .89927\ 7888} = \frac{72\ 2112}{101\ 9418} = .708357.$$

The desired x is therefore

$$x = x_0 + p(x_1 - x_0) = 8.1 + .708357(.1) = 8.17083\ 57$$

To estimate the possible error in this answer, we recall that the maximum error of direct linear interpolation in this table is $\Delta f = 3 \times 10^{-6}$. An approximate value for df/dx is the ratio of the first difference to the argument interval (chapter 25), in this case .010. Hence the maximum error in x is approximately $3 \times 10^{-6}/(.010)$, that is, .0003.

(ii) Subtabulation method. To improve the approximate value of x just obtained, we interpolate directly for $p = .70$, .71 and .72 with the aid of Lagrange's 5-point formula,

x	$xe^x E_1(x)$	δ	δ^2
8.170	.89999 3683		
		1 0151	
8.171	.90000 3834		−2
		1 0149	
8.172	.90001 3983		

Inverse linear interpolation in the new table gives

$$p = \frac{.9 - .89999\ 3683}{.00001\ 0151} = .6223$$

Hence $x = 8.17062\ 23.$

An estimate of the maximum error in this result is

$$\Delta f / \frac{df}{dx} \approx \frac{1 \times 10^{-9}}{.010} = 1 \times 10^{-7}$$

(iii) Aitken's method. This is carried out in the same manner as in direct interpolation.

n	$y_n = xe^x E_1(x)$	x_n	$x_{0,n}$	$x_{0,1,n}$	$x_{0,1,2,n}$	$x_{0,1,2,3,n}$	$y_n - y$
0	.90029 7306	8.2					.00029 7306
1	.89927 7888	8.1	8.17083 5712				−.00072 2112
2	.90129 6033	8.3	8.17023 1505	8.17061 9521			.00129 6033
3	.89823 7113	8.0	8.17113 8043	2 5948	8.17062 2244		−.00176 2887
4	.90227 4695	8.4	8.16992 9437	1 7335	415	8.17062 2318	.00227 4695
5	.89717 4302	7.9	8.17144 0382	2 8142	231	265	−.00282 5698

The estimate of the maximum error in this result is the same as in the subtabulation method. An indication of the error is also provided by the discrepancy in the highest interpolates, in this case $x_{0,1,2,3,4}$ and $x_{0,1,2,3,5}$.

6. Bivariate Interpolation

Bivariate interpolation is generally most simply performed as a sequence of univariate interpolations. We carry out the interpolation in one direction, by one of the methods already described, for several tabular values of the second argument in the neighborhood of its given value. The interpolates are differenced as a check, and interpolation is then carried out in the second direction.

An alternative procedure in the case of functions of a complex variable is to use the Taylor's series expansion, provided that successive derivatives of the function can be computed without much difficulty.

7. Generation of Functions from Recurrence Relations

Many of the special mathematical functions which depend on a parameter, called their index, order or degree, satisfy a linear difference equation (or recurrence relation) with respect to this parameter. Examples are furnished by the Legendre function $P_n(x)$, the Bessel function $J_n(x)$ and the exponential integral $E_n(x)$, for which we have the respective recurrence relations

$$(n+1)P_{n+1} - (2n+1)xP_n + nP_{n-1} = 0$$

$$J_{n+1} - \frac{2n}{x}J_n + J_{n-1} = 0$$

$$nE_{n+1} + xE_n = e^{-x}.$$

Particularly for automatic work, recurrence relations provide an important and powerful computing tool. If the values of $P_n(x)$ or $J_n(x)$ are known for two consecutive values of n, or $E_n(x)$ is known for one value of n, then the function may be computed for other values of n by successive applications of the relation. Since generation is carried out perforce with rounded values, it is vital to know how errors may be propagated in the recurrence process. If the errors do not grow relative to the size of the wanted function, the process is said to be stable. If, however, the relative errors grow and will eventually overwhelm the wanted function, the process is unstable.

It is important to realize that stability may depend on (i) the particular solution of the difference equation being computed; (ii) the values of x or other parameters in the difference equation;

(iii) the direction in which the recurrence is being applied. Examples are as follows.

Stability—increasing n

$P_n(x)$, $P_n^m(x)$

$Q_n(x)$, $Q_n^m(x)$ $(x<1)$

$Y_n(x)$, $K_n(x)$

$J_{-n-\frac{1}{2}}(x)$, $I_{-n-\frac{1}{2}}(x)$

$E_n(x)$ $(n<x)$

Stability—decreasing n

$P_n(x)$, $P_n^m(x)$ $(x<1)$

$Q_n(x)$, $Q_n^m(x)$

$J_n(x)$, $I_n(x)$

$J_{n+\frac{1}{2}}(x)$, $I_{n+\frac{1}{2}}(x)$

$E_n(x)$ $(n>x)$

$F_n(\eta, \rho)$ (Coulomb wave function)

Illustrations of the generation of functions from their recurrence relations are given in the pertinent chapters. It is also shown that even in cases where the recurrence process is unstable, it may still be used when the starting values are known to sufficient accuracy.

Mention must also be made here of a refinement, due to J. C. P. Miller, which enables a recurrence process which is stable for decreasing n to be applied without any knowledge of starting values for large n. Miller's algorithm, which is well-suited to automatic work, is described in **19.28, Example 1.**

8. Acknowledgments

The production of this volume has been the result of the unrelenting efforts of many persons, all of whose contributions have been instrumental in accomplishing the task. The Editor expresses his thanks to each and every one.

The Ad Hoc Advisory Committee individually and together were instrumental in establishing the basic tenets that served as a guide in the formation of the entire work. In particular, special thanks are due to Professor Philip M. Morse for his continuous encouragement and support. Professors J. Todd and A. Erdélyi, panel members of the Conferences on Tables and members of the Advisory Committee have maintained an undiminished interest, offered many suggestions and carefully read all the chapters.

Irene A. Stegun has served effectively as associate editor, sharing in each stage of the planning of the volume. Without her untiring efforts, completion would never have been possible.

Appreciation is expressed for the generous cooperation of publishers and authors in granting permission for the use of their source material. Acknowledgments for tabular material taken wholly or in part from published works are given on the first page of each table. Myrtle R. Kellington corresponded with authors and publishers to obtain formal permission for including their material, maintained uniformity throughout the bibliographic references and assisted in preparing the introductory material.

Valuable assistance in the preparation, checking and editing of the tabular material was received from Ruth E. Capuano, Elizabeth F. Godefroy, David S. Liepman, Kermit Nelson, Bertha H. Walter and Ruth Zucker.

Equally important has been the untiring cooperation, assistance, and patience of the members of the NBS staff in handling the myriad of detail necessarily attending the publication of a volume of this magnitude. Especially appreciated have been the helpful discussions and services from the members of the Office of Technical Information in the areas of editorial format, graphic art layout, printing detail, preprinting reproduction needs, as well as attention to promotional detail and financial support. In addition, the clerical and typing staff of the Applied Mathematics Division merit commendation for their efficient and patient production of manuscript copy involving complicated technical notation.

Finally, the continued support of Dr. E. W. Cannon, chief of the Applied Mathematics Division, and the advice of Dr. F. L. Alt, assistant chief, as well as of the many mathematicians in the Division, is gratefully acknowledged.

M. ABRAMOWITZ.

1. Mathematical Constants

David S. Liepman [1]

Contents

---

[1] National Bureau of Standards.

TABLE 1. 1.   MATHEMATICAL CONSTANTS

n(prime)	$\sqrt{n}$								
2	1. 4142	13562	37309	50488		$10^{1/2}$	3. 1622	77660	16837	93320
3	1. 7320	50807	56887	72935		$10^{1/3}$	2. 1544	34690	03188	37219
5	2. 2360	67977	49978	96964		$10^{1/4}$	1. 7782	79410	03892	28012
7	2. 6457	51311	06459	05905		$10^{1/5}$	1. 5848	93192	46111	34853
11	3. 3166	24790	35539	98491		$100^{1/3}$	4. 6415	88833	61277	88924 *
13	3. 6055	51275	46398	92931		$100^{1/5}$	2. 5118	86431	50958	01112
17	4. 1231	05625	61766	05498		$1000^{1/4}$	5. 6234	13251	90349	08040
19	4. 3588	98943	54067	35522		$1000^{1/5}$	3. 9810	71705	53497	25077 *
23	4. 7958	31523	31271	95416		$2^{1/3}$	1. 2599	21049	89487	31648
29	5. 3851	64807	13450	40313		$3^{1/3}$	1. 4422	49570	30740	83233
31	5. 5677	64362	83002	19221		$2^{1/4}$	1. 1892	07115	00272	10667
37	6. 0827	62530	29821	96890		$3^{1/4}$	1. 3160	74012	95249	24608 *
41	6. 4031	24237	43284	86865		$2^{-1/2}$ (— 1)	7. 0710	67811	86547	52440
43	6. 5574	38524	30200	06523		$3^{-1/2}$ (— 1)	5. 7735	02691	89625	76451
47	6. 8556	54600	40104	41249		$5^{-1/2}$ (— 1)	4. 4721	35954	99957	93928
53	7. 2801	09889	28051	82711						
59	7. 6811	45747	86860	81758						
61	7. 8102	49675	90665	43941		$e^{\pi/2}$	4. 8104	77380	96535	16555
67	8. 1853	52771	87244	99700		$e^{\pi/4}$	2. 1932	80050	73801	54566
71	8. 4261	49773	17635	86306		$e^{-\pi/2}$ (— 1)	2. 0787	95763	50761	90855
73	8. 5440	03745	31753	11679		$e^{-\pi/4}$ (— 1)	4. 5593	81277	65996	23677
79	8. 8881	94417	31558	88501		$e^{1/2}$	1. 6487	21270	70012	81468
83	9. 1104	33579	14429	88819		$e^{-1/2}$ (— 1)	6. 0653	06597	12633	42360
89	9. 4339	81132	05660	38113		$e^{1/3}$	1. 3956	12425	08608	95286
97	9. 8488	57801	79610	47217		$e^{-1/3}$ (— 1)	7. 1653	13105	73789	25043

n	e^n						n	e^{-n}				
1		2. 7182	81828	45904	52353	60287	1	(— 1)	3. 6787	94411	71442	32159 55238
2		7. 3890	56098	93065	02272	30427	2	(— 1)	1. 3533	52832	36612	69189 39995
3	(1)	2. 0085	53692	31876	67740	92853	3	(— 2)	4. 9787	06836	78639	42979 34242
4	(1)	5. 4598	15003	31442	39078	11026	4	(— 2)	1. 8315	63888	87341	80293 71802
5	(2)	1. 4841	31591	02576	60342	11156	5	(— 3)	6. 7379	46999	08546	70966 36048
6	(2)	4. 0342	87934	92735	12260	83872	6	(— 3)	2. 4787	52176	66635	84230 45167
7	(3)	1. 0966	33158	42845	85992	63720	7	(— 4)	9. 1188	19655	54516	20800 31361
8	(3)	2. 9809	57987	04172	82747	43592	8	(— 4)	3. 3546	26279	02511	83882 13891
9	(3)	8. 1030	83927	57538	40077	09997	9	(— 4)	1. 2340	98040	86679	54949 76367
10	(4)	2. 2026	46579	48067	16516	95790	10	(— 5)	4. 5399	92976	24848	51535 59152

n	$e^{n\pi}$						n	$e^{-n\pi}$				
1	(1)	2. 3140	69263	27792	69006		1	(— 2)	4. 3213	91826	37722	49774
2	(2)	5. 3549	16555	24764	73650		2	(— 3)	1. 8674	42731	70798	88144
3	(4)	1. 2391	64780	79166	97482		3	(— 5)	8. 0699	51757	03045	99239
4	(5)	2. 8675	13131	36653	29975		4	(— 6)	3. 4873	42356	20899	54918
5	(6)	6. 6356	23999	34113	42333		5	(— 7)	1. 5070	17275	39006	46107
6	(8)	1. 5355	29353	95446	69392		6	(— 9)	6. 5124	12136	07990	07282
7	(9)	3. 5533	21280	84704	43597		7	(—10)	2. 8142	68457	48555	27211
8	(10)	8. 2226	31558	55949	95275		8	(—11)	1. 2161	55670	94093	08397
9	(12)	1. 9027	73895	29216	12917		9	(—13)	5. 2554	85176	00644	85552
10	(13)	4. 4031	50586	06320	29011		10	(—14)	2. 2711	01068	32409	38387

e^e	(1)	1. 5154	26224	14792	64190		e^{-e}	(— 2)	6. 5988	03584	53125	37077
e^γ		1. 7810	72417	99019	79852		$e^{-\gamma}$	(— 1)	5. 6145	94835	66885	16982

n	$\ln n$						n	$\log_{10} n$				
2	0. 6931	47180	55994	53094	172321		2	(—1) 3. 0102	99956	63981	19521	37389
3	1. 0986	12288	66810	96913	952452		3	(—1) 4. 7712	12547	19662	43729	50279
4	1. 3862	94361	11989	06188	344642		4	(—1) 6. 0205	99913	27962	39042	74778
5	1. 6094	37912	43410	03746	007593		5	(—1) 6. 9897	00043	36018	80478	62611
6	1. 7917	59469	22805	50008	124774		6	(—1) 7. 7815	12503	83643	63250	87668
7	1. 9459	10149	05531	33051	053527		7	(—1) 8. 4509	80400	14256	83071	22163
8	2. 0794	41541	67983	59282	516964		8	(—1) 9. 0308	99869	91943	58564	12167
9	2. 1972	24577	33621	93827	904905		9	(—1) 9. 5424	25094	39324	87459	00558
10	2. 3025	85092	99404	56840	179915		10	1. 0000	00000	00000	00000	00000
11	2. 3978	95272	79837	05440	619436		11	1. 0413	92685	15822	50407	50200
13	2. 5649	49357	46153	67360	534874		13	1. 1139	43352	30683	67692	06505
17	2. 8332	13344	05621	60802	495346		17	1. 2304	48921	37827	39285	40170
19	2. 9444	38979	16644	04600	090274		19	1. 2787	53600	95282	89615	36333
23	3. 1354	94215	92914	96908	067528		23	1. 3617	27836	01759	28788	67777
29	3. 3672	95829	98647	40271	832720		29	1. 4623	97997	89895	60873	32847
31	3. 4339	87204	48514	62459	291643		31	1. 4913	61693	83427	26796	66704
37	3. 6109	17912	64422	44443	680957		37	1. 5682	01724	06699	49968	08451
41	3. 7135	72066	70430	78038	667634		41	1. 6127	83856	71973	54945	09412
43	3. 7612	00115	69356	24234	728425		43	1. 6334	68455	57958	65264	05088

*See page II.

TABLE 1.1. MATHEMATICAL CONSTANTS—Continued

n		$\ln n$				n		$\log_{10} n$			
47	3. 8501	47601	71005	85868	209507	47	1. 6720	97857	93571	74644	14219
53	3. 9702	91913	55212	18341	444691	53	1. 7242	75869	60078	90456	32992
59	4. 0775	37443	90571	94506	160504	59	1. 7708	52011	64214	41902	60656
61	4. 1108	73864	17331	12487	513891	61	1. 7853	29835	01076	70338	85749
67	4. 2046	92619	39096	60596	700720	67	1. 8260	74802	70082	64341	49132
71	4. 2626	79877	04131	54213	294545	71	1. 8512	58348	71907	52860	92829
73	4. 2904	59441	14839	11290	921089	73	1. 8633	22860	12045	59010	74387
79	4. 3694	47852	46702	14941	729455	79	1. 8976	27091	29044	14279	94821
83	4. 4188	40607	79659	79234	754722	83	1. 9190	78092	37607	39038	32760
89	4. 4886	36369	73213	98383	178155	89	1. 9493	90006	64491	27847	23543
97	4. 5747	10978	50338	28221	167216	97	1. 9867	71734	26624	48517	84362
$\ln\pi$	1. 1447	29885	84940	01741	43427	$\log_{10}\pi$	(-1) 4. 9714	98726	94133	85435	12683
$\ln\sqrt{2\pi}$	(-1) 9. 1893	85332	04672	74178	03296	$\log_{10}e$	(-1) 4. 3429	44819	03251	82765	11289

n		$n \ln 10$				n		$n\pi$			
1	2. 3025	85092	99404	56840	17991	1	3. 1415	92653	58979	32384	62643
2	4. 6051	70185	98809	13680	35983	2	6. 2831	85307	17958	64769	25287
3	6. 9077	55278	98213	70520	53974	3	9. 4247	77960	76937	97153	87930
4	9. 2103	40371	97618	27360	71966	4	(1) 1. 2566	37061	43591	72953	85057
5	(1) 1. 1512	92546	49702	28420	08996	5	(1) 1. 5707	96326	79489	66192	31322
6	(1) 1. 3815	51055	79642	74104	10795	6	(1) 1. 8849	55592	15387	59430	77586
7	(1) 1. 6118	09565	09583	19788	12594	7	(1) 2. 1991	14857	51285	52669	23850
8	(1) 1. 8420	68074	39523	65472	14393	8	(1) 2. 5132	74122	87183	45907	70115
9	(1) 2. 0723	26583	69464	11156	16192	9	(1) 2. 8274	33388	23081	39146	16379

n		π^n				n		π^{-n}			
1	3. 1415	92653	58979	32384	62643	1	(-1) 3. 1830	98861	83790	67153	77675
2	9. 8696	04401	08935	86188	34491	2	(-1) 1. 0132	11836	42337	77144	38795
3	(1) 3. 1006	27668	02998	20175	47632	3	(-2) 3. 2251	53443	31994	89184	42205
4	(1) 9. 7409	09103	40024	37236	44033	4	(-2) 1. 0265	98225	46843	35189	15278
5	(2) 3. 0601	96847	85281	45326	27413	5	(-3) 3. 2677	63643	05338	54726	28250
6	(2) 9. 6138	91935	75304	43703	02194	6	(-3) 1. 0401	61473	29585	22960	89838
7	(3) 3. 0202	93227	77679	20675	14206	7	(-4) 3. 3109	36801	77566	76432	59528
8	(3) 9. 4885	31016	07057	40071	28576	8	(-4) 1. 0539	03916	53493	66633	17287
9	(4) 2. 9809	09933	34462	11666	50940	9	(-5) 3. 3546	80357	20886	91287	39854
10	(4) 9. 3648	04747	60830	20973	71669	10	(-5) 1. 0678	27922	68615	33662	04078

$\pi/2$	1. 5707	96326	79489	66192	31322	$3\pi/2$	4. 7123	88980	38468	98576	93965	
$\pi/3$	1. 0471	97551	19659	77461	54214	$4\pi/3$	4. 1887	90204	78639	09846	16858	
$\pi/4$	(-1) 7. 8539	81633	97448	30961	56608	$\pi(2)^{1/2}$	4. 4428	82938	15836	62470	15881	*
$\pi^{1/2}$	1. 7724	53850	90551	60272	98167	$\pi^{-1/2}$	(-1) 5. 6418	95835	47756	28694	80795	
$\pi^{1/3}$	1. 4645	91887	56152	32630	20143	$\pi^{-1/3}$	(-1) 6. 8278	40632	55295	68146	70208	
$\pi^{1/4}$	1. 3313	35363	80038	97127	97535	$\pi^{-1/4}$	(-1) 7. 5112	55444	64942	48285	87030	
$\pi^{2/3}$	2. 1450	29397	11102	56000	77444	$\pi^{-2/3}$	(-1) 4. 6619	40770	35411	61438	19885	
$\pi^{3/4}$	2. 3597	30492	41469	68875	78474	$\pi^{-3/4}$	(-1) 4. 2377	72081	23757	59679	10077	
$\pi^{3/2}$	5. 5683	27996	83170	78452	84818	$\pi^{-3/2}$	(-1) 1. 7958	71221	25166	56168	90820	
π^e	(1) 2. 2459	15771	83610	45473	42715	π^{-e}	(-2) 4. 4525	26726	69229	06151	35273	
$(2\pi)^{1/2}$	2. 5066	28274	63100	05024	15765	$(2\pi)^{-1/2}$	(-1) 3. 9894	22804	01432	67793	99461	
$(\pi/2)^{1/2}$	1. 2533	14137	31550	02512	07883	$(2/\pi)^{1/2}$	(-1) 7. 9788	45608	02865	35587	98921	
$\pi(2)^{-1/2}$	2. 2214	41469	07918	31235	07940	$2^{1/2}/\pi$	(-1) 4. 5015	81580	78553	03477	75996	
$1r$	57. 2957	79513	08232	08767	98155°	$1'$	0. 0002	90888	20866	57215	96154r	
$1°$	0. 0174	53292	51994	32957	69237r	$1''$	0. 0000	04848	13681	10953	59936r	
γ	0. 5772	15664	90153	28606	06512	$\ln\gamma$	$-0. 5495$	39312	98164	48223	37662	
$\Gamma(1/2)$	1. 7724	53850	905516			$1/\Gamma(1/2)$	0. 5641	89583	547756			
$\Gamma(1/3)$	2. 6789	38534	707748			$1/\Gamma(1/3)$	0. 3732	82173	907395			
$\Gamma(2/3)$	1. 3541	17939	426400			$1/\Gamma(2/3)$	0. 7384	88111	621648			
$\Gamma(1/4)$	3. 6256	09908	221908			$1/\Gamma(1/4)$	0. 2758	15662	830209			
$\Gamma(3/4)$	1. 2254	16702	465178			$1/\Gamma(3/4)$	0. 8160	48939	098263			
$\Gamma(4/3)$	0. 8929	79511	569249			$1/\Gamma(4/3)$	1. 1198	46521	722186			
$\Gamma(5/3)$	0. 9027	45292	950934			$1/\Gamma(5/3)$	1. 1077	32167	432472			
$\Gamma(5/4)$	0. 9064	02477	055477			$1/\Gamma(5/4)$	1. 1032	62651	320837			
$\Gamma(7/4)$	0. 9190	62526	848883			$1/\Gamma(7/4)$	1. 0880	65252	131017			
$\ln \Gamma(1/3)$	0. 9854	20646	927767			$\ln \Gamma(4/3)$	$-0. 1131$	91641	740343			
$\ln \Gamma(2/3)$	0. 3031	50275	147523			$\ln \Gamma(5/3)$	$-0. 1023$	14832	960640			
$\ln \Gamma(1/4)$	1. 2880	22524	698077			$\ln \Gamma(5/4)$	$-0. 0982$	71836	421813			
$\ln \Gamma(3/4)$	0. 2032	80951	431296			$\ln \Gamma(7/4)$	$-0. 0844$	01121	020486			

*See page II.

2. Physical Constants and Conversion Factors

A. G. McNish [1]

Contents

[1] National Bureau of Standards.

2. Physical Constants and Conversion Factors

The tables in this chapter supply some of the more commonly needed physical constants and conversion factors.

All scientific measurements in the fields of mechanics and heat are based upon four international arbitrarily adopted units, the magnitudes of which are fixed by four agreed on standards:

Length—the meter—fixed by the vacuum wavelength of radiation corresponding to the transition $2P_{10}-5D_5$ of krypton 86

$$(1 \text{ meter} - 1650763.73\lambda).$$

Mass—the kilogram—fixed by the international kilogram at Sèvres, France.

Time—the second—fixed as 1/31,556,925.9747 of the tropical year 1900 at 12^h ephemeris time, or the duration of 9,192,631,770 cycles of the hyperfine transition frequency of cesium 133.

Temperature—the degree—fixed on a thermodynamic basis by taking the temperature for the triple point of natural water as 273.16 °K. (The Celsius scale is obtained by adding −273.15 to the Kelvin scale.)

Other units are defined in terms of them by assigning the value unity to the proportionality constant in each defining equation. The entire system, including electricity units, is called the Système International d'Unités (SI). Taking the 1/100 part of the meter as the unit of length and the 1/1000 part of the kilogram as the unit of mass, similarly, gives

rise to the CGS system, often used in physics and chemistry.

Table 2.1. Common Units and Conversion Factors

Quantity	SI name	CGS name	SI unit/ CGS unit
Force, F	newton	dyne	10^5
Energy, W	joule	erg	10^7
Power, P	watt	....	10^7

The SI unit of electric current is the ampere defined by the equation $2\Gamma_m I_1 I_2/4\pi = F$ giving the force in vacuo per unit length between two infinitely long parallel conductors of infinitesimal cross-section. If F is in newtons, and Γ_m has the numerical value $4\pi \times 10^{-7}$, then I_1 and I_2 are in amperes. The customary equations define the other electric and magnetic units of SI such as the volt, ohm, farad, henry, etc. The force between electric charges in a vacuum in this system is given by $Q_1 Q_2/4\pi\Gamma_e r^2 = F$, Γ_e having the numerical value $10^7/4\pi c^2$ where c is the speed of light in meters per second ($\Gamma_e = 8.854 \times 10^{-12}$).

The CGS unrationalized system is obtained by deleting 4π in the denominators in these equations and expressing F in dynes, and r in centimeters. Setting Γ_m equal to unity defines the CGS unrationalized electromagnetic system (emu), Γ_e then taking the numerical value of $1/c^2$. Setting Γ_e equal to unity defines the CGS unrationalized electrostatic system (esu), Γ_m then taking the numerical value of $1/c^2$.

Table 2.2. Names and Conversion Factors for Electric and Magnetic Units

Quantity	SI name	emu name	esu name	SI unit/ emu unit	SI unit/ esu unit
Current	ampere	abampere	statampere	10^{-1}	$\sim 3 \times 10^9$
Charge	coulomb	abcoulomb	statcoulomb	10^{-1}	$\sim 3 \times 10^9$
Potential	volt	abvolt	statvolt	10^8	$\sim (1/3) \times 10^{-2}$
Resistance	ohm	abohm	statohm	10^9	$\sim (1/9) \times 10^{-11}$
Inductance	henry	centimeter	- - - - - - - - - - - - -	10^9	$\sim (1/9) \times 10^{-11}$
Capacitance	farad	- - - - - - - - - - - - -	centimeter	10^{-9}	$\sim 9 \times 10^{11}$
Magnetizing force	amp. turns/ meter	oersted	- - - - - - - - - - - - -	$4\pi \times 10^{-3}$*	$\sim 3 \times 10^9$*
Magnetomotive force	amp. turns	gilbert	- - - - - - - - - - - - -	$4\pi \times 10^{-1}$*	$\sim 3/10^6$*
Magnetic flux	weber	maxwell	- - - - - - - - - - - - -	10^8	$\sim (1/3) \times 10^{-2}$
Magnetic flux density	tesla	gauss	- - - - - - - - - - - - -	10^4	$\sim (1/3) \times 10^{-6}$
Electric displacement	- - - - - - - - - - - - -	- - - - - - - - - - - - -	- - - - - - - - - - - - -	10^{-5}*	$\sim 3 \times 10^5$*

Example: If the value assigned to a current is 100 amperes its value in abamperes is $100 \times 10^{-1} = 10$.

*Divide this number by 4π if unrationalized system is involved; other numbers are unchanged.

The values of constants given in Table 2.3 are based on an adjustment by Taylor, Parker, and Langenberg, Rev. Mod. Phys. 41, p.375 (1969). They are being considered for adoption by the Task Group on Fundamental Constants of the Committee on Data for Science and Technology, International Council of Scientific Unions. The uncertainties given are standard errors estimated from the experimental data included in the adjustment. Where applicable, values are based on the unified scale of atomic masses in which the atomic mass unit (u) is defined as 1/12 of the mass of the atom of the ^{12}C nuclide.

Table 2.3. Adjusted Values of Constants

Constant	Symbol	Value	Uncertainty ‡	Systeme International (SI)		Centimeter-gram-second (CGS)	
Speed of light in vacuum	c	2.997 925 0	±10	$\times 10^8$	m/s	$\times 10^{10}$	cm/s
Elementary charge	e	1.602 191 7	70	10^{-19}	C	10^{-20}	$cm^{1/2}g^{1/2}$ *
		4.803 250	21			10^{-10}	$cm^{3/2}g^{1/2}s^{-1}$ †
Avogadro constant	N_A	6.022 169	40	10^{23}	mol^{-1}	10^{23}	mol^{-1}
Atomic mass unit	u	1.660 531	11	10^{-27}	kg	10^{-24}	g
Electron rest mass	m_e	9.109 558	54	10^{-31}	kg	10^{-28}	g
		5.485 930	34	10^{-4}	u	10^{-4}	u
Proton rest mass	m_p	1.672 614	11	10^{-27}	kg	10^{-24}	g
		1.007 276 61	8	10^0	u	10^0	u
Neutron rest mass	m_n	1.674 920	11	10^{-27}	kg	10^{-24}	g
		1.008 665 20	10	10^0	u	10^0	u
Faraday constant	F	9.648 670	54	10^4	C/mol	10^3	$cm^{1/2}g^{1/2}mol^{-1}$*
		2.892 599	16			10^{14}	$cm^{3/2}g^{1/2}s^{-1}mol^{-1}$ †
Planck constant	h	6.626 196	50	10^{-34}	J · s	10^{-27}	erg · s
	$\hbar$	1.054 591 9	80	10^{-34}	J · s	10^{-27}	erg · s
Fine structure constant	a	7.297 351	11	10^{-3}		10^{-3}	
	$1/a$	1.370 360 2	21	10^2		10^2	
Charge to mass ratio for electron	e/m_e	1.758 802 8	54	10^{11}	C/kg	10^7	$cm^{1/2}/g^{1/2}$ *
		5.272 759	16			10^{17}	$cm^{3/2}g^{-1/2}s^{-1}$ †
Quantum-charge ratio	h/e	4.135 708	14	10^{-15}	J · s/C	10^{-7}	$cm^{3/2}g^{1/2}s^{-1}$ *
		1.379 523 4	46			10^{-17}	$cm^{1/2}g^{1/2}$ †
Compton wavelength of electron	λ_C	2.426 309 6	74	10^{-12}	m	10^{-10}	cm
	$\lambda_C/2\pi$	3.861 592	12	10^{-13}	m	10^{-11}	cm
Compton wavelength of proton	$\lambda_{C,p}$	1.321 440 9	90	10^{-15}	m	10^{-13}	cm
	$\lambda_{C,p}/2\pi$	2.103 139	14	10^{-16}	m	10^{-14}	cm
Rydberg constant	$R\infty$	1.097 373 12	11	10^7	m^{-1}	10^5	cm^{-1}
Bohr radius	a_0	5.291 771 5	81	10^{-11}	m	10^{-9}	cm
Electron radius	r_e	2.817 939	13	10^{-15}	m	10^{-13}	cm
Gyromagnetic ratio of proton	γ	2.675 196 5	82	10^8	rad · s^{-1}T^{-1}	10^4	rad · s^{-1}G^{-1} *
	$\gamma/2\pi$	4.257 707	13	10^7	Hz/T	10^3	s^{-1}G^{-1} *
(uncorrected for diamagnetism, H_2O)	γ'	2.675 127 0	82	10^8	rad · s^{-1}T^{-1}	10^4	rad · s^{-1}G^{-1} *
	$\gamma'/2\pi$	4.257 597	13	10^7	Hz/T	10^3	s^{-1}G^{-1} *
Bohr magneton	μ_B	9.274 096	65	10^{-24}	J/T	10^{-21}	erg/G *
Nuclear magneton	μ_N	5.050 951	50	10^{-27}	J/T	10^{-24}	erg/G *
Proton moment	μ_p	1.410 620 3	99	10^{-26}	J/T	10^{-23}	erg/G *
	μ_p/μ_N	2.792 782	17	10^0		10^0	
(uncorrected for diamagnetism, H_2O)	μ'_p/μ_N	2.792 709	17	10^0		10^0	
Gas constant	R	8.314 34	35	10^0	J · K^{-1} mol^{-1}	10^7	erg · K^{-1} mol^{-1}
Normal volume perfect gas	V_0	2.241 36	39	10^{-2}	m^3/mol	10^4	cm^3/mol
Boltzmann constant	k	1.380 622	59	10^{-23}	J/K	10^{-16}	erg/K
First radiation constant ($8\pi hc$)	c_1	4.992 579	38	10^{-24}	J · m	10^{-15}	erg · cm
Second radiation constant	c_2	1.438 833	61	10^{-2}	m · K	10^0	cm · K
Stefan-Boltzmann constant	σ	5.669 61	96	10^{-8}	W · m^{-2}K^{-4}	10^{-5}	erg · cm^{-2}s^{-1}K^{-4}
Gravitational constant	G	6.673 2	31	10^{-11}	N · m^2/kg^2	10^{-8}	dyn · cm^2/g^2

‡Based on 1 std. dev; applies to last digits in preceding column.　　　*Electromagnetic system.　　　†Electrostatic system.

Table 2.4. Miscellaneous Conversion Factors

Standard gravity, g_0	$= 9.806\ 65$ meters per second per second*
Standard atmospheric pressure, P_0	$= 1.013\ 25 \times 10^5$ newtons per square meter*
	$= 1.013\ 25 \times 10^6$ dynes per square centimeter*
1 thermodynamic calorie,[1] cal_c	$= 4.1840$ joules*
1 IT calorie[2], cal_s	$= 4.1868$ joules*
1 liter, l	$= 10^{-3}$ cubic meter*
1 angstrom unit, Å	$= 10^{-10}$ meter*
1 bar	$= 10^5$ newtons per square meter*
	$= 10^6$ dynes per square centimeter*
1 gal	$= 10^{-2}$ meter per second per second*
	$= 1$ centimeter per second per second*
1 astronomical unit, AU	$= 1.496 \times 10^{11}$ meters
1 light year	$= 9.46\ \times 10^{15}$ meters
1 parsec	$= 3.08 \times 10^{16}$ meters
	$= 3.26$ light years

1 curie, the quantity of radioactive material undergoing 3.7×10^{10} disintegrations per second*.

1 roentgen, the exposure of x- or gamma radiation which produces together with its secondaries 2.082×10^9 electron-ion pairs in 0.001 293 gram of air.

The index of refraction of the atmosphere for radio waves of frequency less than 3×10^{10} Hz is given by $(n - 1)10^6 = (77.6/t)\ (p + 4810e/t)$, where n is the refractive index; t, temperature in kelvins; p, total pressure in millibars; e, water vapor partial pressure in millibars.

Factors for converting the customary United States units to units of the metric system are given in Table 2.5.

Geodetic constants for the international (Hayford) spheroid are given in Table 2.6. The gravity values are on the basis of the revised Potsdam value. They are about 14 parts per million smaller than previous values. They are calculated for the surface of the geoid by the international formula.

Table 2.5. Factors for Converting Customary U.S. Units to SI Units

1 yard	0.914 4 meter*
1 foot	0.304 8 meter*
1 inch	0.025 4 meter*
1 statute mile	1 609.344 meters*
1 nautical mile (international)	1 852 meters*
1 pound (avdp.)	0.453 592 37 kilogram*
1 oz. (avdp.)	0.028 349 52 kilogram
1 pound force	4.448 22 newtons
1 slug	14.593 9 kilograms
1 poundal	0.138 255 newtons
1 foot pound	1.355 82 joules
Temperature (Fahrenheit)	32 + (9/5) Celsius temperature*
1 British thermal unit[3]	1055 joules

Table 2.6. Geodetic Constants
$a = 6\ 378\ 388$ m; $f = 1/297$; $b = 6\ 356\ 912$ m

Latitude	Length of 1' of longitude	Length of 1' of latitude	g
	Meters	*Meters*	*m/s²*
0°	1 855.398	1 842.925	9.780 350
15	1 792.580	1 844.170	9.783 800
30	1 608.174	1 847.580	9.793 238
45	1 314.175	1 852.256	9.806 154
60	930.047	1 856.951	9.819 099
75	481.725	1 860.401	9.828 593
90	0	1 861.666	9.832 072

[1] Used principally by chemists.

[2] Used principally by engineers.

[3] Various definitions are given for the British thermal unit. This represents a rounded mean value differing from none of the more important definitions by more than 3 in 10^4.

* Exact value.

3. Elementary Analytical Methods

Milton Abramowitz [1]

Contents

$$n^k, k=1(1)10, 24, 1/2, 1/3, 1/4, 1/5$$
$$n=2(1)999, \text{ Exact or 10S}$$

The author acknowledges the assistance of Peter J. O'Hara and Kermit C. Nelson in the preparation and checking of the table of powers and roots.

[1] National Bureau of Standards. (Deceased.)

3. Elementary Analytical Methods

3.1. Binomial Theorem and Binomial Coefficients; Arithmetic and Geometric Progressions; Arithmetic, Geometric, Harmonic and Generalized Means

Binomial Theorem

3.1.1

$$(a+b)^n = a^n + \binom{n}{1} a^{n-1}b + \binom{n}{2} a^{n-2}b^2$$
$$+ \binom{n}{3} a^{n-3}b^3 + \ldots + b^n$$

$$(n \text{ a positive integer})$$

Binomial Coefficients (see chapter 24)

3.1.2

$$* \quad \binom{n}{k} = {}_nC_k = \frac{n(n-1)\ldots(n-k+1)}{k!} = \frac{n!}{(n-k)!k!}$$

3.1.3 $\quad \binom{n}{k} = \binom{n}{n-k} = (-1)^k \binom{k-n-1}{k}$

3.1.4 $\quad \binom{n+1}{k} = \binom{n}{k} + \binom{n}{k-1}$

3.1.5 $\quad \binom{n}{0} = \binom{n}{n} = 1$

3.1.6 $\quad 1 + \binom{n}{1} + \binom{n}{2} + \ldots + \binom{n}{n} = 2^n$

3.1.7 $\quad 1 - \binom{n}{1} + \binom{n}{2} - \ldots + (-1)^n \binom{n}{n} = 0$

Table of Binomial Coefficients $\binom{n}{k}$

3.1.8

n \ k	0	1	2	3	4	5	6	7	8	9	10	11	12
1	1	1											
2	1	2	1										
3	1	3	3	1									
4	1	4	6	4	1								
5	1	5	10	10	5	1							
6	1	6	15	20	15	6	1						
7	1	7	21	35	35	21	7	1					
8	1	8	28	56	70	56	28	8	1				
9	1	9	36	84	126	126	84	36	9	1			
10	1	10	45	120	210	252	210	120	45	10	1		
11	1	11	55	165	330	462	462	330	165	55	11	1	
12	1	12	66	220	495	792	924	792	495	220	66	12	1

For a more extensive table see chapter **24.**

*See page II.

3.1.9

Sum of Arithmetic Progression to n Terms

$$a + (a+d) + (a+2d) + \ldots + (a+(n-1)d)$$
$$= na + \frac{1}{2}n(n-1)d = \frac{n}{2}(a+l),$$

$$\text{last term in series} = l = a + (n-1)d$$

Sum of Geometric Progression to n Terms

3.1.10

$$s_n = a + ar + ar^2 + \ldots + ar^{n-1} = \frac{a(1-r^n)}{1-r}$$

$$\lim_{n\to\infty} s_n = a/(1-r) \qquad (-1 < r < 1)$$

Arithmetic Mean of n Quantities A

3.1.11 $\qquad A = \frac{a_1 + a_2 + \ldots + a_n}{n}$

Geometric Mean of n Quantities G

3.1.12 $\quad G = (a_1 a_2 \ldots a_n)^{1/n} \qquad (a_k > 0, k = 1, 2, \ldots, n)$

Harmonic Mean of n Quantities H

3.1.13

$$\frac{1}{H} = \frac{1}{n}\left(\frac{1}{a_1} + \frac{1}{a_2} + \ldots + \frac{1}{a_n}\right) \qquad (a_k > 0, k = 1, 2, \ldots, n)$$

Generalized Mean

3.1.14 $\qquad M(t) = \left(\frac{1}{n} \sum_{k=1}^{n} a_k^t\right)^{1/t}$

3.1.15 $\qquad M(t) = 0 \, (t < 0, \text{ some } a_k \text{ zero})$

3.1.16 $\quad \lim_{t\to\infty} M(t) = \text{max.} \qquad (a_1, a_2, \ldots, a_n) = \text{max. } a$

3.1.17 $\quad \lim_{t\to-\infty} M(t) = \text{min.} \qquad (a_1, a_2, \ldots, a_n) = \text{min. } a$

3.1.18 $\qquad \lim_{t\to 0} M(t) = G$

3.1.19 $\qquad M(1) = A$

3.1.20 $\qquad M(-1) = H$

3.2. Inequalities

Relation Between Arithmetic, Geometric, Harmonic and Generalized Means

3.2.1

$$A \geq G \geq H, \text{ equality if and only if } a_1 = a_2 = \ldots = a_n$$

3.2.2 $\qquad \text{min. } a < M(t) < \text{max. } a$

3.2.3
$$\text{min. } a < G < \text{max. } a$$

equality holds if all a_k are equal, or $t<0$ and an a_k is zero

3.2.4 $M(t) < M(s)$ if $t<s$ unless all a_k are equal, or $s<0$ and an a_k is zero.

Triangle Inequalities

3.2.5
$$|a_1| - |a_2| \leq |a_1 + a_2| \leq |a_1| + |a_2|$$

3.2.6
$$\left| \sum_{k=1}^{n} a_k \right| \leq \sum_{k=1}^{n} |a_k|$$

Chebyshev's Inequality

If $a_1 \geq a_2 \geq a_3 \geq \ldots \geq a_n$
$b_1 \geq b_2 \geq b_3 \geq \ldots \geq b_n$

3.2.7
$$n \sum_{k=1}^{n} a_k b_k \geq \left(\sum_{k=1}^{n} a_k \right) \left(\sum_{k=1}^{n} b_k \right)$$

Hölder's Inequality for Sums

If $\dfrac{1}{p} + \dfrac{1}{q} = 1, p>1, q>1$

3.2.8
$$\sum_{k=1}^{n} |a_k b_k| \leq \left(\sum_{k=1}^{n} |a_k|^p \right)^{1/p} \left(\sum_{k=1}^{n} |b_k|^q \right)^{1/q};$$

equality holds if and only if $|b_k| = c|a_k|^{p-1}$ (c=constant>0). If $p=q=2$ we get

Cauchy's Inequality

3.2.9
$$\left[\sum_{k=1}^{n} a_k b_k \right]^2 \leq \sum_{k=1}^{n} a_k^2 \sum_{k=1}^{n} b_k^2 \text{ (equality for } a_k = cb_k,$$
c constant).

Hölder's Inequality for Integrals

If $\dfrac{1}{p} + \dfrac{1}{q} = 1, p>1, q>1$

3.2.10
$$\int_a^b |f(x)g(x)| dx \leq \left[\int_a^b |f(x)|^p dx \right]^{1/p} \left[\int_a^b |g(x)|^q dx \right]^{1/q}$$

equality holds if and only if $|g(x)| = c|f(x)|^{p-1}$ (c=constant>0).
If $p=q=2$ we get

Schwarz's Inequality

3.2.11
$$\left[\int_a^b f(x)g(x)dx \right]^2 \leq \int_a^b [f(x)]^2 dx \int_a^b [g(x)]^2 dx$$

Minkowski's Inequality for Sums

If $p>1$ and $a_k, b_k > 0$ for all k,

3.2.12
$$\left(\sum_{k=1}^{n} (a_k + b_k)^p \right)^{1/p} \leq \left(\sum_{k=1}^{n} a_k^p \right)^{1/p} + \left(\sum_{k=1}^{n} b_k^p \right)^{1/p},$$

equality holds if and only if $b_k = ca_k$ (c=constant>0).

Minkowski's Inequality for Integrals

If $p>1$,

3.2.13
$$\left(\int_a^b |f(x) + g(x)|^p dx \right)^{1/p} \leq \left(\int_a^b |f(x)|^p dx \right)^{1/p}$$
$$+ \left(\int_a^b |g(x)|^p dx \right)^{1/p}$$

equality holds if and only if $g(x) = cf(x)$ (c=constant>0).

3.3. Rules for Differentiation and Integration
Derivatives

3.3.1
$$\frac{d}{dx}(cu) = c\frac{du}{dx}, c \text{ constant}$$

3.3.2
$$\frac{d}{dx}(u+v) = \frac{du}{dx} + \frac{dv}{dx}$$

3.3.3
$$\frac{d}{dx}(uv) = u\frac{dv}{dx} + v\frac{du}{dx}$$

3.3.4
$$\frac{d}{dx}(u/v) = \frac{v\,du/dx - u\,dv/dx}{v^2}$$

3.3.5
$$\frac{d}{dx}u(v) = \frac{du}{dv}\frac{dv}{dx}$$

3.3.6
$$\frac{d}{dx}(u^v) = u^v\left(\frac{v}{u}\frac{du}{dx} + \ln u \frac{dv}{dx} \right)$$

Leibniz's Theorem for Differentiation of an Integral

3.3.7
$$\frac{d}{dc} \int_{a(c)}^{b(c)} f(x,c)dx$$
$$= \int_{a(c)}^{b(c)} \frac{\partial}{\partial c} f(x,c)dx + f(b,c)\frac{db}{dc} - f(a,c)\frac{da}{dc}$$

Leibniz's Theorem for Differentiation of a Product

3.3.8

$$\frac{d^n}{dx^n}(uv)=\frac{d^nu}{dx^n}v+\binom{n}{1}\frac{d^{n-1}u}{dx^{n-1}}\frac{dv}{dx}+\binom{n}{2}\frac{d^{n-2}u}{dx^{n-2}}\frac{d^2v}{dx^2}$$

$$+\cdots+\binom{n}{r}\frac{d^{n-r}u}{dx^{n-r}}\frac{d^rv}{dx^r}+\cdots+u\frac{d^nv}{dx^n}$$

3.3.9
$$\frac{dx}{dy}=1\bigg/\frac{dy}{dx}$$

3.3.10
$$\frac{d^2x}{dy^2}=\frac{-d^2y}{dx^2}\left(\frac{dy}{dx}\right)^{-3}$$

3.3.11
$$\frac{d^3x}{dy^3}=-\left[\frac{d^3y}{dx^3}\frac{dy}{dx}-3\left(\frac{d^2y}{dx^2}\right)^2\right]\left(\frac{dy}{dx}\right)^{-5}$$

Integration by Parts

3.3.12
$$\int u\,dv=uv-\int v\,du$$

3.3.13
$$\int uv\,dx=\left(\int u\,dx\right)v-\int\left(\int u\,dx\right)\frac{dv}{dx}\,dx$$

Integrals of Rational Algebraic Functions

(Integration constants are omitted)

3.3.14
$$\int(ax+b)^n\,dx=\frac{(ax+b)^{n+1}}{a(n+1)}\qquad(n\neq-1)$$

3.3.15
$$\int\frac{dx}{ax+b}=\frac{1}{a}\ln|ax+b|$$

The following formulas are useful for evaluating $\int\frac{P(x)\,dx}{(ax^2+bx+c)^n}$ where $P(x)$ is a polynomial and $n>1$ is an integer.

3.3.16
$$\int\frac{dx}{(ax^2+bx+c)}=\frac{2}{(4ac-b^2)^{\frac{1}{2}}}\arctan\frac{2ax+b}{(4ac-b^2)^{\frac{1}{2}}}$$
$$(b^2-4ac<0)$$

3.3.17
$$=\frac{1}{(b^2-4ac)^{\frac{1}{2}}}\ln\left|\frac{2ax+b-(b^2-4ac)^{\frac{1}{2}}}{2ax+b+(b^2-4ac)^{\frac{1}{2}}}\right|$$
$$(b^2-4ac>0)$$

3.3.18
$$=\frac{-2}{2ax+b}\qquad(b^2-4ac=0)$$

3.3.19
$$\int\frac{x\,dx}{ax^2+bx+c}=\frac{1}{2a}\ln|ax^2+bx+c|-\frac{b}{2a}\int\frac{dx}{ax^2+bx+c}$$

3.3.20
$$\int\frac{dx}{(a+bx)(c+dx)}=\frac{1}{ad-bc}\ln\left|\frac{c+dx}{a+bx}\right|\qquad(ad\neq bc)$$

3.3.21
$$\int\frac{dx}{a^2+b^2x^2}=\frac{1}{ab}\arctan\frac{bx}{a}$$

3.3.22
$$\int\frac{x\,dx}{a^2+b^2x^2}=\frac{1}{2b^2}\ln|a^2+b^2x^2|$$

3.3.23
$$\int\frac{dx}{a^2-b^2x^2}=\frac{1}{2ab}\ln\left|\frac{a+bx}{a-bx}\right|$$

3.3.24
$$\int\frac{dx}{(x^2+a^2)^2}=\frac{1}{2a^3}\arctan\frac{x}{a}+\frac{x}{2a^2(x^2+a^2)}$$

3.3.25
$$\int\frac{dx}{(x^2-a^2)^2}=\frac{-x}{2a^2(x^2-a^2)}+\frac{1}{4a^3}\ln\left|\frac{a+x}{a-x}\right|$$

Integrals of Irrational Algebraic Functions

3.3.26
$$\int\frac{dx}{[(a+bx)(c+dx)]^{1/2}}=\frac{2}{(-bd)^{1/2}}\arctan\left[\frac{-d(a+bx)}{b(c+dx)}\right]^{1/2}\qquad(bd<0)$$

3.3.27
$$=\frac{-1}{(-bd)^{1/2}}\arcsin\left(\frac{2bdx+ad+bc}{bc-ad}\right)\qquad(b>0,\ d<0)$$

3.3.28
$$=\frac{2}{(bd)^{1/2}}\ln\left|[bd(a+bx)]^{1/2}+b(c+dx)^{1/2}\right|\qquad(bd>0)$$

3.3.29
$$\int\frac{dx}{(a+bx)^{1/2}(c+dx)}=\frac{2}{[d(bc-ad)]^{1/2}}\arctan\left[\frac{d(a+bx)}{(bc-ad)}\right]^{1/2}\qquad(d(ad-bc)<0)$$

3.3.30
$$=\frac{1}{[d(ad-bc)]^{1/2}}\ln\left|\frac{d(a+bx)^{1/2}-[d(ad-bc)]^{1/2}}{d(a+bx)^{1/2}+[d(ad-bc)]^{1/2}}\right|\qquad(d(ad-bc)>0)$$

3.3.31

$$\int [(a+bx)(c+dx)]^{1/2}dx$$
$$=\frac{(ad-bc)+2b(c+dx)}{4bd}[(a+bx)(c+dx)]^{1/2}$$
$$-\frac{(ad-bc)^2}{8bd}\int\frac{dx}{[(a+bx)(c+dx)]^{1/2}}$$

3.3.32

$$\int\left[\frac{c+dx}{a+bx}\right]^{1/2}dx=\frac{1}{b}[(a+bx)(c+dx)]^{1/2}$$
$$-\frac{(ad-bc)}{2b}\int\frac{dx}{[(a+bx)(c+dx)]^{1/2}}$$

3.3.33

$$\int\frac{dx}{(ax^2+bx+c)^{1/2}}$$
$$=a^{-1/2}\ln|2a^{1/2}(ax^2+bx+c)^{1/2}+2ax+b|\,(a>0)$$

3.3.34
$$=a^{-1/2}\operatorname{arcsinh}\frac{(2ax+b)}{(4ac-b^2)^{1/2}}$$
$$(a>0,\ 4ac>b^2)$$

3.3.35
$$=a^{-1/2}\ln|2ax+b|\,(a>0,\ b^2=4ac)$$

3.3.36
$$=-(-a)^{-1/2}\arcsin\frac{(2ax+b)}{(b^2-4ac)^{1/2}}$$
$$(a<0,\ b^2>4ac,\ |2ax+b|<(b^2-4ac)^{1/2})$$

3.3.37

$$\int(ax^2+bx+c)^{1/2}dx=\frac{2ax+b}{4a}(ax^2+bx+c)^{1/2}$$
$$+\frac{4ac-b^2}{8a}\int\frac{dx}{(ax^2+bx+c)^{1/2}}$$

3.3.38

$$\int\frac{dx}{x(ax^2+bx+c)^{1/2}}=-\int\frac{dt}{(a+bt+ct^2)^{1/2}}\text{ where }t=1/x$$

3.3.39

$$\int\frac{xdx}{(ax^2+bx+c)^{1/2}}$$
$$=\frac{1}{a}(ax^2+bx+c)^{1/2}-\frac{b}{2a}\int\frac{dx}{(ax^2+bx+c)^{1/2}}$$

3.3.40
$$\int\frac{dx}{(x^2\pm a^2)^{\frac{1}{2}}}=\ln|x+(x^2\pm a^2)^{\frac{1}{2}}|$$

3.3.41

$$\int(x^2\pm a^2)^{\frac{1}{2}}dx=\frac{x}{2}(x^2\pm a^2)^{\frac{1}{2}}\pm\frac{a^2}{2}\ln|x+(x^2\pm a^2)^{\frac{1}{2}}|$$

3.3.42
$$\int\frac{dx}{x(x^2+a^2)^{\frac{1}{2}}}=-\frac{1}{a}\ln\left|\frac{a+(x^2+a^2)^{\frac{1}{2}}}{x}\right|$$

3.3.43
$$\int\frac{dx}{x(x^2-a^2)^{\frac{1}{2}}}=\frac{1}{a}\arccos\frac{a}{x}$$

3.3.44
$$\int\frac{dx}{(a^2-x^2)^{\frac{1}{2}}}=\arcsin\frac{x}{a}$$

3.3.45
$$\int(a^2-x^2)^{\frac{1}{2}}dx=\frac{x}{2}(a^2-x^2)^{\frac{1}{2}}+\frac{a^2}{2}\arcsin\frac{x}{a}$$

3.3.46
$$\int\frac{dx}{x(a^2-x^2)^{\frac{1}{2}}}=-\frac{1}{a}\ln\left|\frac{a+(a^2-x^2)^{\frac{1}{2}}}{x}\right|$$

3.3.47
$$\int\frac{dx}{(2ax-x^2)^{\frac{1}{2}}}=\arcsin\frac{x-a}{a}$$

3.3.48

$$\int(2ax-x^2)^{\frac{1}{2}}dx=\frac{(x-a)}{2}(2ax-x^2)^{\frac{1}{2}}+\frac{a^2}{2}\arcsin\frac{x-a}{a}$$

3.3.49

$$\int\frac{dx}{(ax^2+b)(cx^2+d)^{\frac{1}{2}}}$$
$$=\frac{1}{[b(ad-bc)]^{\frac{1}{2}}}\arctan\frac{x(ad-bc)^{\frac{1}{2}}}{[b(cx^2+d)]^{\frac{1}{2}}}\quad(ad>bc)$$

3.3.50

$$=\frac{1}{2[b(bc-ad)]^{\frac{1}{2}}}\ln\left|\frac{[b(cx^2+d)]^{\frac{1}{2}}+x(bc-ad)^{\frac{1}{2}}}{[b(cx^2+d)]^{\frac{1}{2}}-x(bc-ad)^{\frac{1}{2}}}\right|$$
$$(bc>ad)$$

3.4. Limits, Maxima and Minima

Indeterminate Forms (L'Hospital's Rule)

3.4.1 Let $f(x)$ and $g(x)$ be differentiable on an interval $a\leq x<b$ for which $g'(x)\neq0$.

If
$$\lim_{x\to b-}f(x)=0\text{ and }\lim_{x\to b-}g(x)=0$$

or if
$$\lim_{x\to b-}f(x)=\infty\text{ and }\lim_{x\to b-}g(x)=\infty$$

and if
$$\lim_{x\to b-}\frac{f'(x)}{g'(x)}=l\text{ then }\lim_{x\to b-}\frac{f(x)}{g(x)}=l.$$

Both b and l may be finite or infinite.

Maxima and Minima

3.4.2 (1) *Functions of One Variable*

The function $y=f(x)$ has a maximum at $x=x_0$ if $f'(x_0)=0$ and $f''(x_0)<0$, and a minimum at $x=x_0$ if $f'(x_0)=0$ and $f''(x_0)>0$. Points x_0 for which $f'(x_0)=0$ are called stationary points.

3.4.3 (2) *Functions of Two Variables*

The function $f(x, y)$ has a maximum or minimum for those values of (x_0, y_0) for which

$$\frac{\partial f}{\partial x}=0, \frac{\partial f}{\partial y}=0,$$

and for which $\begin{vmatrix} \partial^2 f/\partial x \partial y & \partial^2 f/\partial x^2 \\ \partial^2 f/\partial y^2 & \partial^2 f/\partial x \partial y \end{vmatrix}<0$;

(a) $f(x, y)$ has a maximum

$$\text{if } \frac{\partial^2 f}{\partial x^2}<0 \text{ and } \frac{\partial^2 f}{\partial y^2}<0 \text{ at } (x_0, y_0),$$

(b) $f(x, y)$ has a minimum

$$\text{if } \frac{\partial^2 f}{\partial x^2}>0 \text{ and } \frac{\partial^2 f}{\partial y^2}>0 \text{ at } (x_0, y_0).$$

3.5. Absolute and Relative Errors

(1) If x_0 is an approximation to the true value of x, then

3.5.1 (a) the *absolute error* of x_0 is $\Delta x=x_0-x$, $x-x_0$ is the correction to x.

3.5.2 (b) the *relative error* of x_0 is $\delta x=\dfrac{\Delta x}{x}\approx\dfrac{\Delta x}{x_0}$

3.5.3 (c) the *percentage error* is 100 times the relative error.

3.5.4 (2) The absolute error of the sum or difference of several numbers is at most equal to the sum of the absolute errors of the individual numbers.

3.5.5 (3) If $f(x_1, x_2, \ldots, x_n)$ is a function of $x_1, x_2, \ldots, x_n$ and the absolute error in x_i ($i=1, 2, \ldots n$) is Δx_i, then the absolute error in f is

$$\Delta f\approx\frac{\partial f}{\partial x_1}\Delta x_1+\frac{\partial f}{\partial x_2}\Delta x_2+\ldots+\frac{\partial f}{\partial x_n}\Delta x_n$$

3.5.6 (4) The relative error of the product or quotient of several factors is at most equal to the sum of the relative errors of the individual factors.

3.5.7

(5) If $y=f(x)$, the relative error $\delta y=\dfrac{\Delta y}{y}\approx\dfrac{f'(x)}{f(x)}\Delta x$

Approximate Values

If $|\epsilon|<<1, |\eta|<<1, b<<a,$

3.5.8 $$(a+b)^k\approx a^k+ka^{k-1}b$$

3.5.9 $$(1+\epsilon)(1+\eta)\approx 1+\epsilon+\eta$$

3.5.10 $$\frac{1+\epsilon}{1+\eta}\approx 1+\epsilon-\eta$$

3.6. Infinite Series

Taylor's Formula for a Single Variable

3.6.1

$$f(x+h)=f(x)+hf'(x)+\frac{h^2}{2!}f''(x)$$
$$+\ldots+\frac{h^{n-1}}{(n-1)!}f^{(n-1)}(x)+R_n$$

3.6.2

$$R_n=\frac{h^n}{n!}f^{(n)}(x+\theta_1 h)=\frac{h^n}{(n-1)!}(1-\theta_2)^{n-1}f^{(n)}(x+\theta_2 h)$$
$$(0<\theta_{1,2}(x)<1)$$

3.6.3

$$=\frac{h^n}{(n-1)!}\int_0^1(1-t)^{n-1}f^{(n)}(x+th)dt$$

3.6.4

$$f(x)=f(a)+\frac{(x-a)}{1!}f'(a)+\frac{(x-a)^2}{2!}f''(a)+$$
$$\ldots+\frac{(x-a)^{n-1}}{(n-1)!}f^{(n-1)}(a)+R_n$$

3.6.5 $$R_n=\frac{(x-a)^n}{n!}f^{(n)}(\xi) \qquad (a<\xi<x)$$

Lagrange's Expansion

If $y=f(x)$, $y_0=f(x_0)$, $f'(x_0)\neq 0$, then

3.6.6

$$x=x_0+\sum_{k=1}^{\infty}\frac{(y-y_0)^k}{k!}\left[\frac{d^{k-1}}{dx^{k-1}}\left\{\frac{x-x_0}{f(x)-y_0}\right\}^k\right]_{x=x_0}$$

3.6.7

$$g(x)=g(x_0)$$

$$+\sum_{k=1}^{\infty}\frac{(y-y_0)^k}{k!}\left[\frac{d^{k-1}}{dx^{k-1}}\left(g'(x)\left\{\frac{x-x_0}{f(x)-y_0}\right\}^k\right)\right]_{x=x_0}$$

where $g(x)$ is any function indefinitely differentiable.

Binomial Series

3.6.8

$$(1+x)^{\alpha}=\sum_{k=0}^{\infty}\binom{\alpha}{k}x^k \qquad (-1<x<1)$$

3.6.9

$$(1+x)^\alpha = 1 + \alpha x + \frac{\alpha(\alpha-1)}{2!} x^2 + \frac{\alpha(\alpha-1)(\alpha-2)}{3!} x^3 + \cdots,$$

3.6.10

$$(1+x)^{-1} = 1 - x + x^2 - x^3 + x^4 - \cdots \qquad (-1 < x < 1)$$

3.6.11

$$(1+x)^{\frac{1}{2}} = 1 + \frac{x}{2} - \frac{x^2}{8} + \frac{x^3}{16} - \frac{5x^4}{128} + \frac{7x^5}{256} - \frac{21x^6}{1024} + \cdots$$
$$(-1 < x < 1)$$

3.6.12

$$(1+x)^{-\frac{1}{2}} = 1 - \frac{x}{2} + \frac{3x^2}{8} - \frac{5x^3}{16} + \frac{35x^4}{128} - \frac{63x^5}{256}$$
$$+ \frac{231x^6}{1024} - \cdots \qquad (-1 < x < 1)$$

3.6.13

$$(1+x)^{\frac{1}{3}} = 1 + \frac{1}{3} x - \frac{1}{9} x^2 + \frac{5}{81} x^3 - \frac{10}{243} x^4$$
$$+ \frac{22}{729} x^5 - \frac{154}{6561} x^6 + \cdots \qquad (-1 < x < 1)$$

3.6.14

$$(1+x)^{-\frac{1}{3}} = 1 - \frac{1}{3} x + \frac{2}{9} x^2 - \frac{14}{81} x^3 + \frac{35}{243} x^4$$
$$- \frac{91}{729} x^5 + \frac{728}{6561} x^6 - \cdots \qquad (-1 < x < 1)$$

Asymptotic Expansions

3.6.15 A series $\sum_{k=0}^{\infty} a_k x^{-k}$ is said to be an asymptotic expansion of a function $f(x)$ if

$$f(x) - \sum_{k=0}^{n-1} a_k x^{-k} = O(x^{-n}) \text{ as } x \to \infty$$

for every $n = 1, 2, \ldots$. We write

$$f(x) \sim \sum_{k=0}^{\infty} a_k x^{-k}.$$

The series itself may be either convergent **or** divergent.

Operations With Series

$$\text{Let } s_1 = 1 + a_1 x + a_2 x^2 + a_3 x^3 + a_4 x^4 + \cdots$$
$$s_2 = 1 + b_1 x + b_2 x^2 + b_3 x^3 + b_4 x^4 + \cdots$$
$$s_3 = 1 + c_1 x + c_2 x^2 + c_3 x^3 + c_4 x^4 + \cdots$$

	Operation	c_1	c_2	c_3	c_4
3.6.16	$s_3 = s_1^{-1}$	$-a_1$	$a_1^2 - a_2$	$2a_1 a_2 - a_3 - a_1^3$	$2a_1 a_3 - 3a_1^2 a_2 - a_4 + a_2^2 + a_1^4$
3.6.17	$s_3 = s_1^{-2}$	$-2a_1$	$3a_1^2 - 2a_2$	$6a_1 a_2 - 2a_3 - 4a_1^3$	$6a_1 a_3 + 3a_2^2 - 2a_4 - 12a_1^2 a_2 + 5a_1^4$
3.6.18	$s_3 = s_1^{\frac{1}{2}}$	$\frac{1}{2} a_1$	$\frac{1}{2} a_2 - \frac{1}{8} a_1^2$	$\frac{1}{2} a_3 - \frac{1}{4} a_1 a_2 + \frac{1}{16} a_1^3$	$\frac{1}{2} a_4 - \frac{1}{4} a_1 a_3 - \frac{1}{8} a_2^2 + \frac{3}{16} a_1^2 a_2 - \frac{5}{128} a_1^4$
3.6.19	$s_3 = s_1^{-\frac{1}{2}}$	$-\frac{1}{2} a_1$	$\frac{3}{8} a_1^2 - \frac{1}{2} a_2$	$\frac{3}{4} a_1 a_2 - \frac{1}{2} a_3 - \frac{5}{16} a_1^3$	$\frac{3}{4} a_1 a_3 + \frac{3}{8} a_2^2 - \frac{1}{2} a_4 - \frac{15}{16} a_1^2 a_2 + \frac{35}{128} a_1^4$
3.6.20	$s_3 = s_1^n$	na_1	$\frac{1}{2}(n-1)c_1 a_1 + na_2$ *	$c_1 a_2 (n-1) + \frac{1}{6} c_1 a_1^2 (n-1)(n-2) + na_3$ *	$na_4 + c_1 a_3 (n-1) + \frac{1}{2} n(n-1) a_2^2 + \frac{1}{2}(n-1)(n-2) c_1 a_1 a_2 + \frac{1}{24}(n-1)(n-2)(n-3) c_1 a_1^3$
3.6.21	$s_3 = s_1 s_2$	$a_1 + b_1$	$b_2 + a_1 b_1 + a_2$	$b_3 + a_1 b_2 + a_2 b_1 + a_3$	$b_4 + a_1 b_3 + a_2 b_2 + a_3 b_1 + a_4$
3.6.22	$s_3 = s_1 / s_2$	$a_1 - b_1$	$a_2 - (b_1 c_1 + b_2)$	$a_3 - (b_1 c_2 + b_2 c_1 + b_3)$	$a_4 - (b_1 c_3 + b_2 c_2 + b_3 c_1 + b_4)$
3.6.23	$s_3 = \exp (s_1 - 1)$	a_1	$a_2 + \frac{1}{2} a_1^2$	$a_3 + a_1 a_2 + \frac{1}{6} a_1^3$	$a_4 + a_1 a_3 + \frac{1}{2} a_2^2 + \frac{1}{2} a_2 a_1^2 + \frac{1}{24} a_1^4$
3.6.24	$s_3 = 1 + \ln s_1$	a_1	$a_2 - \frac{1}{2} a_1 c_1$	$a_3 - \frac{1}{3}(a_2 c_1 + 2a_1 c_2)$	$a_4 - \frac{1}{4}(a_3 c_1 + 2a_2 c_2 + 3a_1 c_3)$ *

*See page II.

Reversion of Series

3.6.25 Given

$$y = ax + bx^2 + cx^3 + dx^4 + ex^5 + fx^6 + gx^7 + \cdots$$

then

$$x = Ay + By^2 + Cy^3 + Dy^4 + Ey^5 + Fy^6 + Gy^7 + \cdots$$

where

$$aA = 1$$
$$a^3 B = -b$$
$$a^5 C = 2b^2 - ac$$
$$a^7 D = 5abc - a^2 d - 5b^3$$
$$a^9 E = 6a^2 bd + 3a^2 c^2 + 14b^4 - a^3 e - 21ab^2 c$$
$$a^{11} F = 7a^3 be + 7a^3 cd + 84ab^3 c - a^4 f \\ - 28a^2 bc^2 - 42b^5 - 28a^2 b^2 d$$
$$a^{13} G = 8a^4 bf + 8a^4 ce + 4a^4 d^2 + 120a^2 b^3 d \\ + 180a^2 b^2 c^2 + 132b^6 - a^5 g - 36a^3 b^2 e \\ - 72a^3 bcd - 12a^3 c^3 - 330ab^4 c$$

Kummer's Transformation of Series

3.6.26 Let $\sum_{k=0}^{\infty} a_k = s$ be a given convergent series and $\sum_{k=0}^{\infty} c_k = c$ be a given convergent series with known sum c such that $\lim_{k \to \infty} \frac{a_k}{c_k} = \lambda \neq 0$. Then

$$s = \lambda c + \sum_{k=0}^{\infty} \left(1 - \lambda \frac{c_k}{a_k}\right) a_k.$$

Euler's Transformation of Series

3.6.27 If $\sum_{k=0}^{\infty} (-1)^k a_k = a_0 - a_1 + a_2 - \cdots$ is a convergent series with sum s then

$$s = \sum_{k=0}^{\infty} \frac{(-1)^k \Delta^k a_0}{2^{k+1}}, \quad \Delta^k a_0 = \sum_{m=0}^{k} (-1)^m \binom{k}{m} a_{k-m}$$

Euler-Maclaurin Summation Formula

3.6.28

$$\sum_{k=1}^{n-1} f_k = \int_0^n f(k)dk - \frac{1}{2}[f(0) + f(n)] + \frac{1}{12}[f'(n) - f'(0)]$$
$$- \frac{1}{720}[f'''(n) - f'''(0)] + \frac{1}{30240}[f^{(V)}(n) - f^{(V)}(0)]$$
$$- \frac{1}{1209600}[f^{(VII)}(n) - f^{(VII)}(0)] + \cdots$$

3.7. Complex Numbers and Functions

Cartesian Form

3.7.1
$$z = x + iy$$

Polar Form

3.7.2
$$z = re^{i\theta} = r(\cos\theta + i\sin\theta)$$

3.7.3
$$\text{Modulus: } |z| = (x^2 + y^2)^{\frac{1}{2}} = r$$

3.7.4 *Argument:* $\arg z = \arctan(y/x) = \theta$ (other notations for $\arg z$ are am z and ph z).

3.7.5
$$\text{Real Part: } x = \mathscr{R}z = r\cos\theta$$

3.7.6
$$\text{Imaginary Part: } y = \mathscr{I}z = r\sin\theta$$

Complex Conjugate of z

3.7.7
$$\bar{z} = x - iy$$

3.7.8
$$|\bar{z}| = |z|$$

3.7.9
$$\arg \bar{z} = -\arg z$$

Multiplication and Division

If $z_1 = x_1 + iy_1$, $z_2 = x_2 + iy_2$, then

3.7.10
$$z_1 z_2 = x_1 x_2 - y_1 y_2 + i(x_1 y_2 + x_2 y_1)$$

3.7.11
$$|z_1 z_2| = |z_1||z_2|$$

3.7.12
$$\arg(z_1 z_2) = \arg z_1 + \arg z_2$$

3.7.13
$$\frac{z_1}{z_2} = \frac{z_1 \bar{z}_2}{|z_2|^2} = \frac{x_1 x_2 + y_1 y_2 + i(x_2 y_1 - x_1 y_2)}{x_2^2 + y_2^2}$$

3.7.14
$$\left|\frac{z_1}{z_2}\right| = \frac{|z_1|}{|z_2|}$$

3.7.15
$$\arg\left(\frac{z_1}{z_2}\right) = \arg z_1 - \arg z_2$$

Powers

3.7.16
$$z^n = r^n e^{in\theta}$$

3.7.17
$$= r^n \cos n\theta + ir^n \sin n\theta \\ (n = 0, \pm1, \pm2, \ldots)$$

3.7.18
$$z^2 = x^2 - y^2 + i(2xy)$$

3.7.19
$$z^3 = x^3 - 3xy^2 + i(3x^2 y - y^3)$$

3.7.20
$$z^4 = x^4 - 6x^2 y^2 + y^4 + i(4x^3 y - 4xy^3)$$

3.7.21
$$z^5 = x^5 - 10x^3 y^2 + 5xy^4 + i(5x^4 y - 10x^2 y^3 + y^5)$$

3.7.22

$$z^n = \left[x^n - \binom{n}{2}x^{n-2}y^2 + \binom{n}{4}x^{n-4}y^4 - \cdots\right]$$
$$+ i\left[\binom{n}{1}x^{n-1}y - \binom{n}{3}x^{n-3}y^3 + \cdots\right],$$
$$(n = 1, 2, \ldots)$$

If $z^n = u_n + iv_n$, then $z^{n+1} = u_{n+1} + iv_{n+1}$ where

3.7.23 $\quad u_{n+1} = xu_n - yv_n;\ v_{n+1} = xv_n + yu_n$

$\mathscr{R}\, z^n$ and $\mathscr{I}\, z^n$ are called harmonic polynomials.

3.7.24
$$\frac{1}{z} = \frac{\bar{z}}{|z|^2} = \frac{x - iy}{x^2 + y^2}$$

3.7.25
$$\frac{1}{z^n} = \frac{\bar{z}^n}{|z|^{2n}} = (z^{-1})^n$$

Roots

3.7.26 $\quad z^{\frac{1}{2}} = \sqrt{z} = r^{\frac{1}{2}} e^{\frac{1}{2}i\theta} = r^{\frac{1}{2}} \cos \frac{1}{2}\theta + ir^{\frac{1}{2}} \sin \frac{1}{2}\theta$

If $-\pi < \theta \le \pi$ this is the principal root. The other root has the opposite sign. The principal root is given by

3.7.27 $\quad z^{\frac{1}{2}} = [\frac{1}{2}(r+x)]^{\frac{1}{2}} \pm i[\frac{1}{2}(r-x)]^{\frac{1}{2}} = u \pm iv$ where $2uv = y$ and where the ambiguous sign is taken to be the same as the sign of y.

3.7.28 $\quad z^{1/n} = r^{1/n} e^{i\theta/n}$, (principal root if $-\pi < \theta \le \pi$). Other roots are $r^{1/n} e^{i(\theta + 2\pi k)/n}$ $(k = 1, 2, 3, \ldots, n-1)$.

Inequalities

3.7.29
$$\left| |z_1| - |z_2| \right| \le |z_1 \pm z_2| \le |z_1| + |z_2|$$

Complex Functions, Cauchy-Riemann Equations

$f(z) = f(x + iy) = u(x, y) + iv(x, y)$ where $u(x, y), v(x, y)$ are real, is *analytic* at those points $z = x + iy$ at which

3.7.30
$$\frac{\partial u}{\partial x} = \frac{\partial v}{\partial y},\ \frac{\partial u}{\partial y} = -\frac{\partial v}{\partial x}$$

If $z = re^{i\theta}$,

3.7.31
$$\frac{\partial u}{\partial r} = \frac{1}{r}\frac{\partial v}{\partial \theta},\ \frac{1}{r}\frac{\partial u}{\partial \theta} = -\frac{\partial v}{\partial r}$$

Laplace's Equation

The functions $u(x, y)$ and $v(x, y)$ are called harmonic functions and satisfy Laplace's equation:

Cartesian Coordinates

3.7.32
$$\frac{\partial^2 u}{\partial x^2} + \frac{\partial^2 u}{\partial y^2} = \frac{\partial^2 v}{\partial x^2} + \frac{\partial^2 v}{\partial y^2} = 0$$

Polar Coordinates

3.7.33 $\quad r\frac{\partial}{\partial r}\left(r\frac{\partial u}{\partial r}\right) + \frac{\partial^2 u}{\partial \theta^2} = r\frac{\partial}{\partial r}\left(r\frac{\partial v}{\partial r}\right) + \frac{\partial^2 v}{\partial \theta^2} = 0$

3.8. Algebraic Equations

Solution of Quadratic Equations

3.8.1 Given $az^2 + bz + c = 0$,

$$z_{1,2} = -\left(\frac{b}{2a}\right) \pm \frac{1}{2a} q^{\frac{1}{2}}, q = b^2 - 4ac,$$

$$z_1 + z_2 = -b/a,\ z_1 z_2 = c/a$$

If $q > 0$, two real roots,
$q = 0$, two equal roots,
$q < 0$, pair of complex conjugate roots.

Solution of Cubic Equations

3.8.2 Given $z^3 + a_2 z^2 + a_1 z + a_0 = 0$, let

$$q = \frac{1}{3} a_1 - \frac{1}{9} a_2^2;\ r = \frac{1}{6}(a_1 a_2 - 3a_0) - \frac{1}{27} a_2^3.$$

If $q^3 + r^2 > 0$, one real root and a pair of complex conjugate roots,

$q^3 + r^2 = 0$, all roots real and at least two are equal,

$q^3 + r^2 < 0$, all roots real (irreducible case).

Let

$$s_1 = [r + (q^3 + r^2)^{\frac{1}{2}}]^{\frac{1}{3}},\ s_2 = [r - (q^3 + r^2)^{\frac{1}{2}}]^{\frac{1}{3}}$$

then

$$z_1 = (s_1 + s_2) - \frac{a_2}{3}$$

$$z_2 = -\frac{1}{2}(s_1 + s_2) - \frac{a_2}{3} + \frac{i\sqrt{3}}{2}(s_1 - s_2)$$

$$z_3 = -\frac{1}{2}(s_1 + s_2) - \frac{a_2}{3} - \frac{i\sqrt{3}}{2}(s_1 - s_2).$$

If z_1, z_2, z_3 are the roots of the cubic equation

$$z_1 + z_2 + z_3 = -a_2$$

$$z_1 z_2 + z_1 z_3 + z_2 z_3 = a_1$$

$$z_1 z_2 z_3 = -a_0$$

Solution of Quartic Equations

3.8.3 Given $z^4 + a_3 z^3 + a_2 z^2 + a_1 z + a_0 = 0$, find the real root u_1 of the cubic equation

$$u^3 - a_2 u^2 + (a_1 a_3 - 4a_0)u - (a_1^2 + a_0 a_3^2 - 4a_0 a_2) = 0$$

and determine the four roots of the quartic as solutions of the two quadratic equations

$$v^2 + \left[\frac{a_3}{2} \mp \left(\frac{a_3^2}{4} + u_1 - a_2\right)^{\frac{1}{2}}\right] v + \frac{u_1}{2} \mp \left[\left(\frac{u_1}{2}\right)^2 - a_0\right]^{\frac{1}{2}} = 0$$

If all roots of the cubic equation are real, use the value of u_1 which gives real coefficients in the *quadratic equation and select signs so that if

$$z^4+a_3z^3+a_2z^2+a_1z+a_0=(z^2+p_1z+q_1)(z^2+p_2z+q_2),$$

then

$$p_1+p_2=a_3, \; p_1p_2+q_1+q_2=a_2, \; p_1q_2+p_2q_1=a_1, \; q_1q_2=a_0.$$

If z_1, z_2, z_3, z_4 are the roots,

$$\Sigma z_i=-a_3, \; \Sigma z_iz_jz_k=-a_1,$$

$$\Sigma z_iz_j=a_2, \; z_1z_2z_3z_4=a_0.$$

3.9. Successive Approximation Methods

General Comments

3.9.1 Let $x=x_1$ be an approximation to $x=\xi$ where $f(\xi)=0$ and both x_1 and ξ are in the interval $a\leq x\leq b$. We define

$$x_{n+1}=x_n+c_nf(x_n) \qquad (n=1, 2, \ldots).$$

Then, if $f'(x)\geq0$ and the constants c_n are negative and bounded, the sequence x_n converges monotonically to the root ξ.

If $c_n=c=\text{constant}<0$ and $f'(x)>0$, then the process converges but not necessarily monotonically.

Degree of Convergence of an Approximation Process

3.9.2 Let x_1, x_2, x_3, $\ldots$ be an infinite sequence of approximations to a number ξ. Then, if

$$|x_{n+1}-\xi|<A|x_n-\xi|^k, \qquad (n=1, 2, \ldots)$$

where A and k are independent of n, the sequence is said to have convergence of at most the kth degree (or order or index) to ξ. If $k=1$ and $A<1$ the convergence is linear; if $k=2$ the convergence is quadratic.

Regula Falsi (False Position)

3.9.3 Given $y=f(x)$ to find ξ such that $f(\xi)=0$, choose x_0 and x_1 such that $f(x_0)$ and $f(x_1)$ have opposite signs and compute

$$x_2=x_1-\frac{(x_1-x_0)}{(f_1-f_0)} \, f_1=\frac{f_1x_0-f_0x_1}{f_1-f_0}.$$

Then continue with x_2 and either of x_0 or x_1 for which $f(x_0)$ or $f(x_1)$ is of opposite sign to $f(x_2)$.

Regula falsi is equivalent to inverse linear interpolation.

Method of Iteration (Successive Substitution)

3.9.4 The iteration scheme $x_{k+1}=F(x_k)$ will converge to a zero of $x=F(x)$ if

(1) $\quad |F'(x)|\leq q<1$ for $a\leq x\leq b$,

(2) $\quad a\leq x_0\pm\dfrac{|F(x_0)-x_0|}{1-q}\leq b$.

Newton's Method of Successive Approximations

3.9.5

Newton's Rule

If $x=x_k$ is an approximation to the solution $x=\xi$ of $f(x)=0$ then the sequence

$$x_{k+1}=x_k-\frac{f(x_k)}{f'(x_k)}$$

will converge quadratically to $x=\xi$: (if instead of the condition (2) above),

(1) *Monotonic convergence*, $f(x_0)f''(x_0)>0$ and $f'(x)$, $f''(x)$ do not change sign in the interval (x_0, ξ), or

(2) *Oscillatory convergence*, $f(x_0)f''(x_0)<0$ and $f'(x)$, $f''(x)$ do not change sign in the interval (x_0, x_1), $x_0\leq\xi\leq x_1$.

Newton's Method Applied to Real nth Roots

3.9.6 Given $x^n=N$, if x_k is an approximation $x=N^{1/n}$ then the sequence

$$x_{k+1}=\frac{1}{n}\left[\frac{N}{x_k^{n-1}}+(n-1)x_k\right]$$

will converge quadratically to x.

If $n=2$, $x_{k+1}=\dfrac{1}{2}\left(\dfrac{N}{x_k}+x_k\right)$,

If $n=3$, $x_{k+1}=\dfrac{1}{3}\left(\dfrac{N}{x_k^2}+2x_k\right)$.

Aitken's δ^2-Process for Acceleration of Sequences

3.9.7 If x_k, x_{k+1}, x_{k+2} are three successive iterates in a sequence converging with an error which is approximately in geometric progression, then

$$\bar{x}_k=x_k-\frac{(x_k-x_{k+1})^2}{\Delta^2 x_k}=\frac{x_kx_{k+2}-x_{k+1}^2}{\Delta^2 x_k};$$

$$\Delta^2 x_k=x_k-2x_{k+1}+x_{k+2}$$

is an improved estimate of x. In fact, if $x_k=x+$ *$O(\lambda^k)$ then $\bar{x}=x+O(\lambda^k)$, $|\lambda|<1$.

---

*See page II.

3.10. Theorems on Continued Fractions

Definitions

3.10.1

(1)   Let
$$f=b_0+\cfrac{a_1}{b_1+\cfrac{a_2}{b_2+\cfrac{a_3}{b_3+}\cdots}}$$

$$=b_0+\cfrac{a_1}{b_1+}\ \cfrac{a_2}{b_2+}\ \cfrac{a_3}{b_3+}\cdots$$

If the number of terms is finite, f is called a terminating continued fraction. If the number of terms is infinite, f is called an infinite continued fraction and the terminating fraction

$$f_n=\frac{A_n}{B_n}=b_0+\cfrac{a_1}{b_1+}\ \cfrac{a_2}{b_2+}\cdots\cfrac{a_n}{b_n}$$

is called the nth convergent of f.

(2)   If $\lim\limits_{n\to\infty}\dfrac{A_n}{B_n}$ exists, the infinite continued fraction f is said to be convergent. If $a_i=1$ and the b_i are integers there is always convergence.

Theorems

(1)   If a_i and b_i are positive then $f_{2n}<f_{2n+2}$, $f_{2n-1}>f_{2n+1}$.

(2)   If $f_n=\dfrac{A_n}{B_n}$,
$$A_n=b_nA_{n-1}+a_nA_{n-2}$$
$$B_n=b_nB_{n-1}+a_nB_{n-2}$$
where $A_{-1}=1$, $A_0=b_0$, $B_{-1}=0$, $B_0=1$.

(3)
$$\begin{bmatrix}A_n\\B_n\end{bmatrix}=\begin{bmatrix}A_{n-1}&A_{n-2}\\B_{n-1}&B_{n-2}\end{bmatrix}\begin{bmatrix}b_n\\a_n\end{bmatrix}$$

(4)
$$A_nB_{n-1}-A_{n-1}B_n=(-1)^{n-1}\prod_{k=1}^{n}a_k$$

(5)   For every $n\ge0$,
$$f_n=b_0+\cfrac{c_1a_1}{c_1b_1+}\ \cfrac{c_1c_2a_2}{c_2b_2+}\ \cfrac{c_2c_3a_3}{c_3b_3+}\cdots\cfrac{c_{n-1}c_na_n}{c_nb_n}.$$

(6) $1+b_2+b_2b_3+\ldots+b_2b_3\ldots b_n$
$$=\cfrac{1}{1-}\ \cfrac{b_2}{b_2+1-}\ \cfrac{b_3}{b_3+1-}\cdots\cfrac{b_n}{-b_n+1}$$

$$\frac{1}{u_1}+\frac{1}{u_2}+\ldots+\frac{1}{u_n}=\cfrac{1}{u_1-}\ \cfrac{u_1^2}{u_1+u_2-}\cdots\cfrac{u_{n-1}^2}{-u_{n-1}+u_n}$$

$$\frac{1}{a_0}-\frac{x}{a_0a_1}+\frac{x^2}{a_0a_1a_2}\ldots+(\ 1)^n\frac{x^n}{a_0a_1a_2\ldots a_n}$$

$$=\cfrac{1}{a_0+}\ \cfrac{a_0x}{a_1-x+}\ \cfrac{a_1x}{a_2-x+}\cdots\cfrac{a_{n-1}x}{+a_n-x}$$

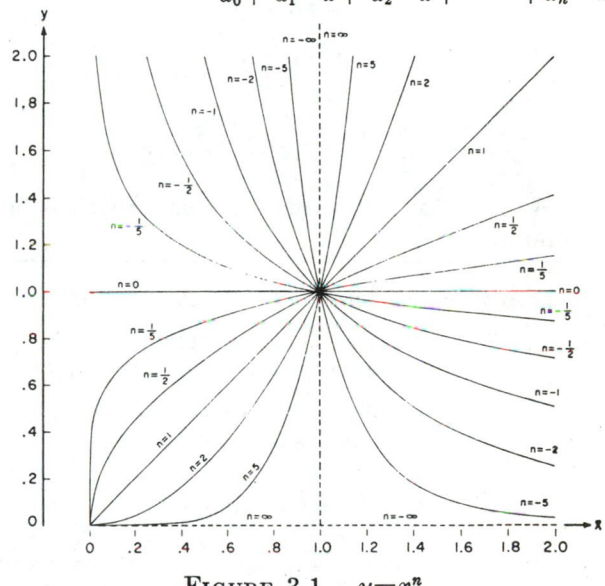

FIGURE 3.1. $y=x^n$.
$$\pm n=0,\ \tfrac{1}{5},\ \tfrac{1}{2},\ 1,\ 2,\ 5.$$

Numerical Methods

3.11. Use and Extension of the Tables

Example 1. Compute x^{19} and x^{47} for $x=29$ using **Table 3.1**.

$x^{19}=x^9\cdot x^{10}$
$\quad=(1.45071\ 4598\cdot10^{13})(4.20707\ 2333\cdot10^{14})$
$\quad=6.10326\ 1248\cdot10^{27}$

$x^{47}=(x^{24})^2/x$
$\quad=(1.25184\ 9008\cdot10^{35})^2/29$
$\quad=5.40388\ 2547\cdot10^{68}$

Example 2.   Compute $x^{-3/4}$ for $x=9.19826$.

$(9.19826)^{1/4}=(919.826/100)^{1/4}=(919.826)^{1/4}/10^{1/2}$

Linear interpolation in **Table 3.1** gives $(919.826)^{1/4}\approx5.507144$.

By Newton's method for fourth roots with $N=919.826$,

$$\frac{1}{4}\left[\frac{919.826}{(5.507144)^3}+3(5.507144)\right]=5.50714\ 3845$$

Repetition yields the same result.   Thus,

$x^{1/4}=5.50714\ 3845/10^{1/2}=1.74151\ 1796,$
$\qquad\qquad x^{-3/4}=x^{1/4}/x=.18933\ 05683.$

3.12. Computing Techniques

Example 3. Solve the quadratic equation $x^2-18.2x+.056$ given the coefficients as $18.2\pm.1$,

.056 ± .001. From **3.8.1** the solution is

$$x = \frac{1}{2}(18.\underline{2} \pm [(18.\underline{2})^2 - 4(.05\underline{6})]^{\frac{1}{2}})$$
$$= \frac{1}{2}(18.\underline{2} \pm [33\underline{1}.016]^{\frac{1}{2}}) = \frac{1}{2}(18.\underline{2} \pm 18.\underline{1939})$$
$$= 18.1\underline{969}, .00\underline{3}$$

The smaller root may be obtained more accurately from

* $.05\underline{6}/18.1\underline{969} = .003\underline{1} \pm .0001.$

Example 4. Compute $(-3 + .0076i)^{\frac{1}{2}}$.

From **3.7.26**, $(-3 + .0076i)^{\frac{1}{2}} = u + iv$ where

$$u = \frac{y}{2v}, \quad v = \left(\frac{r-x}{2}\right)^{\frac{1}{2}}, \quad r = (x^2 + y^2)^{\frac{1}{2}}$$

Thus

$$r = [(-3)^2 + (.0076)^2]^{\frac{1}{2}} = (9.00005776)^{\frac{1}{2}} = 3.00000\ 9627$$

$$v = \left[\frac{3.00000\ 9627 - (-3)}{2}\right]^{\frac{1}{2}} = 1.73205\ 2196$$

$$u = \frac{y}{2v} = \frac{.0076}{2(1.73205\ 2196)} = .00219\ 392926$$

We note that the principal square root has been computed.

Example 5. Solve the cubic equation $x^3 - 18.1x - 34.8 = 0$.

To use Newton's method we first form the table of $f(x) = x^3 - 18.1x - 34.8$

x	$f(x)$
4	-43.2
5	$- .3$
6	72.6
7	181.5

We obtain by linear inverse interpolation:

$$x_0 = 5 + \frac{0 - (-.3)}{72.6 - (-.3)} = 5.004.$$

Using Newton's method, $f'(x) = 3x^2 - 18.1$ we get

$$x_1 \approx x_0 - f(x_0)/f'(x_0)$$

$$\approx 5.004 - \frac{(-.07215\ 9936)}{57.020048} \approx 5.00526.$$

Repetition yields $x_1 = 5.00526\ 5097$. Dividing $f(x)$ by $x - 5.00526\ 5097$ gives $x^2 + 5.00526\ 5097x + 6.95267\ 869$ the zeros of which are $-2.50263\ 2549 \pm .83036\ 800i$.

Example 6. Solve the quartic equation

$$x^4 - 2.37752\ 4922x^3 + 6.07350\ 5741x^2 - 11.17938\ 023x + 9.05265\ 5259 = 0.$$

Resolution Into Quadratic Factors
$$(x^2 + p_1x + q_1)(x^2 + p_2x + q_2)$$
by Inverse Interpolation

Starting with the trial value $q_1 = 1$ we compute successively

q_1	$q_2 = \dfrac{a_0}{q_1}$	$p_1 = \dfrac{a_1 - a_3q_1}{q_2 - q_1}$	$p_2 = a_3 - p_1$	$y(q_1) = q_1 + q_2 + p_1p_2 - a_2$
1	9.053	-1.093	-1.284	5.383
2	4.526	-2.543	.165	.032
2.2	4.115	-3.106	.729	-2.023

We seek that value of q_1 for which $y(q_1) = 0$. Inverse interpolation in $y(q_1)$ gives $y(q_1) \approx 0$ for $q_1 \approx 2.003$. Then,

q_1	q_2	p_1	p_2	$y(q_1)$
2.003	4.520	-2.550	.172	.011

Inverse interpolation between $q_1 = 2.2$ and $q_1 = 2.003$ gives $q_1 = 2.0041$, and thus,

q_1	q_2	p_1	p_2	$y(q_1)$
2.0041	4.51706\ 7640	$-2.55259\ 257$	.17506\ 765	.00078\ 552
2.0042	4.51684\ 2260	$-2.55282\ 851$	.17530\ 358	.00001\ 655
2.0043	4.51661\ 6903	$-2.55306\ 447$	.17553\ 955	$-.00075\ 263$

Inverse interpolation gives $q_1 = 2.00420\ 2152$, and we get finally,

q_1	q_2	p_1	p_2	$y(q_1)$
2.00420\ 2152	4.51683\ 7410	$-2.55283\ 358$	.17530\ 8659	$-.00000\ 0011$

---

*See page II.

Double Precision Multiplication and Division on a Desk Calculator

Example 7. Multiply $M=20243\ 97459\ 71664\ 32102$ by $m=69732\ 82428\ 43662\ 95023$ on a $10\times10\times20$ desk calculating machine.

Let $M_0=20243\ 97459$, $M_1=71664\ 32102$, $m_0=69732\ 82428$, $m_1=43662\ 95023$. Then $Mm=M_0m_010^{20}+(M_0m_1+M_1m_0)10^{10}+M_1m_1$.

(1) Multiply $M_1m_1=31290\ 75681\ 96300\ 28346$ and record the digits $96300\ 28346$ appearing in positions 1 to 10 of the product dial.

(2) Transfer the digits $31290\ 75681$ from positions 11 to 20 of the product dial to positions 1 to 10 of the product dial.

(3) Multiply cumulatively $M_1m_0+M_0m_1+31290\ 75681=58812\ 67160\ 12663\ 25894$ and record the digits $12663\ 25894$ in positions 1 to 10.

(4) Transfer the digits $58812\ 67160$ from positions 11 to 20 to positions 1 to 10.

(5) Multiply cumulatively $M_0m_0+58812\ 67160=14116\ 69523\ 40138\ 17612$. The results as obtained are shown below,

$$96300\ 28346$$
$$12663\ 25894$$
$$14116\ 69523\ 40138\ 17612$$
$$\overline{14116\ 69523\ 40138\ 17612\ 12663\ 25894\ 96300\ 28346}$$

If the product Mm is wanted to 20 digits, only the result obtained in step 5 need be recorded. Further, if the allowable error in the 20th place is a unit, the operation M_1m_1 may be omitted. When either of the factors M or m contains less than 20 digits it is convenient to position the numbers as if they both had 20 digits. This multiplication process may be extended to any higher accuracy desired.

Example 8. Divide $N=14116\ 69523\ 40138\ 17612$ by $d=20243\ 97459\ 71664\ 32102$.

Method (1)—*linear interpolation.*

$$N/20243\ 97459\cdot10^{10}=.69732\ 82430\ 90519\ 39054$$
$$N/20243\ 97460\cdot10^{10}=.69732\ 82427\ 46057\ 26941$$
$$\overline{\text{Difference}=3\ 44462\ 12113.}$$

Difference $\times.71664\ 32102=24685\ 644028\cdot10^{-20}$ (note this is an 11×10 multiplication).

Quotient$=$
$(69732\ 82430\ 90519\ 39054-246856\ 44028)\cdot10^{-20}$
$$=.69732\ 82428\ 43662\ 95026$$

There is an error of 3 units in the 20th place due to neglect of the contribution from second differences.

Method (2)—If N and d are numbers each not more than 19 digits let $N=N_1+N_010^9$, $d=d_1+d_010^9$ where N_0 and d_0 contain 10 digits and N_1 and d_1 not more than 9 digits. Then

$$\frac{N}{d}=\frac{N_010^9+N_1}{d_010^9+d_1}\approx\frac{1}{d_010^9}\left[N-\frac{N_0d_1}{d_0}\right]$$

Here

$N=14116\ 69523\ 40138\ 1761,$
$\qquad\qquad d=20243\ 97459\ 71664\ 3210$
$N_0=14116\ 69523,\ d_0=20243\ 97459,$
$\qquad\qquad\qquad d_1=71664\ 3210$

(1) $N_0d_1=10116\ 63378\ 42188\ 8830$ (product dial).

(2) $(N_0d_1)/d_0=49973\ 55504$ (quotient dial).

(3) $N-(N_0d_1)/d_0=14116\ 69522\ 90164\ 62106$ (product dial).

(4) $[N-(N_0d_1)/d_0]/d_010^9=.69732\ 82428=$ first 10 digits of quotient in quotient dial. Remainder $=r=08839\ 11654$, in positions 1 to 10 of product dial.

(5) $r/(d_010^9)=.43662\ 9502\cdot10^{-10}=$ next 9 digits of quotient. $N/d=.69732\ 82428\ 43662\ 9502$. This method may be modified to give the quotient of 20 digit numbers. Method (1) may be extended to quotients of numbers containing more than 20 digits by employing higher order interpolation.

Example 9. Sum the series $S=1-\frac{1}{2}+\frac{1}{3}-\frac{1}{4}+\ \ldots$ to 5D using the Euler transform.

The sum of the first 8 terms is .634524 to 6D. If $u_n=1/n$ we get

n	u_n	Δu_n	$\Delta^2 u_n$	$\Delta^3 u_n$	$\Delta^4 u_n$
9	.111111				
		-11111			
10	.100000		2020		
		-9091		-505	
11	.090909		1515		156
		-7576		-349	
12	.083333		1166		
		-6410			
13	.076923				

From **3.6.27** we then obtain

$$S=.634524+\frac{.111111}{2}-\frac{(-.011111)}{2^2}+\frac{.002020}{2^3}$$
$$-\frac{(-.000505)}{2^4}+\frac{.000156}{2^5}$$

$$=.634524+.055556+.002778+.000253$$
$$+.000032+.000005$$
$$=.693148$$

$(S=\ln 2=.6931472$ to 7D$)$.

Example 10. Evaluate the integral $\int_0^\infty \dfrac{\sin x}{x}\,dx$ $=\dfrac{\pi}{2}$ to 4D using the Euler transform.

$$\int_0^\infty \frac{\sin x}{x}\,dx = \sum_{k=0}^\infty \int_{k\pi}^{(k+1)\pi} \frac{\sin x}{x}\,dx$$

$$= \sum_{k=0}^\infty \int_0^\pi \frac{\sin(k\pi+t)}{k\pi+t}\,dt = \sum_{k=0}^\infty (-1)^k \int_0^\pi \frac{\sin t}{k\pi+t}\,dt.$$

Evaluating the integrals in the last sum by numerical integration we get

k	$\int_0^\pi \dfrac{\sin t}{k\pi+t}\,dt$	Δ	Δ^2	Δ^3	Δ^4
0	1.85194				
1	.43379				
2	.25661				
3	.18260				
4	.14180				
		-2587			
5	.11593		799		
		-1788		-321	
6	.09805		478		153
		-1310		-168	
7	.08495		310		
		-1000			
8	.07495				

The sum to $k=3$ is 1.49216. Applying the Euler transform to the remainder we obtain

$$\frac{1}{2}\,(.14180) - \frac{1}{2^2}\,(-.02587) + \frac{1}{2^3}\,(.00799)$$

$$-\frac{1}{2^4}\,(-.00321) + \frac{1}{2^5}\,(.00153)$$

$$= .07090 + .00647 + .00100 + .00020$$
$$+ .00005$$

$$= .07862$$

We obtain the value of the integral as 1.570<u>78</u> as compared with 1.57080.

Example 11. Sum the series $\sum_{k=1}^\infty k^{-2} = \dfrac{\pi^2}{6}$ using the Euler-Maclaurin summation formula.

From **3.6.28** we have for $n = \infty$,

$$\sum_{k=1}^\infty k^{-2} = \sum_{k=1}^{10} k^{-2} + \sum_{k=1}^\infty (k+10)^{-2}$$

$$= \sum_{k=1}^{10} k^{-2} + \int_0^\infty f(k)\,dk - \frac{1}{2} f_0 - \frac{1}{12} f_0'$$

$$+ \frac{1}{720} f_0''' - \cdots$$

where $f(k) = (k+10)^{-2}$. Thus,

$$\sum_{k=1}^\infty k^{-2} = 1.54976\ 7731 + .1$$
$$-.005 + .00016\ 6667 - .00000\ 0333$$
$$= 1.64493\ 4065,$$

as compared with $\dfrac{\pi^2}{6} = 1.64493\ 4067$.

Example 12. Compute

$$\arctan x = \frac{x}{1+}\ \frac{x^2}{3+}\ \frac{4x^2}{5+}\ \frac{9x^2}{7+}\ \cdots$$

to 5D for $x = .2$. Here $a_1 = x$, $a_n = (n-1)^2 x^2$ for $n>1$, $b_0 = 0$, $b_n = 2n-1$, $A_{-1} = 1$, $B_{-1} = 0$, $A_0 = 0$, $B_0 = 1$.

For $n \geq 1$

$$\begin{bmatrix} A_n \\ B_n \end{bmatrix} = \begin{vmatrix} A_{n-1} A_{n-2} \\ B_{n-1} B_{n-2} \end{vmatrix} \begin{bmatrix} 2n-1 \\ (n-1)^2 x^2 \end{bmatrix} \qquad \frac{A_0}{B_0} = 0$$

$$\begin{bmatrix} A_1 \\ B_1 \end{bmatrix} = \begin{vmatrix} 0 & 1 \\ 1 & 0 \end{vmatrix} \begin{Vmatrix} 1 \\ .2 \end{Vmatrix} = \begin{vmatrix} .2 \\ 1 \end{vmatrix} \qquad \frac{A_1}{B_1} = .2$$

$$\begin{bmatrix} A_2 \\ B_2 \end{bmatrix} = \begin{vmatrix} .2 & 0 \\ 1 & 1 \end{vmatrix} \begin{Vmatrix} 3 \\ .04 \end{Vmatrix} = \begin{vmatrix} .6 \\ 3.04 \end{vmatrix} \qquad \frac{A_2}{B_2} = .197368$$

$$\begin{bmatrix} A_3 \\ B_3 \end{bmatrix} = \begin{vmatrix} .6 & .2 \\ 3.04 & 1 \end{vmatrix} \begin{Vmatrix} 5 \\ .16 \end{Vmatrix} = \begin{vmatrix} 3.032 \\ 15.36 \end{vmatrix} \qquad \frac{A_3}{B_3} = .197396$$

$$\begin{bmatrix} A_4 \\ B_4 \end{bmatrix} = \begin{vmatrix} 3.032 & .6 \\ 15.36 & 3.04 \end{vmatrix} \begin{Vmatrix} 7 \\ .36 \end{Vmatrix} = \begin{vmatrix} 21.440 \\ 108.6144 \end{vmatrix} \qquad \frac{A_4}{B_4} = .197396$$

Note that in carrying out the recurrence method for computing continued fractions the numerators A_n and the denominators B_n must be used as originally computed. The numerators and denominators obtained by reducing A_n/B_n to lower terms must not be used.

References

Texts

[3.1] R. A. Buckingham, Numerical methods (Pitman Publishing Corp., New York, N.Y., 1957).

[3.2] T. Fort, Finite differences (Clarendon Press, Oxford, England, 1948).

[3.3] L. Fox, The use and construction of mathematical tables, Mathematical Tables, vol. 1, National Physical Laboratory (Her Majesty's Stationery Office, London, England, 1956).

[3.4] G. H. Hardy, A course of pure mathematics, 9th ed. (Cambridge Univ. Press, Cambridge, England, and The Macmillan Co., New York, N.Y., 1947).

[3.5] D. R. Hartree, Numerical analysis (Clarendon Press, Oxford, England, 1952).

[3.6] F. B. Hildebrand, Introduction to numerical analysis (McGraw-Hill Book Co., Inc., New York, N.Y., 1956).

[3.7] A. S. Householder, Principles of numerical analysis (McGraw-Hill Book Co., Inc., New York, N.Y., 1953).

[3.8] L. V. Kantorowitsch and V. I. Krylow, Näherungsmethoden der Höheren Analysis (VEB Deutscher Verlag der Wissenschaften, Berlin, Germany, 1956; translated from Russian, Moscow, U.S.S.R., 1952).

[3.9] K. Knopp, Theory and application of infinite series (Blackie and Son, Ltd., London, England, 1951).

[3.10] Z. Kopal, Numerical analysis (John Wiley & Sons, Inc., New York, N.Y., 1955).

[3.11] G. Kowalewski, Interpolation und genäherte Quadratur (B. G. Teubner, Leipzig, Germany, 1932).

[3.12] K. S. Kunz, Numerical analysis (McGraw-Hill Book Co., Inc., New York, N.Y., 1957).

[3.13] C. Lanczos, Applied analysis (Prentice-Hall, Inc., Englewood Cliffs, N.J., 1956).

[3.14] I. M. Longman, Note on a method for computing infinite integrals of oscillatory functions, Proc. Cambridge Philos. Soc. 52, 764 (1956).

[3.15] S. E. Mikeladze, Numerical methods of mathematical analysis (Russian) (Gos. Izdat. Tehn.-Teor. Lit., Moscow, U.S.S.R., 1953).

[3.16] W. E. Milne, Numerical calculus (Princeton Univ. Press, Princeton, N.J., 1949).

[3.17] L. M. Milne-Thomson, The calculus of finite differences (Macmillan and Co., Ltd., London, England, 1951).

[3.18] H. Mineur, Techniques de calcul numérique (Librairie Polytechnique Ch. Béranger, Paris, France, 1952).

[3.19] National Physical Laboratory, Modern computing methods, Notes on Applied Science No. 16 (Her Majesty's Stationery Office, London, England, 1957).

[3.20] J. B. Rosser, Transformations to speed the convergence of series, J. Research NBS 46, 56–64 (1951).

[3.21] J. B. Scarborough, Numerical mathematical analysis, 3d ed. (The Johns Hopkins Press, Baltimore, Md.; Oxford Univ. Press, London, England, 1955).

[3.22] J. F. Steffensen, Interpolation (Chelsea Publishing Co., New York, N.Y., 1950).

[3.23] H. S. Wall, Analytic theory of continued fractions (D. Van Nostrand Co., Inc., New York, N.Y., 1948).

[3.24] E. T. Whittaker and G. Robinson, The calculus of observations, 4th ed. (Blackie and Son, Ltd., London, England, 1944).

[3.25] R. Zurmühl, Praktische Mathematik (Springer-Verlag, Berlin, Germany, 1953).

Mathematical Tables and Collections of Formulas

[3.26] E. P. Adams, Smithsonian mathematical formulae and tables of elliptic functions, 3d reprint (The Smithsonian Institution, Washington, D.C., 1957).

[3.27] L. J. Comrie, Barlow's tables of squares, cubes, square roots, cube roots and reciprocals of all integers up to 12,500 (Chemical Publishing Co., Inc., New York, N.Y., 1954).

[3.28] H. B. Dwight, Tables of integrals and other mathematical data, 3d ed. (The Macmillan Co., New York, N.Y., 1957).

[3.29] Gt. Britain H.M. Nautical Almanac Office, Interpolation and allied tables (Her Majesty's Stationery Office, London, England, 1956).

[3.30] B. O. Peirce, A short table of integrals, 4th ed. (Ginn and Co., Boston, Mass., 1956).

[3.31] G. Schulz, Formelsammlung zur praktischen Mathematik (de Gruyter and Co., Berlin, Germany, 1945).

Table 3.1   **POWERS AND ROOTS** n^k

See **Examples 1–5** for use of the table.

Floating decimal notation:

$9^{10} = 34867\ 84401$

$= (9)3.4867\ 84401$

k		2	3	4
1	$n^1=$	2	3	4
2	$n^2=$	4	9	16
3	$n^3=$	8	27	64
4	$n^4=$	16	81	256
5	$n^5=$	32	243	1024
6	$n^6=$	64	729	4096
7	$n^7=$	128	2187	16384
8	$n^8=$	256	6561	65536
9	$n^9=$	512	19683	2 62144
10	$n^{10}=$	1024	59049	10 48576
24	$n^{24}=$	167 77216	(11)2.8242 95365	(14)2.8147 49767
1/2	$n^{1/2}=$	1.4142 13562	1.7320 50808	2.0000 00000
1/3	$n^{1/3}=$	1.2599 21050	1.4422 49570	1.5874 01052
1/4	$n^{1/4}=$	1.1892 07115	1.3160 74013	1.4142 13562
1/5	$n^{1/5}=$	1.1486 98355	1.2457 30940	1.3195 07911

k	5	6	7	8	9
1	5	6	7	8	9
2	25	36	49	64	81
3	125	216	343	512	729
4	625	1296	2401	4096	6561
5	3125	7776	16807	32768	59049
6	15625	46656	1 17649	2 62144	5 31441
7	78125	2 79936	8 23543	20 97152	47 82969
8	3 90625	16 79616	57 64801	167 77216	430 46721
9	19 53125	100 77696	403 53607	1342 17728	3874 20489
10	97 65625	604 66176	2824 75249	(9)1.0737 41824	(9)3.4867 84401
24	(16)5.9604 64478	(18)4.7383 81338	(20)1.9158 12314	(21)4.7223 66483	(22)7.9766 44308
1/2	2.2360 67977	2.4494 89743	2.6457 51311	2.8284 27125	3.0000 00000
1/3	1.7099 75947	1.8171 20593	1.9129 31183	2.0000 00000	2.0800 83823
1/4	1.4953 48781	1.5650 84580	1.6265 76562	1.6817 92831	1.7320 50808
1/5	1.3797 29662	1.4309 69081	1.4757 73162	1.5157 16567	1.5518 45574

k	10	11	12	13	14
1	10	11	12	13	14
2	100	121	144	169	196
3	1000	1331	1728	2197	2744
4	10000	14641	20736	28561	38416
5	1 00000	1 61051	2 48832	3 71293	5 37824
6	10 00000	17 71561	29 85984	48 26809	75 29536
7	100 00000	194 87171	358 31808	627 48517	1054 13504
8	1000 00000	2143 58881	4299 81696	8157 30721	(9)1.4757 89056
9	(9)1.0000 00000	(9)2.3579 47691	(9)5.1597 80352	(10)1.0604 49937	(10)2.0661 04678
10	(10)1.0000 00000	(10)2.5937 42460	(10)6.1917 36422	(11)1.3785 84918	(11)2.8925 46550
24	(24)1.0000 00000	(24)9.8497 32676	(25)7.9496 84720	(26)5.4280 07704	(27)3.2141 99700
1/2	3.1622 77660	3.3166 24790	3.4641 01615	3.6055 51275	3.7416 57387
1/3	2.1544 34690	2.2239 80091	2.2894 28485	2.3513 34688	2.4101 42264
1/4	1.7782 79410	1.8211 60287	1.8612 09718	1.8988 28922	1.9343 36420
1/5	1.5848 93192	1.6153 94266	1.6437 51830	1.6702 77652	1.6952 18203

k	15	16	17	18	19
1	15	16	17	18	19
2	225	256	289	324	361
3	3375	4096	4913	5832	6859
4	50625	65536	83521	1 04976	1 30321
5	7 59375	10 48576	14 19857	18 89568	24 76099
6	113 90625	167 77216	241 37569	340 12224	470 45881
7	1708 59375	2684 35456	4103 38673	6122 20032	8938 71739
8	(9)2.5628 90625	(9)4.2949 67296	(9)6.9757 57441	(10)1.1019 96058	(10)1.6983 56304
9	(10)3.8443 35938	(10)6.8719 47674	(11)1.1858 78765	(11)1.9835 92904	(11)3.2268 76978
10	(11)5.7665 03906	(12)1.0995 11628	(12)2.0159 93900	(12)3.5704 67227	(12)6.1310 66258
24	(28)1.6834 11220	(28)7.9228 16251	(29)3.3944 86713	(30)1.3382 58845	(30)4.8987 62931
1/2	3.8729 83346	4.0000 00000	4.1231 05626	4.2426 40687	4.3588 98944
1/3	2.4662 12074	2.5198 42100	2.5712 81591	2.6207 41394	2.6684 01649
1/4	1.9679 89671	2.0000 00000	2.0305 43185	2.0597 67144	2.0877 97630
1/5	1.7187 71928	1.7411 01127	1.7623 40348	1.7826 02458	1.8019 83127

k	20	21	22	23	24
1	20	21	22	23	24
2	400	441	484	529	576
3	8000	9261	10648	12167	13824
4	1 60000	1 94481	2 34256	2 79841	3 31776
5	32 00000	40 84101	51 53632	64 36343	79 62624
6	640 00000	857 66121	1133 79904	1480 35889	1911 02976
7	(9)1.2800 00000	(9)1.8010 88541	(9)2.4943 57888	(9)3.4048 25447	(9)4.5864 71424
8	(10)2.5600 00000	(10)3.7822 85936	(10)5.4875 87354	(10)7.8310 98528	(11)1.1007 53142
9	(11)5.1200 00000	(11)7.9428 00466	(12)1.2072 69218	(12)1.8011 52661	(12)2.6418 07540
10	(13)1.0240 00000	(13)1.6679 88098	(13)2.6559 92279	(13)4.1426 51121	(13)6.3403 38097
24	(31)1.6777 21600	(31)5.4108 19838	(32)1.6525 10926	(32)4.8025 07640	(33)1.3337 35777
1/2	4.4721 35955	4.5825 75695	4.6904 15760	4.7958 31523	4.8989 79486
1/3	2.7144 17617	2.7589 24176	2.8020 39331	2.8438 66980	2.8844 99141
1/4	2.1147 42527	2.1406 95143	2.1657 36771	2.1899 38703	2.2133 63839
1/5	1.8205 64203	1.8384 16287	1.8556 00736	1.8721 71231	1.8881 75023

POWERS AND ROOTS n^k Table 3.1

k	25	26	27	28	29
1	25	26	27	28	29
2	625	676	729	784	841
3	15625	17576	19683	21952	24389
4	3 90625	4 56976	5 31441	6 14656	7 07281
5	97 65625	118 81376	143 48907	172 10368	205 11149
6	2441 40625	3089 15776	3874 20489	4818 90304	5948 23321
7	(9)6.1035 15625	(9)8.0318 10176	(10)1.0460 35320	(10)1.3492 92851	(10)1.7249 87631
8	(11)1.5258 78906	(11)2.0882 70646	(11)2.8242 95365	(11)3.7780 19983	(11)5.0024 64130
9	(12)3.8146 97266	(12)5.4295 03679	(12)7.6255 97485	(13)1.0578 45595	(13)1.4507 14598
10	(14)9.5367 43164	(14)1.4116 70957	(14)2.0589 11321	(14)2.9619 67667	(14)4.2070 72333
24	(33)3.5527 13679	(33)9.1066 85770	(34)2.2528 39954	(34)5.3925 32264	(35)1.2518 49008
1/2	5.0000 00000	5.0990 19514	5.1961 52423	5.2915 02622	5.3851 64807
1/3	2.9240 17738	2.9624 96068	3.0000 00000	3.0365 88972	3.0723 16826
1/4	2.2360 67977	2.2581 00864	2.2795 07057	2.3003 26634	2.3205 95787
1/5	1.9036 53939	1.9186 45192	1.9331 82045	1.9472 94361	1.9610 09057

k	30	31	32	33	34
1	30	31	32	33	34
2	900	961	1024	1089	1156
3	27000	29791	32768	35937	39304
4	8 10000	9 23521	10 48576	11 85921	13 36336
5	243 00000	286 29151	335 54432	391 35393	454 35424
6	7290 00000	8875 03681	(9)1.0737 41824	(9)1.2914 67969	(9)1.5448 04416
7	(10)2.1870 00000	(10)2.7512 61411	(10)3.4359 73837	(10)4.2618 44298	(10)5.2523 35014
8	(11)6.5610 00000	(11)8.5289 10374	(12)1.0995 11628	(12)1.4064 08618	(12)1.7857 93905
9	(13)1.9683 00000	(13)2.6439 62216	(13)3.5184 37209	(13)4.6411 48440	(13)6.0716 99277
10	(14)5.9049 00000	(14)8.1962 82870	(15)1.1258 99907	(15)1.5315 78985	(15)2.0643 77754
24	(35)2.8242 95365	(35)6.2041 26610	(36)1.3292 27996	(36)2.7818 55434	(36)5.6950 03680
1/2	5.4772 25575	5.5677 64363	5.6568 54249	5.7445 62647	5.8309 51895
1/3	3.1072 32506	3.1413 80652	3.1748 02104	3.2075 34330	3.2396 11801
1/4	2.3403 47319	2.3596 11062	2.3784 14230	2.3967 81727	2.4147 36403
1/5	1.9743 50486	1.9873 40755	2.0000 00000	2.0123 46617	2.0243 97459

k	35	36	37	38	39
1	35	36	37	38	39
2	1225	1296	1369	1444	1521
3	42875	46656	50653	54872	59319
4	15 00625	16 79616	18 74161	20 85136	23 13441
5	525 21875	604 66176	693 43957	792 35168	902 24199
6	(9)1.8382 65625	(9)2.1767 82336	(9)2.5657 26409	(9)3.0109 36384	(9)3.5187 43761
7	(10)6.4339 29688	(10)7.8364 16410	(10)9.4931 87713	(11)1.1441 55826	(11)1.3723 10067
8	(12)2.2518 75391	(12)2.8211 09907	(12)3.5124 79454	(12)4.3477 92138	(12)5.3520 09260
9	(13)7.8815 63867	(14)1.0155 99567	(14)1.2996 17398	(14)1.6521 61013	(14)2.0872 83612
10	(15)2.7585 47354	(15)3.6561 58440	(15)4.8085 84372	(15)6.2782 11848	(15)8.1404 06085
24	(37)1.1419 13124	(37)2.2452 25725	(37)4.3335 25711	(37)8.2187 60383	(38)1.5330 29700
1/2	5.9160 79783	6.0000 00000	6.0827 62530	6.1644 14003	6.2449 97998
1/3	3.2710 66310	3.3019 27249	3.3322 21852	3.3619 75407	3.3912 11443
1/4	2.4322 99279	2.4494 89743	2.4663 25715	2.4828 23796	2.4989 99399
1/5	2.0361 68005	2.0476 72511	2.0589 24137	2.0699 35054	2.0807 16549

k	40	41	42	43	44
1	40	41	42	43	44
2	1600	1681	1764	1849	1936
3	64000	68921	74088	79507	85184
4	25 60000	28 25761	31 11696	34 18801	37 48096
5	1024 00000	1158 56201	1306 91232	1470 08443	1649 16224
6	(9)4.0960 00000	(9)4.7501 04241	(9)5.4890 31744	(9)6.3213 63049	(9)7.2563 13856
7	(11)1.6384 00000	(11)1.9475 42739	(11)2.3053 93332	(11)2.7181 86111	(11)3.1927 78097
8	(12)6.5536 00000	(12)7.9849 25229	(12)9.6826 51996	(13)1.1688 20028	(13)1.4048 22363
9	(14)2.6214 40000	(14)3.2738 19344	(14)4.0667 13838	(14)5.0259 26119	(14)6.1812 18395
10	(16)1.0485 76000	(16)1.3422 65931	(16)1.7080 19812	(16)2.1611 48231	(16)2.7197 36094
24	(38)2.8147 49767	(38)5.0911 10945	(38)9.0778 49315	(39)1.5967 72093	(39)2.7724 53276
1/2	6.3245 55320	6.4031 24237	6.4807 40698	6.5574 38524	6.6332 49581
1/3	3.4199 51893	3.4482 17240	3.4760 26645	3.5033 98060	3.5303 48335
1/4	2.5148 66859	2.5304 39534	2.5457 29895	2.5607 49602	2.5755 09577
1/5	2.0912 79105	2.1016 32478	2.1117 85765	2.1217 47461	2.1315 25513

k	45	46	47	48	49
1	45	46	47	48	49
2	2025	2116	2209	2304	2401
3	91125	97336	1 03823	1 10592	1 17649
4	41 00625	44 77456	48 79681	53 08416	57 64801
5	1845 28125	2059 62976	2293 45007	2548 03968	2824 75249
6	(9)8.3037 65625	(9)9.4742 96896	(10)1.0779 21533	(10)1.2230 59046	(10)1.3841 28720
7	(11)3.7366 94531	(11)4.3581 76572	(11)5.0662 31205	(11)5.8706 83423	(11)6.7822 30728
8	(13)1.6815 12539	(13)2.0047 61223	(13)2.3811 28666	(13)2.8179 28043	(13)3.3232 93057
9	(14)7.5668 06426	(14)9.2219 01627	(15)1.1191 30473	(15)1.3526 05461	(15)1.6284 13598
10	(16)3.4050 62892	(16)4.2420 74748	(16)5.2599 13224	(16)6.4925 06211	(16)7.9792 26630
24	(39)4.7544 50505	(39)8.0572 70802	(40)1.3500 46075	(40)2.2376 37322	(40)3.6703 36822
1/2	6.7082 03932	6.7823 29983	6.8556 54600	6.9282 03230	7.0000 00000
1/3	3.5568 93304	3.5830 47871	3.6088 26080	3.6342 41186	3.6593 05710
1/4	2.5900 20064	2.6042 90687	2.6183 30499	2.6321 48026	2.6457 51311
1/5	2.1411 27368	2.1505 60013	2.1598 30012	2.1689 43542	2.1779 06425

$$n^{\frac{1}{2}}\left[(-4)\frac{3}{8}\right] \qquad n^{\frac{1}{3}}\left[(-4)\frac{1}{7}\right] \qquad n^{\frac{1}{4}}\left[(-5)\frac{9}{8}\right] \qquad n^{\frac{1}{5}}\left[(-5)\frac{7}{6}\right]$$

The numbers in square brackets at the bottom of the page mean that the maximum error in a linear interpolate is $a \times 10^{-p}$ (p in parentheses), and that to interpolate to the full tabular accuracy m points must be used in Lagrange's and Aitkens methods for the respective functions $n^{1/r}$.   *

*See page II.

Table 3.1   **POWERS AND ROOTS n^k**

k	50	51	52	53	54
1	50	51	52	53	54
2	2500	2601	2704	2809	2916
3	1 25000	1 32651	1 40608	1 48877	1 57464
4	62 50000	67 65201	73 11616	78 90481	85 03056
5	3125 00000	3450 25251	3802 04032	4181 95493	4591 65024
6	(10)1.5625 00000	(10)1.7596 28780	(10)1.9770 60966	(10)2.2164 36113	(10)2.4794 91130
7	(11)7.8125 00000	(11)8.9741 06779	(12)1.0280 71703	(12)1.1747 11140	(12)1.3389 25210
8	(13)3.9062 50000	(13)4.5767 94457	(13)5.3459 72853	(13)6.2259 69041	(13)7.2301 96134
9	(15)1.9531 25000	(15)2.3341 65173	(15)2.7799 05884	(15)3.2997 63592	(15)3.9043 05912
10	(16)9.7656 25000	(17)1.1904 24238	(17)1.4455 51059	(17)1.7488 74704	(17)2.1083 25193
24	(40)5.9604 64478	(40)9.5870 33090	(41)1.5278 48342	(41)2.4133 53110	(41)3.7796 38253
1/2	7.0710 67812	7.1414 28429	7.2111 02551	7.2801 09889	7.3484 69228
1/3	3.6840 31499	3.7084 29769	3.7325 11157	3.7562 85754	3.7797 63150
1/4	2.6591 47948	2.6723 45118	2.6853 49614	2.6981 67876	2.7108 06011
1/5	2.1867 24148	2.1954 01897	2.2039 44575	2.2123 56822	2.2206 43035

k	55	56	57	58	59
1	55	56	57	58	59
2	3025	3136	3249	3364	3481
3	1 66375	1 75616	1 85193	1 95112	2 05379
4	91 50625	98 34496	105 56001	113 16496	121 17361
5	5032 84375	5507 31776	6016 92057	6563 56768	7149 24299
6	(10)2.7680 64063	(10)3.0840 97946	(10)3.4296 44725	(10)3.8068 69254	(10)4.2180 53364
7	(12)1.5224 35234	(12)1.7270 94850	(12)1.9548 97493	(12)2.2079 84168	(12)2.4886 51485
8	(13)8.3733 93789	(13)9.6717 31157	(14)1.1142 91571	(14)1.2806 30817	(14)1.4683 04376
9	(15)4.6053 66584	(15)5.4161 69448	(15)6.3514 61955	(15)7.4276 58740	(15)8.6629 95819
10	(17)2.5329 51621	(17)3.0330 54891	(17)3.6203 33315	(17)4.3080 42069	(17)5.1111 67533
24	(41)5.8708 98173	(41)9.0471 67858	(42)1.3835 55344	(42)2.1002 54121	(42)3.1655 43453
1/2	7.4161 98487	7.4833 14774	7.5498 34435	7.6157 73106	7.6811 45748
1/3	3.8029 52461	3.8258 62366	3.8485 01131	3.8708 76641	3.8929 96416
1/4	2.7232 69815	2.7355 64800	2.7476 96205	2.7596 69021	2.7714 88002
1/5	2.2288 07384	2.2368 53829	2.2447 86134	2.2526 07878	2.2603 22470

k	60	61	62	63	64
1	60	61	62	63	64
2	3600	3721	3844	3969	4096
3	2 16000	2 26981	2 38328	2 50047	2 62144
4	129 60000	138 45841	147 76336	157 52961	167 77216
5	7776 00000	8445 96301	9161 32832	9924 36543	(9)1.0737 41824
6	(10)4.6656 00000	(10)5.1520 37436	(10)5.6800 23558	(10)6.2523 50221	(10)6.8719 47674
7	(12)2.7993 60000	(12)3.1427 42836	(12)3.5216 14606	(12)3.9389 80639	(12)4.3980 46511
8	(14)1.6796 16000	(14)1.9170 73130	(14)2.1834 01056	(14)2.4815 57803	(14)2.8147 49767
9	(16)1.0077 69600	(16)1.1694 14609	(16)1.3537 08655	(16)1.5633 81416	(16)1.8014 39851
10	(17)6.0466 17600	(17)7.1334 29117	(17)8.3929 93659	(17)9.8493 02919	(18)1.1529 21505
24	(42)4.7383 81338	(42)7.0455 68477	(43)1.0408 79722	(43)1.5281 75339	(43)2.2300 74520
1/2	7.7459 66692	7.8102 49676	7.8740 07874	7.9372 53953	8.0000 00000
1/3	3.9148 67641	3.9364 97183	3.9578 91610	3.9790 57208	4.0000 00000
1/4	2.7831 57684	2.7946 82393	2.8060 66263	2.8173 13247	2.8284 27125
1/5	2.2679 33155	2.2754 43032	2.2828 55056	2.2901 72049	2.2973 96710

k	65	66	67	68	69
1	65	66	67	68	69
2	4225	4356	4489	4624	4761
3	2 74625	2 87496	3 00763	3 14432	3 28509
4	178 50625	189 74736	201 51121	213 81376	226 67121
5	(9)1.1602 90625	(9)1.2523 32576	(9)1.3501 25107	(9)1.4539 33568	(9)1.5640 31349
6	(10)7.5418 89063	(10)8.2653 95002	(10)9.0458 38217	(10)9.8867 48262	(11)1.0791 81631
7	(12)4.9022 27891	(12)5.4551 60701	(12)6.0607 11605	(12)6.7229 88818	(12)7.4463 53253
8	(14)3.1864 48129	(14)3.6004 06063	(14)4.0606 76776	(14)4.5716 32397	(14)5.1379 83744
9	(16)2.0711 91284	(16)2.3762 68001	(16)2.7206 53440	(16)3.1087 10030	(16)3.5452 08784
10	(18)1.3462 74334	(18)1.5683 36881	(18)1.8228 37805	(18)2.1139 22820	(18)2.4461 94061
24	(43)3.2353 44710	(43)4.6671 78950	(43)6.6956 88867	(43)9.5546 30685	(44)1.3563 70007
1/2	8.0622 57748	8.1240 38405	8.1853 52772	8.2462 11251	8.3066 23863
1/3	4.0207 25759	4.0412 40021	4.0615 48100	4.0816 55102	4.1015 65930
1/4	2.8394 11514	2.8502 69883	2.8610 05553	2.8716 21711	2.8821 21417
1/5	2.3045 31620	2.3115 79249	2.3185 41963	2.3254 22030	2.3322 21626

k	70	71	72	73	74
1	70	71	72	73	74
2	4900	5041	5184	5329	5476
3	3 43000	3 57911	3 73248	3 89017	4 05224
4	240 10000	254 11681	268 73856	283 98241	299 86576
5	(9)1.6807 00000	(9)1.8042 29351	(9)1.9349 17632	(9)2.0730 71593	(9)2.2190 06624
6	(11)1.1764 90000	(11)1.2810 02839	(11)1.3931 40695	(11)1.5133 42263	(11)1.6420 64902
7	(12)8.2354 30000	(12)9.0951 20158	(13)1.0030 61300	(13)1.1047 39852	(13)1.2151 28027
8	(14)5.7648 01000	(14)6.4575 35312	(14)7.2220 41363	(14)8.0646 00919	(14)8.9919 47402
9	(16)4.0353 60700	(16)4.5848 50072	(16)5.1998 69781	(16)5.8871 58671	(16)6.6540 41078
10	(18)2.8247 52490	(18)3.2552 43551	(18)3.7439 06243	(18)4.2976 25830	(18)4.9239 90397
24	(44)1.9158 12314	(44)2.6927 76876	(44)3.7668 63772	(44)5.2450 38047	(44)7.2704 49690
1/2	8.3666 00265	8.4261 49773	8.4852 81374	8.5440 03745	8.6023 25267
1/3	4.1212 85300	4.1408 17749	4.1601 67646	4.1793 39196	4.1983 36454
1/4	2.8925 07608	2.9027 83108	2.9129 50630	2.9230 12786	2.9329 72088
1/5	2.3389 42837	2.3455 87669	2.3521 58045	2.3586 55818	2.3650 82769

$$n^{\frac{1}{2}}\begin{bmatrix}(-5)9\\6\end{bmatrix} \qquad n^{\frac{1}{3}}\begin{bmatrix}(-5)4\\6\end{bmatrix} \qquad n^{\frac{1}{4}}\begin{bmatrix}(-5)3\\5\end{bmatrix} \qquad n^{\frac{1}{5}}\begin{bmatrix}(-5)2\\5\end{bmatrix}$$

POWERS AND ROOTS n^k — Table 3.1

k	75	76	77	78	79
1	75	76	77	78	79
2	5625	5776	5929	6084	6241
3	4 21875	4 38976	4 56533	4'74552	4 93039
4	316 40625	333 62176	351 53041	370 15056	389 50081
5	(9)2.3730 46875	(9)2.5355 25376	(9)2.7067 84157	(9)2.8871 74368	(9)3.0770 76399
6	(11)1.7797 85156	(11)1.9269 99286	(11)2.0842 23801	(11)2.2519 96007	(11)2.4308 74555
7	(13)1.3348 38867	(13)1.4645 19457	(13)1.6048 52327	(13)1.7565 56885	(13)1.9203 90899
8	(15)1.0011 29150	(15)1.1130 34787	(15)1.2357 36292	(15)1.3701 14371	(15)1.5171 08810
9	(16)7.5084 68628	(16)8.4590 64385	(16)9.5151 69445	(17)1.0686 89209	(17)1.1985 15960
10	(18)5.6313 51471	(18)6.4288 88932	(18)7.3266 80473	(18)8.3357 75831	(18)9.4682 76083
24	(45)1.0033 91278	(45)1.3788 79182	(45)1.8870 23915	(45)2.5719 97041	(45)3.4918 06676
1/2	8.6602 54038	8.7177 97887	8.7749 64387	8.8317 60866	8.8881 94417
1/3	4.2171 63326	4.2358 23584	4.2543 20865	4.2726 58682	4.2908 40427
1/4	2.9428 30956	2.9525 91724	2.9622 56638	2.9718 27866	2.9813 07501
1/5	2.3714 40610	2.3777 30992	2.3839 55503	2.3901 15677	2.3962 12991

k	80	81	82	83	84
1	80	81	82	83	84
2	6400	6561	6724	6889	7056
3	5 12000	5 31441	5 51368	5 71787	5 92704
4	409 60000	430 46721	452 12176	474 58321	497 87136
5	(9)3.2768 00000	(9)3.4867 84401	(9)3.7073 98432	(9)3.9390 40643	(9)4.1821 19424
6	(11)2.6214 40000	(11)2.8242 95365	(11)3.0400 66714	(11)3.2694 03734	(11)3.5129 80316
7	(13)2.0971 52000	(13)2.2876 79245	(13)2.4928 54706	(13)2.7136 05099	(13)2.9509 03466
8	(15)1.6777 21600	(15)1.8530 20189	(15)2.0441 40859	(15)2.2522 92232	(15)2.4787 58911
9	(17)1.3421 77280	(17)1.5009 46353	(17)1.6761 95504	(17)1.8694 02553	(17)2.0821 57485
10	(19)1.0737 41824	(19)1.2157 66546	(19)1.3744 80313	(19)1.5516 04119	(19)1.7490 12288
24	(45)4.7223 66483	(45)6.3626 85441	(45)8.5414 66801	(46)1.1425 47375	(46)1.5230 10388
1/2	8.9442 71910	9.0000 00000	9.0553 85138	9.1104 33579	9.1651 51390
1/3	4.3088 69380	4.3267 48711	4.3444 81486	4.3620 70671	4.3795 19140
1/4	2.9906 97562	3.0000 00000	3.0092 16698	3.0183 49479	3.0274 00104
1/5	2.4022 48868	2.4082 24685	2.4141 41771	2.4200 01407	2.4258 04834

k	85	86	87	88	89
1	85	86	87	88	89
2	7225	7396	7569	7744	7921
3	6 14125	6 36056	6 58503	6 81472	7 04969
4	522 00625	547 00816	572 89761	599 69536	627 42241
5	(9)4.4370 53125	(9)4.7042 70176	(9)4.9842 09207	(9)5.2773 19168	(9)5.5840 59449
6	(11)3.7714 95156	(11)4.0456 72351	(11)4.3362 62010	(11)4.6440 40868	(11)4.9698 12910
7	(13)3.2057 70883	(13)3.4792 78222	(13)3.7725 47949	(13)4.0867 55964	(13)4.4231 33490
8	(15)2.7249 05250	(15)2.9921 79271	(15)3.2821 16715	(15)3.5963 45248	(15)3.9365 88806
9	(17)2.3161 69463	(17)2.5732 74173	(17)2.8554 41542	(17)3.1647 83818	(17)3.5035 64037
10	(19)1.9687 44043	(19)2.2130 15789	(19)2.4842 34142	(19)2.7850 09760	(19)3.1181 71993
24	(46)2.0232 71747	(46)2.6789 39031	(46)3.5355 91351	(46)4.6514 04745	(46)6.1004 25945
1/2	9.2195 44457	9.2736 18495	9.3273 79053	9.3808 31520	9.4339 81132
1/3	4.3968 29672	4.4140 04962	4.4310 47622	4.4479 60181	4.4647 45096
1/4	3.0363 70277	3.0452 61646	3.0540 75810	3.0628 14314	3.0714 78656
1/5	2.4315 53252	2.4372 47818	2.4428 89656	2.4484 79851	2.4540 19455

k	90	91	92	93	94
1	90	91	92	93	94
2	8100	8281	8464	8649	8836
3	7 29000	7 53571	7 78688	8 04357	8 30584
4	656 10000	685 74961	716 39296	748 05201	780 74896
5	(9)5.9049 00000	(9)6.2403 21451	(9)6.5908 15232	(9)6.9568 83693	(9)7.3390 40224
6	(11)5.3144 10000	(11)5.6786 92520	(11)6.0635 50013	(11)6.4699 01834	(11)6.8986 97811
7	(13)4.7829 69000	(13)5.1676 10194	(13)5.5784 66012	(13)6.0170 08706	(13)6.4847 75942
8	(15)4.3046 72100	(15)4.7025 25276	(15)5.1321 88731	(15)5.5958 18097	(15)6.0956 89385
9	(17)3.8742 04890	(17)4.2792 98001	(17)4.7216 13633	(17)5.2041 10830	(17)5.7299 48022
10	(19)3.4867 84401	(19)3.8941 61181	(19)4.3438 84542	(19)4.8398 23072	(19)5.3861 51141
24	(46)7.9766 44308	(47)1.0399 04400	(47)1.3517 85726	(47)1.7522 28603	(47)2.2650 01461
1/2	9.4868 32981	9.5393 92014	9.5916 63047	9.6436 50761	9.6953 59715
1/3	4.4814 04747	4.4979 41445	4.5143 57435	4.5306 54896	4.5468 35944
1/4	3.0800 70288	3.0885 90619	3.0970 41015	3.1054 22799	3.1137 37258
1/5	2.4595 09486	2.4649 50932	2.4703 44749	2.4756 91866	2.4809 93182

k	95	96	97	98	99
1	95	96	97	98	99
2	9025	9216	9409	9604	9801
3	8 57375	8 84736	9 12673	9 41192	9 70299
4	814 50625	849 34656	885 29281	922 36816	960 59601
5	(9)7.7378 09375	(9)8.1537 26976	(9)8.5873 40257	(9)9.0392 07968	(9)9.5099 00499
6	(11)7.3509 18906	(11)7.8275 77897	(11)8.3297 20049	(11)8.8584 23809	(11)9.4148 01494
7	(13)6.9833 72961	(13)7.5144 74781	(13)8.0798 28448	(13)8.6812 55332	(13)9.3206 53479
8	(15)6.6342 04313	(15)7.2138 95790	(15)7.8374 33594	(15)8.5076 30226	(15)9.2274 46944
9	(17)6.3024 94097	(17)6.9253 39958	(17)7.6023 10587	(17)8.3374 77621	(17)9.1351 72475
10	(19)5.9873 69392	(19)6.6483 26360	(19)7.3742 41269	(19)8.1707 28069	(19)9.0438 20750
24	(47)2.9198 90243	(47)3.7541 32467	(47)4.8141 72219	(47)6.1578 03365	(47)7.8567 81408
1/2	9.7467 94345	9.7979 58971	9.8488 57802	9.8994 94937	9.9498 74371
1/3	4.5629 02635	4.5788 56970	4.5947 00892	4.6104 36292	4.6260 65009
1/4	3.1219 85641	3.1301 69160	3.1382 88993	3.1463 46284	3.1543 42146
1/5	2.4862 49570	2.4914 61879	2.4966 30932	2.5017 57527	2.5068 42442

$$n^{\frac{1}{2}}\left[\begin{matrix}(-5)5\\5\end{matrix}\right] \qquad n^{\frac{1}{3}}\left[\begin{matrix}(-5)2\\5\end{matrix}\right] \qquad n^{\frac{1}{4}}\left[\begin{matrix}(-5)1\\5\end{matrix}\right] \qquad n^{\frac{1}{5}}\left[\begin{matrix}(-6)9\\5\end{matrix}\right]$$

Table 3.1   **POWERS AND ROOTS** n^k

k	100	101	102	103	104
1	100	101	102	103	104
2	10000	10201	10404	10609	10816
3	10 00000	10 30301	10 61208	10 92727	11 24864
4	1000 00000	1040 60401	1082 43216	1125 50881	1169 85856
5	(10)1.0000 00000	(10)1.0510 10050	(10)1.1040 80803	(10)1.1592 74074	(10)1.2166 52902
6	(12)1.0000 00000	(12)1.0615 20151	(12)1.1261 62419	(12)1.1940 52297	(12)1.2653 19018
7	(14)1.0000 00000	(14)1.0721 35352	(14)1.1486 85668	(14)1.2298 73865	(14)1.3159 31779
8	(16)1.0000 00000	(16)1.0828 56706	(16)1.1716 59381	(16)1.2667 70081	(16)1.3685 69050
9	(18)1.0000 00000	(18)1.0936 85273	(18)1.1950 92569	(18)1.3047 73184	(18)1.4233 11812
10	(20)1.0000 00000	(20)1.1046 22125	(20)1.2189 94420	(20)1.3439 16379	(20)1.4802 44285
24	(48)1.0000 00000	(48)1.2697 34649	(48)1.6084 37249	(48)2.0327 94106	(48)2.5633 04165
1/2	(1)1.0000 00000	(1)1.0049 87562	(1)1.0099 50494	(1)1.0148 89157	(1)1.0198 03903
1/3	4.6415 88834	4.6570 09508	4.6723 28728	4.6875 48148	4.7026 69375
1/4	3.1622 77660	3.1701 53880	3.1779 71828	3.1857 32501	3.1934 36868
1/5	2.5118 86432	2.5168 90229	2.5218 54548	2.5267 80083	2.5316 67508

k	105	106	107	108	109
1	105	106	107	108	109
2	11025	11236	11449	11664	11881
3	11 57625	11 91016	12 25043	12 59712	12 95029
4	1215 50625	1262 47696	1310 79601	1360 48896	1411 58161
5	(10)1.2762 81563	(10)1.3382 25578	(10)1.4025 51731	(10)1.4693 28077	(10)1.5386 23955
6	(12)1.3400 95641	(12)1.4185 19112	(12)1.5007 30352	(12)1.5868 74323	(12)1.6771 00111
7	(14)1.4071 00423	(14)1.5036 30259	(14)1.6057 81476	(14)1.7138 24269	(14)1.8280 39121
8	(16)1.4774 55444	(16)1.5938 48075	(16)1.7181 86180	(16)1.8509 30210	(16)1.9925 62642
9	(18)1.5513 28216	(18)1.6894 78959	(18)1.8384 59212	(18)1.9990 04627	(18)2.1718 93279
10	(20)1.6288 94627	(20)1.7908 47697	(20)1.9671 51357	(20)2.1589 24997	(20)2.3673 63675
24	(48)3.2250 99944	(48)4.0489 34641	(48)5.0723 66953	(48)6.3411 80737	(48)7.9110 83175
1/2	(1)1.0246 95077	(1)1.0295 63014	(1)1.0344 08043	(1)1.0392 30485	(1)1.0440 30651
1/3	4.7176 93980	4.7326 23491	4.7474 59398	4.7622 03156	4.7768 56181
1/4	3.2010 85873	3.2086 80436	3.2162 21453	3.2237 09795	3.2311 46315
1/5	2.5365 17482	2.5413 30642	2.5461 07613	2.5508 49001	2.5555 55397

k	110	111	112	113	114
1	110	111	112	113	114
2	12100	12321	12544	12769	12996
3	13 31000	13 67631	14 04928	14 42897	14 81544
4	1464 10000	1518 07041	1573 51936	1630 47361	1688 96016
5	(10)1.6105 10000	(10)1.6850 58155	(10)1.7623 41683	(10)1.8424 35179	(10)1.9254 14582
6	(12)1.7715 61000	(12)1.8704 14552	(12)1.9738 22685	(12)2.0819 51753	(12)2.1949 72624
7	(14)1.9487 17100	(14)2.0761 60153	(14)2.2106 81407	(14)2.3526 05480	(14)2.5022 68791
8	(16)2.1435 88810	(16)2.3045 37770	(16)2.4759 63176	(16)2.6584 44193	(16)2.8525 86422
9	(18)2.3579 47691	(18)2.5580 3692?	(18)2.7730 78757	(18)3.0040 41938	(18)3.2519 48521
10	(20)2.5937 42460	(20)2.8394 20986	(20)3.1058 48208	(20)3.3945 67390	(20)3.7072 21314
24	(48)9.8497 32676	(49)1.2239 15658	(49)1.5178 62893	(49)1.8788 09051	(49)2.3212 20685
1/2	(1)1.0488 08848	(1)1.0535 65375	(1)1.0583 00524	(1)1.0630 14581	(1)1.0677 07825
1/3	4.7914 19857	4.8058 95534	4.8202 84528	4.8345 88127	4.8488 07586
1/4	3.2385 31840	3.2458 67180	3.2531 53123	3.2603 90439	3.2675 79877
1/5	2.5602 27376	2.5648 65499	2.5694 70314	2.5740 42354	2.5785 82140

k	115	116	117	118	119
1	115	116	117	118	119
2	13225	13456	13689	13924	14161
3	15 20875	15 60896	16 01613	16 43032	16 85159
4	1749 00625	1810 63936	1873 88721	1938 77776	2005 33921
5	(10)2.0113 57188	(10)2.1003 41658	(10)2.1924 48036	(10)2.2877 57757	(10)2.3863 53660
6	(12)2.3130 60766	(12)2.4363 96323	(12)2.5651 64202	(12)2.6995 54153	(12)2.8397 60855
7	(14)2.6600 19880	(14)2.8262 19734	(14)3.0012 42116	(14)3.1854 73901	(14)3.3793 15418
8	(16)3.0590 22863	(16)3.2784 14892	(16)3.5114 53276	(16)3.7588 59203	(16)4.0213 85347
9	(18)3.5178 76292	(18)3.8029 61275	(18)4.1084 00333	(18)4.4354 53859	(18)4.7854 48563
10	(20)4.0455 57736	(20)4.4114 35079	(20)4.8068 28389	(20)5.2338 35554	(20)5.6946 83790
24	(49)2.8625 17619	(49)3.5236 41704	(49)4.3297 28675	(49)5.3109 00627	(49)6.5031 99444
1/2	(1)1.0723 80529	(1)1.0770 32961	(1)1.0816 65383	(1)1.0862 78049	(1)1.0908 71211
1/3	4.8629 44131	4.8769 98961	4.8909 73246	4.9048 68131	4.9186 84734
1/4	3.2747 22171	3.2818 18035	3.2888 68168	3.2958 73252	3.3028 33952
1/5	2.5830 90178	2.5875 66964	2.5920 12982	2.5964 28703	2.6008 14587

k	120	121	122	123	124
1	120	121	122	123	124
2	14400	14641	14884	15129	15376
3	17 28000	17 71561	18 15848	18 60867	19 06624
4	2073 60000	2143 58881	2215 33456	2288 86641	2364 21376
5	(10)2.4883 20000	(10)2.5937 42460	(10)2.7027 08163	(10)2.8153 05684	(10)2.9316 25062
6	(12)2.9859 84000	(12)3.1384 28377	(12)3.2973 03959	(12)3.4628 25992	(12)3.6352 15077
7	(14)3.5831 80800	(14)3.7974 98336	(14)4.0227 10830	(14)4.2592 75970	(14)4.5076 66696
8	(16)4.2998 16960	(16)4.5949 72986	(16)4.9077 07213	(16)5.2389 09443	(16)5.5895 06703
9	(18)5.1597 80352	(18)5.5599 17313	(18)5.9874 02800	(18)6.4438 58615	(18)6.9309 88312
10	(20)6.1917 36422	(20)6.7274 99949	(20)7.3046 31415	(20)7.9259 46096	(20)8.5944 25506
24	(49)7.9496 84720	(49)9.7017 23378	(50)1.1820 50242	(50)1.4378 80104	(50)1.7463 06393
1/2	(1)1.0954 45115	(1)1.1000 00000	(1)1.1045 36102	(1)1.1090 53651	(1)1.1135 52873
1/3	4.9324 24149	4.9460 87443	4.9596 75664	4.9731 89833	4.9866 30952
1/4	3.3097 50920	3.3166 24790	3.3234 56186	3.3302 45713	3.3369 93965
1/5	2.6051 71085	2.6094 98635	2.6137 97668	2.6180 68602	2.6223 11847

$$n^{\frac{1}{2}}\left[{(-5)3\atop 4}\right] \qquad n^{\frac{1}{3}}\left[{(-5)1\atop 5}\right] \qquad n^{\frac{1}{4}}\left[{(-6)8\atop 5}\right] \qquad n^{\frac{1}{5}}\left[{(-6)5\atop 4}\right]$$

POWERS AND ROOTS n^k Table 3.1

k	125	126	127	128	129
1	125	126	127	128	129
2	15625	15876	16129	16384	16641
3	19 53125	20 00376	20 48383	20 97152	21 46689
4	2441 40625	2520 47376	2601 44641	2684 35456	2769 22881
5	(10)3.0517 57813	(10)3.1757 96938	(10)3.3038 36941	(10)3.4359 73837	(10)3.5723 05165
6	(12)3.8146 97266	(12)4.0015 04141	(12)4.1958 72915	(12)4.3980 46511	(12)4.6082 73663
7	(14)4.7683 71582	(14)5.0418 95218	(14)5.3287 58602	(14)5.6294 99534	(14)5.9446 73025
8	(16)5.9604 64478	(16)6.3527 87975	(16)6.7675 23424	(16)7.2057 59404	(16)7.6686 28202
9	(18)7.4505 80597	(18)8.0045 12848	(18)8.5947 54749	(18)9.2233 72037	(18)9.8925 30381
10	(20)9.3132 25746	(21)1.0085 68619	(21)1.0915 33853	(21)1.1805 91621	(21)1.2761 36419
24	(50)2.1175 82368	(50)2.5638 52774	(50)3.0994 83316	(50)3.7414 44192	(50)4.5097 56022
1/2	(1)1.1180 33989	(1)1.1224 97216	(1)1.1269 42767	(1)1.1313 70850	(1)1.1357 81669
1/3	5.0000 00000	5.0132 97935	5.0265 25695	5.0396 84200	5.0527 74347
1/4	3.3437 01525	3.3503 68959	3.3569 96823	3.3635 85661	3.3701 36005
1/5	2.6265 27804	2.6307 16865	2.6348 79413	2.6390 15822	2.6431 26458

k	130	131	132	133	134
1	130	131	132	133	134
2	16900	17161	17424	17689	17956
3	21 97000	22 40091	22 99968	23 52637	24 06104
4	2856 10000	2944 99921	3035 95776	3129 00721	3224 17936
5	(10)3.7129 30000	(10)3.8579 48965	(10)4.0074 64243	(10)4.1615 79589	(10)4.3204 00342
6	(12)4.8268 09000	(12)5.0539 13144	(12)5.2898 52801	(12)5.5349 00854	(12)5.7893 36459
7	(14)6.2748 51700	(14)6.6206 26219	(14)6.9826 05697	(14)7.3614 18136	(14)7.7577 10855
8	(16)8.1573 07210	(16)8.6730 20347	(16)9.2170 39521	(16)9.7906 86120	(17)1.0395 33255
9	(19)1.0604 49937	(19)1.1361 65665	(19)1.2166 49217	(19)1.3021 61254	(19)1.3929 74561
10	(21)1.3785 84918	(21)1.4883 77022	(21)1.6059 76966	(21)1.7318 74468	(21)1.8665 85912
24	(50)5.4280 07704	(50)6.5239 57088	(50)7.8302 26935	(50)9.3851 10346	(51)1.1233 50184
1/2	(1)1.1401 75425	(1)1.1445 52314	(1)1.1489 12529	(1)1.1532 56259	(1)1.1575 83690
1/3	5.0657 97019	5.0787 53078	5.0916 43370	5.1044 68722	5.1172 29947
1/4	3.3766 48375	3.3831 23282	3.3895 61224	3.3959 62690	3.4023 28159
1/5	2.6472 11681	2.6512 71840	2.6553 07280	2.6593 18337	2.6633 05339

k	135	136	137	138	139
1	135	136	137	138	139
2	18225	18496	18769	19044	19321
3	24 60375	25 15456	25 71353	26 28072	26 85619
4	3321 50625	3421 02016	3522 75361	3626 73936	3733 01041
5	(10)4.4840 33438	(10)4.6525 87418	(10)4.8261 72446	(10)5.0049 00317	(10)5.1888 84470
6	(12)6.0534 45141	(12)6.3275 18888	(12)6.6118 56251	(12)6.9067 62437	(12)7.2125 49413
7	(14)8.1721 50940	(14)8.6054 25688	(14)9.0582 43063	(14)9.5313 32163	(15)1.0025 44368
8	(17)1.1032 40377	(17)1.1703 37894	(17)1.2409 79300	(17)1.3153 23839	(17)1.3935 36672
9	(19)1.4893 74509	(19)1.5916 59535	(19)1.7001 41641	(19)1.8151 46897	(19)1.9370 15974
10	(21)2.0106 55587	(21)2.1646 56968	(21)2.3291 94048	(21)2.5049 02718	(21)2.6924 52204
24	(51)1.3427 97252	(51)1.6030 01028	(51)1.9111 44882	(51)2.2756 11258	(51)2.7061 70815
1/2	(1)1.1618 95004	(1)1.1661 90379	(1)1.1704 69991	(1)1.1747 34012	(1)1.1789 82612
1/3	5.1299 27840	5.1425 63181	5.1551 36735	5.1676 49252	5.1801 01467
1/4	3.4086 58099	3.4149 52970	3.4212 13222	3.4274 39296	3.4336 31623
1/5	2.6672 68608	2.6712 08461	2.6751 25206	2.6790 19145	2.6828 90577

k	140	141	142	143	144
1	140	141	142	143	144
2	19600	19881	20164	20449	20736
3	27 44000	28 03221	28 63288	29 24207	29 85984
4	3841 60000	3952 54161	4065 86896	4181 61601	4299 81696
5	(10)5.3782 40000	(10)5.5730 83670	(10)5.7735 33923	(10)5.9797 10894	(10)6.1917 36422
6	(12)7.5295 36000	(12)7.8580 47975	(12)8.1984 18171	(12)8.5509 86579	(12)8.9161 00448
7	(15)1.0541 35040	(15)1.1079 84764	(15)1.1641 75380	(15)1.2227 91081	(15)1.2839 18465
8	(17)1.4757 89056	(17)1.5622 58518	(17)1.6531 29040	(17)1.7485 91246	(17)1.8488 42589
9	(19)2.0661 04678	(19)2.2027 84510	(19)2.3474 43237	(19)2.5004 85481	(19)2.6623 33328
10	(21)2.8925 46550	(21)3.1059 26159	(21)3.3333 69396	(21)3.5756 94238	(21)3.8337 59992
24	(51)3.2141 99700	(51)3.8129 28871	(51)4.5177 29930	(51)5.3464 42484	(51)6.3197 48715
1/2	(1)1.1832 15957	(1)1.1874 34209	(1)1.1916 37529	(1)1.1958 26074	(1)1.2000 00000
1/3	5.1924 94102	5.2048 27863	5.2171 03446	5.2293 21532	5.2414 82788
1/4	3.4397 90628	3.4459 16727	3.4520 10326	3.4580 71824	3.4641 01615
1/5	2.6867 39790	2.6905 67070	2.6943 72696	2.6981 56943	2.7019 20077

k	145	146	147	148	149
1	145	146	147	148	149
2	21025	21316	21609	21904	22201
3	30 48625	31 12136	31 76523	32 41792	33 07949
4	4420 50625	4543 71856	4669 48881	4797 85216	4928 84401
5	(10)6.4097 34063	(10)6.6338 29098	(10)6.8641 48551	(10)7.1008 21197	(10)7.3439 77575
6	(12)9.2941 14391	(12)9.6853 90482	(13)1.0090 29837	(13)1.0509 21537	(13)1.0942 52659
7	(15)1.3476 46587	(15)1.4140 67010	(15)1.4832 73860	(15)1.5553 63875	(15)1.6304 36461
8	(17)1.9540 87551	(17)2.0645 37835	(17)2.1804 12575	(17)2.3019 38535	(17)2.4293 50327
9	(19)2.8334 26948	(19)3.0142 25239	(19)3.2052 06485	(19)3.4068 69032	(19)3.6197 31988
10	(21)4.1084 69075	(21)4.4007 68850	(21)4.7116 53533	(21)5.0421 66167	(21)5.3934 00662
24	(51)7.4616 01544	(51)8.7997 13625	(52)1.0366 11527	(52)1.2197 79049	(52)1.4337 40132
1/2	(1)1.2041 59458	(1)1.2083 04597	(1)1.2124 35565	(1)1.2165 52506	(1)1.2206 55562
1/3	5.2535 87872	5.2656 37428	5.2776 32088	5.2895 72473	5.3014 59192
1/4	3.4701 00082	3.4760 67602	3.4820 04545	3.4879 11275	3.4937 88147
1/5	2.7056 62363	2.7093 84058	2.7130 85417	2.7167 66686	2.7204 28110

$$n^{\frac{1}{2}}\left[{(-5)2 \atop 4}\right] \qquad n^{\frac{1}{3}}\left[{(-6)9 \atop 5}\right] \qquad n^{\frac{1}{4}}\left[{(-6)5 \atop 4}\right] \qquad n^{\frac{1}{5}}\left[{(-6)3 \atop 4}\right]$$

Table 3.1

POWERS AND ROOTS n^k

k	150	151	152	153	154
1	150	151	152	153	154
2	22500	22801	23104	23409	23716
3	33 75000	34 42951	35 11808	35 81577	36 52264
4	5062 50000	5198 85601	5337 94816	5479 81281	5624 48656
5	(10)7.5937 50000	(10)7.8502 72575	(10)8.1136 81203	(10)8.3841 13599	(10)8.6617 09302
6	(13)1.1390 62500	(13)1.1853 91159	(13)1.2332 79543	(13)1.2827 69381	(13)1.3339 03233
7	(15)1.7085 93750	(15)1.7899 40650	(15)1.8745 84905	(15)1.9626 37152	(15)2.0542 10978
8	(17)2.5628 90625	(17)2.7028 10381	(17)2.8493 69056	(17)3.0028 34843	(17)3.1634 84906
9	(19)3.8443 35938	(19)4.0812 43676	(19)4.3310 40965	(19)4.5943 37310	(19)4.8717 66756
10	(21)5.7665 03906	(21)6.1626 77950	(21)6.5831 82267	(21)7.0293 36085	(21)7.5025 20804
24	(52)1.6834 11220	(52)1.9744 52704	(52)2.3133 75387	(52)2.7076 61312	(52)3.1659 00782
1/2	(1)1.2247 44871	(1)1.2288 20573	(1)1.2328 82801	(1)1.2369 31688	(1)1.2409 67365
1/3	5.3132 92846	5.3250 74022	5.3368 03297	5.3484 81241	5.3601 08411
1/4	3.4996 35512	3.5054 53712	3.5112 43086	3.5170 03963	3.5227 36670
1/5	2.7240 69927	2.7276 92374	2.7312 95679	2.7348 80069	2.7384 45765

k	155	156	157	158	159
1	155	156	157	158	159
2	24025	24336	24649	24964	25281
3	37 23875	37 96416	38 69893	39 44312	40 19679
4	5772 00625	5922 40896	6075 73201	6232 01296	6391 28961
5	(10)8.9466 09688	(10)9.2389 57978	(10)9.5388 99256	(10)9.8465 80477	(11)1.0162 15048
6	(13)1.3867 24502	(13)1.4412 77445	(13)1.4976 07183	(13)1.5557 59715	(13)1.6157 81926
7	(15)2.1494 22977	(15)2.2483 92813	(15)2.3512 43278	(15)2.4581 00350	(15)2.5690 93263
8	(17)3.3316 05615	(17)3.5074 92789	(17)3.6914 51946	(17)3.8837 98553	(17)4.0848 58288
9	(19)5.1639 88703	(19)5.4716 88751	(19)5.7955 79555	(19)6.1364 01714	(19)6.4949 24678
10	(21)8.0041 82490	(21)8.5358 34451	(21)9.0990 59901	(21)9.6955 14709	(22)1.0326 93024
24	(52)3.6979 47627	(52)4.3150 94990	(52)5.0302 74186	(52)5.8582 79483	(52)6.8160 22003
1/2	(1)1.2449 89960	(1)1.2489 99600	(1)1.2529 96409	(1)1.2569 80509	(1)1.2609 52021
1/3	5.3716 85355	5.3832 12612	5.3946 90712	5.4061 20176	5.4175 01515
1/4	3.5284 41525	3.5341 18843	3.5397 68931	3.5453 92093	3.5509 88625
1/5	2.7419 92987	2.7455 21947	2.7490 32856	2.7525 25920	2.7560 01343

k	160	161	162	163	164
1	160	161	162	163	164
2	25600	25921	26244	26569	26896
3	40 96000	41 73281	42 51528	43 30747	44 10944
4	6553 60000	6718 98241	6887 47536	7059 11761	7233 94816
5	(11)1.0485 76000	(11)1.0817 56168	(11)1.1157 71008	(11)1.1506 36170	(11)1.1863 67498
6	(13)1.6777 21600	(13)1.7416 27430	(13)1.8075 49033	(13)1.8755 36958	(13)1.9456 42697
7	(15)2.6843 54560	(15)2.8040 20163	(15)2.9282 29434	(15)3.0571 25241	(15)3.1908 54023
8	(17)4.2949 67296	(17)4.5144 72463	(17)4.7437 31683	(17)4.9831 14143	(17)5.2330 00598
9	(19)6.8719 47674	(19)7.2683 00665	(19)7.6848 45327	(19)8.1224 76053	(19)8.5821 20981
10	(22)1.0995 11628	(22)1.1701 96407	(22)1.2449 44943	(22)1.3239 63597	(22)1.4074 67841
24	(52)7.9228 16251	(52)9.2007 03274	(53)1.0674 81480	(53)1.2373 78329	(53)1.4330 20335
1/2	(1)1.2649 11064	(1)1.2688 57754	(1)1.2727 92206	(1)1.2767 14533	(1)1.2806 24847
1/3	5.4288 35233	5.4401 21825	5.4513 61778	5.4625 55571	5.4737 03675
1/4	3.5565 58820	3.5621 02966	3.5676 21345	3.5731 14235	3.5785 81908
1/5	2.7594 59323	2.7629 00056	2.7663 23734	2.7697 30547	2.7731 20681

k	165	166	167	168	169
1	165	166	167	168	169
2	27225	27556	27889	28224	28561
3	44 92125	45 74296	46 57463	47 41632	48 26809
4	7412 00625	7593 33136	7777 96321	7965 94176	8157 30721
5	(11)1.2229 81031	(11)1.2604 93006	(11)1.2989 19856	(11)1.3382 78216	(11)1.3785 84918
6	(13)2.0179 18702	(13)2.0924 18390	(13)2.1691 96160	(13)2.2483 07402	(13)2.3298 08512
7	(15)3.3295 65858	(15)3.4734 14527	(15)3.6225 57587	(15)3.7771 56436	(15)3.9373 76386
8	(17)5.4937 83665	(17)5.7658 68114	(17)6.0496 71170	(17)6.3456 22812	(17)6.6541 66092
9	(19)9.0647 43047	(19)9.5713 41070	(20)1.0102 95085	(20)1.0660 64632	(20)1.1245 54070
10	(22)1.4956 82603	(22)1.5888 42618	(22)1.6871 92792	(22)1.7909 88583	(22)1.9004 96377
24	(53)1.6581 15050	(53)1.9168 76411	(53)2.2140 90189	(53)2.5551 87425	(53)2.9463 26763
1/2	(1)1.2845 23258	(1)1.2884 09873	(1)1.2922 84798	(1)1.2961 48140	(1)1.3000 00000
1/3	5.4848 06552	5.4958 64660	5.5068 78446	5.5178 48353	5.5287 74814
1/4	3.5840 24634	3.5894 42676	3.5948 36294	3.6002 05744	3.6055 51275
1/5	2.7764 94317	2.7798 51635	2.7831 92813	2.7865 18023	2.7898 27436

k	170	171	172	173	174
1	170	171	172	173	174
2	28900	29241	29584	29929	30276
3	49 13000	50 00211	50 88448	51 77717	52 68024
4	8352 10000	8550 36081	8752 13056	8957 45041	9166 36176
5	(11)1.4198 57000	(11)1.4621 11699	(11)1.5053 66456	(11)1.5496 38921	(11)1.5949 46946
6	(13)2.4137 56900	(13)2.5002 11004	(13)2.5892 30305	(13)2.6808 75333	(13)2.7752 07686
7	(15)4.1033 86730	(15)4.2753 60818	(15)4.4534 76124	(15)4.6379 14326	(15)4.8288 61374
8	(17)6.9757 57441	(17)7.3108 66998	(17)7.6599 78934	(17)8.0235 91785	(17)8.4022 18792
9	(20)1.1858 78765	(20)1.2501 58257	(20)1.3175 16377	(20)1.3880 81379	(20)1.4619 86070
10	(22)2.0159 93900	(22)2.1377 70619	(22)2.2661 28168	(22)2.4013 80785	(22)2.5438 55761
24	(53)3.3944 86713	(53)3.9075 68945	(53)4.4945 13878	(53)5.1654 29935	(53)5.9317 37979
1/2	(1)1.3038 40481	(1)1.3076 69683	(1)1.3114 87705	(1)1.3152 94644	(1)1.3190 90596
1/3	5.5396 58257	5.5504 99103	5.5612 97766	5.5720 54656	5.5827 70172
1/4	3.6108 73137	3.6161 71571	3.6214 46817	3.6266 99110	3.6319 28683
1/5	2.7931 21220	2.7963 99540	2.7996 62559	2.8029 10436	2.8061 43329

$$n^{\frac{1}{2}}\begin{bmatrix}(-5)2\\4\end{bmatrix} \qquad n^{\frac{1}{3}}\begin{bmatrix}(-6)7\\4\end{bmatrix} \qquad n^{\frac{1}{4}}\begin{bmatrix}(-6)4\\4\end{bmatrix} \qquad n^{\frac{1}{5}}\begin{bmatrix}(-6)3\\4\end{bmatrix}$$

POWERS AND ROOTS n^k Table 3.1

k	175	176	177	178	179
1	175	176	177	178	179
2	30625	30976	31329	31684	32041
3	53 59375	54 51776	55 45233	56 39752	57 35339
4	9378 90625	9595 12576	9815 06241	(9)1.0038 75856	(9)1.0266 25681
5	(11)1.6413 08594	(11)1.6887 42134	(11)1.7372 66047	(11)1.7868 99024	(11)1.8376 59969
6	(13)2.8722 90039	(13)2.9721 86155	(13)3.0749 60902	(13)3.1806 80262	(13)3.2894 11344
7	(15)5.0265 07568	(15)5.2310 47634	(15)5.4426 80797	(15)5.6616 10867	(15)5.8880 46307
8	(17)8.7963 88245	(17)9.2066 43835	(17)9.6335 45011	(18)1.0077 66734	(18)1.0539 60289
9	(20)1.5393 67943	(20)1.6203 69315	(20)1.7051 37467	(20)1.7938 24787	(20)1.8865 88917
10	(22)2.6938 93900	(22)2.8518 49994	(22)3.0180 93317	(22)3.1930 08121	(22)3.3769 74162
24	(53)6.8063 32613	(53)7.8037 62212	(53)8.9404 29702	(54)1.0234 81638	(54)1.1707 73122
1/2	(1)1.3228 75656	(1)1.3266 49916	(1)1.3304 13470	(1)1.3341 66406	(1)1.3379 08816
1/3	5.5934 44710	5.6040 78661	5.6146 72408	5.6252 26328	5.6357 40794
1/4	3.6371 35763	3.6423 20574	3.6474 83337	3.6526 24271	3.6577 43589
1/5	2.8093 61392	2.8125 64777	2.8157 53634	2.8189 28111	2.8220 88352

k	180	181	182	183	184
1	180	181	182	183	184
2	32400	32761	33124	33489	33856
3	58 32000	59 29741	60 28568	61 28407	62 29504
4	(9)1.0497 60000	(9)1.0732 83121	(9)1.0971 99376	(9)1.1215 13121	(9)1.1462 28736
5	(11)1.8895 68000	(11)1.9426 42449	(11)1.9969 02864	(11)2.0523 69011	(11)2.1090 60874
6	(13)3.4012 22400	(13)3.5161 82833	(13)3.6343 63213	(13)3.7558 35291	(13)3.8806 72009
7	(15)6.1222 00320	(15)6.3642 90927	(15)6.6145 41048	(15)6.8731 78582	(15)7.1404 36496
8	(18)1.1019 96058	(18)1.1519 36658	(18)1.2038 46471	(18)1.2577 91681	(18)1.3138 40315
9	(20)1.9835 92904	(20)2.0850 05351	(20)2.1910 00577	(20)2.3017 58775	(20)2.4174 66180
10	(22)3.5704 67227	(22)3.7738 59685	(22)3.9876 21050	(22)4.2122 18559	(22)4.4481 37771
24	(54)1.3382 58845	(54)1.5285 71637	(54)1.7446 70074	(54)1.9898 76639	(54)2.2679 20111
1/2	(1)1.3416 40786	(1)1.3453 62405	(1)1.3490 73756	(1)1.3527 74926	(1)1.3564 65997
1/3	5.6462 16173	5.6566 52826	5.6670 51108	5.6774 11371	5.6877 33960
1/4	3.6628 41501	3.6679 18217	3.6729 73940	3.6780 08871	3.6830 23210
1/5	2.8252 34501	2.8283 66697	2.8314 85080	2.8345 89786	2.8376 80950

k	185	186	187	188	189
1	185	186	187	188	189
2	34225	34596	34969	35344	35721
3	63 31625	64 34856	65 39203	66 44672	67 51269
4	(9)1.1713 50625	(9)1.1968 83216	(9)1.2228 30961	(9)1.2491 98336	(9)1.2759 89841
5	(11)2.1669 98656	(11)2.2262 02782	(11)2.2866 93897	(11)2.3484 92872	(11)2.4116 20799
6	(13)4.0089 47514	(13)4.1407 37174	(13)4.2761 17588	(13)4.4151 66599	(13)4.5579 63311
7	(15)7.4165 52901	(15)7.7017 71144	(15)7.9963 39889	(15)8.3005 13206	(15)8.6145 50658
8	(18)1.3720 62287	(18)1.4325 29433	(18)1.4953 15559	(18)1.5604 96483	(18)1.6281 50074
9	(20)2.5383 15230	(20)2.6645 04745	(20)2.7962 40096	(20)2.9337 33387	(20)3.0772 03640
10	(22)4.6958 83176	(22)4.9559 78826	(22)5.2289 68979	(22)5.5154 18768	(22)5.8159 14881
24	(54)2.5829 82606	(54)2.9397 51775	(54)3.3434 78670	(54)3.8000 41874	(54)4.3160 18526
1/2	(1)1.3601 47051	(1)1.3638 18170	(1)1.3674 79433	(1)1.3711 30920	(1)1.3747 72708
1/3	5.6980 19215	5.7082 67473	5.7184 79065	5.7286 54316	5.7387 93548
1/4	3.6880 17151	3.6929 90888	3.6979 44609	3.7028 78502	3.7077 92751
1/5	2.8407 58702	2.8438 23174	2.8468 74493	2.8499 12786	2.8529 38178

k	190	191	192	193	194
1	190	191	192	193	194
2	36100	36481	36864	37249	37636
3	68 59000	69 67871	70 77888	71 89057	73 01384
4	(9)1.3032 10000	(9)1.3308 63361	(9)1.3589 54496	(9)1.3874 88001	(9)1.4164 68496
5	(11)2.4760 99000	(11)2.5419 49020	(11)2.6091 92632	(11)2.6778 51842	(11)2.7479 48882
6	(13)4.7045 88100	(13)4.8551 22627	(13)5.0096 49854	(13)5.1682 54055	(13)5.3310 20832
7	(15)8.9387 17390	(15)9.2732 84218	(15)9.6185 27720	(15)9.9747 30326	(16)1.0342 18041
8	(18)1.6983 56304	(18)1.7711 97286	(18)1.8467 57322	(18)1.9251 22953	(18)2.0063 83000
9	(20)3.2268 76978	(20)3.3829 86816	(20)3.5457 74059	(20)3.7154 87299	(20)3.8923 83020
10	(22)6.1310 66258	(22)6.4615 04818	(22)6.8078 86193	(22)7.1708 90487	(22)7.5512 23059
24	(54)4.8987 62931	(54)5.5564 93542	(54)6.2983 89130	(54)7.1346 95065	(54)8.0768 40718
1/2	(1)1.3784 04875	(1)1.3820 27496	(1)1.3856 40646	(1)1.3892 44399	(1)1.3928 38828
1/3	5.7488 97079	5.7589 65220	5.7689 98281	5.7789 96565	5.7889 60372
1/4	3.7126 87538	3.7175 63041	3.7224 19436	3.7272 56899	3.7320 75599
1/5	2.8559 50791	2.8589 50746	2.8619 38162	2.8649 13156	2.8678 75844

k	195	196	197	198	199
1	195	196	197	198	199
2	38025	38416	38809	39204	39601
3	74 14875	75 29536	76 45373	77 62392	78 80599
4	(9)1.4459 00625	(9)1.4757 89056	(9)1.5061 38481	(9)1.5369 53616	(9)1.5682 39201
5	(11)2.8195 06219	(11)2.8925 46550	(11)2.9670 92808	(11)3.0431 68160	(11)3.1207 96010
6	(13)5.4980 37127	(13)5.6693 91238	(13)5.8451 72831	(13)6.0254 72956	(13)6.2103 84060
7	(16)1.0721 17240	(16)1.1112 00683	(16)1.1514 99048	(16)1.1930 43645	(16)1.2358 66428
8	(18)2.0906 28617	(18)2.1779 53338	(18)2.2684 53124	(18)2.3622 26418	(18)2.4593 74192
9	(20)4.0767 25804	(20)4.2687 88542	(20)4.4688 52654	(20)4.6772 08307	(20)4.8941 54641
10	(22)7.9496 15318	(22)8.3668 25543	(22)8.8036 39729	(22)9.2608 72448	(22)9.7393 67736
24	(54)9.1375 69069	(55)1.0331 07971	(55)1.1673 18660	(55)1.3181 49187	(55)1.4875 57746
1/2	(1)1.3964 24004	(1)1.4000 00000	(1)1.4035 66885	(1)1.4071 24728	(1)1.4106 73598
1/3	5.7988 89998	5.8087 85734	5.8186 47867	5.8284 76683	5.8382 72461
1/4	3.7368 75706	3.7416 57387	3.7464 20805	3.7511 66123	3.7558 93499
1/5	2.8708 26340	2.8737 64756	2.8766 91203	2.8796 05790	2.8825 08624

$$n^{\frac{1}{2}}\left[\begin{matrix}(-5)1\\4\end{matrix}\right] \qquad n^{\frac{1}{3}}\left[\begin{matrix}(-6)5\\4\end{matrix}\right] \qquad t^{\frac{1}{4}}\left[\begin{matrix}(-6)3\\4\end{matrix}\right] \qquad n^{\frac{1}{5}}\left[\begin{matrix}(-6)2\\4\end{matrix}\right]$$

Table 3.1   **POWERS AND ROOTS** n^k

k	200	201	202	203	204
1	200	201	202	203	204
2	40000	40401	40804	41209	41616
3	80 00000	81 20601	82 42408	83 65427	84 89664
4	(9)1.6000 00000	(9)1.6322 40801	(9)1.6649 66416	(9)1.6981 81681	(9)1.7318 91456
5	(11)3.2000 00000	(11)3.2808 04010	(11)3.3632 32160	(11)3.4473 08812	(11)3.5330 58570
6	(13)6.4000 00000	(13)6.5944 16060	(13)6.7937 28964	(13)6.9980 36889	(13)7.2074 39483
7	(16)1.2800 00000	(16)1.3254 77628	(16)1.3723 33251	(16)1.4206 01489	(16)1.4703 17655
8	(18)2.5600 00000	(18)2.6642 10032	(18)2.7721 13166	(18)2.8838 21022	(18)2.9994 48015
9	(20)5.1200 00000	(20)5.3550 62165	(20)5.5996 68596	(20)5.8541 56674	(20)6.1188 73951
10	(23)1.0240 00000	(23)1.0763 67495	(23)1.1311 33056	(23)1.1883 93805	(23)1.2482 50286
24	(55)1.6777 21600	(55)1.8910 60303	(55)2.1302 61246	(55)2.3983 07745	(55)2.6985 09916
1/2	(1)1.4142 13562	(1)1.4177 44688	(1)1.4212 67040	(1)1.4247 80685	(1)1.4282 85686
1/3	5.8480 35476	5.8577 66003	5.8674 64308	5.8771 30659	5.8867 65317
1/4	3.7606 03093	3.7652 95059	3.7699 69549	3.7746 26716	3.7792 66709
1/5	2.8853 99812	2.8882 79458	2.8911 47666	2.8940 04537	2.8968 50171

k	205	206	207	208	209
1	205	206	207	208	209
2	42025	42436	42849	43264	43681
3	86 15125	87 41816	88 69743	89 98912	91 29329
4	(9)1.7661 00625	(9)1.8008 14096	(9)1.8360 36801	(9)1.8717 73696	(9)1.9080 29761
5	(11)3.6205 06281	(11)3.7096 77038	(11)3.8005 96178	(11)3.8932 89288	(11)3.9877 82200
6	(13)7.4220 37877	(13)7.6419 34698	(13)7.8672 34089	(13)8.0980 41718	(13)8.3344 64799
7	(16)1.5215 17765	(16)1.5742 38548	(16)1.6285 17456	(16)1.6843 92677	(16)1.7419 03143
8	(18)3.1191 11418	(18)3.2429 31408	(18)3.3710 31135	(18)3.5035 36769	(18)3.6405 77569
9	(20)6.3941 78406	(20)6.6804 38701	(20)6.9780 34449	(20)7.2873 56480	(20)7.6088 07119
10	(23)1.3108 06573	(23)1.3761 70372	(23)1.4444 53131	(23)1.5157 70148	(23)1.5902 40688
24	(55)3.0345 38594	(55)3.4104 62581	(55)3.8307 89523	(55)4.3005 10765	(55)4.8251 50531
1/2	(1)1.4317 82106	(1)1.4352 70009	(1)1.4387 49457	(1)1.4422 20510	(1)1.4456 83229
1/3	5.8963 68540	5.9059 40584	5.9154 81700	5.9249 92137	5.9344 72140
1/4	3.7838 89674	3.7884 95756	3.7930 85099	3.7976 57844	3.8022 14131
1/5	2.8996 84668	2.9025 08125	2.9053 20638	2.9081 22302	2.9109 13212

k	210	211	212	213	214
1	210	211	212	213	214
2	44100	44521	44944	45369	45796
3	92 61000	93 93931	95 28128	96 63597	98 00344
4	(9)1.9448 10000	(9)1.9821 19441	(9)2.0199 63136	(9)2.0583 46161	(9)2.0972 73616
5	(11)4.0841 01000	(11)4.1822 72021	(11)4.2823 21848	(11)4.3842 77323	(11)4.4881 65538
6	(13)8.5766 12100	(13)8.8245 93963	(13)9.0785 22318	(13)9.3385 10698	(13)9.6046 74252
7	(16)1.8010 88541	(16)1.8619 89326	(16)1.9246 46732	(16)1.9891 02779	(16)2.0554 00290
8	(18)3.7822 85936	(18)3.9287 97478	(18)4.0802 51071	(18)4.2367 88919	(18)4.3985 56620
9	(20)7.9428 00466	(20)8.2897 62679	(20)8.6501 32270	(20)9.0243 60396	(20)9.4129 11168
10	(23)1.6679 88098	(23)1.7491 39925	(23)1.8338 28041	(23)1.9221 88764	(23)2.0143 62990
24	(55)5.4108 19838	(55)6.0642 75557	(55)6.7929 85105	(55)7.6051 97251	(55)8.5100 19601
1/2	(1)1.4491 37675	(1)1.4525 83905	(1)1.4560 21978	(1)1.4594 51952	(1)1.4628 73884
1/3	5.9439 21953	5.9533 41813	5.9627 31958	5.9720 92620	5.9814 24030
1/4	3.8067 54096	3.8112 77876	3.8157 85604	3.8202 77414	3.8247 53435
1/5	2.9136 93459	2.9164 63134	2.9192 22328	2.9219 71130	2.9247 09627

k	215	216	217	218	219
1	215	216	217	218	219
2	46225	46656	47089	47524	47961
3	99 38375	100 77696	102 18313	103 60232	105 03459
4	(9)2.1367 50625	(9)2.1767 82336	(9)2.2173 73921	(9)2.2585 30576	(9)2.3002 57521
5	(11)4.5940 13844	(11)4.7018 49846	(11)4.8117 01409	(11)4.9235 96656	(11)5.0375 63971
6	(13)9.8771 29764	(14)1.0155 99567	(14)1.0441 39206	(14)1.0733 44071	(14)1.1032 26510
7	(16)2.1235 82899	(16)2.1936 95064	(16)2.2657 82076	(16)2.3398 90075	(16)2.4160 66056
8	(18)4.5657 03233	(18)4.7383 81338	(18)4.9167 47106	(18)5.1009 60363	(18)5.2911 84663
9	(20)9.8162 61952	(21)1.0234 90369	(21)1.0669 34122	(21)1.1120 09359	(21)1.1587 69441
10	(23)2.1104 96320	(23)2.2107 39197	(23)2.3152 47045	(23)2.4241 80403	(23)2.5377 05076
24	(55)9.5175 03342	(56)1.0638 73589	(56)1.1885 94216	(56)1.3272 59512	(56)1.4813 53665
1/2	(1)1.4662 87830	(1)1.4696 93846	(1)1.4730 91986	(1)1.4764 82306	(1)1.4798 64859
1/3	5.9907 26415	6.0000 00000	6.0092 45007	6.0184 61655	6.0276 50160
1/4	3.8292 13796	3.8336 58625	3.8380 88048	3.8425 02187	3.8469 01167
1/5	2.9274 37906	2.9301 56052	2.9328 64149	2.9355 62280	2.9382 50529

k	220	221	222	223	224
1	220	221	222	223	224
2	48400	48841	49284	49729	50176
3	106 48000	107 93861	109 41048	110 89567	112 39424
4	(9)2.3425 60000	(9)2.3854 43281	(9)2.4289 12656	(9)2.4729 73441	(9)2.5176 30976
5	(11)5.1536 32000	(11)5.2718 29651	(11)5.3921 86096	(11)5.5147 30773	(11)5.6394 93386
6	(14)1.1337 99040	(14)1.1650 74353	(14)1.1970 65313	(14)1.2297 84962	(14)1.2632 46519
7	(16)2.4943 57888	(16)2.5748 14320	(16)2.6574 84996	(16)2.7424 20466	(16)2.8296 72201
8	(18)5.4875 87354	(18)5.6903 39647	(18)5.8996 16690	(18)6.1155 97640	(18)6.3384 65731
9	(21)1.2072 69218	(21)1.2575 65062	(21)1.3097 14905	(21)1.3637 78274	(21)1.4198 16324
10	(23)2.6559 92279	(23)2.7792 18787	(23)2.9075 67090	(23)3.0412 25550	(23)3.1803 88565
24	(56)1.6525 10926	(56)1.8425 30003	(56)2.0533 89736	(56)2.2872 66205	(56)2.5465 51362
1/2	(1)1.4832 39697	(1)1.4866 06875	(1)1.4899 66443	(1)1.4933 18452	(1)1.4966 62955
1/3	6.0368 10737	6.0459 43596	6.0550 48947	6.0641 26994	6.0731 77944
1/4	3.8512 85107	3.8556 54127	3.8600 08345	3.8643 47878	3.8686 72841
1/5	2.9409 28975	2.9435 97699	2.9462 56780	2.9489 06295	2.9515 46323

$$n^{\frac{1}{2}}\left[\begin{matrix}(-5)1\\4\end{matrix}\right] \qquad n^{\frac{1}{3}}\left[\begin{matrix}(-6)5\\4\end{matrix}\right] \qquad n^{\frac{1}{4}}\left[\begin{matrix}(-6)2\\4\end{matrix}\right] \qquad n^{\frac{1}{5}}\left[\begin{matrix}(-6)2\\4\end{matrix}\right]$$

POWERS AND ROOTS n^k Table 3.1

k	225	226	227	228	229
1	225	226	227	228	229
2	50625	51076	51529	51984	52441
3	113 90625	115 43176	116 97083	118 52352	120 08989
4	(9)2.5628 90625	(9)2.6087 57776	(9)2.6552 37841	(9)2.7023 36256	(9)2.7500 58481
5	(11)5.7665 03906	(11)5.8957 92574	(11)6.0273 89899	(11)6.1613 26664	(11)6.2976 33921
6	(14)1.2974 63379	(14)1.3324 49122	(14)1.3682 17507	(14)1.4047 82479	(14)1.4421 58168
7	(16)2.9192 92603	(16)3.0113 35015	(16)3.1058 53741	(16)3.2029 04053	(16)3.3025 42205
8	(18)6.5684 08356	(18)6.8056 17134	(18)7.0502 87992	(18)7.3026 21240	(18)7.5628 21649
9	(21)1.4778 91880	(21)1.5380 69472	(21)1.6004 15374	(21)1.6649 97643	(21)1.7318 86158
10	(23)3.3252 56730	(23)3.4760 37007	(23)3.6329 42900	(23)3.7961 94626	(23)3.9660 19301
24	(56)2.8338 73334	(56)3.1521 18526	(56)3.5044 55686	(56)3.8943 62082	(56)4.3256 51988
1/2	(1)1.5000 00000	(1)1.5033 29638	(1)1.5066 51917	(1)1.5099 66887	(1)1.5132 74595
1/3	6.0822 01996	6.0911 99349	6.1001 70200	6.1091 14744	6.1180 33173
1/4	3.8729 83346	3.8772 79507	3.8815 61435	3.8858 29238	3.8900 83026
1/5	2.9541 76939	2.9567 98218	2.9594 10235	2.9620 13062	2.9646 06773

k	230	231	232	233	234
1	230	231	232	233	234
2	52900	53361	53824	54289	54756
3	121 67000	123 26391	124 87168	126 49337	128 12904
4	(9)2.7984 10000	(9)2.8473 96321	(9)2.8970 22976	(9)2.9472 95521	(9)2.9982 19536
5	(11)6.4363 43000	(11)6.5774 85502	(11)6.7210 93304	(11)6.8671 98564	(11)7.0158 33714
6	(14)1.4803 58890	(14)1.5193 99151	(14)1.5592 93647	(14)1.6000 57265	(14)1.6417 05089
7	(16)3.4048 25447	(16)3.5098 12038	(16)3.6175 61260	(16)3.7281 33428	(16)3.8415 89909
8	(18)7.8310 98528	(18)8.1076 65809	(18)8.3927 42123	(18)8.6865 50888	(18)8.9893 20386
9	(21)1.8011 52661	(21)1.8728 70802	(21)1.9471 16173	(21)2.0239 66357	(21)2.1035 00970
10	(23)4.1426 51121	(23)4.3263 31552	(23)4.5173 09521	(23)4.7158 41612	(23)4.9221 92271
24	(56)4.8025 07640	(56)5.3295 12896	(56)5.9116 89798	(56)6.5545 38287	(56)7.2640 79321
1/2	(1)1.5165 75089	(1)1.5198 68415	(1)1.5231 54621	(1)1.5264 33752	(1)1.5297 05854
1/3	6.1269 25675	6.1357 92440	6.1446 33651	6.1534 49494	6.1622 40148
1/4	3.8943 22905	3.8985 48980	3.9027 61357	3.9069 60138	3.9111 45426
1/5	2.9671 91438	2.9697 67129	2.9723 33915	2.9748 91866	2.9774 41049

k	235	236	237	238	239
1	235	236	237	238	239
2	55225	55696	56169	56644	57121
3	129 77875	131 44256	133 12053	134 81272	136 51919
4	(9)3.0498 00625	(9)3.1020 44416	(9)3.1549 56561	(9)3.2085 42736	(9)3.2628 08641
5	(11)7.1670 31469	(11)7.3208 24822	(11)7.4772 47050	(11)7.6363 31712	(11)7.7981 12652
6	(14)1.6842 52395	(14)1.7277 14658	(14)1.7721 07551	(14)1.8174 46947	(14)1.8637 48924
7	(16)3.9579 93129	(16)4.0774 06593	(16)4.1998 94895	(16)4.3255 23735	(16)4.4543 59928
8	(18)9.3012 83852	(18)9.6226 79559	(18)9.9537 50902	(19)1.0294 74649	(19)1.0645 92023
9	(21)2.1858 01705	(21)2.2709 52376	(21)2.3590 38964	(21)2.4501 49664	(21)2.5443 74934
10	(23)5.1366 34007	(23)5.3594 47607	(23)5.5909 22344	(23)5.8313 56201	(23)6.0810 56093
24	(56)8.0469 01671	(56)8.9102 12697	(56)9.8618 93410	(57)1.0910 55818	(57)1.2065 61943
1/2	(1)1.5329 70972	(1)1.5362 29150	(1)1.5394 80432	(1)1.5427 24862	(1)1.5459 62483
1/3	6.1710 05793	6.1797 46606	6.1884 62762	6.1971 54435	6.2058 21795
1/4	3.9153 17320	3.9194 75921	3.9236 21327	3.9277 53635	3.9318 72942
1/5	2.9799 81531	2.9825 13380	2.9850 36660	2.9875 51438	2.9900 57776

k	240	241	242	243	244
1	240	241	242	243	244
2	57600	58081	58564	59049	59536
3	138 24000	139 97521	141 72488	143 48907	145 26784
4	(9)3.3177 60000	(9)3.3734 02561	(9)3.4297 42096	(9)3.4867 84401	(9)3.5445 35296
5	(11)7.9626 24000	(11)8.1299 00172	(11)8.2999 75872	(11)8.4728 86094	(11)8.6486 66122
6	(14)1.9110 29760	(14)1.9593 05941	(14)2.0085 94161	(14)2.0589 11321	(14)2.1102 74534
7	(16)4.5864 71424	(16)4.7219 27319	(16)4.8607 97870	(16)5.0031 54510	(16)5.1490 69863
8	(19)1.1007 53142	(19)1.1379 84484	(19)1.1763 13085	(19)1.2157 66546	(19)1.2563 73046
9	(21)2.6418 07540	(21)2.7425 42606	(21)2.8466 77065	(21)2.9543 12707	(21)3.0655 50233
10	(23)6.3403 38097	(23)6.6095 27681	(23)6.8889 59948	(23)7.1789 79877	(23)7.4799 42569
24	(57)1.3337 35777	(57)1.4736 99791	(57)1.6276 79087	(57)1.7970 10300	(57)1.9831 51223
1/2	(1)1.5491 93338	(1)1.5524 17470	(1)1.5556 34919	(1)1.5588 45727	(1)1.5620 49935
1/3	6.2144 65012	6.2230 84253	6.2316 79684	6.2402 51469	6.2487 99770
1/4	3.9359 79343	3.9400 72930	3.9441 53798	3.9482 22039	3.9522 77742
1/5	2.9925 55740	2.9950 45390	2.9975 26790	3.0000 00000	3.0024 65081

k	245	246	247	248	249
1	245	246	247	248	249
2	60025	60516	61009	61504	62001
3	147 06125	148 86936	150 69223	152 52992	154 38249
4	(9)3.6030 00625	(9)3.6621 86256	(9)3.7220 98081	(9)3.7827 42016	(9)3.8441 24001
5	(11)8.8273 51531	(11)9.0089 78190	(11)9.1935 82260	(11)9.3812 00200	(11)9.5718 68762
6	(14)2.1627 01125	(14)2.2162 08635	(14)2.2708 14818	(14)2.3265 37650	(14)2.3833 95322
7	(16)5.2986 17757	(16)5.4518 73241	(16)5.6089 12601	(16)5.7698 13371	(16)5.9346 54351
8	(19)1.2981 61350	(19)1.3411 60817	(19)1.3854 01412	(19)1.4309 13716	(19)1.4777 28934
9	(21)3.1804 95308	(21)3.2992 55611	(21)3.4219 41489	(21)3.5486 66016	(21)3.6795 45044
10	(23)7.7922 13506	(23)8.1161 68802	(23)8.4521 95477	(23)8.8006 91719	(23)9.1620 67161
24	(57)2.1876 91225	(57)2.4123 62509	(57)2.6590 52293	(57)2.9298 15956	(57)3.2268 91257
1/2	(1)1.5652 47584	(1)1.5684 38714	(1)1.5716 23365	(1)1.5748 01575	(1)1.5779 73384
1/3	6.2573 24746	6.2658 26556	6.2743 05357	6.2827 61305	6.2911 94552
1/4	3.9563 20998	3.9603 51896	3.9643 70523	3.9683 76966	3.9723 71312
1/5	3.0049 22094	3.0073 71096	3.0098 12147	3.0122 45305	3.0146 70627

$$n^{\frac{1}{2}}\left[\begin{matrix}(-6)9\\4\end{matrix}\right] \qquad n^{\frac{1}{3}}\left[\begin{matrix}(-6)3\\4\end{matrix}\right] \qquad n^{\frac{1}{4}}\left[\begin{matrix}(-6)2\\4\end{matrix}\right] \qquad n^{\frac{1}{5}}\left[\begin{matrix}(-6)1\\4\end{matrix}\right]$$

Table 3.1   **POWERS AND ROOTS** n^k

k	250	251	252	253	254
1	250	251	252	253	254
2	62500	63001	63504	64009	64516
3	156 25000	158 13251	160 03008	161 94277	163 87064
4	(9)3.9062 50000	(9)3.9691 26001	(9)4.0327 58016	(9)4.0971 52081	(9)4.1623 14256
5	(11)9.7656 25000	(11)9.9625 06263	(12)1.0162 55020	(12)1.0365 79476	(12)1.0572 27821
6	(14)2.4414 06250	(14)2.5005 89072	(14)2.5609 62650	(14)2.6225 46076	(14)2.6853 58665
7	(16)6.1035 15625	(16)6.2764 78570	(16)6.4536 25879	(16)6.6350 41571	(16)6.8208 11010
8	(19)1.5258 78906	(19)1.5753 96121	(19)1.6263 13472	(19)1.6786 65517	(19)1.7324 85997
9	(21)3.8146 97266	(21)3.9542 44264	(21)4.0983 10578	(21)4.2470 23759	(21)4.4005 14431
10	(23)9.5367 43164	(23)9.9251 53103	(24)1.0327 74266	(24)1.0744 97011	(24)1.1177 30666
24	(57)3.5527 13679	(57)3.9099 33001	(57)4.3014 31179	(57)4.7303 41643	(57)5.2000 70108
1/2	(1)1.5811 38830	(1)1.5842 97952	(1)1.5874 50787	(1)1.5905 97372	(1)1.5937 37745
1/3	6.2996 05249	6.3079 93549	6.3163 59598	6.3247 03543	6.3330 25531
1/4	3.9763 53644	3.9803 24047	3.9842 82604	3.9882 29397	3.9921 64507
1/5	3.0170 88168	3.0194 97986	3.0219 00136	3.0242 94671	3.0266 81647

k	255	256	257	258	259
1	255	256	257	258	259
2	65025	65536	66049	66564	67081
3	165 81375	167 77216	169 74593	171 73512	173 73979
4	(9)4.2282 50625	(9)4.2949 67296	(9)4.3624 70401	(9)4.4307 66096	(9)4.4998 60561
5	(12)1.0782 03909	(12)1.0995 11628	(12)1.1211 54893	(12)1.1431 37653	(12)1.1654 63886
6	(14)2.7494 19969	(14)2.8147 49767	(14)2.8813 68075	(14)2.9492 95144	(14)3.0185 51463
7	(16)7.0110 20921	(16)7.2057 59404	(16)7.4051 15953	(16)7.6091 81472	(16)7.8180 48289
8	(19)1.7878 10335	(19)1.8446 74407	(19)1.9031 14800	(19)1.9631 68820	(19)2.0248 74507
9	(21)4.5589 16354	(21)4.7223 66483	(21)4.8910 03036	(21)5.0649 75555	(21)5.2444 24973
10	(24)1.1625 23670	(24)1.2089 25820	(24)1.2569 88294	(24)1.3067 63693	(24)1.3583 06068
24	(57)5.7143 17018	(57)6.2771 01735	(57)6.8927 88615	(57)7.5661 15089	(57)8.3022 21920
1/2	(1)1.5968 71942	(1)1.6000 00000	(1)1.6031 21954	(1)1.6062 37840	(1)1.6093 47694
1/3	6.3413 25705	6.3496 04208	6.3578 61180	6.3660 96760	6.3743 11088
1/4	3.9960 88015	4.0000 00000	4.0039 00541	4.0077 89716	4.0116 67601
1/5	3.0290 61117	3.0314 33133	3.0337 97748	3.0361 55014	3.0385 04982

k	260	261	262	263	264
1	260	261	262	263	264
2	67600	68121	68644	69169	69696
3	175 76000	177 79581	179 84728	181 91447	183 99744
4	(9)4.5697 60000	(9)4.6404 70641	(9)4.7119 98736	(9)4.7843 50561	(9)4.8575 32416
5	(12)1.1881 37600	(12)1.2111 62837	(12)1.2345 43669	(12)1.2582 84198	(12)1.2823 88558
6	(14)3.0891 57760	(14)3.1611 35005	(14)3.2345 04412	(14)3.3092 87440	(14)3.3855 05793
7	(16)8.0318 10176	(16)8.2505 62364	(16)8.4744 01560	(16)8.7034 25966	(16)8.9377 35293
8	(19)2.0882 70646	(19)2.1533 96777	(19)2.2202 93209	(19)2.2890 01029	(19)2.3595 62117
9	(21)5.4295 03679	(21)5.6203 65588	(21)5.8171 68207	(21)6.0200 72706	(21)6.2292 43990
10	(24)1.4116 70957	(24)1.4669 15418	(24)1.5240 98070	(24)1.5832 79122	(24)1.6445 20413
24	(57)9.1066 85770	(57)9.9855 54265	(58)1.0945 38372	(58)1.1993 27974	(58)1.3136 94086
1/2	(1)1.6124 51550	(1)1.6155 49442	(1)1.6186 41406	(1)1.6217 27474	(1)1.6248 07681
1/3	6.3825 04299	6.3906 76528	6.3988 27910	6.4069 58577	6.4150 68660
1/4	4.0155 34273	4.0193 89807	4.0232 34278	4.0270 67760	4.0308 90325
1/5	3.0408 47703	3.0431 83226	3.0455 11602	3.0478 32879	3.0501 47105

k	265	266	267	268	269
1	265	266	267	268	269
2	70225	70756	71289	71824	72361
3	186 09625	188 21096	190 34163	192 48832	194 65109
4	(9)4.9315 50625	(9)5.0064 11536	(9)5.0821 21521	(9)5.1586 86976	(9)5.2361 14321
5	(12)1.3068 60916	(12)1.3317 05469	(12)1.3569 26446	(12)1.3825 28110	(12)1.4085 14752
6	(14)3.4631 81426	(14)3.5423 36546	(14)3.6229 93611	(14)3.7051 75334	(14)3.7889 04684
7	(16)9.1774 30780	(16)9.4226 15213	(16)9.6733 92947	(16)9.9298 69894	(17)1.0192 15360
8	(19)2.4320 19157	(19)2.5064 15647	(19)2.5827 95915	(19)2.6612 05132	(19)2.7416 89318
9	(21)6.4448 50765	(21)6.6670 65620	(21)6.8960 65094	(21)7.1320 29753	(21)7.3751 44266
10	(24)1.7078 85453	(24)1.7734 39455	(24)1.8412 49380	(24)1.9113 83974	(24)1.9839 13808
24	(58)1.4384 70548	(58)1.5745 60235	(58)1.7229 40472	(58)1.8846 88868	(58)2.0608 89564
1/2	(1)1.6278 82060	(1)1.6309 50643	(1)1.6340 13464	(1)1.6370 70554	(1)1.6401 21947
1/3	6.4231 58289	6.4312 27591	6.4392 76696	6.4473 05727	6.4553 14811
1/4	4.0347 02045	4.0385 02994	4.0422 93240	4.0460 72854	4.0498 41906
1/5	3.0524 54329	3.0547 54599	3.0570 47961	3.0593 34462	3.0616 14147

k	270	271	272	273	274
1	270	271	272	273	274
2	72900	73441	73984	74529	75076
3	196 83000	199 02511	201 23648	203 46417	205 70824
4	(9)5.3144 10000	(9)5.3935 80481	(9)5.4736 32256	(9)5.5545 71841	(9)5.6364 05776
5	(12)1.4348 90700	(12)1.4616 60310	(12)1.4888 27974	(12)1.5163 98113	(12)1.5443 75183
6	(14)3.8742 04890	(14)3.9610 99441	(14)4.0496 12088	(14)4.1397 66847	(14)4.2315 88000
7	(17)1.0460 35320	(17)1.0734 57949	(17)1.1014 94488	(17)1.1301 56349	(17)1.1594 55112
8	(19)2.8242 95365	(19)2.9090 71041	(19)2.9960 65007	(19)3.0853 26834	(19)3.1769 07007
9	(21)7.6255 97485	(21)7.8835 82520	(21)8.1492 96820	(21)8.4229 42256	(21)8.7047 25200
10	(24)2.0589 11321	(24)2.1364 50863	(24)2.2166 08735	(24)2.2994 63236	(24)2.3850 94705
24	(58)2.2528 39954	(58)2.4618 57897	(58)2.6893 89450	(58)2.9369 97176	(58)3.2063 69049
1/2	(1)1.6431 67673	(1)1.6462 07763	(1)1.6492 42250	(1)1.6522 71164	(1)1.6552 94536
1/3	6.4633 04070	6.4712 73627	6.4792 23603	6.4871 54117	6.4950 65288
1/4	4.0536 00464	4.0573 48596	4.0610 86370	4.0648 13851	4.0685 31106
1/5	3.0638 87063	3.0661 53254	3.0684 12765	3.0706 65640	3.0729 11923

$$n^{\frac{1}{2}}\left[\begin{matrix}(-6)8\\4\end{matrix}\right] \qquad n^{\frac{1}{3}}\left[\begin{matrix}(-6)3\\4\end{matrix}\right] \qquad n^{\frac{1}{4}}\left[\begin{matrix}(-6)2\\4\end{matrix}\right] \qquad n^{\frac{1}{5}}\left[\begin{matrix}(-6)1\\4\end{matrix}\right]$$

POWERS AND ROOTS n^k Table 3.1

k	275	276	277	278	279
1	275	276	277	278	279
2	75625	76176	76729	77284	77841
3	207 96875	210 24576	212 53933	214 84952	217 17639
4	(9)5.7191 40625	(9)5.8027 82976	(9)5.8873 39441	(9)5.9728 16656	(9)6.0592 21281
5	(12)1.5727 63672	(12)1.6015 68101	(12)1.6307 93025	(12)1.6604 43030	(12)1.6905 22737
6	(14)4.3251 00098	(14)4.4203 27960	(14)4.5172 96680	(14)4.6160 31624	(14)4.7165 58437
7	(17)1.1894 02527	(17)1.2200 10517	(17)1.2512 91180	(17)1.2832 56792	(17)1.3159 19804
8	(19)3.2708 56949	(19)3.3672 29027	(19)3.4660 76569	(19)3.5674 53881	(19)3.6714 16253
9	(21)8.9948 56609	(21)9.2935 52114	(21)9.6010 32097	(21)9.9175 21788	(22)1.0243 25135
10	(24)2.4735 85568	(24)2.5650 20383	(24)2.6594 85891	(24)2.7570 71057	(24)2.8578 67126
24	(58)3.4993 28001	(58)3.8178 42160	(58)4.1640 35828	(58)4.5402 01230	(58)4.9488 11121
1/2	(1)1.6583 12395	(1)1.6613 24773	(1)1.6643 31698	(1)1.6673 33200	(1)1.6703 29309
1/3	6.5029 57234	6.5108 30071	6.5186 83915	6.5265 18879	6.5343 35077
1/4	4.0722 38199	4.0759 35196	4.0796 22161	4.0832 99156	4.0869 66245
1/5	3.0751 51657	3.0773 84885	3.0796 11650	3.0818 31992	3.0840 45954

k	280	281	282	283	284
1	280	281	282	283	284
2	78400	78961	79524	80089	80656
3	219 52000	221 88041	224 25768	226 65187	229 06304
4	(9)6.1465 60000	(9)6.2348 39521	(9)6.3240 66576	(9)6.4142 47921	(9)6.5053 90336
5	(12)1.7210 36800	(12)1.7519 89905	(12)1.7833 86774	(12)1.8152 32162	(12)1.8475 30855
6	(14)4.8189 03040	(14)4.9230 91634	(14)5.0291 50704	(14)5.1371 07017	(14)5.2469 87629
7	(17)1.3492 92851	(17)1.3833 88749	(17)1.4182 20498	(17)1.4538 01286	(17)1.4901 44487
8	(19)3.7780 19983	(19)3.8873 22385	(19)3.9993 81806	(19)4.1142 57639	(19)4.2320 10342
9	(22)1.0578 45595	(22)1.0923 37590	(22)1.1278 25669	(22)1.1643 34912	(22)1.2018 90937
10	(24)2.9619 67667	(24)3.0694 68629	(24)3.1804 68387	(24)3.2950 67801	(24)3.4133 70262
24	(58)5.3925 32264	(58)5.8742 39885	(58)6.3970 33126	(58)6.9642 51599	(58)7.5794 93086
1/2	(1)1.6733 20053	(1)1.6763 05461	(1)1.6792 85562	(1)1.6822 60384	(1)1.6852 29955
1/3	6.5421 32620	6.5499 11620	6.5576 72186	6.5654 14427	6.5731 38451
1/4	4.0906 23489	4.0942 70950	4.0979 08689	4.1015 36766	4.1051 55240
1/5	3.0862 53577	3.0884 54901	3.0906 49967	3.0928 38815	3.0950 21484

k	285	286	287	288	289
1	285	286	287	288	289
2	81225	81796	82369	82944	83521
3	231 49125	233 93656	236 39903	238 87872	241 37569
4	(9)6.5975 00625	(9)6.6905 85616	(9)6.7846 52161	(9)6.8797 07136	(9)6.9757 57441
5	(12)1.8802 87678	(12)1.9135 07486	(12)1.9471 95170	(12)1.9813 55655	(12)2.0159 93900
6	(14)5.3588 19883	(14)5.4726 31410	(14)5.5884 50138	(14)5.7063 04287	(14)5.8262 22372
7	(17)1.5272 63667	(17)1.5651 72583	(17)1.6038 85190	(17)1.6434 15635	(17)1.6837 78266
8	(19)4.3527 01450	(19)4.4763 93589	(19)4.6031 50495	(19)4.7330 37028	(19)4.8661 19188
9	(22)1.2405 19913	(22)1.2802 48566	(22)1.3211 04192	(22)1.3631 14664	(22)1.4063 08445
10	(24)3.5354 81753	(24)3.6615 10900	(24)3.7915 69031	(24)3.9257 70232	(24)4.0642 31407
24	(58)8.2466 32480	(58)8.9698 42039	(58)9.7536 13040	(59)1.0602 77893	(59)1.1522 54005
1/2	(1)1.6881 94302	(1)1.6911 53453	(1)1.6941 07435	(1)1.6970 56275	(1)1.7000 00000
1/3	6.5808 44365	6.5885 32275	6.5962 02284	6.6038 54498	6.6114 89018
1/4	4.1087 64171	4.1123 63618	4.1159 53637	4.1195 34288	4.1231 05626
1/5	3.0971 98013	3.0993 68441	3.1015 32807	3.1036 91148	3.1058 43502

k	290	291	292	293	294
1	290	291	292	293	294
2	84100	84681	85264	85849	86436
3	243 89000	246 42171	248 97088	251 53757	254 12184
4	(9)7.0728 10000	(9)7.1708 71761	(9)7.2699 49696	(9)7.3700 50801	(9)7.4711 82096
5	(12)2.0511 14900	(12)2.0867 23682	(12)2.1228 25311	(12)2.1594 24885	(12)2.1965 27536
6	(14)5.9482 33210	(14)6.0723 65916	(14)6.1986 49909	(14)6.3271 14912	(14)6.4577 90956
7	(17)1.7249 87631	(17)1.7670 58482	(17)1.8100 05773	(17)1.8538 44669	(17)1.8985 90541
8	(19)5.0024 64130	(19)5.1421 40181	(19)5.2852 16858	(19)5.4317 64881	(19)5.5818 56191
9	(22)1.4507 14598	(22)1.4963 62793	(22)1.5432 83323	(22)1.5915 07110	(22)1.6410 65720
10	(24)4.2070 72333	(24)4.3544 15727	(24)4.5063 87302	(24)4.6631 15833	(24)4.8247 33217
24	(59)1.2518 49008	(59)1.3596 64428	(59)1.4763 46962	(59)1.6025 91698	(59)1.7391 45550
1/2	(1)1.7029 38637	(1)1.7058 72211	(1)1.7088 00749	(1)1.7117 24277	(1)1.7146 42820
1/3	6.6191 05948	6.6267 05387	6.6342 87437	6.6418 52195	6.6493 99761
1/4	4.1266 67707	4.1302 20588	4.1337 64325	4.1372 98970	4.1408 24580
1/5	3.1079 89906	3.1101 30396	3.1122 65011	3.1143 93785	3.1165 16755

k	295	296	297	298	299
1	295	296	297	298	299
2	87025	87616	88209	88804	89401
3	256 72375	259 34336	261 98073	264 63592	267 30899
4	(9)7.5733 50625	(9)7.6765 63456	(9)7.7808 27681	(9)7.8861 50416	(9)7.9925 38801
5	(12)2.2341 38434	(12)2.2722 62783	(12)2.3109 05821	(12)2.3500 72824	(12)2.3897 69101
6	(14)6.5907 08381	(14)6.7258 97838	(14)6.8633 90289	(14)7.0032 17015	(14)7.1454 09613
7	(17)1.9442 58973	(17)1.9908 65760	(17)2.0384 26916	(17)2.0869 58671	(17)2.1364 77474
8	(19)5.7355 63969	(19)5.8929 62649	(19)6.0541 27940	(19)6.2191 36838	(19)6.3880 67649
9	(22)1.6919 91371	(22)1.7443 16944	(22)1.7980 75998	(22)1.8533 02778	(22)1.9100 32227
10	(24)4.9913 74544	(24)5.1631 78155	(24)5.3402 85715	(24)5.5228 42278	(24)5.7109 96358
24	(59)1.8868 10930	(59)2.0464 49657	(59)2.2189 87131	(59)2.4054 16789	(59)2.6068 04847
1/2	(1)1.7175 56404	(1)1.7204 65053	(1)1.7233 68794	(1)1.7262 67650	(1)1.7291 61647
1/3	6.6569 30232	6.6644 43703	6.6719 40272	6.6794 20032	6.6868 83077
1/4	4.1443 41207	4.1478 48904	4.1513 47726	4.1548 37723	4.1583 18947
1/5	3.1186 33956	3.1207 45423	3.1228 51191	3.1249 51295	3.1270 45768

$$n^{\frac{1}{2}}\begin{bmatrix}(-6)7\\4\end{bmatrix} \qquad n^{\frac{1}{3}}\begin{bmatrix}(-6)2\\4\end{bmatrix} \qquad n^{\frac{1}{4}}\begin{bmatrix}(-6)1\\4\end{bmatrix} \qquad n^{\frac{1}{5}}\begin{bmatrix}(-7)8\\4\end{bmatrix}$$

Table 3.1   **POWERS AND ROOTS** n^k

k	300	301	302	303	304
1	300	301	302	303	304
2	90000	90601	91204	91809	92416
3	270 00000	272 70901	275 43608	278 18127	280 94464
4	(9)8.1000 00000	(9)8.2085 41201	(9)8.3181 69616	(9)8.4288 92481	(9)8.5407 17056
5	(12)2.4300 00000	(12)2.4707 70902	(12)2.5120 87224	(12)2.5539 54422	(12)2.5963 77985
6	(14)7.2900 00000	(14)7.4370 20414	(14)7.5865 03417	(14)7.7384 81898	(14)7.8929 89074
7	(17)2.1870 00000	(17)2.2385 43144	(17)2.2911 24032	(17)2.3447 60015	(17)2.3994 68679
8	(19)6.5610 00000	(19)6.7380 14865	(19)6.9191 94576	(19)7.1046 22846	(19)7.2943 84783
9	(22)1.9683 00000	(22)2.0281 42474	(22)2.0895 96762	(22)2.1527 00722	(22)2.2174 92974
10	(24)5.9049 00000	(24)6.1047 08848	(24)6.3105 82221	(24)6.5226 83188	(24)6.7411 78641
24	(59)2.8242 95365	(59)3.0591 15639	(59)3.3125 81949	(59)3.5861 05682	(59)3.8811 99856
1/2	(1)1.7320 50808	(1)1.7349 35157	(1)1.7378 14720	(1)1.7406 89519	(1)1.7435 59577
1/3	6.6943 29501	6.7017 59395	6.7091 72852	6.7165 69962	6.7239 50814
1/4	4.1617 91450	4.1652 55283	4.1687 10496	4.1721 57138	4.1755 95260
1/5	3.1291 34645	3.1312 17958	3.1332 95743	3.1353 68030	3.1374 34853

k	305	306	307	308	309
1	305	306	307	308	309
2	93025	93636	94249	94864	95481
3	283 72625	286 52616	289 34443	292 18112	295 03629
4	(9)8.6536 50625	(9)8.7677 00496	(9)8.8828 74001	(9)8.9991 78496	(9)9.1166 21361
5	(12)2.6393 63441	(12)2.6829 16352	(12)2.7270 42318	(12)2.7717 46977	(12)2.8170 36001
6	(14)8.0500 58494	(14)8.2097 24036	(14)8.3720 19917	(14)8.5369 80688	(14)8.7046 41242
7	(17)2.4552 67841	(17)2.5121 75555	(17)2.5702 10115	(17)2.6293 90052	(17)2.6897 34144
8	(19)7.4885 66914	(19)7.6872 57199	(19)7.8905 45052	(19)8.0985 21360	(19)8.3112 78504
9	(22)2.2840 12909	(22)2.3523 00703	(22)2.4223 97331	(22)2.4943 44579	(22)2.5681 85058
10	(24)6.9662 39372	(24)7.1980 40151	(24)7.4367 59806	(24)7.6825 81303	(24)7.9356 91828
24	(59)4.1994 86063	(59)4.5427 01868	(59)4.9127 08679	(59)5.3115 00125	(59)5.7412 10972
1/2	(1)1.7464 24920	(1)1.7492 85568	(1)1.7521 41547	(1)1.7549 92877	(1)1.7578 39583
1/3	6.7313 15497	6.7386 64101	6.7459 96712	6.7533 13417	6.7606 14302
1/4	4.1790 24910	4.1824 46136	4.1858 58988	4.1892 63512	4.1926 59756
1/5	3.1394 96244	3.1415 52236	3.1436 02859	3.1456 48146	3.1476 88127

k	310	311	312	313	314
1	310	311	312	313	314
2	96100	96721	97344	97969	98596
3	297 91000	300 80231	303 71328	306 64297	309 59144
4	(9)9.2352 10000	(9)9.3549 51841	(9)9.4758 54336	(9)9.5979 24961	(9)9.7211 71216
5	(12)2.8629 15100	(12)2.9093 90023	(12)2.9564 66553	(12)3.0041 50513	(12)3.0524 47762
6	(14)8.8750 36810	(14)9.0482 02970	(14)9.2241 75645	(14)9.4029 91105	(14)9.5846 85972
7	(17)2.7512 61411	(17)2.8139 91124	(17)2.8779 42801	(17)2.9431 36216	(17)3.0095 91395
8	(19)8.5289 10374	(19)8.7515 12395	(19)8.9791 81540	(19)9.2120 16356	(19)9.4501 16981
9	(22)2.6439 62216	(22)2.7217 20355	(22)2.8015 04640	(22)2.8833 61119	(22)2.9673 36732
10	(24)8.1962 82870	(24)8.4645 50303	(24)8.7406 94478	(24)9.0249 20304	(24)9.3174 37339
24	(59)6.2041 26610	(59)6.7026 93132	(59)7.2395 28072	(59)7.8174 31800	(59)8.4393 99655
1/2	(1)1.7606 81686	(1)1.7635 19209	(1)1.7663 52173	(1)1.7691 80601	(1)1.7720 04515
1/3	6.7678 99452	6.7751 68952	6.7824 22886	6.7896 61336	6.7968 84386
1/4	4.1960 47767	4.1994 27591	4.2027 99273	4.2061 62861	4.2095 18398
1/5	3.1497 22833	3.1517 52295	3.1537 76544	3.1557 95609	3.1578 09519

k	315	316	317	318	319
1	315	316	317	318	319
2	99225	99856	1 00489	1 01124	1 01761
3	312 55875	315 54496	318 55013	321 57432	324 61759
4	(9)9.8456 00625	(9)9.9712 20736	(10)1.0098 03912	(10)1.0226 06338	(10)1.0355 30112
5	(12)3.1013 64197	(12)3.1509 05753	(12)3.2010 78401	(12)3.2518 88154	(12)3.3033 41058
6	(14)9.7692 97220	(14)9.9568 62178	(15)1.0147 41853	(15)1.0341 00433	(15)1.0537 65797
7	(17)3.0773 28624	(17)3.1463 68448	(17)3.2167 31675	(17)3.2884 39376	(17)3.3615 12894
8	(19)9.6935 85167	(19)9.9425 24297	(20)1.0197 03941	(20)1.0457 23722	(20)1.0723 22613
9	(22)3.0534 79328	(22)3.1418 37678	(22)3.2324 61493	(22)3.3254 01435	(22)3.4207 09136
10	(24)9.6184 59882	(24)9.9282 07062	(25)1.0240 90293	(25)1.0574 77656	(25)1.0912 06214
24	(59)9.1086 34822	(59)9.8285 62028	(60)1.0602 84208	(60)1.1435 38734	(60)1.2330 37808
1/2	(1)1.7748 23935	(1)1.7776 38883	(1)1.7804 49381	(1)1.7832 55450	(1)1.7860 57110
1/3	6.8040 92116	6.8112 84608	6.8184 61941	6.8256 24197	6.8327 71452
1/4	4.2128 65931	4.2162 05502	4.2195 37156	4.2228 60938	4.2261 76889
1/5	3.1598 18306	3.1618 21997	3.1638 20622	3.1658 14209	3.1678 02787

k	320	321	322	323	324
1	320	321	322	323	324
2	1 02400	1 03041	1 03684	1 04329	1 04976
3	327 68000	330 76161	333 86248	336 98267	340 12224
4	(10)1.0485 76000	(10)1.0617 44768	(10)1.0750 37186	(10)1.0884 54024	(10)1.1019 96058
5	(12)3.3554 43200	(12)3.4082 00706	(12)3.4616 19738	(12)3.5157 06498	(12)3.5704 67227
6	(15)1.0737 41824	(15)1.0940 32426	(15)1.1146 41556	(15)1.1355 73199	(15)1.1568 31381
7	(17)3.4359 73837	(17)3.5118 44089	(17)3.5891 45809	(17)3.6679 01432	(17)3.7481 33676
8	(20)1.0995 11628	(20)1.1273 01953	(20)1.1557 04950	(20)1.1847 32163	(20)1.2143 95311
9	(22)3.5184 37209	(22)3.6186 39268	(22)3.7213 69940	(22)3.8266 84885	(22)3.9346 40808
10	(25)1.1258 99907	(25)1.1615 83205	(25)1.1982 81121	(25)1.2360 19218	(25)1.2748 23622
24	(60)1.3292 27996	(60)1.4325 86248	(60)1.5436 21862	(60)1.6628 78568	(60)1.7909 36736
1/2	(1)1.7888 54382	(1)1.7916 47287	(1)1.7944 35844	(1)1.7972 20076	(1)1.8000 00000
1/3	6.8399 03787	6.8470 21278	6.8541 24002	6.8612 12036	6.8682 85455
1/4	4.2294 85054	4.2327 85474	4.2360 78192	4.2393 63249	4.2426 40687
1/5	3.1697 86385	3.1717 65030	3.1737 38749	3.1757 07571	3.1776 71523

$$n^{\frac{1}{2}}\left[\begin{matrix}(-6)6\\4\end{matrix}\right] \qquad n^{\frac{1}{3}}\left[\begin{matrix}(-6)2\\4\end{matrix}\right] \qquad n^{\frac{1}{4}}\left[\begin{matrix}(-6)1\\4\end{matrix}\right] \qquad n^{\frac{1}{5}}\left[\begin{matrix}(-7)7\\4\end{matrix}\right]$$

POWERS AND ROOTS n^k Table 3.1

k	325	326	327	328	329
1	325	326	327	328	329
2	1 05625	1 06276	1 06929	1 07584	1 08241
3	343 28125	346 45976	349 65783	352 87552	356 11289
4	(10)1.1156 64063	(10)1.1294 58818	(10)1.1433 81104	(10)1.1574 31706	(10)1.1716 11408
5	(12)3.6259 08203	(12)3.6820 35745	(12)3.7388 56210	(12)3.7963 75994	(12)3.8546 01533
6	(15)1.1784 20166	(15)1.2003 43653	(15)1.2226 05981	(15)1.2452 11326	(15)1.2681 63904
7	(17)3.8298 65540	(17)3.9131 20309	(17)3.9979 21557	(17)4.0842 93150	(17)4.1722 59245
8	(20)1.2447 06300	(20)1.2756 77221	(20)1.3073 20349	(20)1.3396 48153	(20)1.3726 73292
9	(22)4.0452 95476	(22)4.1587 07739	(22)4.2749 37542	(22)4.3940 45942	(22)4.5160 95129
10	(25)1.3147 21030	(25)1.3557 38723	(25)1.3979 04576	(25)1.4412 47069	(25)1.4857 95298
24	(60)1.9284 15722	(60)2.0759 76350	(60)2.2343 23554	(60)2.4042 09169	(60)2.5864 34894
1/2	(1)1.8027 75638	(1)1.8055 47009	(1)1.8083 14132	(1)1.8110 77028	(1)1.8138 35715
1/3	6.8753 44335	6.8823 88750	6.8894 18774	6.8964 34481	6.9034 35942
1/4	4.2459 10547	4.2491 72871	4.2524 27697	4.2556 75067	4.2589 15020
1/5	3.1796 30632	3.1815 84924	3.1835 34426	3.1854 79164	3.1874 19165

k	330	331	332	333	334
1	330	331	332	333	334
2	1 08900	1 09561	1 10224	1 10889	1 11556
3	359 37000	362 64691	365 94368	369 26037	372 59704
4	(10)1.1859 21000	(10)1.2003 61272	(10)1.2149 33018	(10)1.2296 37032	(10)1.2444 74114
5	(12)3.9135 39300	(12)3.9731 95811	(12)4.0335 77618	(12)4.0946 91317	(12)4.1565 43539
6	(15)1.2914 67969	(15)1.3151 27813	(15)1.3391 47769	(15)1.3635 32209	(15)1.3882 85542
7	(17)4.2618 44298	(17)4.3530 73062	(17)4.4459 70594	(17)4.5405 62254	(17)4.6368 73711
8	(20)1.4064 08618	(20)1.4408 67184	(20)1.4760 62237	(20)1.5120 07231	(20)1.5487 15819
9	(22)4.6411 48440	(22)4.7692 70378	(22)4.9005 26628	(22)5.0349 84078	(22)5.1727 10837
10	(25)1.5315 78985	(25)1.5786 28495	(25)1.6269 74840	(25)1.6766 49698	(25)1.7276 85420
24	(60)2.7818 55434	(60)2.9913 81825	(60)3.2159 84959	(60)3.4566 99320	(60)3.7146 26935
1/2	(1)1.8165 90212	(1)1.8193 40540	(1)1.8220 86716	(1)1.8248 28759	(1)1.8275 66688
1/3	6.9104 23230	6.9173 96417	6.9243 55573	6.9313 00768	6.9382 32074
1/4	4.2621 47595	4.2653 72832	4.2685 90770	4.2718 01446	4.2750 04899
1/5	3.1893 54454	3.1912 85058	3.1932 11001	3.1951 32308	3.1970 49006

k	335	336	337	338	339
1	335	336	337	338	339
2	1 12225	1 12896	1 13569	1 14244	1 14921
3	375 95375	379 33056	382 72753	386 14472	389 58219
4	(10)1.2594 45063	(10)1.2745 50682	(10)1.2897 91776	(10)1.3051 69154	(10)1.3206 83624
5	(12)4.2191 40959	(12)4.2824 90290	(12)4.3465 98285	(12)4.4114 71739	(12)4.4771 17486
6	(15)1.4134 12221	(15)1.4389 16737	(15)1.4648 03622	(15)1.4910 77448	(15)1.5177 42828
7	(17)4.7349 30942	(17)4.8347 60238	(17)4.9363 88207	(17)5.0398 41774	(17)5.1451 48186
8	(20)1.5862 01865	(20)1.6244 79440	(20)1.6635 62826	(20)1.7034 66520	(20)1.7442 05235
9	(22)5.3137 76249	(22)5.4582 50918	(22)5.6062 06723	(22)5.7577 16836	(22)5.9128 55747
10	(25)1.7801 15044	(25)1.8339 72309	(25)1.8892 91666	(25)1.9461 08291	(25)2.0044 58098
24	(60)3.9909 41565	(60)4.2868 93134	(60)4.6038 12427	(60)4.9431 16051	(60)5.3063 11693
1/2	(1)1.8303 00522	(1)1.8330 30278	(1)1.8357 55975	(1)1.8384 77631	(1)1.8411 95264
1/3	6.9451 49558	6.9520 53290	6.9589 43337	6.9658 19768	6.9726 82649
1/4	4.2782 01166	4.2813 90286	4.2845 72295	4.2877 47230	4.2909 15128
1/5	3.1989 61118	3.2008 68669	3.2027 71684	3.2046 70186	3.2065 64201

k	340	341	342	343	344
1	340	341	342	343	344
2	1 15600	1 16281	1 16964	1 17649	1 18336
3	393 04000	396 51821	400 01688	403 53607	407 07584
4	(10)1.3363 36000	(10)1.3521 27096	(10)1.3680 57730	(10)1.3841 28720	(10)1.4003 40890
5	(12)4.5435 42400	(12)4.6107 53398	(12)4.6787 57435	(12)4.7475 61510	(12)4.8171 72660
6	(15)1.5448 04416	(15)1.5722 66909	(15)1.6001 35043	(15)1.6284 13598	(15)1.6571 07395
7	(17)5.2523 35014	(17)5.3614 30158	(17)5.4724 61847	(17)5.5854 58641	(17)5.7004 49439
8	(20)1.7857 93905	(20)1.8282 47684	(20)1.8715 81952	(20)1.9158 12314	(20)1.9609 54607
9	(22)6.0716 99277	(22)6.2343 24602	(22)6.4008 10274	(22)6.5712 36236	(22)6.7456 83848
10	(25)2.0643 77754	(25)2.1259 04689	(25)2.1890 77114	(25)2.2539 34029	(25)2.3205 15244
24	(60)5.6950 03680	(60)6.1108 98859	(60)6.5558 12822	(60)7.0316 76479	(60)7.5405 43015
1/2	(1)1.8439 08891	(1)1.8466 18531	(1)1.8493 24201	(1)1.8520 25918	(1)1.8547 23699
1/3	6.9795 32047	6.9863 68028	6.9931 90657	7.0000 00000	7.0067 96121
1/4	4.2940 76026	4.2972 29958	4.3003 76961	4.3035 17071	4.3066 50321
1/5	3.2084 53751	3.2103 38860	3.2122 19552	3.2140 95850	3.2159 67776

k	345	346	347	348	349
1	345	346	347	348	349
2	1 19025	1 19716	1 20409	1 21104	1 21801
3	410 63625	414 21736	417 81923	421 44192	425 08549
4	(10)1.4166 95063	(10)1.4331 92066	(10)1.4498 32728	(10)1.4666 17882	(10)1.4835 48360
5	(12)4.8875 97966	(12)4.9588 44547	(12)5.0309 19567	(12)5.1038 30228	(12)5.1775 83777
6	(15)1.6862 21298	(15)1.7157 60213	(15)1.7457 29090	(15)1.7761 32919	(15)1.8069 76738
7	(17)5.8174 63479	(17)5.9365 30338	(17)6.0576 79941	(17)6.1809 42559	(17)6.3063 48816
8	(20)2.0070 24900	(20)2.0540 39497	(20)2.1020 14939	(20)2.1509 68011	(20)2.2009 15737
9	(22)6.9242 35905	(22)7.1069 76659	(22)7.2939 91840	(22)7.4853 68677	(22)7.6811 95921
10	(25)2.3888 61387	(25)2.4590 13924	(25)2.5310 15168	(25)2.6049 08300	(25)2.6807 37377
24	(60)8.0845 95243	(60)8.6661 53376	(60)9.2876 83235	(60)9.9518 04932	(61)1.0661 30203
1/2	(1)1.8574 17562	(1)1.8601 07524	(1)1.8627 93601	(1)1.8654 75811	(1)1.8681 54169
1/3	7.0135 79083	7.0203 48952	7.0271 05788	7.0338 49656	7.0405 80617
1/4	4.3097 76748	4.3128 96386	4.3160 09269	4.3191 15431	4.3222 14906
1/5	3.2178 35355	3.2196 98608	3.2215 57557	3.2234 12226	3.2252 62636

$$n^{\frac{1}{2}}\left[\begin{matrix}(-6)5\\4\end{matrix}\right] \qquad n^{\frac{1}{3}}\left[\begin{matrix}(-6)2\\4\end{matrix}\right] \qquad n^{\frac{1}{4}}\left[\begin{matrix}(-6)1\\4\end{matrix}\right] \qquad n^{\frac{1}{5}}\left[\begin{matrix}(-7)6\\4\end{matrix}\right]$$

Table 3.1   **POWERS AND ROOTS** n^k

k	350	351	352	353	354
1					
2	1 22500	1 23201	1 23904	1 24609	1 25316
3	428 75000	432 43551	436 14208	439 86977	443 61864
4	(10)1.5006 25000	(10)1.5178 48640	(10)1.5352 20122	(10)1.5527 40288	(10)1.5704 09986
5	(12)5.2521 87500	(12)5.3276 48727	(12)5.4039 74828	(12)5.4811 73217	(12)5.5592 51349
6	(15)1.8382 65625	(15)1.8700 04703	(15)1.9021 99139	(15)1.9348 54146	(15)1.9679 74978
7	(17)6.4339 29688	(17)6.5637 16508	(17)6.6957 40971	(17)6.8300 35134	(17)6.9666 31421
8	(20)2.2518 75391	(20)2.3038 64494	(20)2.3569 00822	(20)2.4110 02402	(20)2.4661 87523
9	(22)7.8815 63867	(22)8.0865 64375	(22)8.2962 90893	(22)8.5108 38480	(22)8.7303 03831
10	(25)2.7585 47354	(25)2.8383 84096	(25)2.9202 94394	(25)3.0043 25983	(25)3.0905 27556
24	(61)1.1419 13124	(61)1.2228 43263	(61)1.3092 54042	(61)1.4014 99442	(61)1.4999 55202
1/2	(1)1.8708 28693	(1)1.8734 99400	(1)1.8761 66304	(1)1.8788 29423	(1)1.8814 88772
1/3	7.0472 98732	7.0540 04063	7.0606 96671	7.0673 76615	7.0740 43955
1/4	4.3253 07727	4.3283 93928	4.3314 73541	4.3345 46600	4.3376 13137
1/5	3.2271 08809	3.2289 50768	3.2307 88532	3.2326 22125	3.2344 51567

k	355	356	357	358	359
1					
2	1 26025	1 26736	1 27449	1 28164	1 28881
3	447 38875	451 18016	454 99293	458 82712	462 68279
4	(10)1.5882 30063	(10)1.6062 01370	(10)1.6243 24760	(10)1.6426 01090	(10)1.6610 31216
5	(12)5.6382 16722	(12)5.7180 76876	(12)5.7988 39394	(12)5.8805 11901	(12)5.9631 02066
6	(15)2.0015 66936	(15)2.0356 35368	(15)2.0701 85663	(15)2.1052 23260	(15)2.1407 53642
7	(17)7.1055 62624	(17)7.2468 61909	(17)7.3905 62819	(17)7.5366 99273	(17)7.6853 05573
8	(20)2.5224 74731	(20)2.5798 82840	(20)2.6384 30926	(20)2.6981 38340	(20)2.7590 24701
9	(22)8.9547 85297	(22)9.1843 82909	(22)9.4191 98407	(22)9.6593 35256	(22)9.9048 98676
10	(25)3.1789 48780	(25)3.2696 40316	(25)3.3626 53831	(25)3.4580 42022	(25)3.5558 58625
24	(61)1.6050 20092	(61)1.7171 17251	(61)1.8366 95605	(61)1.9642 31355	(61)2.1002 29556
1/2	(1)1.8841 44368	(1)1.8867 96226	(1)1.8894 44363	(1)1.8920 88793	(1)1.8947 29532
1/3	7.0806 98751	7.0873 41061	7.0939 70945	7.1005 88459	7.1071 93661
1/4	4.3406 73183	4.3437 26771	4.3467 73933	4.3498 14700	4.3528 49104
1/5	3.2362 76880	3.2380 98084	3.2399 15199	3.2417 28247	3.2435 37249

k	360	361	362	363	364
1					
2	1 29600	1 30321	1 31044	1 31769	1 32496
3	466 56000	470 45881	474 37928	478 32147	482 28544
4	(10)1.6796 16000	(10)1.6983 56304	(10)1.7172 52994	(10)1.7363 06936	(10)1.7555 19002
5	(12)6.0466 17600	(12)6.1310 66258	(12)6.2164 55837	(12)6.3027 94178	(12)6.3900 89166
6	(15)2.1767 82336	(15)2.2133 14919	(15)2.2503 57013	(15)2.2879 14287	(15)2.3259 92456
7	(17)7.8364 16410	(17)7.9900 66858	(17)8.1462 92387	(17)8.3051 28860	(17)8.4666 12541
8	(20)2.8211 09907	(20)2.8844 14136	(20)2.9489 57680	(20)3.0147 61776	(20)3.0818 46965
9	(23)1.0155 99567	(23)1.0412 73503	(23)1.0675 22740	(23)1.0943 58525	(23)1.1217 92295
10	(25)3.6561 58440	(25)3.7589 97346	(25)3.8644 32317	(25)3.9725 21445	(25)4.0833 23955
24	(61)2.2452 25771	(61)2.3997 87825	(61)2.5645 17652	(61)2.7400 53237	(61)2.9270 70667
1/2	(1)1.8973 66596	(1)1.9000 00000	(1)1.9026 29759	(1)1.9052 55888	(1)1.9078 78403
1/3	7.1137 86609	7.1203 67359	7.1269 35967	7.1334 92490	7.1400 36982
1/4	4.3558 77175	4.3588 98944	4.3619 14441	4.3649 23697	4.3679 26743
1/5	3.2453 42223	3.2471 43191	3.2489 40172	3.2507 33187	3.2525 22254

k	365	366	367	368	369
1					
2	1 33225	1 33956	1 34689	1 35424	1 36161
3	486 27125	490 27896	494 30863	498 36032	502 43409
4	(10)1.7748 90063	(10)1.7944 20994	(10)1.8141 12672	(10)1.8339 65978	(10)1.8539 81792
5	(12)6.4783 48728	(12)6.5675 80837	(12)6.6577 93507	(12)6.7489 94798	(12)6.8411 92813
6	(15)2.3645 97286	(15)2.4037 34586	(15)2.4434 10217	(15)2.4836 30086	(15)2.5244 00148
7	(17)8.6307 80093	(17)8.7976 68585	(17)8.9673 15496	(17)9.1397 58715	(17)9.3150 36546
8	(20)3.1502 34734	(20)3.2199 46702	(20)3.2910 04787	(20)3.3634 31207	(20)3.4372 48485
9	(23)1.1498 35678	(23)1.1785 00493	(23)1.2077 98757	(23)1.2377 42684	(23)1.2683 44691
10	(25)4.1969 00224	(25)4.3133 11804	(25)4.4326 21438	(25)4.5548 93078	(25)4.6801 91910
24	(61)3.1262 86296	(61)3.3384 59019	(61)3.5643 92671	(61)3.8049 38558	(61)4.0609 98114
1/2	(1)1.9104 97317	(1)1.9131 12647	(1)1.9157 24406	(1)1.9183 32609	(1)1.9209 37271
1/3	7.1465 69499	7.1530 90095	7.1595 98825	7.1660 95742	7.1725 80900
1/4	4.3709 23607	4.3739 14319	4.3768 98909	4.3798 77406	4.3828 49839
1/5	3.2543 07394	3.2560 88625	3.2578 65967	3.2596 39439	3.2614 09059

k	370	371	372	373	374
1					
2	1 36900	1 37641	1 38384	1 39129	1 39876
3	506 53000	510 64811	514 78848	518 95117	523 13624
4	(10)1.8741 61000	(10)1.8945 04488	(10)1.9150 13146	(10)1.9356 87864	(10)1.9565 29538
5	(12)6.9343 95700	(12)7.0286 11651	(12)7.1238 48902	(12)7.2201 15733	(12)7.3174 20471
6	(15)2.5657 26409	(15)2.6076 14922	(15)2.6500 71791	(15)2.6931 03168	(15)2.7367 15256
7	(17)9.4931 87713	(17)9.6742 51362	(17)9.8582 67064	(18)1.0045 27482	(18)1.0235 31506
8	(20)3.5124 79454	(20)3.5891 47255	(20)3.6672 75348	(20)3.7468 87507	(20)3.8280 07832
9	(23)1.2996 17398	(23)1.3315 73632	(23)1.3642 26429	(23)1.3975 89040	(23)1.4316 74929
10	(25)4.8085 84372	(25)4.9401 38174	(25)5.0749 22317	(25)5.2130 07120	(25)5.3544 64234
24	(61)4.3335 25711	(61)4.6235 31606	(61)4.9320 85051	(61)5.2603 17567	(61)5.6094 26383
1/2	(1)1.9235 38406	(1)1.9261 36028	(1)1.9287 30152	(1)1.9313 20792	(1)1.9339 07961
1/3	7.1790 54352	7.1855 16151	7.1919 66348	7.1984 04996	7.2048 32147
1/4	4.3858 16237	4.3887 76627	4.3917 31039	4.3946 79501	4.3976 22040
1/5	3.2631 74848	3.2649 36822	3.2666 95001	3.2684 49404	3.2702 00047

$$n^{\frac{1}{2}}\begin{bmatrix}(-6)5\\4\end{bmatrix} \qquad n^{\frac{1}{3}}\begin{bmatrix}(-6)2\\4\end{bmatrix} \qquad n^{\frac{1}{4}}\begin{bmatrix}(-7)8\\4\end{bmatrix} \qquad n^{\frac{1}{5}}\begin{bmatrix}(-7)5\\4\end{bmatrix}$$

POWERS AND ROOTS n^k Table 3.1

k	375	376	377	378	379
1	375	376	377	378	379
2	1 40625	1 41376	1 42129	1 42884	1 43641
3	527 34375	531 57376	535 82633	540 10152	544 39939
4	(10)1.9775 39063	(10)1.9987 17338	(10)2.0200 65264	(10)2.0415 83746	(10)2.0632 73688
5	(12)7.4157 71484	(12)7.5151 77189	(12)7.6156 46046	(12)7.7171 86558	(12)7.8198 07278
6	(15)2.7809 14307	(15)2.8257 06623	(15)2.8710 98559	(15)2.9170 96519	(15)2.9637 06958
7	(18)1.0428 42865	(18)1.0624 65690	(18)1.0824 04157	(18)1.1026 62484	(18)1.1232 44937
8	(20)3.9106 60744	(20)3.9948 70996	(20)4.0806 63671	(20)4.1680 64190	(20)4.2570 98312
9	(23)1.4664 97779	(23)1.5020 71494	(23)1.5384 10204	(23)1.5755 28264	(23)1.6134 40260
10	(25)5.4993 66671	(25)5.6477 88819	(25)5.7998 06469	(25)5.9554 96838	(25)6.1149 38586
24	(61)5.9806 78067	(61)6.3754 12334	(61)6.7950 46060	(61)7.2410 77507	(61)7.7150 90756
1/2	(1)1.9364 91673	(1)1.9390 71943	(1)1.9416 48784	(1)1.9442 22210	(1)1.9467 92233
1/3	7.2112 47852	7.2176 52160	7.2240 45124	7.2304 26792	7.2367 97216
1/4	4.4005 58684	4.4034 89461	4.4064 14397	4.4093 33520	4.4122 46858
1/5	3.2719 46950	3.2736 90130	3.2754 29605	3.2771 65392	3.2788 97510

k	380	381	382	383	384
1	380	381	382	383	384
2	1 44400	1 45161	1 45924	1 46689	1 47456
3	548 72000	553 06341	557 42968	561 81887	566 23104
4	(10)2.0851 36000	(10)2.1071 71592	(10)2.1293 81378	(10)2.1517 66272	(10)2.1743 27194
5	(12)7.9235 16800	(12)8.0283 23766	(12)8.1342 36862	(12)8.2412 64822	(12)8.3494 16423
6	(15)3.0109 36384	(15)3.0587 91355	(15)3.1072 78481	(15)3.1564 04427	(15)3.2061 75907
7	(18)1.1441 55826	(18)1.1653 99506	(18)1.1869 80380	(18)1.2089 02895	(18)1.2311 71548
8	(20)4.3477 92138	(20)4.4401 72119	(20)4.5342 65051	(20)4.6300 98090	(20)4.7276 98745
9	(23)1.6521 61013	(23)1.6917 05577	(23)1.7320 89250	(23)1.7733 27568	(23)1.8154 36318
10	(25)6.2782 11848	(25)6.4453 98249	(25)6.6165 80933	(25)6.7918 44587	(25)6.9712 75461
24	(61)8.2187 60383	(61)8.7538 56362	(61)9.3222 49236	(61)9.9259 15535	(62)1.0566 94349
1/2	(1)1.9493 58869	(1)1.9519 22130	(1)1.9544 82029	(1)1.9570 38579	(1)1.9595 91794
1/3	7.2431 56443	7.2495 04524	7.2558 41507	7.2621 67440	7.2684 82371
1/4	4.4151 54436	4.4180 56280	4.4209 52418	4.4238 42876	4.4267 27679
1/5	3.2806 25976	3.2823 50807	3.2840 72019	3.2857 89631	3.2875 03659

k	385	386	387	388	389
1	385	386	387	388	389
2	1 48225	1 48996	1 49769	1 50544	1 51321
3	570 66625	575 12456	579 60603	584 11072	588 63869
4	(10)2.1970 65063	(10)2.2199 80802	(10)2.2430 75336	(10)2.2663 49594	(10)2.2898 04504
5	(12)8.4587 00491	(12)8.5691 25894	(12)8.6807 01551	(12)8.7934 36423	(12)8.9073 39521
6	(15)3.2565 99689	(15)3.3076 82595	(15)3.3594 31500	(15)3.4118 53332	(15)3.4649 55074
7	(18)1.2537 90880	(18)1.2767 65482	(18)1.3000 99991	(18)1.3237 99093	(18)1.3478 67524
8	(20)4.8270 94889	(20)4.9283 14759	(20)5.0313 86963	(20)5.1363 40480	(20)5.2432 04667
9	(23)1.8584 31532	(23)1.9023 29497	(23)1.9471 46755	(23)1.9929 00106	(23)2.0396 06615
10	(25)7.1549 61399	(25)7.3429 91859	(25)7.5354 57941	(25)7.7324 52413	(25)7.9340 69734
24	(62)1.1247 53901	(62)1.1970 03202	(62)1.2736 88303	(62)1.3550 69013	(62)1.4414 19629
1/2	(1)1.9621 41687	(1)1.9646 88270	(1)1.9672 31557	(1)1.9697 71560	(1)1.9723 08292
1/3	7.2747 86349	7.2810 79420	7.2873 61631	7.2936 33030	7.2998 93662
1/4	4.4296 06853	4.4324 80423	4.4353 48416	4.4382 10856	4.4410 67768
1/5	3.2892 14120	3.2909 21030	3.2926 24406	3.2943 24265	3.2960 20622

k	390	391	392	393	394
1	390	391	392	393	394
2	1 52100	1 52881	1 53664	1 54449	1 55236
3	593 19000	597 76471	602 36288	606 98457	611 62984
4	(10)2.3134 41000	(10)2.3372 60016	(10)2.3612 62490	(10)2.3854 49360	(10)2.4098 21570
5	(12)9.0224 19900	(12)9.1386 86663	(12)9.2561 48959	(12)9.3748 15985	(12)9.4946 96984
6	(15)3.5187 43761	(15)3.5732 26485	(15)3.6284 10392	(15)3.6843 02682	(15)3.7409 10612
7	(18)1.3723 10067	(18)1.3971 31556	(18)1.4223 36874	(18)1.4479 30954	(18)1.4739 18781
8	(20)5.3520 09260	(20)5.4627 84383	(20)5.5755 60545	(20)5.6903 68650	(20)5.8072 39997
9	(23)2.0872 83612	(23)2.1359 48694	(23)2.1856 19734	(23)2.2363 14879	(23)2.2880 52559
10	(25)8.1404 06085	(25)8.3515 59392	(25)8.5756 29356	(25)8.7887 17476	(25)9.0149 27082
24	(62)1.5330 29700	(62)1.6302 04837	(62)1.7332 67559	(62)1.8425 58176	(62)1.9584 35730
1/2	(1)1.9748 41766	(1)1.9773 71993	(1)1.9798 98987	(1)1.9824 22760	(1)1.9849 43324
1/3	7.3061 43574	7.3123 82812	7.3186 11420	7.3248 29445	7.3310 36930
1/4	4.4439 19178	4.4467 65109	4.4496 05586	4.4524 40634	4.4552 70277
1/5	3.2977 13494	3.2994 02898	3.3010 88848	3.3027 71361	3.3044 50453

k	395	396	397	398	399
1	395	396	397	398	399
2	1 56025	1 56816	1 57609	1 58404	1 59201
3	616 29875	620 99136	625 70773	630 44792	635 21199
4	(10)2.4343 80063	(10)2.4591 25786	(10)2.4840 59688	(10)2.5091 82722	(10)2.5344 95840
5	(12)9.6158 01247	(12)9.7381 38111	(12)9.8617 16962	(12)9.9865 47232	(13)1.0112 63840
6	(15)3.7982 41493	(15)3.8563 02692	(15)3.9151 01634	(15)3.9746 45798	(15)4.0349 42722
7	(18)1.5003 05390	(18)1.5270 95866	(18)1.5542 95349	(18)1.5819 09028	(18)1.6099 42146
8	(20)5.9262 06289	(20)6.0472 99629	(20)6.1705 52534	(20)6.2959 97930	(20)6.4236 69163
9	(23)2.3408 51484	(23)2.3947 30653	(23)2.4497 09356	(23)2.5058 07176	(23)2.5630 43996
10	(25)9.2463 63362	(25)9.4831 33387	(25)9.7253 46143	(25)9.9731 12562	(26)1.0226 54554
24	(62)2.0812 78965	(62)2.2114 87364	(62)2.3494 82217	(62)2.4957 07762	(62)2.6506 32365
1/2	(1)1.9874 60691	(1)1.9899 74874	(1)1.9924 85885	(1)1.9949 93734	(1)1.9974 98436
1/3	7.3372 33921	7.3434 20462	7.3495 96597	7.3557 62368	7.3619 17821
1/4	4.4580 94538	4.4609 13443	4.4637 27013	4.4665 35273	4.4693 38246
1/5	3.3061 26138	3.3077 98433	3.3094 67354	3.3111 32914	3.3127 95131

$$n^{\frac{1}{2}}\left[\begin{matrix}(-6)4\\4\end{matrix}\right] \qquad n^{\frac{1}{3}}\left[\begin{matrix}(-6)2\\4\end{matrix}\right] \qquad n^{\frac{1}{4}}\left[\begin{matrix}(-7)8\\4\end{matrix}\right] \qquad n^{\frac{1}{5}}\left[\begin{matrix}(-7)5\\4\end{matrix}\right]$$

Table 3.1   **POWERS AND ROOTS n^k**

k	400	401	402	403	404
1	400	401	402	403	404
2	1 60000	1 60801	1 61604	1 62409	1 63216
3	640 00000	644 81201	649 64808	654 50827	659 39264
4	(10)2.5600 00000	(10)2.5856 96160	(10)2.6115 85282	(10)2.6376 68328	(10)2.6639 46266
5	(13)1.0240 00000	(13)1.0368 64160	(13)1.0498 57283	(13)1.0629 80336	(13)1.0762 34291
6	(15)4.0960 00000	(15)4.1578 25282	(15)4.2204 26278	(15)4.2838 10755	(15)4.3479 86537
7	(18)1.6384 00000	(18)1.6672 87938	(18)1.6966 11364	(18)1.7263 75734	(18)1.7565 86561
8	(20)6.5536 00000	(20)6.6858 24632	(20)6.8203 77683	(20)6.9572 94209	(20)7.0966 09706
9	(23)2.6214 40000	(23)2.6810 15678	(23)2.7417 91829	(23)2.8037 89566	(23)2.8670 30321
10	(26)1.0485 76000	(26)1.0750 87287	(26)1.1022 00315	(26)1.1299 27195	(26)1.1582 80250
24	(62)2.8147 49767	(62)2.9885 80393	(62)3.1726 72718	(62)3.3676 04703	(62)3.5739 85306
1/2	(1)2.0000 00000	(1)2.0024 98439	(1)2.0049 93766	(1)2.0074 85990	(1)2.0099 75124
1/3	7.3680 62997	7.3741 97940	7.3803 22692	7.3864 37295	7.3925 41792
1/4	4.4721 35955	4.4749 28423	4.4777 15674	4.4804 97729	4.4832 74611
1/5	3.3144 54017	3.3161 09590	3.3177 61862	3.3194 10850	3.3210 56568

k	405	406	407	408	409
1	405	406	407	408	409
2	1 64025	1 64836	1 65649	1 66464	1 67281
3	664 30125	669 23416	674 19143	679 17312	684 17929
4	(10)2.6904 20063	(10)2.7170 90690	(10)2.7439 59120	(10)2.7710 26330	(10)2.7982 93296
5	(13)1.0896 20125	(13)1.1031 38820	(13)1.1167 91362	(13)1.1305 78742	(13)1.1445 01958
6	(15)4.4129 61508	(15)4.4787 43609	(15)4.5453 04843	(15)4.6127 61269	(15)4.6810 13009
7	(18)1.7872 49411	(18)1.8183 69905	(18)1.8499 53723	(18)1.8820 06598	(18)1.9145 34321
8	(20)7.2383 60113	(20)7.3825 81816	(20)7.5293 11653	(20)7.6785 86919	(20)7.8304 45371
9	(23)2.9315 35846	(23)2.9973 28217	(23)3.0644 29843	(23)3.1328 63463	(23)3.2026 52157
10	(26)1.1872 72017	(26)1.2169 15256	(26)1.2472 22946	(26)1.2782 08293	(26)1.3098 84732
24	(62)3.7924 56055	(62)4.0236 92707	(62)4.2684 06980	(62)4.5273 48373	(62)4.8013 06073
1/2	(1)2.0124 61180	(1)2.0149 44168	(1)2.0174 24100	(1)2.0199 00988	(1)2.0223 74842
1/3	7.3986 36223	7.4047 20630	7.4107 95055	7.4168 59539	7.4229 14120
1/4	4.4860 46344	4.4888 12948	4.4915 74446	4.4943 30860	4.4970 82211
1/5	3.3226 99030	3.3243 38251	3.3259 74245	3.3276 07026	3.3292 36609

k	410	411	412	413	414
1	410	411	412	413	414
2	1 68100	1 68921	1 69744	1 70569	1 71396
3	689 21000	694 26531	699 34528	704 44997	709 57944
4	(10)2.8257 61000	(10)2.8534 30424	(10)2.8813 02554	(10)2.9093 78376	(10)2.9376 58882
5	(13)1.1585 62010	(13)1.1727 59904	(13)1.1870 96652	(13)1.2015 73269	(13)1.2161 90777
6	(15)4.7501 04241	(15)4.8200 43207	(15)4.8908 38207	(15)4.9624 97602	(15)5.0350 29817
7	(18)1.9475 42739	(18)1.9810 37758	(18)2.0150 25341	(18)2.0495 11510	(18)2.0845 02344
8	(20)7.9849 25229	(20)8.1420 65185	(20)8.3019 04405	(20)8.4644 82535	(20)8.6298 39705
9	(23)3.2738 19344	(23)3.3463 88791	(23)3.4203 84615	(23)3.4958 31287	(23)3.5727 53638
10	(26)1.3422 65931	(26)1.3753 65793	(26)1.4091 98461	(26)1.4437 78322	(26)1.4791 20006
24	(62)5.0911 10945	(62)5.3976 37632	(62)5.7218 06738	(62)6.0645 87127	(62)6.4269 98328
1/2	(1)2.0248 45673	(1)2.0273 13493	(1)2.0297 78313	(1)2.0322 40143	(1)2.0346 98995
1/3	7.4289 58841	7.4349 93742	7.4410 18861	7.4470 34238	7.4530 39914
1/4	4.4998 28522	4.5025 69814	4.5053 06108	4.5080 37426	4.5107 63788
1/5	3.3308 63008	3.3324 86236	3.3341 06308	3.3357 23237	3.3373 37037

k	415	416	417	418	419
1	415	416	417	418	419
2	1 72225	1 73056	1 73889	1 74724	1 75561
3	714 73375	719 91296	725 11713	730 34632	735 60059
4	(10)2.9661 45063	(10)2.9948 37914	(10)3.0237 38432	(10)3.0528 47618	(10)3.0821 66472
5	(13)1.2309 50201	(13)1.2458 52572	(13)1.2608 98926	(13)1.2760 90304	(13)1.2914 27752
6	(15)5.1084 43334	(15)5.1827 46700	(15)5.2579 48522	(15)5.3340 57471	(15)5.4110 82280
7	(18)2.1200 03984	(18)2.1560 22627	(18)2.1925 64534	(18)2.2296 36023	(18)2.2672 43475
8	(20)8.7980 16532	(20)8.9690 54129	(20)9.1429 94106	(20)9.3198 78576	(20)9.4997 50162
9	(23)3.6511 76861	(23)3.7311 26518	(23)3.8126 28542	(23)3.8957 09245	(23)3.9803 95318
10	(26)1.5152 38397	(26)1.5521 48631	(26)1.5898 66102	(26)1.6284 06464	(26)1.6677 85638
24	(62)6.8101 13045	(62)7.2150 59801	(62)7.6430 25690	(62)8.0952 59269	(62)8.5730 73581
1/2	(1)2.0371 54879	(1)2.0396 07805	(1)2.0420 57786	(1)2.0445 04830	(1)2.0469 48949
1/3	7.4590 35926	7.4650 22314	7.4709 99115	7.4769 66370	7.4829 24114
1/4	4.5134 85215	4.5162 01729	4.5189 13349	4.5216 20097	4.5243 21992
1/5	3.3389 47722	3.3405 55305	3.3421 59799	3.3437 61218	3.3453 59575

k	420	421	422	423	424
1	420	421	422	423	424
2	1 76400	1 77241	1 78084	1 78929	1 79776
3	740 88000	746 18461	751 51448	756 86967	762 25024
4	(10)3.1116 96000	(10)3.1414 37208	(10)3.1713 91106	(10)3.2015 58704	(10)3.2319 41018
5	(13)1.3069 12320	(13)1.3225 45065	(13)1.3383 27047	(13)1.3542 59332	(13)1.3703 42991
6	(15)5.4890 31744	(15)5.5679 14722	(15)5.6477 40136	(15)5.7285 16974	(15)5.8102 54284
7	(18)2.3053 93332	(18)2.3440 92098	(18)2.3833 46338	(18)2.4231 62680	(18)2.4635 47816
8	(20)9.6826 51996	(20)9.8686 27732	(21)1.0057 72154	(21)1.0249 97814	(21)1.0445 44274
9	(23)4.0667 13838	(23)4.1546 92275	(23)4.2443 58492	(23)4.3357 40751	(23)4.4288 67722
10	(26)1.7080 19812	(26)1.7491 25448	(26)1.7911 19284	(26)1.8340 18338	(26)1.8778 39914
24	(62)9.0778 49315	(62)9.6110 38126	(63)1.0174 16609	(63)1.0768 83734	(63)1.1396 73784
1/2	(1)2.0493 90153	(1)2.0518 28453	(1)2.0542 63858	(1)2.0566 96380	(1)2.0591 26028
1/3	7.4888 72387	7.4948 11226	7.5007 40668	7.5066 60749	7.5125 71508
1/4	4.5270 19056	4.5297 11307	4.5323 98767	4.5350 81455	4.5377 59390
1/5	3.3469 54883	3.3485 47155	3.3501 36405	3.3517 22644	3.3533 05887

$$n^{\frac{1}{2}}\begin{bmatrix}(-6)4\\4\end{bmatrix} \qquad n^{\frac{1}{3}}\begin{bmatrix}(-6)1\\4\end{bmatrix} \qquad n^{\frac{1}{4}}\begin{bmatrix}(-7)7\\4\end{bmatrix} \qquad n^{\frac{1}{5}}\begin{bmatrix}(-7)4\\4\end{bmatrix}$$

POWERS AND ROOTS n^k

Table 3.1

k	425	426	427	428	429
1	425	426	427	428	429
2	1 80625	1 81476	1 82329	1 83184	1 84041
3	767 65625	773 08776	778 54483	784 02752	789 53589
4	(10) 3.2625 39063	(10) 3.2933 53858	(10) 3.3243 86424	(10) 3.3556 37786	(10) 3.3871 08968
5	(13) 1.3865 79102	(13) 1.4029 68743	(13) 1.4195 13003	(13) 1.4362 12972	(13) 1.4530 69747
6	(15) 5.8929 61182	(15) 5.9766 46847	(15) 6.0613 20523	(15) 6.1469 91521	(15) 6.2336 69216
7	(18) 2.5045 08502	(18) 2.5460 51557	(18) 2.5881 83863	(18) 2.6309 12371	(18) 2.6742 44094
8	(21) 1.0644 16113	(21) 1.0846 17963	(21) 1.1051 54510	(21) 1.1260 30495	(21) 1.1472 50716
9	(23) 4.5237 68482	(23) 4.6204 72523	(23) 4.7190 09756	(23) 4.8194 10518	(23) 4.9217 05572
10	(26) 1.9226 01605	(26) 1.9683 21295	(26) 2.0150 17166	(26) 2.0627 07702	(26) 2.1114 11691
24	(63) 1.2059 63938	(63) 1.2759 40370	(63) 1.3497 98685	(63) 1.4277 44370	(63) 1.5099 93273
1/2	(1) 2.0615 52813	(1) 2.0639 76744	(1) 2.0663 97832	(1) 2.0688 16087	(1) 2.0712 31518
1/3	7.5184 72981	7.5243 65204	7.5302 48212	7.5361 22043	7.5419 86732
1/4	4.5404 32593	4.5431 01082	4.5457 64877	4.5484 23998	4.5510 78463
1/5	3.3548 86145	3.3564 63431	3.3580 37758	3.3596 09138	3.3611 77583

k	430	431	432	433	434
1	430	431	432	433	434
2	1 84900	1 85761	1 86624	1 87489	1 88356
3	795 07000	800 62991	806 21568	811 82737	817 46504
4	(10) 3.4188 01000	(10) 3.4507 14912	(10) 3.4828 51738	(10) 3.5152 12512	(10) 3.5477 98274
5	(13) 1.4700 84430	(13) 1.4872 58127	(13) 1.5045 91951	(13) 1.5220 87018	(13) 1.5397 44451
6	(15) 6.3213 63049	(15) 6.4100 82528	(15) 6.4998 37227	(15) 6.5906 36787	(15) 6.6824 90916
7	(18) 2.7181 86111	(18) 2.7627 45570	(18) 2.8079 29682	(18) 2.8537 45729	(18) 2.9002 01058
8	(21) 1.1688 20028	(21) 1.1907 43340	(21) 1.2130 25623	(21) 1.2356 71901	(21) 1.2586 87259
9	(23) 5.0259 26119	(23) 5.1321 03797	(23) 5.2402 70690	(23) 5.3504 59329	(23) 5.4627 02704
10	(26) 2.1611 48231	(26) 2.2119 36737	(26) 2.2637 96938	(26) 2.3167 48890	(26) 2.3708 12974
24	(63) 1.5967 72093	(63) 1.6883 18906	(63) 1.7848 83700	(63) 1.8867 28946	(63) 1.9941 30189
1/2	(1) 2.0736 44135	(1) 2.0760 53949	(1) 2.0784 60969	(1) 2.0808 65205	(1) 2.0832 66666
1/3	7.5478 42314	7.5536 88825	7.5595 26299	7.5653 54772	7.5711 74278
1/4	4.5537 28292	4.5563 73502	4.5590 14114	4.5616 50145	4.5642 81614
1/5	3.3627 43107	3.3643 05720	3.3658 65436	3.3674 22267	3.3689 76223

k	435	436	437	438	439
1	435	436	437	438	439
2	1 89225	1 90096	1 90969	1 91844	1 92721
3	823 12875	828 81856	834 53453	840 27672	846 04519
4	(10) 3.5806 10063	(10) 3.6136 48922	(10) 3.6469 15896	(10) 3.6804 12034	(10) 3.7141 38384
5	(13) 1.5575 65377	(13) 1.5755 50930	(13) 1.5937 02247	(13) 1.6120 20471	(13) 1.6305 06751
6	(15) 6.7754 09391	(15) 6.8694 02054	(15) 6.9644 78818	(15) 7.0606 49662	(15) 7.1579 24635
7	(18) 2.9473 03085	(18) 2.9950 59296	(18) 3.0434 77243	(18) 3.0925 64552	(18) 3.1423 28915
8	(21) 1.2820 76842	(21) 1.3058 45853	(21) 1.3299 99555	(21) 1.3545 43274	(21) 1.3794 82394
9	(23) 5.5770 34263	(23) 5.6934 87918	(23) 5.8120 98057	(23) 5.9328 99539	(23) 6.0559 27708
10	(26) 2.4260 09904	(26) 2.4823 60732	(26) 2.5398 86851	(26) 2.5986 09998	(26) 2.6585 52264
24	(63) 2.1073 76666	(63) 2.2267 71952	(63) 2.3526 34640	(63) 2.4852 99040	(63) 2.6251 15920
1/2	(1) 2.0856 65361	(1) 2.0880 61302	(1) 2.0904 54496	(1) 2.0928 44954	(1) 2.0952 32684
1/3	7.5769 84852	7.5827 86527	7.5885 79338	7.5943 63318	7.6001 38502
1/4	4.5669 08540	4.5695 30941	4.5721 48834	4.5747 62238	4.5773 71171
1/5	3.3705 27318	3.3720 75562	3.3736 20969	3.3751 63549	3.3767 03314

k	440	441	442	443	444
1	440	441	442	443	444
2	1 93600	1 94481	1 95364	1 96249	1 97136
3	851 84000	857 66121	863 50888	869 38307	875 28384
4	(10) 3.7480 96000	(10) 3.7822 85936	(10) 3.8167 09250	(10) 3.8513 67000	(10) 3.8862 60250
5	(13) 1.6491 62240	(13) 1.6679 88098	(13) 1.6869 85488	(13) 1.7061 55581	(13) 1.7254 99551
6	(15) 7.2563 13856	(15) 7.3558 27511	(15) 7.4564 75858	(15) 7.5582 69224	(15) 7.6612 18006
7	(18) 3.1927 78097	(18) 3.2439 19933	(18) 3.2957 62329	(18) 3.3483 13266	(18) 3.4015 80795
8	(21) 1.4048 22363	(21) 1.4305 68690	(21) 1.4567 26950	(21) 1.4833 02777	(21) 1.5103 01873
9	(23) 6.1812 18395	(23) 6.3088 07924	(23) 6.4387 33117	(23) 6.5710 33102	(23) 6.7057 40315
10	(26) 2.7197 36094	(26) 2.7821 84294	(26) 2.8459 20038	(26) 2.9109 66867	(26) 2.9773 48700
24	(63) 2.7724 53276	(63) 2.9276 97132	(63) 3.0912 52385	(63) 3.2635 43677	(63) 3.4450 16313
1/2	(1) 2.0976 17696	(1) 2.1000 00000	(1) 2.1023 79604	(1) 2.1047 56518	(1) 2.1071 30751
1/3	7.6059 04922	7.6116 62611	7.6174 11603	7.6231 51930	7.6288 83626
1/4	4.5799 75651	4.5825 75695	4.5851 71321	4.5877 62546	4.5903 49388
1/5	3.3782 40276	3.3797 74445	3.3813 05834	3.3828 34454	3.3843 60316

k	445	446	447	448	449
1	445	446	447	448	449
2	1 98025	1 98916	1 99809	2 00704	2 01601
3	881 21125	887 16536	893 14623	899 15392	905 18849
4	(10) 3.9213 90063	(10) 3.9567 57506	(10) 3.9923 63648	(10) 4.0282 09562	(10) 4.0642 96320
5	(13) 1.7450 18578	(13) 1.7647 13847	(13) 1.7845 86551	(13) 1.8046 37884	(13) 1.8248 69048
6	(15) 7.7653 32671	(15) 7.8706 23760	(15) 7.9771 01882	(15) 8.0847 77719	(15) 8.1936 62024
7	(18) 3.4555 73039	(18) 3.5102 98197	(18) 3.5657 64541	(18) 3.6219 80418	(18) 3.6789 54249
8	(21) 1.5377 30002	(21) 1.5655 92996	(21) 1.5938 96750	(21) 1.6226 47227	(21) 1.6518 50458
9	(23) 6.8428 98510	(23) 6.9825 44761	(23) 7.1247 18472	(23) 7.2694 59578	(23) 7.4168 08555
10	(26) 3.0450 89837	(26) 3.1142 14964	(26) 3.1847 49157	(26) 3.2567 17891	(26) 3.3301 47041
24	(63) 3.6361 37215	(63) 3.8373 95917	(63) 4.0493 05610	(63) 4.2724 04226	(63) 4.5072 55570
1/2	(1) 2.1095 02311	(1) 2.1118 71208	(1) 2.1142 37451	(1) 2.1166 01049	(1) 2.1189 62010
1/3	7.6346 06721	7.6403 21250	7.6460 27242	7.6517 24731	7.6574 13748
1/4	4.5929 31864	4.5955 09991	4.5980 83787	4.6006 53268	4.6032 18450
1/5	3.3858 83431	3.3874 03811	3.3889 21465	3.3904 36406	3.3919 48644

$$n^{\frac{1}{2}}\begin{bmatrix}(-6)4\\4\end{bmatrix} \qquad n^{\frac{1}{3}}\begin{bmatrix}(-6)1\\4\end{bmatrix} \qquad n^{\frac{1}{4}}\begin{bmatrix}(-7)6\\4\end{bmatrix} \qquad n^{\frac{1}{5}}\begin{bmatrix}(-7)4\\4\end{bmatrix}$$

Table 3.1   **POWERS AND ROOTS** n^k

k					
1	450	451	452	453	454
2	2 02500	2 03401	2 04304	2 05209	2 06116
3	911 25000	917 33851	923 45408	929 59677	935 76664
4	(10)4.1006 25000	(10)4.1371 96680	(10)4.1740 12442	(10)4.2110 73368	(10)4.2483 80546
5	(13)1.8452 81250	(13)1.8658 75703	(13)1.8866 53624	(13)1.9076 16236	(13)1.9287 64768
6	(15)8.3037 65625	(15)8.4150 99419	(15)8.5276 74379	(15)8.6415 01548	(15)8.7565 92045
7	(18)3.7366 94531	(18)3.7952 09838	(18)3.8545 08819	(18)3.9146 00201	(18)3.9754 92789
8	(21)1.6815 12539	(21)1.7116 39637	(21)1.7422 37986	(21)1.7733 13891	(21)1.8048 73726
9	(23)7.5668 06426	(23)7.7194 94763	(23)7.8749 15698	(23)8.0331 11927	(23)8.1941 26716
10	(26)3.4050 62892	(26)3.4814 92138	(26)3.5594 61895	(26)3.6389 99703	(26)3.7201 33529
24	(63)4.7544 50505	(63)5.0146 08183	(63)5.2883 77338	(63)5.5764 37619	(63)5.8795 01000
1/2	(1)2.1213 20344	(1)2.1236 76058	(1)2.1260 29163	(1)2.1283 79665	(1)2.1307 27575
1/3	7.6630 94324	7.6687 66491	7.6744 30279	7.6800 85719	7.6857 32843
1/4	4.6057 79352	4.6083 35988	4.6108 88377	4.6134 36534	4.6159 80476
1/5	3.3934 58190	3.3949 65055	3.3964 69249	3.3979 70784	3.3994 69669

k					
1	455	456	457	458	459
2	2 07025	2 07936	2 08849	2 09764	2 10681
3	941 96375	948 18816	954 43993	960 71912	967 02579
4	(10)4.2859 35063	(10)4.3237 38010	(10)4.3617 90480	(10)4.4000 93570	(10)4.4386 48376
5	(13)1.9501 00453	(13)1.9716 24532	(13)1.9933 38249	(13)2.0152 42855	(13)2.0373 39605
6	(15)8.8729 57063	(15)8.9906 07868	(15)9.1095 55800	(15)9.2298 12275	(15)9.3513 88785
7	(18)4.0371 95464	(18)4.0997 17188	(18)4.1630 67001	(18)4.2272 54022	(18)4.2922 87452
8	(21)1.8369 23936	(21)1.8694 71038	(21)1.9025 21619	(21)1.9360 82342	(21)1.9701 59941
9	(23)8.3580 03909	(23)8.5247 87931	(23)8.6945 23800	(23)8.8672 57127	(23)9.0430 34128
10	(26)3.8028 91778	(26)3.8873 03297	(26)3.9733 97377	(26)4.0612 03764	(26)4.1507 52665
24	(63)6.1983 13235	(63)6.5336 55383	(63)6.8863 45396	(63)7.2572 39774	(63)7.6472 35292
1/2	(1)2.1330 72901	(1)2.1354 15650	(1)2.1377 55833	(1)2.1400 93456	(1)2.1424 28529
1/3	7.6913 71681	7.6970 02263	7.7026 24618	7.7082 38778	7.7138 44772
1/4	4.6185 20218	4.6210 55778	4.6235 87171	4.6261 14413	4.6286 37519
1/5	3.4009 65915	3.4024 59532	3.4039 50532	3.4054 38923	3.4069 24718

k					
1	460	461	462	463	464
2	2 11600	2 12521	2 13444	2 14369	2 15296
3	973 36000	979 72181	986 11128	992 52847	998 97344
4	(10)4.4774 56000	(10)4.5165 17544	(10)4.5558 34114	(10)4.5954 06816	(10)4.6352 36762
5	(13)2.0596 29760	(13)2.0821 14588	(13)2.1047 95360	(13)2.1276 73356	(13)2.1507 49857
6	(15)9.4742 96896	(15)9.5985 48250	(15)9.7241 54565	(15)9.8511 27638	(15)9.9794 79338
7	(18)4.3581 76572	(18)4.4249 30743	(18)4.4925 59409	(18)4.5610 72096	(18)4.6304 78413
8	(21)2.0047 61223	(21)2.0398 93073	(21)2.0755 62447	(21)2.1117 76381	(21)2.1485 41984
9	(23)9.2219 01627	(23)9.4039 07065	(23)9.5890 98505	(23)9.7775 24642	(23)9.9692 34804
10	(26)4.2420 74748	(26)4.3352 01157	(26)4.4301 63510	(26)4.5269 93909	(26)4.6257 24949
24	(63)8.0572 70802	(63)8.4883 29103	(63)8.9414 38903	(63)9.4176 76852	(63)9.9181 69666
1/2	(1)2.1447 61059	(1)2.1470 91055	(1)2.1494 18526	(1)2.1517 43479	(1)2.1540 65923
1/3	7.7194 42629	7.7250 32380	7.7306 14052	7.7361 87677	7.7417 53281
1/4	4.6311 56507	4.6336 71390	4.6361 82186	4.6386 88909	4.6411 91574
1/5	3.4084 07924	3.4098 88554	3.4113 66616	3.4128 42121	3.4143 15079

k					
1	465	466	467	468	469
2	2 16225	2 17156	2 18089	2 19024	2 19961
3	1005 44625	1011 94696	1018 47563	1025 03232	1031 61709
4	(10)4.6753 25063	(10)4.7156 72834	(10)4.7562 81192	(10)4.7971 51258	(10)4.8382 84152
5	(13)2.1740 26154	(13)2.1975 03540	(13)2.2211 83317	(13)2.2450 66789	(13)2.2691 55267
6	(16)1.0109 22162	(16)1.0240 36650	(16)1.0372 92609	(16)1.0506 91257	(16)1.0642 33820
7	(18)4.7007 88052	(18)4.7720 10788	(18)4.8441 56484	(18)4.9172 35083	(18)4.9912 56618
8	(21)2.1858 66444	(21)2.2237 57027	(21)2.2622 21078	(21)2.3012 66019	(21)2.3408 99354
9	(24)1.0164 27896	(24)1.0362 70775	(24)1.0564 57243	(24)1.0769 92497	(24)1.0978 81797
10	(26)4.7263 89719	(26)4.8290 21810	(26)4.9336 55326	(26)5.0403 24885	(26)5.1490 65627
24	(64)1.0444 09634	(64)1.0996 69046	(64)1.1577 24259	(64)1.2187 10278	(64)1.2827 68318
1/2	(1)2.1563 85865	(1)2.1587 03314	(1)2.1610 18278	(1)2.1633 30765	(1)2.1656 40783
1/3	7.7473 10895	7.7528 60547	7.7584 02264	7.7639 36077	7.7694 62012
1/4	4.6436 90198	4.6461 84795	4.6486 75380	4.6511 61968	4.6536 44575
1/5	3.4157 85500	3.4172 53393	3.4187 18768	3.4201 81635	3.4216 42003

k					
1	470	471	472	473	474
2	2 20900	2 21841	2 22784	2 23729	2 24676
3	1038 23000	1044 87111	1051 54048	1058 23817	1064 96424
4	(10)4.8796 81000	(10)4.9213 42928	(10)4.9632 71066	(10)5.0054 66544	(10)5.0479 30498
5	(13)2.2934 50070	(13)2.3179 52519	(13)2.3426 63943	(13)2.3675 85675	(13)2.3927 19056
6	(16)1.0779 21533	(16)1.0917 55637	(16)1.1057 37381	(16)1.1198 68024	(16)1.1341 48832
7	(18)5.0662 31205	(18)5.1421 69048	(18)5.2190 80439	(18)5.2969 75756	(18)5.3758 65466
8	(21)2.3811 28666	(21)2.4219 61622	(21)2.4634 05967	(21)2.5054 69532	(21)2.5481 60231
9	(24)1.1191 30473	(24)1.1407 43924	(24)1.1627 27616	(24)1.1850 87089	(24)1.2078 27949
10	(26)5.2599 13224	(26)5.3729 03881	(26)5.4880 74350	(26)5.6054 61930	(26)5.7251 04480
24	(64)1.3500 46075	(64)1.4206 98007	(64)1.4948 85630	(64)1.5727 77826	(64)1.6545 51159
1/2	(1)2.1679 48339	(1)2.1702 53441	(1)2.1725 56098	(1)2.1748 56317	(1)2.1771 54106
1/3	7.7749 80097	7.7804 90361	7.7859 92832	7.7914 87536	7.7969 74500
1/4	4.6561 23215	4.6585 97902	4.6610 68652	4.6635 35480	4.6659 98399
1/5	3.4230 99883	3.4245 55283	3.4260 08213	3.4274 58683	3.4289 06701

$$n^{\frac{1}{2}}\left[(-6)\begin{matrix}3\\4\end{matrix}\right] \qquad n^{\frac{1}{3}}\left[(-6)\begin{matrix}1\\4\end{matrix}\right] \qquad n^{\frac{1}{4}}\left[(-7)\begin{matrix}5\\4\end{matrix}\right] \qquad n^{\frac{1}{5}}\left[(-7)\begin{matrix}3\\3\end{matrix}\right]$$

POWERS AND ROOTS n^k — Table 3.1

k	475	476	477	478	479
1	475	476	477	478	479
2	2 25625	2 26576	2 27529	2 28484	2 29441
3	1071 71875	1078 50176	1085 31333	1092 15352	1099 02239
4	(10)5.0906 64063	(10)5.1336 68378	(10)5.1769 44584	(10)5.2204 93826	(10)5.2643 17248
5	(13)2.4180 65430	(13)2.4436 26148	(13)2.4694 02567	(13)2.4953 96049	(13)2.5216 07962
6	(16)1.1485 81079	(16)1.1631 66046	(16)1.1779 05024	(16)1.1927 99311	(16)1.2078 50214
7	(18)5.4557 60126	(18)5.5366 70380	(18)5.6186 06966	(18)5.7015 80708	(18)5.7856 02524
8	(21)2.5914 86060	(21)2.6354 55101	(21)2.6800 75523	(21)2.7253 55578	(21)2.7713 03609
9	(24)1.2309 55878	(24)1.2544 76628	(24)1.2783 96024	(24)1.3027 19966	(24)1.3274 54429
10	(26)5.8470 40422	(26)5.9713 08750	(26)6.0979 49036	(26)6.2270 01440	(26)6.3585 06713
24	(64)1.7403 90207	(64)1.8304 87912	(64)1.9250 45935	(64)2.0242 75033	(64)2.1283 95451
1/2	(1)2.1794 49472	(1)2.1817 42423	(1)2.1840 32967	(1)2.1863 21111	(1)2.1886 06863
1/3	7.8024 53753	7.8079 25322	7.8133 89232	7.8188 45511	7.8242 94186
1/4	4.6684 57424	4.6709 12569	4.6733 63849	4.6758 11278	4.6782 54870
1/5	3.4303 52278	3.4317 95422	3.4332 36143	3.4346 74449	3.4361 10350

k	480	481	482	483	484
1	480	481	482	483	484
2	2 30400	2 31361	2 32324	2 33289	2 34256
3	1105 92000	1112 84641	1119 80168	1126 78587	1133 79904
4	(10)5.3084 16000	(10)5.3527 91232	(10)5.3974 44098	(10)5.4423 75752	(10)5.4875 87354
5	(13)2.5480 39680	(13)2.5746 92583	(13)2.6015 68055	(13)2.6286 67488	(13)2.6559 92279
6	(16)1.2230 59046	(16)1.2384 27132	(16)1.2539 55803	(16)1.2696 46397	(16)1.2855 00263
7	(18)5.8706 83423	(18)5.9568 34506	(18)6.0440 66968	(18)6.1323 92097	(18)6.2218 21273
8	(21)2.8179 28043	(21)2.8652 37397	(21)2.9132 40279	(21)2.9619 45383	(21)3.0113 61496
9	(24)1.3526 05461	(24)1.3781 79188	(24)1.4041 81814	(24)1.4306 19620	(24)1.4574 98964
10	(26)6.4925 06211	(26)6.6290 41895	(26)6.7681 56345	(26)6.9098 92764	(26)7.0542 94987
24	(64)2.2376 37322	(64)2.3522 41094	(64)2.4724 57971	(64)2.5985 50361	(64)2.7307 92362
1/2	(1)2.1908 90230	(1)2.1931 71220	(1)2.1954 49840	(1)2.1977 26098	(1)2.2000 00000
1/3	7.8297 35282	7.8351 68827	7.8405 94846	7.8460 13365	7.8514 24411
1/4	4.6806 94639	4.6831 30598	4.6855 62762	4.6879 91145	4.6904 15760
1/5	3.4375 43855	3.4389 74973	3.4404 03713	3.4418 30083	3.4432 54092

k	485	486	487	488	489
1	485	486	487	488	489
2	2 35225	2 36196	2 37169	2 38144	2 39121
3	1140 84125	1147 91256	1155 01303	1162 14272	1169 30169
4	(10)5.5330 80063	(10)5.5788 55042	(10)5.6249 13456	(10)5.6712 56474	(10)5.7178 85264
5	(13)2.6835 43830	(13)2.7113 23550	(13)2.7393 32853	(13)2.7675 73159	(13)2.7960 45894
6	(16)1.3015 18758	(16)1.3177 03245	(16)1.3340 55099	(16)1.3505 75702	(16)1.3672 66442
7	(18)6.3123 65975	(18)6.4040 37773	(18)6.4968 48334	(18)6.5908 09424	(18)6.6859 32903
8	(21)3.0614 97498	(21)3.1123 62358	(21)3.1639 65139	(21)3.2163 14999	(21)3.2694 21189
9	(24)1.4848 26286	(24)1.5126 08106	(24)1.5408 51023	(24)1.5695 61719	(24)1.5987 46962
10	(26)7.2014 07489	(26)7.3512 75394	(26)7.5039 44480	(26)7.6594 61191	(26)7.8178 72642
24	(64)2.8694 70250	(64)3.0148 82996	(64)3.1673 42798	(64)3.3271 75643	(64)3.4947 21879
1/2	(1)2.2022 71555	(1)2.2045 40769	(1)2.2068 07649	(1)2.2090 72203	(1)2.2113 34439
1/3	7.8568 28008	7.8622 24183	7.8676 12960	7.8729 94366	7.8783 68425
1/4	4.6928 36620	4.6952 53740	4.6976 67133	4.7000 76812	4.7024 82790
1/5	3.4446 75750	3.4460 95065	3.4475 12045	3.4489 26700	3.4503 39037

k	490	491	492	493	494
1	490	491	492	493	494
2	2 40100	2 41081	2 42064	2 43049	2 44036
3	1176 49000	1183 70771	1190 95488	1198 23157	1205 53784
4	(10)5.7648 01000	(10)5.8120 04856	(10)5.8594 98010	(10)5.9072 81640	(10)5.9553 56930
5	(13)2.8247 52490	(13)2.8536 94384	(13)2.8828 73021	(13)2.9122 89849	(13)2.9419 46323
6	(16)1.3841 28720	(16)1.4011 63943	(16)1.4183 73526	(16)1.4357 58895	(16)1.4533 21484
7	(18)6.7822 30728	(18)6.8797 14959	(18)6.9783 97749	(18)7.0782 91354	(18)7.1794 08129
8	(21)3.3232 93057	(21)3.3779 40045	(21)3.4333 71692	(21)3.4895 97638	(21)3.5466 27616
9	(24)1.6284 13598	(24)1.6585 68562	(24)1.6892 18873	(24)1.7203 71635	(24)1.7520 34042
10	(26)7.9792 26630	(26)8.1435 71639	(26)8.3109 56854	(26)8.4814 32162	(26)8.6550 48169
24	(64)3.6703 36822	(64)3.8543 91376	(64)4.0472 72689	(64)4.2493 84825	(64)4.4611 49467
1/2	(1)2.2135 94362	(1)2.2158 51981	(1)2.2181 07301	(1)2.2203 60331	(1)2.2226 11077
1/3	7.8837 35163	7.8890 94604	7.8944 46773	7.8997 91695	7.9051 29393
1/4	4.7048 85081	4.7072 83697	4.7096 78653	4.7120 69960	4.7144 57633
1/5	3.4517 49066	3.4531 56794	3.4545 62231	3.4559 65384	3.4573 66263

k	495	496	497	498	499
1	495	496	497	498	499
2	2 45025	2 46016	2 47009	2 48004	2 49001
3	1212 87375	1220 23936	1227 63473	1235 05992	1242 51499
4	(10)6.0037 25063	(10)6.0523 87226	(10)6.1013 44608	(10)6.1505 98402	(10)6.2001 49800
5	(13)2.9718 43906	(13)3.0019 84064	(13)3.0323 68270	(13)3.0629 98004	(13)3.0938 74750
6	(16)1.4710 62733	(16)1.4889 84096	(16)1.5070 87030	(16)1.5253 73006	(16)1.5438 43500
7	(18)7.2817 60531	(18)7.3853 61115	(18)7.4902 22541	(18)7.5963 57570	(18)7.7037 79067
8	(21)3.6044 71463	(21)3.6631 39113	(21)3.7226 40603	(21)3.7829 86070	(21)3.8441 85754
9	(24)1.7842 13374	(24)1.8169 17000	(24)1.8501 52380	(24)1.8839 27063	(24)1.9182 48691
10	(26)8.8318 56201	(26)9.0119 08320	(26)9.1952 57326	(26)9.3819 56772	(26)9.5720 60970
24	(64)4.6830 06649	(64)4.9154 15513	(64)5.1588 55098	(64)5.4138 25162	(64)5.6808 47029
1/2	(1)2.2248 59546	(1)2.2271 05745	(1)2.2293 49681	(1)2.2315 91360	(1)2.2338 30790
1/3	7.9104 59893	7.9157 83219	7.9210 99395	7.9264 08444	7.9317 10391
1/4	4.7168 41683	4.7192 22124	4.7215 98967	4.7239 72227	4.7263 41916
1/5	3.4587 64874	3.4601 61227	3.4615 55329	3.4629 47190	3.4643 36816

$$n^{\frac{1}{2}}\left[\begin{matrix}(-6)3\\3\end{matrix}\right] \qquad n^{\frac{1}{3}}\left[\begin{matrix}(-6)1\\4\end{matrix}\right] \qquad n^{\frac{1}{4}}\left[\begin{matrix}(-7)5\\4\end{matrix}\right] \qquad n^{\frac{1}{5}}\left[\begin{matrix}(-7)3\\3\end{matrix}\right]$$

Table 3.1   **POWERS AND ROOTS** n^k

k					
1	500	501	502	503	504
2	2 50000	2 51001	2 52004	2 53009	2 54016
3	1250 00000	1257 51501	1265 06008	1272 63527	1280 24064
4	(10)6.2500 0000	(10)6.3001 50200	(10)6.3506 01602	(10)6.4013 55408	(10)6.4524 12826
5	(13)3.1250 0000	(13)3.1563 75250	(13)3.1880 02004	(13)3.2198 81770	(13)3.2520 16064
6	(16)1.5625 0000	(16)1.5813 44000	(16)1.6003 77006	(16)1.6196 00530	(16)1.6390 16096
7	(18)7.8125 0000	(18)7.9225 33442	(18)8.0338 92570	(18)8.1465 90668	(18)8.2606 41125
8	(21)3.9062 5000	(21)3.9691 89254	(21)4.0330 14070	(21)4.0977 35106	(21)4.1633 63127
9	(24)1.9531 2500	(24)1.9885 63816	(24)2.0245 73063	(24)2.0611 60758	(24)2.0983 35016
10	(26)9.7656 2500	(26)9.9627 04720	(27)1.0163 35678	(27)1.0367 63861	(27)1.0575 60848
24	(64)5.9604 64478	(64)6.2532 44659	(64)6.5597 79050	(64)6.8806 84448	(64)7.2166 04000
1/2	(1)2.2360 67977	(1)2.2383 02929	(1)2.2405 35650	(1)2.2427 66149	(1)2.2449 94432
1/3	7.9370 05260	7.9422 93073	7.9475 73855	7.9528 47628	7.9581 14416
1/4	4.7287 08045	4.7310 70628	4.7334 29676	4.7357 85203	4.7381 37221
1/5	3.4657 24216	3.4671 09398	3.4684 92370	3.4698 73139	3.4712 51715

k					
1	505	506	507	508	509
2	2 55025	2 56036	2 57049	2 58064	2 59081
3	1287 87625	1295 54216	1303 23843	1310 96512	1318 72229
4	(10)6.5037 75063	(10)6.5554 43330	(10)6.6074 18840	(10)6.6597 02810	(10)6.7122 96456
5	(13)3.2844 06407	(13)3.3170 54325	(13)3.3499 61352	(13)3.3831 29027	(13)3.4165 58896
6	(16)1.6586 25235	(16)1.6784 29488	(16)1.6984 30405	(16)1.7186 29546	(16)1.7390 28478
7	(18)8.3760 57438	(18)8.4928 53211	(18)8.6110 42156	(18)8.7306 38093	(18)8.8516 54954
8	(21)4.2299 09006	(21)4.2973 83725	(21)4.3657 98373	(21)4.4351 64151	(21)4.5054 92371
9	(24)2.1361 04048	(24)2.1744 76165	(24)2.2134 59775	(24)2.2530 63389	(24)2.2932 95617
10	(27)1.0787 32544	(27)1.1002 84939	(27)1.1222 24106	(27)1.1445 56202	(27)1.1672 87469
24	(64)7.5682 08268	(64)7.9361 96349	(64)8.3212 97020	(64)8.7242 69942	(64)9.1459 06897
1/2	(1)2.2472 20505	(1)2.2494 44376	(1)2.2516 66050	(1)2.2538 85534	(1)2.2561 02835
1/3	7.9633 74242	7.9686 27129	7.9738 73099	7.9791 12176	7.9843 44383
1/4	4.7404 85740	4.7428 30775	4.7451 72336	4.7475 10436	4.7498 45086
1/5	3.4726 28104	3.4740 02314	3.4753 74353	3.4767 44229	3.4781 11950

k					
1	510	511	512	513	514
2	2 60100	2 61121	2 62144	2 63169	2 64196
3	1326 51000	1334 32831	1342 17728	1350 05697	1357 96744
4	(10)6.7652 01000	(10)6.8184 17664	(10)6.8719 47674	(10)6.9257 92256	(10)6.9799 52642
5	(13)3.4502 52510	(13)3.4842 11426	(13)3.5184 37209	(13)3.5529 31427	(13)3.5876 95658
6	(16)1.7596 28780	(16)1.7804 32039	(16)1.8014 39851	(16)1.8226 53822	(16)1.8440 75568
7	(18)8.9741 06779	(18)9.0980 07719	(18)9.2233 72037	(18)9.3502 14108	(18)9.4785 48420
8	(21)4.5767 94457	(21)4.6490 81944	(21)4.7223 66483	(21)4.7966 59837	(21)4.8719 73888
9	(24)2.3341 65173	(24)2.3756 80873	(24)2.4178 51639	(24)2.4606 86497	(24)2.5041 94578
10	(27)1.1904 24238	(27)1.2139 72926	(27)1.2379 40039	(27)1.2623 32173	(27)1.2871 56013
24	(64)9.5870 33090	(65)1.0048 50848	(65)1.0531 22917	(65)1.1036 12886	(65)1.1564 18034
1/2	(1)2.2583 17958	(1)2.2605 30911	(1)2.2627 41700	(1)2.2649 50331	(1)2.2671 56810
1/3	7.9895 69740	7.9947 88272	8.0000 00000	8.0052 04946	8.0104 03133
1/4	4.7521 76299	4.7545 04087	4.7568 28460	4.7591 49431	4.7614 67011
1/5	3.4794 77522	3.4808 40954	3.4822 02253	3.4835 61427	3.4849 18483

k					
1	515	516	517	518	519
2	2 65225	2 66256	2 67289	2 68324	2 69361
3	1365 90875	1373 88096	1381 88413	1389 91832	1397 98359
4	(10)7.0344 30063	(10)7.0892 25754	(10)7.1443 40952	(10)7.1997 76898	(10)7.2555 34832
5	(13)3.6227 31482	(13)3.6580 40489	(13)3.6936 24272	(13)3.7294 84433	(13)3.7656 22578
6	(16)1.8657 06713	(16)1.8875 48892	(16)1.9096 03749	(16)1.9318 72936	(16)1.9543 58118
7	(18)9.6083 89574	(18)9.7397 52284	(18)9.8726 51381	(19)1.0007 10181	(19)1.0143 11863
8	(21)4.9483 20630	(21)5.0257 12179	(21)5.1041 60764	(21)5.1836 78738	(21)5.2642 78570
9	(24)2.5483 85125	(24)2.5932 67484	(24)2.6388 51115	(24)2.6851 45586	(24)2.7321 60578
10	(27)1.3124 18339	(27)1.3381 26022	(27)1.3642 86026	(27)1.3909 05414	(27)1.4179 91340
24	(65)1.2116 39706	(65)1.2693 83471	(65)1.3297 59294	(65)1.3928 81704	(65)1.4588 69982
1/2	(1)2.2693 61144	(1)2.2715 63338	(1)2.2737 63400	(1)2.2759 61335	(1)2.2781 57150
1/3	8.0155 94581	8.0207 79314	8.0259 57353	8.0311 28718	8.0362 93433
1/4	4.7637 81212	4.7660 92045	4.7683 99522	4.7707 03654	4.7730 04452
1/5	3.4862 73428	3.4876 26271	3.4889 77017	3.4903 25675	3.4916 72252

k					
1	520	521	522	523	524
2	2 70400	2 71441	2 72484	2 73529	2 74576
3	1406 08000	1414 20761	1422 36648	1430 55667	1438 77824
4	(10)7.3116 16000	(10)7.3680 21648	(10)7.4247 53026	(10)7.4818 11384	(10)7.5391 97978
5	(13)3.8020 40320	(13)3.8387 39279	(13)3.8757 21079	(13)3.9129 87354	(13)3.9505 39740
6	(16)1.9770 60966	(16)1.9999 83164	(16)2.0231 26403	(16)2.0464 92386	(16)2.0700 82824
7	(19)1.0280 71703	(19)1.0419 91229	(19)1.0560 71983	(19)1.0703 15518	(19)1.0847 23400
8	(21)5.3459 72853	(21)5.4287 74301	(21)5.5126 95749	(21)5.5977 50159	(21)5.6839 50615
9	(24)2.7799 05884	(24)2.8283 91411	(24)2.8776 27181	(24)2.9276 23333	(24)2.9783 90122
10	(27)1.4455 51059	(27)1.4735 91925	(27)1.5021 21389	(27)1.5311 47003	(27)1.5606 76424
24	(65)1.5278 48342	(65)1.5999 46126	(65)1.6752 98008	(65)1.7540 44200	(65)1.8363 30669
1/2	(1)2.2803 50850	(1)2.2825 42442	(1)2.2847 31932	(1)2.2869 19325	(1)2.2891 04628
1/3	8.0414 51517	8.0466 02993	8.0517 47881	8.0568 86203	8.0620 17979
1/4	4.7753 01928	4.7775 96092	4.7798 86957	4.7821 74532	4.7844 58829
1/5	3.4930 16754	3.4943 59190	3.4956 99566	3.4970 37889	3.4983 74167

$$n^{\frac{1}{2}}\left[\begin{matrix}(-6)3\\3\end{matrix}\right] \qquad n^{\frac{1}{3}}\left[\begin{matrix}(-7)9\\4\end{matrix}\right] \qquad n^{\frac{1}{4}}\left[\begin{matrix}(-7)5\\3\end{matrix}\right] \qquad n^{\frac{1}{5}}\left[\begin{matrix}(-7)3\\3\end{matrix}\right]$$

POWERS AND ROOTS n^k Table 3.1

k					
1	525	526	527	528	529
2	2 75625	2 76676	2 77729	2 78784	2 79841
3	1447 03125	1455 31576	1463 63183	1471 97952	1480 35889
4	(10)7.5969 14063	(10)7.6549 60898	(10)7.7133 39744	(10)7.7720 51866	(10)7.8310 98528
5	(13)3.9883 79883	(13)4.0265 09432	(13)4.0649 30045	(13)4.1036 43385	(13)4.1426 51121
6	(16)2.0938 99438	(16)2.1179 43961	(16)2.1422 18134	(16)2.1667 23707	(16)2.1914 62443
7	(19)1.0992 97205	(19)1.1140 38524	(19)1.1289 48957	(19)1.1440 30117	(19)1.1592 83632
8	(21)5.7713 10327	(21)5.8598 42634	(21)5.9495 61001	(21)6.0404 79020	(21)6.1326 10416
9	(24)3.0299 37922	(24)3.0822 77226	(24)3.1354 18647	(24)3.1893 72923	(24)3.2441 50910
10	(27)1.5907 17409	(27)1.6212 77821	(27)1.6523 65627	(27)1.6839 88903	(27)1.7161 55831
24	(65)1.9223 09365	(65)2.0121 38448	(65)2.1059 82534	(65)2.2040 12944	(65)2.3064 07963
1/2	(1)2.2912 87847	(1)2.2934 68988	(1)2.2956 48057	(1)2.2978 25059	(1)2.3000 00000
1/3	8.0671 43230	8.0722 61977	8.0773 74241	8.0824 80041	8.0875 79399
1/4	4.7867 39859	4.7890 17632	4.7912 92160	4.7935 63454	4.7958 31523
1/5	3.4997 08406	3.5010 40614	3.5023 70797	3.5036 98962	3.5050 25117

k					
1	530	531	532	533	534
2	2 80900	2 81961	2 83024	2 84089	2 85156
3	1488 77000	1497 21291	1505 68768	1514 19437	1522 73304
4	(10)7.8904 81000	(10)7.9502 00552	(10)8.0102 58458	(10)8.0706 55992	(10)8.1313 94434
5	(13)4.1819 54930	(13)4.2215 56493	(13)4.2614 57499	(13)4.3016 59644	(13)4.3421 64628
6	(16)2.2164 36113	(16)2.2416 46498	(16)2.2670 95390	(16)2.2927 84590	(16)2.3187 15911
7	(19)1.1747 11140	(19)1.1903 14290	(19)1.2060 94747	(19)1.2220 54187	(19)1.2381 94297
8	(21)6.2259 69041	(21)6.3205 68882	(21)6.4164 24056	(21)6.5135 48814	(21)6.6119 57543
9	(24)3.2997 63592	(24)3.3562 22076	(24)3.4135 37598	(24)3.4717 21518	(24)3.5307 85328
10	(27)1.7488 74704	(27)1.7821 53922	(27)1.8160 02002	(27)1.8504 27569	(27)1.8854 39365
24	(65)2.4133 53110	(65)2.5250 41417	(65)2.6416 73716	(65)2.7634 58943	(65)2.8906 14446
1/2	(1)2.3021 72887	(1)2.3043 43724	(1)2.3065 12519	(1)2.3086 79276	(1)2.3108 44002
1/3	8.0926 72335	8.0977 58868	8.1028 39019	8.1079 12808	8.1129 80255
1/4	4.7980 96379	4.8003 58033	4.8026 16494	4.8048 71774	4.8071 23882
1/5	3.5063 49267	3.5076 71420	3.5089 91583	3.5103 09762	3.5116 25964

k					
1	535	536	537	538	539
2	2 86225	2 87296	2 88369	2 89444	2 90521
3	1531 30375	1539 90656	1548 54153	1557 20872	1565 90819
4	(10)8.1924 75063	(10)8.2538 99162	(10)8.3156 94016	(10)8.3777 82914	(10)8.4402 45144
5	(13)4.3829 74158	(13)4.4240 89951	(13)4.4655 13725	(13)4.5072 47208	(13)4.5492 92133
6	(16)2.3448 91175	(16)2.3713 12214	(16)2.3979 80870	(16)2.4248 98998	(16)2.4520 68460
7	(19)1.2545 16778	(19)1.2710 23346	(19)1.2877 15727	(19)1.3045 95661	(19)1.3216 64900
8	(21)6.7116 64765	(21)6.8126 85137	(21)6.9150 33455	(21)7.0187 24655	(21)7.1237 73809
9	(24)3.5907 40649	(24)3.6515 99233	(24)3.7133 72966	(24)3.7760 73864	(24)3.8397 14083
10	(27)1.9210 46247	(27)1.9572 57189	(27)1.9940 81282	(27)2.0315 27739	(27)2.0696 05891
24	(65)3.0233 66304	(65)3.1619 49669	(65)3.3066 09101	(65)3.4575 98937	(65)3.6151 83652
1/2	(1)2.3130 06701	(1)2.3151 67381	(1)2.3173 26045	(1)2.3194 82701	(1)2.3216 37353
1/3	8.1180 41379	8.1230 96201	8.1281 44739	8.1331 87014	8.1382 23044
1/4	4.8093 72829	4.8116 18626	4.8138 61283	4.8161 00810	4.8183 37217
1/5	3.5129 40196	3.5142 52463	3.5155 62774	3.5168 71134	3.5181 77550

k					
1	540	541	542	543	544
2	2 91600	2 92681	2 93764	2 94849	2 95936
3	1574 64000	1583 40421	1592 20088	1601 03007	1609 89184
4	(10)8.5030 56000	(10)8.5662 16776	(10)8.6297 28770	(10)8.6935 93280	(10)8.7578 11610
5	(13)4.5916 50240	(13)4.6343 23276	(13)4.6773 12993	(13)4.7206 21151	(13)4.7642 49516
6	(16)2.4794 91130	(16)2.5071 68892	(16)2.5351 03642	(16)2.5632 97285	(16)2.5917 51736
7	(19)1.3389 25210	(19)1.3563 78371	(19)1.3740 26174	(19)1.3918 70426	(19)1.4099 12945
8	(21)7.2301 96134	(21)7.3380 06986	(21)7.4472 21864	(21)7.5578 56412	(21)7.6699 26419
9	(24)3.9043 05912	(24)3.9698 61779	(24)4.0363 94250	(24)4.1039 16032	(24)4.1724 39972
10	(27)2.1083 25193	(27)2.1476 95223	(27)2.1877 25684	(27)2.2284 26405	(27)2.2698 07345
24	(65)3.7796 38253	(65)3.9512 48669	(65)4.1303 12169	(65)4.3171 37789	(65)4.5120 46770
1/2	(1)2.3237 90008	(1)2.3259 40670	(1)2.3280 89345	(1)2.3302 36040	(1)2.3323 80758
1/3	8.1432 52850	8.1482 76449	8.1532 93862	8.1583 05107	8.1633 10204
1/4	4.8205 70514	4.8228 00711	4.8250 27819	4.8272 51847	4.8294 72806
1/5	3.5194 82029	3.5207 84576	3.5220 85199	3.5233 83903	3.5246 80696

k					
1	545	546	547	548	549
2	2 97025	2 98116	2 99209	3 00304	3 01401
3	1618 78625	1627 71336	1636 67323	1645 66592	1654 69149
4	(10)8.8223 85063	(10)8.8873 14946	(10)8.9526 02568	(10)9.0182 49242	(10)9.0842 56280
5	(13)4.8081 99859	(13)4.8524 73960	(13)4.8970 73605	(13)4.9420 00584	(13)4.9872 56698
6	(16)2.6204 68923	(16)2.6494 50782	(16)2.6786 99262	(16)2.7082 16320	(16)2.7380 03927
7	(19)1.4281 55563	(19)1.4466 00127	(19)1.4652 48496	(19)1.4841 02543	(19)1.5031 64156
8	(21)7.7834 47819	(21)7.8984 36694	(21)8.0149 09274	(21)8.1328 81938	(21)8.2523 71216
9	(24)4.2419 79061	(24)4.3125 46435	(24)4.3841 55373	(24)4.4568 19302	(24)4.5305 51798
10	(27)2.3118 78588	(27)2.3546 50354	(27)2.3981 32989	(27)2.4423 36978	(27)2.4872 72937
24	(65)4.7153 73024	(65)4.9274 63602	(65)5.1486 79188	(65)5.3793 94612	(65)5.6199 99369
1/2	(1)2.3345 23506	(1)2.3366 64289	(1)2.3388 03113	(1)2.3409 39982	(1)2.3430 74903
1/3	8.1683 09170	8.1733 02026	8.1782 88788	8.1832 69477	8.1882 44110
1/4	4.8316 90704	4.8339 05553	4.8361 17361	4.8383 26138	4.8405 31895
1/5	3.5259 75582	3.5272 68570	3.5285 59664	3.5298 48871	3.5311 36198

$$n^{\frac{1}{2}}\left[\begin{matrix}(-6)3\\3\end{matrix}\right] \qquad n^{\frac{1}{3}}\left[\begin{matrix}(-7)8\\4\end{matrix}\right] \qquad n^{\frac{1}{4}}\left[\begin{matrix}(-7)4\\3\end{matrix}\right] \qquad n^{\frac{1}{5}}\left[\begin{matrix}(-7)3\\3\end{matrix}\right]$$

Table 3.1   **POWERS AND ROOTS** n^k

k					
1	550	551	552	553	554
2	3 02500	3 03601	3 04704	3 05809	3 06916
3	1663 75000	1672 84151	1681 96608	1691 12377	1700 31464
4	(10)9.1506 25000	(10)9.2173 56720	(10)9.2844 52762	(10)9.3519 14448	(10)9.4197 43106
5	(13)5.0328 43750	(13)5.0787 63553	(13)5.1250 17924	(13)5.1716 08690	(13)5.2185 37681
6	(16)2.7680 64063	(16)2.7983 98718	(16)2.8290 09894	(16)2.8598 99605	(16)2.8910 69875
7	(19)1.5224 35234	(19)1.5419 17693	(19)1.5616 13462	(19)1.5815 24482	(19)1.6016 52711
8	(21)8.3733 93789	(21)8.4959 66491	(21)8.6201 06308	(21)8.7458 30384	(21)8.8731 56018
9	(24)4.6053 66584	(24)4.6812 77536	(24)4.7582 98682	(24)4.8364 44203	(24)4.9157 28434
10	(27)2.5329 51621	(27)2.5793 83922	(27)2.6265 80873	(27)2.6745 53644	(27)2.7233 13552
24	(65)5.8708 98173	(65)6.1325 11516	(65)6.4052 76258	(65)6.6896 46227	(65)6.9860 92851
1/2	(1)2.3452 07880	(1)2.3473 38919	(1)2.3494 68025	(1)2.3515 95203	(1)2.3537 20459
1/3	8.1932 12706	8.1981 75283	8.2031 31859	8.2080 82453	8.2130 27082
1/4	4.8427 34641	4.8449 34384	4.8471 31136	4.8493 24905	4.8515 15700
1/5	3.5324 21650	3.5337 05234	3.5349 86956	3.5362 66821	3.5375 44836

k					
1	555	556	557	558	559
2	3 08025	3 09136	3 10249	3 11364	3 12481
3	1709 53875	1718 79616	1728 08693	1737 41112	1746 76879
4	(10)9.4879 40063	(10)9.5565 06650	(10)9.6254 44200	(10)9.6947 54050	(10)9.7644 37536
5	(13)5.2658 06735	(13)5.3134 17697	(13)5.3613 72419	(13)5.4096 72760	(13)5.4583 20583
6	(16)2.9225 22738	(16)2.9542 60240	(16)2.9862 84438	(16)3.0185 97400	(16)3.0512 01206
7	(19)1.6220 00119	(19)1.6425 68693	(19)1.6633 60432	(19)1.6843 77349	(19)1.7056 21474
8	(21)9.0021 00663	(21)9.1326 81934	(21)9.2649 17605	(21)9.3988 25608	(21)9.5344 24040
9	(24)4.9961 65868	(24)5.0777 71156	(24)5.1605 59106	(24)5.2445 44689	(24)5.3297 43038
10	(27)2.7728 72057	(27)2.8232 40762	(27)2.8744 31422	(27)2.9264 55937	(27)2.9793 26358
24	(65)7.2951 05803	(65)7.6171 93672	(65)7.9528 84664	(65)8.3027 27311	(65)8.6672 91224
1/2	(1)2.3558 43798	(1)2.3579 65225	(1)2.3600 84744	(1)2.3622 02362	(1)2.3643 18084
1/3	8.2179 65765	8.2228 98519	8.2278 25361	8.2327 46311	8.2376 61384
1/4	4.8537 03532	4.8558 88409	4.8580 70341	4.8602 49337	4.8624 25407
1/5	3.5388 21007	3.5400 95340	3.5413 67840	3.5426 38514	3.5439 07368

k					
1	560	561	562	563	564
2	3 13600	3 14721	3 15844	3 16969	3 18096
3	1756 16000	1765 58481	1775 04328	1784 53547	1794 06144
4	(10)9.8344 96000	(10)9.9049 30784	(10)9.9757 43234	(11)1.0046 93470	(11)1.0118 50652
5	(13)5.5073 17760	(13)5.5566 66170	(13)5.6063 67697	(13)5.6564 24234	(13)5.7068 37678
6	(16)3.0840 97946	(16)3.1172 89721	(16)3.1507 78646	(16)3.1845 66844	(16)3.2186 56450
7	(19)1.7270 94850	(19)1.7487 99534	(19)1.7707 37599	(19)1.7929 11133	(19)1.8153 22238
8	(21)9.6717 31157	(21)9.8107 65384	(21)9.9515 45306	(22)1.0094 08968	(22)1.0238 41742
9	(24)5.4161 69448	(24)5.5038 39380	(24)5.5927 68462	(24)5.6829 72489	(24)5.7744 67426
10	(27)3.0330 54891	(27)3.0876 53892	(27)3.1431 35876	(27)3.1995 13511	(27)3.2567 99629
24	(65)9.0471 67858	(65)9.4429 71309	(65)9.8553 39138	(66)1.0284 93323	(66)1.0732 44065
1/2	(1)2.3664 31913	(1)2.3685 43856	(1)2.3706 53918	(1)2.3727 62104	(1)2.3748 68417
1/3	8.2425 70600	8.2474 73974	8.2523 71525	8.2572 63270	8.2621 49226
1/4	4.8645 98558	4.8667 68801	4.8689 36145	4.8711 00598	4.8732 62170
1/5	3.5451 74407	3.5464 39637	3.5477 03064	3.5489 64695	3.5502 24533

k					
1	565	566	567	568	569
2	3 19225	3 20356	3 21489	3 22624	3 23761
3	1803 62125	1813 21496	1822 84263	1832 50432	1842 20009
4	(11)1.0190 46006	(11)1.0262 79667	(11)1.0335 51771	(11)1.0408 62454	(11)1.0482 11851
5	(13)5.7576 09935	(13)5.8087 42917	(13)5.8602 38543	(13)5.9120 98737	(13)5.9643 25433
6	(16)3.2530 49613	(16)3.2877 48491	(16)3.3227 55254	(16)3.3580 72083	(16)3.3937 01172
7	(19)1.8379 73032	(19)1.8608 65646	(19)1.8840 02229	(19)1.9073 84943	(19)1.9310 15967
8	(22)1.0384 54763	(22)1.0532 49956	(22)1.0682 29264	(22)1.0833 94648	(22)1.0987 48085
9	(24)5.8672 69410	(24)5.9613 94749	(24)6.0568 59926	(24)6.1536 81599	(24)6.2518 76604
10	(27)3.3150 07217	(27)3.3741 49428	(27)3.4342 39578	(27)3.4952 91148	(27)3.5573 17788
24	(66)1.1198 57461	(66)1.1684 07534	(66)1.2189 71112	(66)1.2716 27927	(66)1.3264 60719
1/2	(1)2.3769 72865	(1)2.3790 75451	(1)2.3811 76180	(1)2.3832 75058	(1)2.3853 72088
1/3	8.2670 29409	8.2719 03838	8.2767 72529	8.2816 35499	8.2864 92764
1/4	4.8754 20869	4.8775 76704	4.8797 29685	4.8818 79820	4.8840 27117
1/5	3.5514 82586	3.5527 38859	3.5539 93358	3.5552 46087	3.5564 97054

k					
1	570	571	572	573	574
2	3 24900	3 26041	3 27184	3 28329	3 29476
3	1851 93000	1861 69411	1871 49248	1881 32517	1891 19224
4	(11)1.0556 00100	(11)1.0630 27337	(11)1.0704 93699	(11)1.0779 99322	(11)1.0855 44346
5	(13)6.0169 20570	(13)6.0698 86093	(13)6.1232 23956	(13)6.1769 36117	(13)6.2310 24545
6	(16)3.4296 44725	(16)3.4659 04959	(16)3.5024 84103	(16)3.5393 84395	(16)3.5766 08089
7	(19)1.9548 97493	(19)1.9790 31732	(19)2.0034 20907	(19)2.0280 67258	(19)2.0529 73043
8	(22)1.1142 91571	(22)1.1300 27119	(22)1.1459 56759	(22)1.1620 82539	(22)1.1784 06527
9	(24)6.3514 61955	(24)6.4524 54848	(24)6.5548 72660	(24)6.6587 32949	(24)6.7640 53463
10	(27)3.6203 33315	(27)3.6843 51718	(27)3.7493 87161	(27)3.8154 53980	(27)3.8825 66688
24	(66)1.3835 55344	(66)1.4430 00887	(66)1.5048 89774	(66)1.5693 17896	(66)1.6363 84728
1/2	(1)2.3874 67277	(1)2.3895 60629	(1)2.3916 52149	(1)2.3937 41841	(1)2.3958 29710
1/3	8.2913 44342	8.2961 90248	8.3010 30501	8.3058 65115	8.3106 94107
1/4	4.8861 71586	4.8883 13236	4.8904 52074	4.8925 88109	4.8947 21351
1/5	3.5577 46263	3.5589 93720	3.5602 39430	3.5614 83400	3.5627 25633

$$n^{\frac{1}{2}}\left[\begin{matrix}(-6)2\\3\end{matrix}\right] \qquad n^{\frac{1}{3}}\left[\begin{matrix}(-7)8\\4\end{matrix}\right] \qquad n^{\frac{1}{4}}\left[\begin{matrix}(-7)4\\3\end{matrix}\right] \qquad n^{\frac{1}{5}}\left[\begin{matrix}(-7)2\\3\end{matrix}\right]$$

POWERS AND ROOTS n^k — Table 3.1

k	575	576	577	578	579
1	575	576	577	578	579
2	3 30625	3 31776	3 32929	3 34084	3 35241
3	1901 09375	1911 02976	1921 00033	1931 00552	1941 04539
4	(11)1.0931 28906	(11)1.1007 53142	(11)1.1084 17190	(11)1.1161 21191	(11)1.1238 65281
5	(13)6.2854 91211	(13)6.3403 38097	(13)6.3955 67189	(13)6.4511 80481	(13)6.5071 79976
6	(16)3.6141 57446	(16)3.6520 34744	(16)3.6902 42268	(16)3.7287 82318	(16)3.7676 57206
7	(19)2.0781 40532	(19)2.1035 72012	(19)2.1292 69789	(19)2.1552 36180	(19)2.1814 73522
8	(22)1.1949 30806	(22)1.2116 57479	(22)1.2285 88668	(22)1.2457 26512	(22)1.2630 73169
9	(24)6.8708 52133	(24)6.9791 47080	(24)7.0889 56614	(24)7.2002 99239	(24)7.3131 93651
10	(27)3.9507 39976	(27)4.0199 88718	(27)4.0903 27966	(27)4.1617 72960	(27)4.2343 39124
24	(66)1.7061 93459	(66)1.7788 51122	(66)1.8544 68735	(66)1.9331 61432	(66)2.0150 48620
1/2	(1)2.3979 15762	(1)2.4000 00000	(1)2.4020 82430	(1)2.4041 63056	(1)2.4062 41883
1/3	8.3155 17494	8.3203 35292	8.3251 47517	8.3299 54185	8.3347 55313
1/4	4.8968 51807	4.8989 79486	4.9011 04396	4.9032 26546	4.9053 45944
1/5	3.5639 66137	3.5652 04916	3.5664 41976	3.5676 77321	3.5689 10958

k	580	581	582	583	584
1	580	581	582	583	584
2	3 36400	3 37561	3 38724	3 39889	3 41056
3	1951 12000	1961 22941	1971 37368	1981 55287	1991 76704
4	(11)1.1316 49600	(11)1.1394 74287	(11)1.1473 39482	(11)1.1552 45323	(11)1.1631 91951
5	(13)6.5635 67680	(13)6.6203 45609	(13)6.6775 15784	(13)6.7350 80234	(13)6.7930 40996
6	(16)3.8068 69254	(16)3.8464 20799	(16)3.8863 14186	(16)3.9265 51777	(16)3.9671 35942
7	(19)2.2079 84168	(19)2.2347 70484	(19)2.2618 34856	(19)2.2891 79686	(19)2.3168 07390
8	(22)1.2806 30817	(22)1.2984 01651	(22)1.3163 87886	(22)1.3345 91757	(22)1.3530 15516
9	(24)7.4276 58740	(24)7.5437 13594	(24)7.6613 77499	(24)7.7806 69942	(24)7.9016 10612
10	(27)4.3080 42069	(27)4.3828 97598	(27)4.4589 21704	(27)4.5361 30576	(27)4.6145 40597
24	(66)2.1002 54121	(66)2.1889 06331	(66)2.2811 38380	(66)2.3770 88299	(66)2.4768 99188
1/2	(1)2.4083 18916	(1)2.4103 94159	(1)2.4124 67616	(1)2.4145 39294	(1)2.4166 09195
1/3	8.3395 50915	8.3443 41009	8.3491 25609	8.3539 04732	8.3586 78393
1/4	4.9074 62599	4.9095 76518	4.9116 87710	4.9137 96184	4.9159 01946
1/5	3.5701 42892	3.5713 73127	3.5726 01670	3.5738 28526	3.5750 53698

k	585	586	587	588	589
1	585	586	587	588	589
2	3 42225	3 43396	3 44569	3 45744	3 46921
3	2002 01625	2012 30056	2022 62003	2032 97472	2043 36469
4	(11)1.1711 79506	(11)1.1792 08128	(11)1.1872 77958	(11)1.1953 89135	(11)1.2035 41802
5	(13)6.8514 00112	(13)6.9101 59631	(13)6.9693 21611	(13)7.0288 88116	(13)7.0888 61216
6	(16)4.0080 69065	(16)4.0493 53544	(16)4.0909 91786	(16)4.1329 86212	(16)4.1753 39256
7	(19)2.3447 20403	(19)2.3729 21177	(19)2.4014 12178	(19)2.4301 95893	(19)2.4592 74822
8	(22)1.3716 61436	(22)1.3905 31810	(22)1.4096 28949	(22)1.4289 55185	(22)1.4485 12870
9	(24)8.0242 19400	(24)8.1485 16404	(24)8.2745 21928	(24)8.4022 56487	(24)8.5317 40805
10	(27)4.6941 68349	(27)4.7750 30613	(27)4.8571 44372	(27)4.9405 26815	(27)5.0251 95334
24	(66)2.5807 19397	(66)2.6887 02707	(66)2.8010 08521	(66)2.9178 02055	(66)3.0392 54545
1/2	(1)2.4186 77324	(1)2.4207 43687	(1)2.4228 08288	(1)2.4248 71131	(1)2.4269 32220
1/3	8.3634 46607	8.3682 09391	8.3729 66760	8.3777 18728	8.3824 65312
1/4	4.9180 05007	4.9201 05372	4.9222 03051	4.9242 98052	4.9263 90382
1/5	3.5762 77194	3.5774 99018	3.5787 19175	3.5799 37670	3.5811 54508

k	590	591	592	593	594
1	590	591	592	593	594
2	3 48100	3 49281	3 50464	3 51649	3 52836
3	2053 79000	2064 25071	2074 74688	2085 27857	2095 84584
4	(11)1.2117 36100	(11)1.2199 72170	(11)1.2282 50153	(11)1.2365 70192	(11)1.2449 32429
5	(13)7.1492 42990	(13)7.2100 35522	(13)7.2712 40906	(13)7.3328 61239	(13)7.3948 98628
6	(16)4.2180 53364	(16)4.2611 30994	(16)4.3045 74616	(16)4.3483 86715	(16)4.3925 69785
7	(19)2.4886 51485	(19)2.5183 28417	(19)2.5483 08173	(19)2.5785 93322	(19)2.6091 86452
8	(22)1.4683 04376	(22)1.4883 32095	(22)1.5085 98438	(22)1.5291 05840	(22)1.5498 56753
9	(24)8.6629 95819	(24)8.7960 42679	(24)8.9309 02754	(24)9.0675 97630	(24)9.2061 49111
10	(27)5.1111 67533	(27)5.1984 61223	(27)5.2870 94431	(27)5.3770 85394	(27)5.4684 52572
24	(66)3.1655 43453	(66)3.2968 52680	(66)3.4333 72793	(66)3.5753 01250	(66)3.7228 42640
1/2	(1)2.4289 91560	(1)2.4310 49156	(1)2.4331 05012	(1)2.4351 59132	(1)2.4372 11521
1/3	8.3872 06527	8.3919 42387	8.3966 72908	8.4013 98104	8.4061 17992
1/4	4.9284 80050	4.9305 67063	4.9326 51429	4.9347 33156	4.9368 12252
1/5	3.5823 69695	3.5835 83235	3.5847 95134	3.5860 05396	3.5872 14026

k	595	596	597	598	599
1	595	596	597	598	599
2	3 54025	3 55216	3 56409	3 57604	3 58801
3	2106 44875	2117 08736	2127 76173	2138 47192	2149 21799
4	(11)1.2533 37006	(11)1.2617 84067	(11)1.2702 73753	(11)1.2788 06208	(11)1.2873 81576
5	(13)7.4573 55187	(13)7.5202 33037	(13)7.5835 34304	(13)7.6472 61125	(13)7.7114 15640
6	(16)4.4371 26336	(16)4.4820 58890	(16)4.5273 69980	(16)4.5730 62153	(16)4.6191 37969
7	(19)2.6400 90170	(19)2.6713 07098	(19)2.7028 39878	(19)2.7346 91167	(19)2.7668 63643
8	(22)1.5708 53651	(22)1.5920 99031	(22)1.6135 95407	(22)1.6353 45318	(22)1.6573 51322
9	(24)9.3465 79225	(24)9.4889 10223	(24)9.6331 64580	(24)9.7793 65002	(24)9.9275 34420
10	(27)5.5612 14639	(27)5.6553 90493	(27)5.7509 99254	(27)5.8480 60271	(27)5.9465 93118
24	(66)3.8762 08928	(66)4.0356 19703	(66)4.2013 02448	(66)4.3734 92798	(66)4.5524 34829
1/2	(1)2.4392 62184	(1)2.4413 11123	(1)2.4433 58345	(1)2.4454 03852	(1)2.4474 47650
1/3	8.4108 32585	8.4155 41899	8.4202 45948	8.4249 44747	8.4296 38310
1/4	4.9388 88725	4.9409 62581	4.9430 33830	4.9451 02478	4.9471 68534
1/5	3.5884 21030	3.5896 26411	3.5908 30176	3.5920 32329	3.5932 32875

$$n^{\frac{1}{2}}\begin{bmatrix}(-6)2\\3\end{bmatrix} \qquad n^{\frac{1}{3}}\begin{bmatrix}(-7)7\\4\end{bmatrix} \qquad n^{\frac{1}{4}}\begin{bmatrix}(-7)4\\3\end{bmatrix} \qquad n^{\frac{1}{5}}\begin{bmatrix}(-7)2\\3\end{bmatrix}$$

Table 3.1   **POWERS AND ROOTS n^k**

k	600	601	602	603	604
1	600	601	602	603	604
2	3 60000	3 61201	3 62404	3 63609	3 64816
3	2160 00000	2170 81801	2181 67208	2192 56227	2203 48864
4	(11)1.2960 61624	(11)1.3046 61624	(11)1.3133 66592	(11)1.3221 15049	(11)1.3309 07139
5	(13)7.7760 00000	(13)7.8410 16360	(13)7.9064 66885	(13)7.9723 53744	(13)8.0386 79117
6	(16)4.6656 00000	(16)4.7124 50833	(16)4.7596 93065	(16)4.8073 29308	(16)4.8553 62187
7	(19)2.7993 60000	(19)2.8321 82950	(19)2.8653 35225	(19)2.8988 19573	(19)2.9326 38761
8	(22)1.6796 16000	(22)1.7021 41953	(22)1.7249 31805	(22)1.7479 88202	(22)1.7713 13811
9	(25)1.0077 69600	(25)1.0229 87314	(25)1.0384 08947	(25)1.0540 36886	(25)1.0698 73542
10	(27)6.0466 17600	(27)6.1481 53756	(27)6.2512 21860	(27)6.3558 42422	(27)6.4620 36194
24	(66)4.7383 81338	(66)4.9315 94142	(66)5.1323 44384	(66)5.3409 12849	(66)5.5575 90288
1/2	(1)2.4494 89743	(1)2.4515 30134	(1)2.4535 68829	(1)2.4556 05832	(1)2.4576 41145
1/3	8.4343 26653	8.4390 09789	8.4436 87734	8.4483 60500	8.4530 28104
1/4	4.9492 32004	4.9512 92896	4.9533 51218	4.9554 06978	4.9574 60182
1/5	3.5944 31819	3.5956 29165	3.5968 24918	3.5980 19083	3.5992 11665

k	605	606	607	608	609
1	605	606	607	608	609
2	3 66025	3 67236	3 68449	3 69664	3 70881
3	2214 45125	2225 45016	2236 48543	2247 55712	2258 66529
4	(11)1.3397 43006	(11)1.3486 22797	(11)1.3575 46656	(11)1.3665 14729	(11)1.3755 27162
5	(13)8.1054 45188	(13)8.1726 54150	(13)8.2403 08202	(13)8.3084 09552	(13)8.3769 60414
6	(16)4.9037 94339	(16)4.9526 28415	(16)5.0018 67079	(16)5.0515 13008	(16)5.1015 68892
7	(19)2.9667 95575	(19)3.0012 92819	(19)3.0361 33317	(19)3.0713 19909	(19)3.1068 55455
8	(22)1.7949 11323	(22)1.8187 83448	(22)1.8429 32923	(22)1.8673 62504	(22)1.8920 74972
9	(25)1.0859 21350	(25)1.1021 82770	(25)1.1186 60284	(25)1.1353 56403	(25)1.1522 73658
10	(27)6.5698 24169	(27)6.6792 27585	(27)6.7902 67926	(27)6.9029 66929	(27)7.0173 46578
24	(66)5.7826 77757	(66)6.0164 86963	(66)6.2593 40623	(66)6.5115 72833	(66)6.7735 29447
1/2	(1)2.4596 74775	(1)2.4617 06725	(1)2.4637 36999	(1)2.4657 65601	(1)2.4677 92536
1/3	8.4576 90558	8.4623 47878	8.4670 00076	8.4716 47168	8.4762 89168
1/4	4.9595 10838	4.9615 58954	4.9636 04536	4.9656 47592	4.9676 88130
1/5	3.6004 02669	3.6015 92098	3.6027 79959	3.6039 66255	3.6051 50991

k	610	611	612	613	614
1	610	611	612	613	614
2	3 72100	3 73321	3 74544	3 75769	3 76996
3	2269 81000	2280 99131	2292 20928	2303 46397	2314 75544
4	(11)1.3845 84100	(11)1.3936 85690	(11)1.4028 32079	(11)1.4120 23414	(11)1.4212 59840
5	(13)8.4459 63010	(13)8.5154 19568	(13)8.5853 32326	(13)8.6557 03525	(13)8.7265 35419
6	(16)5.1520 37436	(16)5.2029 21356	(16)5.2542 23383	(16)5.3059 46261	(16)5.3580 92747
7	(19)3.1427 42836	(19)3.1789 84949	(19)3.2155 84711	(19)3.2525 45058	(19)3.2898 68947
8	(22)1.9170 73130	(22)1.9423 59804	(22)1.9679 37843	(22)1.9938 10121	(22)2.0199 79533
9	(25)1.1694 14609	(25)1.1867 81840	(25)1.2043 77960	(25)1.2222 05604	(25)1.2402 67433
10	(27)7.1334 29117	(27)7.2512 37043	(27)7.3707 93114	(27)7.4921 20352	(27)7.6152 42041
24	(66)7.0455 68477	(66)7.3280 60494	(66)7.6213 89047	(66)7.9259 51097	(66)8.2421 57465
1/2	(1)2.4698 17807	(1)2.4718 41419	(1)2.4738 63375	(1)2.4758 83681	(1)2.4779 02339
1/3	8.4809 26088	8.4855 57944	8.4901 84749	8.4948 06516	8.4994 23260
1/4	4.9697 26156	4.9717 61679	4.9737 94704	4.9758 25239	4.9778 53291
1/5	3.6063 34171	3.6075 15802	3.6086 95885	3.6098 74428	3.6110 51433

k	615	616	617	618	619
1	615	616	617	618	619
2	3 78225	3 79456	3 80689	3 81924	3 83161
3	2326 08375	2337 44896	2348 85113	2360 29032	2371 76659
4	(11)1.4305 41506	(11)1.4398 68559	(11)1.4492 41147	(11)1.4586 59418	(11)1.4681 23519
5	(13)8.7978 30263	(13)8.8695 90326	(13)8.9418 17878	(13)9.0145 15202	(13)9.0876 84584
6	(16)5.4106 65612	(16)5.4636 67641	(16)5.5171 01631	(16)5.5709 70395	(16)5.6252 76757
7	(19)3.3275 59351	(19)3.3656 19267	(19)3.4040 51706	(19)3.4428 59704	(19)3.4820 46313
8	(22)2.0464 49001	(22)2.0732 21468	(22)2.1002 99903	(22)2.1276 87297	(22)2.1553 86668
9	(25)1.2585 66136	(25)1.2771 04424	(25)1.2958 85040	(25)1.3149 10750	(25)1.3341 84347
10	(27)7.7401 81734	(27)7.8669 63254	(27)7.9956 10697	(27)8.1261 48432	(27)8.2586 01110
24	(66)8.5704 33286	(66)8.9112 18488	(66)9.2649 68280	(66)9.6321 53659	(67)1.0013 26192
1/2	(1)2.4799 19354	(1)2.4819 34729	(1)2.4839 48470	(1)2.4859 60579	(1)2.4879 71061
1/3	8.5040 34993	8.5086 41730	8.5132 43484	8.5178 40269	8.5224 32097
1/4	4.9798 78868	4.9819 01975	4.9839 22621	4.9859 40813	4.9879 56556
1/5	3.6122 26906	3.6134 00850	3.6145 73271	3.6157 44173	3.6169 13560

k	620	621	622	623	624
1	620	621	622	623	624
2	3 84400	3 85641	3 86884	3 88129	3 89376
3	2383 28000	2394 83061	2406 41848	2418 04367	2429 70624
4	(11)1.4776 33600	(11)1.4871 89809	(11)1.4967 92295	(11)1.5064 41206	(11)1.5161 36694
5	(13)9.1613 28320	(13)9.2354 48713	(13)9.3100 48072	(13)9.3851 28716	(13)9.4606 92969
6	(16)5.6800 23558	(16)5.7352 13651	(16)5.7908 49901	(16)5.8469 35190	(16)5.9034 72413
7	(19)3.5216 14606	(19)3.5615 67677	(19)3.6019 08638	(19)3.6426 40623	(19)3.6837 66786
8	(22)2.1834 01056	(22)2.2117 33527	(22)2.2403 87173	(22)2.2693 65108	(22)2.2986 70474
9	(25)1.3537 08655	(25)1.3734 86521	(25)1.3935 20822	(25)1.4138 14463	(25)1.4343 70376
10	(27)8.3929 93659	(27)8.5293 51293	(27)8.6676 99511	(27)8.8080 64101	(27)8.9504 71145
24	(67)1.0408 79722	(67)1.0819 28109	(67)1.1245 25305	(67)1.1687 27115	(67)1.2145 91262
1/2	(1)2.4899 79920	(1)2.4919 87159	(1)2.4939 92783	(1)2.4959 96795	(1)2.4979 99199
1/3	8.5270 18983	8.5316 00940	8.5361 77980	8.5407 50116	8.5453 17363
1/4	4.9899 69859	4.9919 80728	4.9939 89170	4.9959 95191	4.9979 98799
1/5	3.6180 81437	3.6192 47808	3.6204 12677	3.6215 76049	3.6227 37928

$$n^{\frac{1}{2}}\begin{bmatrix}(-6)2\\3\end{bmatrix} \qquad n^{\frac{1}{3}}\begin{bmatrix}(-7)7\\4\end{bmatrix} \qquad n^{\frac{1}{4}}\begin{bmatrix}(-7)3\\3\end{bmatrix} \qquad n^{\frac{1}{5}}\begin{bmatrix}(-7)2\\3\end{bmatrix}$$

POWERS AND ROOTS n^k

Table 3.1

k	625	626	627	628	629
1	625	626	627	628	629
2	3 90625	3 91876	3 93129	3 94384	3 95641
3	2441 40625	2453 14376	2464 91883	2476 73152	2488 58189
4	(11)1.5258 78906	(11)1.5356 67994	(11)1.5455 04106	(11)1.5553 87395	(11)1.5653 18009
5	(13)9.5367 43164	(13)9.6132 81641	(13)9.6903 10747	(13)9.7678 32838	(13)9.8458 50275
6	(16)5.9604 64478	(16)6.0179 14307	(16)6.0758 24838	(16)6.1341 99022	(16)6.1930 39823
7	(19)3.7252 90298	(19)3.7672 14356	(19)3.8095 42174	(19)3.8522 76986	(19)3.8954 22049
8	(22)2.3283 06437	(22)2.3582 76187	(22)2.3885 82943	(22)2.4192 29947	(22)2.4502 20469
9	(25)1.4551 91523	(25)1.4762 80893	(25)1.4976 41505	(25)1.5192 76407	(25)1.5411 88675
10	(27)9.0949 47018	(27)9.2415 18391	(27)9.3902 12238	(27)9.5410 55835	(27)9.6940 76765
24	(67)1.2621 77448	(67)1.3115 47419	(67)1.3627 65028	(67)1.4158 96309	(67)1.4710 09545
1/2	(1)2.5000 00000	(1)2.5019 99201	(1)2.5039 96805	(1)2.5059 92817	(1)2.5079 87241
1/3	8.5498 79733	8.5544 37239	8.5589 89894	8.5635 37711	8.5680 80703
1/4	5.0000 00000	5.0019 98801	5.0039 95209	5.0059 89230	5.0079 80871
1/5	3.6238 98318	3.6250 57224	3.6262 14650	3.6273 70600	3.6285 25079

k	630	631	632	633	634
1	630	631	632	633	634
2	3 96900	3 98161	3 99424	4 00689	4 01956
3	2500 47000	2512 39591	2524 35968	2536 36137	2548 40104
4	(11)1.5752 96100	(11)1.5853 21819	(11)1.5953 95318	(11)1.6055 16747	(11)1.6156 86259
5	(13)9.9243 65430	(14)1.0003 38068	(14)1.0082 89841	(14)1.0162 92101	(14)1.0243 45088
6	(16)6.2523 50221	(16)6.3121 33209	(16)6.3723 91794	(16)6.4331 28999	(16)6.4943 47861
7	(19)3.9389 80639	(19)3.9829 56055	(19)4.0273 51614	(19)4.0721 70657	(19)4.1174 16544
8	(22)2.4815 57803	(22)2.5132 45270	(22)2.5452 86220	(22)2.5776 84026	(22)2.6104 42089
9	(25)1.5633 81416	(25)1.5858 57766	(25)1.6086 20891	(25)1.6316 73988	(25)1.6550 20284
10	(27)9.8493 02919	(28)1.0006 76250	(28)1.0166 48403	(28)1.0328 49635	(28)1.0492 82860
24	(67)1.5281 75339	(67)1.5874 66692	(67)1.6489 59081	(67)1.7127 30535	(67)1.7788 61719
1/2	(1)2.5099 80080	(1)2.5119 71337	(1)2.5139 61018	(1)2.5159 49125	(1)2.5179 35662
1/3	8.5726 18882	8.5771 52262	8.5816 80854	8.5862 04672	8.5907 23728
1/4	5.0099 70139	5.0119 57040	5.0139 41581	5.0159 23768	5.0179 03608
1/5	3.6296 78090	3.6308 29638	3.6319 79727	3.6331 28361	3.6342 75544

k	635	636	637	638	639
1	635	636	637	638	639
2	4 03225	4 04496	4 05769	4 07044	4 08321
3	2560 47875	2572 59456	2584 74853	2596 94072	2609 17119
4	(11)1.6259 04006	(11)1.6361 70140	(11)1.6464 84814	(11)1.6568 48179	(11)1.6672 60390
5	(14)1.0324 49044	(14)1.0406 04209	(14)1.0488 10826	(14)1.0570 69138	(14)1.0653 79389
6	(16)6.5560 51429	(16)6.6182 42770	(16)6.6809 24963	(16)6.7441 01103	(16)6.8077 74299
7	(19)4.1630 92658	(19)4.2092 02402	(19)4.2557 49202	(19)4.3027 36504	(19)4.3501 67777
8	(22)2.6435 63838	(22)2.6770 52728	(22)2.7109 12241	(22)2.7451 45889	(22)2.7797 57209
9	(25)1.6786 63037	(25)1.7026 05535	(25)1.7268 51098	(25)1.7514 03077	(25)1.7762 64857
10	(28)1.0659 51028	(28)1.0828 57120	(28)1.1000 04149	(28)1.1173 95163	(28)1.1350 33244
24	(67)1.8474 36020	(67)1.9185 39634	(67)1.9922 61654	(67)2.0686 94164	(67)2.1479 32334
1/2	(1)2.5199 20634	(1)2.5219 04043	(1)2.5238 85893	(1)2.5258 66188	(1)2.5278 44932
1/3	8.5952 38034	8.5997 47604	8.6042 52449	8.6087 52582	8.6132 48015
1/4	5.0198 81108	5.0218 56273	5.0238 29110	5.0257 99626	5.0277 67827
1/5	3.6354 21280	3.6365 65574	3.6377 08430	3.6388 49851	3.6399 89842

k	640	641	642	643	644
1	640	641	642	643	644
2	4 09600	4 10881	4 12164	4 13449	4 14736
3	2621 44000	2633 74721	2646 09288	2658 47707	2670 89984
4	(11)1.6777 21600	(11)1.6882 31962	(11)1.6987 91629	(11)1.7094 00756	(11)1.7200 59497
5	(14)1.0737 41824	(14)1.0821 56687	(14)1.0906 24226	(14)1.0991 44686	(14)1.1077 18316
6	(16)6.8719 47674	(16)6.9366 24366	(16)7.0018 07530	(16)7.0675 00332	(16)7.1337 05955
7	(19)4.3980 46511	(19)4.4463 76219	(19)4.4951 60434	(19)4.5444 02713	(19)4.5941 06635
8	(22)2.8147 49767	(22)2.8501 27156	(22)2.8858 92999	(22)2.9220 50945	(22)2.9586 04673
9	(25)1.8014 39851	(25)1.8269 31507	(25)1.8527 43305	(25)1.8788 78757	(25)1.9053 41409
10	(28)1.1529 21505	(28)1.1710 63096	(28)1.1894 61202	(28)1.2081 19041	(28)1.2270 39868
24	(67)2.2300 74520	(67)2.3152 22362	(67)2.4034 80891	(67)2.4949 58638	(67)2.5897 67740
1/2	(1)2.5298 22128	(1)2.5317 97780	(1)2.5337 71892	(1)2.5357 44467	(1)2.5377 15508
1/3	8.6177 38760	8.6222 24830	8.6267 06237	8.6311 82992	8.6356 55108
1/4	5.0297 33719	5.0316 97308	5.0336 58602	5.0356 17605	5.0375 74325
1/5	3.6411 28406	3.6422 65548	3.6434 01272	3.6445 35581	3.6456 68481

k	645	646	647	648	649
1	645	646	647	648	649
2	4 16025	4 17316	4 18609	4 19904	4 21201
3	2683 36125	2695 86136	2708 40023	2720 97792	2733 59449
4	(11)1.7307 68006	(11)1.7415 26439	(11)1.7523 34949	(11)1.7631 93692	(11)1.7741 02824
5	(14)1.1163 45364	(14)1.1250 26079	(14)1.1337 60712	(14)1.1425 49513	(14)1.1513 92733
6	(16)7.2004 27598	(16)7.2676 68472	(16)7.3354 31806	(16)7.4037 20841	(16)7.4725 38836
7	(19)4.6442 75801	(19)4.6949 13833	(19)4.7460 24378	(19)4.7976 11105	(19)4.8496 77704
8	(22)2.9955 57891	(22)3.0329 14336	(22)3.0706 77773	(22)3.1088 51996	(22)3.1474 40830
9	(25)1.9321 34840	(25)1.9592 62661	(25)1.9867 28519	(25)2.0145 36093	(25)2.0426 89099
10	(28)1.2462 26972	(28)1.2656 83679	(28)1.2854 13352	(28)1.3054 19389	(28)1.3257 05225
24	(67)2.6880 24057	(67)2.7898 47292	(67)2.8953 61105	(67)3.0046 93247	(67)3.1179 75679
1/2	(1)2.5396 85020	(1)2.5416 53005	(1)2.5436 19468	(1)2.5455 84412	(1)2.5475 47841
1/3	8.6401 22598	8.6445 85472	8.6490 43742	8.6534 97422	8.6579 46522
1/4	5.0395 28767	5.0414 80939	5.0434 30845	5.0453 78492	5.0473 23886
1/5	3.6467 99973	3.6479 30063	3.6490 58755	3.6501 86051	3.6513 11957

$$n^{\frac{1}{2}}\begin{bmatrix}(-6)2\\3\end{bmatrix} \qquad n^{\frac{1}{3}}\begin{bmatrix}(-7)6\\4\end{bmatrix} \qquad n^{\frac{1}{4}}\begin{bmatrix}(-7)3\\3\end{bmatrix} \qquad n^{\frac{1}{5}}\begin{bmatrix}(-7)2\\3\end{bmatrix}$$

Table 3.1 **POWERS AND ROOTS n^k**

k	650	651	652	653	654
1	650	651	652	653	654
2	4 22500	4 23801	4 25104	4 26409	4 27716
3	2746 25000	2758 94451	2771 67808	2784 45077	2797 26264
4	(11)1.7850 62500	(11)1.7960 72876	(11)1.8071 34108	(11)1.8182 46353	(11)1.8294 09767
5	(14)1.1602 90625	(14)1.1692 43442	(14)1.1782 51439	(14)1.1873 14868	(14)1.1964 33987
6	(16)7.5418 89063	(16)7.6117 74809	(16)7.6821 99379	(16)7.7531 66091	(16)7.8246 78277
7	(19)4.9022 27891	(19)4.9552 65401	(19)5.0087 93995	(19)5.0628 17457	(19)5.1173 39593
8	(22)3.1864 48129	(22)3.2258 77776	(22)3.2657 33685	(22)3.3060 19800	(22)3.3467 40094
9	(25)2.0711 91284	(25)2.1000 46432	(25)2.1292 58363	(25)2.1588 30929	(25)2.1887 68021
10	(28)1.3462 74334	(28)1.3671 30227	(28)1.3882 76452	(28)1.4097 16597	(28)1.4314 54286
24	(67)3.2353 44710	(67)3.3569 41134	(67)3.4829 10364	(67)3.6134 02582	(67)3.7485 72888
1/2	(1)2.5495 09757	(1)2.5514 70164	(1)2.5534 29067	(1)2.5553 86468	(1)2.5573 42371
1/3	8.6623 91053	8.6668 31029	8.6712 66460	8.6756 97359	8.6801 23736
1/4	5.0492 67033	5.0512 07939	5.0531 46611	5.0550 83054	5.0570 17274
1/5	3.6524 36476	3.6535 59612	3.6546 81368	3.6558 01749	3.6569 20758

k	655	656	657	658	659
1	655	656	657	658	659
2	4 29025	4 30336	4 31649	4 32964	4 34281
3	2810 11375	2823 00416	2835 93393	2848 90312	2861 91179
4	(11)1.8406 24506	(11)1.8518 90729	(11)1.8632 08592	(11)1.8745 78253	(11)1.8859 99870
5	(14)1.2056 09052	(14)1.2148 40318	(14)1.2241 28045	(14)1.2334 72490	(14)1.2428 73914
6	(16)7.8967 39288	(16)7.9693 52487	(16)8.0425 21255	(16)8.1162 48987	(16)8.1905 39094
7	(19)5.1723 64234	(19)5.2278 95232	(19)5.2839 36465	(19)5.3404 91834	(19)5.3975 65263
8	(22)3.3878 98573	(22)3.4294 99272	(22)3.4715 46257	(22)3.5140 43626	(22)3.5569 55508
9	(25)2.2190 73565	(25)2.2497 51522	(25)2.2808 05891	(25)2.3122 40706	(25)2.3440 60040
10	(28)1.4534 93185	(28)1.4758 36999	(28)1.4984 89470	(28)1.5214 54385	(28)1.5447 35566
24	(67)3.8885 81447	(67)4.0335 93654	(67)4.1837 80288	(67)4.3393 17689	(67)4.5003 87920
1/2	(1)2.5592 96778	(1)2.5612 49695	(1)2.5632 01124	(1)2.5651 51068	(1)2.5670 99531
1/3	8.6845 45603	8.6889 62971	8.6933 75853	8.6977 84260	8.7021 88202
1/4	5.0589 49277	5.0608 79069	5.0628 06656	5.0647 32044	5.0666 55239
1/5	3.6580 38399	3.6591 54676	3.6602 69592	3.6613 83152	3.6624 95358

k	660	661	662	663	664
1	660	661	662	663	664
2	4 35600	4 36921	4 38244	4 39569	4 40896
3	2874 96000	2888 04781	2901 17528	2914 34247	2927 54944
4	(11)1.8974 73600	(11)1.9089 99602	(11)1.9205 78035	(11)1.9322 09058	(11)1.9438 92828
5	(14)1.2523 32576	(14)1.2618 48737	(14)1.2714 22659	(14)1.2810 54605	(14)1.2907 44838
6	(16)8.2653 95002	(16)8.3408 20153	(16)8.4168 18005	(16)8.4933 92032	(16)8.5705 45724
7	(19)5.4551 60701	(19)5.5132 82121	(19)5.5719 33519	(19)5.6311 18918	(19)5.6908 42360
8	(22)3.6004 06063	(22)3.6442 79482	(22)3.6886 19990	(22)3.7334 31842	(22)3.7787 19327
9	(25)2.3762 68001	(25)2.4088 68738	(25)2.4418 66433	(25)2.4752 65311	(25)2.5090 69633
10	(28)1.5683 36881	(28)1.5922 62236	(28)1.6165 15579	(28)1.6411 00901	(28)1.6660 22237
24	(67)4.6671 78950	(67)4.8398 84834	(67)5.0187 05901	(67)5.2038 48947	(67)5.3955 27431
1/2	(1)2.5690 46516	(1)2.5709 92026	(1)2.5729 36066	(1)2.5748 78638	(1)2.5768 19745
1/3	8.7065 87691	8.7109 82739	8.7153 73356	8.7197 59553	8.7241 41343
1/4	5.0685 76246	5.0704 95071	5.0724 11720	5.0743 26200	5.0762 38514
1/5	3.6636 06215	3.6647 15727	3.6658 23896	3.6669 30727	3.6680 36224

k	665	666	667	668	669
1	665	666	667	668	669
2	4 42225	4 43556	4 44889	4 46224	4 47561
3	2940 79625	2954 08296	2967 40963	2980 77632	2994 18309
4	(11)1.9556 29506	(11)1.9674 19251	(11)1.9792 62223	(11)1.9911 58582	(12)2.0031 08487
5	(14)1.3004 93622	(14)1.3103 01221	(14)1.3201 67903	(14)1.3300 93933	(14)1.3400 79578
6	(16)8.6482 82584	(16)8.7266 06135	(16)8.8055 19912	(16)8.8850 27470	(16)8.9651 32376
7	(19)5.7511 07918	(19)5.8119 19686	(19)5.8732 81781	(19)5.9351 98350	(19)5.9976 73560
8	(22)3.8244 86766	(22)3.8707 38511	(22)3.9174 78948	(22)3.9647 12498	(22)4.0124 43612
9	(25)2.5432 83699	(25)2.5779 11848	(25)2.6129 58458	(25)2.6484 27948	(25)2.6843 24776
10	(28)1.6912 83660	(28)1.7168 89291	(28)1.7428 43292	(28)1.7691 49870	(28)1.7958 13275
24	(67)5.5939 61683	(67)5.7993 79113	(67)6.0120 14426	(67)6.2321 09844	(67)6.4599 15340
1/2	(1)2.5787 59392	(1)2.5806 97580	(1)2.5826 34314	(1)2.5845 69597	(1)2.5865 03431
1/3	8.7285 18735	8.7328 91741	8.7372 60372	8.7416 24639	8.7459 84552
1/4	5.0781 48670	5.0800 56673	5.0819 62528	5.0838 66242	5.0857 67819
1/5	3.6691 40389	3.6702 43226	3.6713 44740	3.6724 44934	3.6735 43810

k	670	671	672	673	674
1	670	671	672	673	674
2	4 48900	4 50241	4 51584	4 52929	4 54276
3	3007 63000	3021 11711	3034 64448	3048 21217	3061 82024
4	(11)2.0151 12100	(11)2.0271 69581	(11)2.0392 81091	(11)2.0514 46790	(11)2.0636 66842
5	(14)1.3501 25107	(14)1.3602 30789	(14)1.3703 96893	(14)1.3806 23690	(14)1.3909 11451
6	(16)9.0458 38217	(16)9.1271 48592	(16)9.2090 67120	(16)9.2915 97433	(16)9.3747 43182
7	(19)6.0607 11605	(19)6.1243 16705	(19)6.1884 93105	(19)6.2532 45073	(19)6.3185 76905
8	(22)4.0606 76776	(22)4.1094 16509	(22)4.1586 67366	(22)4.2084 33934	(22)4.2587 20834
9	(25)2.7206 53440	(25)2.7574 18478	(25)2.7946 24470	(25)2.8322 76038	(25)2.8703 77842
10	(28)1.8228 37805	(28)1.8502 27799	(28)1.8779 87644	(28)1.9061 21773	(28)1.9346 34665
24	(67)6.6956 88867	(67)6.9396 96605	(67)7.1922 13208	(67)7.4535 22063	(67)7.7239 15552
1/2	(1)2.5884 35821	(1)2.5903 66769	(1)2.5922 96279	(1)2.5942 24354	(1)2.5961 50997
1/3	8.7503 40123	8.7546 91362	8.7590 38280	8.7633 80887	8.7677 19196
1/4	5.0876 67266	5.0895 64588	5.0914 59790	5.0933 52878	5.0952 43858
1/5	3.6746 41374	3.6757 37627	3.6768 32575	3.6779 26219	3.6790 18565

$$n^{\frac{1}{2}}\begin{bmatrix}(-6)2\\3\end{bmatrix} \qquad n^{\frac{1}{3}}\begin{bmatrix}(-7)6\\4\end{bmatrix} \qquad n^{\frac{1}{4}}\begin{bmatrix}(-7)3\\3\end{bmatrix} \qquad n^{\frac{1}{5}}\begin{bmatrix}(-7)2\\3\end{bmatrix}$$

POWERS AND ROOTS n^k Table 3.1

k					
1	675	676	677	678	679
2	4 55625	4 56976	4 58329	4 59684	4 61041
3	3075 46875	3089 15776	3102 88733	3116 65752	3130 46839
4	(11)2.0759 41406	(11)2.0882 70646	(11)2.1006 54722	(11)2.1130 93799	(11)2.1255 88037
5	(14)1.4012 60449	(14)1.4116 70957	(14)1.4221 43247	(14)1.4326 77595	(14)1.4432 74277
6	(16)9.4585 08032	(16)9.5428 95666	(16)9.6279 09783	(16)9.7135 54097	(16)9.7998 32341
7	(19)6.3844 92922	(19)6.4509 97470	(19)6.5180 94923	(19)6.5857 89678	(19)6.6540 86159
8	(22)4.3095 32722	(22)4.3608 74290	(22)4.4127 50263	(22)4.4651 65402	(22)4.5181 24502
9	(25)2.9089 34587	(25)2.9479 51020	(25)2.9874 31928	(25)3.0273 82142	(25)3.0678 06537
10	(28)1.9635 30847	(28)1.9928 14890	(28)2.0224 91415	(28)2.0525 65092	(28)2.0830 40639
24	(67)8.0036 95322	(67)8.2931 72571	(67)8.5926 68325	(67)8.9025 13744	(67)9.2230 50418
1/2	(1)2.5980 76211	(1)2.6000 00000	(1)2.6019 22366	(1)2.6038 43313	(1)2.6057 62844
1/3	8.7720 53215	8.7763 82955	8.7807 08428	8.7850 29644	8.7893 46612
1/4	5.0971 32735	5.0990 19514	5.1009 04200	5.1027 86801	5.1046 67319
1/5	3.6801 09614	3.6811 99371	3.6822 87840	3.6833 75023	3.6844 60923

k					
1	680	681	682	683	684
2	4 62400	4 63761	4 65124	4 66489	4 67856
3	3144 32000	3158 21241	3172 14568	3186 11987	3200 13504
4	(11)2.1381 37600	(11)2.1507 42651	(11)2.1634 03354	(11)2.1761 19871	(11)2.1888 92367
5	(14)1.4539 33568	(14)1.4646 55745	(14)1.4754 41087	(14)1.4862 89872	(14)1.4972 02379
6	(16)9.8867 48262	(16)9.9743 05627	(17)1.0062 50822	(17)1.0151 35983	(17)1.0240 86427
7	(19)6.7229 88818	(19)6.7925 02132	(19)6.8626 30603	(19)6.9333 78761	(19)7.0047 51164
8	(22)4.5716 32397	(22)4.6256 93952	(22)4.6803 14071	(22)4.7354 97694	(22)4.7912 49796
9	(25)3.1087 10030	(25)3.1500 97581	(25)3.1919 74196	(25)3.2343 44925	(25)3.2772 14860
10	(28)2.1139 22820	(28)2.1452 16453	(28)2.1769 26402	(28)2.2090 57584	(28)2.2416 14965
24	(67)9.5546 30685	(67)9.8976 17949	(68)1.0252 38701	(68)1.0619 32441	(68)1.0998 82878
1/2	(1)2.6076 80962	(1)2.6095 97670	(1)2.6115 12971	(1)2.6134 26869	(1)2.6153 39366
1/3	8.7936 59344	8.7979 67850	8.8022 72141	8.8065 72225	8.8108 68115
1/4	5.1065 45762	5.1084 22134	5.1102 96441	5.1121 68688	5.1140 38880
1/5	3.6855 45546	3.6866 28893	3.6877 10968	3.6887 91774	3.6898 71315

k					
1	685	686	687	688	689
2	4 69225	4 70596	4 71969	4 73344	4 74721
3	3214 19125	3228 28856	3242 42703	3256 60672	3270 82769
4	(11)2.2017 21006	(11)2.2146 05952	(11)2.2275 47370	(11)2.2405 45423	(11)2.2536 00728
5	(14)1.5081 78889	(14)1.5192 19683	(14)1.5303 25043	(14)1.5414 95251	(14)1.5527 30592
6	(17)1.0331 02539	(17)1.0421 84703	(17)1.0513 33304	(17)1.0605 48733	(17)1.0698 31378
7	(19)7.0767 52393	(19)7.1493 87060	(19)7.2226 59802	(19)7.2965 75282	(19)7.3711 38193
8	(22)4.8475 75389	(22)4.9044 79523	(22)4.9619 67284	(22)5.0200 43794	(22)5.0787 14215
9	(25)3.3205 89142	(25)3.3644 72953	(25)3.4088 71524	(25)3.4537 90130	(25)3.4992 34094
10	(28)2.2746 03562	(28)2.3080 28446	(28)2.3418 94737	(28)2.3762 07610	(28)2.4109 72291
24	(68)1.1391 31118	(68)1.1797 19551	(68)1.2216 91886	(68)1.2650 93189	(68)1.3099 69927
1/2	(1)2.6172 50466	(1)2.6191 60171	(1)2.6210 68484	(1)2.6229 75410	(1)2.6248 80950
1/3	8.8151 59819	8.8194 47349	8.8237 30714	8.8280 09925	8.8322 84991
1/4	5.1159 07022	5.1177 73120	5.1196 37179	5.1214 99204	5.1233 59200
1/5	3.6909 49595	3.6920 26615	3.6931 02381	3.6941 76894	3.6952 50159

k					
1	690	691	692	693	694
2	4 76100	4 77481	4 78864	4 80249	4 81636
3	3285 09000	3299 39371	3313 73888	3328 12557	3342 55384
4	(11)2.2667 12100	(11)2.2798 81054	(11)2.2931 07305	(11)2.3063 91020	(11)2.3197 32365
5	(14)1.5640 31349	(14)1.5753 97808	(14)1.5868 30255	(14)1.5983 28977	(14)1.6098 94261
6	(17)1.0791 81631	(17)1.0885 99885	(17)1.0980 86536	(17)1.1076 41981	(17)1.1172 66617
7	(19)7.4463 53253	(19)7.5222 25208	(19)7.5987 58832	(19)7.6759 58928	(19)7.7538 30324
8	(22)5.1379 83744	(22)5.1978 57619	(22)5.2583 41112	(22)5.3194 39537	(22)5.3811 58245
9	(25)3.5452 08784	(25)3.5917 19614	(25)3.6387 72050	(25)3.6863 71599	(25)3.7345 23822
10	(28)2.4461 94061	(28)2.4818 78254	(28)2.5180 30258	(28)2.5546 55518	(28)2.5917 59533
24	(68)1.3563 70007	(68)1.4043 42816	(68)1.4539 39271	(68)1.5052 11857	(68)1.5582 14678
1/2	(1)2.6267 85107	(1)2.6286 87886	(1)2.6305 89288	(1)2.6324 89316	(1)2.6343 87974
1/3	8.8365 55922	8.8408 22729	8.8450 85422	8.8493 44010	8.8535 98503
1/4	5.1252 17173	5.1270 73128	5.1289 27069	5.1307 79001	5.1326 28931
1/5	3.6963 22179	3.6973 92956	3.6984 62494	3.6995 30796	3.7005 97866

k					
1	695	696	697	698	699
2	4 83025	4 84416	4 85809	4 87204	4 88601
3	3357 02375	3371 53536	3386 08873	3400 68392	3415 32099
4	(11)2.3331 31506	(11)2.3465 88611	(11)2.3601 03845	(11)2.3736 77376	(11)2.3873 09372
5	(14)1.6215 26397	(14)1.6332 25673	(14)1.6449 92380	(14)1.6568 26809	(14)1.6687 29251
6	(17)1.1269 60846	(17)1.1367 25068	(17)1.1465 59689	(17)1.1564 65112	(17)1.1664 41746
7	(19)7.8323 77878	(19)7.9116 06476	(19)7.9915 21031	(19)8.0721 26484	(19)8.1534 27808
8	(22)5.4435 02625	(22)5.5064 78107	(22)5.5700 90158	(22)5.6343 44286	(22)5.6992 46038
9	(25)3.7832 34325	(25)3.8325 08763	(25)3.8823 52840	(25)3.9327 72312	(25)3.9837 72980
10	(28)2.6293 47856	(28)2.6674 26099	(28)2.7059 99930	(28)2.7450 75074	(28)2.7846 57313
24	(68)1.6130 03502	(68)1.6696 35809	(68)1.7281 70846	(68)1.7886 69670	(68)1.8511 95210
1/2	(1)2.6362 85265	(1)2.6381 81192	(1)2.6400 75756	(1)2.6419 68963	(1)2.6438 60813
1/3	8.8578 48911	8.8620 95243	8.8663 37511	8.8705 75722	8.8748 09888
1/4	5.1344 76863	5.1363 22801	5.1381 66751	5.1400 08719	5.1418 48708
1/5	3.7016 63707	3.7027 28321	3.7037 91713	3.7048 53884	3.7059 14839

$$n^{\frac{1}{2}}\left[\begin{matrix}(-6)2\\3\end{matrix}\right] \qquad n^{\frac{1}{3}}\left[\begin{matrix}(-7)5\\4\end{matrix}\right] \qquad n^{\frac{1}{4}}\left[\begin{matrix}(-7)3\\3\end{matrix}\right] \qquad n^{\frac{1}{5}}\left[\begin{matrix}(-7)2\\3\end{matrix}\right]$$

Table 3.1　　　　　　　　**POWERS AND ROOTS** n^k

k	700	701	702	703	704
1	700	701	702	703	704
2	4 90000	4 91401	4 92804	4 94209	4 95616
3	3430 00000	3444 72101	3459 48408	3474 28927	3489 13664
4	(11)2.4010 00000	(11)2.4147 49428	(11)2.4285 57824	(11)2.4424 25357	(11)2.4563 52195
5	(14)1.6807 00000	(14)1.6927 39349	(14)1.7048 47593	(14)1.7170 25026	(14)1.7292 71945
6	(17)1.1764 90000	(17)1.1866 10284	(17)1.1968 03010	(17)1.2070 68593	(17)1.2174 07449
7	(19)8.2354 30000	(19)8.3181 38089	(19)8.4015 57130	(19)8.4856 92210	(19)8.5705 48443
8	(22)5.7648 01000	(22)5.8310 14800	(22)5.8978 93105	(22)5.9654 41624	(22)6.0336 66104
9	(25)4.0353 60700	(25)4.0875 41375	(25)4.1403 20960	(25)4.1937 05461	(25)4.2477 00937
10	(28)2.8247 52490	(28)2.8653 66504	(28)2.9065 05314	(28)2.9481 74939	(28)2.9903 81460
24	(68)1.9158 12314	(68)1.9825 87808	(68)2.0515 90555	(68)2.1228 91511	(68)2.1965 63787
1/2	(1)2.6457 51311	(1)2.6476 40459	(1)2.6495 28260	(1)2.6514 14717	(1)2.6532 99832
1/3	8.8790 40017	8.8832 66120	8.8874 88205	8.8917 06283	8.8959 20362
1/4	5.1436 86724	5.1455 22771	5.1473 56856	5.1491 88981	5.1510 19154
1/5	3.7069 74581	3.7080 33112	3.7090 90435	3.7101 46554	3.7112 01473

k	705	706	707	708	709
1	705	706	707	708	709
2	4 97025	4 98436	4 99849	5 01264	5 02681
3	3504 02625	3518 95816	3533 93243	3548 94912	3564 00829
4	(11)2.4703 38506	(11)2.4843 84461	(11)2.4984 90228	(11)2.5126 55977	(11)2.5268 81878
5	(14)1.7415 88647	(14)1.7539 75429	(14)1.7664 32591	(14)1.7789 60432	(14)1.7915 59251
6	(17)1.2278 19996	(17)1.2383 06653	(17)1.2488 67842	(17)1.2595 03986	(17)1.2702 15509
7	(19)8.6561 30972	(19)8.7424 44971	(19)8.8294 95643	(19)8.9172 88218	(19)9.0058 27960
8	(22)6.1025 72335	(22)6.1721 66150	(22)6.2424 53419	(22)6.3134 40059	(22)6.3851 32023
9	(25)4.3023 13497	(25)4.3575 49302	(25)4.4134 14568	(25)4.4699 15561	(25)4.5270 58605
10	(28)3.0331 31015	(28)3.0764 29807	(28)3.1202 84099	(28)3.1647 00218	(28)3.2096 84551
24	(68)2.2726 82709	(68)2.3513 25887	(68)2.4325 73275	(68)2.5165 07242	(68)2.6032 12640
1/2	(1)2.6551 83609	(1)2.6570 66051	(1)2.6589 47160	(1)2.6608 26939	(1)2.6627 05391
1/3	8.9001 30453	8.9043 36564	8.9085 38706	8.9127 36887	8.9169 31117
1/4	5.1528 47377	5.1546 73657	5.1564 97998	5.1583 20404	5.1601 40881
1/5	3.7122 55193	3.7133 07718	3.7143 59051	3.7154 09195	3.7164 58153

k	710	711	712	713	714
1	710	711	712	713	714
2	5 04100	5 05521	5 06944	5 08369	5 09796
3	3579 11000	3594 25431	3609 44128	3624 67097	3639 94344
4	(11)2.5411 68100	(11)2.5555 14814	(11)2.5699 22191	(11)2.5843 90402	(11)2.5989 19616
5	(14)1.8042 29351	(14)1.8169 71033	(14)1.8297 84600	(14)1.8426 70356	(14)1.8556 28606
6	(17)1.2810 02839	(17)1.2918 66404	(17)1.3028 06635	(17)1.3138 23964	(17)1.3249 18825
7	(19)9.0951 20158	(19)9.1851 70136	(19)9.2759 83244	(19)9.3675 64864	(19)9.4599 20408
8	(22)6.4575 35312	(22)6.5306 55967	(22)6.6045 00070	(22)6.6790 73748	(22)6.7543 83171
9	(25)4.5848 50072	(25)4.6432 96392	(25)4.7024 04050	(25)4.7621 79582	(25)4.8226 29584
10	(28)3.2552 43551	(28)3.3013 83735	(28)3.3481 11683	(28)3.3954 34042	(28)3.4433 57523
24	(68)2.6927 76876	(68)2.7852 89985	(68)2.8808 44702	(68)2.9795 36544	(68)3.0814 63889
1/2	(1)2.6645 82519	(1)2.6664 58325	(1)2.6683 32813	(1)2.6702 05985	(1)2.6720 77843
1/3	8.9211 21404	8.9253 07760	8.9294 90191	8.9336 68708	8.9378 43321
1/4	5.1619 59433	5.1637 76065	5.1655 90782	5.1674 03588	5.1692 14489
1/5	3.7175 05928	3.7185 52523	3.7195 97942	3.7206 42186	3.7216 85260

k	715	716	717	718	719
1	715	716	717	718	719
2	5 11225	5 12656	5 14089	5 15524	5 16961
3	3655 25875	3670 61696	3686 01813	3701 46232	3716 94959
4	(11)2.6135 10006	(11)2.6281 61743	(11)2.6428 74999	(11)2.6576 49946	(11)2.6724 86755
5	(14)1.8686 59654	(14)1.8817 63808	(14)1.8949 41374	(14)1.9081 92661	(14)1.9215 17977
6	(17)1.3360 91653	(17)1.3473 42887	(17)1.3586 72965	(17)1.3700 82331	(17)1.3815 71425
7	(19)9.5530 55319	(19)9.6469 75069	(19)9.7416 85162	(19)9.8371 91134	(19)9.9334 98549
8	(22)6.8304 34553	(22)6.9072 34149	(22)6.9847 88261	(22)7.0631 03234	(22)7.1421 85457
9	(25)4.8837 60705	(25)4.9455 79651	(25)5.0080 93183	(25)5.0713 08122	(25)5.1352 31343
10	(28)3.4918 88904	(28)3.5410 35030	(28)3.5908 02813	(28)3.6411 99232	(28)3.6922 31336
24	(68)3.1867 28051	(68)3.2954 33372	(68)3.4076 87302	(68)3.5236 00491	(68)3.6432 86875
1/2	(1)2.6739 48391	(1)2.6758 17632	(1)2.6776 85568	(1)2.6795 52201	(1)2.6814 17536
1/3	8.9420 14037	8.9461 80866	8.9503 43817	8.9545 02899	8.9586 58122
1/4	5.1710 23488	5.1728 30591	5.1746 35801	5.1764 39125	5.1782 40566
1/5	3.7227 27105	3.7237 67905	3.7248 07483	3.7258 45902	3.7268 83164

k	720	721	722	723	724
1	720	721	722	723	724
2	5 18400	5 19841	5 21284	5 22729	5 24176
3	3732 48000	3748 05361	3763 67048	3779 33067	3795 03424
4	(11)2.6873 85600	(11)2.7023 46653	(11)2.7173 70087	(11)2.7324 56074	(11)2.7476 04790
5	(14)1.9349 17632	(14)1.9483 91937	(14)1.9619 41202	(14)1.9755 65742	(14)1.9892 65868
6	(17)1.3931 40695	(17)1.4047 90586	(17)1.4165 21448	(17)1.4283 34031	(17)1.4402 28488
7	(20)1.0030 61300	(20)1.0128 54013	(20)1.0227 28558	(20)1.0326 85505	(20)1.0427 25426
8	(22)7.2220 41363	(22)7.3026 77432	(22)7.3841 00187	(22)7.4663 16199	(22)7.5493 32081
9	(25)5.1998 69781	(25)5.2652 30428	(25)5.3313 20335	(25)5.3981 46612	(25)5.4657 16426
10	(28)3.7439 06243	(28)3.7962 31139	(28)3.8492 13282	(28)3.9028 60000	(28)3.9571 78693
24	(68)3.7668 63772	(68)3.8944 51981	(68)4.0261 75870	(68)4.1621 63488	(68)4.3025 46659
1/2	(1)2.6832 81573	(1)2.6851 44316	(1)2.6870 05769	(1)2.6888 65932	(1)2.6907 24809
1/3	8.9628 09493	8.9669 57022	8.9711 00718	8.9752 40590	8.9793 76646
1/4	5.1800 40128	5.1818 37817	5.1836 33637	5.1854 27593	5.1872 19688
1/5	3.7279 19273	3.7289 54232	3.7299 88042	3.7310 20708	3.7320 52232

$$n^{\frac{1}{2}}\left[\begin{matrix}(-6)2\\3\end{matrix}\right] \qquad n^{\frac{1}{3}}\left[\begin{matrix}(-7)5\\4\end{matrix}\right] \qquad n^{\frac{1}{4}}\left[\begin{matrix}(-7)2\\3\end{matrix}\right] \qquad n^{\frac{1}{5}}\left[\begin{matrix}(-7)2\\3\end{matrix}\right]$$

POWERS AND ROOTS n^k Table 3.1

k	725	726	727	728	729
1	725	726	727	728	729
2	5 25625	5 27076	5 28529	5 29984	5 31441
3	3810 78125	3826 57176	3842 40583	3858 28352	3874 20489
4	(11)2.7628 16406	(11)2.7780 91098	(11)2.7934 29038	(11)2.8088 30403	(11)2.8242 95365
5	(14)2.0030 41895	(14)2.0168 94137	(14)2.0308 22911	(14)2.0448 28533	(14)2.0589 11321
6	(17)1.4522 05374	(17)1.4642 65143	(17)1.4764 08256	(17)1.4886 35172	(17)1.5009 46353
7	(20)1.0528 48896	(20)1.0630 56494	(20)1.0733 48802	(20)1.0837 26405	(20)1.0941 89891
8	(22)7.6331 54495	(22)7.7177 90147	(22)7.8032 45793	(22)7.8895 28230	(22)7.9766 44308
9	(25)5.5340 37009	(25)5.6031 15647	(25)5.6729 59691	(25)5.7435 76552	(25)5.8149 73700
10	(28)4.0121 76831	(28)4.0678 61960	(28)4.1242 41696	(28)4.1813 23730	(28)4.2391 15828
24	(68)4.4474 61095	(68)4.5970 46501	(68)4.7514 46686	(68)4.9108 09683	(68)5.0752 87861
1/2	(1)2.6925 82404	(1)2.6944 38717	(1)2.6962 93753	(1)2.6981 47513	(1)2.7000 00000
1/3	8.9835 08896	8.9876 37347	8.9917 62009	8.9958 82891	9.0000 00000
1/4	5.1890 09928	5.1907 98317	5.1925 84860	5.1943 69560	5.1961 52423
1/5	3.7330 82616	3.7341 11864	3.7351 39979	3.7361 66963	3.7371 92819

k	730	731	732	733	734
1	730	731	732	733	734
2	5 32900	5 34361	5 35824	5 37289	5 38756
3	3890 17000	3906 17891	3922 23168	3938 32837	3954 46904
4	(11)2.8398 24100	(11)2.8554 16783	(11)2.8710 78255	(11)2.8867 94695	(11)2.9025 80275
5	(14)2.0730 71593	(14)2.0873 09669	(14)2.1016 25868	(14)2.1160 20512	(14)2.1304 93922
6	(17)1.5133 42263	(17)1.5258 23368	(17)1.5383 90135	(17)1.5510 43035	(17)1.5637 82539
7	(20)1.1047 39852	(20)1.1153 76882	(20)1.1261 01579	(20)1.1369 14545	(20)1.1478 16384
8	(22)8.0646 00919	(22)8.1534 05006	(22)8.2430 63558	(22)8.3335 83612	(22)8.4249 72255
9	(25)5.8871 58671	(25)5.9601 39059	(25)6.0339 22524	(25)6.1085 16788	(25)6.1839 29635
10	(28)4.2976 25830	(28)4.3568 61652	(28)4.4168 31288	(28)4.4775 42805	(28)4.5390 04352
24	(68)5.2450 38047	(68)5.4202 21655	(68)5.6010 04807	(68)5.7875 58467	(68)5.9800 58576
1/2	(1)2.7018 51217	(1)2.7037 01167	(1)2.7055 49852	(1)2.7073 97274	(1)2.7092 43437
1/3	9.0041 13346	9.0082 22937	9.0123 28782	9.0164 30890	9.0205 29268
1/4	5.1979 33452	5.1997 12653	5.2014 90029	5.2032 65584	5.2050 39324
1/5	3.7382 17550	3.7392 41158	3.7402 63647	3.7412 85019	3.7423 05277

k	735	736	737	738	739
1	735	736	737	738	739
2	5 40225	5 41696	5 43169	5 44644	5 46121
3	3970 65375	3986 88256	4003 15553	4019 47272	4035 83419
4	(11)2.9184 30506	(11)2.9343 45564	(11)2.9503 25626	(11)2.9663 70867	(11)2.9824 81466
5	(14)2.1450 46422	(14)2.1596 78335	(14)2.1743 89986	(14)2.1891 81700	(14)2.2040 53804
6	(17)1.5766 09120	(17)1.5895 23255	(17)1.6025 25402	(17)1.6156 16095	(17)1.6287 95761
7	(20)1.1588 07703	(20)1.1698 89115	(20)1.1810 61234	(20)1.1923 24678	(20)1.2036 80067
8	(22)8.5172 36620	(22)8.6103 83890	(22)8.7044 21297	(22)8.7993 56123	(22)8.8951 95697
9	(25)6.2601 68916	(25)6.3372 42543	(25)6.4151 58496	(25)6.4939 24819	(25)6.5735 49620
10	(28)4.6012 24153	(28)4.6642 10512	(28)4.7279 71812	(28)4.7925 16516	(28)4.8578 53170
24	(68)6.1786 86185	(68)6.3836 27605	(68)6.5950 74542	(68)6.8132 24254	(68)7.0382 79698
1/2	(1)2.7110 88342	(1)2.7129 31993	(1)2.7147 74392	(1)2.7166 15541	(1)2.7184 55444
1/3	9.0246 23926	9.0287 14871	9.0328 02112	9.0368 85658	9.0409 65517
1/4	5.2068 11253	5.2085 81374	5.2103 49693	5.2121 16213	5.2138 80938
1/5	3.7433 24423	3.7443 42461	3.7453 59393	3.7463 75222	3.7473 89950

k	740	741	742	743	744
1	740	741	742	743	744
2	5 47600	5 49081	5 50564	5 52049	5 53536
3	4052 24000	4068 69021	4085 18488	4101 72407	4118 30784
4	(11)2.9986 57600	(11)3.0148 99446	(11)3.0312 07181	(11)3.0475 80984	(11)3.0640 21033
5	(14)2.2190 06624	(14)2.2340 40489	(14)2.2491 55728	(14)2.2643 52671	(14)2.2796 31649
6	(17)1.6420 64902	(17)1.6554 24002	(17)1.6688 73550	(17)1.6824 14035	(17)1.6960 45947
7	(20)1.2151 28027	(20)1.2266 69186	(20)1.2383 04174	(20)1.2500 33628	(20)1.2618 58184
8	(22)8.9919 47402	(22)9.0896 18667	(22)9.1882 16974	(22)9.2877 49854	(22)9.3882 24890
9	(25)6.6540 41078	(25)6.7354 07432	(25)6.8176 56995	(25)6.9007 98142	(25)6.9848 39318
10	(28)4.9239 90397	(28)4.9909 36907	(28)5.0587 01490	(28)5.1272 93019	(28)5.1967 20453
24	(68)7.2704 49690	(68)7.5099 49065	(68)7.7569 98844	(68)8.0118 26396	(68)8.2746 65623
1/2	(1)2.7202 94102	(1)2.7221 31518	(1)2.7239 67694	(1)2.7258 02634	(1)2.7276 36339
1/3	9.0450 41696	9.0491 14206	9.0531 83053	9.0572 48245	9.0613 09792
1/4	5.2156 43874	5.2174 05023	5.2191 64391	5.2209 21982	5.2226 77799
1/5	3.7484 03580	3.7494 16115	3.7504 27557	3.7514 37909	3.7524 47174

k	745	746	747	748	749
1	745	746	747	748	749
2	5 55025	5 56516	5 58009	5 59504	5 61001
3	4134 93625	4151 60936	4168 32723	4185 08992	4201 89749
4	(11)3.0805 27506	(11)3.0971 00583	(11)3.1137 40441	(11)3.1304 47260	(11)3.1472 21220
5	(14)2.2949 92992	(14)2.3104 37035	(14)2.3259 64109	(14)2.3415 74551	(14)2.3572 68694
6	(17)1.7097 69779	(17)1.7235 86028	(17)1.7374 95190	(17)1.7514 97764	(17)1.7655 94252
7	(20)1.2737 78485	(20)1.2857 95177	(20)1.2979 08907	(20)1.3101 20327	(20)1.3224 30094
8	(22)9.4896 49717	(22)9.5920 32018	(22)9.6953 79533	(22)9.7997 00049	(22)9.9050 01408
9	(25)7.0697 89039	(25)7.1556 55886	(25)7.2424 48511	(25)7.3301 75636	(25)7.4188 46054
10	(28)5.2669 92834	(28)5.3381 19291	(28)5.4101 09038	(28)5.4829 71376	(28)5.5567 15695
24	(68)8.5457 57129	(68)8.8253 48404	(68)9.1136 94019	(68)9.4110 55807	(68)9.7177 03069
1/2	(1)2.7294 68813	(1)2.7313 00057	(1)2.7331 30074	(1)2.7349 58866	(1)2.7367 86437
1/3	9.0653 67701	9.0694 21981	9.0734 72639	9.0775 19683	9.0815 63122
1/4	5.2244 31847	5.2261 84131	5.2279 34653	5.2296 83419	5.2314 30432
1/5	3.7534 55355	3.7544 62453	3.7554 68472	3.7564 73415	3.7574 77282

$$n^{\frac{1}{2}}\left[\begin{matrix}(-6)2\\3\end{matrix}\right] \qquad n^{\frac{1}{3}}\left[\begin{matrix}(-7)5\\4\end{matrix}\right] \qquad n^{\frac{1}{4}}\left[\begin{matrix}(-7)2\\3\end{matrix}\right] \qquad n^{\frac{1}{5}}\left[\begin{matrix}(-7)1\\3\end{matrix}\right]$$

Table 3.1　　　　　　　　　**POWERS AND ROOTS** n^k

k	750	751	752	753	754
1	750	751	752	753	754
2	5 62500	5 64001	5 65504	5 67009	5 68516
3	4218 75000	4235 64751	4252 59008	4269 57777	4286 61064
4	(11)3.1640 62500	(11)3.1809 71280	(11)3.1979 47740	(11)3.2149 92061	(11)3.2321 04423
5	(14)2.3730 46875	(14)2.3889 09431	(14)2.4048 56701	(14)2.4208 89022	(14)2.4370 06735
6	(17)1.7797 85156	(17)1.7940 70983	(17)1.8084 52239	(17)1.8229 29433	(17)1.8375 03078
7	(20)1.3348 38867	(20)1.3473 47308	(20)1.3599 56084	(20)1.3726 65863	(20)1.3854 77321
8	(23)1.0011 29150	(23)1.0118 57828	(23)1.0226 86975	(23)1.0336 17395	(23)1.0446 49900
9	(25)7.5084 68628	(25)7.5990 52291	(25)7.6906 06051	(25)7.7831 38985	(25)7.8766 60245
10	(28)5.6313 51471	(28)5.7068 88271	(28)5.7833 35750	(28)5.8607 03656	(28)5.9390 01825
24	(69)1.0033 91278	(69)1.0359 96977	(69)1.0696 16698	(69)1.1042 80565	(69)1.1400 19555
1/2	(1)2.7386 12788	(1)2.7404 37921	(1)2.7422 61840	(1)2.7440 84547	(1)2.7459 06044
1/3	9.0856 02964	9.0896 39217	9.0936 71888	9.0977 00985	9.1017 26517
1/4	5.2331 75697	5.2349 19217	5.2366 60997	5.2384 01041	5.2401 39353
1/5	3.7584 80079	3.7594 81806	3.7604 82467	3.7614 82064	3.7624 80599

k	755	756	757	758	759
1	755	756	757	758	759
2	5 70025	5 71536	5 73049	5 74564	5 76081
3	4303 68875	4320 81216	4337 98093	4355 19512	4372 45479
4	(11)3.2492 85006	(11)3.2665 33993	(11)3.2838 51564	(11)3.3012 37901	(11)3.3186 93186
5	(14)2.4532 10180	(14)2.4694 99699	(14)2.4858 75634	(14)2.5023 38329	(14)2.5188 88128
6	(17)1.8521 73686	(17)1.8669 41772	(17)1.8818 07855	(17)1.8967 72453	(17)1.9118 36089
7	(20)1.3983 91133	(20)1.4114 07980	(20)1.4245 28546	(20)1.4377 53520	(20)1.4510 83592
8	(23)1.0557 85305	(23)1.0670 24433	(23)1.0783 68109	(23)1.0898 17168	(23)1.1013 72446
9	(25)7.9711 79054	(25)8.0667 04711	(25)8.1632 46588	(25)8.2608 14132	(25)8.3594 16865
10	(28)6.0182 40186	(28)6.0984 28762	(28)6.1795 77667	(28)6.2616 97112	(28)6.3447 97401
24	(69)1.1768 65520	(69)1.2148 51214	(69)1.2540 10313	(69)1.2943 77441	(69)1.3359 88198
1/2	(1)2.7477 26333	(1)2.7495 45417	(1)2.7513 63298	(1)2.7531 79980	(1)2.7549 95463
1/3	9.1057 48491	9.1097 66916	9.1137 81798	9.1177 93146	9.1218 00968
1/4	5.2418 75936	5.2436 10795	5.2453 43934	5.2470 75356	5.2488 05067
1/5	3.7634 78075	3.7644 74495	3.7654 69862	3.7664 64176	3.7674 57442

k	760	761	762	763	764
1	760	761	762	763	764
2	5 77600	5 79121	5 80644	5 82169	5 83696
3	4389 76000	4407 11081	4424 50728	4441 94947	4459 43744
4	(11)3.3362 17600	(11)3.3538 11326	(11)3.3714 74547	(11)3.3892 07446	(11)3.4070 10204
5	(14)2.5355 25376	(14)2.5522 50419	(14)2.5690 63605	(14)2.5859 65281	(14)2.6029 55796
6	(17)1.9269 99286	(17)1.9422 62569	(17)1.9576 26467	(17)1.9730 91509	(17)1.9886 58228
7	(20)1.4645 19457	(20)1.4780 61815	(20)1.4917 11368	(20)1.5054 68822	(20)1.5193 34886
8	(23)1.1130 34787	(23)1.1248 05041	(23)1.1366 84062	(23)1.1486 72711	(23)1.1607 71853
9	(25)8.4590 64385	(25)8.5597 66364	(25)8.6615 32555	(25)8.7643 72784	(25)8.8682 96958
10	(28)6.4288 88932	(28)6.5139 82203	(28)6.6000 87807	(28)6.6872 16435	(28)6.7753 78876
24	(69)1.3788 79182	(69)1.4230 88020	(69)1.4686 53390	(69)1.5156 15056	(69)1.5640 13890
1/2	(1)2.7568 09750	(1)2.7586 22845	(1)2.7604 34748	(1)2.7622 45463	(1)2.7640 54992
1/3	9.1258 05271	9.1298 06063	9.1338 03351	9.1377 97144	9.1417 87449
1/4	5.2505 33069	5.2522 59366	5.2539 83963	5.2557 06863	5.2574 28071
1/5	3.7684 49662	3.7694 40838	3.7704 30972	3.7714 20068	3.7724 08126

k	765	766	767	768	769
1	765	766	767	768	769
2	5 85225	5 86756	5 88289	5 89824	5 91361
3	4476 97125	4494 55096	4512 17663	4529 84832	4547 56609
4	(11)3.4248 83006	(11)3.4428 26035	(11)3.4608 39475	(11)3.4789 23510	(11)3.4970 78323
5	(14)2.6200 35500	(14)2.6372 04743	(14)2.6544 63877	(14)2.6718 13255	(14)2.6892 53231
6	(17)2.0043 27157	(17)2.0200 98833	(17)2.0359 73794	(17)2.0519 52580	(17)2.0680 35734
7	(20)1.5333 10275	(20)1.5473 95706	(20)1.5615 91900	(20)1.5758 99582	(20)1.5903 19480
8	(23)1.1729 82361	(23)1.1853 05111	(23)1.1977 40987	(23)1.2102 90879	(23)1.2229 55680
9	(25)8.9733 15059	(25)9.0794 37150	(25)9.1866 73373	(25)9.2950 33948	(25)9.4045 29178
10	(28)6.8645 86020	(28)6.9548 48857	(28)7.0461 78477	(28)7.1385 86072	(28)7.2320 82938
24	(69)1.6138 91907	(69)1.6652 92289	(69)1.7182 59425	(69)1.7728 38934	(69)1.8290 77701
1/2	(1)2.7658 63337	(1)2.7676 70501	(1)2.7694 76485	(1)2.7712 81292	(1)2.7730 84925
1/3	9.1457 74274	9.1497 57625	9.1537 37512	9.1577 13940	9.1616 86919
1/4	5.2591 47590	5.2608 65424	5.2625 81576	5.2642 96052	5.2660 08854
1/5	3.7733 95151	3.7743 81144	3.7753 66108	3.7763 50045	3.7773 32958

k	770	771	772	773	774
1	770	771	772	773	774
2	5 92900	5 94441	5 95984	5 97529	5 99076
3	4565 33000	4583 14011	4600 99648	4618 89917	4636 84824
4	(11)3.5153 04100	(11)3.5336 01025	(11)3.5519 69283	(11)3.5704 09058	(11)3.5889 20538
5	(14)2.7067 84157	(14)2.7244 06390	(14)2.7421 20286	(14)2.7599 26202	(14)2.7778 24496
6	(17)2.0842 23801	(17)2.1005 17327	(17)2.1169 16861	(17)2.1334 22954	(17)2.1500 36160
7	(20)1.6048 52327	(20)1.6194 98859	(20)1.6342 59817	(20)1.6491 35944	(20)1.6641 27988
8	(23)1.2357 36292	(23)1.2486 33620	(23)1.2616 48578	(23)1.2747 82084	(23)1.2880 35063
9	(25)9.5151 69445	(25)9.6269 65212	(25)9.7399 27025	(25)9.8540 65513	(25)9.9693 91385
10	(28)7.3266 80473	(28)7.4223 90179	(28)7.5192 23664	(28)7.6171 92641	(28)7.7163 08932
24	(69)1.8870 23915	(69)1.9467 27094	(69)2.0082 38127	(69)2.0716 09310	(69)2.1368 94378
1/2	(1)2.7748 87385	(1)2.7766 88675	(1)2.7784 88798	(1)2.7802 87755	(1)2.7820 85549
1/3	9.1656 56454	9.1696 22555	9.1735 85227	9.1775 44479	9.1815 00317
1/4	5.2677 19986	5.2694 29452	5.2711 37257	5.2728 43403	5.2745 47894
1/5	3.7783 14849	3.7792 95720	3.7802 75573	3.7812 54412	3.7822 32239

$$n^{\frac{1}{2}}\left[\begin{matrix}(-6)2\\3\end{matrix}\right] \qquad n^{\frac{1}{3}}\left[\begin{matrix}(-7)5\\3\end{matrix}\right] \qquad n^{\frac{1}{4}}\left[\begin{matrix}(-7)2\\3\end{matrix}\right] \qquad n^{\frac{1}{5}}\left[\begin{matrix}(-7)1\\3\end{matrix}\right]$$

POWERS AND ROOTS n^k

Table 3.1

k		775		776		777		778		779
1		775		776		777		778		779
2	6	00625	6	02176	6	03729	6	05284	6	06841
3		4654 84375		4672 88576		4690 97433		4709 10952		4727 29139
4	(11)3.	6075 03906	(11)3.	6261 59350	(11)3.	6448 87054	(11)3.	6636 87207	(11)3.	6825 59993
5	(14)2.	7958 15527	(14)2.	8138 99655	(14)2.	8320 77241	(14)2.	8503 48647	(14)2.	8687 14234
6	(17)2.	1667 57034	(17)2.	1835 86133	(17)2.	2005 24016	(17)2.	2175 71247	(17)2.	2347 28389
7	(20)1.	6792 36701	(20)1.	6944 62839	(20)1.	7098 07161	(20)1.	7252 70430	(20)1.	7408 53415
8	(23)1.	3014 08443	(23)1.	3149 03163	(23)1.	3285 20164	(23)1.	3422 60395	(23)1.	3561 24810
9	(26)1.	0085 91544	(26)1.	0203 64854	(26)1.	0322 60167	(26)1.	0442 78587	(26)1.	0564 21227
10	(28)7.	8165 84463	(28)7.	9180 31271	(28)8.	0206 61501	(28)8.	1244 87408	(28)8.	2295 21359
24	(69)2.	2041 48547	(69)2.	2734 28553	(69)2.	3447 92689	(69)2.	4183 00846	(69)2.	4940 14558
1/2	(1)2.	7838 82181	(1)2.	7856 77655	(1)2.	7874 71973	(1)2.	7892 65136	(1)2.	7910 57147
1/3	9.	1854 52750	9.	1894 01784	9.	1933 47428	9.	1972 89687	9.	2012 28569
1/4	5.	2762 50735	5.	2779 51928	5.	2796 51478	5.	2813 49388	5.	2830 45663
1/5	3.	7832 09055	3.	7841 84864	3.	7851 59667	3.	7861 33467	3.	7871 06266

k		780		781		782		783		784
1		780		781		782		783		784
2	6	08400	6	09961	6	11524	6	13009	6	14656
3		4745 52000		4763 79541		4782 11768		4800 48687		4818 90304
4	(11)3.	7015 05600	(11)3.	7205 24215	(11)3.	7396 16026	(11)3.	7587 81219	(11)3.	7780 19983
5	(14)2.	8871 74368	(14)2.	9057 29412	(14)2.	9243 79732	(14)2.	9431 25695	(14)2.	9619 67667
6	(17)2.	2519 96007	(17)2.	2693 74671	(17)2.	2868 64951	(17)2.	3044 67419	(17)2.	3221 82651
7	(20)1.	7565 56885	(20)1.	7723 81618	(20)1.	7883 28391	(20)1.	8043 97989	(20)1.	8205 91198
8	(23)1.	3701 14371	(23)1.	3842 30044	(23)1.	3984 72802	(23)1.	4128 43625	(23)1.	4273 43499
9	(26)1.	0686 89209	(26)1.	0810 83664	(26)1.	0936 05731	(26)1.	1062 56559	(26)1.	1190 37304
10	(28)8.	3357 75831	(28)8.	4432 63416	(28)8.	5519 96818	(28)8.	6619 88854	(28)8.	7732 52460
24	(69)2.	5719 97041	(69)2.	6523 13239	(69)2.	7350 29868	(69)2.	8202 15463	(69)2.	9079 40422
1/2	(1)2.	7928 48009	(1)2.	7946 37722	(1)2.	7964 26291	(1)2.	7982 13716	(1)2.	8000 00000
1/3	9.	2051 64083	9.	2090 96233	9.	2130 25029	9.	2169 50477	9.	2208 72584
1/4	5.	2847 40305	5.	2864 33318	5.	2881 24706	5.	2898 14473	5.	2915 02622
1/5	3.	7880 78066	3.	7890 48871	3.	7900 18681	3.	7909 87500	3.	7919 55329

k		785		786		787		788		789
1		785		786		787		788		789
2	6	16225	6	17796	6	19369	6	20944	6	22521
3		4837 36625		4855 87656		4874 43403		4893 03872		4911 69069
4	(11)3.	7973 32506	(11)3.	8167 18976	(11)3.	8361 79582	(11)3.	8557 14511	(11)3.	8753 23954
5	(14)2.	9809 06017	(14)2.	9999 41115	(14)3.	0190 73331	(14)3.	0383 03035	(14)3.	0576 30600
6	(17)2.	3400 11224	(17)2.	3579 53717	(17)2.	3760 10711	(17)2.	3941 82792	(17)2.	4124 70543
7	(20)1.	8369 08811	(20)1.	8533 51621	(20)1.	8699 20430	(20)1.	8866 16040	(20)1.	9034 39259
8	(23)1.	4419 73416	(23)1.	4567 34374	(23)1.	4716 27378	(23)1.	4866 53439	(23)1.	5018 13575
9	(26)1.	1319 49132	(26)1.	1449 93218	(26)1.	1581 70747	(26)1.	1714 82910	(26)1.	1849 30911
10	(28)8.	8858 00685	(28)8.	9996 46695	(28)9.	1148 03776	(28)9.	2312 85332	(28)9.	3491 04886
24	(69)2.	9982 77060	(69)3.	0912 99652	(69)3.	1870 84488	(69)3.	2857 09926	(69)3.	3872 56439
1/2	(1)2.	8017 85145	(1)2.	8035 69154	(1)2.	8053 52028	(1)2.	8071 33770	(1)2.	8089 14381
1/3	9.	2247 91357	9.	2287 06804	9.	2326 18931	9.	2365 27746	9.	2404 33255
1/4	5.	2931 89157	5.	2948 74081	5.	2965 57399	5.	2982 39113	5.	2999 19227
1/5	3.	7929 22172	3.	7938 88029	3.	7948 52904	3.	7958 16799	3.	7967 79716

k		790		791		792		793		794
1		790		791		792		793		794
2	6	24100	6	25681	6	27264	6	28849	6	30436
3		4930 39000		4949 13671		4967 93088		4986 77257		5005 66184
4	(11)3.	8950 08100	(11)3.	9147 67138	(11)3.	9346 01257	(11)3.	9545 10648	(11)3.	9744 95501
5	(14)3.	0770 56399	(14)3.	0965 80806	(14)3.	1162 04196	(14)3.	1359 26944	(14)3.	1557 49428
6	(17)2.	4308 74555	(17)2.	4493 95417	(17)2.	4680 33723	(17)2.	4867 90066	(17)2.	5056 65046
7	(20)1.	9203 90899	(20)1.	9374 71775	(20)1.	9546 82708	(20)1.	9720 24523	(20)1.	9894 98046
8	(23)1.	5171 08810	(23)1.	5325 40174	(23)1.	5481 08705	(23)1.	5638 15447	(23)1.	5796 61449
9	(26)1.	1985 15960	(26)1.	2122 39278	(26)1.	2261 02094	(26)1.	2401 05649	(26)1.	2542 51190
10	(28)9.	4682 76083	(28)9.	5888 12687	(28)9.	7107 28588	(28)9.	8340 37797	(28)9.	9587 54451
24	(69)3.	4918 06676	(69)3.	5994 45514	(69)3.	7102 60118	(69)3.	8243 39997	(69)3.	9417 77065
1/2	(1)2.	8106 93865	(1)2.	8124 72222	(1)2.	8142 49456	(1)2.	8160 25568	(1)2.	8178 00561
1/3	9.	2443 35465	9.	2482 34384	9.	2521 30018	9.	2560 22375	9.	2599 11460
1/4	5.	3015 97745	5.	3032 74670	5.	3049 50005	5.	3066 23755	5.	3082 95923
1/5	3.	7977 41656	3.	7987 02623	3.	7996 62619	3.	8006 21646	3.	8015 79705

k		795		796		797		798		799
1		795		796		797		798		799
2	6	32025	6	33616	6	35209	6	36804	6	38401
3		5024 59875		5043 58336		5062 61573		5081 69592		5100 82399
4	(11)3.	9945 56006	(11)4.	0146 92355	(11)4.	0349 04737	(11)4.	0551 93344	(11)4.	0755 58368
5	(14)3.	1756 72025	(14)3.	1956 95114	(14)3.	2158 19075	(14)3.	2360 44289	(14)3.	2563 71136
6	(17)2.	5246 59260	(17)2.	5437 73311	(17)2.	5630 07803	(17)2.	5823 63342	(17)2.	6018 40538
7	(20)2.	0071 04112	(20)2.	0248 43555	(20)2.	0427 17219	(20)2.	0607 25947	(20)2.	0788 70590
8	(23)1.	5956 47769	(23)1.	6117 75470	(23)1.	6280 45624	(23)1.	6444 59306	(23)1.	6610 17601
9	(26)1.	2685 39976	(26)1.	2829 73274	(26)1.	2975 52362	(26)1.	3122 78526	(26)1.	3271 53063
10	(29)1.	0084 89281	(29)1.	0212 46726	(29)1.	0341 49232	(29)1.	0471 98264	(29)1.	0603 95298
24	(69)4.	0626 65702	(69)4.	1871 02820	(69)4.	3151 87922	(69)4.	4470 23172	(69)4.	5827 13463
1/2	(1)2.	8195 74436	(1)2.	8213 47196	(1)2.	8231 18843	(1)2.	8248 89378	(1)2.	8266 58805
1/3	9.	2637 97282	9.	2676 79846	9.	2715 59160	9.	2754 35230	9.	2793 08064
1/4	5.	3099 66512	5.	3116 35526	5.	3133 02968	5.	3149 68841	5.	3166 33150
1/5	3.	8025 36800	3.	8034 92932	3.	8044 48104	3.	8054 02317	3.	8063 55574

$$n^{\frac{1}{2}}\begin{bmatrix}(-6)2\\3\end{bmatrix} \qquad n^{\frac{1}{3}}\begin{bmatrix}(-7)4\\3\end{bmatrix} \qquad n^{\frac{1}{4}}\begin{bmatrix}(-7)2\\3\end{bmatrix} \qquad n^{\frac{1}{5}}\begin{bmatrix}(-7)1\\3\end{bmatrix}$$

Table 3.1　　　　　　　POWERS AND ROOTS n^k

k	800	801	802	803	804
1	800	801	802	803	804
2	6 40000	6 41601	6 43204	6 44809	6 46416
3	5120 00000	5139 22401	5158 49608	5177 81627	5197 18464
4	(11)4.0960 00000	(11)4.1165 18432	(11)4.1371 13856	(11)4.1577 86465	(11)4.1785 36451
5	(14)3.2768 00000	(14)3.2973 31264	(14)3.3179 65313	(14)3.3387 02531	(14)3.3595 43306
6	(17)2.6214 40000	(17)2.6411 62342	(17)2.6610 08181	(17)2.6809 78133	(17)2.7010 72818
7	(20)2.0971 52000	(20)2.1155 71036	(20)2.1341 28561	(20)2.1528 25440	(20)2.1716 62546
8	(23)1.6777 21600	(23)1.6945 72400	(23)1.7115 71106	(23)1.7287 18829	(23)1.7460 16687
9	(26)1.3421 77280	(26)1.3573 52492	(26)1.3726 80027	(26)1.3881 61219	(26)1.4037 97416
10	(29)1.0737 41824	(29)1.0872 39346	(29)1.1008 89382	(29)1.1146 93459	(29)1.1286 53123
24	(69)4.7223 66483	(69)4.8660 92789	(69)5.0140 05879	(69)5.1662 22264	(69)5.3228 61548
1/2	(1)2.8284 27125	(1)2.8301 94340	(1)2.8319 07949	(1)2.8337 25463	(1)2.8354 89376
1/3	9.2831 77667	9.2870 44047	9.2909 07211	9.2947 67164	9.2986 23915
1/4	5.3182 95897	5.3199 57086	5.3216 16720	5.3232 74803	5.3249 31338
1/5	3.8073 07877	3.8082 59229	3.8092 09631	3.8101 59085	3.8111 07593

k	805	806	807	808	809
1	805	806	807	808	809
2	6 48025	6 49636	6 51249	6 52864	6 54481
3	5216 60125	5236 06616	5255 57943	5275 14112	5294 75129
4	(11)4.1993 64006	(11)4.2202 69325	(11)4.2412 52600	(11)4.2623 14025	(11)4.2834 53794
5	(14)3.3804 88025	(14)3.4015 37076	(14)3.4226 90848	(14)3.4439 49732	(14)3.4653 14119
6	(17)2.7212 92860	(17)2.7416 38883	(17)2.7621 11515	(17)2.7827 11384	(17)2.8034 39122
7	(20)2.1906 40752	(20)2.2097 60940	(20)2.2290 23992	(20)2.2484 30798	(20)2.2679 82250
8	(23)1.7634 65806	(23)1.7810 67318	(23)1.7988 22362	(23)1.8167 32085	(23)1.8347 97640
9	(26)1.4195 89974	(26)1.4355 40258	(26)1.4516 49646	(26)1.4679 19524	(26)1.4843 51291
10	(29)1.1427 69929	(29)1.1570 45448	(29)1.1714 81264	(29)1.1860 78976	(29)1.2008 40194
24	(69)5.4840 46503	(69)5.6499 03151	(69)5.8205 60843	(69)5.9961 52346	(69)6.1768 13927
1/2	(1)2.8372 52192	(1)2.8390 13913	(1)2.8407 74542	(1)2.8425 34081	(1)2.8442 92531
1/3	9.3024 77468	9.3063 27832	9.3101 75012	9.3140 19016	9.3178 59849
1/4	5.3265 86329	5.3282 39778	5.3298 91690	5.3315 42067	5.3331 90912
1/5	3.8120 55159	3.8130 01783	3.8139 47468	3.8148 92216	3.8158 36029

k	810	811	812	813	814
1	810	811	812	813	814
2	6 56100	6 57721	6 59344	6 60969	6 62596
3	5314 41000	5334 11731	5353 87328	5373 67797	5393 53144
4	(11)4.3046 72100	(11)4.3259 49138	(11)4.3473 45103	(11)4.3688 00190	(11)4.3903 34592
5	(14)3.4867 84401	(14)3.5083 60971	(14)3.5300 44224	(14)3.5518 34554	(14)3.5737 32358
6	(17)2.8242 95365	(17)2.8452 80748	(17)2.8663 95910	(17)2.8876 41493	(17)2.9090 18139
7	(20)2.2876 79245	(20)2.3075 22686	(20)2.3275 13479	(20)2.3476 52533	(20)2.3679 40765
8	(23)1.8530 20189	(23)1.8714 00899	(23)1.8899 40945	(23)1.9086 41510	(23)1.9275 03783
9	(26)1.5009 46353	(26)1.5177 06129	(26)1.5346 32047	(26)1.5517 25547	(26)1.5689 88079
10	(29)1.2157 66546	(29)1.2308 59670	(29)1.2461 21222	(29)1.2615 52870	(29)1.2771 56297
24	(69)6.3626 85441	(69)6.5539 10420	(69)6.7506 36166	(69)6.9530 13847	(69)7.1611 98588
1/2	(1)2.8460 49894	(1)2.8478 06173	(1)2.8495 61370	(1)2.8513 15486	(1)2.8530 68524
1/3	9.3216 97518	9.3255 32030	9.3293 63391	9.3331 91608	9.3370 16687
1/4	5.3348 38230	5.3364 84023	5.3381 28295	5.3397 71049	5.3414 12288
1/5	3.8167 78910	3.8177 20859	3.8186 61880	3.8196 01974	3.8205 41144

k	815	816	817	818	819
1	815	816	817	818	819
2	6 64225	6 65856	6 67489	6 69124	6 70761
3	5413 43375	5433 38496	5453 38513	5473 43432	5493 53259
4	(11)4.4119 48506	(11)4.4336 42127	(11)4.4554 15651	(11)4.4772 69274	(11)4.4992 03191
5	(14)3.5957 38033	(14)3.6178 51976	(14)3.6400 74587	(14)3.6624 06266	(14)3.6848 47414
6	(17)2.9305 26497	(17)2.9521 67212	(17)2.9739 40938	(17)2.9958 48326	(17)3.0178 90032
7	(20)2.3883 79095	(20)2.4089 68445	(20)2.4297 09746	(20)2.4506 03930	(20)2.4716 51936
8	(23)1.9465 28962	(23)1.9657 18251	(23)1.9850 72863	(23)2.0045 94015	(23)2.0242 82936
9	(26)1.5864 21104	(26)1.6040 26093	(26)1.6218 04529	(26)1.6397 57904	(26)1.6578 87724
10	(29)1.2929 33200	(29)1.3088 85292	(29)1.3250 14300	(29)1.3413 21966	(29)1.3578 10046
24	(69)7.3753 49576	(69)7.5956 30157	(69)7.8222 07941	(69)8.0552 54907	(69)8.2949 47511
1/2	(1)2.8548 20485	(1)2.8565 71371	(1)2.8583 21186	(1)2.8600 69929	(1)2.8618 17604
1/3	9.3408 38634	9.3446 57457	9.3484 73160	9.3522 85752	9.3560 95237
1/4	5.3430 52016	5.3446 90236	5.3463 26950	5.3479 62163	5.3495 95877
1/5	3.8214 79391	3.8224 16717	3.8233 53125	3.8242 88616	3.8252 23193

k	820	821	822	823	824
1	820	821	822	823	824
2	6 72400	6 74041	6 75684	6 77329	6 78976
3	5513 68000	5533 87661	5554 12248	5574 41767	5594 76224
4	(11)4.5212 17600	(11)4.5433 12697	(11)4.5654 88679	(11)4.5877 45742	(11)4.6100 84086
5	(14)3.7073 98432	(14)3.7300 59724	(14)3.7528 31694	(14)3.7757 14746	(14)3.7987 09287
6	(17)3.0400 66714	(17)3.0623 79033	(17)3.0848 27652	(17)3.1074 13236	(17)3.1301 36452
7	(20)2.4928 54706	(20)2.5142 13186	(20)2.5357 28330	(20)2.5574 01093	(20)2.5792 32437
8	(23)2.0441 40859	(23)2.0641 69026	(23)2.0843 68687	(23)2.1047 41100	(23)2.1252 87528
9	(26)1.6761 95504	(26)1.6946 82770	(26)1.7133 51061	(26)1.7322 01925	(26)1.7512 36923
10	(29)1.3744 80313	(29)1.3913 34555	(29)1.4083 74572	(29)1.4256 02184	(29)1.4430 19224
24	(69)8.5414 66801	(69)8.7949 98523	(69)9.0557 33244	(69)9.3238 66467	(69)9.5995 98755
1/2	(1)2.8635 64213	(1)2.8653 09756	(1)2.8670 54237	(1)2.8687 97658	(1)2.8705 40019
1/3	9.3599 01623	9.3637 04916	9.3675 05121	9.3713 02245	9.3750 96295
1/4	5.3512 28095	5.3528 58822	5.3544 88059	5.3561 15810	5.3577 42079
1/5	3.8261 56858	3.8270 89612	3.8280 21458	3.8289 52397	3.8298 82432

$$n^{\frac{1}{2}}\left[\begin{matrix}(-6)1\\3\end{matrix}\right] \qquad n^{\frac{1}{3}}\left[\begin{matrix}(-7)4\\3\end{matrix}\right] \qquad n^{\frac{1}{4}}\left[\begin{matrix}(-7)2\\3\end{matrix}\right] \qquad n^{\frac{1}{5}}\left[\begin{matrix}(-7)1\\3\end{matrix}\right]$$

POWERS AND ROOTS n^k Table 3.1

k	825	826	827	828	829
1	825	826	827	828	829
2	6 80625	6 82276	6 83929	6 85584	6 87241
3	5615 15625	5635 59976	5656 09283	5676 63552	5697 22789
4	(11)4.6325 03906	(11)4.6550 05402	(11)4.6775 88770	(11)4.7002 54211	(11)4.7230 01921
5	(14)3.8218 15723	(14)3.8450 34462	(14)3.8683 65913	(14)3.8918 10486	(14)3.9153 68592
6	(17)3.1529 97971	(17)3.1759 98465	(17)3.1991 38610	(17)3.2224 19083	(17)3.2458 40563
7	(20)2.6012 23326	(20)2.6233 74732	(20)2.6456 87631	(20)2.6681 63000	(20)2.6908 01827
8	(23)2.1460 09244	(23)2.1669 07529	(23)2.1879 83671	(23)2.2092 38964	(23)2.2306 74714
9	(26)1.7704 57626	(26)1.7898 65619	(26)1.8094 62496	(26)1.8292 49863	(26)1.8492 29338
10	(29)1.4606 27542	(29)1.4784 29001	(29)1.4964 25484	(29)1.5146 18886	(29)1.5330 11121
24	(69)9.8831 35853	(70)1.0174 68882	(70)1.0474 47415	(70)1.0782 71392	(70)1.1099 63591
1/2	(1)2.8722 81323	(1)2.8740 21573	(1)2.8757 60769	(1)2.8774 98914	(1)2.8792 36010
1/3	9.3788 87277	9.3826 75196	9.3864 60060	9.3902 41873	9.3940 20643
1/4	5.3593 66869	5.3609 90182	5.3626 12021	5.3642 32391	5.3658 51293
1/5	3.8308 11564	3.8317 39795	3.8326 67128	3.8335 93565	3.8345 19107

k	830	831	832	833	834
1	830	831	832	833	834
2	6 88900	6 90561	6 92224	6 93889	6 95556
3	5717 87000	5738 56191	5759 30368	5780 09537	5800 93704
4	(11)4.7458 32100	(11)4.7687 44947	(11)4.7917 40642	(11)4.8148 19443	(11)4.8379 81491
5	(14)3.9390 40643	(14)3.9628 27051	(14)3.9867 28231	(14)4.0107 44596	(14)4.0348 76564
6	(17)3.2694 03734	(17)3.2931 09279	(17)3.3169 57888	(17)3.3409 50249	(17)3.3650 87054
7	(20)2.7136 05099	(20)2.7365 73811	(20)2.7597 08963	(20)2.7830 11557	(20)2.8064 82603
8	(23)2.2522 92232	(23)2.2740 92837	(23)2.2960 77857	(23)2.3182 48627	(23)2.3406 06491
9	(26)1.8694 02553	(26)1.8897 71148	(26)1.9103 36777	(26)1.9311 01106	(26)1.9520 65814
10	(29)1.5516 04119	(29)1.5703 99824	(29)1.5894 00198	(29)1.6086 07222	(29)1.6280 22889
24	(70)1.1425 47375	(70)1.1760 46709	(70)1.2104 86167	(70)1.2458 90957	(70)1.2822 86929
1/2	(1)2.8809 72058	(1)2.8827 07061	(1)2.8844 41020	(1)2.8861 73938	(1)2.8879 05816
1/3	9.3977 96375	9.4015 69076	9.4053 38751	9.4091 05407	9.4128 69049
1/4	5.3674 68731	5.3690 84709	5.3706 99229	5.3723 12294	5.3739 23907
1/5	3.8354 43756	3.8363 67514	3.8372 90383	3.8382 12366	3.8391 33463

k	835	836	837	838	839
1	835	836	837	838	839
2	6 97225	6 98896	7 00569	7 02244	7 03921
3	5821 82875	5842 77056	5863 76253	5884 80472	5905 89719
4	(11)4.8612 27006	(11)4.8845 56188	(11)4.9079 69238	(11)4.9314 66355	(11)4.9550 47742
5	(14)4.0591 24550	(14)4.0834 88973	(14)4.1079 70252	(14)4.1325 68806	(14)4.1572 85056
6	(17)3.3893 68999	(17)3.4137 96782	(17)3.4383 71101	(17)3.4630 92659	(17)3.4879 62162
7	(20)2.8301 23115	(20)2.8539 34109	(20)2.8779 16611	(20)2.9020 71648	(20)2.9264 00254
8	(23)2.3631 52801	(23)2.3858 88916	(23)2.4088 16204	(23)2.4319 36041	(23)2.4552 49813
9	(26)1.9732 32589	(26)1.9946 03133	(26)2.0161 79163	(26)2.0379 62403	(26)2.0599 54593
10	(29)1.6476 49211	(29)1.6674 88220	(29)1.6875 41959	(29)1.7078 12493	(29)1.7283 01904
24	(70)1.3197 00592	(70)1.3581 59133	(70)1.3976 90431	(70)1.4383 23072	(70)1.4800 86372
1/2	(1)2.8896 36655	(1)2.8913 66459	(1)2.8930 95228	(1)2.8948 22965	(1)2.8965 49672
1/3	9.4166 29685	9.4203 87319	9.4241 41957	9.4278 93606	9.4316 42272
1/4	5.3755 34071	5.3771 42790	5.3787 50067	5.3803 55904	5.3819 60304
1/5	3.8400 53677	3.8409 73010	3.8418 91464	3.8428 09040	3.8437 25741

k	840	841	842	843	844
1	840	841	842	843	844
2	7 05600	7 07281	7 08964	7 10649	7 12336
3	5927 04000	5948 23321	5969 47688	5990 77107	6012 11584
4	(11)4.9787 13600	(11)5.0024 64130	(11)5.0262 99533	(11)5.0502 20012	(11)5.0742 25769
5	(14)4.1821 19424	(14)4.2070 72333	(14)4.2321 44207	(14)4.2573 35470	(14)4.2826 46549
6	(17)3.5129 80316	(17)3.5381 47832	(17)3.5634 65422	(17)3.5889 33801	(17)3.6145 53687
7	(20)2.9509 03466	(20)2.9755 82327	(20)3.0004 37885	(20)3.0254 71195	(20)3.0506 83312
8	(23)2.4787 58911	(23)2.5024 64737	(23)2.5263 68700	(23)2.5504 72217	(23)2.5747 76715
9	(26)2.0821 57485	(26)2.1045 72844	(26)2.1272 02445	(26)2.1500 48079	(26)2.1731 11548
10	(29)1.7490 12288	(29)1.7699 45762	(29)1.7911 04459	(29)1.8124 90531	(29)1.8341 06146
24	(70)1.5230 10388	(70)1.5671 25939	(70)1.6124 64626	(70)1.6590 58848	(70)1.7069 41821
1/2	(1)2.8982 75349	(1)2.9000 00000	(1)2.9017 23626	(1)2.9034 46228	(1)2.9051 67809
1/3	9.4353 87961	9.4391 30677	9.4428 70428	9.4466 07220	9.4503 41057
1/4	5.3835 63271	5.3851 64807	5.3867 64916	5.3883 63600	5.3899 60862
1/5	3.8446 41568	3.8455 56523	3.8464 70609	3.8473 83826	3.8482 96177

k	845	846	847	848	849
1	845	846	847	848	849
2	7 14025	7 15716	7 17409	7 19104	7 20801
3	6033 51125	6054 95736	6076 45423	6098 00192	6119 60049
4	(11)5.0983 17006	(11)5.1224 93927	(11)5.1467 56733	(11)5.1711 05628	(11)5.1955 40816
5	(14)4.3080 77870	(14)4.3336 29862	(14)4.3593 02953	(14)4.3850 97573	(14)4.4110 14153
6	(17)3.6403 25800	(17)3.6662 50863	(17)3.6923 29601	(17)3.7185 62742	(17)3.7449 51016
7	(20)3.0760 75301	(20)3.1016 48230	(20)3.1274 03172	(20)3.1533 41205	(20)3.1794 63412
8	(23)2.5992 83630	(23)2.6239 94403	(23)2.6489 10487	(23)2.6740 33342	(23)2.6993 64437
9	(26)2.1963 54667	(26)2.2198 99265	(26)2.2436 27182	(26)2.2675 80274	(26)2.2917 60407
10	(29)1.8559 53494	(29)1.8780 34778	(29)1.9003 52223	(29)1.9229 08072	(29)1.9457 04586
24	(70)1.7561 47601	(70)1.8067 11101	(70)1.8586 68111	(70)1.9120 55324	(70)1.9669 10351
1/2	(1)2.9068 88371	(1)2.9086 07914	(1)2.9103 26442	(1)2.9120 43956	(1)2.9137 60457
1/3	9.4540 71946	9.4577 99893	9.4615 24903	9.4652 46982	9.4689 66137
1/4	5.3915 56705	5.3931 51133	5.3947 44148	5.3963 35753	5.3979 25951
1/5	3.8492 07664	3.8501 18288	3.8510 28051	3.8519 36956	3.8528 45003

$$n^{\frac{1}{2}}\left[(-6)\begin{smallmatrix}1\\3\end{smallmatrix}\right] \qquad n^{\frac{1}{3}}\left[(-7)\begin{smallmatrix}4\\3\end{smallmatrix}\right] \qquad n^{\frac{1}{4}}\left[(-7)\begin{smallmatrix}2\\3\end{smallmatrix}\right] \qquad n^{\frac{1}{5}}\left[(-7)\begin{smallmatrix}1\\3\end{smallmatrix}\right]$$

Table 3.1   POWERS AND ROOTS n^k

k	850	851	852	853	854
1	850	851	852	853	854
2	7 22500	7 24201	7 25904	7 27609	7 29316
3	6141 25000	6162 95051	6184 70208	6206 50477	6228 35864
4	(11)5.2200 62500	(11)5.2446 70884	(11)5.2693 66207	(11)5.2941 48569	(11)5.3190 18279
5	(14)4.4370 53125	(14)4.4632 14922	(14)4.4894 99979	(14)4.5159 08729	(14)4.5424 41610
6	(17)3.7714 95156	(17)3.7981 95899	(17)3.8250 53982	(17)3.8520 70146	(17)3.8792 45135
7	(20)3.2057 70883	(20)3.2322 64710	(20)3.2589 45993	(20)3.2858 15835	(20)3.3128 75345
8	(23)2.7249 05250	(23)2.7506 57268	(23)2.7766 21986	(23)2.8028 00907	(23)2.8291 95545
9	(26)2.3161 69463	(26)2.3408 09335	(26)2.3656 81932	(26)2.3907 89174	(26)2.4161 32995
10	(29)1.9687 44043	(29)1.9920 28744	(29)2.0155 61006	(29)2.0393 43165	(29)2.0633 77578
24	(70)2.0232 71747	(70)2.0811 79034	(70)2.1406 72719	(70)2.2017 94325	(70)2.2645 86409
1/2	(1)2.9154 75947	(1)2.9171 90429	(1)2.9189 03904	(1)2.9206 16373	(1)2.9223 27839
1/3	9.4726 82372	9.4763 95693	9.4801 06107	9.4838 13619	9.4875 18234
1/4	5.3995 14744	5.4011 02137	5.4026 88131	5.4042 72729	5.4058 55935
1/5	3.8537 52195	3.8546 58534	3.8555 64021	3.8564 68659	3.8573 72448

k	855	856	857	858	859
1	855	856	857	858	859
2	7 31025	7 32736	7 34449	7 36164	7 37881
3	6250 26375	6272 22016	6294 22793	6316 28712	6338 39779
4	(11)5.3439 75506	(11)5.3690 20457	(11)5.3941 53336	(11)5.4193 74349	(11)5.4446 83702
5	(14)4.5690 99058	(14)4.5958 81511	(14)4.6227 89409	(14)4.6498 23191	(14)4.6769 83300
6	(17)3.9065 79694	(17)3.9340 74574	(17)3.9617 30523	(17)3.9895 48298	(17)4.0175 28654
7	(20)3.3401 25639	(20)3.3675 67835	(20)3.3952 03059	(20)3.4230 32440	(20)3.4510 57114
8	(23)2.8558 07421	(23)2.8826 38067	(23)2.9096 89021	(23)2.9369 61833	(23)2.9644 58061
9	(26)2.4417 15345	(26)2.4675 38185	(26)2.4936 03491	(26)2.5199 13253	(26)2.5464 69474
10	(29)2.0876 66620	(29)2.1122 12686	(29)2.1370 18192	(29)2.1620 85571	(29)2.1874 17279
24	(70)2.3290 92589	(70)2.3953 57569	(70)2.4634 27165	(70)2.5333 48329	(70)2.6051 69182
1/2	(1)2.9240 38303	(1)2.9257 47768	(1)2.9274 56234	(1)2.9291 63703	(1)2.9308 70178
1/3	9.4912 19958	9.4949 18797	9.4986 14756	9.5023 07842	9.5059 98059
1/4	5.4074 37751	5.4090 18180	5.4105 97225	5.4121 74889	5.4137 51174
1/5	3.8582 75391	3.8591 77490	3.8600 78746	3.8609 79161	3.8618 78737

k	860	861	862	863	864
1	860	861	862	863	864
2	7 39600	7 41321	7 43044	7 44769	7 46496
3	6360 56000	6382 77381	6405 03928	6427 35647	6449 72544
4	(11)5.4700 81600	(11)5.4955 68250	(11)5.5211 43859	(11)5.5468 08634	(11)5.5725 62780
5	(14)4.7042 70176	(14)4.7316 84264	(14)4.7592 26007	(14)4.7868 95851	(14)4.8146 94242
6	(17)4.0456 72351	(17)4.0739 80151	(17)4.1024 52818	(17)4.1310 91119	(17)4.1598 95825
7	(20)3.4792 78222	(20)3.5076 96910	(20)3.5363 14329	(20)3.5651 31636	(20)3.5941 49993
8	(23)2.9921 79271	(23)3.0201 27039	(23)3.0483 02952	(23)3.0767 08602	(23)3.1053 45594
9	(26)2.5732 74173	(26)2.6003 29381	(26)2.6276 37144	(26)2.6551 99523	(26)2.6830 18593
10	(29)2.2130 15789	(29)2.2388 83597	(29)2.2650 23218	(29)2.2914 37189	(29)2.3181 28064
24	(70)2.6789 39031	(70)2.7547 08410	(70)2.8325 29097	(70)2.9124 54150	(70)2.9945 37938
1/2	(1)2.9325 75660	(1)2.9342 80150	(1)2.9359 83651	(1)2.9376 86164	(1)2.9393 87691
1/3	9.5096 85413	9.5133 69910	9.5170 51555	9.5207 30354	9.5244 06312
1/4	5.4153 26084	5.4168 99621	5.4184 71787	5.4200 42587	5.4216 12022
1/5	3.8627 77475	3.8636 75378	3.8645 72447	3.8654 68684	3.8663 64090

k	865	866	867	868	869
1	865	866	867	868	869
2	7 48225	7 49956	7 51689	7 53424	7 55161
3	6472 14625	6494 61896	6517 14363	6539 72032	6562 34909
4	(11)5.5984 06506	(11)5.6243 40019	(11)5.6503 63527	(11)5.6764 77238	(11)5.7026 81359
5	(14)4.8426 21628	(14)4.8706 78457	(14)4.8988 65178	(14)4.9271 82242	(14)4.9556 30101
6	(17)4.1888 67708	(17)4.2180 07544	(17)4.2473 16109	(17)4.2767 94186	(17)4.3064 42558
7	(20)3.6233 70568	(20)3.6527 94533	(20)3.6824 23067	(20)3.7122 57354	(20)3.7422 98583
8	(23)3.1342 15541	(23)3.1633 20065	(23)3.1926 60799	(23)3.2222 39383	(23)3.2520 57468
9	(26)2.7110 96443	(26)2.7394 35177	(26)2.7680 36913	(26)2.7969 03785	(26)2.8260 37940
10	(29)2.3450 98423	(29)2.3723 50863	(29)2.3998 88003	(29)2.4277 12485	(29)2.4558 26970
24	(70)3.0788 36164	(70)3.1654 05907	(70)3.2543 05644	(70)3.3455 95291	(70)3.4393 36231
1/2	(1)2.9410 88234	(1)2.9427 87794	(1)2.9444 86373	(1)2.9461 83973	(1)2.9478 80595
1/3	9.5280 79435	9.5317 49727	9.5354 17196	9.5390 81845	9.5427 43681
1/4	5.4231 80095	5.4247 46809	5.4263 12167	5.4278 76171	5.4294 38824
1/5	3.8672 58668	3.8681 52418	3.8690 45344	3.8699 37445	3.8708 28725

k	870	871	872	873	874
1	870	871	872	873	874
2	7 56900	7 58641	7 60384	7 62129	7 63876
3	6585 03000	6607 76311	6630 54848	6653 38617	6676 27624
4	(11)5.7289 76100	(11)5.7553 61669	(11)5.7818 38275	(11)5.8084 06126	(11)5.8350 65434
5	(14)4.9842 09207	(14)5.0129 20014	(14)5.0417 62975	(14)5.0707 38548	(14)5.0998 47189
6	(17)4.3362 62010	(17)4.3662 53332	(17)4.3964 17315	(17)4.4267 54753	(17)4.4572 66443
7	(20)3.7725 47949	(20)3.8030 06652	(20)3.8336 75898	(20)3.8645 56899	(20)3.8956 50871
8	(23)3.2821 16715	(23)3.3124 18794	(23)3.3429 65383	(23)3.3737 58173	(23)3.4047 98862
9	(26)2.8554 41542	(26)2.8851 16769	(26)2.9150 65814	(26)2.9452 90885	(26)2.9757 94205
10	(29)2.4842 34142	(29)2.5129 36706	(29)2.5419 37390	(29)2.5712 38943	(29)2.6008 44135
24	(70)3.5355 91351	(70)3.6344 25075	(70)3.7359 03403	(70)3.8400 93943	(70)3.9470 65953
1/2	(1)2.9495 76241	(1)2.9512 70913	(1)2.9529 64612	(1)2.9546 57341	(1)2.9563 49100
1/3	9.5464 02709	9.5500 58934	9.5537 12362	9.5573 62998	9.5610 10846
1/4	5.4310 00130	5.4325 60090	5.4341 18707	5.4356 75984	5.4372 31924
1/5	3.8717 19185	3.8726 08827	3.8734 97651	3.8743 85661	3.8752 72857

$$n^{\frac{1}{2}}\left[(-6)\begin{matrix}1\\3\end{matrix}\right] \qquad n^{\frac{1}{3}}\left[(-7)\begin{matrix}4\\3\end{matrix}\right] \qquad n^{\frac{1}{4}}\left[(-7)\begin{matrix}2\\3\end{matrix}\right] \qquad n^{\frac{1}{5}}\left[(-7)\begin{matrix}1\\3\end{matrix}\right]$$

POWERS AND ROOTS n^k Table 3.1

k					
1	875	876	877	878	879
2	7 65625	7 67376	7 69129	7 70884	7 72641
3	6699 21875	6722 21376	6745 26133	6768 36152	6791 51439
4	(11)5.8618 16406	(11)5.8886 59254	(11)5.9155 94186	(11)5.9426 21415	(11)5.9697 41149
5	(14)5.1290 89355	(14)5.1584 65506	(14)5.1879 76101	(14)5.2176 21602	(14)5.2474 02470
6	(17)4.4879 53186	(17)4.5188 15784	(17)4.5498 55041	(17)4.5810 71767	(17)4.6124 66771
7	(20)3.9269 59038	(20)3.9584 82626	(20)3.9902 22871	(20)4.0221 81011	(20)4.0543 58292
8	(23)3.4360 89158	(23)3.4676 30781	(23)3.4994 25458	(23)3.5314 74928	(23)3.5637 80938
9	(26)3.0065 78013	(26)3.0376 44564	(26)3.0689 96127	(26)3.1006 34987	(26)3.1325 63445
10	(29)2.6307 55762	(29)2.6609 76638	(29)2.6915 09603	(29)2.7223 57518	(29)2.7535 23268
24	(70)4.0568 90376	(70)4.1696 39882	(70)4.2853 88904	(70)4.4042 13682	(70)4.5261 92303
1/2	(1)2.9580 39892	(1)2.9597 29717	(1)2.9614 18579	(1)2.9631 06478	(1)2.9647 93416
1/3	9.5646 55914	9.5682 98205	9.5719 37725	9.5755 74480	9.5792 08475
1/4	5.4387 86530	5.4403 39803	5.4418 91747	5.4434 42365	5.4449 91658
1/5	3.8761 59242	3.8770 44816	3.8779 29583	3.8788 13542	3.8796 96696

k					
1	880	881	882	883	884
2	7 74400	7 76161	7 77924	7 79689	7 81456
3	6814 72000	6837 97841	6861 28968	6884 65387	6908 07104
4	(11)5.9969 53600	(11)6.0242 58979	(11)6.0516 57498	(11)6.0791 49367	(11)6.1067 34799
5	(14)5.2773 19168	(14)5.3073 72161	(14)5.3375 61913	(14)5.3678 88891	(14)5.3983 53563
6	(17)4.6440 40868	(17)4.6757 94874	(17)4.7077 29607	(17)4.7398 45891	(17)4.7721 44549
7	(20)4.0867 55964	(20)4.1193 75284	(20)4.1522 17514	(20)4.1852 83922	(20)4.2185 75782
8	(23)3.5963 45248	(23)3.6291 69625	(23)3.6622 55847	(23)3.6956 05703	(23)3.7292 20991
9	(26)3.1647 83818	(26)3.1972 98440	(26)3.2301 09657	(26)3.2632 19836	(26)3.2966 31356
10	(29)2.7850 09760	(29)2.8168 19925	(29)2.8489 56718	(29)2.8814 23115	(29)2.9142 22119
24	(70)4.6514 04745	(70)4.7799 32920	(70)4.9118 60716	(70)5.0472 74047	(70)5.1862 60897
1/2	(1)2.9664 79395	(1)2.9681 64416	(1)2.9698 48481	(1)2.9715 31592	(1)2.9732 13749
1/3	9.5828 39714	9.5864 68204	9.5900 93948	9.5937 16954	9.5973 37224
1/4	5.4465 39631	5.4480 86284	5.4496 31621	5.4511 75645	5.4527 18358
1/5	3.8805 79047	3.8814 60596	3.8823 41346	3.8832 21296	3.8841 00450

k					
1	885	886	887	888	889
2	7 83225	7 84996	7 86769	7 88544	7 90321
3	6931 54125	6955 06456	6978 64103	7002 27072	7025 95369
4	(11)6.1344 14006	(11)6.1621 87200	(11)6.1900 54594	(11)6.2180 16399	(11)6.2460 72830
5	(14)5.4289 56396	(14)5.4596 97859	(14)5.4905 78425	(14)5.5215 98563	(14)5.5527 58746
6	(17)4.8046 26410	(17)4.8372 92303	(17)4.8701 43063	(17)4.9031 79524	(17)4.9364 02525
7	(20)4.2520 94373	(20)4.2858 40981	(20)4.3198 16896	(20)4.3540 23417	(20)4.3884 61845
8	(23)3.7631 03520	(23)3.7972 55109	(23)3.8316 77587	(23)3.8663 72794	(23)3.9013 42580
9	(26)3.3303 46615	(26)3.3643 68027	(26)3.3986 98020	(26)3.4333 39041	(26)3.4682 93554
10	(29)2.9473 56754	(29)2.9808 30072	(29)3.0146 45144	(29)3.0488 05069	(29)3.0833 12969
24	(70)5.3289 11365	(70)5.4753 17719	(70)5.6255 74442	(70)5.7797 78281	(70)5.9380 28303
1/2	(1)2.9748 94956	(1)2.9765 75213	(1)2.9782 54522	(1)2.9799 32885	(1)2.9816 10303
1/3	9.6009 54766	9.6045 69584	9.6081 81682	9.6117 91067	9.6153 97744
1/4	5.4542 59763	5.4557 99862	5.4573 38658	5.4588 76153	5.4604 12350
1/5	3.8849 78808	3.8858 56373	3.8867 33146	3.8876 09128	3.8884 84321

k					
1	890	891	892	893	894
2	7 92100	7 93881	7 95664	7 97449	7 99236
3	7049 69000	7073 47971	7097 32288	7121 21957	7145 16984
4	(11)6.2742 24100	(11)6.3024 70422	(11)6.3308 12009	(11)6.3592 49076	(11)6.3877 81837
5	(14)5.5840 59449	(14)5.6155 01146	(14)5.6470 84312	(14)5.6788 09425	(14)5.7106 76962
6	(17)4.9698 12910	(17)5.0034 11521	(17)5.0371 99206	(17)5.0711 76816	(17)5.1053 45204
7	(20)4.4231 33490	(20)4.4580 39665	(20)4.4931 81692	(20)4.5285 60897	(20)4.5641 78613
8	(23)3.9365 88806	(23)3.9721 13342	(23)4.0079 18069	(23)4.0440 04881	(23)4.0803 75680
9	(26)3.5035 64037	(26)3.5391 52987	(26)3.5750 62918	(26)3.6112 96359	(26)3.6478 55858
10	(29)3.1181 71993	(29)3.1533 85312	(29)3.1889 56123	(29)3.2248 87648	(29)3.2611 83137
24	(70)6.1004 25945	(70)6.2670 75070	(70)6.4380 82017	(70)6.6135 55666	(70)6.7936 07487
1/2	(1)2.9832 86778	(1)2.9849 62311	(1)2.9866 36905	(1)2.9883 10559	(1)2.9899 83278
1/3	9.6190 01716	9.6226 02990	9.6262 01570	9.6297 97462	9.6333 90671
1/4	5.4619 47252	5.4634 80860	5.4650 13179	5.4665 44210	5.4680 73955
1/5	3.8893 58728	3.8902 32348	3.8911 05185	3.8919 77239	3.8928 48512

k					
1	895	896	897	898	899
2	8 01025	8 02816	8 04609	8 06404	8 08201
3	7169 17375	7193 23136	7217 34273	7241 50792	7265 72699
4	(11)6.4164 10506	(11)6.4451 35299	(11)6.4739 56429	(11)6.5028 74112	(11)6.5318 88564
5	(14)5.7426 87403	(14)5.7748 41228	(14)5.8071 38917	(14)5.8395 80953	(14)5.8721 67819
6	(17)5.1397 05226	(17)5.1742 57740	(17)5.2090 03608	(17)5.2439 43696	(17)5.2790 78869
7	(20)4.6000 36177	(20)4.6361 34935	(20)4.6724 76237	(20)4.7090 61439	(20)4.7458 91904
8	(23)4.1170 32378	(23)4.1539 76902	(23)4.1912 11184	(23)4.2287 37172	(23)4.2665 56821
9	(26)3.6847 43979	(26)3.7219 63304	(26)3.7595 16432	(26)3.7974 05980	(26)3.8356 34582
10	(29)3.2978 45861	(29)3.3348 79120	(29)3.3722 86240	(29)3.4100 70570	(29)3.4482 35490
24	(70)6.9783 51604	(70)7.1679 04854	(70)7.3623 86846	(70)7.5619 20026	(70)7.7666 29743
1/2	(1)2.9916 55060	(1)2.9933 25909	(1)2.9949 95826	(1)2.9966 64813	(1)2.9983 32870
1/3	9.6369 81200	9.6405 69057	9.6441 54244	9.6477 36769	9.6513 16634
1/4	5.4696 02417	5.4711 29599	5.4726 55504	5.4741 80133	5.4757 03489
1/5	3.8937 19006	3.8945 88722	3.8954 57662	3.8963 25828	3.8971 93220

$$n^{\frac{1}{2}}\begin{bmatrix}(-6)1\\3\end{bmatrix} \qquad n^{\frac{1}{3}}\begin{bmatrix}(-7)3\\3\end{bmatrix} \qquad n^{\frac{1}{4}}\begin{bmatrix}(-7)2\\3\end{bmatrix} \qquad n^{\frac{1}{5}}\begin{bmatrix}(-7)1\\3\end{bmatrix}$$

Table 3.1   POWERS AND ROOTS n^k

k	900	901	902	903	904
1	900	901	902	903	904
2	8 10000	8 11801	8 13604	8 15409	8 17216
3	7290 00000	7314 32701	7338 70808	7363 14327	7387 63264
4	(11)6.5610 00000	(11)6.5902 08636	(11)6.6195 14688	(11)6.6489 18373	(11)6.6784 19907
5	(14)5.9049 00000	(14)5.9377 77981	(14)5.9708 02249	(14)6.0039 73291	(14)6.0372 91596
6	(17)5.3144 10000	(17)5.3499 37961	(17)5.3856 63628	(17)5.4215 87881	(17)5.4577 11602
7	(20)4.7829 69000	(20)4.8202 94103	(20)4.8578 68593	(20)4.8956 93857	(20)4.9337 71289
8	(23)4.3046 72100	(23)4.3430 84987	(23)4.3817 97471	(23)4.4208 11553	(23)4.4601 29245
9	(26)3.8742 04890	(26)3.9131 19573	(26)3.9523 81319	(26)3.9919 92832	(26)4.0319 56837
10	(29)3.4867 84401	(29)3.5257 20735	(29)3.5650 47949	(29)3.6047 69527	(29)3.6448 88981
24	(70)7.9766 44308	(70)8.1920 95066	(70)8.4131 16465	(70)8.6398 46120	(70)8.8724 24888
1/2	(1)3.0000 00000	(1)3.0016 66204	(1)3.0033 31484	(1)3.0049 95840	(1)3.0066 59276
1/3	9.6548 93846	9.6584 68409	9.6620 40328	9.6656 09608	9.6691 76254
1/4	5.4772 25575	5.4787 46393	5.4802 65946	5.4817 84235	5.4833 01264
1/5	3.8980 59841	3.8989 25692	3.8997 90774	3.9006 55089	3.9015 18640

k	905	906	907	908	909
1	905	906	907	908	909
2	8 19025	8 20836	8 22649	8 24464	8 26281
3	7412 17625	7436 77416	7461 42643	7486 13312	7510 89429
4	(11)6.7080 19506	(11)6.7377 17389	(11)6.7675 13772	(11)6.7974 08873	(11)6.8274 02910
5	(14)6.0707 57653	(14)6.1043 71954	(14)6.1381 34991	(14)6.1720 47257	(14)6.2061 09245
6	(17)5.4940 35676	(17)5.5305 60991	(17)5.5672 88437	(17)5.6042 18909	(17)5.6413 53304
7	(20)4.9721 02287	(20)5.0106 88258	(20)5.0495 30612	(20)5.0886 30769	(20)5.1279 90153
8	(23)4.4997 52570	(23)4.5396 83561	(23)4.5799 24265	(23)4.6204 76739	(23)4.6613 43049
9	(26)4.0722 76076	(26)4.1129 53307	(26)4.1539 91309	(26)4.1953 92879	(26)4.2371 60832
10	(29)3.6854 09848	(29)3.7263 35696	(29)3.7676 70117	(29)3.8094 16734	(29)3.8515 79196
24	(70)9.1109 96943	(70)9.3557 09844	(70)9.6067 14616	(70)9.8641 65825	(71)1.0128 22166
1/2	(1)3.0083 21791	(1)3.0099 83389	(1)3.0116 44069	(1)3.0133 03835	(1)3.0149 62686
1/3	9.6727 40271	9.6763 01663	9.6798 60436	9.6834 16593	9.6869 70141
1/4	5.4848 17035	5.4863 31551	5.4878 44813	5.4893 56824	5.4908 67587
1/5	3.9023 81426	3.9032 43449	3.9041 04712	3.9049 65216	3.9058 24962

k	910	911	912	913	914
1	910	911	912	913	914
2	8 28100	8 29921	8 31744	8 33569	8 35396
3	7535 71000	7560 58031	7585 50528	7610 48497	7635 51944
4	(11)6.8574 96100	(11)6.8876 88662	(11)6.9179 80815	(11)6.9483 72778	(11)6.9788 64768
5	(14)6.2403 21451	(14)6.2746 84371	(14)6.3091 98504	(14)6.3438 64346	(14)6.3786 82398
6	(17)5.6786 92520	(17)5.7162 37462	(17)5.7539 89035	(17)5.7919 48148	(17)5.8301 15712
7	(20)5.1676 10194	(20)5.2074 92328	(20)5.2476 38000	(20)5.2880 48659	(20)5.3287 25761
8	(23)4.7025 25276	(23)4.7440 25511	(23)4.7858 45856	(23)4.8279 88426	(23)4.8704 55345
9	(26)4.2792 98001	(26)4.3218 07241	(26)4.3646 91421	(26)4.4079 53433	(26)4.4515 96186
10	(29)3.8941 61181	(29)3.9371 66396	(29)3.9805 98576	(29)4.0244 61484	(29)4.0687 58914
24	(71)1.0399 04400	(71)1.0676 79852	(71)1.0961 65476	(71)1.1253 78622	(71)1.1553 37042
1/2	(1)3.0166 20626	(1)3.0182 77655	(1)3.0199 33774	(1)3.0215 88986	(1)3.0232 43292
1/3	9.6905 21083	9.6940 69425	9.6976 15172	9.7011 58327	9.7046 98896
1/4	5.4923 77104	5.4938 85378	5.4953 92410	5.4968 98203	5.4984 02760
1/5	3.9066 83951	3.9075 42186	3.9083 99668	3.9092 56397	3.9101 12376

k	915	916	917	918	919
1	915	916	917	918	919
2	8 37225	8 39056	8 40889	8 42724	8 44561
3	7660 60875	7685 75296	7710 95213	7736 20632	7761 51559
4	(11)7.0094 57006	(11)7.0401 49711	(11)7.0709 43103	(11)7.1018 37402	(11)7.1328 32827
5	(14)6.4136 53161	(14)6.4487 77136	(14)6.4840 54826	(14)6.5194 86735	(14)6.5550 73368
6	(17)5.8684 92642	(17)5.9070 79856	(17)5.9458 78275	(17)5.9848 88823	(17)6.0241 12425
7	(20)5.3696 70767	(20)5.4108 85148	(20)5.4523 70378	(20)5.4941 27939	(20)5.5361 59319
8	(23)4.9132 48752	(23)4.9563 70796	(23)4.9998 23637	(23)5.0436 09448	(23)5.0877 30414
9	(26)4.4956 22608	(26)4.5400 35649	(26)4.5848 38275	(26)4.6300 33473	(26)4.6756 24251
10	(29)4.1134 94687	(29)4.1586 72654	(29)4.2042 96698	(29)4.2503 70729	(29)4.2968 98686
24	(71)1.1860 58902	(71)1.2175 62793	(71)1.2498 67732	(71)1.2829 93183	(71)1.3169 59057
1/2	(1)3.0248 96692	(1)3.0265 49190	(1)3.0282 00786	(1)3.0298 51482	(1)3.0315 01278
1/3	9.7082 36884	9.7117 72294	9.7153 05133	9.7188 35404	9.7223 63112
1/4	5.4999 06083	5.5014 08174	5.5029 09036	5.5044 08671	5.5059 07081
1/5	3.9109 67606	3.9118 22089	3.9126 75826	3.9135 28819	3.9143 81068

k	920	921	922	923	924
1	920	921	922	923	924
2	8 46400	8 48241	8 50084	8 51929	8 53776
3	7786 88000	7812 29961	7837 77448	7863 30467	7888 89024
4	(11)7.1639 29600	(11)7.1951 27941	(11)7.2264 28071	(11)7.2578 30210	(11)7.2893 34582
5	(14)6.5908 15232	(14)6.6267 12833	(14)6.6627 66681	(14)6.6989 77284	(14)6.7353 45154
6	(17)6.0635 50013	(17)6.1032 02520	(17)6.1430 70880	(17)6.1831 56033	(17)6.2234 58922
7	(20)5.5784 66012	(20)5.6210 49521	(20)5.6639 11351	(20)5.7070 53019	(20)5.7504 76044
8	(23)5.1321 88731	(23)5.1769 86608	(23)5.2221 26266	(23)5.2676 09936	(23)5.3134 39864
9	(26)4.7216 13633	(26)4.7680 04666	(26)4.8148 00417	(26)4.8620 03971	(26)4.9096 18435
10	(29)4.3438 84542	(29)4.3913 32298	(29)4.4392 45985	(29)4.4876 29665	(29)4.5364 87434
24	(71)1.3517 85726	(71)1.3874 94035	(71)1.4241 05308	(71)1.4616 41363	(71)1.5001 24518
1/2	(1)3.0331 50178	(1)3.0347 98181	(1)3.0364 45290	(1)3.0380 91506	(1)3.0397 36831
1/3	9.7258 88262	9.7294 10859	9.7329 30906	9.7364 48410	9.7399 63373
1/4	5.5074 04268	5.5089 00236	5.5103 94986	5.5118 88520	5.5133 80842
1/5	3.9152 32576	3.9160 83344	3.9169 33373	3.9177 82664	3.9186 31220

$$n^{\frac{1}{2}}\begin{bmatrix}(-6)1 \\ 3\end{bmatrix} \qquad n^{\frac{1}{3}}\begin{bmatrix}(-7)3 \\ 3\end{bmatrix} \qquad n^{\frac{1}{4}}\begin{bmatrix}(-7)2 \\ 3\end{bmatrix} \qquad n^{\frac{1}{5}}\begin{bmatrix}(-7)1 \\ 3\end{bmatrix}$$

POWERS AND ROOTS n^k

Table 3.1

k	925	926	927	928	929
1	925	926	927	928	929
2	8 55625	8 57476	8 59329	8 61184	8 63041
3	7914 53125	7940 22776	7965 97983	7991 78752	8017 65089
4	(11)7.3209 41406	(11)7.3526 50906	(11)7.3844 63302	(11)7.4163 78819	(11)7.4483 97677
5	(14)6.7718 70801	(14)6.8085 54739	(14)6.8453 97481	(14)6.8823 99544	(14)6.9195 61442
6	(17)6.2639 80491	(17)6.3047 21688	(17)6.3456 83465	(17)6.3868 66776	(17)6.4282 72579
7	(20)5.7941 81954	(20)5.8381 72283	(20)5.8824 48572	(20)5.9270 12369	(20)5.9718 65226
8	(23)5.3596 18307	(23)5.4061 47534	(23)5.4530 29826	(23)5.5002 67478	(23)5.5478 62795
9	(26)4.9576 46934	(26)5.0060 92617	(26)5.0549 58649	(26)5.1042 48220	(26)5.1539 64537
10	(29)4.5858 23414	(29)4.6356 41763	(29)4.6859 46668	(29)4.7367 42348	(29)4.7880 33055
24	(71)1.5395 77607	(71)1.5800 23988	(71)1.6214 87554	(71)1.6639 92748	(71)1.7075 64573
1/2	(1)3.0413 81265	(1)3.0430 24811	(1)3.0446 67470	(1)3.0463 09242	(1)3.0479 50131
1/3	9.7434 75802	9.7469 85700	9.7504 93072	9.7539 97922	9.7575 00256
1/4	5.5148 71952	5.5163 61854	5.5178 50550	5.5193 38042	5.5208 24332
1/5	3.9194 79042	3.9203 26131	3.9211 72488	3.9220 18115	3.9228 63013

k	930	931	932	933	934
1	930	931	932	933	934
2	8 64900	8 66761	8 68624	8 70489	8 72356
3	8043 57000	8069 54491	8095 57568	8121 66237	8147 80504
4	(11)7.4805 20100	(11)7.5127 46311	(11)7.5450 76534	(11)7.5775 10991	(11)7.6100 49907
5	(14)6.9568 83693	(14)6.9943 66816	(14)7.0320 11329	(14)7.0698 17755	(14)7.1077 86613
6	(17)6.4699 01834	(17)6.5117 55505	(17)6.5538 34559	(17)6.5961 39965	(17)6.6386 72697
7	(20)6.0170 08706	(20)6.0624 44376	(20)6.1081 73809	(20)6.1541 98588	(20)6.2005 20299
8	(23)5.5958 18097	(23)5.6441 35714	(23)5.6928 17990	(23)5.7418 67282	(23)5.7912 85959
9	(26)5.2041 10830	(26)5.2546 90349	(26)5.3057 06367	(26)5.3571 62174	(26)5.4090 61086
10	(29)4.8398 23072	(29)4.8921 16715	(29)4.9449 18334	(29)4.9982 32309	(29)5.0520 63054
24	(71)1.7522 28603	(71)1.7980 10997	(71)1.8449 38512	(71)1.8930 38514	(71)1.9423 38996
1/2	(1)3.0495 90136	(1)3.0512 29260	(1)3.0528 67504	(1)3.0545 04870	(1)3.0561 41358
1/3	9.7610 00077	9.7644 97390	9.7679 92199	9.7714 84510	9.7749 74326
1/4	5.5223 09423	5.5237 93317	5.5252 76015	5.5267 57521	5.5282 37837
1/5	3.9237 07185	3.9245 50630	3.9253 93351	3.9262 35348	3.9270 76625

k	935	936	937	938	939
1	935	936	937	938	939
2	8 74225	8 76096	8 77969	8 79844	8 81721
3	8174 00375	8200 25856	8226 56953	8252 93672	8279 36019
4	(11)7.6426 93506	(11)7.6754 42012	(11)7.7082 95650	(11)7.7412 54643	(11)7.7743 19218
5	(14)7.1459 18428	(14)7.1842 13723	(14)7.2226 73024	(14)7.2612 96855	(14)7.3000 85746
6	(17)6.6814 33731	(17)6.7244 24045	(17)6.7676 44623	(17)6.8110 96450	(17)6.8547 80516
7	(20)6.2471 40538	(20)6.2940 60906	(20)6.3412 83012	(20)6.3888 08471	(20)6.4366 38904
8	(23)5.8410 76403	(23)5.8912 41008	(23)5.9417 82182	(23)5.9927 02345	(23)6.0440 03931
9	(26)5.4614 06437	(26)5.5142 01584	(26)5.5674 49905	(26)5.6211 54800	(26)5.6753 19691
10	(29)5.1064 15018	(29)5.1612 92682	(29)5.2167 00561	(29)5.2726 43202	(29)5.3291 25190
24	(71)1.9928 68584	(71)2.0446 56558	(71)2.0977 32860	(71)2.1521 28115	(71)2.2078 73640
1/2	(1)3.0577 76970	(1)3.0594 11708	(1)3.0610 45573	(1)3.0626 78566	(1)3.0643 10689
1/3	9.7784 61652	9.7819 46493	9.7854 28852	9.7889 08735	9.7923 86145
1/4	5.5297 16964	5.5311 94905	5.5326 71663	5.5341 47239	5.5356 21636
1/5	3.9279 17180	3.9287 57017	3.9295 96137	3.9304 34540	3.9312 72229

k	940	941	942	943	944
1	940	941	942	943	944
2	8 83600	8 85481	8 87364	8 89249	8 91136
3	8305 84000	8332 37621	8358 96888	8385 61807	8412 32384
4	(11)7.8074 89600	(11)7.8407 66014	(11)7.8741 48685	(11)7.9076 37840	(11)7.9412 33705
5	(14)7.3390 40224	(14)7.3781 60819	(14)7.4174 48061	(14)7.4569 02483	(14)7.4965 24617
6	(17)6.8986 97811	(17)6.9428 49330	(17)6.9872 36074	(17)7.0318 59042	(17)7.0767 19239
7	(20)6.4847 75942	(20)6.5332 21220	(20)6.5819 76381	(20)6.6310 43076	(20)6.6804 22962
8	(23)6.0956 89385	(23)6.1477 61168	(23)6.2002 21751	(23)6.2530 73621	(23)6.3063 19276
9	(26)5.7299 48022	(26)5.7850 43259	(26)5.8406 08890	(26)5.8966 48424	(26)5.9531 65396
10	(29)5.3861 51141	(29)5.4437 25707	(29)5.5018 53574	(29)5.5605 39464	(29)5.6197 88134
24	(71)2.2650 01461	(71)2.3235 44328	(71)2.3835 35733	(71)2.4450 09921	(71)2.5080 01911
1/2	(1)3.0659 41943	(1)3.0675 72330	(1)3.0692 01851	(1)3.0708 30507	(1)3.0724 58299
1/3	9.7958 61087	9.7993 33566	9.8028 03585	9.8062 71149	9.8097 36263
1/4	5.5370 94855	5.5385 66899	5.5400 37771	5.5415 07472	5.5429 76005
1/5	3.9321 09204	3.9329 45467	3.9337 81020	3.9346 15863	3.9354 49998

k	945	946	947	948	949
1	945	946	947	948	949
2	8 93025	8 94916	8 96809	8 98704	9 00601
3	8439 08625	8465 90536	8492 78123	8519 71392	8546 70349
4	(11)7.9749 36506	(11)8.0087 46471	(11)8.0426 63825	(11)8.0766 88796	(11)8.1108 21612
5	(14)7.5363 14998	(14)7.5762 74161	(14)7.6164 02642	(14)7.6567 00979	(14)7.6971 69710
6	(17)7.1218 17673	(17)7.1671 55356	(17)7.2127 33302	(17)7.2585 52528	(17)7.3046 14055
7	(20)6.7301 17701	(20)6.7801 28967	(20)6.8304 58437	(20)6.8811 07796	(20)6.9320 78738
8	(23)6.3599 61228	(23)6.4140 02003	(23)6.4684 44140	(23)6.5232 90191	(23)6.5785 42722
9	(26)6.0101 63360	(26)6.0676 45895	(26)6.1256 16600	(26)6.1840 79101	(26)6.2430 37043
10	(29)5.6796 04376	(29)5.7399 93016	(29)5.8009 58921	(29)5.8625 06988	(29)5.9246 42154
24	(71)2.5725 47511	(71)2.6386 83331	(71)2.7064 46809	(71)2.7758 76218	(71)2.8470 10693
1/2	(1)3.0740 85230	(1)3.0757 11300	(1)3.0773 36511	(1)3.0789 60864	(1)3.0805 84360
1/3	9.8131 98931	9.8166 59156	9.8201 16944	9.8235 72299	9.8270 25224
1/4	5.5444 43371	5.5459 09574	5.5473 74614	5.5488 38494	5.5503 01217
1/5	3.9362 83427	3.9371 16151	3.9379 48170	3.9387 79487	3.9396 10103

$$n^{\frac{1}{2}}\left[\begin{matrix}(-6)1\\3\end{matrix}\right] \quad n^{\frac{1}{3}}\left[\begin{matrix}(-7)3\\3\end{matrix}\right] \quad n^{\frac{1}{4}}\left[\begin{matrix}(-7)2\\3\end{matrix}\right] \quad n^{\frac{1}{5}}\left[\begin{matrix}(-8)9\\3\end{matrix}\right]$$

Table 3.1   **POWERS AND ROOTS** n^k

k	950	951	952	953	954
1	950	951	952	953	954
2	9 02500	9 04401	9 06304	9 08209	9 10116
3	8573 75000	8600 85351	8628 01408	8655 23177	8682 50664
4	(11)8.1450 62500	(11)8.1794 11688	(11)8.2138 69404	(11)8.2484 35877	(11)8.2831 11335
5	(14)7.7378 09375	(14)7.7786 20515	(14)7.8196 03673	(14)7.8607 59391	(14)7.9020 88213
6	(17)7.3509 18906	(17)7.3974 68110	(17)7.4442 62696	(17)7.4913 03699	(17)7.5385 92155
7	(20)6.9833 72961	(20)7.0349 92173	(20)7.0869 38087	(20)7.1392 12425	(20)7.1918 16916
8	(23)6.6342 04313	(23)6.6902 77556	(23)6.7467 65059	(23)6.8036 69441	(23)6.8609 93338
9	(26)6.3024 94097	(26)6.3624 53956	(26)6.4229 20336	(26)6.4838 96978	(26)6.5453 87645
10	(29)5.9873 69392	(29)6.0506 93712	(29)6.1146 20160	(29)6.1791 53820	(29)6.2442 99813
24	(71)2.9198 90243	(71)2.9945 55775	(71)3.0710 49109	(71)3.1494 12996	(71)3.2296 91146
1/2	(1)3.0822 07001	(1)3.0838 28789	(1)3.0854 49724	(1)3.0870 69808	(1)3.0886 89042
1/3	9.8304 75725	9.8339 23805	9.8373 69469	9.8408 12721	9.8442 53565
1/4	5.5517 62784	5.5532 23198	5.5546 82461	5.5561 40574	5.5575 97541
1/5	3.9404 40019	3.9412 69236	3.9420 97756	3.9429 25580	3.9437 52709

k	955	956	957	958	959
1	955	956	957	958	959
2	9 12025	9 13936	9 15849	9 17764	9 19681
3	8709 83875	8737 22816	8764 67493	8792 17912	8819 74079
4	(11)8.3178 96006	(11)8.3527 90121	(11)8.3877 93908	(11)8.4229 07597	(11)8.4581 31418
5	(14)7.9435 90686	(14)7.9852 67356	(14)8.0271 18770	(14)8.0691 45478	(14)8.1113 48029
6	(17)7.5861 29105	(17)7.6339 15592	(17)7.6819 52663	(17)7.7302 41368	(17)7.7787 82760
7	(20)7.2447 53295	(20)7.2980 23306	(20)7.3516 28698	(20)7.4055 71230	(20)7.4598 52667
8	(23)6.9187 39397	(23)6.9769 10280	(23)7.0355 08664	(23)7.0945 37239	(23)7.1539 98708
9	(26)6.6073 96124	(26)6.6699 26228	(26)6.7329 81792	(26)6.7965 66675	(26)6.8606 84761
10	(29)6.3100 63299	(29)6.3764 49474	(29)6.4434 63575	(29)6.5111 10874	(29)6.5793 96686
24	(71)3.3119 28238	(71)3.3961 69948	(71)3.4824 62966	(71)3.5708 55021	(71)3.6613 94899
1/2	(1)3.0903 07428	(1)3.0919 24967	(1)3.0935 41660	(1)3.0951 57508	(1)3.0967 72513
1/3	9.8476 92005	9.8511 28046	9.8545 61691	9.8579 92945	9.8614 21813
1/4	5.5590 53362	5.5605 08040	5.5619 61578	5.5634 13977	5.5648 65240
1/5	3.9445 79145	3.9454 04889	3.9462 29943	3.9470 54307	3.9478 77983

k	960	961	962	963	964
1	960	961	962	963	964
2	9 21600	9 23521	9 25444	9 27369	9 29296
3	8847 36000	8875 03681	8902 77128	8930 56347	8958 41344
4	(11)8.4934 65600	(11)8.5289 10374	(11)8.5644 65971	(11)8.6001 32622	(11)8.6359 10556
5	(14)8.1537 26976	(14)8.1962 82870	(14)8.2390 16264	(14)8.2819 27715	(14)8.3250 17776
6	(17)7.8275 77897	(17)7.8766 27838	(17)7.9259 33646	(17)7.9754 96389	(17)8.0253 17136
7	(20)7.5144 74781	(20)7.5694 39352	(20)7.6247 48168	(20)7.6804 03023	(20)7.7364 05719
8	(23)7.2138 95790	(23)7.2742 31217	(23)7.3350 07737	(23)7.3962 28111	(23)7.4578 95113
9	(26)6.9253 39958	(26)6.9905 36200	(26)7.0562 77443	(26)7.1225 67671	(26)7.1894 10889
10	(29)6.6483 26360	(29)6.7179 05288	(29)6.7881 38901	(29)6.8590 32667	(29)6.9305 92097
24	(71)3.7541 32467	(71)3.8491 18699	(71)3.9464 05693	(71)4.0460 46699	(71)4.1480 96142
1/2	(1)3.0983 86677	(1)3.1000 00000	(1)3.1016 12484	(1)3.1032 24130	(1)3.1048 34939
1/3	9.8648 48297	9.8682 72403	9.8716 94135	9.8751 13495	9.8785 30490
1/4	5.5663 15367	5.5677 64363	5.5692 12228	5.5706 58964	5.5721 04575
1/5	3.9487 00972	3.9495 23275	3.9503 44894	3.9511 65831	3.9519 86085

k	965	966	967	968	969
1	965	966	967	968	969
2	9 31225	9 33156	9 35089	9 37024	9 38961
3	8986 32125	9014 28696	9042 31063	9070 39232	9098 53209
4	(11)8.6718 00006	(11)8.7078 01203	(11)8.7439 14379	(11)8.7801 39766	(11)8.8164 77595
5	(14)8.3682 87006	(14)8.4117 35962	(14)8.4553 65205	(14)8.4991 75293	(14)8.5431 66790
6	(17)8.0753 96961	(17)8.1257 36940	(17)8.1763 38153	(17)8.2272 01684	(17)8.2783 28619
7	(20)7.7927 58067	(20)7.8494 61884	(20)7.9065 18994	(20)7.9639 31230	(20)8.0217 00432
8	(23)7.5200 11535	(23)7.5825 80180	(23)7.6456 03867	(23)7.7090 85431	(23)7.7730 27719
9	(26)7.2568 11131	(26)7.3247 72454	(26)7.3932 98939	(26)7.4623 94697	(26)7.5320 63859
10	(29)7.0028 22742	(29)7.0757 30190	(29)7.1493 20074	(29)7.2235 98067	(29)7.2985 69880
24	(71)4.2526 09649	(71)4.3596 44069	(71)4.4692 57504	(71)4.5815 09331	(71)4.6964 60232
1/2	(1)3.1064 44913	(1)3.1080 54054	(1)3.1096 62361	(1)3.1112 69837	(1)3.1128 76483
1/3	9.8819 45122	9.8853 57396	9.8887 67316	9.8921 74886	9.8955 80110
1/4	5.5735 49061	5.5749 92425	5.5764 34668	5.5778 75794	5.5793 15803
1/5	3.9528 05659	3.9536 24554	3.9544 42771	3.9552 60312	3.9560 77177

k	970	971	972	973	974
1	970	971	972	973	974
2	9 40900	9 42841	9 44784	9 46729	9 48676
3	9126 73000	9154 98611	9183 30048	9211 67317	9240 10424
4	(11)8.8529 28100	(11)8.8894 91513	(11)8.9261 68067	(11)8.9629 57994	(11)8.9998 61530
5	(14)8.5873 40257	(14)8.6316 96259	(14)8.6762 35361	(14)8.7209 58129	(14)8.7658 65130
6	(17)8.3297 20049	(17)8.3813 77067	(17)8.4333 00771	(17)8.4854 92259	(17)8.5379 52637
7	(20)8.0798 28448	(20)8.1383 17132	(20)8.1971 68349	(20)8.2563 83968	(20)8.3159 65868
8	(23)7.8374 33594	(23)7.9023 05936	(23)7.9676 47635	(23)8.0334 61601	(23)8.0997 50755
9	(26)7.6023 10587	(26)7.6731 39063	(26)7.7445 53501	(26)7.8165 58138	(26)7.8891 57236
10	(29)7.3742 41269	(29)7.4506 18031	(29)7.5277 06003	(29)7.6055 11068	(29)7.6840 39148
24	(71)4.8141 72219	(71)4.9347 08664	(71)5.0581 34323	(71)5.1845 15371	(71)5.3139 19427
1/2	(1)3.1144 82300	(1)3.1160 87290	(1)3.1176 91454	(1)3.1192 94792	(1)3.1208 97307
1/3	9.8989 82992	9.9023 83537	9.9057 81747	9.9091 77627	9.9125 71181
1/4	5.5807 54698	5.5821 92482	5.5836 29155	5.5850 64719	5.5864 99178
1/5	3.9568 93368	3.9577 08886	3.9585 23732	3.9593 37908	3.9601 51415

$$n^{\frac{1}{2}}\left[\begin{matrix}(-6)1\\3\end{matrix}\right] \qquad n^{\frac{1}{3}}\left[\begin{matrix}(-7)3\\3\end{matrix}\right] \qquad n^{\frac{1}{4}}\left[\begin{matrix}(-7)2\\3\end{matrix}\right] \qquad n^{\frac{1}{5}}\left[\begin{matrix}(-8)9\\3\end{matrix}\right]$$

POWERS AND ROOTS n^k

Table 3.1

k	975	976	977	978	979
1	975	976	977	978	979
2	9 50625	9 52576	9 54529	9 56484	9 58441
3	9268 59375	9297 14176	9325 74833	9354 41352	9383 13739
4	(11)9.0368 78906	(11)9.0740 10358	(11)9.1112 56118	(11)9.1486 16423	(11)9.1860 91505
5	(14)8.8109 56934	(14)8.8562 34109	(14)8.9016 97228	(14)8.9473 46861	(14)8.9931 83583
6	(17)8.5906 83010	(17)8.6436 84491	(17)8.6969 58191	(17)8.7505 05230	(17)8.8043 26728
7	(20)8.3759 15935	(20)8.4362 36063	(20)8.4969 28153	(20)8.5579 94115	(20)8.6194 35867
8	(23)8.1665 18037	(23)8.2337 66397	(23)8.3014 98806	(23)8.3697 18245	(23)8.4384 27713
9	(26)7.9623 55086	(26)8.0361 56004	(26)8.1105 64333	(26)8.1855 84443	(26)8.2612 20731
10	(29)7.7632 96209	(29)7.8432 88260	(29)7.9240 21353	(29)8.0055 01586	(29)8.0877 35096
24	(71)5.4464 15584	(71)5.5820 74443	(71)5.7209 68141	(71)5.8631 70383	(71)6.0087 56477
1/2	(1)3.1224 98999	(1)3.1240 99870	(1)3.1256 99922	(1)3.1272 99154	(1)3.1288 97569
1/3	9.9159 62413	9.9193 51328	9.9227 37928	9.9261 22218	9.9295 04202
1/4	5.5879 32533	5.5893 64785	5.5907 95938	5.5922 25992	5.5936 54950
1/5	3.9609 64254	3.9617 76427	3.9625 87934	3.9633 98776	3.9642 08956

k	980	981	982	983	984
1	980	981	982	983	984
2	9 60400	9 62361	9 64324	9 66289	9 68256
3	9411 92000	9440 76141	9469 66168	9498 62087	9527 63904
4	(11)9.2236 81600	(11)9.2613 86943	(11)9.2992 07770	(11)9.3371 44315	(11)9.3751 96815
5	(14)9.0392 07968	(14)9.0854 20591	(14)9.1318 22030	(14)9.1784 12862	(14)9.2251 93666
6	(17)8.8584 23809	(17)8.9127 97600	(17)8.9674 49233	(17)9.0223 79843	(17)9.0775 90568
7	(20)8.6812 55332	(20)8.7434 54446	(20)8.8060 35147	(20)8.8689 99386	(20)8.9323 49119
8	(23)8.5076 30226	(23)8.5773 28811	(23)8.6475 26515	(23)8.7182 26396	(23)8.7894 31533
9	(26)8.3374 77621	(26)8.4143 59564	(26)8.4918 71037	(26)8.5700 16548	(26)8.6488 00628
10	(29)8.1707 28069	(29)8.2544 86732	(29)8.3390 17359	(29)8.4243 26266	(29)8.5104 19818
24	(71)6.1578 03365	(71)6.3103 89657	(71)6.4665 95666	(71)6.6265 03443	(71)6.7901 96812
1/2	(1)3.1304 95168	(1)3.1320 91953	(1)3.1336 87923	(1)3.1352 83081	(1)3.1368 77428
1/3	9.9328 83884	9.9362 61267	9.9396 36356	9.9430 09155	9.9463 79667
1/4	5.5950 82813	5.5965 09584	5.5979 35265	5.5993 59857	5.6007 83363
1/5	3.9650 18474	3.9658 27331	3.9666 35529	3.9674 43069	3.9682 49952

k	985	986	987	988	989
1	985	986	987	988	989
2	9 70225	9 72196	9 74169	9 76144	9 78121
3	9556 71625	9585 85256	9615 04803	9644 30272	9673 61669
4	(11)9.4133 65506	(11)9.4516 50624	(11)9.4900 52406	(11)9.5285 71087	(11)9.5672 06906
5	(14)9.2721 65024	(14)9.3193 27515	(14)9.3666 81724	(14)9.4142 28234	(14)9.4619 67630
6	(17)9.1330 82548	(17)9.1888 56930	(17)9.2449 14862	(17)9.3012 57495	(17)9.3578 85987
7	(20)8.9960 86310	(20)9.0602 12933	(20)9.1247 30969	(20)9.1896 42406	(20)9.2549 49241
8	(23)8.8611 45015	(23)8.9333 69952	(23)9.0061 09466	(23)9.0793 66697	(23)9.1531 44799
9	(26)8.7282 27840	(26)8.8083 02773	(26)8.8890 30043	(26)8.9704 14296	(26)9.0524 60206
10	(29)8.5973 04423	(29)8.6849 86534	(29)8.7734 72653	(29)8.8627 69325	(29)8.9528 83144
24	(71)6.9577 61406	(71)7.1292 84708	(71)7.3048 56083	(71)7.4845 66822	(71)7.6685 10178
1/2	(1)3.1384 70965	(1)3.1400 63694	(1)3.1416 55614	(1)3.1432 46729	(1)3.1448 37039
1/3	9.9497 47896	9.9531 13846	9.9564 77521	9.9598 38925	9.9631 98061
1/4	5.6022 05785	5.6036 27123	5.6050 47381	5.6064 66560	5.6078 84662
1/5	3.9690 56179	3.9698 61752	3.9706 66671	3.9714 70939	3.9722 74555

k	990	991	992	993	994
1	990	991	992	993	994
2	9 80100	9 82081	9 84064	9 86049	9 88036
3	9702 99000	9732 42271	9761 91488	9791 46657	9821 07784
4	(11)9.6059 60100	(11)9.6448 30906	(11)9.6838 19561	(11)9.7229 26304	(11)9.7621 51373
5	(14)9.5099 00499	(14)9.5580 27427	(14)9.6063 49004	(14)9.6548 65820	(14)9.7035 78465
6	(17)9.4148 01494	(17)9.4720 05181	(17)9.5294 98212	(17)9.5872 81759	(17)9.6453 56994
7	(20)9.3206 53479	(20)9.3867 57134	(20)9.4532 62227	(20)9.5201 70787	(20)9.5874 84852
8	(23)9.2274 46944	(23)9.3022 76320	(23)9.3776 36129	(23)9.4535 29591	(23)9.5299 59943
9	(26)9.1351 72475	(26)9.2185 55833	(26)9.3026 15040	(26)9.3873 54884	(26)9.4727 80183
10	(29)9.0438 20750	(29)9.1355 88830	(29)9.2281 94120	(29)9.3216 43400	(29)9.4159 43502
24	(71)7.8567 81408	(71)8.0494 77813	(71)8.2466 98779	(71)8.4485 45822	(71)8.6551 22630
1/2	(1)3.1464 26545	(1)3.1480 15248	(1)3.1496 03150	(1)3.1511 90251	(1)3.1527 76554
1/3	9.9665 54934	9.9699 09547	9.9732 61904	9.9766 12009	9.9799 59866
1/4	5.6093 01690	5.6107 17644	5.6121 32527	5.6135 46340	5.6149 59086
1/5	3.9730 77521	3.9738 79839	3.9746 81509	3.9754 82534	3.9762 82913

k	995	996	997	998	999
1	995	996	997	998	999
2	9 90025	9 92016	9 94009	9 96004	9 98001
3	9850 74875	9880 47936	9910 26973	9940 11992	9970 02999
4	(11)9.8014 95006	(11)9.8409 57443	(11)9.8805 38921	(11)9.9202 39680	(11)9.9600 59960
5	(14)9.7524 87531	(14)9.8015 93613	(14)9.8508 97304	(14)9.9003 99201	(14)9.9500 99900
6	(17)9.7037 25094	(17)9.7623 87238	(17)9.8213 44612	(17)9.8805 98402	(17)9.9401 49800
7	(20)9.6552 06468	(20)9.7233 37689	(20)9.7918 80578	(20)9.8608 37206	(20)9.9302 09650
8	(23)9.6069 30436	(23)9.6844 44339	(23)9.7625 04937	(23)9.8411 15531	(23)9.9202 79441
9	(26)9.5588 95784	(26)9.6457 06561	(26)9.7332 17422	(26)9.8214 33300	(26)9.9103 59161
10	(29)9.5111 01305	(29)9.6071 23735	(29)9.7040 17769	(29)9.8017 90434	(29)9.9004 48802
24	(71)8.8665 35105	(71)9.0828 91413	(71)9.3043 02025	(71)9.5308 79767	(71)9.7627 39866
1/2	(1)3.1543 62059	(1)3.1559 46768	(1)3.1575 30681	(1)3.1591 13800	(1)3.1606 96126
1/3	9.9833 05478	9.9866 48849	9.9899 89983	9.9933 28884	9.9966 65555
1/4	5.6163 70767	5.6177 81384	5.6191 90939	5.6205 99434	5.6220 06871
1/5	3.9770 82648	3.9778 81740	3.9786 80191	3.9794 78001	3.9802 75173

$$n^{\frac{1}{2}}\left[\begin{matrix}(-6)1\\3\end{matrix}\right] \qquad n^{\frac{1}{3}}\left[\begin{matrix}(-7)3\\3\end{matrix}\right] \qquad n^{\frac{1}{4}}\left[\begin{matrix}(-7)1\\3\end{matrix}\right] \qquad n^{\frac{1}{5}}\left[\begin{matrix}(-8)8\\3\end{matrix}\right]$$

4. Elementary Transcendental Functions
Logarithmic, Exponential, Circular and Hyperbolic Functions

RUTH ZUCKER [1]

Contents

---

[1] National Bureau of Standards.

The author acknowledges the assistance of Lois K. Cherwinski and Elizabeth F. Godefroy in the preparation and checking of the tables.

4. Elementary Transcendental Functions
Logarithmic, Exponential, Circular and Hyperbolic Functions

Mathematical Properties

4.1. Logarithmic Function

Integral Representation

4.1.1
$$\ln z = \int_1^z \frac{dt}{t}$$

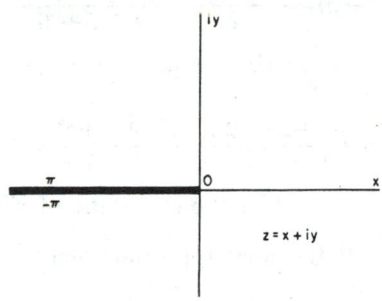

FIGURE 4.1.   *Branch cut for* $\ln z$ *and* z^a.
(a not an integer or zero.)

where the path of integration does not pass through the origin or cross the negative real axis. $\ln z$ is a single-valued function, regular in the z-plane cut along the negative real axis, real when z is positive.

$$z = x + iy = re^{i\theta}.$$

4.1.2 $\ln z = \ln r + i\theta \quad (-\pi < \theta \leq \pi)$.

4.1.3 $r = (x^2 + y^2)^{\frac{1}{2}}, \quad x = r \cos \theta, \quad y = r \sin \theta,$

$$\theta = \arctan \frac{y}{x}.$$

The general logarithmic function is the many-valued function $\mathrm{Ln}\, z$ defined by

4.1.4
$$\mathrm{Ln}\, z = \int_1^z \frac{dt}{t}$$

where the path does not pass through the origin.

4.1.5
$$\mathrm{Ln}\, (re^{i\theta}) = \ln (re^{i\theta}) + 2k\pi i = \ln r + i(\theta + 2k\pi),$$

k being an arbitrary integer. $\ln z$ is said to be the *principal branch* of $\mathrm{Ln}\, z$.

Logarithmic Identities

4.1.6 $\mathrm{Ln}\, (z_1 z_2) = \mathrm{Ln}\, z_1 + \mathrm{Ln}\, z_2.$

(i.e., every value of $\mathrm{Ln}\, (z_1 z_2)$ is one of the values of $\mathrm{Ln}\, z_1 + \mathrm{Ln}\, z_2$.)

4.1.7 $\ln (z_1 z_2) = \ln z_1 + \ln z_2$
$$(-\pi < \arg z_1 + \arg z_2 \leq \pi)$$

4.1.8 $\mathrm{Ln}\, \dfrac{z_1}{z_2} = \mathrm{Ln}\, z_1 - \mathrm{Ln}\, z_2$

4.1.9 $\ln \dfrac{z_1}{z_2} = \ln z_1 - \ln z_2$
$$(-\pi < \arg z_1 - \arg z_2 \leq \pi)$$

4.1.10 $\mathrm{Ln}\, z^n = n\, \mathrm{Ln}\, z \qquad (n \text{ integer})$

4.1.11 $\ln z^n = n \ln z$
$$(n \text{ integer}, \quad -\pi < n \arg z \leq \pi)$$

Special Values   (see chapter 1)

4.1.12 $\ln 1 = 0$

4.1.13 $\ln 0 = -\infty$

4.1.14 $\ln (-1) = \pi i$

4.1.15 $\ln (\pm i) = \pm \tfrac{1}{2}\pi i$

4.1.16 $\ln e = 1, \quad e$ is the real number such that
$$\int_1^e \frac{dt}{t} = 1$$

4.1.17 $e = \lim\limits_{n \to \infty} \left(1 + \dfrac{1}{n}\right)^n = 2.71828\ 18284\ldots$
$$(\text{see } \mathbf{4.2.21})$$

Logarithms to General Base

4.1.18 $\log_a z = \ln z / \ln a$

4.1.19 $\log_a z = \dfrac{\log_b z}{\log_b a}$

4.1.20 $\log_a b = \dfrac{1}{\log_b a}$

4.1.21 $\log_e z = \ln z$

4.1.22 $\log_{10} z = \ln z / \ln 10 = \log_{10} e \ln z$
$$= (.43429\ 44819\ldots) \ln z$$

4.1.23 $\ln z = \ln 10 \log_{10} z = (2.30258\ 50929\ldots)\log_{10} z$

($\log_e x = \ln x$, called natural, Napierian, or hyperbolic logarithms; $\log_{10} x$, called common or Briggs logarithms.)

Series Expansions

4.1.24 $\ln (1+z) = z - \tfrac{1}{2}z^2 + \tfrac{1}{3}z^3 - \cdots$

$$(|z| \leq 1 \text{ and } z \neq -1)$$

4.1.25

$$\ln z = \left(\frac{z-1}{z}\right) + \frac{1}{2}\left(\frac{z-1}{z}\right)^2 + \frac{1}{3}\left(\frac{z-1}{z}\right)^3 + \cdots$$

$$(\mathscr{R}z \geq \tfrac{1}{2})$$

4.1.26

$$\ln z = (z-1) - \tfrac{1}{2}(z-1)^2 + \tfrac{1}{3}(z-1)^3 - \cdots$$

$$(|z-1| \leq 1, \quad z \neq 0)$$

4.1.27

$$\ln z = 2\left[\left(\frac{z-1}{z+1}\right) + \frac{1}{3}\left(\frac{z-1}{z+1}\right)^3 + \frac{1}{5}\left(\frac{z-1}{z+1}\right)^5 + \cdots\right]$$

$$(\mathscr{R}z \geq 0, \quad z \neq 0)$$

4.1.28 $\ln\left(\dfrac{z+1}{z-1}\right) = 2\left(\dfrac{1}{z} + \dfrac{1}{3z^3} + \dfrac{1}{5z^5} + \cdots\right)$

$$(|z| \geq 1, \quad z \neq \pm 1)$$

4.1.29

$$\ln (z+a) = \ln a + 2\left[\left(\frac{z}{2a+z}\right) + \frac{1}{3}\left(\frac{z}{2a+z}\right)^3 \right.$$
$$\left. + \frac{1}{5}\left(\frac{z}{2a+z}\right)^5 + \cdots\right]$$
$$(a > 0, \quad \mathscr{R}z \geq -a \neq z)$$

Limiting Values

4.1.30 $\displaystyle\lim_{x \to \infty} x^{-\alpha} \ln x = 0$

$$(\alpha \text{ constant}, \quad \mathscr{R}\alpha > 0)$$

4.1.31 $\displaystyle\lim_{x \to 0} x^{\alpha} \ln x = 0$

$$(\alpha \text{ constant}, \quad \mathscr{R}\alpha > 0)$$

4.1.32

$$\lim_{m \to \infty}\left(\sum_{k=1}^{m} \frac{1}{k} - \ln m\right) = \gamma \text{ (Euler's constant)}$$
$$= .57721\ 56649\ldots$$

(see chapters **1**, **6** and **23**)

Inequalities

4.1.33 $\dfrac{x}{1+x} < \ln (1+x) < x$

$$(x > -1, \quad x \neq 0)$$

4.1.34 $x < -\ln (1-x) < \dfrac{x}{1-x}$

$$(x < 1, \quad x \neq 0)$$

4.1.35 $|\ln (1-x)| < \dfrac{3x}{2}$ $(0 < x \leq .5828)$

4.1.36 $\ln x \leq x - 1$ $(x > 0)$

4.1.37 $\ln x \leq n(x^{1/n} - 1)$ for any positive n

$$(x > 0)$$

4.1.38 $|\ln (1+z)| \leq -\ln (1-|z|)$ $(|z| < 1)$

Continued Fractions

4.1.39

$$\ln (1+z) = \frac{z}{1+} \frac{z}{2+} \frac{z}{3+} \frac{4z}{4+} \frac{4z}{5+} \frac{9z}{6+} \cdots$$

(z in the plane cut from -1 to $-\infty$)

4.1.40

$$\ln\left(\frac{1+z}{1-z}\right) = \frac{2z}{1-} \frac{z^2}{3-} \frac{4z^2}{5-} \frac{9z^2}{7-} \cdots$$

(z in the cut plane of Figure 4.7.)

Polynomial Approximations [2]

4.1.41 $\dfrac{1}{\sqrt{10}} \leq x \leq \sqrt{10}$

$$\log_{10} x = a_1 t + a_3 t^3 + \epsilon(x), \quad t = (x-1)/(x+1)$$
$$|\epsilon(x)| \leq 6 \times 10^{-4}$$
$$a_1 = .86304 \qquad a_3 = .36415$$

4.1.42 $\dfrac{1}{\sqrt{10}} \leq x \leq \sqrt{10}$

$$\log_{10} x = a_1 t + a_3 t^3 + a_5 t^5 + a_7 t^7 + a_9 t^9 + \epsilon(x)$$
$$t = (x-1)/(x+1)$$
$$|\epsilon(x)| \leq 10^{-7}$$
$$a_1 = .86859\ 1718 \qquad a_7 = .09437\ 6476$$
$$a_3 = .28933\ 5524 \qquad a_9 = .19133\ 7714$$
$$a_5 = .17752\ 2071$$

4.1.43 $0 \leq x \leq 1$

$$\ln (1+x) = a_1 x + a_2 x^2 + a_3 x^3 + a_4 x^4 + a_5 x^5 + \epsilon(x)$$
$$|\epsilon(x)| \leq 1 \times 10^{-5}$$
$$a_1 = .99949\ 556 \qquad a_4 = -.13606\ 275$$
$$a_2 = -.49190\ 896 \qquad a_5 = .03215\ 845$$
$$a_3 = .28947\ 478$$

---

[2] The approximations **4.1.41** to **4.1.44** are from C. Hastings, Jr., Approximations for digital computers. Princeton Univ. Press, Princeton, N.J., 1955 (with permission).

4.1.44 $\qquad 0 \le x \le 1$

$$\ln(1+x) = a_1 x + a_2 x^2 + a_3 x^3 + a_4 x^4 + a_5 x^5 + a_6 x^6$$
$$+ a_7 x^7 + a_8 x^8 + \epsilon(x)$$

$$|\epsilon(x)| \le 3 \times 10^{-8}$$

$a_1 =$	.99999 64239	$a_5 =$	.16765 40711
$a_2 = -$	.49987 41238	$a_6 = -$	.09532 93897
$a_3 =$	.33179 90258	$a_7 =$	.03608 84937
$a_4 = -$	.24073 38084	$a_8 = -$	.00645 35442

Approximation in Terms of Chebyshev Polynomials [3]

4.1.45 $\qquad 0 \le x \le 1$

$$T_n^*(x) = \cos n\theta, \quad \cos \theta = 2x - 1 \quad \text{(see chapter 22)}$$

$$\ln(1+x) = \sum_{n=0}^{\infty} A_n T_n^*(x)$$

n	A_n	n	A_n
0	.37645 2813	6	$-$.00000 8503
1	.34314 5750	7	.00000 1250
2	$-$.02943 7252	8	$-$.00000 0188
3	.00336 7089	9	.00000 0029
4	$-$.00043 3276	10	$-$.00000 0004
5	.00005 9471	11	.00000 0001

Differentiation Formulas

4.1.46
$$\frac{d}{dz} \ln z = \frac{1}{z}$$

4.1.47
$$\frac{d^n}{dz^n} \ln z = (-1)^{n-1}(n-1)! z^{-n}$$

Integration Formulas

4.1.48
$$\int \frac{dz}{z} = \ln z$$

4.1.49
$$\int \ln z \, dz = z \ln z - z$$

4.1.50
$$\int z^n \ln z \, dz = \frac{z^{n+1}}{n+1} \ln z - \frac{z^{n+1}}{(n+1)^2}$$
$$(n \ne -1, \quad n \text{ integer})$$

4.1.51
$$\int z^n (\ln z)^m \, dz = \frac{z^{n+1}(\ln z)^m}{n+1} - \frac{m}{n+1} \int z^n (\ln z)^{m-1} dz$$
$$(n \ne -1)$$

[3] The approximation **4.1.45** is from C. W. Clenshaw, Polynomial approximations to elementary functions, Math. Tables Aids Comp. **8**, 143–147 (1954) (with permission).

4.1.52
$$\int \frac{dz}{z \ln z} = \ln \ln z$$

4.1.53
$$\int \ln[z + (z^2 \pm 1)^{\frac{1}{2}}] dz = z \ln[z + (z^2 \pm 1)^{\frac{1}{2}}] - (z^2 \pm 1)^{\frac{1}{2}}$$

4.1.54
$$\int z^n \ln[z + (z^2 \pm 1)^{\frac{1}{2}}] dz = \frac{z^{n+1}}{n+1} \ln[z + (z^2 \pm 1)^{\frac{1}{2}}]$$
$$- \frac{1}{n+1} \int \frac{z^{n+1}}{(z^2 \pm 1)^{\frac{1}{2}}} dz \quad (n \ne -1)$$

Definite Integrals

4.1.55
$$\int_0^1 \frac{\ln t}{1-t} dt = -\pi^2/6$$

4.1.56
$$\int_0^1 \frac{\ln t}{1+t} dt = -\pi^2/12$$

4.1.57
$$\int_0^x \frac{dt}{\ln t} = li(x) \quad \text{(see 5.1.3)}$$

4.2. Exponential Function

Series Expansion

4.2.1
$$e^z = \exp z = 1 + \frac{z}{1!} + \frac{z^2}{2!} + \frac{z^3}{3!} + \cdots \quad (z = x + iy)$$

where e is the real number defined in **4.1.16**

Fundamental Properties

4.2.2 $\quad \mathrm{Ln}(\exp z) = z + 2k\pi i \quad (k \text{ any integer})$

4.2.3 $\quad \ln(\exp z) = z \quad (-\pi < \mathscr{I}z \le \pi)$

4.2.4 $\quad \exp(\ln z) = \exp(\mathrm{Ln}\, z) = z$

4.2.5
$$\frac{d}{dz} \exp z = \exp z$$

Definition of General Powers

4.2.6 $\quad$ If $N = a^z$, then $z = \mathrm{Log}_a N$

4.2.7 $\qquad a^z = \exp(z \ln a)$

4.2.8 $\quad$ If $a = |a| \exp(i \arg a) \quad (-\pi < \arg a \le \pi)$

4.2.9 $\qquad |a^z| = |a|^x e^{-y \arg a}$

4.2.10 $\qquad \arg(a^z) = y \ln|a| + x \arg a$

4.2.11
$\mathrm{Ln}\, a^z = z \ln a \quad$ for one of the values of $\mathrm{Ln}\, a^z$

4.2.12 $\quad \ln a^x = x \ln a \quad (a \text{ real and positive})$

4.2.13 $\qquad |e^z| = e^x$

4.2.14 $$\arg (e^z)=y$$

4.2.15 $$a^{z_1}a^{z_2}=a^{z_1+z_2}$$

4.2.16 $$a^z b^z=(ab)^z \qquad (-\pi<\arg a+\arg b\leq\pi)$$

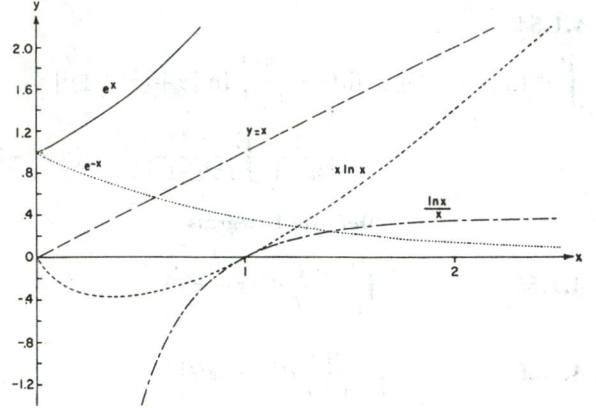

FIGURE 4.2. *Logarithmic and exponential functions.*

Periodic Property

4.2.17 $$e^{z+2\pi ki}=e^z \qquad (k \text{ any integer})$$

Exponential Identities

4.2.18 $$e^{z_1}e^{z_2}=e^{z_1+z_2}$$

4.2.19 $$(e^{z_1})^{z_2}=e^{z_1 z_2} \qquad (-\pi<\mathscr{I}z_1\leq\pi)$$

The restriction $(-\pi<\mathscr{I}z_1\leq\pi)$ can be removed if z_2 is an integer.

Limiting Values

4.2.20
$$\lim_{|z|\to\infty} z^\alpha e^{-z}=0 \quad (|\arg z|\leq\tfrac{1}{2}\pi-\epsilon<\tfrac{1}{2}\pi, \quad \alpha \text{ constant})$$

4.2.21
$$\lim_{m\to\infty}\left(1+\frac{z}{m}\right)^m=e^z$$

Special Values (see chapter 1)

4.2.22 $$e=2.71828\ 18284\ .\ .$$

4.2.23 $$e^0=1$$

4.2.24 $$e^\infty=\infty$$

4.2.25 $$e^{-\infty}=0$$

4.2.26 $$e^{\pm\pi i}=-1$$

4.2.27 $$e^{\pm\frac{\pi i}{2}}=\pm i$$

4.2.28 $$e^{2\pi ki}=1 \quad (k \text{ any integer})$$

Exponential Inequalities

If x is real and different from zero

4.2.29 $$e^{-\frac{x}{1-x}}<1-x<e^{-x} \quad (x<1)$$

4.2.30 $$e^x>1+x$$

4.2.31 $$e^x<\frac{1}{1-x} \quad (x<1)$$

4.2.32 $$\frac{x}{1+x}<(1-e^{-x})<x \quad (x>-1)$$

4.2.33 $$x<(e^x-1)<\frac{x}{1-x} \quad (x<1)$$

4.2.34 $$1+x>e^{\frac{x}{1+x}} \quad (x>-1)$$

4.2.35 $$e^x>1+\frac{x^n}{n!} \quad (n>0, \quad x>0)$$

4.2.36 $$e^x>\left(1+\frac{x}{y}\right)^y>e^{\frac{xy}{x+y}} \quad (x>0, \quad y>0)$$

4.2.37 $$e^{-x}<1-\frac{x}{2} \quad (0<x\leq 1.5936)$$

4.2.38 $$\frac{1}{4}|z|<|e^z-1|<\frac{7}{4}|z| \quad (0<|z|<1)$$

4.2.39 $$|e^z-1|\leq e^{|z|}-1\leq|z|e^{|z|} \quad (\text{all } z)$$

Continued Fractions

4.2.40 $$e^z=\frac{1}{1-}\ \frac{z}{1+}\ \frac{z}{2-}\ \frac{z}{3+}\ \frac{z}{2-}\ \frac{z}{5+}\ \frac{z}{2-}\ \cdots \qquad (|z|<\infty)$$

$$=1+\frac{z}{1-}\ \frac{z}{2+}\ \frac{z}{3-}\ \frac{z}{2+}\ \frac{z}{5-}\ \frac{z}{2+}\ \frac{z}{7-}\ \cdots \qquad (|z|<\infty)$$

$$=1+\frac{z}{(1-z/2)+}\ \frac{z^2/4\cdot 3}{1+}\ \frac{z^2/4\cdot 15}{1+}\ \frac{z^2/4\cdot 35}{1+}\ \cdots\ \frac{z^2/4(4n^2-1)}{1+}\ \cdots\ (|z|<\infty)$$

4.2.41 $$e^z-e_{n-1}(z)=\frac{z^n}{n!-}\ \frac{n!z}{(n+1)+}\ \frac{z}{(n+2)-}\ \frac{(n+1)z}{(n+3)+}\ \frac{2z}{(n+4)-}\ \frac{(n+2)z}{(n+5)+}\ \frac{3z}{(n+6)-}\ \cdots\ (|z|<\infty)$$

(For $e_n(z)$ see **6.5.11**)

4.2.42

$$e^{2a \arctan \frac{1}{z}} = 1 + \frac{2a}{z-a+} \frac{a^2+1}{3z+} \frac{a^2+4}{5z+} \frac{a^2+9}{7z+} \cdots$$

(z in the cut plane of Figure 4.4.)

Polynomial Approximations [4]

4.2.43
$$0 \leq x \leq \ln 2 = .693 \ldots$$

$$e^{-x} = 1 + a_1 x + a_2 x^2 + \epsilon(x)$$

$$|\epsilon(x)| \leq 3 \times 10^{-3}$$

$$a_1 = -.9664 \qquad a_2 = .3536$$

4.2.44
$$0 \leq x \leq \ln 2$$

$$e^{-x} = 1 + a_1 x + a_2 x^2 + a_3 x^3 + a_4 x^4 + \epsilon(x)$$

$$|\epsilon(x)| \leq 3 \times 10^{-5}$$

$$a_1 = -.99986\,84 \qquad a_3 = -.15953\,32$$
$$a_2 = .49829\,26 \qquad a_4 = .02936\,41$$

4.2.45
$$0 \leq x \leq \ln 2$$

$$e^{-x} = 1 + a_1 x + a_2 x^2 + a_3 x^3 + a_4 x^4 + a_5 x^5$$
$$+ a_6 x^6 + a_7 x^7 + \epsilon(x)$$

$$|\epsilon(x)| \leq 2 \times 10^{-10}$$

$$a_1 = -.99999\,99995 \qquad a_5 = -.00830\,13598$$
$$a_2 = .49999\,99206 \qquad a_6 = .00132\,98820$$
$$a_3 = -.16666\,53019 \qquad a_7 = -.00014\,13161$$
$$a_4 = .04165\,73475$$

4.2.46 [5]
$$0 \leq x \leq 1$$

$$10^x = (1 + a_1 x + a_2 x^2 + a_3 x^3 + a_4 x^4)^2 + \epsilon(x)$$

$$|\epsilon(x)| \leq 7 \times 10^{-4}$$

$$a_1 = 1.14991\,96 \qquad a_3 = .20800\,30$$
$$a_2 = .67743\,23 \qquad a_4 = .12680\,89$$

4.2.47
$$0 \leq x \leq 1$$

$$10^x = (1 + a_1 x + a_2 x^2 + a_3 x^3 + a_4 x^4 + a_5 x^5$$
$$+ a_6 x^6 + a_7 x^7)^2 + \epsilon(x)$$

$$|\epsilon(x)| < 5 \times 10^{-8}$$

$$a_1 = 1.15129\,277603 \qquad a_5 = .01742\,111988$$
$$a_2 = .66273\,088429 \qquad a_6 = .00255\,491796$$
$$a_3 = .25439\,357484 \qquad a_7 = .00093\,264267$$
$$a_4 = .07295\,173666$$

[4] The approximations **4.2.43** to **4.2.45** are from B. Carlson, M. Goldstein, Rational approximation of functions, Los Alamos Scientific Laboratory LA–1943, Los Alamos, N. Mex., 1955 (with permission).

[5] The approximations **4.2.46** to **4.2.47** are from C. Hastings, Jr., Approximations for digital computers. Princeton Univ. Press, Princeton, N.J., 1955 (with permission).

Approximations in Terms of Chebyshev Polynomials [6]

4.2.48
$$0 \leq x \leq 1$$

$$T_n^*(x) = \cos n\theta, \quad \cos\theta = 2x - 1 \text{ (see chapter 22)}$$

$$e^x = \sum_{n=0}^{\infty} A_n T_n^*(x) \qquad e^{-x} = \sum_{n=0}^{\infty} A_n T_n^*(x)$$

n	A_n	n	A_n
0	1.75338 7654	0	.64503 5270
1	.85039 1654	1	$-$.31284 1606
2	.10520 8694	2	.03870 4116
3	.00872 2105	3	$-$.00320 8683
4	.00054 3437	4	.00019 9919
5	.00002 7115	5	$-$.00000 9975
6	.00000 1128	6	.00000 0415
7	.00000 0040	7	$-$.00000 0015
8	.00000 0001		

Differentiation Formulas

4.2.49
$$\frac{d}{dz} e^z = e^z$$

4.2.50
$$\frac{d^n}{dz^n} e^{az} = a^n e^{az}$$

4.2.51
$$\frac{d}{dz} a^z = a^z \ln a$$

4.2.52
$$\frac{d}{dz} z^a = a z^{a-1}$$

4.2.53
$$\frac{d}{dz} z^z = (1 + \ln z) z^z$$

Integration Formulas

4.2.54
$$\int e^{az} dz = e^{az}/a$$

4.2.55
$$\int z^n e^{az} dz = \frac{e^{az}}{a^{n+1}} [(az)^n - n(az)^{n-1} + n(n-1)(az)^{n-2}$$
$$+ \ldots + (-1)^{n-1} n!(az) + (-1)^n n!] \quad (n \geq 0)$$

4.2.56
$$\int \frac{e^{az}}{z^n} dz = -\frac{e^{az}}{(n-1)z^{n-1}} + \frac{a}{n-1} \int \frac{e^{az}}{z^{n-1}} dz \quad (n > 1)$$

(See chapters **5, 7** and **29** for other integrals involving exponential functions.)

4.3. Circular Functions

Definitions

4.3.1
$$\sin z = \frac{e^{iz} - e^{-iz}}{2i} \qquad (z = x + iy)$$

4.3.2
$$\cos z = \frac{e^{iz} + e^{-iz}}{2}$$

[6] The approximations **4.2.48** are from C. W. Clenshaw, Polynomial approximations to elementary functions, Math. Tables Aids Comp. 8, 143–147 (1954) (with permission).

4.3.3
$$\tan z = \frac{\sin z}{\cos z}$$

4.3.4
$$\csc z = \frac{1}{\sin z}$$

4.3.5
$$\sec z = \frac{1}{\cos z}$$

4.3.6
$$\cot z = \frac{1}{\tan z}$$

Periodic Properties

4.3.7 $\sin (z+2k\pi) = \sin z$ (k any integer)

4.3.8 $\cos (z+2k\pi) = \cos z$

4.3.9 $\tan (z+k\pi) = \tan z$

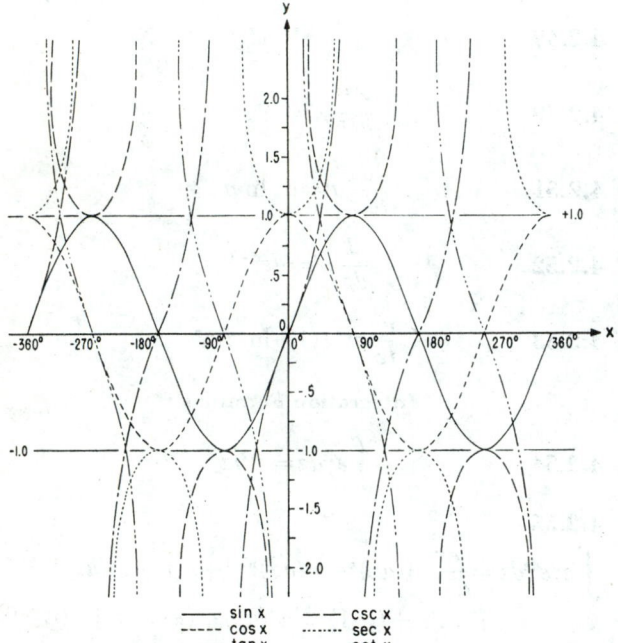

FIGURE 4.3.   *Circular functions.*

Relations Between Circular Functions

4.3.10 $\sin^2 z + \cos^2 z = 1$

4.3.11 $\sec^2 z - \tan^2 z = 1$

4.3.12 $\csc^2 z - \cot^2 z = 1$

Negative Angle Formulas

4.3.13 $\sin (-z) = -\sin z$

4.3.14 $\cos (-z) = \cos z$

4.3.15 $\tan (-z) = -\tan z$

Addition Formulas

4.3.16 $\sin (z_1+z_2) = \sin z_1 \cos z_2 + \cos z_1 \sin z_2$

4.3.17 $\cos (z_1+z_2) = \cos z_1 \cos z_2 - \sin z_1 \sin z_2$

4.3.18 $\tan (z_1+z_2) = \dfrac{\tan z_1 + \tan z_2}{1 - \tan z_1 \tan z_2}$

4.3.19 $\cot (z_1+z_2) = \dfrac{\cot z_1 \cot z_2 - 1}{\cot z_2 + \cot z_1}$

Half-Angle Formulas

4.3.20 $\sin \dfrac{z}{2} = \pm \left(\dfrac{1-\cos z}{2} \right)^{\frac{1}{2}}$

4.3.21 $\cos \dfrac{z}{2} = \pm \left(\dfrac{1+\cos z}{2} \right)^{\frac{1}{2}}$

4.3.22 $\tan \dfrac{z}{2} = \pm \left(\dfrac{1-\cos z}{1+\cos z} \right)^{\frac{1}{2}} = \dfrac{1-\cos z}{\sin z} = \dfrac{\sin z}{1+\cos z}$

The ambiguity in sign may be resolved with the aid of a diagram.

Transformation of Trigonometric Integrals

If $\tan \dfrac{u}{2} = z$ then

4.3.23 $\sin u = \dfrac{2z}{1+z^2}, \quad \cos u = \dfrac{1-z^2}{1+z^2}, \quad du = \dfrac{2}{1+z^2} dz$

Multiple-Angle Formulas

4.3.24 $\sin 2z = 2 \sin z \cos z = \dfrac{2 \tan z}{1+\tan^2 z}$

4.3.25 $\cos 2z = 2 \cos^2 z - 1 = 1 - 2 \sin^2 z$
$$= \cos^2 z - \sin^2 z = \frac{1-\tan^2 z}{1+\tan^2 z}$$

4.3.26 $\tan 2z = \dfrac{2 \tan z}{1-\tan^2 z} = \dfrac{2 \cot z}{\cot^2 z - 1} = \dfrac{2}{\cot z - \tan z}$

4.3.27 $\sin 3z = 3 \sin z - 4 \sin^3 z$

4.3.28 $\cos 3z = -3 \cos z + 4 \cos^3 z$

4.3.29 $\sin 4z = 8 \cos^3 z \sin z - 4 \cos z \sin z$

4.3.30 $\cos 4z = 8 \cos^4 z - 8 \cos^2 z + 1$

Products of Sines and Cosines

4.3.31 $2 \sin z_1 \sin z_2 = \cos (z_1-z_2) - \cos (z_1+z_2)$

4.3.32 $2 \cos z_1 \cos z_2 = \cos (z_1-z_2) + \cos (z_1+z_2)$

4.3.33 $2 \sin z_1 \cos z_2 = \sin (z_1-z_2) + \sin (z_1+z_2)$

Addition and Subtraction of Two Circular Functions

4.3.34
$$\sin z_1 + \sin z_2 = 2 \sin \left(\frac{z_1+z_2}{2} \right) \cos \left(\frac{z_1-z_2}{2} \right)$$

4.3.35
$$\sin z_1 - \sin z_2 = 2 \cos\left(\frac{z_1+z_2}{2}\right) \sin\left(\frac{z_1-z_2}{2}\right)$$

4.3.36
$$\cos z_1 + \cos z_2 = 2 \cos\left(\frac{z_1+z_2}{2}\right) \cos\left(\frac{z_1-z_2}{2}\right)$$

4.3.37
$$\cos z_1 - \cos z_2 = -2 \sin\left(\frac{z_1+z_2}{2}\right) \sin\left(\frac{z_1-z_2}{2}\right)$$

4.3.38
$$\tan z_1 \pm \tan z_2 = \frac{\sin(z_1 \pm z_2)}{\cos z_1 \cos z_2}$$

4.3.39
$$\cot z_1 \pm \cot z_2 = \frac{\sin(z_2 \pm z_1)}{\sin z_1 \sin z_2}$$

Relations Between Squares of Sines and Cosines

4.3.40
$$\sin^2 z_1 - \sin^2 z_2 = \sin(z_1+z_2)\sin(z_1-z_2)$$

4.3.41
$$\cos^2 z_1 - \cos^2 z_2 = -\sin(z_1+z_2)\sin(z_1-z_2)$$

4.3.42
$$\cos^2 z_1 - \sin^2 z_2 = \cos(z_1+z_2)\cos(z_1-z_2)$$

4.3.43

Signs of the Circular Functions in the Four Quadrants

Quadrant	sin csc	cos sec	tan cot
I	+	+	+
II	+	−	−
III	−	−	+
IV	−	+	−

4.3.44

Functions of Angles in Any Quadrant in Terms of Angles in the First Quadrant. $(0 \le \theta \le \frac{\pi}{2},$ k any integer$)$

	$-\theta$	$\frac{\pi}{2}\pm\theta$	$\pi\pm\theta$	$\frac{3\pi}{2}\pm\theta$	$2k\pi\pm\theta$
sin	$-\sin\theta$	$\cos\theta$	$\mp\sin\theta$	$-\cos\theta$	$\pm\sin\theta$
cos	$\cos\theta$	$\mp\sin\theta$	$-\cos\theta$	$\pm\sin\theta$	$+\cos\theta$
tan	$-\tan\theta$	$\mp\cot\theta$	$\pm\tan\theta$	$\mp\cot\theta$	$\pm\tan\theta$
csc	$-\csc\theta$	$+\sec\theta$	$\mp\csc\theta$	$-\sec\theta$	$\pm\csc\theta$
sec	$\sec\theta$	$\mp\csc\theta$	$-\sec\theta$	$\pm\csc\theta$	$+\sec\theta$
cot	$-\cot\theta$	$\mp\tan\theta$	$\pm\cot\theta$	$\mp\tan\theta$	$\pm\cot\theta$

4.3.45 Relations Between Circular (or Inverse Circular) Functions

	$\sin x=a$	$\cos x=a$	$\tan x=a$	$\csc x=a$	$\sec x=a$	$\cot x=a$
sin x	a	$(1-a^2)^{\frac{1}{2}}$	$a(1+a^2)^{-\frac{1}{2}}$	a^{-1}	$a^{-1}(a^2-1)^{\frac{1}{2}}$	$(1+a^2)^{-\frac{1}{2}}$
cos x	$(1-a^2)^{\frac{1}{2}}$	a	$(1+a^2)^{-\frac{1}{2}}$	$a^{-1}(a^2-1)^{\frac{1}{2}}$	a^{-1}	$a(1+a^2)^{-\frac{1}{2}}$
tan x	$a(1-a^2)^{-\frac{1}{2}}$	$a^{-1}(1-a^2)^{\frac{1}{2}}$	a	$(a^2-1)^{-\frac{1}{2}}$	$(a^2-1)^{\frac{1}{2}}$	a^{-1}
csc x	a^{-1}	$(1-a^2)^{-\frac{1}{2}}$	$a^{-1}(1+a^2)^{\frac{1}{2}}$	a	$a(a^2-1)^{-\frac{1}{2}}$	$(1+a^2)^{\frac{1}{2}}$
sec x	$(1-a^2)^{-\frac{1}{2}}$	a^{-1}	$(1+a^2)^{\frac{1}{2}}$	$a(a^2-1)^{-\frac{1}{2}}$	a	$a^{-1}(1+a^2)^{\frac{1}{2}}$
cot x	$a^{-1}(1-a^2)^{\frac{1}{2}}$	$a(1-a^2)^{-\frac{1}{2}}$	a^{-1}	$(a^2-1)^{\frac{1}{2}}$	$(a^2-1)^{-\frac{1}{2}}$	a

$\left(0 \le x \le \frac{\pi}{2}\right)$ Illustration: If $\sin x = a$, $\cot x = a^{-1}(1-a^2)^{\frac{1}{2}}$
$$\operatorname{arcsec} a = \operatorname{arccot}(a^2-1)^{-\frac{1}{2}}$$

4.3.46 Circular Functions for Certain Angles

	0 $0°$	$\pi/12$ $15°$	$\pi/6$ $30°$	$\pi/4$ $45°$	$\pi/3$ $60°$
sin	0	$\frac{\sqrt{2}}{4}(\sqrt{3}-1)$	$1/2$	$\sqrt{2}/2$	$\sqrt{3}/2$
cos	1	$\frac{\sqrt{2}}{4}(\sqrt{3}+1)$	$\sqrt{3}/2$	$\sqrt{2}/2$	$1/2$
tan	0	$2-\sqrt{3}$	$\sqrt{3}/3$	1	$\sqrt{3}$
csc	∞	$\sqrt{2}(\sqrt{3}+1)$	2	$\sqrt{2}$	$2\sqrt{3}/3$
sec	1	$\sqrt{2}(\sqrt{3}-1)$	$2\sqrt{3}/3$	$\sqrt{2}$	2
cot	∞	$2+\sqrt{3}$	$\sqrt{3}$	1	$\sqrt{3}/3$

	$5\pi/12$ $75°$	$\pi/2$ $90°$	$7\pi/12$ $105°$	$2\pi/3$ $120°$
sin	$\frac{\sqrt{2}}{4}(\sqrt{3}+1)$	1	$\frac{\sqrt{2}}{4}(\sqrt{3}+1)$	$\sqrt{3}/2$
cos	$\frac{\sqrt{2}}{4}(\sqrt{3}-1)$	0	$\frac{-\sqrt{2}}{4}(\sqrt{3}-1)$	$-1/2$
tan	$2+\sqrt{3}$	∞	$-(2+\sqrt{3})$	$-\sqrt{3}$
csc	$\sqrt{2}(\sqrt{3}-1)$	1	$\sqrt{2}(\sqrt{3}-1)$	$2\sqrt{3}/3$
sec	$\sqrt{2}(\sqrt{3}+1)$	∞	$-\sqrt{2}(\sqrt{3}+1)$	-2
cot	$2-\sqrt{3}$	0	$-(2-\sqrt{3})$	$-\sqrt{3}/3$

	$3\pi/4$ $135°$	$5\pi/6$ $150°$	$11\pi/12$ $165°$	π $180°$
sin	$\sqrt{2}/2$	$1/2$	$\frac{\sqrt{2}}{4}(\sqrt{3}-1)$	0
cos	$-\sqrt{2}/2$	$-\sqrt{3}/2$	$\frac{-\sqrt{2}}{4}(\sqrt{3}+1)$	-1
tan	-1	$-\sqrt{3}/3$	$-(2-\sqrt{3})$	0
csc	$\sqrt{2}$	2	$\sqrt{2}(\sqrt{3}+1)$	∞
sec	$-\sqrt{2}$	$-2\sqrt{3}/3$	$-\sqrt{2}(\sqrt{3}-1)$	-1
cot	-1	$-\sqrt{3}$	$-(2+\sqrt{3})$	∞

Euler's Formula

4.3.47 $e^z = e^{x+iy} = e^x(\cos y + i \sin y)$

De Moivre's Theorem

4.3.48 $(\cos z + i \sin z)^\nu = \cos \nu z + i \sin \nu z$

$(-\pi < \mathscr{R}z \le \pi$ unless ν is an integer$)$

Relation to Hyperbolic Functions (see 4.5.7 to 4.5.12)

4.3.49 $\sin z = -i \sinh iz$

4.3.50 $\cos z = \cosh iz$

4.3.51 $\tan z = -i \tanh iz$

4.3.52 $\csc z = i \operatorname{csch} iz$

4.3.53 $\sec z = \operatorname{sech} iz$

4.3.54 $\cot z = i \coth iz$

Circular Functions in Terms of Real and Imaginary Parts

4.3.55 $\sin z = \sin x \cosh y + i \cos x \sinh y$

4.3.56 $\cos z = \cos x \cosh y - i \sin x \sinh y$

4.3.57 $\tan z = \dfrac{\sin 2x + i \sinh 2y}{\cos 2x + \cosh 2y}$

4.3.58 $\cot z = \dfrac{\sin 2x - i \sinh 2y}{\cosh 2y - \cos 2x}$

Modulus and Phase (Argument) of Circular Functions

4.3.59 $|\sin z| = (\sin^2 x + \sinh^2 y)^{\frac{1}{2}}$

$= [\tfrac{1}{2}(\cosh 2y - \cos 2x)]^{\frac{1}{2}}$

4.3.60 $\arg \sin z = \arctan(\cot x \tanh y)$

4.3.61 $|\cos z| = (\cos^2 x + \sinh^2 y)^{\frac{1}{2}}$

$= [\tfrac{1}{2}(\cosh 2y + \cos 2x)]^{\frac{1}{2}}$

4.3.62 $\arg \cos z = -\arctan(\tan x \tanh y)$

4.3.63 $|\tan z| = \left(\dfrac{\cosh 2y - \cos 2x}{\cosh 2y + \cos 2x}\right)^{\frac{1}{2}}$

4.3.64 $\arg \tan z = \arctan\left(\dfrac{\sinh 2y}{\sin 2x}\right)$

Series Expansions

4.3.65

$$\sin z = z - \frac{z^3}{3!} + \frac{z^5}{5!} - \frac{z^7}{7!} + \cdots \qquad (|z| < \infty)$$

4.3.66

$$\cos z = 1 - \frac{z^2}{2!} + \frac{z^4}{4!} - \frac{z^6}{6!} + \cdots \qquad (|z| < \infty)$$

4.3.67

$$\tan z = z + \frac{z^3}{3} + \frac{2z^5}{15} + \frac{17z^7}{315} + \cdots$$

$$+ \frac{(-1)^{n-1}2^{2n}(2^{2n}-1)B_{2n}}{(2n)!} z^{2n-1} + \cdots \qquad \left(|z| < \frac{\pi}{2}\right)$$

4.3.68

$$\csc z = \frac{1}{z} + \frac{z}{6} + \frac{7}{360} z^3 + \frac{31}{15120} z^5 + \cdots$$

$$+ \frac{(-1)^{n-1}2(2^{2n-1}-1)B_{2n}}{(2n)!} z^{2n-1} + \cdots \qquad (|z| < \pi)$$

4.3.69

$$\sec z = 1 + \frac{z^2}{2} + \frac{5z^4}{24} + \frac{61z^6}{720} + \cdots$$

$$+ \frac{(-1)^n E_{2n}}{(2n)!} z^{2n} + \cdots \qquad \left(|z| < \frac{\pi}{2}\right)$$

4.3.70

$$\cot z = \frac{1}{z} - \frac{z}{3} - \frac{z^3}{45} - \frac{2z^5}{945} - \cdots$$

$$- \frac{(-1)^{n-1}2^{2n}B_{2n}}{(2n)!} z^{2n-1} - \cdots \qquad (|z| < \pi)$$

4.3.71

$$\ln \frac{\sin z}{z} = \sum_{n=1}^{\infty} \frac{(-1)^n 2^{2n-1} B_{2n}}{n(2n)!} z^{2n} \qquad (|z| < \pi)$$

4.3.72

$$\ln \cos z = \sum_{n=1}^{\infty} \frac{(-1)^n 2^{2n-1}(2^{2n}-1)B_{2n}}{n(2n)!} z^{2n} \qquad (|z| < \tfrac{1}{2}\pi)$$

4.3.73

$$\ln \frac{\tan z}{z} = \sum_{n=1}^{\infty} \frac{(-1)^{n-1}2^{2n}(2^{2n-1}-1)B_{2n}}{n(2n)!} z^{2n}$$

$$(|z| < \tfrac{1}{2}\pi)$$

where B_n and E_n are the Bernoulli and Euler numbers (see chapter **23**).

Limiting Values

4.3.74
$$\lim_{x \to 0} \frac{\sin x}{x} = 1$$

4.3.75
$$\lim_{x \to 0} \frac{\tan x}{x} = 1$$

4.3.76
$$\lim_{n \to \infty} n \sin \frac{x}{n} = x$$

4.3.77
$$\lim_{n \to \infty} n \tan \frac{x}{n} = x$$

4.3.78
$$\lim_{n \to \infty} \cos \frac{x}{n} = 1$$

Inequalities

4.3.79
$$\frac{\sin x}{x} > \frac{2}{\pi} \qquad \left(-\frac{\pi}{2} < x < \frac{\pi}{2}\right)$$

4.3.80
$$\sin x \le x \le \tan x \qquad \left(0 \le x \le \frac{\pi}{2}\right)$$

4.3.81
$$\cos x \le \frac{\sin x}{x} \le 1 \qquad (0 \le x \le \pi)$$

4.3.82
$$\pi < \frac{\sin \pi x}{x(1-x)} \le 4 \qquad (0 < x < 1)$$

4.3.83
$$|\sinh y| \le |\sin z| \le \cosh y$$

4.3.84
$$|\sinh y| \le |\cos z| \le \cosh y$$

4.3.85
$$|\csc z| \le \operatorname{csch}|y|$$

4.3.86
$$|\cos z| \le \cosh|z|$$

4.3.87
$$|\sin z| \le \sinh|z|$$

4.3.88
$$|\cos z| < 2, \quad |\sin z| \le \frac{6}{5}|z| \qquad (|z| < 1)$$

Infinite Products

4.3.89
$$\sin z = z \prod_{k=1}^{\infty} \left(1 - \frac{z^2}{k^2 \pi^2}\right)$$

4.3.90
$$\cos z = \prod_{k=1}^{\infty} \left(1 - \frac{4z^2}{(2k-1)^2 \pi^2}\right)$$

Expansion in Partial Fractions

4.3.91
$$\cot z = \frac{1}{z} + 2z \sum_{k=1}^{\infty} \frac{1}{z^2 - k^2 \pi^2}$$

$$(z \ne 0, \pm \pi, \pm 2\pi, \ldots)$$

4.3.92
$$\csc^2 z = \sum_{k=-\infty}^{\infty} \frac{1}{(z - k\pi)^2}$$

$$(z \ne 0, \pm \pi, \pm 2\pi, \ldots)$$

4.3.93
$$\csc z = \frac{1}{z} + 2z \sum_{k=1}^{\infty} \frac{(-1)^k}{z^2 - k^2 \pi^2}$$

$$(z \ne 0, \pm \pi, \pm 2\pi, \ldots)$$

Continued Fractions

4.3.94
$$\tan z = \frac{z}{1-} \frac{z^2}{3-} \frac{z^2}{5-} \frac{z^2}{7-} \cdots \qquad \left(z \ne \frac{\pi}{2} \pm n\pi\right)$$

4.3.95

$$\tan az = \frac{a \tan z}{1+} \frac{(1-a^2)\tan^2 z}{3+} \frac{(4-a^2)\tan^2 z}{5+}$$

$$\frac{(9-a^2)\tan^2 z}{7+} \cdots \left(-\frac{\pi}{2} < \mathscr{R}\, z < \frac{\pi}{2}, \quad az \ne \frac{\pi}{2} \pm n\pi\right)$$

Polynomial Approximations [7]

4.3.96 $0 \leq x \leq \dfrac{\pi}{2}$

$$\frac{\sin x}{x} = 1 + a_2 x^2 + a_4 x^4 + \epsilon(x)$$

$$|\epsilon(x)| \leq 2 \times 10^{-4}$$

$a_2 = -.16605 \qquad a_4 = .00761$

4.3.97 $0 \leq x \leq \dfrac{\pi}{2}$

$$\frac{\sin x}{x} = 1 + a_2 x^2 + a_4 x^4 + a_6 x^6 + a_8 x^8 + a_{10} x^{10} + \epsilon(x)$$

$$|\epsilon(x)| \leq 2 \times 10^{-9}$$

$a_2 = -.16666\ 66664 \qquad a_8 = \quad .00000\ 27526$

$a_4 = \quad .00833\ 33315 \qquad a_{10} = -.00000\ 00239$

$a_6 = -.00019\ 84090$

4.3.98 $0 \leq x \leq \dfrac{\pi}{2}$

$$\cos x = 1 + a_2 x^2 + a_4 x^4 + \epsilon(x)$$

$$|\epsilon(x)| \leq 9 \times 10^{-4}$$

$a_2 = -.49670 \qquad a_4 = .03705$

4.3.99 $0 \leq x \leq \dfrac{\pi}{2}$

$$\cos x = 1 + a_2 x^2 + a_4 x^4 + a_6 x^6 + a_8 x^8 + a_{10} x^{10} + \epsilon(x)$$

$$|\epsilon(x)| \leq 2 \times 10^{-9}$$

$a_2 = -.49999\ 99963 \qquad a_8 = \quad .00002\ 47609$

$a_4 = \quad .04166\ 66418 \qquad a_{10} = -.00000\ 02605$

$a_6 = -.00138\ 88397$

4.3.100 $0 \leq x \leq \dfrac{\pi}{4}$

$$\frac{\tan x}{x} = 1 + a_2 x^2 + a_4 x^4 + \epsilon(x)$$

$$|\epsilon(x)| \leq 1 \times 10^{-3}$$

$a_2 = .31755 \qquad a_4 = .20330$

4.3.101 $0 \leq x \leq \dfrac{\pi}{4}$

$$\frac{\tan x}{x} = 1 + a_2 x^2 + a_4 x^4 + a_6 x^6 + a_8 x^8 + a_{10} x^{10} + a_{12} x^{12} + \epsilon(x)$$

$$|\epsilon(x)| \leq 2 \times 10^{-8}$$

$a_2 = .33333\ 14036 \qquad a_8 = .02456\ 50893$

$a_4 = .13339\ 23995 \qquad a_{10} = .00290\ 05250$

$a_6 = .05337\ 40603 \qquad a_{12} = .00951\ 68091$

4.3.102 $0 \leq x \leq \dfrac{\pi}{4}$

* $x \cot x = 1 + a_2 x^2 + a_4 x^4 + \epsilon(x)$

$$|\epsilon(x)| \leq 3 \times 10^{-5}$$

$a_2 = -.332867 \qquad a_4 = -.024369$

4.3.103 $0 \leq x \leq \dfrac{\pi}{4}$

$$x \cot x = 1 + a_2 x^2 + a_4 x^4 + a_6 x^6 + a_8 x^8 + a_{10} x^{10} + \epsilon(x)$$

$$|\epsilon(x)| \leq 4 \times 10^{-10}$$

$a_2 = -.33333\ 33410 \qquad a_8 = -.00020\ 78504$

$a_4 = -.02222\ 20287 \qquad a_{10} = -.00002\ 62619$

$a_6 = -.00211\ 77168$

Approximations in Terms of Chebyshev Polynomials [8]

4.3.104 $-1 \leq x \leq 1$

$$T_n^*(x) = \cos n\theta,\ \cos \theta = 2x - 1 \quad \text{(see chapter 22)}$$

$$\sin \tfrac{1}{2}\pi x = x \sum_{n=0}^{\infty} A_n T_n^*(x^2) \qquad \cos \tfrac{1}{2}\pi x = \sum_{n=0}^{\infty} A_n T_n^*(x^2)$$

n	A_n	n	A_n
0	1.27627 8962	0	.47200 1216
1	−.28526 1569	1	−.49940 3258
2	.00911 8016	2	.02799 2080
3	−.00013 6587	3	−.00059 6695
4	.00000 1185	4	.00000 6704
5	−.00000 0007	5	−.00000 0047

[7] The approximations **4.3.96** to **4.3.103** are from B. Carlson, M. Goldstein, Rational approximation of functions, Los Alamos Scientific Laboratory LA–1943, Los Alamos, N. Mex., 1955 (with permission).

[8] The approximations **4.3.104** are from C. W. Clenshaw, Polynomial approximations to elementary functions, Math. Tables Aids Comp. **8**, 143–147 (1954) (with permission).
*See page II.

Differentiation Formulas

4.3.105 $\dfrac{d}{dz}\sin z = \cos z$

4.3.106 $\dfrac{d}{dz}\cos z = -\sin z$

4.3.107 $\dfrac{d}{dz}\tan z = \sec^2 z$

4.3.108 $\dfrac{d}{dz}\csc z = -\csc z \cot z$

4.3.109 $\dfrac{d}{dz}\sec z = \sec z \tan z$

4.3.110 $\dfrac{d}{dz}\cot z = -\csc^2 z$

4.3.111 $\dfrac{d^n}{dz^n}\sin z = \sin\left(z+\dfrac{1}{2}n\pi\right)$

4.3.112 $\dfrac{d^n}{dz^n}\cos z = \cos\left(z+\dfrac{1}{2}n\pi\right)$

Integration Formulas

4.3.113 $\displaystyle\int \sin z\, dz = -\cos z$

4.3.114 $\displaystyle\int \cos z\, dz = \sin z$

4.3.115 $\displaystyle\int \tan z\, dz = -\ln \cos z = \ln \sec z$

4.3.116
$$\int \csc z\, dz = \ln \tan \frac{z}{2} = \ln(\csc z - \cot z) = \frac{1}{2}\ln\frac{1-\cos z}{1+\cos z}$$

4.3.117
$$\int \sec z\, dz = \ln(\sec z + \tan z) = \ln \tan\left(\frac{\pi}{4}+\frac{z}{2}\right) = \mathrm{gd}^{-1}(z)$$
$$= \text{Inverse Gudermannian Function}$$
$$\mathrm{gd}\, z = 2 \arctan e^z - \frac{\pi}{2}$$

4.3.118 $\displaystyle\int \cot z\, dz = \ln \sin z = -\ln \csc z$

4.3.119
$$\int z^n \sin z\, dz = -z^n \cos z + n\int z^{n-1}\cos z\, dz$$

4.3.120
$$\int \frac{\sin z}{z^n}dz = \frac{-\sin z}{(n-1)z^{n-1}} + \frac{1}{n-1}\int \frac{\cos z}{z^{n-1}}dz \quad (n>1)$$

4.3.121 $\displaystyle\int \frac{z}{\sin^2 z}dz = -z \cot z + \ln \sin z$

4.3.122
$$\int \frac{z\, dz}{\sin^n z} = \frac{-z \cos z}{(n-1)\sin^{n-1} z} - \frac{1}{(n-1)(n-2)\sin^{n-2} z}$$
$$+\frac{(n-2)}{(n-1)}\int \frac{z\, dz}{\sin^{n-2} z} \quad (n>2)$$

4.3.123
$$\int z^n \cos z\, dz = z^n \sin z - n\int z^{n-1}\sin z\, dz$$

4.3.124
$$\int \frac{\cos z}{z^n}dz = -\frac{\cos z}{(n-1)z^{n-1}} - \frac{1}{n-1}\int \frac{\sin z}{z^{n-1}}dz \quad (n>1)$$

4.3.125 $\displaystyle\int \frac{z}{\cos^2 z}dz = z \tan z + \ln \cos z$

4.3.126
$$\int \frac{z\, dz}{\cos^n z} = \frac{z \sin z}{(n-1)\cos^{n-1} z} - \frac{1}{(n-1)(n-2)\cos^{n-2} z}$$
$$+\frac{(n-2)}{(n-1)}\int \frac{z\, dz}{\cos^{n-2} z} \quad (n>2)$$

4.3.127
$$\int \sin^m z \cos^n z\, dz = \frac{\sin^{m+1} z \cos^{n-1} z}{m+n}$$
$$+\frac{(n-1)}{(m+n)}\int \sin^m z \cos^{n-2} z\, dz$$
$$=-\frac{\sin^{m-1} z \cos^{n+1} z}{m+n}$$
$$+\frac{(m-1)}{(m+n)}\int \sin^{m-2} z \cos^n z\, dz$$
$$(m \neq -n)$$

4.3.128
$$\int \frac{dz}{\sin^m z \cos^n z} = \frac{1}{(n-1)\sin^{m-1} z \cos^{n-1} z}$$
$$+\frac{m+n-2}{n-1}\int \frac{dz}{\sin^m z \cos^{n-2} z}$$
$$(n>1)$$
$$=\frac{-1}{(m-1)\sin^{m-1} z \cos^{n-1} z}$$
$$+\frac{m+n-2}{m-1}\int \frac{dz}{\sin^{m-2} z \cos^n z}$$
$$(m>1)$$

4.3.129 $\displaystyle\int \tan^n z\, dz = \frac{\tan^{n-1} z}{n-1} - \int \tan^{n-2} z\, dz \quad (n \neq 1)$

4.3.130 $\displaystyle\int \cot^n z\, dz = -\frac{\cot^{n-1} z}{n-1} - \int \cot^{n-2} z\, dz \quad (n \neq 1)$

4.3.131

$$\int \frac{dz}{a+b\sin z} = \frac{2}{(a^2-b^2)^{\frac{1}{2}}}\arctan\frac{a\tan\left(\frac{z}{2}\right)+b}{(a^2-b^2)^{\frac{1}{2}}} \quad (a^2>b^2)$$

$$= \frac{1}{(b^2-a^2)^{\frac{1}{2}}}\ln\left[\frac{a\tan\left(\frac{z}{2}\right)+b-(b^2-a^2)^{\frac{1}{2}}}{a\tan\left(\frac{z}{2}\right)+b+(b^2-a^2)^{\frac{1}{2}}}\right]$$

$$(b^2>a^2)$$

4.3.132

$$\int \frac{dz}{1\pm\sin z} = \mp\tan\left(\frac{\pi}{4}\mp\frac{z}{2}\right)$$

4.3.133

$$\int \frac{dz}{a+b\cos z} = \frac{2}{(a^2-b^2)^{\frac{1}{2}}}\arctan\frac{(a-b)\tan\frac{z}{2}}{(a^2-b^2)^{\frac{1}{2}}} \quad (a^2>b^2)$$

$$= \frac{1}{(b^2-a^2)^{\frac{1}{2}}}\ln\left[\frac{(b-a)\tan\frac{z}{2}+(b^2-a^2)^{\frac{1}{2}}}{(b-a)\tan\frac{z}{2}-(b^2-a^2)^{\frac{1}{2}}}\right]$$

$$(b^2>a^2)$$

4.3.134

$$\int \frac{dz}{1+\cos z} = \tan\frac{z}{2}$$

4.3.135

$$\int \frac{dz}{1-\cos z} = -\cot\frac{z}{2}$$

4.3.136

$$\int e^{az}\sin bz\,dz = \frac{e^{az}}{a^2+b^2}(a\sin bz-b\cos bz)$$

4.3.137

$$\int e^{az}\cos bz\,dz = \frac{e^{az}}{a^2+b^2}(a\cos bz+b\sin bz)$$

4.3.138

$$\int e^{az}\sin^n bz\,dz = \frac{e^{az}\sin^{n-1}bz}{a^2+n^2b^2}(a\sin bz-nb\cos bz)$$

$$+ \frac{n(n-1)b^2}{a^2+n^2b^2}\int e^{az}\sin^{n-2}bz\,dz$$

4.3.139

$$\int e^{az}\cos^n bz\,dz = \frac{e^{az}\cos^{n-1}bz}{a^2+n^2b^2}(a\cos bz+nb\sin bz)$$

$$+ \frac{n(n-1)b^2}{a^2+n^2b^2}\int e^{az}\cos^{n-2}bz\,dz$$

Definite Integrals

4.3.140

$$\int_0^\pi \sin mt\sin nt\,dt = 0$$

$$(m\neq n, \quad m\text{ and }n\text{ integers})$$

$$\int_0^\pi \cos mt\cos nt\,dt = 0$$

4.3.141

$$\int_0^\pi \sin^2 nt\,dt = \int_0^\pi \cos^2 nt\,dt = \frac{\pi}{2}$$

$$(n\text{ an integer}, \quad n\neq 0)$$

4.3.142

$$\int_0^\infty \frac{\sin mt}{t}\,dt = \frac{\pi}{2} \quad (m>0)$$

$$= 0 \quad (m=0)$$

$$= -\frac{\pi}{2} \quad (m<0)$$

4.3.143

$$\int_0^\infty \frac{\cos at-\cos bt}{t}\,dt = \ln(b/a)$$

4.3.144

$$\int_0^\infty \sin t^2\,dt = \int_0^\infty \cos t^2\,dt = \frac{1}{2}\sqrt{\frac{\pi}{2}}$$

4.3.145

$$\int_0^{\pi/2} \ln\sin t\,dt = \int_0^{\pi/2} \ln\cos t\,dt = -\frac{\pi}{2}\ln 2$$

4.3.146

$$\int_0^\infty \frac{\cos mt}{1+t^2}\,dt = \frac{\pi}{2}e^{-m}$$

(See chapters **5** and **7** for other integrals involving circular functions.)
(See [5.3] for Fourier transforms.)

4.3.147

Formulas for Solution of Plane Right Triangles

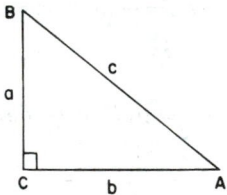

If A, B and C are the vertices (C the right angle), and a, b and c the sides opposite respectively,

$$\sin A = \frac{a}{c} = \frac{1}{\csc A}$$

$$\cos A = \frac{b}{c} = \frac{1}{\sec A}$$

$$\tan A = \frac{a}{b} = \frac{1}{\cot A}$$

$$\text{versine } A = \text{vers } A = 1-\cos A$$

$$\text{coversine } A = \text{covers } A = 1-\sin A$$

$$\text{haversine } A = \text{hav } A = \tfrac{1}{2}\text{ vers } A$$

$$\text{exsecant } A = \text{exsec } A = \sec A-1$$

4.3.148

Formulas for Solution of Plane Triangles

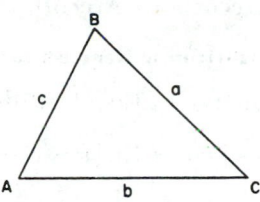

In a triangle with angles A, B and C and sides opposite a, b and c respectively,

$$\frac{a}{\sin A}=\frac{b}{\sin B}=\frac{c}{\sin C}$$

$$\cos A=\frac{c^2+b^2-a^2}{2bc}$$

$$a=b\cos C+c\cos B$$

$$\frac{a+b}{a-b}=\frac{\tan\frac{1}{2}(A+B)}{\tan\frac{1}{2}(A-B)}$$

$$\text{area}=\frac{bc\sin A}{2}=[s(s-a)(s-b)(s-c)]^{\frac{1}{2}}$$

$$s=\frac{1}{2}(a+b+c)$$

4.3.149

Formulas for Solution of Spherical Triangles

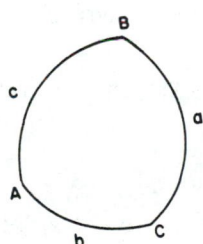

If A, B and C are the three angles and a, b and c the opposite sides,

$$\frac{\sin A}{\sin a}=\frac{\sin B}{\sin b}=\frac{\sin C}{\sin c}$$

$$\cos a=\cos b\cos c+\sin b\sin c\cos A$$
$$=\frac{\cos b\cos(c\pm\theta)}{\cos\theta}$$

where $\tan\theta=\tan b\cos A$

$$\cos A=-\cos B\cos C+\sin B\sin C\cos a$$

4.4. Inverse Circular Functions

Definitions

4.4.1

$$\arcsin z=\int_0^z\frac{dt}{(1-t^2)^{\frac{1}{2}}}\qquad(z=x+iy)$$

4.4.2

$$\arccos z=\int_z^1\frac{dt}{(1-t^2)^{\frac{1}{2}}}=\frac{\pi}{2}-\arcsin z$$

4.4.3

$$\arctan z=\int_0^z\frac{dt}{1+t^2}$$

The path of integration must not cross the real axis in the case of **4.4.1** and **4.4.2** and the imaginary axis in the case of **4.4.3** except possibly inside the unit circle. Each function is single-valued and regular in the z-plane cut along the real axis from $-\infty$ to -1 and $+1$ to $+\infty$ in the case of **4.4.1** and **4.4.2** and along the imaginary axis from i to $i\infty$ and $-i$ to $-i\infty$ in the case of **4.4.3**.

Inverse circular functions are also written $\arcsin z=\sin^{-1}z$, $\arccos z=\cos^{-1}z$, $\arctan z=\tan^{-1}z,\ldots$.

When $-1\le x\le1$, $\arcsin x$ and $\arccos x$ are real and

4.4.4 $\quad-\frac{1}{2}\pi\le\arcsin x\le\frac{1}{2}\pi,\qquad0<\arccos x\le\pi$

4.4.5 $\quad\arctan z+\text{arccot}\,z=\pm\frac{1}{2}\pi\quad\begin{matrix}Rz\ge0\\Rz<0\end{matrix}$

4.4.6 $\quad\text{arccsc}\,z=\arcsin 1/z$

4.4.7 $\quad\text{arcsec}\,z=\arccos 1/z$

4.4.8 $\quad\text{arccot}\,z=\arctan 1/z$

4.4.9 $\quad\text{arcsec}\,z+\text{arccsc}\,z=\frac{1}{2}\pi$

(see **4.3.45**)

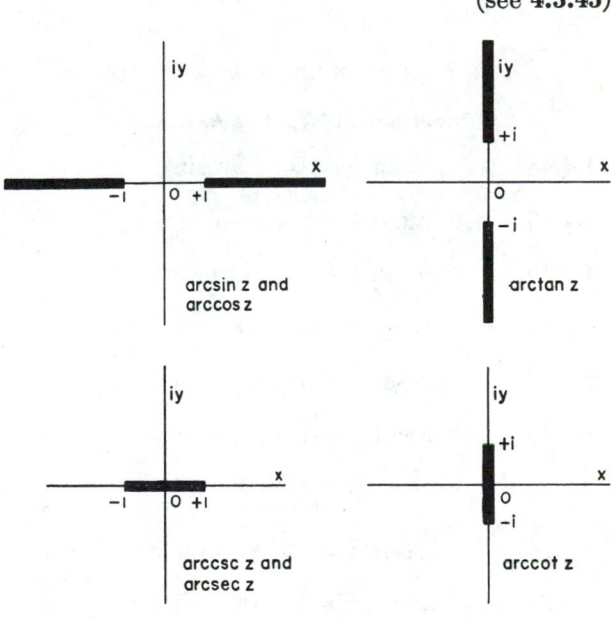

FIGURE 4.4. *Branch cuts for inverse circular functions.*

Fundamental Property

The general solutions of the equations

$$\sin t = z$$

$$\cos t = z$$

$$\tan t = z$$

are respectively

4.4.10 $t = \text{Arcsin } z = (-1)^k \arcsin z + k\pi$

4.4.11 $t = \text{Arccos } z = \pm \arccos z + 2k\pi$

4.4.12
$$t = \text{Arctan } z = \arctan z + k\pi \quad (z^2 \neq -1)$$

where k is an arbitrary integer.

4.4.13   Interval containing principal value

y	x positive or zero	x negative
arcsin x and arctan x	$0 \leq y \leq \pi/2$	$-\pi/2 \leq y < 0$
arccos x and arcsec x	$0 \leq y \leq \pi/2$	$\pi/2 < y \leq \pi$
arccot x and arccsc x	$0 \leq y \leq \pi/2$	$-\pi/2 \leq y < 0$

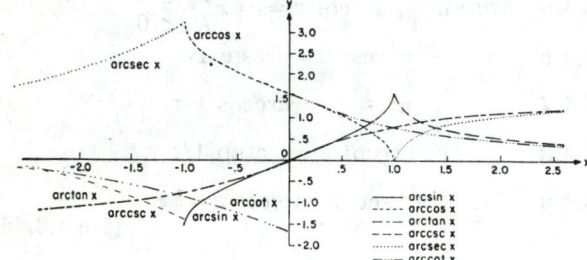

FIGURE 4.5.   *Inverse circular functions.*

Functions of Negative Arguments

4.4.14 $\arcsin(-z) = -\arcsin z$

4.4.15 $\arccos(-z) = \pi - \arccos z$

4.4.16 $\arctan(-z) = -\arctan z$

4.4.17 $\text{arccsc}(-z) = -\text{arccsc } z$

4.4.18 $\text{arcsec}(-z) = \pi - \text{arcsec } z$

4.4.19 $\text{arccot}(-z) = -\text{arccot } z$

Relation to Inverse Hyperbolic Functions (see 4.6.14 to 4.6.19)

4.4.20 $\text{Arcsin } z = -i \text{ Arcsinh } iz$

4.4.21 $\text{Arccos } z = \pm i \text{ Arccosh } z$

4.4.22 $\text{Arctan } z = -i \text{ Arctanh } iz \quad (z^2 \neq -1)$

4.4.23 $\text{Arccsc } z = i \text{ Arccsch } iz$

4.4.24 $\text{Arcsec } z = \pm i \text{ Arcsech } z$

4.4.25 $\text{Arccot } z = i \text{ Arccoth } iz$

Logarithmic Representations

4.4.26 $\text{Arcsin } x = -i \text{ Ln}[(1-x^2)^{\frac{1}{2}} + ix] \quad (x^2 \leq 1)$

4.4.27 $\text{Arccos } x = -i \text{ Ln}[x + i(1-x^2)^{\frac{1}{2}}] \quad (x^2 \leq 1)$

4.4.28 $\text{Arctan } x = \dfrac{i}{2} \text{ Ln} \dfrac{1-i}{1+i}\dfrac{x}{x} = \dfrac{i}{2} \text{ Ln} \dfrac{i+x}{i-x}$

(x real)

4.4.29 $\text{Arccsc } x = -i \text{ Ln}\left[\dfrac{(x^2-1)^{\frac{1}{2}} + i}{x}\right] \quad (x^2 \geq 1)$

4.4.30 $\text{Arcsec } x = -i \text{ Ln}\left[\dfrac{1 + i(x^2-1)^{\frac{1}{2}}}{x}\right] \quad (x^2 \geq 1)$

4.4.31 $\text{Arccot } x = \dfrac{i}{2} \text{ Ln}\left(\dfrac{ix+1}{ix-1}\right) = \dfrac{i}{2} \text{ Ln}\left(\dfrac{x-i}{x+i}\right)$

(x real)

Addition and Subtraction of Two Inverse Circular Functions

4.4.32

$\text{Arcsin } z_1 \pm \text{Arcsin } z_2$
$$= \text{Arcsin }[z_1(1-z_2^2)^{\frac{1}{2}} \pm z_2(1-z_1^2)^{\frac{1}{2}}]$$

4.4.33

$\text{Arccos } z_1 \pm \text{Arccos } z_2$
$$= \text{Arccos }\{z_1 z_2 \mp [(1-z_1^2)(1-z_2^2)]^{\frac{1}{2}}\}$$

4.4.34
$$\text{Arctan } z_1 \pm \text{Arctan } z_2 = \text{Arctan}\left(\frac{z_1 \pm z_2}{1 \mp z_1 z_2}\right)$$

4.4.35

$\text{Arcsin } z_1 \pm \text{Arccos } z_2$
$$= \text{Arcsin}\{z_1 z_2 \pm [(1-z_1^2)(1-z_2^2)]^{\frac{1}{2}}\}$$
$$= \text{Arccos }[z_2(1-z_1^2)^{\frac{1}{2}} \mp z_1(1-z_2^2)^{\frac{1}{2}}]$$

4.4.36

$\text{Arctan } z_1 \pm \text{Arccot } z_2$
$$= \text{Arctan}\left(\frac{z_1 z_2 \pm 1}{z_2 \mp z_1}\right) = \text{Arccot}\left(\frac{z_2 \mp z_1}{z_1 z_2 \pm 1}\right)$$

Inverse Circular Functions in Terms of Real and Imaginary Parts

4.4.37

$\text{Arcsin } z = k\pi + (-1)^k \arcsin \beta$
$$+ (-1)^k i \ln[\alpha + (\alpha^2-1)^{\frac{1}{2}}]$$

4.4.38

$\text{Arccos } z = 2k\pi \pm \{\arccos \beta - i \ln[\alpha + (\alpha^2-1)^{\frac{1}{2}}]\}$

4.4.39

$$\text{Arctan } z = k\pi + \frac{1}{2}\arctan\left(\frac{2x}{1-x^2-y^2}\right)$$

$$+ \frac{i}{4}\ln\left[\frac{x^2+(y+1)^2}{x^2+(y-1)^2}\right]\ (z^2 \neq -1)$$

where k is an integer or zero and

$$\alpha = \frac{1}{2}[(x+1)^2+y^2]^{\frac{1}{2}}+\frac{1}{2}[(x-1)^2+y^2]^{\frac{1}{2}}$$
$$\beta = \frac{1}{2}[(x+1)^2+y^2]^{\frac{1}{2}}-\frac{1}{2}[(x-1)^2+y^2]^{\frac{1}{2}}$$

Series Expansions

4.4.40

$$\arcsin z = z + \frac{z^3}{2\cdot3} + \frac{1\cdot3z^5}{2\cdot4\cdot5} + \frac{1\cdot3\cdot5z^7}{2\cdot4\cdot6\cdot7} + \dots \quad (|z|<1)$$

4.4.41

$$\arcsin(1-z) = \frac{\pi}{2} - (2z)^{\frac{1}{2}}\left[1 + \sum_{k=1}^{\infty}\frac{1\cdot3\cdot5\dots(2k-1)}{2^{2k}(2k+1)k!}z^k\right]$$
$$(|z|<2)$$

4.4.42

$$\arctan z = z - \frac{z^3}{3} + \frac{z^5}{5} - \frac{z^7}{7} + \dots \quad (|z|\leq1 \text{ and } z^2 \neq -1)$$

$$= \frac{\pi}{2} - \frac{1}{z} + \frac{1}{3z^3} - \frac{1}{5z^5} + \dots (|z|>1 \text{ and } z^2 \neq -1)$$

$$= \frac{z}{1+z^2}\left[1 + \frac{2}{3}\frac{z^2}{1+z^2} + \frac{2\cdot4}{3\cdot5}\left(\frac{z^2}{1+z^2}\right)^2 + \dots\right]$$
$$(z^2 \neq -1)$$

Continued Fractions

4.4.43 $\arctan z = \dfrac{z}{1+}\dfrac{z^2}{3+}\dfrac{4z^2}{5+}\dfrac{9z^2}{7+}\dfrac{16z^2}{9+}\dots$

(z in the cut plane of Figure 4.4.)

4.4.44 $\dfrac{\arcsin z}{\sqrt{1-z^2}} = \dfrac{z}{1-}\dfrac{1\cdot2z^2}{3-}\dfrac{1\cdot2z^2}{5-}\dfrac{3\cdot4z^2}{7-}\dfrac{3\cdot4z^2}{9-}\dots$

(z in the cut plane of Figure 4.4.)

Polynomial Approximations [9]

4.4.45 $\qquad 0 \leq x \leq 1$

$$\arcsin x = \frac{\pi}{2} - (1-x)^{\frac{1}{2}}(a_0+a_1x+a_2x^2+a_3x^3) + \epsilon(x)$$

$$|\epsilon(x)| \leq 5\times10^{-5}$$

$a_0=$ 1.57072 88	$a_2=$ .07426 10
$a_1=-.21211 44$	$a_3=-.01872 93$

4.4.46 $\qquad 0 \leq x \leq 1$

$$\arcsin x = \frac{\pi}{2} - (1-x)^{\frac{1}{2}}(a_0+a_1x+a_2x^2+a_3x^3$$
$$+a_4x^4+a_5x^5+a_6x^6+a_7x^7) + \epsilon(x)$$

$$|\epsilon(x)| \leq 2\times10^{-8}$$

$a_0=$ 1.57079 63050	$a_4=$ .03089 18810
$a_1=-.21459 88016$	$a_5=-.01708 81256$
$a_2=$ .08897 89874	$a_6=$ .00667 00901
$a_3=-.05017 43046$	$a_7=-.00126 24911$

4.4.47 $\qquad -1 \leq x \leq 1$

$$\arctan x = a_1x+a_3x^3+a_5x^5+a_7x^7+a_9x^9 + \epsilon(x)$$

$$|\epsilon(x)| \leq 10^{-5}$$

$a_1=$ .99986 60	$a_7=-.08513 30$
$a_3=-.33029 95$	$a_9=$ .02083 51
$a_5=$ .18014 10	

4.4.48 [10] $\qquad -1 \leq x \leq 1$

$$\arctan x = \frac{x}{1+.28x^2} + \epsilon(x)$$

$$|\epsilon(x)| \leq 5\times10^{-3}$$

4.4.49 [11] $\qquad 0 \leq x \leq 1$

$$\frac{\arctan x}{x} = 1 + \sum_{k=1}^{8}a_{2k}x^{2k} + \epsilon(x)$$

$$|\epsilon(x)| \leq 2\times10^{-8}$$

$a_2=-.33333 14528$	$a_{10}=-.07528 96400$
$a_4=$ .19993 55085	$a_{12}=$ .04290 96138
$a_6=-.14208 89944$	$a_{14}=-.01616 57367$
$a_8=$ .10656 26393	$a_{16}=$ .00286 62257

---

[10] The approximation **4.4.48** is from C. Hastings, Jr., Note 143, Math. Tables Aids Comp. **6**, 68 (1953) (with permission).

[11] The approximation **4.4.49** is from B. Carlson, M. Goldstein, Rational approximation of functions, Los Alamos Scientific Laboratory LA-1943, Los Alamos, N. Mex., 1955 (with permission).

---

[9] The approximations **4.4.45** to **4.4.47** are from C. Hastings, Jr., Approximations for digital computers. Princeton Univ. Press, Princeton, N.J., 1955 (with permission).

Approximations in Terms of Chebyshev Polynomials [12]

4.4.50 $\qquad -1 \le x \le 1$

$$T_n^*(x) = \cos n\theta, \qquad \cos\theta = 2x-1 \qquad \text{(see chapter 22)}$$

$$\arctan x = x \sum_{n=0}^{\infty} A_n T_n^*(x^2)$$

n	A_n	n	A_n
0	.88137 3587	6	.00000 3821
1	−.10589 2925	7	−.00000 0570
2	.01113 5843	8	.00000 0086
3	−.00138 1195	9	−.00000 0013
4	.00018 5743	10	.00000 0002
5	−.00002 6215		

For $x > 1$, use $\arctan x = \frac{1}{2}\pi - \arctan (1/x)$

4.4.51 $\qquad -\frac{1}{2}\sqrt{2} \le x \le \frac{1}{2}\sqrt{2}$

$$\arcsin x = x \sum_{n=0}^{\infty} A_n T_n^*(2x^2)$$

$$0 \le x \le \frac{1}{2}\sqrt{2}$$

$$\arccos x = \frac{1}{2}\pi - x \sum_{n=0}^{\infty} A_n T_n^*(2x^2)$$

n	A_n	n	A_n
0	1.05123 1959	5	.00000 5881
1	.05494 6487	6	.00000 0777
2	.00408 0631	7	.00000 0107
3	.00040 7890	8	.00000 0015
4	.00004 6985	9	.00000 0002

For $\frac{1}{2}\sqrt{2} \le x \le 1$, use $\arcsin x = \arccos(1-x^2)^{\frac{1}{2}}$, arccos $x = \arcsin (1-x^2)^{\frac{1}{2}}$.

Differentiation Formulas

4.4.52 $\qquad \dfrac{d}{dz} \arcsin z = (1-z^2)^{-\frac{1}{2}}$

4.4.53 $\qquad \dfrac{d}{dz} \arccos z = -(1-z^2)^{-\frac{1}{2}}$

4.4.54 $\qquad \dfrac{d}{dz} \arctan z = \dfrac{1}{1+z^2}$

4.4.55 $\qquad \dfrac{d}{dz} \operatorname{arccot} z = \dfrac{-1}{1+z^2}$

4.4.56 $\qquad \dfrac{d}{dz} \operatorname{arcsec} z = \dfrac{1}{z(z^2-1)^{\frac{1}{2}}}$

[12] The approximations **4.4.50** to **4.4.51** are from C. W. Clenshaw, Polynomial approximations to elementary functions, Math. Tables Aids Comp. **8**, 143–147 (1954) (with permission).

4.4.57 $\qquad \dfrac{d}{dz} \operatorname{arccsc} z = -\dfrac{1}{z(z^2-1)^{\frac{1}{2}}}$

Integration Formulas

4.4.58 $\quad \displaystyle\int \arcsin z\, dz = z \arcsin z + (1-z^2)^{\frac{1}{2}}$

4.4.59 $\quad \displaystyle\int \arccos z\, dz = z \arccos z - (1-z^2)^{\frac{1}{2}}$

4.4.60 $\quad \displaystyle\int \arctan z\, dz = z \arctan z - \frac{1}{2}\ln (1+z^2)$

4.4.61

$$\int \operatorname{arccsc} z\, dz = z \operatorname{arccsc} z \pm \ln [z+(z^2-1)^{\frac{1}{2}}]$$

$$\begin{bmatrix} 0 < \operatorname{arccsc} z < \dfrac{\pi}{2} \\[2mm] -\dfrac{\pi}{2} < \operatorname{arccsc} z < 0 \end{bmatrix}$$

4.4.62

$$\int \operatorname{arcsec} z\, dz = z \operatorname{arcsec} z \mp \ln [z+(z^2-1)^{\frac{1}{2}}]$$

$$\begin{bmatrix} 0 < \operatorname{arcsec} z < \dfrac{\pi}{2} \\[2mm] \dfrac{\pi}{2} < \operatorname{arcsec} z < \pi \end{bmatrix}$$

4.4.63

$$\int \operatorname{arccot} z\, dz = z \operatorname{arccot} z + \frac{1}{2}\ln (1+z^2)$$

4.4.64

$$\int z \arcsin z\, dz = \left(\frac{z^2}{2} - \frac{1}{4}\right) \arcsin z + \frac{z}{4}(1-z^2)^{\frac{1}{2}}$$

4.4.65

$$\int z^n \arcsin z\, dz = \frac{z^{n+1}}{n+1} \arcsin z - \frac{1}{n+1} \int \frac{z^{n+1}}{(1-z^2)^{\frac{1}{2}}}\, dz$$
$$(n \ne -1)$$

4.4.66

$$\int z \arccos z\, dz = \left(\frac{z^2}{2} - \frac{1}{4}\right) \arccos z - \frac{z}{4}(1-z^2)^{\frac{1}{2}}$$

4.4.67

$$\int z^n \arccos z\, dz = \frac{z^{n+1}}{n+1} \arccos z + \frac{1}{n+1} \int \frac{z^{n+1}}{(1-z^2)^{\frac{1}{2}}}\, dz$$
$$(n \ne -1)$$

4.4.68

$$\int z \arctan z\, dz = \frac{1}{2}(1+z^2) \arctan z - \frac{z}{2}$$

4.4.69

$$\int z^n \arctan z \, dz = \frac{z^{n+1}}{n+1} \arctan z - \frac{1}{n+1} \int \frac{z^{n+1}}{1+z^2} \, dz$$

$$(n \neq -1)$$

4.4.70

$$\int z \, \text{arccot } z \, dz = \frac{1}{2} (1+z^2) \, \text{arccot } z + \frac{z}{2}$$

4.4.71

$$\int z^n \, \text{arccot } z \, dz = \frac{z^{n+1}}{n+1} \, \text{arccot } z + \frac{1}{n+1} \int \frac{z^{n+1}}{1+z^2} \, dz$$

$$(n \neq -1)$$

4.5. Hyperbolic Functions

Definitions

4.5.1 $\qquad \sinh z = \dfrac{e^z - e^{-z}}{2} \qquad (z = x + iy)$

4.5.2 $\qquad \cosh z = \dfrac{e^z + e^{-z}}{2}$

4.5.3 $\qquad \tanh z = \sinh z / \cosh z$

4.5.4 $\qquad \text{csch } z = 1/\sinh z$

4.5.5 $\qquad \text{sech } z = 1/\cosh z$

4.5.6 $\qquad \coth z = 1/\tanh z$

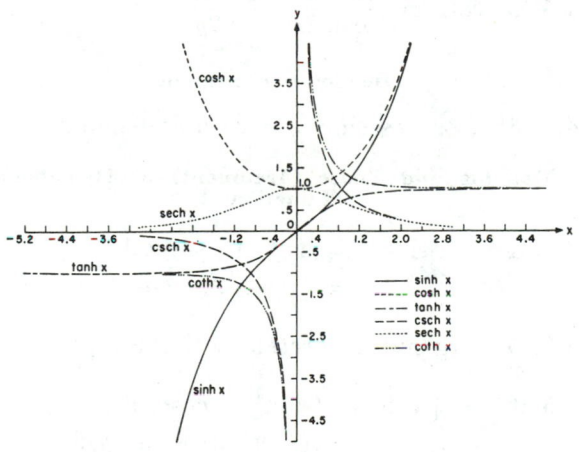

FIGURE 4.6. *Hyperbolic functions.*

Relation to Circular Functions (see **4.3.49** to **4.3.54**)

Hyperbolic formulas can be derived from trigonometric identities by replacing z by iz

4.5.7 $\qquad \sinh z = -i \sin iz$

4.5.8 $\qquad \cosh z = \cos iz$

4.5.9 $\qquad \tanh z = -i \tan iz$

4.5.10 $\qquad \text{csch } z = i \csc iz$

4.5.11 $\qquad \text{sech } z = \sec iz$

4.5.12 $\qquad \coth z = i \cot iz$

Periodic Properties

4.5.13 $\qquad \sinh (z + 2k\pi i) = \sinh z$

$$(k \text{ any integer})$$

4.5.14 $\qquad \cosh (z + 2k\pi i) = \cosh z$

4.5.15 $\qquad \tanh (z + k\pi i) = \tanh z$

Relations Between Hyperbolic Functions

4.5.16 $\qquad \cosh^2 z - \sinh^2 z = 1$

4.5.17 $\qquad \tanh^2 z + \text{sech}^2 z = 1$

4.5.18 $\qquad \coth^2 z - \text{csch}^2 z = 1$

4.5.19 $\qquad \cosh z + \sinh z = e^z$

4.5.20 $\qquad \cosh z - \sinh z = e^{-z}$

Negative Angle Formulas

4.5.21 $\qquad \sinh (-z) = -\sinh z$

4.5.22 $\qquad \cosh (-z) = \cosh z$

4.5.23 $\qquad \tanh (-z) = -\tanh z$

Addition Formulas

4.5.24 $\quad \sinh (z_1 + z_2) = \sinh z_1 \cosh z_2$

$$+ \cosh z_1 \sinh z_2$$

4.5.25 $\quad \cosh (z_1 + z_2) = \cosh z_1 \cosh z_2$

$$+ \sinh z_1 \sinh z_2$$

4.5.26 $\quad \tanh (z_1 + z_2) = (\tanh z_1 + \tanh z_2)/$

$$(1 + \tanh z_1 \tanh z_2)$$

4.5.27 $\quad \coth (z_1 + z_2) = (\coth z_1 \coth z_2 + 1)/$

$$(\coth z_2 + \coth z_1)$$

Half-Angle Formulas

4.5.28

$$\sinh \frac{z}{2} = \left(\frac{\cosh z - 1}{2} \right)^{\frac{1}{2}}$$

4.5.29
$$\cosh \frac{z}{2}=\left(\frac{\cosh z+1}{2}\right)^{\frac{1}{2}}$$

4.5.30
$$\tanh \frac{z}{2}=\left(\frac{\cosh z-1}{\cosh z+1}\right)^{\frac{1}{2}}=\frac{\cosh z-1}{\sinh z}=\frac{\sinh z}{\cosh z+1}$$

Multiple-Angle Formulas

4.5.31 $\quad \sinh 2z=2 \sinh z \cosh z=\dfrac{2 \tanh z}{1-\tanh^2 z}$

4.5.32 $\quad \cosh 2z=2 \cosh^2 z-1=2 \sinh^2 z+1$
$$=\cosh^2 z+\sinh^2 z$$

4.5.33 $\quad \tanh 2z=\dfrac{2 \tanh z}{1+\tanh^2 z}$

4.5.34 $\quad \sinh 3z=3 \sinh z+4 \sinh^3 z$

4.5.35 $\quad \cosh 3z=-3 \cosh z+4 \cosh^3 z$

4.5.36 $\quad \sinh 4z=4 \sinh^3 z \cosh z+4 \cosh^3 z \sinh z$

4.5.37 $\quad \cosh 4z=\cosh^4 z+6 \sinh^2 z \cosh^2 z+\sinh^4 z$

Products of Hyperbolic Sines and Cosines

4.5.38 $\quad 2 \sinh z_1 \sinh z_2=\cosh (z_1+z_2)$
$$-\cosh (z_1-z_2)$$

4.5.39 $\quad 2 \cosh z_1 \cosh z_2=\cosh (z_1+z_2)$
$$+\cosh (z_1-z_2)$$

4.5.40 $\quad 2 \sinh z_1 \cosh z_2=\sinh (z_1+z_2)$
$$+\sinh (z_1-z_2)$$

Addition and Subtraction of Two Hyperbolic Functions

4.5.41
$$\sinh z_1+\sinh z_2=2 \sinh \left(\frac{z_1+z_2}{2}\right) \cosh \left(\frac{z_1-z_2}{2}\right)$$

4.5.42
$$\sinh z_1-\sinh z_2=2 \cosh \left(\frac{z_1+z_2}{2}\right) \sinh \left(\frac{z_1-z_2}{2}\right)$$

4.5.43
$$\cosh z_1+\cosh z_2=2 \cosh \left(\frac{z_1+z_2}{2}\right) \cosh \left(\frac{z_1-z_2}{2}\right)$$

4.5.44
$$\cosh z_1-\cosh z_2=2 \sinh \left(\frac{z_1+z_2}{2}\right) \sinh \left(\frac{z_1-z_2}{2}\right)$$

4.5.45
$$\tanh z_1+\tanh z_2=\frac{\sinh (z_1+z_2)}{\cosh z_1 \cosh z_2}$$

4.5.46
$$\coth z_1+\coth z_2=\frac{\sinh (z_1+z_2)}{\sinh z_1 \sinh z_2}$$

Relations Between Squares of Hyperbolic Sines and Cosines

4.5.47
$$\sinh^2 z_1-\sinh^2 z_2=\sinh (z_1+z_2) \sinh (z_1-z_2)$$
$$=\cosh^2 z_1-\cosh^2 z_2$$

4.5.48
$$\sinh^2 z_1+\cosh^2 z_2=\cosh (z_1+z_2) \cosh (z_1-z_2)$$
$$=\cosh^2 z_1+\sinh^2 z_2$$

Hyperbolic Functions in Terms of Real and Imaginary Parts

$$(z=x+iy)$$

4.5.49 $\quad \sinh z=\sinh x \cos y+i \cosh x \sin y$

4.5.50 $\quad \cosh z=\cosh x \cos y+i \sinh x \sin y$

4.5.51 $\quad \tanh z=\dfrac{\sinh 2x+i \sin 2y}{\cosh 2x+\cos 2y}$

4.5.52 $\quad \coth z=\dfrac{\sinh 2x-i \sin 2y}{\cosh 2x-\cos 2y}$

De Moivre's Theorem

4.5.53 $\quad (\cosh z+\sinh z)^n=\cosh nz+\sinh nz$

Modulus and Phase (Argument) of Hyperbolic Functions

4.5.54 $\quad |\sinh z|=(\sinh^2 x+\sin^2 y)^{\frac{1}{2}}$
$$=[\tfrac{1}{2}(\cosh 2x-\cos 2y)]^{\frac{1}{2}}$$

4.5.55 $\quad \arg \sinh z=\arctan (\coth x \tan y)$

4.5.56 $\quad |\cosh z|=(\sinh^2 x+\cos^2 y)^{\frac{1}{2}}$
$$=[\tfrac{1}{2}(\cosh 2x+\cos 2y)]^{\frac{1}{2}}$$

4.5.57 $\quad \arg \cosh z=\arctan (\tanh x \tan y)$

4.5.58 $\quad |\tanh z|=\left(\dfrac{\cosh 2x-\cos 2y}{\cosh 2x+\cos 2y}\right)^{\frac{1}{2}}$

4.5.59 $\quad \arg \tanh z=\arctan \left(\dfrac{\sin 2y}{\sinh 2x}\right)$

4.5.60 Relations Between Hyperbolic (or Inverse Hyperbolic) Functions

	$\sinh x = a$	$\cosh x = a$	$\tanh x = a$	$\operatorname{csch} x = a$	$\operatorname{sech} x = a$	$\coth x = a$
$\sinh x$	a	$(a^2-1)^{\frac12}$	$a(1-a^2)^{-\frac12}$	a^{-1}	$a^{-1}(1-a^2)^{\frac12}$	$(a^2-1)^{-\frac12}$
$\cosh x$	$(1+a^2)^{\frac12}$	a	$(1-a^2)^{-\frac12}$	$a^{-1}(1+a^2)^{\frac12}$	a^{-1}	$a(a^2-1)^{-\frac12}$
$\tanh x$	$a(1+a^2)^{-\frac12}$	$a^{-1}(a^2-1)^{\frac12}$	a	$(1+a^2)^{-\frac12}$	$(1-a^2)^{\frac12}$	a^{-1}
$\operatorname{csch} x$	a^{-1}	$(a^2-1)^{-\frac12}$	$a^{-1}(1-a^2)^{\frac12}$	a	$a(1-a^2)^{-\frac12}$	$(a^2-1)^{\frac12}$
$\operatorname{sech} x$	$(1+a^2)^{-\frac12}$	a^{-1}	$(1-a^2)^{\frac12}$	$a(1+a^2)^{-\frac12}$	a	$a^{-1}(a^2-1)^{\frac12}$
$\coth x$	$a^{-1}(a^2+1)^{\frac12}$	$a(a^2-1)^{-\frac12}$	a^{-1}	$(1+a^2)^{\frac12}$	$(1-a^2)^{\frac12}$	a

Illustration: If $\sinh x = a$, $\coth x = a^{-1}(a^2+1)^{\frac12}$

$\operatorname{arcsech} a = \operatorname{arccoth} (1-a^2)^{-\frac12}$

4.5.61 Special Values of the Hyperbolic Functions

z	0	$\dfrac{\pi}{2} i$	πi	$\dfrac{3\pi}{2} i$	∞
$\sinh z$	0	i	0	$-i$	∞
$\cosh z$	1	0	-1	0	∞
$\tanh z$	0	∞i	0	$-\infty i$	1
$\operatorname{csch} z$	∞	$-i$	∞	i	0
$\operatorname{sech} z$	1	∞	-1	∞	0
$\coth z$	∞	0	∞	0	1

Series Expansions

4.5.62 $\quad \sinh z = z + \dfrac{z^3}{3!} + \dfrac{z^5}{5!} + \dfrac{z^7}{7!} + \dots \quad (|z| < \infty)$

4.5.63 $\quad \cosh z = 1 + \dfrac{z^2}{2!} + \dfrac{z^4}{4!} + \dfrac{z^6}{6!} + \dots \quad (|z| < \infty)$

4.5.64 $\quad \tanh z = z - \dfrac{z^3}{3} + \dfrac{2}{15} z^5 - \dfrac{17}{315} z^7$

$\qquad + \dots + \dfrac{2^{2n}(2^{2n}-1)B_{2n}}{(2n)!} z^{2n-1} + \dots$

$$\left(|z| < \frac{\pi}{2}\right)$$

4.5.65

$\operatorname{csch} z = \dfrac{1}{z} - \dfrac{z}{6} + \dfrac{7}{360} z^3 - \dfrac{31}{15120} z^5 + \dots$

$\qquad - \dfrac{2(2^{2n-1}-1)B_{2n}}{(2n)!} z^{2n-1} + \dots$

$$(|z| < \pi)$$

4.5.66

$\operatorname{sech} z = 1 - \dfrac{z^2}{2} + \dfrac{5}{24} z^4 - \dfrac{61}{720} z^6 + \dots + \dfrac{E_{2n}}{(2n)!} z^{2n} + \dots$

$$\left(|z| < \frac{\pi}{2}\right)$$

4.5.67

$\coth z = \dfrac{1}{z} + \dfrac{z}{3} - \dfrac{z^3}{45} + \dfrac{2}{945} z^5 - \dots + \dfrac{2^{2n}B_{2n}}{(2n)!} z^{2n-1} + \dots$

$$(|z| < \pi)$$

where B_n and E_n are the nth Bernoulli and Euler numbers, see chapter **23**.

Infinite Products

4.5.68 $\quad \sinh z = z \prod_{k=1}^{\infty} \left(1 + \dfrac{z^2}{k^2\pi^2}\right)$

4.5.69 $\quad \cosh z = \prod_{k=1}^{\infty} \left[1 + \dfrac{4z^2}{(2k-1)^2\pi^2}\right]$

Continued Fraction

4.5.70 $\quad \tanh z = \dfrac{z}{1+} \dfrac{z^2}{3+} \dfrac{z^2}{5+} \dfrac{z^2}{7+} \dots$

$$\left(z \neq \frac{\pi}{2} i \pm n\pi i\right)$$

Differentiation Formulas

4.5.71 $\quad \dfrac{d}{dz} \sinh z = \cosh z$

4.5.72 $\quad \dfrac{d}{dz} \cosh z = \sinh z$

4.5.73 $\quad \dfrac{d}{dz} \tanh z = \operatorname{sech}^2 z$

4.5.74 $\quad \dfrac{d}{dz} \operatorname{csch} z = -\operatorname{csch} z \coth z$

*See page II.

4.5.75 $\quad \dfrac{d}{dz} \operatorname{sech} z = -\operatorname{sech} z \tanh z$

4.5.76 $\quad \dfrac{d}{dz} \coth z = -\operatorname{csch}^2 z$

Integration Formulas

4.5.77 $\quad \displaystyle\int \sinh z \, dz = \cosh z$

4.5.78 $\quad \displaystyle\int \cosh z \, dz = \sinh z$

4.5.79 $\quad \displaystyle\int \tanh z \, dz = \ln \cosh z$

4.5.80 $\quad \displaystyle\int \operatorname{csch} z \, dz = \ln \tanh \dfrac{z}{2}$

4.5.81 $\quad \displaystyle\int \operatorname{sech} z \, dz = \arctan (\sinh z)$

4.5.82 $\quad \displaystyle\int \coth z \, dz = \ln \sinh z$

4.5.83
$$\int z^n \sinh z \, dz = z^n \cosh z - n \int z^{n-1} \cosh z \, dz$$

4.5.84
$$\int z^n \cosh z \, dz = z^n \sinh z - n \int z^{n-1} \sinh z \, dz$$

4.5.85
$$\int \sinh^m z \cosh^n z \, dz = \frac{1}{m+n} \sinh^{m+1} z \cosh^{n-1} z$$
$$+ \frac{n-1}{m+n} \int \sinh^m z \cosh^{n-2} z \, dz$$
$$= \frac{1}{m+n} \sinh^{m-1} z \cosh^{n+1} z$$
$$- \frac{m-1}{m+n} \int \sinh^{m-2} z \cosh^n z \, dz$$
$$(m+n \neq 0)$$

4.5.86 $\displaystyle\int \dfrac{dz}{\sinh^m z \cosh^n z} = \dfrac{-1}{m-1} \dfrac{1}{\sinh^{m-1} z \cosh^{n-1} z}$
$$- \frac{m+n-2}{m-1} \int \frac{dz}{\sinh^{m-2} z \cosh^n z} \quad (m \neq 1)$$
$$= \frac{1}{n-1} \frac{1}{\sinh^{m-1} z \cosh^{n-1} z}$$
$$+ \frac{m+n-2}{n-1} \int \frac{dz}{\sinh^m z \cosh^{n-2} z} \quad (n \neq 1)$$

4.5.87
$$\int \tanh^n z \, dz = -\frac{\tanh^{n-1} z}{n-1} + \int \tanh^{n-2} z \, dz$$
$$(n \neq 1)$$

4.5.88
$$\int \coth^n z \, dz = -\frac{\coth^{n-1} z}{n-1} + \int \coth^{n-2} z \, dz$$
$$(n \neq 1)$$

(See chapters **5** and **7** for other integrals involving hyperbolic functions.)

4.6. Inverse Hyperbolic Functions

Definitions

4.6.1 $\quad \operatorname{arcsinh} z = \displaystyle\int_0^z \dfrac{dt}{(1+t^2)^{\frac{1}{2}}} \quad (z = x+iy)$

4.6.2 $\quad \operatorname{arccosh} z = \displaystyle\int_1^z \dfrac{dt}{(t^2-1)^{\frac{1}{2}}}$

4.6.3 $\quad \operatorname{arctanh} z = \displaystyle\int_0^z \dfrac{dt}{1-t^2}$

The paths of integration must not cross the following cuts.

4.6.1 imaginary axis from $-i\infty$ to $-i$ and i to $i\infty$

4.6.2 real axis from $-\infty$ to $+1$

4.6.3 real axis from $-\infty$ to -1 and $+1$ to $+\infty$

Inverse hyperbolic functions are also written $\sinh^{-1} z$, arsinh z, $\mathscr{A}r$ sinh z, etc.

4.6.4 $\quad \operatorname{arccsch} z = \operatorname{arcsinh} 1/z$

4.6.5 $\quad \operatorname{arcsech} z = \operatorname{arccosh} 1/z$

4.6.6 $\quad \operatorname{arccoth} z = \operatorname{arctanh} 1/z$

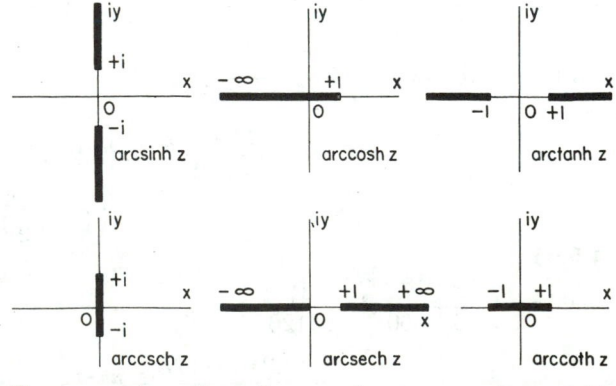

FIGURE 4.7. *Branch cuts for inverse hyperbolic functions.*

4.6.7 $\quad$ arctanh $z =$ arccoth $z \pm \frac{1}{2}\pi i$

(see **4.5.60**) $\qquad$ (according as $\mathscr{I}z \gtrless 0$)

Fundamental Property

The general solutions of the equations

$$z = \sinh t$$
$$z = \cosh t$$
$$z = \tanh t$$

are respectively

4.6.8 $\quad t = $ Arcsinh $z = (-1)^k$ arcsinh $z + k\pi i$

4.6.9 $\quad t = $ Arccosh $z = \pm$ arccosh $z + 2k\pi i$

4.6.10 $t = $ Arctanh $z = $ arctanh $z + k\pi i$

$$(k, \text{ integer})$$

Functions of Negative Arguments

4.6.11 $\quad$ arcsinh $(-z) = -$ arcsinh z

*4.6.12 $\quad$ arccosh $(-z) = \pi i -$ arccosh z

4.6.13 $\quad$ arctanh $(-z) = -$ arctanh z

Relation to Inverse Circular Functions (see **4.4.20** to **4.4.25**)

Hyperbolic identities can be derived from trigonometric identities by replacing z by iz.

4.6.14 $\quad$ Arcsinh $z = -i$ Arcsin iz

4.6.15 $\quad$ Arccosh $z = \pm i$ Arccos z

4.6.16 $\quad$ Arctanh $z = -i$ Arctan iz

4.6.17 $\quad$ Arccsch $z = i$ Arccsc iz

4.6.18 $\quad$ Arcsech $z = \pm i$ Arcsec z

4.6.19 $\quad$ Arccoth $z = i$ Arccot iz

Logarithmic Representations

4.6.20 $\quad$ arcsinh $x = \ln [x + (x^2 + 1)^{\frac{1}{2}}]$

4.6.21 $\quad$ arccosh $x = \ln [x + (x^2 - 1)^{\frac{1}{2}}]$ $\quad (x \geq 1)$

4.6.22 $\quad$ arctanh $x = \frac{1}{2} \ln \dfrac{1+x}{1-x}$ $\quad (0 \leq x^2 < 1)$

4.6.23 $\quad$ arccsch $x = \ln \left[\dfrac{1}{x} + \left(\dfrac{1}{x^2} + 1 \right)^{\frac{1}{2}} \right]$ $\quad (x \neq 0)$

4.6.24 $\quad$ arcsech $x = \ln \left[\dfrac{1}{x} + \left(\dfrac{1}{x^2} - 1 \right)^{\frac{1}{2}} \right]$ $\quad (0 < x \leq 1)$

4.6.25 $\quad$ arccoth $x = \dfrac{1}{2} \ln \dfrac{x+1}{x-1}$ $\qquad (x^2 > 1)$

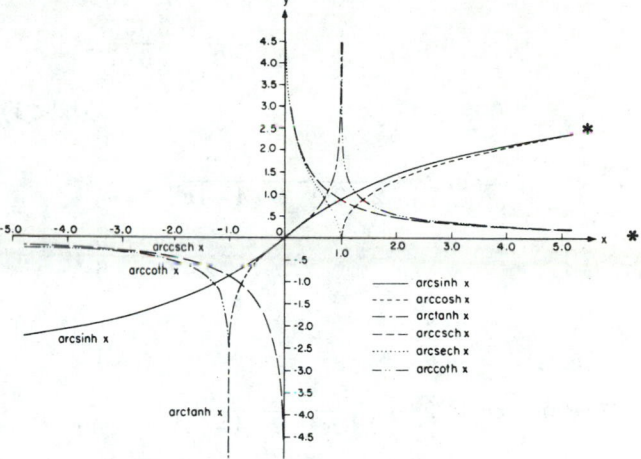

FIGURE 4.8.   *Inverse hyperbolic functions.*

Addition and Subtraction of Two Inverse Hyperbolic Functions

4.6.26

Arcsinh $z_1 \pm$ Arcsinh z_2

$$= \text{Arcsinh } [z_1(1+z_2^2)^{\frac{1}{2}} \pm z_2(1+z_1^2)^{\frac{1}{2}}]$$

4.6.27

Arccosh $z_1 \pm$ Arccosh z_2

$$= \text{Arccosh } \{ z_1 z_2 \pm [(z_1^2 - 1)(z_2^2 - 1)]^{\frac{1}{2}} \}$$

4.6.28

$$\text{Arctanh } z_1 \pm \text{Arctanh } z_2 = \text{Arctanh } \left(\frac{z_1 \pm z_2}{1 \pm z_1 z_2} \right)$$

4.6.29

Arcsinh $z_1 \pm$ Arccosh z_2

$$= \text{Arcsinh} \{ z_1 z_2 \pm [(1+z_1^2)(z_2^2 - 1)]^{\frac{1}{2}} \}$$

$$= \text{Arccosh } [z_2(1+z_1^2)^{\frac{1}{2}} \pm z_1(z_2^2 - 1)^{\frac{1}{2}}]$$

4.6.30

$$\text{Arctanh } z_1 \pm \text{Arccoth } z_2 = \text{Arctanh } \left(\frac{z_1 z_2 \pm 1}{z_2 \pm z_1} \right)$$

$$= \text{Arccoth } \left(\frac{z_2 \pm z_1}{z_1 z_2 \pm 1} \right)$$

Series Expansions

4.6.31
$$\operatorname{arcsinh} z = z - \frac{1}{2 \cdot 3} z^3 + \frac{1 \cdot 3}{2 \cdot 4 \cdot 5} z^5$$
$$- \frac{1 \cdot 3 \cdot 5}{2 \cdot 4 \cdot 6 \cdot 7} z^7 + \cdots$$
$$(|z| < 1)$$
$$= \ln 2z + \frac{1}{2 \cdot 2z^2} - \frac{1 \cdot 3}{2 \cdot 4 \cdot 4z^4}$$
$$+ \frac{1 \cdot 3 \cdot 5}{2 \cdot 4 \cdot 6 \cdot 6z^6} - \cdots$$
$$(|z| > 1)$$

4.6.32
$$\operatorname{arccosh} z = \ln 2z - \frac{1}{2 \cdot 2z^2} - \frac{1 \cdot 3}{2 \cdot 4 \cdot 4z^4}$$
$$- \frac{1 \cdot 3 \cdot 5}{2 \cdot 4 \cdot 6 \cdot 6z^6} - \cdots$$
$$(|z| > 1)$$

4.6.33 $\operatorname{arctanh} z = z + \dfrac{z^3}{3} + \dfrac{z^5}{5} + \dfrac{z^7}{7} + \cdots$ $(|z| < 1)$

4.6.34 $\operatorname{arccoth} z = \dfrac{1}{z} + \dfrac{1}{3z^3} + \dfrac{1}{5z^5} + \dfrac{1}{7z^7} + \cdots$
$$(|z| > 1)$$

Continued Fractions

4.6.35 $\operatorname{arctanh} z = \dfrac{z}{1-} \dfrac{z^2}{3-} \dfrac{4z^2}{5-} \dfrac{9z^2}{7-} \cdots$

(z in the cut plane of Figure 4.7.)

4.6.36
$$\frac{\operatorname{arcsinh} z}{\sqrt{1+z^2}} = \frac{z}{1+} \frac{1 \cdot 2z^2}{3+} \frac{1 \cdot 2z^2}{5+} \frac{3 \cdot 4z^2}{7+} \frac{3 \cdot 4z^2}{9+} \cdots$$

Differentiation Formulas

4.6.37 $\dfrac{d}{dz} \operatorname{arcsinh} z = (1+z^2)^{-\frac{1}{2}}$

4.6.38 $\dfrac{d}{dz} \operatorname{arccosh} z = (z^2-1)^{-\frac{1}{2}}$

4.6.39 $\dfrac{d}{dz} \operatorname{arctanh} z = (1-z^2)^{-1}$

4.6.40 $\dfrac{d}{dz} \operatorname{arccsch} z = \mp \dfrac{1}{z(1+z^2)^{\frac{1}{2}}}$

(according as $\mathscr{R}z \gtrless 0$)

4.6.41 $\dfrac{d}{dz} \operatorname{arcsech} z = \mp \dfrac{1}{z(1-z^2)^{\frac{1}{2}}}$

4.6.42 $\dfrac{d}{dz} \operatorname{arccoth} z = (1-z^2)^{-1}$

Integration Formulas

4.6.43 $\displaystyle\int \operatorname{arcsinh} z \, dz = z \operatorname{arcsinh} z - (1+z^2)^{\frac{1}{2}}$

4.6.44 $\displaystyle\int \operatorname{arccosh} z \, dz = z \operatorname{arccosh} z - (z^2-1)^{\frac{1}{2}}$

4.6.45 $\displaystyle\int \operatorname{arctanh} z \, dz = z \operatorname{arctanh} z + \tfrac{1}{2} \ln (1-z^2)$

4.6.46 $\displaystyle\int \operatorname{arccsch} z \, dz = z \operatorname{arccsch} z \pm \operatorname{arcsinh} z$ *

(according as $\mathscr{R}z \gtrless 0$)

4.6.47 $\displaystyle\int \operatorname{arcsech} z \, dz = z \operatorname{arcsech} z \pm \arcsin z$ *

4.6.48 $\displaystyle\int \operatorname{arccoth} z \, dz = z \operatorname{arccoth} z + \tfrac{1}{2} \ln (z^2-1)$

4.6.49
$$\int z \operatorname{arcsinh} z \, dz = \frac{2z^2+1}{4} \operatorname{arcsinh} z - \frac{z}{4}(z^2+1)^{\frac{1}{2}}$$

4.6.50
$$\int z^n \operatorname{arcsinh} z \, dz = \frac{z^{n+1}}{n+1} \operatorname{arcsinh} z - \frac{1}{n+1} \int \frac{z^{n+1}}{(1+z^2)^{\frac{1}{2}}} dz$$
$$(n \neq -1)$$

4.6.51
$$\int z \operatorname{arccosh} z \, dz = \frac{2z^2-1}{4} \operatorname{arccosh} z - \frac{z}{4}(z^2-1)^{\frac{1}{2}}$$

4.6.52
$$\int z^n \operatorname{arccosh} z \, dz = \frac{z^{n+1}}{n+1} \operatorname{arccosh} z - \frac{1}{n+1} \int \frac{z^{n+1}}{(z^2-1)^{\frac{1}{2}}} dz$$
$$(n \neq -1)$$

4.6.53
$$\int z \operatorname{arctanh} z \, dz = \frac{z^2-1}{2} \operatorname{arctanh} z + \frac{z}{2}$$

4.6.54
$$\int z^n \operatorname{arctanh} z \, dz = \frac{z^{n+1}}{n+1} \operatorname{arctanh} z - \frac{1}{n+1} \int \frac{z^{n+1}}{1-z^2} dz$$
$$(n \neq -1)$$

4.6.55
$$\int z \operatorname{arccsch} z \, dz = \frac{z^2}{2} \operatorname{arccsch} z \pm \frac{1}{2}(1+z^2)^{\frac{1}{2}}$$ *

(according as $\mathscr{R}z \gtrless 0$)

4.6.56
$$\int z^n \operatorname{arccsch} z \, dz = \frac{z^{n+1}}{n+1} \operatorname{arccsch} z \pm \frac{1}{n+1} \int \frac{z^n}{(z^2+1)^{\frac{1}{2}}} dz$$ *
$$(n \neq -1)$$

---

*See page II.

4.6.57

$$\int z \operatorname{arcsech} z \, dz = \frac{z^2}{2} \operatorname{arcsech} z \mp \frac{1}{2}(1-z^2)^{\frac{1}{2}}$$

(according as $\mathscr{R}z \gtrless 0$)

4.6.58

$$\int z^n \operatorname{arcsech} z \, dz = \frac{z^{n+1}}{n+1} \operatorname{arcsech} z \pm \frac{1}{n+1}\int \frac{z^n}{(1-z^2)^{\frac{1}{2}}} dz$$

$$(n \neq -1)$$

4.6.59

$$\int z \operatorname{arccoth} z \, dz = \frac{z^2-1}{2} \operatorname{arccoth} z + \frac{z}{2}$$

4.6.60

$$\int z^n \operatorname{arccoth} z \, dz = \frac{z^{n+1}}{n+1} \operatorname{arccoth} z + \frac{1}{n+1}\int \frac{z^{n+1}}{z^2-1} dz$$

$$(n \neq -1)$$

Numerical Methods

4.7. Use and Extension of the Tables

NOTE: In the examples given it is assumed that the arguments are exact.

Example 1. Computation of Common Logarithms.

To compute common logarithms, the number must be expressed in the form $x \cdot 10^q$, $(1 \leq x < 10, -\infty \leq q \leq \infty)$. The common logarithm of $x \cdot 10^q$ consists of an integral part which is called the characteristic and a decimal part which is called the mantissa. **Table 4.1** gives the common logarithm of x.

x	$x \cdot 10^q$	$\log_{10} x \cdot 10^q$
.009836	$9.836 \cdot 10^{-3}$	$\bar{3}.99281\ 85 = (-2.00718\ 15)$
.09836	$9.836 \cdot 10^{-2}$	$\bar{2}.99281\ 85 = (-1.00718\ 15)$
.9836	$9.836 \cdot 10^{-1}$	$\bar{1}.99281\ 85 = (-0.00718\ 15)$
9.836	$9.836 \cdot 10^{0}$	$0.99281\ 85$
98.36	$9.836 \cdot 10^{1}$	$1.99281\ 85$
983.6	$9.836 \cdot 10^{2}$	$2.99281\ 85$

Interpolation in **Table 4.1** between 983 and 984 gives .99281 85 as the mantissa of 9836.

Note that $\bar{3}.99281\ 85 = -3 + .99281\ 85$. When q is negative the common logarithm can be expressed in the alternative forms

$$\log_{10}(.009836) = \bar{3}.99281\ 85 = 7.99281\ 85 - 10$$
$$= -2.00718\ 15.$$

The last form is convenient for conversion from common logarithms to natural logarithms.

The inverse of $\log_{10} x$ is called the antilogarithm of x, and is written antilog x or $\log^{-1} x$. The logarithm of the reciprocal of a number is called the cologarithm, written colog.

Example 2.

Compute $x^{-3/4}$ for $x = 9.19826$ to 10D using the Table of Common Logarithms.

From **Table 4.1**, four-point Lagrangian interpolation gives $\log_{10}(9.19826) = .96370\ 56812$. Then,

$$-\frac{3}{4}\log_{10}(x) = -.72277\ 92609 = 9.27722\ 07391 - 10.$$

Linear inverse interpolation in **Table 4.1** yields antilog $(\bar{1}.27722) = .18933$. For 10 place accuracy subtabulation with 4-point Lagrangian interpolants produces the table

N	$\log_{10} N$	Δ	Δ^2
.18933	.27721 94350		
		2 29379	
.18934	.27724 23729		-13
		2 29366	
.18935	.27726 53095		

By linear inverse interpolation

$$x^{-3/4} = .18933\ 05685.$$

Example 3.

Convert $\log_{10} x$ to $\ln x$ for $x = .009836$.
Using **4.1.23** and **Table 4.1**, $\ln(.009836) = \ln 10 \log_{10}(.009836) = 2.30258\ 5093\ (-2.00718\ 15) = -4.62170\ 62$.

Example 4.

Compute $\ln x$ for $x = .00278$ to 6D.
Using **4.1.7, 4.1.11** and **Table 4.2**, $\ln(.00278) = \ln(.278 \cdot 10^{-2}) = \ln(.278) - 2 \ln 10 = -5.885304$.
Linear interpolation between $x = .002$ and $x = .003$ would give $\ln(.00278) = -5.898$. To obtain 5 decimal place accuracy with linear interpolation it is necessary that $x > .175$.

Example 5.

Compute $\ln x$ for $x = 1131.718$ to 8D.
Using **4.1.7, 4.1.11** and **Table 4.2**

$$\ln 1131.718 = \ln\left(\frac{1131.718}{1131} 1131\right)$$

$$= \ln \frac{1131.718}{1131} + \ln 1.131 + \ln 10^3$$

$$= \ln(1.00063\ 4836) + \ln 1.131 + 3 \ln 10.$$

*See page II.

Then from **4.1.24**

ln 1131.718$=(.00063\ 4836)-\tfrac{1}{2}(.00063\ 4836)^2$

$+$ln 1.131$+3$ ln 10$=.00063\ 4836-.00000\ 0202$

$+.12310\ 2197+6.90775\ 5279=7.03149\ 211.$

Example 6.

Compute ln x working with 16D for
$$x=1.38967\ 12458\ 179231.$$

Since $\dfrac{x}{1.389}=1.00048\ 32583\ 282384=1+a$, using

4.1.24 and **Table 4.2** we compute successively

$$a=\quad .00048\ 32583\ 282384$$

$$-\frac{a^2}{2}=-.\qquad\qquad 1167\ 693059$$

$$\frac{a^3}{3}=\quad .\qquad\qquad\qquad 376199$$

$$-\frac{a^4}{4}=-.\qquad\qquad\qquad\qquad 136$$

$$\ln(1+a)=\quad .00048\ 31415\ 965388$$

$$\ln 1.389=\quad .32858\ 40637\ 722067$$
$$\ln x=\quad .32906\ 72053\ 687455.$$

Example 7.

Compute the principal value of ln $(\pm2\pm3i)$.
From **4.1.2**, **4.1.3** and **Tables 4.2** and **4.14**.

$$\ln(2+3i)=\frac{1}{2}\ln(2^2+3^2)+i\arctan\frac{3}{2}$$
$$=1.282475+i(.982794)$$

$$\ln(-2+3i)=\frac{1}{2}\ln 13+i\left(\pi-\arctan\frac{3}{2}\right)$$
$$=1.282475+i(2.158799)$$

$$\ln(-2-3i)=\frac{1}{2}\ln 13+i\left(-\pi+\arctan\frac{3}{2}\right)$$
$$=1.282475-i(2.158799)$$

$$\ln(2-3i)=\frac{1}{2}\ln 13+i\left(-\arctan\frac{3}{2}\right)$$
$$=1.282475-i(.982794).$$

Example 8.

Compute $(.227)^{.69}$ to 7D.
Using **4.2.7** and **Tables 4.2** and **4.4**,

$$(.227)^{.69}=e^{.69\ \ln(.227)}=e^{.69(-1.48280\ 5262)}$$
$$=e^{-1.02313\ 5631}=.35946\ 60.$$

Example 9.

Compute $e^{4.99728\ 69}$ to 7S.
Using **4.2.18** and **Table 4.4**,

$$e^{4.99728\ 69}=e^{4.9}e^{.09728\ 69}.$$

Linear interpolation gives $e^{.09728\ 69}=1.10217\ 67$ with an error of 1×10^{-7},

$$e^{4.99728\ 69}=(134.28978)(1.10217\ 67)=148.0111.$$

Example 10.

Compute e^x to 18D for
$$x=.86725\ 13489\ 24685\ 12693.$$

Let $a=x-.867$. Using **4.2.1**, compute successively

$$1.00000\ 00000\ 00000\ 00000$$
$$a=\ .00025\ 13489\ 24685\ 12693$$
$$\frac{a^2}{2!}=\ .\qquad 315\ 88140\ 97019$$
$$\frac{a^3}{3!}=\ .\qquad\quad 2646\ 54842$$
$$\frac{a^4}{4!}=\ .\qquad\qquad\quad 16630$$

$$e^a=1.00025\ 13805\ 15472\ 81184$$
$$e^{.867}=2.37976\ 08513\ 29496\ 863\ \text{from Table 4.4}$$
$$e^a e^{.867}=e^x=2.38035\ 90768\ 39006\ 089.$$

Example 11.

Compute e^{648} to 7S.

Let $n=\dfrac{x}{\ln 10}$ and $d=$ the decimal part of $\dfrac{x}{\ln 10}$. Then

$$\exp x=\exp\left(\frac{x}{\ln 10}\ln 10\right)=\exp[(n+d)\ln 10]$$
$$=\exp(\ln 10^n)\exp(d\ln 10)$$
$$=10^n\exp(d\ln 10)$$

From **Table 4.4**

$$e^{648}=\exp\left(\frac{648}{\ln 10}\ln 10\right)=\exp(281.42282\ 42\ln 10)$$
$$=10^{281}\exp(.42282\ 42\ln 10)=10^{281}\exp .97358\ 87$$
$$=10^{281}(2.647428)=(281)2.647428.$$

Example 12.

Compute e^{-x} for $x=.75$ using the expansion in Chebyshev polynomials.

Following the procedure in [4.3] we have from **4.2.48**

$$e^{-x} = \sum_{k=0}^{7} A_k T_k^*(x)$$

where $T_k^*(x)$ are the Chebyshev polynomials defined in chapter **22.** Assuming $b_8 = b_9 = 0$ we generate b_k, $k = 7, 6, 5, \ldots 0$ from the recurrence relation

$$b_k = (4x-2)b_{k+1} - b_{k+2} + A_k$$

k	b_k
7	$-.00000\ 0015$
6	$.00000\ 0400$
5	$-.00000\ 9560$
4	$.00018\ 9959$
3	$-.00300\ 9164$
2	$.03550\ 4993$
1	$-.27432\ 7449$
0	$.33520\ 2828$

since $f(x) = b_0 - (2x-1)b_1$,

$$e^{-.75} = .33520\ 2828 - (.5)(-.27432\ 7449)$$
$$= .47236\ 6553.$$

Example 13.

Express $38°42'32''$ in radians to 6D.

$$1° = .01745\ 32925\ 19943\ 29577\ r$$
$$1' = .00029\ 08882\ 08665\ 72159\ 62\ r$$
$$1'' = .00000\ 48481\ 36811\ 09535\ 9936\ r$$

Therefore

$$38° = .66322\ 51\ r$$
$$42' = .01221\ 73\ r$$
$$32'' = .00015\ 51\ r$$
$$38°42'32'' = .675598\ r.$$

Example 14.

Express $x = 1.6789$ radians in degrees, minutes and seconds to the nearest tenth of a second.

From **Table 1.1** giving the mathematical constants we have

$$1\ r = \frac{180°}{\pi} = 57.29577\ 95130° \ldots$$
$$1.6789\ r = 96.19388°$$
$$.19388° \times 60 = 11.633'$$
$$.633' \times 60 = 38.0''$$
$$1.6789\ r = 96°11'38.0''.$$

Example 15.

Compute $\sin x$ and $\cos x$ for $x = 2.317$ to 7D. From **4.3.44** and **Table 4.6**

$$\sin (2.317) = \sin (\pi - 2.317) = \sin (.82459\ 2654)$$
$$= .73427\ 12$$

$$\cos (2.317) = \cos (\pi - 2.317) = -\cos (.82459\ 2654)$$
$$= -.67885\ 60.$$

Linear interpolation for $x = .82459\ 2654$ gives an error of 9×10^{-8}.

Example 16.

Compute $\sin x$ for $x = 12.867$ to 8D. From **4.3.16** and **Tables 4.6** and **4.8**

$$\sin (12.867) = \sin 12 \cos .867 + \cos 12 \sin .867$$
$$= .29612\ 142.$$

The method of reduction to an angle in the first quadrant which was given in **Example 15** may also be used.

Example 17.

Compute $\sin x$ to 19D for

$$x = .86725\ 13489\ 24685\ 12693.$$

Let $\alpha = .867$, $\beta = x - \alpha$. From **4.3.16** and **Table 4.6**

$$\sin (\alpha + \beta) = \sin \alpha \cos \beta + \cos \alpha \sin \beta$$
$$\sin \alpha = .76239\ 10208\ 07866\ 22598$$
$$\cos \alpha = .64711\ 66288\ 94312\ 75010$$

With the series expansions for $\sin \beta$ and $\cos \beta$ we compute successively

	$1.00000\ \ 00000\ \ 00000\ \ 00000$
$-\dfrac{\beta^2}{2!} = -\ \cdot$	$315\ \ 88140\ \ 97019$
$\dfrac{\beta^4}{4!} = \ \cdot$	16630
$\cos \beta =$	$.99999\ \ 99684\ \ 11859\ \ 19611$
$\beta =$	$.00025\ \ 13489\ \ 24685\ \ 12693$
$-\dfrac{\beta^3}{3!} = -\ \cdot$	$2646\ \ 54842$
$\dfrac{\beta^5}{5!} = \ \cdot$	1
$\sin \beta =$	$.00025\ \ 13489\ \ 22038\ \ 57852$
$\sin \alpha \cos \beta =$	$.76239\ \ 09967\ \ 25351\ \ 31308$
$\cos \alpha \sin \beta =$	$.00016\ \ 26520\ \ 67105\ \ 82436$
$\sin x =$	$.76255\ \ 36487\ \ 92457\ \ 1374$

This procedure is equivalent to interpolation with Taylor's formula **3.6.4.**

Example 18.

In the plane triangle ABC, $a=123$, $B=29°16'$, $c=321$; find A, b.

$$b^2=a^2+c^2-2ac \cos B=(123)^2+(321)^2$$
$$-2(123)(321) \cos 29°16'$$

$$b=221.99934\ 00$$

$$\sin A=\frac{a \sin B}{b}=\frac{(123)(.48887\ 50196)}{221.99934\ 00}=.27086\ 39918$$

$$A=15°42'56.469''.$$

Example 19.

In the plane triangle ABC, $a=4$, $b=7$, $c=9$, find A, B, and C.

$$\cos A=\frac{c^2+b^2-a^2}{2bc}=\frac{81+49-16}{2\cdot 7\cdot 9}=\frac{114}{126}=.90476\ 1905$$

$$A=.43997\ 5954=25°12'31.6''$$

$$\sin A=.42591\ 7709$$

$$\sin B=\frac{7(.42591\ 7709)}{4}, \quad B=.84106\ 8670$$
$$=48°11'22.9''$$

$$\sin C=\frac{9(.42591\ 7709)}{4}, \quad C=1.86054\ 803$$
$$=106°36'5.6''$$

where the supplementary angle must be chosen for C. As a check we get $A+B+C=180°00'.1''$.

Example 20.

Compute $\cot x$ for $x=.4589$ to 6D.
Since $x<.5$, using **Table 4.9** with interpolation in $(x^{-1}-\cot x)$, we find $\frac{1}{.4589}-\cot(.4589)=$ .155159. Therefore $\cot (.4589)=2.179124-$ .155159$=2.023965$.

Example 21.

Compute $\arcsin x$ for $x=.99511$.
For $x>.95$, using **Table 4.14** with interpolation in the auxiliary function $f(x)$ we find

$$\arcsin x=\frac{\pi}{2}-[2(1-x)]^{\frac{1}{2}}f(x)$$

$$\arcsin (.99511)=\frac{\pi}{2}-[2(.00489)]^{\frac{1}{2}}f(.99511)$$

$$=1.57079\ 6327-(.09889\ 388252)$$
$$(1.00040\ 7951)$$
$$=1.47186\ 2100.$$

Example 22.

Compute $\arctan 20$ and $\operatorname{arccot} 20$ to 9D.
Using **4.4.5**, **4.4.8**, and **Table 4.14**   *

$$\arctan 20=\frac{\pi}{2}-\arctan 1/20=1.52083\ 7931$$

$$\operatorname{arccot} 20=\frac{\pi}{2}-\arctan 20=\arctan .05=.04995\ 8396.$$

Example 23.

Express $z=3+9i$ in polar form.

$z=x+iy=re^{i\theta}$, where $r=(x^2+y^2)^{\frac{1}{2}}$,

$\theta=\arctan \frac{y}{x}+2\pi k$, k is   an   integer.   For $k=0$,
$r=(3^2+9^2)^{\frac{1}{2}}=\sqrt{90}=9.486833$

$$\theta=\arctan 9/3=\arctan 3=1.24904\ 58.$$

Thus $3+9i=9.486833 \exp (1.24904\ 58i)$.

Example 24.

Compute $\arctan x$ for $x=1/3$ to 12D.
From **4.4.34** and **4.4.42** we have

$\arctan x=\arctan (x_0+h)$

$$=\arctan x_0+\arctan \frac{h}{1+x_0h+x_0^2}$$

$$=\arctan x_0+\left(\frac{h}{1+x_0h+x_0^2}\right)-\frac{1}{3}\left(\frac{h}{1+x_0h+x_0^2}\right)^3+\cdots.$$

We have
$x=\frac{1}{3}=.33333\ 33333\ 33$ so that $h=.00033\ 33333\ 33$ and, from **Table 4.14**, $\arctan x_0=\arctan .333$ $=.32145 \quad 05244 \quad 03.$ Since $\frac{h}{1+x_0h+x_0^2}=$ .00030 00300 03 we get

$$\arctan x=.32145\ 05244\ 03+.00030\ 00300\ 03$$
$$-.00000\ 00000\ 09$$
$$=.32175\ 05543\ 97.$$

If x is given in the form b/a it is convenient to use **4.4.34** in the form

$$\arctan \frac{b}{a}=\arctan x_0+\arctan \frac{b-ax_0}{a+bx_0}.$$

In the present example we get

$$\arctan \frac{1}{3}=\arctan .333+\arctan \frac{1}{3333}.$$

Example 25.

Compute arcsec 2.8 to 5D.
Using **4.3.45** and **Table 4.14**

$$\text{arcsec } z = \arcsin \frac{(z^2-1)^{\frac{1}{2}}}{z}$$

$$\text{arcsec } 2.8 = \arcsin \frac{[(2.8)^2-1]^{\frac{1}{2}}}{2.8}$$

$$= \arcsin .93404\ 97735$$

$$= 1.20559$$

or using **4.3.45** and **Table 4.14**

$$\text{arcsec } z = \arctan\ (z^2-1)^{\frac{1}{2}}$$

$$\text{arcsec } 2.8 = \arctan 2.61533\ 9366$$

$$= \frac{\pi}{2} - \arctan .38235\ 95564,$$

from **4.4.3** and **4.4.8**

$$= 1.570796 - .365207$$

$$= 1.20559.$$

Example 26.

Compute arctanh x for $x = .96035$ to 6D.
From **4.6.22** and **Table 4.2**

$$\text{arctanh } .96035 = \tfrac{1}{2} \ln \frac{1+.96035}{1-.96035} = \tfrac{1}{2} \ln \frac{1.96035}{.03965}$$

$$= \tfrac{1}{2} \ln 49.44136\ 191$$

$$= \tfrac{1}{2}(3.90078\ 7359) = 1.950394.$$

Example 27.

Compute arccosh x for $x = 1.5368$ to 6D.
Using **Table 4.17**

$$\frac{\text{arccosh } x}{(x^2-1)^{\frac{1}{2}}} = \frac{\text{arccosh } 1.5368}{[(1.5368)^2-1]^{\frac{1}{2}}} = .852346$$

$$\text{arccosh } 1.5368 = (.852346)(1.361754)^{\frac{1}{2}}$$

$$= (.852346)(1.166942)$$

$$= .994638.$$

Example 28.

Compute arccosh x for $x = 31.2$ to 5D.
Using **Tables 4.2** and **4.17** with $1/x = 1/31.2 = .03205\ 128205$

$$\text{arccosh } 31.2 - \ln 31.2 = .692886$$

$$\text{arccosh } 31.2 = .692886 + 3.440418 = 4.13330.$$

References

Texts

[4.1] B. Carlson, M. Goldstein, Rational approximation of functions, Los Alamos Scientific Laboratory LA–1943 (Los Alamos, N. Mex., 1955).

[4.2] C. W. Clenshaw, Polynomial approximations to elementary functions, Math. Tables Aids Comp. **8**, 143–147 (1954).

[4.3] C. W. Clenshaw, A note on the summation of Chebyshev series, Math. Tables Aids Comp. **9**, 118–120 (1955).

[4.4] G. H. Hardy, A course of pure mathematics, 9th ed. (Cambridge Univ. Press, Cambridge, England, and The Macmillan Co., New York, N.Y., 1947).

[4.5] C. Hastings, Jr., Approximations for digital computers (Princeton Univ. Press, Princeton, N.J., 1955).

[4.6] C. Hastings, Jr., Note 143, Math. Tables Aids Comp. **6**, 68 (1953).

[4.7] E. W. Hobson, A treatise on plane trigonometry, 4th ed. (Cambridge Univ. Press, Cambridge, England, 1918).

[4.8] H. S. Wall, Analytic theory of continued fractions (D. Van Nostrand Co., Inc., New York, N.Y., 1948).

Tables

[4.9] E. P. Adams, Smithsonian mathematical formulae and tables of elliptic functions, 3d reprint (The Smithsonian Institution, Washington, D.C., 1957).

[4.10] H. Andoyer, Nouvelles tables trigonométriques fondamentales (Hermann et fils, Paris, France, 1916).

[4.11] British Association for the Advancement of Science, Mathematical Tables, vol. I. Circular and hyperbolic functions, exponential, sine and cosine integrals, factorial function and allied functions, Hermitian probability functions, 3d ed. (Cambridge Univ. Press, Cambridge, England, 1951).

[4.12] Chemical Rubber Company, Standard mathematical tables, 12th ed. (Chemical Rubber Publ. Co., Cleveland, Ohio, 1959).

[4.13] L. J. Comrie, Chambers' six-figure mathematical tables, vol. 2 (W. R. Chambers, Ltd., London, England, 1949).

[4.14] H. B. Dwight, Tables of integrals and other mathematical data, 3d rev. ed. (The Macmillan Co., New York, N.Y., 1957).

[4.15] W. Gröbner and N. Hofreiter, Integraltafel, unbestimmte und bestimmte Integrale (Springer-Verlag, Wien und Innsbruck, Austria, 1949-1950).

[4.16] Harvard Computation Laboratory, Tables of the function arcsin z (Harvard Univ. Press, Cambridge, Mass., 1956). $z = x + iy$, $0 \leq x \leq 475$, $0 \leq y \leq 475$, 6D, varying intervals.

[4.17] Harvard Computation Laboratory, Tables of inverse hyperbolic functions (Harvard Univ. Press, Cambridge, Mass., 1949). arctanh x, $0 \leq x < 1$; arcsinh x, $0 \leq x < 3.5$; arccosh x, $1 \leq x < 3.5$; arcsinh x, arccosh x, $3.5 \leq x \leq 22980$, 9D, varying intervals.

[4.18] National Bureau of Standards, Tables of 10^x, Applied Math. Series 27 (U.S. Government Printing Office, Washington, D.C., 1953). $x = 0(.00001)1$, 10 D. Radix table of $10^{n \cdot 10^{-p}}$, $n = 1(1)999$, $p = 3(3)15$, 15D.

[4.19] National Bureau of Standards, Table of natural logarithms for arguments between zero and five to sixteen decimal places, 2d ed., Applied Math. Series 31 (U.S. Government Printing office, Washington, D.C., 1953). $x=0(.0001)5$, 16 D.

[4.20] National Bureau of Standards, Tables of the exponential function e^x, 3d ed., Applied Math. Series 14 (U.S. Government Printing Office, Washington, D.C., 1951). $x=-2.4999(.0001).9999$, 18D, $x=1(.0001)2.4999$, 15D, $x=2.5(.001)4.999$, 15D, $x=5(.01)9.99$, 12D, $x=-.000099(.000001).000099$, 18D, $x=-100(1)100$, 19S, $x=-9\times10^{-n}(10^{-n})9\times10^{-n}$, $n=10, 9, 8, 7$, 18D; values of e and $1/e$, 2556D.

[4.21] National Bureau of Standards, Table of the descending exponential, $x=2.5$ to $x=10$, Applied Math. Series 46 (U.S. Government Printing Office, Washington, D.C., 1955). $x=2.5(.001)10$, 20D.

[4.22] National Bureau of Standards, Tables of sines and cosines for radian arguments, 2d ed., Applied Math. Series 43 (U.S. Government Printing Office, Washington, D.C., 1955). $\sin x$, $\cos x$, $x=0(.001)25.2$, $0(1)100$, 8D, $x=10^{-n}(10^{-n})9\times10^{-n}$, $n=5, 4, 3, 2, 1$, 15D, $x=0(.00001).01$, 12D.

[4.23] National Bureau of Standards, Tables of circular and hyperbolic sines and cosines for radian arguments, 2d ed., Applied Math. Series 36 (U.S. Government Printing Office, Washington, D.C., 1953). $\sin x$, $\cos x$, $\sinh x$, $\cosh x$, $x=0(.0001)1.9999$, $0(.1)10$, 9D.

[4.24] National Bureau of Standards, Table of circular and hyperbolic tangents and cotangents for radian arguments, 2d printing (Columbia Univ. Press, New York, N.Y., 1947). $\tan x$, $\cot x$, $\tanh x$, $\coth x$, $x=0(.0001)2$, 8D or 8S, $x=0(.1)10$, 10D.

[4.25] National Bureau of Standards, Table of sines and cosines to fifteen decimal places at hundredths of a degree, Applied Math. Series 5 (U.S. Government Printing Office, Washington, D.C., 1949). $\sin x$, $\cos x$, $x=0°(.01°)90°$, 15D; supplementary table of $\sin x$, $\cos x$, $x=1°(1°)89°$, 30 D.

[4.26] National Bureau of Standards, Table of secants and cosecants to nine significant figures at hundredths of a degree, Applied Math. Series 40 (U.S. Government Printing Office, Washington, D.C., 1954).

[4.27] National Bureau of Standards, Tables of functions and of zeros of functions, Collected short tables of the Computation Laboratory, Applied Math. Series 37 (U.S. Government Printing Office, Washington, D.C., 1954).

[4.28] National Bureau of Standards, Table of arcsin x (Columbia Univ. Press, New York, N.Y., 1945). arcsin x, $x=0(.0001).989(.00001)1$, 12D; auxiliary table of $f(v)=[\frac{1}{2}\pi-\arcsin(1-v)]/(2v)^{\frac{1}{2}}$, $v=0(.00001).0005$, 13D.

[4.29] National Bureau of Standards, Tables of arctan x, 2d ed., Applied Math. Series 26 (U.S. Government Printing Office, Washington D.C., 1953). $x=0(.001)7(.01)50(.1)300(1)2000(10)10000$, 12D.

[4.30] National Bureau of Standards, Table of hyperbolic sines and cosines, $x=2$ to $x=10$, Applied Math. Series 45 (U.S. Government Printing Office, Washington, D.C., 1955). $x=2(.001)10$, 9S.

[4.31] B. O. Peirce, A short table of integrals, 4th ed. (Ginn and Co., Boston, Mass., 1956).

[4.32] J. Peters, Ten-place logarithm table, vols. 1, 2 (together with an appendix of mathematical tables), (Berlin, 1922; rev. ed., Frederick Ungar Publ. Co., New York, N.Y., 1957).

[4.33] J. Peters, Seven-place values of trigonometric functions for every thousandth of a degree (Berlin-Friedenau, 1918; D. Van Nostrand Co., Inc., New York, N.Y., 1942).

[4.34] L. W. Pollak, Rechentafeln zur harmonischen Analyse (Johann Ambrosius Barth, Leipzig, Germany, 1926).

[4.35] A. J. Thompson, Standard table of logarithms to twenty decimal places, Tracts for Computers, No. 22 (Cambridge Univ. Press, Cambridge, England, and New York, N.Y., 1952).

[4.36] J. Todd, Table of arctangents of rational numbers, NBS Applied Math. Series 11 (U.S. Government Printing Office, Washington, D.C., 1951). arctan m/n and arccot m/n, $0<m<n\leq100$, 12D; reductions of arctan m/n, $0<m<n\leq100$; reductions of arctan n for reducible $n\leq2089$.

[4.37] U.S. Department of Commerce, Coast and Geodetic Survey, Natural sines and cosines to eight decimal places, Special Publication No. 231 (U.S. Government Printing Office, Washington, D.C., 1942).

[4.38] C. E. Van Ostrand, Tables of the exponential function and of the circular sine and cosine to radian arguments, Memoirs of the National Academy of Sciences 14, 5th Memoir (U.S. Government Printing Office, Washington, D.C., 1921).

[4.39] B. V. Vega, Logarithmic tables of numbers and trigonometrical functions (G. E. Stechert & Co., New York, N.Y., 1905); $\log_{10} x$, $x=1(1)100000$; logarithms of the trigonometrical functions for every ten seconds.

COMMON LOGARITHMS

Table 4.1

x	$\log_{10} x$	x	$\log_{10} x$	x	$\log_{10} x$	x	$\log_{10} x$	x	$\log_{10} x$
100	00000 00000	150	17609 12591	200	30102 99957	250	39794 00087	300	47712 12547
101	00432 13738	151	17897 69473	201	30319 60574	251	39967 37215	301	47856 64956
102	00860 01718	152	18184 35879	202	30535 13694	252	40140 05408	302	48000 69430
103	01283 72247	153	18469 14308	203	30749 60379	253	40312 05212	303	48144 26285
104	01703 33393	154	18752 07208	204	30963 01674	254	40483 37166	304	48287 35836
105	02118 92991	155	19033 16982	205	31175 38611	255	40654 01804	305	48429 98393
106	02530 58653	156	19312 45984	206	31386 72204	256	40823 99653	306	48572 14265
107	02938 37777	157	19589 96524	207	31597 03455	257	40993 31233	307	48713 83755
108	03342 37555	158	19865 70870	208	31806 33350	258	41161 97060	308	48855 07165
109	03742 64979	159	20139 71243	209	32014 62861	259	41329 97641	309	48995 84794
110	04139 26852	160	20411 99827	210	32221 92947	260	41497 33480	310	49136 16938
111	04532 29788	161	20682 58760	211	32428 24553	261	41664 05073	311	49276 03890
112	04921 80227	162	20951 50145	212	32633 58609	262	41830 12913	312	49415 45940
113	05307 84435	163	21218 76044	213	32837 96034	263	41995 57485	313	49554 43375
114	05690 48513	164	21484 38480	214	33041 37733	264	42160 39269	314	49692 96481
115	06069 78404	165	21748 39442	215	33243 84599	265	42324 58739	315	49831 05538
116	06445 79892	166	22010 80880	216	33445 37512	266	42488 16366	316	49968 70826
117	06818 58617	167	22271 64711	217	33645 97338	267	42651 12614	317	50105 92622
118	07188 20073	168	22530 92817	218	33845 64936	268	42813 47940	318	50242 71200
119	07554 69614	169	22788 67046	219	34044 41148	269	42975 22800	319	50379 06831
120	07918 12460	170	23044 89214	220	34242 26808	270	43136 37642	320	50514 99783
121	08278 53703	171	23299 61104	221	34439 22737	271	43296 92909	321	50650 50324
122	08635 98307	172	23552 84469	222	34635 29745	272	43456 89040	322	50785 58717
123	08990 51114	173	23804 61031	223	34830 48630	273	43616 26470	323	50920 25223
124	09342 16852	174	24054 92483	224	35024 80183	274	43775 05628	324	51054 50102
125	09691 00130	175	24303 80487	225	35218 25181	275	43933 26938	325	51188 33610
126	10037 05451	176	24551 26678	226	35410 84391	276	44090 90821	326	51321 76001
127	10380 37210	177	24797 32664	227	35602 58572	277	44247 97691	327	51454 77527
128	10720 99696	178	25042 00023	228	35793 48470	278	44404 47959	328	51587 38437
129	11058 97103	179	25285 30310	229	35983 54823	279	44560 42033	329	51719 58979
130	11394 33523	180	25527 25051	230	36172 78360	280	44715 80313	330	51851 39399
131	11727 12957	181	25767 85749	231	36361 19799	281	44870 63199	331	51982 79938
132	12057 39312	182	26007 13880	232	36548 79849	282	45024 91083	332	52113 80837
133	12385 16410	183	26245 10897	233	36735 59210	283	45178 64355	333	52244 42335
134	12710 47984	184	26481 78230	234	36921 58574	284	45331 83400	334	52374 64668
135	13033 37685	185	26717 17284	235	37106 78623	285	45484 48600	335	52504 48070
136	13353 89084	186	26951 29442	236	37291 20030	286	45636 60331	336	52633 92774
137	13672 05672	187	27184 16065	237	37474 83460	287	45788 18967	337	52762 99009
138	13987 90864	188	27415 78493	238	37657 69571	288	45939 24878	338	52891 67003
139	14301 48003	189	27646 18042	239	37839 79009	289	46089 78428	339	53019 96982
140	14612 80357	190	27875 36010	240	38021 12417	290	46239 79979	340	53147 89170
141	14921 91127	191	28103 33672	241	38201 70426	291	46389 29890	341	53275 43790
142	15228 83444	192	28330 12287	242	38381 53660	292	46538 28514	342	53402 61061
143	15533 60375	193	28555 73090	243	38560 62736	293	46686 76204	343	53529 41200
144	15836 24921	194	28780 17299	244	38738 98263	294	46834 73304	344	53655 84426
145	16136 80022	195	29003 46114	245	38916 60844	295	46982 20160	345	53781 90951
146	16435 28558	196	29225 60714	246	39093 51071	296	47129 17111	346	53907 60988
147	16731 73347	197	29446 62262	247	39269 69533	297	47275 64493	347	54032 94748
148	17026 17154	198	29666 51903	248	39445 16808	298	47421 62641	348	54157 92439
149	17318 62684	199	29885 30764	249	39619 93471	299	47567 11883	349	54282 54270
150	17609 12591	200	30102 99957	250	39794 00087	300	47712 12547	350	54406 80444

$$\begin{bmatrix} (-6)6 \\ 5 \end{bmatrix} \qquad \begin{bmatrix} (-6)2 \\ 5 \end{bmatrix} \qquad \begin{bmatrix} (-6)1 \\ 4 \end{bmatrix} \qquad \begin{bmatrix} (-7)9 \\ 4 \end{bmatrix} \qquad \begin{bmatrix} (-7)6 \\ 4 \end{bmatrix}$$

For use of common logarithms see **Examples 1–3.** For $100 < x < 135$ interpolate in the range $1000 < x < 1350$. Compiled from A. J. Thompson, Standard table of logarithms to twenty decimal places, Tracts for Computers, No. 22. Cambridge Univ. Press, Cambridge, England, 1952 (with permission).

Table 4.1   **COMMON LOGARITHMS**

x	$\log_{10} x$	x	$\log_{10} x$	x	$\log_{10} x$	x	$\log_{10} x$	x	$\log_{10} x$
350	54406 80444	400	60205 99913	450	65321 25138	500	69897 00043	550	74036 26895
351	54530 71165	401	60314 43726	451	65417 65419	501	69983 77259	551	74115 15989
352	54654 26635	402	60422 60531	452	65513 84348	502	70070 37171	552	74193 90777
353	54777 47054	403	60530 50461	453	65609 82020	503	70156 79851	553	74272 51313
354	54900 32620	404	60638 13651	454	65705 58529	504	70243 05364	554	74350 97647
355	55022 83531	405	60745 50232	455	65801 13967	505	70329 13781	555	74429 29831
356	55144 99980	406	60852 60336	456	65896 48427	506	70415 05168	556	74507 47916
357	55266 82161	407	60959 44092	457	65991 62001	507	70500 79593	557	74585 51952
358	55388 30266	408	61066 01631	458	66086 54780	508	70586 37123	558	74663 41989
359	55509 44486	409	61172 33080	459	66181 26855	509	70671 77823	559	74741 18079
360	55630 25008	410	61278 38567	460	66275 78317	510	70757 01761	560	74818 80270
361	55750 72019	411	61384 18219	461	66370 09254	511	70842 09001	561	74896 28613
362	55870 85705	412	61489 72160	462	66464 19756	512	70926 99610	562	74973 63156
363	55990 66250	413	61595 00517	463	66558 09910	513	71011 73651	563	75050 83949
364	56110 13836	414	61700 03411	464	66651 79806	514	71096 31190	564	75127 91040
365	56229 28645	415	61804 80967	465	66745 29529	515	71180 72290	565	75204 84478
366	56348 10854	416	61909 33306	466	66838 59167	516	71264 97016	566	75281 64312
367	56466 60643	417	62013 60550	467	66931 68806	517	71349 05431	567	75358 30589
368	56584 78187	418	62117 62818	468	67024 58531	518	71432 97597	568	75434 83357
369	56702 63662	419	62221 40230	469	67117 28427	519	71516 73578	569	75511 22664
370	56820 17241	420	62324 92904	470	67209 78579	520	71600 33436	570	75587 48557
371	56937 39096	421	62428 20958	471	67302 09071	521	71683 77233	571	75663 61082
372	57054 29399	422	62531 24510	472	67394 19986	522	71767 05030	572	75739 60288
373	57170 88318	423	62634 03674	473	67486 11407	523	71850 16889	573	75815 46220
374	57287 16022	424	62736 58566	474	67577 83417	524	71933 12870	574	75891 18924
375	57403 12677	425	62838 89301	475	67669 36096	525	72015 93034	575	75966 78447
376	57518 78449	426	62940 95991	476	67760 69527	526	72098 57442	576	76042 24834
377	57634 13502	427	63042 78750	477	67851 83790	527	72181 06152	577	76117 58132
378	57749 17998	428	63144 37690	478	67942 78966	528	72263 39225	578	76192 78384
379	57863 92100	429	63245 72922	479	68033 55134	529	72345 56720	579	76267 85637
380	57978 35966	430	63346 84556	480	68124 12374	530	72427 58696	580	76342 79936
381	58092 49757	431	63447 72702	481	68214 50764	531	72509 45211	581	76417 61324
382	58206 33629	432	63548 37468	482	68304 70382	532	72591 16323	582	76492 29846
383	58319 87740	433	63648 78964	483	68394 71308	533	72672 72090	583	76566 85548
384	58433 12244	434	63748 97295	484	68484 53616	534	72754 12570	584	76641 28471
385	58546 07295	435	63848 92570	485	68574 17386	535	72835 37820	585	76715 58661
386	58658 73047	436	63948 64893	486	68663 62693	536	72916 47897	586	76789 76160
387	58771 09650	437	64048 14370	487	68752 89612	537	72997 42857	587	76863 81012
388	58883 17256	438	64147 41105	488	68841 98220	538	73078 22757	588	76937 73261
389	58994 96013	439	64246 45202	489	68930 88591	539	73158 87652	589	77011 52948
390	59106 46070	440	64345 26765	490	69019 60800	540	73239 37598	590	77085 20116
391	59217 67574	441	64443 85895	491	69108 14921	541	73319 72651	591	77158 74809
392	59328 60670	442	64542 22693	492	69196 51028	542	73399 92865	592	77232 17067
393	59439 25504	443	64640 37262	493	69284 69193	543	73479 98296	593	77305 46934
394	59549 62218	444	64738 29701	494	69372 69489	544	73559 88997	594	77378 64450
395	59659 70956	445	64836 00110	495	69460 51989	545	73639 65023	595	77451 69657
396	59769 51859	446	64933 48587	496	69548 16765	546	73719 26427	596	77524 62597
397	59879 05068	447	65030 75231	497	69635 63887	547	73798 73263	597	77597 43311
398	59988 30721	448	65127 80140	498	69722 93428	548	73878 05585	598	77670 11840
399	60097 28957	449	65224 63410	499	69810 05456	549	73957 23445	599	77742 68224
400	60205 99913	450	65321 25138	500	69897 00043	550	74036 26895	600	77815 12504
	$\begin{bmatrix} (-7)4 \\ 4 \end{bmatrix}$		$\begin{bmatrix} (-7)3 \\ 4 \end{bmatrix}$		$\begin{bmatrix} (-7)3 \\ 4 \end{bmatrix}$		$\begin{bmatrix} (-7)2 \\ 4 \end{bmatrix}$		$\begin{bmatrix} (-7)2 \\ 4 \end{bmatrix}$

COMMON LOGARITHMS

Table 4.1

x	$\log_{10} x$	x	$\log_{10} x$	x	$\log_{10} x$	x	$\log_{10} x$	x	$\log_{10} x$
600	77815 12504	650	81291 33566	700	84509 80400	750	87506 12634	800	90308 99870
601	77887 44720	651	81358 09886	701	84571 80180	751	87563 99370	801	90363 25161
602	77959 64913	652	81424 75957	702	84633 71121	752	87621 78406	802	90417 43683
603	78031 73121	653	81491 31813	703	84695 53250	753	87679 49762	803	90471 55453
604	78103 69386	654	81557 77483	704	84757 26591	754	87737 13459	804	90525 60487
605	78175 53747	655	81624 13000	705	84818 91170	755	87794 69516	805	90579 58804
606	78247 26242	656	81690 38394	706	84880 47011	756	87852 17955	806	90633 50418
607	78318 86911	657	81756 53696	707	84941 94138	757	87909 58795	807	90687 35347
608	78390 35793	658	81822 58936	708	85003 32577	758	87966 92056	808	90741 13608
609	78461 72926	659	81888 54146	709	85064 62352	759	88024 17759	809	90794 85216
610	78532 98350	660	81954 39355	710	85125 83487	760	88081 35923	810	90848 50189
611	78604 12102	661	82020 14595	711	85186 96007	761	88138 46568	811	90902 08542
612	78675 14221	662	82085 79894	712	85247 99936	762	88195 49713	812	90955 60292
613	78746 04745	663	82151 35284	713	85308 95299	763	88252 45380	813	91009 05456
614	78816 83711	664	82216 80794	714	85369 82118	764	88309 33586	814	91062 44049
615	78887 51158	665	82282 16453	715	85430 60418	765	88366 14352	815	91115 76087
616	78958 07122	666	82347 42292	716	85491 30223	766	88422 87696	816	91169 01588
617	79028 51640	667	82412 58339	717	85551 91557	767	88479 53639	817	91222 20565
618	79098 84751	668	82477 64625	718	85612 44442	768	88536 12200	818	91275 33037
619	79169 06490	669	82542 61178	719	85672 88904	769	88592 63398	819	91328 39018
620	79239 16895	670	82607 48027	720	85733 24964	770	88649 07252	820	91381 38524
621	79309 16002	671	82672 25202	721	85793 52647	771	88705 43781	821	91434 31571
622	79379 03847	672	82736 92731	722	85853 71976	772	88761 73003	822	91487 18175
623	79448 80467	673	82801 50642	723	85913 82973	773	88817 94939	823	91539 98352
624	79518 45897	674	82865 98965	724	85973 85662	774	88874 09607	824	91592 72117
625	79588 00173	675	82930 37728	725	86033 80066	775	88930 17025	825	91645 39485
626	79657 43332	676	82994 66959	726	86093 66207	776	88986 17213	826	91698 00473
627	79726 75408	677	83058 86687	727	86153 44109	777	89042 10188	827	91750 55096
628	79795 96437	678	83122 96939	728	86213 13793	778	89097 95970	828	91803 03368
629	79865 06454	679	83186 97743	729	86272 75283	779	89153 74577	829	91855 45306
630	79934 05495	680	83250 89127	730	86332 28601	780	89209 46027	830	91907 80924
631	80002 93592	681	83314 71119	731	86391 73770	781	89265 10339	831	91960 10238
632	80071 70783	682	83378 43747	732	86451 10811	782	89320 67531	832	92012 33263
633	80140 37100	683	83442 07037	733	86510 39746	783	89376 17621	833	92064 50014
634	80208 92579	684	83505 61017	734	86569 60599	784	89431 60627	834	92116 60506
635	80277 37253	685	83569 05715	735	86628 73391	785	89486 96567	835	92168 64755
636	80345 71156	686	83632 41157	736	86687 78143	786	89542 25460	836	92220 62774
637	80413 94323	687	83695 67371	737	86746 74879	787	89597 47324	837	92272 54580
638	80482 06787	688	83758 84382	738	86805 63618	788	89652 62175	838	92324 40186
639	80550 08582	689	83821 92219	739	86864 44384	789	89707 70032	839	92376 19608
640	80617 99740	690	83884 90907	740	86923 17197	790	89762 70913	840	92427 92861
641	80685 80295	691	83947 80474	741	86981 82080	791	89817 64835	841	92479 59958
642	80753 50281	692	84010 60945	742	87040 39053	792	89872 51816	842	92531 20915
643	80821 09729	693	84073 32346	743	87098 88138	793	89927 31873	843	92582 75746
644	80888 58674	694	84135 94705	744	87157 29355	794	89982 05024	844	92634 24466
645	80955 97146	695	84198 48046	745	87215 62727	795	90036 71287	845	92685 67089
646	81023 25180	696	84260 92396	746	87273 88275	796	90091 30677	846	92737 03630
647	81090 42807	697	84323 27781	747	87332 06018	797	90145 83214	847	92788 34103
648	81157 50059	698	84385 54226	748	87390 15979	798	90200 28914	848	92839 58523
649	81224 46968	699	84447 71757	749	87448 18177	799	90254 67793	849	92890 76902
650	81291 33566	700	84509 80400	750	87506 12634	800	90308 99870	850	92941 89257
	$\begin{bmatrix} (-7)2 \\ 4 \end{bmatrix}$		$\begin{bmatrix} (-7)1 \\ 4 \end{bmatrix}$		$\begin{bmatrix} (-7)1 \\ 4 \end{bmatrix}$		$\begin{bmatrix} (-7)1 \\ 4 \end{bmatrix}$		$\begin{bmatrix} (-8)8 \\ 4 \end{bmatrix}$

Table 4.1 **COMMON LOGARITHMS**

x	$\log_{10} x$	x	$\log_{10} x$	x	$\log_{10} x$	x	$\log_{10} x$	x	$\log_{10} x$
850	92941 89257	900	95424 25094	950	97772 36053	1000	00000 00000	1050	02118 92991
851	92992 95601	901	95472 47910	951	97818 05169	1001	00043 40775	1051	02160 27160
852	93043 95948	902	95520 65375	952	97863 69484	1002	00086 77215	1052	02201 57398
853	93094 90312	903	95568 77503	953	97909 29006	1003	00130 09330	1053	02242 83712
854	93145 78707	904	95616 84305	954	97954 83747	1004	00173 37128	1054	02284 06109
855	93196 61147	905	95664 85792	955	98000 33716	1005	00216 60618	1055	02325 24596
856	93247 37647	906	95712 81977	956	98045 78923	1006	00259 79807	1056	02366 39182
857	93298 08219	907	95760 72871	957	98091 19378	1007	00302 94706	1057	02407 49873
858	93348 72878	908	95808 58485	958	98136 55091	1008	00346 05321	1058	02448 56677
859	93399 31638	909	95856 38832	959	98181 86072	1009	00389 11662	1059	02489 59601
860	93449 84512	910	95904 13923	960	98227 12330	1010	00432 13738	1060	02530 58653
861	93500 31515	911	95951 83770	961	98272 33877	1011	00475 11556	1061	02571 53839
862	93550 72658	912	95999 48383	962	98317 50720	1012	00518 05125	1062	02612 45167
863	93601 07957	913	96047 07775	963	98362 62871	1013	00560 94454	1063	02653 32645
364	93651 37425	914	96094 61957	964	98407 70339	1014	00603 79550	1064	02694 16280
865	93701 61075	915	96142 10941	965	98452 73133	1015	00646 60422	1065	02734 96078
866	93751 78920	916	96189 54737	966	98497 71264	1016	00689 37079	1066	02775 72047
867	93801 90975	917	96236 93357	967	98542 64741	1017	00732 09529	1067	02816 44194
868	93851 97252	918	96284 26812	968	98587 53573	1018	00774 77780	1068	02857 12527
869	93901 97764	919	96331 55114	969	98632 37771	1019	00817 41840	1069	02897 77052
870	93951 92526	920	96378 78273	970	98677 17343	1020	00860 01718	1070	02938 37777
871	94001 81550	921	96425 96302	971	98721 92299	1021	00902 57421	1071	02978 94708
872	94051 64849	922	96473 09211	972	98766 62649	1022	00945 08958	1072	03019 47854
873	94101 42437	923	96520 17010	973	98811 28403	1023	00987 56337	1073	03059 97220
874	94151 14326	924	96567 19712	974	98855 89569	1024	01029 99566	1074	03100 42814
875	94200 80530	925	96614 17327	975	98900 46157	1025	01072 38654	1075	03140 84643
876	94250 41062	926	96661 09867	976	98944 98177	1026	01114 73608	1076	03181 22713
877	94299 95934	927	96707 97341	977	98989 45637	1027	01157 04436	1077	03221 57033
878	94349 45159	928	96754 79762	978	99033 88548	1028	01199 31147	1078	03261 87609
879	94398 88751	929	96801 57140	979	99078 26918	1029	01241 53748	1079	03302 14447
880	94448 26722	930	96848 29486	980	99122 60757	1030	01283 72247	1080	03342 37555
881	94497 59084	931	96894 96810	981	99166 90074	1031	01325 86653	1081	03382 56940
882	94546 85851	932	96941 59124	982	99211 14878	1032	01367 96973	1082	03422 72608
883	94596 07036	933	96988 16437	983	99255 35178	1033	01410 03215	1083	03462 84566
884	94645 22650	934	97034 68762	984	99299 50984	1034	01452 05388	1084	03502 92822
885	94694 32707	935	97081 16109	985	99343 62305	1035	01494 03498	1085	03542 97382
886	94743 37219	936	97127 58487	986	99387 69149	1036	01535 97554	1086	03582 98253
887	94792 36198	937	97173 95909	987	99431 71527	1037	01577 87564	1087	03622 95441
888	94841 29658	938	97220 28384	988	99475 69446	1038	01619 73535	1088	03662 88954
889	94890 17610	939	97266 55923	989	99519 62916	1039	01661 55476	1089	03702 78798
890	94939 00066	940	97312 78536	990	99563 51946	1040	01703 33393	1090	03742 64979
891	94987 77040	941	97358 96234	991	99607 36545	1041	01745 07295	1091	03782 47506
892	95036 48544	942	97405 09028	992	99651 16722	1042	01786 77190	1092	03822 26384
893	95085 14589	943	97451 16927	993	99694 92485	1043	01828 43084	1093	03862 01619
894	95133 75188	944	97497 19943	994	99738 63844	1044	01870 04987	1094	03901 73220
895	95182 30353	945	97543 18085	995	99782 30807	1045	01911 62904	1095	03941 41192
896	95230 80097	946	97589 11364	996	99825 93384	1046	01953 16845	1096	03981 05541
897	95279 24430	947	97634 99790	997	99869 51583	1047	01994 66817	1097	04020 66276
898	95327 63367	948	97680 83373	998	99913 05413	1048	02036 12826	1098	04060 23401
899	95375 96917	949	97726 62124	999	99956 54882	1049	02077 54882	1099	04099 76924
900	95424 25094	950	97772 36053	1000	00000 00000	1050	02118 92991	1100	04139 26852
	$\begin{bmatrix} (-8)8 \\ 4 \end{bmatrix}$		$\begin{bmatrix} (-8)7 \\ 4 \end{bmatrix}$		$\begin{bmatrix} (-8)6 \\ 3 \end{bmatrix}$		$\begin{bmatrix} (-8)5 \\ 3 \end{bmatrix}$		$\begin{bmatrix} (-8)5 \\ 3 \end{bmatrix}$

COMMON LOGARITHMS

Table 4.1

x	$\log_{10} x$	x	$\log_{10} x$	x	$\log_{10} x$	x	$\log_{10} x$	x	$\log_{10} x$
1100	04139 26852	1150	06069 78404	1200	07918 12460	1250	09691 00130	1300	11394 33523
1101	04178 73190	1151	06107 53236	1201	07954 30074	1251	09725 73097	1301	11427 72966
1102	04218 15945	1152	06145 24791	1202	07990 44677	1252	09760 43289	1302	11461 09842
1103	04257 55124	1153	06182 93073	1203	08026 56273	1253	09795 10710	1303	11494 44157
1104	04296 90734	1154	06220 58088	1204	08062 64869	1254	09829 75365	1304	11527 75914
1105	04336 22780	1155	06258 19842	1205	08098 70469	1255	09864 37258	1305	11561 05117
1106	04375 51270	1156	06295 78341	1206	08134 73078	1256	09898 96394	1306	11594 31769
1107	04414 76209	1157	06333 33590	1207	08170 72701	1257	09933 52777	1307	11627 55876
1108	04453 97604	1158	06370 85594	1208	08206 69343	1258	09968 06411	1308	11660 77440
1109	04493 15461	1159	06408 34360	1209	08242 63009	1259	10002 57301	1309	11693 96466
1110	04532 29788	1160	06445 79892	1210	08278 53703	1260	10037 05451	1310	11727 12957
1111	04571 40589	1161	06483 22197	1211	08314 41431	1261	10071 50866	1311	11760 26917
1112	04610 47872	1162	06520 61281	1212	08350 26198	1262	10105 93549	1312	11793 38350
1113	04649 51643	1163	06557 97147	1213	08386 08009	1263	10140 33506	1313	11826 47261
1114	04688 51908	1164	06595 29803	1214	08421 86867	1264	10174 70739	1314	11859 53652
1115	04727 48674	1165	06632 59254	1215	08457 62779	1265	10209 05255	1315	11892 57528
1116	04766 41946	1166	06669 85504	1216	08493 35749	1266	10243 37057	1316	11925 58893
1117	04805 31731	1167	06707 08560	1217	08529 05782	1267	10277 66149	1317	11958 57750
1118	04844 18036	1168	06744 28428	1218	08564 72883	1268	10311 92535	1318	11991 54103
1119	04883 00865	1169	06781 45112	1219	08600 37056	1269	10346 16221	1319	12024 47955
1120	04921 80227	1170	06818 58617	1220	08635 98307	1270	10380 37210	1320	12057 39312
1121	04960 56126	1171	06855 68951	1221	08671 56639	1271	10414 55506	1321	12090 28176
1122	04999 28569	1172	06892 76117	1222	08707 12059	1272	10448 71113	1322	12123 14551
1123	05037 97563	1173	06929 80121	1223	08742 64570	1273	10482 84037	1323	12155 98442
1124	05076 63112	1174	06966 80969	1224	08778 14178	1274	10516 94280	1324	12188 79851
1125	05115 25224	1175	07003 78666	1225	08813 60887	1275	10551 01848	1325	12221 58783
1126	05153 83905	1176	07040 73217	1226	08849 04702	1276	10585 06744	1326	12254 35241
1127	05192 39160	1177	07077 64628	1227	08884 45627	1277	10619 08973	1327	12287 09229
1128	05230 90996	1178	07114 52905	1228	08919 83668	1278	10653 08538	1328	12319 80750
1129	05269 39419	1179	07151 38051	1229	08955 18829	1279	10687 05445	1329	12352 49809
1130	05307 84435	1180	07188 20073	1230	08990 51114	1280	10720 99696	1330	12385 16410
1131	05346 26049	1181	07224 98976	1231	09025 80529	1281	10754 91297	1331	12417 80555
1132	05384 64269	1182	07261 74765	1232	09061 07078	1282	10788 80252	1332	12450 42248
1133	05422 99099	1183	07298 47446	1233	09096 30766	1283	10822 66564	1333	12483 01494
1134	05461 30546	1184	07335 17024	1234	09131 51597	1284	10856 50237	1334	12515 58296
1135	05499 58615	1185	07371 83503	1235	09166 69576	1285	10890 31277	1335	12548 12657
1136	05537 83314	1186	07408 46890	1236	09201 84708	1286	10924 09686	1336	12580 64581
1137	05576 04647	1187	07445 07190	1237	09236 96996	1287	10957 85469	1337	12613 14073
1138	05614 22621	1188	07481 64406	1238	09272 06447	1288	10991 58630	1338	12645 61134
1139	05652 37241	1189	07518 18546	1239	09307 13064	1289	11025 29174	1339	12678 05770
1140	05690 48513	1190	07554 69614	1240	09342 16852	1290	11058 97103	1340	12710 47984
1141	05728 56444	1191	07591 17615	1241	09377 17815	1291	11092 62423	1341	12742 87779
1142	05766 61039	1192	07627 62554	1242	09412 15958	1292	11126 25137	1342	12775 25158
1143	05804 62304	1193	07664 04437	1243	09447 11286	1293	11159 85249	1343	12807 60127
1144	05842 60245	1194	07700 43268	1244	09482 03804	1294	11193 42763	1344	12839 92687
1145	05880 54867	1195	07736 79053	1245	09516 93514	1295	11226 97684	1345	12872 22843
1146	05918 46176	1196	07773 11797	1246	09551 80423	1296	11260 50015	1346	12904 50599
1147	05956 34179	1197	07809 41504	1247	09586 64535	1297	11293 99761	1347	12936 75957
1148	05994 18881	1198	07845 68181	1248	09621 45853	1298	11327 46925	1348	12968 98922
1149	06032 00287	1199	07881 91831	1249	09656 24384	1299	11360 91511	1349	13001 19497
1150	06069 78404	1200	07918 12460	1250	09691 00130	1300	11394 33523	1350	13033 37685
	$\begin{bmatrix} (-8)5 \\ 3 \end{bmatrix}$		$\begin{bmatrix} (-8)4 \\ 3 \end{bmatrix}$		$\begin{bmatrix} (-8)4 \\ 3 \end{bmatrix}$		$\begin{bmatrix} (-8)3 \\ 3 \end{bmatrix}$		$\begin{bmatrix} (-8)3 \\ 3 \end{bmatrix}$

Table 4.2   **NATURAL LOGARITHMS**

x	$\ln x$	x	$\ln x$	x	$\ln x$
0.000	$-\infty$	0.050	-2.99573 22735 539910	0.100	-2.30258 50929 940457
0.001	-6.90775 52789 821371	0.051	-2.97592 96462 578113	0.101	-2.29263 47621 408776
0.002	-6.21460 80984 221917	0.052	-2.95651 15604 007097	0.102	-2.28278 24656 978660
0.003	-5.80914 29903 140274	0.053	-2.93746 33654 300152	0.103	-2.27302 62907 525013
0.004	-5.52146 09178 622464	0.054	-2.91877 12324 178627	0.104	-2.26336 43798 407644
0.005	-5.29831 73665 480367	0.055	-2.90042 20937 496661	0.105	-2.25379 49288 246137
0.006	-5.11599 58097 540821	0.056	-2.88240 35882 469878	0.106	-2.24431 61848 700699
0.007	-4.96184 51299 268237	0.057	-2.86470 40111 475869	0.107	-2.23492 64445 202309
0.008	-4.82831 37373 023011	0.058	-2.84731 22684 357177	0.108	-2.22562 40518 579174
0.009	-4.71053 07016 459177	0.059	-2.83021 78350 764176	0.109	-2.21640 73967 529934
0.010	-4.60517 01859 880914	0.060	-2.81341 07167 600364	0.110	-2.20727 49131 897208
0.011	-4.50986 00061 837665	0.061	-2.79688 14148 088258	0.111	-2.19822 50776 698029
0.012	-4.42284 86291 941367	0.062	-2.78062 08939 370455	0.112	-2.18925 64076 870425
0.013	-4.34280 59215 206003	0.063	-2.76462 05525 906044	0.113	-2.18036 74602 697965
0.014	-4.26869 79493 668784	0.064	-2.74887 21956 224652	0.114	-2.17155 68305 876416
0.015	-4.19970 50778 799270	0.065	-2.73336 80090 864999	0.115	-2.16282 31506 188870
0.016	-4.13516 65567 423558	0.066	-2.71810 05369 557115	0.116	-2.15416 50878 757724
0.017	-4.07454 19349 259210	0.067	-2.70306 26595 911710	0.117	-2.14558 13441 843809
0.018	-4.01738 35210 859724	0.068	-2.68824 75738 060304	0.118	-2.13707 06545 164723
0.019	-3.96331 62998 156966	0.069	-2.67364 87743 848777	0.119	-2.12863 17858 706077
0.020	-3.91202 30054 281461	0.070	-2.65926 00369 327781	0.120	-2.12026 35362 000911
0.021	-3.86323 28412 587141	0.071	-2.64507 54019 408216	0.121	-2.11196 47333 853960
0.022	-3.81671 28256 238212	0.072	-2.63108 91599 660817	0.122	-2.10373 42342 488805
0.023	-3.77226 10630 529874	0.073	-2.61729 58378 337459	0.123	-2.09557 09236 097196
0.024	-3.72970 14486 341914	0.074	-2.60369 01857 779673	0.124	-2.08747 37133 771002
0.025	-3.68887 94541 139363	0.075	-2.59026 71654 458266	0.125	-2.07944 15416 798359
0.026	-3.64965 87409 606550	0.076	-2.57702 19386 958060	0.126	-2.07147 33720 306591
0.027	-3.61191 84129 778080	0.077	-2.56394 98571 284532	0.127	-2.06356 81925 235458
0.028	-3.57555 07688 069331	0.078	-2.55104 64522 925453	0.128	-2.05572 50150 625199
0.029	-3.54045 94489 956630	0.079	-2.53830 74265 151156	0.129	-2.04794 28746 204649
0.030	-3.50655 78973 199817	0.080	-2.52572 86443 082554	0.130	-2.04022 08285 265546
0.031	-3.47376 80744 969908	0.081	-2.51330 61243 096983	0.131	-2.03255 79557 809855
0.032	-3.44201 93761 824105	0.082	-2.50103 60317 178839	0.132	-2.02495 33563 957662
0.033	-3.41124 77175 156568	0.083	-2.48891 46711 855391	0.133	-2.01740 61507 603833
0.034	-3.38139 47543 659757	0.084	-2.47693 84801 388234	0.134	-2.00991 54790 312257
0.035	-3.35240 72174 927234	0.085	-2.46510 40224 918206	0.135	-2.00248 05005 437076
0.036	-3.32423 63405 260271	0.086	-2.45340 79827 286293	0.136	-1.99510 03932 460850
0.037	-3.29683 73663 379126	0.087	-2.44184 71603 275533	0.137	-1.98777 43531 540121
0.038	-3.27016 91192 557513	0.088	-2.43041 84645 039306	0.138	-1.98050 15938 249324
0.039	-3.24419 36328 524906	0.089	-2.41911 89092 499972	0.139	-1.97328 13458 514453
0.040	-3.21887 58248 682007	0.090	-2.40794 56086 518720	0.140	-1.96611 28563 728328
0.041	-3.19418 32122 778292	0.091	-2.39689 57724 652870	0.141	-1.95899 53886 039688
0.042	-3.17008 56606 987687	0.092	-2.38596 67019 330967	0.142	-1.95192 82213 808763
0.043	-3.14655 51632 885746	0.093	-2.37515 57858 288811	0.143	-1.94491 06487 222298
0.044	-3.12356 56450 638759	0.094	-2.36446 04967 121332	0.144	-1.93794 19794 061364
0.045	-3.10109 27892 118173	0.095	-2.35387 83873 815962	0.145	-1.93102 15365 615627
0.046	-3.07911 38824 930421	0.096	-2.34340 70875 143008	0.146	-1.92414 86572 738006
0.047	-3.05760 76772 720785	0.097	-2.33304 43004 787542	0.147	-1.91732 26922 034008
0.048	-3.03655 42680 742461	0.098	-2.32278 78003 115651	0.148	-1.91054 30052 180220
0.049	-3.01593 49808 715104	0.099	-2.31263 54288 475471	0.149	-1.90380 89730 366779
0.050	-2.99573 22735 539910	0.100	-2.30258 50929 940457	0.150	-1.89711 99848 858813
			$\begin{bmatrix} (-5)5 \\ 12 \end{bmatrix}$		$\begin{bmatrix} (-5)1 \\ 9 \end{bmatrix}$

For use of natural logarithms see **Examples 4–7**.

$$\ln 10 = 2.30258\ 50929\ 940457$$

NATURAL LOGARITHMS

Table 4.2

x	$\ln x$	x	$\ln x$	x	$\ln x$
0.150	-1.89711 99848 858813	0.200	-1.60943 79124 341004	0.250	-1.38629 43611 198906
0.151	-1.89047 54421 672127	0.201	-1.60445 03709 230613	0.251	-1.38230 23398 503532
0.152	-1.88387 47581 358607	0.202	-1.59948 75815 809323	0.252	-1.37832 61914 707137
0.153	-1.87731 73575 897016	0.203	-1.59454 92999 403497	0.253	-1.37436 57902 546168
0.154	-1.87080 26765 685079	0.204	-1.58963 52851 379207	0.254	-1.37042 10119 636005
0.155	-1.86433 01620 628904	0.205	-1.58474 52998 437289	0.255	-1.36649 17338 237109
0.156	-1.85789 92717 326000	0.206	-1.57987 91101 925560	0.256	-1.36257 78345 025746
0.157	-1.85150 94736 338290	0.207	-1.57503 64857 167680	0.257	-1.35867 91940 869173
0.158	-1.84516 02459 551702	0.208	-1.57021 71992 808191	0.258	-1.35479 56940 605196
0.159	-1.83885 10767 619055	0.209	-1.56542 10270 173260	0.259	-1.35092 72172 825993
0.160	-1.83258 14637 483101	0.210	-1.56064 77482 646684	0.260	-1.34707 36479 666093
0.161	-1.82635 09139 976741	0.211	-1.55589 71455 060706	0.261	-1.34323 48716 594436
0.162	-1.82015 89437 497530	0.212	-1.55116 90043 101246	0.262	-1.33941 07752 210402
0.163	-1.81400 50781 753747	0.213	-1.54646 31132 727119	0.263	-1.33560 12468 043725
0.164	-1.80788 88511 579386	0.214	-1.54177 92639 602856	0.264	-1.33180 61758 358209
0.165	-1.80180 98050 815564	0.215	-1.53711 72508 544743	0.265	-1.32802 54529 959148
0.166	-1.79576 74906 255938	0.216	-1.53247 68712 979720	0.266	-1.32425 89702 004380
0.167	-1.78976 14665 653819	0.217	-1.52785 79254 416775	0.267	-1.32050 66205 818875
0.168	-1.78379 12995 788781	0.218	-1.52326 02161 930480	0.268	-1.31676 82984 712804
0.169	-1.77785 65640 590636	0.219	-1.51868 35491 656362	0.269	-1.31304 38993 802979
0.170	-1.77195 68419 318753	0.220	-1.51412 77326 297755	0.270	-1.30933 33199 837623
0.171	-1.76609 17224 794772	0.221	-1.50959 25774 643842	0.271	-1.30563 64581 024362
0.172	-1.76026 08021 686840	0.222	-1.50507 78971 098576	0.272	-1.30195 32126 861397
0.173	-1.75446 36844 843581	0.223	-1.50058 35075 220183	0.273	-1.29828 34837 971773
0.174	-1.74869 99797 676080	0.224	-1.49610 92271 270972	0.274	-1.29462 71725 940668
0.175	-1.74296 93050 586230	0.225	-1.49165 48767 777169	0.275	-1.29098 41813 155658
0.176	-1.73727 12839 439853	0.226	-1.48722 02797 098512	0.276	-1.28735 44132 649871
0.177	-1.73160 55464 083079	0.227	-1.48280 52615 007344	0.277	-1.28373 77727 947986
0.178	-1.72597 17286 900519	0.228	-1.47840 96500 276963	0.278	-1.28013 41652 915000
0.179	-1.72036 94731 413821	0.229	-1.47403 32754 278974	0.279	-1.27654 34971 607714
0.180	-1.71479 84280 919267	0.230	-1.46967 59700 589417	0.280	-1.27296 56758 128874
0.181	-1.70925 82477 163113	0.231	-1.46533 75684 603435	0.281	-1.26940 06096 483913
0.182	-1.70374 85919 053417	0.232	-1.46101 79073 158271	0.282	-1.26584 82080 440235
0.183	-1.69826 91261 407161	0.233	-1.45671 68254 164365	0.283	-1.26230 83813 388994
0.184	-1.69281 95213 731514	0.234	-1.45243 41636 244356	0.284	-1.25878 10408 209310
0.185	-1.68739 94539 038122	0.235	-1.44816 97648 379781	0.285	-1.25526 60987 134865
0.186	-1.68200 86052 689358	0.236	-1.44392 34739 565270	0.286	-1.25176 34681 622845
0.187	-1.67664 66621 275504	0.237	-1.43969 51378 470059	0.287	-1.24827 30632 225159
0.188	-1.67131 33161 521878	0.238	-1.43548 46053 106624	0.288	-1.24479 47988 461911
0.189	-1.66600 82639 224947	0.239	-1.43129 17270 506264	0.289	-1.24132 85908 697049
0.190	-1.66073 12068 216509	0.240	-1.42711 63556 401457	0.290	-1.23787 43560 016173
0.191	-1.65548 18509 355072	0.241	-1.42295 83454 914821	0.291	-1.23443 20118 106445
0.192	-1.65025 99069 543555	0.242	-1.41881 75528 254507	0.292	-1.23100 14767 138553
0.193	-1.64506 50900 772515	0.243	-1.41469 38356 415886	0.293	-1.22758 26699 650697
0.194	-1.63989 71199 188089	0.244	-1.41058 70536 889352	0.294	-1.22417 55116 434554
0.195	-1.63475 57204 183903	0.245	-1.40649 70684 374101	0.295	-1.22077 99226 423172
0.196	-1.62964 06197 516198	0.246	-1.40242 37430 497742	0.296	-1.21739 58246 580767
0.197	-1.62455 15502 441485	0.247	-1.39836 69423 541599	0.297	-1.21402 31401 794374
0.198	-1.61948 82482 876018	0.248	-1.39432 65328 171549	0.298	-1.21066 17924 767326
0.199	-1.61445 04542 576447	0.249	-1.39030 23825 174294	0.299	-1.20731 17055 914506
0.200	-1.60943 79124 341004	0.250	-1.38629 43611 198906	0.300	-1.20397 28043 259360

$$\begin{bmatrix} (-6)5 \\ 8 \end{bmatrix} \qquad \begin{bmatrix} (-6)3 \\ 8 \end{bmatrix} \qquad \begin{bmatrix} (-6)2 \\ 7 \end{bmatrix}$$

$$\ln 10 = 2.30258\ 50929\ 940457$$

Table 4.2

NATURAL LOGARITHMS

x	$\ln x$	x	$\ln x$	x	$\ln x$
0.300	-1.20397 28043 259360	0.350	-1.04982 21244 986777	0.400	-0.91629 07318 741551
0.301	-1.20064 50142 332613	0.351	-1.04696 90555 162712	0.401	-0.91379 38516 755679
0.302	-1.19732 82616 072674	0.352	-1.04412 41033 840400	0.402	-0.91130 31903 631160
0.303	-1.19402 24734 727679	0.353	-1.04128 72220 488403	0.403	-0.90881 87170 354541
0.304	-1.19072 75775 759154	0.354	-1.03845 83658 483626	0.404	-0.90634 04010 209870
0.305	-1.18744 35023 747254	0.355	-1.03563 74895 067213	0.405	-0.90386 82118 755979
0.306	-1.18417 01770 297563	0.356	-1.03282 45481 301066	0.406	-0.90140 21193 804044
0.307	-1.18090 75313 949399	0.357	-1.03001 94972 024980	0.407	-0.89894 20935 395421
0.308	-1.17765 54960 085626	0.358	-1.02722 22925 814367	0.408	-0.89648 81045 779754
0.309	-1.17441 40020 843916	0.359	-1.02443 28904 938582	0.409	-0.89404 01229 393353
0.310	-1.17118 29815 029451	0.360	-1.02165 12475 319814	0.410	-0.89159 81192 837836
0.311	-1.16796 23668 029029	0.361	-1.01887 73206 492561	0.411	-0.88916 20644 859024
0.312	-1.16475 20911 726547	0.362	-1.01611 10671 563660	0.412	-0.88673 19296 326107
0.313	-1.16155 20884 419838	0.363	-1.01335 24447 172863	0.413	-0.88430 76860 211043
0.314	-1.15836 22930 738837	0.364	-1.01060 14113 453964	0.414	-0.88188 93051 568227
0.315	-1.15518 26401 565040	0.365	-1.00785 79253 996455	0.415	-0.87947 67587 514388
0.316	-1.15201 30653 952249	0.366	-1.00512 19455 807708	0.416	-0.87707 00187 208738
0.317	-1.14885 35051 048564	0.367	-1.00239 34309 275668	0.417	-0.87466 90571 833356
0.318	-1.14570 38962 019602	0.368	-0.99967 23408 132061	0.418	-0.87227 38464 573807
0.319	-1.14256 41761 972925	0.369	-0.99695 86349 416099	0.419	-0.86988 43590 599993
0.320	-1.13943 42831 883648	0.370	-0.99425 22733 438669	0.420	-0.86750 05677 047231
0.321	-1.13631 41558 521212	0.371	-0.99155 32163 747019	0.421	-0.86512 24452 997556
0.322	-1.13320 37334 377287	0.372	-0.98886 14247 089905	0.422	-0.86274 99649 461252
0.323	-1.13010 29557 594805	0.373	-0.98617 68593 383215	0.423	-0.86038 30999 358591
0.324	-1.12701 17631 898077	0.374	-0.98349 94815 676051	0.424	-0.85802 18237 501793
0.325	-1.12393 00966 523996	0.375	-0.98082 92530 117262	0.425	-0.85566 61100 577202
0.326	-1.12085 78976 154294	0.376	-0.97816 61355 922425	0.426	-0.85331 59327 127666
0.327	-1.11779 51080 848837	0.377	-0.97551 00915 341263	0.427	-0.85097 12657 535125
0.328	-1.11474 16705 979933	0.378	-0.97286 10833 625494	0.428	-0.84863 20834 003403
0.329	-1.11169 75282 167652	0.379	-0.97021 90738 997107	0.429	-0.84629 83600 541201
0.330	-1.10866 26245 216111	0.380	-0.96758 40262 617056	0.430	-0.84397 00702 945289
0.331	-1.10563 69036 050742	0.381	-0.96495 59038 554361	0.431	-0.84164 71888 783893
0.332	-1.10262 03100 656485	0.382	-0.96233 46703 755619	0.432	-0.83932 96907 380267
0.333	-1.09961 27890 016932	0.383	-0.95972 02898 014911	0.433	-0.83701 75509 796472
0.334	-1.09661 42860 054366	0.384	-0.95711 27263 944102	0.434	-0.83471 07448 817322
0.335	-1.09362 47471 570706	0.385	-0.95451 19446 943528	0.435	-0.83240 92478 934530
0.336	-1.09064 41190 189328	0.386	-0.95191 79095 173062	0.436	-0.83011 30356 331027
0.337	-1.08767 23486 297753	0.387	-0.94933 05859 523552	0.437	-0.82782 20838 865469
0.338	-1.08470 93834 991183	0.388	-0.94674 99393 588636	0.438	-0.82553 63686 056909
0.339	-1.08175 51716 016868	0.389	-0.94417 59353 636908	0.439	-0.82325 58659 069657
0.340	-1.07880 96613 719300	0.390	-0.94160 85398 584449	0.440	-0.82098 05520 698302
0.341	-1.07587 28016 986203	0.391	-0.93904 77189 967713	0.441	-0.81871 04035 352911
0.342	-1.07294 45419 195319	0.392	-0.93649 34391 916745	0.442	-0.81644 53969 044389
0.343	-1.07002 48318 161971	0.393	-0.93394 56671 128758	0.443	-0.81418 55089 370014
0.344	-1.06711 36216 087387	0.394	-0.93140 43696 842032	0.444	-0.81193 07165 499123
0.345	-1.06421 08619 507773	0.395	-0.92886 95140 810152	0.445	-0.80968 09968 158968
0.346	-1.06131 65039 244128	0.396	-0.92634 10677 276565	0.446	-0.80743 63269 620730
0.347	-1.05843 04990 352779	0.397	-0.92381 89982 949466	0.447	-0.80519 66843 685682
0.348	-1.05555 27992 076627	0.398	-0.92130 32736 976993	0.448	-0.80296 20465 671519
0.349	-1.05268 33567 797099	0.399	-0.91879 38620 922736	0.449	-0.80073 23912 398828
0.350	-1.04982 21244 986777	0.400	-0.91629 07318 741551	0.450	-0.79850 76962 177716

$$\begin{bmatrix} (-6)1 \\ 7 \end{bmatrix} \qquad \begin{bmatrix} (-6)1 \\ 7 \end{bmatrix} \qquad * \begin{bmatrix} (-7)8 \\ 7 \end{bmatrix}$$

$$\ln 10 = 2.30258 \ 50929 \ 940457$$

*See page II.

NATURAL LOGARITHMS

Table 4.2

x	$\ln x$	x	$\ln x$	x	$\ln x$
0.450	−0.79850 76962 177716	0.500	−0.69314 71805 599453	0.550	−0.59783 70007 556204
0.451	−0.79628 79394 794587	0.501	−0.69114 91778 972723	0.551	−0.59602 04698 292226
0.452	−0.79407 30991 499059	0.502	−0.68915 51592 904079	0.552	−0.59420 72327 050417
0.453	−0.79186 31534 991030	0.503	−0.68716 51088 823978	0.553	−0.59239 72774 598023
0.454	−0.78965 80809 407891	0.504	−0.68517 90109 107684	0.554	−0.59059 05922 348532
0.455	−0.78745 78600 311866	0.505	−0.68319 68497 067772	0.555	−0.58878 71652 357025
0.456	−0.78526 24694 677510	0.506	−0.68121 86096 946715	0.556	−0.58698 69847 315547
0.457	−0.78307 18880 879324	0.507	−0.67924 42753 909539	0.557	−0.58519 00390 548530
0.458	−0.78088 60948 679521	0.508	−0.67727 38314 036552	0.558	−0.58339 63166 008261
0.459	−0.77870 50689 215919	0.509	−0.67530 72624 316143	0.559	−0.58160 58058 270379
0.460	−0.77652 87894 989964	0.510	−0.67334 45532 637656	0.560	−0.57981 84952 529421
0.461	−0.77435 72359 854885	0.511	−0.67138 56887 784326	0.561	−0.57803 43734 594407
0.462	−0.77219 03879 003982	0.512	−0.66943 06539 426293	0.562	−0.57625 34290 884460
0.463	−0.77002 82248 959030	0.513	−0.66747 94338 113675	0.563	−0.57447 56508 424467
0.464	−0.76787 07267 558818	0.514	−0.66553 20135 269719	0.564	−0.57270 10274 840782
0.465	−0.76571 78733 947807	0.515	−0.66358 83783 184009	0.565	−0.57092 95478 356961
0.466	−0.76356 96448 564912	0.516	−0.66164 85135 005743	0.566	−0.56916 12007 789541
0.467	−0.76142 60213 132397	0.517	−0.65971 24044 737079	0.567	−0.56739 59752 543850
0.468	−0.75928 69830 644903	0.518	−0.65778 00367 226540	0.568	−0.56563 38602 609857
0.469	−0.75715 25105 358577	0.519	−0.65585 13958 162484	0.569	−0.56387 48448 558061
0.470	−0.75502 25842 780328	0.520	−0.65392 64674 066640	0.570	−0.56211 89181 535412
0.471	−0.75289 71849 657193	0.521	−0.65200 52372 287701	0.571	−0.56036 60693 261268
0.472	−0.75077 62933 965817	0.522	−0.65008 76910 994983	0.572	−0.55861 62876 023392
0.473	−0.74865 98904 902041	0.523	−0.64817 38149 172142	0.573	−0.55686 95622 673975
0.474	−0.74654 79572 870606	0.524	−0.64626 35946 610949	0.574	−0.55512 58826 625706
0.475	−0.74444 04749 474958	0.525	−0.64435 70163 905133	0.575	−0.55338 52381 847866
0.476	−0.74233 74247 507170	0.526	−0.64245 40662 444272	0.576	−0.55164 76182 862458
0.477	−0.74023 87880 937958	0.527	−0.64055 47304 407747	0.577	−0.54991 30124 740375
0.478	−0.73814 45464 906811	0.528	−0.63865 89952 758756	0.578	−0.54818 14103 097596
0.479	−0.73605 46815 712218	0.529	−0.63676 68471 238377	0.579	−0.54645 28014 091418
0.480	−0.73396 91750 802004	0.530	−0.63487 82724 359695	0.580	−0.54472 71754 416720
0.481	−0.73188 80088 763759	0.531	−0.63299 32577 401982	0.581	−0.54300 45221 302258
0.482	−0.72981 11649 315367	0.532	−0.63111 17896 404927	0.582	−0.54128 48312 506992
0.483	−0.72773 86253 295644	0.533	−0.62923 38548 162925	0.583	−0.53956 80926 316447
0.484	−0.72567 03722 655053	0.534	−0.62735 94400 219422	0.584	−0.53785 42961 539100
0.485	−0.72360 63880 446539	0.535	−0.62548 85320 861305	0.585	−0.53614 34317 502806
0.486	−0.72154 66550 816433	0.536	−0.62362 11179 113351	0.586	−0.53443 54894 051244
0.487	−0.71949 11558 995473	0.537	−0.62175 71844 732724	0.587	−0.53273 04591 540406
0.488	−0.71743 98731 289899	0.538	−0.61989 67188 203526	0.588	−0.53102 83310 835101
0.489	−0.71539 27895 072650	0.539	−0.61803 97080 731399	0.589	−0.52932 90953 305503
0.490	−0.71334 98878 774648	0.540	−0.61618 61394 238170	0.590	−0.52763 27420 823719
0.491	−0.71131 11511 876165	0.541	−0.61433 60001 356555	0.591	−0.52593 92615 760389
0.492	−0.70927 65624 898289	0.542	−0.61248 92775 424908	0.592	−0.52424 86440 981314
0.493	−0.70724 61049 394469	0.543	−0.61064 59590 482016	0.593	−0.52256 08799 844116
0.494	−0.70521 97617 942145	0.544	−0.60880 60321 261944	0.594	−0.52087 59596 194921
0.495	−0.70319 75164 134468	0.545	−0.60696 94843 188930	0.595	−0.51919 38734 365073
0.496	−0.70117 93522 572096	0.546	−0.60513 63032 372320	0.596	−0.51751 46119 167873
0.497	−0.69916 52528 855083	0.547	−0.60330 64765 601558	0.597	−0.51583 81655 895350
0.498	−0.69715 52019 574841	0.548	−0.60147 99920 341215	0.598	−0.51416 45250 315053
0.499	−0.69514 91832 306184	0.549	−0.59965 68374 726064	0.599	−0.51249 36808 666877
0.500	−0.69314 71805 599453	0.550	−0.59783 70007 556204	0.600	−0.51082 56237 659907

$$\begin{bmatrix} (-7)6 \\ 7 \end{bmatrix} \qquad \begin{bmatrix} (-7)5 \\ 6 \end{bmatrix} \qquad \begin{bmatrix} (-7)4 \\ 6 \end{bmatrix}$$

$\ln 10 = 2.30258\ 50929\ 940457$

Table 4.2　　　　　　　　　　　**NATURAL LOGARITHMS**

x	$\ln x$	x	$\ln x$	x	$\ln x$
0.600	−0.51082 56237 659907	0.650	−0.43078 29160 924543	0.700	−0.35667 49439 387324
0.601	−0.50916 03444 469295	0.651	−0.42924 56367 735678	0.701	−0.35524 73919 475470
0.602	−0.50749 78336 733160	0.652	−0.42771 07170 554841	0.702	−0.35382 18749 563259
0.603	−0.50583 80822 549516	0.653	−0.42617 81497 057060	0.703	−0.35239 83871 714721
0.604	−0.50418 10810 473221	0.654	−0.42464 79275 249384	0.704	−0.35097 69228 240947
0.605	−0.50252 68209 512956	0.655	−0.42312 00433 468851	0.705	−0.34955 74761 698684
0.606	−0.50087 52929 128226	0.656	−0.42159 44900 380480	0.706	−0.34814 00414 888950
0.607	−0.49922 64879 226388	0.657	−0.42007 12604 975265	0.707	−0.34672 46130 855643
0.608	−0.49758 03970 159700	0.658	−0.41855 03476 568199	0.708	−0.34531 11852 884173
0.609	−0.49593 70112 722400	0.659	−0.41703 17444 796298	0.709	−0.34389 97524 500096
0.610	−0.49429 63218 147801	0.660	−0.41551 54439 616658	0.710	−0.34249 03089 467759
0.611	−0.49265 83198 105417	0.661	−0.41400 14391 304508	0.711	−0.34108 28491 788962
0.612	−0.49102 29964 698110	0.662	−0.41248 97230 451288	0.712	−0.33967 73675 701613
0.613	−0.48939 03430 459257	0.663	−0.41098 02887 962745	0.713	−0.33827 38585 678411
0.614	−0.48776 03508 349946	0.664	−0.40947 31295 057032	0.714	−0.33687 23166 425527
0.615	−0.48613 30111 756192	0.665	−0.40796 82383 262829	0.715	−0.33547 27362 881294
0.616	−0.48450 83154 486173	0.666	−0.40646 56084 417479	0.716	−0.33407 51120 214914
0.617	−0.48288 62550 767492	0.667	−0.40496 52330 665133	0.717	−0.33267 94383 825167
0.618	−0.48126 68215 244463	0.668	−0.40346 71054 454913	0.718	−0.33128 57099 339129
0.619	−0.47965 00062 975409	0.669	−0.40197 12188 539086	0.719	−0.32989 39212 610904
0.620	−0.47803 58009 429998	0.670	−0.40047 75665 971253	0.720	−0.32850 40669 720361
0.621	−0.47642 41970 486583	0.671	−0.39898 61420 104553	0.721	−0.32711 61416 971880
0.622	−0.47481 51862 429576	0.672	−0.39749 69384 589875	0.722	−0.32573 01400 893108
0.623	−0.47320 87601 946839	0.673	−0.39600 99493 374092	0.723	−0.32434 60568 233724
0.624	−0.47160 49106 127094	0.674	−0.39452 51680 698300	0.724	−0.32296 38865 964207
0.625	−0.47000 36292 457356	0.675	−0.39304 25881 096072	0.725	−0.32158 36241 274623
0.626	−0.46840 49078 820385	0.676	−0.39156 22029 391730	0.726	−0.32020 52641 573410
0.627	−0.46680 87383 492164	0.677	−0.39008 40060 698621	0.727	−0.31882 88014 486177
0.628	−0.46521 51125 139384	0.678	−0.38860 79910 417415	0.728	−0.31745 42307 854511
0.629	−0.46362 40222 816965	0.679	−0.38713 41514 234409	0.729	−0.31608 15469 734789
0.630	−0.46203 54595 965587	0.680	−0.38566 24808 119847	0.730	−0.31471 07448 397002
0.631	−0.46044 94164 409239	0.681	−0.38419 29728 326247	0.731	−0.31334 18192 323585
0.632	−0.45886 58848 352796	0.682	−0.38272 56211 386750	0.732	−0.31197 47650 208255
0.633	−0.45728 48568 379609	0.683	−0.38126 04194 113470	0.733	−0.31060 95770 954856
0.634	−0.45570 63245 449111	0.684	−0.37979 73613 595866	0.734	−0.30924 62503 676215
0.635	−0.45413 02800 894454	0.685	−0.37833 64407 199118	0.735	−0.30788 47797 693004
0.636	−0.45255 67156 420149	0.686	−0.37687 76512 562518	0.736	−0.30652 51602 532608
0.637	−0.45098 56234 099737	0.687	−0.37542 09867 597877	0.737	−0.30516 73867 928004
0.638	−0.44941 69956 373472	0.688	−0.37396 64410 487934	0.738	−0.30381 14543 816646
0.639	−0.44785 08246 046022	0.689	−0.37251 40079 684785	0.739	−0.30245 73580 339353
0.640	−0.44628 71026 284195	0.690	−0.37106 36813 908320	0.740	−0.30110 50927 839216
0.641	−0.44472 58220 614670	0.691	−0.36961 54552 144672	0.741	−0.29975 46536 860502
0.642	−0.44316 69752 921759	0.692	−0.36816 93233 644675	0.742	−0.29840 60358 147566
0.643	−0.44161 05547 445177	0.693	−0.36672 52797 922338	0.743	−0.29705 92342 643779
0.644	−0.44005 65528 777834	0.694	−0.36528 33184 753326	0.744	−0.29571 42441 490452
0.645	−0.43850 49621 863646	0.695	−0.36384 34334 173449	0.745	−0.29437 10606 025775
0.646	−0.43695 57751 995352	0.696	−0.36240 56186 477174	0.746	−0.29302 96787 783762
0.647	−0.43540 89844 812365	0.697	−0.36096 98682 216132	0.747	−0.29169 00938 493197
0.648	−0.43386 45826 298624	0.698	−0.35953 61762 197646	0.748	−0.29035 23010 076598
0.649	−0.43232 25622 780471	0.699	−0.35810 45367 483268	0.749	−0.28901 62954 649176
0.650	−0.43078 29160 924543	0.700	−0.35667 49439 387324	0.750	−0.28768 20724 517809
	$\begin{bmatrix} (-7)3 \\ 6 \end{bmatrix}$		$\begin{bmatrix} (-7)3 \\ 6 \end{bmatrix}$		$\begin{bmatrix} (-7)3 \\ 6 \end{bmatrix}$

$$\ln 10 = 2.30258\ 50929\ 940457$$

NATURAL LOGARITHMS

Table 4.2

x	$\ln x$	x	$\ln x$	x	$\ln x$
0.750	−0.28768 20724 517809	0.800	−0.22314 35513 142098	0.850	−0.16251 89294 977749
0.751	−0.28634 96272 180023	0.801	−0.22189 43319 137778	0.851	−0.16134 31504 087629
0.752	−0.28501 89550 322973	0.802	−0.22064 66711 156226	0.852	−0.16016 87521 528213
0.753	−0.28369 00511 822435	0.803	−0.21940 05650 353754	0.853	−0.15899 57314 904579
0.754	−0.28236 29109 741810	0.804	−0.21815 60098 031707	0.854	−0.15782 40851 935672
0.755	−0.28103 75297 331123	0.805	−0.21691 30015 635737	0.855	−0.15665 38100 453768
0.756	−0.27971 39028 026041	0.806	−0.21567 15364 755088	0.856	−0.15548 49028 403950
0.757	−0.27839 20255 446883	0.807	−0.21443 16107 121883	0.857	−0.15431 73603 843573
0.758	−0.27707 18933 397654	0.808	−0.21319 32204 610417	0.858	−0.15315 11794 941748
0.759	−0.27575 35015 865071	0.809	−0.21195 63619 236454	0.859	−0.15198 63569 978817
0.760	−0.27443 68457 017603	0.810	−0.21072 10313 156526	0.860	−0.15082 28897 345836
0.761	−0.27312 19211 204512	0.811	−0.20948 72248 667241	0.861	−0.14966 07745 544063
0.762	−0.27180 87232 954908	0.812	−0.20825 49388 204591	0.862	−0.14850 00083 184440
0.763	−0.27049 72476 976800	0.813	−0.20702 41694 343265	0.863	−0.14734 05878 987091
0.764	−0.26918 74898 156166	0.814	−0.20579 49129 795968	0.864	−0.14618 25101 780814
0.765	−0.26787 94451 556012	0.815	−0.20456 71657 412743	0.865	−0.14502 57720 502577
0.766	−0.26657 31092 415458	0.816	−0.20334 09240 180300	0.866	−0.14387 03704 197019
0.767	−0.26526 84776 148809	0.817	−0.20211 61841 221342	0.867	−0.14271 63022 015952
0.768	−0.26396 55458 344649	0.818	−0.20089 29423 793900	0.868	−0.14156 35643 217869
0.769	−0.26266 43094 764931	0.819	−0.19967 11951 290676	0.869	−0.14041 21537 167450
0.770	−0.26136 47641 344075	0.820	−0.19845 09387 238383	0.870	−0.13926 20673 335076
0.771	−0.26006 69054 188076	0.821	−0.19723 21695 297088	0.871	−0.13811 33021 296343
0.772	−0.25877 07289 573609	0.822	−0.19601 48839 259571	0.872	−0.13696 58550 731574
0.773	−0.25747 62303 947151	0.823	−0.19479 90783 050672	0.873	−0.13581 97231 425348
0.774	−0.25618 34053 924099	0.824	−0.19358 47490 726654	0.874	−0.13467 49033 266016
0.775	−0.25489 22496 287901	0.825	−0.19237 18926 474561	0.875	−0.13353 13926 245226
0.776	−0.25360 27587 989183	0.826	−0.19116 05054 611590	0.876	−0.13238 91880 457456
0.777	−0.25231 49286 144896	0.827	−0.18995 05839 584457	0.877	−0.13124 82866 099540
0.778	−0.25102 87548 037454	0.828	−0.18874 21245 968774	0.878	−0.13010 86853 470204
0.779	−0.24974 42331 113888	0.829	−0.18753 51238 468421	0.879	−0.12897 03812 969601
0.780	−0.24846 13592 984996	0.830	−0.18632 95781 914934	0.880	−0.12783 33715 098849
0.781	−0.24718 01291 424511	0.831	−0.18512 54841 266889	0.881	−0.12669 76530 459575
0.782	−0.24590 05384 368260	0.832	−0.18392 28381 609285	0.882	−0.12556 32229 753457
0.783	−0.24462 25829 913340	0.833	−0.18272 16368 152944	0.883	−0.12443 00783 781770
0.784	−0.24334 62586 317292	0.834	−0.18152 18766 233903	0.884	−0.12329 82163 444936
0.785	−0.24207 15611 997286	0.835	−0.18032 35541 312816	0.885	−0.12216 76339 742075
0.786	−0.24079 84865 529305	0.836	−0.17912 66658 974354	0.886	−0.12103 83283 770561
0.787	−0.23952 70305 647338	0.837	−0.17793 12084 926617	0.887	−0.11991 02966 725576
0.788	−0.23825 71891 242579	0.838	−0.17673 71785 000540	0.888	−0.11878 35359 899670
0.789	−0.23698 89581 362628	0.839	−0.17554 45725 149309	0.889	−0.11765 80434 682325
0.790	−0.23572 23335 210699	0.840	−0.17435 33871 447778	0.890	−0.11653 38162 559515
0.791	−0.23445 73112 144832	0.841	−0.17316 36190 091890	0.891	−0.11541 08515 113277
0.792	−0.23319 38871 677112	0.842	−0.17197 52647 398103	0.892	−0.11428 91464 021277
0.793	−0.23193 20573 472891	0.843	−0.17078 83209 802816	0.893	−0.11316 86981 056380
0.794	−0.23067 18177 350013	0.844	−0.16960 27843 861799	0.894	−0.11204 95038 086229
0.795	−0.22941 31643 278052	0.845	−0.16841 86516 249632	0.895	−0.11093 15607 072817
0.796	−0.22815 60931 377540	0.846	−0.16723 59193 759138	0.896	−0.10981 48660 072066
0.797	−0.22690 06001 919220	0.847	−0.16605 45843 300827	0.897	−0.10869 94169 233409
0.798	−0.22564 66815 323283	0.848	−0.16487 46431 902340	0.898	−0.10758 52106 799374
0.799	−0.22439 43332 158624	0.849	−0.16369 60926 707897	0.899	−0.10647 22445 105168
0.800	−0.22314 35513 142098	0.850	−0.16251 89294 977749	0.900	−0.10536 05156 578263
	$\left[\begin{smallmatrix}(-7)2\\6\end{smallmatrix}\right]$		$\left[\begin{smallmatrix}(-7)2\\6\end{smallmatrix}\right]$		$\left[\begin{smallmatrix}(-7)2\\6\end{smallmatrix}\right]$

$$\ln 10 = 2.30258\ 50929\ 940457$$

Table 4.2   **NATURAL LOGARITHMS**

x	$\ln x$	x	$\ln x$	x	$\ln x$
0.900	−0.10536 05156 578263	0.950	−0.05129 32943 875505	1.000	0.00000 00000 000000
0.901	−0.10425 00213 737991	0.951	−0.05024 12164 367467	1.001	0.00099 95003 330835
0.902	−0.10314 07589 195134	0.952	−0.04919 02441 907717	1.002	0.00199 80026 626731
0.903	−0.10203 27255 651516	0.953	−0.04814 03753 279349	1.003	0.00299 55089 797985
0.904	−0.10092 59185 899606	0.954	−0.04709 16075 338505	1.004	0.00399 20212 695375
0.905	−0.09982 03352 822109	0.955	−0.04604 39385 014068	1.005	0.00498 75415 110391
0.906	−0.09871 59729 391577	0.956	−0.04499 73659 307358	1.006	0.00598 20716 775475
0.907	−0.09761 28288 670004	0.957	−0.04395 18875 291828	1.007	0.00697 56137 364252
0.908	−0.09651 09003 808438	0.958	−0.04290 75010 112765	1.008	0.00796 81696 491769
0.909	−0.09541 01848 046582	0.959	−0.04186 42040 986988	1.009	0.00895 97413 714719
0.910	−0.09431 06794 712413	0.960	−0.04082 19945 202551	1.010	0.00995 03308 531681
0.911	−0.09321 23817 221787	0.961	−0.03978 08700 118446	1.011	0.01093 99400 383344
0.912	−0.09211 52889 078057	0.962	−0.03874 08283 164306	1.012	0.01192 85708 652738
0.913	−0.09101 93983 871686	0.963	−0.03770 18671 840115	1.013	0.01291 62252 665463
0.914	−0.08992 47075 279870	0.964	−0.03666 39843 715914	1.014	0.01390 29051 689914
0.915	−0.08883 12137 066157	0.965	−0.03562 71776 431511	1.015	0.01488 86124 937507
0.916	−0.08773 89143 080068	0.966	−0.03459 14447 696191	1.016	0.01587 33491 562901
0.917	−0.08664 78067 256722	0.967	−0.03355 67835 288427	1.017	0.01685 71170 664229
0.918	−0.08555 78883 616466	0.968	−0.03252 31917 055600	1.018	0.01783 99181 283310
0.919	−0.08446 91566 264500	0.969	−0.03149 06670 913708	1.019	0.01882 17542 405878
0.920	−0.08338 16089 390511	0.970	−0.03045 92074 847085	1.020	0.01980 26272 961797
0.921	−0.08229 52427 268302	0.971	−0.02942 88106 908121	1.021	0.02078 25391 825285
0.922	−0.08121 00554 255432	0.972	−0.02839 94745 216980	1.022	0.02176 14917 815127
0.923	−0.08012 60444 792849	0.973	−0.02737 11967 961320	1.023	0.02273 94869 694894
0.924	−0.07904 32073 404529	0.974	−0.02634 39753 396020	1.024	0.02371 65266 173160
0.925	−0.07796 15414 697119	0.975	−0.02531 78079 842899	1.025	0.02469 26125 903715
0.926	−0.07688 10443 359577	0.976	−0.02429 26925 690446	1.026	0.02566 77467 485778
0.927	−0.07580 17134 162819	0.977	−0.02326 86269 393543	1.027	0.02664 19309 464212
0.928	−0.07472 35461 959365	0.978	−0.02224 56089 473197	1.028	0.02761 51670 329734
0.929	−0.07364 65401 682985	0.979	−0.02122 36364 516267	1.029	0.02858 74568 519126
0.930	−0.07257 06928 348354	0.980	−0.02020 27073 175194	1.030	0.02955 88022 415444
0.931	−0.07149 60017 050700	0.981	−0.01918 28194 167740	1.031	0.03052 92050 348229
0.932	−0.07042 24642 965459	0.982	−0.01816 39706 276712	1.032	0.03149 86670 593710
0.933	−0.06935 00781 347932	0.983	−0.01714 61588 349705	1.033	0.03246 71901 375015
0.934	−0.06827 88407 532944	0.984	−0.01612 93819 298836	1.034	0.03343 47760 862374
0.935	−0.06720 87496 934501	0.985	−0.01511 36378 100482	1.035	0.03440 14267 173324
0.936	−0.06613 98025 045450	0.986	−0.01409 89243 795016	1.036	0.03536 71438 372913
0.937	−0.06507 19967 437149	0.987	−0.01308 52395 486555	1.037	0.03633 19292 473903
0.938	−0.06400 53299 759124	0.988	−0.01207 25812 342692	1.038	0.03729 57847 436969
0.939	−0.06293 97997 738741	0.989	−0.01106 09473 594249	1.039	0.03825 87121 170903
0.940	−0.06187 54037 180875	0.990	−0.01005 03358 535014	1.040	0.03922 07131 532813
0.941	−0.06081 21393 967574	0.991	−0.00904 07446 521491	1.041	0.04018 17896 328318
0.942	−0.05975 00044 057740	0.992	−0.00803 21716 972643	1.042	0.04114 19433 311752
0.943	−0.05868 89963 486796	0.993	−0.00702 46149 369645	1.043	0.04210 11760 186354
0.944	−0.05762 91128 366364	0.994	−0.00601 80723 255630	1.044	0.04305 94894 604470
0.945	−0.05657 03514 883943	0.995	−0.00501 25418 235443	1.045	0.04401 68854 167743
0.946	−0.05551 27099 302588	0.996	−0.00400 80213 975388	1.046	0.04497 33656 427312
0.947	−0.05445 61857 960588	0.997	−0.00300 45090 202987	1.047	0.04592 89318 883998
0.948	−0.05340 07767 271152	0.998	−0.00200 20026 706731	1.048	0.04688 35858 988504
0.949	−0.05234 64803 722092	0.999	−0.00100 05003 335835	1.049	0.04783 73294 141601
0.950	−0.05129 32943 875505	1.000	0.00000 00000 000000	1.050	0.04879 01641 694320

$$\begin{bmatrix} (-7)2 \\ 6 \end{bmatrix} \qquad \begin{bmatrix} (-7)1 \\ 6 \end{bmatrix} \qquad \begin{bmatrix} (-7)1 \\ 6 \end{bmatrix}$$

$\ln 10 = 2.30258\ 50929\ 940457$

NATURAL LOGARITHMS

Table 4.2

x	$\ln x$	x	$\ln x$	x	$\ln x$
1.050	0.04879 01641 694320	1.100	0.09531 01798 043249	1.150	0.13976 19423 751587
1.051	0.04974 20918 948141	1.101	0.09621 88577 405429	1.151	0.14063 11297 397456
1.052	0.05069 31143 155181	1.102	0.09712 67107 307227	1.152	0.14149 95622 736995
1.053	0.05164 32331 518384	1.103	0.09803 37402 713654	1.153	0.14236 72412 869220
1.054	0.05259 24501 191706	1.104	0.09893 99478 549036	1.154	0.14323 41680 859078
1.055	0.05354 07669 280298	1.105	0.09984 53349 697161	1.155	0.14410 03439 737569
1.056	0.05448 81852 840697	1.106	0.10074 99031 001431	1.156	0.14496 57702 501857
1.057	0.05543 47068 881006	1.107	0.10165 36537 264998	1.157	0.14583 04482 115395
1.058	0.05638 03334 361076	1.108	0.10255 65883 250921	1.158	0.14669 43791 508035
1.059	0.05732 50666 192694	1.109	0.10345 87083 682300	1.159	0.14755 75643 576147
1.060	0.05826 89081 239758	1.110	0.10436 00153 242428	1.160	0.14842 00051 182733
1.061	0.05921 18596 318461	1.111	0.10526 05106 574929	1.161	0.14928 17027 157544
1.062	0.06015 39228 197471	1.112	0.10616 01958 283906	1.162	0.15014 26584 297195
1.063	0.06109 50993 598109	1.113	0.10705 90722 934078	1.163	0.15100 28735 365274
1.064	0.06203 53909 194526	1.114	0.10795 71415 050923	1.164	0.15186 23493 092461
1.065	0.06297 47991 613884	1.115	0.10885 44049 120821	1.165	0.15272 10870 176639
1.066	0.06391 33257 436528	1.116	0.10975 08639 591192	1.166	0.15357 90879 283006
1.067	0.06485 09723 196163	1.117	0.11064 65200 870637	1.167	0.15443 63533 044189
1.068	0.06578 77405 380031	1.118	0.11154 13747 329074	1.168	0.15529 28844 060353
1.069	0.06672 36320 429082	1.119	0.11243 54293 297882	1.169	0.15614 86824 899314
1.070	0.06765 86484 738148	1.120	0.11332 86853 070032	1.170	0.15700 37488 096648
1.071	0.06859 27914 656117	1.121	0.11422 11440 900229	1.171	0.15785 80846 155803
1.072	0.06952 60626 486102	1.122	0.11511 28071 005046	1.172	0.15871 16911 548209
1.073	0.07045 84636 485614	1.123	0.11600 36757 563061	1.173	0.15956 45696 713384
1.074	0.07138 99960 866729	1.124	0.11689 37514 714993	1.174	0.16041 67214 059047
1.075	0.07232 06615 796261	1.125	0.11778 30356 563835	1.175	0.16126 81475 961223
1.076	0.07325 04617 395927	1.126	0.11867 15297 174986	1.176	0.16211 88494 764352
1.077	0.07417 93981 742515	1.127	0.11955 92350 576392	1.177	0.16296 88282 781397
1.078	0.07510 74724 868054	1.128	0.12044 61530 758672	1.178	0.16381 80852 293950
1.079	0.07603 46862 759976	1.129	0.12133 22851 675250	1.179	0.16466 66215 552339
1.080	0.07696 10411 361283	1.130	0.12221 76327 242492	1.180	0.16551 44384 775734
1.081	0.07788 65386 570712	1.131	0.12310 21971 339834	1.181	0.16636 15372 152253
1.082	0.07881 11804 242898	1.132	0.12398 59797 809912	1.182	0.16720 79189 839065
1.083	0.07973 49680 188536	1.133	0.12486 89820 458693	1.183	0.16805 35849 962497
1.084	0.08065 79030 174545	1.134	0.12575 12053 055603	1.184	0.16889 85364 618139
1.085	0.08157 99869 924229	1.135	0.12663 26509 333660	1.185	0.16974 27745 870945
1.086	0.08250 12215 117437	1.136	0.12751 33202 989596	1.186	0.17058 63005 755337
1.087	0.08342 16081 390724	1.137	0.12839 32147 683990	1.187	0.17142 91156 275310
1.088	0.08434 11484 337509	1.138	0.12927 23357 041392	1.188	0.17227 12209 404532
1.089	0.08525 98439 508234	1.139	0.13015 06844 650451	1.189	0.17311 26177 086448
1.090	0.08617 76962 410523	1.140	0.13102 82624 064041	1.190	0.17395 33071 234380
1.091	0.08709 47068 509338	1.141	0.13190 50708 799386	1.191	0.17479 32903 731631
1.092	0.08801 08773 227133	1.142	0.13278 11112 338185	1.192	0.17563 25686 431580
1.093	0.08892 62091 944015	1.143	0.13365 63848 126736	1.193	0.17647 11431 157791
1.094	0.08984 07039 997895	1.144	0.13453 08929 576062	1.194	0.17730 90149 704103
1.095	0.09075 43632 684641	1.145	0.13540 46370 062030	1.195	0.17814 61853 834740
1.096	0.09166 71885 258238	1.146	0.13627 76182 925478	1.196	0.17898 26555 284400
1.097	0.09257 91812 930932	1.147	0.13714 98381 472336	1.197	0.17981 84265 758361
1.098	0.09349 03430 873389	1.148	0.13802 12978 973747	1.198	0.18065 34996 932576
1.099	0.09440 06754 214843	1.149	0.13889 19988 666186	1.199	0.18148 78760 453772
1.100	0.09531 01798 043249	1.150	0.13976 19423 751587	1.200	0.18232 15567 939546
	$\left[\begin{smallmatrix}(-7)1\\6\end{smallmatrix}\right]$		$\left[\begin{smallmatrix}(-7)1\\6\end{smallmatrix}\right]$		$\left[\begin{smallmatrix}(-8)9\\6\end{smallmatrix}\right]$

$$\ln 10 = 2.30258\ 50929\ 940457$$

Table 4.2   **NATURAL LOGARITHMS**

x	$\ln x$	x	$\ln x$	x	$\ln x$
1.200	0.18232 15567 939546	1.250	0.22314 35513 142098	1.300	0.26236 42644 674911
1.201	0.18315 45430 978465	1.251	0.22394 32314 847741	1.301	0.26313 31995 303682
1.202	0.18398 68361 130158	1.252	0.22474 22726 779068	1.302	0.26390 15437 863775
1.203	0.18481 84369 925418	1.253	0.22554 06759 139312	1.303	0.26466 92981 427081
1.204	0.18564 93468 866293	1.254	0.22633 84422 107290	1.304	0.26543 64635 044612
1.205	0.18647 95669 426183	1.255	0.22713 55725 837472	1.305	0.26620 30407 746567
1.206	0.18730 90983 049937	1.256	0.22793 20680 460069	1.306	0.26696 90308 542393
1.207	0.18813 79421 153944	1.257	0.22872 79296 081104	1.307	0.26773 44346 420849
1.208	0.18896 60995 126232	1.258	0.22952 31582 782488	1.308	0.26849 92530 350070
1.209	0.18979 35716 326556	1.259	0.23031 77550 622101	1.309	0.26926 34869 277629
1.210	0.19062 03596 086497	1.260	0.23111 17209 633866	1.310	0.27002 71372 130602
1.211	0.19144 64645 709552	1.261	0.23190 50569 827825	1.311	0.27079 02047 815628
1.212	0.19227 18876 471227	1.262	0.23269 77641 190214	1.312	0.27155 26905 218973
1.213	0.19309 66299 619131	1.263	0.23348 98433 683541	1.313	0.27231 45953 206591
1.214	0.19392 06926 373065	1.264	0.23428 12957 246657	1.314	0.27307 59200 624188
1.215	0.19474 40767 925118	1.265	0.23507 21221 794836	1.315	0.27383 66656 297279
1.216	0.19556 67835 439753	1.266	0.23586 23237 219844	1.316	0.27459 68329 031255
1.217	0.19638 88140 053901	1.267	0.23665 19013 390020	1.317	0.27535 64227 611440
1.218	0.19721 01692 877053	1.268	0.23744 08560 150342	1.318	0.27611 54360 803155
1.219	0.19803 08504 991345	1.269	0.23822 91887 322506	1.319	0.27687 38737 351775
1.220	0.19885 08587 451652	1.270	0.23901 69004 704999	1.320	0.27763 17365 982795
1.221	0.19967 01951 285676	1.271	0.23980 39922 073170	1.321	0.27838 90255 401883
1.222	0.20048 88607 494036	1.272	0.24059 04649 179304	1.322	0.27914 57414 294945
1.223	0.20130 68567 050353	1.273	0.24137 63195 752695	1.323	0.27990 18851 328186
1.224	0.20212 41840 901343	1.274	0.24216 15571 499716	1.324	0.28065 74575 148165
1.225	0.20294 08439 966903	1.275	0.24294 61786 103895	1.325	0.28141 24594 381855
1.226	0.20375 68375 140197	1.276	0.24373 01849 225981	1.326	0.28216 68917 636708
1.227	0.20457 21657 287744	1.277	0.24451 35770 504022	1.327	0.28292 07553 500705
1.228	0.20538 68297 249507	1.278	0.24529 63559 553431	1.328	0.28367 40510 542421
1.229	0.20620 08305 838978	1.279	0.24607 85225 967056	1.329	0.28442 67797 311083
1.230	0.20701 41693 843261	1.280	0.24686 00779 315258	1.330	0.28517 89422 336624
1.231	0.20782 68472 023165	1.281	0.24764 10229 145972	1.331	0.28593 05394 129746
1.232	0.20863 88651 113280	1.282	0.24842 13584 984783	1.332	0.28668 15721 181974
1.233	0.20945 02241 822072	1.283	0.24920 10856 334994	1.333	0.28743 20411 965716
1.234	0.21026 09254 831961	1.284	0.24998 02052 677694	1.334	0.28818 19474 934320
1.235	0.21107 09700 799405	1.285	0.25075 87183 471831	1.335	0.28893 12918 522129
1.236	0.21188 03590 354990	1.286	0.25153 66258 154276	1.336	0.28968 00751 144540
1.237	0.21268 90934 103508	1.287	0.25231 39286 139896	1.337	0.29042 82981 198061
1.238	0.21349 71742 624044	1.288	0.25309 06276 821619	1.338	0.29117 59617 060367
1.239	0.21430 46026 470054	1.289	0.25386 67239 570503	1.339	0.29192 30667 090355
1.240	0.21511 13796 169455	1.290	0.25464 22183 735807	1.340	0.29266 96139 628200
1.241	0.21591 75062 224702	1.291	0.25541 71118 645054	1.341	0.29341 56042 995415
1.242	0.21672 29835 112870	1.292	0.25619 14053 604101	1.342	0.29416 10385 494901
1.243	0.21752 78125 285741	1.293	0.25696 50997 897204	1.343	0.29490 59175 411005
1.244	0.21833 19943 169877	1.294	0.25773 81960 787088	1.344	0.29565 02421 009578
1.245	0.21913 55299 166709	1.295	0.25851 06951 515011	1.345	0.29639 40130 538024
1.246	0.21993 84203 652614	1.296	0.25928 25979 300830	1.346	0.29713 72312 225361
1.247	0.22074 06666 978994	1.297	0.26005 39053 343068	1.347	0.29787 98974 282269
1.248	0.22154 22699 472359	1.298	0.26082 46182 818983	1.348	0.29862 20124 901153
1.249	0.22234 32311 434406	1.299	0.26159 47376 884625	1.349	0.29936 35772 256188
1.250	0.22314 35513 142098	1.300	0.26236 42644 674911	1.350	0.30010 45924 503381

$$\begin{bmatrix} (-8)9 \\ 6 \end{bmatrix} \qquad\qquad \begin{bmatrix} (-8)8 \\ 6 \end{bmatrix} \qquad\qquad \begin{bmatrix} (-8)7 \\ 6 \end{bmatrix}$$

ln 10 = 2.30258 50929 940457

NATURAL LOGARITHMS

Table 4.2

x	$\ln x$	x	$\ln x$	x	$\ln x$
1.350	0.30010 45924 503381	1.400	0.33647 22366 212129	1.450	0.37156 35564 324830
1.351	0.30084 50589 780618	1.401	0.33718 62673 548700	1.451	0.37225 29739 020508
1.352	0.30158 49776 207723	1.402	0.33789 97886 123983	1.452	0.37294 19164 026043
1.353	0.30232 43491 886510	1.403	0.33861 28011 203239	1.453	0.37363 03845 881459
1.354	0.30306 31744 900833	1.404	0.33932 53056 036194	1.454	0.37431 83791 113276
1.355	0.30380 14543 316642	1.405	0.34003 73027 857091	1.455	0.37500 59006 234558
1.356	0.30453 91895 182038	1.406	0.34074 87933 884732	1.456	0.37569 29497 744942
1.357	0.30527 63808 527321	1.407	0.34145 97781 322520	1.457	0.37637 95272 130678
1.358	0.30601 30291 365044	1.408	0.34217 02577 358507	1.458	0.37706 56335 864664
1.359	0.30674 91351 690067	1.409	0.34288 02329 165432	1.459	0.37775 12695 406486
1.360	0.30748 46997 479606	1.410	0.34358 97043 900769	1.460	0.37843 64357 202451
1.361	0.30821 97236 693290	1.411	0.34429 86728 706770	1.461	0.37912 11327 685624
1.362	0.30895 42077 273206	1.412	0.34500 71390 710503	1.462	0.37980 53613 275868
1.363	0.30968 81527 143956	1.413	0.34571 51037 023904	1.463	0.38048 91220 379873
1.364	0.31042 15594 212704	1.414	0.34642 25674 743810	1.464	0.38117 24155 391198
1.365	0.31115 44286 369231	1.415	0.34712 95310 952009	1.465	0.38185 52424 690306
1.366	0.31188 67611 485983	1.416	0.34783 59952 715280	1.466	0.38253 76034 644597
1.367	0.31261 85577 418125	1.417	0.34854 19607 085434	1.467	0.38321 94991 608447
1.368	0.31334 98192 003587	1.418	0.34924 74281 099358	1.468	0.38390 09301 923238
1.369	0.31408 05463 063118	1.419	0.34995 23981 779056	1.469	0.38458 18971 917403
1.370	0.31481 07398 400335	1.420	0.35065 68716 131694	1.470	0.38526 24007 906449
1.371	0.31554 04005 801773	1.421	0.35136 08491 149636	1.471	0.38594 24416 193005
1.372	0.31626 95293 036935	1.422	0.35206 43313 810491	1.472	0.38662 20203 066845
1.373	0.31699 81267 858340	1.423	0.35276 73191 077153	1.473	0.38730 11374 804932
1.374	0.31772 61938 001576	1.424	0.35346 98129 897840	1.474	0.38797 97937 671449
1.375	0.31845 37311 185346	1.425	0.35417 18137 206138	1.475	0.38865 79897 917831
1.376	0.31918 07395 111519	1.426	0.35487 33219 921042	1.476	0.38933 57261 782808
1.377	0.31990 72197 465178	1.427	0.35557 43384 946994	1.477	0.39001 30035 492427
1.378	0.32063 31725 914668	1.428	0.35627 48639 173926	1.478	0.39068 98225 260100
1.379	0.32135 85988 111648	1.429	0.35697 48989 477304	1.479	0.39136 61837 286627
1.380	0.32208 34991 691133	1.430	0.35767 44442 718159	1.480	0.39204 20877 760237
1.381	0.32280 78744 271551	1.431	0.35837 35005 743139	1.481	0.39271 75352 856617
1.382	0.32353 17253 454782	1.432	0.35907 20685 384539	1.482	0.39339 25268 738951
1.383	0.32425 50526 826212	1.433	0.35977 01488 460348	1.483	0.39406 70631 557950
1.384	0.32497 78571 954778	1.434	0.36046 77421 774286	1.484	0.39474 11447 451887
1.385	0.32570 01396 393018	1.435	0.36116 48492 115844	1.485	0.39541 47722 546629
1.386	0.32642 19007 677115	1.436	0.36186 14706 260324	1.486	0.39608 79462 955674
1.387	0.32714 31413 326945	1.437	0.36255 76070 968879	1.487	0.39676 06674 780180
1.388	0.32786 38620 846128	1.438	0.36325 32592 988549	1.488	0.39743 29364 109001
1.389	0.32858 40637 722067	1.439	0.36394 84279 052308	1.489	0.39810 47537 018719
1.390	0.32930 37471 426004	1.440	0.36464 31135 879093	1.490	0.39877 61199 573678
1.391	0.33002 29129 413059	1.441	0.36533 73170 173850	1.491	0.39944 70357 826014
1.392	0.33074 15619 122279	1.442	0.36603 10388 627573	1.492	0.40011 75017 815691
1.393	0.33145 96947 976686	1.443	0.36672 42797 917338	1.493	0.40078 75185 570533
1.394	0.33217 73123 383321	1.444	0.36741 70404 706345	1.494	0.40145 70867 106256
1.395	0.33289 44152 733290	1.445	0.36810 93215 643955	1.495	0.40212 62068 426497
1.396	0.33361 10043 401807	1.446	0.36880 11237 365729	1.496	0.40279 48795 522855
1.397	0.33432 70802 748248	1.447	0.36949 24476 493468	1.497	0.40346 31054 374913
1.398	0.33504 26438 116185	1.448	0.37018 32939 635246	1.498	0.40413 08850 950277
1.399	0.33575 76956 833441	1.449	0.37087 36633 385453	1.499	0.40479 82191 204607
1.400	0.33647 22366 212129	1.450	0.37156 35564 324830	1.500	0.40546 51081 081644
	$\left[\begin{matrix}(-8)7\\6\end{matrix}\right]$		$\left[\begin{matrix}(-8)6\\6\end{matrix}\right]$		$\left[\begin{matrix}(-8)6\\5\end{matrix}\right]$

$$\ln 10 = 2.30258\ 50929\ 940457$$

Table 4.2   **NATURAL LOGARITHMS**

x	$\ln x$	x	$\ln x$	x	$\ln x$
1.500	0.40546 51081 081644	1.550	0.43825 49309 311553	1.600	0.47000 36292 457356
1.501	0.40613 15526 513249	1.551	0.43889 98841 944018	1.601	0.47062 84340 145776
1.502	0.40679 75533 419430	1.552	0.43954 44217 610270	1.602	0.47125 28486 461675
1.503	0.40746 31107 708374	1.553	0.44018 85441 665500	1.603	0.47187 68736 274159
1.504	0.40812 82255 276481	1.554	0.44083 22519 454557	1.604	0.47250 05094 443228
1.505	0.40879 28982 008391	1.555	0.44147 55456 311975	1.605	0.47312 37565 819792
1.506	0.40945 71293 777018	1.556	0.44211 84257 561999	1.606	0.47374 66155 245699
1.507	0.41012 09196 443584	1.557	0.44276 08928 518613	1.607	0.47436 90867 553755
1.508	0.41078 42695 857643	1.558	0.44340 29474 485565	1.608	0.47499 11707 567746
1.509	0.41144 71797 857118	1.559	0.44404 45900 756395	1.609	0.47561 28680 102462
1.510	0.41210 96508 268330	1.560	0.44468 58212 614457	1.610	0.47623 41789 963716
1.511	0.41277 16832 906025	1.561	0.44532 66415 332950	1.611	0.47685 51041 948373
1.512	0.41343 32777 573413	1.562	0.44596 70514 174942	1.612	0.47747 56440 844365
1.513	0.41409 44348 062189	1.563	0.44660 70514 393396	1.613	0.47809 57991 430718
1.514	0.41475 51550 152570	1.564	0.44724 66421 231193	1.614	0.47871 55698 477571
1.515	0.41541 54389 613325	1.565	0.44788 58239 921165	1.615	0.47933 49566 746199
1.516	0.41607 52872 201799	1.566	0.44852 45975 686114	1.616	0.47995 39600 989036
1.517	0.41673 47003 663952	1.567	0.44916 29633 738838	1.617	0.48057 25805 949698
1.518	0.41739 36789 734382	1.568	0.44980 09219 282161	1.618	0.48119 08186 362999
1.519	0.41805 22236 136358	1.569	0.45043 84737 508955	1.619	0.48180 86746 954981
1.520	0.41871 03348 581850	1.570	0.45107 56193 602167	1.620	0.48242 61492 442927
1.521	0.41936 80132 771558	1.571	0.45171 23592 734841	1.621	0.48304 32427 535391
1.522	0.42002 52594 394941	1.572	0.45234 86940 070148	1.622	0.48365 99556 932212
1.523	0.42068 20739 130248	1.573	0.45298 46240 761408	1.623	0.48427 62885 324542
1.524	0.42133 84572 644545	1.574	0.45362 01499 952115	1.624	0.48489 22417 394862
1.525	0.42199 44100 593749	1.575	0.45425 52722 775964	1.625	0.48550 78157 817008
1.526	0.42264 99328 622653	1.576	0.45488 99914 356874	1.626	0.48612 30111 256188
1.527	0.42330 50262 364954	1.577	0.45552 43079 809013	1.627	0.48673 78282 369007
1.528	0.42395 96907 443287	1.578	0.45615 82224 236825	1.628	0.48735 22675 803486
1.529	0.42461 39269 469252	1.579	0.45679 17352 735050	1.629	0.48796 63296 199081
1.530	0.42526 77354 043441	1.580	0.45742 48470 388754	1.630	0.48858 00148 186710
1.531	0.42592 11166 755467	1.581	0.45805 75582 273350	1.631	0.48919 33236 388768
1.532	0.42657 40713 183996	1.582	0.45868 98693 454621	1.632	0.48980 62565 419153
1.533	0.42722 65998 896771	1.583	0.45932 17808 988751	1.633	0.49041 88139 883281
1.534	0.42787 87029 450644	1.584	0.45995 32933 922341	1.634	0.49103 09964 378111
1.535	0.42853 03810 391605	1.585	0.46058 44073 292439	1.635	0.49164 28043 492167
1.536	0.42918 16347 254804	1.586	0.46121 51232 126562	1.636	0.49225 42381 805553
1.537	0.42983 24645 564588	1.587	0.46184 54415 442720	1.637	0.49286 52983 889979
1.538	0.43048 28710 834522	1.588	0.46247 53628 249440	1.638	0.49347 59854 308777
1.539	0.43113 28548 567422	1.589	0.46310 48875 545789	1.639	0.49408 62997 616926
1.540	0.43178 24164 255378	1.590	0.46373 40162 321402	1.640	0.49469 62418 361071
1.541	0.43243 15563 379787	1.591	0.46436 27493 556498	1.641	0.49530 58121 079538
1.542	0.43308 02751 411377	1.592	0.46499 10874 221913	1.642	0.49591 50110 302365
1.543	0.43372 85733 810238	1.593	0.46561 90309 279115	1.643	0.49652 38390 551310
1.544	0.43437 64516 025844	1.594	0.46624 65803 680233	1.644	0.49713 22966 339882
1.545	0.43502 39103 497088	1.595	0.46687 37362 368079	1.645	0.49774 03842 173352
1.546	0.43567 09501 652302	1.596	0.46750 04990 276170	1.646	0.49834 81022 548781
1.547	0.43631 75715 909291	1.597	0.46812 68692 328754	1.647	0.49895 54511 955033
1.548	0.43696 37751 675354	1.598	0.46875 28473 440829	1.648	0.49956 24314 872800
1.549	0.43760 95614 347316	1.599	0.46937 84338 518172	1.649	0.50016 90435 774619
1.550	0.43825 49309 311553	1.600	0.47000 36292 457356	1.650	0.50077 52879 124892

$$\begin{bmatrix} (-8)6 \\ 5 \end{bmatrix} \qquad \begin{bmatrix} (-8)5 \\ 5 \end{bmatrix} \qquad \begin{bmatrix} (-8)5 \\ 5 \end{bmatrix}$$

$$\ln 10 = 2.30258\ 50929\ 940457$$

NATURAL LOGARITHMS

Table 4.2

x	$\ln x$	x	$\ln x$	x	$\ln x$
1.650	0.50077 52879 124892	1.700	0.53062 82510 621704	1.750	0.55961 57879 354227
1.651	0.50138 11649 379910	1.701	0.53121 63134 137247	1.751	0.56018 70533 037148
1.652	0.50198 66750 987863	1.702	0.53180 40301 511824	1.752	0.56075 79925 141997
1.653	0.50259 18188 388871	1.703	0.53239 14016 805512	1.753	0.56132 86059 390974
1.654	0.50319 65966 014996	1.704	0.53297 84284 071240	1.754	0.56189 88939 499913
1.655	0.50380 10088 290262	1.705	0.53356 51107 354801	1.755	0.56246 88569 178291
1.656	0.50440 50559 630679	1.706	0.53415 14490 694874	1.756	0.56303 84952 129249
1.657	0.50500 87384 444259	1.707	0.53473 74438 123036	1.757	0.56360 78092 049601
1.658	0.50561 20567 131032	1.708	0.53532 30953 663781	1.758	0.56417 67992 629853
1.659	0.50621 50112 083074	1.709	0.53590 84041 334538	1.759	0.56474 54657 554211
1.660	0.50681 76023 684519	1.710	0.53649 33705 145685	1.760	0.56531 38090 500604
1.661	0.50741 98306 311578	1.711	0.53707 79949 100564	1.761	0.56588 18295 140691
1.662	0.50802 16964 332564	1.712	0.53766 22777 195504	1.762	0.56644 95275 139878
1.663	0.50862 32002 107906	1.713	0.53824 62193 419829	1.763	0.56701 69034 157332
1.664	0.50922 43423 990168	1.714	0.53882 98201 755880	1.764	0.56758 39575 845996
1.665	0.50982 51234 324071	1.715	0.53941 30806 179032	1.765	0.56815 06903 852601
1.666	0.51042 55437 446509	1.716	0.53999 60010 657705	1.766	0.56871 71021 817683
1.667	0.51102 56037 686569	1.717	0.54057 85819 153385	1.767	0.56928 31933 375593
1.668	0.51162 53039 365550	1.718	0.54116 08235 620636	1.768	0.56984 89642 154517
1.669	0.51222 46446 796980	1.719	0.54174 27264 007122	1.769	0.57041 44151 776482
1.670	0.51282 36264 286637	1.720	0.54232 42908 253617	1.770	0.57097 95465 857378
1.671	0.51342 22496 132567	1.721	0.54290 55172 294024	1.771	0.57154 43588 006965
1.672	0.51402 05146 625099	1.722	0.54348 64060 055391	1.772	0.57210 88521 828892
1.673	0.51461 84220 046869	1.723	0.54406 69575 457926	1.773	0.57267 30270 920708
1.674	0.51521 59720 672836	1.724	0.54464 71722 415014	1.774	0.57323 68838 873877
1.675	0.51581 31652 770298	1.725	0.54522 70504 833231	1.775	0.57380 04229 273791
1.676	0.51641 00020 598913	1.726	0.54580 65926 612362	1.776	0.57436 36445 699783
1.677	0.51700 64828 410718	1.727	0.54638 57991 645415	1.777	0.57492 65491 725143
1.678	0.51760 26080 450144	1.728	0.54696 46703 818639	1.778	0.57548 91370 917128
1.679	0.51819 83780 954038	1.729	0.54754 32067 011534	1.779	0.57605 14086 836981
1.680	0.51879 37934 151676	1.730	0.54812 14085 096876	1.780	0.57661 33643 039938
1.681	0.51938 88544 264786	1.731	0.54869 92761 940722	1.781	0.57717 50043 075246
1.682	0.51998 35615 507563	1.732	0.54927 68101 402434	1.782	0.57773 63290 486176
1.683	0.52057 79152 086690	1.733	0.54985 40107 334690	1.783	0.57829 73388 810034
1.684	0.52117 19158 201350	1.734	0.55043 08783 583501	1.784	0.57885 80341 578176
1.685	0.52176 55638 043250	1.735	0.55100 74133 988225	1.785	0.57941 84152 316024
1.686	0.52235 88595 796637	1.736	0.55158 36162 381584	1.786	0.57997 84824 543073
1.687	0.52295 18035 638312	1.737	0.55215 94872 589679	1.787	0.58053 82361 772910
1.688	0.52354 43961 737654	1.738	0.55273 50268 432003	1.788	0.58109 76767 513224
1.689	0.52413 66378 256630	1.739	0.55331 02353 721460	1.789	0.58165 68045 265821
1.690	0.52472 85289 349821	1.740	0.55388 51132 264377	1.790	0.58221 56198 526636
1.691	0.52532 00699 164432	1.741	0.55445 96607 860520	1.791	0.58277 41230 785747
1.692	0.52591 12611 840315	1.742	0.55503 38784 303111	1.792	0.58333 23145 527387
1.693	0.52650 21031 509983	1.743	0.55560 77665 378839	1.793	0.58389 01946 229958
1.694	0.52709 25962 298627	1.744	0.55618 13254 867879	1.794	0.58444 77636 366044
1.695	0.52768 27408 324136	1.745	0.55675 45556 543905	1.795	0.58500 50219 402422
1.696	0.52827 25373 697113	1.746	0.55732 74574 174105	1.796	0.58556 19698 800079
1.697	0.52886 19862 520893	1.747	0.55790 00311 519195	1.797	0.58611 86078 014220
1.698	0.52945 10878 891556	1.748	0.55847 22772 333437	1.798	0.58667 49360 494285
1.699	0.53003 98426 897950	1.749	0.55904 41960 364650	1.799	0.58723 09549 683961
1.700	0.53062 82510 621704	1.750	0.55961 57879 354227	1.800	0.58778 66649 021190
	$\left[\begin{smallmatrix}(-8)5\\5\end{smallmatrix}\right]$		$\left[\begin{smallmatrix}(-8)4\\5\end{smallmatrix}\right]$		$\left[\begin{smallmatrix}(-8)4\\5\end{smallmatrix}\right]$

$$\ln 10 = 2.30258\ 50929\ 940457$$

Table 4.2   **NATURAL LOGARITHMS**

x	$\ln x$	x	$\ln x$	x	$\ln x$
1.800	0.58778 66649 021190	1.850	0.61518 56390 902335	1.900	0.64185 38861 723948
1.801	0.58834 20661 938190	1.851	0.61572 60335 913605	1.901	0.64238 00635 062921
1.802	0.58889 71591 861462	1.852	0.61626 61362 239876	1.902	0.64290 59641 231986
1.803	0.58945 19442 211802	1.853	0.61680 59473 032227	1.903	0.64343 15883 140124
1.804	0.59000 64216 404319	1.854	0.61734 54671 436634	1.904	0.64395 69363 691736
1.805	0.59056 05917 848442	1.855	0.61788 46960 593985	1.905	0.64448 20085 786643
1.806	0.59111 44549 947937	1.856	0.61842 36343 640088	1.906	0.64500 68052 320104
1.807	0.59166 80116 100914	1.857	0.61896 22823 705687	1.907	0.64553 13266 182820
1.808	0.59222 12619 699848	1.858	0.61950 06403 916468	1.908	0.64605 55730 260948
1.809	0.59277 42064 131581	1.859	0.62003 87087 393070	1.909	0.64657 95447 436106
1.810	0.59332 68452 777344	1.860	0.62057 64877 251099	1.910	0.64710 32420 585385
1.811	0.59387 91789 012763	1.861	0.62111 39776 601137	1.911	0.64762 66652 581360
1.812	0.59443 12076 207876	1.862	0.62165 11788 548753	1.912	0.64814 98146 292095
1.813	0.59498 29317 727140	1.863	0.62218 80916 194514	1.913	0.64867 26904 581158
1.814	0.59553 43516 929449	1.864	0.62272 47162 633994	1.914	0.64919 52930 307625
1.815	0.59608 54677 168141	1.865	0.62326 10530 957789	1.915	0.64971 76226 326093
1.816	0.59663 62801 791016	1.866	0.62379 71024 251521	1.916	0.65023 96795 486688
1.817	0.59718 67894 140341	1.867	0.62433 28645 595856	1.917	0.65076 14640 635074
1.818	0.59773 69957 552871	1.868	0.62486 83398 066509	1.918	0.65128 29764 612465
1.819	0.59828 68995 359852	1.869	0.62540 35284 734258	1.919	0.65180 42170 255629
1.820	0.59883 65010 887040	1.870	0.62593 84308 664953	1.920	0.65232 51860 396902
1.821	0.59938 58007 454709	1.871	0.62647 30472 919526	1.921	0.65284 58837 864196
1.822	0.59993 47988 377666	1.872	0.62700 73780 554003	1.922	0.65336 63105 481007
1.823	0.60048 34956 965260	1.873	0.62754 14234 619515	1.923	0.65388 64666 066427
1.824	0.60103 18916 521396	1.874	0.62807 51838 162304	1.924	0.65440 63522 435147
1.825	0.60157 99870 344548	1.875	0.62860 86594 223741	1.925	0.65492 59677 397475
1.826	0.60212 77821 727767	1.876	0.62914 18505 840329	1.926	0.65544 53133 759338
1.827	0.60267 52773 958697	1.877	0.62967 47576 043718	1.927	0.65596 43894 322293
1.828	0.60322 24730 319583	1.878	0.63020 73807 860712	1.928	0.65648 31961 883539
1.829	0.60376 93694 087286	1.879	0.63073 97204 313283	1.929	0.65700 17339 235920
1.830	0.60431 59668 533296	1.880	0.63127 17768 418578	1.930	0.65752 00029 167942
1.831	0.60486 22656 923737	1.881	0.63180 35503 188933	1.931	0.65803 80034 463774
1.832	0.60540 82662 519385	1.882	0.63233 50411 631879	1.932	0.65855 57357 903263
1.833	0.60595 39688 575680	1.883	0.63286 62496 750154	1.933	0.65907 32002 261938
1.834	0.60649 93738 342731	1.884	0.63339 71761 541713	1.934	0.65959 03970 311026
1.835	0.60704 44815 065336	1.885	0.63392 78208 999741	1.935	0.66010 73264 817451
1.836	0.60758 92921 982987	1.886	0.63445 81842 112658	1.936	0.66062 39888 543853
1.837	0.60813 38062 329886	1.887	0.63498 82663 864132	1.937	0.66114 03844 248588
1.838	0.60867 80239 334953	1.888	0.63551 80677 233089	1.938	0.66165 65134 685745
1.839	0.60922 19456 221840	1.889	0.63604 75885 193725	1.939	0.66217 23762 605148
1.840	0.60976 55716 208943	1.890	0.63657 68290 715510	1.940	0.66268 79730 752368
1.841	0.61030 89022 509408	1.891	0.63710 57896 763204	1.941	0.66320 33041 868732
1.842	0.61085 19378 331151	1.892	0.63763 44706 296865	1.942	0.66371 83698 691332
1.843	0.61139 46786 876862	1.893	0.63816 28722 271858	1.943	0.66423 31703 953030
1.844	0.61193 71251 344021	1.894	0.63869 09947 638865	1.944	0.66474 77060 382473
1.845	0.61247 92774 924905	1.895	0.63921 88385 343897	1.945	0.66526 19770 704096
1.846	0.61302 11360 806604	1.896	0.63974 64038 328301	1.946	0.66577 59837 638133
1.847	0.61356 27012 171029	1.897	0.64027 36909 528772	1.947	0.66628 97263 900626
1.848	0.61410 39732 194924	1.898	0.64080 07001 877361	1.948	0.66680 32052 203434
1.849	0.61464 49524 049878	1.899	0.64132 74318 301488	1.949	0.66731 64205 254238
1.850	0.61518 56390 902335	1.900	0.64185 38861 723948	1.950	0.66782 93725 756554
	$\left[\begin{array}{c}(-8)4 \\ 5\end{array}\right]$		$\left[\begin{array}{c}(-8)4 \\ 5\end{array}\right]$		$\left[\begin{array}{c}(-8)3 \\ 5\end{array}\right]$

$$\ln 10 = 2.30258 \ 50929 \ 940457$$

NATURAL LOGARITHMS

Table 4.2

x	$\ln x$	x	$\ln x$	x	$\ln x$
1.950	0.66782 93725 756554	2.000	0.69314 71805 599453	2.050	0.71783 97931 503168
1.951	0.66834 20616 409742	2.001	0.69364 70556 015964	2.051	0.71832 74790 902436
1.952	0.66885 44879 909007	2.002	0.69414 66808 930288	2.052	0.71881 49273 085231
1.953	0.66936 66518 945419	2.003	0.69464 60566 836812	2.053	0.71930 21380 367965
1.954	0.66987 85536 205910	2.004	0.69514 51832 226184	2.054	0.71978 91115 063665
1.955	0.67039 01934 373291	2.005	0.69564 40607 585325	2.055	0.72027 58479 481979
1.956	0.67090 15716 126256	2.006	0.69614 26895 397438	2.056	0.72076 23475 929187
1.957	0.67141 26884 139392	2.007	0.69664 10698 142011	2.057	0.72124 86106 708201
1.958	0.67192 35441 083186	2.008	0.69713 92018 294828	2.058	0.72173 46374 118579
1.959	0.67243 41389 624037	2.009	0.69763 70858 327974	2.059	0.72222 04280 456524
1.960	0.67294 44732 424259	2.010	0.69813 47220 709844	2.060	0.72270 59828 014897
1.961	0.67345 45472 142092	2.011	0.69863 21107 905150	2.061	0.72319 13019 083220
1.962	0.67396 43611 431713	2.012	0.69912 92522 374928	2.062	0.72367 63855 947682
1.963	0.67447 39152 943240	2.013	0.69962 61466 576544	2.063	0.72416 12340 891148
1.964	0.67498 32099 322741	2.014	0.70012 27942 963706	2.064	0.72464 58476 193163
1.965	0.67549 22453 212246	2.015	0.70061 91953 986463	2.065	0.72513 02264 129961
1.966	0.67600 10217 249748	2.016	0.70111 53502 091222	2.066	0.72561 43706 974468
1.967	0.67650 95394 069220	2.017	0.70161 12589 720747	2.067	0.72609 82806 996312
1.968	0.67701 77986 300617	2.018	0.70210 69219 314172	2.068	0.72658 19566 461827
1.969	0.67752 57996 569885	2.019	0.70260 23393 307004	2.069	0.72706 53987 634060
1.970	0.67803 35427 498971	2.020	0.70309 75114 131134	2.070	0.72754 86072 772777
1.971	0.67854 10281 705832	2.021	0.70359 24384 214840	2.071	0.72803 15824 134471
1.972	0.67904 82561 804437	2.022	0.70408 71205 982797	2.072	0.72851 43243 972366
1.973	0.67955 52270 404783	2.023	0.70458 15581 856084	2.073	0.72899 68334 536425
1.974	0.68006 19410 112898	2.024	0.70507 57514 252191	2.074	0.72947 91098 073356
1.975	0.68056 83983 530852	2.025	0.70556 97005 585025	2.075	0.72996 11536 826616
1.976	0.68107 45993 256761	2.026	0.70606 34058 264916	2.076	0.73044 29653 036422
1.977	0.68158 05441 884799	2.027	0.70655 68674 698630	2.077	0.73092 45448 939753
1.978	0.68208 62332 005204	2.028	0.70705 00857 289367	2.078	0.73140 58926 770357
1.979	0.68259 16666 204287	2.029	0.70754 30608 436777	2.079	0.73188 70088 758759
1.980	0.68309 68447 064439	2.030	0.70803 57930 536960	2.080	0.73236 78937 132266
1.981	0.68360 17677 164139	2.031	0.70852 82825 982476	2.081	0.73284 85474 114974
1.982	0.68410 64359 077962	2.032	0.70902 05297 162355	2.082	0.73332 89701 927771
1.983	0.68461 08495 376589	2.033	0.70951 25346 462096	2.083	0.73380 91622 788349
1.984	0.68511 50088 626811	2.034	0.71000 42976 263682	2.084	0.73428 91238 911205
1.985	0.68561 89141 391537	2.035	0.71049 58188 945583	2.085	0.73476 88552 507648
1.986	0.68612 25656 229808	2.036	0.71098 70986 882763	2.086	0.73524 83565 785807
1.987	0.68662 59635 696798	2.037	0.71147 81372 446688	2.087	0.73572 76280 950637
1.988	0.68712 91082 343823	2.038	0.71196 89348 005331	2.088	0.73620 66700 203923
1.989	0.68763 19998 718351	2.039	0.71245 94915 923181	2.089	0.73668 54825 744287
1.990	0.68813 46387 364010	2.040	0.71294 98078 561250	2.090	0.73716 40659 767196
1.991	0.68863 70250 820592	2.041	0.71343 98838 277077	2.091	0.73764 24204 464965
1.992	0.68913 91591 624065	2.042	0.71392 97197 424738	2.092	0.73812 05462 026765
1.993	0.68964 10412 306577	2.043	0.71441 93158 354850	2.093	0.73859 84434 638627
1.994	0.69014 26715 396466	2.044	0.71490 86723 414580	2.094	0.73907 61124 483451
1.995	0.69064 40503 418268	2.045	0.71539 77894 947651	2.095	0.73955 35533 741011
1.996	0.69114 51778 892722	2.046	0.71588 66675 294347	2.096	0.74003 07664 587957
1.997	0.69164 60544 336782	2.047	0.71637 53066 791525	2.097	0.74050 77519 197829
1.998	0.69214 66802 263618	2.048	0.71686 37071 772614	2.098	0.74098 45099 741054
1.999	0.69264 70555 182630	2.049	0.71735 18692 567627	2.099	0.74146 10408 384959
2.000	0.69314 71805 599453	2.050	0.71783 97931 503168	2.100	0.74193 73447 293773

$$\left[\begin{matrix}(-8)3\\5\end{matrix}\right] \qquad \left[\begin{matrix}(-8)3\\5\end{matrix}\right] \qquad \left[\begin{matrix}(-8)3\\5\end{matrix}\right]$$

For $x>2.1$ see **Example 5.** $\ln 10 = 2.30258\ 50929\ 940457$

Table 4.3   **RADIX TABLE OF NATURAL LOGARITHMS**

x	n	$\ln\,(1+x10^{-n})$	$-\ln\,(1-x10^{-n})$
1	10	0. 00000 00000 99999 99999 50000	0. 00000 00001 00000 00000 50000
2	10	0. 00000 00001 99999 99998 00000	0. 00000 00002 00000 00002 00000
3	10	0. 00000 00002 99999 99995 50000	0. 00000 00003 00000 00004 50000
4	10	0. 00000 00003 99999 99992 00000	0. 00000 00004 00000 00008 00000
5	10	0. 00000 00004 99999 99987 50000	0. 00000 00005 00000 00012 50000
6	10	0. 00000 00005 99999 99982 00000	0. 00000 00006 00000 00018 00000
7	10	0. 00000 00006 99999 99975 50000	0. 00000 00007 00000 00024 50000
8	10	0. 00000 00007 99999 99968 00000	0. 00000 00008 00000 00032 00000
9	10	0. 00000 00008 99999 99959 50000	0. 00000 00009 00000 00040 50000
1	9	0. 00000 00009 99999 99950 00000	0. 00000 00010 00000 00050 00000
2	9	0. 00000 00019 99999 99800 00000	0. 00000 00020 00000 00200 00000
3	9	0. 00000 00029 99999 99550 00000	0. 00000 00030 00000 00450 00000
4	9	0. 00000 00039 99999 99200 00000	0. 00000 00040 00000 00800 00000
5	9	0. 00000 00049 99999 98750 00000	0. 00000 00050 00000 01250 00000
6	9	0. 00000 00059 99999 98200 00001	0. 00000 00060 00000 01800 00001
7	9	0. 00000 00069 99999 97550 00001	0. 00000 00070 00000 02450 00001
8	9	0. 00000 00079 99999 96800 00002	0. 00000 00080 00000 03200 00002
9	9	0. 00000 00089 99999 95950 00002	0. 00000 00090 00000 04050 00002
1	8	0. 00000 00099 99999 95000 00003	0. 00000 00100 00000 05000 00003
2	8	0. 00000 00199 99999 80000 00027	0. 00000 00200 00000 20000 00027
3	8	0. 00000 00299 99999 55000 00090	0. 00000 00300 00000 45000 00090
4	8	0. 00000 00399 99999 20000 00213	0. 00000 00400 00000 80000 00213
5	8	0. 00000 00499 99998 75000 00417	0. 00000 00500 00001 25000 00417
6	8	0. 00000 00599 99998 20000 00720	0. 00000 00600 00001 80000 00720
7	8	0. 00000 00699 99997 55000 01143	0. 00000 00700 00002 45000 01143
8	8	0. 00000 00799 99996 80000 01707	0. 00000 00800 00003 20000 01707
9	8	0. 00000 00899 99995 95000 02430	0. 00000 00900 00004 05000 02430
1	7	0. 00000 00999 99995 00000 03333	0. 00000 01000 00005 00000 03333
2	7	0. 00000 01999 99980 00000 26667	0. 00000 02000 00020 00000 26667
3	7	0. 00000 02999 99955 00000 90000	0. 00000 03000 00045 00000 90000
4	7	0. 00000 03999 99920 00002 13333	0. 00000 04000 00080 00002 13333
5	7	0. 00000 04999 99875 00004 16667	0. 00000 05000 00125 00004 16667
6	7	0. 00000 05999 99820 00007 20000	0. 00000 06000 00180 00007 20000
7	7	0. 00000 06999 99755 00011 43333	0. 00000 07000 00245 00011 43334
8	7	0. 00000 07999 99680 00017 06666	0. 00000 08000 00320 00017 06668
9	7	0. 00000 08999 99595 00024 29998	0. 00000 09000 00405 00024 30002
1	6	0. 00000 09999 99500 00033 33331	0. 00000 10000 00500 00033 33336
2	6	0. 00000 19999 98000 00266 66627	0. 00000 20000 02000 00266 66707
3	6	0. 00000 29999 95500 00899 99798	0. 00000 30000 04500 00900 00203
4	6	0. 00000 39999 92000 02133 32693	0. 00000 40000 08000 02133 33973
5	6	0. 00000 49999 87500 04166 65104	0. 00000 50000 12500 04166 68229
6	6	0. 00000 59999 82000 07199 96760	0. 00000 60000 18000 07200 03240
7	6	0. 00000 69999 75500 11433 27331	0. 00000 70000 24500 11433 39336
8	6	0. 00000 79999 68000 17066 56427	0. 00000 80000 32000 17066 76907
9	6	0. 00000 89999 59500 24299 83598	0. 00000 90000 40500 24300 16403

For $n>10$, $\ln\,(1\pm x10^{-n}) = \pm x10^{-n} - \dfrac{1}{2}x^2 10^{-2n}$ to 25 D.

RADIX TABLE OF NATURAL LOGARITHMS

Table 4.3

x	n	$\ln\,(1+x10^{-n})$	$-\ln\,(1-x10^{-n})$
1	5	0. 00000 99999 50000 33333 08334	0. 00001 00000 50000 33333 58334
2	5	0. 00001 99998 00002 66662 66673	0. 00002 00002 00002 66670 66673
3	5	0. 00002 99995 50008 99979 75049	0. 00003 00004 50009 00020 25049
4	5	0. 00003 99992 00021 33269 33538	0. 00004 00008 00021 33397 33538
5	5	0. 00004 99987 50041 66510 42292	0. 00005 00012 50041 66822 92292
6	5	0. 00005 99982 00071 99676 01555	0. 00006 00018 00072 00324 01555
7	5	0. 00006 99975 50114 32733 11695	0. 00007 00024 50114 33933 61695
8	5	0. 00007 99968 00170 65642 73220	0. 00008 00032 00170 67690 73221
9	5	0. 00008 99959 50242 98359 86809	0. 00009 00040 50243 01640 36811
1	4	0. 00009 99950 00333 30833 53332	0. 00010 00050 00333 35833 53335
2	4	0. 00019 99800 02666 26673 06560	0. 00020 00200 02667 06673 06773
3	4	0. 00029 99550 08997 97548 58785	0. 00030 00450 09002 02548 61215
4	4	0. 00039 99200 21326 93538 06509	0. 00040 00800 21339 73538 20162
5	4	0. 00049 98750 41651 04791 40636	0. 00050 01250 41682 29791 92719
6	4	0. 00059 98200 71967 61554 42280	0. 00060 01800 72032 41555 97800
7	4	0. 00069 97551 14273 34192 77369	0. 00070 02451 14393 39196 69533
8	4	0. 00079 96801 70564 33215 90059	0. 00080 03201 70769 13224 63873
9	4	0. 00089 95952 42836 09300 94948	0. 00090 04052 43164 14318 66419
1	3	0. 00099 95003 33083 53316 68094	0. 00100 05003 33583 53350 01430
2	3	0. 00199 80026 62673 05601 82538	0. 00200 20026 70673 07735 16511
3	3	0. 00299 55089 79798 47881 16106	0. 00300 45090 20298 72181 32509
4	3	0. 00399 20212 69537 45299 90751	0. 00400 80213 97538 81834 87927
5	3	0. 00498 75415 11039 07361 21022	0. 00501 25418 23544 28204 30937
6	3	0. 00598 20716 77547 46378 20189	0. 00601 80723 25563 01620 19350
7	3	0. 00697 56137 36425 24209 95222	0. 00702 46149 36964 45987 41123
8	3	0. 00796 81696 49176 87351 07973	0. 00803 21716 97264 25903 86494
9	3	0. 00895 97413 71471 90444 31465	0. 00904 07446 52149 06220 55241
1	2	0. 00995 03308 53168 08284 82154	0. 01005 03358 53501 44118 35489
2	2	0. 01980 26272 96179 71302 60291	0. 02020 27073 17519 44840 80453
3	2	0. 02955 88022 41544 40273 26194	0. 03045 92074 84708 54591 92613
4	2	0. 03922 07131 53281 29626 92009	0. 04082 19945 20255 12955 45771
5	2	0. 04879 01641 69432 00306 53744	0. 05129 32943 87550 53342 61961
6	2	0. 05826 89081 23975 77552 57184	0. 06187 54037 18087 47179 78001
7	2	0. 06765 86484 73814 80526 84159	0. 07257 06928 34835 43071 15733
8	2	0. 07696 10411 36128 32498 42170	0. 08338 16089 39051 05839 47658
9	2	0. 08617 76962 41052 33234 13335	0. 09431 06794 71241 32687 71427
1	1	0. 09531 01798 04324 86004 39521	0. 10536 05156 57826 30122 75010
2	1	0. 18232 15567 93954 62621 17180	0. 22314 35513 14209 75576 62951
3	1	0. 26236 42644 67491 05203 54960	0. 35667 49439 38732 37891 26387
4	1	0. 33647 22366 21212 93050 45934	0. 51082 56237 65990 68320 55141
5	1	0. 40546 51081 08164 38197 80131	0. 69314 71805 59945 30941 72321
6	1	0. 47000 36292 45735 55365 09370	0. 91629 07318 74155 06518 35272
7	1	0. 53062 82510 62170 39623 15432	1. 20397 28043 25935 99262 27462
8	1	0. 58778 66649 02119 00818 97311	1. 60943 79124 34100 37460 07593
9	1	0. 64185 38861 72394 77599 10360	2. 30258 50929 94045 68401 79915
1	0	0. 69314 71805 59945 30941 72321	∞

Table 4.4   **EXPONENTIAL FUNCTION**

x	e^x	e^{-x}
0.000	1.00000 00000 00000 000	1.00000 00000 00000 000
0.001	1.00100 05001 66708 342	0.99900 04998 33374 992
0.002	1.00200 20013 34000 267	0.99800 19986 67333 067
0.003	1.00300 45045 03377 026	0.99700 44955 03372 976
0.004	1.00400 80106 77341 872	0.99600 79893 43991 472
0.005	1.00501 25208 59401 063	0.99501 24791 92682 313
0.006	1.00601 80360 54064 865	0.99401 79640 53935 265
0.007	1.00702 45572 66848 555	0.99302 44429 33235 105
0.008	1.00803 20855 04273 431	0.99203 19148 37060 630
0.009	1.00904 06217 73867 814	0.99104 03787 72883 662
0.010	1.01005 01670 84168 058	0.99004 ,8337 49168 054
0.011	1.01106 07224 44719 556	0.98906 02787 75368 698
0.012	1.01207 22888 66077 754	0.98807 17128 61930 540
0.013	1.01308 48673 59809 158	0.98708 41350 20287 583
0.014	1.01409 84589 38492 345	0.98609 75442 62861 903
0.015	1.01511 30646 15718 979	0.98511 19396 03062 661
0.016	1.01612 86854 06094 822	0.98412 73200 55285 115
0.017	1.01714 53223 25240 748	0.98314 36846 34909 635
0.018	1.01816 29763 89793 761	0.98216 10323 58300 718
0.019	1.01918 16486 17408 011	0.98117 93622 42806 006
0.020	1.02020 13400 26755 810	0.98019 86733 06755 302
0.021	1.02122 20516 37528 653	0.97921 89645 69459 588
0.022	1.02224 37844 70438 235	0.97824 02350 51210 045
0.023	1.02326 65395 47217 475	0.97726 24837 73277 073
0.024	1.02429 03178 90621 534	0.97628 57097 57909 314
0.025	1.02531 51205 24428 841	0.97530 99120 28332 669
0.026	1.02634 09484 73442 115	0.97433 50896 08749 328
0.027	1.02736 78027 63489 392	0.97336 12415 24336 791
0.028	1.02839 56844 21425 045	0.97238 83668 01246 891
0.029	1.02942 45944 75130 820	0.97141 64644 66604 825
0.030	1.03045 45339 53516 856	0.97044 55335 48508 177
0.031	1.03148 55038 86522 716	0.96947 55730 76025 948
0.032	1.03251 75053 05118 420	0.96850 65820 79197 585
0.033	1.03355 05392 41305 472	0.96753 85595 89032 009
0.034	1.03458 46067 28117 894	0.96657 15046 37506 651
0.035	1.03561 97087 99623 260	0.96560 54162 57566 478
0.036	1.03665 58464 90932 727	0.96464 02934 83123 030
0.037	1.03769 30208 38157 074	0.96367 61353 49053 452
0.038	1.03873 12328 78497 733	0.96271 29408 91199 529
0.039	1.03977 04836 50157 831	0.96175 07091 46366 723
0.040	1.04081 07741 92388 227	0.96078 94391 52323 209
0.041	1.04185 21055 45479 549	0.95982 91299 47798 914
0.042	1.04289 44787 50763 238	0.95886 97805 72484 552
0.043	1.04393 78948 50612 586	0.95791 13900 67030 669
0.044	1.04498 23548 88443 779	0.95695 39574 73046 678
0.045	1.04602 78599 08716 943	0.95599 74818 33099 907
0.046	1.04707 44109 56937 184	0.95504 19621 90714 635
0.047	1.04812 20090 79655 638	0.95408 73975 90371 141
0.048	1.04917 06553 24470 516	0.95313 37870 77504 745
0.049	1.05022 03507 40028 148	0.95218 11296 98504 853
0.050	1.05127 10963 76024 040	0.95122 94245 00714 009
	$\begin{bmatrix} (-7)1 \\ 6 \end{bmatrix}$	$\begin{bmatrix} (-7)1 \\ 6 \end{bmatrix}$

For use and extension of the table see **Examples 8–11**.

See **Table 7.1** for values of $\frac{2}{\sqrt{\pi}} e^{-x^2}$ and **Table 26.1** for $\frac{1}{\sqrt{2\pi}} e^{-\frac{x^2}{2}}$.

EXPONENTIAL FUNCTION

Table 4.4

x	e^x	e^{-x}
0.050	1.05127 10963 76024 040	0.95122 94245 00714 009
0.051	1.05232 28932 83203 913	0.95027 86705 32426 935
0.052	1.05337 57425 13364 763	0.94932 88668 42889 583
0.053	1.05442 96451 19355 907	0.94838 00124 82298 184
0.054	1.05548 46021 55080 041	0.94743 21065 01798 300
0.055	1.05654 06146 75494 286	0.94648 51479 53483 869
0.056	1.05759 76837 36611 252	0.94553 91358 90396 267
0.057	1.05865 58103 95500 087	0.94459 40693 66523 349
0.058	1.05971 49957 10287 540	0.94364 99474 36798 514
0.059	1.06077 52407 40159 012	0.94270 67691 57099 754
0.060	1.06183 65465 45359 622	0.94176 45335 84248 710
0.061	1.06289 89141 87195 264	0.94082 32397 76009 730
0.062	1.06396 23447 28033 669	0.93988 28867 91088 928
0.063	1.06502 68392 31305 464	0.93894 34736 89133 241
0.064	1.06609 23987 61505 244	0.93800 49995 30729 488
0.065	1.06715 90243 84192 625	0.93706 74633 77403 433
0.066	1.06822 67171 65993 321	0.93613 08642 91618 844
0.067	1.06929 54781 74600 202	0.93519 52013 36776 558
0.068	1.07036 53084 78774 366	0.93426 04735 77213 542
0.069	1.07143 62091 48346 205	0.93332 66800 78201 958
0.070	1.07250 81812 54216 479	0.93239 38199 05948 229
0.071	1.07358 12258 68357 383	0.93146 18921 27592 106
0.072	1.07465 53440 63813 620	0.93053 08958 11205 732
0.073	1.07573 05369 14703 476	0.92960 08300 25792 713
0.074	1.07680 68054 96219 891	0.92867 16938 41287 187
0.075	1.07788 41508 84631 536	0.92774 34863 28552 892
0.076	1.07896 25741 57283 889	0.92681 62065 59382 237
0.077	1.08004 20763 92600 313	0.92588 98536 06495 377
0.078	1.08112 26586 70083 133	0.92496 44265 43539 280
0.079	1.08220 43220 70314 717	0.92403 99244 45086 807
0.080	1.08328 70676 74958 554	0.92311 63463 86635 783
0.081	1.08437 08965 66760 341	0.92219 36914 44608 072
0.082	1.08545 58098 29549 059	0.92127 19586 96348 654
0.083	1.08654 18085 48238 061	0.92035 11472 20124 706
0.084	1.08762 88938 08826 156	0.91943 12560 95124 674
0.085	1.08871 70666 98398 696	0.91851 22844 01457 356
0.086	1.08980 63283 05128 660	0.91759 42312 20150 982
0.087	1.09089 66797 18277 747	0.91667 70956 33152 295
0.088	1.09198 81220 28197 460	0.91576 08767 23325 631
0.089	1.09308 06563 26330 201	0.91484 55735 74452 003
0.090	1.09417 42837 05210 358	0.91393 11852 71228 187
0.091	1.09526 90052 58465 401	0.91301 77108 99265 803
0.092	1.09636 48220 80816 975	0.91210 51495 45090 403
0.093	1.09746 17352 68081 994	0.91119 35002 96140 557
0.094	1.09855 97459 17173 736	0.91028 27622 40766 940
0.095	1.09965 88551 26102 942	0.90937 29344 68231 420
0.096	1.10075 90639 93978 912	0.90846 40160 68706 150
0.097	1.10186 03736 21010 606	0.90755 60061 33272 654
0.098	1.10296 27851 08507 743	0.90664 89037 53920 921
0.099	1.10406 62995 58881 902	0.90574 27080 23548 496
0.100	1.10517 09180 75647 625	0.90483 74180 35959 573
	$\begin{bmatrix} (-7)1 \\ 6 \end{bmatrix}$	$\begin{bmatrix} (-7)1 \\ 6 \end{bmatrix}$

Table 4.4   EXPONENTIAL FUNCTION

x	e^x	e^{-x}
0.100	1.10517 09180 75647 625	0.90483 74180 35959 573
0.101	1.10627 66417 63423 521	0.90393 30328 85864 089
0.102	1.10738 34717 27933 371	0.90302 95516 68876 819
0.103	1.10849 14090 76007 230	0.90212 69734 81516 470
0.104	1.10960 04549 15582 540	0.90122 52974 21204 780
0.105	1.11071 06103 55705 232	0.90032 45225 86265 613
0.106	1.11182 18765 06530 839	0.89942 46480 75924 059
0.107	1.11293 42544 79325 605	0.89852 56729 90305 534
0.108	1.11404 77453 86467 594	0.89762 75964 30434 876
0.109	1.11516 23503 41447 807	0.89673 04174 98235 450
0.110	1.11627 80704 58871 292	0.89583 41352 96528 251
0.111	1.11739 49068 54458 258	0.89493 87489 29031 000
0.112	1.11851 28606 45045 196	0.89404 42575 00357 257
0.113	1.11963 19329 48585 987	0.89315 06601 16015 519
0.114	1.12075 21248 84153 031	0.89225 79558 82408 325
0.115	1.12187 34375 71938 354	0.89136 61439 06831 368
0.116	1.12299 58721 33254 738	0.89047 52232 97472 599
0.117	1.12411 94296 90536 839	0.88958 51931 63411 334
0.118	1.12524 41113 67342 307	0.88869 60526 14617 364
0.119	1.12636 99182 88352 913	0.88780 78007 61950 067
0.120	1.12749 68515 79375 671	0.88692 04367 17157 516
0.121	1.12862 49123 67343 967	0.88603 39595 92875 591
0.122	1.12975 41017 80318 682	0.88514 83685 02627 096
0.123	1.13088 44209 47489 324	0.88426 36625 60820 866
0.124	1.13201 58709 99175 153	0.88337 98408 82750 886
0.125	1.13314 84530 66826 317	0.88249 69025 84595 403
0.126	1.13428 21682 83024 976	0.88161 48467 83416 046
0.127	1.13541 70177 81486 442	0.88073 36725 97156 940
0.128	1.13655 30026 97060 307	0.87985 33791 44643 827
0.129	1.13769 01241 65731 582	0.87897 39655 45583 178
0.130	1.13882 83833 24621 831	0.87809 54309 20561 324
0.131	1.13996 77813 11990 306	0.87721 77743 91043 564
0.132	1.14110 83192 67235 091	0.87634 09950 79373 297
0.133	1.14224 99983 30894 235	0.87546 50921 08771 138
0.134	1.14339 28196 44646 898	0.87459 00646 03334 043
0.135	1.14453 67843 51314 488	0.87371 59116 88034 434
0.136	1.14568 18935 94861 807	0.87284 26324 88719 322
0.137	1.14682 81485 20398 195	0.87197 02261 32109 436
0.138	1.14797 55502 74178 672	0.87109 86917 45798 347
0.139	1.14912 41000 03605 088	0.87022 80284 58251 595
0.140	1.15027 37988 57227 268	0.86935 82353 98805 820
0.141	1.15142 46479 84744 161	0.86848 93116 97667 890
0.142	1.15257 66485 37004 992	0.86762 12564 85914 032
0.143	1.15372 98016 66010 407	0.86675 40688 95488 962
0.144	1.15488 41085 24913 632	0.86588 77480 59205 017
0.145	1.15603 95702 68021 623	0.86502 22931 10741 288
0.146	1.15719 61880 50796 218	0.86415 77031 84642 755
0.147	1.15835 39630 29855 297	0.86329 39774 16319 421
0.148	1.15951 28963 62973 936	0.86243 11149 42045 443
0.149	1.16067 29892 09085 563	0.86156 91148 98958 277
0.150	1.16183 42427 28283 123	0.86070 79764 25057 807
	$\left[\begin{smallmatrix}(-7)1\\6\end{smallmatrix}\right]$	$\left[\begin{smallmatrix}(-7)1\\6\end{smallmatrix}\right]$

EXPONENTIAL FUNCTION   Table 4.4

x	e^x	e^{-x}
0.150	1.16183 42427 28283 123	0.86070 79764 25057 807
0.151	1.16299 66580 81820 230	0.85984 76986 59205 488
0.152	1.16416 02364 32112 335	0.85898 82807 41123 482
0.153	1.16532 49789 42737 886	0.85812 97218 11393 800
0.154	1.16649 08867 78439 490	0.85727 20210 11457 440
0.155	1.16765 79611 05125 080	0.85641 51774 83613 531
0.156	1.16882 62030 89869 080	0.85555 91903 71018 473
0.157	1.16999 56139 00913 572	0.85470 40588 17685 083
0.158	1.17116 61947 07669 465	0.85384 97819 68481 735
0.159	1.17233 79466 80717 662	0.85299 63589 69131 511
0.160	1.17351 00709 91810 235	0.85214 37889 66211 338
0.161	1.17468 49688 13871 592	0.85129 20711 07151 144
0.162	1.17586 02413 20999 654	0.85044 12045 40232 998
0.163	1.17703 66896 88467 025	0.84959 11884 14590 263
0.164	1.17821 43150 92722 171	0.84874 20218 80206 741
0.165	1.17939 31187 11390 594	0.84789 37040 87915 828
0.166	1.18057 31017 23276 011	0.84704 62341 89399 660
0.167	1.18175 42653 08361 533	0.84619 96113 37188 270
0.168	1.18293 66106 47810 843	0.84535 38346 84658 733
0.169	1.18412 01389 23969 378	0.84450 89033 86034 326
0.170	1.18530 48513 20365 514	0.84366 48165 96383 682
0.171	1.18649 07490 21711 746	0.84282 15734 71619 939
0.172	1.18767 78332 13905 874	0.84197 91731 68499 904
0.173	1.18886 61050 84032 188	0.84113 76148 44623 201
0.174	1.19005 55658 20362 660	0.84029 68976 58431 438
0.175	1.19124 62166 12358 122	0.83945 70207 69207 358
0.176	1.19243 80586 50669 468	0.83861 79833 37074 003
0.177	1.19363 10931 27138 834	0.83777 97845 22993 869
0.178	1.19482 53212 34800 796	0.83694 24234 88768 073
0.179	1.19602 07441 67883 563	0.83610 58993 97035 511
0.180	1.19721 73631 21810 165	0.83527 02114 11272 021
0.181	1.19841 51792 93199 657	0.83443 53586 95789 549
0.182	1.19961 41938 79868 311	0.83360 13404 15735 309
0.183	1.20081 44080 80830 812	0.83276 81557 37090 951
0.184	1.20201 58230 96301 462	0.83193 58038 26671 728
0.185	1.20321 84401 27695 376	0.83110 42838 52125 659
0.186	1.20442 22603 77629 686	0.83027 35949 81932 701
0.187	1.20562 72850 49924 742	0.82944 37363 85403 915
0.188	1.20683 35153 49605 317	0.82861 47072 32680 634
0.189	1.20804 09524 82901 811	0.82778 65066 94733 637
0.190	1.20924 95976 57251 458	0.82695 91339 43362 318
0.191	1.21045 94520 81299 533	0.82613 25881 51193 854
0.192	1.21167 05169 64900 562	0.82530 68684 91682 387
0.193	1.21288 27935 19119 527	0.82448 19741 39108 186
0.194	1.21409 62829 56233 085	0.82365 79042 68576 832
0.195	1.21531 09864 89730 774	0.82283 46580 56018 384
0.196	1.21652 69053 34316 229	0.82201 22346 78186 562
0.197	1.21774 40407 05908 396	0.82119 06333 12657 919
0.198	1.21896 23938 21642 747	0.82036 98531 37831 021
0.199	1.22018 19658 99872 499	0.81954 98933 32925 626
0.200	1.22140 27581 60169 834	0.81873 07530 77981 859

$$\left[\begin{matrix}(-7)1\\6\end{matrix}\right] \qquad\qquad\qquad \left[\begin{matrix}(-7)1\\6\end{matrix}\right]$$

Table 4.4   EXPONENTIAL FUNCTION

x	e^x	e^{-x}
0.200	1.22140 27581 60169 834	0.81873 07530 77981 859
0.201	1.22262 47718 23327 112	0.81791 24315 53859 397
0.202	1.22384 80081 11358 099	0.81709 49279 42236 649
0.203	1.22507 24682 47499 185	0.81627 82414 25609 934
0.204	1.22629 81534 56210 607	0.81546 23711 87292 668
0.205	1.22752 50649 63177 678	0.81464 73164 11414 545
0.206	1.22875 32039 95312 005	0.81383 30762 82920 720
0.207	1.22998 25717 80752 723	0.81301 96499 87570 998
0.208	1.23121 31695 48867 721	0.81220 70367 11939 015
0.209	1.23244 49985 30254 869	0.81139 52356 43411 427
0.210	1.23367 80599 56743 251	0.81058 42459 70187 100
0.211	1.23491 23550 61394 396	0.80977 40668 81276 291
0.212	1.23614 78850 78503 512	0.80896 46975 66499 845
0.213	1.23738 46512 43600 719	0.80815 61372 16488 379
0.214	1.23862 26547 93452 285	0.80734 83850 22681 475
0.215	1.23986 18969 66061 862	0.80654 14401 77326 874
0.216	1.24110 23790 00671 728	0.80573 53018 73479 662
0.217	1.24234 41021 37764 020	0.80492 99693 05001 467
0.218	1.24358 70676 19061 978	0.80412 54416 66559 655
0.219	1.24483 12766 87531 187	0.80332 17181 53626 521
0.220	1.24607 67305 87380 820	0.80251 87979 62478 483
0.221	1.24732 34305 64064 879	0.80171 66802 90195 284
0.222	1.24857 13778 64283 447	0.80091 53643 34659 186
0.223	1.24982 05737 35983 926	0.80011 48492 94554 165
0.224	1.25107 10194 28362 294	0.79931 51343 69365 114
0.225	1.25232 27161 91864 345	0.79851 62187 59377 043
0.226	1.25357 56652 78186 948	0.79771 81016 65674 274
0.227	1.25482 98679 40279 295	0.79692 07822 90139 647
0.228	1.25608 53254 32344 151	0.79612 42598 35453 721
0.229	1.25734 20390 09839 113	0.79532 85335 05093 973
0.230	1.25860 00099 29477 863	0.79453 36025 03334 008
0.231	1.25985 92394 49231 426	0.79373 94660 35242 758
0.232	1.26111 97288 28329 426	0.79294 61233 06683 687
0.233	1.26238 14793 27261 349	0.79215 35735 24314 003
0.234	1.26364 44922 07777 797	0.79136 18158 95583 855
0.235	1.26490 87687 32891 756	0.79057 08496 28735 550
0.236	1.26617 43101 66879 857	0.78978 06739 32802 754
0.237	1.26744 11177 75283 640	0.78899 12880 17609 706
0.238	1.26870 91928 24910 818	0.78820 26910 93770 426
0.239	1.26997 85365 83836 547	0.78741 48823 72687 922
0.240	1.27124 91503 21404 692	0.78662 78610 66553 409
0.241	1.27252 10353 08229 095	0.78584 16263 88345 515
0.242	1.27379 41928 16194 849	0.78505 61775 51829 496
0.243	1.27506 86241 18459 570	0.78427 15137 71556 451
0.244	1.27634 43304 89454 665	0.78348 76342 62862 532
0.245	1.27762 13132 04886 611	0.78270 45382 41868 168
0.246	1.27889 95735 41738 230	0.78192 22249 25477 270
0.247	1.28017 91127 78269 966	0.78114 06935 31376 458
0.248	1.28145 99321 94021 162	0.78035 99432 78034 273
0.249	1.28274 20330 69811 341	0.77957 99733 84700 396
0.250	1.28402 54166 87741 484	0.77880 07830 71404 868
	$\left[\begin{array}{c}(-7)2 \\ 6\end{array}\right]$	$\left[\begin{array}{c}(-7)1 \\ 6\end{array}\right]$

EXPONENTIAL FUNCTION

Table 4.4

x	e^x	e^{-x}
0.250	1.28402 54166 87741 484	0.77880 07830 71404 868
0.251	1.28531 00843 31195 317	0.77802 23715 58957 312
0.252	1.28659 60372 84840 591	0.77724 47380 68946 150
0.253	1.28788 32768 34630 366	0.77646 78818 23737 828
0.254	1.28917 18042 67804 299	0.77569 18020 46476 034
0.255	1.29046 16208 72889 931	0.77491 64979 61080 928
0.256	1.29175 27279 39703 974	0.77414 19687 92248 360
0.257	1.29304 51267 59353 603	0.77336 82137 65449 096
0.258	1.29433 88186 24237 745	0.77259 52321 06928 045
0.259	1.29563 38048 28048 373	0.77182 30230 43703 483
0.260	1.29693 00866 65771 798	0.77105 15858 03566 284
0.261	1.29822 76654 33689 967	0.77028 09196 15079 142
0.262	1.29952 65424 29381 755	0.76951 10237 07575 806
0.263	1.30082 67189 51724 266	0.76874 18973 11160 303
0.264	1.30212 81963 00894 131	0.76797 35396 56706 173
0.265	1.30343 09757 78368 808	0.76720 59499 75855 698
0.266	1.30473 50586 86927 883	0.76643 91275 01019 133
0.267	1.30604 04463 30654 372	0.76567 30714 65373 938
0.268	1.30734 71400 14936 028	0.76490 77811 02864 015
0.269	1.30865 51410 46466 646	0.76414 32556 48198 937
0.270	1.30996 44507 33247 364	0.76337 94943 36853 186
0.271	1.31127 50703 84587 979	0.76261 64964 05065 386
0.272	1.31258 70013 11108 252	0.76185 42610 89837 543
0.273	1.31390 02448 24739 218	0.76109 27876 28934 278
0.274	1.31521 48022 38724 500	0.76033 20752 60882 066
0.275	1.31653 06748 67621 623	0.75957 21232 24968 476
0.276	1.31784 78640 27303 324	0.75881 29307 61241 409
0.277	1.31916 63710 34958 873	0.75805 44971 10508 337
0.278	1.32048 61972 09095 387	0.75729 68215 14335 547
0.279	1.32180 73438 69539 151	0.75653 99032 15047 380
0.280	1.32312 98123 37436 936	0.75578 37414 55725 472
0.281	1.32445 36039 35257 318	0.75502 83354 80208 002
0.282	1.32577 87199 86792 007	0.75427 36845 33088 932
0.283	1.32710 51618 17157 164	0.75351 97878 59717 250
0.284	1.32843 29307 52794 731	0.75276 66447 06196 222
0.285	1.32976 20281 21473 753	0.75201 42543 19382 630
0.286	1.33109 24552 52291 710	0.75126 26159 46886 026
0.287	1.33242 42134 75675 843	0.75051 17288 37067 974
0.288	1.33375 73041 23384 488	0.74976 15922 39041 301
0.289	1.33509 17285 28508 403	0.74901 22054 02669 348
0.290	1.33642 74880 25472 103	0.74826 35675 78565 215
0.291	1.33776 45839 50035 199	0.74751 56780 18091 016
0.292	1.33910 30176 39293 724	0.74676 85359 73357 128
0.293	1.34044 27904 31681 481	0.74602 21406 97221 444
0.294	1.34178 39036 66971 373	0.74527 64914 43288 626
0.295	1.34312 63586 86276 747	0.74453 15874 65909 357
0.296	1.34447 01568 32052 735	0.74378 74280 20179 599
0.297	1.34581 52994 48097 594	0.74304 40123 61939 843
0.298	1.34716 17878 79554 052	0.74230 13397 47774 369
0.299	1.34850 96234 72910 654	0.74155 94094 35010 502
0.300	1.34985 88075 76003 104	0.74081 82206 81717 866
	$\left[\dfrac{(-7)2}{6}\right]$	$\left[\dfrac{(-7)1}{6}\right]$

Table 4.4   EXPONENTIAL FUNCTION

x	e^x	e^{-x}
0.300	1.34985 88075 76003 104	0.74081 82206 81717 866
0.301	1.35120 93415 38015 618	0.74007 77727 46707 647
0.302	1.35256 12267 09482 272	0.73933 80648 89531 848
0.303	1.35391 44644 42288 348	0.73859 90963 70482 549
0.304	1.35526 90560 89671 692	0.73786 08664 50591 171
0.305	1.35662 50030 06224 066	0.73712 33743 91627 732
0.306	1.35798 23065 47892 497	0.73638 66194 56100 112
0.307	1.35934 09680 71980 642	0.73565 06009 07253 313
0.308	1.36070 09889 37150 137	0.73491 53180 09068 726
0.309	1.36206 23705 03421 961	0.73418 07700 26263 391
0.310	1.36342 51141 32177 794	0.73344 69562 24289 264
0.311	1.36478 92211 86161 378	0.73271 38758 69332 482
0.312	1.36615 46930 29479 880	0.73198 15282 28312 628
0.313	1.36752 15310 27605 258	0.73124 99125 68882 001
0.314	1.36888 97365 47375 624	0.73051 90281 59424 881
0.315	1.37025 93109 56996 611	0.72978 88742 69056 797
0.316	1.37163 02556 26042 743	0.72905 94501 67623 797
0.317	1.37300 25719 25458 804	0.72833 07551 25701 720
0.318	1.37437 62612 27561 208	0.72760 27884 14595 463
0.319	1.37575 13249 06039 370	0.72687 55493 06338 254
0.320	1.37712 77643 35957 085	0.72614 90370 73690 925
0.321	1.37850 55808 93753 895	0.72542 32509 90141 181
0.322	1.37988 47759 57246 476	0.72469 81903 29902 880
0.323	1.38126 53509 05630 003	0.72397 38543 67915 300
0.324	1.38264 73071 19479 542	0.72325 02423 79842 419
0.325	1.38403 06459 80751 421	0.72252 73536 42072 189
0.326	1.38541 53688 72784 617	0.72180 51874 31715 812
0.327	1.38680 14771 80302 136	0.72108 37430 26607 016
0.328	1.38818 89722 89412 403	0.72036 30197 05301 338
0.329	1.38957 78555 87610 642	0.71964 30167 47075 395
0.330	1.39096 81284 63780 266	0.71892 37334 31926 170
0.331	1.39235 97923 08194 268	0.71820 51690 40570 286
0.332	1.39375 28485 12516 609	0.71748 73228 54443 294
0.333	1.39514 72984 69803 608	0.71677 01941 55698 947
0.334	1.39654 31435 74505 339	0.71605 37822 27208 486
0.335	1.39794 03852 22467 023	0.71533 80863 52559 924
0.336	1.39933 90248 10930 424	0.71462 31058 16057 326
0.337	1.40073 90637 38535 249	0.71390 88399 02720 095
0.338	1.40214 05034 05320 540	0.71319 52878 98282 260
0.339	1.40354 33452 12726 081	0.71248 24490 89191 756
0.340	1.40494 75905 63593 797	0.71177 03227 62609 715
0.341	1.40635 32408 62169 155	0.71105 89082 06409 751
0.342	1.40776 02975 14102 572	0.71034 82047 09177 248
0.343	1.40916 87619 26450 817	0.70963 82115 60208 649
0.344	1.41057 86355 07678 418	0.70892 89280 49510 748
0.345	1.41198 99196 67659 075	0.70822 03534 67799 973
0.346	1.41340 26158 17677 066	0.70751 24871 06501 685
0.347	1.41481 67253 70428 658	0.70680 53282 57749 463
0.348	1.41623 22497 40023 522	0.70609 88762 14384 398
0.349	1.41764 91903 41986 146	0.70539 31302 69954 390
0.350	1.41906 75485 93257 248	0.70468 80897 18713 434
	$\left[\begin{smallmatrix}(-7)2\\6\end{smallmatrix}\right]$	$\left[\begin{smallmatrix}(-8)9\\6\end{smallmatrix}\right]$

EXPONENTIAL FUNCTION

Table 4.4

x	e^x	e^{-x}
0.350	1.41906 75485 93257 248	0.70468 80897 18713 434
0.351	1.42048 73259 12195 200	0.70398 37538 55620 921
0.352	1.42190 85237 18577 438	0.70328 01219 76340 929
0.353	1.42333 11434 33601 886	0.70257 71933 77241 521
0.354	1.42475 51864 79888 380	0.70187 49673 55394 037
0.355	1.42618 06542 81480 082	0.70117 34432 08572 398
0.356	1.42760 75482 63844 915	0.70047 26202 35252 399
0.357	1.42903 58698 53876 979	0.69977 24977 34611 008
0.358	1.43046 56204 79897 983	0.69907 30750 06525 666
0.359	1.43189 68015 71658 672	0.69837 43513 51573 587
0.360	1.43332 94145 60340 258	0.69767 63260 71031 057
0.361	1.43476 34608 78555 848	0.69697 89984 66872 738
0.362	1.43619 89419 60351 880	0.69628 23678 41770 967
0.363	1.43763 58592 41209 556	0.69558 64334 99095 062
0.364	1.43907 42141 58046 276	0.69489 11947 42910 621
0.365	1.44051 40081 49217 078	0.69419 66508 77978 831
0.366	1.44195 52426 54516 071	0.69350 28012 09755 768
0.367	1.44339 79191 15177 881	0.69280 96450 44391 707
0.368	1.44484 20389 73879 090	0.69211 71816 88730 425
0.369	1.44628 76036 74739 677	0.69142 54104 50308 508
0.370	1.44773 46146 63324 462	0.69073 43306 37354 660
0.371	1.44918 30733 86644 554	0.69004 39415 58789 010
0.372	1.45063 29812 93158 799	0.68935 42425 24222 423
0.373	1.45208 43398 32775 223	0.68866 52328 43955 806
0.374	1.45353 71504 56852 487	0.68797 69118 28979 422
0.375	1.45499 14146 18201 336	0.68728 92787 90972 199
0.376	1.45644 71337 71086 052	0.68660 23330 42301 040
0.377	1.45790 43093 71225 910	0.68591 60738 96020 141
0.378	1.45936 29428 75796 632	0.68523 05006 65870 297
0.379	1.46082 30357 43431 842	0.68454 56126 66278 222
0.380	1.46228 45894 34224 532	0.68386 14092 12355 858
0.381	1.46374 76054 09728 512	0.68317 78896 19899 696
0.382	1.46521 20851 32959 881	0.68249 50532 05390 084
0.383	1.46667 80300 68398 485	0.68181 28992 85990 553
0.384	1.46814 54416 81989 380	0.68113 14271 79547 125
0.385	1.46961 43214 41144 302	0.68045 06362 04587 638
0.386	1.47108 46708 14743 133	0.67977 05256 80321 060
0.387	1.47255 64912 73135 370	0.67909 10949 26636 810
0.388	1.47402 97842 88141 592	0.67841 23432 64104 077
0.389	1.47550 45513 33054 939	0.67773 42700 13971 142
0.390	1.47698 07938 82642 577	0.67705 68744 98164 700
0.391	1.47845 85134 13147 180	0.67638 01560 39289 177
0.392	1.47993 77114 02288 401	0.67570 41139 60626 058
0.393	1.48141 83893 29264 352	0.67502 87475 86133 209
0.394	1.48290 05486 74753 084	0.67435 40562 40444 198
0.395	1.48438 41909 20914 066	0.67368 00392 48867 624
0.396	1.48586 93175 51389 667	0.67300 66959 37386 438
0.397	1.48735 59300 51306 642	0.67233 40256 32657 274
0.398	1.48884 40299 07277 615	0.67166 20276 62009 771
0.399	1.49033 36186 07402 565	0.67099 07013 53445 901
0.400	1.49182 46976 41270 318	0.67032 00460 35639 301
	$\begin{bmatrix} (-7)2 \\ 6 \end{bmatrix}$	$\begin{bmatrix} (-8)9 \\ 6 \end{bmatrix}$

Table 4.4   **EXPONENTIAL FUNCTION**

x	e^x	e^{-x}
0.400	1.49182 46976 41270 318	0.67032 00460 35639 301
0.401	1.49331 72684 99960 030	0.66965 00610 37934 596
0.402	1.49481 13326 76042 686	0.66898 07456 90346 733
0.403	1.49630 68916 63582 585	0.66831 20993 23560 309
0.404	1.49780 39469 58138 840	0.66764 41212 68928 902
0.405	1.49930 25000 56766 870	0.66697 68108 58474 400
0.406	1.50080 25524 58019 898	0.66631 01674 24886 338
0.407	1.50230 41056 61950 452	0.66564 41903 01521 227
0.408	1.50380 71611 70111 860	0.66497 88788 22401 888
0.409	1.50531 17204 85559 754	0.66431 42323 22216 786
0.410	1.50681 77851 12853 578	0.66365 02501 36319 366
0.411	1.50832 53565 58058 082	0.66298 69316 00727 386
0.412	1.50983 44363 28744 838	0.66232 42760 52122 256
0.413	1.51134 50259 33993 742	0.66166 22828 27848 372
0.414	1.51285 71268 84394 526	0.66100 09512 65912 454
0.415	1.51437 07406 92048 265	0.66034 02807 04982 886
0.416	1.51588 58688 70568 894	0.65968 02704 84389 050
0.417	1.51740 25129 35084 718	0.65902 09199 44120 673
0.418	1.51892 06744 02239 927	0.65836 22284 24827 158
0.419	1.52044 03547 90196 115	0.65770 41952 67816 932
0.420	1.52196 15556 18633 796	0.65704 68198 15056 782
0.421	1.52348 42784 08753 926	0.65639 01014 09171 201
0.422	1.52500 85246 83279 422	0.65573 40393 93441 728
0.423	1.52653 42959 66456 685	0.65507 86331 11806 293
0.424	1.52806 15937 84057 126	0.65442 38819 08858 560
0.425	1.52959 04196 63378 690	0.65376 97851 29847 271
0.426	1.53112 07751 33247 382	0.65311 63421 20675 593
0.427	1.53265 26617 24018 802	0.65246 35522 27900 462
0.428	1.53418 60809 67579 666	0.65181 14147 98731 930
0.429	1.53572 10343 97349 347	0.65115 99291 81032 515
0.430	1.53725 75235 48281 402	0.65050 90947 23316 545
0.431	1.53879 55499 56865 110	0.64985 89107 74749 506
0.432	1.54033 51151 61127 008	0.64920 93766 85147 398
0.433	1.54187 62207 00632 428	0.64856 04918 04976 075
0.434	1.54341 88681 16487 038	0.64791 22554 85350 604
0.435	1.54496 30589 51338 384	0.64726 46670 78034 611
0.436	1.54650 87947 49377 427	0.64661 77259 35439 635
0.437	1.54805 60770 56340 096	0.64597 14314 10624 479
0.438	1.54960 49074 19508 826	0.64532 57828 57294 565
0.439	1.55115 52873 87714 108	0.64468 07796 29801 285
0.440	1.55270 72185 11336 042	0.64403 64210 83141 359
0.441	1.55426 07023 42305 879	0.64339 27065 72956 185
0.442	1.55581 57404 34107 580	0.64274 96354 55531 200
0.443	1.55737 23343 41779 367	0.64210 72070 87795 233
0.444	1.55893 04856 21915 277	0.64146 54208 27319 863
0.445	1.56049 01958 32666 719	0.64082 42760 32318 776
0.446	1.56205 14665 33744 035	0.64018 37720 61647 123
0.447	1.56361 42992 86418 055	0.63954 39082 74800 880
0.448	1.56517 86956 53521 663	0.63890 46840 31916 208
0.449	1.56674 46571 99451 356	0.63826 60986 93768 809
0.450	1.56831 21854 90168 811	0.63762 81516 21773 293
	$\left[\dfrac{(-7)2}{6}\right]$	$\left[\dfrac{(-8)9}{6}\right]$

EXPONENTIAL FUNCTION

Table 4.4

x	e^x	e^{-x}
0.450	1.56831 21854 90168 811	0.63762 81516 21773 293
0.451	1.56988 12820 93202 449	0.63699 08421 77982 535
0.452	1.57145 19485 77649 003	0.63635 41697 25087 037
0.453	1.57302 41865 14175 089	0.63571 81336 26414 293
0.454	1.57459 79974 75018 775	0.63508 27332 45928 153
0.455	1.57617 33830 33991 152	0.63444 79679 48228 182
0.456	1.57775 03447 66477 911	0.63381 38370 98549 030
0.457	1.57932 88842 49440 916	0.63318 03400 62759 794
0.458	1.58090 90030 61419 781	0.63254 74762 07363 387
0.459	1.58249 07027 82533 449	0.63191 52448 99495 898
0.460	1.58407 39849 94401 775	0.63128 36455 06925 969
0.461	1.58565 88512 80547 101	0.63065 26773 98054 154
0.462	1.58724 53032 25595 846	0.63002 23399 41912 291
0.463	1.58883 33424 16080 087	0.62939 26325 08162 872
0.464	1.59042 29704 40039 147	0.62876 35544 67098 411
0.465	1.59201 41888 87101 182	0.62813 51051 89640 814
0.466	1.59360 69993 48484 772	0.62750 72840 47340 750
0.467	1.59520 14034 17000 511	0.62688 00904 12377 027
0.468	1.59679 74026 87052 601	0.62625 35236 57555 956
0.469	1.59839 49987 54640 444	0.62562 75831 56310 730
0.470	1.59999 41932 17360 241	0.62500 22682 82700 796
0.471	1.60159 49876 74406 589	0.62437 75784 11411 229
0.472	1.60319 73837 26574 077	0.62375 35129 17752 104
0.473	1.60480 13829 76258 891	0.62313 00711 77657 876
0.474	1.60640 69870 27460 416	0.62250 72525 67686 754
0.475	1.60801 41974 85782 835	0.62188 50564 65020 075
0.476	1.60962 30159 58436 741	0.62126 34822 47461 685
0.477	1.61123 34440 54240 740	0.62064 25292 93437 314
0.478	1.61284 54833 83623 064	0.62002 21969 81993 957
0.479	1.61445 91355 58623 174	0.61940 24846 92799 250
0.480	1.61607 44021 92893 382	0.61878 33918 06140 853
0.481	1.61769 12849 01700 456	0.61816 49177 02925 827
0.482	1.61930 97853 01927 238	0.61754 70617 64680 018
0.483	1.62092 99050 12074 265	0.61692 98233 73547 436
0.484	1.62255 16456 52261 382	0.61631 32019 12289 639
0.485	1.62417 50088 44229 364	0.61569 71967 64285 113
0.486	1.62579 99962 11341 538	0.61508 18073 13528 659
0.487	1.62742 66093 78585 406	0.61446 70329 44630 776
0.488	1.62905 48499 72574 272	0.61385 28730 42817 043
0.489	1.63068 47196 21548 865	0.61323 93269 93927 508
0.490	1.63231 62199 55378 970	0.61262 63941 84416 069
0.491	1.63394 93526 05565 057	0.61201 40740 01349 867
0.492	1.63558 41192 05239 912	0.61140 23658 32408 668
0.493	1.63722 05213 89170 270	0.61079 12690 65884 251
0.494	1.63885 85607 93758 453	0.61018 07830 90679 799
0.495	1.64049 82390 57044 002	0.60957 09072 96309 287
0.496	1.64213 95578 18705 315	0.60896 16410 72896 868
0.497	1.64378 25187 20061 292	0.60835 29838 11176 269
0.498	1.64542 71234 04072 971	0.60774 49349 02490 178
0.499	1.64707 33735 15345 173	0.60713 74937 38789 634
0.500	1.64872 12707 00128 147	0.60653 06597 12633 424
	$\begin{bmatrix} (-7)2 \\ 6 \end{bmatrix}$	$\begin{bmatrix} (-8)8 \\ 6 \end{bmatrix}$

Table 4.4   **EXPONENTIAL FUNCTION**

x	e^x	e^{-x}
0.500	1.64872 12707 00128 147	0.60653 06597 12633 424
0.501	1.65037 08166 06319 214	0.60592 44322 17187 470
0.502	1.65202 20128 83464 418	0.60531 88106 46224 228
0.503	1.65367 48611 82760 175	0.60471 37943 94122 075
0.504	1.65532 93631 57054 920	0.60410 93828 55864 709
0.505	1.65698 55204 60850 766	0.60350 55754 27040 541
0.506	1.65864 33347 50305 156	0.60290 23715 03842 093
0.507	1.66030 28076 83232 516	0.60229 97704 83065 390
0.508	1.66196 39409 19105 918	0.60169 77717 62109 362
0.509	1.66362 67361 19058 736	0.60109 63747 38975 237
0.510	1.66529 11949 45886 308	0.60049 55788 12265 943
0.511	1.66695 73190 64047 601	0.59989 53833 81185 502
0.512	1.66862 51101 39666 871	0.59929 57878 45538 434
0.513	1.67029 45698 40535 333	0.59869 67916 05729 153
0.514	1.67196 56998 36112 826	0.59809 83940 62761 369
0.515	1.67363 85017 97529 486	0.59750 05946 18237 489
0.516	1.67531 29773 97587 414	0.59690 33926 74358 019
0.517	1.67698 91283 10762 348	0.59630 67876 33920 965
0.518	1.67866 69562 13205 342	0.59571 07789 00321 238
0.519	1.68034 64627 82744 439	0.59511 53658 77550 053
0.520	1.68202 76496 98886 347	0.59452 05479 70194 339
0.521	1.68371 05186 42818 123	0.59392 63245 83436 138
0.522	1.68539 50712 97408 851	0.59333 26951 23052 015
0.523	1.68708 13093 47211 326	0.59273 96589 95412 460
0.524	1.68876 92344 78463 738	0.59214 72156 07481 294
0.525	1.69045 88483 79091 359	0.59155 53643 66815 082
0.526	1.69215 01527 38708 232	0.59096 41046 81562 533
0.527	1.69384 31492 48618 855	0.59037 34359 60463 912
0.528	1.69553 78396 01819 881	0.58978 33576 12850 450
0.529	1.69723 42254 93001 803	0.58919 38690 48643 749
0.530	1.69893 23086 18550 654	0.58860 49696 78355 196
0.531	1.70063 20906 76549 702	0.58801 66589 13085 372
0.532	1.70233 35733 66781 146	0.58742 89361 64523 463
0.533	1.70403 67583 90727 817	0.58684 18008 44946 670
0.534	1.70574 16474 51574 883	0.58625 52523 67219 626
0.535	1.70744 82422 54211 545	0.58566 92901 44793 803
0.536	1.70915 65445 05232 748	0.58508 39135 91706 932
0.537	1.71086 65559 12940 887	0.58449 91221 22582 409
0.538	1.71257 82781 87347 510	0.58391 49151 52628 716
0.539	1.71429 17130 40175 036	0.58333 12920 97638 836
0.540	1.71600 68621 84858 460	0.58274 82523 73989 665
0.541	1.71772 37273 36547 069	0.58216 57953 98641 430
0.542	1.71944 23102 12106 159	0.58158 39205 89137 107
0.543	1.72116 26125 30118 747	0.58100 26273 63601 839
0.544	1.72288 46360 10887 296	0.58042 19151 40742 351
0.545	1.72460 83823 76435 429	0.57984 17833 39846 373
0.546	1.72633 38533 50509 656	0.57926 22313 80782 055
0.547	1.72806 10506 58581 095	0.57868 32586 83997 389
0.548	1.72978 99760 27847 197	0.57810 48646 70519 631
0.549	1.73152 06311 87233 477	0.57752 70487 61954 718
0.550	1.73325 30178 67395 237	0.57694 98103 80486 695
	$\left[\dfrac{(-7)2}{6}\right]$	$\left[\dfrac{(-8)8}{6}\right]$

EXPONENTIAL FUNCTION

Table 4.4

x	e^x	e^{-x}
0.550	1.73325 30178 67395 237	0.57694 98103 80486 695
0.551	1.73498 71378 00719 302	0.57637 31489 48877 132
0.552	1.73672 29927 21325 750	0.57579 70638 90464 548
0.553	1.73846 05843 65069 647	0.57522 15546 29163 839
0.554	1.74019 99144 69542 780	0.57464 66205 89465 693
0.555	1.74194 09847 74075 399	0.57407 22611 96436 024
0.556	1.74368 37970 19737 955	0.57349 84758 75715 391
0.557	1.74542 83529 49342 837	0.57292 52640 53518 425
0.558	1.74717 46543 07446 121	0.57235 26251 56633 257
0.559	1.74892 27028 40349 310	0.57178 05586 12420 941
0.560	1.75067 25002 96101 083	0.57120 90638 48814 886
0.561	1.75242 40484 24499 041	0.57063 81402 94320 280
0.562	1.75417 73489 77091 459	0.57006 77873 78013 522
0.563	1.75593 24037 07179 036	0.56949 80045 29541 648
0.564	1.75768 92143 69816 648	0.56892 87911 79121 761
0.565	1.75944 77827 21815 104	0.56836 01467 57540 464
0.566	1.76120 81105 21742 902	0.56779 20706 96153 288
0.567	1.76297 01995 29927 989	0.56722 45624 26884 125
0.568	1.76473 40515 08459 520	0.56665 76213 82224 657
0.569	1.76649 96682 21189 621	0.56609 12469 95233 792
0.570	1.76826 70514 33735 152	0.56552 54386 99537 097
0.571	1.77003 62029 13479 471	0.56496 01959 29326 229
0.572	1.77180 71244 29574 208	0.56439 55181 19358 370
0.573	1.77357 98177 52941 024	0.56383 14047 04955 664
0.574	1.77535 42846 56273 392	0.56326 78551 22004 648
0.575	1.77713 05269 14038 362	0.56270 48688 06955 693
0.576	1.77890 85463 02478 341	0.56214 24451 96822 437
0.577	1.78068 83445 99612 864	0.56158 05837 29181 224
0.578	1.78246 99235 85240 377	0.56101 92838 42170 538
0.579	1.78425 32850 40940 016	0.56045 85449 74490 445
0.580	1.78603 84307 50073 382	0.55989 83665 65402 033
0.581	1.78782 53624 97786 336	0.55933 87480 54726 843
0.582	1.78961 40820 71010 772	0.55877 96888 82846 320
0.583	1.79140 45912 58466 414	0.55822 11884 90701 245
0.584	1.79319 68918 50662 599	0.55766 32463 19791 179
0.585	1.79499 09856 39900 067	0.55710 58618 12173 905
0.586	1.79678 68744 20272 757	0.55654 90344 10464 868
0.587	1.79858 45599 87669 600	0.55599 27635 57836 621
0.588	1.80038 40441 39776 313	0.55543 70486 98018 264
0.589	1.80218 53286 76077 198	0.55488 18892 75294 892
0.590	1.80398 84153 97856 940	0.55432 72847 34507 035
0.591	1.80579 33061 08202 413	0.55377 32345 21050 107
0.592	1.80760 00026 12004 477	0.55321 97380 80873 848
0.593	1.80940 85067 15959 787	0.55266 67948 60481 771
0.594	1.81121 88202 28572 596	0.55211 44043 06930 610
0.595	1.81303 09449 60156 569	0.55156 25658 67829 766
0.596	1.81484 48827 22836 588	0.55101 12789 91340 753
0.597	1.81666 06353 30550 566	0.55046 05431 26176 649
0.598	1.81847 82045 99051 264	0.54991 03577 21601 542
0.599	1.82029 75923 45908 101	0.54936 07222 27429 984
0.600	1.82211 88003 90508 975	0.54881 16360 94026 433
	$\begin{bmatrix} (-7)2 \\ 6 \end{bmatrix}$	$\begin{bmatrix} (-8)7 \\ 6 \end{bmatrix}$

Table 4.4   EXPONENTIAL FUNCTION

x	e^x	e^{-x}
0.600	1.82211 88003 90508 975	0.54881 16360 94026 433
0.601	1.82394 18305 54062 083	0.54826 30987 72304 710
0.602	1.82576 66846 59597 740	0.54771 51097 13727 448
0.603	1.82759 33645 31970 203	0.54716 76683 70305 543
0.604	1.82942 18719 97859 499	0.54662 07741 94597 605
0.605	1.83125 22088 85773 244	0.54607 44266 39709 413
0.606	1.83308 43770 26048 479	0.54552 86251 59293 368
0.607	1.83491 83782 50853 497	0.54498 33692 07547 943
0.608	1.83675 42143 94189 676	0.54443 86582 39217 140
0.609	1.83859 18872 91893 312	0.54389 44917 09589 946
0.610	1.84043 13987 81637 455	0.54335 08690 74499 787
0.611	1.84227 27507 02933 750	0.54280 77897 90323 981
0.612	1.84411 59448 97134 270	0.54226 52533 13983 200
0.613	1.84596 09832 07433 364	0.54172 32591 02940 922
0.614	1.84780 78674 78869 496	0.54118 18066 15202 890
0.615	1.84965 65995 58327 090	0.54064 08953 09316 571
0.616	1.85150 71812 94538 381	0.54010 05246 44370 616
0.617	1.85335 96145 38085 258	0.53956 06940 79994 313
0.618	1.85521 39011 41401 120	0.53902 14030 76357 053
0.619	1.85707 00429 58772 725	0.53848 26510 94167 789
0.620	1.85892 80418 46342 044	0.53794 44375 94674 492
0.621	1.86078 78996 62108 121	0.53740 67620 39663 618
0.622	1.86264 96182 65928 925	0.53686 96238 91459 568
0.623	1.86451 31995 19523 215	0.53633 30226 12924 149
0.624	1.86637 86452 86472 402	0.53579 69576 67456 037
0.625	1.86824 59574 32222 407	0.53526 14285 18990 242
0.626	1.87011 51378 24085 530	0.53472 64346 31997 571
0.627	1.87198 61883 31242 321	0.53419 19754 71484 093
0.628	1.87385 91108 24743 442	0.53365 80505 02990 602
0.629	1.87573 39071 77511 543	0.53312 46591 92592 086
0.630	1.87761 05792 64343 132	0.53259 18010 06897 190
0.631	1.87948 91289 61910 454	0.53205 94754 13047 683
0.632	1.88136 95581 48763 361	0.53152 76818 78717 927
0.633	1.88325 18687 05331 198	0.53099 64198 72114 344
0.634	1.88513 60625 13924 678	0.53046 56888 61974 883
0.635	1.88702 21414 58737 766	0.52993 54883 17568 489
0.636	1.88891 01074 25849 565	0.52940 58177 08694 574
0.637	1.89079 99623 03226 199	0.52887 66765 05682 485
0.638	1.89269 17079 80722 703	0.52834 80641 79390 975
0.639	1.89458 53463 50084 912	0.52781 99802 01207 673
0.640	1.89648 08793 04951 353	0.52729 24240 43048 557
0.641	1.89837 83087 40855 140	0.52676 53951 77357 426
0.642	1.90027 76365 55225 865	0.52623 88930 77105 369
0.643	1.90217 88646 47391 502	0.52571 29172 15790 242
0.644	1.90408 19949 18580 301	0.52518 74670 67436 140
0.645	1.90598 70292 71922 692	0.52466 25421 06592 872
0.646	1.90789 39696 12453 188	0.52413 81418 08335 432
0.647	1.90980 28178 47112 287	0.52361 42656 48263 478
0.648	1.91171 35758 84748 384	0.52309 09131 02500 807
0.649	1.91362 62456 36119 674	0.52256 80836 47694 830
0.650	1.91554 08290 13896 070	0.52204 57767 61016 048
	$\left[\dfrac{(-7)2}{6}\right]$	$\left[\dfrac{(-8)7}{6}\right]$

EXPONENTIAL FUNCTION

Table 4.4

x	e^x	e^{-x}
0.650	1.91554 08290 13896 070	0.52204 57767 61016 048
0.651	1.91745 73279 32661 108	0.52152 39919 20157 530
0.652	1.91937 57443 08913 867	0.52100 27286 03334 394
0.653	1.92129 60800 61070 883	0.52048 19862 89283 277
0.654	1.92321 83371 09468 067	0.51996 17644 57261 823
0.655	1.92514 25173 76362 630	0.51944 20625 87048 156
0.656	1.92706 86227 85934 997	0.51892 28801 58940 364
0.657	1.92899 66552 64290 740	0.51840 42166 53755 974
0.658	1.93092 66167 39462 496	0.51788 60715 52831 438
0.659	1.93285 85091 41411 902	0.51736 84443 38021 612
0.660	1.93479 23344 02031 522	0.51685 13344 91699 238
0.661	1.93672 80944 55146 776	0.51633 47414 96754 426
0.662	1.93866 57912 36517 879	0.51581 86648 36594 140
0.663	1.94060 54266 83841 774	0.51530 31039 95141 674
0.664	1.94254 70027 36754 070	0.51478 80584 56836 146
0.665	1.94449 05213 36830 982	0.51427 35277 06631 974
0.666	1.94643 59844 27591 272	0.51375 95112 29998 365
0.667	1.94838 33939 54498 192	0.51324 60085 12918 798
0.668	1.95033 27518 64961 432	0.51273 30190 41890 516
0.669	1.95228 40601 08339 065	0.51222 05423 03924 002
0.670	1.95423 73206 35939 496	0.51170 85777 86542 478
0.671	1.95619 25354 01023 417	0.51119 71249 77781 383
0.672	1.95814 97063 58805 754	0.51068 61833 66187 865
0.673	1.96010 88354 66457 630	0.51017 57524 40820 271
0.674	1.96206 99246 83108 314	0.50966 58316 91247 632
0.675	1.96403 29759 69847 187	0.50915 64206 07549 157
0.676	1.96599 79912 89725 700	0.50864 75186 80313 718
0.677	1.96796 49726 07759 335	0.50813 91254 00639 348
0.678	1.96993 39218 90929 575	0.50763 12402 60132 723
0.679	1.97190 48411 08185 868	0.50712 38627 50908 661
0.680	1.97387 77322 30447 594	0.50661 69923 65589 610
0.681	1.97585 25972 30606 040	0.50611 06285 97305 142
0.682	1.97782 94380 83526 371	0.50560 47709 39691 448
0.683	1.97980 82567 66049 605	0.50509 94188 86890 827
0.684	1.98178 90552 56994 589	0.50459 45719 33551 185
0.685	1.98377 18355 37159 979	0.50409 02295 74825 526
0.686	1.98575 65995 89326 220	0.50358 63913 06371 449
0.687	1.98774 33493 98257 531	0.50308 30566 24350 644
0.688	1.98973 20869 50703 885	0.50258 02250 25428 387
0.689	1.99172 28142 35403 001	0.50207 .3960 06773 037
0.690	1.99371 55332 43082 329	0.50157 60690 66055 534
0.691	1.99571 02459 66461 043	0.50107 47437 01448 895
0.692	1.99770 69544 00252 033	0.50057 39194 11627 713
0.693	1.99970 56605 41163 899	0.50007 35956 95767 658
0.694	2.00170 63663 87902 948	0.49957 37720 53544 971
0.695	2.00370 90739 41175 193	0.49907 44479 85135 969
0.696	2.00571 37852 03688 356	0.49857 56229 91216 541
0.697	2.00772 05021 80153 865	0.49807 72965 72961 653
0.698	2.00972 92268 77288 865	0.49757 94682 32044 844
0.699	2.01173 99613 03818 219	0.49708 21374 70637 732
0.700	2.01375 27074 70476 522	0.49658 53037 91409 515
	$\begin{bmatrix} (-7)2 \\ 6 \end{bmatrix}$	$\begin{bmatrix} (-8)6 \\ 6 \end{bmatrix}$

Table 4.4   **EXPONENTIAL FUNCTION**

x	e^x	e^{-x}
0.700	2.01375 27074 70476 522	0.49658 53037 91409 515
0.701	2.01576 74673 90010 108	0.49608 89666 97526 471
0.702	2.01778 42430 77179 065	0.49559 31256 92651 465
0.703	2.01980 30365 48759 247	0.49509 77802 80943 451
0.704	2.02182 38498 23544 296	0.49460 29299 67056 976
0.705	2.02384 66849 22347 653	0.49410 85742 56141 685
0.706	2.02587 15438 68004 586	0.49361 47126 53841 826
0.707	2.02789 84286 85374 210	0.49312 13446 66295 756
0.708	2.02992 73414 01341 511	0.49262 84698 00135 445
0.709	2.03195 82840 44819 374	0.49213 60875 62485 987
0.710	2.03399 12586 46750 612	0.49164 41974 60965 102
0.711	2.03602 62672 40109 996	0.49115 27990 03682 649
0.712	2.03806 33118 59906 288	0.49066 18916 99240 129
0.713	2.04010 23945 43184 280	0.49017 14750 56730 197
0.714	2.04214 35173 29026 822	0.48968 15485 85736 169
0.715	2.04418 66822 58556 873	0.48919 21117 96331 534
0.716	2.04623 18913 74939 531	0.48870 31641 99079 460
0.717	2.04827 91467 23384 083	0.48821 47053 05032 312
0.718	2.05032 84503 51146 049	0.48772 67346 25731 153
0.719	2.05237 98043 07529 226	0.48723 92516 73205 263
0.720	2.05443 32106 43887 743	0.48675 22559 59971 650
0.721	2.05648 86714 13628 106	0.48626 57469 99034 560
0.722	2.05854 61886 72211 257	0.48577 97243 03884 990
0.723	2.06060 57644 77154 626	0.48529 41873 88500 207
0.724	2.06266 74008 88034 189	0.48480 91357 67343 253
0.725	2.06473 10999 66486 529	0.48432 45689 55362 467
0.726	2.06679 68637 76210 896	0.48384 04864 67990 997
0.727	2.06886 46943 82971 273	0.48335 68878 21146 315
0.728	2.07093 45938 54598 438	0.48287 37725 31229 734
0.729	2.07300 65642 60992 036	0.48239 11401 15125 923
0.730	2.07508 06076 74122 645	0.48190 89900 90202 427
0.731	2.07715 67261 68033 852	0.48142 73219 74309 180
0.732	2.07923 49218 18844 323	0.48094 61352 85778 027
0.733	2.08131 51967 04749 882	0.48046 54295 43422 238
0.734	2.08339 75529 06025 589	0.47998 52042 66536 031
0.735	2.08548 19925 05027 819	0.47950 54589 74894 090
0.736	2.08756 85175 86196 344	0.47902 61931 88751 082
0.737	2.08965 71302 36056 419	0.47854 74064 28841 182
0.738	2.09174 78325 43220 868	0.47806 90982 16377 589
0.739	2.09384 06265 98392 173	0.47759 12680 73052 052
0.740	2.09593 55144 94364 563	0.47711 39155 21034 388
0.741	2.09803 24983 26026 109	0.47663 70400 82972 004
0.742	2.10013 15801 90360 816	0.47616 06412 81989 423
0.743	2.10223 27621 86450 725	0.47568 47186 41687 803
0.744	2.10433 60464 15478 007	0.47520 92716 86144 466
0.745	2.10644 14349 80727 065	0.47473 42999 39912 416
0.746	2.10854 89299 87586 641	0.47425 98029 28019 867
0.747	2.11065 85335 43551 917	0.47378 57801 75969 767
0.748	2.11277 02477 58226 625	0.47331 22312 09739 326
0.749	2.11488 40747 43325 155	0.47283 91555 55779 537
0.750	2.11700 00166 12674 669	0.47236 65527 41014 707
	$\begin{bmatrix} (-7)3 \\ 6 \end{bmatrix}$	$\begin{bmatrix} (-8)6 \\ 6 \end{bmatrix}$

EXPONENTIAL FUNCTION

Table 4.4

x	e^x	e^{-x}
0.750	2.11700 00166 12674 669	0.47236 65527 41014 707
0.751	2.11911 80754 82217 212	0.47189 44222 92841 982
0.752	2.12123 82534 70011 830	0.47142 27637 39130 875
0.753	2.12336 05526 96236 688	0.47095 15766 08222 791
0.754	2.12548 49752 83191 190	0.47048 08604 28930 562
0.755	2.12761 15233 55298 098	0.47001 06147 30537 969
0.756	2.12974 01990 39105 663	0.46954 08390 42799 274
0.757	2.13187 10044 63289 745	0.46907 15328 95938 749
0.758	2.13400 39417 58655 946	0.46860 26958 20650 211
0.759	2.13613 90130 58141 739	0.46813 43273 48096 543
0.760	2.13827 62204 96810 602	0.46766 64270 09909 234
0.761	2.14041 55662 11894 152	0.46719 89943 38187 907
0.762	2.14255 70523 42714 282	0.46673 20288 65499 852
0.763	2.14470 06810 30765 301	0.46626 55301 24879 557
0.764	2.14684 64544 19676 075	0.46579 94976 49828 242
0.765	2.14899 43746 55220 173	0.46533 39309 74313 393
0.766	2.15114 44438 85318 010	0.46486 88296 32768 297
0.767	2.15329 66642 60038 993	0.46440 41931 60091 573
0.768	2.15545 10379 31603 678	0.46394 00210 91646 708
0.769	2.15760 75670 54385 916	0.46347 63129 63261 598
0.770	2.15976 62537 84915 008	0.46301 30683 11228 073
0.771	2.16192 71002 81877 866	0.46255 02866 72301 444
0.772	2.16409 01087 06121 167	0.46208 79675 83700 034
0.773	2.16625 52812 20653 514	0.46162 61105 83104 714
0.774	2.16842 26199 90647 604	0.46116 47152 08658 446
0.775	2.17059 21271 83442 386	0.46070 37809 98965 818
0.776	2.17276 38049 68545 234	0.46024 33074 93092 580
0.777	2.17493 76555 17634 114	0.45978 32942 30565 189
0.778	2.17711 36810 04559 757	0.45932 37407 51370 344
0.779	2.17929 18836 05347 830	0.45886 46465 95954 527
0.780	2.18147 22654 98201 117	0.45840 60113 05223 545
0.781	2.18365 48288 63501 691	0.45794 78344 20542 069
0.782	2.18583 95758 83813 099	0.45749 01154 83733 175
0.783	2.18802 65087 43882 545	0.45703 28540 37077 890
0.784	2.19021 56296 30643 070	0.45657 60496 23314 727
0.785	2.19240 69407 33215 744	0.45611 97017 85639 236
0.786	2.19460 04442 42911 852	0.45566 38100 67703 540
0.787	2.19679 61423 53235 086	0.45520 83740 13615 885
0.788	2.19899 40372 59883 740	0.45475 33931 67940 176
0.789	2.20119 41311 60752 903	0.45429 88670 75695 532
0.790	2.20339 64262 55936 659	0.45384 47952 82355 822
0.791	2.20560 09247 47730 288	0.45339 11773 33849 215
0.792	2.20780 76288 40632 465	0.45293 80127 76557 724
0.793	2.21001 65407 41347 466	0.45248 53011 57316 754
0.794	2.21222 76626 58787 377	0.45203 30420 23414 649
0.795	2.21444 09968 04074 299	0.45158 12349 22592 237
0.796	2.21665 65453 90542 561	0.45112 98794 03042 379
0.797	2.21887 43106 33740 936	0.45067 89750 13409 518
0.798	2.22109 42947 51434 850	0.45022 85213 02789 227
0.799	2.22331 64999 63608 607	0.44977 85178 20727 758
0.800	2.22554 09284 92467 605	0.44932 89641 17221 591

$$\begin{bmatrix}(-7)3 \\ 6\end{bmatrix} \qquad \begin{bmatrix}(-8)6 \\ 6\end{bmatrix}$$

Table 4.4   **EXPONENTIAL FUNCTION**

x	e^x	e^{-x}
0.800	2.22554 09284 92467 605	0.44932 89641 17221 591
0.801	2.22776 75825 62440 556	0.44887 98597 42716 986
0.802	2.22999 64644 00181 717	0.44843 12042 48109 530
0.803	2.23222 75762 34573 111	0.44798 29971 84743 691
0.804	2.23446 09202 96726 759	0.44753 52381 04412 369
0.805	2.23669 64988 19986 909	0.44708 79265 59356 447
0.806	2.23893 43140 39932 270	0.44664 10621 02264 340
0.807	2.24117 43681 94378 249	0.44619 46442 86271 556
0.808	2.24341 66635 23379 186	0.44574 86726 64960 242
0.809	2.24566 12022 69230 599	0.44530 31467 92358 738
0.810	2.24790 79866 76471 419	0.44485 80662 22941 134
0.811	2.25015 70189 91886 242	0.44441 34305 11626 826
0.812	2.25240 83014 64507 569	0.44396 92392 13780 063
0.813	2.25466 18363 45618 061	0.44352 54918 85209 512
0.814	2.25691 76258 88752 788	0.44308 21880 82167 806
0.815	2.25917 56723 49701 480	0.44263 93273 61351 106
0.816	2.26143 59779 86510 786	0.44219 69092 79898 654
0.817	2.26369 85450 59486 532	0.44175 49333 95392 332
0.818	2.26596 33758 31195 979	0.44131 33992 65856 218
0.819	2.26823 04725 66470 087	0.44087 23064 49756 146
0.820	2.27049 98375 32405 781	0.44043 16545 05999 263
0.821	2.27277 14729 98368 215	0.43999 14429 93933 588
0.822	2.27504 53812 35993 046	0.43955 16714 73347 574
0.823	2.27732 15645 19188 700	0.43911 23395 04469 662
0.824	2.27960 00251 24138 650	0.43867 34466 47967 847
0.825	2.28188 07653 29303 690	0.43823 49924 64949 237
0.826	2.28416 37874 15424 217	0.43779 69765 16959 611
0.827	2.28644 90936 65522 506	0.43735 93983 65982 985
0.828	2.28873 66863 64904 998	0.43692 22575 74441 171
0.829	2.29102 65678 01164 583	0.43648 55537 05193 342
0.830	2.29331 87402 64182 888	0.43604 92863 21535 593
0.831	2.29561 32060 46132 567	0.43561 34549 87200 502
0.832	2.29790 99674 41479 593	0.43517 80592 66356 699
0.833	2.30020 90267 46985 553	0.43474 30987 23608 428
0.834	2.30251 03862 61709 945	0.43430 85729 23995 109
0.835	2.30481 40482 87012 474	0.43387 44814 32990 906
0.836	2.30712 00151 26555 358	0.43344 08238 16504 293
0.837	2.30942 82890 86305 628	0.43300 75996 40877 616
0.838	2.31173 88724 74537 437	0.43257 48084 72886 664
0.839	2.31405 17676 01834 366	0.43214 24498 79740 233
0.840	2.31636 69767 81091 734	0.43171 05234 29079 693
0.841	2.31868 45023 27518 913	0.43127 90286 88978 558
0.842	2.32100 43465 58641 644	0.43084 79652 27942 052
0.843	2.32332 65117 94304 351	0.43041 73326 14906 679
0.844	2.32565 10003 56672 462	0.42998 71304 19239 788
0.845	2.32797 78145 70234 734	0.42955 73582 10739 148
0.846	2.33030 69567 61805 575	0.42912 80155 59632 516
0.847	2.33263 84292 60527 370	0.42869 91020 36577 204
0.848	2.33497 22343 97872 812	0.42827 06172 12659 654
0.849	2.33730 83745 07647 233	0.42784 25606 59395 005
0.850	2.33964 68519 25990 937	0.42741 49319 48726 670
	$\left[\begin{matrix} (-7)3 \\ 6 \end{matrix} \right]$	$\left[\begin{matrix} (-8)6 \\ 6 \end{matrix} \right]$

EXPONENTIAL FUNCTION

Table 4.4

x	e^x	e^{-x}
0.850	2.33964 68519 25990 937	0.42741 49319 48726 670
0.851	2.34198 76689 91381 538	0.42698 77306 53025 901
0.852	2.34433 08280 44636 295	0.42656 09563 45091 367
0.853	2.34667 63314 28914 459	0.42613 46085 98148 726
0.854	2.34902 41814 89719 607	0.42570 86869 85850 193
0.855	2.35137 43805 74901 997	0.42528 31910 82274 123
0.856	2.35372 69310 34660 911	0.42485 81204 61924 574
0.857	2.35608 18352 21547 002	0.42443 34746 99730 893
0.858	2.35843 90954 90464 656	0.42400 92533 71047 281
0.859	2.36079 87141 98674 336	0.42358 54560 51652 373
0.860	2.36316 06937 05794 948	0.42316 20823 17748 817
0.861	2.36552 50363 73806 196	0.42273 91317 45962 841
0.862	2.36789 17445 67050 946	0.42231 66039 13343 840
0.863	2.37026 08206 52237 586	0.42189 44983 97363 945
0.864	2.37263 22669 98442 400	0.42147 28147 75917 606
0.865	2.37500 60859 77111 933	0.42105 15526 27321 165
0.866	2.37738 22799 62065 359	0.42063 07115 30312 439
0.867	2.37976 08513 29496 863	0.42021 02910 64050 296
0.868	2.38214 18024 57978 010	0.41979 02908 08114 234
0.869	2.38452 51357 28460 126	0.41937 07103 42503 963
0.870	2.38691 08535 24276 682	0.41895 15492 47638 983
0.871	2.38929 89582 31145 671	0.41853 28071 04358 162
0.872	2.39168 94522 37171 999	0.41811 44834 93919 324
0.873	2.39408 23379 32849 872	0.41769 65779 97998 822
0.874	2.39647 76177 11065 184	0.41727 90901 98691 126
0.875	2.39887 52939 67097 915	0.41686 20196 78508 403
0.876	2.40127 53690 98624 518	0.41644 53660 20380 096
0.877	2.40367 78455 05720 327	0.41602 91288 07652 513
0.878	2.40608 27255 90861 947	0.41561 33076 24088 408
0.879	2.40849 00117 58929 666	0.41519 79020 53866 560
0.880	2.41089 97064 17209 851	0.41478 29116 81581 367
0.881	2.41331 18119 75397 361	0.41436 83360 92242 420
0.882	2.41572 63308 45597 956	0.41395 41748 71274 097
0.883	2.41814 32654 42330 708	0.41354 04276 04515 140
0.884	2.42056 26181 82530 413	0.41312 70938 78218 250
0.885	2.42298 43914 85550 015	0.41271 41732 79049 666
0.886	2.42540 85877 73163 018	0.41230 16653 94088 753
0.887	2.42783 52094 69565 911	0.41188 95698 10827 593
0.888	2.43026 42590 01380 593	0.41147 78861 17170 568
0.889	2.43269 57387 97656 799	0.41106 66139 01433 949
0.890	2.43512 96512 89874 527	0.41065 57527 52345 488
0.891	2.43756 59989 11946 472	0.41024 53022 59044 001
0.892	2.44000 47841 00220 460	0.40983 52620 11078 959
0.893	2.44244 60092 93481 882	0.40942 56315 98410 082
0.894	2.44488 96769 32956 134	0.40901 64106 11406 922
0.895	2.44733 57894 62311 060	0.40860 75986 40848 458
0.896	2.44978 43493 27659 394	0.40819 91952 77922 685
0.897	2.45223 53589 77561 203	0.40779 12001 14226 207
0.898	2.45468 88208 63026 343	0.40738 36127 41763 826
0.899	2.45714 47374 37516 904	0.40697 64327 52948 135
0.900	2.45960 31111 56949 664	0.40656 96597 40599 112
	$\begin{bmatrix}(-7)3\\6\end{bmatrix}$	$\begin{bmatrix}(-8)5\\6\end{bmatrix}$

Table 4.4   EXPONENTIAL FUNCTION

x	e^x	e^{-x}
0.900	2.45960 31111 56949 664	0.40656 96597 40599 112
0.901	2.46206 39444 79698 548	0.40616 32932 97943 710
0.902	2.46452 72398 66597 083	0.40575 73330 18615 453
0.903	2.46699 29997 80940 863	0.40535 17784 96654 028
0.904	2.46946 12266 88490 006	0.40494 66293 26504 879
0.905	2.47193 19230 57471 626	0.40454 18851 03018 802
0.906	2.47440 50913 58582 298	0.40413 75454 21451 540
0.907	2.47688 07340 64990 529	0.40373 36098 77463 377
0.908	2.47935 88536 52339 232	0.40333 00780 67118 736
0.909	2.48183 94525 98748 200	0.40292 69495 86885 773
0.910	2.48432 25333 84816 587	0.40252 42240 33635 975
0.911	2.48680 80984 93625 386	0.40212 19010 04643 753
0.912	2.48929 61504 10739 912	0.40171 99800 97586 047
0.913	2.49178 66916 24212 291	0.40131 84609 10541 915
0.914	2.49427 97246 24583 942	0.40091 73430 41992 136
0.915	2.49677 52519 04888 075	0.40051 66260 90818 809
0.916	2.49927 32759 60652 177	0.40011 63096 56304 950
0.917	2.50177 37992 89900 513	0.39971 63933 38134 089
0.918	2.50427 68243 93156 620	0.39931 68767 36389 877
0.919	2.50678 23537 73445 810	0.39891 77594 51555 677
0.920	2.50929 03899 36297 671	0.39851 90410 84514 173
0.921	2.51180 09353 89748 577	0.39812 07212 36546 962
0.922	2.51431 39926 44344 189	0.39772 27995 09334 165
0.923	2.51682 95642 13141 971	0.39732 52755 04954 021
0.924	2.51934 76526 11713 703	0.39692 81488 25882 492
0.925	2.52186 82603 58147 991	0.39653 14190 74992 866
0.926	2.52439 13899 73052 794	0.39613 50858 55555 360
0.927	2.52691 70439 79557 936	0.39573 91487 71236 720
0.928	2.52944 52249 03317 633	0.39534 36074 26099 830
0.929	2.53197 59352 72513 022	0.39494 84614 24603 311
0.930	2.53450 91776 17854 680	0.39455 37103 71601 130
0.931	2.53704 49544 72585 166	0.39415 93538 72342 199
0.932	2.53958 32683 72481 544	0.39376 53915 32469 987
0.933	2.54212 41218 55857 927	0.39337 18229 58022 122
0.934	2.54466 75174 63568 010	0.39297 86477 55429 996
0.935	2.54721 34577 39007 611	0.39258 58655 31518 373
0.936	2.54976 19452 28117 220	0.39219 34758 93504 997
0.937	2.55231 29824 79384 537	0.39180 14784 49000 198
0.938	2.55486 65720 43847 026	0.39140 98728 06006 497
0.939	2.55742 27164 75094 464	0.39101 86585 72918 221
0.940	2.55998 14183 29271 496	0.39062 78353 58521 102
0.941	2.56254 26801 65080 189	0.39023 74027 71991 894
0.942	2.56510 65045 43782 593	0.38984 73604 22897 977
0.943	2.56767 28940 29203 299	0.38945 77079 21196 971
0.944	2.57024 18511 87732 007	0.38906 84448 77236 341
0.945	2.57281 33785 88326 089	0.38867 95709 01753 010
0.946	2.57538 74788 02513 161	0.38829 10856 05872 971
0.947	2.57796 41544 04393 651	0.38790 29886 01110 896
0.948	2.58054 34079 70643 376	0.38751 52794 99369 747
0.949	2.58312 52420 80516 117	0.38712 79579 12940 390
0.950	2.58570 96593 15846 199	0.38674 10234 54501 207
	$\left[\dfrac{(-7)3}{6}\right]$	$\left[\dfrac{(-8)5}{6}\right]$

EXPONENTIAL FUNCTION

Table 4.4

x	e^x	e^{-x}
0.950	2.58570 96593 15846 199	0.38674 10234 54501 207
0.951	2.58829 66622 61051 072	0.38635 44757 37117 707
0.952	2.59088 62535 03133 898	0.38596 83143 74242 140
0.953	2.59347 84356 31686 135	0.38558 25389 79713 111
0.954	2.59607 32112 38890 126	0.38519 71491 67755 194
0.955	2.59867 05829 19521 695	0.38481 21445 52978 545
0.956	2.60127 05532 70952 740	0.38442 75247 50378 516
0.957	2.60387 31248 93153 828	0.38404 32893 75335 273
0.958	2.60647 83003 88696 799	0.38365 94380 43613 409
0.959	2.60908 60823 62757 366	0.38327 59703 71361 560
0.960	2.61169 64734 23117 718	0.38289 28859 75112 023
0.961	2.61430 94761 80169 136	0.38251 01844 71780 368
0.962	2.61692 50932 46914 592	0.38212 78654 78665 061
0.963	2.61954 33272 38971 373	0.38174 59286 13447 076
0.964	2.62216 41807 74573 688	0.38136 43734 94189 517
0.965	2.62478 76564 74575 291	0.38098 31997 39337 233
0.966	2.62741 37569 62452 101	0.38060 24069 67716 437
0.967	2.63004 24848 64304 825	0.38022 19947 98534 325
0.968	2.63267 38428 08861 583	0.37984 19628 51378 697
0.969	2.63530 78334 27480 539	0.37946 23107 46217 574
0.970	2.63794 44593 54152 532	0.37908 30381 03398 818
0.971	2.64058 37232 25503 708	0.37870 41445 43649 757
0.972	2.64322 56276 80798 158	0.37832 56296 88076 798
0.973	2.64587 01753 61940 558	0.37794 74931 58165 054
0.974	2.64851 73689 13478 808	0.37756 97345 75777 964
0.975	2.65116 72109 82606 682	0.37719 23535 63156 913
0.976	2.65381 97042 19166 470	0.37681 53497 42920 859
0.977	2.65647 48512 75651 628	0.37643 87227 38065 949
0.978	2.65913 26548 07209 434	0.37606 24721 71965 147
0.979	2.66179 31174 71643 642	0.37568 65976 68367 855
0.980	2.66445 62419 29417 138	0.37531 10988 51399 539
0.981	2.66712 20308 43654 602	0.37493 59753 45561 350
0.982	2.66979 04868 80145 169	0.37456 12267 75729 751
0.983	2.67246 16127 07345 099	0.37418 68527 67156 142
0.984	2.67513 54109 96380 441	0.37381 28529 45466 482
0.985	2.67781 18844 21049 708	0.37343 92269 36660 918
0.986	2.68049 10356 57826 547	0.37306 59743 67113 412
0.987	2.68317 28673 85862 418	0.37269 30948 63571 361
0.988	2.68585 73822 86989 272	0.37232 05880 53155 231
0.989	2.68854 45830 45722 235	0.37194 84535 63358 181
0.990	2.69123 44723 49262 289	0.37157 66910 22045 691
0.991	2.69392 70528 87498 962	0.37120 53000 57455 187
0.992	2.69662 23273 53013 016	0.37083 42802 98195 674
0.993	2.69932 02984 41079 142	0.37046 36313 73247 362
0.994	2.70202 09688 49668 652	0.37009 33529 11961 296
0.995	2.70472 43412 79452 181	0.36972 34445 44058 983
0.996	2.70743 04184 33802 382	0.36935 39058 99632 024
0.997	2.71013 92030 18796 637	0.36898 47366 09141 744
0.998	2.71285 06977 43219 755	0.36861 59363 03418 822
0.999	2.71556 49053 18566 687	0.36824 75046 13662 921
1.000	2.71828 18284 59045 235	0.36787 94411 71442 322
	$\left[\dfrac{(-7)3}{6}\right]$	$\left[\dfrac{(-8)5}{6}\right]$

Table 4.4   **EXPONENTIAL FUNCTION**

x	e^x	e^{-x}
0.0	1.00000 00000 00000	1.00000 00000 00000 00000
0.1	1.10517 09180 75648	0.90483 74180 35959 57316
0.2	1.22140 27581 60170	0.81873 07530 77981 85867
0.3	1.34985 88075 76003	0.74081 82206 81717 86607
0.4	1.49182 46976 41270	0.67032 00460 35639 30074
0.5	1.64872 12707 00128	0.60653 06597 12633 42360
0.6	1.82211 88003 90509	0.54881 16360 94026 43263
0.7	2.01375 27074 70477	0.49658 53037 91409 51470
0.8	2.22554 09284 92468	0.44932 89641 17221 59143
0.9	2.45960 31111 56950	0.40656 96597 40599 11188
1.0	2.71828 18284 59045	0.36787 94411 71442 32160
1.1	3.00416 60239 46433	0.33287 10836 98079 55329
1.2	3.32011 69227 36547	0.30119 42119 12202 09664
1.3	3.66929 66676 19244	0.27253 17930 34012 60312
1.4	4.05519 99668 44675	0.24659 69639 41606 47694
1.5	4.48168 90703 38065	0.22313 01601 48429 82893
1.6	4.95303 24243 95115	0.20189 65179 94655 40849
1.7	5.47394 73917 27200	0.18268 35240 52734 65022
1.8	6.04964 74644 12946	0.16529 88882 21586 53830
1.9	6.68589 44422 79269	0.14956 86192 22635 05264
2.0	7.38905 60989 30650	0.13533 52832 36612 69189
2.1	8.16616 99125 67650	0.12245 64282 52981 91022
2.2	9.02501 34994 34121	0.11080 31583 62333 88333
2.3	9.97418 24548 14721	0.10025 88437 22803 73373
2.4	11.02317 63806 41602	0.09071 79532 89412 50338
2.5	12.18249 39607 03473	0.08208 49986 23898 79517
2.6	13.46373 80350 01690	0.07427 35782 14333 88043
2.7	14.87973 17248 72834	0.06720 55127 39749 76513
2.8	16.44464 67710 97050	0.06081 00626 25217 96500
2.9	18.17414 53694 43061	0.05502 32200 56407 22903
3.0	20.08553 69231 87668	0.04978 70683 67863 94298
3.1	22.19795 12814 41633	0.04504 92023 93557 80607
3.2	24.53253 01971 09349	0.04076 22039 78366 21517
3.3	27.11263 89206 57887	0.03688 31674 01240 00545
3.4	29.96410 00473 97013	0.03337 32699 60326 07948
3.5	33.11545 19586 92314	0.03019 73834 22318 50074
3.6	36.59823 44436 77988	0.02732 37224 47292 56080
3.7	40.44730 43600 67391	0.02472 35264 70339 39120
3.8	44.70118 44933 00823	0.02237 07718 56165 59578
3.9	49.40244 91055 30174	0.02024 19114 45804 38847
4.0	54.59815 00331 44239	0.01831 56388 88734 18029
4.1	60.34028 75973 61969	0.01657 26754 01761 24754
4.2	66.68633 10409 25142	0.01499 55768 20477 70621
4.3	73.69979 36995 95797	0.01356 85590 12200 93176
4.4	81.45086 86649 68117	0.01227 73399 03068 44118
4.5	90.01713 13005 21814	0.01110 89965 38242 30650
4.6	99.48431 56419 33809	0.01005 18357 44633 58164
4.7	109.94717 24521 23499	0.00909 52771 01695 81709
4.8	121.51041 75187 34881	0.00822 97470 49020 02884
4.9	134.28977 96849 35485	0.00744 65830 70924 34052
5.0	148.41315 91025 76603	0.00673 79469 99085 46710

From C. E. Van Orstrand, Tables of the exponential function and of the circular sine and cosine to radian arguments, Memoirs of the National Academy of Sciences, vol. 14, Fifth Memoir. U.S. Government Printing Office, Washington, D.C., 1921 (with permission) for e^{-x}, $x \leq 2.4$.

EXPONENTIAL FUNCTION

Table 4.4

x	e^x	e^{-x}
5.0	148.41315 91025 77	0.00673 79469 99085 46710
5.1	164.02190 72999 02	0.00609 67465 65515 63611
5.2	181.27224 18751 51	0.00551 65644 20760 77242
5.3	200.33680 99747 92	0.00499 15939 06910 21621
5.4	221.40641 62041 87	0.00451 65809 42612 66798
5.5	244.69193 22642 20	0.00408 67714 38464 06699
5.6	270.42640 74261 53	0.00369 78637 16482 93082
5.7	298.86740 09670 60	0.00334 59654 57471 27277
5.8	330.29955 99096 49	0.00302 75547 45375 81475
5.9	365.03746 78653 29	0.00273 94448 18768 36923
6.0	403.42879 34927 35	0.00247 87521 76666 35842
6.1	445.85777 00825 17	0.00224 28677 19485 80247
6.2	492.74904 10932 56	0.00202 94306 36295 73436
6.3	544.57191 01259 29	0.00183 63047 77028 90683
6.4	601.84503 78720 82	0.00166 15572 73173 93450
6.5	665.14163 30443 62	0.00150 34391 92977 57245
6.6	735.09518 92419 73	0.00136 03680 37547 89342
6.7	812.40582 51675 43	0.00123 09119 02673 48118
6.8	897.84729 16504 18	0.00111 37751 47844 80308
6.9	992.27471 56050 26	0.00100 77854 29048 51076
7.0	1096.63315 84284 59	0.00091 18819 65554 51621
7.1	1211.96707 44925 77	0.00082 51049 23265 90427
7.2	1339.43076 43944 18	0.00074 65858 08376 67937
7.3	1480.29992 75845 45	0.00067 55387 75193 84424
7.4	1635.98442 99959 27	0.00061 12527 61129 57256
7.5	1808.04241 44560 63	0.00055 30843 70147 83358
7.6	1998.19589 51041 18	0.00050 04514 33440 61070
7.7	2208.34799 18872 09	0.00045 28271 82886 79706
7.8	2440.60197 76244 99	0.00040 97349 78979 78671
7.9	2697.28232 82685 09	0.00037 07435 40459 08837
8.0	2980.95798 70417 28	0.00033 54626 27902 51184
8.1	3294.46807 52838 41	0.00030 35391 38078 86666
8.2	3640.95030 73323 55	0.00027 46535 69972 14233
8.3	4023.87239 38223 10	0.00024 85168 27107 95202
8.4	4447.06674 76998 56	0.00022 48673 24178 84827
8.5	4914.76884 02991 34	0.00020 34683 69010 64417
8.6	5431.65959 13629 80	0.00018 41057 93667 57912
8.7	6002.91221 72610 22	0.00016 65858 10987 63341
8.8	6634.24400 62778 85	0.00015 07330 75095 47660
8.9	7331.97353 91559 93	0.00013 63889 26482 01145
9.0	8103.08392 75753 84	0.00012 34098 04086 67955
9.1	8955.29270 34825 12	0.00011 16658 08490 11474
9.2	9897.12905 87439 16	0.00010 10394 01837 09335
9.3	10938.01920 81651 84	0.00009 14242 31478 17334
9.4	12088.38073 02169 84	0.00008 27240 65556 63226
9.5	13359.72682 96618 72	0.00007 48518 29887 70059
9.6	14764.78156 55772 73	0.00006 77287 36490 85387
9.7	16317.60719 80154 32	0.00006 12834 95053 22210
9.8	18033.74492 78285 11	0.00005 54515 99432 17698
9.9	19930.37043 82302 89	0.00005 01746 82056 17530
10.0	22026.46579 48067 17	0.00004 53999 29762 48485

Table 4.4   **EXPONENTIAL FUNCTION**

x	e^x	e^{-x}
0	(0)1.00000 00000 00000 000	(0)1.00000 00000 00000 000
1	(0)2.71828 18284 59045 235	(− 1)3.67879 44117 14423 216
2	(0)7.38905 60989 30650 227	(− 1)1.35335 28323 66126 919
3	(1)2.00855 36923 18766 774	(− 2)4.97870 68367 86394 298
4	(1)5.45981 50033 14423 908	(− 2)1.83156 38888 73418 029
5	(2)1.48413 15910 25766 034	(− 3)6.73794 69990 85467 097
6	(2)4.03428 79349 27351 226	(− 3)2.47875 21766 66358 423
7	(3)1.09663 31584 28458 599	(− 4)9.11881 96555 45162 080
8	(3)2.98095 79870 41728 275	(− 4)3.35462 62790 25118 388
9	(3)8.10308 39275 75384 008	(− 4)1.23409 80408 66795 495
10	(4)2.20264 65794 80671 652	(− 5)4.53999 29762 48485 154
11	(4)5.98741 41715 19781 846	(− 5)1.67017 00790 24565 931
12	(5)1.62754 79141 90039 208	(− 6)6.14421 23533 28209 759
13	(5)4.42413 39200 89205 033	(− 6)2.26032 94069 81054 326
14	(6)1.20260 42841 64776 778	(− 7)8.31528 71910 35678 841
15	(6)3.26901 73724 72110 639	(− 7)3.05902 32050 18257 884
16	(6)8.88611 05205 07872 637	(− 7)1.12535 17471 92591 145
17	(7)2.41549 52753 57529 821	(− 8)4.13993 77187 85166 660
18	(7)6.56599 69137 33051 114	(− 8)1.52299 79744 71262 844
19	(8)1.78482 30096 31872 608	(− 9)5.60279 64375 37267 540
20	(8)4.85165 19540 97902 780	(− 9)2.06115 36224 38557 828
21	(9)1.31881 57344 83214 697	(−10)7.58256 04279 11906 728
22	(9)3.58491 28461 31591 562	(−10)2.78946 80928 68924 808
23	(9)9.74480 34462 48902 600	(−10)1.02618 79631 70189 030
24	(10)2.64891 22129 84347 229	(−11)3.77513 45442 79097 752
25	(10)7.20048 99337 38587 252	(−11)1.38879 43864 96402 059
26	(11)1.95729 60942 88387 643	(−12)5.10908 90280 63324 720
27	(11)5.32048 24060 17986 167	(−12)1.87952 88165 39083 295
28	(12)1.44625 70642 91475 174	(−13)6.91440 01069 40203 009
29	(12)3.93133 42971 44042 074	(−13)2.54366 56473 76922 910
30	(13)1.06864 74581 52446 215	(−14)9.35762 29688 40174 605
31	(13)2.90488 49665 24742 523	(−14)3.44247 71084 69976 458
32	(13)7.89629 60182 68069 516	(−14)1.26641 65549 09417 572
33	(14)2.14643 57978 59160 646	(−15)4.65888 61451 03397 364
34	(14)5.83461 74252 74548 814	(−15)1.71390 84315 42012 966
35	(15)1.58601 34523 13430 728	(−16)6.30511 67601 46989 386
36	(15)4.31123 15471 15195 227	(−16)2.31952 28302 43569 388
37	(16)1.17191 42372 80261 131	(−17)8.53304 76257 44065 794
38	(16)3.18559 31757 11375 622	(−17)3.13913 27920 48029 629
39	(16)8.65934 00423 99374 695	(−17)1.15482 24173 01578 599
40	(17)2.35385 26683 70199 854	(−18)4.24835 42552 91588 995
41	(17)6.39843 49353 00549 492	(−18)1.56288 21893 34988 768
42	(18)1.73927 49415 20501 047	(−19)5.74952 22642 93559 807
43	(18)4.72783 94682 29346 561	(−19)2.11513 10375 91080 487
44	(19)1.28516 00114 35930 828	(−20)7.78113 22411 33796 516
45	(19)3.49342 71057 48509 535	(−20)2.86251 85805 49393 644
46	(19)9.49611 94206 02448 875	(−20)1.05306 17357 55381 238
47	(20)2.58131 28861 90067 396	(−21)3.87399 76286 87187 113
48	(20)7.01673 59120 97631 739	(−21)1.42516 40827 40935 106
49	(21)1.90734 65724 95099 691	(−22)5.24288 56633 63463 937
50	(21)5.18470 55285 87072 464	(−22)1.92874 98479 63917 783

EXPONENTIAL FUNCTION

Table 4.4

x	e^x	e^{-x}
50	(21) 5.18470 55285 87072 464	(−22) 1.92874 98479 63917 783
51	(22) 1.40934 90824 26938 796	(−23) 7.09547 41622 84704 139
52	(22) 3.83100 80007 16576 849	(−23) 2.61027 90696 67704 805
53	(23) 1.04137 59433 02908 780	(−24) 9.60268 00545 08676 030
54	(23) 2.83075 33032 74693 900	(−24) 3.53262 85722 00807 030
55	(23) 7.69478 52651 42017 138	(−24) 1.29958 14250 07503 074
56	(24) 2.09165 94960 12996 154	(−25) 4.78089 28838 85469 081
57	(24) 5.68571 99993 35932 223	(−25) 1.75879 22024 24311 649
58	(25) 1.54553 89355 90103 930	(−26) 6.47023 49256 45460 326
59	(25) 4.20121 04037 90514 255	(−26) 2.38026 64086 94400 606
60	(26) 1.14200 73898 15684 284	(−27) 8.75651 07626 96520 338
61	(26) 3.10429 79357 01919 909	(−27) 3.22134 02859 92516 089
62	(26) 8.43835 66687 41454 489	(−27) 1.18506 48642 33981 006
63	(27) 2.29378 31594 69609 879	(−28) 4.35961 00000 63080 974
64	(27) 6.23514 90808 11616 883	(−28) 1.60381 08905 48637 853
65	(28) 1.69488 92444 10333 714	(−29) 5.90009 05415 97061 391
66	(28) 4.60718 66343 31291 543	(−29) 2.17052 20113 03639 412
67	(29) 1.25236 31708 42213 781	(−30) 7.98490 42456 86978 808
68	(29) 3.40427 60499 31740 521	(−30) 2.93748 21117 10802 947
69	(29) 9.25378 17255 87787 600	(−30) 1.08063 92777 07278 495
70	(30) 2.51543 86709 19167 006	(−31) 3.97544 97359 08646 808
71	(30) 6.83767 12297 62743 867	(−31) 1.46248 62272 51230 947
72	(31) 1.85867 17452 84127 980	(−32) 5.38018 61600 21138 414
73	(31) 5.05239 36302 76104 195	(−32) 1.97925 98779 46904 554
74	(32) 1.37338 29795 40176 188	(−33) 7.28129 01783 21643 834
75	(32) 3.73324 19967 99001 640	(−33) 2.67863 69618 08077 944
76	(33) 1.01480 03881 13888 728	(−34) 9.85415 46861 11258 029
77	(33) 2.75851 34545 23170 206	(−34) 3.62514 09191 43559 224
78	(33) 7.49841 69969 90120 435	(−34) 1.33361 48155 02261 341
79	(34) 2.03828 10665 12668 767	(−35) 4.90609 47306 49280 566
80	(34) 5.54062 23843 93510 053	(−35) 1.80485 13878 45415 172
81	(35) 1.50609 73145 85030 548	(−36) 6.63967 71995 80734 401
82	(35) 4.09399 69621 27454 697	(−36) 2.44260 07377 40527 679
83	(36) 1.11286 37547 91759 412	(−37) 8.98582 59440 49380 670
84	(36) 3.02507 73222 01142 338	(−37) 3.30570 06267 60734 298
85	(36) 8.22301 27146 22913 510	(−37) 1.21609 92992 52825 564
86	(37) 2.23524 66037 34715 047	(−38) 4.47377 93061 81120 735
87	(37) 6.07603 02250 56872 150	(−38) 1.64581 14310 82273 651
88	(38) 1.65163 62549 94001 856	(−39) 6.05460 18954 01185 885
89	(38) 4.48961 28191 74345 246	(−39) 2.22736 35617 95743 739
90	(39) 1.22040 32943 17840 802	(−40) 8.19401 26239 90515 430
91	(39) 3.31740 00983 35742 626	(−40) 3.01440 87850 65374 553
92	(39) 9.01762 84050 34298 931	(−40) 1.10893 90193 12136 379
93	(40) 2.45124 55429 20085 786	(−41) 4.07955 86671 77560 158
94	(40) 6.66317 62164 10895 834	(−41) 1.50078 57627 07394 888
95	(41) 1.81123 90828 89023 282	(−42) 5.52108 22770 28532 732
96	(41) 4.92345 82860 12058 400	(−42) 2.03109 26627 34810 926
97	(42) 1.33833 47192 04269 500	(−43) 7.47197 23373 42990 161
98	(42) 3.63797 09476 08804 579	(−43) 2.74878 50079 10214 930
99	(42) 9.88903 03193 46946 771	(−43) 1.01122 14926 10448 530
100	(43) 2.68811 71418 16135 448	(−44) 3.72007 59760 20835 963

For $|x| > 100$ see **Example 11**.

Table 4.5   **RADIX TABLE OF THE EXPONENTIAL FUNCTION**

x	n	$e^{x10^{-n}}$	$e^{-x10^{-n}}$
1	10	1. 00000 00001 00000 00000 50000	0. 99999 99999 00000 00000 50000
2	10	1. 00000 00002 00000 00002 00000	0. 99999 99998 00000 00002 00000
3	10	1. 00000 00003 00000 00004 50000	0. 99999 99997 00000 00004 50000
4	10	1. 00000 00004 00000 00008 00000	0. 99999 99996 00000 00008 00000
5	10	1. 00000 00005 00000 00012 50000	0. 99999 99995 00000 00012 50000
6	10	1. 00000 00006 00000 00018 00000	0. 99999 99994 00000 00018 00000
7	10	1. 00000 00007 00000 00024 50000	0. 99999 99993 00000 00024 50000
8	10	1. 00000 00008 00000 00032 00000	0. 99999 99992 00000 00032 00000
9	10	1. 00000 00009 00000 00040 50000	0. 99999 99991 00000 00040 50000
1	9	1. 00000 00010 00000 00050 00000	0. 99999 99990 00000 00050 00000
2	9	1. 00000 00020 00000 00200 00000	0. 99999 99980 00000 00200 00000
3	9	1. 00000 00030 00000 00450 00000	0. 99999 99970 00000 00450 00000
4	9	1. 00000 00040 00000 00800 00000	0. 99999 99960 00000 00800 00000
5	9	1. 00000 00050 00000 01250 00000	0. 99999 99950 00000 01250 00000
6	9	1. 00000 00060 00000 01800 00000	0. 99999 99940 00000 01800 00000
7	9	1. 00000 00070 00000 02450 00001	0. 99999 99930 00000 02449 99999
8	9	1. 00000 00080 00000 03200 00001	0. 99999 99920 00000 03199 99999
9	9	1. 00000 00090 00000 04050 00001	0. 99999 99910 00000 04049 99999
1	8	1. 00000 00100 00000 05000 00002	0. 99999 99900 00000 04999 99998
2	8	1. 00000 00200 00000 20000 00013	0. 99999 99800 00000 19999 99987
3	8	1. 00000 00300 00000 45000 00045	0. 99999 99700 00000 44999 99955
4	8	1. 00000 00400 00000 80000 00107	0. 99999 99600 00000 79999 99893
5	8	1. 00000 00500 00001 25000 00208	0. 99999 99500 00001 24999 99792
6	8	1. 00000 00600 00001 80000 00360	0. 99999 99400 00001 79999 99640
7	8	1. 00000 00700 00002 45000 00572	0. 99999 99300 00002 44999 99428
8	8	1. 00000 00800 00003 20000 00853	0. 99999 99200 00003 19999 99147
9	8	1. 00000 00900 00004 05000 01215	0. 99999 99100 00004 04999 98785
1	7	1. 00000 01000 00005 00000 01667	0. 99999 99000 00004 99999 98333
2	7	1. 00000 02000 00020 00000 13333	0. 99999 98000 00019 99999 86667
3	7	1. 00000 03000 00045 00000 45000	0. 99999 97000 00044 99999 55000
4	7	1. 00000 04000 00080 00001 06667	0. 99999 96000 00079 99998 93333
5	7	1. 00000 05000 00125 00002 08333	0. 99999 95000 00124 99997 91667
6	7	1. 00000 06000 00180 00003 60000	0. 99999 94000 00179 99996 40000
7	7	1. 00000 07000 00245 00005 71667	0. 99999 93000 00244 99994 28333
8	7	1. 00000 08000 00320 00008 53334	0. 99999 92000 00319 99991 46667
9	7	1. 00000 09000 00405 00012 15000	0. 99999 91000 00404 99987 85000
1	6	1. 00000 10000 00500 00016 66667	0. 99999 90000 00499 99983 33334
2	6	1. 00000 20000 02000 00133 33340	0. 99999 80000 01999 99866 66673
3	6	1. 00000 30000 04500 00450 00034	0. 99999 70000 04499 99550 00034
4	6	1. 00000 40000 08000 01066 66773	0. 99999 60000 07999 98933 33440
5	6	1. 00000 50000 12500 02083 33594	0. 99999 50000 12499 97916 66927
6	6	1. 00000 60000 18000 03600 00540	0. 99999 40000 17999 96400 00540
7	6	1. 00000 70000 24500 05716 67667	0. 99999 30000 24499 94283 34334
8	6	1. 00000 80000 32000 08533 35040	0. 99999 20000 31999 91466 68373
9	6	1. 00000 90000 40500 12150 02734	0. 99999 10000 40499 87850 02734

For $n>10$, $e^{\pm x10^{-n}}=1\pm x10^{-n}+\frac{1}{2}\,x^2 10^{-2n}$ to 25D.

Compiled from C. E. Van Orstrand, Tables of the exponential function and of the circular sine and cosine to radian arguments, Memoirs of the National Academy of Sciences, vol. 14, Fifth Memoir. U.S. Government Printing Office, Washington, D.C., 1921 (with permission).

RADIX TABLE OF THE EXPONENTIAL FUNCTION   Table 4.5

x	n	$e^{x10^{-n}}$	$e^{-x10^{-n}}$
1	5	1.00001 00000 50000 16666 70833	0.99999 00000 49999 83333 37500
2	5	1.00002 00002 00001 33334 00000	0.99998 00001 99998 66667 33333
3	5	1.00003 00004 50004 50003 37502	0.99997 00004 49995 50003 37498
4	5	1.00004 00008 00010 66677 33342	0.99996 00007 99989 33343 99991
5	5	1.00005 00012 50020 83359 37526	0.99995 00012 49979 16692 70807
6	5	1.00006 00018 00036 00054 00065	0.99994 00017 99964 00053 99935
7	5	1.00007 00024 50057 16766 70973	0.99993 00024 49942 83433 37360
8	5	1.00008 00032 00085 33504 00273	0.99992 00031 99914 66837 33060
9	5	1.00009 00040 50121 50273 37992	0.99991 00040 49878 50273 37008
1	4	1.00010 00050 00166 67083 34167	0.99990 00049 99833 33749 99167
2	4	1.00020 00200 01333 40000 26668	0.99980 00199 98666 73333 06668
3	4	1.00030 00450 04500 33752 02510	0.99970 00449 95500 33747 97510
4	4	1.00040 00800 10667 73341 86724	0.99960 00799 89334 39991 46724
5	4	1.00050 01250 20835 93776 04384	0.99950 01249 79169 27057 29384
6	4	1.00060 01800 36005 40064 80648	0.99940 01799 64005 39935 20648
7	4	1.00070 02450 57176 67223 40801	0.99930 02449 42843 33609 95801
8	4	1.00080 03200 85350 40273 10308	0.99920 03199 14683 73060 30307
9	4	1.00090 04051 21527 34242 14882	0.99910 04048 78527 33257 99880
1	3	1.00100 05001 66708 34166 80558	0.99900 04998 33374 99166 80554
2	3	1.00200 20013 34000 26675 55810	0.99800 19986 67333 06675 55302
3	3	1.00300 45045 03377 02601 29341	0.99700 44955 03372 97601 20662
4	3	1.00400 80106 77341 87235 88080	0.99600 79893 43991 47235 23064
5	3	1.00501 25208 59401 06338 35662	0.99501 24791 92682 31335 25642
6	3	1.00601 80360 54064 86485 55845	0.99401 79640 53935 26474 44988
7	3	1.00702 45572 66848 55523 16000	0.99302 44429 33235 10490 47970
8	3	1.00803 20855 04273 43117 20736	0.99203 19148 37060 63033 98697
9	3	1.00904 06217 73867 81406 25705	0.99104 03787 72883 66216 45648
1	2	1.01005 01670 84168 05754 21655	0.99004 98337 49168 05357 39060
2	2	1.02020 13400 26755 81016 01439	0.98019 86733 06755 30222 08141
3	2	1.03045 45339 53516 85561 24400	0.97044 55335 48508 17693 25284
4	2	1.04081 07741 92388 22675 70448	0.96078 94391 52323 20943 92107
5	2	1.05127 10963 76024 03969 75176	0.95122 94245 00714 00909 14253
6	2	1.06183 65465 45359 62222 46849	0.94176 45335 84248 70953 71528
7	2	1.07250 81812 54216 47905 31039	0.93239 38199 05948 22885 79726
8	2	1.08328 70676 74958 55443 59878	0.92311 63463 86635 78291 07598
9	2	1.09417 42837 05210 35787 28976	0.91393 11852 71228 18674 73535
1	1	1.10517 09180 75647 62481 17078	0.90483 74180 35959 57316 42491
2	1	1.22140 27581 60169 83392 10720	0.81873 07530 77981 85866 99355
3	1	1.34985 88075 76003 10398 37443	0.74081 82206 81717 86606 68738
4	1	1.49182 46976 41270 31782 48530	0.67032 00460 35639 30074 44329
5	1	1.64872 12707 00128 14684 86508	0.60653 06597 12633 42360 37995
6	1	1.82211 88003 90508 97487 53677	0.54881 16360 94026 43262 84589
7	1	2.01375 27074 70476 52162 45494	0.49658 53037 91409 51470 48001
8	1	2.22554 09284 92467 60457 95375	0.44932 89641 17221 59143 01024
9	1	2.45960 31111 56949 66380 01266	0.40656 96597 40599 11188 34542
1	0	2.71828 18284 59045 23536 02875	0.36787 94411 71442 32159 55238

Table 4.6   CIRCULAR SINES AND COSINES FOR RADIAN ARGUMENTS

x	$\sin x$	$\cos x$
0.000	0.00000 00000 00000 00000 000	1.00000 00000 00000 00000 000
0.001	0.00099 99998 33333 34166 667	0.99999 95000 00041 66666 528
0.002	0.00199 99986 66666 93333 331	0.99999 80000 00666 66657 778
0.003	0.00299 99955 00002 02499 957	0.99999 55000 03374 99898 750
0.004	0.00399 99893 33341 86666 342	0.99999 20000 10666 66097 778
0.005	0.00499 99791 66692 70831 783	0.99998 75000 26041 64496 529
0.006	0.00599 99640 00064 79994 446	0.99998 20000 53999 93520 004
0.007	0.00699 99428 33473 39150 327	0.99997 55001 00041 50326 542
0.008	0.00799 99146 66939 73291 723	0.99996 80001 70666 30257 819
0.009	0.00899 98785 00492 07405 100	0.99995 95002 73374 26188 857
0.010	0.00999 98333 34166 66468 254	0.99995 00004 16665 27778 026
0.011	0.01099 97781 68008 75446 684	0.99993 95006 10039 20617 059
0.012	0.01199 97120 02073 59289 053	0.99992 80008 63995 85281 066
0.013	0.01299 96338 36427 42921 659	0.99991 55011 90034 96278 551
0.014	0.01399 95426 71148 51241 801	0.99990 20016 00656 20901 438
0.015	0.01499 94375 06328 09109 944	0.99988 75021 09359 17975 106
0.016	0.01599 93173 42071 41340 585	0.99987 20027 30643 36508 430
0.017	0.01699 91811 78498 72691 726	0.99985 55034 80008 14243 829
0.018	0.01799 90280 15746 27852 832	0.99983 80043 73952 76107 331
0.019	0.01899 88568 53967 31431 205	0.99981 95054 29976 32558 650
0.020	0.01999 86666 93333 07936 649	0.99980 00066 66577 77841 270
0.021	0.02099 84565 34033 81764 335	0.99977 95081 03255 88132 556
0.022	0.02199 82253 76279 77175 771	0.99975 80097 60509 19593 878
0.023	0.02299 79722 20302 18277 769	0.99973 55116 59836 06320 750
0.024	0.02399 76960 66354 28999 311	0.99971 20138 23734 58193 002
0.025	0.02499 73959 14712 33066 217	0.99968 75162 75702 58624 967
0.026	0.02599 70707 65676 53973 517	0.99966 20190 40237 62215 698
0.027	0.02699 67196 19572 14955 411	0.99963 55221 42836 92299 214
0.028	0.02799 63414 76750 38952 746	0.99960 80256 09997 38394 779
0.029	0.02899 59353 37589 48577 881	0.99957 95294 69215 53557 207
0.030	0.02999 55002 02495 66076 853	0.99955 00337 48987 51627 216
0.031	0.03099 50350 71904 13288 752	0.99951 95384 78809 04381 810
0.032	0.03199 45389 46280 11602 188	0.99948 80436 89175 38584 710
0.033	0.03299 40108 26119 81908 762	0.99945 55494 11581 32936 824
0.034	0.03399 34497 11951 44553 435	0.99942 20556 78521 14926 773
0.035	0.03499 28546 04336 19281 702	0.99938 75625 23488 57581 460
0.036	0.03599 22245 03869 25183 461	0.99935 20699 80976 76116 700
0.037	0.03699 15584 11180 80633 489	0.99931 55780 86478 24487 902
0.038	0.03799 08553 26937 03228 414	0.99927 80868 76484 91840 819
0.039	0.03899 01142 51841 09720 085	0.99923 95963 88487 98862 358
0.040	0.03998 93341 86634 15945 255	0.99920 01066 60977 94031 457
0.041	0.04098 85141 32096 36751 449	0.99915 96177 33444 49770 040
0.042	0.04198 76530 89047 85918 946	0.99911 81296 46376 58494 043
0.043	0.04298 67500 58349 76078 755	0.99907 56424 41262 28564 524
0.044	0.04398 58040 40905 18626 492	0.99903 21561 60588 80138 853
0.045	0.04498 48140 37660 23632 066	0.99898 76708 47842 40921 992
0.046	0.04598 37790 49604 99745 054	0.99894 21865 47508 41817 869
0.047	0.04698 26980 77774 54095 689	0.99889 57033 05071 12480 849
0.048	0.04798 15701 23249 92191 340	0.99884 82211 67013 76767 299
0.049	0.04898 03941 87159 17808 403	0.99879 97401 80818 48087 272
0.050	0.04997 91692 70678 32879 487	0.99875 02603 94966 24656 287
	$\begin{bmatrix} (-9)6 \\ 7 \end{bmatrix}$	$\begin{bmatrix} (-7)1 \\ 7 \end{bmatrix}$

For conversion from degrees to radians see **Example 13**.

For use and extension of the table see **Examples 15–17**.

From C. E. Van Orstrand, Tables of the exponential function and of the circular sine and cosine to radian arguments, Memoirs of the National Academy of Sciences, vol. 14, Fifth Memoir. U.S. Government Printing Office, Washington, D.C., 1921 (with permission). Known errors have been corrected.

CIRCULAR SINES AND COSINES FOR RADIAN ARGUMENTS   Table 4.6

x	$\sin x$	$\cos x$
0.050	0.04997 91692 70678 32879 487	0.99875 02603 94966 24656 287
0.051	0.05097 78943 75032 37375 800	0.99869 97818 58936 84647 237
0.052	0.05197 65685 01496 29184 649	0.99864 83046 23208 81242 407
0.053	0.05297 51906 51396 03981 925	0.99859 58287 39259 37585 623
0.054	0.05397 37598 26109 55099 505	0.99854 23542 59564 41634 531
0.055	0.05497 22750 27067 73387 446	0.99848 78812 37598 40913 005
0.056	0.05597 07352 55755 47070 891	0.99843 24097 27834 37163 704
0.057	0.05696 91395 13712 61601 567	0.99837 59397 85743 80900 770
0.058	0.05796 74868 02534 99503 794	0.99831 84714 67796 65862 676
0.059	0.05896 57761 23875 40214 896	0.99826 00048 31461 23365 235
0.060	0.05996 40064 79444 59919 909	0.99820 05399 35204 16554 766
0.061	0.06096 21768 71012 31380 500	0.99814 00768 38490 34561 437
0.062	0.06196 02863 00408 23757 982	0.99807 86156 01782 86552 769
0.063	0.06295 83337 69523 02430 343	0.99801 61562 86542 95687 334
0.064	0.06395 63182 80309 28803 166	0.99795 26989 55229 92968 628
0.065	0.06495 42388 34782 60114 361	0.99788 82436 71301 10999 144
0.066	0.06595 20944 35022 49232 601	0.99782 27904 99211 77634 635
0.067	0.06694 98840 83173 44449 361	0.99775 63395 04415 09538 592
0.068	0.06794 76067 81445 89264 458	0.99768 88907 53362 05636 926
0.069	0.06894 52615 32117 22165 004	0.99762 04443 13501 40472 866
0.070	0.06994 28473 37532 76397 655	0.99755 10002 53279 57462 091
0.071	0.07094 03632 00106 79734 071	0.99748 05586 42140 62048 084
0.072	0.07193 78081 22323 54229 480	0.99740 91195 50526 14757 726
0.073	0.07293 51811 06738 15974 250	0.99733 66830 49875 24157 139
0.074	0.07393 24811 55977 74838 360	0.99726 32492 12624 39707 777
0.075	0.07492 97072 72742 34208 684	0.99718 88181 12207 44522 774
0.076	0.07592 68584 59805 90718 980	0.99711 33898 23055 48023 568
0.077	0.07692 39337 20017 33972 485	0.99703 69644 20596 78496 785
0.078	0.07792 09320 56301 46257 015	0.99695 95419 81256 75551 417
0.079	0.07891 78524 71660 02252 478	0.99688 11225 82457 82476 279
0.080	0.07991 46939 69172 68730 688	0.99680 17063 02619 38497 771
0.081	0.08091 14555 51998 04247 389	0.99672 12932 21157 70937 933
0.082	0.08190 81362 23374 58826 394	0.99663 98834 18485 87272 823
0.083	0.08290 47349 86621 73635 718	0.99655 74769 76013 67091 212
0.084	0.08390 12508 45140 80655 638	0.99647 40739 76147 53953 598
0.085	0.08489 76828 02416 02338 544	0.99638 96745 02290 47151 570
0.086	0.08589 40298 62015 51260 514	0.99630 42786 38841 93367 506
0.087	0.08689 02910 27592 29764 492	0.99621 78864 71197 78234 626
0.088	0.08788 64653 02885 29594 973	0.99613 04980 85750 17797 412
0.089	0.08888 25516 91720 31524 112	0.99604 21135 69887 49872 388
0.090	0.08987 85491 98011 04969 125	0.99595 27330 11994 25309 284
0.091	0.09087 44568 25760 07600 919	0.99586 23565 01450 99152 586
0.092	0.09187 02735 79059 84943 819	0.99577 09841 28634 21703 483
0.093	0.09286 59984 62093 69966 323	0.99567 86159 84916 29482 217
0.094	0.09386 16304 79136 82662 751	0.99558 52521 62665 36090 844
0.095	0.09485 71686 34557 29625 724	0.99549 08927 55245 22976 426
0.096	0.09585 26119 32817 03609 347	0.99539 55378 57015 30094 649
0.097	0.09684 79593 78472 83083 006	0.99529 91875 63330 46473 881
0.098	0.09784 32099 76177 31775 683	0.99520 18419 70541 00679 686
0.099	0.09883 83627 30679 98210 683	0.99510 35011 75992 51179 796
0.100	0.09983 34166 46828 15230 681	0.99500 41652 78025 76609 556
	$\begin{bmatrix} (-8)1 \\ 7 \end{bmatrix}$	$\begin{bmatrix} (-7)1 \\ 7 \end{bmatrix}$

Table 4.6   **CIRCULAR SINES AND COSINES FOR RADIAN ARGUMENTS**

x	sin x	cos x
0.100	0.09983 34166 46828 15230 681	0.99500 41652 78025 76609 556
0.101	0.10082 83707 29567 99512 975	0.99490 38343 75976 65937 840
0.102	0.10182 32239 83945 51074 864	0.99480 25085 70176 08533 469
0.103	0.10281 79754 15107 52769 040	0.99470 01879 61949 84132 117
0.104	0.10381 26240 28302 69768 897	0.99459 68726 53618 52703 737
0.105	0.10480 71688 28882 49043 655	0.99449 25627 48497 44220 501
0.106	0.10580 16088 22302 18823 209	0.99438 72583 50896 48325 268
0.107	0.10679 59430 14121 88052 588	0.99428 09595 66120 03900 596
0.108	0.10779 01704 10007 45835 941	0.99417 36665 00466 88538 307
0.109	0.10878 42900 15731 60869 939	0.99406 53792 61230 07909 607
0.110	0.10977 83008 37174 80866 495	0.99395 60979 56696 85035 784
0.111	0.11077 22018 80326 31964 714	0.99384 58226 96148 49459 483
0.112	0.11176 59921 51285 18131 952	0.99373 45535 89860 26316 578
0.113	0.11275 96706 56261 20553 909	0.99362 22907 49101 25308 652
0.114	0.11375 32364 01575 97013 636	0.99350 90342 86134 29576 080
0.115	0.11474 66883 93663 81259 372	0.99339 47843 14215 84471 755
0.116	0.11574 00256 39072 82361 097	0.99327 95409 47595 86235 439
0.117	0.11673 32471 44465 84055 722	0.99316 33043 01517 70568 768
0.118	0.11772 63519 16621 44080 790	0.99304 60744 92218 01110 921
0.119	0.11871 93389 62434 93496 613	0.99292 78516 36926 57814 950
0.120	0.11971 22072 88919 35996 735	0.99280 86358 53866 25224 810
0.121	0.12070 49559 03206 47206 615	0.99268 84272 62252 80653 067
0.122	0.12169 75838 12547 73970 447	0.99256 72259 82294 82259 329
0.123	0.12269 00900 24315 33626 003	0.99244 50321 35193 57029 382
0.124	0.12368 24735 46003 13267 407	0.99232 18458 43142 88655 070
0.125	0.12467 47333 85227 68995 744	0.99219 76672 29329 05314 910
0.126	0.12566 68685 49729 25157 389	0.99207 24964 17930 67355 462
0.127	0.12665 88780 47372 73569 978	0.99194 63335 34118 54873 474
0.128	0.12765 07608 86148 72735 909	0.99181 91787 04055 55198 803
0.129	0.12864 25160 74174 47043 273	0.99169 10320 54896 50278 123
0.130	0.12963 41426 19694 85954 121	0.99156 18937 14788 03959 451
0.131	0.13062 56395 31083 43179 968	0.99143 17638 12868 49177 481
0.132	0.13161 70058 16843 35844 433	0.99130 06424 79267 75039 751
0.133	0.13260 82404 85608 43632 907	0.99116 85298 45107 13813 659
0.134	0.13359 93425 46144 07929 171	0.99103 54260 42499 27814 325
0.135	0.13459 03110 07348 30938 844	0.99090 13312 04547 96193 339
0.136	0.13558 11448 78252 74799 575	0.99076 62454 65348 01628 375
0.137	0.13657 18431 68023 60677 867	0.99063 01689 59985 16913 714
0.138	0.13756 24048 85962 67852 453	0.99049 31018 24535 91451 667
0.139	0.13855 28290 41508 32784 107	0.99035 50441 96067 37644 937
0.140	0.13954 31146 44236 48171 799	0.99021 59962 12637 17189 895
0.141	0.14053 32607 03861 61995 092	0.99007 59580 13293 27270 829
0.142	0.14152 32662 30237 76542 691	0.98993 49421 38073 86655 145
0.143	0.14251 31302 33359 47427 025	0.98979 29115 28007 21689 546
0.144	0.14350 28517 23362 82584 791	0.98964 99035 25111 52197 214
0.145	0.14449 24297 10526 41263 332	0.98950 59058 72394 77275 984
0.146	0.14548 18632 05272 32992 773	0.98936 09187 13854 60997 551
0.147	0.14647 11512 18167 16543 800	0.98921 49421 94478 18007 704
0.148	0.14746 02927 59922 98870 997	0.98906 79764 60241 99027 617
0.149	0.14844 92868 41398 34041 627	0.98892 00216 58111 76256 193
0.150	0.14943 81324 73599 22149 773	0.98877 10779 36042 28673 498
	$\begin{bmatrix} (-8)2 \\ 7 \end{bmatrix}$	$\begin{bmatrix} (-7)1 \\ 7 \end{bmatrix}$

CIRCULAR SINES AND COSINES FOR RADIAN ARGUMENTS   Table 4.6

x	sin x	cos x
0.150	0.14943 81324 73599 22149 773	0.98877 10779 36042 28673 498
0.151	0.15042 68286 67680 08215 725	0.98862 11454 42977 27245 283
0.152	0.15141 53744 34944 81070 532	0.98847 02243 28849 20028 611
0.153	0.15240 37687 86847 72225 604	0.98831 83147 44579 17178 614
0.154	0.15339 20107 34994 54727 267	0.98816 54168 42076 75856 382
0.155	0.15438 00992 91143 41996 190	0.98801 15307 74239 85038 006
0.156	0.15536 80334 67205 86651 555	0.98785 66566 94954 50224 794
0.157	0.15635 58122 75247 79319 902	0.98770 07947 59094 78054 663
0.158	0.15734 34347 27490 47428 529	0.98754 39451 22522 60814 736
0.159	0.15833 08998 36311 53983 354	0.98738 61079 42087 60855 150
0.160	0.15931 82066 14245 96331 146	0.98722 72833 75626 94904 095
0.161	0.16030 53540 73987 04906 020	0.98706 74715 81965 18284 099
0.162	0.16129 23412 28387 41960 095	0.98690 66727 20914 09029 574
0.163	0.16227 91670 90460 00278 226	0.98674 48869 53272 51905 638
0.164	0.16326 58306 73379 01876 705	0.98658 21144 40826 22328 234
0.165	0.16425 23309 90480 96685 825	0.98641 83553 46347 70185 554
0.166	0.16523 86670 55265 61216 228	0.98625 36098 33596 03560 791
0.167	0.16622 48378 81396 97208 916	0.98608 78780 67316 72356 233
0.168	0.16721 08424 82704 30268 843	0.98592 11602 13241 51818 712
0.169	0.16819 66798 73183 08481 981	0.98575 34564 38088 25966 434
0.170	0.16918 23490 66996 01015 762	0.98558 47669 09560 70917 193
0.171	0.17016 78490 78473 96702 805	0.98541 50917 96348 38117 998
0.172	0.17115 31789 22117 02607 812	0.98524 44312 68126 37476 124
0.173	0.17213 83376 12595 42577 560	0.98507 27854 95555 20391 598
0.174	0.17312 33241 64750 55773 865	0.98490 01546 50280 62691 158
0.175	0.17410 81375 93595 95189 433	0.98472 65389 04933 47463 670
0.176	0.17509 27769 14318 26146 505	0.98455 19384 33129 47797 052
0.177	0.17607 72411 42278 24778 176	0.98437 63534 09469 09416 699
0.178	0.17706 15292 93011 76492 317	0.98419 97840 09537 33225 443
0.179	0.17804 56403 82230 74417 975	0.98402 22304 09903 57745 046
0.180	0.17902 95734 25824 17834 180	0.98384 36927 88121 41459 272
0.181	0.18001 33274 39859 10581 029	0.98366 41713 22728 45058 522
0.182	0.18099 69014 40581 59452 980	0.98348 36661 93246 13586 083
0.183	0.18198 02944 44417 72574 233	0.98330 21775 80179 58485 974
0.184	0.18296 35054 67974 57756 116	0.98311 97056 65017 39552 448
0.185	0.18394 65335 28041 20836 370	0.98293 62506 30231 46781 122
0.186	0.18492 93776 41589 64000 231	0.98275 18126 59276 82121 799
0.187	0.18591 20368 25775 84083 224	0.98256 63919 36591 41132 959
0.188	0.18689 45100 97940 70855 554	0.98237 99886 47595 94537 971
0.189	0.18787 67964 75611 05288 013	0.98219 26029 78693 69683 022
0.190	0.18885 88949 76500 57799 285	0.98200 42351 17270 31896 788
0.191	0.18984 08046 18510 86484 571	0.98181 48852 51693 65751 875
0.192	0.19082 25244 19732 35325 424	0.98162 45535 71313 56228 034
0.193	0.19180 40533 98445 32380 691	0.98143 32402 66461 69777 178
0.194	0.19278 53905 73120 87958 485	0.98124 09455 28451 35290 214
0.195	0.19376 65349 62421 92769 058	0.98104 76695 49577 24965 723
0.196	0.19474 74855 85204 16058 510	0.98085 34125 23115 35080 479
0.197	0.19572 82414 60517 03723 204	0.98065 81746 43322 66661 867
0.198	0.19670 88016 07604 76404 820	0.98046 19561 05437 06062 170
0.199	0.19768 91650 45907 27565 917	0.98026 47571 05677 05434 796
0.200	0.19866 93307 95061 21545 941	0.98006 65778 41241 63112 420
	$\begin{bmatrix} (-8)2 \\ 7 \end{bmatrix}$	$\begin{bmatrix} (-7)1 \\ 7 \end{bmatrix}$

Table 4.6   **CIRCULAR SINES AND COSINES FOR RADIAN ARGUMENTS**

x	$\sin x$	$\cos x$
0.200	0.19866 93307 95061 21545 941	0.98006 65778 41241 63112 420
0.201	0.19964 92978 74900 91597 545	0.97986 74185 10310 03887 090
0.202	0.20062 90653 05459 37903 151	0.97966 72793 12041 59192 306
0.203	0.20160 86321 06969 25571 640	0.97946 61604 46575 47187 084
0.204	0.20258 79972 99863 82615 083	0.97926 40621 15030 52742 047
0.205	0.20356 71599 04777 97905 397	0.97906 09845 19505 07327 536
0.206	0.20454 61189 42549 19110 856	0.97885 69278 63076 68803 784
0.207	0.20552 48734 34218 50612 330	0.97865 18923 49802 01113 156
0.208	0.20650 34224 01031 51399 175	0.97844 58781 84716 53874 491
0.209	0.20748 17648 64439 32944 665	0.97823 88855 73834 41879 553
0.210	0.20845 98998 46099 57060 871	0.97803 09147 24148 24491 614
0.211	0.20943 78263 67877 33732 895	0.97782 19658 43628 84946 201
0.212	0.21041 55434 51846 18932 346	0.97761 20391 41225 09554 014
0.213	0.21139 30501 20289 12409 982	0.97740 11348 26863 66806 039
0.214	0.21237 03453 95699 55467 398	0.97718 92531 11448 86380 882
0.215	0.21334 74283 00782 28707 677	0.97697 63942 06862 38054 344
0.216	0.21432 42978 58454 49764 905	0.97676 25583 25963 10511 247
0.217	0.21530 09530 91846 71012 439	0.97654 77456 82586 90059 555
0.218	0.21627 73930 24303 77249 851	0.97633 19564 91546 39246 782
0.219	0.21725 36166 79385 83368 434	0.97611 51909 68630 75378 736
0.220	0.21822 96230 80869 31995 179	0.97589 74493 30605 48940 602
0.221	0.21920 54112 52747 91115 124	0.97567 87317 95212 21920 392
0.222	0.22018 09802 19233 51671 977	0.97545 90385 81168 46034 788
0.223	0.22115 63290 04757 25146 920	0.97523 83699 08167 40857 388
0.224	0.22213 14566 33970 41115 484	0.97501 67259 96877 71849 392
0.225	0.22310 63621 31745 44782 417	0.97479 41070 68943 28292 737
0.226	0.22408 10445 23176 94494 428	0.97457 05133 46983 01125 708
0.227	0.22505 55028 33582 59230 720	0.97434 59450 54590 60681 052
0.228	0.22602 97360 88504 16071 214	0.97412 04024 16334 34326 607
0.229	0.22700 37433 13708 47642 363	0.97389 38856 57756 84008 477
0.230	0.22797 75235 35188 39540 462	0.97366 63950 05374 83696 773
0.231	0.22895 10757 79163 77732 354	0.97343 79306 86678 96733 940
0.232	0.22992 43990 72082 45933 437	0.97320 84929 30133 53085 695
0.233	0.23089 74924 40621 22962 869	0.97297 80819 65176 26494 602
0.234	0.23187 03549 11686 80075 884	0.97274 66980 22218 11536 294
0.235	0.23284 29855 12416 78273 112	0.97251 43413 32643 00578 389
0.236	0.23381 53832 70180 65586 809	0.97228 10121 28807 60642 091
0.237	0.23478 75472 12580 74343 904	0.97204 67106 44041 10166 529
0.238	0.23575 94763 67453 18405 752	0.97181 14371 12644 95675 843
0.239	0.23673 11697 62868 90384 520	0.97157 51917 69892 68349 034
0.240	0.23770 26264 27134 58836 079	0.97133 79748 52029 60492 618
0.241	0.23867 38453 88793 65429 334	0.97109 97865 96272 61916 095
0.242	0.23964 48256 76627 22091 869	0.97086 06272 40809 96210 262
0.243	0.24061 55663 19655 08131 828	0.97062 04970 24800 96928 391
0.244	0.24158 60663 47136 67335 933	0.97037 93961 88375 83670 294
0.245	0.24255 63247 88572 05043 522	0.97013 73249 72635 38069 313
0.246	0.24352 63406 73702 85196 546	0.96989 42836 19650 79682 233
0.247	0.24449 61130 32513 27365 389	0.96965 02723 72463 41782 166
0.248	0.24546 56408 95231 03750 445	0.96940 52914 75084 47054 425
0.249	0.24643 49232 92328 36159 337	0.96915 93411 72494 83195 397
0.250	0.24740 39592 54522 92959 685	0.96891 24217 10644 78414 459
	$\begin{bmatrix}(-8)3\\7\end{bmatrix}$	$\begin{bmatrix}(-7)1\\7\end{bmatrix}$

CIRCULAR SINES AND COSINES FOR RADIAN ARGUMENTS   Table 4.6

x	sin x	cos x
0.250	0.24740 39592 54522 92959 685	0.96891 24217 10644 78414 459
0.251	0.24837 27478 12778 86007 332	0.96866 45333 36453 76838 955
0.252	0.24934 12879 98307 67549 922	0.96841 56762 97810 13822 250
0.253	0.25030 95788 42569 27105 742	0.96816 58508 43570 91154 897
0.254	0.25127 76193 77272 88317 722	0.96791 50572 23561 52178 941
0.255	0.25224 54086 34378 05782 506	0.96766 32956 88575 56805 375
0.256	0.25321 29456 46095 61854 486	0.96741 05664 90374 56434 780
0.257	0.25418 02294 44888 63424 714	0.96715 68698 81687 68781 180
0.258	0.25514 72590 63473 38674 587	0.96690 22061 16211 52599 126
0.259	0.25611 40335 34820 33804 209	0.96664 65754 48609 82314 035
0.260	0.25708 05518 92155 09735 339	0.96638 99781 34513 22555 822
0.261	0.25804 68131 68959 38788 820	0.96613 24144 30519 02595 835
0.262	0.25901 28163 98972 01336 401	0.96587 38845 94190 90687 131
0.263	0.25997 85606 16189 82426 844	0.96561 43888 84058 68308 107
0.264	0.26094 40448 54868 68386 239	0.96535 39275 59618 04309 520
0.265	0.26190 92681 49524 43392 399	0.96509 25008 81330 28964 923
0.266	0.26287 42295 34933 86023 278	0.96483 01091 10622 07924 537
0.267	0.26383 89280 46135 65779 278	0.96456 67525 09885 16072 584
0.268	0.26480 33627 18431 39579 372	0.96430 24313 42476 11288 118
0.269	0.26576 75325 87386 48230 942	0.96403 71458 72716 08109 368
0.270	0.26673 14366 88831 12873 229	0.96377 08963 65890 51301 623
0.271	0.26769 50740 58861 31394 301	0.96350 36830 88248 89328 696
0.272	0.26865 84437 33839 74821 451	0.96323 55063 07004 47727 972
0.273	0.26962 15447 50396 83684 915	0.96296 63662 90334 02389 084
0.274	0.27058 43761 45431 64354 828	0.96269 62633 07377 52736 246
0.275	0.27154 69369 56112 85351 302	0.96242 51976 28237 94814 248
0.276	0.27250 92262 19879 73627 557	0.96215 31695 23980 94278 169
0.277	0.27347 12429 74443 10825 981	0.96188 01792 66634 59286 807
0.278	0.27443 29862 57786 29507 043	0.96160 62271 29189 13299 879
0.279	0.27539 44551 08166 09350 952	0.96133 13133 85596 67778 997
0.280	0.27635 56485 64113 73331 967	0.96105 54383 10770 94792 459
0.281	0.27731 65656 64435 83865 270	0.96077 86021 80586 99523 878
0.282	0.27827 72054 48215 38926 293	0.96050 08052 71880 92684 682
0.283	0.27923 75669 54812 68142 411	0.96022 20478 62449 62830 504
0.284	0.28019 76492 23866 28856 909	0.95994 23302 31050 48581 495
0.285	0.28115 74512 95294 02165 110	0.95966 16526 57401 10746 590
0.286	0.28211 69722 09293 88922 591	0.95938 00154 22179 04351 746
0.287	0.28307 62110 06345 05725 374	0.95909 74188 07021 50572 193
0.288	0.28403 51667 27208 80861 997	0.95881 38630 94525 08568 713
0.289	0.28499 38384 12929 50237 384	0.95852 93485 68245 47227 984
0.290	0.28595 22251 04835 53268 394	0.95824 38755 12697 16807 013
0.291	0.28691 03258 44540 28750 981	0.95795 74442 13353 20481 688
0.292	0.28786 81396 73943 10698 841	0.95767 00549 56644 85799 478
0.293	0.28882 56656 35230 24153 475	0.95738 17080 29961 36036 308
0.294	0.28978 29027 70875 80965 551	0.95709 24037 21649 61457 636
0.295	0.29073 98501 23642 75547 489	0.95680 21423 21013 90483 768
0.296	0.29169 65067 36583 80597 155	0.95651 09241 18315 60759 429
0.297	0.29265 28716 53042 42792 582	0.95621 87494 04772 90127 632
0.298	0.29360 89439 16653 78457 616	0.95592 56184 72560 47507 858
0.299	0.29456 47225 71345 69198 389	0.95563 15316 14809 23678 590
0.300	0.29552 02066 61339 57510 532	0.95533 64891 25606 01964 231
	$\begin{bmatrix} (-8)4 \\ 7 \end{bmatrix}$	$\begin{bmatrix} (-7)1 \\ 7 \end{bmatrix}$

Table 4.6   CIRCULAR SINES AND COSINES FOR RADIAN ARGUMENTS

x	$\sin x$	$\cos x$
0.300	0.29552 02066 61339 57510 532	0.95533 64891 25606 01964 231
0.301	0.29647 53952 31151 42357 025	0.95504 04912 99993 28826 414
0.302	0.29743 02873 25592 74716 586	0.95474 35384 33968 84359 763
0.303	0.29838 48819 89771 53102 518	0.95444 56308 24485 52692 116
0.304	0.29933 91782 69093 19051 897	0.95414 67687 69450 92289 242
0.305	0.30029 31752 09261 52585 026	0.95384 69525 67727 06164 084
0.306	0.30124 68718 56279 67635 045	0.95354 61825 19130 11990 559
0.307	0.30220 02672 56451 07447 613	0.95324 44589 24430 12121 945
0.308	0.30315 33604 56380 39950 549	0.95294 17820 85350 63513 878
0.309	0.30410 61505 02974 53093 365	0.95263 81523 04568 47552 001
0.310	0.30505 86364 43443 50156 564	0.95233 35698 85713 39784 281
0.311	0.30601 08173 25301 45030 632	0.95202 80351 33367 79558 038
0.312	0.30696 26921 96367 57464 615	0.95172 15483 53066 39561 711
0.313	0.30791 42601 04767 08284 189	0.95141 41098 51295 95271 383
0.314	0.30886 55200 98932 14579 138	0.95110 57199 35494 94302 111
0.315	0.30981 64712 27602 84860 120	0.95079 63789 14053 25664 080
0.316	0.31076 71125 39828 14184 658	0.95048 60870 96311 88923 617
0.317	0.31171 74430 84966 79252 234	0.95017 48447 92562 63269 094
0.318	0.31266 74619 12688 33468 402	0.94986 26523 14047 76481 749
0.319	0.31361 71680 72974 01977 833	0.94954 95099 72959 73811 467
0.320	0.31456 65606 16117 76666 176	0.94923 54180 82440 86757 531
0.321	0.31551 56385 92727 11130 659	0.94892 03769 56583 01754 395
0.322	0.31646 44010 53724 15619 332	0.94860 43869 10427 28762 501
0.323	0.31741 28470 50346 51938 844	0.94828 74482 59963 69764 173
0.324	0.31836 09756 34148 28330 674	0.94796 95613 22130 87164 613
0.325	0.31930 87858 57000 94315 718	0.94765 07264 14815 72098 048
0.326	0.32025 62767 71094 35507 128	0.94733 09438 56853 12639 034
0.327	0.32120 34474 28937 68391 319	0.94701 02139 68025 61918 976
0.328	0.32215 02968 83360 35077 048	0.94668 85370 69063 06147 877
0.329	0.32309 68241 87512 98012 460	0.94636 59134 81642 32541 351
0.330	0.32404 30283 94868 34670 020	0.94604 23435 28386 97152 941
0.331	0.32498 89085 59222 32199 224	0.94571 78275 32866 92611 768
0.332	0.32593 44637 34694 82047 011	0.94539 23658 19598 15765 535
0.333	0.32687 96929 75730 74545 756	0.94506 59587 14042 35228 939
0.334	0.32782 45953 37100 93468 777	0.94473 86065 42606 58837 502
0.335	0.32876 91698 73903 10553 241	0.94441 03096 32643 01006 864
0.336	0.32971 34156 41562 79990 386	0.94408 10683 12448 49997 577
0.337	0.33065 73316 95834 32882 957	0.94375 08829 11264 35085 413
0.338	0.33160 09170 92801 71669 766	0.94341 97537 59275 93637 243
0.339	0.33254 41708 88879 64517 288	0.94308 76811 87612 38092 499
0.340	0.33348 70921 40814 39678 177	0.94275 46655 28346 22850 264
0.341	0.33442 96799 05684 79816 635	0.94242 07071 14493 11062 025
0.342	0.33537 19332 40903 16300 519	0.94208 58062 80011 41330 105
0.343	0.33631 38512 04216 23460 104	0.94174 99633 59801 94311 834
0.344	0.33725 54328 53706 12813 399	0.94141 31786 89707 59229 468
0.345	0.33819 66772 47791 27257 928	0.94107 54526 06513 00285 905
0.346	0.33913 75834 45227 35228 880	0.94073 67854 47944 22986 218
0.347	0.34007 81505 05108 24823 531	0.94039 71775 52668 40365 059
0.348	0.34101 83774 86866 97891 850	0.94005 66292 60293 39119 944
0.349	0.34195 82634 50276 64093 188	0.93971 51409 11367 45650 473
0.350	0.34289 78074 55451 34918 963	0.93937 27128 47378 92003 503
	$\begin{bmatrix} (-8)4 \\ 7 \end{bmatrix}$	$\begin{bmatrix} (-7)1 \\ 7 \end{bmatrix}$

CIRCULAR SINES AND COSINES FOR RADIAN ARGUMENTS   Table 4.6

x	$\sin x$	$\cos x$
0.350	0.34289 78074 55451 34918 963	0.93937 27128 47378 92003 503
0.351	0.34383 70085 62847 17681 237	0.93902 93454 10755 81724 321
0.352	0.34477 58658 33263 09467 102	0.93868 50389 44865 55613 841
0.353	0.34571 43783 27841 91058 778	0.93833 97937 94014 57391 869
0.354	0.34665 25451 08071 20819 319	0.93799 36103 03447 99266 461
0.355	0.34759 03652 35784 28543 852	0.93764 64888 19349 27409 412
0.356	0.34852 78377 73161 09276 237	0.93729 84296 88839 87337 915
0.357	0.34946 49617 82729 17091 064	0.93694 94332 59978 89202 418
0.358	0.35040 17363 27364 58840 891	0.93659 94998 81762 72980 716
0.359	0.35133 81604 70292 87868 632	0.93624 86299 04124 73578 312
0.360	0.35227 42332 75089 97684 991	0.93589 68236 77934 85835 091
0.361	0.35320 99538 05683 15610 866	0.93554 40815 54999 29438 322
0.362	0.35414 53211 26351 96384 608	0.93519 04038 88060 13742 042
0.363	0.35508 03343 01729 15734 065	0.93483 57910 30795 02492 855
0.364	0.35601 49923 96801 63913 294	0.93448 02433 37816 78462 165
0.365	0.35694 92944 76911 39203 863	0.93412 37611 64673 07984 897
0.366	0.35788 32396 07756 41380 647	0.93376 63448 67846 05404 739
0.367	0.35881 68268 55391 65142 021	0.93340 79948 04751 97425 922
0.368	0.35975 00552 86229 93504 354	0.93304 87113 33740 87371 606
0.369	0.36068 29239 67042 91160 721	0.93268 84948 14096 19348 871
0.370	0.36161 54319 64961 97803 729	0.93232 73456 06034 42320 381
0.371	0.36254 75783 47479 21412 373	0.93196 52640 70704 74082 737
0.372	0.36347 93621 82448 31502 813	0.93160 22505 70188 65151 560
0.373	0.36441 07825 38085 52343 006	0.93123 83054 67499 62553 347
0.374	0.36534 18384 82970 56131 067	0.93087 34291 26582 73524 125
0.375	0.36627 25290 86047 56137 291	0.93050 76219 12314 29114 948
0.376	0.36720 28534 16625 99809 733	0.93014 08841 90501 47704 265
0.377	0.36813 28105 44381 61843 251	0.92977 32163 27881 98417 211
0.378	0.36906 23995 39357 37211 926	0.92940 46186 92123 64451 836
0.379	0.36999 16194 71964 34164 758	0.92903 50916 51824 06312 328
0.380	0.37092 04694 12982 67184 549	0.92866 46355 76510 24949 253
0.381	0.37184 89484 33562 49909 881	0.92829 32508 36638 24806 858
0.382	0.37277 70556 05224 88020 096	0.92792 09378 03592 76777 471
0.383	0.37370 47899 99862 72083 184	0.92754 76968 49686 81063 030
0.384	0.37463 21506 89741 70366 479	0.92717 35283 48161 29943 792
0.385	0.37555 91367 47501 21610 089	0.92679 84326 73184 70454 235
0.386	0.37648 57472 46155 27762 945	0.92642 24101 99852 66966 223
0.387	0.37741 19812 59093 46681 397	0.92604 54613 04187 63679 438
0.388	0.37833 78378 60081 84790 240	0.92566 75863 63138 47019 143
0.389	0.37926 33161 23263 89706 110	0.92528 87857 54580 07941 297
0.390	0.38018 84151 23161 42823 118	0.92490 90598 57313 04145 068
0.391	0.38111 31339 34675 51860 671	0.92452 84090 51063 22192 776
0.392	0.38203 74716 33087 43373 349	0.92414 68337 16481 39537 314
0.393	0.38296 14272 94059 55222 774	0.92376 43342 35142 86457 070
0.394	0.38388 49999 93636 29011 366	0.92338 09109 89547 07898 401
0.395	0.38480 81888 08245 02477 888	0.92299 65643 63117 25225 693
0.396	0.38573 09928 14697 01854 707	0.92261 12947 40199 97879 040
0.397	0.38665 34110 90188 34186 658	0.92222 51025 06064 84939 589
0.398	0.38757 54427 12300 79611 426	0.92183 79880 46904 06602 584
0.399	0.38849 70867 59002 83601 363	0.92144 99517 49832 05558 150
0.400	0.38941 83423 08650 49166 631	0.92106 09940 02885 08279 853
	$\begin{bmatrix}(-8)5\\7\end{bmatrix}$	$\begin{bmatrix}(-7)1\\7\end{bmatrix}$

Table 4.6   **CIRCULAR SINES AND COSINES FOR RADIAN ARGUMENTS**

x	$\sin x$	$\cos x$
0.400	0.38941 83423 08650 49166 631	0.92106 09940 02885 08279 853
0.401	0.39033 92084 39988 29019 595	0.92067 11151 95020 86221 075
0.402	0.39125 96842 32150 17700 358	0.92028 03157 16118 16919 248
0.403	0.39217 97687 64660 43663 363	0.91988 85959 56976 45007 979
0.404	0.39309 94611 17434 61324 955	0.91949 59563 09315 43137 110
0.405	0.39401 87603 70780 43071 820	0.91910 23971 65774 72800 745
0.406	0.39493 76656 05398 71230 202	0.91870 79189 19913 45073 295
0.407	0.39585 61759 02384 29995 816	0.91831 25219 66209 81253 568
0.408	0.39677 42903 43226 97324 356	0.91791 62067 00060 73416 956
0.409	0.39769 20080 09812 36782 508	0.91751 89735 17781 44875 737
0.410	0.39860 93279 84422 89359 380	0.91712 08228 16605 10547 564
0.411	0.39952 62493 49738 65238 251	0.91672 17549 94682 37232 150
0.412	0.40044 27711 88838 35528 558	0.91632 17704 51081 03796 202
0.413	0.40135 88925 85200 23958 010	0.91592 08695 85785 61266 649
0.414	0.40227 46126 22702 98524 766	0.91551 90527 99696 92832 194
0.415	0.40318 99303 85626 63109 550	0.91511 63204 94631 73753 232
0.416	0.40410 48449 58653 49047 645	0.91471 26730 73322 31180 180
0.417	0.40501 93554 26869 06660 654	0.91430 81109 39416 03880 251
0.418	0.40593 34608 75762 96747 939	0.91390 26344 97475 01872 722
0.419	0.40684 71603 91229 82037 655	0.91349 62441 52975 65972 725
0.420	0.40776 04530 59570 18597 279	0.91308 89403 12308 27243 609
0.421	0.40867 33379 67491 47203 546	0.91268 07233 82776 66357 915
0.422	0.40958 58142 02108 84671 703	0.91227 15937 72597 72866 996
0.423	0.41049 78808 50946 15143 980	0.91186 15518 90901 04379 332
0.424	0.41140 95370 01936 81337 201	0.91145 05981 47728 45647 576
0.425	0.41232 07817 43424 75749 435	0.91103 87329 54033 67564 373
0.426	0.41323 16141 64165 31825 593	0.91062 59567 21681 86066 990
0.427	0.41414 20333 53326 15081 889	0.91021 22698 63449 20950 808
0.428	0.41505 20384 00488 14189 067	0.90979 76727 93022 54591 701
0.429	0.41596 16283 95646 32014 301	0.90938 21659 24998 90577 360
0.430	0.41687 08024 29210 76621 692	0.90896 57496 74885 12247 591
0.431	0.41777 95595 92007 52231 243	0.90854 84244 59097 41143 638
0.432	0.41868 78989 75279 50136 257	0.90813 01906 94960 95366 563
0.433	0.41959 58196 70687 39579 028	0.90771 10488 00709 47844 729
0.434	0.42050 33207 70310 58584 774	0.90729 09991 95484 84510 435
0.435	0.42141 04013 66648 04753 684	0.90687 00422 99336 62385 731
0.436	0.42231 70605 52619 26011 018	0.90644 81785 33221 67577 465
0.437	0.42322 32974 21565 11315 146	0.90602 54083 19003 73181 601
0.438	0.42412 91110 67248 81323 456	0.90560 17320 79452 97096 848
0.439	0.42503 45005 83856 79016 027	0.90517 71502 38245 59747 647
0.440	0.42593 94650 65999 60276 972	0.90475 16632 19963 41716 554
0.441	0.42684 40036 08712 84433 381	0.90432 52714 50093 41286 061
0.442	0.42774 81153 07458 04751 750	0.90389 79753 55027 31889 904
0.443	0.42865 17992 58123 58891 823	0.90346 97753 62061 19473 892
0.444	0.42955 50545 57025 59317 745	0.90304 06718 99394 99766 305
0.445	0.43045 78803 00908 83666 443	0.90261 06653 96132 15457 899
0.446	0.43136 02755 86947 65073 141	0.90217 97562 82279 13291 573
0.447	0.43226 22395 12746 82453 917	0.90174 79449 88745 01061 718
0.448	0.43316 37711 76342 50745 219	0.90131 52319 47341 04523 319
0.449	0.43406 48696 76203 11100 244	0.90088 16175 90780 24210 832
0.450	0.43496 55341 11230 21042 084	0.90044 71023 52676 92166 884
	$\left[\begin{smallmatrix}(-8)5\\7\end{smallmatrix}\right]$	$\left[\begin{smallmatrix}(-7)1\\7\end{smallmatrix}\right]$

CIRCULAR SINES AND COSINES FOR RADIAN ARGUMENTS   Table 4.6

x	$\sin x$	$\cos x$
0.450	0.43496 55341 11230 21042 084	0.90044 71023 52676 92166 884
0.451	0.43586 57635 80759 44573 567	0.90001 16866 67546 28580 847
0.452	0.43676 55571 84561 42243 681	0.89957 53709 70803 98337 319
0.453	0.43766 49140 22842 61170 507	0.89913 81556 98765 67474 569
0.454	0.43856 38331 96246 25020 568	0.89870 00412 88646 59552 965
0.455	0.43946 23138 05853 23944 492	0.89826 10281 78561 11933 463
0.456	0.44036 03549 53183 04468 918	0.89782 11168 07522 31966 167
0.457	0.44125 79557 40194 59344 542	0.89738 03076 15441 53089 030
0.458	0.44215 51152 69287 17350 215	0.89693 86010 43127 90836 721
0.459	0.44305 18326 43301 33053 008	0.89649 59975 32287 98759 714
0.460	0.44394 81069 65519 76524 151	0.89605 24975 25525 24253 639
0.461	0.44484 39373 39668 23010 752	0.89560 81014 66339 64298 937
0.462	0.44573 93228 69916 42563 218	0.89516 28097 99127 21110 867
0.463	0.44663 42626 60878 89618 275	0.89471 66229 69179 57699 908
0.464	0.44752 87558 17615 92537 506	0.89426 95414 22683 53342 602
0.465	0.44842 28014 45634 43101 319	0.89382 15656 06720 58962 873
0.466	0.44931 63986 50888 85958 244	0.89337 26959 69266 52423 883
0.467	0.45020 95465 39782 08029 479	0.89292 29329 59190 93730 459
0.468	0.45110 22442 19166 27868 603	0.89247 22770 26256 80142 134
0.469	0.45199 44907 96343 84976 342	0.89202 07286 21120 01196 857
0.470	0.45288 62853 79068 29070 327	0.89156 82881 95328 93645 402
0.471	0.45377 76270 75545 09309 736	0.89111 49562 01323 96296 541
0.472	0.45466 85149 94432 63474 735	0.89066 07330 92437 04773 005
0.473	0.45555 89482 44843 07100 635	0.89020 56193 22891 26178 292
0.474	0.45644 89259 36343 22566 671	0.88974 96153 47800 33674 367
0.475	0.45733 84471 78955 48139 307	0.88929 27216 23168 20970 288
0.476	0.45822 75110 83158 66969 994	0.88883 49386 05888 56721 822
0.477	0.45911 61167 59888 96047 279	0.88837 62667 53744 38842 074
0.478	0.46000 42633 20540 75103 180	0.88791 67065 25407 48723 197
0.479	0.46089 19498 76967 55473 739	0.88745 62583 80438 05369 212
0.480	0.46177 91755 41482 88913 664	0.88699 49227 79284 19439 995
0.481	0.46266 59394 26861 16364 968	0.88653 27001 83281 47206 469
0.482	0.46355 22406 46338 56679 522	0.88606 95910 54652 44417 051
0.483	0.46443 80783 13613 95295 430	0.88560 55958 56506 20075 401
0.484	0.46532 34515 42849 72867 132	0.88514 07150 52837 90129 517
0.485	0.46620 83594 48672 73849 162	0.88467 49491 08528 31072 223
0.486	0.46709 28011 46175 15033 451	0.88420 82984 89343 33453 094
0.487	0.46797 67757 50915 34040 104	0.88374 07636 61933 55301 874
0.488	0.46886 02823 78918 77761 558	0.88327 23450 93833 75463 416
0.489	0.46974 33201 46678 90760 024	0.88280 30432 53462 46844 214
0.490	0.47062 58881 71158 03618 136	0.88233 28586 10121 49570 547
0.491	0.47150 79855 69788 21242 715	0.88186 17916 33995 44058 307
0.492	0.47238 96114 60472 11121 556	0.88138 98427 96151 23994 541
0.493	0.47327 07649 61583 91533 149	0.88091 70125 68537 69230 763
0.494	0.47415 14451 91970 19709 261	0.88044 33014 23984 98588 075
0.495	0.47503 16512 70950 79950 264	0.87996 87098 36204 22574 157
0.496	0.47591 13823 18319 71693 150	0.87949 32382 79786 96012 154
0.497	0.47679 06374 54345 97532 118	0.87901 68872 30204 70581 529
0.498	0.47766 94157 99774 51191 668	0.87853 96571 63808 47270 917
0.499	0.47854 77164 75827 05452 099	0.87806 15485 57828 28743 023
0.500	0.47942 55386 04203 00027 329 $\begin{bmatrix}(-8)6\\7\end{bmatrix}$	0.87758 25618 90372 71611 628 $\begin{bmatrix}(-7)1\\7\end{bmatrix}$

Table 4.6   **CIRCULAR SINES AND COSINES FOR RADIAN ARGUMENTS**

x	$\sin x$	$\cos x$
0.500	0.47942 55386 04203 00027 329	0.87758 25618 90372 71611 628
0.501	0.48030 28813 07080 29394 947	0.87710 26976 40428 38630 733
0.502	0.48117 97437 07116 30578 414	0.87662 19562 87859 50795 903
0.503	0.48205 61249 27448 70881 314	0.87614 03383 13407 39357 847
0.504	0.48293 20240 91696 35573 583	0.87565 78441 98689 97748 295
0.505	0.48380 74403 23960 15529 617	0.87517 44744 26201 33418 203
0.506	0.48468 23727 48823 94818 170	0.87469 02294 79311 19588 355
0.507	0.48555 68204 91355 38243 967	0.87420 51098 42264 46912 391
0.508	0.48643 07826 77106 78840 928	0.87371 91160 00180 75052 318
0.509	0.48730 42584 32116 05316 931	0.87323 22484 39053 84166 561
0.510	0.48817 72468 82907 49450 013	0.87274 45076 45751 26310 581
0.511	0.48904 97471 56492 73435 934	0.87225 58941 08013 76750 129
0.512	0.48992 17583 80371 57187 006	0.87176 64083 14454 85187 176
0.513	0.49079 32796 82532 85582 104	0.87127 60507 54560 26898 565
0.514	0.49166 43101 91455 35667 778	0.87078 48219 18687 53787 441
0.515	0.49253 48490 36108 63810 364	0.87029 27222 98065 45347 504
0.516	0.49340 48953 45953 92799 025	0.86979 97523 84793 59540 132
0.517	0.49427 44482 50944 98899 617	0.86930 59126 71841 83584 429
0.518	0.49514 35068 81528 98859 309	0.86881 12036 53049 84660 240
0.519	0.49601 20703 68647 36861 855	0.86831 56258 23126 60524 189
0.520	0.49688 01378 43736 71433 446	0.86781 91796 77649 90038 785
0.521	0.49774 77084 38729 62299 043	0.86732 18657 13065 83614 647
0.522	0.49861 47812 86055 57189 109	0.86682 36844 26688 33565 898
0.523	0.49948 13555 18641 78596 658	0.86632 46363 16698 64378 779
0.524	0.50034 74302 69914 10484 518	0.86582 47218 82144 82893 524
0.525	0.50121 30046 73797 84942 748	0.86532 39416 22941 28399 561
0.526	0.50207 80778 64718 68796 092	0.86482 22960 39868 22644 077
0.527	0.50294 26489 77603 50161 411	0.86431 97856 34571 19753 996
0.528	0.50380 67171 47881 24954 981	0.86381 64109 09560 56071 436
0.529	0.50467 02815 11483 83349 596	0.86331 21723 68210 99902 671
0.530	0.50553 33412 04846 96181 366	0.86280 70705 14761 01180 670
0.531	0.50639 58953 64911 01306 143	0.86230 11058 54312 41041 248
0.532	0.50725 79431 29121 89905 473	0.86179 42788 92829 81312 894
0.533	0.50811 94836 35431 92741 999	0.86128 65901 37140 13920 311
0.534	0.50898 05160 22300 66364 220	0.86077 80400 94932 10201 726
0.535	0.50984 10394 28695 79260 534	0.86026 86292 74755 70140 025
0.536	0.51070 10529 94093 97962 456	0.85975 83581 86021 71507 760
0.537	0.51156 05558 58481 73096 946	0.85924 72273 39001 18926 068
0.538	0.51241 95471 62356 25387 754	0.85873 52372 44824 92837 581
0.539	0.51327 80260 46726 31605 686	0.85822 23884 15482 98393 339
0.540	0.51413 59916 53113 10467 728	0.85770 86813 63824 14253 797
0.541	0.51499 34431 23551 08484 914	0.85719 41166 03555 41303 947
0.542	0.51585 03796 00588 85758 874	0.85667 86946 49241 51282 623
0.543	0.51670 68002 27290 01726 969	0.85616 24160 16304 35326 032
0.544	0.51756 27041 47234 00855 920	0.85564 52812 21022 52425 567
0.545	0.51841 80905 04516 98283 861	0.85512 72907 80530 77799 957
0.546	0.51927 29584 43752 65410 714	0.85460 84452 12819 51181 787
0.547	0.52012 73071 10073 15436 812	0.85408 87450 36734 25018 472
0.548	0.52098 11356 49129 88849 675	0.85356 81907 71975 12587 703
0.549	0.52183 44432 07094 38858 868	0.85304 67829 39096 36027 442
0.550	0.52268 72289 30659 16778 838	0.85252 45220 59505 74280 498
	$\begin{bmatrix} (-8)7 \\ 7 \end{bmatrix}$	$\begin{bmatrix} (-7)1 \\ 7 \end{bmatrix}$

CIRCULAR SINES AND COSINES FOR RADIAN ARGUMENTS   Table 4.6

x	$\sin x$	$\cos x$
0.550	0.52268 72289 30659 16778 838	0.85252 45220 59505 74280 498
0.551	0.52353 94919 67038 57359 653	0.85200 14086 55464 10953 761
0.552	0.52439 12314 63969 64065 565	0.85147 74432 50084 82092 114
0.553	0.52524 24465 69712 94301 297	0.85095 26263 67333 23867 110
0.554	0.52609 31364 33053 44585 976	0.85042 69585 32026 20180 431
0.555	0.52694 33002 03301 35674 635	0.84990 04402 69831 50182 218
0.556	0.52779 29370 30292 97627 180	0.84937 30721 07267 35704 287
0.557	0.52864 20460 64391 54824 757	0.84884 48545 71701 88608 318
0.558	0.52949 06264 56488 10933 415	0.84831 57881 91352 58049 047
0.559	0.53033 86773 58002 33815 002	0.84778 58734 95285 77652 517
0.560	0.53118 61979 20883 40385 187	0.84725 51110 13416 12609 452
0.561	0.53203 31872 97610 81418 533	0.84672 35012 76506 06683 799
0.562	0.53287 96446 41195 26300 543	0.84619 10448 16165 29136 481
0.563	0.53372 55691 05179 47726 585	0.84565 77421 64850 21564 438
0.564	0.53457 09598 43639 06347 607	0.84512 35938 55863 44654 991
0.565	0.53541 58160 11183 35362 572	0.84458 86004 23353 24855 579
0.566	0.53626 01367 62956 25057 521	0.84405 27624 02313 00958 945
0.567	0.53710 39212 54637 07291 168	0.84351 60803 28580 70603 796
0.568	0.53794 71686 42441 39926 969	0.84297 85547 38838 36691 011
0.569	0.53878 98780 83121 91211 553	0.84244 01861 70611 53715 445
0.570	0.53963 20487 33969 24099 446	0.84190 09751 62268 74013 376
0.571	0.54047 36797 52812 80524 005	0.84136 09222 53020 93925 658
0.572	0.54131 47702 98021 65614 465	0.84082 00279 82920 99876 632
0.573	0.54215 53195 28505 31859 028	0.84027 82928 92863 14368 839
0.574	0.54299 53266 03714 63213 905	0.83973 57175 24582 41893 605
0.575	0.54383 47906 83642 59158 222	0.83919 23024 20654 14757 543
0.576	0.54467 37109 28825 18694 718	0.83864 80481 24493 38825 019
0.577	0.54551 20865 00342 24296 136	0.83810 29551 80354 39176 658
0.578	0.54634 99165 59818 25797 231	0.83755 70241 33330 05683 918
0.579	0.54718 72002 69423 24232 321	0.83701 02555 29351 38499 807
0.580	0.54802 39367 91873 55618 270	0.83646 26499 15186 93465 789
0.581	0.54886 01252 90432 74682 851	0.83591 42078 38442 27434 927
0.582	0.54969 57649 28912 38538 382	0.83536 49298 47559 43511 337
0.583	0.55053 08548 71672 90300 563	0.83481 48164 91816 36205 988
0.584	0.55136 53942 83624 42652 424	0.83426 38683 21326 36508 907
0.585	0.55219 93823 30227 61353 309	0.83371 20858 87037 56877 861
0.586	0.55303 28181 77494 48692 799	0.83315 94697 40732 36143 543
0.587	0.55386 57009 91989 26889 504	0.83260 60204 35026 84331 337
0.588	0.55469 80299 40829 21434 637	0.83205 17385 23370 27399 720
0.589	0.55552 98041 91685 44380 278	0.83149 66245 60044 51895 332
0.590	0.55636 10229 12783 77572 254	0.83094 06791 00163 49524 800
0.591	0.55719 16852 72905 55827 556	0.83038 39026 99672 61643 346
0.592	0.55802 17904 41388 50056 192	0.82982 62959 15348 23660 255
0.593	0.55885 13375 88127 50327 409	0.82926 78593 04797 09361 243
0.594	0.55968 03258 83575 48880 201	0.82870 85934 26455 75147 786
0.595	0.56050 87544 98744 23078 004	0.82814 84988 39590 04193 468
0.596	0.56133 66226 05205 18307 516	0.82758 75761 04294 50517 407
0.597	0.56216 39293 75090 30821 541	0.82702 58257 81491 82974 799
0.598	0.56299 06739 81092 90525 792	0.82646 32484 32932 29164 660
0.599	0.56381 68555 96468 43709 545	0.82589 98446 21193 19254 799
0.600	0.56464 24733 95035 35720 095	0.82533 56149 09678 29724 095
	$\begin{bmatrix}(-8)7\\7\end{bmatrix}$	$\begin{bmatrix}(-7)1\\7\end{bmatrix}$

Table 4.6   CIRCULAR SINES AND COSINES FOR RADIAN ARGUMENTS

x	$\sin x$	$\cos x$
0.600	0.56464 24733 95035 35720 095	0.82533 56149 09678 29724 095
0.601	0.56546 75265 51175 93580 897	0.82477 05598 62617 27022 123
0.602	0.56629 20142 39837 08553 336	0.82420 46800 45065 11146 193
0.603	0.56711 59356 36531 18642 028	0.82363 79760 22901 59135 858
0.604	0.56793 92899 17336 91043 574	0.82307 04483 62830 68484 934
0.605	0.56876 20762 58900 04538 687	0.82250 20976 32380 00471 116
0.606	0.56958 42938 38434 31827 607	0.82193 29243 99900 23403 216
0.607	0.57040 59418 33722 21808 719	0.82136 29292 34564 55786 102
0.608	0.57122 70194 23115 81800 299	0.82079 21127 06368 09403 380
0.609	0.57204 75257 85537 59705 300	0.82022 04753 86127 32317 893
0.610	0.57286 74601 00481 26119 098	0.81964 80178 45479 51790 075
0.611	0.57368 68215 48012 56380 111	0.81907 47406 56882 17114 225
0.612	0.57450 56093 08770 12563 221	0.81850 06443 93612 42372 770
0.613	0.57532 38225 63966 25415 904	0.81792 57296 29766 49108 549
0.614	0.57614 14604 95387 76236 989	0.81734 99969 40259 08915 198
0.615	0.57695 85222 85396 78697 975	0.81677 34469 00822 85945 685
0.616	0.57777 50071 16931 60606 809	0.81619 60800 88007 79339 051
0.617	0.57859 09141 73507 45614 047	0.81561 78970 79180 65565 411
0.618	0.57940 62426 39217 34861 330	0.81503 88984 52524 40689 288
0.619	0.58022 09916 98732 88572 073	0.81445 90847 87037 62551 318
0.620	0.58103 51605 37305 07584 296	0.81387 84566 62533 92868 400
0.621	0.58184 87483 40765 14825 522	0.81329 70146 59641 39252 335
0.622	0.58266 17542 95525 36729 641	0.81271 47593 59801 97147 027
0.623	0.58347 41775 88579 84595 681	0.81213 16913 45270 91684 290
0.624	0.58428 60174 07505 35888 387	0.81154 78111 99116 19458 331
0.625	0.58509 72729 40462 15480 540	0.81096 31195 05217 90218 953
0.626	0.58590 79433 76194 76836 923	0.81037 76168 48267 68483 556
0.627	0.58671 80279 04032 83139 861	0.80979 13038 13768 15067 973
0.628	0.58752 75257 13891 88356 252	0.80920 41809 88032 28536 214
0.629	0.58833 64359 96274 18246 006	0.80861 62489 58182 86569 178
0.630	0.58914 47579 42269 51311 811	0.80802 75083 12151 87252 371
0.631	0.58995 24907 43555 99690 151	0.80743 79596 38679 90282 722
0.632	0.59075 96335 92400 89983 484	0.80684 76035 27315 58094 522
0.633	0.59156 61856 81661 44033 509	0.80625 64405 68414 96904 569
0.634	0.59237 21462 04785 59635 440	0.80566 44713 53140 97676 566
0.635	0.59317 75143 55812 91193 198	0.80507 16964 73462 77004 837
0.636	0.59398 22893 29375 30315 454	0.80447 81165 22155 17917 411
0.637	0.59478 64703 20697 86352 425	0.80388 37320 92798 10598 548
0.638	0.59559 00565 25599 66873 364	0.80328 85437 79775 93030 752
0.639	0.59639 30471 40494 58084 641	0.80269 25521 78276 91556 338
0.640	0.59719 54413 62392 05188 355	0.80209 57578 84292 61358 611
0.641	0.59799 72383 88897 92681 375	0.80149 81614 94617 26862 715
0.642	0.59879 84374 18215 24594 757	0.80089 97636 06847 22056 216
0.643	0.59959 90376 49145 04673 426	0.80030 05648 19380 30729 469
0.644	0.60039 90382 81087 16496 070	0.79970 05657 31415 26635 842
0.645	0.60119 84385 14041 03535 151	0.79909 97669 42951 13571 848
0.646	0.60199 72375 48606 49156 949	0.79849 81690 54786 65377 243
0.647	0.60279 54345 85984 56561 576	0.79789 57726 68519 65855 159
0.648	0.60359 30288 27978 28662 868	0.79729 25783 86546 48612 327
0.649	0.60439 00194 76993 47908 070	0.79668 85868 12061 36819 444
0.650	0.60518 64057 36039 56037 252	0.79608 37985 49055 82891 760

$$\left[\begin{smallmatrix}(-8)8\\7\end{smallmatrix}\right] \qquad \left[\begin{smallmatrix}(-7)1\\7\end{smallmatrix}\right]$$

CIRCULAR SINES AND COSINES FOR RADIAN ARGUMENTS   Table 4.6

x	sin x	cos x
0.650	0.60518 64057 36039 56037 252	0.79608 37985 49055 82891 760
0.651	0.60598 21868 08730 33782 358	0.79547 82142 02318 08089 927
0.652	0.60677 73618 99284 80505 818	0.79487 18343 77432 42041 183
0.653	0.60757 19302 12527 93778 646	0.79426 46596 80778 62180 929
0.654	0.60836 58909 53891 48897 929	0.79365 66907 19531 33114 757
0.655	0.60915 92433 29414 78343 652	0.79304 79281 01659 45900 987
0.656	0.60995 19865 45745 51174 755	0.79243 83724 35925 57253 785
0.657	0.61074 41198 10140 52364 359	0.79182 80243 31885 28666 909
0.658	0.61153 56423 30466 62074 073	0.79121 68843 99886 65458 154
0.659	0.61232 65533 15201 34867 307	0.79060 49532 51069 55734 550
0.660	0.61311 68519 73433 78861 515	0.78999 22314 97365 09278 382
0.661	0.61390 65375 14865 34819 272	0.78937 87197 51494 96354 080
0.662	0.61469 56091 49810 55178 137	0.78876 44186 26970 86436 061
0.663	0.61548 40660 89197 83019 186	0.78814 93287 38093 86857 558
0.664	0.61627 19075 44570 30974 165	0.78753 34506 99953 81380 523
0.665	0.61705 91327 28086 60071 171	0.78691 67851 28428 68686 643
0.666	0.61784 57408 52521 58518 785	0.78629 93326 40184 00789 551
0.667	0.61863 17311 31267 20428 576	0.78568 10938 52672 21368 279
0.668	0.61941 71027 78333 24475 901	0.78506 20693 84132 04022 017
0.669	0.62020 18550 08348 12498 919	0.78444 22598 53587 90446 244
0.670	0.62098 59870 36559 68035 744	0.78382 16658 80849 28530 294
0.671	0.62176 94980 78835 94799 654	0.78320 02880 86510 10376 414
0.672	0.62255 23873 51665 95092 281	0.78257 81270 91948 10240 374
0.673	0.62333 46540 72160 48154 700	0.78195 51835 19324 22393 698
0.674	0.62411 62974 58052 88456 349	0.78133 14579 91581 98907 578
0.675	0.62489 73167 27699 83921 682	0.78070 69511 32446 87358 526
0.676	0.62567 77111 00082 14094 496	0.78008 16635 66425 68455 830
0.677	0.62645 74797 94805 48239 849	0.77945 55959 18805 93590 877
0.678	0.62723 66220 32101 23383 477	0.77882 87488 15655 22308 414
0.679	0.62801 51370 32827 22288 658	0.77820 11228 83820 59699 786
0.680	0.62879 30240 18468 51370 418	0.77757 27187 50927 93718 239
0.681	0.62957 02822 11138 18547 018	0.77694 35370 45381 32416 339
0.682	0.63034 69108 33578 11028 644	0.77631 35783 96362 41105 566
0.683	0.63112 29091 09159 73043 207	0.77568 28434 33829 79438 156
0.684	0.63189 82762 61884 83499 197	0.77505 13327 88518 38411 247
0.685	0.63267 30115 16386 33585 498	0.77441 90470 91938 77293 390
0.686	0.63344 71140 97929 04308 084	0.77378 59869 76376 60473 500
0.687	0.63422 05832 32410 43963 542	0.77315 21530 74891 94232 293
0.688	0.63499 34181 46361 45549 306	0.77251 75460 21318 63436 286
0.689	0.63576 56180 66947 24110 566	0.77188 21664 50263 68154 418
0.690	0.63653 71822 21967 94023 743	0.77124 60149 97106 60197 354
0.691	0.63730 81098 39859 46216 467	0.77060 90922 97998 79579 541
0.692	0.63807 84001 49694 25323 984	0.76997 13989 89862 90904 069
0.693	0.63884 80523 81182 06781 899	0.76933 29357 10392 19670 418
0.694	0.63961 70657 64670 73855 200	0.76869 37030 98049 88505 132
0.695	0.64038 54395 31146 94603 464	0.76805 37017 92068 53315 502
0.696	0.64115 31729 12236 98782 185	0.76741 29324 32449 39366 321
0.697	0.64192 02651 40207 54680 136	0.76677 13956 59961 77279 757
0.698	0.64268 67154 47966 45892 698	0.76612 90921 16142 38958 434
0.699	0.64345 25230 69063 48031 063	0.76548 60224 43294 73431 759
0.700	0.64421 76872 37691 05367 261	0.76484 21872 84488 42625 586
	$\begin{bmatrix} (-8)8 \\ 7 \end{bmatrix}$	$\begin{bmatrix} (-7)1 \\ 7 \end{bmatrix}$

ELEMENTARY TRANSCENDENTAL FUNCTIONS

Table 4.6   CIRCULAR SINES AND COSINES FOR RADIAN ARGUMENTS

x	$\sin x$	$\cos x$
0.700	0.64421 76872 37691 05367 261	0.76484 21872 84488 42625 586
0.701	0.64498 22071 88685 07414 902	0.76419 75872 83558 57055 252
0.702	0.64574 60821 57525 65445 583	0.76355 22230 85105 11442 075
0.703	0.64650 93113 80337 88940 870	0.76290 60953 34492 20253 368
0.704	0.64727 18940 93892 61979 783	0.76225 92046 77847 53166 023
0.705	0.64803 38295 35607 19561 705	0.76161 15517 62061 70453 752
0.706	0.64879 51169 43546 23864 641	0.76096 31372 34787 58298 030
0.707	0.64955 57555 56422 40438 747	0.76031 39617 44439 64022 815
0.708	0.65031 57446 13597 14335 062	0.75966 40259 40193 31253 107
0.709	0.65107 50833 55081 46169 354	0.75901 33304 71984 34997 406
0.710	0.65183 37710 21536 68121 013	0.75836 18759 90508 16654 146
0.711	0.65259 18068 54275 19866 915	0.75770 96631 47219 18942 159
0.712	0.65334 91900 95261 24450 173	0.75705 66925 94330 20755 235
0.713	0.65410 59199 87111 64083 709	0.75640 29649 84811 71940 852
0.714	0.65486 19957 73096 55888 565	0.75574 84809 72391 28003 128
0.715	0.65561 74166 97140 27566 883	0.75509 32412 11552 84730 074
0.716	0.65637 21820 03821 93009 463	0.75443 72463 57536 12745 203
0.717	0.65712 62909 38376 27837 851	0.75378 04970 66335 91983 563
0.718	0.65787 97427 46694 44880 853	0.75312 29939 94701 46092 263
0.719	0.65863 25366 75324 69585 417	0.75246 47378 00135 76755 558
0.720	0.65938 46719 71473 15361 800	0.75180 57291 40894 97944 549
0.721	0.66013 61478 83004 58862 952	0.75114 59686 75987 70091 576
0.722	0.66088 69636 58443 15198 027	0.75048 54570 65174 34189 363
0.723	0.66163 71185 46973 13079 967	0.74982 41949 68966 45814 983
0.724	0.66238 66117 98439 69907 065	0.74916 21830 48626 09078 707
0.725	0.66313 54426 63349 66778 441	0.74849 94219 66165 10497 806
0.726	0.66388 36103 92872 23443 354	0.74783 59123 84344 52795 369
0.727	0.66463 11142 38839 73184 280	0.74717 16549 66673 88624 209
0.728	0.66537 79534 53748 37633 666	0.74650 66503 77410 54215 910
0.729	0.66612 41272 90759 01524 309	0.74584 08992 81559 02955 103
0.730	0.66686 96350 03697 87373 259	0.74517 44023 44870 38879 013
0.731	0.66761 44758 47057 30099 195	0.74450 71602 33841 50102 364
0.732	0.66835 86490 75996 51573 181	0.74383 91736 15714 42167 693
0.733	0.66910 21539 46342 35102 739	0.74317 04431 58475 71321 153
0.734	0.66984 49897 14589 99849 159	0.74250 09695 30855 77713 862
0.735	0.67058 71556 37903 75177 973	0.74183 07534 02328 18528 866
0.736	0.67132 86509 74117 74942 523	0.74115 97954 43109 01033 791
0.737	0.67206 94749 81736 71700 537	0.74048 80963 24156 15559 237
0.738	0.67280 96269 19936 70863 650	0.73981 56567 17168 68402 998
0.739	0.67354 91060 48565 84779 796	0.73914 24772 94586 14660 158
0.740	0.67428 79116 28145 06748 388	0.73846 85587 29587 90979 142
0.741	0.67502 60429 19868 84968 216	0.73779 39016 96092 48243 787
0.742	0.67576 34991 85605 96417 996	0.73711 85068 68756 84181 492
0.743	0.67650 02796 87900 20669 485	0.73644 23749 22975 75897 532
0.744	0.67723 63836 89971 13633 096	0.73576 55065 34881 12335 582
0.745	0.67797 18104 55714 81235 936	0.73508 79023 81341 26664 537
0.746	0.67870 65592 49704 53032 193	0.73440 95631 39960 28591 681
0.747	0.67944 06293 37191 55745 803	0.73373 04894 89077 36602 285
0.748	0.68017 40199 84105 86745 313	0.73305 06821 07766 10125 695
0.749	0.68090 67304 57056 87450 880	0.73237 01416 75833 81627 975
0.750	0.68163 87600 23334 16673 324	0.73168 88688 73820 88631 184
	$\begin{bmatrix} (-8)9 \\ 7 \end{bmatrix}$	$\begin{bmatrix} (-7)1 \\ 7 \end{bmatrix}$

CIRCULAR SINES AND COSINES FOR RADIAN ARGUMENTS   Table 4.6

x	sin x	cos x
0.750	0.68163 87600 23334 16673 324	0.73168 88688 73820 88631 184
0.751	0.68237 01079 50908 23885 163	0.73100 68643 83000 05659 342
0.752	0.68310 07735 08431 22423 554	0.73032 41288 85375 76111 160
0.753	0.68383 07559 65237 62625 080	0.72964 06630 63683 44059 608
0.754	0.68456 00545 91345 04892 285	0.72895 64676 01388 85978 367
0.755	0.68528 86686 57454 92691 917	0.72827 15431 82687 42395 268
0.756	0.68601 65974 34953 25484 772	0.72758 58904 92503 49472 750
0.757	0.68674 38401 95911 31587 089	0.72689 95102 16489 70515 436
0.758	0.68747 03962 13086 40963 419	0.72621 24030 41026 27404 867
0.759	0.68819 62647 59922 57950 885	0.72552 45696 53220 31961 494
0.760	0.68892 14451 10551 33914 776	0.72483 60107 40905 17233 969
0.761	0.68964 59365 39792 39835 383	0.72414 67269 92639 68715 814
0.762	0.69036 97383 23154 38826 030	0.72345 67190 97707 55489 548
0.763	0.69109 28497 36835 58582 200	0.72276 59877 46116 61298 318
0.764	0.69181 52700 57724 63761 700	0.72207 45336 28598 15545 123
0.765	0.69253 69985 63401 28295 794	0.72138 23574 36606 24219 693
0.766	0.69325 80345 32137 07631 223	0.72068 94598 62317 00753 084
0.767	0.69397 83772 42896 10903 039	0.71999 58415 98627 96800 072
0.768	0.69469 80259 75335 73038 195	0.71930 15033 39157 32949 410
0.769	0.69541 69800 09807 26789 802	0.71860 64457 78243 29362 010
0.770	0.69613 52386 27356 74701 988	0.71791 06696 10943 36337 129
0.771	0.69685 28011 09725 61005 296	0.71721 41755 33033 64806 626
0.772	0.69756 96667 39351 43442 524	0.71651 69642 41008 16757 355
0.773	0.69828 58347 99368 65024 972	0.71581 90364 32078 15581 770
0.774	0.69900 13045 73609 25718 983	0.71512 03928 04171 36356 807
0.775	0.69971 60753 46603 54062 747	0.71442 10340 55931 36051 117
0.776	0.70043 01464 03580 78713 256	0.71372 09608 86716 83660 709
0.777	0.70114 35170 30469 99923 379	0.71302 01739 96600 90273 093
0.778	0.70185 61865 13900 60948 949	0.71231 86740 86370 39059 972
0.779	0.70256 81541 41203 19385 818	0.71161 64618 57525 15198 564
0.780	0.70327 94192 00410 18436 790	0.71091 35380 12277 35721 626
0.781	0.70398 99809 80256 58108 374	0.71020 99032 53550 79296 239
0.782	0.70469 98387 70180 66337 280	0.70950 55582 84980 15931 435
0.783	0.70540 89918 60324 70046 581	0.70880 05038 10910 36614 737
0.784	0.70611 74395 41535 66131 480	0.70809 47405 36395 82877 671
0.785	0.70682 51811 05365 92374 614	0.70738 82691 67199 76290 330
0.786	0.70753 22158 44073 98290 801	0.70668 10904 09793 47885 059
0.787	0.70823 85430 50625 15901 193	0.70597 32049 71355 67509 330
0.788	0.70894 41620 18692 30436 730	0.70526 46135 59771 73107 880
0.789	0.70964 90720 42656 50970 857	0.70455 53168 83632 99934 173
0.790	0.71035 32724 17607 80981 403	0.70384 53156 52236 09691 278
0.791	0.71105 67624 39345 88841 574	0.70313 46105 75582 19602 208
0.792	0.71175 95414 04380 78239 979	0.70242 32023 64376 31409 812
0.793	0.71246 16086 09933 58529 620	0.70171 10917 30026 60306 275
0.794	0.71316 29633 53937 15005 776	0.70099 82793 84643 63792 314
0.795	0.71386 36049 35036 79112 713	0.70028 47660 41039 70466 123
0.796	0.71456 35326 52590 98579 148	0.69957 05524 12728 08742 151
0.797	0.71526 27458 06672 07482 391	0.69885 56392 13922 35499 779
0.798	0.71596 12436 98066 96241 109	0.69814 00271 59535 64661 971
0.799	0.71665 90256 28277 81536 630	0.69742 37169 65179 95703 964
0.800	0.71735 60908 99522 76162 718	0.69670 67093 47165 42092 075
	$\begin{bmatrix} (-8)9 \\ 7 \end{bmatrix}$	$\begin{bmatrix} (-8)9 \\ 7 \end{bmatrix}$

Table 4.6 CIRCULAR SINES AND COSINES FOR RADIAN ARGUMENTS

x	$\sin x$	$\cos x$
0.800	0.71735 60908 99522 76162 718	0.69670 67093 47165 42092 075
0.801	0.71805 24388 14736 58803 753	0.69598 90050 22499 59652 695
0.802	0.71874 80686 77571 43741 255	0.69527 06047 08886 74871 538
0.803	0.71944 29797 92397 50488 651	0.69455 15091 24727 13123 218
0.804	0.72013 71714 64303 73354 263	0.69383 17189 89116 26831 236
0.805	0.72083 06429 99098 50932 396	0.69311 12350 21844 23558 425
0.806	0.72152 33937 03310 35522 503	0.69239 00579 43394 94027 956
0.807	0.72221 54228 84188 62476 322	0.69166 81884 74945 40074 951
0.808	0.72290 67298 49704 19472 935	0.69094 56273 38365 02528 784
0.809	0.72359 73139 08550 15721 677	0.69022 23752 56214 89026 151
0.810	0.72428 71743 70142 51092 818	0.68949 84329 51747 01754 964
0.811	0.72497 63105 44620 85175 959	0.68877 38011 48903 65129 158
0.812	0.72566 47217 42849 06266 069	0.68804 84805 72316 53394 472
0.813	0.72635 24072 76416 00277 085	0.68732 24719 47306 18165 280
0.814	0.72703 93664 57636 19583 027	0.68659 57759 99881 15892 545
0.815	0.72772 55985 99550 51786 534	0.68586 83934 56737 35262 969
0.816	0.72841 11030 15926 88414 775	0.68514 03250 45257 24529 414
0.817	0.72909 58790 21260 93542 651	0.68441 15714 93509 18772 652
0.818	0.72977 99259 30776 72343 223	0.68368 21335 30246 67094 544
0.819	0.73046 32430 60427 39565 302	0.68295 20118 84907 59742 692
0.820	0.73114 58297 26895 87938 131	0.68222 12072 87613 55166 656
0.821	0.73182 76852 47595 56503 084	0.68148 97204 69169 07005 802
0.822	0.73250 88089 40670 98872 320	0.68075 75521 61060 91008 857
0.823	0.73318 92001 24998 51414 329	0.68002 47030 95457 31885 232
0.824	0.73386 88581 20187 01366 283	0.67929 11740 05207 30088 213
0.825	0.73454 77822 46578 54873 150	0.67855 69656 23839 88530 058
0.826	0.73522 59718 25249 04953 477	0.67782 20786 85563 39229 106
0.827	0.73590 34261 78008 99391 793	0.67708 65139 25264 69888 949
0.828	0.73658 01446 27404 08557 557	0.67635 02720 78508 50409 750
0.829	0.73725 61264 96715 93150 579	0.67561 33538 81536 59331 781
0.830	0.73793 13711 09962 71872 858	0.67487 57600 71267 10211 246
0.831	0.73860 58777 91899 89026 752	0.67413 74913 85293 77928 481
0.832	0.73927 96458 68020 82039 434	0.67339 85485 61885 24928 580
0.833	0.73995 26746 64557 48913 544	0.67265 89323 39984 27394 537
0.834	0.74062 49635 08481 15603 989	0.67191 86434 59207 01352 983
0.835	0.74129 65117 27503 03320 808	0.67117 76826 59842 28712 570
0.836	0.74196 73186 50074 95758 049	0.67043 60506 82850 83235 098
0.837	0.74263 73836 05390 06248 576	0.66969 37482 69864 56439 445
0.838	0.74330 67059 23383 44844 755	0.66895 07761 63185 83438 385
0.839	0.74397 52849 34732 85324 932	0.66820 71351 05786 68708 357
0.840	0.74464 31199 70859 32125 657	0.66746 28258 41308 11792 267
0.841	0.74531 02103 63927 87199 577	0.66671 78491 14059 32935 396
0.842	0.74597 65554 46848 16798 923	0.66597 22056 69016 98654 482
0.843	0.74664 21545 53275 18184 539	0.66522 58962 51824 47240 065
0.844	0.74730 70070 17609 86260 385	0.66447 89216 08791 14192 152
0.845	0.74797 11121 74999 80133 429	0.66373 12824 86891 57589 286
0.846	0.74863 44693 61339 89598 886	0.66298 29796 33764 83391 100
0.847	0.74929 70779 13273 01550 724	0.66223 40137 97713 70674 409
0.848	0.74995 89371 68190 66317 368	0.66148 43857 27703 96802 946
0.849	0.75062 00464 64233 63922 547	0.66073 40961 73363 62530 783
0.850	0.75128 04051 40292 70271 207	0.65998 31458 84982 17039 542

$$\begin{bmatrix} (-8)9 \\ 7 \end{bmatrix} \qquad\qquad\qquad \begin{bmatrix} (-8)9 \\ 7 \end{bmatrix}$$

CIRCULAR SINES AND COSINES FOR RADIAN ARGUMENTS   Table 4.6

x	sin x	cos x
0.850	0.75128 04051 40292 70271 207	0.65998 31458 84982 17039 542
0.851	0.75194 00125 36009 23260 432	0.65923 15356 13509 82909 449
0.852	0.75259 88679 91775 88815 295	0.65847 92661 10556 81024 321
0.853	0.75325 69708 48737 26849 594	0.65772 63381 28392 55410 547
0.854	0.75391 43204 48790 57151 380	0.65697 27524 19944 98010 152
0.855	0.75457 09161 34586 25193 237	0.65621 85097 38799 73388 013
0.856	0.75522 67572 49528 67867 227	0.65546 36108 39199 43373 300
0.857	0.75588 18431 37776 79144 450	0.65470 80564 76042 91635 218
0.858	0.75653 61731 44244 75659 143	0.65395 18474 04884 48193 134
0.859	0.75718 97466 14602 62217 260	0.65319 49843 81933 13861 148
0.860	0.75784 25628 95276 97229 459	0.65243 74681 64051 84627 203
0.861	0.75849 46213 33451 58068 441	0.65167 92995 08756 75466 794
0.862	0.75914 59212 77068 06350 566	0.65092 04791 74216 47091 357
0.863	0.75979 64620 74826 53141 684	0.65016 10079 19251 25131 418
0.864	0.76044 62430 76186 24087 122	0.64940 08865 03332 29254 574
0.865	0.76109 52636 31366 24465 750	0.64864 01156 86580 94718 373
0.866	0.76174 35230 91346 04168 073	0.64787 86962 29767 96858 196
0.867	0.76239 10208 07866 22598 272	0.64711 66288 94312 75010 176
0.868	0.76303 77561 33429 13500 144	0.64635 39144 42282 56369 276
0.869	0.76368 37284 21299 49706 858	0.64559 05536 36391 79782 561
0.870	0.76432 89370 25505 07814 480	0.64482 65472 40001 19477 766
0.871	0.76497 33813 00837 32779 191	0.64406 18960 17117 08727 234
0.872	0.76561 70606 02852 02438 134	0.64329 66007 32390 63447 280
0.873	0.76625 99742 87869 91953 834	0.64253 06621 51117 05733 091
0.874	0.76690 21217 12977 38182 114	0.64176 40810 39234 87329 202
0.875	0.76754 35022 36027 03963 458	0.64099 68581 63325 13035 656
0.876	0.76818 41152 15638 42337 736	0.64022 89942 90610 64049 903
0.877	0.76882 39600 11198 60682 252	0.63946 04901 88955 21244 528
0.878	0.76946 30359 82862 84773 027	0.63869 13466 26862 88380 872
0.879	0.77010 13424 91555 22769 271	0.63792 15643 73477 15258 639
0.880	0.77073 88788 98969 29120 965	0.63715 11441 98580 20801 550
0.881	0.77137 56445 67568 68399 506	0.63638 00868 72592 16079 131
0.882	0.77201 16388 60587 79051 337	0.63560 83931 66570 27264 710
0.883	0.77264 68611 42032 37074 497	0.63483 60638 52208 18529 695
0.884	0.77328 13107 76680 19618 049	0.63406 30997 01835 14874 218
0.885	0.77391 49871 30081 68504 290	0.63328 95014 88415 24894 213
0.886	0.77454 78895 68560 53673 706	0.63251 52699 85546 63485 020
0.887	0.77518 00174 59214 36552 600	0.63174 04059 67460 74481 571
0.888	0.77581 13701 69915 33343 321	0.63096 49102 09021 53235 256
0.889	0.77644 19470 69310 78237 045	0.63018 87834 85724 69127 530
0.890	0.77707 17475 26823 86549 033	0.62941 20265 73696 88020 355
0.891	0.77770 07709 12654 17776 316	0.62863 46402 49694 94643 540
0.892	0.77832 90165 97778 38577 722	0.62785 66252 91105 14919 057
0.893	0.77895 64839 53950 85676 211	0.62707 79824 75942 38222 428
0.894	0.77958 31723 53704 28683 432	0.62629 87125 82849 39581 242
0.895	0.78020 90811 70350 32846 443	0.62551 88163 91096 01810 880
0.896	0.78083 42097 77980 21716 548	0.62473 82946 80578 37587 545
0.897	0.78145 85575 51465 39740 163	0.62395 71482 31818 11458 656
0.898	0.78208 21238 66458 14771 667	0.62317 53778 25961 61790 683
0.899	0.78270 49080 99392 20508 171	0.62239 29842 44779 22654 524
0.900	0.78332 69096 27483 38846 138	0.62160 99682 70664 45648 472
	$\begin{bmatrix} (-7)1 \\ 7 \end{bmatrix}$	$\begin{bmatrix} (-8)8 \\ 7 \end{bmatrix}$

Table 4.6   CIRCULAR SINES AND COSINES FOR RADIAN ARGUMENTS

x	$\sin x$	$\cos x$
0.900	0.78332 69096 27483 38846 138	0.62160 99682 70664 45648 472
0.901	0.78394 81278 28730 22159 796	0.62082 63306 86633 21658 870
0.902	0.78456 85620 81914 55501 279	0.62004 20722 76323 02558 530
0.903	0.78518 82117 66602 18722 439	0.61925 71938 23992 22842 983
0.904	0.78580 70762 63143 48518 260	0.61847 16961 14519 21204 658
0.905	0.78642 51549 52674 00391 817	0.61768 55799 33401 62045 040
0.906	0.78704 24472 17115 10540 713	0.61689 88460 66755 56924 921
0.907	0.78765 89524 39174 57664 940	0.61611 14953 01314 85952 792
0.908	0.78827 46700 02347 24696 094	0.61532 35284 24430 19111 466
0.909	0.78888 95992 90915 60447 888	0.61453 49462 24068 37523 020
0.910	0.78950 37396 89950 41187 896	0.61374 57494 88811 54652 118
0.911	0.79011 70905 85311 32130 474	0.61295 59390 07856 37447 803
0.912	0.79072 96513 63647 48850 789	0.61216 55155 71013 27423 839
0.913	0.79134 14214 12398 18619 897	0.61137 44799 68705 61677 674
0.914	0.79195 24001 19793 41660 812	0.61058 28329 91968 93848 110
0.915	0.79256 25868 74854 52325 499	0.60979 05754 32450 15011 758
0.916	0.79317 19810 67394 80192 738	0.60899 77080 82406 74518 350
0.917	0.79378 05820 88020 11086 785	0.60820 42317 34706 00764 999
0.918	0.79438 83893 28129 48016 785	0.60741 01471 82824 21909 476
0.919	0.79499 54021 79915 72036 860	0.60661 54552 20845 86522 589
0.920	0.79560 16200 36366 03026 828	0.60582 01566 43462 84179 741
0.921	0.79620 70422 91262 60393 471	0.60502 42522 45973 65991 745
0.922	0.79681 16683 39183 23692 319	0.60422 77428 24282 65074 984
0.923	0.79741 54975 75501 93169 858	0.60343 06291 74899 16960 980
0.924	0.79801 85293 96389 50226 129	0.60263 29120 94936 79945 468
0.925	0.79862 07631 98814 17797 639	0.60183 45923 82112 55377 043
0.926	0.79922 21983 80542 20660 537	0.60103 56708 34746 07885 466
0.927	0.79982 28343 40138 45653 978	0.60023 61482 51758 85549 703
0.928	0.80042 26704 76967 01823 638	0.59943 60254 32673 40005 791
0.929	0.80102 17061 91191 80485 294	0.59863 53031 77612 46494 584
0.930	0.80161 99408 83777 15208 432	0.59783 39822 87298 23849 491
0.931	0.80221 73739 56488 41719 806	0.59703 20635 63051 54424 260
0.932	0.80281 40048 11892 57726 899	0.59622 95478 06791 03960 905
0.933	0.80340 98328 53358 82661 218	0.59542 64358 21032 41397 846
0.934	0.80400 48574 85059 17341 371	0.59462 27284 08887 58618 345
0.935	0.80459 90781 11969 03555 863	0.59381 84263 74063 90139 324
0.936	0.80519 24941 39867 83565 545	0.59301 35305 20863 32740 634
0.937	0.80578 51049 75339 59525 671	0.59220 80416 54181 65034 867
0.938	0.80637 69100 25773 52827 488	0.59140 19605 79507 66977 785
0.939	0.80696 79086 99364 63359 313	0.59059 52881 02922 39319 443
0.940	0.80755 81004 05114 28687 022	0.58978 80250 31098 22996 099
0.941	0.80814 74845 52830 83153 915	0.58898 01721 71298 18462 976
0.942	0.80873 60605 53130 16899 872	0.58817 17303 31375 04967 973
0.943	0.80932 38278 17436 34799 758	0.58736 27003 19770 59766 388
0.944	0.80991 07857 57982 15321 017	0.58655 30829 45514 77276 748
0.945	0.81049 69337 87809 69300 383	0.58574 28790 18224 88177 827
0.946	0.81108 22713 20770 98639 669	0.58493 20893 48104 78446 913
0.947	0.81166 67977 71528 54920 560	0.58412 07147 45944 08339 436
0.948	0.81225 05125 55555 97938 351	0.58330 87560 23117 31310 012
0.949	0.81283 34150 89138 54154 591	0.58249 62139 91583 12874 994
0.950	0.81341 55047 89373 75068 542	0.58168 30894 63883 49416 618
	$\begin{bmatrix} (-7)1 \\ 7 \end{bmatrix}$	$\begin{bmatrix} (-8)8 \\ 7 \end{bmatrix}$

CIRCULAR SINES AND COSINES FOR RADIAN ARGUMENTS   Table 4.6

x	$\sin x$	$\cos x$
0.950	0.81341 55047 89373 75068 542	0.58168 30894 63883 49416 618
0.951	0.81399 67810 74171 95507 433	0.58086 93832 53142 86928 810
0.952	0.81457 72433 62256 91835 411	0.58005 50961 73067 39704 748
0.953	0.81515 68910 73166 40081 165	0.57924 02290 37944 08966 253
0.954	0.81573 57236 27252 73984 145	0.57842 47826 62640 01435 096
0.955	0.81631 37404 45683 42959 322	0.57760 87578 62601 47846 300
0.956	0.81689 09409 50441 69980 433	0.57679 21554 53853 21403 511
0.957	0.81746 73245 64327 09381 654	0.57597 49762 52997 56176 536
0.958	0.81804 28907 10956 04577 644	0.57515 72210 77213 65441 113
0.959	0.81861 76388 14762 45701 891	0.57433 88907 44256 59961 007
0.960	0.81919 15683 00998 27163 322	0.57351 99860 72456 66212 505
0.961	0.81976 46785 95734 05121 101	0.57270 05078 80718 44551 395
0.962	0.82033 69691 25859 54877 569	0.57188 04569 88520 07322 513
0.963	0.82090 84393 19084 28189 263	0.57105 98342 15912 36911 940
0.964	0.82147 90886 03938 10495 962	0.57023 86403 83518 03741 923
0.965	0.82204 89164 09771 78067 694	0.56941 68763 12530 84208 614
0.966	0.82261 79221 66757 55069 656	0.56859 45428 24714 78562 699
0.967	0.82318 61053 05889 70544 986	0.56777 16407 42403 28733 004
0.968	0.82375 34652 58985 15315 328	0.56694 81708 88498 36093 162
0.969	0.82432 00014 58683 98799 136	0.56612 41340 86469 79171 417
0.970	0.82488 57133 38450 05747 662	0.56529 95311 60354 31303 653
0.971	0.82545 06003 32571 52898 564	0.56447 43629 34754 78229 727
0.972	0.82601 46618 76161 45547 087	0.56364 86302 34839 35633 190
0.973	0.82657 78974 05158 34034 750	0.56282 23338 86340 66624 480
0.974	0.82714 03063 56326 70155 495	0.56199 54747 15554 99167 663
0.975	0.82770 18881 67257 63479 226	0.56116 80535 49341 43450 813
0.976	0.82826 26422 76369 37592 699	0.56034 00712 15121 09200 110
0.977	0.82882 25681 22907 86257 689	0.55951 15285 40876 22937 736
0.978	0.82938 16651 46947 29486 397	0.55868 24263 55149 45183 654
0.979	0.82993 99327 89390 69534 022	0.55785 27654 87042 87601 358
0.980	0.83049 73704 91970 46808 453	0.55702 25467 66217 30087 666
0.981	0.83105 39776 97248 95697 028	0.55619 17710 22891 37806 645
0.982	0.83160 97538 48619 00310 290	0.55536 04390 87840 78167 757
0.983	0.83216 46983 90304 50142 703	0.55452 85517 92397 37748 295
0.984	0.83271 88107 67360 95650 254	0.55369 61099 68448 39160 207
0.985	0.83327 20904 25676 03744 902	0.55286 31144 48435 57861 376
0.986	0.83382 45368 11970 13205 801	0.55202 95660 65354 38911 453
0.987	0.83437 61493 73796 79007 262	0.55119 54656 52753 13672 322
0.988	0.83492 69275 59543 82563 379	0.55036 08140 44732 16453 272
0.989	0.83547 68708 18432 76889 279	0.54952 56120 75943 01100 969
0.990	0.83602 59786 00520 51678 926	0.54868 98605 81587 57534 313
0.991	0.83657 42503 56699 33299 444	0.54785 35603 97417 28224 252
0.992	0.83712 16855 38697 50701 883	0.54701 67123 59732 24618 647
0.993	0.83766 82835 99079 90248 385	0.54617 93173 05380 43512 268
0.994	0.83821 40439 91248 50455 694	0.54534 13760 71756 83362 006
0.995	0.83875 89661 69442 96654 953	0.54450 28894 96802 60547 375
0.996	0.83930 30495 88741 15567 733	0.54366 38584 19004 25576 412
0.997	0.83984 62937 05059 69798 245	0.54282 42836 77392 79237 026
0.998	0.84038 86979 75154 52241 668	0.54198 41661 11542 88693 907
0.999	0.84093 02618 56621 40408 555	0.54114 35065 61572 03531 067
1.000	0.84147 09848 07896 50665 250	0.54030 23058 68139 71740 094
	$\begin{bmatrix} (-7)1 \\ 7 \end{bmatrix}$	$\begin{bmatrix} (-8)7 \\ 7 \end{bmatrix}$

Table 4.6   CIRCULAR SINES AND COSINES FOR RADIAN ARGUMENTS

x	$\sin x$	$\cos x$
1.000	0. 84147 09848 07896 50665 250	0. 54030 23058 68139 71740 094
1.001	0. 84201 08662 88256 92390 268	0. 53946 05648 72446 55654 214
1.002	0. 84254 99057 57821 22046 578	0. 53861 82844 16233 47828 237
1.003	0. 84308 81026 77549 97169 747	0. 53777 54653 41780 86864 465
1.004	0. 84362 54565 09246 30271 873	0. 53693 21084 91907 73184 669
1.005	0. 84416 19667 15556 42661 273	0. 53608 82147 09970 84748 188
1.006	0. 84469 76327 59970 18177 851	0. 53524 37848 39863 92716 262
1.007	0. 84523 24541 06821 56844 116	0. 53439 88197 26016 77062 668
1.008	0. 84576 64302 21289 28431 774	0. 53355 33202 13394 42130 747
1.009	0. 84629 95605 69397 25943 853	0. 53270 72871 47496 32136 904
1.010	0. 84683 18446 18015 19012 310	0. 53186 07213 74355 46620 673
1.011	0. 84736 32818 34859 07211 051	0. 53101 36237 40537 55841 426
1.012	0. 84789 38716 88491 73284 331	0. 53016 59950 93140 16121 808
1.013	0. 84842 36136 48323 36290 466	0. 52931 78362 79791 85137 984
1.014	0. 84895 25071 84612 04660 810	0. 52846 91481 48651 37156 798
1.015	0. 84948 05517 68464 29173 940	0. 52761 99315 48406 78219 896
1.016	0. 85000 77468 71835 55845 003	0. 52677 01873 28274 61274 932
1.017	0. 85053 40919 67530 78730 164	0. 52591 99163 37999 01253 921
1.018	0. 85105 95865 29204 92646 111	0. 52506 91194 27850 90098 832
1.019	0. 85158 42300 31363 45804 549	0. 52421 77974 48627 11734 503
1.020	0. 85210 80219 49362 92361 655	0. 52336 59512 51649 56988 961
1.021	0. 85263 09617 59411 44882 415	0. 52251 35816 88764 38461 245
1.022	0. 85315 30489 38569 26719 808	0. 52166 06896 12341 05336 792
1.023	0. 85367 42829 64749 24308 778	0. 52080 72758 75271 58150 502
1.024	0. 85419 46633 16717 39374 945	0. 51995 33413 30969 63497 542
1.025	0. 85471 41894 74093 41057 997	0. 51909 88868 33369 68691 985
1.026	0. 85523 28609 17351 17949 715	0. 51824 39132 36926 16373 373
1.027	0. 85575 06771 27819 30046 586	0. 51738 84213 96612 59061 276
1.028	0. 85626 76375 87681 60616 931	0. 51653 24121 67920 73657 956
1.029	0. 85678 37417 79977 67982 525	0. 51567 58864 06859 75899 186
1.030	0. 85729 89891 88603 37214 627	0. 51481 88449 69955 34753 350
1.031	0. 85781 33792 98311 31744 398	0. 51396 12887 14248 86768 878
1.032	0. 85832 69115 94711 44887 626	0. 51310 32184 97296 50370 116
1.033	0. 85883 95855 64271 51283 734	0. 51224 46351 77168 40101 715
1.034	0. 85935 14006 94317 58248 998	0. 51138 55396 12447 80821 625
1.035	0. 85986 23564 73034 57043 938	0. 51052 59326 62230 21842 776
1.036	0. 86037 24523 89466 74054 819	0. 50966 58151 86122 51023 535
1.037	0. 86088 16879 33518 21889 224	0. 50880 51880 44242 08807 028
1.038	0. 86139 00625 95953 50385 634	0. 50794 40520 97216 02209 404
1.039	0. 86189 75758 68397 97536 975	0. 50708 24082 06180 18757 138
1.040	0. 86240 42272 43338 40328 079	0. 50622 02572 32778 40373 447
1.041	0. 86291 00162 14123 45486 997	0. 50535 76000 39161 57213 919
1.042	0. 86341 49422 74964 20150 131	0. 50449 44374 87986 81451 427
1.043	0. 86391 90049 20934 62441 124	0. 50363 07704 42416 61010 426
1.044	0. 86442 22036 47972 11963 456	0. 50276 65997 66117 93250 711
1.045	0. 86492 45379 52878 00206 699	0. 50190 19263 23261 38600 728
1.046	0. 86542 60073 33318 00866 385	0. 50103 67509 78520 34140 520
1.047	0. 86592 66112 87822 80077 424	0. 50017 10745 97070 07134 396
1.048	0. 86642 63493 15788 46561 037	0. 49930 48980 44586 88513 415
1.049	0. 86692 52209 17477 01685 140	0. 49843 82221 87247 26307 756
1.050	0. 86742 32255 94016 89438 141	0. 49757 10478 91726 99029 085

$$\begin{bmatrix} (-7)1 \\ 7 \end{bmatrix} \qquad\qquad\qquad\qquad \begin{bmatrix} (-8)7 \\ 7 \end{bmatrix}$$

CIRCULAR SINES AND COSINES FOR RADIAN ARGUMENTS   Table 4.6

x	$\sin x$	$\cos x$
1.050	0.86742 32255 94016 89438 141	0.49757 10478 91726 99029 085
1.051	0.86792 03628 47403 46316 092	0.49670 33760 25200 29002 975
1.052	0.86841 66321 80499 51123 146	0.49583 52074 55338 95651 499
1.053	0.86891 20330 97035 74685 276	0.49496 65430 50311 48726 051
1.054	0.86940 65651 01611 29477 198	0.49409 73836 78782 21490 510
1.055	0.86990 02276 99694 19162 460	0.49322 77302 09910 43854 806
1.056	0.87039 30203 97621 88046 624	0.49235 75835 13349 55459 008
1.057	0.87088 49427 02601 70443 529	0.49148 69444 59246 18707 979
1.058	0.87137 59941 22711 39954 543	0.49061 58139 18239 31756 732
1.059	0.87186 61741 66899 58660 794	0.48974 41927 61459 41446 534
1.060	0.87235 54823 44986 26228 295	0.48887 20818 60527 56191 864
1.061	0.87284 39181 67663 28925 947	0.48799 94820 87554 58818 317
1.062	0.87333 14811 46494 88556 345	0.48712 63943 15140 19351 528
1.063	0.87381 81707 93918 11299 356	0.48625 28194 16372 07757 202
1.064	0.87430 39866 23243 36468 402	0.48537 87582 64825 06632 362
1.065	0.87478 89281 48654 85179 424	0.48450 42117 34560 23847 867
1.066	0.87527 29948 85211 08932 453	0.48362 91807 00124 05142 311
1.067	0.87575 61863 48845 38105 753	0.48275 36660 36547 46667 387
1.068	0.87623 85020 56366 30362 492	0.48187 76686 19345 07484 800
1.069	0.87671 99415 25458 18969 874	0.48100 11893 24514 22014 811
1.070	0.87720 05042 74681 61030 706	0.48012 42290 28534 12436 509
1.071	0.87768 01898 23473 85627 336	0.47924 67886 08365 01039 904
1.072	0.87815 89976 92149 41877 919	0.47836 88689 41447 22529 904
1.073	0.87863 69274 01900 46904 963	0.47749 04709 05700 36282 289
1.074	0.87911 39784 74797 33716 111	0.47661 15953 79522 38551 762
1.075	0.87959 01504 33788 98997 101	0.47573 22432 41788 74632 160
1.076	0.88006 54428 02703 50816 869	0.47485 24153 71851 50968 911
1.077	0.88053 98551 06248 56244 731	0.47397 21126 49538 47223 840
1.078	0.88101 33868 70011 88879 619	0.47309 13359 55152 28292 396
1.079	0.88148 60376 20461 76291 297	0.47221 00861 69469 56273 392
1.080	0.88195 78068 84947 47373 533	0.47132 83641 73740 02391 353
1.081	0.88242 86941 91699 79609 169	0.47044 61708 49685 58871 547
1.082	0.88289 86990 69831 46247 031	0.46956 35070 79499 50767 810
1.083	0.88336 78210 49337 63390 660	0.46868 03737 45845 47743 217
1.084	0.88383 60596 61096 36998 790	0.46779 67717 31856 75803 727
1.085	0.88430 34144 36869 09797 534	0.46691 27019 21135 28984 862
1.086	0.88476 98849 09301 08104 243	0.46602 81651 97750 80991 522
1.087	0.88523 54706 11921 88562 972	0.46514 31624 46239 96791 014
1.088	0.88570 01710 79145 84791 522	0.46425 76945 51605 44159 401
1.089	0.88616 39858 46272 53940 000	0.46337 17623 99315 05181 235
1.090	0.88662 69144 49487 23160 860	0.46248 53668 75300 87702 790
1.091	0.88708 89564 25861 35990 371	0.46159 85088 65958 36738 852
1.092	0.88755 01113 13352 98641 470	0.46071 11892 58145 45833 190
1.093	0.88801 03786 50807 26207 951	0.45982 34089 39181 68372 764
1.094	0.88846 97579 77956 88779 948	0.45893 51687 96847 28855 783
1.095	0.88892 82488 35422 57470 660	0.45804 64697 19382 34113 686
1.096	0.88938 58507 64713 50354 274	0.45715 73125 95485 84487 142
1.097	0.88984 25633 08227 78315 047	0.45626 76983 14314 84956 158
1.098	0.89029 83860 09252 90807 488	0.45537 76277 65483 56224 382
1.099	0.89075 33184 11966 21527 609	0.45448 71018 39062 45757 688
1.100	0.89120 73600 61435 33995 180	0.45359 61214 25577 38777 137

$$\begin{bmatrix} (-7)1 \\ 7 \end{bmatrix} \qquad\qquad\qquad \begin{bmatrix} (-8)6 \\ 7 \end{bmatrix}$$

Table 4.6   CIRCULAR SINES AND COSINES FOR RADIAN ARGUMENTS

x	$\sin x$	$\cos x$
1.100	0.89120 73600 61435 33995 180	0.45359 61214 25577 38777 137
1.101	0.89166 05105 03618 67046 971	0.45270 46874 16008 69206 400
1.102	0.89211 27692 85365 80240 901	0.45181 28007 01790 30573 730
1.103	0.89256 41359 54417 99171 080	0.45092 04621 74808 86868 576
1.104	0.89301 46100 59408 60693 678	0.45002 76727 27402 83352 928
1.105	0.89346 41911 49863 58063 585	0.44913 44332 52361 57327 478
1.106	0.89391 28787 76201 85981 812	0.44824 07446 42924 48852 689
1.107	0.89436 06724 89735 85553 594	0.44734 66077 92780 11424 866
1.108	0.89480 75718 42671 89157 146	0.44645 20235 96065 22607 305
1.109	0.89525 35763 88110 65223 027	0.44555 69929 47363 94616 628
1.110	0.89569 86856 80047 62924 063	0.44466 15167 41706 84864 374
1.111	0.89614 28992 73373 56775 801	0.44376 55958 74570 06453 951
1.112	0.89658 62167 23874 91147 427	0.44286 92312 41874 38633 030
1.113	0.89702 86375 88234 24683 120	0.44197 24237 39984 37201 474
1.114	0.89747 01614 24030 74633 785	0.44107 51742 65707 44874 890
1.115	0.89791 07877 89740 61099 138	0.44017 74837 16293 01603 891
1.116	0.89835 05162 44737 51180 079	0.43927 93529 89431 54849 166
1.117	0.89878 93463 49293 03041 321	0.43838 07829 83253 69812 438
1.118	0.89922 72776 64577 09884 230	0.43748 17745 96329 39623 410
1.119	0.89966 43097 52658 43829 826	0.43658 23287 27666 95482 777
1.120	0.90010 04421 76504 99711 910	0.43568 24462 76712 16761 399
1.121	0.90053 56744 99984 38780 263	0.43478 21281 43347 41055 736
1.122	0.90097 00062 87864 32313 880	0.43388 13752 27890 74199 612
1.123	0.90140 34371 05813 05144 201	0.43298 01884 31095 00232 420
1.124	0.90183 59665 20399 79088 276	0.43207 85686 54146 91323 845
1.125	0.90226 75940 99095 16291 842	0.43117 65167 98666 17655 197
1.126	0.90269 83194 10271 62482 258	0.43027 40337 66704 57257 452
1.127	0.90312 81420 23203 90131 256	0.42937 11204 60745 05806 078
1.128	0.90355 70615 08069 41527 464	0.42846 77777 83700 86372 749
1.129	0.90398 50774 35948 71758 658	0.42756 40066 38914 59134 030
1.130	0.90441 21893 78825 91603 708	0.42665 98079 30157 31037 122
1.131	0.90483 83969 09589 10334 160	0.42575 51825 61627 65422 763
1.132	0.90526 36996 02030 78425 425	0.42485 01314 37950 91605 376
1.133	0.90568 80970 30848 30177 523	0.42394 46554 64178 14410 540
1.134	0.90611 15887 71644 26245 348	0.42303 87555 45785 23669 902
1.135	0.90653 41744 00926 96078 401	0.42213 24325 88672 03673 585
1.136	0.90695 58534 96110 80269 960	0.42122 56874 99161 42580 219
1.137	0.90737 66256 35516 72815 632	0.42031 85211 83998 41784 656
1.138	0.90779 64903 98372 63281 260	0.41941 09345 50349 25243 478
1.139	0.90821 54473 64813 78880 126	0.41850 29285 05800 48758 379
1.140	0.90863 34961 15883 26459 422	0.41759 45039 58358 09217 519
1.141	0.90905 06362 33532 34395 940	0.41668 56618 16446 53794 933
1.142	0.90946 68673 00620 94400 939	0.41577 64029 88907 89108 094
1.143	0.90988 21889 00918 03234 153	0.41486 67283 85000 90333 707
1.144	0.91029 66006 19102 04326 885	0.41395 66389 14400 10281 852
1.145	0.91071 01020 40761 29314 164	0.41304 61354 87194 88428 529
1.146	0.91112 26927 52394 39475 912	0.41213 52190 13888 59906 732
1.147	0.91153 43723 41410 67087 073	0.41122 38904 05397 64456 120
1.148	0.91194 51403 96130 56676 684	0.41031 21505 73050 55331 381
1.149	0.91235 49965 05786 06195 821	0.40940 00004 28587 08169 395
1.150	0.91276 39402 60521 08094 403	0.40848 74408 84157 29815 258

$$\left[\begin{matrix} (-7)1 \\ 7 \end{matrix} \right] \qquad\qquad \left[\begin{matrix} (-8)6 \\ 7 \end{matrix} \right]$$

CIRCULAR SINES AND COSINES FOR RADIAN ARGUMENTS   Table 4.6

x	$\sin x$	$\cos x$
1.150	0.91276 39402 60521 08094 403	0.40848 74408 84157 29815 258
1.151	0.91317 19712 51391 90306 792	0.40757 44728 52320 67107 284
1.152	0.91357 90890 70367 57146 165	0.40666 10972 46045 15621 071
1.153	0.91398 52933 10330 30107 602	0.40574 73149 78706 28372 706
1.154	0.91439 05835 65075 88579 865	0.40483 31269 64086 24481 224
1.155	0.91479 49594 29314 10465 816	0.40391 85341 16372 97790 397
1.156	0.91519 84204 98669 12711 431	0.40300 35373 50159 25449 945
1.157	0.91560 09663 69679 91743 383	0.40208 81375 80441 76456 266
1.158	0.91600 25966 39800 63815 143	0.40117 23357 22620 20152 779
1.159	0.91640 33109 07401 05261 556	0.40025 61326 92496 34689 958
1.160	0.91680 31087 71766 92661 866	0.39933 95294 06273 15445 164
1.161	0.91720 19898 33100 42911 136	0.39842 25267 80553 83402 355
1.162	0.91759 99536 92520 53200 023	0.39750 51257 32340 93491 775
1.163	0.91799 69999 52063 40902 883	0.39658 73271 79035 42889 706
1.164	0.91839 31282 14682 83374 147	0.39566 91320 38435 79278 377
1.165	0.91878 83380 84250 57652 941	0.39475 05412 28737 09066 125
1.166	0.91918 26291 65556 80075 906	0.39383 15556 68530 05567 898
1.167	0.91957 60010 64310 45798 178	0.39291 21762 76800 17146 187
1.168	0.91996 84533 87139 68222 492	0.39199 24039 72926 75312 486
1.169	0.92035 99857 41592 18336 360	0.39107 22396 76682 02789 366
1.170	0.92075 05977 36135 63957 301	0.39015 16843 08230 21533 266
1.171	0.92114 02889 80158 08886 071	0.38923 07387 88126 60718 072
1.172	0.92152 90590 83968 31967 851	0.38830 94040 37316 64679 599
1.173	0.92191 69076 58796 26061 369	0.38738 76809 77135 00821 054
1.174	0.92230 38343 16793 36915 902	0.38646 55705 29304 67479 575
1.175	0.92268 98386 71033 01956 127	0.38554 30736 15936 01753 942
1.176	0.92307 49203 35510 88974 783	0.38462 01911 59525 87293 547
1.177	0.92345 90789 25145 34733 097	0.38369 69240 82956 62048 718
1.178	0.92384 23140 55777 83468 944	0.38277 32733 09495 25982 487
1.179	0.92422 46253 44173 25312 701	0.38184 92397 62792 48743 902
1.180	0.92460 60124 08020 34610 754	0.38092 48243 66881 77302 960
1.181	0.92498 64748 65932 08156 619	0.38000 00280 46178 43547 271
1.182	0.92536 60123 37446 03329 642	0.37907 48517 25478 71840 534
1.183	0.92574 46244 43024 76141 242	0.37814 92963 29958 86542 917
1.184	0.92612 23108 04056 19188 645	0.37722 33627 85174 19493 444
1.185	0.92649 90710 42853 99516 095	0.37629 70520 17058 17454 471
1.186	0.92687 49047 82657 96383 480	0.37537 03649 51921 49518 342
1.187	0.92724 98116 47634 38942 352	0.37444 33025 16451 14476 334
1.188	0.92762 37912 62876 43819 290	0.37351 58656 37709 48149 962
1.189	0.92799 68432 54404 52606 588	0.37258 80552 43133 30684 752
1.190	0.92836 89672 49166 69260 202	0.37165 98722 60532 93806 568
1.191	0.92874 01628 75038 97404 950	0.37073 13176 18091 28040 589
1.192	0.92911 04297 60825 77546 899	0.36980 23922 44362 89893 026
1.193	0.92947 97675 36260 24192 928	0.36887 30970 68273 08995 672
1.194	0.92984 81758 32004 62877 403	0.36794 34330 19116 95213 382
1.195	0.93021 56542 79650 67095 956	0.36701 34010 26558 45714 570
1.196	0.93058 22025 11719 15146 303	0.36608 30020 20629 52004 819
1.197	0.93094 78201 61664 26876 083	0.36515 22369 31729 06923 698
1.198	0.93131 25068 63866 00337 679	0.36422 11066 90622 11604 876
1.199	0.93167 62622 53638 48349 974	0.36328 96122 28438 82399 631
1.200	0.93203 90859 67226 34967 013	0.36235 77544 76673 57763 837
	$\begin{bmatrix} (-7)1 \\ 7 \end{bmatrix}$	$\begin{bmatrix} (-8)5 \\ 7 \end{bmatrix}$

Table 4.6　　　CIRCULAR SINES AND COSINES FOR RADIAN ARGUMENTS

x	$\sin x$	$\cos x$
1.200	0.93203 90859 67226 34967 013	0.36235 77544 76673 57763 837
1.201	0.93240 09776 41805 91853 542	0.36142 55343 67184 05108 539
1.202	0.93276 19369 15485 54567 367	0.36049 29528 32190 27614 189
1.203	0.93312 19634 27305 98748 519	0.35956 00108 04273 71008 651
1.204	0.93348 10568 17240 76215 175	0.35862 67092 16376 30309 065
1.205	0.93383 92167 26196 50966 302	0.35769 30490 01799 56527 660
1.206	0.93419 64427 96013 35090 992	0.35675 90310 94203 63341 607
1.207	0.93455 27346 69465 24584 444	0.35582 46564 27606 33727 018
1.208	0.93490 80919 90260 35070 567	0.35488 99259 36382 26557 166
1.209	0.93526 25144 03041 37431 162	0.35395 48405 55261 83165 039
1.210	0.93561 60015 53385 93341 646	0.35301 94012 19330 33870 301
1.211	0.93596 85530 87806 90713 291	0.35208 36088 64027 04470 775
1.212	0.93632 01686 53752 79041 926	0.35114 74644 25144 22698 521
1.213	0.93667 08478 99608 04663 095	0.35021 09688 38826 24640 616
1.214	0.93702 05904 74693 45913 598	0.34927 41230 41568 61124 730
1.215	0.93736 93960 29266 48199 416	0.34833 69279 70217 04069 578
1.216	0.93771 72642 14521 58969 959	0.34739 93845 61966 52800 358
1.217	0.93806 41946 82590 62598 617	0.34646 14937 54360 40329 260
1.218	0.93841 01870 86543 15169 574	0.34552 32564 85289 39601 140
1.219	0.93875 52410 80386 79170 848	0.34458 46736 92990 69704 455
1.220	0.93909 93563 19067 58093 524	0.34364 57463 16047 02047 552
1.221	0.93944 25324 58470 30937 151	0.34270 64752 93385 66500 405
1.222	0.93978 47691 55418 86621 257	0.34176 68615 64277 57501 890
1.223	0.94012 60660 67676 58302 957	0.34082 69060 68336 40132 702
1.224	0.94046 64228 53946 57600 622	0.33988 66097 45517 56153 996
1.225	0.94080 58391 73872 08723 559	0.33894 59735 36117 30011 855
1.226	0.94114 43146 88036 82507 685	0.33800 49983 80771 74807 668
1.227	0.94148 18490 57965 30357 157	0.33706 36852 20455 98234 533
1.228	0.94181 84419 46123 18091 912	0.33612 20349 96483 08479 750
1.229	0.94215 40930 15917 59701 104	0.33518 00486 50503 20093 523
1.230	0.94248 88019 31697 51002 382	0.33423 77271 24502 59823 955
1.231	0.94282 25683 58754 03206 998	0.33329 50713 60802 72418 427
1.232	0.94315 53919 63320 76390 684	0.33235 20822 02059 26391 462
1.233	0.94348 72724 12574 12870 299	0.33140 87608 91261 19759 164
1.234	0.94381 82093 74633 70486 175	0.33046 51080 71729 85740 328
1.235	0.94414 82025 18562 55790 164	0.32952 11247 87117 98424 316
1.236	0.94447 72515 14367 57139 322	0.32857 68119 81408 78405 786
1.237	0.94480 53560 32999 77695 223	0.32763 21705 98914 98386 387
1.238	0.94513 25157 46354 68328 851	0.32668 72015 84277 88743 487
1.239	0.94545 87303 27272 60431 046	0.32574 19058 82466 43066 054
1.240	0.94578 39994 49538 98628 471	0.32479 62844 38776 23657 769
1.241	0.94610 83227 87884 73405 063	0.32385 03381 98828 67007 475
1.242	0.94643 17000 17986 53628 942	0.32290 40681 08569 89227 042
1.243	0.94675 41308 16467 18984 738	0.32195 74751 14269 91456 764
1.244	0.94707 56148 60895 92311 309	0.32101 05601 62521 65238 364
1.245	0.94739 61518 29788 71844 815	0.32006 33242 00239 97855 712
1.246	0.94771 57414 02608 63367 118	0.31911 57681 74660 77643 341
1.247	0.94803 43832 59766 12259 472	0.31816 78930 33339 99262 871
1.248	0.94835 20770 82619 35461 479	0.31721 96997 24152 68947 423
1.249	0.94866 88225 53474 53335 262	0.31627 11891 95292 09714 116
1.250	0.94898 46193 55586 21434 849	0.31532 23623 95268 66544 754
	$\begin{bmatrix} (-7)1 \\ 7 \end{bmatrix}$	$\begin{bmatrix} (-8)5 \\ 7 \end{bmatrix}$

CIRCULAR SINES AND COSINES FOR RADIAN ARGUMENTS   Table 4.6

x	$\sin x$	$\cos x$
1.250	0.94898 46193 55586 21434 849	0.31532 23623 95268 66544 754
1.251	0.94929 94671 73157 62180 713	0.31437 32202 72909 11534 791
1.252	0.94961 33656 91340 96439 444	0.31342 37637 77355 49010 665
1.253	0.94992 63145 96237 75008 528	0.31247 39938 58064 20615 601
1.254	0.95023 83135 74899 10006 196	0.31152 39114 64805 10363 979
1.255	0.95054 93623 15326 06166 303	0.31057 35175 47660 49664 355
1.256	0.95085 94605 06469 92038 225	0.30962 28130 57024 22311 242
1.257	0.95116 86078 38232 51091 729	0.30867 17989 43600 69445 729
1.258	0.95147 68040 01466 52726 783	0.30772 04761 58403 94485 052
1.259	0.95178 40486 87975 83188 287	0.30676 88456 52756 68021 196
1.260	0.95209 03415 90515 76385 682	0.30581 69083 78289 32688 634
1.261	0.95239 56824 02793 44617 416	0.30486 46652 86939 08001 291
1.262	0.95270 00708 19468 09200 227	0.30391 21173 30948 95158 833
1.263	0.95300 35065 36151 31003 222	0.30295 92654 62866 81822 373
1.264	0.95330 59892 49407 40886 709	0.30200 61106 35544 46859 693
1.265	0.95360 75186 56753 70045 767	0.30105 26538 02136 65060 070
1.266	0.95390 80944 56660 80258 512	0.30009 88959 16100 11818 814
1.267	0.95420 77163 48552 94039 032	0.29914 48379 31192 67791 595
1.268	0.95450 63840 32808 24694 963	0.29819 04808 01472 23518 675
1.269	0.95480 40972 10759 06289 671	0.29723 58254 81295 84019 121
1.270	0.95510 08555 84692 23509 018	0.29628 08729 25318 73355 114
1.271	0.95539 66588 57849 41432 673	0.29532 56240 88493 39166 425
1.272	0.95569 15067 34427 35209 944	0.29437 00799 26068 57175 182
1.273	0.95598 53989 19578 19640 104	0.29341 42413 93588 35661 000
1.274	0.95627 83351 19409 78657 170	0.29245 81094 46891 19906 579
1.275	0.95657 03150 40985 94719 118	0.29150 16850 42108 96613 869
1.276	0.95686 13383 92326 78101 497	0.29054 49691 35665 98290 890
1.277	0.95715 14048 82408 96095 419	0.28958 79626 84278 07609 308
1.278	0.95744 05142 21166 02109 886	0.28863 06666 44951 61732 860
1.279	0.95772 86661 19488 64678 437	0.28767 30819 74982 56616 726
1.280	0.95801 58602 89224 96370 075	0.28671 52096 31955 51277 939
1.281	0.95830 20964 43180 82604 453	0.28575 70505 73742 72036 934
1.282	0.95858 73742 95120 10371 286	0.28479 86057 58503 16730 332
1.283	0.95887 16935 59764 96853 962	0.28383 98761 44681 58895 050
1.284	0.95915 50539 52796 17957 320	0.28288 08626 91007 51923 831
1.285	0.95943 74551 90853 36739 577	0.28192 15663 56494 33192 303
1.286	0.95971 88969 91535 31748 357	0.28096 19881 00438 28157 651
1.287	0.95999 93790 73400 25260 814	0.28000 21288 82417 54428 993
1.288	0.96027 89011 55966 11427 805	0.27904 19896 62291 25809 577
1.289	0.96055 74629 59710 84322 094	0.27808 15714 00198 56310 871
1.290	0.96083 50642 06072 65890 556	0.27712 08750 56557 64138 661
1.291	0.96111 17046 17450 33810 354	0.27615 99015 92064 75651 234
1.292	0.96138 73839 17203 49249 056	0.27519 86519 67693 29289 769
1.293	0.96166 21018 29652 84528 675	0.27423 71271 44692 79480 997
1.294	0.96193 58580 80080 50693 590	0.27327 53280 84588 00512 263
1.295	0.96220 86523 94730 24982 339	0.27231 32557 49177 90379 053
1.296	0.96248 04845 00807 78203 231	0.27135 09111 00534 74605 108
1.297	0.96275 13541 26481 02013 782	0.27038 82951 01003 10035 206
1.298	0.96302 12610 00880 36103 915	0.26942 54087 13198 88600 711
1.299	0.96329 02048 54098 95282 920	0.26846 22529 00008 41057 992
1.300	0.96355 81854 17192 96470 135	0.26749 88286 24587 40699 798

$$\left[\begin{array}{c}(-7)1 \\ 7\end{array}\right] \qquad \left[\begin{array}{c}(-8)4 \\ 7\end{array}\right]$$

Table 4.6   CIRCULAR SINES AND COSINES FOR RADIAN ARGUMENTS

x	$\sin x$	$\cos x$
1.300	0.96355 81854 17192 96470 135	0.26749 88286 24587 40699 798
1.301	0.96382 52024 22181 85589 331	0.26653 51368 50360 07039 695
1.302	0.96409 12556 02048 64366 761	0.26557 11785 41018 09469 650
1.303	0.96435 63446 90740 17032 855	0.26460 69546 60519 70890 877
1.304	0.96462 04694 23167 36927 537	0.26364 24661 73088 71318 016
1.305	0.96488 36295 35205 53009 126	0.26267 77140 43213 51456 761
1.306	0.96514 58247 63694 56266 806	0.26171 26992 35646 16255 031
1.307	0.96540 70548 46439 26036 635	0.26074 74227 15401 38427 774
1.308	0.96566 73195 22209 56221 061	0.25978 18854 47755 61955 494
1.309	0.96592 66185 30740 81411 924	0.25881 60883 98246 05556 626
1.310	0.96618 49516 12734 02916 926	0.25785 00325 32669 66133 818
1.311	0.96644 23185 09856 14689 520	0.25688 37188 17082 22194 242
1.312	0.96669 87189 64740 29162 218	0.25591 71482 17797 37244 030
1.313	0.96695 41527 20986 02983 276	0.25495 03217 01385 63156 911
1.314	0.96720 86195 23159 62656 736	0.25398 32402 34673 43517 173
1.315	0.96746 21191 16794 30085 794	0.25301 59047 84742 16937 022
1.316	0.96771 46512 48390 48019 478	0.25204 83163 18927 20348 457
1.317	0.96796 62156 65416 05402 607	0.25108 04758 04816 92269 738
1.318	0.96821 68121 16306 62628 991	0.25011 23842 10251 76046 556
1.319	0.96846 64403 50465 76697 879	0.24914 40425 03323 23067 996
1.320	0.96871 51001 18265 26273 590	0.24817 54516 52372 95957 398
1.321	0.96896 27911 71045 36648 340	0.24720 66126 25991 71738 199
1.322	0.96920 95132 61115 04608 211	0.24623 75263 93018 44974 865
1.323	0.96945 52661 41752 23202 252	0.24526 81939 22539 30889 004
1.324	0.96970 00495 67204 06414 685	0.24429 86161 83886 68450 760
1.325	0.96994 38632 92687 13740 188	0.24332 87941 46638 23445 582
1.326	0.97018 67070 74387 74662 236	0.24235 87287 80615 91516 463
1.327	0.97042 85806 69462 13034 465	0.24138 84210 55885 01181 759
1.328	0.97066 94838 36036 71365 051	0.24041 78719 42753 16828 662
1.329	0.97090 94163 33208 35004 060	0.23944 70824 11769 41682 448
1.330	0.97114 83779 21044 56233 768	0.23847 60534 33723 20751 578
1.331	0.97138 63683 60583 78261 900	0.23750 47859 79643 43748 768
1.332	0.97162 33874 13835 59117 786	0.23653 32810 20797 47988 097
1.333	0.97185 94348 43780 95451 405	0.23556 15395 28690 21258 288
1.334	0.97209 45104 14372 46235 282	0.23458 95624 75063 04672 221
1.335	0.97232 86138 90534 56369 230	0.23361 73508 31892 95492 805
1.336	0.97256 17450 38163 80187 900	0.23264 49055 71391 49935 286
1.337	0.97279 39036 24129 04871 129	0.23167 22276 66003 85946 099
1.338	0.97302 50894 16271 73757 046	0.23069 93180 88407 85958 358
1.339	0.97325 53021 83406 09557 931	0.22972 61778 11512 99624 085
1.340	0.97348 45416 95319 37478 787	0.22875 28078 08459 46523 264
1.341	0.97371 28077 22772 08238 616	0.22777 92090 52617 18849 831
1.342	0.97394 01000 37498 20994 365	0.22680 53825 17584 84074 691
1.343	0.97416 64184 12205 46167 522	0.22583 13291 77188 87585 859
1.344	0.97439 17626 20575 48173 349	0.22485 70500 05482 55305 819
1.345	0.97461 61324 37264 08052 713	0.22388 25459 76744 96286 212
1.346	0.97483 95276 37901 46006 501	0.22290 78180 65480 05279 929
1.347	0.97506 19479 99092 43832 603	0.22193 28672 46415 65290 729
1.348	0.97528 33932 98416 67265 423	0.22095 76944 94502 50100 463
1.349	0.97550 38633 14428 88217 916	0.21998 23007 84913 26774 007
1.350	0.97572 33578 26659 06926 111	0.21900 66870 93041 58142 002
	$\begin{bmatrix} (-7)1 \\ 7 \end{bmatrix}$	$\begin{bmatrix} (-8)3 \\ 7 \end{bmatrix}$

CIRCULAR SINES AND COSINES FOR RADIAN ARGUMENTS   Table 4.6

x	$\sin x$	$\cos x$
1.350	0.97572 33578 26659 06926 111	0.21900 66870 93041 58142 002
1.351	0.97594 18766 15612 73996 110	0.21803 08543 94501 05261 504
1.352	0.97615 94194 62771 12353 536	0.21705 48036 65124 29854 627
1.353	0.97637 59861 50591 39095 407	0.21607 85358 80961 96725 291
1.354	0.97659 15764 62506 87244 418	0.21510 20520 18281 76154 163
1.355	0.97680 61901 82927 27405 609	0.21412 53530 53567 46271 899
1.356	0.97701 98270 97238 89325 386	0.21314 84399 63517 95410 772
1.357	0.97723 24869 91804 83352 894	0.21217 13137 25046 24434 790
1.358	0.97744 41696 53965 21803 706	0.21119 39753 15278 49048 406
1.359	0.97765 48748 72037 40225 805	0.21021 64257 11553 02083 908
1.360	0.97786 46024 35316 18567 849	0.20923 86658 91419 35767 598
1.361	0.97807 33521 34074 02249 690	0.20826 06968 32637 23964 842
1.362	0.97828 11237 59561 23135 125	0.20728 25195 13175 64404 112
1.363	0.97848 79171 04006 20406 864	0.20630 41349 11211 80880 089
1.364	0.97869 37319 60615 61343 685	0.20532 55440 05130 25435 952
1.365	0.97889 85681 23574 61999 774	0.20434 67477 73521 80524 932
1.366	0.97910 24253 88047 07786 196	0.20336 77471 95182 61151 240
1.367	0.97930 53035 50175 73954 516	0.20238 85432 49113 16990 457
1.368	0.97950 72024 07082 45982 521	0.20140 91369 14517 34489 495
1.369	0.97970 81217 56868 39862 027	0.20042 95291 70801 38946 217
1.370	0.97990 80613 98614 22288 769	0.19944 97209 97572 96568 820
1.371	0.98010 70211 32380 30754 328	0.19846 97133 74640 16515 079
1.372	0.98030 50007 59206 93540 094	0.19748 95072 82010 52911 545
1.373	0.98050 20000 81114 49613 233	0.19650 91036 99890 06852 798
1.374	0.98069 80189 01103 68424 652	0.19552 85036 08682 28380 853
1.375	0.98089 30570 23155 69608 920	0.19454 77079 88987 18444 822
1.376	0.98108 71142 52232 42586 155	0.19356 67178 21600 30840 918
1.377	0.98128 01903 94276 66065 826	0.19258 55340 87511 74132 912
1.378	0.98147 22852 56212 27452 479	0.19160 41577 67905 13553 129
1.379	0.98166 33986 45944 42153 343	0.19062 25898 44156 72884 094
1.380	0.98185 35303 72359 72787 813	0.18964 08312 97834 36320 915
1.381	0.98204 26802 45326 48298 791	0.18865 88831 10696 50314 508
1.382	0.98223 08480 75694 82965 850	0.18767 67462 64691 25395 757
1.383	0.98241 80336 75296 95320 221	0.18669 44217 41955 37980 715
1.384	0.98260 42368 56947 26961 571	0.18571 19105 24813 32156 930
1.385	0.98278 94574 34442 61276 561	0.18472 92135 95776 21451 016
1.386	0.98297 36952 22562 42059 162	0.18374 63319 37540 90577 542
1.387	0.98315 69500 37068 92032 708	0.18276 32665 32988 97169 360
1.388	0.98333 92216 94707 31273 673	0.18178 00183 65185 73489 451
1.389	0.98352 05100 13205 95537 148	0.18079 65884 17379 28124 404
1.390	0.98370 08148 11276 54484 004	0.17981 29776 72999 47659 616
1.391	0.98388 01359 08614 29809 722	0.17882 91871 15656 98336 311
1.392	0.98405 84731 25898 13274 870	0.17784 52177 29142 27690 484
1.393	0.98423 58262 84790 84637 207	0.17686 10704 97424 66173 860
1.394	0.98441 21952 07939 29485 405	0.17587 67464 04651 28756 976
1.395	0.98458 75797 18974 56974 360	0.17489 22464 35146 16514 467
1.396	0.98476 19796 42512 17462 083	0.17390 75715 73409 18192 681
1.397	0.98493 53948 04152 20048 145	0.17292 27228 04115 11759 690
1.398	0.98510 78250 30479 50013 670	0.17193 77011 12112 65937 830
1.399	0.98527 92701 49063 86162 846	0.17095 25074 82423 41718 833
1.400	0.98544 97299 88460 18065 947	0.16996 71429 00240 93861 675

$$\begin{bmatrix} (-7)1 \\ 7 \end{bmatrix} \qquad\qquad\qquad \begin{bmatrix} (-8)3 \\ 7 \end{bmatrix}$$

Table 4.6   CIRCULAR SINES AND COSINES FOR RADIAN ARGUMENTS

x	$\sin x$	$\cos x$
1.400	0.98544 97299 88460 18065 947	0.16996 71429 00240 93861 675
1.401	0.98561 92043 78208 63203 840	0.16898 16083 50929 72373 233
1.402	0.98578 76931 48834 84013 966	0.16799 59048 20024 23971 842
1.403	0.98595 51961 31850 04837 776	0.16701 00332 93227 93533 854
1.404	0.98612 17131 59751 28769 609	0.16602 39947 56412 25523 303
1.405	0.98628 72440 66021 54406 982	0.16503 77901 95615 65404 770
1.406	0.98645 17886 85129 92502 294	0.16405 14205 97042 61039 544
1.407	0.98661 53468 52531 82515 912	0.16306 48869 47062 64065 184
1.408	0.98677 79184 04669 09070 631	0.16207 81902 32209 31258 571
1.409	0.98693 95031 78970 18307 486	0.16109 13314 39179 25882 568
1.410	0.98710 01010 13850 34142 909	0.16010 43115 54831 19016 356
1.411	0.98725 97117 48711 74427 198	0.15911 71315 66184 90869 577
1.412	0.98741 83352 23943 67004 304	0.15812 97924 60420 32080 359
1.413	0.98757 59712 80922 65672 895	0.15714 22952 24876 44997 336
1.414	0.98773 26197 62012 66048 706	0.15615 46408 47050 44945 751
1.415	0.98788 82805 10565 21328 142	0.15516 68303 14596 61477 752
1.416	0.98804 29533 70919 57953 120	0.15417 88646 15325 39606 967
1.417	0.98819 66381 88402 91177 144	0.15319 07447 37202 41027 471
1.418	0.98834 93348 09330 40532 586	0.15220 24716 68347 45317 231
1.419	0.98850 10430 81005 45199 170	0.15121 40463 97033 51126 135
1.420	0.98865 17628 51719 79273 627	0.15022 54699 11685 77348 698
1.421	0.98880 14939 70753 66940 521	0.14923 67432 00880 64281 559
1.422	0.98895 02362 88375 97544 222	0.14824 78672 53344 74765 840
1.423	0.98909 79896 55844 40562 021	0.14725 88430 57953 95314 499
1.424	0.98924 47539 25405 60478 351	0.14626 96716 03732 37224 747
1.425	0.98939 05289 50295 31560 129	0.14528 03538 79851 37675 648
1.426	0.98953 53145 84738 52533 174	0.14429 08908 75628 60810 986
1.427	0.98967 91106 83949 61159 714	0.14330 12835 80526 98807 514
1.428	0.98982 19171 04132 48716 941	0.14231 15329 84153 72928 666
1.429	0.98996 37337 02480 74376 619	0.14132 16400 76259 34563 848
1.430	0.99010 45603 37177 79485 729	0.14033 16058 46736 66253 390
1.431	0.99024 43968 67397 01748 121	0.13934 14312 85619 82699 275
1.432	0.99038 32431 53301 89307 176	0.13835 11173 83083 31761 733
1.433	0.99052 10990 56046 14729 460	0.13736 06651 29440 95441 799
1.434	0.99065 79644 37773 88889 346	0.13637 00755 15144 90849 940
1.435	0.99079 38391 61619 74754 605	0.13537 93495 30784 71160 849
1.436	0.99092 87230 91709 01072 941	0.13438 84881 67086 26554 495
1.437	0.99106 26160 93157 75959 459	0.13339 74924 14910 85143 546
1.438	0.99119 55180 32073 00385 060	0.13240 63632 65254 13887 244
1.439	0.99132 74287 75552 81565 735	0.13141 51017 09245 19491 852
1.440	0.99145 83481 91686 46252 760	0.13042 37087 38145 49297 752
1.441	0.99158 82761 49554 53923 766	0.12943 21853 43347 92153 306
1.442	0.99171 72125 19229 09874 676	0.12844 05325 16375 79275 576
1.443	0.99184 51571 71773 78212 505	0.12744 87512 48881 85098 002
1.444	0.99197 21099 79243 94748 990	0.12645 68425 32647 28105 135
1.445	0.99209 80708 14686 79795 055	0.12546 48073 59580 71654 525
1.446	0.99222 30395 52141 50856 088	0.12447 26467 21717 24785 871
1.447	0.99234 70160 66639 35228 024	0.12348 03616 11217 43017 513
1.448	0.99247 00002 34203 82494 216	0.12248 79530 20366 29130 391
1.449	0.99259 19919 31850 76923 086	0.12149 54219 41572 33939 548
1.450	0.99271 29910 37588 49766 535	0.12050 27693 67366 57053 287
	$\begin{bmatrix} (-7)1 \\ 7 \end{bmatrix}$	$\begin{bmatrix} (-8)2 \\ 7 \end{bmatrix}$

CIRCULAR SINES AND COSINES FOR RADIAN ARGUMENTS   Table 4.6

x	$\sin x$	$\cos x$
1.450	0.99271 29910 37588 49766 535	0.12050 27693 67366 57053 287
1.451	0.99283 29974 30417 91459 118	0.11950 99962 90401 47620 080
1.452	0.99295 20109 90332 63717 946	0.11851 71037 03450 05063 327
1.453	0.99307 00315 98319 11543 325	0.11752 40925 99404 79804 068
1.454	0.99318 70591 36356 75120 114	0.11653 09639 71276 73971 735
1.455	0.99330 30934 87418 01619 777	0.11553 77188 12194 42103 061
1.456	0.99341 81345 35468 56903 143	0.11454 43581 15402 91829 237
1.457	0.99353 21821 65467 37123 830	0.11355 08828 74262 84551 407
1.458	0.99364 52362 63366 80232 355	0.11255 72940 82249 36104 618
1.459	0.99375 72967 16112 77380 893	0.11156 35927 32951 17410 313
1.460	0.99386 83634 11644 84228 683	0.11056 97798 20069 55117 465
1.461	0.99397 84362 38896 32148 075	0.10957 58563 37417 32232 463
1.462	0.99408 75150 87794 39331 194	0.10858 18232 78917 88737 835
1.463	0.99419 55998 49260 21797 223	0.10758 76816 38604 22199 915
1.464	0.99430 26904 15209 04300 286	0.10659 34324 10617 88365 556
1.465	0.99440 87866 78550 31137 923	0.10559 90765 89208 01747 983
1.466	0.99451 38885 33187 76860 141	0.10460 46151 68730 36201 884
1.467	0.99461 79958 74019 56879 043	0.10361 00491 43646 25487 846
1.468	0.99472 11085 96938 37979 012	0.10261 53795 08521 63826 230
1.469	0.99482 32265 98831 48727 437	0.10162 06072 58026 06440 584
1.470	0.99492 43497 77580 89785 993	0.10062 57333 86931 70090 698
1.471	0.99502 44780 32063 44122 430	0.09963 07588 90112 33595 391
1.472	0.99512 36112 62150 87122 898	0.09863 56847 62542 38345 147
1.473	0.99522 17493 68709 96604 762	0.09764 05119 99295 88804 678
1.474	0.99531 88922 53602 62729 932	0.09664 52415 95545 53005 525
1.475	0.99541 50398 19685 97818 664	0.09564 98745 46561 63028 806
1.476	0.99551 01919 70812 46063 854	0.09465 44118 47711 15478 186
1.477	0.99560 43486 11829 93145 787	0.09365 88544 94456 71943 189
1.478	0.99569 75096 48581 75747 356	0.09266 32034 82355 59452 948
1.479	0.99578 96749 87906 90969 720	0.09166 74598 07058 70920 484
1.480	0.99588 08445 37640 05648 408	0.09067 16244 64309 65577 623
1.481	0.99597 10182 06611 65569 851	0.08967 56984 49943 69400 641
1.482	0.99606 01959 04648 04588 337	0.08867 96827 59886 75526 752
1.483	0.99614 83775 42571 53643 374	0.08768 35783 90154 44661 519
1.484	0.99623 55630 32200 49677 461	0.08668 73863 36851 05477 303
1.485	0.99632 17522 86349 44454 246	0.08569 11075 96168 55002 845
1.486	0.99640 69452 18829 13277 079	0.08469 47431 64385 59004 070
1.487	0.99649 11417 44446 43607 933	0.08369 82940 37866 52356 240
1.488	0.99657 43417 79005 43586 693	0.08270 17612 13060 39407 518
1.489	0.99665 65452 39305 50450 815	0.08170 51456 86499 94334 076
1.490	0.99673 77520 43143 38855 320	0.08070 84484 54800 61486 832
1.491	0.99681 79621 09312 29093 143	0.07971 16705 14659 55729 907
1.492	0.99689 71753 57602 15215 811	0.07871 48128 62854 62770 926
1.493	0.99697 53917 08799 73054 448	0.07771 78764 96243 39483 234
1.494	0.99705 26110 84688 68141 099	0.07672 08624 11762 14220 152
1.495	0.99712 88334 08049 63530 364	0.07572 37716 06424 87121 354
1.496	0.99720 40586 02660 27521 334	0.07472 66050 77322 30411 478
1.497	0.99727 82865 93295 41279 821	0.07372 93638 21620 88691 060
1.498	0.99735 15173 05727 06360 877	0.07273 20488 36561 79219 898
1.499	0.99742 37506 66724 52131 595	0.07173 46611 19459 92192 943
1.500	0.99749 49866 04054 43094 172	0.07073 72016 67702 91008 819

$$\begin{bmatrix} (-7)1 \\ 7 \end{bmatrix} \qquad\qquad\qquad \begin{bmatrix} (-8)2 \\ 7 \end{bmatrix}$$

Table 4.6   CIRCULAR SINES AND COSINES FOR RADIAN ARGUMENTS

x	$\sin x$	$\cos x$
1.500	0.99749 49866 04054 43094 172	0.07073 72016 67702 91008 819
1.501	0.99756 52250 46480 86109 251	0.06973 96714 78750 12531 065
1.502	0.99763 44659 23765 37519 509	0.06874 20715 50131 67342 208
1.503	0.99770 27091 66667 10173 501	0.06774 44028 79447 39990 761
1.504	0.99776 99547 06942 80349 750	0.06674 66664 64365 89231 245
1.505	0.99783 62024 77346 94581 063	0.06574 88633 02623 48257 343
1.506	0.99790 14524 11631 76379 092	0.06475 09943 92023 24928 268
1.507	0.99796 57044 44547 32859 104	0.06375 30607 30434 01988 470
1.508	0.99802 89585 11841 61264 976	0.06275 50633 15789 37280 758
1.509	0.99809 12145 50260 55394 397	0.06175 70031 46086 63952 953
1.510	0.99815 24724 97548 11924 274	0.06075 88812 19385 90658 160
1.511	0.99821 27322 92446 36636 332	0.05976 06985 33809 01748 769
1.512	0.99827 19938 74695 50542 912	0.05876 24560 87538 57464 281
1.513	0.99833 02571 85033 95912 947	0.05776 41548 78816 94113 053
1.514	0.99838 75221 65198 42198 118	0.05676 57959 05945 24248 072
1.515	0.99844 37887 57923 91859 188	0.05576 73801 67282 36836 851
1.516	0.99849 90569 06943 86092 495	0.05476 89086 61243 97425 545
1.517	0.99855 33265 56990 10456 612	0.05377 03823 86301 48297 399
1.518	0.99860 65976 53793 00399 163	0.05277 18023 40981 08625 609
1.519	0.99865 88701 44081 46683 784	0.05177 31695 23862 74620 716
1.520	0.99871 01439 75583 00717 231	0.05077 44849 33579 19672 613
1.521	0.99876 04190 97023 79776 634	0.04977 57495 68814 94487 284
1.522	0.99880 96954 58128 72136 872	0.04877 69644 28305 27218 360
1.523	0.99885 79730 09621 42098 089	0.04777 81305 10835 23593 598
1.524	0.99890 52517 03224 34913 328	0.04677 92488 15238 67036 388
1.525	0.99895 15314 91658 81616 285	0.04578 03203 40397 18782 371
1.526	0.99899 68123 28645 03749 180	0.04478 13460 85239 17991 291
1.527	0.99904 10941 68902 17990 729	0.04378 23270 48738 81854 166
1.528	0.99908 43769 68148 40684 234	0.04278 32642 29915 05695 871
1.529	0.99912 66606 83100 92265 762	0.04178 41586 27830 63073 262
1.530	0.99916 79452 71476 01592 427	0.04078 50112 41591 05868 899
1.531	0.99920 82306 91989 10170 755	0.03978 58230 70343 64380 513
1.532	0.99924 75169 04354 76285 152	0.03878 65951 13276 47406 277
1.533	0.99928 58038 69286 79026 436	0.03778 73283 69617 42326 008
1.534	0.99932 30915 48498 22220 463	0.03678 80238 38633 15178 390
1.535	0.99935 93799 04701 38256 819	0.03578 86825 19628 10734 312
1.536	0.99939 46689 01607 91817 592	0.03478 93054 11943 52566 435
1.537	0.99942 89585 03928 83506 202	0.03378 98935 14956 43115 073
1.538	0.99946 22486 77374 53376 306	0.03279 04478 28078 63750 505
1.539	0.99949 45393 88654 84360 752	0.03179 09693 50755 74831 796
1.540	0.99952 58306 05479 05600 596	0.03079 14590 82466 15762 248
1.541	0.99955 61222 96555 95674 180	0.02979 19180 22720 05041 568
1.542	0.99958 54144 31593 85726 242	0.02879 23471 71058 40314 858
1.543	0.99961 37069 81300 62497 095	0.02779 27475 27051 98418 526
1.544	0.99964 09999 17383 71251 832	0.02679 31200 90300 35423 217
1.545	0.99966 72932 12550 18609 586	0.02579 34658 60430 86673 867
1.546	0.99969 25868 40506 75272 821	0.02479 37858 37097 66826 971
1.547	0.99971 68807 75959 78656 660	0.02379 40810 19980 69885 184
1.548	0.99974 01749 94615 35418 249	0.02279 43524 08784 69229 328
1.549	0.99976 24694 73179 23886 150	0.02179 46010 03238 17647 934
1.550	0.99978 37641 89356 96389 761	0.02079 48278 03092 47364 391

$$\begin{bmatrix} (-7)1 \\ 7 \end{bmatrix} \qquad\qquad\qquad \begin{bmatrix} (-9)9 \\ 7 \end{bmatrix}$$

CIRCULAR SINES AND COSINES FOR RADIAN ARGUMENTS Table 4.6

x	$\sin x$	$\cos x$
1.550	0.99978 37641 89356 96389 761	0.02079 48278 03092 47364 391
1.551	0.99980 40591 21853 81488 767	0.01979 50338 08120 70061 827
1.552	0.99982 33542 50374 86102 606	0.01879 52200 18116 76905 802
1.553	0.99984 16495 55624 97539 966	0.01779 53874 32894 38564 929
1.554	0.99985 89450 19308 85428 298	0.01679 55370 52286 05229 507
1.555	0.99987 52406 24131 03543 342	0.01579 56698 76142 06628 284
1.556	0.99989 05363 53795 91538 676	0.01479 57869 04329 52043 433
1.557	0.99990 48321 93007 76575 277	0.01379 58891 36731 30323 849
1.558	0.99991 81281 27470 74851 093	0.01279 59775 73245 09896 874
1.559	0.99993 04241 43888 93030 623	0.01179 60532 13782 38778 533
1.560	0.99994 17202 29966 29574 517	0.01079 61170 58267 44582 392
1.561	0.99995 20163 74406 75969 172	0.00979 61701 06636 34527 146
1.562	0.99996 13125 66914 17856 344	0.00879 62133 58835 95443 014
1.563	0.99996 96087 98192 36062 758	0.00779 62478 14822 93777 062
1.564	0.99997 69050 59945 07529 731	0.00679 62744 74562 75597 546
1.565	0.99998 32013 44876 06142 794	0.00579 62943 38028 66597 372
1.566	0.99998 84976 46689 03461 318	0.00479 63084 05200 72096 784
1.567	0.99999 27939 60087 69348 142	0.00379 63176 76064 77045 359
1.568	0.99999 60902 80775 72499 201	0.00279 63231 50611 46023 436
1.569	0.99999 83866 05456 80873 162	0.00179 63258 28835 23243 059
1.570	0.99999 96829 31834 62021 053	+0.00079 63267 10733 32548 541
1.571	0.99999 99792 58612 83315 895	−0.00020 36732 03695 22583 254
1.572	0.99999 92755 85495 12082 337	−0.00120 36729 14450 59042 804
1.573	0.99999 75719 13185 15626 285	−0.00220 36714 21533 14087 901
1.574	0.99999 48682 43386 61164 539	−0.00320 36677 24944 45343 613
1.575	0.99999 11645 78803 15654 423	−0.00420 36608 24688 30802 109
1.576	0.99998 64609 23138 45523 419	−0.00520 36497 20771 68822 280
1.577	0.99998 07572 81096 16298 798	−0.00620 36334 13205 78129 029
1.578	0.99997 40536 58379 92137 261	−0.00720 36109 02006 97812 142
1.579	0.99996 63500 61693 35254 568	−0.00820 35811 87197 87324 647
1.580	0.99995 76464 98740 05255 179	−0.00920 35432 68808 26480 539
1.581	0.99994 79429 78223 58361 895	−0.01020 34961 46876 15451 796
1.582	0.99993 72395 09847 46545 499	−0.01120 34388 21448 74764 568
1.583	0.99992 55361 04315 16554 408	−0.01220 33702 92583 45294 454
1.584	0.99991 28327 73330 08844 324	−0.01320 32895 60348 88260 743
1.585	0.99989 91295 29595 56407 893	−0.01420 31956 24825 85219 553
1.586	0.99988 44263 86814 83504 374	−0.01520 30874 86108 38055 737
1.587	0.99986 87233 59691 04289 313	−0.01620 29641 44304 68973 475
1.588	0.99985 20204 63927 21344 232	−0.01720 28245 99538 20485 440
1.589	0.99983 43177 16226 24106 322	−0.01820 26678 51948 55400 452
1.590	0.99981 56151 34290 87198 158	−0.01920 24929 01692 56809 503
1.591	0.99979 59127 36823 68657 422	−0.02020 22987 48945 28070 065
1.592	0.99977 52105 43527 08066 646	−0.02120 20843 93900 92788 583
1.593	0.99975 35085 75103 24582 972	−0.02220 18488 36773 94801 039
1.594	0.99973 08068 53254 14867 933	−0.02320 15910 77799 98151 502
1.595	0.99970 71054 00681 50917 259	−0.02420 13101 17236 87068 552
1.596	0.99968 24042 41086 77790 702	−0.02520 10049 55365 65939 492
1.597	0.99965 67033 99171 11241 891	−0.02620 06745 92491 59282 234
1.598	0.99963 00029 00635 35248 219	−0.02720 03180 28945 11714 764
1.599	0.99960 23027 72179 99440 759	−0.02819 99342 65082 87922 093
1.600	0.99957 36030 41505 16434 211	−0.02919 95223 01288 72620 577
	$\begin{bmatrix}(-7)1 \\ 7\end{bmatrix}$	$\begin{bmatrix}(-9)3 \\ 7\end{bmatrix}$

For $x>1.6$ see **Example 16**.

$\dfrac{\pi}{2}$=1.57079 63267 94896 61923 132 π=3.14159 26535 89793 23846 264

Table 4.7   RADIX TABLE OF CIRCULAR SINES AND COSINES

x	n	$\sin x 10^{-n}$	$\cos x 10^{-n}$
1	10	0. 00000 00001 00000 00000 00000	0. 99999 99999 99999 99999 50000
2	10	0. 00000 00002 00000 00000 00000	0. 99999 99999 99999 99998 00000
3	10	0. 00000 00003 00000 00000 00000	0. 99999 99999 99999 99995 50000
4	10	0. 00000 00004 00000 00000 00000	0. 99999 99999 99999 99992 00000
5	10	0. 00000 00005 00000 00000 00000	0. 99999 99999 99999 99987 50000
6	10	0. 00000 00006 00000 00000 00000	0. 99999 99999 99999 99982 00000
7	10	0. 00000 00007 00000 00000 00000	0. 99999 99999 99999 99975 50000
8	10	0. 00000 00008 00000 00000 00000	0. 99999 99999 99999 99968 00000
9	10	0. 00000 00009 00000 00000 00000	0. 99999 99999 99999 99959 50000
1	9	0. 00000 00010 00000 00000 00000	0. 99999 99999 99999 99950 00000
2	9	0. 00000 00020 00000 00000 00000	0. 99999 99999 99999 99800 00000
3	9	0. 00000 00030 00000 00000 00000	0. 99999 99999 99999 99550 00000
4	9	0. 00000 00040 00000 00000 00000	0. 99999 99999 99999 99200 00000
5	9	0. 00000 00050 00000 00000 00000	0. 99999 99999 99999 98750 00000
6	9	0. 00000 00060 00000 00000 00000	0. 99999 99999 99999 98200 00000
7	9	0. 00000 00069 99999 99999 99999	0. 99999 99999 99999 97550 00000
8	9	0. 00000 00079 99999 99999 99999	0. 99999 99999 99999 96800 00000
9	9	0. 00000 00089 99999 99999 99999	0. 99999 99999 99999 95950 00000
1	8	0. 00000 00099 99999 99999 99998	0. 99999 99999 99999 95000 00000
2	8	0. 00000 00199 99999 99999 99987	0. 99999 99999 99999 80000 00000
3	8	0. 00000 00299 99999 99999 99955	0. 99999 99999 99999 55000 00000
4	8	0. 00000 00399 99999 99999 99893	0. 99999 99999 99999 20000 00000
5	8	0. 00000 00499 99999 99999 99792	0. 99999 99999 99998 75000 00000
6	8	0. 00000 00599 99999 99999 99640	0. 99999 99999 99998 20000 00000
7	8	0. 00000 00699 99999 99999 99428	0. 99999 99999 99997 55000 00000
8	8	0. 00000 00799 99999 99999 99147	0. 99999 99999 99996 80000 00000
9	8	0. 00000 00899 99999 99999 98785	0. 99999 99999 99995 95000 00000
1	7	0. 00000 00999 99999 99999 98333	0. 99999 99999 99995 00000 00000
2	7	0. 00000 01999 99999 99999 86667	0. 99999 99999 99980 00000 00000
3	7	0. 00000 02999 99999 99999 55000	0. 99999 99999 99955 00000 00000
4	7	0. 00000 03999 99999 99998 93333	0. 99999 99999 99920 00000 00000
5	7	0. 00000 04999 99999 99997 91667	0. 99999 99999 99875 00000 00000
6	7	0. 00000 05999 99999 99996 40000	0. 99999 99999 99820 00000 00000
7	7	0. 00000 06999 99999 99994 28333	0. 99999 99999 99755 00000 00000
8	7	0. 00000 07999 99999 99991 46667	0. 99999 99999 99680 00000 00000
9	7	0. 00000 08999 99999 99987 85000	0. 99999 99999 99595 00000 00000
1	6	0. 00000 09999 99999 99983 33333	0. 99999 99999 99500 00000 00000
2	6	0. 00000 19999 99999 99866 66667	0. 99999 99999 98000 00000 00007
3	6	0. 00000 29999 99999 99550 00000	0. 99999 99999 95500 00000 00034
4	6	0. 00000 39999 99999 98933 33333	0. 99999 99999 92000 00000 00107
5	6	0. 00000 49999 99999 97916 66667	0. 99999 99999 87500 00000 00260
6	6	0. 00000 59999 99999 96400 00000	0. 99999 99999 82000 00000 00540
7	6	0. 00000 69999 99999 94283 33333	0. 99999 99999 75500 00000 01000
8	6	0. 00000 79999 99999 91466 66667	0. 99999 99999 68000 00000 01707
9	6	0. 00000 89999 99999 87850 00000	0. 99999 99999 59500 00000 02734
1	5	0. 00000 99999 99999 83333 33333	0. 99999 99999 50000 00000 04167
2	5	0. 00001 99999 99998 66666 66667	0. 99999 99998 00000 00000 66667
3	5	0. 00002 99999 99995 50000 00002	0. 99999 99995 50000 00003 37500
4	5	0. 00003 99999 99989 33333 33342	0. 99999 99992 00000 00010 66667
5	5	0. 00004 99999 99979 16666 66693	0. 99999 99987 50000 00026 04167
6	5	0. 00005 99999 99964 00000 00065	0. 99999 99982 00000 00054 00000
7	5	0. 00006 99999 99942 83333 33473	0. 99999 99975 50000 00100 04167
8	5	0. 00007 99999 99914 66666 66940	0. 99999 99968 00000 00170 66667
9	5	0. 00008 99999 99878 50000 00492	0. 99999 99959 50000 00273 37500
1	4	0. 00009 99999 99833 33333 34167	0. 99999 99950 00000 00416 66667
2	4	0. 00019 99999 98666 66666 93333	0. 99999 99800 00000 06666 66666
3	4	0. 00029 99999 95500 00002 02500	0. 99999 99550 00000 33749 99990
4	4	0. 00039 99999 89333 33341 86667	0. 99999 99200 00001 06666 66610
5	4	0. 00049 99999 79166 66692 70833	0. 99999 98750 00002 60416 66450
6	4	0. 00059 99999 64000 00064 80000	0. 99999 98200 00005 39999 99352
7	4	0. 00069 99999 42833 33473 39167	0. 99999 97550 00010 04166 65033
8	4	0. 00079 99999 14666 66939 73333	0. 99999 96800 00017 06666 63026
9	4	0. 00089 99998 78500 00492 07499	0. 99999 95950 00027 33749 92619
1	3	0. 00099 99998 33333 34166 66665	0. 99999 95000 00041 66666 52778

For $n > 10$, $\sin x 10^{-n} = x 10^{-n}$; $\cos x 10^{-n} = 1 - \frac{1}{2} x^2 10^{-2n}$; to 25D.

From C. E. Van Orstrand, Tables of the exponential function and of the circular sine and cosine to radian arguments, Memoirs of the National Academy of Sciences, vol. 14, Fifth Memoir. U.S. Government Printing Office, Washington, D.C., 1921 (with permission).

CIRCULAR SINES AND COSINES FOR LARGE RADIAN ARGUMENTS   Table 4.8

x	$\sin x$	$\cos x$
0	0. 00000 00000 00000 00000 000	1. 00000 00000 00000 00000 000
1	+0. 84147 09848 07896 50665 250	+0. 54030 23058 68139 71740 094
2	+0. 90929 74268 25681 69539 602	−0. 41614 68365 47142 38699 757
3	+0. 14112 00080 59867 22210 074	−0. 98999 24966 00445 45727 157
4	−0. 75680 24953 07928 25137 264	−0. 65364 36208 63611 91463 917
5	−0. 95892 42746 63138 46889 315	+0. 28366 21854 63226 26446 664
6	−0. 27941 54981 98925 87281 156	+0. 96017 02866 50366 02054 565
7	+0. 65698 65987 18789 09039 700	+0. 75390 22543 43304 63814 120
8	+0. 98935 82466 23381 77780 812	−0. 14550 00338 08613 52586 884
9	+0. 41211 84852 41756 56975 627	−0. 91113 02618 84676 98836 829
10	−0. 54402 11108 89369 81340 475	−0. 83907 15290 76452 45225 886
11	−0. 99999 02065 50703 45705 156	+0. 00442 56979 88050 78574 836
12	−0. 53657 29180 00434 97166 537	+0. 84385 39587 32492 10465 396
13	+0. 42016 70368 26640 92186 896	+0. 90744 67814 50196 21385 269
14	+0. 99060 73556 94870 30787 535	+0. 13673 72182 07833 59424 893
15	+0. 65028 78401 57116 86582 974	−0. 75968 79128 58821 27384 815
16	−0. 28790 33166 65065 29478 446	−0. 95765 94803 23384 64189 964
17	−0. 96139 74918 79556 85726 164	−0. 27516 33380 51596 92222 034
18	−0. 75098 72467 71676 10375 016	+0. 66031 67082 44080 14481 610
19	+0. 14987 72096 62952 32975 424	+0. 98870 46181 86669 25289 835
20	+0. 91294 52507 27627 65437 610	+0. 40808 20618 13391 98606 227
21	+0. 83665 56385 36056 03186 648	−0. 54772 92602 24268 42138 427
22	−0. 00885 13092 90403 87592 169	−0. 99996 08263 94637 12645 417
23	−0. 84622 04041 75170 63524 133	−0. 53283 30203 33397 55521 576
24	−0. 90557 83620 06623 84513 579	+0. 42417 90073 36996 97593 705
25	−0. 13235 17500 97773 02890 201	+0. 99120 28118 63473 59808 329
26	+0. 76255 84504 79602 73751 582	+0. 64691 93223 28640 34272 138
27	+0. 95637 59284 04503 01343 234	−0. 29213 88087 33836 19337 140
28	+0. 27090 57883 07869 01998 634	−0. 96260 58663 13566 60197 545
29	−0. 66363 38842 12967 50215 117	−0. 74805 75296 89000 35176 519
30	−0. 98803 16240 92861 78998 775	+0. 15425 14498 87584 05071 866
31	−0. 40403 76453 23065 00604 877	+0. 91474 23578 04531 27896 244
32	+0. 55142 66812 41690 55066 156	+0. 83422 33605 06510 27221 553
33	+0. 99991 18601 07267 14572 808	−0. 01327 67472 23059 47891 522
34	+0. 52908 26861 20023 82083 249	−0. 84857 02747 84605 18659 997
35	−0. 42818 26694 96151 00440 675	−0. 90369 22050 91506 75984 730
36	−0. 99177 88534 43115 73683 529	−0. 12796 36896 27404 68102 833
37	−0. 64353 81333 56999 46068 567	+0. 76541 40519 45343 35649 108
38	+0. 29636 85787 09385 31739 230	+0. 95507 36440 47294 85758 654
39	+0. 96379 53862 84087 75326 066	+0. 26664 29323 59937 25152 683
40	+0. 74511 31604 79348 78698 771	−0. 66693 80616 52261 84438 409
41	−0. 15862 26688 04708 98710 332	−0. 98733 92775 23826 45822 883
42	−0. 91652 15479 15633 78589 899	−0. 39998 53149 88351 29395 471
43	−0. 83177 47426 28598 28820 958	+0. 55511 33015 20625 67704 483
44	+0. 01770 19251 05413 57780 795	+0. 99984 33086 47691 22006 901
45	+0. 85090 35245 34118 42486 238	+0. 52532 19888 17729 69604 746
46	+0. 90178 83476 48809 18503 329	−0. 43217 79448 84778 29495 278
47	+0. 12357 31227 45224 00406 153	−0. 99233 54691 50928 71827 975
48	−0. 76825 46613 23666 79904 497	−0. 64014 43394 69199 73131 294
49	−0. 95375 26527 59471 81836 042	+0. 30059 25437 43637 08368 703
50	−0. 26237 48537 03928 78591 439	+0. 96496 60284 92113 27406 896

From C. E. Van Orstrand, Tables of the exponential function and of the circular sine and cosine to radian arguments, Memoirs of the National Academy of Sciences, vol. 14, Fifth Memoir. U.S. Government Printing Office, Washington, D.C., 1921 (with permission) for $x \leq 100$.

Table 4.8   CIRCULAR SINES AND COSINES FOR LARGE RADIAN ARGUMENTS

x	$\sin x$	$\cos x$
50	−0.26237 48537 03928 78591 439	+0.96496 60284 92113 27406 896
51	+0.67022 91758 43374 73449 435	+0.74215 41968 13782 53946 738
52	+0.98662 75920 40485 29658 757	−0.16299 07807 95705 48100 333
53	+0.39592 51501 81834 18150 339	−0.91828 27862 12118 89119 973
54	−0.55878 90488 51616 24581 787	−0.82930 98328 63150 14772 785
55	−0.99975 51733 58619 83659 863	+0.02212 67562 61955 73456 356
56	−0.52155 10020 86911 88018 741	+0.85322 01077 22584 11396 968
57	+0.43616 47552 47824 95908 053	+0.89986 68269 69193 78650 300
58	+0.99287 26480 84537 11816 509	+0.11918 01354 48819 28543 584
59	+0.63673 80071 39137 88077 123	−0.77108 02229 75845 22938 744
60	−0.30481 06211 02216 70562 565	−0.95241 29804 15156 29269 382
61	−0.96611 77700 08392 94701 829	−0.25810 16359 38267 44570 121
62	−0.73918 06966 49222 86727 602	+0.67350 71623 23586 25288 783
63	+0.16735 57003 02806 92152 784	+0.98589 65815 82549 69743 864
64	+0.92002 60381 96790 68335 154	+0.39185 72304 29550 00516 171
65	+0.82682 86794 90103 46771 021	−0.56245 38512 38172 03106 212
66	−0.02655 11540 23966 79446 384	−0.99964 74559 66349 96483 045
67	−0.85551 99789 75322 25899 683	−0.51776 97997 89505 06565 339
68	−0.89792 76806 89291 26040 073	+0.44014 30224 96040 70593 105
69	−0.11478 48137 83187 22054 507	+0.99339 03797 22271 63756 155
70	+0.77389 06815 57889 09778 733	+0.63331 92030 86299 83233 201
71	+0.95105 46532 54374 63665 657	−0.30902 27281 66070 70291 749
72	+0.25382 33627 62036 27306 903	−0.96725 05882 73882 48729 171
73	−0.67677 19568 87307 62215 498	−0.73619 27182 27315 96016 815
74	−0.98514 62604 68247 37085 189	+0.17171 73418 30777 55609 845
75	−0.38778 16354 09430 43773 094	+0.92175 12697 24749 31639 230
76	+0.56610 76368 98180 32361 028	+0.82433 13311 07557 75991 501
77	+0.99952 01585 80731 24386 610	−0.03097 50317 31216 45752 196
78	+0.51397 84559 87535 21169 609	−0.85780 30932 44987 85540 835
79	−0.44411 26687 07508 36850 760	−0.89597 09467 90963 14833 703
80	−0.99388 86539 23375 18973 081	−0.11038 72438 39047 55811 787
81	−0.62988 79942 74453 87856 521	+0.77668 59820 21631 15768 342
82	+0.31322 87824 33085 15263 353	+0.94967 76978 82543 20471 326
83	+0.96836 44611 00185 40435 015	+0.24954 01179 73338 12437 735
84	+0.73319 03200 73292 16636 321	−0.68002 34955 87338 79542 720
85	−0.17607 56199 48587 07696 212	−0.98437 66433 94041 89491 821
86	−0.92345 84470 04059 80260 163	−0.38369 84449 49741 84477 893
87	−0.82181 78366 30822 54487 211	+0.56975 03342 65311 92000 851
88	+0.03539 83027 33660 68362 543	+0.99937 32836 95124 65698 442
89	+0.86006 94058 12453 22683 685	+0.51017 70449 41668 89902 379
90	+0.89399 66636 00557 89051 827	−0.44807 36161 29170 15236 548
91	+0.10598 75117 51156 85002 021	−0.99436 74609 28201 52610 672
92	−0.77946 60696 15804 68855 400	−0.62644 44479 10339 06880 027
93	−0.94828 21412 69947 23213 104	+0.31742 87015 19701 64974 551
94	−0.24525 19854 67654 32522 044	+0.96945 93666 69987 60380 439
95	+0.68326 17147 36120 98369 958	+0.73017 35609 94819 66479 352
96	+0.98358 77454 34344 85760 773	−0.18043 04492 91083 95011 850
97	+0.37960 77390 27521 69648 192	−0.92514 75365 96413 89170 475
98	−0.57338 18719 90422 88494 922	−0.81928 82452 91459 25267 566
99	−0.99920 68341 86353 69443 272	+0.03982 08803 93138 89816 180
100	−0.50636 56411 09758 79365 656	+0.86231 88722 87683 93410 194

CIRCULAR SINES AND COSINES FOR LARGE RADIAN ARGUMENTS   Table 4.8

x	$\sin x$	$\cos x$	x	$\sin x$	$\cos x$
100	−0.50636 564	+0.86231 887	150	−0.71487 643	+0.69925 081
101	+0.45202 579	+0.89200 487	151	+0.20214 988	+0.97935 460
102	+0.99482 679	+0.10158 570	152	+0.93332 052	+0.35904 429
103	+0.62298 863	−0.78223 089	153	+0.80640 058	−0.59136 968
104	−0.32162 240	−0.94686 801	154	−0.06192 034	−0.99808 109
105	−0.97053 528	−0.24095 905	155	−0.87331 198	−0.48716 135
106	−0.72714 250	+0.68648 655	156	−0.88178 462	+0.47165 229
107	+0.18478 174	+0.98277 958	157	−0.07954 854	+0.99683 099
108	+0.92681 851	+0.37550 960	158	+0.79582 410	+0.60552 787
109	+0.81674 261	−0.57700 218	159	+0.93951 973	−0.34249 478
110	−0.04424 268	−0.99902 081	160	+0.21942 526	−0.97562 931
111	−0.86455 145	−0.50254 432	161	−0.70240 779	−0.71177 476
112	−0.88999 560	+0.45596 910	162	−0.97845 035	−0.20648 223
113	−0.09718 191	+0.99526 664	163	−0.35491 018	+0.93490 040
114	+0.78498 039	+0.61952 061	164	+0.59493 278	+0.80377 546
115	+0.94543 533	−0.32580 981	165	+0.99779 728	−0.06633 694
116	+0.23666 139	−0.97159 219	166	+0.48329 156	−0.87545 946
117	−0.68969 794	−0.72409 720	167	−0.47555 019	−0.87968 859
118	−0.98195 217	+0.18912 942	168	−0.99717 329	−0.07513 609
119	−0.37140 410	+0.92847 132	169	−0.60199 987	+0.79849 619
120	+0.58061 118	+0.81418 097	170	+0.34664 946	+0.93799 475
121	+0.99881 522	−0.04866 361	171	+0.97659 087	+0.21510 527
122	+0.49871 315	−0.86676 709	172	+0.70865 914	−0.70555 101
123	−0.45990 349	−0.88796 891	173	−0.21081 053	−0.97752 694
124	−0.99568 699	−0.09277 620	174	−0.93646 197	−0.35076 911
125	−0.61604 046	+0.78771 451	175	−0.80113 460	+0.59848 422
126	+0.32999 083	+0.94398 414	176	+0.07075 224	+0.99749 392
127	+0.97263 007	+0.23235 910	177	+0.87758 979	+0.47941 231
128	+0.72103 771	−0.69289 582	178	+0.87757 534	−0.47943 877
129	−0.19347 339	−0.98110 552	179	+0.07072 217	−0.99749 605
130	−0.93010 595	−0.36729 133	180	−0.80115 264	−0.59846 007
131	−0.81160 339	+0.58420 882	181	−0.93645 140	+0.35079 734
132	+0.05308 359	+0.99859 007	182	−0.21078 107	+0.97753 329
133	+0.86896 576	+0.49487 222	183	+0.70868 041	+0.70552 964
134	+0.88592 482	−0.46382 887	184	+0.97658 438	−0.21513 471
135	+0.08836 869	−0.99608 784	185	+0.34662 118	−0.93800 520
136	−0.79043 321	−0.61254 824	186	−0.60202 394	−0.79847 804
137	−0.94251 445	+0.33416 538	187	−0.99717 102	+0.07516 615
138	−0.22805 226	+0.97364 889	188	−0.47552 367	+0.87970 293
139	+0.69608 013	+0.71796 410	189	+0.48331 795	+0.87544 489
140	+0.98023 966	−0.19781 357	190	+0.99779 928	+0.06630 686
141	+0.36317 137	−0.93172 236	191	+0.59490 855	−0.80379 339
142	−0.58779 501	−0.80900 991	192	−0.35493 836	−0.93488 971
143	−0.99834 536	+0.05750 253	193	−0.97845 657	−0.20645 273
144	−0.49102 159	+0.87114 740	194	−0.70238 633	+0.71179 593
145	+0.46774 516	+0.88386 337	195	+0.21945 467	+0.97562 270
146	+0.99646 917	+0.08395 944	196	+0.93953 006	+0.34246 646
147	+0.60904 402	−0.79313 642	197	+0.79580 584	−0.60555 186
148	−0.33833 339	−0.94102 631	198	−0.07957 859	−0.99682 859
149	−0.97464 865	−0.22374 095	199	−0.88179 884	−0.47162 571
150	−0.71487 643	+0.69925 081	200	−0.87329 730	+0.48718 768

Table 4.8   **CIRCULAR SINES AND COSINES FOR LARGE RADIAN ARGUMENTS**

x	$\sin x$	$\cos x$	x	$\sin x$	$\cos x$
200	−0.87329 730	+0.48718 768	250	−0.97052 802	+0.24098 831
201	−0.06189 025	+0.99808 296	251	−0.32159 386	+0.94687 771
202	+0.80641 841	+0.59134 538	252	+0.62301 221	+0.78221 211
203	+0.93330 970	−0.35907 242	253	+0.99482 373	−0.10161 569
204	+0.20212 036	−0.97936 069	254	+0.45199 890	−0.89201 850
205	−0.71489 751	−0.69922 926	255	−0.50639 163	−0.86230 361
206	−0.97464 190	+0.22377 033	256	−0.99920 803	−0.03979 076
207	−0.33830 503	+0.94103 651	257	−0.57335 717	+0.81930 553
208	+0.60906 793	+0.79311 806	258	+0.37963 563	+0.92513 609
209	+0.99646 664	−0.08398 947	259	+0.98359 318	+0.18040 080
210	+0.46771 852	−0.88387 747	260	+0.68323 970	−0.73019 416
211	−0.49104 785	−0.87113 260	261	−0.24528 121	−0.96945 197
212	−0.99834 709	−0.05747 243	262	−0.94829 171	−0.31740 012
213	−0.58777 062	+0.80902 763	263	−0.77944 719	+0.62646 794
214	+0.36319 945	+0.93171 141	264	+0.10601 749	+0.99436 427
215	+0.98024 562	+0.19778 403	265	+0.89401 017	+0.44804 667
216	+0.69605 849	−0.71798 508	266	+0.86005 403	−0.51020 297
217	−0.22808 161	−0.97364 202	267	+0.03536 818	−0.99937 435
218	−0.94252 453	−0.33413 697	268	−0.82183 501	−0.56972 556
219	−0.79041 474	+0.61257 207	269	−0.92344 688	+0.38372 628
220	+0.08839 871	+0.99608 517	270	−0.17604 595	+0.98438 195
221	+0.88593 880	+0.46380 216	271	+0.73321 082	+0.68000 139
222	+0.86895 084	−0.49489 841	272	+0.96835 694	−0.24956 931
223	+0.05305 349	−0.99859 167	273	+0.31320 015	−0.94968 714
224	−0.81162 100	−0.58418 435	274	−0.62991 141	−0.77666 699
225	−0.93009 488	+0.36731 937	275	−0.99388 533	+0.11041 720
226	−0.19344 382	+0.98111 135	276	−0.44408 566	+0.89598 433
227	+0.72105 860	+0.69287 409	277	+0.51400 431	+0.85778 760
228	+0.97262 306	−0.23238 842	278	+0.99952 109	+0.03094 490
229	+0.32996 237	−0.94399 409	279	+0.56608 279	−0.82434 840
230	−0.61606 420	−0.78769 594	280	−0.38780 942	−0.92173 958
231	−0.99568 419	+0.09280 622	281	−0.98515 144	−0.17168 765
232	−0.45987 672	+0.88798 277	282	−0.67674 976	+0.73621 312
233	+0.49873 928	+0.86675 206	283	+0.25385 252	+0.96724 294
234	+0.99881 669	+0.04863 350	284	+0.95106 397	+0.30899 406
235	+0.58058 664	−0.81419 847	285	+0.77387 159	−0.63334 253
236	−0.37143 209	−0.92846 012	286	−0.11481 476	−0.99338 692
237	−0.98195 787	−0.18909 982	287	−0.89794 095	−0.44011 595
238	−0.68967 611	+0.72411 799	288	−0.85550 437	+0.51779 559
239	+0.23669 068	+0.97158 506	289	−0.02652 102	+0.99964 826
240	+0.94544 515	+0.32578 131	290	+0.82684 563	+0.56242 893
241	+0.78496 171	−0.61954 428	291	+0.92001 423	−0.39188 496
242	−0.09721 191	−0.99526 371	292	+0.16732 598	−0.98590 163
243	−0.89000 935	−0.45594 228	293	−0.73920 100	−0.67348 488
244	−0.86453 630	+0.50257 038	294	−0.96610 999	+0.25813 076
245	−0.04421 256	+0.99902 215	295	−0.30478 191	+0.95242 217
246	+0.81676 000	+0.57697 756	296	+0.63676 125	+0.77106 103
247	+0.92680 719	−0.37553 754	297	+0.99286 906	−0.11921 006
248	+0.18475 212	−0.98278 515	298	+0.43613 763	−0.89987 997
249	−0.72716 319	−0.68646 463	299	−0.52157 672	−0.85320 439
250	−0.97052 802	+0.24098 831	300	−0.99975 584	−0.02209 662

CIRCULAR SINES AND COSINES FOR LARGE RADIAN ARGUMENTS   Table 4.8

x	$\sin x$	$\cos x$	x	$\sin x$	$\cos x$
300	−0.99975 584	−0.02209 662	350	−0.95893 283	−0.28363 328
301	−0.55876 405	+0.82932 668	351	−0.75678 279	+0.65366 643
302	+0.39595 283	+0.91827 085	352	+0.14114 985	+0.98998 824
303	+0.98663 250	+0.16296 104	353	+0.90930 997	+0.41611 943
304	+0.67020 680	−0.74217 440	354	+0.84145 470	−0.54032 767
305	−0.26240 394	−0.96495 812	355	−0.00003 014	−1.00000 000
306	−0.95376 171	−0.30056 379	356	−0.84148 727	−0.54027 694
307	−0.76823 536	+0.64016 750	357	−0.90928 488	+0.41617 425
308	+0.12360 304	+0.99233 174	358	−0.14109 017	+0.98999 675
309	+0.90180 137	+0.43215 076	359	+0.75682 220	+0.65362 081
310	+0.85088 769	−0.52534 764	360	+0.95891 572	−0.28369 109
311	+0.01767 179	−0.99984 384	361	+0.27938 655	−0.96017 871
312	−0.83179 148	−0.55508 823	362	−0.65700 932	−0.75388 245
313	−0.91650 949	+0.40001 294	363	−0.98935 386	+0.14552 986
314	−0.15859 291	+0.98734 406	364	−0.41209 102	+0.91114 268
315	+0.74513 326	+0.66691 560	365	+0.54404 640	+0.83905 513
316	+0.96378 735	−0.26667 199	366	+0.99999 007	−0.00445 584
317	+0.29633 979	−0.95508 258	367	+0.53654 748	−0.84387 013
318	−0.64356 121	−0.76539 465	368	−0.42019 439	−0.90743 412
319	−0.99177 500	+0.12799 359	369	−0.99061 148	−0.13670 736
320	−0.42815 543	+0.90370 511	370	−0.65026 494	+0.75970 752
321	+0.52910 827	+0.84855 433	371	+0.28793 218	+0.95765 080
322	+0.99991 226	+0.01324 661	372	+0.96140 579	+0.27513 436
323	+0.55140 153	−0.83423 998	373	+0.75096 734	−0.66033 935
324	−0.40406 522	−0.91473 018	374	−0.14990 701	−0.98870 010
325	−0.98803 627	−0.15422 167	375	−0.91295 755	−0.40805 454
326	−0.66361 133	+0.74807 753	376	−0.83663 913	+0.54775 448
327	+0.27093 481	+0.96259 770	377	+0.00888 145	+0.99996 056
328	+0.95638 473	+0.29210 998	378	+0.84623 647	+0.53280 751
329	+0.76253 895	−0.64694 231	379	+0.90556 557	−0.42420 631
330	−0.13238 163	−0.99119 882	380	+0.13232 187	−0.99120 680
331	−0.90559 115	−0.42415 171	381	−0.76257 795	−0.64689 634
332	−0.84620 434	+0.53285 853	382	−0.95636 712	+0.29216 764
333	−0.00882 117	+0.99996 109	383	−0.27087 677	+0.96261 403
334	+0.83667 215	+0.54770 404	384	+0.66365 643	+0.74803 752
335	+0.91293 295	−0.40810 958	385	+0.98802 697	−0.15428 123
336	+0.14984 741	−0.98870 914	386	+0.40401 007	−0.91475 454
337	−0.75100 715	−0.66029 407	387	−0.55145 183	−0.83420 674
338	−0.96138 920	+0.27519 232	388	−0.99991 146	+0.01330 689
339	−0.28787 445	+0.95766 816	389	−0.52905 711	+0.84858 622
340	+0.65031 074	+0.75966 831	390	+0.42820 991	+0.90367 930
341	+0.99060 323	−0.13676 708	391	+0.99178 271	+0.12793 379
342	+0.42013 968	−0.90745 945	392	+0.64351 506	−0.76543 345
343	−0.53659 836	−0.84383 778	393	−0.29639 737	−0.95506 471
344	−0.99999 034	−0.00439 555	394	−0.96380 342	−0.26661 388
345	−0.54399 582	+0.83908 793	395	−0.74509 306	+0.66696 052
346	+0.41214 595	+0.91111 784	396	+0.15865 243	+0.98733 450
347	+0.98936 263	+0.14547 021	397	+0.91653 361	+0.39995 769
348	+0.65696 387	−0.75392 206	398	+0.83175 801	−0.55513 837
349	−0.27944 444	−0.96016 186	399	−0.01773 206	−0.99984 277
350	−0.95893 283	−0.28363 328	400	−0.85091 936	−0.52529 634

Table 4.8   CIRCULAR SINES AND COSINES FOR LARGE RADIAN ARGUMENTS

x	$\sin x$	$\cos x$	x	$\sin x$	$\cos x$
400	−0.85091 936	−0.52529 634	450	−0.68328 373	−0.73015 296
401	−0.90177 532	+0.43220 513	451	−0.98358 231	+0.18046 010
402	−0.12354 321	+0.99233 919	452	−0.37957 985	+0.92515 898
403	+0.76827 396	+0.64012 118	453	+0.57340 657	+0.81927 096
404	+0.95374 359	−0.30062 129	454	+0.99920 563	−0.03985 100
405	+0.26234 577	−0.96497 394	455	+0.50633 965	−0.86233 414
406	−0.67025 155	−0.74213 399	456	−0.45205 268	−0.89199 124
407	−0.98662 268	+0.16302 052	457	−0.99482 985	−0.10155 572
408	−0.39589 747	+0.91829 472	458	−0.62296 505	+0.78224 967
409	+0.55881 405	+0.82929 299	459	+0.32165 095	+0.94685 832
410	+0.99975 451	−0.02215 689	460	+0.97054 255	+0.24092 979
411	+0.52152 528	−0.85323 583	461	+0.72712 181	−0.68650 847
412	−0.43619 188	−0.89985 368	462	−0.18481 137	−0.98277 401
413	−0.99287 624	−0.11915 021	463	−0.92682 982	−0.37548 166
414	−0.63671 476	+0.77109 942	464	−0.81672 521	+0.57702 680
415	+0.30483 933	+0.95240 379	465	+0.04427 279	+0.99901 948
416	+0.96612 555	+0.25807 251	466	+0.86456 660	+0.50251 826
417	+0.73916 039	−0.67352 944	467	+0.88998 186	−0.45599 593
418	−0.16738 542	−0.98589 154	468	+0.09715 190	−0.99526 957
419	−0.92003 785	−0.39182 950	469	−0.78499 906	−0.61949 695
420	−0.82681 172	+0.56247 878	470	−0.94542 551	+0.32583 830
421	+0.02658 129	+0.99964 666	471	−0.23663 211	+0.97159 932
422	+0.85553 559	+0.51774 401	472	+0.68971 977	+0.72407 641
423	+0.89791 441	−0.44017 009	473	+0.98194 647	−0.18915 902
424	+0.11475 487	−0.99339 384	474	+0.37137 611	−0.92848 252
425	−0.77390 977	−0.63329 587	475	−0.58063 573	−0.81416 347
426	−0.95104 534	+0.30905 140	476	−0.99881 376	+0.04869 372
427	−0.25379 421	+0.96725 824	477	−0.49868 703	+0.86678 212
428	+0.67679 415	+0.73617 232	478	+0.45993 026	+0.88795 504
429	+0.98514 108	−0.17174 704	479	+0.99568 978	+0.09274 619
430	+0.38775 385	−0.92176 296	480	+0.61601 671	−0.78773 308
431	−0.56613 249	−0.82431 427	481	−0.33001 928	−0.94397 419
432	−0.99951 922	+0.03100 516	482	−0.97263 707	−0.23232 978
433	−0.51395 260	+0.85781 859	483	−0.72101 682	+0.69291 756
434	+0.44413 968	+0.89595 756	484	+0.19350 297	+0.98109 969
435	+0.99389 198	+0.11035 728	485	+0.93011 702	+0.36726 329
436	+0.62986 458	−0.77670 497	486	+0.81158 578	−0.58423 328
437	−0.31325 741	−0.94966 826	487	−0.05311 369	−0.99858 847
438	−0.96837 198	−0.24951 093	488	−0.86898 067	−0.49484 603
439	−0.73316 982	+0.68004 560	489	−0.88591 083	+0.46385 557
440	+0.17610 529	+0.98437 134	490	−0.08833 866	+0.99609 050
441	+0.92347 001	+0.38367 061	491	+0.79045 167	+0.61252 441
442	+0.82180 066	−0.56977 511	492	+0.94250 438	−0.33419 379
443	−0.03542 843	−0.99937 222	493	+0.22802 291	−0.97365 577
444	−0.86008 478	−0.51015 112	494	−0.69610 177	−0.71794 312
445	−0.89398 316	+0.44810 056	495	−0.98023 370	+0.19784 312
446	−0.10595 754	+0.99437 066	496	−0.36314 328	+0.93173 331
447	+0.77948 495	+0.62642 095	497	+0.58781 939	+0.80899 219
448	+0.94827 257	−0.31745 729	498	+0.99834 363	−0.05753 262
449	+0.24522 276	−0.96946 676	499	+0.49099 533	−0.87116 220
450	−0.68328 373	−0.73015 296	500	−0.46777 181	−0.88384 927

CIRCULAR SINES AND COSINES FOR LARGE RADIAN ARGUMENTS   Table 4.8

x	$\sin x$	$\cos x$	x	$\sin x$	$\cos x$
500	−0.46777 181	−0.88384 927	550	−0.21948 408	−0.97561 608
501	−0.99647 170	−0.08392 940	551	−0.93954 038	−0.34243 814
502	−0.60902 011	+0.79315 478	552	−0.79578 759	+0.60557 585
503	+0.33836 176	+0.94101 611	553	+0.07960 864	+0.99682 620
504	+0.97465 539	+0.22371 157	554	+0.88181 305	+0.47159 913
505	+0.71485 535	−0.69927 236	555	+0.87328 261	−0.48721 400
506	−0.20217 940	−0.97934 850	556	+0.06186 016	−0.99808 483
507	−0.93333 135	−0.35901 615	557	−0.80643 623	−0.59132 107
508	−0.80638 275	+0.59139 399	558	−0.93329 888	+0.35910 055
509	+0.06195 042	+0.99807 923	559	−0.20209 084	+0.97936 678
510	+0.87332 667	+0.48713 502	560	+0.71491 859	+0.69920 771
511	+0.88177 040	−0.47167 887	561	+0.97463 516	−0.22379 971
512	+0.07951 849	−0.99683 339	562	+0.33827 666	−0.94104 671
513	−0.79584 235	−0.60550 389	563	−0.60909 184	−0.79309 970
514	−0.93950 941	+0.34252 310	564	−0.99646 411	+0.08401 951
515	−0.21939 585	+0.97563 593	565	−0.46769 187	+0.88389 157
516	+0.70242 924	+0.71175 358	566	+0.49107 411	+0.87111 780
517	+0.97844 413	−0.20651 172	567	+0.99834 883	+0.05744 234
518	+0.35488 199	−0.93491 110	568	+0.58774 623	−0.80904 534
519	−0.59495 701	−0.80375 753	569	−0.36322 754	−0.93170 046
520	−0.99779 528	+0.06636 701	570	−0.98025 158	−0.19775 448
521	−0.48326 517	+0.87547 403	571	−0.69603 684	+0.71800 607
522	+0.47557 670	+0.87967 426	572	+0.22811 096	+0.97363 514
523	+0.99717 555	+0.07510 603	573	+0.94253 460	+0.33410 856
524	+0.60197 580	−0.79851 433	574	+0.79039 628	−0.61259 589
525	−0.34667 773	−0.93798 430	575	−0.08842 874	−0.99608 251
526	−0.97659 735	−0.21507 583	576	−0.88595 278	−0.46377 546
527	−0.70863 787	+0.70557 237	577	−0.86893 592	+0.49492 461
528	+0.21084 000	+0.97752 059	578	−0.05302 338	+0.99859 327
529	+0.93647 255	+0.35074 088	579	+0.81163 861	+0.58415 989
530	+0.80111 655	−0.59850 837	580	+0.93008 380	−0.36734 740
531	−0.07078 230	−0.99749 179	581	+0.19341 424	−0.98111 719
532	−0.87760 424	−0.47938 586	582	−0.72107 948	−0.69285 235
533	−0.87756 088	+0.47946 522	583	−0.97261 606	+0.23241 774
534	−0.07069 210	+0.99749 818	584	−0.32993 391	+0.94400 403
535	+0.80117 068	+0.59843 592	585	+0.61608 795	+0.78767 737
536	+0.93644 083	−0.35082 557	586	+0.99568 139	−0.09283 623
537	+0.21075 160	−0.97753 965	587	+0.45984 996	−0.88799 663
538	−0.70870 168	−0.70550 828	588	−0.49876 541	−0.86673 702
539	−0.97657 790	+0.21516 415	589	−0.99881 816	−0.04860 339
540	−0.34659 290	+0.93801 565	590	−0.58056 210	+0.81421 597
541	+0.60204 801	+0.79845 989	591	+0.37146 008	+0.92844 893
542	+0.99716 876	−0.07519 621	592	+0.98196 357	+0.18907 022
543	+0.47549 715	−0.87971 726	593	+0.68965 428	−0.72413 878
544	−0.48334 434	−0.87543 032	594	−0.23671 997	−0.97157 792
545	−0.99780 128	−0.06627 678	595	−0.94545 497	−0.32575 281
546	−0.59488 432	+0.80381 133	596	−0.78494 304	+0.61956 794
547	+0.35496 654	+0.93487 901	597	+0.09724 191	+0.99526 078
548	+0.97846 280	+0.20642 324	598	+0.89002 309	+0.45591 545
549	+0.70236 487	−0.71181 710	599	+0.86452 115	−0.50259 644
550	−0.21948 408	−0.97561 608	600	+0.04418 245	−0.99902 348

Table 4.8   CIRCULAR SINES AND COSINES FOR LARGE RADIAN ARGUMENTS

x	$\sin x$	$\cos x$	x	$\sin x$	$\cos x$
600	+0.04418 245	−0.99902 348	650	+0.30475 320	−0.95243 136
601	−0.81677 739	−0.57695 294	651	−0.63678 449	−0.77104 183
602	−0.92679 586	+0.37556 547	652	−0.99286 546	+0.11923 999
603	−0.18472 249	+0.98279 072	653	−0.43611 050	+0.89989 312
604	+0.72718 389	+0.68644 271	654	+0.52160 244	+0.85318 866
605	+0.97052 075	−0.24101 756	655	+0.99975 651	+0.02206 648
606	+0.32156 532	−0.94688 740	656	+0.55873 905	−0.82934 352
607	−0.62303 579	−0.78219 333	657	−0.39598 051	−0.91825 891
608	−0.99482 067	+0.10164 568	658	−0.98663 742	−0.16293 130
609	−0.45197 201	+0.89203 212	659	−0.67018 443	+0.74219 460
610	+0.50641 763	+0.86228 834	660	+0.26243 303	+0.96495 021
611	+0.99920 923	+0.03976 064	661	+0.95377 077	+0.30053 504
612	+0.57333 248	−0.81932 281	662	+0.76821 607	−0.64019 066
613	−0.37966 351	−0.92512 465	663	−0.12363 295	−0.99232 802
614	−0.98359 862	−0.18037 115	664	−0.90181 440	−0.43212 358
615	−0.68321 769	+0.73021 475	665	−0.85087 185	+0.52537 329
616	+0.24531 043	+0.96944 458	666	−0.01764 165	+0.99984 437
617	+0.94830 128	+0.31737 153	667	+0.83180 821	+0.55506 315
618	+0.77942 830	−0.62649 144	668	+0.91649 743	−0.40004 057
619	−0.10604 746	−0.99436 107	669	+0.15856 314	−0.98734 884
620	−0.89402 368	−0.44801 972	670	−0.74515 337	−0.66689 314
621	−0.86003 865	+0.51022 890	671	−0.96377 931	+0.26670 104
622	−0.03533 805	+0.99937 542	672	−0.29631 100	+0.95509 151
623	+0.82185 218	+0.56970 079	673	+0.64358 428	+0.76537 525
624	+0.92343 531	−0.38375 412	674	+0.99177 114	−0.12802 348
625	+0.17601 627	−0.98438 726	675	+0.42812 819	−0.90371 802
626	−0.73323 132	−0.67997 929	676	−0.52913 384	−0.84853 838
627	−0.96834 941	+0.24959 850	677	−0.99991 266	−0.01321 646
628	−0.31317 153	+0.94969 658	678	−0.55137 639	+0.83425 660
629	+0.62993 482	+0.77664 801	679	+0.40409 279	+0.91471 800
630	+0.99388 200	−0.11044 716	680	+0.98804 092	+0.15419 188
631	+0.44405 865	−0.89599 772	681	+0.66358 878	−0.74809 754
632	−0.51403 017	−0.85777 210	682	−0.27096 382	−0.96258 953
633	−0.99952 202	−0.03091 477	683	−0.95639 354	−0.29208 115
634	−0.56605 794	+0.82436 546	684	−0.76251 945	+0.64696 529
635	+0.38783 721	+0.92172 789	685	+0.13241 151	+0.99119 483
636	+0.98515 661	+0.17165 795	686	+0.90560 393	+0.42412 441
637	+0.67672 757	−0.73623 352	687	+0.84618 828	−0.53288 404
638	−0.25388 168	−0.96723 528	688	+0.00879 102	−0.99996 136
639	−0.95107 328	−0.30896 539	689	−0.83668 866	−0.54767 882
640	−0.77385 250	+0.63336 586	690	−0.91292 065	+0.40813 710
641	+0.11484 470	+0.99338 346	691	−0.14981 760	+0.98871 365
642	+0.89795 421	+0.44008 889	692	+0.75102 706	+0.66027 143
643	+0.85548 876	−0.51782 138	693	+0.96138 090	−0.27522 130
644	+0.02649 089	−0.99964 905	694	+0.28784 558	−0.95767 684
645	−0.82686 259	−0.56240 400	695	−0.65033 364	−0.75964 871
646	−0.92000 241	+0.39191 270	696	−0.99059 911	+0.13679 694
647	−0.16729 626	+0.98590 667	697	−0.42011 233	+0.90747 211
648	+0.73922 130	+0.67346 260	698	+0.53662 379	+0.84382 161
649	+0.96610 221	−0.25815 988	699	+0.99999 047	+0.00436 541
650	+0.30475 320	−0.95243 136	700	+0.54397 052	−0.83910 433

CIRCULAR SINES AND COSINES FOR LARGE RADIAN ARGUMENTS   Table 4.8

x	$\sin x$	$\cos x$	x	$\sin x$	$\cos x$
700	+0.54397 052	−0.83910 433	750	+0.74507 295	−0.66698 298
701	−0.41217 342	−0.91110 541	751	−0.15868 219	−0.98732 971
702	−0.98936 702	−0.14544 039	752	−0.91654 566	−0.39993 006
703	−0.65694 115	+0.75394 186	753	−0.83174 127	+0.55516 345
704	+0.27947 339	+0.96015 344	754	+0.01776 220	+0.99984 224
705	+0.95894 137	+0.28360 437	755	+0.85093 519	+0.52527 069
706	+0.75676 309	−0.65368 925	756	+0.90176 229	−0.43223 231
707	−0.14117 969	−0.98998 399	757	+0.12351 330	−0.99234 292
708	−0.90932 251	−0.41609 202	758	−0.76829 325	−0.64009 802
709	−0.84143 841	+0.54035 304	759	−0.95373 453	+0.30065 004
710	+0.00006 029	+1.00000 000	760	−0.26231 668	+0.96498 184
711	+0.84150 356	+0.54025 157	761	+0.67027 392	+0.74211 379
712	+0.90927 234	−0.41620 166	762	+0.98661 776	−0.16305 026
713	+0.14106 032	−0.99000 100	763	+0.39586 979	−0.91830 665
714	−0.75684 190	−0.65359 799	764	−0.55883 905	−0.82927 614
715	−0.95890 717	+0.28372 000	765	−0.99975 384	+0.02218 703
716	−0.27935 761	+0.96018 713	766	−0.52149 956	+0.85325 155
717	+0.65703 205	+0.75386 264	767	+0.43621 901	+0.89984 053
718	+0.98934 947	−0.14555 968	768	+0.99287 983	+0.11912 028
719	+0.41206 355	−0.91115 511	769	+0.63669 152	−0.77111 861
720	−0.54407 170	−0.83903 873	770	−0.30486 804	−0.95239 460
721	−0.99998 994	+0.00448 599	771	−0.96613 333	−0.25804 339
722	−0.53652 204	+0.84388 631	772	−0.73914 009	+0.67355 173
723	+0.42022 174	+0.90742 145	773	+0.16741 514	+0.98588 649
724	+0.99061 560	+0.13667 750	774	+0.92004 966	+0.39180 176
725	+0.65024 204	−0.75972 712	775	+0.82679 477	−0.56250 370
726	−0.28796 105	−0.95764 212	776	−0.02661 142	−0.99964 585
727	−0.96141 408	−0.27510 538	777	−0.85555 119	−0.51771 822
728	−0.75094 744	+0.66036 198	778	−0.89790 114	+0.44019 716
729	+0.14993 682	+0.98869 558	779	−0.11472 492	+0.99339 730
730	+0.91296 985	+0.40802 702	780	+0.77392 886	+0.63327 255
731	+0.83662 262	−0.54777 970	781	+0.95103 602	−0.30908 007
732	−0.00891 160	−0.99996 029	782	+0.25376 505	−0.96726 589
733	−0.84625 253	−0.53278 200	783	−0.67681 634	−0.73615 192
734	−0.90555 279	+0.42423 360	784	−0.98513 591	+0.17177 673
735	−0.13229 199	+0.99121 079	785	−0.38772 606	+0.92177 465
736	+0.76259 745	+0.64687 335	786	+0.56615 733	+0.82429 720
737	+0.95635 831	−0.29219 647	787	+0.99951 829	−0.03103 529
738	+0.27084 775	−0.96262 220	788	+0.51392 674	−0.85783 408
739	−0.66367 898	−0.74801 752	789	−0.44416 668	−0.89594 417
740	−0.98802 232	+0.15431 102	790	−0.99389 531	−0.11032 732
741	−0.40398 250	+0.91476 672	791	−0.62984 117	+0.77672 396
742	+0.55147 697	+0.83419 011	792	+0.31328 604	+0.94965 881
743	+0.99991 106	−0.01333 703	793	+0.96837 950	+0.24948 174
744	+0.52903 153	−0.84860 217	794	+0.73314 932	−0.68006 770
745	−0.42823 715	−0.90366 639	795	−0.17613 497	−0.98436 603
746	−0.99178 657	−0.12790 390	796	−0.92348 158	−0.38364 277
747	−0.64349 199	+0.76545 285	797	−0.82178 349	+0.56979 988
748	+0.29642 616	+0.95505 577	798	+0.03545 855	+0.99937 115
749	+0.96381 146	+0.26658 483	799	+0.86010 016	+0.51012 519
750	+0.74507 295	−0.66698 298	800	+0.89396 965	−0.44812 751

Table 4.8   CIRCULAR SINES AND COSINES FOR LARGE RADIAN ARGUMENTS

x	$\sin x$	$\cos x$	x	$\sin x$	$\cos x$
800	+0.89396 965	−0.44812 751	850	+0.98022 773	−0.19787 267
801	+0.10592 756	−0.99437 385	851	+0.36311 519	−0.93174 426
802	−0.77950 384	−0.62639 745	852	−0.58784 378	−0.80897 447
803	−0.94826 300	+0.31748 587	853	−0.99834 189	+0.05756 271
804	−0.24519 354	+0.96947 415	854	−0.49096 907	+0.87117 700
805	+0.68330 573	+0.73013 237	855	+0.46779 845	+0.88383 517
806	+0.98357 687	−0.18048 975	856	+0.99647 423	+0.08389 936
807	+0.37955 196	−0.92517 042	857	+0.60899 620	−0.79317 314
808	−0.57343 126	−0.81925 368	858	−0.33839 013	−0.94100 591
809	−0.99920 443	+0.03988 112	859	−0.97466 214	−0.22368 219
810	−0.50631 365	+0.86234 940	860	−0.71483 427	+0.69929 390
811	+0.45207 956	+0.89197 762	861	+0.20220 893	+0.97934 241
812	+0.99483 291	+0.10152 573	862	+0.93334 217	+0.35898 802
813	+0.62294 147	−0.78226 845	863	+0.80636 493	−0.59141 830
814	−0.32167 949	−0.94684 862	864	−0.06198 051	−0.99807 736
815	−0.97054 981	−0.24090 054	865	−0.87334 135	−0.48710 870
816	−0.72710 111	+0.68653 039	866	−0.88175 618	+0.47170 545
817	+0.18484 099	+0.98276 844	867	−0.07948 845	+0.99683 579
818	+0.92684 114	+0.37545 372	868	+0.79586 060	+0.60547 989
819	+0.81670 782	−0.57705 142	869	+0.93949 908	−0.34255 142
820	−0.04430 291	−0.99901 814	870	+0.21936 644	−0.97564 254
821	−0.86458 174	−0.50249 220	871	−0.70245 070	−0.71173 241
822	−0.88996 811	+0.45602 276	872	−0.97843 790	+0.20654 122
823	−0.09712 190	+0.99527 249	873	−0.35485 381	+0.93492 180
824	+0.78501 774	+0.61947 329	874	+0.59498 124	+0.80373 959
825	+0.94541 569	−0.32586 680	875	+0.99779 328	−0.06639 709
826	+0.23660 282	−0.97160 646	876	+0.48323 878	−0.87548 859
827	−0.68974 159	−0.72405 561	877	−0.47560 322	−0.87965 992
828	−0.98194 076	+0.18918 862	878	−0.99717 782	−0.07507 597
829	−0.37134 812	+0.92849 371	879	−0.60195 173	+0.79853 248
830	+0.58066 027	+0.81414 596	880	+0.34670 601	+0.93797 385
831	+0.99881 229	−0.04872 383	881	+0.97660 383	+0.21504 639
832	+0.49866 090	−0.86679 716	882	+0.70861 660	−0.70559 373
833	−0.45995 702	−0.88794 118	883	−0.21086 947	−0.97751 423
834	−0.99569 258	−0.09271 618	884	−0.93648 312	−0.35071 265
835	−0.61599 297	+0.78775 165	885	−0.80109 851	+0.59853 252
836	+0.33004 774	+0.94396 424	886	+0.07081 237	+0.99748 965
837	+0.97264 407	+0.23230 046	887	+0.87761 869	+0.47935 940
838	+0.72099 594	−0.69293 929	888	+0.87754 643	−0.47949 167
839	−0.19353 254	−0.98109 386	889	+0.07066 203	−0.99750 031
840	−0.93012 809	−0.36723 525	890	−0.80118 871	−0.59841 177
841	−0.81156 816	+0.58425 775	891	−0.93643 025	+0.35085 380
842	+0.05314 379	+0.99858 687	892	−0.21072 213	+0.97754 600
843	+0.86899 559	+0.49481 983	893	+0.70872 294	+0.70548 692
844	+0.88589 685	−0.46388 228	894	+0.97657 141	−0.21519 358
845	+0.08830 863	−0.99609 316	895	+0.34656 463	−0.93802 610
846	−0.79047 014	−0.61250 058	896	−0.60207 208	−0.79844 174
847	−0.94249 431	+0.33422 221	897	−0.99716 649	+0.07522 627
848	−0.22799 356	+0.97366 264	898	−0.47547 063	+0.87973 159
849	+0.69612 342	+0.71792 213	899	+0.48337 073	+0.87541 575
850	+0.98022 773	−0.19787 267	900	+0.99780 327	+0.06624 670

CIRCULAR SINES AND COSINES FOR LARGE RADIAN ARGUMENTS   Table 4.8

x	$\sin x$	$\cos x$	x	$\sin x$	$\cos x$
900	+0.99780 327	+0.06624 670	950	+0.94546 479	+0.32572 431
901	+0.59486 009	−0.80382 926	951	+0.78492 436	−0.61959 160
902	−0.35499 472	−0.93486 831	952	−0.09727 191	−0.99525 784
903	−0.97846 902	−0.20639 374	953	−0.89003 684	−0.45588 862
904	−0.70234 341	+0.71183 827	954	−0.86450 600	+0.50262 250
905	+0.21951 349	+0.97560 947	955	−0.04415 233	+0.99902 481
906	+0.93955 070	+0.34240 981	956	+0.81679 478	+0.57692 832
907	+0.79576 933	−0.60559 984	957	+0.92678 454	−0.37559 341
908	−0.07963 869	−0.99682 380	958	+0.18469 287	−0.98279 629
909	−0.88182 727	−0.47157 255	959	−0.72720 458	−0.68642 079
910	−0.87326 792	+0.48724 032	960	−0.97051 349	+0.24104 682
911	−0.06183 008	+0.99808 669	961	−0.32153 677	+0.94689 709
912	+0.80645 406	+0.59129 676	962	+0.62305 937	+0.78217 455
913	+0.93328 805	−0.35912 869	963	+0.99481 760	−0.10167 567
914	+0.20206 131	−0.97937 287	964	+0.45194 512	−0.89204 574
915	−0.71493 966	−0.69918 616	965	−0.50644 362	−0.86227 308
916	−0.97462 841	+0.22382 909	966	−0.99921 043	−0.03973 052
917	−0.33824 829	+0.94105 690	967	−0.57330 778	+0.81934 009
918	+0.60911 575	+0.79308 134	968	+0.37969 140	+0.92511 320
919	+0.99646 158	−0.08404 955	969	+0.98360 406	+0.18034 150
920	+0.46766 523	−0.88390 567	970	+0.68319 568	−0.73023 535
921	−0.49110 037	−0.87110 299	971	−0.24533 966	−0.96943 718
922	−0.99835 056	−0.05741 224	972	−0.94831 084	−0.31734 294
923	−0.58772 184	+0.80906 306	973	−0.77940 942	+0.62651 493
924	+0.36325 562	+0.93168 952	974	+0.10607 744	+0.99435 787
925	+0.98025 754	+0.19772 493	975	+0.89403 718	+0.44799 277
926	+0.69601 520	−0.71802 705	976	+0.86002 327	−0.51025 482
927	−0.22814 031	−0.97362 827	977	+0.03530 793	−0.99937 648
928	−0.94254 467	−0.33408 015	978	−0.82186 936	−0.56967 601
929	−0.79037 781	+0.61261 972	979	−0.92342 374	+0.38378 195
930	+0.08845 877	+0.99607 984	980	−0.17598 660	+0.98439 256
931	+0.88596 676	+0.46374 875	981	+0.73325 181	+0.67995 719
932	+0.86892 100	−0.49495 080	982	+0.96834 189	−0.24962 769
933	+0.05299 328	−0.99859 487	983	+0.31314 290	−0.94970 602
934	−0.81165 622	−0.58413 542	984	−0.62995 823	−0.77662 902
935	−0.93007 273	+0.36737 544	985	−0.99387 867	+0.11047 712
936	−0.19338 467	+0.98112 302	986	−0.44403 164	+0.89601 111
937	+0.72110 037	+0.69283 061	987	+0.51405 603	+0.85775 661
938	+0.97260 905	−0.23244 706	988	+0.99952 296	+0.03088 464
939	+0.32990 546	−0.94401 398	989	+0.56603 309	−0.82438 252
940	−0.61611 169	−0.78765 880	990	−0.38786 499	−0.92171 620
941	−0.99567 859	+0.09286 625	991	−0.98516 179	−0.17162 825
942	−0.45982 319	+0.88801 049	992	−0.67670 538	+0.73625 392
943	+0.49879 154	+0.86672 199	993	+0.25391 083	+0.96722 763
944	+0.99881 962	+0.04857 328	994	+0.95108 260	+0.30893 672
945	+0.58053 755	−0.81423 347	995	+0.77383 341	−0.63338 919
946	−0.37148 806	−0.92843 773	996	−0.11487 465	−0.99338 000
947	−0.98196 927	−0.18904 062	997	−0.89796 748	−0.44006 182
948	−0.68963 246	+0.72415 957	998	−0.85547 315	+0.51784 716
949	+0.23674 926	+0.97157 078	999	−0.02646 075	+0.99964 985
950	+0.94546 479	+0.32572 431	1000	+0.82687 954	+0.56237 908

For $x > 1000$ see **Example 16**.

Table 4.9

CIRCULAR TANGENTS, COTANGENTS, SECANTS AND COSECANTS FOR RADIAN ARGUMENTS

x	$\tan x$	$\cot x$	$\sec x$	$\csc x$	$x^{-1} - \cot x$	$\csc x - x^{-1}$
0.00	0.00000 0000	∞	1.00000 00	∞	0.00000 000	0.00000 000
0.01	0.01000 0333	99.99666 66	1.00005 00	100.00166 67	0.00333 335	0.00166 668
0.02	0.02000 2667	49.99333 32	1.00020 00	50.00333 35	0.00666 684	0.00333 349
0.03	0.03000 9003	33.32333 27	1.00045 02	33.33833 39	0.01000 060	0.00500 053
0.04	0.04002 1347	24.98666 52	1.00080 05	25.00666 79	0.01333 476	0.00666 791
0.05	0.05004 1708	19.98333 06	1.00125 13	20.00833 58	0.01666 944	0.00833 576
0.06	0.06007 2104	16.64666 19	1.00180 27	16.67667 09	0.02000 480	0.01000 420
0.07	0.07011 4558	14.26237 33	1.00245 50	14.29738 76	0.02334 096	0.01167 334
0.08	0.08017 1105	12.47332 19	1.00320 86	12.51334 32	0.02667 805	0.01334 330
0.09	0.09024 3790	11.08109 49	1.00406 37	11.12612 53	0.03001 621	0.01501 419
0.10	0.10033 467	9.96664 44	1.00502 09	10.01668 61	0.03335 558	0.01668 614
0.11	0.11044 582	9.05421 28	1.00608 07	9.10926 83	0.03669 628	0.01835 925
0.12	0.12057 934	8.29329 49	1.00724 35	8.35336 70	0.04003 845	0.02003 365
0.13	0.13073 732	7.64892 55	1.00850 99	7.71401 72	0.04338 223	0.02170 946
0.14	0.14092 189	7.09612 94	1.00988 07	7.16624 39	0.04672 776	0.02338 680
0.15	0.15113 522	6.61659 15	1.01135 64	6.69173 24	0.05007 516	0.02506 578
0.16	0.16137 946	6.19657 54	1.01293 80	6.27674 65	0.05342 458	0.02674 653
0.17	0.17165 682	5.82567 68	1.01462 61	5.91078 21	0.05677 615	0.02842 915
0.18	0.18196 953	5.49542 56	1.01642 16	5.58566 93	0.06013 000	0.03011 379
0.19	0.19231 984	5.19967 16	1.01832 55	5.29495 84	0.06348 628	0.03180 054
0.20	0.20271 004	4.93315 49	1.02033 88	5.03348 95	0.06684 512	0.03348 955
0.21	0.21314 244	4.69169 81	1.02246 26	4.79708 57	0.07020 667	0.03518 092
0.22	0.22361 942	4.47188 35	1.02469 78	4.58232 93	0.07357 105	0.03687 477
0.23	0.23414 336	4.27088 77	1.02704 58	4.38639 73	0.07693 841	0.03857 124
0.24	0.24471 670	4.08635 78	1.02950 78	4.20693 71	0.08030 889	0.04027 044
0.25	0.25534 192	3.91631 74	1.03208 50	4.04197 25	0.08368 264	0.04197 250
0.26	0.26602 154	3.75909 41	1.03477 89	3.88983 14	0.08705 978	0.04367 754
0.27	0.27675 814	3.61326 32	1.03759 10	3.74908 94	0.09044 046	0.04538 569
0.28	0.28755 433	3.47760 37	1.04052 27	3.61852 56	0.09382 483	0.04709 707
0.29	0.29841 279	3.35106 28	1.04357 57	3.49708 77	0.09721 302	0.04881 181
0.30	0.30933 625	3.23272 81	1.04675 16	3.38386 34	0.10060 519	0.05053 003
0.31	0.32032 751	3.12180 50	1.05005 22	3.27805 83	0.10400 147	0.05225 186
0.32	0.33138 941	3.01759 80	1.05347 94	3.17897 74	0.10740 202	0.05397 744
0.33	0.34252 487	2.91949 61	1.05703 51	3.08600 99	0.11080 697	0.05570 689
0.34	0.35373 688	2.82696 00	1.06072 13	2.99861 68	0.11421 648	0.05744 034
0.35	0.36502 849	2.73951 22	1.06454 02	2.91632 08	0.11763 070	0.05917 792
0.36	0.37640 285	2.65672 80	1.06849 38	2.83869 75	0.12104 976	0.06091 976
0.37	0.38786 316	2.57822 89	1.07258 47	2.76536 87	0.12447 383	0.06266 601
0.38	0.39941 272	2.50367 59	1.07681 50	2.69599 57	0.12790 306	0.06441 678
0.39	0.41105 492	2.43276 50	1.08118 74	2.63027 48	0.13133 759	0.06617 222
0.40	0.42279 322	2.36522 24	1.08570 44	2.56793 25	0.13477 758	0.06793 246
0.41	0.43463 120	2.30080 12	1.09036 89	2.50872 20	0.13822 318	0.06969 763
0.42	0.44657 255	2.23927 78	1.09518 36	2.45242 03	0.14167 456	0.07146 789
0.43	0.45862 102	2.18044 95	1.10015 15	2.39882 48	0.14513 185	0.07324 336
0.44	0.47078 053	2.12413 20	1.10527 57	2.34775 15	0.14859 524	0.07502 418
0.45	0.48305 507	2.07015 74	1.11055 94	2.29903 27	0.15206 486	0.07681 051
0.46	0.49544 877	2.01837 22	1.11600 60	2.25251 55	0.15554 089	0.07860 247
0.47	0.50796 590	1.96863 61	1.12161 91	2.20805 98	0.15902 348	0.08040 022
0.48	0.52061 084	1.92082 05	1.12740 22	2.16553 72	0.16251 280	0.08220 390
0.49	0.53338 815	1.87480 73	1.13335 91	2.12483 00	0.16600 901	0.08401 366
0.50	0.54630 249	1.83048 77	1.13949 39	2.08582 96	0.16951 228	0.08582 964
	$\begin{bmatrix} (-5)2 \\ 4 \end{bmatrix}$		$\begin{bmatrix} (-5)2 \\ 4 \end{bmatrix}$		$\begin{bmatrix} (-7)9 \\ 4 \end{bmatrix}$	$\begin{bmatrix} (-7)8 \\ 4 \end{bmatrix}$

Compilation of $\tan x$ and $\cot x$ from National Bureau of Standards, Table of circular and hyperbolic tangents and cotangents for radian arguments, 2d printing. Columbia Univ. Press, New York, N.Y., 1947 (with permission).

CIRCULAR TANGENTS, COTANGENTS, SECANTS AND COSECANTS Table 4.9

FOR RADIAN ARGUMENTS

x	$\tan x$	$\cot x$	$\sec x$	$\csc x$
0.50	0.54630 249	1.83048 772	1.13949 39	2.08582 96
0.51	0.55935 872	1.78776 154	1.14581 07	2.04843 63
0.52	0.57256 183	1.74653 626	1.15231 38	2.01255 78
0.53	0.58591 701	1.70672 634	1.15900 77	1.97810 89
0.54	0.59942 962	1.66825 255	1.16589 70	1.94501 07
0.55	0.61310 521	1.63104 142	1.17298 68	1.91319 00
0.56	0.62694 954	1.59502 471	1.18028 21	1.88257 90
0.57	0.64096 855	1.56013 894	1.18778 81	1.85311 45
0.58	0.65516 845	1.52632 503	1.19551 06	1.82473 78
0.59	0.66955 565	1.49352 784	1.20345 53	1.79739 41
0.60	0.68413 681	1.46169 595	1.21162 83	1.77103 22
0.61	0.69891 886	1.43078 125	1.22003 59	1.74560 45
0.62	0.71390 901	1.40073 873	1.22868 47	1.72106 62
0.63	0.72911 473	1.37152 626	1.23758 16	1.69737 57
0.64	0.74454 382	1.34310 429	1.24673 39	1.67449 37
0.65	0.76020 440	1.31543 569	1.25614 92	1.65238 34
0.66	0.77610 491	1.28848 559	1.26583 52	1.63101 05
0.67	0.79225 417	1.26222 118	1.27580 04	1.61034 23
0.68	0.80866 138	1.23661 155	1.28605 34	1.59034 84
0.69	0.82533 611	1.21162 759	1.29660 31	1.57100 01
0.70	0.84228 838	1.18724 183	1.30745 93	1.55227 03
0.71	0.85952 867	1.16342 833	1.31863 17	1.53413 35
0.72	0.87706 790	1.14016 258	1.33013 09	1.51656 54
0.73	0.89491 753	1.11742 140	1.34196 77	1.49954 35
0.74	0.91308 953	1.09518 285	1.35415 38	1.48304 60
0.75	0.93159 646	1.07342 615	1.36670 11	1.46705 27
0.76	0.95045 146	1.05213 158	1.37962 24	1.45154 43
0.77	0.96966 833	1.03128 046	1.39293 10	1.43650 25
0.78	0.98926 154	1.01085 503	1.40664 08	1.42190 99
0.79	1.00924 629	0.99083 842	1.42076 67	1.40775 03
0.80	1.02963 857	0.97121 460	1.43532 42	1.39400 78
0.81	1.05045 514	0.95196 830	1.45032 96	1.38066 78
0.82	1.07171 372	0.93308 500	1.46580 02	1.36771 62
0.83	1.09343 292	0.91455 085	1.48175 42	1.35513 96
0.84	1.11563 235	0.89635 264	1.49821 08	1.34292 52
0.85	1.13833 271	0.87847 778	1.51519 02	1.33106 09
0.86	1.16155 586	0.86091 426	1.53271 39	1.31953 53
0.87	1.18532 486	0.84365 058	1.55080 46	1.30833 72
0.88	1.20966 412	0.82667 575	1.56948 63	1.29745 63
0.89	1.23459 946	0.80997 930	1.58878 44	1.28688 25
0.90	1.26015 822	0.79355 115	1.60872 58	1.27660 62
0.91	1.28636 938	0.77738 169	1.62933 92	1.26661 84
0.92	1.31326 370	0.76146 169	1.65065 49	1.25691 05
0.93	1.34087 383	0.74578 232	1.67270 52	1.24747 40
0.94	1.36923 448	0.73033 510	1.69552 44	1.23830 10
0.95	1.39838 259	0.71511 188	1.71914 92	1.22938 40
0.96	1.42835 749	0.70010 485	1.74361 84	1.22071 57
0.97	1.45920 113	0.68530 649	1.76897 37	1.21228 91
0.98	1.49095 827	0.67070 959	1.79525 95	1.20409 77
0.99	1.52367 674	0.65630 719	1.82252 32	1.19613 51
1.00	1.55740 772	0.64209 262	1.85081 57	1.18839 51
	$*\begin{bmatrix}(-4)1\\5\end{bmatrix}$	$\begin{bmatrix}(-4)2\\6\end{bmatrix}$	$\begin{bmatrix}(-4)1\\5\end{bmatrix}$	$\begin{bmatrix}(-4)2\\5\end{bmatrix}$

*See page II.

ELEMENTARY TRANSCENDENTAL FUNCTIONS

Table 4.9   CIRCULAR TANGENTS, COTANGENTS, SECANTS AND COSECANTS
FOR RADIAN ARGUMENTS

x	$\tan x$	$\cot x$	$\sec x$	$\csc x$
1.00	1.55740 77	0.64209 262	1.85081 57	1.18839 51
1.01	1.59220 60	0.62805 942	1.88019 15	1.18087 20
1.02	1.62813 04	0.61420 141	1.91070 89	1.17356 01
1.03	1.66524 40	0.60051 260	1.94243 08	1.16645 42
1.04	1.70361 46	0.58698 722	1.97542 47	1.15954 90
1.05	1.74331 53	0.57361 970	2.00976 32	1.15283 98
1.06	1.78442 48	0.56040 467	2.04552 49	1.14632 17
1.07	1.82702 82	0.54733 693	2.08279 43	1.13999 02
1.08	1.87121 73	0.53441 147	2.12166 31	1.13384 11
1.09	1.91709 18	0.52162 342	2.16223 06	1.12787 01
1.10	1.96475 97	0.50896 811	2.20460 44	1.12207 33
1.11	2.01433 82	0.49644 096	2.24890 16	1.11644 69
1.12	2.06595 53	0.48403 759	2.29524 97	1.11098 71
1.13	2.11975 01	0.47175 371	2.34378 77	1.10569 05
1.14	2.17587 51	0.45958 520	2.39466 75	1.10055 37
1.15	2.23449 69	0.44752 802	2.44805 57	1.09557 35
1.16	2.29579 85	0.43557 829	2.50413 48	1.09074 67
1.17	2.35998 11	0.42373 221	2.56310 57	1.08607 04
1.18	2.42726 64	0.41198 610	2.62518 99	1.08154 17
1.19	2.49789 94	0.40033 638	2.69063 21	1.07715 79
1.20	2.57215 16	0.38877 957	2.75970 36	1.07291 64
1.21	2.65032 46	0.37731 227	2.83270 55	1.06881 46
1.22	2.73275 42	0.36593 119	2.90997 35	1.06485 01
1.23	2.81981 57	0.35463 310	2.99188 25	1.06102 06
1.24	2.91192 99	0.34341 486	3.07885 30	1.05732 39
1.25	3.00956 97	0.33227 342	3.17135 77	1.05375 79
1.26	3.11326 91	0.32120 577	3.26993 04	1.05032 05
1.27	3.22363 32	0.31020 899	3.37517 57	1.04700 98
1.28	3.34135 00	0.29928 023	3.48778 15	1.04382 41
1.29	3.46720 57	0.28841 670	3.60853 36	1.04076 14
1.30	3.60210 24	0.27761 565	3.73833 41	1.03782 00
1.31	3.74708 10	0.26687 440	3.87822 33	1.03499 85
1.32	3.90334 78	0.25619 034	4.02940 74	1.03229 53
1.33	4.07230 98	0.24556 088	4.19329 31	1.02970 88
1.34	4.25561 79	0.23498 350	4.37153 10	1.02723 77
1.35	4.45522 18	0.22445 572	4.56607 06	1.02488 07
1.36	4.67344 12	0.21397 509	4.77923 14	1.02263 65
1.37	4.91305 81	0.20353 922	5.01379 49	1.02050 39
1.38	5.17743 74	0.19314 574	5.27312 60	1.01848 18
1.39	5.47068 86	0.18279 234	5.56133 39	1.01656 93
1.40	5.79788 37	0.17247 673	5.88349 01	1.01476 51
1.41	6.16535 61	0.16219 663	6.24592 80	1.01306 85
1.42	6.58111 95	0.15194 983	6.65666 08	1.01147 85
1.43	7.05546 38	0.14173 413	7.12597 85	1.00999 43
1.44	7.60182 61	0.13154 734	7.66731 76	1.00861 52
1.45	8.23809 28	0.12138 732	8.29856 45	1.00734 05
1.46	8.98860 76	0.11125 194	9.04406 25	1.00616 95
1.47	9.88737 49	0.10113 908	9.93781 58	1.00510 15
1.48	10.98337 93	0.09104 6660	11.02880 87	1.00413 62
1.49	12.34985 64	0.08097 2601	12.39027 66	1.00327 29
1.50	14.10141 99	0.07091 4844	14.13683 29	1.00251 13
1.51	16.42809 17	0.06087 1343	16.45849 92	1.00185 09
1.52	19.66952 78	0.05084 0061	19.69493 14	1.00129 15
1.53	24.49841 04	0.04081 8975	24.51881 14	1.00083 27
1.54	32.46113 89	0.03080 6066	32.47653 83	1.00047 44
1.55	48.07848 25	0.02079 9325	48.08888 10	1.00021 63
1.56	92.62049 63	0.01079 6746	92.62589 45	1.00005 83
1.57	+1255.76559 15	+ 0.00079 6327	+1255.76598 97	1.00000 03
1.58	− 108.64920 36	− 0.00920 3933	− 108.65380 55	1.00004 24
1.59	− 52.06696 96	− 0.01920 6034	− 52.07657 18	1.00018 44
1.60	− 34.23253 27	− 0.02921 1978	− 34.24713 56	1.00042 66

For $x > 1.6$, use 4.3.44.

$\begin{bmatrix} (-5)2 \\ 5 \end{bmatrix}$

$\begin{bmatrix} (-5)3 \\ 4 \end{bmatrix}$

CIRCULAR SINES AND COSINES TO TENTHS OF A DEGREE   Table 4.10

θ	$\sin \theta$	$\cos \theta$	$90° - \theta$
0.0°	0.00000 00000 00000	1.00000 00000 00000	90.0°
0.1	0.00174 53283 65898	0.99999 84769 13288	89.9
0.2	0.00349 06514 15224	0.99999 39076 57790	89.8
0.3	0.00523 59638 31420	0.99998 62922 47427	89.7
0.4	0.00698 12602 97962	0.99997 56307 05395	89.6
0.5	0.00872 65354 98374	0.99996 19230 64171	89.5
0.6	0.01047 17841 16246	0.99994 51693 65512	89.4
0.7	0.01221 70008 35247	0.99992 53696 60452	89.3
0.8	0.01396 21803 39145	0.99990 25240 09304	89.2
0.9	0.01570 73173 11821	0.99987 66324 81661	89.1
1.0	0.01745 24064 37284	0.99984 76951 56391	89.0
1.1	0.01919 74423 99690	0.99981 57121 21644	88.9
1.2	0.02094 24198 83357	0.99978 06834 74845	88.8
1.3	0.02268 73335 72781	0.99974 26093 22698	88.7
1.4	0.02443 21781 52653	0.99970 14897 81183	88.6
1.5	0.02617 69483 07873	0.99965 73249 75557	88.5
1.6	0.02792 16387 23569	0.99961 01150 40354	88.4
1.7	0.02966 62440 85111	0.99955 98601 19384	88.3
1.8	0.03141 07590 78128	0.99950 65603 65732	88.2
1.9	0.03315 51783 88526	0.99945 02159 41757	88.1
2.0	0.03489 94967 02501	0.99939 08270 19096	88.0
2.1	0.03664 37087 06556	0.99932 83937 78656	87.9
2.2	0.03838 78090 87520	0.99926 29164 10621	87.8
2.3	0.04013 17925 32560	0.99919 43951 14446	87.7
2.4	0.04187 56537 29200	0.99912 28300 98858	87.6
2.5	0.04361 93873 65336	0.99904 82215 81858	87.5
2.6	0.04536 29881 29254	0.99897 05697 90715	87.4
2.7	0.04710 64507 09643	0.99888 98749 61970	87.3
2.8	0.04884 97697 95613	0.99880 61373 41434	87.2
2.9	0.05059 29400 76713	0.99871 93571 84186	87.1
3.0	0.05233 59562 42944	0.99862 95347 54574	87.0
3.1	0.05407 88129 84775	0.99853 66703 26212	86.9
3.2	0.05582 15049 93164	0.99844 07641 81981	86.8
3.3	0.05756 40269 59567	0.99834 18166 14028	86.7
3.4	0.05930 63735 75962	0.99823 98279 23765	86.6
3.5	0.06104 85395 34857	0.99813 47984 21867	86.5
3.6	0.06279 05195 29313	0.99802 67284 28272	86.4
3.7	0.06453 23082 52958	0.99791 56182 72179	86.3
3.8	0.06627 39004 00000	0.99780 14682 92050	86.2
3.9	0.06801 52906 65248	0.99768 42788 35605	86.1
4.0	0.06975 64737 44125	0.99756 40502 59824	86.0
4.1	0.07149 74443 32686	0.99744 07829 30944	85.9
4.2	0.07323 81971 27632	0.99731 44772 24458	85.8
4.3	0.07497 87268 26328	0.99718 51335 25116	85.7
4.4	0.07671 90281 26819	0.99705 27522 26920	85.6
4.5	0.07845 90957 27845	0.99691 73337 33128	85.5
4.6	0.08019 89243 28859	0.99677 88784 56247	85.4
4.7	0.08193 85086 30041	0.99663 73868 18037	85.3
4.8	0.08367 78433 32315	0.99649 28592 49504	85.2
4.9	0.08541 69231 37367	0.99634 52961 90906	85.1
5.0	0.08715 57427 47658	0.99619 46980 91746	85.0
$90° - \theta$	$\cos \theta$	$\sin \theta$	θ
	$*\begin{bmatrix}(-8)3 \\ 5\end{bmatrix}$	$\begin{bmatrix}(-7)4 \\ 5\end{bmatrix}$	

For conversion from radians to degrees see **Example 14**.

*See page II.

Table 4.10   CIRCULAR SINES AND COSINES TO TENTHS OF A DEGREE

θ	$\sin \theta$	$\cos \theta$	$90° - \theta$
5.0°	0.08715 57427 47658	0.99619 46980 91746	85.0°
5.1	0.08889 42968 66442	0.99604 10654 10770	84.9
5.2	0.09063 25801 97780	0.99588 43986 15970	84.8
5.3	0.09237 05874 46562	0.99572 46981 84582	84.7
5.4	0.09410 83133 18514	0.99556 19646 03080	84.6
5.5	0.09584 57525 20224	0.99539 61983 67179	84.5
5.6	0.09758 28997 59149	0.99522 73999 81831	84.4
5.7	0.09931 97497 43639	0.99505 55699 61226	84.3
5.8	0.10105 62971 82946	0.99488 07088 28788	84.2
5.9	0.10279 25367 87247	0.99470 28171 17174	84.1
6.0	0.10452 84632 67653	0.99452 18953 68273	84.0
6.1	0.10626 40713 36233	0.99433 79441 33205	83.9
6.2	0.10799 93557 06023	0.99415 09639 72315	83.8
6.3	0.10973 43110 91045	0.99396 09554 55180	83.7
6.4	0.11146 89322 06325	0.99376 79191 60596	83.6
6.5	0.11320 32137 67907	0.99357 18556 76587	83.5
6.6	0.11493 71504 92867	0.99337 27656 00396	83.4
6.7	0.11667 07370 99333	0.99317 06495 38486	83.3
6.8	0.11840 39683 06501	0.99296 55081 06537	83.2
6.9	0.12013 68388 34647	0.99275 73419 29446	83.1
7.0	0.12186 93434 05147	0.99254 61516 41322	83.0
7.1	0.12360 14767 40493	0.99233 19378 85489	82.9
7.2	0.12533 32335 64304	0.99211 47013 14478	82.8
7.3	0.12706 46086 01350	0.99189 44425 90030	82.7
7.4	0.12879 55965 77563	0.99167 11623 83090	82.6
7.5	0.13052 61922 20052	0.99144 48613 73810	82.5
7.6	0.13225 63902 57122	0.99121 55402 51542	82.4
7.7	0.13398 61854 18292	0.99098 31997 14836	82.3
7.8	0.13571 55724 34304	0.99074 78404 71444	82.2
7.9	0.13744 45460 37147	0.99050 94632 38309	82.1
8.0	0.13917 31009 60065	0.99026 80687 41570	82.0
8.1	0.14090 12319 37583	0.99002 36577 16558	81.9
8.2	0.14262 89337 05512	0.98977 62309 07789	81.8
8.3	0.14435 62010 00973	0.98952 57890 68969	81.7
8.4	0.14608 30285 62412	0.98927 23329 62988	81.6
8.5	0.14780 94111 29611	0.98901 58633 61917	81.5
8.6	0.14953 53434 43710	0.98875 63810 47006	81.4
8.7	0.15126 08202 47219	0.98849 38868 08684	81.3
8.8	0.15298 58362 84038	0.98822 83814 46553	81.2
8.9	0.15471 03862 99468	0.98795 98657 69389	81.1
9.0	0.15643 44650 40231	0.98768 83405 95138	81.0
9.1	0.15815 80672 54484	0.98741 38067 50911	80.9
9.2	0.15988 11876 91835	0.98713 62650 72988	80.8
9.3	0.16160 38211 03361	0.98685 57164 06807	80.7
9.4	0.16332 59622 41622	0.98657 21616 06969	80.6
9.5	0.16504 76058 60678	0.98628 56015 37231	80.5
9.6	0.16676 87467 16102	0.98599 60370 70505	80.4
9.7	0.16848 93795 65003	0.98570 34690 88854	80.3
9.8	0.17020 94991 66033	0.98540 78984 83490	80.2
9.9	0.17192 91002 79410	0.98510 93261 54774	80.1
10.0	0.17364 81776 66930	0.98480 77530 12208	80.0
$90° - \theta$	$\cos \theta$	$\sin \theta$	θ

$$* \begin{bmatrix} (-8)7 \\ 5 \end{bmatrix} \qquad \begin{bmatrix} (-7)4 \\ 5 \end{bmatrix}$$

---

*See page II.

CIRCULAR SINES AND COSINES TO TENTHS OF A DEGREE   Table 4.10

θ	$\sin \theta$	$\cos \theta$	$90° - \theta$
10.0°	0.17364 81776 66930	0.98480 77530 12208	80.0°
10.1	0.17536 67260 91987	0.98450 31799 74437	79.9
10.2	0.17708 47403 19583	0.98419 56079 69242	79.8
10.3	0.17880 22151 16350	0.98388 50379 33542	79.7
10.4	0.18051 91452 50560	0.98357 14708 13386	79.6
10.5	0.18223 55254 92147	0.98325 49075 63955	79.5
10.6	0.18395 13506 12720	0.98293 53491 49554	79.4
10.7	0.18566 66153 85577	0.98261 27965 43615	79.3
10.8	0.18738 13145 85725	0.98228 72507 28689	79.2
10.9	0.18909 54429 89891	0.98195 87126 96444	79.1
11.0	0.19080 89953 76545	0.98162 71834 47664	79.0
11.1	0.19252 19665 25907	0.98129 26639 92245	78.9
11.2	0.19423 43512 19972	0.98095 51553 49192	78.8
11.3	0.19594 61442 42518	0.98061 46585 46613	78.7
11.4	0.19765 73403 79126	0.98027 11746 21722	78.6
11.5	0.19936 79344 17197	0.97992 47046 20830	78.5
11.6	0.20107 79211 45965	0.97957 52495 99344	78.4
11.7	0.20278 72953 56512	0.97922 28106 21766	78.3
11.8	0.20449 60518 41790	0.97886 73887 61685	78.2
11.9	0.20620 41853 96630	0.97850 89851 01778	78.1
12.0	0.20791 16908 17759	0.97814 76007 33806	78.0
12.1	0.20961 85629 03822	0.97778 32367 58606	77.9
12.2	0.21132 47964 55389	0.97741 58942 86096	77.8
12.3	0.21303 03862 74977	0.97704 55744 35264	77.7
12.4	0.21473 53271 67063	0.97667 22783 34168	77.6
12.5	0.21643 96139 38103	0.97629 60071 19933	77.5
12.6	0.21814 32413 96543	0.97591 67619 38747	77.4
12.7	0.21984 62043 52838	0.97553 45439 45857	77.3
12.8	0.22154 84976 19467	0.97514 93543 05563	77.2
12.9	0.22325 01160 10951	0.97476 11941 91222	77.1
13.0	0.22495 10543 43865	0.97437 00647 85235	77.0
13.1	0.22665 13074 36855	0.97397 59672 79052	76.9
13.2	0.22835 08701 10656	0.97357 89028 73160	76.8
13.3	0.23004 97371 88104	0.97317 88727 77088	76.7
13.4	0.23174 79034 94157	0.97277 58782 09397	76.6
13.5	0.23344 53638 55905	0.97236 99203 97677	76.5
13.6	0.23514 21131 02590	0.97196 10005 78546	76.4
13.7	0.23683 81460 65619	0.97154 91199 97646	76.3
13.8	0.23853 34575 78581	0.97113 42799 09636	76.2
13.9	0.24022 80424 77264	0.97071 64815 78191	76.1
14.0	0.24192 18955 99668	0.97029 57262 75996	76.0
14.1	0.24361 50117 86023	0.96987 20152 84747	75.9
14.2	0.24530 73858 78803	0.96944 53498 95139	75.8
14.3	0.24699 90127 22743	0.96901 57314 06870	75.7
14.4	0.24868 98871 64855	0.96858 31611 28631	75.6
14.5	0.25038 00040 54441	0.96814 76403 78108	75.5
14.6	0.25206 93582 43114	0.96770 91704 81971	75.4
14.7	0.25375 79445 84806	0.96726 77527 75877	75.3
14.8	0.25544 57579 35791	0.96682 33886 04459	75.2
14.9	0.25713 27931 54696	0.96637 60793 21329	75.1
15.0	0.25881 90451 02521	0.96592 58262 89068	75.0
$90° - \theta$	$\cos \theta$	$\sin \theta$	θ

$$* \begin{bmatrix} (-7)1 \\ 5 \end{bmatrix} \qquad \begin{bmatrix} (-7)4 \\ 5 \end{bmatrix}$$

*See page II.

Table 4.10　　**CIRCULAR SINES AND COSINES TO TENTHS OF A DEGREE**

θ	$\sin \theta$	$\cos \theta$	$90° - \theta$
15.0°	0.25881 90451 02521	0.96592 58262 89068	75.0°
15.1	0.26050 45086 42648	0.96547 26308 79225	74.9
15.2	0.26218 91786 40865	0.96501 64944 72311	74.8
15.3	0.26387 30499 65373	0.96455 74184 57798	74.7
15.4	0.26555 61174 86809	0.96409 54042 34110	74.6
15.5	0.26723 83760 78257	0.96363 04532 08623	74.5
15.6	0.26891 98206 15266	0.96316 25667 97658	74.4
15.7	0.27060 04459 75864	0.96269 17464 26479	74.3
15.8	0.27228 02470 40574	0.96221 79935 29285	74.2
15.9	0.27395 92186 92432	0.96174 13095 49211	74.1
16.0	0.27563 73558 16999	0.96126 16959 38319	74.0
16.1	0.27731 46533 02378	0.96077 91541 57594	73.9
16.2	0.27899 11060 39229	0.96029 36856 76943	73.8
16.3	0.28066 67089 20788	0.95980 52919 75187	73.7
16.4	0.28234 14568 42876	0.95931 39745 40058	73.6
16.5	0.28401 53447 03923	0.95881 97348 68193	73.5
16.6	0.28568 83674 04974	0.95832 25744 65133	73.4
16.7	0.28736 05198 49712	0.95782 24948 45315	73.3
16.8	0.28903 17969 44472	0.95731 94975 32067	73.2
16.9	0.29070 21935 98252	0.95681 35840 57607	73.1
17.0	0.29237 17047 22737	0.95630 47559 63035	73.0
17.1	0.29404 03252 32304	0.95579 30147 98330	72.9
17.2	0.29570 80500 44047	0.95527 83621 22344	72.8
17.3	0.29737 48740 77786	0.95476 07995 02797	72.7
17.4	0.29904 07922 56087	0.95424 03285 16277	72.6
17.5	0.30070 57995 04273	0.95371 69507 48227	72.5
17.6	0.30236 98907 50445	0.95319 06677 92947	72.4
17.7	0.30403 30609 25490	0.95266 14812 53586	72.3
17.8	0.30569 53049 63106	0.95212 93927 42139	72.2
17.9	0.30735 66177 99807	0.95159 44038 79438	72.1
18.0	0.30901 69943 74947	0.95105 65162 95154	72.0
18.1	0.31067 64296 30732	0.95051 57316 27784	71.9
18.2	0.31233 49185 12233	0.94997 20515 24653	71.8
18.3	0.31399 24559 67405	0.94942 54776 41904	71.7
18.4	0.31564 90369 47102	0.94887 60116 44497	71.6
18.5	0.31730 46564 05092	0.94832 36552 06199	71.5
18.6	0.31895 93092 98070	0.94776 84100 09586	71.4
18.7	0.32061 29905 85676	0.94721 02777 46029	71.3
18.8	0.32226 56952 30511	0.94664 92601 15696	71.2
18.9	0.32391 74181 98149	0.94608 53588 27545	71.1
19.0	0.32556 81544 57157	0.94551 85755 99317	71.0
19.1	0.32721 78989 79104	0.94494 89121 57531	70.9
19.2	0.32886 66467 38583	0.94437 63702 37481	70.8
19.3	0.33051 43927 13223	0.94380 09515 83229	70.7
19.4	0.33216 11318 83703	0.94322 26579 47601	70.6
19.5	0.33380 68592 33771	0.94264 14910 92178	70.5
19.6	0.33545 15697 50255	0.94205 74527 87297	70.4
19.7	0.33709 52584 23082	0.94147 05448 12038	70.3
19.8	0.33873 79202 45291	0.94088 07689 54225	70.2
19.9	0.34037 95502 13050	0.94028 81270 10419	70.1
20.0	0.34202 01433 25669	0.93969 26207 85908	70.0
$90° - \theta$	$\cos \theta$	$\sin \theta$	θ
	$\ast \begin{bmatrix} (-7)1 \\ 5 \end{bmatrix}$	$\begin{bmatrix} (-7)4 \\ 5 \end{bmatrix}$	

*See page II.

CIRCULAR SINES AND COSINES TO TENTHS OF A DEGREE   Table 4.10

θ	$\sin \theta$	$\cos \theta$	$90° - \theta$
20.0°	0.34202 01433 25669	0.93969 26207 85908	70.0°
20.1	0.34365 96945 85616	0.93909 42520 94709	69.9
20.2	0.34529 81989 98535	0.93849 30227 59556	69.8
20.3	0.34693 56515 73256	0.93788 89346 11898	69.7
20.4	0.34857 20473 21815	0.93728 19894 91892	69.6
20.5	0.35020 73812 59467	0.93667 21892 48398	69.5
20.6	0.35184 16484 04702	0.93605 95357 38973	69.4
20.7	0.35347 48437 79257	0.93544 40308 29867	69.3
20.8	0.35510 69624 08137	0.93482 56763 96014	69.2
20.9	0.35673 79993 19625	0.93420 44743 21030	69.1
21.0	0.35836 79495 45300	0.93358 04264 97202	69.0
21.1	0.35999 68081 20051	0.93295 35348 25489	68.9
21.2	0.36162 45700 82092	0.93232 38012 15512	68.8
21.3	0.36325 12304 72978	0.93169 12275 85549	68.7
21.4	0.36487 67843 37620	0.93105 58158 62528	68.6
21.5	0.36650 12267 24297	0.93041 75679 82025	68.5
21.6	0.36812 45526 84678	0.92977 64858 88251	68.4
21.7	0.36974 67572 73829	0.92913 25715 34056	68.3
21.8	0.37136 78355 50235	0.92848 58268 80914	68.2
21.9	0.37298 77825 75809	0.92783 62538 98920	68.1
22.0	0.37460 65934 15912	0.92718 38545 66787	68.0
22.1	0.37622 42631 39366	0.92652 86308 71837	67.9
22.2	0.37784 07868 18467	0.92587 05848 09995	67.8
22.3	0.37945 61595 29005	0.92520 97183 85782	67.7
22.4	0.38107 03763 50274	0.92454 60336 12313	67.6
22.5	0.38268 34323 65090	0.92387 95325 11287	67.5
22.6	0.38429 53226 59804	0.92321 02171 12981	67.4
22.7	0.38590 60423 24319	0.92253 80894 56246	67.3
22.8	0.38751 55864 52103	0.92186 31515 88501	67.2
22.9	0.38912 39501 40206	0.92118 54055 65721	67.1
23.0	0.39073 11284 89274	0.92050 48534 52440	67.0
23.1	0.39233 71166 03561	0.91982 14973 21738	66.9
23.2	0.39394 19095 90951	0.91913 53392 55234	66.8
23.3	0.39554 55025 62965	0.91844 63813 43087	66.7
23.4	0.39714 78906 34781	0.91775 46256 83981	66.6
23.5	0.39874 90689 25246	0.91706 00743 85124	66.5
23.6	0.40034 90325 56895	0.91636 27295 62240	66.4
23.7	0.40194 77766 55960	0.91566 25933 39561	66.3
23.8	0.40354 52963 52390	0.91495 96678 49825	66.2
23.9	0.40514 15867 79863	0.91425 39552 34264	66.1
24.0	0.40673 66430 75800	0.91354 54576 42601	66.0
24.1	0.40833 04603 81385	0.91283 41772 33043	65.9
24.2	0.40992 30338 41573	0.91212 01161 72273	65.8
24.3	0.41151 43586 05109	0.91140 32766 35445	65.7
24.4	0.41310 44298 24542	0.91068 36608 06177	65.6
24.5	0.41469 32426 56239	0.90996 12708 76543	65.5
24.6	0.41628 07922 60401	0.90923 61090 47069	65.4
24.7	0.41786 70738 01077	0.90850 81775 26722	65.3
24.8	0.41945 20824 46177	0.90777 74785 32909	65.2
24.9	0.42103 58133 67491	0.90704 40142 91465	65.1
25.0	0.42261 82617 40699	0.90630 77870 36650	65.0
$90° - \theta$	$\cos \theta$	$\sin \theta$	θ

$$* \quad \begin{bmatrix} (-7)2 \\ 5 \end{bmatrix} \qquad \begin{bmatrix} (-7)4 \\ 5 \end{bmatrix}$$

*See page II.

Table 4.10   **CIRCULAR SINES AND COSINES TO TENTHS OF A DEGREE**

θ	$\sin\theta$	$\cos\theta$	$90°-\theta$
25.0°	0.42261 82617 40699	0.90630 77870 36650	65.0°
25.1	0.42419 94227 45390	0.90556 87990 11140	64.9
25.2	0.42577 92915 65073	0.90482 70524 66020	64.8
25.3	0.42735 78633 87192	0.90408 25496 60778	64.7
25.4	0.42893 51334 03146	0.90333 52928 63301	64.6
25.5	0.43051 10968 08295	0.90258 52843 49861	64.5
25.6	0.43208 57488 01982	0.90183 25264 05114	64.4
25.7	0.43365 90845 87544	0.90107 70213 22092	64.3
25.8	0.43523 10993 72328	0.90031 87714 02194	64.2
25.9	0.43680 17883 67702	0.89955 77789 55180	64.1
26.0	0.43837 11467 89077	0.89879 40462 99167	64.0
26.1	0.43993 91698 55915	0.89802 75757 60616	63.9
26.2	0.44150 58527 91745	0.89725 83696 74328	63.8
26.3	0.44307 11908 24180	0.89648 64303 83441	63.7
26.4	0.44463 51791 84927	0.89571 17602 39413	63.6
26.5	0.44619 78131 09809	0.89493 43616 02025	63.5
26.6	0.44775 90878 38770	0.89415 42368 39368	63.4
26.7	0.44931 89986 15897	0.89337 13883 27838	63.3
26.8	0.45087 75406 89431	0.89258 58184 52125	63.2
26.9	0.45243 47093 11783	0.89179 75296 05214	63.1
27.0	0.45399 04997 39547	0.89100 65241 88368	63.0
27.1	0.45554 49072 33516	0.89021 28046 11127	62.9
27.2	0.45709 79270 58694	0.88941 63732 91298	62.8
27.3	0.45864 95544 84315	0.88861 72326 54949	62.7
27.4	0.46019 97847 83852	0.88781 53851 36401	62.6
27.5	0.46174 86132 35034	0.88701 08331 78222	62.5
27.6	0.46329 60351 19862	0.88620 35792 31215	62.4
27.7	0.46484 20457 24620	0.88539 36257 54416	62.3
27.8	0.46638 66403 39891	0.88458 09752 15084	62.2
27.9	0.46792 98142 60573	0.88376 56300 88693	62.1
28.0	0.46947 15627 85891	0.88294 75928 58927	62.0
28.1	0.47101 18812 19410	0.88212 68660 17668	61.9
28.2	0.47255 07648 69054	0.88130 34520 64992	61.8
28.3	0.47408 82090 47116	0.88047 73535 09162	61.7
28.4	0.47562 42090 70275	0.87964 85728 66617	61.6
28.5	0.47715 87602 59608	0.87881 71126 61965	61.5
28.6	0.47869 18579 40607	0.87798 29754 27981	61.4
28.7	0.48022 34974 43189	0.87714 61637 05589	61.3
28.8	0.48175 36741 01715	0.87630 66800 43864	61.2
28.9	0.48328 23832 55002	0.87546 45270 00018	61.1
29.0	0.48480 96202 46337	0.87461 97071 39396	61.0
29.1	0.48633 53804 23490	0.87377 22230 35465	60.9
29.2	0.48785 96591 38733	0.87292 20772 69810	60.8
29.3	0.48938 24517 48846	0.87206 92724 32121	60.7
29.4	0.49090 37536 15141	0.87121 38111 20189	60.6
29.5	0.49242 35601 03467	0.87035 56959 39900	60.5
29.6	0.49394 18665 84231	0.86949 49295 05219	60.4
29.7	0.49545 86684 32408	0.86863 15144 38191	60.3
29.8	0.49697 39610 27555	0.86776 54533 68928	60.2
29.9	0.49848 77397 53830	0.86689 67489 35603	60.1
30.0	0.50000 00000 00000	0.86602 54037 84439	60.0
$90°-\theta$	$\cos\theta$	$\sin\theta$	θ

$$* \quad \begin{bmatrix} (-7)2 \\ 5 \end{bmatrix} \qquad \begin{bmatrix} (-7)4 \\ 5 \end{bmatrix}$$

---

*See page II.

CIRCULAR SINES AND COSINES TO TENTHS OF A DEGREE   Table 4.10

θ	$\sin \theta$	$\cos \theta$	$90° - \theta$
30.0°	0.50000 00000 00000	0.86602 54037 84439	60.0°
30.1	0.50151 07371 59457	0.86515 14205 69704	59.9
30.2	0.50301 99466 30235	0.86427 48019 53705	59.8
30.3	0.50452 76238 15019	0.86339 55506 06772	59.7
30.4	0.50603 37641 21164	0.86251 36692 07257	59.6
30.5	0.50753 83629 60704	0.86162 91604 41526	59.5
30.6	0.50904 14157 50371	0.86074 20270 03944	59.4
30.7	0.51054 29179 11606	0.85985 22715 96873	59.3
30.8	0.51204 28648 70572	0.85895 98969 30664	59.2
30.9	0.51354 12520 58170	0.85806 49057 23645	59.1
31.0	0.51503 80749 10054	0.85716 73007 02112	59.0
31.1	0.51653 33288 66642	0.85626 70846 00328	58.9
31.2	0.51802 70093 73130	0.85536 42601 60507	58.8
31.3	0.51951 91118 79509	0.85445 88301 32807	58.7
31.4	0.52100 96318 40576	0.85355 07972 75327	58.6
31.5	0.52249 85647 15949	0.85264 01643 54092	58.5
31.6	0.52398 59059 70079	0.85172 69341 43048	58.4
31.7	0.52547 16510 72268	0.85081 11094 24051	58.3
31.8	0.52695 57954 96678	0.84989 26929 86864	58.2
31.9	0.52843 83347 22347	0.84897 16876 29141	58.1
32.0	0.52991 92642 33205	0.84804 80961 56426	58.0
32.1	0.53139 85795 18083	0.84712 19213 82137	57.9
32.2	0.53287 62760 70730	0.84619 31661 27564	57.8
32.3	0.53435 23493 89826	0.84526 18332 21856	57.7
32.4	0.53582 67949 78997	0.84432 79255 02015	57.6
32.5	0.53729 96083 46824	0.84339 14458 12886	57.5
32.6	0.53877 07850 06863	0.84245 23970 07148	57.4
32.7	0.54024 03204 77655	0.84151 07819 45306	57.3
32.8	0.54170 82102 82740	0.84056 66034 95684	57.2
32.9	0.54317 44499 50671	0.83961 98645 34413	57.1
33.0	0.54463 90350 15027	0.83867 05679 45424	57.0
33.1	0.54610 19610 14429	0.83771 87166 20439	56.9
33.2	0.54756 32234 92550	0.83676 43134 58962	56.8
33.3	0.54902 28179 98132	0.83580 73613 68270	56.7
33.4	0.55048 07400 84996	0.83484 78632 63407	56.6
33.5	0.55193 69853 12058	0.83388 58220 67168	56.5
33.6	0.55339 15492 43344	0.83292 12407 10099	56.4
33.7	0.55484 44274 47999	0.83195 41221 30483	56.3
33.8	0.55629 56155 00305	0.83098 44692 74328	56.2
33.9	0.55774 51089 79690	0.83001 22850 95367	56.1
34.0	0.55919 29034 70747	0.82903 75725 55042	56.0
34.1	0.56063 89945 63242	0.82806 03346 22494	55.9
34.2	0.56208 33778 52131	0.82708 05742 74562	55.8
34.3	0.56352 60489 37571	0.82609 82944 95764	55.7
34.4	0.56496 70034 24938	0.82511 34982 78295	55.6
34.5	0.56640 62369 24833	0.82412 61886 22016	55.5
34.6	0.56784 37450 53101	0.82313 63685 34442	55.4
34.7	0.56927 95234 30844	0.82214 40410 30737	55.3
34.8	0.57071 35676 84432	0.82114 92091 33704	55.2
34.9	0.57214 58734 45516	0.82015 18758 73772	55.1
35.0	0.57357 64363 51046	0.81915 20442 88992	55.0
$90° - \theta$	$\cos \theta$	$\sin \theta$	θ

$$* \begin{bmatrix} (-7)2 \\ 5 \end{bmatrix} \qquad \begin{bmatrix} (-7)3 \\ 5 \end{bmatrix}$$

*See page II.

Table 4.10   CIRCULAR SINES AND COSINES TO TENTHS OF A DEGREE

θ	$\sin \theta$	$\cos \theta$	$90° - \theta$
35.0°	0.57357 64363 51046	0.81915 20442 88992	55.0°
35.1	0.57500 52520 43279	0.81814 97174 25023	54.9
35.2	0.57643 23161 69793	0.81714 48983 35129	54.8
35.3	0.57785 76243 83505	0.81613 75900 80160	54.7
35.4	0.57928 11723 42679	0.81512 77957 28554	54.6
35.5	0.58070 29557 10940	0.81411 55183 56319	54.5
35.6	0.58212 29701 57289	0.81310 07610 47028	54.4
35.7	0.58354 12113 56118	0.81208 35268 91806	54.3
35.8	0.58495 76749 87215	0.81106 38189 89327	54.2
35.9	0.58637 23567 35789	0.81004 16404 45796	54.1
36.0	0.58778 52522 92473	0.80901 69943 74947	54.0
36.1	0.58919 63573 53342	0.80798 98838 98031	53.9
36.2	0.59060 56676 19925	0.80696 03121 43802	53.8
36.3	0.59201 31787 99220	0.80592 82822 48516	53.7
36.4	0.59341 88866 03701	0.80489 37973 55914	53.6
36.5	0.59482 27867 51341	0.80385 68606 17217	53.5
36.6	0.59622 48749 65616	0.80281 74751 91115	53.4
36.7	0.59762 51469 75521	0.80177 56442 43754	53.3
36.8	0.59902 35985 15586	0.80073 13709 48733	53.2
36.9	0.60042 02253 25884	0.79968 46584 87091	53.1
37.0	0.60181 50231 52048	0.79863 55100 47293	53.0
37.1	0.60320 79877 45282	0.79758 39288 25229	52.9
37.2	0.60459 91148 62375	0.79652 99180 24196	52.8
37.3	0.60598 84002 65711	0.79547 34808 54896	52.7
37.4	0.60737 58397 23287	0.79441 46205 35418	52.6
37.5	0.60876 14290 08721	0.79335 33402 91235	52.5
37.6	0.61014 51639 01268	0.79228 96433 55191	52.4
37.7	0.61152 70401 85831	0.79122 35329 67490	52.3
37.8	0.61290 70536 52976	0.79015 50123 75690	52.2
37.9	0.61428 52000 98943	0.78908 40848 34691	52.1
38.0	0.61566 14753 25658	0.78801 07536 06722	52.0
38.1	0.61703 58751 40749	0.78693 50219 61337	51.9
38.2	0.61840 83953 57554	0.78585 68931 75402	51.8
38.3	0.61977 90317 95140	0.78477 63705 33083	51.7
38.4	0.62114 77802 78310	0.78369 34573 25840	51.6
38.5	0.62251 46366 37620	0.78260 81568 52414	51.5
38.6	0.62387 95967 09386	0.78152 04724 18819	51.4
38.7	0.62524 26563 35705	0.78043 04073 38330	51.3
38.8	0.62660 38113 64461	0.77933 79649 31474	51.2
38.9	0.62796 30576 49338	0.77824 31485 26021	51.1
39.0	0.62932 03910 49837	0.77714 59614 56971	51.0
39.1	0.63067 58074 31286	0.77604 64070 66546	50.9
39.2	0.63202 93026 64851	0.77494 44887 04180	50.8
39.3	0.63338 08726 27550	0.77384 02097 26506	50.7
39.4	0.63473 05132 02268	0.77273 35734 97351	50.6
39.5	0.63607 82202 77764	0.77162 45833 87720	50.5
39.6	0.63742 39897 48690	0.77051 32427 75789	50.4
39.7	0.63876 78175 15598	0.76939 95550 46895	50.3
39.8	0.64010 96994 84955	0.76828 35235 93523	50.2
39.9	0.64144 96315 69158	0.76716 51518 15300	50.1
40.0	0.64278 76096 86539	0.76604 44431 18978	50.0
$90° - \theta$	$\cos \theta$	$\sin \theta$	θ

$$* \begin{bmatrix} (-7)2 \\ 5 \end{bmatrix} \qquad \begin{bmatrix} (-7)3 \\ 5 \end{bmatrix}$$

---

*See page II.

CIRCULAR SINES AND COSINES TO TENTHS OF A DEGREE — Table 4.10

θ	$\sin \theta$	$\cos \theta$	$90° - \theta$
40.0°	0.64278 76096 86539	0.76604 44431 18978	50.0°
40.1	0.64412 36297 61387	0.76492 14009 18432	49.9
40.2	0.64545 76877 23951	0.76379 60286 34642	49.8
40.3	0.64678 97795 10460	0.76266 83296 95688	49.7
40.4	0.64811 99010 63131	0.76153 83075 36737	49.6
40.5	0.64944 80483 30184	0.76040 59656 00031	49.5
40.6	0.65077 42172 65851	0.75927 13073 34881	49.4
40.7	0.65209 84038 30392	0.75813 43361 97652	49.3
40.8	0.65342 06039 90105	0.75699 50556 51756	49.2
40.9	0.65474 08137 17340	0.75585 34691 67640	49.1
41.0	0.65605 90289 90507	0.75470 95802 22772	49.0
41.1	0.65737 52457 94096	0.75356 33923 01638	48.9
41.2	0.65868 94601 18680	0.75241 49088 95724	48.8
41.3	0.66000 16679 60937	0.75126 41335 03511	48.7
41.4	0.66131 18653 23652	0.75011 10696 30460	48.6
41.5	0.66262 00482 15737	0.74895 57207 89002	48.5
41.6	0.66392 62126 52242	0.74779 80904 98532	48.4
41.7	0.66523 03546 54361	0.74663 81822 85391	48.3
41.8	0.66653 24702 49452	0.74547 59996 82862	48.2
41.9	0.66783 25554 71047	0.74431 15462 31154	48.1
42.0	0.66913 06063 58858	0.74314 48254 77394	48.0
42.1	0.67042 66189 58799	0.74197 58409 75616	47.9
42.2	0.67172 05893 22990	0.74080 45962 86750	47.8
42.3	0.67301 25135 09773	0.73963 10949 78610	47.7
42.4	0.67430 23875 83723	0.73845 53406 25884	47.6
42.5	0.67559 02076 15660	0.73727 73368 10124	47.5
42.6	0.67687 59696 82661	0.73609 70871 19734	47.4
42.7	0.67815 96698 68071	0.73491 45951 49960	47.3
42.8	0.67944 13042 61517	0.73372 98645 02876	47.2
42.9	0.68072 08689 58918	0.73254 28987 87379	47.1
43.0	0.68199 83600 62499	0.73135 37016 19170	47.0
43.1	0.68327 37736 80799	0.73016 22766 20752	46.9
43.2	0.68454 71059 28689	0.72896 86274 21412	46.8
43.3	0.68581 83529 27376	0.72777 27576 57210	46.7
43.4	0.68708 75108 04423	0.72657 46709 70976	46.6
43.5	0.68835 45756 93754	0.72537 43710 12288	46.5
43.6	0.68961 95437 35670	0.72417 18614 37468	46.4
43.7	0.69088 24110 76858	0.72296 71459 09568	46.3
43.8	0.69214 31738 70407	0.72176 02280 98362	46.2
43.9	0.69340 18282 75813	0.72055 11116 80330	46.1
44.0	0.69465 83704 58997	0.71933 98003 38651	46.0
44.1	0.69591 27965 92314	0.71812 62977 63189	45.9
44.2	0.69716 51028 54565	0.71691 06076 50483	45.8
44.3	0.69841 52854 31006	0.71569 27337 03736	45.7
44.4	0.69966 33405 13365	0.71447 26796 32803	45.6
44.5	0.70090 92642 99851	0.71325 04491 54182	45.5
44.6	0.70215 30529 95162	0.71202 60459 90996	45.4
44.7	0.70339 47028 10504	0.71079 94738 72992	45.3
44.8	0.70463 42099 63595	0.70957 07365 36521	45.2
44.9	0.70587 15706 78681	0.70833 98377 24529	45.1
45.0	0.70710 67811 86548	0.70710 67811 86548	45.0
$90° - \theta$	$\cos \theta$	$\sin \theta$	θ

$$* \begin{bmatrix} (-7)3 \\ 5 \end{bmatrix} \qquad\qquad \begin{bmatrix} (-7)3 \\ 5 \end{bmatrix}$$

---

*See page II.

**Table 4.11   CIRCULAR TANGENTS, COTANGENTS, SECANTS AND COSECANTS
TO FIVE TENTHS OF A DEGREE**

θ	$\tan \theta$	$\cot \theta$	$\sec \theta$	$\csc \theta$	$90° - \theta$
0.0°	0.00000 00000 00000	∞	1.00000 000	∞	90.0°
0.5	0.00872 68677 90759	114.58865 01293 09608	1.00003 808	114.59301 348	89.5
1.0	0.01745 50649 28217	57.28996 16307 59424	1.00015 233	57.29868 850	89.0
1.5	0.02618 59215 69187	38.18845 92970 25609	1.00034 279	38.20155 001	88.5
2.0	0.03492 07694 91747	28.63625 32829 15603	1.00060 954	28.65370 835	88.0
2.5	0.04366 09429 08512	22.90376 55484 31198	1.00095 269	22.92558 563	87.5
3.0	0.05240 77792 83041	19.08113 66877 28211	1.00137 235	19.10732 261	87.0
3.5	0.06116 26201 50484	16.34985 54760 99672	1.00186 869	16.38040 824	86.5
4.0	0.06992 68119 43510	14.30066 62567 11928	1.00244 190	14.33558 703	86.0
4.5	0.07870 17068 24618	12.70620 47361 74704	1.00309 220	12.74549 484	85.5
5.0	0.08748 86635 25924	11.43005 23027 61343	1.00381 984	11.47371 325	85.0
5.5	0.09628 90481 97538	10.38539 70801 38159	1.00462 509	10.43343 052	84.5
6.0	0.10510 42352 65676	9.51436 44542 22585	1.00550 828	9.56677 223	84.0
6.5	0.11393 56083 01645	8.77688 73568 69956	1.00646 973	8.83367 147	83.5
7.0	0.12278 45609 02904	8.14434 64279 74594	1.00750 983	8.20550 905	83.0
7.5	0.13165 24975 87396	7.59575 41127 25150	1.00862 896	7.66129 758	82.5
8.0	0.14054 08347 02391	7.11536 97223 84209	1.00982 757	7.18529 653	82.0
8.5	0.14945 10013 49128	6.69115 62383 17409	1.01110 613	6.76546 908	81.5
9.0	0.15838 44403 24536	6.31375 15146 75043	1.01246 513	6.39245 322	81.0
9.5	0.16734 26090 81419	5.97576 43644 33065	1.01390 510	6.05885 796	80.5
10.0	0.17632 69807 08465	5.67128 18196 17709	1.01542 661	5.75877 049	80.0
10.5	0.18533 90449 31534	5.39551 71743 19137	1.01703 027	5.48740 427	79.5
11.0	0.19438 03091 37718	5.14455 40159 70310	1.01871 670	5.24084 307	79.0
11.5	0.20345 22994 23699	4.91515 70310 71205	1.02048 657	5.01585 174	78.5
12.0	0.21255 65616 70022	4.70463 01094 78454	1.02234 059	4.80973 435	78.0
12.5	0.22169 46626 42940	4.51070 85036 62057	1.02427 951	4.62022 632	77.5
13.0	0.23086 81911 25563	4.33147 58742 84155	1.02630 411	4.44541 148	77.0
13.5	0.24007 87590 80116	4.16529 97700 90417	1.02841 519	4.28365 757	76.5
14.0	0.24932 80028 43180	4.01078 09335 35844	1.03061 363	4.13356 550	76.0
14.5	0.25861 75843 55890	3.86671 30948 98738	1.03290 031	3.99392 916	75.5
15.0	0.26794 91924 31122	3.73205 08075 68877	1.03527 618	3.86370 331	75.0
15.5	0.27732 45440 59838	3.60588 35087 60874	1.03774 221	3.74197 754	74.5
16.0	0.28674 53857 58808	3.48741 44438 40908	1.04029 944	3.62795 528	74.0
16.5	0.29621 34949 62080	3.37594 34225 91246	1.04294 891	3.52093 652	73.5
17.0	0.30573 06814 58660	3.27085 26184 84141	1.04569 176	3.42030 362	73.0
17.5	0.31529 87888 78983	3.17159 48023 63212	1.04852 913	3.32550 952	72.5
18.0	0.32491 96962 32906	3.07768 35371 75253	1.05146 222	3.23606 798	72.0
18.5	0.33459 53195 02073	2.98868 49627 42893	1.05449 231	3.15154 530	71.5
19.0	0.34432 76132 89665	2.90421 08776 75823	1.05762 068	3.07155 349	71.0
19.5	0.35411 85725 30698	2.82391 28856 00801	1.06084 870	2.99574 431	70.5
20.0	0.36397 02342 66202	2.74747 74194 54622	1.06417 777	2.92380 440	70.0
20.5	0.37388 46794 84804	2.67462 14939 26824	1.06760 936	2.85545 095	69.5
21.0	0.38386 40350 35416	2.60508 90646 93801	1.07114 499	2.79042 811	69.0
21.5	0.39391 04756 14942	2.53864 78956 64307	1.07478 624	2.72850 383	68.5
22.0	0.40402 62258 35157	2.47508 68534 16296	1.07853 474	2.66946 716	68.0
22.5	0.41421 35623 73095	2.41421 35623 73095	1.08239 220	2.61312 593	67.5
$90° - \theta$	$\cot \theta$	$\tan \theta$	$\csc \theta$	$\sec \theta$	θ

$$\begin{bmatrix} (-5)1 \\ 8 \end{bmatrix} \qquad \begin{bmatrix} (-5)1 \\ 4 \end{bmatrix}$$

CIRCULAR TANGENTS, COTANGENTS, SECANTS AND COSECANTS
TO FIVE TENTHS OF A DEGREE

Table 4.11

θ	tan θ	cot θ	sec θ	csc θ	$90°-\theta$
22.5°	0.41421 35623 73095	2.41421 35623 73095	1.08239 220	2.61312 593	67.5°
23.0	0.42447 48162 09604	2.35585 23658 23753	1.08636 038	2.55930 467	67.0
23.5	0.43481 23749 60933	2.29984 25472 36257	1.09044 110	2.50784 285	66.5
24.0	0.44522 86853 08536	2.24603 67739 04216	1.09463 628	2.45859 334	66.0
24.5	0.45572 62555 32584	2.19429 97311 65038	1.09894 787	2.41142 102	65.5
25.0	0.46630 76581 54998	2.14450 69205 09558	1.10337 792	2.36620 158	65.0
25.5	0.47697 55326 98160	2.09654 35990 88174	1.10792 854	2.32282 050	64.5
26.0	0.48773 25885 65861	2.05030 38415 79296	1.11260 194	2.28117 203	64.0
26.5	0.49858 16080 53431	2.00568 97082 59020	1.11740 038	2.24115 845	63.5
27.0	0.50952 54494 94429	1.96261 05055 05150	1.12232 624	2.20268 926	63.0
27.5	0.52056 70505 51746	1.92090 21269 71166	1.12738 195	2.16568 057	62.5
28.0	0.53170 94316 61479	1.88072 64653 46332	1.13257 005	2.13005 447	62.0
28.5	0.54295 56996 38437	1.84177 08860 33458	1.13789 318	2.09573 853	61.5
29.0	0.55430 90514 52769	1.80404 77552 71424	1.14335 407	2.06266 534	61.0
29.5	0.56577 27781 87770	1.76749 40162 42891	1.14895 554	2.03077 204	60.5
30.0	0.57735 02691 89626	1.73205 08075 68877	1.15470 054	2.00000 000	60.0
30.5	0.58904 50164 20551	1.69766 31193 26089	1.16059 210	1.97029 441	59.5
31.0	0.60086 06190 27560	1.66427 94823 50518	1.16663 340	1.94160 403	59.0
31.5	0.61280 07881 39932	1.63185 16871 28789	1.17282 770	1.91388 086	58.5
32.0	0.62486 93519 09327	1.60033 45290 41050	1.17917 840	1.88707 991	58.0
32.5	0.63707 02608 07493	1.56968 55771 17490	1.18568 905	1.86115 900	57.5
33.0	0.64940 75931 97510	1.53986 49638 14583	1.19236 329	1.83607 846	57.0
33.5	0.66188 55611 95691	1.51083 51936 14901	1.19920 494	1.81180 103	56.5
34.0	0.67450 85168 42426	1.48256 09685 12740	1.20621 795	1.78829 165	56.0
34.5	0.68728 09586 01613	1.45500 90286 72445	1.21340 641	1.76551 728	55.5
35.0	0.70020 75382 09710	1.42814 80067 42114	1.22077 459	1.74344 680	55.0
35.5	0.71329 30678 97005	1.40194 82944 76336	1.22832 691	1.72205 082	54.5
36.0	0.72654 25280 05361	1.37638 19204 71173	1.23606 798	1.70130 162	54.0
36.5	0.73996 10750 28487	1.35142 24379 45808	1.24400 257	1.68117 299	53.5
37.0	0.75355 40501 02794	1.32704 48216 20410	1.25213 566	1.66164 014	53.0
37.5	0.76732 69879 78960	1.30322 53728 41206	1.26047 241	1.64267 963	52.5
38.0	0.78128 56265 06717	1.27994 16321 93079	1.26901 822	1.62426 925	52.0
38.5	0.79543 59166 67828	1.25717 22989 18954	1.27777 866	1.60638 793	51.5
39.0	0.80978 40331 95007	1.23489 71565 35051	1.28675 957	1.58901 573	51.0
39.5	0.82433 63858 17495	1.21309 70040 92932	1.29596 700	1.57213 369	50.5
40.0	0.83909 96311 77280	1.19175 35925 94210	1.30540 729	1.55572 383	50.0
40.5	0.85408 06854 63466	1.17084 95661 12539	1.31508 700	1.53976 904	49.5
41.0	0.86928 67378 16226	1.15036 84072 21009	1.32501 299	1.52425 309	49.0
41.5	0.88472 52645 55944	1.13029 43863 61753	1.33519 242	1.50916 050	48.5
42.0	0.90040 40442 97840	1.11061 25148 29193	1.34563 273	1.49447 655	48.0
42.5	0.91633 11740 17423	1.09130 85010 69271	1.35634 170	1.48018 723	47.5
43.0	0.93251 50861 37661	1.07236 87100 24682	1.36732 746	1.46627 919	47.0
43.5	0.94896 45667 14880	1.05378 01252 80962	1.37859 847	1.45273 967	46.5
44.0	0.96568 87748 07074	1.03553 03137 90569	1.39016 359	1.43955 654	46.0
44.5	0.98269 72631 15690	1.01760 73929 72125	1.40203 206	1.42671 819	45.5
45.0	1.00000 00000 00000	1.00000 00000 00000	1.41421 356	1.41421 356	45.0
$90°-\theta$	cot θ	tan θ	csc θ	sec θ	θ
	$\begin{bmatrix}(-5)4\\9\end{bmatrix}$	$\begin{bmatrix}(-4)3\end{bmatrix}$	$\begin{bmatrix}(-5)4\\5\end{bmatrix}$	$\begin{bmatrix}(-4)3\\6\end{bmatrix}$	

Table 4.12　　　　CIRCULAR FUNCTIONS FOR THE ARGUMENT $\frac{\pi}{2}x$

x	$\sin\frac{\pi}{2}x$	$\cos\frac{\pi}{2}x$	$\tan\frac{\pi}{2}x$	$1-x$
0.00	0.00000 00000 00000 00000	1.00000 00000 00000 00000	0.00000 00000 00000 00000	1.00
0.01	0.01570 73173 11820 67575	0.99987 66324 81660 59864	0.01570 92553 23664 91632	0.99
0.02	0.03141 07590 78128 29384	0.99950 65603 65731 55700	0.03142 62660 43351 14782	0.98
0.03	0.04710 64507 09642 66090	0.99888 98749 61969 97264	0.04715 88028 77480 47448	0.97
0.04	0.06279 05195 29313 37607	0.99802 67284 28271 56195	0.06291 46672 53649 75722	0.96
0.05	0.07845 90957 27844 94503	0.99691 73337 33127 97620	0.07870 17068 24618 44806	0.95
0.06	0.09410 83133 18514 31847	0.99556 19646 03080 01290	0.09452 78311 79282 04901	0.94
0.07	0.10973 43110 91045 26802	0.99396 09554 55179 68775	0.11040 10278 15818 94497	0.93
0.08	0.12533 32335 64304 24537	0.99211 47013 14477 83105	0.12632 93784 46108 17478	0.92
0.09	0.14090 12319 37582 66116	0.99002 36577 16557 56725	0.14232 10757 02942 94229	0.91
0.10	0.15643 44650 40230 86901	0.98768 83405 95137 72619	0.15838 44403 24536 29384	0.90
0.11	0.17192 91002 79409 54661	0.98510 93261 54773 91802	0.17452 79388 94365 08461	0.89
0.12	0.18738 13145 85724 63054	0.98228 72507 28688 68108	0.19076 02022 18566 74856	0.88
0.13	0.20278 72953 56512 48344	0.97922 28106 21765 78086	0.20709 00444 27938 70402	0.87
0.14	0.21814 32413 96542 55202	0.97591 67619 38747 39896	0.22352 64828 97149 10184	0.86
0.15	0.23344 53638 55905 41177	0.97236 99203 97676 60183	0.24007 87590 80116 03926	0.85
0.16	0.24868 98871 64854 78824	0.96858 31611 28631 11949	0.25675 63603 67726 78332	0.84
0.17	0.26387 30499 65372 89696	0.96455 74184 57798 09366	0.27356 90430 82237 23655	0.83
0.18	0.27899 11060 39229 25185	0.96029 36856 76943 07175	0.29052 68567 31916 45432	0.82
0.19	0.29404 03252 32303 95777	0.95579 30147 98330 12664	0.30764 01696 59898 29067	0.81
0.20	0.30901 69943 74947 42410	0.95105 65162 95153 57211	0.32491 96962 32906 32615	0.80
0.21	0.32391 74181 98149 41440	0.94608 53588 27545 31853	0.34237 65257 28683 05965	0.79
0.22	0.33873 79202 45291 38122	0.94088 07689 54225 47232	0.36002 21530 95756 62634	0.78
0.23	0.35347 48437 79257 12472	0.93544 40308 29867 32518	0.37786 85117 75820 93670	0.77
0.24	0.36812 45526 84677 95915	0.92977 64858 88251 40366	0.39592 80087 97721 26049	0.76
0.25	0.38268 34323 65089 77173	0.92387 95325 11286 75613	0.41421 35623 73095 04880	0.75
0.26	0.39714 78906 34780 61375	0.91775 46256 83981 14114	0.43273 86422 47425 93197	0.74
0.27	0.41151 43586 05108 77405	0.91140 32766 35445 24821	0.45151 73130 86983 28945	0.73
0.28	0.42577 92915 65072 64886	0.90482 70524 66019 52771	0.47056 42812 12251 49308	0.72
0.29	0.43993 91698 55915 14083	0.89802 75757 60615 63093	0.48989 49450 22477 05270	0.71
0.30	0.45399 04997 39546 79156	0.89100 65241 88367 86236	0.50952 54494 94428 81051	0.70
0.31	0.46792 98142 60573 37723	0.88376 56300 88693 42432	0.52947 27451 82014 63252	0.69
0.32	0.48175 36741 01715 27498	0.87630 66800 43863 58731	0.54975 46521 92770 07429	0.68
0.33	0.49545 86684 32407 53805	0.86863 15144 38191 24777	0.57038 99296 73294 88698	0.67
0.34	0.50904 14157 50371 30028	0.86074 20270 03943 63716	0.59139 83513 99471 09817	0.66
0.35	0.52249 85647 15948 86499	0.85264 01643 54092 22152	0.61280 07881 39931 99664	0.65
0.36	0.53582 67949 78996 61827	0.84432 79255 02015 07855	0.63461 92975 44148 10071	0.64
0.37	0.54902 28179 98131 74352	0.83580 73613 68270 25847	0.65687 72224 01279 37691	0.63
0.38	0.56208 33778 52130 60010	0.82708 05742 74561 82492	0.67959 92982 24526 52184	0.62
0.39	0.57500 52520 43278 56590	0.81814 97174 25023 43213	0.70281 17712 40357 33761	0.61
0.40	0.58778 52522 92473 12917	0.80901 69943 74947 42410	0.72654 25280 05360 88589	0.60
0.41	0.60042 02253 25884 04976	0.79968 46584 87090 53868	0.75082 12380 38764 68575	0.59
0.42	0.61290 70536 52976 49336	0.79015 50123 75690 36516	0.77567 95110 49613 10378	0.58
0.43	0.62524 26563 35705 17290	0.78043 04073 38329 73585	0.80115 10705 58751 23382	0.57
0.44	0.63742 39897 48689 71017	0.77051 32427 75789 23080	0.82727 19459 72475 63403	0.56
0.45	0.64944 80483 30183 65572	0.76040 59656 00030 93817	0.85408 06854 63466 63752	0.55
0.46	0.66131 18653 23651 87657	0.75011 10696 30459 54151	0.88161 85923 63189 11465	0.54
0.47	0.67301 25135 09773 33872	0.73963 10949 78609 69747	0.90992 99881 77737 46579	0.53
0.48	0.68454 71059 28688 67373	0.72896 86274 21411 52314	0.93906 25058 17492 35255	0.52
0.49	0.69591 27965 92314 32549	0.71812 62977 63188 83037	0.96906 74171 93793 27618	0.51
0.50	0.70710 67811 86547 52440	0.70710 67811 86547 52440	1.00000 00000 00000 00000	0.50

$1-x$	$\cos\frac{\pi}{2}x$	$\sin\frac{\pi}{2}x$	$\cot\frac{\pi}{2}x$	x
	$\left[\dfrac{(-5)2}{10}\right]$	$\left[\dfrac{(-5)3}{10}\right]$	$\left[\dfrac{(-4)1}{10}\right]$	

CIRCULAR FUNCTIONS FOR THE ARGUMENT $\frac{\pi}{2}x$

Table 4.12

x	$\cot \frac{\pi}{2} x$	$\sec \frac{\pi}{2} x$	$\csc \frac{\pi}{2} x$	$1-x$
0.00	∞	1.00000 00000 00000 00000	∞	1.00
0.01	63.65674 11628 71580 99500	1.00012 33827 39761 81169	63.66459 53060 00564 58546	0.99
0.02	31.82051 59537 73958 03934	1.00049 36832 37144 42400	31.83622 52090 97622 95566	0.98
0.03	21.20494 87896 88751 52283	1.00111 13587 85243 76109	21.22851 50958 16816 17580	0.97
0.04	15.89454 48438 65303 44576	1.00197 71730 71142 10978	15.92597 11099 08654 59358	0.96
0.05	12.70620 47361 74704 64602	1.00309 21984 82825 50283	12.74549 48431 82374 28619	0.95
0.06	10.57889 49934 05635 52417	1.00445 78193 57019 51480	10.62605 37962 83115 99865	0.94
0.07	9.05788 66862 38928 19329	1.00607 57361 86291 90575	9.11292 00161 49841 72675	0.93
0.08	7.91581 50883 05826 84427	1.00794 79708 09297 28943	7.97872 97555 59476 60149	0.92
0.09	7.02636 62290 41380 19848	1.01007 68726 13784 19104	7.09717 00264 69225 38129	0.91
0.10	6.31375 15146 75043 09898	1.01246 51257 88002 93136	6.39245 32214 99661 54704	0.90
0.11	5.72974 16467 24314 86192	1.01511 57576 62501 87437	5.81635 10329 24944 03199	0.89
0.12	5.24218 35811 13176 73758	1.01803 21481 91042 38259	5.33671 14122 92458 78659	0.88
0.13	4.82881 73521 92759 97818	1.02121 80406 26567 47910	4.93127 53949 49859 96253	0.87
0.14	4.47374 28292 11554 62415	1.02467 75534 55900 33566	4.58414 38570 27373 56913	0.86
0.15	4.16529 97700 90417 20387	1.02841 51936 65208 54585	4.28365 75697 31185 03924	0.85
0.16	3.89474 28549 29859 33474	1.03243 58714 17339 88710	4.02107 22333 75967 50952	0.84
0.17	3.65538 43546 52259 73004	1.03674 49162 32016 53065	3.78970 11465 59780 81919	0.83
0.18	3.44202 25766 69218 62809	1.04134 80947 70681 14007	3.58434 36523 72161 57038	0.82
0.19	3.25055 08012 99836 37634	1.04625 16303 39647 78848	3.40089 40753 61802 31848	0.81
0.20	3.07768 35371 75253 40257	1.05146 22242 38267 21205	3.23606 79774 99789 69641	0.80
0.21	2.92076 09892 98816 40048	1.05698 70790 93232 61183	3.08720 66268 08416 38088	0.79
0.22	2.77760 68539 14974 88865	1.06283 39243 36113 96396	2.95213 47928 09339 97327	0.78
0.23	2.64642 32102 86631 86514	1.06901 10439 98926 01199	2.82905 56388 91501 64260	0.77
0.24	2.52571 16894 47304 99451	1.07552 73070 22247 78234	2.71647 18916 65871 74307	0.76
0.25	2.41421 35623 73095 04880	1.08239 22002 92393 96880	2.61312 59297 52753 05571	0.75
0.26	2.31086 36538 82410 63708	1.08961 58646 48705 30888	2.51795 36983 10349 34110	0.74
0.27	2.21475 44978 13361 51875	1.09720 91341 29537 26252	2.43004 88648 55296 52041	0.73
0.28	2.12510 81731 57202 76115	1.10518 35787 56399 59380	2.34863 46560 54351 86300	0.72
0.29	2.04125 39671 21703 26026	1.11355 15511 90413 37268	2.27304 15214 61957 72361	0.71
0.30	1.96261 05055 05150 58230	1.12232 62376 34360 80715	2.20268 92645 85266 62156	0.70
0.31	1.88867 13416 31067 67620	1.13152 17133 97749 42882	2.13707 26325 27611 85837	0.69
0.32	1.81899 32472 81066 27571	1.14115 30035 92241 17245	2.07574 96076 48793 05903	0.68
0.33	1.75318 66324 72237 08332	1.15123 61494 81376 51287	2.01833 18280 89559 43676	0.67
0.34	1.69090 76557 85011 24674	1.16178 82810 72765 98515	1.96447 66988 67248 48330	0.66
0.35	1.63185 16871 28789 61767	1.17282 76966 14008 94955	1.91388 08554 30942 72280	0.65
0.36	1.57574 78599 68651 08688	1.18437 39497 36918 17500	1.86627 47167 00567 54120	0.64
0.37	1.52235 45068 96131 24085	1.19644 79214 89806 17366	1.82141 79214 74081 38479	0.63
0.38	1.47145 53158 19969 04283	1.20907 20434 06541 15436	1.77909 54854 79867 33350	0.62
0.39	1.42285 60774 31870 59031	1.22227 01770 86068 14117	1.73911 45497 30640 74960°	0.61
0.40	1.37638 19204 71173 53820	1.23606 79774 99789 69641	1.70130 16167 04079 86436	0.60
0.41	1.33187 49515 02597 59439	1.25049 29154 09784 85573	1.66550 01910 65749 08074	0.59
0.42	1.28919 22317 85066 67042	1.26557 44560 72090 15648	1.63156 87575 13749 73007	0.58
0.43	1.24820 40363 53049 43751	1.28134 42308 20677 31999	1.59937 90408 68062 88301	0.57
0.44	1.20879 23504 09609 13115	1.29783 62271 84727 12712	1.56881 45035 05365 75750	0.56
0.45	1.17084 95661 12539 22520	1.31508 69998 90784 80424	1.53976 90432 22366 30748	0.55
0.46	1.13427 73492 55405 46422	1.33313 59504 50172 40410	1.51214 58610 31226 40092	0.54
0.47	1.09898 56505 36301 56382	1.35202 53634 40027 12805	1.48585 64735 81717 76608	0.53
0.48	1.06489 18403 24791 86700	1.37180 11480 64918 28453	1.46081 98491 22513 12750	0.52
0.49	1.03191 99492 80495 57182	1.39251 27141 49012 49662	1.43696 16493 57094 20394	0.51
0.50	1.00000 00000 00000 00000	1.41421 35623 73095 04880	1.41421 35623 73095 04880	0.50
$1-x$	$\tan \frac{\pi}{2} x$	$\csc \frac{\pi}{2} x$	$\sec \frac{\pi}{2} x$	x

$$\left[(-4)1\right]$$

Table 4.13   HARMONIC ANALYSIS

r	$\sin\dfrac{2\pi r}{s}$	$\cos\dfrac{2\pi r}{s}$	$\sin\dfrac{2\pi r}{s}$	$\cos\dfrac{2\pi r}{s}$	$\sin\dfrac{2\pi r}{s}$	$\cos\dfrac{2\pi r}{s}$
	$s=3$		$s=4$		$s=5$	
1	0.86602 54038	−0.50000 00000	1.00000 00000	+0.00000 00000	0.95105 65163	+0.30901 69944
2			0.00000 00000	−1.00000 00000	0.58778 52523	−0.80901 69944
	$s=6$		$s=7$		$s=8$	
1	0.86602 54038	+0.50000 00000	0.78183 14824	+0.62348 98019	0.70710 67812	0.70710 67812
2	0.86602 54038	−0.50000 00000	0.97492 79122	−0.22252 09340	1.00000 00000	+0.00000 00000
3	0.00000 00000	−1.00000 00000	0.43388 37391	−0.90096 88679	0.70710 67812	−0.70710 67812
4					0.00000 00000	−1.00000 00000
	$s=9$		$s=10$		$s=11$	
1	0.64278 76097	0.76604 44431	0.58778 52523	0.80901 69944	0.54064 08174	0.84125 35328
2	0.98480 77530	+0.17364 81777	0.95105 65163	+0.30901 69944	0.90963 19953	+0.41541 50130
3	0.86602 54038	−0.50000 00000	0.95105 65163	−0.30901 69944	0.98982 14419	−0.14231 48383
4	0.34202 01433	−0.93969 26208	0.58778 52523	−0.80901 69944	0.75574 95743	−0.65486 07340
5			0.00000 00000	−1.00000 00000	0.28173 25568	−0.95949 29736
	$s=12$		$s=13$		$s=14$	
1	0.50000 00000	0.86602 54038	0.46472 31720	0.88545 60257	0.43388 37391	0.90096 88679
2	0.86602 54038	0.50000 00000	0.82298 38659	0.56806 47468	0.78183 14825	0.62348 98019
3	1.00000 00000	+0.00000 00000	0.99270 88741	+0.12053 66803	0.97492 79122	+0.22252 09340
4	0.86602 54038	−0.50000 00000	0.93501 62427	−0.35460 48871	0.97492 79122	−0.22252 09340
5	0.50000 00000	−0.86602 54038	0.66312 26582	−0.74851 07482	0.78183 14825	−0.62348 98019
6	0.00000 00000	−1.00000 00000	0.23931 56643	−0.97094 18174	0.43388 37391	−0.90096 88679
7					0.00000 00000	−1.00000 00000
	$s=15$		$s=16$		$s=17$	
1	0.40673 66431	0.91354 54576	0.38268 34324	0.92387 95325	0.36124 16662	0.93247 22294
2	0.74314 48255	0.66913 06064	0.70710 67812	0.70710 67812	0.67369 56436	0.73900 89172
3	0.95105 65163	+0.30901 69944	0.92387 95325	0.38268 34324	0.89516 32913	0.44573 83558
4	0.99452 18954	−0.10452 84633	1.00000 00000	+0.00000 00000	0.99573 41763	+0.09226 83595
5	0.86602 54038	−0.50000 00000	0.92387 95325	−0.38268 34324	0.96182 56432	−0.27366 29901
6	0.58778 52523	−0.80901 69944	0.70710 67812	−0.70710 67812	0.79801 72273	−0.60263 46364
7	0.20791 16908	−0.97814 76007	0.38268 34324	−0.92387 95325	0.52643 21629	−0.85021 71357
8			0.00000 00000	−1.00000 00000	0.18374 95178	−0.98297 30997
	$s=18$		$s=19$		$s=20$	
1	0.34202 01433	0.93969 26208	0.32469 94692	0.94581 72417	0.30901 69944	0.95105 65163
2	0.64278 76097	0.76604 44431	0.61421 27127	0.78914 05094	0.58778 52523	0.80901 69944
3	0.86602 54038	0.50000 00000	0.83716 64782	0.54694 81581	0.80901 69944	0.58778 52523
4	0.98480 77530	+0.17364 81777	0.96940 02659	+0.24548 54872	0.95105 65163	0.30901 69944
5	0.98480 77530	−0.17364 81777	0.99658 44930	−0.08257 93455	1.00000 00000	+0.00000 00000
6	0.86602 54038	−0.50000 00000	0.91577 33266	−0.40169 54247	0.95105 65163	−0.30901 69944
7	0.64278 76097	−0.76604 44431	0.73572 39107	−0.67728 15716	0.80901 69944	−0.58778 52523
8	0.34202 01433	−0.93969 26208	0.47594 73930	−0.87947 37512	0.58778 52523	−0.80901 69944
9	0.00000 00000	−1.00000 00000	0.16459 45903	−0.98636 13034	0.30901 69944	−0.95105 65163
10					0.00000 00000	−1.00000 00000
	$s=21$		$s=22$		$s=23$	
1	0.29475 51744	0.95557 28058	0.28173 25568	0.95949 29736	0.26979 67711	0.96291 72874
2	0.56332 00580	0.82623 87743	0.54064 08174	0.84125 35328	0.51958 39500	0.85441 94046
3	0.78183 14825	0.62348 98019	0.75574 95743	0.65486 07340	0.73083 59643	0.68255 31432
4	0.93087 37486	0.36534 10244	0.90963 19953	0.41541 50130	0.88788 52184	0.46006 50378
5	0.99720 37972	+0.07473 00936	0.98982 14419	+0.14231 48383	0.97908 40877	+0.20345 60131
6	0.97492 79122	−0.22252 09340	0.98982 14419	−0.14231 48383	0.99766 87692	−0.06824 24134
7	0.86602 54038	−0.50000 00000	0.90963 19953	−0.41541 50130	0.94226 09221	−0.33487 96122
8	0.68017 27378	−0.73305 18718	0.75574 95743	−0.65486 07340	0.81696 98930	−0.57668 03221
9	0.43388 37391	−0.90096 88679	0.54064 08174	−0.84125 35328	0.63108 79443	−0.77571 12907
10	0.14904 22662	−0.98883 08262	0.28173 25568	−0.95949 29736	0.39840 10898	−0.91721 13015
11			0.00000 00000	−1.00000 00000	0.13616 66491	−0.99068 59460
	$s=24$		$s=25$			
1	0.25881 90451	0.96592 58263	0.24868 98872	0.96858 31611		
2	0.50000 00000	0.86602 54038	0.48175 36741	0.87630 66801		
3	0.70710 67812	0.70710 67812	0.68454 71059	0.72896 86274		
4	0.86602 54038	0.50000 00000	0.84432 79255	0.53582 67950		
5	0.96592 58263	0.25881 90451	0.95105 65163	0.30901 69944		
6	1.00000 00000	+0.00000 00000	0.99802 67284	+0.06279 05196		
7	0.96592 58263	−0.25881 90451	0.98228 72507	−0.18738 13146		
8	0.86602 54038	−0.50000 00000	0.90482 70525	−0.42577 92916		
9	0.70710 67812	−0.70710 67812	0.77051 32428	−0.63742 39898		
10	0.50000 00000	−0.86602 54038	0.58778 52523	−0.80901 69944		
11	0.25881 90451	−0.96592 58263	0.36812 45527	−0.92977 64859		
12	0.00000 00000	−1.00000 00000	0.12533 32336	−0.99211 47013		

INVERSE CIRCULAR SINES AND TANGENTS   Table 4.14

x	arcsin x	arctan x	x	arcsin x	arctan x
0.000	0.00000 00000 00	0.00000 00000 00	0.050	0.05002 08568 06	0.04995 83957 22
0.001	0.00100 00001 67	0.00099 99996 67	0.051	0.05102 21344 17	0.05095 58518 77
0.002	0.00200 00013 33	0.00199 99973 33	0.052	0.05202 34632 28	0.05195 32065 61
0.003	0.00300 00045 00	0.00299 99910 00	0.053	0.05302 48442 51	0.05295 04578 05
0.004	0.00400 00106 67	0.00399 99786 67	0.054	0.05402 62784 97	0.05394 76036 42
0.005	0.00500 00208 34	0.00499 99583 34	0.055	0.05502 77669 81	0.05494 46421 07
0.006	0.00600 00360 01	0.00599 99280 02	0.056	0.05602 93107 15	0.05594 15712 34
0.007	0.00700 00571 68	0.00699 98856 70	0.057	0.05703 09107 14	0.05693 83890 60
0.008	0.00800 00853 36	0.00799 98293 40	0.058	0.05803 25679 92	0.05793 50936 23
0.009	0.00900 01215 04	0.00899 97570 12	0.059	0.05903 42835 64	0.05893 16829 64
0.010	0.01000 01666 74	0.00999 96666 87	0.060	0.06003 60584 45	0.05992 81551 21
0.011	0.01100 02218 45	0.01099 95563 66	0.061	0.06103 78936 52	0.06092 45081 38
0.012	0.01200 02880 19	0.01199 94240 50	0.062	0.06203 97902 01	0.06192 07400 58
0.013	0.01300 03661 95	0.01299 92677 41	0.063	0.06304 17491 09	0.06291 68489 26
0.014	0.01400 04573 74	0.01399 90854 41	0.064	0.06404 37713 94	0.06391 28327 89
0.015	0.01500 05625 57	0.01499 88751 52	0.065	0.06504 58580 75	0.06490 86896 93
0.016	0.01600 06827 45	0.01599 86348 76	0.066	0.06604 80101 69	0.06590 44176 90
0.017	0.01700 08189 40	0.01699 83626 17	0.067	0.06705 02286 97	0.06690 00148 29
0.018	0.01800 09721 42	0.01799 80563 78	0.068	0.06805 25146 79	0.06789 54791 63
0.019	0.01900 11433 52	0.01899 77141 62	0.069	0.06905 48691 36	0.06889 08087 46
0.020	0.02000 13335 73	0.01999 73339 73	0.070	0.07005 72930 88	0.06988 60016 35
0.021	0.02100 15438 06	0.02099 69138 17	0.071	0.07105 97875 58	0.07088 10558 85
0.022	0.02200 17750 53	0.02199 64516 97	0.072	0.07206 23535 68	0.07187 59695 56
0.023	0.02300 20283 16	0.02299 59456 20	0.073	0.07306 49921 42	0.07287 07407 09
0.024	0.02400 23045 97	0.02399 53935 92	0.074	0.07406 77043 03	0.07386 53674 06
0.025	0.02500 26048 99	0.02499 47936 19	0.075	0.07507 04910 77	0.07485 98477 11
0.026	0.02600 29302 25	0.02599 41437 08	0.076	0.07607 33534 87	0.07585 41796 89
0.027	0.02700 32815 77	0.02699 34418 68	0.077	0.07707 62925 62	0.07684 83614 08
0.028	0.02800 36599 58	0.02799 26861 07	0.078	0.07807 93093 26	0.07784 23909 37
0.029	0.02900 40663 72	0.02899 18744 33	0.079	0.07908 24048 07	0.07883 62663 48
0.030	0.03000 45018 23	0.02999 10048 57	0.080	0.08008 55800 34	0.07982 99857 12
0.031	0.03100 49673 15	0.03099 00753 89	0.081	0.08108 88360 35	0.08082 35471 05
0.032	0.03200 54638 51	0.03198 90840 39	0.082	0.08209 21738 40	0.08181 69486 04
0.033	0.03300 59924 37	0.03298 80288 21	0.083	0.08309 55944 79	0.08281 01882 86
0.034	0.03400 65540 77	0.03398 69077 46	0.084	0.08409 90989 83	0.08380 32642 31
0.035	0.03500 71497 75	0.03498 57188 29	0.085	0.08510 26883 84	0.08479 61745 23
0.036	0.03600 77805 38	0.03598 44600 82	0.086	0.08610 63637 15	0.08578 89172 45
0.037	0.03700 84473 72	0.03698 31295 22	0.087	0.08711 01260 09	0.08678 14904 84
0.038	0.03800 91512 81	0.03798 17251 64	0.088	0.08811 39763 00	0.08777 38923 27
0.039	0.03900 98932 73	0.03898 02450 25	0.089	0.08911 79156 23	0.08876 61208 65
0.040	0.04001 06743 54	0.03997 86871 23	0.090	0.09012 19450 15	0.08975 81741 90
0.041	0.04101 14955 31	0.04097 70494 77	0.091	0.09112 60655 11	0.09075 00503 96
0.042	0.04201 23578 12	0.04197 53301 05	0.092	0.09213 02781 49	0.09174 17475 79
0.043	0.04301 32622 04	0.04297 35270 30	0.093	0.09313 45839 68	0.09273 32638 38
0.044	0.04401 42097 16	0.04397 16382 71	0.094	0.09413 89840 07	0.09372 45972 74
0.045	0.04501 52013 56	0.04496 96618 52	0.095	0.09514 34793 06	0.09471 57459 88
0.046	0.04601 62381 33	0.04596 75957 97	0.096	0.09614 80709 05	0.09570 67080 87
0.047	0.04701 73210 57	0.04696 54381 30	0.097	0.09715 27598 48	0.09669 74816 76
0.048	0.04801 84511 37	0.04796 31868 77	0.098	0.09815 75471 75	0.09768 80648 65
0.049	0.04901 96293 83	0.04896 08400 65	0.099	0.09916 24339 32	0.09867 84557 66
0.050	0.05002 08568 06	0.04995 83957 22	0.100	0.10016 74211 62	0.09966 86524 91
	$\begin{bmatrix}(-9)6\\4\end{bmatrix}$	$\begin{bmatrix}(-8)1\\4\end{bmatrix}$		$\begin{bmatrix}(-8)1\\4\end{bmatrix}$	$\begin{bmatrix}(-8)2\\4\end{bmatrix}$

For use and extension of the table see **Examples 21–25**. For other inverse functions see **4.4** and **4.3.45**.

$$\frac{\pi}{2}=1.57079\ 63267\ 95$$

Compilation of arcsin x from National Bureau of Standards, Table of arcsin x. Columbia Univ. Press, New York, N.Y., 1945 (with permission).

Table 4.14   INVERSE CIRCULAR SINES AND TANGENTS

x	arcsin x	arctan x	x	arcsin x	arctan x
0.100	0.10016 74211 62	0.09966 86524 91	0.150	0.15056 82727 77	0.14888 99476 09
0.101	0.10117 25099 11	0.10065 86531 58	0.151	0.15157 97940 40	0.14986 77989 58
0.102	0.10217 77012 25	0.10164 84558 83	0.152	0.15259 14716 20	0.15084 53616 21
0.103	0.10318 29961 53	0.10263 80587 89	0.153	0.15360 33066 23	0.15182 26338 59
0.104	0.10418 83957 41	0.10362 74599 97	0.154	0.15461 53001 61	0.15279 96139 37
0.105	0.10519 39010 40	0.10461 66576 33	0.155	0.15562 74533 44	0.15377 63001 20
0.106	0.10619 95131 00	0.10560 56498 23	0.156	0.15663 97672 86	0.15475 26906 78
0.107	0.10720 52329 72	0.10659 44346 99	0.157	0.15765 22431 01	0.15572 87838 86
0.108	0.10821 10617 08	0.10758 30103 93	0.158	0.15866 48819 05	0.15670 45780 19
0.109	0.10921 70003 62	0.10857 13750 39	0.159	0.15967 76848 15	0.15768 00713 58
0.110	0.11022 30499 88	0.10955 95267 74	0.160	0.16069 06529 52	0.15865 52621 86
0.111	0.11122 92116 41	0.11054 74637 38	0.161	0.16170 37874 35	0.15963 01487 91
0.112	0.11223 54863 77	0.11153 51840 74	0.162	0.16271 70893 88	0.16060 47294 61
0.113	0.11324 18752 55	0.11252 26859 25	0.163	0.16373 05599 34	0.16157 90024 91
0.114	0.11424 83793 32	0.11350 99674 40	0.164	0.16474 42001 99	0.16255 29661 78
0.115	0.11525 49996 68	0.11449 70267 67	0.165	0.16575 80113 10	0.16352 66188 21
0.116	0.11626 17373 23	0.11548 38620 60	0.166	0.16677 19943 96	0.16449 99587 25
0.117	0.11726 85933 61	0.11647 04714 73	0.167	0.16778 61505 87	0.16547 29841 97
0.118	0.11827 55688 42	0.11745 68531 63	0.168	0.16880 04810 17	0.16644 56935 49
0.119	0.11928 26648 32	0.11844 30052 90	0.169	0.16981 49868 19	0.16741 80850 93
0.120	0.12028 98823 95	0.11942 89260 18	0.170	0.17082 96691 29	0.16839 01571 48
0.121	0.12129 72225 97	0.12041 46135 12	0.171	0.17184 45290 84	0.16936 19080 34
0.122	0.12230 46865 07	0.12140 00659 40	0.172	0.17285 95678 23	0.17033 33360 78
0.123	0.12331 22751 92	0.12238 52814 72	0.173	0.17387 47864 87	0.17130 44396 07
0.124	0.12431 99897 22	0.12337 02582 82	0.174	0.17489 01862 19	0.17227 52169 54
0.125	0.12532 78311 68	0.12435 49945 47	0.175	0.17590 57681 64	0.17324 56664 52
0.126	0.12633 58006 02	0.12533 94884 45	0.176	0.17692 15334 66	0.17421 57864 43
0.127	0.12734 38990 98	0.12632 37381 58	0.177	0.17793 74832 75	0.17518 55752 68
0.128	0.12835 21277 29	0.12730 77418 71	0.178	0.17895 36187 40	0.17615 50312 74
0.129	0.12936 04875 72	0.12829 14977 71	0.179	0.17996 99410 13	0.17712 41528 10
0.130	0.13036 89797 03	0.12927 50040 48	0.180	0.18098 64512 47	0.17809 29382 31
0.131	0.13137 76052 01	0.13025 82588 96	0.181	0.18200 31505 97	0.17906 13858 94
0.132	0.13238 63651 45	0.13124 12605 10	0.182	0.18302 00402 20	0.18002 94941 59
0.133	0.13339 52606 16	0.13222 40070 89	0.183	0.18403 71212 76	0.18099 72613 91
0.134	0.13440 42926 95	0.13320 64968 35	0.184	0.18505 43949 25	0.18196 46859 59
0.135	0.13541 34624 67	0.13418 87279 52	0.185	0.18607 18623 31	0.18293 17662 35
0.136	0.13642 27710 15	0.13517 06986 49	0.186	0.18708 95246 57	0.18389 85005 94
0.137	0.13743 22194 25	0.13615 24071 35	0.187	0.18810 73830 71	0.18486 48874 16
0.138	0.13844 18087 85	0.13713 38516 25	0.188	0.18912 54387 40	0.18583 09250 85
0.139	0.13945 15401 83	0.13811 50303 34	0.189	0.19014 36928 36	0.18679 66119 87
0.140	0.14046 14147 10	0.13909 59414 82	0.190	0.19116 21465 31	0.18776 19465 14
0.141	0.14147 14334 56	0.14007 65832 92	0.191	0.19218 08009 99	0.18872 69270 59
0.142	0.14248 15975 13	0.14105 69539 90	0.192	0.19319 96574 17	0.18969 15520 22
0.143	0.14349 19079 77	0.14203 70518 03	0.193	0.19421 87169 63	0.19065 58198 05
0.144	0.14450 23659 42	0.14301 68749 65	0.194	0.19523 79808 18	0.19161 97288 15
0.145	0.14551 29725 04	0.14399 64217 09	0.195	0.19625 74501 64	0.19258 32774 60
0.146	0.14652 37287 64	0.14497 56902 74	0.196	0.19727 71261 85	0.19354 64641 55
0.147	0.14753 46358 19	0.14595 46789 00	0.197	0.19829 70100 69	0.19450 92873 18
0.148	0.14854 56947 71	0.14693 33858 33	0.198	0.19931 71030 03	0.19547 17453 71
0.149	0.14955 69067 22	0.14791 18093 19	0.199	0.20033 74061 80	0.19643 38367 38
0.150	0.15056 82727 77	0.14888 99476 09	0.200	0.20135 79207 90	0.19739 55598 50
	$\begin{bmatrix} (-8)2 \\ 4 \end{bmatrix}$	$\begin{bmatrix} (-8)4 \\ 4 \end{bmatrix}$		$\begin{bmatrix} (-8)3 \\ 4 \end{bmatrix}$	$\begin{bmatrix} (-8)5 \\ 4 \end{bmatrix}$

$$\frac{\pi}{2} = 1.57079\ 63267\ 95$$

INVERSE CIRCULAR SINES AND TANGENTS

Table 4.14

x	$\arcsin x$	$\arctan x$	x	$\arcsin x$	$\arctan x$
0.200	0.20135 79207 90	0.19739 55598 50	0.250	0.25268 02551 42	0.24497 86631 27
0.201	0.20237 86480 31	0.19835 69131 40	0.251	0.25371 31886 28	0.24591 96179 19
0.202	0.20339 95890 97	0.19931 78950 44	0.252	0.25474 63988 49	0.24686 01284 51
0.203	0.20442 07451 90	0.20027 85040 06	0.253	0.25577 98871 33	0.24780 01933 77
0.204	0.20544 21175 10	0.20123 87384 69	0.254	0.25681 36548 08	0.24873 98113 53
0.205	0.20646 37072 61	0.20219 85968 83	0.255	0.25784 77032 07	0.24967 89810 38
0.206	0.20748 55156 48	0.20315 80777 01	0.256	0.25888 20336 66	0.25061 77010 99
0.207	0.20850 75438 81	0.20411 71793 81	0.257	0.25991 66475 22	0.25155 59702 05
0.208	0.20952 97931 68	0.20507 59003 83	0.258	0.26095 15461 18	0.25249 37870 29
0.209	0.21055 22647 22	0.20603 42391 73	0.259	0.26198 67307 97	0.25343 11502 51
0.210	0.21157 49597 58	0.20699 21942 20	0.260	0.26302 22029 08	0.25436 80585 53
0.211	0.21259 78794 93	0.20794 97639 97	0.261	0.26405 79638 02	0.25530 45106 23
0.212	0.21362 10251 46	0.20890 69469 83	0.262	0.26509 40148 31	0.25624 05051 53
0.213	0.21464 43979 39	0.20986 37416 57	0.263	0.26613 03573 53	0.25717 60408 40
0.214	0.21566 79990 96	0.21082 01465 06	0.264	0.26716 69927 28	0.25811 11163 83
0.215	0.21669 18298 42	0.21177 61600 20	0.265	0.26820 39223 20	0.25904 57304 89
0.216	0.21771 58914 06	0.21273 17806 92	0.266	0.26924 11474 95	0.25997 98818 68
0.217	0.21874 01850 19	0.21368 70070 19	0.267	0.27027 86696 22	0.26091 35692 33
0.218	0.21976 47119 15	0.21464 18375 04	0.268	0.27131 64900 75	0.26184 67913 04
0.219	0.22078 94733 28	0.21559 62706 53	0.269	0.27235 46102 31	0.26277 95468 05
0.220	0.22181 44704 97	0.21655 03049 76	0.270	0.27339 30314 67	0.26371 18344 62
0.221	0.22283 97046 62	0.21750 39389 87	0.271	0.27443 17551 69	0.26464 36530 10
0.222	0.22386 51770 66	0.21845 71712 05	0.272	0.27547 07827 21	0.26557 50011 84
0.223	0.22489 08889 55	0.21941 00001 53	0.273	0.27651 01155 13	0.26650 58777 27
0.224	0.22591 68415 75	0.22036 24243 57	0.274	0.27754 97549 38	0.26743 62813 84
0.225	0.22694 30361 79	0.22131 44423 48	0.275	0.27858 97023 92	0.26836 62109 06
0.226	0.22796 94740 17	0.22226 60526 61	0.276	0.27962 99592 75	0.26929 56650 49
0.227	0.22899 61563 45	0.22321 72538 37	0.277	0.28067 05269 90	0.27022 46425 71
0.228	0.23002 30844 22	0.22416 80444 19	0.278	0.28171 14069 43	0.27115 31422 39
0.229	0.23105 02595 07	0.22511 84229 53	0.279	0.28275 26005 45	0.27208 11628 19
0.230	0.23207 76828 63	0.22606 83879 94	0.280	0.28379 41092 08	0.27300 87030 87
0.231	0.23310 53557 56	0.22701 79380 96	0.281	0.28483 59343 51	0.27393 57618 19
0.232	0.23413 32794 53	0.22796 70718 22	0.282	0.28587 80773 93	0.27486 23377 99
0.233	0.23516 14552 26	0.22891 57877 34	0.283	0.28692 05397 58	0.27578 84298 14
0.234	0.23618 98843 48	0.22986 40844 03	0.284	0.28796 33228 75	0.27671 40366 55
0.235	0.23721 85680 94	0.23081 19604 03	0.285	0.28900 64281 74	0.27763 91571 20
0.236	0.23824 75077 44	0.23175 94143 10	0.286	0.29004 98570 89	0.27856 37900 08
0.237	0.23927 67045 78	0.23270 64447 07	0.287	0.29109 36110 61	0.27948 79341 26
0.238	0.24030 61598 80	0.23365 30501 80	0.288	0.29213 76915 30	0.28041 15882 83
0.239	0.24133 58749 37	0.23459 92293 19	0.289	0.29318 20999 43	0.28133 47512 95
0.240	0.24236 58510 39	0.23554 49807 21	0.290	0.29422 68377 49	0.28225 74219 81
0.241	0.24339 60894 77	0.23649 03029 83	0.291	0.29527 19064 01	0.28317 95991 65
0.242	0.24442 65915 47	0.23743 51947 10	0.292	0.29631 73073 57	0.28410 12816 76
0.243	0.24545 73585 45	0.23837 96545 10	0.293	0.29736 30420 76	0.28502 24683 46
0.244	0.24648 83917 73	0.23932 36809 95	0.294	0.29840 91120 25	0.28594 31580 14
0.245	0.24751 96925 34	0.24026 72727 81	0.295	0.29945 55186 70	0.28686 33495 23
0.246	0.24855 12621 33	0.24121 04284 90	0.296	0.30050 22634 85	0.28778 30417 18
0.247	0.24958 31018 81	0.24215 31467 47	0.297	0.30154 93479 45	0.28870 22334 53
0.248	0.25061 52130 88	0.24309 54261 82	0.298	0.30259 67735 30	0.28962 09235 83
0.249	0.25164 75970 69	0.24403 72654 29	0.299	0.30364 45417 24	0.29053 91109 69
0.250	0.25268 02551 42	0.24497 86631 27	0.300	0.30469 26540 15	0.29145 67944 78
	$\left[\begin{smallmatrix}(-8)4\\4\end{smallmatrix}\right]$	$\left[\begin{smallmatrix}(-8)6\\4\end{smallmatrix}\right]$		$\left[\begin{smallmatrix}(-8)4\\4\end{smallmatrix}\right]$	$\left[\begin{smallmatrix}(-8)6\\4\end{smallmatrix}\right]$

$$\frac{\pi}{2} = 1.57079\ 63267\ 95$$

Table 4.14   **INVERSE CIRCULAR SINES AND TANGENTS**

x	arcsin x	arctan x	x	arcsin x	arctan x
0.300	0.30469 26540 15	0.29145 67944 78	0.350	0.35757 11036 46	0.33667 48193 87
0.301	0.30574 11118 95	0.29237 39729 79	0.351	0.35863 88378 55	0.33756 54100 58
0.302	0.30678 99168 60	0.29329 06453 47	0.352	0.35970 69995 85	0.33845 54442 85
0.303	0.30783 90704 09	0.29420 68104 62	0.353	0.36077 55905 70	0.33934 49211 81
0.304	0.30888 85740 46	0.29512 24672 09	0.354	0.36184 46125 51	0.34023 38398 61
0.305	0.30993 84292 78	0.29603 76144 75	0.355	0.36291 40672 71	0.34112 21994 49
0.306	0.31098 86376 19	0.29695 22511 55	0.356	0.36398 39564 82	0.34200 99990 70
0.307	0.31203 92005 83	0.29786 63761 46	0.357	0.36505 42819 39	0.34289 72378 56
0.308	0.31309 01196 91	0.29877 99883 52	0.358	0.36612 50454 05	0.34378 39149 42
0.309	0.31414 13964 68	0.29969 30866 80	0.359	0.36719 62486 46	0.34467 00294 69
0.310	0.31519 30324 41	0.30060 56700 42	0.360	0.36826 78934 37	0.34555 55805 82
0.311	0.31624 50291 43	0.30151 77373 55	0.361	0.36933 99815 54	0.34644 05674 30
0.312	0.31729 73881 12	0.30242 92875 41	0.362	0.37041 25147 84	0.34732 49891 68
0.313	0.31835 01108 88	0.30334 03195 25	0.363	0.37148 54949 16	0.34820 88449 54
0.314	0.31940 31990 18	0.30425 08322 38	0.364	0.37255 89237 46	0.34909 21339 52
0.315	0.32045 66540 50	0.30516 08246 16	0.365	0.37363 28030 75	0.34997 48553 30
0.316	0.32151 04775 38	0.30607 02955 99	0.366	0.37470 71347 12	0.35085 70082 60
0.317	0.32256 46710 42	0.30697 92441 31	0.367	0.37578 19204 71	0.35173 85919 21
0.318	0.32361 92361 24	0.30788 76691 62	0.368	0.37685 71621 69	0.35261 96054 93
0.319	0.32467 41743 51	0.30879 55696 46	0.369	0.37793 28616 34	0.35350 00481 64
0.320	0.32572 94872 95	0.30970 29445 42	0.370	0.37900 90206 96	0.35437 99191 23
0.321	0.32678 51765 31	0.31060 97928 14	0.371	0.38008 56411 93	0.35525 92175 68
0.322	0.32784 12436 42	0.31151 61134 29	0.372	0.38116 27249 69	0.35613 79426 98
0.323	0.32889 76902 11	0.31242 19053 60	0.373	0.38224 02738 73	0.35701 60937 18
0.324	0.32995 45178 29	0.31332 71675 84	0.374	0.38331 82897 61	0.35789 36698 38
0.325	0.33101 17280 89	0.31423 18990 84	0.375	0.38439 67744 96	0.35877 06702 71
0.326	0.33206 93225 91	0.31513 60988 47	0.376	0.38547 57299 45	0.35964 70942 35
0.327	0.33312 73029 38	0.31603 97658 63	0.377	0.38655 51579 83	0.36052 29409 56
0.328	0.33418 56707 38	0.31694 28991 30	0.378	0.38763 50604 92	0.36139 82096 58
0.329	0.33524 44276 04	0.31784 54976 47	0.379	0.38871 54393 57	0.36227 28995 76
0.330	0.33630 35751 54	0.31874 75604 21	0.380	0.38979 62964 74	0.36314 70099 46
0.331	0.33736 31150 09	0.31964 90864 60	0.381	0.39087 76337 42	0.36402 05400 09
0.332	0.33842 30487 98	0.32055 00747 81	0.382	0.39195 94530 68	0.36489 34890 12
0.333	0.33948 33781 50	0.32145 05244 03	0.383	0.39304 17563 64	0.36576 58562 04
0.334	0.34054 41047 05	0.32235 04343 49	0.384	0.39412 45455 51	0.36663 76408 40
0.335	0.34160 52301 02	0.32324 98036 48	0.385	0.39520 78225 54	0.36750 88421 81
0.336	0.34266 67559 88	0.32414 86123 34	0.386	0.39629 15893 06	0.36837 94594 90
0.337	0.34372 86840 15	0.32504 69164 46	0.387	0.39737 58477 48	0.36924 94920 36
0.338	0.34479 10158 39	0.32594 46580 25	0.388	0.39846 05998 24	0.37011 89390 92
0.339	0.34585 37531 21	0.32684 18551 19	0.389	0.39954 58474 89	0.37098 77999 35
0.340	0.34691 68975 27	0.32773 85067 81	0.390	0.40063 15927 01	0.37185 60738 49
0.341	0.34798 04507 29	0.32863 46120 66	0.391	0.40171 78374 28	0.37272 37601 18
0.342	0.34904 44144 03	0.32953 01700 37	0.392	0.40280 45836 44	0.37359 08580 36
0.343	0.35010 87902 30	0.33042 51797 60	0.393	0.40389 18333 27	0.37445 73668 96
0.344	0.35117 35798 98	0.33131 96403 04	0.394	0.40497 95884 67	0.37532 32860 01
0.345	0.35223 87850 97	0.33221 35507 47	0.395	0.40606 78510 57	0.37618 86146 53
0.346	0.35330 44075 25	0.33310 69101 67	0.396	0.40715 66231 00	0.37705 33521 62
0.347	0.35437 04488 84	0.33399 97176 49	0.397	0.40824 59066 02	0.37791 74978 43
0.348	0.35543 69108 81	0.33489 19722 83	0.398	0.40933 57035 81	0.37878 10510 12
0.349	0.35650 37952 29	0.33578 36731 63	0.399	0.41042 60160 60	0.37964 40109 93
0.350	0.35757 11036 46	0.33667 48193 87	0.400	0.41151 68460 67	0.38050 63771 12
	$\begin{bmatrix}(-8)5\\4\end{bmatrix}$	$\begin{bmatrix}(-8)7\\4\end{bmatrix}$		$\begin{bmatrix}(-8)6\\4\end{bmatrix}$	$\begin{bmatrix}(-8)8\\4\end{bmatrix}$

$$\frac{\pi}{2} = 1.57079\ 63267\ 95$$

INVERSE CIRCULAR SINES AND TANGENTS

Table 4.14

x	arcsin x	arctan x	x	arcsin x	arctan x
0.400	0.41151 68460 67	0.38050 63771 12	0.450	0.46676 53390 47	0.42285 39261 33
0.401	0.41260 81956 42	0.38136 81487 02	0.451	0.46788 54404 09	0.42368 52156 87
0.402	0.41370 00668 29	0.38222 93250 97	0.452	0.46900 61761 03	0.42451 58823 89
0.403	0.41479 24616 80	0.38308 99056 39	0.453	0.47012 75486 20	0.42534 59257 92
0.404	0.41588 53822 54	0.38394 98896 72	0.454	0.47124 95604 59	0.42617 53454 56
0.405	0.41697 88306 20	0.38480 92765 46	0.455	0.47237 22141 29	0.42700 41409 43
0.406	0.41807 28088 50	0.38566 80656 14	0.456	0.47349 55121 50	0.42783 23118 21
0.407	0.41916 73190 29	0.38652 62562 34	0.457	0.47461 94570 53	0.42865 98576 60
0.408	0.42026 23632 45	0.38738 38477 69	0.458	0.47574 40513 79	0.42948 67780 36
0.409	0.42135 79435 96	0.38824 08395 85	0.459	0.47686 92976 80	0.43031 30725 28
0.410	0.42245 40621 87	0.38909 72310 55	0.460	0.47799 51985 19	0.43113 87407 19
0.411	0.42355 07211 31	0.38995 30215 54	0.461	0.47912 17564 68	0.43196 37821 96
0.412	0.42464 79225 49	0.39080 82104 62	0.462	0.48024 89741 12	0.43278 81965 51
0.413	0.42574 56685 70	0.39166 27971 64	0.463	0.48137 68540 46	0.43361 19833 80
0.414	0.42684 39613 30	0.39251 67810 48	0.464	0.48250 53988 75	0.43443 51422 81
0.415	0.42794 28029 74	0.39337 01615 09	0.465	0.48363 46112 18	0.43525 76728 58
0.416	0.42904 21956 53	0.39422 29379 43	0.466	0.48476 44937 02	0.43607 95747 19
0.417	0.43014 21415 30	0.39507 51097 52	0.467	0.48589 50489 67	0.43690 08474 74
0.418	0.43124 26427 72	0.39592 66763 44	0.468	0.48702 62796 64	0.43772 14907 40
0.419	0.43234 37015 57	0.39677 76371 29	0.469	0.48815 81884 55	0.43854 15041 36
0.420	0.43344 53200 70	0.39762 79915 22	0.470	0.48929 07780 14	0.43936 08872 85
0.421	0.43454 75005 03	0.39847 77389 43	0.471	0.49042 40510 26	0.44017 96398 14
0.422	0.43565 02450 60	0.39932 68788 14	0.472	0.49155 80101 88	0.44099 77613 55
0.423	0.43675 35559 49	0.40017 54105 66	0.473	0.49269 26582 08	0.44181 52515 43
0.424	0.43785 74353 90	0.40102 33336 29	0.474	0.49382 79978 07	0.44263 21100 17
0.425	0.43896 18856 10	0.40187 06474 40	0.475	0.49496 40317 17	0.44344 83364 20
0.426	0.44006 69088 44	0.40271 73514 42	0.476	0.49610 07626 82	0.44426 39303 99
0.427	0.44117 25073 36	0.40356 34450 79	0.477	0.49723 81934 59	0.44507 88916 06
0.428	0.44227 86833 39	0.40440 89278 00	0.478	0.49837 63268 16	0.44589 32196 95
0.429	0.44338 54391 16	0.40525 37990 60	0.479	0.49951 51655 34	0.44670 69143 24
0.430	0.44449 27769 36	0.40609 80583 18	0.480	0.50065 47124 05	0.44751 99751 57
0.431	0.44560 06990 78	0.40694 17050 34	0.481	0.50179 49702 34	0.44833 24018 60
0.432	0.44670 92078 31	0.40778 47386 77	0.482	0.50293 59418 39	0.44914 41941 03
0.433	0.44781 83054 92	0.40862 71587 18	0.483	0.50407 76300 52	0.44995 53515 61
0.434	0.44892 79943 67	0.40946 89646 31	0.484	0.50522 00377 13	0.45076 58739 11
0.435	0.45003 82767 71	0.41031 01558 96	0.485	0.50636 31676 79	0.45157 57608 36
0.436	0.45114 91550 28	0.41115 07319 97	0.486	0.50750 70228 19	0.45238 50120 20
0.437	0.45226 06314 71	0.41199 06924 22	0.487	0.50865 16060 14	0.45319 36271 55
0.438	0.45337 27084 44	0.41283 00366 64	0.488	0.50979 69201 57	0.45400 16059 33
0.439	0.45448 53882 99	0.41366 87642 17	0.489	0.51094 29681 57	0.45480 89480 51
0.440	0.45559 86733 96	0.41450 68745 85	0.490	0.51208 97529 34	0.45561 56532 11
0.441	0.45671 25661 07	0.41534 43672 70	0.491	0.51323 72774 22	0.45642 17211 17
0.442	0.45782 70688 11	0.41618 12417 83	0.492	0.51438 55445 69	0.45722 71514 78
0.443	0.45894 21838 99	0.41701 74976 36	0.493	0.51553 45573 34	0.45803 19440 06
0.444	0.46005 79137 71	0.41785 31343 48	0.494	0.51668 43186 93	0.45883 60984 16
0.445	0.46117 42608 35	0.41868 81514 38	0.495	0.51783 48316 32	0.45963 96144 30
0.446	0.46229 12275 10	0.41952 25484 34	0.496	0.51898 60991 55	0.46044 24917 71
0.447	0.46340 88162 25	0.42035 63248 66	0.497	0.52013 81242 77	0.46124 47301 65
0.448	0.46452 70294 19	0.42118 94802 67	0.498	0.52129 09100 26	0.46204 63293 45
0.449	0.46564 58695 40	0.42202 20141 75	0.499	0.52244 44594 47	0.46284 72890 44
0.450	0.46676 53390 47	0.42285 39261 33	0.500	0.52359 87755 98	0.46364 76090 01
	$\begin{bmatrix} (-8)8 \\ 4 \end{bmatrix}$	$\begin{bmatrix} (-8)8 \\ 4 \end{bmatrix}$		$\begin{bmatrix} (-7)1 \\ 4 \end{bmatrix}$	$\begin{bmatrix} (-8)8 \\ 4 \end{bmatrix}$

$$\frac{\pi}{2} = 1.57079\ 63267\ 95$$

Table 4.14　　　　　　INVERSE CIRCULAR SINES AND TANGENTS

x	arcsin x	arctan x	x	arcsin x	arctan x
0.500	0.52359 87755 98	0.46364 76090 01	0.550	0.58236 42378 69	0.50284 32109 28
0.501	0.52475 38615 51	0.46444 72889 58	0.551	0.58356 20792 89	0.50361 06410 37
0.502	0.52590 97203 91	0.46524 63286 62	0.552	0.58476 08688 33	0.50437 74226 73
0.503	0.52706 63552 20	0.46604 47278 61	0.553	0.58596 06104 84	0.50514 35557 57
0.504	0.52822 37691 54	0.46684 24863 09	0.554	0.58716 13082 43	0.50590 90402 12
0.505	0.52938 19653 22	0.46763 96037 63	0.555	0.58836 29661 37	0.50667 38759 68
0.506	0.53054 09468 69	0.46843 60799 83	0.556	0.58956 55882 10	0.50743 80629 53
0.507	0.53170 07169 56	0.46923 19147 34	0.557	0.59076 91785 32	0.50820 16011 02
0.508	0.53286 12787 56	0.47002 71077 82	0.558	0.59197 37411 92	0.50896 44903 52
0.509	0.53402 26354 61	0.47082 16589 00	0.559	0.59317 92803 04	0.50972 67306 43
0.510	0.53518 47902 76	0.47161 55678 62	0.560	0.59438 58000 01	0.51048 83219 17
0.511	0.53634 77464 20	0.47240 88344 48	0.561	0.59559 33044 41	0.51124 92641 21
0.512	0.53751 15071 30	0.47320 14584 38	0.562	0.59680 17978 05	0.51200 95572 04
0.513	0.53867 60756 57	0.47399 34396 20	0.563	0.59801 12842 95	0.51276 92011 19
0.514	0.53984 14552 69	0.47478 47777 82	0.564	0.59922 17681 37	0.51352 81958 22
0.515	0.54100 76492 49	0.47557 54727 17	0.565	0.60043 32535 81	0.51428 65412 69
0.516	0.54217 46608 96	0.47636 55242 22	0.566	0.60164 57448 99	0.51504 42374 25
0.517	0.54334 24935 25	0.47715 49320 97	0.567	0.60285 92463 89	0.51580 12842 52
0.518	0.54451 11504 67	0.47794 36961 45	0.568	0.60407 37623 71	0.51655 76817 18
0.519	0.54568 06350 69	0.47873 18161 73	0.569	0.60528 92971 89	0.51731 34297 96
0.520	0.54685 09506 96	0.47951 92919 93	0.570	0.60650 58552 13	0.51806 85284 57
0.521	0.54802 21007 28	0.48030 61234 17	0.571	0.60772 34408 36	0.51882 29776 79
0.522	0.54919 40885 61	0.48109 23102 64	0.572	0.60894 20584 75	0.51957 67774 41
0.523	0.55036 69176 11	0.48187 78523 54	0.573	0.61016 17125 74	0.52032 99277 27
0.524	0.55154 05913 07	0.48266 27495 12	0.574	0.61138 24076 01	0.52108 24285 22
0.525	0.55271 51130 97	0.48344 70015 67	0.575	0.61260 41480 49	0.52183 42798 14
0.526	0.55389 04864 46	0.48423 06083 50	0.576	0.61382 69384 37	0.52258 54815 96
0.527	0.55506 67148 37	0.48501 35696 94	0.577	0.61505 07833 09	0.52333 60338 62
0.528	0.55624 38017 69	0.48579 58854 40	0.578	0.61627 56872 37	0.52408 59366 09
0.529	0.55742 17507 59	0.48657 75554 29	0.579	0.61750 16548 17	0.52483 51898 38
0.530	0.55860 05653 43	0.48735 85795 05	0.580	0.61872 86906 72	0.52558 37935 52
0.531	0.55978 02490 72	0.48813 89575 18	0.581	0.61995 67994 52	0.52633 17477 57
0.532	0.56096 08055 18	0.48891 86893 19	0.582	0.62118 59858 34	0.52707 90524 63
0.533	0.56214 22382 69	0.48969 77747 65	0.583	0.62241 62545 21	0.52782 57076 82
0.534	0.56332 45509 33	0.49047 62137 12	0.584	0.62364 76102 44	0.52857 17134 28
0.535	0.56450 77471 34	0.49125 40060 25	0.585	0.62488 00577 61	0.52931 70697 19
0.536	0.56569 18305 17	0.49203 11515 68	0.586	0.62611 36018 60	0.53006 17765 76
0.537	0.56687 68047 44	0.49280 76502 10	0.587	0.62734 82473 54	0.53080 58340 23
0.538	0.56806 26734 97	0.49358 35018 23	0.588	0.62858 39990 87	0.53154 92420 86
0.539	0.56924 94404 76	0.49435 87062 83	0.589	0.62982 08619 28	0.53229 20007 93
0.540	0.57043 71094 00	0.49513 32634 68	0.590	0.63105 88407 78	0.53303 41101 77
0.541	0.57162 56840 08	0.49590 71732 62	0.591	0.63229 79405 66	0.53377 55702 73
0.542	0.57281 51680 58	0.49668 04355 48	0.592	0.63353 81662 50	0.53451 63811 18
0.543	0.57400 55653 28	0.49745 30502 17	0.593	0.63477 95228 17	0.53525 65427 53
0.544	0.57519 68796 15	0.49822 50171 59	0.594	0.63602 20152 84	0.53599 60552 20
0.545	0.57638 91147 36	0.49899 63362 71	0.595	0.63726 56487 00	0.53673 49185 66
0.546	0.57758 22745 29	0.49976 70074 50	0.596	0.63851 04281 42	0.53747 31328 39
0.547	0.57877 63628 51	0.50053 70305 98	0.597	0.63975 63587 17	0.53821 06980 90
0.548	0.57997 13835 79	0.50130 64056 22	0.598	0.64100 34455 66	0.53894 76143 74
0.549	0.58116 73406 12	0.50207 51324 28	0.599	0.64225 16938 57	0.53968 38817 48
0.550	0.58236 42378 69	0.50284 32109 28	0.600	0.64350 11087 93	0.54041 95002 71
	$\begin{bmatrix} (-7)1 \\ 4 \end{bmatrix}$	$\begin{bmatrix} (-8)8 \\ 4 \end{bmatrix}$		$\begin{bmatrix} (-7)2 \\ 5 \end{bmatrix}$	$\begin{bmatrix} (-8)8 \\ 4 \end{bmatrix}$

$$\frac{\pi}{2} = 1.57079\ 63267\ 95$$

INVERSE CIRCULAR SINES AND TANGENTS

Table 4.14

x	arcsin x	arctan x	x	arcsin x	arctan x
0.600	0.64350 11087 93	0.54041 95002 71	0.650	0.70758 44367 25	0.57637 52205 91
0.601	0.64475 16956 07	0.54115 44700 04	0.651	0.70890 10818 82	0.57707 78870 95
0.602	0.64600 34595 63	0.54188 87910 15	0.652	0.71021 92154 53	0.57777 99113 37
0.603	0.64725 64059 60	0.54262 24633 69	0.653	0.71153 88447 93	0.57848 12935 07
0.604	0.64851 05401 26	0.54335 54871 37	0.654	0.71285 99773 14	0.57918 20337 94
0.605	0.64976 58674 24	0.54408 78623 92	0.655	0.71418 26204 76	0.57988 21323 94
0.606	0.65102 23932 51	0.54481 95892 10	0.656	0.71550 67817 97	0.58058 15895 01
0.607	0.65228 01230 34	0.54555 06676 70	0.657	0.71683 24688 45	0.58128 04053 13
0.608	0.65353 90622 38	0.54628 10978 51	0.658	0.71815 96892 45	0.58197 85800 31
0.609	0.65479 92163 58	0.54701 08798 38	0.659	0.71948 84506 75	0.58267 61138 57
0.610	0.65606 05909 25	0.54774 00137 16	0.660	0.72081 87608 70	0.58337 30069 94
0.611	0.65732 31915 05	0.54846 84995 75	0.661	0.72215 06276 21	0.58406 92596 49
0.612	0.65858 70237 00	0.54919 63375 05	0.662	0.72348 40587 76	0.58476 48720 31
0.613	0.65985 20931 44	0.54992 35276 01	0.663	0.72481 90622 40	0.58545 98443 49
0.614	0.66111 84055 09	0.55065 00699 59	0.664	0.72615 56459 74	0.58615 41768 17
0.615	0.66238 59665 02	0.55137 59646 79	0.665	0.72749 38180 01	0.58684 78696 50
0.616	0.66365 47818 67	0.55210 12118 61	0.666	0.72883 35864 02	0.58754 09230 63
0.617	0.66492 48573 84	0.55282 58116 10	0.667	0.73017 49593 16	0.58823 33372 77
0.618	0.66619 61988 69	0.55354 97640 33	0.668	0.73151 79449 44	0.58892 51125 11
0.619	0.66746 88121 78	0.55427 30692 38	0.669	0.73286 25515 49	0.58961 62489 89
0.620	0.66874 27032 02	0.55499 57273 39	0.670	0.73420 87874 53	0.59030 67469 35
0.621	0.67001 78778 71	0.55571 77384 48	0.671	0.73555 66610 44	0.59099 66065 77
0.622	0.67129 43421 53	0.55643 91026 82	0.672	0.73690 61807 69	0.59168 58281 44
0.623	0.67257 21020 54	0.55715 98201 62	0.673	0.73825 73551 41	0.59237 44118 66
0.624	0.67385 11636 20	0.55787 98910 07	0.674	0.73961 01927 39	0.59306 23579 77
0.625	0.67513 15329 37	0.55859 93153 44	0.675	0.74096 47022 03	0.59374 96667 11
0.626	0.67641 32161 29	0.55931 80932 97	0.676	0.74232 08922 43	0.59443 63383 05
0.627	0.67769 62193 62	0.56003 62249 97	0.677	0.74367 87716 32	0.59512 23729 99
0.628	0.67898 05488 41	0.56075 37105 74	0.678	0.74503 83492 13	0.59580 77710 32
0.629	0.68026 62108 12	0.56147 05501 63	0.679	0.74639 96338 96	0.59649 25326 49
0.630	0.68155 32115 63	0.56218 67439 00	0.680	0.74776 26346 60	0.59717 66580 93
0.631	0.68284 15574 24	0.56290 22919 24	0.681	0.74912 73605 52	0.59786 01476 11
0.632	0.68413 12547 66	0.56361 71943 75	0.682	0.75049 38206 91	0.59854 30014 52
0.633	0.68542 23100 04	0.56433 14513 97	0.683	0.75186 20242 68	0.59922 52198 66
0.634	0.68671 47295 93	0.56504 50631 37	0.684	0.75323 19805 42	0.59990 68031 06
0.635	0.68800 85200 35	0.56575 80297 42	0.685	0.75460 36988 49	0.60058 77514 26
0.636	0.68930 36878 74	0.56647 03513 63	0.686	0.75597 71885 95	0.60126 80650 81
0.637	0.69060 02396 97	0.56718 20281 53	0.687	0.75735 24592 63	0.60194 77443 31
0.638	0.69189 81821 37	0.56789 30602 67	0.688	0.75872 95204 10	0.60262 67894 35
0.639	0.69319 75218 73	0.56860 34478 63	0.689	0.76010 83816 68	0.60330 52006 54
0.640	0.69449 82656 27	0.56931 31911 01	0.690	0.76148 90527 48	0.60398 29782 53
0.641	0.69580 04201 68	0.57002 22901 42	0.691	0.76287 15434 36	0.60466 01224 96
0.642	0.69710 39923 13	0.57073 07451 52	0.692	0.76425 58636 00	0.60533 66336 52
0.643	0.69840 89889 23	0.57143 85562 98	0.693	0.76564 20231 84	0.60601 25119 88
0.644	0.69971 54169 09	0.57214 57237 47	0.694	0.76703 00322 15	0.60668 77577 76
0.645	0.70102 32832 27	0.57285 22476 73	0.695	0.76841 99008 00	0.60736 23712 89
0.646	0.70233 25948 84	0.57355 81282 48	0.696	0.76981 16391 29	0.60803 63528 01
0.647	0.70364 33589 34	0.57426 33656 48	0.697	0.77120 52574 75	0.60870 97025 88
0.648	0.70495 55824 80	0.57496 79600 51	0.698	0.77260 07661 95	0.60938 24209 28
0.649	0.70626 92726 76	0.57567 19116 38	0.699	0.77399 81757 30	0.61005 45081 01
0.650	0.70758 44367 25	0.57637 52205 91	0.700	0.77539 74966 11	0.61072 59643 89
	$\begin{bmatrix} (-7)2 \\ 5 \end{bmatrix}$	$\begin{bmatrix} (-8)8 \\ 4 \end{bmatrix}$		$\begin{bmatrix} (-7)2 \\ 5 \end{bmatrix}$	$\begin{bmatrix} (-8)8 \\ 4 \end{bmatrix}$

$$\frac{\pi}{2} = 1.57079\ 63267\ 95$$

Table 4.14 INVERSE CIRCULAR SINES AND TANGENTS

x	arcsin x	arctan x	x	arcsin x	arctan x
0.700	0.77539 74966 11	0.61072 59643 89	0.750	0.84806 20789 81	0.64350 11087 93
0.701	0.77679 87394 52	0.61139 67900 75	0.751	0.84957 52355 56	0.64414 08016 53
0.702	0.77820 19149 57	0.61206 69854 44	0.752	0.85109 10007 70	0.64477 98804 75
0.703	0.77960 70339 20	0.61273 65507 83	0.753	0.85260 93916 63	0.64541 83456 20
0.704	0.78101 41072 23	0.61340 54863 79	0.754	0.85413 04254 45	0.64605 61974 52
0.705	0.78242 31458 43	0.61407 37925 25	0.755	0.85565 41195 04	0.64669 34363 37
0.706	0.78383 41608 47	0.61474 14695 10	0.756	0.85718 04914 02	0.64733 00626 40
0.707	0.78524 71633 95	0.61540 85176 29	0.757	0.85870 95588 84	0.64796 60767 30
0.708	0.78666 21647 44	0.61607 49371 78	0.758	0.86024 13398 74	0.64860 14789 75
0.709	0.78807 91762 45	0.61674 07284 52	0.759	0.86177 58524 85	0.64923 62697 45
0.710	0.78949 82093 46	0.61740 58917 52	0.760	0.86331 31150 16	0.64987 04494 12
0.711	0.79091 92755 96	0.61807 04273 76	0.761	0.86485 31459 55	0.65050 40183 48
0.712	0.79234 23866 39	0.61873 43356 27	0.762	0.86639 59639 86	0.65113 69769 28
0.713	0.79376 75542 24	0.61939 76168 09	0.763	0.86794 15879 89	0.65176 93255 25
0.714	0.79519 47901 99	0.62006 02712 26	0.764	0.86949 00370 42	0.65240 10645 18
0.715	0.79662 41065 16	0.62072 22991 86	0.765	0.87104 13304 26	0.65303 21942 83
0.716	0.79805 55152 32	0.62138 37009 97	0.766	0.87259 54876 26	0.65366 27151 99
0.717	0.79948 90285 08	0.62204 44769 70	0.767	0.87415 25283 38	0.65429 26276 46
0.718	0.80092 46586 13	0.62270 46274 14	0.768	0.87571 24724 65	0.65492 19320 05
0.719	0.80236 24179 26	0.62336 41526 45	0.769	0.87727 53401 29	0.65555 06286 59
0.720	0.80380 23189 33	0.62402 30529 77	0.770	0.87884 11516 69	0.65617 87179 91
0.721	0.80524 43742 33	0.62468 13287 26	0.771	0.88040 99276 42	0.65680 62003 87
0.722	0.80668 85965 35	0.62533 89802 10	0.772	0.88198 16888 33	0.65743 30762 31
0.723	0.80813 49986 66	0.62599 60077 48	0.773	0.88355 64562 55	0.65805 93459 11
0.724	0.80958 35935 64	0.62665 24116 63	0.774	0.88513 42511 51	0.65868 50098 15
0.725	0.81103 43942 88	0.62730 81922 76	0.775	0.88671 50950 00	0.65931 00683 33
0.726	0.81248 74140 11	0.62796 33499 11	0.776	0.88829 90095 19	0.65993 45218 55
0.727	0.81394 26660 28	0.62861 78848 95	0.777	0.88988 60166 70	0.66055 83707 72
0.728	0.81540 01637 58	0.62927 17975 54	0.778	0.89147 61386 58	0.66118 16154 79
0.729	0.81685 99207 37	0.62992 50882 17	0.779	0.89306 93979 43	0.66180 42563 67
0.730	0.81832 19506 32	0.63057 77572 15	0.780	0.89466 58172 34	0.66242 62938 33
0.731	0.81978 62672 31	0.63122 98048 79	0.781	0.89626 54195 03	0.66304 77282 73
0.732	0.82125 28844 52	0.63188 12315 41	0.782	0.89786 82279 83	0.66366 85600 83
0.733	0.82272 18163 44	0.63253 20375 38	0.783	0.89947 42661 72	0.66428 87896 62
0.734	0.82419 30770 85	0.63318 22232 04	0.784	0.90108 35578 41	0.66490 84174 09
0.735	0.82566 66809 86	0.63383 17888 78	0.785	0.90269 61270 38	0.66552 74437 26
0.736	0.82714 26424 94	0.63448 07348 99	0.786	0.90431 19980 87	0.66614 58690 12
0.737	0.82862 09761 92	0.63512 90616 06	0.787	0.90593 11956 01	0.66676 36936 71
0.738	0.83010 16968 01	0.63577 67693 42	0.788	0.90755 37444 80	0.66738 09181 07
0.739	0.83158 48191 83	0.63642 38584 50	0.789	0.90917 96699 17	0.66799 75427 24
0.740	0.83307 03583 42	0.63707 03292 76	0.790	0.91080 89974 07	0.66861 35679 28
0.741	0.83455 83294 24	0.63771 61821 64	0.791	0.91244 17527 48	0.66922 89941 25
0.742	0.83604 87477 24	0.63836 14174 63	0.792	0.91407 79620 46	0.66984 38217 24
0.743	0.83754 16286 83	0.63900 60355 21	0.793	0.91571 76517 23	0.67045 80511 32
0.744	0.83903 69878 93	0.63965 00366 89	0.794	0.91736 08485 19	0.67107 16827 61
0.745	0.84053 48410 98	0.64029 34213 19	0.795	0.91900 75795 02	0.67168 47170 20
0.746	0.84203 52041 95	0.64093 61897 63	0.796	0.92065 78720 67	0.67229 71543 22
0.747	0.84353 80932 39	0.64157 83423 76	0.797	0.92231 17539 49	0.67290 89950 79
0.748	0.84504 35244 42	0.64221 98795 14	0.798	0.92396 92532 24	0.67352 02397 05
0.749	0.84655 15141 77	0.64286 08015 33	0.799	0.92563 03983 15	0.67413 08886 15
0.750	0.84806 20789 81	0.64350 11087 93	0.800	0.92729 52180 02	0.67474 09422 24
	$\begin{bmatrix} (-7)3 \\ 5 \end{bmatrix}$	$\begin{bmatrix} (-8)8 \\ 4 \end{bmatrix}$		$\begin{bmatrix} (-7)5 \\ 5 \end{bmatrix}$	$\begin{bmatrix} (-8)8 \\ 4 \end{bmatrix}$

$$\frac{\pi}{2} = 1.57079\ 63267\ 95$$

INVERSE CIRCULAR SINES AND TANGENTS

Table 4.14

x	arcsin x	arctan x	x	arcsin x	arctan x
0.800	0.92729 52180 02	0.67474 09422 24	0.850	1.01598 52938 15	0.70449 40642 42
0.801	0.92896 37414 22	0.67535 04009 49	0.851	1.01788 65272 25	0.70507 43293 58
0.802	0.93063 59980 83	0.67595 92652 08	0.852	1.01979 36361 62	0.70565 40219 63
0.303	0.93231 20178 64	0.67656 75354 19	0.853	1.02170 66824 41	0.70623 31425 16
0.804	0.93399 18310 25	0.67717 52120 01	0.854	1.02362 57289 29	0.70681 16914 73
0.805	0.93567 54682 12	0.67778 22953 77	0.855	1.02555 08395 76	0.70738 96692 96
0.806	0.93736 29604 66	0.67838 87859 65	0.856	1.02748 20794 40	0.70796 70764 42
0.807	0.93905 43392 28	0.67899 46841 90	0.857	1.02941 95147 10	0.70854 39133 73
0.808	0.94074 96363 49	0.67959 99904 74	0.858	1.03136 32127 41	0.70912 01805 50
0.809	0.94244 88840 95	0.68020 47052 41	0.859	1.03331 32420 77	0.70969 58784 34
0.810	0.94415 21151 54	0.68080 88289 16	0.860	1.03526 96724 81	0.71027 10074 87
0.811	0.94585 93626 48	0.68141 23619 25	0.861	1.03723 25749 68	0.71084 55681 72
0.812	0.94757 06601 38	0.68201 53046 96	0.862	1.03920 20218 39	0.71141 95609 52
0.813	0.94928 60416 29	0.68261 76576 55	0.863	1.04117 80867 05	0.71199 29862 92
0.814	0.95100 55415 87	0.68321 94212 31	0.864	1.04316 08445 30	0.71256 58446 55
0.815	0.95272 91949 40	0.68382 05958 54	0.865	1.04515 03716 61	0.71313 81365 07
0.816	0.95445 70370 88	0.68442 11819 54	0.866	1.04714 67458 63	0.71370 98623 14
0.817	0.95618 91039 18	0.68502 11799 62	0.867	1.04915 00463 62	0.71428 10225 41
0.818	0.95792 54318 04	0.68562 05903 10	0.868	1.05116 03538 76	0.71485 16176 56
0.819	0.95966 60576 23	0.68621 94134 31	0.869	1.05317 77506 61	0.71542 16481 25
0.820	0.96141 10187 64	0.68681 76497 59	0.870	1.05520 23205 49	0.71599 11144 16
0.821	0.96316 03531 36	0.68741 52997 28	0.871	1.05723 41489 91	0.71656 00169 99
0.822	0.96491 40991 79	0.68801 23637 73	0.872	1.05927 33231 01	0.71712 83563 41
0.823	0.96667 22958 76	0.68860 88423 31	0.873	1.06131 99317 03	0.71769 61329 12
0.824	0.96843 49827 60	0.68920 47358 39	0.874	1.06337 40653 78	0.71826 33471 82
0.825	0.97020 21999 29	0.68980 00447 34	0.875	1.06543 58165 11	0.71882 99996 22
0.826	0.97197 39880 56	0.69039 47694 55	0.876	1.06750 52793 43	0.71939 60907 02
0.827	0.97375 03884 00	0.69098 89104 41	0.877	1.06958 25500 24	0.71996 16208 94
0.828	0.97553 14428 17	0.69158 24681 33	0.878	1.07166 77266 67	0.72052 65906 70
0.829	0.97731 71937 77	0.69217 54429 71	0.879	1.07376 09094 07	0.72109 10005 03
0.830	0.97910 76843 68	0.69276 78353 97	0.880	1.07586 22004 54	0.72165 48508 65
0.831	0.98090 29583 19	0.69335 96458 54	0.881	1.07797 17041 59	0.72221 81422 30
0.832	0.98270 30600 05	0.69395 08747 85	0.882	1.08008 95270 75	0.72278 08750 71
0.833	0.98450 80344 64	0.69454 15226 33	0.883	1.08221 57780 22	0.72334 30498 64
0.834	0.98631 79274 13	0.69513 15898 44	0.884	1.08435 05681 59	0.72390 46670 83
0.835	0.98813 27852 56	0.69572 10768 63	0.885	1.08649 40110 49	0.72446 57272 04
0.836	0.98995 26551 06	0.69630 99841 36	0.886	1.08864 62227 36	0.72502 62307 01
0.837	0.99177 75847 95	0.69689 83121 11	0.887	1.09080 73218 22	0.72558 61780 53
0.838	0.99360 76228 94	0.69748 60612 34	0.888	1.09297 74295 43	0.72614 55697 34
0.839	0.99544 28187 22	0.69807 32319 55	0.889	1.09515 66698 56	0.72670 44062 23
0.840	0.99728 32223 72	0.69865 98247 21	0.890	1.09734 51695 23	0.72726 26879 97
0.841	0.99912 88847 18	0.69924 58399 85	0.891	1.09954 30581 99	0.72782 04155 34
0.842	1.00097 98574 39	0.69983 12781 94	0.892	1.10175 04685 30	0.72837 75893 12
0.843	1.00283 61930 35	0.70041 61398 02	0.893	1.10396 75362 43	0.72893 42098 11
0.844	1.00469 79448 46	0.70100 04252 59	0.894	1.10619 44002 56	0.72949 02775 09
0.845	1.00656 51670 67	0.70158 41350 19	0.895	1.10843 12027 75	0.73004 57928 87
0.846	1.00843 79147 75	0.70216 72695 35	0.896	1.11067 80894 12	0.73060 07564 24
0.847	1.01031 62439 41	0.70274 98292 60	0.897	1.11293 52092 94	0.73115 51686 02
0.848	1.01220 02114 56	0.70333 18146 49	0.898	1.11520 27151 85	0.73170 90299 00
0.849	1.01408 98751 50	0.70391 32261 58	0.899	1.11748 07636 13	0.73226 23408 01
0.850	1.01598 52938 15	0.70449 40642 42	0.900	1.11976 95149 99	0.73281 51017 87
	$\left[\begin{matrix}(-7)7\\5\end{matrix}\right]$	$\left[\begin{matrix}(-8)7\\4\end{matrix}\right]$		$\left[\begin{matrix}(-6)1\\6\end{matrix}\right]$	$\left[\begin{matrix}(-8)7\\4\end{matrix}\right]$

$$\frac{\pi}{2} = 1.57079\ 63267\ 95$$

Table 4.14   INVERSE CIRCULAR SINES AND TANGENTS

x	arcsin x	arctan x	x	arcsin x	arctan x	$f(x)$
0.900	1.11976 95149 99	0.73281 51017 87	0.950	1.25323 58975 03	0.75976 27548 76	1.00421 42513 02
0.901	1.12206 91337 93	0.73336 73133 38	0.951	1.25645 42223 06	0.76028 81166 70	1.00412 90197 55
0.902	1.12437 97886 21	0.73391 89759 38	0.952	1.25970 47250 03	0.76081 29540 28	1.00404 38274 04
0.903	1.12670 16524 29	0.73447 00900 70	0.953	1.26298 84259 28	0.76133 72674 43	1.00395 86742 15
0.904	1.12903 49026 45	0.73502 06562 16	0.954	1.26630 64000 67	0.76186 10574 14	1.00387 35601 52
0.905	1.13137 97213 39	0.73557 06748 62	0.955	1.26965 97812 42	0.76238 43244 37	1.00378 84851 78
0.906	1.13373 62953 96	0.73612 01464 89	0.956	1.27304 97667 20	0.76290 70690 08	1.00370 34492 58
0.907	1.13610 48166 99	0.73666 90715 84	0.957	1.27647 76222 92	0.76342 92916 23	1.00361 84523 57
0.908	1.13848 54823 12	0.73721 74506 30	0.958	1.27994 46878 88	0.76395 09927 81	1.00353 34944 39
0.909	1.14087 84946 83	0.73776 52841 13	0.959	1.28345 23838 00	0.76447 21729 78	1.00344 85754 69
0.910	1.14328 40618 50	0.73831 25725 17	0.960	1.28700 22175 87	0.76499 28327 11	1.00336 36954 10
0.911	1.14570 23976 58	0.73885 93163 30	0.961	1.29059 57917 69	0.76551 29724 78	1.00327 88542 28
0.912	1.14813 37219 91	0.73940 55160 36	0.962	1.29423 48124 14	0.76603 25927 75	1.00319 40518 88
0.913	1.15057 82610 10	0.73995 11721 22	0.963	1.29792 10987 43	0.76655 16941 02	1.00310 92883 53
0.914	1.15303 62474 12	0.74049 62850 76	0.964	1.30165 65939 20	0.76707 02769 55	1.00302 45635 89
0.915	1.15550 79206 90	0.74104 08553 83	0.965	1.30544 33771 97	0.76758 83418 33	1.00293 98775 61
0.916	1.15799 35274 19	0.74158 48835 32	0.966	1.30928 36776 35	0.76810 58892 33	1.00285 52302 33
0.917	1.16049 33215 50	0.74212 83700 10	0.967	1.31317 98896 52	0.76862 29196 53	1.00277 06215 71
0.918	1.16300 75647 25	0.74267 13153 04	0.968	1.31713 45907 19	0.76913 94335 92	1.00268 60515 39
0.919	1.16553 65266 04	0.74321 37199 05	0.969	1.32115 05615 54	0.76965 54315 49	1.00260 15201 02
0.920	1.16808 04852 14	0.74375 55842 99	0.970	1.32523 08092 80	0.77017 09140 20	1.00251 70272 25
0.921	1.17063 97273 16	0.74429 69089 76	0.971	1.32937 85940 93	0.77068 58815 06	1.00243 25728 74
0.922	1.17321 45487 95	0.74483 76944 25	0.972	1.33359 74601 02	0.77120 03345 05	1.00234 81570 13
0.923	1.17580 52550 71	0.74537 79411 35	0.973	1.33789 12711 79	0.77171 42735 14	1.00226 37796 07
0.924	1.17841 21615 31	0.74591 76495 97	0.974	1.34226 42528 47	0.77222 76990 34	1.00217 94406 23
0.925	1.18103 55939 97	0.74645 68203 00	0.975	1.34672 10414 93	0.77274 06115 63	1.00209 51400 25
0.926	1.18367 58892 09	0.74699 54537 35	0.976	1.35126 67425 45	0.77325 30116 01	1.00201 08777 78
0.927	1.18633 33953 44	0.74753 35503 92	0.977	1.35590 69996 85	0.77376 48996 45	1.00192 66538 49
0.928	1.18900 84725 71	0.74807 11107 62	0.978	1.36064 80777 70	0.77427 62761 95	1.00184 24682 01
0.929	1.19170 14936 35	0.74860 81353 36	0.979	1.36549 69629 42	0.77478 71417 51	1.00175 83208 02
0.930	1.19441 28444 77	0.74914 46246 06	0.980	1.37046 14844 72	0.77529 74968 12	1.00167 42116 16
0.931	1.19714 29249 00	0.74968 05790 63	0.981	1.37555 04644 29	0.77580 73418 77	1.00159 01406 08
0.932	1.19989 21492 75	0.75021 59991 99	0.982	1.38077 39033 32	0.77631 66774 45	1.00150 61077 45
0.933	1.20266 09472 92	0.75075 08855 06	0.983	1.38614 32129 70	0.77682 55040 17	1.00142 21129 93
0.934	1.20544 97647 69	0.75128 52384 76	0.984	1.39167 15119 16	0.77733 38220 91	1.00133 81563 16
0.935	1.20825 90645 07	0.75181 90586 03	0.985	1.39737 40056 99	0.77784 16321 67	1.00125 42376 80
0.936	1.21108 93272 10	0.75235 23463 79	0.986	1.40326 84832 96	0.77834 89347 44	1.00117 03570 52
0.937	1.21394 10524 70	0.75288 51022 96	0.987	1.40937 59766 46	0.77885 57303 23	1.00108 65143 98
0.938	1.21681 47598 22	0.75341 73268 49	0.988	1.41572 16538 31	0.77936 20194 04	1.00100 27096 82
0.939	1.21971 09898 74	0.75394 90205 30	0.989	1.42233 60557 98	0.77986 78024 85	1.00091 89428 72
0.940	1.22263 03055 22	0.75448 01838 34	0.990	1.42925 68534 70	0.78037 30800 67	1.00083 52139 33
0.941	1.22557 32932 59	0.75501 08172 55	0.991	1.43653 14207 77	0.78087 78526 49	1.00075 15228 31
0.942	1.22854 05645 81	0.75554 09212 86	0.992	1.44422 07408 32	0.78138 21207 32	1.00066 78695 32
0.943	1.23153 27575 05	0.75607 04964 22	0.993	1.45240 56012 67	0.78188 58848 15	1.00058 42540 02
0.944	1.23455 05382 02	0.75659 95431 57	0.994	1.46119 69689 63	0.78238 91453 98	1.00050 06762 08
0.945	1.23759 46027 74	0.75712 80619 86	0.995	1.47075 46131 83	0.78289 19029 81	1.00041 71361 15
0.946	1.24066 56791 62	0.75765 60534 05	0.996	1.48132 37665 90	0.78339 41580 64	1.00033 36336 91
0.947	1.24376 45292 24	0.75818 35179 08	0.997	1.49331 72818 71	0.78389 59111 47	1.00025 01689 01
0.948	1.24689 19509 90	0.75871 04559 90	0.998	1.50754 02279 20	0.78439 71627 31	1.00016 67417 11
0.949	1.25004 87811 06	0.75923 68681 48	0.999	1.52607 12396 26	0.78489 79133 14	1.00008 33520 89
0.950	1.25323 58975 03	0.75976 27548 76	1.000	1.57079 63267 95	0.78539 81633 97	1.00000 00000 00
	$\begin{bmatrix} (-6)4 \\ 6 \end{bmatrix}$	$\begin{bmatrix} (-8)7 \\ 4 \end{bmatrix}$			$\begin{bmatrix} (-8)7 \\ 4 \end{bmatrix}$	$\begin{bmatrix} (-9)5 \\ 4 \end{bmatrix}$

For arctan x, $x>1$ see **Example 22.**

$$\text{arcsin } x = \frac{\pi}{2} - [2(1-x)]^{\frac{1}{2}} f(x) \qquad \frac{\pi}{2} = 1.57079\ 63267\ 95$$

HYPERBOLIC FUNCTIONS   Table 4.15

x	$\sinh x$	$\cosh x$	$\tanh x$	$\coth x$
0.00	0.00000 0000	1.00000 0000	0.00000 000	∞
0.01	0.01000 0167	1.00005 0000	0.00999 967	100.00333 33
0.02	0.02000 1333	1.00020 0007	0.01999 733	50.00666 65
0.03	0.03000 4500	1.00045 0034	0.02999 100	33.34333 27
0.04	0.04001 0668	1.00080 0107	0.03997 868	25.01333 19
0.05	0.05002 0836	1.00125 0260	0.04995 838	20.01666 39
0.06	0.06003 6006	1.00180 0540	0.05992 810	16.68666 19
0.07	0.07005 7181	1.00245 1001	0.06988 589	14.30904 00
0.08	0.08008 5361	1.00320 1707	0.07982 977	12.52665 53
0.09	0.09012 1549	1.00405 2734	0.08975 779	11.14109 49
0.10	0.10016 6750	1.00500 4168	0.09966 800	10.03331 11
0.11	0.11022 1968	1.00605 6103	0.10955 847	9.12754 62
0.12	0.12028 8207	1.00720 8644	0.11942 730	8.37329 50
0.13	0.13036 6476	1.00846 1907	0.12927 258	7.73559 23
0.14	0.14045 7782	1.00981 6017	0.13909 245	7.18946 29
0.15	0.15056 3133	1.01127 1110	0.14888 503	6.71659 18
0.16	0.16068 3541	1.01282 7330	0.15864 850	6.30324 25
0.17	0.17082 0017	1.01448 4834	0.16838 105	5.93891 07
0.18	0.18097 3576	1.01624 3787	0.17808 087	5.61542 64
0.19	0.19114 5232	1.01810 4366	0.18774 621	5.32633 93
0.20	0.20133 6003	1.02006 6756	0.19737 532	5.06648 96
0.21	0.21154 6907	1.02213 1153	0.20696 650	4.83169 98
0.22	0.22177 8966	1.02429 7764	0.21651 806	4.61855 23
0.23	0.23203 3204	1.02656 6806	0.22602 835	4.42422 37
0.24	0.24231 0645	1.02893 8506	0.23549 575	4.24636 11
0.25	0.25261 2317	1.03141 3100	0.24491 866	4.08298 82
0.26	0.26293 9250	1.03399 0836	0.25429 553	3.93243 24
0.27	0.27329 2478	1.03667 1973	0.26362 484	3.79326 93
0.28	0.28367 3035	1.03945 6777	0.27290 508	3.66427 77
0.29	0.29408 1960	1.04234 5528	0.28213 481	3.54440 49
0.30	0.30452 0293	1.04533 8514	0.29131 261	3.43273 84
0.31	0.31498 9079	1.04843 6035	0.30043 710	3.32848 38
0.32	0.32548 9364	1.05163 8401	0.30950 692	3.23094 55
0.33	0.33602 2198	1.05494 5931	0.31852 078	3.13951 26
0.34	0.34658 8634	1.05835 8957	0.32747 740	3.05364 59
0.35	0.35718 9729	1.06187 7819	0.33637 554	2.97286 77
0.36	0.36782 6544	1.06550 2870	0.34521 403	2.89675 36
0.37	0.37850 0142	1.06923 4473	0.35399 171	2.82492 49
0.38	0.38921 1590	1.07307 2999	0.36270 747	2.75704 28
0.39	0.39996 1960	1.07701 8834	0.37136 023	2.69280 32
0.40	0.41075 2326	1.08107 2372	0.37994 896	2.63193 24
0.41	0.42158 3767	1.08523 4018	0.38847 268	2.57418 36
0.42	0.43245 7368	1.08950 4188	0.39693 043	2.51933 32
0.43	0.44337 4214	1.09388 3309	0.40532 131	2.46717 85
0.44	0.45433 5399	1.09837 1820	0.41364 444	2.41753 52
0.45	0.46534 2017	1.10297 0169	0.42189 901	2.37023 55
0.46	0.47639 5170	1.10767 8815	0.43008 421	2.32512 60
0.47	0.48749 5962	1.11249 8231	0.43819 932	2.28206 66
0.48	0.49864 5505	1.11742 8897	0.44624 361	2.24092 84
0.49	0.50984 4913	1.12247 1307	0.45421 643	2.20159 36
0.50	0.52109 5305	1.12762 5965	0.46211 716	2.16395 34
	$\begin{bmatrix} (-6)6 \\ 4 \end{bmatrix}$	$\begin{bmatrix} (-5)1 \\ 4 \end{bmatrix}$	$\begin{bmatrix} (-6)9 \\ 4 \end{bmatrix}$	

For coth x, $x \leq .1$ use 4.5.67.

Compilation of tanh x and coth x from National Bureau of Standards, Table of circular and hyperbolic tangents and cotangents for radian arguments, 2d printing. Columbia Univ. Press, New York, N.Y., 1947 (with permission).

Table 4.15　　　　　　　　　　**HYPERBOLIC FUNCTIONS**

x	$\sinh x$	$\cosh x$	$\tanh x$	$\coth x$
0.50	0.52109 5305	1.12762 5965	0.46211 716	2.16395 34
0.51	0.53239 7808	1.13289 3387	0.46994 520	2.12790 77
0.52	0.54375 3551	1.13827 4099	0.47770 001	2.09336 40
0.53	0.55516 3669	1.14376 8639	0.48538 109	2.06023 68
0.54	0.56662 9305	1.14937 7557	0.49298 797	2.02844 71
0.55	0.57815 1604	1.15510 1414	0.50052 021	1.99792 13
0.56	0.58973 1718	1.16094 0782	0.50797 743	1.96859 14
0.57	0.60137 0806	1.16689 6245	0.51535 928	1.94039 39
0.58	0.61307 0032	1.17296 8399	0.52266 543	1.91326 98
0.59	0.62483 0565	1.17915 7850	0.52989 561	1.88716 42
0.60	0.63665 3582	1.18546 5218	0.53704 957	1.86202 55
0.61	0.64854 0265	1.19189 1134	0.54412 710	1.83780 59
0.62	0.66049 1802	1.19843 6240	0.55112 803	1.81446 04
0.63	0.67250 9389	1.20510 1190	0.55805 222	1.79194 70
0.64	0.68459 4228	1.21188 6652	0.56489 955	1.77022 62
0.65	0.69674 7526	1.21879 3303	0.57166 997	1.74926 10
0.66	0.70897 0500	1.22582 1834	0.57836 341	1.72901 67
0.67	0.72126 4371	1.23297 2949	0.58497 988	1.70946 05
0.68	0.73363 0370	1.24024 7362	0.59151 940	1.69056 16
0.69	0.74606 9732	1.24764 5801	0.59798 200	1.67229 11
0.70	0.75858 3702	1.25516 9006	0.60436 778	1.65462 16
0.71	0.77117 3531	1.26281 7728	0.61067 683	1.63752 73
0.72	0.78384 0477	1.27059 2733	0.61690 930	1.62098 38
0.73	0.79658 5809	1.27849 4799	0.62306 535	1.60496 81
0.74	0.80941 0799	1.28652 4715	0.62914 516	1.58945 83
0.75	0.82231 6732	1.29468 3285	0.63514 895	1.57443 38
0.76	0.83530 4897	1.30297 1324	0.64107 696	1.55987 51
0.77	0.84837 6593	1.31138 9661	0.64692 945	1.54576 36
0.78	0.86153 3127	1.31993 9138	0.65270 671	1.53208 17
0.79	0.87477 5815	1.32862 0611	0.65840 904	1.51881 27
0.80	0.88810 5982	1.33743 4946	0.66403 677	1.50594 07
0.81	0.90152 4960	1.34638 3026	0.66959 026	1.49345 06
0.82	0.91503 4092	1.35546 5746	0.67506 987	1.48132 81
0.83	0.92863 4727	1.36468 4013	0.68047 601	1.46955 95
0.84	0.94232 8227	1.37403 8750	0.68580 906	1.45813 18
0.85	0.95611 5960	1.38353 0892	0.69106 947	1.44703 25
0.86	0.96999 9306	1.39316 1388	0.69625 767	1.43624 99
0.87	0.98397 9652	1.40293 1201	0.70137 413	1.42577 26
0.88	0.99805 8397	1.41284 1309	0.70641 932	1.41558 98
0.89	1.01223 6949	1.42289 2702	0.71139 373	1.40569 13
0.90	1.02651 6726	1.43308 6385	0.71629 787	1.39606 73
0.91	1.04089 9155	1.44342 3379	0.72113 225	1.38670 82
0.92	1.05538 5674	1.45390 4716	0.72589 742	1.37760 51
0.93	1.06997 7734	1.46453 1444	0.73059 390	1.36874 95
0.94	1.08467 6791	1.47530 4627	0.73522 225	1.36013 29
0.95	1.09948 4318	1.48622 5341	0.73978 305	1.35174 76
0.96	1.11440 1794	1.49729 4680	0.74427 687	1.34358 60
0.97	1.12943 0711	1.50851 3749	0.74870 429	1.33564 08
0.98	1.14457 2572	1.51988 3670	0.75306 591	1.32790 50
0.99	1.15982 8891	1.53140 5582	0.75736 232	1.32037 20
1.00	1.17520 1194 $\left[\begin{smallmatrix}(-5)1\\4\end{smallmatrix}\right]$	1.54308 0635 $\left[\begin{smallmatrix}(-5)2\\4\end{smallmatrix}\right]$	0.76159 416 $\left[\begin{smallmatrix}(-6)9\\4\end{smallmatrix}\right]$	1.31303 53 $\left[\begin{smallmatrix}(-4)2\\5\end{smallmatrix}\right]$

HYPERBOLIC FUNCTIONS

Table 4.15

x	$\sinh x$	$\cosh x$	$\tanh x$	$\coth x$
1.00	1.17520 1194	1.54308 0635	0.76159 416	1.31303 53
1.01	1.19069 1018	1.55490 9997	0.76576 202	1.30588 87
1.02	1.20629 9912	1.56689 4852	0.76986 654	1.29892 64
1.03	1.22202 9437	1.57903 6398	0.77390 834	1.29214 27
1.04	1.23788 1166	1.59133 5848	0.77788 807	1.28553 20
1.05	1.25385 6684	1.60379 4434	0.78180 636	1.27908 91
1.06	1.26995 7589	1.61641 3400	0.78566 386	1.27280 90
1.07	1.28618 5491	1.62919 4009	0.78946 122	1.26668 67
1.08	1.30254 2013	1.64213 7538	0.79319 910	1.26071 75
1.09	1.31902 8789	1.65524 5283	0.79687 814	1.25489 70
1.10	1.33564 7470	1.66851 8554	0.80049 902	1.24922 08
1.11	1.35239 9717	1.68195 8678	0.80406 239	1.24368 46
1.12	1.36928 7204	1.69556 6999	0.80756 892	1.23828 44
1.13	1.38631 1622	1.70934 4878	0.81101 926	1.23301 63
1.14	1.40347 4672	1.72329 3694	0.81441 409	1.22787 66
1.15	1.42077 8070	1.73741 4840	0.81775 408	1.22286 15
1.16	1.43822 3548	1.75170 9728	0.82103 988	1.21796 76
1.17	1.45581 2849	1.76617 9790	0.82427 217	1.21319 15
1.18	1.47354 7732	1.78082 6471	0.82745 161	1.20852 99
1.19	1.49142 9972	1.79565 1236	0.83057 887	1.20397 96
1.20	1.50946 1355	1.81065 5567	0.83365 461	1.19953 75
1.21	1.52764 3687	1.82584 0966	0.83667 949	1.19520 08
1.22	1.54597 8783	1.84120 8950	0.83965 418	1.19096 65
1.23	1.56446 8479	1.85676 1057	0.84257 933	1.18683 19
1.24	1.58311 4623	1.87249 8841	0.84545 560	1.18279 42
1.25	1.60191 9080	1.88842 3877	0.84828 364	1.17885 10
1.26	1.62088 3730	1.90453 7757	0.85106 411	1.17499 96
1.27	1.64001 0470	1.92084 2092	0.85379 765	1.17123 77
1.28	1.65930 1213	1.93733 8513	0.85648 492	1.16756 29
1.29	1.67875 7886	1.95402 8669	0.85912 654	1.16397 29
1.30	1.69838 2437	1.97091 4230	0.86172 316	1.16046 55
1.31	1.71817 6828	1.98799 6884	0.86427 541	1.15703 86
1.32	1.73814 3038	2.00527 8340	0.86678 393	1.15369 01
1.33	1.75828 3063	2.02276 0324	0.86924 933	1.15041 79
1.34	1.77859 8918	2.04044 4587	0.87167 225	1.14722 02
1.35	1.79909 2635	2.05833 2896	0.87405 329	1.14409 50
1.36	1.81976 6262	2.07642 7039	0.87639 307	1.14104 05
1.37	1.84062 1868	2.09472 8828	0.87869 219	1.13805 50
1.38	1.86166 1537	2.11324 0090	0.88095 127	1.13513 66
1.39	1.88288 7374	2.13196 2679	0.88317 089	1.13228 37
1.40	1.90430 1501	2.15089 8465	0.88535 165	1.12949 47
1.41	1.92590 6060	2.17004 9344	0.88749 413	1.12676 80
1.42	1.94770 3212	2.18941 7229	0.88959 892	1.12410 21
1.43	1.96969 5135	2.20900 4057	0.89166 660	1.12149 54
1.44	1.99188 4029	2.22881 1788	0.89369 773	1.11894 66
1.45	2.01427 2114	2.24884 2402	0.89569 287	1.11645 41
1.46	2.03686 1627	2.26909 7902	0.89765 260	1.11401 67
1.47	2.05965 4828	2.28958 0313	0.89957 745	1.11163 30
1.48	2.08265 3996	2.31029 1685	0.90146 799	1.10930 17
1.49	2.10586 1432	2.33123 4087	0.90332 474	1.10702 16
1.50	2.12927 9455	2.35240 9615	0.90514 825	1.10479 14
	$\left[\begin{matrix}(-5)3\\4\end{matrix}\right]$	$\left[\begin{matrix}(-5)3\\5\end{matrix}\right]$	$\left[\begin{matrix}(-6)8\\4\end{matrix}\right]$	$\left[\begin{matrix}(-5)2\\4\end{matrix}\right]$

Table 4.15   **HYPERBOLIC FUNCTIONS**

x	sinh x	cosh x	tanh x	coth x
1.50	2.12927 9455	2.35240 9615	0.90514 825	1.10479 14
1.51	2.15291 0408	2.37382 0386	0.90693 905	1.10260 99
1.52	2.17675 6654	2.39546 8541	0.90869 766	1.10047 60
1.53	2.20082 0577	2.41735 6245	0.91042 459	1.09838 86
1.54	2.22510 4585	2.43948 5686	0.91212 037	1.09634 65
1.55	2.24961 1104	2.46185 9078	0.91378 549	1.09434 87
1.56	2.27434 2587	2.48447 8658	0.91542 046	1.09239 42
1.57	2.29930 1506	2.50734 6688	0.91702 576	1.09048 19
1.58	2.32449 0357	2.53046 5455	0.91860 189	1.08861 09
1.59	2.34991 1658	2.55383 7270	0.92014 933	1.08678 01
1.60	2.37556 7953	2.57746 4471	0.92166 855	1.08498 87
1.61	2.40146 1807	2.60134 9421	0.92316 003	1.08323 58
1.62	2.42759 5809	2.62549 4508	0.92462 422	1.08152 04
1.63	2.45397 2572	2.64990 2146	0.92606 158	1.07984 18
1.64	2.48059 4735	2.67457 4777	0.92747 257	1.07819 90
1.65	2.50746 4959	2.69951 4868	0.92885 762	1.07659 13
1.66	2.53458 5932	2.72472 4912	0.93021 718	1.07501 78
1.67	2.56196 0366	2.75020 7431	0.93155 168	1.07347 77
1.68	2.58959 0998	2.77596 4974	0.93286 155	1.07197 04
1.69	2.61748 0591	2.80200 0115	0.93414 721	1.07049 51
1.70	2.64563 1934	2.82831 5458	0.93540 907	1.06905 10
1.71	2.67404 7843	2.85491 3635	0.93664 754	1.06763 75
1.72	2.70273 1158	2.88179 7306	0.93786 303	1.06625 38
1.73	2.73168 4749	2.90896 9159	0.93905 593	1.06489 93
1.74	2.76091 1511	2.93643 1912	0.94022 664	1.06357 34
1.75	2.79041 4366	2.96418 8310	0.94137 554	1.06227 53
1.76	2.82019 6265	2.99224 1129	0.94250 301	1.06100 46
1.77	2.85026 0186	3.02059 3175	0.94360 942	1.05976 05
1.78	2.88060 9136	3.04924 7283	0.94469 516	1.05854 25
1.79	2.91124 6148	3.07820 6318	0.94576 057	1.05735 01
1.80	2.94217 4288	3.10747 3176	0.94680 601	1.05618 26
1.81	2.97339 6648	3.13705 0785	0.94783 185	1.05503 95
1.82	3.00491 6349	3.16694 2100	0.94883 842	1.05392 02
1.83	3.03673 6545	3.19715 0113	0.94982 608	1.05282 43
1.84	3.06886 0417	3.22767 7844	0.95079 514	1.05175 13
1.85	3.10129 1178	3.25852 8344	0.95174 596	1.05070 05
1.86	3.13403 2071	3.28970 4701	0.95267 884	1.04967 17
1.87	3.16708 6369	3.32121 0031	0.95359 412	1.04866 42
1.88	3.20045 7378	3.35304 7484	0.95449 211	1.04767 76
1.89	3.23414 8436	3.38522 0245	0.95537 312	1.04671 15
1.90	3.26816 2912	3.41773 1531	0.95623 746	1.04576 53
1.91	3.30250 4206	3.45058 4593	0.95708 542	1.04483 88
1.92	3.33717 5754	3.48378 2716	0.95791 731	1.04393 14
1.93	3.37218 1022	3.51732 9220	0.95873 341	1.04304 28
1.94	3.40752 3510	3.55122 7460	0.95953 401	1.04217 25
1.95	3.44320 6754	3.58548 0826	0.96031 939	1.04132 02
1.96	3.47923 4322	3.62009 2743	0.96108 983	1.04048 55
1.97	3.51560 9816	3.65506 6672	0.96184 561	1.03966 79
1.98	3.55233 6874	3.69040 6111	0.96258 698	1.03886 72
1.99	3.58941 9168	3.72611 4594	0.96331 422	1.03808 29
2.00	3.62686 0408	3.76219 5691	0.96402 758	1.03731 47
	$\begin{bmatrix} (-5)4 \\ 5 \end{bmatrix}$	$\begin{bmatrix} (-5)5 \\ 5 \end{bmatrix}$	$\begin{bmatrix} (-6)4 \\ 4 \end{bmatrix}$	$\begin{bmatrix} (-6)6 \\ 3 \end{bmatrix}$

HYPERBOLIC FUNCTIONS

Table 4.15

x	$\sinh x$	$\cosh x$	$\tanh x$	$\coth x$
2.0	3. 62686 0408	3. 76219 5691	0. 96402 75801	1. 03731 47207
2.1	4. 02185 6742	4. 14431 3170	0. 97045 19366	1. 03044 77350
2.2	4. 45710 5171	4. 56790 8329	0. 97574 31300	1. 02485 98932
2.3	4. 93696 1806	5. 03722 0649	0. 98009 63963	1. 02030 78022
2.4	5. 46622 9214	5. 55694 7167	0. 98367 48577	1. 01659 60756
2.5	6. 05020 4481	6. 13228 9480	0. 98661 42982	1. 01356 73098
2.6	6. 69473 2228	6. 76900 5807	0. 98902 74022	1. 01109 43314
2.7	7. 40626 3106	7. 47346 8619	0. 99100 74537	1. 00907 41460
2.8	8. 19191 8354	8. 25272 8417	0. 99263 15202	1. 00742 31773
2.9	9. 05956 1075	9. 11458 4295	0. 99396 31674	1. 00607 34973
3.0	10. 01787 4927	10. 06766 1996	0. 99505 47537	1. 00496 98233
3.1	11. 07645 1040	11. 12150 0242	0. 99594 93592	1. 00406 71152
3.2	12. 24588 3997	12. 28664 6201	0. 99668 23978	1. 00332 86453
3.3	13. 53787 7877	13. 57476 1044	0. 99728 29601	1. 00272 44423
3.4	14. 96536 3389	14. 99873 6659	0. 99777 49279	1. 00223 00341
3.5	16. 54262 7288	16. 57282 4671	0. 99817 78976	1. 00182 54285
3.6	18. 28545 5361	18. 31277 9083	0. 99850 79423	1. 00149 42872
3.7	20. 21129 0417	20. 23601 3943	0. 99877 82413	1. 00122 32532
3.8	22. 33940 6861	22. 36177 7633	0. 99899 95978	1. 00100 14040
3.9	24. 69110 3597	24. 71134 5508	0. 99918 08657	1. 00081 98059
4.0	27. 28991 7197	27. 30823 2836	0. 99932 92997	1. 00067 11504
4.1	30. 16185 7461	30. 17843 0136	0. 99945 08437	1. 00054 94581
4.2	33. 33566 7732	33. 35066 3309	0. 99955 03665	1. 00044 98358
4.3	36. 84311 2570	36. 85668 1129	0. 99963 18562	1. 00036 82794
4.4	40. 71929 5663	40. 73157 3002	0. 99969 85793	1. 00030 15116
4.5	45. 00301 1152	45. 01412 0149	0. 99975 32108	1. 00024 68501
4.6	49. 73713 1903	49. 74718 3739	0. 99979 79416	1. 00020 20992
4.7	54. 96903 8588	54. 97813 3865	0. 99983 45656	1. 00016 54618
4.8	60. 75109 3886	60. 75932 3633	0. 99986 45517	1. 00013 54666
4.9	67. 14116 6551	67. 14861 3134	0. 99988 91030	1. 00011 09093
5.0	74. 20321 0578	74. 20994 8525	0. 99990 92043	1. 00009 08040
5.1	82. 00790 5277	82. 01400 2023	0. 99992 56621	1. 00007 43434
5.2	90. 63336 2655	90. 63887 9220	0. 99993 91369	1. 00006 08668
5.3	100. 16590 9190	100. 17090 0784	0. 99995 01692	1. 00004 98333
5.4	110. 70094 9812	110. 70546 6393	0. 99995 92018	1. 00004 07998
5.5	122. 34392 2746	122. 34800 9518	0. 99996 65972	1. 00003 34040
5.6	135. 21135 4781	135. 21505 2645	0. 99997 26520	1. 00002 73488
5.7	149. 43202 7501	149. 43537 3466	0. 99997 76093	1. 00002 23912
5.8	165. 14826 6177	165. 15129 3732	0. 99998 16680	1. 00001 83323
5.9	182. 51736 4210	182. 52010 3655	0. 99998 49910	1. 00001 50092
6.0	201. 71315 7370	201. 71563 6122	0. 99998 77117 $\begin{bmatrix}(-4)1\\6\end{bmatrix}$	1. 00001 22885 $\begin{bmatrix}(-4)2\\9\end{bmatrix}$

Table 4.15   **HYPERBOLIC FUNCTIONS**

x	$\sinh x$	$\cosh x$	$\tanh x$	$\coth x$
6.0	201.71315 7370	201.71563 6122	0.99998 77117	1.00001 22885
6.1	222.92776 3607	222.93000 6475	0.99998 99391	1.00001 00610
6.2	246.37350 5831	246.37553 5262	0.99999 17629	1.00000 82372
6.3	272.28503 6911	272.28687 3215	0.99999 32560	1.00000 67441
6.4	300.92168 8157	300.92334 9715	0.99999 44785	1.00000 55216
6.5	332.57006 4803	332.57156 8242	0.99999 54794	1.00000 45207
6.6	367.54691 4437	367.54827 4805	0.99999 62988	1.00000 37012
6.7	406.20229 7128	406.20352 8040	0.99999 69697	1.00000 30303
6.8	448.92308 8938	448.92420 2713	0.99999 75190	1.00000 24810
6.9	496.13685 3910	496.13786 1695	0.99999 79687	1.00000 20313
7.0	548.31612 3273	548.31703 5155	0.99999 83369	1.00000 16631
7.1	605.98312 4694	605.98394 9799	0.99999 86384	1.00000 13616
7.2	669.71500 8904	669.71575 5490	0.99999 88852	1.00000 11148
7.3	740.14962 6023	740.15030 1562	0.99999 90873	1.00000 09127
7.4	817.99190 9372	817.99252 0624	0.99999 92527	1.00000 07473
7.5	904.02093 0686	904.02148 3770	0.99999 93882	1.00000 06118
7.6	999.09769 7326	999.09819 7778	0.99999 94991	1.00000 05009
7.7	1104.17376 9530	1104.17422 2357	0.99999 95899	1.00000 04101
7.8	1220.30078 3945	1220.30119 3680	0.99999 96642	1.00000 03358
7.9	1348.64097 8762	1348.64134 9506	0.99999 97251	1.00000 02749
8.0	1490.47882 5790	1490.47916 1252	0.99999 97749	1.00000 02251
8.1	1647.23388 5872	1647.23418 9411	0.99999 98157	1.00000 01843
8.2	1820.47501 6339	1820.47529 0993	0.99999 98491	1.00000 01509
8.3	2011.93607 2653	2011.93632 1170	0.99999 98765	1.00000 01235
8.4	2223.53326 1416	2223.53348 6284	0.99999 98989	1.00000 01011
8.5	2457.38431 8415	2457.38452 1884	0.99999 99172	1.00000 00828
8.6	2715.82970 3629	2715.82988 7734	0.99999 99322	1.00000 00678
8.7	3001.45602 5338	3001.45619 1923	0.99999 99445	1.00000 00555
8.8	3317.12192 7772	3317.12207 8505	0.99999 99546	1.00000 00454
8.9	3665.98670 1384	3665.98683 7772	0.99999 99628	1.00000 00372
9.0	4051.54190 2083	4051.54202 5493	0.99999 99695	1.00000 00305
9.1	4477.64629 5908	4477.64640 7574	0.99999 99751	1.00000 00249
9.2	4948.56447 8852	4948.56457 9892	0.99999 99796	1.00000 00204
9.3	5469.00955 8370	5469.00964 9795	0.99999 99833	1.00000 00167
9.4	6044.19032 3746	6044.19040 6471	0.99999 99863	1.00000 00137
9.5	6679.86337 7405	6679.86345 2257	0.99999 99888	1.00000 00112
9.6	7382.39074 8924	7382.39081 6653	0.99999 99908	1.00000 00092
9.7	8158.80356 8366	8158.80362 9649	0.99999 99925	1.00000 00075
9.8	9016.87243 6188	9016.87249 1640	0.99999 99939	1.00000 00061
9.9	9965.18519 4028	9965.18524 4202	0.99999 99950	1.00000 00050
10.0	11013.23287 4703	11013.23292 0103	0.99999 99959	1.00000 00041
			$*\begin{bmatrix}(-8)5\\5\end{bmatrix}$	$\begin{bmatrix}(-8)7\\5\end{bmatrix}$

For $x \gg 0$, $\sinh x \sim \cosh x \sim \frac{1}{2} e^x$.   For $x > 10$, $\tanh x \sim 1 - 2e^{-2x}$, $\coth x \sim 1 + 2e^{-2x}$ to 10D.

---

*See page II.

EXPONENTIAL AND HYPERBOLIC FUNCTIONS FOR THE ARGUMENT πx — Table 4.16

x	$e^{\pi x}$	$e^{-\pi x}$	$\sinh \pi x$	$\cosh \pi x$	$\tanh \pi x$
0.00	1.00000 00000	1.00000 00000	0.00000 00000	1.00000 00000	0.00000 00000
0.01	1.03191 46153	0.96907 24263	0.03142 10945	1.00049 35208	0.03140 55952
0.02	1.06484 77733	0.93910 13674	0.06287 32029	1.00197 45704	0.06274 93000
0.03	1.09883 19803	0.91005 72407	0.09438 73698	1.00444 46105	0.09396 97111
0.04	1.13390 07803	0.88191 13783	0.12599 47010	1.00790 60793	0.12500 63906
0.05	1.17008 87875	0.85463 59992	0.15772 63942	1.01236 23933	0.15580 03292
0.06	1.20743 17210	0.82820 41813	0.18961 37699	1.01781 79512	0.18629 43856
0.07	1.24596 64399	0.80258 98355	0.22168 83022	1.02427 81377	0.21643 36952
0.08	1.28573 09795	0.77776 76792	0.25398 16502	1.03174 93294	0.24616 60434
0.09	1.32676 45892	0.75371 32120	0.28652 56886	1.04023 89006	0.27544 21974
0.10	1.36910 77706	0.73040 26910	0.31935 25398	1.04975 52308	0.30421 61929
0.11	1.41280 23184	0.70781 31080	0.35249 46052	1.06030 77132	0.33244 55730
0.12	1.45789 13610	0.68592 21659	0.38598 45975	1.07190 67634	0.36009 15776
0.13	1.50441 94029	0.66470 82576	0.41985 55727	1.08456 38303	0.38711 92833
0.14	1.55243 23694	0.64415 04440	0.45414 09627	1.09829 14067	0.41349 76928
0.15	1.60197 76513	0.62422 84336	0.48887 46088	1.11310 30425	0.43919 97777
0.16	1.65310 41518	0.60492 25628	0.52409 07945	1.12901 33573	0.46420 24748
0.17	1.70586 23348	0.58621 37756	0.55982 42796	1.14603 80552	0.48848 66406
0.18	1.76030 42750	0.56808 36059	0.59611 03346	1.16419 39405	0.51203 69673
0.19	1.81648 37088	0.55051 41583	0.63298 47753	1.18349 89335	0.53484 18637
0.20	1.87445 60876	0.53348 80911	0.67048 39982	1.20397 20893	0.55689 33069
0.21	1.93427 86325	0.51698 85988	0.70864 50169	1.22563 36157	0.57818 66683
0.22	1.99601 03910	0.50099 93958	0.74750 54976	1.24850 48934	0.59872 05188
0.23	2.05971 22948	0.48550 47001	0.78710 37973	1.27260 84975	0.61849 64181
0.24	2.12544 72203	0.47048 92177	0.82747 90013	1.29796 82190	0.63751 86920
0.25	2.19328 00507	0.45593 81278	0.86867 09615	1.32460 90893	0.65579 42026
0.26	2.26327 77398	0.44183 70677	0.91072 03361	1.35255 74038	0.67333 21140
0.27	2.33550 93782	0.42817 21192	0.95366 86295	1.38184 07487	0.69014 36583
0.28	2.41004 62616	0.41492 97945	0.99755 82336	1.41248 80280	0.70624 19035
0.29	2.48696 19609	0.40209 70227	1.04243 24691	1.44452 94918	0.72164 15276
0.30	2.56633 23952	0.38966 11374	1.08833 56289	1.47799 67663	0.73635 85995
0.31	2.64823 59064	0.37760 98638	1.13531 30213	1.51292 28851	0.75041 03695
0.32	2.73275 33366	0.36593 13069	1.18341 10148	1.54934 23218	0.76381 50706
0.33	2.81996 81081	0.35461 39395	1.23267 70843	1.58729 10238	0.77659 17313
0.34	2.90996 63054	0.34364 65907	1.28315 98573	1.62680 64481	0.78876 00021
0.35	3.00283 67606	0.33301 84355	1.33490 91626	1.66792 75980	0.80033 99933
0.36	3.09867 11407	0.32271 89833	1.38797 60787	1.71069 50620	0.81135 21279
0.37	3.19756 40381	0.31273 80681	1.44241 29850	1.75515 10531	0.82181 70068
0.38	3.29961 30643	0.30306 58385	1.49827 36129	1.80133 94514	0.83175 52873
0.39	3.40491 89460	0.29369 27474	1.55561 30993	1.84930 58467	0.84118 75743
0.40	3.51358 56243	0.28460 95433	1.61448 80405	1.89909 75838	0.85013 43239
0.41	3.62572 03579	0.27580 72607	1.67495 65486	1.95076 38093	0.85861 57589
0.42	3.74143 38283	0.26727 72113	1.73707 83085	2.00435 55198	0.86665 17947
0.43	3.86084 02496	0.25901 09757	1.80091 46370	2.05992 56127	0.87426 19762
0.44	3.98405 74810	0.25100 03946	1.86652 85432	2.11752 89378	0.88146 54241
0.45	4.11120 71429	0.24323 75614	1.93398 47907	2.17722 23522	0.88828 07899
0.46	4.24241 47373	0.23571 48138	2.00334 99617	2.23906 47756	0.89472 62194
0.47	4.37780 97717	0.22842 47266	2.07469 25226	2.30311 72491	0.90081 93236
0.48	4.51752 58864	0.22136 01040	2.14808 28912	2.36944 29952	0.90657 71557
0.49	4.66170 09873	0.21451 39731	2.22359 35071	2.43810 74802	0.91201 61950
0.50	4.81047 73810	0.20787 95764	2.30129 89023	2.50917 84787	0.91715 23357
	$\begin{bmatrix}(-4)6\\6\end{bmatrix}$	$\begin{bmatrix}(-4)1\\6\end{bmatrix}$	$\begin{bmatrix}(-4)3\\6\end{bmatrix}$	$\begin{bmatrix}(-4)3\\6\end{bmatrix}$	$\begin{bmatrix}(-5)9\\7\end{bmatrix}$

Compiled from British Association for the Advancement of Science, Mathematical Tables, vol. I. Circular and hyperbolic functions, exponential, sine and cosine integrals, factorial function and allied functions, Hermitian probability functions, 3d ed. Cambridge Univ. Press, Cambridge, England, 1951 (with permission). Known errors have been corrected.

Table 4.16   **EXPONENTIAL AND HYPERBOLIC FUNCTIONS FOR THE ARGUMENT** πx

x	$e^{\pi x}$	$e^{-\pi x}$	$\sinh \pi x$	$\cosh \pi x$	$\tanh \pi x$
0.50	4. 81047 73810	0. 20787 95764	2. 30129 89023	2. 50917 84787	0. 91715 23357
0.51	4. 96400 19160	0. 20145 03654	2. 38127 57753	2. 58272 61407	0. 92200 08803
0.52	5. 12242 61276	0. 19521 99944	2. 46360 30666	2. 65882 30610	0. 92657 65378
0.53	5. 28590 63869	0. 18918 23136	2. 54836 20366	2. 73754 43503	0. 93089 34251
0.54	5. 45460 40558	0. 18333 13637	2. 63563 63461	2. 81896 77098	0. 93496 50714
0.55	5. 62868 56460	0. 17766 13694	2. 72551 21383	2. 90317 35077	0. 93880 44259
0.56	5. 80832 29831	0. 17216 67343	2. 81807 81244	2. 99024 48587	0. 94242 38675
0.57	5. 99369 33767	0. 16684 20350	2. 91342 56709	3. 08026 77058	0. 94583 52160
0.58	6. 18497 97951	0. 16168 20156	3. 01164 88897	3. 17333 09054	0. 94904 97460
0.59	6. 38237 10460	0. 15668 15832	3. 11284 47314	3. 26952 63146	0. 95207 82009
0.60	6. 58606 19627	0. 15183 58020	3. 21711 30804	3. 36894 88823	0. 95493 08086
0.61	6. 79625 35967	0. 14713 98890	3. 32455 68538	3. 47169 67428	0. 95761 72978
0.62	7. 01315 34158	0. 14258 92093	3. 43528 21032	3. 57787 13125	0. 96014 69151
0.63	7. 23697 55091	0. 13817 92710	3. 54939 81191	3. 68757 73901	0. 96252 84417
0.64	7. 46794 07985	0. 13390 57214	3. 66701 75386	3. 80092 32600	0. 96477 02118
0.65	7. 70627 72563	0. 12976 43423	3. 78825 64570	3. 91802 07993	0. 96688 01293
0.66	7. 95222 01304	0. 12575 10461	3. 91323 45422	4. 03898 55883	0. 96886 56859
0.67	8. 20601 21768	0. 12186 18713	4. 04207 51527	4. 16393 70240	0. 97073 39783
0.68	8. 46790 38986	0. 11809 29793	4. 17490 54597	4. 29299 84390	0. 97249 17255
0.69	8. 73815 37941	0. 11444 06500	4. 31185 65720	4. 42629 72220	0. 97414 52857
0.70	9. 01702 86109	0. 11090 12784	4. 45306 36663	4. 56396 49447	0. 97570 06726
0.71	9. 30480 36103	0. 10747 13709	4. 59866 61197	4. 70613 74906	0. 97716 35718
0.72	9. 60176 28381	0. 10414 75422	4. 74880 76480	4. 85295 51901	0. 97853 93563
0.73	9. 90819 94054	0. 10092 65114	4. 90363 64470	5. 00456 29584	0. 97983 31019
0.74	10. 22441 57779	0. 09780 50993	5. 06330 53393	5. 16111 04386	0. 98104 96015
0.75	10. 55072 40742	0. 09478 02248	5. 22797 19247	5. 32275 21495	0. 98219 33800
0.76	10. 88744 63743	0. 09184 89025	5. 39779 87359	5. 48964 76384	0. 98326 87071
0.77	11. 23491 50371	0. 08900 82388	5. 57295 33992	5. 66196 16379	0. 98427 96111
0.78	11. 59347 30285	0. 08625 54299	5. 75360 87993	5. 83986 42292	0. 98522 98912
0.79	11. 96347 42604	0. 08358 77587	5. 93994 32508	6. 02353 10095	0. 98612 31297
0.80	12. 34528 39392	0. 08100 25922	6. 13214 06735	6. 21314 32657	0. 98696 27033
0.81	12. 73927 89270	0. 07849 73785	6. 33039 07743	6. 40888 81528	0. 98775 17946
0.82	13. 14584 81133	0. 07606 96451	6. 53488 92341	6. 61095 88792	0. 98849 34022
0.83	13. 56539 27988	0. 07371 69955	6. 74583 79017	6. 81955 48972	0. 98919 03509
0.84	13. 99832 70916	0. 07143 71077	6. 96344 49919	7. 03488 20996	0. 98984 53014
0.85	14. 44507 83157	0. 06922 77313	7. 18792 52922	7. 25715 30235	0. 99046 07591
0.86	14. 90608 74333	0. 06708 66855	7. 41950 03739	7. 48658 70594	0. 99103 90830
0.87	15. 38180 94795	0. 06501 18571	7. 65839 88112	7. 72341 06683	0. 99158 24938
0.88	15. 87271 40119	0. 06300 11981	7. 90485 64069	7. 96785 76050	0. 99209 30818
0.89	16. 37928 55735	0. 06105 27239	8. 15911 64248	8. 22016 91487	0. 99257 28142
0.90	16. 90202 41717	0. 05916 45113	8. 42142 98302	8. 48059 43415	0. 99302 35419
0.91	17. 44144 57711	0. 05733 46965	8. 69205 55373	8. 74939 02338	0. 99344 70066
0.92	17. 99808 28034	0. 05556 14735	8. 97126 06650	9. 02682 21384	0. 99384 48468
0.93	18. 57248 46925	0. 05384 30919	9. 25932 08003	9. 31316 38922	0. 99421 86036
0.94	19. 16521 83968	0. 05217 78557	9. 55652 02706	9. 60869 81263	0. 99456 97268
0.95	19. 77686 89693	0. 05056 41212	9. 86315 24240	9. 91371 65453	0. 99489 95797
0.96	20. 40804 01345	0. 04900 02956	10. 17951 99195	10. 22852 02151	0. 99520 94443
0.97	21. 05935 48847	0. 04748 48354	10. 50593 50247	10. 55341 98601	0. 99550 05263
0.98	21. 73145 60946	0. 04601 62446	10. 84271 99250	10. 88873 61696	0. 99577 39591
0.99	22. 42500 71560	0. 04459 30738	11. 19020 70411	11. 23480 01149	0. 99603 08084
1.00	23. 14069 26328	0. 04321 39183	11. 54873 93573	11. 59195 32755	0. 99627 20762
	$\left[\begin{smallmatrix}(-3)3\\6\end{smallmatrix}\right]$	$\left[\begin{smallmatrix}(-5)3\\5\end{smallmatrix}\right]$	$\left[\begin{smallmatrix}(-3)1\\6\end{smallmatrix}\right]$	$\left[\begin{smallmatrix}(-3)1\\6\end{smallmatrix}\right]$	$\left[\begin{smallmatrix}(-5)4\\6\end{smallmatrix}\right]$

INVERSE HYPERBOLIC FUNCTIONS

Table 4.17

x	arcsinh x	arctanh x	x	arcsinh x	arctanh x
0.00	0.00000 0000	0.00000 0000	0.50	0.48121 1825	0.54930 6144
0.01	0.00999 9833	0.01000 0333	0.51	0.49013 8161	0.56272 9769
0.02	0.01999 8667	0.02000 2667	0.52	0.49902 8444	0.57633 9754
0.03	0.02999 5502	0.03000 9004	0.53	0.50788 2413	0.59014 5160
0.04	0.03998 9341	0.04002 1353	0.54	0.51669 9824	0.60415 5603
0.05	0.04997 9190	0.05004 1729	0.55	0.52548 0448	0.61838 1313
0.06	0.05996 4058	0.06007 2156	0.56	0.53422 4074	0.63283 3186
0.07	0.06994 2959	0.07011 4671	0.57	0.54293 0505	0.64752 2844
0.08	0.07991 4912	0.08017 1325	0.58	0.55159 9562	0.66246 2707
0.09	0.08987 8941	0.09024 4188	0.59	0.56023 1077	0.67766 6068
0.10	0.09983 4079	0.10033 5347	0.60	0.56882 4899	0.69314 7180
0.11	0.10977 9366	0.11044 6915	0.61	0.57738 0892	0.70892 1359
0.12	0.11971 3851	0.12058 1028	0.62	0.58589 8932	0.72500 5087
0.13	0.12963 6590	0.13073 9850	0.63	0.59437 8911	0.74141 6144
0.14	0.13954 6654	0.14092 5576	0.64	0.60282 0733	0.75817 3745
0.15	0.14944 3120	0.15114 0436	0.65	0.61122 4314	0.77529 8706
0.16	0.15932 5080	0.16138 6696	0.66	0.61958 9584	0.79281 3631
0.17	0.16919 1636	0.17166 6663	0.67	0.62791 6485	0.81074 3125
0.18	0.17904 1904	0.18198 2689	0.68	0.63620 4970	0.82911 4038
0.19	0.18887 5015	0.19233 7169	0.69	0.64445 5005	0.84795 5755
0.20	0.19869 0110	0.20273 2554	0.70	0.65266 6566	0.86730 0527
0.21	0.20848 6350	0.21317 1346	0.71	0.66083 9641	0.88718 3863
0.22	0.21826 2908	0.22365 6109	0.72	0.66897 4227	0.90764 4983
0.23	0.22801 8972	0.23418 9466	0.73	0.67707 0332	0.92872 7364
0.24	0.23775 3749	0.24477 4112	0.74	0.68512 7974	0.95047 9381
0.25	0.24746 6462	0.25541 2812	0.75	0.69314 7181	0.97295 5074
0.26	0.25715 6349	0.26610 8407	0.76	0.70112 7988	0.99621 5082
0.27	0.26682 2667	0.27686 3823	0.77	0.70907 0441	1.02032 7758
0.28	0.27646 4691	0.28768 2072	0.78	0.71697 4594	1.04537 0548
0.29	0.28608 1715	0.29856 6264	0.79	0.72484 0509	1.07143 1684
0.30	0.29567 3048	0.30951 9604	0.80	0.73266 8256	1.09861 2289
0.31	0.30523 8020	0.32054 5409	0.81	0.74045 7912	1.12702 9026
0.32	0.31477 5980	0.33164 7108	0.82	0.74820 9563	1.15681 7465
0.33	0.32428 6295	0.34282 8254	0.83	0.75592 3300	1.18813 6404
0.34	0.33376 8352	0.35409 2528	0.84	0.76359 9222	1.22117 3518
0.35	0.34322 1555	0.36544 3754	0.85	0.77123 7433	1.25615 2811
0.36	0.35264 5330	0.37688 5901	0.86	0.77883 8046	1.29334 4672
0.37	0.36203 9121	0.38842 3100	0.87	0.78640 1177	1.33307 9629
0.38	0.37140 2391	0.40005 9650	0.88	0.79392 6950	1.37576 7657
0.39	0.38073 4624	0.41180 0034	0.89	0.80141 5491	1.42192 5871
0.40	0.39003 5320	0.42364 8930	0.90	0.80886 6936	1.47221 9490
0.41	0.39930 4001	0.43561 1223	0.91	0.81628 1421	1.52752 4425
0.42	0.40854 0208	0.44769 2023	0.92	0.82365 9091	1.58902 6915
0.43	0.41774 3500	0.45989 6681	0.93	0.83100 0091	1.65839 0020
0.44	0.42691 3454	0.47223 0804	0.94	0.83830 4575	1.73804 9345
0.45	0.43604 9669	0.48470 0279	0.95	0.84557 2697	1.83178 0823
0.46	0.44515 1759	0.49731 1288	0.96	0.85280 4617	1.94591 0149
0.47	0.45421 9359	0.51007 0337	0.97	0.86000 0498	2.09229 5720
0.48	0.46325 2120	0.52298 4278	0.98	0.86716 0507	2.29755 9925
0.49	0.47224 9713	0.53606 0337	0.99	0.87428 4812	2.64665 2412
0.50	0.48121 1825	0.54930 6144	1.00	0.88137 3587	∞
	$\begin{bmatrix}(-6)5\\4\end{bmatrix}$	$\begin{bmatrix}(-5)2\\5\end{bmatrix}$		$\begin{bmatrix}(-6)5\\4\end{bmatrix}$	

For use of the table see **Examples 26–28**.

$Q_0(x)$ (Legendre Function—Second Kind)$=$arctanh $x(|x|<1)$
$=$arccoth $x(|x|>1)$

Compiled from Harvard Computation Laboratory, Tables of inverse hyperbolic functions. Harvard Univ. Press, Cambridge, Mass., 1949 (with permission).

Table 4.17   **INVERSE HYPERBOLIC FUNCTIONS**

x	arcsinh x	$\dfrac{\text{arccosh } x}{(x^2-1)^{\frac{1}{2}}}$	x	arcsinh x	$\dfrac{\text{arccosh } x}{(x^2-1)^{\frac{1}{2}}}$
1.00	0.88137 3587	1.00000 000	1.50	1.19476 3217	0.86081 788
1.01	0.88842 7007	0.99667 995	1.51	1.20029 7449	0.85849 554
1.02	0.89544 5249	0.99338 621	1.52	1.20580 6263	0.85618 806
1.03	0.90242 8496	0.99011 848	1.53	1.21128 9840	0.85389 528
1.04	0.90937 6928	0.98687 641	1.54	1.21674 8362	0.85161 706
1.05	0.91629 0732	0.98365 968	1.55	1.22218 2008	0.84935 324
1.06	0.92317 0094	0.98046 798	1.56	1.22759 0958	0.84710 368
1.07	0.93001 5204	0.97730 099	1.57	1.23297 5390	0.84486 823
1.08	0.93682 6251	0.97415 841	1.58	1.23833 5478	0.84264 676
1.09	0.94360 3429	0.97103 994	1.59	1.24367 1400	0.84043 913
1.10	0.95034 6930	0.96794 529	1.60	1.24898 3328	0.83824 520
1.11	0.95705 6950	0.96487 415	1.61	1.25427 1436	0.83606 483
1.12	0.96373 3684	0.96182 625	1.62	1.25953 5895	0.83389 788
1.13	0.97037 7331	0.95880 131	1.63	1.26477 6877	0.83174 424
1.14	0.97698 8088	0.95579 904	1.64	1.26999 4549	0.82960 376
1.15	0.98356 6154	0.95281 918	1.65	1.27518 9081	0.82747 632
1.16	0.99011 1729	0.94986 146	1.66	1.28036 0639	0.82536 179
1.17	0.99662 5013	0.94692 561	1.67	1.28550 9389	0.82326 005
1.18	1.00310 6208	0.94401 139	1.68	1.29063 5495	0.82117 097
1.19	1.00955 5514	0.94111 853	1.69	1.29573 9120	0.81909 443
1.20	1.01597 3134	0.93824 678	1.70	1.30082 0427	0.81703 032
1.21	1.02235 9270	0.93539 589	1.71	1.30587 9576	0.81497 850
1.22	1.02871 4123	0.93256 563	1.72	1.31091 6727	0.81293 888
1.23	1.03503 7896	0.92975 576	1.73	1.31593 2038	0.81091 132
1.24	1.04133 0792	0.92696 604	1.74	1.32092 5666	0.80889 572
1.25	1.04759 3013	0.92419 624	1.75	1.32589 7767	0.80689 197
1.26	1.05382 4760	0.92144 613	1.76	1.33084 8496	0.80489 994
1.27	1.06002 6237	0.91871 550	1.77	1.33577 8006	0.80291 954
1.28	1.06619 7645	0.91600 411	1.78	1.34068 6450	0.80095 066
1.29	1.07233 9185	0.91331 175	1.79	1.34557 3978	0.79899 318
1.30	1.07845 1059	0.91063 821	1.80	1.35044 0740	0.79704 701
1.31	1.08453 3467	0.90798 328	1.81	1.35528 6886	0.79511 203
1.32	1.09058 6610	0.90534 676	1.82	1.36011 2562	0.79318 816
1.33	1.09661 0688	0.90272 843	1.83	1.36491 7914	0.79127 527
1.34	1.10260 5899	0.90012 810	1.84	1.36970 3089	0.78937 328
1.35	1.10857 2442	0.89754 557	1.85	1.37446 8228	0.78748 209
1.36	1.11451 0515	0.89498 064	1.86	1.37921 3477	0.78560 160
1.37	1.12042 0317	0.89243 313	1.87	1.38393 8975	0.78373 170
1.38	1.12630 2042	0.88990 284	1.88	1.38864 4863	0.78187 231
1.39	1.13215 5887	0.88738 959	1.89	1.39333 1280	0.78002 334
1.40	1.13798 2046	0.88489 320	1.90	1.39799 8365	0.77818 468
1.41	1.14378 0715	0.88241 348	1.91	1.40264 6254	0.77635 625
1.42	1.14955 2086	0.87995 026	1.92	1.40727 5083	0.77453 796
1.43	1.15529 6351	0.87750 336	1.93	1.41188 4987	0.77272 971
1.44	1.16101 3703	0.87507 261	1.94	1.41647 6099	0.77093 142
1.45	1.16670 4331	0.87265 784	1.95	1.42104 8552	0.76914 300
1.46	1.17236 8425	0.87025 888	1.96	1.42560 2476	0.76736 437
1.47	1.17800 6174	0.86787 557	1.97	1.43013 8002	0.76559 544
1.48	1.18361 7765	0.86550 774	1.98	1.43465 5259	0.76383 612
1.49	1.18920 3384	0.86315 523	1.99	1.43915 4374	0.76208 633
1.50	1.19476 3217	0.86081 788	2.00	1.44363 5475	0.76034 600
	$\left[\begin{matrix}(-6)4\\4\end{matrix}\right]$	$\left[\begin{matrix}(-6)3\\4\end{matrix}\right]$		$\left[\begin{matrix}(-6)3\\4\end{matrix}\right]$	$\left[\begin{matrix}(-6)2\\4\end{matrix}\right]$

INVERSE HYPERBOLIC FUNCTIONS

Table 4.17

x^{-1}	arcsinh $x-\ln x$	arccosh $x-\ln x$	$\langle x \rangle$	x^{-1}	arcsinh $x-\ln x$	arccosh $x-\ln x$	$\langle x \rangle$
0.50	0.75048 82946	0.62381 07164	2	0.25	0.70841 81861	0.67714 27078	4
0.49	0.74839 16011	0.62685 90940	2	0.24	0.70724 57326	0.67842 57947	4
0.48	0.74632 48341	0.62981 77884	2	0.23	0.70611 72820	0.67965 18411	4
0.47	0.74428 85962	0.63268 90778	2	0.22	0.70503 32895	0.68082 14660	5
0.46	0.74228 34908	0.63547 51194	2	0.21	0.70399 41963	0.68193 52541	5
0.45	0.74031 01215	0.63817 79566	2	0.20	0.70300 04288	0.68299 37571	5
0.44	0.73836 90921	0.64079 95268	2	0.19	0.70205 23983	0.68399 74947	5
0.43	0.73646 10057	0.64334 16670	2	0.18	0.70115 05002	0.68494 69555	6
0.42	0.73458 64641	0.64580 61207	2	0.17	0.70029 51134	0.68584 25981	6
0.41	0.73274 60676	0.64819 45429	2	0.16	0.69948 66000	0.68668 48518	6
0.40	0.73094 04145	0.65050 85051	3	0.15	0.69872 53043	0.68747 41175	7
0.39	0.72917 01001	0.65274 95004	3	0.14	0.69801 15527	0.68821 07683	7
0.38	0.72743 57167	0.65491 89477	3	0.13	0.69734 56533	0.68889 51504	8
0.37	0.72573 78524	0.65701 81952	3	0.12	0.69672 78946	0.68952 75836	8
0.36	0.72407 70912	0.65904 85249	3	0.11	0.69615 85462	0.69010 83616	9
0.35	0.72245 40117	0.66101 11555	3	0.10	0.69563 78573	0.69063 77531	10
0.34	0.72086 91873	0.66290 72458	3	0.09	0.69516 60572	0.69111 60018	11
0.33	0.71932 31846	0.66473 78974	3	0.08	0.69474 33542	0.69154 33269	13
0.32	0.71781 65636	0.66650 41577	3	0.07	0.69436 99357	0.69191 99235	14
0.31	0.71634 98766	0.66820 70226	3	0.06	0.69404 59680	0.69224 59631	17
0.30	0.71492 36678	0.66984 74382	3	0.05	0.69377 15954	0.69252 15938	20
0.29	0.71353 84725	0.67142 63038	3	0.04	0.69354 69408	0.69274 69403	25
0.28	0.71219 48165	0.67294 44732	4	0.03	0.69337 21047	0.69292 21046	33
0.27	0.71089 32154	0.67440 27575	4	0.02	0.69324 71656	0.69304 71656	50
0.26	0.70963 41742	0.67580 19258	4	0.01	0.69317 21796	0.69312 21796	100
0.25	0.70841 81861	0.67714 27078	4	0.00	0.69314 71806	0.69314 71806	∞

$$\begin{bmatrix} (-6)5 \\ 5 \end{bmatrix} \qquad \begin{bmatrix} (-5)1 \\ 6 \end{bmatrix} \qquad \begin{bmatrix} (-6)6 \\ 5 \end{bmatrix} \qquad * \begin{bmatrix} (-6)7 \\ 5 \end{bmatrix}$$

$\langle x \rangle$ = nearest integer to x.

ROOTS x_n OF $\cos x_n \cosh x_n = 1$

Table 4.18

n	x_n
1	4.73004 07
2	7.85320 46
3	10.99560 78
4	14.13716 55
5	17.27875 96

For $n \geq 5$, $x_n = \frac{1}{2}[2n+1]\pi$

ROOTS x_n OF $\cos x_n \cosh x_n = -1$

n	x_n
1	1.87510 41
2	4.69409 11
3	7.85475 74
4	10.99554 07
5	14.13716 84

For $n > 5$, $x_n = \frac{1}{2}[2n-1]\pi$

*See page II.

Table 4.19   ROOTS x_n OF $\tan x_n = \lambda x_n$

$-\lambda$	x_1	x_2	x_3	x_4	x_5	x_6	x_7	x_8	x_9
0.00	3.14159	6.28319	9.42478	12.56637	15.70796	18.84956	21.99115	25.13274	28.27433
0.05	2.99304	5.99209	9.00185	12.02503	15.06247	18.11361	21.17717	24.25156	27.33519
0.10	2.86277	5.76056	8.70831	11.70268	14.73347	17.79083	20.86724	23.95737	27.05755
0.15	2.75032	5.58578	8.51805	11.52018	14.56638	17.64009	20.73148	23.83468	26.94607
0.20	2.65366	5.45435	8.39135	11.40863	14.46987	17.55621	20.65782	23.76928	26.88740
0.25	2.57043	5.35403	8.30293	11.33482	14.40797	17.50343	20.61203	23.72894	26.85142
0.30	2.49840	5.27587	8.23845	11.28284	14.36517	17.46732	20.58092	23.70166	26.82716
0.35	2.43566	5.21370	8.18965	11.24440	14.33391	17.44113	20.55844	23.68201	26.80971
0.40	2.38064	5.16313	8.15156	11.21491	14.31012	17.42129	20.54146	23.66719	26.79656
0.45	2.33208	5.12176	8.12108	11.19159	14.29142	17.40574	20.52818	23.65561	26.78631
0.50	2.28893	5.08698	8.09616	11.17271	14.27635	17.39324	20.51752	23.64632	26.77809
0.55	2.25037	5.05750	8.07544	11.15712	14.26395	17.38298	20.50877	23.63871	26.77135
0.60	2.21571	5.03222	8.05794	11.14403	14.25357	17.37439	20.50147	23.63235	26.76572
0.65	2.18440	5.01031	8.04298	11.13289	14.24475	17.36711	20.49528	23.62697	26.76096
0.70	2.15598	4.99116	8.03004	11.12330	14.23717	17.36086	20.48996	23.62235	26.75688
0.75	2.13008	4.97428	8.01875	11.11496	14.23059	17.35543	20.48535	23.61834	26.75333
0.80	2.10638	4.95930	8.00881	11.10764	14.22482	17.35068	20.48131	23.61483	26.75023
0.85	2.08460	4.94592	7.99999	11.10116	14.21971	17.34648	20.47774	23.61173	26.74749
0.90	2.06453	4.93389	7.99212	11.09538	14.21517	17.34274	20.47457	23.60897	26.74506
0.95	2.04597	4.92303	7.98505	11.09021	14.21110	17.33939	20.47172	23.60651	26.74288
1.00	2.02876	4.91318	7.97867	11.08554	14.20744	17.33638	20.46917	23.60428	26.74092

λ^{-1}	x_1	x_2	x_3	x_4	x_5	x_6	x_7	x_8	x_9	$\langle\lambda\rangle$
-1.00	2.02876	4.91318	7.97867	11.08554	14.20744	17.33638	20.46917	23.60428	26.74092	- 1
-0.95	2.01194	4.90375	7.97258	11.08110	14.20395	17.33351	20.46673	23.60217	26.73905	- 1
-0.90	1.99465	4.89425	7.96648	11.07665	14.20046	17.33064	20.46430	23.60006	26.73718	- 1
-0.85	1.97687	4.88468	7.96036	11.07219	14.19697	17.32777	20.46187	23.59795	26.73532	- 1
-0.80	1.95857	4.87504	7.95422	11.06773	14.19347	17.32490	20.45943	23.59584	26.73345	- 1
-0.75	1.93974	4.86534	7.94807	11.06326	14.18997	17.32203	20.45700	23.59372	26.73159	- 1
-0.70	1.92035	4.85557	7.94189	11.05879	14.18647	17.31915	20.45456	23.59161	26.72972	- 1
-0.65	1.90036	4.84573	7.93571	11.05431	14.18296	17.31628	20.45212	23.58949	26.72785	- 2
-0.60	1.87976	4.83583	7.92950	11.04982	14.17946	17.31340	20.44968	23.58738	26.72598	- 2
-0.55	1.85852	4.82587	7.92329	11.04533	14.17594	17.31052	20.44724	23.58526	26.72411	- 2
-0.50	1.83660	4.81584	7.91705	11.04083	14.17243	17.30764	20.44480	23.58314	26.72225	- 2
-0.45	1.81396	4.80575	7.91080	11.03633	14.16892	17.30476	20.44236	23.58102	26.72038	- 2
-0.40	1.79058	4.79561	7.90454	11.03182	14.16540	17.30187	20.43992	23.57891	26.71851	- 3
-0.35	1.76641	4.78540	7.89827	11.02730	14.16188	17.29899	20.43748	23.57679	26.71664	- 3
-0.30	1.74140	4.77513	7.89198	11.02278	14.15835	17.29610	20.43503	23.57467	26.71477	- 3
-0.25	1.71551	4.76481	7.88567	11.01826	14.15483	17.29321	20.43259	23.57255	26.71290	- 4
-0.20	1.68868	4.75443	7.87936	11.01373	14.15130	17.29033	20.43014	23.57043	26.71102	- 5
-0.15	1.66087	4.74400	7.87303	11.00920	14.14777	17.28744	20.42769	23.56831	26.70915	- 7
-0.10	1.63199	4.73351	7.86669	11.00466	14.14424	17.28454	20.42525	23.56619	26.70728	-10
-0.05	1.60200	4.72298	7.86034	11.00012	14.14070	17.28165	20.42280	23.56407	26.70541	-20
0.00	1.57080	4.71239	7.85398	10.99557	14.13717	17.27875	20.42035	23.56194	26.70354	∞
0.05	1.53830	4.70176	7.84761	10.99102	14.13363	17.27586	20.41790	23.55982	26.70166	20
0.10	1.50442	4.69108	7.84123	10.98647	14.13009	17.27297	20.41545	23.55770	26.69979	10
0.15	1.46904	4.68035	7.83484	10.98192	14.12655	17.27007	20.41300	23.55558	26.69792	7
0.20	1.43203	4.66958	7.82844	10.97736	14.12301	17.26718	20.41055	23.55345	26.69604	5
0.25	1.39325	4.65878	7.82203	10.97279	14.11946	17.26428	20.40810	23.55133	26.69417	4
0.30	1.35252	4.64793	7.81562	10.96823	14.11592	17.26138	20.40565	23.54921	26.69230	3
0.35	1.30965	4.63705	7.80919	10.96366	14.11237	17.25848	20.40320	23.54708	26.69042	3
0.40	1.26440	4.62614	7.80276	10.95909	14.10882	17.25558	20.40075	23.54496	26.68855	3
0.45	1.21649	4.61519	7.79633	10.95452	14.10527	17.25268	20.39829	23.54283	26.68668	2
0.50	1.16556	4.60422	7.78988	10.94994	14.10172	17.24978	20.39584	23.54071	26.68480	2
0.55	1.11118	4.59321	7.78344	10.94537	14.09817	17.24688	20.39339	23.53858	26.68293	2
0.60	1.05279	4.58219	7.77698	10.94079	14.09462	17.24398	20.39094	23.53646	26.68105	2
0.65	0.98966	4.57114	7.77053	10.93621	14.09107	17.24108	20.38848	23.53433	26.67918	2
0.70	0.92079	4.56007	7.76407	10.93163	14.08752	17.23817	20.38603	23.53221	26.67730	1
0.75	0.84473	4.54899	7.75760	10.92704	14.08396	17.23527	20.38357	23.53008	26.67543	1
0.80	0.75931	4.53789	7.75114	10.92246	14.08041	17.23237	20.38112	23.52796	26.67355	1
0.85	0.66086	4.52678	7.74467	10.91788	14.07686	17.22946	20.37867	23.52583	26.67168	1
0.90	0.54228	4.51566	7.73820	10.91329	14.07330	17.22656	20.37621	23.52370	26.66980	1
0.95	0.38537	4.50454	7.73172	10.90871	14.06975	17.22366	20.37376	23.52158	26.66793	1
1.00	0.00000	4.49341	7.72525	10.90412	14.06619	17.22075	20.37130	23.51945	26.66605	1

For $\lambda = 0$, see $j_{\frac{1}{2},s}$ of **Table 10.6**. $\langle\lambda\rangle$ = nearest integer to λ.

ROOTS x_n OF $\cot x_n = \lambda x_n$ — Table 4.20

λ	x_1	x_2	x_3	x_4	x_5	x_6	x_7	x_8	x_9
0.00	1.57080	4.71239	7.85398	10.99557	14.13717	17.27876	20.42035	23.56194	26.70354
0.05	1.49613	4.49148	7.49541	10.51167	13.54198	16.58639	19.64394	22.71311	25.79232
0.10	1.42887	4.30580	7.22811	10.20026	13.21418	16.25936	19.32703	22.41085	25.50638
0.15	1.36835	4.15504	7.04126	10.01222	13.03901	16.10053	19.18401	22.28187	25.38952
0.20	1.31384	4.03357	6.90960	9.89275	12.93522	16.01066	19.10552	22.21256	25.32765
0.25	1.26459	3.93516	6.81401	9.81188	12.86775	15.95363	19.05645	22.16965	25.28961
0.30	1.21995	3.85460	6.74233	9.75407	12.82073	15.91443	19.02302	22.14058	25.26392
0.35	1.17933	3.78784	6.68698	9.71092	12.78621	15.88591	18.99882	22.11960	25.24544
0.40	1.14223	3.73184	6.64312	9.67758	12.75985	15.86426	18.98052	22.10377	25.23150
0.45	1.10820	3.68433	6.60761	9.65109	12.73907	15.84728	18.96619	22.09140	25.22062
0.50	1.07687	3.64360	6.57833	9.62956	12.72230	15.83361	18.95468	22.08147	25.21190
0.55	1.04794	3.60834	6.55380	9.61173	12.70847	15.82237	18.94523	22.07333	25.20475
0.60	1.02111	3.57756	6.53297	9.59673	12.69689	15.81297	18.93734	22.06653	25.19878
0.65	0.99617	3.55048	6.51508	9.58394	12.68704	15.80500	18.93065	22.06077	25.19373
0.70	0.97291	3.52649	6.49954	9.57292	12.67857	15.79814	18.92490	22.05583	25.18939
0.75	0.95116	3.50509	6.48593	9.56331	12.67121	15.79219	18.91991	22.05154	25.18563
0.80	0.93076	3.48590	6.47392	9.55486	12.66475	15.78698	18.91554	22.04778	25.18234
0.85	0.91158	3.46859	6.46324	9.54738	12.65904	15.78237	18.91168	22.04447	25.17943
0.90	0.89352	3.45292	6.45368	9.54072	12.65395	15.77827	18.90825	22.04151	25.17684
0.95	0.87647	3.43865	6.44508	9.53473	12.64939	15.77459	18.90518	22.03887	25.17453
1.00	0.86033	3.42562	6.43730	9.52933	12.64529	15.77128	18.90241	22.03650	25.17245

λ^{-1}	x_1	x_2	x_3	x_4	x_5	x_6	x_7	x_8	x_9	$\langle\lambda\rangle$
1.00	0.86033	3.42562	6.43730	9.52933	12.64529	15.77128	18.90241	22.03650	25.17245	1
0.95	0.84426	3.41306	6.42987	9.52419	12.64138	15.76814	18.89978	22.03424	25.17047	1
0.90	0.82740	3.40034	6.42241	9.51904	12.63747	15.76499	18.89715	22.03197	25.16848	1
0.85	0.80968	3.38744	6.41492	9.51388	12.63355	15.76184	18.89451	22.02971	25.16650	1
0.80	0.79103	3.37438	6.40740	9.50871	12.62963	15.75868	18.89188	22.02745	25.16452	1
0.75	0.77136	3.36113	6.39984	9.50353	12.62570	15.75553	18.88924	22.02519	25.16254	1
0.70	0.75056	3.34772	6.39226	9.49834	12.62177	15.75237	18.88660	22.02292	25.16055	1
0.65	0.72851	3.33413	6.38464	9.49314	12.61784	15.74921	18.88396	22.02066	25.15857	2
0.60	0.70507	3.32037	6.37700	9.48793	12.61390	15.74605	18.88132	22.01839	25.15659	2
0.55	0.68006	3.30643	6.36932	9.48271	12.60996	15.74288	18.87868	22.01612	25.15460	2
0.50	0.65327	3.29231	6.36162	9.47749	12.60601	15.73972	18.87604	22.01386	25.15262	2
0.45	0.62444	3.27802	6.35389	9.47225	12.60206	15.73655	18.87339	22.01159	25.15063	2
0.40	0.59324	3.26355	6.34613	9.46700	12.59811	15.73338	18.87075	22.00932	25.14864	3
0.35	0.55922	3.24891	6.33835	9.46175	12.59415	15.73021	18.86810	22.00705	25.14666	3
0.30	0.52179	3.23409	6.33054	9.45649	12.59019	15.72704	18.86546	22.00478	25.14467	3
0.25	0.48009	3.21910	6.32270	9.45122	12.58623	15.72386	18.86281	22.00251	25.14268	4
0.20	0.43284	3.20393	6.31485	9.44595	12.58226	15.72068	18.86016	22.00024	25.14070	5
0.15	0.37788	3.18860	6.30696	9.44067	12.57829	15.71751	18.85751	21.99797	25.13871	7
0.10	0.31105	3.17310	6.29906	9.43538	12.57432	15.71433	18.85486	21.99569	25.13672	10
0.05	0.22176	3.15743	6.29113	9.43008	12.57035	15.71114	18.85221	21.99342	25.13473	20
0.00	0.00000	3.14159	6.28319	9.42478	12.56637	15.70796	18.84956	21.99115	25.13274	∞
*		$\begin{bmatrix}(-5)2\\3\end{bmatrix}$	$\begin{bmatrix}(-5)1\\2\end{bmatrix}$	$\begin{bmatrix}(-5)1\\2\end{bmatrix}$	$\begin{bmatrix}(-5)1\\2\end{bmatrix}$	$\begin{bmatrix}(-5)1\\2\end{bmatrix}$	$\begin{bmatrix}(-5)1\\2\end{bmatrix}$	$\begin{bmatrix}(-5)1\\2\end{bmatrix}$	$\begin{bmatrix}(-5)1\\2\end{bmatrix}$	

$\langle\lambda\rangle$ = nearest integer to λ.

For $\lambda^{-1} > .20$, the maximum error in linear interpolation is $(-4)7$; five-point interpolation gives 5D.

For $\lambda^{-1} \leq .20$,

$$x_1 \approx \frac{1}{\sqrt{\lambda}}\left[1 - \frac{1}{6\lambda} + \frac{11}{360\lambda}2 - \frac{1}{432\lambda}3 + \ldots\right].$$

*See page II.

5. Exponential Integral and Related Functions

Walter Gautschi [1] and William F. Cahill [2]

Contents

The authors acknowledge the assistance of David S. Liepman in the preparation and checking of the tables, Robert L. Durrah for the computation of **Table 5.2,** and Alfred E. Beam for the computation of **Table 5.6.**

[1] Guest worker, National Bureau of Standards, from the American University. (Presently Purdue University.)

[2] National Bureau of Standards. (Presently NASA.)

5. Exponential Integral and Related Functions

Mathematical Properties

5.1. Exponential Integral

Definitions

5.1.1
$$E_1(z) = \int_z^\infty \frac{e^{-t}}{t}\, dt \qquad (|\arg z| < \pi)$$

5.1.2
$$\mathrm{Ei}(x) = -\!\!\int_{-x}^\infty \frac{e^{-t}}{t}\, dt = \int_{-\infty}^x \frac{e^t}{t}\, dt \qquad (x > 0)$$

5.1.3
$$\mathrm{li}(x) = \int_0^x \frac{dt}{\ln t} = \mathrm{Ei}(\ln x) \qquad (x > 1)$$

5.1.4
$$E_n(z) = \int_1^\infty \frac{e^{-zt}}{t^n}\, dt \qquad (n = 0, 1, 2, \ldots;\ \mathscr{R}\, z > 0)$$

5.1.5
$$\alpha_n(z) = \int_1^\infty t^n e^{-zt}\, dt \qquad (n = 0, 1, 2, \ldots;\ \mathscr{R}\, z > 0)$$

5.1.6
$$\beta_n(z) = \int_{-1}^1 t^n e^{-zt}\, dt \qquad (n = 0, 1, 2, \ldots)$$

In **5.1.1** it is assumed that the path of integration excludes the origin and does not cross the negative real axis.

Analytic continuation of the functions in **5.1.1**, **5.1.2**, and **5.1.4** for $n > 0$ yields multi-valued functions with branch points at $z = 0$ and $z = \infty$.[3] They are single-valued functions in the z-plane cut along the negative real axis.[4] The function $\mathrm{li}(z)$, the logarithmic integral, has an additional branch point at $z = 1$.

Interrelations

5.1.7
$$E_1(-x \pm i0) = -\mathrm{Ei}(x) \mp i\pi,$$
$$-\mathrm{Ei}(x) = \tfrac{1}{2}[E_1(-x+i0) + E_1(-x-i0)] \qquad (x > 0)$$

---

[3] Some authors [5.14], [5.16] use the entire function $\int_0^z (1 - e^{-t})\, dt/t$ as the basic function and denote it by $\mathrm{Ein}(z)$. We have $\mathrm{Ein}(z) = E_1(z) + \ln z + \gamma$.

[4] Various authors define the integral $\int_{-\infty}^z (e^t/t)\, dt$ in the z-plane cut along the positive real axis and denote it also by $\mathrm{Ei}(z)$. For $z = x > 0$ additional notations such as $\overline{\mathrm{Ei}}(x)$ (e.g., in [5.10], [5.25]), $E^*(x)$ (in [5.2]), $\mathrm{Ei}^*(x)$ (in [5.6]) are then used to designate the principal value of the integral. Correspondingly, $E_1(x)$ is often denoted by $-\mathrm{Ei}(-x)$.

Explicit Expressions for $\alpha_n(z)$ and $\beta_n(z)$

5.1.8
$$\alpha_n(z) = n!\, z^{-n-1} e^{-z}\left(1 + z + \frac{z^2}{2!} + \ldots + \frac{z^n}{n!}\right)$$

5.1.9
$$\beta_n(z) = n!\, z^{-n-1}\left\{ e^z\left[1 - z + \frac{z^2}{2!} - \ldots + (-1)^n \frac{z^n}{n!}\right] \right.$$
$$\left. - e^{-z}\left(1 + z + \frac{z^2}{2!} + \ldots + \frac{z^n}{n!}\right) \right\}$$

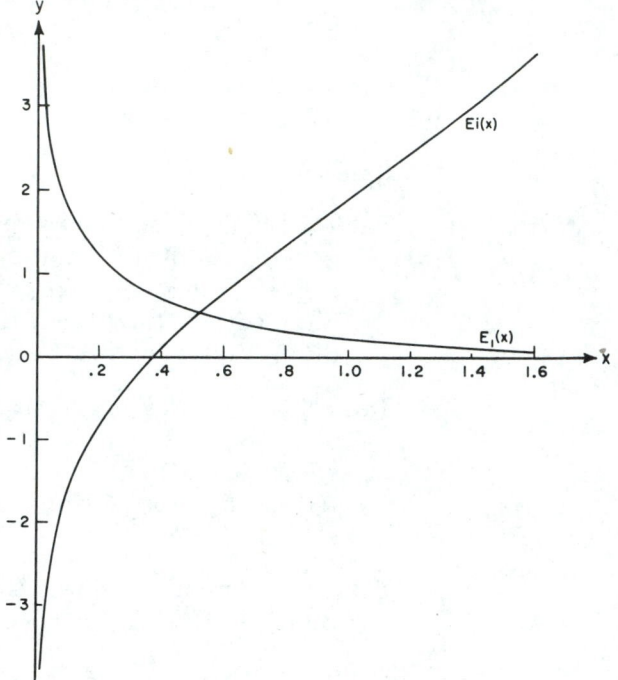

FIGURE 5.1. $y = \mathrm{Ei}(x)$ and $y = E_1(x)$.

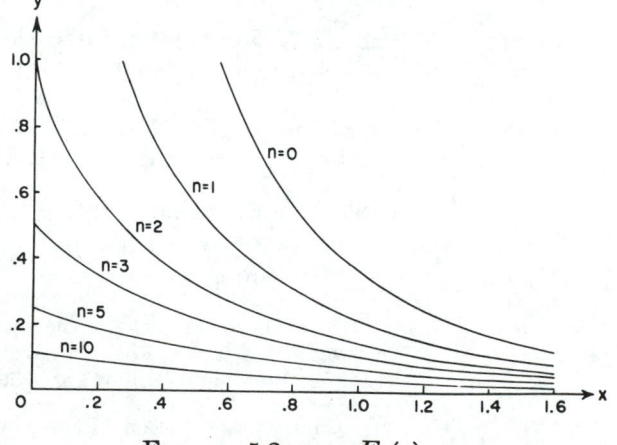

FIGURE 5.2. $y = E_n(x)$
$n = 0, 1, 2, 3, 5, 10$

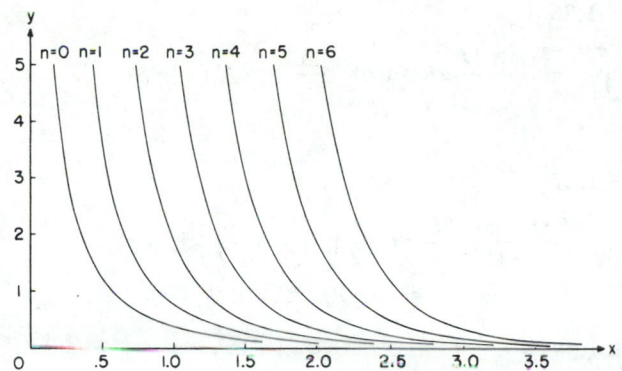

FIGURE 5.3. $y=\alpha_n(x)$
$n=0(1)6$

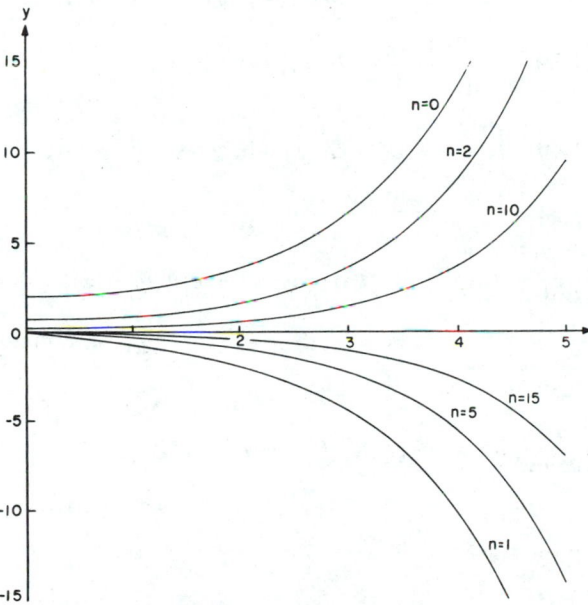

FIGURE 5.4. $y=\beta_n(x)$
$n=0, 1, 2, 5, 10, 15$

Series Expansions

5.1.10 $\qquad \mathrm{Ei}(x)=\gamma+\ln x+\sum_{n=1}^{\infty}\frac{x^n}{nn!} \qquad (x>0)$

5.1.11

$$E_1(z)=-\gamma-\ln z-\sum_{n=1}^{\infty}\frac{(-1)^n z^n}{nn!} \qquad (|\arg z|<\pi)$$

5.1.12

$$E_n(z)=\frac{(-z)^{n-1}}{(n-1)!}[-\ln z+\psi(n)]-\sum_{\substack{m=0\\m\neq n-1}}^{\infty}{}'\frac{(-z)^m}{(m-n+1)m!}$$
$$(|\arg z|<\pi)$$

$$\psi(1)=-\gamma, \ \psi(n)=-\gamma+\sum_{m=1}^{n-1}\frac{1}{m} \qquad (n>1)$$

$\gamma=.57721\ 56649\ \ldots$ is Euler's constant.

Symmetry Relation

5.1.13 $\qquad E_n(\bar{z})=\overline{E_n(z)}$

Recurrence Relations

5.1.14

$$E_{n+1}(z)=\frac{1}{n}\left[e^{-z}-zE_n(z)\right] \quad (n=1,2,3,\ldots)$$

5.1.15 $\quad z\alpha_n(z)=e^{-z}+n\alpha_{n-1}(z) \qquad (n=1,2,3,\ldots)$

5.1.16

$$z\beta_n(z)=(-1)^n e^z-e^{-z}+n\beta_{n-1}(z) \qquad (n=1,2,3,\ldots)$$

Inequalities [5.8], [5.4]

5.1.17

$$\frac{n-1}{n}E_n(x)<E_{n+1}(x)<E_n(x) \quad (x>0; n=1,2,3,\ldots)$$

5.1.18

$$E_n^2(x)<E_{n-1}(x)E_{n+1}(x) \qquad (x>0; n=1,2,3,\ldots)$$

5.1.19

$$\frac{1}{x+n}<e^x E_n(x)\leq\frac{1}{x+n-1} \qquad (x>0; n=1,2,3,\ldots)$$

5.1.20

$$\tfrac{1}{2}\ln\left(1+\frac{2}{x}\right)<e^x E_1(x)<\ln\left(1+\frac{1}{x}\right) \qquad (x>0)$$

5.1.21

$$\frac{d}{dx}\left[\frac{E_n(x)}{E_{n-1}(x)}\right]>0 \qquad (x>0; n=1,2,3,\ldots)$$

Continued Fraction

5.1.22
$$E_n(z)=e^{-z}\left(\frac{1}{z+}\ \frac{n}{1+}\ \frac{1}{z+}\ \frac{n+1}{1+}\ \frac{2}{z+}\cdots\right) \quad (|\arg z|<\pi)$$

Special Values

5.1.23 $\qquad E_n(0)=\frac{1}{n-1} \qquad (n>1)$

5.1.24 $\qquad E_0(z)=\frac{e^{-z}}{z}$

5.1.25 $\qquad \alpha_0(z)=\frac{e^{-z}}{z}, \ \beta_0(z)=\frac{2}{z}\sinh z$

Derivatives

5.1.26 $\quad \dfrac{dE_n(z)}{dz} = -E_{n-1}(z) \qquad (n=1,2,3,\ldots)$

5.1.27

$$\frac{d^n}{dz^n}[e^z E_1(z)] = \frac{d^{n-1}}{dz^{n-1}}[e^z E_1(z)]$$
$$+ \frac{(-1)^n(n-1)!}{z^n} \quad (n=1,2,3,\ldots)$$

Definite and Indefinite Integrals

(For more extensive tables of integrals see [5.3], [5.6], [5.11], [5.12], [5.13]. For integrals involving $E_n(x)$ see [5.9].)

5.1.28 $\quad \displaystyle\int_0^\infty \frac{e^{-at}}{b+t}\,dt = e^{ab}E_1(ab)$

5.1.29

$$\int_0^\infty \frac{e^{iat}}{b+t}\,dt = e^{-iab}E_1(-iab) \qquad (a>0, b>0)$$

5.1.30

$$\int_0^\infty \frac{t-ib}{t^2+b^2}\,e^{iat}dt = e^{ab}E_1(ab) \qquad (a>0, b>0)$$

5.1.31

$$\int_0^\infty \frac{t+ib}{t^2+b^2}\,e^{iat}dt = e^{-ab}(-\mathrm{Ei}(ab)+i\pi)$$
$$(a>0, b>0)$$

5.1.32 $\quad \displaystyle\int_0^\infty \frac{e^{-at}-e^{-bt}}{t}\,dt = \ln \frac{b}{a}$

5.1.33 $\quad \displaystyle\int_0^\infty E_1^2(t)\,dt = 2\ln 2$

5.1.34

$$\int_0^\infty e^{-at}E_n(t)\,dt =$$
$$\frac{(-1)^{n-1}}{a^n}[\ln(1+a)+\sum_{k=1}^{n-1}\frac{(-1)^k a^k}{k}] \qquad (a>-1)$$

5.1.35

$$\int_0^1 \frac{e^{at}\sin bt}{t}\,dt = \pi - \arctan\frac{b}{a}+\mathscr{I}E_1(-a+ib)$$
$$(a>0, b>0)$$

5.1.36

$$\int_0^1 \frac{e^{-at}\sin bt}{t}\,dt = \arctan\frac{b}{a}+\mathscr{I}E_1(a+ib)$$
$$(a>0, b\text{ real})$$

5.1.37

$$\int_0^1 \frac{e^{at}(1-\cos bt)}{t}\,dt = \frac{1}{2}\ln\left(1+\frac{b^2}{a^2}\right)+\mathrm{Ei}(a)$$
$$+\mathscr{R}E_1(-a+ib) \quad (a>0, b\text{ real})$$

5.1.38

$$\int_0^1 \frac{e^{-at}(1-\cos bt)}{t}\,dt = \frac{1}{2}\ln\left(1+\frac{b^2}{a^2}\right)-E_1(a)$$
$$+\mathscr{R}E_1(a+ib) \qquad (a>0, b\text{ real})$$

5.1.39 $\quad \displaystyle\int_0^z \frac{1-e^{-t}}{t}\,dt = E_1(z)+\ln z+\gamma$

5.1.40 $\quad \displaystyle\int_0^x \frac{e^t-1}{t}\,dt = \mathrm{Ei}(x)-\ln x-\gamma \qquad (x>0)$

5.1.41

$$\int \frac{e^{ix}}{a^2+x^2}\,dx = \frac{i}{2a}[e^{-a}E_1(-a-ix)-e^a E_1(a-ix)]$$
$$+\mathrm{const.}$$

5.1.42

$$\int \frac{xe^{ix}}{a^2+x^2}\,dx = -\frac{1}{2}[e^{-a}E_1(-a-ix)+e^a E_1(a-ix)]$$
$$+\mathrm{const.}$$

5.1.43

$$\int \frac{e^x}{a^2+x^2}\,dx = -\frac{1}{a}\,\mathscr{I}(e^{ia}E_1(-x+ia))+\mathrm{const.} \quad (a>0)$$

5.1.44

$$\int \frac{xe^x}{a^2+x^2}\,dx = -\mathscr{R}(e^{ia}E_1(-x+ia))+\mathrm{const.} \qquad (a>0)$$

Relation to Incomplete Gamma Function (see 6.5)

5.1.45 $\qquad E_n(z) = z^{n-1}\Gamma(1-n, z)$

5.1.46 $\qquad \alpha_n(z) = z^{-n-1}\Gamma(n+1, z)$

5.1.47 $\quad \beta_n(z) = z^{-n-1}[\Gamma(n+1, -z)-\Gamma(n+1, z)]$

Relation to Spherical Bessel Functions (see 10.2)

5.1.48 $\quad \alpha_0(z) = \sqrt{\dfrac{2}{\pi z}}\,K_{\frac{1}{2}}(z), \quad \beta_0(z) = \sqrt{\dfrac{2\pi}{z}}\,I_{\frac{1}{2}}(z)$

5.1.49 $\quad \alpha_1(z) = \sqrt{\dfrac{2}{\pi z}}\,K_{3/2}(z), \quad \beta_1(z) = -\sqrt{\dfrac{2\pi}{z}}\,I_{3/2}(z)$

Number-Theoretic Significance of $\mathrm{li}\,(x)$

(Assuming Riemann's hypothesis that all non-real zeros of $\zeta(z)$ have a real part of $\frac{1}{2}$)

5.1.50 $\mathrm{li}\,(x) - \pi(x) = O(\sqrt{x}\,\ln x)$ $(x \to \infty)$

$\pi(x)$ is the number of primes less than or equal to x.

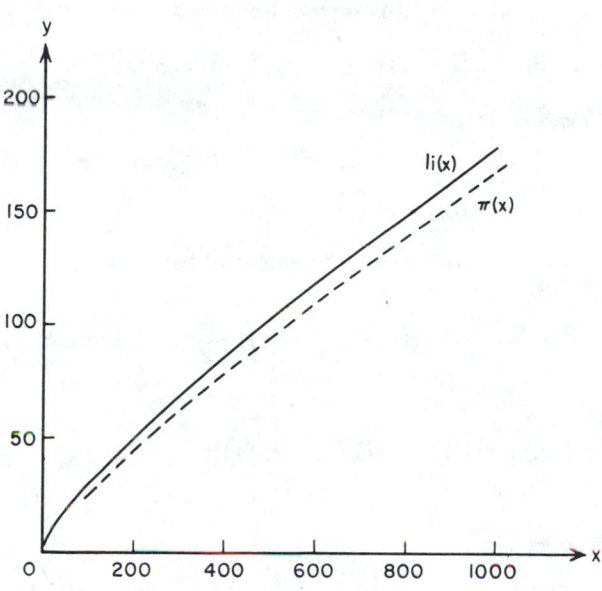

FIGURE 5.5. $y = \mathrm{li}(x)$ and $y = \pi(x)$

Asymptotic Expansion

5.1.51

$$E_n(z) \sim \frac{e^{-z}}{z}\left\{1 - \frac{n}{z} + \frac{n(n+1)}{z^2} - \frac{n(n+1)(n+2)}{z^3} + \ldots\right\}$$

$$(|\arg z| < \tfrac{3}{2}\pi)$$

Representation of $E_n(x)$ for Large n

5.1.52

$$E_n(x) = \frac{e^{-x}}{x+n}\left\{1 + \frac{n}{(x+n)^2} + \frac{n(n-2x)}{(x+n)^4}\right.$$

$$\left. + \frac{n(6x^2 - 8nx + n^2)}{(x+n)^6} + R(n,x)\right\}$$

$$-.36n^{-4} \leq R(n,x) \leq \left(1 + \frac{1}{x+n-1}\right)n^{-4} \quad (x>0)$$

Polynomial and Rational Approximations [5]

5.1.53 $0 \leq x \leq 1$

$$E_1(x) + \ln x = a_0 + a_1 x + a_2 x^2 + a_3 x^3 + a_4 x^4 + a_5 x^5 + \epsilon(x)$$
$$|\epsilon(x)| < 2 \times 10^{-7}$$

$$
\begin{aligned}
a_0 &= -.57721\ 566 & a_3 &= .05519\ 968 \\
a_1 &= .99999\ 193 & a_4 &= -.00976\ 004 \\
a_2 &= -.24991\ 055 & a_5 &= .00107\ 857
\end{aligned}
$$

5.1.54 $1 \leq x < \infty$

$$xe^x E_1(x) = \frac{x^2 + a_1 x + a_2}{x^2 + b_1 x + b_2} + \epsilon(x)$$

$$|\epsilon(x)| < 5 \times 10^{-5}$$

$$
\begin{aligned}
a_1 &= 2.334733 & b_1 &= 3.330657 \\
a_2 &= .250621 & b_2 &= 1.681534
\end{aligned}
$$

5.1.55 $10 \leq x < \infty$

$$xe^x E_1(x) = \frac{x^2 + a_1 x + a_2}{x^2 + b_1 x + b_2} + \epsilon(x)$$

$$|\epsilon(x)| < 10^{-7}$$

$$
\begin{aligned}
a_1 &= 4.03640 & b_1 &= 5.03637 \\
a_2 &= 1.15198 & b_2 &= 4.19160
\end{aligned}
$$

5.1.56 $1 \leq x < \infty$

$$xe^x E_1(x) = \frac{x^4 + a_1 x^3 + a_2 x^2 + a_3 x + a_4}{x^4 + b_1 x^3 + b_2 x^2 + b_3 x + b_4} + \epsilon(x)$$

$$|\epsilon(x)| < 2 \times 10^{-8}$$

$$
\begin{aligned}
a_1 &= 8.57332\ 87401 & b_1 &= 9.57332\ 23454 \\
a_2 &= 18.05901\ 69730 & b_2 &= 25.63295\ 61486 \\
a_3 &= 8.63476\ 08925 & b_3 &= 21.09965\ 30827 \\
a_4 &= .26777\ 37343 & b_4 &= 3.95849\ 69228
\end{aligned}
$$

5.2. Sine and Cosine Integrals

Definitions

5.2.1 $\displaystyle \mathrm{Si}\,(z) = \int_0^z \frac{\sin t}{t}\,dt$

5.2.2 [6] $\displaystyle \mathrm{Ci}\,(z) = \gamma + \ln z + \int_0^z \frac{\cos t - 1}{t}\,dt$ $(|\arg z| < \pi)$

5.2.3 [7] $\displaystyle \mathrm{Shi}\,(z) = \int_0^z \frac{\sinh t}{t}\,dt$

5.2.4 [7] $\displaystyle \mathrm{Chi}\,(z) = \gamma + \ln z + \int_0^z \frac{\cosh t - 1}{t}\,dt$ $(|\arg z| < \pi)$

---

[5] The approximation **5.1.53** is from E. E. Allen, Note **169**, MTAC **8**, 240 (1954); approximations **5.1.54** and **5.1.56** are from C. Hastings, Jr., Approximations for digital computers, Princeton Univ. Press, Princeton, N.J., 1955; approximation **5.1.55** is from C. Hastings, Jr., Note 143, MTAC **7**, 68 (1953) (with permission).

[6] Some authors [5.14], [5.16] use the entire function $\int_0^z (1 - \cos t)\,dt/t$ as the basic function and denote it by $\mathrm{Cin}(z)$. We have
$$\mathrm{Cin}(z) = -\mathrm{Ci}(z) + \ln z + \gamma.$$

[7] The notations $\mathrm{Sih}(z) = \int_0^z \sinh t\,dt/t$, $\mathrm{Cinh}(z) = \int_0^z (\cosh t - 1)\,dt/t$ have also been proposed [5.14.]

5.2.5
$$\text{si}(z) = \text{Si}(z) - \frac{\pi}{2}$$

Auxiliary Functions

5.2.6
$$f(z) = \text{Ci}(z)\sin z - \text{si}(z)\cos z$$

5.2.7
$$g(z) = -\text{Ci}(z)\cos z - \text{si}(z)\sin z$$

Sine and Cosine Integrals in Terms of Auxiliary Functions

5.2.8
$$\text{Si}(z) = \frac{\pi}{2} - f(z)\cos z - g(z)\sin z$$

5.2.9
$$\text{Ci}(z) = f(z)\sin z - g(z)\cos z$$

Integral Representations

5.2.10
$$\text{si}(z) = -\int_0^{\frac{\pi}{2}} e^{-z\cos t}\cos\,(z\sin t)dt$$

5.2.11
$$\text{Ci}(z) + E_1(z) = \int_0^{\frac{\pi}{2}} e^{-z\cos t}\sin\,(z\sin t)dt$$

5.2.12
$$f(z) = \int_0^\infty \frac{\sin t}{t+z}dt = \int_0^\infty \frac{e^{-zt}}{t^2+1}dt \quad (\mathscr{R}z>0)$$

5.2.13
$$g(z) = \int_0^\infty \frac{\cos t}{t+z}dt = \int_0^\infty \frac{te^{-zt}}{t^2+1}dt \quad (\mathscr{R}z>0)$$

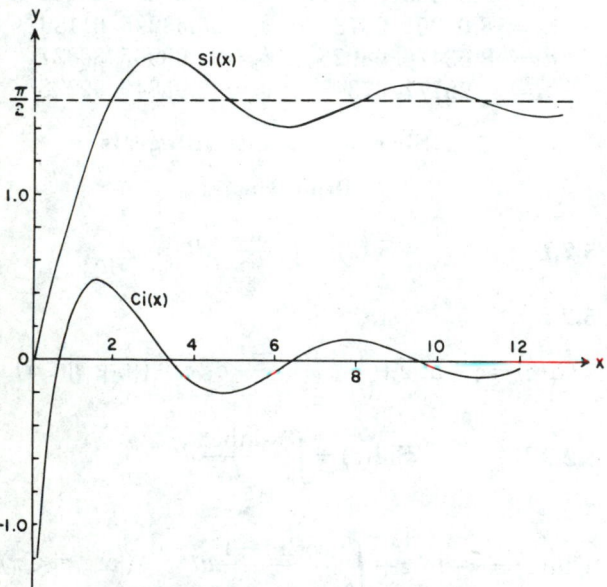

FIGURE 5.6. $y = \text{Si}(x)$ and $y = \text{Ci}(x)$

Series Expansions

5.2.14
$$\text{Si}(z) = \sum_{n=0}^\infty \frac{(-1)^n z^{2n+1}}{(2n+1)(2n+1)!}$$

5.2.15
$$\text{Si}(z) = \pi \sum_{n=0}^\infty J_{n+\frac{1}{2}}^2\left(\frac{z}{2}\right)$$

5.2.16
$$\text{Ci}(z) = \gamma + \ln z + \sum_{n=1}^\infty \frac{(-1)^n z^{2n}}{2n(2n)!}$$

5.2.17
$$\text{Shi}(z) = \sum_{n=0}^\infty \frac{z^{2n+1}}{(2n+1)(2n+1)!}$$

5.2.18
$$\text{Chi}(z) = \gamma + \ln z + \sum_{n=1}^\infty \frac{z^{2n}}{2n(2n)!}$$

Symmetry Relations

5.2.19
$$\text{Si}(-z) = -\text{Si}(z),\ \text{Si}(\bar{z}) = \overline{\text{Si}(z)}$$

5.2.20
$$\text{Ci}(-z) = \text{Ci}(z) - i\pi \quad (0 < \arg z < \pi)$$
$$\text{Ci}(\bar{z}) = \overline{\text{Ci}(z)}$$

Relation to Exponential Integral

5.2.21
$$\text{Si}(z) = \frac{1}{2i}[E_1(iz) - E_1(-iz)] + \frac{\pi}{2} \quad (|\arg z| < \frac{\pi}{2})$$

5.2.22
$$\text{Si}(ix) = \frac{i}{2}[\text{Ei}(x) + E_1(x)] \quad (x>0)$$

5.2.23
$$\text{Ci}(z) = -\frac{1}{2}[E_1(iz) + E_1(-iz)] \quad (|\arg z| < \frac{\pi}{2})$$

5.2.24
$$\text{Ci}(ix) = \frac{1}{2}[\text{Ei}(x) - E_1(x)] + i\frac{\pi}{2} \quad (x>0)$$

Value at Infinity

5.2.25
$$\lim_{x\to\infty} \text{Si}(x) = \frac{\pi}{2}$$

Integrals

(For more extensive tables of integrals see [5.3], [5.6], [5.11], [5.12], [5.13].)

5.2.26
$$\int_z^\infty \frac{\sin t}{t}dt = -\text{si}\,(z) \quad (|\arg z| < \pi)$$

5.2.27
$$\int_z^\infty \frac{\cos t}{t}dt = -\text{Ci}\,(z) \quad (|\arg z| < \pi)$$

5.2.28
$$\int_0^\infty e^{-at}\text{Ci}\,(t)dt = -\frac{1}{2a}\ln\,(1+a^2) \quad (\mathscr{R}a>0)*$$

5.2.29
$$\int_0^\infty e^{-at}\text{si}\,(t)dt = -\frac{1}{a}\arctan a \quad (\mathscr{R}a>0)$$

5.2.30
$$\int_0^\infty \cos t\,\text{Ci}\,(t)dt = \int_0^\infty \sin t\,\text{si}\,(t)dt = -\frac{\pi}{4}$$

---

*See page II.

5.2.31 $$\int_0^\infty \text{Ci}^2(t)\,dt = \int_0^\infty \text{si}^2(t)\,dt = \frac{\pi}{2}$$

5.2.32* $$\int_0^\infty \text{Ci}(t)\,\text{si}(t)\,dt = \ln 2$$

5.2.33

$$\int_0^1 \frac{(1-e^{-at})\cos bt}{t}\,dt = \frac{1}{2}\ln\left(1+\frac{a^2}{b^2}\right) + \text{Ci}(b)$$
$$+ \mathscr{R}E_1(a+ib) \quad (a\text{ real}, b>0)$$

Asymptotic Expansions

5.2.34

$$f(z) \sim \frac{1}{z}\left(1 - \frac{2!}{z^2} + \frac{4!}{z^4} - \frac{6!}{z^6} + \dots\right) \quad (|\arg z| < \pi)$$

5.2.35

$$g(z) \sim \frac{1}{z^2}\left(1 - \frac{3!}{z^2} + \frac{5!}{z^4} - \frac{7!}{z^6} + \dots\right) \quad (|\arg z| < \pi)$$

Rational Approximations [8]

5.2.36 $$1 \le x < \infty$$

$$f(x) = \frac{1}{x}\left(\frac{x^4 + a_1 x^2 + a_2}{x^4 + b_1 x^2 + b_2}\right) + \epsilon(x)$$

$$|\epsilon(x)| < 2 \times 10^{-4}$$

$a_1 = 7.241163 \qquad b_1 = 9.068580$
$a_2 = 2.463936 \qquad b_2 = 7.157433$

5.2.37 $$1 \le x < \infty$$

$$g(x) = \frac{1}{x^2}\left(\frac{x^4 + a_1 x^2 + a_2}{x^4 + b_1 x^2 + b_2}\right) + \epsilon(x)$$

$$|\epsilon(x)| < 10^{-4}$$

$a_1 = 7.547478 \qquad b_1 = 12.723684 \qquad *$
$a_2 = 1.564072 \qquad b_2 = 15.723606 \qquad *$

5.2.38 $$1 \le x < \infty$$

$$f(x) = \frac{1}{x}\left(\frac{x^8 + a_1 x^6 + a_2 x^4 + a_3 x^2 + a_4}{x^8 + b_1 x^6 + b_2 x^4 + b_3 x^2 + b_4}\right) + \epsilon(x)$$

$$|\epsilon(x)| < 5 \times 10^{-7}$$

$a_1 = 38.027264 \qquad b_1 = 40.021433$
$a_2 = 265.187033 \qquad b_2 = 322.624911$
$a_3 = 335.677320 \qquad b_3 = 570.236280$
$a_4 = 38.102495 \qquad b_4 = 157.105423$

5.2.39 $$1 \le x < \infty$$

$$g(x) = \frac{1}{x^2}\left(\frac{x^8 + a_1 x^6 + a_2 x^4 + a_3 x^2 + a_4}{x^8 + b_1 x^6 + b_2 x^4 + b_3 x^2 + b_4}\right) + \epsilon(x)$$

$$|\epsilon(x)| < 3 \times 10^{-7}$$

$a_1 = 42.242855 \qquad b_1 = 48.196927$
$a_2 = 302.757865 \qquad b_2 = 482.485984$
$a_3 = 352.018498 \qquad b_3 = 1114.978885$
$a_4 = 21.821899 \qquad b_4 = 449.690326$

Numerical Methods

5.3. Use and Extension of the Tables

Example 1. Compute Ci (.25) to 5D. From **Tables 5.1** and **4.2** we have

$$\frac{\text{Ci}(.25) - \ln(.25) - \gamma}{(.25)^2} = -.249350,$$

Ci $(.25) = (.25)^2(-.249350) + (-1.38629)$
$$+ .577216 = -.82466.$$

Example 2. Compute Ei (8) to 5S.
From **Table 5.1** we have $xe^{-x}\text{Ei}(x) = 1.18185$ for $x=8$. From **Table 4.4**, $e^8 = 2.98096 \times 10^3$. Thus Ei $(8) = 440.38$.

*See page II.
[8] From C. Hastings, Jr., Approximations for digital computers, Princeton Univ. Press, Princeton, N.J., 1955 (with permission).

Example 3. Compute Si (20) to 5D.
Since $1/20 = .05$ from **Table 5.2** we find $f(20) = .049757$, $g(20) = .002464$. From **Table 4.8**, $\sin 20 = .912945$, $\cos 20 = .408082$. Using **5.2.8**

$$\text{Si}(20) = \frac{\pi}{2} - f(20)\cos 20 - g(20)\sin 20$$
$$= 1.570796 - .022555 = 1.54824.$$

Example 4. Compute $E_n(x)$, $n = 1(1)N$, to 5S for $x = 1.275$, $N = 10$.

If x is less than about five, the recurrence relation **5.1.14** can be used in increasing order of n without serious loss of accuracy.

By quadratic interpolation in **Table 5.1** we get $E_1(1.275) = .1408099$, and from **Table 4.4**, $e^{-1.275} = .2794310$. The recurrence formula **5.1.14** then yields

n	$E_n(1.275)$		$E_n(1.275)$
1	.1408099	6	.0430168
2	.0998984	7	.0374307
3	.0760303	8	.0331009
4	.0608307	9	.0296534
5	.0504679	10	.0268469

Interpolating directly in **Table 5.4** for $n=10$ we get $E_{10}(1.275)=.0268470$ as a check.

Example 5. Compute $E_n(x)$, $n=1(1)N$, to 5S for $x=10$, $N=10$.

If, as in this example, x is appreciably larger than five and $N \leq x$, then the recurrence relation **5.1.14** may be safely used in decreasing order of n ([5.5]). From **Table 5.5** for $x^{-1}=.1$ we get $(x+10)e^x E_{10}(x)=1.02436$ so that $E_{10}(10)=2.32529 \times 10^{-6}$. Using this as the initial value we obtain column (2).

n	$10^5 E_n(10)$ (1)	$10^5 E_n(10)$ (2)
1	.41570	.41570
2	.38300	.38302
3	.35500̲	.35488
4	.33000̲	.33041
5	.31000̲	.30898
6	.28800̲	.29005
7	.27667̲	.27325
8	.25333̲	.25822
9	.25084̲	.24472
10	.22573̲	.23253

From **Table 5.2** we get $xe^x E_1(x)=.915633$ so that $E_1(10)=4.15697\times 10^{-6}$ as a check. Forward recurrence starting with $E_1(10)=4.1570\times 10^{-6}$ yields the values in column (1). The underlined figures are in error.

Example 6. Compute $E_n(x)$, $n=1(1)N$, to 5S for $x=12.3$, $N=20$.

If N is appreciably larger than x, and x appreciably larger than five, then the recurrence relation **5.1.14** should be used in the backward direction to generate $E_n(x)$ for $n<n_0$, and in the forward direction to generate $E_n(x)$ for $n>n_0$, where $n_0=\langle x \rangle$.

From **5.1.52**, with $n_0=12$, $x=12.3$, we have

$$E_{n_0}(x)=\frac{e^{-12.3}}{24.3}(1+.02032-.00043-.00001)$$
$$=1.91038\times 10^{-7}.$$

Using the recurrence relation **5.1.14**, as indicated, we get

n	$10^6 E_n(12.3)$	$10^6 E_n(12.3)$	n
12	.191038	.191038	12
11	.199213	.183498	13
10	.208098	.176516	14
9	.217793	.170042	15
8	.228406	.164015	16
7	.240073	.158397	17
6	.252951	.153144	18
5	.267234	.148226	19
4	.283155	.143608	20
3	.300998		
2	.321117		
1	.343953		

From **Tables 5.2** and **5.5** we find $E_1(12.3)=.343953 \times 10^{-6}$, $E_{20}(12.3)=.143609\times 10^{-6}$ as a check.

Example 7. Compute $\alpha_n(2)$ to 6S for $n=1(1)5$.

The recurrence formula **5.1.15** can be used for all $x>0$ in increasing order of n without loss of accuracy. From **5.1.25** we have $\alpha_0(2)=\frac{1}{2}e^{-2}=.0676676$, so we get

n	$\alpha_n(2)$
0	.0676676
1	.101501
2	.169169
3	.321421
4	.710510
5	1.84394

Independent calculation with **5.1.8** yields the same result for $\alpha_5(2)$.

The functions $\alpha_0(x)$ and $\alpha_1(x)$ can be obtained from **Table 10.8** using **5.1.48**, **5.1.49**.

Example 8. Compute $\beta_n(x)$, $n=0(1)N$ to 6S for $x=1$, $N=5$.

Use the recurrence relation **5.1.16** in increasing order of n if

$$x>.368N+.184 \ln N+.821$$

and in decreasing order of n otherwise [5.5].

From **5.1.9** with $n=5$ we get $\beta_5(1)=-.324297$ correctly rounded to 6D. Using the recurrence formula **5.1.16** in decreasing order of n and carrying 9D we get the values in column (2).

n	$\beta_n(1)$ (1)	$\beta_n(1)$ (2)
0	2.35040 2	2.35040 2389
1	−.73575 9269̲	−.73575 8880̲
2	.87888 3849̲	.87888 4629̲
3	−.44950 9722̲	−.44950 7383̲
4	.55236 3499̲	.55237 2854̲
5	−.32434 3774̲	−.32429 7

Using forward recurrence instead, starting with

$\beta_0(1) = 2 \sinh 1 = 2.350402$ and again carrying 9D, we obtain column (1). The underlined figures are in error. The above shows that three significant figures are lost in forward recurrence, whereas about three significant figures are gained in backward recurrence!

An alternative procedure is to start with an arbitrary value for n sufficiently large (see also [5.1]). To illustrate, starting with the value zero at $n=11$ we get

n	$\beta_n(1)$	n	$\beta_n(1)$
11	0.	5	$-.324297$
10	.280560	4	.552373
9	$-.206984$	3	$-.449507$
8	.319908	2	.878885
7	$-.253812$	1	$-.735759$
6	.404621	0	2.350402

The functions $\beta_0(x)$ and $\beta_1(x)$ can be obtained from **Table 10.8** using **5.1.48, 5.1.49**.

Example 9. Compute $E_1(z)$ for $z = 3.2578 + 6.8943i$.

From **Table 5.6** we have for $z_0 = x_0 + iy_0 = 3 + 7i$

$$z_0 e^{z_0} E_1(z_0) = .934958 + .095598i,$$

$$e^{z_0} E_1(z_0) = .059898 - .107895i.$$

From Taylor's formula with $f(z) = e^z E_1(z)$ we have

$$f(z) = f(z_0 + \Delta z) = f(z_0) + \frac{f'(z_0)}{1!} \Delta z$$

$$+ \frac{f''(z_0)}{2!} (\Delta z)^2 + \cdots$$

with $\Delta z = z - z_0 = .2578 - .1057i$. Thus with **5.1.27** we get

k	$f^{(k)}(z_0)/k!$		$(\Delta z)^k f^{(k)}(z_0)/k!$	
0	.059898	$-.107895i$	.059898	$-.107895i$
1	.008174	$+.012795i$	.003460	$+.002435i$
2	$-.001859$	$+.000155i$	$-.000094$	$+.000110i$
3	.000088	$-.000212i$	$-.000003$	$-.000004i$

$$f(z) = .063261 \quad -.105354i$$
$$e^{-z} = .031510 \quad -.022075i$$
$$E_1(z) = -.000332 \quad -.004716i$$

Repeating the calculation with $z_0 = 3 + 6i$ and $\Delta z = .2578 + .8943i$ we get the same result.

An alternative procedure is to perform bivariate interpolation in the real and imaginary parts of $z e^z E_1(z)$.

Example 10. Compute $E_1(z)$ for $z = -4.2 + 12.7i$.

Using the formula at the bottom of **Table 5.6**

$$e^z E_1(z) \approx \frac{.711093}{-3.784225 + 12.7i}$$

$$+ \frac{.278518}{-1.90572 + 12.7i} + \frac{.010389}{2.0900 + 12.7i}$$

$$= -.0184106 - .0736698i$$
$$E_1(z) \approx -1.87133 - 4.70540i.$$

References

Texts

[5.1] F. J. Corbató, On the computation of auxiliary functions for two-center integrals by means of a high-speed computer, J. Chem. Phys. **24**, 452–453 (1956).

[5.2] A. Erdélyi et al., Higher transcendental functions, vol. 2 (McGraw-Hill Book Co., Inc., New York, N.Y., 1953).

[5.3] A. Erdélyi et al., Tables of integral transforms, vols. 1, 2 (McGraw-Hill Book Co., Inc., New York, N.Y., 1954).

[5.4] W. Gautschi, Some elementary inequalities relating to the gamma and incomplete gamma function, J. Math. Phys. **38**, 77–81 (1959).

[5.5] W. Gautschi, Recursive computation of certain integrals, J. Assoc. Comput. Mach. **8**, 21–40 (1961).

[5.6] W. Gröbner and N. Hofreiter, Integraltafel (Springer-Verlag, Wien and Innsbruck, Austria, 1949–50).

[5.7] C. Hastings, Jr., Approximations for digital computers (Princeton Univ. Press, Princeton, N.J., 1955).

[5.8] E. Hopf, Mathematical problems of radiative equilibrium, Cambridge Tracts in Mathematics and Mathematical Physics, No. 31 (Cambridge Univ. Press, Cambridge, England, 1934).

[5.9] V. Kourganoff, Basic methods in transfer problems (Oxford Univ. Press, London, England, 1952).

[5.10] F. Lösch and F. Schoblik, Die Fakultät und verwandte Funktionen (B. G. Teubner, Leipzig, Germany, 1951).

[5.11] N. Nielsen, Theorie des Integrallogarithmus (B. G. Teubner, Leipzig, Germany, 1906).

[5.12] F. Oberhettinger, Tabellen zur Fourier Transformation (Springer-Verlag, Berlin, Göttingen, Heidelberg, Germany, 1957).

[5.13] I. M. Ryshik and I. S. Gradstein, Tables of series, products and integrals (VEB Deutscher Verlag der Wissenschaften, Berlin, Germany, 1957).

[5.14] S. A. Schelkunoff, Proposed symbols for the modified cosine and exponential integral, Quart. Appl. Math. **2**, 90 (1944).

[5.15] J. Todd, Evaluation of the exponential integral for large complex arguments, J. Research NBS **52**, 313–317 (1954) RP 2508.

[5.16] F. G. Tricomi, Funzioni ipergeometriche confluenti (Edizioni Cremonese, Rome, Italy, 1954).

Tables

[5.17] British Association for the Advancement of Science, Mathematical Tables, vol. I. Circular and hyperbolic functions, exponential, sine and cosine integrals, etc., 3d ed. (Cambridge Univ. Press, Cambridge, England, 1951). $\mathrm{Ei}(x) - \ln x$, $-E_1(x) - \ln x$, $\mathrm{Ci}(x) - \ln x$, $\mathrm{Si}(x)$, $x = 0(.1)5$, 11D; $\mathrm{Ei}(x)$, $x = 5(.1)15$, 10–11S; $E_1(x)$, $x = 5(.1)15$, 13–14D; $\mathrm{Si}(x)$, $\mathrm{Ci}(x)$, $x = 5(.1)20(.2)40$, 10D.

[5.18] L. Fox, Tables of Weber parabolic cylinder functions and other functions for large arguments, Mathematical Tables, vol. 4, National Physical Laboratory (Her Majesty's Stationery Office, London, England, 1960). $e^{-x}\mathrm{Ei}(x)$, $e^x E_1(x)$, $x^{-1} = 0(.001).1$, 10D; $f(x)$, $g(x)$, $x^{-1} = 0(.001).1$, 10D.

[5.19] J. W. L. Glaisher, Tables of the numerical values of the sine-integral, cosine-integral and exponential integral, Philos. Trans. Roy. Soc. London **160**, 367–388 (1870). $\mathrm{Si}(x)$, $\mathrm{Ci}(x)$, $\mathrm{Ei}(x)$, $-E_1(x)$, $x = 0(.01)1$, 18D, $x = 1(.1)5(1)15$, 11D.

[5.20] B. S. Gourary and M. E. Lynam, Tables of the auxiliary molecular integrals $A_n(x)$ and the auxiliary functions $C_n(x)$, The Johns Hopkins Univ. Applied Physics Laboratory, CM Report 905, Baltimore, Md. (1957). $\alpha_n(x)$, $n!e_n(x)$, $x = .05(.05)15$, $n = 0(1)18$, 9S.

[5.21] F. E. Harris, Tables of the exponential integral $\mathrm{Ei}(x)$, Math. Tables Aids Comp. **11**, 9–16 (1957). $E_1(x)$, $e^x E_1(x)$, $\mathrm{Ei}(x)$, $e^{-x}\mathrm{Ei}(x)$, $x = 1(1)4(.4)8(1)50$, 18–19S.

[5.22] Harvard University, The Annals of the Computation Laboratory, vols. 18, 19; Tables of the generalized sine- and cosine-integral functions, parts I, II (Harvard Univ. Press, Cambridge, Mass., 1949). $S(a, x) = \int_0^x \frac{\sin u}{u} dx$, $C(a, x) = \int_0^x \frac{1 - \cos u}{u} dx$, 6D; $\mathrm{Ss}(a, x) = \int_0^x \frac{\sin u}{u} \sin x dx$, $\mathrm{Sc}(a, x) = \int_0^x \frac{\sin u}{u} \cos x dx$, 6D; $\mathrm{Cs}(a, x) = \int_0^x \frac{\cos u}{u} \sin x dx$, $\mathrm{Cc}(a, x) = \int_0^x \frac{\cos u}{u}(1 - \cos x) dx$, 6D; $u = \sqrt{x^2 + a^2}$, $0 \leq a < 25$, $0 \leq x \leq 25$.

[5.23] Harvard University, The Annals of the Computation Laboratory, vol. 21; Tables of the generalized exponential-integral functions (Harvard Univ. Press, Cambridge, Mass., 1949). $E(a, x) = \int_0^x \frac{1 - e^{-u}}{u} dx$, $\mathrm{Es}(a, x) = \int_0^x \frac{e^{-u} \sin u}{u} dx$, $\mathrm{Ec}(a, x) = \int_0^x \frac{1 - e^{-u} \cos u}{u} dx$, 6D; $u = \sqrt{x^2 + a^2}$, $0 \leq a < 10$, $0 \leq x < 10$.

[5.24] A. V. Hershey, Computing programs for the complex exponential integral, U.S. Naval Proving Ground, Dahlgren, Va., NPG Report No. 1646 (1959). $-E_1(-z)$, $z = x + iy$, $x = -20(1)20$, $y = 0(1)20$, 13S.

[5.25] E. Jahnke and F. Emde, Tables of functions, 4th ed. (Dover Publications, Inc., New York, N.Y., 1945). $-E_1(x)$, $\mathrm{Ei}(x)$, $x = 0(.01)1(.1)5(1)15$, 4–6S; $\mathrm{Si}(x)$, $\mathrm{Ci}(x)$, $x = 0(.01)1(.1)5(1)15(5)100(10)200(100)10^{3(1)7}$, generally 4–5S; maxima and minima of $\mathrm{Ci}(x)$ and $\mathrm{si}(x)$, $0 < x < 16$, 5S.

[5.26] K. A. Karpov and S. N. Razumovskiĭ, Tablitsy integral' nogo logarifma (Izdat. Akad. Nauk SSSR., Moscow, U.S.S.R., 1956). $\mathrm{li}(x)$, $x = 0(.0001)2.5(.001)20(.01)200(.1)500(1)10000(10)25000$, 7S; $\mathrm{li}(x) - \ln |1 - x|$, $x = .95(.0001)1.05$, 6D.

[5.27] M. Kotani, A. Amemiya, E. Ishiguro, T. Kimura, Table of molecular integrals (Maruzen Co., Ltd., Tokyo, Japan, 1955). $\alpha_n(x)$, $x = .25(.25)9(.5)19(1)25$, $n = 0(1)15$, 11S; $\beta_n(x)$, $x = .25(.25)8(.5)19(1)25$, $n = 0(1)8$, 11S.

[5.28] M. Mashiko, Tables of generalized exponential-, sine- and cosine-integrals, Numerical Computation Bureau Report No. 7, Tokyo, Japan (1953). $E_1(z) + \ln|z| = C_\alpha(\xi) + \ln \xi - iS_\alpha(\xi)$, $z = \xi e^{i\alpha}$, $\xi = 0(.05)5$, $\alpha = 0°(2°)60°(1°)90°$, 6D; $ze^z E_1(z) = A_\alpha(\eta)$ exp $[i\Phi_\alpha(\eta)]$, $z = \frac{1}{\eta} e^{i\alpha}$, $\eta = .01(.01).2$, $\alpha = 0°(2°)60°(1°)90°$, 5–6D.

[5.29] G. F. Miller, Tables of generalized exponential integrals, Mathematical Tables, vol. 3, National Physical Laboratory (Her Majesty's Stationery Office, London, England, 1958). $(x + n)e^x E_n(x)$, $x = 0(.01)1$, $n = 1(1)8$, $x = 0(.1)20$, $n = 1(1)24$, $x^{-1} = 0(.001).05$, $n = 1(1)24$; 8D.

[5.30] J. Miller, J. M. Gerhauser, and F. A. Matsen, Quantum chemistry integrals and tables (Univ. of Texas Press, Austin, Tex., 1959). $\alpha_n(x)$, $x = .125(.125)25$, $n = 0(1)16$, 14S; $\beta_n(x)$, $x = 0(.125)24.875$, $n = 0(1)16$, 12–14S.

[5.31] J. Miller and R. P. Hurst, Simplified calculation of the exponential integral, Math. Tables Aids Comp. **12**, 187–193 (1958). $e^{-x}\mathrm{Ei}(x)$, $\mathrm{Ei}(x)$, $e^x E_1(x)$, $E_1(x)$, $x = .2(.05)5(.1)10(.2)20(.5)50(1)80$; 16S.

[5.32] National Bureau of Standards, Tables of sine, cosine and exponential integrals, vol. I (U.S. Government Printing Office, Washington, D.C., 1940). $\mathrm{Si}(x)$, $\mathrm{Ci}(x)$, $\mathrm{Ei}(x)$, $E_1(x)$, $x = 0(.0001)2$, $x = 0(.1)10$; 9D.

[5.33] National Bureau of Standards, Tables of sine, cosine and exponential integrals, vol. II (U.S. Government Printing Office, Washington, D.C., 1940). $\mathrm{Si}(x)$, $\mathrm{Ci}(x)$, $\mathrm{Ei}(x)$, $E_1(x)$, $x = 0(.001)10$, 9–10 D or S; $\mathrm{Si}(x)$, $\mathrm{Ci}(x)$, $x = 10(.1)40$, 10D; $\mathrm{Ei}(x)$, $E_1(x)$, $x = 10(.1)15$, 7–11S.

[5.34] National Bureau of Standards, Table of sine and cosine integrals for arguments from 10 to 100, Applied Math. Series 32 (U.S. Government Printing Office, Washington, D.C., 1954). $Si(x)$, $Ci(x)$, $x = 10(.01)100$, 10D.

[5.35] National Bureau of Standards, Tables of functions and of zeros of functions, Collected short tables of the Computation Laboratory, Applied Math. Series 37 (U.S. Government Printing Office, Washington, D.C., 1954). $E_n(x)$, $n = 0(1)20$, $x = 0(.01)2(.1)10$, 4–9S; $E_2(x) - x \ln x$, $x = 0(.01).5$, 7S; $E_3(x) + \frac{1}{2}x^2 \ln x$, $x = 0(.01).1$, 7S.

[5.36] National Bureau of Standards, Tables of the exponential integral for complex arguments, Applied Math. Series 51 (U.S. Government Printing Office, Washington, D.C., 1958). $E_1(z) + \ln z$, 6D, $x = 0(.02)1$, $y = 0(.02)1$, $x = -1(.1)0$, $y = 0(.1)1$; $E_1(z)$, 6D, $x = 0(.02)4$, $y = 0(.02)3(.05)10$, $x = 0(1)20$, $y = 0(1)20$, $x = -3.1(.1)0$, $y = 0(.1)3.1$, $x = -4.5(.5)0$, $y = 0(.1)4(.5)10$, $x = -10(.5)-4.5$, $y = 0(.5)10$, $x = -20(1)0$, $y = 0(1)20$; $e^z E_1(z)$, 6D, $x = 4(.1)10$, $y = 0(.5)10$.

[5.37] S. Oberländer, Tabellen von Exponentialfunktionen und-integralen zur Anwendung auf Gebieten der Thermodynamik, Halbleitertheorie und Gaskinetik (Akademie-Verlag, Berlin, Germany, 1959). $\frac{\Delta E}{kT}$, $\frac{kT}{\Delta E}$, $\exp\left(\frac{-\Delta E}{kT}\right)$, $\frac{kT}{\Delta E}\exp\left(\frac{-\Delta E}{kT}\right)$, $E_1\left(\frac{\Delta E}{kT}\right)$, $\frac{k}{\Delta E}\int_0^T \exp\left(\frac{-\Delta E}{kT}\right)dT$, $\frac{\Delta E}{kT}\exp\left(\frac{\Delta E}{kT}\right)\times$ $E_1\left(\frac{\Delta E}{kT}\right)$, $1 - \frac{\Delta E}{kT}\exp\left(\frac{\Delta E}{kT}\right)E_1\left(\frac{\Delta E}{kT}\right)$; $\Delta E = .2(.2)2$, $T = 25(25)1000$, $T = 150(10)390$, 3–4S; x^{-1}, $\exp(-x^{-1})$, $x \exp(-x^{-1})$, $E_1(x^{-1})$, $\int_0^x \exp(-t^{-1})dt$, $x^{-1}\exp(x^{-1})E_1(x^{-1})$, $1 - x^{-1}\exp(x^{-1})E_1(x^{-1})$; $x = .01(.0001).1$, 5–6S.

[5.38] V. I. Pagurova, Tables of the exponential integral $E_\nu(x) = \int_1^\infty e^{-xu}u^{-\nu}du$. Translated from the Russian by D. G. Fry (Pergamon Press, New York, N.Y.; Oxford, London, England; Paris, France, 1961). $E_n(x)$, $n = 0(1)20$, $x = 0(.01)2(.1)10$, 4–9S; $E_2(x) - x \ln x$, $x = 0(.01)5$, 7S; $E_3(x) + \frac{1}{2}x^2 \ln x$, $x = 0(.01).1$, 7S; $e^x E_n(x)$, $n = 2(1)10$, $x = 10(.1)20$, 7D; $e^x E_\nu(x)$, $\nu = 0(.1)1$, $x = .01(.01)7(.05)12(.1)20$, 7 S or D.

[5.39] Tablitsy integral'nogo sinusa i kosinusa (Izdat. Akad. Nauk SSSR., Moscow, U.S.S.R., 1954). $Si(x)$, $Ci(x)$, $x = 0(.0001)2(.001)10(.005)100$, 7D; $Ci(x) - \ln x$, $x = 0(.0001).01$, 7D.

[5.40] Tablitsy integral'noĭ pokazatel'noĭ funktsii (Izdat. Akad. Nauk SSSR., Moscow, U.S.S.R., 1954). $Ei(x)$, $E_1(x)$, $x = 0(.0001)1.3(.001)3(.0005)10(.1)15$, 7D.

[5.41] D. K. Trubey, A table of three exponential integrals, Oak Ridge National Laboratory Report 2750, Oak Ridge, Tenn. (June 1959). $E_1(x)$, $E_2(x)$, $E_3(x)$, $x = 0(.0005).1(.001)2(.01)10(.1)20$, 6S.

Table 5.1   **SINE, COSINE AND EXPONENTIAL INTEGRALS**

x	$x^{-1}\mathrm{Si}(x)$	$x^{-2}[\mathrm{Ci}(x)-\ln x-\gamma]$	$x^{-1}[\mathrm{Ei}(x)-\ln x-\gamma]$	$x^{-1}[E_1(x)+\ln x+\gamma]$
0.00	1.00000 00000	−0.25000 00000	1.00000 0000	1.00000 00000
0.01	0.99999 44444	−0.24999 89583	1.00250 5566	0.99750 55452
0.02	0.99997 77781	−0.24999 58333	1.00502 2306	0.99502 21392
0.03	0.99995 00014	−0.24999 06250	1.00755 0283	0.99254 97201
0.04	0.99991 11154	−0.24998 33339	1.01008 9560	0.99008 82265
0.05	0.99986 11215	−0.24997 39598	1.01264 0202	0.98763 75971
0.06	0.99980 00216	−0.24996 25030	1.01520 2272	0.98519 77714
0.07	0.99972 78178	−0.24994 89639	1.01777 5836	0.98276 86889
0.08	0.99964 45127	−0.24993 33429	1.02036 0958	0.98035 02898
0.09	0.99955 01094	−0.24991 56402	1.02295 7705	0.97794 25142
0.10	0.99944 46111	−0.24989 58564	1.02556 6141	0.97554 53033
0.11	0.99932 80218	−0.24987 39923	1.02818 6335	0.97315 85980
0.12	0.99920 03455	−0.24985 00480	1.03081 8352	0.97078 23399
0.13	0.99906 15870	−0.24982 40244	1.03346 2259	0.96841 64710
0.14	0.99891 17512	−0.24979 59223	1.03611 8125	0.96606 09336
0.15	0.99875 08435	−0.24976 57422	1.03878 6018	0.96371 56702
0.16	0.99857 88696	−0.24973 34850	1.04146 6006	0.96138 06240
0.17	0.99839 58357	−0.24969 91516	1.04415 8158	0.95905 57383
0.18	0.99820 17486	−0.24966 27429	1.04686 2544	0.95674 09569
0.19	0.99799 66151	−0.24962 42598	1.04957 9234	0.95443 62237
0.20	0.99778 04427	−0.24958 37035	1.05230 8298	0.95214 14833
0.21	0.99755 32390	−0.24954 10749	1.05504 9807	0.94985 66804
0.22	0.99731 50122	−0.24949 63752	1.05780 3833	0.94758 17603
0.23	0.99706 57709	−0.24944 96056	1.06057 0446	0.94531 66684
0.24	0.99680 55242	−0.24940 07674	1.06334 9719	0.94306 13506
0.25	0.99653 42813	−0.24934 98618	1.06614 1726	0.94081 57528
0.26	0.99625 20519	−0.24929 68902	1.06894 6539	0.93857 98221
0.27	0.99595 88464	−0.24924 18540	1.07176 4232	0.93635 35046
0.28	0.99565 46750	−0.24918 47546	1.07459 4879	0.93413 67481
0.29	0.99533 95489	−0.24912 55938	1.07743 8555	0.93192 94997
0.30	0.99501 34793	−0.24906 43727	1.08029 5334	0.92973 17075
0.31	0.99467 64779	−0.24900 10933	1.08316 5293	0.92754 33196
0.32	0.99432 85570	−0.24893 57573	1.08604 8507	0.92536 42845
0.33	0.99396 97288	−0.24886 83662	1.08894 5053	0.92319 45510
0.34	0.99360 00064	−0.24879 89219	1.09185 5008	0.92103 40684
0.35	0.99321 94028	−0.24872 74263	1.09477 8451	0.91888 27858
0.36	0.99282 79320	−0.24865 38813	1.09771 5458	0.91674 06533
0.37	0.99242 56078	−0.24857 82887	1.10066 6108	0.91460 76209
0.38	0.99201 24449	−0.24850 06507	1.10363 0481	0.91248 36388
0.39	0.99158 84579	−0.24842 09693	1.10660 8656	0.91036 86582
0.40	0.99115 36619	−0.24833 92466	1.10960 0714	0.90826 26297
0.41	0.99070 80728	−0.24825 54849	1.11260 6735	0.90616 55048
0.42	0.99025 17063	−0.24816 96860	1.11562 6800	0.90407 72350
0.43	0.98978 45790	−0.24808 18528	1.11866 0991	0.90199 77725
0.44	0.98930 67074	−0.24799 19870	1.12170 9391	0.89992 70693
0.45	0.98881 81089	−0.24790 00913	1.12477 2082	0.89786 50778
0.46	0.98831 88008	−0.24780 61685	1.12784 9147	0.89581 17511
0.47	0.98780 88010	−0.24771 02206	1.13094 0671	0.89376 70423
0.48	0.98728 81278	−0.24761 22500	1.13404 6738	0.89173 09048
0.49	0.98675 67998	−0.24751 22600	1.13716 7432	0.88970 32920
0.50	0.98621 48361	−0.24741 02526	1.14030 2841	0.88768 41584
	$\begin{bmatrix}(-6)1\\4\end{bmatrix}$	$\begin{bmatrix}(-7)3\\4\end{bmatrix}$	$\begin{bmatrix}(-6)2\\4\end{bmatrix}$	$\begin{bmatrix}(-6)2\\4\end{bmatrix}$

See **Examples 1–2**.

$$\gamma = 0.57721\ 56649$$

SINE, COSINE AND EXPONENTIAL INTEGRALS   Table 5.1

x	$Si(x)$	$Ci(x)$	$Ei(x)$	$E_1(x)$
0.50	0.49310 74180	−0.17778 40788	0.45421 9905	0.55977 3595
0.51	0.50268 77506	−0.16045 32390	0.48703 2167	0.54782 2352
0.52	0.51225 15212	−0.14355 37358	0.51953 0633	0.53621 9798
0.53	0.52179 84228	−0.12707 07938	0.55173 0445	0.52495 1510
0.54	0.53132 81492	−0.11099 04567	0.58364 5931	0.51400 3886
0.55	0.54084 03951	−0.09529 95274	0.61529 0657	0.50336 4081
0.56	0.55033 48563	−0.07998 55129	0.64667 7490	0.49301 9959
0.57	0.55981 12298	−0.06503 65744	0.67781 8642	0.48296 0034
0.58	0.56926 92137	−0.05044 14815	0.70872 5720	0.47317 3433
0.59	0.57870 85069	−0.03618 95707	0.73940 9764	0.46364 9849
0.60	0.58812 88096	−0.02227 07070	0.76988 1290	0.45437 9503
0.61	0.59752 98233	−0.00067 52486	0.80015 0320	0.44535 3112
0.62	0.60691 12503	+0.00460 59849	0.83022 6417	0.43656 1854
0.63	0.61627 27944	0.01758 17424	0.86011 8716	0.42799 7338
0.64	0.62561 41603	0.03026 03686	0.88983 5949	0.41965 1581
0.65	0.63493 50541	0.04264 98293	0.91938 6468	0.41151 6976
0.66	0.64423 51831	0.05475 77343	0.94877 8277	0.40358 6275
0.67	0.65351 42557	0.06659 13594	0.97801 9042	0.39585 2563
0.68	0.66277 19817	0.07815 76659	1.00711 6121	0.38830 9243
0.69	0.67200 80721	0.08946 33195	1.03607 6576	0.38095 0010
0.70	0.68122 22391	0.10051 47070	1.06490 7195	0.37376 8843
0.71	0.69041 41965	0.11131 79525	1.09361 4501	0.36675 9981
0.72	0.69958 36590	0.12187 89322	1.12220 4777	0.35991 7914
0.73	0.70873 03430	0.13220 32879	1.15068 4069	0.35323 7364
0.74	0.71785 39660	0.14229 64404	1.17905 8208	0.34671 3279
0.75	0.72695 42472	0.15216 36010	1.20733 2816	0.34034 0813
0.76	0.73603 09067	0.16180 97827	1.23551 3319	0.33411 5321
0.77	0.74508 36664	0.17123 98110	1.26360 4960	0.32803 2346
0.78	0.75411 22494	0.18045 83335	1.29161 2805	0.32208 7610
0.79	0.76311 63804	0.18946 98290	1.31954 1753	0.31627 7004
0.80	0.77209 57855	0.19827 86160	1.34739 6548	0.31059 6579
0.81	0.78105 01921	0.20688 88610	1.37518 1783	0.30504 2539
0.82	0.78997 93293	0.21530 45859	1.40290 1910	0.29961 1236
0.83	0.79888 29277	0.22352 96752	1.43056 1245	0.29429 9155
0.84	0.80776 07191	0.23156 78824	1.45816 3978	0.28910 2918
0.85	0.81661 24372	0.23942 28368	1.48571 4176	0.28401 9269
0.86	0.82543 78170	0.24709 80486	1.51321 5791	0.27904 5070
0.87	0.83423 65953	0.25459 69153	1.54067 2664	0.27417 7301
0.88	0.84300 85102	0.26192 27264	1.56808 8534	0.26941 3046
0.89	0.85175 33016	0.26907 86687	1.59546 7036	0.26474 9496
0.90	0.86047 07107	0.27606 78305	1.62281 1714	0.26018 3939
0.91	0.86916 04808	0.28289 32065	1.65012 6019	0.25571 3758
0.92	0.87782 23564	0.28955 77018	1.67741 3317	0.25133 6425
0.93	0.88645 60839	0.29606 41358	1.70467 6891	0.24704 9501
0.94	0.89506 14112	0.30241 52458	1.73191 9946	0.24285 0627
0.95	0.90363 80880	0.30861 36908	1.75914 5612	0.23873 7524
0.96	0.91218 58656	0.31466 20547	1.78635 6947	0.23470 7988
0.97	0.92070 44970	0.32056 28495	1.81355 6941	0.23075 9890
0.98	0.92919 37370	0.32631 85183	1.84074 8519	0.22689 1167
0.99	0.93765 33420	0.33193 14382	1.86793 4543	0.22309 9826
1.00	0.94608 30704	0.33740 39229	1.89511 7816	0.21938 3934
	$\begin{bmatrix} (-6)4 \\ 4 \end{bmatrix}$	$\begin{bmatrix} (-5)5 \\ 6 \end{bmatrix}$	$\begin{bmatrix} (-5)4 \\ 5 \end{bmatrix}$	$\begin{bmatrix} (-5)4 \\ 5 \end{bmatrix}$

Table 5.1   **SINE, COSINE AND EXPONENTIAL INTEGRALS**

x	$Si(x)$	$Ci(x)$	$Ei(x)$	$E_1(x)$
1.00	0.94608 30704	0.33740 39229	1.89511 7816	0.21938 3934
1.01	0.95448 26820	0.34273 82254	1.92230 1085	0.21574 1624
1.02	0.96285 19387	0.34793 65405	1.94948 7042	0.21217 1083
1.03	0.97119 06039	0.35300 10067	1.97667 8325	0.20867 0559
1.04	0.97949 84431	0.35793 37091	2.00387 7525	0.20523 8352
1.05	0.98777 52233	0.36273 66810	2.03108 7184	0.20187 2813
1.06	0.99602 07135	0.36741 19060	2.05830 9800	0.19857 2347
1.07	1.00423 46846	0.37196 13201	2.08554 7825	0.19533 5403
1.08	1.01241 69091	0.37638 68132	2.11280 3672	0.19216 0479
1.09	1.02056 71617	0.38069 02312	2.14007 9712	0.18904 6118
1.10	1.02868 52187	0.38487 33774	2.16737 8280	0.18599 0905
1.11	1.03677 08583	0.38893 80142	2.19470 1672	0.18299 3465
1.12	1.04482 38608	0.39288 58645	2.22205 2152	0.18005 2467
1.13	1.05284 40082	0.39671 86134	2.24943 1949	0.17716 6615
1.14	1.06083 10845	0.40043 79090	2.27684 3260	0.17433 4651
1.15	1.06878 48757	0.40404 53647	2.30428 8252	0.17155 5354
1.16	1.07670 51696	0.40754 25593	2.33176 9062	0.16882 7535
1.17	1.08459 17561	0.41093 10390	2.35928 7800	0.16615 0040
1.18	1.09244 44270	0.41421 23185	2.38684 6549	0.16352 1748
1.19	1.10026 29760	0.41738 78816	2.41444 7367	0.16094 1567
1.20	1.10804 71990	0.42045 91829	2.44209 2285	0.15840 8437
1.21	1.11579 68937	0.42342 76482	2.46978 3315	0.15592 1324
1.22	1.12351 18599	0.42629 46760	2.49752 2442	0.15347 9226
1.23	1.13119 18994	0.42906 16379	2.52531 1634	0.15108 1164
1.24	1.13883 68160	0.43172 98802	2.55315 2836	0.14872 6188
1.25	1.14644 64157	0.43430 07240	2.58104 7974	0.14641 3373
1.26	1.15402 05063	0.43677 54665	2.60899 8956	0.14414 1815
1.27	1.16155 88978	0.43915 53815	2.63700 7673	0.14191 0639
1.28	1.16906 14023	0.44144 17205	2.66507 5997	0.13971 8989
1.29	1.17652 78340	0.44363 57130	2.69320 5785	0.13756 6032
1.30	1.18395 80091	0.44573 85675	2.72139 8880	0.13545 0958
1.31	1.19135 17459	0.44775 14723	2.74965 7110	0.13337 2975
1.32	1.19870 88649	0.44967 55955	2.77798 2287	0.13133 1314
1.33	1.20602 91886	0.45151 20863	2.80637 6214	0.12932 5224
1.34	1.21331 25418	0.45326 20753	2.83484 0677	0.12735 3972
1.35	1.22055 87513	0.45492 66752	2.86337 7453	0.12541 6844
1.36	1.22776 76460	0.45650 69811	2.89198 8308	0.12351 3146
1.37	1.23493 90571	0.45800 40711	2.92067 4997	0.12164 2198
1.38	1.24207 28180	0.45941 90071	2.94943 9263	0.11980 3337
1.39	1.24916 87640	0.46075 28349	2.97828 2844	0.11799 5919
1.40	1.25622 67328	0.46200 65851	3.00720 7464	0.11621 9313
1.41	1.26324 65642	0.46318 12730	3.03621 4843	0.11447 2903
1.42	1.27022 81004	0.46427 78995	3.06530 6691	0.11275 6090
1.43	1.27717 11854	0.46529 74513	3.09448 4712	0.11106 8287
1.44	1.28407 56658	0.46624 09014	3.12375 0601	0.10940 8923
1.45	1.29094 13902	0.46710 92094	3.15310 6049	0.10777 7440
1.46	1.29776 82094	0.46790 33219	3.18255 2741	0.10617 3291
1.47	1.30455 59767	0.46862 41732	3.21209 2355	0.10459 5946
1.48	1.31130 45473	0.46927 26848	3.24172 6566	0.10304 4882
1.49	1.31801 37788	0.46984 97667	3.27145 7042	0.10151 9593
1.50	1.32468 35312	0.47035 63172	3.30128 5449	0.10001 9582
	$\begin{bmatrix} (-6)5 \\ 4 \end{bmatrix}$	$\begin{bmatrix} (-5)2 \\ 5 \end{bmatrix}$	$\begin{bmatrix} (-5)1 \\ 5 \end{bmatrix}$	$\begin{bmatrix} (-6)9 \\ 5 \end{bmatrix}$

SINE, COSINE AND EXPONENTIAL INTEGRALS

Table 5.1

x	$Si(x)$	$Ci(x)$	$Ei(x)$	$E_1(x)$
1.50	1.32468 35312	0.47035 63172	3.30128 5449	0.10001 9582
1.51	1.33131 36664	0.47079 32232	3.33121 3449	0.09854 4365
1.52	1.33790 40489	0.47116 13608	3.36124 2701	0.09709 3466
1.53	1.34445 45453	0.47146 15952	3.39137 4858	0.09566 6424
1.54	1.35096 50245	0.47169 47815	3.42161 1576	0.09426 2786
1.55	1.35743 53577	0.47186 17642	3.45195 4503	0.09288 2108
1.56	1.36386 54183	0.47196 33785	3.48240 5289	0.09152 3960
1.57	1.37025 50823	0.47200 04495	3.51296 5580	0.09018 7917
1.58	1.37660 42275	0.47197 37932	3.54363 7024	0.08887 3566
1.59	1.38291 27345	0.47188 42164	3.57442 1266	0.08758 0504
1.60	1.38918 04859	0.47173 25169	3.60531 9949	0.08630 8334
1.61	1.39540 73666	0.47151 94840	3.63633 4719	0.08505 6670
1.62	1.40159 32640	0.47124 58984	3.66746 7221	0.08382 5133
1.63	1.40773 80678	0.47091 25325	3.69871 9099	0.08261 3354
1.64	1.41384 16698	0.47052 01507	3.73009 1999	0.08142 0970
1.65	1.41990 39644	0.47006 95096	3.76158 7569	0.08024 7627
1.66	1.42592 48482	0.46956 13580	3.79320 7456	0.07909 2978
1.67	1.43190 42202	0.46899 64372	3.82495 3310	0.07795 6684
1.68	1.43784 19816	0.46837 54812	3.85682 6783	0.07683 8412
1.69	1.44373 80361	0.46769 92169	3.88882 9528	0.07573 7839
1.70	1.44959 22897	0.46696 83642	3.92096 3201	0.07465 4644
1.71	1.45540 46507	0.46618 36359	3.95322 9462	0.07358 8518
1.72	1.46117 50299	0.46534 57385	3.98562 9972	0.07253 9154
1.73	1.46690 33404	0.46445 53716	4.01816 6395	0.07150 6255
1.74	1.47258 94974	0.46351 32286	4.05084 0400	0.07048 9527
1.75	1.47823 34189	0.46251 99967	4.08365 3659	0.06948 8685
1.76	1.48383 50249	0.46147 63568	4.11660 7847	0.06850 3447
1.77	1.48939 42379	0.46038 29839	4.14970 4645	0.06753 3539
1.78	1.49491 09830	0.45924 05471	4.18294 5736	0.06657 8691
1.79	1.50038 51872	0.45804 97097	4.21633 2809	0.06563 8641
1.80	1.50581 67803	0.45681 11294	4.24986 7557	0.06471 3129
1.81	1.51120 56942	0.45552 54585	4.28355 1681	0.06380 1903
1.82	1.51655 18633	0.45419 33436	4.31738 6883	0.06290 4715
1.83	1.52185 52243	0.45281 54262	4.35137 4872	0.06202 1320
1.84	1.52711 57165	0.45139 23427	4.38551 7364	0.06115 1482
1.85	1.53233 32813	0.44992 47241	4.41981 6080	0.06029 4967
1.86	1.53750 78626	0.44841 31966	4.45427 2746	0.05945 1545
1.87	1.54263 94066	0.44685 83813	4.48888 9097	0.05862 0994
1.88	1.54772 78621	0.44526 08948	4.52366 6872	0.05780 3091
1.89	1.55277 31800	0.44362 13486	4.55860 7817	0.05699 7623
1.90	1.55777 53137	0.44194 03497	4.59371 3687	0.05620 4378
1.91	1.56273 42192	0.44021 85005	4.62898 6242	0.05542 3149
1.92	1.56764 98545	0.43845 63991	4.66442 7249	0.05465 3731
1.93	1.57252 21801	0.43665 46388	4.70003 8485	0.05389 5927
1.94	1.57735 11591	0.43481 38088	4.73582 1734	0.05314 9540
1.95	1.58213 67567	0.43293 44941	4.77177 8785	0.05241 4380
1.96	1.58687 89407	0.43101 72752	4.80791 1438	0.05169 0257
1.97	1.59157 76810	0.42906 27288	4.84422 1501	0.05097 6988
1.98	1.59623 29502	0.42707 14273	4.88071 0791	0.05027 4392
1.99	1.60084 47231	0.42504 39391	4.91738 1131	0.04958 2291
2.00	1.60541 29768	0.42298 08288	4.95423 4356	0.04890 0511
	$\left[\begin{smallmatrix}(-6)5\\4\end{smallmatrix}\right]$	$\left[\begin{smallmatrix}(-6)9\\5\end{smallmatrix}\right]$	$\left[\begin{smallmatrix}(-5)2\\4\end{smallmatrix}\right]$	$\left[\begin{smallmatrix}(-6)3\\4\end{smallmatrix}\right]$

Table 5.1   **SINE, COSINE AND EXPONENTIAL INTEGRALS**

x	$Si(x)$	$Ci(x)$	$xe^{-x}Ei(x)$	$xe^{x}E_1(x)$
2.0	1.60541 29768	0.42298 08288	1.34096 5420	0.72265 7234
2.1	1.64869 86362	0.40051 19878	1.37148 6802	0.73079 1502
2.2	1.68762 48272	0.37507 45990	1.39742 1992	0.73843 1132
2.3	1.72220 74818	0.34717 56175	1.41917 1534	0.74562 2149
2.4	1.75248 55008	0.31729 16174	1.43711 8315	0.75240 4829
2.5	1.77852 01734	0.28587 11964	1.45162 5159	0.75881 4592
2.6	1.80039 44505	0.25333 66161	1.46303 3397	0.76488 2722
2.7	1.81821 20765	0.22008 48786	1.47166 2153	0.77063 6987
2.8	1.83209 65891	0.18648 83896	1.47780 8187	0.77610 2123
2.9	1.84219 01946	0.15289 53242	1.48174 6162	0.78130 0252
3.0	1.84865 25280	0.11962 97860	1.48372 9204	0.78625 1221
3.1	1.85165 93077	0.08699 18312	1.48398 9691	0.79097 2900
3.2	1.85140 08970	0.05525 74117	1.48274 0191	0.79548 1422
3.3	1.84808 07828	+0.02467 82846	1.48017 4491	0.79979 1408
3.4	1.84191 39833	−0.00451 80779	1.47646 8706	0.80391 6127
3.5	1.83312 53987	−0.03212 85485	1.47178 2389	0.80786 7661
3.6	1.82194 81156	−0.05797 43519	1.46625 9659	0.81165 7037
3.7	1.80862 16809	−0.08190 10013	1.46003 0313	0.81529 4342
3.8	1.79339 03548	−0.10377 81504	1.45321 0902	0.81878 8821
3.9	1.77650 13604	−0.12349 93492	1.44590 5765	0.82214 8967
4.0	1.75820 31389	−0.14098 16979	1.43820 8032	0.82538 2600
4.1	1.73874 36265	−0.15616 53918	1.43020 0557	0.82849 6926
4.2	1.71836 85637	−0.16901 31568	1.42195 6813	0.83149 8602
4.3	1.69731 98507	−0.17950 95725	1.41354 1719	0.83439 3794
4.4	1.67583 39594	−0.18766 02868	1.40501 2424	0.83718 8207
4.5	1.65414 04144	−0.19349 11221	1.39641 9030	0.83988 7144
4.6	1.63246 03525	−0.19704 70797	1.38780 5263	0.84249 5539
4.7	1.61100 51718	−0.19839 12468	1.37920 9093	0.84501 7971
4.8	1.58997 52782	−0.19760 36133	1.37066 3313	0.84745 8721
4.9	1.56955 89381	−0.19477 98060	1.36219 6054	0.84982 1778
5.0	1.54993 12449	−0.19002 97497	1.35383 1278	0.85211 0880
5.1	1.53125 32047	−0.18347 62632	1.34558 9212	0.85432 9519
5.2	1.51367 09468	−0.17525 36023	1.33748 6755	0.85648 0958
5.3	1.49731 50636	−0.16550 59586	1.32953 7845	0.85856 8275
5.4	1.48230 00826	−0.15438 59262	1.32175 3788	0.86059 4348
5.5	1.46872 40727	−0.14205 29476	1.31414 3566	0.86256 1885
5.6	1.45666 83847	−0.12867 17494	1.30671 4107	0.86447 3436
5.7	1.44619 75285	−0.11441 07808	1.29947 0536	0.86633 1399
5.8	1.43735 91823	−0.09944 06647	1.29241 6395	0.86813 8040
5.9	1.43018 43341	−0.08393 26741	1.28555 3849	0.86989 5494
6.0	1.42468 75513	−0.06805 72439	1.27888 3860	0.87160 5775
6.1	1.42086 73734	−0.05198 25290	1.27240 6357	0.87327 0793
6.2	1.41870 68241	−0.03587 30193	1.26612 0373	0.87489 2347
6.3	1.41817 40348	−0.01988 82206	1.26002 4184	0.87647 2150
6.4	1.41922 29740	−0.00418 14110	1.25411 5417	0.87801 1816
6.5	1.42179 42744	+0.01110 15195	1.24839 1155	0.87951 2881
6.6	1.42581 61486	0.02582 31381	1.24284 8032	0.88097 6797
6.7	1.43120 53853	0.03985 54400	1.23748 2309	0.88240 4955
6.8	1.43786 84161	0.05308 07167	1.23228 9952	0.88379 8662
6.9	1.44570 24427	0.06539 23140	1.22726 6684	0.88515 9176
7.0	1.45459 66142	0.07669 52785	1.22240 8053	0.88648 7675
	$\begin{bmatrix} (-4)5 \\ 7 \end{bmatrix}$	$\begin{bmatrix} (-4)4 \\ 8 \end{bmatrix}$	$\begin{bmatrix} (-4)6 \\ 7 \end{bmatrix}$	$\begin{bmatrix} (-5)6 \\ 6 \end{bmatrix}$

SINE, COSINE AND EXPONENTIAL INTEGRALS

Table 5.1

x	$\mathrm{Si}(x)$	$\mathrm{Ci}(x)$	$xe^{-x}\mathrm{Ei}(x)$	$xe^{x}E_1(x)$
7.0	1.45459 66142	0.07669 52785	1.22240 8053	0.88648 7675
7.1	1.46443 32441	0.08690 68881	1.21770 9472	0.88778 5294
7.2	1.47508 90554	0.09595 70643	1.21316 6264	0.88905 3119
7.3	1.48643 64451	0.10378 86664	1.20877 3699	0.89029 2173
7.4	1.49834 47533	0.11035 76658	1.20452 7026	0.89150 3440
7.5	1.51068 15309	0.11563 32032	1.20042 1500	0.89268 7854
7.6	1.52331 37914	0.11959 75293	1.19645 2401	0.89384 6312
7.7	1.53610 92381	0.12224 58319	1.19261 5063	0.89497 9666
7.8	1.54893 74581	0.12358 59542	1.18890 4881	0.89608 8737
7.9	1.56167 10702	0.12363 80071	1.18531 7334	0.89717 4302
8.0	1.57418 68217	0.12243 38825	1.18184 7987	0.89823 7113
8.1	1.58636 66225	0.12001 66733	1.17849 2509	0.89927 7888
8.2	1.59809 85106	0.11644 00055	1.17524 6676	0.90029 7306
8.3	1.60927 75419	0.11176 72931	1.17210 6376	0.90129 6033
8.4	1.61980 65968	0.10607 09196	1.16906 7617	0.90227 4695
8.5	1.62959 70996	0.09943 13586	1.16612 6526	0.90323 3900
8.6	1.63856 96454	0.09193 62396	1.16327 9354	0.90417 4228
8.7	1.64665 45309	0.08367 93696	1.16052 2476	0.90509 6235
8.8	1.65379 21861	0.07475 97196	1.15785 2390	0.90600 0459
8.9	1.65993 35052	0.06528 03850	1.15526 5719	0.90688 7415
9.0	1.66504 00758	0.05534 75313	1.15275 9209	0.90775 7602
9.1	1.66908 43056	0.04506 93325	1.15032 9724	0.90861 1483
9.2	1.67204 94480	0.03455 49134	1.14797 4251	0.90944 9530
9.3	1.67392 95283	0.02391 33045	1.14568 9889	0.91027 2177
9.4	1.67472 91725	0.01325 24187	1.14347 3855	0.91107 9850
9.5	1.67446 33423	+0.00267 80588	1.14132 3476	0.91187 2958
9.6	1.67315 69801	−0.00770 70361	1.13923 6185	0.91265 1897
9.7	1.67084 45697	−0.01780 40977	1.13720 9523	0.91341 7043
9.8	1.66756 96169	−0.02751 91811	1.13524 1130	0.91416 8766
9.9	1.66338 40566	−0.03676 39563	1.13332 8746	0.91490 7418
10.0	1.65834 75942	−0.04545 64330	1.13147 0205	0.91563 3339
	$\begin{bmatrix}(-4)1\\7\end{bmatrix}$	$\begin{bmatrix}(-4)2\\7\end{bmatrix}$	$\begin{bmatrix}(-5)2\\5\end{bmatrix}$	$\begin{bmatrix}(-6)4\\4\end{bmatrix}$

Table 5.2

SINE, COSINE AND EXPONENTIAL INTEGRALS FOR LARGE ARGUMENTS

x^{-1}	$xf(x)$	$x^2g(x)$	$xe^{-x}\mathrm{Ei}(x)$	$xe^{x}E_1(x)$	$\langle x\rangle$
0.100	0.98191 0351	0.94885 39	1.13147 021	0.91563 33394	10
0.095	0.98353 4427	0.95323 18	1.12249 671	0.91925 68286	11
0.090	0.98509 9171	0.95748 44	1.11389 377	0.92293 15844	11
0.085	0.98660 1776	0.96160 17	1.10564 739	0.92665 90998	12
0.080	0.98803 9405	0.96557 23	1.09773 775	0.93044 09399	13
0.075	0.98940 9188	0.96938 56	1.09014 087	0.93427 87466	13
0.070	0.99070 8244	0.97302 98	1.08283 054	0.93817 42450	14
0.065	0.99193 3695	0.97649 35	1.07578 038	0.94212 92486	15
0.060	0.99308 2682	0.97976 47	1.06896 548	0.94614 56670	17
0.055	0.99415 2385	0.98283 17	1.06236 365	0.95022 55126	18
0.050	0.99514 0052	0.98568 24	1.05595 591	0.95437 09099	20
0.045	0.99604 3013	0.98830 52	1.04972 640	0.95858 41038	22
0.040	0.99685 8722	0.99068 81	1.04366 194	0.96286 74711	25
0.035	0.99758 4771	0.99282 12	1.03775 135	0.96722 35311	29
0.030	0.99821 8937	0.99469 37	1.03198 503	0.97165 49596	33
0.025	0.99875 9204	0.99629 57	1.02635 451	0.97616 46031	40
0.020	0.99920 3795	0.99761 89	1.02085 228	0.98075 54965	50
0.015	0.99955 1207	0.99865 60	1.01547 157	0.98543 08813	67
0.010	0.99980 0239	0.99940 12	1.01020 625	0.99019 42287	100
0.005	0.99995 0015	0.99985 01	1.00505 077	0.99504 92646	200
0.000	1.00000 0000	1.00000 00	1.00000 000	1.00000 00000	∞
	$\begin{bmatrix}(-5)1\\5\end{bmatrix}$	$\begin{bmatrix}(-5)4\\4\end{bmatrix}$	$\begin{bmatrix}(-5)5\\6\end{bmatrix}$	$\begin{bmatrix}(-5)1\\6\end{bmatrix}$	

$$\mathrm{Si}(x)=\frac{\pi}{2}-f(x)\cos x-g(x)\sin x \qquad \mathrm{Ci}(x)=f(x)\sin x-g(x)\cos x$$

$$\frac{\pi}{2}=1.57079\ 63268$$

$$\langle x\rangle=\text{nearest integer to } x.$$

See Example 3.

Table 5.3   **SINE AND COSINE INTEGRALS FOR ARGUMENTS** πx

x	$\text{Si}(\pi x)$	$\text{Cin}(\pi x)$	x	$\text{Si}(\pi x)$	$\text{Cin}(\pi x)$
0.0	0.00000 00	0.00000 00	5.0	1.63396 48	3.32742 23
0.1	0.31244 18	0.02457 28	5.1	1.63088 98	3.36670 50
0.2	0.61470 01	0.09708 67	5.2	1.62211 92	3.40335 81
0.3	0.89718 92	0.21400 75	5.3	1.60871 21	3.43582 68
0.4	1.15147 74	0.36970 10	5.4	1.59212 99	3.46297 82
0.5	1.37076 22	0.55679 77	5.5	1.57408 24	3.48419 47
0.6	1.55023 35	0.76666 63	5.6	1.55635 75	3.49941 45
0.7	1.68729 94	0.98995 93	5.7	1.54064 82	3.50911 89
0.8	1.78166 12	1.21719 42	5.8	1.52839 53	3.51426 89
0.9	1.83523 65	1.43932 68	5.9	1.52065 96	3.51619 81
1.0	1.85193 70	1.64827 75	6.0	1.51803 39	3.51647 44
1.1	1.83732 28	1.83737 48	6.1	1.52060 20	3.51674 38
1.2	1.79815 90	2.00168 51	6.2	1.52794 77	3.51857 25
1.3	1.74191 10	2.13821 22	6.3	1.53921 04	3.52330 06
1.4	1.67621 68	2.24595 41	6.4	1.55318 17	3.53192 30
1.5	1.60837 27	2.32581 82	6.5	1.56843 12	3.54500 55
1.6	1.54487 36	2.38040 96	6.6	1.58344 97	3.56264 55
1.7	1.49103 51	2.41370 98	6.7	1.59679 62	3.58447 72
1.8	1.45072 37	2.43067 75	6.8	1.60723 30	3.60972 10
1.9	1.42621 05	2.43680 30	6.9	1.61383 85	3.63727 15
2.0	1.41815 16	2.43765 34	7.0	1.61608 55	3.66581 26
2.1	1.42569 13	2.43844 23	7.1	1.61388 08	3.69395 05
2.2	1.44667 38	2.44365 73	7.2	1.60756 18	3.72034 97
2.3	1.47794 03	2.45676 95	7.3	1.59785 21	3.74385 98
2.4	1.51568 40	2.48004 47	7.4	1.58578 13	3.76362 13
2.5	1.55583 10	2.51446 40	7.5	1.57257 88	3.77914 01
2.6	1.59441 60	2.55975 53	7.6	1.55954 96	3.79032 64
2.7	1.62792 16	2.61452 59	7.7	1.54794 81	3.79749 22
2.8	1.65355 62	2.67647 93	7.8	1.53885 84	3.80131 21
2.9	1.66945 05	2.74269 41	7.9	1.53309 50	3.80274 91
3.0	1.67476 18	2.80993 76	8.0	1.53113 13	3.80295 56
3.1	1.66968 11	2.87498 49	8.1	1.53306 26	3.80315 83
3.2	1.65535 02	2.93491 77	8.2	1.53860 67	3.80453 88
3.3	1.63369 82	2.98737 63	8.3	1.54713 99	3.80812 16
3.4	1.60721 88	3.03074 73	8.4	1.55776 52	3.81467 97
3.5	1.57870 92	3.06427 25	8.5	1.56940 54	3.82466 68
3.6	1.55099 62	3.08807 51	8.6	1.58091 06	3.83818 15
3.7	1.52667 49	3.10310 38	8.7	1.59117 06	3.85496 61
3.8	1.50788 19	3.11100 53	8.8	1.59922 11	3.87444 05
3.9	1.49612 20	3.11393 95	8.9	1.60433 29	3.89576 52
4.0	1.49216 12	3.11435 65	9.0	1.60607 69	3.91792 84
4.1	1.49599 24	3.11475 82	9.1	1.60435 85	3.93984 77
4.2	1.50687 40	3.11746 60	9.2	1.59942 00	3.96047 61
4.3	1.52343 40	3.12441 61	9.3	1.59180 91	3.97890 22
4.4	1.54382 74	3.13699 91	9.4	1.58232 00	3.99443 58
4.5	1.56593 04	3.15595 79	9.5	1.57191 16	4.00666 94
4.6	1.58755 15	3.18134 84	9.6	1.56161 12	4.01551 22
4.7	1.60664 04	3.21256 74	9.7	1.55241 46	4.02119 22
4.8	1.62147 45	3.24843 85	9.8	1.54519 00	4.02422 80
4.9	1.63080 69	3.28734 92	9.9	1.54059 74	4.02537 29
5.0	1.63396 48	3.32742 23	10.0	1.53902 91	4.02553 78
	$\begin{bmatrix}(-3)5\\8\end{bmatrix}$	$\begin{bmatrix}(-3)6\\8\end{bmatrix}$		$\begin{bmatrix}(-4)7\\7\end{bmatrix}$	$\begin{bmatrix}(-4)7\\7\end{bmatrix}$

$$\text{Ci}(\pi x) = \gamma + \ln \pi + \ln x - \text{Cin}(\pi x) \qquad\qquad \gamma + \ln \pi = 1.72194\ 55508$$

$\text{Si}(n\pi)$ are maximum values of $\text{Si}(x)$ if $n>0$ is odd, and minimum values if $n>0$ is even.

$\text{Ci}\left[\left(n+\frac{1}{2}\right)\pi\right]$ are maximum values of $\text{Ci}(x)$ if $n>0$ is even, and minimum values if $n>0$ is odd. We have

$$\text{Si}(n\pi) \sim \frac{\pi}{2} - \frac{(-1)^n}{n\pi}\left[1 - \frac{2!}{n^2\pi^2} + \frac{4!}{n^4\pi^4} - \cdots\right] \quad (n\to\infty)$$

$$\text{Ci}\left[\left(n+\frac{1}{2}\right)\pi\right] \sim \frac{(-1)^n}{\left(n+\frac{1}{2}\right)\pi}\left[1 - \frac{2!}{\left(n+\frac{1}{2}\right)^2\pi^2} + \frac{4!}{\left(n+\frac{1}{2}\right)^4\pi^4} - \cdots\right] \quad (n\to\infty)$$

EXPONENTIAL INTEGRALS $E_n(x)$ Table 5.4

x	$E_2(x) - x \ln x$	$E_3(x)$	$E_4(x)$	$E_{10}(x)$	$E_{20}(x)$
0.00	1.00000 00	0.50000 00	0.33333 33	0.11111 11	0.05263 16
0.01	0.99572 22	0.49027 66	0.32838 24	0.10986 82	0.05207 90
0.02	0.99134 50	0.48096 83	0.32352 64	0.10863 95	0.05153 21
0.03	0.98686 87	0.47199 77	0.31876 19	0.10742 46	0.05099 11
0.04	0.98229 39	0.46332 39	0.31408 55	0.10622 36	0.05045 58
0.05	0.97762 11	0.45491 88	0.30949 45	0.10503 63	0.04992 60
0.06	0.97285 08	0.44676 09	0.30498 63	0.10386 24	0.04940 19
0.07	0.96798 34	0.43883 27	0.30055 85	0.10270 18	0.04888 33
0.08	0.96301 94	0.43111 97	0.29620 89	0.10155 44	0.04837 02
0.09	0.95795 93	0.42360 96	0.29193 54	0.10042 00	0.04786 24
0.10	0.95280 35	0.41629 15	0.28773 61	0.09929 84	0.04736 00
0.11	0.94755 26	0.40915 57	0.28360 90	0.09818 96	0.04686 29
0.12	0.94220 71	0.40219 37	0.27955 24	0.09709 34	0.04637 10
0.13	0.93676 72	0.39539 77	0.27556 46	0.09600 95	0.04588 43
0.14	0.93123 36	0.38876 07	0.27164 39	0.09493 80	0.04540 27
0.15	0.92560 67	0.38227 61	0.26778 89	0.09387 86	0.04492 62
0.16	0.91988 70	0.37593 80	0.26399 79	0.09283 12	0.04445 47
0.17	0.91407 48	0.36974 08	0.26026 96	0.09179 56	0.04398 82
0.18	0.90817 06	0.36367 95	0.25660 26	0.09077 18	0.04352 66
0.19	0.90217 50	0.35774 91	0.25299 56	0.08975 95	0.04306 98
0.20	0.89608 82	0.35194 53	0.24944 72	0.08875 87	0.04261 79
0.21	0.88991 09	0.34626 38	0.24595 63	0.08776 93	0.04217 07
0.22	0.88364 33	0.34070 05	0.24252 16	0.08679 10	0.04172 82
0.23	0.87728 60	0.33525 18	0.23914 19	0.08582 38	0.04129 03
0.24	0.87083 93	0.32991 42	0.23581 62	0.08486 75	0.04085 71
0.25	0.86430 37	0.32468 41	0.23254 32	0.08392 20	0.04042 85
0.26	0.85767 97	0.31955 85	0.22932 21	0.08298 72	0.04000 43
0.27	0.85096 76	0.31453 43	0.22615 17	0.08206 30	0.03958 46
0.28	0.84416 78	0.30960 86	0.22303 11	0.08114 92	0.03916 93
0.29	0.83728 08	0.30477 87	0.21995 93	0.08024 57	0.03875 84
0.30	0.83030 71	0.30004 18	0.21693 52	0.07935 24	0.03835 18
0.31	0.82324 69	0.29539 56	0.21395 81	0.07846 93	0.03794 95
0.32	0.81610 07	0.29083 74	0.21102 70	0.07759 60	0.03755 15
0.33	0.80886 90	0.28636 52	0.20814 11	0.07673 27	0.03715 76
0.34	0.80155 21	0.28197 65	0.20529 94	0.07587 90	0.03676 78
0.35	0.79415 04	0.27766 93	0.20250 13	0.07503 50	0.03638 22
0.36	0.78666 44	0.27344 16	0.19974 58	0.07420 06	0.03600 06
0.37	0.77909 43	0.26929 13	0.19703 22	0.07337 55	0.03562 31
0.38	0.77144 07	0.26521 65	0.19435 97	0.07255 97	0.03524 95
0.39	0.76370 39	0.26121 55	0.19172 76	0.07175 31	0.03487 98
0.40	0.75588 43	0.25728 64	0.18913 52	0.07095 57	0.03451 40
0.41	0.74798 23	0.25342 76	0.18658 16	0.07016 71	0.03415 21
0.42	0.73999 82	0.24963 73	0.18406 64	0.06938 75	0.03379 39
0.43	0.73193 24	0.24591 41	0.18158 87	0.06861 67	0.03343 96
0.44	0.72378 54	0.24225 63	0.17914 79	0.06785 45	0.03308 89
0.45	0.71555 75	0.23866 25	0.17674 33	0.06710 09	0.03274 20
0.46	0.70724 91	0.23513 13	0.17437 44	0.06635 58	0.03239 87
0.47	0.69886 05	0.23166 12	0.17204 05	0.06561 91	0.03205 90
0.48	0.69039 21	0.22825 08	0.16974 10	0.06489 07	0.03172 29
0.49	0.68184 43	0.22489 90	0.16747 53	0.06417 04	0.03139 03
0.50	0.67321 75	0.22160 44	0.16524 28	0.06345 83	0.03106 12
	$\begin{bmatrix} (-5)1 \\ 3 \end{bmatrix}$	$\begin{bmatrix} (-5)5 \\ 6 \end{bmatrix}$	$\begin{bmatrix} (-5)1 \\ 4 \end{bmatrix}$	$\begin{bmatrix} (-6)2 \\ 3 \end{bmatrix}$	$\begin{bmatrix} (-7)7 \\ 3 \end{bmatrix}$

See Examples 4—6.

Table 5.4   **EXPONENTIAL INTEGRALS** $E_n(x)$

x	$E_2(x)$	$E_3(x)$	$E_4(x)$	$E_{10}(x)$	$E_{20}(x)$
0.50	0.32664 39	0.22160 44	0.16524 28	0.06345 83	0.03106 12
0.51	0.32110 62	0.21836 57	0.16304 30	0.06275 42	0.03073 56
0.52	0.31568 63	0.21518 18	0.16087 53	0.06205 80	0.03041 34
0.53	0.31038 07	0.21205 16	0.15873 92	0.06136 96	0.03009 46
0.54	0.30518 62	0.20897 39	0.15663 41	0.06068 89	0.02977 91
0.55	0.30009 96	0.20594 75	0.15455 96	0.06001 59	0.02946 70
0.56	0.29511 79	0.20297 15	0.15251 50	0.05935 05	0.02915 81
0.57	0.29023 82	0.20004 48	0.15050 00	0.05869 25	0.02885 25
0.58	0.28545 78	0.19716 64	0.14851 39	0.05804 19	0.02855 01
0.59	0.28077 39	0.19433 53	0.14655 65	0.05739 86	0.02825 08
0.60	0.27618 39	0.19155 06	0.14462 71	0.05676 26	0.02795 48
0.61	0.27168 55	0.18881 14	0.14272 53	0.05613 36	0.02766 18
0.62	0.26727 61	0.18611 66	0.14085 07	0.05551 18	0.02737 19
0.63	0.26295 35	0.18346 56	0.13900 28	0.05489 69	0.02708 50
0.64	0.25871 54	0.18085 73	0.13718 13	0.05428 89	0.02680 12
0.65	0.25455 97	0.17829 10	0.13538 55	0.05368 77	0.02652 04
0.66	0.25048 44	0.17576 58	0.13361 53	0.05309 33	0.02624 25
0.67	0.24648 74	0.17328 10	0.13187 01	0.05250 55	0.02596 75
0.68	0.24256 67	0.17083 58	0.13014 95	0.05192 43	0.02569 54
0.69	0.23872 06	0.16842 94	0.12845 33	0.05134 97	0.02542 62
0.70	0.23494 71	0.16606 12	0.12678 08	0.05078 15	0.02515 98
0.71	0.23124 46	0.16373 03	0.12513 19	0.05021 96	0.02489 62
0.72	0.22761 14	0.16143 60	0.12350 61	0.04966 40	0.02463 53
0.73	0.22404 57	0.15917 78	0.12190 31	0.04911 47	0.02437 72
0.74	0.22054 61	0.15695 49	0.12032 24	0.04857 15	0.02412 19
0.75	0.21711 09	0.15476 67	0.11876 38	0.04803 44	0.02386 92
0.76	0.21373 88	0.15261 25	0.11722 70	0.04750 33	0.02361 91
0.77	0.21042 82	0.15049 17	0.11571 15	0.04697 81	0.02337 17
0.78	0.20717 77	0.14840 37	0.11421 70	0.04645 88	0.02312 69
0.79	0.20398 60	0.14634 79	0.11274 33	0.04594 53	0.02288 46
0.80	0.20085 17	0.14432 38	0.11129 00	0.04543 76	0.02264 49
0.81	0.19777 36	0.14233 07	0.10985 67	0.04493 56	0.02240 78
0.82	0.19475 04	0.14036 81	0.10844 33	0.04443 91	0.02217 31
0.83	0.19178 10	0.13843 55	0.10704 93	0.04394 82	0.02194 08
0.84	0.18886 41	0.13653 24	0.10567 44	0.04346 28	0.02171 11
0.85	0.18599 86	0.13465 81	0.10431 85	0.04298 29	0.02148 37
0.86	0.18318 33	0.13281 22	0.10298 12	0.04250 82	0.02125 87
0.87	0.18041 73	0.13099 43	0.10166 22	0.04203 89	0.02103 61
0.88	0.17769 94	0.12920 37	0.10036 12	0.04157 49	0.02081 58
0.89	0.17502 87	0.12744 01	0.09907 80	0.04111 60	0.02059 78
0.90	0.17240 41	0.12570 30	0.09781 23	0.04066 22	0.02038 21
0.91	0.16982 47	0.12399 19	0.09656 39	0.04021 35	0.02016 87
0.92	0.16728 95	0.12230 63	0.09533 24	0.03976 98	0.01995 75
0.93	0.16479 77	0.12064 59	0.09411 77	0.03933 11	0.01974 86
0.94	0.16234 82	0.11901 02	0.09291 94	0.03889 73	0.01954 18
0.95	0.15994 04	0.11739 88	0.09173 74	0.03846 83	0.01933 72
0.96	0.15757 32	0.11581 13	0.09057 13	0.03804 41	0.01913 47
0.97	0.15524 59	0.11424 72	0.08942 11	0.03762 46	0.01893 44
0.98	0.15295 78	0.11270 63	0.08828 63	0.03720 98	0.01873 62
0.99	0.15070 79	0.11118 80	0.08716 69	0.03679 96	0.01854 01
1.00	0.14849 55	0.10969 20	0.08606 25	0.03639 40	0.01834 60
	$\left[\begin{smallmatrix}(-5)1\\4\end{smallmatrix}\right]$	$\left[\begin{smallmatrix}(-6)7\\3\end{smallmatrix}\right]$	$\left[\begin{smallmatrix}(-6)4\\3\end{smallmatrix}\right]$	$\left[\begin{smallmatrix}(-6)1\\3\end{smallmatrix}\right]$	$\left[\begin{smallmatrix}(-7)4\\3\end{smallmatrix}\right]$

EXPONENTIAL INTEGRALS $E_n(x)$ Table 5.4

x	$E_2(x)$	$E_3(x)$	$E_4(x)$	$E_{10}(x)$	$E_{20}(x)$
1.00	0.14849 55	0.10969 20	0.08606 25	0.03639 40	0.01834 60
1.01	0.14631 99	0.10821 79	0.08497 30	0.03599 29	0.01815 39
1.02	0.14418 04	0.10676 54	0.08389 81	0.03559 63	0.01796 39
1.03	0.14207 63	0.10533 42	0.08283 76	0.03520 41	0.01777 59
1.04	0.14000 68	0.10392 38	0.08179 13	0.03481 63	0.01758 98
1.05	0.13797 13	0.10253 39	0.08075 90	0.03443 28	0.01740 57
1.06	0.13596 91	0.10116 43	0.07974 06	0.03405 35	0.01722 35
1.07	0.13399 96	0.09981 45	0.07873 57	0.03367 85	0.01704 33
1.08	0.13206 22	0.09848 42	0.07774 42	0.03330 77	0.01686 49
1.09	0.13015 62	0.09717 31	0.07676 59	0.03294 10	0.01668 84
1.10	0.12828 11	0.09588 09	0.07580 07	0.03257 84	0.01651 37
1.11	0.12643 62	0.09460 74	0.07484 83	0.03221 98	0.01634 09
1.12	0.12462 10	0.09335 21	0.07390 85	0.03186 52	0.01616 99
1.13	0.12283 50	0.09211 49	0.07298 12	0.03151 45	0.01600 07
1.14	0.12107 75	0.09089 53	0.07206 61	0.03116 78	0.01583 33
1.15	0.11934 81	0.08969 32	0.07116 32	0.03082 49	0.01566 76
1.16	0.11764 62	0.08850 83	0.07027 22	0.03048 58	0.01550 37
1.17	0.11597 14	0.08734 02	0.06939 30	0.03015 05	0.01534 14
1.18	0.11432 31	0.08618 88	0.06852 53	0.02981 89	0.01518 09
1.19	0.11270 08	0.08505 37	0.06766 91	0.02949 10	0.01502 21
1.20	0.11110 41	0.08393 47	0.06682 42	0.02916 68	0.01486 49
1.21	0.10953 25	0.08283 15	0.06599 04	0.02884 61	0.01470 94
1.22	0.10798 55	0.08174 39	0.06516 75	0.02852 90	0.01455 55
1.23	0.10646 27	0.08067 17	0.06435 55	0.02821 55	0.01440 32
1.24	0.10496 37	0.07961 46	0.06355 40	0.02790 54	0.01425 26
1.25	0.10348 81	0.07857 23	0.06276 31	0.02759 88	0.01410 35
1.26	0.10203 53	0.07754 47	0.06198 25	0.02729 55	0.01395 59
1.27	0.10060 51	0.07653 16	0.06121 22	0.02699 57	0.01381 00
1.28	0.09919 70	0.07553 26	0.06045 19	0.02669 91	0.01366 55
1.29	0.09781 06	0.07454 76	0.05970 15	0.02640 59	0.01352 26
1.30	0.09644 55	0.07357 63	0.05896 09	0.02611 59	0.01338 11
1.31	0.09510 15	0.07261 86	0.05822 99	0.02582 91	0.01324 12
1.32	0.09377 80	0.07167 42	0.05750 85	0.02554 55	0.01310 27
1.33	0.09247 47	0.07074 29	0.05679 64	0.02526 51	0.01296 57
1.34	0.09119 13	0.06982 46	0.05609 36	0.02498 78	0.01283 01
1.35	0.08992 75	0.06891 91	0.05539 98	0.02471 35	0.01269 59
1.36	0.08868 29	0.06802 60	0.05471 51	0.02444 23	0.01256 31
1.37	0.08745 71	0.06714 53	0.05403 93	0.02417 41	0.01243 17
1.38	0.08624 99	0.06627 68	0.05337 22	0.02390 88	0.01230 17
1.39	0.08506 10	0.06542 03	0.05271 37	0.02364 65	0.01217 31
1.40	0.08388 99	0.06457 55	0.05206 37	0.02338 72	0.01204 58
1.41	0.08273 65	0.06374 24	0.05142 22	0.02313 06	0.01191 98
1.42	0.08160 04	0.06292 07	0.05078 89	0.02287 70	0.01179 52
1.43	0.08048 13	0.06211 04	0.05016 37	0.02262 61	0.01167 19
1.44	0.07937 89	0.06131 11	0.04954 66	0.02237 80	0.01154 99
1.45	0.07829 30	0.06052 27	0.04893 74	0.02213 27	0.01142 91
1.46	0.07722 33	0.05974 52	0.04833 61	0.02189 01	0.01130 96
1.47	0.07616 94	0.05897 82	0.04774 25	0.02165 01	0.01119 14
1.48	0.07513 13	0.05822 17	0.04715 65	0.02141 28	0.01107 44
1.49	0.07410 85	0.05747 55	0.04657 80	0.02117 82	0.01095 86
1.50	0.07310 08	0.05673 95	0.04600 70	0.02094 61	0.01084 40
1.51	0.07210 80	0.05601 35	0.04544 32	0.02071 67	0.01073 07
1.52	0.07112 98	0.05529 73	0.04488 67	0.02048 97	0.01061 85
1.53	0.07016 60	0.05459 08	0.04433 72	0.02026 53	0.01050 75
1.54	0.06921 64	0.05389 39	0.04379 48	0.02004 33	0.01039 77
1.55	0.06828 07	0.05320 64	0.04325 93	0.01982 38	0.01028 90
1.56	0.06735 87	0.05252 83	0.04273 07	0.01960 67	0.01018 15
1.57	0.06645 02	0.05185 92	0.04220 87	0.01939 21	0.01007 50
1.58	0.06555 49	0.05119 92	0.04169 35	0.01917 98	0.00996 97
1.59	0.06467 26	0.05054 81	0.04118 47	0.01896 98	0.00986 56
1.60	0.06380 32	0.04990 57	0.04068 25	0.01876 22	0.00976 24
	$\begin{bmatrix} (-6)5 \\ 3 \end{bmatrix}$	$\begin{bmatrix} (-6)3 \\ 3 \end{bmatrix}$	$\begin{bmatrix} (-6)2 \\ 3 \end{bmatrix}$	$\begin{bmatrix} (-7)6 \\ 3 \end{bmatrix}$	$\begin{bmatrix} (-7)3 \\ 3 \end{bmatrix}$

Table 5.4   **EXPONENTIAL INTEGRALS $E_n(x)$**

x	$E_2(x)$	$E_3(x)$	$E_4(x)$	$E_{10}(x)$	$E_{20}(x)$
1.60	0.06380 32	0.04990 57	0.04068 25	0.01876 22	0.00976 24
1.61	0.06294 64	0.04927 20	0.04018 66	0.01855 68	0.00966 04
1.62	0.06210 20	0.04864 67	0.03969 70	0.01835 38	0.00955 95
1.63	0.06126 98	0.04802 99	0.03921 36	0.01815 30	0.00945 96
1.64	0.06044 97	0.04742 13	0.03873 64	0.01795 43	0.00936 07
1.65	0.05964 13	0.04682 09	0.03826 52	0.01775 79	0.00926 29
1.66	0.05884 46	0.04622 84	0.03779 99	0.01756 37	0.00916 61
1.67	0.05805 94	0.04564 39	0.03734 06	0.01737 16	0.00907 03
1.68	0.05728 54	0.04506 72	0.03688 70	0.01718 16	0.00897 56
1.69	0.05652 26	0.04449 82	0.03643 92	0.01699 37	0.00888 18
1.70	0.05577 06	0.04393 67	0.03599 70	0.01680 79	0.00878 90
1.71	0.05502 94	0.04338 27	0.03556 04	0.01662 42	0.00869 72
1.72	0.05429 88	0.04283 61	0.03512 93	0.01644 24	0.00860 63
1.73	0.05357 86	0.04229 67	0.03470 37	0.01626 27	0.00851 64
1.74	0.05286 86	0.04176 45	0.03428 34	0.01608 50	0.00842 74
1.75	0.05216 87	0.04123 93	0.03386 84	0.01590 92	0.00833 94
1.76	0.05147 88	0.04072 11	0.03345 86	0.01573 54	0.00825 22
1.77	0.05079 86	0.04020 97	0.03305 39	0.01556 34	0.00816 60
1.78	0.05012 81	0.03970 51	0.03265 44	0.01539 34	0.00808 07
1.79	0.04946 70	0.03920 71	0.03225 98	0.01522 53	0.00799 63
1.80	0.04881 53	0.03871 57	0.03187 02	0.01505 90	0.00791 28
1.81	0.04817 27	0.03823 08	0.03148 55	0.01489 45	0.00783 02
1.82	0.04753 92	0.03775 22	0.03110 56	0.01473 18	0.00774 84
1.83	0.04691 46	0.03728 00	0.03073 04	0.01457 10	0.00766 74
1.84	0.04629 87	0.03681 39	0.03035 99	0.01441 19	0.00758 74
1.85	0.04569 15	0.03635 40	0.02999 41	0.01425 46	0.00750 81
1.86	0.04509 28	0.03590 01	0.02963 28	0.01409 90	0.00742 97
1.87	0.04450 24	0.03545 21	0.02927 61	0.01394 51	0.00735 21
1.88	0.04392 03	0.03501 00	0.02892 38	0.01379 29	0.00727 53
1.89	0.04334 63	0.03457 37	0.02857 59	0.01364 24	0.00719 93
1.90	0.04278 03	0.03414 30	0.02823 23	0.01349 35	0.00712 42
1.91	0.04222 22	0.03371 80	0.02789 30	0.01334 63	0.00704 98
1.92	0.04167 18	0.03329 86	0.02755 79	0.01320 07	0.00697 62
1.93	0.04112 91	0.03288 46	0.02722 70	0.01305 67	0.00690 33
1.94	0.04059 38	0.03247 59	0.02690 02	0.01291 43	0.00683 12
1.95	0.04006 60	0.03207 27	0.02657 75	0.01277 34	0.00675 99
1.96	0.03954 55	0.03167 46	0.02625 87	0.01263 41	0.00668 93
1.97	0.03903 22	0.03128 17	0.02594 40	0.01249 64	0.00661 95
1.98	0.03852 59	0.03089 39	0.02563 31	0.01236 01	0.00655 04
1.99	0.03802 67	0.03051 12	0.02532 61	0.01222 54	0.00648 20
2.00	0.03753 43	0.03013 34	0.02502 28	0.01209 21	0.00641 43
	$\left[\begin{matrix}(-6)2\\3\end{matrix}\right]$	$\left[\begin{matrix}(-6)1\\3\end{matrix}\right]$	$\left[\begin{matrix}(-7)8\\3\end{matrix}\right]$	$\left[\begin{matrix}(-7)3\\3\end{matrix}\right]$	$\left[\begin{matrix}(-7)1\\3\end{matrix}\right]$

Table 5.5   **EXPONENTIAL INTEGRALS $E_n(x)$ FOR LARGE ARGUMENTS**

x^{-1}	$(x+2)e^x E_2(x)$	$(x+3)e^x E_3(x)$	$(x+4)e^x E_4(x)$	$(x+10)e^x E_{10}(x)$	$(x+20)e^x E_{20}(x)$	$\langle x\rangle$
0.50	1.10937	1.11329	1.10937	1.07219	1.04270	2
0.45	1.09750	1.10285	1.10071	1.06926	1.04179	2
0.40	1.08533	1.09185	1.09136	1.06586	1.04067	3
0.35	1.07292	1.08026	1.08125	1.06187	1.03932	3
0.30	1.06034	1.06808	1.07031	1.05712	1.03762	3
0.25	1.04770	1.05536	1.05850	1.05138	1.03543	4
0.20	1.03522	1.04222	1.04584	1.04432	1.03249	5
0.15	1.02325	1.02895	1.03247	1.03550	1.02837	7
0.10	1.01240	1.01617	1.01889	1.02436	1.02222	10
0.09	1.01045	1.01377	1.01624	1.02182	1.02060	11
0.08	1.00861	1.01147	1.01366	1.01917	1.01883	13
0.07	1.00688	1.00927	1.01116	1.01642	1.01688	14
0.06	1.00528	1.00721	1.00878	1.01360	1.01472	17
0.05	1.00384	1.00531	1.00654	1.01074	1.01234	20
0.04	1.00258	1.00361	1.00451	1.00790	1.00973	25
0.03	1.00152	1.00217	1.00275	1.00516	1.00692	33
0.02	1.00071	1.00103	1.00133	1.00271	1.00401	50
0.01	1.00019	1.00027	1.00036	1.00081	1.00137	100
0.00	1.00000	1.00000	1.00000	1.00000	1.00000	∞
	$\left[\begin{matrix}(-4)1\\4\end{matrix}\right]$	$\left[\begin{matrix}(-5)7\\4\end{matrix}\right]$	$\left[\begin{matrix}(-4)1\\4\end{matrix}\right]$	$\left[\begin{matrix}(-4)3\\4\end{matrix}\right]$	$\left[\begin{matrix}(-4)3\\4\end{matrix}\right]$	

$\langle x\rangle$=nearest integer to x.

EXPONENTIAL INTEGRAL FOR COMPLEX ARGUMENTS   Table 5.6

$$ze^z E_1(z)$$

$y\backslash x$	$\mathscr{R}$ -19	$\mathscr{I}$	$\mathscr{R}$ -18	$\mathscr{I}$	$\mathscr{R}$ -17	$\mathscr{I}$	$\mathscr{R}$ -16	$\mathscr{I}$	$\mathscr{R}$ -15	$\mathscr{I}$
0	1.059305	0.000000	1.063087	0.000001	1.067394	0.000002	1.072345	0.000006	1.078103	0.000014
1	1.059090	0.003539	1.062827	0.004010	1.067073	0.004584	1.071942	0.005296	1.077584	0.006195
2	1.058456	0.007000	1.062061	0.007918	1.066135	0.009032	1.070774	0.010403	1.076102	0.012118
3	1.057431	0.010310	1.060829	0.011633	1.064636	0.013226	1.068925	0.015172	1.073783	0.017579
4	1.056058	0.013410	1.059190	0.015079	1.062657	0.017075	1.066508	0.019486	1.070793	0.022432
5	1.054391	0.016252	1.057215	0.018202	1.060297	0.020512	1.063659	0.023272	1.067318	0.026598
6	1.052490	0.018806	1.054981	0.020969	1.057655	0.023505	1.060510	0.026499	1.063538	0.030055
7	1.050413	0.021055	1.052565	0.023364	1.054829	0.026044	1.057187	0.029167	1.059610	0.032823
8	1.048217	0.022996	1.050037	0.025391	1.051905	0.028141	1.053795	0.031306	1.055664	0.034957
9	1.045956	0.024637	1.047458	0.027066	1.048958	0.029824	1.050421	0.032960	1.051797	0.036527
10	1.043672	0.025993	1.044880	0.028412	1.046045	0.031130	1.047129	0.034183	1.048081	0.037609
11	1.041402	0.027086	1.042345	0.029461	1.043212	0.032102	1.043967	0.035034	1.044559	0.038282
12	1.039177	0.027940	1.039882	0.030245	1.040490	0.032781	1.040965	0.035567	1.041259	0.038616
13	1.037018	0.028581	1.037515	0.030796	1.037901	0.033211	1.038140	0.035036	1.038192	0.038677
14	1.034942	0.029034	1.035259	0.031148	1.035456	0.033431	1.035501	0.035888	1.035359	0.038520
15	1.032959	0.029326	1.033123	0.031330	1.033162	0.033476	1.033049	0.035765	1.032754	0.038193
16	1.031076	0.029477	1.031110	0.031368	1.031019	0.033377	1.030780	0.035502	1.030365	0.037735
17	1.029296	0.029511	1.029222	0.031288	1.029025	0.033162	1.028685	0.035129	1.028180	0.037179
18	1.027620	0.029445	1.027456	0.031110	1.027174	0.032855	1.026756	0.034672	1.026183	0.036552
19	1.026046	0.029296	1.025809	0.030854	1.025459	0.032474	1.024981	0.034150	1.024360	0.035873
20	1.024570	0.029080	1.024275	0.030534	1.023872	0.032037	1.023349	0.033582	1.022695	0.035160

$y\backslash x$	-14		-13		-12		-11		-10	
0	1.084892	0.000037	1.093027	0.000092	1.102975	0.000232	1.115431	0.000577	1.131470	0.001426
1	1.084200	0.007359	1.092067	0.008913	1.101566	0.011063	1.113230	0.014169	1.127796	0.018879
2	1.082276	0.014306	1.089498	0.017161	1.098025	0.020981	1.108170	0.026241	1.120286	0.033700
3	1.079313	0.020604	1.085635	0.024471	1.092873	0.029507	1.101137	0.036189	1.110462	0.045218
4	1.075560	0.026075	1.080853	0.030637	1.086686	0.036422	1.093013	0.043843	1.099666	0.053451
5	1.071279	0.030642	1.075522	0.035599	1.079985	0.041724	1.084526	0.049336	1.088877	0.058817
6	1.066708	0.034303	1.069960	0.039405	1.073185	0.045552	1.076197	0.052967	1.078701	0.061886
7	1.062046	0.037117	1.064412	0.042169	1.066578	0.048115	1.068350	0.055093	1.069450	0.063225
8	1.057448	0.039174	1.059054	0.044041	1.060352	0.049644	1.061159	0.056057	1.061235	0.063322
9	1.053021	0.040580	1.053997	0.045176	1.054606	0.050359	1.054687	0.056158	1.054046	0.062566
10	1.048834	0.041444	1.049303	0.045719	1.049380	0.050452	1.048933	0.055640	1.047807	0.061249
11	1.044928	0.041867	1.044997	0.045801	1.044674	0.050084	1.043853	0.054695	1.042417	0.059584
12	1.041320	0.041938	1.041080	0.045531	1.040464	0.049384	1.039389	0.053465	1.037766	0.057719
13	1.038010	0.041734	1.037537	0.044999	1.036713	0.048452	1.035473	0.052056	1.033752	0.055758
14	1.034989	0.041321	1.034344	0.044277	1.033378	0.047365	1.032040	0.050547	1.030282	0.053773
15	1.032241	0.040751	1.031474	0.043422	1.030414	0.046180	1.029026	0.048991	1.027274	0.051808
16	1.029747	0.040066	1.028895	0.042477	1.027781	0.044941	1.026377	0.047428	1.024658	0.049894
17	1.027486	0.039301	1.026579	0.041475	1.025438	0.043679	1.024043	0.045883	1.022375	0.048049
18	1.025437	0.038481	1.024499	0.040444	1.023352	0.042417	1.021981	0.044374	1.020375	0.046282
19	1.023580	0.037629	1.022628	0.039401	1.021489	0.041170	1.020155	0.042912	1.018617	0.044599
20	1.021896	0.036759	1.020942	0.038361	1.019824	0.039950	1.018533	0.041505	1.017066	0.043001

$y\backslash x$	-9		-8		-7		-6		-5	
0	1.152759	0.003489	1.181848	0.008431	1.222408	0.020053	1.278884	0.046723	1.353831	0.105839
1	1.146232	0.026376	1.169677	0.038841	1.199049	0.060219	1.233798	0.097331	1.268723	0.160826
2	1.134679	0.044579	1.151385	0.060814	1.169639	0.085335	1.186778	0.122162	1.196351	0.175646
3	1.120694	0.057595	1.131255	0.074701	1.140733	0.098259	1.146266	0.130005	1.142853	0.170672
4	1.106249	0.065948	1.111968	0.082156	1.115404	0.102861	1.114273	0.128444	1.105376	0.158134
5	1.092564	0.070592	1.094818	0.085055	1.094475	0.102411	1.089952	0.122397	1.079407	0.143879
6	1.080246	0.072520	1.080188	0.084987	1.077672	0.099188	1.071684	0.114638	1.061236	0.130280
7	1.069494	0.072580	1.067987	0.083120	1.064339	0.094618	1.057935	0.106568	1.048279	0.118116
8	1.060276	0.071425	1.057920	0.080250	1.053778	0.089537	1.047493	0.098840	1.038838	0.107508
9	1.052450	0.069523	1.049645	0.076885	1.045382	0.084405	1.039464	0.091717	1.031806	0.098337
10	1.045832	0.067197	1.042834	0.073340	1.038659	0.079462	1.033205	0.085271	1.026459	0.090413
11	1.040241	0.064664	1.037210	0.069803	1.033231	0.074821	1.028260	0.079488	1.022317	0.083544
12	1.035508	0.062063	1.032539	0.066381	1.028808	0.070524	1.024300	0.074315	1.019052	0.077561
13	1.031490	0.059482	1.028638	0.063128	1.025171	0.066576	1.021090	0.069688	1.016439	0.072320
14	1.028065	0.056975	1.025359	0.060070	1.022152	0.062962	1.018458	0.065542	1.014319	0.067702
15	1.025132	0.054573	1.022583	0.057215	1.019626	0.059658	1.016277	0.061817	1.012577	0.063610
16	1.022608	0.052291	1.020219	0.054559	1.017494	0.056638	1.014452	0.058460	1.011130	0.059962
17	1.020426	0.050135	1.018192	0.052094	1.015681	0.053874	1.012912	0.055424	1.009915	0.056694
18	1.018530	0.048106	1.016444	0.049806	1.014129	0.051341	1.011600	0.052670	1.008887	0.053752
19	1.016874	0.046201	1.014929	0.047684	1.012790	0.049015	1.010476	0.050161	1.008009	0.051092
20	1.015422	0.044413	1.013607	0.045714	1.011629	0.046875	1.009505	0.047870	1.007254	0.048675

For $|z|>4$, linear interpolation will yield about four decimals, eight-point interpolation will yield about six decimals.

See Examples 9 – 10.

Table 5.6

EXPONENTIAL INTEGRAL FOR COMPLEX ARGUMENTS
$$ze^z E_1(z)$$

	$\mathscr{R}$	$\mathscr{I}$	$\mathscr{R}$	$\mathscr{I}$	$\mathscr{R}$	$\mathscr{I}$	$\mathscr{R}$	$\mathscr{I}$	$\mathscr{R}$	$\mathscr{I}$
$y\backslash x$	−4		−3		−2		−1		0	
0	1.438208	0.230161	1.483729	0.469232	1.340965	0.850337	0.697175	1.155727	0.577216	0.000000
1	1.287244	0.263705	1.251069	0.410413	1.098808	0.561916	0.813486	0.578697	0.621450	0.343378
2	1.185758	0.247356	1.136171	0.328439	1.032990	0.388428	0.896419	0.378838	0.798042	0.289091
3	1.123282	0.217835	1.080316	0.262814	1.013205	0.289366	0.936283	0.280906	0.875873	0.237665
4	1.085153	0.189003	1.051401	0.215118	1.006122	0.228399	0.957446	0.222612	0.916770	0.198713
5	1.061263	0.164466	1.035185	0.180487	1.003172	0.187857	0.969809	0.183963	0.940714	0.169481
6	1.045719	0.144391	1.025396	0.154746	1.001788	0.159189	0.977582	0.156511	0.955833	0.147129
7	1.035205	0.128073	1.019109	0.135079	1.001077	0.137939	0.982756	0.136042	0.965937	0.129646
8	1.027834	0.114732	1.014861	0.119660	1.000684	0.121599	0.986356	0.120218	0.972994	0.115678
9	1.022501	0.103711	1.011869	0.107294	1.000454	0.108665	0.988955	0.107634	0.978103	0.104303
10	1.018534	0.094502	1.009688	0.097181	1.000312	0.098184	0.990887	0.097396	0.981910	0.094885
11	1.015513	0.086718	1.008052	0.088770	1.000221	0.089525	0.992361	0.088911	0.984819	0.086975
12	1.013163	0.080069	1.006795	0.081673	1.000161	0.082255	0.993508	0.081769	0.987088	0.080245
13	1.011303	0.074333	1.005809	0.075609	1.000119	0.076067	0.994418	0.075676	0.988891	0.074457
14	1.009806	0.069340	1.005022	0.070371	1.000090	0.070738	0.995151	0.070419	0.990345	0.069429
15	1.008585	0.064959	1.004384	0.065803	1.000070	0.066102	0.995751	0.065838	0.991534	0.065024
16	1.007577	0.061086	1.003859	0.061786	1.000055	0.062032	0.996246	0.061812	0.992518	0.061135
17	1.006735	0.057640	1.003423	0.058227	1.000043	0.058432	0.996661	0.058246	0.993342	0.057677
18	1.006025	0.054555	1.003057	0.055052	1.000035	0.055224	0.997011	0.055066	0.994038	0.054583
19	1.005420	0.051779	1.002747	0.052202	1.000028	0.052349	0.997309	0.052214	0.994631	0.051801
20	1.004902	0.049267	1.002481	0.049631	1.000023	0.049757	0.997565	0.049640	0.995140	0.049284

	$\mathscr{R}$	$\mathscr{I}$	$\mathscr{R}$	$\mathscr{I}$	$\mathscr{R}$	$\mathscr{I}$	$\mathscr{R}$	$\mathscr{I}$	$\mathscr{R}$	$\mathscr{I}$
$y\backslash x$	1		2		3		4		5	
0	0.596347	0.000000	0.722657	0.000000	0.786251	0.000000	0.825383	0.000000	0.852111	0.000000
1	0.673321	0.147864	0.747012	0.075661	0.797036	0.045686	0.831126	0.030619	0.855544	0.021985
2	0.777514	0.186570	0.796965	0.118228	0.823055	0.078753	0.846097	0.055494	0.864880	0.040999
3	0.847460	0.181226	0.844361	0.132252	0.853176	0.096659	0.865521	0.072180	0.877860	0.055341
4	0.891460	0.165207	0.881036	0.131686	0.880584	0.103403	0.885308	0.081408	0.892143	0.064825
5	0.919826	0.148271	0.907873	0.125136	0.903152	0.103577	0.903231	0.085187	0.906058	0.070209
6	0.938827	0.132986	0.927384	0.116656	0.921006	0.100357	0.918527	0.085460	0.918708	0.072544
7	0.952032	0.119807	0.941722	0.107990	0.934958	0.095598	0.931209	0.083666	0.929765	0.072792
8	0.961512	0.108589	0.952435	0.099830	0.945868	0.090303	0.941594	0.080755	0.939221	0.071700
9	0.968512	0.099045	0.960582	0.092408	0.954457	0.084986	0.950072	0.077313	0.947219	0.069799
10	0.973810	0.090888	0.966885	0.085758	0.961283	0.079898	0.957007	0.073688	0.953955	0.067447
11	0.977904	0.083871	0.971842	0.079836	0.966766	0.075147	0.962708	0.070080	0.959626	0.064878
12	0.981127	0.077790	0.975799	0.074567	0.971216	0.070769	0.967423	0.066599	0.964412	0.062242
13	0.983706	0.072484	0.979000	0.069873	0.974865	0.066762	0.971351	0.063300	0.968464	0.059630
14	0.985799	0.067822	0.981621	0.065679	0.977888	0.063104	0.974646	0.060206	0.971911	0.057096
15	0.987519	0.063698	0.983791	0.061921	0.980414	0.059767	0.977430	0.057322	0.974858	0.054671
16	0.988949	0.060029	0.985606	0.058539	0.982544	0.056723	0.979799	0.054644	0.977391	0.052371
17	0.990149	0.056745	0.987138	0.055485	0.984353	0.053941	0.981827	0.052162	0.979579	0.050200
18	0.991167	0.053792	0.988442	0.052717	0.985902	0.051394	0.983574	0.049861	0.981478	0.048160
19	0.992036	0.051122	0.989561	0.050199	0.987237	0.049057	0.985089	0.047728	0.983135	0.046245
20	0.992784	0.048699	0.990527	0.047900	0.988395	0.046909	0.986410	0.045749	0.984587	0.044449

	$\mathscr{R}$	$\mathscr{I}$	$\mathscr{R}$	$\mathscr{I}$	$\mathscr{R}$	$\mathscr{I}$	$\mathscr{R}$	$\mathscr{I}$	$\mathscr{R}$	$\mathscr{I}$
$y\backslash x$	6		7		8		9		10	
0	0.871606	0.000000	0.886488	0.000000	0.898237	0.000000	0.907758	0.000000	0.915633	0.000000
1	0.873827	0.016570	0.888009	0.012947	0.899327	0.010401	0.908565	0.008543	0.916249	0.007143
2	0.880023	0.031454	0.892327	0.024866	0.902453	0.020140	0.910901	0.016639	0.918040	0.013975
3	0.889029	0.043517	0.898793	0.034995	0.907236	0.028693	0.914531	0.023921	0.920856	0.020230
4	0.899484	0.052380	0.906591	0.042967	0.913167	0.035755	0.919127	0.030145	0.924479	0.025717
5	0.910242	0.058259	0.914952	0.048780	0.919729	0.041242	0.924336	0.035208	0.928664	0.030334
6	0.920534	0.061676	0.923283	0.052667	0.926481	0.045242	0.929836	0.039123	0.933175	0.034063
7	0.929945	0.063220	0.931193	0.054971	0.933096	0.047942	0.935365	0.041986	0.937807	0.036944
8	0.938313	0.063425	0.938469	0.056047	0.939359	0.049570	0.940731	0.043936	0.942398	0.039060
9	0.945629	0.062714	0.945023	0.056211	0.945154	0.050349	0.945812	0.045128	0.946833	0.040514
10	0.951965	0.061408	0.950850	0.055725	0.950427	0.050481	0.950535	0.045711	0.951035	0.041413
11	0.957427	0.059735	0.955987	0.054790	0.955176	0.050135	0.954870	0.045818	0.954959	0.041861
12	0.962128	0.057855	0.960495	0.053560	0.959421	0.049444	0.958814	0.045563	0.958586	0.041948
13	0.966178	0.055877	0.964444	0.052146	0.963201	0.048514	0.962379	0.045038	0.961913	0.041755
14	0.969673	0.053874	0.967903	0.050627	0.966559	0.047425	0.965591	0.044319	0.964949	0.041347
15	0.972699	0.051894	0.970935	0.049062	0.969539	0.046236	0.968477	0.043463	0.967710	0.040780
16	0.975326	0.049966	0.973577	0.047489	0.972185	0.044992	0.971067	0.042516	0.970214	0.040095
17	0.977617	0.048109	0.975940	0.045935	0.974538	0.043724	0.973393	0.041512	0.972484	0.039329
18	0.979622	0.046332	0.978009	0.044419	0.976632	0.042456	0.975481	0.040477	0.974540	0.038508
19	0.981384	0.044641	0.979839	0.042951	0.978500	0.041205	0.977357	0.039431	0.976402	0.037653
20	0.982938	0.043036	0.981465	0.041538	0.980169	0.039980	0.979047	0.038388	0.978090	0.036781

*   If $x>10$ or $y>10$ then (see [5.15])

$$e^z E_1(z) = \frac{0.711093}{z+0.415775} + \frac{0.278518}{z+2.29428} + \frac{0.010389}{z+6.2900} + \epsilon, |\epsilon| < 3 \times 10^{-6}.$$

$$E_1(iy) = -\mathrm{Ci}(y) + i\,\mathrm{si}(y) \quad (y \text{ real})$$

EXPONENTIAL INTEGRAL FOR COMPLEX ARGUMENTS   Table 5.6

$$ze^z E_1(z)$$

y\x	11 $\mathscr{R}$	11 $\mathscr{I}$	12 $\mathscr{R}$	12 $\mathscr{I}$	13 $\mathscr{R}$	13 $\mathscr{I}$	14 $\mathscr{R}$	14 $\mathscr{I}$	15 $\mathscr{R}$	15 $\mathscr{I}$
0	0.922260	0.000000	0.927914	0.000000	0.932796	0.000000	0.937055	0.000000	0.940804	0.000000
1	0.922740	0.006063	0.928295	0.005212	0.933105	0.004528	0.937308	0.003972	0.941014	0.003512
2	0.924143	0.011902	0.929416	0.010258	0.934013	0.008932	0.938055	0.007847	0.941636	0.006949
3	0.926370	0.017321	0.931205	0.014991	0.935473	0.013098	0.939261	0.011540	0.942643	0.010242
4	0.929270	0.022171	0.933560	0.019295	0.937408	0.016934	0.940870	0.014974	0.943994	0.013331
5	0.932672	0.026361	0.936356	0.023091	0.939729	0.020373	0.942816	0.018095	0.945640	0.016169
6	0.936400	0.029857	0.939462	0.026339	0.942338	0.023378	0.945024	0.020867	0.947522	0.018725
7	0.940297	0.032670	0.942757	0.029036	0.945140	0.025934	0.947419	0.023273	0.949582	0.020980
8	0.944229	0.034847	0.946132	0.031205	0.948047	0.028052	0.949933	0.025315	0.951765	0.022931
9	0.948093	0.036453	0.949500	0.032887	0.950985	0.029756	0.952502	0.027004	0.954018	0.024582
10	0.951816	0.037566	0.952792	0.034134	0.953895	0.031081	0.955075	0.028365	0.956296	0.025949
11	0.955347	0.038261	0.955958	0.035004	0.956729	0.032068	0.957610	0.029426	0.958563	0.027052
12	0.958659	0.038612	0.958968	0.035552	0.959454	0.032761	0.960073	0.030221	0.960787	0.027915
13	0.961739	0.038604	0.961800	0.035833	0.962049	0.033201	0.962443	0.030781	0.962947	0.028564
14	0.964583	0.038534	0.964447	0.035893	0.964499	0.033128	0.964702	0.031140	0.965026	0.029024
15	0.967199	0.038211	0.966907	0.035775	0.966799	0.033479	0.966843	0.031327	0.967011	0.029320
16	0.969597	0.037756	0.969184	0.035515	0.968947	0.033384	0.968860	0.031370	0.968897	0.029476
17	0.971789	0.037200	0.971285	0.035144	0.970946	0.033172	0.970752	0.031293	0.970680	0.029512
18	0.973792	0.036572	0.973220	0.034687	0.972802	0.032865	0.972521	0.031117	0.972359	0.029448
19	0.975621	0.035893	0.974999	0.034166	0.974521	0.032485	0.974172	0.030862	0.973936	0.029301
20	0.977290	0.035179	0.976634	0.033597	0.976112	0.032049	0.975709	0.030542	0.975414	0.029086

y\x	16 $\mathscr{R}$	16 $\mathscr{I}$	17 $\mathscr{R}$	17 $\mathscr{I}$	18 $\mathscr{R}$	18 $\mathscr{I}$	19 $\mathscr{R}$	19 $\mathscr{I}$	20 $\mathscr{R}$	20 $\mathscr{I}$
0	0.944130	0.000000	0.947100	0.000000	0.949769	0.000000	0.952181	0.000000	0.954371	0.000000
1	0.944306	0.003128	0.947250	0.002804	0.949897	0.002527	0.952291	0.002290	0.954467	0.002085
2	0.944829	0.006196	0.947693	0.005560	0.950277	0.005016	0.952619	0.004549	0.954752	0.004144
3	0.945678	0.009150	0.948416	0.008223	0.950898	0.007430	0.953156	0.006745	0.955219	0.006151
4	0.946824	0.011940	0.949395	0.010754	0.951741	0.009735	0.953887	0.008853	0.955856	0.008084
5	0.948226	0.014529	0.950600	0.013121	0.952782	0.011904	0.954793	0.010847	0.956650	0.009922
6	0.949842	0.016886	0.951995	0.015296	0.953995	0.013916	0.955853	0.012709	0.957581	0.011649
7	0.951624	0.018994	0.953545	0.017265	0.955349	0.015753	0.957043	0.014425	0.958631	0.013253
8	0.953527	0.020847	0.955212	0.019019	0.956815	0.017409	0.958337	0.015986	0.959779	0.014723
9	0.955509	0.022445	0.956960	0.020555	0.958363	0.018878	0.959712	0.017387	0.961004	0.016056
10	0.957530	0.023797	0.958758	0.021878	0.959966	0.020163	0.961144	0.018628	0.962288	0.017250
11	0.959559	0.024917	0.960576	0.022998	0.961598	0.021270	0.962612	0.019712	0.963611	0.018305
12	0.961568	0.025823	0.962391	0.023927	0.963238	0.022207	0.964097	0.020645	0.964956	0.019227
13	0.963534	0.026534	0.964181	0.024679	0.964868	0.022984	0.965582	0.021436	0.966310	0.020021
14	0.965443	0.027070	0.965931	0.025271	0.966472	0.023616	0.967052	0.022094	0.967658	0.020694
15	0.967280	0.027453	0.967628	0.025720	0.968039	0.024114	0.968496	0.022629	0.968990	0.021255
16	0.969038	0.027700	0.969264	0.026041	0.969558	0.024493	0.969906	0.023052	0.970297	0.021712
17	0.970712	0.027831	0.970832	0.026249	0.971023	0.024765	0.971273	0.023375	0.971571	0.022075
18	0.972300	0.027862	0.972328	0.026361	0.972430	0.024943	0.972594	0.023607	0.972808	0.022352
19	0.973800	0.027809	0.973751	0.026388	0.973775	0.025038	0.973863	0.023760	0.974004	0.022552
20	0.975215	0.027685	0.975099	0.026343	0.975057	0.025062	0.975079	0.023842	0.975155	0.022684

EXPONENTIAL INTEGRAL FOR SMALL COMPLEX ARGUMENTS   Table 5.7

$$e^z E_1(z)$$

y\x	-4.0 $\mathscr{R}$	-4.0 $\mathscr{I}$	-3.5 $\mathscr{R}$	-3.5 $\mathscr{I}$	-3.0 $\mathscr{R}$	-3.0 $\mathscr{I}$	-2.5 $\mathscr{R}$	-2.5 $\mathscr{I}$	-2.0 $\mathscr{R}$	-2.0 $\mathscr{I}$
0.0	-0.359552	-0.057540	-0.420509	-0.094868	-0.494576	-0.156411	-0.580650	-0.257878	-0.670483	-0.425168
0.2	-0.347179	-0.078283	-0.400596	-0.119927	-0.462493	-0.185573	-0.528987	-0.289009	-0.587558	-0.451225
0.4	-0.333373	-0.096648	-0.379278	-0.141221	-0.429554	-0.208800	-0.478303	-0.310884	-0.510543	-0.463193
0.6	-0.318556	-0.112633	-0.357202	-0.158890	-0.396730	-0.226575	-0.429978	-0.324774	-0.441128	-0.464163
0.8	-0.303109	-0.126301	-0.334923	-0.173169	-0.364785	-0.239500	-0.384941	-0.332047	-0.380013	-0.457088
1.0	-0.287369	-0.137768	-0.312894	-0.184355	-0.334280	-0.248231	-0.343719	-0.334043	-0.327140	-0.444528

$$E_1(z)+\ln z$$

y\x	-2.0 $\mathscr{R}$	-2.0 $\mathscr{I}$	-1.5 $\mathscr{R}$	-1.5 $\mathscr{I}$	-1.0 $\mathscr{R}$	-1.0 $\mathscr{I}$	-0.5 $\mathscr{R}$	-0.5 $\mathscr{I}$	0 $\mathscr{R}$	0 $\mathscr{I}$
0.0	-4.261087	0.000000	-2.895820	0.000000	-1.895118	0.000000	-1.147367	0.000000	-0.577216	0.000000
0.2	-4.219228	0.636779	-2.867070	0.462804	-1.875155	0.342700	-1.133341	0.258840	-0.567232	0.199556
0.4	-4.094686	1.260867	-2.781497	0.917127	-1.815717	0.679691	-1.091560	0.513806	-0.537482	0.396461
0.6	-3.890531	1.859922	-2.641121	1.354712	-1.718135	1.005410	-1.022911	0.761122	-0.488555	0.588128
0.8	-3.611783	2.422284	-2.449241	1.767748	-1.584591	1.314586	-0.928842	0.997200	-0.421423	0.772095
1.0	-3.265262	2.937296	-2.210344	2.149077	-1.418052	1.602372	-0.811327	1.218731	-0.337404	0.946083

y\x	0.5 $\mathscr{R}$	0.5 $\mathscr{I}$	1.0 $\mathscr{R}$	1.0 $\mathscr{I}$	1.5 $\mathscr{R}$	1.5 $\mathscr{I}$	2.0 $\mathscr{R}$	2.0 $\mathscr{I}$	2.5 $\mathscr{R}$	2.5 $\mathscr{I}$
0.0	-0.133374	0.000000	0.219384	0.000000	0.505485	0.000000	0.742048	0.000000	0.941206	0.000000
0.2	-0.126168	0.157081	0.224661	0.126210	0.509410	0.103432	0.745014	0.086359	0.943484	0.073355
0.4	-0.104687	0.312331	0.240402	0.251143	0.521123	0.205962	0.753871	0.172075	0.950289	0.146246
0.6	-0.069328	0.463961	0.266336	0.373547	0.540441	0.306707	0.768490	0.256515	0.961532	0.218215
0.8	-0.020743	0.610264	0.302022	0.492229	0.567061	0.404823	0.788664	0.339075	0.977068	0.288822
1.0	+0.040177	0.749655	0.346856	0.606074	0.600568	0.499516	0.814107	0.419185	0.996699	0.357653

6. Gamma Function and Related Functions

Philip J. Davis[1]

Contents

---

[1] National Bureau of Standards.

The author acknowledges the assistance of Mary Orr in the preparation and checking of the tables; and the assistance of Patricia Farrant in checking the formulas.

6. Gamma Function and Related Functions

Mathematical Properties

6.1. Gamma (Factorial) Function

Euler's Integral

6.1.1
$$\Gamma(z) = \int_0^\infty t^{z-1} e^{-t}\, dt \qquad (\mathscr{R} z > 0)$$

$$= k^z \int_0^\infty t^{z-1} e^{-kt}\, dt \qquad (\mathscr{R} z > 0,\ \mathscr{R} k > 0)$$

Euler's Formula

6.1.2
$$\Gamma(z) = \lim_{n\to\infty} \frac{n!\, n^z}{z(z+1)\cdots(z+n)} \qquad (z \neq 0, -1, -2, \ldots)$$

Euler's Infinite Product

6.1.3
$$\frac{1}{\Gamma(z)} = z e^{\gamma z} \prod_{n=1}^{\infty}\left[\left(1+\frac{z}{n}\right) e^{-z/n}\right] \qquad (|z| < \infty)$$

$$\gamma = \lim_{m\to\infty}\left[1 + \frac{1}{2} + \frac{1}{3} + \frac{1}{4} + \ldots + \frac{1}{m} - \ln m\right]$$

$$= .57721\ 56649\ldots$$

γ is known as Euler's constant and is given to 25 decimal places in chapter 1. $\Gamma(z)$ is single valued and analytic over the entire complex plane, save for the points $z = -n (n = 0, 1, 2, \ldots)$ where it possesses simple poles with residue $(-1)^n/n!$. Its reciprocal $1/\Gamma(z)$ is an entire function possessing simple zeros at the points $z = -n (n = 0, 1, 2, \ldots)$.

Hankel's Contour Integral

6.1.4
$$\frac{1}{\Gamma(z)} = \frac{i}{2\pi} \int_C (-t)^{-z} e^{-t}\, dt \qquad (|z| < \infty)$$

The path of integration C starts at $+\infty$ on the real axis, circles the origin in the counterclockwise direction and returns to the starting point.

Factorial and Π Notations

6.1.5
$$\Pi(z) = z! = \Gamma(z+1)$$

Integer Values

6.1.6
$$\Gamma(n+1) = 1 \cdot 2 \cdot 3 \cdots (n-1)n = n!$$

6.1.7
$$\lim_{z\to n}\frac{1}{\Gamma(-z)} = 0 = \frac{1}{(-n-1)!} \qquad (n = 0, 1, 2, \ldots)$$

Fractional Values

6.1.8
$$\Gamma(\tfrac{1}{2}) = 2\int_0^\infty e^{-t^2}\, dt = \pi^{\frac{1}{2}} = 1.77245\ 38509\ldots = (-\tfrac{1}{2})!$$

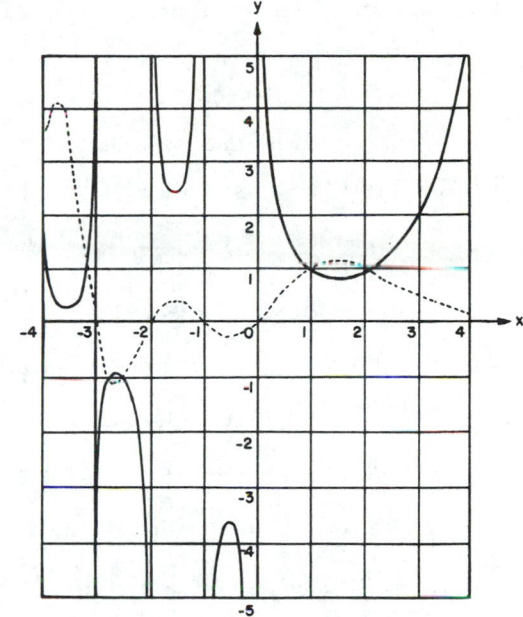

FIGURE 6.1. *Gamma function.* *

————, $y = \Gamma(x)$, $----$, $y = 1/\Gamma(x)$

6.1.9 $\Gamma(3/2) = \tfrac{1}{2}\pi^{\frac{1}{2}} = .88622\ 69254\ldots = (\tfrac{1}{2})!$

6.1.10 $\Gamma(n+\tfrac{1}{4}) = \dfrac{1 \cdot 5 \cdot 9 \cdot 13 \cdots (4n-3)}{4^n}\, \Gamma(\tfrac{1}{4})$

$$\Gamma(\tfrac{1}{4}) = 3.62560\ 99082\ldots$$

6.1.11 $\Gamma(n+\tfrac{1}{3}) = \dfrac{1 \cdot 4 \cdot 7 \cdot 10 \cdots (3n-2)}{3^n}\, \Gamma(\tfrac{1}{3})$

$$\Gamma(\tfrac{1}{3}) = 2.67893\ 85347\ldots$$

6.1.12 $\Gamma(n+\tfrac{1}{2}) = \dfrac{1 \cdot 3 \cdot 5 \cdot 7 \cdots (2n-1)}{2^n}\, \Gamma(\tfrac{1}{2})$

6.1.13 $\Gamma(n+\tfrac{2}{3}) = \dfrac{2 \cdot 5 \cdot 8 \cdot 11 \cdots (3n-1)}{3^n}\, \Gamma(\tfrac{2}{3})$

$$\Gamma(\tfrac{2}{3}) = 1.35411\ 79394\ldots$$

6.1.14 $\Gamma(n+\tfrac{3}{4}) = \dfrac{3 \cdot 7 \cdot 11 \cdot 15 \cdots (4n-1)}{4^n}\, \Gamma(\tfrac{3}{4})$

$$\Gamma(\tfrac{3}{4}) = 1.22541\ 67024\ldots$$

Recurrence Formulas

6.1.15 $\Gamma(z+1)=z\Gamma(z)=z!=z(z-1)!$

6.1.16
$$\Gamma(n+z)=(n-1+z)(n-2+z)\ldots(1+z)\Gamma(1+z)$$
$$=(n-1+z)!$$
$$=(n-1+z)(n-2+z)\ldots(1+z)z!$$

Reflection Formula

6.1.17 $\Gamma(z)\Gamma(1-z)=-z\Gamma(-z)\Gamma(z)=\pi\csc\pi z$
$$=\int_0^\infty\frac{t^{z-1}}{1+t}dt\qquad(0<\mathscr{R}z<1)$$

Duplication Formula

6.1.18 $\Gamma(2z)=(2\pi)^{-\frac{1}{2}}2^{2z-\frac{1}{2}}\Gamma(z)\Gamma(z+\tfrac{1}{2})$

Triplication Formula

6.1.19 $\Gamma(3z)=(2\pi)^{-1}3^{3z-\frac{1}{2}}\Gamma(z)\Gamma(z+\tfrac{1}{3})\Gamma(z+\tfrac{2}{3})$

Gauss' Multiplication Formula

6.1.20 $\Gamma(nz)=(2\pi)^{\frac{1}{2}(1-n)}n^{nz-\frac{1}{2}}\prod_{k=0}^{n-1}\Gamma\left(z+\frac{k}{n}\right)$

Binomial Coefficient

6.1.21 $\binom{z}{w}=\frac{z!}{w!(z-w)!}=\frac{\Gamma(z+1)}{\Gamma(w+1)\Gamma(z-w+1)}$

Pochhammer's Symbol

6.1.22
$(z)_0=1,$
$(z)_n=z(z+1)(z+2)\ldots(z+n-1)=\frac{\Gamma(z+n)}{\Gamma(z)}$

Gamma Function in the Complex Plane

6.1.23 $\Gamma(\bar{z})=\overline{\Gamma(z)};\ \ln\Gamma(\bar{z})=\overline{\ln\Gamma(z)}$

6.1.24 $\arg\Gamma(z+1)=\arg\Gamma(z)+\arctan\frac{y}{x}$

6.1.25 $\left|\frac{\Gamma(x+iy)}{\Gamma(x)}\right|^2=\prod_{n=0}^\infty\left[1+\frac{y^2}{(x+n)^2}\right]^{-1}$

6.1.26 $|\Gamma(x+iy)|\le|\Gamma(x)|$

6.1.27
$$\arg\Gamma(x+iy)=y\psi(x)+\sum_{n=0}^\infty\left(\frac{y}{x+n}-\arctan\frac{y}{x+n}\right)$$
$$(x+iy\ne0,-1,-2,\ldots)$$

where $\psi(z)=\Gamma'(z)/\Gamma(z)$

6.1.28 $\Gamma(1+iy)=iy\,\Gamma(iy)$

6.1.29 $\Gamma(iy)\Gamma(-iy)=|\Gamma(iy)|^2=\dfrac{\pi}{y\sinh\pi y}$

6.1.30 $\Gamma(\tfrac{1}{2}+iy)\Gamma(\tfrac{1}{2}-iy)=|\Gamma(\tfrac{1}{2}+iy)|^2=\dfrac{\pi}{\cosh\pi y}$

6.1.31 $\Gamma(1+iy)\Gamma(1-iy)=|\Gamma(1+iy)|^2=\dfrac{\pi y}{\sinh\pi y}$

6.1.32 $\Gamma(\tfrac{1}{4}+iy)\Gamma(\tfrac{3}{4}-iy)=\dfrac{\pi\sqrt{2}}{\cosh\pi y+i\sinh\pi y}$

Power Series

6.1.33
$$\ln\Gamma(1+z)=-\ln(1+z)+z(1-\gamma)$$
$$+\sum_{n=2}^\infty(-1)^n[\zeta(n)-1]z^n/n\quad(|z|<2)$$

$\zeta(n)$ is the Riemann Zeta Function (see chapter 23).

Series Expansion [2] for $1/\Gamma(z)$

6.1.34 $\dfrac{1}{\Gamma(z)}=\sum\limits_{k=1}^\infty c_k z^k\qquad(|z|<\infty)$

k	c_k
1	1. 00000 00000 000000
2	0. 57721 56649 015329
3	—0. 65587 80715 202538
4	—0. 04200 26350 340952
5	0. 16653 86113 822915
6	—0. 04219 77345 555443
7	—0. 00962 19715 278770
8	0. 00721 89432 466630
9	—0. 00116 51675 918591
10	—0. 00021 52416 741149
11	0. 00012 80502 823882
12	—0. 00002 01348 547807
13	—0. 00000 12504 934821
14	0. 00000 11330 272320
15	—0. 00000 02056 338417
16	0. 00000 00061 160950
17	0. 00000 00050 020075
18	—0. 00000 00011 812746
19	0. 00000 00001 043427
20	0. 00000 00000 077823
21	—0. 00000 00000 036968
22	0. 00000 00000 005100
23	—0. 00000 00000 000206
24	—0. 00000 00000 000054
25	0. 00000 00000 000014
26	0. 00000 00000 000001

[2] The coefficients c_k are from H. T. Davis, Tables of higher mathematical functions, 2 vols., Principia Press, Bloomington, Ind., 1933, 1935 (with permission); with corrections due to H. E. Salzer.

Polynomial Approximations[3]

6.1.35 $0 \leq x \leq 1$

$$\Gamma(x+1)=x!=1+a_1x+a_2x^2+a_3x^3+a_4x^4+a_5x^5+\epsilon(x)$$

$$|\epsilon(x)| \leq 5 \times 10^{-5}$$

$a_1=-.57486\ 46 \qquad a_4=\ \ .42455\ 49$

$a_2=\ \ .95123\ 63 \qquad a_5=-.10106\ 78$

$a_3=-.69985\ 88$

6.1.36 $0 \leq x \leq 1$

$$\Gamma(x+1)=x!=1+b_1x+b_2x^2+\ .\ .\ .\ +b_8x^8+\epsilon(x)$$

$$|\epsilon(x)| \leq 3 \times 10^{-7}$$

$b_1=-.57719\ 1652 \qquad b_5=-.75670\ 4078$

$b_2=\ \ .98820\ 5891 \qquad b_6=\ \ .48219\ 9394$

$b_3=-.89705\ 6937 \qquad b_7=-.19352\ 7818$

$b_4=\ \ .91820\ 6857 \qquad b_8=\ \ .03586\ 8343$

Stirling's Formula

6.1.37

$$\Gamma(z) \sim e^{-z}z^{z-\frac{1}{2}}(2\pi)^{\frac{1}{2}}\left[1+\frac{1}{12z}+\frac{1}{288z^2}-\frac{139}{51840z^3}\right.$$

$$\left.-\frac{571}{2488320z^4}+\cdots\right] \qquad (z\to\infty \text{ in } |\arg z|<\pi)$$

6.1.38

$$x!=\sqrt{2\pi}\ x^{x+\frac{1}{2}}\exp\left(-x+\frac{\theta}{12x}\right) \qquad (x>0,\ 0<\theta<1)$$

Asymptotic Formulas

6.1.39

$$\Gamma(az+b) \sim \sqrt{2\pi}\ e^{-az}(az)^{az+b-\frac{1}{2}} \qquad (|\arg z|<\pi,\ a>0)$$

6.1.40

$$\ln \Gamma(z) \sim (z-\tfrac{1}{2})\ \ln\ z-z+\tfrac{1}{2}\ \ln\ (2\pi)$$

$$+\sum_{m=1}^{\infty}\frac{B_{2m}}{2m(2m-1)z^{2m-1}} \qquad (z\to\infty \text{ in } |\arg z|<\pi)$$

For B_n see chapter **23**

6.1.41

$$\ln \Gamma(z) \sim (z-\tfrac{1}{2})\ \ln\ z-z+\tfrac{1}{2}\ \ln\ (2\pi)+\frac{1}{12z}-\frac{1}{360z^3}$$

$$+\frac{1}{1260z^5}-\frac{1}{1680z^7}+\cdots \qquad (z\to\infty \text{ in } |\arg z|<\pi)$$

---

[3] From C. Hastings, Jr., Approximations for digital computers, Princeton Univ. Press, Princeton, N.J., 1955 (with permission).

Error Term for Asymptotic Expansion

6.1.42

If

$$R_n(z)= \ln \Gamma\ (z)-(z-\tfrac{1}{2})\ \ln\ z+z-\tfrac{1}{2}\ \ln\ (2\pi)$$

$$-\sum_{m=1}^{n}\frac{B_{2m}}{2m(2m-1)z^{2m-1}}$$

then

$$|R_n(z)| \leq \frac{|B_{2n+2}|K(z)}{(2n+1)(2n+2)|z|^{2n+1}}$$

where

$$K(z)=\underset{u\geq 0}{\text{upper bound}}|z^2/(u^2+z^2)|$$

For z real and positive, R_n is less in absolute value than the first term neglected and has the same sign.

6.1.43

$$\mathscr{R}\ln \Gamma(iy)=\mathscr{R}\ln \Gamma(-iy)$$

$$=\tfrac{1}{2}\ \ln\left(\frac{\pi}{y \sinh\ \pi y}\right)$$

$$\sim\tfrac{1}{2}\ \ln\ (2\pi)-\tfrac{1}{2}\pi y-\tfrac{1}{2}\ln\ y, \qquad (y\to+\infty)$$

6.1.44

$$\mathscr{I}\ln \Gamma(iy)=\arg \Gamma(iy)=-\arg \Gamma(-iy)$$

$$=-\mathscr{I}\ln \Gamma(-iy)$$

$$\sim y \ln y-y-\tfrac{1}{4}\pi-\sum_{n=1}^{\infty}\frac{(-1)^{n-1}\ B_{2n}}{(2n-1)(2n)y^{2n-1}}$$

$$(y\to+\infty)$$

6.1.45 $\displaystyle\lim_{|y|\to\infty}(2\pi)^{-\frac{1}{2}}|\Gamma(x+iy)|e^{\frac{1}{2}\pi|y|}|y|^{\frac{1}{2}-x}=1$

6.1.46 $\displaystyle\lim_{n\to\infty} n^{b-a}\frac{\Gamma(n+a)}{\Gamma(n+b)}=1$

6.1.47

$$z^{b-a}\frac{\Gamma(z+a)}{\Gamma(z+b)} \sim 1+\frac{(a-b)(a+b-1)}{2z}$$

$$+\frac{1}{12}\binom{a-b}{2}\left(3(a+b-1)^2-a+b-1\right)\frac{1}{z^2}+\cdots$$

as $z\to\infty$ along any curve joining $z=0$ and $z=\infty$, providing $z\neq -a,\ -a-1,\ \ldots$; $z\neq -b,\ -b-1,$ . . . .

Continued Fraction

6.1.48

$$\ln \Gamma(z)+z-(z-\tfrac{1}{2}) \ln z-\tfrac{1}{2} \ln (2\pi)$$

$$=\frac{a_0}{z+}\frac{a_1}{z+}\frac{a_2}{z+}\frac{a_3}{z+}\frac{a_4}{z+}\frac{a_5}{z+}\ldots \qquad (\mathscr{R}\, z > 0)$$

$$a_0=\frac{1}{12},\ a_1=\frac{1}{30},\ a_2=\frac{53}{210},\ a_3=\frac{195}{371},$$

$$a_4=\frac{22999}{22737},\ a_5=\frac{29944523}{19733142},\ a_6=\frac{109535241009}{48264275462}$$

Wallis' Formula [4]

6.1.49

$$\frac{2}{\pi}\int_0^{\pi/2}\binom{\sin}{\cos}^{2n} x\,dx=\frac{1\cdot 3\cdot 5\,\ldots\,(2n-1)}{2\cdot 4\cdot 6\,\ldots\,(2n)}$$

$$=\frac{(2n)!}{2^{2n}(n!)^2}=\frac{1}{2^{2n}}\binom{2n}{n}=\frac{\Gamma(n+\tfrac{1}{2})}{\pi^{\frac{1}{2}}\Gamma(n+1)}$$

$$\sim \frac{1}{\pi^{\frac{1}{2}}n^{\frac{1}{2}}}\left[1-\frac{1}{8n}+\frac{1}{128n^2}-\ldots\right]$$

$$(n\to\infty)$$

Some Definite Integrals

6.1.50

$$\ln \Gamma(z)=\int_0^{\infty}\left[(z-1)\,e^{-t}-\frac{e^{-t}-e^{-zt}}{1-e^{-t}}\right]\frac{dt}{t} \qquad (\mathscr{R}\,z > 0)$$

$$=(z-\tfrac{1}{2})\,\ln z-z+\tfrac{1}{2}\ln 2\pi$$

$$+2\int_0^{\infty}\frac{\arctan\,(t/z)}{e^{2\pi t}-1}\,dt \qquad (\mathscr{R}\,z > 0)$$

6.2. Beta Function

6.2.1

$$B(z,w)=\int_0^1 t^{z-1}\,(1-t)^{w-1}\,dt=\int_0^{\infty}\frac{t^{z-1}}{(1+t)^{z+w}}\,dt$$

$$=2\int_0^{\pi/2}(\sin t)^{2z-1}\,(\cos t)^{2w-1}\,dt$$

$$(\mathscr{R}\,z > 0,\ \mathscr{R}\,w > 0)$$

6.2.2 $$B(z,w)=\frac{\Gamma(z)\,\Gamma(w)}{\Gamma(z+w)}=B(w,z)$$

6.3. Psi (Digamma) Function [5]

6.3.1 $$\psi(z)=d[\ln \Gamma(z)]/dz=\Gamma'(z)/\Gamma(z)$$

---

[4] Some authors employ the special double factorial notation as follows:

$$(2n)\,!\,!=2\cdot 4\cdot 6\,\ldots\,(2n)=2^n n\,!$$
$$(2n-1)\,!\,!=1\cdot 3\cdot 5\,\ldots\,(2n-1)=\pi^{-\frac{1}{2}}\,2^n\,\Gamma(n+\tfrac{1}{2})$$

[5] Some authors write $\psi(z)=\dfrac{d}{dz}\ln\Gamma(z+1)$ and similarly for the polygamma functions.

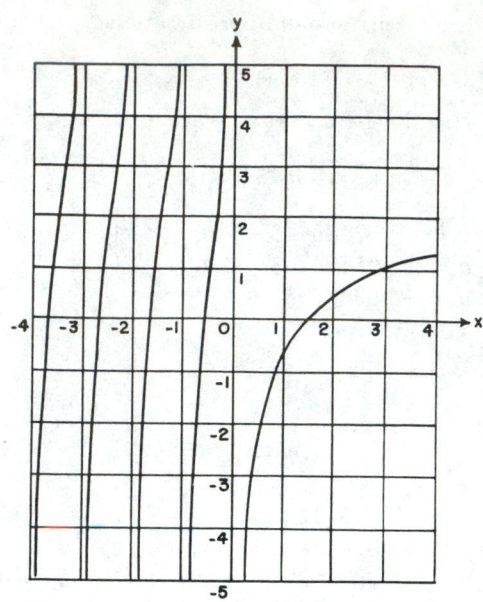

FIGURE 6.2. *Psi function.*

$$y=\psi(x)=d\ln\Gamma(x)/dx$$

Integer Values

6.3.2 $$\psi(1)=-\gamma,\ \psi(n)=-\gamma+\sum_{k=1}^{n-1}k^{-1} \qquad (n\geq 2)$$

Fractional Values

6.3.3

$$\psi(\tfrac{1}{2})=-\gamma-2\ln 2=-1.96351\ 00260\ 21423\,\ldots$$

6.3.4

$$\psi(n+\tfrac{1}{2})=-\gamma-2\ln 2+2\left(1+\frac{1}{3}+\cdots+\frac{1}{2n-1}\right)$$

$$(n\geq 1)$$

Recurrence Formulas

6.3.5 $$\psi(z+1)=\psi(z)+\frac{1}{z}$$

6.3.6

$$\psi(n+z)=\frac{1}{(n-1)+z}+\frac{1}{(n-2)+z}+\cdots$$

$$+\frac{1}{2+z}+\frac{1}{1+z}+\psi(1+z)$$

Reflection Formula

6.3.7
$$\psi(1-z)=\psi(z)+\pi \cot \pi z$$

Duplication Formula

6.3.8
$$\psi(2z)=\tfrac{1}{2}\psi(z)+\tfrac{1}{2}\psi(z+\tfrac{1}{2})+\ln 2$$

Psi Function in the Complex Plane

6.3.9
$$\psi(\bar{z})=\overline{\psi(z)}$$

6.3.10
$$\mathscr{R}\psi(iy)=\mathscr{R}\psi(-iy)=\mathscr{R}\psi(1+iy)=\mathscr{R}\psi(1-iy)$$

6.3.11
$$\mathscr{I}\psi(iy)=\tfrac{1}{2}y^{-1}+\tfrac{1}{2}\pi \coth \pi y$$

6.3.12
$$\mathscr{I}\psi(\tfrac{1}{2}+iy)=\tfrac{1}{2}\pi \tanh \pi y$$

6.3.13
$$\mathscr{I}\psi(1+iy)=-\frac{1}{2y}+\tfrac{1}{2}\pi \coth \pi y$$
$$=y\sum_{n=1}^{\infty}(n^2+y^2)^{-1}$$

Series Expansions

6.3.14 $\quad \psi(1+z)=-\gamma+\sum_{n=2}^{\infty}(-1)^n \zeta(n)z^{n-1} \quad (|z|<1)$

6.3.15
$$\psi(1+z)=\tfrac{1}{2}z^{-1}-\tfrac{1}{2}\pi \cot \pi z-(1-z^2)^{-1}+1-\gamma$$
$$-\sum_{n=1}^{\infty}[\zeta(2n+1)-1]z^{2n} \quad (|z|<2)$$

6.3.16
$$\psi(1+z)=-\gamma+\sum_{n=1}^{\infty}\frac{z}{n(n+z)} \quad (z\neq -1,-2,-3,\ldots)$$

6.3.17
$$\mathscr{R}\psi(1+iy)=1-\gamma-\frac{1}{1+y^2}$$
$$+\sum_{n=1}^{\infty}(-1)^{n+1}[\zeta(2n+1)-1]y^{2n}$$
$$(|y|<2)$$
$$=-\gamma+y^2\sum_{n=1}^{\infty}n^{-1}(n^2+y^2)^{-1}$$
$$(-\infty<y<\infty)$$

Asymptotic Formulas

6.3.18
$$\psi(z)\sim\ln z-\frac{1}{2z}-\sum_{n=1}^{\infty}\frac{B_{2n}}{2nz^{2n}}$$
$$=\ln z-\frac{1}{2z}-\frac{1}{12z^2}+\frac{1}{120z^4}-\frac{1}{252z^6}+\cdots$$
$$(z\to\infty \text{ in } |\arg z|<\pi)$$

6.3.19
$$\mathscr{R}\psi(1+iy)\sim\ln y+\sum_{n=1}^{\infty}\frac{(-1)^{n-1}B_{2n}}{2ny^{2n}}$$
$$=\ln y+\frac{1}{12y^2}+\frac{1}{120y^4}+\frac{1}{252y^6}+\cdots$$
$$(y\to\infty)$$

Extrema[6] of $\Gamma(x)$ — Zeros of $\psi(x)$

$$\Gamma'(x_n)=\psi(x_n)=0$$

n	x_n	$\Gamma(x_n)$
0	$+1.462$	$+0.886$
1	-0.504	-3.545
2	-1.573	$+2.302$
3	-2.611	-0.888
4	-3.635	$+0.245$
5	-4.653	-0.053
6	-5.667	$+0.009$
7	-6.678	-0.001

$$x_0=1.46163 \quad 21449 \quad 68362$$
$$\Gamma(x_0)=.88560 \quad 31944 \quad 10889$$

6.3.20 $\quad x_n=-n+(\ln n)^{-1}+o[(\ln n)^{-2}]$

Definite Integrals

6.3.21
$$\psi(z)=\int_0^\infty\left[\frac{e^{-t}}{t}-\frac{e^{-zt}}{1-e^{-t}}\right]dt \quad (\mathscr{R}z>0)$$
$$=\int_0^\infty\left[e^{-t}-\frac{1}{(1+t)^z}\right]\frac{dt}{t}$$
$$=\ln z-\frac{1}{2z}-2\int_0^\infty\frac{t\,dt}{(t^2+z^2)(e^{2\pi t}-1)}$$
$$\left(|\arg z|<\frac{\pi}{2}\right)$$

6.3.22
$$\psi(z)+\gamma=\int_0^\infty\frac{e^{-t}-e^{-zt}}{1-e^{-t}}dt=\int_0^1\frac{1-t^{z-1}}{1-t}dt$$
$$\gamma=\int_0^\infty\left(\frac{1}{e^t-1}-\frac{1}{te^t}\right)dt$$
$$=\int_0^\infty\left(\frac{1}{1+t}-e^{-t}\right)\frac{dt}{t}$$

---

[6] From W. Sibagaki, Theory and applications of the gamma function, Iwanami Syoten, Tokyo, Japan, 1952 (with permission).

6.4. Polygamma Functions [7]

6.4.1

$$\psi^{(n)}(z) = \frac{d^n}{dz^n}\psi(z) = \frac{d^{n+1}}{dz^{n+1}}\ln\Gamma(z)$$

$$(n = 1, 2, 3, \dots)$$

$$* \qquad = (-1)^{n+1}\int_0^\infty \frac{t^n e^{-zt}}{1-e^{-t}}dt \qquad (\mathscr{R}z > 0)$$

$\psi^{(n)}(z), (n = 0, 1, \dots)$, is a single valued analytic function over the entire complex plane save at the points $z = -m (m = 0, 1, 2, \dots)$ where it possesses poles of order $(n+1)$.

Integer Values

6.4.2

$$\psi^{(n)}(1) = (-1)^{n+1}n!\zeta(n+1) \qquad (n = 1, 2, 3, \dots)$$

6.4.3

$$\psi^{(m)}(n+1) = (-1)^m m!\left[-\zeta(m+1) + 1 + \frac{1}{2^{m+1}} + \dots + \frac{1}{n^{m+1}}\right]$$

Fractional Values

6.4.4

$$\psi^{(n)}(\tfrac{1}{2}) = (-1)^{n+1}n!(2^{n+1}-1)\zeta(n+1) \qquad (n = 1, 2, \dots)$$

6.4.5

$$\psi'(n+\tfrac{1}{2}) = \tfrac{1}{2}\pi^2 - 4\sum_{k=1}^n (2k-1)^{-2}$$

Recurrence Formula

6.4.6

$$\psi^{(n)}(z+1) = \psi^{(n)}(z) + (-1)^n n! z^{-n-1}$$

Reflection Formula

6.4.7

$$\psi^{(n)}(1-z) + (-1)^{n+1}\psi^{(n)}(z) = (-1)^n \pi \frac{d^n}{dz^n}\cot \pi z$$

Multiplication Formula

6.4.8

$$* \qquad \psi^{(n)}(mz) = \delta\ln m + \frac{1}{m^{n+1}}\sum_{k=0}^{m-1}\psi^{(n)}\left(z + \frac{k}{m}\right)$$

$$\delta = 1, \quad n = 0$$
$$\delta = 0, \quad n > 0$$

---

[7] ψ' is known as the trigamma function. ψ'', $\psi^{(3)}$, $\psi^{(4)}$ are the tetra-, penta-, and hexagamma functions respectively. Some authors write $\psi(z) = d[\ln\Gamma(z+1)]/dz$, and similarly for the polygamma functions.

*See page II.

Series Expansions

6.4.9

$$\psi^{(n)}(1+z) = (-1)^{n+1}\left[n!\zeta(n+1)\right.$$
$$\left. -\frac{(n+1)!}{1!}\zeta(n+2)z + \frac{(n+2)!}{2!}\zeta(n+3)z^2 - \dots\right]$$

$$(|z| < 1)$$

6.4.10

$$\psi^{(n)}(z) = (-1)^{n+1}n!\sum_{k=0}^\infty (z+k)^{-n-1}$$

$$(z \neq 0, -1, -2, \dots)$$

Asymptotic Formulas

6.4.11

$$\psi^{(n)}(z) \sim (-1)^{n-1}\left[\frac{(n-1)!}{z^n} + \frac{n!}{2z^{n+1}}\right.$$
$$\left. + \sum_{k=1}^\infty B_{2k}\frac{(2k+n-1)!}{(2k)!z^{2k+n}}\right] \quad (z \to \infty \text{ in } |\arg z| < \pi)$$

6.4.12

$$\psi'(z) \sim \frac{1}{z} + \frac{1}{2z^2} + \frac{1}{6z^3} - \frac{1}{30z^5} + \frac{1}{42z^7} - \frac{1}{30z^9} + \dots$$

$$(z \to \infty \text{ in } |\arg z| < \pi)$$

6.4.13

$$\psi''(z) \sim -\frac{1}{z^2} - \frac{1}{z^3} - \frac{1}{2z^4} + \frac{1}{6z^6} - \frac{1}{6z^8} + \frac{3}{10z^{10}} - \frac{5}{6z^{12}} + \dots$$

$$(z \to \infty \text{ in } |\arg z| < \pi)$$

6.4.14

$$\psi^{(3)}(z) \sim \frac{2}{z^3} + \frac{3}{z^4} + \frac{2}{z^5} - \frac{1}{z^7} + \frac{4}{3z^9} - \frac{3}{z^{11}} + \frac{10}{z^{13}} - \dots$$

$$(z \to \infty \text{ in } |\arg z| < \pi)$$

6.5. Incomplete Gamma Function
(see also 26.4)

6.5.1

$$P(a, x) = \frac{1}{\Gamma(a)}\int_0^x e^{-t}t^{a-1}dt \qquad (\mathscr{R}a > 0)$$

6.5.2

$$\gamma(a, x) = P(a, x)\Gamma(a) = \int_0^x e^{-t}t^{a-1}dt \qquad (\mathscr{R}a > 0)$$

6.5.3

$$\Gamma(a, x) = \Gamma(a) - \gamma(a, x) = \int_x^\infty e^{-t}t^{a-1}dt$$

6.5.4

$$\gamma^*(a, x) = x^{-a}P(a, x) = \frac{x^{-a}}{\Gamma(a)}\gamma(a, x)$$

γ^* is a single valued analytic function of a and x possessing no finite singularities.

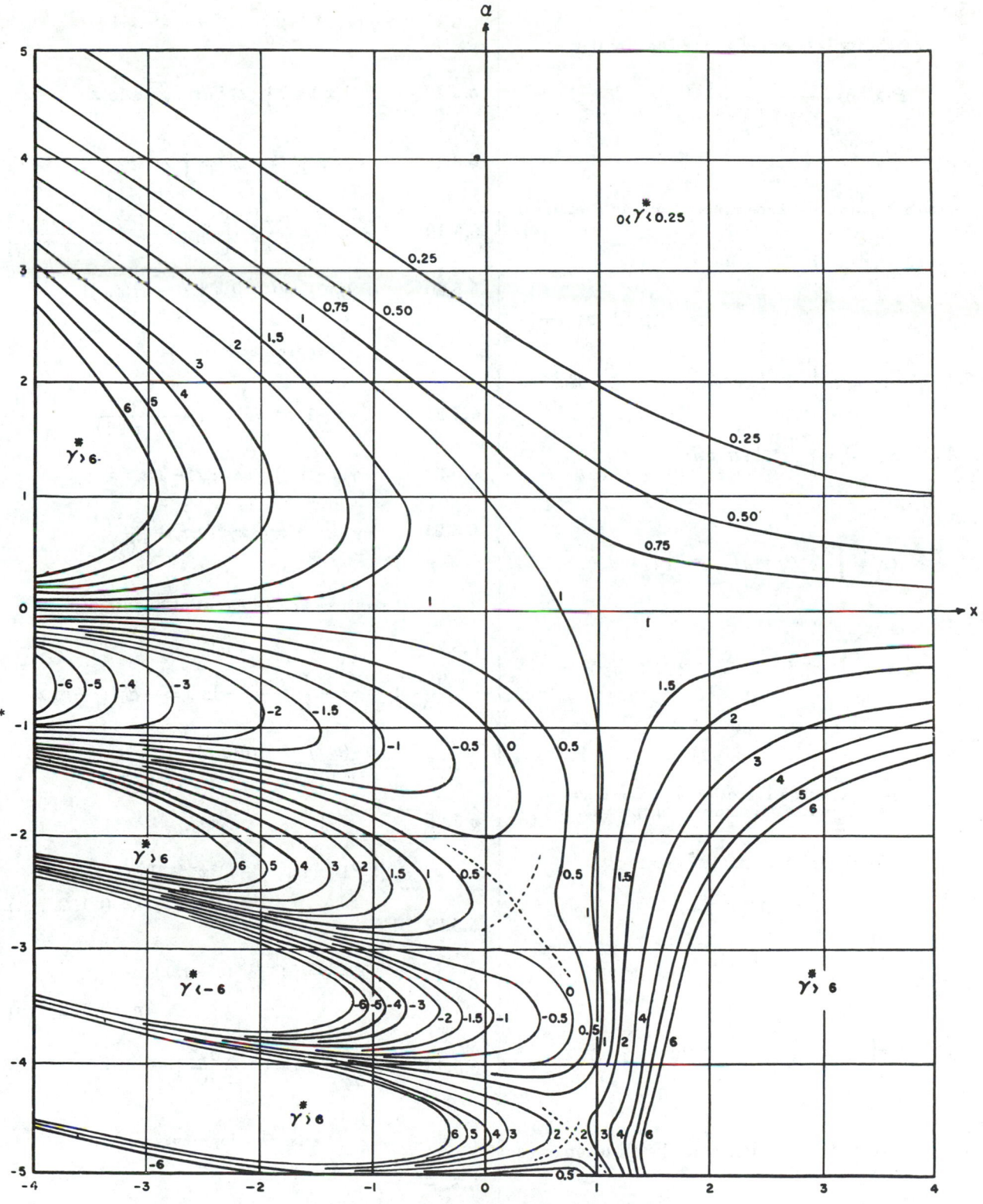

FIGURE 6.3. *Incomplete gamma function.*

$$\gamma^*(a,x) = \frac{x^{-a}}{\Gamma(a)} \int_0^x e^{-t}t^{a-1}dt$$

From F. G. Tricomi, Sulla funzione gamma incompleta, Annali di Matematica, IV, 33, 1950 (with permission).

*See page II.

6.5.5

Probability Integral of the χ^2-Distribution

$$P(\chi^2|\nu) = \frac{1}{2^{\frac{1}{2}\nu}\Gamma\left(\frac{\nu}{2}\right)} \int_0^{\chi^2} t^{\frac{1}{2}\nu-1} e^{-\frac{t}{2}} dt$$

6.5.6

(Pearson's Form of the Incomplete Gamma Function)

$$I(u, p) = \frac{1}{\Gamma(p+1)} \int_0^{u\sqrt{p+1}} e^{-t} t^p \, dt$$

$$= P(p+1, u\sqrt{p+1})$$

6.5.7
$$C(x,a) = \int_x^\infty t^{a-1} \cos t \, dt \qquad (\mathscr{R}a < 1)$$

6.5.8
$$S(x,a) = \int_x^\infty t^{a-1} \sin t \, dt \qquad (\mathscr{R}a < 1)$$

6.5.9
$$E_n(x) = \int_1^\infty e^{-xt} t^{-n} dt = x^{n-1} \Gamma(1-n,x)$$

6.5.10
$$\alpha_n(x) = \int_1^\infty e^{-xt} t^n dt = x^{-n-1} \Gamma(1+n,x)$$

6.5.11
$$e_n(x) = \sum_{j=0}^n \frac{x^j}{j!}$$

Incomplete Gamma Function as a Confluent Hypergeometric Function (see chapter 13)

6.5.12
$$\gamma(a,x) = a^{-1} x^a e^{-x} M(1, 1+a, x)$$

$$= a^{-1} x^a M(a, 1+a, -x)$$

Special Values

6.5.13

$$P(n,x) = 1 - \left(1 + x + \frac{x^2}{2!} + \ldots + \frac{x^{n-1}}{(n-1)!}\right) e^{-x}$$

$$= 1 - e_{n-1}(x) e^{-x}$$

For relation to the Poisson distribution, see **26.4**.

6.5.14
$$\gamma^*(-n, x) = x^n$$

6.5.15
$$\Gamma(0, x) = \int_x^\infty e^{-t} t^{-1} dt = E_1(x)$$

6.5.16
$$\gamma\left(\tfrac{1}{2}, x^2\right) = 2 \int_0^x e^{-t^2} dt = \sqrt{\pi} \, \mathrm{erf} \, x$$

6.5.17
$$\Gamma(\tfrac{1}{2}, x^2) = 2 \int_x^\infty e^{-t^2} dt = \sqrt{\pi} \, \mathrm{erfc} \, x$$

6.5.18
$$\tfrac{1}{2}\sqrt{\pi} \, x \, \gamma^*(\tfrac{1}{2}, -x^2) = \int_0^x e^{t^2} dt$$

6.5.19
$$\Gamma(-n,x) = \frac{(-1)^n}{n!} \left[E_1(x) - e^{-x} \sum_{j=0}^{n-1} \frac{(-1)^j j!}{x^{j+1}} \right]$$

6.5.20
$$\Gamma(a,ix) = e^{\frac{1}{2}\pi i a} [C(x,a) - iS(x,a)]$$

Recurrence Formulas

6.5.21
$$P(a+1, x) = P(a, x) - \frac{x^a e^{-x}}{\Gamma(a+1)}$$

6.5.22
$$\gamma(a+1,x) = a\gamma(a,x) - x^a e^{-x}$$

6.5.23
$$\gamma^*(a-1,x) = x\gamma^*(a,x) + \frac{e^{-x}}{\Gamma(a)}$$

Derivatives and Differential Equations

6.5.24
$$\left(\frac{\partial \gamma^*}{\partial \alpha}\right)_{\alpha=0} = -\int_x^\infty \frac{e^{-t} dt}{t} - \ln x = -E_1(x) - \ln x$$

6.5.25
$$\frac{\partial \gamma(a,x)}{\partial x} = -\frac{\partial \Gamma(a,x)}{\partial x} = x^{a-1} e^{-x}$$

6.5.26
$$\frac{\partial^n}{\partial x^n} [x^{-a}\Gamma(a,x)] = (-1)^n x^{-a-n} \Gamma(a+n,x)$$
$$(n=0, 1, 2, \ldots)$$

6.5.27
$$\frac{\partial^n}{\partial x^n} [e^x x^a \gamma^*(a,x)] = e^x x^{a-n} \gamma^*(a-n, x)$$
$$(n=0, 1, 2, \ldots)$$

6.5.28
$$x\frac{\partial^2 \gamma^*}{\partial x^2} + (a+1+x)\frac{\partial \gamma^*}{\partial x} + a\gamma^* = 0$$

Series Developments

6.5.29

$$\gamma^*(a,z) = e^{-z} \sum_{n=0}^\infty \frac{z^n}{\Gamma(a+n+1)} = \frac{1}{\Gamma(a)} \sum_{n=0}^\infty \frac{(-z)^n}{(a+n)n!}$$

$$(|z| < \infty)$$

6.5.30

$\gamma(a, x+y) - \gamma(a, x)$

$= e^{-x} x^{a-1} \sum_{n=0}^{\infty} \frac{(a-1)(a-2)\dots(a-n)}{x^n} [1 - e^{-y} e_n(y)]$

$(|y| < |x|)$

Continued Fraction

6.5.31

$\Gamma(a,x) = e^{-x} x^a \left(\frac{1}{x+} \frac{1-a}{1+} \frac{1}{x+} \frac{2-a}{1+} \frac{2}{x+} \dots \right)$

$(x>0, |a|<\infty)$

Asymptotic Expansions

6.5.32

$\Gamma(a, z) \sim z^{a-1} e^{-z} \left[1 + \frac{a-1}{z} + \frac{(a-1)(a-2)}{z^2} + \dots \right]$

$\left(z \to \infty \text{ in } |\arg z| < \frac{3\pi}{2} \right)$

Suppose $R_n(a,z) = u_{n+1}(a,z) + \dots$ is the remainder after n terms in this series. Then if a, z are real, we have for $n > a-2$

$|R_n(a,z)| \le |u_{n+1}(a,z)|$

and sign $R_n(a,z) = $ sign $u_{n+1}(a,z)$.

6.5.33 $\gamma(a,z) \sim \sum_{n=0}^{\infty} \frac{(-1)^n z^{a+n}}{(a+n)n!}$ $(a \to +\infty)$

6.5.34 $\lim_{n \to \infty} \frac{e_n(\alpha n)}{e^{\alpha n}} = \begin{cases} 0 \text{ for } \alpha > 1 \\ \frac{1}{2} \text{ for } \alpha = 1 \\ 1 \text{ for } 0 \le \alpha < 1 \end{cases}$

6.5.35

$\Gamma(z+1, z) \sim e^{-z} z^z \left(\sqrt{\frac{\pi}{2}} z^{\frac{1}{2}} + \frac{2}{3} + \frac{\sqrt{2\pi}}{24} \frac{1}{z^{\frac{1}{2}}} + \dots \right)$

$(z \to \infty \text{ in } | \arg z| < \frac{1}{2}\pi)$

Definite Integrals

6.5.36

$\int_0^{\infty} e^{-at} \Gamma(b, ct)\, dt = \frac{\Gamma(b)}{a} \left[1 - \frac{c^b}{(a+c)^b} \right]$ *

$(\mathscr{R}(a+c) > 0, \mathscr{R} b > -1)$

6.5.37

$\int_0^{\infty} t^{a-1} \Gamma(b, t)\, dt = \frac{\Gamma(a+b)}{a}$

$(\mathscr{R}(a+b) > 0, \quad \mathscr{R} a > 0)$

6.6. Incomplete Beta Function

6.6.1 $B_x(a,b) = \int_0^x t^{a-1} (1-t)^{b-1} dt$

6.6.2 $I_x(a,b) = B_x(a,b)/B(a,b)$

For statistical applications, see **26.5**.

Symmetry

6.6.3 $I_x(a,b) = 1 - I_{1-x}(b,a)$

Relation to Binomial Expansion

6.6.4 $I_p(a, n-a+1) = \sum_{j=a}^{n} \binom{n}{j} p^j (1-p)^{n-j}$

For binomial distribution, see **26.1**.

Recurrence Formulas

6.6.5 $I_x(a,b) = x I_x(a-1,b) + (1-x) I_x(a, b-1)$

6.6.6 $(a+b-ax) I_x(a,b)$ *

$= a(1-x) I_x(a+1, b-1) + b I_x(a, b+1)$

6.6.7 $(a+b) I_x(a,b) = a I_x(a+1, b) + b I_x(a, b+1)$

Relation to Hypergeometric Function

6.6.8 $B_x(a,b) = a^{-1} x^a F(a, 1-b; a+1; x)$

Numerical Methods

6.7. Use and Extension of the Tables

Example 1. Compute $\Gamma(6.38)$ to 8S. Using the recurrence relation **6.1.16** and **Table 6.1** we have,

$\Gamma(6.38) = [(5.38)(4.38)(3.38)(2.38)(1.38)] \Gamma(1.38)$

$= 232.43671$.

Example 2. Compute $\ln \Gamma(56.38)$, using **Table 6.4** and linear interpolation in f_2. We have

$\ln \Gamma(56.38) = (56.38 - \frac{1}{2}) \ln (56.38) - (56.38)$

$+ f_2(56.38)$

The error of linear interpolation in the table of the function f_2 is smaller than 10^{-7} in this region. Hence, $f_2(56.38) = .92041\ 67$ and $\ln \Gamma(56.38) = 169.85497\ 42$.

Direct interpolation in **Table 6.4** of $\log_{10} \Gamma(n)$ eliminates the necessity of employing logarithms. However, the error of linear interpolation is .002 so that $\log_{10} \Gamma(n)$ is obtained with a relative error of 10^{-5}.

*See page II.

Example 3. Compute $\psi(6.38)$ to 8S. Using the recurrence relation **6.3.6** and **Table 6.1**.

$$\psi(6.38) = \frac{1}{5.38} + \frac{1}{4.38} + \frac{1}{3.38} + \frac{1}{2.38} + \frac{1}{1.38} + \psi(1.38)$$

$$= 1.77275\ 59.$$

Example 4. Compute $\psi(56.38)$. Using **Table 6.3** we have $\psi(56.38) = \ln 56.38 - f_3(56.38)$.

The error of linear interpolation in the table of the function f_3 is smaller than 8×10^{-7} in this region. Hence, $f_3(56.38) = .00889\ 53$ and $\psi(56.38) = 4.023219$.

Example 5. Compute $\ln \Gamma(1-i)$. From the reflection principle **6.1.23** and **Table 6.7**, $\ln \Gamma(1-i) = \overline{\ln \Gamma(1+i)} = -.6509 + .3016i$.

Example 6. Compute $\ln \Gamma(\frac{1}{2} + \frac{1}{2}i)$. Taking the logarithm of the recurrence relation **6.1.15** we have,

$$\ln \Gamma(\tfrac{1}{2} + \tfrac{1}{2}i) = \ln \Gamma(\tfrac{3}{2} + \tfrac{1}{2}i) - \ln(\tfrac{1}{2} + \tfrac{1}{2}i)$$
$$= -.23419 + .03467i$$
$$- (\tfrac{1}{2}\ln\tfrac{1}{2} + i \arctan 1)$$
$$= .11239 - .75073i$$

The logarithms of complex numbers are found from **4.1.2**.

Example 7. Compute $\ln \Gamma(3+7i)$ using the duplication formula **6.1.18**. Taking the logarithm of **6.1.18**, we have

$$-\tfrac{1}{2}\ln 2\pi = -\ \ .91894$$
$$(\tfrac{5}{2} + 7i)\ln 2 = \ \ 1.73287 + \ \ 4.85203i$$
$$\ln \Gamma(\tfrac{3}{2} + \tfrac{7}{2}i) = -3.31598 + \ \ 2.32553i$$
$$\ln \Gamma(2 + \tfrac{7}{2}i) = -2.66047 + \ \ 2.93869i$$
$$\overline{}$$
$$\ln \Gamma(3 + 7i) = -5.16252 + 10.11625i$$

Example 8. Compute $\ln \Gamma(3+7i)$ to 5D using the asymptotic formula **6.1.41**. We have

$$\ln(3+7i) = 2.03022\ 15 + 1.16590\ 45i.$$

Then,

$$(2.5 + 7i)\ln(3+7i) = -3.0857779 + 17.1263119i$$
$$-(3+7i) = -3.0000000 - 7.0000000i$$
$$\tfrac{1}{2}\ln(2\pi) = \ \ .9189385$$
$$[12(3+7i)]^{-1} = \ \ .0043103 - \ \ .0100575i$$
$$-[360(3+7i)^3]^{-1} = \ \ .0000059 - \ \ .0000022i$$
$$\overline{}$$
$$\ln \Gamma(3+7i) = -5.16252 \ \ +10.11625i$$

6.8. Summation of Rational Series by Means of Polygamma Functions

An infinite series whose general term is a rational function of the index may always be reduced to a finite series of psi and polygamma functions. The method will be illustrated by writing the explicit formula when the denominator contains a triple root.

Let the general term of an infinite series have the form

$$u_n = \frac{p(n)}{d_1(n)d_2(n)d_3(n)}$$

where

$$d_1(n) = (n+\alpha_1)(n+\alpha_2)\ \ .\ .\ .\ (n+\alpha_m)$$
$$d_2(n) = (n+\beta_1)^2(n+\beta_2)^2\ .\ .\ .\ (n+\beta_r)^2$$
$$d_3(n) = (n+\gamma_1)^3(n+\gamma_2)^3\ .\ .\ .\ (n+\gamma_s)^3$$

where $p(n)$ is a polynomial of degree $m+2r+3s-2$ at most and where the constants α_i, β_i, and γ_i are distinct. Expand u_n in partial fractions as follows

$$u_n = \sum_{k=1}^{m} \frac{a_k}{(n+\alpha_k)} + \sum_{k=1}^{r} \frac{b_{1k}}{(n+\beta_k)} + \frac{b_{2k}}{(n+\beta_k)^2}$$

$$+ \sum_{k=1}^{s} \frac{c_{1k}}{(n+\gamma_k)} + \frac{c_{2k}}{(n+\gamma_k)^2} + \frac{c_{3k}}{(n+\gamma_k)^3}$$

$$\sum_{k=1}^{m} a_k + \sum_{k=1}^{r} b_{1k} + \sum_{k=1}^{s} c_{1k} = 0.$$

Then, we may express $\sum_{n=1}^{\infty} u_n$ in terms of the constants appearing in this partial fraction expansion as follows

$$\sum_{n=1}^{\infty} u_n = -\sum_{j=1}^{m} a_j \psi(1+\alpha_j)$$

$$- \sum_{j=1}^{r} b_{1j} \psi(1+\beta_j) + \sum_{j=1}^{r} b_{2j} \psi'(1+\beta_j)$$

$$- \sum_{j=1}^{s} c_{1j} \psi(1+\gamma_j) + \sum_{j=1}^{s} c_{2j} \psi'(1+\gamma_j)$$

$$- \sum_{j=1}^{s} \frac{c_{3j}}{2!} \psi''(1+\gamma_j).$$

Higher order repetitions in the denominator are handled similarly. If the denominator contains only simple or double roots, omit the corresponding lines.

Example 9. Find

$$s=\sum_{n=1}^{\infty}\frac{1}{(n+1)(2n+1)(4n+1)}.$$

Since

$$\frac{1}{(n+1)(2n+1)(4n+1)}=\frac{\frac{1}{3}}{n+1}-\frac{1}{n+\frac{1}{2}}+\frac{\frac{2}{3}}{n+\frac{1}{4}},$$

we have

$$\alpha_1=1,\ \alpha_2=\tfrac{1}{2},\ \alpha_3=\tfrac{1}{4},\ a_1=\tfrac{1}{3},\ a_2=-1,\ a_3=\tfrac{2}{3}.$$

Thus,

$$s=-\tfrac{1}{3}\psi(2)+\psi(1\tfrac{1}{2})-\tfrac{2}{3}\psi(1\tfrac{1}{4})=.047198.$$

Example 10.

Find $s=\sum_{n=1}^{\infty}\frac{1}{n^2(8n+1)^2}.$

Since $\frac{1}{n^2(8n+1)^2}=-\frac{16}{n}+\frac{16}{n+\frac{1}{8}}+\frac{1}{n^2}+\frac{1}{(n+\frac{1}{8})^2},$

we have,

$$\beta_1=0,\ \beta_2=\tfrac{1}{8},\ b_{11}=-16,\ b_{12}=16,\ b_{21}=1,\ b_{22}=1.$$

Therefore

$$s=16\psi(1)-16\psi(1\tfrac{1}{8})+\psi'(1)+\psi'(1\tfrac{1}{8})=.013499.$$

Example 11.

Evaluate $s=\sum_{n=1}^{\infty}\frac{1}{(n^2+1)(n^2+4)}$ (see also **6.3.13**).

We have, $\frac{1}{(n^2+1)(n^2+4)}=\frac{i}{6}\left(\frac{1}{n+i}-\frac{1}{n-i}\right)$

$$-\frac{i}{12}\left(\frac{1}{n+2i}-\frac{1}{n-2i}\right).$$

Hence, $a_1=\frac{i}{6},\ a_2=\frac{-i}{6},\ a_3=\frac{-i}{12},\ a_4=\frac{i}{12},$

$$\alpha_1=i,\ \alpha_2=-i,\ \alpha_3=2i,\ \alpha_4=-2i,$$

and therefore

$$s=\frac{-i}{6}[\psi(1+i)-\psi(1-i)]+\frac{i}{12}[\psi(1+2i)-\psi(1-2i)].$$

By **6.3.9**, this reduces to

$$s=\frac{1}{3}\mathscr{I}\psi(1+i)-\frac{1}{6}\mathscr{I}\psi(1+2i).$$

From **Table 6.8**, $s=.13876.$

References

Texts

[6.1] E. Artin, Einführung in die Theorie der Gammafunktion (Leipzig, Germany, 1931).

[6.2] P. E. Böhmer, Differenzengleichungen und bestimmte Integrale, chs. 3, 4, 5 (K. F. Koehler, Leipzig, Germany, 1939).

[6.3] G. Doetsch, Handbuch der Laplace-Transformation, vol. II, pp. 52–61 (Birkhauser, Basel, Switzerland, 1955).

[6.4] A. Erdélyi et al., Higher transcendental functions, vol. 1, ch. 1, ch. 2, sec. 5; vol. 2, ch. 9 (McGraw-Hill Book Co., Inc., New York, N.Y., 1953).

[6.5] C. Hastings, Jr., Approximations for digital computers (Princeton Univ. Press, Princeton, N.J., 1955).

[6.6] F. Lösch and F. Schoblik, Die Fakultät und verwandte Funktionen (B. G. Teubner, Leipzig, Germany, 1951).

[6.7] W. Sibagaki, Theory and applications of the gamma function (Iwanami Syoten, Tokyo, Japan, 1952).

[6.8] E. T. Whittaker and G. N. Watson, A course of modern analysis, ch. 12, 4th ed. (Cambridge Univ. Press, Cambridge, England, 1952).

Tables

[6.9] A. Abramov, Tables of ln $\Gamma(z)$ for complex argument. Translated from the Russian by D. G. Fry (Pergamon Press, New York, N.Y., 1960). ln $\Gamma(x+iy)$, $x=0(.01)10$, $y=0(.01)4$, 6D.

[6.10] Ballistic Research Laboratory, A table of the factorial numbers and their reciprocals from 1! through 1000! to 20 significant digits. Technical Note No. 381, Aberdeen Proving Ground, Md., 1951.

[6.11] British Association for the Advancement of Science, Mathematical tables, vol. 1, 3d ed., pp. 40–59 (Cambridge Univ. Press, Cambridge, England, 1951). The gamma and polygamma functions. Also $1+\int_0^x \log_{10}(t)!\,dt$, $x=0(.01)1$, 10D.

[6.12] H. T. Davis, Tables of the higher mathematical functions, 2 vols. (Principia Press, Bloomington, Ind., 1933, 1935). Extensive, many place tables of the gamma and polygamma functions up to $\psi^{(4)}(x)$ and of their logarithms.

[6.13] F. J. Duarte, Nouvelles tables de $\log_{10} n!$ à 33 décimales depuis $n=1$ jusqu'à $n=3000$ (Kundig, Geneva, Switzerland; Index Generalis, Paris, France, 1927).

[6.14] National Bureau of Standards, Tables of $n!$ and $\Gamma(n+\tfrac{1}{2})$ for the first thousand values of n, Applied Math. Series 16 (U.S. Government Printing Office, Washington, D.C., 1951). $n!, 16S; \Gamma(n+\tfrac{1}{2})$, 8S.

[6.15] National Bureau of Standards, Table of Coulomb wave functions, vol. I, pp. 114–135, Applied Math. Series 17 (U.S. Government Printing Office, Washington, D.C., 1952).

$\mathscr{R}[\Gamma'(1+i\eta)/\Gamma(1+i\eta], \eta=0(.005)2\,(.01)6\,(.02)10(.1)$

$20\,(.2)60(.5)110, 10D; \arg\Gamma(1+i\eta), \eta=0(.01)1(.02)$

$3\,(.05)10(.2)20(.4)30(.5)85,\ 8D.$

[6.16] National Bureau of Standards, Table of the gamma function for complex arguments, Applied Math. Series 34 (U.S. Government Printing Office, Washington, D.C., 1954).

$\ln\ \Gamma(x+iy),\ x=0(.1)10,\ y=0(.1)10,\quad 12D.$

Contains an extensive bibliography.

[6.17] National Physical Laboratory, Tables of Weber parabolic cylinder functions, pp. 226–233 (Her Majesty's Stationery Office, London, England, 1955).

Real and imaginary parts of $\ln\ \Gamma(\tfrac{1}{4}k+\tfrac{1}{2}ia)$, $k=0(1)3$, $a=0(.1)5(.2)20,\ \ 8D;\ \ (|\Gamma(\tfrac{3}{4}+\tfrac{1}{2}ia)/\Gamma(\tfrac{1}{4}+\tfrac{1}{2}ia)|)^{-1/2}$

$a=0(.02)1(.1)5(.2)20,\ \ 8D.$

[6.18] E. S. Pearson, Table of the logarithms of the complete Γ-function, arguments 2 to 1200, Tracts for Computers No. VIII (Cambridge Univ. Press, Cambridge, England, 1922). $\mathrm{Log}_{10}\ \Gamma(p),\ p=2(.1)$ $5(.2)70(1)1200,\ 10D.$

[6.19] J. Peters, Ten-place logarithm tables, vol. I, Appendix, pp. 58–68 (Frederick Ungar Publ. Co., New York, N.Y., 1957). $n!,\ n=1(1)60$, exact; $(n!)^{-1},\ n=1(1)43,\ 54D;\ \mathrm{Log}_{10}(n!),\ n=1(1)1200$, 18D.

[6.20] J. P. Stanley and M. V. Wilkes, Table of the reciprocal of the gamma function for complex argument (Univ. of Toronto Press, Toronto, Canada, 1950). $x=-.5(.01).5,\ y=0(.01)1,\ 6D.$

[6.21] M. Zyczkowski, Tablice funkcyj eulera i pokrewnych (Panstwowe Wydawnictwo Naukowe, Warsaw, Poland, 1954). Extensive tables of integrals involving gamma and beta functions.

For references to tabular material on the incomplete gamma and incomplete beta functions, see the references in chapter **26**.

GAMMA, DIGAMMA AND TRIGAMMA FUNCTIONS   Table 6.1

x	$\Gamma(x)$	$\ln \Gamma(x)$	$\psi(x)$	$\psi'(x)$	
1.000	1.00000 00000	0.00000 00000	-0.57721 56649	1.64493 40668	0.000
1.005	0.99713 85354	-0.00286 55666	-0.56902 09113	1.63299 41567	0.005
1.010	0.99432 58512	-0.00569 03079	-0.56088 54579	1.62121 35283	0.010
1.015	0.99156 12888	-0.00847 45187	-0.55280 85156	1.60958 91824	0.015
1.020	0.98884 42033	-0.01121 84893	-0.54478 93105	1.59811 81919	0.020
1.025	0.98617 39633	-0.01392 25067	-0.53682 70828	1.58679 76993	0.025
1.030	0.98354 99506	-0.01658 68539	-0.52892 10873	1.57562 49154	0.030
1.035	0.98097 15606	-0.01921 18101	-0.52107 05921	1.56459 71163	0.035
1.040	0.97843 82009	-0.02179 76511	-0.51327 48789	1.55371 16426	0.040
1.045	0.97594 92919	-0.02434 46490	-0.50553 32428	1.54296 58968	0.045
1.050	0.97350 42656	-0.02685 30725	-0.49784 49913	1.53235 73421	0.050
1.055	0.97110 25663	-0.02932 31868	-0.49020 94448	1.52188 35001	0.055
1.060	0.96874 36495	-0.03175 52537	-0.48262 59358	1.51154 19500	0.060
1.065	0.96642 69823	-0.03414 95318	-0.47509 38088	1.50133 03259	0.065
1.070	0.96415 20425	-0.03650 62763	-0.46761 24199	1.49124 63164	0.070
1.075	0.96191 83189	-0.03882 57395	-0.46018 11367	1.48128 76622	0.075
1.080	0.95972 53107	-0.04110 81702	-0.45279 93380	1.47145 21556	0.080
1.085	0.95757 25273	-0.04335 38143	-0.44546 64135	1.46173 76377	0.085
1.090	0.95545 94882	-0.04556 29148	-0.43818 17635	1.45214 19988	0.090
1.095	0.95338 57227	-0.04773 57114	-0.43094 47988	1.44266 31755	0.095
1.100	0.95135 07699	-0.04987 24413	-0.42375 49404	1.43329 91508	0.100
1.105	0.94935 41778	-0.05197 33384	-0.41661 16193	1.42404 79514	0.105
1.110	0.94739 55040	-0.05403 86341	-0.40951 42761	1.41490 76482	0.110
1.115	0.94547 43149	-0.05606 85568	-0.40246 23611	1.40587 63535	0.115
1.120	0.94359 01856	-0.05806 33325	-0.39545 53339	1.39695 22213	0.120
1.125	0.94174 26997	-0.06002 31841	-0.38849 26633	1.38813 34449	0.125
1.130	0.93993 14497	-0.06194 83322	-0.38157 38268	1.37941 82573	0.130
1.135	0.93815 60356	-0.06383 89946	-0.37469 83110	1.37080 49288	0.135
1.140	0.93641 60657	-0.06569 53867	-0.36786 56106	1.36229 17670	0.140
1.145	0.93471 11562	-0.06751 77212	-0.36107 52291	1.35387 71152	0.145
1.150	0.93304 09311	-0.06930 62087	-0.35432 66780	1.34555 93520	0.150
1.155	0.93140 50217	-0.07106 10569	-0.34761 94768	1.33733 68900	0.155
1.160	0.92980 30666	-0.07278 24716	-0.34095 31528	1.32920 81752	0.160
1.165	0.92823 47120	-0.07447 06558	-0.33432 72413	1.32117 16859	0.165
1.170	0.92669 96106	-0.07612 58106	-0.32774 12847	1.31322 59322	0.170
1.175	0.92519 74225	-0.07774 81345	-0.32119 48332	1.30536 94548	0.175
1.180	0.92372 78143	-0.07933 78240	-0.31468 74438	1.29760 08248	0.180
1.185	0.92229 04591	-0.08089 50733	-0.30821 86809	1.28991 86421	0.185
1.190	0.92088 50371	-0.08242 00745	-0.30178 81156	1.28232 15358	0.190
1.195	0.91951 12341	-0.08391 30174	-0.29539 53259	1.27480 81622	0.195
1.200	0.91816 87424	-0.08537 40900	-0.28903 98966	1.26737 72054	0.200
1.205	0.91685 72606	-0.08680 34780	-0.28272 14187	1.26002 73755	0.205
1.210	0.91557 64930	-0.08820 13651	-0.27643 94897	1.25275 74090	0.210
1.215	0.91432 61500	-0.08956 79331	-0.27019 37135	1.24556 60671	0.215
1.220	0.91310 59475	-0.09090 33619	-0.26398 37000	1.23845 21360	0.220
1.225	0.91191 56071	-0.09220 78291	-0.25780 90652	1.23141 44258	0.225
1.230	0.91075 48564	-0.09348 15108	-0.25166 94307	1.22445 17702	0.230
1.235	0.90962 34274	-0.09472 45811	-0.24556 44243	1.21756 30254	0.235
1.240	0.90852 10583	-0.09593 72122	-0.23949 36791	1.21074 70707	0.240
1.245	0.90744 74922	-0.09711 95744	-0.23345 68341	1.20400 28063	0.245
1.250	0.90640 24771	-0.09827 18364	-0.22745 35334	1.19732 91545	0.250
	$y!$	$\ln y!$	$\dfrac{d}{dy} \ln y!$	$\dfrac{d^2}{dy^2} \ln y!$	y
	$\begin{bmatrix} (-6)6 \\ 5 \end{bmatrix}$	$\begin{bmatrix} (-6)5 \\ 5 \end{bmatrix}$	$\begin{bmatrix} (-6)7 \\ 5 \end{bmatrix}$	$\begin{bmatrix} (-5)2 \\ 5 \end{bmatrix}$	

For $x>2$ see **Examples 1–4**. $\log_{10} e = 0.43429\ 44819$

Compiled from H. T. Davis, Tables of the higher mathematical functions, 2 vols. (Principia Press, Bloomington, Ind., 1933, 1935) (with permission). Known error has been corrected.

Table 6.1   **GAMMA, DIGAMMA AND TRIGAMMA FUNCTIONS**

x	$\Gamma(x)$	$\ln \Gamma(x)$	$\psi(x)$	$\psi'(x)$	
1.250	0.90640 24771	−0.09827 18364	−0.22745 35334	1.19732 91545	0.250
1.255	0.90538 57663	−0.09939 41651	−0.22148 34266	1.19072 50579	0.255
1.260	0.90439 71178	−0.10048 67254	−0.21554 61686	1.18418 94799	0.260
1.265	0.90343 62946	−0.10154 96809	−0.20964 14193	1.17772 14030	0.265
1.270	0.90250 30645	−0.10258 31932	−0.20376 88437	1.17131 98301	0.270
1.275	0.90159 71994	−0.10358 74224	−0.19792 81118	1.16498 37821	0.275
1.280	0.90071 84765	−0.10456 25269	−0.19211 88983	1.15871 22990	0.280
1.285	0.89986 66769	−0.10550 86634	−0.18634 08828	1.15250 44385	0.285
1.290	0.89904 15863	−0.10642 59872	−0.18059 37494	1.14635 92764	0.290
1.295	0.89824 29947	−0.10731 46519	−0.17487 71870	1.14027 59053	0.295
1.300	0.89747 06963	−0.10817 48095	−0.16919 08889	1.13425 34350	0.300
1.305	0.89672 44895	−0.10900 66107	−0.16353 45526	1.12829 09915	0.305
1.310	0.89600 41767	−0.10981 02045	−0.15790 78803	1.12238 77175	0.310
1.315	0.89530 95644	−0.11058 57384	−0.15231 05782	1.11654 27706	0.315
1.320	0.89464 04630	−0.11133 33587	−0.14674 23568	1.11075 53246	0.320
1.325	0.89399 66866	−0.11205 32100	−0.14120 29305	1.10502 45678	0.325
1.330	0.89337 80535	−0.11274 54356	−0.13569 20180	1.09934 97037	0.330
1.335	0.89278 43850	−0.11341 01772	−0.13020 93416	1.09372 99497	0.335
1.340	0.89221 55072	−0.11404 75756	−0.12475 46279	1.08816 45379	0.340
1.345	0.89167 12485	−0.11465 77697	−0.11932 76069	1.08265 27136	0.345
1.350	0.89115 14420	−0.11524 08974	−0.11392 80127	1.07719 37361	0.350
1.355	0.89065 59235	−0.11579 70951	−0.10855 55827	1.07178 68773	0.355
1.360	0.89018 45324	−0.11632 64980	−0.10321 00582	1.06643 14226	0.360
1.365	0.88973 71116	−0.11682 92401	−0.09789 11840	1.06112 66696	0.365
1.370	0.88931 35074	−0.11730 54539	−0.09259 87082	1.05587 19286	0.370
1.375	0.88891 35692	−0.11775 52707	−0.08733 23825	1.05066 65216	0.375
1.380	0.88853 71494	−0.11817 88209	−0.08209 19619	1.04550 97829	0.380
1.385	0.88818 41041	−0.11857 62331	−0.07687 72046	1.04040 10578	0.385
1.390	0.88785 42918	−0.11894 76353	−0.07168 78723	1.03533 97036	0.390
1.395	0.88754 75748	−0.11929 31538	−0.06652 37297	1.03032 50881	0.395
1.400	0.88726 38175	−0.11961 29142	−0.06138 45446	1.02535 65905	0.400
1.405	0.88700 28884	−0.11990 70405	−0.05627 00879	1.02043 36002	0.405
1.410	0.88676 46576	−0.12017 56559	−0.05118 01337	1.01555 55173	0.410
1.415	0.88654 89993	−0.12041 88823	−0.04611 44589	1.01072 17518	0.415
1.420	0.88635 57896	−0.12063 68406	−0.04107 28433	1.00593 17241	0.420
1.425	0.88618 49081	−0.12082 96505	−0.03605 50697	1.00118 48640	0.425
1.430	0.88603 62361	−0.12099 74307	−0.03106 09237	0.99648 06113	0.430
1.435	0.88590 96587	−0.12114 02987	−0.02609 01935	0.99181 84147	0.435
1.440	0.88580 50635	−0.12125 83713	−0.02114 26703	0.98719 77326	0.440
1.445	0.88572 23397	−0.12135 17638	−0.01621 81479	0.98261 80318	0.445
1.450	0.88566 13803	−0.12142 05907	−0.01131 64226	0.97807 87886	0.450
1.455	0.88562 20800	−0.12146 49657	−0.00643 72934	0.97357 94874	0.455
1.460	0.88560 43364	−0.12148 50010	−0.00158 05620	0.96911 96215	0.460
1.465	0.88560 80495	−0.12148 08083	+0.00325 39677	0.96469 86921	0.465
1.470	0.88563 31217	−0.12145 24980	0.00806 64890	0.96031 62091	0.470
1.475	0.88567 94575	−0.12140 01797	0.01285 71930	0.95597 16896	0.475
1.480	0.88574 69646	−0.12132 39621	0.01762 62684	0.95166 46592	0.480
1.485	0.88583 55520	−0.12122 39528	0.02237 39013	0.94739 46509	0.485
1.490	0.88594 51316	−0.12110 02585	0.02710 02758	0.94316 12052	0.490
1.495	0.88607 56174	−0.12095 29852	0.03180 55736	0.93896 38700	0.495
1.500	0.88622 69255	−0.12078 22376	0.03648 99740	0.93480 22005	0.500
	$y!$	$\ln y!$	$* \dfrac{d}{dy} \ln y!$	$* \dfrac{d^2}{dy^2} \ln y!$	y
	$\begin{bmatrix} (-6)4 \\ 5 \end{bmatrix}$	$\begin{bmatrix} (-6)4 \\ 4 \end{bmatrix}$	$\begin{bmatrix} (-6)4 \\ 5 \end{bmatrix}$	$\begin{bmatrix} (-6)9 \\ 5 \end{bmatrix}$	

$$\log_{10} e = 0.43429 \ 44819$$

*See page II.

GAMMA, DIGAMMA AND TRIGAMMA FUNCTIONS — Table 6.1

x	$\Gamma(x)$	$\ln \Gamma(x)$	$\psi(x)$	$\psi'(x)$	
1.500	0.88622 69255	−0.12078 22376	0.03648 99740	0.93480 22005	0.500
1.505	0.88639 89744	−0.12058 81200	0.04115 36543	0.93067 57588	0.505
1.510	0.88659 16850	−0.12037 07353	0.04579 67896	0.92658 41142	0.510
1.515	0.88680 49797	−0.12013 01860	0.05041 95527	0.92252 68425	0.515
1.520	0.88703 87833	−0.11986 65735	0.05502 21146	0.91850 35265	0.520
1.525	0.88729 30231	−0.11957 99983	0.05960 46439	0.91451 37552	0.525
1.530	0.88756 76278	−0.11927 05601	0.06416 73074	0.91055 71245	0.530
1.535	0.88786 25287	−0.11893 83580	0.06871 02697	0.90663 32361	0.535
1.540	0.88817 76586	−0.11858 34900	0.07323 36936	0.90274 16984	0.540
1.545	0.88851 29527	−0.11820 60534	0.07773 77400	0.89888 21253	0.545
1.550	0.88886 83478	−0.11780 61446	0.08222 25675	0.89505 41371	0.550
1.555	0.88924 37830	−0.11738 38595	0.08668 83334	0.89125 73596	0.555
1.560	0.88963 91990	−0.11693 92928	0.09113 51925	0.88749 14249	0.560
1.565	0.89005 45387	−0.11647 25388	0.09556 32984	0.88375 59699	0.565
1.570	0.89048 97463	−0.11598 36908	0.09997 28024	0.88005 06378	0.570
1.575	0.89094 47686	−0.11547 28415	0.10436 38544	0.87637 50766	0.575
1.580	0.89141 95537	−0.11494 00828	0.10873 66023	0.87272 89402	0.580
1.585	0.89191 40515	−0.11438 55058	0.11309 11923	0.86911 18871	0.585
1.590	0.89242 82141	−0.11380 92009	0.11742 77690	0.86552 35815	0.590
1.595	0.89296 19949	−0.11321 12579	0.12174 64754	0.86196 36921	0.595
1.600	0.89351 53493	−0.11259 17657	0.12604 74528	0.85843 18931	0.600
1.605	0.89408 82342	−0.11195 08127	0.13033 08407	0.85492 78630	0.605
1.610	0.89468 06085	−0.11128 84864	0.13459 67772	0.85145 12856	0.610
1.615	0.89529 24327	−0.11060 48737	0.13884 53988	0.84800 18488	0.615
1.620	0.89592 36685	−0.10990 00610	0.14307 68404	0.84457 92455	0.620
1.625	0.89657 42800	−0.10917 41338	0.14729 12354	0.84118 31730	0.625
1.630	0.89724 42326	−0.10842 71769	0.15148 87158	0.83781 33330	0.630
1.635	0.89793 34930	−0.10765 92746	0.15566 94120	0.83446 94315	0.635
1.640	0.89864 20302	−0.10687 05105	0.15983 34529	0.83115 11790	0.640
1.645	0.89936 98138	−0.10606 09676	0.16398 09660	0.82785 82897	0.645
1.650	0.90011 68163	−0.10523 07282	0.16811 20776	0.82459 04826	0.650
1.655	0.90088 30104	−0.10437 98739	0.17222 69122	0.82134 74802	0.655
1.660	0.90166 83712	−0.10350 84860	0.17632 55933	0.81812 90092	0.660
1.665	0.90247 28748	−0.10261 66447	0.18040 82427	0.81493 48001	0.665
1.670	0.90329 64995	−0.10170 44301	0.18447 49813	0.81176 45875	0.670
1.675	0.90413 92243	−0.10077 19212	0.18852 59282	0.80861 81094	0.675
1.680	0.90500 10302	−0.09981 91969	0.19256 12015	0.80549 51079	0.680
1.685	0.90588 18996	−0.09884 63351	0.19658 09180	0.80239 53282	0.685
1.690	0.90678 18160	−0.09785 34135	0.20058 51931	0.79931 85198	0.690
1.695	0.90770 07650	−0.09684 05088	0.20457 41410	0.79626 44350	0.695
1.700	0.90863 87329	−0.09580 76974	0.20854 78749	0.79323 28302	0.700
1.705	0.90959 57079	−0.09475 50552	0.21250 65064	0.79022 34645	0.705
1.710	0.91057 16796	−0.09368 26573	0.21645 01462	0.78723 61012	0.710
1.715	0.91156 66390	−0.09259 05785	0.22037 89037	0.78427 05060	0.715
1.720	0.91258 05779	−0.09147 88929	0.22429 28871	0.78132 64486	0.720
1.725	0.91361 34904	−0.09034 76741	0.22819 22037	0.77840 37011	0.725
1.730	0.91466 53712	−0.08919 69951	0.23207 69593	0.77550 20396	0.730
1.735	0.91573 62171	−0.08802 69286	0.23594 72589	0.77262 12424	0.735
1.740	0.91682 60252	−0.08683 75466	0.23980 32061	0.76976 10915	0.740
1.745	0.91793 47950	−0.08562 89203	0.24364 49038	0.76692 13714	0.745
1.750	0.91906 25268	−0.08440 11210	0.24747 24535	0.76410 18699	0.750
	$y!$	$\ln y!$	$\dfrac{d}{dy} \ln y!$	$\dfrac{d^2}{dy^2} \ln y!$	y
	$\begin{bmatrix}(-6)3 \\ 4\end{bmatrix}$	$\begin{bmatrix}(-6)3 \\ 4\end{bmatrix}$	$\begin{bmatrix}(-6)3 \\ 4\end{bmatrix}$	$\begin{bmatrix}(-6)4 \\ 5\end{bmatrix}$	

$\log_{10} e = 0.43429\ 44819$

Table 6.1　　　　GAMMA, DIGAMMA AND TRIGAMMA FUNCTIONS

x	$\Gamma(x)$	$\ln \Gamma(x)$	$\psi(x)$	$\psi'(x)$	
1.750	0.91906 25268	−0.08440 11210	0.24747 24535	0.76410 18699	0.750
1.755	0.92020 92224	−0.08315 42192	0.25128 59559	0.76130 23773	0.755
1.760	0.92137 48846	−0.08188 82847	0.25508 55103	0.75852 26870	0.760
1.765	0.92255 95178	−0.08060 33871	0.25887 12154	0.75576 25950	0.765
1.770	0.92376 31277	−0.07929 95955	0.26264 31686	0.75302 19003	0.770
1.775	0.92498 57211	−0.07797 69782	0.26640 14664	0.75030 04040	0.775
1.780	0.92622 73062	−0.07663 56034	0.27014 62043	0.74759 79107	0.780
1.785	0.92748 78926	−0.07527 55386	0.27387 74769	0.74491 42268	0.785
1.790	0.92876 74904	−0.07389 68509	0.27759 53776	0.74224 91617	0.790
1.795	0.93006 61123	−0.07249 96070	0.28129 99992	0.73960 25271	0.795
1.800	0.93138 37710	−0.07108 38729	0.28499 14333	0.73697 41375	0.800
1.805	0.93272 04811	−0.06964 97145	0.28866 97707	0.73436 38093	0.805
1.810	0.93407 62585	−0.06819 71969	0.29233 51012	0.73177 13620	0.810
1.815	0.93545 11198	−0.06672 63850	0.29598 75138	0.72919 66166	0.815
1.820	0.93684 50832	−0.06523 73431	0.29962 70966	0.72663 93972	0.820
1.825	0.93825 81682	−0.06373 01353	0.30325 39367	0.72409 95297	0.825
1.830	0.93969 03951	−0.06220 48248	0.30686 81205	0.72157 68426	0.830
1.835	0.94114 17859	−0.06066 14750	0.31046 97335	0.71907 11662	0.835
1.840	0.94261 23634	−0.05910 01483	0.31405 88602	0.71658 23333	0.840
1.845	0.94410 21519	−0.05752 09071	0.31763 55846	0.71411 01788	0.845
1.850	0.94561 11764	−0.05592 38130	0.32119 99895	0.71165 45396	0.850
1.855	0.94713 94637	−0.05430 89276	0.32475 21572	0.70921 52546	0.855
1.860	0.94868 70417	−0.05267 63117	0.32829 21691	0.70679 21650	0.860
1.865	0.95025 39389	−0.05102 60260	0.33182 01056	0.70438 51138	0.865
1.870	0.95184 01855	−0.04935 81307	0.33533 60467	0.70199 39461	0.870
1.875	0.95344 58127	−0.04767 26854	0.33884 00713	0.69961 85089	0.875
1.880	0.95507 08530	−0.04596 97497	0.34233 22577	0.69725 86512	0.880
1.885	0.95671 53398	−0.04424 93824	0.34581 26835	0.69491 42236	0.885
1.890	0.95837 93077	−0.04251 16423	0.34928 14255	0.69258 50790	0.890
1.895	0.96006 27927	−0.04075 65875	0.35273 85596	0.69027 10717	0.895
1.900	0.96176 58319	−0.03898 42759	0.35618 41612	0.68797 20582	0.900
1.905	0.96348 84632	−0.03719 47650	0.35961 83049	0.68568 78965	0.905
1.910	0.96523 07261	−0.03538 81118	0.36304 10646	0.68341 84465	0.910
1.915	0.96699 26608	−0.03356 43732	0.36645 25136	0.68116 35696	0.915
1.920	0.96877 43090	−0.03172 36054	0.36985 27244	0.67892 31293	0.920
1.925	0.97057 57134	−0.02986 58646	0.37324 17688	0.67669 69903	0.925
1.930	0.97239 69178	−0.02799 12062	0.37661 97179	0.67448 50194	0.930
1.935	0.97423 79672	−0.02609 96858	0.37998 66424	0.67228 70846	0.935
1.940	0.97609 89075	−0.02419 13581	0.38334 26119	0.67010 30559	0.940
1.945	0.97797 97861	−0.02226 62778	0.38668 76959	0.66793 28044	0.945
1.950	0.97988 06513	−0.02032 44991	0.39002 19627	0.66577 62034	0.950
1.955	0.98180 15524	−0.01836 60761	0.39334 54805	0.66363 31270	0.955
1.960	0.98374 25404	−0.01639 10621	0.39665 83163	0.66150 34514	0.960
1.965	0.98570 36664	−0.01439 95106	0.39996 05371	0.65938 70538	0.965
1.970	0.98768 49838	−0.01239 14744	0.40325 22088	0.65728 38134	0.970
1.975	0.98968 65462	−0.01036 70060	0.40653 33970	0.65519 36104	0.975
1.980	0.99170 84087	−0.00832 61578	0.40980 41664	0.65311 63266	0.980
1.985	0.99375 06274	−0.00626 89816	0.41306 45816	0.65105 18450	0.985
1.990	0.99581 32598	−0.00419 55291	0.41631 47060	0.64900 00505	0.990
1.995	0.99789 63643	−0.00210 58516	0.41955 46030	0.64696 08286	0.995
2.000	1.00000 00000	0.00000 00000	0.42278 43351	0.64493 40668	1.000

$y!$	$\ln y!$	$\dfrac{d}{dy} \ln y!$	$\dfrac{d^2}{dy^2} \ln y!$	y
$\begin{bmatrix} (-6)2 \\ 4 \end{bmatrix}$	$\begin{bmatrix} (-6)2 \\ 4 \end{bmatrix}$	$\begin{bmatrix} (-6)2 \\ 4 \end{bmatrix}$	$\begin{bmatrix} (-6)2 \\ 4 \end{bmatrix}$	

$\log_{10} e = 0.43429\ 44819$

TETRAGAMMA AND PENTAGAMMA FUNCTIONS

Table 6.2

x	$\psi''(x)$	$\psi^{(3)}(x)$	x	x	$\psi''(x)$	$\psi^{(3)}(x)$	x
1.00	−2.40411 38063	6.49393 94023	0.00	1.50	−0.82879 66442	1.40909 10340	0.50
1.01	−2.34039 86771	6.25106 18729	0.01	1.51	−0.81487 76121	1.37489 70527	0.51
1.02	−2.27905 42052	6.01969 49890	0.02	1.52	−0.80129 51399	1.34177 21104	0.52
1.03	−2.21996 85963	5.79918 38573	0.03	1.53	−0.78803 87419	1.30967 56244	0.53
1.04	−2.16303 63855	5.58891 68399	0.04	1.54	−0.77509 83287	1.27856 88154	0.54
1.05	−2.10815 80219	5.38832 23132	0.05	1.55	−0.76246 41904	1.24841 46160	0.55
1.06	−2.05523 94833	5.19686 56970	0.06	1.56	−0.75012 69793	1.21917 75841	0.56
1.07	−2.00419 19194	5.01404 67303	0.07	1.57	−0.73807 76946	1.19082 38216	0.57
1.08	−1.95493 13213	4.83939 69702	0.08	1.58	−0.72630 76669	1.16332 08979	0.58
1.09	−1.90737 82154	4.67247 74947	0.09	1.59	−0.71480 85441	1.13663 77770	0.59
1.10	−1.86145 73783	4.51287 67903	0.10	1.60	−0.70357 22779	1.11074 47490	0.60
1.11	−1.81709 75731	4.36020 88083	0.11	1.61	−0.69259 11105	1.08561 33658	0.61
1.12	−1.77423 13035	4.21411 11755	0.12	1.62	−0.68185 75627	1.06121 63792	0.62
1.13	−1.73279 45852	4.07424 35447	0.13	1.63	−0.67136 44220	1.03752 76035	0.63
1.14	−1.69272 67342	3.94028 60737	0.14	1.64	−0.66110 47316	1.01452 22608	0.64
1.15	−1.65397 01677	3.81193 80220	0.15	1.65	−0.65107 17793	0.99217 61290	0.65
1.16	−1.61647 02206	3.68891 64540	0.16	1.66	−0.64125 90881	0.97046 62927	0.66
1.17	−1.58017 49731	3.57095 50416	0.17	1.67	−0.63166 04061	0.94937 06973	0.67
1.18	−1.54503 50903	3.45780 29554	0.18	1.68	−0.62226 96973	0.92886 81843	0.68
1.19	−1.51100 36723	3.34922 38402	0.19	1.69	−0.61308 11332	0.90893 84502	0.69
1.20	−1.47803 61144	3.24499 48647	0.20	1.70	−0.60408 90841	0.88956 20066	0.70
1.21	−1.44608 99765	3.14490 58422	0.21	1.71	−0.59528 81112	0.87072 01433	0.71
1.22	−1.41512 48602	3.04875 84139	0.22	1.72	−0.58667 29593	0.85239 48922	0.72
1.23	−1.38510 22950	2.95636 52925	0.23	1.73	−0.57823 85490	0.83456 89940	0.73
1.24	−1.35598 56308	2.86754 95589	0.24	1.74	−0.56997 99702	0.81722 58660	0.74
1.25	−1.32773 99375	2.78214 40092	0.25	1.75	−0.56189 24756	0.80034 95719	0.75
1.26	−1.30033 19112	2.69999 05478	0.26	1.76	−0.55397 14738	0.78392 47929	0.76
1.27	−1.27372 97857	2.62093 96227	0.27	1.77	−0.54621 25238	0.76793 68005	0.77
1.28	−1.24790 32496	2.54484 97000	0.28	1.78	−0.53861 13291	0.75237 14300	0.78
1.29	−1.22282 33691	2.47158 67746	0.29	1.79	−0.53116 37320	0.73721 50564	0.79
1.30	−1.19846 25147	2.40102 39143	0.30	1.80	−0.52386 57084	0.72245 45705	0.80
1.31	−1.17479 42923	2.33304 08348	0.31	1.81	−0.51671 33630	0.70807 73565	0.81
1.32	−1.15179 34794	2.26752 35032	0.32	1.82	−0.50970 29242	0.69407 12710	0.82
1.33	−1.12943 59642	2.20436 37678	0.33	1.83	−0.50283 07396	0.68042 46226	0.83
1.34	−1.10769 86881	2.14345 90132	0.34	1.84	−0.49609 32712	0.66712 61527	0.84
1.35	−1.08655 95925	2.08471 18367	0.35	1.85	−0.48948 70921	0.65416 50169	0.85
1.36	−1.06599 75682	2.02802 97472	0.36	1.86	−0.48300 88813	0.64153 07680	0.86
1.37	−1.04599 24073	1.97332 48830	0.37	1.87	−0.47665 54207	0.62921 33389	0.87
1.38	−1.02652 47586	1.92051 37473	0.38	1.88	−0.47042 35909	0.61720 30270	0.88
1.39	−1.00757 60850	1.86951 69616	0.39	1.89	−0.46431 03677	0.60549 04793	0.89
1.40	−0.98912 86236	1.82025 90339	0.40	1.90	−0.45831 28188	0.59406 66772	0.90
1.41	−0.97116 53479	1.77266 81419	0.41	1.91	−0.45242 81007	0.58292 29238	0.91
1.42	−0.95366 99322	1.72667 59295	0.42	1.92	−0.44665 34549	0.57205 08299	0.92
1.43	−0.93662 67177	1.68221 73161	0.43	1.93	−0.44098 62055	0.56144 23020	0.93
1.44	−0.92002 06808	1.63923 03178	0.44	1.94	−0.43542 37563	0.55108 95304	0.94
1.45	−0.90383 74031	1.59765 58792	0.45	1.95	−0.42996 35876	0.54098 49774	0.95
1.46	−0.88806 30426	1.55743 77157	0.46	1.96	−0.42460 32537	0.53112 13668	0.96
1.47	−0.87268 43070	1.51852 21649	0.47	1.97	−0.41934 03805	0.52149 16733	0.97
1.48	−0.85768 84281	1.48085 80478	0.48	1.98	−0.41417 26631	0.51208 91127	0.98
1.49	−0.84306 31376	1.44439 65370	0.49	1.99	−0.40909 78630	0.50290 71324	0.99
1.50	−0.82879 66442	1.40909 10340	0.50	2.00	−0.40411 38063	0.49393 94023	1.00

$*$

$\dfrac{d^3}{dy^3}\ln y!$	$\dfrac{d^4}{dy^4}\ln y!$	y	$\dfrac{d^3}{dy^3}\ln y!$	$\dfrac{d^4}{dy^4}\ln y!$	y
$\begin{bmatrix}(-4)3\\7\end{bmatrix}$	$\begin{bmatrix}(-3)1\\7\end{bmatrix}$		$\begin{bmatrix}(-5)4\\6\end{bmatrix}$	$\begin{bmatrix}(-4)1\\6\end{bmatrix}$	

$*$

Compiled from H. T. Davis, Tables of the higher mathematical functions, 2 vols. (Principia Press, Bloomington, Ind., 1933, 1935) (with permission).

*See page II.

Table 6.3 GAMMA AND DIGAMMA FUNCTIONS FOR INTEGER AND HALF-INTEGER VALUES

n	$\Gamma(n)$	$1/\Gamma(n)$	$\Gamma(n+\tfrac{1}{2})$	$\psi(n)$	$f_1(n)$	$f_3(n)$
1	(0)1.00000 00000	(0)1.00000 000	(−1)8.86226 93	−0.57721 56649	1.08443 755	0.57721 566
2	(0)1.00000 00000	(0)1.00000 000	(0)1.32934 04	+0.42278 43351	1.04220 712	0.27036 285
3	(0)2.00000 00000	(− 1)5.00000 000	(0)3.32335 10	0.92278 43351	1.02806 452	0.17582 795
4	(0)6.00000 00000	(− 1)1.66666 667	(1)1.16317 28	1.25611 76684	1.02100 830	0.13017 669
5	(1)2.40000 00000	(− 2)4.16666 667	(1)5.23427 78	1.50611 76684	1.01678 399	0.10332 024
6	(2)1.20000 00000	(− 3)8.33333 333	(2)2.87885 28	1.70611 76684	1.01397 285	0.08564 180
7	(2)7.20000 00000	(− 3)1.38888 889	(3)1.87125 43	1.87278 43351	1.01196 776	0.07312 581
8	(3)5.04000 00000	(− 4)1.98412 698	(4)1.40344 07	2.01564 14780	1.01046 565	0.06380 006
9	(4)4.03200 00000	(− 5)2.48015 873	(5)1.19292 46	2.14064 14780	1.00929 843	0.05658 310
10	(5)3.62880 00000	(− 6)2.75573 192	(6)1.13327 84	2.25175 25891	1.00836 536	0.05083 250
11	(6)3.62880 00000	(− 7)2.75573 192	(7)1.18994 23	2.35175 25891	1.00760 243	0.04614 268
12	(7)3.99168 00000	(− 8)2.50521 084	(8)1.36843 37	2.44266 16800	1.00696 700	0.04224 497
13	(8)4.79001 60000	(− 9)2.08767 570	(9)1.71054 21	2.52599 50133	1.00642 958	0.03895 434
14	(9)6.22702 08000	(−10)1.60590 438	(10)2.30923 18	2.60291 80902	1.00596 911	0.03613 924
15	(10)8.71782 91200	(−11)1.14707 456	(11)3.34838 61	2.67434 66617	1.00557 019	0.03370 354
16	(12)1.30767 43680	(−13)7.64716 373	(12)5.18999 85	2.74101 33283	1.00522 124	0.03157 539
17	(13)2.09227 89888	(−14)4.77947 733	(13)8.56349 74	2.80351 33283	1.00491 343	0.02970 002
18	(14)3.55687 42810	(−15)2.81145 725	(15)1.49861 21	2.86233 68577	1.00463 988	0.02803 490
19	(15)6.40237 37057	(−16)1.56192 070	(16)2.77243 23	2.91789 24133	1.00439 519	0.02654 657
20	(17)1.21645 10041	(−18)8.22063 525	(17)5.40624 30	2.97052 39922	1.00417 501	0.02520 828
21	(18)2.43290 20082	(−19)4.11031 762	(19)1.10827 98	3.02052 39922	1.00397 584	0.02399 845
22	(19)5.10909 42172	(−20)1.95729 411	(20)2.38280 16	3.06814 30399	1.00379 480	0.02289 941
23	(21)1.12400 07278	(−22)8.89679 139	(21)5.36130 36	3.11359 75853	1.00362 953	0.02189 663
24	(22)2.58520 16739	(−23)3.86817 017	(23)1.25990 63	3.15707 58462	1.00347 806	0.02097 798
25	(23)6.20448 40173	(−24)1.61173 757	(24)3.08677 05	3.19874 25129	1.00333 872	0.02013 331
26	(25)1.55112 10043	(−26)6.44695 029	(25)7.87126 49	3.23874 25129	1.00321 011	0.01935 403
27	(26)4.03291 46113	(−27)2.47959 626	(27)2.08588 52	3.27720 40513	1.00309 105	0.01863 281
28	(28)1.08888 69450	(−29)9.18368 986	(28)5.73618 43	3.31424 10884	1.00298 050	0.01796 342
29	(29)3.04888 34461	(−30)3.27988 924	(30)1.63481 25	3.34995 53741	1.00287 758	0.01734 046
30	(30)8.84176 19937	(−31)1.13099 629	(31)4.82269 69	3.38443 81327	1.00278 154	0.01675 925
31	(32)2.65252 85981	(−33)3.76998 763	(33)1.47092 26	3.41777 14660	1.00269 170	0.01621 574
32	(33)8.22283 86542	(−34)1.21612 504	(34)4.63340 61	3.45002 95305	1.00260 748	0.01570 637
33	(35)2.63130 83693	(−36)3.80039 076	(36)1.50585 70	3.48127 95305	1.00252 837	0.01522 803
34	(36)8.68331 76188	(−37)1.15163 356	(37)5.04462 09	3.51158 25608	1.00245 392	0.01477 796
35	(38)2.95232 79904	(−39)3.38715 754	(39)1.74039 42	3.54099 43255	1.00238 372	0.01435 374
36	(40)1.03331 47966	(−41)9.67759 296	(40)6.17839 94	3.56956 57541	1.00231 744	0.01395 318
37	(41)3.71993 32679	(−42)2.68822 027	(42)2.25511 58	3.59734 35319	1.00225 474	0.01357 438
38	(43)1.37637 53091	(−44)7.26546 018	(43)8.45668 42	3.62437 05589	1.00219 534	0.01321 560
39	(44)5.23022 61747	(−45)1.91196 320	(45)3.25582 34	3.65068 63484	1.00213 899	0.01287 530
40	(46)2.03978 82081	(−47)4.90246 976	(47)1.28605 02	3.67632 73740	1.00208 546	0.01255 208
41	(47)8.15915 28325	(−48)1.22561 744	(48)5.20850 35	3.70132 73740	1.00203 455	0.01224 469
42	(49)3.34525 26613	(−50)2.98931 083	(50)2.16152 90	3.72571 76179	1.00198 606	0.01195 200
43	(51)1.40500 61178	(−52)7.11740 673	(51)9.18649 81	3.74952 71417	1.00193 983	0.01167 297
44	(52)6.04152 63063	(−53)1.65521 087	(53)3.99612 67	3.77278 29557	1.00189 570	0.01140 668
45	(54)2.65827 15748	(−55)3.76184 288	(55)1.77827 64	3.79551 02284	1.00185 354	0.01115 226
46	(56)1.19622 22087	(−57)8.35965 084	(56)8.09115 74	3.81773 24506	1.00181 321	0.01090 895
47	(57)5.50262 21598	(−58)1.81731 540	(58)3.76238 82	3.83947 15811	1.00177 460	0.01067 602
48	(59)2.58623 24151	(−60)3.86662 851	(60)1.78713 44	3.86074 81768	1.00173 759	0.01045 283
49	(61)1.24139 15593	(−62)8.05547 607	(61)8.66760 18	3.88158 15102	1.00170 210	0.01023 879
50	(62)6.08281 86403	(−63)1.64397 471	(63)4.29046 29	3.90198 96734	1.00166 803	0.01003 333
51	(64)3.04140 93202	(−65)3.28794 942	(65)2.16668 38	3.92198 96734	1.00163 530	0.00983 596
	$(n-1)!$	$1/(n-1)!$	$(n-\tfrac{1}{2})!$	$\dfrac{d}{dn} \, ln(n-1)!$ *		

$$n! = (2\pi)^{\frac{1}{2}} n^{n+\frac{1}{2}} e^{-n} f_1(n) \qquad \Gamma(n) = (2\pi)^{\frac{1}{2}} n^{n-\frac{1}{2}} e^{-n} f_1(n) \qquad \psi(n) = \ln n - f_3(n) \qquad (2\pi)^{\frac{1}{2}} = 2.50662\ 82746\ 31001$$

$\psi(n)$ compiled from H. T. Davis, Tables of the higher mathematical functions, 2 vols. (Principia Press, Bloomington, Ind., 1933, 1935) (with permission).

---

*See page II.

GAMMA AND DIGAMMA FUNCTIONS FOR INTEGER AND HALF-INTEGER VALUES   Table 6.3

n	$\Gamma(n)$	$1/\Gamma(n)$	$\Gamma(n+\tfrac{1}{2})$	$\psi(n)$	$f_1(n)$	$f_3(n)$
51	(64)3.04140 93202	(− 65)3.28794 942	(65)2.16668 38	3.92198 96734	1.00163 530	0.00983 596
52	(66)1.55111 87533	(− 67)6.44695 964	(67)1.11584 21	3.94159 75166	1.00160 383	0.00964 620
53	(67)8.06581 75171	(− 68)1.23979 993	(68)5.85817 12	3.96082 82858	1.00157 355	0.00946 363
54	(69)4.27488 32841	(− 70)2.33924 515	(70)3.13412 16	3.97969 62103	1.00154 438	0.00928 784
55	(71)2.30843 69734	(− 72)4.33193 547	(72)1.70809 63	3.99821 47288	1.00151 628	0.00911 846
56	(73)1.26964 03354	(− 74)7.87624 631	(73)9.47993 44	4.01639 65470	1.00148 919	0.00895 514
57	(74)7.10998 58780	(− 75)1.40647 255	(75)5.35616 29	4.03425 36899	1.00146 304	0.00879 758
58	(76)4.05269 19505	(− 77)2.46749 571	(77)3.07979 37	4.05179 75495	1.00143 780	0.00864 546
59	(78)2.35056 13313	(− 79)4.25430 295	(79)1.80167 93	4.06903 89288	1.00141 341	0.00849 852
60	(80)1.38683 11855	(− 81)7.21068 296	(81)1.07199 92	4.08598 80814	1.00138 984	0.00835 648
61	(81)8.32090 71127	(− 82)1.20178 049	(82)6.48559 51	4.10265 47481	1.00136 704	0.00821 912
62	(83)5.07580 21388	(− 84)1.97013 196	(84)3.98864 10	4.11904 81907	1.00134 498	0.00808 619
63	(85)3.14699 73260	(− 86)3.17763 219	(86)2.49290 06	4.13517 72229	1.00132 362	0.00795 750
64	(87)1.98260 83154	(− 88)5.04386 062	(88)1.58299 19	4.15105 02388	1.00130 292	0.00783 284
65	(89)1.26886 93219	(− 90)7.88103 221	(90)1.02102 98	4.16667 52388	1.00128 286	0.00771 203
66	(90)8.24765 05921	(− 91)1.21246 649	(91)6.68774 50	4.18205 98542	1.00126 341	0.00759 489
67	(92)5.44344 93908	(− 93)1.83707 044	(93)4.44735 04	4.19721 13693	1.00124 455	0.00748 125
68	(94)3.64711 10918	(− 95)2.74189 619	(95)3.00196 15	4.21213 67425	1.00122 623	0.00737 096
69	(96)2.48003 55424	(− 97)4.03220 028	(97)2.05634 36	4.22684 26248	1.00120 845	0.00726 388
70	(98)1.71122 45243	(− 99)5.84376 852	(99)1.42915 88	4.24133 53785	1.00119 118	0.00715 986
71	(100)1.19785 71670	(−101)8.34824 074	(101)1.00755 70	4.25562 10927	1.00117 439	0.00705 878
72	(101)8.50478 58857	(−102)1.17580 856	(102)7.20403 24	4.26970 55998	1.00115 807	0.00696 052
73	(103)6.12344 58377	(−104)1.63306 744	(104)5.22292 35	4.28359 44887	1.00114 220	0.00686 495
74	(105)4.47011 54615	(−106)2.23707 868	(106)3.83884 87	4.29729 31188	1.00112 675	0.00677 197
75	(107)3.30788 54415	(−108)3.02307 930	(108)2.85994 23	4.31080 66323	1.00111 172	0.00668 148
76	(109)2.48091 40811	(−110)4.03077 240	(110)2.15925 64	4.32413 99657	1.00109 709	0.00659 337
77	(111)1.88549 47017	(−112)5.30364 789	(112)1.65183 12	4.33729 78604	1.00108 283	0.00650 756
78	(113)1.45183 09203	(−114)6.88785 441	(114)1.28016 92	4.35028 48734	1.00106 894	0.00642 395
79	(115)1.13242 81178	(−116)8.83058 257	(116)1.00493 28	4.36310 53862	1.00105 540	0.00634 247
80	(116)8.94618 21308	(−117)1.11779 526	(117)7.98921 57	4.37576 36140	1.00104 220	0.00626 302
81	(118)7.15694 57046	(−119)1.39724 408	(119)6.43131 87	4.38826 36140	1.00102 933	0.00618 554
82	(120)5.79712 60207	(−121)1.72499 269	(121)5.24152 47	4.40060 92931	1.00101 677	0.00610 995
83	(122)4.75364 33370	(−123)2.10364 962	(123)4.32425 79	4.41280 44150	1.00100 452	0.00603 619
84	(124)3.94552 39697	(−125)2.53451 761	(125)3.61075 53	4.42485 26078	1.00099 255	0.00596 419
85	(126)3.31424 01346	(−127)3.01728 287	(127)3.05108 83	4.43675 73697	1.00098 087	0.00589 389
86	(128)2.81710 41144	(−129)3.54974 456	(129)2.60868 05	4.44852 20756	1.00096 946	0.00582 522
87	(130)2.42270 95384	(−131)4.12760 995	(131)2.25650 86	4.46014 99825	1.00095 831	0.00575 814
88	(132)2.10775 72984	(−133)4.74437 926	(133)1.97444 50	4.47164 42354	1.00094 741	0.00569 258
89	(134)1.85482 64226	(−135)5.39134 006	(135)1.74738 38	4.48300 78718	1.00093 676	0.00562 850
90	(136)1.65079 55161	(−137)6.05768 546	(137)1.56390 85	4.49424 38268	1.00092 635	0.00556 584
91	(138)1.48571 59645	(−139)6.73076 163	(139)1.41533 72	4.50535 49379	1.00091 617	0.00550 457
92	(140)1.35200 15277	(−141)7.39644 134	(141)1.29503 36	4.51634 39489	1.00090 620	0.00544 463
93	(142)1.24384 14055	(−143)8.03961 016	(143)1.19790 60	4.52721 35142	1.00089 646	0.00538 598
94	(144)1.15677 25071	(−145)8.64474 211	(145)1.12004 22	4.53796 62023	1.00088 691	0.00532 858
95	(146)1.08736 61567	(−147)9.19653 415	(147)1.05843 98	4.54860 45002	1.00087 757	0.00527 239
96	(148)1.03299 78488	(−149)9.68056 227	(149)1.01081 00	4.55913 08160	1.00086 843	0.00521 738
97	(149)9.91677 93487	(−150)1.00839 190	(150)9.75431 69	4.56954 74827	1.00085 947	0.00516 350
98	(151)9.61927 59682	(−152)1.03957 928	(152)9.51045 90	4.57985 67610	1.00085 070	0.00511 072
99	(153)9.42689 04489	(−154)1.06079 519	(154)9.36780 21	4.59006 08426	1.00084 210	0.00505 901
100	(155)9.33262 15444	(−156)1.07151 029	(156)9.32096 31	4.60016 18527	1.00083 368	0.00500 833
101	(157)9.33262 15444	(−158)1.07151 029	(158)9.36756 79	4.61016 18527	1.00082 542	0.00495 866
	$(n-1)!$	$1/(n-1)!$	$(n-\tfrac{1}{2})!$	$*\,\dfrac{d}{dn}\ln(n-1)!$	$\begin{bmatrix}(-7)2\\3\end{bmatrix}$	$\begin{bmatrix}(-6)1\\4\end{bmatrix}$

$$n!=(2\pi)^{\frac{1}{2}}n^{n+\frac{1}{2}}e^{-n}f_1(n) \qquad \Gamma(n)=(2\pi)^{\frac{1}{2}}n^{n-\frac{1}{2}}e^{-n}f_1(n) \qquad \psi(n)=\ln n-f_3(n) \qquad (2\pi)^{\frac{1}{2}}=2.50662\ 82746\ 31001$$

*See page II.

Table 6.4 **LOGARITHMS OF THE GAMMA FUNCTION**

n	$\log_{10} \Gamma(n)$	$\log_{10} \Gamma(n+\tfrac{1}{3})$	$\log_{10} \Gamma(n+\tfrac{1}{2})$	$\log_{10} \Gamma(n+\tfrac{2}{3})$	$f_2(n)$
1	0.00000 000	-0.04915 851	-0.05245 506	-0.04443 477	1.00000 000
2	0.00000 000	+0.07578 023	+0.12363 620	+0.17741 398	0.96027 923
3	0.30103 000	0.44375 702	0.52157 621	0.60338 271	0.94661 646
4	0.77815 125	0.96663 576	1.06564 43	1.16765 41	0.93972 921
5	1.38021 12	1.60345 79	1.71885 68	1.83666 09	0.93558 323
6	2.07918 12	2.33045 66	2.45921 95	2.58998 86	0.93281 466
7	2.85733 25	3.13208 89	3.27213 28	3.41389 73	0.93083 524
8	3.70243 05	3.99739 04	4.14719 41	4.29850 39	0.92934 980
9	4.60552 05	4.91820 91	5.07661 30	5.23635 60	0.92819 400
10	5.55976 30	5.88824 59	6.05433 66	6.22163 27	0.92726 910
11	6.55976 30	6.90248 63	7.07552 59	7.24966 15	0.92651 221
12	7.60115 57	7.95684 40	8.13622 37	8.31660 83	0.92588 137
13	8.68033 70	9.04792 45	9.23313 38	9.41927 06	0.92534 753
14	9.79428 03	10.17286 3	10.36346 8	10.55493 3	0.92488 990
15	10.94040 8	11.32921 0	11.52483 6	11.72126 5	0.92449 327
16	12.116500	12.514847	12.715167	12.916241	0.92414 619
17	13.320620	13.727922	13.932651	14.138090	0.92383 993
18	14.551069	14.966804	15.175689	15.385245	0.92356 769
19	15.806341	16.230045	16.442861	16.656311	0.92332 409
20	17.085095	17.516352	17.732896	17.950042	0.92310 485
21	18.386125	18.824561	19.044649	19.265313	0.92290 649
22	19.708344	20.153619	20.377088	20.601105	0.92272 615
23	21.050767	21.502573	21.729270	21.956492	0.92256 149
24	22.412494	22.870550	23.100338	23.330629	0.92241 055
25	23.792706	24.256751	24.489504	24.722740	0.92227 169
26	25.190646	25.660444	25.896045	26.132109	0.92214 350
27	26.605619	27.080949	27.319290	27.558078	0.92202 481
28	28.036983	28.517642	28.758623	29.000035	0.92191 460
29	29.484141	29.969940	30.213468	30.457412	0.92181 198
30	30.946539	31.437301	31.683290	31.929681	0.92171 621
31	32.423660	32.919221	33.167590	33.416347	0.92162 661
32	33.915022	34.415228	34.665900	34.916950	0.92154 262
33	35.420172	35.924878	36.177784	36.431055	0.92146 371
34	36.938686	37.447757	37.702829	37.958255	0.92138 944
35	38.470165	38.983473	39.240648	39.498167	0.92131 942
36	40.014233	40.531658	40.790876	41.050429	0.92125 329
37	41.570535	42.091963	42.353169	42.614701	0.92119 073
38	43.138737	43.664060	43.927200	44.190658	0.92113 146
39	44.718520	45.247636	45.512661	45.777995	0.92107 524
40	46.309585	46.842397	47.109258	47.376420	0.92102 182
41	47.911645	48.448061	48.716713	48.985659	0.92097 101
42	49.524429	50.064362	50.334761	50.605448	0.92092 262
43	51.147678	51.691044	51.963150	52.235536	0.92087 648
44	52.781147	53.327866	53.601639	53.875686	0.92083 244
45	54.424599	54.974597	55.249999	55.525670	0.92079 035
46	56.077812	56.631014	56.908011	57.185269	0.92075 010
47	57.740570	58.296908	58.575464	58.854276	0.92071 156
48	59.412668	59.972075	60.252157	60.532491	0.92067 462
49	61.093909	61.656322	61.937899	62.219723	0.92063 919
50	62.784105	63.349462	63.632504	63.915788	0.92060 518
51	64.483075	65.051318	65.335796	65.620510	0.92057 250
	$\log_{10}(n-1)!$	$\log_{10}(n-\tfrac{2}{3})!$	$\log_{10}(n-\tfrac{1}{2})!$	$\log_{10}(n-\tfrac{1}{3})!$	

$$\ln \Gamma(n) = \ln (n-1)! = (n-\tfrac{1}{2}) \ln n - n + f_2(n) \qquad \ln 10 = 2.30258\ 509299$$

$\log_{10} \Gamma(n)$ compiled from E. S. Pearson, Table of the logarithms of the complete Γ-function, arguments 2 to 1200. Tracts for Computers No. VIII (Cambridge Univ. Press, Cambridge, England, 1922) (with permission).

LOGARITHMS OF THE GAMMA FUNCTION

Table 6.4

n	$\log_{10} \Gamma(n)$	$\log_{10} \Gamma(n+\frac{1}{3})$	$\log_{10} \Gamma(n+\frac{1}{2})$	$\log_{10} \Gamma(n+\frac{2}{3})$	$f_2(n)$
51	64.483075	65.051318	65.335796	65.620510	0.92057 250
52	66.190645	66.761717	67.047603	67.333720	0.92054 108
53	67.906648	68.480496	68.767762	69.055256	0.92051 084
54	69.630924	70.207494	70.496116	70.784961	0.92048 173
55	71.363318	71.942561	72.232512	72.522683	0.92045 367
56	73.103681	73.685548	73.976805	74.268279	0.92042 661
57	74.851869	75.436313	75.728854	76.021606	0.92040 051
58	76.607744	77.194720	77.488522	77.782531	0.92037 530
59	78.371172	78.960637	79.255677	79.550922	0.92035 095
60	80.142024	80.733936	81.030194	81.326654	0.92032 741
61	81.920175	82.514493	82.811950	83.109604	0.92030 464
62	83.705505	84.302190	84.600825	84.899655	0.92028 261
63	85.497896	86.096910	86.396705	86.696691	0.92026 127
64	87.297237	87.898542	88.199479	88.500604	0.92024 061
65	89.103417	89.706978	90.009038	90.311284	0.92022 057
66	90.916330	91.522113	91.825280	92.128629	0.92020 115
67	92.735874	93.343845	93.648101	93.952538	0.92018 231
68	94.561949	95.172075	95.477405	95.782913	0.92016 401
69	96.394458	97.006708	97.313096	97.619659	0.92014 625
70	98.233307	98.847650	99.155080	99.462684	0.92012 900
71	100.07841	100.69481	101.00327	101.31190	0.92011 223
72	101.92966	102.54810	102.85758	103.16722	0.92009 593
73	103.78700	104.40744	104.71791	105.02855	0.92008 008
74	105.65032	106.27274	106.58420	106.89582	0.92006 465
75	107.51955	108.14393	108.45636	108.76895	0.92004 964
76	109.39461	110.02091	110.33430	110.64785	0.92003 502
77	111.27543	111.90363	112.21797	112.53246	0.92002 078
78	113.16192	113.79200	114.10727	114.42269	0.92000 690
79	115.05401	115.68594	116.00214	116.31848	0.91999 338
80	116.95164	117.58540	117.90250	118.21976	0.91998 019
81	118.85473	119.49029	119.80830	120.12646	0.91996 733
82	120.76321	121.40056	121.71946	122.03850	0.91995 479
83	122.67703	123.31614	123.63591	123.95583	0.91994 254
84	124.59610	125.23696	125.55760	125.87838	0.91993 059
85	126.52038	127.16296	127.48445	127.80610	0.91991 892
86	128.44980	129.09407	129.41642	129.73891	0.91990 752
87	130.38430	131.03025	131.35344	131.67676	0.91989 638
88	132.32382	132.97143	133.29545	133.61959	0.91988 550
89	134.26830	134.91756	135.24239	135.56735	0.91987 486
90	136.21769	136.86857	137.19421	137.51999	0.91986 446
91	138.17194	138.82442	139.15086	139.47743	0.91985 428
92	140.13098	140.78505	141.11228	141.43964	0.91984 433
93	142.09477	142.75041	143.07842	143.40657	0.91983 459
94	144.06325	144.72044	145.04923	145.37815	0.91982 505
95	146.03638	146.69511	147.02467	147.35435	0.91981 572
96	148.01410	148.67435	149.00467	149.33511	0.91980 659
97	149.99637	150.65813	150.98920	151.32039	0.91979 764
98	151.98314	152.64639	152.97820	153.31013	0.91978 887
99	153.97437	154.63909	154.97164	155.30430	0.91978 028
100	155.97000	156.63619	156.96946	157.30285	0.91977 186
101	157.97000	158.63763	158.97163	159.30574	0.91976 361
	$\log_{10}(n-1)!$	$\log_{10}(n-\frac{2}{3})!$	$\log_{10}(n-\frac{1}{2})!$	$\log_{10}(n-\frac{1}{3})!$	$\left[\begin{smallmatrix}(-7)2\\3\end{smallmatrix}\right]$

$$\ln \Gamma(n)=\ln(n-1)!=(n-\tfrac{1}{2})\ln n-n+f_2(n) \qquad \ln 10=2.30258\ 509299$$

Table 6.5   AUXILIARY FUNCTIONS FOR GAMMA AND DIGAMMA FUNCTIONS

x^{-1}	$f_1(x)$	$f_2(x)$	$f_3(x)$	$\langle x \rangle$
0.015	1.00125 077	0.92018 852	0.00751 875	67
0.014	1.00116 735	0.92010 519	0.00701 633	71
0.013	1.00108 391	0.92002 186	0.00651 408	77
0.012	1.00100 050	0.91993 853	0.00601 200	83
0.011	1.00091 708	0.91985 520	0.00551 008	91
0.010	1.00083 368	0.91977 186	0.00500 833	100
0.009	1.00075 028	0.91968 853	0.00450 675	111
0.008	1.00066 689	0.91960 520	0.00400 533	125
0.007	1.00058 350	0.91952 187	0.00350 408	143
0.006	1.00050 012	0.91943 853	0.00300 300	167
0.005	1.00041 675	0.91935 520	0.00250 208	200
0.004	1.00033 339	0.91927 187	0.00200 133	250
0.003	1.00025 003	0.91918 853	0.00150 075	333
0.002	1.00016 668	0.91910 520	0.00100 033	500
0.001	1.00008 334	0.91902 187	0.00050 008	1000
0.000	1.00000 000 $\begin{bmatrix} (-8)1 \\ 2 \end{bmatrix}$	0.91893 853 $\begin{bmatrix} (-8)1 \\ 2 \end{bmatrix}$	0.00000 000 $\begin{bmatrix} (-8)2 \\ 3 \end{bmatrix}$	∞

$$x! = (2\pi)^{\frac{1}{2}} x^{x+\frac{1}{2}} e^{-x} f_1(x)$$

$$\Gamma(x) = (2\pi)^{\frac{1}{2}} x^{x-\frac{1}{2}} e^{-x} f_1(x)$$

$$\ln \Gamma(x) = \ln (x-1)! = (x-\tfrac{1}{2}) \ln x - x + f_2(x)$$

$$\psi(x) = \ln x - f_3(x)$$

$$(2\pi)^{\frac{1}{2}} = 2.50662\ 82746\ 31001$$

$$\langle x \rangle = \text{nearest integer to } x.$$

Table 6.6   FACTORIALS FOR LARGE ARGUMENTS

n	$n!$	n	$n!$
100	(157)9.3326 21544 39441 52682	600	(1408)1.2655 72316 22543 07425
200	(374)7.8865 78673 64790 50355	700	(1689)2.4220 40124 75027 21799
300	(614)3.0605 75122 16440 63604	800	(1976)7.7105 30113 35386 00414
400	(868)6.4034 52284 66238 95262	900	(2269)6.7526 80220 96458 41584
500	(1134)1.2201 36825 99111 00687	1000	(2567)4.0238 72600 77093 77354
	$\Gamma(n+1)$		$\Gamma(n+1)$

Compiled from Ballistic Research Laboratory,   A table of the factorial numbers and their reciprocals from 1! to 1000! to 20 significant digits, Technical Note No. 381, Aberdeen Proving Ground, Md.(1951) (with permission).

GAMMA FUNCTION FOR COMPLEX ARGUMENTS

Table 6.7

$$x=1.0$$

y	$\mathscr{R}\ln\Gamma(z)$	$\mathscr{I}\ln\Gamma(z)$	y	$\mathscr{R}\ln\Gamma(z)$	$\mathscr{I}\ln\Gamma(z)$
0.0	0.00000 00000 00	0.00000 00000 00	5.0	− 6.13032 41445 53	3.81589 85746 15
0.1	− 0.00819 77805 65	− 0.05732 29404 17	5.1	− 6.27750 24635 84	3.97816 38691 88
0.2	− 0.03247 62923 18	− 0.11230 22226 44	5.2	− 6.42487 30533 35	4.14237 74050 86
0.3	− 0.07194 62509 00	− 0.16282 06721 68	5.3	− 6.57242 85885 29	4.30850 21885 83
0.4	− 0.12528 93748 21	− 0.20715 58263 16	5.4	− 6.72016 21547 03	4.47650 25956 68
0.5	− 0.19094 54991 87	− 0.24405 82989 05	5.5	− 6.86806 72180 48	4.64634 42978 70
0.6	− 0.26729 00682 14	− 0.27274 38104 91	5.6	− 7.01613 75979 76	4.81799 41933 05
0.7	− 0.35276 86908 60	− 0.29282 63511 87	5.7	− 7.16436 74421 06	4.99142 03424 89
0.8	− 0.44597 87835 49	− 0.30422 56029 76	5.8	− 7.31275 12034 30	5.16659 19085 37
0.9	− 0.54570 51286 05	− 0.30707 43756 42	5.9	− 7.46128 36194 29	5.34347 91013 53
1.0	− 0.65092 31993 02	− 0.30164 03204 68	6.0	− 7.60995 96929 51	5.52205 31255 15
1.1	− 0.76078 39588 41	− 0.28826 66142 39	6.1	− 7.75877 46746 55	5.70228 61315 35
1.2	− 0.87459 04638 95	− 0.26733 05805 81	6.2	− 7.90772 40468 98	5.88415 11702 39
1.3	− 0.99177 27669 59	− 0.23921 67844 65	6.3	− 8.05680 35089 04	6.06762 21500 13
1.4	− 1.11186 45664 26	− 0.20430 07241 49	6.4	− 8.20600 89631 00	6.25267 37967 05
1.5	− 1.23448 30515 47	− 0.16293 97694 80	6.5	− 8.35533 65025 11	6.43928 16159 76
1.6	− 1.35931 22484 65	− 0.11546 87935 89	6.6	− 8.50478 23991 25	6.62742 18579 12
1.7	− 1.48608 96127 57	− 0.06219 86983 29	6.7	− 8.65434 30931 23	6.81707 14837 44
1.8	− 1.61459 53960 00	− 0.00341 66314 77	6.8	− 8.80401 51829 10	7.00820 81345 02
1.9	− 1.74464 42761 74	+ 0.06061 28742 95	6.9	− 8.95379 54158 79	7.20081 01014 93
2.0	− 1.87607 87864 31	0.12964 63163 10	7.0	− 9.10368 06798 32	7.39485 62984 36
2.1	− 2.00876 41504 71	0.20345 94738 33	7.1	− 9.25366 79950 15	7.59032 62351 84
2.2	− 2.14258 42092 96	0.28184 56584 26	7.2	− 9.40375 45067 08	7.78719 99928 77
2.3	− 2.27743 81922 04	0.36461 40489 50	7.3	− 9.55393 74783 21	7.98545 82004 68
2.4	− 2.41323 81411 84	0.45158 81524 41	7.4	− 9.70421 42849 72	8.18508 20125 03
2.5	− 2.54990 68424 95	0.54260 44058 52	7.5	− 9.85458 24074 86	8.38605 30880 89
2.6	− 2.68737 61537 50	0.63751 09190 46	7.6	−10.00503 94267 90	8.58835 35709 62
2.7	− 2.82558 56411 91	0.73616 63516 79	7.7	−10.15558 30186 86	8.79196 60705 87
2.8	− 2.96448 14617 89	0.83843 89130 96	7.8	−10.30621 09489 48	8.99687 36442 29
2.9	− 3.10401 54399 01	0.94420 54730 39	7.9	−10.45692 10687 39	9.20305 97799 25
3.0	− 3.24414 42995 90	1.05335 07710 69	8.0	−10.60771 13103 15	9.41050 83803 12
3.1	− 3.38482 90223 77	1.16576 67132 86	8.1	−10.75857 96829 95	9.61920 37472 42
3.2	− 3.52603 43067 09	1.28135 17459 32	8.2	−10.90952 42693 78	9.82913 05671 62
3.3	− 3.66772 81104 88	1.40001 02965 76	8.3	−11.06054 32217 92	10.04027 38971 80
3.4	− 3.80988 12618 23	1.52165 22746 73	8.4	−11.21163 47589 48	10.25261 91518 09
3.5	− 3.95246 71261 89	1.64619 26242 69	8.5	−11.36279 71628 04	10.46615 20903 24
3.6	− 4.09546 13204 51	1.77355 09225 91	8.6	−11.51402 87756 02	10.68085 88047 12
3.7	− 4.23884 14660 71	1.90365 10190 19	8.7	−11.66532 79970 81	10.89672 57081 77
3.8	− 4.38258 69752 28	2.03642 07096 93	8.8	−11.81669 32818 48	11.13373 95241 57
3.9	− 4.52667 88647 16	2.17179 14436 05	8.9	−11.96812 31369 01	11.33188 72758 53
4.0	− 4.67109 95934 09	2.30969 80565 73	9.0	−12.11961 61192 81	11.55115 62762 02
4.1	− 4.81583 29197 96	2.45007 85299 47	9.1	−12.27117 08338 67	11.77153 41183 09
4.2	− 4.96086 37766 87	2.59287 37713 19	9.2	−12.42278 59312 81	11.99300 86662 85
4.3	− 5.10617 81606 63	2.73802 74148 20	9.3	−12.57446 01059 08	12.21556 80464 79
4.4	− 5.25176 30342 30	2.88548 56389 27	9.4	−12.72619 20940 29	12.43920 06390 90
4.5	− 5.39760 62389 84	3.03519 69999 22	9.5	−12.87798 06720 44	12.66389 50701 28
4.6	− 5.54369 64183 04	3.18711 22793 89	9.6	−13.02982 46547 89	12.88964 02037 08
4.7	− 5.69002 29483 73	3.34118 43443 27	9.7	−13.18172 28939 51	13.11642 51346 66
4.8	− 5.83657 58764 54	3.49736 80186 15	9.8	−13.33367 42765 47	13.34423 91814 77
4.9	− 5.98334 58655 32	3.65561 99647 12	9.9	−13.48567 77234 95	13.57307 18794 55
5.0	− 6.13032 41445 53	3.81589 85746 15	10.0	−13.63773 21882 47	13.80291 29742 30

Linear interpolation will yield about three figures; eight-point interpolation will yield about eight figures.

For z outside the range of the table, see **Examples 5–8.**

$$\mathscr{R}\ln\Gamma(z)=\ln|\Gamma(z)| \qquad\qquad \mathscr{I}\ln\Gamma(z)=\arg\Gamma(z)$$

Table 6.7   GAMMA FUNCTION FOR COMPLEX ARGUMENTS

$$x=1.1$$

y	$\mathscr{R} \ln \Gamma(z)$	$\mathscr{I} \ln \Gamma(z)$	y	$\mathscr{R} \ln \Gamma(z)$	$\mathscr{I} \ln \Gamma(z)$
0.0	− 0.04987 24412 60	0.00000 00000 00	5.0	− 5.96893 91493 52	3.96198 63258 60
0.1	− 0.05702 02290 38	− 0.04206 65443 76	5.1	− 6.11415 43840 05	4.12446 68364 90
0.2	− 0.07824 35801 68	− 0.08230 97383 98	5.2	− 6.25959 93585 61	4.28888 73284 80
0.3	− 0.11291 43470 17	− 0.11905 06275 18	5.3	− 6.40526 53566 40	4.45521 12743 47
0.4	− 0.16008 21257 99	− 0.15086 79240 09	5.4	− 6.55114 41480 20	4.62340 34819 04
0.5	− 0.21858 96764 09	− 0.17666 11398 43	5.5	− 6.69722 79531 89	4.79343 00232 04
0.6	− 0.28718 99839 43	− 0.19566 16788 64	5.6	− 6.84350 94110 69	4.96525 81683 67
0.7	− 0.36464 38731 53	− 0.20740 35526 60	5.7	− 6.98998 15495 70	5.13885 63238 91
0.8	− 0.44978 83131 87	− 0.21167 10325 55	5.8	− 7.13663 77586 96	5.31419 39750 77
0.9	− 0.54157 54093 11	− 0.20843 91333 00	5.9	− 7.28347 17659 19	5.49124 16322 40
1.0	− 0.63908 78153 48	− 0.19781 78257 67	6.0	− 7.43047 76136 25	5.66997 07803 94
1.1	− 0.74153 80620 74	− 0.18000 55175 74	6.1	− 7.57764 96383 95	5.85035 38321 46
1.2	− 0.84825 85646 26	− 0.15525 33222 12	6.2	− 7.72498 24519 72	6.03236 40835 50
1.3	− 0.95868 73364 97	− 0.12383 93047 38	6.3	− 7.87247 09237 38	6.21597 56726 90
1.4	− 1.07235 26519 67	− 0.08605 08957 00	6.4	− 8.02011 01645 61	6.40116 35407 92
1.5	− 1.18885 84815 22	− 0.04217 34907 11	6.5	− 8.16789 55118 88	6.58790 33956 67
1.6	− 1.30787 15575 95	+ 0.00751 65191 79	6.6	− 8.31582 25159 69	6.77617 16773 32
1.7	− 1.42911 03402 04	0.06275 56777 30	6.7	− 8.46388 69271 17	6.96594 55256 30
1.8	− 1.55233 58336 11	0.12329 53847 15	6.8	− 8.61208 46838 95	7.15720 27497 24
1.9	− 1.67734 40572 49	0.18890 25358 69	6.9	− 8.76041 19021 72	7.34992 17993 20
2.0	− 1.80395 99248 63	0.25935 93780 23	7.0	− 8.90886 48649 60	7.54408 17375 09
2.1	− 1.93203 22878 13	0.33446 29085 79	7.1	− 9.05744 00129 63	7.73966 22151 13
2.2	− 2.06142 99239 46	0.41402 40321 50	7.2	− 9.20613 39357 92	7.93664 34464 25
2.3	− 2.19203 82866 29	0.49786 66085 82	7.3	− 9.35494 33637 73	8.13500 61862 70
2.4	− 2.32375 68617 01	0.58582 64745 04	7.4	− 9.50386 51603 25	8.33473 17082 71
2.5	− 2.45649 70097 26	0.67775 04868 09	7.5	− 9.65289 63148 29	8.53580 17842 76
2.6	− 2.59018 01959 43	0.77349 56148 91	7.6	− 9.80203 39359 83	8.73819 86648 33
2.7	− 2.72473 65306 67	0.87292 80949 66	7.7	− 9.95127 52455 81	8.94190 50606 84
2.8	− 2.86010 35591 81	0.97592 26515 07	7.8	−10.10061 75726 94	9.14690 41251 84
2.9	− 2.99622 52529 98	1.08236 17859 08	7.9	−10.25005 83482 21	9.35317 94376 01
3.0	− 3.13305 11644 50	1.19213 51297 05	8.0	−10.39959 50997 80	9.56071 49872 49
3.1	− 3.27053 57144 30	1.30513 88581 77	8.1	−10.54922 54469 17	9.76949 51583 85
3.2	− 3.40863 75892 32	1.42127 51595 43	8.2	−10.69894 70966 06	9.97950 47158 43
3.3	− 3.54731 92273 03	1.54045 17547 76	8.3	−10.84875 78390 24	10.19072 87913 49
3.4	− 3.68654 63804 17	1.66258 14631 94	8.4	−10.99865 55435 72	10.40315 28704 84
3.5	− 3.82628 77368 25	1.78758 18092 68	8.5	−11.14863 81551 38	10.61676 27802 52
3.6	− 3.96651 45962 20	1.91537 46664 26	8.6	−11.29870 36905 72	10.83154 46772 22
3.7	− 4.10720 05882 64	2.04588 59340 24	8.7	−11.44885 02353 71	11.04748 50362 14
3.8	− 4.24832 14278 81	2.17904 52440 32	8.8	−11.59907 59405 42	11.26457 06394 86
3.9	− 4.38985 47017 40	2.31478 56943 26	8.9	−11.74937 90196 53	11.48278 85664 18
4.0	− 4.53177 96812 84	2.45304 36058 25	9.0	−11.89975 77460 43	11.70212 61836 32
4.1	− 4.67407 71584 70	2.59375 83010 13	9.1	−12.05021 04501 83	11.92257 11355 62
4.2	− 4.81672 93009 83	2.73687 19016 54	9.2	−12.20073 55171 88	12.14411 13354 15
4.3	− 4.95971 95242 44	2.88232 91437 48	9.3	−12.35133 13844 58	12.36673 49565 33
4.4	− 5.10303 23779 21	3.03007 72080 09	9.4	−12.50199 65394 43	12.59043 04241 06
4.5	− 5.24665 34450 28	3.18006 55643 29	9.5	−12.65272 95175 33	12.81518 64072 43
4.6	− 5.39056 92519 72	3.33224 58288 43	9.6	−12.80352 89000 52	13.04099 18113 65
4.7	− 5.53476 71881 64	3.48657 16324 07	9.7	−12.95439 33123 60	13.26783 57709 12
4.8	− 5.67923 54339 89	3.64299 84993 84	9.8	−13.10532 14220 44	13.49570 76423 49
4.9	− 5.82396 28961 29	3.80148 37357 79	9.9	−13.25631 19372 14	13.72459 69974 44
5.0	− 5.96893 91493 52	3.96198 63258 60	10.0	−13.40736 36048 74	13.95449 36168 27

GAMMA FUNCTION FOR COMPLEX ARGUMENTS

Table 6.7

$x=1.2$

y	$\mathscr{R} \ln \Gamma(z)$	$\mathscr{I} \ln \Gamma(z)$	y	$\mathscr{R} \ln \Gamma(z)$	$\mathscr{I} \ln \Gamma(z)$
0.0	− 0.08537 40900 03	0.00000 00000 00	5.0	− 5.80731 52672 85	4.10609 64053 70
0.1	− 0.09169 75124 13	− 0.02865 84973 21	5.1	− 5.95057 66519 39	4.26883 00575 53
0.2	− 0.11050 89067 86	− 0.05586 39903 67	5.2	− 6.09410 47211 91	4.43349 40204 01
0.3	− 0.14135 09532 62	− 0.08025 91592 09	5.3	− 6.23788 94064 81	4.60005 23089 91
0.4	− 0.18352 07443 57	− 0.10066 05658 03	5.4	− 6.38192 11972 10	4.76847 02339 50
0.5	− 0.23614 32688 51	− 0.11610 77219 87	5.5	− 6.52619 11003 82	4.93871 43339 56
0.6	− 0.29824 98509 35	− 0.12588 00935 13	5.6	− 6.67069 06038 24	5.11075 23127 64
0.7	− 0.36884 83560 49	− 0.12948 68069 28	5.7	− 6.81541 16425 98	5.28455 29803 68
0.8	− 0.44697 73864 90	− 0.12663 80564 16	5.8	− 6.96034 65682 97	5.46008 61980 02
0.9	− 0.53174 22756 96	− 0.11720 77278 71	5.9	− 7.10548 81209 15	5.63732 28266 55
1.0	− 0.62233 46814 87	− 0.10119 48344 90	6.0	− 7.25082 94030 54	5.81623 46788 41
1.1	− 0.71803 95313 44	− 0.07868 85726 52	6.1	− 7.39636 38562 29	5.99679 44733 73
1.2	− 0.81823 34133 20	− 0.04983 92764 14	6.2	− 7.54208 52390 70	6.17897 57929 16
1.3	− 0.92237 79303 78	− 0.01483 57562 65	6.3	− 7.68798 76072 47	6.36275 30441 11
1.4	− 1.03001 06294 86	+ 0.02611 15201 47	6.4	− 7.83406 52949 57	6.54810 14200 83
1.5	− 1.14073 52341 62	0.07278 23932 61	6.5	− 7.98031 28978 26	6.73499 68651 55
1.6	− 1.25421 22047 39	0.12495 51937 38	6.6	− 8.12672 52570 99	6.92341 60416 24
1.7	− 1.37015 01536 37	0.18241 21090 01	6.7	− 8.27329 74450 10	7.11333 62984 34
1.8	− 1.48829 83245 09	0.24494 25273 48	6.8	− 8.42002 47512 17	7.30473 56416 32
1.9	− 1.60844 01578 57	0.31234 49712 35	6.9	− 8.56690 26702 20	7.49759 27064 69
2.0	− 1.73038 78680 93	0.38442 80719 73	7.0	− 8.71392 68896 74	7.69188 67310 43
2.1	− 1.85397 79144 87	0.46101 09100 87	7.1	− 8.86109 32795 24	7.88759 75313 86
2.2	− 1.97906 72374 32	0.54192 29484 31	7.2	− 9.00839 78818 89	8.08470 54778 77
2.3	− 2.10553 01371 17	0.62700 37140 16	7.3	− 9.15583 69016 37	8.28319 14729 22
2.4	− 2.23325 56848 33	0.71610 23338 39	7.4	− 9.30340 66975 98	8.48303 69297 94
2.5	− 2.36214 55727 43	0.80907 69945 69	7.5	− 9.45110 37743 60	8.68422 37525 82
2.6	− 2.49211 23232 46	0.90579 43715 71	7.6	− 9.59892 47746 01	8.88673 43171 55
2.7	− 2.62307 77928 95	1.00612 90561 43	7.7	− 9.74686 64719 23	9.09055 14530 96
2.8	− 2.75497 19177 39	1.10996 29987 33	7.8	− 9.89492 57641 38	9.29565 84265 39
2.9	− 2.88773 16568 77	1.21718 49784 62	7.9	−10.04309 96669 84	9.50203 89238 50
3.0	− 3.02130 00992 07	1.32769 01044 18	8.0	−10.19138 53082 31	9.70967 70361 08
3.1	− 3.15562 57049 65	1.44137 93510 29	8.1	−10.33977 99221 46	9.91855 72443 36
3.2	− 3.29066 16590 00	1.55815 91278 68	8.2	−10.48828 08443 04	10.12866 44054 34
3.3	− 3.42636 53170 56	1.67794 08829 56	8.3	−10.63688 55067 01	10.33998 37387 77
3.4	− 3.56269 77297 54	1.80064 07379 67	8.4	−10.78559 14331 66	10.55250 08134 40
3.5	− 3.69962 32317 85	1.92617 91533 49	8.5	−10.93439 62350 38	10.76620 15360 05
3.6	− 3.83710 90860 24	2.05448 06211 84	8.6	−11.08329 76070 93	10.98107 21389 38
3.7	− 3.97512 51741 07	2.18547 33836 08	8.7	−11.23229 33237 11	11.19709 91694 76
3.8	− 4.11364 37264 61	2.31908 91746 67	8.8	−11.38138 12352 53	11.41426 94790 19
3.9	− 4.25263 90859 57	2.45526 29835 70	8.9	−11.53055 92646 46	11.63257 02129 90
4.0	− 4.39208 75003 42	2.59393 28374 55	9.0	−11.67982 54041 57	11.85198 88011 32
4.1	− 4.53196 69393 70	2.73503 96019 03	9.1	−11.82917 77123 44	12.07251 29482 35
4.2	− 4.67225 69332 23	2.87852 67976 01	9.2	−11.97861 43111 70	12.29413 06252 48
4.3	− 4.81293 84293 30	3.02434 04316 86	9.3	−12.12813 33832 78	12.51683 00607 77
4.4	− 4.95399 36651 50	3.17242 88424 26	9.4	−12.27773 31694 04	12.74059 97329 36
4.5	− 5.09540 60548 36	3.32274 25560 43	9.5	−12.42741 19659 29	12.96542 83615 35
4.6	− 5.23716 00880 20	3.47523 41545 72	9.6	−12.57716 81225 64	13.19130 49005 92
4.7	− 5.37924 12391 93	3.62985 81537 79	9.7	−12.72700 00401 42	13.41821 85311 47
4.8	− 5.52163 58863 97	3.78657 08902 31	9.8	−12.87690 61685 35	13.64615 86543 64
4.9	− 5.66433 12381 00	3.94533 04167 32	9.9	−13.02688 50046 68	13.87511 48849 16
5.0	− 5.80731 52672 85	4.10609 64053 70	10.0	−13.17693 50906 38	14.10507 70446 23

Table 6.7   **GAMMA FUNCTION FOR COMPLEX ARGUMENTS**

$$x=1.3$$

y	$\mathscr{R} \ln \Gamma(z)$	$\mathscr{I} \ln \Gamma(z)$	y	$\mathscr{R} \ln \Gamma(z)$	$\mathscr{I} \ln \Gamma(z)$
0.0	− 0.10817 48095 08	0.00000 00000 00	5.0	− 5.64541 41381 33	4.24823 90621 27
0.1	− 0.11383 61080 85	− 0.01671 99199 34	5.1	− 5.78673 23355 37	4.41126 31957 95
0.2	− 0.13070 20636 90	− 0.03225 84033 35	5.2	− 5.92835 35606 66	4.57620 66023 67
0.3	− 0.15843 10081 49	− 0.04549 95427 81	5.3	− 6.07026 64370 51	4.74303 39118 17
0.4	− 0.19649 12771 78	− 0.05544 82296 06	5.4	− 6.21246 02140 03	4.91171 10050 12
0.5	− 0.24420 93680 45	− 0.06126 78750 55	5.5	− 6.35492 47217 66	5.08220 49501 77
0.6	− 0.30082 34434 02	− 0.06229 79103 48	5.6	− 6.49765 03305 97	5.25448 39434 72
0.7	− 0.36553 39002 19	− 0.05805 28252 04	5.7	− 6.64062 79133 72	5.42851 72533 50
0.8	− 0.43754 53407 27	− 0.04820 73993 35	5.8	− 6.78384 88113 55	5.60427 51684 12
0.9	− 0.51609 74046 40	− 0.03257 37450 94	5.9	− 6.92730 48028 21	5.78172 89485 09
1.0	− 0.60048 45154 05	− 0.01107 52190 48	6.0	− 7.07098 80742 52	5.96085 07788 45
1.1	− 0.69006 62005 12	+ 0.01627 90894 04	6.1	− 7.21489 11938 62	6.14161 37268 52
1.2	− 0.78427 03001 02	0.04941 70710 23	6.2	− 7.35900 70872 13	6.32399 17016 49
1.3	− 0.88259 13601 03	0.08822 25250 96	6.3	− 7.50332 90147 58	6.50795 94158 99
1.4	− 0.98458 61322 90	0.13255 01649 50	6.4	− 7.64785 05510 98	6.69349 23498 81
1.5	− 1.08986 76158 16	0.18223 70479 17	6.5	− 7.79256 55658 27	6.88056 67176 38
1.6	− 1.19809 86148 04	0.23711 09920 47	6.6	− 7.93746 82058 02	7.06915 94350 45
1.7	− 1.30898 54162 82	0.29699 65855 44	6.7	− 8.08255 28787 24	7.25924 80896 76
1.8	− 1.42227 19237 14	0.36171 93463 93	6.8	− 8.22781 42379 13	7.45081 09123 38
1.9	− 1.53773 44011 63	0.43110 85022 51	6.9	− 8.37324 71681 76	7.64382 67501 64
2.0	− 1.65517 68709 10	0.50499 87656 67	7.0	− 8.51884 67726 68	7.83827 50411 67
2.1	− 1.77442 71431 91	0.58323 13926 09	7.1	− 8.66460 83606 78	8.03413 57901 50
2.2	− 1.89533 34239 28	0.66565 47394 67	7.2	− 8.81052 74362 48	8.23138 95458 91
2.3	− 2.01776 14331 34	0.75212 44759 30	7.3	− 8.95659 96875 66	8.43001 73795 19
2.4	− 2.14159 19646 87	0.84250 35670 42	7.4	− 9.10282 09770 73	8.63000 08640 04
2.5	− 2.26671 88222 04	0.93666 21049 03	7.5	− 9.24918 73322 19	8.83132 20546 97
2.6	− 2.39304 70725 18	1.03447 70464 53	7.6	− 9.39569 49368 29	9.03396 34708 43
2.7	− 2.52049 15659 37	1.13583 18965 15	7.7	− 9.54234 01230 14	9.23790 80780 23
2.8	− 2.64897 56799 18	1.24061 63628 56	7.8	− 9.68911 93636 11	9.44313 92714 58
2.9	− 2.77843 02497 03	1.34872 60013 87	7.9	− 9.83602 92650 88	9.64964 08601 22
3.0	− 2.90879 26554 06	1.46006 18633 96	8.0	− 9.98306 65608 89	9.85739 70516 25
3.1	− 3.04000 60402 26	1.57453 01525 07	8.1	−10.13022 81051 96	10.06639 24378 12
3.2	− 3.17201 86387 60	1.69204 18960 57	8.2	−10.27751 08670 60	10.27661 19810 47
3.3	− 3.30478 31979 94	1.81251 26335 69	8.3	−10.42491 19248 88	10.48804 10011 24
3.4	− 3.43825 64765 05	1.93586 21235 97	8.4	−10.57242 84612 54	10.70066 51627 91
3.5	− 3.57239 88099 07	2.06201 40693 37	8.5	−10.72005 77580 15	10.91447 04638 39
3.6	− 3.70717 37325 19	2.19089 58627 45	8.6	−10.86779 71917 09	11.12944 32237 30
3.7	− 3.84254 76469 59	2.32243 83465 44	8.7	−11.01564 42292 16	11.34557 00727 24
3.8	− 3.97848 95346 95	2.45657 55932 86	8.8	−11.16359 64236 64	11.56283 79415 00
3.9	− 4.11497 07016 98	2.59324 47004 59	8.9	−11.31165 14105 63	11.78123 40512 20
4.0	− 4.25196 45543 18	2.73238 56006 34	9.0	−11.45980 69041 59	12.00074 59040 23
4.1	− 4.38944 64012 12	2.87394 08855 80	9.1	−11.60806 06939 74	12.22136 12739 31
4.2	− 4.52739 32778 30	3.01785 56433 48	9.2	−11.75641 06415 49	12.44306 81981 38
4.3	− 4.66578 37904 84	3.16407 73073 22	9.3	−11.90485 46773 52	12.66585 49686 64
4.4	− 4.80459 79774 65	3.31255 55163 23	9.4	−12.05339 07978 49	12.88971 01243 51
4.5	− 4.94381 71850 33	3.46324 19848 78	9.5	−12.20201 70627 34	13.11462 24431 99
4.6	− 5.08342 39564 42	3.61609 03828 59	9.6	−12.35073 15923 02	13.34058 09350 03
4.7	− 5.22340 19323 94	3.77105 62237 32	9.7	−12.49953 25649 49	13.56757 48342 95
4.8	− 5.36373 57615 52	3.92809 67607 19	9.8	−12.64841 82148 10	13.79559 35935 62
4.9	− 5.50441 10199 31	4.08717 08902 55	9.9	−12.79738 68295 12	14.02462 68767 33
5.0	− 5.64541 41381 33	4.24823 90621 27	10.0	−12.94643 67480 34	14.25466 45529 28

GAMMA FUNCTION FOR COMPLEX ARGUMENTS

Table 6.7

$$x=1.4$$

y	$\mathscr{R} \ln \Gamma(z)$	$\mathscr{I} \ln \Gamma(z)$	y	$\mathscr{R} \ln \Gamma(z)$	$\mathscr{I} \ln \Gamma(z)$
0.0	− 0.11961 29141 72	0.00000 00000 00	5.0	− 5.48319 80511 50	4.38842 59888 87
0.1	− 0.12473 21357 76	− 0.00597 40017 43	5.1	− 5.62258 51037 75	4.55177 72808 10
0.2	− 0.14000 01552 88	− 0.01097 08056 66	5.2	− 5.76231 08530 59	4.71703 54898 14
0.3	− 0.16515 59551 89	− 0.01405 93840 03	5.3	− 5.90236 26637 68	4.88416 59286 80
0.4	− 0.19978 93616 12	− 0.01439 47989 49	5.4	− 6.04272 85898 90	5.05313 51119 86
0.5	− 0.24337 34438 09	− 0.01124 72025 18	5.5	− 6.18339 73257 62	5.22391 06968 84
0.6	− 0.29530 16779 62	− 0.00401 77865 38	5.6	− 6.32435 81614 11	5.39646 14275 35
0.7	− 0.35492 46161 10	+ 0.00775 78473 84	5.7	− 6.46560 09417 01	5.57075 70829 41
0.8	− 0.42158 20669 55	0.02441 65124 32	5.8	− 6.60711 60288 99	5.74676 84279 33
0.9	− 0.49462 85345 46	0.04618 11610 42	5.9	− 6.74889 42683 24	5.92446 71670 92
1.0	− 0.57345 12921 03	0.07317 82199 73	6.0	− 6.89092 69567 80	6.10382 59013 94
1.1	− 0.65748 16506 41	0.10545 58409 92	6.1	− 7.03320 58135 18	6.28481 80874 01
1.2	− 0.74620 06322 98	0.14300 11986 37	6.2	− 7.17572 29534 78	6.46741 79988 09
1.3	− 0.83914 04638 04	0.18575 57618 52	6.3	− 7.31847 08625 98	6.65160 06901 96
1.4	− 0.93588 32199 21	0.23362 80933 40	6.4	− 7.46144 23750 25	6.83734 19628 28
1.5	− 1.03605 77156 27	0.28650 41540 26	6.5	− 7.60463 06520 25	7.02461 83323 73
1.6	− 1.13933 54742 88	0.34425 53337 92	6.6	− 7.74802 91624 64	7.21340 69984 03
1.7	− 1.24542 63479 49	0.40674 45404 87	6.7	− 7.89163 16647 23	7.40368 58155 67
1.8	− 1.35407 41615 64	0.47383 07041 21	6.8	− 8.03543 21899 02	7.59543 32663 20
1.9	− 1.46505 26007 14	0.54537 20299 26	6.9	− 8.17942 50262 34	7.78862 84351 12
2.0	− 1.57816 14562 85	0.62122 82885 81	7.0	− 8.32360 47045 82	7.98325 09839 40
2.1	− 1.69322 32702 19	0.70126 23803 49	7.1	− 8.46796 59849 44	8.17928 11291 83
2.2	− 1.81008 03838 54	0.78534 13608 50	7.2	− 8.61250 38438 82	8.37669 96196 29
2.3	− 1.92859 23663 09	0.87333 70735 61	7.3	− 8.75721 34627 90	8.57548 77156 28
2.4	− 2.04863 37884 08	0.96512 64991 00	7.4	− 8.90209 02169 54	8.77562 71692 98
2.5	− 2.17009 23032 73	1.06059 19035 92	7.5	− 9.04712 96653 17	8.97710 02057 23
2.6	− 2.29286 69947 17	1.15962 08468 95	7.6	− 9.19232 75409 21	9.17988 95050 80
2.7	− 2.41686 69570 58	1.26210 60952 18	7.7	− 9.33767 97419 53	9.38397 81856 34
2.8	− 2.54201 00734 84	1.36794 54704 02	7.8	− 9.48318 23233 58	9.58934 97875 68
2.9	− 2.66822 19640 86	1.47704 16591 47	7.9	− 9.62883 14889 78	9.79598 82575 76
3.0	− 2.79543 50784 95	1.58930 19987 43	8.0	− 9.77462 35841 76	10.00387 79341 91
3.1	− 2.92358 79116 75	1.70463 82510 60	8.1	− 9.92055 50889 05	10.21300 35337 97
3.2	− 3.05262 43245 92	1.82296 63729 35	8.2	−10.06662 26112 05	10.42335 01372 94
3.3	− 3.18249 29542 71	1.94420 62885 89	8.3	−10.21282 28810 76	10.63490 31773 72
3.4	− 3.31314 67001 61	2.06828 16678 10	8.4	−10.35915 27447 20	10.84764 84263 58
3.5	− 3.44454 22757 38	2.19511 97123 13	8.5	−10.50560 91591 10	11.06157 19846 19
3.6	− 3.57663 98160 21	2.32465 09517 70	8.6	−10.65218 91868 81	11.27666 02694 74
3.7	− 3.70940 25331 00	2.45680 90502 77	8.7	−10.79888 99915 05	11.49290 00045 92
3.8	− 3.84279 64130 02	2.59153 06235 98	8.8	−10.94570 88327 39	11.71027 82098 57
3.9	− 3.97678 99482 49	2.72875 50671 88	8.9	−11.09264 30623 27	11.92878 21916 70
4.0	− 4.11135 39012 79	2.86842 43947 56	9.0	−11.23969 01199 39	12.14839 95336 59
4.1	− 4.24646 10946 69	3.01048 30870 18	9.1	−11.38684 75293 27	12.36911 80877 89
4.2	− 4.38208 62246 51	3.15487 79501 77	9.2	−11.53411 28946 97	12.59092 59658 40
4.3	− 4.51820 56949 47	3.30155 79836 24	9.3	−11.68148 38972 65	12.81381 15312 39
4.4	− 4.65479 74683 75	3.45047 42563 18	9.4	−11.82895 82920 01	13.03776 33912 29
4.5	− 4.79184 09340 18	3.60157 97913 33	9.5	−11.97653 39045 38	13.26277 03893 53
4.6	− 4.92931 67880 70	3.75482 94580 13	9.6	−12.12420 86282 47	13.48882 15982 45
4.7	− 5.06720 69267 30	3.91017 98712 52	9.7	−12.27198 04214 52	13.71590 63127 03
4.8	− 5.20549 43497 23	4.06758 92973 81	9.8	−12.41984 73048 02	13.94401 40430 46
4.9	− 5.34416 30732 30	4.22701 75662 27	9.9	−12.56780 73587 55	14.17313 45087 16
5.0	− 5.48319 80511 50	4.38842 59888 87	10.0	−12.71585 87212 03	14.40325 76321 42

Table 6.7

GAMMA FUNCTION FOR COMPLEX ARGUMENTS

$$x = 1.5$$

y	$\mathscr{R} \ln \Gamma(z)$	$\mathscr{I} \ln \Gamma(z)$	y	$\mathscr{R} \ln \Gamma(z)$	$\mathscr{I} \ln \Gamma(z)$
0.0	− 0.12078 22376 35	0.00000 00000 00	5.0	− 5.32063 00229 09	4.52667 02683 19
0.1	− 0.12545 03928 11	0.00378 68415 10	5.1	− 5.45809 92990 12	4.69038 46594 51
0.2	− 0.13938 53175 79	0.00839 39012 17	5.2	− 5.59594 21987 69	4.85599 23475 89
0.3	− 0.16238 37050 76	0.01460 80536 11	5.3	− 5.73414 48816 77	5.02345 93914 30
0.4	− 0.19412 35254 45	0.02315 34211 15	5.4	− 5.87269 42552 05	5.19275 29984 42
0.5	− 0.23418 63474 70	0.03466 89612 75	5.5	− 6.01157 79223 61	5.36384 14702 24
0.6	− 0.28208 36136 63	0.04969 46638 36	5.6	− 6.15078 41337 33	5.53669 41510 65
0.7	− 0.33728 34790 33	0.06866 64150 66	5.7	− 6.29030 17435 55	5.71128 13794 95
0.8	− 0.39923 54301 20	0.09191 83319 43	5.8	− 6.43012 01693 96	5.88757 44426 18
0.9	− 0.46739 08704 08	0.11969 06415 60	5.9	− 6.57022 93551 39	6.06554 55330 63
1.0	− 0.54121 88685 47	0.15214 09934 52	6.0	− 6.71061 97369 14	6.24516 77083 65
1.1	− 0.62021 70896 71	0.18935 73091 01	6.1	− 6.85128 22117 36	6.42641 48526 40
1.2	− 0.70391 84698 97	0.23137 07067 73	6.2	− 6.99220 81085 67	6.60926 16403 83
1.3	− 0.79189 44573 28	0.27816 75270 32	6.3	− 7.13338 91616 09	6.79368 35022 65
1.4	− 0.88375 56946 74	0.32969 99180 52	6.4	− 7.27481 74856 07	6.97965 65928 01
1.5	− 0.97915 09391 81	0.38589 47712 67	6.5	− 7.41648 55529 97	7.16715 77597 60
1.6	− 1.07776 48736 47	0.44666 10201 49	6.6	− 7.55838 61727 29	7.35616 45152 22
1.7	− 1.17931 53061 81	0.51189 54441 75	6.7	− 7.70051 24706 26	7.54665 50081 65
1.8	− 1.28355 01134 19	0.58148 71805 09	6.8	− 7.84285 78711 49	7.73860 79984 87
1.9	− 1.39024 41643 92	0.65532 11610 93	6.9	− 7.98541 60804 40	7.93200 28323 86
2.0	− 1.49919 63725 85	0.73328 06816 91	7.0	− 8.12818 10705 51	8.12681 94190 02
2.1	− 1.61022 69592 23	0.81524 92850 60	7.1	− 8.27114 70647 52	8.32303 82082 45
2.2	− 1.72317 49667 28	0.90111 21116 92	7.2	− 8.41430 85238 40	8.52064 01697 48
2.3	− 1.83789 60327 96	0.99075 68430 94	7.3	− 8.55766 01333 52	8.71960 67728 67
2.4	− 1.95426 04180 71	1.08407 43370 92	7.4	− 8.70119 67916 34	8.91991 99676 60
2.5	− 2.07215 12706 83	1.18095 90329 08	7.5	− 8.84491 35986 81	9.12156 21668 12
2.6	− 2.19146 31061 38	1.28130 91860 05	7.6	− 8.98880 58456 98	9.32451 62284 17
2.7	− 2.31210 04795 77	1.38502 69784 97	7.7	− 9.13286 90053 22	9.52876 54395 97
2.8	− 2.43397 68277 27	1.49201 85397 98	7.8	− 9.27709 87224 65	9.73429 35008 92
2.9	− 2.55701 34593 17	1.60219 39035 70	7.9	− 9.42149 08057 13	9.94108 45113 82
3.0	− 2.68113 86746 74	1.71546 69204 67	8.0	− 9.56604 12192 67	10.14912 29545 01
3.1	− 2.80628 69972 89	1.83175 51411 18	8.1	− 9.71074 60753 60	10.35839 36845 06
3.2	− 2.93239 85022 62	1.95097 96800 61	8.2	− 9.85560 16271 36	10.56888 19135 53
3.3	− 3.05941 82284 63	2.07306 50684 28	8.3	−10.00060 42619 46	10.78057 31993 69
3.4	− 3.18729 56630 57	2.19793 91011 06	8.4	−10.14575 04950 41	10.99345 34334 60
3.5	− 3.31598 42885 64	2.32553 26824 38	8.5	−10.29103 69636 22	11.20750 88298 51
3.6	− 3.44544 11840 65	2.45577 96733 92	8.6	−10.43646 04212 40	11.42272 59143 12
3.7	− 3.57562 66733 10	2.58861 67421 82	8.7	−10.58201 77325 09	11.63909 15140 53
3.8	− 3.70650 40135 44	2.72398 32197 35	8.8	−10.72770 58681 09	11.85659 27478 60
3.9	− 3.83803 91197 27	2.86182 09608 36	8.9	−10.87352 19000 77	12.07521 70166 56
4.0	− 3.97020 03195 93	3.00207 42115 08	9.0	−11.01946 29973 44	12.29495 19944 46
4.1	− 4.10295 81356 26	3.14468 94828 47	9.1	−11.16552 64215 28	12.51578 56196 58
4.2	− 4.23628 50905 75	3.28961 54314 23	9.2	−11.31170 95229 33	12.73770 60868 20
4.3	− 4.37015 55336 09	3.43680 27461 51	9.3	−11.45800 97367 84	12.96070 18385 99
4.4	− 4.50454 54845 89	3.58620 40415 07	9.4	−11.60442 45796 38	13.18476 15581 47
4.5	− 4.63943 24943 00	3.73777 37568 62	9.5	−11.75095 16459 94	13.40987 41617 61
4.6	− 4.77479 55187 51	3.89146 80616 79	9.6	−11.89758 86050 76	13.63602 87918 31
4.7	− 4.91061 48059 11	4.04724 47663 05	9.7	−12.04433 31977 78	13.86321 48100 75
4.8	− 5.04687 17934 63	4.20506 32380 55	9.8	−12.19118 32337 59	14.09142 17910 27
4.9	− 5.18354 90163 32	4.36488 43223 09	9.9	−12.33813 65886 95	14.32063 95157 82
5.0	− 5.32063 00229 09	4.52667 02683 19	10.0	−12.48519 12016 51	14.55085 79659 84

GAMMA FUNCTION FOR COMPLEX ARGUMENTS

Table 6.7

$$x=1.6$$

y	$\mathscr{R}\ln\Gamma(z)$	$\mathscr{I}\ln\Gamma(z)$	y	$\mathscr{R}\ln\Gamma(z)$	$\mathscr{I}\ln\Gamma(z)$
0.0	− 0.11259 17656 97	0.00000 00000 00	5.0	− 5.15767 38696 89	4.66298 63139 40
0.1	− 0.11687 93076 07	0.01272 17953 11	5.1	− 5.29324 00046 70	4.82709 89421 23
0.2	− 0.12968 70233 13	0.02614 08547 67	5.2	− 5.42921 38858 50	4.99309 00410 26
0.3	− 0.15085 38452 14	0.04092 98346 69	5.3	− 5.56558 05247 67	5.16092 64732 77
0.4	− 0.18012 29875 82	0.05771 47266 93	5.4	− 5.70232 57347 10	5.33057 61938 29
0.5	− 0.21715 76591 72	0.07705 74009 90	5.5	− 5.83943 60752 49	5.50200 82001 33
0.6	− 0.26155 99560 50	0.09944 39491 75	5.6	− 5.97689 88014 04	5.67519 24850 30
0.7	− 0.31289 07142 69	0.12527 90746 90	5.7	− 6.11470 18170 24	5.85009 99922 08
0.8	− 0.37068 83847 40	0.15488 59553 99	5.8	− 6.25283 36319 59	6.02670 25740 71
0.9	− 0.43448 55339 80	0.18851 04588 87	5.9	− 6.39128 33226 66	6.20497 29518 79
1.0	− 0.50382 21960 58	0.22632 83631 44	6.0	− 6.53004 04959 33	6.38488 46780 37
1.1	− 0.57825 58588 66	0.26845 42738 89	6.1	− 6.66909 52554 28	6.56641 21003 90
1.2	− 0.65736 82809 44	0.31495 11405 00	6.2	− 6.80843 81708 20	6.74953 03284 11
1.3	− 0.74076 95833 61	0.36583 95580 78	6.3	− 6.94806 02492 33	6.93421 52011 79
1.4	− 0.82810 01661 20	0.42110 63293 75	6.4	− 7.08795 29088 41	7.12044 32570 25
1.5	− 0.91903 10002 05	0.48071 20031 31	6.5	− 7.22810 79544 00	7.30819 17047 52
1.6	− 1.01326 27864 52	0.54459 72874 22	6.6	− 7.36851 75545 64	7.49743 83963 44
1.7	− 1.11052 43845 66	0.61268 83586 73	6.7	− 7.50917 42208 19	7.68816 18010 64
1.8	− 1.21057 08228 70	0.68490 11588 51	6.8	− 7.65007 07879 17	7.88034 09808 67
1.9	− 1.31318 11150 50	0.76114 48080 60	6.9	− 7.79120 03956 68	8.07395 55670 43
2.0	− 1.41815 60399 85	0.84132 42695 09	7.0	− 7.93255 64719 90	8.26898 57380 27
2.1	− 1.52531 59861 47	0.92534 23984 61	7.1	− 8.07413 27171 08	8.46541 21983 05
2.2	− 1.63449 89215 98	1.01310 14934 56	7.2	− 8.21592 30888 20	8.66321 61583 45
2.3	− 1.74555 85219 99	1.10450 44515 88	7.3	− 8.35792 17887 32	8.86237 93155 10
2.4	− 1.85836 24696 22	1.19945 56127 07	7.4	− 8.50012 32493 99	9.06288 38358 78
2.5	− 1.97279 09238 15	1.29786 13618 36	7.5	− 8.64252 21222 97	9.26471 23369 30
2.6	− 2.08873 51557 24	1.39963 05453 39	7.6	− 8.78511 32665 62	9.46784 78710 61
2.7	− 2.20609 63358 10	1.50467 47448 81	7.7	− 8.92789 17384 38	9.67227 39098 48
2.8	− 2.32478 44606 95	1.61290 84436 93	7.8	− 9.07085 27813 87	9.87797 43290 61
2.9	− 2.44471 74052 94	1.72424 91120 48	7.9	− 9.21399 18168 02	10.08493 33943 44
3.0	− 2.56582 00865 46	1.83861 72327 21	8.0	− 9.35730 44352 92	10.29313 57475 61
3.1	− 2.68802 37258 40	1.95593 62824 65	8.1	− 9.50078 63884 89	10.50256 63937 51
3.2	− 2.81126 51983 53	2.07613 26817 55	8.2	− 9.64443 35813 39	10.71321 06886 60
3.3	− 2.93548 64586 59	2.19913 57221 55	8.3	− 9.78824 20648 48	10.92505 43268 31
3.4	− 3.06063 40331 69	2.32487 74784 17	8.4	− 9.93220 80292 58	11.13808 33302 08
3.5	− 3.18665 85710 48	2.45329 27106 82	8.5	−10.07632 77975 98	11.35228 40372 42
3.6	− 3.31351 44463 00	2.58431 87608 00	8.6	−10.22059 78196 20	11.56764 30924 55
3.7	− 3.44115 94046 31	2.71789 54457 96	8.7	−10.36501 46660 67	11.78414 74364 58
3.8	− 3.56955 42495 22	2.85396 49506 80	8.8	−10.50957 50232 55	12.00178 42963 80
3.9	− 3.69866 25626 62	2.99247 17222 46	8.9	−10.65427 56879 66	12.22054 11767 06
4.0	− 3.82845 04545 47	3.13336 23649 89	9.0	−10.79911 35626 11	12.44040 58504 89
4.1	− 3.95888 63415 67	3.27658 55399 89	9.1	−10.94408 56506 53	12.66136 63509 22
4.2	− 4.08994 07464 23	3.42209 18672 73	9.2	−11.08918 90522 76	12.88341 09632 56
4.3	− 4.22158 61190 90	3.56983 38320 36	9.3	−11.23442 09602 86	13.10652 82170 40
4.4	− 4.35379 66759 32	3.71976 56948 92	9.4	−11.37977 86562 21	13.33070 68786 75
4.5	− 4.48654 82548 65	3.87184 34062 62	9.5	−11.52525 95066 64	13.55593 59442 57
4.6	− 4.61981 81847 38	4.02602 45248 92	9.6	−11.67086 09597 45	13.78220 46327 06
4.7	− 4.75358 51673 33	4.18226 81404 46	9.7	−11.81658 05418 21	14.00950 23791 60
4.8	− 4.88782 91705 81	4.34053 48000 81	9.8	−11.96241 58543 24	14.23781 88286 23
4.9	− 5.02253 13317 74	4.50078 64388 72	9.9	−12.10836 45707 60	14.46714 38298 57
5.0	− 5.15767 38696 89	4.66298 63139 40	10.0	−12.25442 44338 60	14.69746 74295 03

Table 6.7   **GAMMA FUNCTION FOR COMPLEX ARGUMENTS**

$$x = 1.7$$

y	$\mathscr{R} \ln \Gamma(z)$	$\mathscr{I} \ln \Gamma(z)$	y	$\mathscr{R} \ln \Gamma(z)$	$\mathscr{I} \ln \Gamma(z)$
0.0	− 0.09580 76974 07	0.00000 00000 00	5.0	− 4.99429 42740 24	4.79738 98064 85
0.1	− 0.09977 01624 55	0.02095 53101 47	5.1	− 5.12797 31077 01	4.96193 49448 28
0.2	− 0.11161 35203 43	0.04250 99781 99	5.2	− 5.26209 29486 79	5.12834 25830 88
0.3	− 0.13120 82417 20	0.06524 48506 20	5.3	− 5.39663 77210 79	5.29658 04404 97
0.4	− 0.15834 67099 43	0.08970 54480 34	5.4	− 5.53159 21994 12	5.46661 72692 91
0.5	− 0.19275 44989 43	0.11638 82473 83	5.5	− 5.66694 19505 53	5.63842 28098 55
0.6	− 0.23410 41754 11	0.14573 09476 06	5.6	− 5.80267 32805 14	5.81196 77481 03
0.7	− 0.28203 01468 30	0.17810 70108 82	5.7	− 5.93877 31855 28	5.98722 36749 88
0.8	− 0.33614 32007 35	0.21382 42284 85	5.8	− 6.07522 93070 61	6.16416 30480 45
0.9	− 0.39604 36829 33	0.25312 66649 29	5.9	− 6.21202 98903 76	6.34275 91548 66
1.0	− 0.46133 26441 19	0.29619 91243 57	6.0	− 6.34916 37463 25	6.52298 60784 05
1.1	− 0.53162 06562 78	0.34317 32455 42	6.1	− 6.48662 02160 75	6.70481 86640 24
1.2	− 0.60653 43029 30	0.39413 44205 39	6.2	− 6.62438 91385 04	6.88823 24881 89
1.3	− 0.68572 05552 37	0.44912 88915 80	6.3	− 6.76246 08200 42	7.07320 38287 20
1.4	− 0.76884 93610 19	0.50817 05624 82	6.4	− 6.90082 60067 27	7.25970 96365 25
1.5	− 0.85561 48134 32	0.57124 72307 84	6.5	− 7.03947 58582 98	7.44772 75087 22
1.6	− 0.94573 52538 42	0.63832 60866 03	6.6	− 7.17840 19241 47	7.63723 56630 84
1.7	− 1.03895 26210 76	0.70935 84280 02	6.7	− 7.31759 61209 77	7.82821 29137 39
1.8	− 1.13503 13039 83	0.78428 36123 89	6.8	− 7.45705 07120 18	8.02063 86480 35
1.9	− 1.23375 66975 90	0.86303 23052 04	6.9	− 7.59675 82876 82	8.21449 28045 37
2.0	− 1.33493 36116 09	0.94552 91079 51	7.0	− 7.73671 17475 34	8.40975 58520 62
2.1	− 1.43838 46369 05	1.03169 46541 37	7.1	− 7.87690 42834 81	8.60640 87697 25
2.2	− 1.54394 85411 53	1.12144 72591 94	7.2	− 8.01732 93640 69	8.80443 30279 13
2.3	− 1.65147 87389 10	1.21470 42030 73	7.3	− 8.15798 07198 22	9.00381 05701 63
2.4	− 1.76084 18623 15	1.31138 27144 41	7.4	− 8.29885 23295 23	9.20452 37958 73
2.5	− 1.87191 64452 44	1.41140 07152 26	7.5	− 8.43993 84073 80	9.40655 55438 14
2.6	− 1.98459 17246 80	1.51467 73744 45	7.6	− 8.58123 33910 02	9.60988 90763 93
2.7	− 2.09876 65571 99	1.62113 35114 76	7.7	− 8.72273 19301 22	9.81450 80646 38
2.8	− 2.21434 84448 82	1.73069 18813 34	7.8	− 8.86442 88760 30	10.02039 65738 46
2.9	− 2.33125 26629 53	1.84327 73680 71	7.9	− 9.00631 92716 38	10.22753 90498 84
3.0	− 2.44940 14805 61	1.95881 71071 34	8.0	− 9.14839 83421 51	10.43592 03060 85
3.1	− 2.56872 34658 89	2.07724 05531 98	8.1	− 9.29066 14862 98	10.64552 55107 28
3.2	− 2.68915 28670 03	2.19847 95064 74	8.2	− 9.43310 42680 75	10.85634 01750 59
3.3	− 2.81062 90603 59	2.32246 81077 41	8.3	− 9.57572 24089 73	11.06835 01418 23
3.4	− 2.93309 60594 79	2.44914 28100 87	8.4	− 9.71851 17806 54	11.28154 15743 00
3.5	− 3.05650 20770 24	2.57844 23336 16	8.5	− 9.86146 83980 47	11.49590 09457 89
3.6	− 3.18079 91341 33	2.71030 76079 67	8.6	−10.00458 84128 32	11.71141 50295 52
3.7	− 3.30594 27115 93	2.84468 17064 22	8.7	−10.14786 81072 85	11.92807 08891 58
3.8	− 3.43189 14379 84	2.98150 97744 80	8.8	−10.29130 38884 74	12.14585 58692 46
3.9	− 3.55860 68105 24	3.12073 89551 42	8.9	−10.43489 22827 58	12.36475 75866 47
4.0	− 3.68605 29448 47	3.26231 83125 99	9.0	−10.57862 99305 96	12.58476 39218 81
4.1	− 3.81419 63503 82	3.40619 87555 93	9.1	−10.72251 35816 27	12.80586 30109 93
4.2	− 3.94300 57284 13	3.55233 29614 33	9.2	−10.86654 00900 14	13.02804 32377 08
4.3	− 4.07245 17902 59	3.70067 53013 46	9.3	−11.01070 64100 32	13.25129 32259 06
4.4	− 4.20250 70933 22	3.85118 17677 02	9.4	−11.15500 95918 83	13.47560 18323 86
4.5	− 4.33314 58930 01	4.00380 99034 45	9.5	−11.29944 67777 28	13.70095 81399 16
4.6	− 4.46434 40087 52	4.15851 87339 90	9.6	−11.44401 51979 25	13.92735 14505 47
4.7	− 4.59607 87027 47	4.31526 87017 23	9.7	−11.58871 21674 47	14.15477 12791 90
4.8	− 4.72832 85697 79	4.47402 16031 94	9.8	−11.73353 50824 91	14.38320 73474 23
4.9	− 4.86107 34372 26	4.63474 05290 18	9.9	−11.87848 14172 43	14.61264 95775 51
5.0	− 4.99429 42740 24	4.79738 98064 85	10.0	−12.02354 87208 09	14.84308 80868 68

GAMMA FUNCTION FOR COMPLEX ARGUMENTS

Table 6.7

$$x=1.8$$

y	$\mathscr{R} \ln \Gamma(z)$	$\mathscr{I} \ln \Gamma(z)$	y	$\mathscr{R} \ln \Gamma(z)$	$\mathscr{I} \ln \Gamma(z)$
0.0	− 0.07108 38729 14	0.00000 00000 00	5.0	− 4.83045 68451 13	4.92989 76263 84
0.1	− 0.07476 57386 86	0.02858 63331 36	5.1	− 4.96226 53555 54	5.09490 86275 80
0.2	− 0.08577 55297 09	0.05769 29209 31	5.2	− 5.09454 72216 70	5.26176 50781 04
0.3	− 0.10400 76857 32	0.08782 58538 91	5.3	− 5.22728 53433 89	5.43043 56009 62
0.4	− 0.12929 22486 30	0.11946 40495 57	5.4	− 5.36046 35143 73	5.60088 97905 12
0.5	− 0.16140 31015 52	0.15304 83729 82	5.5	− 5.49406 63619 68	5.77309 81726 78
0.6	− 0.20006 82029 53	0.18897 35429 70	5.6	− 5.62807 92920 13	5.94703 21669 16
0.7	− 0.24498 08149 51	0.22758 31014 17	5.7	− 5.76248 84380 56	6.12266 40498 86
0.8	− 0.29581 07721 71	0.26916 73612 58	5.8	− 5.89728 06145 63	6.29996 69207 68
0.9	− 0.35221 50054 25	0.31396 39650 50	5.9	− 6.03244 32737 64	6.47891 46681 58
1.0	− 0.41384 67690 74	0.36216 05120 09	6.0	− 6.16796 44658 02	6.65948 19384 99
1.1	− 0.48036 32669 52	0.41389 86472 00	6.1	− 6.30383 28019 05	6.84164 41059 65
1.2	− 0.55143 15880 74	0.46927 90315 88	6.2	− 6.44003 74202 92	7.02537 72437 42
1.3	− 0.62673 30272 43	0.52836 66950 54	6.3	− 6.57656 79546 04	7.21065 80966 53
1.4	− 0.70596 59713 03	0.59119 63857 23	6.4	− 6.71341 45046 23	7.39746 40550 43
1.5	− 0.78884 75850 80	0.65777 76436 65	6.5	− 6.85056 76090 92	7.58577 31298 85
1.6	− 0.87511 45440 57	0.72809 94297 11	6.6	− 6.98801 82204 65	7.77556 39290 39
1.7	− 0.96452 30468 26	0.80213 42229 48	6.7	− 7.12575 76814 17	7.96681 56346 11
1.8	− 1.05684 83111 80	0.87984 15616 08	6.8	− 7.26377 77029 87	8.15950 79813 46
1.9	− 1.15188 37223 02	0.96117 10434 30	6.9	− 7.40207 03441 98	8.35362 12360 30
2.0	− 1.24943 97659 29	1.04606 48267 65	7.0	− 7.54062 79930 63	8.54913 61778 15
2.1	− 1.34934 28469 99	1.13445 96865 98	7.1	− 7.67944 33488 49	8.74603 40794 54
2.2	− 1.45143 40669 35	1.22628 86841 72	7.2	− 7.81850 94055 06	8.94429 66893 74
2.3	− 1.55556 80105 11	1.32148 25078 65	7.3	− 7.95781 94361 78	9.14390 62145 64
2.4	− 1.66161 15761 22	1.41997 05387 49	7.4	− 8.09736 69787 03	9.34484 53042 25
2.5	− 1.76944 28703 84	1.52168 16884 90	7.5	− 8.23714 58220 35	9.54709 70341 42
2.6	− 1.87895 01786 38	1.62654 50508 69	7.6	− 8.37714 99935 16	9.75064 48917 54
2.7	− 1.99003 10163 61	1.73449 04020 35	7.7	− 8.51737 37469 39	9.95547 27618 74
2.8	− 2.10259 12619 95	1.84544 85788 28	7.8	− 8.65781 15513 42	10.16156 49130 30
2.9	− 2.21654 43688 12	1.95935 17594 45	7.9	− 8.79845 80804 75	10.36890 59844 02
3.0	− 2.33181 06516 27	2.07613 66663 29	8.0	− 8.93930 82029 08	10.57748 09733 12
3.1	− 2.44831 66432 13	2.19572 97074 49	8.1	− 9.08035 69727 14	10.78727 52232 56
3.2	− 2.56599 45147 78	2.31807 70690 52	8.2	− 9.22159 96207 08	10.99827 44124 32
3.3	− 2.68478 15548 41	2.44311 47704 17	8.3	− 9.36303 15461 81	11.21046 45427 62
3.4	− 2.80461 97009 53	2.57078 36890 62	8.4	− 9.50464 83091 20	11.42383 19293 59
3.5	− 2.92545 51190 19	2.70102 65631 50	8.5	− 9.64644 56228 63	11.63836 31904 38
3.6	− 3.04723 78253 42	2.83378 79764 90	8.6	− 9.78841 93471 63	11.85404 52376 37
3.7	− 3.16992 13469 31	2.96901 43304 05	8.7	− 9.93056 54816 43	12.07086 52667 34
3.8	− 3.29346 24159 89	3.10665 38058 79	8.8	−10.07288 01596 06	12.28881 07487 37
3.9	− 3.41782 06949 39	3.24665 63186 51	8.9	−10.21535 96421 85	12.50786 94213 31
4.0	− 3.54295 85286 89	3.38897 34693 93	9.0	−10.35800 03128 01	12.72802 92806 69
4.1	− 3.66884 07212 13	3.53355 84906 21	9.1	−10.50079 86719 24	12.94927 85734 79
4.2	− 3.79543 43338 26	3.68036 61916 47	9.2	−10.64375 13321 05	13.17160 57894 90
4.3	− 3.92270 85028 21	3.82935 29025 75	9.3	−10.78685 50132 67	13.39499 96541 43
4.4	− 4.05063 42744 24	3.98047 64181 31	9.4	−10.93010 65382 43	13.61944 91215 87
4.5	− 4.17918 44552 05	4.13369 59419 14	9.5	−11.07350 28285 39	13.84494 33679 42
4.6	− 4.30833 34763 48	4.28897 20315 17	9.6	−11.21704 09003 12	14.07147 17848 17
4.7	− 4.43805 72703 06	4.44626 65448 66	9.7	−11.36071 78605 47	14.29902 39730 75
4.8	− 4.56833 31585 96	4.60554 25879 92	9.8	−11.50453 09034 33	14.52758 97368 21
4.9	− 4.69913 97495 61	4.76676 44644 38	9.9	−11.64847 73069 06	14.75715 90776 29
5.0	− 4.83045 68451 13	4.92989 76263 84	10.0	−11.79255 44293 69	14.98772 21889 61

Table 6.7   **GAMMA FUNCTION FOR COMPLEX ARGUMENTS**

$$x = 1.9$$

y	$\mathscr{R} \ln \Gamma(z)$	$\mathscr{I} \ln \Gamma(z)$	y	$\mathscr{R} \ln \Gamma(z)$	$\mathscr{I} \ln \Gamma(z)$
0.0	− 0.03898 42759 23	0.00000 00000 00	5.0	− 4.66612 81728 77	5.06052 77830 38
0.1	− 0.04242 16648 18	0.03569 47077 36	5.1	− 4.79608 44074 24	5.22603 70297 75
0.2	− 0.05270 43596 13	0.07184 49288 73	5.2	− 4.92654 53878 64	5.39337 36626 27
0.3	− 0.06974 53071 16	0.10889 51730 33	5.3	− 5.05749 30552 47	5.56250 72499 47
0.4	− 0.09340 38158 25	0.14726 87453 39	5.4	− 5.18891 02823 51	5.73340 82679 93
0.5	− 0.12349 16727 26	0.18735 90383 60	5.5	− 5.32078 08121 05	5.90604 80662 49
0.6	− 0.15978 08372 30	0.22952 28050 02	5.6	− 5.45308 92008 98	6.08039 88340 38
0.7	− 0.20201 20244 82	0.27407 56544 06	5.7	− 5.58582 07663 21	6.25643 35684 02
0.8	− 0.24990 35004 09	0.32128 97690 64	5.8	− 5.71896 15389 41	6.43412 60432 49
0.9	− 0.30315 95035 34	0.37139 36389 55	5.9	− 5.85249 82177 50	6.61345 07797 49
1.0	− 0.36147 78527 10	0.42457 34706 81	6.0	− 5.98641 81289 78	6.79438 30179 35
1.1	− 0.42455 64621 11	0.48097 58618 37	6.1	− 6.12070 91879 56	6.97689 86894 96
1.2	− 0.49209 86372 39	0.54071 13247 70	6.2	− 6.25535 98637 85	7.16097 43917 16
1.3	− 0.56381 71504 20	0.60385 82827 52	6.3	− 6.39035 91465 66	7.34658 73625 14
1.4	− 0.63943 71834 98	0.67046 72268 81	6.4	− 6.52569 65169 71	7.53371 54565 59
1.5	− 0.71869 82795 42	0.74056 47971 47	6.5	− 6.66136 19179 75	7.72233 71224 13
1.6	− 0.80135 54698 30	0.81415 76239 52	6.6	− 6.79734 57285 54	7.91243 13806 57
1.7	− 0.88717 97447 03	0.89123 58296 55	6.7	− 6.93363 87392 01	8.10397 78029 64
1.8	− 0.97595 80247 42	0.97177 61401 47	6.8	− 7.07023 21291 12	8.29695 64920 80
1.9	− 1.06749 27687 53	1.05574 45936 43	6.9	− 7.20711 74449 04	8.49134 80626 65
2.0	− 1.16160 13318 68	1.14309 88592 34	7.0	− 7.34428 65807 56	8.68713 36229 72
2.1	− 1.25811 51641 83	1.23379 01934 57	7.1	− 7.48173 17598 49	8.88429 47573 07
2.2	− 1.35687 89195 14	1.32776 50714 39	7.2	− 7.61944 55170 18	9.08281 35092 45
2.3	− 1.45774 95259 72	1.42496 65323 75	7.3	− 7.75742 06825 11	9.28267 23655 74
2.4	− 1.56059 52554 63	1.52533 52787 28	7.4	− 7.89565 03667 87	9.48385 42409 11
2.5	− 1.66529 48176 11	1.62881 05662 06	7.5	− 8.03412 79462 62	9.68634 24629 88
2.6	− 1.77173 64947 51	1.73533 09179 80	7.6	− 8.17284 70499 43	9.89012 07585 45
2.7	− 1.87981 73280 00	1.84483 46926 69	7.7	− 8.31180 15468 79	10.09517 32398 33
2.8	− 1.98944 23595 80	1.95726 05315 67	7.8	− 8.45098 55343 75	10.30148 43916 76
2.9	− 2.10052 39332 16	2.07254 77068 08	7.9	− 8.59039 33269 14	10.50903 90590 64
3.0	− 2.21298 10520 42	2.19063 63887 13	8.0	− 8.73001 94457 32	10.71782 24352 78
3.1	− 2.32673 87919 77	2.31146 78475 36	8.1	− 8.86985 86090 10	10.92782 00504 91
3.2	− 2.44172 77675 72	2.43498 46022 00	8.2	− 9.00990 57226 31	11.13901 77608 39
3.3	− 2.55788 36468 15	2.56113 05263 98	8.3	− 9.15015 58714 69	11.35140 17379 39
3.4	− 2.67514 67111 48	2.68985 09205 60	8.4	− 9.29060 43111 75	11.56495 84588 29
3.5	− 2.79346 14569 24	2.82109 25566 19	8.5	− 9.43124 64604 23	11.77967 46963 13
3.6	− 2.91277 62346 38	2.95480 37012 40	8.6	− 9.57207 78935 85	11.99553 75096 87
3.7	− 3.03304 29224 14	3.09093 41220 91	8.7	− 9.71309 43338 13	12.21253 42358 42
3.8	− 3.15421 66305 10	3.22943 50808 91	8.8	− 9.85429 16464 97	12.43065 24807 06
3.9	− 3.27625 54337 96	3.37025 93162 16	8.9	− 9.99566 58330 75	12.64988 01110 27
4.0	− 3.39912 01294 42	3.51336 10185 24	9.0	−10.13721 30251 72	12.87020 52464 75
4.1	− 3.52277 40173 08	3.65869 57993 21	9.1	−10.27892 94790 52	13.09161 62520 42
4.2	− 3.64718 27007 49	3.80622 06560 50	9.2	−10.42081 15703 58	13.31410 17307 41
4.3	− 3.77231 39057 84	3.95589 39339 63	9.3	−10.56285 57891 26	13.53765 05165 78
4.4	− 3.89813 73167 71	4.10767 52859 66	9.4	−10.70505 87350 54	13.76225 16677 85
4.5	− 4.02462 44269 53	4.26152 56312 41	9.5	−10.84741 71130 08	13.98789 44603 16
4.6	− 4.15174 84023 59	4.41740 71132 72	9.6	−10.98992 77287 64	14.21456 83815 73
4.7	− 4.27948 39577 56	4.57528 30577 67	9.7	−11.13258 74849 48	14.44226 31243 75
4.8	− 4.40780 72434 44	4.73511 79308 60	9.8	−11.27539 33771 93	14.67096 85811 36
4.9	− 4.53669 57418 38	4.89687 72979 01	9.9	−11.41834 24904 66	14.90067 48382 65
5.0	− 4.66612 81728 77	5.06052 77830 38	10.0	−11.56143 19955 88	15.13137 21707 60

GAMMA FUNCTION FOR COMPLEX ARGUMENTS

Table 6.7

$$x=2.0$$

y	$\mathscr{R} \ln \Gamma(z)$	$\mathscr{I} \ln \Gamma(z)$	y	$\mathscr{R} \ln \Gamma(z)$	$\mathscr{I} \ln \Gamma(z)$
0.0	0.00000 00000 00	0.00000 00000 00	5.0	− 4.50127 58755 42	5.18929 93415 60
0.1	− 0.00322 26151 39	0.04234 57120 74	5.1	− 4.62939 88796 82	5.35533 82031 27
0.2	− 0.01286 59357 41	0.08509 33372 06	5.2	− 4.75805 70222 52	5.52318 54439 62
0.3	− 0.02885 74027 79	0.12863 61223 10	5.3	− 4.88723 13522 76	5.69281 16137 11
0.4	− 0.05107 93722 62	0.17335 05507 97	5.4	− 5.01690 38831 33	5.86418 81052 00
0.5	− 0.07937 37235 30	0.21958 93100 95	5.5	− 5.14705 75299 57	6.03728 71248 73
0.6	− 0.11354 77183 40	0.26767 56897 80	5.6	− 5.27767 60518 81	6.21208 16640 30
0.7	− 0.15338 06308 81	0.31789 96132 02	5.7	− 5.40874 39987 03	6.38854 54709 43
0.8	− 0.19863 06626 31	0.37051 53392 47	5.8	− 5.54024 66615 82	6.56665 30238 56
0.9	− 0.24904 17059 66	0.42574 07261 44	5.9	− 5.67217 00274 24	6.74637 95048 97
1.0	− 0.30434 96090 22	0.48375 78429 30	6.0	− 5.80450 07366 29	6.92770 07748 95
1.1	− 0.36428 77010 76	0.54471 46524 35	6.1	− 5.93722 60439 25	7.11059 33491 13
1.2	− 0.42859 14442 42	0.60872 74700 17	6.2	− 6.07033 37820 31	7.29503 43738 76
1.3	− 0.49700 21701 52	0.67588 39160 88	6.3	− 6.20381 23278 98	7.48100 16040 81
1.4	− 0.56926 99322 58	0.74624 61166 63	6.4	− 6.33765 05713 36	7.66847 33815 76
1.5	− 0.64515 55533 76	0.81985 39537 67	6.5	− 6.47183 78858 22	7.85742 86143 76
1.6	− 0.72443 19760 33	0.89672 82178 63	6.6	− 6.60636 41013 16	8.04784 67567 00
1.7	− 0.80688 50339 42	0.97687 35612 07	6.7	− 6.74121 94789 19	8.23970 77898 07
1.8	− 0.89231 37613 78	1.06028 11909 26	6.8	− 6.87639 46872 45	8.43299 22035 86
1.9	− 0.98053 03476 69	1.14693 12720 53	6.9	− 7.01188 07803 50	8.62768 09788 99
2.0	− 1.07135 98302 14	1.23679 50341 04	7.0	− 7.14766 91771 18	8.82375 55706 27
2.1	− 1.16463 96040 42	1.32983 65907 26	7.1	− 7.28375 16419 82	9.02119 78914 05
2.2	− 1.26021 88108 76	1.42601 44920 94	7.2	− 7.42012 02668 81	9.21999 02960 14
2.3	− 1.35795 76568 48	1.52528 30352 04	7.3	− 7.55676 74543 62	9.42011 55664 09
2.4	− 1.45772 66961 57	1.62759 33595 36	7.4	− 7.69368 59017 46	9.62155 68973 45
2.5	− 1.55940 61080 61	1.73289 43555 35	7.5	− 7.83086 85862 69	9.82429 78825 87
2.6	− 1.66288 49866 52	1.84113 34120 22	7.6	− 7.96830 87511 38	10.02832 25016 83
2.7	− 1.76806 06566 17	1.95225 70264 63	7.7	− 8.10599 98924 36	10.23361 51072 54
2.8	− 1.87483 80234 65	2.06621 12994 71	7.8	− 8.24393 57468 08	10.44016 04128 09
2.9	− 1.98312 89631 02	2.18294 23322 91	7.9	− 8.38211 02798 83	10.64794 34810 35
3.0	− 2.09285 17530 93	2.30239 65434 67	8.0	− 8.52051 76753 67	10.85694 97125 60
3.1	− 2.20393 05460 64	2.42452 09185 18	8.1	− 8.65915 23247 82	11.06716 48351 59
3.2	− 2.31629 48844 77	2.54926 32043 52	8.2	− 8.79800 88177 87	11.27857 48933 86
3.3	− 2.42987 92551 37	2.67657 20582 60	8.3	− 8.93708 19330 47	11.49116 62386 10
3.4	− 2.54462 26813 03	2.80639 71597 50	8.4	− 9.07636 66296 28	11.70492 55194 45
3.5	− 2.66046 83499 73	2.93868 92920 59	8.5	− 9.21585 80388 55	11.91983 96725 52
3.6	− 2.77736 32717 84	3.07340 03990 47	8.6	− 9.35555 14566 37	12.13589 59137 86
3.7	− 2.89525 79709 78	3.21048 36221 88	8.7	− 9.49544 23361 92	12.35308 17297 01
3.8	− 3.01410 62029 30	3.34989 33215 16	8.8	− 9.63552 62811 84	12.57138 48693 62
3.9	− 3.13386 46968 42	3.49158 50837 57	8.9	− 9.77579 90392 11	12.79079 33364 76
4.0	− 3.25449 29213 81	3.63551 57202 41	9.0	− 9.91625 64956 49	13.01129 53818 23
4.1	− 3.37595 28711 45	3.78164 32567 78	9.1	−10.05689 46678 12	13.23287 94959 63
4.2	− 3.49820 88720 59	3.92992 69172 45	9.2	−10.19770 96994 20	13.45553 44022 19
4.3	− 3.62122 74039 03	4.08032 71023 23	9.3	−10.33869 78553 49	13.67924 90499 21
4.4	− 3.74497 69383 89	4.23280 53645 81	9.4	−10.47985 55166 49	13.90401 26078 95
4.5	− 3.86942 77912 99	4.38732 43808 43	9.5	−10.62117 91758 12	14.12981 44581 93
4.6	− 3.99455 19873 65	4.54384 79226 20	9.6	−10.76266 54322 81	14.35664 41900 46
4.7	− 4.12032 31366 90	4.70234 08252 48	9.7	−10.90431 09881 75	14.58449 15940 42
4.8	− 4.24671 63216 20	4.86276 89562 20	9.8	−11.04611 26442 29	14.81334 66565 09
4.9	− 4.37370 79930 87	5.02509 91831 32	9.9	−11.18806 72959 27	15.04319 95540 92
5.0	− 4.50127 58755 42	5.18929 93415 60	10.0	−11.33017 19298 27	15.27404 06485 34

Table 6.8   DIGAMMA FUNCTION FOR COMPLEX ARGUMENTS
$x=1.0$

y	$\mathscr{R}\psi(z)$	$\mathscr{I}\psi(z)$	y	$\mathscr{R}\psi(z)$	$\mathscr{I}\psi(z)$
0.0	−0.57721 56649	0.00000	5.0	1.61278 48446	1.47080
0.1	−0.56529 77902	0.16342	5.1	1.63245 69889	1.47276
0.2	−0.53073 04055	0.32064	5.2	1.65175 20861	1.47464
0.3	−0.47675 48934	0.46653	5.3	1.67068 42228	1.47646
0.4	−0.40786 79442	0.59770	5.4	1.68926 67162	1.47820
0.5	−0.32888 63572	0.71269	5.5	1.70751 21687	1.47989
0.6	−0.24419 65809	0.81160	5.6	1.72543 25175	1.48151
0.7	−0.15733 61258	0.89563	5.7	1.74303 90807	1.48308
0.8	−0.07088 34022	0.96655	5.8	1.76034 25988	1.48459
0.9	+0.01345 20154	1.02628	5.9	1.77735 32733	1.48605
1.0	0.09465 03206	1.07667	6.0	1.79408 08018	1.48746
1.1	0.17219 05426	1.11938	6.1	1.81053 44105	1.48883
1.2	0.24588 65515	1.15580	6.2	1.82672 28842	1.49015
1.3	0.31576 20906	1.18707	6.3	1.84265 45939	1.49143
1.4	0.38196 28134	1.21413	6.4	1.85833 75219	1.49267
1.5	0.44469 79402	1.23772	6.5	1.87377 92858	1.49387
1.6	0.50420 34618	1.25843	6.6	1.88898 71602	1.49504
1.7	0.56072 00645	1.27675	6.7	1.90396 80964	1.49617
1.8	0.61448 06554	1.29306	6.8	1.91872 87422	1.49727
1.9	0.66570 39172	1.30766	6.9	1.93327 54582	1.49833
2.0	0.71459 15154	1.32081	7.0	1.94761 43346	1.49937
2.1	0.76132 74328	1.33271	7.1	1.96175 12062	1.50037
2.2	0.80607 84807	1.34353	7.2	1.97569 16663	1.50135
2.3	0.84899 54079	1.35341	7.3	1.98944 10799	1.50230
2.4	0.89021 42662	1.36246	7.4	2.00300 45959	1.50323
2.5	0.92985 78387	1.37080	7.5	2.01638 71585	1.50413
2.6	0.96803 70243	1.37849	7.6	2.02959 35177	1.50501
2.7	1.00485 21252	1.38561	7.7	2.04262 82397	1.50586
2.8	1.04039 40175	1.39222	7.8	2.05549 57159	1.50669
2.9	1.07474 51976	1.39838	7.9	2.06820 01717	1.50751
3.0	1.10798 07107	1.40413	8.0	2.08074 56749	1.50830
3.1	1.14016 89703	1.40951	8.1	2.09313 61434	1.50907
3.2	1.17137 24783	1.41455	8.2	2.10537 53524	1.50982
3.3	1.20164 84581	1.41928	8.3	2.11746 69410	1.51056
3.4	1.23104 94107	1.42374	8.4	2.12941 44191	1.51127
3.5	1.25962 36033	1.42794	8.5	2.14122 11731	1.51197
3.6	1.28741 54995	1.43191	8.6	2.15289 04718	1.51266
3.7	1.31446 61381	1.43566	8.7	2.16442 54716	1.51332
3.8	1.34081 34679	1.43922	8.8	2.17582 92217	1.51398
3.9	1.36649 26435	1.44259	8.9	2.18710 46687	1.51462
4.0	1.39153 62879	1.44580	9.0	2.19825 46616	1.51524
4.1	1.41597 47255	1.44885	9.1	2.20928 19555	1.51585
4.2	1.43983 61892	1.45175	9.2	2.22018 92160	1.51645
4.3	1.46314 70060	1.45452	9.3	2.23097 90229	1.51703
4.4	1.48593 17620	1.45716	9.4	2.24165 38740	1.51760
4.5	1.50821 34505	1.45969	9.5	2.25221 61882	1.51816
4.6	1.53001 36052	1.46210	9.6	2.26266 83093	1.51871
4.7	1.55135 24197	1.46441	9.7	2.27301 25085	1.51925
4.8	1.57224 88550	1.46663	9.8	2.28325 09877	1.51978
4.9	1.59272 07370	1.46876	9.9	2.29338 58823	1.52029
5.0	1.61278 48446	1.47080	10.0	2.30341 92637	1.52080
	$\begin{bmatrix}(-3)2\\5\end{bmatrix}$			$\begin{bmatrix}(-5)5\\6\end{bmatrix}$	$\begin{bmatrix}(-5)1\\2\end{bmatrix}$

$$\mathscr{I}\psi(1+iy)=\tfrac{1}{2}\pi\coth\pi y-\frac{1}{2y}$$

$\psi(z)$ to 5D, computed by M. Goldstein, Los Alamos Scientific Laboratory.

AUXILIARY FUNCTION FOR $\mathscr{R}\psi(1+iy)$

y^{-1}	$f_4(y)$	$\langle y\rangle$	y^{-1}	$f_4(y)$	$\langle y\rangle$
0.11	0.00100 956	9	0.05	0.00020 839	20
0.10	0.00083 417	10	0.04	0.00013 335	25
0.09	0.00067 555	11	0.03	0.00007 501	33
0.08	0.00053 368	13	0.02	0.00003 333	50
0.07	0.00040 853	14	0.01	0.00000 833	100
0.06	0.00030 011	17	0.00	0.00000 000	∞
	$\begin{bmatrix}(-6)2\\3\end{bmatrix}$			$\begin{bmatrix}(-6)2\\3\end{bmatrix}$	

$$\mathscr{R}\psi(1+iy)=\ln y+f_4(y)$$
$\langle y\rangle$=nearest integer to y.

DIGAMMA FUNCTION FOR COMPLEX ARGUMENTS

Table 6.8

$x = 1.1$ $\qquad x = 1.2$

y	$\mathscr{R}\psi(z)$	$\mathscr{I}\psi(z)$	y	$\mathscr{R}\psi(z)$	$\mathscr{I}\psi(z)$	y	$\mathscr{R}\psi(z)$	$\mathscr{I}\psi(z)$	y	$\mathscr{R}\psi(z)$	$\mathscr{I}\psi(z)$
0.0	−0.42375	0.00000	5.0	1.61498	1.45097	0.0	−0.28904	0.00000	5.0	1.61756	1.43125
0.1	−0.41451	0.14258	5.1	1.63457	1.45332	0.1	−0.28169	0.12620	5.1	1.63705	1.43396
0.2	−0.38753	0.28082	5.2	1.65378	1.45557	0.2	−0.26014	0.24926	5.2	1.65617	1.43658
0.3	−0.34490	0.41099	5.3	1.67264	1.45774	0.3	−0.22578	0.36640	5.3	1.67494	1.43910
0.4	−0.28961	0.53042	5.4	1.69115	1.45983	0.4	−0.18064	0.47552	5.4	1.69336	1.44152
0.5	−0.22498	0.63764	5.5	1.70933	1.46184	0.5	−0.12710	0.57530	5.5	1.71146	1.44386
0.6	−0.15426	0.73229	5.6	1.72718	1.46378	0.6	−0.06753	0.66517	5.6	1.72924	1.44612
0.7	−0.08023	0.81484	5.7	1.74473	1.46565	0.7	−0.00412	0.74519	5.7	1.74672	1.44829
0.8	−0.00509	0.88630	5.8	1.76197	1.46746	0.8	+0.06130	0.81589	5.8	1.76390	1.45039
0.9	+0.06954	0.94792	5.9	1.77893	1.46921	0.9	0.12730	0.87806	5.9	1.78079	1.45243
1.0	0.14255	1.00102	6.0	1.79561	1.47090	1.0	0.19280	0.93260	6.0	1.79740	1.45439
1.1	0.21327	1.04687	6.1	1.81201	1.47253	1.1	0.25707	0.98046	6.1	1.81375	1.45629
1.2	0.28131	1.08660	6.2	1.82815	1.47411	1.2	0.31960	1.02252	6.2	1.82983	1.45813
1.3	0.34649	1.12119	6.3	1.84404	1.47565	1.3	0.38012	1.05960	6.3	1.84567	1.45991
1.4	0.40880	1.15146	6.4	1.85968	1.47713	1.4	0.43846	1.09240	6.4	1.86126	1.46164
1.5	0.46829	1.17810	6.5	1.87508	1.47857	1.5	0.49459	1.12153	6.5	1.87661	1.46331
1.6	0.52507	1.20169	6.6	1.89025	1.47996	1.6	0.54851	1.14752	6.6	1.89173	1.46493
1.7	0.57930	1.22269	6.7	1.90519	1.48132	1.7	0.60028	1.17082	6.7	1.90663	1.46651
1.8	0.63111	1.24148	6.8	1.91992	1.48263	1.8	0.64999	1.19179	6.8	1.92132	1.46803
1.9	0.68067	1.25839	6.9	1.93443	1.48391	1.9	0.69774	1.21074	6.9	1.93579	1.46952
2.0	0.72813	1.27368	7.0	1.94874	1.48515	2.0	0.74362	1.22794	7.0	1.95006	1.47096
2.1	0.77363	1.28755	7.1	1.96284	1.48635	2.1	0.78775	1.24362	7.1	1.96413	1.47236
2.2	0.81730	1.30021	7.2	1.97675	1.48752	2.2	0.83022	1.25796	7.2	1.97800	1.47372
2.3	0.85928	1.31179	7.3	1.99047	1.48866	2.3	0.87114	1.27112	7.3	1.99169	1.47505
2.4	0.89967	1.32243	7.4	2.00401	1.48977	2.4	0.91060	1.28323	7.4	2.00519	1.47634
2.5	0.93858	1.33224	7.5	2.01736	1.49085	2.5	0.94868	1.29442	7.5	2.01852	1.47760
2.6	0.97610	1.34131	7.6	2.03054	1.49190	2.6	0.98546	1.30478	7.6	2.03167	1.47882
2.7	1.01234	1.34972	7.7	2.04356	1.49292	2.7	1.02103	1.31441	7.7	2.04465	1.48001
2.8	1.04736	1.35753	7.8	2.05640	1.49392	2.8	1.05546	1.32337	7.8	2.05746	1.48117
2.9	1.08124	1.36482	7.9	2.06908	1.49489	2.9	1.08881	1.33173	7.9	2.07012	1.48230
3.0	1.11405	1.37162	8.0	2.08160	1.49584	3.0	1.12113	1.33955	8.0	2.08262	1.48341
3.1	1.14586	1.37800	8.1	2.09397	1.49676	3.1	1.15250	1.34688	8.1	2.09496	1.48448
3.2	1.17671	1.38398	8.2	2.10619	1.49767	3.2	1.18295	1.35377	8.2	2.10716	1.48553
3.3	1.20667	1.38960	8.3	2.11826	1.49855	3.3	1.21254	1.36024	8.3	2.11921	1.48656
3.4	1.23578	1.39489	8.4	2.13019	1.49940	3.4	1.24132	1.36635	8.4	2.13111	1.48756
3.5	1.26409	1.39989	8.5	2.14198	1.50024	3.5	1.26932	1.37211	8.5	2.14288	1.48853
3.6	1.29164	1.40461	8.6	2.15363	1.50106	3.6	1.29659	1.37756	8.6	2.15451	1.48949
3.7	1.31847	1.40907	8.7	2.16515	1.50186	3.7	1.32315	1.38272	8.7	2.16601	1.49042
3.8	1.34461	1.41331	8.8	2.17654	1.50265	3.8	1.34905	1.38761	8.8	2.17738	1.49133
3.9	1.37010	1.41732	8.9	2.18780	1.50341	3.9	1.37432	1.39226	8.9	2.18862	1.49222
4.0	1.39496	1.42114	9.0	2.19893	1.50416	4.0	1.39898	1.39667	9.0	2.19973	1.49310
4.1	1.41924	1.42478	9.1	2.20995	1.50489	4.1	1.42306	1.40088	9.1	2.21073	1.49395
4.2	1.44294	1.42824	9.2	2.22084	1.50561	4.2	1.44659	1.40489	9.2	2.22160	1.49478
4.3	1.46611	1.43154	9.3	2.23161	1.50631	4.3	1.46959	1.40871	9.3	2.23236	1.49560
4.4	1.48876	1.43469	9.4	2.24228	1.50699	4.4	1.49209	1.41236	9.4	2.24301	1.49640
4.5	1.51092	1.43771	9.5	2.25283	1.50766	4.5	1.51410	1.41586	9.5	2.25354	1.49718
4.6	1.53261	1.44059	9.6	2.26326	1.50832	4.6	1.53565	1.41920	9.6	2.26397	1.49794
4.7	1.55384	1.44335	9.7	2.27360	1.50896	4.7	1.55676	1.42240	9.7	2.27429	1.49869
4.8	1.57463	1.44600	9.8	2.28382	1.50960	4.8	1.57743	1.42547	9.8	2.28450	1.49943
4.9	1.59501	1.44854	9.9	2.29395	1.51021	4.9	1.59769	1.42842	9.9	2.29461	1.50015
5.0	1.61498 $\left[\begin{smallmatrix}(-3)2\\5\end{smallmatrix}\right]$	1.45097 $\left[\begin{smallmatrix}(-3)2\\5\end{smallmatrix}\right]$	10.0	2.30397 $\left[\begin{smallmatrix}(-5)5\\3\end{smallmatrix}\right]$	1.51082 $\left[\begin{smallmatrix}(-5)1\\2\end{smallmatrix}\right]$	5.0	1.61756 $\left[\begin{smallmatrix}(-3)1\\5\end{smallmatrix}\right]$	1.43125 $\left[\begin{smallmatrix}(-3)1\\5\end{smallmatrix}\right]$	10.0	2.30462 $\left[\begin{smallmatrix}(-5)5\\3\end{smallmatrix}\right]$	1.50085 $\left[\begin{smallmatrix}(-5)1\\2\end{smallmatrix}\right]$

Table 6.8

DIGAMMA FUNCTION FOR COMPLEX ARGUMENTS

	$x=1.3$						$x=1.4$				
y	$\mathscr{R}\psi(z)$	$\mathscr{I}\psi(z)$	y	$\mathscr{R}\psi(z)$	$\mathscr{I}\psi(z)$	y	$\mathscr{R}\psi(z)$	$\mathscr{I}\psi(z)$	y	$\mathscr{R}\psi(z)$	$\mathscr{I}\psi(z)$
0.0	−0.16919	0.00000	5.0	1.62052	1.41163	0.0	−0.06138	0.00000	5.0	1.62386	1.39213
0.1	−0.16323	0.11303	5.1	1.63990	1.41472	0.1	−0.05646	0.10223	5.1	1.64311	1.39559
0.2	−0.14567	0.22372	5.2	1.65891	1.41769	0.2	−0.04192	0.20269	5.2	1.66200	1.39891
0.3	−0.11748	0.32997	5.3	1.67758	1.42055	0.3	−0.01844	0.29974	5.3	1.68055	1.40211
0.4	−0.08009	0.43011	5.4	1.69591	1.42331	0.4	+0.01295	0.39204	5.4	1.69878	1.40519
0.5	−0.03520	0.52298	5.5	1.71392	1.42597	0.5	0.05100	0.47862	5.5	1.71668	1.40817
0.6	+0.01541	0.60796	5.6	1.73161	1.42853	0.6	0.09436	0.55886	5.6	1.73428	1.41103
0.7	0.07003	0.68491	5.7	1.74900	1.43101	0.7	0.14171	0.63250	5.7	1.75158	1.41380
0.8	0.12718	0.75404	5.8	1.76611	1.43340	0.8	0.19183	0.69957	5.8	1.76860	1.41648
0.9	0.18561	0.81582	5.9	1.78292	1.43571	0.9	0.24367	0.76033	5.9	1.78533	1.41907
1.0	0.24434	0.87085	6.0	1.79947	1.43794	1.0	0.29635	0.81517	6.0	1.80180	1.42157
1.1	0.30262	0.91983	6.1	1.81575	1.44011	1.1	0.34918	0.86457	6.1	1.81800	1.42399
1.2	0.35994	0.96341	6.2	1.83177	1.44220	1.2	0.40163	0.90903	6.2	1.83395	1.42634
1.3	0.41593	1.00227	6.3	1.84754	1.44423	1.3	0.45331	0.94907	6.3	1.84966	1.42861
1.4	0.47035	1.03698	6.4	1.86308	1.44619	1.4	0.50395	0.98517	6.4	1.86513	1.43081
1.5	0.52310	1.06809	6.5	1.87837	1.44810	1.5	0.55336	1.01778	6.5	1.88036	1.43294
1.6	0.57409	1.09605	6.6	1.89344	1.44995	1.6	0.60144	1.04730	6.6	1.89537	1.43502
1.7	0.62333	1.12126	6.7	1.90829	1.45174	1.7	0.64811	1.07409	6.7	1.91017	1.43702
1.8	0.67084	1.14409	6.8	1.92293	1.45348	1.8	0.69337	1.09849	6.8	1.92475	1.43898
1.9	0.71667	1.16483	6.9	1.93735	1.45517	1.9	0.73722	1.12075	6.9	1.93912	1.44087
2.0	0.76087	1.18373	7.0	1.95158	1.45681	2.0	0.77968	1.14113	7.0	1.95330	1.44271
2.1	0.80353	1.20102	7.1	1.96560	1.45841	2.1	0.82078	1.15984	7.1	1.96727	1.44450
2.2	0.84470	1.21688	7.2	1.97944	1.45996	2.2	0.86058	1.17707	7.2	1.98106	1.44625
2.3	0.88447	1.23148	7.3	1.99309	1.46147	2.3	0.89913	1.19296	7.3	1.99467	1.44794
2.4	0.92290	1.24495	7.4	2.00655	1.46294	2.4	0.93647	1.20768	7.4	2.00809	1.44959
2.5	0.96007	1.25743	7.5	2.01984	1.46438	2.5	0.97265	1.22133	7.5	2.02134	1.45119
2.6	0.99604	1.26900	7.6	2.03296	1.46577	2.6	1.00775	1.23402	7.6	2.03442	1.45276
2.7	1.03088	1.27976	7.7	2.04591	1.46713	2.7	1.04179	1.24585	7.7	2.04733	1.45428
2.8	1.06464	1.28980	7.8	2.05869	1.46845	2.8	1.07484	1.25689	7.8	2.06008	1.45576
2.9	1.09739	1.29918	7.9	2.07131	1.46974	2.9	1.10693	1.26723	7.9	2.07267	1.45721
3.0	1.12917	1.30797	8.0	2.08378	1.47100	3.0	1.13813	1.27693	8.0	2.08510	1.45862
3.1	1.16004	1.31621	8.1	2.09610	1.47223	3.1	1.16846	1.28604	8.1	2.09739	1.46000
3.2	1.19005	1.32396	8.2	2.10827	1.47342	3.2	1.19797	1.29461	8.2	2.10952	1.46134
3.3	1.21923	1.33126	8.3	2.12029	1.47459	3.3	1.22670	1.30269	8.3	2.12151	1.46266
3.4	1.24763	1.33814	8.4	2.13217	1.47573	3.4	1.25469	1.31032	8.4	2.13337	1.46394
3.5	1.27529	1.34464	8.5	2.14391	1.47685	3.5	1.28196	1.31753	8.5	2.14508	1.46519
3.6	1.30223	1.35080	8.6	2.15552	1.47794	3.6	1.30855	1.32436	8.6	2.15666	1.46641
3.7	1.32851	1.35663	8.7	2.16700	1.47900	3.7	1.33450	1.33084	8.7	2.16811	1.46760
3.8	1.35413	1.36216	8.8	2.17834	1.48004	3.8	1.35983	1.33699	8.8	2.17943	1.46877
3.9	1.37915	1.36742	8.9	2.18956	1.48106	3.9	1.38456	1.34283	8.9	2.19063	1.46991
4.0	1.40357	1.37242	9.0	2.20066	1.48205	4.0	1.40873	1.34840	9.0	2.20170	1.47103
4.1	1.42744	1.37718	9.1	2.21163	1.48302	4.1	1.43235	1.35370	9.1	2.21265	1.47212
4.2	1.45077	1.38172	9.2	2.22249	1.48397	4.2	1.45546	1.35876	9.2	2.22349	1.47319
4.3	1.47358	1.38606	9.3	2.23323	1.48490	4.3	1.47806	1.36359	9.3	2.23421	1.47423
4.4	1.49590	1.39020	9.4	2.24386	1.48582	4.4	1.50019	1.36821	9.4	2.24481	1.47525
4.5	1.51775	1.39416	9.5	2.25437	1.48671	4.5	1.52185	1.37263	9.5	2.25531	1.47626
4.6	1.53914	1.39795	9.6	2.26478	1.48758	4.6	1.54307	1.37686	9.6	2.26570	1.47724
4.7	1.56010	1.40158	9.7	2.27508	1.48844	4.7	1.56387	1.38092	9.7	2.27598	1.47820
4.8	1.58064	1.40507	9.8	2.28528	1.48927	4.8	1.58425	1.38481	9.8	2.28616	1.47914
4.9	1.60078	1.40841	9.9	2.29537	1.49010	4.9	1.60425	1.38854	9.9	2.29623	1.48006
5.0	1.62052	1.41163	10.0	2.30537	1.49090	5.0	1.62386	1.39213	10.0	2.30621	1.48096
	$\begin{bmatrix}(-3)2\\5\end{bmatrix}$	$\begin{bmatrix}(-3)1\\5\end{bmatrix}$		$\begin{bmatrix}(-5)5\\3\end{bmatrix}$	$\begin{bmatrix}(-5)2\\3\end{bmatrix}$		$\begin{bmatrix}(-3)1\\4\end{bmatrix}$	$\begin{bmatrix}(-4)8\\4\end{bmatrix}$		$\begin{bmatrix}(-5)5\\3\end{bmatrix}$	$\begin{bmatrix}(-5)2\\3\end{bmatrix}$

DIGAMMA FUNCTION FOR COMPLEX ARGUMENTS

Table 6.8

$x = 1.5$

$x = 1.6$

y	$\mathscr{R}\psi(z)$	$\mathscr{I}\psi(z)$	y	$\mathscr{R}\psi(z)$	$\mathscr{I}\psi(z)$	y	$\mathscr{R}\psi(z)$	$\mathscr{I}\psi(z)$	y	$\mathscr{R}\psi(z)$	$\mathscr{I}\psi(z)$
0.0	0.03649	0.00000	5.0	1.62756	1.37278	0.0	0.12605	0.00000	5.0	1.63162	1.35357
0.1	0.04062	0.09325	5.1	1.64667	1.37658	0.1	0.12955	0.08566	5.1	1.65057	1.35773
0.2	0.05284	0.18511	5.2	1.66543	1.38025	0.2	0.13995	0.17023	5.2	1.66919	1.36173
0.3	0.07266	0.27432	5.3	1.68386	1.38378	0.3	0.15687	0.25268	5.3	1.68748	1.36558
0.4	0.09932	0.35978	5.4	1.70196	1.38719	0.4	0.17976	0.33214	5.4	1.70546	1.36930
0.5	0.13189	0.44066	5.5	1.71976	1.39047	0.5	0.20790	0.40789	5.5	1.72313	1.37289
0.6	0.16935	0.51640	5.6	1.73725	1.39364	0.6	0.24050	0.47942	5.6	1.74051	1.37635
0.7	0.21064	0.58668	5.7	1.75445	1.39670	0.7	0.27674	0.54642	5.7	1.75760	1.37969
0.8	0.25479	0.65144	5.8	1.77137	1.39965	0.8	0.31581	0.60875	5.8	1.77441	1.38293
0.9	0.30091	0.71078	5.9	1.78801	1.40251	0.9	0.35697	0.66642	5.9	1.79095	1.38605
1.0	0.34824	0.76494	6.0	1.80439	1.40528	1.0	0.39957	0.71957	6.0	1.80724	1.38908
1.1	0.39614	0.81424	6.1	1.82051	1.40796	1.1	0.44305	0.76840	6.1	1.82327	1.39200
1.2	0.44411	0.85907	6.2	1.83638	1.41055	1.2	0.48692	0.81319	6.2	1.83906	1.39484
1.3	0.49175	0.89980	6.3	1.85201	1.41306	1.3	0.53082	0.85423	6.3	1.85460	1.39759
1.4	0.53878	0.93684	6.4	1.86741	1.41549	1.4	0.57445	0.89183	6.4	1.86992	1.40025
1.5	0.58497	0.97054	6.5	1.88258	1.41786	1.5	0.61757	0.92629	6.5	1.88501	1.40284
1.6	0.63018	1.00127	6.6	1.89752	1.42015	1.6	0.66001	0.95790	6.6	1.89989	1.40534
1.7	0.67432	1.02932	6.7	1.91225	1.42237	1.7	0.70167	0.98693	6.7	1.91455	1.40778
1.8	0.71732	1.05500	6.8	1.92677	1.42453	1.8	0.74244	1.01363	6.8	1.92900	1.41014
1.9	0.75916	1.07855	6.9	1.94109	1.42663	1.9	0.78228	1.03824	6.9	1.94326	1.41244
2.0	0.79983	1.10020	7.0	1.95521	1.42866	2.0	0.82115	1.06096	7.0	1.95731	1.41467
2.1	0.83935	1.12015	7.1	1.96914	1.43065	2.1	0.85905	1.08197	7.1	1.97118	1.41684
2.2	0.87772	1.13857	7.2	1.98287	1.43257	2.2	0.89597	1.10144	7.2	1.98487	1.41895
2.3	0.91499	1.15563	7.3	1.99643	1.43445	2.3	0.93193	1.11953	7.3	1.99837	1.42101
2.4	0.95118	1.17146	7.4	2.00981	1.43628	2.4	0.96694	1.13635	7.4	2.01169	1.42301
2.5	0.98634	1.18618	7.5	2.02301	1.43805	2.5	1.00102	1.15204	7.5	2.02485	1.42496
2.6	1.02050	1.19990	7.6	2.03604	1.43978	2.6	1.03421	1.16668	7.6	2.03784	1.42686
2.7	1.05370	1.21271	7.7	2.04891	1.44147	2.7	1.06653	1.18039	7.7	2.05066	1.42871
2.8	1.08598	1.22469	7.8	2.06162	1.44312	2.8	1.09801	1.19324	7.8	2.06332	1.43051
2.9	1.11738	1.23592	7.9	2.07417	1.44472	2.9	1.12867	1.20530	7.9	2.07583	1.43227
3.0	1.14794	1.24647	8.0	2.08657	1.44628	3.0	1.15856	1.21664	8.0	2.08819	1.43398
3.1	1.17769	1.25639	8.1	2.09882	1.44781	3.1	1.18770	1.22733	8.1	2.10040	1.43565
3.2	1.20667	1.26574	8.2	2.11092	1.44930	3.2	1.21611	1.23741	8.2	2.11246	1.43728
3.3	1.23491	1.27457	8.3	2.12288	1.45075	3.3	1.24383	1.24693	8.3	2.12439	1.43888
3.4	1.26245	1.28290	8.4	2.13470	1.45217	3.4	1.27089	1.25594	8.4	2.13617	1.44043
3.5	1.28931	1.29080	8.5	2.14638	1.45355	3.5	1.29731	1.26448	8.5	2.14782	1.44195
3.6	1.31552	1.29828	8.6	2.15794	1.45491	3.6	1.32311	1.27257	8.6	2.15934	1.44344
3.7	1.34112	1.30537	8.7	2.16936	1.45623	3.7	1.34833	1.28026	8.7	2.17073	1.44489
3.8	1.36612	1.31212	8.8	2.18065	1.45753	3.8	1.37297	1.28757	8.8	2.18199	1.44631
3.9	1.39055	1.31853	8.9	2.19182	1.45879	3.9	1.39707	1.29454	8.9	2.19313	1.44770
4.0	1.41443	1.32464	9.0	2.20286	1.46003	4.0	1.42065	1.30117	9.0	2.20415	1.44905
4.1	1.43779	1.33047	9.1	2.21379	1.46124	4.1	1.44373	1.30750	9.1	2.21504	1.45038
4.2	1.46065	1.33603	9.2	2.22460	1.46242	4.2	1.46632	1.31354	9.2	2.22583	1.45168
4.3	1.48302	1.34134	9.3	2.23530	1.46358	4.3	1.48844	1.31932	9.3	2.23650	1.45295
4.4	1.50493	1.34642	9.4	2.24588	1.46471	4.4	1.51012	1.32485	9.4	2.24706	1.45420
4.5	1.52639	1.35128	9.5	2.25635	1.46582	4.5	1.53136	1.33014	9.5	2.25751	1.45542
4.6	1.54742	1.35594	9.6	2.26672	1.46691	4.6	1.55219	1.33522	9.6	2.26785	1.45661
4.7	1.56804	1.36041	9.7	2.27698	1.46798	4.7	1.57262	1.34009	9.7	2.27809	1.45778
4.8	1.58826	1.36470	9.8	2.28714	1.46902	4.8	1.59265	1.34476	9.8	2.28822	1.45892
4.9	1.60810	1.36882	9.9	2.29720	1.47004	4.9	1.61232	1.34925	9.9	2.29826	1.46005
5.0	1.62756	1.37278	10.0	2.30716	1.47105	5.0	1.63162	1.35357	10.0	2.30820	1.46115
	$\begin{bmatrix}(-3)1\\4\end{bmatrix}$	$\begin{bmatrix}(-4)7\\4\end{bmatrix}$		$\begin{bmatrix}(-5)4\\3\end{bmatrix}$	$\begin{bmatrix}(-5)2\\3\end{bmatrix}$		$\begin{bmatrix}(-4)9\\4\end{bmatrix}$	$\begin{bmatrix}(-4)6\\4\end{bmatrix}$		$\begin{bmatrix}(-5)4\\3\end{bmatrix}$	$\begin{bmatrix}(-5)2\\3\end{bmatrix}$

$$\mathscr{I}\psi(1.5+iy) = \tfrac{1}{2}\pi \tanh \pi y - \frac{4y}{4y^2+1}$$

Table 6.8   **DIGAMMA FUNCTION FOR COMPLEX ARGUMENTS**

	$x=1.7$						$x=1.8$				
y	$\mathscr{R}\psi(z)$	$\mathscr{I}\psi(z)$	y	$\mathscr{R}\psi(z)$	$\mathscr{I}\psi(z)$	y	$\mathscr{R}\psi(z)$	$\mathscr{I}\psi(z)$	y	$\mathscr{R}\psi(z)$	$\mathscr{I}\psi(z)$
0.0	0.20855	0.00000	5.0	1.63603	1.33453	0.0	0.28499	0.00000	5.0	1.64078	1.31566
0.1	0.21156	0.07918	5.1	1.65482	1.33902	0.1	0.28760	0.07358	5.1	1.65939	1.32048
0.2	0.22050	0.15747	5.2	1.67328	1.34335	0.2	0.29537	0.14644	5.2	1.67769	1.32513
0.3	0.23511	0.23407	5.3	1.69142	1.34752	0.3	0.30809	0.21792	5.3	1.69567	1.32961
0.4	0.25494	0.30824	5.4	1.70926	1.35154	0.4	0.32541	0.28740	5.4	1.71336	1.33393
0.5	0.27945	0.37937	5.5	1.72680	1.35543	0.5	0.34693	0.35437	5.5	1.73076	1.33810
0.6	0.30803	0.44701	5.6	1.74405	1.35918	0.6	0.37215	0.41842	5.6	1.74787	1.34213
0.7	0.34001	0.51086	5.7	1.76102	1.36280	0.7	0.40053	0.47928	5.7	1.76472	1.34603
0.8	0.37474	0.57074	5.8	1.77772	1.36630	0.8	0.43155	0.53675	5.8	1.78130	1.34979
0.9	0.41161	0.62661	5.9	1.79416	1.36969	0.9	0.46469	0.59076	5.9	1.79762	1.35344
1.0	0.45005	0.67852	6.0	1.81034	1.37297	1.0	0.49947	0.64131	6.0	1.81369	1.35697
1.1	0.48957	0.72661	6.1	1.82627	1.37614	1.1	0.53546	0.68847	6.1	1.82952	1.36038
1.2	0.52973	0.77107	6.2	1.84196	1.37922	1.2	0.57226	0.73237	6.2	1.84511	1.36369
1.3	0.57018	0.81211	6.3	1.85742	1.38220	1.3	0.60955	0.77316	6.3	1.86047	1.36690
1.4	0.61063	0.84996	6.4	1.87266	1.38509	1.4	0.64706	0.81103	6.4	1.87561	1.37001
1.5	0.65085	0.88488	6.5	1.88767	1.38789	1.5	0.68455	0.84617	6.5	1.89053	1.37303
1.6	0.69065	0.91710	6.6	1.90246	1.39061	1.6	0.72184	0.87877	6.6	1.90525	1.37596
1.7	0.72990	0.94685	6.7	1.91705	1.39326	1.7	0.75879	0.90903	6.7	1.91975	1.37881
1.8	0.76849	0.97436	6.8	1.93143	1.39582	1.8	0.79528	0.93713	6.8	1.93406	1.38158
1.9	0.80636	0.99982	6.9	1.94562	1.39832	1.9	0.83122	0.96326	6.9	1.94817	1.38426
2.0	0.84345	1.02342	7.0	1.95961	1.40074	2.0	0.86655	0.98757	7.0	1.96210	1.38688
2.1	0.87973	1.04533	7.1	1.97342	1.40310	2.1	0.90123	1.01022	7.1	1.97583	1.38942
2.2	0.91519	1.06570	7.2	1.98704	1.40539	2.2	0.93523	1.03136	7.2	1.98939	1.39189
2.3	0.94981	1.08468	7.3	2.00048	1.40762	2.3	0.96853	1.05110	7.3	2.00277	1.39430
2.4	0.98362	1.10238	7.4	2.01375	1.40980	2.4	1.00111	1.06957	7.4	2.01598	1.39664
2.5	1.01661	1.11893	7.5	2.02685	1.41191	2.5	1.03299	1.08687	7.5	2.02903	1.39892
2.6	1.04879	1.13441	7.6	2.03979	1.41398	2.6	1.06416	1.10310	7.6	2.04191	1.40115
2.7	1.08020	1.14893	7.7	2.05256	1.41599	2.7	1.09463	1.11836	7.7	2.05463	1.40332
2.8	1.11084	1.16257	7.8	2.06518	1.41794	2.8	1.12442	1.13270	7.8	2.06719	1.40543
2.9	1.14075	1.17539	7.9	2.07764	1.41986	2.9	1.15353	1.14622	7.9	2.07960	1.40749
3.0	1.16993	1.18747	8.0	2.08996	1.42172	3.0	1.18200	1.15898	8.0	2.09187	1.40950
3.1	1.19842	1.19886	8.1	2.10212	1.42354	3.1	1.20982	1.17103	8.1	2.10399	1.41146
3.2	1.22625	1.20962	8.2	2.11415	1.42531	3.2	1.23703	1.18243	8.2	2.11597	1.41338
3.3	1.25342	1.21981	8.3	2.12603	1.42704	3.3	1.26363	1.19322	8.3	2.12781	1.41525
3.4	1.27997	1.22945	8.4	2.13778	1.42874	3.4	1.28965	1.20345	8.4	2.13952	1.41708
3.5	1.30592	1.23859	8.5	2.14939	1.43039	3.5	1.31511	1.21317	8.5	2.15109	1.41886
3.6	1.33129	1.24727	8.6	2.16087	1.43200	3.6	1.34003	1.22241	8.6	2.16253	1.42061
3.7	1.35610	1.25553	8.7	2.17222	1.43358	3.7	1.36441	1.23119	8.7	2.17385	1.42231
3.8	1.38037	1.26338	8.8	2.18345	1.43513	3.8	1.38829	1.23956	8.8	2.18504	1.42398
3.9	1.40413	1.27087	8.9	2.19456	1.43664	3.9	1.41168	1.24754	8.9	2.19611	1.42561
4.0	1.42738	1.27800	9.0	2.20555	1.43811	4.0	1.43459	1.25516	9.0	2.20707	1.42720
4.1	1.45015	1.28481	9.1	2.21642	1.43956	4.1	1.45704	1.26243	9.1	2.21790	1.42876
4.2	1.47246	1.29132	9.2	2.22717	1.44097	4.2	1.47904	1.26939	9.2	2.22862	1.43029
4.3	1.49432	1.29755	9.3	2.23781	1.44235	4.3	1.50062	1.27605	9.3	2.23923	1.43178
4.4	1.51574	1.30351	9.4	2.24834	1.44371	4.4	1.52178	1.28242	9.4	2.24974	1.43324
4.5	1.53675	1.30922	9.5	2.25877	1.44503	4.5	1.54254	1.28854	9.5	2.26013	1.43468
4.6	1.55736	1.31470	9.6	2.26908	1.44633	4.6	1.56292	1.29440	9.6	2.27042	1.43608
4.7	1.57758	1.31996	9.7	2.27930	1.44760	4.7	1.58291	1.30004	9.7	2.28061	1.43745
4.8	1.59742	1.32501	9.8	2.28941	1.44885	4.8	1.60255	1.30545	9.8	2.29069	1.43880
4.9	1.61690	1.32986	9.9	2.29942	1.45007	4.9	1.62183	1.31065	9.9	2.30068	1.44012
5.0	1.63603	1.33453	10.0	2.30933	1.45127	5.0	1.64078	1.31566	10.0	2.31057	1.44142
	$\begin{bmatrix}(-4)7\\4\end{bmatrix}$	$\begin{bmatrix}(-4)5\\4\end{bmatrix}$		$\begin{bmatrix}(-5)4\\3\end{bmatrix}$	$\begin{bmatrix}(-5)2\\3\end{bmatrix}$		$\begin{bmatrix}(-4)6\\4\end{bmatrix}$	$\begin{bmatrix}(-4)4\\4\end{bmatrix}$		$\begin{bmatrix}(-5)4\\3\end{bmatrix}$	$\begin{bmatrix}(-5)2\\3\end{bmatrix}$

*

*See page II.

DIGAMMA FUNCTION FOR COMPLEX ARGUMENTS

Table 6.8

$x = 1.9$ $x = 2.0$

y	$\mathscr{R}\psi(z)$	$\mathscr{I}\psi(z)$	y	$\mathscr{R}\psi(z)$	$\mathscr{I}\psi(z)$	y	$\mathscr{R}\psi(z)$	$\mathscr{I}\psi(z)$	y	$\mathscr{R}\psi(z)$	$\mathscr{I}\psi(z)$
0.0	0.35618	0.00000	5.0	1.64585	1.29698	0.0	0.42278	0.00000	5.0	1.65125	1.27849
0.1	0.35847	0.06870	5.1	1.66428	1.30212	0.1	0.42480	0.06441	5.1	1.66948	1.28394
0.2	0.36528	0.13681	5.2	1.68240	1.30707	0.2	0.43081	0.12833	5.2	1.68742	1.28919
0.3	0.37644	0.20377	5.3	1.70022	1.31185	0.3	0.44068	0.19130	5.3	1.70506	1.29426
0.4	0.39169	0.26908	5.4	1.71775	1.31647	0.4	0.45420	0.25288	5.4	1.72242	1.29916
0.5	0.41071	0.33229	5.5	1.73500	1.32092	0.5	0.47111	0.31269	5.5	1.73951	1.30389
0.6	0.43309	0.39306	5.6	1.75197	1.32522	0.6	0.49110	0.37042	5.6	1.75633	1.30846
0.7	0.45842	0.45110	5.7	1.76868	1.32938	0.7	0.51380	0.42583	5.7	1.77290	1.31288
0.8	0.48625	0.50624	5.8	1.78513	1.33341	0.8	0.53887	0.47874	5.8	1.78921	1.31715
0.9	0.51614	0.55838	5.9	1.80133	1.33730	0.9	0.56594	0.52904	5.9	1.80528	1.32129
1.0	0.54770	0.60749	6.0	1.81728	1.34107	1.0	0.59465	0.57667	6.0	1.82111	1.32530
1.1	0.58053	0.65359	6.1	1.83300	1.34473	1.1	0.62468	0.62165	6.1	1.83671	1.32918
1.2	0.61431	0.69677	6.2	1.84848	1.34827	1.2	0.65572	0.66400	6.2	1.85208	1.33295
1.3	0.64872	0.73714	6.3	1.86374	1.35170	1.3	0.68751	0.70380	6.3	1.86723	1.33660
1.4	0.68351	0.77483	6.4	1.87878	1.35503	1.4	0.71980	0.74116	6.4	1.88217	1.34015
1.5	0.71846	0.80999	6.5	1.89361	1.35826	1.5	0.75239	0.77618	6.5	1.89690	1.34358
1.6	0.75338	0.84278	6.6	1.90824	1.36140	1.6	0.78510	0.80899	6.6	1.91143	1.34692
1.7	0.78814	0.87335	6.7	1.92266	1.36445	1.7	0.81779	0.83973	6.7	1.92576	1.35017
1.8	0.82261	0.90188	6.8	1.93688	1.36741	1.8	0.85033	0.86853	6.8	1.93990	1.35332
1.9	0.85669	0.92851	6.9	1.95092	1.37029	1.9	0.88262	0.89551	6.9	1.95385	1.35639
2.0	0.89031	0.95338	7.0	1.96476	1.37308	2.0	0.91459	0.92081	7.0	1.96761	1.35937
2.1	0.92342	0.97664	7.1	1.97843	1.37581	2.1	0.94617	0.94454	7.1	1.98120	1.36227
2.2	0.95598	0.99840	7.2	1.99192	1.37846	2.2	0.97731	0.96681	7.2	1.99462	1.36509
2.3	0.98795	1.01879	7.3	2.00523	1.38104	2.3	1.00798	0.98775	7.3	2.00786	1.36784
2.4	1.01932	1.03792	7.4	2.01838	1.38355	2.4	1.03814	1.00743	7.4	2.02094	1.37052
2.5	1.05008	1.05588	7.5	2.03136	1.38599	2.5	1.06779	1.02597	7.5	2.03385	1.37313
2.6	1.08022	1.07278	7.6	2.04418	1.38838	2.6	1.09690	1.04344	7.6	2.04661	1.37567
2.7	1.10975	1.08868	7.7	2.05684	1.39070	2.7	1.12548	1.05992	7.7	2.05921	1.37815
2.8	1.13867	1.10367	7.8	2.06935	1.39297	2.8	1.15352	1.07548	7.8	2.07167	1.38056
2.9	1.16698	1.11782	7.9	2.08171	1.39518	2.9	1.18102	1.09020	7.9	2.08397	1.38292
3.0	1.19470	1.13119	8.0	2.09393	1.39734	3.0	1.20798	1.10413	8.0	2.09613	1.38522
3.1	1.22184	1.14384	8.1	2.10600	1.39944	3.1	1.23442	1.11733	8.1	2.10815	1.38746
3.2	1.24841	1.15583	8.2	2.11793	1.40149	3.2	1.26034	1.12985	8.2	2.12003	1.38966
3.3	1.27442	1.16719	8.3	2.12973	1.40350	3.3	1.28575	1.14174	8.3	2.13178	1.39180
3.4	1.29990	1.17798	8.4	2.14139	1.40546	3.4	1.31067	1.15304	8.4	2.14339	1.39389
3.5	1.32485	1.18823	8.5	2.15292	1.40738	3.5	1.33510	1.16379	8.5	2.15487	1.39593
3.6	1.34929	1.19798	8.6	2.16432	1.40925	3.6	1.35905	1.17403	8.6	2.16623	1.39793
3.7	1.37324	1.20727	8.7	2.17560	1.41108	3.7	1.38254	1.18379	8.7	2.17746	1.39988
3.8	1.39670	1.21613	8.8	2.18675	1.41286	3.8	1.40558	1.19310	8.8	2.18858	1.40179
3.9	1.41970	1.22458	8.9	2.19778	1.41461	3.9	1.42818	1.20200	8.9	2.19957	1.40366
4.0	1.44226	1.23265	9.0	2.20870	1.41632	4.0	1.45036	1.21050	9.0	2.21045	1.40548
4.1	1.46437	1.24037	9.1	2.21950	1.41800	4.1	1.47212	1.21864	9.1	2.22121	1.40727
4.2	1.48606	1.24775	9.2	2.23019	1.41964	4.2	1.49348	1.22643	9.2	2.23187	1.40902
4.3	1.50734	1.25482	9.3	2.24077	1.42124	4.3	1.51446	1.23389	9.3	2.24241	1.41074
4.4	1.52822	1.26160	9.4	2.25124	1.42281	4.4	1.53505	1.24105	9.4	2.25284	1.41241
4.5	1.54872	1.26810	9.5	2.26160	1.42435	4.5	1.55527	1.24792	9.5	2.26318	1.41406
4.6	1.56885	1.27434	9.6	2.27186	1.42586	4.6	1.57514	1.25452	9.6	2.27340	1.41566
4.7	1.58861	1.28033	9.7	2.28202	1.42733	4.7	1.59466	1.26086	9.7	2.28353	1.41724
4.8	1.60803	1.28610	9.8	2.29207	1.42878	4.8	1.61385	1.26696	9.8	2.29356	1.41879
4.9	1.62710	1.29164	9.9	2.30203	1.43020	4.9	1.63270	1.27283	9.9	2.30349	1.42030
5.0	1.64585	1.29698	10.0	2.31190	1.43159	5.0	1.65125	1.27849	10.0	2.31332	1.42179
	$\begin{bmatrix}(-4)6\\4\end{bmatrix}$	$\begin{bmatrix}(-4)4\\4\end{bmatrix}$		$\begin{bmatrix}(-5)4\\3\end{bmatrix}$	$\begin{bmatrix}(-5)2\\3\end{bmatrix}$		$\begin{bmatrix}(-4)5\\4\end{bmatrix}$	$\begin{bmatrix}(-4)3\\4\end{bmatrix}$		$\begin{bmatrix}(-5)4\\3\end{bmatrix}$	$\begin{bmatrix}(-5)3\\3\end{bmatrix}$

$$\mathscr{I}\psi(2+iy) = \tfrac{1}{2}\pi \coth \pi y - \frac{1+3y^2}{2y(1+y^2)}$$

7. Error Function and Fresnel Integrals

Walter Gautschi[1]

Contents

Table 7.1. Error Function and its Derivative $(0 \leq x \leq 2)$ 310

$$(2/\sqrt{\pi})\,e^{-x^2}, \text{ erf } x = (2/\sqrt{\pi}) \int_0^x e^{-t^2} dt, \quad x = 0(.01)2, \quad 10\text{D}$$

Table 7.2. Derivative of the Error Function $(2 \leq x \leq 10)$ 312

$$(2/\sqrt{\pi})\,e^{-x^2}, \quad x = 2(.01)10, \quad 8\text{S}$$

Table 7.3. Complementary Error Function $(2 \leq x \leq \infty)$. 316

$$xe^{x^2} \text{ erfc } x = (2/\sqrt{\pi})xe^{x^2} \int_x^\infty e^{-t^2} dt, \quad x^{-2} = .25(-.005)0, \quad 7\text{D}$$

$$\text{erfc } \sqrt{n\pi}, \quad n = 1(1)10, \quad 15\text{D}$$

Table 7.4. Repeated Integrals of the Error Function $(0 \leq x \leq 5)$ . . . 317

$$2^n \Gamma\left(\frac{n}{2}+1\right) i^n \text{ erfc } x = 2^{n+1} \Gamma\left(\frac{n}{2}+1\right) \sqrt{\frac{1}{\pi}} \int_x^\infty \frac{(t-x)^n}{n!} e^{-t^2} dt$$

$$x = 0(.1)5, n = 1(1)6, 10, 11, \quad 6\text{S}$$

Table 7.5. Dawson's Integral $(0 \leq x \leq \infty)$ 319

$$e^{-x^2} \int_0^x e^{t^2} dt, \quad x = 0(.02)2, \quad 10\text{D}$$

$$xe^{-x^2} \int_0^x e^{t^2} dt, \quad x^{-2} = .25(-.005)0, \quad 9\text{D}$$

---

[1] Guest worker, National Bureau of Standards, from The American University. (Presently Purdue University.)

The author acknowledges the assistance of Alfred E. Beam, Ruth E. Capuano, Lois K. Cherwinski, Elizabeth F. Godefroy, David S. Liepman, Mary Orr, Bertha H. Walter, and Ruth Zucker in the preparation and checking of the tables.

7. Error Function and Fresnel Integrals

Mathematical Properties

7.1. Error Function

Definitions

7.1.1
$$\operatorname{erf} z=\frac{2}{\sqrt{\pi}}\int_0^z e^{-t^2}dt$$

7.1.2
$$\operatorname{erfc} z=\frac{2}{\sqrt{\pi}}\int_z^\infty e^{-t^2}dt=1-\operatorname{erf} z$$

7.1.3
$$w(z)=e^{-z^2}\left(1+\frac{2i}{\sqrt{\pi}}\int_0^z e^{t^2}dt\right)=e^{-z^2}\operatorname{erfc}(-iz)$$

In **7.1.2** the path of integration is subject to the restriction $\arg t\to\alpha$ with $|\alpha|<\frac{\pi}{4}$ as $t\to\infty$ along the path. ($\alpha=\frac{\pi}{4}$ is permissible if $\mathscr{R}t^2$ remains bounded to the left.)

Integral Representation

7.1.4
$$w(z)=\frac{i}{\pi}\int_{-\infty}^\infty\frac{e^{-t^2}dt}{z-t}=\frac{2iz}{\pi}\int_0^\infty\frac{e^{-t^2}dt}{z^2-t^2}\qquad(\mathscr{I}z>0)$$

Series Expansions

7.1.5
$$\operatorname{erf} z=\frac{2}{\sqrt{\pi}}\sum_{n=0}^\infty\frac{(-1)^n z^{2n+1}}{n!(2n+1)}$$

7.1.6
$$=\frac{2}{\sqrt{\pi}}\,e^{-z^2}\sum_{n=0}^\infty\frac{2^n}{1\cdot3\ldots(2n+1)}z^{2n+1}$$

7.1.7
$$=\sqrt{2}\sum_{n=0}^\infty(-1)^n[I_{2n+1/2}(z^2)-I_{2n+3/2}(z^2)]$$

7.1.8
$$w(z)=\sum_{n=0}^\infty\frac{(iz)^n}{\Gamma\left(\frac{n}{2}+1\right)}$$

For $I_{n-\frac{1}{2}}(x)$, see chapter **10**.

Symmetry Relations

7.1.9
$$\operatorname{erf}(-z)=-\operatorname{erf} z$$

7.1.10
$$\operatorname{erf}\bar{z}=\overline{\operatorname{erf} z}$$

7.1.11
$$w(-z)=2e^{-z^2}-w(z)$$

7.1.12
$$w(\bar{z})=\overline{w(-z)}$$

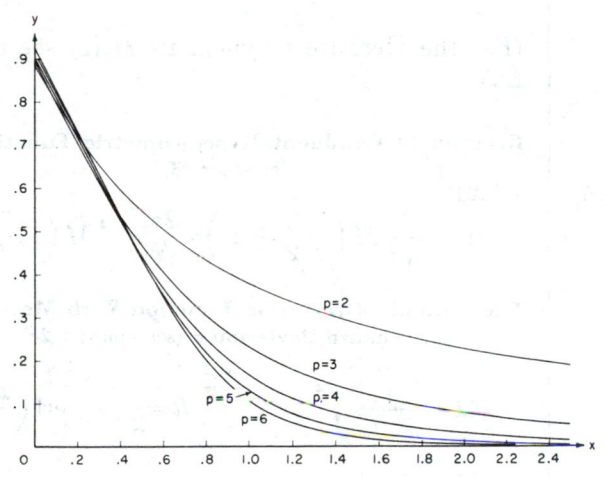

FIGURE 7.1. $\quad y=e^{x^p}\int_x^\infty e^{-t^p}dt.$

$p=2(1)6$

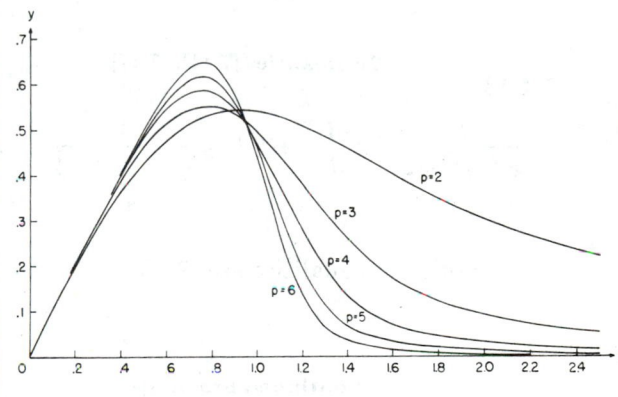

FIGURE 7.2. $\quad y=e^{-x^p}\int_0^x e^{t^p}dt.$

$p=2(1)6$

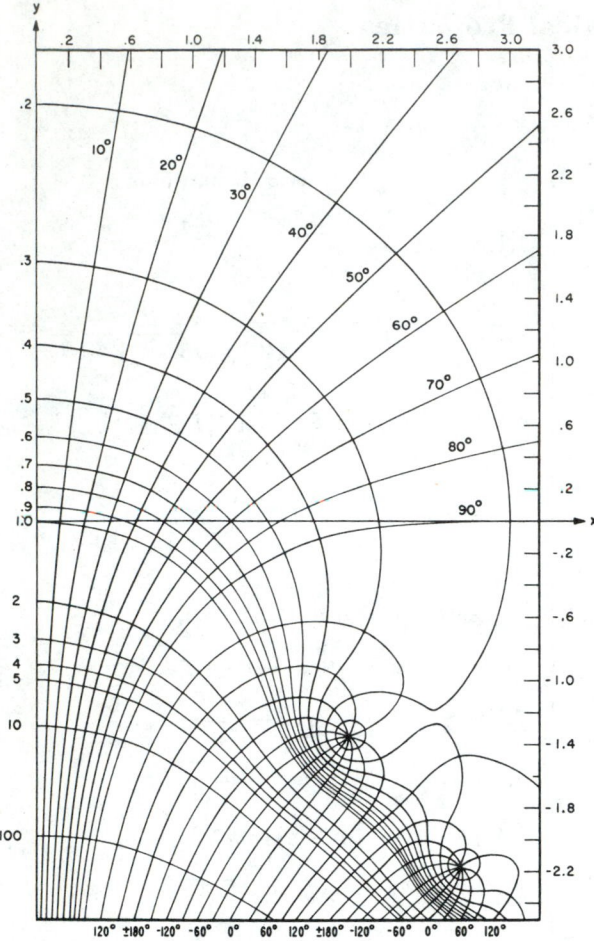

FIGURE 7.3. *Altitude Chart of $w(z)$.*

Inequalities [7.11], [7.17]

7.1.13

$$\frac{1}{x+\sqrt{x^2+2}}<e^{x^2}\int_x^\infty e^{-t^2}dt\leq\frac{1}{x+\sqrt{x^2+\frac{4}{\pi}}}\qquad(x\geq0)$$

(For other inequalities see [7.2].)

Continued Fractions

7.1.14

$$2e^{z^2}\int_z^\infty e^{-t^2}dt=\frac{1}{z+}\frac{1/2}{z+}\frac{1}{z+}\frac{3/2}{z+}\frac{2}{z+}\cdots\ (\mathscr{R}z>0)$$

7.1.15

$$\frac{1}{\sqrt{\pi}}\int_{-\infty}^\infty\frac{e^{-t^2}dt}{z-t}=\frac{1}{z-}\frac{1/2}{z-}\frac{1}{z-}\frac{3/2}{z-}\frac{2}{z-}\cdots$$

$$=\frac{1}{\sqrt{\pi}}\lim_{n\to\infty}\sum_{k=1}^n\frac{H_k^{(n)}}{z-x_k^{(n)}}\qquad(\mathscr{I}z\neq0)$$

$x_k^{(n)}$ and $H_k^{(n)}$ are the zeros and weight factors of the Hermite polynomials. For numerical values see chapter **25**.

Value at Infinity

7.1.16 $\operatorname{erf} z\to1\left(z\to\infty\text{ in }|\arg z|<\frac{\pi}{4}\right)$

Maximum and Inflection Points for Dawson's Integral [7.31]

$$F(x)=e^{-x^2}\int_0^x e^{t^2}dt$$

7.1.17 $F(.92413\ 88730\ldots)=.54104\ 42246\ldots$

7.1.18 $F(1.50197\ 52682\ldots)=.42768\ 66160\ldots$

Derivatives

7.1.19

$$\frac{d^{n+1}}{dz^{n+1}}\operatorname{erf} z=(-1)^n\frac{2}{\sqrt{\pi}}H_n(z)e^{-z^2}\qquad(n=0,1,2,\ldots)$$

7.1.20

$$w^{(n+2)}(z)+2zw^{(n+1)}(z)+2(n+1)w^{(n)}(z)=0$$
$$(n=0,1,2,\ldots)$$

$$w^{(0)}(z)=w(z),\qquad w'(z)=-2zw(z)+\frac{2i}{\sqrt{\pi}}$$

(For the Hermite polynomials $H_n(z)$ see chapter **22**.)

Relation to Confluent Hypergeometric Function (see chapter **13**)

7.1.21

$$\operatorname{erf} z=\frac{2z}{\sqrt{\pi}}M\left(\frac{1}{2},\frac{3}{2},-z^2\right)=\frac{2z}{\sqrt{\pi}}e^{-z^2}M\left(1,\frac{3}{2},z^2\right)$$

The Normal Distribution Function With Mean m and Standard Deviation σ (see chapter **26**)

7.1.22 $\dfrac{1}{\sigma\sqrt{2\pi}}\displaystyle\int_{-\infty}^x e^{-\frac{(t-m)^2}{2\sigma^2}}dt=\dfrac{1}{2}\left(1+\operatorname{erf}\left(\dfrac{x-m}{\sigma\sqrt{2}}\right)\right)$

Asymptotic Expansion

7.1.23

$$\sqrt{\pi}ze^{z^2}\operatorname{erfc} z\sim1+\sum_{m=1}^\infty(-1)^m\frac{1\cdot3\ldots(2m-1)}{(2z^2)^m}$$

$$\left(z\to\infty,\ |\arg z|<\frac{3\pi}{4}\right)$$

If $R_n(z)$ is the remainder after n terms then

7.1.24

$$R_n(z)=(-1)^n\,\frac{1\cdot 3\,\ldots\,(2n-1)}{(2z^2)^n}\,\theta,$$

$$\theta=\int_0^\infty e^{-t}\left(1+\frac{t}{z^2}\right)^{-n-\frac{1}{2}}dt\qquad\left(|\arg z|<\frac{\pi}{2}\right)$$

$$|\theta|<1\qquad\left(|\arg z|<\frac{\pi}{4}\right)$$

For x real, $R_n(x)$ is less in absolute value than the first neglected term and of the same sign.

Rational Approximations [2] ($0\le x<\infty$)

7.1.25

$$\text{erf } x=1-(a_1t+a_2t^2+a_3t^3)\,e^{-x^2}+\epsilon(x),\qquad t=\frac{1}{1+px}$$

$$|\epsilon(x)|\le 2.5\times10^{-5}$$

$p=.47047\qquad a_1=.34802\ 42\qquad a_2=-.09587\ 98$
$$a_3=.74785\ 56$$

7.1.26

$$\text{erf } x=1-(a_1t+a_2t^2+a_3t^3+a_4t^4+a_5t^5)\,e^{-x^2}+\epsilon(x),$$

$$t=\frac{1}{1+px}$$

$$|\epsilon(x)|\le 1.5\times10^{-7}$$

$p=.32759\ 11\qquad a_1=.25482\ 9592$
$a_2=-.28449\ 6736\qquad a_3=1.42141\ 3741$
$a_4=-1.45315\ 2027\qquad a_5=1.06140\ 5429$

7.1.27

$$\text{erf } x=1-\frac{1}{[1+a_1x+a_2x^2+a_3x^3+a_4x^4]^4}+\epsilon(x)$$

$$|\epsilon(x)|\le 5\times10^{-4}$$

$a_1=.278393\qquad a_2=.230389$
$a_3=.000972\qquad a_4=.078108$

7.1.28

$$\text{erf } x=1-\frac{1}{[1+a_1x+a_2x^2+\cdots+a_6x^6]^{16}}+\epsilon(x)$$

$$|\epsilon(x)|\le 3\times10^{-7}$$

$a_1=.07052\ 30784\qquad a_2=.04228\ 20123$
$a_3=.00927\ 05272\qquad a_4=.00015\ 20143$
$a_5=.00027\ 65672\qquad a_6=.00004\ 30638$

---

[2] Approximations **7.1.25–7.1.28** are from C. Hastings, Jr., Approximations for digital computers. Princeton Univ. Press, Princeton, N. J., 1955 (with permission).

Infinite Series Approximation for Complex Error Function [7.19]

7.1.29

$$\text{erf }(x+iy)=\text{erf } x+\frac{e^{-x^2}}{2\pi x}\left[(1-\cos 2xy)+i\sin 2xy\right]$$

$$+\frac{2}{\pi}\,e^{-x^2}\sum_{n=1}^\infty\frac{e^{-\frac{1}{4}n^2}}{n^2+4x^2}[f_n(x,y)+ig_n(x,y)]+\epsilon(x,y)$$

where

$$f_n(x,y)=2x-2x\cosh ny\cos 2xy+n\sinh ny\sin 2xy$$
$$g_n(x,y)=2x\cosh ny\sin 2xy+n\sinh ny\cos 2xy$$
$$|\epsilon(x,y)|\approx10^{-16}|\text{ erf }(x+iy)|$$

7.2. Repeated Integrals of the Error Function

Definition

7.2.1

$$\text{i}^n\text{ erfc } z=\int_z^\infty\text{i}^{n-1}\text{ erfc } t\,dt\qquad(n=0,1,2,\ldots)$$

$$\text{i}^{-1}\text{ erfc } z=\frac{2}{\sqrt\pi}\,e^{-z^2},\ \text{i}^0\text{ erfc } z=\text{erfc } z$$

Differential Equation

7.2.2

$$\frac{d^2y}{dz^2}+2z\,\frac{dy}{dz}-2ny=0$$

$$y=A\text{i}^n\text{ erfc } z+B\text{i}^n\text{ erfc }(-z)$$

(A and B are constants.)

Expression as a Single Integral

7.2.3

$$\text{i}^n\text{ erfc } z=\frac{2}{\sqrt\pi}\int_z^\infty\frac{(t-z)^n}{n!}\,e^{-t^2}dt$$

Power Series [3]

7.2.4

$$\text{i}^n\text{ erfc } z=\sum_{k=0}^\infty\frac{(-1)^kz^k}{2^{n-k}k!\,\Gamma\left(1+\dfrac{n-k}{2}\right)}$$

Recurrence Relations

7.2.5

$$\text{i}^n\text{ erfc } z=-\frac{z}{n}\text{i}^{n-1}\text{ erfc } z+\frac{1}{2n}\text{i}^{n-2}\text{ erfc } z$$

$$(n=1,2,3,\ldots)$$

7.2.6

$$2(n+1)(n+2)\text{i}^{n+2}\text{ erfc } z$$
$$=(2n+1+2z^2)\text{i}^n\text{ erfc } z-\frac{1}{2}\text{i}^{n-2}\text{ erfc } z$$

$$(n=1,2,3,\ldots)$$

---

[3] The terms in this series corresponding to $k=n+2$, $n+4$, $n+6$, $\ldots$ are understood to be zero.

Value at Zero

7.2.7

$$i^n \operatorname{erfc} 0 = \frac{1}{2^n \Gamma\left(\dfrac{n}{2}+1\right)} \qquad (n=-1,0,1,2,\ldots)$$

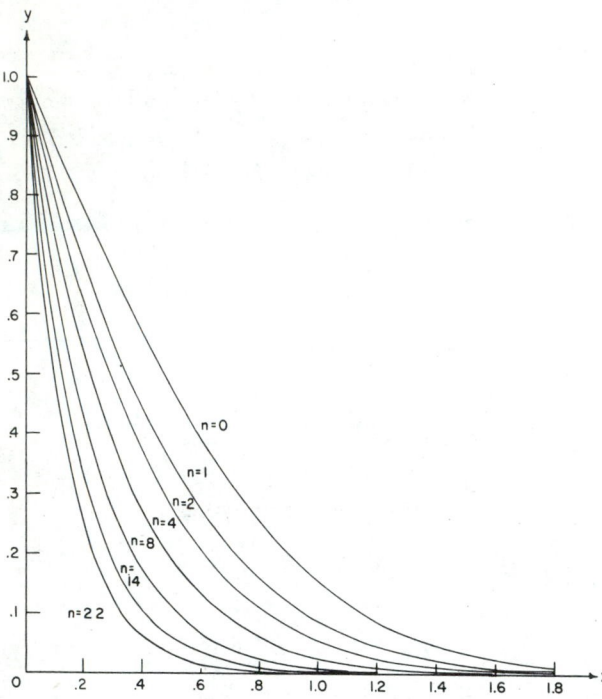

FIGURE 7.4. *Repeated Integrals of the Error Function.*

$$y = 2^n \Gamma\left(\frac{n}{2}+1\right) i^n \operatorname{erfc} x$$

$$n = 0, 1, 2, 4, 8, 14, 22$$

Derivatives

7.2.8 $\quad \dfrac{d}{dz} i^n \operatorname{erfc} z = -i^{n-1} \operatorname{erfc} z \qquad (n=0,1,2,\ldots)$

7.2.9

$$\frac{d^n}{dz^n}\left(e^{z^2} \operatorname{erfc} z\right) = (-1)^n 2^n n! \, e^{z^2} i^n \operatorname{erfc} z$$

$$(n=0,1,2,\ldots)$$

Relation to $Hh_n(z)$ (see 19.14)

7.2.10 $\qquad i^n \operatorname{erfc} z = \dfrac{1}{(2^{n-1}\pi)^{\frac{1}{2}}} Hh_n(\sqrt{2}z)$

Relation to Hermite Polynomials (see chapter 22)

7.2.11 $\quad (-1)^n i^n \operatorname{erfc} z + i^n \operatorname{erfc}(-z) = \dfrac{i^{-n}}{2^{n-1}n!} H_n(iz)$

Relation to the Confluent Hypergeometric Function (see chapter 13)

7.2.12

$$i^n \operatorname{erfc} z = e^{-z^2}\left[\frac{1}{2^n \Gamma\left(\dfrac{n}{2}+1\right)} M\left(\frac{n+1}{2},\frac{1}{2}, z^2\right) \right.$$

$$\left. - \frac{z}{2^{n-1}\Gamma\left(\dfrac{n+1}{2}\right)} M\left(\frac{n}{2}+1,\frac{3}{2}, z^2\right) \right]$$

Relation to Parabolic Cylinder Functions (see chapter 19)

7.2.13 $\qquad i^n \operatorname{erfc} z = \dfrac{e^{-\frac{1}{2}z^2}}{(2^{n-1}\pi)^{\frac{1}{2}}} D_{-n-1}(z\sqrt{2})$

Asymptotic Expansion

7.2.14

$$i^n \operatorname{erfc} z \sim \frac{2}{\sqrt{\pi}} \frac{e^{-z^2}}{(2z)^{n+1}} \sum_{m=0}^{\infty} \frac{(-1)^m (2m+n)!}{n! \, m! \, (2z)^{2m}}$$

$$\left(z \to \infty, \, |\arg z| < \frac{3\pi}{4}\right)$$

7.3. Fresnel Integrals

Definition

7.3.1 $\qquad C(z) = \displaystyle\int_0^z \cos\left(\frac{\pi}{2} t^2\right) dt$

7.3:2 $\qquad S(z) = \displaystyle\int_0^z \sin\left(\frac{\pi}{2} t^2\right) dt$

The following functions are also in use:

7.3.3

$$C_1(x) = \sqrt{\frac{2}{\pi}} \int_0^x \cos t^2 \, dt, \quad C_2(x) = \frac{1}{\sqrt{2\pi}} \int_0^x \frac{\cos t}{\sqrt{t}} \, dt$$

7.3.4

$$S_1(x) = \sqrt{\frac{2}{\pi}} \int_0^x \sin t^2 \, dt, \quad S_2(x) = \frac{1}{\sqrt{2\pi}} \int_0^x \frac{\sin t}{\sqrt{t}} \, dt$$

Auxiliary Functions

7.3.5

$$f(z) = \left[\frac{1}{2} - S(z)\right] \cos\left(\frac{\pi}{2} z^2\right) - \left[\frac{1}{2} - C(z)\right] \sin\left(\frac{\pi}{2} z^2\right)$$

7.3.6

$$g(z) = \left[\frac{1}{2} - C(z)\right] \cos\left(\frac{\pi}{2} z^2\right) + \left[\frac{1}{2} - S(z)\right] \sin\left(\frac{\pi}{2} z^2\right)$$

Interrelations

7.3.7 $\qquad C(x) = C_1\left(x\sqrt{\frac{\pi}{2}}\right) = C_2\left(\frac{\pi}{2} x^2\right)$

7.3.8
$$S(x)=S_1\left(x\sqrt{\frac{\pi}{2}}\right)=S_2\left(\frac{\pi}{2}x^2\right)$$

7.3.9
$$C(z)=\frac{1}{2}+f(z)\sin\left(\frac{\pi}{2}z^2\right)-g(z)\cos\left(\frac{\pi}{2}z^2\right)$$

7.3.10
$$S(z)=\frac{1}{2}-f(z)\cos\left(\frac{\pi}{2}z^2\right)-g(z)\sin\left(\frac{\pi}{2}z^2\right)$$

Series Expansions

7.3.11
$$C(z)=\sum_{n=0}^{\infty}\frac{(-1)^n(\pi/2)^{2n}}{(2n)!(4n+1)}z^{4n+1}$$

7.3.12
$$C(z)=\cos\left(\frac{\pi}{2}z^2\right)\sum_{n=0}^{\infty}\frac{(-1)^n\pi^{2n}}{1\cdot3\ldots(4n+1)}z^{4n+1}$$
$$+\sin\left(\frac{\pi}{2}z^2\right)\sum_{n=0}^{\infty}\frac{(-1)^n\pi^{2n+1}}{1\cdot3\ldots(4n+3)}z^{4n+3}$$

7.3.13
$$S(z)=\sum_{n=0}^{\infty}\frac{(-1)^n(\pi/2)^{2n+1}}{(2n+1)!(4n+3)}z^{4n+3}$$

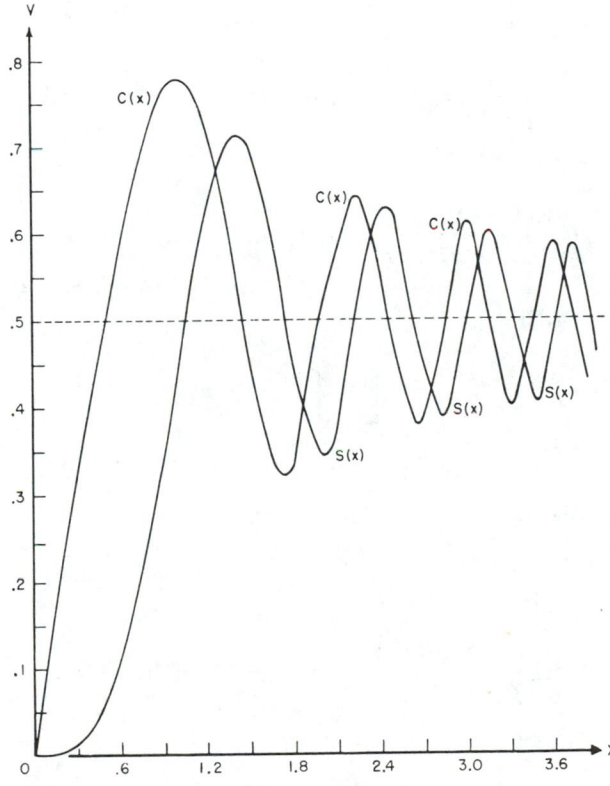

7.3.14
$$S(z)=-\cos\left(\frac{\pi}{2}z^2\right)\sum_{n=0}^{\infty}\frac{(-1)^n\pi^{2n+1}}{1\cdot3\ldots(4n+3)}z^{4n+3}$$
$$+\sin\left(\frac{\pi}{2}z^2\right)\sum_{n=0}^{\infty}\frac{(-1)^n\pi^{2n}}{1\cdot3\ldots(4n+1)}z^{4n+1}$$

7.3.15 $\quad C_2(z)=J_{1/2}(z)+J_{5/2}(z)+J_{9/2}(z)+\ldots$

7.3.16 $\quad S_2(z)=J_{3/2}(z)+J_{7/2}(z)+J_{11/2}(z)+\ldots$

For Bessel functions $J_{n+1/2}(z)$ see chapter **10.**

Symmetry Relations

7.3.17 $\quad C(-z)=-C(z),\quad S(-z)=-S(z)$

7.3.18 $\quad C(iz)=iC(z),\quad S(iz)=-iS(z)$

7.3.19 $\quad C(\bar{z})=\overline{C(z)},\quad S(\bar{z})=\overline{S(z)}$

Value at Infinity

7.3.20 $\quad C(x)\to\frac{1}{2},\qquad S(x)\to\frac{1}{2}\qquad(x\to\infty)$

Derivatives

7.3.21 $\quad \dfrac{df(x)}{dx}=-\pi xg(x),\qquad \dfrac{dg(x)}{dx}=\pi xf(x)-1$

Relation to Error Function (see 7.1.1, 7.1.3)

7.3.22
$$C(z)+iS(z)=\frac{1+i}{2}\text{ erf}\left[\frac{\sqrt{\pi}}{2}(1-i)z\right]$$
$$=\frac{1+i}{2}\left\{1-e^{i\frac{\pi}{2}z^2}w\left[\frac{\sqrt{\pi}}{2}(1+i)z\right]\right\}$$

7.3.23 $\quad g(x)=\mathscr{R}\left\{\dfrac{1+i}{2}w\left[\dfrac{\sqrt{\pi}}{2}(1+i)x\right]\right\}$

7.3.24 $\quad f(x)=\mathscr{I}\left\{\dfrac{1+i}{2}w\left[\dfrac{\sqrt{\pi}}{2}(1+i)x\right]\right\}$

Relation to Confluent Hypergeometric Function (see chapter 13)

7.3.25
$$C(z)+iS(z)=zM\left(\frac{1}{2},\frac{3}{2},i\frac{\pi}{2}z^2\right)$$
$$=ze^{i\frac{\pi}{2}z^2}M\left(1,\frac{3}{2},-i\frac{\pi}{2}z^2\right)$$

Relation to Spherical Bessel Functions (see chapter 10)

7.3.26 $\quad C_2(z)=\dfrac{1}{2}\displaystyle\int_0^z J_{-\frac{1}{2}}(t)dt,\ S_2(z)=\dfrac{1}{2}\displaystyle\int_0^z J_{\frac{1}{2}}(t)dt$

FIGURE 7.5. *Fresnel Integrals.*
$y=C(x),\ y=S(x)$

Asymptotic Expansions

7.3.27

$$\pi z f(z) \sim 1 + \sum_{m=1}^{\infty} (-1)^m \frac{1 \cdot 3 \ldots (4m-1)}{(\pi z^2)^{2m}}$$

$$\left(z \to \infty, \, |\arg z| < \frac{\pi}{2} \right)$$

7.3.28

$$\pi z g(z) \sim \sum_{m=0}^{\infty} (-1)^m \frac{1 \cdot 3 \ldots (4m+1)}{(\pi z^2)^{2m+1}}$$

$$\left(z \to \infty, \, |\arg z| < \frac{\pi}{2} \right)$$

If $R_n^{(f)}(z)$, $R_n^{(g)}(z)$ are the remainders after n terms in **7.3.27**, **7.3.28**, respectively, then

7.3.29

$$R_n^{(f)}(z) = (-1)^n \frac{1 \cdot 3 \ldots (4n-1)}{(\pi z^2)^{2n}} \theta^{(f)},$$

$$\theta^{(f)} = \frac{1}{\Gamma(2n+\frac{1}{2})} \int_0^{\infty} \frac{e^{-t} t^{2n-\frac{1}{2}}}{1 + \left(\frac{2t}{\pi z^2}\right)^2} dt \left(|\arg z| < \frac{\pi}{4} \right)$$

7.3.30

$$R_n^{(g)}(z) = (-1)^n \frac{1 \cdot 3 \ldots (4n+1)}{(\pi z^2)^{2n}} \theta^{(g)},$$

$$\theta^{(g)} = \frac{1}{\Gamma(2n+\frac{3}{2})} \int_0^{\infty} \frac{e^{-t} t^{2n+\frac{1}{2}}}{1 + \left(\frac{2t}{\pi z^2}\right)^2} dt \left(|\arg z| < \frac{\pi}{4} \right)$$

7.3.31 $|\theta^{(f)}| < 1, \, |\theta^{(g)}| < 1$ $\left(|\arg z| \leq \frac{\pi}{8} \right)$

For x real, $R_n^{(f)}(x)$ and $R_n^{(g)}(x)$ are less in absolute value than the first neglected term and of the same sign.

Rational Approximations [4] ($0 \leq x \leq \infty$)

7.3.32

$$f(x) = \frac{1 + .926x}{2 + 1.792x + 3.104x^2} + \epsilon(x) |\epsilon(x)| \leq 2 \times 10^{-3}$$

7.3.33

$$g(x) = \frac{1}{2 + 4.142x + 3.492x^2 + 6.670x^3} + \epsilon(x)$$

$$|\epsilon(x)| \leq 2 \times 10^{-3}$$

(For more accurate approximations see [7.1].)

7.4. Definite and Indefinite Integrals

For a more extensive list of integrals see [7.5], [7 8], [7.15].

7.4.1 $\int_0^{\infty} e^{-t^2} dt = \frac{\sqrt{\pi}}{2}$

---

[4] Approximations **7.3.32**, **7.3.33** are based on those given in C. Hastings, Jr., Approximations for calculating Fresnel integrals, Approximation Newsletter, April 1956, Note 10. [See also MTAC **10**, 173, 1956.]

7.4.2

$$\int_0^{\infty} e^{-(at^2 + 2bt + c)} dt = \frac{1}{2} \sqrt{\frac{\pi}{a}} e^{\frac{b^2 - ac}{a}} \operatorname{erfc} \frac{b}{\sqrt{a}}   (\mathscr{R}a > 0)$$

7.4.3

$$\int_0^{\infty} e^{-at^2 - \frac{b}{t^2}} dt = \frac{1}{2} \sqrt{\frac{\pi}{a}} e^{-2\sqrt{ab}}   (\mathscr{R}a > 0, \mathscr{R}b > 0)$$

7.4.4

$$\int_0^{\infty} t^{2n} e^{-at^2} dt = \frac{1 \cdot 3 \ldots (2n-1)}{2^{n+1} a^n} \sqrt{\frac{\pi}{a}}$$

$$= \frac{\Gamma(n+\frac{1}{2})}{2a^{n+\frac{1}{2}}}   (\mathscr{R}a > 0; n = 0, 1, 2, \ldots)$$

7.4.5

$$\int_0^{\infty} t^{2n+1} e^{-at^2} dt = \frac{n!}{2a^{n+1}}   (\mathscr{R}a > 0; n = 0, 1, 2, \ldots)$$

7.4.6

$$\int_0^{\infty} e^{-at^2} \cos(2xt) dt = \frac{1}{2} \sqrt{\frac{\pi}{a}} e^{-\frac{x^2}{a}}   (\mathscr{R}a > 0)$$

7.4.7

$$\int_0^{\infty} e^{-at^2} \sin(2xt) dt = \frac{1}{\sqrt{a}} e^{-x^2/a} \int_0^{x/\sqrt{a}} e^{t^2} dt$$

$$(\mathscr{R}a > 0)$$

7.4.8

$$\int_0^{\infty} \frac{e^{-at} dt}{\sqrt{t+z^2}} = \sqrt{\frac{\pi}{a}} e^{az^2} \operatorname{erfc} \sqrt{az}   (\mathscr{R}a > 0, \mathscr{R}z > 0)$$

7.4.9

$$\int_0^{\infty} \frac{e^{-at} dt}{\sqrt{t}(t+z)} = \frac{\pi}{\sqrt{z}} e^{az} \operatorname{erfc} \sqrt{az}$$

$$(\mathscr{R}a > 0, z \neq 0, |\arg z| < \pi)$$

7.4.10

$$\int_0^{\infty} \frac{e^{-at^2} dt}{t+x} = e^{-ax^2} \left[\sqrt{\pi} \int_0^{\sqrt{ax}} e^{t^2} dt - \frac{1}{2} \operatorname{Ei}(ax^2) \right]   *$$

$$(a > 0, x > 0)$$

7.4.11

$$\int_0^{\infty} \frac{e^{-at^2} dt}{t^2 + x^2} = \frac{\pi}{2x} e^{ax^2} \operatorname{erfc} \sqrt{ax}   (a > 0, x > 0)$$

7.4.12 $\int_0^1 \frac{e^{-at^2} dt}{t^2 + 1} = \frac{\pi}{4} e^a [1 - (\operatorname{erf} \sqrt{a})^2]   (a > 0)$

7.4.13

$$\int_{-\infty}^{\infty} \frac{y e^{-t^2} dt}{(x-t)^2 + y^2} = \pi \mathscr{R} w(x+iy)   (x \text{ real}, y > 0)$$

*See page II.

7.4.14

$$\int_{-\infty}^{\infty} \frac{(x-t)e^{-t^2}dt}{(x-t)^2+y^2} = \pi \mathscr{I} w(x+iy) \qquad (x \text{ real}, y>0)$$

7.4.15

$$\int_0^{\infty} \frac{[t^2-(x^2-y^2)]e^{-t^2}dt}{t^4-2(x^2-y^2)t^2+(x^2+y^2)^2} = \frac{\pi}{2} \mathscr{R} \frac{w(x+iy)}{y-ix}$$

$$(x \text{ real}, y>0)$$

7.4.16

$$\int_0^{\infty} \frac{2xye^{-t^2}dt}{t^4-2(x^2-y^2)t^2+(x^2+y^2)^2} = \frac{\pi}{2} \mathscr{I} \frac{w(x+iy)}{y-ix}$$

$$(x \text{ real}, y>0)$$

7.4.17

$$\int_0^{\infty} e^{-at} \operatorname{erf} bt \, dt = \frac{1}{a} e^{\frac{a^2}{4b^2}} \operatorname{erfc} \frac{a}{2b}$$

$$\left(\mathscr{R} a>0, |\arg b|<\frac{\pi}{4} \right)$$

7.4.18

$$\int_0^{\infty} \sin(2at) \operatorname{erfc} bt \, dt = \frac{1}{2a}[1-e^{-(a/b)^2}](a>0, \mathscr{R}b>0)$$

7.4.19

$$\int_0^{\infty} e^{-at} \operatorname{erf} \sqrt{bt} \, dt = \frac{1}{a}\sqrt{\frac{b}{a+b}} \qquad (\mathscr{R}(a+b)>0)$$

7.4.20

$$\int_0^{\infty} e^{-at} \operatorname{erfc} \sqrt{\frac{b}{t}} \, dt = \frac{1}{a} e^{-2\sqrt{ab}} \qquad (\mathscr{R}a>0, \mathscr{R}b>0)$$

7.4.21

$$\int_0^{\infty} e^{(a-b)t} \operatorname{erfc} \left(\sqrt{at}+\sqrt{\frac{c}{t}}\right) dt = \frac{e^{-2(\sqrt{ac}+\sqrt{bc})}}{\sqrt{b}(\sqrt{a}+\sqrt{b})}$$

$$(\mathscr{R}b>0, \mathscr{R}c>0)$$

7.4.22

$$\int_0^{\infty} e^{-at} \cos(t^2) dt = \sqrt{\frac{\pi}{2}} \left\{ \left[\frac{1}{2}-S\left(\frac{a}{2}\sqrt{\frac{2}{\pi}}\right)\right] \cos\left(\frac{a^2}{4}\right) \right.$$

$$\left. -\left[\frac{1}{2}-C\left(\frac{a}{2}\sqrt{\frac{2}{\pi}}\right)\right] \sin\left(\frac{a^2}{4}\right) \right\} \qquad (\mathscr{R}a>0)$$

7.4.23

$$\int_0^{\infty} e^{-at} \sin(t^2) dt = \sqrt{\frac{\pi}{2}} \left\{ \left[\frac{1}{2}-C\left(\frac{a}{2}\sqrt{\frac{2}{\pi}}\right)\right] \cos\left(\frac{a^2}{4}\right) \right.$$

$$\left. +\left[\frac{1}{2}-S\left(\frac{a}{2}\sqrt{\frac{2}{\pi}}\right)\right] \sin\left(\frac{a^2}{4}\right) \right\} \qquad (\mathscr{R}a>0)$$

7.4.24

$$\int_0^{\infty} e^{-at} \frac{\sin(t^2)}{t} dt = \frac{\pi}{2} \left[\frac{1}{2}-C\left(\frac{a}{2}\sqrt{\frac{2}{\pi}}\right)\right]^2$$

$$+\frac{\pi}{2}\left[\frac{1}{2}-S\left(\frac{a}{2}\sqrt{\frac{2}{\pi}}\right)\right]^2 \qquad (\mathscr{R}a>0)$$

7.4.25

$$\int_0^{\infty} \frac{e^{-at}\sqrt{t}}{t^2+b^2} dt = \pi\sqrt{\frac{2}{b}} \left\{ \left[\frac{1}{2}-C\left(\sqrt{\frac{2ab}{\pi}}\right)\right] \cos(ab) \right.$$

$$\left. +\left[\frac{1}{2}-S\left(\sqrt{\frac{2ab}{\pi}}\right)\right] \sin(ab) \right\} \qquad (\mathscr{R}a>0, \mathscr{R}b>0)$$

7.4.26

$$\int_0^{\infty} \frac{e^{-at}dt}{\sqrt{t}(t^2+b^2)} = \frac{\pi}{b}\sqrt{\frac{2}{b}} \left\{ \left[\frac{1}{2}-S\left(\sqrt{\frac{2ab}{\pi}}\right)\right] \cos(ab) \right.$$

$$\left. -\left[\frac{1}{2}-C\left(\sqrt{\frac{2ab}{\pi}}\right)\right] \sin(ab) \right\} \qquad (\mathscr{R}a>0, \mathscr{R}b>0)$$

7.4.27

$$\int_0^{\infty} e^{-at} C(t) dt = \frac{1}{a} \left\{ \left[\frac{1}{2}-S\left(\frac{a}{\pi}\right)\right] \cos\left(\frac{a^2}{2\pi}\right) \right.$$

$$\left. -\left[\frac{1}{2}-C\left(\frac{a}{\pi}\right)\right] \sin\left(\frac{a^2}{2\pi}\right) \right\} \qquad (\mathscr{R}a>0)$$

7.4.28

$$\int_0^{\infty} e^{-at} S(t) dt = \frac{1}{a} \left\{ \left[\frac{1}{2}-C\left(\frac{a}{\pi}\right)\right] \cos\left(\frac{a^2}{2\pi}\right) \right.$$

$$\left. +\left[\frac{1}{2}-S\left(\frac{a}{\pi}\right)\right] \sin\left(\frac{a^2}{2\pi}\right) \right\} \qquad (\mathscr{R}a>0)$$

7.4.29

$$\int_0^{\infty} e^{-at} C\left(\sqrt{\frac{2t}{\pi}}\right) dt = \frac{1}{2a(\sqrt{a^2+1}-a)^{\frac{1}{2}}\sqrt{a^2+1}}$$

$$(\mathscr{R}a>0)$$

7.4.30

$$\int_0^{\infty} e^{-at} S\left(\sqrt{\frac{2t}{\pi}}\right) dt = \frac{1}{2a(\sqrt{a^2+1}+a)^{\frac{1}{2}}\sqrt{a^2+1}}$$

$$(\mathscr{R}a>0)$$

7.4.31 $\quad \displaystyle\int_0^{\infty} \left\{ \left[\frac{1}{2}-C(t)\right]^2 + \left[\frac{1}{2}-S(t)\right]^2 \right\} dt = \frac{1}{\pi}$

7.4.32

$$\int e^{-(ax^2+2bx+c)} dx = \frac{1}{2}\sqrt{\frac{\pi}{a}} e^{\frac{b^2-ac}{a}} \operatorname{erf}\left(\sqrt{a}x+\frac{b}{\sqrt{a}}\right)+\text{const.}$$

$$(a \neq 0)$$

7.4.33

$$\int e^{-a^2x^2-\frac{b^2}{x^2}}dx=\frac{\sqrt{\pi}}{4a}\left[e^{2ab}\operatorname{erf}\left(ax+\frac{b}{x}\right)\right.$$
$$\left.+e^{-2ab}\operatorname{erf}\left(ax-\frac{b}{x}\right)\right]+\text{const.}\qquad(a\neq0)$$

7.4.34

$$\int e^{-a^2x^2+\frac{b^2}{x^2}}dx=-\frac{\sqrt{\pi}}{4a}e^{-a^2x^2+\frac{b^2}{x^2}}\left[w\left(\frac{b}{x}+iax\right)\right.$$
$$\left.+w\left(-\frac{b}{x}+iax\right)\right]+\text{const.}\qquad(a\neq0)$$

7.4.35 $\quad\int\operatorname{erf}x\,dx=x\operatorname{erf}x+\frac{1}{\sqrt{\pi}}e^{-x^2}+\text{const.}$

7.4.36

$$\int e^{ax}\operatorname{erf}bx\,dx=\frac{1}{a}\left[e^{ax}\operatorname{erf}bx-e^{\frac{a^2}{4b^2}}\operatorname{erf}\left(bx-\frac{a}{2b}\right)\right]$$
$$+\text{const.}\qquad(a\neq0)$$

7.4.37

$$\int e^{ax}\operatorname{erf}\sqrt{\frac{b}{x}}\,dx=\frac{1}{a}\left\{e^{ax}\operatorname{erf}\sqrt{\frac{b}{x}}\right.$$
$$+\frac{1}{2}e^{ax-\frac{b}{x}}\left[w\left(\sqrt{ax}+i\sqrt{\frac{b}{x}}\right)+w\left(-\sqrt{ax}+i\sqrt{\frac{b}{x}}\right)\right]\right\}$$
$$+\text{const.}\qquad(a\neq0)$$

7.4.38

$$\int\cos(ax^2+2bx+c)dx$$
$$=\sqrt{\frac{\pi}{2a}}\left\{\cos\left(\frac{b^2-ac}{a}\right)C\left[\sqrt{\frac{2}{a\pi}}(ax+b)\right]\right.$$
$$\left.+\sin\left(\frac{b^2-ac}{a}\right)S\left[\sqrt{\frac{2}{a\pi}}(ax+b)\right]\right\}+\text{const.}$$

7.4.39

$$\int\sin(ax^2+2bx+c)dx$$
$$=\sqrt{\frac{\pi}{2a}}\left\{\cos\left(\frac{b^2-ac}{a}\right)S\left[\sqrt{\frac{2}{a\pi}}(ax+b)\right]\right.$$
$$\left.-\sin\left(\frac{b^2-ac}{a}\right)C\left[\sqrt{\frac{2}{a\pi}}(ax+b)\right]\right\}+\text{const.}$$

7.4.40 $\int C(x)dx=xC(x)-\frac{1}{\pi}\sin\left(\frac{\pi}{2}x^2\right)+\text{const.}$

7.4.41 $\int S(x)dx=xS(x)+\frac{1}{\pi}\cos\left(\frac{\pi}{2}x^2\right)+\text{const.}$

Numerical Methods

7.5. Use and Extension of the Tables

Example 1. Compute erf .745 and $e^{-(.745)^2}$ using Taylor's series.

With the aid of Taylor's theorem and **7.1.19** it can be shown that

$$\operatorname{erf}(x_0+ph)=\operatorname{erf}x_0$$
$$+\frac{2}{\sqrt{\pi}}e^{-x_0^2}ph\left[1-phx_0+\frac{1}{3}p^2h^2(2x_0^2-1)\right]+\epsilon$$

$$e^{-(x_0+ph)^2}=e^{-x_0^2}\left[1-2phx_0+p^2h^2(2x_0^2-1)\right.$$
$$\left.-\frac{2}{3}p^3h^3x_0(2x_0^2-3)\right]+\eta$$

where $|\epsilon|<1.2\times10^{-10}$, $|\eta|<3.2\times10^{-10}$ if $h=10^{-2}$, $|p|\leq\frac{1}{2}$. With $x_0=.74$, $p=.5$ and using **Table 7.1**

erf .745$=.70467\ 80779+(.5)(.00652\ 58247)\times$
$$[1-(.005)(.74)+(.00000\ 83333)(.0952)]$$
$$=.70792\ 8920$$

$$e^{-(.745)^2}=\frac{\sqrt{\pi}}{2}(.65258\ 24665)[1-.0074$$
$$+(.000025)(.0952)+(.00000\ 00833)(.74)(1.9048)]$$
$$=.57405\ 7910.$$

As a check the computation was repeated with $x_0=.75$, $p=-.5$.

Example 2. Compute erfc x to 5S for $x=4.8$.

We have $1/x^2=.0434028$. With **Table 7.2** and linear interpolation in **Table 7.3**, we obtain

erfc 4.8$=\frac{1}{4.8}(1.11253)(10^{-10})(.552669)\frac{\sqrt{\pi}}{2}$
$$=(1.1352)10^{-11}.$$

Example 3. Compute $e^{-x^2}\int_0^x e^{t^2}dt$ to 5S for $x=6.5$.

With $1/x^2=.0236686$ and linear interpolation in **Table 7.5**

$$e^{-(6.5)^2}\int_0^{6.5} e^{t^2}dt = (.506143)/(6.5) = .077868.$$

Example 4. Compute i^2 erfc 1.72 using the recurrence relation and **Table 7.1**.

By **7.2.1**, using **Table 7.1**,

$$i^{-1}\text{erfc } 1.72 = .05856\ 50.$$

Using the recurrence relation **7.2.5** and **Table 7.1**

$$i\ \text{erfc } 1.72 = -(1.72)(.01499\ 72)+(.5)(.05856\ 50)$$
$$= .0034873$$
$$i^2\ \text{erfc } 1.72 = -(.86)(.0034873)+(.25)(.01499\ 72)$$
$$= .0007502.$$

Note the loss of two significant digits.

Example 5. Compute i^k erfc 1.72 for $k=1, 2, 3$ by backward recurrence.

Let the sequence $w_\mu^m(x)(\mu=m, m-1, \ldots, 1, 0, -1)$ be generated by backward use of the recurrence relation **7.2.5** starting with $w_{m+2}^m=0$, $w_{m+1}^m=1$. Then, for any fixed k, (see [7.7]),

$$\lim_{m\to\infty} \frac{w_k^m(x)}{w_{-1}^m(x)} = \frac{\sqrt{\pi}}{2}\ e^{x^2}i^k\ \text{erfc } x \qquad (x>0).$$

With $x=1.72$, $m=15$ we obtain

μ	$w_\mu^{15}(1.72)$	μ	$w_\mu^{15}(1.72)$	μ	$w_\mu^{15}(1.72)$	μ	$w_\mu^{15}(1.72)$
17	0	12	(3) 2.1011	7	(7) 2.5879	2	(11) 1.2920
16	1	11	(4) 1.3831	6	(8) 1.5569	1	(11) 6.0064
15	3.44	10	(4) 9.8005	5	(8) 8.9787	0	(12) 2.5830
14	(1) 4.3834	9	(5) 6.4143	4	(9) 4.9570	−1	(13) 1.0087
13	(2) 2.5399	8	(6) 4.1666	3	(10) 2.6031		

From **Table 7.1** we have $\frac{2}{\sqrt{\pi}}\ e^{-(1.72)^2}=.058565$. Thus,

$$i\ \text{erfc } 1.72 \approx (.058565)(6.0064\times10^{11})/1.0087\times10^{13}$$
$$= 3.4873\times10^{-3}$$
$$i^2\ \text{erfc } 1.72 \approx (.058565)(1.2920\times10^{11})/1.0087\times10^{13}$$
$$= 7.5013\times10^{-4}$$
$$i^3\ \text{erfc } 1.72 \approx (.058565)(2.6031\times10^{10})/1.0087\times10^{13}$$
$$= 1.5114\times10^{-4}.$$

Example 6. Compute $C(8.65)$ using **Table 7.8**. With $x=8.65$, $1/x=.115607$ we have from **Table 7.8** by linear interpolation

$$f(8.65)=.036797, \quad g(8.65)=.000159.$$

From **Table 4.6**

$$\sin\left(\frac{\pi}{2}\ x^2\right)=-.961382, \quad \cos\left(\frac{\pi}{2}\ x^2\right)=-.275218.$$

Using **7.3.9**

$$C(8.65)=.5+(.036797)(-.961382)$$
$$-(.000159)(-.275218)=.46467.$$

Example 7. Compute $S_1(1.1)$ to 10D. Using **7.3.8** and **7.3.10** we obtain by 6-pt interpolation in **Table 7.8**

$$S_1(1.1)=S\left(1.1\ \sqrt{\frac{2}{\pi}}\right)$$
$$=S(.87767\ 30169)=.31865\ 57172.$$

Example 8. Compute $S_2(5.24)$ to 6D. Enter **Table 7.7** in the column headed by u. Using Aitken's scheme of interpolation

u	$S_2(u)$					
5.20310 58	.43280 06	.03689 42				
5.31808 80	.41573 97	−.07808 80	.42732 63			
5.08938 01	.45093 88	.15061 99	691 63	.42718 63		
5.43432 70	.39999 44	−.19432 70	756 60	6 52	.42717 71	
4.97691 11	.46990 94	.26308 89	674 79	9 39	61	.42717 67

$$S_2(5.24)=.427177$$

Example 9. Compute $S_2(5.24)$ using Taylor's series and **Table 7.8**. Using **7.3.21** we can write Taylor's series for $f_2(u)=f\left(\sqrt{\frac{2u}{\pi}}\right)$ and $g_2(u)=g\left(\sqrt{\frac{2u}{\pi}}\right)$ in the form

$$f_2(u)=c_0+c_1(u-u_0)+\frac{c_2}{2!}(u-u_0)^2+\frac{c_3}{3!}(u-u_0)^3+\ldots,$$

$$g_2(u)=-\left[c_1+c_2(u-u_0)\right.$$
$$\left.+\frac{c_3}{2!}(u-u_0)^2+\frac{c_4}{3!}(u-u_0)^3+\ldots\right],$$

where

$$c_0=f_2(u_0),\quad c_1=-g_2(u_0),$$
$$c_{k+2}=-c_k+(-1)^k\frac{1\cdot3\ldots(2k-1)}{\sqrt{2\pi u_0}(2u_0)^k}$$

$$(k=0,\,1,\,2,\,\ldots).$$

Consulting **Table 7.8** we chose $u_0=1/.185638=5.386819$, thus having $u-u_0=5.24-5.386819=-.146819$. From **Table 7.8**

$$f_2(u_0)=.168270,\quad g_2(u_0)=.014483.$$

Hence, applying the series above,

$$f_2(5.24)=.170436,\quad g_2(5.24)=.015030.$$

Using the 4th formula at the bottom of **Table 7.8**

$$S_2(5.24)=.5-(.170436)(.503471)$$
$$-(.015030)(-.864012)=.42718.$$

Example 10. Compute $S_2(2)$ using **7.3.16**. Generating the values of $J_{n+\frac{1}{2}}(2)$ as described in chapter **10** we find

$$S_2(2)=J_{3/2}(2)+J_{7/2}(2)+J_{11/2}(2)+J_{15/2}(2)+\ldots$$
$$=.49129+.06852+.00297+.00006=.56284.$$

Example 11. Compute $\int_1^\infty\frac{Y_0(t)}{t}\,dt$ by numerical integration using **Tables 9.1** and **7.8**. [$Y_0(t)$ is the Bessel function of the second kind defined in **9.1.16**.] We decompose the integral into three parts

$$\int_1^\infty Y_0(t)\frac{dt}{t}=\int_1^{10}Y_0(t)\frac{dt}{t}+\int_{10}^\infty[Y_0(t)-\tilde{Y}_0(t)]\frac{dt}{t}$$
$$+\int_{10}^\infty\tilde{Y}_0(t)\frac{dt}{t}$$

where

$$\tilde{Y}_0(t)=\left(1-\frac{9}{128t^2}\right)\frac{\sin\left(t-\frac{\pi}{4}\right)}{\sqrt{\frac{1}{2}\pi t}}$$
$$-\left(1-\frac{75}{128t^2}\right)\frac{\cos\left(t-\frac{\pi}{4}\right)}{8t\sqrt{\frac{1}{2}\pi t}}$$

represents the first two terms of the asymptotic expansion **9.2.2**.

By numerical integration, using **Table 9.1**,

$$\int_1^{10}Y_0(t)\frac{dt}{t}=.41826\ 00.$$

Using the fact that the remainder terms of the asymptotic expansion are less in absolute value than the first neglected terms, we can estimate

$$\left|\int_{10}^\infty[Y_0(t)-\tilde{Y}_0(t)]\frac{dt}{t}\right|\leq\sqrt{\frac{2}{\pi}}\int_{10}^\infty\left[\frac{3^2\cdot5^2\cdot7^2}{2^{12}\cdot4!}t^{-11/2}\right.$$
$$\left.+\frac{3^2\cdot5^2\cdot7^2\cdot9^2}{2^{15}\cdot5!}t^{-13/2}\right]dt=7.33\times10^{-7}.$$

Finally,

$$\int_{10}^\infty\tilde{Y}_0(t)\frac{dt}{t}=\frac{14659}{6720}\sqrt{2}[1-C_2(10)-S_2(10)]$$
$$-\frac{5953819}{2688000}\frac{\cos10-\sin10}{\sqrt{10\pi}}$$
$$-\frac{23107}{2150400}\frac{\cos10+\sin10}{\sqrt{10\pi}}=-.02298\ 78,$$

using **Tables 7.8** and **4.8**. Hence

$$\int_1^\infty Y_0(t)\frac{dt}{t}=.41826\ 00-.02298\ 78=.39527\ 22.$$

The answer correct to 8D is .39527 290 (**Table 11.2**).

Example 12. Compute $w(.44+.67i)$ using bivariate linear interpolation.

By linear interpolation in **Table 7.9** along the x-direction at $y=.6$ and $y=.7$

$$w(.44+.6i)\approx.6(.522246+.167880i)+.4(.498591$$
$$+.202666i)=.512784+.181794i$$

$$w(.44+.7i)\approx.6(.487556+.147975i)+.4(.467521$$
$$+.179123i)=.479542+.160434i.$$

By linear interpolation along the y-direction at $x=.44$

$$w(.44+.67i)\approx.3(.512784+.181794i)+.7(.479542$$
$$+.160434i)=.489515+.166842i.$$

The correct answer is $.489557+.166889i$.

Example 13. Compute $\mathscr{R}w(z)$ for $z=.44+.61i$. Bivariate linear interpolation, as described in **Example 12,** is most accurate if z lies near the center or along a diagonal of one of the squares of the tabular grid [7.6]. It is not as accurate for z near the midpoint of a side of a square, as in this example. However, we may introduce an auxiliary square (see diagram) which contains z close to its center. Bivariate linear interpolation can then be applied within this auxiliary square.

The values of $w(z)$ needed at $z=\zeta_1$, and $z=\zeta_2$ are easily approximated by the average of the four neighboring tabular values. Furthermore the parts to be used are given by

$$\frac{|z_0-\lambda_1|}{|z_0-\zeta_1|}=p_1+p_2, \quad \frac{|z_0-\lambda_2|}{|z_0-\zeta_2|}=p_1-p_2$$

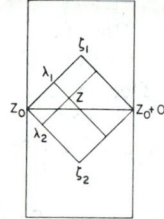

where $z=z_0+.1(p_1+ip_2)$. Thus, with $z_0=.4+.6i$, $\zeta_1=.45+.65i$, $\zeta_2=.45+.55i$, $p_1=.4$, $p_2=.1$, we get from **Table 7.9**

$$\mathscr{R}w(\zeta_1) \approx \tfrac{1}{4}(.522246+.498591+.487556+.467521)$$
$$=.493979$$

$$\mathscr{R}w(\zeta_2) \approx \tfrac{1}{4}(.522246+.498591+.561252+.533157)$$
$$=.528812$$

$$\mathscr{R}w(z) \approx [1-(.4+.1)]\{[1-(.4-.1)].522246$$
$$+(.4-.1).528812\}+(.4+.1)\times$$
$$\{[1-(.4-.1)]\ .493979+(.4-.1).498591\}=.509789.$$

The correct answer is .509756. Straightforward bivariate interpolation gives .509460.

Example 14. Compute $\mathscr{I}w(.39+.61i)$ to 6D using Taylor's series.

Let $z=.39+.61i$, $z_0=.4+.6i$. From **7.1.20**, and using **Table 7.9**, we have

$$w(z_0)=.522246+.167880i$$

$$w'(z_0)=-.21634+.36738i, \quad z-z_0=(-1+i)10^{-2}$$
$$\tfrac{1}{2}w''(z_0)=-.215-.185i, \quad (z-z_0)^2=-2i\times10^{-4}$$

$$\mathscr{I}w(z)=.167880-.0021634-.0036738$$
$$+.0000430=.162086.$$

Example 15. Compute $w(.4-1.3i)$.
From **7.1.11, 7.1.12**

$$w(.4-1.3i)=\overline{w(-.4-1.3i)}=2e^{-(.4-1.3i)^2}$$
$$-\overline{w(.4+1.3i)}.$$

Using **Tables 7.9, 4.4** and **4.6**

$$w(.4-1.3i)=4.33342+8.04201i.$$

Example 16. Compute $w(7+2i)$.
Using the second formula at the end of **Table 7.9**

$$w(7+2i)=(-2+7i)\left(\frac{.5124242}{44.72474+28i}\right.$$
$$\left.+\frac{.05176536}{42.27525+28i}\right)=.021853+.075010i.$$

Example 17. Compute erf $(2+i)$.
From **7.1.3, 7.1.12** we have

$$\text{erf } z=1-e^{-z^2}w(iz)=1-e^{y^2-x^2}(\cos 2xy$$
$$-i\sin 2xy)\overline{w(y+ix)} \qquad (z=x+iy).$$

Using **Tables 7.9, 4.4, 4.6**

$$\text{erf } (2+i)=1-e^{-3}(\cos 4-i\sin 4)\overline{w(1+2i)}$$
$$=1.003606-.0112590i.$$

Example 18. Compute $S_1\!\left(\left(\tfrac{1}{2}+i\right)\sqrt{2}\right)$.
From **7.3.22, 7.3.8, 7.3.18** we have

$$S_1(z)=\frac{1}{2}-\frac{1-i}{4}\,e^{iz^2}w\left[(1+i)\,\frac{z}{\sqrt{2}}\right]$$
$$-\frac{1+i}{4}\,e^{-iz^2}w\left[(i-1)\,\frac{z}{\sqrt{2}}\right].$$

Setting $z=\left(\tfrac{1}{2}+i\right)\sqrt{2}$ and making use of **7.1.11, 7.1.12**, and **Table 7.9**

$$S_1\!\left(\left(\tfrac{1}{2}+i\right)\sqrt{2}\right)=$$
$$-\frac{i}{2}-\frac{1-i}{4}\,e^{-2}\left(\cos\frac{3}{2}-i\sin\frac{3}{2}\right)\overline{w\left(\frac{1}{2}+\frac{3}{2}i\right)}$$
$$+\frac{1+i}{4}\,e^2\left(\cos\frac{3}{2}+i\sin\frac{3}{2}\right)w\left(\frac{3}{2}+\frac{1}{2}i\right)$$
$$=-.990734-.681619i.$$

Example 19. Compute $\displaystyle\int_0^\infty e^{-(1/4)t^2-3t}\cos(2t)dt$ using **Table 7.9**.
Setting $b=y+ix$, $c=0$ in **7.4.2** and using **7.1.3, 7.1.12** we find

$$\int_0^\infty e^{-at^2-2yt}\cos(2xt)dt=\frac{1}{2}\sqrt{\frac{\pi}{a}}\,\mathscr{R}w\left(\frac{x+iy}{\sqrt{a}}\right)$$
$$(a>0, x, y \text{ real}).$$

Hence from **Table 7.9**

$$\int_0^\infty e^{-(1/4)t^2-3t}\cos(2t)dt=\sqrt{\pi}\,\mathscr{R}w(2+3i)=.231761.$$

References

Texts

[7.1] J. Boersma, Computation of Fresnel integrals, Math. Comp. **14**, 380 (1960).

[7.2] A. V. Boyd, Inequalities for Mills' ratio, Rep. Statist. Appl. Res. Un. Jap. Sci. Engrs. **6**, 44–46 (1959).

[7.3] O. Emersleben, Numerische Werte des Fehlerintegrals für $\sqrt{n\pi}$, Z. Angew. Math. Mech. **31**, 393–394 (1951).

[7.4] A. Erdélyi et al., Higher transcendental functions, vol. 2 (McGraw-Hill Book Co., Inc., New York, N.Y., Toronto, Canada, London, England, 1953).

[7.5] A. Erdélyi et al., Tables of integral transforms, vol. 1 (McGraw-Hill Book Co., Inc., New York, N.Y., Toronto, Canada, London, England, 1954).

[7.6] W. Gautschi, Note on bivariate linear interpolation for analytic functions, Math. Tables Aids Comp. **13**, 91–96 (1959).

[7.7] W. Gautschi, Recursive computation of the repeated integrals of the error function, Math. Comp. **15**, 227–232 (1961).

[7.8] W. Gröbner and N. Hofreiter, Integraltafel (Springer-Verlag, Wien and Innsbruck, Austria, 1949–50).

[7.9] D. R. Hartree, Some properties and applications of the repeated integrals of the error function, Mem. Proc. Manchester Lit. Philos. Soc. **80**, 85–102 (1936).

[7.10] C. Hastings, Jr., Approximations for digital computers (Princeton Univ. Press, Princeton, N.J., 1955).

[7.11] Y. Komatu, Elementary inequalities for Mills' ratio, Rep. Statist. Appl. Res. Un. Jap. Sci. Engrs. **4**, 69–70 (1955–57).

[7.12] E. Kreyszig, On the zeros of the Fresnel integrals, Canad. J. Math. **9**, 118–131 (1957).

[7.13] Th. Laible, Höhenkarte des Fehlerintegrals, Z. Angew. Math. Phys. **2**, 484–486 (1951).

[7.14] F. Lösch and F. Schoblik, Die Fakultät (B. G. Teubner, Leipzig, Germany, 1951).

[7.15] F. Oberhettinger, Tabellen zur Fourier Transformation (Springer-Verlag, Berlin, Göttingen, Heidelberg, Germany, 1957).

[7.16] J. R. Philip, The function inv erfc θ, Austral. J. Phys. **13**, 13–20 (1960).

[7.17] H. O. Pollak, A remark on "Elementary inequalities for Mills' ratio" by Yûsaku Komatu, Rep. Statist. Appl. Res. Un. Jap. Sci. Engrs. **4**, 110 (1955–57).

[7.18] J. B. Rosser, Theory and application of $\int_0^z e^{-x^2}dx$ and $\int_0^z e^{-p^2v^2}dy \int_0^v e^{-x^2}dx$ (Mapleton House, Brooklyn, N.Y., 1948).

[7.19] H. E. Salzer, Formulas for calculating the error function of a complex variable, Math. Tables Aids Comp. **5**, 67–70 (1951).

[7.20] H. E. Salzer, Complex zeros of the error function, J. Franklin Inst. **260**, 209–211 (1955).

[7.21] F. G. Tricomi, Funzioni ipergeometriche confluenti (Edizioni Cremonese, Rome, Italy, 1954).

[7.22] G. N. Watson, A treatise on the theory of Bessel functions, 2d ed. (Cambridge Univ. Press, London, England, 1958).

Tables

[7.23] M. Abramowitz, Table of the integral $\int_0^z e^{-u^3}du$, J. Math. Phys. **30**, 162–163 (1951). $x=0(.01)2.5$, 8D.

[7.24] P. C. Clemmow and Cara M. Munford, A table of $\sqrt{(\tfrac{1}{2}\pi)}e^{\frac{1}{2}i\pi\rho^2}\int_\rho^\infty e^{-\frac{1}{2}i\pi\lambda^2}d\lambda$ for complex values of ρ, Philos. Trans. Roy. Soc. London $\{A\}$, **245**, 189–211 (1952). $|\rho|=0(.01).8$, arg $\rho=0°(1°)45°$, 4D.

[7.25] R. B. Dingle, Doreen Arndt and S. K. Roy, The integrals

$$C_p(x)=(p!)^{-1}\int_0^\infty \epsilon^p(\epsilon^2+x^2)^{-1}e^{-\epsilon}d\epsilon$$

and

$$D_p(x)=(p!)^{-1}\int_0^\infty \epsilon^p(\epsilon^2+x^2)^{-2}e^{-\epsilon}d\epsilon$$

and their tabulation, Appl. Sci. Res. B **6**, 155–164 (1956). $C(x)$, $S(x)$, $x=0(1)20$, 12D.

[7.26] V. N. Faddeeva and N. M. Terent'ev, Tables of values of the function $w(z)=e^{-z^2}\left(1+\dfrac{2i}{\sqrt{\pi}}\int_0^z e^{t^2}dt\right)$ for complex argument. Translated from the Russian by D. G. Fry (Pergamon Press, New York, N.Y., 1961). $w(z), z=x+iy$; $x,y=0(.02)3$; $x=3(.1)5$, $y=0(.1)3$; $x=0(.1)5$, $y=3(.1)5$; 6D.

[7.27] B. D. Fried and S. D. Conte, The plasma dispersion function (Academic Press, New York, N.Y. and London, England, 1961). $i\sqrt{\pi}w(z)$, $i\sqrt{\pi}w'(z)$, $z=x+iy$; $x=0(.1)9.9$, $y=-9.1(.1)10$; $x=$var. $(.1)9.9$, $y=-10(.1)-9.2$; 6S.

[7.28] K. A. Karpov, Tablitsy funktsii $w(z)=e^{-z^2}\int_0^z e^{x^2}dx$ v kompleksnoi oblasti (Izdat. Akad. Nauk SSSR., Moscow, U.S.S.R., 1954). $z=x$; $x=0(.001)2(.01)10$; 5D; $z=\rho e^{i\theta}$; $\theta=2.5°(2.5°)30°(1.25°)35°(.625°)40°$; $\rho=\rho_\theta(.001)\rho'_\theta(.01)\rho''_\theta(.0002)5$, $0\le\rho_\theta\le\rho'_\theta\le\rho''_\theta\le5$, 5D; $z=iy$; $y=0(.001)3(.0002)5$, 5S.

[7.29] K. A. Karpov, Tablitsy funktsii $F(z)=\int_0^z e^{x^2}dx$ v kompleksnoi oblasti (Izdat. Akad. Nauk SSSR., Moscow, U.S.S.R., 1958). $z=\rho e^{i\theta}$; $\theta=45°(.3125°)48.75°(.625°)55°(1.25°)65°(2.5°)90°$, $\rho=\rho_\theta(.001)\rho'_\theta(.01)\rho''_\theta$, $0\le\rho_\theta<\rho'_\theta\le\rho''_\theta\le5$, 5D; $z=x$; $x=0(.001)10$, 5S.

[7.30] J. Kaye, A table of the first eleven repeated integrals of the error function, J. Math. Phys. **34**, 119–125 (1955). $i^n\mathrm{erfc}\,x$, $x=0(.01).2(.05)1(.1)3$, $n=-1(1)11$, 6D.

[7.31] B. Lohmander and S. Rittsten, Table of the function $y=e^{-x^2}\int_0^x e^{t^2}dt$, Kungl. Fysiogr. Sällsk. i Lund Förh. **28**, 45–52 (1958). $x=0(.01)3(.02)5$, $x^{-1}=0(.005).2$, 10D; $x=.5(.5)10$, 20D. Contains also 20D values for maximum and inflection points.

[7.32] W. Lash Miller and A. R. Gordon, Numerical evaluation of infinite series and integrals which arise in certain problems of linear heat flow, electrochemical diffusion, etc., J. Phys. Chem. **35**, 2785–2884 (1931).

$$F(x) = e^{-x^2} \int_0^x e^{t^2} dt; \quad x = 0(.01)1.99, \quad 6D;$$

$$x = 2(.01)4(.05)7.5(.1)10(.2)12, \quad 8S.$$

[7.33] National Bureau of Standards, Tables of the error function and its derivative, Applied Math. Series 41, 2d ed. (U.S. Government Printing Office, Washington, D.C., 1954).

$$(2/\sqrt{\pi})e^{-x^2}, \text{ erf } x, \quad x = 0(.0001)1(.001)5.6, \quad 15D;$$

$$(2/\sqrt{\pi})e^{-x^2}, \text{ erfc } x, \quad x = 4(.01)10, \quad 8S.$$

[7.34] T. Pearcey, Table of the Fresnel integral (Cambridge Univ. Press, London, England, 1956).

$$C\left(\sqrt{\frac{2x}{\pi}}\right), \ S\left(\sqrt{\frac{2x}{\pi}}\right), \ x = 0(.01)50, \quad 6\text{–}7D.$$

[7.35] Tablitsy integralov Frenelya (Izdat. Akad. Nauk SSSR., Moscow, U.S.S.R., 1953). $C(x)$, $S(x)$, $x = 0(.001)25$, 7D; $S(x)$, $x = 0(.001) .58$, 7S; $C(x)$, $x = 0(.001) .101$, 7S.

[7.36] A. van Wijngaarden and W. L. Scheen, Table of Fresnel integrals, Verh. Nederl. Akad. Wetensch., Afd. Natuurk. Sec. **I**, 19, No. 4, 1–26 (1949). $C(x)$, $S(x)$, $x = 0(.01)20$, 5D. (Also contains numerical values of the coefficients in Taylor and asymptotic expansions.)

Table 7.1   ERROR FUNCTION AND ITS DERIVATIVE

x	$\frac{2}{\sqrt{\pi}} e^{-x^2}$	erf x	x	$\frac{2}{\sqrt{\pi}} e^{-x^2}$	erf x
0.00	1.12837 91671	0.00000 00000	0.50	0.87878 25789	0.52049 98778
0.01	1.12826 63348	0.01128 34156	0.51	0.86995 15467	0.52924 36198
0.02	1.12792 79057	0.02256 45747	0.52	0.86103 70343	0.53789 86305
0.03	1.12736 40827	0.03384 12223	0.53	0.85204 34444	0.54646 40969
0.04	1.12657 52040	0.04511 11061	0.54	0.84297 51813	0.55493 92505
0.05	1.12556 17424	0.05637 19778	0.55	0.83383 66473	0.56332 33663
0.06	1.12432 43052	0.06762 15944	0.56	0.82463 22395	0.57161 57638
0.07	1.12286 36333	0.07885 77198	0.57	0.81536 63461	0.57981 58062
0.08	1.12118 06004	0.09007 81258	0.58	0.80604 33431	0.58792 29004
0.09	1.11927 62126	0.10128 05939	0.59	0.79666 75911	0.59593 64972
0.10	1.11715 16068	0.11246 29160	0.60	0.78724 34317	0.60385 60908
0.11	1.11480 80500	0.12362 28962	0.61	0.77777 51846	0.61168 12189
0.12	1.11224 69379	0.13475 83518	0.62	0.76826 71442	0.61941 14619
0.13	1.10946 97934	0.14586 71148	0.63	0.75872 35764	0.62704 64433
0.14	1.10647 82654	0.15694 70331	0.64	0.74914 87161	0.63458 58291
0.15	1.10327 41267	0.16799 59714	0.65	0.73954 67634	0.64202 93274
0.16	1.09985 92726	0.17901 18132	0.66	0.72992 18814	0.64937 66880
0.17	1.09623 57192	0.18999 24612	0.67	0.72027 81930	0.65662 77023
0.18	1.09240 56008	0.20093 58390	0.68	0.71061 97784	0.66378 22027
0.19	1.08837 11683	0.21183 98922	0.69	0.70095 06721	0.67084 00622
0.20	1.08413 47871	0.22270 25892	0.70	0.69127 48604	0.67780 11938
0.21	1.07969 89342	0.23352 19230	0.71	0.68159 62792	0.68466 55502
0.22	1.07506 61963	0.24429 59116	0.72	0.67191 88112	0.69143 31231
0.23	1.07023 92672	0.25502 25996	0.73	0.66224 62838	0.69810 39429
0.24	1.06522 09449	0.26570 00590	0.74	0.65258 24665	0.70467 80779
0.25	1.06001 41294	0.27632 63902	0.75	0.64293 10692	0.71115 56337
0.26	1.05462 18194	0.28689 97232	0.76	0.63329 57399	0.71753 67528
0.27	1.04904 71098	0.29741 82185	0.77	0.62368 00626	0.72382 16140
0.28	1.04329 31885	0.30788 00680	0.78	0.61408 75556	0.73001 04313
0.29	1.03736 33334	0.31828 34959	0.79	0.60452 16696	0.73610 34538
0.30	1.03126 09096	0.32862 67595	0.80	0.59498 57863	0.74210 09647
0.31	1.02498 93657	0.33890 81503	0.81	0.58548 32161	0.74800 32806
0.32	1.01855 22310	0.34912 59948	0.82	0.57601 71973	0.75381 07509
0.33	1.01195 31119	0.35927 86550	0.83	0.56659 08944	0.75952 37569
0.34	1.00519 56887	0.36936 45293	0.84	0.55720 73967	0.76514 27115
0.35	0.99828 37121	0.37938 20536	0.85	0.54786 97173	0.77066 80576
0.36	0.99122 10001	0.38932 97011	0.86	0.53858 07918	0.77610 02683
0.37	0.98401 14337	0.39920 59840	0.87	0.52934 34773	0.78143 98455
0.38	0.97665 89542	0.40900 94534	0.88	0.52016 05514	0.78668 73192
0.39	0.96916 75592	0.41873 87001	0.89	0.51103 47116	0.79184 32468
0.40	0.96154 12988	0.42839 23550	0.90	0.50196 85742	0.79690 82124
0.41	0.95378 42727	0.43796 90902	0.91	0.49296 46742	0.80188 28258
0.42	0.94590 06256	0.44746 76184	0.92	0.48402 54639	0.80676 77215
0.43	0.93789 45443	0.45688 66945	0.93	0.47515 33132	0.81156 35586
0.44	0.92977 02537	0.46622 51153	0.94	0.46635 05090	0.81627 10190
0.45	0.92153 20130	0.47548 17198	0.95	0.45761 92546	0.82089 08073
0.46	0.91318 41122	0.48465 53900	0.96	0.44896 16700	0.82542 36496
0.47	0.90473 08685	0.49374 50509	0.97	0.44037 97913	0.82987 02930
0.48	0.89617 66223	0.50274 96707	0.98	0.43187 55710	0.83423 15043
0.49	0.88752 57337	0.51166 82612	0.99	0.42345 08779	0.83850 80696
0.50	0.87878 25789	0.52049 98778	1.00	0.41510 74974	0.84270 07929
	$\begin{bmatrix}(-5)3\\5\end{bmatrix}$	$\begin{bmatrix}(-5)1\\5\end{bmatrix}$		$\begin{bmatrix}(-5)1\\5\end{bmatrix}$	$\begin{bmatrix}(-5)1\\5\end{bmatrix}$

See Example 1.

$$\text{erf } x = \frac{2}{\sqrt{\pi}} \int_0^x e^{-t^2} dt \qquad\qquad \frac{\sqrt{\pi}}{2} = 0.88622 \ 69255$$

ERROR FUNCTION AND ITS DERIVATIVE

Table 7.1

x	$\frac{2}{\sqrt{\pi}}\,e^{-x^2}$	erf x	x	$\frac{2}{\sqrt{\pi}}\,e^{-x^2}$	erf x
1.00	0.41510 74974	0.84270 07929	1.50	0.11893 02892	0.96610 51465
1.01	0.40684 71315	0.84681 04962	1.51	0.11540 38270	0.96727 67481
1.02	0.39867 13992	0.85083 80177	1.52	0.11195 95356	0.96841 34969
1.03	0.39058 18368	0.85478 42115	1.53	0.10859 63195	0.96951 62091
1.04	0.38257 98986	0.85864 99465	1.54	0.10531 30683	0.97058 56899
1.05	0.37466 69570	0.86243 61061	1.55	0.10210 86576	0.97162 27333
1.06	0.36684 43034	0.86614 35866	1.56	0.09898 19506	0.97262 81220
1.07	0.35911 31488	0.86977 32972	1.57	0.09593 17995	0.97360 26275
1.08	0.35147 46245	0.87332 61584	1.58	0.09295 70461	0.97454 70093
1.09	0.34392 97827	0.87680 31019	1.59	0.09005 65239	0.97546 20158
1.10	0.33647 95978	0.88020 50696	1.60	0.08722 90586	0.97634 83833
1.11	0.32912 49667	0.88353 30124	1.61	0.08447 34697	0.97720 68366
1.12	0.32186 67103	0.88678 78902	1.62	0.08178 85711	0.97803 80884
1.13	0.31470 55742	0.88997 06704	1.63	0.07917 31730	0.97884 28397
1.14	0.30764 22299	0.89308 23276	1.64	0.07662 60821	0.97962 17795
1.15	0.30067 72759	0.89612 38429	1.65	0.07414 61034	0.98037 55850
1.16	0.29381 12389	0.89909 62029	1.66	0.07173 20405	0.98110 49213
1.17	0.28704 45748	0.90200 03990	1.67	0.06938 26972	0.98181 04416
1.18	0.28037 76702	0.90483 74269	1.68	0.06709 68781	0.98249 27870
1.19	0.27381 08437	0.90760 82860	1.69	0.06487 33895	0.98315 25869
1.20	0.26734 43470	0.91031 39782	1.70	0.06271 10405	0.98379 04586
1.21	0.26097 83664	0.91295 55080	1.71	0.06060 86436	0.98440 70075
1.22	0.25471 30243	0.91553 38810	1.72	0.05856 50157	0.98500 28274
1.23	0.24854 83805	0.91805 01041	1.73	0.05657 89788	0.98557 84998
1.24	0.24248 44335	0.92050 51843	1.74	0.05464 93607	0.98613 45950
1.25	0.23652 11224	0.92290 01283	1.75	0.05277 49959	0.98667 16712
1.26	0.23065 83281	0.92523 59418	1.76	0.05095 47262	0.98719 02752
1.27	0.22489 58748	0.92751 36293	1.77	0.04918 74012	0.98769 09422
1.28	0.21923 35317	0.92973 41930	1.78	0.04747 18791	0.98817 41959
1.29	0.21367 10145	0.93189 86327	1.79	0.04580 70274	0.98864 05487
1.30	0.20820 79868	0.93400 79449	1.80	0.04419 17233	0.98909 05016
1.31	0.20284 40621	0.93606 31228	1.81	0.04262 48543	0.98952 45446
1.32	0.19757 88048	0.93806 51551	1.82	0.04110 53185	0.98994 31565
1.33	0.19241 17326	0.94001 50262	1.83	0.03963 20255	0.99034 68051
1.34	0.18734 23172	0.94191 37153	1.84	0.03820 38966	0.99073 59476
1.35	0.18236 99865	0.94376 21961	1.85	0.03681 98653	0.99111 10301
1.36	0.17749 41262	0.94556 14366	1.86	0.03547 88774	0.99147 24883
1.37	0.17271 40811	0.94731 23980	1.87	0.03417 98920	0.99182 07476
1.38	0.16802 91568	0.94901 60353	1.88	0.03292 18811	0.99215 62228
1.39	0.16343 86216	0.95067 32958	1.89	0.03170 38307	0.99247 93184
1.40	0.15894 17077	0.95228 51198	1.90	0.03052 47404	0.99279 04292
1.41	0.15453 76130	0.95385 24394	1.91	0.02938 36241	0.99308 99398
1.42	0.15022 55027	0.95537 61786	1.92	0.02827 95101	0.99337 82251
1.43	0.14600 45107	0.95685 72531	1.93	0.02721 14412	0.99365 56502
1.44	0.14187 37413	0.95829 65696	1.94	0.02617 84752	0.99392 25709
1.45	0.13783 22708	0.95969 50256	1.95	0.02517 96849	0.99417 93336
1.46	0.13387 91486	0.96105 35095	1.96	0.02421 41583	0.99442 62755
1.47	0.13001 33993	0.96237 28999	1.97	0.02328 09986	0.99466 37246
1.48	0.12623 40239	0.96365 40654	1.98	0.02237 93244	0.99489 20004
1.49	0.12254 00011	0.96489 78648	1.99	0.02150 82701	0.99511 14132
1.50	0.11893 02892	0.96610 51465	2.00	0.02066 69854	0.99532 22650
	$\left[\begin{smallmatrix}(-5)1\\5\end{smallmatrix}\right]$	$\left[\begin{smallmatrix}(-5)1\\5\end{smallmatrix}\right]$		$\left[\begin{smallmatrix}(-5)1\\5\end{smallmatrix}\right]$	$\left[\begin{smallmatrix}(-6)4\\5\end{smallmatrix}\right]$

$$\frac{\sqrt{\pi}}{2}=0.88622\ 69255$$

Table 7.2   **DERIVATIVE OF THE ERROR FUNCTION**

x	$\frac{2}{\sqrt{\pi}}e^{-x^2}$	x	$\frac{2}{\sqrt{\pi}}e^{-x^2}$	x	$\frac{2}{\sqrt{\pi}}e^{-x^2}$	x	$\frac{2}{\sqrt{\pi}}e^{-x^2}$
2.00	(− 2) 2.0666 985	2.50	(− 3) 2.1782 842	3.00	(− 4) 1.3925 305	3.50	(− 6) 5.3994 268
2.01	(− 2) 1.9854 636	2.51	(− 3) 2.0718 409	3.01	(− 4) 1.3113 047	3.51	(− 6) 5.0338 887
2.02	(− 2) 1.9070 402	2.52	(− 3) 1.9702 048	3.02	(− 4) 1.2345 698	3.52	(− 6) 4.6921 589
2.03	(− 2) 1.8313 482	2.53	(− 3) 1.8731 800	3.03	(− 4) 1.1620 929	3.53	(− 6) 4.3727 530
2.04	(− 2) 1.7583 088	2.54	(− 3) 1.7805 771	3.04	(− 4) 1.0936 521	3.54	(− 6) 4.0742 749
2.05	(− 2) 1.6878 448	2.55	(− 3) 1.6922 136	3.05	(− 4) 1.0290 362	3.55	(− 6) 3.7954 113
2.06	(− 2) 1.6198 806	2.56	(− 3) 1.6079 137	3.06	(− 5) 9.6804 434	3.56	(− 6) 3.5349 275
2.07	(− 2) 1.5543 422	2.57	(− 3) 1.5275 078	3.07	(− 5) 9.1048 542	3.57	(− 6) 3.2916 626
2.08	(− 2) 1.4911 571	2.58	(− 3) 1.4508 325	3.08	(− 5) 8.5617 765	3.58	(− 6) 3.0645 257
2.09	(− 2) 1.4302 545	2.59	(− 3) 1.3777 304	3.09	(− 5) 8.0494 817	3.59	(− 6) 2.8524 914
2.10	(− 2) 1.3715 650	2.60	(− 3) 1.3080 500	3.10	(− 5) 7.5663 267	3.60	(− 6) 2.6545 968
2.11	(− 2) 1.3150 207	2.61	(− 3) 1.2416 455	3.11	(− 5) 7.1107 499	3.61	(− 6) 2.4699 374
2.12	(− 2) 1.2605 554	2.62	(− 3) 1.1783 764	3.12	(− 5) 6.6812 674	3.62	(− 6) 2.2976 636
2.13	(− 2) 1.2081 043	2.63	(− 3) 1.1181 075	3.13	(− 5) 6.2764 699	3.63	(− 6) 2.1369 782
2.14	(− 2) 1.1576 041	2.64	(− 3) 1.0607 090	3.14	(− 5) 5.8950 187	3.64	(− 6) 1.9871 328
2.15	(− 2) 1.1089 930	2.65	(− 3) 1.0060 558	3.15	(− 5) 5.5356 429	3.65	(− 6) 1.8474 250
2.16	(− 2) 1.0622 108	2.66	(− 4) 9.5402 778	3.16	(− 5) 5.1971 360	3.66	(− 6) 1.7171 961
2.17	(− 2) 1.0171 986	2.67	(− 4) 9.0450 949	3.17	(− 5) 4.8783 532	3.67	(− 6) 1.5958 281
2.18	(− 3) 9.7389 910	2.68	(− 4) 8.5738 992	3.18	(− 5) 4.5782 082	3.68	(− 6) 1.4827 416
2.19	(− 3) 9.3225 623	2.69	(− 4) 8.1256 247	3.19	(− 5) 4.2956 707	3.69	(− 6) 1.3773 933
2.20	(− 3) 8.9221 551	2.70	(− 4) 7.6992 476	3.20	(− 5) 4.0297 636	3.70	(− 6) 1.2792 741
2.21	(− 3) 8.5372 378	2.71	(− 4) 7.2937 850	3.21	(− 5) 3.7795 604	3.71	(− 6) 1.1879 068
2.22	(− 3) 8.1672 930	2.72	(− 4) 6.9082 932	3.22	(− 5) 3.5441 831	3.72	(− 6) 1.1028 445
2.23	(− 3) 7.8118 164	2.73	(− 4) 6.5418 671	3.23	(− 5) 3.3227 997	3.73	(− 6) 1.0236 686
2.24	(− 3) 7.4703 176	2.74	(− 4) 6.1936 378	3.24	(− 5) 3.1146 217	3.74	(− 7) 9.4998 679
2.25	(− 3) 7.1423 190	2.75	(− 4) 5.8627 725	3.25	(− 5) 2.9189 025	3.75	(− 7) 8.8143 219
2.26	(− 3) 6.8273 562	2.76	(− 4) 5.5484 722	3.26	(− 5) 2.7349 351	3.76	(− 7) 8.1766 120
2.27	(− 3) 6.5249 776	2.77	(− 4) 5.2499 713	3.27	(− 5) 2.5620 500	3.77	(− 7) 7.5835 232
2.28	(− 3) 6.2347 440	2.78	(− 4) 4.9665 360	3.28	(− 5) 2.3996 135	3.78	(− 7) 7.0320 473
2.29	(− 3) 5.9562 287	2.79	(− 4) 4.6974 632	3.29	(− 5) 2.2470 263	3.79	(− 7) 6.5193 709
2.30	(− 3) 5.6890 172	2.80	(− 4) 4.4420 794	3.30	(− 5) 2.1037 210	3.80	(− 7) 6.0428 629
2.31	(− 3) 5.4327 069	2.81	(− 4) 4.1997 400	3.31	(− 5) 1.9691 613	3.81	(− 7) 5.6000 632
2.32	(− 3) 5.1869 067	2.82	(− 4) 3.9698 274	3.32	(− 5) 1.8428 397	3.82	(− 7) 5.1886 725
2.33	(− 3) 4.9512 374	2.83	(− 4) 3.7517 508	3.33	(− 5) 1.7242 768	3.83	(− 7) 4.8065 419
2.34	(− 3) 4.7253 306	2.84	(− 4) 3.5449 449	3.34	(− 5) 1.6130 192	3.84	(− 7) 4.4516 637
2.35	(− 3) 4.5088 292	2.85	(− 4) 3.3488 688	3.35	(− 5) 1.5086 387	3.85	(− 7) 4.1221 624
2.36	(− 3) 4.3013 869	2.86	(− 4) 3.1630 053	3.36	(− 5) 1.4107 306	3.86	(− 7) 3.8162 867
2.37	(− 3) 4.1026 681	2.87	(− 4) 2.9868 598	3.37	(− 5) 1.3189 127	3.87	(− 7) 3.5324 013
2.38	(− 3) 3.9123 473	2.88	(− 4) 2.8199 597	3.38	(− 5) 1.2328 243	3.88	(− 7) 3.2689 796
2.39	(− 3) 3.7301 092	2.89	(− 4) 2.6618 533	3.39	(− 5) 1.1521 246	3.89	(− 7) 3.0245 971
2.40	(− 3) 3.5556 487	2.90	(− 4) 2.5121 089	3.40	(− 5) 1.0764 921	3.90	(− 7) 2.7979 245
2.41	(− 3) 3.3886 700	2.91	(− 4) 2.3703 144	3.41	(− 5) 1.0056 235	3.91	(− 7) 2.5877 218
2.42	(− 3) 3.2288 871	2.92	(− 4) 2.2360 761	3.42	(− 6) 9.3923 243	3.92	(− 7) 2.3928 327
2.43	(− 3) 3.0760 230	2.93	(− 4) 2.1090 184	3.43	(− 6) 8.7704 910	3.93	(− 7) 2.2121 788
2.44	(− 3) 2.9298 098	2.94	(− 4) 1.9887 824	3.44	(− 6) 8.1881 894	3.94	(− 7) 2.0447 548
2.45	(− 3) 2.7899 886	2.95	(− 4) 1.8750 262	3.45	(− 6) 7.6430 199	3.95	(− 7) 1.8896 240
2.46	(− 3) 2.6563 089	2.96	(− 4) 1.7674 231	3.46	(− 6) 7.1327 211	3.96	(− 7) 1.7459 135
2.47	(− 3) 2.5285 285	2.97	(− 4) 1.6656 619	3.47	(− 6) 6.6551 620	3.97	(− 7) 1.6128 098
2.48	(− 3) 2.4064 136	2.98	(− 4) 1.5694 459	3.48	(− 6) 6.2083 353	3.98	(− 7) 1.4895 557
2.49	(− 3) 2.2897 383	2.99	(− 4) 1.4784 919	3.49	(− 6) 5.7903 503	3.99	(− 7) 1.3754 458
2.50	(− 3) 2.1782 842	3.00	(− 4) 1.3925 305	3.50	(− 6) 5.3994 268	4.00	(− 7) 1.2698 235

$$\frac{\sqrt{\pi}}{2} = 0.88622\ 69255$$

DERIVATIVE OF THE ERROR FUNCTION

Table 7.2

x	$\frac{2}{\sqrt{\pi}}e^{-x^2}$	x	$\frac{2}{\sqrt{\pi}}e^{-x^2}$	x	$\frac{2}{\sqrt{\pi}}e^{-x^2}$	x	$\frac{2}{\sqrt{\pi}}e^{-x^2}$
4.00	(− 7)1.2698 235	4.50	(− 9)1.8113 059	5.00	(−11)1.5670 866	5.50	(−14)8.2233 160
4.01	(− 7)1.1720 776	4.51	(− 9)1.6552 434	5.01	(−11)1.4178 169	5.51	(−14)7.3659 906
4.02	(− 7)1.0816 394	4.52	(− 9)1.5123 248	5.02	(−11)1.2825 089	5.52	(−14)6.5967 265
4.03	(− 8)9.9797 993	4.53	(− 9)1.3814 699	5.03	(−11)1.1598 820	5.53	(−14)5.9066 187
4.04	(− 8)9.2060 694	4.54	(− 9)1.2616 849	5.04	(−11)1.0487 702	5.54	(−14)5.2876 480
4.05	(− 8)8.4906 281	4.55	(− 9)1.1520 559	5.05	(−12)9.4811 285	5.55	(−14)4.7325 943
4.06	(− 8)7.8292 207	4.56	(− 9)1.0517 423	5.06	(−12)8.5694 483	5.56	(−14)4.2349 585
4.07	(− 8)7.2178 923	4.57	(−10)9.5997 127	5.07	(−12)7.7438 839	5.57	(−14)3.7888 917
4.08	(− 8)6.6529 674	4.58	(−10)8.7603 264	5.08	(−12)6.9964 533	5.58	(−14)3.3891 310
4.09	(− 8)6.1310 313	4.59	(−10)7.9927 363	5.09	(−12)6.3198 998	5.59	(−14)3.0309 422
4.10	(− 8)5.6489 121	4.60	(−10)7.2909 450	5.10	(−12)5.7076 270	5.60	(−14)2.7100 675
4.11	(− 8)5.2036 639	4.61	(−10)6.6494 435	5.11	(−12)5.1536 405	5.61	(−14)2.4226 780
4.12	(− 8)4.7925 517	4.62	(−10)6.0631 724	5.12	(−12)4.6524 937	5.62	(−14)2.1653 317
4.13	(− 8)4.4130 364	4.63	(−10)5.5274 864	5.13	(−12)4.1992 391	5.63	(−14)1.9349 346
4.14	(− 8)4.0627 618	4.64	(−10)5.0381 209	5.14	(−12)3.7893 835	5.64	(−14)1.7287 067
4.15	(− 8)3.7395 414	4.65	(−10)4.5911 621	5.15	(−12)3.4188 470	5.65	(−14)1.5441 499
4.16	(− 8)3.4413 471	4.66	(−10)4.1830 187	5.16	(−12)3.0839 257	5.66	(−14)1.3790 206
4.17	(− 8)3.1662 977	4.67	(−10)3.8103 962	5.17	(−12)2.7812 580	5.67	(−14)1.2313 037
4.18	(− 8)2.9126 490	4.68	(−10)3.4702 727	5.18	(−12)2.5077 937	5.68	(−14)1.0991 900
4.19	(− 8)2.6787 841	4.69	(−10)3.1598 772	5.19	(−12)2.2607 652	5.69	(−15)9.8105 529
4.20	(− 8)2.4632 041	4.70	(−10)2.8766 694	5.20	(−12)2.0376 626	5.70	(−15)8.7544 193
4.21	(− 8)2.2645 204	4.71	(−10)2.6183 207	5.21	(−12)1.8362 094	5.71	(−15)7.8104 192
4.22	(− 8)2.0814 463	4.72	(−10)2.3826 973	5.22	(−12)1.6543 420	5.72	(−15)6.9668 183
4.23	(− 8)1.9127 901	4.73	(−10)2.1678 441	5.23	(−12)1.4901 896	5.73	(−15)6.2130 917
4.24	(− 8)1.7574 484	4.74	(−10)1.9719 702	5.24	(−12)1.3420 568	5.74	(−15)5.5398 013
4.25	(− 8)1.6143 994	4.75	(−10)1.7934 357	5.25	(−12)1.2084 075	5.75	(−15)4.9384 851
4.26	(− 8)1.4826 974	4.76	(−10)1.6307 388	5.26	(−12)1.0878 501	5.76	(−15)4.4015 583
4.27	(− 8)1.3614 673	4.77	(−10)1.4825 049	5.27	(−13)9.7912 433	5.77	(−15)3.9222 232
4.28	(− 8)1.2498 993	4.78	(−10)1.3474 759	5.28	(−13)8.8108 899	5.78	(−15)3.4943 893
4.29	(− 8)1.1472 445	4.79	(−10)1.2245 007	5.29	(−13)7.9271 093	5.79	(−15)3.1126 008
4.30	(− 8)1.0528 102	4.80	(−10)1.1125 261	5.30	(−13)7.1305 505	5.80	(−15)2.7719 710
4.31	(− 9)9.6595 598	4.81	(−10)1.0105 888	5.31	(−13)6.4127 516	5.81	(−15)2.4681 247
4.32	(− 9)8.8608 977	4.82	(−11)9.1780 821	5.32	(−13)5.7660 568	5.82	(−15)2.1971 447
4.33	(− 9)8.1266 442	4.83	(−11)8.3337 894	5.33	(−13)5.1835 412	5.83	(−15)1.9555 249
4.34	(− 9)7.4517 438	4.84	(−11)7.5656 500	5.34	(−13)4.6589 423	5.84	(−15)1.7401 279
4.35	(− 9)6.8315 260	4.85	(−11)6.8669 377	5.35	(−13)4.1865 979	5.85	(−15)1.5481 468
4.36	(− 9)6.2616 772	4.86	(−11)6.2315 074	5.36	(−13)3.7613 895	5.86	(−15)1.3770 708
4.37	(− 9)5.7382 144	4.87	(−11)5.6537 456	5.37	(−13)3.3786 913	5.87	(−15)1.2246 543
4.38	(− 9)5.2574 603	4.88	(−11)5.1285 259	5.38	(−13)3.0343 233	5.88	(−15)1.0888 898
4.39	(− 9)4.8160 210	4.89	(−11)4.6511 675	5.39	(−13)2.7245 096	5.89	(−16)9.6798 241
4.40	(− 9)4.4107 647	4.90	(−11)4.2173 976	5.40	(−13)2.4458 396	5.90	(−16)8.6032 817
4.41	(− 9)4.0388 018	4.91	(−11)3.8233 166	5.41	(−13)2.1952 336	5.91	(−16)7.6449 380
4.42	(− 9)3.6974 673	4.92	(−11)3.4653 660	5.42	(−13)1.9699 112	5.92	(−16)6.7919 883
4.43	(− 9)3.3843 033	4.93	(−11)3.1402 998	5.43	(−13)1.7673 627	5.93	(−16)6.0329 959
4.44	(− 9)3.0970 439	4.94	(−11)2.8451 570	5.44	(−13)1.5853 234	5.94	(−16)5.3577 479
4.45	(− 9)2.8336 002	4.95	(−11)2.5772 379	5.45	(−13)1.4217 499	5.95	(−16)4.7571 261
4.46	(− 9)2.5920 474	4.96	(−11)2.3340 811	5.46	(−13)1.2747 989	5.96	(−16)4.2229 913
4.47	(− 9)2.3706 118	4.97	(−11)2.1134 428	5.47	(−13)1.1428 081	5.97	(−16)3.7480 801
4.48	(− 9)2.1676 596	4.98	(−11)1.9132 785	5.48	(−13)1.0242 785	5.98	(−16)3.3259 113
4.49	(− 9)1.9816 862	4.99	(−11)1.7317 254	5.49	(−14)9.1785 895	5.99	(−16)2.9507 038
4.50	(− 9)1.8113 059	5.00	(−11)1.5670 866	5.50	(−14)8.2233 160	6.00	(−16)2.6173 012

$$\frac{\sqrt{\pi}}{2} = 0.88622\ 69255$$

Table 7.2   **DERIVATIVE OF THE ERROR FUNCTION**

x	$\frac{2}{\sqrt{\pi}}e^{-x^2}$	x	$\frac{2}{\sqrt{\pi}}e^{-x^2}$	x	$\frac{2}{\sqrt{\pi}}e^{-x^2}$	x	$\frac{2}{\sqrt{\pi}}e^{-x^2}$
6.00	(−16) 2.6173 012	6.50	(−19) 5.0525 800	7.00	(−22) 5.9159 630	7.50	(−25) 4.2013 654
6.01	(−16) 2.3211 058	6.51	(−19) 4.4362 038	7.01	(−22) 5.1425 768	7.51	(−25) 3.6157 871
6.02	(−16) 2.0580 187	6.52	(−19) 3.8942 418	7.02	(−22) 4.4694 005	7.52	(−25) 3.1112 033
6.03	(−16) 1.8243 864	6.53	(−19) 3.4178 066	7.03	(−22) 3.8835 679	7.53	(−25) 2.6764 989
6.04	(−16) 1.6169 533	6.54	(−19) 2.9990 603	7.04	(−22) 3.3738 492	7.54	(−25) 2.3020 719
6.05	(−16) 1.4328 188	6.55	(−19) 2.6310 921	7.05	(−22) 2.9304 450	7.55	(−25) 1.9796 292
6.06	(−16) 1.2693 992	6.56	(−19) 2.3078 100	7.06	(−22) 2.5448 057	7.56	(−25) 1.7020 094
6.07	(−16) 1.1243 934	6.57	(−19) 2.0238 447	7.07	(−22) 2.2094 736	7.57	(−25) 1.4630 299
6.08	(−17) 9.9575 277	6.58	(−19) 1.7744 651	7.08	(−22) 1.9179 450	7.58	(−25) 1.2573 541
6.09	(−17) 8.8165 340	6.59	(−19) 1.5555 031	7.09	(−22) 1.6645 491	7.59	(−25) 1.0803 765
6.10	(−17) 7.8047 211	6.60	(−19) 1.3632 874	7.10	(−22) 1.4443 426	7.60	(−26) 9.2812 353
6.11	(−17) 6.9076 453	6.61	(−19) 1.1945 852	7.11	(−22) 1.2530 171	7.61	(−26) 7.9716 752
6.12	(−17) 6.1124 570	6.62	(−19) 1.0465 500	7.12	(−22) 1.0868 181	7.62	(−26) 6.8455 216
6.13	(−17) 5.4077 268	6.63	(−20) 9.1667 618	7.13	(−23) 9.4247 516	7.63	(−26) 5.8772 834
6.14	(−17) 4.7832 911	6.64	(−20) 8.0275 879	7.14	(−23) 8.1713 928	7.64	(−26) 5.0449 849
6.15	(−17) 4.2301 135	6.65	(−20) 7.0285 758	7.15	(−23) 7.0832 963	7.65	(−26) 4.3296 844
6.16	(−17) 3.7401 616	6.66	(−20) 6.1526 575	7.16	(−23) 6.1388 620	7.66	(−26) 3.7150 594
6.17	(−17) 3.3062 970	6.67	(−20) 5.3848 212	7.17	(−23) 5.3192 876	7.67	(−26) 3.1870 466
6.18	(−17) 2.9221 768	6.68	(−20) 4.7118 664	7.18	(−23) 4.6082 095	7.68	(−26) 2.7335 323
6.19	(−17) 2.5821 666	6.69	(−20) 4.1221 880	7.19	(−23) 3.9913 893	7.69	(−26) 2.3440 839
6.20	(−17) 2.2812 620	6.70	(−20) 3.6055 852	7.20	(−23) 3.4564 408	7.70	(−26) 2.0097 185
6.21	(−17) 2.0150 194	6.71	(−20) 3.1530 937	7.21	(−23) 2.9925 904	7.71	(−26) 1.7227 031
6.22	(−17) 1.7794 936	6.72	(−20) 2.7568 372	7.22	(−23) 2.5904 701	7.72	(−26) 1.4763 822
6.23	(−17) 1.5711 830	6.73	(−20) 2.4098 972	7.23	(−23) 2.2419 351	7.73	(−26) 1.2650 285
6.24	(−17) 1.3869 801	6.74	(−20) 2.1061 973	7.24	(−23) 1.9399 057	7.74	(−26) 1.0837 147
6.25	(−17) 1.2241 281	6.75	(−20) 1.8404 021	7.25	(−23) 1.6782 295	7.75	(−27) 9.2820 251
6.26	(−17) 1.0801 812	6.76	(−20) 1.6078 278	7.26	(−23) 1.4515 608	7.76	(−27) 7.9484 723
6.27	(−18) 9.5297 064	6.77	(−20) 1.4043 634	7.27	(−23) 1.2552 558	7.77	(−27) 6.8051 505
6.28	(−18) 8.4057 325	6.78	(−20) 1.2264 013	7.28	(−23) 1.0852 815	7.78	(−27) 5.8251 209
6.29	(−18) 7.4128 421	6.79	(−20) 1.0707 765	7.29	(−24) 9.3813 574	7.79	(−27) 4.9852 310
6.30	(−18) 6.5359 252	6.80	(−21) 9.3471 286	7.30	(−24) 8.1077 830	7.80	(−27) 4.2655 868
6.31	(−18) 5.7615 925	6.81	(−21) 8.1577 565	7.31	(−24) 7.0057 026	7.81	(−27) 3.6490 970
6.32	(−18) 5.0779 819	6.82	(−21) 7.1183 018	7.32	(−24) 6.0522 159	7.82	(−27) 3.1210 820
6.33	(−18) 4.4745 863	6.83	(−21) 6.2100 515	7.33	(−24) 5.2274 546	7.83	(−27) 2.6689 356
6.34	(−18) 3.9421 013	6.84	(−21) 5.4166 048	7.34	(−24) 4.5141 841	7.84	(−27) 2.2818 346
6.35	(−18) 3.4722 886	6.85	(−21) 4.7235 904	7.35	(−24) 3.8974 577	7.85	(−27) 1.9504 883
6.36	(−18) 3.0578 557	6.86	(−21) 4.1184 183	7.36	(−24) 3.3643 153	7.86	(−27) 1.6669 236
6.37	(−18) 2.6923 486	6.87	(−21) 3.5900 610	7.37	(−24) 2.9035 220	7.87	(−27) 1.4242 990
6.38	(−18) 2.3700 568	6.88	(−21) 3.1288 615	7.38	(−24) 2.5053 400	7.88	(−27) 1.2167 456
6.39	(−18) 2.0859 281	6.89	(−21) 2.7263 649	7.39	(−24) 2.1613 315	7.89	(−27) 1.0392 297
6.40	(−18) 1.8354 945	6.90	(−21) 2.3751 704	7.40	(−24) 1.8641 859	7.90	(−28) 8.8743 478
6.41	(−18) 1.6148 045	6.91	(−21) 2.0688 010	7.41	(−24) 1.6075 712	7.91	(−28) 7.5766 022
6.42	(−18) 1.4203 650	6.92	(−21) 1.8015 892	7.42	(−24) 1.3860 036	7.92	(−28) 6.4673 396
6.43	(−18) 1.2490 883	6.93	(−21) 1.5685 776	7.43	(−24) 1.1947 351	7.93	(−28) 5.5193 762
6.44	(−18) 1.0982 455	6.94	(−21) 1.3654 297	7.44	(−24) 1.0296 557	7.94	(−28) 4.7094 204
6.45	(−19) 9.6542 574	6.95	(−21) 1.1883 540	7.45	(−25) 8.8720 826	7.95	(−28) 4.0175 202
6.46	(−19) 8.4849 924	6.96	(−21) 1.0340 356	7.46	(−25) 7.6431 480	7.96	(−28) 3.4265 874
6.47	(−19) 7.4558 503	6.97	(−22) 8.9957 684	7.47	(−25) 6.5831 250	7.97	(−28) 2.9219 899
6.48	(−19) 6.5502 224	6.98	(−22) 7.8244 565	7.48	(−25) 5.6689 820	7.98	(−28) 2.4912 008
6.49	(−19) 5.7534 461	6.99	(−22) 6.8042 967	7.49	(−25) 4.8808 021	7.99	(−28) 2.1234 982
6.50	(−19) 5.0525 800	7.00	(−22) 5.9159 630	7.50	(−25) 4.2013 654	8.00	(−28) 1.8097 068

$$\frac{\sqrt{\pi}}{2} = 0.88622\ 69255$$

DERIVATIVE OF THE ERROR FUNCTION

Table 7.2

x	$\frac{2}{\sqrt{\pi}}e^{-x^2}$	x	$\frac{2}{\sqrt{\pi}}e^{-x^2}$	x	$\frac{2}{\sqrt{\pi}}e^{-x^2}$	x	$\frac{2}{\sqrt{\pi}}e^{-x^2}$
8.00	(−28)1.8097 068	8.50	(−32)4.7280 139	9.00	(−36)7.4920 734	9.50	(−40)7.2007 555
8.01	(−28)1.5419 762	8.51	(−32)3.9884 601	9.01	(−36)6.2572 800	9.51	(−40)5.9541 351
8.02	(−28)1.3135 913	8.52	(−32)3.3639 141	9.02	(−36)5.2249 519	9.52	(−40)4.9223 495
8.03	(−28)1.1188 091	8.53	(−32)2.8365 973	9.03	(−36)4.3620 651	9.53	(−40)4.0685 471
8.04	(−29)9.5271 911	8.54	(−32)2.3914 628	9.04	(−36)3.6409 535	9.54	(−40)3.3621 678
8.05	(−29)8.1112 334	8.55	(−32)2.0157 780	9.05	(−36)3.0384 441	9.55	(−40)2.7778 742
8.06	(−29)6.9043 382	8.56	(−32)1.6987 713	9.06	(−36)2.5351 317	9.56	(−40)2.2946 629
8.07	(−29)5.8758 453	8.57	(−32)1.4313 316	9.07	(−36)2.1147 690	9.57	(−40)1.8951 272
8.08	(−29)4.9995 601	8.58	(−32)1.2057 541	9.08	(−36)1.7637 559	9.58	(−40)1.5648 437
8.09	(−29)4.2531 077	8.59	(−32)1.0155 245	9.09	(−36)1.4707 105	9.59	(−40)1.2918 638
8.10	(−29)3.6173 797	8.60	(−33)8.5513 598	9.10	(−36)1.2261 088	9.60	(−40)1.0662 907
8.11	(−29)3.0760 612	8.61	(−33)7.1993 468	9.11	(−36)1.0219 837	9.61	(−41)8.7992 901
8.12	(−29)2.6152 245	8.62	(−33)6.0598 819	9.12	(−37)8.5167 148	9.62	(−41)7.2599 363
8.13	(−29)2.2229 829	8.63	(−33)5.0997 438	9.13	(−37)7.0959 960	9.63	(−41)5.9886 802
8.14	(−29)1.8891 933	8.64	(−33)4.2908 734	9.14	(−37)5.9110 925	9.64	(−41)4.9390 403
8.15	(−29)1.6052 025	8.65	(−33)3.6095 760	9.15	(−37)4.9230 619	9.65	(−41)4.0725 570
8.16	(−29)1.3636 296	8.66	(−33)3.0358 465	9.16	(−37)4.0993 592	9.66	(−41)3.3574 141
8.17	(−29)1.1581 801	8.67	(−33)2.5527 988	9.17	(−37)3.4127 918	9.67	(−41)2.7672 971
8.18	(−30)9.8348 778	8.68	(−33)2.1461 817	9.18	(−37)2.8406 437	9.68	(−41)2.2804 460
8.19	(−30)8.3497 786	8.69	(−33)1.8039 709	9.19	(−37)2.3639 423	9.69	(−41)1.8788 710
8.20	(−30)7.0875 167	8.70	(−33)1.5160 228	9.20	(−37)1.9668 449	9.70	(−41)1.5477 017
8.21	(−30)6.0148 717	8.71	(−33)1.2737 818	9.21	(−37)1.6361 251	9.71	(−41)1.2746 493
8.22	(−30)5.1035 431	8.72	(−33)1.0700 339	9.22	(−37)1.3607 427	9.72	(−41)1.0495 600
8.23	(−30)4.3294 262	8.73	(−34)8.9869 668	9.23	(−37)1.1314 847	9.73	(−42)8.6404 628
8.24	(−30)3.6719 947	8.74	(−34)7.5464 360	9.24	(−38)9.4066 395	9.74	(−42)7.1118 055
8.25	(−30)3.1137 725	8.75	(−34)6.3355 422	9.25	(−38)7.8186 802	9.75	(−42)5.8524 252
8.26	(−30)2.6398 841	8.76	(−34)5.3178 836	9.26	(−38)6.4974 888	9.76	(−42)4.8150 968
8.27	(−30)2.2376 697	8.77	(−34)4.4627 957	9.27	(−38)5.3984 710	9.77	(−42)3.9608 401
8.28	(−30)1.8963 577	8.78	(−34)3.7444 525	9.28	(−38)4.4844 496	9.78	(−42)3.2574 873
8.29	(−30)1.6067 846	8.79	(−34)3.1411 074	9.29	(−38)3.7244 373	9.79	(−42)2.6784 979
8.30	(−30)1.3611 569	8.80	(−34)2.6344 525	9.30	(−38)3.0926 112	9.80	(−42)2.2019 782
8.31	(−30)1.1528 476	8.81	(−34)2.2090 784	9.31	(−38)2.5674 566	9.81	(−42)1.8098 720
8.32	(−31)9.7622 228	8.82	(−34)1.8520 172	9.32	(−38)2.1310 520	9.82	(−42)1.4872 907
8.33	(−31)8.2649 206	8.83	(−34)1.5523 585	9.33	(−38)1.7684 718	9.83	(−42)1.2219 600
8.34	(−31)6.9958 710	8.84	(−34)1.3009 248	9.34	(−38)1.4672 880	9.84	(−42)1.0037 632
8.35	(−31)5.9204 954	8.85	(−34)1.0899 975	9.35	(−38)1.2171 545	9.85	(−43)8.2436 338
8.36	(−31)5.0094 199	8.86	(−35)9.1308 655	9.36	(−38)1.0094 602	9.86	(−43)6.7689 179
8.37	(−31)4.2376 977	8.87	(−35)7.6473 600	9.37	(−39)8.3703 932	9.87	(−43)5.5569 047
8.38	(−31)3.5841 456	8.88	(−35)6.4036 010	9.38	(−39)6.9392 997	9.88	(−43)4.5609 970
8.39	(−31)3.0307 803	8.89	(−35)5.3610 534	9.39	(−39)5.7517 311	9.89	(−43)3.7428 271
8.40	(−31)2.5623 380	8.90	(−35)4.4873 418	9.40	(−39)4.7664 456	9.90	(−43)3.0708 096
8.41	(−31)2.1658 657	8.91	(−35)3.7552 711	9.41	(−39)3.9491 520	9.91	(−43)2.5189 477
8.42	(−31)1.8303 736	8.92	(−35)3.1420 030	9.42	(−39)3.2713 439	9.92	(−43)2.0658 489
8.43	(−31)1.5465 399	8.93	(−35)2.6283 611	9.43	(−39)2.7093 286	9.93	(−43)1.6939 130
8.44	(−31)1.3064 586	8.94	(−35)2.1982 476	9.44	(−39)2.2434 186	9.94	(−43)1.3886 628
8.45	(−31)1.1034 263	8.95	(−35)1.8381 516	9.45	(−39)1.8572 574	9.95	(−43)1.1381 922
8.46	(−32)9.3176 012	8.96	(−35)1.5367 357	9.46	(−39)1.5372 589	9.96	(−44)9.3271 204
8.47	(−32)7.8664 369	8.97	(−35)1.2844 884	9.47	(−39)1.2721 404	9.97	(−44)7.6417 477
8.48	(−32)6.6399 552	8.98	(−35)1.0734 315	9.48	(−39)1.0525 343	9.98	(−44)6.2596 629
8.49	(−32)5.6035 774	8.99	(−36)8.9687 435	9.49	(−40)8.7066 400	9.99	(−44)5.1265 162
8.50	(−32)4.7280 139	9.00	(−36)7.4920 734	9.50	(−40)7.2007 555	10.00	(−44)4.1976 562

$$\frac{\sqrt{\pi}}{2} = 0.88622\ 69255$$

Table 7.3   **COMPLEMENTARY ERROR FUNCTION**

x^{-2}	$xe^{x^2} \operatorname{erfc} x$	$\langle x \rangle$	x^{-2}	$xe^{x^2} \operatorname{erfc} x$	$\langle x \rangle$
0.250	0.51079 14	2	0.125	0.53406 72	3
0.245	0.51163 07	2	0.120	0.53511 47	3
0.240	0.51247 67	2	0.115	0.53617 29	3
0.235	0.51332 94	2	0.110	0.53724 20	3
0.230	0.51418 90	2	0.105	0.53832 23	3
0.225	0.51505 55	2	0.100	0.53941 41	3
0.220	0.51592 92	2	0.095	0.54051 76	3
0.215	0.51681 01	2	0.090	0.54163 32	3
0.210	0.51769 83	2	0.085	0.54276 11	3
0.205	0.51859 40	2	0.080	0.54390 16	4
0.200	0.51949 74	2	0.075	0.54505 51	4
0.195	0.52040 85	2	0.070	0.54622 19	4
0.190	0.52132 75	2	0.065	0.54740 24	4
0.185	0.52225 45	2	0.060	0.54859 69	4
0.180	0.52318 98	2	0.055	0.54980 58	4
0.175	0.52413 33	2	0.050	0.55102 95	4
0.170	0.52508 55	2	0.045	0.55226 85	5
0.165	0.52604 63	2	0.040	0.55352 32	5
0.160	0.52701 59	3	0.035	0.55479 41	5
0.155	0.52799 46	3	0.030	0.55608 17	6
0.150	0.52898 25	3	0.025	0.55738 65	6
0.145	0.52997 98	3	0.020	0.55870 90	7
0.140	0.53098 67	3	0.015	0.56005 00	8
0.135	0.53200 35	3	0.010	0.56140 99	10
0.130	0.53303 02	3	0.005	0.56278 96	14
0.125	0.53406 72	3	0.000	0.56418 96	∞

$$\begin{bmatrix} (-6)1 \\ 3 \end{bmatrix} \qquad\qquad\qquad \begin{bmatrix} (-6)3 \\ 3 \end{bmatrix}$$

See **Example 2.**

$$\langle x \rangle = \text{nearest integer to } x.$$

n	$\operatorname{erfc} \sqrt{n\pi}$	n	$\operatorname{erfc} \sqrt{n\pi}$
1	0.01218 88821 84803	6	0.00000 00008 25422
2	0.00039 27505 88282	7	0.00000 00000 33136
3	0.00001 41444 02689	8	0.00000 00000 01343
4	0.00000 05351 64662	9	0.00000 00000 00055
5	0.00000 00208 26552	10	0.00000 00000 00002

$$\operatorname{erfc} x = \frac{2}{\sqrt{\pi}} \int_x^\infty e^{-t^2} dt = 1 - \operatorname{erf} x$$

$\operatorname{erfc} \sqrt{n\pi}$ compiled from O. Emersleben, Numerische Werte des Fehlerintegrals für $\sqrt{n\pi}$, Z. Angew. Math. Mech. **31**, 393–394, 1951 (with permission).

REPEATED INTEGRALS OF THE ERROR FUNCTION

Table 7.4

$$2^n \Gamma\left(\frac{n}{2}+1\right) i^n \operatorname{erfc} x$$

x	$n=1$	$n=2$	$n=3$	$n=4$
0.0	1.00000	1.00000	1.00000	1.00000
0.1	(− 1)8.32738	(− 1)7.93573	(− 1)7.62409	(− 1)7.36220
0.2	(− 1)6.85245	(− 1)6.22654	(− 1)5.74882	(− 1)5.36163
0.3	(− 1)5.56938	(− 1)4.82842	(− 1)4.28565	(− 1)3.86125
0.4	(− 1)4.46884	(− 1)3.69906	(− 1)3.15756	(− 1)2.74894
0.5	(− 1)3.53855	(− 1)2.79859	(− 1)2.29846	(− 1)1.93408
0.6	(− 1)2.76388	(− 1)2.09021	(− 1)1.65244	(− 1)1.34438
0.7	(− 1)2.12869	(− 1)1.54061	(− 1)1.17295	(− 2)9.22962
0.8	(− 1)1.61601	(− 1)1.12021	(− 2)8.21802	(− 2)6.25650
0.9	(− 1)1.20884	(− 2)8.03288	(− 2)5.68138	(− 2)4.18643
1.0	(− 2)8.90739	(− 2)5.67901	(− 2)3.87449	(− 2)2.76442
1.1	(− 2)6.46332	(− 2)3.95711	(− 2)2.60573	(− 2)1.80092
1.2	(− 2)4.61706	(− 2)2.71686	(− 2)1.72776	(− 2)1.15720
1.3	(− 2)3.24613	(− 2)1.83748	(− 2)1.12918	(− 3)7.33229
1.4	(− 2)2.24570	(− 2)1.22388	(− 3)7.27211	(− 3)4.58017
1.5	(− 2)1.52836	(− 3)8.02626	(− 3)4.61400	(− 3)2.81992
1.6	(− 2)1.02305	(− 3)5.18140	(− 3)2.88347	(− 3)1.71085
1.7	(− 3)6.73408	(− 3)3.29192	(− 3)1.77452	(− 3)1.02261
1.8	(− 3)4.35805	(− 3)2.05795	(− 3)1.07519	(− 4)6.02074
1.9	(− 3)2.77245	(− 3)1.26566	(− 4)6.41281	(− 4)3.49094
2.0	(− 3)1.73350	(− 4)7.65644	(− 4)3.76431	(− 4)1.99301
2.1	(− 3)1.06515	(− 4)4.55498	(− 4)2.17431	(− 4)1.12014
2.2	(− 4)6.43074	(− 4)2.66457	(− 4)1.23562	(− 5)6.19670
2.3	(− 4)3.81436	(− 4)1.53245	(− 5)6.90731	(− 5)3.37364
2.4	(− 4)2.22250	(− 5)8.66372	(− 5)3.79773	(− 5)1.80727
2.5	(− 4)1.27195	(− 5)4.81417	(− 5)2.05339	(− 6)9.52500
2.6	(− 5)7.14929	(− 5)2.62896	(− 5)1.09167	(− 6)4.93818
2.7	(− 5)3.94619	(− 5)1.41072	(− 6)5.70591	(− 6)2.51807
2.8	(− 5)2.13882	(− 6)7.43784	(− 6)2.93172	(− 6)1.26274
2.9	(− 5)1.13820	(− 6)3.85260	(− 6)1.48058	(− 7)6.22654
3.0	(− 6)5.94664	(− 6)1.96029	(− 7)7.34867	(− 7)3.01870
3.1	(− 6)3.05003	(− 7)9.79725	(− 7)3.58429	(− 7)1.43874
3.2	(− 6)1.53562	(− 7)4.80916	(− 7)1.71780	(− 8)6.74044
3.3	(− 7)7.58899	(− 7)2.31835	(− 8)8.08871	(− 8)3.10379
3.4	(− 7)3.68109	(− 7)1.09748	(− 8)3.74180	(− 8)1.40460
3.5	(− 7)1.75241	(− 8)5.10148	(− 8)1.70036	(− 9)6.24636
3.6	(− 8)8.18726	(− 8)2.32831	(− 9)7.58967	(− 9)2.72947
3.7	(− 8)3.75373	(− 8)1.04329	(− 9)3.32733	(− 9)1.17184
3.8	(− 8)1.68883	(− 9)4.58945	(− 9)1.43260	(−10)4.94271
3.9	(− 9)7.45575	(− 9)1.98190	(−10)6.05736	(−10)2.04800
4.0	(− 9)3.22966	(−10)8.40124	(−10)2.51501	(−11)8.33554
4.1	(− 9)1.37267	(−10)3.49560	(−10)1.02533	(−11)3.33230
4.2	(−10)5.72405	(−10)1.42757	(−11)4.10427	(−11)1.30837
4.3	(−10)2.34181	(−11)5.72196	(−11)1.61297	(−12)5.04508
4.4	(−11)9.39929	(−11)2.25085	(−12)6.22316	(−12)1.91041
4.5	(−11)3.70102	(−12)8.68930	(−12)2.35705	(−13)7.10366
4.6	(−11)1.42960	(−12)3.29184	(−13)8.76348	(−13)2.59364
4.7	(−12)5.41708	(−12)1.22375	(−13)3.19826	(−14)9.29786
4.8	(−12)2.01353	(−13)4.46407	(−13)1.14567	(−14)3.27252
4.9	(−13)7.34149	(−13)1.59785	(−14)4.02809	(−14)1.13080
5.0	(−13)2.62561	(−14)5.61169	(−14)1.38998	(−15)3.83592

$$\left[2^n \Gamma\left(\frac{n}{2}+1\right)\right]^{-1}$$

(−1)5.64189 58355	(−1)2.50000 00000	(−2)9.40315 97258	(−2)3.12500

See **Examples 4** and **5**.

Table 7.4   **REPEATED INTEGRALS OF THE ERROR FUNCTION**

$$2^n \Gamma\left(\frac{n}{2}+1\right) i^n \operatorname{erfc} x$$

x	$n=5$	$n=6$	$n=10$	$n=11$
0.0	1.00000	1.00000	1.00000	1.00000
0.1	(− 1)7.13475	(− 1)6.93283	(− 1)6.28971	(− 1)6.15727
0.2	(− 1)5.03608	(− 1)4.75548	(− 1)3.91490	(− 1)3.75188
0.3	(− 1)3.51572	(− 1)3.22652	(− 1)2.41089	(− 1)2.26201
0.4	(− 1)2.42671	(− 1)2.16478	(− 1)1.46861	(− 1)1.34906
0.5	(− 1)1.65569	(− 1)1.43588	(− 2)8.84744	(− 2)7.95749
0.6	(− 1)1.11630	(− 2)9.41309	(− 2)5.27007	(− 2)4.64127
0.7	(− 2)7.43528	(− 2)6.09742	(− 2)3.10323	(− 2)2.67626
0.8	(− 2)4.89121	(− 2)3.90166	(− 2)1.80600	(− 2)1.52533
0.9	(− 2)3.17704	(− 2)2.46567	(− 2)1.03859	(− 3)8.59126
1.0	(− 2)2.03707	(− 2)1.53850	(− 3)5.90062	(− 3)4.78106
1.1	(− 2)1.28901	(− 3)9.47623	(− 3)3.31130	(− 3)2.62835
1.2	(− 3)8.04765	(− 3)5.76033	(− 3)1.83510	(− 3)1.42708
1.3	(− 3)4.95614	(− 3)3.45489	(− 3)1.00415	(− 4)7.65146
1.4	(− 3)3.01008	(− 3)2.04411	(− 4)5.42413	(− 4)4.05030
1.5	(− 3)1.80252	(− 3)1.19278	(− 4)2.89186	(− 4)2.11641
1.6	(− 3)1.06403	(− 4)6.86307	(− 4)1.52145	(− 4)1.09146
1.7	(− 4)6.19032	(− 4)3.89303	(− 5)7.89765	(− 5)5.55435
1.8	(− 4)3.54870	(− 4)2.17663	(− 5)4.04407	(− 5)2.78871
1.9	(− 4)2.00419	(− 4)1.19930	(− 5)2.04244	(− 5)1.38116
2.0	(− 4)1.11492	(− 5)6.51088	(− 5)1.01722	(− 6)6.74666
2.1	(− 5)6.10810	(− 5)3.48211	(− 6)4.99509	(− 6)3.24987
2.2	(− 5)3.29497	(− 5)1.83427	(− 6)2.41807	(− 6)1.54350
2.3	(− 5)1.74988	(− 6)9.51547	(− 6)1.15378	(− 7)7.22681
2.4	(− 6)9.14767	(− 6)4.86044	(− 7)5.42553	(− 7)3.33519
2.5	(− 6)4.70641	(− 6)2.44418	(− 7)2.51397	(− 7)1.51693
2.6	(− 6)2.38278	(− 6)1.20988	(− 7)1.14766	(− 8)6.79864
2.7	(− 6)1.18695	(− 7)5.89435	(− 8)5.16116	(− 8)3.00212
2.8	(− 7)5.81672	(− 7)2.82592	(− 8)2.28612	(− 8)1.30595
2.9	(− 7)2.80391	(− 7)1.33308	(− 9)9.97266	(− 9)5.59577
3.0	(− 7)1.32935	(− 8)6.18684	(− 9)4.28380	(− 9)2.36143
3.1	(− 8)6.19798	(− 8)2.82454	(− 9)1.81176	(−10)9.81330
3.2	(− 8)2.84151	(− 8)1.26835	(−10)7.54345	(−10)4.01541
3.3	(− 8)1.28082	(− 9)5.60145	(−10)3.09165	(−10)1.61759
3.4	(− 9)5.67576	(− 9)2.43265	(−10)1.24712	(−11)6.41479
3.5	(− 9)2.47236	(− 9)1.03880	(−11)4.95086	(−11)2.50393
3.6	(− 9)1.05855	(−10)4.36132	(−11)1.93401	(−12)9.61928
3.7	(−10)4.45435	(−10)1.80009	(−12)7.43354	(−12)3.63661
3.8	(−10)1.84200	(−11)7.30331	(−12)2.81094	(−12)1.35283
3.9	(−11)7.48503	(−11)2.91245	(−12)1.04564	(−13)4.95149
4.0	(−11)2.98854	(−11)1.14149	(−13)3.82601	(−13)1.78294
4.1	(−11)1.17234	(−12)4.39668	(−13)1.37691	(−14)6.31544
4.2	(−12)4.51802	(−12)1.66412	(−14)4.87328	(−14)2.20038
4.3	(−12)1.71044	(−13)6.18894	(−14)1.69612	(−15)7.54020
4.4	(−13)6.36069	(−13)2.26147	(−15)5.80461	(−15)2.54109
4.5	(−13)2.32332	(−14)8.11851	(−15)1.95316	(−16)8.42124
4.6	(−14)8.33482	(−14)2.86315	(−16)6.46126	(−16)2.74419
4.7	(−14)2.93656	(−15)9.91898	(−16)2.10125	(−17)8.79230
4.8	(−14)1.01604	(−15)3.37534	(−17)6.71719	(−17)2.76954
4.9	(−15)3.45215	(−15)1.12815	(−17)2.11065	(−18)8.57626
5.0	(−15)1.15173	(−16)3.70336	(−18)6.51829	(−18)2.61062

$$\left[2^n \Gamma\left(\frac{n}{2}+1\right)\right]^{-1}$$

(−3)9.40315 97258	(−3)2.60416 66667	(−6)8.13802 08333	(−6)1.69609 66316

DAWSON'S INTEGRAL

Table 7.5

x	$e^{-x^2}\int_0^x e^{t^2}dt$	x	$e^{-x^2}\int_0^x e^{t^2}dt$	x^{-2}	$xe^{-x^2}\int_0^x e^{t^2}dt$	$\langle x\rangle$
0.00	0.00000 00000	1.00	0.53807 95069	0.250	0.60268 0777	2
0.02	0.01999 46675	1.02	0.53637 44359	0.245	0.60046 6027	2
0.04	0.03995 73606	1.04	0.53431 71471	0.240	0.59819 8606	2
0.06	0.05985 62071	1.06	0.53192 50787	0.235	0.59588 1008	2
0.08	0.07965 95389	1.08	0.52921 57454	0.230	0.59351 6018	2
0.10	0.09933 59924	1.10	0.52620 66800	0.225	0.59110 6724	2
0.12	0.11885 46083	1.12	0.52291 53777	0.220	0.58865 6517	2
0.14	0.13818 49287	1.14	0.51935 92435	0.215	0.58616 9107	2
0.16	0.15729 70920	1.16	0.51555 55409	0.210	0.58364 8516	2
0.18	0.17616 19254	1.18	0.51152 13448	0.205	0.58109 9080	2
0.20	0.19475 10334	1.20	0.50727 34964	0.200	0.57852 5444	2
0.22	0.21303 68833	1.22	0.50282 85611	0.195	0.57593 2550	2
0.24	0.23099 28865	1.24	0.49820 27897	0.190	0.57332 5618	2
0.26	0.24859 34747	1.26	0.49341 20827	0.185	0.57071 0126	2
0.28	0.26581 41727	1.28	0.48847 19572	0.180	0.56809 1778	2
0.30	0.28263 16650	1.30	0.48339 75174	0.175	0.56547 6462	2
0.32	0.29902 38575	1.32	0.47820 34278	0.170	0.56287 0205	2
0.34	0.31496 99336	1.34	0.47290 38898	0.165	0.56027 9114	2
0.36	0.33045 04051	1.36	0.46751 26208	0.160	0.55770 9305	3
0.38	0.34544 71562	1.38	0.46204 28368	0.155	0.55516 6829	3
0.40	0.35994 34819	1.40	0.45650 72375	0.150	0.55265 7582	3
0.42	0.37392 41210	1.42	0.45091 79943	0.145	0.55018 7208	3
0.44	0.38737 52812	1.44	0.44528 67410	0.140	0.54776 0994	3
0.46	0.40028 46599	1.46	0.43962 45670	0.135	0.54538 3766	3
0.48	0.41264 14572	1.48	0.43394 20135	0.130	0.54305 9774	3
0.50	0.42443 63835	1.50	0.42824 90711	0.125	0.54079 2591	3
0.52	0.43566 16609	1.52	0.42255 51804	0.120	0.53858 5013	3
0.54	0.44631 10184	1.54	0.41686 92347	0.115	0.53643 8983	3
0.56	0.45637 96813	1.56	0.41119 95842	0.110	0.53435 5529	3
0.58	0.46586 43551	1.58	0.40555 40424	0.105	0.53233 4747	3
0.60	0.47476 32037	1.60	0.39993 98943	0.100	0.53037 5810	3
0.62	0.48307 58219	1.62	0.39436 39058	0.095	0.52847 7031	3
0.64	0.49080 32040	1.64	0.38883 23346	0.090	0.52663 5967	3
0.66	0.49794 77064	1.66	0.38335 09429	0.085	0.52484 9575	3
0.68	0.50451 30066	1.68	0.37792 50103	0.080	0.52311 4393	4
0.70	0.51050 40576	1.70	0.37255 93490	0.075	0.52142 6749	4
0.72	0.51592 70382	1.72	0.36725 83182	0.070	0.51978 2972	4
0.74	0.52078 93010	1.74	0.36202 58410	0.065	0.51817 9571	4
0.76	0.52509 93152	1.76	0.35686 54206	0.060	0.51661 3369	4
0.78	0.52886 66089	1.78	0.35178 01580	0.055	0.51508 1573	4
0.80	0.53210 17071	1.80	0.34677 27691	0.050	0.51358 1788	4
0.82	0.53481 60684	1.82	0.34184 56029	0.045	0.51211 1971	5
0.84	0.53702 20202	1.84	0.33700 06597	0.040	0.51067 0372	5
0.86	0.53873 26921	1.86	0.33223 96091	0.035	0.50925 5466	5
0.88	0.53996 19480	1.88	0.32756 38080	0.030	0.50786 5903	6
0.90	0.54072 43187	1.90	0.32297 43193	0.025	0.50650 0473	6
0.92	0.54103 49328	1.92	0.31847 19293	0.020	0.50515 8078	7
0.94	0.54090 94485	1.94	0.31405 71655	0.015	0.50383 7717	8
0.96	0.54036 39857	1.96	0.30973 03141	0.010	0.50253 8471	10
0.98	0.53941 50580	1.98	0.30549 14372	0.005	0.50125 9494	14
1.00	0.53807 95069	2.00	0.30134 03889	0.000	0.50000 0000	∞
	$\begin{bmatrix}(-5)7\\4\end{bmatrix}$		$\begin{bmatrix}(-5)4\\4\end{bmatrix}$		$\begin{bmatrix}(-6)8\\6\end{bmatrix}$	

See Example 3. $\langle x\rangle$ = nearest integer to x.

Compiled from J. B. Rosser, Theory and application of $\int_0^z e^{-x^2}dx$ and $\int_0^z e^{-p^2y^2}dy\int_0^y e^{-x^2}dx$. Mapleton House, Brooklyn, N.Y., 1948; and B. Lohmander and S. Rittsten, Table of the function $y=e^{-x^2}\int_0^x e^{t^2}dt$, Kungl. Fysiogr. Sällsk. i Lund Förh. **28**, 45–52, 1958 (with permission).

Table 7.6

$$\frac{3}{\Gamma\left(\frac{1}{3}\right)}\int_0^x e^{-t^3}dt$$

x	$\dfrac{3}{\Gamma\left(\frac{1}{3}\right)}\displaystyle\int_0^x e^{-t^3}dt$	x	$\dfrac{3}{\Gamma\left(\frac{1}{3}\right)}\displaystyle\int_0^x e^{-t^3}dt$	x	$\dfrac{3}{\Gamma\left(\frac{1}{3}\right)}\displaystyle\int_0^x e^{-t^3}dt$
0.00	0.00000 00	0.70	0.72276 69	1.40	0.98973 54
0.02	0.02239 69	0.72	0.73842 49	1.42	0.99109 36
0.04	0.04479 31	0.74	0.75360 34	1.44	0.99229 70
0.06	0.06718 72	0.76	0.76829 12	1.46	0.99335 97
0.08	0.08957 63	0.78	0.78247 88	1.48	0.99429 49
0.10	0.11195 67	0.80	0.79615 78	1.50	0.99511 49
0.12	0.13432 36	0.82	0.80932 16	1.52	0.99583 14
0.14	0.15667 11	0.84	0.82196 48	1.54	0.99645 52
0.16	0.17899 22	0.86	0.83408 41	1.56	0.99699 62
0.18	0.20127 90	0.88	0.84567 73	1.58	0.99746 38
0.20	0.22352 24	0.90	0.85674 42	1.60	0.99786 63
0.22	0.24571 24	0.92	0.86728 62	1.62	0.99821 16
0.24	0.26783 80	0.94	0.87730 62	1.64	0.99850 65
0.26	0.28988 71	0.96	0.88680 89	1.66	0.99875 75
0.28	0.31184 70	0.98	0.89580 05	1.68	0.99897 03
0.30	0.33370 37	1.00	0.90428 86	1.70	0.99914 99
0.32	0.35544 26	1.02	0.91228 25		
0.34	0.37704 82	1.04	0.91979 27		
0.36	0.39850 45	1.06	0.92683 11		
0.38	0.41979 45	1.08	0.93341 06		
0.40	0.44090 07	1.10	0.93954 56	1.70	0.99914 99
0.42	0.46180 52	1.12	0.94525 09	1.74	0.99942 75
0.44	0.48248 96	1.14	0.95054 27	1.78	0.99962 05
0.46	0.50293 51	1.16	0.95543 76	1.82	0.99975 26
0.48	0.52312 25	1.18	0.95995 30	1.86	0.99984 14
0.50	0.54303 28	1.20	0.96410 64	1.90	0.99990 01
0.52	0.56264 66	1.22	0.96791 62	1.94	0.99993 82
0.54	0.58194 46	1.24	0.97140 05	1.98	0.99996 24
0.56	0.60090 80	1.26	0.97457 79	2.02	0.99997 76
0.58	0.61951 78	1.28	0.97746 66	2.06	0.99998 69
0.60	0.63775 57	1.30	0.98008 48	2.10	0.99999 25
0.62	0.65560 39	1.32	0.98245 07	2.14	0.99999 57
0.64	0.67304 52	1.34	0.98458 18	2.18	0.99999 77
0.66	0.69006 30	1.36	0.98649 52	2.22	0.99999 87
0.68	0.70664 18	1.38	0.98820 77	2.26	0.99999 93
0.70	0.72276 69	1.40	0.98973 54	2.30	0.99999 97
	$\begin{bmatrix}(-5)6\\4\end{bmatrix}$		$\begin{bmatrix}(-5)7\\5\end{bmatrix}$		$\begin{bmatrix}(-5)1\\5\end{bmatrix}$

$$\frac{\Gamma\left(\frac{1}{3}\right)}{3}=0.89297\ 95116$$

Compiled from M. Abramowitz, Table of the integral $\int_0^x e^{-u^3}du$, J. Math. Phys. **30**, 162–163, 1951 (with permission).

FRESNEL INTEGRALS

Table 7.7

$$C(x)=\int_0^x \cos\left(\frac{\pi}{2}t^2\right)dt \qquad C_2(u)=\frac{1}{\sqrt{2\pi}}\int_0^u \frac{\cos t}{\sqrt{t}}\,dt=C\left(\sqrt{\frac{2u}{\pi}}\right)$$

$$S(x)=\int_0^x \sin\left(\frac{\pi}{2}t^2\right)dt \qquad S_2(u)=\frac{1}{\sqrt{2\pi}}\int_0^u \frac{\sin t}{\sqrt{t}}\,dt=S\left(\sqrt{\frac{2u}{\pi}}\right)$$

x	$u=\frac{\pi}{2}x^2$	$C(x)=C_2(u)$	$S(x)=S_2(u)$	x	$u=\frac{\pi}{2}x^2$	$C(x)=C_2(u)$	$S(x)=S_2(u)$
0.00	0.00000 00	0.00000 00	0.00000 00	1.00	1.57079 63	0.77989 34	0.43825 91
0.02	0.00062 83	0.02000 00	0.00000 42	1.02	1.63425 65	0.77926 11	0.45824 58
0.04	0.00251 33	0.04000 00	0.00003 35	1.04	1.69897 33	0.77735 01	0.47815 08
0.06	0.00565 49	0.05999 98	0.00011 31	1.06	1.76494 68	0.77414 34	0.49788 84
0.08	0.01005 31	0.07999 92	0.00026 81	1.08	1.83217 68	0.76963 03	0.51736 86
0.10	0.01570 80	0.09999 75	0.00052 36	1.10	1.90066 36	0.76380 67	0.53649 79
0.12	0.02261 95	0.11999 39	0.00090 47	1.12	1.97040 69	0.75667 60	0.55517 92
0.14	0.03078 76	0.13998 67	0.00143 67	1.14	2.04140 69	0.74824 94	0.57331 28
0.16	0.04021 24	0.15997 41	0.00214 44	1.16	2.11366 35	0.73854 68	0.59079 66
0.18	0.05089 38	0.17995 34	0.00305 31	1.18	2.18717 68	0.72759 68	0.60752 74
0.20	0.06283 19	0.19992 11	0.00418 76	1.20	2.26194 67	0.71543 77	0.62340 09
0.22	0.07602 65	0.21987 29	0.00557 30	1.22	2.33797 33	0.70211 76	0.63831 34
0.24	0.09047 79	0.23980 36	0.00723 40	1.24	2.41525 64	0.68769 47	0.65216 19
0.26	0.10618 58	0.25970 70	0.00919 54	1.26	2.49379 62	0.67223 78	0.66484 56
0.28	0.12315 04	0.27957 56	0.01148 16	1.28	2.57359 27	0.65582 63	0.67626 72
0.30	0.14137 17	0.29940 10	0.01411 70	1.30	2.65464 58	0.63855 05	0.68633 33
0.32	0.16084 95	0.31917 31	0.01712 56	1.32	2.73695 55	0.62051 11	0.69495 62
0.34	0.18158 41	0.33888 06	0.02053 11	1.34	2.82052 19	0.60181 95	0.70205 50
0.36	0.20357 52	0.35851 09	0.02435 68	1.36	2.90534 49	0.58259 73	0.70755 67
0.38	0.22682 30	0.37804 96	0.02862 55	1.38	2.99142 45	0.56297 59	0.71139 77
0.40	0.25132 74	0.39748 08	0.03335 94	1.40	3.07876 08	0.54309 58	0.71352 51
0.42	0.27708 85	0.41678 68	0.03858 02	1.42	3.16735 37	0.52310 58	0.71389 77
0.44	0.30410 62	0.43594 82	0.04430 85	1.44	3.25720 33	0.50316 23	0.71248 78
0.46	0.33238 05	0.45494 40	0.05056 42	1.46	3.34830 95	0.48342 80	0.70928 16
0.48	0.36191 15	0.47375 10	0.05736 63	1.48	3.44067 23	0.46407 05	0.70428 12
0.50	0.39269 91	0.49234 42	0.06473 24	1.50	3.53429 17	0.44526 12	0.69750 50
0.52	0.42474 33	0.51069 69	0.07267 89	1.52	3.62916 78	0.42717 32	0.68898 88
0.54	0.45804 42	0.52878 01	0.08122 06	1.54	3.72530 06	0.40997 99	0.67878 67
0.56	0.49260 17	0.54656 30	0.09037 08	1.56	3.82268 99	0.39385 29	0.66697 13
0.58	0.52841 59	0.56401 31	0.10014 09	1.58	3.92133 60	0.37895 96	0.65363 46
0.60	0.56548 67	0.58109 54	0.11054 02	1.60	4.02123 86	0.36546 17	0.63888 77
0.62	0.60381 41	0.59777 37	0.12157 59	1.62	4.12239 79	0.35351 20	0.62286 07
0.64	0.64339 82	0.61400 94	0.13325 28	1.64	4.22481 38	0.34325 29	0.60570 26
0.66	0.68423 89	0.62976 25	0.14557 29	1.66	4.32848 64	0.33481 32	0.58758 04
0.68	0.72633 62	0.64499 12	0.15853 54	1.68	4.43341 56	0.32830 61	0.56867 83
0.70	0.76969 02	0.65965 24	0.17213 65	1.70	4.53960 14	0.32382 69	0.54919 60
0.72	0.81430 08	0.67370 12	0.18636 89	1.72	4.64704 39	0.32145 02	0.52934 73
0.74	0.86016 81	0.68709 20	0.20122 21	1.74	4.75574 30	0.32122 83	0.50935 84
0.76	0.90729 20	0.69977 79	0.21668 16	1.76	4.86569 87	0.32318 87	0.48946 49
0.78	0.95567 25	0.71171 13	0.23272 88	1.78	4.97691 11	0.32733 25	0.46990 94
0.80	1.00530 96	0.72284 42	0.24934 14	1.80	5.08938 01	0.33363 29	0.45093 88
0.82	1.05620 35	0.73312 83	0.26649 22	1.82	5.20310 58	0.34203 39	0.43280 06
0.84	1.10835 39	0.74251 54	0.28414 98	1.84	5.31808 80	0.35244 96	0.41573 97
0.86	1.16176 10	0.75095 79	0.30227 80	1.86	5.43432 70	0.36476 35	0.39999 44
0.88	1.21642 47	0.75840 90	0.32083 55	1.88	5.55182 25	0.37882 93	0.38579 25
0.90	1.27234 50	0.76482 30	0.33977 63	1.90	5.67057 47	0.39447 05	0.37334 73
0.92	1.32952 20	0.77015 63	0.35904 93	1.92	5.79058 36	0.41148 24	0.36285 37
0.94	1.38795 56	0.77436 72	0.37859 81	1.94	5.91184 91	0.42963 33	0.35448 37
0.96	1.44764 59	0.77741 68	0.39836 12	1.96	6.03437 12	0.44866 69	0.34838 30
0.98	1.50859 28	0.77926 95	0.41827 21	1.98	6.15814 99	0.46830 56	0.34466 65
1.00	1.57079 63	0.77989 34	0.43825 91	2.00	6.28318 53	0.48825 34	0.34341 57
	$\begin{bmatrix}(-4)2\\3\end{bmatrix}$	$\begin{bmatrix}(-4)2\\5\end{bmatrix}$	$\begin{bmatrix}(-5)8\\5\end{bmatrix}$		$\begin{bmatrix}(-4)2\\3\end{bmatrix}$	$\begin{bmatrix}(-4)3\\5\end{bmatrix}$	$\begin{bmatrix}(-4)3\\6\end{bmatrix}$

See Example 8.

For $x\to 0$: $C(x)\approx x-\frac{\pi^2}{40}x^5 \qquad S(x)\approx \frac{\pi}{6}x^3-\frac{\pi^3}{336}x^7$

Table 7.7 **FRESNEL INTEGRALS**

$$C(x)=\int_0^x \cos\left(\frac{\pi}{2}t^2\right)dt \qquad S(x)=\int_0^x \sin\left(\frac{\pi}{2}t^2\right)dt$$

x	$C(x)$	$S(x)$	x	$C(x)$	$S(x)$	x	$C(x)$	$S(x)$
2.00	0.48825 34	0.34341 57	3.00	0.60572 08	0.49631 30	4.00	0.49842 60	0.42051 58
2.02	0.50820 04	0.34467 48	3.02	0.60383 73	0.51619 42	4.02	0.51821 54	0.42301 99
2.04	0.52782 73	0.34844 87	3.04	0.59823 78	0.53536 29	4.04	0.53675 05	0.43039 00
2.06	0.54681 06	0.35470 04	3.06	0.58910 11	0.55311 95	4.06	0.55284 04	0.44217 81
2.08	0.56482 79	0.36334 98	3.08	0.57674 01	0.56880 28	4.08	0.56543 47	0.45764 45
2.10	0.58156 41	0.37427 34	3.10	0.56159 39	0.58181 59	4.10	0.57369 56	0.47579 83
2.12	0.59671 75	0.38730 37	3.12	0.54421 58	0.59165 11	4.12	0.57705 88	0.49545 71
2.14	0.61000 60	0.40223 09	3.14	0.52525 53	0.59791 29	4.14	0.57527 76	0.51532 14
2.16	0.62117 32	0.41880 45	3.16	0.50543 56	0.60033 66	4.16	0.56844 74	0.53405 87
2.18	0.62999 53	0.43673 63	3.18	0.48552 76	0.59880 34	4.18	0.55700 75	0.55039 41
2.20	0.63628 60	0.45570 46	3.20	0.46632 03	0.59334 95	4.20	0.54171 92	0.56319 89
2.22	0.63990 31	0.47535 85	3.22	0.44858 96	0.58416 97	4.22	0.52362 06	0.57157 23
2.24	0.64075 25	0.49532 41	3.24	0.43306 55	0.57161 47	4.24	0.50396 08	0.57491 03
2.26	0.63879 28	0.51521 11	3.26	0.42040 05	0.55618 06	4.26	0.48411 63	0.57295 47
2.28	0.63403 83	0.53462 03	3.28	0.41113 97	0.53849 35	4.28	0.46549 61	0.56582 05
2.30	0.62656 17	0.55315 16	3.30	0.40569 44	0.51928 61	4.30	0.44944 12	0.55399 59
2.32	0.61649 45	0.57041 28	3.32	0.40431 99	0.49936 95	4.32	0.43712 50	0.53831 55
2.34	0.60402 69	0.58602 84	3.34	0.40709 96	0.47960 04	4.34	0.42946 40	0.51990 77
2.36	0.58940 65	0.59964 89	3.36	0.41393 66	0.46084 46	4.36	0.42704 39	0.50011 73
2.38	0.57293 44	0.61095 96	3.38	0.42455 18	0.44393 82	4.38	0.43006 79	0.48041 08
2.40	0.55496 14	0.61969 00	3.40	0.43849 17	0.42964 95	4.40	0.43833 29	0.46226 80
2.42	0.53588 11	0.62562 11	3.42	0.45514 37	0.41864 11	4.42	0.45123 59	0.44707 06
2.44	0.51612 29	0.62859 38	3.44	0.47375 96	0.41143 69	4.44	0.46781 05	0.43599 33
2.46	0.49614 28	0.62851 43	3.46	0.49348 70	0.40839 28	4.46	0.48679 41	0.42990 86
2.48	0.47641 35	0.62535 98	3.48	0.51340 62	0.40967 54	4.48	0.50671 95	0.42931 16
2.50	0.45741 30	0.61918 18	3.50	0.53257 24	0.41524 80	4.50	0.52602 59	0.43427 30
2.52	0.43961 32	0.61010 76	3.52	0.55006 11	0.42486 72	4.52	0.54318 11	0.44442 34
2.54	0.42346 72	0.59834 06	3.54	0.56501 32	0.43808 83	4.54	0.55680 46	0.45897 36
2.56	0.40939 65	0.58415 75	3.56	0.57668 02	0.45428 17	4.56	0.56578 27	0.47676 89
2.58	0.39777 91	0.56790 42	3.58	0.58446 43	0.47265 92	4.58	0.56936 57	0.49637 56
2.60	0.38893 75	0.54998 93	3.60	0.58795 33	0.49230 95	4.60	0.56723 67	0.51619 23
2.62	0.38312 73	0.53087 53	3.62	0.58694 64	0.51224 12	4.62	0.55954 81	0.53457 97
2.64	0.38052 80	0.51106 79	3.64	0.58147 10	0.53143 21	4.64	0.54691 86	0.54999 67
2.66	0.38123 50	0.49110 35	3.66	0.57178 75	0.54888 15	4.66	0.53039 13	0.56113 28
2.68	0.38525 32	0.47153 52	3.68	0.55838 18	0.56366 38	4.68	0.51135 38	0.56702 44
2.70	0.39249 40	0.45291 75	3.70	0.54194 57	0.57498 04	4.70	0.49142 65	0.56714 55
2.72	0.40277 39	0.43578 98	3.72	0.52334 49	0.58220 56	4.72	0.47232 71	0.56146 19
2.74	0.41581 68	0.42066 03	3.74	0.50357 70	0.58492 61	4.74	0.45572 30	0.55044 52
2.76	0.43125 85	0.40798 90	3.76	0.48371 94	0.58296 92	4.76	0.44308 30	0.53504 16
2.78	0.44865 46	0.39817 24	3.78	0.46487 19	0.57641 91	4.78	0.43554 28	0.51659 82
2.80	0.46749 17	0.39152 84	3.80	0.44809 49	0.56561 87	4.80	0.43379 66	0.49675 02
2.82	0.48720 04	0.38828 41	3.82	0.43434 86	0.55115 74	4.82	0.43802 47	0.47728 00
2.84	0.50717 21	0.38856 43	3.84	0.42443 43	0.53384 32	4.84	0.44786 69	0.45995 75
2.86	0.52677 66	0.39238 50	3.86	0.41894 43	0.51466 22	4.86	0.46244 40	0.44637 74
2.88	0.54538 21	0.39964 80	3.88	0.41822 16	0.49472 45	4.88	0.48042 90	0.43780 82
2.90	0.56237 64	0.41014 06	3.90	0.42233 27	0.47520 24	4.90	0.50016 10	0.43506 74
2.92	0.57718 78	0.42353 87	3.92	0.43105 68	0.45726 13	4.92	0.51979 51	0.43843 48
2.94	0.58930 60	0.43941 39	3.94	0.44389 17	0.44198 92	4.94	0.53747 34	0.44761 56
2.96	0.59830 19	0.45724 45	3.96	0.46007 70	0.43032 79	4.96	0.55150 25	0.46175 67
2.98	0.60384 56	0.47643 06	3.98	0.47863 51	0.42301 17	4.98	0.56051 94	0.47951 78
3.00	0.60572 08	0.49631 30	4.00	0.49842 60	0.42051 58	5.00	0.56363 12	0.49919 14
	$\begin{bmatrix}(-4)5\\6\end{bmatrix}$	$\begin{bmatrix}(-4)4\\6\end{bmatrix}$		$\begin{bmatrix}(-4)6\\7\end{bmatrix}$	$\begin{bmatrix}(-4)6\\7\end{bmatrix}$		$\begin{bmatrix}(-4)7\\7\end{bmatrix}$	$\begin{bmatrix}(-4)8\\7\end{bmatrix}$

$$\text{For } x>5 \quad \frac{C(x)}{S(x)}=0.5\pm\left(0.3183099-\frac{0.0968}{x^4}\right)\frac{\sin\left(\frac{\pi}{2}x^2\right)}{\cos\left(\frac{\pi}{2}x^2\right)}\cdot\frac{1}{x}-\left(0.10132-\frac{0.154}{x^4}\right)\frac{\cos\left(\frac{\pi}{2}x^2\right)}{\sin\left(\frac{\pi}{2}x^2\right)}\cdot\frac{1}{x^3}+\epsilon(x) \qquad \epsilon(x)|<3\times10^{-7}$$

$$\text{For } u>39 \quad \frac{C_2(u)}{S_2(u)}=0.5\pm\left(0.3989423-\frac{0.3}{u^2}\right)\frac{\sin(u)}{\cos(u)}\cdot\frac{1}{\sqrt{u}}-\left(0.19947-\frac{0.748}{u^2}\right)\frac{\cos(u)}{\sin(u)}\cdot\frac{1}{u\sqrt{u}}+\epsilon(u) \qquad \epsilon(u)|<3\times10^{-7}$$

AUXILIARY FUNCTIONS

Table 7.8

x	$u=\frac{\pi}{2}x^2$	$f(x)=f_2(u)$	$g(x)=g_2(u)$
0.00	0.00000 00000 00000	0.50000 00000 00000	0.50000 00000 00000
0.02	0.00062 83185 30718	0.49969 41196 39303	0.48031 40626 54163
0.04	0.00251 32741 22872	0.49880 88057 20520	0.46125 51239 79101
0.06	0.00565 48667 76462	0.49739 07811 66949	0.44281 99356 00196
0.08	0.01005 30964 91487	0.49548 44294 00553	0.42500 33536 38036
0.10	0.01570 79632 67949	0.49313 18256 06624	0.40779 85545 29930
0.12	0.02261 94671 05847	0.49037 27777 82254	0.39119 72364 96391
0.14	0.03078 76080 05180	0.48724 48761 11561	0.37518 98069 99885
0.16	0.04021 23859 65949	0.48378 35493 31728	0.35976 55566 09573
0.18	0.05089 38009 88155	0.48002 21268 70713	0.34491 28197 39391
0.20	0.06283 18530 71796	0.47599 19056 49140	0.33061 91227 69034
0.22	0.07602 65422 16873	0.47172 22205 45221	0.31687 13200 89318
0.24	0.09047 78684 23386	0.46724 05176 22164	0.30365 57186 36191
0.26	0.10618 58316 91335	0.46257 24293 12303	0.29095 81914 92531
0.28	0.12315 04320 20720	0.45774 18508 40978	0.27876 42811 44593
0.30	0.14137 16694 11541	0.45277 10172 56087	0.26705 92929 81728
0.32	0.16084 95438 63797	0.44768 05805 06203	0.25582 83796 24420
0.34	0.18158 40553 77490	0.44248 96860 81319	0.24505 66166 57772
0.36	0.20357 52039 52619	0.43721 60487 95888	0.23472 90703 35799
0.38	0.22682 29895 89183	0.43187 60273 53913	0.22483 08578 07150
0.40	0.25132 74122 87183	0.42648 46973 90789	0.21534 72003 95520
0.42	0.27708 84720 46620	0.42105 59227 36507	0.20626 34704 48744
0.44	0.30410 61688 67492	0.41560 24246 90070	0.19756 52322 49727
0.46	0.33238 05027 49800	0.41013 58491 35691	0.18923 82774 60398
0.48	0.36191 14736 93544	0.40466 68313 67950	0.18126 86555 47172
0.50	0.39269 90816 98724	0.39920 50585 25702	0.17364 26996 13238
0.52	0.42474 33267 65340	0.39375 93295 63563	0.16634 70480 39628
0.54	0.45804 42088 93392	0.38833 76127 15400	0.15936 86623 13733
0.56	0.49260 17280 82880	0.38294 71004 26771	0.15269 48414 00876
0.58	0.52841 58843 33803	0.37759 42617 52882	0.14631 32329 91905
0.60	0.56548 66776 46163	0.37228 48922 35620	0.14021 18419 37684
0.62	0.60381 41080 19958	0.36702 41612 87842	0.13437 90361 59907
0.64	0.64339 81754 55190	0.36181 66571 25476	0.12880 35503 06985
0.66	0.68423 88799 51857	0.35666 64292 98472	0.12347 44874 03863
0.68	0.72633 62215 09960	0.35157 70288 80259	0.11838 13187 25611
0.70	0.76969 02001 29499	0.34655 15463 82434	0.11351 38821 06517
0.72	0.81430 08158 10474	0.34159 26474 67053	0.10886 23788 79214
0.74	0.86016 80685 52885	0.33670 26065 33192	0.10441 73696 22082
0.76	0.90729 19583 56732	0.33188 33382 57734	0.10016 97688 77848
0.78	0.95567 24852 22015	0.32713 64271 72503	0.09611 08389 91866
0.80	1.00530 96491 48734	0.32246 31553 61284	0.09223 21832 05037
0.82	1.05620 34501 36888	0.31786 45283 60796	0.08852 57381 23702
0.84	1.10835 38881 86479	0.31334 12993 49704	0.08498 37656 77045
0.86	1.16176 09632 97506	0.30889 39917 09068	0.08159 88446 61614
0.88	1.21642 46754 69968	0.30452 29200 36579	0.07836 38619 62362
0.90	1.27234 50247 03866	0.30022 82096 95385	0.07527 20035 30280
0.92	1.32952 20109 99200	0.29600 98149 76518	0.07231 67451 87932
0.94	1.38795 56343 55971	0.29186 75359 51781	0.06949 18433 26312
0.96	1.44764 58947 74177	0.28780 10340 91658	0.06679 13255 49021
0.98	1.50859 27922 53819	0.28380 98467 20271	0.06420 94813 13093
1.00	1.57079 63267 94897	0.27989 34003 76823	0.06174 08526 09645
	$\left[\begin{matrix}(-4)2\\3\end{matrix}\right]$	$\left[\begin{matrix}(-5)7\\10\end{matrix}\right]$	$\left[\begin{matrix}(-5)8\\10\end{matrix}\right]$

See **Examples 6, 7,** and **9.**

$$C(x)=\frac{1}{2}+f(x)\ \sin\left(\frac{\pi}{2}x^2\right)-g(x)\ \cos\left(\frac{\pi}{2}x^2\right) \qquad C_2(u)=\frac{1}{2}+f_2(u)\ \sin u-g_2(u)\ \cos u$$

$$S(x)=\frac{1}{2}-f(x)\ \cos\left(\frac{\pi}{2}x^2\right)-g(x)\ \sin\left(\frac{\pi}{2}x^2\right) \qquad S_2(u)=\frac{1}{2}-f_2(u)\ \cos u-g_2(u)\ \sin u$$

Table 7.8   **AUXILIARY FUNCTIONS**

x^{-1}	$u^{-1}=\dfrac{2}{\pi x^2}$	$f(x)=f_2(u)$	$g(x)=g_2(u)$	$<x>$	$<u>$
1.00	0.63661 97723 67581	0.27989 34003 76823	0.06174 08526 09645	1	2
0.98	0.61140 96293 81825	0.27597 33733 36442	0.05933 31378 64174	1	2
0.96	0.58670 87822 13963	0.27197 11505 76851	0.05693 89827 01255	1	2
0.94	0.56251 72308 63995	0.26788 56989 47656	0.05456 06112 91100	1	2
0.92	0.53883 49753 31921	0.26371 60682 37287	0.05220 03510 52931	1	2
0.90	0.51566 20156 17741	0.25946 14023 65674	0.04986 06317 93636	1	2
0.88	0.49299 83517 21455	0.25512 09512 80091	0.04754 39838 94725	1	2
0.86	0.47084 39836 43063	0.25069 40835 25766	0.04525 30354 03048	1	2
0.84	0.44919 89113 82565	0.24618 02994 44393	0.04299 05078 69390	1	2
0.82	0.42806 31349 39962	0.24157 92449 31459	0.04075 92107 68723	1	2
0.80	0.40743 66543 15252	0.23689 07256 57089	0.03856 20343 27312	1	2
0.78	0.38731 94695 08436	0.23211 47216 24632	0.03640 19405 75704	1	3
0.76	0.36771 15805 19515	0.22725 14019 06110	0.03428 19524 44132	1	3
0.74	0.34861 29873 48488	0.22230 11393 53995	0.03220 51407 19129	1	3
0.72	0.33002 36899 95354	0.21726 45250 44609	0.03017 46086 88637	1	3
0.70	0.31194 36884 60115	0.21214 23821 60229	0.02819 34743 19381	1	3
0.68	0.29437 29827 42770	0.20693 57789 65521	0.02626 48498 36510	1	3
0.66	0.27731 15728 43318	0.20164 60404 80635	0.02439 18186 13588	2	4
0.64	0.26075 94587 61761	0.19627 47584 00004	0.02257 74093 32978	2	4
0.62	0.24471 66404 98098	0.19082 37987 55563	0.02082 45674 44482	2	4
0.60	0.22918 31180 52329	0.18529 53067 79209	0.01913 61240 35536	2	4
0.58	0.21415 88914 24454	0.17969 17083 86674	0.01751 47623 30357	2	5
0.56	0.19964 39606 14474	0.17401 57076 89207	0.01596 29821 58470	2	5
0.54	0.18563 83256 22387	0.16827 02799 47273	0.01448 30628 73722	2	5
0.52	0.17214 19864 48194	0.16245 86594 19322	0.01307 70253 60097	2	6
0.50	0.15915 49430 91895	0.15658 43216 36302	0.01174 65939 24659	2	6
0.48	0.14667 71955 53491	0.15065 09597 56320	0.01049 31590 42015	2	7
0.46	0.13470 87438 32980	0.14466 24548 29603	0.00931 77420 66589	2	7
0.44	0.12324 95879 30364	0.13862 28400 34552	0.00822 09631 52815	2	8
0.42	0.11229 97278 45641	0.13253 62592 29647	0.00720 30137 00215	2	9
0.40	0.10185 91635 78813	0.12640 69204 94864	0.00626 36346 49122	3	10
0.38	0.09192 78951 29879	0.12023 90456 93806	0.00540 21018 72942	3	11
0.36	0.08250 59224 98839	0.11403 68174 47880	0.00461 72197 27002	3	12
0.34	0.07359 32456 85692	0.10780 43252 41741	0.00390 73235 12822	3	14
0.32	0.06518 98646 90440	0.10154 55126 32988	0.00327 02912 03254	3	15
0.30	0.05729 57795 13082	0.09526 41276 74844	0.00270 35642 68526	3	17
0.28	0.04991 09901 53618	0.08896 36786 39974	0.00220 41768 84885	4	20
0.26	0.04303 54966 12048	0.08264 73969 33180	0.00176 87922 53708	4	23
0.24	0.03666 92988 88373	0.07631 82087 00913	0.00139 37442 77909	4	27
0.22	0.03081 23969 82591	0.06997 87161 16730	0.00107 50825 02743	5	32
0.20	0.02546 47908 94703	0.06363 11887 04012	0.00080 86180 82883	5	39
0.18	0.02062 64806 24710	0.05727 75644 30652	0.00058 99686 10701	6	48
0.16	0.01629 74661 72610	0.05091 94597 59575	0.00041 45999 18234	6	61
0.14	0.01247 77475 38405	0.04455 81874 32960	0.00027 78633 97799	7	80
0.12	0.00916 73247 22093	0.03819 47805 44642	0.00017 50279 00844	8	109
0.10	0.00636 61977 23676	0.03183 00214 15118	0.00010 13057 94484	10	157
0.08	0.00407 43665 43153	0.02546 44738 95252	0.00005 18732 17470	13	245
0.06	0.00229 18311 80523	0.01909 85179 38105	0.00002 18849 44630	17	436
0.04	0.00101 85916 35788	0.01273 23855 39770	0.00000 64845 30524	25	982
0.02	0.00025 46479 08947	0.00636 61974 14061	0.00000 08105 69272	50	3927
0.00	0.00000 00000 00000	0.00000 00000 00000	0.00000 00000 00000	∞	∞
	$\begin{bmatrix}(-5)6\\3\end{bmatrix}$	$\begin{bmatrix}(-5)1\\12\end{bmatrix}$	$\begin{bmatrix}(-5)1\\12\end{bmatrix}$		

$$C(x)=\tfrac{1}{2}+f(x)\,\sin\left(\tfrac{\pi}{2}x^2\right)-g(x)\,\cos\left(\tfrac{\pi}{2}x^2\right) \qquad C_2(u)=\tfrac{1}{2}+f_2(u)\,\sin u-g_2(u)\,\cos u$$

$$S(x)=\tfrac{1}{2}-f(x)\,\cos\left(\tfrac{\pi}{2}x^2\right)-g(x)\,\sin\left(\tfrac{\pi}{2}x^2\right) \qquad S_2(u)=\tfrac{1}{2}-f_2(u)\,\cos u-g_2(u)\,\sin u$$

$$<x>=\text{nearest integer to } x.$$

ERROR FUNCTION FOR COMPLEX ARGUMENTS Table 7.9

$$w(z)=e^{-z^2}\operatorname{erfc}(-iz) \qquad z=x+iy$$

y	$\mathscr{R}w(z)$ $\mathscr{I}w(z)$ $x=0$	$\mathscr{R}w(z)$ $\mathscr{I}w(z)$ $x=0.1$	$\mathscr{R}w(z)$ $\mathscr{I}w(z)$ $x=0.2$	$\mathscr{R}w(z)$ $\mathscr{I}w(z)$ $x=0.3$	$\mathscr{R}w(z)$ $\mathscr{I}w(z)$ $x=0.4$
0.0	1.000000 0.000000	0.990050 0.112089	0.960789 0.219753	0.913931 0.318916	0.852144 0.406153
0.1	0.896457 0.000000	0.888479 0.094332	0.864983 0.185252	0.827246 0.269600	0.777267 0.344688
0.2	0.809020 0.000000	0.802567 0.080029	0.783538 0.157403	0.752895 0.229653	0.712146 0.294653
0.3	0.734599 0.000000	0.729337 0.068410	0.713801 0.134739	0.688720 0.197037	0.655244 0.253613
0.4	0.670788 0.000000	0.666463 0.058897	0.653680 0.116147	0.632996 0.170203	0.605295 0.219706
0.5	0.615690 0.000000	0.612109 0.051048	0.601513 0.100782	0.584333 0.147965	0.561252 0.191500
0.6	0.567805 0.000000	0.564818 0.044524	0.555974 0.087993	0.541605 0.129408	0.522246 0.167880
0.7	0.525930 0.000000	0.523423 0.039064	0.515991 0.077275	0.503896 0.113821	0.487556 0.147975
0.8	0.489101 0.000000	0.486982 0.034465	0.480697 0.068235	0.470452 0.100647	0.456579 0.131101
0.9	0.456532 0.000000	0.454731 0.030566	0.449383 0.060563	0.440655 0.089444	0.428808 0.116714
1.0	0.427584 0.000000	0.426044 0.027242	0.421468 0.054014	0.413989 0.079864	0.403818 0.104380
1.1	0.401730 0.000000	0.400406 0.024392	0.396470 0.048393	0.390028 0.071628	0.381250 0.093752
1.2	0.378557 0.000000	0.377393 0.021934	0.373989 0.043542	0.368412 0.064510	0.360799 0.084547
1.3	0.357643 0.000000	0.356649 0.019805	0.353691 0.039336	0.348839 0.058329	0.342206 0.076538
1.4	0.338744 0.000000	0.337876 0.017951	0.335294 0.035671	0.331054 0.052930	0.325248 0.069538
1.5	0.321585 0.000000	0.320825 0.016329	0.318561 0.032463	0.314839 0.048210	0.309736 0.063393
1.6	0.305953 0.000000	0.305284 0.014905	0.303290 0.029643	0.300009 0.044051	0.295506 0.057978
1.7	0.291663 0.000000	0.291072 0.013648	0.289309 0.027154	0.286406 0.040377	0.282417 0.053186
1.8	0.278560 0.000000	0.278035 0.012536	0.276470 0.024948	0.273892 0.037118	0.270346 0.048931
1.9	0.266509 0.000000	0.266042 0.011547	0.264648 0.022987	0.262350 0.034217	0.259186 0.045139
2.0	0.255396 0.000000	0.254978 0.010664	0.253732 0.021236	0.251677 0.031626	0.248844 0.041748
2.1	0.245119 0.000000	0.244745 0.009874	0.243628 0.019669	0.241783 0.029304	0.239239 0.038706
2.2	0.235593 0.000000	0.235256 0.009165	0.234251 0.018260	0.232592 0.027217	0.230300 0.035968
2.3	0.226742 0.000000	0.226438 0.008526	0.225531 0.016991	0.224033 0.025335	0.221963 0.033498
2.4	0.218499 0.000000	0.218224 0.007949	0.217404 0.015845	0.216047 0.023633	0.214172 0.031263
2.5	0.210806 0.000000	0.210557 0.007427	0.209813 0.014806	0.208582 0.022090	0.206879 0.029234
2.6	0.203613 0.000000	0.203387 0.006952	0.202710 0.013862	0.201589 0.020687	0.200039 0.027389
2.7	0.196874 0.000000	0.196668 0.006520	0.196050 0.013002	0.195028 0.019409	0.193613 0.025706
2.8	0.190549 0.000000	0.190360 0.006125	0.189796 0.012216	0.188861 0.018241	0.187566 0.024168
2.9	0.184602 0.000000	0.184429 0.005764	0.183912 0.011498	0.183056 0.017172	0.181868 0.022759
3.0	0.179001 0.000000	0.178842 0.005433	0.178368 0.010839	0.177581 0.016192	0.176491 0.021466

y	$x=0.5$	$x=0.6$	$x=0.7$	$x=0.8$	$x=0.9$
0.0	0.778801 0.478925	0.697676 0.535713	0.612626 0.576042	0.527292 0.600412	0.444858 0.610142
0.1	0.717588 0.408474	0.651076 0.459665	0.580698 0.497744	0.509299 0.522932	0.439421 0.536087
0.2	0.663223 0.350751	0.608322 0.396852	0.549739 0.432442	0.489710 0.457569	0.430271 0.472773
0.3	0.614852 0.303124	0.569238 0.344645	0.520192 0.377688	0.469480 0.402194	0.418736 0.418491
0.4	0.571717 0.263563	0.533581 0.300989	0.492289 0.331535	0.449244 0.355082	0.405763 0.371813
0.5	0.533157 0.230488	0.501079 0.264268	0.466127 0.292432	0.429418 0.314828	0.392021 0.331544
0.6	0.498591 0.202666	0.471453 0.233206	0.441712 0.259136	0.410264 0.280290	0.377977 0.296692
0.7	0.467521 0.179123	0.444434 0.206787	0.418998 0.230646	0.391936 0.250532	0.363957 0.266427
0.8	0.439512 0.159087	0.419766 0.184200	0.397906 0.206155	0.374518 0.224789	0.350182 0.240057
0.9	0.414191 0.141945	0.397216 0.164793	0.378341 0.185005	0.358043 0.202429	0.336799 0.217004
1.0	0.391234 0.127202	0.376571 0.148036	0.360200 0.166660	0.342511 0.182932	0.323899 0.196783
1.1	0.370363 0.114460	0.357637 0.133501	0.343375 0.150681	0.327900 0.165868	0.311537 0.178990
1.2	0.351335 0.103395	0.340241 0.120838	0.327766 0.136706	0.314176 0.150787	0.299741 0.163281
1.3	0.333942 0.093744	0.324229 0.109759	0.313273 0.124435	0.301294 0.137661	0.288519 0.149370
1.4	0.318001 0.085288	0.309463 0.100026	0.299804 0.113620	0.289208 0.125971	0.277865 0.137012
1.5	0.303355 0.077851	0.295820 0.091443	0.287274 0.104054	0.277869 0.115594	0.267766 0.126002
1.6	0.289866 0.071283	0.283192 0.083845	0.275602 0.095563	0.267228 0.106355	0.258203 0.116164
1.7	0.277412 0.065461	0.271479 0.077096	0.264718 0.088001	0.257237 0.098103	0.249151 0.107348
1.8	0.265890 0.060283	0.260598 0.071081	0.254554 0.081245	0.247851 0.090710	0.240586 0.099427
1.9	0.255205 0.055661	0.250469 0.065701	0.245050 0.075190	0.239027 0.084068	0.232482 0.092291
2.0	0.245276 0.051521	0.241025 0.060876	0.236152 0.069748	0.230724 0.078085	0.224813 0.085845
2.1	0.236031 0.047804	0.232204 0.056534	0.227810 0.064842	0.222905 0.072680	0.217552 0.080009
2.2	0.227407 0.044454	0.223952 0.052617	0.219978 0.060409	0.215535 0.067785	0.210676 0.074712
2.3	0.219347 0.041428	0.216219 0.049073	0.212616 0.056391	0.208581 0.063342	0.204160 0.069894
2.4	0.211800 0.038686	0.208961 0.045859	0.205686 0.052741	0.202013 0.059298	0.197982 0.065500
2.5	0.204723 0.036196	0.202139 0.042936	0.199155 0.049417	0.195804 0.055610	0.192120 0.061486
2.6	0.198074 0.033929	0.195717 0.040271	0.192992 0.046384	0.189928 0.052238	0.186554 0.057811
2.7	0.191818 0.031859	0.189664 0.037836	0.187170 0.043608	0.184362 0.049150	0.181265 0.054439
2.8	0.185924 0.029966	0.183950 0.035607	0.181662 0.041064	0.179084 0.046315	0.176237 0.051339
2.9	0.180361 0.028231	0.178549 0.033561	0.176447 0.038728	0.174074 0.043708	0.171452 0.048485
3.0	0.175105 0.026636	0.173437 0.031680	0.171502 0.036577	0.169315 0.041306	0.166895 0.045851

See Examples 12–19. $\qquad w(x)=e^{-x^2}+\dfrac{2i}{\sqrt\pi}e^{-x^2}\int_0^x e^{t^2}\,dt$

$$w(-x+iy)=\overline{w(x+iy)} \qquad w(x-iy)=2e^{y^2-x^2}(\cos 2xy+i\sin 2xy)-\overline{w(x+iy)}$$

$$w(iy)=e^{y^2}\operatorname{erfc}y \qquad w[(1+i)u]=e^{-2iu^2}\left\{1+(i-1)\left[C\!\left(\dfrac{2u}{\sqrt\pi}\right)+iS\!\left(\dfrac{2u}{\sqrt\pi}\right)\right]\right\}$$

Table 7.9   **ERROR FUNCTION FOR COMPLEX ARGUMENTS**

$$w(z)=e^{-z^2}\,\text{erfc}\,(-iz)\qquad z=x+iy$$

y	$\mathscr{R}w(z)$ $x=1.0$	$\mathscr{I}w(z)$	$\mathscr{R}w(z)$ $x=1.1$	$\mathscr{I}w(z)$	$\mathscr{R}w(z)$ $x=1.2$	$\mathscr{I}w(z)$	$\mathscr{R}w(z)$ $x=1.3$	$\mathscr{I}w(z)$	$\mathscr{R}w(z)$ $x=1.4$	$\mathscr{I}w(z)$
0.0	0.367879	0.607158	0.298197	0.593761	0.236928	0.572397	0.184520	0.545456	0.140858	0.515113
0.1	0.373170	0.538555	0.312136	0.532009	0.257374	0.518283	0.209431	0.499216	0.168407	0.476535
0.2	0.373153	0.478991	0.319717	0.477439	0.270928	0.469488	0.227362	0.456555	0.189247	0.440005
0.3	0.369386	0.427225	0.322586	0.429275	0.279199	0.425667	0.239793	0.417491	0.204662	0.405823
0.4	0.363020	0.382166	0.321993	0.386777	0.283443	0.386412	0.247908	0.381908	0.215711	0.374110
0.5	0.354900	0.342872	0.318884	0.349266	0.284638	0.351299	0.252654	0.349611	0.223262	0.344868
0.6	0.345649	0.308530	0.313978	0.316128	0.283540	0.319910	0.254784	0.320368	0.228026	0.318022
0.7	0.335721	0.278445	0.307816	0.286815	0.280740	0.291851	0.254895	0.293927	0.230578	0.293453
0.8	0.325446	0.252024	0.300807	0.260847	0.276693	0.266757	0.253461	0.270040	0.231385	0.271015
0.9	0.315064	0.228759	0.293259	0.237800	0.271752	0.244295	0.250858	0.248462	0.230826	0.250549
1.0	0.304744	0.208219	0.285402	0.217306	0.266189	0.224168	0.247381	0.228967	0.229205	0.231897
1.1	0.294606	0.190036	0.277407	0.199046	0.260213	0.206108	0.243266	0.211343	0.226767	0.214902
1.2	0.284731	0.173896	0.269401	0.182742	0.253985	0.189878	0.238695	0.195398	0.223710	0.199416
1.3	0.275174	0.159531	0.261476	0.168151	0.247628	0.175271	0.233813	0.180957	0.220192	0.185299
1.4	0.265967	0.146712	0.253697	0.155066	0.241233	0.162100	0.228733	0.167863	0.216340	0.172423
1.5	0.257128	0.135242	0.246112	0.143305	0.234870	0.150205	0.223542	0.155975	0.212253	0.160668
1.6	0.248665	0.124954	0.238752	0.132711	0.228592	0.139441	0.218309	0.145167	0.208014	0.149927
1.7	0.240578	0.115702	0.231635	0.123147	0.222436	0.129684	0.213086	0.135326	0.203684	0.140103
1.8	0.232861	0.107361	0.224775	0.114495	0.216428	0.120822	0.207912	0.126353	0.199315	0.131106
1.9	0.225503	0.099824	0.218176	0.106650	0.210587	0.112760	0.202818	0.118158	0.194947	0.122858
2.0	0.218493	0.092998	0.211839	0.099523	0.204926	0.105411	0.197827	0.110662	0.190608	0.115286
2.1	0.211816	0.086801	0.205760	0.093035	0.199452	0.098700	0.192953	0.103795	0.186324	0.108325
2.2	0.205457	0.081162	0.199935	0.087116	0.194166	0.092562	0.188208	0.097495	0.182112	0.101919
2.3	0.199402	0.076021	0.194356	0.081706	0.189072	0.086936	0.183599	0.091706	0.177985	0.096015
2.4	0.193634	0.071324	0.189014	0.076753	0.184165	0.081773	0.179131	0.086378	0.173954	0.090567
2.5	0.188139	0.067024	0.183901	0.072208	0.179444	0.077024	0.174805	0.081467	0.170024	0.085532
2.6	0.182903	0.063080	0.179008	0.068031	0.174903	0.072651	0.170623	0.076933	0.166201	0.080873
2.7	0.177910	0.059456	0.174324	0.064186	0.170538	0.068617	0.166582	0.072742	0.162487	0.076557
2.8	0.173147	0.056118	0.169840	0.060639	0.166342	0.064890	0.162681	0.068863	0.158883	0.072553
2.9	0.168602	0.053041	0.165546	0.057363	0.162310	0.061440	0.158916	0.065266	0.155389	0.068834
3.0	0.164261	0.050197	0.161434	0.054331	0.158435	0.058243	0.155285	0.061926	0.152005	0.065375

y	$\mathscr{R}w(z)$ $x=1.5$	$\mathscr{I}w(z)$	$\mathscr{R}w(z)$ $x=1.6$	$\mathscr{I}w(z)$	$\mathscr{R}w(z)$ $x=1.7$	$\mathscr{I}w(z)$	$\mathscr{R}w(z)$ $x=1.8$	$\mathscr{I}w(z)$	$\mathscr{R}w(z)$ $x=1.9$	$\mathscr{I}w(z)$
0.0	0.105399	0.483227	0.077305	0.451284	0.055576	0.420388	0.039164	0.391291	0.027052	0.364437
0.1	0.134049	0.451763	0.105843	0.426168	0.083112	0.400743	0.065099	0.376214	0.051038	0.353066
0.2	0.156521	0.421076	0.128895	0.400837	0.105929	0.380161	0.087090	0.359721	0.071811	0.340004
0.3	0.173865	0.391665	0.147272	0.375911	0.124612	0.359313	0.105522	0.342479	0.089592	0.325873
0.4	0.186984	0.363828	0.161702	0.351803	0.139717	0.338676	0.120793	0.324985	0.104641	0.311161
0.5	0.196636	0.337720	0.172820	0.328777	0.151751	0.318584	0.133288	0.307609	0.117233	0.296240
0.6	0.203461	0.313397	0.181177	0.306990	0.161171	0.299261	0.143369	0.290613	0.127644	0.281392
0.7	0.207990	0.290847	0.187245	0.286517	0.168379	0.280846	0.151366	0.274180	0.136134	0.266823
0.8	0.210664	0.270016	0.191423	0.267738	0.173725	0.263418	0.157578	0.258431	0.142949	0.252681
0.9	0.211846	0.250823	0.194049	0.249556	0.177513	0.247012	0.162268	0.243439	0.148310	0.239067
1.0	0.211837	0.233171	0.195407	0.233009	0.180002	0.231630	0.165667	0.229244	0.152418	0.226046
1.1	0.210881	0.216954	0.195734	0.217678	0.181414	0.217253	0.167977	0.215857	0.155452	0.213656
1.2	0.209182	0.202067	0.195228	0.203494	0.181938	0.203847	0.169373	0.203272	0.157569	0.201914
1.3	0.206902	0.188403	0.194053	0.190384	0.181733	0.191366	0.170003	0.191471	0.158906	0.190821
1.4	0.204177	0.175862	0.192347	0.178275	0.180933	0.179762	0.169997	0.180425	0.159585	0.180367
1.5	0.201115	0.164349	0.190222	0.167092	0.179651	0.168980	0.169465	0.170099	0.159709	0.170534
1.6	0.197806	0.153773	0.187772	0.156765	0.177983	0.158969	0.168500	0.160457	0.159369	0.161300
1.7	0.194320	0.144054	0.185073	0.147226	0.176008	0.149674	0.167183	0.151458	0.158641	0.152637
1.8	0.190717	0.135113	0.182189	0.138412	0.173792	0.141045	0.165579	0.143063	0.157593	0.144516
1.9	0.187043	0.126883	0.179172	0.130262	0.171390	0.133033	0.163746	0.135234	0.156282	0.136908
2.0	0.183335	0.119298	0.176064	0.122723	0.168849	0.125590	0.161733	0.127931	0.154757	0.129781
2.1	0.179623	0.112302	0.172901	0.115744	0.166206	0.118674	0.159580	0.121118	0.153059	0.123108
2.2	0.175930	0.105842	0.169710	0.109277	0.163493	0.112243	0.157320	0.114761	0.151224	0.116858
2.3	0.172276	0.099870	0.166513	0.103280	0.160737	0.106260	0.154982	0.108827	0.149281	0.111003
2.4	0.168674	0.094343	0.163330	0.097713	0.157958	0.100689	0.152591	0.103285	0.147256	0.105519
2.5	0.165136	0.089222	0.160175	0.092541	0.155175	0.095499	0.150165	0.098107	0.145172	0.100378
2.6	0.161669	0.084472	0.157060	0.087732	0.152402	0.090660	0.147752	0.093265	0.143045	0.095558
2.7	0.158281	0.080061	0.153993	0.083254	0.149649	0.086143	0.145274	0.088735	0.140892	0.091037
2.8	0.154975	0.075960	0.150981	0.079082	0.146927	0.081925	0.142834	0.084493	0.138725	0.086794
2.9	0.151753	0.072142	0.148030	0.075191	0.144243	0.077982	0.140411	0.080519	0.136555	0.082809
3.0	0.148618	0.068585	0.145144	0.071558	0.141602	0.074293	0.138012	0.076794	0.134391	0.079065

See Examples 12–19. $w(x)=e^{-x^2}+\dfrac{2i}{\sqrt{\pi}}e^{-x^2}\int_0^x e^{t^2}\,dt$

$$w(-x+iy)=\overline{w(x+iy)}\qquad\qquad w(x-iy)=2e^{y^2-x^2}(\cos 2xy+i\sin 2xy)-\overline{w(x+iy)}$$

$$w(iy)=e^{y^2}\,\text{erfc}\,y\qquad\qquad w[(1+i)u]=e^{-2iu^2}\left\{1+(i-1)\left[C\left(\tfrac{2u}{\sqrt{\pi}}\right)+iS\left(\tfrac{2u}{\sqrt{\pi}}\right)\right]\right\}$$

ERROR FUNCTION FOR COMPLEX ARGUMENTS   Table 7.9

$$w(z)=e^{-z^2}\operatorname{erfc}(-iz) \qquad z=x+iy$$

y	$\mathscr{R}w(z)$ $x=2.0$	$\mathscr{I}w(z)$	$\mathscr{R}w(z)$ $x=2.1$	$\mathscr{I}w(z)$	$\mathscr{R}w(z)$ $x=2.2$	$\mathscr{I}w(z)$	$\mathscr{R}w(z)$ $x=2.3$	$\mathscr{I}w(z)$	$\mathscr{R}w(z)$ $x=2.4$	$\mathscr{I}w(z)$
0.0	0.018316	0.340026	0.012155	0.318073	0.007907	0.298468	0.005042	0.281026	0.003151	0.265522
0.1	0.040201	0.331583	0.031936	0.311886	0.025678	0.293982	0.020958	0.277795	0.017397	0.263201
0.2	0.059531	0.321332	0.049726	0.303894	0.041927	0.287771	0.035728	0.272968	0.030792	0.259435
0.3	0.076396	0.309831	0.065521	0.294574	0.056586	0.280232	0.049248	0.266865	0.043211	0.254478
0.4	0.090944	0.297529	0.079385	0.284327	0.069655	0.271710	0.061473	0.259775	0.054585	0.248566
0.5	0.103359	0.284786	0.091422	0.273482	0.081182	0.262499	0.072408	0.251953	0.064890	0.241914
0.6	0.113836	0.271881	0.101765	0.262308	0.091245	0.252844	0.082092	0.243617	0.074132	0.234714
0.7	0.122574	0.259031	0.110558	0.251016	0.099943	0.242947	0.090585	0.234952	0.082345	0.227129
0.8	0.129768	0.246396	0.117948	0.239772	0.107383	0.232968	0.097963	0.226111	0.089576	0.219302
0.9	0.135600	0.234096	0.124081	0.228703	0.113679	0.223037	0.104309	0.217219	0.095884	0.211349
1.0	0.140240	0.222213	0.129097	0.217904	0.118941	0.213253	0.109709	0.208376	0.101336	0.203368
1.1	0.143840	0.210805	0.133125	0.207442	0.123277	0.203692	0.114251	0.199660	0.105999	0.195438
1.2	0.146541	0.199904	0.136286	0.197366	0.126788	0.194410	0.118019	0.191133	0.109942	0.187620
1.3	0.148466	0.189529	0.138689	0.187705	0.129570	0.185446	0.121092	0.182840	0.113232	0.179965
1.4	0.149725	0.179687	0.140432	0.178478	0.131709	0.176827	0.123548	0.174814	0.115935	0.172510
1.5	0.150415	0.170371	0.141604	0.169691	0.133284	0.168569	0.125454	0.167078	0.118109	0.165281
1.6	0.150622	0.161572	0.142283	0.161343	0.134367	0.160680	0.126877	0.159646	0.119812	0.158299
1.7	0.150418	0.153274	0.142540	0.153429	0.135021	0.153161	0.127873	0.152526	0.121096	0.151576
1.8	0.149870	0.145457	0.142434	0.145938	0.135305	0.146009	0.128495	0.145721	0.122010	0.145120
1.9	0.149032	0.138100	0.142021	0.138855	0.135269	0.139217	0.128792	0.139229	0.122597	0.138933
2.0	0.147953	0.131180	0.141347	0.132164	0.134959	0.132773	0.128805	0.133045	0.122897	0.133015
2.1	0.146675	0.124674	0.140453	0.125849	0.134414	0.126667	0.128574	0.127161	0.122945	0.127363
2.2	0.145234	0.118558	0.139375	0.119891	0.133669	0.120885	0.128130	0.121569	0.122773	0.121972
2.3	0.143660	0.112810	0.138145	0.114272	0.132755	0.115413	0.127506	0.116258	0.122411	0.116834
2.4	0.141982	0.107408	0.136789	0.108973	0.131699	0.110236	0.126726	0.111218	0.121884	0.111942
2.5	0.140220	0.102329	0.135331	0.103977	0.130524	0.105339	0.125814	0.106436	0.121215	0.107286
2.6	0.138395	0.097554	0.133791	0.099265	0.129252	0.100709	0.124792	0.101901	0.120424	0.102858
2.7	0.136523	0.093062	0.132187	0.094822	0.127900	0.096330	0.123676	0.097601	0.119530	0.098648
2.8	0.134619	0.088837	0.130533	0.090631	0.126483	0.092189	0.122484	0.093523	0.118548	0.094646
2.9	0.132693	0.084859	0.128842	0.086677	0.125016	0.088273	0.121229	0.089658	0.117492	0.090842
3.0	0.130757	0.081113	0.127125	0.082944	0.123510	0.084568	0.119922	0.085992	0.116375	0.087227

y	$x=2.5$		$x=2.6$		$x=2.7$		$x=2.8$		$x=2.9$	
0.0	0.001930	0.251723	0.001159	0.239403	0.000682	0.228355	0.000394	0.218399	0.000223	0.209377
0.1	0.014698	0.250050	0.012635	0.238187	0.011037	0.227458	0.009778	0.217722	0.008769	0.208854
0.2	0.026841	0.247092	0.023653	0.235838	0.021057	0.225569	0.018918	0.216181	0.017134	0.207577
0.3	0.038226	0.243042	0.034087	0.232504	0.030626	0.222800	0.027707	0.213858	0.025225	0.205607
0.4	0.048773	0.238092	0.043849	0.228337	0.039656	0.219268	0.036064	0.210843	0.032967	0.203014
0.5	0.058437	0.232420	0.052885	0.223482	0.048090	0.215093	0.043930	0.207232	0.040304	0.199873
0.6	0.067205	0.226190	0.061167	0.218077	0.055890	0.210387	0.051264	0.203119	0.047194	0.196262
0.7	0.075088	0.219546	0.068691	0.212247	0.063043	0.205258	0.058046	0.198594	0.053611	0.192256
0.8	0.082112	0.212614	0.075467	0.206103	0.069548	0.199804	0.064266	0.193741	0.059543	0.187927
0.9	0.088317	0.205504	0.081521	0.199744	0.075416	0.194111	0.069927	0.188638	0.064986	0.183344
1.0	0.093751	0.198307	0.086885	0.193255	0.080670	0.188258	0.075043	0.183354	0.069944	0.178568
1.1	0.098466	0.191099	0.091598	0.186707	0.085338	0.182311	0.079632	0.177950	0.074431	0.173654
1.2	0.102518	0.183943	0.095702	0.180163	0.089451	0.176328	0.083718	0.172480	0.078462	0.168651
1.3	0.105960	0.176889	0.099243	0.173670	0.093044	0.170357	0.087328	0.166990	0.082059	0.163603
1.4	0.108848	0.169977	0.102264	0.167270	0.096155	0.164438	0.090492	0.161519	0.085245	0.158547
1.5	0.111233	0.163237	0.104811	0.160996	0.098820	0.158604	0.093239	0.156099	0.088044	0.153515
1.6	0.113165	0.156692	0.106925	0.154872	0.101076	0.152882	0.095601	0.150758	0.090482	0.148534
1.7	0.114690	0.150359	0.108647	0.148918	0.102957	0.147292	0.097608	0.145518	0.092584	0.143625
1.8	0.115851	0.144249	0.110016	0.143147	0.104498	0.141851	0.099288	0.140395	0.094376	0.138807
1.9	0.116689	0.138368	0.111067	0.137569	0.105730	0.136571	0.100671	0.135403	0.095882	0.134094
2.0	0.117239	0.132720	0.111834	0.132191	0.106683	0.131459	0.101783	0.130553	0.097127	0.129498
2.1	0.117534	0.127305	0.112347	0.127015	0.107386	0.126522	0.102649	0.125851	0.098133	0.125027
2.2	0.117606	0.122121	0.112635	0.122042	0.107864	0.121762	0.103293	0.121303	0.098922	0.120688
2.3	0.117481	0.117164	0.112723	0.117271	0.108140	0.117180	0.103737	0.116911	0.099513	0.116484
2.4	0.117184	0.112428	0.112633	0.112699	0.108238	0.112775	0.104002	0.112676	0.099925	0.112419
2.5	0.116737	0.107909	0.112389	0.108322	0.108177	0.108546	0.104105	0.108597	0.100177	0.108493
2.6	0.116160	0.103597	0.112008	0.104136	0.107975	0.104489	0.104066	0.104674	0.100284	0.104707
2.7	0.115471	0.099487	0.111508	0.100133	0.107648	0.100601	0.103898	0.100905	0.100261	0.101058
2.8	0.114685	0.095570	0.110904	0.096309	0.107213	0.096876	0.103617	0.097284	0.100122	0.097546
2.9	0.113816	0.091838	0.110210	0.092657	0.106682	0.093310	0.103236	0.093810	0.099879	0.094168
3.0	0.112878	0.088283	0.109439	0.089170	0.106067	0.089898	0.102767	0.090479	0.099544	0.090921

See Examples 12–19. $\qquad w(x)=e^{-x^2}+\dfrac{2i}{\sqrt{\pi}}e^{-x^2}\int_0^x e^{t^2}\,dt$

$$w(-x+iy)=\overline{w(x+iy)} \qquad w(x-iy)=2e^{y^2-x^2}(\cos 2xy+i\sin 2xy)-\overline{w(x+iy)}$$

$$w(iy)=e^{y^2}\operatorname{erfc}y \qquad w[(1+i)u]=e^{-2iu^2}\left\{1+(i-1)\left[C\!\left(\dfrac{2u}{\sqrt{\pi}}\right)+iS\!\left(\dfrac{2u}{\sqrt{\pi}}\right)\right]\right\}$$

Table 7.9   **ERROR FUNCTION FOR COMPLEX ARGUMENTS**

$$w(z)=e^{-z^2}\operatorname{erfc}(-iz) \qquad z=x+iy$$

y	$\mathcal{R}w(z)$ $\mathcal{I}w(z)$ $x=3.0$	$x=3.1$	$x=3.2$	$x=3.3$	$x=3.4$
0.0	0.000123 0.201157	0.000067 0.193630	0.000036 0.186704	0.000019 0.180302	0.000010 0.174362
0.1	0.007943 0.200742	0.007254 0.193292	0.006670 0.186421	0.006167 0.180061	0.005728 0.174152
0.2	0.015627 0.199669	0.014338 0.192376	0.013225 0.185630	0.012252 0.179369	0.011394 0.173542
0.3	0.023095 0.197980	0.021250 0.190915	0.019639 0.184354	0.018222 0.178245	0.016966 0.172545
0.4	0.030279 0.195732	0.027929 0.188951	0.025862 0.182626	0.024032 0.176715	0.022403 0.171181
0.5	0.037126 0.192984	0.034328 0.186532	0.031849 0.180484	0.029643 0.174808	0.027670 0.169475
0.6	0.043598 0.189798	0.040407 0.183709	0.037565 0.177970	0.035022 0.172560	0.032738 0.167455
0.7	0.049665 0.186239	0.046141 0.180534	0.042983 0.175128	0.040144 0.170006	0.037582 0.165151
0.8	0.055311 0.182368	0.051509 0.177061	0.048083 0.172003	0.044989 0.167184	0.042185 0.162596
0.9	0.060529 0.178243	0.056501 0.173340	0.052854 0.168637	0.049544 0.164132	0.046532 0.159821
1.0	0.065318 0.173918	0.061114 0.169418	0.057289 0.165072	0.053801 0.160886	0.050615 0.156858
1.1	0.069685 0.169445	0.065350 0.165339	0.061387 0.161349	0.057757 0.157480	0.054428 0.153738
1.2	0.073641 0.164866	0.069216 0.161145	0.065151 0.157502	0.061413 0.153948	0.057971 0.150490
1.3	0.077202 0.160223	0.072722 0.156872	0.068589 0.153567	0.064773 0.150320	0.061246 0.147141
1.4	0.080385 0.155551	0.075883 0.152553	0.071711 0.149572	0.067844 0.146623	0.064258 0.143717
1.5	0.083210 0.150880	0.078712 0.148217	0.074529 0.145545	0.070636 0.142882	0.067012 0.140239
1.6	0.085697 0.146236	0.081229 0.143888	0.077055 0.141510	0.073158 0.139120	0.069518 0.136731
1.7	0.087870 0.141640	0.083450 0.139588	0.079306 0.137488	0.075423 0.135357	0.071785 0.133209
1.8	0.089749 0.137113	0.085394 0.135335	0.081297 0.133495	0.077445 0.131609	0.073823 0.129691
1.9	0.091355 0.132667	0.087080 0.131146	0.083044 0.129548	0.079236 0.127892	0.075646 0.126192
2.0	0.092711 0.128317	0.088525 0.127031	0.084562 0.125660	0.080811 0.124219	0.077263 0.122723
2.1	0.093835 0.124071	0.089749 0.123003	0.085867 0.121840	0.082182 0.120600	0.078687 0.119296
2.2	0.094748 0.119936	0.090767 0.119068	0.086974 0.118099	0.083364 0.117045	0.079930 0.115919
2.3	0.095467 0.115919	0.091597 0.115233	0.087900 0.114442	0.084370 0.113560	0.081004 0.112602
2.4	0.096010 0.112023	0.092255 0.111503	0.088657 0.110875	0.085213 0.110153	0.081921 0.109349
2.5	0.096393 0.108249	0.092754 0.107881	0.089259 0.107403	0.085905 0.106827	0.082690 0.106166
2.6	0.096632 0.104600	0.093110 0.104370	0.089719 0.104027	0.086458 0.103586	0.083324 0.103057
2.7	0.096739 0.101076	0.093336 0.100969	0.090050 0.100751	0.086883 0.100433	0.083832 0.100026
2.8	0.096729 0.097674	0.093442 0.097680	0.090263 0.097575	0.087190 0.097369	0.084225 0.097073
2.9	0.096613 0.094395	0.093442 0.094502	0.090368 0.094499	0.087391 0.094396	0.084511 0.094202
3.0	0.096402 0.091236	0.093345 0.091434	0.090375 0.091523	0.087493 0.091513	0.084700 0.091413

y	$x=3.5$	$x=3.6$	$x=3.7$	$x=3.8$	$x=3.9$
0.0	0.000005 0.168830	0.000002 0.163662	0.000001 0.158821	0.000001 0.154273	0.000000 0.149992
0.1	0.005340 0.168645	0.004995 0.163498	0.004685 0.158673	0.004406 0.154140	0.004153 0.149871
0.2	0.010633 0.168102	0.009952 0.163011	0.009339 0.158235	0.008786 0.153743	0.008282 0.149510
0.3	0.015846 0.167212	0.014841 0.162211	0.013935 0.157513	0.013115 0.153088	0.012368 0.148913
0.4	0.020944 0.165990	0.019632 0.161111	0.018446 0.156516	0.017370 0.152183	0.016389 0.148088
0.5	0.025897 0.164456	0.024297 0.159725	0.022847 0.155260	0.021529 0.151040	0.020326 0.147044
0.6	0.030677 0.162633	0.028812 0.158075	0.027118 0.153760	0.025574 0.149672	0.024162 0.145792
0.7	0.035263 0.160548	0.033158 0.156181	0.031239 0.152034	0.029486 0.148094	0.027880 0.144346
0.8	0.039637 0.158227	0.037316 0.154066	0.035195 0.150102	0.033253 0.146324	0.031469 0.142721
0.9	0.043785 0.155698	0.041274 0.151755	0.038974 0.147985	0.036861 0.144380	0.034916 0.140931
1.0	0.047698 0.152988	0.045023 0.149271	0.042565 0.145703	0.040301 0.142279	0.038212 0.138993
1.1	0.051370 0.150124	0.048556 0.146637	0.045962 0.143277	0.043567 0.140039	0.041352 0.136922
1.2	0.054798 0.147132	0.051869 0.143878	0.049161 0.140727	0.046653 0.137680	0.044328 0.134735
1.3	0.057984 0.144038	0.054962 0.141014	0.052159 0.138074	0.049558 0.135218	0.047139 0.132448
1.4	0.060928 0.140862	0.057835 0.138067	0.054958 0.135336	0.052279 0.132671	0.049783 0.130076
1.5	0.063637 0.137628	0.060491 0.135056	0.057557 0.132530	0.054819 0.130054	0.052260 0.127633
1.6	0.066116 0.134354	0.062936 0.131999	0.059962 0.129674	0.057179 0.127384	0.054572 0.125133
1.7	0.068374 0.131058	0.065176 0.128913	0.062177 0.126782	0.059362 0.124673	0.056720 0.122591
1.8	0.070419 0.127755	0.067217 0.125812	0.064206 0.123869	0.061374 0.121935	0.058708 0.120016
1.9	0.072260 0.124460	0.069068 0.122709	0.066058 0.120947	0.063219 0.119182	0.060540 0.117422
2.0	0.073908 0.121185	0.070736 0.119617	0.067738 0.118027	0.064903 0.116425	0.062222 0.114817
2.1	0.075373 0.117940	0.072232 0.116545	0.069254 0.115120	0.066433 0.113673	0.063759 0.112212
2.2	0.076666 0.114735	0.073563 0.113503	0.070615 0.112234	0.067815 0.110935	0.065156 0.109614
2.3	0.077796 0.111578	0.074739 0.110500	0.071829 0.109377	0.069058 0.108218	0.066420 0.107031
2.4	0.078774 0.108474	0.075770 0.107540	0.072902 0.106556	0.070166 0.105530	0.067556 0.104469
2.5	0.079611 0.105431	0.076664 0.104631	0.073845 0.103777	0.071149 0.102875	0.068572 0.101935
2.6	0.080316 0.102451	0.077430 0.101777	0.074663 0.101044	0.072013 0.100260	0.069474 0.099433
2.7	0.080898 0.099538	0.078076 0.098981	0.075366 0.098362	0.072764 0.097688	0.070267 0.096968
2.8	0.081366 0.096696	0.078612 0.096247	0.075961 0.095734	0.073411 0.095163	0.070959 0.094543
2.9	0.081730 0.093927	0.079044 0.093577	0.076455 0.093162	0.073959 0.092688	0.071555 0.092162
3.0	0.081996 0.091230	0.079381 0.090973	0.076855 0.090649	0.074415 0.090265	0.072061 0.089826

If $x>3.9$ or $y>3$ $\quad w(z)=iz\left(\dfrac{0.4613135}{z^2-0.1901635}+\dfrac{0.09999216}{z^2-1.7844927}+\dfrac{0.002883894}{z^2-5.5253437}\right)+\epsilon(z)\quad |\epsilon(z)|<2\times10^{-6}$

If $x>6$ or $y>6$ $\quad w(z)=iz\left(\dfrac{0.5124242}{z^2-0.2752551}+\dfrac{0.05176536}{z^2-2.724745}\right)+\eta(z)\quad |\eta(z)|<10^{-6}$ *

*See page II.

COMPLEX ZEROS OF THE ERROR FUNCTION Table 7.10

$$\text{erf } z_n = 0 \qquad z_n = x_n + iy_n$$

n	x_n	y_n	n	x_n	y_n
1	1.45061 616	1.88094 300	6	4.15899 840	4.43557 144
2	2.24465 928	2.61657 514	7	4.51631 940	4.78044 764
3	2.83974 105	3.17562 810	8	4.84797 031	5.10158 804
4	3.33546 074	3.64617 438	9	5.15876 791	5.40333 264
5	3.76900 557	4.06069 723	10	5.45219 220	5.68883 744

$$\text{erf } z_n = \text{erf}(-z_n) = \text{erf } \bar{z}_n = \text{erf}(-\bar{z}_n) = 0$$

$$\begin{matrix} x_n \\ y_n \end{matrix} \approx \frac{1}{2}\sqrt{\pi\left(4n - \frac{1}{2}\right)} \mp \frac{\ln\left(\pi\sqrt{2n - \frac{1}{4}}\right)}{2\sqrt{\pi\left(4n - \frac{1}{2}\right)}} \qquad (n > 0)$$

From H. E. Salzer, Complex zeros of the error function, J. Franklin Inst. **260**, 209–211, 1955 (with permission).

COMPLEX ZEROS OF FRESNEL INTEGRALS Table 7.11

$$C(z_n) = 0 \qquad z_n = x_n + iy_n$$
$$S(z_n^*) = 0 \qquad z_n^* = x_n^* + iy_n^*$$

n	x_n	y_n	x_n^*	y_n^*
0	0.0000	0.0000	0.0000	0.0000
1	1.7437	0.3057	2.0093	0.2886
2	2.6515	0.2529	2.8335	0.2443
3	3.3208	0.2239	3.4675	0.2185
4	3.8759	0.2047	4.0026	0.2008
5	4.3611	0.1909	4.4742	0.1877

$$\frac{C}{S}(z_n) = \frac{C}{S}(-z_n) = \frac{C}{S}(\bar{z}_n) = \frac{C}{S}(-\bar{z}_n) = \frac{C}{S}(iz_n) = \frac{C}{S}(-iz_n) = \frac{C}{S}(-i\bar{z}_n) = \frac{C}{S}(i\bar{z}_n) = 0$$

$$x_n \approx \sqrt{4n-1} - \frac{\ln(\pi\sqrt{4n-1})}{\pi^2(4n-1)^{3/2}} \qquad y_n \approx \frac{\ln(\pi\sqrt{4n-1})}{\pi\sqrt{4n-1}}$$

$$(n > 0)$$

$$x_n^* \approx 2\sqrt{n} - \frac{\ln(2\pi\sqrt{n})}{8\pi^2 n^{3/2}} \qquad y_n^* \approx \frac{\ln(2\pi\sqrt{n})}{2\pi\sqrt{n}}$$

MAXIMA AND MINIMA OF FRESNEL INTEGRALS Table 7.12

$$M_n = C\left(\sqrt{4n+1}\right) \qquad m_n = C\left(\sqrt{4n+3}\right) \qquad M_n^* = S\left(\sqrt{4n+2}\right) \qquad m_n^* = S\left(\sqrt{4n+4}\right)$$

n	M_n	m_n	M_n^*	m_n^*
0	0.779893	0.321056	0.713972	0.343415
1	0.640807	0.380389	0.628940	0.387969
2	0.605721	0.404260	0.600361	0.408301
3	0.588128	0.417922	0.584942	0.420516
4	0.577121	0.427036	0.574957	0.428877
5	0.569413	0.433666	0.567822	0.435059

$$M_n \sim \frac{1}{2} + \frac{\pi^2(4n+1)^2 - 3}{\pi^3(4n+1)^{5/2}} \qquad m_n \sim \frac{1}{2} - \frac{\pi^2(4n+3)^2 - 3}{\pi^3(4n+3)^{5/2}}$$

$$(n \to \infty)$$

$$M_n^* \sim \frac{1}{2} + \frac{\pi^2(4n+2)^2 - 3}{\pi^3(4n+2)^{5/2}} \qquad m_n^* \sim \frac{1}{2} - \frac{16\pi^2(n+1)^2 - 3}{32\pi^3(n+1)^{5/2}}$$

From G. N. Watson, A treatise on the theory of Bessel functions, 2d ed. Cambridge Univ. Press, Cambridge, England, 1958 (with permission).

8. Legendre Functions

Irene A. Stegun [1]

Contents

The author acknowledges the assistance of Ruth E. Capuano, Elizabeth F. Godefroy, David S. Liepman, and Bertha H. Walter in the preparation and checking of the tables and examples.

[1] National Bureau of Standards.

8. Legendre Functions

Mathematical Properties

Notation

The conventions used are $z=x+iy$, x, y real, and in particular, x always means a real number in the interval $-1\leq x\leq +1$ with $\cos\theta=x$ where θ is likewise a real number; n and m are positive integers or zero; ν and μ are unrestricted except where otherwise indicated.

Other notations are:

$$P^n(x)\ \text{for}\ \frac{n!P_n(x)}{(2n-1)!!}$$

$$P_{nm}(x)\ \text{for}\ (-1)^m P_n^m(x)$$

$$T_n^m(x)\ \text{for}\ (-1)^m P_n^m(x)$$

$$\overline{P_n^m}(x)\ \text{for}\ (-1)^m\sqrt{\frac{(2n+1)(n-m)!}{2(n+m)!}}\ P_n^m(x)$$

$$\mathfrak{P}_\nu^\mu(z)\ \text{for}\ P_\nu^\mu(z),\ \mathfrak{Q}_\nu^\mu(z)\ \text{for}\ Q_\nu^\mu(z)\qquad(\mathcal{R}z>1)$$

$$\mathfrak{Q}_\nu^\mu(z)\ \text{for}\ e^{\mu\pi i}Q_\nu^\mu(z)$$

$$\mathsf{Q}_\nu^\mu(z)\ \text{for}\ \frac{\sin(\nu+u)\pi}{\sin\nu\pi}\ Q_\nu^\mu(z)$$

Various other definitions of the functions occur as well as mixing of definitions.

8.1. Differential Equation

8.1.1

$$(1-z^2)\frac{d^2w}{dz^2}-2z\frac{dw}{dz}+[\nu(\nu+1)-\frac{\mu^2}{1-z^2}]w=0$$

Solutions

(Degree ν and order μ with singularities at $z=\pm 1$, ∞ as ordinary branch points—μ, ν arbitrary complex constants.)

$P_\nu^\mu(z)$, $Q_\nu^\mu(z)$ —**Associated Legendre Functions (Spherical Harmonics) of the First and Second Kinds** [2]

$$|\arg(z\pm 1)|<\pi,\qquad|\arg z|<\pi$$

$$(z^2-1)^{\frac{1}{2}\mu}=(z-1)^{\frac{1}{2}\mu}(z+1)^{\frac{1}{2}\mu}$$

(For $P_\nu^\mu(z)$, $\mu=0$, Legendre polynomials, see chapter **22**.)

8.1.2

$$P_\nu^\mu(z)=\frac{1}{\Gamma(1-\mu)}\left[\frac{z+1}{z-1}\right]^{\frac{1}{2}\mu}F\left(-\nu,\nu+1;1-\mu;\frac{1-z}{2}\right)$$
$$(|1-z|<2)$$

(For $F(a,b;c;z)$ see chapter **15**.)

8.1.3 $Q_\nu^\mu(z)=e^{i\mu\pi}2^{-\nu-1}\pi^{\frac{1}{2}}\dfrac{\Gamma(\nu+\mu+1)}{\Gamma(\nu+\frac{3}{2})}\ z^{-\nu-\mu-1}(z^2-1)^{\frac{1}{2}\mu}F\left(1+\dfrac{\nu}{2}+\dfrac{\mu}{2},\dfrac{1}{2}+\dfrac{\nu}{2}+\dfrac{\mu}{2};\nu+\dfrac{3}{2};\dfrac{1}{z^2}\right)$ $(|z|>1)$

Alternate Forms

(Additional forms may be obtained by means of the transformation formulas of the hypergeometric function, see [8.1].)

8.1.4 $P_\nu^\mu(z)=2^\mu\pi^{\frac{1}{2}}(z^2-1)^{-\frac{1}{2}\mu}\left\{\dfrac{F\left(-\dfrac{\nu}{2}-\dfrac{\mu}{2},\dfrac{1}{2}+\dfrac{\nu}{2}-\dfrac{\mu}{2};\dfrac{1}{2};z^2\right)}{\Gamma\left(\dfrac{1}{2}-\dfrac{\nu}{2}-\dfrac{\mu}{2}\right)\Gamma\left(1+\dfrac{\nu}{2}-\dfrac{\mu}{2}\right)}-2z\dfrac{F\left(\dfrac{1}{2}-\dfrac{\nu}{2}-\dfrac{\mu}{2},1+\dfrac{\nu}{2}-\dfrac{\mu}{2};\dfrac{3}{2};z^2\right)}{\Gamma\left(\dfrac{1}{2}+\dfrac{\nu}{2}-\dfrac{\mu}{2}\right)\Gamma\left(-\dfrac{\nu}{2}-\dfrac{\mu}{2}\right)}\right\}$ $(|z^2|<1)$

8.1.5 $P_\nu^\mu(z)=\dfrac{2^{-\nu-1}\pi^{-\frac{1}{2}}\Gamma(-\frac{1}{2}-\nu)z^{-\nu+\mu-1}}{(z^2-1)^{\mu/2}\Gamma(-\nu-\mu)}\ F\left(\dfrac{1}{2}+\dfrac{\nu}{2}-\dfrac{\mu}{2},1+\dfrac{\nu}{2}-\dfrac{\mu}{2};\nu+\dfrac{3}{2};z^{-2}\right)$

$$+\frac{2^\nu\Gamma(\frac{1}{2}+\nu)z^{\nu+\mu}}{(z^2-1)^{\mu/2}\Gamma(1+\nu-\mu)}\ F\left(-\frac{\nu}{2}-\frac{\mu}{2},\frac{1}{2}-\frac{\nu}{2}-\frac{\mu}{2};\frac{1}{2}-\nu;z^{-2}\right)\qquad(|z^{-2}|<1)$$

8.1.6 $e^{-i\mu\pi}Q_\nu^\mu(z)=\dfrac{\Gamma(1+\nu+\mu)\Gamma(-\mu)(z-1)^{\frac{1}{2}\mu}(z+1)^{-\frac{1}{2}\mu}}{2\Gamma(1+\nu-\mu)}\ F\left(-\nu,1+\nu;1+\mu;\dfrac{1-z}{2}\right)$

$$+\frac{1}{2}\Gamma(\mu)(z+1)^{\frac{1}{2}\mu}(z-1)^{-\frac{1}{2}\mu}F\left(-\nu,1+\nu;1-\mu;\frac{1-z}{2}\right)\qquad(|1-z|<2)\qquad *$$

---

[2] The functions $Y_n^m(\theta,\varphi)=\begin{Bmatrix}\cos m\varphi\\\sin m\varphi\end{Bmatrix}P_n^m(\cos\theta)$ called surface harmonics of the first kind, tesseral for $m<n$ and sectoral for $m=n$. With $0\leq\theta\leq\pi$, $0\leq\varphi\leq 2\pi$, they are everywhere one valued and continuous functions on the surface of the unit sphere $x^2+y^2+z^2=1$ where $x=\sin\theta\cos\varphi$, $y=\sin\theta\sin\varphi$ and $z=\cos\theta$.

*See page II.

8.1.7 $\quad e^{-i\mu\pi}Q_\nu^\mu(z)=\pi^{\frac{1}{2}}2^\mu(z^2-1)^{-\frac{1}{2}\mu}\left\{\dfrac{\Gamma\left(\dfrac{1}{2}+\dfrac{\nu}{2}+\dfrac{\mu}{2}\right)}{2\Gamma\left(1+\dfrac{\nu}{2}-\dfrac{\mu}{2}\right)}e^{\pm i\frac{1}{2}\pi(\mu-\nu-1)}F\left(-\dfrac{\nu}{2}-\dfrac{\mu}{2},\dfrac{1}{2}+\dfrac{\nu}{2}-\dfrac{\mu}{2};\dfrac{1}{2};z^2\right)\right.$

$$+\left.\frac{z\Gamma\left(1+\dfrac{\nu}{2}+\dfrac{\mu}{2}\right)e^{\pm i\frac{1}{2}\pi(\mu-\nu)}}{\Gamma\left(\dfrac{1}{2}+\dfrac{\nu}{2}-\dfrac{\mu}{2}\right)}F\left(\dfrac{1}{2}-\dfrac{\nu}{2}-\dfrac{\mu}{2},1+\dfrac{\nu}{2}-\dfrac{\mu}{2};\dfrac{3}{2};z^2\right)\right\} \qquad (|z^2|<1)$$

(Upper and lower signs according as $\mathscr{I}z\gtrless 0$.)

Wronskian

8.1.8

$$W\{P_\nu^\mu(z),Q_\nu^\mu(z)\}=\frac{e^{i\mu\pi}2^{2\mu}\Gamma\left(\dfrac{\nu+\mu+2}{2}\right)\Gamma\left(\dfrac{\nu+\mu+1}{2}\right)}{(1-z^2)\Gamma\left(\dfrac{\nu-\mu+2}{2}\right)\Gamma\left(\dfrac{\nu-\mu+1}{2}\right)}$$

8.1.9 $\qquad W\{P_n(z),Q_n(z)\}=-(z^2-1)^{-1}$

8.2. Relations Between Legendre Functions

Negative Degree

8.2.1 $\qquad P_{-\nu-1}^\mu(z)=P_\nu^\mu(z)$

8.2.2

$$Q_{-\nu-1}^\mu(z)=\{-\pi e^{i\mu\pi}\cos\nu\pi P_\nu^\mu(z)$$
$$+Q_\nu^\mu(z)\sin[\pi(\nu+\mu)]\}/\sin[\pi(\nu-\mu)]$$

Negative Argument ($\mathscr{I}z\gtrless 0$)

8.2.3

$$P_\nu^\mu(-z)=e^{\mp i\nu\pi}P_\nu^\mu(z)-\frac{2}{\pi}e^{-i\mu\pi}\sin[\pi(\nu+\mu)]Q_\nu^\mu(z)$$

8.2.4 $\qquad Q_\nu^\mu(-z)=-e^{\pm i\nu\pi}Q_\nu^\mu(z)$

Negative Order

8.2.5

$$P_\nu^{-\mu}(z)=\frac{\Gamma(\nu-\mu+1)}{\Gamma(\nu+\mu+1)}\left[P_\nu^\mu(z)-\frac{2}{\pi}e^{-i\mu\pi}\sin(\mu\pi)Q_\nu^\mu(z)\right]$$

8.2.6 $\qquad Q_\nu^{-\mu}(z)=e^{-2i\mu\pi}\dfrac{\Gamma(\nu-\mu+1)}{\Gamma(\nu+\mu+1)}Q_\nu^\mu(z)$

Degree $\mu+\frac{1}{2}$ and Order $\nu+\frac{1}{2}$ *
$$\mathscr{R}z>0$$

8.2.7 $\quad P_{-\mu-\frac{1}{2}}^{-\nu-\frac{1}{2}}\left(\dfrac{z}{(z^2-1)^{1/2}}\right)=\dfrac{(z^2-1)^{1/4}e^{-i\mu\pi}Q_\nu^\mu(z)}{(\frac{1}{2}\pi)^{1/2}\Gamma(\nu+\mu+1)}$ *

8.2.8
$$Q_{-\mu-\frac{1}{2}}^{-\nu-\frac{1}{2}}\left(\frac{z}{(z^2-1)^{1/2}}\right) \quad *$$
$$=-i(\tfrac{1}{2}\pi)^{1/2}\Gamma(-\nu-\mu)(z^2-1)^{1/4}e^{-i\nu\pi}P_\nu^\mu(z)$$

8.3. Values on the Cut
$$(-1<x<1)$$

8.3.1
$$P_\nu^\mu(x)=\tfrac{1}{2}[e^{\frac{1}{2}i\mu\pi}P_\nu^\mu(x+i0)+e^{-\frac{1}{2}i\mu\pi}P_\nu^\mu(x-i0)]$$

8.3.2
$$P_\nu^\mu(x)=e^{\pm\frac{1}{2}i\mu\pi}P_\nu^\mu(x\pm i0) \quad *$$

8.3.3
$$=i\pi^{-1}e^{-i\mu\pi}[e^{-\frac{1}{2}i\mu\pi}Q_\nu^\mu(x+i0) \quad *$$
$$-e^{\frac{1}{2}i\mu\pi}Q_\nu^\mu(x-i0)] \quad *$$

8.3.4
$$Q_\nu^\mu(x)=\tfrac{1}{2}e^{-i\mu\pi}[e^{-\frac{1}{2}i\mu\pi}Q_\nu^\mu(x+i0)+e^{\frac{1}{2}i\mu\pi}Q_\nu^\mu(x-i0)]$$

(Formulas for $P_\nu^\mu(x)$ and $Q_\nu^\mu(x)$ are obtained with the replacement of $z-1$ by $(1-x)e^{\pm i\pi}$, (z^2-1) by $(1-x^2)e^{\pm i\pi}$, $z+1$ by $x+1$ for $z=x\pm i0$.)

8.4. Explicit Expressions
$$(x=\cos\theta)$$

8.4.1 $\qquad P_0(z)=1 \qquad P_0(x)=1$

8.4.2
$$Q_0(z)=\frac{1}{2}\ln\left(\frac{z+1}{z-1}\right) \qquad Q_0(x)=\frac{1}{2}\ln\left(\frac{1+x}{1-x}\right)$$
$$=xF(\tfrac{1}{2},1;\tfrac{3}{2};x^2)$$

8.4.3 $\qquad P_1(z)=z \qquad P_1(x)=x=\cos\theta$

8.4.4
$$Q_1(z)=\frac{z}{2}\ln\left(\frac{z+1}{z-1}\right)-1 \qquad Q_1(x)=\frac{x}{2}\ln\left(\frac{1+x}{1-x}\right)-1$$

8.4.5
$$P_2(z)=\tfrac{1}{2}(3z^2-1) \qquad P_2(x)=\tfrac{1}{2}(3x^2-1)$$
$$=\tfrac{1}{4}(3\cos 2\theta+1)$$

8.4.6
$$Q_2(z)=\frac{1}{2}P_2(z)\ln\left(\frac{z+1}{z-1}\right) \qquad Q_2(x)=$$
$$-\frac{3z}{2} \qquad\qquad \left(\frac{3x^2-1}{4}\right)\ln\left(\frac{1+x}{1-x}\right)-\frac{3x}{2}$$

8.5. Recurrence Relations

(Both P_ν^μ and Q_ν^μ satisfy the same recurrence relations.)

Varying Order

8.5.1
$$P_\nu^{\mu+1}(z)=(z^2-1)^{-\frac{1}{2}}\{(\nu-\mu)zP_\nu^\mu(z)-(\nu+\mu)P_{\nu-1}^\mu(z)\}$$

*See page II.

8.5.2

$$(z^2-1)\frac{dP_\nu^\mu(z)}{dz}=(\nu+\mu)(\nu-\mu+1)(z^2-1)^{\frac12}P_\nu^{\mu-1}(z)$$
$$-\mu z P_\nu^\mu(z)$$

Varying Degree

8.5.3

$$(\nu-\mu+1)P_{\nu+1}^\mu(z)=(2\nu+1)zP_\nu^\mu(z)-(\nu+\mu)P_{\nu-1}^\mu(z)$$

8.5.4 $(z^2-1)\dfrac{dP_\nu^\mu(z)}{dz}=\nu z P_\nu^\mu(z)-(\nu+\mu)P_{\nu-1}^\mu(z)$

Varying Order and Degree

8.5.5 $P_{\nu+1}^\mu(z)=P_{\nu-1}^\mu(z)+(2\nu+1)(z^2-1)^{\frac12}P_\nu^{\mu-1}(z)$

8.6. Special Values

$$x=0$$

8.6.1

$$P_\nu^\mu(0)$$
$$=2^\mu\pi^{-\frac12}\cos\left[\tfrac12\pi(\nu+\mu)\right]\Gamma(\tfrac12\nu+\tfrac12\mu+\tfrac12)/\Gamma(\tfrac12\nu-\tfrac12\mu+1)$$

8.6.2

$$Q_\nu^\mu(0)=$$
$$-2^{\mu-1}\pi^{\frac12}\sin\left[\tfrac12\pi(\nu+\mu)\right]\Gamma(\tfrac12\nu+\tfrac12\mu+\tfrac12)/\Gamma(\tfrac12\nu-\tfrac12\mu+1)$$

8.6.3

$$\left[\frac{dP_\nu^\mu(x)}{dx}\right]_{x=0}=$$
$$2^{\mu+1}\pi^{-\frac12}\sin\left[\tfrac12\pi(\nu+\mu)\right]\Gamma(\tfrac12\nu+\tfrac12\mu+1)/\Gamma(\tfrac12\nu-\tfrac12\mu+\tfrac12)$$

8.6.4

$$\left[\frac{dQ_\nu^\mu(x)}{dx}\right]_{x=0}=$$
$$2^\mu\pi^{\frac12}\cos\left[\tfrac12\pi(\nu+\mu)\right]\Gamma(\tfrac12\nu+\tfrac12\mu+1)/\Gamma(\tfrac12\nu-\tfrac12\mu+\tfrac12)$$

8.6.5

$$W\{P_\nu^\mu(x),Q_\nu^\mu(x)\}_{x=0}=\frac{2^{2\mu}\Gamma(\tfrac12\nu+\tfrac12\mu+1)\Gamma(\tfrac12\nu+\tfrac12\mu+\tfrac12)}{\Gamma(\tfrac12\nu-\tfrac12\mu+1)\Gamma(\tfrac12\nu-\tfrac12\mu+\tfrac12)}$$

$$\mu=m=1,2,3,\ldots$$

8.6.6

$$P_\nu^m(z)=(z^2-1)^{\frac12 m}\frac{d^m P_\nu(z)}{dz^m},$$

$$P_\nu^m(x)=(-1)^m(1-x^2)^{\frac12 m}\frac{d^m P_\nu(x)}{dx^m}$$

8.6.7

$$Q_\nu^m(z)=(z^2-1)^{\frac12 m}\frac{d^m Q_\nu(z)}{dz^m},$$

$$Q_\nu^m(x)=(-1)^m(1-x^2)^{\frac12 m}\frac{d^m Q_\nu(x)}{dx^m}$$

$$\mu=\pm\tfrac12$$

8.6.8

$$P_\nu^{\frac12}(z)=(z^2-1)^{-1/4}(2\pi)^{-1/2}\{[z+(z^2-1)^{1/2}]^{\nu+\frac12}$$
$$+[z+(z^2-1)^{1/2}]^{-\nu-\frac12}\}$$

8.6.9

$$P_\nu^{-\frac12}(z)=\left(\frac{2}{\pi}\right)^{1/2}\frac{(z^2-1)^{-1/4}}{2\nu+1}\{[z+(z^2-1)^{1/2}]^{\nu+\frac12}$$
$$-[z+(z^2-1)^{1/2}]^{-\nu-\frac12}\}$$

8.6.10

$$Q_\nu^{\frac12}(z)=i(\tfrac12\pi)^{1/2}(z^2-1)^{-1/4}[z+(z^2-1)^{1/2}]^{-\nu-\frac12}$$

8.6.11

$$Q_\nu^{-\frac12}(z)=-i(2\pi)^{1/2}\frac{(z^2-1)^{-1/4}}{2\nu+1}[z+(z^2-1)^{1/2}]^{-\nu-\frac12} \quad *$$

8.6.12

$$P_\nu^{\frac12}(\cos\theta)=(\tfrac12\pi)^{-\frac12}(\sin\theta)^{-\frac12}\cos\left[(\nu+\tfrac12)\theta\right]$$

8.6.13

$$Q_\nu^{\frac12}(\cos\theta)=-(\tfrac12\pi)^{\frac12}(\sin\theta)^{-\frac12}\sin\left[(\nu+\tfrac12)\theta\right]$$

8.6.14

$$P_\nu^{-\frac12}(\cos\theta)=(\tfrac12\pi)^{-\frac12}(\nu+\tfrac12)^{-1}(\sin\theta)^{-\frac12}\sin\left[(\nu+\tfrac12)\theta\right]$$

8.6.15

$$Q_\nu^{-\frac12}(\cos\theta)=(2\pi)^{\frac12}(2\nu+1)^{-1}(\sin\theta)^{-\frac12}\cos\left[(\nu+\tfrac12)\theta\right] \quad *$$

$$\mu=-\nu$$

8.6.16 $P_\nu^{-\nu}(z)=\dfrac{2^{-\nu}(z^2-1)^{\frac12\nu}}{\Gamma(\nu+1)}$

8.6.17 $P_\nu^{-\nu}(\cos\theta)=\dfrac{2^{-\nu}(\sin\theta)^\nu}{\Gamma(\nu+1)}$

$$\mu=0,\ \nu=n$$

(Rodrigues' Formula)

8.6.18 $P_n(z)=\dfrac{1}{2^n n!}\dfrac{d^n(z^2-1)^n}{dz^n}$

8.6.19 $Q_n(x)=\dfrac12 P_n(x)\ln\dfrac{1+x}{1-x}-W_{n-1}(x)$

where

$$W_{n-1}(x)=\frac{2n-1}{1\cdot n}P_{n-1}(x)+\frac{2n-5}{3(n-1)}P_{n-3}(x)$$
$$+\frac{2n-9}{5(n-2)}P_{n-5}(x)+\ldots$$
$$=\sum_{m=1}^n\frac{1}{m}P_{m-1}(x)P_{n-m}(x)$$

$$W_{-1}(x)=0$$

$$\nu = 0, 1$$

8.6.20 $\quad \left[\dfrac{\partial P_\nu(\cos\theta)}{\partial\nu}\right]_{\nu=0} = 2\ln(\cos\tfrac{1}{2}\theta)$

8.6.21 $\quad \left[\dfrac{\partial P_\nu^{-1}(\cos\theta)}{\partial\nu}\right]_{\nu=0} = -\tan\tfrac{1}{2}\theta - 2\cot\tfrac{1}{2}\theta\,\ln(\cos\tfrac{1}{2}\theta)$

8.6.22 $\quad \left[\dfrac{\partial P_\nu^{-1}(\cos\theta)}{\partial\nu}\right]_{\nu=1} = -\tfrac{1}{2}\tan\tfrac{1}{2}\theta\,\sin^2\tfrac{1}{2}\theta + \sin\theta\,\ln(\cos\tfrac{1}{2}\theta)$

8.7. Trigonometric Expansions $(0 < \theta < \pi)$

8.7.1 $\quad P_\nu^\mu(\cos\theta) = \pi^{-1/2}2^{\mu+1}(\sin\theta)^\mu\,\dfrac{\Gamma(\nu+\mu+1)}{\Gamma(\nu+\frac{3}{2})}\sum_{k=0}^{\infty}\dfrac{(\mu+\frac{1}{2})_k(\nu+\mu+1)_k}{k!\,(\nu+\frac{3}{2})_k}\sin[(\nu+\mu+2k+1)\theta]$

8.7.2 $\quad Q_\nu^\mu(\cos\theta) = \pi^{1/2}2^\mu(\sin\theta)^\mu\,\dfrac{\Gamma(\nu+\mu+1)}{\Gamma(\nu+\frac{3}{2})}\sum_{k=0}^{\infty}\dfrac{(\mu+\frac{1}{2})_k(\nu+\mu+1)_k}{k!(\nu+\frac{3}{2})_k}\cos[(\nu+\mu+2k+1)\theta]$

8.7.3 $\quad P_n(\cos\theta) = \dfrac{2^{2n+2}(n!)^2}{\pi(2n+1)!}\left[\sin(n+1)\theta + \dfrac{n+1}{2n+3}\sin(n+3)\theta + \dfrac{1\cdot3}{2!}\dfrac{(n+1)(n+2)}{(2n+3)(2n+5)}\sin(n+5)\theta + \ldots\right]$

8.7.4 $\quad Q_n(\cos\theta) = \dfrac{2^{2n+1}(n!)^2}{(2n+1)!}\left[\cos(n+1)\theta + \dfrac{n+1}{2n+3}\cos(n+3)\theta + \dfrac{1\cdot3}{2!}\dfrac{(n+1)(n+2)}{(2n+3)(2n+5)}\cos(n+5)\theta + \ldots\right]$

8.8. Integral Representations

(z not on the real axis between -1 and $-\infty$)   *

8.8.1 $\quad P_\nu^\mu(z) = \dfrac{2^{-\nu}(z^2-1)^{-\frac{1}{2}\mu}}{\Gamma(-\nu-\mu)\Gamma(\nu+1)}\int_0^\infty(z+\cosh t)^{\mu-\nu-1}(\sinh t)^{2\nu+1}dt \qquad (\mathscr{R}(-\mu) > \mathscr{R}\nu > -1)$

8.8.2 $\quad Q_\nu^\mu(z) = \dfrac{e^{i\mu\pi}\sqrt{\pi}2^{-\mu}}{\Gamma(\mu+\frac{1}{2})}\dfrac{\Gamma(\nu+\mu+1)}{\Gamma(\nu-\mu+1)}(z^2-1)^{\frac{1}{2}\mu}\int_0^\infty[z+(z^2-1)^{\frac{1}{2}}\cosh t]^{-\nu-\mu-1}(\sinh t)^{2\mu}dt \quad (\mathscr{R}(\nu\pm\mu+1)>0)$ *

8.8.3 $\quad Q_n(z) = \dfrac{1}{2}\int_{-1}^1(z-t)^{-1}P_n(t)\,dt = (-1)^{n+1}Q_n(-z)$

(For other integral representations see [8.2].)

8.9. Summation Formulas

8.9.1 $\quad (\xi-z)\sum_{m=0}^{n}(2m+1)P_m(z)P_m(\xi) = (n+1)[P_{n+1}(\xi)P_n(z) - P_n(\xi)P_{n+1}(z)]$

8.9.2 $\quad (\xi-z)\sum_{m=0}^{n}(2m+1)P_m(z)Q_m(\xi) = 1 - (n+1)[P_{n+1}(z)Q_n(\xi) - P_n(z)Q_{n+1}(\xi)]$

8.10. Asymptotic Expansions

For fixed z and ν and $\mathscr{R}\mu \to \infty$, **8.10.1–8.10.3** are asymptotic expansions if z is not on the real axis between $-\infty$ and -1 and $+\infty$ and $+1$. (Upper or lower signs according as $\mathscr{I}z \gtrless 0$.)

8.10.1 $\quad P_\nu^\mu(z) = \dfrac{\Gamma(\nu+\mu+1)\Gamma(\mu-\nu)}{\pi\Gamma(\mu+1)}\left(\dfrac{z+1}{z-1}\right)^{\frac{1}{2}\mu}\sin\mu\pi\left[F\left(-\nu,\nu+1;1+\mu;\tfrac{1}{2}+\tfrac{1}{2}z\right)\right.$

$$\left. -\dfrac{\sin\nu\pi}{\sin\mu\pi}e^{\mp i\mu\pi}\left(\dfrac{z-1}{z+1}\right)^\mu F\left(-\nu,\nu+1;1+\mu;\tfrac{1}{2}-\tfrac{1}{2}z\right)\right]$$

8.10.2 $\quad Q_\nu^\mu(z) = \tfrac{1}{2}e^{i\mu\pi}\dfrac{\Gamma(\nu+\mu+1)}{\Gamma(\mu+1)}\left(\dfrac{z+1}{z-1}\right)^{\frac{1}{2}\mu}\Gamma(\mu-\nu)\left[F\left(-\nu,\nu+1;1+\mu;\tfrac{1}{2}+\tfrac{1}{2}z\right)\right.$

$$\left. -e^{\mp i\nu\pi}\left(\dfrac{z-1}{z+1}\right)^\mu F\left(-\nu,\nu+1;1+\mu;\tfrac{1}{2}-\tfrac{1}{2}z\right)\right]$$

*See page II.

8.10.3 $Q_\nu^{-\mu}(z) = \dfrac{e^{-i\mu\pi}\csc[\pi(\nu-\mu)]}{2\pi\Gamma(1+\mu)}\left[e^{\mp i\nu\pi}\left(\dfrac{z+1}{z-1}\right)^{-\frac{1}{2}\mu}F(-\nu,\nu+1;1+\mu;\tfrac{1}{2}-\tfrac{1}{2}z)\right.$

$$\left.-\left(\dfrac{z-1}{z+1}\right)^{-\frac{1}{2}\mu}F(-\nu,\nu+1;1+\mu;\tfrac{1}{2}+\tfrac{1}{2}z)\right]$$

With μ replaced by $-\mu$, **8.1.2** is an asymptotic expansion for $P_\nu^{-\mu}(z)$ for fixed z and ν and $\mathscr{R}\,\mu\to\infty$ if z is not on the real axis between $-\infty$ and -1.

For fixed z and μ and $\mathscr{R}\nu\to\infty$, **8.10.4** and **8.10.6** are asymptotic expansions if z is not on the real axis between $-\infty$ and -1 and $+\infty$ and $+1$; **8.10.5** if z is not on the real axis between $-\infty$ and $+1$.

8.10.4 $P_\nu^\mu(z) = (2\pi)^{-\frac{1}{2}}(z^2-1)^{-1/4}\dfrac{\Gamma(\nu+\mu+1)}{\Gamma(\nu+\frac{3}{2})}\left\{[z+(z^2-1)^{\frac{1}{2}}]^{\nu+\frac{1}{2}}F(\tfrac{1}{2}+\mu,\tfrac{1}{2}-\mu;\tfrac{3}{2}+\nu;\dfrac{z+(z^2-1)^{\frac{1}{2}}}{2(z^2-1)^{\frac{1}{2}}})\right.$

$$\left.+ie^{-i\mu\pi}[z-(z^2-1)^{\frac{1}{2}}]^{\nu+\frac{1}{2}}F(\tfrac{1}{2}+\mu,\tfrac{1}{2}-\mu;\tfrac{3}{2}+\nu;\dfrac{-z+(z^2-1)^{\frac{1}{2}}}{2(z^2-1)^{\frac{1}{2}}})\right\}$$

8.10.5 $Q_\nu^\mu(z) = e^{i\mu\pi}(\tfrac{1}{2}\pi)^{\frac{1}{2}}(z^2-1)^{-1/4}\dfrac{\Gamma(\nu+\mu+1)}{\Gamma(\nu+\frac{3}{2})}[z-(z^2-1)^{\frac{1}{2}}]^{\nu+\frac{1}{2}}F(\tfrac{1}{2}+\mu,\tfrac{1}{2}-\mu;\tfrac{3}{2}+\nu;\dfrac{-z+(z^2-1)^{\frac{1}{2}}}{2(z^2-1)^{\frac{1}{2}}})$

8.10.6 $Q_{-\nu}^\mu(z) = \dfrac{e^{i\mu\pi}(\tfrac{1}{2}\pi)^{\frac{1}{2}}(z^2-1)^{-1/4}}{\sin[\pi(\mu-\nu)]}\dfrac{\Gamma(\mu+\nu)}{\Gamma(\frac{1}{2}-\mu)}\left\{\cos\nu\pi[z+(z^2-1)^{\frac{1}{2}}]^{\nu-\frac{1}{2}}F(\tfrac{1}{2}+\mu,\tfrac{1}{2}-\mu;\tfrac{1}{2}+\nu;\dfrac{z+(z^2-1)^{\frac{1}{2}}}{2(z^2-1)^{\frac{1}{2}}})\right.$

$$\left.+ie^{i\nu\pi}\cos\mu\pi[z-(z^2-1)^{\frac{1}{2}}]^{\nu-\frac{1}{2}}F(\tfrac{1}{2}+\mu,\tfrac{1}{2}-\mu;\tfrac{1}{2}+\nu;\dfrac{-z+(z^2-1)^{\frac{1}{2}}}{2(z^2-1)^{\frac{1}{2}}})\right\}$$

The related asymptotic expansion for $P_{-\nu}^\mu(z)$ may be derived from **8.10.4** together with **8.2.1**.

8.10.7 $P_\nu^\mu(\cos\theta) = \dfrac{\Gamma(\nu+\mu+1)}{\Gamma(\nu+\frac{3}{2})}(\tfrac{1}{2}\pi\,\sin\theta)^{-\frac{1}{2}}\cos[(\nu+\tfrac{1}{2})\theta-\dfrac{\pi}{4}+\dfrac{\mu\pi}{2}]+O(\nu^{-1})$

8.10.8 $Q_\nu^\mu(\cos\theta) = \dfrac{\Gamma(\nu+\mu+1)}{\Gamma(\nu+\frac{3}{2})}\left(\dfrac{\pi}{2\sin\theta}\right)^{\frac{1}{2}}\cos[(\nu+\tfrac{1}{2})\theta+\dfrac{\pi}{4}+\dfrac{\mu\pi}{2}]+O(\nu^{-1})$ $(\epsilon<\theta<\pi-\epsilon,\ \epsilon>0)$

For other asymptotic expansions, see [8.7] and [8.9].

8.11. Toroidal Functions (or Ring Functions)

(Only special properties are given; other properties and representations follow from the earlier sections.)

8.11.1 $P_{\nu-\frac{1}{2}}^\mu(\cosh\eta) = [\Gamma(1-\mu)]^{-1}2^{2\mu}(1-e^{-2\eta})^{-\mu}e^{-(\nu+\frac{1}{2})\eta}F(\tfrac{1}{2}-\mu,\tfrac{1}{2}+\nu-\mu;1-2\mu;1-e^{-2\eta})$

8.11.2 $P_{n-\frac{1}{2}}^m(\cosh\eta) = \dfrac{\Gamma(n+m+\frac{1}{2})(\sinh\eta)^m}{\Gamma(n-m+\frac{1}{2})2^m\sqrt{\pi}\Gamma(m+\frac{1}{2})}\displaystyle\int_0^\pi\dfrac{(\sin\varphi)^{2m}d\varphi}{(\cosh\eta+\cos\varphi\sinh\eta)^{n+m+\frac{1}{2}}}$

8.11.3 $Q_{\nu-\frac{1}{2}}^\mu(\cosh\eta) = [\Gamma(1+\nu)]^{-1}\sqrt{\pi}e^{i\mu\pi}\Gamma(\tfrac{1}{2}+\nu+\mu)(1-e^{-2\eta})^\mu e^{-(\nu+\frac{1}{2})\eta}F(\tfrac{1}{2}+\mu,\tfrac{1}{2}+\nu+\mu;1+\nu;e^{-2\eta})$ *

8.11.4 $Q_{n-\frac{1}{2}}^m(\cosh\eta) = \dfrac{(-1)^m\Gamma(n+\frac{1}{2})}{\Gamma(n-m+\frac{1}{2})}\displaystyle\int_0^\infty\dfrac{\cosh mt\,dt}{(\cosh\eta+\cosh t\sinh\eta)^{n+\frac{1}{2}}}$ * $(n>m)$

---

*See page II.

8.12. Conical Functions

$$(P^\mu_{-\frac{1}{2}+i\lambda}(\cos\theta),\ Q^\mu_{-\frac{1}{2}+i\lambda}(\cos\theta))$$

(Only special properties are given as other properties and representations follow from earlier sections with $\nu=-\frac{1}{2}+i\lambda$ (λ, a real parameter) and $z=\cos\theta$.)

8.12.1

$$P_{-\frac{1}{2}+i\lambda}(\cos\theta)=1+\frac{4\lambda^2+1^2}{2^2}\sin^2\frac{\theta}{2}$$

$$+\frac{(4\lambda^2+1^2)(4\lambda^2+3^2)}{2^24^2}\sin^4\frac{\theta}{2}+\ \ldots\qquad(0\le\theta<\pi)$$

8.12.2 $\qquad P_{-\frac{1}{2}+i\lambda}(\cos\theta)=P_{-\frac{1}{2}-i\lambda}(\cos\theta)$

8.12.3 $\quad P_{-\frac{1}{2}+i\lambda}(\cos\theta)=\dfrac{2}{\pi}\displaystyle\int_0^\theta\frac{\cosh\lambda t\,dt}{\sqrt{2}\,(\cos t-\cos\theta)}$

8.12.4

$$Q_{-\frac{1}{2}\mp i\lambda}(\cos\theta)=\pm i\sinh\lambda\pi\int_0^\infty\frac{\cos\lambda t\,dt}{\sqrt{2(\cosh t+\cos\theta)}}$$

$$+\int_0^\infty\frac{\cosh\lambda t\,dt}{\sqrt{2(\cosh t-\cos\theta)}}$$

8.12.5

$$P_{-\frac{1}{2}+i\lambda}(-\cos\theta)$$

$$=\frac{\cosh\lambda\pi}{\pi}\left[Q_{-\frac{1}{2}+i\lambda}(\cos\theta)+Q_{-\frac{1}{2}-i\lambda}(\cos\theta)\right]$$

* 8.13. Relation to Elliptic Integrals
(see chapter 17) ($\mathscr{R}\eta>0$)

8.13.1 $\qquad P_{-\frac{1}{2}}(z)=\dfrac{2}{\pi}\sqrt{\dfrac{2}{z+1}}\,K\left(\sqrt{\dfrac{z-1}{z+1}}\right)$

8.13.2 $\quad P_{-\frac{1}{2}}(\cosh\eta)=\left[\dfrac{\pi}{2}\cosh\dfrac{\eta}{2}\right]^{-1}K\left(\tanh\dfrac{\eta}{2}\right)$

8.13.3 $\qquad Q_{-\frac{1}{2}}(z)=\sqrt{\dfrac{2}{z+1}}\,K\left(\sqrt{\dfrac{2}{z+1}}\right)$

8.13.4 $\qquad Q_{-\frac{1}{2}}(\cosh\eta)=2e^{-\eta/2}K(e^{-\eta})$

8.13.5

$$P_{\frac{1}{2}}(z)=\frac{2}{\pi}\,(z+\sqrt{z^2-1})^{\frac{1}{2}}E\left(\sqrt{\frac{2(z^2-1)^{1/2}}{z+(z^2-1)^{1/2}}}\right)$$

8.13.6 $\quad P_{\frac{1}{2}}(\cosh\eta)=\dfrac{2}{\pi}\,e^{\eta/2}E\left(\sqrt{1-e^{-2\eta}}\right)$

8.13.7

$$Q_{\frac{1}{2}}(z)=z\,\sqrt{\frac{2}{z+1}}\,K\left(\sqrt{\frac{2}{z+1}}\right)$$

$$-[2(z+1)]^{\frac{1}{2}}E\left(\sqrt{\frac{2}{z+1}}\right)\qquad *$$

$$(-1<x<1)$$

8.13.8 $\qquad P_{-\frac{1}{2}}(x)=\dfrac{2}{\pi}K\left(\sqrt{\dfrac{1-x}{2}}\right)$

8.13.9 $\qquad P_{-\frac{1}{2}}(\cos\theta)=\dfrac{2}{\pi}K\left(\sin\dfrac{\theta}{2}\right)$

8.13.10 $\qquad Q_{-\frac{1}{2}}(x)=K\left(\sqrt{\dfrac{1+x}{2}}\right)\ *$

8.13.11 $\quad P_{\frac{1}{2}}(x)=\dfrac{2}{\pi}\left[2E\left(\sqrt{\dfrac{1-x}{2}}\right)-K\left(\sqrt{\dfrac{1-x}{2}}\right)\right]$

8.13.12 $\quad Q_{\frac{1}{2}}(x)=K\left(\sqrt{\dfrac{1+x}{2}}\right)-2E\left(\sqrt{\dfrac{1+x}{2}}\right)\ *$

8.14. Integrals

8.14.1 $\displaystyle\int_1^\infty P_\nu(x)Q_\rho(x)dx=[(\rho-\nu)(\rho+\nu+1)]^{-1}\qquad\qquad(\mathscr{R}\rho>\mathscr{R}\nu>0)$

8.14.2 $\displaystyle\int_1^\infty Q_\nu(x)Q_\rho(x)dx=[(\rho-\nu)(\rho+\nu+1)]^{-1}[\psi(\rho+1)-\psi(\nu+1)]\qquad(\mathscr{R}(\rho+\nu)>-1,\ \rho+\nu+1\neq0;$

$$\nu,\ \rho\neq-1,\ -2,\ -3,\ \ldots)$$

8.14.3 $\displaystyle\int_1^\infty[Q_\nu(x)]^2dx=(2\nu+1)^{-1}\psi'(\nu+1)\qquad\qquad(\mathscr{R}\nu>-\tfrac{1}{2})$

8.14.4 $\displaystyle\int_{-1}^1 P_\nu(x)P_\rho(x)dx=\frac{2}{\pi^2}\,[(\rho-\nu)(\rho+\nu+1)]^{-1}\{2\sin\pi\nu\sin\pi\rho[\psi(\nu+1)-\psi(\rho+1)]+\pi\sin(\pi\rho-\pi\nu)\}$

$$(\rho+\nu+1\neq0)$$

8.14.5 $\displaystyle\int_{-1}^1[P_\nu(x)]^2dx=\frac{\pi^2-2(\sin\pi\nu)^2\psi'(\nu+1)}{\pi^2(\nu+\frac{1}{2})}\qquad *$

8.14.6 $\displaystyle\int_{-1}^1 Q_\nu(x)Q_\rho(x)dx=[(\rho-\nu)(\rho+\nu+1)]^{-1}\{[\psi(\nu+1)-\psi(\rho+1)][1+\cos\rho\pi\cos\nu\pi]-\tfrac{1}{2}\pi\sin(\nu\pi-\rho\pi)\}\ *$

$$(\rho+\nu+1\neq0;\ \nu,\ \rho\neq-1,\ -2,\ -3,\ \ldots)$$

8.14.7 $\displaystyle\int_{-1}^1[Q_\nu(x)]^2dx=(2\nu+1)^{-1}\{\tfrac{1}{2}\pi^2-\psi'(\nu+1)[1+(\cos\nu\pi)^2]\}\qquad\qquad(\nu\neq-1,\ -2,\ -3,\ \ldots)$

8.14.8 $\int_{-1}^{1} P_\nu(x) Q_\rho(x) dx = [(\nu-\rho)(\rho+\nu+1)]^{-1} \left\{ 1 - \cos(\rho\pi - \nu\pi) - \frac{2}{\pi}\sin \pi\nu \cos \pi\nu[\psi(\nu+1) - \psi(\rho+1)] \right\}$

$$(\mathscr{R}\nu>0,\ \mathscr{R}\rho>0,\ \rho\neq\nu)$$

8.14.9 $\int_{-1}^{1} P_\nu(x) Q_\nu(x) dx = -\frac{1}{\pi}(2\nu+1)^{-1}\sin 2\nu\pi \psi'(\nu+1)$ $\qquad (\mathscr{R}\nu>0)$

$$(m,\ n,\ l\ \text{positive integers})$$

8.14.10

$$\int_{-1}^{1} Q_n^m(x) P_l^m(x) dx = (-1)^m \frac{1-(-1)^{l+n}(n+m)!}{(l-n)(l+n+1)(n-m)!}$$

8.14.11 $\int_{-1}^{1} P_n^m(x) P_l^m(x) dx = 0$ $\qquad (l\neq n)$

8.14.12 $\int_{-1}^{1} P_n^m(x) P_n^l(x)(1-x^2)^{-1} dx = 0$ $\qquad (l\neq m)$

8.14.13 $\int_{-1}^{1} [P_n^m(x)]^2 dx = (n+\tfrac{1}{2})^{-1}(n+m)!/(n-m)!$

8.14.14

$$\int_{-1}^{1} (1-x^2)^{-1}[P_n^m(x)]^2 dx = (n+m)!/m(n-m)!$$

8.14.15

$$\int_0^1 P_\nu(x) x^\rho dx = \frac{\pi^{\frac{1}{2}}2^{-\rho-1}\Gamma(1+\rho)}{\Gamma(1+\tfrac{1}{2}\rho-\tfrac{1}{2}\nu)\Gamma(\tfrac{1}{2}\rho+\tfrac{1}{2}\nu+\tfrac{3}{2})}$$

$$(\mathscr{R}\rho>-1)$$

8.14.16

$$\int_0^\pi (\sin t)^{\alpha-1} P_\nu^{-\mu}(\cos t) dt = \frac{2^{-\mu}\pi\Gamma(\tfrac{1}{2}\alpha+\tfrac{1}{2}\mu)\Gamma(\tfrac{1}{2}\alpha-\tfrac{1}{2}\mu)}{\Gamma(\tfrac{1}{2}+\tfrac{1}{2}\alpha+\tfrac{1}{2}\nu)\Gamma(\tfrac{1}{2}\alpha-\tfrac{1}{2}\nu)\Gamma(\tfrac{1}{2}\mu+\tfrac{1}{2}\nu+1)\Gamma(\tfrac{1}{2}\mu-\tfrac{1}{2}\nu+\tfrac{1}{2})} \qquad (\mathscr{R}(\alpha\pm\mu)>0)$$

8.14.17

$$P_\nu^{-m}(z) = (z^2-1)^{-\frac{1}{2}m}\int_1^z \cdots \int_1^z P_\nu(z)(dz)^m$$

8.14.18

$$Q_\nu^{-m}(z) = (-1)^m(z^2-1)^{-\frac{1}{2}m}\int_z^\infty \cdots \int_z^\infty Q_\nu(z)(dz)^m$$

For other integrals, see [8.2], [8.4] and chapter **22.**

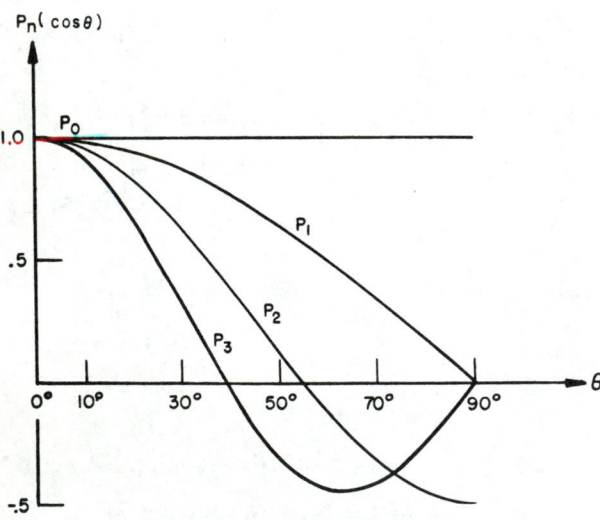

FIGURE 8.1. $P_n(\cos\theta)$. $n=0(1)3$.

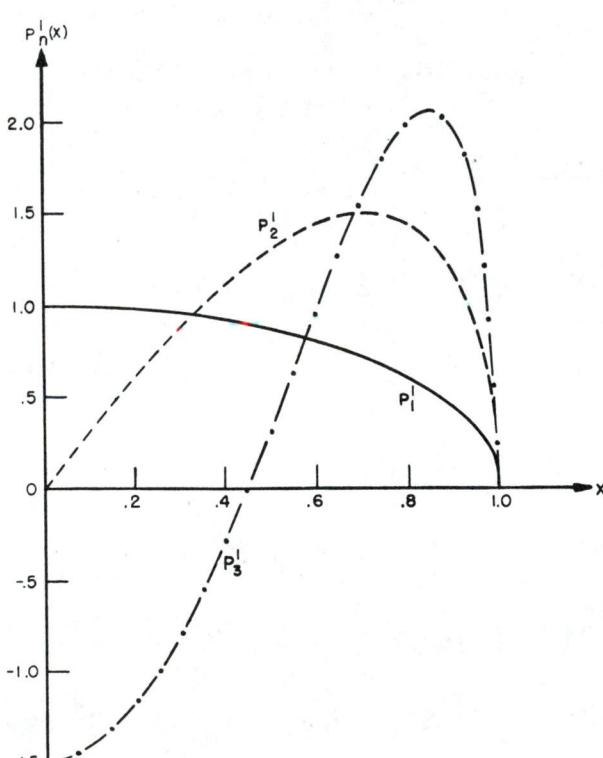

FIGURE 8.2. $P_n^1(x)$. $n=1(1)3,\ x\leq1$.

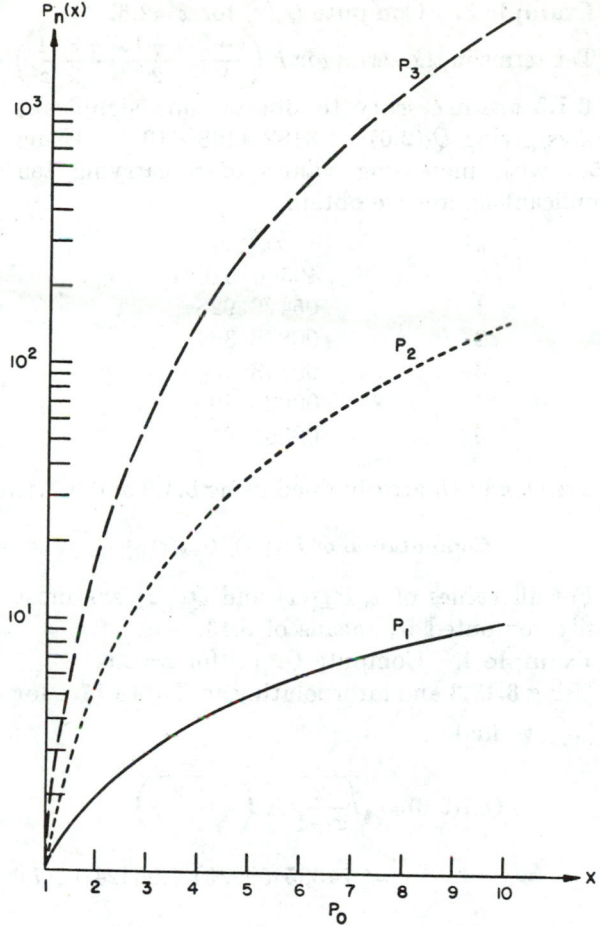

FIGURE 8.3. $P_n(x)$. $n=0(1)3, x\geq 1$.

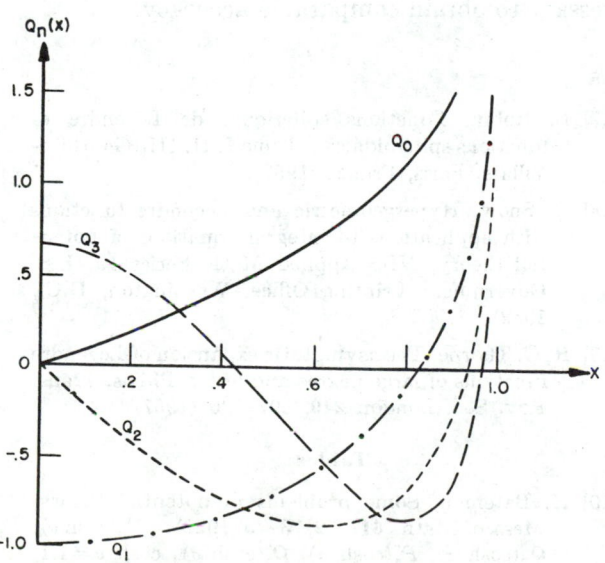

FIGURE 8.4. $Q_n(x)$. $n=0(1)3, x<1$.

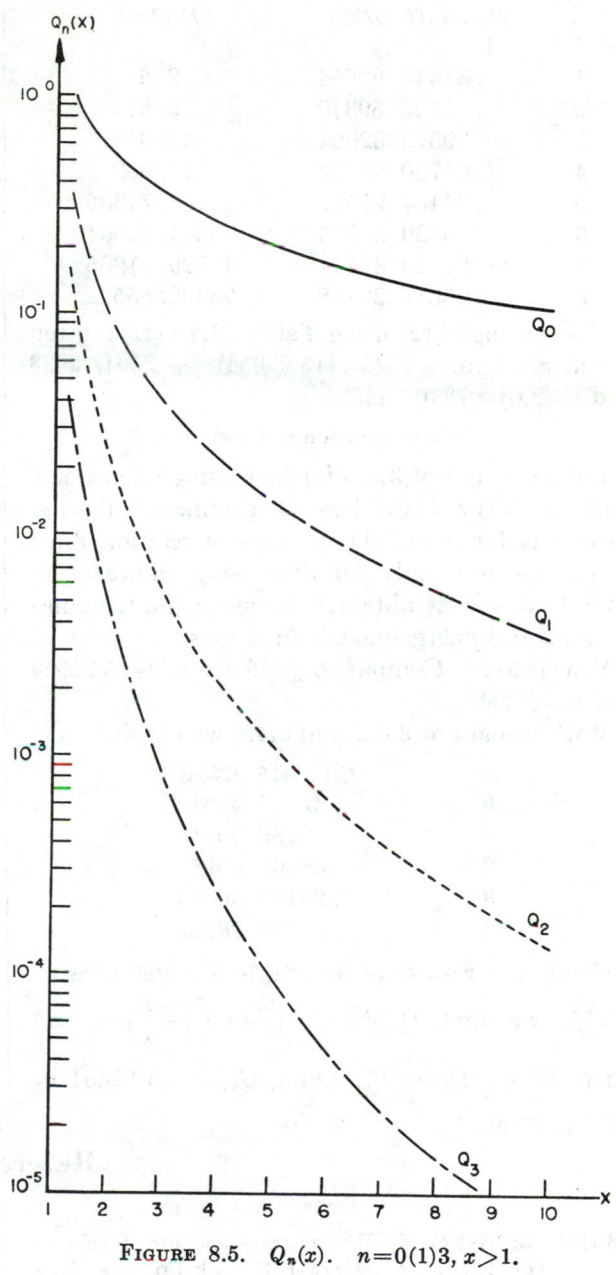

FIGURE 8.5. $Q_n(x)$. $n=0(1)3, x>1$.

Numerical Methods

8.15. Use and Extension of the Tables

Computation of $P_n(x)$

For all values of x there is very little loss of significant figures (except at zeros) in using the recurrence relation **8.5.3** for increasing values of n.

Example 1. Compute $P_n(x)$ for $x=.31415\ 92654$ and $x=2.6$ for $n=2(1)8$.

n	$P_n(.31415\ 92654)$	$P_n(2.6)$
0	1	1
1	.31415 92654	2.6
2	−.35195 59340	9.64
3	−.39372 32064	40.04
4	.04750 63122	174.952
5	.34184 27517	786.74336
6	.15729 86975	3604.350016
7	−.20123 39354	16729.51005
8	−.25617 29328	78402.55522

Computing $P_8(x)$ using **Table 22.9** carrying ten significant figures, $P_8(.31415\ 92654) = -.25617\ 2933$ and $P_8(2.6) = 78402.55526$.

Computation of $Q_n(x)$

For $x < 1$, use of **8.5.3** for increasing values of n leads to very little loss of significant figures. However, for $x > 1$, the recurrence relation **8.5.3** should be used only for decreasing values of n, after having first obtained Q_n using the formulas in terms of hypergeometric functions.

Example 2. Compute $Q_n(x)$ for $x = .31415\ 92654$ and $n = 0(1)4$.

With the aid of **8.4.2** and **8.4.4** we obtain

n	$Q_n(.31415\ 92654)$
0	.32515 34813
1	−.89785 00212
2	−.58567 85953
3	.29190 60854
4	.59974 26989

Using the results of **Example 1** together with **8.6.19**, we find $Q_4(x) = \frac{1}{2}P_4(x)\ln\left(\frac{1+x}{1-x}\right) - W_3(x)$ where $W_3 = \frac{7}{4}\,P_3 + \frac{1}{3}\,P_1$, giving $Q_4(.31415\ 92654) = .59974\ 26989$.

Example 3. Compute $Q_5(x)$ for $x = 2.6$.

Ten terms in the series for $F\left(\frac{\nu+2}{2}, \frac{\nu+1}{2}; \nu+\frac{3}{2}; \frac{1}{z^2}\right)$ of **8.1.3** are necessary to obtain nine significant figures giving $Q_5(2.6) = 4.8182\ 4468 \times 10^{-5}$. Using **8.5.3** with increasing values of n carrying ten significant figures we obtain

n	$Q_n(2.6)$
0	.40546 51081
1	.05420 928
2	.00868 364
3	.00148 95
4	.00026 49
5	.00004 81

where Q_0 and Q_1 are obtained using **8.4.2** and **8.4.4**.

Computation of $P_{\pm\frac{1}{2}}(x)$, $Q_{\pm\frac{1}{2}}(x)$

For all values of x, $P_{\pm\frac{1}{2}}(x)$ and $Q_{\pm\frac{1}{2}}(x)$ are most easily computed by means of **8.13**.

Example 4. Compute $Q_{-\frac{1}{2}}(x)$ for $x = 2.6$.

Using **8.13.3** and interpolating in **Table 17.1** for $K(.5)$, we find

$$Q_{-\frac{1}{2}}(2.6) = \sqrt{\frac{2}{x+1}}\,K\left(\sqrt{\frac{2}{x+1}}\right)$$

$$= (.74535\ 59925)(1.90424\ 1417)$$

$$= 1.41933\ 7751.$$

On the other hand, at least nine terms in the expansion of $F\left(\frac{\nu+2}{2}, \frac{\nu+1}{2}; \nu+\frac{3}{2}; \frac{1}{z^2}\right)$ of **8.1.3** are necessary to obtain comparable accuracy.

References

Texts

[8.1] A. Erdélyi et al., Higher transcendental functions, vol. 1, ch. 3 (McGraw-Hill Book Co., Inc., New York, N.Y., 1953).

[8.2] E. W. Hobson, The theory of spherical and ellipsoidal harmonics (Chelsea Publishing Co., New York, N.Y., 1955).

[8.3] J. Lense, Kugelfunktionen (Akademische Verlagsgesellschaft, Leipzig, Germany, 1950).

[8.4] T. M. MacRobert, Spherical harmonics, 2d rev. ed. (Dover Publications, Inc., New York, N.Y., 1948).

[8.5] W. Magnus and F. Oberhettinger, Formulas and theorems for the special functions of mathematical physics (Chelsea Publishing Co., New York, N.Y., 1949).

[8.6] G. Prasad, A treatise on spherical harmonics and the functions of Bessel and Lamé, Part II (Advanced) (Mahamandal Press, Benares City, India, 1932).

[8.7] L. Robin, Fonctions sphériques de Legendre e fonctions sphéroidales. Tome I, II, III (Gauthier-Villars, Paris, France, 1957).

[8.8] C. Snow, Hypergeometric and Legendre functions with applications to integral equations of potential theory, NBS Applied Math. Series 19 (U.S. Government Printing Office, Washington, D.C., 1952).

[8.9] R. C. Thorne, The asymptotic expansion of Legendre functions of large degree and order, Philos. Trans. Roy. Soc. London **249**, 597–620 (1957).

Tables

[8.10] H. Bateman, Some problems in potential theory, Mess. of Math., **617**, 52, 73–75 (1922). $P_n(\cosh\sigma)$, $Q_n(\cosh\sigma)$, $P'_n(\cosh\sigma)$, $Q'_n(\cosh\sigma)$; $\cosh\sigma = 1.1$, $n = 0(1)20$, 10D; $\cosh\sigma = 1.2$, 2, 3; $n = 0(1)10$, exact or 10D.

[8.11] Centre National d'Études des Télécommunications, Tables des fonctions de Legendre associées. Fonction associée de première espèce $P_n^m(\cos \theta)$ (Éditions de La Revue d'Optique, Paris, France, 1952). $n = -\frac{1}{2}(.1)10$, $m = 0(1)5$, $\theta = 0°(1°)90°$ (variable number of figures).

[8.12] Centre National d'Études des Télécommunications, Tables numérique des fonctions associées de Legendre. Fonctions associées de première espèce $P_n^m(\cos \theta)$ (Éditions de La Revue d'Optique, Paris, France, 1959). $n = -\frac{1}{2}(.1)10$, $m = 0(1)2$, $\theta = 0°(1°)180°$ (variable number of figures).

[8.13] G. C. Clark and S. W. Churchill, Table of Legendre polynomials $P_n(\cos \theta)$ for $n = 0(1)80$ and $\theta = 0°(1°)180°$, Engineering Research Institute Publications (Univ. of Michigan Press, Ann Arbor, Mich., 1957).

[8.14] R. O. Gumprecht and G. M. Sliepcevich, Tables of functions of the first and second partial derivatives of Legendre polynomials (Univ. of Michigan Press, Ann Arbor, Mich., 1951). Values of $[x\pi_n - (1-x^2)\pi_n'] \cdot 10^4$ and $\pi_n 10^4$ for $\gamma = 0°(10°)170°(1°)180°$, $n = 1(1)420$, 5S.

[8.15] M. E. Lynam, Table of Legendre functions for complex arguments TG–323, The Johns Hopkins Univ. Applied Physics Laboratory, Baltimore, Md. (1958).

[8.16] National Bureau of Standards, Tables of associated Legendre functions (Columbia Univ. Press, New York, N.Y., 1945). $P_n^m(\cos \theta)$, $\frac{d}{d\theta} P_n^m(\cos \theta)$, $n = 1(1)10$, $m(\leq n) = 0(1)4$, $\theta = 0°(1°)90°$, 6S; $P_n^m(x)$, $\frac{d}{dx} P_n^m(x)$, $n = 1(1)10$, $(-1)^m Q_n^m(x)$, $(-1)^{m+1}\frac{d}{dx} Q_n^m(x)$, $n = 0(1)10$, $m(\leq n) = 0(1)4$, $x = 1(.1)10$, 6S or exact; $i^{-n}P_n^m(ix)$, $i^{-n}\frac{d}{dx} P_n^m(ix)$, $n = 1(1)10$, $i^{n+2m+1}Q_n^m(ix)$, $i^{n+2m-1}\frac{d}{dx} Q_n^m(ix)$, $n = 0(1)10$, $m(\leq n) = 0(1)4$, $x = 0(.1)10$, 6S; $P_{n+\frac{1}{2}}^m(x)$, $\frac{d}{dx} P_{n-\frac{1}{2}}^m(x)$, $(-1)^m Q_{n-\frac{1}{2}}^m(x)$, $(-1)^{m+1}\frac{d}{dx} Q_{n+\frac{1}{2}}^m$, $n = -1(1)4$, $m = 0(1)4$, $x = 1(.1)10$, 4–6S.

[8.17] G. Prevost, Tables des fonctions sphériques et de leurs intégrales (Gauthier-Villars, Bordeaux and Paris, France, 1933). $P_n(x), \int_0^x P_n(t)\,dt$, $n = 1(1)10$; $P_n^j(x), \int_0^x P_n^j(t)\,dt$, $n = 0(1)8$, $j = 0(1)n$, $x = 0(.01)1$, 5S.

[8.18] H. Tallqvist, Sechsstellige Tafeln der 32 ersten Kugelfunktionen $P_n(\cos \theta)$, Acta Soc. Sci. Fenn., Nova Series A, II, 11 (1938). $P_n(\cos \theta)$, $n = 1(1)32$, $0°(10')90°$; 6D.

[8.19] H. Tallqvist, Acta Soc. Sci. Fenn., Nova Series A, II, 4(1937). $P_n(x)$, $n = 1(1)16$, $x = 0(.001)1$, 6D.

[8.20] H. Tallqvist, Tafeln der Kugelfunktionen $P_{25}(\cos \theta)$ bis $P_{32}(\cos \theta)$, Soc. Sci. Fenn. Comment. Phys.-Math., VI, 10(1932). $P_n(\cos \theta)$, $n = 25(1)32$, $\theta = 0°(1°)90°$, 5D.

[8.21] H. Tallqvist, Tafeln der 24 ersten Kugelfunktionen $P_n(\cos \theta)$, Soc. Sci. Fenn. Comment. Phys.-Math., VI, 3(1932). $P_n(\cos \theta)$, $n = 1(1)24$, $\theta = 0°(1°)90°$. 5D.

Table 8.1   **LEGENDRE FUNCTION—FIRST KIND $P_n(x)$**

$$P_0(x) = 1 \qquad\qquad P_1(x) = x$$

x	arccos x	$P_2(x)$	$P_3(x)$	$P_9(x)$	$P_{10}(x)$
0.00	90.00000 00	−0.50000	0.00000 00	0.00000 000	−0.24609 37
0.01	89.42703 26	−0.49985	−0.01499 75	0.02457 330	−0.24474 14
0.02	88.85400 80	−0.49940	−0.02998 00	0.04893 045	−0.24069 84
0.03	88.28086 87	−0.49865	−0.04493 25	0.07285 701	−0.23400 69
0.04	87.70755 72	−0.49760	−0.05984 00	0.09614 188	−0.22473 64
0.05	87.13401 60	−0.49625	−0.07468 75	0.11857 899	−0.21298 35
0.06	86.56018 72	−0.49460	−0.08946 00	0.13996 890	−0.19887 11
0.07	85.98601 28	−0.49265	−0.10414 25	0.16012 040	−0.18254 68
0.08	85.41143 43	−0.49040	−0.11872 00	0.17885 206	−0.16418 20
0.09	84.83639 29	−0.48785	−0.13317 75	0.19599 366	−0.14397 02
0.10	84.26082 95	−0.48500	−0.14750 00	0.21138 764	−0.12212 50
0.11	83.68468 44	−0.48185	−0.16167 25	0.22489 042	−0.09887 86
0.12	83.10789 74	−0.47840	−0.17568 00	0.23637 363	−0.07447 93
0.13	82.53040 77	−0.47465	−0.18950 75	0.24572 526	−0.04918 90
0.14	81.95215 37	−0.47060	−0.20314 00	0.25285 070	−0.02328 12
0.15	81.37307 34	−0.46625	−0.21656 25	0.25767 367	+0.00296 18
0.16	80.79310 38	−0.46160	−0.22976 00	0.26013 706	0.02925 20
0.17	80.21218 10	−0.45665	−0.24271 75	0.26020 358	0.05529 81
0.18	79.63024 02	−0.45140	−0.25542 00	0.25785 632	0.08080 85
0.19	79.04721 58	−0.44585	−0.26785 25	0.25309 918	0.10549 42
0.20	78.46304 10	−0.44000	−0.28000 00	0.24595 712	0.12907 20
0.21	77.87764 77	−0.43385	−0.29184 75	0.23647 631	0.15126 74
0.22	77.29096 70	−0.42740	−0.30338 00	0.22472 407	0.17181 75
0.23	76.70292 82	−0.42065	−0.31458 25	0.21078 870	0.19047 36
0.24	76.11345 96	−0.41360	−0.32544 00	0.19477 914	0.20700 49
0.25	75.52248 78	−0.40625	−0.33593 75	0.17682 442	0.22120 02
0.26	74.92993 79	−0.39860	−0.34606 00	0.15707 305	0.23287 14
0.27	74.33573 31	−0.39065	−0.35579 25	0.13569 215	0.24185 52
0.28	73.73979 53	−0.38240	−0.36512 00	0.11286 642	0.24801 62
0.29	73.14204 40	−0.37385	−0.37402 75	0.08879 707	0.25124 81
0.30	72.54239 69	−0.36500	−0.38250 00	0.06370 038	0.25147 63
0.31	71.94076 95	−0.35585	−0.39052 25	0.03780 634	0.24865 91
0.32	71.33707 51	−0.34640	−0.39808 00	+0.01135 691	0.24278 89
0.33	70.73122 45	−0.33665	−0.40515 75	−0.01539 566	0.23389 37
0.34	70.12312 59	−0.32660	−0.41174 00	−0.04219 085	0.22203 73
0.35	69.51268 49	−0.31625	−0.41781 25	−0.06876 185	0.20732 00
0.36	68.89980 39	−0.30560	−0.42336 00	−0.09483 780	0.18987 83
0.37	68.28438 27	−0.29465	−0.42836 75	−0.12014 608	0.16988 48
0.38	67.66631 73	−0.28340	−0.43282 00	−0.14441 472	0.14754 72
0.39	67.04550 06	−0.27185	−0.43670 25	−0.16737 489	0.12310 73
0.40	66.42182 15	−0.26000	−0.44000 00	−0.18876 356	0.09683 91
0.41	65.79516 52	−0.24785	−0.44269 75	−0.20832 609	0.06904 71
0.42	65.16541 25	−0.23540	−0.44478 00	−0.22581 900	0.04006 39
0.43	64.53243 99	−0.22265	−0.44623 25	−0.24101 269	+0.01024 69
0.44	63.89611 88	−0.20960	−0.44704 00	−0.25369 426	−0.02002 45
0.45	63.25631 61	−0.19625	−0.44718 75	−0.26367 022	−0.05035 30
0.46	62.61289 25	−0.18260	−0.44666 00	−0.27076 932	−0.08032 72
0.47	61.96570 35	−0.16865	−0.44544 25	−0.27484 521	−0.10952 64
0.48	61.31459 80	−0.15440	−0.44352 00	−0.27577 908	−0.13752 51
0.49	60.65941 84	−0.13985	−0.44087 75	−0.27348 225	−0.16389 87
0.50	60.00000 00	−0.12500	−0.43750 00	−0.26789 856	−0.18822 86
		$\begin{bmatrix} (-4)5 \\ 5 \end{bmatrix}$	$\begin{bmatrix} (-5)4 \\ 3 \end{bmatrix}$	$\begin{bmatrix} (-5)9 \\ 4 \end{bmatrix}$	$\begin{bmatrix} (-4)4 \\ 6 \end{bmatrix}$	$\begin{bmatrix} (-4)4 \\ 6 \end{bmatrix}$

$$P_2(x) = \tfrac{1}{2}(-1 + 3x^2) \qquad P_3(x) = \frac{x}{2}(-3 + 5x^2)$$

$$P_9(x) = \frac{x}{512}(1260 - 18480x^2 + 72072x^4 - 102960x^6 + 48620x^8)$$

$$P_{10}(x) = \frac{1}{1024}(-252 + 13860x^2 - 120120x^4 + 360360x^6 - 437580x^8 + 184756x^{10})$$

$$(n+1)P_{n+1}(x) = (2n+1)xP_n(x) - nP_{n-1}(x)$$

For coefficients of other polynomials, see chapter **22**.

LEGENDRE FUNCTION—FIRST KIND $P_n(x)$ Table 8.1

$$P_0(x) = 1 \qquad P_1(x) = x$$

x	arccos x	$P_2(x)$	$P_3(x)$	$P_9(x)$	$P_{10}(x)$
0.50	60.00000 00	−0.12500	−0.43750 00	−0.26789 856	−0.18822 86
0.51	59.33617 03	−0.10985	−0.43337 25	−0.25900 667	−0.21010 83
0.52	58.66774 85	−0.09440	−0.42848 00	−0.24682 215	−0.22914 92
0.53	57.99454 51	−0.07865	−0.42280 75	−0.23139 939	−0.24498 73
0.54	57.31636 11	−0.06260	−0.41634 00	−0.21283 321	−0.25728 92
0.55	56.63298 70	−0.04625	−0.40906 25	−0.19126 025	−0.26575 85
0.56	55.94420 22	−0.02960	−0.40096 00	−0.16686 000	−0.27014 28
0.57	55.24977 42	−0.01265	−0.39201 75	−0.13985 552	−0.27023 97
0.58	54.54945 74	+0.00460	−0.38222 00	−0.11051 366	−0.26590 30
0.59	53.84299 18	0.02215	−0.37155 25	−0.07914 497	−0.25704 92
0.60	53.13010 24	0.04000	−0.36000 00	−0.04610 304	−0.24366 27
0.61	52.41049 70	0.05815	−0.34754 75	−0.01178 332	−0.22580 16
0.62	51.68386 55	0.07660	−0.33418 00	+0.02337 862	−0.20360 19
0.63	50.94987 75	0.09535	−0.31988 25	0.05890 951	−0.17728 16
0.64	50.20818 05	0.11440	−0.30464 00	0.09430 141	−0.14714 41
0.65	49.45839 81	0.13375	−0.28843 75	0.12901 554	−0.11358 05
0.66	48.70012 72	0.15340	−0.27126 00	0.16248 693	−0.07707 01
0.67	47.93293 52	0.17335	−0.25309 25	0.19412 981	−0.03818 08
0.68	47.15635 69	0.19360	−0.23392 00	0.22334 410	+0.00243 30
0.69	46.36989 11	0.21415	−0.21372 75	0.24952 270	0.04403 37
0.70	45.57299 60	0.23500	−0.19250 00	0.27205 993	0.08580 58
0.71	44.76508 47	0.25615	−0.17022 25	0.29036 111	0.12686 31
0.72	43.94551 96	0.27760	−0.14688 00	0.30385 323	0.16625 89
0.73	43.11360 59	0.29935	−0.12245 75	0.31199 698	0.20299 76
0.74	42.26858 44	0.32140	−0.09694 00	0.31430 004	0.23605 08
0.75	41.40962 21	0.34375	−0.07031 25	0.31033 185	0.26437 45
0.76	40.53580 21	0.36640	−0.04256 00	0.29973 981	0.28693 19
0.77	39.64611 11	0.38935	−0.01366 75	0.28226 712	0.30271 79
0.78	38.73942 46	0.41260	+0.01638 00	0.25777 224	0.31078 93
0.79	37.81448 85	0.43615	0.04759 75	0.22625 012	0.31029 79
0.80	36.86989 76	0.46000	0.08000 00	0.18785 528	0.30052 98
0.81	35.90406 86	0.48415	0.11360 25	0.14292 678	0.28094 87
0.82	34.91520 62	0.50860	0.14842 00	0.09201 529	0.25124 52
0.83	33.90126 20	0.53335	0.18446 75	+0.03591 226	0.21139 19
0.84	32.85988 04	0.55840	0.22176 00	−0.02431 874	0.16170 50
0.85	31.78833 06	0.58375	0.26031 25	−0.08730 820	0.10291 23
0.86	30.68341 71	0.60940	0.30014 00	−0.15134 456	+0.03622 91
0.87	29.54136 05	0.63535	0.34125 75	−0.21433 544	−0.03655 86
0.88	28.35763 66	0.66160	0.38368 00	−0.27376 627	−0.11300 29
0.89	27.12675 31	0.68815	0.42742 25	−0.32665 610	−0.18989 29
0.90	25.84193 28	0.71500	0.47250 00	−0.36951 049	−0.26314 56
0.91	24.49464 85	0.74215	0.51892 75	−0.39827 146	−0.32768 58
0.92	23.07391 81	0.76960	0.56672 00	−0.40826 421	−0.37731 58
0.93	21.56518 50	0.79735	0.61589 25	−0.39414 060	−0.40457 43
0.94	19.94844 36	0.82540	0.66646 00	−0.34981 919	−0.40058 29
0.95	18.19487 23	0.85375	0.71843 75	−0.26842 182	−0.35488 03
0.96	16.26020 47	0.88240	0.77184 00	−0.14220 642	−0.25524 34
0.97	14.06986 77	0.91135	0.82668 25	+0.03750 397	−0.08749 40
0.98	11.47834 09	0.94060	0.88298 00	0.28039 609	+0.16470 81
0.99	8.10961 44	0.97015	0.94074 75	0.59724 553	0.52008 90
1.00	0.00000 00	1.00000	1.00000 00	1.00000 000	1.00000 00
		$\begin{bmatrix}(-5)4\\3\end{bmatrix}$	$\begin{bmatrix}(-4)2\\4\end{bmatrix}$	$\begin{bmatrix}(-2)1\\7\end{bmatrix}$	$\begin{bmatrix}(-2)2\\7\end{bmatrix}$

$$P_2(x) = \tfrac{1}{2}(-1+3x^2) \qquad P_3(x) = \tfrac{x}{2}(-3+5x^2)$$

$$P_9(x) = \tfrac{x}{512}(1260 - 18480x^2 + 72072x^4 - 102960x^6 + 48620x^8)$$

$$P_{10}(x) = \tfrac{1}{1024}(-252 + 13860x^2 - 120120x^4 + 360360x^6 - 437580x^8 + 184756x^{10})$$

$$(n+1)P_{n+1}(x) = (2n+1)xP_n(x) - nP_{n-1}(x)$$

For coefficients of other polynomials, see chapter 22.

Table 8.2   **DERIVATIVE OF THE LEGENDRE FUNCTION—FIRST KIND $P_n'(x)$**

$$P_1'(x) = 1 \qquad P_2'(x) = 3x$$

x	$P_3'(x)$	$P_4'(x)$	$P_9'(x)$	$P_{10}'(x)$
0.00	−1.50000	0.00000 00	2.46093 75	0.00000 00
0.01	−1.49925	−0.07498 25	2.45011 64	0.27023 41
0.02	−1.49700	−0.14986 00	2.41773 75	0.53765 93
0.03	−1.49325	−0.22452 75	2.36405 34	0.79949 17
0.04	−1.48800	−0.29888 00	2.28948 35	1.05299 82
0.05	−1.48125	−0.37281 25	2.19461 13	1.29552 05
0.06	−1.47300	−0.44622 00	2.08018 11	1.52449 98
0.07	−1.46325	−0.51899 75	1.94709 32	1.73750 05
0.08	−1.45200	−0.59104 00	1.79639 87	1.93223 25
0.09	−1.43925	−0.66224 25	1.62929 31	2.10657 29
0.10	−1.42500	−0.73250 00	1.44710 87	2.25858 73
0.11	−1.40925	−0.80170 75	1.25130 64	2.38654 80
0.12	−1.39200	−0.86976 00	1.04346 68	2.48895 24
0.13	−1.37325	−0.93655 25	0.82528 00	2.56453 90
0.14	−1.35300	−1.00198 00	0.59853 47	2.61230 18
0.15	−1.33125	−1.06593 75	0.36510 73	2.63150 28
0.16	−1.30800	−1.12832 00	+0.12694 88	2.62168 25
0.17	−1.28325	−1.18902 25	−0.11392 76	2.58266 81
0.18	−1.25700	−1.24794 00	−0.35546 01	2.51458 04
0.19	−1.22925	−1.30496 75	−0.59555 27	2.41783 68
0.20	−1.20000	−1.36000 00	−0.83208 96	2.29315 33
0.21	−1.16925	−1.41293 25	−1.06295 03	2.14154 35
0.22	−1.13700	−1.46366 00	−1.28602 54	1.96431 51
0.23	−1.10325	−1.51207 75	−1.49923 18	1.76306 37
0.24	−1.06800	−1.55808 00	−1.70052 94	1.53966 43
0.25	−1.03125	−1.60156 25	−1.88793 72	1.29625 99
0.26	−0.99300	−1.64242 00	−2.05954 92	1.03524 77
0.27	−0.95325	−1.68054 75	−2.21355 15	0.75926 26
0.28	−0.91200	−1.71584 00	−2.34823 78	0.47115 77
0.29	−0.86925	−1.74819 25	−2.46202 63	+0.17398 30
0.30	−0.82500	−1.77750 00	−2.55347 51	−0.12903 87
0.31	−0.77925	−1.80365 75	−2.62129 80	−0.43453 90
0.32	−0.73200	−1.82656 00	−2.66437 95	−0.73903 23
0.33	−0.68325	−1.84610 25	−2.68178 96	−1.03894 72
0.34	−0.63300	−1.86218 00	−2.67279 74	−1.33065 96
0.35	−0.58125	−1.87468 75	−2.63688 47	−1.61052 81
0.36	−0.52800	−1.88352 00	−2.57375 82	−1.87493 10
0.37	−0.47325	−1.88857 25	−2.48336 07	−2.12030 43
0.38	−0.41700	−1.88974 00	−2.36588 14	−2.34318 21
0.39	−0.35925	−1.88691 75	−2.22176 52	−2.54023 74
0.40	−0.30000	−1.88000 00	−2.05172 01	−2.70832 36
0.41	−0.23925	−1.86888 25	−1.85672 35	−2.84451 75
0.42	−0.17700	−1.85346 00	−1.63802 69	−2.94616 13
0.43	−0.11325	−1.83362 75	−1.39715 86	−3.01090 51
0.44	−0.04800	−1.80928 00	−1.13592 50	−3.03674 96
0.45	+0.01875	−1.78031 25	−0.85640 91	−3.02208 63
0.46	0.08700	−1.74662 00	−0.56096 76	−2.96573 83
0.47	0.15675	−1.70809 75	−0.25222 53	−2.86699 80
0.48	0.22800	−1.66464 00	+0.06693 30	−2.72566 30
0.49	0.30075	−1.61614 25	0.39337 29	−2.54206 98
0.50	0.37500	−1.56250 00	0.72372 44	−2.31712 34
	$\begin{bmatrix}(-4)2\\3\end{bmatrix}$	$\begin{bmatrix}(-4)6\\4\end{bmatrix}$	$\begin{bmatrix}(-3)3\\6\end{bmatrix}$	$\begin{bmatrix}(-3)5\\6\end{bmatrix}$

$$P_3'(x) = \tfrac{1}{2}(-3 + 15x^2) \qquad P_4'(x) = \frac{x}{8}(-60 + 140x^2)$$

$$P_9'(x) = \frac{1}{512}(1260 - 55440x^2 + 360360x^4 - 720720x^6 + 437580x^8)$$

$$P_{10}'(x) = \frac{x}{1024}(27720 - 480480x^2 + 2162160x^4 - 3500640x^6 + 1847560x^8)$$

$$P_n'(x) = \frac{n+1}{1-x^2}[xP_n(x) - P_{n+1}(x)]$$

DERIVATIVE OF THE LEGENDRE FUNCTION—FIRST KIND $P'_n(x)$

Table 8.2

$$P'_1(x) = 1 \qquad P'_2(x) = 3x$$

x	$P'_3(x)$	$P'_4(x)$	$P'_9(x)$	$P'_{10}(x)$
0.50	0.37500	$-$ 1.56250 00	0.72372 44	$-$ 2.31712 34
0.51	0.45075	$-$ 1.50360 75	1.05439 75	$-$ 2.05232 40
0.52	0.52800	$-$ 1.43936 00	1.38160 24	$-$ 1.74978 82
0.53	0.60675	$-$ 1.36965 25	1.70137 21	$-$ 1.41226 67
0.54	0.68700	$-$ 1.29438 00	2.00958 86	$-$ 1.04315 43
0.55	0.76875	$-$ 1.21343 75	2.30201 29	$-$ 0.64649 54
0.56	0.85200	$-$ 1.12672 00	2.57431 87	$-$ 0.22698 16
0.57	0.93675	$-$ 1.03412 25	2.82213 05	$+$ 0.21005 92
0.58	1.02300	$-$ 0.93554 00	3.04106 49	0.65868 10
0.59	1.11075	$-$ 0.83086 75	3.22677 77	1.11234 92
0.60	1.20000	$-$ 0.72000 00	3.37501 44	1.56397 82
0.61	1.29075	$-$ 0.60283 25	3.48166 60	2.00598 31
0.62	1.38300	$-$ 0.47926 00	3.54283 00	2.43034 08
0.63	1.47675	$-$ 0.34917 75	3.55487 57	2.82866 68
0.64	1.57200	$-$ 0.21248 00	3.51451 63	3.19230 45
0.65	1.66875	$-$ 0.06906 25	3.41888 50	3.51243 07
0.66	1.76700	$+$ 0.08118 00	3.26561 84	3.78017 74
0.67	1.86675	0.23835 25	3.05294 51	3.98677 13
0.68	1.96800	0.40256 00	2.77978 03	4.12369 16
0.69	2.07075	0.57390 75	2.44582 82	4.18284 84
0.70	2.17500	0.75250 00	2.05168 93	4.15678 18
0.71	2.28075	0.93844 25	1.59897 66	4.03888 45
0.72	2.38800	1.13184 00	1.09043 73	3.82364 72
0.73	2.49675	1.33279 75	$+$ 0.53008 28	3.50693 03
0.74	2.60700	1.54142 00	$-$ 0.07667 36	3.08626 20
0.75	2.71875	1.75781 25	$-$ 0.72287 14	2.56116 49
0.76	2.83200	1.98208 00	$-$ 1.39984 93	1.93351 26
0.77	2.94675	2.21432 75	$-$ 2.09708 32	1.20791 71
0.78	3.06300	2.45466 00	$-$ 2.80201 52	$+$ 0.39215 05
0.79	3.18075	2.70318 25	$-$ 3.49987 45	$-$ 0.50239 96
0.80	3.30000	2.96000 00	$-$ 4.17348 81	$-$ 1.46023 77
0.81	3.42075	3.22521 75	$-$ 4.80308 26	$-$ 2.46122 91
0.82	3.54300	3.49894 00	$-$ 5.36607 64	$-$ 3.48002 97
0.83	3.66675	3.78127 25	$-$ 5.83686 10	$-$ 4.48547 21
0.84	3.79200	4.07232 00	$-$ 6.18657 35	$-$ 5.43990 91
0.85	3.91875	4.37218 75	$-$ 6.38285 68	$-$ 6.29851 03
0.86	4.04700	4.68098 00	$-$ 6.38961 06	$-$ 7.00851 07
0.87	4.17675	4.99880 25	$-$ 6.16672 97	$-$ 7.50840 93
0.88	4.30800	5.32576 00	$-$ 5.66983 23	$-$ 7.72711 51
0.89	4.44075	5.66195 75	$-$ 4.84997 54	$-$ 7.58303 90
0.90	4.57500	6.00750 00	$-$ 3.65335 89	$-$ 6.98312 79
0.91	4.71075	6.36249 25	$-$ 2.02101 73	$-$ 5.82184 03
0.92	4.84800	6.72704 00	$+$ 0.11150 20	$-$ 3.98006 04
0.93	4.98675	7.10124 75	2.81447 18	$-$ 1.32394 73
0.94	5.12700	7.48522 00	6.16433 35	$+$ 2.29628 14
0.95	5.26875	7.87906 25	10.24405 70	7.04763 58
0.96	5.41200	8.28288 00	15.14351 59	13.11571 11
0.97	5.55675	8.69677 75	20.95987 66	20.70612 01
0.98	5.70300	9.12086 00	27.79800 16	30.04600 25
0.99	5.85075	9.55523 25	35.77086 77	41.38561 43
1.00	6.00000	10.00000 00	45.00000 00	55.00000 00
	$\begin{bmatrix} (-4)2 \\ 3 \end{bmatrix}$	$\begin{bmatrix} (-3)1 \\ 4 \end{bmatrix}$	$\begin{bmatrix} (-1)2 \\ 7 \end{bmatrix}$	$\begin{bmatrix} (-1)3 \\ 7 \end{bmatrix}$

$$P'_3(x) = \tfrac{1}{2}(-3+15x^2) \qquad P'_4(x) = \frac{x}{8}(-60+140x^2)$$

$$P'_9(x) = \frac{1}{512}(1260 - 55440x^2 + 360360x^4 - 720720x^6 + 437580x^8)$$

$$P'_{10}(x) = \frac{x}{1024}(27720 - 480480x^2 + 2162160x^4 - 3500640x^6 + 1847560x^8)$$

$$P'_n(x) = \frac{n+1}{1-x^2}[xP_n(x) - P_{n+1}(x)]$$

Table 8.3   **LEGENDRE FUNCTION—SECOND KIND $Q_n(x)$**

x	$Q_0(x)$	$Q_1(x)$	$Q_2(x)$	$Q_3(x)$	$Q_9(x)$	$Q_{10}(x)$
0.00	0.00000 000	−1.00000 000	0.00000 000	0.66666 667	−0.40634 921	0.00000 000
0.01	0.01000 033	−0.99990 000	−0.01999 867	0.66626 669	−0.40452 191	−0.04056 181
0.02	0.02000 267	−0.99959 995	−0.03998 933	0.66506 699	−0.39905 538	−0.08068 584
0.03	0.03000 900	−0.99909 973	−0.05996 399	0.66306 829	−0.38999 553	−0.11993 860
0.04	0.04002 135	−0.99839 915	−0.07991 463	0.66027 179	−0.37741 852	−0.15789 513
0.05	0.05004 173	−0.99749 791	−0.09983 321	0.65667 917	−0.36143 026	−0.19414 321
0.06	0.06007 216	−0.99639 567	−0.11971 169	0.65229 261	−0.34216 562	−0.22828 745
0.07	0.07011 467	−0.99509 197	−0.13954 199	0.64711 475	−0.31978 750	−0.25995 321
0.08	0.08017 133	−0.99358 629	−0.15931 602	0.64114 873	−0.29448 565	−0.28879 038
0.09	0.09024 419	−0.99187 802	−0.17902 563	0.63439 817	−0.26647 538	−0.31447 701
0.10	0.10033 535	−0.98996 647	−0.19866 264	0.62686 720	−0.23599 595	−0.33672 259
0.11	0.11044 692	−0.98785 084	−0.21821 885	0.61856 044	−0.20330 891	−0.35527 122
0.12	0.12058 103	−0.98553 028	−0.23768 596	0.60948 299	−0.16869 616	−0.36990 435
0.13	0.13073 985	−0.98300 382	−0.25705 567	0.59964 048	−0.13245 792	−0.38044 330
0.14	0.14092 558	−0.98027 042	−0.27631 958	0.58903 905	−0.09491 050	−0.38675 142
0.15	0.15114 044	−0.97732 893	−0.29546 923	0.57768 532	−0.05638 395	−0.38873 587
0.16	0.16138 670	−0.97417 813	−0.31449 610	0.56558 646	−0.01721 959	−0.38634 905
0.17	0.17166 666	−0.97081 667	−0.33339 158	0.55275 016	+0.02223 260	−0.37958 962
0.18	0.18198 269	−0.96724 312	−0.35214 699	0.53918 465	0.06161 670	−0.36850 308
0.19	0.19233 717	−0.96345 594	−0.37075 353	0.52489 868	0.10057 361	−0.35318 198
0.20	0.20273 255	−0.95945 349	−0.38920 232	0.50990 155	0.13874 395	−0.33376 565
0.21	0.21317 135	−0.95523 402	−0.40748 439	0.49420 314	0.17577 093	−0.31043 947
0.22	0.22365 611	−0.95079 566	−0.42559 062	0.47781 388	0.21130 336	−0.28343 378
0.23	0.23418 947	−0.94613 642	−0.44351 180	0.46074 476	0.24499 861	−0.25302 221
0.24	0.24477 411	−0.94125 421	−0.46123 857	0.44300 738	0.27652 557	−0.21951 969
0.25	0.25541 281	−0.93614 680	−0.47876 145	0.42461 393	0.30556 765	−0.18327 994
0.26	0.26610 841	−0.93081 181	−0.49607 081	0.40557 719	0.33182 571	−0.14469 251
0.27	0.27686 382	−0.92524 677	−0.51315 685	0.38591 059	0.35502 089	−0.10417 949
0.28	0.28768 207	−0.91944 902	−0.53000 962	0.36562 819	0.37489 746	−0.06219 173
0.29	0.29856 626	−0.91341 578	−0.54661 900	0.34474 467	0.39122 551	−0.01920 468
0.30	0.30951 960	−0.90714 412	−0.56297 466	0.32327 542	0.40380 351	+0.02428 610
0.31	0.32054 541	−0.90063 092	−0.57906 608	0.30123 647	0.41246 080	0.06776 975
0.32	0.33164 711	−0.89387 293	−0.59488 256	0.27864 459	0.41705 981	0.11072 534
0.33	0.34282 825	−0.88686 668	−0.61041 313	0.25551 723	0.41749 822	0.15262 723
0.34	0.35409 253	−0.87960 854	−0.62564 662	0.23187 261	0.41371 084	0.19295 076
0.35	0.36544 375	−0.87209 469	−0.64057 159	0.20772 970	0.40567 128	0.23117 811
0.36	0.37688 590	−0.86432 108	−0.65517 633	0.18310 825	0.39339 336	0.26680 432
0.37	0.38842 310	−0.85628 345	−0.66944 887	0.15802 883	0.37693 227	0.29934 337
0.38	0.40005 965	−0.84797 733	−0.68337 690	0.13251 285	0.35638 546	0.32833 437
0.39	0.41180 003	−0.83939 799	−0.69694 784	0.10658 256	0.33189 317	0.35334 774
0.40	0.42364 893	−0.83054 043	−0.71014 872	0.08026 114	0.30363 867	0.37399 123
0.41	0.43561 122	−0.82139 940	−0.72296 624	0.05357 267	0.27184 811	0.38991 596
0.42	0.44769 202	−0.81196 935	−0.73538 670	+0.02654 221	0.23679 006	0.40082 218
0.43	0.45989 668	−0.80224 443	−0.74739 600	−0.00080 418	0.19877 461	0.40646 477
0.44	0.47223 080	−0.79221 845	−0.75897 958	−0.02843 939	0.15815 208	0.40665 845
0.45	0.48470 028	−0.78188 487	−0.77012 243	−0.05633 524	0.11531 136	0.40128 259
0.46	0.49731 129	−0.77123 681	−0.78080 904	−0.08446 239	0.07067 773	0.39028 551
0.47	0.51007 034	−0.76026 694	−0.79102 336	−0.11279 034	+0.02471 030	0.37368 827
0.48	0.52298 428	−0.74896 755	−0.80074 877	−0.14128 732	−0.02210 100	0.35158 779
0.49	0.53606 034	−0.73733 044	−0.80996 804	−0.16992 027	−0.06923 897	0.32415 933
0.50	0.54930 614	−0.72534 693	−0.81866 327	−0.19865 477	−0.11616 303	0.29165 814
	$\begin{bmatrix} (-5)2 \\ 5 \end{bmatrix}$	$\begin{bmatrix} (-5)4 \\ 5 \end{bmatrix}$	$\begin{bmatrix} (-5)7 \\ 5 \end{bmatrix}$	$\begin{bmatrix} (-4)1 \\ 5 \end{bmatrix}$	$\begin{bmatrix} (-4)5 \\ 6 \end{bmatrix}$	$\begin{bmatrix} (-4)7 \\ 6 \end{bmatrix}$

$$Q_0(x) = \tfrac{1}{2} \ln\left(\frac{1+x}{1-x}\right) \qquad Q_1(x) = \frac{x}{2} \ln\left(\frac{1+x}{1-x}\right) - 1$$

$$Q_2(x) = \frac{3x^2-1}{4} \ln\left(\frac{1+x}{1-x}\right) - \frac{3x}{2} \qquad Q_3(x) = \frac{x}{4}(5x^2-3) \ln\left(\frac{1+x}{1-x}\right) - \frac{5x^2}{2} + \frac{2}{3}$$

$$(n+1)Q_{n+1}(x) = (2n+1)xQ_n(x) - nQ_{n-1}(x)$$

$Q_0(x) = \operatorname{arctanh} x$ (**Table 4.17**) is included here for completeness.

LEGENDRE FUNCTION—SECOND KIND $Q_n(x)$

Table 8.3

x	$Q_0(x)$	$Q_1(x)$	$Q_2(x)$	$Q_3(x)$	$Q_9(x)$	$Q_{10}(x)$
0.50	0.54930 614	−0.72534 693	−0.81866 327	−0.19865 477	−0.11616 303	+0.29165 814
0.51	0.56272 977	−0.71300 782	−0.82681 587	−0.22745 494	−0.16231 372	0.25442 027
0.52	0.57633 975	−0.70030 333	−0.83440 647	−0.25628 339	−0.20711 759	0.21286 243
0.53	0.59014 516	−0.68722 307	−0.84141 492	−0.28510 113	−0.24999 263	0.16748 087
0.54	0.60415 560	−0.67375 597	−0.84782 014	−0.31386 748	−0.29035 406	0.11884 913
0.55	0.61838 131	−0.65989 028	−0.85360 014	−0.34253 994	−0.32762 069	0.06761 470
0.56	0.63283 319	−0.64561 342	−0.85873 186	−0.37107 413	−0.36122 172	+0.01449 441
0.57	0.64752 284	−0.63091 198	−0.86319 116	−0.39942 362	−0.39060 386	−0.03973 144
0.58	0.66246 271	−0.61577 163	−0.86695 267	−0.42753 983	−0.41523 901	−0.09422 630
0.59	0.67766 607	−0.60017 702	−0.86998 970	−0.45537 186	−0.43463 218	−0.14810 594
0.60	0.69314 718	−0.58411 169	−0.87227 411	−0.48286 632	−0.44832 986	−0.20044 847
0.61	0.70892 136	−0.56755 797	−0.87377 622	−0.50996 718	−0.45592 864	−0.25030 577
0.62	0.72500 509	−0.55049 685	−0.87446 461	−0.53661 553	−0.45708 410	−0.29671 648
0.63	0.74141 614	−0.53290 783	−0.87430 597	−0.56274 938	−0.45151 989	−0.33872 031
0.64	0.75817 374	−0.51476 880	−0.87326 492	−0.58830 338	−0.43903 693	−0.37537 391
0.65	0.77529 871	−0.49605 584	−0.87130 380	−0.61320 855	−0.41952 271	−0.40576 815
0.66	0.79281 363	−0.47674 300	−0.86838 239	−0.63739 196	−0.39296 048	−0.42904 673
0.67	0.81074 313	−0.45680 211	−0.86445 768	−0.66077 634	−0.35943 834	−0.44442 606
0.68	0.82911 404	−0.43620 245	−0.85948 352	−0.68327 969	−0.31915 810	−0.45121 636
0.69	0.84795 576	−0.41491 053	−0.85341 027	−0.70481 480	−0.27244 363	−0.44884 377
0.70	0.86730 053	−0.39288 963	−0.84618 438	−0.72528 868	−0.21974 878	−0.43687 329
0.71	0.88718 386	−0.37009 946	−0.83774 785	−0.74460 199	−0.16166 443	−0.41503 236
0.72	0.90764 498	−0.34649 561	−0.82803 775	−0.76264 823	−0.09892 467	−0.38323 471
0.73	0.92872 736	−0.32202 902	−0.81698 546	−0.77931 296	−0.03241 178	−0.34160 431
0.74	0.95047 938	−0.29664 526	−0.80451 593	−0.79447 280	+0.03684 038	−0.29049 884
0.75	0.97295 507	−0.27028 369	−0.79054 669	−0.80799 424	0.10764 474	−0.23053 218
0.76	0.99621 508	−0.24287 654	−0.77498 679	−0.81973 225	0.17866 149	−0.16259 543
0.77	1.02032 776	−0.21434 763	−0.75773 539	−0.82952 866	0.24840 151	−0.08787 565
0.78	1.04537 055	−0.18461 097	−0.73868 011	−0.83721 016	0.31523 275	−0.00787 146
0.79	1.07143 168	−0.15356 897	−0.71769 507	−0.84258 586	0.37739 063	+0.07559 560
0.80	1.09861 229	−0.12111 017	−0.69463 835	−0.84544 435	0.43299 312	0.16037 522
0.81	1.12702 903	−0.08710 649	−0.66934 890	−0.84555 002	0.48006 146	0.24398 961
0.82	1.15681 746	−0.05140 968	−0.64164 264	−0.84263 849	0.51654 781	0.32364 357
0.83	1.18813 640	−0.01384 678	−0.61130 745	−0.83641 078	0.54037 123	0.39624 661
0.84	1.22117 352	+0.02578 575	−0.57809 671	−0.82652 589	0.54946 418	0.45844 913
0.85	1.25615 281	0.06772 989	−0.54172 080	−0.81259 105	0.54183 191	0.50669 726
0.86	1.29334 467	0.11227 642	−0.50183 576	−0.79414 886	0.51562 828	0.53731 190
0.87	1.33307 963	0.15977 928	−0.45802 786	−0.77065 991	0.46925 273	0.54659 757
0.88	1.37576 766	0.21067 554	−0.40979 212	−0.74147 880	0.40147 508	0.53099 253
0.89	1.42192 587	0.26551 403	−0.35650 171	−0.70582 022	0.31159 776	0.48727 156
0.90	1.47221 949	0.32499 754	−0.29736 306	−0.66270 962	0.19967 037	0.41282 291
0.91	1.52752 443	0.39004 723	−0.23134 775	−0.61090 890	+0.06677 934	0.30602 901
0.92	1.58902 692	0.46190 476	−0.15708 489	−0.54880 000	−0.08454 828	+0.16680 029
0.93	1.65839 002	0.54230 272	−0.07268 272	−0.47419 336	−0.24975 925	−0.00265 428
0.94	1.73804 934	0.63376 638	+0.02458 593	−0.38399 297	−0.42137 701	−0.19666 273
0.95	1.83178 082	0.74019 178	0.13888 288	−0.27356 330	−0.58752 240	−0.40421 502
0.96	1.94591 015	0.86807 374	0.27707 112	−0.13540 204	−0.72921 201	−0.60564 435
0.97	2.09229 572	1.02952 685	0.45181 370	+0.04408 092	−0.81464 729	−0.76587 179
0.98	2.29755 993	1.25160 873	0.69108 487	0.29436 613	−0.78406 452	−0.81720 735
0.99	2.64665 241	1.62018 589	1.08264 984	0.70624 831	−0.48875 677	−0.59305 105
1.00	∞	∞	∞	∞	∞	∞

$$Q_0(x) = \tfrac{1}{2} \ln \left(\frac{1+x}{1-x}\right) \qquad Q_1(x) = \frac{x}{2} \ln \left(\frac{1+x}{1-x}\right) - 1$$

$$Q_2(x) = \frac{3x^2-1}{4} \ln \left(\frac{1+x}{1-x}\right) - \frac{3x}{2} \qquad Q_3(x) = \frac{x}{4}(5x^2-3) \ln \left(\frac{1+x}{1-x}\right) - \frac{5x^2}{2} + \frac{2}{3}$$

$$(n+1)Q_{n+1}(x) = (2n+1)xQ_n(x) - nQ_{n-1}(x)$$

Table 8.4   DERIVATIVE OF THE LEGENDRE FUNCTION—SECOND KIND $Q'_n(x)$

x	$Q'_0(x)$	$Q'_1(x)$	$Q'_2(x)$	$Q'_3(x)$	$Q'_9(x)$	$Q'_{10}(x)$
0.00	1.00000 000	0.00000 000	−2.00000 000	0.00000 000	0.00000 00	−4.06349 21
0.01	1.00010 001	0.02000 133	−1.99959 998	−0.07999 200	0.36520 25	−4.04156 71
0.02	1.00040 016	0.04001 067	−1.99839 968	−0.15993 599	0.72733 83	−3.97600 70
0.03	1.00090 081	0.06003 603	−1.99639 838	−0.23978 392	1.08336 24	−3.86745 44
0.04	1.00160 256	0.08008 546	−1.99359 487	−0.31948 767	1.43027 23	−3.71697 43
0.05	1.00250 627	0.10016 704	−1.98998 747	−0.39899 900	1.76512 98	−3.52604 61
0.06	1.00361 301	0.12028 894	−1.98557 401	−0.47826 951	2.08508 14	−3.29655 13
0.07	1.00492 413	0.14045 936	−1.98035 179	−0.55725 060	2.38737 90	−3.03075 84
0.08	1.00644 122	0.16068 662	−1.97431 766	−0.63589 347	2.66939 94	−2.73130 45
0.09	1.00816 615	0.18097 914	−1.96746 792	−0.71414 899	2.92866 44	−2.40117 40
0.10	1.01010 101	0.20134 545	−1.95979 839	−0.79196 777	3.16285 86	−2.04367 37
0.11	1.01224 820	0.22179 422	−1.95130 431	−0.86930 001	3.36984 76	−1.66240 59
0.12	1.01461 039	0.24233 428	−1.94198 044	−0.94609 554	3.54769 49	−1.26123 82
0.13	1.01719 052	0.26297 462	−1.93182 094	−1.02230 373	3.69467 78	−0.84427 11
0.14	1.01999 184	0.28372 443	−1.92081 942	−1.09787 345	3.80930 18	−0.41580 27
0.15	1.02301 790	0.30459 312	−1.90896 890	−1.17275 302	3.89031 48	+0.01970 77
0.16	1.02627 258	0.32559 031	−1.89626 181	−1.24689 019	3.93671 92	0.45767 92
0.17	1.02976 007	0.34672 587	−1.88268 994	−1.32023 203	3.94778 25	0.89344 90
0.18	1.03348 491	0.36800 997	−1.86824 444	−1.39272 496	3.92304 76	1.32231 56
0.19	1.03745 202	0.38945 305	−1.85291 580	−1.46431 458	3.86234 02	1.73958 08
0.20	1.04166 667	0.41106 589	−1.83669 380	−1.53494 573	3.76577 54	2.14059 45
0.21	1.04613 453	0.43285 960	−1.81956 752	−1.60456 234	3.63376 26	2.52079 94
0.22	1.05086 171	0.45484 568	−1.80152 526	−1.67310 742	3.46700 84	2.87577 54
0.23	1.05585 471	0.47703 605	−1.78255 455	−1.74052 294	3.26651 77	3.20128 51
0.24	1.06112 054	0.49944 304	−1.76264 210	−1.80674 982	3.03359 33	3.49331 81
0.25	1.06666 667	0.52207 948	−1.74177 372	−1.87172 780	2.76983 31	3.74813 48
0.26	1.07250 107	0.54495 869	−1.71993 437	−1.93539 537	2.47712 56	3.96230 97
0.27	1.07863 229	0.56809 454	−1.69710 801	−1.99768 972	2.15764 35	4.13277 26
0.28	1.08506 944	0.59150 152	−1.67327 761	−2.05854 661	1.81383 48	4.25684 84
0.29	1.09182 225	0.61519 472	−1.64842 510	−2.11790 027	1.44841 22	4.33229 46
0.30	1.09890 110	0.63918 993	−1.62253 126	−2.17568 334	1.06434 02	4.35733 72
0.31	1.10631 707	0.66350 370	−1.59557 570	−2.23182 672	0.66482 02	4.33070 22
0.32	1.11408 200	0.68815 335	−1.56753 678	−2.28625 944	+0.25327 32	4.25164 55
0.33	1.12220 851	0.71315 706	−1.53839 152	−2.33890 860	−0.16667 95	4.11997 79
0.34	1.13071 009	0.73853 396	−1.50811 553	−2.38969 914	−0.59123 78	3.93608 76
0.35	1.13960 114	0.76430 415	−1.47668 292	−2.43855 378	−1.01644 63	3.70095 66
0.36	1.14889 706	0.79048 884	−1.44406 617	−2.48539 281	−1.43822 04	3.41617 42
0.37	1.15861 430	0.81711 039	−1.41023 606	−2.53013 394	−1.85237 43	3.08394 42
0.38	1.16877 045	0.84419 242	−1.37516 155	−2.57269 210	−2.25465 05	2.70708 74
0.39	1.17938 436	0.87175 994	−1.33880 960	−2.61297 926	−2.64075 25	2.28903 82
0.40	1.19047 619	0.89983 941	−1.30114 509	−2.65090 420	−3.00637 81	1.83383 54
0.41	1.20206 756	0.92845 892	−1.26213 064	−2.68637 229	−3.34725 61	1.34610 61
0.42	1.21418 164	0.95764 831	−1.22172 641	−2.71928 520	−3.65918 35	0.83104 35
0.43	1.22684 333	0.98743 931	−1.17988 995	−2.74954 067	−3.93806 51	+0.29437 81
0.44	1.24007 937	1.01786 572	−1.13657 597	−2.77703 216	−4.17995 45	−0.25765 92
0.45	1.25391 850	1.04896 360	−1.09173 613	−2.80164 855	−4.38109 69	−0.81838 00
0.46	1.26839 168	1.08077 146	−1.04531 874	−2.82327 375	−4.53797 26	−1.38069 01
0.47	1.28353 228	1.11333 051	−0.99726 854	−2.84178 630	−4.64734 21	−1.93714 78
0.48	1.29937 630	1.14668 490	−0.94752 634	−2.85705 896	−4.70629 25	−2.48003 04
0.49	1.31596 263	1.18088 202	−0.89602 868	−2.86895 817	−4.71228 35	−3.00140 86
0.50	1.33333 333	1.21597 281	−0.84270 745	−2.87734 353	−4.66319 54	−3.49322 79
	$\begin{bmatrix} (-4)1 \\ 5 \end{bmatrix}$	$\begin{bmatrix} (-4)1 \\ 5 \end{bmatrix}$	$\begin{bmatrix} (-4)2 \\ 5 \end{bmatrix}$	$\begin{bmatrix} (-4)4 \\ 5 \end{bmatrix}$	$\begin{bmatrix} (-3)7 \\ 6 \end{bmatrix}$	$\begin{bmatrix} (-3)6 \\ 6 \end{bmatrix}$

DERIVATIVE OF THE LEGENDRE FUNCTION—SECOND KIND $Q_n'(x)$ Table 8.4

x	$Q_0'(x)$	$Q_1'(x)$	$Q_2'(x)$	$Q_3'(x)$	$Q_9'(x)$	$Q_{10}'(x)$
0.50	1.33333 333	1.21597 281	− 0.84270 74	− 2.87734 35	− 4.66319 54	− 3.493228
0.51	1.35153 399	1.25201 210	− 0.78748 95	− 2.88206 72	− 4.55737 62	− 3.947399
0.52	1.37061 403	1.28905 905	− 0.73029 59	− 2.88297 33	− 4.39368 94	− 4.355894
0.53	1.39062 717	1.32717 756	− 0.67104 20	− 2.87989 70	− 4.17156 11	− 4.710854
0.54	1.41163 185	1.36643 680	− 0.60963 61	− 2.87266 39	− 3.91102 65	− 5.004695
0.55	1.43369 176	1.40691 178	− 0.54597 91	− 2.86108 89	− 3.55277 54	− 5.230233
0.56	1.45687 646	1.44868 400	− 0.47996 38	− 2.84497 53	− 3.15819 61	− 5.380807
0.57	1.48126 204	1.49184 220	− 0.41147 39	− 2.82411 36	− 2.70941 73	− 5.450406
0.58	1.50693 189	1.53648 320	− 0.34038 30	− 2.79828 02	− 2.20934 79	− 5.433812
0.59	1.53397 760	1.58271 285	− 0.26655 35	− 2.76723 56	− 1.66171 26	− 5.326732
0.60	1.56250 000	1.63064 718	− 0.18983 51	− 2.73072 34	− 1.07108 51	− 5.125950
0.61	1.59261 029	1.68041 364	− 0.11006 36	− 2.68846 75	− 0.44291 60	− 4.829465
0.62	1.62443 145	1.73215 259	− 0.02705 91	− 2.64017 05	+ 0.21644 47	− 4.436645
0.63	1.65809 982	1.78601 903	+ 0.05937 63	− 2.58551 08	0.89973 10	− 3.948368
0.64	1.69376 694	1.84218 458	0.14946 05	− 2.52414 00	1.59875 12	− 3.367169
0.65	1.73160 173	1.90083 983	0.24343 42	− 2.45567 92	2.30438 77	− 2.697375
0.66	1.77179 305	1.96219 705	0.34156 40	− 2.37971 49	3.00660 55	− 1.945245
0.67	1.81455 271	2.02649 344	0.44414 64	− 2.29579 49	3.69447 22	− 1.119087
0.68	1.86011 905	2.09399 499	0.55151 17	− 2.20342 26	4.35619 14	− 0.229371
0.69	1.90876 121	2.16500 099	0.66402 96	− 2.10205 04	4.97914 99	+ 0.711177
0.70	1.96078 431	2.23984 955	0.78211 54	− 1.99107 23	5.54998 34	1.687501
0.71	2.01653 559	2.31892 413	0.90623 72	− 1.86981 51	6.05466 05	2.682165
0.72	2.07641 196	2.40266 159	1.03692 51	− 1.73752 72	6.47859 09	3.675339
0.73	2.14086 919	2.49156 187	1.17478 21	− 1.59336 54	6.80675 90	4.644816
0.74	2.21043 324	2.58619 998	1.32049 75	− 1.43637 96	7.02388 88	5.566082
0.75	2.28571 429	2.68724 079	1.47486 32	− 1.26549 27	7.11464 51	6.412431
0.76	2.36742 424	2.79545 751	1.63879 46	− 1.07947 65	7.06387 68	7.155161
0.77	2.45639 892	2.91175 493	1.81335 60	− 0.87692 20	6.85691 02	7.763836
0.78	2.55362 615	3.03719 894	1.99979 32	− 0.65620 16	6.47990 33	8.206652
0.79	2.66028 199	3.17305 446	2.19957 51	− 0.41542 09	5.92027 14	8.450921
0.80	2.77777 778	3.32083 451	2.41444 73	− 0.15235 72	5.16720 18	8.463693
0.81	2.90782 204	3.48236 488	2.64650 26	+ 0.13562 04	4.21227 67	8.212559
0.82	3.05250 305	3.65986 997	2.89827 40	0.45165 68	3.05023 28	7.666669
0.83	3.21440 051	3.85608 883	3.17286 02	0.79955 16	1.67989 36	6.798024
0.84	3.39673 913	4.07443 439	3.47409 64	1.18395 08	+ 0.10532 57	5.583115
0.85	3.60360 360	4.31921 588	3.80679 33	1.61061 19	− 1.66270 85	4.005017
0.86	3.84024 578	4.59595 604	4.17707 50	2.08677 72	− 3.60489 91	+ 2.056070
0.87	4.11353 352	4.91185 380	4.59287 14	2.62171 45	− 5.69098 02	− 0.258625
0.88	4.43262 411	5.27647 688	5.06465 07	3.22751 63	− 7.87652 81	− 2.916594
0.89	4.81000 481	5.70283 015	5.60654 69	3.92032 16	−10.09858 18	− 5.871760
0.90	5.26315 789	6.20906 159	6.23815 05	4.72224 63	−12.26944 98	− 9.045801
0.91	5.81733 566	6.82129 988	6.98747 73	5.66456 11	−14.26758 89	−12.315713
0.92	6.51041 667	7.57861 025	7.89613 09	6.79318 58	−15.92348 54	−15.495090
0.93	7.40192 450	8.54217 980	9.02883 27	8.17876 62	−16.99643 22	−18.304274
0.94	8.59106 529	9.81365 072	10.49236 44	9.93658 04	−17.13329 84	−20.319071
0.95	10.25641 026	11.57537 057	12.47698 56	12.26978 50	−15.78782 62	−20.873659
0.96	12.75510 204	14.19080 811	15.35932 33	15.57616 37	−12.04072 38	−18.851215
0.97	16.92047 377	18.50515 528	20.00905 43	20.76422 38	− 4.11777 87	−12.140718
0.98	25.25252 525	27.04503 467	29.00735 14	30.50045 90	+12.32933 89	+ 4.242107
0.99	50.25125 628	52.39539 613	55.11181 39	57.80864 53	54.86521 05	49.428990
1.00	∞	∞	∞	∞	∞	∞

Table 8.5   **LEGENDRE FUNCTION—FIRST KIND** $P_n(x)$

$$P_0(x) = 1 \qquad P_1(x) = x$$

x	$P_2(x)$	$P_3(x)$	$P_4(x)$	$P_5(x)$	$P_9(x)$	$P_{10}(x)$
1.0	1.00	1.00	1.00000	1.00000	1.00000	1.00000
1.2	1.66	2.52	4.04700	6.72552	(1)6.02754	(2)1.06544
1.4	2.44	4.76	9.83200	(1)2.09686	(2)5.03668	(3)1.13789
1.6	3.34	7.84	(1)1.94470	(1)4.97354	(3)2.45973	(3)6.65436
1.8	4.36	11.88	(1)3.41520	(2)1.01148	(3)8.97882	(4)2.81110
2.0	5.50	17.00	(1)5.53750	(2)1.85750	(4)2.71007	(4)9.60605
2.2	6.76	23.32	(1)8.47120	(2)3.16804	(4)7.13591	(5)2.81929
2.4	8.14	30.96	(2)1.23927	(2)5.10597	(5)1.69353	(5)7.37020
2.6	9.64	40.04	(2)1.74952	(2)7.86743	(5)3.70173	(6)1.75809
2.8	11.26	50.68	(2)2.39887	(3)1.16849	(5)7.56647	(6)3.89219
3.0	13.00	63.00	(2)3.21000	(3)1.68300	(6)1.46256	(6)8.09745
3.2	14.86	77.12	(2)4.20727	(3)2.36169	(6)2.69625	(7)1.59814
3.4	16.84	93.16	(2)5.41672	(3)3.24050	(6)4.77208	(7)3.01437
3.6	18.94	111.24	(2)6.86607	(3)4.36022	(6)8.15181	(7)5.46578
3.8	21.16	131.48	(2)8.58472	(3)5.76676	(7)1.34978	(7)9.57313
4.0	23.50	154.00	(3)1.06038	(3)7.51150	(7)2.17406	(8)1.62597
4.2	25.96	178.92	(3)1.29559	(3)9.65154	(7)3.41632	(8)2.68690
4.4	28.54	206.36	(3)1.56757	(4)1.22500	(7)5.25060	(8)4.33189
4.6	31.24	236.44	(3)1.87991	(4)1.53765	(7)7.90944	(8)6.82993
4.8	34.06	269.28	(3)2.23641	(4)1.91071	(8)1.16994	(9)1.05524
5.0	37.00	305.00	(3)2.64100	(4)2.35250	(8)1.70196	(9)1.60047
5.2	40.06	343.72	(3)3.09781	(4)2.87205	(8)2.43839	(9)2.38657
5.4	43.24	385.56	(3)3.61111	(4)3.47916	(8)3.44472	(9)3.50362
5.6	46.54	430.64	(3)4.18537	(4)4.18440	(8)4.80363	(9)5.06985
5.8	49.96	479.08	(3)4.82519	(4)4.99917	(8)6.61853	(9)7.23884
6.0	53.50	531.00	(3)5.53538	(4)5.93572	(8)9.01781	(10)1.02082
6.2	57.16	586.52	(3)6.32087	(4)7.00717	(9)1.21596	(10)1.42299
6.4	60.94	645.76	(3)7.18681	(4)8.22754	(9)1.62372	(10)1.96229
6.6	64.84	708.84	(3)8.13847	(4)9.61180	(9)2.14858	(10)2.67872
6.8	68.86	775.88	(3)9.18133	(5)1.11759	(9)2.81890	(10)3.62216
7.0	73.00	847.00	(4)1.03210	(5)1.29367	(9)3.66876	(10)4.85435
7.2	77.26	922.32	(4)1.15633	(5)1.49122	(9)4.73885	(10)6.45123
7.4	81.64	1001.96	(4)1.29142	(5)1.71215	(9)6.07749	(10)8.50564
7.6	86.14	1086.04	(4)1.43797	(5)1.95846	(9)7.74185	(11)1.11305
7.8	90.76	1174.68	(4)1.59663	(5)2.23227	(9)9.79919	(11)1.44623
8.0	95.50	1268.00	(4)1.76804	(5)2.53583	(10)1.23283	(11)1.86653
8.2	100.36	1366.12	(4)1.95286	(5)2.87149	(10)1.54212	(11)2.39363
8.4	105.34	1469.16	(4)2.15176	(5)3.24171	(10)1.91848	(11)3.05098
8.6	110.44	1577.24	(4)2.36546	(5)3.64912	(10)2.37430	(11)3.86641
8.8	115.66	1690.48	(4)2.59466	(5)4.09643	(10)2.92387	(11)4.87282
9.0	121.00	1809.00	(4)2.84010	(5)4.58649	(10)3.58363	(11)6.10897
9.2	126.46	1932.92	(4)3.10252	(5)5.12230	(10)4.37243	(11)7.62030
9.4	132.04	2062.36	(4)3.38268	(5)5.70699	(10)5.31184	(11)9.45994
9.6	137.74	2197.44	(4)3.68137	(5)6.34383	(10)6.42640	(12)1.16898
9.8	143.56	2338.28	(4)3.99938	(5)7.03621	(10)7.74404	(12)1.43817
10.0	149.50	2485.00	(4)4.33754	(5)7.78769	(10)9.29640	(12)1.76188

From National Bureau of Standards, Tables of associated Legendre functions. Columbia Univ. Press, New York, N.Y., 1945 (with permission).

DERIVATIVE OF THE LEGENDRE FUNCTION—FIRST KIND $P'_n(x)$　　　Table 8.6

$$P'_1(x)=1 \qquad P'_2(x)=3x$$

x	$P'_3(x)$	$P'_4(x)$	$P'_5(x)$	$P'_9(x)$	$P'_{10}(x)$
1.0	6.000	(1)1.00000	(1)1.50000	(1)4.50000	(1)5.50000
1.2	9.300	(1)2.12400	(1)4.57230	(2)7.77587	(3)1.53586
1.4	(1)1.320	(1)3.75200	(2)1.01688	(3)4.50787	(4)1.13477
1.6	(1)1.770	(1)5.96800	(2)1.92723	(4)1.74282	(4)5.24824
1.8	(1)2.280	(1)8.85600	(2)3.30168	(4)5.33445	(5)1.85808
2.0	(1)2.850	(2)1.25000	(2)5.26875	(5)1.39531	(5)5.50068
2.2	(1)3.480	(2)1.69840	(2)7.97208	(5)3.25362	(6)1.42939
2.4	(1)4.170	(2)2.23920	(3)1.15704	(5)6.94480	(6)3.36028
2.6	(1)4.920	(2)2.88080	(3)1.62377	(6)1.38132	(6)7.29317
2.8	(1)5.730	(2)3.63160	(3)2.21628	(6)2.59296	(7)1.48267
3.0	(1)6.600	(2)4.50000	(3)2.95500	(6)4.63721	(7)2.85372
3.2	(1)7.530	(2)5.49440	(3)3.86184	(6)7.95819	(7)5.24287
3.4	(1)8.520	(2)6.62320	(3)4.96025	(7)1.31805	(7)9.25345
3.6	(1)9.570	(2)7.89480	(3)6.27516	(7)2.11632	(8)1.57706
3.8	(2)1.068	(2)9.31760	(3)7.83305	(7)3.30652	(8)2.60626
4.0	(2)1.185	(3)1.09000	(3)9.66187	(7)5.04229	(8)4.19097
4.2	(2)1.308	(3)1.26504	(4)1.17911	(7)7.52431	(8)6.57653
4.4	(2)1.437	(3)1.45772	(4)1.42518	(8)1.10110	(9)1.00955
4.6	(2)1.572	(3)1.66888	(4)1.70764	(8)1.58313	(9)1.51918
4.8	(2)1.713	(3)1.89936	(4)2.02990	(8)2.23988	(9)2.24508
5.0	(2)1.860	(3)2.15000	(4)2.39550	(8)3.12290	(9)3.26340
5.2	(2)2.013	(3)2.42164	(4)2.80816	(8)4.29574	(9)4.67217
5.4	(2)2.172	(3)2.71512	(4)3.27172	(8)5.83620	(9)6.59627
5.6	(2)2.337	(3)3.03128	(4)3.79020	(8)7.83868	(9)9.19329
5.8	(2)2.508	(3)3.37096	(4)4.36775	(9)1.04169	(10)1.26604
6.0	(2)2.685	(3)3.73500	(4)5.00869	(9)1.37071	(10)1.72421
6.2	(2)2.868	(3)4.12424	(4)5.71746	(9)1.78712	(10)2.32397
6.4	(2)3.057	(3)4.53952	(4)6.49870	(9)2.31006	(10)3.10217
6.6	(2)3.252	(3)4.98168	(4)7.35714	(9)2.96206	(10)4.10354
6.8	(2)3.453	(3)5.45156	(4)8.29772	(9)3.76947	(10)5.38214
7.0	(2)3.660	(3)5.95000	(4)9.32550	(9)4.76295	(10)7.00283
7.2	(2)3.873	(3)6.47784	(5)1.04457	(9)5.97809	(10)9.04307
7.4	(2)4.092	(3)7.03592	(5)1.16637	(9)7.45591	(11)1.15949
7.6	(2)4.317	(3)7.62508	(5)1.29849	(9)9.24362	(11)1.47670
7.8	(2)4.548	(3)8.24616	(5)1.44152	(10)1.13953	(11)1.86875
8.0	(2)4.785	(3)8.90000	(5)1.59602	(10)1.39725	(11)2.35063
8.2	(2)5.028	(3)9.58744	(5)1.76260	(10)1.70455	(11)2.93985
8.4	(2)5.277	(4)1.03093	(5)1.94187	(10)2.06937	(11)3.65675
8.6	(2)5.532	(4)1.10665	(5)2.13445	(10)2.50070	(11)4.52490
8.8	(2)5.793	(4)1.18598	(5)2.34099	(10)3.00866	(11)5.57149
9.0	(2)6.060	(4)1.26900	(5)2.56215	(10)3.60463	(11)6.82780
9.2	(2)6.333	(4)1.35580	(5)2.79860	(10)4.30137	(11)8.32969
9.4	(2)6.612	(4)1.44647	(5)3.05102	(10)5.11311	(12)1.01182
9.6	(2)6.897	(4)1.54109	(5)3.32013	(10)6.05576	(12)1.22399
9.8	(2)7.188	(4)1.63974	(5)3.60663	(10)7.14698	(12)1.47481
10.0	(2)7.485	(4)1.74250	(5)3.91127	(10)8.40642	(12)1.77028

From National Bureau of Standards, Tables of associated Legendre functions. Columbia Univ. Press, New York, N.Y., 1945 (with permission).

Table 8.7   **LEGENDRE FUNCTION—SECOND KIND $Q_n(x)$**

x	$Q_0(x)$	$Q_1(x)$	$Q_2(x)$	$Q_3(x)$	$Q_9(x)$	$Q_{10}(x)$
1.0	∞	∞	∞	∞	∞	∞
1.2	1.19895	(−1)4.38737	(−1)1.90253	(−2)8.80147	(− 3)1.32079	(− 4)6.75615
1.4	(−1)8.95880	(−1)2.54232	(−2)8.59466	(−2)3.10542	(− 4)1.06810	(− 5)4.27633
1.6	(−1)7.33169	(−1)1.73070	(−2)4.87829	(−2)1.47080	(− 5)1.71471	(− 6)5.73368
1.8	(−1)6.26381	(−1)1.27487	(−2)3.10233	(−3)8.07870	(− 6)3.91902	(− 6)1.13241
2.0	(−1)5.49306	(−2)9.86123	(−2)2.11838	(−3)4.87112	(− 6)1.12179	(− 7)2.86313
2.2	(−1)4.90415	(−2)7.89122	(−2)1.52029	(−3)3.13576	(− 7)3.76522	(− 8)8.62195
2.4	(−1)4.43652	(−2)6.47638	(−2)1.13240	(−3)2.12013	(− 7)1.42488	(− 8)2.96212
2.6	(−1)4.05465	(−2)5.42093	(−3)8.68364	(−3)1.48960	(− 8)5.92566	(− 8)1.12879
2.8	(−1)3.73607	(−2)4.61002	(−3)6.81708	(−3)1.07961	(− 8)2.66020	(− 9)4.67876
3.0	(−1)3.46574	(−2)3.97208	(−3)5.45667	(−4)8.02854	(− 8)1.27252	(− 9)2.07945
3.2	(−1)3.23314	(−2)3.46035	(−3)4.43984	(−4)6.10146	(− 9)6.42269	(−10)9.80358
3.4	(−1)3.03068	(−2)3.04309	(−3)3.66347	(−4)4.72397	(− 9)3.39441	(−10)4.86183
3.6	(−1)2.85272	(−2)2.69807	(−3)3.05981	(−4)3.71695	(− 9)1.86714	(−10)2.51945
3.8	(−1)2.69498	(−2)2.40934	(−3)2.58298	(−4)2.96625	(− 9)1.06372	(−10)1.35695
4.0	(−1)2.55413	(−2)2.16512	(−3)2.20108	(−4)2.39697	(−10)6.25130	(−11)7.56235
4.2	(−1)2.42754	(−2)1.95664	(−3)1.89145	(−4)1.95866	(−10)3.77701	(−11)4.34493
4.4	(−1)2.31312	(−2)1.77717	(−3)1.63766	(−4)1.61661	(−10)2.33956	(−11)2.56563
4.6	(−1)2.20916	(−2)1.62153	(−3)1.42759	(−4)1.34641	(−10)1.48213	(−11)1.55290
4.8	(−1)2.11428	(−2)1.48564	(−3)1.25217	(−4)1.13061	(−11)9.58309	(−12)9.61271
5.0	(−1)2.02733	(−2)1.36628	(−3)1.10450	(−5)9.56532	(−11)6.31274	(−12)6.07362
5.2	(−1)1.94732	(−2)1.26084	(−4)9.79278	(−5)8.14823	(−11)4.23006	(−12)3.91025
5.4	(−1)1.87347	(−2)1.16723	(−4)8.72377	(−5)6.98500	(−11)2.87937	(−12)2.56132
5.6	(−1)1.80507	(−2)1.08374	(−4)7.80551	(−5)6.02278	(−11)1.98859	(−12)1.70471
5.8	(−1)1.74153	(−2)1.00894	(−4)7.01223	(−5)5.22117	(−11)1.39197	(−12)1.15147
6.0	(−1)1.68236	(−3)9.41671	(−4)6.32330	(−5)4.54896	(−12)9.86572	(−13)7.88519
6.2	(−1)1.62711	(−3)8.80944	(−4)5.72204	(−5)3.98181	(−12)7.07418	(−13)5.46920
6.4	(−1)1.57541	(−3)8.25935	(−4)5.19491	(−5)3.50058	(−12)5.12787	(−13)3.83900
6.6	(−1)1.52691	(−3)7.75944	(−4)4.73078	(−5)3.09006	(−12)3.75499	(−13)2.72499
6.8	(−1)1.48133	(−3)7.30377	(−4)4.32050	(−5)2.73812	(−12)2.77600	(−13)1.95462
7.0	(−1)1.43841	(−3)6.88725	(−4)3.95644	(−5)2.43500	(−12)2.07071	(−13)1.41592
7.2	(−1)1.39792	(−3)6.50550	(−4)3.63228	(−5)2.17277	(−12)1.55770	(−13)1.03525
7.4	(−1)1.35967	(−3)6.15475	(−4)3.34266	(−5)1.94497	(−12)1.18115	(−14)7.63577
7.6	(−1)1.32346	(−3)5.83171	(−4)3.08311	(−5)1.74631	(−13)9.02383	(−14)5.67877
7.8	(−1)1.28915	(−3)5.53353	(−4)2.84980	(−5)1.57242	(−13)6.94338	(−14)4.25654
8.0	(−1)1.25657	(−3)5.25771	(−4)2.63950	(−5)1.41968	(−13)5.37876	(−14)3.21427
8.2	(−1)1.22561	(−3)5.00208	(−4)2.44944	(−5)1.28507	(−13)4.19350	(−14)2.44439
8.4	(−1)1.19615	(−3)4.76469	(−4)2.27723	(−5)1.16606	(−13)3.28941	(−14)1.87141
8.6	(−1)1.16807	(−3)4.54386	(−4)2.12082	(−5)1.06054	(−13)2.59524	(−14)1.44191
8.8	(−1)1.14129	(−3)4.33807	(−4)1.97844	(−6)9.66707	(−13)2.05891	(−14)1.11775
9.0	(−1)1.11572	(−3)4.14598	(−4)1.84855	(−6)8.83037	(−13)1.64205	(−15)8.71513
9.2	(−1)1.09127	(−3)3.96640	(−4)1.72979	(−6)8.08237	(−13)1.31620	(−15)6.83294
9.4	(−1)1.06787	(−3)3.79827	(−4)1.62102	(−6)7.41202	(−13)1.06011	(−15)5.38569
9.6	(−1)1.04546	(−3)3.64063	(−4)1.52119	(−6)6.80982	(−14)8.57794	(−15)4.26656
9.8	(−1)1.02397	(−3)3.49262	(−4)1.42940	(−6)6.26763	(−14)6.97159	(−15)3.39644
10.0	(−1)1.00335	(−3)3.35348	(−4)1.34486	(−6)5.77839	(−14)5.69010	(−15)2.71639

From National Bureau of Standards, Tables of associated Legendre functions. Columbia Univ. Press, New York, N.Y., 1945 (with permission).

DERIVATIVE OF THE LEGENDRE FUNCTION—SECOND KIND $Q'_n(x)$ · Table 8.8

x	$-Q'_0(x)$	$-Q'_1(x)$	$-Q'_2(x)$	$-Q'_3(x)$	$-Q'_9(x)$	$-Q'_{10}(x)$
1.0	∞	∞	∞	∞	∞	∞
1.2	2.27273	1.52833	(−1)9.56516	(−1)5.77060	(− 2)2.06667	(− 2)1.15922
1.4	1.04167	(−1)5.62454	(−1)2.78972	(−1)1.32721	(− 3)1.11220	(− 4)4.88977
1.6	(−1)6.41026	(−1)2.92472	(−1)1.21817	(−2)4.85580	(− 4)1.39114	(− 5)5.11106
1.8	(−1)4.46429	(−1)1.77190	(−2)6.39686	(−2)2.20736	(− 5)2.64367	(− 6)8.39591
2.0	(−1)3.33333	(−1)1.17361	(−2)3.74965	(−2)1.14416	(− 6)6.52419	(− 6)1.83053
2.2	(−1)2.60417	(−2)8.25020	(−2)2.36801	(−3)6.48766	(− 6)1.93263	(− 7)4.86561
2.4	(−1)2.10084	(−2)6.05501	(−2)1.57925	(−3)3.93006	(− 7)6.56197	(− 7)1.49994
2.6	(−1)1.73611	(−2)4.59238	(−2)1.09833	(−3)2.50557	(− 7)2.47880	(− 8)5.19235
2.8	(−1)1.46199	(−2)3.57495	(−3)7.89834	(−3)1.66411	(− 7)1.02057	(− 8)1.97390
3.0	(−1)1.25000	(−2)2.84264	(−3)5.83769	(−3)1.14304	(− 8)4.51200	(− 9)8.10849
3.2	(−1)1.08225	(−2)2.30068	(−3)4.41472	(−4)8.07587	(− 8)2.11821	(− 9)3.55578
3.4	(−2)9.46970	(−2)1.89018	(−3)3.40437	(−4)5.84465	(− 8)1.04686	(− 9)1.64904
3.6	(−2)8.36120	(−2)1.57309	(−3)2.66980	(−4)4.31867	(− 9)5.40951	(−10)8.02794
3.8	(−2)7.44048	(−2)1.32398	(−3)2.12471	(−4)3.24956	(− 9)2.90659	(−10)4.07799
4.0	(−2)6.66667	(−2)1.12539	(−3)1.71292	(−4)2.48459	(− 9)1.61660	(−10)2.15091
4.2	(−2)6.00962	(−3)9.64994	(−3)1.39691	(−4)1.92694	(−10)9.27220	(−10)1.17316
4.4	(−2)5.44662	(−3)8.33966	(−3)1.15099	(−4)1.51364	(−10)5.46705	(−11)6.59413
4.6	(−2)4.96032	(−3)7.25823	(−4)9.57184	(−4)1.20274	(−10)3.30481	(−11)3.80849
4.8	(−2)4.53721	(−3)6.35742	(−4)8.02725	(−5)9.65712	(−10)2.04345	(−11)2.25453
5.0	(−2)4.16667	(−3)5.60078	(−4)6.78356	(−5)7.82792	(−10)1.28985	(−11)1.36497
5.2	(−2)3.84025	(−3)4.96040	(−4)5.77277	(−5)6.40058	(−11)8.29696	(−12)8.43598
5.4	(−2)3.55114	(−3)4.41464	(−4)4.94423	(−5)5.27543	(−11)5.43056	(−12)5.31340
5.6	(−2)3.29381	(−3)3.94656	(−4)4.25974	(−5)4.38019	(−11)3.61188	(−12)3.40566
5.8	(−2)3.06373	(−3)3.54273	(−4)3.69015	(−5)3.66172	(−11)2.43819	(−12)2.21848
6.0	(−2)2.85714	(−3)3.19245	(−4)3.21299	(−5)3.08050	(−11)1.66874	(−12)1.46703
6.2	(−2)2.67094	(−3)2.88709	(−4)2.81078	(−5)2.60683	(−11)1.15686	(−13)9.83782
6.4	(−2)2.50250	(−3)2.61964	(−4)2.46977	(−5)2.21813	(−12)8.11673	(−13)6.68395
6.6	(−2)2.34962	(−3)2.38436	(−4)2.17910	(−5)1.89709	(−12)5.75903	(−13)4.59703
6.8	(−2)2.21043	(−3)2.17655	(−4)1.93008	(−5)1.63035	(−12)4.12938	(−13)3.19817
7.0	(−2)2.08333	(−3)1.99230	(−4)1.71573	(−5)1.40747	(−12)2.99029	(−13)2.24909
7.2	(−2)1.96696	(−3)1.82834	(−4)1.53040	(−5)1.22023	(−12)2.18566	(−13)1.59779
7.4	(−2)1.86012	(−3)1.68195	(−4)1.36949	(−5)1.06216	(−12)1.61163	(−13)1.14602
7.6	(−2)1.76180	(−3)1.55083	(−4)1.22923	(−6)9.28073	(−12)1.19826	(−14)8.29452
7.8	(−2)1.67112	(−3)1.43304	(−4)1.10651	(−6)8.13829	(−13)8.97939	(−14)6.05494
8.0	(−2)1.58730	(−3)1.32691	(−5)9.98765	(−6)7.16078	(−13)6.77915	(−14)4.45610
8.2	(−2)1.50966	(−3)1.23104	(−5)9.03846	(−6)6.32104	(−13)5.15433	(−14)3.30480
8.4	(−2)1.43761	(−3)1.14421	(−5)8.19960	(−6)5.59691	(−13)3.94535	(−14)2.46898
8.6	(−2)1.37061	(−3)1.06538	(−5)7.45601	(−6)4.97021	(−13)3.03931	(−14)1.85745
8.8	(−2)1.30822	(−4)9.93646	(−5)6.79498	(−6)4.42597	(−13)2.35565	(−14)1.40670
9.0	(−2)1.25000	(−4)9.28224	(−5)6.20573	(−6)3.95179	(−13)1.83641	(−14)1.07211
9.2	(−2)1.19560	(−4)8.68435	(−5)5.67908	(−6)3.53736	(−13)1.43959	(−15)8.22064
9.4	(−2)1.14469	(−4)8.13682	(−5)5.20722	(−6)3.17406	(−13)1.13452	(−15)6.33995
9.6	(−2)1.09697	(−4)7.63447	(−5)4.78344	(−6)2.85468	(−14)8.98657	(−15)4.91668
9.8	(−2)1.05219	(−4)7.17272	(−5)4.40196	(−6)2.57314	(−14)7.15298	(−15)3.83321
10.0	(−2)1.01010	(−4)6.74753	(−5)4.05782	(−6)2.32430	(−14)5.72014	(−15)3.00374

From National Bureau of Standards, Tables of associated Legendre functions. Columbia Univ. Press, New York, N.Y., 1945 (with permission).

9. Bessel Functions of Integer Order

F. W. J. Olver [1]

Contents

[1] National Bureau of Standards, on leave from the National Physical Laboratory.

The author acknowledges the assistance of Alfred E. Beam, Ruth E. Capuano, Lois K. Cherwinski, Elizabeth F. Godefroy, David S. Liepman, Mary Orr, Bertha H. Walter, and Ruth Zucker of the National Bureau of Standards, and N. F. Bird, C. W. Clenshaw, and Joan M. Felton of the National Physical Laboratory in the preparation and checking of the tables and graphs.

9. Bessel Functions of Integer Order

Mathematical Properties

Notation

The tables in this chapter are for Bessel functions of integer order; the text treats general orders. The conventions used are:

$z = x + iy$; x, y real.

n is a positive integer or zero.

ν, μ are unrestricted except where otherwise indicated; ν is supposed real in the sections devoted to Kelvin functions **9.9**, **9.10**, and **9.11**.

The notation used for the Bessel functions is that of Watson [9.15] and the British Association and Royal Society Mathematical Tables. The function $Y_\nu(z)$ is often denoted $N_\nu(z)$ by physicists and European workers.

Other notations are those of:

Aldis, Airey:

$$G_n(z) \text{ for } -\tfrac{1}{2}\pi Y_n(z), K_n(z) \text{ for } (-)^n K_n(z).$$

Clifford:

$$C_n(x) \text{ for } x^{-\frac{1}{2}n} J_n(2\sqrt{x}).$$

Gray, Mathews and MacRobert [9.9]:

$$Y_n(z) \text{ for } \tfrac{1}{2}\pi Y_n(z) + (\ln 2 - \gamma) J_n(z),$$

$$\overline{Y}_\nu(z) \text{ for } \pi e^{\nu\pi i} \sec(\nu\pi) Y_\nu(z),$$

$$G_\nu(z) \text{ for } \tfrac{1}{2}\pi i H_\nu^{(1)}(z).$$

Jahnke, Emde and Lösch [9.32]:

$$\Lambda_\nu(z) \text{ for } \Gamma(\nu+1)(\tfrac{1}{2}z)^{-\nu} J_\nu(z).$$

Jeffreys:

$$Hs_\nu(z) \text{ for } H_\nu^{(1)}(z), \ Hi_\nu(z) \text{ for } H_\nu^{(2)}(z),$$

$$Kh_\nu(z) \text{ for } (2/\pi) K_\nu(z).$$

Heine:

$$K_n(z) \text{ for} -\tfrac{1}{2}\pi Y_n(z).$$

Neumann:

$$Y^n(z) \text{ for } \tfrac{1}{2}\pi Y_n(z) + (\ln 2 - \gamma) J_n(z).$$

Whittaker and Watson [9.18]:

$$K_\nu(z) \text{ for } \cos(\nu\pi) K_\nu(z).$$

Bessel Functions J and Y

9.1. Definitions and Elementary Properties

Differential Equation

9.1.1
$$z^2 \frac{d^2 w}{dz^2} + z \frac{dw}{dz} + (z^2 - \nu^2) w = 0$$

Solutions are the Bessel functions of the first kind $J_{\pm\nu}(z)$, of the second kind $Y_\nu(z)$ (also called Weber's function) and of the third kind $H_\nu^{(1)}(z)$, $H_\nu^{(2)}(z)$ (also called the Hankel functions). Each is a regular (holomorphic) function of z throughout the z-plane cut along the negative real axis, and for fixed $z(\neq 0)$ each is an entire (integral) function of ν. When $\nu = \pm n$, $J_\nu(z)$ has no branch point and is an entire (integral) function of z.

Important features of the various solutions are as follows: $J_\nu(z)(\mathscr{R}\nu \geq 0)$ is bounded as $z \to 0$ in any bounded range of arg z. $J_\nu(z)$ and $J_{-\nu}(z)$ are linearly independent except when ν is an integer. $J_\nu(z)$ and $Y_\nu(z)$ are linearly independent for all values of ν.

$H_\nu^{(1)}(z)$ tends to zero as $|z| \to \infty$ in the sector $0 < \arg z < \pi$; $H_\nu^{(2)}(z)$ tends to zero as $|z| \to \infty$ in the sector $-\pi < \arg z < 0$. For all values of ν, $H_\nu^{(1)}(z)$ and $H_\nu^{(2)}(z)$ are linearly independent.

Relations Between Solutions

9.1.2
$$Y_\nu(z) = \frac{J_\nu(z) \cos(\nu\pi) - J_{-\nu}(z)}{\sin(\nu\pi)}$$

The right of this equation is replaced by its limiting value if ν is an integer or zero.

9.1.3

$$H_\nu^{(1)}(z) = J_\nu(z) + i Y_\nu(z)$$
$$= i \csc(\nu\pi) \{ e^{-\nu\pi i} J_\nu(z) - J_{-\nu}(z) \}$$

9.1.4

$$H_\nu^{(2)}(z) = J_\nu(z) - i Y_\nu(z)$$
$$= i \csc(\nu\pi) \{ J_{-\nu}(z) - e^{\nu\pi i} J_\nu(z) \}$$

9.1.5 $\quad J_{-n}(z) = (-)^n J_n(z) \qquad Y_{-n}(z) = (-)^n Y_n(z)$

9.1.6 $\quad H_{-\nu}^{(1)}(z) = e^{\nu\pi i} H_\nu^{(1)}(z) \qquad H_{-\nu}^{(2)}(z) = e^{-\nu\pi i} H_\nu^{(2)}(z)$

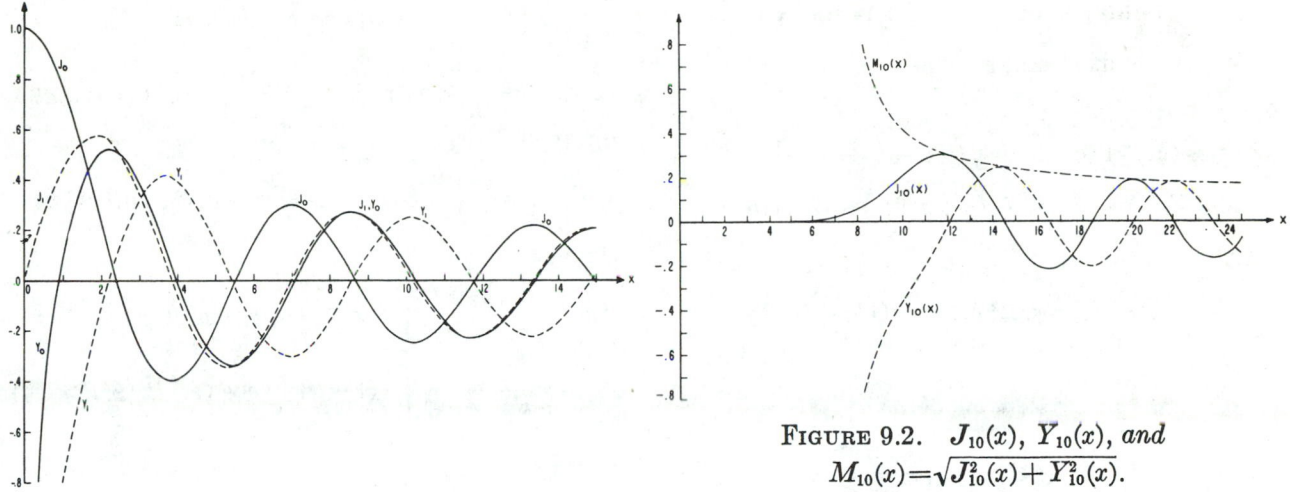

FIGURE 9.2. $J_{10}(x)$, $Y_{10}(x)$, and $M_{10}(x)=\sqrt{J_{10}^2(x)+Y_{10}^2(x)}$.

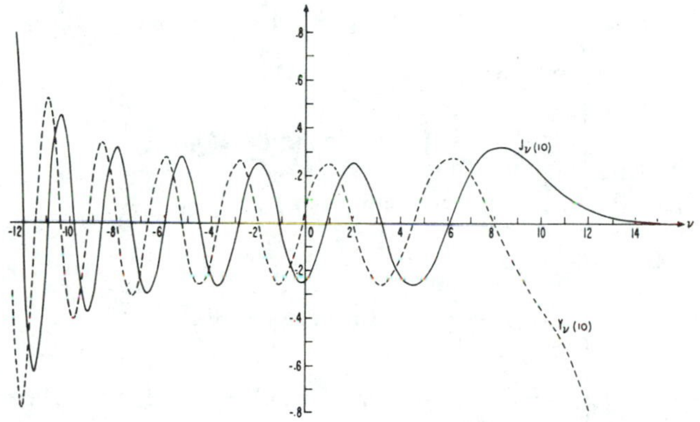

FIGURE 9.1. $J_0(x)$, $Y_0(x)$, $J_1(x)$, $Y_1(x)$.

FIGURE 9.3. $J_\nu(10)$ and $Y_\nu(10)$.

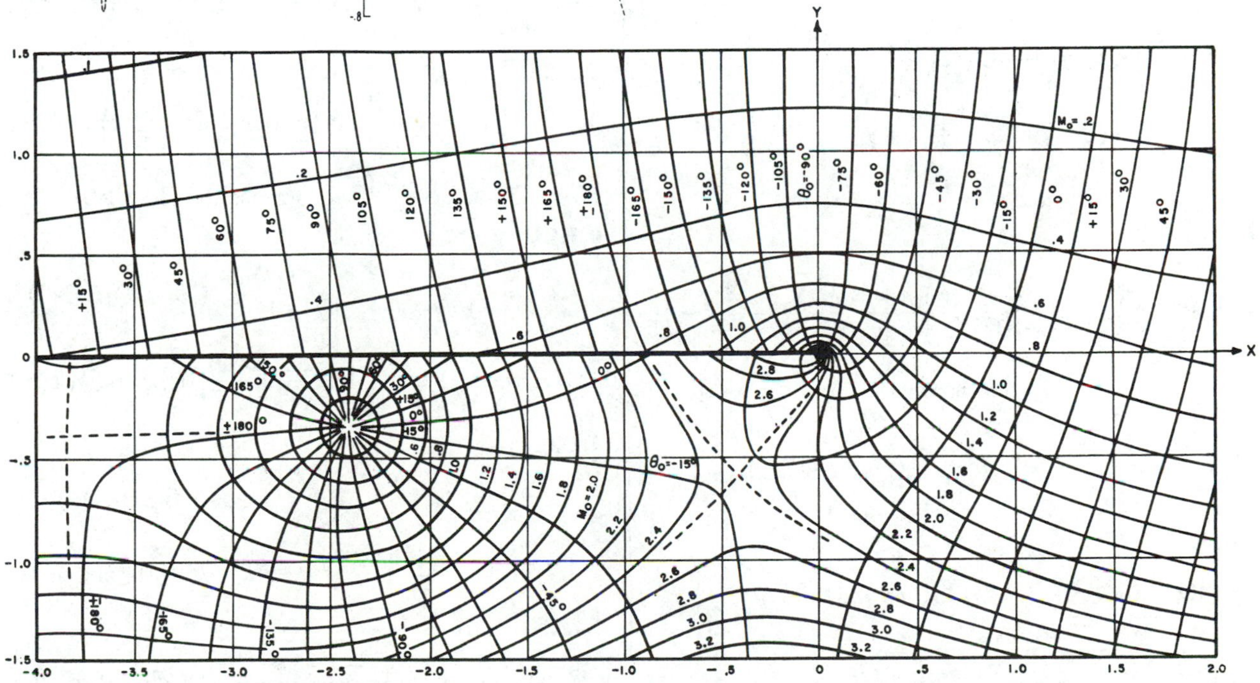

FIGURE 9.4. *Contour lines of the modulus and phase of the Hankel Function $H_0^{(1)}(x+iy)=M_0 e^{i\theta_0}$.* From E. Jahnke, F. Emde, and F. Lösch, Tables of higher functions, McGraw-Hill Book Co., Inc., New York, N.Y., 1960 (with permission).

Limiting Forms for Small Arguments

When ν is fixed and $z \to 0$

9.1.7

$$J_\nu(z) \sim (\tfrac{1}{2}z)^\nu / \Gamma(\nu+1) \qquad (\nu \neq -1, -2, -3, \ldots)$$

9.1.8 $Y_0(z) \sim -iH_0^{(1)}(z) \sim iH_0^{(2)}(z) \sim (2/\pi) \ln z$

9.1.9

$$Y_\nu(z) \sim -iH_\nu^{(1)}(z) \sim iH_\nu^{(2)}(z) \sim -(1/\pi)\Gamma(\nu)(\tfrac{1}{2}z)^{-\nu}$$
$$(\mathscr{R}\nu > 0)$$

Ascending Series

9.1.10 $J_\nu(z) = (\tfrac{1}{2}z)^\nu \sum_{k=0}^{\infty} \dfrac{(-\tfrac{1}{4}z^2)^k}{k!\,\Gamma(\nu+k+1)}$

9.1.11

$$Y_n(z) = -\frac{(\tfrac{1}{2}z)^{-n}}{\pi} \sum_{k=0}^{n-1} \frac{(n-k-1)!}{k!}(\tfrac{1}{4}z^2)^k$$
$$+ \frac{2}{\pi} \ln(\tfrac{1}{2}z) J_n(z)$$
$$- \frac{(\tfrac{1}{2}z)^n}{\pi} \sum_{k=0}^{\infty} \{\psi(k+1)+\psi(n+k+1)\} \frac{(-\tfrac{1}{4}z^2)^k}{k!\,(n+k)!}$$

where $\psi(n)$ is given by **6.3.2.**

9.1.12 $J_0(z) = 1 - \dfrac{\tfrac{1}{4}z^2}{(1!)^2} + \dfrac{(\tfrac{1}{4}z^2)^2}{(2!)^2} - \dfrac{(\tfrac{1}{4}z^2)^3}{(3!)^2} + \cdots$

9.1.13

$$Y_0(z) = \frac{2}{\pi}\{\ln(\tfrac{1}{2}z)+\gamma\}J_0(z) + \frac{2}{\pi}\left\{\frac{\tfrac{1}{4}z^2}{(1!)^2}\right.$$
$$\left. -(1+\tfrac{1}{2})\frac{(\tfrac{1}{4}z^2)^2}{(2!)^2} + (1+\tfrac{1}{2}+\tfrac{1}{3})\frac{(\tfrac{1}{4}z^2)^3}{(3!)^2} - \cdots\right\}$$

9.1.14

$$J_\nu(z) J_\mu(z) =$$
$$(\tfrac{1}{2}z)^{\nu+\mu} \sum_{k=0}^{\infty} \frac{(-)^k \Gamma(\nu+\mu+2k+1)(\tfrac{1}{4}z^2)^k}{\Gamma(\nu+k+1)\Gamma(\mu+k+1)\Gamma(\nu+\mu+k+1)\,k!}$$

Wronskians

9.1.15

$$W\{J_\nu(z), J_{-\nu}(z)\} = J_{\nu+1}(z)J_{-\nu}(z) + J_\nu(z)J_{-(\nu+1)}(z)$$
$$= -2\sin(\nu\pi)/(\pi z)$$

9.1.16

$$W\{J_\nu(z), Y_\nu(z)\} = J_{\nu+1}(z)Y_\nu(z) - J_\nu(z)Y_{\nu+1}(z)$$
$$= 2/(\pi z)$$

9.1.17

$$W\{H_\nu^{(1)}(z), H_\nu^{(2)}(z)\} = H_{\nu+1}^{(1)}(z)H_\nu^{(2)}(z) - H_\nu^{(1)}(z)H_{\nu+1}^{(2)}(z)$$
$$= -4i/(\pi z)$$

Integral Representations

9.1.18

$$J_0(z) = \frac{1}{\pi}\int_0^\pi \cos(z\sin\theta)\,d\theta = \frac{1}{\pi}\int_0^\pi \cos(z\cos\theta)\,d\theta$$

9.1.19

$$Y_0(z) = \frac{4}{\pi^2}\int_0^{\frac{1}{2}\pi} \cos(z\cos\theta)\{\gamma+\ln(2z\sin^2\theta)\}\,d\theta$$

9.1.20

$$J_\nu(z) = \frac{(\tfrac{1}{2}z)^\nu}{\pi^{\frac{1}{2}}\Gamma(\nu+\tfrac{1}{2})}\int_0^\pi \cos(z\cos\theta)\sin^{2\nu}\theta\,d\theta$$
$$= \frac{2(\tfrac{1}{2}z)^\nu}{\pi^{\frac{1}{2}}\Gamma(\nu+\tfrac{1}{2})}\int_0^1 (1-t^2)^{\nu-\frac{1}{2}}\cos(zt)\,dt \quad (\mathscr{R}\nu > -\tfrac{1}{2})$$

9.1.21

$$J_n(z) = \frac{1}{\pi}\int_0^\pi \cos(z\sin\theta - n\theta)\,d\theta$$
$$= \frac{i^{-n}}{\pi}\int_0^\pi e^{iz\cos\theta}\cos(n\theta)\,d\theta$$

9.1.22

$$J_\nu(z) = \frac{1}{\pi}\int_0^\pi \cos(z\sin\theta - \nu\theta)\,d\theta$$
$$- \frac{\sin(\nu\pi)}{\pi}\int_0^\infty e^{-z\sinh t - \nu t}\,dt \quad (|\arg z| < \tfrac{1}{2}\pi)$$

$$Y_\nu(z) = \frac{1}{\pi}\int_0^\pi \sin(z\sin\theta - \nu\theta)\,d\theta$$
$$- \frac{1}{\pi}\int_0^\infty \{e^{\nu t} + e^{-\nu t}\cos(\nu\pi)\}e^{-z\sinh t}\,dt \quad (|\arg z| < \tfrac{1}{2}\pi)$$

9.1.23

$$J_0(x) = \frac{2}{\pi}\int_0^\infty \sin(x\cosh t)\,dt \quad (x > 0)$$
$$Y_0(x) = -\frac{2}{\pi}\int_0^\infty \cos(x\cosh t)\,dt \quad (x > 0)$$

9.1.24

$$J_\nu(x) = \frac{2(\tfrac{1}{2}x)^{-\nu}}{\pi^{\frac{1}{2}}\Gamma(\tfrac{1}{2}-\nu)}\int_1^\infty \frac{\sin(xt)\,dt}{(t^2-1)^{\nu+\frac{1}{2}}} \quad (|\mathscr{R}\nu| < \tfrac{1}{2}, x > 0)$$

$$Y_\nu(x) = -\frac{2(\tfrac{1}{2}x)^{-\nu}}{\pi^{\frac{1}{2}}\Gamma(\tfrac{1}{2}-\nu)}\int_1^\infty \frac{\cos(xt)\,dt}{(t^2-1)^{\nu+\frac{1}{2}}} \quad (|\mathscr{R}\nu| < \tfrac{1}{2}, x > 0)$$

9.1.25

$$H_\nu^{(1)}(z) = \frac{1}{\pi i}\int_{-\infty}^{\infty+\pi i} e^{z\sinh t - \nu t}\,dt \quad (|\arg z| < \tfrac{1}{2}\pi)$$

$$H_\nu^{(2)}(z) = -\frac{1}{\pi i}\int_{-\infty}^{\infty-\pi i} e^{z\sinh t - \nu t}\,dt \quad (|\arg z| < \tfrac{1}{2}\pi)$$

9.1.26

$$J_\nu(x) = \frac{1}{2\pi i}\int_{-i\infty}^{i\infty} \frac{\Gamma(-t)(\tfrac{1}{2}x)^{\nu+2t}}{\Gamma(\nu+t+1)}\,dt \quad (\mathscr{R}\nu > 0, x > 0)$$

In the last integral the path of integration must lie to the left of the points $t = 0, 1, 2, \ldots$.

Recurrence Relations

9.1.27

$$\mathscr{C}_{\nu-1}(z)+\mathscr{C}_{\nu+1}(z)=\frac{2\nu}{z}\mathscr{C}_{\nu}(z)$$

$$\mathscr{C}_{\nu-1}(z)-\mathscr{C}_{\nu+1}(z)=2\mathscr{C}'_{\nu}(z)$$

$$\mathscr{C}'_{\nu}(z)=\mathscr{C}_{\nu-1}(z)-\frac{\nu}{z}\mathscr{C}_{\nu}(z)$$

$$\mathscr{C}'_{\nu}(z)=-\mathscr{C}_{\nu+1}(z)+\frac{\nu}{z}\mathscr{C}_{\nu}(z)$$

$\mathscr{C}$ denotes $J, Y, H^{(1)}, H^{(2)}$ or any linear combination of these functions, the coefficients in which are independent of z and ν.

9.1.28 $\quad J'_0(z)=-J_1(z) \qquad Y'_0(z)=-Y_1(z)$

If $f_\nu(z)=z^p\mathscr{C}_\nu(\lambda z^q)$ where p, q, λ are independent of ν, then

9.1.29

$$f_{\nu-1}(z)+f_{\nu+1}(z)=(2\nu/\lambda)z^{-q}f_\nu(z)$$

$$(p+\nu q)f_{\nu-1}(z)+(p-\nu q)f_{\nu+1}(z)=(2\nu/\lambda)z^{1-q}f'_\nu(z)$$

$$zf'_\nu(z)=\lambda qz^q f_{\nu-1}(z)+(p-\nu q)f_\nu(z)$$

$$zf'_\nu(z)=-\lambda qz^q f_{\nu+1}(z)+(p+\nu q)f_\nu(z)$$

Formulas for Derivatives

9.1.30

$$\left(\frac{1}{z}\frac{d}{dz}\right)^k\{z^\nu\mathscr{C}_\nu(z)\}=z^{\nu-k}\mathscr{C}_{\nu-k}(z)$$

$$\left(\frac{1}{z}\frac{d}{dz}\right)^k\{z^{-\nu}\mathscr{C}_\nu(z)\}=(-)^k z^{-\nu-k}\mathscr{C}_{\nu+k}(z)$$

$$(k=0,1,2,\ldots)$$

9.1.31

$$\mathscr{C}_\nu^{(k)}(z)=\frac{1}{2^k}\left\{\mathscr{C}_{\nu-k}(z)-\binom{k}{1}\mathscr{C}_{\nu-k+2}(z)\right.$$

$$\left.+\binom{k}{2}\mathscr{C}_{\nu-k+4}(z)-\ldots+(-)^k\mathscr{C}_{\nu+k}(z)\right\}$$

$$(k=0,1,2,\ldots)$$

Recurrence Relations for Cross-Products

If

9.1.32

$$p_\nu=J_\nu(a)Y_\nu(b)-J_\nu(b)Y_\nu(a)$$

$$q_\nu=J_\nu(a)Y'_\nu(b)-J'_\nu(b)Y_\nu(a)$$

$$r_\nu=J'_\nu(a)Y_\nu(b)-J_\nu(b)Y'_\nu(a)$$

$$s_\nu=J'_\nu(a)Y'_\nu(b)-J'_\nu(b)Y'_\nu(a)$$

then

9.1.33

$$p_{\nu+1}-p_{\nu-1}=-\frac{2\nu}{a}q_\nu-\frac{2\nu}{b}r_\nu$$

$$q_{\nu+1}+r_\nu=\frac{\nu}{a}p_\nu-\frac{\nu+1}{b}p_{\nu+1}$$

$$r_{\nu+1}+q_\nu=\frac{\nu}{b}p_\nu-\frac{\nu+1}{a}p_{\nu+1}$$

$$s_\nu=\frac{1}{2}p_{\nu+1}+\frac{1}{2}p_{\nu-1}-\frac{\nu^2}{ab}p_\nu$$

and

9.1.34

$$p_\nu s_\nu-q_\nu r_\nu=\frac{4}{\pi^2 ab}$$

Analytic Continuation

In **9.1.35** to **9.1.38**, m is an integer.

9.1.35 $\quad J_\nu(ze^{m\pi i})=e^{m\nu\pi i}J_\nu(z)$

9.1.36

$$Y_\nu(ze^{m\pi i})=e^{-m\nu\pi i}Y_\nu(z)+2i\sin(m\nu\pi)\cot(\nu\pi)J_\nu(z)$$

9.1.37

$$\sin(\nu\pi)H_\nu^{(1)}(ze^{m\pi i})=-\sin\{(m-1)\nu\pi\}H_\nu^{(1)}(z)$$

$$-e^{-\nu\pi i}\sin(m\nu\pi)H_\nu^{(2)}(z)$$

9.1.38

$$\sin(\nu\pi)H_\nu^{(2)}(ze^{m\pi i})=\sin\{(m+1)\nu\pi\}H_\nu^{(2)}(z)$$

$$+e^{\nu\pi i}\sin(m\nu\pi)H_\nu^{(1)}(z)$$

9.1.39

$$H_\nu^{(1)}(ze^{\pi i})=-e^{-\nu\pi i}H_\nu^{(2)}(z)$$

$$H_\nu^{(2)}(ze^{-\pi i})=-e^{\nu\pi i}H_\nu^{(1)}(z)$$

9.1.40

$$J_\nu(\bar{z})=\overline{J_\nu(z)} \qquad Y_\nu(\bar{z})=\overline{Y_\nu(z)}$$

$$H_\nu^{(1)}(\bar{z})=\overline{H_\nu^{(2)}(z)} \qquad H_\nu^{(2)}(\bar{z})=\overline{H_\nu^{(1)}(z)} \qquad (\nu\text{ real})$$

Generating Function and Associated Series

9.1.41 $\quad e^{\frac{1}{2}z(t-1/t)}=\sum_{k=-\infty}^{\infty}t^k J_k(z) \qquad (t\neq 0)$

9.1.42 $\quad \cos(z\sin\theta)=J_0(z)+2\sum_{k=1}^{\infty}J_{2k}(z)\cos(2k\theta)$

9.1.43 $\quad \sin(z\sin\theta)=2\sum_{k=0}^{\infty}J_{2k+1}(z)\sin\{(2k+1)\theta\}$

9.1.44

$$\cos(z\cos\theta)=J_0(z)+2\sum_{k=1}^{\infty}(-)^k J_{2k}(z)\cos(2k\theta)$$

9.1.45

$$\sin(z\cos\theta)=2\sum_{k=0}^{\infty}(-)^k J_{2k+1}(z)\cos\{(2k+1)\theta\}$$

9.1.46 $\quad 1=J_0(z)+2J_2(z)+2J_4(z)+2J_6(z)+\ldots$

9.1.47

$$\cos z=J_0(z)-2J_2(z)+2J_4(z)-2J_6(z)+\ldots$$

9.1.48 $\quad \sin z=2J_1(z)-2J_3(z)+2J_5(z)-\ldots$

Other Differential Equations

9.1.49 $w'' + \left(\lambda^2 - \frac{\nu^2 - \frac{1}{4}}{z^2}\right)w = 0, \qquad w = z^{\frac{1}{2}}\mathscr{C}_\nu(\lambda z)$

9.1.50 $w'' + \left(\frac{\lambda^2}{4z} - \frac{\nu^2 - 1}{4z^2}\right)w = 0, \qquad w = z^{\frac{1}{2}}\mathscr{C}_\nu(\lambda z^{\frac{1}{2}})$

9.1.51 $w'' + \lambda^2 z^{p-2}w = 0, \qquad w = z^{\frac{1}{2}}\mathscr{C}_{1/p}(2\lambda z^{\frac{1}{2}p}/p)$

9.1.52

$$w'' - \frac{2\nu-1}{z}w' + \lambda^2 w = 0, \qquad w = z^\nu \mathscr{C}_\nu(\lambda z)$$

9.1.53

$$z^2 w'' + (1-2p)zw' + (\lambda^2 q^2 z^{2q} + p^2 - \nu^2 q^2)w = 0,$$
$$w = z^p \mathscr{C}_\nu(\lambda z^q)$$

9.1.54

$$w'' + (\lambda^2 e^{2z} - \nu^2)w = 0, \qquad w = \mathscr{C}_\nu(\lambda e^z)$$

9.1.55

$$z^2(z^2 - \nu^2)w'' + z(z^2 - 3\nu^2)w'$$
$$+ \{(z^2 - \nu^2)^2 - (z^2 + \nu^2)\}w = 0, \qquad w = \mathscr{C}'_\nu(z)$$

9.1.56

$$w^{(2n)} = (-)^n \lambda^{2n} z^{-n} w, \qquad w = z^{\frac{1}{2}n}\mathscr{C}_n(2\lambda\alpha z^{\frac{1}{2}})$$

where α is any of the $2n$ roots of unity.

Differential Equations for Products

In the following $\vartheta \equiv z\frac{d}{dz}$ and $\mathscr{C}_\nu(z)$, $\mathscr{D}_\mu(z)$ are any cylinder functions of orders ν, μ respectively.

9.1.57

$$\{\vartheta^4 - 2(\nu^2 + \mu^2)\vartheta^2 + (\nu^2 - \mu^2)^2\}w$$
$$+ 4z^2(\vartheta + 1)(\vartheta + 2)w = 0, \qquad w = \mathscr{C}_\nu(z)\mathscr{D}_\mu(z)$$

9.1.58

$$\vartheta(\vartheta^2 - 4\nu^2)w + 4z^2(\vartheta + 1)w = 0, \qquad w = \mathscr{C}_\nu(z)\mathscr{D}_\nu(z)$$

9.1.59

$$z^3 w''' + z(4z^2 + 1 - 4\nu^2)w' + (4\nu^2 - 1)w = 0,$$
$$w = z\mathscr{C}_\nu(z)\mathscr{D}_\nu(z)$$

Upper Bounds

9.1.60 $|J_\nu(x)| \leq 1 \ (\nu \geq 0), \ |J_\nu(x)| \leq 1/\sqrt{2} \qquad (\nu \geq 1)$

9.1.61 $0 < J_\nu(\nu) < \dfrac{2^{\frac{1}{3}}}{3^{\frac{1}{3}}\Gamma(\frac{2}{3})\nu^{\frac{1}{3}}} \qquad (\nu > 0)$

9.1.62 $|J_\nu(z)| \leq \dfrac{|\frac{1}{2}z|^\nu e^{|\mathscr{I}z|}}{\Gamma(\nu+1)} \qquad (\nu \geq -\frac{1}{2}) \qquad *$

9.1.63 $|J_n(nz)| \leq \left|\dfrac{z^n \exp\{n\sqrt{(1-z^2)}\}}{\{1 + \sqrt{(1-z^2)}\}^n}\right|$

Derivatives With Respect to Order

9.1.64

$$\frac{\partial}{\partial\nu}J_\nu(z) = J_\nu(z)\ln(\tfrac{1}{2}z)$$
$$- (\tfrac{1}{2}z)^\nu \sum_{k=0}^\infty (-)^k \frac{\psi(\nu+k+1)}{\Gamma(\nu+k+1)}\frac{(\frac{1}{4}z^2)^k}{k!}$$

9.1.65

$$\frac{\partial}{\partial\nu}Y_\nu(z) = \cot(\nu\pi)\{\frac{\partial}{\partial\nu}J_\nu(z) - \pi Y_\nu(z)\}$$
$$- \csc(\nu\pi)\frac{\partial}{\partial\nu}J_{-\nu}(z) - \pi J_\nu(z)$$
$$(\nu \neq 0, \pm 1, \pm 2, \ldots)$$

9.1.66

$$\left[\frac{\partial}{\partial\nu}J_\nu(z)\right]_{\nu=n} = \frac{\pi}{2}Y_n(z) + \frac{n!(\frac{1}{2}z)^{-n}}{2}\sum_{k=0}^{n-1}\frac{(\frac{1}{2}z)^k J_k(z)}{(n-k)k!}$$

9.1.67

$$\left[\frac{\partial}{\partial\nu}Y_\nu(z)\right]_{\nu=n} = -\frac{\pi}{2}J_n(z) + \frac{n!(\frac{1}{2}z)^{-n}}{2}\sum_{k=0}^{n-1}\frac{(\frac{1}{2}z)^k Y_k(z)}{(n-k)k!}$$

9.1.68

$$\left[\frac{\partial}{\partial\nu}J_\nu(z)\right]_{\nu=0} = \frac{\pi}{2}Y_0(z), \left[\frac{\partial}{\partial\nu}Y_\nu(z)\right]_{\nu=0} = -\frac{\pi}{2}J_0(z)$$

Expressions in Terms of Hypergeometric Functions

9.1.69

$$J_\nu(z) = \frac{(\frac{1}{2}z)^\nu}{\Gamma(\nu+1)}\,_0F_1(\nu+1; -\tfrac{1}{4}z^2)$$
$$= \frac{(\frac{1}{2}z)^\nu e^{-iz}}{\Gamma(\nu+1)}M(\nu+\tfrac{1}{2}, 2\nu+1, 2iz)$$

9.1.70

$$J_\nu(z) = \frac{(\frac{1}{2}z)^\nu}{\Gamma(\nu+1)}\lim F\left(\lambda, \mu; \nu+1; -\frac{z^2}{4\lambda\mu}\right)$$

as λ, $\mu \to \infty$ through real or complex values; z, ν being fixed.

($_0F_1$ is the generalized hypergeometric function. For $M(a, b, z)$ and $F(a, b; c; z)$ see chapters **13** and **15**.)

Connection With Legendre Functions

If μ and x are fixed and $\nu \to \infty$ through real positive values

9.1.71

$$\lim\{\nu^\mu P_\nu^{-\mu}\left(\cos\frac{x}{\nu}\right)\} = J_\mu(x) \qquad (x > 0)$$

9.1.72

$$\lim \{\nu^\mu Q_\nu^{-\mu}\left(\cos\frac{x}{\nu}\right)\}=-\tfrac{1}{2}\pi\, Y_\mu(x) \qquad (x>0)$$

For $P_\nu^{-\mu}$ and $Q_\nu^{-\mu}$, see chapter **8**.

Continued Fractions

9.1.73

$$\frac{J_\nu(z)}{J_{\nu-1}(z)}=\frac{1}{2\nu z^{-1}-}\ \frac{1}{2(\nu+1)z^{-1}-}\ \frac{1}{2(\nu+2)z^{-1}-}\cdots$$

$$=\frac{\tfrac{1}{2}z/\nu}{1-}\ \frac{\tfrac{1}{4}z^2/\{\nu(\nu+1)\}}{1-}\ \frac{\tfrac{1}{4}z^2/\{(\nu+1)(\nu+2)\}}{1-}\cdots$$

Multiplication Theorem

9.1.74

$$\mathscr{C}_\nu(\lambda z)=\lambda^{\pm\nu}\sum_{k=0}^{\infty}\frac{(\mp)^k(\lambda^2-1)^k(\tfrac{1}{2}z)^k}{k!}\,\mathscr{C}_{\nu\pm k}(z)$$

$$(|\lambda^2-1|<1)$$

If $\mathscr{C}=J$ and the upper signs are taken, the restriction on λ is unnecessary.

This theorem will furnish expansions of $\mathscr{C}_\nu(re^{i\theta})$ in terms of $\mathscr{C}_{\nu\pm k}(r)$.

Addition Theorems

Neumann's

9.1.75 $\quad\mathscr{C}_\nu(u\pm v)=\sum_{k=-\infty}^{\infty}\mathscr{C}_{\nu\mp k}(u)J_k(v) \qquad (|v|<|u|)$

The restriction $|v|<|u|$ is unnecessary when $\mathscr{C}=J$ and ν is an integer or zero. Special cases are

9.1.76

$$1=J_0^2(z)+2\sum_{k=1}^{\infty}J_k^2(z)$$

9.1.77

$$0=\sum_{k=0}^{2n}(-)^k J_k(z)J_{2n-k}(z)+2\sum_{k=1}^{\infty}J_k(z)J_{2n+k}(z)\quad(n\geq1)$$

9.1.78

$$J_n(2z)=\sum_{k=0}^{n}J_k(z)J_{n-k}(z)+2\sum_{k=1}^{\infty}(-)^k J_k(z)J_{n+k}(z)$$

Graf's

9.1.79

$$\mathscr{C}_\nu(w)\,{\cos\atop\sin}\,\nu\chi=\sum_{k=-\infty}^{\infty}\mathscr{C}_{\nu+k}(u)J_k(v)\,{\cos\atop\sin}\,k\alpha\,(|ve^{\pm i\alpha}|<|u|)$$

Gegenbauer's

9.1.80

$$\frac{\mathscr{C}_\nu(w)}{w^\nu}=2^\nu\Gamma(\nu)\sum_{k=0}^{\infty}(\nu+k)\frac{\mathscr{C}_{\nu+k}(u)}{u^\nu}\frac{J_{\nu+k}(v)}{v^\nu}\,C_k^{(\nu)}(\cos\alpha)$$

$$(\nu\neq0,-1,\ldots,|ve^{\pm i\alpha}|<|u|)$$

In **9.1.79** and **9.1.80**,

$$w=\surd(u^2+v^2-2uv\cos\alpha),$$

$$u-v\cos\alpha=w\cos\chi,\ v\sin\alpha=w\sin\chi$$

the branches being chosen so that $w\to u$ and $\chi\to0$ as $v\to0$. $C_k^{(\nu)}(\cos\alpha)$ is Gegenbauer's polynomial (see chapter **22**).

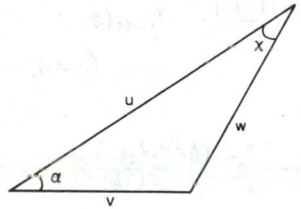

Gegenbauer's addition theorem.

If u,v are real and positive and $0\leq\alpha\leq\pi$, then w,χ are real and non-negative, and the geometrical relationship of the variables is shown in the diagram.

The restrictions $|ve^{\pm i\alpha}|<|u|$ are unnecessary in **9.1.79** when $\mathscr{C}=J$ and ν is an integer or zero, and in **9.1.80** when $\mathscr{C}=J$.

Degenerate Form $(u=\infty)$:

9.1.81

$$e^{iv\cos\alpha}=\Gamma(\nu)(\tfrac{1}{2}v)^{-\nu}\sum_{k=0}^{\infty}(\nu+k)i^k J_{\nu+k}(v)C_k^{(\nu)}(\cos\alpha)$$

$$(\nu\neq0,-1,\ldots)$$

Neumann's Expansion of an Arbitrary Function in a Series of Bessel Functions

9.1.82 $\quad f(z)=a_0 J_0(z)+2\sum_{k=1}^{\infty}a_k J_k(z) \qquad (|z|<c)$

where c is the distance of the nearest singularity of $f(z)$ from $z=0$,

9.1.83 $\quad a_k=\frac{1}{2\pi i}\int_{|z|=c'}f(t)O_k(t)dt \qquad (0<c'<c)$

and $O_k(t)$ is Neumann's polynomial. The latter is defined by the generating function

9.1.84

$$\frac{1}{t-z}=J_0(z)O_0(t)+2\sum_{k=1}^{\infty}J_k(z)O_k(t) \qquad (|z|<|t|)$$

$O_n(t)$ is a polynomial of degree $n+1$ in $1/t$; $O_0(t)=1/t$,

9.1.85

$$O_n(t)=\frac{1}{4}\sum_{k=0}^{\leq\frac{1}{2}n}\frac{n(n-k-1)!}{k!}\left(\frac{2}{t}\right)^{n-2k+1} \qquad (n=1,2,\ldots)$$

The more general form of expansion

9.1.86 $\quad f(z)=a_0 J_\nu(z)+2\sum_{k=1}^{\infty}a_k J_{\nu+k}(z)$

also called a Neumann expansion, is investigated in [9.7] and [9.15] together with further generalizations. Examples of Neumann expansions are **9.1.41** to **9.1.48** and the Addition Theorems. Other examples are

9.1.87

$$(\tfrac{1}{2}z)^\nu = \sum_{k=0}^\infty \frac{(\nu+2k)\,\Gamma(\nu+k)}{k!}\,J_{\nu+2k}(z)$$

$$(\nu \neq 0, -1, -2, \ldots)$$

9.1.88

$$Y_n(z) = -\frac{n!(\tfrac{1}{2}z)^{-n}}{\pi}\sum_{k=0}^{n-1}\frac{(\tfrac{1}{2}z)^k J_k(z)}{(n-k)k!}$$

$$+\frac{2}{\pi}\{\ln(\tfrac{1}{2}z) - \psi(n+1)\}J_n(z)$$

$$-\frac{2}{\pi}\sum_{k=1}^\infty (-)^k\frac{(n+2k)J_{n+2k}(z)}{k(n+k)}$$

where $\psi(n)$ is given by **6.3.2**.

9.1.89

$$Y_0(z) = \frac{2}{\pi}\{\ln(\tfrac{1}{2}z)+\gamma\}J_0(z) - \frac{4}{\pi}\sum_{k=1}^\infty (-)^k\frac{J_{2k}(z)}{k}$$

9.2. Asymptotic Expansions for Large Arguments

Principal Asymptotic Forms

When ν is fixed and $|z| \to \infty$

9.2.1

$$J_\nu(z) = \sqrt{2/(\pi z)}\{\cos(z - \tfrac{1}{2}\nu\pi - \tfrac{1}{4}\pi) + e^{|\mathscr{I}z|}O(|z|^{-1})\}$$

$$(|\arg z| < \pi)$$

9.2.2

$$Y_\nu(z) = \sqrt{2/(\pi z)}\{\sin(z - \tfrac{1}{2}\nu\pi - \tfrac{1}{4}\pi) + e^{|\mathscr{I}z|}O(|z|^{-1})\}$$

$$(|\arg z| < \pi)$$

9.2.3

$$H_\nu^{(1)}(z) \sim \sqrt{2/(\pi z)}\,e^{i(z-\tfrac{1}{2}\nu\pi-\tfrac{1}{4}\pi)} \qquad (-\pi < \arg z < 2\pi)$$

9.2.4

$$H_\nu^{(2)}(z) \sim \sqrt{2/(\pi z)}\,e^{-i(z-\tfrac{1}{2}\nu\pi-\tfrac{1}{4}\pi)} \qquad (-2\pi < \arg z < \pi)$$

Hankel's Asymptotic Expansions

When ν is fixed and $|z| \to \infty$

9.2.5

$$J_\nu(z) = \sqrt{2/(\pi z)}\{P(\nu, z)\cos\chi - Q(\nu, z)\sin\chi\}$$

$$(|\arg z| < \pi)$$

9.2.6

$$Y_\nu(z) = \sqrt{2/(\pi z)}\{P(\nu, z)\sin\chi + Q(\nu, z)\cos\chi\}$$

$$(|\arg z| < \pi)$$

9.2.7

$$H_\nu^{(1)}(z) = \sqrt{2/(\pi z)}\{P(\nu, z)+iQ(\nu, z)\}e^{i\chi}$$

$$(-\pi < \arg z < 2\pi)$$

9.2.8

$$H_\nu^{(2)}(z) = \sqrt{2/(\pi z)}\{P(\nu, z)-iQ(\nu, z)\}e^{-i\chi}$$

$$(-2\pi < \arg z < \pi)$$

where $\chi = z - (\tfrac{1}{2}\nu + \tfrac{1}{4})\pi$ and, with $4\nu^2$ denoted by μ,

9.2.9

$$P(\nu, z) \sim \sum_{k=0}^\infty (-)^k\frac{(\nu, 2k)}{(2z)^{2k}} = 1 - \frac{(\mu-1)(\mu-9)}{2!(8z)^2}$$

$$+\frac{(\mu-1)(\mu-9)(\mu-25)(\mu-49)}{4!(8z)^4} - \cdots$$

9.2.10

$$Q(\nu, z) \sim \sum_{k=0}^\infty (-)^k\frac{(\nu, 2k+1)}{(2z)^{2k+1}}$$

$$= \frac{\mu-1}{8z} - \frac{(\mu-1)(\mu-9)(\mu-25)}{3!(8z)^3} + \cdots$$

If ν is real and non-negative and z is positive, the remainder after k terms in the expansion of $P(\nu, z)$ does not exceed the $(k+1)$th term in absolute value and is of the same sign, provided that $k > \tfrac{1}{2}\nu - \tfrac{1}{4}$. The same is true of $Q(\nu,z)$ provided that $k > \tfrac{1}{2}\nu - \tfrac{3}{4}$.

Asymptotic Expansions of Derivatives

With the conditions and notation of the preceding subsection

9.2.11

$$J_\nu'(z) = \sqrt{2/(\pi z)}\{-R(\nu, z)\sin\chi - S(\nu, z)\cos\chi\}$$

$$(|\arg z| < \pi)$$

9.2.12

$$Y_\nu'(z) = \sqrt{2/(\pi z)}\{R(\nu, z)\cos\chi - S(\nu, z)\sin\chi\}$$

$$(|\arg z| < \pi)$$

9.2.13

$$H_\nu^{(1)\prime}(z) = \sqrt{2/(\pi z)}\{iR(\nu, z)-S(\nu, z)\}e^{i\chi}$$

$$(-\pi < \arg z < 2\pi)$$

9.2.14

$$H_\nu^{(2)\prime}(z) = \sqrt{2/(\pi z)}\{-iR(\nu, z)-S(\nu, z)\}e^{-i\chi}$$

$$(-2\pi < \arg z < \pi)$$

9.2.15

$$R(\nu, z) \sim \sum_{k=0}^{\infty} (-)^k \frac{4\nu^2 + 16k^2 - 1}{4\nu^2 - (4k-1)^2} \frac{(\nu, 2k)}{(2z)^{2k}}$$

$$= 1 - \frac{(\mu-1)(\mu+15)}{2!(8z)^2} + \cdots$$

9.2.16

$$S(\nu, z) \sim \sum_{k=0}^{\infty} (-)^k \frac{4\nu^2 + 4(2k+1)^2 - 1}{4\nu^2 - (4k+1)^2} \frac{(\nu, 2k+1)}{(2z)^{2k+1}}$$

$$= \frac{\mu+3}{8z} - \frac{(\mu-1)(\mu-9)(\mu+35)}{3!(8z)^3} + \cdots$$

Modulus and Phase

For real ν and positive x

9.2.17

$$M_\nu = |H_\nu^{(1)}(x)| = \sqrt{\{J_\nu^2(x) + Y_\nu^2(x)\}}$$
$$\theta_\nu = \arg H_\nu^{(1)}(x) = \arctan\{Y_\nu(x)/J_\nu(x)\}$$

9.2.18

$$N_\nu = |H_\nu^{(1)\prime}(x)| = \sqrt{\{J_\nu^{\prime 2}(x) + Y_\nu^{\prime 2}(x)\}}$$
$$\varphi_\nu = \arg H_\nu^{(1)\prime}(x) = \arctan\{Y_\nu'(x)/J_\nu'(x)\}$$

9.2.19 $\quad J_\nu(x) = M_\nu \cos\theta_\nu, \qquad Y_\nu(x) = M_\nu \sin\theta_\nu,$

9.2.20 $\quad J_\nu'(x) = N_\nu \cos\varphi_\nu, \qquad Y_\nu'(x) = N_\nu \sin\varphi_\nu.$

In the following relations, primes denote differentiations with respect to x.

9.2.21 $\quad M_\nu^2\theta_\nu' = 2/(\pi x) \qquad N_\nu^2\varphi_\nu' = 2(x^2-\nu^2)/(\pi x^3)$

9.2.22 $\quad N_\nu^2 = M_\nu'^2 + M_\nu^2\theta_\nu'^2 = M_\nu'^2 + 4/(\pi x M_\nu)^2$

9.2.23 $\quad (x^2-\nu^2)M_\nu M_\nu' + x^2 N_\nu N_\nu' + x N_\nu^2 = 0$

9.2.24

$$\tan(\varphi_\nu - \theta_\nu) = M_\nu \theta_\nu'/M_\nu' = 2/(\pi x M_\nu M_\nu')$$
$$M_\nu N_\nu \sin(\varphi_\nu - \theta_\nu) = 2/(\pi x)$$

9.2.25 $\quad x^2 M_\nu'' + x M_\nu' + (x^2-\nu^2)M_\nu - 4/(\pi^2 M_\nu^3) = 0$

9.2.26

$$x^3 w''' + x(4x^2 + 1 - 4\nu^2)w' + (4\nu^2 - 1)w = 0, \quad w = x M_\nu^2$$

9.2.27 $\quad \theta_\nu'^2 + \frac{1}{2}\frac{\theta_\nu'''}{\theta_\nu'} - \frac{3}{4}\left(\frac{\theta_\nu''}{\theta_\nu'}\right)^2 = 1 - \frac{\nu^2 - \frac{1}{4}}{x^2}$

Asymptotic Expansions of Modulus and Phase

When ν is fixed, x is large and positive, and $\mu = 4\nu^2$

9.2.28

$$M_\nu^2 \sim \frac{2}{\pi x}\left\{1 + \frac{1}{2}\frac{\mu-1}{(2x)^2} + \frac{1\cdot3}{2\cdot4}\frac{(\mu-1)(\mu-9)}{(2x)^4}\right.$$
$$\left. + \frac{1\cdot3\cdot5}{2\cdot4\cdot6}\frac{(\mu-1)(\mu-9)(\mu-25)}{(2x)^6} + \cdots\right\}$$

9.2.29

$$\theta_\nu \sim x - (\tfrac{1}{2}\nu + \tfrac{1}{4})\pi + \frac{\mu-1}{2(4x)}$$
$$+ \frac{(\mu-1)(\mu-25)}{6(4x)^3} + \frac{(\mu-1)(\mu^2 - 114\mu + 1073)}{5(4x)^5}$$
$$+ \frac{(\mu-1)(5\mu^3 - 1535\mu^2 + 54703\mu - 375733)}{14(4x)^7} + \cdots$$

9.2.30

$$N_\nu^2 \sim \frac{2}{\pi x}\left\{1 - \frac{1}{2}\frac{\mu-3}{(2x)^2} - \frac{1\cdot1}{2\cdot4}\frac{(\mu-1)(\mu-45)}{(2x)^4} - \cdots\right\}$$

The general term in the last expansion is given by

$$-\frac{1\cdot1\cdot3\ldots(2k-3)}{2\cdot4\cdot6\ldots(2k)}$$

$$\times\frac{(\mu-1)(\mu-9)\ldots\{\mu-(2k-3)^2\}\{\mu-(2k+1)(2k-1)^2\}}{(2x)^{2k}}$$ *

9.2.31

$$\phi_\nu \sim x - (\tfrac{1}{2}\nu - \tfrac{1}{4})\pi + \frac{\mu+3}{2(4x)} + \frac{\mu^2 + 46\mu - 63}{6(4x)^3}$$
$$+ \frac{\mu^3 + 185\mu^2 - 2053\mu + 1899}{5(4x)^5} + \cdots$$

If $\nu \geq 0$, the remainder after k terms in **9.2.28** does not exceed the $(k+1)$th term in absolute value and is of the same sign, provided that $k > \nu - \frac{1}{2}$.

9.3. Asymptotic Expansions for Large Orders

Principal Asymptotic Forms

In the following equations it is supposed that $\nu \to \infty$ through real positive values, the other variables being fixed.

9.3.1

$$J_\nu(z) \sim \frac{1}{\sqrt{2\pi\nu}}\left(\frac{ez}{2\nu}\right)^\nu$$

$$Y_\nu(z) \sim -\sqrt{\frac{2}{\pi\nu}}\left(\frac{ez}{2\nu}\right)^{-\nu}$$

9.3.2

$$J_\nu(\nu\,\mathrm{sech}\,\alpha) \sim \frac{e^{\nu(\tanh\alpha - \alpha)}}{\sqrt{2\pi\nu\tanh\alpha}} \qquad (\alpha > 0)$$

$$Y_\nu(\nu\,\mathrm{sech}\,\alpha) \sim -\frac{e^{\nu(\alpha - \tanh\alpha)}}{\sqrt{\frac{1}{2}\pi\nu\tanh\alpha}} \qquad (\alpha > 0)$$

*See page II.

9.3.3

$$J_\nu(\nu \sec \beta) =$$
$$\sqrt{2/(\pi\nu \tan \beta)} \{\cos(\nu \tan \beta - \nu\beta - \tfrac{1}{4}\pi) + O(\nu^{-1})\}$$
$$(0 < \beta < \tfrac{1}{2}\pi)$$

$$Y_\nu(\nu \sec \beta) =$$
$$\sqrt{2/(\pi\nu \tan \beta)} \{\sin(\nu \tan \beta - \nu\beta - \tfrac{1}{4}\pi) + O(\nu^{-1})\}$$
$$(0 < \beta < \tfrac{1}{2}\pi)$$

9.3.4

$$J_\nu(\nu + z\nu^{1/3}) = 2^{1/3}\nu^{-1/3} \operatorname{Ai}(-2^{1/3}z) + O(\nu^{-1})$$

$$Y_\nu(\nu + z\nu^{1/3}) = -2^{1/3}\nu^{-1/3} \operatorname{Bi}(-2^{1/3}z) + O(\nu^{-1})$$

9.3.5
$$J_\nu(\nu) \sim \frac{2^{1/3}}{3^{2/3}\Gamma(\tfrac{2}{3})} \frac{1}{\nu^{1/3}}$$

$$Y_\nu(\nu) \sim -\frac{2^{1/3}}{3^{1/6}\Gamma(\tfrac{2}{3})} \frac{1}{\nu^{1/3}}$$

9.3.6

$$J_\nu(\nu z) = \left(\frac{4\zeta}{1-z^2}\right)^{1/4}\Big\{\frac{\operatorname{Ai}(\nu^{2/3}\zeta)}{\nu^{1/3}}$$
$$+ \frac{\exp(-\tfrac{2}{3}\nu\zeta^{3/2})}{1+\nu^{1/6}|\zeta|^{1/4}} O\left(\frac{1}{\nu^{1/3}}\right)\Big\} \quad (|\arg z| < \pi)$$

$$Y_\nu(\nu z) = -\left(\frac{4\zeta}{1-z^2}\right)^{1/4}\Big\{\frac{\operatorname{Bi}(\nu^{2/3}\zeta)}{\nu^{1/3}}$$
$$+ \frac{\exp|\mathscr{R}(\tfrac{2}{3}\nu\zeta^{3/2})|}{1+\nu^{1/6}|\zeta|^{1/4}} O\left(\frac{1}{\nu^{1/3}}\right)\Big\} \quad (|\arg z| < \pi)$$

In the last two equations ζ is given by **9.3.38** and **9.3.39** below.

Debye's Asymptotic Expansions

(i) If α is fixed and positive and ν is large and positive

9.3.7

$$J_\nu(\nu \operatorname{sech} \alpha) \sim \frac{e^{\nu(\tanh \alpha - \alpha)}}{\sqrt{2\pi\nu \tanh \alpha}} \Big\{1 + \sum_{k=1}^{\infty} \frac{u_k(\coth \alpha)}{\nu^k}\Big\}$$

9.3.8

$$Y_\nu(\nu \operatorname{sech} \alpha) \sim$$
$$-\frac{e^{\nu(\alpha - \tanh \alpha)}}{\sqrt{\tfrac{1}{2}\pi\nu \tanh \alpha}} \Big\{1 + \sum_{k=1}^{\infty} (-)^k \frac{u_k(\coth \alpha)}{\nu^k}\Big\}$$

where

9.3.9

$$u_0(t) = 1$$
$$u_1(t) = (3t - 5t^3)/24$$
$$u_2(t) = (81t^2 - 462t^4 + 385t^6)/1152$$
$$u_3(t) = (30375t^3 - 3\,69603t^5 + 7\,65765t^7$$
$$\qquad\qquad\qquad - 4\,25425t^9)/4\,14720$$
$$u_4(t) = (44\,65125t^4 - 941\,21676t^6 + 3499\,22430t^8$$
$$\qquad - 4461\,85740t^{10} + 1859\,10725t^{12})/398\,13120$$

For $u_5(t)$ and $u_6(t)$ see [9.4] or [9.21].

9.3.10

$$u_{k+1}(t) = \tfrac{1}{2}t^2(1-t^2)u_k'(t) + \frac{1}{8}\int_0^t (1-5t^2)u_k(t)dt$$
$$(k = 0, 1, \ldots)$$

Also

9.3.11

$$J_\nu'(\nu \operatorname{sech} \alpha) \sim$$
$$\sqrt{\frac{\sinh 2\alpha}{4\pi\nu}} e^{\nu(\tanh \alpha - \alpha)} \Big\{1 + \sum_{k=1}^{\infty} \frac{v_k(\coth \alpha)}{\nu^k}\Big\}$$

9.3.12

$$Y_\nu'(\nu \operatorname{sech} \alpha)$$
$$\sim \sqrt{\frac{\sinh 2\alpha}{\pi\nu}} e^{\nu(\alpha - \tanh \alpha)} \Big\{1 + \sum_{k=1}^{\infty} (-)^k \frac{v_k(\coth \alpha)}{\nu^k}\Big\}$$

where

9.3.13

$$v_0(t) = 1$$
$$v_1(t) = (-9t + 7t^3)/24$$
$$v_2(t) = (-135t^2 + 594t^4 - 455t^6)/1152$$
$$v_3(t) = (-42525t^3 + 4\,51737t^5 - 8\,83575t^7$$
$$\qquad\qquad\qquad + 4\,75475t^9)/4\,14720$$

9.3.14

$$v_k(t) = u_k(t) + t(t^2-1)\{\tfrac{1}{2}u_{k-1}(t) + tu_{k-1}'(t)\}$$
$$(k = 1, 2, \ldots)$$

(ii) If β is fixed, $0 < \beta < \tfrac{1}{2}\pi$ and ν is large and positive

9.3.15

$$J_\nu(\nu \sec \beta) = \sqrt{2/(\pi\nu \tan \beta)}\{L(\nu, \beta) \cos \Psi + M(\nu, \beta) \sin \Psi\}$$

9.3.16

$$Y_\nu(\nu \sec \beta) = \sqrt{2/(\pi\nu \tan \beta)}\{L(\nu, \beta) \sin \Psi - M(\nu, \beta) \cos \Psi\}$$

where $\Psi = \nu(\tan \beta - \beta) - \tfrac{1}{4}\pi$

9.3.17

$$L(\nu, \beta) \sim \sum_{k=0}^{\infty} \frac{u_{2k}(i \cot \beta)}{\nu^{2k}}$$
$$= 1 - \frac{81 \cot^2 \beta + 462 \cot^4 \beta + 385 \cot^6 \beta}{1152\nu^2} + \ldots$$

9.3.18

$$M(\nu, \beta) \sim -i \sum_{k=0}^{\infty} \frac{u_{2k+1}(i \cot \beta)}{\nu^{2k+1}}$$
$$= \frac{3 \cot \beta + 5 \cot^3 \beta}{24\nu} - \cdots$$

Also

9.3.19

$$J_{\nu}'(\nu \sec \beta) = \sqrt{(\sin 2\beta)/(\pi\nu)} \{ -N(\nu, \beta) \sin \Psi - O(\nu, \beta) \cos \Psi \}$$

9.3.20

$$Y_{\nu}'(\nu \sec \beta) = \sqrt{(\sin 2\beta)/(\pi\nu)} \{ N(\nu, \beta) \cos \Psi - O(\nu, \beta) \sin \Psi \}$$

where

9.3.21

$$N(\nu, \beta) \sim \sum_{k=0}^{\infty} \frac{v_{2k}(i \cot \beta)}{\nu^{2k}}$$
$$= 1 + \frac{135 \cot^2 \beta + 594 \cot^4 \beta + 455 \cot^6 \beta}{1152\nu^2} - \cdots$$

9.3.22

$$O(\nu, \beta) \sim i \sum_{k=0}^{\infty} \frac{v_{2k+1}(i \cot \beta)}{\nu^{2k+1}} = \frac{9 \cot \beta + 7 \cot^3 \beta}{24\nu} - \cdots$$

Asymptotic Expansions in the Transition Regions

When z is fixed, $|\nu|$ is large and $|\arg \nu| < \frac{1}{2}\pi$

9.3.23

$$J_{\nu}(\nu + z\nu^{1/3}) \sim \frac{2^{1/3}}{\nu^{1/3}} \text{Ai}(-2^{1/3}z) \{ 1 + \sum_{k=1}^{\infty} \frac{f_k(z)}{\nu^{2k/3}} \}$$
$$+ \frac{2^{2/3}}{\nu} \text{Ai}'(-2^{1/3}z) \sum_{k=0}^{\infty} \frac{g_k(z)}{\nu^{2k/3}}$$

9.3.24

$$Y_{\nu}(\nu + z\nu^{1/3}) \sim -\frac{2^{1/3}}{\nu^{1/3}} \text{Bi}(-2^{1/3}z) \{ 1 + \sum_{k=1}^{\infty} \frac{f_k(z)}{\nu^{2k/3}} \}$$
$$- \frac{2^{2/3}}{\nu} \text{Bi}'(-2^{1/3}z) \sum_{k=0}^{\infty} \frac{g_k(z)}{\nu^{2k/3}}$$

where

9.3.25

$$f_1(z) = -\frac{1}{5} z$$

$$f_2(z) = -\frac{9}{100} z^5 + \frac{3}{35} z^2$$

$$f_3(z) = \frac{957}{7000} z^6 - \frac{173}{3150} z^3 - \frac{1}{225}$$

$$f_4(z) = \frac{27}{20000} z^{10} - \frac{23573}{147000} z^7 + \frac{5903}{138600} z^4 + \frac{947}{346500} z$$

9.3.26

$$g_0(z) = \frac{3}{10} z^2$$

$$g_1(z) = -\frac{17}{70} z^3 + \frac{1}{70}$$

$$g_2(z) = -\frac{9}{1000} z^7 + \frac{611}{3150} z^4 - \frac{37}{3150} z$$

$$g_3(z) = \frac{549}{28000} z^8 - \frac{110767}{693000} z^5 + \frac{79}{12375} z^2$$

The corresponding expansions for $H_{\nu}^{(1)}(\nu + z\nu^{1/3})$ and $H_{\nu}^{(2)}(\nu + z\nu^{1/3})$ are obtained by use of **9.1.3** and **9.1.4**; they are valid for $-\frac{1}{2}\pi < \arg \nu < \frac{3}{2}\pi$ and $-\frac{3}{2}\pi < \arg \nu < \frac{1}{2}\pi$, respectively.

9.3.27

$$J_{\nu}'(\nu + z\nu^{1/3}) \sim -\frac{2^{2/3}}{\nu^{2/3}} \text{Ai}'(-2^{1/3}z) \{ 1 + \sum_{k=1}^{\infty} \frac{h_k(z)}{\nu^{2k/3}} \}$$
$$+ \frac{2^{1/3}}{\nu^{4/3}} \text{Ai}(-2^{1/3}z) \sum_{k=0}^{\infty} \frac{l_k(z)}{\nu^{2k/3}}$$

9.3.28

$$Y_{\nu}'(\nu + z\nu^{1/3}) \sim \frac{2^{2/3}}{\nu^{2/3}} \text{Bi}'(-2^{1/3}z) \{ 1 + \sum_{k=1}^{\infty} \frac{h_k(z)}{\nu^{2k/3}} \}$$
$$- \frac{2^{1/3}}{\nu^{4/3}} \text{Bi}(-2^{1/3}z) \sum_{k=0}^{\infty} \frac{l_k(z)}{\nu^{2k/3}}$$

where

9.3.29

$$h_1(z) = -\frac{4}{5} z$$

$$h_2(z) = -\frac{9}{100} z^5 + \frac{57}{70} z^2$$

$$h_3(z) = \frac{699}{3500} z^6 - \frac{2617}{3150} z^3 + \frac{23}{3150}$$

$$h_4(z) = \frac{27}{20000} z^{10} - \frac{46631}{147000} z^7 + \frac{3889}{4620} z^4 - \frac{1159}{115500} z$$

9.3.30

$$l_0(z) = \frac{3}{5} z^3 - \frac{1}{5}$$

$$l_1(z) = -\frac{131}{140} z^4 + \frac{1}{5} z$$

$$l_2(z) = -\frac{9}{500} z^8 + \frac{5437}{4500} z^5 - \frac{593}{3150} z^2$$

$$l_3(z) = \frac{369}{7000} z^9 - \frac{999443}{693000} z^6 + \frac{31727}{173250} z^3 + \frac{947}{346500}$$

9.3.31 $J_\nu(\nu) \sim \dfrac{a}{\nu^{1/3}}\{1+\sum\limits_{k=1}^{\infty}\dfrac{\alpha_k}{\nu^{2k}}\} - \dfrac{b}{\nu^{5/3}}\sum\limits_{k=0}^{\infty}\dfrac{\beta_k}{\nu^{2k}}$

9.3.32 $Y_\nu(\nu) \sim -\dfrac{3^{1/2}a}{\nu^{1/3}}\{1+\sum\limits_{k=1}^{\infty}\dfrac{\alpha_k}{\nu^{2k}}\} - \dfrac{3^{1/2}b}{\nu^{5/3}}\sum\limits_{k=0}^{\infty}\dfrac{\beta_k}{\nu^{2k}}$

9.3.33 $J_\nu'(\nu) \sim \dfrac{b}{\nu^{2/3}}\{1+\sum\limits_{k=1}^{\infty}\dfrac{\gamma_k}{\nu^{2k}}\} - \dfrac{a}{\nu^{4/3}}\sum\limits_{k=0}^{\infty}\dfrac{\delta_k}{\nu^{2k}}$

9.3.34 $Y_\nu'(\nu) \sim \dfrac{3^{1/2}b}{\nu^{2/3}}\{1+\sum\limits_{k=1}^{\infty}\dfrac{\gamma_k}{\nu^{2k}}\} + \dfrac{3^{1/2}a}{\nu^{4/3}}\sum\limits_{k=0}^{\infty}\dfrac{\delta_k}{\nu^{2k}}$

where

$$a=\frac{2^{1/3}}{3^{2/3}\Gamma(\frac{2}{3})}=.44730\ 73184,\qquad 3^{\frac14}a=.77475\ 90021$$

$$b=\frac{2^{2/3}}{3^{1/3}\Gamma(\frac{1}{3})}=.41085\ 01939,\qquad 3^{\frac14}b=.71161\ 34101$$

$$\alpha_0=1,\qquad \alpha_1=-\frac{1}{225}=-.00\dot{4},$$

$$\alpha_2=.00069\ 3735\ldots,\qquad \alpha_3=-.00035\ 38\ldots$$

$$\beta_0=\frac{1}{70}=.01428\ 57143\ldots,$$

$$\beta_1=-\frac{1213}{10\ 23750}=-.00118\ 48596\ldots,$$

$$\beta_2=.00043\ 78\ldots,\qquad \beta_3=-.00038\ldots$$

$$\gamma_0=1,\qquad \gamma_1=\frac{23}{3150}=.00730\ 15873\ldots,$$

$$\gamma_2=-.00093\ 7300\ldots,\qquad \gamma_3=.00044\ 40\ldots$$

$$\delta_0=\frac{1}{5},\quad \delta_1=-\frac{947}{3\ 46500}=-.00273\ 30447\ldots,$$

$$\delta_2=.00060\ 47\ldots,\qquad \delta_3=-.00038\ldots$$

Uniform Asymptotic Expansions

These are more powerful than the previous expansions of this section, save for **9.3.31** and **9.3.32**, but their coefficients are more complicated. They reduce to **9.3.31** and **9.3.32** when the argument equals the order.

9.3.35

$$J_\nu(\nu z) \sim \left(\frac{4\zeta}{1-z^2}\right)^{1/4}\{\frac{\mathrm{Ai}\,(\nu^{2/3}\zeta)}{\nu^{1/3}}\sum\limits_{k=0}^{\infty}\frac{a_k(\zeta)}{\nu^{2k}}$$
$$+\frac{\mathrm{Ai}\,'(\nu^{2/3}\zeta)}{\nu^{5/3}}\sum\limits_{k=0}^{\infty}\frac{b_k(\zeta)}{\nu^{2k}}\}$$

9.3.36

$$Y_\nu(\nu z) \sim -\left(\frac{4\zeta}{1-z^2}\right)^{1/4}\{\frac{\mathrm{Bi}\,(\nu^{2/3}\zeta)}{\nu^{1/3}}\sum\limits_{k=0}^{\infty}\frac{a_k(\zeta)}{\nu^{2k}}$$
$$+\frac{\mathrm{Bi}\,'(\nu^{2/3}\zeta)}{\nu^{5/3}}\sum\limits_{k=0}^{\infty}\frac{b_k(\zeta)}{\nu^{2k}}\}$$

9.3.37

$$H_\nu^{(1)}(\nu z) \sim 2e^{-\pi i/3}\left(\frac{4\zeta}{1-z^2}\right)^{1/4}\{\frac{\mathrm{Ai}\,(e^{2\pi i/3}\nu^{2/3}\zeta)}{\nu^{1/3}}\sum\limits_{k=0}^{\infty}\frac{a_k(\zeta)}{\nu^{2k}}$$
$$+\frac{e^{2\pi i/3}\mathrm{Ai}\,'(e^{2\pi i/3}\nu^{2/3}\zeta)}{\nu^{5/3}}\sum\limits_{k=0}^{\infty}\frac{b_k(\zeta)}{\nu^{2k}}\}$$

When $\nu\to+\infty$, these expansions hold uniformly with respect to z in the sector $|\arg z|\le\pi-\epsilon$, where ϵ is an arbitrary positive number. The corresponding expansion for $H_\nu^{(2)}(\nu z)$ is obtained by changing the sign of i in **9.3.37**.

Here

9.3.38

$$\frac{2}{3}\zeta^{3/2}=\int_z^1\frac{\sqrt{1-t^2}}{t}\,dt=\ln\frac{1+\sqrt{1-z^2}}{z}-\sqrt{1-z^2}$$

equivalently,

9.3.39

$$\frac{2}{3}(-\zeta)^{3/2}=\int_1^z\frac{\sqrt{t^2-1}}{t}\,dt=\sqrt{z^2-1}-\arccos\left(\frac{1}{z}\right)$$

the branches being chosen so that ζ is real when z is positive. The coefficients are given by

9.3.40

$$a_k(\zeta)=\sum\limits_{s=0}^{2k}\mu_s\zeta^{-3s/2}u_{2k-s}\{(1-z^2)^{-\frac12}\}$$

$$b_k(\zeta)=-\zeta^{-\frac12}\sum\limits_{s=0}^{2k+1}\lambda_s\zeta^{-3s/2}u_{2k-s+1}\{(1-z^2)^{-\frac12}\}$$

where u_k is given by **9.3.9** and **9.3.10**, $\lambda_0=\mu_0=1$ and

9.3.41

$$\lambda_s=\frac{(2s+1)(2s+3)\ldots(6s-1)}{s!\,(144)^s},\qquad \mu_s=-\frac{6s+1}{6s-1}\lambda_s$$

Thus $a_0(\zeta)=1$,

9.3.42

$$b_0(\zeta)=-\frac{5}{48\zeta^2}+\frac{1}{\zeta^{\frac12}}\{\frac{5}{24(1-z^2)^{3/2}}-\frac{1}{8(1-z^2)^{\frac12}}\}$$
$$=-\frac{5}{48\zeta^2}+\frac{1}{(-\zeta)^{\frac12}}\{\frac{5}{24(z^2-1)^{3/2}}+\frac{1}{8(z^2-1)^{\frac12}}\}$$

Tables of the early coefficients are given below. For more extensive tables of the coefficients and for bounds on the remainder terms in **9.3.35** and **9.3.36** see [9.38].

Uniform Expansions of the Derivatives

With the conditions of the preceding subsection

9.3.43

$$J'_\nu(\nu z) \sim -\frac{2}{z}\left(\frac{1-z^2}{4\zeta}\right)^{\frac{1}{4}}\left\{\frac{\text{Ai}\,(\nu^{2/3}\zeta)}{\nu^{4/3}}\sum_{k=0}^{\infty}\frac{c_k(\zeta)}{\nu^{2k}}\right.$$
$$\left.+\frac{\text{Ai}'\,(\nu^{2/3}\zeta)}{\nu^{2/3}}\sum_{k=0}^{\infty}\frac{d_k(\zeta)}{\nu^{2k}}\right\}$$

9.3.44

$$Y'_\nu(\nu z) \sim \frac{2}{z}\left(\frac{1-z^2}{4\zeta}\right)^{\frac{1}{4}}\left\{\frac{\text{Bi}\,(\nu^{2/3}\zeta)}{\nu^{4/3}}\sum_{k=0}^{\infty}\frac{c_k(\zeta)}{\nu^{2k}}\right.$$
$$\left.+\frac{\text{Bi}'\,(\nu^{2/3}\zeta)}{\nu^{2/3}}\sum_{k=0}^{\infty}\frac{d_k(\zeta)}{\nu^{2k}}\right\}$$

9.3.45

$$H_\nu^{(1)\prime}(\nu z) \sim \frac{4e^{2\pi i/3}}{z}\left(\frac{1-z^2}{4\zeta}\right)^{\frac{1}{4}}\left\{\frac{\text{Ai}\,(e^{2\pi i/3}\nu^{2/3}\zeta)}{\nu^{4/3}}\sum_{k=0}^{\infty}\frac{c_k(\zeta)}{\nu^{2k}}\right.$$
$$\left.+\frac{e^{2\pi i/3}\,\text{Ai}'\,(e^{2\pi i/3}\nu^{2/3}\zeta)}{\nu^{2/3}}\sum_{k=0}^{\infty}\frac{d_k(\zeta)}{\nu^{2k}}\right\}$$

where

9.3.46

$$c_k(\zeta) = -\zeta^{\frac{1}{2}}\sum_{s=0}^{2k+1}\mu_s\zeta^{-3s/2}v_{2k-s+1}\{(1-z^2)^{-\frac{1}{2}}\}$$

$$d_k(\zeta) = \sum_{s=0}^{2k}\lambda_s\zeta^{-3s/2}v_{2k-s}\{(1-z^2)^{-\frac{1}{2}}\}$$

and v_k is given by **9.3.13** and **9.3.14**. For bounds on the remainder terms in **9.3.43** and **9.3.44** see [9.38].

ζ	$b_0(\zeta)$	$a_1(\zeta)$	$c_0(\zeta)$	$d_1(\zeta)$
0	0. 0180	−0. 004	0. 1587	0. 007
1	. 0278	−. 004	. 1785	. 009
2	. 0351	−. 001	. 1862	. 007
3	. 0366	+. 002	. 1927	. 005
4	. 0352	. 003	. 2031	. 004
5	. 0331	. 004	. 2155	. 003
6	. 0311	. 004	. 2284	. 003
7	. 0294	. 004	. 2413	. 003
8	. 0278	. 004	. 2539	. 003
9	. 0265	. 004	. 2662	. 003
10	. 0253	. 004	. 2781	. 003

$-\zeta$	$b_0(\zeta)$	$a_1(\zeta)$	$c_0(\zeta)$	$d_1(\zeta)$
0	0. 0180	−0. 004	0. 1587	0. 007
1	. 0109	−. 003	. 1323	. 004
2	. 0067	−. 002	. 1087	. 002
3	. 0044	−. 001	. 0903	. 001
4	. 0031	−. 001	. 0764	. 001
5	. 0022	−. 000	. 0658	. 000
6	. 0017	−. 000	. 0576	. 000
7	. 0013	−. 000	. 0511	. 000
8	. 0011	−. 000	. 0459	. 000
9	. 0009	−. 000	. 0415	. 000
10	. 0007	−. 000	. 0379	. 000

For $\zeta > 10$ use

$$b_0(\zeta) \sim \frac{1}{12}\zeta^{-\frac{1}{2}}-.104\zeta^{-2}, \qquad a_1(\zeta)=.003,$$

$$c_0(\zeta) \sim \frac{1}{12}\zeta^{\frac{1}{2}}+.146\zeta^{-1}, \qquad d_1(\zeta)=.003.$$

For $\zeta < -10$ use

$$b_0(\zeta) \sim \frac{1}{12}\zeta^{-2}, \qquad a_1(\zeta)=.000,$$

$$c_0(\zeta) \sim -\frac{5}{12}\zeta^{-1}-1.33(-\zeta)^{-5/2}, \qquad d_1(\zeta)=.000.$$

Maximum values of higher coefficients:

$$|b_1(\zeta)|=.003, \qquad |a_2(\zeta)|=.0008, \qquad |d_2(\zeta)|=.001$$

$$|c_1(\zeta)|=.008\ (\zeta<10), \quad c_1(\zeta)\sim-.003\zeta^{\frac{1}{2}}\ \text{as}\ \zeta\to+\infty.$$

9.4. Polynomial Approximations [2]

9.4.1 $\qquad -3 \leq x \leq 3$

$$J_0(x) = 1-2.24999\,97(x/3)^2+1.26562\,08(x/3)^4$$
$$-.31638\,66(x/3)^6+.04444\,79(x/3)^8$$
$$-.00394\,44(x/3)^{10}+.00021\,00(x/3)^{12}+\epsilon$$

$$|\epsilon|<5\times10^{-8}$$

9.4.2 $\qquad 0 < x \leq 3$

$$Y_0(x) = (2/\pi)\ln(\tfrac{1}{2}x)J_0(x)+.36746\,691$$
$$+.60559\,366(x/3)^2-.74350\,384(x/3)^4$$
$$+.25300\,117(x/3)^6-.04261\,214(x/3)^8$$
$$+.00427\,916(x/3)^{10}-.00024\,846(x/3)^{12}+\epsilon$$

$$|\epsilon|<1.4\times10^{-8}$$

9.4.3 $\qquad 3 \leq x < \infty$

$$J_0(x)=x^{-\frac{1}{2}}f_0\cos\theta_0 \qquad Y_0(x)=x^{-\frac{1}{2}}f_0\sin\theta_0$$

$$f_0=.79788\,456-.00000\,077(3/x)-.00552\,740(3/x)^2$$
$$-.00009\,512(3/x)^3+.00137\,237(3/x)^4$$
$$-.00072\,805(3/x)^5+.00014\,476(3/x)^6+\epsilon$$

$$|\epsilon|<1.6\times10^{-8}$$

---

[2] Equations **9.4.1** to **9.4.6** and **9.8.1** to **9.8.8** are taken from E. E. Allen, Analytical approximations, Math. Tables Aids Comp. **8**, 240–241 (1954), and Polynomial approximations to some modified Bessel functions, Math. Tables Aids Comp. **10**, 162–164 (1956) (with permission). They were checked at the National Physical Laboratory by systematic tabulation; new bounds for the errors, ϵ, given here were obtained as a result.

$$\theta_0 = x - .78539\ 816 - .04166\ 397(3/x)$$
$$- .00003\ 954(3/x)^2 + .00262\ 573(3/x)^3$$
$$- .00054\ 125(3/x)^4 - .00029\ 333(3/x)^5$$
$$+ .00013\ 558(3/x)^6 + \epsilon$$

$$|\epsilon| < 7 \times 10^{-8}$$

9.4.4 $\qquad -3 \leq x \leq 3$

$$x^{-1} J_1(x) = \tfrac{1}{2} - .56249\ 985(x/3)^2 + .21093\ 573(x/3)^4$$
$$- .03954\ 289(x/3)^6 + .00443\ 319(x/3)^8$$
$$- .00031\ 761(x/3)^{10} + .00001\ 109(x/3)^{12} + \epsilon$$

$$|\epsilon| < 1.3 \times 10^{-8}$$

9.4.5 $\qquad 0 < x \leq 3$

$$x Y_1(x) = (2/\pi) x \ln(\tfrac{1}{2}x) J_1(x) - .63661\ 98$$
$$+ .22120\ 91(x/3)^2 + 2.16827\ 09(x/3)^4$$
$$- 1.31648\ 27(x/3)^6 + .31239\ 51(x/3)^8$$
$$- .04009\ 76(x/3)^{10} + .00278\ 73(x/3)^{12} + \epsilon$$

$$|\epsilon| < 1.1 \times 10^{-7}$$

9.4.6 $\qquad 3 \leq x < \infty$

$$J_1(x) = x^{-\frac{1}{2}} f_1 \cos \theta_1, \qquad Y_1(x) = x^{-\frac{1}{2}} f_1 \sin \theta_1$$

$$f_1 = .79788\ 456 + .00000\ 156(3/x) + .01659\ 667(3/x)^2$$
$$+ .00017\ 105(3/x)^3 - .00249\ 511(3/x)^4$$
$$+ .00113\ 653(3/x)^5 - .00020\ 033(3/x)^6 + \epsilon$$

$$|\epsilon| < 4 \times 10^{-8}$$

$$\theta_1 = x - 2.35619\ 449 + .12499\ 612(3/x)$$
$$+ .00005\ 650(3/x)^2 - .00637\ 879(3/x)^3$$
$$+ .00074\ 348(3/x)^4 + .00079\ 824(3/x)^5$$
$$- .00029\ 166(3/x)^6 + \epsilon$$

$$|\epsilon| < 9 \times 10^{-8}$$

For expansions of $J_0(x)$, $Y_0(x)$, $J_1(x)$, and $Y_1(x)$ in series of Chebyshev polynomials for the ranges $0 \leq x \leq 8$ and $0 \leq 8/x \leq 1$, see [9.37].

9.5. Zeros

Real Zeros

When ν is real, the functions $J_\nu(z)$, $J'_\nu(z)$, $Y_\nu(z)$ and $Y'_\nu(z)$ each have an infinite number of real zeros, all of which are simple with the possible exception of $z=0$. For *non-negative* ν the sth positive zeros of these functions are denoted by $j_{\nu,s}$, $j'_{\nu,s}$, $y_{\nu,s}$ and $y'_{\nu,s}$ respectively, except that $z=0$ is counted as the first zero of $J'_0(z)$. Since $J'_0(z) = -J_1(z)$, it follows that

9.5.1 $\quad j'_{0,1} = 0, \qquad j'_{0,s} = j_{1,s-1} \qquad (s=2,\ 3,\ \ldots)$

The zeros interlace according to the inequalities

9.5.2
$$j_{\nu,1} < j_{\nu+1,1} < j_{\nu,2} < j_{\nu+1,2} < j_{\nu,3} < \ldots$$
$$y_{\nu,1} < y_{\nu+1,1} < y_{\nu,2} < y_{\nu+1,2} < y_{\nu,3} < \ldots$$
$$\nu \leq j'_{\nu,1} < y_{\nu,1} < y'_{\nu,1} < j_{\nu,1} < j'_{\nu,2}$$
$$< y_{\nu,2} < y'_{\nu,2} < j_{\nu,2} < j'_{\nu,3} < \ldots$$

The positive zeros of any two real distinct cylinder functions of the same order are interlaced, as are the positive zeros of any real cylinder function $\mathscr{C}_\nu(z)$, defined as in **9.1.27,** and the contiguous function $\mathscr{C}_{\nu+1}(z)$.

If ρ_ν is a zero of the cylinder function

9.5.3 $\quad \mathscr{C}_\nu(z) = J_\nu(z) \cos(\pi t) + Y_\nu(z) \sin(\pi t)$

where t is a parameter, then

9.5.4 $\qquad \mathscr{C}'_\nu(\rho_\nu) = \mathscr{C}_{\nu-1}(\rho_\nu) = -\mathscr{C}_{\nu+1}(\rho_\nu)$

If σ_ν is a zero of $\mathscr{C}'_\nu(z)$ then

9.5.5 $\quad \mathscr{C}_\nu(\sigma_\nu) = \dfrac{\sigma_\nu}{\nu} \mathscr{C}_{\nu-1}(\sigma_\nu) = \dfrac{\sigma_\nu}{\nu} \mathscr{C}_{\nu+1}(\sigma_\nu)$

The parameter t may be regarded as a continuous variable and ρ_ν, σ_ν as functions $\rho_\nu(t)$, $\sigma_\nu(t)$ of t. If these functions are fixed by

9.5.6 $\qquad \rho_\nu(0) = 0, \qquad \sigma_\nu(0) = j'_{\nu,1}$

then

9.5.7
$$j_{\nu,s} = \rho_\nu(s), \qquad y_{\nu,s} = \rho_\nu(s - \tfrac{1}{2}) \qquad (s=1,2,\ldots)$$

9.5.8
$$j'_{\nu,s} = \sigma_\nu(s-1), \qquad y'_{\nu,s} = \sigma_\nu(s - \tfrac{1}{2}) \qquad (s=1,2,\ldots)$$

9.5.9 $\quad \mathscr{C}'_\nu(\rho_\nu) = \left(\dfrac{\rho_\nu}{2} \dfrac{d\rho_\nu}{dt}\right)^{-\frac{1}{2}}, \ \mathscr{C}_\nu(\sigma_\nu) = \left(\dfrac{\sigma_\nu^2 - \nu^2}{2\sigma_\nu} \dfrac{d\sigma_\nu}{dt}\right)^{-\frac{1}{2}}$

Infinite Products

9.5.10 $\qquad J_\nu(z) = \dfrac{(\tfrac{1}{2}z)^\nu}{\Gamma(\nu+1)} \prod_{s=1}^{\infty} \left(1 - \dfrac{z^2}{j_{\nu,s}^2}\right)$

9.5.11 $\quad J'_\nu(z) = \dfrac{(\tfrac{1}{2}z)^{\nu-1}}{2\Gamma(\nu)} \prod_{s=1}^{\infty} \left(1 - \dfrac{z^2}{j_{\nu,s}'^2}\right) \qquad (\nu > 0)$

McMahon's Expansions for Large Zeros

When ν is fixed, $s \gg \nu$ and $\mu = 4\nu^2$

9.5.12

$$j_{\nu,s}, y_{\nu,s} \sim \beta - \frac{\mu-1}{8\beta} - \frac{4(\mu-1)(7\mu-31)}{3(8\beta)^3} - \frac{32(\mu-1)(83\mu^2-982\mu+3779)}{15(8\beta)^5}$$

$$- \frac{64(\mu-1)(6949\mu^3 - 1\,53855\mu^2 + 15\,85743\mu - 62\,77237)}{105(8\beta)^7} - \cdots$$

where $\beta = (s + \frac{1}{2}\nu - \frac{1}{4})\pi$ for $j_{\nu,s}$, $\beta = (s + \frac{1}{2}\nu - \frac{3}{4})\pi$ for $y_{\nu,s}$. With $\beta = (t + \frac{1}{2}\nu - \frac{1}{4})\pi$, the right of **9.5.12** is the asymptotic expansion of $\rho_\nu(t)$ for large t.

9.5.13

$$j'_{\nu,s}, y'_{\nu,s} \sim \beta' - \frac{\mu+3}{8\beta'} - \frac{4(7\mu^2+82\mu-9)}{3(8\beta')^3} - \frac{32(83\mu^3+2075\mu^2-3039\mu+3537)}{15(8\beta')^5}$$

$$- \frac{64(6949\mu^4 + 2\,96492\mu^3 - 12\,48002\mu^2 + 74\,14380\mu - 58\,53627)}{105(8\beta')^7} - \cdots$$

where $\beta' = (s + \frac{1}{2}\nu - \frac{3}{4})\pi$ for $j'_{\nu,s}$, $\beta' = (s + \frac{1}{2}\nu - \frac{1}{4})\pi$ for $y'_{\nu,s}$, $\beta' = (t + \frac{1}{2}\nu + \frac{1}{4})\pi$ for $\sigma_\nu(t)$. For higher terms in **9.5.12** and **9.5.13** see [9.4] or [9.40].

Asymptotic Expansions of Zeros and Associated Values for Large Orders

9.5.14

$$j_{\nu,1} \sim \nu + 1.85575\ 71\nu^{1/3} + 1.03315\ 0\nu^{-1/3}$$
$$- .00397\nu^{-1} - .0908\nu^{-5/3} + .043\nu^{-7/3} + \cdots$$

9.5.15

$$y_{\nu,1} \sim \nu + .93157\ 68\nu^{1/3} + .26035\ 1\nu^{-1/3}$$
$$+ .01198\nu^{-1} - .0060\nu^{-5/3} - .001\nu^{-7/3} + \cdots$$

9.5.16

$$j'_{\nu,1} \sim \nu + .80861\ 65\nu^{1/3} + .07249\ 0\nu^{-1/3}$$
$$- .05097\nu^{-1} + .0094\nu^{-5/3} + \cdots$$

9.5.17

$$y'_{\nu,1} \sim \nu + 1.82109\ 80\nu^{1/3} + .94000\ 7\nu^{-1/3}$$
$$- .05808\nu^{-1} - .0540\nu^{-5/3} + \cdots$$

9.5.18

$$J'_\nu(j_{\nu,1}) \sim -1.11310\ 28\nu^{-2/3}/(1 + 1.48460\ 6\nu^{-2/3}$$
$$+ .43294\nu^{-4/3} - .1943\nu^{-2} + .019\nu^{-8/3} + \cdots)$$

9.5.19

$$Y'_\nu(y_{\nu,1}) \sim .95554\ 86\nu^{-2/3}/(1 + .74526\ 1\nu^{-2/3}$$
$$+ .10910\nu^{-4/3} - .0185\nu^{-2} - .003\nu^{-8/3} + \cdots)$$

9.5.20

$$J_\nu(j'_{\nu,1}) \sim .67488\ 51\nu^{-1/3}(1 - .16172\ 3\nu^{-2/3}$$
$$+ .02918\nu^{-4/3} - .0068\nu^{-2} + \cdots)$$

9.5.21

$$Y_\nu(y'_{\nu,1}) \sim .57319\ 40\nu^{-1/3}(1 - .36422\ 0\nu^{-2/3}$$
$$+ .09077\nu^{-4/3} + .0237\nu^{-2} + \cdots)$$

Corresponding expansions for $s = 2, 3$ are given in [9.40]. These expansions become progressively weaker as s increases; those which follow do not suffer from this defect.

Uniform Asymptotic Expansions of Zeros and Associated Values for Large Orders

9.5.22 $\quad j_{\nu,s} \sim \nu z(\zeta) + \sum_{k=1}^{\infty} \frac{f_k(\zeta)}{\nu^{2k-1}}$ with $\zeta = \nu^{-2/3}a_s$

9.5.23

$$J'_\nu(j_{\nu,s}) \sim -\frac{2}{\nu^{2/3}} \frac{\mathrm{Ai}'(a_s)}{z(\zeta)h(\zeta)} \left\{ 1 + \sum_{k=1}^{\infty} \frac{F_k(\zeta)}{\nu^{2k}} \right\}$$

$$\text{with } \zeta = \nu^{-2/3}a_s$$

9.5.24 $\quad j'_{\nu,s} \sim \nu z(\zeta) + \sum_{k=1}^{\infty} \frac{g_k(\zeta)}{\nu^{2k-1}}$ with $\zeta = \nu^{-2/3}a'_s$

9.5.25

$$J_\nu(j'_{\nu,s}) \sim \mathrm{Ai}(a'_s) \frac{h(\zeta)}{\nu^{1/3}} \left\{ 1 + \sum_{k=1}^{\infty} \frac{G_k(\zeta)}{\nu^{2k}} \right\} \text{ with } \zeta = \nu^{-2/3}a'_s$$

where a_s, a'_s are the sth negative zeros of $\mathrm{Ai}(z)$, $\mathrm{Ai}'(z)$ (see **10.4**), $z = z(\zeta)$ is the inverse function defined implicitly by **9.3.39**, and

9.5.26

$$h(\zeta) = \{4\zeta/(1-z^2)\}^{\frac{1}{4}}$$
$$f_1(\zeta) = \tfrac{1}{2}z(\zeta)\{h(\zeta)\}^2 b_0(\zeta)$$
$$g_1(\zeta) = \tfrac{1}{2}\zeta^{-1}z(\zeta)\{h(\zeta)\}^2 c_0(\zeta)$$

where $b_0(\zeta)$, $c_0(\zeta)$ appear in **9.3.42** and **9.3.46**. Tables of the leading coefficients follow. More extensive tables are given in [9.40].

The expansions of $y_{\nu,s}$, $Y'_\nu(y_{\nu,s})$, $y'_{\nu,s}$ and $Y_\nu(y'_{\nu,s})$ corresponding to **9.5.22** to **9.5.25** are obtained by changing the symbols j, J, Ai, Ai', a_s and a'_s to y, Y, $-\mathrm{Bi}$, $-\mathrm{Bi}'$, b_s and b'_s respectively.

$-\zeta$	$z(\zeta)$	$h(\zeta)$	$f_1(\zeta)$	$F_1(\zeta)$	$(-\zeta)g_1(\zeta)$	$(-\zeta)^3 g_2(\zeta)$	$(-\zeta)^2 G_1(\zeta)$
0. 0	1. 000000	1. 25992	0. 0143	−0. 007	−0. 1260	−0. 010	0. 000
0. 2	1. 166284	1. 22076	. 0142	−. 005	−. 1335	−. 010	. 002
0. 4	1. 347557	1. 18337	. 0139	−. 004	−. 1399	−. 009	. 004
0. 6	1. 543615	1. 14780	. 0135	−. 003	−. 1453	−. 009	. 005
0. 8	1. 754187	1. 11409	. 0131	−. 003	−. 1498	−. 008	. 006
1. 0	1. 978963	1. 08220	0. 0126	−0. 002	−0. 1533	−0. 008	0. 006

$-\zeta$	$z(\zeta)$	$h(\zeta)$	$f_1(\zeta)$	$F_1(\zeta)$	$g_1(\zeta)$	$g_2(\zeta)$	$G_1(\zeta)$
1. 0	1. 978963	1. 08220	0. 0126	−0. 002	−0. 1533	−0. 008	0. 006
1. 2	2. 217607	1. 05208	. 0121	−. 002	−. 1301	−. 004	. 004
1. 4	2. 469770	1. 02367	. 0115	−. 001	−. 1130	−. 002	. 003
1. 6	2. 735103	0. 99687	. 0110	−. 001	−. 0998	−. 001	. 002
1. 8	3. 013256	. 97159	. 0105	−. 001	−. 0893	−. 001	. 002
2. 0	3. 303889	0. 94775	0. 0100	−0. 001	−0. 0807	−0. 001	0. 001
2. 2	3. 606673	. 92524	. 0095	−0. 001	−. 0734		. 001
2. 4	3. 921292	. 90397	. 0091		−. 0673		. 001
2. 6	4. 247441	. 88387	. 0086		−. 0619		. 001
2. 8	4. 584833	. 86484	. 0082		−. 0573		0. 001
3. 0	4. 933192	0. 84681	0. 0078		−0. 0533		
3. 2	5. 292257	. 82972	. 0075		−. 0497		
3. 4	5. 661780	. 81348	. 0071		−. 0464		
3. 6	6. 041525	. 79806	. 0068		−. 0436		
3. 8	6. 431269	. 78338	. 0065		−. 0410		
4. 0	6. 830800	0. 76939	0. 0062		−0. 0386		
4. 2	7. 239917	. 75605	. 0060		−. 0365		
4. 4	7. 658427	. 74332	. 0057		−. 0345		
4. 6	8. 086150	. 73115	. 0055		−. 0328		
4. 8	8. 522912	. 71951	. 0052		−. 0311		
5. 0	8. 968548	0. 70836	0. 0050		−0. 0296		
5. 2	9. 422900	. 69768	. 0048		−. 0282		
5. 4	9. 885820	. 68742	. 0047		−. 0270		
5. 6	10. 357162	. 67758	. 0045		−. 0258		
5. 8	10. 836791	. 66811	. 0043		−. 0246		
6. 0	11. 324575	0. 65901	0. 0042		−0. 0236		
6. 2	11. 820388	. 65024	. 0040		−. 0227		
6. 4	12. 324111	. 64180	. 0039		−. 0218		
6. 6	12. 835627	. 63366	. 0037		−. 0209		
6. 8	13. 354826	. 62580	. 0036		−. 0201		
7. 0	13. 881601	0. 61821	0. 0035		−0. 0194		

$(-\zeta)^{-\frac{1}{2}}$	$z(\zeta) - \frac{2}{3}(-\zeta)^{\frac{3}{2}}$	$(-\zeta)^{\frac{1}{4}}h(\zeta)$	$f_1(\zeta)$	$g_1(\zeta)$
0. 40	1. 528915	1. 62026	0. 0040	−0. 0224
. 35	1. 541532	1. 65351	. 0029	−. 0158
. 30	1. 551741	1. 68067	. 0020	−. 0104
. 25	1. 559490	1. 70146	. 0012	−. 0062
. 20	1. 564907	1. 71607	. 0006	−. 0033
0. 15	1. 568285	1. 72523	0. 0003	−0. 0014
. 10	1. 570048	1. 73002	. 0001	−. 0004
. 05	1. 570703	1. 73180	. 0000	−. 0001
. 00	1. 570796	1. 73205	. 0000	−. 0000

Maximum Values of Higher Coefficients

$$|f_2(\zeta)| = .001, \quad |F_2(\zeta)| = .0004 \qquad (0 \leq -\zeta < \infty)$$

$$|g_3(\zeta)| = .001, \quad |G_2(\zeta)| = .0007 \qquad (1 \leq -\zeta < \infty)$$

$$|(-\zeta)^5 g_3(\zeta)| = .002, \quad |(-\zeta)^4 G_2(\zeta)| = .0007$$

$$(0 \leq -\zeta \leq 1)$$

Complex Zeros of $J_\nu(z)$

When $\nu \geq -1$ the zeros of $J_\nu(z)$ are all real. If $\nu < -1$ and ν is not an integer the number of complex zeros of $J_\nu(z)$ is twice the integer part of $(-\nu)$; if the integer part of $(-\nu)$ is odd two of these zeros lie on the imaginary axis.

If $\nu \geq 0$, all zeros of $J'_\nu(z)$ are real.

Complex Zeros of $Y_\nu(z)$

When ν is real the pattern of the complex zeros of $Y_\nu(z)$ and $Y'_\nu(z)$ depends on the non-integer part of ν. Attention is confined here to the case $\nu = n$, a positive integer or zero.

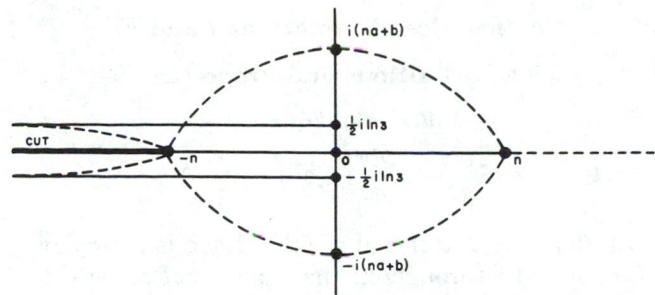

FIGURE 9.5.   *Zeros of* $Y_n(z)$ *and* $Y_n'(z)$ . . .

$$|\arg z| \leq \pi.$$

Figure 9.5 shows the approximate distribution of the complex zeros of $Y_n(z)$ in the region $|\arg z| \leq \pi$. The figure is symmetrical about the real axis. The two curves on the left extend to infinity, having the asymptotes

$$\mathscr{I}z = \pm \tfrac{1}{2}\ln 3 = \pm.54931 \ldots$$

There are an infinite number of zeros near each of these curves.

The two curves extending from $z=-n$ to $z=n$ and bounding an eye-shaped domain intersect the imaginary axis at the points $\pm i(na+b)$, where

$$a = \sqrt{t_0^2 - 1} = .66274 \ldots$$

$$b = \tfrac{1}{2}\sqrt{1 - t_0^{-2}} \ln 2 = .19146 \ldots$$

and $t_0 = 1.19968 \ldots$ is the positive root of $\coth t = t$. There are n zeros near each of these curves. Asymptotic expansions of these zeros for large n

are given by the right of **9.5.22** with $\nu = n$ and $\zeta = n^{-2/3}\beta_s$ or $n^{-2/3}\bar{\beta}_s$, where β_s, $\bar{\beta}_s$ are the complex zeros of $\text{Bi}(z)$ (see **10.4**).

Figure 9.5 is also applicable to the zeros of $Y_n'(z)$. There are again an infinite number near the infinite curves, and n near each of the finite curves. Asymptotic expansions of the latter for large n are given by the right of **9.5.24** with $\nu = n$ and $\zeta = n^{-2/3}\beta_s'$ or $n^{-2/3}\bar{\beta}_s'$; where β_s' and $\bar{\beta}_s'$ are the complex zeros of $\text{Bi}'(z)$.

Numerical values of the three smallest complex zeros of $Y_0(z)$, $Y_1(z)$ and $Y_1'(z)$ in the region $0 < \arg z < \pi$ are given below.

For further details see [9.36] and [9.13]. The latter reference includes tables to facilitate computation.

Complex Zeros of the Hankel Functions

The approximate distribution of the zeros of $H_n^{(1)}(z)$ and its derivative in the region $|\arg z| \leq \pi$ is indicated in a similar manner on **Figure 9.6**.

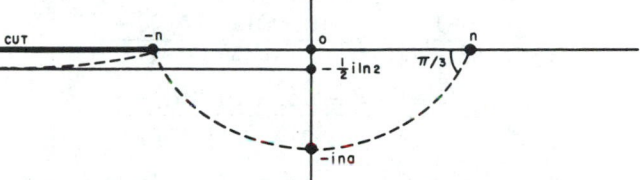

FIGURE 9.6.   *Zeros of* $H_n^{(1)}(z)$ *and* $H_n^{(1)\prime}(z)$ . . .

$$|\arg z| \leq \pi.$$

The asymptote of the solitary infinite curve is given by

$$\mathscr{I}z = -\tfrac{1}{2}\ln 2 = -.34657 \ldots$$

Zeros of $Y_0(z)$ *and Values of* $Y_1(z)$ *at the Zeros*[3]

Zero		Y_1	
Real	*Imag.*	*Real*	*Imag.*
$-2.40301\ 6632$	$+.53988\ 2313$	$+.10074\ 7689$	$-.88196\ 7710$
$-5.51987\ 6702$	$+.54718\ 0011$	$-.02924\ 6418$	$+.58716\ 9503$
$-8.65367\ 2403$	$+.54841\ 2067$	$+.01490\ 8063$	$-.46945\ 8752$

Zeros of $Y_1(z)$ *and Values of* $Y_0(z)$ *at the Zeros*

Zero		Y_0	
Real	*Imag.*	*Real*	*Imag.*
$-0.50274\ 3273$	$+.78624\ 3714$	$-.45952\ 7684$	$+1.31710\ 1937$
$-3.83353\ 5193$	$+.56235\ 6538$	$+.04830\ 1909$	$-0.69251\ 2884$
$-7.01590\ 3683$	$+.55339\ 3046$	$-.02012\ 6949$	$+0.51864\ 2833$

Zeros of $Y_1'(z)$ *and Values of* $Y_1(z)$ *at the Zeros*

Zero		Y_1	
Real	*Imag.*	*Real*	*Imag.*
$+0.57678\ 5129$	$+.90398\ 4792$	$-.76349\ 7088$	$+.58924\ 4865$
$-1.94047\ 7342$	$+.72118\ 5919$	$+.16206\ 4006$	$-.95202\ 7886$
$-5.33347\ 8617$	$+.56721\ 9637$	$-.03179\ 4008$	$+.59685\ 3673$

[3] From National Bureau of Standards, Tables of the Bessel functions $Y_0(z)$ and $Y_1(z)$ for complex arguments, Columbia Univ. Press, New York, N.Y., 1950 (with permission).

There are n zeros of each function near the finite curve extending from $z=-n$ to $z=n$; the asymptotic expansions of these zeros for large n are given by the right side of **9.5.22** or **9.5.24** with $\nu=n$ and $\zeta=e^{-2\pi i/3}n^{-2/3}a_s$ or $\zeta=e^{-2\pi i/3}n^{-2/3}a_s'$.

Zeros of Cross-Products

If ν is real and λ is positive, the zeros of the function

9.5.27 $\qquad J_\nu(z)Y_\nu(\lambda z)-J_\nu(\lambda z)Y_\nu(z)$

are real and simple. If $\lambda>1$, the asymptotic expansion of the sth zero is

9.5.28 $\qquad \beta+\dfrac{p}{\beta}+\dfrac{q-p^2}{\beta^3}+\dfrac{r-4pq+2p^3}{\beta^5}+\cdots$

where with $4\nu^2$ denoted by μ,

9.5.29
$$\beta=s\pi/(\lambda-1)$$
$$p=\frac{\mu-1}{8\lambda}, \qquad q=\frac{(\mu-1)(\mu-25)(\lambda^3-1)}{6(4\lambda)^3(\lambda-1)}$$
$$r=\frac{(\mu-1)(\mu^2-114\mu+1073)(\lambda^5-1)}{5(4\lambda)^5(\lambda-1)}$$

The asymptotic expansion of the large positive zeros (not necessarily the sth) of the function

9.5.30 $\qquad J_\nu'(z)Y_\nu'(\lambda z)-J_\nu'(\lambda z)Y_\nu'(z) \qquad (\lambda>1)$

is given by **9.5.28** with the same value of β, but instead of **9.5.29** we have

9.5.31
$$p=\frac{\mu+3}{8\lambda}, \qquad q=\frac{(\mu^2+46\mu-63)(\lambda^3-1)}{6(4\lambda)^3(\lambda-1)}$$
$$r=\frac{(\mu^3+185\mu^2-2053\mu+1899)(\lambda^5-1)}{5(4\lambda)^5(\lambda-1)}$$

The asymptotic expansion of the large positive zeros of the function

9.5.32 $\qquad J_\nu'(z)Y_\nu(\lambda z)-Y_\nu'(z)J_\nu(\lambda z)$

is given by **9.5.28** with

9.5.33
$$\beta=(s-\tfrac{1}{2})\pi/(\lambda-1)$$
$$p=\frac{(\mu+3)\lambda-(\mu-1)}{8\lambda(\lambda-1)}$$
$$q=\frac{(\mu^2+46\mu-63)\lambda^3-(\mu-1)(\mu-25)}{6(4\lambda)^3(\lambda-1)}$$
$$5(4\lambda)^5(\lambda-1)r=(\mu^3+185\mu^2-2053\mu+1899)\lambda^5$$
$$-(\mu-1)(\mu^2-114\mu+1073)$$

Modified Bessel Functions I and K

9.6. Definitions and Properties

Differential Equation

9.6.1 $\qquad z^2\dfrac{d^2w}{dz^2}+z\dfrac{dw}{dz}-(z^2+\nu^2)w=0$

Solutions are $I_{\pm\nu}(z)$ and $K_\nu(z)$. Each is a regular function of z throughout the z-plane cut along the negative real axis, and for fixed $z(\neq0)$ each is an entire function of ν. When $\nu=\pm n$, $I_\nu(z)$ is an entire function of z.

$I_\nu(z)$ ($\mathscr{R}\nu\geq0$) is bounded as $z\to0$ in any bounded range of arg z. $I_\nu(z)$ and $I_{-\nu}(z)$ are linearly independent except when ν is an integer. $K_\nu(z)$ tends to zero as $|z|\to\infty$ in the sector $|\arg z|<\tfrac{1}{2}\pi$, and for all values of ν, $I_\nu(z)$ and $K_\nu(z)$ are linearly independent. $I_\nu(z)$, $K_\nu(z)$ are real and positive when $\nu>-1$ and $z>0$.

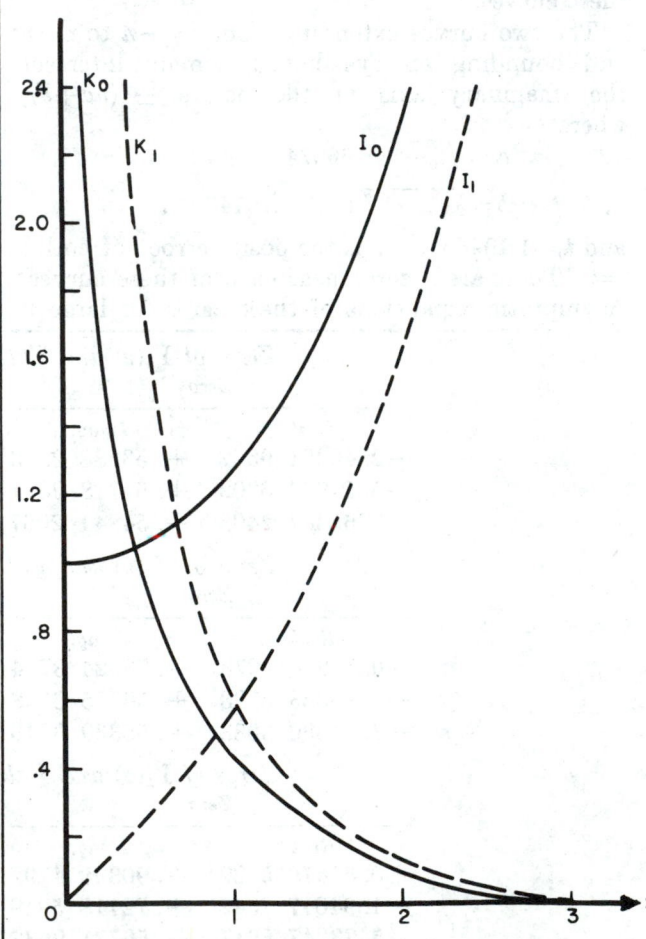

FIGURE 9.7. $I_0(x)$, $K_0(x)$, $I_1(x)$ and $K_1(x)$.

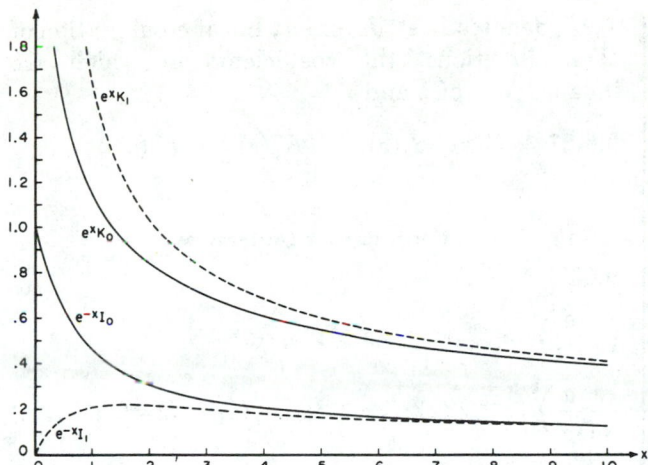

FIGURE 9.8. $e^{-x}I_0(x), e^{-x}I_1(x), e^x K_0(x)$ and $e^x K_1(x)$.

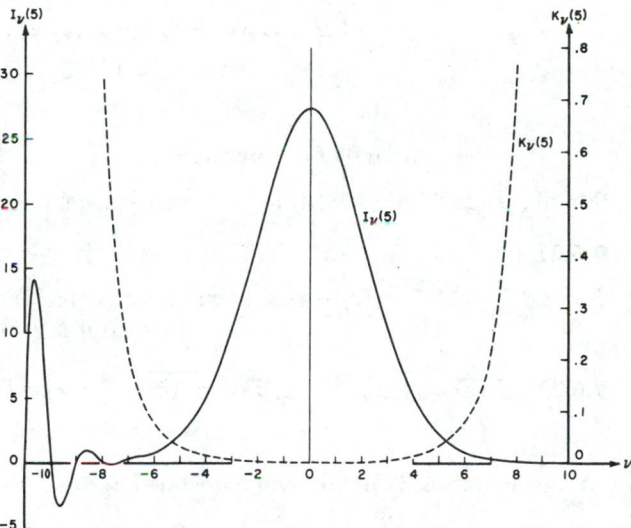

FIGURE 9.9. $I_\nu(5)$ and $K_\nu(5)$.

Relations Between Solutions

9.6.2
$$K_\nu(z) = \tfrac{1}{2}\pi \frac{I_{-\nu}(z) - I_\nu(z)}{\sin(\nu\pi)}$$

The right of this equation is replaced by its limiting value if ν is an integer or zero.

9.6.3
$$I_\nu(z) = e^{-\frac{1}{2}\nu\pi i} J_\nu(ze^{\frac{1}{2}\pi i}) \qquad (-\pi < \arg z \le \tfrac{1}{2}\pi)$$
$$I_\nu(z) = e^{3\nu\pi i/2} J_\nu(ze^{-3\pi i/2}) \qquad (\tfrac{1}{2}\pi < \arg z \le \pi)$$

9.6.4
$$K_\nu(z) = \tfrac{1}{2}\pi i e^{\frac{1}{2}\nu\pi i} H_\nu^{(1)}(ze^{\frac{1}{2}\pi i}) \qquad (-\pi < \arg z \le \tfrac{1}{2}\pi)$$
$$K_\nu(z) = -\tfrac{1}{2}\pi i e^{-\frac{1}{2}\nu\pi i} H_\nu^{(2)}(ze^{-\frac{1}{2}\pi i}) \quad (-\tfrac{1}{2}\pi < \arg z \le \pi)$$

9.6.5
$$Y_\nu(ze^{\frac{1}{2}\pi i}) = e^{\frac{1}{2}(\nu+1)\pi i} I_\nu(z) - (2/\pi)e^{-\frac{1}{2}\nu\pi i} K_\nu(z)$$
$$(-\pi < \arg z \le \tfrac{1}{2}\pi)$$

9.6.6 $\qquad I_{-n}(z) = I_n(z), \quad K_{-\nu}(z) = K_\nu(z)$

Most of the properties of modified Bessel functions can be deduced immediately from those of ordinary Bessel functions by application of these relations.

Limiting Forms for Small Arguments

When ν is fixed and $z \to 0$

9.6.7
$$I_\nu(z) \sim (\tfrac{1}{2}z)^\nu / \Gamma(\nu+1) \qquad (\nu \ne -1, -2, \ldots)$$

9.6.8 $\qquad K_0(z) \sim -\ln z$

9.6.9 $\qquad K_\nu(z) \sim \tfrac{1}{2}\Gamma(\nu)(\tfrac{1}{2}z)^{-\nu} \qquad (\mathscr{R}\nu > 0)$

Ascending Series

9.6.10 $\qquad I_\nu(z) = (\tfrac{1}{2}z)^\nu \sum_{k=0}^\infty \dfrac{(\tfrac{1}{4}z^2)^k}{k!\,\Gamma(\nu+k+1)}$

9.6.11
$$K_n(z) = \tfrac{1}{2}(\tfrac{1}{2}z)^{-n} \sum_{k=0}^{n-1} \frac{(n-k-1)!}{k!}(-\tfrac{1}{4}z^2)^k$$
$$+ (-)^{n+1}\ln(\tfrac{1}{2}z) I_n(z)$$
$$+ (-)^n \tfrac{1}{2}(\tfrac{1}{2}z)^n \sum_{k=0}^\infty \{\psi(k+1) + \psi(n+k+1)\} \frac{(\tfrac{1}{4}z^2)^k}{k!(n+k)!}$$

where $\psi(n)$ is given by **6.3.2**.

9.6.12 $\quad I_0(z) = 1 + \dfrac{\tfrac{1}{4}z^2}{(1!)^2} + \dfrac{(\tfrac{1}{4}z^2)^2}{(2!)^2} + \dfrac{(\tfrac{1}{4}z^2)^3}{(3!)^2} + \cdots$

9.6.13
$$K_0(z) = -\{\ln(\tfrac{1}{2}z) + \gamma\} I_0(z) + \frac{\tfrac{1}{4}z^2}{(1!)^2}$$
$$+ (1+\tfrac{1}{2})\frac{(\tfrac{1}{4}z^2)^2}{(2!)^2} + (1+\tfrac{1}{2}+\tfrac{1}{3})\frac{(\tfrac{1}{4}z^2)^3}{(3!)^2} + \cdots$$

Wronskians

9.6.14
$$W\{I_\nu(z), I_{-\nu}(z)\} = I_\nu(z)I_{-(\nu+1)}(z) - I_{\nu+1}(z)I_{-\nu}(z)$$
$$= -2\sin(\nu\pi)/(\pi z)$$

9.6.15
$$W\{K_\nu(z), I_\nu(z)\} = I_\nu(z)K_{\nu+1}(z) + I_{\nu+1}(z)K_\nu(z) = 1/z$$

Integral Representations

9.6.16
$$I_0(z) = \frac{1}{\pi} \int_0^\pi e^{\pm z \cos \theta} \, d\theta = \frac{1}{\pi} \int_0^\pi \cosh(z \cos \theta) \, d\theta$$

9.6.17 $\quad K_0(z) = -\frac{1}{\pi} \int_0^\pi e^{\pm z \cos \theta} \{ \gamma + \ln(2z \sin^2 \theta) \} \, d\theta$

9.6.18
$$I_\nu(z) = \frac{(\frac{1}{2}z)^\nu}{\pi^{\frac{1}{2}} \Gamma(\nu + \frac{1}{2})} \int_0^\pi e^{\pm z \cos \theta} \sin^{2\nu} \theta \, d\theta$$
$$= \frac{(\frac{1}{2}z)^\nu}{\pi^{\frac{1}{2}} \Gamma(\nu + \frac{1}{2})} \int_{-1}^1 (1-t^2)^{\nu-\frac{1}{2}} e^{\pm zt} \, dt \qquad (\mathscr{R}\nu > -\frac{1}{2})$$

9.6.19 $\qquad I_n(z) = \frac{1}{\pi} \int_0^\pi e^{z \cos \theta} \cos(n\theta) \, d\theta$

9.6.20
$$I_\nu(z) = \frac{1}{\pi} \int_0^\pi e^{z \cos \theta} \cos(\nu \theta) \, d\theta$$
$$- \frac{\sin(\nu\pi)}{\pi} \int_0^\infty e^{-z \cosh t - \nu t} \, dt \qquad (|\arg z| < \tfrac{1}{2}\pi)$$

9.6.21
$$K_0(x) = \int_0^\infty \cos(x \sinh t) \, dt = \int_0^\infty \frac{\cos(xt)}{\sqrt{t^2+1}} \, dt$$
$$(x > 0)$$

9.6.22
$$K_\nu(x) = \sec(\tfrac{1}{2}\nu\pi) \int_0^\infty \cos(x \sinh t) \cosh(\nu t) \, dt$$
$$= \csc(\tfrac{1}{2}\nu\pi) \int_0^\infty \sin(x \sinh t) \sinh(\nu t) \, dt$$
$$(|\mathscr{R}\nu| < 1, \, x > 0)$$

9.6.23
$$K_\nu(z) = \frac{\pi^{\frac{1}{2}}(\frac{1}{2}z)^\nu}{\Gamma(\nu + \frac{1}{2})} \int_0^\infty e^{-z \cosh t} \sinh^{2\nu} t \, dt$$
$$= \frac{\pi^{\frac{1}{2}}(\frac{1}{2}z)^\nu}{\Gamma(\nu + \frac{1}{2})} \int_1^\infty e^{-zt} (t^2-1)^{\nu-\frac{1}{2}} \, dt$$
$$(\mathscr{R}\nu > -\tfrac{1}{2}, \, |\arg z| < \tfrac{1}{2}\pi)$$

9.6.24 $\quad K_\nu(z) = \int_0^\infty e^{-z \cosh t} \cosh(\nu t) \, dt \quad (|\arg z| < \tfrac{1}{2}\pi)$

9.6.25
$$K_\nu(xz) = \frac{\Gamma(\nu + \frac{1}{2})(2z)^\nu}{\pi^{\frac{1}{2}} x^\nu} \int_0^\infty \frac{\cos(xt) \, dt}{(t^2 + z^2)^{\nu + \frac{1}{2}}}$$
$$(\mathscr{R}\nu > -\tfrac{1}{2}, \, x > 0, \, |\arg z| < \tfrac{1}{2}\pi)^*$$

Recurrence Relations

9.6.26
$$\mathscr{Z}_{\nu-1}(z) - \mathscr{Z}_{\nu+1}(z) = \frac{2\nu}{z} \mathscr{Z}_\nu(z)$$
$$\mathscr{Z}'_\nu(z) = \mathscr{Z}_{\nu-1}(z) - \frac{\nu}{z} \mathscr{Z}_\nu(z)$$
$$\mathscr{Z}_{\nu-1}(z) + \mathscr{Z}_{\nu+1}(z) = 2\mathscr{Z}'_\nu(z)$$
$$\mathscr{Z}'_\nu(z) = \mathscr{Z}_{\nu+1}(z) + \frac{\nu}{z} \mathscr{Z}_\nu(z)$$

$\mathscr{Z}_\nu$ denotes I_ν, $e^{\nu\pi i} K_\nu$, or any linear combination of these functions, the coefficients in which are independent of z and ν.

9.6.27 $\quad I_0'(z) = I_1(z), \qquad K_0'(z) = -K_1(z)$

Formulas for Derivatives

9.6.28
$$\left(\frac{1}{z} \frac{d}{dz} \right)^k \{ z^\nu \mathscr{Z}_\nu(z) \} = z^{\nu-k} \mathscr{Z}_{\nu-k}(z)$$
$$\left(\frac{1}{z} \frac{d}{dz} \right)^k \{ z^{-\nu} \mathscr{Z}_\nu(z) \} = z^{-\nu-k} \mathscr{Z}_{\nu+k}(z) \qquad (k=0,1,2,\ldots)$$

9.6.29
$$\mathscr{Z}_\nu^{(k)}(z) = \frac{1}{2^k} \left\{ \mathscr{Z}_{\nu-k}(z) + \binom{k}{1} \mathscr{Z}_{\nu-k+2}(z) \right.$$
$$\left. + \binom{k}{2} \mathscr{Z}_{\nu-k+4}(z) + \ldots + \mathscr{Z}_{\nu+k}(z) \right\}$$
$$(k=0,1,2,\ldots)$$

Analytic Continuation

9.6.30 $\quad I_\nu(ze^{m\pi i}) = e^{m\nu\pi i} I_\nu(z) \qquad (m \text{ an integer})$

9.6.31
$$K_\nu(ze^{m\pi i}) = e^{-m\nu\pi i} K_\nu(z) - \pi i \sin(m\nu\pi) \csc(\nu\pi) I_\nu(z)$$
$$(m \text{ an integer})$$

9.6.32 $\quad I_\nu(\bar{z}) = \overline{I_\nu(z)}, \qquad K_\nu(\bar{z}) = \overline{K_\nu(z)} \qquad (\nu \text{ real})$

Generating Function and Associated Series

9.6.33 $\qquad e^{\frac{1}{2}z(t+1/t)} = \sum_{k=-\infty}^\infty t^k I_k(z) \qquad (t \neq 0)$

9.6.34 $\qquad e^{z \cos \theta} = I_0(z) + 2 \sum_{k=1}^\infty I_k(z) \cos(k\theta)$

9.6.35
$$e^{z \sin \theta} = I_0(z) + 2 \sum_{k=0}^\infty (-)^k I_{2k+1}(z) \sin\{(2k+1)\theta\}$$
$$+ 2 \sum_{k=1}^\infty (-)^k I_{2k}(z) \cos(2k\theta)$$

9.6.36 $\quad 1 = I_0(z) - 2I_2(z) + 2I_4(z) - 2I_6(z) + \cdots$

9.6.37 $\quad e^z = I_0(z) + 2I_1(z) + 2I_2(z) + 2I_3(z) + \cdots$

9.6.38 $\quad e^{-z} = I_0(z) - 2I_1(z) + 2I_2(z) - 2I_3(z) + \cdots$

9.6.39
$$\cosh z = I_0(z) + 2I_2(z) + 2I_4(z) + 2I_6(z) + \cdots$$

9.6.40 $\quad \sinh z = 2I_1(z) + 2I_3(z) + 2I_5(z) + \cdots$

Other Differential Equations

The quantity λ^2 in equations **9.1.49** to **9.1.54** and **9.1.56** can be replaced by $-\lambda^2$ if at the same time the symbol $\mathscr{C}$ in the given solutions is replaced by $\mathscr{L}$.

9.6.41

$$z^2 w'' + z(1 \pm 2z)w' + (\pm z - \nu^2)w = 0, \qquad w = e^{\mp z}\mathscr{L}_\nu(z)$$

Differential equations for products may be obtained from **9.1.57** to **9.1.59** by replacing z by iz.

Derivatives With Respect to Order

9.6.42

$$\frac{\partial}{\partial \nu} I_\nu(z) = I_\nu(z) \ln\left(\tfrac{1}{2}z\right) - \left(\tfrac{1}{2}z\right)^\nu \sum_{k=0}^\infty \frac{\psi(\nu+k+1)}{\Gamma(\nu+k+1)} \frac{\left(\tfrac{1}{4}z^2\right)^k}{k!}$$

9.6.43

$$\frac{\partial}{\partial \nu} K_\nu(z) = \tfrac{1}{2}\pi \csc(\nu\pi)\left\{\frac{\partial}{\partial \nu} I_{-\nu}(z) - \frac{\partial}{\partial \nu} I_\nu(z)\right\}$$
$$- \pi \cot(\nu\pi)K_\nu(z) \qquad (\nu \neq 0, \pm 1, \pm 2, \ldots)$$

9.6.44

$$(-)^n \left[\frac{\partial}{\partial \nu} I_\nu(z)\right]_{\nu=n} =$$
$$-K_n(z) + \frac{n!\left(\tfrac{1}{2}z\right)^{-n}}{2} \sum_{k=0}^{n-1} (-)^k \frac{\left(\tfrac{1}{2}z\right)^k I_k(z)}{(n-k)k!}$$

9.6.45

$$\left[\frac{\partial}{\partial \nu} K_\nu(z)\right]_{\nu=n} = \frac{n!\left(\tfrac{1}{2}z\right)^{-n}}{2} \sum_{k=0}^{n-1} \frac{\left(\tfrac{1}{2}z\right)^k K_k(z)}{(n-k)k!}$$

9.6.46

$$\left[\frac{\partial}{\partial \nu} I_\nu(z)\right]_{\nu=0} = -K_0(z), \qquad \left[\frac{\partial}{\partial \nu} K_\nu(z)\right]_{\nu=0} = 0$$

Expressions in Terms of Hypergeometric Functions

9.6.47

$$I_\nu(z) = \frac{\left(\tfrac{1}{2}z\right)^\nu}{\Gamma(\nu+1)} \; {}_0F_1\left(\nu+1; \tfrac{1}{4}z^2\right)$$
$$= \frac{\left(\tfrac{1}{2}z\right)^\nu e^{-z}}{\Gamma(\nu+1)} M\left(\nu+\tfrac{1}{2}, 2\nu+1, 2z\right) = \frac{z^{-\frac{1}{2}}M_{0,\nu}(2z)}{2^{2\nu+\frac{1}{2}}\Gamma(\nu+1)}$$

9.6.48

$$K_\nu(z) = \left(\frac{\pi}{2z}\right)^{\frac{1}{2}} W_{0,\nu}(2z)$$

(${}_0F_1$ is the generalized hypergeometric function. For $M(a, b, z)$, $M_{0,\nu}(z)$ and $W_{0,\nu}(z)$ see chapter **13**.)

Connection With Legendre Functions

If μ and z are fixed, $\mathscr{R}z > 0$, and $\nu \to \infty$ through real positive values

9.6.49

$$\lim\left\{\nu^\mu P_\nu^{-\mu}\left(\cosh\frac{z}{\nu}\right)\right\} = I_\mu(z)$$

9.6.50

$$\lim\left\{\nu^{-\mu}e^{-\mu\pi i} Q_\nu^\mu\left(\cosh\frac{z}{\nu}\right)\right\} = K_\mu(z)$$

For the definition of $P_\nu^{-\mu}$ and Q_ν^μ, see chapter **8**.

Multiplication Theorems

9.6.51

$$\mathscr{L}_\nu(\lambda z) = \lambda^{\pm\nu} \sum_{k=0}^\infty \frac{(\lambda^2-1)^k \left(\tfrac{1}{2}z\right)^k}{k!} \mathscr{L}_{\nu\pm k}(z) \qquad (|\lambda^2-1| < 1)$$

If $\mathscr{L} = I$ and the upper signs are taken, the restriction on λ is unnecessary.

9.6.52

$$I_\nu(z) = \sum_{k=0}^\infty \frac{z^k}{k!} J_{\nu+k}(z), \qquad J_\nu(z) = \sum_{k=0}^\infty (-)^k \frac{z^k}{k!} I_{\nu+k}(z)$$

Neumann Series for $K_n(z)$

9.6.53

$$K_n(z) = (-)^{n-1}\left\{\ln\left(\tfrac{1}{2}z\right) - \psi(n+1)\right\} I_n(z)$$
$$+ \frac{n!\left(\tfrac{1}{2}z\right)^{-n}}{2} \sum_{k=0}^{n-1} (-)^k \frac{\left(\tfrac{1}{2}z\right)^k I_k(z)}{(n-k)k!}$$
$$+ (-)^n \sum_{k=1}^\infty \frac{(n+2k)I_{n+2k}(z)}{k(n+k)}$$

9.6.54 $\quad K_0(z) = -\left\{\ln\left(\tfrac{1}{2}z\right) + \gamma\right\} I_0(z) + 2 \sum_{k=1}^\infty \frac{I_{2k}(z)}{k}$

Zeros

Properties of the zeros of $I_\nu(z)$ and $K_\nu(z)$ may be deduced from those of $J_\nu(z)$ and $H_\nu^{(1)}(z)$ respectively, by application of the transformations **9.6.3** and **9.6.4**.

For example, if ν is real the zeros of $I_\nu(z)$ are all complex unless $-2k < \nu < -(2k-1)$ for some positive integer k, in which event $I_\nu(z)$ has two real zeros.

The approximate distribution of the zeros of $K_n(z)$ in the region $-\tfrac{3}{2}\pi \leq \arg z \leq \tfrac{1}{2}\pi$ is obtained on rotating **Figure 9.6** through an angle $-\tfrac{1}{2}\pi$ so that the cut lies along the positive imaginary axis. The zeros in the region $-\tfrac{1}{2}\pi \leq \arg z \leq \tfrac{3}{2}\pi$ are their conjugates. $K_n(z)$ has no zeros in the region $|\arg z| \leq \tfrac{1}{2}\pi$; this result remains true when n is replaced by any real number ν.

9.7. Asymptotic Expansions

Asymptotic Expansions for Large Arguments

When ν is fixed, $|z|$ is large and $\mu = 4\nu^2$

9.7.1

$$I_\nu(z) \sim \frac{e^z}{\sqrt{2\pi z}}\left\{1 - \frac{\mu-1}{8z} + \frac{(\mu-1)(\mu-9)}{2!(8z)^2}\right.$$
$$\left. - \frac{(\mu-1)(\mu-9)(\mu-25)}{3!(8z)^3} + \ldots\right\} \qquad (|\arg z| < \tfrac{1}{2}\pi)$$

9.7.2

$$K_\nu(z) \sim \sqrt{\frac{\pi}{2z}}\, e^{-z}\{1 + \frac{\mu-1}{8z} + \frac{(\mu-1)(\mu-9)}{2!(8z)^2}$$
$$+ \frac{(\mu-1)(\mu-9)(\mu-25)}{3!(8z)^3} + \ldots\} \quad (|\arg z| < \tfrac{3}{2}\pi)$$

9.7.3

$$I'_\nu(z) \sim \frac{e^z}{\sqrt{2\pi z}}\{1 - \frac{\mu+3}{8z} + \frac{(\mu-1)(\mu+15)}{2!(8z)^2}$$
$$- \frac{(\mu-1)(\mu-9)(\mu+35)}{3!(8z)^3} + \ldots\} \quad (|\arg z| < \tfrac{1}{2}\pi)$$

9.7.4

$$K'_\nu(z) \sim -\sqrt{\frac{\pi}{2z}}e^{-z}\{1 + \frac{\mu+3}{8z} + \frac{(\mu-1)(\mu+15)}{2!(8z)^2}$$
$$+ \frac{(\mu-1)(\mu-9)(\mu+35)}{3!(8z)^3} + \ldots\} \quad (|\arg z| < \tfrac{3}{2}\pi)$$

The general terms in the last two expansions can be written down by inspection of **9.2.15** and **9.2.16**.

If ν is real and non-negative and z is positive the remainder after k terms in the expansion **9.7.2** does not exceed the $(k+1)$th term in absolute value and is of the same sign, provided that $k \geq \nu - \tfrac{1}{2}$.

9.7.5

$$I_\nu(z)K_\nu(z) \sim \frac{1}{2z}\{1 - \frac{1}{2}\frac{\mu-1}{(2z)^2}$$
$$+ \frac{1\cdot 3}{2\cdot 4}\frac{(\mu-1)(\mu-9)}{(2z)^4} - \ldots\}$$
$$(|\arg z| < \tfrac{1}{2}\pi)$$

9.7.6

$$I'_\nu(z)K'_\nu(z) \sim -\frac{1}{2z}\{1 + \frac{1}{2}\frac{\mu-3}{(2z)^2}$$
$$- \frac{1\cdot 1}{2\cdot 4}\frac{(\mu-1)(\mu-45)}{(2z)^4} + \ldots\}$$
$$(|\arg z| < \tfrac{1}{2}\pi)$$

The general terms can be written down by inspection of **9.2.28** and **9.2.30**.

Uniform Asymptotic Expansions for Large Orders

9.7.7 $I_\nu(\nu z) \sim \dfrac{1}{\sqrt{2\pi\nu}}\dfrac{e^{\nu\eta}}{(1+z^2)^{1/4}}\{1 + \sum_{k=1}^\infty \dfrac{u_k(t)}{\nu^k}\}$

9.7.8

$$K_\nu(\nu z) \sim \sqrt{\frac{\pi}{2\nu}}\frac{e^{-\nu\eta}}{(1+z^2)^{1/4}}\{1 + \sum_{k=1}^\infty (-)^k \frac{u_k(t)}{\nu^k}\}$$

9.7.9 $I'_\nu(\nu z) \sim \dfrac{1}{\sqrt{2\pi\nu}}\dfrac{(1+z^2)^{1/4}}{z}\, e^{\nu\eta}\{1 + \sum_{k=1}^\infty \dfrac{v_k(t)}{\nu^k}\}$

9.7.10

$$K'_\nu(\nu z) \sim -\sqrt{\frac{\pi}{2\nu}}\frac{(1+z^2)^{1/4}}{z}\, e^{-\nu\eta}\{1 + \sum_{k=1}^\infty (-)^k \frac{v_k(t)}{\nu^k}\}$$

When $\nu \to +\infty$, these expansions hold uniformly with respect to z in the sector $|\arg z| \leq \tfrac{1}{2}\pi - \epsilon$, where ϵ is an arbitrary positive number. Here

9.7.11 $t = 1/\sqrt{1+z^2}, \qquad \eta = \sqrt{1+z^2} + \ln\dfrac{z}{1+\sqrt{1+z^2}}$

and $u_k(t)$, $v_k(t)$ are given by **9.3.9**, **9.3.10**, **9.3.13** and **9.3.14**. See [9.38] for tables of η, $u_k(t)$, $v_k(t)$, and also for bounds on the remainder terms in **9.7.7** to **9.7.10**.

9.8. Polynomial Approximations [4]

In equations **9.8.1** to **9.8.4**, $t = x/3.75$.

9.8.1 $-3.75 \leq x \leq 3.75$

$$I_0(x) = 1 + 3.5156229t^2 + 3.0899424t^4 + 1.2067492t^6$$
$$+ .2659732t^8 + .0360768t^{10} + .0045813t^{12} + \epsilon$$
$$|\epsilon| < 1.6 \times 10^{-7}$$

9.8.2 $3.75 \leq x < \infty$

$$x^{\frac{1}{2}}e^{-x}I_0(x) = .39894228 + .01328592t^{-1}$$
$$+ .00225319t^{-2} - .00157565t^{-3}$$
$$+ .00916281t^{-4} - .02057706t^{-5}$$
$$+ .02635537t^{-6} - .01647633t^{-7}$$
$$+ .00392377t^{-8} + \epsilon$$
$$|\epsilon| < 1.9 \times 10^{-7}$$

9.8.3 $-3.75 \leq x \leq 3.75$

$$x^{-1}I_1(x) = \tfrac{1}{2} + .87890594t^2 + .51498869t^4$$
$$+ .15084934t^6 + .02658733t^8$$
$$+ .00301532t^{10} + .00032411t^{12} + \epsilon$$
$$|\epsilon| < 8 \times 10^{-9}$$

9.8.4 $3.75 \leq x < \infty$

$$x^{\frac{1}{2}}e^{-x}I_1(x) = .39894228 - .03988024t^{-1}$$
$$- .00362018t^{-2} + .00163801t^{-3}$$
$$- .01031555t^{-4} + .02282967t^{-5}$$
$$- .02895312t^{-6} + .01787654t^{-7}$$
$$- .00420059t^{-8} + \epsilon$$
$$|\epsilon| < 2.2 \times 10^{-7}$$

---

[4] See footnote 2, section **9.4**.

9.8.5 $\qquad\qquad 0 < x \le 2$

$$K_0(x) = -\ln\,(x/2) I_0(x) - .57721\ 566$$
$$+ .42278\ 420(x/2)^2 + .23069\ 756(x/2)^4$$
$$+ .03488\ 590(x/2)^6 + .00262\ 698(x/2)^8$$
$$+ .00010\ 750(x/2)^{10} + .00000\ 740(x/2)^{12} + \epsilon$$

$$|\epsilon| < 1 \times 10^{-8}$$

9.8.6 $\qquad\qquad 2 \le x < \infty$

$$x^{\frac{1}{2}} e^x K_0(x) = 1.25331\ 414 - .07832\ 358(2/x)$$
$$+ .02189\ 568(2/x)^2 - .01062\ 446(2/x)^3$$
$$+ .00587\ 872(2/x)^4 - .00251\ 540(2/x)^5$$
$$+ .00053\ 208(2/x)^6 + \epsilon$$

$$|\epsilon| < 1.9 \times 10^{-7}$$

9.8.7 $\qquad\qquad 0 < x \le 2$

$$x K_1(x) = x \ln\,(x/2) I_1(x) + 1 + .15443\ 144(x/2)^2$$
$$- .67278\ 579(x/2)^4 - .18156\ 897(x/2)^6$$
$$- .01919\ 402(x/2)^8 - .00110\ 404(x/2)^{10}$$
$$- .00004\ 686(x/2)^{12} + \epsilon$$

$$|\epsilon| < 8 \times 10^{-9}$$

9.8.8 $\qquad\qquad 2 \le x < \infty$

$$x^{\frac{1}{2}} e^x K_1(x) = 1.25331\ 414 + .23498\ 619(2/x)$$
$$- .03655\ 620(2/x)^2 + .01504\ 268(2/x)^3$$
$$- .00780\ 353(2/x)^4 + .00325\ 614(2/x)^5$$
$$- .00068\ 245(2/x)^6 + \epsilon$$

$$|\epsilon| < 2.2 \times 10^{-7}$$

For expansions of $I_0(x)$, $K_0(x)$, $I_1(x)$, and $K_1(x)$ in series of Chebyshev polynomials for the ranges $0 \le x \le 8$ and $0 \le 8/x \le 1$, see [9.37].

Kelvin Functions

9.9. Definitions and Properties

In this and the following section ν is real, x is real and non-negative, and n is again a positive integer or zero.

Definitions

9.9.1

$$\text{ber}_\nu\,x + i\,\text{bei}_\nu\,x = J_\nu(xe^{3\pi i/4}) = e^{\nu\pi i} J_\nu(xe^{-\pi i/4})$$
$$= e^{\frac{1}{2}\nu\pi i} I_\nu(xe^{\pi i/4}) = e^{3\nu\pi i/2} I_\nu(xe^{-3\pi i/4})$$

9.9.2

$$\text{ker}_\nu\,x + i\,\text{kei}_\nu\,x = e^{-\frac{1}{2}\nu\pi i} K_\nu(xe^{\pi i/4})$$
$$= \tfrac{1}{2}\pi i H_\nu^{(1)}(xe^{3\pi i/4}) = -\tfrac{1}{2}\pi i e^{-\nu\pi i} H_\nu^{(2)}(xe^{-\pi i/4})$$

When $\nu = 0$, suffices are usually suppressed.

Differential Equations

9.9.3

$$x^2 w'' + x w' - (ix^2 + \nu^2) w = 0,$$
$$w = \text{ber}_\nu\,x + i\,\text{bei}_\nu\,x, \qquad \text{ber}_{-\nu}\,x + i\,\text{bei}_{-\nu}\,x,$$
$$\text{ker}_\nu\,x + i\,\text{kei}_\nu\,x, \quad \text{ker}_{-\nu}\,x + i\,\text{kei}_{-\nu}\,x$$

9.9.4

$$x^4 w^{iv} + 2x^3 w''' - (1 + 2\nu^2)(x^2 w'' - x w')$$
$$+ (\nu^4 - 4\nu^2 + x^4) w = 0,$$
$$w = \text{ber}_{\pm\nu}\,x, \text{ bei}_{\pm\nu}\,x, \text{ ker}_{\pm\nu}\,x, \text{ kei}_{\pm\nu}\,x$$

Relations Between Solutions

9.9.5

$$\text{ber}_{-\nu}\,x = \cos(\nu\pi)\,\text{ber}_\nu\,x + \sin(\nu\pi)\,\text{bei}_\nu\,x$$
$$+ (2/\pi)\sin(\nu\pi)\,\text{ker}_\nu\,x$$

$$\text{bei}_{-\nu}\,x = -\sin(\nu\pi)\,\text{ber}_\nu\,x + \cos(\nu\pi)\,\text{bei}_\nu\,x$$
$$+ (2/\pi)\sin(\nu\pi)\,\text{kei}_\nu\,x$$

9.9.6

$$\text{ker}_{-\nu}\,x = \cos(\nu\pi)\,\text{ker}_\nu\,x - \sin(\nu\pi)\,\text{kei}_\nu\,x$$
$$\text{kei}_{-\nu}\,x = \sin(\nu\pi)\,\text{ker}_\nu\,x + \cos(\nu\pi)\,\text{kei}_\nu\,x$$

9.9.7 $\quad \text{ber}_{-n}\,x = (-)^n\,\text{ber}_n\,x, \quad \text{bei}_{-n}\,x = (-)^n\,\text{bei}_n\,x$

9.9.8 $\quad \text{ker}_{-n}\,x = (-)^n\,\text{ker}_n\,x, \quad \text{kei}_{-n}\,x = (-)^n\,\text{kei}_n\,x$

Ascending Series

9.9.9

$$\text{ber}_\nu\,x = (\tfrac{1}{2}x)^\nu \sum_{k=0}^\infty \frac{\cos\{(\tfrac{3}{4}\nu + \tfrac{1}{2}k)\pi\}}{k!\,\Gamma(\nu+k+1)}(\tfrac{1}{4}x^2)^k$$

$$\text{bei}_\nu\,x = (\tfrac{1}{2}x)^\nu \sum_{k=0}^\infty \frac{\sin\{(\tfrac{3}{4}\nu + \tfrac{1}{2}k)\pi\}}{k!\,\Gamma(\nu+k+1)}(\tfrac{1}{4}x^2)^k$$

9.9.10

$$\text{ber}\,x = 1 - \frac{(\tfrac{1}{4}x^2)^2}{(2!)^2} + \frac{(\tfrac{1}{4}x^2)^4}{(4!)^2} - \cdots$$

$$\text{bei}\,x = \tfrac{1}{4}x^2 - \frac{(\tfrac{1}{4}x^2)^3}{(3!)^2} + \frac{(\tfrac{1}{4}x^2)^5}{(5!)^2} - \cdots$$

9.9.11

$$\text{ker}_n\,x = \tfrac{1}{2}(\tfrac{1}{2}x)^{-n} \sum_{k=0}^{n-1} \cos\{(\tfrac{3}{4}n + \tfrac{1}{2}k)\pi\}$$

$$\times \frac{(n-k-1)!}{k!}(\tfrac{1}{4}x^2)^k - \ln(\tfrac{1}{2}x)\,\text{ber}_n\,x + \tfrac{1}{4}\pi\,\text{bei}_n\,x$$

$$+ \tfrac{1}{2}(\tfrac{1}{2}x)^n \sum_{k=0}^\infty \cos\{(\tfrac{3}{4}n + \tfrac{1}{2}k)\pi\}$$

$$\times \frac{\{\psi(k+1) + \psi(n+k+1)\}}{k!\,(n+k)!}(\tfrac{1}{4}x^2)^k$$

$$\text{kei}_n\, x = -\tfrac{1}{2}(\tfrac{1}{2}x)^{-n} \sum_{k=0}^{n-1} \sin\{(\tfrac{3}{4}n+\tfrac{1}{2}k)\pi\}$$

$$\times \frac{(n-k-1)!}{k!}(\tfrac{1}{4}x^2)^k - \ln(\tfrac{1}{2}x)\,\text{bei}_n\, x - \tfrac{1}{4}\pi\,\text{ber}_n\, x$$

$$+\tfrac{1}{2}(\tfrac{1}{2}x)^n \sum_{k=0}^{\infty} \sin\{(\tfrac{3}{4}n+\tfrac{1}{2}k)\pi\}$$

$$\times \frac{\{\psi(k+1)+\psi(n+k+1)\}}{k!(n+k)!}(\tfrac{1}{4}x^2)^k$$

where $\psi(n)$ is given by **6.3.2.**

9.9.12

$$\text{ker}\, x = -\ln(\tfrac{1}{2}x)\,\text{ber}\, x + \tfrac{1}{4}\pi\,\text{bei}\, x$$

$$+\sum_{k=0}^{\infty}(-)^k\frac{\psi(2k+1)}{\{(2k)!\}^2}(\tfrac{1}{4}x^2)^{2k}$$

$$\text{kei}\, x = -\ln(\tfrac{1}{2}x)\,\text{bei}\, x - \tfrac{1}{4}\pi\,\text{ber}\, x$$

$$+\sum_{k=0}^{\infty}(-)^k\frac{\psi(2k+2)}{\{(2k+1)!\}^2}(\tfrac{1}{4}x^2)^{2k+1}$$

Functions of Negative Argument

In general Kelvin functions have a branch point at $x=0$ and individual functions with arguments $xe^{\pm\pi i}$ are complex. The branch point is absent however in the case of ber$_\nu$ and bei$_\nu$ when ν is an integer, and

9.9.13

$$\text{ber}_n(-x)=(-)^n\,\text{ber}_n\, x, \qquad \text{bei}_n(-x)=(-)^n\,\text{bei}_n\, x$$

Recurrence Relations

9.9.14

$$f_{\nu+1}+f_{\nu-1}=-\frac{\nu\sqrt{2}}{x}(f_\nu-g_\nu)$$

$$f_\nu'=\frac{1}{2\sqrt{2}}(f_{\nu+1}+g_{\nu+1}-f_{\nu-1}-g_{\nu-1})$$

$$f_\nu'-\frac{\nu}{x}f_\nu=\frac{1}{\sqrt{2}}(f_{\nu+1}+g_{\nu+1})$$

$$f_\nu'+\frac{\nu}{x}f_\nu=-\frac{1}{\sqrt{2}}(f_{\nu-1}+g_{\nu-1})$$

where

9.9.15

$$\left.\begin{aligned}f_\nu&=\text{ber}_\nu\, x\\g_\nu&=\text{bei}_\nu\, x\end{aligned}\right\} \qquad \left.\begin{aligned}f_\nu&=\text{bei}_\nu\, x\\g_\nu&=-\text{ber}_\nu\, x\end{aligned}\right\}$$

$$\left.\begin{aligned}f_\nu&=\text{ker}_\nu\, x\\g_\nu&=\text{kei}_\nu\, x\end{aligned}\right\} \qquad \left.\begin{aligned}f_\nu&=\text{kei}_\nu\, x\\g_\nu&=-\text{ker}_\nu\, x\end{aligned}\right\}$$

9.9.16

$$\sqrt{2}\,\text{ber}'\, x=\text{ber}_1\, x+\text{bei}_1\, x$$

$$\sqrt{2}\,\text{bei}'\, x=-\text{ber}_1\, x+\text{bei}_1\, x$$

9.9.17

$$\sqrt{2}\,\text{ker}'\, x=\text{ker}_1\, x+\text{kei}_1\, x$$

$$\sqrt{2}\,\text{kei}'\, x=-\text{ker}_1\, x+\text{kei}_1\, x$$

Recurrence Relations for Cross-Products

If

9.9.18

$$p_\nu=\text{ber}_\nu^2\, x+\text{bei}_\nu^2\, x$$

$$q_\nu=\text{ber}_\nu\, x\,\text{bei}_\nu'\, x-\text{ber}_\nu'\, x\,\text{bei}_\nu\, x$$

$$r_\nu=\text{ber}_\nu\, x\,\text{ber}_\nu'\, x+\text{bei}_\nu\, x\,\text{bei}_\nu'\, x$$

$$s_\nu=\text{ber}_\nu'^2\, x+\text{bei}_\nu'^2\, x$$

then

9.9.19

$$p_{\nu+1}=p_{\nu-1}-\frac{4\nu}{x}r_\nu$$

$$q_{\nu+1}=-\frac{\nu}{x}p_\nu+r_\nu=-q_{\nu-1}+2r_\nu$$

$$r_{\nu+1}=-\frac{(\nu+1)}{x}p_{\nu+1}+q_\nu$$

$$s_\nu=\frac{1}{2}p_{\nu+1}+\frac{1}{2}p_{\nu-1}-\frac{\nu^2}{x^2}p_\nu$$

and

9.9.20 $$\qquad p_\nu s_\nu=r_\nu^2+q_\nu^2$$

The same relations hold with ber, bei replaced throughout by ker, kei, respectively.

Indefinite Integrals

In the following f_ν, g_ν are any one of the pairs given by equations **9.9.15** and f_ν^*, g_ν^* are either the same pair or any other pair.

9.9.21

$$\int x^{1+\nu}f_\nu dx=-\frac{x^{1+\nu}}{\sqrt{2}}(f_{\nu+1}-g_{\nu+1})=-x^{1+\nu}\left(\frac{\nu}{x}g_\nu-g_\nu'\right)$$

9.9.22

$$\int x^{1-\nu}f_\nu dx=\frac{x^{1-\nu}}{\sqrt{2}}(f_{\nu-1}-g_{\nu-1})=x^{1-\nu}\left(\frac{\nu}{x}g_\nu+g_\nu'\right)$$

9.9.23

$$\int x(f_\nu g_\nu^*-g_\nu f_\nu^*)dx=\frac{x}{2\sqrt{2}}\{f_\nu^*(f_{\nu+1}+g_{\nu+1})$$

$$-g_\nu^*(f_{\nu+1}-g_{\nu+1})-f_\nu(f_{\nu+1}^*+g_{\nu+1}^*)+g_\nu(f_{\nu+1}^*-g_{\nu+1}^*)\}$$

$$=\frac{1}{2}x(f_\nu'f_\nu^*-f_\nu f_\nu^{*\prime}+g_\nu'g_\nu^*-g_\nu g_\nu^{*\prime})$$

9.9.24

$$\int x(f_\nu g_\nu^* + g_\nu f_\nu^*) dx = \frac{1}{4} x^2 (2f_\nu g_\nu^* - f_{\nu-1} g_{\nu+1}^*$$
$$- f_{\nu+1} g_{\nu-1}^* + 2g_\nu f_\nu^* - g_{\nu-1} f_{\nu+1}^* - g_{\nu+1} f_{\nu-1}^*)$$

9.9.25

$$\int x(f_\nu^2 + g_\nu^2) dx = x(f_\nu g_\nu' - f_\nu' g_\nu)$$
$$= -(x/\sqrt{2})(f_\nu f_{\nu+1} + g_\nu g_{\nu+1} - f_\nu g_{\nu+1} + f_{\nu+1} g_\nu)$$

9.9.26

$$\int x f_\nu g_\nu dx = \frac{1}{4} x^2 (2f_\nu g_\nu - f_{\nu-1} g_{\nu+1} - f_{\nu+1} g_{\nu-1})$$

9.9.27

$$\int x(f_\nu^2 - g_\nu^2) dx = \frac{1}{2} x^2 (f_\nu^2 - f_{\nu-1} f_{\nu+1} - g_\nu^2 + g_{\nu-1} g_{\nu+1})$$

Ascending Series for Cross-Products

9.9.28

$$\text{ber}_\nu^2 x + \text{bei}_\nu^2 x =$$
$$(\tfrac{1}{2} x)^{2\nu} \sum_{k=0}^{\infty} \frac{1}{\Gamma(\nu+k+1)\Gamma(\nu+2k+1)} \frac{(\tfrac{1}{4} x^2)^{2k}}{k!}$$

9.9.29

$$\text{ber}_\nu x \, \text{bei}_\nu' x - \text{ber}_\nu' x \, \text{bei}_\nu x$$
$$= (\tfrac{1}{2} x)^{2\nu+1} \sum_{k=0}^{\infty} \frac{1}{\Gamma(\nu+k+1)\Gamma(\nu+2k+2)} \frac{(\tfrac{1}{4} x^2)^{2k}}{k!}$$

9.9.30

$$\text{ber}_\nu x \, \text{ber}_\nu' x + \text{bei}_\nu x \, \text{bei}_\nu' x$$
$$= \tfrac{1}{2} (\tfrac{1}{2} x)^{2\nu-1} \sum_{k=0}^{\infty} \frac{1}{\Gamma(\nu+k+1)\Gamma(\nu+2k)} \frac{(\tfrac{1}{4} x^2)^{2k}}{k!}$$

9.9.31

$$\text{ber}_\nu'^2 x + \text{bei}_\nu'^2 x$$
$$= (\tfrac{1}{2} x)^{2\nu-2} \sum_{k=0}^{\infty} \frac{(2k^2 + 2\nu k + \tfrac{1}{4}\nu^2)}{\Gamma(\nu+k+1)\Gamma(\nu+2k+1)} \frac{(\tfrac{1}{4} x^2)^{2k}}{k!}$$

Expansions in Series of Bessel Functions

9.9.32

$$\text{ber}_\nu x + i \, \text{bei}_\nu x = \sum_{k=0}^{\infty} \frac{e^{(3\nu+k)\pi i/4} x^k J_{\nu+k}(x)}{2^{\frac{1}{2}k} k!}$$
$$= \sum_{k=0}^{\infty} \frac{e^{(3\nu+3k)\pi i/4} x^k I_{\nu+k}(x)}{2^{\frac{1}{2}k} k!}$$

9.9.33

$$\text{ber}_n(x\sqrt{2}) = \sum_{k=-\infty}^{\infty} (-)^{n+k} J_{n+2k}(x) I_{2k}(x)$$

$$\text{bei}_n(x\sqrt{2}) = \sum_{k=-\infty}^{\infty} (-)^{n+k} J_{n+2k+1}(x) I_{2k+1}(x)$$

Zeros of Functions of Order Zero [5]

	ber x	bei x	ker x	kei x
1st zero	2. 84892	5. 02622	1. 71854	3. 91467
2nd zero	7. 23883	9. 45541	6. 12728	8. 34422
3rd zero	11. 67396	13. 89349	10. 56294	12. 78256
4th zero	16. 11356	18. 33398	15. 00269	17. 22314
5th zero	20. 55463	22. 77544	19. 44381	21. 66464

	ber' x	bei' x	ker' x	kei' x
1st zero	6. 03871	3. 77320	2. 66584	4. 93181
2nd zero	10. 51364	8. 28099	7. 17212	9. 40405
3rd zero	14. 96844	12. 74215	11. 63218	13. 85827
4th zero	19. 41758	17. 19343	16. 08312	18. 30717
5th zero	23. 86430	21. 64114	20. 53068	22. 75379

9.10. Asymptotic Expansions

Asymptotic Expansions for Large Arguments

When ν is fixed and x is large

9.10.1

$$\text{ber}_\nu x = \frac{e^{x/\sqrt{2}}}{\sqrt{2\pi x}} \{f_\nu(x) \cos \alpha + g_\nu(x) \sin \alpha\}$$
$$- \frac{1}{\pi} \{\sin (2\nu\pi) \, \text{ker}_\nu x + \cos (2\nu\pi) \, \text{kei}_\nu x\}$$

9.10.2

$$\text{bei}_\nu x = \frac{e^{x/\sqrt{2}}}{\sqrt{2\pi x}} \{f_\nu(x) \sin \alpha - g_\nu(x) \cos \alpha\}$$
$$+ \frac{1}{\pi} \{\cos (2\nu\pi) \, \text{ker}_\nu x - \sin (2\nu\pi) \, \text{kei}_\nu x\}$$

9.10.3

$$\text{ker}_\nu x = \sqrt{\pi/(2x)} \, e^{-x/\sqrt{2}} \{f_\nu(-x) \cos \beta - g_\nu(-x) \sin \beta\}$$

9.10.4

$$\text{kei}_\nu x = \sqrt{\pi/(2x)} \, e^{-x/\sqrt{2}} \{-f_\nu(-x) \sin \beta - g_\nu(-x) \cos \beta\}$$

where

9.10.5

$$\alpha = (x/\sqrt{2}) + (\tfrac{1}{2}\nu - \tfrac{1}{8})\pi, \quad \beta = (x/\sqrt{2}) + (\tfrac{1}{2}\nu + \tfrac{1}{8})\pi = \alpha + \tfrac{1}{4}\pi$$

and, with $4\nu^2$ denoted by μ,

9.10.6

$$f_\nu(\pm x)$$
$$\sim 1 + \sum_{k=1}^{\infty} (\mp)^k \frac{(\mu-1)(\mu-9)\ldots\{\mu-(2k-1)^2\}}{k!(8x)^k} \cos\left(\frac{k\pi}{4}\right)$$

[5] From British Association for the Advancement of Science, Annual Report (J. R. Airey), 254 (1927) with permission. This reference also gives 5-decimal values of the next five zeros of each function.

9.10.7

$$g_\nu(\pm x)$$

$$\sim \sum_{k=1}^{\infty} (\mp)^k \frac{(\mu-1)(\mu-9)\ldots\{\mu-(2k-1)^2\}}{k!(8x)^k} \sin\left(\frac{k\pi}{4}\right)$$

The terms[6] in $\ker_\nu x$ and $\kei_\nu x$ in equations **9.10.1** and **9.10.2** are asymptotically negligible compared with the other terms, but their inclusion in numerical calculations yields improved accuracy.

The corresponding series for $\ber'_\nu x$, $\bei'_\nu x$, $\ker'_\nu x$ and $\kei'_\nu x$ can be derived from **9.2.11** and **9.2.13** with $z=xe^{3\pi i/4}$; the extra terms in the expansions of $\ber'_\nu x$ and $\bei'_\nu x$ are respectively

$$-(1/\pi)\{\sin(2\nu\pi)\ker'_\nu x + \cos(2\nu\pi)\kei'_\nu x\}$$

and

$$(1/\pi)\{\cos(2\nu\pi)\ker'_\nu x - \sin(2\nu\pi)\kei'_\nu x\}.$$

Modulus and Phase

9.10.8

$$M_\nu = \sqrt{(\ber_\nu^2 x + \bei_\nu^2 x)}, \qquad \theta_\nu = \arctan(\bei_\nu x/\ber_\nu x)$$

9.10.9 $\quad \ber_\nu x = M_\nu \cos \theta_\nu, \qquad \bei_\nu x = M_\nu \sin \theta_\nu$

9.10.10 $\qquad M_{-n} = M_n, \qquad \theta_{-n} = \theta_n - n\pi$

9.10.11

$$\ber'_\nu x = \tfrac{1}{2} M_{\nu+1} \cos(\theta_{\nu+1} - \tfrac{1}{4}\pi) - \tfrac{1}{2} M_{\nu-1} \cos(\theta_{\nu-1} - \tfrac{1}{4}\pi)$$
$$= (\nu/x) M_\nu \cos \theta_\nu + M_{\nu+1} \cos(\theta_{\nu+1} - \tfrac{1}{4}\pi)$$
$$= -(\nu/x) M_\nu \cos \theta_\nu - M_{\nu-1} \cos(\theta_{\nu-1} - \tfrac{1}{4}\pi)$$

9.10.12

$$\bei'_\nu x = \tfrac{1}{2} M_{\nu+1} \sin(\theta_{\nu+1} - \tfrac{1}{4}\pi) - \tfrac{1}{2} M_{\nu-1} \sin(\theta_{\nu-1} - \tfrac{1}{4}\pi)$$
$$= (\nu/x) M_\nu \sin \theta_\nu + M_{\nu+1} \sin(\theta_{\nu+1} - \tfrac{1}{4}\pi)$$
$$= -(\nu/x) M_\nu \sin \theta_\nu - M_{\nu-1} \sin(\theta_{\nu-1} - \tfrac{1}{4}\pi)$$

9.10.13

$$\ber' x = M_1 \cos(\theta_1 - \tfrac{1}{4}\pi), \qquad \bei' x = M_1 \sin(\theta_1 - \tfrac{1}{4}\pi)$$

9.10.14

$$M'_\nu = (\nu/x) M_\nu + M_{\nu+1} \cos(\theta_{\nu+1} - \theta_\nu - \tfrac{1}{4}\pi)$$
$$= -(\nu/x) M_\nu - M_{\nu-1} \cos(\theta_{\nu-1} - \theta_\nu - \tfrac{1}{4}\pi)$$

9.10.15

$$\theta'_\nu = (M_{\nu+1}/M_\nu) \sin(\theta_{\nu+1} - \theta_\nu - \tfrac{1}{4}\pi)$$
$$= -(M_{\nu-1}/M_\nu) \sin(\theta_{\nu-1} - \theta_\nu - \tfrac{1}{4}\pi)$$

---

[6] The coefficients of these terms given in [9.17] are incorrect. The present results are due to Mr. G. F. Miller.

9.10.16

$$M'_0 = M_1 \cos(\theta_1 - \theta_0 - \tfrac{1}{4}\pi)$$
$$\theta'_0 = (M_1/M_0) \sin(\theta_1 - \theta_0 - \tfrac{1}{4}\pi)$$

9.10.17

$$d(xM_\nu^2\theta'_\nu)/dx = \lambda M_\nu^2, \qquad x^2 M''_\nu + x M'_\nu - \nu^2 M_\nu = x^2 M_\nu \theta'^2_\nu$$

9.10.18

$$N_\nu = \sqrt{(\ker_\nu^2 x + \kei_\nu^2 x)}, \qquad \phi_\nu = \arctan(\kei_\nu x/\ker_\nu x)$$

9.10.19 $\quad \ker_\nu x = N_\nu \cos \phi_\nu, \qquad \kei_\nu x = N_\nu \sin \phi_\nu$

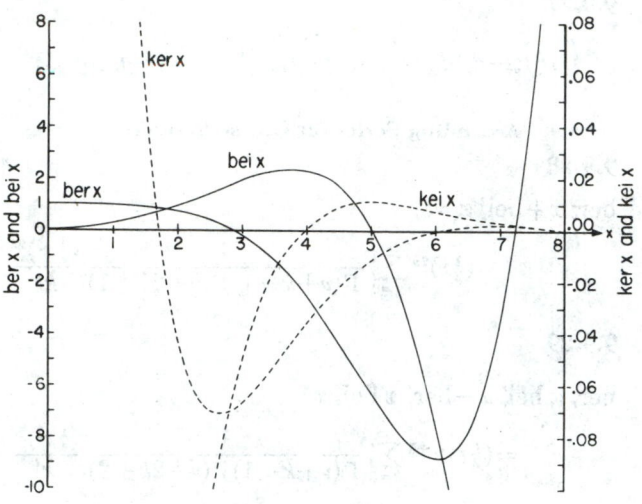

FIGURE 9.10. $\ber x$, $\bei x$, $\ker x$ and $\kei x$.

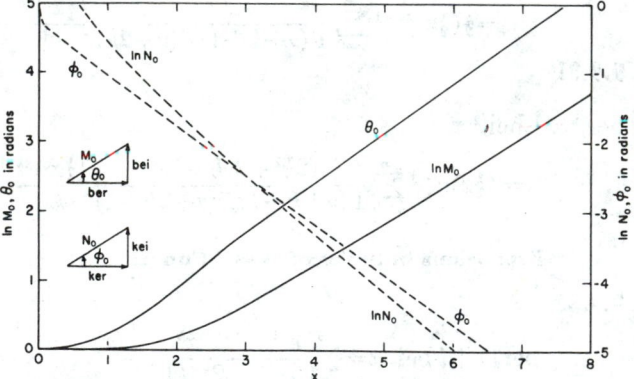

FIGURE 9.11. $\ln M_0(x)$, $\theta_0(x)$, $\ln N_0(x)$ and $\phi_0(x)$.

Equations **9.10.11** to **9.10.17** hold with the symbols b, M, θ replaced throughout by k, N, ϕ, respectively. In place of **9.10.10**

9.10.20 $\qquad N_{-\nu} = N_\nu, \qquad \phi_{-\nu} = \phi_\nu + \nu\pi$

Asymptotic Expansions of Modulus and Phase

When ν is fixed, x is large and $\mu=4\nu^2$

9.10.21

$$M_\nu=\frac{e^{x/\sqrt2}}{\sqrt{2\pi x}}\Big\{1-\frac{\mu-1}{8\sqrt2}\frac{1}{x}+\frac{(\mu-1)^2}{256}\frac{1}{x^2}$$
$$-\frac{(\mu-1)(\mu^2+14\mu-399)}{6144\sqrt2}\frac{1}{x^3}+O\Big(\frac{1}{x^4}\Big)\Big\}$$

9.10.22

$$\ln M_\nu=\frac{x}{\sqrt2}-\tfrac12\ln(2\pi x)-\frac{\mu-1}{8\sqrt2}\frac1x-\frac{(\mu-1)(\mu-25)}{384\sqrt2}\frac{1}{x^3}$$
$$-\frac{(\mu-1)(\mu-13)}{128}\frac{1}{x^4}+O\Big(\frac{1}{x^5}\Big)$$

9.10.23

$$\theta_\nu=\frac{x}{\sqrt2}+\Big(\frac12\nu-\frac18\Big)\pi+\frac{\mu-1}{8\sqrt2}\frac1x+\frac{\mu-1}{16}\frac{1}{x^2}$$
$$-\frac{(\mu-1)(\mu-25)}{384\sqrt2}\frac{1}{x^3}+O\Big(\frac{1}{x^5}\Big)$$

9.10.24

$$N_\nu=\sqrt{\frac{\pi}{2x}}\,e^{-x/\sqrt2}\Big\{1+\frac{\mu-1}{8\sqrt2}\frac1x+\frac{(\mu-1)^2}{256}\frac{1}{x^2}$$
$$+\frac{(\mu-1)(\mu^2+14\mu-399)}{6144\sqrt2}\frac{1}{x^3}+O\Big(\frac{1}{x^4}\Big)\Big\}$$

9.10.25

$$\ln N_\nu=-\frac{x}{\sqrt2}+\tfrac12\ln\Big(\frac{\pi}{2x}\Big)+\frac{\mu-1}{8\sqrt2}\frac1x+\frac{(\mu-1)(\mu-25)}{384\sqrt2}\frac{1}{x^3}$$
$$-\frac{(\mu-1)(\mu-13)}{128}\frac{1}{x^4}+O\Big(\frac{1}{x^5}\Big)$$

9.10.26

$$\phi_\nu=-\frac{x}{\sqrt2}-\Big(\frac12\nu+\frac18\Big)\pi-\frac{\mu-1}{8\sqrt2}\frac1x+\frac{\mu-1}{16}\frac{1}{x^2}$$
$$+\frac{(\mu-1)(\mu-25)}{384\sqrt2}\frac{1}{x^3}+O\Big(\frac{1}{x^5}\Big)$$

Asymptotic Expansions of Cross-Products

If x is large

9.10.27

$$\mathrm{ber}^2\,x+\mathrm{bei}^2\,x\sim\frac{e^{x\sqrt2}}{2\pi x}\Big(1+\frac{1}{4\sqrt2}\frac1x+\frac{1}{64}\frac{1}{x^2}$$
$$-\frac{33}{256\sqrt2}\frac{1}{x^3}-\frac{1797}{8192}\frac{1}{x^4}+\cdots\Big)$$

9.10.28

$$\mathrm{ber}\,x\,\mathrm{bei}'\,x-\mathrm{ber}'\,x\,\mathrm{bei}\,x\sim\frac{e^{x\sqrt2}}{2\pi x}\Big(\frac{1}{\sqrt2}+\frac18\frac1x$$
$$+\frac{9}{64\sqrt2}\frac{1}{x^2}+\frac{39}{512}\frac{1}{x^3}+\frac{75}{8192\sqrt2}\frac{1}{x^4}+\cdots\Big)$$

9.10.29

$$\mathrm{ber}\,x\,\mathrm{ber}'\,x+\mathrm{bei}\,x\,\mathrm{bei}'\,x\sim\frac{e^{x\sqrt2}}{2\pi x}\Big(\frac{1}{\sqrt2}-\frac38\frac1x$$
$$-\frac{15}{64\sqrt2}\frac{1}{x^2}-\frac{45}{512}\frac{1}{x^3}+\frac{315}{8192\sqrt2}\frac{1}{x^4}+\cdots\Big)$$

9.10.30

$$\mathrm{ber}'^2\,x+\mathrm{bei}'^2\,x\sim\frac{e^{x\sqrt2}}{2\pi x}\Big(1-\frac{3}{4\sqrt2}\frac1x+\frac{9}{64}\frac{1}{x^2}$$
$$+\frac{75}{256\sqrt2}\frac{1}{x^3}+\frac{2475}{8192}\frac{1}{x^4}+\cdots\Big)$$

9.10.31

$$\mathrm{ker}^2\,x+\mathrm{kei}^2\,x\sim\frac{\pi}{2x}e^{-x\sqrt2}\Big(1-\frac{1}{4\sqrt2}\frac1x+\frac{1}{64}\frac{1}{x^2}$$
$$+\frac{33}{256\sqrt2}\frac{1}{x^3}-\frac{1797}{8192}\frac{1}{x^4}+\cdots\Big)$$

9.10.32

$$\mathrm{ker}\,x\,\mathrm{kei}'\,x-\mathrm{ker}'\,x\,\mathrm{kei}\,x\sim-\frac{\pi}{2x}\,e^{-x\sqrt2}\Big(\frac{1}{\sqrt2}-\frac18\frac1x$$
$$+\frac{9}{64\sqrt2}\frac{1}{x^2}-\frac{39}{512}\frac{1}{x^3}+\frac{75}{8192\sqrt2}\frac{1}{x^4}+\cdots\Big)$$

9.10.33

$$\mathrm{ker}\,x\,\mathrm{ker}'\,x+\mathrm{kei}\,x\,\mathrm{kei}'\,x\sim-\frac{\pi}{2x}\,e^{-x\sqrt2}\Big(\frac{1}{\sqrt2}+\frac38\frac1x$$
$$-\frac{15}{64\sqrt2}\frac{1}{x^2}+\frac{45}{512}\frac{1}{x^3}+\frac{315}{8192\sqrt2}\frac{1}{x^4}+\cdots\Big)$$

9.10.34

$$\mathrm{ker}'^2\,x+\mathrm{kei}'^2\,x\sim\frac{\pi}{2x}\,e^{-x\sqrt2}\Big(1+\frac{3}{4\sqrt2}\frac1x+\frac{9}{64}\frac{1}{x^2}$$
$$-\frac{75}{256\sqrt2}\frac{1}{x^3}+\frac{2475}{8192}\frac{1}{x^4}+\cdots\Big)$$

Asymptotic Expansions of Large Zeros

Let

9.10.35

$$f(\delta)=\frac{\mu-1}{16\delta}+\frac{\mu-1}{32\delta^2}+\frac{(\mu-1)(5\mu+19)}{1536\delta^3}+\frac{3(\mu-1)^2}{512\delta^4}+\cdots$$

where $\mu=4\nu^2$. Then if s is a large positive integer

9.10.36

Zeros of $\mathrm{ber}_\nu\,x\sim\sqrt2\{\delta-f(\delta)\}$, $\qquad\delta=(s-\tfrac12\nu-\tfrac38)\pi$

Zeros of $\mathrm{bei}_\nu\,x\sim\sqrt2\{\delta-f(\delta)\}$, $\qquad\delta=(s-\tfrac12\nu+\tfrac18)\pi$

Zeros of $\mathrm{ker}_\nu\,x\sim\sqrt2\{\delta+f(-\delta)\}$, $\qquad\delta=(s-\tfrac12\nu-\tfrac58)\pi$

Zeros of $\mathrm{kei}_\nu\,x\sim\sqrt2\{\delta+f(-\delta)\}$, $\qquad\delta=(s-\tfrac12\nu-\tfrac18)\pi$

For $\nu=0$ these expressions give the sth zero of each function; for other values of ν the zeros represented may not be the sth.

Uniform Asymptotic Expansions for Large Orders

When ν is large and positive

9.10.37

$$\text{ber}_\nu(\nu x)+i \text{ bei}_\nu(\nu x) \sim$$

$$\frac{e^{\nu\xi}}{\sqrt{2\pi\nu\xi}}\left(\frac{xe^{3\pi i/4}}{1+\xi}\right)^\nu \left\{1+\sum_{k=1}^{\infty}\frac{u_k(\xi^{-1})}{\nu^k}\right\}$$

9.10.38

$$\text{ker}_\nu(\nu x)+i \text{ kei}_\nu(\nu x)$$

$$\sim \sqrt{\frac{\pi}{2\nu\xi}}\, e^{-\nu\xi}\left(\frac{xe^{3\pi i/4}}{1+\xi}\right)^{-\nu}\left\{1+\sum_{k=1}^{\infty}(-)^k\frac{u_k(\xi^{-1})}{\nu^k}\right\}$$

9.10.39

$$\text{ber}'_\nu(\nu x)+i \text{ bei}'_\nu(\nu x)$$

$$\sim \sqrt{\frac{\xi}{2\pi\nu}}\,\frac{e^{\nu\xi}}{x}\left(\frac{xe^{3\pi i/4}}{1+\xi}\right)^\nu\left\{1+\sum_{k=1}^{\infty}\frac{v_k(\xi^{-1})}{\nu^k}\right\}$$

9.10.40

$$\text{ker}'_\nu(\nu x)+i \text{ kei}'_\nu(\nu x)$$

$$\sim -\sqrt{\frac{\pi\xi}{2\nu}}\,\frac{e^{-\nu\xi}}{x}\left(\frac{xe^{3\pi i/4}}{1+\xi}\right)^{-\nu}\left\{1+\sum_{k=1}^{\infty}(-)^k\frac{v_k(\xi^{-1})}{\nu^k}\right\}$$

where

9.10.41

$$\xi=\sqrt{1+ix^2}$$

and $u_k(t)$, $v_k(t)$ are given by **9.3.9** and **9.3.13**. All fractional powers take their principal values.

9.11. Polynomial Approximations

9.11.1 $-8\leq x\leq 8$

$$\text{ber }x=1-64(x/8)^4+113.77777\ 774(x/8)^8$$
$$-32.36345\ 652(x/8)^{12}+2.64191\ 397(x/8)^{16}$$
$$-.08349\ 609(x/8)^{20}+.00122\ 552(x/8)^{24}$$
$$-.00000\ 901(x/8)^{28}+\epsilon$$

$$|\epsilon|<1\times10^{-9}$$

9.11.2 $-8\leq x\leq 8$

$$\text{bei }x=16(x/8)^2-113.77777\ 774(x/8)^6$$
$$+72.81777\ 742(x/8)^{10}-10.56765\ 779(x/8)^{14}$$
$$+.52185\ 615(x/8)^{18}-.01103\ 667(x/8)^{22}$$
$$+.00011\ 346(x/8)^{26}+\epsilon$$

$$|\epsilon|<6\times10^{-9}$$

9.11.3 $0<x\leq 8$

$$\text{ker }x=-\ln(\tfrac{1}{2}x)\,\text{ber }x+\tfrac{1}{4}\pi\,\text{bei }x-.57721\ 566$$
$$-59.05819\ 744(x/8)^4+171.36272\ 133(x/8)^8$$
$$-60.60977\ 451(x/8)^{12}+5.65539\ 121(x/8)^{16}$$
$$-.19636\ 347(x/8)^{20}+.00309\ 699(x/8)^{24}$$
$$-.00002\ 458(x/8)^{28}+\epsilon$$

$$|\epsilon|<1\times10^{-8}$$

9.11.4 $0<x\leq 8$

$$\text{kei }x=-\ln(\tfrac{1}{2}x)\text{bei }x-\tfrac{1}{4}\pi\,\text{ber }x+6.76454\ 936(x/8)^2$$
$$-142.91827\ 687(x/8)^6+124.23569\ 650(x/8)^{10}$$
$$-21.30060\ 904(x/8)^{14}+1.17509\ 064(x/8)^{18}$$
$$-.02695\ 875(x/8)^{22}+.00029\ 532(x/8)^{26}+\epsilon$$

$$|\epsilon|<3\times10^{-9}$$

9.11.5 $-8\leq x\leq 8$

$$\text{ber}'\,x=x[-4(x/8)^2+14.22222\ 222(x/8)^6$$
$$-6.06814\ 810(x/8)^{10}+.66047\ 849(x/8)^{14}$$
$$-.02609\ 253(x/8)^{18}+.00045\ 957(x/8)^{22}$$
$$-.00000\ 394(x/8)^{26}]+\epsilon$$

$$|\epsilon|<2.1\times10^{-8}$$

9.11.6 $-8\leq x\leq 8$

$$\text{bei}'\,x=x[\tfrac{1}{2}-10.66666\ 666(x/8)^4$$
$$+11.37777\ 772(x/8)^8-2.31167\ 514(x/8)^{12}$$
$$+.14677\ 204(x/8)^{16}-.00379\ 386(x/8)^{20}$$
$$+.00004\ 609(x/8)^{24}]+\epsilon$$

$$|\epsilon|<7\times10^{-8}$$

9.11.7 $0<x\leq 8$

$$\text{ker}'\,x=-\ln(\tfrac{1}{2}x)\,\text{ber}'\,x-x^{-1}\,\text{ber }x+\tfrac{1}{4}\pi\,\text{bei}'\,x$$
$$+x[-3.69113\ 734(x/8)^2+21.42034\ 017(x/8)^6$$
$$-11.36433\ 272(x/8)^{10}+1.41384\ 780(x/8)^{14}$$
$$-.06136\ 358(x/8)^{18}+.00116\ 137(x/8)^{22}$$
$$-.00001\ 075(x/8)^{26}]+\epsilon$$

$$|\epsilon|<8\times10^{-8}$$

9.11.8 $$0 < x \leq 8$$

$$\text{kei}' \, x = -\ln \, (\tfrac{1}{2}x) \, \text{bei}' \, x - x^{-1} \, \text{bei} \, x - \tfrac{1}{4}\pi \, \text{ber}' \, x$$
$$+ x[.21139 \, 217 - 13.39858 \, 846(x/8)^4$$
$$+ 19.41182 \, 758(x/8)^8 - 4.65950 \, 823(x/8)^{12}$$
$$+ .33049 \, 424(x/8)^{16} - .00926 \, 707(x/8)^{20}$$
$$+ .00011 \, 997(x/8)^{24}] + \epsilon$$
$$|\epsilon| < 7 \times 10^{-8}$$

9.11.9 $$8 \leq x < \infty$$

$$\text{ker} \, x + i \, \text{kei} \, x = f(x)(1 + \epsilon_1)$$
$$f(x) = \sqrt{\frac{\pi}{2x}} \exp\left[-\frac{1+i}{\sqrt{2}} \, x + \theta(-x)\right]$$
$$|\epsilon_1| < 1 \times 10^{-7}$$

9.11.10 $$8 \leq x < \infty$$

$$\text{ber} \, x + i \, \text{bei} \, x - \frac{i}{\pi} \, (\text{ker} \, x + i \, \text{kei} \, x) = g(x)(1 + \epsilon_2)$$
$$g(x) = \frac{1}{\sqrt{2\pi x}} \exp\left[\frac{1+i}{\sqrt{2}} \, x + \theta(x)\right]$$
$$|\epsilon_2| < 3 \times 10^{-7}$$

where

9.11.11

$$\theta(x) = (.00000 \, 00 - .39269 \, 91i)$$
$$+ (.01104 \, 86 - .01104 \, 85i)(8/x)$$
$$+ (.00000 \, 00 - .00097 \, 65i)(8/x)^2$$
$$+ (-.00009 \, 06 - .00009 \, 01i)(8/x)^3$$
$$+ (-.00002 \, 52 + .00000 \, 00i)(8/x)^4$$
$$+ (-.00000 \, 34 + .00000 \, 51i)(8/x)^5$$
$$+ (.00000 \, 06 + .00000 \, 19i)(8/x)^6$$

9.11.12 $$8 \leq x < \infty$$

$$\text{ker}' \, x + i \, \text{kei}' \, x = -f(x)\phi(-x)(1 + \epsilon_3)$$
$$|\epsilon_3| < 2 \times 10^{-7}$$

9.11.13 $$8 \leq x < \infty$$

$$\text{ber}' \, x + i \, \text{bei}' \, x - \frac{i}{\pi} \, (\text{ker}' \, x + i \, \text{kei}' \, x) = g(x)\phi(x)(1 + \epsilon_4)$$
$$|\epsilon_4| < 3 \times 10^{-7}$$

where

9.11.14

$$\phi(x) = (.70710 \, 68 + .70710 \, 68i)$$
$$+ (-.06250 \, 01 - .00000 \, 01i)(8/x)$$
$$+ (-.00138 \, 13 + .00138 \, 11i)(8/x)^2$$
$$+ (.00000 \, 05 + .00024 \, 52i)(8/x)^3$$
$$+ (.00003 \, 46 + .00003 \, 38i)(8/x)^4$$
$$+ (.00001 \, 17 - .00000 \, 24i)(8/x)^5$$
$$+ (.00000 \, 16 - .00000 \, 32i)(8/x)^6$$

Numerical Methods

9.12. Use and Extension of the Tables

Example 1. To evaluate $J_n(1.55)$, $n = 0, 1, 2, . . .$, each to 5 decimals.

The recurrence relation

$$J_{n-1}(x) + J_{n+1}(x) = (2n/x)J_n(x)$$

can be used to compute $J_0(x)$, $J_1(x)$, $J_2(x)$, . . ., successively provided that $n < x$, otherwise severe accumulation of rounding errors will occur. Since, however, $J_n(x)$ is a decreasing function of n when $n > x$, recurrence can always be carried out in the direction of decreasing n.

Inspection of **Table 9.2** shows that $J_n(1.55)$ vanishes to 5 decimals when $n > 7$. Taking arbitrary values zero for J_9 and unity for J_8, we compute by recurrence the entries in the second column of the following table, rounding off to the nearest integer at each step.

n	*Trial values*	$J_n(1.55)$
9	0	.00000
8	1	.00000
7	10	.00003
6	89	.00028
5	679	.00211
4	4292	.01331
3	21473	.06661
2	78829	.24453
1	181957	.56442
0	155954	.48376

We normalize the results by use of the equation **9.1.46**, namely

$$J_0(x) + 2J_2(x) + 2J_4(x) + . . . = 1$$

This yields the normalization factor

$$1/322376 = .00000 \, 31019 \, 7$$

and multiplying the trial values by this factor we obtain the required results, given in the third column. As a check we may verify the value of $J_0(1.55)$ by interpolation in **Table 9.1.**

Remarks. (i) In this example it was possible to estimate immediately the value of $n=N$, say, at which to begin the recurrence. This may not always be the case and an arbitrary value of N may have to be taken. The number of correct significant figures in the final values is the same as the number of digits in the respective trial values. If the chosen N is too small the trial values will have too few digits and insufficient accuracy is obtained in the results. The calculation must then be repeated taking a higher value. On the other hand if N were too large unnecessary effort would be expended. This could be offset to some extent by discarding significant figures in the trial values which are in excess of the number of decimals required in J_n.

(ii) If we had required, say, $J_0(1.55)$, $J_1(1.55)$, . . ., $J_{10}(1.55)$, each to 5 significant figures, we would have found the values of $J_{10}(1.55)$ and $J_{11}(1.55)$ to 5 significant figures by interpolation in **Table 9.3** and then computed by recurrence $J_9, J_8, \ldots, J_0$, no normalization being required. Alternatively, we could begin the recurrence at a higher value of N and retain only 5 significant figures in the trial values for $n \leq 10$.

(iii) Exactly similar methods can be used to compute the modified Bessel function $I_n(x)$ by means of the relations **9.6.26** and **9.6.36.** If x is large, however, considerable cancellation will take place in using the latter equation, and it is preferable to normalize by means of **9.6.37.**

Example 2. To evaluate $Y_n(1.55)$, $n=0, 1, 2,$ . . ., 10, each to 5 significant figures.

The recurrence relation

$$Y_{n-1}(x) + Y_{n+1}(x) = (2n/x) Y_n(x)$$

can be used to compute $Y_n(x)$ in the direction of increasing n both for $n<x$ and $n>x$, because in the latter event $Y_n(x)$ is a numerically increasing function of n.

We therefore compute $Y_0(1.55)$ and $Y_1(1.55)$ by interpolation in **Table 9.1,** generate $Y_2(1.55)$, $Y_3(1.55)$, . . ., $Y_{10}(1.55)$ by recurrence and check $Y_{10}(1.55)$ by interpolation in **Table 9.3.**

n	$Y_n(1.55)$	n	$Y_n(1.55)$
0	$+0.40225$	6	-1.9917×10^2
1	-0.37970	7	-1.5100×10^3
2	-0.89218	8	-1.3440×10^4
3	-1.9227	9	-1.3722×10^5
4	-6.5505	10	-1.5801×10^6
5	-31.886		

Remarks. (i) An alternative way of computing $Y_0(x)$, should $J_0(x)$, $J_2(x)$, $J_4(x)$, . . ., be available (see **Example 1**), is to use formula **9.1.89.** The other starting value for the recurrence, $Y_1(x)$, can then be found from the Wronskian relation $J_1(x) Y_0(x) - J_0(x) Y_1(x) = 2/(\pi x)$. This is a convenient procedure for use with an automatic computer.

(ii) Similar methods can be used to compute the modified Bessel function $K_n(x)$ by means of the recurrence relation **9.6.26** and the relation **9.6.54,** except that if x is large severe cancellation will occur in the use of **9.6.54** and other methods for evaluating $K_0(x)$ may be preferable, for example, use of the asymptotic expansion **9.7.2** or the polynomial approximation **9.8.6.**

Example 3. To evaluate $J_0(.36)$ and $Y_0(.36)$ each to 5 decimals, using the multiplication theorem.

From **9.1.74** we have

$$\mathscr{C}_0(\lambda z) = \sum_{k=0}^{\infty} a_k \mathscr{C}_k(z), \text{ where } a_k = \frac{(-)^k(\lambda^2-1)^k(\frac{1}{2}z)^k}{k!}.$$

We take $z=.4$. Then $\lambda=.9$, $(\lambda^2-1)(\frac{1}{2}z)=-.038$, and extracting the necessary values of $J_k(.4)$ and $Y_k(.4)$ from **Tables 9.1** and **9.2,** we compute the required results as follows:

k	a_k	$a_k J_k(.4)$	$a_k Y_k(.4)$
0	$+1.0$	$+.96040$	$-.60602$
1	$+0.038$	$+.00745$	$-.06767$
2	$+0.7220 \times 10^{-3}$	$+.00001$	$-.00599$
3	$+0.914 \times 10^{-5}$		$-.00074$
4	$+0.87 \times 10^{-7}$		$-.00011$
5	$+0.7 \times 10^{-9}$		$-.00002$

$$J_0(.36) = +.96786 \qquad Y_0(.36) = -.68055$$

Remark. This procedure is equivalent to interpolating by means of the Taylor series

$$\mathscr{C}_0(z+h) = \sum_{k=0}^{\infty} \frac{h^k}{k!} \mathscr{C}_0^{(k)}(z)$$

at $z=.4$, and expressing the derivatives $\mathscr{C}_0^{(k)}(z)$ in terms of $\mathscr{C}_k(z)$ by means of the recurrence relations and differential equation for the Bessel functions.

Example 4. To evaluate $J_\nu(x)$, $J_\nu'(x)$, $Y_\nu(x)$ and $Y_\nu'(x)$ for $\nu=50$, $x=75$, each to 6 decimals.

We use the asymptotic expansions **9.3.35, 9.3.36, 9.3.43,** and **9.3.44.** Here $z=x/\nu=3/2$. From **9.3.39** we find

$$\frac{2}{3}(-\zeta)^{3/2} = \frac{1}{2}\sqrt{5} - \arccos\frac{2}{3} = +.2769653.$$

Hence

$$\zeta = -.5567724 \text{ and } \left(\frac{4\zeta}{1-z^2}\right)^{1/4} = +1.155332.$$

Next,

$$\nu^{1/3} = 3.684031, \qquad \nu^{2/3}\zeta = -7.556562.$$

Interpolating in **Table 10.11**, we find that

$$\text{Ai}(\nu^{2/3}\zeta) = +.299953, \qquad \text{Ai}'(\nu^{2/3}\zeta) = +.451441,$$

$$\text{Bi}(\nu^{2/3}\zeta) = -.160565, \qquad \text{Bi}'(\nu^{2/3}\zeta) = +.819542.$$

As a check on the interpolation, we may verify that $\text{Ai Bi}' - \text{Ai}'\text{Bi} = 1/\pi$.

Interpolating in the table following **9.3.46** we obtain

$$b_0(\zeta) = +.0136, \qquad c_0(\zeta) = +.1442.$$

The contributions of the terms involving $a_1(\zeta)$ and $d_1(\zeta)$ are negligible, and substituting in the asymptotic expansions we find that

$$J_{50}(75) = +1.155332(50^{-1/3} \times .299953$$
$$+ 50^{-5/3} \times .451441 \times .0136) = +.094077,$$

$$J_{50}'(75) = -(4/3)(1.155332)^{-1}(50^{-4/3} \times .299953$$
$$\times .1442 + 50^{-2/3} \times .451441) = -.038658,$$

$$Y_{50}(75) = -1.155332(-50^{-1/3} \times .160565$$
$$+ 50^{-5/3} \times .819542 \times .0136) = +.050335,$$

$$Y_{50}'(75) = +(4/3)(1.155332)^{-1}(-50^{-4/3} \times .160565$$
$$\times .1442 + 50^{-2/3} \times .819542) = +.069543.$$

As a check we may verify that

$$JY' - J'Y = 2/(75\pi).$$

Remarks. This example may also be computed using the Debye expansions **9.3.15, 9.3.16, 9.3.19,** and **9.3.20.** Four terms of each of these series are required, compared with two in the computations above. The closer the argument-order ratio is to unity, the less effective the Debye expansions become. In the neighborhood of unity the expansions **9.3.23, 9.3.24, 9.3.27,** and **9.3.28** will furnish results of moderate accuracy; for high-accuracy work the uniform expansions should again be used.

Example 5. To evaluate the 5th positive zero of $J_{10}(x)$ and the corresponding value of $J_{10}'(x)$, each to 5 decimals.

We use the asymptotic expansions **9.5.22** and **9.5.23** setting $\nu = 10$, $s = 5$. From **Table 10.11**

we find

$$a_5 = -7.944134, \qquad \text{Ai}'(a_5) = +.947336.$$

Hence

$$\zeta = 10^{-2/3}a_5 = .21544347a_5 = -1.7115118.$$

Interpolating in the table following **9.5.26** we obtain

$$z(\zeta) = +2.888631, \qquad h(\zeta) = +.98259,$$
$$f_1(\zeta) = +.0107, \qquad F_1(\zeta) = -.001.$$

The bounds given at the foot of the table show that the contributions of higher terms to the asymptotic series are negligible. Hence

$$j_{10,5} = 28.88631 + .00107 + \ldots = 28.88738,$$

$$J_{10}'(j_{10,5}) = -\frac{2}{10^{2/3}} \frac{.947336}{2.888631 \times .98259}$$
$$\times (1 - .00001 + \ldots) = -.14381.$$

Example 6. To evaluate the first root of $J_0(x)Y_0(\lambda x) - Y_0(x)J_0(\lambda x) = 0$ for $\lambda = \frac{3}{2}$ to 4 significant figures.

Let $\alpha_\lambda^{(1)}$ denote the root. Direct interpolation in **Table 9.7** is impracticable owing to the divergence of the differences. Inspection of **9.5.28** suggests that a smoother function is $(\lambda - 1)\alpha_\lambda^{(1)}$. Using **Table 9.7** we compute the following values

$1/\lambda$	$(\lambda-1)\alpha_\lambda^{(1)}$	δ	δ^2
0.4	3.110		
		+21	
0.6	3.131		−12
		+9	
0.8	3.140		−7
		+2	
1.0	3.142 (π)		

Interpolating for $1/\lambda = .667$, we obtain $(\lambda - 1)\alpha_\lambda^{(1)} = 3.134$ and thence the required root $\alpha_{1.5}^{(1)} = 6.268$.

Example 7. To evaluate $\text{ber}_n 1.55$, $\text{bei}_n 1.55$, $n = 0, 1, 2, \ldots$, each to 5 decimals.

We use the recurrence relation

$$J_{n-1}(xe^{3\pi i/4}) + J_{n+1}(xe^{3\pi i/4})$$
$$= -\frac{n\sqrt{2}}{x}(1+i)J_n(xe^{3\pi i/4}),$$

taking arbitrary values zero for $J_9(xe^{3\pi i/4})$ and $1 + 0i$ for $J_8(xe^{3\pi i/4})$ (see **Example 1**).

n	Real trial values	Imag. trial values	$\mathrm{ber}_n x$	$\mathrm{bei}_n x$
9	0	0	.00000	.00000
8	+1	0	.00000	.00000
7	−7	−7	−.00002	−.00003
6	−1	+89	−.00003	+.00030
5	+500	−475	+.00181	−.00148
4	−4447	−203	−.01494	−.00180
3	+14989	+17446	+.04614	+.06258
2	+11172	−88578	+.05994	−.29580
1	−197012	+123804	−.69531	+.36781
0	+281539	+155373	+.91004	+.59461
Σ	+106734	+207449	+.30763	+.72619

The values of $\mathrm{ber}_n x$ and $\mathrm{bei}_n x$ are computed by multiplication of the trial values by the normalizing factor

$$1/(294989 - 22011i) = (.337119 + .025155i) \times 10^{-5},$$

obtained from the relation

$$J_0(xe^{3\pi i/4}) + 2J_2(xe^{3\pi i/4}) + 2J_4(xe^{3\pi i/4}) + \ldots = 1.$$

Adequate checks are furnished by interpolating in **Table 9.12** for ber 1.55 and bei 1.55, and the use of a simple sum check on the normalization.

Should $\mathrm{ker}_n x$ and $\mathrm{kei}_n x$ be required they can be computed by forward recurrence using formulas **9.9.14**, taking the required starting values for $n = 0$ and 1 from **Table 9.12** (see **Example 2**). If an independent check on the recurrence is required the asymptotic expansion **9.10.38** can be used.

References

Texts

[9.1] E. E. Allen, Analytical approximations, Math. Tables Aids Comp. **8**, 240–241 (1954).

[9.2] E. E. Allen, Polynomial approximations to some modified Bessel functions, Math. Tables Aids Comp. **10**, 162–164 (1956).

[9.3] H. Bateman and R. C. Archibald, A guide to tables of Bessel functions, Math. Tables Aids Comp. **1**, 205–308 (1944).

[9.4] W. G. Bickley, Bessel functions and formulae (Cambridge Univ. Press, Cambridge, England, 1953). This is a straight reprint of part of the preliminaries to [9.21].

[9.5] H. S. Carslaw and J. C. Jaeger, Conduction of heat in solids (Oxford Univ. Press, London, England, 1947).

[9.6] E. T. Copson, An introduction to the theory of functions of a complex variable (Oxford Univ. Press, London, England, 1935).

[9.7] A. Erdélyi et al., Higher transcendental functions, vol 2, ch. 7 (McGraw-Hill Book Co., Inc., New York, N.Y., 1953).

[9.8] E. T. Goodwin, Recurrence relations for cross-products of Bessel functions, Quart. J. Mech. Appl. Math. **2**, 72–74 (1949).

[9.9] A. Gray, G. B. Mathews and T. M. MacRobert, A treatise on the theory of Bessel functions, 2d ed. (Macmillan and Co., Ltd., London, England, 1931).

[9.10] W. Magnus and F. Oberhettinger, Formeln und Sätze für die speziellen Funktionen der mathematischen Physik, 2d ed. (Springer-Verlag, Berlin, Germany, 1948).

[9.11] N. W. McLachlan, Bessel functions for engineers, 2d ed. (Clarendon Press, Oxford, England, 1955).

[9.12] F. W. J. Olver, Some new asymptotic expansions for Bessel functions of large orders. Proc. Cambridge Philos. Soc. **48**, 414–427 (1952).

[9.13] F. W. J. Olver, The asymptotic expansion of Bessel functions of large order. Philos. Trans. Roy. Soc. London **A247**, 328–368 (1954).

[9.14] G. Petiau, La théorie des fonctions de Bessel (Centre National de la Recherche Scientifique, Paris, France, 1955).

[9.15] G. N. Watson, A treatise on the theory of Bessel functions, 2d ed. (Cambridge Univ. Press, Cambridge, England, 1958).

[9.16] R. Weyrich, Die Zylinderfunktionen und ihre Anwendungen (B. G. Teubner, Leipzig, Germany, 1937).

[9.17] C. S. Whitehead, On a generalization of the functions ber x, bei x, ker x, kei x. Quart. J. Pure Appl. Math. **42**, 316–342 (1911).

[9.18] E. T. Whittaker and G. N. Watson, A course of modern analysis, 4th ed. (Cambridge Univ. Press, Cambridge, England, 1952).

Tables

[9.19] J. F. Bridge and S. W. Angrist, An extended table of roots of $J_n'(x) Y_n'(\beta x) - J_n'(\beta x) Y_n'(x) = 0$, Math. Comp. **16**, 198–204 (1962).

[9.20] British Association for the Advancement of Science, Bessel functions, Part I. Functions of orders zero and unity, Mathematical Tables, vol. VI (Cambridge Univ. Press, Cambridge, England, 1950).

[9.21] British Association for the Advancement of Science, Bessel functions, Part II. Functions of positive integer order, Mathematical Tables, vol. X (Cambridge Univ. Press, Cambridge, England, 1952).

[9.22] British Association for the Advancement of Science, Annual Report (J. R. Airey), 254 (1927).

[9.23] E. Cambi, Eleven- and fifteen-place tables of Bessel functions of the first kind, to all significant orders (Dover Publications, Inc., New York, N.Y., 1948).

[9.24] E. A. Chistova, Tablitsy funktsii Besselya ot deistvitel'nogo argumenta i integralov ot nikh (Izdat. Akad. Nauk SSSR, Moscow, U.S.S.R., 1958). (Table of Bessel functions with real argument and their integrals).

[9.25] H. B. Dwight, Tables of integrals and other mathematical data (The Macmillan Co., New York, N.Y., 1957).
This includes formulas for, and tables of Kelvin functions.

[9.26] H. B. Dwight, Table of roots for natural frequencies in coaxial type cavities, J. Math. Phys. **27**, 84–89 (1948).
This gives zeros of the functions **9.5.27** and **9.5.30** for $n=0,1,2,3$.

[9.27] V. N. Faddeeva and M. K. Gavurin, Tablitsy funktsii Besselia $J_n(x)$ tselykh nomerov ot 0 do 120 (Izdat. Akad. Nauk SSSR, Moscow, U.S.S.R., 1950). (Table of $J_n(x)$ for orders 0 to 120).

[9.28] L. Fox, A short table for Bessel functions of integer orders and large arguments. Royal Society Shorter Mathematical Tables No. 3 (Cambridge Univ. Press, Cambridge, England, 1954).

[9.29] E. T. Goodwin and J. Staton, Table of $J_0(j_{0,n}r)$, Quart. J. Mech. Appl. Math. **1**, 220–224 (1948).

[9.30] Harvard Computation Laboratory, Tables of the Bessel functions of the first kind of orders 0 through 135, vols. 3–14 (Harvard Univ. Press, Cambridge, Mass., 1947–1951).

[9.31] K. Hayashi, Tafeln der Besselschen, Theta, Kugel- und anderer Funktionen (Springer, Berlin, Germany, 1930).

[9.32] E. Jahnke, F. Emde, and F. Lösch, Tables of higher functions, ch. IX, 6th ed. (McGraw-Hill Book Co., Inc., New York, N.Y., 1960).

[9.33] L. N. Karmazina and E. A. Chistova, Tablitsy funktsii Besselya ot mnimogo argumenta i integralov ot nikh (Izdat. Akad. Nauk SSSR, Moscow, U.S.S.R., 1958). (Tables of Bessel functions with imaginary argument and their integrals).

[9.34] Mathematical Tables Project, Table of $f_n(x)=n!(\tfrac{1}{2}x)^{-n}J_n(x)$. J. Math. Phys. **23**, 45–60 (1944).

[9.35] National Bureau of Standards, Table of the Bessel functions $J_0(z)$ and $J_1(z)$ for complex arguments, 2d ed. (Columbia Univ. Press, New York, N.Y., 1947).

[9.36] National Bureau of Standards, Tables of the Bessel functions $Y_0(z)$ and $Y_1(z)$ for complex arguments (Columbia Univ. Press, New York, N.Y., 1950).

[9.37] National Physical Laboratory Mathematical Tables, vol. 5, Chebyshev series for mathematical functions, by C. W. Clenshaw (Her Majesty's Stationery Office, London, England, 1902).

[9.38] National Physical Laboratory Mathematical Tables, vol. 6, Tables for Bessel functions of moderate or large orders, by F. W. J. Olver (Her Majesty's Stationery Office, London, England, 1962).

[9.39] L. N. Nosova, Tables of Thomson (Kelvin) functions and their first derivatives, translated from the Russian by P. Basu (Pergamon Press, New York, N.Y., 1961).

[9.40] Royal Society Mathematical Tables, vol. 7, Bessel functions, Part III. Zeros and associated values, edited by F. W. J. Olver (Cambridge Univ. Press, Cambridge, England, 1960).
The introduction includes many formulas connected with zeros.

[9.41] Royal Society Mathematical Tables, vol. 10, Bessel functions, Part IV. Kelvin functions, by A. Young and A. Kirk (Cambridge Univ. Press, Cambridge, England, 1963).
The introduction includes many formulas for Kelvin functions.

[9.42] W. Sibagaki, 0.01 % tables of modified Bessel functions, with the account of the methods used in the calculation (Baifukan, Tokyo, Japan, 1955).

Table 9.1   BESSEL FUNCTIONS—ORDERS 0, 1 AND 2

x	$J_0(x)$	$J_1(x)$	$J_2(x)$
0.0	1.00000 00000 00000	0.00000 00000	0.00000 00000
0.1	0.99750 15620 66040	0.04993 75260	0.00124 89587
0.2	0.99002 49722 39576	0.09950 08326	0.00498 33542
0.3	0.97762 62465 38296	0.14831 88163	0.01116 58619
0.4	0.96039 82266 59563	0.19602 65780	0.01973 46631
0.5	0.93846 98072 40813	0.24226 84577	0.03060 40235
0.6	0.91200 48634 97211	0.28670 09881	0.04366 50967
0.7	0.88120 08886 07405	0.32899 57415	0.05878 69444
0.8	0.84628 73527 50480	0.36884 20461	0.07581 77625
0.9	0.80752 37981 22545	0.40594 95461	0.09458 63043
1.0	0.76519 76865 57967	0.44005 05857	0.11490 34849
1.1	0.71962 20185 27511	0.47090 23949	0.13656 41540
1.2	0.67113 27442 64363	0.49828 90576	0.15934 90183
1.3	0.62008 59895 61509	0.52202 32474	0.18302 66988
1.4	0.56685 51203 74289	0.54194 77139	0.20735 58995
1.5	0.51182 76717 35918	0.55793 65079	0.23208 76721
1.6	0.45540 21676 39381	0.56989 59353	0.25696 77514
1.7	0.39798 48594 46109	0.57776 52315	0.28173 89424
1.8	0.33998 64110 42558	0.58151 69517	0.30614 35353
1.9	0.28181 85593 74385	0.58115 70727	0.32992 57277
2.0	0.22389 07791 41236	0.57672 48078	0.35283 40286
2.1	0.16660 69803 31990	0.56829 21358	0.37462 36252
2.2	0.11036 22669 22174	0.55596 30498	0.39505 86875
2.3	0.05553 97844 45602	0.53987 25326	0.41391 45917
2.4	+0.00250 76832 97244	0.52018 52682	0.43098 00402
2.5	−0.04838 37764 68198	0.49709 41025	0.44605 90584
2.6	−0.09680 49543 97038	0.47081 82665	0.45897 28517
2.7	−0.14244 93700 46012	0.44160 13791	0.46956 15027
2.8	−0.18503 60333 64387	0.40970 92469	0.47768 54954
2.9	−0.22431 15457 91968	0.37542 74818	0.48322 70505
3.0	−0.26005 19549 01933	0.33905 89585	0.48609 12606
3.1	−0.29206 43476 50698	0.30092 11331	0.48620 70142
3.2	−0.32018 81696 57123	0.26134 32488	0.48352 77001
3.3	−0.34429 62603 98885	0.22066 34530	0.47803 16865
3.4	−0.36429 55967 62000	0.17922 58517	0.46972 25683
3.5	−0.38012 77399 87263	0.13737 75274	0.45862 91842
3.6	−0.39176 89837 00798	0.09546 55472	0.44480 53988
3.7	−0.39923 02033 71191	0.05383 39877	0.42832 96562
3.8	−0.40255 64101 78564	+0.01282 10029	0.40930 43065
3.9	−0.40182 60148 87640	−0.02724 40396	0.38785 47125
4.0	−0.39714 98098 63847	−0.06604 33280	0.36412 81459
4.1	−0.38866 96798 35854	−0.10327 32577	0.33829 24809
4.2	−0.37655 70543 67568	−0.13864 69421	0.31053 47010
4.3	−0.36101 11172 36535	−0.17189 65602	0.28105 92288
4.4	−0.34225 67900 03886	−0.20277 55219	0.25008 60982
4.5	−0.32054 25089 85121	−0.23106 04319	0.21784 89837
4.6	−0.29613 78165 74141	−0.25655 28361	0.18459 31052
4.7	−0.26933 07894 19753	−0.27908 07358	0.15057 30295
4.8	−0.24042 53272 91183	−0.29849 98581	0.11605 03864
4.9	−0.20973 83275 85326	−0.31469 46710	0.08129 15231
5.0	−0.17759 67713 14338	−0.32757 91376	0.04656 51163
	$\begin{bmatrix} (-4)6 \\ 11 \end{bmatrix}$	$\begin{bmatrix} (-4)5 \\ 8 \end{bmatrix}$	$\begin{bmatrix} (-4)3 \\ 7 \end{bmatrix}$

$$J_{n+1}(x) = \frac{2n}{x} J_n(x) - J_{n-1}(x)$$

Compiled from British Association for the Advancement of Science, Bessel functions, Part II. Functions of positive integer order, Mathematical Tables, vol. X (Cambridge Univ. Press, Cambridge, England, 1952) and Harvard Computation Laboratory, Tables of the Bessel functions of the first kind of orders 0 through 135, vols. 3–14 (Harvard Univ. Press, Cambridge, Mass., 1947–1951) (with permission).

BESSEL FUNCTIONS—ORDERS 0, 1 AND 2   Table 9.1

x	$Y_0(x)$	$Y_1(x)$	$Y_2(x)$
0.0	$-\infty$	$-\infty$	$-\infty$
0.1	-1.53423 86514	-6.45895 10947	-127.64478 324
0.2	-1.08110 53224	-3.32382 49881	- 32.15714 456
0.3	-0.80727 35778	-2.29310 51384	- 14.48009 401
0.4	-0.60602 45684	-1.78087 20443	- 8.29833 565
0.5	-0.44451 87335	-1.47147 23927	- 5.44137 084
0.6	-0.30850 98701	-1.26039 13472	- 3.89279 462
0.7	-0.19066 49293	-1.10324 98719	- 2.96147 756
0.8	-0.08680 22797	-0.97814 41767	- 2.35855 816
0.9	+0.00562 83066	-0.87312 65825	- 1.94590 960
1.0	0.08025 69642	-0.78121 28213	- 1.65068 261
1.1	0.16216 32029	-0.69811 95601	- 1.43147 149
1.2	0.22808 35032	-0.62113 63797	- 1.26331 080
1.3	0.28653 53572	-0.54851 97300	- 1.13041 186
1.4	0.33789 51297	-0.47914 69742	- 1.02239 081
1.5	0.38244 89238	-0.41230 86270	- 0.93219 376
1.6	0.42042 68964	-0.34757 80083	- 0.85489 941
1.7	0.45202 70002	-0.28472 62451	- 0.78699 905
1.8	0.47743 17149	-0.22366 48682	- 0.72594 824
1.9	0.49681 99713	-0.16440 57723	- 0.66987 868
2.0	0.51037 56726	-0.10703 24315	- 0.61740 810
2.1	0.51829 37375	-0.05167 86121	- 0.56751 146
2.2	0.52078 42854	+0.00148 77893	- 0.51943 175
2.3	0.51807 53962	0.05227 73158	- 0.47261 686
2.4	0.51041 47487	0.10048 89383	- 0.42667 397
2.5	0.49807 03596	0.14591 81380	- 0.38133 585
2.6	0.48133 05906	0.18836 35444	- 0.33643 556
2.7	0.46050 35491	0.22763 24459	- 0.29188 692
2.8	0.43591 59856	0.26354 53936	- 0.24766 928
2.9	0.40791 17692	0.29594 00546	- 0.20381 518
3.0	0.37685 00100	0.32467 44248	- 0.16040 039
3.1	0.34310 28894	0.34962 94823	- 0.11753 548
3.2	0.30705 32501	0.37071 13384	- 0.07535 866
3.3	0.26909 19951	0.38785 29310	- 0.03402 961
3.4	0.22961 53372	0.40101 52921	+ 0.00627 601
3.5	0.18902 19439	0.41018 84179	0.04537 144
3.6	0.14771 00126	0.41539 17621	0.08306 319
3.7	0.10607 43153	0.41667 43727	0.11915 508
3.8	0.06450 32467	0.41411 46893	0.15345 185
3.9	+0.02337 59082	0.40782 00193	0.18576 256
4.0	-0.01694 07393	0.39792 57106	0.21590 359
4.1	-0.05609 46266	0.38459 40348	0.24370 147
4.2	-0.09375 12013	0.36801 28079	0.26899 540
4.3	-0.12959 59029	0.34839 37583	0.29163 951
4.4	-0.16333 64628	0.32597 06708	0.31150 495
4.5	-0.19470 50086	0.30099 73231	0.32848 160
4.6	-0.22345 99526	0.27374 52415	0.34247 962
4.7	-0.24938 76472	0.24450 12968	0.35343 075
4.8	-0.27230 37945	0.21356 51673	0.36128 928
4.9	-0.29205 45942	0.18124 66920	0.36603 284
5.0	-0.30851 76252	0.14786 31434	0.36766 288

$$Y_{n+1}(x) = \frac{2n}{x} Y_n(x) - Y_{n-1}(x)$$

Table 9.1   BESSEL FUNCTIONS—ORDERS 0, 1 AND 2

x	$J_0(x)$	$J_1(x)$	$J_2(x)$
5.0	−0.17759 67713 14338	−0.32757 91376	0.04656 51163
5.1	−0.14433 47470 60501	−0.33709 72020	+0.01213 97659
5.2	−0.11029 04397 90987	−0.34322 30059	−0.02171 84086
5.3	−0.07580 31115 85584	−0.34596 08338	−0.05474 81465
5.4	−0.04121 01012 44991	−0.34534 47908	−0.08669 53768
5.5	−0.00684 38694 17819	−0.34143 82154	−0.11731 54816
5.6	+0.02697 08846 85114	−0.33433 28363	−0.14637 54691
5.7	0.05992 00097 24037	−0.32414 76802	−0.17365 60379
5.8	0.09170 25675 74816	−0.31102 77443	−0.19895 35139
5.9	0.12203 33545 92823	−0.29514 24447	−0.22208 16409
6.0	0.15064 52572 50997	−0.27668 38581	−0.24287 32100
6.1	0.17729 14222 42744	−0.25586 47726	−0.26118 15116
6.2	0.20174 72229 48904	−0.23291 65671	−0.27688 15994
6.3	0.22381 20061 32191	−0.20808 69402	−0.28987 13522
6.4	0.24331 06048 23407	−0.18163 75090	−0.30007 23264
6.5	0.26009 46055 81606	−0.15384 13014	−0.30743 03906
6.6	0.27404 33606 24146	−0.12498 01652	−0.31191 61379
6.7	0.28506 47377 10576	−0.09534 21180	−0.31352 50715
6.8	0.29309 56031 04273	−0.06521 86634	−0.31227 75629
6.9	0.29810 20354 04820	−0.03490 20961	−0.30821 85850
7.0	0.30007 92705 19556	−0.00468 28235	−0.30141 72201
7.1	0.29905 13805 01550	+0.02515 32743	−0.29196 59511
7.2	0.29507 06914 00958	0.05432 74202	−0.27997 97413
7.3	0.28821 69476 35014	0.08257 04305	−0.26559 49119
7.4	0.27859 62326 57478	0.10962 50949	−0.24896 78286
7.5	0.26633 96578 80378	0.13524 84276	−0.23027 34105
7.6	0.25160 18338 49976	0.15921 37684	−0.20970 34737
7.7	0.23455 91395 86464	0.18131 27153	−0.18746 49278
7.8	0.21540 78077 46263	0.20135 68728	−0.16377 78404
7.9	0.19436 18448 41278	0.21917 93999	−0.13887 33892
8.0	0.17165 08071 37554	0.23463 63469	−0.11299 17204
8.1	0.14751 74540 44378	0.24760 77670	−0.08637 97338
8.2	0.12221 53017 84138	0.25799 85976	−0.05928 88146
8.3	0.09600 61008 95010	0.26573 93020	−0.03197 25341
8.4	0.06915 72616 56985	0.27078 62683	−0.00468 43406
8.5	0.04193 92518 42935	0.27312 19637	+0.02232 47396
8.6	+0.01462 29912 78741	0.27275 48445	0.04880 83679
8.7	−0.01252 27324 49665	0.26971 90241	0.07452 71058
8.8	−0.03923 38031 76542	0.26407 37032	0.09925 05539
8.9	−0.06525 32468 51244	0.25590 23714	0.12275 93977
9.0	−0.09033 36111 82876	0.24531 17866	0.14484 73415
9.1	−0.11423 92326 83199	0.23243 07450	0.16532 29129
9.2	−0.13674 83707 64864	0.21740 86550	0.18401 11218
9.3	−0.15765 51899 43403	0.20041 39278	0.20075 49594
9.4	−0.17677 15727 51508	0.18163 22040	0.21541 67225
9.5	−0.19392 87476 87422	0.16126 44308	0.22787 91542
9.6	−0.20897 87183 68872	0.13952 48117	0.23804 63875
9.7	−0.22179 54820 31723	0.11663 86479	0.24584 46878
9.8	−0.23227 60275 79367	0.09284 00911	0.25122 29849
9.9	−0.24034 11055 34760	0.06836 98323	0.25415 31929
10.0	−0.24593 57644 51348	0.04347 27462	0.25463 03137
	$\begin{bmatrix} (-4)4 \\ 11 \end{bmatrix}$	$\begin{bmatrix} (-4)4 \\ 8 \end{bmatrix}$	$\begin{bmatrix} (-4)4 \\ 7 \end{bmatrix}$

$$J_{n+1}(x) = \frac{2n}{x} J_n(x) - J_{n-1}(x)$$

BESSEL FUNCTIONS—ORDERS 0, 1 AND 2   Table 9.1

x	$Y_0(x)$	$Y_1(x)$	$Y_2(x)$
5.0	−0.30851 76252	0.14786 31434	0.36766 288
5.1	−0.32160 24491	0.11373 64420	0.36620 498
5.2	−0.33125 09348	0.07919 03430	0.36170 876
5.3	−0.33743 73011	0.04454 76191	0.35424 772
5.4	−0.34016 78783	+0.01012 72667	0.34391 872
5.5	−0.33948 05929	−0.02375 82390	0.33084 123
5.6	−0.33544 41812	−0.05680 56144	0.31515 646
5.7	−0.32815 71408	−0.08872 33405	0.29702 614
5.8	−0.31774 64300	−0.11923 41135	0.27663 122
5.9	−0.30436 59300	−0.14807 71525	0.25417 029
6.0	−0.28819 46840	−0.17501 03443	0.22985 790
6.1	−0.26943 49304	−0.19981 22045	0.20392 273
6.2	−0.24830 99505	−0.22228 36406	0.17660 555
6.3	−0.22506 17496	−0.24224 95005	0.14815 715
6.4	−0.19994 85953	−0.25955 98934	0.11883 613
6.5	−0.17324 24349	−0.27409 12740	0.08890 666
6.6	−0.14522 62172	−0.28574 72791	0.05863 613
6.7	−0.11619 11427	−0.29445 93130	+0.02829 284
6.8	−0.08643 38683	−0.30018 68758	−0.00185 639
6.9	−0.05625 36922	−0.30291 76343	−0.03154 852
7.0	−0.02594 97440	−0.30266 72370	−0.06052 661
7.1	+0.00418 17932	−0.29947 88746	−0.08854 204
7.2	0.03385 04048	−0.29342 25939	−0.11535 668
7.3	0.06277 38864	−0.28459 43719	−0.14074 495
7.4	0.09068 08802	−0.27311 49598	−0.16449 573
7.5	0.11731 32861	−0.25912 85105	−0.18641 422
7.6	0.14242 85247	−0.24280 10021	−0.20632 353
7.7	0.16580 16324	−0.22431 84743	−0.22406 617
7.8	0.18722 71733	−0.20388 50954	−0.23950 540
7.9	0.20652 09481	−0.18172 10773	−0.25252 628
8.0	0.22352 14894	−0.15806 04617	−0.26303 660
8.1	0.23809 13287	−0.13314 87960	−0.27096 757
8.2	0.25011 80276	−0.10724 07223	−0.27627 430
8.3	0.25951 49638	−0.08059 75035	−0.27893 605
8.4	0.26622 18674	−0.05348 45084	−0.27895 627
8.5	0.27020 51054	−0.02616 86794	−0.27636 244
8.6	0.27145 77123	+0.00108 39918	−0.27120 562
8.7	0.26999 91703	0.02801 09592	−0.26355 987
8.8	0.26587 49418	0.05435 55633	−0.25352 140
8.9	0.25915 57617	0.07986 93974	−0.24120 758
9.0	0.24593 66983	0.10431 45752	−0.22675 568
9.1	0.23833 59921	0.12746 58820	−0.21032 151
9.2	0.22449 36870	0.14911 27879	−0.19207 786
9.3	0.20857 00676	0.16906 13071	−0.17221 280
9.4	0.19074 39189	0.18713 56847	−0.15092 782
9.5	0.17121 06262	0.20317 98994	−0.12843 591
9.6	0.15018 01353	0.21705 89660	−0.10495 952
9.7	0.12787 47920	0.22866 00298	−0.08072 839
9.8	0.10452 70840	0.23789 32421	−0.05597 744
9.9	0.08037 73052	0.24469 24113	−0.03094 449
10.0	0.05567 11673 $\left[\dfrac{(-4)4}{8}\right]$	0.24901 54242 $\left[\dfrac{(-4)4}{8}\right]$	−0.00586 808 $\left[\dfrac{(-4)4}{6}\right]$

$$Y_{n+1}(x) = \frac{2n}{x} Y_n(x) - Y_{n-1}(x)$$

Table 9.1 BESSEL FUNCTIONS—ORDERS 0, 1 AND 2

x	$J_0(x)$	$J_1(x)$	$J_2(x)$
10.0	−0.24593 57644 51348	0.04347 27462	0.25463 03137
10.1	−0.24902 96505 80910	+0.01839 55155	0.25267 23269
10.2	−0.24961 70698 54127	−0.00661 57433	0.24831 98653
10.3	−0.24771 68134 82244	−0.03131 78295	0.24163 56815
10.4	−0.24337 17507 14207	−0.05547 27618	0.23270 39119
10.5	−0.23664 81944 62347	−0.07885 00142	0.22162 91441
10.6	−0.22763 50476 20693	−0.10122 86626	0.20853 53000
10.7	−0.21644 27399 23818	−0.12239 94239	0.19356 43429
10.8	−0.20320 19671 12039	−0.14216 65683	0.17687 48248
10.9	−0.18806 22459 63342	−0.16034 96867	0.15864 02851
11.0	−0.17119 03004 07196	−0.17678 52990	0.13904 75188
11.1	−0.15276 82954 35677	−0.19132 82878	0.11829 47301
11.2	−0.13299 19368 59575	−0.20385 31459	0.09658 95894
11.3	−0.11206 84561 09807	−0.21425 50262	0.07414 72125
11.4	−0.09021 45002 47520	−0.22245 05864	0.05118 80816
11.5	−0.06765 39481 11665	−0.22837 86207	0.02793 59271
11.6	−0.04461 56740 94438	−0.23200 04746	+0.00461 55923
11.7	−0.02133 12813 88500	−0.23330 02408	−0.01854 91017
11.8	+0.00196 71733 06740	−0.23228 47343	−0.04133 74673
11.9	0.02504 94416 99590	−0.22898 32497	−0.06353 40215
12.0	0.04768 93107 96834	−0.22344 71045	−0.08493 04949
12.1	0.06966 67736 06807	−0.21574 89734	−0.10532 77609
12.2	0.09077 01231 70505	−0.20598 20217	−0.12453 76677
12.3	0.11079 79503 07585	−0.19425 88480	−0.14238 47549
12.4	0.12956 10265 17502	−0.18071 02469	−0.15870 78405
12.5	0.14688 40547 00421	−0.16548 38046	−0.17336 14634
12.6	0.16260 72717 45511	−0.14874 23434	−0.18621 71675
12.7	0.17658 78885 61499	−0.13066 22290	−0.19716 46175
12.8	0.18870 13547 80683	−0.11143 15593	−0.20611 25359
12.9	0.19884 24371 36331	−0.09124 82522	−0.21298 94530
13.0	0.20692 61023 77068	−0.07031 80521	−0.21774 42642
13.1	0.21288 81975 22060	−0.04885 24733	−0.22034 65904
13.2	0.21668 59222 58564	−0.02706 67028	−0.22078 69378
13.3	0.21829 80903 19277	−0.00517 74806	−0.21907 66588
13.4	0.21772 51787 31184	+0.01659 90199	−0.21524 77131
13.5	0.21498 91658 80401	0.03804 92921	−0.20935 22337
13.6	0.21013 31613 69248	0.05896 45572	−0.20146 19030
13.7	0.20322 08326 33007	0.07914 27651	−0.19166 71443
13.8	0.19433 56352 15629	0.09839 05167	−0.18007 61400
13.9	0.18357 98554 57870	0.11652 48904	−0.16681 36842
14.0	0.17107 34761 10459	0.13337 51547	−0.15201 98826
14.1	0.15695 28770 32601	0.14878 43513	−0.13584 87137
14.2	0.14136 93846 57129	0.16261 07342	−0.11846 64643
14.3	0.12448 76852 83919	0.17472 90520	−0.10005 00556
14.4	0.10648 41184 90342	0.18503 16616	−0.08078 52766
14.5	0.08754 48680 10376	0.19342 94636	−0.06086 49420
14.6	0.06786 40683 23379	0.19985 26514	−0.04048 69928
14.7	0.04764 18459 01522	0.20425 12683	−0.01985 25577
14.8	0.02708 23145 85872	0.20659 55672	+0.00083 60053
14.9	+0.00639 15448 90853	0.20687 61718	0.02137 70688
15.0	−0.01422 44728 26781	0.20510 40386	0.04157 16780
	$\begin{bmatrix} (-4)3 \\ 11 \end{bmatrix}$	$\begin{bmatrix} (-4)3 \\ 7 \end{bmatrix}$	$\begin{bmatrix} (-4)3 \\ 7 \end{bmatrix}$

$$J_{n+1}(x) = \frac{2n}{x} J_n(x) - J_{n-1}(x)$$

BESSEL FUNCTIONS—ORDERS 0, 1 AND 2   Table 9.1

x	$Y_0(x)$	$Y_1(x)$	$Y_2(x)$
10.0	0.05567 11673	0.24901 54242	−0.00586 808
10.1	0.03065 73806	0.25084 44363	+0.01901 478
10.2	+0.00558 52273	0.25018 58292	0.04347 082
10.3	−0.01929 78497	0.24706 99395	0.06727 260
10.4	−0.04374 86190	0.24155 05610	0.09020 065
10.5	−0.06753 03725	0.23370 42284	0.11204 546
10.6	−0.09041 51548	0.22362 92892	0.13260 936
10.7	−0.11218 58897	0.21144 47763	0.15170 828
10.8	−0.13263 83844	0.19728 90905	0.16917 340
10.9	−0.15158 31932	0.18131 85097	0.18485 264
11.0	−0.16884 73239	0.16370 55374	0.19861 197
11.1	−0.18427 57716	0.14463 71102	0.21033 651
11.2	−0.19773 28675	0.12431 26795	0.21993 156
11.3	−0.20910 34295	0.10294 21889	0.22732 329
11.4	−0.21829 37073	0.08074 39654	0.23245 932
11.5	−0.22523 21117	0.05794 25471	0.23530 908
11.6	−0.22986 97260	0.03476 64663	0.23586 394
11.7	−0.23218 05930	+0.01144 60113	0.23413 718
11.8	−0.23216 17790	−0.01178 90120	0.23016 364
11.9	−0.22983 32139	−0.03471 14983	0.22399 935
12.0	−0.22523 73126	−0.05709 92183	0.21572 078
12.1	−0.21843 83806	−0.07873 69315	0.20542 401
12.2	−0.20952 18128	−0.09941 84171	0.19322 371
12.3	−0.19859 30946	−0.11894 84033	0.17925 189
12.4	−0.18577 66153	−0.13714 43766	0.16365 655
12.5	−0.17121 43068	−0.15383 82565	0.14660 019
12.6	−0.15506 41238	−0.16887 79186	0.12825 810
12.7	−0.13749 83780	−0.18212 85528	0.10881 672
12.8	−0.11870 19463	−0.19347 38454	0.08847 166
12.9	−0.09887 03702	−0.20281 69743	0.06742 588
13.0	−0.07820 78645	−0.21008 14084	0.04588 765
13.1	−0.05692 52568	−0.21521 15060	0.02406 854
13.2	−0.03523 78771	−0.21817 29066	+0.00218 138
13.3	−0.01336 34191	−0.21895 27145	−0.01956 180
13.4	+0.00848 02072	−0.21755 94728	−0.04095 177
13.5	0.03007 70090	−0.21402 29303	−0.06178 411
13.6	0.05121 50115	−0.20839 36044	−0.08186 113
13.7	0.07168 83040	−0.20074 21453	−0.10099 373
13.8	0.09129 90143	−0.19115 85095	−0.11900 315
13.9	0.10985 91895	−0.17975 09511	−0.13572 264
14.0	0.12719 25686	−0.16664 48419	−0.15099 897
14.1	0.14313 62286	−0.15198 13335	−0.16469 386
14.2	0.15754 20895	−0.13591 58742	−0.17668 517
14.3	0.17027 82640	−0.11861 65967	−0.18686 800
14.4	0.18123 02411	−0.10026 25924	−0.19515 560
14.5	0.19030 18912	−0.08104 20909	−0.20148 011
14.6	0.19741 62858	−0.06115 05609	−0.20579 307
14.7	0.20251 63238	−0.04078 87536	−0.20806 581
14.8	0.20556 51604	−0.02016 07059	−0.20828 958
14.9	0.20654 64347	+0.00052 82751	−0.20647 553
15.0	0.20546 42960	0.02107 36280	−0.20265 448
	$\begin{bmatrix}(-4)3\\8\end{bmatrix}$	$\begin{bmatrix}(-4)3\\8\end{bmatrix}$	$\begin{bmatrix}(-4)3\\6\end{bmatrix}$

$$Y_{n+1}(x) = \frac{2n}{x} Y_n(x) - Y_{n-1}(x)$$

Table 9.1 **BESSEL FUNCTIONS—ORDERS 0, 1 AND 2**

x	$J_0(x)$	$J_1(x)$	$J_2(x)$
15.0	−0.01422 44728 26781	0.20510 40386	0.04157 16780
15.1	−0.03456 18514 55565	0.20131 02204	0.06122 54568
15.2	−0.05442 07968 44039	0.19554 54359	0.08015 04595
15.3	−0.07360 75449 51123	0.18787 94498	0.09816 69502
15.4	−0.09193 62278 62321	0.17840 02717	0.11510 50943
15.5	−0.10923 06509 00050	0.16721 31804	0.13080 65451
15.6	−0.12532 59640 22481	0.15443 95871	0.14512 59111
15.7	−0.14007 02118 29049	0.14021 57469	0.15793 20904
15.8	−0.15332 57477 60686	0.12469 13334	0.16910 94608
15.9	−0.16497 04994 85671	0.10802 78901	0.17855 89133
16.0	−0.17489 90739 83629	0.09039 71757	0.18619 87209
16.1	−0.18302 36924 65310	0.07197 94186	0.19196 52352
16.2	−0.18927 49469 77945	0.05296 14991	0.19581 34037
16.3	−0.19360 23723 28377	0.03353 50765	0.19771 71056
16.4	−0.19597 48287 91007	+0.01389 46807	0.19766 93020
16.5	−0.19638 06929 36861	−0.00576 42137	0.19568 20004
16.6	−0.19482 78558 05566	−0.02524 71116	0.19178 60351
16.7	−0.19134 35295 25189	−0.04436 24008	0.18603 06671
16.8	−0.18597 38653 47601	−0.06292 32177	0.17848 30061
16.9	−0.17878 33878 91219	−0.08074 92543	0.16922 72631
17.0	−0.16985 42521 51184	−0.09766 84928	0.15836 38412
17.1	−0.15928 53315 32265	−0.11351 88483	0.14600 82733
17.2	−0.14719 11467 66030	−0.12814 97057	0.13229 00182
17.3	−0.13370 06470 75764	−0.14142 33355	0.11735 11285
17.4	−0.11895 58563 36348	−0.15321 61760	0.10134 48016
17.5	−0.10311 03982 28686	−0.16341 99694	0.08443 38303
	$\begin{bmatrix} (-4)2 \\ 11 \end{bmatrix}$	$\begin{bmatrix} (-4)2 \\ 7 \end{bmatrix}$	$\begin{bmatrix} (-4)2 \\ 7 \end{bmatrix}$

$$J_{n+1}(x)=\frac{2n}{x}J_n(x)-J_{n-1}(x)$$

Table 9.1 **BESSEL FUNCTIONS—MODULUS AND PHASE OF ORDERS 0, 1 AND 2**

$$J_n(x)=M_n(x)\cos\theta_n(x) \qquad Y_n(x)=M_n(x)\sin\theta_n(x)$$

x^{-1}	$x^{\frac{1}{2}}M_0(x)$	$\theta_0(x)-x$	$x^{\frac{1}{2}}M_1(x)$	$\theta_1(x)-x$	$x^{\frac{1}{2}}M_2(x)$	$\theta_2(x)-x$	$\langle x\rangle$
0.10	0.79739 375	−0.79783 499	0.79936 575	−2.31885 508	0.80542 555	−3.73985 605	10
0.09	0.79748 584	−0.79660 186	0.79908 654	−2.32256 201	0.80398 367	−3.75850 527	11
0.08	0.79756 868	−0.79536 548	0.79883 586	−2.32627 732	0.80269 711	−3.77717 539	13
0.07	0.79764 214	−0.79412 617	0.79861 398	−2.33000 016	0.80156 472	−3.79586 377	14
0.06	0.79770 609	−0.79288 426	0.79842 116	−2.33372 965	0.80058 549	−3.81456 786	17
0.05	0.79776 040	−0.79164 009	0.79825 761	−2.33746 488	0.79975 851	−3.83328 521	20
0.04	0.79780 498	−0.79039 402	0.79812 353	−2.34120 495	0.79900 299	−3.85201 346	25
0.03	0.79783 975	−0.78914 641	0.79801 908	−2.34494 891	0.79855 829	−3.87075 034	33
0.02	0.79786 463	−0.78789 764	0.79794 438	−2.34869 580	0.79818 387	−3.88949 363	50
0.01	0.79787 957	−0.78664 810	0.79789 952	−2.35244 465	0.79795 937	−3.90824 117	100
0.00	0.79788 456	−0.78539 816	0.79788 456	−2.35619 449	0.79788 456	−3.92699 082	∞
	$\begin{bmatrix} (-6)1 \\ 4 \end{bmatrix}$	$\begin{bmatrix} (-7)4 \\ 4 \end{bmatrix}$	$\begin{bmatrix} (-6)4 \\ 4 \end{bmatrix}$	$\begin{bmatrix} (-6)1 \\ 4 \end{bmatrix}$	$\begin{bmatrix} (-5)2 \\ 4 \end{bmatrix}$	$\begin{bmatrix} (-6)3 \\ 4 \end{bmatrix}$	

$\langle x\rangle$=nearest integer to x.

BESSEL FUNCTIONS—ORDERS 0, 1 AND 2 Table 9.1

x	$Y_0(x)$	$Y_1(x)$	$Y_2(x)$
15.0	0.20546 42960	0.02107 36280	−0.20265 448
15.1	0.20234 32292	0.04127 35340	−0.19687 654
15.2	0.19722 76821	0.06093 08736	−0.18921 046
15.3	0.19018 15001	0.07985 51269	−0.17974 292
15.4	0.18128 71741	0.09786 41973	−0.16857 754
15.5	0.17064 49112	0.11478 61425	−0.15583 380
15.6	0.15837 15368	0.13046 07959	−0.14164 579
15.7	0.14459 92412	0.14474 12638	−0.12616 086
15.8	0.12947 41833	0.15749 52835	−0.10953 807
15.9	0.11315 49657	0.16860 64314	−0.09194 661
16.0	0.09581 09971	0.17797 51689	−0.07356 410
16.1	0.07762 07587	0.18551 97173	−0.05457 483
16.2	0.05876 99918	0.19117 67538	−0.03516 792
16.3	0.03944 98249	0.19490 19240	−0.01553 548
16.4	0.01985 48596	0.19667 01648	+0.00412 931
16.5	+0.00018 12325	0.19647 58378	0.02363 402
16.6	−0.01937 53254	0.19433 26715	0.04278 890
16.7	−0.03862 14147	0.19027 35142	0.06140 866
16.8	−0.05736 78596	0.18434 99015	0.07931 428
16.9	−0.07543 15476	0.17663 14431	0.09633 468
17.0	−0.09263 71984	0.16720 50361	0.11230 838
17.1	−0.10881 90473	0.15617 39131	0.12708 500
17.2	−0.12382 24237	0.14365 65362	0.14052 667
17.3	−0.13750 52134	0.12978 53467	0.15250 930
17.4	−0.14973 91883	0.11470 53859	0.16292 372
17.5	−0.16041 11925	0.09857 27987	0.17167 666
	$\begin{bmatrix} (-4)2 \\ 7 \end{bmatrix}$	$\begin{bmatrix} (-4)2 \\ 7 \end{bmatrix}$	$\begin{bmatrix} (-4)2 \\ 6 \end{bmatrix}$

$$Y_{n+1}(x) = \frac{2n}{x} Y_n(x) - Y_{n-1}(x)$$

Table 9.1

BESSEL FUNCTIONS—AUXILIARY TABLE FOR SMALL ARGUMENTS

x	$f_1(x)$	$f_2(x)$	x	$f_1(x)$	$f_2(x)$
0.0	−0.07380 430	−0.63661 977	1.0	0.08825 696	−0.78121 282
0.1	−0.07202 984	−0.63857 491	1.1	0.11849 917	−0.79936 142
0.2	−0.06672 574	−0.64437 529	1.2	0.15018 546	−0.81476 705
0.3	−0.05794 956	−0.65382 684	1.3	0.18296 470	−0.82642 473
0.4	−0.04579 663	−0.66660 964	1.4	0.21647 200	−0.83332 875
0.5	−0.03039 904	−0.68228 315	1.5	0.25033 233	−0.83449 074
0.6	−0.01192 435	−0.70029 342	1.6	0.28416 437	−0.82895 780
0.7	+0.00942 612	−0.71998 221	1.7	0.31758 436	−0.81583 036
0.8	0.03341 927	−0.74059 789	1.8	0.35020 995	−0.79427 978
0.9	0.05979 263	−0.76130 792	1.9	0.38166 415	−0.76356 508
1.0	0.08825 696	−0.78121 282	2.0	0.41157 912	−0.72304 896
	$\begin{bmatrix} (-4)4 \\ 6 \end{bmatrix}$	$\begin{bmatrix} (-4)5 \\ 7 \end{bmatrix}$		$\begin{bmatrix} (-4)2 \\ 6 \end{bmatrix}$	$\begin{bmatrix} (-3)1 \\ 6 \end{bmatrix}$

$$Y_0(x) = f_1(x) + \frac{2}{\pi} J_0(x) \ln x \qquad Y_1(x) = \frac{1}{x} f_2(x) + \frac{2}{\pi} J_1(x) \ln x$$

Table 9.2   **BESSEL FUNCTIONS—ORDERS 3–9**

x	$J_3(x)$	$J_4(x)$	$J_5(x)$	$J_6(x)$	$J_7(x)$	$J_8(x)$	$J_9(x)$
0.0	0.0000	0.0000	0.0000	0.0000	0.0000	0.0000	0.0000
0.2	(−4)1.6625	(−6)4.1583	(−8)8.3195	(−9)1.3869	(−11)1.9816	(−13)2.4774	(−15)2.7530
0.4	(−3)1.3201	(−5)6.6135	(−6)2.6489	(−8)8.8382	(−9)2.5270	(−11)6.3210	(−12)1.4053
0.6	(−3)4.3997	(−4)3.3147	(−5)1.9948	(−7)9.9956	(−8)4.2907	(−9)1.6110	(−11)5.3755
0.8	(−2)1.0247	(−3)1.0330	(−5)8.3084	(−6)5.5601	(−7)3.1864	(−8)1.5967	(−10)7.1092
1.0	(−2)1.9563	(−3)2.4766	(−4)2.4976	(−5)2.0938	(−6)1.5023	(−8)9.4223	(−9)5.2493
1.2	(−2)3.2874	(−3)5.0227	(−4)6.1010	(−5)6.1541	(−6)5.3093	(−7)4.0021	(−8)2.6788
1.4	(−2)5.0498	(−3)9.0629	(−3)1.2901	(−4)1.5231	(−5)1.5366	(−6)1.3538	(−7)1.0587
1.6	(−2)7.2523	(−2)1.4995	(−3)2.4524	(−4)3.3210	(−5)3.8397	(−6)3.8744	(−7)3.4687
1.8	(−2)9.8802	(−2)2.3197	(−3)4.2936	(−4)6.5690	(−5)8.5712	(−6)9.7534	(−7)9.8426
2.0	0.12894	(−2)3.3996	(−3)7.0396	(−3)1.2024	(−4)1.7494	(−5)2.2180	(−6)2.4923
2.2	0.16233	(−2)4.7647	(−2)1.0937	(−3)2.0660	(−4)3.3195	(−5)4.6434	(−6)5.7535
2.4	0.19811	(−2)6.4307	(−2)1.6242	(−3)3.3669	(−4)5.9274	(−5)9.0756	(−5)1.2300
2.6	0.23529	(−2)8.4013	(−2)2.3207	(−3)5.2461	(−3)1.0054	(−4)1.6738	(−5)2.4647
2.8	0.27270	(−1)1.0667	(−2)3.2069	(−3)7.8634	(−3)1.6314	(−4)2.9367	(−5)4.6719
3.0	0.30906	0.13203	(−2)4.3028	(−2)1.1394	(−3)2.5473	(−4)4.9344	(−5)8.4395
3.2	0.34307	0.15972	(−2)5.6238	(−2)1.6022	(−3)3.8446	(−4)7.9815	(−4)1.4615
3.4	0.37339	0.18920	(−2)7.1785	(−2)2.1934	(−3)5.6301	(−3)1.2482	(−4)2.4382
3.6	0.39876	0.21980	(−2)8.9680	(−2)2.9311	(−3)8.0242	(−3)1.8940	(−4)3.9339
3.8	0.41803	0.25074	(−1)1.0984	(−2)3.8316	(−2)1.1159	(−3)2.7966	(−4)6.1597
4.0	0.43017	0.28113	0.13209	(−2)4.9088	(−2)1.5176	(−3)4.0287	(−4)9.3860
4.2	0.43439	0.31003	0.15614	(−2)6.1725	(−2)2.0220	(−3)5.6739	(−3)1.3952
4.4	0.43013	0.33645	0.18160	(−2)7.6279	(−2)2.6433	(−3)7.8267	(−3)2.0275
4.6	0.41707	0.35941	0.20799	(−2)9.2745	(−2)3.3953	(−2)1.0591	(−3)2.8852
4.8	0.39521	0.37796	0.23473	(−1)1.1105	(−2)4.2901	(−2)1.4079	(−3)4.0270
5.0	0.36483	0.39123	0.26114	0.13105	(−2)5.3376	(−2)1.8405	(−3)5.5203
5.2	0.32652	0.39847	0.28651	0.15252	(−2)6.5447	(−2)2.3689	(−3)7.4411
5.4	0.28113	0.39906	0.31007	0.17515	(−2)7.9145	(−2)3.0044	(−3)9.8734
5.6	0.22978	0.39257	0.33103	0.19856	(−2)9.4455	(−2)3.7577	(−2)1.2907
5.8	0.17382	0.37877	0.34862	0.22230	(−1)1.1131	(−2)4.6381	(−2)1.6639
6.0	0.11477	0.35764	0.36209	0.24584	0.12959	(−2)5.6532	(−2)2.1165
6.2	+0.05428	0.32941	0.37077	0.26860	0.14910	(−2)6.8077	(−2)2.6585
6.4	−0.00591	0.29453	0.37408	0.28996	0.16960	(−2)8.1035	(−2)3.2990
6.6	−0.06406	0.25368	0.37155	0.30928	0.19077	(−2)9.5385	(−2)4.0468
6.8	−0.11847	0.20774	0.36288	0.32590	0.21224	(−1)1.1107	(−2)4.9093
7.0	−0.16756	0.15780	0.34790	0.33920	0.23358	0.12797	(−2)5.8921
7.2	−0.20987	0.10509	0.32663	0.34857	0.25432	0.14594	(−2)6.9987
7.4	−0.24420	+0.05097	0.29930	0.35349	0.27393	0.16476	(−2)8.2300
7.6	−0.26958	−0.00313	0.26629	0.35351	0.29188	0.18417	(−2)9.5839
7.8	−0.28535	−0.05572	0.22820	0.34828	0.30762	0.20385	(−1)1.1054
8.0	−0.29113	−0.10536	0.18577	0.33758	0.32059	0.22345	0.12632
8.2	−0.28692	−0.15065	0.13994	0.32131	0.33027	0.24257	0.14303
8.4	−0.27302	−0.19033	0.09175	0.29956	0.33619	0.26075	0.16049
8.6	−0.25005	−0.22326	+0.04237	0.27253	0.33790	0.27755	0.17847
8.8	−0.21896	−0.24854	−0.00699	0.24060	0.33508	0.29248	0.19670
9.0	−0.18094	−0.26547	−0.05504	0.20432	0.32746	0.30507	0.21488
9.2	−0.13740	−0.27362	−0.10053	0.16435	0.31490	0.31484	0.23266
9.4	−0.08997	−0.27284	−0.14224	0.12152	0.29737	0.32138	0.24965
9.6	−0.04034	−0.26326	−0.17904	0.07676	0.27499	0.32427	0.26546
9.8	+0.00970	−0.24528	−0.20993	+0.03107	0.24797	0.32318	0.27967
10.0	0.05838	−0.21960	−0.23406	−0.01446	0.21671	0.31785	0.29186

Compiled from British Association for the Advancement of Science, Bessel functions, Part II.   Functions of positive integer order, Mathematical Tables, vol. X   (Cambridge Univ. Press, Cambridge, England, 1952) and Mathematical Tables Project, Table of $f_n(x) = n!(\frac{1}{2}x)^{-n}J_n(x)$. J. Math. Phys. **23**, 45–60 (1944) (with permission).

BESSEL FUNCTIONS—ORDERS 3–9   Table 9.2

x	$Y_3(x)$	$Y_4(x)$	$Y_5(x)$	$Y_6(x)$	$Y_7(x)$	$Y_8(x)$	$Y_9(x)$
0.0	$-\infty$	$-\infty$	$-\infty$	$-\infty$	$-\infty$	$-\infty$	$-\infty$
0.2	(2) -6.3982	(4) -1.9162	(5) -7.6586	(7) -3.8274	(9) -2.2957	(11) -1.6066	(13) -1.2850
0.4	(1) -8.1202	(3) -1.2097	(4) -2.4114	(5) -6.0163	(7) -1.8025	(8) -6.3027	(10) -2.5193
0.6	(1) -2.4692	(2) -2.4302	(3) -3.2156	(4) -5.3351	(6) -1.0638	(7) -2.4769	(8) -6.5943
0.8	(1) -1.0815	(1) -7.8751	(2) -7.7670	(3) -9.6300	(5) -1.4367	(6) -2.5046	(7) -4.9949
1.0	-5.8215	(1) -3.3278	(2) -2.6041	(3) -2.5708	(4) -3.0589	(5) -4.2567	(6) -6.7802
1.2	-3.5899	(1) -1.6686	(2) -1.0765	(2) -8.8041	(3) -8.6964	(5) -1.0058	(6) -1.3323
1.4	-2.4420	-9.4432	(1) -5.1519	(2) -3.5855	(3) -3.0218	(4) -2.9859	(5) -3.3823
1.6	-1.7897	-5.8564	(1) -2.7492	(2) -1.6597	(3) -1.2173	(4) -1.0485	(5) -1.0364
1.8	-1.3896	-3.9059	(1) -1.5970	(1) -8.4816	(2) -5.4947	(3) -4.1889	(4) -3.6685
2.0	-1.1278	-2.7659	-9.9360	(1) -4.6914	(2) -2.7155	(3) -1.8539	(4) -1.4560
2.2	-0.94591	-2.0603	-6.5462	(1) -2.7695	(2) -1.4452	(2) -8.9196	(3) -6.3425
2.4	-0.81161	-1.6024	-4.5296	(1) -1.7271	(1) -8.1825	(2) -4.6004	(3) -2.9851
2.6	-0.70596	-1.2927	-3.2716	(1) -1.1290	(1) -4.8837	(2) -2.5168	(3) -1.5000
2.8	-0.61736	-1.0752	-2.4548	-7.6918	(1) -3.0510	(2) -1.4486	(2) -7.9725
3.0	-0.53854	-0.91668	-1.9059	-5.4365	(1) -1.9840	(1) -8.7150	(2) -4.4496
3.2	-0.46491	-0.79635	-1.5260	-3.9723	(1) -1.3370	(1) -5.4522	(2) -2.5924
3.4	-0.39363	-0.70092	-1.2556	-2.9920	-9.3044	(1) -3.5320	(2) -1.5691
3.6	-0.32310	-0.62156	-1.0581	-2.3177	-6.6677	(1) -2.3612	(1) -9.8275
3.8	-0.25259	-0.55227	-0.91009	-1.8427	-4.9090	(1) -1.6243	(1) -6.3483
4.0	-0.18202	-0.48894	-0.79585	-1.5007	-3.7062	(1) -1.1471	(1) -4.2178
4.2	-0.11183	-0.42875	-0.70484	-1.2494	-2.8650	-8.3005	(1) -2.8756
4.4	-0.04278	-0.36985	-0.62967	-1.0612	-2.2645	-6.1442	(1) -2.0078
4.6	$+0.02406$	-0.31109	-0.56509	-0.91737	-1.8281	-4.6463	(1) -1.4333
4.8	0.08751	-0.25190	-0.50735	-0.80507	-1.5053	-3.5855	(1) -1.0446
5.0	0.14627	-0.19214	-0.45369	-0.71525	-1.2629	-2.8209	-7.7639
5.2	0.19905	-0.13204	-0.40218	-0.64139	-1.0780	-2.2608	-5.8783
5.4	0.24463	-0.07211	-0.35146	-0.57874	-0.93462	-1.8444	-4.5302
5.6	0.28192	-0.01310	-0.30063	-0.52375	-0.82168	-1.5304	-3.5510
5.8	0.31001	$+0.04407$	-0.24922	-0.47377	-0.73099	-1.2907	-2.8295
6.0	0.32825	0.09839	-0.19706	-0.42683	-0.65659	-1.1052	-2.2907
6.2	0.33622	0.14877	-0.14426	-0.38145	-0.59403	-0.95990	-1.8831
6.4	0.33383	0.19413	-0.09117	-0.33658	-0.53992	-0.84450	-1.5713
6.6	0.32128	0.23344	-0.03833	-0.29151	-0.49169	-0.75147	-1.3301
6.8	0.29909	0.26576	$+0.01357$	-0.24581	-0.44735	-0.67521	-1.1414
7.0	0.26808	0.29031	0.06370	-0.19931	-0.40537	-0.61144	-0.99220
7.2	0.22934	0.30647	0.11119	-0.15204	-0.36459	-0.55689	-0.87293
7.4	0.18420	0.31385	0.15509	-0.10426	-0.32416	-0.50902	-0.77643
7.6	0.13421	0.31228	0.19450	-0.05635	-0.28348	-0.46585	-0.69726
7.8	0.08106	0.30186	0.22854	-0.00886	-0.24217	-0.42581	-0.63128
8.0	$+0.02654$	0.28294	0.25640	$+0.03756$	-0.20006	-0.38767	-0.57528
8.2	-0.02753	0.25613	0.27741	0.08218	-0.15716	-0.35049	-0.52673
8.4	-0.07935	0.22228	0.29104	0.12420	-0.11361	-0.31355	-0.48363
8.6	-0.12723	0.18244	0.29694	0.16284	-0.06973	-0.27635	-0.44440
8.8	-0.16959	0.13789	0.29495	0.19728	-0.02593	-0.23853	-0.40777
9.0	-0.20509	0.09003	0.28512	0.22677	$+0.01724$	-0.19995	-0.37271
9.2	-0.23262	$+0.04037$	0.26773	0.25064	0.05920	-0.16056	-0.33843
9.4	-0.25136	-0.00951	0.24326	0.26830	0.09925	-0.12048	-0.30433
9.6	-0.26079	-0.05804	0.21243	0.27932	0.13672	-0.07994	-0.26995
9.8	-0.26074	-0.10366	0.17612	0.28338	0.17087	-0.03928	-0.23499
10.0	-0.25136	-0.14495	0.13540	0.28035	0.20102	$+0.00108$	-0.19930

Table 9.2   **BESSEL FUNCTIONS—ORDERS 3–9**

x	$J_3(x)$	$J_4(x)$	$J_5(x)$	$J_6(x)$	$J_7(x)$	$J_8(x)$	$J_9(x)$
10.0	0.05838	-0.21960	-0.23406	-0.01446	0.21671	0.31785	0.29186
10.2	0.10400	-0.18715	-0.25078	-0.05871	0.18170	0.30811	0.30161
10.4	0.14497	-0.14906	-0.25964	-0.10059	0.14358	0.29386	0.30852
10.6	0.17992	-0.10669	-0.26044	-0.13901	0.10308	0.27515	0.31224
10.8	0.20768	-0.06150	-0.25323	-0.17297	0.06104	0.25210	0.31244
11.0	0.22735	-0.01504	-0.23829	-0.20158	+0.01838	0.22497	0.30886
11.2	0.23835	+0.03110	-0.21614	-0.22408	-0.02395	0.19414	0.30130
11.4	0.24041	0.07534	-0.18754	-0.23985	-0.06494	0.16010	0.28964
11.6	0.23359	0.11621	-0.15345	-0.24849	-0.10361	0.12344	0.27388
11.8	0.21827	0.15232	-0.11500	-0.24978	-0.13901	0.08485	0.25407
12.0	0.19514	0.18250	-0.07347	-0.24372	-0.17025	0.04510	0.23038
12.2	0.16515	0.20576	-0.03023	-0.23053	-0.19653	+0.00501	0.20310
12.4	0.12951	0.22138	+0.01331	-0.21064	-0.21716	-0.03453	0.17260
12.6	0.08963	0.22890	0.05571	-0.18469	-0.23160	-0.07264	0.13935
12.8	0.04702	0.22815	0.09557	-0.15349	-0.23947	-0.10843	0.10393
13.0	+0.00332	0.21928	0.13162	-0.11803	-0.24057	-0.14105	0.06698
13.2	-0.03984	0.20268	0.16267	-0.07944	-0.23489	-0.16969	+0.02921
13.4	-0.08085	0.17905	0.18774	-0.03894	-0.22261	-0.19364	-0.00860
13.6	-0.11822	0.14931	0.20605	+0.00220	-0.20411	-0.21231	-0.04567
13.8	-0.15059	0.11460	0.21702	0.04266	-0.17993	-0.22520	-0.08117
14.0	-0.17681	0.07624	0.22038	0.08117	-0.15080	-0.23197	-0.11431
14.2	-0.19598	+0.03566	0.21607	0.11650	-0.11762	-0.23246	-0.14432
14.4	-0.20747	-0.00566	0.20433	0.14756	-0.08136	-0.22666	-0.17048
14.6	-0.21094	-0.04620	0.18563	0.17335	-0.04315	-0.21472	-0.19216
14.8	-0.20637	-0.08450	0.16069	0.19308	-0.00415	-0.19700	-0.20883
15.0	-0.19402	-0.11918	0.13046	0.20615	+0.03446	-0.17398	-0.22005
15.2	-0.17445	-0.14901	0.09603	0.21219	0.07149	-0.14634	-0.22553
15.4	-0.14850	-0.17296	0.05865	0.21105	0.10580	-0.11487	-0.22514
15.6	-0.11723	-0.19021	+0.01968	0.20283	0.13634	-0.08047	-0.21888
15.8	-0.08188	-0.20020	-0.01949	0.18787	0.16217	-0.04417	-0.20690
16.0	-0.04385	-0.20264	-0.05747	0.16672	0.18251	-0.00702	-0.18953
16.2	-0.00461	-0.19752	-0.09293	0.14016	0.19675	+0.02987	-0.16725
16.4	+0.03432	-0.18511	-0.12462	0.10913	0.20447	0.06542	-0.14065
16.6	0.07146	-0.16596	-0.15144	0.07473	0.20546	0.09855	-0.11047
16.8	0.10542	-0.14083	-0.17248	0.03817	0.19974	0.12829	-0.07756
17.0	0.13493	-0.11074	-0.18704	+0.00072	0.18755	0.15374	-0.04286
17.2	0.15891	-0.07685	-0.19466	-0.03632	0.16932	0.17414	-0.00733
17.4	0.17651	-0.04048	-0.19512	-0.07166	0.14570	0.18889	+0.02799
17.6	0.18712	-0.00300	-0.18848	-0.10410	0.11751	0.19757	0.06210
17.8	0.19041	+0.03417	-0.17505	-0.13251	0.08571	0.19993	0.09400
18.0	0.18632	0.06964	-0.15537	-0.15596	0.05140	0.19593	0.12276
18.2	0.17510	0.10209	-0.13022	-0.17364	+0.01573	0.18574	0.14756
18.4	0.15724	0.13033	-0.10058	-0.18499	-0.02007	0.16972	0.16766
18.6	0.13351	0.15334	-0.06756	-0.18966	-0.05481	0.14841	0.18247
18.8	0.10487	0.17031	-0.03240	-0.18755	-0.08731	0.12253	0.19159
19.0	0.07249	0.18065	+0.00357	-0.17877	-0.11648	0.09294	0.19474
19.2	0.03764	0.18403	0.03904	-0.16370	-0.14135	0.06063	0.19187
19.4	+0.00170	0.18039	0.07269	-0.14292	-0.16110	+0.02667	0.18309
19.6	-0.03395	0.16994	0.10331	-0.11723	-0.17508	-0.00783	0.16869
19.8	-0.06791	0.15313	0.12978	-0.08759	-0.18287	-0.04171	0.14916
20.0	-0.09890	0.13067	0.15117	-0.05509	-0.18422	-0.07387	0.12513
	$\begin{bmatrix} (-3)1 \\ 5 \end{bmatrix}$	$\begin{bmatrix} (-3)1 \\ 5 \end{bmatrix}$	$\begin{bmatrix} (-3)1 \\ 5 \end{bmatrix}$	$\begin{bmatrix} (-4)9 \\ 5 \end{bmatrix}$	$\begin{bmatrix} (-4)8 \\ 5 \end{bmatrix}$	$\begin{bmatrix} (-4)8 \\ 5 \end{bmatrix}$	$\begin{bmatrix} (-4)8 \\ 5 \end{bmatrix}$

BESSEL FUNCTIONS—ORDERS 3–9 Table 9.2

x	$Y_3(x)$	$Y_4(x)$	$Y_5(x)$	$Y_6(x)$	$Y_7(x)$	$Y_8(x)$	$Y_9(x)$
10.0	−0.25136	−0.14495	0.13540	0.28035	0.20102	0.00108	−0.19930
10.2	−0.23314	−0.18061	0.09148	0.27030	0.22652	0.04061	−0.16282
10.4	−0.20686	−0.20954	+0.04567	0.25346	0.24678	0.07874	−0.12563
10.6	−0.17359	−0.23087	−0.00065	0.23025	0.26131	0.11488	−0.08791
10.8	−0.13463	−0.24397	−0.04609	0.20130	0.26975	0.14838	−0.04993
11.0	−0.09148	−0.24851	−0.08925	0.16737	0.27184	0.17861	−0.01205
11.2	−0.04577	−0.24445	−0.12884	0.12941	0.26750	0.20496	+0.02530
11.4	+0.00082	−0.23203	−0.16365	0.08848	0.25678	0.22687	0.06163
11.6	0.04657	−0.21178	−0.19262	0.04573	0.23992	0.24384	0.09640
11.8	0.08981	−0.18450	−0.21489	+0.00238	0.21732	0.25545	0.12906
12.0	0.12901	−0.15122	−0.22982	−0.04030	0.18952	0.26140	0.15902
12.2	0.16277	−0.11317	−0.23698	−0.08107	0.15724	0.26151	0.18573
12.4	0.18994	−0.07175	−0.23623	−0.11875	0.12130	0.25571	0.20865
12.6	0.20959	−0.02845	−0.22766	−0.15223	0.08268	0.24409	0.22728
12.8	0.22112	+0.01518	−0.21163	−0.18052	0.04240	0.22689	0.24122
13.0	0.22420	0.05759	−0.18876	−0.20279	+0.00157	0.20448	0.25010
13.2	0.21883	0.09729	−0.15987	−0.21840	−0.03868	0.17738	0.25369
13.4	0.20534	0.13289	−0.12600	−0.22692	−0.07722	0.14625	0.25184
13.6	0.18432	0.16318	−0.08833	−0.22813	−0.11296	0.11185	0.24454
13.8	0.15666	0.18712	−0.04819	−0.22204	−0.14489	0.07505	0.23190
14.0	0.12350	0.20393	−0.00697	−0.20891	−0.17209	+0.03682	0.21417
14.2	0.08615	0.21308	+0.03390	−0.18921	−0.19380	−0.00186	0.19170
14.4	0.04605	0.21434	0.07303	−0.16363	−0.20939	−0.03994	0.16501
14.6	+0.00477	0.20775	0.10907	−0.13305	−0.21842	−0.07640	0.13470
14.8	−0.03613	0.19364	0.14080	−0.09850	−0.22067	−0.11024	0.10149
15.0	−0.07511	0.17261	0.16717	−0.06116	−0.21610	−0.14053	0.06620
15.2	−0.11072	0.14550	0.18730	−0.02228	−0.20489	−0.16644	+0.02969
15.4	−0.14165	0.11339	0.20055	+0.01684	−0.18743	−0.18723	−0.00710
15.6	−0.16678	0.07750	0.20652	0.05489	−0.16430	−0.20234	−0.04322
15.8	−0.18523	+0.03920	0.20507	0.09059	−0.13627	−0.21134	−0.07775
16.0	−0.19637	−0.00007	0.19633	0.12278	−0.10425	−0.21399	−0.10975
16.2	−0.19986	−0.03885	0.18067	0.15038	−0.06928	−0.21025	−0.13838
16.4	−0.19566	−0.07571	0.15873	0.17250	−0.03251	−0.20025	−0.16286
16.6	−0.18402	−0.10930	0.13135	0.18843	+0.00487	−0.18432	−0.18253
16.8	−0.16547	−0.13841	0.09956	0.19767	0.04164	−0.16297	−0.19685
17.0	−0.14078	−0.16200	0.06455	0.19996	0.07660	−0.13688	−0.20543
17.2	−0.11098	−0.17924	+0.02761	0.19529	0.10864	−0.10686	−0.20805
17.4	−0.07725	−0.18956	−0.00990	0.18387	0.13671	−0.07387	−0.20464
17.6	−0.04094	−0.19265	−0.04663	0.16616	0.15991	−0.03895	−0.19533
17.8	−0.00347	−0.18846	−0.08123	0.14282	0.17752	−0.00320	−0.18039
18.0	+0.03372	−0.17722	−0.11249	0.11472	0.18897	+0.03225	−0.16030
18.2	0.06920	−0.15942	−0.13928	0.08289	0.19393	0.06629	−0.13566
18.4	0.10163	−0.13580	−0.16067	0.04848	0.19229	0.09782	−0.10722
18.6	0.12977	−0.10731	−0.17593	+0.01272	0.18414	0.12587	−0.07586
18.8	0.15261	−0.07506	−0.18455	−0.02310	0.16980	0.14955	−0.04252
19.0	0.16930	−0.04031	−0.18628	−0.05773	0.14982	0.16812	−0.00824
19.2	0.17927	−0.00440	−0.18111	−0.08993	0.12490	0.18100	+0.02593
19.4	0.18221	+0.03131	−0.16930	−0.11857	0.09595	0.18782	0.05895
19.6	0.17805	0.06546	−0.15134	−0.14267	0.06399	0.18838	0.08979
19.8	0.16705	0.09678	−0.12794	−0.16139	+0.03013	0.18270	0.11750
20.0	0.14967	0.12409	−0.10004	−0.17411	−0.00443	0.17101	0.14124
	$\begin{bmatrix}(-3)1\\5\end{bmatrix}$	$\begin{bmatrix}(-3)1\\5\end{bmatrix}$	$\begin{bmatrix}(-3)1\\5\end{bmatrix}$	$\begin{bmatrix}(-4)9\\5\end{bmatrix}$	$\begin{bmatrix}(-4)9\\5\end{bmatrix}$	$\begin{bmatrix}(-4)8\\5\end{bmatrix}$	$\begin{bmatrix}(-4)8\\5\end{bmatrix}$

Table 9.3   **BESSEL FUNCTIONS—ORDERS 10, 11, 20 AND 21**

x	$10^{10}x^{-10}J_{10}(x)$	$10^{11}x^{-11}J_{11}(x)$	$10^{-9}x^{10}Y_{10}(x)$	$10^{25}x^{-20}J_{20}(x)$	$10^{27}x^{-21}J_{21}(x)$	$10^{-23}x^{20}Y_{20}(x)$
0.0	2.69114 446	1.22324 748	−0.11828 049	3.91990	9.33311	−0.406017
0.1	2.69053 290	1.22299 266	−0.11831 335	3.91944	9.33205	−0.406071
0.2	2.68869 898	1.22222 850	−0.11841 200	3.91804	9.32886	−0.406231
0.3	2.68564 500	1.22095 588	−0.11857 661	3.91571	9.32357	−0.406499
0.4	2.68137 477	1.21917 626	−0.11880 750	3.91244	9.31615	−0.406873
0.5	2.67589 362	1.21689 169	−0.11910 510	3.90825	9.30663	−0.407355
0.6	2.66920 838	1.21410 481	−0.11946 998	3.90314	9.29500	−0.407945
0.7	2.66132 738	1.21081 883	−0.11990 282	3.89710	9.28128	−0.408644
0.8	2.65226 043	1.20703 750	−0.12040 444	3.89015	9.26546	−0.409452
0.9	2.64201 878	1.20276 518	−0.12097 581	3.88228	9.24758	−0.410369
1.0	2.63061 512	1.19800 675	−0.12161 801	3.87350	9.22762	−0.411397
1.1	2.61806 358	1.19276 764	−0.12233 229	3.86383	9.20562	−0.412536
1.2	2.60437 963	1.18705 385	−0.12312 002	3.85325	9.18157	−0.413788
1.3	2.58958 012	1.18087 185	−0.12398 273	3.84179	9.15550	−0.415153
1.4	2.57368 323	1.17422 867	−0.12492 212	3.82945	9.12743	−0.416632
1.5	2.55670 842	1.16713 182	−0.12594 004	3.81624	9.09737	−0.418228
1.6	2.53867 639	1.15958 931	−0.12703 852	3.80216	9.06534	−0.419940
1.7	2.51960 907	1.15160 961	−0.12821 977	3.78723	9.03137	−0.421771
1.8	2.49952 955	1.14320 168	−0.12948 616	3.77146	8.99546	−0.423722
1.9	2.47846 207	1.13437 488	−0.13084 030	3.75485	8.95766	−0.425795
2.0	2.45643 192	1.12513 904	−0.13228 497	3.73742	8.91797	−0.427992
2.1	2.43346 545	1.11550 438	−0.13382 319	3.71918	8.87643	−0.430315
2.2	2.40959 000	1.10548 152	−0.13545 821	3.70015	8.83306	−0.432764
2.3	2.38483 384	1.09508 144	−0.13719 351	3.68032	8.78790	−0.435344
2.4	2.35922 612	1.08431 551	−0.13903 284	3.65973	8.74096	−0.438056
2.5	2.33279 682	1.07319 540	−0.14098 022	3.63837	8.69228	−0.440902
2.6	2.30557 673	1.06173 312	−0.14303 997	3.61627	8.64189	−0.443885
2.7	2.27759 732	1.04994 098	−0.14521 672	3.59344	8.58981	−0.447007
2.8	2.24889 074	1.03783 155	−0.14751 543	3.56989	8.53609	−0.450272
2.9	2.21948 976	1.02541 767	−0.14994 141	3.54564	8.48076	−0.453682
3.0	2.18942 770	1.01271 242	−0.15250 037	3.52071	8.42385	−0.457241
3.1	2.15873 836	0.99972 906	−0.15519 840	3.49510	8.36539	−0.460951
3.2	2.12745 598	0.98648 108	−0.15804 206	3.46885	8.30542	−0.464816
3.3	2.09561 517	0.97298 213	−0.16103 836	3.44195	8.24397	−0.468840
3.4	2.06325 085	0.95924 599	−0.16419 482	3.41444	8.18110	−0.473027
3.5	2.03039 820	0.94528 659	−0.16751 951	3.38633	8.11682	−0.477379
3.6	1.99709 260	0.93111 794	−0.17102 110	3.35763	8.05119	−0.481902
3.7	1.96336 956	0.91675 415	−0.17470 889	3.32837	7.98424	−0.486600
3.8	1.92926 467	0.90220 939	−0.17859 286	3.29855	7.91600	−0.491476
3.9	1.89481 352	0.88749 785	−0.18268 376	3.26821	7.84653	−0.496537
4.0	1.86005 168	0.87263 375	−0.18699 314	3.23736	7.77586	−0.501786
4.1	1.82501 462	0.85763 130	−0.19153 346	3.20601	7.70403	−0.507229
4.2	1.78973 765	0.84250 469	−0.19631 812	3.17419	7.63108	−0.512872
4.3	1.75425 588	0.82726 806	−0.20136 159	3.14192	7.55707	−0.518719
4.4	1.71860 416	0.81193 548	−0.20667 950	3.10921	7.48202	−0.524777
4.5	1.68281 701	0.79652 093	−0.21228 873	3.07608	7.40598	−0.531051
4.6	1.64692 860	0.78103 829	−0.21820 757	3.04256	7.32900	−0.537549
4.7	1.61097 267	0.76550 130	−0.22445 582	3.00866	7.25112	−0.544276
4.8	1.57498 249	0.74992 357	−0.23105 498	2.97440	7.17238	−0.551240
4.9	1.53899 084	0.73431 852	−0.23802 840	2.93981	7.09282	−0.558448
5.0	1.50302 991	0.71869 942	−0.24540 147	2.90490	7.01250	−0.565907
	$\begin{bmatrix}(-4)2\\5\end{bmatrix}$	$\begin{bmatrix}(-5)6\\5\end{bmatrix}$	$\begin{bmatrix}(-5)5\\5\end{bmatrix}$	$\begin{bmatrix}(-4)1\\3\end{bmatrix}$	$\begin{bmatrix}(-4)3\\3\end{bmatrix}$	$\begin{bmatrix}(-5)3\\3\end{bmatrix}$

$$J_{n+1}(x)=\frac{2n}{x}J_n(x)-J_{n-1}(x) \qquad\qquad Y_{n+1}(x)=\frac{2n}{x}Y_n(x)-Y_{n-1}(x)$$

Compiled from British Association for the Advancement of Science, Bessel functions, Part II. Functions of positive integer order, Mathematical Tables, vol. X (Cambridge Univ. Press, Cambridge, England, 1952), L. Fox, A short table for Bessel functions of integer orders and large arguments. Royal Society Shorter Mathematical Tables No. 3 (Cambridge Univ. Press, Cambridge, England, 1954), and Mathematical Tables Project, Table of $f_n(x)=n!(\tfrac{1}{2}x)^{-n}J_n(x)$. J. Math. Phys. **23**, 45–60 (1944) (with permission).

BESSEL FUNCTIONS—ORDERS 10, 11, 20 AND 21   Table 9.3

x	$10^{10}x^{-10}J_{10}(x)$	$10^{11}x^{-11}J_{11}(x)$	$10^{-9}x^{10}Y_{10}(x)$	$10^{25}x^{-20}J_{20}(x)$	$10^{27}x^{-21}J_{21}(x)$	$10^{-23}x^{20}Y_{20}(x)$
5.0	1.50302 991	0.71869 942	−0.24540 147	2.90490	7.01250	−0.565907
5.1	1.46713 132	0.70307 931	−0.25320 186	2.86969	6.93145	−0.573626
5.2	1.43132 603	0.68747 104	−0.26145 975	2.83421	6.84971	−0.581612
5.3	1.39564 431	0.67188 722	−0.27020 813	2.79846	6.76734	−0.589875
5.4	1.36011 571	0.65634 019	−0.27948 304	2.76248	6.68437	−0.598423
5.5	1.32476 904	0.64084 205	−0.28932 400	2.72628	6.60085	−0.607266
5.6	1.28963 229	0.62540 463	−0.29977 431	2.68988	6.51682	−0.616414
5.7	1.25473 264	0.61003 945	−0.31088 154	2.65330	6.43233	−0.625876
5.8	1.22009 642	0.59475 774	−0.32269 795	2.61656	6.34742	−0.635663
5.9	1.18574 907	0.57957 041	−0.33528 105	2.57967	6.26213	−0.645788
6.0	1.15171 513	0.56448 805	−0.34869 413	2.54267	6.17651	−0.656261
6.1	1.11801 822	0.54952 091	−0.36300 693	2.50556	6.09059	−0.667094
6.2	1.08468 098	0.53467 890	−0.37829 631	2.46837	6.00443	−0.678301
6.3	1.05172 510	0.51997 158	−0.39464 698	2.43111	5.91806	−0.689895
6.4	1.01917 129	0.50540 814	−0.41215 232	2.39381	5.83152	−0.701890
6.5	0.98703 926	0.49099 740	−0.43091 524	2.35647	5.74485	−0.714300
6.6	0.95534 769	0.47674 781	−0.45104 907	2.31913	5.65810	−0.727140
6.7	0.92411 427	0.46266 745	−0.47267 855	2.28179	5.57131	−0.740427
6.8	0.89335 563	0.44876 400	−0.49594 084	2.24448	5.48451	−0.754178
6.9	0.86308 740	0.43504 477	−0.52098 648	2.20721	5.39775	−0.768410
7.0	0.83332 414	0.42151 665	−0.54798 051	2.17000	5.31106	−0.783140
7.1	0.80407 941	0.40818 616	−0.57710 346	2.13286	5.22448	−0.798389
7.2	0.77536 570	0.39505 943	−0.60855 234	2.09582	5.13805	−0.814177
7.3	0.74719 450	0.38214 216	−0.64254 159	2.05888	5.05181	−0.830524
7.4	0.71957 626	0.36943 970	−0.67930 390	2.02206	4.96579	−0.847452
7.5	0.69252 040	0.35695 696	−0.71909 088	1.98539	4.88002	−0.864985
7.6	0.66603 536	0.34469 850	−0.76217 356	1.94887	4.79455	−0.883147
7.7	0.64012 854	0.33266 845	−0.80884 258	1.91252	4.70940	−0.901963
7.8	0.61480 640	0.32087 058	−0.85940 807	1.87635	4.62461	−0.921460
7.9	0.59007 439	0.30930 826	−0.91419 914	1.84038	4.54021	−0.941665
8.0	0.56593 704	0.29798 448	−0.97356 279	1.80462	4.45624	−0.962608
8.1	0.54239 791	0.28690 187	−1.03786 231	1.76908	4.37272	−0.984319
8.2	0.51945 967	0.27606 265	−1.10747 485	1.73378	4.28968	−1.006831
8.3	0.49712 408	0.26546 873	−1.18278 826	1.69874	4.20716	−1.030178
8.4	0.47539 201	0.25512 162	−1.26419 685	1.66395	4.12518	−1.054394
8.5	0.45426 352	0.24502 250	−1.35209 608	1.62944	4.04377	−1.079518
8.6	0.43373 779	0.23517 220	−1.44687 598	1.59521	3.96296	−1.105589
8.7	0.41381 323	0.22557 121	−1.54891 312	1.56128	3.88277	−1.132647
8.8	0.39448 748	0.21621 969	−1.65856 097	1.52765	3.80323	−1.160736
8.9	0.37575 740	0.20711 750	−1.77613 854	1.49434	3.72436	−1.189902
9.0	0.35761 917	0.19826 418	−1.90191 706	1.46136	3.64619	−1.220192
9.1	0.34006 823	0.18965 897	−2.03610 452	1.42872	3.56873	−1.251657
9.2	0.32309 939	0.18130 082	−2.17882 801	1.39641	3.49201	−1.284351
9.3	0.30670 683	0.17318 839	−2.33011 366	1.36447	3.41606	−1.318328
9.4	0.29088 411	0.16532 010	−2.48986 396	1.33288	3.34088	−1.353647
9.5	0.27562 422	0.15769 409	−2.65783 251	1.30166	3.26651	−1.390372
9.6	0.26091 963	0.15030 825	−2.83359 602	1.27082	3.19294	−1.428567
9.7	0.24676 227	0.14316 025	−3.01652 353	1.24036	3.12022	−1.468301
9.8	0.23314 362	0.13624 751	−3.20574 283	1.21029	3.04834	−1.509646
9.9	0.22005 470	0.12956 726	−3.40010 421	1.18061	2.97733	−1.552680
10.0	0.20748 611	0.12311 653	−3.59814 152	1.15134	2.90720	−1.597484
	$\begin{bmatrix}(-5)8\\5\end{bmatrix}$	$\begin{bmatrix}(-5)3\\5\end{bmatrix}$	$\begin{bmatrix}(-3)1\\7\end{bmatrix}$	$\begin{bmatrix}(-5)5\\3\end{bmatrix}$	$\begin{bmatrix}(-4)1\\3\end{bmatrix}$	$\begin{bmatrix}(-4)2\\4\end{bmatrix}$

Table 9.3   BESSEL FUNCTIONS—ORDERS 10, 11, 20 AND 21

x	$J_{10}(x)$	$J_{11}(x)$	$Y_{10}(x)$	$10^{25}x^{-20}J_{20}(x)$	$10^{27}x^{-21}J_{21}(x)$	$10^{-23}x^{20}Y_{20}(x)$
10.0	0.20748 611	0.12311 653	-0.35981 415	1.151337	2.907199	- 1.59748
10.1	0.21587 417	0.13041 285	-0.34383 078	1.122469	2.837961	- 1.64414
10.2	0.22413 707	0.13787 866	-0.32793 809	1.094012	2.769629	- 1.69275
10.3	0.23223 256	0.14549 509	-0.31207 433	1.065970	2.702215	- 1.74339
10.4	0.24011 699	0.15324 123	-0.29618 615	1.038347	2.635729	- 1.79618
10.5	0.24774 554	0.16109 407	-0.28022 819	1.011148	2.570182	- 1.85121
10.6	0.25507 240	0.16902 861	-0.26416 276	0.984374	2.505582	- 1.90861
10.7	0.26205 109	0.17701 780	-0.24795 949	0.958030	2.441939	- 1.96848
10.8	0.26863 466	0.18503 266	-0.23159 513	0.932118	2.379259	- 2.03097
10.9	0.27477 603	0.19304 230	-0.21505 324	0.906639	2.317550	- 2.09619
11.0	0.28042 823	0.20101 401	-0.19832 403	0.881596	2.256817	- 2.16430
11.1	0.28554 479	0.20891 340	-0.18140 409	0.856989	2.197065	- 2.23544
11.2	0.29007 999	0.21670 446	-0.16429 620	0.832821	2.138299	- 2.30977
11.3	0.29398 925	0.22434 974	-0.14700 917	0.809092	2.080523	- 2.38746
11.4	0.29722 944	0.23181 048	-0.12955 753	0.785801	2.023738	- 2.46870
11.5	0.29975 923	0.23904 680	-0.11196 142	0.762950	1.967947	- 2.55367
11.6	0.30153 946	0.24601 789	-0.09424 628	0.740539	1.913152	- 2.64257
11.7	0.30253 345	0.25268 218	-0.07644 263	0.718565	1.859352	- 2.73563
11.8	0.30270 737	0.25899 761	-0.05858 580	0.697029	1.806548	- 2.83307
11.9	0.30203 061	0.26492 183	-0.04071 566	0.675930	1.754740	- 2.93513
12.0	0.30047 604	0.27041 248	-0.02287 631	0.655266	1.703925	- 3.04208
12.1	0.29802 036	0.27542 744	-0.00511 577	0.635035	1.654102	- 3.15419
12.2	0.29464 445	0.27992 508	+0.01251 441	0.615236	1.605267	- 3.27175
12.3	0.29033 357	0.28386 459	0.02995 946	0.595866	1.557418	- 3.39509
12.4	0.28507 771	0.28720 623	0.04716 182	0.576923	1.510551	- 3.52453
12.5	0.27887 175	0.28991 166	0.06406 154	0.558403	1.464660	- 3.66044
12.6	0.27171 575	0.29194 422	0.08059 668	0.540305	1.419743	- 3.80321
12.7	0.26361 509	0.29326 923	0.09670 381	0.522625	1.375791	- 3.95323
12.8	0.25458 064	0.29385 431	0.11231 845	0.505359	1.332800	- 4.11095
12.9	0.24462 889	0.29366 968	0.12737 554	0.488504	1.290762	- 4.27684
13.0	0.23378 201	0.29268 843	0.14180 995	0.472056	1.249671	- 4.45140
13.1	0.22206 793	0.29088 684	0.15555 698	0.456011	1.209520	- 4.63518
13.2	0.20952 032	0.28824 464	0.16855 286	0.440365	1.170299	- 4.82874
13.3	0.19617 859	0.28474 526	0.18073 529	0.425114	1.132001	- 5.03272
13.4	0.18208 776	0.28037 612	0.19204 392	0.410252	1.094617	- 5.24778
13.5	0.16729 840	0.27512 884	0.20242 090	0.395776	1.058137	- 5.47464
13.6	0.15186 646	0.26899 942	0.21181 137	0.381681	1.022552	- 5.71407
13.7	0.13585 302	0.26198 851	0.22016 393	0.367961	0.987853	- 5.96691
13.8	0.11932 411	0.25410 149	0.22743 118	0.354612	0.954028	- 6.23405
13.9	0.10235 036	0.24534 866	0.23357 014	0.341628	0.921067	- 6.51646
14.0	0.08500 671	0.23574 535	0.23854 273	0.329005	0.888960	- 6.81520
14.1	0.06737 200	0.22531 197	0.24231 614	0.316736	0.857694	- 7.13138
14.2	0.04952 862	0.21407 407	0.24486 329	0.304816	0.827260	- 7.46624
14.3	0.03156 199	0.20206 238	0.24616 313	0.293240	0.797644	- 7.82110
14.4	+0.01356 013	0.18931 275	0.24620 100	0.282001	0.768835	- 8.19739
14.5	-0.00438 689	0.17586 611	0.24496 888	0.271095	0.740821	- 8.59667
14.6	-0.02218 745	0.16176 836	0.24246 568	0.260516	0.713590	- 9.02062
14.7	-0.03974 898	0.14707 028	0.23869 741	0.250257	0.687129	- 9.47109
14.8	-0.05697 854	0.13182 729	0.23367 730	0.240312	0.661426	- 9.95006
14.9	-0.07378 344	0.11609 931	0.22742 597	0.230676	0.636467	-10.45971
15.0	-0.09007 181	0.09995 048	0.21997 141	0.221343	0.612240	-11.00239
	$\begin{bmatrix}(-4)1\\6\end{bmatrix}$	$\begin{bmatrix}(-4)1\\5\end{bmatrix}$	$\begin{bmatrix}(-4)2\\6\end{bmatrix}$	$\begin{bmatrix}(-5)6\\3\end{bmatrix}$	$\begin{bmatrix}(-4)1\\4\end{bmatrix}$	$\begin{bmatrix}(-3)4\\4\end{bmatrix}$

BESSEL FUNCTIONS—ORDERS 10, 11, 20 AND 21

Table 9.3

x	$J_{10}(x)$	$J_{11}(x)$	$Y_{10}(x)$	$10^{25}x^{-20}J_{20}(x)$	$10^{27}x^{-21}J_{21}(x)$	$10^{-23}x^{20}Y_{20}(x)$
15.0	−0.09007 181	0.09995 048	0.21997 141	0.22134 33	0.61224 04	− 11.0024
15.1	−0.10575 330	0.08344 886	0.21134 904	0.21230 71	0.58873 25	− 11.5807
15.2	−0.12073 964	0.06666 618	0.20160 159	0.20356 16	0.56593 06	− 12.1974
15.3	−0.13494 535	0.04967 738	0.19077 902	0.19510 08	0.54382 12	− 12.8555
15.4	−0.14828 828	0.03256 035	0.17893 834	0.18691 87	0.52239 14	− 13.5585
15.5	−0.16069 032	+0.01539 539	0.16614 338	0.17900 91	0.50162 76	− 14.3098
15.6	−0.17207 791	−0.00173 513	0.15246 453	0.17136 62	0.48151 66	− 15.1136
15.7	−0.18238 269	−0.01874 731	0.13797 838	0.16398 38	0.46204 52	− 15.9742
15.8	−0.19154 204	−0.03555 621	0.12276 733	0.15685 60	0.44319 99	− 16.8962
15.9	−0.19949 958	−0.05207 632	0.10691 918	0.14997 67	0.42496 74	− 17.8849
16.0	−0.20620 569	−0.06822 215	0.09052 660	0.14334 00	0.40733 43	− 18.9460
16.1	−0.21161 797	−0.08390 874	0.07368 666	0.13694 00	0.39028 75	− 20.0855
16.2	−0.21570 160	−0.09905 224	0.05650 016	0.13077 08	0.37381 35	− 21.3104
16.3	−0.21842 977	−0.11357 046	0.03907 110	0.12482 65	0.35789 93	− 22.6279
16.4	−0.21978 394	−0.12738 344	0.02150 600	0.11910 14	0.34253 16	− 24.0462
16.5	−0.21975 411	−0.14041 403	+0.00391 319	0.11358 96	0.32769 75	− 25.5740
16.6	−0.21833 905	−0.15258 841	−0.01359 786	0.10828 55	0.31338 39	− 27.2209
16.7	−0.21554 637	−0.16383 668	−0.03091 729	0.10318 34	0.29957 78	− 28.9975
16.8	−0.21139 267	−0.17409 338	−0.04793 557	0.09827 77	0.28626 66	− 30.9150
16.9	−0.20590 350	−0.18329 797	−0.06454 431	0.09356 30	0.27343 76	− 32.9859
17.0	−0.19911 332	−0.19139 539	−0.08063 696	0.08903 37	0.26107 81	− 35.2237
17.1	−0.19106 538	−0.19833 646	−0.09610 960	0.08468 45	0.24917 57	− 37.6429
17.2	−0.18181 155	−0.20407 831	−0.11086 170	0.08051 02	0.23771 82	− 40.2594
17.3	−0.17141 203	−0.20858 485	−0.12479 683	0.07650 53	0.22669 32	− 43.0904
17.4	−0.15993 505	−0.21182 701	−0.13782 343	0.07266 49	0.21608 89	− 46.1543
17.5	−0.14745 649	−0.21378 318	−0.14985 544	0.06898 37	0.20589 33	− 49.4711
17.6	−0.13405 943	−0.21443 935	−0.16081 304	0.06545 69	0.19609 48	− 53.0622
17.7	−0.11983 363	−0.21378 944	−0.17062 321	0.06207 96	0.18668 17	− 56.9506
17.8	−0.10487 499	−0.21183 538	−0.17922 038	0.05884 68	0.17764 27	− 61.1611
17.9	−0.08928 492	−0.20858 727	−0.18654 691	0.05575 39	0.16896 66	− 65.7197
18.0	−0.07316 966	−0.20406 341	−0.19255 365	0.05279 63	0.16064 24	− 70.6543
18.1	−0.05663 961	−0.19829 032	−0.19720 030	0.04996 93	0.15265 91	− 75.9946
18.2	−0.03980 852	−0.19130 265	−0.20045 582	0.04726 85	0.14500 62	− 81.7717
18.3	−0.02279 278	−0.18314 307	−0.20229 875	0.04468 96	0.13767 32	− 88.0182
18.4	−0.00571 052	−0.17386 213	−0.20271 742	0.04222 83	0.13064 97	− 94.7683
18.5	+0.01131 917	−0.16351 793	−0.20171 011	0.03988 04	0.12392 57	−102.0574
18.6	0.02817 711	−0.15217 591	−0.19928 520	0.03764 17	0.11749 14	−109.9219
18.7	0.04474 490	−0.13990 845	−0.19546 113	0.03550 84	0.11133 69	−118.3992
18.8	0.06090 579	−0.12679 446	−0.19026 637	0.03347 64	0.10545 28	−127.5270
18.9	0.07654 556	−0.11291 893	−0.18373 930	0.03154 21	0.09982 98	−137.3432
19.0	0.09155 333	−0.09837 240	−0.17592 797	0.02970 16	0.09445 89	−147.8850
19.1	0.10582 247	−0.08325 039	−0.16688 985	0.02795 15	0.08933 10	−159.1885
19.2	0.11925 134	−0.06765 283	−0.15669 143	0.02628 80	0.08443 76	−171.2882
19.3	0.13174 416	−0.05168 334	−0.14540 785	0.02470 79	0.07977 01	−184.2155
19.4	0.14321 168	−0.03544 863	−0.13312 231	0.02320 78	0.07532 03	−197.9980
19.5	0.15357 193	−0.01905 771	−0.11992 560	0.02178 44	0.07108 01	−212.6582
19.6	0.16275 089	−0.00262 120	−0.10591 538	0.02043 46	0.06704 16	−228.2122
19.7	0.17068 305	+0.01374 948	−0.09119 555	0.01915 54	0.06319 71	−244.6678
19.8	0.17731 198	0.02994 285	−0.07587 548	0.01794 37	0.05953 92	−262.0226
19.9	0.18259 079	0.04584 818	−0.06006 922	0.01679 67	0.05606 06	−280.2622
20.0	0.18648 256	0.06135 630	−0.04389 465	0.01571 16	0.05275 42	−299.3574
	$\begin{bmatrix} (-4)2 \\ 6 \end{bmatrix}$	$\begin{bmatrix} (-4)2 \\ 6 \end{bmatrix}$	$\begin{bmatrix} (-4)2 \\ 6 \end{bmatrix}$	$\begin{bmatrix} (-5)4 \\ 4 \end{bmatrix}$	$\begin{bmatrix} (-5)9 \\ 4 \end{bmatrix}$	$\begin{bmatrix} (-1)1 \\ 5 \end{bmatrix}$

Table 9.3

BESSEL FUNCTIONS—MODULUS AND PHASE OF ORDERS 10, 11, 20 AND 21

$$J_n(x) = M_n(x) \cos \theta_n(x) \qquad\qquad Y_n(x) = M_n(x) \sin \theta_n(x)$$

x^{-1}	$x^{\frac{1}{2}} M_{10}(x)$	$\theta_{10}(x) - x$	$x^{\frac{1}{2}} M_{11}(x)$	$\theta_{11}(x) - x$	$\langle x \rangle$
0.050	0.85676 701	−13.94798 864	0.87222 790	−14.96758 686	20
0.048	0.85136 682	−14.05389 581	0.86513 271	−15.09771 672	21
0.046	0.84633 336	−14.15926 984	0.85857 314	−15.22701 466	22
0.044	0.84164 245	−14.26413 968	0.85250 587	−15.35552 901	23
0.042	0.83727 251	−14.36853 333	0.84689 281	−15.48330 635	24
0.040	0.83320 419	−14.47247 807	0.84170 044	−15.61039 144	25
0.038	0.82942 012	−14.57600 035	0.83689 917	−15.73682 771	26
0.036	0.82590 472	−14.67912 589	0.83246 283	−15.86265 679	28
0.034	0.82264 403	−14.78187 967	0.82836 826	−15.98791 896	29
0.032	0.81962 546	−14.88428 611	0.82459 496	−16.11265 291	31
0.030	0.81683 775	−14.98636 880	0.82112 469	−16.23689 620	33
0.028	0.81427 076	−15.08815 085	0.81794 133	−16.36068 504	36
0.026	0.81191 546	−15.18965 477	0.81503 056	−16.48405 469	38
0.024	0.80976 370	−15.29090 253	0.81237 970	−16.60703 912	42
0.022	0.80780 825	−15.39191 569	0.80997 751	−16.72967 149	45
0.020	0.80604 267	−15.49271 527	0.80781 410	−16.85198 406	50
0.018	0.80446 127	−15.59332 192	0.80588 079	−16.97400 835	56
0.016	0.80305 902	−15.69375 598	0.80416 997	−17.09577 505	63
0.014	0.80183 156	−15.79403 741	0.80267 505	−17.21731 438	71
0.012	0.80077 512	−15.89418 589	0.80139 036	−17.33865 590	83
0.010	0.79988 647	−15.99422 093	0.80031 114	−17.45982 880	100
0.008	0.79916 297	−16.09416 168	0.79943 341	−17.58086 166	125
0.006	0.79860 244	−16.19402 726	0.79875 398	−17.70178 301	167
0.004	0.79820 323	−16.29383 652	0.79827 039	−17.82262 084	250
0.002	0.79796 417	−16.39360 832	0.79798 093	−17.94340 316	500
0.000	0.79788 456	−16.49336 143	0.79788 456	−18.06415 776	∞
	$\begin{bmatrix} (-5)5 \\ 5 \end{bmatrix}$	$\begin{bmatrix} (-5)7 \\ 5 \end{bmatrix}$	$\begin{bmatrix} (-5)7 \\ 6 \end{bmatrix}$	$\begin{bmatrix} (-4)1 \\ 6 \end{bmatrix}$	

x^{-1}	$x^{\frac{1}{2}} M_{20}(x)$	$\theta_{20}(x) - x$	$x^{\frac{1}{2}} M_{21}(x)$	$\theta_{21}(x) - x$	$\langle x \rangle$
0.050	1.474083	−21.047407	1.791133	−21.290925	20
0.048	1.320938	−21.606130	1.525581	−21.927545	21
0.046	1.211667	−22.149524	1.347435	−22.550082	22
0.044	1.131459	−22.676802	1.224460	−23.154248	23
0.042	1.070845	−23.188535	1.136653	−23.738936	24
0.040	1.023762	−23.685951	1.071741	−24.304948	25
0.038	0.986284	−24.170500	1.022171	−24.853951	26
0.036	0.955823	−24.643620	0.983229	−25.387848	28
0.034	0.930635	−25.106640	0.951902	−25.908478	29
0.032	0.909513	−25.560748	0.926211	−26.417500	31
0.030	0.891605	−26.006988	0.904821	−26.916369	33
0.028	0.876293	−26.446280	0.886799	−27.406346	36
0.026	0.863121	−26.879433	0.871483	−27.888527	38
0.024	0.851743	−27.307159	0.858385	−28.363869	42
0.022	0.841895	−27.730098	0.847145	−28.833211	45
0.020	0.833375	−28.148822	0.837487	−29.297299	50
0.018	0.826019	−28.563847	0.829198	−29.756800	56
0.016	0.819702	−28.975650	0.822114	−30.212318	63
0.014	0.814321	−29.384666	0.816105	−30.664405	71
0.012	0.809796	−29.791303	0.811069	−31.113569	83
0.010	0.806062	−30.195941	0.806925	−31.560285	100
0.008	0.803071	−30.598942	0.803612	−32.005000	125
0.006	0.800781	−31.000652	0.801081	−32.448139	167
0.004	0.799165	−31.401404	0.799297	−32.890109	250
0.002	0.798204	−31.801522	0.798237	−33.331307	500
0.000	0.797885	−32.201325	0.797885	−33.772121	∞
	$\begin{bmatrix} (-3)5 \\ 7 \end{bmatrix}$	$\begin{bmatrix} (-3)2 \\ 7 \end{bmatrix}$	$\begin{bmatrix} (-2)1 \\ 8 \end{bmatrix}$	$\begin{bmatrix} (-3)2 \\ 7 \end{bmatrix}$	

$\langle x \rangle$ = nearest integer to x.

Compiled from L. Fox, A short table for Bessel functions of integer orders and large arguments. Royal Society Shorter Mathematical Tables No. 3 (Cambridge Univ. Press, Cambridge, England, 1954) (with permission).

BESSEL FUNCTIONS—VARIOUS ORDERS   Table 9.4

n	$J_n(1)$	$J_n(2)$	$J_n(5)$
0	(− 1)7.65197 6866	(− 1)2.23890 7791	(− 1)−1.77596 7713
1	(− 1)4.40050 5857	(− 1)5.76724 8078	(− 1)−3.27579 1376
2	(− 1)1.14903 4849	(− 1)3.52834 0286	(− 2)+4.65651 1628
3	(− 2)1.95633 5398	(− 1)1.28943 2495	(− 1) 3.64831 2306
4	(− 3)2.47663 8964	(− 2)3.39957 1981	(− 1) 3.91232 3605
5	(− 4)2.49757 7302	(− 3)7.03962 9756	(− 1) 2.61140 5461
6	(− 5)2.09383 3800	(− 3)1.20242 8972	(− 1) 1.31048 7318
7	(− 6)1.50232 5817	(− 4)1.74944 0749	(− 2) 5.33764 1016
8	(− 8)9.42234 4173	(− 5)2.21795 5229	(− 2) 1.84052 1665
9	(− 9)5.24925 0180	(− 6)2.49234 3435	(− 3) 5.52028 3139
10	(− 10)2.63061 5124	(− 7)2.51538 6283	(− 3) 1.46780 2647
11	(− 11)1.19800 6746	(− 8)2.30428 4758	(− 4) 3.50927 4498
12	(− 13)4.99971 8179	(− 9)1.93269 5149	(− 5) 7.62781 3166
13	(− 14)1.92561 6764	(− 10)1.49494 2010	(− 5) 1.52075 8221
14	(− 16)6.88540 8200	(− 11)1.07294 6448	(− 6) 2.80129 5810
15	(− 17)2.29753 1532	(− 13)7.18301 6356	(− 7) 4.79674 3278
16	(− 19)7.18639 6587	(− 14)4.50600 5896	(− 8) 7.67501 5694
17	(− 20)2.11537 5568	(− 15)2.65930 7805	(− 8) 1.15266 7666
18	(− 22)5.88034 4574	(− 16)1.48173 7249	(− 9) 1.63124 4339
19	(− 23)1.54847 8441	(− 18)7.81924 3273	(− 10) 2.18282 5842
20	(− 25)3.87350 3009	(− 19)3.91897 2805	(− 11) 2.77033 0052
30	(− 42)3.48286 9794	(− 33)3.65025 6266	(− 21) 2.67117 7278
40	(− 60)1.10791 5851	(− 48)1.19607 7458	(− 33) 8.70224 1617
50	(− 80)2.90600 4948	(− 65)3.22409 5839	(− 45) 2.29424 7616
100	(−189)8.43182 8790	(−158)1.06095 3112	(−119) 6.26778 9396

n	$J_n(10)$	$J_n(50)$	$J_n(100)$
0	(− 1)−2.45935 7645	(− 2)+5.58123 2767	(−2)+1.99858 5030
1	(− 2)+4.34727 4617	(− 2)−9.75118 2813	(−2)−7.71453 5201
2	(− 1)+2.54630 3137	(− 2)−5.97128 0079	(−2)−2.15287 5734
3	(− 2)+5.83793 7931	(− 2)+9.27348 0406	(−2)+7.62842 0172
4	(− 1)−2.19602 6861	(− 2)+7.08409 7728	(−2)+2.61058 0945
5	(− 1)−2.34061 5282	(− 2)−8.14002 4770	(−2)−7.41957 3696
6	(− 2)−1.44588 4208	(− 2)−8.71210 2682	(−2)−3.35253 8314
7	(− 1)+2.16710 9177	(− 2)+6.04912 0126	(−2)+7.01726 9099
8	(− 1) 3.17854 1268	(− 1)+1.04058 5632	(−2)+4.33495 5988
9	(− 1) 2.91855 6853	(− 2)−2.71924 6104	(−2)−6.32367 6141
10	(− 1) 2.07486 1066	(− 1)−1.13847 8491	(−2)−5.47321 7694
11	(− 1) 1.23116 5280	(− 2)−1.83466 7862	(−2)+5.22903 2602
12	(− 2) 6.33702 5497	(− 1)+1.05775 3106	(−2)+6.62360 4866
13	(− 2) 2.89720 8393	(− 2)+6.91188 2768	(−2)−3.63936 7434
14	(− 2) 1.19571 6324	(− 2)−6.98335 2016	(−2)−7.56984 0399
15	(− 3) 4.50797 3144	(− 1)−1.08225 5990	(−2)+1.51981 2122
16	(− 3) 1.56675 6192	(− 3)+4.89816 0778	(−2)+8.02578 4036
17	(− 4) 5.05646 6697	(− 1)+1.11360 4219	(−2)+1.04843 8769
18	(− 4) 1.52442 4853	(− 2)+7.08269 2610	(−2)−7.66931 4854
19	(− 5) 4.31462 7752	(− 2)−6.03650 3508	(−2)−3.80939 2116
20	(− 5) 1.15133 6925	(− 1)−1.16704 3528	(−2)+6.22174 5850
30	(−12) 1.55109 6078	(− 2)+4.84342 5725	(−2)+8.14601 2958
40	(−21) 6.03089 5312	(− 1)−1.38176 2812	(−2)+7.27017 5482
50	(−30) 1.78451 3608	(− 1)+1.21409 0219	(−2)−3.86983 3973
100	(−89) 6.59731 6064	(−21)+1.11592 7368	(−2)+9.63666 7330

Table 9.4   **BESSEL FUNCTIONS—VARIOUS ORDERS**

n	$Y_n(1)$	$Y_n(2)$	$Y_n(5)$
0	$(-2)+8.82569\ 6420$	$(-1)+5.10375\ 6726$	$(-1)-3.08517\ 6252$
1	$(-1)-7.81212\ 8213$	$(-1)-1.07032\ 4315$	$(-1)+1.47863\ 1434$
2	$(0)-1.65068\ 2607$	$(-1)-6.17408\ 1042$	$(-1)+3.67662\ 8826$
3	$(0)-5.82151\ 7606$	$(0)-1.12778\ 3777$	$(-1)+1.46267\ 1627$
4	$(1)-3.32784\ 2303$	$(0)-2.76594\ 3226$	$(-1)-1.92142\ 2874$
5	$(2)-2.60405\ 8666$	$(0)-9.93598\ 9128$	$(-1)-4.53694\ 8225$
6	$(3)-2.57078\ 0243$	$(1)-4.69140\ 0242$	$(-1)-7.15247\ 3576$
7	$(4)-3.05889\ 5705$	$(2)-2.71548\ 0254$	$(0)-1.26289\ 8836$
8	$(5)-4.25674\ 6185$	$(3)-1.85392\ 2175$	$(0)-2.82086\ 9383$
9	$(6)-6.78020\ 4939$	$(4)-1.45598\ 2938$	$(0)-7.76388\ 3188$
10	$(8)-1.21618\ 0143$	$(5)-1.29184\ 5422$	$(1)-2.51291\ 1010$
11	$(9)-2.42558\ 0081$	$(6)-1.27728\ 5593$	$(1)-9.27525\ 5719$
12	$(10)-5.32411\ 4376$	$(7)-1.39209\ 5698$	$(2)-3.82982\ 1416$
13	$(12)-1.27536\ 1870$	$(8)-1.65774\ 1981$	$(3)-1.74556\ 1722$
14	$(13)-3.31061\ 6748$	$(9)-2.14114\ 3619$	$(3)-8.69393\ 8814$
15	$(14)-9.25697\ 3276$	$(10)-2.98102\ 3646$	$(4)-4.69404\ 9564$
16	$(16)-2.77378\ 1366$	$(11)-4.45012\ 4034$	$(5)-2.72949\ 0350$
17	$(17)-8.86684\ 3398$	$(12)-7.09038\ 8217$	$(6)-1.69993\ 3328$
18	$(19)-3.01195\ 2974$	$(14)-1.20091\ 5873$	$(7)-1.12865\ 9760$
19	$(21)-1.08341\ 6386$	$(15)-2.15455\ 8183$	$(7)-7.95635\ 6938$
20	$(22)-4.11397\ 0315$	$(16)-4.08165\ 1389$	$(8)-5.93396\ 5297$
30	$(39)-3.04812\ 8783$	$(30)-2.91322\ 3848$	$(18)-4.02856\ 8418$
40	$(57)-7.18487\ 4797$	$(45)-6.66154\ 1235$	$(29)-9.21681\ 6571$
50	$(77)-2.19114\ 2813$	$(62)-1.97615\ 0576$	$(42)-2.78883\ 7017$
100	$(185)-3.77528\ 7810$	$(155)-3.00082\ 6049$	$(115)-5.08486\ 3915$

n	$Y_n(10)$	$Y_n(50)$	$Y_n(100)$
0	$(-2)+5.56711\ 6730$	$(-2)-9.80649\ 9547$	$(-2)-7.72443\ 1337$
1	$(-1)+2.49015\ 4242$	$(-2)-5.67956\ 6856$	$(-2)-2.03723\ 1200$
2	$(-3)-5.86808\ 2460$	$(-2)+9.57931\ 6873$	$(-2)+7.68368\ 6713$
3	$(-1)-2.51362\ 6572$	$(-2)+6.44591\ 2206$	$(-2)+2.34457\ 8669$
4	$(-1)-1.44949\ 5119$	$(-2)-8.80580\ 7408$	$(-2)-7.54301\ 1992$
5	$(-1)+1.35403\ 0477$	$(-2)-7.85484\ 1391$	$(-2)-2.94801\ 9628$
6	$(-1)+2.80352\ 5596$	$(-2)+7.23483\ 9130$	$(-2)+7.24821\ 0030$
7	$(-1)+2.01020\ 0238$	$(-2)+9.59120\ 2782$	$(-2)+3.81780\ 4832$
8	$(-3)+1.07547\ 3712$	$(-2)-4.54930\ 2351$	$(-2)-6.71371\ 7353$
9	$(-1)-1.99299\ 2658$	$(-1)-1.10469\ 7953$	$(-2)-4.89199\ 9608$
10	$(-1)-3.59814\ 1522$	$(-3)+5.72389\ 7182$	$(-2)+5.83315\ 7424$
11	$(-1)-5.20329\ 0386$	$(-1)+1.12759\ 3542$	$(-2)+6.05863\ 1093$
12	$(-1)-7.84909\ 7327$	$(-2)+4.38902\ 1867$	$(-2)-4.50025\ 8583$
13	$(0)-1.36345\ 4320$	$(-2)-9.16920\ 4926$	$(-2)-7.13869\ 3153$
14	$(0)-2.76007\ 1499$	$(-2)-9.15700\ 8429$	$(-2)+2.64419\ 8363$
15	$(0)-6.36474\ 5877$	$(-2)+4.04128\ 0205$	$(-2)+7.87906\ 8695$
16	$(1)-1.63341\ 6613$	$(-1)+1.15817\ 7655$	$(-3)-2.80477\ 7550$
17	$(1)-4.59045\ 8575$	$(-2)+3.37105\ 6788$	$(-2)-7.96882\ 1576$
18	$(2)-1.39741\ 4254$	$(-2)-9.28945\ 7936$	$(-2)-2.42892\ 1581$
19	$(2)-4.57164\ 5457$	$(-1)-1.00594\ 6650$	$(-2)+7.09440\ 9807$
20	$(3)-1.59748\ 3848$	$(-2)+1.64426\ 3395$	$(-2)+5.12479\ 7308$
30	$(9)-7.25614\ 2316$	$(-1)-1.16457\ 2349$	$(-3)+6.13883\ 9212$
40	$(18)-1.36280\ 3297$	$(-2)-4.53080\ 1120$	$(-2)+4.07468\ 5217$
50	$(27)-3.64106\ 6502$	$(-1)-2.10316\ 5546$	$(-2)+7.65052\ 6394$
100	$(85)-4.84914\ 8271$	$(+18)-3.29380\ 0188$	$(-1)-1.66921\ 4114$

Table 9.5

ZEROS AND ASSOCIATED VALUES OF BESSEL FUNCTIONS AND THEIR DERIVATIVES

s	$j_{0,s}$	$J'_0(j_{0,s})$	$j_{1,s}$	$J'_1(j_{1,s})$	$j_{2,s}$	$J'_2(j_{2,s})$
1	2.40482 55577	−0.51914 74973	3.83171	−0.40276	5.13562	−0.33967
2	5.52007 81103	+0.34026 48065	7.01559	+0.30012	8.41724	+0.27138
3	8.65372 79129	−0.27145 22999	10.17347	−0.24970	11.61984	−0.23244
4	11.79153 44391	+0.23245 98314	13.32369	+0.21836	14.79595	+0.20654
5	14.93091 77086	−0.20654 64331	16.47063	−0.19647	17.95982	−0.18773
6	18.07106 39679	+0.18772 88030	19.61586	+0.18006	21.11700	+0.17326
7	21.21163 66299	−0.17326 58942	22.76008	−0.16718	24.27011	−0.16170
8	24.35247 15308	+0.16170 15507	25.90367	+0.15672	27.42057	+0.15218
9	27.49347 91320	−0.15218 12138	29.04683	−0.14801	30.56920	−0.14417
10	30.63460 64684	+0.14416 59777	32.18968	+0.14061	33.71652	+0.13730
11	33.77582 02136	−0.13729 69434	35.33231	−0.13421	36.86286	−0.13132
12	36.91709 83537	+0.13132 46267	38.47477	+0.12862	40.00845	+0.12607
13	40.05842 57646	−0.12606 94971	41.61709	−0.12367	43.15345	−0.12140
14	43.19979 17132	+0.12139 86248	44.75932	+0.11925	46.29800	+0.11721
15	46.34118 83717	−0.11721 11989	47.90146	−0.11527	49.44216	−0.11343
16	49.48260 98974	+0.11342 91926	51.04354	+0.11167	52.58602	+0.10999
17	52.62405 18411	−0.10999 11430	54.18555	−0.10839	55.72963	−0.10685
18	55.76551 07550	+0.10684 78883	57.32753	+0.10537	58.87302	+0.10396
19	58.90698 39261	−0.10395 95729	60.46946	−0.10260	62.01622	−0.10129
20	62.04846 91902	+0.10129 34989	63.61136	+0.10004	65.15927	+0.09882

s	$j_{3,s}$	$J'_3(j_{3,s})$	$j_{4,s}$	$J'_4(j_{4,s})$	$j_{5,s}$	$J'_5(j_{5,s})$
1	6.38016	−0.29827	7.58834	−0.26836	8.77148	−0.24543
2	9.76102	+0.24942	11.06471	+0.23188	12.33860	+0.21743
3	13.01520	−0.21828	14.37254	−0.20636	15.70017	−0.19615
4	16.22347	+0.19644	17.61597	+0.18766	18.98013	+0.17993
5	19.40942	−0.18005	20.82693	−0.17323	22.21780	−0.16712
6	22.58273	+0.16718	24.01902	+0.16168	25.43034	+0.15669
7	25.74817	−0.15672	27.19909	−0.15217	28.62662	−0.14799
8	28.90835	+0.14801	30.37101	+0.14416	31.81172	+0.14059
9	32.06485	−0.14060	33.53714	−0.13729	34.98878	−0.13420
10	35.21867	+0.13421	36.69900	+0.13132	38.15987	+0.12861
11	38.37047	−0.12862	39.85763	−0.12607	41.32638	−0.12366
12	41.52072	+0.12367	43.01374	+0.12140	44.48932	+0.11925
13	44.66974	−0.11925	46.16785	−0.11721	47.64940	−0.11527
14	47.81779	+0.11527	49.32036	+0.11343	50.80717	+0.11167
15	50.96503	−0.11167	52.47155	−0.10999	53.96303	−0.10838
16	54.11162	+0.10839	55.62165	+0.10685	57.11730	+0.10537
17	57.25765	−0.10537	58.77084	−0.10396	60.27025	−0.10260
18	60.40322	+0.10260	61.91925	+0.10129	63.42205	+0.10003
19	63.54840	−0.10004	65.06700	−0.09882	66.57289	−0.09765
20	66.69324	+0.09765	68.21417	+0.09652	69.72289	+0.09543

s	$j_{6,s}$	$J'_6(j_{6,s})$	$j_{7,s}$	$J'_7(j_{7,s})$	$j_{8,s}$	$J'_8(j_{8,s})$
1	9.93611	−0.22713	11.08637	−0.21209	12.22509	−0.19944
2	13.58929	+0.20525	14.82127	+0.19479	16.03777	+0.18569
3	17.00382	−0.18726	18.28758	−0.17942	19.55454	−0.17244
4	20.32079	+0.17305	21.64154	+0.16688	22.94517	+0.16130
5	23.58608	−0.16159	24.93493	−0.15657	26.26681	−0.15196
6	26.82015	+0.15212	28.19119	+0.14792	29.54566	+0.14404
7	30.03372	−0.14413	31.42279	−0.14055	32.79580	−0.13722
8	33.23304	+0.13727	34.63709	+0.13418	36.02562	+0.13127
9	36.42202	−0.13131	37.83872	−0.12859	39.24045	−0.12603
10	39.60324	+0.12606	41.03077	+0.12365	42.44389	+0.12137
11	42.77848	−0.12139	44.21541	−0.11924	45.63844	−0.11719
12	45.94902	+0.11721	47.39417	+0.11526	48.82593	+0.11342
13	49.11577	−0.11343	50.56818	−0.11166	52.00769	−0.10998
14	52.27945	+0.10999	53.73833	+0.10838	55.18475	+0.10684
15	55.44059	−0.10685	56.90525	−0.10537	58.35789	−0.10395
16	58.59961	+0.10396	60.06948	+0.10260	61.52774	+0.10129
17	61.75682	−0.10129	63.23142	−0.10003	64.69478	−0.09882
18	64.91251	+0.09882	66.39141	+0.09765	67.85943	+0.09652
19	68.06689	−0.09652	69.54971	−0.09543	71.02200	−0.09438
20	71.22013	+0.09438	72.70655	+0.09336	74.18277	+0.09237

Table 9.5

ZEROS AND ASSOCIATED VALUES OF BESSEL FUNCTIONS AND THEIR DERIVATIVES

s	$y_{0,s}$	$Y'_0(y_{0,s})$	$y_{1,s}$	$Y'_1(y_{1,s})$	$y_{2,s}$	$Y'_2(y_{2,s})$
1	0.89357 697	+0.87942 080	2.19714	+0.52079	3.38424	+0.39921
2	3.95767 842	−0.40254 267	5.42968	−0.34032	6.79381	−0.29992
3	7.08605 106	+0.30009 761	8.59601	+0.27146	10.02348	+0.24967
4	10.22234 504	−0.24970 124	11.74915	−0.23246	13.20999	−0.21835
5	13.36109 747	+0.21835 830	14.89744	+0.20655	16.37897	+0.19646
6	16.50092 244	−0.19646 494	18.04340	−0.18773	19.53904	−0.18006
7	19.64130 970	+0.18006 318	21.18807	+0.17327	22.69396	+0.16718
8	22.78202 805	−0.16718 450	24.33194	−0.16170	25.84561	−0.15672
9	25.92295 765	+0.15672 493	27.47529	+0.15218	28.99508	+0.14801
10	29.06403 025	−0.14801 108	30.61829	−0.14417	32.14300	−0.14061
11	32.20520 412	+0.14060 578	33.76102	+0.13730	35.28979	+0.13421
12	35.34645 231	−0.13421 123	36.90356	−0.13132	38.43573	−0.12862
13	38.48775 665	+0.12861 661	40.04594	+0.12607	41.58101	+0.12367
14	41.62910 447	−0.12366 795	43.18822	−0.12140	44.72578	−0.11925
15	44.77048 661	+0.11924 981	46.33040	+0.11721	47.87012	+0.11527
16	47.91189 633	−0.11527 369	49.47251	−0.11343	51.01413	−0.11167
17	51.05332 855	+0.11167 049	52.61455	+0.10999	54.15785	+0.10839
18	54.19477 936	−0.10838 535	55.75654	−0.10685	57.30135	−0.10537
19	57.33624 570	+0.10537 405	58.89850	+0.10396	60.44464	+0.10260
20	60.47772 516	−0.10260 057	62.04041	−0.10129	63.58777	−0.10004

s	$y_{3,s}$	$Y'_3(y_{3,s})$	$y_{4,s}$	$Y'_4(y_{4,s})$	$y_{5,s}$	$Y'_5(y_{5,s})$
1	4.52702	+0.33256	5.64515	+0.28909	6.74718	+0.25795
2	8.09755	−0.27080	9.36162	−0.24848	10.59718	−0.23062
3	11.39647	+0.23232	12.73014	+0.21805	14.03380	+0.20602
4	14.62308	−0.20650	15.99963	−0.19635	17.34709	−0.18753
5	17.81846	+0.18771	19.22443	+0.18001	20.60290	+0.17317
6	20.99728	−0.17326	22.42481	−0.16716	23.82654	−0.16165
7	24.16624	+0.16170	25.61027	+0.15671	27.03013	+0.15215
8	27.32880	−0.15218	28.78589	−0.14800	30.22034	−0.14415
9	30.48699	+0.14416	31.95469	+0.14060	33.40111	+0.13729
10	33.64205	−0.13730	35.11853	−0.13421	36.57497	−0.13132
11	36.79479	+0.13132	38.27867	+0.12861	39.74363	+0.12606
12	39.94577	−0.12607	41.43596	−0.12367	42.90825	−0.12140
13	43.09537	+0.12140	44.59102	+0.11925	46.06968	+0.11721
14	46.24387	−0.11721	47.74429	−0.11527	49.22854	−0.11343
15	49.39150	+0.11343	50.89611	+0.11167	52.38531	+0.10999
16	52.53840	−0.10999	54.04673	−0.10838	55.54035	−0.10685
17	55.68470	+0.10685	57.19635	+0.10537	58.69393	+0.10396
18	58.83049	−0.10396	60.34513	−0.10260	61.84628	−0.10129
19	61.97586	+0.10129	63.49320	+0.10003	64.99759	+0.09882
20	65.12086	−0.09882	66.64065	−0.09765	68.14799	−0.09652

s	$y_{6,s}$	$Y'_6(y_{6,s})$	$y_{7,s}$	$Y'_7(y_{7,s})$	$y_{8,s}$	$Y'_8(y_{8,s})$
1	7.83774	+0.23429	8.91961	+0.21556	9.99463	+0.20027
2	11.81104	−0.21591	13.00771	−0.20352	14.19036	−0.19289
3	15.31362	+0.19571	16.57392	+0.18672	17.81789	+0.17880
4	18.67070	−0.17975	19.97434	−0.17283	21.26093	−0.16662
5	21.95829	+0.16703	23.29397	+0.16148	24.61258	+0.15643
6	25.20621	−0.15664	26.56676	−0.15206	27.91052	−0.14785
7	28.42904	+0.14796	29.80953	+0.14449	31.17370	+0.14051
8	31.63488	−0.14058	33.03177	−0.13725	34.41286	−0.13415
9	34.82864	+0.13419	36.23927	+0.13130	37.63465	+0.12857
10	38.01347	−0.12860	39.43579	−0.12605	40.84342	−0.12364
11	41.19152	+0.12366	42.62391	+0.12138	44.04215	+0.11923
12	44.36427	−0.11924	45.80544	−0.11720	47.23298	−0.11526
13	47.53282	+0.11527	48.98171	+0.11342	50.41746	+0.11166
14	50.69796	−0.11167	52.15369	−0.10999	53.59675	−0.10838
15	53.86031	+0.10838	55.32215	+0.10684	56.77177	+0.10537
16	57.02034	−0.10537	58.48767	−0.10396	59.94319	−0.10260
17	60.17842	+0.10260	61.65071	+0.10129	63.11158	+0.10003
18	63.33485	−0.10003	64.81164	−0.09882	66.27738	−0.09765
19	66.48986	+0.09765	67.97075	+0.09652	69.44095	+0.09543
20	69.64364	−0.09543	71.12830	−0.09438	72.60259	−0.09336

Table 9.5
ZEROS AND ASSOCIATED VALUES OF BESSEL FUNCTIONS AND THEIR DERIVATIVES

s	$j'_{0,s}$	$J_0(j'_{0,s})$	$j'_{1,s}$	$J_1(j'_{1,s})$	$j'_{2,s}$	$J_2(j'_{2,s})$
1	0.00000 00000	+1.00000 00000	1.84118	+0.58187	3.05424	+0.48650
2	3.83170 59702	−0.40275 93957	5.33144	−0.34613	6.70613	−0.31353
3	7.01558 66698	+0.30011 57525	8.53632	+0.27330	9.96947	+0.25474
4	10.17346 81351	−0.24970 48771	11.70600	−0.23330	13.17037	−0.22088
5	13.32369 19363	+0.21835 94072	14.86359	+0.20701	16.34752	+0.19794
6	16.47063 00509	−0.19646 53715	18.01553	−0.18802	19.51291	−0.18101
7	19.61585 85105	+0.18006 33753	21.16437	+0.17346	22.67158	+0.16784
8	22.76008 43806	−0.16718 46005	24.31133	−0.16184	25.82604	−0.15720
9	25.90367 20876	+0.15672 49863	27.45705	+0.15228	28.97767	+0.14836
10	29.04682 85349	−0.14801 11100	30.60192	−0.14424	32.12733	−0.14088
11	32.18967 99110	+0.14060 57982	33.74618	+0.13736	35.27554	+0.13443
12	35.33230 75501	−0.13421 12403	36.88999	−0.13137	38.42265	−0.12879
13	38.47476 62348	+0.12861 66221	40.03344	+0.12611	41.56893	+0.12381
14	41.61709 42128	−0.12366 79608	43.17663	−0.12143	44.71455	−0.11937
15	44.75931 89977	+0.11924 98120	46.31960	+0.11724	47.85964	+0.11537
16	47.90146 08872	−0.11527 36941	49.46239	−0.11345	51.00430	−0.11176
17	51.04353 51836	+0.11167 04969	52.60504	+0.11001	54.14860	+0.10846
18	54.18555 36411	−0.10838 53489	55.74757	−0.10687	57.29260	−0.10544
19	57.32752 54379	+0.10537 40554	58.89000	+0.10397	60.43635	+0.10266
20	60.46945 78453	−0.10260 05671	62.03235	−0.10131	63.57989	−0.10008

s	$j'_{3,s}$	$J_3(j'_{3,s})$	$j'_{4,s}$	$J_4(j'_{4,s})$	$j'_{5,s}$	$J_5(j'_{5,s})$
1	4.20119	+0.43439	5.31755	+0.39965	6.41562	+0.37409
2	8.01524	−0.29116	9.28240	−0.27438	10.51986	−0.26109
3	11.34592	+0.24074	12.68191	+0.22959	13.98719	+0.22039
4	14.58585	−0.21097	15.96411	−0.20276	17.31284	−0.19580
5	17.78875	+0.19042	19.19603	+0.18403	20.57551	+0.17849
6	20.97248	−0.17505	22.40103	−0.16988	23.80358	−0.16533
7	24.14490	+0.16295	25.58976	+0.15866	27.01031	+0.15482
8	27.31006	−0.15310	28.76784	−0.14945	30.20285	−0.14616
9	30.47027	+0.14487	31.93854	+0.14171	33.38544	+0.13885
10	33.62695	−0.13784	35.10392	−0.13509	36.56078	−0.13256
11	36.78102	+0.13176	38.26532	+0.12932	39.73064	+0.12707
12	39.93311	−0.12643	41.42367	−0.12425	42.89627	−0.12223
13	43.08365	+0.12169	44.57962	+0.11973	46.05857	+0.11790
14	46.23297	−0.11746	47.73367	−0.11568	49.21817	−0.11402
15	49.38130	+0.11364	50.88616	+0.11202	52.37559	+0.11049
16	52.52882	−0.11017	54.03737	−0.10868	55.53120	−0.10728
17	55.67567	+0.10700	57.18752	+0.10563	58.68528	+0.10434
18	58.82195	−0.10409	60.33677	−0.10283	61.83809	−0.10163
19	61.96775	+0.10141	63.48526	+0.10023	64.98980	+0.09912
20	65.11315	−0.09893	66.63309	−0.09783	68.14057	−0.09678

s	$j'_{6,s}$	$J_6(j'_{6,s})$	$j'_{7,s}$	$J_7(j'_{7,s})$	$j'_{8,s}$	$J_8(j'_{8,s})$
1	7.50127	+0.35414	8.57784	+0.33793	9.64742	+0.32438
2	11.73494	−0.25017	12.93239	−0.24096	14.11552	−0.23303
3	15.26818	+0.21261	16.52937	+0.20588	17.77401	+0.19998
4	18.63744	−0.18978	19.94185	−0.18449	21.22906	−0.17979
5	21.93172	+0.17363	23.26805	+0.16929	24.58720	+0.16539
6	25.18393	−0.16127	26.54503	−0.15762	27.88927	−0.15431
7	28.40978	+0.15137	29.79075	+0.14823	31.15533	+0.14537
8	31.61788	−0.14317	33.01518	−0.14044	34.39663	−0.13792
9	34.81339	+0.13623	36.22438	+0.13381	37.62008	+0.13158
10	37.99964	−0.13024	39.42227	−0.12808	40.83018	−0.12608
11	41.17885	+0.12499	42.61152	+0.12305	44.03001	+0.12124
12	44.35258	−0.12035	45.79400	−0.11859	47.22176	−0.11695
13	47.52196	+0.11620	48.97107	+0.11460	50.40702	+0.11309
14	50.68782	−0.11246	52.14375	−0.11099	53.58700	−0.10960
15	53.85079	+0.10906	55.31282	+0.10771	56.76260	+0.10643
16	57.01138	−0.10596	58.47887	−0.10471	59.93454	−0.10352
17	60.16995	+0.10311	61.64239	+0.10195	63.10340	+0.10084
18	63.32681	−0.10049	64.80374	−0.09940	66.26961	−0.09837
19	66.48221	+0.09805	67.96324	+0.09704	69.43356	+0.09607
20	69.63635	−0.09579	71.12113	−0.09484	72.59554	−0.09393

Table 9.5

ZEROS AND ASSOCIATED VALUES OF BESSEL FUNCTIONS AND THEIR DERIVATIVES

s	$y'_{0,s}$	$Y_0(y'_{0,s})$	$y'_{1,s}$	$Y_1(y'_{1,s})$	$y'_{2,s}$	$Y_2(y'_{2,s})$
1	2.19714 133	+0.52078 641	3.68302	+0.41673	5.00258	+0.36766
2	5.42968 104	−0.34031 805	6.94150	−0.30317	8.35072	−0.27928
3	8.59600 587	+0.27145 988	10.12340	+0.25091	11.57420	+0.23594
4	11.74915 483	−0.23246 177	13.28576	−0.21897	14.76091	−0.20845
5	14.89744 213	+0.20654 711	16.44006	+0.19683	17.93129	+0.18890
6	18.04340 228	−0.18772 909	19.59024	−0.18030	21.09289	−0.17405
7	21.18806 893	+0.17326 604	22.73803	+0.16735	24.24923	+0.16225
8	24.33194 257	−0.16170 163	25.88431	−0.15684	27.40215	−0.15259
9	27.47529 498	+0.15218 126	29.02958	+0.14810	30.55271	+0.14448
10	30.61828 649	−0.14416 600	32.17412	−0.14067	33.70159	−0.13754
11	33.76101 780	+0.13729 696	35.31813	+0.13427	36.84921	+0.13152
12	36.90355 532	−0.13132 464	38.46175	−0.12866	39.99589	−0.12623
13	40.04594 464	+0.12606 951	41.60507	+0.12370	43.14182	+0.12153
14	43.18821 810	−0.12139 863	44.74814	−0.11928	46.28716	−0.11732
15	46.33039 925	+0.11721 120	47.89101	+0.11530	49.43202	+0.11352
16	49.47250 568	−0.11342 920	51.03373	−0.11169	52.57649	−0.11007
17	52.61455 077	+0.10999 115	54.17632	+0.10840	55.72063	+0.10692
18	55.75654 488	−0.10684 789	57.31880	−0.10539	58.86450	−0.10402
19	58.89849 617	+0.10395 957	60.46118	+0.10261	62.00814	+0.10135
20	62.04041 115	−0.10129 350	63.60349	−0.10005	65.15159	−0.09887

s	$y'_{3,s}$	$Y_3(y'_{3,s})$	$y'_{4,s}$	$Y_4(y'_{4,s})$	$y'_{5,s}$	$Y_5(y'_{5,s})$
1	6.25363	+0.33660	7.46492	+0.31432	8.64956	+0.29718
2	9.69879	−0.26195	11.00517	−0.24851	12.28087	−0.23763
3	12.97241	+0.22428	14.33172	+0.21481	15.66080	+0.20687
4	16.19045	−0.19987	17.58444	−0.19267	18.94974	−0.18650
5	19.38239	+0.18223	20.80106	+0.17651	22.19284	+0.17151
6	22.55979	−0.16867	23.99700	−0.16397	25.40907	−0.15980
7	25.72821	+0.15779	27.17989	+0.15384	28.60804	+0.15030
8	28.89068	−0.14881	30.35396	−0.14543	31.79520	−0.14236
9	32.04898	+0.14122	33.52180	+0.13828	34.97389	+0.13559
10	35.20427	−0.13470	36.68505	−0.13211	38.14631	−0.12973
11	38.35728	+0.12901	39.84483	+0.12671	41.31392	+0.12458
12	41.50855	−0.12399	43.00191	−0.12193	44.47779	−0.12001
13	44.65845	+0.11952	46.15686	+0.11765	47.63867	+0.11591
14	47.80725	−0.11550	49.31009	−0.11380	50.79713	−0.11221
15	50.95515	+0.11186	52.46191	+0.11031	53.95360	+0.10885
16	54.10232	−0.10855	55.61257	−0.10712	57.10841	−0.10578
17	57.24887	+0.10552	58.76225	+0.10420	60.26183	+0.10295
18	60.39491	−0.10273	61.91110	−0.10151	63.41407	−0.10035
19	63.54050	+0.10015	65.05925	+0.09901	66.56530	+0.09793
20	66.68571	−0.09775	68.20679	−0.09669	69.71565	−0.09568

s	$y'_{6,s}$	$Y_6(y'_{6,s})$	$y'_{7,s}$	$Y_7(y'_{7,s})$	$y'_{8,s}$	$Y_8(y'_{8,s})$
1	9.81480	+0.28339	10.96515	+0.27194	12.10364	+0.26220
2	13.53281	−0.22854	14.76569	−0.22077	15.98284	−0.21402
3	16.96553	+0.20007	18.25012	+0.19414	19.51773	+0.18891
4	20.29129	−0.18111	21.61275	−0.17634	22.91696	−0.17207
5	23.56186	+0.16708	24.91131	+0.16311	26.24370	+0.15953
6	26.79950	−0.15607	28.17105	−0.15269	29.52596	−0.14962
7	30.01567	+0.14709	31.40518	+0.14417	32.77857	+0.14149
8	33.21697	−0.13957	34.62140	−0.13700	36.01026	−0.13463
9	36.40752	+0.13313	37.82455	+0.13085	39.22658	+0.12874
10	39.59002	−0.12753	41.01785	−0.12549	42.43122	−0.12359
11	42.76632	+0.12260	44.20351	+0.12076	45.62678	+0.11904
12	45.93775	−0.11822	47.38314	−0.11654	48.81512	−0.11497
13	49.10528	+0.11428	50.55791	+0.11275	51.99761	+0.11131
14	52.26963	−0.11072	53.72870	−0.10931	55.17529	−0.10798
15	55.43136	+0.10748	56.89619	+0.10618	58.34899	+0.10494
16	58.59089	−0.10451	60.06092	−0.10330	61.51933	−0.10216
17	61.74857	+0.10177	63.22331	+0.10065	64.68681	+0.09958
18	64.90468	−0.09925	66.38370	−0.09820	67.85185	−0.09720
19	68.05943	+0.09690	69.54237	+0.09592	71.01478	+0.09498
20	71.21301	−0.09471	72.69955	−0.09379	74.17587	−0.09291

BESSEL FUNCTIONS—$J_0(j_{0,s}x)$ Table 9.6

x	$J_0(j_{0,1}x)$	$J_0(j_{0,2}x)$	$J_0(j_{0,3}x)$	$J_0(j_{0,4}x)$	$J_0(j_{0,5}x)$
0.00	1.00000	1.00000	1.00000	1.00000	1.00000
0.02	0.99942	0.99696	0.99253	0.98614	0.97783
0.04	0.99769	0.98785	0.97027	0.94515	0.91280
0.06	0.99480	0.97276	0.93373	0.87872	0.80920
0.08	0.99077	0.95184	0.88372	0.78961	0.67388
0.10	0.98559	0.92526	0.82136	0.68146	0.51568
0.12	0.97929	0.89328	0.74804	0.55871	0.34481
0.14	0.97186	0.85617	0.66537	0.42632	0.17211
0.16	0.96333	0.81429	0.57518	0.28958	+0.00827
0.18	0.95370	0.76800	0.47943	0.15386	−0.13693
0.20	0.94300	0.71773	0.38020	+0.02438	−0.25533
0.22	0.93124	0.66392	0.27960	−0.09404	−0.34090
0.24	0.91844	0.60706	0.17976	−0.19716	−0.39013
0.26	0.90463	0.54766	+0.08277	−0.28155	−0.40225
0.28	0.88982	0.48623	−0.00942	−0.34466	−0.37917
0.30	0.87405	0.42333	−0.09498	−0.38498	−0.32527
0.32	0.85734	0.35950	−0.17226	−0.40207	−0.24698
0.34	0.83972	0.29529	−0.23986	−0.39653	−0.15223
0.36	0.82122	0.23126	−0.29664	−0.36998	−0.04980
0.38	0.80187	0.16795	−0.34171	−0.32493	+0.05137
0.40	0.78171	0.10590	−0.37453	−0.26467	0.14293
0.42	0.76077	+0.04562	−0.39482	−0.19304	0.21767
0.44	0.73908	−0.01240	−0.40264	−0.11431	0.27011
0.46	0.71669	−0.06769	−0.39835	−0.03289	0.29684
0.48	0.69362	−0.11983	−0.38259	+0.04684	0.29671
0.50	0.66993	−0.16840	−0.35628	0.12078	0.27086
0.52	0.64565	−0.21306	−0.32056	0.18527	0.22252
0.54	0.62081	−0.25349	−0.27678	0.23725	0.15667
0.56	0.59547	−0.28941	−0.22648	0.27445	+0.07960
0.58	0.56967	−0.32062	−0.17130	0.29541	−0.00168
0.60	0.54345	−0.34692	−0.11295	0.29959	−0.08007
0.62	0.51685	−0.36821	−0.05320	0.28731	−0.14891
0.64	0.48992	−0.38441	+0.00622	0.25977	−0.20259
0.66	0.46270	−0.39551	0.06363	0.21892	−0.23697
0.68	0.43524	−0.40152	0.11745	0.16735	−0.24965
0.70	0.40758	−0.40255	0.16625	0.10814	−0.24019
0.72	0.37977	−0.39871	0.20878	+0.04470	−0.21003
0.74	0.35186	−0.39019	0.24399	−0.01945	−0.16237
0.76	0.32389	−0.37721	0.27107	−0.08082	−0.10179
0.78	0.29591	−0.36003	0.28945	−0.13618	−0.03389
0.80	0.26796	−0.33896	0.29882	−0.18270	+0.03525
0.82	0.24009	−0.31433	0.29915	−0.21808	0.09960
0.84	0.21234	−0.28652	0.29063	−0.24067	0.15369
0.86	0.18476	−0.25591	0.27374	−0.24957	0.19306
0.88	0.15739	−0.22293	0.24914	−0.24461	0.21464
0.90	0.13027	−0.18800	0.21774	−0.22637	0.21694
0.92	0.10346	−0.15157	0.18059	−0.19613	0.20021
0.94	0.07698	−0.11411	0.13891	−0.15580	0.16630
0.96	0.05089	−0.07605	0.09399	−0.10779	0.11854
0.98	0.02521	−0.03787	0.04722	−0.05486	0.06138
1.00	0.00000	0.00000	0.00000	0.00000	0.00000
	$\begin{bmatrix} (-4)1 \\ 3 \end{bmatrix}$	$\begin{bmatrix} (-4)8 \\ 4 \end{bmatrix}$	$\begin{bmatrix} (-3)2 \\ 5 \end{bmatrix}$	$\begin{bmatrix} (-3)3 \\ 5 \end{bmatrix}$	$\begin{bmatrix} (-3)5 \\ 6 \end{bmatrix}$

From E. T. Goodwin and J. Staton, Table of $J_0(j_{0,n}r)$, Quart. J. Mech. Appl. Math. **1**, 220–224 (1948) (with permission).

Table 9.7 BESSEL FUNCTIONS—MISCELLANEOUS ZEROS

s^{th} Zero of $xJ_1(x) - \lambda J_0(x)$

$\lambda \backslash s$	1	2	3	4	5
0.00	0.0000	3.8317	7.0156	10.1735	13.3237
0.02	0.1995	3.8369	7.0184	10.1754	13.3252
0.04	0.2814	3.8421	7.0213	10.1774	13.3267
0.06	0.3438	3.8473	7.0241	10.1794	13.3282
0.08	0.3960	3.8525	7.0270	10.1813	13.3297
0.10	0.4417	3.8577	7.0298	10.1833	13.3312
0.20	0.6170	3.8835	7.0440	10.1931	13.3387
0.40	0.8516	3.9344	7.0723	10.2127	13.3537
0.60	1.0184	3.9841	7.1004	10.2322	13.3686
0.80	1.1490	4.0325	7.1282	10.2516	13.3835
1.00	1.2558	4.0795	7.1558	10.2710	13.3984

$\lambda^{-1} \backslash s$	1	2	3	4	5	$\langle\lambda\rangle$
1.00	1.2558	4.0795	7.1558	10.2710	13.3984	1
0.80	1.3659	4.1361	7.1898	10.2950	13.4169	1
0.60	1.5095	4.2249	7.2453	10.3346	13.4476	2
0.40	1.7060	4.3818	7.3508	10.4118	13.5079	3
0.20	1.9898	4.7131	7.6177	10.6223	13.6786	5
0.10	2.1795	5.0332	7.9569	10.9363	13.9580	10
0.08	2.2218	5.1172	8.0624	11.0477	14.0666	13
0.06	2.2656	5.2085	8.1852	11.1864	14.2100	17
0.04	2.3108	5.3068	8.3262	11.3575	14.3996	25
0.02	2.3572	5.4112	8.4840	11.5621	14.6433	50
0.00	2.4048	5.5201	8.6537	11.7915	14.9309	∞

s^{th} Zero of $J_1(x) - \lambda x J_0(x)$

$\lambda \backslash s$	1	2	3	4	5
0.5	0.0000	5.1356	8.4172	11.6198	14.7960
0.6	1.1231	5.2008	8.4569	11.6486	14.8185
0.7	1.4417	5.2476	8.4853	11.6691	14.8346
0.8	1.6275	5.2826	8.5066	11.6845	14.8467
0.9	1.7517	5.3098	8.5231	11.6964	14.8561
1.0	1.8412	5.3314	8.5363	11.7060	14.8636

$\lambda^{-1} \backslash s$	1	2	3	4	5	$\langle\lambda\rangle$
1.00	1.8412	5.3314	8.5363	11.7060	14.8636	1
0.80	1.9844	5.3702	8.5600	11.7232	14.8771	1
0.60	2.1092	5.4085	8.5836	11.7404	14.8906	2
0.40	2.2192	5.4463	8.6072	11.7575	14.9041	3
0.20	2.3171	5.4835	8.6305	11.7745	14.9175	5
0.10	2.3621	5.5019	8.6421	11.7830	14.9242	10
0.08	2.3709	5.5055	8.6445	11.7847	14.9256	13
0.06	2.3795	5.5092	8.6468	11.7864	14.9269	17
0.04	2.3880	5.5128	8.6491	11.7881	14.9282	25
0.02	2.3965	5.5165	8.6514	11.7898	14.9296	50
0.00	2.4048	5.5201	8.6537	11.7915	14.9309	∞

$\langle\lambda\rangle$ = nearest integer to λ.

Compiled from H. S. Carslaw and J. C. Jaeger, Conduction of heat in solids (Oxford Univ. Press, London, England, 1947) and British Association for the Advancement of Science, Bessel functions, Part I. Functions of orders zero and unity, Mathematical Tables, vol. VI (Cambridge Univ. Press, Cambridge, England, 1950) (with permission).

BESSEL FUNCTIONS—MISCELLANEOUS ZEROS   Table 9.7

s^{th} Zero of $J_0(x)Y_0(\lambda x) - Y_0(x)J_0(\lambda x)$

$\lambda^{-1}\backslash s$	1	2	3	4	5	$\langle\lambda\rangle$
* 0.80	12.55847 031	25.12877	37.69646	50.26349	62.83026	1
0.60	4.69706 410	9.41690	14.13189	18.84558	23.55876	2
0.40	2.07322 886	4.17730	6.27537	8.37167	10.46723	3
0.20	0.76319 127	1.55710	2.34641	3.13403	3.92084	5
0.10	0.33139 387	0.68576	1.03774	1.38864	1.73896	10
0.08	0.25732 649	0.53485	0.81055	1.08536	1.35969	13 *
0.06	0.18699 458	0.39079	0.59334	0.79522	0.99673	17
0.04	0.12038 637	0.25340	0.38570	0.51759	0.64923	25
0.02	0.05768 450	0.12272	0.18751	0.25214	0.31666	50
0.00	0.00000 000	0.00000	0.00000	0.00000	0.00000	∞

s^{th} Zero of $J_1(x)Y_1(\lambda x) - Y_1(x)J_1(\lambda x)$

$\lambda^{-1}\backslash s$	1	2	3	4	5	$\langle\lambda\rangle$
* 0.80	12.59004 151	25.14465	37.70706	50.27145	62.83662	1
0.60	4.75805 426	9.44837	14.15300	18.86146	23.57148	2
0.40	2.15647 249	4.22309	6.30658	8.39528	10.48619	3
0.20	0.84714 961	1.61108	2.38532	3.16421	3.94541	5
0.10	0.39409 416	0.73306	1.07483	1.41886	1.76433	10
0.08	0.31223 576	0.57816	0.84552	1.11441	1.38440	13 *
0.06	0.23235 256	0.42843	0.62483	0.82207	1.02001	17
0.04	0.15400 729	0.28296	0.41157	0.54044	0.66961	25
0.02	0.07672 788	0.14062	0.20409	0.26752	0.33097	50
0.00	0.00000 000	0.00000	0.00000	0.00000	0.00000	∞

s^{th} Zero of $J_1(x)Y_0(\lambda x) - Y_1(x)J_0(\lambda x)$

$\lambda^{-1}\backslash s$	1	2	3	4	5	$\langle\lambda\rangle$
* * 0.80	6.56973 310	18.94971	31.47626	44.02544	56.58224	1
0.60	2.60328 138	7.16213	11.83783	16.53413	21.23751	2
0.40	1.24266 626	3.22655	5.28885	7.36856	9.45462	3
0.20	0.51472 663	1.24657	2.00959	2.78326	3.56157	5
0.10	0.24481 004	0.57258	0.90956	1.25099	1.59489	10
0.08	0.19461 772	0.45251	0.71635	0.98327	1.25203	13 *
0.06	0.14523 798	0.33597	0.53005	0.72594	0.92301	17
0.04	0.09647 602	0.22226	0.34957	0.47768	0.60634	25
0.02	0.04813 209	0.11059	0.17353	0.23666	0.29991	50
0.00	0.00000 000	0.00000	0.00000	0.00000	0.00000	∞

$\langle\lambda\rangle$ =nearest integer to λ.

Compiled from British Association for the Advancement of Science, Bessel functions, Part I. Functions of orders zero and unity, Mathematical Tables, vol. VI (Cambridge Univ. Press, Cambridge, England, 1950) (with permission).

*See page II.

Table 9.8   MODIFIED BESSEL FUNCTIONS—ORDERS 0, 1 AND 2

x	$e^{-x}I_0(x)$	$e^{-x}I_1(x)$	$x^{-2}I_2(x)$
0.0	1.00000 00000	0.00000 00000	0.12500 00000
0.1	0.90710 09258	0.04529 84468	0.12510 41992
0.2	0.82693 85516	0.08228 31235	0.12541 71878
0.3	0.75758 06252	0.11237 75606	0.12594 01407
0.4	0.69740 21705	0.13676 32243	0.12667 50222
0.5	0.64503 52706	0.15642 08032	0.12762 45967
0.6	0.59932 72031	0.17216 44195	0.12879 24416
0.7	0.55930 55265	0.18466 99828	0.13018 29658
0.8	0.52414 89420	0.19449 86933	0.13180 14318
0.9	0.49316 29662	0.20211 65309	0.13365 39819
1.0	0.46575 96077	0.20791 04154	0.13574 76698
1.1	0.44144 03776	0.21220 16132	0.13809 04952
1.2	0.41978 20789	0.21525 68594	0.14069 14455
1.3	0.40042 49127	0.21729 75878	0.14356 05405
1.4	0.38306 25154	0.21850 75923	0.14670 88837
1.5	0.36743 36090	0.21903 93874	0.15014 87192
1.6	0.35331 49978	0.21901 94899	0.15389 34944
1.7	0.34051 56880	0.21855 28066	0.15795 79288
1.8	0.32887 19497	0.21772 62788	0.16235 80900
1.9	0.31824 31629	0.21661 19112	0.16711 14772
2.0	0.30850 83225	0.21526 92892	0.17223 71119
2.1	0.29956 30945	0.21374 76721	0.17775 56370
2.2	0.29131 73331	0.21208 77328	0.18368 94251
2.3	0.28369 29857	0.21032 30051	0.19006 26964
2.4	0.27662 23231	0.20848 10887	0.19690 16460
2.5	0.27004 64416	0.20658 46495	0.20423 45837
2.6	0.26391 39957	0.20465 22544	0.21209 20841
2.7	0.25818 01238	0.20269 90640	0.22050 71509
2.8	0.25280 55337	0.20073 74113	0.22951 53938
2.9	0.24775 57304	0.19877 72816	0.23915 52213
3.0	0.24300 03542	0.19682 67133	0.24946 80490
3.1	0.23851 26187	0.19489 21309	0.26049 85252
3.2	0.23426 88316	0.19297 86229	0.27229 47757
3.3	0.23024 79845	0.19109 01727	0.28490 86686
3.4	0.22643 14011	0.18922 98511	0.29839 61010
3.5	0.22280 24380	0.18739 99766	0.31281 73100
3.6	0.21934 62245	0.18560 22484	0.32823 72078
3.7	0.21604 94417	0.18383 78580	0.34472 57467
3.8	0.21290 01308	0.18210 75810	0.36235 83128
3.9	0.20988 75279	0.18041 18543	0.38121 61528
4.0	0.20700 19211	0.17875 08394	0.40138 68359
4.1	0.20423 45274	0.17712 44763	0.42296 47539
4.2	0.20157 73840	0.17553 25260	0.44605 16629
4.3	0.19902 32571	0.17397 46091	0.47075 72701
4.4	0.19656 55589	0.17245 02337	0.49719 98689
4.5	0.19419 82777	0.17095 88223	0.52550 70272
4.6	0.19191 59151	0.16949 97311	0.55581 63319
4.7	0.18971 34330	0.16807 22681	0.58827 61978
4.8	0.18758 62042	0.16667 57058	0.62304 67409
4.9	0.18552 99721	0.16530 92936	0.66030 07270
5.0	0.18354 08126	0.16397 22669	0.70022 45988
	$\begin{bmatrix} (-3)2 \\ 9 \end{bmatrix}$	$\begin{bmatrix} (-3)1 \\ 9 \end{bmatrix}$	$\begin{bmatrix} (-4)3 \\ 7 \end{bmatrix}$

$$I_{n+1}(x) = -\frac{2n}{x}I_n(x) + I_{n-1}(x)$$

Compiled from British Association for the Advancement of Science, Bessel functions, Part I. Functions of orders zero and unity, Mathematical Tables, vol. VI , Part II. Functions of positive integer order, Mathematical Tables, vol. X (Cambridge Univ. Press, Cambridge, England, 1950, 1952) and L. Fox, A short table for Bessel functions of integer orders and large arguments. Royal Society Shorter Mathematical Tables No. 3 (Cambridge Univ. Press, Cambridge, England, 1954) (with permission).

MODIFIED BESSEL FUNCTIONS—ORDERS 0, 1 AND 2   Table 9.8

x	$e^x K_0(x)$	$e^x K_1(x)$	$x^2 K_2(x)$
0.0	∞	∞	2.00000 0000
0.1	2.68232 61023	10.89018 2683	1.99503 9646
0.2	2.14075 73233	5.83338 6037	1.98049 7172
0.3	1.85262 73007	4.12515 7762	1.95711 6625
0.4	1.66268 20891	3.25867 3880	1.92580 8202
0.5	1.52410 93857	2.73100 97082	1.88754 5888
0.6	1.41673 76214	2.37392 00376	1.84330 9881
0.7	1.33012 36562	2.11501 13128	1.79405 1681
0.8	1.25820 31216	1.91793 02990	1.74067 2762
0.9	1.19716 33803	1.76238 82197	1.68401 1992
1.0	1.14446 30797	1.63615 34863	1.62483 8899
1.1	1.09833 02828	1.53140 37541	1.56385 0953
1.2	1.05748 45322	1.44289 75522	1.50167 3576
1.3	1.02097 31613	1.36698 72841	1.43886 2011
1.4	0.98806 99961	1.30105 37400	1.37590 4446
1.5	0.95821 00533	1.24316 58736	1.31322 5917
1.6	0.93094 59808	1.19186 75654	1.25119 2681
1.7	0.90591 81386	1.14603 92462	1.19011 6819
1.8	0.88283 35270	1.10480 53726	1.13026 0897
1.9	0.86145 06168	1.06747 09298	1.07184 2567
2.0	0.84156 82151	1.03347 68471	1.01503 9018
2.1	0.82301 71525	1.00236 80527	0.95999 1226
2.2	0.80565 39812	0.97377 01679	0.90680 7952
2.3	0.78935 61312	0.94737 22250	0.85556 9487
2.4	0.77401 81407	0.92291 36650	0.80633 1113
2.5	0.75954 86903	0.90017 44239	0.75912 6289
2.6	0.74586 82430	0.87896 72806	0.71396 9565
2.7	0.73290 71515	0.85913 18867	0.67085 9227
2.8	0.72060 41251	0.84053 00604	0.62977 9698
2.9	0.70890 49774	0.82304 20403	0.59070 3688
3.0	0.69776 15980	0.80656 34800	0.55359 4126
3.1	0.68713 11010	0.79100 30157	0.51840 5885
3.2	0.67697 51139	0.77628 02824	0.48508 7306
3.3	0.66725 91831	0.76232 42864	0.45358 1550
3.4	0.65795 22725	0.74907 20613	0.42382 7789
3.5	0.64902 63377	0.73646 75480	0.39576 2241
3.6	0.64045 59647	0.72446 06608	0.36931 9074
3.7	0.63221 80591	0.71300 65010	0.34443 1194
3.8	0.62429 15812	0.70206 46931	0.32103 0914
3.9	0.61665 73147	0.69159 88206	0.29905 0529
4.0	0.60929 76693	0.68157 59452	0.27842 2808
4.1	0.60219 65064	0.67196 61952	0.25908 1398
4.2	0.59533 89889	0.66274 24110	0.24096 1165
4.3	0.58871 14486	0.65387 98395	0.22399 8474
4.4	0.58230 12704	0.64535 58689	0.20813 1411
4.5	0.57609 67897	0.63714 97988	0.19329 9963
4.6	0.57008 72022	0.62924 26383	0.17944 6150
4.7	0.56426 24840	0.62161 69312	0.16651 4127
4.8	0.55861 33194	0.61425 66003	0.15445 0249
4.9	0.55313 10397	0.60714 68131	0.14320 3117
5.0	0.54780 75643	0.60027 38587	0.13272 3593 $\begin{bmatrix} (-3)1 \\ 11 \end{bmatrix}$

$$K_{n+1}(x) = \frac{2n}{x} K_n(x) + K_{n-1}(x)$$

Table 9.8   MODIFIED BESSEL FUNCTIONS—ORDERS 0, 1 AND 2

x	$e^{-x}I_0(x)$	$e^{-x}I_1(x)$	$e^{-x}I_2(x)$
5.0	0.18354 08126	0.16397 22669	0.11795 1906
5.1	0.18161 51021	0.16266 38546	0.11782 5355
5.2	0.17974 94883	0.16138 32850	0.11767 8994
5.3	0.17794 08646	0.16012 97913	0.11751 4528
5.4	0.17618 63475	0.15890 26150	0.11733 3527
5.5	0.17448 32564	0.15770 10090	0.11713 7435
5.6	0.17282 90951	0.15652 42405	0.11692 7581
5.7	0.17122 15362	0.15537 15922	0.11670 5188
5.8	0.16965 84061	0.15424 23641	0.11647 1384
5.9	0.16813 76726	0.15313 58742	0.11622 7207
6.0	0.16665 74327	0.15205 14593	0.11597 3613
6.1	0.16521 59021	0.15098 84754	0.11571 1484
6.2	0.16381 14064	0.14994 62978	0.11544 1633
6.3	0.16244 23718	0.14892 43212	0.11516 4809
6.4	0.16110 73175	0.14792 19595	0.11488 1705
6.5	0.15980 48490	0.14693 86457	0.11459 2958
6.6	0.15853 36513	0.14597 38314	0.11429 9157
6.7	0.15729 24831	0.14502 69866	0.11400 0845
6.8	0.15608 01720	0.14409 75991	0.11369 8525
6.9	0.15489 56090	0.14318 51745	0.11339 2660
7.0	0.15373 77447	0.14228 92347	0.11308 3678
7.1	0.15260 55844	0.14140 93186	0.11277 1974
7.2	0.15149 81855	0.14054 49809	0.11245 7913
7.3	0.15041 46530	0.13969 57915	0.11214 1833
7.4	0.14935 41371	0.13886 13353	0.11182 4046
7.5	0.14831 58301	0.13804 12115	0.11150 4840
7.6	0.14729 89636	0.13723 50333	0.11118 4481
7.7	0.14630 28062	0.13644 24270	0.11086 3215
7.8	0.14532 66611	0.13566 30318	0.11054 1268
7.9	0.14436 98642	0.13489 64995	0.11021 8852
8.0	0.14343 17818	0.13414 24933	0.10989 6158
8.1	0.14251 18095	0.13340 06883	0.10957 3368
8.2	0.14160 93695	0.13267 07705	0.10925 0645
8.3	0.14072 39098	0.13195 24362	0.10892 8142
8.4	0.13985 49027	0.13124 53923	0.10860 6000
8.5	0.13900 18430	0.13054 93551	0.10828 4348
8.6	0.13816 42474	0.12986 40505	0.10796 3305
8.7	0.13734 16526	0.12918 92134	0.10764 2983
8.8	0.13653 36147	0.12852 45873	0.10732 3481
8.9	0.13573 97082	0.12786 99242	0.10700 4894
9.0	0.13495 95247	0.12722 49839	0.10668 7306
9.1	0.13419 26720	0.12658 95342	0.10637 0796
9.2	0.13343 87740	0.12596 33501	0.10605 5437
9.3	0.13269 74691	0.12534 62139	0.10574 1294
9.4	0.13196 84094	0.12473 79145	0.10542 8428
9.5	0.13125 12609	0.12413 82477	0.10511 6893
9.6	0.13054 57016	0.12354 70154	0.10480 6740
9.7	0.12985 14223	0.12296 40258	0.10449 8015
9.8	0.12916 81248	0.12238 90929	0.10419 0759
9.9	0.12849 55220	0.12182 20364	0.10388 5010
10.0	0.12783 33371	0.12126 26814	0.10358 0801
	$\begin{bmatrix} (-6)8 \\ 6 \end{bmatrix}$	$\begin{bmatrix} (-6)3 \\ 5 \end{bmatrix}$	$\begin{bmatrix} (-6)2 \\ 5 \end{bmatrix}$

MODIFIED BESSEL FUNCTIONS—ORDERS 0, 1 AND 2   Table 9.8

x	$e^x K_0(x)$	$e^x K_1(x)$	$e^x K_2(x)$
5.0	0.54780 75643	0.60027 38587	0.78791 711
5.1	0.54263 53519	0.59362 50463	0.77542 949
5.2	0.53760 73540	0.58718 86062	0.76344 913
5.3	0.53271 69744	0.58095 36085	0.75194 475
5.4	0.52795 80329	0.57490 98871	0.74088 762
5.5	0.52332 47316	0.56904 79741	0.73025 127
5.6	0.51881 16252	0.56335 90393	0.72001 128
5.7	0.51441 35938	0.55783 48348	0.71014 511
5.8	0.51012 58183	0.55246 76495	0.70063 190
5.9	0.50594 37583	0.54725 02639	0.69145 232
6.0	0.50186 31309	0.54217 59104	0.68258 843
6.1	0.49787 98929	0.53723 82386	0.67402 358
6.2	0.49399 02237	0.53243 12833	0.66574 225
6.3	0.49019 05093	0.52774 94344	0.65773 001
6.4	0.48647 73291	0.52318 74101	0.64997 339
6.5	0.48284 74413	0.51874 02336	0.64245 982
6.6	0.47929 77729	0.51440 32108	0.63517 753
6.7	0.47582 54066	0.51017 19097	0.62811 553
6.8	0.47242 75723	0.50604 21421	0.62126 350
6.9	0.46910 16370	0.50200 99471	0.61461 177
7.0	0.46584 50959	0.49807 15749	0.60815 126
7.1	0.46265 55657	0.49422 34737	0.60187 345
7.2	0.45953 07756	0.49046 22755	0.59577 030
7.3	0.45646 85618	0.48678 47842	0.58983 426
7.4	0.45346 68594	0.48318 79648	0.58405 820
7.5	0.45052 36991	0.47966 89336	0.57843 541
7.6	0.44763 71996	0.47622 49486	0.57295 955
7.7	0.44480 55636	0.47285 33995	0.56762 463
7.8	0.44202 70724	0.46955 18010	0.56242 497
7.9	0.43930 00819	0.46631 77847	0.55735 522
8.0	0.43662 30185	0.46314 90928	0.55241 029
8.1	0.43399 43754	0.46004 35709	0.54758 538
8.2	0.43141 27084	0.45699 91615	0.54287 592
8.3	0.42887 66329	0.45401 39001	0.53827 757
8.4	0.42638 48214	0.45108 59089	0.53378 623
8.5	0.42393 59993	0.44821 33915	0.52939 797
8.6	0.42152 89433	0.44539 46295	0.52510 909
8.7	0.41916 24781	0.44262 79775	0.52091 604
8.8	0.41683 54743	0.43991 18594	0.51681 544
8.9	0.41454 68462	0.43724 47648	0.51280 410
9.0	0.41229 55493	0.43462 52454	0.50887 894
9.1	0.41008 05783	0.43205 19116	0.50503 704
9.2	0.40790 09662	0.42952 34301	0.50127 562
9.3	0.40575 57809	0.42703 85204	0.49759 202
9.4	0.40364 41245	0.42459 59520	0.49398 369
9.5	0.40156 51322	0.42219 45430	0.49044 819
9.6	0.39951 79693	0.41983 31565	0.48698 321
9.7	0.39750 18313	0.41751 06989	0.48358 651
9.8	0.39551 59416	0.41522 61179	0.48025 597
9.9	0.39355 95506	0.41297 84003	0.47698 953
10.0	0.39163 19344	0.41076 65704	0.47378 525
	$\begin{bmatrix} (-5)2 \\ 6 \end{bmatrix}$	$\begin{bmatrix} (-5)3 \\ 6 \end{bmatrix}$	$\begin{bmatrix} (-5)6 \\ 5 \end{bmatrix}$

Table 9.8   MODIFIED BESSEL FUNCTIONS—ORDERS 0, 1 AND 2

x	$e^{-x}I_0(x)$	$e^{-x}I_1(x)$	$e^{-x}I_2(x)$
10.0	0.12783 33371	0.12126 26814	0.10358 0801
10.2	0.12653 91639	0.12016 64024	0.10297 7124
10.4	0.12528 35822	0.11909 89584	0.10237 9936
10.6	0.12406 47082	0.11805 91273	0.10178 9401
10.8	0.12288 07840	0.11704 57564	0.10120 5644
11.0	0.12173 01682	0.11605 77582	0.10062 8758
11.2	0.12061 13250	0.11509 41055	0.10005 8806
11.4	0.11952 28165	0.11415 38276	0.09949 5829
11.6	0.11846 32942	0.11323 60059	0.09893 9845
11.8	0.11743 14923	0.11233 97710	0.09839 0853
12.0	0.11642 62212	0.11146 42993	0.09784 8838
12.2	0.11544 63616	0.11060 88096	0.09731 3770
12.4	0.11449 08594	0.10977 25611	0.09678 5608
12.6	0.11355 87206	0.10895 48501	0.09626 4300
12.8	0.11264 90074	0.10815 50080	0.09574 9787
13.0	0.11176 08338	0.10737 23993	0.09524 2003
13.2	0.11089 33621	0.10660 64190	0.09474 0874
13.4	0.11004 57995	0.10585 64916	0.09424 6323
13.6	0.10921 73954	0.10512 20685	0.09375 8268
13.8	0.10840 74378	0.10440 26267	0.09327 6622
14.0	0.10761 52517	0.10369 76675	0.09280 1299
14.2	0.10684 01959	0.10300 67148	0.09233 2208
14.4	0.10608 16613	0.10232 93142	0.09186 9257
14.6	0.10533 90688	0.10166 50311	0.09141 2352
14.8	0.10461 18671	0.10101 34506	0.09096 1401
15.0	0.10389 95314	0.10037 41751	0.09051 6308
15.2	0.10320 15618	0.09974 68245	0.09007 6980
15.4	0.10251 74813	0.09913 10348	0.08964 3321
15.6	0.10184 68351	0.09852 64572	0.08921 5238
15.8	0.10118 91887	0.09793 27574	0.08879 2637
16.0	0.10054 41273	0.09734 96147	0.08837 5426
16.2	0.09991 12544	0.09677 67216	0.08796 3511
16.4	0.09929 01906	0.09621 37828	0.08755 6802
16.6	0.09868 05729	0.09566 05145	0.08715 5210
16.8	0.09808 20539	0.09511 66444	0.08675 8644
17.0	0.09749 43005	0.09458 19107	0.08636 7017
17.2	0.09691 69938	0.09405 60614	0.08598 0242
17.4	0.09634 98277	0.09353 88542	0.08559 8235
17.6	0.09579 25085	0.09303 00560	0.08522 0911
17.8	0.09524 47546	0.09252 94423	0.08484 8188
18.0	0.09470 62952	0.09203 67968	0.08447 9984
18.2	0.09417 68703	0.09155 19113	0.08411 6221
18.4	0.09365 62299	0.09107 45848	0.08375 6819
18.6	0.09314 41336	0.09060 46237	0.08340 1701
18.8	0.09264 03503	0.09014 18411	0.08305 0793
19.0	0.09214 46572	0.08968 60569	0.08270 4020
19.2	0.09165 68400	0.08923 70968	0.08236 1309
19.4	0.09117 66923	0.08879 47929	0.08202 2590
19.6	0.09070 40151	0.08835 89829	0.08168 7792
19.8	0.09023 86167	0.08792 95099	0.08135 6848
20.0	0.08978 03119	0.08750 62222	0.08102 9690
	$\begin{bmatrix} (-6)5 \\ 6 \end{bmatrix}$	$\begin{bmatrix} (-6)4 \\ 6 \end{bmatrix}$	$\begin{bmatrix} (-7)9 \\ 5 \end{bmatrix}$

MODIFIED BESSEL FUNCTIONS—ORDERS 0, 1 AND 2   Table 9.8

x	$e^x K_0(x)$	$e^x K_1(x)$	$e^x K_2(x)$
10.0	0.39163 19344	0.41076 65704	0.47378 525
10.2	0.38786 02539	0.40644 68479	0.46755 571
10.4	0.38419 55846	0.40225 98277	0.46155 324
10.6	0.38063 29549	0.39819 88825	0.45576 482
10.8	0.37716 77125	0.39425 78391	0.45017 842
11.0	0.37379 54971	0.39043 09362	0.44478 294
11.2	0.37051 22156	0.38671 27920	0.43956 807
11.4	0.36731 40243	0.38309 83725	0.43452 427
11.6	0.36419 73076	0.37958 29618	0.42964 265
11.8	0.36115 86616	0.37616 21391	0.42491 496
12.0	0.35819 48784	0.37283 17534	0.42033 350
12.2	0.35530 29318	0.36958 79032	0.41589 111
12.4	0.35247 99643	0.36642 69191	0.41158 108
12.6	0.34972 32746	0.36334 53438	0.40739 714
12.8	0.34703 03081	0.36033 99192	0.40333 342
13.0	0.34439 86455	0.35740 75702	0.39938 443
13.2	0.34182 59943	0.35454 53922	0.39554 499
13.4	0.33931 01806	0.35175 06397	0.39181 028
13.6	0.33684 91405	0.34902 07143	0.38817 572
13.8	0.33444 09142	0.34635 31558	0.38463 702
14.0	0.33208 36383	0.34374 56322	0.38119 016
14.2	0.32977 55402	0.34119 59314	0.37783 131
14.4	0.32751 49332	0.33870 19539	0.37455 687
14.6	0.32530 02091	0.33626 17039	0.37136 346
14.8	0.32312 98364	0.33387 32858	0.36824 785
15.0	0.32100 23534	0.33153 48949	0.36520 701
15.2	0.31891 63655	0.32924 48132	0.36223 805
15.4	0.31687 05405	0.32700 14043	0.35933 826
15.6	0.31486 36051	0.32480 31080	0.35650 503
15.8	0.31289 43424	0.32264 84361	0.35373 592
16.0	0.31096 15880	0.32053 59682	0.35102 858
16.2	0.30906 42269	0.31846 43471	0.34838 081
16.4	0.30720 11919	0.31643 22766	0.34579 049
16.6	0.30537 14592	0.31443 85164	0.34325 562
16.8	0.30357 40487	0.31248 18807	0.34077 427
17.0	0.30180 80193	0.31056 12340	0.33834 464
17.2	0.30007 24678	0.30867 54888	0.33596 497
17.4	0.29836 65276	0.30682 36027	0.33363 361
17.6	0.29668 93657	0.30500 45765	0.33134 898
17.8	0.29504 01817	0.30321 74518	0.32910 956
18.0	0.29341 82062	0.30146 13089	0.32691 391
18.2	0.29182 26987	0.29973 52642	0.32476 064
18.4	0.29025 29472	0.29803 84697	0.32264 843
18.6	0.28870 82654	0.29637 01096	0.32057 602
18.8	0.28718 79933	0.29472 94003	0.31854 218
19.0	0.28569 14944	0.29311 55877	0.31654 577
19.2	0.28421 81554	0.29152 79458	0.31458 565
19.4	0.28276 73848	0.28996 57766	0.31266 076
19.6	0.28133 86117	0.28842 84068	0.31077 008
19.8	0.27993 12862	0.28691 51886	0.30891 262
20.0	0.27854 48766	0.28542 54970	0.30708 743
	$\begin{bmatrix} (-5)1 \\ 6 \end{bmatrix}$	$\begin{bmatrix} (-5)2 \\ 6 \end{bmatrix}$	$\begin{bmatrix} (-5)3 \\ 5 \end{bmatrix}$

Table 9.8   **MODIFIED BESSEL FUNCTIONS—AUXILIARY TABLE FOR LARGE ARGUMENTS**

x^{-1}	$x^{\frac{1}{2}}e^{-x}I_0(x)$	$x^{\frac{1}{2}}e^{-x}I_1(x)$	$x^{\frac{1}{2}}e^{-x}I_2(x)$	$\pi^{-1}x^{\frac{1}{2}}e^{x}K_0(x)$	$\pi^{-1}x^{\frac{1}{2}}e^{x}K_1(x)$	$\pi^{-1}x^{\frac{1}{2}}e^{x}K_2(x)$	$\langle x\rangle$
0.050	0.40150 9761	0.39133 9722	0.36237 579	0.39651 5620	0.40631 0355	0.43714 666	20
0.048	0.40140 4058	0.39164 8743	0.36380 578	0.39661 0241	0.40601 9771	0.43558 814	21
0.046	0.40129 8619	0.39195 7336	0.36523 854	0.39670 5057	0.40572 8854	0.43403 211	22
0.044	0.40119 3443	0.39226 5502	0.36667 408	0.39680 0069	0.40543 7604	0.43247 858	23
0.042	0.40108 8526	0.39257 3245	0.36811 237	0.39689 5278	0.40514 6017	0.43092 754	24
0.040	0.40098 3868	0.39288 0567	0.36955 342	0.39699 0686	0.40485 4094	0.42937 901	25
0.038	0.40087 9466	0.39318 7470	0.37099 722	0.39708 6293	0.40456 1832	0.42783 299	26
0.036	0.40077 5319	0.39349 3958	0.37244 375	0.39718 2101	0.40426 9230	0.42628 949	28
0.034	0.40067 1424	0.39380 0032	0.37389 302	0.39727 8110	0.40397 6286	0.42474 850	29
0.032	0.40056 7781	0.39410 5695	0.37534 502	0.39737 4322	0.40368 2998	0.42321 003	31
0.030	0.40046 4387	0.39441 0950	0.37679 973	0.39747 0738	0.40338 9365	0.42167 410	33
0.028	0.40036 1241	0.39471 5798	0.37825 716	0.39756 7359	0.40309 5386	0.42014 070	36
0.026	0.40025 8340	0.39502 0243	0.37971 729	0.39766 4186	0.40280 1058	0.41860 984	38
0.024	0.40015 5684	0.39532 4286	0.38118 012	0.39776 1221	0.40250 6380	0.41708 153	42
0.022	0.40005 3270	0.39562 7929	0.38264 564	0.39785 8465	0.40221 1349	0.41555 576	45
0.020	0.39995 1098	0.39593 1176	0.38411 385	0.39795 5918	0.40191 5965	0.41403 256	50
0.018	0.39984 9164	0.39623 4028	0.38558 474	0.39805 3583	0.40162 0226	0.41251 191	56
0.016	0.39974 7469	0.39653 6487	0.38705 830	0.39815 1460	0.40132 4130	0.41099 383	63
0.014	0.39964 6009	0.39683 8556	0.38853 453	0.39824 9551	0.40102 7674	0.40947 833	71
0.012	0.39954 4785	0.39714 0236	0.39001 342	0.39834 7857	0.40073 0858	0.40796 540	83
0.010	0.39944 3793	0.39744 1530	0.39149 496	0.39844 6379	0.40043 3679	0.40645 505	100
0.008	0.39934 3033	0.39774 2440	0.39297 915	0.39854 5119	0.40013 6136	0.40494 730	125
0.006	0.39924 2503	0.39804 2968	0.39446 599	0.39864 4077	0.39983 8226	0.40344 214	167
0.004	0.39914 2202	0.39834 3116	0.39595 546	0.39874 3256	0.39953 9949	0.40193 958	250
0.002	0.39904 2128	0.39864 2886	0.39744 756	0.39884 2657	0.39924 1300	0.40043 962	500
0.000	0.39894 2280	0.39894 2280	0.39894 228	0.39894 2280	0.39894 2280	0.39894 228	∞
	$\left[\begin{array}{c}(-8)3\\3\end{array}\right]$	$\left[\begin{array}{c}(-8)5\\3\end{array}\right]$	$\left[\begin{array}{c}(-7)3\\3\end{array}\right]$	$\left[\begin{array}{c}(-8)3\\3\end{array}\right]$	$\left[\begin{array}{c}(-8)5\\3\end{array}\right]$	$\left[\begin{array}{c}(-7)3\\3\end{array}\right]$	

For interpolating near $x^{-1}=0$ note that if $f_n(x^{-1})=x^{\frac{1}{2}}e^{-x}I_n(x)$ then $f_n(-x^{-1})=\pi^{-1}x^{\frac{1}{2}}e^{x}K_n(x)$.

$\langle x\rangle$ = nearest integer to x.

Compiled from L. Fox, A short table for Bessel functions of integer orders and large arguments. Royal Society Shorter Mathematical Tables No. 3 (Cambridge Univ. Press, Cambridge, England, 1954) (with permission).

MODIFIED BESSEL FUNCTIONS—AUXILIARY TABLE FOR SMALL ARGUMENTS

x	$K_0(x)+I_0(x)\ln x$	$x[K_1(x)-I_1(x)\ln x]$	x	$K_0(x)+I_0(x)\ln x$	$x[K_1(x)-I_1(x)\ln x]$
0.0	0.11593 152	1.00000 000	1.0	0.42102 444	0.60190 723
0.1	0.11872 387	0.99691 180	1.1	0.49199 896	0.49390 093
0.2	0.12713 128	0.98754 448	1.2	0.57261 444	0.36514 944
0.3	0.14124 511	0.97158 819	1.3	0.66373 364	0.21236 381
0.4	0.16121 862	0.94852 090	1.4	0.76632 938	+0.03176 677
0.5	0.18726 857	0.91759 992	1.5	0.88149 436	-0.18096 553
0.6	0.21967 734	0.87784 980	1.6	1.01045 200	-0.43076 964
0.7	0.25879 579	0.82804 659	1.7	1.15456 879	-0.72326 976
0.8	0.30504 682	0.76669 810	1.8	1.31536 786	-1.06486 242
0.9	0.35892 957	0.69201 997	1.9	1.49454 429	-1.46281 214
1.0	0.42102 444	0.60190 723	2.0	1.69398 200	-1.92535 914
	$\left[\begin{array}{c}(-3)1\\6\end{array}\right]$	$\left[\begin{array}{c}(-3)2\\7\end{array}\right]$		$\left[\begin{array}{c}(-3)3\\7\end{array}\right]$	$\left[\begin{array}{c}(-3)8\\7\end{array}\right]$

MODIFIED BESSEL FUNCTIONS—ORDERS 3–9 — Table 9.9

x	$e^{-x}I_3(x)$	$e^{-x}I_4(x)$	$e^{-x}I_5(x)$	$e^{-x}I_6(x)$	$e^{-x}I_7(x)$	$e^{-x}I_8(x)$	$e^{-x}I_9(x)$
0.0	0.0000	0.0000	0.0000	0.0000	0.0000	0.0000	0.0000
0.2	(−4)1.3680	(−6)3.4182	(−8)6.8341	* (−9)1.1388	(−11)1.6265	(−13)2.0328	(−15)2.2585
0.4	(−4)9.0273	(−5)4.5047	(−6)1.7995	(−8)5.9925	(−9)1.7109	(−11)4.2750	(−13)9.4957
0.6	(−3)2.5257	(−4)1.8858	(−5)1.1281	(−7)5.6286	(−8)2.4084	(−10)9.0201	(−11)3.0037
0.8	(−3)4.9877	(−4)4.9483	(−5)3.9377	(−6)2.6152	(−7)1.4902	(−9)7.4343	(−10)3.2983
1.0	(−3)8.1553	(−3)1.0069	(−5)9.9866	(−6)8.2731	(−7)5.8832	(−8)3.6643	(−9)2.0301
1.2	(−2)1.1855	(−3)1.7471	(−4)2.0719	(−5)2.0544	(−6)1.7497	(−7)1.3058	(−9)8.6707
1.4	(−2)1.5911	(−3)2.7189	(−4)3.7459	(−5)4.3203	(−6)4.2831	(−7)3.7225	(−8)2.8797
1.6	(−2)2.0168	(−3)3.9110	(−4)6.1288	(−5)8.0504	(−6)9.0974	(−7)9.0178	(−8)7.9596
1.8	(−2)2.4495	(−3)5.3023	(−4)9.2978	(−4)1.3686	(−5)1.7349	(−6)1.9302	(−7)1.9131
2.0	(−2)2.8791	(−3)6.8654	(−3)1.3298	(−4)2.1656	(−5)3.0402	(−6)3.7487	(−7)4.1199
2.2	(−2)3.2978	(−3)8.5701	(−3)1.8142	(−4)3.2349	(−5)4.9776	(−6)6.7325	(−7)8.1206
2.4	(−2)3.7001	(−2)1.0386	(−3)2.3819	(−4)4.6097	(−5)7.7080	(−5)1.1339	(−6)1.4883
2.6	(−2)4.0823	(−2)1.2283	(−3)3.0293	(−4)6.3166	(−4)1.1395	(−5)1.8099	(−6)2.5669
2.8	(−2)4.4421	(−2)1.4234	(−3)3.7511	(−4)8.3747	(−4)1.6197	(−5)2.7609	(−6)4.2048
3.0	(−2)4.7783	(−2)1.6216	(−3)4.5409	(−3)1.0796	(−4)2.2265	(−5)4.0512	(−6)6.5905
3.2	(−2)5.0907	(−2)1.8206	(−3)5.3913	(−3)1.3584	(−4)2.9735	(−5)5.7482	(−6)9.9425
3.4	(−2)5.3795	(−2)2.0188	(−3)6.2947	(−3)1.6738	(−4)3.8725	(−5)7.9208	(−5)1.4507
3.6	(−2)5.6454	(−2)2.2145	(−3)7.2431	(−3)2.0249	(−4)4.9334	(−4)1.0638	(−5)2.0556
3.8	(−2)5.8893	(−2)2.4065	(−3)8.2288	(−3)2.4106	(−4)6.1640	(−4)1.3965	(−5)2.8380
4.0	(−2)6.1124	(−2)2.5940	(−3)9.2443	(−3)2.8291	(−4)7.5698	(−4)1.7968	(−5)3.8284
4.2	(−2)6.3161	(−2)2.7761	(−2)1.0283	(−3)3.2785	(−4)9.1545	(−4)2.2703	(−5)5.0587
4.4	(−2)6.5015	(−2)2.9523	(−2)1.1337	(−3)3.7566	(−3)1.0919	(−4)2.8224	(−5)6.5607
4.6	(−2)6.6699	(−2)3.1221	(−2)1.2402	(−3)4.2609	(−3)1.2864	(−4)3.4578	(−5)8.3667
4.8	(−2)6.8227	(−2)3.2854	(−2)1.3471	(−3)4.7890	(−3)1.4986	(−4)4.1806	(−4)1.0508
5.0	(−2)6.9611	(−2)3.4419	(−2)1.4540	(−3)5.3384	(−3)1.7282	(−4)4.9939	(−4)1.3015
5.2	(−2)7.0861	(−2)3.5916	(−2)1.5605	(−3)5.9065	(−3)1.9747	(−4)5.9005	(−4)1.5916
5.4	(−2)7.1989	(−2)3.7346	(−2)1.6662	(−3)6.4909	(−3)2.2374	(−4)6.9020	(−4)1.9240
5.6	(−2)7.3005	(−2)3.8708	(−2)1.7707	(−3)7.0892	(−3)2.5157	(−4)7.9996	(−4)2.3010
5.8	(−2)7.3917	(−2)4.0005	(−2)1.8738	(−3)7.6990	(−3)2.8087	(−4)9.1937	(−4)2.7249
6.0	(−2)7.4736	(−2)4.1238	(−2)1.9752	(−3)8.3181	(−3)3.1156	(−3)1.0484	(−4)3.1978
6.2	(−2)7.5468	(−2)4.2408	(−2)2.0747	(−3)8.9445	(−3)3.4355	(−3)1.1870	(−4)3.7214
6.4	(−2)7.6121	(−2)4.3518	(−2)2.1723	(−3)9.5763	(−3)3.7674	(−3)1.3351	(−4)4.2971
6.6	(−2)7.6702	(−2)4.4570	(−2)2.2677	(−2)1.0212	(−3)4.1105	(−3)1.4924	(−4)4.9261
6.8	(−2)7.7216	(−2)4.5567	(−2)2.3608	(−2)1.0849	(−3)4.4637	(−3)1.6587	(−4)5.6094
7.0	(−2)7.7670	(−2)4.6509	(−2)2.4516	(−2)1.1486	(−3)4.8261	(−3)1.8337	(−4)6.3475
7.2	(−2)7.8068	(−2)4.7401	(−2)2.5401	(−2)1.2122	(−3)5.1969	(−3)2.0172	(−4)7.1409
7.4	(−2)7.8416	(−2)4.8244	(−2)2.6261	(−2)1.2756	(−3)5.5750	(−3)2.2089	(−4)7.9897
7.6	(−2)7.8717	(−2)4.9040	(−2)2.7096	(−2)1.3387	(−3)5.9596	(−3)2.4084	(−4)8.8937
7.8	(−2)7.8975	(−2)4.9791	(−2)2.7907	(−2)1.4012	(−3)6.3499	(−3)2.6152	(−4)9.8527
8.0	(−2)7.9194	(−2)5.0500	(−2)2.8694	(−2)1.4633	(−3)6.7449	(−3)2.8292	(−3)1.0866
8.2	(−2)7.9378	(−2)5.1169	(−2)2.9456	(−2)1.5247	(−3)7.1440	(−3)3.0497	(−3)1.1933
8.4	(−2)7.9528	(−2)5.1800	(−2)3.0195	(−2)1.5854	(−3)7.5464	(−3)3.2766	(−3)1.3053
8.6	(−2)7.9649	(−2)5.2395	(−2)3.0909	(−2)1.6453	(−3)7.9513	(−3)3.5093	(−3)1.4224
8.8	(−2)7.9741	(−2)5.2954	(−2)3.1601	(−2)1.7045	(−3)8.3582	(−3)3.7475	(−3)1.5446
9.0	(−2)7.9808	(−2)5.3482	(−2)3.2269	(−2)1.7627	(−3)8.7663	(−3)3.9907	(−3)1.6716
9.2	(−2)7.9852	(−2)5.3978	(−2)3.2915	(−2)1.8201	(−3)9.1750	(−3)4.2386	(−3)1.8035
9.4	(−2)7.9875	(−2)5.4445	(−2)3.3539	(−2)1.8765	(−3)9.5839	(−3)4.4908	(−3)1.9399
9.6	(−2)7.9878	(−2)5.4883	(−2)3.4141	(−2)1.9319	(−3)9.9924	(−3)4.7470	(−3)2.0808
9.8	(−2)7.9862	(−2)5.5296	(−2)3.4723	(−2)1.9864	(−2)1.0400	(−3)5.0066	(−3)2.2260
10.0	(−2)7.9830	(−2)5.5683	(−2)3.5284	(−2)2.0398	(−2)1.0806	(−3)5.2694	(−3)2.3753
10.5	(−2)7.9687	(−2)5.6549	(−2)3.6602	(−2)2.1690	(−2)1.1814	(−3)5.9380	(−3)2.7653
11.0	(−2)7.9465	(−2)5.7284	(−2)3.7804	(−2)2.2916	(−2)1.2805	(−3)6.6192	(−3)3.1769
11.5	(−2)7.9182	(−2)5.7905	(−2)3.8900	(−2)2.4078	(−2)1.3775	(−3)7.3082	(−3)3.6073
12.0	(−2)7.8848	(−2)5.8425	(−2)3.9898	(−2)2.5176	(−2)1.4722	(−3)8.0010	(−3)4.0537
12.5	(−2)7.8474	(−2)5.8857	(−2)4.0805	(−2)2.6212	(−2)1.5642	(−3)8.6939	(−3)4.5134
13.0	(−2)7.8067	(−2)5.9211	(−2)4.1630	(−2)2.7188	(−2)1.6533	(−3)9.3836	(−3)4.9837
13.5	(−2)7.7635	(−2)5.9497	(−2)4.2378	(−2)2.8106	(−2)1.7394	(−2)1.0068	(−3)5.4622
14.0	(−2)7.7183	(−2)5.9723	(−2)4.3056	(−2)2.8969	(−2)1.8225	(−2)1.0744	(−3)5.9469
14.5	(−2)7.6716	(−2)5.9896	(−2)4.3670	(−2)2.9779	(−2)1.9025	(−2)1.1410	(−3)6.4354
15.0	(−2)7.6236	(−2)6.0022	(−2)4.4225	(−2)3.0538	(−2)1.9794	(−2)1.2064	(−3)6.9260
15.5	(−2)7.5749	(−2)6.0106	(−2)4.4726	(−2)3.1251	(−2)2.0532	(−2)1.2705	(−3)7.4171
16.0	(−2)7.5256	(−2)6.0155	(−2)4.5179	(−2)3.1918	(−2)2.1240	(−2)1.3333	(−3)7.9071
16.5	(−2)7.4759	(−2)6.0170	(−2)4.5585	(−2)3.2543	(−2)2.1918	(−2)1.3946	(−3)8.3947
17.0	(−2)7.4260	(−2)6.0158	(−2)4.5951	(−2)3.3128	(−2)2.2567	(−2)1.4543	(−3)8.8788
17.5	(−2)7.3761	(−2)6.0119	(−2)4.6278	(−2)3.3675	(−2)2.3187	(−2)1.5125	(−3)9.3584
18.0	(−2)7.3263	(−2)6.0059	(−2)4.6571	(−2)3.4186	(−2)2.3780	(−2)1.5691	(−3)9.8324
18.5	(−2)7.2768	(−2)5.9978	(−2)4.6831	(−2)3.4664	(−2)2.4346	(−2)1.6240	(−2)1.0300
19.0	(−2)7.2275	(−2)5.9880	(−2)4.7062	(−2)3.5111	(−2)2.4886	(−2)1.6774	(−2)1.0761
19.5	(−2)7.1785	(−2)5.9767	(−2)4.7266	(−2)3.5528	(−2)2.5402	(−2)1.7291	(−2)1.1215
20.0	(−2)7.1300	(−2)5.9640	(−2)4.7444	(−2)3.5917	(−2)2.5894	(−2)1.7792	(−2)1.1661

Compiled from British Association for the Advancement of Science, Bessel functions, Part II. Functions of positive integer order, Mathematical Tables, vol. X (Cambridge Univ. Press, Cambridge, England, 1952) (with permission).

*See page II.

Table 9.9   **MODIFIED BESSEL FUNCTIONS—ORDERS 3–9**

x	$e^x K_3(x)$	$e^x K_4(x)$	$e^x K_5(x)$	$e^x K_6(x)$	$e^x K_7(x)$	$e^x K_8(x)$	$e^x K_9(x)$
0.0	∞	∞	∞	∞	∞	∞	∞
0.2	(3)1.2153	(4)3.6520	(6)1.4620	(7)7.3138	(9)4.3897	(11)3.0735	(13)2.4593
0.4	(2)1.8282	(3)2.7602	(4)5.5388	(6)1.3875	(7)4.1679	(9)1.4602	(10)5.8448
0.6	(1)6.4573	(2)6.5506	(3)8.7987	(5)1.4730	(6)2.9548	(7)6.9092	(9)1.8454
0.8	(1)3.2183	(2)2.4743	(3)2.5064	(4)3.1578	(5)4.7618	(6)8.3647	(8)1.6777
1.0	(1)1.9303	(2)1.2024	(2)9.8119	(3)9.9322	(5)1.2017	(6)1.6923	(7)2.7197
1.2	(1)1.2984	(1)6.8382	(2)4.6886	(3)3.9756	(4)4.0225	(5)4.7326	(6)6.3504
1.4	(0)9.4345	(1)4.3280	(2)2.5675	(3)1.8772	(4)1.6347	(5)1.6535	(6)1.9061
1.6	(0)7.2438	(1)2.9585	(2)1.5517	(2)9.9939	(3)7.6506	(4)6.7942	(5)6.8707
1.8	(0)5.7946	(1)2.1426	(2)1.0102	(2)5.8265	(3)3.9853	(4)3.1580	(5)2.8469
2.0	(0)4.7836	(1)1.6226	(1)6.9687	(2)3.6466	(3)2.2576	(4)1.6168	(5)1.3160
2.2	(0)4.0481	(1)1.2731	(1)5.0344	(2)2.4157	(3)1.3680	(3)8.9469	(4)6.6436
2.4	(0)3.4948	(1)1.0280	(1)3.7762	(2)1.6762	(2)8.7586	(3)5.2768	(4)3.6055
2.6	(0)3.0667	(0)8.4989	(1)2.9217	(2)1.2087	(2)5.8709	(3)3.2821	(4)2.0785
2.8	(0)2.7276	(0)7.1659	(1)2.3202	(1)9.0029	(2)4.0904	(3)2.1352	(4)1.2610
3.0	(0)2.4539	(0)6.1432	(1)1.8836	(1)6.8929	(2)2.9455	(3)1.4435	(3)7.9932
3.2	(0)2.2290	(0)5.3415	(1)1.5583	(1)5.4037	(2)2.1822	(3)1.0088	(3)5.2620
3.4	(0)2.0415	(0)4.7013	(1)1.3103	(1)4.3240	(2)1.6572	(2)7.2560	(3)3.5803
3.6	(0)1.8833	(0)4.1817	(1)1.1176	(1)3.5226	(2)1.2860	(2)5.3532	(3)2.5078
3.8	(0)1.7482	(0)3.7541	(0)9.6515	(1)2.9153	(2)1.0171	(2)4.0388	(3)1.8023
4.0	(0)1.6317	(0)3.3976	(0)8.4268	(1)2.4465	(1)8.1821	(2)3.1084	(3)1.3252
4.2	(0)1.5303	(0)3.0971	(0)7.4295	(1)2.0786	(1)6.6819	(2)2.4352	(2)9.9450
4.4	(0)1.4414	(0)2.8412	(0)6.6072	(1)1.7858	(1)5.5310	(2)1.9384	(2)7.6019
4.6	(0)1.3629	(0)2.6213	(0)5.9217	(1)1.5495	(1)4.6342	(2)1.5654	(2)5.9082
4.8	(0)1.2931	(0)2.4309	(0)5.3445	(1)1.3565	(1)3.9258	(2)1.2807	(2)4.6615
5.0	(0)1.2306	(0)2.2646	(0)4.8540	(1)1.1973	(1)3.3589	(2)1.0602	(2)3.7285
5.2	(0)1.1745	(0)2.1186	(0)4.4338	(1)1.0645	(1)2.9000	(1)8.8721	(2)3.0199
5.4	(0)1.1237	(0)1.9895	(0)4.0711	(0)9.5285	(1)2.5245	(1)7.4980	(2)2.4741
5.6	(0)1.0777	(0)1.8746	(0)3.7557	(0)8.5813	(1)2.2144	(1)6.3942	(2)2.0483
5.8	(0)1.0357	(0)1.7720	(0)3.4798	(0)7.7717	(1)1.9559	(1)5.4983	(2)1.7124
6.0	(–1)9.9723	(0)1.6798	(0)3.2370	(0)7.0748	(1)1.7387	(1)4.7644	(2)1.4444
6.2	(–1)9.6194	(0)1.5967	(0)3.0221	(0)6.4711	(1)1.5547	(1)4.1577	(2)1.2284
6.4	(–1)9.2942	(0)1.5213	(0)2.8311	(0)5.9448	(1)1.3978	(1)3.6521	(2)1.0528
6.6	(–1)8.9936	(0)1.4528	(0)2.6603	(0)5.4835	(1)1.2630	(1)3.2275	(1)9.0873
6.8	(–1)8.7149	(0)1.3902	(0)2.5071	(0)5.0771	(1)1.1467	(1)2.8685	(1)7.8960
7.0	(–1)8.4559	(0)1.3329	(0)2.3689	(0)4.7171	(1)1.0455	(1)2.5628	(1)6.9034
7.2	(–1)8.2145	(0)1.2803	(0)2.2440	(0)4.3970	(0)9.5723	(1)2.3010	(1)6.0705
7.4	(–1)7.9890	(0)1.2318	(0)2.1306	(0)4.1110	(0)8.7970	(1)2.0754	(1)5.3671
7.6	(–1)7.7778	(0)1.1870	(0)2.0273	(0)3.8544	(0)8.1132	(1)1.8800	(1)4.7692
7.8	(–1)7.5797	(0)1.1455	(0)1.9328	(0)3.6235	(0)7.5074	(1)1.7098	(1)4.2581
8.0	(–1)7.3935	(0)1.1069	(0)1.8463	(0)3.4148	(0)6.9684	(1)1.5610	(1)3.8188
8.2	(–1)7.2182	(0)1.0710	(0)1.7667	(0)3.2256	(0)6.4871	(1)1.4301	(1)3.4392
8.4	(–1)7.0527	(0)1.0376	(0)1.6934	(0)3.0535	(0)6.0556	(1)1.3146	(1)3.1096
8.6	(–1)6.8963	(0)1.0062	(0)1.6257	(0)2.8966	(0)5.6674	(1)1.2123	(1)2.8221
8.8	(–1)6.7483	(–1)9.7693	(0)1.5629	(0)2.7530	(0)5.3170	(1)1.1212	(1)2.5702
9.0	(–1)6.6079	(–1)9.4941	(0)1.5047	(0)2.6213	(0)4.9998	(1)1.0399	(1)2.3486
9.2	(–1)6.4746	(–1)9.2354	(0)1.4505	(0)2.5002	(0)4.7117	(0)9.6702	(1)2.1529
9.4	(–1)6.3480	(–1)8.9918	(0)1.4001	(0)2.3886	(0)4.4493	(0)9.0153	(1)1.9794
9.6	(–1)6.2274	(–1)8.7620	(0)1.3529	(0)2.2855	(0)4.2098	(0)8.4247	(1)1.8251
9.8	(–1)6.1125	(–1)8.5449	(0)1.3088	(0)2.1900	(0)3.9904	(0)7.8906	(1)1.6873
10.0	(–1)6.0028	(–1)8.3395	(0)1.2674	(0)2.1014	(0)3.7891	(0)7.4062	(1)1.5639
10.5	(–1)5.7493	(–1)7.8717	(0)1.1747	(0)1.9059	(0)3.3529	(0)6.3764	(1)1.3069
11.0	(–1)5.5217	(–1)7.4597	(0)1.0947	(0)1.7411	(0)2.9941	(0)5.5518	(1)1.1070
11.5	(–1)5.3161	(–1)7.0942	(0)1.0251	(0)1.6008	(0)2.6956	(0)4.8824	(0)9.4885
12.0	(–1)5.1294	(–1)6.7680	(–1)9.6415	(0)1.4803	(0)2.4444	(0)4.3321	(0)8.2205
12.5	(–1)4.9591	(–1)6.4751	(–1)9.1031	(0)1.3758	(0)2.2310	(0)3.8745	(0)7.1904
13.0	(–1)4.8030	(–1)6.2106	(–1)8.6249	(0)1.2845	(0)2.0482	(0)3.4902	(0)6.3439
13.5	(–1)4.6593	(–1)5.9706	(–1)8.1974	(0)1.2043	(0)1.8902	(0)3.1645	(0)5.6407
14.0	(–1)4.5266	(–1)5.7519	(–1)7.8133	(0)1.1333	(0)1.7527	(0)2.8860	(0)5.0510
14.5	(–1)4.4036	(–1)5.5517	(–1)7.4666	(0)1.0701	(0)1.6323	(0)2.6461	(0)4.5521
15.0	(–1)4.2892	(–1)5.3678	(–1)7.1520	(0)1.0136	(0)1.5261	(0)2.4379	(0)4.1265
15.5	(–1)4.1826	(–1)5.1982	(–1)6.8656	(–1)9.6276	(0)1.4319	(0)2.2561	(0)3.7608
16.0	(–1)4.0829	(–1)5.0414	(–1)6.6036	(–1)9.1686	(0)1.3480	(0)2.0964	(0)3.4444
16.5	(–1)3.9895	(–1)4.8959	(–1)6.3633	(–1)8.7524	(0)1.2729	(0)1.9552	(0)3.1689
17.0	(–1)3.9017	(–1)4.7605	(–1)6.1420	(–1)8.3734	(0)1.2053	(0)1.8299	(0)2.9275
17.5	(–1)3.8191	(–1)4.6343	(–1)5.9376	(–1)8.0272	(0)1.1442	(0)1.7181	(0)2.7150
18.0	(–1)3.7411	(–1)4.5162	(–1)5.7483	(–1)7.7097	(0)1.0888	(0)1.6178	(0)2.5269
18.5	(–1)3.6674	(–1)4.4055	(–1)5.5725	(–1)7.4176	(0)1.0384	(0)1.5276	(0)2.3595
19.0	(–1)3.5976	(–1)4.3015	(–1)5.4087	(–1)7.1482	(–1)9.9234	(0)1.4460	(0)2.2100
19.5	(–1)3.5313	(–1)4.2037	(–1)5.2559	(–1)6.8990	(–1)9.5015	(0)1.3721	(0)2.0759
20.0	(–1)3.4684	(–1)4.1114	(–1)5.1130	(–1)6.6679	(–1)9.1137	(0)1.3048	(0)1.9552

MODIFIED BESSEL FUNCTIONS—ORDERS 10, 11, 20 AND 21   Table 9.10

x	$10^9 x^{-10} I_{10}(x)$	$10^{11} x^{-11} I_{11}(x)$	$10^{-8} x^{10} K_{10}(x)$	$10^{24} x^{-20} I_{20}(x)$	$10^{26} x^{-21} I_{21}(x)$	$10^{-22} x^{20} K_{20}(x)$
0.0	0.26911 445	1.22324 748	1.85794 560	0.391990	0.933311	6.37771
0.2	0.26935 920	1.22426 724	1.85588 251	0.392177	0.933736	6.37435
0.4	0.27009 468	1.22733 125	1.84970 867	0.392738	0.935008	6.36429
0.6	0.27132 457	1.23245 366	1.83947 021	0.393674	0.937136	6.34757
0.8	0.27305 504	1.23965 820	1.82524 326	0.394988	0.940123	6.32424
1.0	0.27529 480	1.24897 831	1.80713 290	0.396684	0.943974	6.29437
1.2	0.27805 517	1.26045 740	1.78527 169	0.398766	0.948703	6.25807
1.4	0.28135 012	1.27414 918	1.75981 781	0.401239	0.954321	6.21545
1.6	0.28519 648	1.29011 798	1.73095 297	0.404112	0.960843	6.16665
1.8	0.28961 396	1.30843 932	1.69887 992	0.407392	0.968285	6.11184
2.0	0.29462 538	1.32920 036	1.66381 982	0.411087	0.976669	6.05118
2.2	0.30025 682	1.35250 061	1.62600 944	0.415209	0.986016	5.98488
2.4	0.30653 784	1.37845 262	1.58569 822	0.419768	0.996351	5.91314
2.6	0.31350 170	1.40718 285	1.54314 529	0.424778	1.007703	5.83620
2.8	0.32118 565	1.43883 260	1.49861 645	0.430253	1.020101	5.75428
3.0	0.32963 121	1.47355 907	1.45238 126	0.436209	1.033581	5.66764
3.2	0.33888 455	1.51153 657	1.40471 020	0.442662	1.048178	5.57655
3.4	0.34899 681	1.55295 782	1.35587 192	0.449632	1.063935	5.48128
3.6	0.36002 459	1.59803 551	1.30613 075	0.457139	1.080893	5.38210
3.8	0.37203 039	1.64700 388	1.25574 432	0.465205	1.099102	5.27932
4.0	0.38508 316	1.70012 064	1.20496 150	0.473853	1.118613	5.17321
4.2	0.39925 889	1.75766 896	1.15402 052	0.483111	1.139481	5.06408
4.4	0.41464 125	1.81995 978	1.10314 736	0.493006	1.161768	4.95224
4.6	0.43132 237	1.88733 435	1.05255 442	0.503569	1.185538	4.83797
4.8	0.44940 362	1.96016 700	1.00243 944	0.514832	1.210861	4.72159
5.0	0.46899 655	2.03886 82	0.95298 465	0.526830	1.237813	4.60339
5.2	0.49022 387	2.12388 83	0.90435 626	0.539601	1.266475	4.48367
5.4	0.51322 061	2.21572 08	0.85670 405	0.553186	1.296933	4.36272
5.6	0.53813 536	2.31490 71	0.81016 129	0.567630	1.329281	4.24084
5.8	0.56513 169	2.42204 09	0.76484 483	0.582979	1.363622	4.11830
6.0	0.59438 965	2.53777 36	0.72085 532	0.599284	1.400061	3.99537
6.2	0.62610 759	2.66282 00	0.67827 767	0.616599	1.438715	3.87234
6.4	0.66050 400	2.79796 48	0.63718 161	0.634984	1.479709	3.74945
6.6	0.69781 972	2.94406 93	0.59762 235	0.654501	1.523176	3.62695
6.8	0.73832 033	3.10208 00	0.55964 137	0.675219	1.569259	3.50507
7.0	0.78229 881	3.27303 69	0.52326 729	0.697210	1.618113	3.38405
7.2	0.83007 854	3.45808 34	0.48851 672	0.720554	1.669904	3.26411
7.4	0.88201 663	3.65847 74	0.45539 529	0.745333	1.724808	3.14542
7.6	0.93850 764	3.87560 29	0.42389 854	0.771639	1.783016	3.02821
7.8	0.99998 773	4.11098 38	0.39401 295	0.799570	1.844734	2.91264
8.0	1.06693 936	4.36629 90	0.36571 690	0.829231	1.910180	2.79887
8.2	1.13989 641	4.64339 88	0.33898 159	0.860735	1.979593	2.68705
8.4	1.21945 007	4.94432 35	0.31377 202	0.894204	2.053225	2.57733
8.6	1.30625 534	5.27132 42	0.29004 783	0.929769	2.131351	2.46983
8.8	1.40103 829	5.62688 64	0.26776 418	0.967571	2.214264	2.36466
9.0	1.50460 429	6.01375 48	0.24687 251	1.007764	2.302281	2.26193
9.2	1.61784 713	6.43496 31	0.22732 134	1.050510	2.395741	2.16172
9.4	1.74175 933	6.89386 57	0.20905 690	1.095988	2.495011	2.06411
9.6	1.87744 369	7.39417 36	0.19202 382	1.144389	2.600488	1.96916
9.8	2.02612 620	7.93999 51	0.17616 568	1.195919	2.712593	1.87692
10.0	2.18917 062	8.53588 02	0.16142 553	1.250800	2.831786	1.78744
	$\begin{bmatrix} (-3)2 \\ 7 \end{bmatrix}$	$\begin{bmatrix} (-3)6 \\ 6 \end{bmatrix}$	$\begin{bmatrix} (-4)5 \\ 6 \end{bmatrix}$	$\begin{bmatrix} (-4)4 \\ 4 \end{bmatrix}$	$\begin{bmatrix} (-4)9 \\ 5 \end{bmatrix}$	$\begin{bmatrix} (-4)8 \\ 4 \end{bmatrix}$

$$I_{n+1}(x) = -\frac{2n}{x} I_n(x) + I_{n-1}(x) \qquad\qquad K_{n+1}(x) = \frac{2n}{x} K_n(x) + K_{n-1}(x)$$

Compiled from British Association for the Advancement of Science, Bessel functions, Part II. Functions of positive integer order, Mathematical Tables, vol. X (Cambridge Univ. Press, Cambridge, England, 1952) and L. Fox, A short table for Bessel functions of integer orders and large arguments. Royal Society Shorter Mathematical Tables No. 3 (Cambridge Univ. Press, Cambridge, England, 1954) (with permission).

Table 9.10   **MODIFIED BESSEL FUNCTIONS—ORDERS 10, 11, 20 AND 21**

x	$e^{-x}I_{10}(x)$	$e^{-x}I_{11}(x)$	$e^{x}K_{10}(x)$	$10^{24}x^{-20}I_{20}(x)$	$10^{26}x^{-21}I_{21}(x)$	$10^{-22}x^{20}K_{20}(x)$
10.0	0.00099 38819	0.00038 75284	35.55633 91	1.25080	2.83179	1.787443
10.2	0.00107 29935	0.00042 45861	32.60759 68	1.30927	2.95856	1.700753
10.4	0.00115 52835	0.00046 37417	29.98423 91	1.37160	3.09345	1.616873
10.6	0.00124 06973	0.00050 50080	27.64297 29	1.43806	3.23703	1.535814
10.8	0.00132 91744	0.00054 83934	25.54714 23	1.50895	3.38992	1.457578
11.0	0.00142 06490	0.00059 39013	23.66558 79	1.58462	3.55278	1.382160
11.2	0.00151 50508	0.00064 15309	21.97172 20	1.66540	3.72634	1.309546
11.4	0.00161 23051	0.00069 12768	20.44277 46	1.75169	3.91139	1.239714
11.6	0.00171 23339	0.00074 31298	19.05917 72	1.84390	4.10876	1.172637
11.8	0.00181 50559	0.00079 70766	17.80405 56	1.94249	4.31937	1.108279
12.0	0.00192 03870	0.00085 31003	16.66281 24	2.04795	4.54421	1.046601
12.2	0.00202 82412	0.00091 11805	15.62277 97	2.16080	4.78434	0.987556
12.4	0.00213 85303	0.00097 12937	14.67293 16	2.28162	5.04093	0.931095
12.6	0.00225 11650	0.00103 34132	13.80364 34	2.41105	5.31521	0.877164
12.8	0.00236 60548	0.00109 75097	13.00649 01	2.54975	5.60856	0.825703
13.0	0.00248 31086	0.00116 35512	12.27407 71	2.69846	5.92244	0.776652
13.2	0.00260 22347	0.00123 15035	11.59989 74	2.85799	6.25845	0.729947
13.4	0.00272 33415	0.00130 13301	10.97821 07	3.02921	6.61832	0.685520
13.6	0.00284 63375	0.00137 29926	10.40394 07	3.21306	7.00393	0.643305
13.8	0.00297 11314	0.00144 64509	9.87258 79	3.41058	7.41731	0.603230
14.0	0.00309 76327	0.00152 16634	9.38015 52	3.62289	7.86068	0.565225
14.2	0.00322 57518	0.00159 85870	8.92308 36	3.85121	8.33644	0.529218
14.4	0.00335 53999	0.00167 71776	8.49819 79	4.09686	8.84722	0.495137
14.6	0.00348 64894	0.00175 73898	8.10265 95	4.36131	9.39585	0.462910
14.8	0.00361 89341	0.00183 91776	7.73392 53	4.64613	9.98543	0.432464
15.0	0.00375 26491	0.00192 24942	7.38971 31	4.95305	10.61932	0.403728
15.2	0.00388 75510	0.00200 72921	7.06797 04	5.28394	11.30119	0.376630
15.4	0.00402 35583	0.00209 35235	6.76684 87	5.64087	12.03503	0.351101
15.6	0.00416 05908	0.00218 11403	6.48467 94	6.02608	12.82520	0.327070
15.8	0.00429 85705	0.00227 00942	6.21995 46	6.44202	13.67643	0.304470
16.0	0.00443 74209	0.00236 03366	5.97130 87	6.89137	14.59389	0.283235
16.2	0.00457 70675	0.00245 18192	5.73750 35	7.37705	15.58322	0.263299
16.4	0.00471 74378	0.00254 44936	5.51741 43	7.90228	16.65059	0.244598
16.6	0.00485 84612	0.00263 83118	5.31001 78	8.47055	17.80271	0.227071
16.8	0.00500 00690	0.00273 32259	5.11438 19	9.08571	19.04691	0.210658
17.0	0.00514 21947	0.00282 91884	4.92965 63	9.75197	20.39124	0.195301
17.2	0.00528 47735	0.00292 61523	4.75506 40	10.47392	21.84444	0.180944
17.4	0.00542 77427	0.00302 40709	4.58989 42	11.25663	23.41611	0.167532
17.6	0.00557 10418	0.00312 28982	4.43349 60	12.10562	25.11674	0.155012
17.8	0.00571 46119	0.00322 25887	4.28527 20	13.02697	26.95781	0.143336
18.0	0.00585 83964	0.00332 30977	4.14467 40	14.02734	28.95188	0.132454
18.2	0.00600 23403	0.00342 43808	4.01119 75	15.11406	31.11272	0.122321
18.4	0.00614 63909	0.00352 63948	3.88437 85	16.29515	33.45541	0.112891
18.6	0.00629 04971	0.00362 90969	3.76378 89	17.57946	35.99648	0.104124
18.8	0.00643 46098	0.00373 24450	3.64903 41	18.97668	38.75407	0.095978
19.0	0.00657 86817	0.00383 63982	3.53974 93	20.49749	41.74804	0.088414
19.2	0.00672 26672	0.00394 09161	3.43559 74	22.15363	45.00024	0.081397
19.4	0.00686 65226	0.00404 59590	3.33626 62	23.95803	48.53460	0.074892
19.6	0.00701 02059	0.00415 14885	3.24146 65	25.92489	52.37745	0.068865
19.8	0.00715 36768	0.00425 74667	3.15093 00	28.06989	56.55768	0.063285
20.0	0.00729 68965	0.00436 38567	3.06440 75	30.41029	61.10706	0.058124
	$\begin{bmatrix}(-7)4\\5\end{bmatrix}$	$\begin{bmatrix}(-7)3\\5\end{bmatrix}$	$\begin{bmatrix}(-2)4\\8\end{bmatrix}$	$\begin{bmatrix}(-2)2\\5\end{bmatrix}$	$\begin{bmatrix}(-2)5\\5\end{bmatrix}$	$\begin{bmatrix}(-4)4\\4\end{bmatrix}$

MODIFIED BESSEL FUNCTIONS—AUXILIARY TABLE FOR LARGE ARGUMENTS   Table 9.10

x^{-1}	$\ln\left[x^{\frac{1}{2}}e^{-x}I_{10}(x)\right]$	$\ln\left[x^{\frac{1}{2}}e^{-x}I_{11}(x)\right]$	$\ln\left[\pi^{-1}x^{\frac{1}{2}}e^{x}K_{10}(x)\right]$	$\ln\left[x^{\frac{1}{2}}e^{-x}I_{20}(x)\right]$	$\ln\left[x^{\frac{1}{2}}e^{-x}I_{21}(x)\right]$	$\ln\left[\pi^{-1}x^{\frac{1}{2}}e^{x}K_{20}(x)\right]$	$\langle x\rangle$
0.050	−3.42244 002	−3.93653 292	1.47299 048	−10.434749	−11.346341	8.250182	20
0.049	−3.37318 689	−3.87762 888	1.42771 939	−10.263511	−11.160467	8.088946	20
0.048	−3.32386 306	−3.81861 524	1.38232 785	−10.091302	−10.973471	7.926737	21
0.047	−3.27447 055	−3.75949 454	1.33681 644	− 9.918126	−10.785351	7.763551	21
0.046	−3.22501 139	−3.70026 938	1.29118 575	− 9.743983	−10.596108	7.599386	22
0.045	−3.17548 766	−3.64094 242	1.24543 642	− 9.568876	−10.405744	7.434240	22
0.044	−3.12590 147	−3.58151 639	1.19956 910	− 9.392809	−10.214259	7.268110	23
0.043	−3.07625 496	−3.52199 408	1.15358 449	− 9.215785	−10.021658	7.100994	23
0.042	−3.02655 033	−3.46237 835	1.10748 332	− 9.037810	− 9.827944	6.932893	24
0.041	−2.97678 979	−3.40267 211	1.06126 635	− 8.858889	− 9.633121	6.763806	24
0.040	−2.92697 559	−3.34287 833	1.01493 437	− 8.679029	− 9.437195	6.593733	25
0.039	−2.87711 002	−3.28300 006	0.96848 822	− 8.498236	− 9.240173	6.422673	26
0.038	−2.82719 539	−3.22304 039	0.92192 874	− 8.316519	− 9.042063	6.250630	26
0.037	−2.77723 405	−3.16300 246	0.87525 686	− 8.133888	− 8.842873	6.077603	27
0.036	−2.72722 837	−3.10288 949	0.82847 349	− 7.950352	− 8.642612	5.903597	28
0.035	−2.67718 076	−3.04270 472	0.78157 961	− 7.765923	− 8.441293	5.728614	29
0.034	−2.62709 365	−2.98245 146	0.73457 624	− 7.580613	− 8.238927	5.552659	29
0.033	−2.57696 948	−2.92213 308	0.68746 441	− 7.394434	− 8.035529	5.375732	30
0.032	−2.52681 074	−2.86175 298	0.64024 520	− 7.207403	− 7.831113	5.197843	31
0.031	−2.47661 992	−2.80131 461	0.59291 975	− 7.019533	− 7.625695	5.018998	32
0.030	−2.42639 955	−2.74082 147	0.54548 920	− 6.830842	− 7.419294	4.839203	33
0.029	−2.37615 216	−2.68027 709	0.49795 475	− 6.641348	− 7.211929	4.658466	34
0.028	−2.32588 032	−2.61968 504	0.45031 764	− 6.451070	− 7.003620	4.476796	36
0.027	−2.27558 659	−2.55904 894	0.40257 915	− 6.260027	− 6.794389	4.294202	37
0.026	−2.22527 356	−2.49837 243	0.35474 059	− 6.068243	− 6.584261	4.110696	38
0.025	−2.17494 384	−2.43765 918	0.30680 331	− 5.875738	− 6.373261	3.926290	40
0.024	−2.12460 002	−2.37691 291	0.25876 871	− 5.682539	− 6.161416	3.740995	42
0.023	−2.07424 475	−2.31613 733	0.21063 822	− 5.488669	− 5.948754	3.554826	43
0.022	−2.02388 063	−2.25533 620	0.16241 332	− 5.294155	− 5.735305	3.367799	45
0.021	−1.97351 031	−2.19451 329	0.11409 551	− 5.099025	− 5.521102	3.179929	48
0.020	−1.92313 643	−2.13367 239	0.06568 636	− 4.903309	− 5.306177	2.991233	50
0.019	−1.87276 162	−2.07281 731	+0.01718 745	− 4.707035	− 5.090565	2.801730	53
0.018	−1.82238 853	−2.01195 186	−0.03139 959	− 4.510235	− 4.874302	2.611440	56
0.017	−1.77201 979	−1.95107 986	−0.08007 306	− 4.312943	− 4.657427	2.420383	59
0.016	−1.72165 806	−1.89020 514	−0.12883 128	− 4.115190	− 4.439978	2.228582	63
0.015	−1.67130 595	−1.82933 153	−0.17767 247	− 3.917011	− 4.221995	2.036059	67
0.014	−1.62096 610	−1.76846 286	−0.22659 485	− 3.718443	− 4.003521	1.842840	71
0.013	−1.57064 113	−1.70760 295	−0.27559 659	− 3.519520	− 3.784599	1.648949	77
0.012	−1.52033 365	−1.64675 564	−0.32467 581	− 3.320281	− 3.565272	1.454415	83
0.011	−1.47004 626	−1.58592 472	−0.37383 061	− 3.120763	− 3.345586	1.259264	91
0.010	−1.41978 154	−1.52511 400	−0.42305 904	− 2.921004	− 3.125587	1.063526	100
0.009	−1.36954 207	−1.46432 725	−0.47235 911	− 2.721043	− 2.905322	0.867231	111
0.008	−1.31933 040	−1.40356 824	−0.52172 881	− 2.520921	− 2.684838	0.670412	125
0.007	−1.26914 908	−1.34284 072	−0.57116 608	− 2.320676	− 2.464184	0.473099	143
0.006	−1.21900 063	−1.28214 841	−0.62066 881	− 2.120350	− 2.243408	0.275328	167
0.005	−1.16888 754	−1.22149 499	−0.67023 489	− 1.919982	− 2.022558	+0.077133	200
0.004	−1.11881 229	−1.16088 414	−0.71986 215	− 1.719613	− 1.801685	−0.121451	250
0.003	−1.06877 735	−1.10031 949	−0.76954 839	− 1.519284	− 1.580838	−0.320388	333
0.002	−1.01878 514	−1.03980 463	−0.81929 138	− 1.319036	− 1.360065	−0.519640	500
0.001	−0.96883 808	−0.97934 314	−0.86908 886	− 1.118907	− 1.139416	−0.719170	1000
0.000	−0.91893 853	−0.91893 853	−0.91893 853	− 0.918939	− 0.918939	−0.918939	∞
	$\left[\begin{smallmatrix}(-6)9\\4\end{smallmatrix}\right]$	$\left[\begin{smallmatrix}(-5)1\\4\end{smallmatrix}\right]$	$\left[\begin{smallmatrix}(-5)2\\4\end{smallmatrix}\right]$	$\left[\begin{smallmatrix}(-4)1\\4\end{smallmatrix}\right]$	$\left[\begin{smallmatrix}(-4)1\\4\end{smallmatrix}\right]$	$\left[\begin{smallmatrix}(-4)1\\4\end{smallmatrix}\right]$	

$\langle x\rangle$ = nearest integer to x.

Compiled from L. Fox, A short table for Bessel functions of integer orders and large arguments. Royal Society Shorter Mathematical Tables No. 3 (Cambridge Univ. Press, Cambridge, England, 1954) (with permission).

Table 9.11   **MODIFIED BESSEL FUNCTIONS—VARIOUS ORDERS**

n	$I_n(1)$	$I_n(2)$	$I_n(5)$
0	(0)1.26606 5878	(0)2.27958 5302	(1)2.72398 7182
1	(− 1)5.65159 1040	(0)1.59063 6855	(1)2.43356 4214
2	(− 1)1.35747 6698	(− 1)6.88948 4477	(1)1.75056 1497
3	(− 2)2.21684 2492	(− 1)2.12739 9592	(1)1.03311 5017
4	(− 3)2.73712 0221	(− 2)5.07285 6998	(0)5.10823 4764
5	(− 4)2.71463 1560	(− 3)9.82567 9323	(0)2.15797 4547
6	(− 5)2.24886 6148	(− 3)1.60017 3364	(− 1)7.92285 6690
7	(− 6)1.59921 8231	(− 4)2.24639 1420	(− 1)2.56488 9417
8	(− 8)9.96062 4033	(− 5)2.76993 6951	(− 2)7.41166 3216
9	(− 9)5.51838 5863	(− 6)3.04418 5903	(− 2)1.93157 1882
10	(− 10)2.75294 8040	(− 7)3.01696 3879	(− 3)4.58004 4419
11	(− 11)1.24897 8308	(− 8)2.72220 2336	(− 4)9.95541 1401
12	(− 13)5.19576 1153	(− 9)2.25413 0978	(− 4)1.99663 4027
13	(− 14)1.99563 1678	(− 10)1.72451 6264	(− 5)3.71568 0720
14	(− 16)7.11879 0054	(− 11)1.22598 3451	(− 6)6.44800 5272
15	(− 17)2.37046 3051	(− 13)8.13943 2531	(− 6)1.04797 7675
16	(− 19)7.40090 0286	(− 14)5.06857 1401	(− 7)1.60139 2190
17	(− 20)2.17495 9747	(− 15)2.97182 8970	(− 8)2.30866 7371
18	(− 22)6.03714 4636	(− 16)1.64621 5204	(− 9)3.14983 7806
19	(− 23)1.58767 8369	(− 18)8.64160 3385	(− 10)4.07841 5017
20	(− 25)3.96683 5986	(− 19)4.31056 0576	(− 11)5.02423 9358
30	(− 42)3.53950 0588	(− 33)3.89351 9664	(− 21)3.99784 4971
40	(− 60)1.12150 9741	(− 48)1.25586 9192	(− 32)1.18042 6980
50	(− 80)2.93463 5309	(− 65)3.35304 2830	(− 45)2.93146 9647
100	(−189)8.47367 4008	(−158)1.08217 1475	(−119)7.09355 1489

n	$I_n(10)$	$I_n(50)$	$I_n(100)$
0	(3)2.81571 6628	(20)2.93255 378	(42)1.07375 171
1	(3)2.67098 8304	(20)2.90307 859	(42)1.06836 939
2	(3)2.28151 8968	(20)2.81643 064	(42)1.05238 432
3	(3)1.75838 0717	(20)2.67776 414	(42)1.02627 402
4	(3)1.22649 0538	(20)2.49509 894	(41)9.90807 878
5	(2)7.77188 2864	(20)2.27854 831	(41)9.47009 387
6	(2)4.49302 2514	(20)2.03938 928	(41)8.96106 940
7	(2)2.38025 5848	(20)1.78909 488	(41)8.39476 555
8	(2)1.16066 4327	(20)1.53844 272	(41)7.78580 222
9	(1)5.23192 9250	(20)1.29679 321	(41)7.14903 719
10	(1)2.18917 0616	(20)1.07159 716	(41)6.49897 552
11	(0)8.53588 0176	(19)8.68154 347	(41)5.84924 209
12	(0)3.11276 9776	(19)6.89609 247	(41)5.21214 227
13	(0)1.06523 2713	(19)5.37141 909	(41)4.59832 794
14	(− 1)3.43164 7223	(19)4.10295 454	(41)4.01657 700
15	(− 1)1.04371 4907	(19)3.07376 455	(41)3.47368 638
16	(− 2)3.00502 5016	(19)2.25869 581	(41)2.97447 109
17	(− 3)8.21069 0206	(19)1.62819 923	(41)2.52185 563
18	(− 3)2.13390 3457	(19)1.15152 033	(41)2.11704 017
19	(− 4)5.28637 7589	(18)7.99104 593	(41)1.75972 117
20	(− 4)1.25079 9736	(18)5.44200 840	(41)1.44834 613
30	(−12)7.78756 9783	(16)4.27499 365	(40)1.20615 487
40	(−20)2.04212 3274	(13)6.00717 897	(38)3.84170 550
50	(−30)4.75689 4561	(+10)1.76508 024	(36)4.82195 809
100	(−88)1.08234 4202	(−16)2.72788 795	(21)4.64153 494

MODIFIED BESSEL FUNCTIONS—VARIOUS ORDERS Table 9.11

n	$K_n(1)$	$K_n(2)$	$K_n(5)$
0	(− 1)4.21024 4382	(−1)1.13893 8728	(−3)3.69109 8334
1	(− 1)6.01907 2302	(−1)1.39865 8818	(−3)4.04461 3445
2	(0)1.62483 8899	(−1)2.53759 7546	(−3)5.30894 3712
3	(0)7.10126 2825	(−1)6.47385 3909	(−3)8.29176 8415
4	(1)4.42324 1585	(0)2.19591 5927	(−2)1.52590 6581
5	(2)3.60960 5896	(0)9.43104 9101	(−2)3.27062 7371
6	(3)3.65383 8312	(1)4.93511 6143	(−2)8.06716 1323
7	(4)4.42070 2033	(2)3.05538 0177	(−1)2.26318 1455
8	(5)6.22552 1230	(3)2.18811 7285	(−1)7.14362 4206
9	(7)1.00050 4099	(4)1.78104 7630	(0)2.51227 7891
10	(8)1.80713 2899	(5)1.62482 4040	(0)9.75856 2829
11	(9)3.62427 0839	(6)1.64263 4516	(1)4.15465 2921
12	(10)7.99146 7175	(7)1.82314 6208	(2)1.92563 2913
13	(12)1.92157 6393	(8)2.20420 1795	(2)9.65850 3277
14	(13)5.00409 0088	(9)2.88369 3795	(3)5.21498 4995
15	(15)1.40306 6801	(10)4.05921 3332	(4)3.01697 6630
16	(16)4.21420 4494	(11)6.11765 6935	(5)1.86233 5828
17	(18)1.34994 8505	(12)9.82884 3230	(6)1.22206 4696
18	(19)4.59403 9121	(14)1.67702 1006	(6)8.49627 3517
19	(21)1.65520 4032	(15)3.02846 6654	(7)6.23952 3402
20	(22)6.29436 9360	(16)5.77085 6853	(8)4.82700 0521
30	(39)4.70614 5527	(30)4.27112 5755	(18)4.11213 2063
40	(58)1.11422 0651	(45)9.94083 9886	(30)1.05075 6722
50	(77)3.40689 6854	(62)2.97998 1740	(42)3.39432 2243
100	(185)5.90033 3184	(155)4.61941 5978	(115)7.03986 0193

n	$K_n(10)$	$K_n(50)$	$K_n(100)$
0	(−5)1.77800 6232	(−23)3.41016 774	(−45)4.65662 823
1	(−5)1.86487 7345	(−23)3.44410 222	(−45)4.67985 373
2	(−5)2.15098 1701	(−23)3.54793 183	(−45)4.75022 530
3	(−5)2.72527 0026	(−23)3.72793 677	(−45)4.86986 274
4	(−5)3.78614 3716	(−23)3.99528 424	(−45)5.04241 707
5	(−5)5.75418 4999	(−23)4.36718 224	(−45)5.27325 611
6	(−5)9.54032 8715	(−23)4.86872 069	(−45)5.56974 268
7	(−4)1.72025 7946	(−23)5.53567 521	(−45)5.94162 523
8	(−4)3.36239 3995	(−23)6.41870 975	(−45)6.40157 021
9	(−4)7.10008 8338	(−23)7.58966 233	(−45)6.96587 646
10	(−3)1.61425 5300	(−23)9.15098 819	(−45)7.65542 797
11	(−3)3.93851 9435	(−22)1.12500 576	(−45)8.49696 206
12	(−2)1.02789 9806	(−22)1.41010 135	(−45)9.52475 963
13	(−2)2.86081 1477	(−22)1.80185 441	(−44)1.07829 044
14	(−2)8.46600 9646	(−22)2.34706 565	(−44)1.23283 148
15	(−1)2.65656 3849	(−22)3.11621 117	(−44)1.42348 325
16	(−1)8.81629 2510	(−22)4.21679 235	(−44)1.65987 645
17	(0)3.08686 9988	(−22)5.81495 828	(−44)1.95464 371
18	(1)1.13769 8721	(−22)8.17096 398	(−44)2.32445 531
19	(1)4.40440 2395	(−21)1.16980 523	(−44)2.79144 763
20	(2)1.78744 2782	(−21)1.70614 838	(−44)3.38520 541
30	(9)2.03024 7813	(−19)2.00581 681	(−43)3.97060 205
40	(17)5.93822 4681	(−16)1.29986 971	(−41)1.20842 080
50	(27)2.06137 3775	(−13)4.00601 347	(−40)9.27452 265
100	(85)4.59667 4084	(+13)1.63940 352	(−25)7.61712 963

Table 9.12   **KELVIN FUNCTIONS—ORDERS 0 AND 1**

x	ber x	bei x	$\text{ber}_1 x$	$\text{bei}_1 x$
0.0	1.00000 00000	0.00000 00000	0.00000 00000	0.00000 00000
0.1	0.99999 84375	0.00249 99996	-0.03539 95148	0.03531 11265
0.2	0.99997 50000	0.00999 99722	-0.07106 36418	0.07035 65360
0.3	0.99987 34379	0.02249 96836	-0.10725 47768	0.10486 83082
0.4	0.99960 00044	0.03999 82222	-0.14423 08645	0.13857 41359
0.5	0.99902 34640	0.06249 32184	-0.18224 31238	0.17119 51797
0.6	0.99797 51139	0.08997 97504	-0.22153 37177	0.20244 39824
0.7	0.99624 88284	0.12244 89390	-0.26233 33470	0.23202 24623
0.8	0.99360 11377	0.15988 62295	-0.30485 87511	0.25962 00070
0.9	0.98975 13567	0.20226 93635	-0.34931 01000	0.28491 16898
1.0	0.98438 17812	0.24956 60400	-0.39586 82610	0.30755 66314
1.1	0.97713 79732	0.30173 12692	-0.44469 19268	0.32719 65305
1.2	0.96762 91558	0.35870 44199	-0.49591 45913	0.34345 43903
1.3	0.95542 87468	0.42040 59656	-0.54964 13636	0.35593 34649
1.4	0.94007 50567	0.48673 39336	-0.60594 56099	0.36421 64560
1.5	0.92107 21835	0.55756 00623	-0.66486 54180	0.36786 49890
1.6	0.89789 11386	0.63272 56770	-0.72639 98786	0.36641 93986
1.7	0.86997 12370	0.71203 72924	-0.79050 51846	0.35939 88584
1.8	0.83672 17942	0.79526 19548	-0.85709 05470	0.34630 18876
1.9	0.79752 41670	0.88212 23406	-0.92601 39357	0.32660 72722
2.0	0.75173 41827	0.97229 16273	-0.99707 76519	0.29977 54370
2.1	0.69868 50014	1.06538 81608	-1.07002 37462	0.26525 03092
2.2	0.63769 04571	1.16096 99438	-1.14452 92997	0.22246 17120
2.3	0.56804 89261	1.25852 89751	-1.22020 15903	0.17082 83322
2.4	0.48904 77721	1.35748 54765	-1.29657 31717	0.10976 13027
2.5	0.39996 84171	1.45718 20442	-1.37309 68976	+0.03866 84440
2.6	0.30009 20903	1.55687 77737	-1.44914 09315	-0.04304 07916
2.7	0.18870 63040	1.65574 24073	-1.52398 37854	-0.13594 96285
2.8	+0.06511 21084	1.75285 05638	-1.59680 94413	-0.24062 74875
2.9	-0.07136 78258	1.84717 61157	-1.66670 26139	-0.35762 26713
3.0	-0.22138 02496	1.93758 67853	-1.73264 42211	-0.48745 41770
3.1	-0.38553 14550	2.02283 90420	-1.79350 71373	-0.63060 25952
3.2	-0.56437 64305	2.10157 33881	-1.84805 23125	-0.78750 00586
3.3	-0.75840 70121	2.17231 01315	-1.89492 53482	-0.95851 04088
3.4	-0.96803 89953	2.23344 57503	-1.93265 36306	-1.14396 11510
3.5	-1.19359 81796	2.28324 99669	-1.95964 41313	-1.34404 23731
3.6	-1.43530 53217	2.31986 36548	-1.97418 19924	-1.55888 06139
3.7	-1.69325 99843	2.34129 77145	-1.97443 00262	-1.78847 96677
3.8	-1.96742 32727	2.34543 30614	-1.95842 92665	-2.03271 31257
3.9	-2.25759 94661	2.33002 18823	-1.92410 07174	-2.29130 70630
4.0	-2.56341 65573	2.29269 03227	-1.86924 84590	-2.56382 16886
4.1	-2.88430 57320	2.23094 27803	-1.79156 42730	-2.84963 19932
4.2	-3.21947 98323	2.14216 79867	-1.68863 39648	-3.14790 74393
4.3	-3.56791 08628	2.02364 70694	-1.55794 55649	-3.45759 07560
4.4	-3.92830 66215	1.87256 37958	-1.39689 95997	-3.77737 59182
4.5	-4.29908 65516	1.68601 72036	-1.20282 16315	-4.10568 54084
4.6	-4.67835 69372	1.46103 68359	-0.97297 72697	-4.44064 68813
4.7	-5.06388 55867	1.19460 07968	-0.70458 98649	-4.78006 93721
4.8	-5.45307 61749	0.88365 68537	-0.39486 10961	-5.12141 92170
4.9	-5.84294 24419	0.52514 68109	-0.04099 46681	-5.46179 58790
5.0	-6.23008 24787	0.11603 43816	+0.35977 66668	-5.79790 79018
	$\left[\dfrac{(-3)2}{8}\right]$	$\left[\dfrac{(-3)6}{8}\right]$	$\left[\dfrac{(-3)6}{8}\right]$	$\left[\dfrac{(-3)2}{8}\right]$

KELVIN FUNCTIONS—AUXILIARY TABLE FOR SMALL ARGUMENTS

x	ker $x+$ber x ln x	kei $x+$bei x ln x	$x(\text{ker}_1 x+\text{ber}_1 x$ ln $x)$	$x(\text{kei}_1 x+\text{bei}_1 x$ ln $x)$
0.0	0.11593 1516	-0.78539 8163	-0.70710 6781	-0.70710 6781
0.1	0.11789 2485	-0.78260 7108	-0.70651 7131	-0.70215 4903
0.2	0.12374 5076	-0.77421 9267	-0.70486 2164	-0.68733 0339
0.3	0.13339 8210	-0.76019 0919	-0.70248 3157	-0.66272 8003
0.4	0.14669 9682	-0.74045 0212	-0.69994 6658	-0.62851 1738
0.5	0.16343 5574	-0.71489 8693	-0.69804 1049	-0.58492 2770
	$\left[\dfrac{(-4)5}{7}\right]$	$\left[\dfrac{(-4)8}{7}\right]$	$\left[\dfrac{(-4)1}{7}\right]$	$\left[\dfrac{(-3)1}{7}\right]$

Compiled from National Bureau of Standards, Tables of the Bessel functions $J_0(z)$ and $J_1(z)$ for complex arguments, 2d ed. (Columbia Univ. Press, New York, N.Y., 1947) and National Bureau of Standards, Tables of the Bessel functions $Y_0(z)$ and $Y_1(z)$ for complex arguments (Columbia Univ. Press, New York, N.Y., 1950) (with permission).

KELVIN FUNCTIONS—ORDERS 0 AND 1   Table 9.12

x	ker x	kei x	$\ker_1 x$	$\kei_1 x$
0.0	∞	−0.78539 8163	− ∞	− ∞
0.1	2.42047 3980	−0.77685 0646	−7.14668 1711	−6.94024 2153
0.2	1.73314 2752	−0.75812 4933	−3.63868 3342	−3.32341 7218
0.3	1.33721 8637	−0.73310 1912	−2.47074 2357	−2.08283 4751
0.4	1.06262 3902	−0.70380 0212	−1.88202 4050	−1.44430 5150
0.5	0.85590 5872	−0.67158 1695	−1.52240 3406	−1.05118 2085
0.6	0.69312 0695	−0.63744 9494	−1.27611 7712	−0.78373 8860
0.7	0.56137 8274	−0.60217 5451	−1.09407 2943	−0.59017 5251
0.8	0.45288 2093	−0.56636 7650	−0.95203 2751	−0.44426 9985
0.9	0.36251 4812	−0.53051 1122	−0.83672 7829	−0.33122 6820
1.0	0.28670 6208	−0.49499 4636	−0.74032 2276	−0.24199 5966
1.1	0.22284 4513	−0.46012 9528	−0.65791 0729	−0.17068 4462
1.2	0.16894 5592	−0.42616 3604	−0.58627 4386	−0.11325 6800
1.3	0.12345 5395	−0.39329 1826	−0.52321 5989	−0.06683 2622
1.4	0.08512 6048	−0.36166 4781	−0.46718 3076	−0.02928 3749
1.5	0.05293 4915	−0.33139 5562	−0.41704 4285	+0.00100 8681
1.6	0.02602 9861	−0.30256 5474	−0.37195 1238	0.02530 6776
1.7	+0.00369 1104	−0.27522 8834	−0.33125 0485	0.04461 5190
1.8	−0.01469 6087	−0.24941 7069	−0.29442 5803	0.05974 7779
1.9	−0.02966 1407	−0.22514 2235	−0.26105 9495	0.07137 3592
2.0	−0.04166 4514	−0.20240 0068	−0.23080 5929	0.08004 9398
2.1	−0.05110 6500	−0.18117 2644	−0.20337 3135	0.08624 3202
2.2	−0.05833 8834	−0.16143 0701	−0.17850 9812	0.09035 1619
2.3	−0.06367 0454	−0.14313 5677	−0.15599 6054	0.09271 2940
2.4	−0.06737 3493	−0.12624 1488	−0.13563 6638	0.09361 7161
2.5	−0.06968 7972	−0.11069 6099	−0.11725 6136	0.09331 3788
2.6	−0.07082 5700	−0.09644 2891	−0.10069 5314	0.09201 8037
2.7	−0.07097 3560	−0.08342 1858	−0.08580 8451	0.08991 5810
2.8	−0.07029 6321	−0.07157 0648	−0.07246 1339	0.08716 7762
2.9	−0.06893 9052	−0.06082 5473	−0.06052 9755	0.08391 2666
3.0	−0.06702 9233	−0.05112 1884	−0.04989 8308	0.08027 0223
3.1	−0.06467 8610	−0.04239 5446	−0.04045 9533	0.07634 3451
3.2	−0.06198 4833	−0.03458 2313	−0.03211 3183	0.07222 0724
3.3	−0.05903 2916	−0.02761 9697	−0.02476 5662	0.06797 7529
3.4	−0.05589 6550	−0.02144 6287	−0.01832 9556	0.06367 7999
3.5	−0.05263 9277	−0.01600 2568	−0.01272 3249	0.05937 6256
3.6	−0.04931 5556	−0.01123 1096	−0.00787 0585	0.05511 7592
3.7	−0.04597 1723	−0.00707 6704	−0.00370 0576	0.05093 9514
3.8	−0.04264 6864	−0.00348 6665	−0.00014 7138	0.04687 2681
3.9	−0.03937 3608	−0.00041 0809	+0.00285 1155	0.04294 1728
4.0	−0.03617 8848	+0.00219 8399	0.00535 1296	0.03916 6011
4.1	−0.03308 4395	0.00438 5818	0.00740 6063	0.03556 0272
4.2	−0.03010 7574	0.00619 3613	0.00906 4226	0.03213 5235
4.3	−0.02726 1764	0.00766 1269	0.01037 0752	0.02889 8142
4.4	−0.02455 6892	0.00882 5624	0.01136 6998	0.02585 3229
4.5	−0.02199 9875	0.00972 0918	0.01209 0904	0.02300 2160
4.6	−0.01959 5024	0.01037 8865	0.01257 7182	0.02034 4409
4.7	−0.01734 4409	0.01082 8725	0.01285 7498	0.01787 7607
4.8	−0.01524 8188	0.01109 7399	0.01296 0651	0.01559 7847
4.9	−0.01330 4899	0.01120 9526	0.01291 2753	0.01349 9960
5.0	−0.01151 1727	0.01118 7587	0.01273 7390	0.01157 7754

KELVIN FUNCTIONS—AUXILIARY TABLE FOR SMALL ARGUMENTS

x	ker x+ber $x \ln x$	kei x+bei $x \ln x$	$x(\ker_1 x$+ber$_1 x \ln x)$	$x(\kei_1 x$+bei$_1 x \ln x)$
0.5	0.16343 5574	−0.71489 8693	−0.69804 1049	−0.58492 2770
0.6	0.18332 9435	−0.68341 3456	−0.69777 1567	−0.53229 1460
0.7	0.20604 1279	−0.64584 9920	−0.70035 3648	−0.47105 2294
0.8	0.23116 6407	−0.60204 5231	−0.70720 4389	−0.40176 2012
0.9	0.25823 4099	−0.55182 2327	−0.71993 1903	−0.32512 0736
1.0	0.28670 6208	−0.49499 4636	−0.74032 2276	−0.24199 5966
	$\begin{bmatrix} (-4)4 \\ 7 \end{bmatrix}$	$\begin{bmatrix} (-4)8 \\ 7 \end{bmatrix}$	$\begin{bmatrix} (-3)1 \\ 7 \end{bmatrix}$	$\begin{bmatrix} (-3)1 \\ 7 \end{bmatrix}$

Table 9.12   KELVIN FUNCTIONS—MODULUS AND PHASE

ber $x = M_0(x) \cos \theta_0(x)$ ber$_1$ $x = M_1(x) \cos \theta_1(x)$

bei $x = M_0(x) \sin \theta_0(x)$ bei$_1$ $x = M_1(x) \sin \theta_1(x)$

x	$M_0(x)$	$\theta_0(x)$	$M_1(x)$	$\theta_1(x)$
0.0	1.000000	0.000000	0.000000	2.356194
0.2	1.000025	0.010000	0.100000	2.361194
0.4	1.000400	0.039993	0.200013	2.376194
0.6	1.002023	0.089919	0.300101	2.401189
0.8	1.006383	0.159548	0.400427	2.436166
1.0	1.015525	0.248294	0.501301	2.481086
1.2	1.031976	0.354999	0.603235	2.535872
1.4	1.058608	0.477755	0.706982	2.600386
1.6	1.098431	0.613860	0.813585	2.674406
1.8	1.154359	0.759999	0.924407	2.757605
2.0	1.229006	0.912639	1.041167	2.849536
2.2	1.324576	1.068511	1.165949	2.949617
2.4	1.442891	1.225011	1.301211	3.057139
2.6	1.585536	1.380379	1.449780	3.171285
2.8	1.754059	1.533667	1.614838	3.291160
3.0	1.950193	1.684559	1.799908	3.415839
3.2	2.176036	1.833156	2.008844	3.544415
3.4	2.434210	1.979784	2.245840	3.676044
3.6	2.727979	2.124854	2.515453	3.809981
3.8	3.061341	2.268771	2.822653	3.945601
4.0	3.439118	2.411887	3.172896	4.082407
4.2	3.867032	2.554483	3.572227	4.220023
4.4	4.351791	2.696771	4.027393	4.358179
4.6	4.901189	2.838893	4.545990	4.496691
4.8	5.524209	2.980942	5.136619	4.635441
5.0	6.231163	3.122970	5.809060	4.774362
5.2	7.033841	3.265002	6.574474	4.913417
5.4	7.945700	3.407044	7.445618	5.052589
5.6	8.982083	3.549094	8.437083	5.191872
5.8	10.160473	3.691142	9.565568	5.331267
6.0	11.500794	3.833179	10.850182	5.470772
6.2	13.025757	3.975197	12.312791	5.610390
6.4	14.761257	4.117190	13.978402	5.750117
6.6	16.736836	4.259152	15.875614	5.889950
6.8	18.986208	4.401083	18.037122	6.029884
7.0	21.547863	4.542982	20.500302	6.169913
	$\begin{bmatrix} (-2)4 \\ 7 \end{bmatrix}$	$\begin{bmatrix} (-3)2 \\ 8 \end{bmatrix}$	$\begin{bmatrix} (-2)4 \\ 6 \end{bmatrix}$	$\begin{bmatrix} (-3)1 \\ 6 \end{bmatrix}$

KELVIN FUNCTIONS—MODULUS AND PHASE FOR LARGE ARGUMENTS

x^{-1}	$x^{\frac{1}{2}}e^{-x/\sqrt2}M_0(x)$	$\theta_0(x)-(x/\sqrt2)$	$x^{\frac{1}{2}}e^{-x/\sqrt2}M_1(x)$	$\theta_1(x)-(x/\sqrt2)$	$\langle x \rangle$
0.15	0.40418	-0.40758	0.38359	1.22254	7
0.14	0.40383	-0.40644	0.38457	1.21922	7
0.13	0.40349	-0.40534	0.38556	1.21598	8
0.12	0.40315	-0.40427	0.38655	1.21280	8
0.11	0.40281	-0.40323	0.38755	1.20968	9
0.10	0.40246	-0.40221	0.38856	1.20660	10
0.09	0.40211	-0.40119	0.38957	1.20356	11
0.08	0.40176	-0.40019	0.39060	1.20057	13
0.07	0.40141	-0.39921	0.39162	1.19762	14
0.06	0.40106	-0.39824	0.39266	1.19471	17
0.05	0.40071	-0.39728	0.39369	1.19184	20
0.04	0.40035	-0.39634	0.39474	1.18901	25
0.03	0.40000	-0.39541	0.39578	1.18622	33
0.02	0.39965	-0.39449	0.39683	1.18348	50
0.01	0.39930	-0.39359	0.39789	1.18077	100
0.00	0.39894	-0.39270	0.39894	1.17810	∞
	$\begin{bmatrix} (-5)1 \\ 2 \end{bmatrix}$	$\begin{bmatrix} (-5)1 \\ 2 \end{bmatrix}$	$\begin{bmatrix} (-6)3 \\ 2 \end{bmatrix}$	$\begin{bmatrix} (-5)1 \\ 2 \end{bmatrix}$	

$\langle x \rangle$ = nearest integer to x.

KELVIN FUNCTIONS—MODULUS AND PHASE   Table 9.12

$$\text{ker } x = N_0(x) \cos \phi_0(x) \qquad\qquad \text{ker}_1 x = N_1(x) \cos \phi_1(x)$$

$$\text{kei } x = N_0(x) \sin \phi_0(x) \qquad\qquad \text{kei}_1 x = N_1(x) \sin \phi_1(x)$$

x	$N_0(x)$	$\phi_0(x)$	$N_1(x)$	$\phi_1(x)$
0.0	∞	0.000000	∞	−2.356194
0.2	1.891702	−0.412350	4.927993	−2.401447
0.4	1.274560	−0.584989	2.372347	−2.487035
0.6	0.941678	−0.743582	1.497572	−2.590827
0.8	0.725172	−0.896284	1.050591	−2.704976
1.0	0.572032	−1.045803	0.778870	−2.825662
1.2	0.458430	−1.193368	0.597114	−2.950763
1.4	0.371548	−1.339631	0.468100	−3.078993
1.6	0.303683	−1.484977	0.372811	−3.209526
1.8	0.249850	−1.629650	0.300427	−3.341804
2.0	0.206644	−1.773013	0.244293	−3.475437
2.2	0.171649	−1.917579	0.200073	−3.610143
2.4	0.143095	−2.061029	0.164807	−3.745715
2.6	0.119656	−2.204225	0.136407	−3.881994
2.8	0.100319	−2.347212	0.113353	−4.018860
3.0	0.084299	−2.490025	0.094515	−4.156217
3.2	0.070979	−2.632692	0.079039	−4.293990
3.4	0.059870	−2.775236	0.066264	−4.432118
3.6	0.050578	−2.917672	0.055677	−4.570551
3.8	0.042789	−3.060017	0.046873	−4.709250
4.0	0.036246	−3.202283	0.039530	−4.848179
4.2	0.030738	−3.344478	0.033389	−4.987312
4.4	0.026095	−3.486612	0.028242	−5.126623
4.6	0.022174	−3.628692	0.023918	−5.266093
4.8	0.018859	−3.770724	0.020280	−5.405705
5.0	0.016052	−3.912712	0.017213	−5.545443
5.2	0.013674	−4.054662	0.014624	−5.685295
5.4	0.011656	−4.196576	0.012435	−5.825250
5.6	0.009942	−4.338460	0.010583	−5.965298
5.8	0.008485	−4.480314	0.009013	−6.105430
6.0	0.007246	−4.622142	0.007682	−6.245638
6.2	0.006191	−4.763947	0.006551	−6.385917
6.4	0.005292	−4.905730	0.005590	−6.526260
6.6	0.004526	−5.047493	0.004773	−6.666662
6.8	0.003872	−5.189238	0.004077	−6.807119
7.0	0.003315	−5.330966	0.003485	−6.947625

KELVIN FUNCTIONS—MODULUS AND PHASE FOR LARGE ARGUMENTS

x^{-1}	$x^{\frac{1}{2}} e^{x/\sqrt{2}} N_0(x)$	$\phi_0(x)+(x/\sqrt{2})$	$x^{\frac{1}{2}} e^{x/\sqrt{2}} N_1(x)$	$\phi_1(x)+(x/\sqrt{2})$	$\langle x \rangle$
0.15	1.23695	−0.38070	1.30377	−1.99943	7
0.14	1.23802	−0.38142	1.30039	−1.99725	7
0.13	1.23909	−0.38217	1.29701	−1.99505	8
0.12	1.24017	−0.38291	1.29363	−1.99281	8
0.11	1.24125	−0.38367	1.29024	−1.99055	9
0.10	1.24233	−0.38444	1.28687	−1.98825	10
0.09	1.24342	−0.38522	1.28349	−1.98592	11
0.08	1.24451	−0.38600	1.28012	−1.98357	13
0.07	1.24560	−0.38680	1.27675	−1.98118	14
0.06	1.24670	−0.38761	1.27339	−1.97876	17
0.05	1.24779	−0.38843	1.27002	−1.97630	20
0.04	1.24889	−0.38926	1.26667	−1.97381	25
0.03	1.25000	−0.39010	1.26332	−1.97128	33
0.02	1.25110	−0.39096	1.25998	−1.96872	50
0.01	1.25221	−0.39182	1.25664	−1.96613	100
0.00	1.25331	−0.39270	1.25331	−1.96350	∞
	$\begin{bmatrix} (-6)1 \\ 2 \end{bmatrix}$	$\begin{bmatrix} (-6)3 \\ 2 \end{bmatrix}$	$\begin{bmatrix} (-6)3 \\ 2 \end{bmatrix}$	$\begin{bmatrix} (-6)5 \\ 2 \end{bmatrix}$	

$\langle x \rangle$ = nearest integer to x.

10. Bessel Functions of Fractional Order

H. A. ANTOSIEWICZ [1]

Contents

---

[1] National Bureau of Standards. (Presently, University of Southern California.)

The author acknowledges the assistance of Bertha H. Walter and Ruth Zucker in the preparation and checking of the tables and graphs.

10. Bessel Functions of Fractional Order

Mathematical Properties

10.1. Spherical Bessel Functions

Definitions

Differential Equation

10.1.1

$$z^2 w'' + 2zw' + [z^2 - n(n+1)]w = 0$$

$$(n = 0, \pm 1, \pm 2, \ldots)$$

Particular solutions are the *Spherical Bessel functions of the first kind*

$$j_n(z) = \sqrt{\tfrac{1}{2}\pi/z}\, J_{n+\frac{1}{2}}(z),$$

the *Spherical Bessel functions of the second kind*

$$y_n(z) = \sqrt{\tfrac{1}{2}\pi/z}\, Y_{n+\frac{1}{2}}(z),$$

and the *Spherical Bessel functions of the third kind*

$$h_n^{(1)}(z) = j_n(z) + iy_n(z) = \sqrt{\tfrac{1}{2}\pi/z}\, H_{n+\frac{1}{2}}^{(1)}(z),$$
$$h_n^{(2)}(z) = j_n(z) - iy_n(z) = \sqrt{\tfrac{1}{2}\pi/z}\, H_{n+\frac{1}{2}}^{(2)}(z).$$

The pairs $j_n(z)$, $y_n(z)$ and $h_n^{(1)}(z)$, $h_n^{(2)}(z)$ are linearly independent solutions for every n. For general properties see the remarks after **9.1.1**.

Ascending Series (See 9.1.2, 9.1.10)

10.1.2

$$j_n(z) = \frac{z^n}{1 \cdot 3 \cdot 5 \ldots (2n+1)} \left\{ 1 - \frac{\frac{1}{2}z^2}{1!(2n+3)} + \frac{(\frac{1}{2}z^2)^2}{2!(2n+3)(2n+5)} - \ldots \right\}$$

10.1.3

$$y_n(z) = -\frac{1 \cdot 3 \cdot 5 \ldots (2n-1)}{z^{n+1}} \left\{ 1 - \frac{\frac{1}{2}z^2}{1!(1-2n)} + \frac{(\frac{1}{2}z^2)^2}{2!(1-2n)(3-2n)} - \ldots \right\}$$

$$(n = 0, 1, 2, \ldots)$$

Limiting Values as $z \to 0$

10.1.4

$$z^{-n} j_n(z) \to \frac{1}{1 \cdot 3 \cdot 5 \ldots (2n+1)}$$

10.1.5

$$z^{n+1} y_n(z) \to -1 \cdot 3 \cdot 5 \ldots (2n-1) \qquad (n = 0, 1, 2, \ldots)$$

Wronskians

10.1.6

$$W\{j_n(z), y_n(z)\} = z^{-2}$$

10.1.7

$$W\{h_n^{(1)}(z), h_n^{(2)}(z)\} = -2iz^{-2} \qquad (n = 0, 1, 2, \ldots)$$

Representations by Elementary Functions

10.1.8

$$j_n(z) = z^{-1}[P(n+\tfrac{1}{2}, z) \sin (z - \tfrac{1}{2}n\pi) + Q(n+\tfrac{1}{2}, z) \cos (z - \tfrac{1}{2}n\pi)]$$

10.1.9

$$y_n(z) = (-1)^{n+1} z^{-1}[P(n+\tfrac{1}{2}, z) \cos (z + \tfrac{1}{2}n\pi) - Q(n+\tfrac{1}{2}, z) \sin (z + \tfrac{1}{2}n\pi)]$$

$$P(n+\tfrac{1}{2}, z) = 1 - \frac{(n+2)!}{2!\,\Gamma(n-1)}(2z)^{-2} + \frac{(n+4)!}{4!\,\Gamma(n-3)}(2z)^{-4} - \ldots$$

$$= \sum_0^{[\frac{1}{2}n]} (-1)^k (n+\tfrac{1}{2}, 2k)(2z)^{-2k}$$

$$Q(n+\tfrac{1}{2}, z) = \frac{(n+1)!}{1!\,\Gamma(n)}(2z)^{-1} - \frac{(n+3)!}{3!\,\Gamma(n-2)}(2z)^{-3} + \frac{(n+5)!}{5!\,\Gamma(n-4)}(2z)^{-5} - \ldots$$

$$= \sum_0^{[\frac{1}{2}(n-1)]} (-1)^k (n+\tfrac{1}{2}, 2k+1)(2z)^{-2k-1}$$

$$(n = 0, 1, 2, \ldots)$$

$$(n+\tfrac{1}{2}, k) = \frac{(n+k)!}{k!\,\Gamma(n-k+1)}$$

n \ k	1	2	3	4	5
1	2				
2	6	12			
3	12	60	120		
4	20	180	840	1680	
5	30	420	3360	15120	30240

10.1.10

$$j_n(z) = f_n(z) \sin z + (-1)^{n+1} f_{-n-1}(z) \cos z$$

$$f_0(z) = z^{-1}, \qquad f_1(z) = z^{-2}$$

$$f_{n-1}(z) + f_{n+1}(z) = (2n+1) z^{-1} f_n(z)$$

$$(n = 0, \pm 1, \pm 2, \ldots)$$

The Functions $j_n(z)$, $y_n(z)$ for $n = 0, 1, 2$

10.1.11

$$j_0(z) = \frac{\sin z}{z}$$

$$j_1(z) = \frac{\sin z}{z^2} - \frac{\cos z}{z}$$

$$j_2(z) = \left(\frac{3}{z^3} - \frac{1}{z}\right) \sin z - \frac{3}{z^2} \cos z \qquad *$$

10.1.12

$$y_0(z) = -j_{-1}(z) = -\frac{\cos z}{z}$$

$$y_1(z) = j_{-2}(z) = -\frac{\cos z}{z^2} - \frac{\sin z}{z}$$

$$y_2(z) = -j_{-3}(z) = \left(-\frac{3}{z^3} + \frac{1}{z}\right) \cos z - \frac{3}{z^2} \sin z \qquad *$$

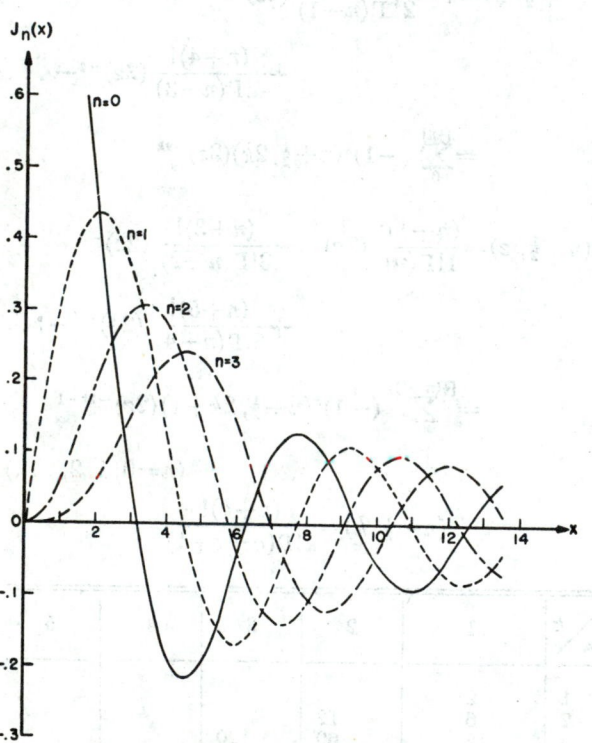

FIGURE 10.1. $j_n(x)$. $n = 0(1)3$.

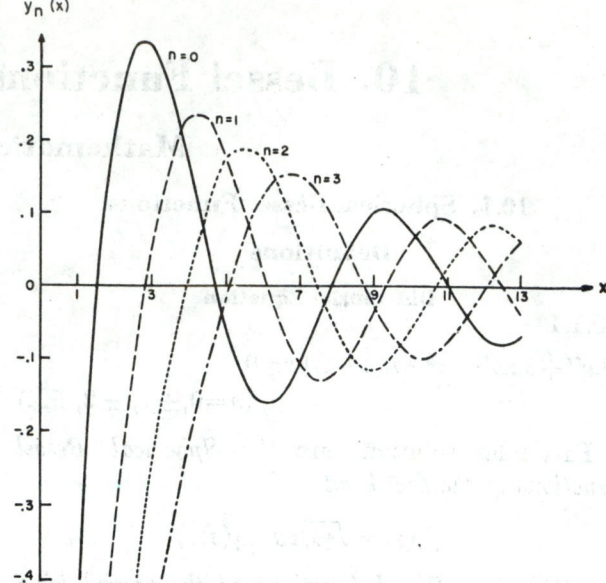

FIGURE 10.2. $y_n(x)$. $n = 0(1)3$.

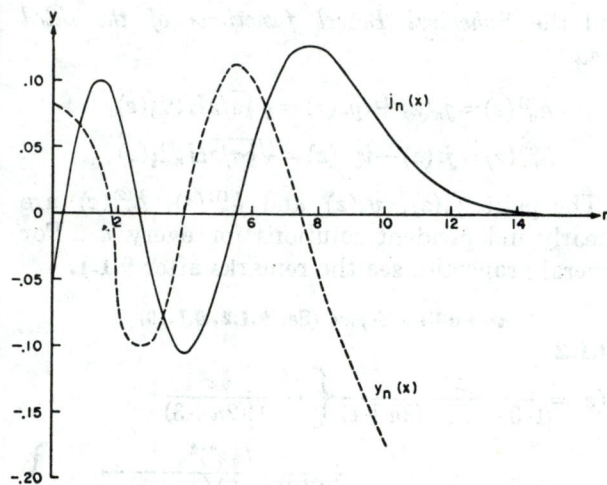

FIGURE 10.3. $j_n(x)$, $y_n(x)$. $x = 10$.

Poisson's Integral and Gegenbauer's Generalization

10.1.13 $\quad j_n(z) = \dfrac{z^n}{2^{n+1} n!} \displaystyle\int_0^\pi \cos(z \cos \theta) \sin^{2n+1} \theta \, d\theta$

(See **9.1.20**.)

10.1.14

$$= \frac{1}{2} (-i)^n \int_0^\pi e^{iz \cos \theta} P_n(\cos \theta) \sin \theta \, d\theta$$

$$(n = 0, 1, 2, \ldots)$$

Spherical Bessel Functions of the Second and Third Kind

10.1.15

$$y_n(z)=(-1)^{n+1}j_{-n-1}(z) \qquad (n=0,\ \pm1,\ \pm2,\ \ldots)$$

10.1.16

$$h_n^{(1)}(z)=i^{-n-1}z^{-1}e^{iz}\sum_0^n\ (n+\tfrac{1}{2},\ k)(-2iz)^{-k}$$

10.1.17

$$h_n^{(2)}(z)=i^{n+1}z^{-1}e^{-iz}\sum_0^n\ (n+\tfrac{1}{2},\ k)\ (2iz)^{-k} \qquad *$$

10.1.18

$$h_{-n-1}^{(1)}(z)=i(-1)^n h_n^{(1)}(z)$$
$$h_{-n-1}^{(2)}(z)=-i(-1)^n h_n^{(2)}(z) \qquad (n=0,\ 1,\ 2,\ \ldots)$$

Elementary Properties
Recurrence Relations

$$f_n(z):j_n(z),\ y_n(z),\ h_n^{(1)}(z),\ h_n^{(2)}(z)$$
$$(n=0,\ \pm1,\ \pm2,\ \ldots)$$

10.1.19 $\quad f_{n-1}(z)+f_{n+1}(z)=(2n+1)z^{-1}f_n(z)$

10.1.20 $\quad nf_{n-1}(z)-(n+1)f_{n+1}(z)=(2n+1)\dfrac{d}{dz}f_n(z)$

10.1.21 $\quad \dfrac{n+1}{z}f_n(z)+\dfrac{d}{dz}f_n(z)=f_{n-1}(z)$

(See **10.1.23.**)

10.1.22 $\quad \dfrac{n}{z}f_n(z)-\dfrac{d}{dz}f_n(z)=f_{n+1}(z)$

(See **10.1.24.**)

Differentiation Formulas

$$f_n(z):j_n(z),\ y_n(z),\ h_n^{(1)}(z),\ h_n^{(2)}(z)$$
$$(n=0,\ \pm1,\ \pm2,\ \ldots)$$

10.1.23 $\quad \left(\dfrac{1}{z}\dfrac{d}{dz}\right)^m[z^{n+1}f_n(z)]=z^{n-m+1}f_{n-m}(z)$

10.1.24 $\quad \left(\dfrac{1}{z}\dfrac{d}{dz}\right)^m[z^{-n}f_n(z)]=(-1)^m z^{-n-m}f_{n+m}(z)$
$$(m=1,\ 2,\ 3,\ \ldots)$$

Rayleigh's Formulas

10.1.25

$$j_n(z)=z^n\left(-\dfrac{1}{z}\dfrac{d}{dz}\right)^n\dfrac{\sin z}{z}$$

10.1.26

$$y_n(z)=-z^n\left(-\dfrac{1}{z}\dfrac{d}{dz}\right)^n\dfrac{\cos z}{z} \qquad (n=0,\ 1,\ 2,\ \ldots)$$

Modulus and Phase

$$j_n(z)=\sqrt{\tfrac{1}{2}\pi/z}\,M_{n+\frac{1}{2}}(z)\,\cos\theta_{n+\frac{1}{2}}(z),$$
$$y_n(z)=\sqrt{\tfrac{1}{2}\pi/z}\,M_{n+\frac{1}{2}}(z)\,\sin\theta_{n+\frac{1}{2}}(z)$$

(See **9.2.17.**)

10.1.27

$$(\tfrac{1}{2}\pi/z)\,M_{n+\frac{1}{2}}^2(z)=\dfrac{1}{z^2}\sum_0^n\dfrac{(2n-k)!(2n-2k)!}{k![(n-k)!]^2}(2z)^{2k-2n}$$

(See **9.2.28.**)

10.1.28 $\quad (\tfrac{1}{2}\pi/z)M_{1/2}^2(z)=j_0^2(z)+y_0^2(z)=z^{-2}$

10.1.29

$$(\tfrac{1}{2}\pi/z)M_{3/2}^2(z)=j_1^2(z)+y_1^2(z)=z^{-2}+z^{-4}$$

10.1.30

$$(\tfrac{1}{2}\pi/z)M_{5/2}^2(z)=j_2^2(z)+y_2^2(z)=z^{-2}+3z^{-4}+9z^{-6}$$

Cross Products

10.1.31 $\quad j_n(z)y_{n-1}(z)-j_{n-1}(z)y_n(z)=z^{-2}$

10.1.32

$$j_{n+1}(z)y_{n-1}(z)-j_{n-1}(z)y_{n+1}(z)=(2n+1)z^{-3}$$

10.1.33

$$j_0(z)j_n(z)+y_0(z)y_n(z)$$
$$=z^{-2}\sum_0^{[\frac{1}{2}n]}(-1)^k 2^{n-2k}\left(k+\tfrac{1}{2}\right)_{n-2k}\binom{n-k}{k}z^{2k-n}$$
$$(n=0,\ 1,\ 2,\ \ldots)$$

Analytic Continuation

10.1.34 $\quad j_n(ze^{m\pi i})=e^{mn\pi i}j_n(z)$

10.1.35 $\quad y_n(ze^{m\pi i})=(-1)^m e^{mn\pi i}y_n(z)$

10.1.36 $\quad h_n^{(1)}(ze^{(2m+1)\pi i})=(-1)^n h_n^{(2)}(z)$

10.1.37 $\quad h_n^{(2)}(ze^{(2m+1)\pi i})=(-1)^n h_n^{(1)}(z)$

10.1.38 $\quad h_n^{(l)}(ze^{2m\pi i})=h_n^{(l)}(z)$
$$(l=1,\ 2;\ m,\ n=0,\ 1,\ 2,\ \ldots)$$

Generating Functions

10.1.39

$$\dfrac{1}{z}\sin\sqrt{z^2+2zt}=\sum_0^\infty\dfrac{(-t)^n}{n!}y_{n-1}(z) \qquad (2|t|<|z|)$$

10.1.40 $\quad \dfrac{1}{z}\cos\sqrt{z^2-2zt}=\sum_0^\infty\dfrac{t^n}{n!}j_{n-1}(z)$

*See page II.

Derivatives With Respect to Order

10.1.41

$$\left[\frac{\partial}{\partial\nu}j_\nu(x)\right]_{\nu=0}=(\tfrac{1}{2}\pi/x)\{\operatorname{Ci}(2x)\sin x-\operatorname{Si}(2x)\cos x\}$$

10.1.42

$$\left[\frac{\partial}{\partial\nu}j_\nu(x)\right]_{\nu=-1}=(\tfrac{1}{2}\pi/x)\{\operatorname{Ci}(2x)\cos x+\operatorname{Si}(2x)\sin x\}$$

10.1.43

$$\left[\frac{\partial}{\partial\nu}y_\nu(x)\right]_{\nu=0}=(\tfrac{1}{2}\pi/x)\{\operatorname{Ci}(2x)\cos x+[\operatorname{Si}(2x)-\pi]\sin x\}$$

10.1.44

$$\left[\frac{\partial}{\partial\nu}y_\nu(x)\right]_{\nu=-1}=$$
$$(\tfrac{1}{2}\pi/x)\{\operatorname{Ci}(2x)\sin x-[\operatorname{Si}(2x)-\pi]\cos x\}$$

Addition Theorems and Degenerate Forms

r,ρ,θ,λ arbitrary complex; $R=\sqrt{(r^2+\rho^2-2r\rho\cos\theta)}$

10.1.45 $\quad\dfrac{\sin\lambda R}{\lambda R}=\sum_0^\infty(2n+1)j_n(\lambda r)j_n(\lambda\rho)P_n(\cos\theta)$

*10.1.46 $-\dfrac{\cos\lambda R}{\lambda R}=\sum_0^\infty(2n+1)j_n(\lambda r)y_n(\lambda\rho)P_n(\cos\theta)$
$$|re^{\pm\iota\theta}|<|\rho|$$

10.1.47 $\quad e^{iz\cos\theta}=\sum_0^\infty(2n+1)e^{\frac{1}{2}n\pi i}j_n(z)P_n(\cos\theta)$

10.1.48

$$J_0(z\sin\theta)=\sum_0^\infty(4n+1)\frac{(2n)!}{2^{2n}(n!)^2}j_{2n}(z)P_{2n}(\cos\theta)$$

Duplication Formula

10.1.49

$j_n(2z)=$

* $\qquad -n!z^{n+1}\sum_0^n\dfrac{2n-2k+1}{k!(2n-k+1)!}j_{n-k}(z)y_{n-k}(z)$

Some Infinite Series Involving $j_n^2(z)$

10.1.50 $\qquad\sum_0^\infty(2n+1)j_n^2(z)=1$

10.1.51 $\qquad\sum_0^\infty(-1)^n(2n+1)j_n^2(z)=\dfrac{\sin 2z}{2z}$

10.1.52 $\qquad\sum_0^\infty j_n^2(z)=\dfrac{\operatorname{Si}(2z)}{2z}$

*See page II.

Fresnel Integrals

10.1.53

$$C(\sqrt{2x/\pi})=\frac{1}{2}\int_0^x J_{-\frac{1}{2}}(t)dt$$
$$=\sqrt{2}[\cos\tfrac{1}{2}x\sum_0^\infty(-1)^n J_{2n+\frac{1}{2}}(\tfrac{1}{2}x)$$
$$+\sin\tfrac{1}{2}x\sum_0^\infty(-1)^n J_{2n+3/2}(\tfrac{1}{2}x)]$$

10.1.54

$$S(\sqrt{2x/\pi})=\frac{1}{2}\int_0^x J_{\frac{1}{2}}(t)dt$$
$$=\sqrt{2}[\sin\tfrac{1}{2}x\sum_0^\infty(-1)^n J_{2n+\frac{1}{2}}(\tfrac{1}{2}x)$$
$$-\cos\tfrac{1}{2}x\sum_0^\infty(-1)^n J_{2n+3/2}(\tfrac{1}{2}x)].$$

(See also 11.1.1, 11.1.2.)

Zeros and Their Asymptotic Expansions

The zeros of $j_n(x)$ and $y_n(x)$ are the same as the zeros of $J_{n+\frac{1}{2}}(x)$ and $Y_{n+\frac{1}{2}}(x)$ and the formulas for $j_{\nu,s}$ and $y_{\nu,s}$ given in **9.5** are applicable with $\nu=n+\frac{1}{2}$. There are, however, no simple relations connecting the zeros of the derivatives. Accordingly, we now give formulas for $a'_{n,s}$, $b'_{n,s}$, the s-th positive zero of $j'_n(z)$, $y'_n(z)$, respectively; $z=0$ is counted as the first zero of $j'_0(z)$.

(Tables of $a'_{n,s}$, $b'_{n,s}$, $j_n(a'_{n,s})$, $y_n(b'_{n,s})$ are given in [10.31].)

Elementary Relations

$$f_n(z)=j_n(z)\cos\pi t+y_n(z)\sin\pi t$$
$$(t\text{ a real parameter, }0\le t\le1)$$

If τ_n is a zero of $f'_n(z)$ then

10.1.55 $\quad f_n(\tau_n)=[\tau_n/(n+1)]f_{n-1}(\tau_n)$
(See **10.1.21**.)

10.1.56 $\qquad=(\tau_n/n)f_{n+1}(\tau_n)$
(See **10.1.22**.)

10.1.57 $\qquad=\left\{\dfrac{1}{\pi}[\tau_n^2-n(n+1)]\dfrac{d\tau_n}{d\tau}\right\}^{-\frac{1}{2}}$

McMahon's Expansions for n Fixed and s Large

10.1.58

$$a'_{n,s}, b'_{n,s} \sim \beta - (\mu+7)(8\beta)^{-1}$$

$$-\frac{4}{3}(7\mu^2 + 154\mu + 95)(8\beta)^{-3}$$

$$-\frac{32}{15}(85\mu^3 + 3535\mu^2 + 3561\mu + 6133)(8\beta)^{-5}$$

$$-\frac{64}{105}(6949\mu^4 + 474908\mu^3 + 330638\mu^2$$

$$+9040780\mu - 5075147)(8\beta)^{-7} - \cdots$$

$$\beta = \pi(s + \tfrac{1}{2}n - \tfrac{1}{2}) \text{ for } a'_{n,s}, \quad \beta = \pi(s + \tfrac{1}{2}n) \text{ for } b'_{n,s};$$

$$\mu = (2n+1)^2$$

Asymptotic Expansions of Zeros and Associated Values for n Large

10.1.59

$$a'_{n,1} \sim (n+\tfrac{1}{2}) + .8086165(n+\tfrac{1}{2})^{1/3} - .236680(n+\tfrac{1}{2})^{-1/3}$$

$$-.20736(n+\tfrac{1}{2})^{-1} + .0233(n+\tfrac{1}{2})^{-5/3} + \cdots$$

10.1.60

$$b'_{n,1} \sim (n+\tfrac{1}{2}) + 1.8210980(n+\tfrac{1}{2})^{1/3}$$

$$+.802728(n+\tfrac{1}{2})^{-1/3} - .11740(n+\tfrac{1}{2})^{-1}$$

$$+.0249(n+\tfrac{1}{2})^{-5/3} + \cdots$$

10.1.61

$$j_n(a'_{n,1}) \sim .8458430(n+\tfrac{1}{2})^{-5/6}\{1 - .566032(n+\tfrac{1}{2})^{-2/3}$$

$$+.38081(n+\tfrac{1}{2})^{-4/3} - .2203(n+\tfrac{1}{2})^{-2} + \cdots\}$$

10.1.62

$$y_n(b'_{n,1}) \sim .7183921(n+\tfrac{1}{2})^{-5/6}\{1 - 1.274769(n+\tfrac{1}{2})^{-2/3}$$

$$+1.23038(n+\tfrac{1}{2})^{-4/3} - 1.0070(n+\tfrac{1}{2})^{-2} + \cdots\}$$

See [10.31] for corresponding expansions for $s=2, 3$.

Uniform Asymptotic Expansions of Zeros and Associated Values for n Large

10.1.63

$$a'_{n,s} \sim (n+\tfrac{1}{2})\{z[(n+\tfrac{1}{2})^{-2/3}a'_s]$$

$$+\sum_{k=1}^{\infty} h_k[(n+\tfrac{1}{2})^{-2/3}a'_s](n+\tfrac{1}{2})^{-2k}\}$$

10.1.64

$$b'_{n,s} \sim (n+\tfrac{1}{2})\{z[(n+\tfrac{1}{2})^{-2/3}b'_s]$$

$$+\sum_{k=1}^{\infty} h_k[(n+\tfrac{1}{2})^{-2/3}b'_s](n+\tfrac{1}{2})^{-2k}\}$$

10.1.65

$$j_n(a'_{n,s}) \sim \sqrt{\tfrac{1}{2}\pi}\operatorname{Ai}(a'_s)(n+\tfrac{1}{2})^{-5/6}$$

$$h[(n+\tfrac{1}{2})^{-2/3}a'_s](z[(n+\tfrac{1}{2})^{-2/3}a'_s])^{-1/2}$$

$$\{1+\sum_{k=1}^{\infty} H_k[(n+\tfrac{1}{2})^{-2/3}a'_s](n+\tfrac{1}{2})^{-2k}\}$$

10.1.66

$$y_n(b'_{n,s}) \sim -\sqrt{\tfrac{1}{2}\pi}\operatorname{Bi}(b'_s)(n+\tfrac{1}{2})^{-5/6}$$

$$h[(n+\tfrac{1}{2})^{-2/3}b'_s](z[(n+\tfrac{1}{2})^{-2/3}b'_s])^{-1/2}$$

$$\{1+\sum_{k=1}^{\infty} H_k[(n+\tfrac{1}{2})^{-2/3}b'_s](n+\tfrac{1}{2})^{-2k}\}$$

$h(\xi)$, $z(\xi)$ are defined as in **9.5.26, 9.3.38, 9.3.39**. a'_s, b'_s s-th (negative) real zero of $\operatorname{Ai}'(z)$, $\operatorname{Bi}'(z)$ (see **10.4.95, 10.4.99**.)

Complex Zeros of $h_n^{(1)}(z)$, $h_n^{(1)\prime}(z)$

$h_n^{(1)}(z)$ and $h_n^{(1)}(ze^{2m\pi i})$, m any integer, have the same zeros.

$h_n^{(1)}(z)$ has n zeros, symmetrically distributed with respect to the imaginary axis and lying approximately on the finite arc joining $z=-n$ and $z=n$ shown in **Figure 9.6**. If n is odd, one zero lies on the imaginary axis.

$h_n^{(1)\prime}(z)$ has $n+1$ zeros lying approximately on the same curve. If n is even, one zero lies on the imaginary axis.

$-\zeta$	$(-\zeta)h_1(\zeta)$	$(-\zeta)h_2(\zeta)$	$(-\zeta)h_3(\zeta)$	$(-\zeta)^2 H_1(\zeta)$	$(-\zeta)^4 H_2(\zeta)$	$(-\zeta)^6 H_3(\zeta)$
0.0	$-.4409724$	$-.122500$	$-.06806$	.000000	.00000	.0000
0.2	$-.4572444$	$-.114201$	$-.05986$	.027518	.00575	.0023
0.4	$-.4702250$	$-.107243$	$-.05279$	.049069	.01118	.0043
0.6	$-.4802184$	$-.101318$	$-.04674$	.065677	.01592	.0061
0.8	$-.4875705$	$-.096159$	$-.04160$	.078255	.01983	.0075
1.0	$-.4926355$	$-.091561$	$-.03725$	.087587	.02290	.0085

$-\zeta$	$h_1(\zeta)$	$h_2(\zeta)$	$h_3(\zeta)$	$H_1(\zeta)$	$H_2(\zeta)$	
1.0	$-.4926355$	$-.09156$	$-.037$	.087587	.0229	
1.2	$-.4131280$	$-.05056$	$-.014$	.065507	.0121	
1.4	$-.3551700$	$-.03043$	$-.006$	.050524	.0070	
1.6	$-.3108548$	$-.01950$	$-.003$	.039890	.0042	
1.8	$-.2757704$	$-.01310$	$-.001$	.032085	.0027	
2.0	$-.2472521$	$-.00914$		.026206	.0018	
2.2	$-.2235898$	$-.00658$		.021682	.0012	
2.4	$-.2036314$	$-.00485$		.018141	.0008	
2.6	$-.1865701$	$-.00366$		.015326	.0006	
2.8	$-.1718217$	$-.00280$		.013061	.0004	
3.0	$-.1589519$	$-.00219$		.011217	.0003	
3.2	$-.1476304$	$-.00173$		.009701	.0002	
3.4	$-.1376005$	$-.00138$		.008443	.0002	
3.6	$-.1286601$	$-.00112$		.007391	.0001	
3.8	$-.1206469$	$-.00091$		.006505	.0001	
4.0	$-.1134296$	$-.00075$		.005753		
4.2	$-.1069004$	$-.00062$		.005111		
4.4	$-.1009699$	$-.00052$		.004560		
4.6	$-.0955634$	$-.00044$		.004085		
4.8	$-.0906180$	$-.00037$		.003672		
5.0	$-.0860804$	$-.00032$		.003313		
5.2	$-.0819049$	$-.00027$		.002998		
5.4	$-.0780523$	$-.00023$		.002722		
5.6	$-.0744888$	$-.00020$		.002478		
5.8	$-.0711850$	$-.00018$		.002262		
6.0	$-.0681152$	$-.00015$		.002070		
6.2	$-.0652570$	$-.00013$		.001899		
6.4	$-.0625905$	$-.00012$		.001746		
6.6	$-.0600985$	$-.00010$		.001609		
6.8	$-.0577653$	$-.00009$		.001486		
7.0	$-.0555773$	$-.00008$		.001375		

$(-\zeta)^{-\frac{1}{2}}$	$h_1(\zeta)$	$h_2(\zeta)$	$H_1(\zeta)$
0.40	$-.0645731$	$-.00013$	.001859
.36	$-.0487592$	$-.00005$	.001056
.32	$-.0352949$	$-.00002$	.000551
.28	$-.0242415$	$-.00001$	.000259
.24	$-.0155683$		.000106
.20	$-.0091416$		.000037
.16	$-.0047276$		.000010
.12	$-.0020068$		.000002
.08	$-.0005965$		
.04	$-.0000747$		
.00	$-.0000000$		

10.2. Modified Spherical Bessel Functions

Definitions

Differential Equation

10.2.1

$$z^2 w'' + 2zw' - [z^2 + n(n+1)]w = 0$$

$$(n = 0, \pm 1, \pm 2, \ldots)$$

Particular solutions are the *Modified Spherical Bessel functions of the first kind,*

10.2.2

$$\sqrt{\tfrac{1}{2}\pi/z}\, I_{n+\frac{1}{2}}(z) = e^{-n\pi i/2} j_n(z e^{\pi i/2}) \quad (-\pi < \arg z \leq \tfrac{1}{2}\pi)$$

$$= e^{3n\pi i/2} j_n(z e^{-3\pi i/2}) \quad (\tfrac{1}{2}\pi < \arg z \leq \pi)$$

of the second kind,

10.2.3

$$\sqrt{\tfrac{1}{2}\pi/z}\, I_{-n-\frac{1}{2}}(z) = e^{3(n+1)\pi i/2} y_n(z e^{\pi i/2})$$

$$(-\pi < \arg z \leq \tfrac{1}{2}\pi)$$

$$= e^{-(n+1)\pi i/2} y_n(z e^{-3\pi i/2})$$

$$(\tfrac{1}{2}\pi < \arg z \leq \pi)$$

of the third kind,

10.2.4

$$\sqrt{\tfrac{1}{2}\pi/z}\, K_{n+\frac{1}{2}}(z) = \tfrac{1}{2}\pi(-1)^{n+1}\sqrt{\tfrac{1}{2}\pi/z}\, [I_{n+\frac{1}{2}}(z) - I_{-n-\frac{1}{2}}(z)]$$

The pairs

$$\sqrt{\tfrac{1}{2}\pi/z}\, I_{n+\frac{1}{2}}(z),\ \sqrt{\tfrac{1}{2}\pi/z}\, I_{-n-\frac{1}{2}}(z)$$

and

$$\sqrt{\tfrac{1}{2}\pi/z}\, I_{n+\frac{1}{2}}(z),\ \sqrt{\tfrac{1}{2}\pi/z}\, K_{n+\frac{1}{2}}(z)$$

are linearly independent solutions for every n.

Most properties of the Modified Spherical Bessel functions can be derived from those of the Spherical Bessel functions by use of the above relations.

Ascending Series

10.2.5

$$\sqrt{\tfrac{1}{2}\pi/z}\, I_{n+\frac{1}{2}}(z) = \frac{z^n}{1 \cdot 3 \cdot 5 \ldots (2n+1)}$$

$$\left\{ 1 + \frac{\tfrac{1}{2}z^2}{1!(2n+3)} + \frac{(\tfrac{1}{2}z^2)^2}{2!(2n+3)(2n+5)} + \ldots \right\}$$

10.2.6

$$\sqrt{\tfrac{1}{2}\pi/z}\, I_{-n-\frac{1}{2}}(z) = \frac{1 \cdot 3 \cdot 5 \ldots (2n-1)}{(-1)^n z^{n+1}}$$

$$\left\{ 1 + \frac{\tfrac{1}{2}z^2}{1!(1-2n)} + \frac{(\tfrac{1}{2}z^2)^2}{2!(1-2n)(3-2n)} + \ldots \right\}$$

$$(n = 0, 1, 2, \ldots)$$

Wronskians

10.2.7

$$W\{\sqrt{\tfrac{1}{2}\pi/z}\, I_{n+\frac{1}{2}}(z),\ \sqrt{\tfrac{1}{2}\pi/z}\, I_{-n-\frac{1}{2}}(z)\} = (-1)^{n+1} z^{-2}$$

10.2.8

$$W\{\sqrt{\tfrac{1}{2}\pi/z}\, I_{n+\frac{1}{2}}(z),\ \sqrt{\tfrac{1}{2}\pi/z}\, K_{n+\frac{1}{2}}(z)\} = -\tfrac{1}{2}\pi z^{-2}$$

Representations by Elementary Functions

10.2.9

$$\sqrt{\tfrac{1}{2}\pi/z}\, I_{n+\frac{1}{2}}(z) = (2z)^{-1}[R(n+\tfrac{1}{2}, -z)e^z - (-1)^n R(n+\tfrac{1}{2}, z)e^{-z}]$$

10.2.10

$$\sqrt{\tfrac{1}{2}\pi/z}\, I_{-n-\frac{1}{2}}(z) = (2z)^{-1}[R(n+\tfrac{1}{2}, -z)e^z + (-1)^n R(n+\tfrac{1}{2}, z)e^{-z}]$$

10.2.11

$$R(n+\tfrac{1}{2},\, z) = 1 + \frac{(n+1)!}{1!\,\Gamma(n)}\,(2z)^{-1}$$

$$+ \frac{(n+2)!}{2!\,\Gamma(n-1)}\,(2z)^{-2} + \ldots$$

$$= \sum_0^n (n+\tfrac{1}{2}, k)(2z)^{-k}$$

$$(n = 0, 1, 2, \ldots)$$

(See **10.1.9**.)

10.2.12

$$\sqrt{\tfrac{1}{2}\pi/z}\, I_{n+\frac{1}{2}}(z) = g_n(z) \sinh z + g_{-n-1}(z) \cosh z$$

$$g_0(z) = z^{-1},\ g_1(z) = -z^{-2}$$

$$g_{n-1}(z) - g_{n+1}(z) = (2n+1) z^{-1} g_n(z)$$

$$(n = 0, \pm 1, \pm 2, \ldots)$$

The Functions $\sqrt{\tfrac{1}{2}\pi/z}\, I_{\pm(n+\frac{1}{2})}(z)$, $n = 0, 1, 2$

10.2.13

$$\sqrt{\tfrac{1}{2}\pi/z}\, I_{1/2}(z) = \frac{\sinh z}{z}$$

$$\sqrt{\tfrac{1}{2}\pi/z}\, I_{3/2}(z) = -\frac{\sinh z}{z^2} + \frac{\cosh z}{z}$$

$$\sqrt{\tfrac{1}{2}\pi/z}\, I_{5/2}(z) = \left(\frac{3}{z^3} + \frac{1}{z}\right) \sinh z - \frac{3}{z^2} \cosh z \qquad *$$

10.2.14

$$\sqrt{\tfrac{1}{2}\pi/z}\, I_{-1/2}(z) = \frac{\cosh z}{z}$$

$$\sqrt{\tfrac{1}{2}\pi/z}\, I_{-3/2}(z) = \frac{\sinh z}{z} - \frac{\cosh z}{z^2}$$

$$\sqrt{\tfrac{1}{2}\pi/z}\, I_{-5/2}(z) = -\frac{3}{z^2} \sinh z + \left(\frac{3}{z^3} + \frac{1}{z}\right) \cosh z \qquad *$$

---

*See page II.

Modified Spherical Bessel Functions of the Third Kind

10.2.15

$$\sqrt{\tfrac{1}{2}\pi/z}\,K_{n+\frac{1}{2}}(z)=\tfrac{1}{2}\pi i e^{(n+1)\pi i/2}h_n^{(1)}(ze^{\frac{1}{2}\pi i})$$
$$(-\pi<\arg z\le\tfrac{1}{2}\pi)$$
$$=-\tfrac{1}{2}\pi i e^{-(n+1)\pi i/2}h_n^{(2)}(ze^{-\frac{1}{2}\pi i})$$
$$(\tfrac{1}{2}\pi<\arg z\le\pi)$$
$$=(\tfrac{1}{2}\pi/z)e^{-z}\sum_0^n\,(n+\tfrac{1}{2},k)(2z)^{-k}$$

10.2.16

$$K_{n+\frac{1}{2}}(z)=K_{-n-\frac{1}{2}}(z)\qquad(n=0,1,2,\ldots)$$

The Functions $\sqrt{\tfrac{1}{2}\pi/z}\,K_{n+\frac{1}{2}}(z),n=0,1,2$

10.2.17 $\quad\sqrt{\tfrac{1}{2}\pi/z}\,K_{1/2}(z)=(\tfrac{1}{2}\pi/z)e^{-z}$

$$\sqrt{\tfrac{1}{2}\pi/z}\,K_{3/2}(z)=(\tfrac{1}{2}\pi/z)e^{-z}(1+z^{-1})$$
$$\sqrt{\tfrac{1}{2}\pi/z}\,K_{5/2}(z)=(\tfrac{1}{2}\pi/z)e^{-z}(1+3z^{-1}+3z^{-2})$$

Elementary Properties

Recurrence Relations

$$f_n(z):\sqrt{\tfrac{1}{2}\pi/z}I_{n+\frac{1}{2}}(z),\ (-1)^{n+1}\sqrt{\tfrac{1}{2}\pi/z}K_{n+\frac{1}{2}}(z)$$
$$(n=0,\pm1,\pm2,\ldots)$$

10.2.18 $\quad f_{n-1}(z)-f_{n+1}(z)=(2n+1)z^{-1}f_n(z)$

10.2.19 $\quad nf_{n-1}(z)+(n+1)f_{n+1}(z)=(2n+1)\dfrac{d}{dz}f_n(z)$

10.2.20 $\quad\dfrac{n+1}{z}f_n(z)+\dfrac{d}{dz}f_n(z)=f_{n-1}(z)$
(See **10.2.22**.)

10.2.21 $\quad-\dfrac{n}{z}f_n(z)+\dfrac{d}{dz}f_n(z)=f_{n+1}(z)$
(See **10.2.23**.)

Differentiation Formulas

$$f_n(z):\sqrt{\tfrac{1}{2}\pi/z}I_{n+\frac{1}{2}}(z),\ (-1)^{n+1}\sqrt{\tfrac{1}{2}\pi/z}K_{n+\frac{1}{2}}(z)$$
$$(n=0,\pm1,\pm2,\ldots)$$

10.2.22 $\quad\left(\dfrac{1}{z}\dfrac{d}{dz}\right)^m[z^{n+1}f_n(z)]=z^{n-m+1}f_{n-m}(z)$

10.2.23 $\quad\left(\dfrac{1}{z}\dfrac{d}{dz}\right)^m[z^{-n}f_n(z)]=z^{-n-m}f_{n+m}(z)$
$$(m=1,2,3,\ldots)$$

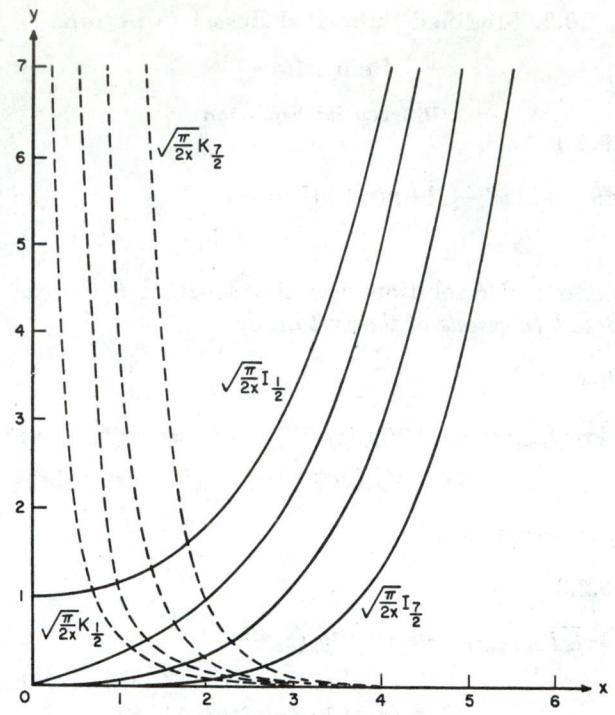

FIGURE 10.4. $\sqrt{\dfrac{\pi}{2x}}\,I_{n+\frac{1}{2}}(x),\ \sqrt{\dfrac{\pi}{2x}}\,K_{n+\frac{1}{2}}(x).\quad n=0(1)3.$

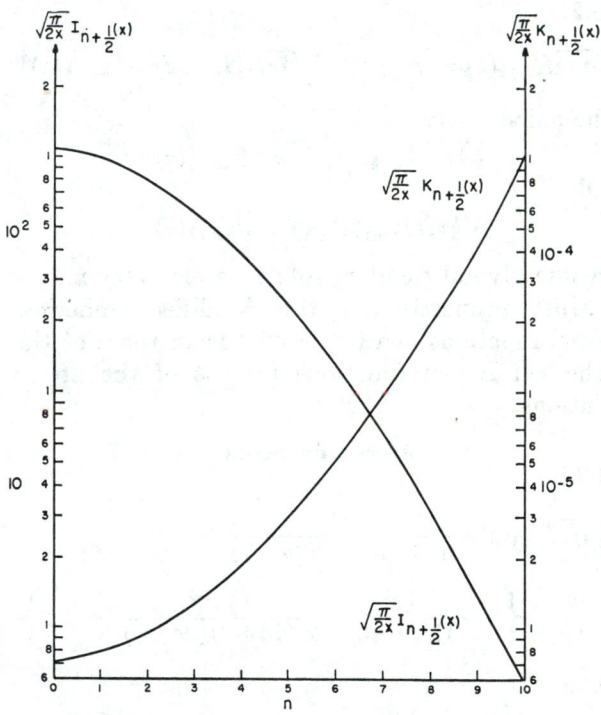

FIGURE 10.5. $\sqrt{\dfrac{\pi}{2x}}\,I_{n+\frac{1}{2}}(x),\ \sqrt{\dfrac{\pi}{2x}}\,K_{n+\frac{1}{2}}(x).\quad x=10.$

Formulas of Rayleigh's Type

10.2.24 $\sqrt{\tfrac{1}{2}\pi/z}\,I_{n+\frac{1}{2}}(z)=z^n\left(\dfrac{1}{z}\dfrac{d}{dz}\right)^n\dfrac{\sinh z}{z}$

10.2.25

$$\sqrt{\tfrac{1}{2}\pi/z}\,I_{-n-\frac{1}{2}}(z)=z^n\left(\dfrac{1}{z}\dfrac{d}{dz}\right)^n\dfrac{\cosh z}{z}$$

$$(n=0,1,2,\ldots)$$

Formulas for $I_{n+\frac{1}{2}}^2(z)-I_{-n-\frac{1}{2}}^2(z)$

10.2.26

$(\tfrac{1}{2}\pi/z)[I_{n+\frac{1}{2}}^2(z)-I_{-n-\frac{1}{2}}^2(z)]$

$$=\frac{1}{z^2}\sum_{0}^{n}(-1)^{k+1}\frac{(2n-k)!\,(2n-2k)!}{k!\,[(n-k)!]^2}(2z)^{2k-2n}$$

$$(n=0,1,2,\ldots)$$

10.2.27 $(\tfrac{1}{2}\pi/z)[I_{1/2}^2(z)-I_{-1/2}^2(z)]=-z^{-2}$

10.2.28 $(\tfrac{1}{2}\pi/z)[I_{3/2}^2(z)-I_{-3/2}^2(z)]=z^{-2}-z^{-4}$

10.2.29

$$(\tfrac{1}{2}\pi/z)[I_{5/2}^2(z)-I_{-5/2}^2(z)]=-z^{-2}+3z^{-4}-9z^{-6}$$

Generating Functions

16.2.30

$$\frac{1}{z}\sinh\sqrt{z^2-2izt}=\sum_{0}^{\infty}\frac{(-it)^n}{n!}\,[\sqrt{\tfrac{1}{2}\pi/z}\,I_{-n+\frac{1}{2}}(z)]$$

$$(2|t|<|z|)$$

10.2.31

$$\frac{1}{z}\cosh\sqrt{z^2+2izt}=\sum_{0}^{\infty}\frac{(it)^n}{n!}\,[\sqrt{\tfrac{1}{2}\pi/z}\,I_{n-\frac{1}{2}}(z)]$$

Derivatives With Respect to Order

10.2.32

$$\left[\frac{\partial}{\partial\nu}I_\nu(x)\right]_{\nu=\frac{1}{2}}=-\frac{1}{2\pi x}\left[\mathrm{Ei}(2x)e^{-x}-E_1(-2x)e^x\right]$$

10.2.33

$$\left[\frac{\partial}{\partial\nu}I_\nu(x)\right]_{\nu=-\frac{1}{2}}=\frac{1}{2\pi x}\left[\mathrm{Ei}(2x)e^{-x}+E_1(-2x)e^x\right]$$

10.2.34 $\left[\dfrac{\partial}{\partial\nu}K_\nu(x)\right]_{\nu=\pm\frac{1}{2}}=\mp\sqrt{\pi/2x}\,\mathrm{Ei}(-2x)e^x$

For $E_1(x)$ and $\mathrm{Ei}(x)$, see **5.1.1, 5.1.2.**

*See page II.

Addition Theorems and Degenerate Forms

r,ρ,θ,λ arbitrary complex; $R=\sqrt{r^2+\rho^2-2r\rho\,\cos\theta}$

10.2.35

$$\frac{e^{-\lambda R}}{\lambda R}=\frac{2}{\pi}\sum_{0}^{\infty}(2n+1)\,[\sqrt{\tfrac{1}{2}\pi/\lambda r}\,I_{n+\frac{1}{2}}(\lambda r)]$$

$$[\sqrt{\tfrac{1}{2}\pi/\lambda\rho}\,K_{n+\frac{1}{2}}(\lambda\rho)]P_n(\cos\theta)$$

10.2.36

$$e^{z\cos\theta}=\sum_{0}^{\infty}(2n+1)\,[\sqrt{\tfrac{1}{2}\pi/z}\,I_{n+\frac{1}{2}}(z)]P_n(\cos\theta)$$

10.2.37

$$e^{-z\cos\theta}=\sum_{0}^{\infty}(-1)^n(2n+1)[\sqrt{\tfrac{1}{2}\pi/z}\,I_{n+\frac{1}{2}}(z)]P_n(\cos\theta)$$

Duplication Formula

10.2.38

$K_{n+\frac{1}{2}}(2z)=$

$$n!\,\pi^{-\frac{1}{2}}z^{n+\frac{1}{2}}\sum_{0}^{n}\frac{(-1)^k(2n-2k+1)}{k!\,(2n-k+1)!}\,K_{n-k+\frac{1}{2}}^2(z)$$

10.3. Riccati-Bessel Functions

Differential Equation

10.3.1

$$z^2w''+[z^2-n(n+1)]w=0$$

$$(n=0,\pm1,\pm2,\ldots)$$

Pairs of linearly independent solutions are

$$zj_n(z),\ zy_n(z)$$
$$zh_n^{(1)}(z),\ zh_n^{(2)}(z)$$

All properties of these functions follow directly from those of the Spherical Bessel functions.

The Functions $zj_n(z),\ zy_n(z),\ n=0,1,2$

10.3.2

$zj_0(z)=\sin z,\qquad zj_1(z)=z^{-1}\sin z-\cos z$

$$zj_2(z)=(3z^{-2}-1)\sin z-3z^{-1}\cos z\quad *$$

10.3.3

$zy_0(z)=-\cos z,\qquad zy_1(z)=-\sin z-z^{-1}\cos z$

$$zy_2(z)=-3z^{-1}\sin z-(3z^{-2}-1)\cos z\quad *$$

Wronskians

10.3.4 $W\{zj_n(z),\ zy_n(z)\}=1$

10.3.5 $W\{zh_n^{(1)}(z),\ zh_n^{(2)}(z)\}=-2i$

$$(n=0,1,2,\ldots)$$

10.4. Airy Functions

Definitions and Elementary Properties

Differential Equation

10.4.1
$$w'' - zw = 0$$

Pairs of linearly independent solutions are

$$\text{Ai }(z), \text{ Bi }(z),$$
$$\text{Ai }(z), \text{ Ai }(ze^{2\pi i/3}),$$
$$\text{Ai }(z), \text{ Ai }(ze^{-2\pi i/3}).$$

Ascending Series

10.4.2 $\text{Ai }(z) = c_1 f(z) - c_2 g(z)$

10.4.3 $\text{Bi }(z) = \sqrt{3}\,[c_1 f(z) + c_2 g(z)]$

$$f(z) = 1 + \frac{1}{3!}\,z^3 + \frac{1\cdot 4}{6!}\,z^6 + \frac{1\cdot 4\cdot 7}{9!}\,z^9 + \cdots$$

$$= \sum_0^\infty 3^k \left(\frac{1}{3}\right)_k \frac{z^{3k}}{(3k)!}$$

$$g(z) = z + \frac{2}{4!}\,z^4 + \frac{2\cdot 5}{7!}\,z^7 + \frac{2\cdot 5\cdot 8}{10!}\,z^{10} + \cdots$$

$$= \sum_0^\infty 3^k \left(\frac{2}{3}\right)_k \frac{z^{3k+1}}{(3k+1)!}$$

$$\left(\alpha + \frac{1}{3}\right)_0 = 1$$

$$3^k \left(\alpha + \frac{1}{3}\right)_k = (3\alpha + 1)(3\alpha + 4)\cdots(3\alpha + 3k - 2)$$

$$(\alpha \text{ arbitrary}; k = 1, 2, 3, \ldots)$$

(See **6.1.22.**)

10.4.4
$$c_1 = \text{Ai }(0) = \text{Bi }(0)/\sqrt{3} = 3^{-2/3}/\Gamma(2/3)$$
$$= .35502\ 80538\ 87817$$

10.4.5
$$c_2 = -\text{Ai}'(0) = \text{Bi}'(0)/\sqrt{3} = 3^{-1/3}/\Gamma(1/3)$$
$$= .25881\ 94037\ 92807$$

Relations Between Solutions

10.4.6 $\text{Bi }(z) = e^{\pi i/6}\,\text{Ai }(ze^{2\pi i/3}) + e^{-\pi i/6}\,\text{Ai }(ze^{-2\pi i/3})$

10.4.7
$$\text{Ai }(z) + e^{2\pi i/3}\,\text{Ai }(ze^{2\pi i/3}) + e^{-2\pi i/3}\,\text{Ai }(ze^{-2\pi i/3}) = 0$$

10.4.8
$$\text{Bi }(z) + e^{2\pi i/3}\,\text{Bi }(ze^{2\pi i/3}) + e^{-2\pi i/3}\,\text{Bi }(ze^{-2\pi i/3}) = 0$$

10.4.9 $\text{Ai }(ze^{\pm 2\pi i/3}) = \tfrac{1}{2}e^{\pm \pi i/3}[\text{Ai }(z) \mp i\,\text{Bi }(z)]$

Wronskians

10.4.10 $\qquad W\{\text{Ai }(z),\ \text{Bi }(z)\} = \pi^{-1}$

10.4.11 $W\{\text{Ai }(z),\ \text{Ai }(ze^{2\pi i/3})\} = \tfrac{1}{2}\pi^{-1}e^{-\pi i/6}$

10.4.12 $W\{\text{Ai }(z),\ \text{Ai }(ze^{-2\pi i/3})\} = \tfrac{1}{2}\pi^{-1}e^{\pi i/6}$

10.4.13 $W\{\text{Ai }(ze^{2\pi i/3}),\ \text{Ai }(ze^{-2\pi i/3})\} = \tfrac{1}{2}i\pi^{-1}$

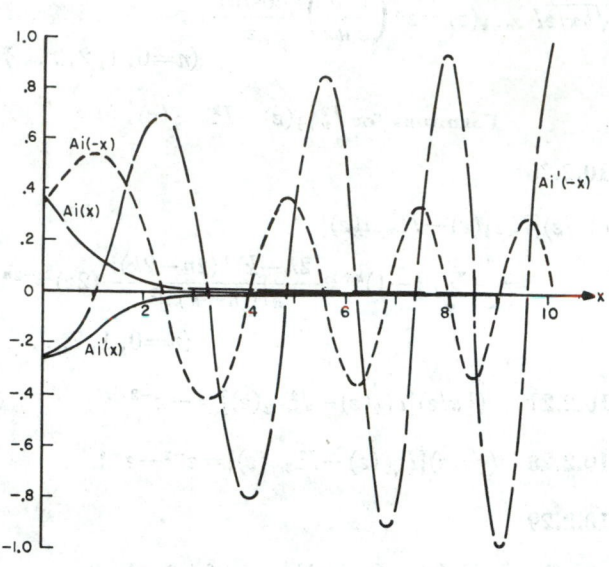

FIGURE 10.6. Ai $(\pm x)$, Ai' $(\pm x)$.

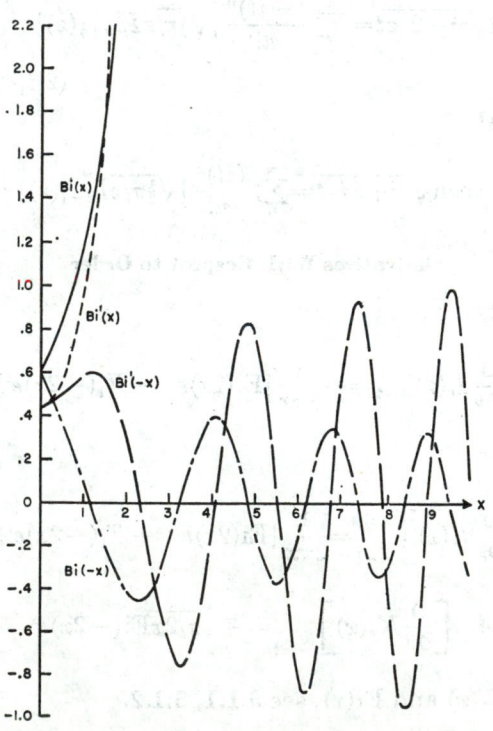

FIGURE 10.7. Bi $(\pm x)$, Bi' $(\pm x)$.

Representations in Terms of Bessel Functions

$$\zeta = \tfrac{2}{3}z^{3/2}$$

10.4.14
$$\text{Ai }(z)=\tfrac{1}{3}\sqrt{z}[I_{-1/3}(\zeta)-I_{1/3}(\zeta)]=\pi^{-1}\sqrt{z/3}K_{1/3}(\zeta)$$

10.4.15
$$\text{Ai }(-z)=\tfrac{1}{3}\sqrt{z}[J_{1/3}(\zeta)+J_{-1/3}(\zeta)]$$
$$=\tfrac{1}{2}\sqrt{z/3}[e^{\pi i/6}H^{(1)}_{1/3}(\zeta)+e^{-\pi i/6}H^{(2)}_{1/3}(\zeta)]$$

10.4.16
$$* - \text{Ai}'(z)=\tfrac{1}{3}z[I_{-2/3}(\zeta)-I_{2/3}(\zeta)]=\pi^{-1}(z/\sqrt{3})K_{2/3}(\zeta)$$

10.4.17
$$\text{Ai}'(-z)=-\tfrac{1}{3}z[J_{-2/3}(\zeta)-J_{2/3}(\zeta)]$$
$$=\tfrac{1}{2}(z/\sqrt{3})[e^{-\pi i/6}H^{(1)}_{2/3}(\zeta)+e^{\pi i/6}H^{(2)}_{2/3}(\zeta)]$$

10.4.18
$$\text{Bi }(z)=\sqrt{z/3}[I_{-1/3}(\zeta)+I_{1/3}(\zeta)]$$

10.4.19
$$\text{Bi }(-z)=\sqrt{z/3}[J_{-1/3}(\zeta)-J_{1/3}(\zeta)]$$
$$=\tfrac{1}{2}i\sqrt{z/3}[e^{\pi i/6}H^{(1)}_{1/3}(\zeta)-e^{-\pi i/6}H^{(2)}_{1/3}(\zeta)]$$

10.4.20
$$\text{Bi}'(z)=(z/\sqrt{3})[I_{-2/3}(\zeta)+I_{2/3}(\zeta)]$$

10.4.21
$$\text{Bi}'(-z)=(z/\sqrt{3})[J_{-2/3}(\zeta)+J_{2/3}(\zeta)]$$
$$=\tfrac{1}{2}i(z/\sqrt{3})[e^{-\pi i/6}H^{(1)}_{2/3}(\zeta)-e^{\pi i/6}H^{(2)}_{2/3}(\zeta)]$$

Representations of Bessel Functions in Terms of Airy Functions

$$z=\left(\tfrac{3}{2}\zeta\right)^{2/3}$$

10.4.22 $\quad J_{\pm1/3}(\zeta)=\tfrac{1}{2}\sqrt{3/z}[\sqrt{3}\text{ Ai }(-z)\mp\text{Bi }(-z)]$

*10.4.23 $\quad H^{(1)}_{\pm1/3}(\zeta)=e^{\mp\pi i/6}\sqrt{3/z}[\text{Ai }(-z)-i\text{ Bi }(-z)]$

10.4.24 $\quad H^{(2)}_{\pm1/3}(\zeta)=e^{\pm\pi i/6}\sqrt{3/z}[\text{Ai }(-z)+i\text{ Bi }(-z)]$

10.4.25 $\quad I_{\pm1/3}(\zeta)=\tfrac{1}{2}\sqrt{3/z}[\mp\sqrt{3}\text{ Ai }(z)+\text{Bi }(z)]$

10.4.26 $\quad K_{\pm1/3}(\zeta)=\pi\sqrt{3/z}\text{ Ai }(z)$

10.4.27 $\quad J_{\pm2/3}(\zeta)=(\sqrt{3}/2z)[\pm\sqrt{3}\text{ Ai}'(-z)+\text{Bi}'(-z)]$

10.4.28
$$H^{(1)}_{2/3}(\zeta)=e^{-2\pi i/3}H^{(1)}_{-2/3}{}'(\zeta)$$
$$=e^{\pi i/6}(\sqrt{3}/z)[\text{Ai}'(-z)-i\text{ Bi}'(-z)]$$

10.4.29
$$H^{(2)}_{2/3}(\zeta)=e^{2\pi i/3}H^{(2)}_{-2/3}(\zeta)$$
$$=e^{-\pi i/6}(\sqrt{3}/z)[\text{Ai}'(-z)+i\text{ Bi}'(-z)]$$

10.4.30 $\quad I_{\pm2/3}(\zeta)=(\sqrt{3}/2z)[\pm\sqrt{3}\text{ Ai}'(z)+\text{Bi}'(z)]$

10.4.31 $\quad K_{\pm2/3}(\zeta)=-\pi(\sqrt{3}/z)\text{ Ai}'(z)$

Integral Representations

10.4.32
$$(3a)^{-1/3}\pi\text{ Ai }[\pm(3a)^{-1/3}x]=\int_0^\infty\cos(at^3\pm xt)dt$$

10.4.33
$$(3a)^{-1/3}\pi\text{ Bi }[\pm(3a)^{-1/3}x]$$
$$=\int_0^\infty[\exp(-at^3\pm xt)+\sin(at^3\pm xt)]dt$$

The Integrals $\int_0^z\text{Ai }(\pm t)dt,\ \int_0^z\text{Bi }(\pm t)dt$

$$\zeta=\tfrac{2}{3}z^{3/2}$$

10.4.34 $\quad\displaystyle\int_0^z\text{Ai }(t)dt=\frac{1}{3}\int_0^\zeta[I_{-1/3}(t)-I_{1/3}(t)]dt$

10.4.35 $\quad\displaystyle\int_0^z\text{Ai }(-t)dt=\frac{1}{3}\int_0^\zeta[J_{-1/3}(t)+J_{1/3}(t)]dt$

10.4.36 $\quad\displaystyle\int_0^z\text{Bi }(t)dt=\frac{1}{\sqrt{3}}\int_0^\zeta[I_{-1/3}(t)+I_{1/3}(t)]dt$

10.4.37 $\quad\displaystyle\int_0^z\text{Bi }(-t)dt=\frac{1}{\sqrt{3}}\int_0^\zeta[J_{-1/3}(t)-J_{1/3}(t)]dt$

Ascending Series for $\int_0^z\text{Ai }(\pm t)dt,\ \int_0^z\text{Bi }(\pm t)dt$

10.4.38 $\quad\displaystyle\int_0^z\text{Ai }(t)dt=c_1F(z)-c_2G(z)$

(See **10.4.2.**)

10.4.39 $\quad\displaystyle\int_0^z\text{Ai }(-t)dt=-c_1F(-z)+c_2G(-z)$

10.4.40 $\quad\displaystyle\int_0^z\text{Bi }(t)dt=\sqrt{3}[c_1F(z)+c_2G(z)]$

(See **10.4.3.**)

10.4.41
$$\int_0^z\text{Bi }(-t)dt=-\sqrt{3}[c_1F(-z)+c_2G(-z)]$$

$$F(z)=z+\frac{1}{4!}z^4+\frac{1\cdot4}{7!}z^7+\frac{1\cdot4\cdot7}{10!}z^{10}+\cdots$$

$$=\sum_0^\infty 3^k\left(\frac{1}{3}\right)_k\frac{z^{3k+1}}{(3k+1)!}$$

$$G(z)=\frac{1}{2!}z^2+\frac{2}{5!}z^5+\frac{2\cdot5}{8!}z^8+\frac{2\cdot5\cdot8}{11!}z^{11}+\cdots$$

$$=\sum_0^\infty 3^k\left(\frac{2}{3}\right)_k\frac{z^{3k+2}}{(3k+2)!}$$

The constants c_1, c_2 are given in **10.4.4, 10.4.5.**

The Functions Gi(z), Hi(z)

10.4.42

$$\text{Gi} (z) = \pi^{-1} \int_0^\infty \sin\left(\frac{1}{3} t^3 + zt\right) dt$$

$$= \frac{1}{3} \text{Bi} (z) + \int_0^z [\text{Ai} (z) \text{Bi} (t) - \text{Ai} (t) \text{Bi} (z)] dt$$

10.4.43

$$\text{Gi}' (z) = \frac{1}{3} \text{Bi}' (z) + \int_0^z [\text{Ai}'(z) \text{Bi} (t) - \text{Ai} (t) \text{Bi}'(z)] dt$$

10.4.44

$$\text{Hi}(z) = \pi^{-1} \int_0^\infty \exp\left(-\frac{1}{3} t^3 + zt\right) dt$$

$$= \frac{2}{3} \text{Bi} (z) + \int_0^z [\text{Ai} (t) \text{Bi} (z) - \text{Ai}(z) \text{Bi} (t)] dt$$

10.4.45

$$\text{Hi}' (z) = \frac{2}{3} \text{Bi}' (z) + \int_0^z [\text{Ai} (t) \text{Bi}'(z) - \text{Ai}'(z) \text{Bi} (t)] dt$$

10.4.46 $\text{Gi} (z) + \text{Hi} (z) = \text{Bi} (z)$

Representations of $\int_0^z \text{Ai}(\pm t) dt$, $\int_0^z \text{Bi}(\pm t) dt$ by Gi $(\pm z)$, Hi $(\pm z)$

10.4.47

$$\int_0^z \text{Ai} (t) dt = \frac{1}{3} + \pi[\text{Ai}' (z) \text{Gi}(z) - \text{Ai} (z) \text{Gi}'(z)]$$

10.4.48

$$= -\frac{2}{3} - \pi[\text{Ai}' (z) \text{Hi} (z) - \text{Ai} (z) \text{Hi}' (z)]$$

10.4.49

$$\int_0^z \text{Ai} (-t) dt = -\frac{1}{3} - \pi[\text{Ai}' (-z) \text{Gi} (-z)$$
$$-\text{Ai} (-z) \text{Gi}' (-z)]$$

10.4.50

$$= \frac{2}{3} + \pi[\text{Ai}' (-z) \text{Hi} (-z)$$
$$-\text{Ai} (-z)\text{Hi}'(-z)]$$

10.4.51

$$\int_0^z \text{Bi} (t) dt = \pi[\text{Bi}' (z) \text{Gi} (z) - \text{Bi} (z) \text{Gi}' (z)]$$

10.4.52 $= -\pi[\text{Bi}' (z) \text{Hi} (z) - \text{Bi} (z) \text{Hi}'(z)]$

10.4.53

$$\int_0^z \text{Bi} (-t) dt = -\pi[\text{Bi}' (-z) \text{Gi} (-z)$$
$$-\text{Bi} (-z) \text{Gi}' (-z)]$$

10.4.54 $= \pi[\text{Bi}' (-z) \text{Hi} (-z)$
$$-\text{Bi} (-z) \text{Hi}' (-z)]$$

Differential Equations for Gi (z), Hi (z)

10.4.55 $w'' - zw = -\pi^{-1}$

$$w(0) = \frac{1}{3} \text{Bi} (0) = \frac{1}{\sqrt{3}} \text{Ai} (0) = .20497\ 55424\ 78$$

$$w'(0) = \frac{1}{3} \text{Bi}' (0) = -\frac{1}{\sqrt{3}} \text{Ai}' (0) = .14942\ 94524\ 49$$

$$w(z) = \text{Gi}(z)$$

10.4.56 $w'' - zw = \pi^{-1}$

$$w(0) = \frac{2}{3} \text{Bi} (0) = \frac{2}{\sqrt{3}} \text{Ai} (0) = .40995\ 10849\ 56$$

$$w'(0) = \frac{2}{3} \text{Bi}' (0) = -\frac{2}{\sqrt{3}} \text{Ai}' (0) = .29885\ 89048\ 98$$

$$w(z) = \text{Hi} (z)$$

Differential Equation for Products of Airy Functions

10.4.57 $w''' - 4zw' - 2w = 0$

Linearly independent solutions are Ai^2 (z), Ai (z) Bi (z), Bi^2 (z).

Wronskian for Products of Airy Functions

10.4.58 $W\{\text{Ai}^2 (z), \text{Ai} (z) \text{Bi} (z), \text{Bi}^2 (z)\} = 2\pi^{-3}$

Asymptotic Expansions for $|z|$ Large

$$c_0 = 1, c_k = \frac{\Gamma(3k + \frac{1}{2})}{54^k k! \Gamma(k + \frac{1}{2})} = \frac{(2k+1)(2k+3) \ldots (6k-1)}{216^k k!},$$

$$d_0 = 1, d_k = -\frac{6k+1}{6k-1} c_k \qquad (k = 1, 2, 3, \ldots)$$

$$\zeta = \frac{2}{3} z^{3/2}$$

10.4.59

$$\text{Ai} (z) \sim \frac{1}{2} \pi^{-1/2} z^{-1/4} e^{-\zeta} \sum_0^\infty (-1)^k c_k \zeta^{-k} \quad (|\arg z| < \pi)$$

10.4.60

$$\text{Ai} (-z) \sim \pi^{-1/2} z^{-1/4} \left[\sin\left(\zeta + \frac{\pi}{4}\right) \sum_0^\infty (-1)^k c_{2k} \zeta^{-2k} \right.$$
$$\left. - \cos\left(\zeta + \frac{\pi}{4}\right) \sum_0^\infty (-1)^k c_{2k+1} \zeta^{-2k-1} \right]$$

$$(|\arg z| < \tfrac{2}{3}\pi)$$

10.4.61

$$\text{Ai}' (z) \sim -\frac{1}{2}\pi^{-1/2} z^{1/4} e^{-\zeta} \sum_0^\infty (-1)^k d_k \zeta^{-k}$$

$$(|\arg z| < \pi)$$

10.4.62

$$\text{Ai}' \, (-z) \sim -\pi^{-\frac12} z^{\frac14} \left[\cos\left(\zeta+\frac{\pi}{4}\right) \sum_0^\infty (-1)^k d_{2k} \zeta^{-2k} \right.$$
$$\left. +\sin\left(\zeta+\frac{\pi}{4}\right) \sum_0^\infty (-1)^k d_{2k+1} \zeta^{-2k-1} \right]$$
$$\left(|\arg z| < \tfrac{2}{3}\pi\right)$$

10.4.63

$$\text{Bi} \, (z) \sim \pi^{-\frac12} z^{-\frac14} e^{\zeta} \sum_0^\infty c_k \zeta^{-k} \qquad \left(|\arg z| < \tfrac{1}{3}\pi\right)$$

10.4.64

$$\text{Bi} \, (-z) \sim \pi^{-\frac12} z^{-\frac14} \left[\cos\left(\zeta+\frac{\pi}{4}\right) \sum_0^\infty (-1)^k c_{2k} \zeta^{-2k} \right.$$
$$\left. +\sin\left(\zeta+\frac{\pi}{4}\right) \sum_0^\infty (-1)^k c_{2k+1} \zeta^{-2k-1} \right]$$
$$\left(|\arg z| < \tfrac{2}{3}\pi\right)$$

10.4.65

$$\text{Bi} \, (ze^{\pm\pi i/3})$$
$$\sim \sqrt{2/\pi}\, e^{\pm\pi i/6} z^{-\frac14} \left[\sin\left(\zeta+\frac{\pi}{4}\mp\frac{i}{2}\ln 2\right) \sum_0^\infty (-1)^k c_{2k} \zeta^{-2k} \right.$$
$$\left. -\cos\left(\zeta+\frac{\pi}{4}\mp\frac{i}{2}\ln 2\right) \sum_0^\infty (-1)^k c_{2k+1} \zeta^{-2k-1} \right]$$
$$\left(|\arg z| < \tfrac{2}{3}\pi\right)$$

10.4.66

$$* \quad \text{Bi}' \, (z) \sim \pi^{-\frac12} z^{\frac14} e^{\zeta} \sum_0^\infty d_k \zeta^{-k} \quad \left(|\arg z| < \tfrac{1}{3}\pi\right)$$

10.4.67

$$\text{Bi}' \, (-z) \sim \pi^{-\frac12} z^{\frac14} \left[\sin\left(\zeta+\frac{\pi}{4}\right) \sum_0^\infty (-1)^k d_{2k} \zeta^{-2k} \right.$$
$$\left. -\cos\left(\zeta+\frac{\pi}{4}\right) \sum_0^\infty (-1)^k d_{2k+1} \zeta^{-2k-1} \right]$$
$$\left(|\arg z| < \tfrac{2}{3}\pi\right)$$

10.4.68

$$\text{Bi}' \, (ze^{\pm\pi i/3})$$
$$\sim \sqrt{2/\pi}\, e^{\mp\pi i/6} z^{\frac14} \left[\cos\left(\zeta+\frac{\pi}{4}\mp\frac{i}{2}\ln 2\right) \sum_0^\infty (-1)^k d_{2k} \zeta^{-2k} \right.$$
$$\left. +\sin\left(\zeta+\frac{\pi}{4}\mp\frac{i}{2}\ln 2\right) \sum_0^\infty (-1)^k d_{2k+1} \zeta^{-2k-1} \right]$$
$$\left(|\arg z| < \tfrac{2}{3}\pi\right)$$

Modulus and Phase

10.4.69

$$\text{Ai} \, (-x) = M(x)\cos\theta(x), \quad \text{Bi} \, (-x) = M(x)\sin\theta(x)$$
$$M(x) = \sqrt{[\text{Ai}^2 \, (-x) + \text{Bi}^2 \, (-x)]},$$
$$\theta(x) = \arctan \, [\text{Bi} \, (-x)/\text{Ai} \, (-x)]$$

10.4.70

$$\text{Ai}' \, (-x) = N(x)\cos\phi(x), \quad \text{Bi}' \, (-x) = N(x)\sin\phi(x)$$
$$N(x) = \sqrt{[\text{Ai}'^2 \, (-x) + \text{Bi}'^2 \, (-x)]},$$
$$\phi(x) = \arctan \, [\text{Bi}' \, (-x)/\text{Ai}' \, (-x)]$$

Differential Equations for Modulus and Phase

Primes denote differentiation with respect to x

10.4.71
$$M^2\theta' = -\pi^{-1}, \quad N^2\phi' = -\pi^{-1}x$$

10.4.72
$$N^2 = M'^2 + M^2\theta'^2 = M'^2 + \pi^{-2}M^{-2} \qquad *$$

10.4.73
$$NN' = -xMM'$$

10.4.74
$$\tan \, (\phi-\theta) = M\theta'/M' = -(\pi MM')^{-1},$$
$$MN\sin \, (\phi-\theta) = \pi^{-1}$$

10.4.75
$$M'' + xM - \pi^{-2}M^{-3} = 0$$

10.4.76
$$(M^2)''' + 4x(M^2)' - 2M^2 = 0$$

10.4.77
$$\theta'^2 + \tfrac12(\theta'''/\theta') - \tfrac34(\theta''/\theta')^2 = x$$

Asymptotic Expansions of Modulus and Phase for Large x

10.4.78
$$M^2(x) \sim \frac{1}{\pi}\, x^{-1/2} \sum_0^\infty \frac{(-1)^k}{12^k k!} 2^{3k} \left(\frac{1}{2}\right)_{3k} (2x)^{-3k}$$

10.4.79
$$\theta(x) \sim \frac14\,\pi - \frac23\,x^{3/2} \left[1 - \frac54\,(2x)^{-3} + \frac{1105}{96}\,(2x)^{-6} \right.$$
$$\left. -\frac{82825}{128}\,(2x)^{-9} + \frac{12820\,31525}{14336}\,(2x)^{-12} - \dots \right]$$

10.4.80
$$N^2(x) \sim \frac{1}{\pi} x^{\frac12} \sum_0^\infty \frac{(-1)^{k+1}}{12^k k!} \frac{6k+1}{6k-1} 2^{3k} \left(\frac{1}{2}\right)_{3k} (2x)^{-3k}$$

10.4.81
$$\phi(x) \sim \frac34\,\pi - \frac23\,x^{3/2} \left[1 + \frac74\,(2x)^{-3} - \frac{1463}{96}\,(2x)^{-6} \right.$$
$$\left. +\frac{495271}{640}\,(2x)^{-9} - \frac{2065\,30429}{2048}\,(2x)^{-12} + \dots \right]$$

Asymptotic Forms of $\int_0^x \text{Ai} \, (\pm t)\,dt, \int_0^x \text{Bi} \, (\pm t)\,dt$ for Large x

10.4.82
$$\int_0^x \text{Ai} \, (t)\,dt \sim \frac13 - \frac12\,\pi^{-1/2}x^{-3/4}\exp\left(-\frac23\,x^{3/2}\right)$$

10.4.83
$$\int_0^x \text{Ai} \, (-t)\,dt \sim \frac23 - \pi^{-1/2}x^{-3/4}\cos\left(\frac23\,x^{3/2} + \frac{\pi}{4}\right)$$

10.4.84 $\displaystyle\int_0^z \mathrm{Bi}\,(t)dt \sim \pi^{-1/2}x^{-3/4}\exp\left(\frac{2}{3}\,x^{3/2}\right)$

10.4.85 $\displaystyle\int_0^z \mathrm{Bi}\,(-t)dt \sim \pi^{-1/2}x^{-3/4}\sin\left(\frac{2}{3}\,x^{3/2}+\frac{\pi}{4}\right)$

Asymptotic Forms of $\mathrm{Gi}\,(\pm x),\ \mathrm{Gi}'\,(\pm x),\ \mathrm{Hi}\,(\pm x),\ \mathrm{Hi}'\,(\pm x)$ **for Large x**

10.4.86 $\qquad \mathrm{Gi}\,(x) \sim \pi^{-1}x^{-1}$

10.4.87 $\quad \mathrm{Gi}\,(-x) \sim \pi^{-1/2}x^{-1/4}\cos\left(\frac{2}{3}\,x^{3/2}+\frac{\pi}{4}\right)$

10.4.88 $\qquad \mathrm{Gi}'\,(x) \sim \frac{7}{96}\,\pi^{-1}x^{-2}$

10.4.89 $\quad \mathrm{Gi}'\,(-x) \sim \pi^{-1/2}x^{1/4}\sin\left(\frac{2}{3}\,x^{3/2}+\frac{\pi}{4}\right)$

10.4.90 $\qquad \mathrm{Hi}\,(x) \sim \pi^{-1/2}x^{-1/4}\exp\left(\frac{2}{3}x^{3/2}\right)$

10.4.91 $\qquad \mathrm{Hi}\,(-x) \sim \pi^{-1}x^{-1}$

10.4.92 $\qquad \mathrm{Hi}'\,(x) \sim \pi^{-1/2}x^{1/4}\exp\left(\frac{2}{3}x^{3/2}\right)$

10.4.93 $\qquad \mathrm{Hi}'\,(-x) \sim -\frac{3}{2}\,\pi^{-1}x^{-2}$

Zeros and Their Asymptotic Expansions

$\mathrm{Ai}\,(z)$, $\mathrm{Ai}'\,(z)$ have zeros on the negative real axis only. $\mathrm{Bi}\,(z)$, $\mathrm{Bi}'\,(z)$ have zeros on the negative real axis and in the sector $\frac{1}{3}\pi<|\arg z|<\frac{1}{2}\pi$. $a_s,\ a_s';\ b_s,\ b_s'$ s-th (real) negative zero of $\mathrm{Ai}\,(z)$, $\mathrm{Ai}'\,(z)$; $\mathrm{Bi}\,(z)$, $\mathrm{Bi}'\,(z)$, respectively. $\beta_s,\ \beta_s';\ \bar\beta_s,\ \bar\beta_s'$ s-th complex zero of $\mathrm{Bi}\,(z)$, $\mathrm{Bi}'\,(z)$ in the sectors $\frac{1}{3}\pi<\arg z<\frac{1}{2}\pi$, $-\frac{1}{2}\pi<\arg z<-\frac{1}{3}\pi$, respectively.

10.4.94 $\qquad a_s = -f[3\pi(4s-1)/8]$

10.4.95 $\qquad a_s' = -g[3\pi(4s-3)/8]$

10.4.96 $\quad \mathrm{Ai}'\,(a_s) = (-1)^{s-1}f_1[3\pi(4s-1)/8]$

10.4.97 $\quad \mathrm{Ai}\,(a_s') = (-1)^{s-1}g_1[3\pi(4s-3)/8]$

10.4.98 $\qquad b_s = -f[3\pi(4s-3)/8]$

10.4.99 $\qquad b_s' = -g[3\pi(4s-1)/8]$

10.4.100 $\quad \mathrm{Bi}'\,(b_s) = (-1)^{s-1}f_1[3\pi(4s-3)/8]$

10.4.101 $\quad \mathrm{Bi}\,(b_s') = (-1)^{s}g_1[3\pi(4s-1)/8]$

10.4.102 $\quad \beta_s = e^{\pi i/3}f\left[\frac{3\pi}{8}\,(4s-1)+\frac{3i}{4}\ln 2\right]$

10.4.103 $\quad \beta_s' = e^{\pi i/3}g\left[\frac{3\pi}{8}\,(4s-3)+\frac{3i}{4}\ln 2\right]$

10.4.104

$$\mathrm{Bi}'\,(\beta_s) = (-1)^{s}\sqrt{2}e^{-\pi i/6}f_1\left[\frac{3\pi}{8}\,(4s-1)+\frac{3i}{4}\ln 2\right]$$

10.4.105

$$\mathrm{Bi}\,(\beta_s') = (-1)^{s-1}\sqrt{2}e^{\pi i/6}g_1\left[\frac{3\pi}{8}\,(4s-3)+\frac{3i}{4}\ln 2\right]$$

$|z|$ sufficiently large

$$f(z) \sim z^{2/3}\left(1+\frac{5}{48}\,z^{-2}-\frac{5}{36}\,z^{-4}+\frac{77125}{82944}\,z^{-6}\right.$$
$$-\frac{1080\,56875}{69\,67296}\,z^{-8}$$
$$\left.+\frac{16\,23755\,96875}{3344\,30208}\,z^{-10}-\ldots\right)$$

$$g(z) \sim z^{2/3}\left(1-\frac{7}{48}\,z^{-2}+\frac{35}{288}\,z^{-4}-\frac{181223}{207360}\,z^{-6}\right.$$
$$+\frac{186\,83371}{12\,44160}\,z^{-8}$$
$$\left.-\frac{9\,11458\,84361}{1911\,02976}\,z^{-10}+\ldots\right)$$

$$f_1(z) \sim \pi^{-1/2}z^{1/6}\left(1+\frac{5}{48}\,z^{-2}-\frac{1525}{4608}\,z^{-4}\right.$$
$$\left.+\frac{23\,97875}{6\,63552}\,z^{-6}-\ldots\right)$$

$$g_1(z) \sim \pi^{-1/2}z^{-1/6}\left(1-\frac{7}{96}\,z^{-2}+\frac{1673}{6144}\,z^{-4}\right. \qquad *$$
$$\left.-\frac{843\,94709}{265\,42080}\,z^{-6}+\ldots\right)$$

Formal and Asymptotic Solutions of Ordinary Differential Equations of Second Order With Turning Points

An equation

10.4.106 $\quad W''+a(z,\lambda)W'+b(z,\lambda)W=0$

in which λ is a real or complex parameter and, for fixed λ, $a(z,\lambda)$ is analytic in z and $b(z,\lambda)$ is continuous in z in some region of the z-plane, may be reduced by the transformation

10.4.107 $\quad W(z)=w(z)\exp\left(-\frac{1}{2}\int^z a(t,\lambda)dt\right)$

to the equation

10.4.108

$w''+\varphi(z,\lambda)w=0$

$$\varphi(z,\lambda)=b(z,\lambda)-\frac{1}{4}\,a^2(z,\lambda)-\frac{1}{2}\,\frac{d}{dz}\,a(z,\lambda).$$

If $\varphi(z, \lambda)$ can be written in the form

10.4.109
$$\varphi(z, \lambda) = \lambda^2 p(z) + q(z, \lambda)$$

where $q(z, \lambda)$ is bounded in a region R of the z-plane, then the zeros of $p(z)$ in R are said to be turning points of the equation **10.4.108**.

The Special Case $w'' + [\lambda^2 z + q(z, \lambda)]w = 0$

Let $\lambda = |\lambda|e^{i\omega}$ vary over a sectorial domain S: $|\lambda| \geq \lambda_0(>0)$, $\omega_1 \leq \omega \leq \omega_2$, and suppose that $q(z, \lambda)$ is continuous in z for $|z| < r$ and λ in S, and $q(z, \lambda)$ $\sim \sum_0^\infty q_n(z)\lambda^{-n}$ as $\lambda \to \infty$ in S.

Formal Series Solution

10.4.110
$$w(z) = u(z) \sum_0^\infty \varphi_n(z)\lambda^{-n} + \lambda^{-1} u'(z) \sum_0^\infty \psi_n(z)\lambda^{-n}$$
$$u'' + \lambda^2 z u = 0$$

$$\varphi_0(z) = c_0, \qquad \psi_0(z) = z^{-\frac{1}{2}} c_1, \qquad c_0, c_1 \text{ constants}$$

$$\varphi_{n+1}(z) = -\frac{1}{2}\psi_n'(z) - \frac{1}{2}\int_0^z \sum_0^n q_{n-k}(t)\psi_k(t)\,dt$$

$$\psi_n(z) = \frac{1}{2} z^{-\frac{1}{2}} \int_0^z t^{-\frac{1}{2}}\left[\varphi_n''(t) + \sum_0^n q_{n-k}(t)\varphi_k(t)\right]dt$$

$$(n = 0, 1, 2, \ldots)$$

Uniform Asymptotic Expansions of Solutions

For z real, i.e. for the equation

10.4.111
$$y'' + [\lambda^2 x + q(x, \lambda)]y = 0$$

where x varies in a bounded interval $a \leq x \leq b$ that includes the origin and where, for each fixed λ in S, $q(x, \lambda)$ is continuous in x for $a \leq x \leq b$, the following asymptotic representations hold.

(i) If λ is real and positive, there are solutions $y_0(x)$, $y_1(x)$ such that, uniformly in x on $a \leq x \leq 0$,

10.4.112
$$y_0(x) = \mathrm{Ai}\,(-\lambda^{2/3}x)[1 + O(\lambda^{-1})] \qquad (\lambda \to \infty)$$
$$y_1(x) = \mathrm{Bi}\,(-\lambda^{2/3}x)[1 + O(\lambda^{-1})]$$

and, uniformly in x on $0 \leq x \leq b$

10.4.113
$$y_0(x) = \mathrm{Ai}\,(-\lambda^{2/3}x)[1 + O(\lambda^{-1})] + \mathrm{Bi}\,(-\lambda^{2/3}x)O(\lambda^{-1}),$$
$$y_1(x) = \mathrm{Bi}\,(-\lambda^{2/3}x)[1 + O(\lambda^{-1})] + \mathrm{Ai}\,(-\lambda^{2/3}x)O(\lambda^{-1})$$
$$(\lambda \to \infty)$$

(ii) If $\mathcal{R}\lambda \geq 0$, $\mathcal{I}\lambda \neq 0$, there are solutions $y_0(x)$, $y_1(x)$ such that, uniformly in x on $a \leq x \leq b$,

10.4.114
$$y_0(x) = \mathrm{Ai}\,(-\lambda^{2/3}x)[1 + O(\lambda^{-1})]$$
$$y_1(x) = \mathrm{Bi}\,(-\lambda^{2/3}x)[1 + O(\lambda^{-1})] \qquad (|\lambda| \to \infty)$$

For further representations and details, we refer to [10.4].

When z is complex (bounded or unbounded), conditions under which the formal series **10.4.110** yields a uniform asymptotic expansion of a solution are given in [10.12] if $q(z, \lambda)$ is independent of λ and $|\lambda| \to \infty$ with fixed ω, and in [10.14] if λ lies in any region of the complex plane. Further references are [10.2; 10.9; 10.10].

The General Case $w'' + [\lambda^2 p(z) + q(z, \lambda)]w = 0$

Let $\lambda = |\lambda|e^{i\omega}$ where $|\lambda| \geq \lambda_0(>0)$ and $-\pi \leq \omega \leq \pi$; suppose that $p(z)$ is analytic in a region R and has a zero $z = z_0$ in R, and that, for fixed λ, $q(z, \lambda)$ is analytic in z for z in R. The transformation $\xi = \xi(z)$, $v = [p(z)/\xi]^{1/4}w(z)$, where ξ is defined as the (unique) solution of the equation

10.4.115
$$\xi\left(\frac{d\xi}{dz}\right)^2 = p(z),$$

yields the special case

10.4.116
$$\frac{d^2v}{d\xi^2} + [\lambda^2\xi + f(\xi, \lambda)]v = 0, \qquad *$$

$$f(\xi, \lambda) = \left(\frac{d\xi}{dz}\right)^{-2} q(z, \lambda) - \left(\frac{d\xi}{dz}\right)^{-\frac{1}{2}} \frac{d^2}{d\xi^2}\left[\left(\frac{d\xi}{dz}\right)^{\frac{1}{2}}\right].$$

Example:

Consider the equation

10.4.117
$$y'' + [\lambda^2 - (\lambda^2 - \tfrac{1}{4})\, x^{-2}]y = 0$$

for which the points $x = 0$, ∞ are singular points and $x = 1$ is a turning point. It has the functions $x^{\frac{1}{2}}J_\lambda(\lambda x)$, $x^{\frac{1}{2}}Y_\lambda(\lambda x)$ as particular solutions (see **9.1.49**).

The equation **10.4.115** becomes

$$\xi\left(\frac{d\xi}{dx}\right)^2 = \frac{x^2 - 1}{x^2}$$

whence

$$\tfrac{2}{3}(-\xi)^{3/2} = -\sqrt{1 - x^2} + \ln x^{-1}(1 + \sqrt{1 - x^2})$$
$$(0 < x \leq 1)$$

$$\tfrac{2}{3}\xi^{3/2} = \sqrt{x^2 - 1} - \arccos x^{-1} \qquad (1 \leq x < \infty).$$

Thus

10.4.118
$$v(\xi) = \left(\frac{x^2 - 1}{x^2\xi}\right)^{1/4} y(x)$$

*See page II.

satisfies the equation

10.4.119 $\quad \dfrac{d^2v}{d\xi^2}+\left[\lambda^2\xi-\dfrac{5}{16\xi^2}+\dfrac{\xi^2}{4}\dfrac{x^2(x^2+4)}{(x^2-1)^3}\right]v=0$

which is of the form **10.4.111** with x replaced by ξ and $q(\xi,\lambda)$ independent of λ.

Suppose $\mathscr{R}\lambda\geq0$, $\mathscr{I}\lambda\neq0$. By the first equation of **10.4.114** there is a solution $v_0(\xi)$ of **10.4.119**, i.e., a solution $y_0(x)$ of **10.4.117** for which the representation

10.4.120

$$v_0(\xi)=\left(\dfrac{x^2-1}{x^2\xi}\right)^{1/4}y_0(x)=\mathrm{Ai}\,(-\lambda^{2/3}\xi)[1+O(\lambda^{-1})]$$

holds uniformly in x on $0<x<\infty$ as $|\lambda|\to\infty$.

To identify $y_0(x)$ in terms of $x^{\frac12}J_\lambda(\lambda x)$, $x^{\frac12}Y_\lambda(\lambda x)$, restrict x to $0<x\leq b<1$ so that by **10.4.118** ξ is negative, and replace the Airy function by its asymptotic representation **10.4.59**. This yields

10.4.121

$$y_0(x)$$
$$=\left(\dfrac{x^2-1}{x^2\xi}\right)^{-1/4}\dfrac{1}{2}\pi^{-1/2}\lambda^{-1/6}(-\xi)^{1/4}\exp\left(\dfrac{2}{3}\lambda(-\xi)^{3/2}\right)$$
$$[1+O(\lambda^{-1})]$$

$$=\dfrac{1}{2}\pi^{-1/2}\lambda^{-1/6}\left(\dfrac{1-x^2}{x^2}\right)^{-1/4}\exp\left(\dfrac{2}{3}\lambda(-\xi)^{3/2}\right)$$
$$[1+O(\lambda^{-1})]$$

Let now λ be fixed and $x\to0$ in **10.4.121**. There results

10.4.122 $\quad y_0(x)\sim\frac12\pi^{-1/2}\lambda^{-1/6}x^{1/2}\left(\frac12 x\right)^\lambda e^\lambda.$

On the other hand, $y_0(x)$ is a solution of **10.4.117** and therefore it can be written in the form

10.4.123 $\quad y_0(x)=x^{1/2}[c_1J_\lambda(\lambda x)+c_2Y_\lambda(\lambda x)]$

where, from **9.1.7** for λ fixed and $x\to0$

$$J_\lambda(\lambda x)\sim\dfrac{(\frac12\lambda x)^\lambda}{\Gamma(\lambda+1)},$$

$$Y_\lambda(\lambda x)\sim\dfrac{(\frac12\lambda x)^\lambda}{\Gamma(\lambda+1)}\cot\lambda\pi-\dfrac{(\frac12\lambda x)^{-\lambda}}{\Gamma(1-\lambda)}\csc\lambda\pi.$$

Thus, letting $x\to0$ in **10.4.123** and comparing the resulting relation with **10.4.122** one finds that $c_2=0$ and

10.4.124 $\quad y_0(x)=\frac12\pi^{-1/2}\lambda^{-\lambda-1/6}e^\lambda\Gamma(\lambda+1)x^{1/2}J_\lambda(\lambda x).$

It follows from **10.4.120** that uniformly in x on $0<x<\infty$

10.4.125

$$J_\lambda(\lambda x)$$
$$=\dfrac{2\pi^{1/2}}{\Gamma(\lambda+1)}\lambda^{\lambda+1/6}e^{-\lambda}\left(\dfrac{x^2-1}{\xi}\right)^{-1/4}\mathrm{Ai}\,(-\lambda^{2/3}\xi)[1+O(\lambda^{-1})]$$
$$(|\lambda|\to\infty)$$

Numerical Methods

10.5. Use and Extension of the Tables

Spherical Bessel Functions

To compute $j_n(x)$, $y_n(x)$, $n=0, 1, 2$, for values of x outside the range of **Table 10.1**, use formulas **10.1.11**, **10.1.12** and obtain values for the circular functions from **Tables 4.6–4.8**.

Example 1. Compute $j_1(x)$ for $x=11.425$.

From **10.1.11**, $j_1(x)=\dfrac{\sin x}{x^2}-\dfrac{\cos x}{x}$. Hence, using **Tables 4.6** and **4.8**,

$$j_1(11.425)=-\dfrac{.90920\,500}{(11.425)^2}-\dfrac{.41634\,873}{11.425}$$
$$=-.00696\,54535-.03644\,1902$$
$$=-.04340\,7356.$$

To compute $j_n(x)$, $11\leq n\leq20$, for a value of x within the range of **Table 10.3**, obtain from **Table 10.3**, directly or possibly by linear interpolation, $j_{21}(x)$, $j_{20}(x)$ and use these as starting values in the recurrence relation **10.1.19** for decreasing n.

An alternative procedure which often yields better accuracy and which also applies to computations of $j_n(x)$ when both n and x are outside the range of **Table 10.1** is the following device essentially due to J. C. P. Miller [9.20].

At some value N larger than the desired value n, assume tentatively $F_{N+1}=0$, $F_N=1$ and use recurrence relation **10.1.19** for decreasing N to obtain the sequence $F_{N-1}, \ldots, F_0$. If N was chosen large enough, each term of this sequence up to F_n is proportional, to a certain number of significant figures, to the corresponding term in the sequence $j_{N-1}(x), \ldots, j_0(x)$ of true values. The factor of proportionality, p, may be obtained by comparing, say, F_0 with the true value $j_0(x)$ computed separately. The terms in the sequence $pF_0, \ldots pF_n$ are then accurate to the number of significant figures present in the tentative values. If the accuracy obtained is not sufficient, the process may be repeated by starting from a larger value N.

Example 2. Compute $j_{15}(x)$ for $x=24.6$.

Interpolation in **Table 10.3** yields for $x=24.6$

$$x^{-21}e^{x^2/86}j_{21}(x)=(-28)3.934616$$
$$x^{-20}e^{x^2/82}j_{20}(x)=(-27)9.48683$$

whence

$$j_{21}(24.6)=.05604\ 29,\quad j_{20}(24.6)=.03896\ 98.$$

From the recurrence relation **10.1.19** there results

$$\begin{aligned}
j_{19}(24.6)=&\quad.00890\ 67660 &&[.00890\ 70]\\
j_{18}(24.6)=&-.02484\ 93173 &&[-.02485\ 90]\\
j_{17}(24.6)=&-.04628\ 17554 &&[-.04628\ 16]\\
j_{16}(24.6)=&-.04099\ 87086 &&[-.04099\ 88]\\
j_{15}(24.6)=&-.00871\ 65122 &&[-.00871\ 67]
\end{aligned}$$

For comparison, the correct values are shown in brackets.

To compute $j_{15}(x)$ for $x=24.6$ by Miller's device, take, for example, $N=39$ and assume $F_{40}=0$, $F_{39}=1$. Using **10.1.19** with decreasing N, i.e., $F_{N-1}=[(2N+1)/x]F_N-F_{N+1}$, $N=39, 38, \ldots,$ 1, 0, generate the sequence $F_{38}, F_{37}, \ldots, F_1, F_0$, compute from **Table 4.6**, $j_0(24.6)=(\sin 24.6)/24.6$ $=-.02064\ 620296$, and obtain the factor of proportionality

$$p=j_0(24.6)/F_0=.00000\ 03839\ 17642.$$

The value pF_{15} equals $j_{15}(24.6)$ to 8 decimals. The final part of the computations is shown in the following table, in which the correct values are given for comparison.

N	F_N	pF_N	$j_N(24.6)$
15	$-22704.\ 71107$	$-.00871\ 67391$	$-.00871\ 674$
14	$+78178.\ 88236$	$+.03001\ 42522$	$+.03001\ 425$
13	$+114866.\ 80811$	$+.04409\ 93941$	$+.04409\ 939$
12	$+47894.\ 44353$	$+.01838\ 75218$	$+.01838\ 752$
11	$-66193.\ 59317$	$-.02541\ 28882$	$-.02541\ 289$
10	$-109782.\ 76234$	$-.04214\ 75392$	$-.04214\ 754$
9	$-27523.\ 39903$	$-.01056\ 67185$	$-.01056\ 672$
8	$+88524.\ 85252$	$+.03398\ 62526$	$+.03398\ 625$
7	$+88699.\ 11017$	$+.03405\ 31532$	$+.03405\ 315$
6	$-34440.\ 02929$	$-.01322\ 21348$	$-.01322\ 213$
5	$-106899.\ 12565$	$-.04104\ 04602$	$-.04104\ 046$
4	$-13360.\ 39272$	$-.00512\ 92905$	$-.00512\ 929$
3	$+102011.\ 17704$	$+.03916\ 38905$	$+.03916\ 389$
2	$+42387.\ 96341$	$+.01627\ 34870$	$+.01627\ 349$
1	$-93395.\ 73728$	$-.03585\ 62712$	$-.03585\ 627$
0	$-53777.\ 68747$	$-.02064\ 62030$	$-.02064\ 620$

It may be observed that the normalization of the sequence $F_N, F_{N-1}, \ldots, F_0$ can also be obtained from formula **10.1.50** by computing the sum $\sigma=\sum_0^\infty (2k+1)F_k^2$ and finding $p=1/\sqrt{\sigma}$. This yields, in the case of the example, $p=1/\sqrt{\sigma}=$ $.00000\ 03839\ 177$.

Modified Spherical Bessel Functions

To compute $\sqrt{\tfrac{1}{2}\pi/x}I_{n+\frac{1}{2}}(x)$, $\sqrt{\tfrac{1}{2}\pi/x}K_{n+\frac{1}{2}}(x)$, $n=0, 1, 2, \ldots$ for values of x outside the range of **Table 10.8**, use formulas **10.2.13**, **10.2.14** together with **10.2.4** and obtain values for the hyperbolic and exponential functions from **Tables 4.4** and **4.15**. In those cases when $\sqrt{\tfrac{1}{2}\pi/x}I_{n+\frac{1}{2}}(x)$ and $\sqrt{\tfrac{1}{2}\pi/x}I_{-n-\frac{1}{2}}(x)$ are nearly equal, i.e., when x is sufficiently large, compute $\sqrt{\tfrac{1}{2}\pi/x}K_{n+\frac{1}{2}}(x)$ from formula **10.2.15**, for which the coefficients $(n+\frac{1}{2}, k)$ are given in **10.1.9**.

Example 3. Compute $\sqrt{\tfrac{1}{2}\pi/x}I_{5/2}(x)$, $\sqrt{\tfrac{1}{2}\pi/x}K_{5/2}(x)$ for $x=16.2$.

From **10.2.13**, $\sqrt{\tfrac{1}{2}\pi/x}I_{5/2}(x)=(3+x^2)$ $\sinh x/x^3-$ $3\cosh x/x^2$; from **Table 4.4**, $\cosh 16.2=(6)5.4267\ 59950$ and this equals the value of $\sinh 16.2$ to the same number of significant figures. Hence

$$\begin{aligned}
\sqrt{\tfrac{1}{2}\pi/16.2}I_{5/2}(16.2)=&(.06243\ 402371\\
&-.01143\ 118427)[(6)5.4267\ 59950]\\
=&338814.4594-62034.29298\\
=&276780.1664.
\end{aligned}$$

To compute $\sqrt{\tfrac{1}{2}\pi/16.2}K_{5/2}(16.2)$ use **10.2.17** and obtain

$$\begin{aligned}
\sqrt{\tfrac{1}{2}\pi/16.2}K_{5/2}(16.2)=&\pi e^{-16.2}\left[\frac{1}{32.4}+\frac{6}{(32.4)^2}+\frac{12}{(32.4)^3}\right]\\
=&(-7)2.8945\ 38069[.036932\ 60400]\\
=&(-8)1.0690\ 28283.
\end{aligned}$$

To compute $\sqrt{\tfrac{1}{2}\pi/x}I_{n+\frac{1}{2}}(x)$, $3\leq n\leq 8$, for a value of x within the range of **Table 10.9**, obtain from **Table 10.9**, $\sqrt{\tfrac{1}{2}\pi/x}I_{19/2}(x)$, $\sqrt{\tfrac{1}{2}\pi/x}I_{21/2}(x)$ for the desired value of x and use these as starting values in the recurrence relation **10.2.18** for decreasing n.

To compute $\sqrt{\tfrac{1}{2}\pi/x}K_{n+\frac{1}{2}}(x)$ for some integer n outside the range of **Table 10.9**, obtain from **10.2.15** or from **Table 10.8**, $\sqrt{\tfrac{1}{2}\pi/x}K_{\frac{1}{2}}(x)$, $\sqrt{\tfrac{1}{2}\pi/x}K_{3/2}(x)$ for the desired value of x and use these as starting values in the recurrence relation **10.2.18** for increasing n. If x lies within the range of **Table 10.9** and $n>10$, the recurrence may be started with $\sqrt{\tfrac{1}{2}\pi/x}K_{19/2}(x)$, $\sqrt{\tfrac{1}{2}\pi/x}K_{21/2}(x)$ obtained from **Table 10.9**.

Example 4. Compute $\sqrt{\tfrac{1}{2}\pi/x}K_{11/2}(x)$ for $x=3.6$. Obtain from **Table 10.8** for $x=3.6$

$$\sqrt{\tfrac{1}{2}\pi/x}K_{1/2}(x)=.01192\ 222$$

$$\sqrt{\tfrac{1}{2}\pi/x}K_{3/2}(x)=.01523\ 3952$$

The recurrence relation **10.2.18** yields successively

$$-\sqrt{\tfrac{1}{2}\pi/3.6}\,K_{5/2}(3.6) = -.01192\ 222$$

$$-\frac{3}{3.6}\,(.01523\ 3952)$$

$$= -.02461\ 718$$

$$\sqrt{\tfrac{1}{2}\pi/3.6}\,K_{7/2}(3.6) = .01523\ 3952$$

$$+\frac{5}{3.6}\,(.02461\ 718)$$

$$= .04942\ 4480$$

$$-\sqrt{\tfrac{1}{2}\pi/3.6}\,K_{9/2}(3.6) = -.02461\ 718$$

$$-\frac{7}{3.6}\,(.04942\ 4480)$$

$$= -.12072\ 034$$

$$\sqrt{\tfrac{1}{2}\pi/3.6}\,K_{11/2}(3.6) = .04942\ 4480$$

$$+\frac{9}{3.6}\,(.12072\ 034)$$

$$= .35122\ 533.$$

As a check, the recurrence can be carried out until $n=9$ and the value of $\sqrt{\tfrac{1}{2}\pi/3.6}\,K_{19/2}(3.6)$ so obtained can be compared with the corresponding value from **Table 10.9.**

To compute $\sqrt{\tfrac{1}{2}\pi/x}\,I_{n+\frac{1}{2}}(x)$ when both n and x are outside the range of **Table 10.9,** use the device described in [9.20].

Airy Functions

To compute $\mathrm{Ai}(x)$, $\mathrm{Bi}(x)$ for values of x beyond 1, use auxiliary functions from **Table 10.11.**

Example 5. Compute $\mathrm{Ai}(x)$ for $x=4.5$.

First, for $x=4.5$,

$$\xi=\tfrac{2}{3}x^{3/2}=6.36396\ 1029,\quad \xi^{-1}=.15713\ 48403.$$

Hence, from **Table 10.11,** $f(-\xi)=.55848\ 24$ and thus

$$\mathrm{Ai}(4.5)$$

$$=\tfrac{1}{2}(4.5)^{-1/4}(.55848\ 24)\ \exp\,(-6.36396\ 1029)$$

$$=\tfrac{1}{2}(.68658\ 905)(.55848\ 24)(.00172\ 25302)$$

$$=.00033\ 02503.$$

To compute the zeros c, c' of a solution $y(x)$ of the equation $y''-xy=0$ and of its derivative

$y'(x)$, respectively, the following formulas may be used, in which d, d' denote approximations to c, c' and $u=y(d)/y'(d)$, $v=y'(d')/d'^2 y(d')$.

$$c=d-u-2d\,\frac{u^3}{3!}+2\,\frac{u^4}{4!}-24d^2\,\frac{u^5}{5!}$$

$$+88d\,\frac{u^6}{6!}-(88+720d^3)\,\frac{u^7}{7!}$$

$$+5856d^2\,\frac{u^8}{8!}-(16640d+40320d^4)\frac{u^9}{9!}+\cdots$$

$$c'=d'\left\{1-v-\frac{v^2}{2!}-(3+2d'^3)\frac{v^3}{3!}-(15+10d'^3)\frac{v^4}{4!}\right.$$

$$-(105+76d'^3+24d'^6)\frac{v^5}{5!}$$

$$\left.-(945+756d'^3+272d'^6)\frac{v^6}{6!}-\cdots\right\}$$

$$y'(c)=y'(d)\left\{1-d\,\frac{u^2}{2!}+\frac{u^3}{3!}-3d^2\,\frac{u^4}{4!}+14d\,\frac{u^5}{5!}\right.$$

$$-(14+45d^3)\frac{u^6}{6!}+471d^2\,\frac{u^7}{7!}$$

$$\left.-(1432d+1575d^4)\frac{u^8}{8!}+\cdots\right\}$$

$$y(c')=y(d')\left\{1-d'^2\,\frac{v^2}{2!}-d'^3\,\frac{v^3}{3!}-(3d'^3+3d'^6)\frac{v^4}{4!}\right.$$

$$-(15d'^3+14d'^6)\frac{v^5}{5!}$$

$$\left.-(105d'^3+101d'^6+45d'^4)\frac{v^6}{6!}-\cdots\right\}$$

Example 6. Compute the zero of $y(x)=\mathrm{Ai}(x)-\mathrm{Bi}(x)$ near $d=-.4$.

From **Table 10.11,**

$$y(-.4)=.02420\ 467,\quad y'(-.4)=-.71276\ 627$$

whence $u=y(-.4)/y'(-.4)=-.03395\ 8776$. From the above formulas

$$c=-.4+.03395\ 8776-.00000\ 5221$$

$$+.00000\ 0111+.00000\ 0001$$

$$=-.36604\ 6333.$$

$$y'(c)=(-.71276\ 627)\{1+.00023\ 0640$$

$$-.00000\ 6527-.00000\ 0027+.00000\ 0002\}$$

$$=(-.71276\ 627)(1.00022\ 4088)$$

$$=-.71292\ 599.$$

References

Texts

[10.1] H. Bateman and R. C. Archibald, A guide to tables of Bessel functions, Math. Tables Aids Comp. **1**, 205–308 (1944), in particular, pp. 229–240.

[10.2] T. M. Cherry, Uniform asymptotic formulae for functions with transition points, Trans. Amer. Math. Soc. **68**, 224–257 (1950).

[10.3] A. Erdélyi et al., Higher transcendental functions, vol. 1, 2 (McGraw-Hill Book Co., Inc., New York, N.Y., 1953).

[10.4] A. Erdélyi, Asymptotic expansions, California Institute of Technology, Dept. of Math., Technical Report No. 3, Pasadena, Calif. (1955).

[10.5] A. Erdélyi, Asymptotic solutions of differential equations with transition points or singularities, J. Mathematical Physics **1**, 16–26 (1960).

[10.6] H. Jeffreys, On certain approximate solutions of linear differential equations of the second order, Proc. London Math. Soc. **23**, 428–436 (1925).

[10.7] H. Jeffreys, The effect on Love waves of heterogeneity in the lower layer, Monthly Nat. Roy. Astr. Soc., Geophys. Suppl. **2**, 101–111 (1928).

[10.8] H. Jeffreys, On the use of asymptotic approximations of Green's type when the coefficient has zeros, Proc. Cambridge Philos. Soc. **52**, 61–66 (1956).

[10.9] R. E. Langer, On the asymptotic solutions of differential equations with an application to the Bessel functions of large complex order, Trans. Amer. Math. Soc. **34**, 447–480 (1932).

[10.10] R. E. Langer, The asymptotic solutions of ordinary linear differential equations of the second order, with special reference to a turning point, Trans. Amer. Math. Soc. **67**, 461–490 (1949).

[10.11] W. Magnus and F. Oberhettinger, Formeln und Sätze für die speziellen Funktionen der mathematischen Physik, 2d ed. (Springer-Verlag, Berlin, Germany, 1948).

[10.12] F. W. J. Olver, The asymptotic solution of linear differential equations of the second order for large values of a parameter, Philos. Trans. Roy. Soc. London [A] **247**, 307–327 (1954–55).

[10.13] F. W. J. Olver, The asymptotic expansion of Bessel functions of large order, Philos. Trans. Roy. Soc. London [A] **247**, 328–368 (1954).

[10.14] F. W. J. Olver, Uniform asymptotic expansions of solutions of linear second-order differential equations for large values of a parameter, Philos. Trans. Roy. Soc. London [A] **250**, 479–517 (1958).

[10.15] W. R. Wasow, Turning point problems for systems of linear differential equations. Part I: The formal theory; Part II: The analytic theory. Comm. Pure Appl. Math. **14**, 657–673 (1961); **15**, 173–187 (1962).

[10.16] G. N. Watson, A treatise on the theory of Bessel functions, 2d ed. (Cambridge Univ. Press, Cambridge, England, 1958).

Tables

[10.17] H. K. Crowder and G. C. Francis, Tables of spherical Bessel functions and ordinary Bessel functions of order half and odd integer of the first and second kind, Ballistic Research Laboratory Memorandum Report No. 1027, Aberdeen Proving Ground, Md. (1956).

[10.18] A. T. Doodson, Bessel functions of half integral order [Riccati-Bessel functions], British Assoc. Adv. Sci. Report, 87–102 (1914).

[10.19] A. T. Doodson, Riccati-Bessel functions, British Assoc. Adv. Sci. Report, 97–107 (1916).

[10.20] A. T. Doodson, Riccati-Bessel functions, British Assoc. Adv. Sci. Report, 263–270 (1922).

[10.21] Harvard University, Tables of the modified Hankel functions of order one-third and of their derivatives (Harvard Univ. Press, Cambridge, Mass., 1945).

[10.22] E. Jahnke and F. Emde, Tables of functions, 4th ed. (Dover Publications, Inc., New York, N.Y., 1945).

[10.23] C. W. Jones, A short table for the Bessel functions $I_{n+\frac{1}{2}}(x)$, $(2/\pi)K_{n+\frac{1}{2}}(x)$ (Cambridge Univ. Press, Cambridge, England, 1952).

[10.24] J. C. P. Miller, The Airy integral, British Assoc. Adv. Sci. Mathematical Tables, Part-vol. B (Cambridge Univ. Press, Cambridge, England, 1946).

[10.25] National Bureau of Standards, Tables of spherical Bessel functions, vols. I, II (Columbia Univ. Press, New York, N.Y., 1947).

[10.26] National Bureau of Standards, Tables of Bessel functions of fractional order, vols. I, II (Columbia Univ. Press, New York, N.Y., 1948–49).

[10.27] National Bureau of Standards, Integrals of Airy functions, Applied Math. Series 52 (U.S. Government Printing Office, Washington, D.C., 1958).

[10.28] J. Proudman, A. T. Doodson and G. Kennedy, Numerical results of the theory of the diffraction of a plane electromagnetic wave by a conducting sphere, Philos. Trans. Roy. Soc. London [A] **217**, 279–314 (1916–18), in particular pp. 284–288.

[10.29] M. Rothman, The problem of an infinite plate under an inclined loading, with tables of the integrals of Ai $(\pm x)$, Bi $(\pm x)$, Quart. J. Mech. Appl. Math. **7**, 1–7 (1954).

[10.30] M. Rothman, Tables of the integrals and differential coefficients of Gi $(+x)$, Hi $(-x)$, Quart. J. Mech. Appl. Math. **7**, 379–384 (1954).

[10.31] Royal Society Mathematical Tables, vol. 7, Bessel functions, Part III. Zeros and associated values (Cambridge Univ. Press, Cambridge, England, 1960).

[10.32] R. S. Scorer, Numerical evaluation of integrals of the form

$$I = \int_{x_1}^{x_2} f(x)\, e^{ip(x)}\, dx$$

and the tabulation of the function

$$\text{Gi}\,(z) = (1/\pi) \int_0^\infty \sin\,(uz + 1/3u^3)\, du,$$

Quart. J. Mech. Appl. Math. **3**, 107–112 (1950).

[10.33] A. D. Smirnov, Tables of Airy functions (and special confluent hypergeometric functions). Translated from the Russian by D. G. Fry (Pergamon Press, New York, N.Y., 1960).

[10.34] I. M. Vinogradov and N. G. Cetaev, Tables of Bessel functions of imaginary argument (Izdat. Akad. Nauk SSSR., Moscow, U.S.S.R., 1950).

[10.35] P. M. Woodward, A. M. Woodward, R. Hensman, H. H. Davies and N. Gamble, Four-figure tables of the Airy functions in the complex plane, Phil. Mag. (7) **37**, 236–261 (1946).

SPHERICAL BESSEL FUNCTIONS—ORDERS 0, 1 AND 2   Table 10.1

x	$j_0(x)$	$j_1(x)$	$j_2(x)$	$y_0(x)$	$y_1(x)$	$y_2(x)$
0.0	1.00000 000	0.00000 0000	0.00000 000000	$-\infty$	$-\infty$	$-\infty$
0.1	0.99833 417	0.03330 0012	0.00066 619061	−9.95004 17	−100.49875	− 3005.0125
0.2	0.99334 665	0.06640 0381	0.00265 90561	−4.90033 29	−25.495011	− 377.52483
0.3	0.98506 736	0.09910 2888	0.00596 15249	−3.18445 50	−11.599917	− 112.81472
0.4	0.97354 586	0.13121 215	0.01054 5302	−2.30265 25	− 6.73017 71	−48.173676
0.5	0.95885 108	0.16253 703	0.01637 1107	−1.75516 51	− 4.46918 13	− 25.059923
0.6	0.94107 079	0.19289 196	0.02338 8995	−1.37555 94	− 3.23366 97	−14.792789
0.7	0.92031 098	0.22209 828	0.03153 8780	−1.09263 17	− 2.48121 34	− 9.54114 00
0.8	0.89669 511	0.24998 551	0.04075 0531	−0.87088 339	− 1.98529 93	− 6.57398 92
0.9	0.87036 323	0.27639 252	0.05094 5155	−0.69067 774	− 1.63778 29	− 4.76859 87
1.0	0.84147 098	0.30116 868	0.06203 5052	−0.54030 231	− 1.38177 33	− 3.60501 76
1.1	0.81018 851	0.32417 490	0.07392 4849	−0.41236 011	− 1.18506 13	− 2.81962 54
1.2	0.77669 924	0.34528 457	0.08651 2186	−0.30196 480	− 1.02833 66	− 2.26887 66
1.3	0.74119 860	0.36438 444	0.09968 8571	−0.20576 833	− 0.89948 193	− 1.86995 92
1.4	0.70389 266	0.38137 537	0.11334 028	−0.12140 510	− 0.79061 059	− 1.57276 05
1.5	0.66499 666	0.39617 297	0.12734 928	−0.04715 8134	− 0.69643 541	− 1.34571 27
1.6	0.62473 350	0.40870 814	0.14159 426	+0.01824 9701	− 0.61332 744	− 1.16823 87
1.7	0.58333 224	0.41892 749	0.15595 157	0.07579 0879	− 0.53874 937	− 1.02652 51
1.8	0.54102 646	0.42679 364	0.17029 628	0.12622 339	− 0.47090 236	− 0.91106 065
1.9	0.49805 268	0.43228 539	0.18450 320	0.17015 240	− 0.40849 878	− 0.81515 048
2.0	0.45464 871	0.43539 778	0.19844 795	0.20807 342	− 0.35061 200	− 0.73399 142
2.1	0.41105 208	0.43614 199	0.21200 791	0.24040 291	− 0.29657 450	− 0.66408 077
2.2	0.36749 837	0.43454 522	0.22506 330	0.26750 051	− 0.24590 723	− 0.60282 854
2.3	0.32421 966	0.43065 030	0.23749 812	0.28968 523	− 0.19826 956	− 0.54829 769
2.4	0.28144 299	0.42451 529	0.24920 113	0.30724 738	− 0.15342 325	− 0.49902 644
2.5	0.23938 886	0.41621 299	0.26006 673	0.32045 745	− 0.11120 588	− 0.45390 450
2.6	0.19826 976	0.40583 020	0.26999 585	0.32957 260	− 0.07151 1067	− 0.41208 537
2.7	0.15828 884	0.39346 703	0.27889 675	0.33484 153	− 0.03427 3462	− 0.37292 316
2.8	0.11963 863	0.37923 606	0.28668 572	0.33650 798	+ 0.00054 2796	− 0.33592 641
2.9	0.08249 9769	0.36326 136	0.29328 784	0.33481 316	0.03295 3045	− 0.30072 380
3.0	0.04704 0003	0.34567 750	0.29863 750	0.32999 750	0.06295 9164	− 0.26703 834
3.1	+0.01341 3117	0.32662 847	0.30267 895	0.32230 166	0.09055 5161	− 0.23466 763
3.2	−0.01824 1920	0.30626 652	0.30536 678	0.31196 712	0.11573 164	− 0.20346 870
3.3	−0.04780 1726	0.28475 092	0.30666 620	0.29923 629	0.13847 939	− 0.17334 594
3.4	−0.07515 9148	0.26224 678	0.30655 336	0.28435 241	0.15879 221	− 0.14424 164
3.5	−0.10022 378	0.23892 369	0.30501 551	0.26755 905	0.17666 922	− 0.11612 829
3.6	−0.12292 235	0.21495 446	0.30205 107	0.24909 956	0.19211 667	− 0.08900 2337
3.7	−0.14319 896	0.19051 380	0.29766 961	0.22921 622	0.20514 929	− 0.06287 8964
3.8	−0.16101 523	0.16577 697	0.29189 179	0.20814 940	0.21579 139	− 0.03778 7773
3.9	−0.17635 030	0.14091 846	0.28474 912	0.18613 649	0.22407 760	− 0.01376 9102
4.0	−0.18920 062	0.11611 075	0.27628 369	0.16341 091	0.23005 335	+ 0.00912 9107
4.1	−0.19957 978	0.09152 2967	0.26654 781	0.14020 096	0.23377 514	0.03085 4018
4.2	−0.20751 804	0.06731 9710	0.25560 355	0.11672 877	0.23531 060	0.05135 0236
4.3	−0.21306 185	0.04365 9843	0.24352 220	0.09320 9110	0.23473 838	0.07056 1855
4.4	−0.21627 320	+0.02069 5380	0.23038 368	0.06984 8380	0.23214 783	0.08843 4232
4.5	−0.21722 892	−0.00142 95812	0.21627 586	0.04684 3511	0.22763 858	0.10491 554
4.6	−0.21601 978	−0.02257 9838	0.20129 380	0.02438 0984	0.22132 000	0.11995 814
4.7	−0.21274 963	−0.04262 9993	0.18553 900	+0.00263 5886	0.21331 046	0.13351 972
4.8	−0.20753 429	−0.06146 5266	0.16911 850	−0.01822 8955	0.20373 659	0.14556 433
4.9	−0.20050 053	−0.07898 2225	0.15214 407	−0.03806 3749	0.19273 242	0.15606 319
5.0	−0.19178 485	−0.09508 9408	0.13473 121	−0.05673 2437	0.18043 837	0.16499 546
	$\begin{bmatrix} (-4)4 \\ 6 \end{bmatrix}$	$\begin{bmatrix} (-4)3 \\ 6 \end{bmatrix}$	$\begin{bmatrix} (-4)2 \\ 6 \end{bmatrix}$			

$$j_n(x) = \sqrt{\tfrac{1}{2}\pi/x}\, J_{n+\frac{1}{2}}(x) \qquad\qquad y_n(x) = \sqrt{\tfrac{1}{2}\pi/x}\, Y_{n+\frac{1}{2}}(x) = (-1)^{n+1}\sqrt{\tfrac{1}{2}\pi/x}\, J_{-(n+\frac{1}{2})}(x)$$

Compiled from National Bureau of Standards, Tables of spherical Bessel functions, vols. I, II.   Columbia Univ. Press, New York, N.Y., 1947 (with permission).

Table 10.1   **SPHERICAL BESSEL FUNCTIONS—ORDERS 0, 1 AND 2**

x	$j_0(x)$	$j_1(x)$	$j_2(x)$	$y_0(x)$	$y_1(x)$	$y_2(x)$
5.0	(−1) −1.9178	(−2) −9.5089	(−1) 1.3473	(−2) −5.6732	(−1) 1.8044	(−1) 1.6500
5.1	(−1) −1.8153	(−1) −1.0971	(−1) 1.1700	(−2) −7.4113	(−1) 1.6700	(−1) 1.7235
5.2	(−1) −1.6990	(−1) −1.2277	(−2) 9.9065	(−2) −9.0099	(−1) 1.5257	(−1) 1.7812
5.3	(−1) −1.5703	(−1) −1.3423	(−2) 8.1054	(−1) −1.0460	(−1) 1.3730	(−1) 1.8231
5.4	(−1) −1.4310	(−1) −1.4404	(−2) 6.3084	(−1) −1.1754	(−1) 1.2134	(−1) 1.8495
5.5	(−1) −1.2828	(−1) −1.5217	(−2) 4.5277	(−1) −1.2885	(−1) 1.0485	(−1) 1.8604
5.6	(−1) −1.1273	(−1) −1.5862	(−2) 2.7749	(−1) −1.3849	(−2) 8.7995	(−1) 1.8563
5.7	(−2) −9.6611	(−1) −1.6339	(−2) +1.0617	(−1) −1.4644	(−2) 7.0920	(−1) 1.8377
5.8	(−2) −8.0104	(−1) −1.6649	(−3) −6.0100	(−1) −1.5268	(−2) 5.3780	(−1) 1.8049
5.9	(−2) −6.3369	(−1) −1.6794	(−2) −2.2024	(−1) −1.5720	(−2) 3.6725	(−1) 1.7587
6.0	(−2) −4.6569	(−1) −1.6779	(−2) −3.7326	(−1) −1.6003	(−2) 1.9898	(−1) 1.6998
6.1	(−2) −2.9863	(−1) −1.6609	(−2) −5.1819	(−1) −1.6119	(−3) +3.4379	(−1) 1.6288
6.2	(−2) −1.3402	(−1) −1.6289	(−2) −6.5418	(−1) −1.6073	(−2) −1.2523	(−1) 1.5467
6.3	(−3) +2.6689	(−1) −1.5828	(−2) −7.8042	(−1) −1.5871	(−2) −2.7861	(−1) 1.4544
6.4	(−2) 1.8211	(−1) −1.5234	(−2) −8.9620	(−1) −1.5519	(−2) −4.2458	(−1) 1.3528
6.5	(−2) 3.3095	(−1) −1.4515	(−1) −1.0009	(−1) −1.5024	(−2) −5.6210	(−1) 1.2430
6.6	(−2) 4.7203	(−1) −1.3682	(−1) −1.0940	(−1) −1.4397	(−2) −6.9018	(−1) 1.1260
6.7	(−2) 6.0425	(−1) −1.2746	(−1) −1.1750	(−1) −1.3648	(−2) −8.0795	(−1) 1.0030
6.8	(−2) 7.2664	(−1) −1.1717	(−1) −1.2435	(−1) −1.2785	(−2) −9.1466	(−2) 8.7500
6.9	(−2) 8.3832	(−1) −1.0607	(−1) −1.2995	(−1) −1.1822	(−1) −1.0097	(−2) 7.4323
7.0	(−2) 9.3855	(−2) −9.4292	(−1) −1.3427	(−1) −1.0770	(−1) −1.0924	(−2) 6.0883
7.1	(−1) 1.0267	(−2) −8.1954	(−1) −1.3730	(−2) −9.6415	(−1) −1.1625	(−2) 4.7295
7.2	(−1) 1.1023	(−2) −6.9183	(−1) −1.3906	(−2) −8.4493	(−1) −1.2197	(−2) 3.3674
7.3	(−1) 1.1650	(−2) −5.6107	(−1) −1.3956	(−2) −7.2065	(−1) −1.2637	(−2) 2.0132
7.4	(−1) 1.2145	(−2) −4.2851	(−1) −1.3882	(−2) −5.9263	(−1) −1.2946	(−3) +6.7812
7.5	(−1) 1.2507	(−2) −2.9542	(−1) −1.3688	(−2) −4.6218	(−1) −1.3123	(−3) −6.2736
7.6	(−1) 1.2736	(−2) −1.6303	(−1) −1.3379	(−2) −3.3061	(−1) −1.3171	(−2) −1.8929
7.7	(−1) 1.2833	(−3) −3.2520	(−1) −1.2960	(−2) −1.9919	(−1) −1.3092	(−2) −3.1089
7.8	(−1) 1.2802	(−3) +9.4953	(−1) −1.2437	(−3) −6.9174	(−1) −1.2891	(−2) −4.2662
7.9	(−1) 1.2645	(−2) 2.1829	(−1) −1.1816	(−3) +5.8231	(−1) −1.2571	(−2) −5.3561
8.0	(−1) 1.2367	(−2) 3.3646	(−1) −1.1105	(−2) 1.8188	(−1) −1.2140	(−2) −6.3711
8.1	(−1) 1.1974	(−2) 4.4850	(−1) −1.0313	(−2) 3.0067	(−1) −1.1603	(−2) −7.3040
8.2	(−1) 1.1472	(−2) 5.5351	(−2) −9.4473	(−2) 4.1360	(−1) −1.0968	(−2) −8.1487
8.3	(−1) 1.0870	(−2) 6.5069	(−2) −8.5177	(−2) 5.1973	(−1) −1.0243	(−2) −8.8997
8.4	(−1) 1.0174	(−2) 7.3932	(−2) −7.5334	(−2) 6.1820	(−2) −9.4378	(−2) −9.5527
8.5	(−2) 9.3940	(−2) 8.1877	(−2) −6.5042	(−2) 7.0825	(−2) −8.5607	(−1) −1.0104
8.6	(−2) 8.5395	(−2) 8.8851	(−2) −5.4401	(−2) 7.8921	(−2) −7.6218	(−1) −1.0551
8.7	(−2) 7.6203	(−2) 9.4810	(−2) −4.3510	(−2) 8.6051	(−2) −6.6312	(−1) −1.0892
8.8	(−2) 6.6468	(−2) 9.9723	(−2) −3.2471	(−2) 9.2170	(−2) −5.5994	(−1) −1.1126
8.9	(−2) 5.6294	(−1) 1.0357	(−2) −2.1385	(−2) 9.7240	(−2) −4.5369	(−1) −1.1253
9.0	(−2) 4.5791	(−1) 1.0632	(−2) −1.0349	(−1) 1.0124	(−2) −3.4542	(−1) −1.1275
9.1	(−2) 3.5066	(−1) 1.0800	(−4) +5.3818	(−1) 1.0415	(−2) −2.3621	(−1) −1.1193
9.2	(−2) 2.4227	(−1) 1.0859	(−2) 1.1184	(−1) 1.0596	(−2) −1.2710	(−1) −1.1011
9.3	(−2) 1.3382	(−1) 1.0813	(−2) 2.1498	(−1) 1.0669	(−3) −1.9101	(−1) −1.0731
9.4	(−3) +2.6357	(−1) 1.0663	(−2) 3.1395	(−1) 1.0635	(−3) +8.6782	(−1) −1.0358
9.5	(−3) −7.9106	(−1) 1.0413	(−2) 4.0795	(−1) 1.0497	(−2) 1.8960	(−2) −9.8978
9.6	(−2) −1.8159	(−1) 1.0068	(−2) 4.9622	(−1) 1.0257	(−2) 2.8844	(−2) −9.3558
9.7	(−2) −2.8017	(−2) 9.6325	(−2) 5.7808	(−2) 9.9213	(−2) 3.8245	(−2) −8.7385
9.8	(−2) −3.7396	(−2) 9.1126	(−2) 6.5291	(−2) 9.4941	(−2) 4.7084	(−2) −8.0528
9.9	(−2) −4.6216	(−2) 8.5149	(−2) 7.2018	(−2) 8.9817	(−2) 5.5288	(−2) −7.3063
10.0	(−2) −5.4402	(−2) 7.8467	(−2) 7.7942	(−2) 8.3907	(−2) 6.2793	(−2) −6.5069

$$j_n(x) = \sqrt{\tfrac{1}{2}\pi/x}\, J_{n+\frac{1}{2}}(x) \qquad\qquad y_n(x) = \sqrt{\tfrac{1}{2}\pi/x}\, Y_{n+\frac{1}{2}}(x) = (-1)^{n+1}\sqrt{\tfrac{1}{2}\pi/x}\, J_{-(n+\frac{1}{2})}(x)$$

SPHERICAL BESSEL FUNCTIONS—ORDERS 3–10 · Table 10.2

x	$j_3(x)$	$j_4(x)$	$j_5(x)$	$j_6(x)$	$j_7(x)$	$j_8(x)$	$10^9 x^{-9} j_9(x)$	$10^{11} x^{-10} j_{10}(x)$
0.0	0.0000	0.0000	0.0000	0.0000	0.0000	0.0000	1.52734 93	7.27309 19
0.1	(−6)9.5185	(−7)1.0577	(−10)9.6163	(−12)7.3975	(−14)4.9319	(−16)2.9012	1.52698 56	7.27151 10
0.2	(−5)7.6021	(−6)1.6900	(−8)3.0737	(−10)4.7297	(−12)6.3072	(−14)7.4212	1.52589 53	7.26677 00
0.3	(−4)2.5586	(−6)8.5364	(−7)2.3296	(−9)5.3784	(−10)1.0761	(−12)1.8995	1.52407 96	7.25887 47
0.4	(−4)6.0413	(−5)2.6894	(−7)9.7904	(−8)3.0149	(−10)8.0448	(−11)1.8938	1.52154 09	7.24783 46
0.5	(−3)1.1740	(−5)6.5390	(−6)2.9775	(−7)1.1467	(−9)3.8259	(−10)1.1261	1.51828 26	7.23366 29
0.6	(−3)2.0163	(−4)1.3491	(−6)7.3776	(−7)3.4113	(−8)1.3665	(−10)4.8282	1.51430 88	7.21637 65
0.7	(−3)3.1787	(−4)2.4847	(−5)1.5866	(−7)8.5649	(−8)4.0046	(−9)1.6515	1.50962 48	7.19599 61
0.8	(−3)4.7053	(−4)4.2098	(−5)3.0755	(−6)1.8989	(−7)1.0153	(−9)4.7873	1.50423 66	7.17254 61
0.9	(−3)6.6361	(−4)6.6912	(−5)5.5059	(−6)3.8277	(−7)2.3040	(−8)1.2228	1.49815 12	7.14605 44
1.0	(−3)9.0066	(−3)1.0110	(−5)9.2561	(−6)7.1569	(−7)4.7901	(−8)2.8265	1.49137 65	7.11655 26
1.1	(−2)1.1847	(−3)1.4661	(−4)1.4786	(−5)1.2590	(−7)9.2769	(−8)6.0254	1.48392 11	7.08407 57
1.2	(−2)1.5183	(−3)2.0546	(−4)2.2643	(−5)2.1058	(−6)1.6942	(−7)1.2013	1.47579 48	7.04866 21
1.3	(−2)1.9033	(−3)2.7976	(−4)3.3461	(−5)3.3756	(−6)2.9451	(−7)2.2640	1.46700 80	7.01035 39
1.4	(−2)2.3411	(−3)3.7164	(−4)4.7963	(−5)5.2181	(−6)4.9082	(−7)4.0669	1.45757 18	6.96919 61
1.5	(−2)2.8325	(−3)4.8324	(−4)6.6962	(−5)7.8174	(−6)7.8875	(−7)7.0086	1.44749 84	6.92523 71
1.6	(−2)3.3774	(−3)6.1667	(−4)9.1354	(−4)1.1395	(−5)1.2279	(−6)1.1649	1.43680 05	6.87852 85
1.7	(−2)3.9754	(−3)7.7397	(−3)1.2212	(−4)1.6212	(−5)1.8587	(−6)1.8756	1.42549 17	6.82912 49
1.8	(−2)4.6252	(−3)9.5709	(−3)1.6031	(−4)2.2577	(−5)2.7444	(−6)2.9356	1.41358 63	6.77708 37
1.9	(−2)5.3249	(−2)1.1679	(−3)2.0705	(−4)3.0840	(−5)3.9632	(−6)4.4800	1.40109 93	6.72246 53
2.0	(−2)6.0722	(−2)1.4079	(−3)2.6352	(−4)4.1404	(−5)5.6097	(−6)6.6832	1.38804 63	6.66533 28
2.1	(−2)6.8639	(−2)1.6788	(−3)3.3094	(−4)5.4720	(−5)7.7975	(−6)9.7670	1.37444 35	6.60575 19
2.2	(−2)7.6962	(−2)1.9817	(−3)4.1059	(−4)7.1289	(−4)1.0661	(−5)1.4009	1.36030 78	6.54379 07
2.3	(−2)8.5650	(−2)2.3176	(−3)5.0375	(−4)9.1665	(−4)1.4358	(−5)1.9754	1.34565 67	6.47951 98
2.4	(−2)9.4654	(−2)2.6872	(−3)6.1171	(−3)1.1645	(−4)1.9071	(−5)2.7420	1.33050 81	6.41301 19
2.5	(−1)1.0392	(−2)3.0911	(−3)7.3576	(−3)1.4630	(−4)2.5009	(−5)3.7516	1.31488 05	6.34434 22
2.6	(−1)1.1339	(−2)3.5292	(−3)8.7717	(−3)1.8192	(−4)3.2410	(−5)5.0647	1.29879 28	6.27358 74
2.7	(−1)1.2301	(−2)4.0014	(−2)1.0372	(−3)2.2404	(−4)4.1542	(−5)6.7532	1.28226 44	6.20082 63
2.8	(−1)1.3270	(−2)4.5071	(−2)1.2169	(−3)2.7345	(−4)5.2705	(−5)8.9013	1.26531 50	6.12613 95
2.9	(−1)1.4241	(−2)5.0454	(−2)1.4174	(−3)3.3096	(−4)6.6231	(−4)1.1607	1.24796 48	6.04960 91
3.0	(−1)1.5205	(−2)5.6150	(−2)1.6397	(−3)3.9744	(−4)8.2484	(−4)1.4983	1.23023 41	5.97131 85
3.1	(−1)1.6156	(−2)6.2142	(−2)1.8848	(−3)4.7374	(−3)1.0187	(−4)1.9160	1.21214 38	5.89135 26
3.2	(−1)1.7087	(−2)6.8409	(−2)2.1532	(−3)5.6074	(−3)1.2481	(−4)2.4283	1.19371 48	5.80979 75
3.3	(−1)1.7989	(−2)7.4929	(−2)2.4457	(−3)6.5935	(−3)1.5177	(−4)3.0520	1.17496 82	5.72674 00
3.4	(−1)1.8857	(−2)8.1673	(−2)2.7626	(−3)7.7045	(−3)1.8326	(−4)3.8056	1.15592 54	5.64226 82
3.5	(−1)1.9681	(−2)8.8610	(−2)3.1042	(−3)8.9491	(−3)2.1980	(−4)4.7098	1.13660 79	5.55647 05
3.6	(−1)2.0456	(−2)9.5706	(−2)3.4705	(−2)1.0356	(−3)2.6195	(−4)5.7875	1.11703 73	5.46943 61
3.7	(−1)2.1174	(−1)1.0292	(−2)3.8614	(−2)1.1873	(−3)3.1030	(−4)7.0639	1.09723 52	5.38125 47
3.8	(−1)2.1829	(−1)1.1022	(−2)4.2765	(−2)1.3569	(−3)3.6544	(−4)8.5665	1.07722 33	5.29201 62
3.9	(−1)2.2414	(−1)1.1756	(−2)4.7151	(−2)1.5429	(−3)4.2801	(−3)1.0325	1.05702 31	5.20181 05
4.0	(−1)2.2924	(−1)1.2489	(−2)5.1766	(−2)1.7462	(−3)4.9865	(−3)1.2372	1.03665 63	5.11072 78
4.1	(−1)2.3354	(−1)1.3217	(−2)5.6596	(−2)1.9673	(−3)5.7801	(−3)1.4743	1.01614 44	5.01885 80
4.2	(−1)2.3697	(−1)1.3935	(−2)6.1630	(−2)2.2065	(−3)6.6676	(−3)1.7473	0.99550 88	4.92629 07
4.3	(−1)2.3951	(−1)1.4637	(−2)6.6851	(−2)2.4645	(−3)7.6554	(−3)2.0603	0.97477 06	4.83311 51
4.4	(−1)2.4110	(−1)1.5319	(−2)7.2242	(−2)2.7413	(−3)8.7501	(−3)2.4174	0.95395 10	4.73942 00
4.5	(−1)2.4174	(−1)1.5976	(−2)7.7780	(−2)3.0371	(−3)9.9581	(−3)2.8229	0.93307 06	4.64529 34
4.6	(−1)2.4138	(−1)1.6602	(−2)8.3444	(−2)3.3520	(−2)1.1286	(−3)3.2814	0.91215 01	4.55082 25
4.7	(−1)2.4001	(−1)1.7193	(−2)8.9207	(−2)3.6857	(−2)1.2739	(−3)3.7976	0.89120 97	4.45609 35
4.8	(−1)2.3763	(−1)1.7743	(−2)9.5043	(−2)4.0381	(−2)1.4322	(−3)4.3763	0.87026 94	4.36119 18
4.9	(−1)2.3423	(−1)1.8247	(−1)1.0092	(−2)4.4086	(−2)1.6042	(−3)5.0226	0.84934 88	4.26620 13
5.0	(−1)2.2982	(−1)1.8702	(−1)1.0681	(−2)4.7967	(−2)1.7903	(−3)5.7414	0.82846 70	4.17120 50

$$j_n(x) = \sqrt{\tfrac{1}{2}\pi/x}\, J_{n+\frac{1}{2}}(x)$$

$$\begin{bmatrix}(-5)9\\4\end{bmatrix} \qquad \begin{bmatrix}(-4)4\\4\end{bmatrix}$$

Compiled from National Bureau of Standards, Tables of spherical Bessel functions, vols. I, II. Columbia Univ. Press, New York, N.Y., 1947 (with permission).

Table 10.2   **SPHERICAL BESSEL FUNCTIONS—ORDERS 3–10**

x	$y_3(x)$	$y_4(x)$	$y_5(x)$	$y_6(x)$	$y_7(x)$	$y_8(x)$	$10^{-8}x^{10}y_9(x)$	$10^{-9}x^{11}y_{10}(x)$
0.0	$-\infty$	$-\infty$	$-\infty$	$-\infty$	$-\infty$	$-\infty$	-0.34459 42	-0.65472 90
0.1	(5)-1.5015	(7)-1.0507	(8)-9.4553	(11)-1.0400	(13)-1.3519	(15)-2.0277	-0.34469 56	-0.65490 14
0.2	(3)-9.4126	(5)-3.2906	(7)-1.4798	(8)-8.1359	(10)-5.2868	(12)-3.9643	-0.34499 99	-0.65541 86
0.3	(3)-1.8686	(4)-4.3489	(6)-1.3028	(7)-4.7726	(9)-2.0668	(11)-1.0329	-0.34550 77	-0.65628 18
0.4	(2)-5.9544	(4)-1.0372	(5)-2.3278	(6)-6.3910	(8)-2.0747	(9)-7.7739	-0.34622 02	-0.65749 23
0.5	(2)-2.4613	(3)-3.4208	(4)-6.1328	(6)-1.3458	(7)-3.4929	(9)-1.0465	-0.34713 86	-0.65905 23
0.6	(2)-1.2004	(3)-1.3857	(4)-2.0665	(5)-3.7747	(6)-8.1579	(8)-2.0357	-0.34826 48	-0.66096 47
0.7	(1)-6.5670	(2)-6.4716	(3)-8.2549	(5)-1.2907	(6)-2.3888	(7)-5.1060	-0.34960 12	-0.66323 28
0.8	(1)-3.9102	(2)-3.3557	(3)-3.7361	(4)-5.1035	(5)-8.2559	(7)-1.5429	-0.35115 04	-0.66586 06
0.9	(1)-2.4854	(2)-1.8854	(3)-1.8606	(4)-2.2552	(5)-3.2389	(6)-5.3756	-0.35291 56	-0.66885 29
1.0	(1)-1.6643	(2)-1.1290	(2)-9.9944	(4)-1.0881	(5)-1.4045	(6)-2.0959	-0.35490 04	-0.67221 50
1.1	(1)-1.1631	(1)-7.1198	(2)-5.7090	(3)-5.6378	(4)-6.6058	(5)-8.9515	-0.35710 89	-0.67595 30
1.2	(0)-8.4253	(1)-4.6879	(2)-3.4317	(3)-3.0988	(4)-3.3227	(5)-4.1224	-0.35954 56	-0.68007 37
1.3	(0)-6.2927	(1)-3.2014	(2)-2.1534	(3)-1.7901	(4)-1.7686	(5)-2.0227	-0.36221 57	-0.68458 47
1.4	(0)-4.8264	(1)-2.2559	(2)-1.4020	(3)-1.0790	(3)-9.8790	(5)-1.0477	-0.36512 46	-0.68949 42
1.5	(0)-3.7893	(1)-1.6338	(1)-9.4236	(2)-6.7473	(3)-5.7534	(4)-5.6859	-0.36827 87	-0.69481 14
1.6	(0)-3.0374	(1)-1.2120	(1)-6.5140	(2)-4.3572	(3)-3.4751	(4)-3.2143	-0.37168 46	-0.70054 60
1.7	(0)-2.4804	(0)-9.1871	(1)-4.6157	(2)-2.8948	(3)-2.1675	(4)-1.8835	-0.37534 96	-0.70670 90
1.8	(0)-2.0598	(0)-7.0994	(1)-3.3437	(2)-1.9724	(3)-1.3911	(4)-1.1395	-0.37928 17	-0.71331 20
1.9	(0)-1.7366	(0)-5.5830	(1)-2.4709	(2)-1.3747	(2)-9.1587	(3)-7.0931	-0.38348 96	-0.72036 75
2.0	(0)-1.4844	(0)-4.4613	(1)-1.8591	(1)-9.7792	(2)-6.1705	(3)-4.5301	-0.38798 26	-0.72788 93
2.1	(0)-1.2846	(0)-3.6178	(1)-1.4220	(1)-7.0870	(2)-4.2450	(3)-2.9613	-0.39277 08	-0.73589 19
2.2	(0)-1.1242	(0)-2.9740	(1)-1.1042	(1)-5.2238	(2)-2.9764	(3)-1.9771	-0.39786 50	-0.74439 11
2.3	(-1)-9.9368	(0)-2.4760	(0)-8.6948	(1)-3.9108	(2)-2.1235	(3)-1.3458	-0.40327 71	-0.75340 38
2.4	(-1)-8.8622	(0)-2.0858	(0)-6.9354	(1)-2.9702	(2)-1.5395	(2)-9.3247	-0.40901 97	-0.76294 81
2.5	(-1)-7.9660	(0)-1.7766	(0)-5.5991	(1)-2.2859	(2)-1.1327	(2)-6.5676	-0.41510 62	-0.77304 34
2.6	(-1)-7.2096	(0)-1.5290	(0)-4.5716	(1)-1.7812	(1)-8.4491	(2)-4.6963	-0.42155 14	-0.78371 06
2.7	(-1)-6.5632	(0)-1.3287	(0)-3.7725	(1)-1.4041	(1)-6.3832	(2)-3.4058	-0.42837 10	-0.79497 18
2.8	(-1)-6.0041	(0)-1.1651	(0)-3.1446	(1)-1.1189	(1)-4.8802	(2)-2.5025	-0.43558 18	-0.80685 08
2.9	(-1)-5.5144	(0)-1.0303	(0)-2.6462	(0)-9.0069	(1)-3.7729	(2)-1.8615	-0.44320 20	-0.81937 31
3.0	(-1)-5.0802	(-1)-9.1835	(0)-2.2470	(0)-7.3207	(1)-2.9476	(2)-1.4006	-0.45125 11	-0.83256 59
3.1	(-1)-4.6905	(-1)-8.2448	(0)-1.9246	(0)-6.0048	(1)-2.3257	(2)-1.0653	-0.45975 01	-0.84645 82
3.2	(-1)-4.3365	(-1)-7.4514	(0)-1.6621	(0)-4.9682	(1)-1.8521	(1)-8.1850	-0.46872 14	-0.86108 11
3.3	(-1)-4.0112	(-1)-6.7752	(0)-1.4467	(0)-4.1447	(1)-1.4881	(1)-6.3496	-0.47818 95	-0.87646 78
3.4	(-1)-3.7091	(-1)-6.1940	(0)-1.2687	(0)-3.4851	(1)-1.2057	(1)-4.9707	-0.48818 03	-0.89265 39
3.5	(-1)-3.4257	(-1)-5.6901	(0)-1.1206	(0)-2.9528	(0)-9.8471	(1)-3.9249	-0.49872 20	-0.90967 72
3.6	(-1)-3.1573	(-1)-5.2492	(-1)-9.9657	(0)-2.5201	(0)-8.1040	(1)-3.1246	-0.50984 49	-0.92757 84
3.7	(-1)-2.9012	(-1)-4.8600	(-1)-8.9204	(0)-2.1660	(0)-6.7182	(1)-2.5070	-0.52158 17	-0.94640 10
3.8	(-1)-2.6551	(-1)-4.5131	(-1)-8.0339	(0)-1.8743	(0)-5.6086	(1)-2.0265	-0.53396 75	-0.96619 15
3.9	(-1)-2.4173	(-1)-4.2011	(-1)-7.2774	(0)-1.6325	(0)-4.7139	(1)-1.6498	-0.54704 05	-0.98699 97
4.0	(-1)-2.1864	(-1)-3.9175	(-1)-6.6280	(0)-1.4310	(0)-3.9878	(1)-1.3523	-0.56084 19	-1.00887 91
4.1	(-1)-1.9615	(-1)-3.6574	(-1)-6.0670	(0)-1.2620	(0)-3.3947	(1)-1.1158	-0.57541 63	-1.03188 69
4.2	(-1)-1.7418	(-1)-3.4165	(-1)-5.5793	(0)-1.1196	(0)-2.9075	(0)-9.2642	-0.59081 20	-1.05608 44
4.3	(-1)-1.5269	(-1)-3.1913	(-1)-5.1525	(-1)-9.9895	(0)-2.5048	(0)-7.7389	-0.60708 14	-1.08153 78
4.4	(-1)-1.3165	(-1)-2.9788	(-1)-4.7765	(-1)-8.9625	(0)-2.1704	(0)-6.5027	-0.62428 15	-1.10831 79
4.5	(-1)-1.1107	(-1)-2.7768	(-1)-4.4430	(-1)-8.0839	(0)-1.8910	(0)-5.4951	-0.64247 43	-1.13650 10
4.6	(-2)-9.0931	(-1)-2.5833	(-1)-4.1450	(-1)-7.3286	(0)-1.6566	(0)-4.6692	-0.66172 73	-1.16616 90
4.7	(-2)-7.1268	(-1)-2.3966	(-1)-3.8766	(-1)-6.6763	(0)-1.4590	(0)-3.9887	-0.68211 42	-1.19741 05
4.8	(-2)-5.2107	(-1)-2.2155	(-1)-3.6331	(-1)-6.1102	(0)-1.2915	(0)-3.4251	-0.70371 55	-1.23032 08
4.9	(-2)-3.3484	(-1)-2.0390	(-1)-3.4102	(-1)-5.6166	(0)-1.1491	(0)-2.9560	-0.72661 94	-1.26500 29
5.0	(-2)-1.5443	(-1)-1.8662	(-1)-3.2047	(-1)-5.1841	(0)-1.0274	(0)-2.5638	-0.75092 23	-1.30156 80

$$y_n(x)=\sqrt{\tfrac{1}{2}\pi/x}\,Y_{n+\frac{1}{2}}(x)=(-1)^{n+1}\sqrt{\tfrac{1}{2}\pi/x}\,J_{-(n+\frac{1}{2})}(x) \qquad \left[\begin{matrix}(-4)2\\5\end{matrix}\right] \qquad \left[\begin{matrix}(-4)2\\5\end{matrix}\right]$$

SPHERICAL BESSEL FUNCTIONS—ORDERS 3–10

Table **10.2**

x	$j_3(x)$	$j_4(x)$	$j_5(x)$	$j_6(x)$	$j_7(x)$	$j_8(x)$	$10^9 x^{-9} j_9(x)$	$10^{11} x^{-10} j_{10}(x)$
5.0	(−1) 2.2982	(−1) 1.8702	(−1) 1.0681	(−2) 4.7967	(−2) 1.7903	(−3) 5.7414	0.82846 70	4.17120 50
5.1	(−1) 2.2441	(−1) 1.9102	(−1) 1.1268	(−2) 5.2015	(−2) 1.9908	(−3) 6.5379	0.80764 29	4.07628 42
5.2	(−1) 2.1803	(−1) 1.9443	(−1) 1.1849	(−2) 5.6221	(−2) 2.2061	(−3) 7.4172	0.78689 50	3.98151 88
5.3	(−1) 2.1069	(−1) 1.9722	(−1) 1.2421	(−2) 6.0573	(−2) 2.4365	(−3) 8.3843	0.76624 10	3.88698 72
5.4	(−1) 2.0245	(−1) 1.9935	(−1) 1.2980	(−2) 6.5057	(−2) 2.6821	(−3) 9.4443	0.74569 86	3.79276 59
5.5	(−1) 1.9335	(−1) 2.0078	(−1) 1.3522	(−2) 6.9660	(−2) 2.9429	(−2) 1.0602	0.72528 47	3.69892 98
5.6	(−1) 1.8340	(−1) 2.0150	(−1) 1.4044	(−2) 7.4364	(−2) 3.2191	(−2) 1.1862	0.70501 58	3.60555 18
5.7	(−1) 1.7270	(−1) 2.0147	(−1) 1.4542	(−2) 7.9151	(−2) 3.5104	(−2) 1.3229	0.68490 78	3.51270 30
5.8	(−1) 1.6131	(−1) 2.0069	(−1) 1.5011	(−2) 8.4000	(−2) 3.8166	(−2) 1.4707	0.66497 60	3.42045 23
5.9	(−1) 1.4928	(−1) 1.9913	(−1) 1.5448	(−2) 8.8889	(−2) 4.1374	(−2) 1.6299	0.64523 54	3.32886 66
6.0	(−1) 1.3669	(−1) 1.9679	(−1) 1.5850	(−2) 9.3796	(−2) 4.4722	(−2) 1.8010	0.62570 01	3.23801 06
6.1	(−1) 1.2361	(−1) 1.9367	(−1) 1.6213	(−2) 9.8696	(−2) 4.8205	(−2) 1.9842	0.60638 37	3.14794 66
6.2	(−1) 1.1014	(−1) 1.8977	(−1) 1.6533	(−1) 1.0356	(−2) 5.1815	(−2) 2.1797	0.58729 93	3.05873 50
6.3	(−2) 9.6346	(−1) 1.8509	(−1) 1.6807	(−1) 1.0837	(−2) 5.5543	(−2) 2.3877	0.56845 94	2.97043 34
6.4	(−2) 8.2324	(−1) 1.7966	(−1) 1.7033	(−1) 1.1309	(−2) 5.9379	(−2) 2.6084	0.54987 57	2.88309 73
6.5	(−2) 6.8161	(−1) 1.7349	(−1) 1.7206	(−1) 1.1769	(−2) 6.3311	(−2) 2.8417	0.53155 94	2.79677 98
6.6	(−2) 5.3947	(−1) 1.6661	(−1) 1.7325	(−1) 1.2214	(−2) 6.7327	(−2) 3.0876	0.51352 10	2.71153 12
6.7	(−2) 3.9773	(−1) 1.5905	(−1) 1.7388	(−1) 1.2642	(−2) 7.1412	(−2) 3.3461	0.49577 04	2.62739 98
6.8	(−2) 2.5729	(−1) 1.5084	(−1) 1.7391	(−1) 1.3049	(−2) 7.5551	(−2) 3.6168	0.47831 68	2.54443 09
6.9	(−2) +1.1905	(−1) 1.4203	(−1) 1.7335	(−1) 1.3432	(−2) 7.9728	(−2) 3.8996	0.46116 89	2.46266 76
7.0	(−3) −1.6120	(−1) 1.3265	(−1) 1.7217	(−1) 1.3789	(−2) 8.3923	(−2) 4.1940	0.44433 45	2.38215 03
7.1	(−2) −1.4736	(−1) 1.2277	(−1) 1.7036	(−1) 1.4117	(−2) 8.8118	(−2) 4.4994	0.42782 11	2.30291 70
7.2	(−2) −2.7385	(−1) 1.1243	(−1) 1.6793	(−1) 1.4412	(−2) 9.2292	(−2) 4.8154	0.41163 52	2.22500 27
7.3	(−2) −3.9479	(−1) 1.0170	(−1) 1.6486	(−1) 1.4672	(−2) 9.6425	(−2) 5.1412	0.39578 30	2.14844 05
7.4	(−2) −5.0945	(−2) 9.0628	(−1) 1.6117	(−1) 1.4895	(−1) 1.0049	(−2) 5.4759	0.38026 97	2.07326 03
7.5	(−2) −6.1713	(−2) 7.9285	(−1) 1.5685	(−1) 1.5077	(−1) 1.0448	(−2) 5.8188	0.36510 02	1.99948 99
7.6	(−2) −7.1719	(−2) 6.7736	(−1) 1.5193	(−1) 1.5217	(−1) 1.0835	(−2) 6.1686	0.35027 86	1.92715 45
7.7	(−2) −8.0904	(−2) 5.6051	(−1) 1.4642	(−1) 1.5312	(−1) 1.1209	(−2) 6.5244	0.33580 85	1.85627 66
7.8	(−2) −8.9217	(−2) 4.4300	(−1) 1.4033	(−1) 1.5360	(−1) 1.1568	(−2) 6.8849	0.32169 28	1.78687 63
7.9	(−2) −9.6613	(−2) 3.2552	(−1) 1.3370	(−1) 1.5361	(−1) 1.1908	(−2) 7.2486	0.30793 39	1.71897 14
8.0	(−1) −1.0305	(−2) 2.0880	(−1) 1.2654	(−1) 1.5312	(−1) 1.2227	(−2) 7.6143	0.29453 36	1.65257 72
8.1	(−1) −1.0851	(−3) +9.3549	(−1) 1.1890	(−1) 1.5212	(−1) 1.2524	(−2) 7.9804	0.28149 30	1.58770 64
8.2	(−1) −1.1296	(−3) −1.9533	(−1) 1.1081	(−1) 1.5060	(−1) 1.2795	(−2) 8.3451	0.26881 29	1.52436 97
8.3	(−1) −1.1638	(−2) −1.2975	(−1) 1.0231	(−1) 1.4857	(−1) 1.3039	(−2) 8.7069	0.25649 33	1.46257 53
8.4	(−1) −1.1877	(−2) −2.3644	(−2) 9.3440	(−1) 1.4601	(−1) 1.3252	(−2) 9.0640	0.24453 39	1.40232 92
8.5	(−1) −1.2014	(−2) −3.3894	(−2) 8.4249	(−1) 1.4292	(−1) 1.3434	(−2) 9.4145	0.23293 38	1.34363 53
8.6	(−1) −1.2048	(−2) −4.3664	(−2) 7.4784	(−1) 1.3932	(−1) 1.3581	(−2) 9.7564	0.22169 16	1.28649 51
8.7	(−1) −1.1982	(−2) −5.2894	(−2) 6.5099	(−1) 1.3520	(−1) 1.3693	(−1) 1.0088	0.21080 54	1.23090 84
8.8	(−1) −1.1817	(−2) −6.1529	(−2) 5.5245	(−1) 1.3059	(−1) 1.3767	(−1) 1.0407	0.20027 29	1.17687 25
8.9	(−1) −1.1558	(−2) −6.9520	(−2) 4.5278	(−1) 1.2548	(−1) 1.3801	(−1) 1.0712	0.19009 14	1.12438 32
9.0	(−1) −1.1207	(−2) −7.6819	(−2) 3.5255	(−1) 1.1991	(−1) 1.3795	(−1) 1.1000	0.18025 78	1.07343 42
9.1	(−1) −1.0770	(−2) −8.3387	(−2) 2.5233	(−1) 1.1389	(−1) 1.3746	(−1) 1.1270	0.17076 84	1.02401 72
9.2	(−1) −1.0252	(−2) −8.9186	(−2) 1.5269	(−1) 1.0744	(−1) 1.3655	(−1) 1.1520	0.16161 93	0.97612 24
9.3	(−2) −9.6572	(−2) −9.4187	(−3) +5.4232	(−1) 1.0060	(−1) 1.3520	(−1) 1.1747	0.15280 62	0.92973 83
9.4	(−2) −8.9931	(−2) −9.8365	(−3) −4.2485	(−2) 9.3394	(−1) 1.3341	(−1) 1.1949	0.14432 46	0.88485 16
9.5	(−2) −8.2662	(−1) −1.0170	(−2) −1.3689	(−2) 8.5853	(−1) 1.3117	(−1) 1.2126	0.13616 93	0.84144 75
9.6	(−2) −7.4836	(−1) −1.0419	(−2) −2.2842	(−2) 7.8016	(−1) 1.2849	(−1) 1.2275	0.12833 53	0.79950 99
9.7	(−2) −6.6527	(−1) −1.0582	(−2) −3.1654	(−2) 6.9921	(−1) 1.2536	(−1) 1.2394	0.12081 68	0.75902 10
9.8	(−2) −5.7814	(−1) −1.0659	(−2) −4.0072	(−2) 6.1608	(−1) 1.2180	(−1) 1.2482	0.11360 83	0.71996 20
9.9	(−2) −4.8776	(−1) −1.0651	(−2) −4.8048	(−2) 5.3120	(−1) 1.1780	(−1) 1.2537	0.10670 35	0.68231 26
10.0	(−2) −3.9496	(−1) −1.0559	(−2) −5.5535	(−2) 4.4501	(−1) 1.1339	(−1) 1.2558	0.10009 64	0.64605 15

$$j_n(x) = \sqrt{\tfrac{1}{2}\pi/x}\, J_{n+\frac{1}{2}}(x)$$

$$\begin{bmatrix} (-5)5 \\ 4 \end{bmatrix} \qquad \begin{bmatrix} (-4)2 \\ 5 \end{bmatrix}$$

Table 10.2   SPHERICAL BESSEL FUNCTIONS—ORDERS 3–10

x	$y_3(x)$	$y_4(x)$	$y_5(x)$	$y_6(x)$	$y_7(x)$	$y_8(x)$	$10^{-8}x^{10}y_9(x)$	$10^{-9}x^{11}y_{10}(x)$
5.0	(−2)−1.5443	(−1)−1.8662	(−1)−3.2047	(−1)−5.1841	(0)−1.0274	(0)−2.5638	−0.75092 23	−1.30156 80
5.1	(−3)+1.9691	(−1)−1.6965	(−1)−3.0134	(−1)−4.8031	(−1)−9.2298	(0)−2.2343	−0.77673 01	−1.34013 68
5.2	(−2) 1.8700	(−1)−1.5295	(−1)−2.8341	(−1)−4.4658	(−1)−8.3305	(0)−1.9564	−0.80415 92	−1.38083 98
5.3	(−2) 3.4698	(−1)−1.3649	(−1)−2.6647	(−1)−4.1656	(−1)−7.5528	(0)−1.7210	−0.83333 74	−1.42381 86
5.4	(−2) 4.9908	(−1)−1.2025	(−1)−2.5033	(−1)−3.8967	(−1)−6.8777	(0)−1.5208	−0.86440 56	−1.46922 70
5.5	(−2) 6.4276	(−1)−1.0424	(−1)−2.3484	(−1)−3.6545	(−1)−6.2895	(0)−1.3499	−0.89751 90	−1.51723 25
5.6	(−2) 7.7750	(−2)−8.8447	(−1)−2.1990	(−1)−3.4349	(−1)−5.7750	(0)−1.2034	−0.93284 85	−1.56801 75
5.7	(−2) 9.0279	(−2)−7.2898	(−1)−2.0538	(−1)−3.2345	(−1)−5.3232	(0)−1.0774	−0.97058 31	−1.62178 08
5.8	(−1) 1.0182	(−2)−5.7610	(−1)−1.9121	(−1)−3.0503	(−1)−4.9248	(−1)−9.6863	−1.01093 09	−1.67873 97
5.9	(−1) 1.1232	(−2)−4.2612	(−1)−1.7732	(−1)−2.8799	(−1)−4.5723	(−1)−8.7446	−1.05412 18	−1.73913 16
6.0	(−1) 1.2175	(−2)−2.7936	(−1)−1.6365	(−1)−2.7210	(−1)−4.2589	(−1)−7.9262	−1.10040 93	−1.80321 67
6.1	(−1) 1.3007	(−2)−1.3619	(−1)−1.5017	(−1)−2.5717	(−1)−3.9791	(−1)−7.2128	−1.15007 32	−1.87128 02
6.2	(−1) 1.3726	(−4)+2.9727	(−1)−1.3683	(−1)−2.4306	(−1)−3.7281	(−1)−6.5889	−1.20342 16	−1.94363 49
6.3	(−1) 1.4329	(−2) 1.3770	(−1)−1.2362	(−1)−2.2961	(−1)−3.5018	(−1)−6.0416	−1.26079 38	−2.02062 45
6.4	(−1) 1.4815	(−2) 2.6754	(−1)−1.1052	(−1)−2.1672	(−1)−3.2969	(−1)−5.5598	−1.32256 26	−2.10262 69
6.5	(−1) 1.5183	(−2) 3.9204	(−2)−9.7544	(−1)−2.0428	(−1)−3.1101	(−1)−5.1344	−1.38913 71	−2.19005 78
6.6	(−1) 1.5432	(−2) 5.1073	(−2)−8.4678	(−1)−1.9220	(−1)−2.9390	(−1)−4.7576	−1.46096 57	−2.28337 46
6.7	(−1) 1.5564	(−2) 6.2315	(−2)−7.1937	(−1)−1.8042	(−1)−2.7813	(−1)−4.4227	−1.53853 78	−2.38308 14
6.8	(−1) 1.5580	(−2) 7.2886	(−2)−5.9337	(−1)−1.6887	(−1)−2.6351	(−1)−4.1239	−1.62238 69	−2.48973 26
6.9	(−1) 1.5482	(−2) 8.2743	(−2)−4.6896	(−1)−1.5751	(−1)−2.4985	(−1)−3.8565	−1.71309 24	−2.60393 95
7.0	(−1) 1.5273	(−2) 9.1846	(−2)−3.4641	(−1)−1.4628	(−1)−2.3703	(−1)−3.6163	−1.81128 11	−2.72637 44
7.1	(−1) 1.4956	(−1) 1.0016	(−2)−2.2599	(−1)−1.3517	(−1)−2.2489	(−1)−3.3996	−1.91762 85	−2.85777 73
7.2	(−1) 1.4535	(−1) 1.0764	(−2)−1.0801	(−1)−1.2414	(−1)−2.1334	(−1)−3.2032	−2.03285 95	−2.99896 17
7.3	(−1) 1.4016	(−1) 1.1427	(−4)+7.1768	(−1)−1.1319	(−1)−2.0228	(−1)−3.0246	−2.15774 75	−3.15082 08
7.4	(−1) 1.3404	(−1) 1.2001	(−2) 1.1922	(−1)−1.0229	(−1)−1.9162	(−1)−2.8613	−2.29311 31	−3.31433 45
7.5	(−1) 1.2705	(−1) 1.2485	(−2) 2.2774	(−2)−9.1449	(−1)−1.8129	(−1)−2.7112	−2.43982 13	−3.49057 53
7.6	(−1) 1.1925	(−1) 1.2877	(−2) 3.3235	(−2)−8.0665	(−1)−1.7122	(−1)−2.5726	−2.59877 67	−3.68071 56
7.7	(−1) 1.1073	(−1) 1.3176	(−2) 4.3267	(−2)−6.9945	(−1)−1.6136	(−1)−2.4439	−2.77091 77	−3.88603 37
7.8	(−1) 1.0156	(−1) 1.3380	(−2) 5.2830	(−2)−5.9299	(−1)−1.5166	(−1)−2.3236	−2.95720 73	−4.10791 96
7.9	(−2) 9.1812	(−1) 1.3491	(−2) 6.1887	(−2)−4.8741	(−1)−1.4209	(−1)−2.2106	−3.15862 24	−4.34788 05
8.0	(−2) 8.1577	(−1) 1.3509	(−2) 7.0400	(−2)−3.8290	(−1)−1.3262	(−1)−2.1038	−3.37613 93	−4.60754 55
8.1	(−2) 7.0941	(−1) 1.3435	(−2) 7.8334	(−2)−2.7968	(−1)−1.2322	(−1)−2.0022	−3.61071 67	−4.88866 85
8.2	(−2) 5.9992	(−1) 1.3270	(−2) 8.5654	(−2)−1.7798	(−1)−1.1387	(−1)−1.9050	−3.86327 49	−5.19312 95
8.3	(−2) 4.8821	(−1) 1.3017	(−2) 9.2329	(−3)−7.8077	(−1)−1.0456	(−1)−1.8115	−4.13466 98	−5.52293 51
8.4	(−2) 3.7517	(−1) 1.2679	(−2) 9.8330	(−3)+1.9747	(−2)−9.5274	(−1)−1.7211	−4.42566 38	−5.88021 45
8.5	(−2) 2.6172	(−1) 1.2259	(−1) 1.0363	(−2) 1.1519	(−2)−8.6015	(−1)−1.6331	−4.73689 09	−6.26721 41
8.6	(−2) 1.4876	(−1) 1.1762	(−1) 1.0821	(−2) 2.0793	(−2)−7.6780	(−1)−1.5471	−5.06881 69	−6.68628 70
8.7	(−3)+3.7160	(−1) 1.1191	(−1) 1.1205	(−2) 2.9765	(−2)−6.7573	(−1)−1.4627	−5.42169 35	−7.13987 95
8.8	(−3)−7.2210	(−1) 1.0551	(−1) 1.1513	(−2) 3.8403	(−2)−5.8403	(−1)−1.3795	−5.79550 68	−7.63051 13
8.9	(−2)−1.7852	(−2) 9.8492	(−1) 1.1745	(−2) 4.6672	(−2)−4.9278	(−1)−1.2973	−6.18991 88	−8.16074 96
9.0	(−2)−2.8097	(−2) 9.0898	(−1) 1.1899	(−2) 5.4540	(−2)−4.0214	(−1)−1.2156	−6.60420 33	−8.73317 65
9.1	(−2)−3.7880	(−2) 8.2794	(−1) 1.1976	(−2) 6.1976	(−2)−3.1227	(−1)−1.1345	−7.03717 50	−9.35034 96
9.2	(−2)−4.7130	(−2) 7.4246	(−1) 1.1976	(−2) 6.8948	(−2)−2.2335	(−1)−1.0536	−7.48710 95	−10.01475 2
9.3	(−2)−5.5782	(−2) 6.5321	(−1) 1.1900	(−2) 7.5427	(−2)−1.3560	(−2)−9.7298	−7.95166 19	−10.72873 2
9.4	(−2)−6.3774	(−2) 5.6089	(−1) 1.1748	(−2) 8.1384	(−3)−4.9250	(−2)−8.9243	−8.42777 38	−11.49443 4
9.5	(−2)−7.1053	(−2) 4.6623	(−1) 1.1522	(−2) 8.6793	(−3)+3.5462	(−2)−8.1193	−8.91157 56	−12.31371 5
9.6	(−2)−7.7572	(−2) 3.6995	(−1) 1.1225	(−2) 9.1630	(−2) 1.1827	(−2)−7.3150	−9.39828 63	−13.18805 0
9.7	(−2)−8.3288	(−2) 2.7280	(−1) 1.0860	(−2) 9.5874	(−2) 1.9892	(−2)−6.5114	−9.88210 58	−14.11841 9
9.8	(−2)−8.8169	(−2) 1.7550	(−1) 1.0429	(−2) 9.9507	(−2) 2.7712	(−2)−5.7090	−10.35610 3	−15.10518 2
9.9	(−2)−9.2189	(−3)+7.8793	(−2) 9.9352	(−1) 1.0251	(−2) 3.5259	(−2)−4.9088	−10.81210 4	−16.14793 9
10.0	(−2)−9.5327	(−3)−1.6599	(−2) 9.3834	(−1) 1.0488	(−2) 4.2506	(−2)−4.1117	−11.24057 9	−17.24536 7

$$y_n(x) = \sqrt{\tfrac{1}{2}\pi/x}\,Y_{n+\frac{1}{2}}(x) = (-1)^{n+1}\sqrt{\tfrac{1}{2}\pi/x}\,J_{-(n+\frac{1}{2})}(x) \qquad \begin{bmatrix}(-3)3\\6\end{bmatrix} \qquad \begin{bmatrix}(-3)7\\6\end{bmatrix}$$

SPHERICAL BESSEL FUNCTIONS—ORDERS 20 AND 21 Table 10.3

x	$10^{26}f_{20}(x)$	$10^{27}f_{21}(x)$	$10^{-24}g_{20}(x)$	$10^{-25}g_{21}(x)$
0.0	7.62597 90	1.77348 35	−0.31983 10	−1.31130 70
0.5	7.62705 91	1.77371 23	−0.31988 11	−1.31149 33
1.0	7.63028 29	1.77439 56	−0.32003 25	−1.31205 61
1.5	7.63560 15	1.77552 32	−0.32028 86	−1.31300 70
2.0	7.64293 25	1.77707 85	−0.32065 49	−1.31436 61
2.5	7.65215 99	1.77903 78	−0.32113 96	−1.31616 11
3.0	7.66313 22	1.78137 03	−0.32175 30	−1.31842 87
3.5	7.67566 19	1.78403 80	−0.32250 82	−1.32121 43
4.0	7.68952 28	1.78699 49	−0.32342 08	−1.32457 29
4.5	7.70444 90	1.79018 73	−0.32450 98	−1.32856 95
5.0	7.72013 23	1.79355 29	−0.32579 69	−1.33328 02
5.5	7.73621 95	1.79702 05	−0.32730 79	−1.33879 33
6.0	7.75231 00	1.80050 95	−0.32907 24	−1.34521 03
6.5	7.76795 28	1.80392 94	−0.33112 44	−1.35264 77
7.0	7.78264 38	1.80717 91	−0.33350 34	−1.36123 89
7.5	7.79582 23	1.81014 64	−0.33625 47	−1.37113 69
8.0	7.80686 80	1.81270 77	−0.33943 07	−1.38251 67
8.5	7.81509 84	1.81472 70	−0.34309 23	−1.39557 96
9.0	7.81976 53	1.81605 56	−0.34731 02	−1.41055 73
9.5	7.82005 32	1.81653 14	−0.35216 70	−1.42771 82
10.0	7.815076	1.815979	−0.35776 04	−1.447374
10.5	7.803876	1.814208	−0.36420 59	−1.469891
11.0	7.785428	1.811016	−0.37164 20	−1.495697
11.5	7.758627	1.806185	−0.38023 59	−1.525305
12.0	7.722309	1.799482	−0.39019 23	−1.559325
12.5	7.675238	1.790664	−0.40176 53	−1.598497
13.0	7.616116	1.779472	−0.41527 46	−1.643728
13.5	7.543601	1.765639	−0.43113 22	−1.696143
14.0	7.456316	1.748885	−0.44987 76	−1.757166
14.5	7.352841	1.728929	−0.47223 40	−1.828625
15.0	7.231764	1.705481	−0.49918 70	−1.912922
15.5	7.091689	1.678251	−0.53209 15	−2.013273
16.0	6.931265	1.646956	−0.57279 98	−2.134049
16.5	6.749220	1.611324	−0.62378 79	−2.281228
17.0	6.544411	1.571096	−0.68821 72	−2.462936
17.5	6.315851	1.526041	−0.76981 49	−2.689957
18.0	6.062784	1.475960	−0.87240 01	−2.975953
18.5	5.784739	1.420698	−0.99883 14	−3.336925
19.0	5.481584	1.360155	−1.149171	−3.789188
19.5	5.153621	1.294299	−1.317987	−4.344958
20.0	4.801647	1.223178	−1.490982	−5.004711
20.5	4.427041	1.146936	−1.641599	−5.745922
21.0	4.031843	1.065826	−1.728777	−6.508927
21.5	3.618830	0.98022 63	−1.697442	−7.182333
22.0	3.191590	0.89065 46	−1.483467	−7.592679
22.5	2.754567	0.79777 92	−1.024223	−7.504782
23.0	2.313103	0.70243 25	−0.274630	−6.640003
23.5	1.873442	0.60561 45	+0.773430	−4.717888
24.0	1.442686	0.50849 80	2.072631	−1.52185
24.5	1.028721	0.41242 27	3.508629	+3.01816
25.0	0.640055	0.31888 30	4.901591	+8.74251
	$\begin{bmatrix}(-3)3\\6\end{bmatrix}$	$\begin{bmatrix}(-4)7\\5\end{bmatrix}$		

$$j_n(x) = f_n x^n \exp\left(-x^2/4n+2\right) \qquad y_n(x) = g_n x^{-(n+1)} \exp\left(x^2/4n+2\right)$$

Compiled from National Bureau of Standards, Tables of spherical Bessel functions, vols. I, II. Columbia Univ. Press, New York, N.Y., 1947 (with permission).

Table 10.4

SPHERICAL BESSEL FUNCTIONS—MODULUS AND PHASE—ORDERS 9, 10, 20 AND 21

$$j_n(x) = \sqrt{\tfrac{1}{2}\pi/x}\, M_{n+\frac{1}{2}}(x)\,\cos\theta_{n+\frac{1}{2}}(x) \qquad y_n(x) = \sqrt{\tfrac{1}{2}\pi/x}\, M_{n+\frac{1}{2}}(x)\,\sin\theta_{n+\frac{1}{2}}(x)$$

x^{-1}	$\sqrt{\tfrac{1}{2}\pi x}\,M_{19\frac{1}{2}}(x)$	$\theta_{19\frac{1}{2}}(x)-x$	$\sqrt{\tfrac{1}{2}\pi x}\,M_{21\frac{1}{2}}(x)$	$\theta_{21\frac{1}{2}}(x)-x$	$\langle x\rangle$
0.100	1.50513 630	1.72311 121	1.84157 799	1.35401 461	10
0.095	1.41043 073	1.44562 029	1.65174 534	1.00196 372	11
0.090	1.33509 121	1.17232 718	1.50947 539	0.65310 249	11
0.085	1.27462 197	0.90378 457	1.40190 550	+0.30984 705	12
0.080	1.22560 809	0.64017 615	1.31955 792	−0.02643 915	13
0.075	1.18548 011	0.38142 613	1.25559 223	−0.35524 574	13
0.070	1.15231 423	+0.12729 416	1.20514 049	−0.67664 889	14
0.065	1.12467 134	−0.12255 277	1.16476 186	−0.99107 278	15
0.060	1.10147 221	−0.36849 087	1.13202 416	−1.29911 571	17
0.055	1.08190 340	−0.61090 826	1.10519 883	−1.60143 947	18
0.050	1.06534 781	−0.85018 673	1.08304 588	−1.89870 678	20
0.045	1.05133 389	−1.08669 229	1.06466 562	−2.19155 009	22
0.040	1.03949 892	−1.32077 114	1.04939 746	−2.48055 907	25
0.035	1.02956 235	−1.55274 891	1.03675 104	−2.76627 814	29
0.030	1.02130 658	−1.78293 175	1.02635 931	−3.04920 936	33
0.025	1.01456 304	−2.01160 832	1.01794 637	−3.32981 737	40
0.020	1.00920 210	−2.23905 224	1.01130 529	−3.60853 532	50
0.015	1.00512 574	−2.46552 469	1.00628 277	−3.88577 070	67
0.010	1.00226 240	−2.69127 701	1.00276 864	−4.16191 106	100
0.005	1.00056 327	−2.91655 326	1.00068 866	−4.43732 935	200
0.000	1.00000 000 $\begin{bmatrix}(-3)2\\9\end{bmatrix}$	−3.14159 265 $\begin{bmatrix}(-4)6\\9\end{bmatrix}$	1.00000 000 $\begin{bmatrix}(-3)6\\9\end{bmatrix}$	−4.71238 898 $\begin{bmatrix}(-4)9\\10\end{bmatrix}$	∞

x^{-1}	$\sqrt{\tfrac{1}{2}\pi x}\,M_{41\frac{1}{2}}(x)$	$\theta_{41\frac{1}{2}}(x)-x$	$\sqrt{\tfrac{1}{2}\pi x}\,M_{43\frac{1}{2}}(x)$	$\theta_{43\frac{1}{2}}(x)-x$	$\langle x\rangle$
0.040	1.31126 605	1.12909 207	1.37979 868	+0.54348 547	25
0.038	1.25741 042	0.61321 135	1.30763 025	−0.04056 472	26
0.036	1.21433 612	+0.11048 098	1.25205 767	−0.60729 830	28
0.034	1.17917 949	−0.38066 745	1.20806 627	−1.15885 172	29
0.032	1.15001 033	−0.86163 915	1.17245 178	−1.69717 688	31
0.030	1.12549 256	−1.33366 819	1.14310 153	−2.22398 514	33
0.028	1.10467 736	−1.79783 172	1.11857 851	−2.74075 480	36
0.026	1.08687 488	−2.25507 118	1.09787 629	−3.24876 024	38
0.024	1.07157 283	−2.70621 373	1.08027 122	−3.74910 503	42
0.022	1.05838 371	−3.15199 149	1.06523 083	−4.24275 239	45
0.020	1.04700 987	−3.59305 805	1.05235 561	−4.73055 105	50
0.018	1.03721 972	−4.03000 220	1.04134 092	−5.21325 651	56
0.016	1.02883 137	−4.46335 928	1.03195 154	−5.69154 843	63
0.014	1.02170 104	−4.89362 072	1.02400 423	−6.16604 479	71
0.012	1.01571 485	−5.32124 187	1.01735 560	−6.63731 350	83
0.010	1.01078 282	−5.74664 872	1.01189 351	−7.10588 196	100
0.008	1.00683 452	−6.17024 356	1.00753 093	−7.57224 522	125
0.006	1.00381 592	−6.59240 995	1.00420 153	−8.03687 285	167
0.004	1.00168 705	−7.01351 707	1.00185 654	−8.50021 498	250
0.002	1.00042 044	−7.43392 365	1.00046 253	−8.96270 770	500
0.000	1.00000 000 $\begin{bmatrix}(-3)1\\9\end{bmatrix}$	−7.85398 164 $\begin{bmatrix}(-3)2\\9\end{bmatrix}$	1.00000 000 $\begin{bmatrix}(-3)2\\10\end{bmatrix}$	−9.42477 796 $\begin{bmatrix}(-3)2\\9\end{bmatrix}$	∞

$\langle x\rangle$ = nearest integer to x.

Compiled from National Bureau of Standards, Tables of spherical Bessel functions, vols. I, II. Columbia Univ. Press, New York, N.Y., 1947 (with permission).

SPHERICAL BESSEL FUNCTIONS—VARIOUS ORDERS

Table 10.5

$$j_n(x)$$

n	$x=1$	$x=2$	$x=5$
0	(– 1)8.41470 9848	(– 1)4.54648 7134	(– 1)-1.91784 8549
1	(– 1)3.01168 6789	(– 1)4.35397 7750	(– 2)-9.50894 0808
2	(– 2)6.20350 5201	(– 1)1.98447 9491	(– 1)+1.34731 2101
3	(– 3)9.00658 1117	(– 2)6.07220 9766	(– 1) 2.29820 6182
4	(– 3)1.01101 5808	(– 2)1.40793 9276	(– 1) 1.87017 6553
5	(– 5)9.25611 5861	(– 3)2.63516 9770	(– 1) 1.06811 1615
6	(– 6)7.15693 6310	(– 4)4.14040 9734	(– 2) 4.79668 9986
7	(– 7)4.79013 4199	(– 5)5.60965 5703	(– 2) 1.79027 7818
8	(– 8)2.82649 8802	(– 6)6.68320 4324	(– 3) 5.74143 4675
9	(– 9)1.49137 6503	(– 7)7.10679 7192	(– 3) 1.61809 9715
10	(– 11)7.11655 2640	(– 8)6.82530 0865	(– 4) 4.07344 2442
11	(– 12)3.09955 1855	(– 9)5.97687 1612	(– 5) 9.27461 1037
12	(– 13)1.24166 2597	(– 10)4.81014 8901	(– 5) 1.92878 6347
13	(– 15)4.60463 7678	(– 11)3.58145 1402	(– 6) 3.69320 6998
14	(– 16)1.58957 5988	(– 12)2.48104 9119	(– 7) 6.55454 3131
15	(– 18)5.13268 6115	(– 13)1.60698 2166	(– 7) 1.08428 0182
16	(– 19)1.55670 8271	(– 15)9.77323 7728	(– 8) 1.67993 9976
17	(– 21)4.45117 7504	(– 16)5.60205 9151	(– 9) 2.44802 0198
18	(– 22)1.20385 5742	(– 17)3.03657 8644	(– 10) 3.36741 6303
19	(– 24)3.08874 2364	(– 18)1.56113 3992	(– 11) 4.38678 6630
20	(– 26)7.53779 5722	(– 20)7.63264 1101	(– 12) 5.42772 6761
30	(– 43)5.56683 1267	(– 34)5.83661 7888	(– 22) 4.28273 0217
40	(– 61)1.53821 0374	(– 49)1.66097 8779	(– 33) 1.21034 7583
50	(– 81)3.61527 4717	(– 66)4.01157 5290	(– 46) 2.85747 9350
100	(–190)7.44472 7742	(–160)9.36783 2591	(–120) 5.53565 0303

n	$x=10$	$x=50$	$x=100$
0	(– 2)-5.44021 1109	(– 3)-5.24749 7074	(–3)-5.06365 6411
1	(– 2)+7.84669 4180	(– 2)-1.94042 7051	(–3)-8.67382 5287
2	(– 2)+7.79421 9363	(– 3)+4.08324 0843	(–3)+4.80344 1652
3	(– 2)-3.94958 4498	(– 2)+1.98125 9460	(–3)+8.91399 7370
4	(– 1)-1.05589 2851	(– 3)-1.30947 7600	(–3)-4.17946 1837
5	(– 2)-5.55345 1162	(– 2)-2.00483 0056	(–3)-9.29014 8935
6	(– 2)+4.45013 2233	(– 3)-3.10114 8524	(–3)+3.15754 5454
7	(– 1) 1.13386 2307	(– 2)+1.92420 0195	(–3)+9.70062 9844
8	(– 1) 1.25578 0236	(– 3)+8.87374 9108	(–3)-1.70245 0977
9	(– 1) 1.00096 4095	(– 2)-1.62249 2725	(–3)-9.99004 6510
10	(– 2) 6.46051 5449	(– 2)-1.50392 2146	(–4)-1.95657 8597
11	(– 2) 3.55744 1489	(– 3)+9.90845 4236	(–3)+9.94895 8359
12	(– 2) 1.72159 9974	(– 2)+1.95971 1041	(–3)+2.48391 8282
13	(– 3) 7.46558 4477	(– 4)-1.09899 0300	(–3)-9.32797 8789
14	(– 3) 2.94107 8342	(– 2)-1.96564 5589	(–3)-5.00247 2555
15	(– 3) 1.06354 2715	(– 2)-1.12908 4539	(–3)+7.87726 1748
16	(– 4) 3.55904 0735	(– 2)+1.26561 3175	(–3)+7.44442 3697
17	(– 4) 1.10940 7280	(– 2)+1.96438 9234	(–3)-5.42060 1928
18	(– 5) 3.23884 7439	(– 3)+1.09459 2888	(–3)-9.34163 4372
19	(– 6) 8.89662 7269	(– 2)-1.88338 9360	(–3)+1.96419 7210
20	(– 6) 2.30837 1961	(– 2)-1.57850 2990	(–2)+1.01076 7128
30	(–13) 2.51205 7385	(– 3)-1.49467 3454	(–3)+8.70062 8514
40	(–22) 8.43567 1634	(– 2)-2.60633 6952	(–2)+1.04341 0851
50	(–31) 2.23069 6023	(– 2)+1.88291 0737	(–4)+5.79714 0882
100	(–90) 5.83204 0182	(–22)+1.01901 2263	(–2)+1.08804 7701

Table 10.5   **SPHERICAL BESSEL FUNCTIONS—VARIOUS ORDERS**

$$y_n(x)$$

n	$x=1$	$x=2$	$x=5$
0	(−1)−5.40302 3059	(−1)+2.08073 4183	(−2)−5.67324 3709
1	(0)−1.38177 3291	(−1)−3.50612 0043	(−1)+1.80438 3675
2	(0)−3.60501 7566	(−1)−7.33991 4247	(−1)+1.64995 4576
3	(1)−1.66433 1454	(0)−1.48436 6557	(−2)−1.54429 0991
4	(2)−1.12898 1842	(0)−4.46129 1526	(−1)−1.86615 5315
5	(2)−9.99440 3434	(1)−1.85914 4531	(−1)−3.20465 0467
6	(4)−1.08809 4559	(1)−9.77916 5769	(−1)−5.18407 5714
7	(5)−1.40452 8524	(2)−6.17054 3296	(0)−1.02739 4639
8	(6)−2.09591 1840	(3)−4.53011 5815	(0)−2.56377 6345
9	(7)−3.54900 4843	(4)−3.78889 3009	(0)−7.68944 4934
10	(8)−6.72215 0083	(5)−3.55414 7201	(1)−2.66561 1441
11	(10)−1.40810 2512	(6)−3.69396 5631	(2)−1.04266 2356
12	(11)−3.23191 3629	(7)−4.21251 9003	(2)−4.52968 5692
13	(12)−8.06570 3047	(8)−5.22870 9098	(3)−2.16057 6611
14	(14)−2.17450 7909	(9)−7.01663 2092	(4)−1.12141 4513
15	(15)−6.29800 7233	(11)−1.01218 2944	(4)−6.28814 6513
16	(17)−1.95020 7734	(12)−1.56186 6932	(5)−3.78650 9387
17	(18)−6.42938 7516	(13)−2.56695 8608	(6)−2.43621 4730
18	(20)−2.24833 5423	(14)−4.47655 8894	(7)−1.66748 5217
19	(21)−8.31241 1677	(15)−8.25596 4368	(8)−1.20957 6913
20	(23)−3.23959 2219	(17)−1.60543 6493	(8)−9.26795 1403
30	(40)−2.94642 8547	(31)−1.40739 3871	(18)−7.76071 7570
40	(58)−8.02845 0851	(46)−3.72092 9322	(30)−2.05575 8716
50	(78)−2.73919 2285	(63)−1.23502 1944	(42)−6.96410 9188
100	(186)−6.68307 9463	(156)−2.65595 5830	(116)−1.79971 3983

n	$x=10$	$x=50$	$x=100$
0	(−2)+8.39071 5291	(−2)−1.92993 2057	(−3)−8.62318 8723
1	(−2)+6.27928 2638	(−3)+4.86151 0663	(−3)+4.97742 4524
2	(−2)−6.50693 0499	(−2)+1.95910 1121	(−3)+8.77251 1459
3	(−2)−9.53274 7888	(−3)−2.90240 9542	(−3)−4.53879 8951
4	(−3)−1.65993 0220	(−2)−1.99973 4855	(−3)−9.09022 7385
5	(−2)+9.38335 4168	(−4)−6.97113 1965	(−3)+3.72067 8486
6	(−1)+1.04876 8261	(−2)+1.98439 8364	(−3)+9.49950 2019
7	(−2)+4.25063 3221	(−3)+5.85654 8943	(−3)−2.48574 3224
8	(−2)−4.11173 2775	(−2)−1.80870 1896	(−3)−9.87236 3502
9	(−1)−1.12405 7894	(−2)−1.20061 3539	(−4)+8.07441 4285
10	(−1)−1.72453 6721	(−2)+1.35246 8751	(−2)+1.00257 7737
11	(−1)−2.49746 9220	(−2)+1.76865 0414	(−3)+1.29797 1820
12	(−1)−4.01964 2485	(−3)−5.38889 5605	(−3)−9.72724 3855
13	(−1)−7.55163 6993	(−2)−2.03809 5195	(−3)−3.72978 2784
14	(0)−1.63697 7739	(−3)−5.61681 8446	(−3)+8.72020 2503
15	(0)−3.99207 1745	(−2)+1.71231 9725	(−3)+6.25864 1510
16	(1)−1.07384 4467	(−2)+1.62332 0074	(−3)−6.78002 3635
17	(1)−3.14447 9567	(−3)−6.40928 4759	(−3)−8.49604 9309
18	(1)−9.93183 4017	(−2)−2.07197 0007	(−3)+3.80640 6377
19	(2)−3.36033 0630	(−3)−8.92329 3294	(−3)+9.90441 9669
20	(3)−1.21121 0605	(−2)+1.37595 3130	(−5)+5.63172 9379
30	(9)−6.90831 8646	(−2)−2.24122 6812	(−3)−5.41292 9349
40	(18)−1.51030 4919	(−5)+4.97879 7221	(−4)−7.04842 0407
50	(27)−4.52822 7272	(−2)−4.19000 0150	(−2)+1.07478 2297
100	(85)−8.57322 6309	(+18)−1.12569 2891	(−2)−2.29838 5049

ZEROS OF BESSEL FUNCTIONS OF HALF-INTEGER ORDER

$$J_\nu(j_{\nu,s})=0 \qquad Y_\nu(y_{\nu,s})=0$$

Table 10.6

ν	s	$j_{\nu,s}$	$J'_\nu(j_{\nu,s})$	$y_{\nu,s}$	$(-1)^{n+1}Y'_\nu(y_{\nu,s})$
1/2	1	3.141593	−0.45015 82	1.570796	−0.63661 98
	2	6.283185	+0.31830 99	4.712389	+0.36755 26
	3	9.424778	−0.25989 89	7.853982	−0.28470 50
	4	12.566370	+0.22507 91	10.995574	+0.24061 97
	5	15.707963	−0.20131 68	14.137167	−0.21220 66
	6	18.849556	+0.18377 63	17.278760	+0.19194 81
	7	21.991149	−0.17014 38	20.420352	−0.17656 66
	8			23.561945	+0.16437 45
3/2	1	4.493409	−0.36741 35	2.798386	+0.44914 84
	2	7.725252	+0.28469 20	6.121250	−0.31827 37
	3	10.904122	−0.24061 69	9.317866	+0.25989 33
	4	14.066194	+0.21220 57	12.486454	−0.22507 76
	5	17.220755	−0.19194 77	15.644128	+0.20131 63
	6	20.371303	+0.17656 64	18.796404	−0.18377 61
	7	23.519452	−0.16437 44	21.945613	+0.17014 37
5/2	1	5.763459	−0.31710 58	3.959528	−0.36184 68
	2	9.095011	+0.25973 30	7.451610	+0.28430 75
	3	12.322941	−0.22503 59	10.715647	−0.24053 93
	4	15.514603	+0.20130 14	13.921686	+0.21218 15
	5	18.689036	−0.18376 96	17.103359	−0.19193 81
	6	21.853874	+0.17014 05	20.272369	+0.17656 19
	7			23.433926	−0.16437 21
7/2	1	6.987932	−0.28223 71	5.088498	+0.30882 36
	2	10.417119	+0.24019 23	8.733710	−0.25896 77
	3	13.698023	−0.21208 02	12.067544	+0.22485 68
	4	16.923621	+0.19189 90	15.315390	−0.20124 01
	5	20.121806	−0.17654 40	18.525210	+0.18374 36
	6	23.304247	+0.16436 28	21.714547	−0.17012 77
	7			24.891503	+0.15914 62
9/2	1	8.182561	−0.25620 49	6.197831	−0.27236 25
	2	11.704907	+0.22432 53	9.982466	+0.23908 76
	3	15.039665	−0.20107 12	13.385287	−0.21179 27
	4	18.301256	+0.18367 64	16.676625	+0.19179 35
	5	21.525418	−0.17009 46	19.916796	−0.17649 69
	6	24.727566	+0.15912 86	23.128642	+0.16433 89
11/2	1	9.355812	−0.23580 60	7.293692	+0.24538 14
	2	12.966530	+0.21109 29	11.206497	−0.22293 49
	3	16.354710	−0.19155 58	14.676387	+0.20067 86
	4	19.653152	+0.17639 49	18.011609	−0.18352 21
	5	22.904551	−0.16428 83	21.283249	+0.17002 38
	6			24.518929	−0.15909 15
13/2	1	10.512835	−0.21926 48	8.379626	−0.22441 70
	2	14.207392	+0.19983 04	12.411301	+0.20946 65
	3	17.647975	−0.18321 82	15.945983	−0.19106 59
	4	20.983463	+0.16988 82	19.324820	+0.17619 60
	5	24.262768	−0.15902 21	22.628417	−0.16419 26

ν	s	$j_{\nu,s}$	$J'_\nu(j_{\nu,s})$	$y_{\nu,s}$	$(-1)^{n+1}Y'_\nu(y_{\nu,s})$
15/2	1	11.657032	−0.20550 46	9.457882	+0.20754 83
	2	15.431289	+0.19008 87	13.600629	−0.19801 01
	3	18.922999	−0.17582 99	17.197777	+0.18264 01
	4	22.295348	+0.16402 38	20.619612	−0.16964 44
	5			23.955267	+0.15890 14
17/2	1	12.790782	−0.19382 82	10.529989	−0.19361 38
	2	16.641003	+0.18155 15	14.777175	+0.18810 92
	3	20.182471	−0.16922 10	18.434529	−0.17517 27
	4	23.591275	+0.15870 04	21.898570	+0.16373 75
19/2	1	13.915823	−0.18376 12	11.597038	+0.18186 42
	2	17.838643	+0.17398 80	15.942945	−0.17744 10
	3	21.428487	−0.16326 17	19.658369	+0.16849 33
	4	24.873214	+0.15383 84	23.163734	−0.15837 45
21/2	1	15.033469	−0.17496 82	12.659840	−0.17179 22
	2	19.025854	+0.16722 59	17.099480	+0.17176 97
	3	22.662721	−0.15785 09	20.870973	−0.16247 13
	4			24.416749	+0.15347 56
23/2	1	16.144743	−0.16720 39	13.719013	+0.16304 06
	2	20.203943	+0.16113 25	18.247994	−0.16491 86
	3	23.886531	−0.15290 87	22.073692	+0.15700 50
25/2	1	17.250455	−0.16028 44	14.775045	−0.15534 97
	2	21.373972	+0.15560 47	19.389462	+0.15875 20
	3			23.267630	−0.15201 34
27/2	1	18.351261	−0.15406 88	15.828325	+0.14852 56
	2	22.536817	+0.15056 00	20.524680	−0.15316 36
	3			24.453705	+0.14743 15
29/2	1	19.447703	−0.14844 69	16.879170	−0.14242 04
	2	23.693208	+0.14593 21	21.654309	+0.14806 91
31/2	1	20.540230	−0.14333 12	17.927842	+0.13691 88
	2	24.843763	+0.14166 70	22.778902	−0.14340 05
33/2	1	21.629221	−0.13865 11	18.974562	−0.13192 99
	2			23.898931	+0.13910 20
35/2	1	22.715002	−0.13434 93	20.019515	+0.12738 05
37/2	1	23.797849	−0.13037 81	21.062860	−0.12321 13
39/2	1	24.878005	−0.12669 81	22.104735	+0.11937 34

Values to greater accuracy and over a wider range are given in [10.31].
From National Bureau of Standards, Tables of spherical Bessel functions, vols. I, II. Columbia Univ. Press, New York, N.Y., 1947 (with permission).

Table 10.7

ZEROS OF THE DERIVATIVE OF BESSEL FUNCTIONS OF HALF-INTEGER ORDER

$$J'_\nu(j'_{\nu,s})=0 \qquad Y'_\nu(y'_{\nu,s})=0$$

ν	s	$j'_{\nu,s}$	$J_\nu(j'_{\nu,s})$	$y'_{\nu,s}$	$(-1)^{n+1}Y_\nu(y'_{\nu,s})$	ν	s	$j'_{\nu,s}$	$J_\nu(j'_{\nu,s})$	$y'_{\nu,s}$	$(-1)^{n+1}Y_\nu(y'_{\nu,s})$
1/2	1	1.165561	+0.679192	2.975086	−0.456186	15/2	1	9.113402	+0.330874	11.535731	+0.266883
	2	4.604217	−0.369672	6.202750	+0.319331		2	13.525575	−0.236854	15.376058	−0.217283
	3	7.789884	+0.285287	9.371475	−0.260267		3	17.153587	+0.202841	18.885886	+0.191447
	4	10.949944	−0.240870	12.526476	+0.225258		4	20.587450	−0.182077	22.266861	−0.174147
	5	14.101725	+0.212340	15.676078	−0.201419		5	23.929631	+0.167294		
	6	17.249782	−0.192029	18.822999	+0.183841						
	7	20.395842	+0.176620	21.968393	−0.170188						
	8	23.540708	−0.164412								
						17/2	1	10.180054	+0.318378	12.669130	−0.257833
							2	14.702493	−0.229449	16.586323	+0.210950
							3	18.390930	+0.197291	20.145940	−0.186505
							4	21.866965	−0.177623	23.563314	+0.170098
3/2	1	2.460536	+0.525338	4.354435	+0.388891						
	2	6.029292	−0.328062	7.655545	−0.290138						
	3	9.261402	+0.263295	10.856531	+0.242910						
	4	12.445260	−0.226711	14.029845	−0.213417	19/2	1	11.241675	+0.307606	13.793646	+0.249935
	5	15.611585	+0.202245	17.191285	+0.192678		2	15.868463	−0.222927	17.784362	−0.205332
	6	18.769469	−0.184363	20.346496	−0.177046		3	19.615227	+0.192335	21.392422	+0.182067
	7	21.922619	+0.170542	23.498023	+0.164709		4	23.132584	−0.173605	24.845689	−0.166427
5/2	1	3.632797	+0.457398	5.634297	−0.350669	21/2	1	12.299124	+0.298179	14.910648	−0.242951
	2	7.367009	−0.301449	9.030902	+0.270006		2	17.025072	−0.217118	18.971857	+0.200296
	3	10.663561	+0.247304	12.278863	−0.229783		3	20.828186	+0.187870	22.627032	−0.178048
	4	13.883370	−0.215670	15.480655	+0.203956		4	24.385974	−0.169950		
	5	17.072849	+0.194015	18.661309	−0.185432						
	6	20.246945	−0.177917	21.830390	+0.171262						
	7	23.412100	+0.165314	24.992411	−0.159953						
						23/2	1	13.353045	+0.289825	16.021196	+0.236710
							2	18.173567	−0.211893	20.150142	−0.195742
							3	22.031181	+0.183813	23.851147	+0.174383
7/2	1	4.762196	+0.415533	6.863232	+0.324651						
	2	8.653134	−0.282237	10.356373	−0.254849						
	3	12.018262	+0.234875	13.656304	+0.219318	25/2	1	14.403937	+0.282348	17.126125	−0.231081
	4	15.279081	−0.206685	16.891400	−0.196124		2	19.314945	−0.207156	21.320300	+0.191594
	5	18.496200	+0.187103	20.095393	+0.179270		3	23.225333	+0.180103		
	6	21.690284	−0.172377	23.281796	−0.166245						
	7	24.870602	+0.160741								
						27/2	1	15.452196	+0.275596	18.226109	+0.225965
							2	20.450018	−0.202830	22.483219	−0.187792
							3	24.411571	+0.176690		
9/2	1	5.868420	+0.386006	8.060030	−0.305246						
	2	9.904306	−0.267385	11.646354	+0.242810						
	3	13.337928	+0.224788	14.999624	−0.210673	29/2	1	16.498138	+0.269455	19.321702	−0.221286
	4	16.641787	−0.199151	18.270330	+0.189472		2	21.579459	−0.198856	23.639641	+0.184287
	5	19.888934	+0.181169	21.500029	−0.173929						
	6	23.105297	−0.167534	24.705942	+0.161826						
						31/2	1	17.542024	+0.263833	20.413362	+0.216981
							2	22.703832	−0.195187	24.790191	−0.181040
11/2	1	6.959746	+0.363557	9.234274	+0.289946						
	2	11.129856	−0.255385	12.909478	−0.232895						
	3	14.630406	+0.216349	16.315912	+0.203344	33/2	1	18.584071	+0.258658	21.501477	−0.213000
	4	17.977886	−0.192692	19.623229	−0.183714		2	23.823614	−0.191783		
	5	21.256291	+0.175987	22.879980	+0.169229						
	6	24.496327	−0.163244								
						35/2	1	19.624460	+0.253871	22.586374	+0.209303
							2	24.939214	−0.188612		
13/2	1	8.040535	+0.345649	10.391621	−0.277420	37/2	1	20.663347	+0.249423	23.668335	−0.205855
	2	12.335631	−0.245384	14.151399	+0.224513						
	3	15.901023	+0.209127	17.610124	−0.197009						
	4	19.291967	−0.187058	20.954335	+0.178651	39/2	1	21.700865	+0.245275	24.747606	+0.202629
	5	22.602185	+0.171399	24.238863	−0.165043						

Values to greater accuracy and over a wider range are given in [10.31].
From National Bureau of Standards, Tables of spherical Bessel functions, vols. I, II. Columbia Univ. Press, New York, N.Y., 1947 (with permission).

MODIFIED SPHERICAL BESSEL FUNCTIONS—ORDERS 0, 1 AND 2   Table 10.8

x	$i_0(x)$	$i_1(x)$	$i_2(x)$	$k_0(x)$	$k_1(x)$	$k_2(x)$
0.0	1.00000 000	0.00000 000	0.00000 0000	∞	∞	∞
0.1	1.00166 750	0.03336 668	0.00066 7143	14.21315 293	156.344682	4704.5536
0.2	1.00668 001	0.06693 370	0.00267 4294	6.43029 630	38.58177 78	585.15696
0.3	1.01506 764	0.10090 290	0.00603 8668	3.87891 513	16.80863 22	171.96524
0.4	1.02688 081	0.13547 889	0.01078 9114	2.63234 067	9.21319 233	71.731283
0.5	1.04219 061	0.17087 071	0.01696 6360	1.90547 226	5.71641 679	36.203973
0.6	1.06108 930	0.20729 319	0.02462 3348	1.43678 550	3.83142 801	20.593926
0.7	1.08369 100	0.24496 858	0.03382 5678	1.11433 482	2.70624 170	12.712514
0.8	1.11013 248	0.28412 808	0.04465 2156	0.88225 536	1.98507 456	8.32628 49
0.9	1.14057 414	0.32501 361	0.05719 5452	0.70959 792	1.49804 005	5.70306 48
1.0	1.17520 119	0.36787 944	0.07156 2871	0.57786 367	1.15572 735	4.04504 57
1.1	1.21422 497	0.41299 416	0.08787 7251	0.47533 880	0.90746 4974	2.95024 33
1.2	1.25788 446	0.46064 259	0.10627 7995	0.39426 230	0.72281 4219	2.20129 78
1.3	1.30644 803	0.51112 785	0.12692 2227	0.32930 149	0.58261 0332	1.67378 69
1.4	1.36021 536	0.56477 365	0.14998 6112	0.27668 115	0.47431 0537	1.29306 09
1.5	1.41951 964	0.62192 665	0.17566 6332	0.23366 136	0.38943 5596	1.01253 25
1.6	1.48472 997	0.68295 906	0.20418 1728	0.19821 144	0.32209 3595	0.80213 693
1.7	1.55625 408	0.74827 140	0.23577 5138	0.16879 918	0.26809 2818	0.64190 415
1.8	1.63454 127	0.81829 550	0.27071 5433	0.14425 049	0.22438 9655	0.51823 325
1.9	1.72008 574	0.89349 778	0.30929 9770	0.12365 360	0.18873 4440	0.42165 535
2.0	1.81343 020	0.97438 274	0.35185 6089	0.10629 208	0.15943 8124	0.34544 927
2.1	1.91516 988	1.06149 681	0.39874 5868	0.09159 719	0.13521 4906	0.28476 135
2.2	2.02595 690	1.15543 247	0.45036 7165	0.07911 327	0.11507 3847	0.23603 215
2.3	2.14650 513	1.25683 283	0.50715 7959	0.06847 227	0.09824 2824	0.19661 508
2.4	2.27759 551	1.36639 653	0.56959 9849	0.05937 476	0.08411 4246	0.16451 757
2.5	2.42008 179	1.48488 308	0.63822 2102	0.05157 553	0.07220 5736	0.13822 241
2.6	2.57489 701	1.61311 877	0.71360 6125	0.04487 256	0.06213 1241	0.11656 246
2.7	2.74306 041	1.75200 304	0.79639 0365	0.03909 858	0.05357 9539	0.09863 140
2.8	2.92568 513	1.90251 546	0.88727 5704	0.03411 437	0.04629 8067	0.08371 944
2.9	3.12398 658	2.06572 335	0.98703 1387	0.02980 354	0.04008 0625	0.07126 626
3.0	3.33929 164	2.24279 012	1.09650 152	0.02606 845	0.03475 7931	0.06082 638
3.1	3.57304 872	2.43498 437	1.21661 224	0.02282 681	0.03019 0302	0.05204 323
3.2	3.82683 875	2.64368 983	1.34837 954	0.02000 910	0.02626 1944	0.04462 967
3.3	4.10238 723	2.87041 631	1.49291 787	0.01755 635	0.02287 6452	0.03835 312
3.4	4.40157 747	3.11681 153	1.65144 965	0.01541 841	0.01995 3243	0.03302 422
3.5	4.72646 494	3.38467 421	1.82531 562	0.01355 255	0.01742 4712	0.02848 802
3.6	5.07929 316	3.67596 831	2.01598 623	0.01192 222	0.01523 3952	0.02461 718
3.7	5.46251 092	3.99283 865	2.22507 418	0.01049 611	0.01333 2903	0.02130 658
3.8	5.87879 128	4.33762 799	2.45434 813	0.00924 735	0.01168 0862	0.01846 908
3.9	6.33105 220	4.71289 572	2.70574 780	0.00815 280	0.01024 3262	0.01603 223
4.0	6.82247 930	5.12143 838	2.98140 051	0.00719 253	0.00899 0668	0.01393 554
4.1	7.35655 060	5.56631 208	3.28363 932	0.00634 934	0.00789 7961	0.01212 834
4.2	7.93706 374	6.05085 704	3.61502 300	0.00560 833	0.00694 3650	0.01056 808
4.3	8.56816 571	6.57872 451	3.97835 791	0.00495 661	0.00610 9316	0.00921 893
4.4	9.25438 538	7.15390 628	4.37672 200	0.00438 300	0.00537 9136	0.00805 059
4.5	10.00066 914	7.78076 689	4.81349 122	0.00387 777	0.00473 9498	0.00703 744
4.6	10.81241 998	8.46407 908	5.29236 840	0.00343 248	0.00417 8666	0.00615 769
4.7	11.69554 012	9.20906 250	5.81741 513	0.00303 975	0.00368 6506	0.00539 284
4.8	12.65647 789	10.02142 620	6.39308 652	0.00269 318	0.00325 4257	0.00472 709
4.9	13.70227 889	10.90741 515	7.02426 961	0.00238 716	0.00287 4331	0.00414 695
5.0	14.84064 212	11.87386 128	7.71632 535	0.00211 679	0.00254 0146	0.00364 088
	$\begin{bmatrix}(-2)1\\7\end{bmatrix}$	$\begin{bmatrix}(-2)1\\7\end{bmatrix}$	$\begin{bmatrix}(-3)8\\7\end{bmatrix}$			

$$i_n(x) = \sqrt{\tfrac{1}{2}\pi/x}\, I_{n+\frac{1}{2}}(x) \qquad\qquad k_n(x) = \sqrt{\tfrac{1}{2}\pi/x}\, K_{n+\frac{1}{2}}(x)$$

Table 10.9   **MODIFIED SPHERICAL BESSEL FUNCTIONS—ORDERS 9 AND 10**

x	$10^9 x^{-9} i_9(x)$	$10^{10} x^{-10} i_{10}(x)$	$10^{-7} x^{10} k_9(x)$	$10^{-9} x^{11} k_{10}(x)$
0.0	1.52734 93	0.72730 92	5.41287 38	1.02844 60
0.1	1.52771 30	0.72746 73	5.41128 21	1.02817 54
0.2	1.52880 46	0.72794 19	5.40650 99	1.02736 41
0.3	1.53062 54	0.72873 35	5.39856 70	1.02601 35
0.4	1.53317 79	0.72984 30	5.38746 92	1.02412 59
0.5	1.53646 54	0.73127 18	5.37323 85	1.02170 47
0.6	1.54049 23	0.73302 17	5.35590 33	1.01875 42
0.7	1.54526 36	0.73509 47	5.33549 79	1.01527 95
0.8	1.55078 57	0.73749 33	5.31206 23	1.01128 67
0.9	1.55706 60	0.74022 04	5.28564 31	1.00678 27
1.0	1.56411 27	0.74327 93	5.25629 13	1.00177 53
1.1	1.57193 49	0.74667 38	5.22406 45	0.99627 31
1.2	1.58054 32	0.75040 79	5.18902 48	0.99028 56
1.3	1.58994 87	0.75448 62	5.15123 93	0.98382 30
1.4	1.60016 42	0.75891 37	5.11078 01	0.97689 61
1.5	1.61120 30	0.76369 58	5.06772 38	0.96951 68
1.6	1.62308 02	0.76883 83	5.02215 07	0.96169 72
1.7	1.63581 13	0.77434 76	4.97414 57	0.95345 03
1.8	1.64941 38	0.78023 05	4.92379 68	0.94478 97
1.9	1.66390 60	0.78649 43	4.87119 57	0.93572 94
2.0	1.67930 73	0.79314 68	4.81643 66	0.92628 41
2.1	1.69563 90	0.80019 63	4.75961 72	0.91646 88
2.2	1.71292 33	0.80765 17	4.70083 65	0.90629 89
2.3	1.73118 39	0.81552 21	4.64019 67	0.89579 04
2.4	1.75044 59	0.82381 79	4.57780 09	0.88495 95
2.5	1.77073 63	0.83254 94	4.51375 41	0.87382 25
2.6	1.79208 32	0.84172 78	4.44816 23	0.86239 63
2.7	1.81451 64	0.85136 49	4.38113 22	0.85069 78
2.8	1.83806 76	0.86147 30	4.31277 10	0.83874 39
2.9	1.86277 03	0.87206 54	4.24318 63	0.82655 20
3.0	1.88865 96	0.88315 57	4.17248 53	0.81413 92
3.1	1.91577 24	0.89475 86	4.10077 50	0.80152 28
3.2	1.94414 79	0.90688 95	4.02816 19	0.78872 01
3.3	1.97382 74	0.91956 42	3.95475 12	0.77574 83
3.4	2.00485 39	0.93279 97	3.88064 76	0.76262 45
3.5	2.03727 33	0.94661 40	3.80595 33	0.74936 56
3.6	2.07113 33	0.96102 55	3.73076 99	0.73598 84
3.7	2.10648 43	0.97605 38	3.65519 70	0.72250 95
3.8	2.14337 94	0.99171 97	3.57933 16	0.70894 53
3.9	2.18187 40	1.00804 44	3.50326 88	0.69531 19
4.0	2.22202 68	1.02505 08	3.42710 13	0.68162 50
4.1	2.26389 90	1.04276 26	3.35091 95	0.66790 02
4.2	2.30755 54	1.06120 45	3.27481 07	0.65415 25
4.3	2.35306 35	1.08040 28	3.19885 96	0.64039 66
4.4	2.40049 43	1.10038 47	3.12314 76	0.62664 70
4.5	2.44992 27	1.12117 91	3.04775 39	0.61291 75
4.6	2.50142 71	1.14281 58	2.97275 34	0.59922 16
4.7	2.55508 99	1.16532 63	2.89821 88	0.58557 24
4.8	2.61099 74	1.18874 39	2.82421 90	0.57198 25
4.9	2.66924 03	1.21310 29	2.75081 98	0.55846 39
5.0	2.72991 40	1.23843 97	2.67808 38	0.54502 82
	$\begin{bmatrix} (-4)3 \\ 5 \end{bmatrix}$	$\begin{bmatrix} (-4)1 \\ 4 \end{bmatrix}$	$\begin{bmatrix} (-4)4 \\ 5 \end{bmatrix}$	$\begin{bmatrix} (-5)7 \\ 4 \end{bmatrix}$

$$i_n(x) = \sqrt{\tfrac{1}{2}\pi/x}\, I_{n+\frac{1}{2}}(x) \qquad\qquad k_n(x) = \sqrt{\tfrac{1}{2}\pi/x}\, K_{n+\frac{1}{2}}(x)$$

Compiled from C. W. Jones, A short table for the Bessel functions $I_{n+\frac{1}{2}}(x)$, $(2/\pi)K_{n+\frac{1}{2}}(x)$. Cambridge Univ. Press, Cambridge, England, 1952 (with permission).

MODIFIED SPHERICAL BESSEL FUNCTIONS—ORDERS 9 AND 10 Table 10.9

x	$e^{-x}I_{\frac{19}{2}}(x)$	$e^{-x}I_{\frac{21}{2}}(x)$	$\frac{2}{\pi}e^{x}K_{\frac{19}{2}}(x)$	$\frac{2}{\pi}e^{x}K_{\frac{21}{2}}(x)$
5.0	(−5)6.40961	(−5)1.45387	(2)4.62276	(3)1.88159
5.1	(−5)7.16216	(−5)1.65403	(2)4.11899	(3)1.64774
5.2	(−5)7.97716	(−5)1.87488	(2)3.68187	(3)1.44818
5.3	(−5)8.85734	(−5)2.11778	(2)3.30123	(3)1.27719
5.4	(−5)9.80541	(−5)2.38413	(2)2.96863	(3)1.13013
5.5	(−4)1.08240	(−5)2.67535	(2)2.67706	(3)1.00320
5.6	(−4)1.19157	(−5)2.99285	(2)2.42066	(2)8.93250
5.7	(−4)1.30831	(−5)3.33809	(2)2.19449	(2)7.97686
5.8	(−4)1.43285	(−5)3.71252	(2)1.99441	(2)7.14360
5.9	(−4)1.56545	(−5)4.11760	(2)1.81692	(2)6.41477
6.0	(−4)1.70632	(−5)4.55480	(2)1.65905	(2)5.77537
6.1	(−4)1.85569	(−5)5.02559	(2)1.51825	(2)5.21281
6.2	(−4)2.01376	(−5)5.53143	(2)1.39236	(2)4.71647
6.3	(−4)2.18075	(−5)6.07377	(2)1.27955	(2)4.27737
6.4	(−4)2.35684	(−5)6.65407	(2)1.17821	(2)3.88791
6.5	(−4)2.54221	(−5)7.27375	(2)1.08697	(2)3.54160
6.6	(−4)2.73703	(−5)7.93423	(2)1.00464	(2)3.23292
6.7	(−4)2.94147	(−5)8.63691	(1)9.30213	(2)2.95714
6.8	(−4)3.15568	(−5)9.38317	(1)8.62775	(2)2.71019
6.9	(−4)3.37978	(−4)1.01743	(1)8.01557	(2)2.48857
7.0	(−4)3.61391	(−4)1.10117	(1)7.45880	(2)2.28926
7.1	(−4)3.85819	(−4)1.18967	(1)6.95148	(2)2.10966
7.2	(−4)4.11271	(−4)1.28304	(1)6.48840	(2)1.94748
7.3	(−4)4.37758	(−4)1.38142	(1)6.06498	(2)1.80076
7.4	(−4)4.65288	(−4)1.48492	(1)5.67717	(2)1.66777
7.5	(−4)4.93867	(−4)1.59365	(1)5.32140	(2)1.54701
7.6	(−4)5.23503	(−4)1.70773	(1)4.99452	(2)1.43717
7.7	(−4)5.54199	(−4)1.82727	(1)4.69371	(2)1.33708
7.8	(−4)5.85960	(−4)1.95236	(1)4.41649	(2)1.24573
7.9	(−4)6.18789	(−4)2.08311	(1)4.16065	(2)1.16223
8.0	(−4)6.52688	(−4)2.21961	(1)3.92420	(2)1.08577
8.1	(−4)6.87657	(−4)2.36195	(1)3.70539	(2)1.01566
8.2	(−4)7.23697	(−4)2.51020	(1)3.50262	(1)9.51284
8.3	(−4)7.60807	(−4)2.66447	(1)3.31448	(1)8.92076
8.4	(−4)7.98985	(−4)2.82481	(1)3.13970	(1)8.37549
8.5	(−4)8.38228	(−4)2.99130	(1)2.97713	(1)7.87266
8.6	(−4)8.78533	(−4)3.16400	(1)2.82574	(1)7.40835
8.7	(−4)9.19895	(−4)3.34298	(1)2.68460	(1)6.97906
8.8	(−4)9.62308	(−4)3.52828	(1)2.55287	(1)6.58165
8.9	(−3)1.00576	(−4)3.71997	(1)2.42979	(1)6.21331
9.0	(−3)1.05026	(−4)3.91809	(1)2.31467	(1)5.87149
9.1	(−3)1.09579	(−4)4.12268	(1)2.20689	(1)5.55393
9.2	(−3)1.14235	(−4)4.33377	(1)2.10586	(1)5.25858
9.3	(−3)1.18991	(−4)4.55140	(1)2.01109	(1)4.98356
9.4	(−3)1.23849	(−4)4.77560	(1)1.92209	(1)4.72722
9.5	(−3)1.28806	(−4)5.00639	(1)1.83843	(1)4.48802
9.6	(−3)1.33861	(−4)5.24378	(1)1.75973	(1)4.26461
9.7	(−3)1.39014	(−4)5.48779	(1)1.68563	(1)4.05572
9.8	(−3)1.44263	(−4)5.73844	(1)1.61578	(1)3.86022
9.9	(−3)1.49607	(−4)5.99571	(1)1.54991	(1)3.67709
10.0	(−3)1.55045	(−4)6.25963	(1)1.48772	(1)3.50537

Table 10.9

MODIFIED SPHERICAL BESSEL FUNCTIONS—ORDERS 9 AND 10

x^{-1}	$f_9(x)$	$f_{10}(x)$	$g_9(x)$	$g_{10}(x)$	$\langle x \rangle$
0.100	1.10630 573	1.21411 149	0.65502 364	0.56777 303	10
0.095	1.08238 951	1.17260 877	0.68557 030	0.60351 931	11
0.090	1.06167 683	1.13650 462	0.71563 676	0.63926 956	11
0.085	1.04394 741	1.10534 464	0.74502 124	0.67473 612	12
0.080	1.02899 406	1.07872 041	0.77352 114	0.70961 813	13
0.075	1.01661 895	1.05626 085	0.80093 667	0.74360 745	13
0.070	1.00662 998	1.03762 412	0.82707 483	0.77639 538	14
0.065	0.99883 728	1.02248 982	0.85175 354	0.80768 018	15
0.060	0.99304 985	1.01055 159	0.87480 587	0.83717 510	17
0.055	0.98907 251	1.00151 009	0.89608 425	0.86461 675	18
0.050	0.98670 320	0.99506 643	0.91546 455	0.88977 340	20
0.045	0.98573 080	0.99091 634	0.93284 978	0.91245 301	22
0.040	0.98593 357	0.98874 519	0.94817 344	0.93251 041	25
0.035	0.98707 842	0.98822 421	0.96140 216	0.94985 358	29
0.030	0.98892 100	0.98900 824	0.97253 769	0.96444 830	33
0.025	0.99120 680	0.99073 519	0.98161 804	0.97632 121	40
0.020	0.99367 323	0.99302 746	0.98871 764	0.98556 077	50
0.015	0.99605 259	0.99549 538	0.99394 654	0.99231 623	67
0.010	0.99807 595	0.99774 259	0.99744 863	0.99679 434	100
0.005	0.99947 760	0.99937 316	0.99939 894	0.99925 415	200
0.000	1.00000 000	1.00000 000	1.00000 000	1.00000 000	∞
	$\begin{bmatrix} (-4)4 \\ 6 \end{bmatrix}$	$\begin{bmatrix} (-4)7 \\ 7 \end{bmatrix}$	$\begin{bmatrix} (-4)3 \\ 6 \end{bmatrix}$	$\begin{bmatrix} (-4)3 \\ 7 \end{bmatrix}$	

$$\sqrt{2\pi x}\, I_{\frac{19}{2}}(x) = f_9(x)\, e^{x - 45x^{-1}}$$

$$\sqrt{2\pi x}\, I_{\frac{21}{2}}(x) = f_{10}(x)\, e^{x - 55x^{-1}}$$

$$\sqrt{2x/\pi}\, K_{\frac{19}{2}}(x) = g_9(x)\, e^{-x + 45x^{-1}}$$

$$\sqrt{2x/\pi}\, K_{\frac{21}{2}}(x) = g_{10}(x)\, e^{-x + 55x^{-1}}$$

$\langle x \rangle = $ nearest integer to x.

Table 10.10

MODIFIED SPHERICAL BESSEL FUNCTIONS—VARIOUS ORDERS

$$\sqrt{\tfrac{1}{2}\,\pi/x}\,I_{n+\frac{1}{2}}(x)$$

n	$x=1$	$x=2$	$x=5$
0	(0)1.17520 1194	(0)1.81343 0204	(1)1.48406 4212
1	(− 1)3.67879 4412	(− 1)9.74382 7436	(1)1.18738 6128
2	(− 2)7.15628 7013	(− 1)3.51856 0886	(0)7.71632 5346
3	(− 2)1.00650 9052	(− 2)9.47425 2220	(0)4.15753 5935
4	(− 3)1.10723 6461	(− 2)2.02572 6087	(0)1.89577 5037
5	(− 5)9.99623 7520	(− 3)3.58484 8301	(− 1)7.45140 8690
6	(− 6)7.65033 3778	(− 4)5.40595 2086	(− 1)2.56465 1251
7	(− 7)5.08036 0873	(− 5)7.09794 4523	(− 2)7.83315 4364
8	(− 8)2.97924 6909	(− 6)8.24936 9394	(− 2)2.14704 9422
9	(− 9)1.56411 2692	(− 7)8.59805 3854	(− 3)5.33186 3294
10	(− 11)7.43279 3549	(− 8)8.12182 3211	(− 3)1.20941 3702
11	(− 12)3.22604 7141	(− 9)7.01394 8275	(− 4)2.52325 7454
12	(− 13)1.28851 2381	(− 10)5.57826 9483	(− 5)4.87152 7330
13	(− 15)4.76618 7751	(− 11)4.11114 2138	(− 6)8.74937 8858
14	(− 16)1.64168 8672	(− 12)2.82275 9636	(− 6)1.46862 7470
15	(− 18)5.29060 2725	(− 13)1.81406 6530	(− 7)2.31339 5316
16	(− 19)1.60182 7153	(− 14)1.09565 1449	(− 8)3.43223 7424
17	(− 21)4.57312 0086	(− 16)6.24163 9390	(− 9)4.81186 1587
18	(− 22)1.23512 2995	(− 17)3.36455 5792	(− 10)6.39343 1309
19	(− 24)3.16500 3796	(− 18)1.72111 7468	(− 11)8.07224 1852
20	(− 26)7.71514 7565	(− 20)8.37672 8478	(− 12)9.70826 6441
30	(− 43)5.65589 8686	(− 34)6.21921 4440	(− 22)6.36889 3001
40	(− 61)1.55685 5122	(− 49)1.74298 6176	(− 33)1.63577 1994
50	(− 81)3.65054 5412	(− 66)4.17042 9214	(− 46)3.64245 9664
100	(−190)7.48149 1755	(−160)9.55425 1030	(−120)6.26113 6933

n	$x=10$	$x=50$	$x=100$
0	(3)1.10132 3287	(19)5.18470 5529	(41)1.34405 8571
1	(2)9.91190 9633	(19)5.08101 1418	(41)1.33061 7985
2	(2)8.03965 9985	(19)4.87984 4844	(41)1.30414 0031
3	(2)5.89207 9640	(19)4.59302 6934	(41)1.26541 0984
4	(2)3.91520 4237	(19)4.23682 1073	(41)1.21556 1262
5	(2)2.36839 5827	(19)3.83039 9141	(41)1.15601 0470
6	(2)1.30996 8827	(19)3.39413 3262	(41)1.08840 0111
7	(1)6.65436 3519	(19)2.94792 4492	(41)1.01451 8456
8	(1)3.11814 2991	(19)2.50975 5914	(40)9.36222 3425
9	(1)1.35352 0435	(19)2.09460 7482	(40)8.55360 6574
10	(0)5.46454 1653	(19)1.71380 5071	(40)7.73703 8176
11	(0)2.05966 6874	(19)1.37480 9352	(40)6.92882 8557
12	(− 1)7.27307 8439	(19)1.08139 2769	(40)6.14340 7607
13	(− 1)2.41397 2641	(18)8.34112 9672	(40)5.39297 6655
14	(− 2)7.55352 3093	(18)6.30971 7670	(40)4.68730 3911
15	(− 2)2.23450 9437	(18)4.68149 3423	(40)4.03365 8521
16	(− 3)6.26543 8379	(18)3.40719 1747	(40)3.43686 9769
17	(− 3)1.66914 7720	(18)2.43274 6870	(40)2.89949 1497
18	(− 4)4.23421 3574	(18)1.70426 8938	(40)2.42204 7745
19	(− 4)1.02488 6979	(18)1.17158 7856	(40)2.00333 3832
20	(− 5)2.37154 3577	(17)7.90430 4104	(40)1.64074 7551
30	(− 12)1.22928 4325	(15)5.67659 3929	(39)1.30147 2327
40	(− 21)2.81471 5830	(12)7.34905 8082	(37)3.95371 9716
50	(− 31)5.88991 6154	(+ 9)2.00489 8633	(35)4.74095 0959
100	(− 90)9.54463 8661	(− 17)2.34189 3740	(20)3.73598 8741

Table 10.10

MODIFIED SPHERICAL BESSEL FUNCTIONS—VARIOUS ORDERS

$$\sqrt{\tfrac{1}{2}\pi/x}\,K_{n+\frac{1}{2}}(x)$$

n	x=1	x=2	x=5
0	(− 1)5.77863 6749	(− 1)1.06292 0829	(− 3)2.11678 8479
1	(0)1.15572 7350	(− 1)1.59438 1243	(− 3)2.54014 6175
2	(0)4.04504 5724	(− 1)3.45449 2694	(− 3)3.64087 6184
3	(1)2.13809 5597	(0)1.02306 1298	(− 3)6.18102 2359
4	(2)1.53711 7375	(0)3.92616 3812	(− 2)1.22943 0749
5	(3)1.40478 6594	(1)1.86907 9845	(− 2)2.83107 7584
6	(4)1.56063 6427	(2)1.06725 5553	(− 2)7.45780 1433
7	(5)2.04287 5221	(2)7.12406 9079	(− 1)2.22213 6131
8	(6)3.07991 9195	(3)5.44977 7364	(− 1)7.41218 8536
9	(7)5.25629 1384	(4)4.70355 1451	(0)2.74235 7715
10	(9)1.00177 5282	(5)4.52287 1652	(1)1.11621 7817
11	(10)2.10898 4384	(6)4.79605 0749	(1)4.96235 0604
12	(11)4.86068 1836	(7)5.56068 7078	(2)2.39430 3059
13	(13)1.21727 9443	(8)6.99881 9354	(3)1.24677 5036
14	(14)3.29151 5179	(9)9.50401 2999	(3)6.97201 5499
15	(15)9.55756 6814	(11)1.38508 0704	(4)4.16844 6493
16	(17)2.96613 7227	(12)2.15637 9105	(5)2.65415 6981
17	(18)9.79781 0417	(13)3.57187 6330	(6)1.79342 8072
18	(20)3.43219 9783	(14)6.27234 7368	(7)1.28194 1220
19	(22)1.27089 3701	(16)1.16395 6139	(7)9.66570 7838
20	(23)4.95991 7633	(17)2.27598 6819	(8)7.66744 6235
30	(40)4.55045 5450	(31)2.06581 6824	(18)7.97979 3303
40	(59)1.24524 3351	(46)5.55624 8963	(30)2.35318 1718
50	(78)4.25947 0196	(63)1.86314 7755	(42)8.49795 8757
100	(87)1.04451 3645	(156)4.08894 4237	(116)2.49323 8041

n	x=10	x=50	x=100
0	(−6)7.13140 4291	(−24)6.05934 6353	(−46)5.84348 1679
1	(−6)7.84454 4720	(−24)6.18053 3280	(−46)5.90191 6495
2	(−6)9.48476 7707	(−24)6.43017 8350	(−46)6.02053 9173
3	(−5)1.25869 2857	(−24)6.82355 1115	(−46)6.20294 3454
4	(−5)1.82956 1771	(−24)7.38547 5506	(−46)6.45474 5215
5	(−5)2.90529 8451	(−24)8.15293 6706	(−46)6.78387 0523
6	(−5)5.02539 0067	(−24)9.17912 1581	(−46)7.20097 0973
7	(−5)9.43830 5538	(−23)1.05395 0832	(−46)7.71999 6750
8	(−4)1.91828 4837	(−23)1.23409 7408	(−46)8.35897 0485
9	(−4)4.20491 4777	(−23)1.47354 3950	(−46)9.14102 1732
10	(−4)9.90762 2914	(−23)1.79404 4109	(−45)1.00957 6461
11	(−3)2.50109 2290	(−23)2.22704 2476	(−45)1.12611 3230
12	(−3)6.74327 4558	(−23)2.81848 3648	(−45)1.26858 2504
13	(−2)1.93592 7868	(−23)3.63628 4300	(−45)1.44325 8856
14	(−2)5.90133 2701	(−23)4.78207 7170	(−45)1.65826 2396
15	(−1)1.90497 9270	(−23)6.40988 9058	(−45)1.92415 4951
16	(−1)6.49556 9007	(−23)8.75620 8386	(−45)2.25475 0430
17	(0)2.33403 5699	(−22)1.21889 8659	(−45)2.66822 2593
18	(0)8.81868 1848	(−22)1.72884 9900	(−45)3.18862 8338
19	(1)3.49631 5854	(−22)2.49824 7585	(−45)3.84801 5078
20	(2)1.45175 0001	(−22)3.67748 3017	(−45)4.68935 4218
30	(9)1.99043 6138	(−20)4.72460 0057	(−44)5.77221 5084
40	(17)6.68871 7408	(−17)3.32175 1557	(−42)1.84121 2999
50	(27)2.59020 6572	(−13)1.10246 0162	(−40)1.47876 1633
100	(85)8.14750 7624	(+12)5.97531 1344	(−25)1.48279 6529

AIRY FUNCTIONS

Table 10.11

x	$Ai(x)$	$Ai'(x)$	$Bi(x)$	$Bi'(x)$	x	$Ai(x)$	$Ai'(x)$	$Bi(x)$	$Bi'(x)$
0.00	0.35502 805	−0.25881 940	0.61492 663	0.44828 836	0.50	0.23169 361	−0.22491 053	0.85427 704	0.54457 256
0.01	0.35243 992	−0.25880 174	0.61940 962	0.44831 926	0.51	0.22945 031	−0.22374 617	0.85974 431	0.54890 049
0.02	0.34985 214	−0.25874 909	0.62389 322	0.44841 254	0.52	0.22721 872	−0.22257 027	0.86525 543	0.55334 239
0.03	0.34726 505	−0.25866 197	0.62837 808	0.44856 911	0.53	0.22499 894	−0.22138 322	0.87081 154	0.55789 959
0.04	0.34467 901	−0.25854 090	0.63286 482	0.44878 987	0.54	0.22279 109	−0.22018 541	0.87641 381	0.56257 345
0.05	0.34209 435	−0.25838 640	0.63735 409	0.44907 570	0.55	0.22059 527	−0.21897 720	0.88206 341	0.56736 532
0.06	0.33951 139	−0.25819 898	0.64184 655	0.44942 752	0.56	0.21841 158	−0.21775 898	0.88776 152	0.57227 662
0.07	0.33693 047	−0.25797 916	0.64634 286	0.44984 622	0.57	0.21624 012	−0.21653 112	0.89350 934	0.57730 873
0.08	0.33435 191	−0.25772 745	0.65084 370	0.45033 270	0.58	0.21408 099	−0.21529 399	0.89930 810	0.58246 311
0.09	0.33177 603	−0.25744 437	0.65534 975	0.45088 787	0.59	0.21193 427	−0.21404 790	0.90515 902	0.58774 120
0.10	0.32920 313	−0.25713 042	0.65986 169	0.45151 263	0.60	0.20980 006	−0.21279 326	0.91106 334	0.59314 448
0.11	0.32663 352	−0.25678 613	0.66438 023	0.45220 789	0.61	0.20767 844	−0.21153 041	0.91702 233	0.59867 447
0.12	0.32406 751	−0.25641 200	0.66890 609	0.45297 457	0.62	0.20556 948	−0.21025 970	0.92303 726	0.60433 267
0.13	0.32150 538	−0.25600 854	0.67343 997	0.45381 357	0.63	0.20347 327	−0.20898 146	0.92910 941	0.61012 064
0.14	0.31894 743	−0.25557 625	0.67798 260	0.45472 582	0.64	0.20138 987	−0.20769 605	0.93524 011	0.61603 997
0.15	0.31639 395	−0.25511 565	0.68253 473	0.45571 223	0.65	0.19931 937	−0.20640 378	0.94143 066	0.62209 226
0.16	0.31384 521	−0.25462 724	0.68709 709	0.45677 373	0.66	0.19726 182	−0.20510 500	0.94768 241	0.62827 912
0.17	0.31130 150	−0.25411 151	0.69167 046	0.45791 125	0.67	0.19521 729	−0.20380 004	0.95399 670	0.63460 222
0.18	0.30876 307	−0.25356 898	0.69625 558	0.45912 572	0.68	0.19318 584	−0.20248 920	0.96037 491	0.64106 324
0.19	0.30623 020	−0.25300 013	0.70085 323	0.46041 808	0.69	0.19116 752	−0.20117 281	0.96681 843	0.64766 389
0.20	0.30370 315	−0.25240 547	0.70546 420	0.46178 928	0.70	0.18916 240	−0.19985 119	0.97332 866	0.65440 592
0.21	0.30118 218	−0.25178 548	0.71008 928	0.46324 026	0.71	0.18717 052	−0.19852 464	0.97990 703	0.66129 109
0.22	0.29866 753	−0.25114 067	0.71472 927	0.46477 197	0.72	0.18519 192	−0.19719 347	0.98655 496	0.66832 121
0.23	0.29615 945	−0.25047 151	0.71938 499	0.46638 539	0.73	0.18322 666	−0.19585 798	0.99327 394	0.67549 810
0.24	0.29365 818	−0.24977 850	0.72405 726	0.46808 147	0.74	0.18127 478	−0.19451 846	1.00006 542	0.68282 363
0.25	0.29116 395	−0.24906 211	0.72874 690	0.46986 119	0.75	0.17933 631	−0.19317 521	1.00693 091	0.69029 970
0.26	0.28867 701	−0.24832 284	0.73345 477	0.47172 554	0.76	0.17741 128	−0.19182 851	1.01387 192	0.69792 824
0.27	0.28619 757	−0.24756 115	0.73818 170	0.47367 549	0.77	0.17549 975	−0.19047 865	1.02088 999	0.70571 121
0.28	0.28372 586	−0.24677 753	0.74292 857	0.47571 205	0.78	0.17360 172	−0.18912 591	1.02798 667	0.71365 062
0.29	0.28126 209	−0.24597 244	0.74769 624	0.47783 623	0.79	0.17171 724	−0.18777 055	1.03516 353	0.72174 849
0.30	0.27880 648	−0.24514 636	0.75248 559	0.48004 903	0.80	0.16984 632	−0.18641 286	1.04242 217	0.73000 690
0.31	0.27635 923	−0.24429 976	0.75729 752	0.48235 148	0.81	0.16798 899	−0.18505 310	1.04976 421	0.73842 795
0.32	0.27392 055	−0.24343 309	0.76213 292	0.48474 462	0.82	0.16614 526	−0.18369 153	1.05719 128	0.74701 380
0.33	0.27149 064	−0.24254 682	0.76699 272	0.48722 948	0.83	0.16431 516	−0.18232 840	1.06470 504	0.75576 663
0.34	0.26906 968	−0.24164 140	0.77187 782	0.48980 713	0.84	0.16249 870	−0.18096 398	1.07230 717	0.76468 865
0.35	0.26665 787	−0.24071 730	0.77678 917	0.49247 861	0.85	0.16069 588	−0.17959 851	1.07999 939	0.77378 215
0.36	0.26425 540	−0.23977 495	0.78172 770	0.49524 501	0.86	0.15890 673	−0.17823 223	1.08778 340	0.78304 942
0.37	0.26186 243	−0.23881 481	0.78669 439	0.49810 741	0.87	0.15713 124	−0.17686 539	1.09566 096	0.79249 282
0.38	0.25947 916	−0.23783 731	0.79169 018	0.50106 692	0.88	0.15536 942	−0.17549 823	1.10363 385	0.80211 473
0.39	0.25710 574	−0.23684 291	0.79671 605	0.50412 463	0.89	0.15362 128	−0.17413 097	1.11170 386	0.81191 759
0.40	0.25474 235	−0.23583 203	0.80177 300	0.50728 168	0.90	0.15188 680	−0.17276 384	1.11987 281	0.82190 389
0.41	0.25238 916	−0.23480 512	0.80686 202	0.51053 920	0.91	0.15016 600	−0.17139 708	1.12814 255	0.83207 615
0.42	0.25004 630	−0.23376 259	0.81198 412	0.51389 833	0.92	0.14845 886	−0.17003 090	1.13651 496	0.84243 695
0.43	0.24771 395	−0.23270 487	0.81714 033	0.51736 025	0.93	0.14676 538	−0.16866 551	1.14499 193	0.85298 891
0.44	0.24539 226	−0.23163 239	0.82233 167	0.52092 614	0.94	0.14508 555	−0.16730 113	1.15357 539	0.86373 470
0.45	0.24308 135	−0.23054 556	0.82755 920	0.52459 717	0.95	0.14341 935	−0.16593 797	1.16226 728	0.87467 704
0.46	0.24078 139	−0.22944 479	0.83282 397	0.52837 457	0.96	0.14176 678	−0.16457 623	1.17106 959	0.88581 871
0.47	0.23849 250	−0.22833 050	0.83812 705	0.53225 956	0.97	0.14012 782	−0.16321 611	1.17998 433	0.89716 253
0.48	0.23621 482	−0.22720 310	0.84346 952	0.53625 338	0.98	0.13850 245	−0.16185 781	1.18901 352	0.90871 137
0.49	0.23394 848	−0.22606 297	0.84885 248	0.54035 729	0.99	0.13689 066	−0.16050 153	1.19815 925	0.92046 818
0.50	0.23169 361	−0.22491 053	0.85427 704	0.54457 256	1.00	0.13529 242	−0.15914 744	1.20742 359	0.93243 593
	$\begin{bmatrix}(-6)1\\4\end{bmatrix}$	$\begin{bmatrix}(-6)4\\4\end{bmatrix}$	$\begin{bmatrix}(-6)5\\4\end{bmatrix}$	$\begin{bmatrix}(-5)1\\4\end{bmatrix}$		$\begin{bmatrix}(-6)2\\4\end{bmatrix}$	$\begin{bmatrix}(-6)1\\4\end{bmatrix}$	$\begin{bmatrix}(-5)1\\4\end{bmatrix}$	$\begin{bmatrix}(-5)3\\4\end{bmatrix}$

AIRY FUNCTIONS—AUXILIARY FUNCTIONS FOR LARGE POSITIVE ARGUMENTS

ζ^{-1}	x	$f(-\zeta)$	$f(\zeta)$	$g(-\zeta)$	$g(\zeta)$	ζ^{-1}	x	$f(-\zeta)$	$f(\zeta)$	$g(-\zeta)$	$g(\zeta)$
1.5	1.000000	0.527027	0.619912	0.619954	0.478728	0.50	2.080084	0.548230	0.593015	0.587245	0.526011
1.4	1.047069	0.528783	0.620335	0.617156	0.479925	0.45	2.231443	0.549584	0.589451	0.585235	0.530678
1.3	1.100099	0.530601	0.620327	0.614275	0.481658	0.40	2.413723	0.550980	0.585855	0.583174	0.535345
1.2	1.160397	0.532488	0.619799	0.611305	0.484018	0.35	2.638450	0.552421	0.582330	0.581056	0.539902
1.1	1.229700	0.534448	0.618649	0.608239	0.487107	0.30	2.924018	0.553912	0.578985	0.578878	0.544235
1.0	1.310371	0.536489	0.616764	0.605068	0.491037	0.25	3.301927	0.555456	0.575908	0.576635	0.548255
0.9	1.405721	0.538618	0.614022	0.601782	0.495921	0.20	3.831547	0.557058	0.573135	0.574320	0.551930
0.8	1.520550	0.540844	0.610309	0.598372	0.501859	0.15	4.641589	0.558724	0.570636	0.571927	0.555296
0.7	1.662119	0.543180	0.605543	0.594823	0.508909	0.10	6.082202	0.560462	0.568343	0.569448	0.558428
0.6	1.842016	0.545636	0.599723	0.591120	0.517032	0.05	9.654894	0.562280	0.566204	0.566873	0.561382
0.5	2.080084	0.548230	0.593015	0.587245	0.526011	0.00	∞	0.564190	0.564190	0.564190	0.564190
	$\begin{bmatrix}(-3)7\\9\end{bmatrix}$	$\begin{bmatrix}(-5)2\\4\end{bmatrix}$	$\begin{bmatrix}(-4)1\\6\end{bmatrix}$	$\begin{bmatrix}(-5)2\\4\end{bmatrix}$	$\begin{bmatrix}(-4)1\\6\end{bmatrix}$			$\begin{bmatrix}(-5)1\\4\end{bmatrix}$	$\begin{bmatrix}(-5)4\\4\end{bmatrix}$	$\begin{bmatrix}(-5)1\\4\end{bmatrix}$	$\begin{bmatrix}(-5)4\\6\end{bmatrix}$

$$Ai(x)=\tfrac{1}{2}x^{-\frac{1}{4}}e^{-\zeta}f(-\zeta) \quad Bi(x)=x^{-\frac{1}{4}}e^{\zeta}f(\zeta) \quad Ai'(x)=-\tfrac{1}{2}x^{\frac{1}{4}}e^{-\zeta}g(-\zeta) \quad Bi'(x)=x^{\frac{1}{4}}e^{\zeta}g(\zeta) \qquad \zeta=\tfrac{2}{3}x^{\frac{3}{2}}$$

From J. C. P. Miller, The Airy integral, British Assoc. Adv. Sci. Mathematical Tables, Part–vol. B. Cambridge Univ. Press, Cambridge, England, 1946 (with permission).

Table 10.11   **AIRY FUNCTIONS**

x	$Ai(-x)$	$Ai'(-x)$	$Bi(-x)$	$Bi'(-x)$	x	$Ai(-x)$	$Ai'(-x)$	$Bi(-x)$	$Bi'(-x)$
0.00	0.35502 805	−0.25881 940	0.61492 663	0.44828 836	0.50	0.47572 809	−0.20408 167	0.38035 266	0.50593 371
0.01	0.35761 619	−0.25880 157	0.61044 364	0.44831 896	0.51	0.47775 692	−0.20167 409	0.37528 379	0.50784 166
0.02	0.36020 397	−0.25874 771	0.60596 005	0.44841 015	0.52	0.47976 138	−0.19920 846	0.37019 579	0.50976 123
0.03	0.36279 102	−0.25865 731	0.60147 524	0.44856 104	0.53	0.48174 089	−0.19668 449	0.36508 853	0.51169 132
0.04	0.36537 699	−0.25852 986	0.59698 863	0.44877 074	0.54	0.48369 487	−0.19410 192	0.35996 193	0.51363 080
0.05	0.36796 149	−0.25836 484	0.59249 963	0.44903 833	0.55	0.48562 274	−0.19146 050	0.35481 589	0.51557 853
0.06	0.37054 416	−0.25816 173	0.58800 767	0.44936 293	0.56	0.48752 389	−0.18875 999	0.34965 033	0.51753 339
0.07	0.37312 460	−0.25792 001	0.58351 218	0.44974 364	0.57	0.48939 774	−0.18600 016	0.34446 520	0.51949 424
0.08	0.37570 243	−0.25763 918	0.57901 261	0.45017 955	0.58	0.49124 369	−0.18318 078	0.33926 043	0.52145 991
0.09	0.37827 725	−0.25731 872	0.57450 841	0.45066 976	0.59	0.49306 115	−0.18030 166	0.33403 599	0.52342 927
0.10	0.38084 867	−0.25695 811	0.56999 904	0.45121 336	0.60	0.49484 953	−0.17736 260	0.32879 184	0.52540 115
0.11	0.38341 628	−0.25655 685	0.56548 397	0.45180 945	0.61	0.49660 821	−0.17436 341	0.32352 796	0.52737 438
0.12	0.38597 967	−0.25611 443	0.56096 268	0.45245 712	0.62	0.49833 659	−0.17130 392	0.31824 435	0.52934 780
0.13	0.38853 843	−0.25563 033	0.55643 466	0.45315 546	0.63	0.50003 408	−0.16818 399	0.31294 101	0.53132 022
0.14	0.39109 213	−0.25510 406	0.55189 940	0.45390 355	0.64	0.50170 007	−0.16500 345	0.30761 795	0.53329 046
0.15	0.39364 037	−0.25453 511	0.54735 642	0.45470 047	0.65	0.50333 395	−0.16176 218	0.30227 521	0.53525 733
0.16	0.39618 269	−0.25392 297	0.54280 523	0.45554 530	0.66	0.50493 511	−0.15846 007	0.29691 282	0.53721 964
0.17	0.39871 868	−0.25326 716	0.53824 536	0.45643 713	0.67	0.50650 295	−0.15509 701	0.29153 084	0.53917 618
0.18	0.40124 789	−0.25256 716	0.53367 634	0.45737 503	0.68	0.50803 685	−0.15167 290	0.28612 932	0.54112 575
0.19	0.40376 987	−0.25182 250	0.52909 771	0.45835 806	0.69	0.50953 620	−0.14818 768	0.28070 835	0.54306 714
0.20	0.40628 419	−0.25103 267	0.52450 903	0.45938 529	0.70	0.51100 040	−0.14464 129	0.27526 801	0.54499 912
0.21	0.40879 038	−0.25019 720	0.51990 986	0.46045 578	0.71	0.51242 882	−0.14103 366	0.26980 840	0.54692 048
0.22	0.41128 798	−0.24931 559	0.51529 977	0.46156 860	0.72	0.51382 087	−0.13736 479	0.26432 964	0.54883 000
0.23	0.41377 653	−0.24838 737	0.51067 835	0.46272 279	0.73	0.51517 591	−0.13363 464	0.25883 185	0.55072 642
0.24	0.41625 557	−0.24741 206	0.50604 518	0.46391 740	0.74	0.51649 336	−0.12984 322	0.25331 516	0.55260 852
0.25	0.41872 461	−0.24638 919	0.50139 987	0.46515 148	0.75	0.51777 258	−0.12599 055	0.24777 973	0.55447 506
0.26	0.42118 319	−0.24531 828	0.49674 203	0.46642 408	0.76	0.51901 296	−0.12207 665	0.24222 571	0.55632 480
0.27	0.42363 082	−0.24419 888	0.49207 127	0.46773 423	0.77	0.52021 390	−0.11810 157	0.23665 329	0.55815 647
0.28	0.42606 701	−0.24303 053	0.48738 722	0.46908 095	0.78	0.52137 479	−0.11406 538	0.23106 265	0.55996 884
0.29	0.42849 126	−0.24181 276	0.48268 953	0.47046 327	0.79	0.52249 501	−0.10996 815	0.22545 398	0.56176 063
0.30	0.43090 310	−0.24054 513	0.47797 784	0.47188 022	0.80	0.52357 395	−0.10580 999	0.21982 751	0.56353 059
0.31	0.43330 200	−0.23922 719	0.47325 181	0.47333 081	0.81	0.52461 101	−0.10159 101	0.21418 345	0.56527 745
0.32	0.43568 747	−0.23785 851	0.46851 112	0.47481 405	0.82	0.52560 557	−0.09731 134	0.20852 204	0.56699 994
0.33	0.43805 900	−0.23643 865	0.46375 543	0.47632 895	0.83	0.52655 703	−0.09297 113	0.20284 354	0.56869 679
0.34	0.44041 607	−0.23496 718	0.45898 443	0.47787 450	0.84	0.52746 479	−0.08857 055	0.19714 820	0.57036 671
0.35	0.44275 817	−0.23344 368	0.45419 784	0.47944 970	0.85	0.52832 824	−0.08410 979	0.19143 630	0.57200 845
0.36	0.44508 477	−0.23186 773	0.44939 534	0.48105 354	0.86	0.52914 678	−0.07958 904	0.18570 813	0.57362 071
0.37	0.44739 535	−0.23023 893	0.44457 667	0.48268 500	0.87	0.52991 982	−0.07500 854	0.17996 399	0.57520 220
0.38	0.44968 937	−0.22855 687	0.43974 156	0.48434 307	0.88	0.53064 676	−0.07036 852	0.17420 419	0.57675 165
0.39	0.45196 631	−0.22682 116	0.43488 973	0.48602 670	0.89	0.53132 700	−0.06566 925	0.16842 906	0.57826 777
0.40	0.45422 561	−0.22503 141	0.43002 094	0.48773 486	0.90	0.53195 995	−0.06091 100	0.16263 895	0.57974 926
0.41	0.45646 675	−0.22318 723	0.42513 495	0.48946 652	0.91	0.53254 502	−0.05609 407	0.15683 420	0.58119 484
0.42	0.45868 918	−0.22128 826	0.42023 153	0.49122 062	0.92	0.53308 163	−0.05121 879	0.15101 518	0.58260 321
0.43	0.46089 233	−0.21933 412	0.41531 047	0.49299 611	0.93	0.53356 920	−0.04628 549	0.14518 226	0.58397 309
0.44	0.46307 567	−0.21732 447	0.41037 154	0.49479 193	0.94	0.53400 715	−0.04129 452	0.13933 585	0.58530 317
0.45	0.46523 864	−0.21525 894	0.40541 457	0.49660 702	0.95	0.53439 490	−0.03624 628	0.13347 634	0.58659 217
0.46	0.46738 066	−0.21313 721	0.40043 934	0.49844 031	0.96	0.53473 189	−0.03114 116	0.12760 415	0.58783 879
0.47	0.46950 119	−0.21095 893	0.39544 570	0.50029 070	0.97	0.53501 754	−0.02597 957	0.12171 971	0.58904 174
0.48	0.47159 965	−0.20872 379	0.39043 348	0.50215 713	0.98	0.53525 129	−0.02076 197	0.11582 346	0.59019 973
0.49	0.47367 548	−0.20643 147	0.38540 251	0.50403 850	0.99	0.53543 259	−0.01548 880	0.10991 587	0.59131 145
0.50	0.47572 809	−0.20408 167	0.38035 266	0.50593 371	1.00	0.53556 088	−0.01016 057	0.10399 739	0.59237 563
	$\begin{bmatrix}(-6)3\\4\end{bmatrix}$	$\begin{bmatrix}(-6)7\\4\end{bmatrix}$	$\begin{bmatrix}(-6)2\\4\end{bmatrix}$	$\begin{bmatrix}(-6)8\\4\end{bmatrix}$		$\begin{bmatrix}(-6)7\\4\end{bmatrix}$	$\begin{bmatrix}(-6)8\\4\end{bmatrix}$	$\begin{bmatrix}(-6)2\\4\end{bmatrix}$	$\begin{bmatrix}(-6)6\\4\end{bmatrix}$

AIRY FUNCTIONS

Table 10.11

x	Ai(−x)	Ai′(−x)	Bi(−x)	Bi′(−x)	x	Ai(−x)	Ai′(−x)	Bi(−x)	Bi′(−x)
1.0	0.53556 088	−0.01016 057	+0.10399 739	0.59237 563	5.5	+0.01778 154	0.86419 722	−0.36781 345	+0.02511 158
1.1	0.53381 051	+0.04602 915	+0.04432 659	0.60011 970	5.6	−0.06833 070	0.85003 256	−0.36017 223	−0.17783 760
1.2	0.52619 437	0.10703 157	−0.01582 137	0.60171 016	5.7	−0.15062 016	0.78781 722	−0.33245 825	−0.37440 903
1.3	0.51227 201	0.17199 181	−0.07576 964	0.59592 975	5.8	−0.22435 192	0.67943 152	−0.28589 021	−0.55300 203
1.4	0.49170 018	0.23981 912	−0.13472 406	0.58165 624	5.9	−0.28512 278	0.52962 857	−0.22282 969	−0.70247 952
1.5	0.46425 658	0.30918 697	−0.19178 486	0.55790 810	6.0	−0.32914 517	0.34593 549	−0.14669 838	−0.81289 879
1.6	0.42986 298	0.37854 219	−0.24596 320	0.52389 354	6.1	−0.35351 168	+0.13836 394	−0.06182 255	−0.87622 530
1.7	0.38860 704	0.44612 455	−0.29620 266	0.47906 134	6.2	−0.35642 107	−0.08106 856	+0.02679 081	−0.88697 896
1.8	0.34076 156	0.50999 763	−0.34140 583	0.42315 137	6.3	−0.33734 765	−0.29899 161	0.11373 701	−0.84276 110
1.9	0.28680 006	0.56809 172	−0.38046 588	0.35624 251	6.4	−0.29713 762	−0.50147 985	0.19354 136	−0.74461 387
2.0	0.22740 743	0.61825 902	−0.41230 259	0.27879 517	6.5	−0.23802 030	−0.67495 249	0.26101 266	−0.59717 067
2.1	0.16348 451	0.65834 069	−0.43590 235	0.19168 563	6.6	−0.16352 646	−0.80711 925	0.31159 995	−0.40856 734
2.2	0.09614 538	0.68624 482	−0.45036 098	+0.09622 919	6.7	−0.07831 247	−0.88790 797	0.34172 774	−0.19009 878
2.3	+0.02670 633	0.70003 366	−0.45492 823	−0.00581 106	6.8	+0.01210 452	−0.91030 401	0.34908 418	+0.04437 678
2.4	−0.04333 414	0.69801 760	−0.44905 228	−0.11223 237	6.9	0.10168 800	−0.87103 106	0.33283 784	0.27926 391
2.5	−0.11232 507	0.67885 273	−0.43242 247	−0.22042 015	7.0	0.18428 084	−0.77100 817	0.29376 207	0.49824 459
2.6	−0.17850 243	0.64163 799	−0.40500 828	−0.32739 717	7.1	0.25403 633	−0.61552 879	0.23425 088	0.68542 058
2.7	−0.24003 811	0.58600 720	−0.36709 211	−0.42989 534	7.2	0.30585 152	−0.41412 428	0.15821 739	0.82650 634
2.8	−0.29509 759	0.51221 098	−0.31929 389	−0.52445 040	7.3	0.33577 037	−0.18009 580	+0.07087 411	0.90998 427
2.9	−0.34190 510	0.42118 281	−0.26258 500	−0.60751 829	7.4	0.34132 375	+0.07027 632	−0.02159 652	0.92812 809
3.0	−0.37881 429	0.31458 377	−0.19828 963	−0.67561 122	7.5	0.32177 572	0.31880 951	−0.11246 349	0.87780 228
3.1	−0.40438 222	0.19482 045	−0.12807 165	−0.72544 957	7.6	0.27825 023	0.54671 882	−0.19493 376	0.76095 509
3.2	−0.41744 342	+0.06503 115	−0.05390 576	−0.75412 455	7.7	0.21372 037	0.73605 242	−0.26267 007	0.58474 045
3.3	−0.41718 094	−0.07096 362	+0.02196 800	−0.75926 518	7.8	0.13285 154	0.87115 540	−0.31030 057	0.36122 930
3.4	−0.40319 048	−0.20874 905	0.09710 619	−0.73920 163	7.9	+0.04170 188	0.94004 300	−0.33387 856	+0.10670 215
3.5	−0.37553 382	−0.34344 343	0.16893 984	−0.69311 628	8.0	−0.05270 505	0.93556 094	−0.33125 158	−0.15945 050
3.6	−0.33477 748	−0.46986 397	0.23486 631	−0.62117 283	8.1	−0.14290 815	0.85621 859	−0.30230 331	−0.41615 664
3.7	−0.28201 306	−0.58272 780	0.29235 261	−0.52461 361	8.2	−0.22159 945	0.70659 870	−0.24904 019	−0.64232 293
3.8	−0.21885 598	−0.67688 257	0.33904 647	−0.40581 592	8.3	−0.28223 176	0.49727 619	−0.17550 556	−0.81860 044
3.9	−0.14741 991	−0.74755 809	0.37289 058	−0.26829 836	8.4	−0.31959 219	+0.24422 089	−0.08751 798	−0.92910 958
4.0	−0.07026 553	−0.79062 858	0.39223 471	−0.11667 057	8.5	−0.33029 024	−0.03231 335	+0.00775 444	−0.96296 917
4.1	+0.00967 698	−0.80287 254	0.39593 974	+0.04347 872	8.6	−0.31311 245	−0.30933 027	0.10235 647	−0.91547 918
4.2	0.08921 076	−0.78221 561	0.38346 736	0.20575 691	8.7	−0.26920 454	−0.56297 685	0.18820 363	−0.78882 623
4.3	0.16499 781	−0.72794 081	0.35494 906	0.36320 468	8.8	−0.20205 445	−0.77061 301	0.25778 240	−0.59221 371
4.4	0.23370 326	−0.64085 018	0.31122 860	0.50858 932	8.9	−0.11726 631	−0.91289 276	0.30483 241	−0.34136 475
4.5	0.29215 278	−0.52336 253	0.25387 266	0.63474 477	9.0	−0.02213 372	−0.97566 398	0.32494 732	−0.05740 051
4.6	0.33749 598	−0.37953 391	0.18514 576	0.73494 444	9.1	+0.07495 989	−0.95149 682	0.31603 471	+0.23484 379
4.7	0.36736 748	−0.21499 018	0.10794 695	0.80328 926	9.2	0.16526 800	−0.84067 107	0.27858 425	0.50894 402
4.8	0.38003 668	−0.03676 510	+0.02570 779	0.83508 976	9.3	0.24047 380	−0.65149 241	0.21570 835	0.73928 028
4.9	0.37453 635	+0.14695 743	−0.05774 655	0.82721 903	9.4	0.29347 756	−0.39986 237	0.13293 876	0.90348 537
5.0	0.35076 101	0.32719 282	−0.13836 913	0.77841 177	9.5	0.31910 325	−0.10809 532	+0.03778 543	0.98471 407
5.1	0.30952 600	0.49458 600	−0.21208 913	0.68948 513	9.6	0.31465 158	+0.19695 044	−0.06091 293	0.97349 918
5.2	0.25258 034	0.63990 517	−0.27502 704	0.56345 898	9.7	0.28023 750	0.48628 629	−0.15379 421	0.86898 388
5.3	0.18256 793	0.75457 542	−0.32371 608	0.40555 694	9.8	0.21886 743	0.73154 486	−0.23186 331	0.67936 774
5.4	0.10293 460	0.83122 307	−0.35531 708	0.22307 496	9.9	0.13623 503	0.90781 333	−0.28738 356	0.42147 209
5.5	0.01778 154	0.86419 722	−0.36781 345	0.02511 158	10.0	0.04024 124	0.99626 504	−0.31467 983	0.11941 411

$$\begin{bmatrix}(-3)2\\8\end{bmatrix} \quad \begin{bmatrix}(-3)5\\8\end{bmatrix} \quad \begin{bmatrix}(-3)2\\8\end{bmatrix} \quad \begin{bmatrix}(-3)5\\9\end{bmatrix} \quad \begin{bmatrix}(-3)4\\9\end{bmatrix} \quad \begin{bmatrix}(-2)1\\10\end{bmatrix} \quad \begin{bmatrix}(-3)4\\9\end{bmatrix} \quad \begin{bmatrix}(-2)1\\10\end{bmatrix}$$

AIRY FUNCTIONS—AUXILIARY FUNCTIONS FOR LARGE NEGATIVE ARGUMENTS

ζ^{-1}	x	$f_1(\zeta)$	$f_2(\zeta)$	$g_1(\zeta)$	$g_2(\zeta)$	$\langle\zeta\rangle$
0.05	9.654894	0.39752 21	0.40028 87	0.40092 31	0.39704 87	20
0.04	11.203512	0.39781 14	0.40002 58	0.40052 06	0.39741 99	25
0.03	13.572088	0.39809 83	0.39975 97	0.40012 11	0.39779 49	33
0.02	17.784467	0.39838 24	0.39949 03	0.39972 48	0.39817 37	50
0.01	28.231081	0.39866 38	0.39921 79	0.39933 19	0.39855 62	100
0.00	∞	0.39894 23	0.39894 23	0.39894 23	0.39894 23	∞

$$\begin{bmatrix}(-7)4\\3\end{bmatrix} \quad \begin{bmatrix}(-7)4\\3\end{bmatrix} \quad \begin{bmatrix}(-7)4\\3\end{bmatrix} \quad \begin{bmatrix}(-7)5\\3\end{bmatrix}$$

$$\mathrm{Ai}(-x)=x^{-\frac{1}{4}}\left[f_1(\zeta)\cos\zeta+f_2(\zeta)\sin\zeta\right] \qquad \mathrm{Bi}(-x)=x^{-\frac{1}{4}}\left[f_2(\zeta)\cos\zeta-f_1(\zeta)\sin\zeta\right]$$

$$\mathrm{Ai}'(-x)=x^{\frac{1}{4}}\left[g_1(\zeta)\sin\zeta-g_2(\zeta)\cos\zeta\right] \qquad \mathrm{Bi}'(-x)=x^{\frac{1}{4}}\left[g_1(\zeta)\cos\zeta+g_2(\zeta)\sin\zeta\right]$$

$$\zeta=\frac{2}{3}x^{\frac{3}{2}} \qquad \langle\zeta\rangle=\text{nearest integer to } \zeta.$$

Table 10.12   INTEGRALS OF AIRY FUNCTIONS

x	$\int_0^x Ai(t)\,dt$	$\int_0^x Ai(-t)\,dt$	$\int_0^x Bi(t)\,dt$	$\int_0^x Bi(-t)\,dt$	x	$\int_0^x Ai(t)\,dt$	$\int_0^x Ai(-t)\,dt$	$\int_0^x Bi(-t)\,dt$
0.0	0.00000 00	0.00000 00	0.00000 00	0.00000 00	5.0	0.33328 76	-0.71788 22	-0.15873 09
0.1	0.03421 01	-0.03679 54	0.06373 67	-0.05924 87	5.1	0.33329 73	-0.75103 62	-0.14113 39
0.2	0.06585 15	-0.07615 70	0.13199 45	-0.11398 10	5.2	0.33330 50	-0.77926 27	-0.11667 30
0.3	0.09497 09	-0.11802 51	0.20487 68	-0.16411 57	5.3	0.33331 11	-0.80111 58	-0.08660 41
0.4	0.12164 06	-0.16229 44	0.28256 70	-0.20952 89	5.4	0.33331 59	-0.81545 49	-0.05250 03
0.5	0.14595 33	-0.20880 95	0.36533 85	-0.25006 28	5.5	0.33331 97	-0.82151 82	-0.01617 86
0.6	0.16801 79	-0.25736 07	0.45356 50	-0.28553 62	5.6	0.33332 27	-0.81897 90	+0.02038 99
0.7	0.18795 52	-0.30768 05	0.54773 36	-0.31575 56	5.7	0.33332 50	-0.80797 96	0.05518 54
0.8	0.20589 45	-0.35944 15	0.64845 82	-0.34052 58	5.8	0.33332 69	-0.78914 06	0.08625 18
0.9	0.22196 97	-0.41225 56	0.75649 64	-0.35966 27	5.9	0.33332 83	-0.76354 19	0.11181 25
1.0	0.23631 73	-0.46567 40	0.87276 91	-0.37300 50	6.0	0.33332 95	-0.73267 53	0.13038 11
1.1	0.24907 33	-0.51918 94	0.99838 41	-0.38042 77	6.1	0.33333 03	-0.69836 93	0.14086 00
1.2	0.26037 12	-0.57224 05	1.13466 38	-0.38185 43	6.2	0.33333 10	-0.66268 96	0.14262 05
1.3	0.27034 09	-0.62421 79	1.28318 00	-0.37726 99	6.3	0.33333 16	-0.62781 93	0.13555 73
1.4	0.27910 66	-0.67447 31	1.44579 42	-0.36673 34	6.4	0.33333 20	-0.59592 62	0.12011 15
1.5	0.28678 67	-0.72232 88	1.62470 81	-0.35038 81	6.5	0.33333 23	-0.56902 35	0.09726 08
1.6	0.29349 24	-0.76709 26	1.82252 33	-0.32847 24	6.6	0.33333 25	-0.54883 59	0.06847 29
1.7	0.29932 75	-0.80807 24	2.04231 52	-0.30132 67	6.7	0.33333 27	-0.53667 65	0.03562 42
1.8	0.30438 82	-0.84459 41	2.28772 12	-0.26939 97	6.8	0.33333 29	-0.53334 74	+0.00088 80
1.9	0.30876 29	-0.87602 06	2.56304 90	-0.23325 04	6.9	0.33333 30	-0.53906 98	-0.03340 40
2.0	0.31253 28	-0.90177 28	2.87340 83	-0.19354 74	7.0	0.33333 31	-0.55345 17	-0.06491 67
2.1	0.31577 11	-0.92135 09	.	-0.15106 46	7.1	0.33333 31	-0.57549 72	-0.09147 36
2.2	0.31854 43	-0.93435 56	.	-0.10667 18	7.2	0.33333 32	-0.60365 96	-0.11121 47
2.3	0.32091 19	-0.94050 97	.	-0.06132 23	7.3	0.33333 32	-0.63593 60	-0.12273 90
2.4	0.32292 74	-0.93967 67	.	-0.01603 45	7.4	0.33333 33	-0.66999 96	-0.12521 80
2.5	0.32463 80	-0.93187 78	.	+0.02812 94	7.5	0.33333 33	-0.70336 19	-0.11847 31
2.6	0.32608 57	-0.91730 54	.	0.07009 01	7.6	.	-0.73355 34	-0.10300 57
2.7	0.32730 74	-0.89633 20	.	0.10878 06	7.7	.	-0.75830 99	-0.07997 85
2.8	0.32833 55	-0.86951 37	.	0.14317 88	7.8	.	-0.77575 13	-0.05114 35
2.9	0.32919 83	-0.83758 77	.	0.17234 20	7.9	.	-0.78453 65	-0.01872 22
3.0	0.32992 04	-0.80146 29	.	0.19544 25	8.0	.	-0.78398 26	+0.01475 64
3.1	0.33052 31	-0.76220 32	.	0.21180 21	8.1	.	-0.77413 57	0.04664 84
3.2	0.33102 49	-0.72100 37	.	0.22092 49	8.2	.	-0.75578 55	0.07440 43
3.3	0.33144 15	-0.67915 91	.	0.22252 61	8.3	.	-0.73041 93	0.09577 87
3.4	0.33178 65	-0.63802 56	.	0.21655 57	8.4	.	-0.70011 70	0.10902 22
3.5	0.33207 15	-0.59897 71	.	0.20321 50	8.5	.	-0.66739 21	0.11303 86
3.6	0.33230 63	-0.56335 61	.	0.18296 47	8.6	.	-0.63499 08	0.10749 35
3.7	0.33249 93	-0.53242 25	.	0.15652 33	8.7	.	-0.60566 32	0.09285 98
3.8	0.33265 76	-0.50730 05	.	0.12485 43	8.8	.	-0.58192 70	0.07039 64
3.9	0.33278 70	-0.48892 77	.	0.08914 28	8.9	.	-0.56584 22	0.04205 63
4.0	0.33289 27	-0.47800 75	.	+0.05076 01	9.0	.	-0.55881 97	+0.01033 04
4.1	0.33297 86	-0.47496 79	.	+0.01121 78	9.1	.	-0.56148 12	-0.02196 26
4.2	0.33304 84	-0.47992 95	.	-0.02788 79	9.2	.	-0.57358 51	-0.05192 24
4.3	0.33310 50	-0.49268 51	.	-0.06494 00	9.3	.	-0.59403 00	-0.07682 93
4.4	0.33315 07	-0.51269 28	.	-0.09837 02	9.4	.	-0.62093 76	-0.09439 87
4.5	0.33318 76	-0.53908 35	.	-0.12673 04	9.5	.	-0.65181 01	-0.10300 27
4.6	0.33321 73	-0.57068 59	.	-0.14876 50	9.6	.	-0.68375 25	-0.10183 70
4.7	0.33324 11	-0.60606 63	.	-0.16347 66	9.7	.	-0.71373 85	-0.09101 44
4.8	0.33326 02	-0.64358 51	.	-0.17018 59	9.8	.	-0.73889 84	-0.07157 33
4.9	0.33327 54	-0.68146 70	.	-0.16857 74	9.9	.	-0.75680 07	-0.04539 57
5.0	0.33328 76	-0.71788 22	.	-0.15873 09	10.0	.	-0.76569 84	-0.01504 04
	$\begin{bmatrix}(-4)3\\5\end{bmatrix}$	$\begin{bmatrix}(-3)1\\7\end{bmatrix}$	$\begin{bmatrix}(-3)4\\7\end{bmatrix}$	$\begin{bmatrix}(-3)1\\6\end{bmatrix}$		$\begin{bmatrix}(-7)3\\3\end{bmatrix}$	$\begin{bmatrix}(-3)1\\7\end{bmatrix}$	$\begin{bmatrix}(-3)1\\7\end{bmatrix}$

Table 10.13   ZEROS AND ASSOCIATED VALUES OF AIRY FUNCTIONS
                         AND THEIR DERIVATIVES

s	a_s	$Ai'(a_s)$	a'_s	$Ai(a'_s)$	b_s	$Bi'(b_s)$	b'_s	$Bi(b'_s)$
1	-2.33810 741	+0.70121 082	-1.01879 297	+0.53565 666	-1.17371 322	+0.60195 789	-2.29443 968	-0.45494 438
2	-4.08794 944	-0.80311 137	-3.24819 758	-0.41901 548	-3.27109 330	-0.76031 014	-4.07315 509	+0.39652 284
3	-5.52055 983	+0.86520 403	-4.82009 921	+0.38040 647	-4.83073 784	+0.83699 101	-5.51239 573	-0.36796 916
4	-6.78670 809	-0.91085 074	-6.16330 736	-0.35790 794	-6.16985 213	-0.88947 990	-6.78129 445	+0.34949 912
5	-7.94413 359	+0.94733 571	-7.37217 726	+0.34230 124	-7.37676 208	+0.92998 364	-7.94017 869	-0.33602 624
6	-9.02265 085	-0.97792 281	-8.48848 673	-0.33047 623	-8.49194 885	-0.96323 443	-9.01958 336	+0.32550 974
7	-10.04017 434	+1.00437 012	-9.53544 905	+0.32102 229	-9.53819 438	+0.99158 637	-10.03769 633	-0.31693 465
8	-11.00852 430	-1.02773 869	-10.52766 040	-0.31318 539	-10.52991 351	-1.01638 966	-11.00646 267	+0.30972 594
9	-11.93601 556	+1.04872 065	-11.47505 663	+0.30651 729	-11.47695 355	+1.03849 429	-11.93426 165	-0.30352 766
10	-12.82877 675	-1.06779 386	-12.38478 837	-0.30073 083	-12.38641 714	-1.05847 184	-12.82725 831	+0.29810 491

AUXILIARY TABLE—COMPLEX ZEROS AND ASSOCIATED VALUES OF
$Bi(z)$ AND $Bi'(z)$

s	$e^{-\pi i/3}\beta_s$ Modulus	Phase	$Bi'(\beta_s)$ Modulus	Phase	$e^{-\pi i/3}\beta'_s$ Modulus	Phase	$Bi(\beta'_s)$ Modulus	Phase
1	2.354	0.095	0.993	+2.641	1.121	0.331	0.750	+0.466
2	4.093	0.042	1.136	-0.513	3.257	0.059	0.592	-2.632
3	5.524	0.027	1.224	+2.625	4.824	0.033	0.538	+0.515
4	6.789	0.020	1.288	-0.519	6.166	0.023	0.506	-2.624
5	7.946	0.015	1.340	+2.622	7.374	0.017	0.484	+0.519

From J. C. P. Miller, The Airy integral, British Assoc. Adv. Sci. Mathematical
Tables Part–vol. B. Cambridge Univ. Press, Cambridge, England, 1946 and
F. W. J. Olver, The asymptotic expansion of Bessel functions of large order.
Philos. Trans. Roy. Soc. London [A] 247, 328–368, 1954 (with permission).

11. Integrals of Bessel Functions

Yudell L. Luke [1]

Contents

$$\left.\begin{array}{l} \displaystyle\int_0^x J_0(t)dt, \int_0^x Y_0(t)dt, 10D \\[2ex] \displaystyle e^{-x}\int_0^x I_0(t)dt,\ e^x\int_x^\infty K_0(t)dt, 7D \end{array}\right\} x=0(.1)10$$

$$\left.\begin{array}{l} \displaystyle\int_0^x \frac{[1-J_0(t)]dt}{t}, \int_x^\infty \frac{Y_0(t)dt}{t}, 8D \\[2ex] \displaystyle e^{-x}\int_0^x \frac{[I_0(t)-1]dt}{t}, 8D;\ xe^x\int_x^\infty \frac{K_0(t)dt}{t}, 6D \end{array}\right\} x=0(.1)5$$

The author acknowledges the assistance of Geraldine Coombs, Betty Kahn, Marilyn Kemp, Betty Ruhlman, and Anna Lee Samuels for checking formulas and developing numerical examples, only a portion of which could be accommodated here.

---

[1] Midwest Research Institute. (Prepared under contract with the National Bureau of Standards.)

11. Integrals of Bessel Functions

Mathematical Properties

11.1. Simple Integrals of Bessel Functions

$$\int_0^z t^\mu J_\nu(t)dt$$

11.1.1

$$\int_0^z t^\mu J_\nu(t)dt = \frac{z^\mu \Gamma\left(\frac{\nu+\mu+1}{2}\right)}{\Gamma\left(\frac{\nu-\mu+1}{2}\right)}$$

$$\times \sum_{k=0}^\infty \frac{(\nu+2k+1)\Gamma\left(\frac{\nu-\mu+1}{2}+k\right)}{\Gamma\left(\frac{\nu+\mu+3}{2}+k\right)} J_{\nu+2k+1}(z)$$

$$(\mathscr{R}(\mu+\nu+1)>0)$$

11.1.2

$$\int_0^z J_\nu(t)dt = 2\sum_{k=0}^\infty J_{\nu+2k+1}(z) \quad (\mathscr{R}\nu>-1)$$

11.1.3 $\quad \int_0^z J_{2n}(t)dt = \int_0^z J_0(t)dt - 2\sum_{k=0}^{n-1} J_{2k+1}(z)$

11.1.4 $\quad \int_0^z J_{2n+1}(t)dt = 1 - J_0(z) - 2\sum_{k=1}^n J_{2k}(z)$

Recurrence Relations

11.1.5

$$\int_0^z J_{n+1}(t)dt = \int_0^z J_{n-1}(t)dt - 2J_n(z) \quad (n>0)$$

11.1.6 $\quad \int_0^z J_1(t)dt = 1 - J_0(z)$

$$\int J_0(t)dt, \int Y_0(t)dt, \int I_0(t)dt, \int K_0(t)dt$$

11.1.7

$$\int_0^z \mathscr{C}_0(t)dt = x\mathscr{C}_0(x) + \frac{1}{2}\pi x\{\mathbf{H}_0(x)\mathscr{C}_1(x) - \mathbf{H}_1(x)\mathscr{C}_0(x)\}$$

$$\mathscr{C}_\nu(x) = AJ_\nu(x) + BY_\nu(x), \nu=0,1$$

A and B are constants.

11.1.8

$$\int_0^z Z_0(t)dt = xZ_0(x) + \frac{1}{2}\pi x\{-\mathbf{L}_0(x)Z_1(x) + \mathbf{L}_1(x)Z_0(x)\}$$

$$Z_\nu(x) = AI_\nu(x) + Be^{i\nu\pi}K_\nu(x), \nu=0,1$$

A and B are constants.

$\mathbf{H}_\nu(x)$ and $\mathbf{L}_\nu(x)$ are Struve functions (see chapter **12**).

11.1.9

$$\int_0^z K_0(t)dt = -\left(\gamma+\ln\frac{x}{2}\right)x\sum_{k=0}^\infty \frac{(x/2)^{2k}}{(k!)^2(2k+1)}$$

$$+x\sum_{k=0}^\infty \frac{(x/2)^{2k}}{(k!)^2(2k+1)^2}$$

$$+x\sum_{k=1}^\infty \frac{(x/2)^{2k}}{(k!)^2(2k+1)}\left(1+\frac{1}{2}+\ldots+\frac{1}{k}\right)$$

γ (Euler's constant) $= .57721\ 56649\ \ldots$

In this and all other integrals of **11.1**, x is real and positive although all the results remain valid for extended portions of the complex plane unless stated to the contrary.

11.1.10

$$\int_0^{-ix} K_0(t)dt = \frac{\pi}{2}\int_0^x J_0(t)dt + i\frac{\pi}{2}\int_0^x Y_0(t)dt$$

Asymptotic Expansions

11.1.11

$$\int_x^\infty [J_0(t)+iY_0(t)]dt \sim \left(\frac{2}{\pi x}\right)^{\frac{1}{2}} e^{i(x-\pi/4)}$$

$$\times\left[\sum_{k=0}^\infty (-)^k a_{2k+1}x^{-2k-1} + i\sum_{k=0}^\infty (-)^k a_{2k}x^{-2k}\right]$$

11.1.12 $\quad a_k = \frac{\Gamma(k+\frac{1}{2})}{\Gamma(\frac{1}{2})}\sum_{s=0}^k \frac{\Gamma(s+\frac{1}{2})}{2^s s!\Gamma(\frac{1}{2})}$

11.1.13

$$2(k+1)a_{k+1} = 3\left(k+\frac{1}{2}\right)\left(k+\frac{5}{6}\right)a_k$$

$$-\left(k+\frac{1}{2}\right)^2\left(k-\frac{1}{2}\right)a_{k-1}$$

11.1.14 $\quad x^{\frac{1}{2}}e^{-x}\int_0^x I_0(t)dt \sim (2\pi)^{-\frac{1}{2}}\sum_{k=0}^{\infty} a_k x^{-k}$

where the a_k are defined as in **11.1.12**.

11.1.15 $\quad x^{\frac{1}{2}}e^x\int_x^{\infty} K_0(t)dt \sim \left(\frac{\pi}{2}\right)^{\frac{1}{2}}\sum_{k=0}^{\infty} (-)^k a_k x^{-k}$

where the a_k are defined as in **11.1.12**.

Polynomial Approximations [2]

11.1.16 $\qquad 8\leq x\leq\infty$

$$\int_x^{\infty} [J_0(t)+iY_0(t)]dt$$

$$= x^{-\frac{1}{2}}e^{i(x-\pi/4)}\left[\sum_{k=0}^{7} (-)^k a_k(x/8)^{-2k-1}\right.$$

$$\left.+i\sum_{k=0}^{7} (-)^k b_k(x/8)^{-2k}+\epsilon(x)\right]$$

$$|\epsilon(x)|\leq 2\times10^{-9}$$

k	a_k	b_k
0	. 06233 47304	. 79788 45600
1	. 00404 03539	. 01256 42405
2	. 00100 89872	. 00178 70944
3	. 00053 66169	. 00067 40148
4	. 00039 92825	. 00041 00676
5	. 00027 55037	. 00025 43955
6	. 00012 70039	. 00011 07299
7	. 00002 68482	. 00002 26238

11.1.17 $\qquad 8\leq x\leq\infty$

$$x^{\frac{1}{2}}e^{-x}\int_0^x I_0(t)dt = \sum_{k=0}^{6} d_k(x/8)^{-k}+\epsilon(x)$$

$$|\epsilon(x)|\leq 2\times10^{-6}$$

k	d_k
0	. 39894 23
1	. 03117 34
2	. 00591 91
3	. 00559 56
4	$-$. 01148 58
5	. 01774 40
6	$-$. 00739 95

[2] Approximation **11.1.16** is from A. J. M. Hitchcock. Polynomial approximations to Bessel functions of order zero and one and to related functions, Math. Tables Aids Comp. **11**, 86–88 (1957) (with permission).

11.1.18 $\qquad 7\leq x\leq\infty$

$$x^{\frac{1}{2}}e^x\int_x^{\infty} K_0(t)dt = \sum_{k=0}^{6} (-)^k e_k(x/7)^{-k}+\epsilon(x)$$

$$|\epsilon(x)|\leq 2\times10^{-7}$$

k	e_k
0	1. 25331 414
1	0. 11190 289
2	. 02576 646
3	. 00933 994
4	. 00417 454
5	. 00163 271
6	. 00033 934

$$\int\frac{J_0(t)dt}{t}, \int\frac{Y_0(t)dt}{t}, \int\frac{K_0(t)dt}{t}$$

11.1.19

$$\int_0^x \frac{1-J_0(t)}{t}\,dt$$

$$= 2x^{-1}\sum_{k=0}^{\infty} (2k+3)[\psi(k+2)-\psi(1)]\,J_{2k+3}(x)$$

$$= 1-2x^{-1}J_1(x)$$

$$+2x^{-1}\sum_{k=0}^{\infty} (2k+5)[\psi(k+3)-\psi(1)-1]J_{2k+5}(x)$$

For $\psi(z)$, see **6.3**.

11.1.20

$$\int_x^{\infty} \frac{J_0(t)dt}{t}+\gamma+\ln\frac{x}{2}=\int_0^x \frac{[1-J_0(t)]dt}{t}$$

$$= -\sum_{k=1}^{\infty} \frac{(-)^k\left(\frac{x}{2}\right)^{2k}}{2k(k!)^2}$$

11.1.21

$$\int_x^{\infty} \frac{Y_0(t)dt}{t}=-\frac{1}{\pi}\left(\ln\frac{x}{2}\right)^2-\frac{2\gamma}{\pi}\left(\ln\frac{x}{2}\right)+\frac{1}{\pi}\left(\frac{\pi^2}{6}-\gamma^2\right)$$

$$+\frac{2}{\pi}\sum_{k=1}^{\infty} \frac{(-)^k\left(\frac{x}{2}\right)^{2k}}{2k(k!)^2}\left\{\psi(k+1)+\frac{1}{2k}-\ln\frac{x}{2}\right\}$$

11.1.22

$$\int_x^{\infty} \frac{K_0(t)dt}{t}=\frac{1}{2}\left(\ln\frac{x}{2}\right)^2+\gamma\ln\frac{x}{2}+\frac{\pi^2}{24}+\frac{\gamma^2}{2}$$

$$-\sum_{k=1}^{\infty} \frac{\left(\frac{x}{2}\right)^{2k}}{2k(k!)^2}\left\{\psi(k+1)+\frac{1}{2k}-\ln\frac{x}{2}\right\}$$

11.1.23

$$\int_{-ix}^{-i\infty} \frac{K_0(t)dt}{t}=\frac{i\pi}{2}\int_x^{\infty} \frac{J_0(t)dt}{t}-\frac{\pi}{2}\int_x^{\infty} \frac{Y_0(t)dt}{t}$$

Asymptotic Expansions

11.1.24 $\displaystyle\int_x^\infty \frac{\mathscr{C}_0(t)dt}{t}=\frac{2g_1(x)\mathscr{C}_0(x)}{x^2}-\frac{g_0(x)\mathscr{C}_1(x)}{x}$

where

$$g_0(x)\sim\sum_{k=0}^\infty (-)^k\left(\frac{x}{2}\right)^{-2k}(k!)^2,$$

$$g_1(x)\sim\sum_{k=0}^\infty (-)^k\left(\frac{x}{2}\right)^{-2k}k!(k+1)!$$

11.1.25 $\displaystyle g_0(x)=2x^2\int_x^\infty \frac{g_1(t)dt}{t^3}$

11.1.26 $\displaystyle x^{3/2}e^x\int_x^\infty \frac{K_0(t)dt}{t}\sim\left(\frac{\pi}{2}\right)^{\frac{1}{2}}\sum_{k=0}^\infty (-)^k c_k x^{-k}$

where

11.1.27 $\displaystyle c_0=1, c_1=\frac{13}{8}$

$$2(k+1)c_{k+1}=\left[3(k+1)^2+\frac{1}{4}\right]c_k-\left(k+\frac{1}{2}\right)^3 c_{k-1}$$

11.1.28 $\displaystyle x^{3/2}e^{-x}\int_0^x \frac{[I_0(t)-1]dt}{t}\sim(2\pi)^{-\frac{1}{2}}\sum_{k=0}^\infty c_k x^{-k}$

where c_k is defined as in **11.1.27**.

Polynomial Approximations

11.1.29 $5\leq x\leq \infty$

$$\int_x^\infty \frac{\mathscr{C}_0(t)dt}{t}=\frac{2g_1(x)\mathscr{C}_0(x)}{x^2}-\frac{g_0(x)\mathscr{C}_1(x)}{x}$$

where

$$g_0(x)=\sum_{k=0}^9 (-)^k a_k(x/5)^{-2k}+\epsilon(x),$$

$$g_1(x)=\sum_{k=0}^9 (-)^k b_k(x/5)^{-2k}+\epsilon(x)$$

$$|\epsilon(x)|\leq 2\times 10^{-7}$$

k	a_k	b_k
0	1. 0	1. 0
1	0. 15999 2815	0. 31998 5629
2	. 10161 9385	. 30485 8155
3	. 13081 1585	. 52324 6341
4	. 20740 4022	1. 03702 0112
5	. 28330 0508	1. 69980 3050
6	. 27902 9488	1. 95320 6413
7	. 17891 5710	1. 43132 5684
8	. 06622 8328	0. 59605 4956
9	. 01070 2234	. 10702 2336

11.1.30 $4\leq x\leq \infty$

$$x^{\frac{3}{2}}e^x\int_x^\infty \frac{K_0(t)dt}{t}=\sum_{k=0}^6 (-)^k d_k\left(\frac{x}{4}\right)^{-k}+\epsilon(x)$$

$$|\epsilon(x)|\leq 6\times 10^{-6}$$

k	d_k
0	1. 25331 41
1	0. 50913 39
2	. 32191 84
3	. 26214 46
4	. 20601 26
5	. 11103 96
6	. 02724 00

11.1.31 $5\leq x\leq \infty$

$$x^{\frac{3}{2}}e^{-x}\int_0^x \frac{[I_0(t)-1]dt}{t}=\sum_{k=0}^{10} f_k\left(\frac{x}{5}\right)^{-k}+\epsilon(x)$$

$$|\epsilon(x)|\leq 1.1\times 10^{-5}$$

k	f_k
0	0. 39893 14
1	. 13320 55
2	−. 04938 43
3	1. 47800 44
4	−8. 65560 13
5	28. 12214 78
6	−48. 05241 15
7	40. 39473 40
8	−11. 90943 95
9	−3. 51950 09
10	2. 19454 64

11.2. Repeated Integrals of $J_n(z)$ and $K_0(z)$

Repeated Integrals of $J_n(z)$

Let

11.2.1

$$f_{0,n}(z)=J_n(z),$$

$$f_{1,n}(z)=\int_0^z J_n(t)dt, \ldots, f_{r,n}(z)=\int_0^z f_{r-1,n}(t)dt$$

11.2.2 $f_{-r,n}(z)=\dfrac{d^r}{dz^r}J_n(z)$

Then

11.2.3

$$f_{r,n}(z)=\frac{1}{\Gamma(r)}\int_0^z (z-t)^{r-1}J_n(t)dt \quad (\mathscr{R}r>0)$$

11.2.4 $f_{r,n}(z)=\dfrac{2^r}{\Gamma(r)}\sum_{k=0}^\infty \dfrac{\Gamma(k+r)}{k!}J_{n+r+2k}(z)$

11.2.5

$$r(r-1)f_{r+1,\,n}(z)=2(r-1)zf_{r,\,n}(z)$$
$$-[(1-r)^2-n^2+z^2]f_{r-1,\,n}(z)$$
$$+(2r-3)zf_{r-2,\,n}(z)-z^2f_{r-3,\,n}(z)$$

11.2.6

$$rf_{r+1,\,0}(z)=zf_{r,\,0}(z)-(r-1)f_{r-1,\,0}(z)+zf_{r-2,\,0}(z)$$

11.2.7 $\quad f_{r+1,\,n+1}(z)=f_{r+1,\,n-1}(z)-2f_{r,\,n}(z)$

Repeated Integrals of $K_0(z)$

Let

11.2.8

$$\mathrm{Ki}_0(z)=K_0(z),$$

$$\mathrm{Ki}_1(z)=\int_z^\infty K_0(t)dt,\ \ldots,\ \mathrm{Ki}_r(z)=\int_z^\infty \mathrm{Ki}_{r-1}(t)dt$$

11.2.9 $\quad \mathrm{Ki}_{-r}(z)=(-)^r\,\dfrac{d^r}{dz^r}\,K_0(z)$

Then

11.2.10

$$\mathrm{Ki}_r(z)=\int_0^\infty \frac{e^{-z\cosh t}\,dt}{\cosh^r t}$$

$$(\mathscr{R}z\geq 0,\ \mathscr{R}r>0,\ \mathscr{R}z>0,\ r=0)$$

11.2.11

$$\mathrm{Ki}_r(z)=\frac{1}{\Gamma(r)}\int_z^\infty (t-z)^{r-1}K_0(t)dt$$

$$(\mathscr{R}z\geq 0,\ \mathscr{R}r>0)$$

11.2.12 $\quad \mathrm{Ki}_{2r}(0)=\dfrac{\Gamma(r)\,\Gamma(\frac{1}{2})}{\Gamma(r+\frac{1}{2})}\qquad (\mathscr{R}r>0)$

11.2.13 $\quad \mathrm{Ki}_{2r+1}(0)=\dfrac{\frac{\pi}{2}\,\Gamma(r+\frac{1}{2})}{\Gamma(\frac{1}{2})\,\Gamma(r+1)}\qquad \left(\mathscr{R}r>-\dfrac{1}{2}\right)$

11.2.14

$$r\mathrm{Ki}_{r+1}(z)=-z\mathrm{Ki}_r(z)+(r-1)\mathrm{Ki}_{r-1}(z)+z\mathrm{Ki}_{r-2}(z)$$

11.3. Reduction Formulas for Indefinite Integrals

Let

11.3.1 $\qquad g_{\mu,\nu}(z)=\int^z e^{-pt}t^\mu Z_\nu(t)dt$

where $Z_\nu(z)$ represents any of the Bessel functions of the first three kinds or the modified Bessel functions. The parameters a and b appearing in the reduction formulae are associated with the particular type of Bessel function as delineated in the following table.

11.3.2

$Z_\nu(z)$	a	b
$J_\nu(z),\ Y_\nu(z),\ H_\nu^{(1)}(z),\ H_\nu^{(2)}(z)$	1	1
$I_\nu(z)$	-1	1
$K_\nu(z)$	1	-1

11.3.3

$$pg_{\mu,\nu}(z)=-e^{-pz}z^\mu Z_\nu(z)$$
$$+(\mu+\nu)g_{\mu-1,\nu}(z)-ag_{\mu,\nu+1}(z)$$

11.3.4

$$pg_{\mu,\nu+1}(z)=-e^{-pz}z^\mu Z_{\nu+1}(z)$$
$$+(\mu-\nu-1)g_{\mu-1,\nu+1}(z)+bg_{\mu,\nu}(z)$$

11.3.5

$$(p^2+ab)g_{\mu,\nu}(z)=ae^{-pz}z^\mu Z_{\nu+1}(z)$$
$$+(\mu-\nu-1)e^{-pz}z^{\mu-1}Z_\nu(z)-pe^{-pz}z^\mu Z_\nu(z)$$
$$+p(2\mu-1)g_{\mu-1,\nu}(z)+[\nu^2-(\mu-1)^2]g_{\mu-2,\nu}(z)$$

11.3.6

$$a(\nu-\mu)g_{\mu,\nu+1}(z)=-2\nu e^{-pz}z^\mu Z_\nu(z)-2\nu pg_{\mu,\nu}(z)$$
$$+b(\mu+\nu)g_{\mu,\nu-1}(z)$$

Case 1: $\qquad p^2+ab=0,\ \nu=\pm(\mu-1)$

11.3.7 $\quad g_{\nu,\nu}(z)=\dfrac{e^{-pz}z^{\nu+1}}{2\nu+1}\left\{Z_\nu(z)-\dfrac{a}{p}Z_{\nu+1}(z)\right\}$

11.3.8 $\quad g_{-\nu,\nu}(z)=-\dfrac{e^{-pz}z^{-\nu+1}}{2\nu-1}\left\{Z_\nu(z)+\dfrac{b}{p}Z_{\nu-1}(z)\right\}$

11.3.9

$$\int_0^z e^{it}t^\nu J_\nu(t)dt=\frac{e^{iz}z^{\nu+1}}{2\nu+1}[J_\nu(z)-iJ_{\nu+1}(z)]$$

$$(\mathscr{R}\nu>-\tfrac{1}{2})$$

11.3.10

$$\int_0^z e^{it}t^{-\nu}J_\nu(t)dt=-\frac{e^{iz}z^{-\nu+1}}{2\nu-1}[J_\nu(z)+iJ_{\nu-1}(z)]$$

$$+\frac{i}{2^{\nu-1}(2\nu-1)\Gamma(\nu)}\qquad (\nu\neq\tfrac{1}{2})$$

11.3.11

$$\int_0^z e^{it}t^\nu Y_\nu(t)dt=\frac{e^{iz}z^{\nu+1}}{2\nu+1}[Y_\nu(z)-iY_{\nu+1}(z)]$$

$$-\frac{i2^{\nu+1}\Gamma(\nu+1)}{\pi(2\nu+1)}\qquad (\mathscr{R}\nu>-\tfrac{1}{2})$$

11.3.12

$$\int_0^z e^{\pm z}t^\nu I_\nu(t)dt=\frac{e^{\pm z}z^{\nu+1}}{2\nu+1}[I_\nu(z)\mp I_{\nu+1}(z)]$$

$$(\mathscr{R}\nu>-\tfrac{1}{2})$$

11.3.13

$$\int_0^z e^{-t}I_n(t)dt=ze^{-z}[I_0(z)+I_1(z)]$$
$$+n[e^{-z}I_0(z)-1]+2e^{-z}\sum_{k=1}^{n-1}(n-k)I_k(z)$$

11.3.14

$$\int_0^z e^{\pm t}t^{-\nu}I_\nu(t)dt=-\frac{e^{\pm z}z^{-\nu+1}}{2\nu-1}[I_\nu(z)\mp I_{\nu-1}(z)]$$
$$\mp\frac{1}{2^{\nu-1}(2\nu-1)\Gamma(\nu)}\qquad(\nu\neq\tfrac{1}{2})$$

11.3.15

$$\int_0^z e^{\pm t}t^\nu K_\nu(t)dt=\frac{e^{\pm z}z^{\nu+1}}{2\nu+1}[K_\nu(z)\pm K_{\nu+1}(z)]$$
$$\mp\frac{2^\nu\Gamma(\nu+1)}{2\nu+1}\qquad(\mathscr{R}\nu>-\tfrac{1}{2})$$

King's integral (see [11.5])

11.3.16 $\int_0^z e^t K_0(t)dt=ze^z[K_0(z)+K_1(z)]-1$

11.3.17

$$\int_z^\infty e^t t^{-\nu}K_\nu(t)dt$$
$$=\frac{e^z z^{-\nu+1}}{2\nu-1}[K_\nu(z)+K_{\nu-1}(z)]\qquad(\mathscr{R}\nu>\tfrac{1}{2})$$

Case 2: $\qquad p=0,\ \mu=\pm\nu$

11.3.18 $\qquad bg_{\nu,\nu-1}(z)=z^\nu Z_\nu(z)$

11.3.19 $\qquad ag_{-\nu,\nu+1}(z)=-z^{-\nu}Z_\nu(z)$

11.3.20 $\int_0^z t^\nu J_{\nu-1}(t)dt=z^\nu J_\nu(z)\qquad(\mathscr{R}\nu>0)$

11.3.21 $\int_0^z t^{-\nu}J_{\nu+1}(t)dt=\frac{1}{2^\nu\Gamma(\nu+1)}-z^{-\nu}J_\nu(z)$

11.3.22

$$2n\int_0^z\frac{J_{2n}(t)}{t}dt=1-\frac{2}{z}\sum_{k=1}^n(2k-1)J_{2k-1}(z)$$
$$=\frac{2}{z}\sum_{k=n+1}^\infty(2k-1)J_{2k-1}(z)\qquad(n>0)$$

11.3.23

$$(2n+1)\int_0^z\frac{J_{2n+1}(t)}{t}dt=\int_0^z J_0(t)dt$$
$$-J_1(z)-\frac{4}{z}\sum_{k=1}^n kJ_{2k}(z)$$

11.3.24

$$\int_0^z t^\nu Y_{\nu-1}(t)dt=z^\nu Y_\nu(z)+\frac{2^\nu\Gamma(\nu)}{\pi}\qquad(\mathscr{R}\nu>0)$$

11.3.25 $\int_0^z t^\nu I_{\nu-1}(t)dt=z^\nu I_\nu(z)\qquad(\mathscr{R}\nu>0)$

11.3.26 $\int_0^z t^{-\nu}I_{\nu+1}(t)dt=z^{-\nu}I_\nu(z)-\frac{1}{2^\nu\Gamma(\nu+1)}$

11.3.27

$$\int_0^z t^\nu K_{\nu-1}(t)dt=-z^\nu K_\nu(z)+2^{\nu-1}\Gamma(\nu)\qquad(\mathscr{R}\nu>0)$$

11.3.28 $\int_z^\infty t^{-\nu}K_{\nu+1}(t)dt=z^{-\nu}K_\nu(z)$

Indefinite Integrals of Products of Bessel Functions

Let $\mathscr{C}_\mu(z)$ and $\mathscr{D}_\nu(z)$ denote any two cylinder functions of orders μ and ν respectively.

11.3.29

$$\int^z\left\{(k^2-l^2)t-\frac{(\mu^2-\nu^2)}{t}\right\}\mathscr{C}_\mu(kt)\mathscr{D}_\nu(lt)dt$$
$$=z\{k\mathscr{C}_{\mu+1}(kz)\mathscr{D}_\nu(lz)-l\mathscr{C}_\mu(kz)\mathscr{D}_{\nu+1}(lz)\}\quad*$$
$$-(\mu-\nu)\mathscr{C}_\mu(kz)\mathscr{D}_\nu(lz)$$

11.3.30

$$\int^z t^{-\mu-\nu-1}\mathscr{C}_{\mu+1}(t)\mathscr{D}_{\nu+1}(t)dt$$
$$=-\frac{z^{-\mu-\nu}}{2(\mu+\nu+1)}\{\mathscr{C}_\mu(z)\mathscr{D}_\nu(z)+\mathscr{C}_{\mu+1}(z)\mathscr{D}_{\nu+1}(z)\}$$

11.3.31

$$\int^z t^{\mu+\nu+1}\mathscr{C}_\mu(t)\mathscr{D}_\nu(t)dt$$
$$=\frac{z^{\mu+\nu+2}}{2(\mu+\nu+1)}\{\mathscr{C}_\mu(z)\mathscr{D}_\nu(z)+\mathscr{C}_{\mu+1}(z)\mathscr{D}_{\nu+1}(z)\}$$

11.3.32

$$\int_0^z tJ_{\nu-1}^2(t)dt=2\sum_{k=0}^\infty(\nu+2k)J_{\nu+2k}^2(z)\qquad(\mathscr{R}\nu>0)$$

11.3.33

$$\int_0^z t[J_{\nu-1}^2(t)-J_{\nu+1}^2(t)]dt=2\nu J_\nu^2(z)\qquad(\mathscr{R}\nu>0)$$

11.3.34 $\int_0^z tJ_0^2(t)dt=\frac{z^2}{2}[J_0^2(z)+J_1^2(z)]$

11.3.35

$$\int_0^z J_n(t)J_{n+1}(t)dt=\frac{1}{2}[1-J_0^2(z)]-\sum_{k=1}^n J_k^2(z)$$
$$=\sum_{k=n+1}^\infty J_k^2(z)$$

---

*See page II.

11.3.36

$$(\mu+\nu)\int^z t^{-1}\mathscr{C}_\mu(t)\,\mathscr{D}_\nu(t)dt$$

$$-(\mu+\nu+2n)\int^z t^{-1}\mathscr{C}_{\mu+n}(t)\,\mathscr{D}_{\nu+n}(t)dt$$

$$=\mathscr{C}_\mu(z)\mathscr{D}_\nu(z)+\mathscr{C}_{\mu+n}(z)\mathscr{D}_{\nu+n}(z)+2\sum_{k=1}^{n-1}\mathscr{C}_{\mu+k}(z)\mathscr{D}_{\nu+k}(z)$$

Convolution Type Integrals

11.3.37

$$\int_0^z J_\mu(t)\,J_\nu(z-t)dt=2\sum_{k=0}^\infty (-)^k J_{\mu+\nu+2k+1}(z)$$

$$(\mathscr{R}\mu>-1,\ \mathscr{R}\nu>-1)$$

11.3.38

$$\int_0^z J_\nu(t)\,J_{1-\nu}(z-t)dt=J_0(z)-\cos z \quad (-1<\mathscr{R}\nu<2)$$

11.3.39

$$\int_0^z J_\nu(t)\,J_{-\nu}(z-t)\,dt=\sin z \quad (|\mathscr{R}\nu|<1)$$

11.3.40

$$\int_0^z t^{-1}J_\mu(t)\,J_\nu(z-t)dt=\frac{J_{\mu+\nu}(z)}{\mu}$$

$$(\mathscr{R}\mu>0,\ \mathscr{R}\nu>-1)$$

11.3.41

$$\int_0^z \frac{J_\mu(t)\,J_\nu(z-t)dt}{t(z-t)}=\frac{(\mu+\nu)\,J_{\mu+\nu}(z)}{\mu\nu z}$$

$$(\mathscr{R}\mu>0,\ \mathscr{R}\nu>0)$$

11.4. Definite Integrals

Orthogonality Properties of Bessel Functions

Let $\mathscr{C}_\nu(z)$ be a cylinder function of order ν. In particular, let

11.4.1
$$\mathscr{C}_\nu(z)=AJ_\nu(z)+BY_\nu(z)$$

where A and B are real constants. Then

11.4.2

$$\int_a^b t\mathscr{C}_\nu(\lambda_m t)\mathscr{C}_\nu(\lambda_n t)dt=0\ (m\neq n)$$

$$=\left[\frac{1}{2}\,t^2\left\{\left(1-\frac{\nu^2}{\lambda_n^2 t^2}\right)\mathscr{C}_\nu^2(\lambda_n t)+\mathscr{C}_\nu'^2(\lambda_n t)\right\}\right]_a^b$$

$$(m=n)\,(0<a<b)$$

provided the following two conditions hold:
1. λ_n is a real zero of

11.4.3
$$h_1\lambda\mathscr{C}_{\nu+1}(\lambda b)-h_2\mathscr{C}_\nu(\lambda b)=0$$

2. There must exist numbers k_1 and k_2 (both not zero) so that for all n

11.4.4
$$k_1\lambda_n\mathscr{C}_{\nu+1}(\lambda_n a)-k_2\mathscr{C}_\nu(\lambda_n a)=0$$

In connection with these formulae, see **11.3.29**. If $a=0$, the above is valid provided $B=0$. This case is covered by the following result.

11.4.5

$$\int_0^1 tJ_\nu(\alpha_m t)J_\nu(\alpha_n t)dt=0 \quad (m\neq n,\nu>-1)$$

$$=\tfrac{1}{2}[J_\nu'(\alpha_n)]^2$$

$$(m=n,b=0,\nu>-1)$$

$$=\frac{1}{2\alpha_n^2}\left[\frac{a^2}{b^2}+\alpha_n^2-\nu^2\right][J_\nu(\alpha_n)]^2$$

$$(m=n,\ b\neq0,\ \nu\geq-1)$$

$\alpha_1,\ \alpha_2,\ \ldots$ are the positive zeros of $aJ_\nu(x)+bxJ_\nu'(x)=0$, where a and b are real constants.

11.4.6

$$\int_0^\infty t^{-1}J_{\nu+2n+1}(t)J_{\nu+2m+1}(t)dt=0 \quad (m\neq n)$$

$$=\frac{1}{2(2n+\nu+1)}$$

$$(m=n)(\nu+n+m>-1)$$

Definite Integrals Over a Finite Range

11.4.7
$$\int_0^{\frac{\pi}{2}} J_{2n}(2z\sin t)dt=\frac{\pi}{2}\,J_n^2(z)$$

11.4.8
$$\int_0^\pi J_0(2z\sin t)\cos 2ntdt=\pi J_n^2(z)$$

11.4.9
$$\int_0^{\frac{\pi}{2}} Y_0(2z\sin t)\cos 2ntdt=\frac{\pi}{2}\,J_n(z)Y_n(z)$$

11.4.10

$$\int_0^{\frac{\pi}{2}} J_\mu(z\sin t)\,\sin^{\mu+1}t\,\cos^{2\nu+1}tdt$$

$$=\frac{2^\nu\Gamma(\nu+1)}{z^{\nu+1}}\,J_{\mu+\nu+1}(z) \quad (\mathscr{R}\mu>-1,\ \mathscr{R}\nu>-1)$$

11.4.11

$$\int_0^{\frac{\pi}{2}} J_\mu(z\sin^2 t)J_\nu(z\cos^2 t)\csc 2tdt$$

$$=\frac{(\mu+\nu)}{4\mu\nu}\,J_{\mu+\nu}(z) \quad (\mathscr{R}\mu>0,\ \mathscr{R}\nu>0)$$

Infinite Integrals

Integrals of the Form $\int_0^\infty e^{-pt}t^\mu Z_\nu(t)dt$

11.4.12

$$\int_0^\infty e^{it}t^{\mu-1}J_\nu(t)dt=\frac{e^{\frac{1}{2}i\pi(\mu+\nu)}\Gamma(\mu+\nu)\,\Gamma(\frac{1}{2}-\mu)}{\Gamma(\frac{1}{2})2^\mu\Gamma(\nu-\mu+1)}$$

$$\left(\mathscr{R}\mu<\frac{1}{2},\,\mathscr{R}(\mu+\nu)>0\right)$$

11.4.13

$$\int_0^\infty e^{-t}t^{\mu-1}I_\nu(t)dt=\frac{\Gamma(\mu+\nu)\Gamma(\frac{1}{2}-\mu)}{\Gamma(\frac{1}{2})2^\mu\Gamma(\nu-\mu+1)}$$

$$\left(\mathscr{R}\mu<\frac{1}{2},\,\mathscr{R}(\mu+\nu)>0\right)$$

11.4.14

$$\int_0^\infty \cos bt\,K_0(t)\,dt=\frac{\frac{1}{2}\pi}{(1+b^2)^{\frac{1}{2}}}\qquad(|\mathscr{I}b|<1)$$

11.4.15

$$\int_0^\infty \sin bt\,K_0(t)dt=\frac{\text{arc sinh }b}{(1+b^2)^{\frac{1}{2}}}\qquad(|\mathscr{I}b|<1)$$

11.4.16 $\displaystyle\int_0^\infty t^\mu J_\nu(t)dt=\frac{2^\mu\Gamma\left(\dfrac{\nu+\mu+1}{2}\right)}{\Gamma\left(\dfrac{\nu-\mu+1}{2}\right)}$

$$\left(\mathscr{R}(\mu+\nu)>-1,\,\mathscr{R}\mu<\frac{1}{2}\right)$$

11.4.17 $\displaystyle\int_0^\infty J_\nu(t)dt=1\qquad(\mathscr{R}\nu>-1)$

11.4.18

$$\int_0^\infty\frac{[1-J_0(t)]dt}{t^\mu}=\frac{\Gamma\left(\dfrac{\mu-1}{2}\right)\Gamma\left(\dfrac{3-\mu}{2}\right)}{2^\mu\left\{\Gamma\left(\dfrac{\mu+1}{2}\right)\right\}^2}\,(1<\mathscr{R}\mu<3)$$

11.4.19

$$\int_0^\infty t^\mu Y_\nu(t)dt=\frac{2^\mu}{\pi}\,\Gamma\left(\frac{\mu+\nu+1}{2}\right)\Gamma\left(\frac{\mu-\nu+1}{2}\right)$$

$$\times\sin\frac{\pi}{2}(\mu-\nu)\left(\mathscr{R}(\mu\pm\nu)>-1,\,\mathscr{R}\mu<\frac{1}{2}\right)$$

11.4.20 $\displaystyle\int_0^\infty Y_\nu(t)dt=-\tan\frac{\nu\pi}{2}\qquad(|\mathscr{R}\nu|<1)$

11.4.21 $\displaystyle\int_0^\infty Y_0(t)dt=0$

11.4.22

$$\int_0^\infty t^\mu K_\nu(t)dt=2^{\mu-1}\Gamma\left(\frac{\mu+\nu+1}{2}\right)\Gamma\left(\frac{\mu-\nu+1}{2}\right)$$

$$(\mathscr{R}(\mu\pm\nu)>-1)$$

11.4.23 $\displaystyle\int_0^\infty K_0(t)dt=\frac{\pi}{2}$

11.4.24 $\displaystyle\int_{-\infty}^\infty e^{-i\omega t}J_n(t)dt=\frac{2(-i)^nT_n(\omega)}{(1-\omega^2)^{\frac{1}{2}}}\,(\omega^2<1)$

$$=0\,(\omega^2>1)$$

where $T_n(\omega)$ is the Chebyshev polynomial of the first kind (see chapter **22**).

11.4.25

$$\int_{-\infty}^\infty t^{-1}e^{-i\omega t}J_n(t)dt$$

$$=\frac{2i}{n}(-i)^n(1-\omega^2)^{\frac{1}{2}}U_{n-1}(\omega)\,(\omega^2<1)$$

$$=0\,(\omega^2>1)$$

where $U_n(\omega)$ is the Chebyshev polynomial of the second kind (see chapter **22**).

11.4.26

$$\int_{-\infty}^\infty t^{-\frac{1}{2}}e^{-i\omega t}J_{n+\frac{1}{2}}(t)dt=(-i)^n(2\pi)^{\frac{1}{2}}P_n(\omega)\,(\omega^2<1)$$

$$=0\,(\omega^2>1)$$

where $P_n(\omega)$ is the Legendre polynomial (see chapter **22**).

11.4.27

$$\int_0^\infty e^{-t}t^{\frac{a}{2}-1}J_a[2(zt)^{\frac{1}{2}}]dt=\frac{\gamma(a,z)}{z^{a/2}}\qquad(\mathscr{R}a>0,\,\mathscr{R}z>0)$$

where $\gamma(a,z)$ is the incomplete gamma function (see chapter **6**).

Integrals of the Form $\int_0^\infty e^{-a^2t^2}t^\mu Z_\nu(bt)dt$

11.4.28

$$\int_0^\infty e^{-a^2t^2}t^{\mu-1}J_\nu(bt)dt$$

$$=\frac{\Gamma\left(\dfrac{1}{2}\nu+\dfrac{1}{2}\mu\right)\left(\dfrac{1}{2}\dfrac{b}{a}\right)^\nu}{2a^\mu\Gamma(\nu+1)}\,M\left(\frac{1}{2}\nu+\frac{1}{2}\mu,\nu+1,-\frac{b^2}{4a^2}\right)$$

$$(\mathscr{R}(\mu+\nu)>0,\,\mathscr{R}a^2>0)$$

where the notation $M(a,b,z)$ stands for the confluent hypergeometric function (see chapter **13**).

11.4.29

$$\int_0^\infty e^{-a^2t^2}t^{\nu+1}J_\nu(bt)dt$$

$$=\frac{b^\nu}{(2a^2)^{\nu+1}}\,e^{\frac{b^2}{4a^2}}\qquad(\mathscr{R}\nu>-1,\,\mathscr{R}a^2>0)$$

11.4.30

$$\int_0^\infty e^{-a^2 t^2} Y_{2\nu}(bt)dt = -\frac{\pi^{\frac{1}{2}}}{2a} e^{-\frac{b^2}{8a^2}} \left[I_\nu \left(\frac{b^2}{8a^2} \right) \tan \nu\pi \right.$$
$$\left. + \frac{1}{\pi} K_\nu \left(\frac{b^2}{8a^2} \right) \sec \nu\pi \right] \quad \left(|\mathscr{R}\nu| < \frac{1}{2}, \mathscr{R}a^2 > 0 \right)$$

11.4.31

$$\int_0^\infty e^{-a^2 t^2} I_\nu(bt)dt = \frac{\pi^{\frac{1}{2}}}{2a} e^{\frac{b^2}{8a^2}} I_{\frac{1}{2}\nu} \left(\frac{b^2}{8a^2} \right)$$
$$(\mathscr{R}\nu > -1, \mathscr{R}a^2 > 0)$$

11.4.32

$$\int_0^\infty e^{-a^2 t^2} K_0(bt)dt = \frac{\pi^{\frac{1}{2}}}{4a} e^{\frac{b^2}{8a^2}} K_0 \left(\frac{b^2}{8a^2} \right) \quad (\mathscr{R}a^2 > 0)$$

Weber-Schafheitlin Type Integrals

11.4.33

$$\int_0^\infty \frac{J_\mu(at)J_\nu(bt)dt}{t^\lambda} = \frac{b^\nu \Gamma \left(\frac{\mu+\nu-\lambda+1}{2} \right)}{2^\lambda a^{\nu-\lambda+1} \Gamma(\nu+1) \Gamma \left(\frac{\mu-\nu+\lambda+1}{2} \right)}$$
$$\times {}_2F_1 \left(\frac{\mu+\nu-\lambda+1}{2}, \frac{\nu-\mu-\lambda+1}{2}; \nu+1; \frac{b^2}{a^2} \right)$$
$$(\mathscr{R}(\mu+\nu-\lambda+1) > 0, \mathscr{R}\lambda > -1, 0 < b < a)$$

11.4.34

$$\int_0^\infty \frac{J_\mu(at)J_\nu(bt)dt}{t^\lambda} = \frac{a^\mu \Gamma \left(\frac{\mu+\nu-\lambda+1}{2} \right)}{2^\lambda b^{\mu-\lambda+1} \Gamma(\mu+1) \Gamma \left(\frac{\nu-\mu+\lambda+1}{2} \right)}$$
$$\times {}_2F_1 \left(\frac{\mu+\nu-\lambda+1}{2}, \frac{\mu-\nu-\lambda+1}{2}; \mu+1; \frac{a^2}{b^2} \right)$$
$$(\mathscr{R}(\mu+\nu-\lambda+1) > 0, \mathscr{R}\lambda > -1, 0 < a < b)$$

For ${}_2F_1$, see chapter **15**.

Special Cases of the Discontinuous Weber-Schafheitlin Integral

11.4.35

$$\int_0^\infty \frac{J_\mu(at) \sin bt\, dt}{t} = \frac{1}{\mu} \sin \left[\mu \arcsin \frac{b}{a} \right] \quad (0 \leq b \leq a)$$
$$= \frac{a^\mu \sin \frac{\pi\mu}{2}}{\mu[b + (b^2-a^2)^{\frac{1}{2}}]^\mu} \quad (b \geq a > 0)$$
$$(\mathscr{R}\mu > -1)$$

11.4.36

$$\int_0^\infty \frac{J_\mu(at) \cos bt\, dt}{t} = \frac{1}{\mu} \cos \left[\mu \arcsin \frac{b}{a} \right] \quad (0 \leq b \leq a)$$
$$= \frac{a^\mu \cos \frac{\pi\mu}{2}}{\mu[b + (b^2-a^2)^{\frac{1}{2}}]^\mu} \quad (b \geq a > 0)$$
$$(\mathscr{R}\mu > 0)$$

11.4.37

$$\int_0^\infty J_\mu(at) \cos bt\, dt = \frac{\cos \left[\mu \arcsin \frac{b}{a} \right]}{(a^2-b^2)^{\frac{1}{2}}} \quad (0 \leq b < a)$$
$$= \frac{-a^\mu \sin \frac{\pi\mu}{2}}{(b^2-a^2)^{\frac{1}{2}}[b + (b^2-a^2)^{\frac{1}{2}}]^\mu}$$
$$(b > a > 0) \quad (\mathscr{R}\mu > -1)$$

11.4.38

$$\int_0^\infty J_\mu(at) \sin bt\, dt = \frac{\sin \left[\mu \arcsin \frac{b}{a} \right]}{(a^2-b^2)^{\frac{1}{2}}} \quad (0 \leq b < a)$$
$$= \frac{a^\mu \cos \frac{\pi\mu}{2}}{(b^2-a^2)^{\frac{1}{2}}[b + (b^2-a^2)^{\frac{1}{2}}]^\mu}$$
$$(b > a > 0) \quad (\mathscr{R}\mu > -2)$$

11.4.39 $\quad \int_0^\infty e^{ibt} J_0(at)dt = \dfrac{1}{(a^2-b^2)^{\frac{1}{2}}} \quad (0 \leq b < a)$
$$= \frac{i}{(b^2-a^2)^{\frac{1}{2}}} \quad (0 < a < b)$$

11.4.40

$$\int_0^\infty e^{ibt} Y_0(at)dt = \frac{2i}{\pi(a^2-b^2)^{\frac{1}{2}}} \arcsin \frac{b}{a} \quad (0 \leq b < a)$$
$$= \frac{-1}{(b^2-a^2)^{\frac{1}{2}}} + \frac{2i}{\pi(b^2-a^2)^{\frac{1}{2}}}$$
$$\times \ln \left\{ \frac{b - (b^2-a^2)^{\frac{1}{2}}}{a} \right\} \quad (0 < a < b)$$

11.4.41

$$\int_0^\infty t^{\mu-\nu+1} J_\mu(at) J_\nu(bt)dt = 0 \quad (0 < b < a)$$
$$= \frac{2^{\mu-\nu+1} a^\mu (b^2-a^2)^{\nu-\mu-1}}{b^\nu \Gamma(\nu-\mu)}$$
$$(b > a > 0)$$
$$(\mathscr{R}\nu > \mathscr{R}\mu > -1)$$

11.4.42 $\quad \int_0^\infty J_\mu(at) J_{\mu-1}(bt)dt = \dfrac{b^{\mu-1}}{a^\mu} \quad (0 < b < a)$
$$= \frac{1}{2b} \quad (0 < b = a)$$
$$= 0 \quad (b > a > 0)$$
$$(\mathscr{R}\mu > 0)$$

11.4.43 $\quad \int_0^\infty \dfrac{J_0(at)}{t} \{1 - J_0(bt)\}dt = 0 \quad (0 < b \leq a)$
$$= \ln \frac{b}{a} \quad (b \geq a > 0)$$

Hankel-Nicholson Type Integrals

11.4.44

$$\int_0^\infty \frac{t^{\nu+1} J_\nu(at) dt}{(t^2+z^2)^{\mu+1}} = \frac{a^\mu z^{\nu-\mu}}{2^\mu \Gamma(\mu+1)} K_{\nu-\mu}(az)$$

$$\left(a>0,\ \mathscr{R}z>0,\ -1<\mathscr{R}\nu<2\mathscr{R}\mu+\frac{3}{2}\right)$$

11.4.45

$$\int_0^\infty \frac{J_\nu(at) dt}{t^\nu(t^2+z^2)} = \frac{\pi}{2z^{\nu+1}}[I_\nu(az) - \mathbf{L}_\nu(az)]$$

$$\left(a>0,\ \mathscr{R}z>0,\ \mathscr{R}\nu>-\frac{5}{2}\right)$$

11.4.46

$$\int_0^\infty \frac{Y_0(at) dt}{t^2+z^2} = -\frac{K_0(az)}{z} \qquad (a>0,\ \mathscr{R}z>0)$$

11.4.47

$$\int_0^\infty \frac{K_\nu(at) dt}{t^\nu(t^2+z^2)} = \frac{\pi^2}{4z^{\nu+1}\cos\nu\pi}[\mathbf{H}_\nu(az) - Y_\nu(az)]$$

$$(\mathscr{R}a>0,\ \mathscr{R}z>0,\ \mathscr{R}\nu<\tfrac{1}{2})$$

11.4.48

$$\int_0^\infty \frac{J_\nu(at) dt}{(t^2+z^2)^{\frac{1}{2}}} = I_{\frac{1}{2}\nu}(\tfrac{1}{2}az) K_{\frac{1}{2}\nu}(\tfrac{1}{2}az)$$

$$(a>0,\ \mathscr{R}z>0,\ \mathscr{R}\nu>-1)$$

11.4.49

$$\int_0^\infty \frac{J_\nu(at) dt}{t^\nu(t^2+z^2)^{\nu+\frac{1}{2}}} = \frac{\left(\frac{2a}{z^2}\right)^\nu \Gamma(\nu+1)}{\Gamma(2\nu+1)} I_\nu(\tfrac{1}{2}az) K_\nu(\tfrac{1}{2}az)$$

$$(a>0,\ \mathscr{R}z>0,\ \mathscr{R}\nu>-\tfrac{1}{2})$$

Numerical Methods

11.5. Use and Extension of the Tables

$$\int_0^x J_0(t)dt,\ \int_0^x Y_0(t)dt,\ \int_0^x I_0(t)dt,\ \int_x^\infty K_0(t)dt$$

For moderate values of x, use **11.1.2** and **11.1.7–11.1.10** as appropriate. For x sufficiently large, use the asymptotic expansions or the polynomial approximations **11.1.11–11.1.18**.

Example 1. Compute $\int_0^{3.05} J_0(t)dt$ to 5D. Using **11.1.2** and interpolating in **Tables 9.1** and **9.2**, we have

$$\int_0^{3.05} J_0(t)dt = 2[.32019\ 09 + .31783\ 69 + .04611\ 52$$
$$+ .00283\ 19 + .00009\ 72 + .00000\ 21]$$
$$= 1.37415$$

Example 2. Compute $\int_0^{3.05} J_0(t)dt$ to 5D by interpolation of **Table 11.1** using Taylor's formula. We have

$$\int_0^{x+h} J_0(t)dt = \int_0^x J_0(t)dt + hJ_0(x) - \frac{h^2}{2}J_1(x)$$
$$+ \frac{h^3}{12}[J_2(x) - J_0(x)] + \frac{h^4}{96}[3J_1(x) - J_3(x)] + \dots$$

Then with $x=3.0$ and $h=.05$,

$$\int_0^{3.05} J_0(t)dt = 1.387567 + (.05)(-.260052)$$
$$- (.00125)(.339059)$$
$$+ (.000010)(.746143) = 1.37415$$

This value is readily checked using $x=3.1$ and $h=-.05$. Now $|J_0(x)|\le 1$ for all x and $|J_n(x)| < 2^{-\frac{1}{2}}$, $n\ge 1$ for all x. In **Table 11.1**, we can always choose $|h|\le .05$. Thus if all terms of $O(h^4)$ and higher are neglected, then a bound for the absolute error is $2^{\frac{1}{2}}h^4/48 < .2\cdot 10^{-6}$ for all x if $|h| \le .05$. Similarly, the absolute error for quadratic interpolation does not exceed

$$h^3(2^{\frac{1}{2}}+2)/24 < .2\cdot 10^{-4}.$$

Example 3. Interpolation of $\int_0^x J_0(t)dt$ using Simpson's rule. We have

$$\int_0^{x+h} J_0(t)dt = \int_0^x J_0(t)dt + \int_x^{x+h} J_0(t)dt$$
$$\int_x^{x+h} J_0(t)dt = \frac{h}{6}\left[J_0(x) + 4J_0\left(x+\frac{h}{2}\right) + J_0(x+h)\right] + R$$
$$R = -\frac{h^5}{2880}J_0^{(4)}(\xi), \qquad x<\xi<x+h$$

Now

$$J_0^{(4)}(x) = \frac{1}{8}[J_4(x) - 4J_2(x) + 3J_0(x)]$$

$$|J_0^{(4)}(x)| < \frac{6+5\sqrt{2}}{16} < .82$$

and with $|h|\le .05$, it follows that

$$|R| < .9\cdot 10^{-10}$$

Thus if $x=3.0$ and $h=.05$

$$\int_0^{3.05} J_0(t)dt = 1.38756\ 72520 + \frac{(.05)}{6}[-.26005\ 19549$$
$$+ 4(-.26841\ 13883) - .27653\ 49599]$$
$$= 1.37414\ 86481$$

which is correct to 10D. The above procedure gives high accuracy though it may be necessary to interpolate twice in $J_0(x)$ to compute $J_0\left(x+\frac{h}{2}\right)$ and $J_0(x+h)$. A similar technique based on the trapezoidal rule is less accurate, but at most only one interpolation of $J_0(x)$ is required.

Example 4. Compute $\int_0^3 J_0(t)dt$ and $\int_0^3 Y_0(t)dt$ to 5D using the representation in terms of Struve functions and the tables in chapters **9** and **12**.

For $x=3$, from **Tables 9.1** and **12.1**

$$J_0=-.260052 \qquad J_1=.339059$$
$$Y_0=.376850 \qquad Y_1=.324674$$
$$\mathbf{H}_0=.574306 \qquad \mathbf{H}_1=1.020110$$

Using **11.1.7**, we have

$$\int_0^3 J_0(t)dt=3(-.260052)+\frac{3\pi}{2}\left[(.574306)(.339059)\right.$$
$$\left.-(1.020110)(-.260052)\right]$$
$$=1.38757$$

Similarly,

$$\int_0^3 Y_0(t)dt=.19766$$

Using **11.1.8** and **Tables 9.8** and **12.1**, one can compute $\int_0^x I_0(t)dt$ and $\int_0^x K_0(t)dt$.

$$\underline{\int_x^\infty \frac{J_0(t)dt}{t}, \int_x^\infty \frac{Y_0(t)dt}{t}, \int_0^x \frac{[I_0(t)-1]dt}{t}, \int_x^\infty \frac{K_0(t)dt}{t}}$$

For moderate values of x, use **11.1.19–11.1.23**. For x sufficiently large, use the asymptotic expansions or the polynomial approximations **11.1.24–11.1.31**.

Repeated Integrals of $J_0(x)$

For moderate values of x and r, use **11.2.4**. If $r=1$, see **Example 1**. For moderate values of x, use the recurrence formula **11.2.5**. If x is large and $x\gg r$, see the discussion below.

Example 5. Compute $f_{r,0}(x)=f_r(x)$ to 5D for $x=2$ and $r=0(1)5$ using **11.2.6**. We have

$$rf_{r+1}(x)=xf_r(x)-(r-1)f_{r-1}(x)+xf_{r-2}(x)$$

$$f_{-1}(x)=-J_1(x), \quad f_0(x)=J_0(x), \quad f_1(x)=\int_0^x J_0(t)\,dt$$

and the terms on this last line are tabulated. Thus for $x=2$,

$$f_{-1}=-.57672\ 48, f_0=.22389\ 08, f_1=1.42577\ 03$$

The recurrence formula gives

$$f_2=2(f_1+f_{-1})=1.69809\ 10$$

Similarly,

$$f_3=1.20909\ 66, f_4=.62451\ 73, f_5=.25448\ 17$$

When $x\gg r$, it is convenient to use the auxiliary function

$$g_r(x)=(r-1)!x^{-r+1}f_r(x)$$

This satisfies the recurrence relation

$$x^2g_{r+1}(x)=x^2g_r-(r-1)^2g_{r-1}(x)$$
$$+(r-1)(r-2)g_{r-2}(x), \quad r\geq3$$

$$g_1(x)=\int_0^x J_0(t)dt, \quad g_2(x)=g_1(x)-J_1(x)$$
$$g_3(x)=[x^2g_2(x)-g_1(x)+xJ_0(x)]/x^2$$

Example 6. Compute $g_r(x)$ to 5D for $x=10$ and $r=0(1)6$. We have for $x=10$,

$$J_0=-.24593\ 58, J_1=.04347\ 27, g_1=1.06701\ 13$$

Thus

$$g_2=1.02353\ 86, g_3=.98827\ 49$$

and the forward recurrence formula gives

$$g_4=.96867\ 36, g_5=.94114\ 12, g_6=.90474\ 64$$

For tables of $2^{-r}f_r(x)$, see [11.16].

Repeated Integrals of $K_0(x)$

For moderate values of x, use the recurrence formula **11.2.14** for all r.

Example 7. Compute $\mathrm{Ki}_r(x)$ to 5D for $x=2$ and $r=0(1)5$. We have

$$r\mathrm{Ki}_{r+1}(x)=-x\mathrm{Ki}_r(x)+(r-1)\mathrm{Ki}_{r-1}(x)+x\mathrm{Ki}_{r-2}(x)$$

$$\mathrm{Ki}_{-1}(x)=K_1(x), \ \mathrm{Ki}_0(x)=K_0(x), \ \mathrm{Ki}_1(x)=\int_x^\infty K_0(t)dt$$

and the functions on this last line are tabulated Thus for $x=2$,

$$K_0=.11389\ 39, \ K_1=.13986\ 59, \ \mathrm{Ki}_1=.09712\ 06$$

and

$$\mathrm{Ki}_2=-2\mathrm{Ki}_1+2K_1=.08549\ 06$$

Similarly,

$$\mathrm{Ki}_3=.07696\ 36, \ \mathrm{Ki}_4=.07043\ 17, \ \mathrm{Ki}_5=.06525\ 22$$

If x/r is not large the formula can still be used provided that the starting values are sufficiently accurate to offset the growth of rounding error.

For tables of $\mathrm{Ki}_r(x)$, see [11.11].

Now

$$f_m(x)=x^{-m}\int_0^x t^m K_0(t)dt$$

$$f_0(x)=\int_0^x K_0(t)dt, f_1(x)=[1-xK_1(x)]/x$$

the latter following from **11.3.27** with $\nu=1$. In **11.3.5**, put $a=1$, $b=-1$, $p=0$ and $\nu=0$. Let $\mu=m$. Then

$$f_m(x)=[(m-1)^2f_{m-2}(x)-x^2K_1(x)$$
$$-x(m-1)K_0(x)]/x^2 \qquad (m>1)$$

Using tabular values of f_0 and f_1, one can compute in succession $f_2, f_3, \ldots$ provided that m/x is not large.

Example 8. Compute $f_m(x)$ to 5D for $x=5$ and $m=0(1)6$. We have, retaining two additional decimals

$$K_0=.00369\ 11 \qquad K_1=.00404\ 46$$
$$f_0=1.56738\ 74 \qquad f_1=.19595\ 54$$

Thus

$$f_2=.05791\ 27, f_4=.01458\ 93, f_6=.00685\ 36$$

Similarly starting with f_1, we can compute f_3 and f_5.

If $m>x$, employ the recurrence formula in backward form and write

$$f_{m-2}(x)=[x^2f_m(x)+x^2K_1(x)+x(m-1)K_0(x)]/(m-1)^2$$

In the latter expression, replace f_m by g_m. Fix x. Take $r>m$ and assume $g_r=0$. Compute g_{r-2}, g_{r-4}, etc. Then

$$\lim_{r\to\infty} g_{r-2k}(x)=f_m(x), \quad m=r-2k$$

Apart from round-off error, the value of r needed to achieve a stated accuracy for given x and m can be determined a priori. Let

$$\epsilon_r=|g_r-f_r|$$

Then

$$\epsilon_{r-2k}=\frac{x^{2k}\epsilon_r}{(r-1)^2(r-3)^2\ldots(r-2k+1)^2}$$

$$\epsilon_r\le[x^2K_1(x)+x(r-1)K_0(x)]/(r-1)^2$$

since for x fixed, $f_r(x)$ is positive and decreases as r increases.

Example 9. Compute $f_m(x)$ to 5D for $x=3$ and $m=0(2)10$. We have

$$K_0=.03473\ 95 \qquad K_1=.04015\ 64$$

If $r=16$,

$$\epsilon_{16}<.86\cdot10^{-2} \qquad \epsilon_{10}<1.4\cdot10^{-6}$$

Taking $g_{16}=0$, we compute the following values of $g_{14}, g_{12}, \ldots, g_0$ by recurrence. Also recorded are the required values of f_m to 5D.

m	g_m	f_m
14	.00855 42	
12	.01061 09	
10	.01325 05	.01325
8	.01751 39	.01751
6	.02548 09	.02548
4	.04447 31	.04447
2	.11936 90	.11937
0	1.53994 71	1.53995

For tables of $f_m(x)$, see [11.21].

References

Texts

[11.1] H. Bateman and R. C. Archibald, A guide to tables of Bessel functions, Math. Tables Aids Comp. **1**, 247–252 (1943). See also Supplements I, II, IV, same journal, **1**, 403–404 (1943); **2**, 59 (1946); **2**, 190 (1946), respectively.

[11.2] A. Erdélyi et al., Higher transcendental functions, vol. 2, ch. 7 (McGraw-Hill Book Co., Inc., New York, N.Y., 1953).

[11.3] A. Erdélyi et al., Tables of integral transforms, vols. 1, 2 (McGraw-Hill Book Co., Inc., New York, N.Y., 1954).

[11.4] W. Gröbner and N. Hofreiter, Integraltafel, II Teil (Springer-Verlag, Wien and Innsbruck, Austria, 1949–1950).

[11.5] L. V. King, On the convection of heat from small cylinders in a stream of fluid, Trans. Roy. Soc. London **214A**, 373–432 (1914).

[11.6] Y. L. Luke, Some notes on integrals involving Bessel functions, J. Math. Phys. **29**, 27–30 (1950).

[11.7] Y. L. Luke, An associated Bessel function, J. Math. Phys. **31**, 131–138 (1952).

[11.8] F. Oberhettinger, On some expansions for Bessel integral functions, J. Research NBS **59**, 197–201 (1957) RP 2786.

[11.9] G. Petiau, La théorie des fonctions de Bessel (Centre National de la Recherche Scientifique, Paris, France, 1955).

[11.10] G. N. Watson, A treatise on the theory of Bessel functions, 2d ed. (Cambridge Univ. Press, Cambridge, England, 1958).

Tables

[11.11] W. G. Bickley and J. Nayler, A short table of the functions $Ki_n(x)$, from $n=1$ to $n=16$. Philos. Mag. **7**, **20**, 343–347 (1935). $Ki_1(x)=\int_x^\infty K_0(t)dt$, $Ki_n(x)=\int_x^\infty Ki_{n-1}(t)dt$, $n=1(1)16$, $x=0(.05).2$ $(.1)2, 3, \quad 9D$.

[11.12] V. R. Bursian and V. Fock, Table of the functions $\int_x^\infty K_0(x)\,dx, \int_0^x I_0(x)\,dx, e^x\int_x^\infty K_0(x)\,dx, e^{-x}\int_0^x I_0(x)\,dx$, Akad. Nauk, Leningrad, Inst. Fiz. Mat., Trudy (Travaux) 2, 6–10 (1931). $\int_x^\infty K_0(t)\,dt$, $x=0(.1)12$, 7D; $e^x\int_x^\infty K_0(t)\,dt$, $x=0(.1)16$, 7D; $\int_0^x I_0(t)\,dt$, $x=0(.1)6$, 7D; $e^{-x}\int_0^x I_0(t)\,dt$, $x=0\,(.1)16$, 7D.

[11.13] E. A. Chistova, Tablitsy funktsii Besselya ot deistvitel' nogo argumenta i integralov ot nikh (Izdat. Akad. Nauk SSSR., Moscow, U.S.S.R., 1958). $J_n(x)$, $Y_n(x)$, $\int_x^\infty \dfrac{J_n(t)}{t}\,dt$, $\int_x^\infty \dfrac{Y_n(t)}{t}\,dt$, $n=0$, 1; $x=0(.001)15(.01)100$, 7D. Also tabulated are auxiliary expressions to facilitate interpolation near the origin.

[11.14] A. J. M. Hitchcock, Polynomial approximations to Bessel functions of order zero and one and to related functions, Math. Tables Aids Comp. 11, 86–88 (1957). Polynomial approximations for $\int_0^x J_0(t)\,dt$ and $\int_x^\infty K_0(t)\,dt$.

[11.15] C. W. Horton, A short table of Struve functions and of some integrals involving Bessel and Struve functions, J. Math. Phys. 29, 56–58 (1950). $C_n(x)=\int_0^x t^n J_n(t)\,dt, n=1(1)4, x=0(.1)10$, 4D; $D_n(x)=\int_0^x t^n \mathbf{H}_n(t)\,dt, n=0(1)4, x=0(.1)10$, 4D, where $\mathbf{H}_n(x)$ is Struve's function; see chapter 12.

[11.16] J. C. Jaeger, Repeated integrals of Bessel functions and the theory of transients in filter circuits, J. Math. Phys. 27, 210–219 (1948). $f_1(x)=\int_0^x J_0(t)\,dt, f_r(x)=\int_0^x f_{r-1}(t)\,dt, 2^{-r}f_r(x), r=1(1)7, x=0(1)24, 8D$. Also $\Phi_n(x)=\int_0^\infty J_0[2(xt)^{1/2}]J_n(t)\,dt$, $\Phi_n(x)$, $\Phi_n'(x)$, $n=1(1)7$, $x=0(1)24$, 4D.

[11.17] L. N. Karmazina and E. A. Chistova, Tablitsy funktsii Besselya ot mnimogo argumenta i integralov ot nikh (Izdat. Akad. Nauk SSSR., Moscow, U.S.S.R., 1958). $e^{-x}I_0(x)$, $e^{-x}I_1(x)$, $e^x K_0(x)$, $e^x K_1(x)$, e^x, $e^{-x}\int_0^x I_0(t)\,dt, e^x\int_x^\infty K_0(t)\,dt$, $x=0(.001)5(.005)15(.01)100$, 7D except for e^x which is 7S. Also tabulated are auxiliary expressions to facilitate interpolation near the origin.

[11.18] H. L. Knudsen, Bidrag til teorien for antennesystemer med hel eller delvis rotations-symmetri. I (Kommission Has Teknisk Forlag, Copenhagen, Denmark, 1953). $\int_0^x J_n(t)\,dt$, $n=0(1)8$, $x=0(.01)10$, 5D. Also $\int_0^x J_n(t)e^{i\alpha}dt, \alpha=t, \alpha=x-t$.

[11.19] Y. L. Luke and D. Ufford, Tables of the function $\overline{K}_0(x)=\int_x^\infty K_0(t)\,dt$. Math. Tables Aids Comp. UMT 129. $\overline{K}_0(x)=-[\gamma+\ln(x/2)]A_1(x)+A_2(x)$, $A_1(x)$, $A_2(x)$. $x=0(.01).5(.05)1$, 8D.

[11.20] C. Mack and M. Castle, Tables of $\int_0^a I_0(x)\,dx$ and $\int_a^\infty K_0(x)\,dx$, Roy. Soc. Unpublished Math. Table File No. 6. $a=0(.02)2(.1)4$, 9D.

[11.21] G. M. Muller, Table of the function $$\mathrm{Kj}_n(x)=x^{-n}\int_0^x u^n K_0(u)\,du,$$ Office of Technical Services, U.S. Department of Commerce, Washington, D.C. (1954). $n=0(1)31, x=0(.01)2(.02)5$, 8S.

[11.22] National Bureau of Standards, Tables of functions and zeros of functions, Applied Math. Series 37 (U.S. Government Printing Office, Washington, D.C., 1954). (1) pp. 21–31: $\int_0^x J_0(t)\,dt, \int_0^x Y_0(t)\,dt$, $x=0(.01)10$, 10D. (2) pp. 33–39: $\int_x^\infty J_0(t)\,dt/t$, $x=0(.1)10(1)22$, 10D; $F(x)=\int_x^\infty J_0(t)\,dt/t$ $+\ln(x/2)$, $x=0(.1)3$, 10D; $F^{(n)}(x)/n!$, $x=10(1)22, n=0(1)13$, 12D.

[11.23] National Physical Laboratory, Integrals of Bessel functions, Roy. Soc. Unpublished Math. Table File No. 17. $\int_0^x J_0(t)\,dt$, $\int_0^x Y_0(t)\,dt$, $x=0(.5)50$, 10D.

[11.24] M. Rothman, Table of $\int_0^x I_0(x)\,dx$ for $0(.1)20(1)25$, Quart. J. Mech. Appl. Math. 2, 212–217 (1949). 8S–9S.

[11.25] P. W. Schmidt, Tables of $\int_0^x J_0(t)\,dt$ for large x, J. Math. Phys. 34, 169–172 (1955). $x=10(.2)40$, 6D.

[11.26] G. N. Watson, A treatise on the theory of Bessel functions, 2d ed. (Cambridge Univ. Press, Cambridge, England, 1958). Table VIII, p. 752: $\frac{1}{2}\int_0^x J_0(t)\,dt, \frac{1}{2}\int_0^x Y_0(t)\,dt$, $x=0(.02)1$, 7D, with the first 16 maxima and minima of the integrals to 7D.

Table 11.1   **INTEGRALS OF BESSEL FUNCTIONS**

x	$\int_0^x J_0(t)\,dt$	$\int_0^x Y_0(t)\,dt$	$e^{-x}\int_0^x I_0(t)\,dt$	$e^x\int_x^\infty K_0(t)\,dt$
0.0	0.00000 00000	0.00000 00000	0.00000 00	1.57079 63
0.1	0.09991 66979	-0.21743 05666	0.09055 92	1.35784 82
0.2	0.19933 43325	-0.34570 88380	0.16429 28	1.25032 54
0.3	0.29775 75802	-0.43928 31758	0.22391 79	1.17280 09
0.4	0.39469 85653	-0.50952 48283	0.27172 46	1.11171 28
0.5	0.48968 05066	-0.56179 54559	0.30964 29	1.06127 17
0.6	0.58224 12719	-0.59927 15570	0.33929 99	1.01836 48
0.7	0.67193 68094	-0.62409 96341	0.36206 71	0.98109 70
0.8	0.75834 44308	-0.63786 88991	0.37910 05	0.94821 80
0.9	0.84106 59149	-0.64184 01770	0.39137 42	0.91885 56
1.0	0.91973 04101	-0.63706 93766	0.39970 88	0.89237 52
1.1	0.99399 71082	-0.62447 91607	0.40479 52	0.86829 97
1.2	1.06355 76711	-0.60490 26964	0.40721 52	0.84626 10
1.3	1.12813 83885	-0.57911 12548	0.40745 78	0.82596 89
1.4	1.18750 20495	-0.54783 19295	0.40593 39	0.80719 04
1.5	1.24144 95144	-0.51175 90340	0.40298 85	0.78973 57
1.6	1.28982 09734	-0.47156 13039	0.39891 09	0.77344 80
1.7	1.33249 68829	-0.42788 62338	0.39394 29	0.75819 62
1.8	1.36939 85727	-0.38136 24134	0.38828 68	0.74386 97
1.9	1.40048 85208	-0.33260 04453	0.38211 11	0.73037 44
2.0	1.42577 02932	-0.28219 28501	0.37555 57	0.71762 95
2.1	1.44528 81525	-0.23071 32490	0.36873 67	0.70556 50
2.2	1.45912 63387	-0.17871 50399	0.36174 98	0.69412 02
2.3	1.46740 80303	-0.12672 97284	0.35467 38	0.68324 16
2.4	1.47029 39949	-0.07526 50420	0.34757 29	0.67288 26
2.5	1.46798 09446	-0.02480 29261	0.34049 93	0.66300 15
2.6	1.46069 96081	+0.02420 24953	0.33349 48	0.65356 16
2.7	1.44871 25408	0.07132 69288	0.32659 30	0.64452 98
2.8	1.43231 16899	0.11617 78353	0.31981 99	0.63587 68
2.9	1.41181 57386	0.15839 62206	0.31319 59	0.62757 60
3.0	1.38756 72520	0.19765 82565	0.30673 62	0.61960 34
3.1	1.35992 96508	0.23367 66986	0.30045 18	0.61193 74
3.2	1.32928 40386	0.26620 20748	0.29435 04	0.60455 84
3.3	1.29602 59125	0.29502 36222	0.28843 67	0.59744 84
3.4	1.26056 17835	0.31996 99576	0.28271 31	0.59059 11
3.5	1.22330 57382	0.34090 94657	0.27718 02	0.58397 14
3.6	1.18467 59706	0.35775 03989	0.27183 70	0.57757 57
3.7	1.14509 13136	0.37044 06831	0.26668 11	0.57139 13
3.8	1.10496 78009	0.37896 74266	0.26170 94	0.56540 66
3.9	1.06471 52877	0.38335 61369	0.25691 78	0.55961 09
4.0	1.02473 41595	0.38366 96479	0.25230 18	0.55399 42
4.1	0.98541 21560	0.38000 67672	0.24785 61	0.54854 72
4.2	0.94712 13375	0.37250 06552	0.24357 56	0.54326 15
4.3	0.91021 52175	0.36131 69475	0.23945 46	0.53812 91
4.4	0.87502 60866	0.34665 16398	0.23548 74	0.53314 27
4.5	0.84186 25481	0.32872 87513	0.23166 83	0.52829 52
4.6	0.81100 72858	0.30779 77892	0.22799 15	0.52358 03
4.7	0.78271 50802	0.28413 10351	0.22445 13	0.51899 19
4.8	0.75721 10902	0.25802 06786	0.22104 21	0.51452 43
4.9	0.73468 94106	0.22977 58227	0.21775 83	0.51017 24
5.0	0.71531 19178	0.19971 93876	0.21459 46	0.50593 10
	$\begin{bmatrix}(-4)7\\7\end{bmatrix}$		$\begin{bmatrix}(-3)2\\6\end{bmatrix}$	

INTEGRALS OF BESSEL FUNCTIONS

Table 11.1

x	$\int_0^x J_0(t)\,dt$	$\int_0^x Y_0(t)\,dt$	$e^{-x}\int_0^x I_0(t)\,dt$	$e^x\int_x^\infty K_0(t)\,dt$
5.0	0.71531 19178	0.19971 93876	0.21459 46	0.50593 10
5.1	0.69920 74098	0.16818 49405	0.21154 58	0.50179 55
5.2	0.68647 10457	0.13551 34784	0.20860 68	0.49776 16
5.3	0.67716 40870	0.10205 01932	0.20577 28	0.49382 50
5.4	0.67131 39407	0.06814 12463	0.20303 89	0.48998 19
5.5	0.66891 44989	0.03413 05806	0.20040 08	0.48622 86
5.6	0.66992 67724	+0.00035 67983	0.19785 40	0.48256 16
5.7	0.67427 98068	−0.03284 98697	0.19539 44	0.47897 75
5.8	0.68187 18713	−0.06517 04775	0.19301 81	0.47547 34
5.9	0.69257 19078	−0.09630 01348	0.19072 13	0.47204 60
6.0	0.70622 12236	−0.12595 06129	0.18850 02	0.46869 29
6.1	0.72263 54100	−0.15385 27646	0.18635 16	0.46541 11
6.2	0.74160 64692	−0.17975 87372	0.18427 20	0.46219 83
6.3	0.76290 51256	−0.20344 39625	0.18225 84	0.45905 20
6.4	0.78628 33012	−0.22470 89068	0.18030 78	0.45596 99
6.5	0.81147 67291	−0.24338 05692	0.17841 74	0.45294 98
6.6	0.83820 76824	−0.25931 37161	0.17658 44	0.44998 97
6.7	0.86618 77897	−0.27239 18447	0.17480 64	0.44708 76
6.8	0.89512 09137	−0.28252 78684	0.17308 09	0.44424 15
6.9	0.92470 60635	−0.28966 45218	0.17140 55	0.44144 97
7.0	0.95464 03155	−0.29377 44843	0.16977 82	0.43871 05
7.1	0.98462 17153	−0.29486 02239	0.16819 68	0.43602 22
7.2	1.01435 21344	−0.29295 35658	0.16665 93	0.43338 34
7.3	1.04354 00558	−0.28811 49927	0.16516 39	0.43079 23
7.4	1.07190 32638	−0.28043 26862	0.16370 89	0.42824 76
7.5	1.09917 14142	−0.27002 13202	0.16229 24	0.42574 81
7.6	1.12508 84628	−0.25702 06208	0.16091 30	0.42329 20
7.7	1.14941 49299	−0.24159 37080	0.15956 91	0.42087 86
7.8	1.17192 99830	−0.22392 52368	0.15825 93	0.41850 63
7.9	1.19243 33198	−0.20421 93575	0.15698 21	0.41617 40
8.0	1.21074 68348	−0.18269 75150	0.15573 64	0.41388 07
8.1	1.22671 60587	−0.15959 61109	0.15452 08	0.41162 52
8.2	1.24021 13565	−0.13516 40494	0.15333 42	0.40940 65
8.3	1.25112 88778	−0.10966 01934	0.15217 55	0.40722 37
8.4	1.25939 12520	−0.08335 07540	0.15104 36	0.40507 56
8.5	1.26494 80240	−0.05650 66385	0.14993 74	0.40296 15
8.6	1.26777 58297	−0.02940 07834	0.14885 61	0.40088 04
8.7	1.26787 83120	−0.00230 54965	0.14779 88	0.39883 15
8.8	1.26528 57796	+0.02451 01664	0.14676 44	0.39681 40
8.9	1.26005 46162	0.05078 29664	0.14575 23	0.39482 69
9.0	1.25226 64460	0.07625 79635	0.14476 16	0.39286 97
9.1	1.24202 70675	0.10069 08937	0.14379 16	0.39094 15
9.2	1.22946 51666	0.12385 04194	0.14284 16	0.38904 17
9.3	1.21473 08237	0.14552 02334	0.14191 08	0.38716 95
9.4	1.19799 38314	0.16550 09969	0.14099 87	0.38532 41
9.5	1.17944 18392	0.18361 20962	0.14010 46	0.38350 53
9.6	1.15927 83464	0.19969 32017	0.13922 78	0.38171 20
9.7	1.13772 05614	0.21360 56169	0.13836 79	0.37994 39
9.8	1.11499 71504	0.22523 34059	0.13752 43	0.37820 03
9.9	1.09134 58985	0.23448 42919	0.13669 65	0.37648 06
10.0	1.06701 13040	0.24129 03183	0.13588 40	0.37478 43
	$\left[\begin{smallmatrix}(-4)4\\7\end{smallmatrix}\right]$	$\left[\begin{smallmatrix}(-4)4\\7\end{smallmatrix}\right]$	$\left[\begin{smallmatrix}(-5)1\\4\end{smallmatrix}\right]$	$\left[\begin{smallmatrix}(-5)1\\4\end{smallmatrix}\right]$

Table 11.2   **INTEGRALS OF BESSEL FUNCTIONS**

x	$\int_0^x \dfrac{1-J_0(t)}{t}\,dt$	$\int_x^\infty \dfrac{Y_0(t)}{t}\,dt$	$e^{-x}\int_0^x \dfrac{I_0(t)-1}{t}\,dt$	$xe^x\int_x^\infty \dfrac{K_0(t)}{t}\,dt$
0.0	0.00000 000	$-\infty$	0.00000 000	0.000000
0.1	0.00124 961	−1.34138 382	0.00113 140	0.368126
0.2	0.00499 375	−0.43423 067	0.00409 877	0.460111
0.3	0.01121 841	−0.05107 832	0.00835 768	0.506394
0.4	0.01990 030	+0.15238 037	0.01347 363	0.532910
0.5	0.03100 699	0.26968 854	0.01910 285	0.548819
0.6	0.04449 711	0.33839 213	0.02497 622	0.558366
0.7	0.06032 057	0.37689 807	0.03088 584	0.563828
0.8	0.07841 882	0.39543 866	0.03667 383	0.566545
0.9	0.09872 519	0.40022 301	0.04222 295	0.567355
1.0	0.12116 525	0.39527 290	0.04744 889	0.566811
1.1	0.14565 721	0.38332 909	0.05229 376	0.565291
1.2	0.17211 240	0.36633 694	0.05672 080	0.563058
1.3	0.20043 570	0.34572 398	0.06070 995	0.560302
1.4	0.23052 610	0.32256 701	0.06425 420	0.557163
1.5	0.26227 724	0.29769 696	0.06735 663	0.553745
1.6	0.29557 796	0.27176 713	0.07002 797	0.550126
1.7	0.33031 288	0.24529 896	0.07228 458	0.546364
1.8	0.36636 308	0.21871 360	0.07414 688	0.542506
1.9	0.40360 666	0.19235 409	0.07563 806	0.538587
2.0	0.44191 940	0.16650 135	0.07678 298	0.534635
2.1	0.48117 541	0.14138 594	0.07760 744	0.530670
2.2	0.52124 775	0.11719 681	0.07813 746	0.526711
2.3	0.56200 913	0.09408 798	0.07839 884	0.522768
2.4	0.60333 248	0.07218 365	0.07841 674	0.518854
2.5	0.64509 164	0.05158 229	0.07821 544	0.514976
2.6	0.68716 194	0.03235 987	0.07781 809	0.511139
2.7	0.72942 081	+0.01457 248	0.07724 664	0.507350
2.8	0.77174 836	−0.00174 144	0.07652 168	0.503610
2.9	0.81402 795	−0.01655 931	0.07566 245	0.499924
3.0	0.85614 669	−0.02987 272	0.07468 681	0.496292
3.1	0.89799 596	−0.04168 613	0.07361 124	0.492717
3.2	0.93947 188	−0.05201 554	0.07245 090	0.489198
3.3	0.98047 571	−0.06088 740	0.07121 963	0.485736
3.4	1.02091 428	−0.06833 756	0.06993 006	0.482332
3.5	1.06070 032	−0.07441 025	0.06859 360	0.478984
3.6	1.09975 277	−0.07915 722	0.06722 060	0.475694
3.7	1.13799 707	−0.08263 683	0.06582 033	0.472459
3.8	1.17536 536	−0.08491 323	0.06440 109	0.469280
3.9	1.21179 667	−0.08605 553	0.06297 029	0.466155
4.0	1.24723 707	−0.08613 706	0.06153 450	0.463085
4.1	1.28163 975	−0.08523 459	0.06009 952	0.460067
4.2	1.31496 504	−0.08342 762	0.05867 042	0.457100
4.3	1.34718 044	−0.08079 769	0.05725 166	0.454185
4.4	1.37826 060	−0.07742 769	0.05584 708	0.451320
4.5	1.40818 716	−0.07340 123	0.05446 000	0.448503
4.6	1.43694 870	−0.06880 199	0.05309 325	0.445734
4.7	1.46454 052	−0.06371 317	0.05174 921	0.443012
4.8	1.49096 446	−0.05821 690	0.05042 989	0.440335
4.9	1.51622 864	−0.05239 371	0.04913 691	0.437703
5.0	1.54034 722	−0.04632 205	0.04787 161	0.435114
	$\begin{bmatrix}(-4)3\\6\end{bmatrix}$		$\begin{bmatrix}(-4)2\\7\end{bmatrix}$	

12. Struve Functions and Related Functions

Milton Abramowitz[1]

Contents

The author acknowledges the assistance of Bertha H. Walter in the preparation and checking of the tables.

---

[1] National Bureau of Standards. (Deceased.)

12. Struve Functions and Related Functions

Mathematical Properties

12.1. Struve Function $\mathbf{H}_\nu(z)$

Differential Equation and General Solution

12.1.1

$$z^2\frac{d^2w}{dz^2}+z\frac{dw}{dz}+(z^2-\nu^2)w=\frac{4(\frac{1}{2}z)^{\nu+1}}{\sqrt{\pi}\,\Gamma(\nu+\frac{1}{2})}$$

The general solution is

12.1.2 $\quad w=aJ_\nu(z)+bY_\nu(z)+\mathbf{H}_\nu(z)\ \ (a,b,\ \text{constants})$

where $z^{-\nu}\mathbf{H}_\nu(z)$ is an entire function of z.

Power Series Expansion

12.1.3

$$\mathbf{H}_\nu(z)=(\tfrac{1}{2}z)^{\nu+1}\sum_{k=0}^\infty\frac{(-1)^k(\frac{1}{2}z)^{2k}}{\Gamma(k+\frac{3}{2})\Gamma(k+\nu+\frac{3}{2})}$$

12.1.4 $\quad \mathbf{H}_0(z)=\frac{2}{\pi}\left[z-\frac{z^3}{1^2\cdot3^2}+\frac{z^5}{1^2\cdot3^2\cdot5^2}-\cdots\right]$

12.1.5

$$\mathbf{H}_1(z)=\frac{2}{\pi}\left[\frac{z^2}{1^2\cdot3}-\frac{z^4}{1^2\cdot3^2\cdot5}+\frac{z^6}{1^2\cdot3^2\cdot5^2\cdot7}-\cdots\right]$$

Integral Representations

If $\mathscr{R}\,\nu>-\frac{1}{2}$,

12.1.6

$$\mathbf{H}_\nu(z)=\frac{2\,(\frac{1}{2}z)^\nu}{\sqrt{\pi}\,\Gamma(\nu+\frac{1}{2})}\int_0^1(1-t^2)^{\nu-\frac{1}{2}}\sin(z\,t)\,dt$$

12.1.7 $\quad =\frac{2(\frac{1}{2}z)^\nu}{\sqrt{\pi}\,\Gamma(\nu+\frac{1}{2})}\int_0^{\frac{\pi}{2}}\sin(z\cos\theta)\sin^{2\nu}\theta\,d\theta$

12.1.8 $\quad =Y_\nu(z)$

$$+\frac{2(\frac{1}{2}z)^\nu}{\sqrt{\pi}\,\Gamma(\nu+\frac{1}{2})}\int_0^\infty e^{-zt}(1+t^2)^{\nu-\frac{1}{2}}dt$$

$$\left(|\arg z|<\frac{\pi}{2}\right)$$

Recurrence Relations

12.1.9 $\quad \mathbf{H}_{\nu-1}+\mathbf{H}_{\nu+1}=\frac{2\nu}{z}\,\mathbf{H}_\nu+\frac{(\frac{1}{2}z)^\nu}{\sqrt{\pi}\,\Gamma(\nu+\frac{3}{2})}$

12.1.10 $\quad \mathbf{H}_{\nu-1}-\mathbf{H}_{\nu+1}=2\mathbf{H}_\nu'-\frac{(\frac{1}{2}z)^\nu}{\sqrt{\pi}\,\Gamma(\nu+\frac{3}{2})}$

12.1.11 $\quad \mathbf{H}_0'=(2/\pi)-\mathbf{H}_1$

12.1.12 $\quad \frac{d}{dz}(z^\nu\mathbf{H}_\nu)=z^\nu\mathbf{H}_{\nu-1}$

12.1.13 $\quad \frac{d}{dz}(z^{-\nu}\mathbf{H}_\nu)=\frac{1}{\sqrt{\pi}\,2^\nu\Gamma(\nu+\frac{3}{2})}-z^{-\nu}\mathbf{H}_{\nu+1}$

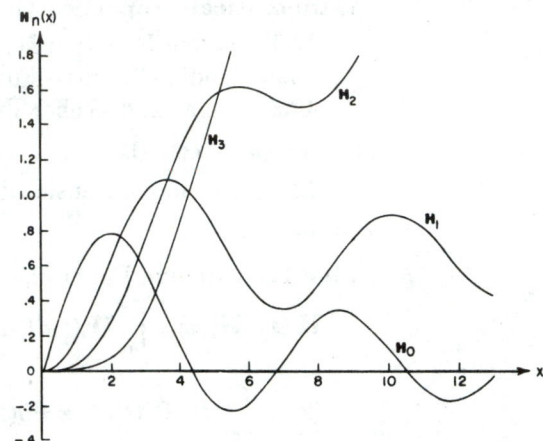

FIGURE 12.1. *Struve functions.*

$\mathbf{H}_n(x),\ n=0(1)3$

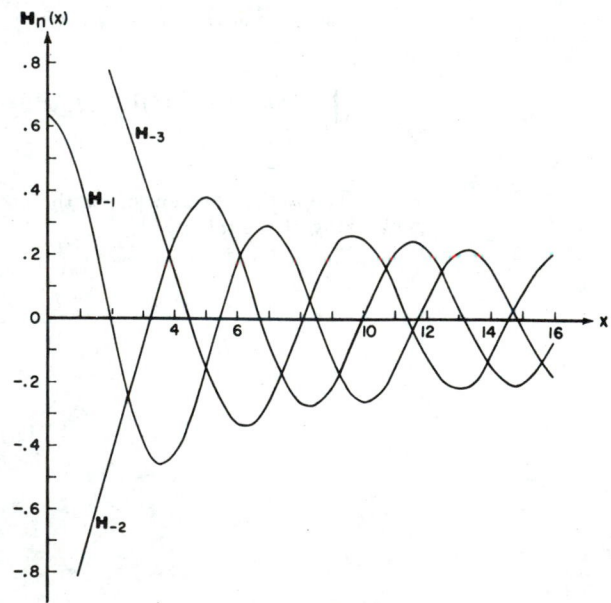

FIGURE 12.2. *Struve functions.*

$\mathbf{H}_n(x),\ -n=1(1)3$

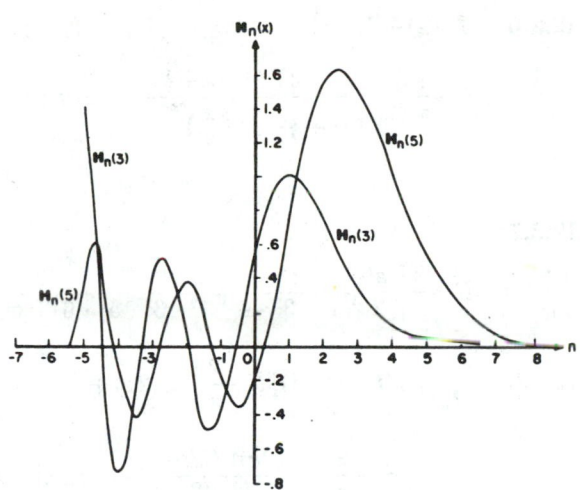

FIGURE 12.3. *Struve functions.*

$$\mathbf{H}_n(x),\ x=3,\ 5$$

Special Properties

12.1.14 $\qquad \mathbf{H}_\nu(x)\geq 0 \qquad\qquad (x>0 \text{ and } \nu\geq \tfrac{1}{2})$

12.1.15
$$\mathbf{H}_{-(n+\frac{1}{2})}(z)=(-1)^n J_{n+\frac{1}{2}}(z) \quad (n \text{ an integer}\geq 0)$$

12.1.16 $\qquad \mathbf{H}_{\frac{1}{2}}(z)=\left(\dfrac{2}{\pi z}\right)^{\frac{1}{2}}(1-\cos z)$

12.1.17
$$\mathbf{H}_{\frac{3}{2}}(z)=\left(\frac{z}{2\pi}\right)^{\frac{1}{2}}\left(1+\frac{2}{z^2}\right)-\left(\frac{2}{\pi z}\right)^{\frac{1}{2}}\left(\sin z+\frac{\cos z}{z}\right)$$

12.1.18 $\quad \mathbf{H}_\nu(ze^{m\pi i})=e^{m(\nu+1)\pi i}\,\mathbf{H}_\nu(z)\ (m \text{ an integer})$

12.1.19 $\qquad \mathbf{H}_0(z)=\dfrac{4}{\pi}\displaystyle\sum_{k=0}^{\infty}\dfrac{J_{2k+1}(z)}{2k+1}$

12.1.20 $\quad \mathbf{H}_1(z)=\dfrac{2}{\pi}-\dfrac{2}{\pi}\,J_0(z)+\dfrac{4}{\pi}\displaystyle\sum_{k=1}^{\infty}\dfrac{J_{2k}(z)}{4k^2-1}$

12.1.21 $\quad \mathbf{H}_\nu(z)=\dfrac{2(z/2)^{\nu+1}}{\sqrt{\pi}\,\Gamma(\nu+\frac{3}{2})}\,{}_1F_2\left(1;\frac{3}{2}+\nu,\frac{3}{2};-\frac{z^2}{4}\right)$

Integrals (See chapter 11)

12.1.22 $\qquad \displaystyle\int_0^\infty t^{-1}\,\mathbf{H}_0(t)dt=\dfrac{\pi}{2}$

12.1.23
$$\int_0^z \mathbf{H}_0(t)dt=\frac{2}{\pi}\left[\frac{z^2}{2}-\frac{z^4}{1^2\cdot 3^2\cdot 4}+\frac{z^6}{1^2\cdot 3^2\cdot 5^2\cdot 6}-\cdots\right]$$

12.1.24 $\quad \displaystyle\int_0^z t^{-\nu}\,\mathbf{H}_{\nu+1}(t)dt=\dfrac{z}{2^\nu\sqrt{\pi}\,\Gamma(\nu+\frac{3}{2})}-z^{-\nu}\,\mathbf{H}_\nu(z)$

Struve's Integral

12.1.25
$$\frac{4}{\pi}\int_z^\infty t^{-2}\mathbf{H}_1(t)dt=\frac{2}{\pi z}\mathbf{H}_1(z)+\frac{2}{\pi}\int_z^\infty t^{-1}\mathbf{H}_0(t)dt$$

12.1.26
$$\frac{2}{\pi}\int_z^\infty t^{-1}\mathbf{H}_0(t)dt=1-\frac{4}{\pi^2}\left[z-\frac{z^3}{1^2\cdot 3^2\cdot 3}+\frac{z^5}{1^2\cdot 3^2\cdot 5^2\cdot 5}-\cdots\right]$$

12.1.27
$$\int_0^\infty t^{\mu-\nu-1}\mathbf{H}_\nu(t)dt=\frac{\Gamma(\frac{1}{2}\mu)2^{\mu-\nu-1}\tan(\frac{1}{2}\pi\mu)}{\Gamma(\nu-\frac{1}{2}\mu+1)}$$
$$(|\mathscr{R}\mu|<1,\ \mathscr{R}\nu>\mathscr{R}\mu-\tfrac{3}{2})$$

$$\text{If } f_\nu(z)=\int_0^z \mathbf{H}_\nu(t)t^\nu dt$$

12.1.28
$$f_{\nu+1}=(2\nu+1)f_\nu(z)-z^{\nu+1}\mathbf{H}_\nu(z)$$
$$+\frac{z^{2\nu+2}}{(\nu+1)2^{\nu+1}\Gamma(\frac{1}{2})\Gamma(\nu+\frac{3}{2})}\ (\mathscr{R}\nu>-\tfrac{1}{2})$$

Asymptotic Expansions for Large |z|

12.1.29
$$\mathbf{H}_\nu(z)-Y_\nu(z)=\frac{1}{\pi}\sum_{k=0}^{m-1}\frac{\Gamma(k+\frac{1}{2})}{\Gamma(\nu+\frac{1}{2}-k)}\left(\frac{z}{2}\right)^{2k-\nu+1}+R_m$$
$$(|\arg z|<\pi)$$

where $R_m=O(|z|^{\nu-2m-1})$. If ν is real, z positive *
and $m+\frac{1}{2}-\nu\geq 0$, the remainder after m terms is
of the same sign and numerically less than the
first term neglected.

12.1.30
$$\mathbf{H}_0(z)-Y_0(z)\sim\frac{2}{\pi}\left[\frac{1}{z}-\frac{1}{z^3}+\frac{1^2\cdot 3^2}{z^5}-\frac{1^2\cdot 3^2\cdot 5^2}{z^7}+\cdots\right]$$
$$(|\arg z|<\pi)$$

12.1.31
$$\mathbf{H}_1(z)-Y_1(z)\sim\frac{2}{\pi}\left[1+\frac{1}{z^2}-\frac{1^2\cdot 3}{z^4}+\frac{1^2\cdot 3^2\cdot 5}{z^6}-\cdots\right]$$
$$(|\arg z|<\pi)$$

12.1.32
$$\int_0^z [\mathbf{H}_0(t)-Y_0(t)]dt-\frac{2}{\pi}[\ln(2z)+\gamma]$$
$$\sim\frac{2}{\pi}\sum_{k=1}^{\infty}\frac{(-1)^{k+1}(2k)!(2k-1)!}{(k!)^2(2z)^{2k}}\quad (|\arg z|<\pi)$$

where $\gamma=.57721\ 56649\ \ldots$ is Euler's constant.

12.1.33
$$\int_z^\infty t^{-1}[\mathbf{H}_0(t)-Y_0(t)]dt\sim\frac{2}{\pi z}\sum_{k=0}^{\infty}\frac{(-1)^k[(2k)!]^2}{(k!)^2(2k+1)(2z)^{2k}}$$
$$(|\arg z|<\pi)$$

Asymptotic Expansions for Large Orders

12.1.34

$$\mathbf{H}_\nu(z) - Y_\nu(z) \sim \frac{2(\tfrac{1}{2}z)^\nu}{\sqrt{\pi}\,\Gamma(\nu+\tfrac{1}{2})} \sum_{k=0}^{\infty} \frac{k!\,b_k}{z^{k+1}}$$

$$(|\arg z| < \tfrac{1}{2}\pi, |\nu| < |z|)$$

$$b_0 = 1,\ b_1 = 2\nu/z,\ b_2 = 6(\nu/z)^2 - \tfrac{1}{2},\ b_3 = 20(\nu/z)^3 - 4(\nu/z)$$

12.1.35

$$\mathbf{H}_\nu(z) + iJ_\nu(z) \sim \frac{2(\tfrac{1}{2}z)^\nu}{\sqrt{\pi}\,\Gamma(\nu+\tfrac{1}{2})} \sum_{k=0}^{\infty} \frac{k!\,b_k}{z^{k+1}} \qquad (|\nu| > |z|)$$

12.2. Modified Struve Function $\mathbf{L}_\nu(z)$

Power Series Expansion

12.2.1 $\quad \mathbf{L}_\nu(z) = -ie^{-\frac{i\nu\pi}{2}} \mathbf{H}_\nu(iz)$

$$= (\tfrac{1}{2}z)^{\nu+1} \sum_{k=0}^{\infty} \frac{(z/2)^{2k}}{\Gamma(k+\tfrac{3}{2})\,\Gamma(k+\nu+\tfrac{3}{2})}$$

Integral Representations

12.2.2 $\quad \mathbf{L}_\nu(z) = \frac{2(z/2)^\nu}{\sqrt{\pi}\,\Gamma(\nu+\tfrac{1}{2})} \int_0^{\frac{\pi}{2}} \sinh(z\cos\theta)\sin^{2\nu}\theta\,d\theta$

$$(\mathscr{R}\nu > -\tfrac{1}{2})$$

12.2.3

$$I_{-\nu}(x) - \mathbf{L}_\nu(x) = \frac{2(x/2)^\nu}{\sqrt{\pi}\,\Gamma(\nu+\tfrac{1}{2})} \int_0^{\infty} \sin(tx)(1+t^2)^{\nu-\frac{1}{2}}\,dt$$

$$(\mathscr{R}\nu < \tfrac{1}{2}, x > 0)$$

Recurrence Relations

12.2.4 $\quad \mathbf{L}_{\nu-1} - \mathbf{L}_{\nu+1} = \frac{2\nu}{z}\,\mathbf{L}_\nu + \frac{(z/2)^\nu}{\sqrt{\pi}\,\Gamma(\nu+\tfrac{3}{2})}$

12.2.5 $\quad \mathbf{L}_{\nu-1} + \mathbf{L}_{\nu+1} = 2\mathbf{L}_\nu' - \frac{(z/2)^\nu}{\sqrt{\pi}\,\Gamma(\nu+\tfrac{3}{2})}$

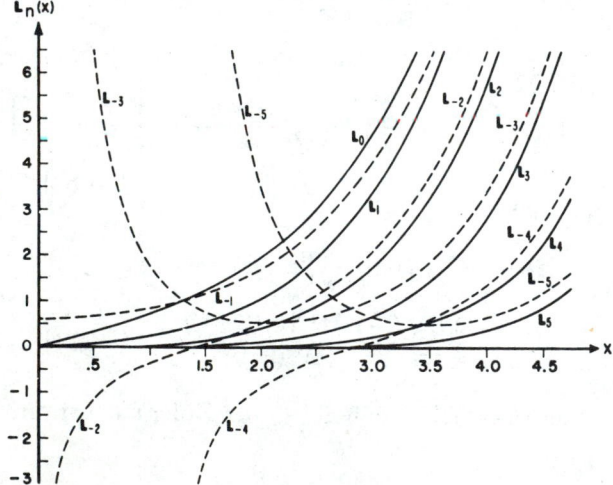

FIGURE 12.4. *Modified Struve functions.*

$$\mathbf{L}_n(x),\ \pm n = 0(1)5$$

Asymptotic Expansion for Large $|z|$

12.2.6 $\quad \mathbf{L}_\nu(z) - I_{-\nu}(z)$

$$\sim \frac{1}{\pi} \sum_{k=0}^{\infty} \frac{(-1)^{k+1}\Gamma(k+\tfrac{1}{2})}{\Gamma(\nu+\tfrac{1}{2}-k)\left(\dfrac{z}{2}\right)^{2k-\nu+1}}\ (|\arg z| < \tfrac{1}{2}\pi)$$

Integrals

12.2.7

$$\int_0^z \mathbf{L}_0(t)\,dt = \frac{2}{\pi}\left[\frac{z^2}{2} + \frac{z^4}{1^2\cdot 3^2\cdot 4} + \frac{z^6}{1^2\cdot 3^2\cdot 5^2\cdot 6} + \cdots\right]$$

12.2.8 $\quad \displaystyle\int_0^z [I_0(t) - \mathbf{L}_0(t)]\,dt - \frac{2}{\pi}[\ln(2z) + \gamma]$

$$\sim -\frac{2}{\pi} \sum_{k=1}^{\infty} \frac{(2k)!\,(2k-1)!}{(k!)^2(2z)^{2k}}\ (|\arg z| < \tfrac{1}{2}\pi)$$

12.2.9 $\quad \displaystyle\int_0^z \mathbf{L}_1(t)\,dt = \mathbf{L}_0(z) - \frac{2}{\pi}\,z$

Relation to Modified Spherical Bessel Function

12.2.10 $\quad \mathbf{L}_{-(n+\frac{1}{2})}(z) = I_{(n+\frac{1}{2})}(z) \qquad (n \text{ an integer} \geq 0)$

12.3. Anger and Weber Functions

Anger's Function

12.3.1 $\quad \mathbf{J}_\nu(z) = \frac{1}{\pi} \int_0^\pi \cos(\nu\theta - z\sin\theta)\,d\theta$

12.3.2 $\quad \mathbf{J}_n(z) = J_n(z) \qquad\qquad (n \text{ an integer})$

Weber's Function

12.3.3 $\quad \mathbf{E}_\nu(z) = \frac{1}{\pi} \int_0^\pi \sin(\nu\theta - z\sin\theta)\,d\theta$

Relations Between Anger's and Weber's Function

12.3.4 $\quad \sin(\nu\pi)\,\mathbf{J}_\nu(z) = \cos(\nu\pi)\,\mathbf{E}_\nu(z) - \mathbf{E}_{-\nu}(z)$

12.3.5 $\quad \sin(\nu\pi)\,\mathbf{E}_\nu(z) = \mathbf{J}_{-\nu}(z) - \cos(\nu\pi)\,\mathbf{J}_\nu(z)$

Relations Between Weber's Function and Struve's Function

If n is a positive integer or zero,

12.3.6 $\quad \mathbf{E}_n(z) = \frac{1}{\pi} \sum_{k=0}^{\left[\frac{n-1}{2}\right]} \frac{\Gamma(k+\tfrac{1}{2})(\tfrac{1}{2}z)^{n-2k-1}}{\Gamma(n+\tfrac{1}{2}-k)} - \mathbf{H}_n(z) \qquad *$

12.3.7

$$\mathbf{E}_{-n}(z) = \frac{(-1)^{n+1}}{\pi} \sum_{k=0}^{\left[\frac{n-1}{2}\right]} \frac{\Gamma(n-k-\tfrac{1}{2})(\tfrac{1}{2}z)^{-n+2k+1}}{\Gamma(k+\tfrac{3}{2})} - \mathbf{H}_{-n}(z) \qquad *$$

12.3.8 $\qquad \mathbf{E}_0(z) = -\mathbf{H}_0(z)$

12.3.9 $\qquad \mathbf{E}_1(z) = \dfrac{2}{\pi} - \mathbf{H}_1(z)$

12.3.10 $\qquad \mathbf{E}_2(z) = \dfrac{2z}{3\pi} - \mathbf{H}_2(z)$

Numerical Methods

12.4. Use and Extension of the Tables

Example 1. Compute $\mathbf{L}_0(2)$ to 6D. From **Table 12.1** $I_0(2) - \mathbf{L}_0(2) = .342152$; from **Table 9.11** we have $I_0(2) = 2.279585$ so that $\mathbf{L}_0(2) = 1.937433$.

Example 2. Compute $\mathbf{H}_0(10)$ to 6D. From **Table 12.2** for $x^{-1} = .1$, $\mathbf{H}_0(10) - Y_0(10) = .063072$; from **Table 9.1** we have $Y_0(10) = .055671$. Thus, $\mathbf{H}_0(10) = .118743$.

Example 3. Compute $\int_0^x \mathbf{H}_0(t)dt$ for $x = 6$ to 5D. Using **Tables 12.2, 11.1** and **4.2**, we have

$$\int_0^6 \mathbf{H}_0(t)dt = \int_0^6 Y_0(t)dt + \frac{2}{\pi}\ln 6 + f_1(6)$$

$$= -.125951 + (.636620)(1.791759)$$
$$+ .816764$$

$$= 1.83148$$

Example 4. Compute $\mathbf{H}_n(x)$ for $x = 4$, $-n = 0(1)8$ to 6S. From **Table 12.1** we have $\mathbf{H}_0(4) = .1350146$, $\mathbf{H}_1(4) = 1.0697267$. Using **12.1.9** we find

$\mathbf{H}_{-1}(4) = -.433107$	$\mathbf{H}_{-5}(4) = .689652$
$\mathbf{H}_{-2}(4) = .240694$	$\mathbf{H}_{-6}(4) = -1.21906$
$\mathbf{H}_{-3}(4) = .152624$	$\mathbf{H}_{-7}(4) = 2.82066$
$\mathbf{H}_{-4}(4) = -.439789$	$\mathbf{H}_{-8}(4) = -8.24933$

Example 5. Compute $\mathbf{H}_n(x)$ for $x = 4$, $n = 0(1)10$ to 7S. Starting with the values of $\mathbf{H}_0(4)$ and $\mathbf{H}_1(4)$ and using **12.1.9** with forward recurrence, we get

$\mathbf{H}_0(4) = .13501\ 46$	$\mathbf{H}_6(4) = .05433\ 54$
$\mathbf{H}_1(4) = 1.06972\ 67$	$\mathbf{H}_7(4) = .01510\ 37$
$\mathbf{H}_2(4) = 1.24867\ 51$	$\mathbf{H}_8(4) = .00367\ 33$
$\mathbf{H}_3(4) = .85800\ 95$	$\mathbf{H}_9(4) = .00080\ 02$
$\mathbf{H}_4(4) = .42637\ 41$	$\mathbf{H}_{10}(4) = .00018\ 25$
$\mathbf{H}_5(4) = .16719\ 87$	

We note that for $n > 6$ there is a rapid loss of significant figures. On the other hand using **12.1.3** for $x = 4$ we find $\mathbf{H}_9(4) = .0007935729$, $\mathbf{H}_{10}(4) = .00015447630$ and backward recurrence with **12.1.9** gives

$\mathbf{H}_8(4) = .00367\ 1495$	$\mathbf{H}_3(4) = .85800\ 94$
$\mathbf{H}_7(4) = .01510\ 315$	$\mathbf{H}_2(4) = 1.24867\ 6$
$\mathbf{H}_6(4) = .05433\ 519$	$\mathbf{H}_1(4) = 1.06972\ 7$
$\mathbf{H}_5(4) = .16719\ 87$	$\mathbf{H}_0(4) = .13501\ 4$
$\mathbf{H}_4(4) = .42637\ 43$	

Example 6. Compute $\mathbf{L}_n(.5)$ for $n = 0(1)5$ to 8S. From **12.2.1** we find $\mathbf{L}_5(.5) = 9.6307462 \times 10^{-7}$, $\mathbf{L}_4(.5) = 2.1212342 \times 10^{-5}$. Then, with **12.2.4** we get

$\mathbf{L}_3(.5) = 3.82465\ 03 \times 10^{-4}$	$\mathbf{L}_1(.5) = .05394\ 2181$
$\mathbf{L}_2(.5) = 5.36867\ 34 \times 10^{-3}$	$\mathbf{L}_0(.5) = .32724\ 068$

Example 7. Compute $\mathbf{L}_n(.5)$ for $-n = 0(1)5$ to 6S. From **Tables 12.1** and **9.8** we find $\mathbf{L}_0(.5) = .327240$, $\mathbf{L}_1(.5) = .053942$. Then employing **12.2.4** with backward recurrence we get

$\mathbf{L}_{-1}(.5) = .690562$	$\mathbf{L}_{-4}(.5) = -75.1418$
$\mathbf{L}_{-2}(.5) = -1.16177$	$\mathbf{L}_{-5}(.5) = 1056.92$
$\mathbf{L}_{-3}(.5) = 7.43824$	

Example 8. Compute $\mathbf{L}_n(x)$ for $x = 6$ and $-n = 0(1)6$ to 8S. From **Tables 12.2** and **9.8** we find $\mathbf{L}_0(6) = 67.124454$, $\mathbf{L}_1(6) = 60.725011$. Using **12.2.4** we get

$\mathbf{L}_{-1}(6) = 61.361631$	$\mathbf{L}_{-4}(6) = 16.626028$
$\mathbf{L}_{-2}(6) = 46.776680$	$\mathbf{L}_{-5}(6) = 7.984089$
$\mathbf{L}_{-3}(6) = 30.159494$	$\mathbf{L}_{-6}(6) = 3.32780$

We note that there is no essential loss of accuracy until $n = -6$. However, if further values were necessary the recurrence procedure becomes unstable. To avoid the instability use the methods described in **Examples 5** and **6**.

References

Texts

[12.1] R. K. Cook, Some properties of Struve functions, J. Washington Acad. Sci. **47,** 11, 365–368 (1957).

[12.2] A. Erdélyi et al., Higher transcendental functions, vol. 2, ch. 7 (McGraw-Hill Book Co., Inc., New York, N.Y., 1953).

[12.3] A. Gray and G. B. Mathews, A treatise on Bessel functions, ch. 14 (The Macmillan Co., New York, N.Y., 1931).

[12.4] N. W. McLachlan, Bessel functions for engineers, 2d ed. ch. 4 (Clarendon Press, Oxford, England, 1955).

[12.5] F. Oberhettinger, On some expansions for Bessel integral functions, J. Research NBS **59** (1957) RP2786.

[12.6] G. Petiau, La théorie des fonctions de Bessel, ch. 10 (Centre National de la Recherche Scientifique, Paris, France, 1955).

[12.7] G. N. Watson, A treatise on the theory of Bessel functions, ch. 10 (Cambridge Univ. Press, Cambridge, England, 1958).

Tables

[12.8] M. Abramowitz, Tables of integrals of Struve functions, J. Math. Phys. **29,** 49–51 (1950).

[12.9] C. W. Horton, On the extension of some Lommel integrals to Struve functions with an application to acoustic radiation, J. Math. Phys. **29,** 31–37 (1950).

[12.10] C. W. Horton, A short table of Struve functions and of some integrals involving Bessel and Struve functions, J. Math. Phys. **29,** 56–58 (1950).

[12.11] Mathematical Tables Project, Table of the Struve functions $\mathbf{L}_\nu(x)$ and $\mathbf{H}_\nu(x)$, J. Math. Phys. **25,** 252–259 (1946).

STRUVE FUNCTIONS

Table 12.1

x	$\mathbf{H}_0(x)$	$\mathbf{H}_1(x)$	$\int_0^x \mathbf{H}_0(t)dt$	$I_0(x)-\mathbf{L}_0(x)$	$I_1(x)-\mathbf{L}_1(x)$	$f_0(x)$	$\frac{2}{\pi}\int_x^\infty \frac{\mathbf{H}_0(t)}{t}dt$
0.0	0.00000 00	0.00000 00	0.000000	1.000000	0.000000	0.00000	1.000000
0.1	0.06359 13	0.00212 07	0.003181	0.938769	0.047939	0.09690	0.959487
0.2	0.12675 90	0.00846 57	0.012704	0.882134	0.091990	0.18791	0.919063
0.3	0.18908 29	0.01898 43	0.028505	0.829724	0.132480	0.27347	0.878819
0.4	0.25014 97	0.03359 25	0.050479	0.781198	0.169710	0.35398	0.838843
0.5	0.30955 59	0.05217 37	0.078480	0.736243	0.203952	0.42982	0.799223
0.6	0.36691 14	0.07457 97	0.112322	0.694573	0.235457	0.50134	0.760044
0.7	0.42184 24	0.10063 17	0.151781	0.655927	0.264454	0.56884	0.721389
0.8	0.47399 44	0.13012 25	0.196597	0.620063	0.291151	0.63262	0.683341
0.9	0.52303 50	0.16281 75	0.246476	0.586763	0.315740	0.69294	0.645976
1.0	0.56865 66	0.19845 73	0.301090	0.555823	0.338395	0.75005	0.609371
1.1	0.61057 87	0.23675 97	0.360084	0.527058	0.359276	0.80418	0.573596
1.2	0.64855 00	0.27742 18	0.423074	0.500300	0.378530	0.85553	0.538719
1.3	0.68235 03	0.32012 31	0.489655	0.475391	0.396290	0.90430	0.504803
1.4	0.71179 25	0.36452 80	0.559399	0.452188	0.412679	0.95066	0.471907
1.5	0.73672 35	0.41028 85	0.631863	0.430561	0.427810	0.99479	0.440086
1.6	0.75702 55	0.45704 72	0.706590	0.410388	0.441783	1.03682	0.409388
1.7	0.77261 68	0.50444 07	0.783111	0.391558	0.454694	1.07691	0.379857
1.8	0.78345 23	0.55210 21	0.860954	0.373970	0.466629	1.11518	0.351533
1.9	0.78952 36	0.59966 45	0.939643	0.357530	0.477666	1.15174	0.324450
2.0	0.79085 88	0.64676 37	1.018701	0.342152	0.487877	1.18672	0.298634
2.1	0.78752 22	0.69304 18	1.097659	0.327756	0.497329	1.22020	0.274109
2.2	0.77961 35	0.73814 96	1.176053	0.314270	0.506083	1.25230	0.250891
2.3	0.76726 65	0.78174 98	1.253434	0.301627	0.514194	1.28309	0.228992
2.4	0.75064 85	0.82351 98	1.329364	0.289765	0.521712	1.31265	0.208417
2.5	0.72995 77	0.86315 42	1.403427	0.278627	0.528685	1.34106	0.189168
2.6	0.70542 23	0.90036 74	1.475227	0.268162	0.535156	1.36840	0.171238
2.7	0.67729 77	0.93489 57	1.544392	0.258319	0.541164	1.39472	0.154618
2.8	0.64586 46	0.96649 98	1.610577	0.249056	0.546746	1.42008	0.139293
2.9	0.61142 64	0.99496 63	1.673465	0.240332	0.551933	1.44455	0.125242
3.0	0.57430 61	1.02010 96	1.732773	0.232107	0.556757	1.46816	0.112439
3.1	0.53484 44	1.04177 30	1.788248	0.224348	0.561246	1.49098	0.100857
3.2	0.49339 57	1.05983 03	1.839675	0.217022	0.565426	1.51305	0.090460
3.3	0.45032 57	1.07418 63	1.886873	0.210099	0.569319	1.53440	0.081212
3.4	0.40600 80	1.08477 74	1.929699	0.203553	0.572948	1.55508	0.073071
3.5	0.36082 08	1.09157 23	1.968046	0.197357	0.576333	1.57512	0.065992
3.6	0.31514 40	1.09457 16	2.001847	0.191488	0.579492	1.59456	0.059928
3.7	0.26935 59	1.09380 77	2.031071	0.185924	0.582442	1.61343	0.054829
3.8	0.22382 98	1.08934 44	2.055726	0.180646	0.585199	1.63176	0.050642
3.9	0.17893 12	1.08127 62	2.075858	0.175634	0.587776	1.64957	0.047311
4.0	0.13501 46	1.06972 67	2.091545	0.170872	0.590187	1.66689	0.044781
4.1	0.09242 08	1.05484 79	2.102905	0.166343	0.592445	1.68375	0.042994
4.2	0.05147 40	1.03681 86	2.110084	0.162032	0.594560	1.70017	0.041891
4.3	+0.01247 93	1.01584 22	2.113265	0.157926	0.596542	1.71616	0.041414
4.4	−0.02427 98	0.99214 51	2.112655	0.154012	0.598402	1.73176	0.041502
4.5	−0.05854 33	0.96597 44	2.108492	0.150279	0.600147	1.74697	0.042096
4.6	−0.09007 71	0.93759 56	2.101037	0.146714	0.601787	1.76182	0.043139
4.7	−0.11867 42	0.90729 01	2.090574	0.143309	0.603328	1.77632	0.044571
4.8	−0.14415 67	0.87535 28	2.077406	0.140053	0.604777	1.79049	0.046335
4.9	−0.16637 66	0.84208 90	2.061852	0.136938	0.606142	1.80434	0.048376
5.0	−0.18521 68	0.80781 19	2.044244	0.133955	0.607426	1.81788	0.050640
	$\left[\begin{smallmatrix}(-4)6\\5\end{smallmatrix}\right]$	$\left[\begin{smallmatrix}(-4)5\\5\end{smallmatrix}\right]$	$\left[\begin{smallmatrix}(-4)8\\5\end{smallmatrix}\right]$	$\left[\begin{smallmatrix}(-4)6\\5\end{smallmatrix}\right]$	$\left[\begin{smallmatrix}(-4)5\\5\end{smallmatrix}\right]$	$\left[\begin{smallmatrix}(-4)7\\4\end{smallmatrix}\right]$	$\left[\begin{smallmatrix}(-4)2\\4\end{smallmatrix}\right]$

$$\int_0^x [I_0(t)-\mathbf{L}_0(t)]dt=f_0(x)$$

$\mathbf{H}_0(x)$, $\mathbf{H}_1(x)$, $\mathbf{L}_0(x)$, $\mathbf{L}_1(x)$, compiled from Mathematical Tables Project, Table of the Struve functions $\mathbf{L}_\nu(x)$ and $\mathbf{H}_\nu(x)$, J. Math. Phys. **25**, 252–259, 1946 (with permission).

$\int_0^x \mathbf{H}_0(t)dt$, $\int_0^x [I_0(t)-\mathbf{L}_0(t)]dt$, $\frac{2}{\pi}\int_x^\infty \frac{\mathbf{H}_0(t)}{t}dt$, compiled from M. Abramowitz, Tables of integrals of Struve functions, J. Math. Phys. **29**, 49–51, 1950 (with permission).

Table 12.2   **STRUVE FUNCTIONS FOR LARGE ARGUMENTS**

x^{-1}	$H_0(x)-Y_0(x)$	$H_1(x)-Y_1(x)$	$f_1(x)$	$I_0(x)-L_0(x)$	$I_1(x)-L_1(x)$	$f_2(x)$	$f_3(x)$	$<x>$
0.20	0.123301	0.659949	0.819924	0.133955	0.607426	0.793280	0.125868	5
0.19	0.117449	0.657819	0.818935	0.126683	0.610467	0.794902	0.119694	5
0.18	0.111556	0.655774	0.817981	0.119468	0.613348	0.796448	0.113505	6
0.17	0.105625	0.653818	0.817062	0.112319	0.616060	0.797910	0.107299	6
0.16	0.099655	0.651952	0.816182	0.105242	0.618598	0.799279	0.101079	6
0.15	0.093647	0.650180	0.815341	0.098241	0.620955	0.800551	0.094843	7
0.14	0.087602	0.648504	0.814541	0.091318	0.623129	0.801721	0.088593	7
0.13	0.081521	0.646927	0.813785	0.084474	0.625119	0.802787	0.082328	8
0.12	0.075404	0.645452	0.813074	0.077706	0.626927	0.803750	0.076051	8
0.11	0.069254	0.644081	0.812411	0.071010	0.628558	0.804611	0.069761	9
0.10	0.063072	0.642817	0.811796	0.064379	0.630018	0.805374	0.063460	10
0.09	0.056860	0.641663	0.811232	0.057805	0.631315	0.806047	0.057147	11
0.08	0.050620	0.640622	0.810722	0.051279	0.632457	0.806634	0.050824	13
0.07	0.044354	0.639696	0.810266	0.044793	0.633450	0.807140	0.044492	14
0.06	0.038064	0.638888	0.809866	0.038340	0.634302	0.807572	0.038152	17
0.05	0.031753	0.638200	0.809525	0.031912	0.635016	0.807933	0.031805	20
0.04	0.025425	0.637634	0.809244	0.025506	0.635596	0.808225	0.025451	25
0.03	0.019082	0.637191	0.809023	0.019116	0.636045	0.808450	0.019093	33
0.02	0.012727	0.636874	0.808865	0.012738	0.636365	0.808611	0.012731	50
0.01	0.006366	0.636683	0.808770	0.006367	0.636556	0.808706	0.006366	100
0.00	0.000000	0.636620	0.808738	0.000000	0.636620	0.808738	0.000000	∞
	$\begin{bmatrix}(-6)5\\3\end{bmatrix}$	$\begin{bmatrix}(-5)2\\3\end{bmatrix}$	$\begin{bmatrix}(-6)8\\3\end{bmatrix}$	$\begin{bmatrix}(-5)1\\3\end{bmatrix}$	$\begin{bmatrix}(-5)2\\3\end{bmatrix}$	$\begin{bmatrix}(-5)1\\3\end{bmatrix}$	$\begin{bmatrix}(-6)2\\3\end{bmatrix}$	

$$\int_0^x [H_0(t)-Y_0(t)]dt = \frac{2}{\pi}\ln x + f_1(x)$$

$$\int_0^x [L_0(t)-I_0(t)]dt = \frac{2}{\pi}\ln x + f_2(x)$$

$$\int_x^\infty \left[\frac{H_0(t)-Y_0(t)}{t}\right]dt = f_3(x)$$

$<x>$ = nearest integer to x.

Starting with $H_0(x)$ and $H_1(x)$, recurrence formula **12.1.9** may be used to generate $H_n(x)$ for $n<0$. As long as $n<x/2$ (approx.), $H_n(x)$ may be generated by forward recurrence. When $n>x/2$, forward recurrence is unstable. To avoid the instability, choose $n>>x$, compute $H_k(x)$ and $H_{k+1}(x)$ with **12.1.3**, and then use backward recurrence with **12.1.9**.

If $n>0$, $L_n(x)$ must be generated by backward recurrence. If $n<0$, $L_n(x)$ may be generated by backward recurrence as long as $L_n(x)$ increases. If $n<0$ and $L_n(x)$ is decreasing, forward recurrence should be used.

See Examples 4–8.

13. Confluent Hypergeometric Functions

Lucy Joan Slater [1]

Contents

The tables were calculated by the author on the electronic calculator EDSACI in the Mathematical Laboratory of Cambridge University, by kind permission of its director, Dr. M. V. Wilkes. The table of $M(a, b, x)$ was recomputed by Alfred E. Beam for uniformity to eight significant figures.

[1] University Mathematical Laboratory, Cambridge. (Prepared under contract with the National Bureau of Standards.)

13. Confluent Hypergeometric Functions

Mathematical Properties

13.1. Definitions of Kummer and Whittaker Functions

Kummer's Equation

13.1.1
$$z \frac{d^2w}{dz^2} + (b-z) \frac{dw}{dz} - aw = 0$$

It has a regular singularity at $z=0$ and an irregular singularity at ∞.

Independent solutions are

Kummer's Function

13.1.2
$$M(a, b, z) = 1 + \frac{az}{b} + \frac{(a)_2 z^2}{(b)_2 2!} + \ldots + \frac{(a)_n z^n}{(b)_n n!} + \ldots$$

where
$$(a)_n = a(a+1)(a+2) \ldots (a+n-1), \quad (a)_0 = 1,$$

and

13.1.3
$$U(a, b, z) = \frac{\pi}{\sin \pi b} \left\{ \frac{M(a, b, z)}{\Gamma(1+a-b)\Gamma(b)} - z^{1-b} \frac{M(1+a-b, 2-b, z)}{\Gamma(a)\Gamma(2-b)} \right\}$$

Parameters $(m, n$ positive integers$)$		$M(a, b, z)$
$b \neq -n$	$a \neq -m$	a convergent series for all values of a, b and z
$b \neq -n$	$a = -m$	a polynomial of degree m in z
$b = -n$	$a \neq -m$	
$b = -n$	$a = -m$, $m > n$	a simple pole at $b = -n$
$b = -n$	$a = -m$, $m \leq n$	undefined

$U(a, b, z)$ is defined even when $b \to \pm n$

As $|z| \to \infty$,

13.1.4
$$M(a, b, z) = \frac{\Gamma(b)}{\Gamma(a)} e^z z^{a-b} [1 + O(|z|^{-1})] \qquad (\mathscr{R} z > 0)$$

and

13.1.5
$$M(a, b, z) = \frac{\Gamma(b)}{\Gamma(b-a)} (-z)^{-a} [1 + O(|z|^{-1})] \qquad (\mathscr{R} z < 0)$$

$U(a, b, z)$ is a many-valued function. Its principal branch is given by $-\pi < \arg z \leq \pi$.

Logarithmic Solution

13.1.6
$$U(a, n+1, z) = \frac{(-1)^{n+1}}{n! \Gamma(a-n)} \left[M(a, n+1, z) \ln z + \sum_{r=0}^{\infty} \frac{(a)_r z^r}{(n+1)_r r!} \{\psi(a+r) - \psi(1+r) - \psi(1+n+r)\} \right] + \frac{(n-1)!}{\Gamma(a)} z^{-n} M(a-n, 1-n, z)_n$$

for $n = 0, 1, 2, \ldots$, where the last function is the sum to n terms. It is to be interpreted as zero when $n = 0$, and $\psi(a) = \Gamma'(a)/\Gamma(a)$.

13.1.7 $\quad U(a, 1-n, z) = z^n U(a+n, 1+n, z)$

As $\mathscr{R} z \to \infty$

13.1.8 $\qquad U(a, b, z) = z^{-a}[1 + O(|z|^{-1})]$

Analytic Continuation

13.1.9
$$U(a, b, ze^{\pm \pi i}) = \frac{\pi}{\sin \pi b} e^{-z} \left\{ \frac{M(b-a, b, z)}{\Gamma(1+a-b)\Gamma(b)} - \frac{e^{\pm \pi i(1-b)} z^{1-b} M(1-a, 2-b, z)}{\Gamma(a)\Gamma(2-b)} \right\}$$

where either upper or lower signs are to be taken throughout.

13.1.10
$$U(a, b, ze^{2\pi in}) = [1 - e^{-2\pi ibn}] \frac{\Gamma(1-b)}{\Gamma(1+a-b)} M(a, b, z) + e^{-2\pi ibn} U(a, b, z)$$

Alternative Notations

$_1F_1(a; b; z)$ or $\Phi(a; b; z)$ for $M(a, b, z)$

$z^{-a} {}_2F_0(a, 1+a-b; ;-1/z)$ or $\Psi(a; b; z)$ for $U(a, b, z)$

Complete Solution

13.1.11 $\quad y = AM(a, b, z) + BU(a, b, z)$

where A and B are arbitrary constants, $b \neq -n$.

Eight Solutions

13.1.12 $\quad y_1 = M(a, b, z)$

13.1.13 $\quad y_2 = z^{1-b} M(1+a-b, 2-b, z)$

13.1.14 $\quad y_3 = e^z M(b-a, b, -z)$

13.1.15 $\quad y_4 = z^{1-b}e^z M(1-a, 2-b, -z)$

13.1.16 $\quad y_5 = U(a, b, z)$

13.1.17 $\quad y_6 = z^{1-b}U(1+a-b, 2-b, z)$

13.1.18 $\quad y_7 = e^z U(b-a, b, -z)$

13.1.19 $\quad y_8 = z^{1-b}e^z U(1-a, 2-b, -z)$

Wronskians

If $W\{m, n\} = y_m y_n' - y_n y_m'$ and

$\epsilon = sgn\ (\mathscr{I}z) = 1$ if $\mathscr{I}z > 0$,

$\qquad\qquad\quad = -1$ if $\mathscr{I}z \leq 0$

13.1.20

$$W\{1, 2\} = W\{3, 4\} = W\{1, 4\} = -W\{2, 3\}$$
$$= (1-b)z^{-b}e^z$$

13.1.21

$$W\{1, 3\} = W\{2, 4\} = W\{5, 6\} = W\{7, 8\} = 0$$

13.1.22 $\quad W\{1, 5\} = -\Gamma(b)z^{-b}e^z/\Gamma(a)$

13.1.23 $\quad W\{1, 7\} = \Gamma(b)e^{\epsilon\pi i b}z^{-b}e^z/\Gamma(b-a)$

13.1.24 $\quad W\{2, 5\} = -\Gamma(2-b)z^{-b}e^z/\Gamma(1+a-b)$

13.1.25 $\quad W\{2, 7\} = -\Gamma(2-b)z^{-b}e^z/\Gamma(1-a)$

13.1.26 $\quad W\{5, 7\} = e^{\epsilon\pi i(b-a)}z^{-b}e^z$

Kummer Transformations

13.1.27 $\quad M(a, b, z) = e^z M(b-a, b, -z)$

13.1.28

$$z^{1-b}M(1+a-b, 2-b, z) = z^{1-b}e^z M(1-a, 2-b, -z)$$

13.1.29 $\quad U(a, b, z) = z^{1-b}U(1+a-b, 2-b, z)$

13.1.30

$$e^z U(b-a, b, -z) = e^{\epsilon\pi i(1-b)}e^z z^{1-b}U(1-a, 2-b, -z)$$

Whittaker's Equation

13.1.31 $\quad \dfrac{d^2w}{dz^2} + [-\dfrac{1}{4} + \dfrac{\kappa}{z} + \dfrac{(\frac{1}{4}-\mu^2)}{z^2}]\,w = 0$

Solutions:

Whittaker's Functions

13.1.32 $\quad M_{\kappa, \mu}(z) = e^{-\frac{1}{2}z}z^{\frac{1}{2}+\mu}M(\frac{1}{2}+\mu-\kappa, 1+2\mu, z)$

13.1.33

$$W_{\kappa, \mu}(z) = e^{-\frac{1}{2}z}z^{\frac{1}{2}+\mu}U(\frac{1}{2}+\mu-\kappa, 1+2\mu, z)$$
$$(-\pi < \arg z \leq \pi, \kappa = \frac{1}{2}b-a, \mu = \frac{1}{2}b-\frac{1}{2})$$

*See page II.

13.1.34

$$W_{\kappa, \mu}(z) = \frac{\Gamma(-2\mu)}{\Gamma(\frac{1}{2}-\mu-\kappa)}\,M_{\kappa, \mu}(z) + \frac{\Gamma(2\mu)}{\Gamma(\frac{1}{2}+\mu-\kappa)}\,M_{\kappa, -\mu}(z)$$

General Confluent Equation

13.1.35

$$w'' + [\frac{2A}{Z} + 2f' + \frac{bh'}{h} - h' - \frac{h''}{h'}]w'$$

$$+ [\left(\frac{bh'}{h} - h' - \frac{h''}{h'}\right)\left(\frac{A}{Z}+f'\right) + \frac{A(A-1)}{Z^2}$$

$$+ \frac{2Af'}{Z} + f'' + f'^2 - \frac{ah'^2}{h}]w = 0$$

Solutions:

13.1.36 $\qquad Z^{-A}e^{-f(Z)}M(a, b, h(Z))$

13.1.37 $\qquad Z^{-A}e^{-f(Z)}U(a, b, h(Z))$

13.2. Integral Representations

$$\mathscr{R}b > \mathscr{R}a > 0$$

13.2.1

$$\frac{\Gamma(b-a)\Gamma(a)}{\Gamma(b)}\,M(a, b, z)$$

$$= \int_0^1 e^{zt}t^{a-1}(1-t)^{b-a-1}dt$$

13.2.2

$$= 2^{1-b}e^{\frac{1}{2}z}\int_{-1}^{+1}e^{-\frac{1}{2}zt}(1+t)^{b-a-1}(1-t)^{a-1}dt$$

13.2.3

$$= 2^{1-b}e^{\frac{1}{2}z}\int_0^\pi e^{-\frac{1}{2}z\cos\theta}\sin^{b-1}\theta\cot^{b-2a}(\tfrac{1}{2}\theta)\,d\theta$$

13.2.4

$$= e^{-Az}\int_A^B e^{zt}(t-A)^{a-1}(B-t)^{b-a-1}dt$$
$$(A = B-1)$$
$$\mathscr{R}a > 0,\ \mathscr{R}z > 0$$

13.2.5

$$\Gamma(a)U(a, b, z) = \int_0^\infty e^{-zt}t^{a-1}(1+t)^{b-a-1}dt$$

13.2.6

$$= e^z \int_1^\infty e^{-zt}(t-1)^{a-1}t^{b-a-1}dt$$

13.2.7

$$= 2^{1-b}e^{\frac{1}{2}z}\int_0^\infty e^{-\frac{1}{2}z\cosh\theta}\sinh^{b-1}\theta\coth^{b-2a}(\tfrac{1}{2}\theta)\,d\theta\ \ *$$

13.2.8 $\Gamma(a)U(a,b,z)$

$$=e^{Az}\int_A^\infty e^{-zt}(t-A)^{a-1}(t+B)^{b-a-1}dt$$
$$(A=1-B)$$

Similar integrals for $M_{\kappa,\mu}(z)$ and $W_{\kappa,\mu}(z)$ can be deduced with the help of **13.1.32** and **13.1.33**.

Barnes-type Contour Integrals

13.2.9

$$\frac{\Gamma(a)}{\Gamma(b)}M(a,b,z)=\frac{1}{2\pi i}\int_{c-i\infty}^{c+i\infty}\frac{\Gamma(-s)\Gamma(a+s)}{\Gamma(b+s)}(-z)^s ds$$

for $|\arg(-z)|<\tfrac{1}{2}\pi$, a, $b\neq 0$, -1, -2, ... The contour must separate the poles of $\Gamma(-s)$ from those of $\Gamma(a+s)$; c is finite.

13.2.10

$$\Gamma(a)\Gamma(1+a-b)z^a U(a,b,z)$$
$$=\frac{1}{2\pi i}\int_{c-i\infty}^{c+i\infty}\Gamma(-s)\Gamma(a+s)\Gamma(1+a-b+s)z^{-s}ds$$

for $|\arg z|<\dfrac{3\pi}{2}$, $a\neq 0$, -1, -2, ..., $b-a\neq 1$, 2, 3, ... The contour must separate the poles of $\Gamma(-s)$ from those of $\Gamma(a+s)$ and $\Gamma(1+a-b+s)$.

13.3. Connections With Bessel Functions
(see chapters **9** and **10**)

Bessel Functions as Limiting Cases

If b and z are fixed,

13.3.1 $\lim\limits_{a\to\infty}\{M(a,b,z/a)/\Gamma(b)\}=z^{\frac{1}{2}-\frac{1}{2}b}I_{b-1}(2\sqrt{z})$

13.3.2 $\lim\limits_{a\to\infty}\{M(a,b,-z/a)/\Gamma(b)\}=z^{\frac{1}{2}-\frac{1}{2}b}J_{b-1}(2\sqrt{z})$

13.3.3

$$\lim_{a\to\infty}\{\Gamma(1+a-b)\,U(a,b,z/a)\}=2z^{\frac{1}{2}-\frac{1}{2}b}K_{b-1}(2\sqrt{z})$$

13.3.4

$$\lim_{a\to\infty}\{\Gamma(1+a-b)U(a,b,-z/a)\}$$
$$=-\pi i e^{\pi i b}z^{\frac{1}{2}-\frac{1}{2}b}H_{b-1}^{(1)}(2\sqrt{z})\quad(\mathscr{I}z>0)$$

13.3.5 $=\pi i e^{-\pi i b}z^{\frac{1}{2}-\frac{1}{2}b}H_{b-1}^{(2)}(2\sqrt{z})\quad(\mathscr{I}z<0)$

Expansions in Series
13.3.6

$$M(a,b,z)=e^{\frac{1}{2}z}\Gamma\,(b-a-\tfrac{1}{2})(\tfrac{1}{4}z)^{a-b+\frac{1}{2}}$$
$$*\quad\cdot\sum_{n=0}^\infty\frac{(2b-2a-1)_n(b-2a)_n(b-a-\tfrac{1}{2}+n)}{n!(b)_n}$$
$$(-1)^n\,I_{b-a-\frac{1}{2}+n}(\tfrac{1}{2}z)\,(b\neq 0,-1,-2,\ldots)$$

13.3.7

$$\frac{M(a,b,z)}{\Gamma(b)}=e^{\frac{1}{2}z}(\tfrac{1}{2}bz-az)^{\frac{1}{2}-\frac{1}{2}b}$$
$$\cdot\sum_{n=0}^\infty A_n(\tfrac{1}{2}z)^{\frac{1}{2}n}(b-2a)^{-\frac{1}{2}n}J_{b-1+n}\left(\sqrt{(2zb-4za)}\right)$$

where
$$A_0=1,\ A_1=0,\ A_2=\tfrac{1}{2}b,$$
$$(n+1)A_{n+1}=(n+b-1)A_{n-1}+(2a-b)A_{n-2},$$
$$(a\text{ real})$$

13.3.8

$$\frac{M(a,b,z)}{\Gamma(b)}$$
$$=e^{hz}\sum_{n=0}^\infty C_n z^n(-az)^{\frac{1}{2}(1-b-n)}J_{b-1+n}(2\sqrt{(-az)})$$

where
$$C_0=1,\ C_1=-bh,\ C_2=-\tfrac{1}{2}(2h-1)a+\tfrac{1}{2}b(b+1)h^2,$$
$$(n+1)C_{n+1}=[(1-2h)n-bh]C_n$$
$$+[(1-2h)a-h(h-1)(b+n-1)]C_{n-1}$$
$$-h(h-1)aC_{n-2}\quad(h\text{ real})$$

13.3.9 $M(a,b,z)=\sum\limits_{n=0}^\infty C_n(a,b)I_n(z)$

where
$$C_0=1,\ C_1(a,b)=2a/b,$$
$$C_{n+1}(a,b)=2aC_n(a+1,b+1)/b-C_{n-1}(a,b)$$

13.4. Recurrence Relations and Differential Properties

13.4.1

$$(b-a)M(a-1,b,z)+(2a-b+z)M(a,b,z)$$
$$-aM(a+1,b,z)=0$$

13.4.2

$$b(b-1)M(a,b-1,z)+b(1-b-z)M(a,b,z)$$
$$+z(b-a)M(a,b+1,z)=0$$

13.4.3

$$(1+a-b)M(a,b,z)-aM(a+1,b,z)$$
$$+(b-1)M(a,b-1,z)=0$$

13.4.4

$$bM(a,b,z)-bM(a-1,b,z)-zM(a,b+1,z)=0$$

13.4.5

$$b(a+z)M(a,b,z)+z(a-b)M(a,b+1,z)$$
$$-abM(a+1,b,z)=0$$

---
*See page II.

13.4.6

$$(a-1+z)M(a, b, z)+(b-a)M(a-1, b, z) \\ +(1-b)M(a, b-1, z)=0$$

13.4.7

$$b(1-b+z)M(a, b, z)+b(b-1)M(a-1, b-1, z) \\ -azM(a+1, b+1, z)=0$$

13.4.8 $\quad M'(a, b, z)=\dfrac{a}{b} M(a+1, b+1, z)$

13.4.9 $\quad \dfrac{d^n}{dz^n} \{M(a, b, z)\}=\dfrac{(a)_n}{(b)_n} M(a+n, b+n, z)$

13.4.10 $\quad aM(a+1, b, z)=aM(a, b, z)+zM'(a, b, z)$

13.4.11

$$(b-a)M(a-1, b, z)=(b-a-z)M(a, b, z) \\ +zM'(a, b, z)$$

13.4.12

$$(b-a)M(a, b+1, z)=bM(a, b, z)-bM'(a, b, z)$$

13.4.13

$$(b-1)M(a, b-1, z)=(b-1)M(a, b, z) \\ +zM'(a, b, z)$$

13.4.14

$$(b-1)M(a-1, b-1, z)=(b-1-z)M(a, b, z) \\ +zM'(a, b, z)$$

13.4.15

$$U(a-1, b, z)+(b-2a-z)U(a, b, z) \\ +a(1+a-b)U(a+1, b, z)=0$$

13.4.16

$$(b-a-1)U(a, b-1, z)+(1-b-z)U(a, b, z) \\ +zU(a, b+1, z)=0$$

13.4.17

$$U(a, b, z)-aU(a+1, b, z)-U(a, b-1, z)=0$$

13.4.18

$$(b-a)U(a, b, z)+U(a-1, b, z) \\ -zU(a, b+1, z)=0$$

13.4.19

$$(a+z)U(a, b, z)-zU(a, b+1, z) \\ +a(b-a-1)U(a+1, b, z)=0$$

13.4.20

$$(a+z-1)U(a, b, z)-U(a-1, b, z) \\ +(1+a-b)U(a, b-1, z)=0$$

13.4.21 $\quad U'(a, b, z)=-aU(a+1, b+1, z)$

13.4.22

$$\dfrac{d^n}{dz^n} \{U(a, b, z)\}=(-1)^n(a)_nU(a+n, b+n, z)$$

13.4.23

$$a(1+a-b)U(a+1, b, z)=aU(a, b, z) \\ +zU'(a, b, z)$$

13.4.24

$$(1+a-b)U(a, b-1, z)=(1-b)U(a, b, z) \\ -zU'(a, b, z)$$

13.4.25 $\quad U(a, b+1, z)=U(a, b, z)-U'(a, b, z)$

13.4.26

$$U(a-1, b, z)=(a-b+z)U(a, b, z)-zU'(a, b, z)$$

13.4.27

$$U(a-1, b-1, z)=(1-b+z)U(a, b, z) \\ -zU'(a, b, z)$$

13.4.28 $\quad 2\mu M_{\kappa-\frac{1}{2}, \mu-\frac{1}{2}}(z)-z^{\frac{1}{2}}M_{\kappa, \mu}(z)=2\mu M_{\kappa+\frac{1}{2}, \mu-\frac{1}{2}}(z)$

13.4.29

$$(1+2\mu+2\kappa)M_{\kappa+1, \mu}(z)-(1+2\mu-2\kappa)M_{\kappa-1, \mu}(z) \\ =2(2\kappa-z)M_{\kappa, \mu}(z)$$

13.4.30

$$W_{\kappa+\frac{1}{2}, \mu}(z)-z^{\frac{1}{2}}W_{\kappa, \mu+\frac{1}{2}}(z)+(\kappa+\mu)W_{\kappa-\frac{1}{2}, \mu}(z)=0$$

13.4.31

$$(2\kappa-z)W_{\kappa, \mu}(z)+W_{\kappa+1, \mu}(z) \\ =(\mu-\kappa+\tfrac{1}{2})(\mu+\kappa-\tfrac{1}{2})W_{\kappa-1, \mu}(z)$$

13.4.32

$$zM'_{\kappa, \mu}(z)=(\tfrac{1}{2}z-\kappa)M_{\kappa, \mu}(z)+(\tfrac{1}{2}+\mu+\kappa)M_{\kappa+1, \mu}(z)$$

13.4.33 $\quad zW'_{\kappa, \mu}(z)=(\tfrac{1}{2}z-\kappa)W_{\kappa, \mu}(z)-W_{\kappa+1, \mu}(z)$

13.5. Asymptotic Expansions and Limiting Forms

For $|z|$ large, $(a, b$ fixed)

13.5.1

$$\frac{M(a, b, z)}{\Gamma(b)}=$$

$$\frac{e^{\pm i\pi a}z^{-a}}{\Gamma(b-a)}\left\{\sum_{n=0}^{R-1}\frac{(a)_n(1+a-b)_n}{n!}(-z)^{-n}+O(|z|^{-R})\right\}$$

$$+\frac{e^z z^{a-b}}{\Gamma(a)}\left\{\sum_{n=0}^{S-1}\frac{(b-a)_n(1-a)_n}{n!}z^{-n}+O(|z|^{-S})\right\}$$

the upper sign being taken if $-\frac{1}{2}\pi<\arg z<\frac{3}{2}\pi$, the lower sign if $-\frac{3}{2}\pi<\arg z\le-\frac{1}{2}\pi$.

13.5.2

$$U(a, b, z)=z^{-a}\left\{\sum_{n=0}^{R-1}\frac{(a)_n(1+a-b)_n}{n!}(-z)^{-n}\right.$$

$$\left.+O(|z|^{-R})\right\}\ (-\tfrac{3}{2}\pi<\arg z<\tfrac{3}{2}\pi)$$

Converging Factors for the Remainders

13.5.3

$$O(|z|^{-R})=\frac{(a)_R(1+a-b)_R}{R!}(-z)^{-R}$$

$$\left[\tfrac{1}{2}+\frac{(\tfrac{1}{8}+\tfrac{1}{4}b-\tfrac{1}{2}a+\tfrac{1}{4}z-\tfrac{1}{4}R)}{z}+O(|z|^{-2})\right]$$

and

13.5.4

$$O(|z|^{-S})=\frac{(b-a)_S(1-a)_S}{S!}z^{-s}$$

$$\left[\tfrac{2}{3}-b+2a+z-S+O(|z|^{-1})\right]$$

where the R'th and S'th terms are the smallest in the expansions **13.5.1** and **13.5.2**.

For small z $(a, b$ fixed)

13.5.5   As $|z|\to0$, $M(a, b, 0)=1$, $b\ne-n$

13.5.6 $U(a, b, z)=\dfrac{\Gamma(b-1)}{\Gamma(a)}z^{1-b}+O(|z|^{\mathscr{R}b-2})$

$$(\mathscr{R}b\ge2, b\ne2)$$

13.5.7 $=\dfrac{\Gamma(b-1)}{\Gamma(a)}z^{1-b}+O(|\ln z|)$

$$(b=2)$$

13.5.8 $=\dfrac{\Gamma(b-1)}{\Gamma(a)}z^{1-b}+O(1)$

$$(1<\mathscr{R}b<2)$$

13.5.9 $=-\dfrac{1}{\Gamma(a)}[\ln z+\psi(a)]+O(|z\ln z|)$

$$(b=1)$$

13.5.10 $U(a, b, z)=\dfrac{\Gamma(1-b)}{\Gamma(1+a-b)}+O(|z|^{1-\mathscr{R}b})$

$$(0<\mathscr{R}b<1)$$

13.5.11 $=\dfrac{1}{\Gamma(1+a)}+O(|z\ln z|)$ $(b=0)$

13.5.12 $=\dfrac{\Gamma(1-b)}{\Gamma(1+a-b)}+O(|z|)$

$$(\mathscr{R}b\le0, b\ne0)$$

For large a $(b, z$ fixed)

13.5.13

$$M(a, b, z)=$$

$$\Gamma(b)e^{\frac{1}{2}z}(\tfrac{1}{2}bz-az)^{\frac{1}{4}-\frac{1}{2}b}J_{b-1}(\sqrt{(2bz-4az)})$$

$$[1+O(|\tfrac{1}{2}b-a|^{-\sigma})]$$

where

$$|z|=\left|\tfrac{1}{2}b-a\right|^\rho \text{ and } \sigma=\min(1-\rho, \tfrac{1}{2}-\tfrac{3}{2}\rho), 0\le\rho<\tfrac{1}{3}.$$

13.5.14

$$M(a, b, x)=\Gamma(b)e^{\frac{1}{2}x}(\tfrac{1}{2}bx-ax)^{\frac{1}{4}-\frac{1}{2}b}\pi^{-\frac{1}{2}}$$

$$\cos(\sqrt{(2bx-4ax)}-\tfrac{1}{2}b\pi+\tfrac{1}{4}\pi)$$

$$[1+O(|\tfrac{1}{2}b-a|^{-\frac{1}{2}})]$$

as $a\to-\infty$ for b bounded, x real.

13.5.15

$$U(a, b, z)=$$

$$\Gamma(\tfrac{1}{2}b-a+\tfrac{1}{2})e^{\frac{1}{2}z}z^{\frac{1}{4}-\frac{1}{2}b}[\cos(a\pi)J_{b-1}(\sqrt{(2bz-4az)})$$

$$-\sin(a\pi)Y_{b-1}(\sqrt{(2bz-4az)})]\ [1+O(|\tfrac{1}{2}b-a|^{-\sigma})]$$

where σ is defined in **13.5.13**.

13.5.16

$$U(a, b, x)=\Gamma(\tfrac{1}{2}b-a+\tfrac{1}{4})\pi^{-\frac{1}{2}}e^{\frac{1}{2}x}x^{\frac{1}{4}-\frac{1}{2}b}$$

$$\cos(\sqrt{(2bx-4ax)}-\tfrac{1}{2}b\pi+a\pi+\tfrac{1}{4}\pi)$$

$$[1+O(|\tfrac{1}{2}b-a|^{-\frac{1}{2}})]$$

as $a\to-\infty$ for b bounded, x real.

For large real a, b, x

If $\cosh^2\theta=x/(2b-4a)$ so that $x>2b-a>1$,

13.5.17

$$M(a, b, x)=\Gamma(b)\sin(a\pi)$$

$$\exp[(b-2a)(\tfrac{1}{2}\sinh2\theta-\theta+\cosh^2\theta)]$$

$$[(b-2a)\cosh\theta]^{1-b}[\pi(\tfrac{1}{2}b-a)\sinh2\theta]^{-\frac{1}{2}}$$

$$[1+O(|\tfrac{1}{2}b-a|^{-1})]$$

13.5.18

$$U(a, b, x)=\exp[(b-2a)(\tfrac{1}{2}\sinh2\theta-\theta+\cosh^2\theta)]$$

$$[(b-2a)\cosh\theta]^{1-b}[(\tfrac{1}{2}b-a)\sinh2\theta]^{-\frac{1}{2}}$$

$$[1+O(|\tfrac{1}{2}b-a|^{-1})]$$

If $x=(2b-4a)[1+t/(b-2a)^{\frac{1}{3}}]$, so that

$$x\sim 2b-4a$$

13.5.19

$$M(a, b, x)=e^{\frac{1}{2}x}(b-2a)^{\frac{1}{3}-b}\Gamma(b)[\mathrm{Ai}(t)\cos(a\pi)$$
$$+\mathrm{Bi}(t)\sin(a\pi)+O(|\tfrac{1}{2}b-a|^{-\frac{2}{3}})]$$

13.5.20

$$U(a, b, x)=e^{\frac{1}{2}x+a-\frac{1}{2}b}\Gamma(\tfrac{1}{3})\pi^{-\frac{1}{2}}x6^{-\frac{1}{3}}$$
$$\{1-t\Gamma(\tfrac{5}{6})(bx-2ax)^{-\frac{1}{2}}3^{\frac{1}{3}}\pi^{-\frac{1}{2}}+O(|\tfrac{1}{2}b-a|^{-\frac{2}{3}})\}$$

If $\cos^2\theta=x/(2b-4a)$ so that $2b-4a>x>0$,

13.5.21

$$M(a, b, x)=\Gamma(b)\exp\{(b-2a)\cos^2\theta\}$$
$$[(b-2a)\cos\theta]^{1-b}[\pi(\tfrac{1}{2}b-a)\sin 2\theta]^{-\frac{1}{2}}$$
$$[\sin(a\pi)+\sin\{(\tfrac{1}{2}b-a)(2\theta-\sin 2\theta)+\tfrac{1}{4}\pi\}$$
$$+O(|\tfrac{1}{2}b-a|^{-1})]$$

13.5.22

$$U(a, b, x)=\exp[(b-2a)\cos^2\theta][(b-2a)\cos\theta]^{1-b}$$
$$[(\tfrac{1}{2}b-a)\sin 2\theta]^{-\frac{1}{2}}\{\sin[(\tfrac{1}{2}b-a)$$
$$(2\theta-\sin 2\theta)+\tfrac{1}{4}\pi]+O(|\tfrac{1}{2}b-a|^{-1})\}$$

13.6. Special Cases

	$M(a, b, z)$			Relation	Function
	a	b	z		
13.6.1	$\nu+\frac{1}{2}$	$2\nu+1$	$2iz$	$\Gamma(1+\nu)e^{iz}(\frac{1}{2}z)^{-\nu}J_\nu(z)$	Bessel
13.6.2	$-\nu+\frac{1}{2}$	$-2\nu+1$	$2iz$	$\Gamma(1-\nu)e^{iz}(\frac{1}{2}z)^{\nu}[\cos(\nu\pi)J_\nu(z)-\sin(\nu\pi)Y_\nu(z)]$	Bessel
13.6.3	$\nu+\frac{1}{2}$	$2\nu+1$	$2z$	$\Gamma(1+\nu)e^{z}(\frac{1}{2}z)^{-\nu}I_\nu(z)$	Modified Bessel
13.6.4	$n+1$	$2n+2$	$2iz$	$\Gamma(\frac{3}{2}+n)e^{iz}(\frac{1}{2}z)^{-n-\frac{1}{2}}J_{n+\frac{1}{2}}(z)$	Spherical Bessel
13.6.5	$-n$	$-2n$	$2iz$	$\Gamma(\frac{1}{2}-n)e^{iz}(\frac{1}{2}z)^{n+\frac{1}{2}}J_{-n-\frac{1}{2}}(z)$	Spherical Bessel
13.6.6	$n+1$	$2n+2$	$2z$	$\Gamma(\frac{3}{2}+n)e^{z}(\frac{1}{2}z)^{-n-\frac{1}{2}}I_{n+\frac{1}{2}}(z)$ *	Spherical Bessel
13.6.7	$n+\frac{1}{2}$	$2n+1$	$-2\sqrt{i}x$	$\Gamma(1+n)e^{-\frac{1}{2}ix\pi}(\frac{1}{2}ix\pi)^{-n}(\mathrm{ber}_n x+i\,\mathrm{bei}_n x)$	Kelvin
13.6.8	$L+1-i\eta$	$2L+2$	$2ix$	$e^{ix}F_L(\eta, x)x^{-L-1}/C_L(\eta)$	Coulomb Wave
13.6.9	$-n$	$\alpha+1$	x	$\dfrac{n!}{(\alpha+1)_n}L_n^{(\alpha)}(x)$	Laguerre
13.6.10	a	$a+1$	$-x$	$ax^{-a}\gamma(a, x)$	Incomplete Gamma
13.6.11	$-n$	$1+\nu-n$	x	$\dfrac{(n!)^{\frac{1}{2}}x^{\frac{1}{2}n}}{(1+\nu-n)_n}\rho_n(\nu, x)$	Poisson-Charlier
13.6.12	a	a	z	e^z	Exponential
13.6.13	1	2	$-2iz$	$\dfrac{e^{-iz}}{z}\sin z$	Trigonometric
13.6.14	1	2	$2z$	$\dfrac{e^z}{z}\sinh z$	Hyperbolic
13.6.15	$-\frac{1}{2}\nu$	$\frac{1}{2}$	$\frac{1}{2}z^2$	$2^{-\frac{1}{2}}\exp(\frac{1}{4}z^2)E_\nu^{(0)}(z)$	Weber or Parabolic Cylinder
13.6.16	$\frac{1}{2}-\frac{1}{2}\nu$	$\frac{3}{2}$	$\frac{1}{2}z^2$	$\dfrac{\exp(\frac{1}{4}z^2)}{2z}E_\nu^{(1)}(z)$	Weber or Parabolic Cylinder
13.6.17	$-n$	$\frac{1}{2}$	$\frac{1}{2}x^2$	$\dfrac{n!}{(2n)!}(-\frac{1}{2})^{-n}He_{2n}(x)$	Hermite
13.6.18	$-n$	$\frac{3}{2}$	$\frac{1}{2}x^2$	$\dfrac{n!}{(2n+1)!}(-\frac{1}{2})^{-n}\dfrac{1}{x}He_{2n+1}(x)$ *	Hermite
13.6.19	$\frac{1}{2}$	$\frac{3}{2}$	$-x^2$	$\dfrac{\pi^{\frac{1}{2}}}{2x}\mathrm{erf}\,x$	Error Integral
13.6.20	$\frac{1}{2}m+\frac{1}{2}$	$1+n$	r^2	$\dfrac{n!r^{-2n+m-1}}{\Gamma(\frac{1}{2}m+\frac{1}{2})}e^{r^2}T(m, n, r)$ *	Toronto

*See page II.

13.6. Special Cases—Continued

	$U(a, b, z)$			Relation	Function
	a	b	z		
13.6.21	$\nu+\frac{1}{2}$	$2\nu+1$	$2z$	$\pi^{-\frac{1}{2}}e^z(2z)^{-\nu}K_\nu(z)$	Modified Bessel
13.6.22	$\nu+\frac{1}{2}$	$2\nu+1$	$-2iz$	$\frac{1}{2}\pi^{\frac{1}{2}}e^{i[\pi(\nu+\frac{1}{2})-z]}(2z)^{-\nu}H_\nu^{(1)}(z)$*	Hankel
13.6.23	$\nu+\frac{1}{2}$	$2\nu+1$	$2iz$	$\frac{1}{2}\pi^{\frac{1}{2}}e^{-i[\pi(\nu+\frac{1}{2})-z]}(2z)^{-\nu}H_\nu^{(2)}(z)$*	Hankel
13.6.24	$n+1$	$2n+2$	$2z$	$\pi^{-\frac{1}{2}}e^z(2z)^{-n-\frac{1}{2}}K_{n+\frac{1}{2}}(z)$	Spherical Bessel
13.6.25	$\frac{5}{6}$	$\frac{5}{3}$	$\frac{4}{3}z^{3/2}$	$\pi^{\frac{1}{2}}z^{-1}\exp(\frac{2}{3}z^{3/2})2^{-2/3}3^{5/6}$ Ai (z)	Airy
13.6.26	$n+\frac{1}{2}$	$2n+1$	$\sqrt{ix}$	$i^n\pi^{-\frac{1}{2}}e^{\sqrt{ix}}(2\sqrt{ix})^{-n}[\text{ker}_n x+i\,\text{kei}_n x]$	Kelvin
13.6.27	$-n$	$\alpha+1$	x	$(-1)^n n! L_n^{(\alpha)}(x)$	Laguerre
13.6.28	$1-a$	$1-a$	x	$e^x\Gamma(a, x)$	Incomplete Gamma
13.6.29	1	1	$-x$	$-e^{-x}$ Ei (x)	Exponential Integral
13.6.30	1	1	x	$e^x E_1(x)$	Exponential Integral
13.6.31	1	1	$-\ln x$	$-\dfrac{1}{x}$ li (x)	Logarithmic Integral
13.6.32	$\frac{1}{2}m-n$	$1+m$	x	$\Gamma(1+n-\frac{1}{2}m)e^{x-\pi i(\frac{1}{2}m-n)}\omega_{n, m}(x)$	Cunningham
13.6.33	$-\frac{1}{2}\nu$	0	$2x$	$\Gamma(1+\frac{1}{2}\nu)e^x k_\nu(x)$ for $x>0$	Bateman
13.6.34	1	1	ix	$e^{ix}[-\frac{1}{2}\pi i+i$ Si $(x)-$ Ci $(x)]$	Sine and Cosine Integral
13.6.35	1	1	$-ix$	$e^{-ix}[\frac{1}{2}\pi i-i$ Si $(x)-$ Ci $(x)]$	Sine and Cosine Integral
13.6.36	$-\frac{1}{2}\nu$	$\frac{1}{2}$	$\frac{1}{2}z^2$	$2^{-\frac{1}{2}\nu}e^{z^2/4}D_\nu(z)$	Weber
13.6.37	$\frac{1}{2}-\frac{1}{2}\nu$	$\frac{3}{2}$	$\frac{1}{2}z^2$	$2^{\frac{1}{2}-\frac{1}{2}\nu}e^{z^2/4}D_\nu(z)/z$ *	or Parabolic Cylinder
13.6.38	$\frac{1}{2}-\frac{1}{2}n$	$\frac{1}{2}$	x^2	$2^{-n}H_n(x)/x$ *	Hermite
13.6.39	$\frac{1}{2}$	$\frac{1}{2}$	x^2	$\sqrt{\pi}\exp(x^2)$ erfc x	Error Integral

13.7. Zeros and Turning Values

If $j_{b-1, r}$ is the r'th positive zero of $J_{b-1}(x)$, then a first approximation X_0 to the r'th positive zero of $M(a, b, x)$ is

13.7.1 $\quad X_0=j_{b-1, r}^2\{1/(2b-4a)+O(1/(\frac{1}{2}b-a)^2)\}$

13.7.2 $\qquad X_0\approx\dfrac{\pi^2(r+\frac{1}{2}b-\frac{3}{4})^2}{2b-4a}$

A closer approximation is given by

13.7.3 $\quad X_1=X_0-M(a, b, X_0)/M'(a, b, X_0)$

For the derivative,

13.7.4

$$M'(a, b, X_1)=$$
$$M'(a, b, X_0)\left\{1+(b-X_0)\frac{M(a, b, X_0)}{M'(a, b, X_0)}\right\}$$

If X_0' is the first approximation to a turning value of $M(a, b, x)$, that is, to a zero of $M'(a, b, x)$ then a better approximation is

13.7.5 $\qquad X_1'=X_0'-\dfrac{X_0'M'(a, b, X_0')}{aM(a, b, X_0')}$

*See page II.

The self-adjoint equation **13.1.1** can also be written

13.7.6
$$\frac{d}{dz}[z^b e^{-z} \frac{dw}{dz}] = az^{b-1} e^{-z} w$$

The Sonine-Polya Theorem

The maxima and minima of $|w|$ form an increasing or decreasing sequence according as

$$-ax^{2b-1}e^{-2x}$$

is an increasing or decreasing function of x, that is, they form an increasing sequence for $M(a, b, x)$ if $a>0$, $x<b-\frac{1}{2}$ or if $a<0$, $x>b-\frac{1}{2}$, and a decreasing sequence if $a>0$ and $x>b-\frac{1}{2}$ or if $a<0$ and $x<b-\frac{1}{2}$.

The turning values of $|w|$ lie near the curves

13.7.7
$$w = \pm \Gamma(b) \pi^{-1/2} e^{x/2} (\tfrac{1}{2}bx - ax)^{\frac{1}{4} - \frac{1}{2}b} \{1 - x/(2b-4a)\}^{-1/4}$$

Numerical Methods

13.8. Use and Extension of the Tables

Calculation of $M(a, b, x)$

Kummer's Transformation

Example 1. Compute $M(.3, .2, -.1)$ to 7S. Using **13.1.27** and **Tables 4.4** and **13.1** we have $a=.3$, $b=.2$ so that

$$M(.3, .2, -.1) = e^{-.1}M(-.1, .2, .1)$$
$$= .85784\ 90.$$

Thus **13.1.27** can be used to extend **Table 13.1** to negative values of x. Kummer's transformation should also be used when a and b are large and nearly equal, for x large or small.

Example 2. Compute $M(17, 16, 1)$ to 7S. Here $a=17$, $b=16$, and

$$M(17, 16, 1) = e^1 M(-1, 16, -1)$$
$$= 2.71828\ 18 \times 1.06250\ 00$$
$$= 2.88817\ 44.$$

Recurrence Relations

Example 3. Compute $M(-1.3, 1.2, .1)$ to 7S. Using **13.4.1** and **Table 13.1** we have $a=-.3$, $b=.2$ so that

$$M(-1.3, .2, .1) = 2[.7\ M(-.3, .2, .1) - .3\ M(.7, .2, .1)]$$
$$= .35821\ 23.$$

By **13.4.5** when $a=-1.3$ and $b=.2$,

$$M(-1.3, 1.2, .1) = [.26\ M(-.3, .2, .1)$$
$$- .24\ M(-1.3, .2, .1)]/.15$$
$$= .89241\ 08.$$

Similarly when $a=-.3$ and $b=.2$

$$M(-.3, 1.2, .1) = .97459\ 52.$$

Check, by **13.4.6**,

$$M(-1.3, 1.2, .1) = [.2\ M(-.3, .2, .1)$$
$$+ 1.2\ M(-.3, 1.2, .1)]/1.5$$
$$= .89241\ 08.$$

In this way **13.4.1–13.4.7** can be used together with **13.1.27** to extend **Table 13.1** to the range

$$-10 \leq a \leq 10, \quad -10 \leq b \leq 10, \quad -10 \leq x \leq 10.$$

This extension of ten units in any direction is possible with the loss of about 1S. All the recurrence relations are stable except i) if $a<0$, $b<0$ and $|a|>|b|$, $x>0$, or ii) $b<a$, $b<0$, $|b-a|>|b|$, $x<0$, when the oscillations may become large, especially if $|x|$ also is large.

Neither interpolation nor the use of recurrence relations should be attempted in the strips $b=-n\pm.1$ where the function is very large numerically. In particular $M(a, b, x)$ cannot be evaluated in the neighborhood of the points $a=-m$, $b=-n$, $m \leq n$, as near these points small changes in a, b or x can produce very large changes in the numerical value of $M(a, b, x)$.

Example 4. At the point $(-1, -1, x)$, $M(a, b, x)$ is undefined.

When $a=-1$, $M(-1, b, x) = 1 - \frac{x}{b}$ for all x.

Hence $\lim_{b \to -1} M(-1, b, x) = 1 + x$. But $M(b, b, x) = e^x$ for all x, when $a=b$. Hence $\lim_{b \to -1} M(b, b, x) = e^x$. In the first case $b \to -1$ along the line $a=-1$, and in the second case $b \to -1$ along the line $a=b$.

Derivatives

Example 5. To evaluate $M'(-.7, -.6, .5)$ to 7S. By **13.4.8**, when $a=-.7$ and $b=-.6$, we have

$$M'(-.7, -.6, .5) = \frac{-.7}{-.6} M(.3, .4, .5)$$
$$= 1.724128.$$

Asymptotic Formulas

For $x \geq 10$, a and b small, $M(a, b, x)$ should be evaluated by **13.5.1** using converging factors **13.5.3** and **13.5.4** to improve the accuracy if necessary.

Example 6. Calculate $M(.9, .1, 10)$ to 7S, using **13.5.1.**

$$M(.9, .1, 10) = \frac{\Gamma(.1)}{\Gamma(-.8)} e^{.9\pi i} 10^{-.9} \sum_{n=0}^{N} \frac{(.9)_n (1.8)_n}{n! (-10)^n}$$

$$+ \frac{\Gamma(.1)}{\Gamma(.9)} e^{10} 10^{.8} \sum_{n=0}^{N} \frac{(-.8)_n (.1)_n}{n! 10^n} + O(10^{-N})$$

$$= -.198(.869) + 1237253(.99190\ 285)$$
$$+ O(1)$$

$$= 1227235.23 - .17 + O(1)$$

$$= 1227235 + O(1)$$

Check, from **Table 13.1,** $M(.9, .1, 10) = 1227235$. To evaluate $M(a, b, x)$ with a large, x small and b small or large **13.5.13–14** should be used.

Example 7. Compute $M(-52.5, .1, 1)$ to 3S, using **13.5.14.**

$$M(-52.5, .1, 1) = \Gamma(.1)e^{.5}(.05 + 52.5)^{.25 - .05}$$

$$.5642 \cos [(.2 - 4(-52.5))^{.5} - .05\pi + .25\pi]$$

$$[1 + O((.05 + 52.5)^{-.5})] = -16.34 + O(.2)$$

By direct application of a recurrence relation, $M(-52.5, .1, 1)$ has been calculated as -16.447. To evaluate $M(a, b, x)$ with x, a and/or b large, **13.5.17, 19** or **21** should be tried.

Example 8. Compute $M(-52.5, .1, 1)$ using **13.5.21** to 3S, $\cos \theta = \sqrt{1/210.2}$.

$$M(-52.5, .1, 1)$$
$$= \Gamma(.1)e^{105.1 \cos^2 \theta}[105.1 \cos \theta]^{1 - .1}.5641$$

$$52.55^{-\frac{1}{4}} \sin 2\theta^{-\frac{1}{2}}[\sin (-52.5\pi)$$

$$+ \sin \{52.55(2\theta - \sin 2\theta) + \tfrac{1}{4}\pi\}$$

$$+ O((52.55)^{-1})] = -16.47 + O(.02)$$

A full range of asymptotic formulas to cover all possible cases is not yet known.

Calculation of $U(a, b, x)$

For $-10 \leq x \leq 10$, $-10 \leq a \leq 10$, $-10 \leq b \leq 10$ this is possible by **13.1.3,** using **Table 13.1** and the recurrence relations **13.4.15–20.**

Example 9. Compute $U(1.1, .2, 1)$ to 5S. Using **Tables 13.1, 4.12** and **6.1** and **13.1.3,** we have

$$U(.1, .2, 1) =$$

$$\frac{\pi}{\sin (.2\pi)} \left\{ \frac{M(.1, .2, 1)}{\Gamma(.9)\Gamma(.2)} - \frac{M(.9, 1.8, 1)}{\Gamma(.1)\Gamma(1.8)} \right\}.$$

But $M(.9, 1.8, 1) = .8[M(.9, .8, 1) - M(-.1, .8, 1)]$
$$= 1.72329, \text{ using } \textbf{13.4.4.}$$

Hence

$$U(.1, .2, 1) = 5.344799(.371765 - .194486)$$
$$= .94752.$$

Similarly

$$U(-.9, .2, 1) = .91272.$$

Hence by **13.4.15**

$$U(1.1, .2, 1) = [U(.1, .2, 1) - U(-.9, .2, 1)]/.09$$
$$= .38664.$$

Example 10. To compute $U'(-.9, -.8, 1)$ to 5S. By **13.4.21**

$$U'(-.9, -.8, 1) = .9U(.1, .2, 1)$$
$$= (.9)(.94752)$$
$$= .85276.$$

Asymptotic Formulas

Example 11. To compute $U(1, .1, 100)$ to 5S. By **13.5.2**

$$U(1, .1, 100) = \frac{1}{100}\left\{1 - \frac{1.9}{100} + \frac{1.9}{100}\frac{2.9}{100}\right.$$

$$\left. - \frac{1.9}{100}\frac{2.9}{100}\frac{3.9}{100} + O(10^{-9})\right\}.$$

$$= .01\{1 - .019 + .000551 - .000021$$
$$+ O(10^{-9})\},$$

$$= .00981\ 53.$$

Example 12. To evaluate $U(.1, .2, .01)$. For x small, **13.5.6–12** should be used.

$$U(.1, .2, .01) = \frac{\Gamma(1 - .2)}{\Gamma(1.1 - .2)} + O((.01)^{1 - .2})$$

$$= \frac{\Gamma(.8)}{\Gamma(.9)} + O((.01)^{.8})$$

$$= 1.09 \text{ to 3S, by } \textbf{13.5.10.}$$

To evaluate $U(a, b, x)$ with a large, x small and b small or large **13.5.15** or **16** should be used.

To evaluate $U(a, b, x)$ with x, a and/or b large **13.5.18, 20** or **22** should be tried. In all these cases the size of the remainder term is the guide to the number of significant figures obtainable.

Calculation of the Whittaker Functions

Example 13. Compute $M_{.0, -.4}(1)$ and $W_{.0, -.4}(1)$ to 5S. By formulas **13.1.32** and **13.1.33** and **Tables 13.1, 4.4**

$$M_{.0, -.4}(1) = e^{-.5}M(.1, .2, 1) = 1.10622,$$
$$W_{.0, -.4}(1) = e^{-.5}U(.1, .2, 1) = .57469.$$

Thus the values of $M_{\kappa,\mu}(x)$ and $W_{\kappa,\mu}(x)$ can always be found if the values of $M(a, b, x)$ and $U(a, b, x)$ are known.

13.9. Calculation of Zeros and Turning Points

Example 14. Compute the smallest positive zero of $M(-4, .6, x)$. This is outside the range of **Table 13.2**. Using **13.7.2** we have, as a first approximation

$$X_0 = \frac{(.55\pi)^2}{17.2} = .174.$$

Using **13.7.3** we have

$$X_1 = X_0 - M(-4, .6, X_0)/M'(-4, .6, X_0).$$

But, by **13.4.8**,

$$M'(-4, .6, X_0) = -(.15)^{-1}M(-3, 1.6, X_0)$$

Hence

$$X_1 = X_0 + .15M(-4, .6, X_0)/M(-3, 1.6, X_0),$$

$$= .174 + (.15)(.030004)$$

$$= .17850 \text{ as a second approximation.}$$

If we repeat this calculation, we find that

$$X_2 = X_1 + .00002\ 99 = .17852\ 99 \text{ to 7S.}$$

Calculation of Maxima and Minima

Example 15. Compute the value of x at which $M(-1.8, -.2, x)$ has a turning value. Using **13.4.8** and **Table 13.2**, we find that $M'(-1.8, -.2, x) = 9M(-.8, .8, x) = 0$ when $x = .94291\ 59$. Also $M''(-1.8, -.2, x) = 9M'(-.8, .8, x) = -9M(.2, 1.8, x)$ and $M(.2, 1.8, .94291\ 59) > 0$. Hence $M(-1.8, -.2, x)$ has a maximum in x when $x = .94291\ 59$.

Example 16. Compute the smallest positive value of x for which $M(-3, .6, x)$ has a turning value, X_1'. This is outside the range of **Table 13.2**. Using **13.4.8** we have

$$M'(-3, .6, x) = -3M(-2, 1.6, x)/.6.$$

By **13.7.2** for $M(-2, 1.6, x)$,

$$X_0 = (1.05\pi)^2/(11.2) = .9715.$$

This is a first approximation to X_0' for $M(-3, .6, x)$. Using **13.7.5** and **13.4.8** we find a second approximation

$$X_1' = X_0'\ [1 - \frac{M'(-3, .6, X_0')}{-3M(-3, .6, X_0')}]$$

$$= X_0'\ [1 - M(-2, 1.6, X_0')/.6M(-3, .6, X_0')]$$

$$= .9715 \times 1.0163 = .9873 \text{ to 4S.}$$

This process can be repeated to give as many significant figures as are required.

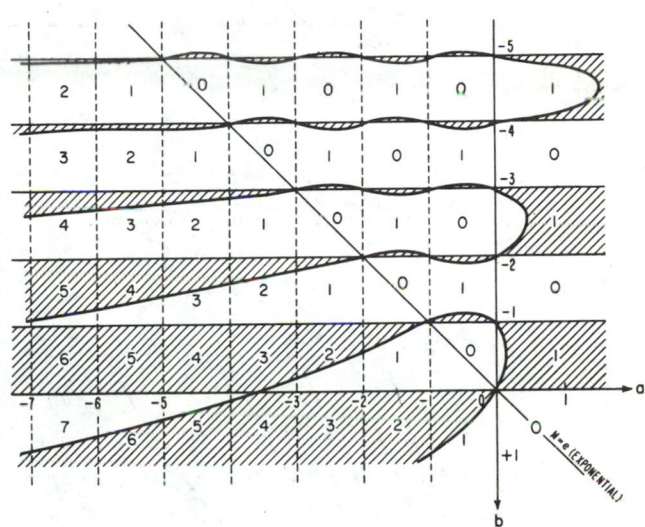

FIGURE 13.1.

Figure 13.1 shows the curves on which $M(a, b, x) = 0$ in the a, b plane when $x = 1$. The function is positive in the unshaded areas, and negative in the shaded areas. The number in each square gives the number of real positive zeros of $M(a, b, x)$ as a function of x in that square. The vertical boundaries to the left are to be included in each square.

13.10. Graphing $M(a, b, x)$

Example 17. Sketch $M(-4.5, 1, x)$. Firstly, from **Figure 13.1** we see that the function has five real positive zeros. From **13.5.1**, we find that $M \to -\infty$, $M' \to -\infty$ as $x \to +\infty$ and that $M \to +\infty$, $M' \to +\infty$ as $x \to -\infty$. By **13.7.2** we have as first approximations to the zeros, .3, 1.5, 3.7, 6.9, 10.6, and by **13.7.2** and **13.4.8** we find as first approximations to the turning values .9, 2.8, 5.8, 9.9. From **13.7.7**, we see that these must lie near the curves

$$y = \pm e^{\frac{1}{2}x}(5x)^{-\frac{1}{4}}(1 - x/11)^{-\frac{1}{4}}\pi^{-\frac{1}{2}}.$$

From these facts we can form a rough graph of the behavior of the function, **Figure 13.2.**

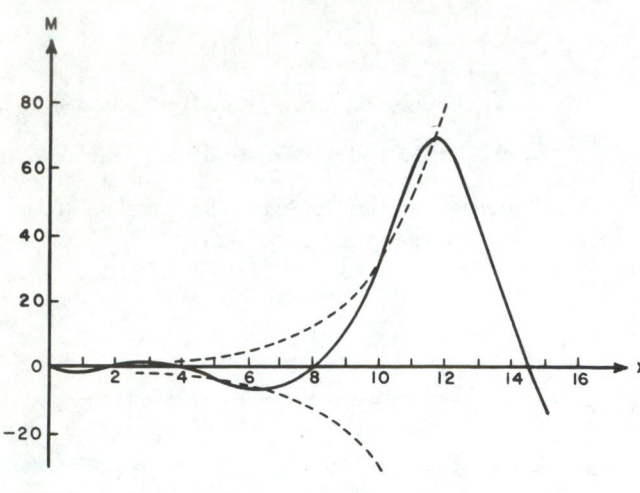

FIGURE 13.2. $M(-4.5, 1, x)$.

(From F. G. Tricomi, Funzioni, ipergeometriche confluenti, Edizioni. Cremonese, Rome, Italy, 1954, with permission.)

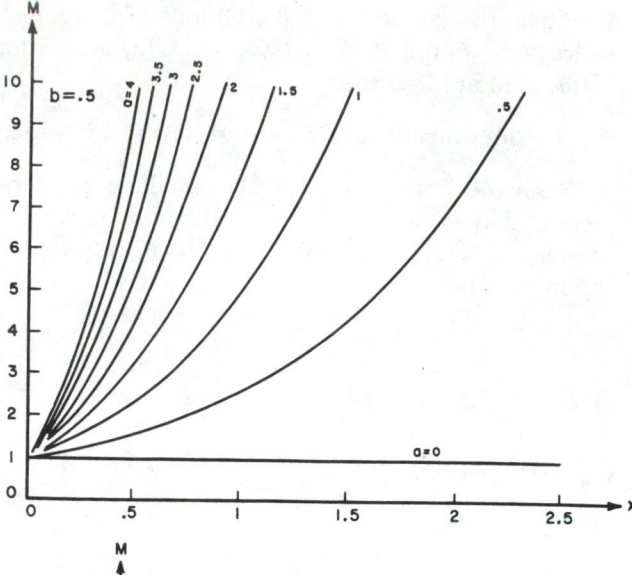

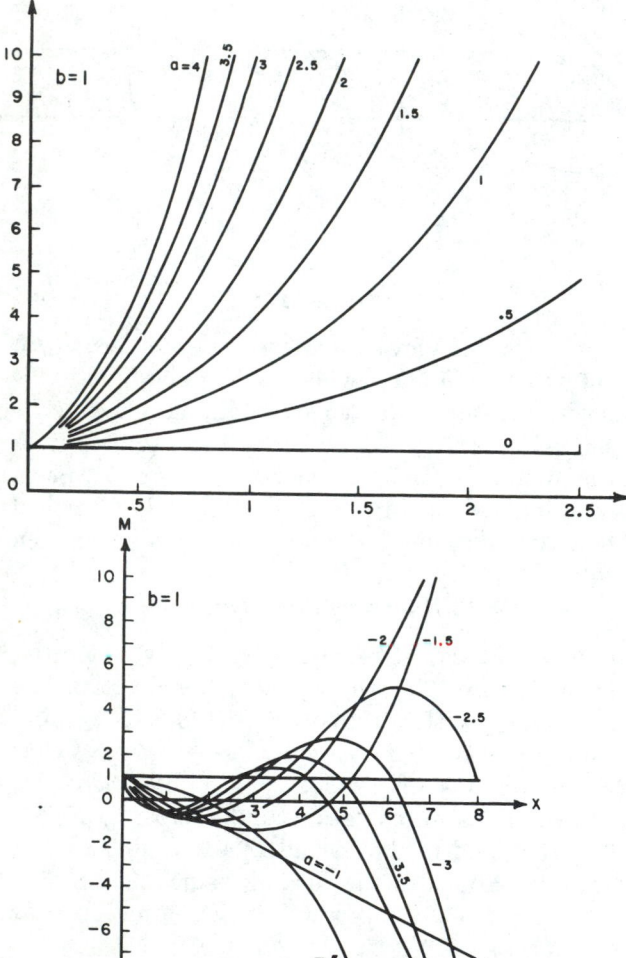

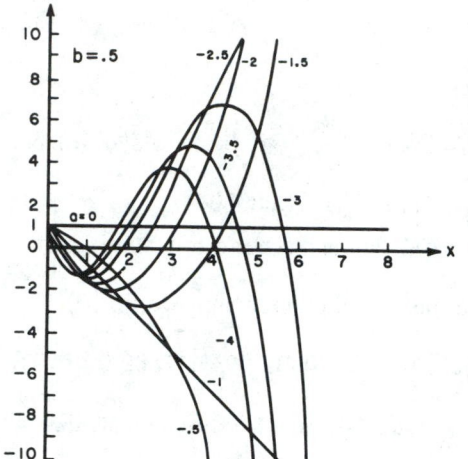

FIGURE 13.4. $M(a, .5, x)$.

(From E. Jahnke and F. Emde, Tables of functions, Dover Publications, Inc., New York, N.Y., 1945, with permission.)

FIGURE 13.3. $M(a, 1, x)$.

(From E. Jahnke and F. Emde, Tables of functions, Dover Publications, Inc., New York, N.Y., 1945, with permission.)

References

Texts

[13.1] H. Buchholz, Die konfluente hypergeometrische Funktion (Springer-Verlag, Berlin, Germany, 1953). On Whittaker functions, with a large bibliography.

[13.2] A. Erdélyi et al., Higher transcendental functions, vol. 1, ch. 6 (McGraw-Hill Book Co., Inc., New York, N.Y., 1953). On Kummer functions.

[13.3] H. Jeffreys and B. S. Jeffreys, Methods of mathematical physics, ch. 23 (Cambridge Univ. Press, Cambridge, England, 1950). On Kummer functions.

[13.4] J. C. P. Miller, Note on the general solutions of the confluent hypergeometric equation, Math. Tables Aids Comp. **9**, 97–99 (1957).

[13.5] L. J. Slater, On the evaluation of the confluent hypergeometric function, Proc. Cambridge Philos. Soc. **49**, 612–622 (1953).

[13.6] L. J. Slater, The evaluation of the basic confluent hypergeometric function, Proc. Cambridge Philos. Soc. **50**, 404–413 (1954).

[13.7] L. J. Slater, The real zeros of the confluent hypergeometric function, Proc. Cambridge Philos. Soc. **52**, 626–635 (1956).

[13.8] C. A. Swanson and A. Erdélyi, Asymptotic forms of confluent hypergeometric functions, Memoir 25, Amer. Math. Soc. (1957).

[13.9] F. G. Tricomi, Funzioni ipergeometriche confluenti (Edizioni Cremonese, Rome, Italy, 1954). On Kummer functions.

[13.10] E. T. Whittaker and G. N. Watson, A course of modern analysis, ch. 16, 4th ed. (Cambridge Univ. Press, Cambridge, England, 1952). On Whittaker functions.

Tables

[13.11] J. R. Airey, The confluent hypergeometric function, British Association Reports, Oxford, 276–294 (1926), and Leeds, 220–244 (1927). $M(a, b, x)$, $a = -4(.5)4$, $b = \frac{1}{2}$, 1, $\frac{3}{2}$, 2, 3, 4, $x = .1(.1)2(.2)3$ $(.5)8$, 5D.

[13.12] J. R. Airey and H. A. Webb, The practical importance of the confluent hypergeometric function, Phil. Mag. **36**, 129–141 (1918). $M(a, b, x)$, $a = -3(.5)4$, $b = 1(1)7$, $x = 1(1)6(2)10$, 4S.

[13.13] E. Jahnke and F. Emde, Tables of functions, ch. 10, 4th ed. (Dover Publications, Inc., New York, N.Y., 1945). Graphs of $M(a, b, x)$ based on the tables of [13.11].

[13.14] P. Nath, Confluent hypergeometric functions, Sankhya J. Indian Statist. Soc. **11**, 153–166 (1951). $M(a, b, x)$, $a = 1(1)40$, $b = 3$, $x = .02(.02)$ $.1(.1)1(1)10(10)50$, 100, 200, 6D.

[13.15] S. Rushton and E. D. Lang, Tables of the confluent hypergeometric function, Sankhya J. Indian Statist. Soc. **13**, 369–411 (1954). $M(a, b, x)$, $a = .5(.5)40$, $b = .5(.5)3.5$, $x = .02(.02).1(.1)1(1)10(10)50$, 100, 200, 7S.

[13.16] L. J. Slater, Confluent hypergeometric functions (Cambridge Univ. Press, Cambridge, England, 1960). $M(a, b, x)$, $a = -1(.1)1$, $b = .1(.1)1$, $x = .1(.1)10$, 8S; $M(a, b, 1)$, $a = -11(.2)2$, $b = -4(.2)1$, 8S; and smallest positive values of x for which $M(a, b, x) = 0$, $a = -4(.1) - .1$, $b = .1(.1)2.5$, 8S.

Table 13.1   CONFLUENT HYPERGEOMETRIC FUNCTION $M(a, b, x)$

$$x = 0.1$$

$a\backslash b$	0.1	0.2	0.3	0.4	0.5
-1.0	0.00000 00	(-1) 5.00000 00	(-1) 6.66666 67	(-1) 7.50000 00	(-1) 8.00000 00
-0.9	(-2) 9.58364 34	(-1) 5.48093 23	(-1) 6.98827 46	(-1) 7.74183 96	(-1) 8.19391 07
-0.8	(-1) 1.92586 25	(-1) 5.96605 00	(-1) 7.31245 77	(-1) 7.98547 23	(-1) 8.38915 99
-0.7	(-1) 2.90253 86	(-1) 6.45537 25	(-1) 7.63922 74	(-1) 8.23090 56	(-1) 8.58575 33
-0.6	(-1) 3.88843 71	(-1) 6.94891 92	(-1) 7.96859 49	(-1) 8.47814 73	(-1) 8.78369 61
-0.5	(-1) 4.88360 25	(-1) 7.44670 94	(-1) 8.30057 19	(-1) 8.72720 49	(-1) 8.98299 40
-0.4	(-1) 5.88807 94	(-1) 7.94876 28	(-1) 8.63516 97	(-1) 8.97808 60	(-1) 9.18365 22
-0.3	(-1) 6.90191 26	(-1) 8.45509 89	(-1) 8.97239 98	(-1) 9.23079 84	(-1) 9.38567 64
-0.2	(-1) 7.92514 70	(-1) 8.96573 73	(-1) 9.31227 38	(-1) 9.48534 97	(-1) 9.58907 21
-0.1	(-1) 8.95782 77	(-1) 9.48069 78	(-1) 9.65480 34	(-1) 9.74174 76	(-1) 9.79384 48
0.0	(0) 1.00000 00	(0) 1.00000 00	(0) 1.00000 00	(0) 1.00000 00	(0) 1.00000 00
0.1	(0) 1.10517 09	(0) 1.05236 64	(0) 1.03478 75	(0) 1.02601 15	(0) 1.02075 43
0.2	(0) 1.21130 01	(0) 1.10517 09	(0) 1.06984 41	(0) 1.05220 99	(0) 1.04164 80
0.3	(0) 1.31839 21	(0) 1.15841 56	(0) 1.10517 09	(0) 1.07859 61	(0) 1.06268 16
0.4	(0) 1.42645 14	(0) 1.21210 24	(0) 1.14076 91	(0) 1.10517 09	(0) 1.08385 58
0.5	(0) 1.53548 28	(0) 1.26623 34	(0) 1.17663 99	(0) 1.13193 51	(0) 1.10517 09
0.6	(0) 1.64549 07	(0) 1.32081 05	(0) 1.21278 44	(0) 1.15888 93	(0) 1.12662 77
0.7	(0) 1.75647 99	(0) 1.37583 59	(0) 1.24920 38	(0) 1.18603 45	(0) 1.14822 66
0.8	(0) 1.86845 49	(0) 1.43131 14	(0) 1.28589 94	(0) 1.21337 14	(0) 1.16996 83
0.9	(0) 1.98142 05	(0) 1.48723 92	(0) 1.32287 23	(0) 1.24090 08	(0) 1.19185 34
1.0	(0) 2.09538 12	(0) 1.54362 12	(0) 1.36012 38	(0) 1.26862 36	(0) 1.21388 22

$a\backslash b$	0.6	0.7	0.8	0.9	1.0
-1.0	(-1) 8.33333 33	(-1) 8.57142 86	(-1) 8.75000 00	(-1) 8.88888 89	(-1) 9.00000 00
-0.9	(-1) 8.49524 54	(-1) 8.71045 21	(-1) 8.87183 35	(-1) 8.99733 47	(-1) 9.09772 21
-0.8	(-1) 8.65820 31	(-1) 8.85031 91	(-1) 8.99436 39	(-1) 9.10636 73	(-1) 9.19594 59
-0.7	(-1) 8.82221 06	(-1) 8.99103 26	(-1) 9.11759 38	(-1) 9.21598 87	(-1) 9.29467 31
-0.6	(-1) 8.98727 18	(-1) 9.13259 59	(-1) 9.24152 56	(-1) 9.32620 11	(-1) 9.39390 52
-0.5	(-1) 9.15339 10	(-1) 9.27501 22	(-1) 9.36616 18	(-1) 9.43700 64	(-1) 9.49364 42
-0.4	(-1) 9.32057 22	(-1) 9.41828 47	(-1) 9.49150 52	(-1) 9.54840 68	(-1) 9.59389 16
-0.3	(-1) 9.48881 96	(-1) 9.56241 64	(-1) 9.61755 81	(-1) 9.66040 42	(-1) 9.69464 91
-0.2	(-1) 9.65813 72	(-1) 9.70741 08	(-1) 9.74432 32	(-1) 9.77300 09	(-1) 9.79591 86
-0.1	(-1) 9.82852 93	(-1) 9.85327 09	(-1) 9.87180 29	(-1) 9.88619 88	(-1) 9.89770 16
0.0	(0) 1.00000 00	(0) 1.00000 00	(0) 1.00000 00	(0) 1.00000 00	(0) 1.00000 00
0.1	(0) 1.01725 53	(0) 1.01476 01	(0) 1.01289 17	(0) 1.01144 07	(0) 1.01028 15
0.2	(0) 1.03461 94	(0) 1.02960 78	(0) 1.02585 56	(0) 1.02294 21	(0) 1.02061 50
0.3	(0) 1.05209 25	(0) 1.04454 34	(0) 1.03889 21	(0) 1.03450 45	(0) 1.03100 04
0.4	(0) 1.06967 52	(0) 1.05956 71	(0) 1.05200 13	(0) 1.04612 80	(0) 1.04143 81
0.5	(0) 1.08736 79	(0) 1.07467 94	(0) 1.06518 35	(0) 1.05781 30	(0) 1.05192 82
0.6	(0) 1.10517 09	(0) 1.08988 06	(0) 1.07843 90	(0) 1.06955 95	(0) 1.06247 09
0.7	(0) 1.12308 48	(0) 1.10517 09	(0) 1.09176 81	(0) 1.08136 79	(0) 1.07306 64
0.8	(0) 1.14110 98	(0) 1.12055 08	(0) 1.10517 09	(0) 1.09323 83	(0) 1.08371 47
0.9	(0) 1.15924 65	(0) 1.13602 05	(0) 1.11864 79	(0) 1.10517 09	(0) 1.09441 62
1.0	(0) 1.17749 53	(0) 1.15158 03	(0) 1.13219 91	(0) 1.11716 60	(0) 1.10517 09

For $0 \leq x \leq 1$, linear interpolation in a, b or x provides 3–4S. Lagrange four-point interpolation gives 7S in a, b or x over most of the table, but the Lagrange six-point formula is needed over the range $1 \leq x \leq 10$. Any interpolation formula can be reapplied to give two dimensional interpolates in a and b, a and x or b and x. This calculation can be checked by being repeated in a different order.

CONFLUENT HYPERGEOMETRIC FUNCTION $M(a, b, x)$ Table 13.1

$$x = 0.2$$

$a\backslash b$	0.1	0.2	0.3	0.4	0.5
−1.0	(0) −1.00000 00	0.00000 00	(−1) 3.33333 33	(−1) 5.00000 00	(−1) 6.00000 00
−0.9	(−1) −8.16955 02	(−2) 9.22415 48	(−1) 3.95232 64	(−1) 5.46684 38	(−1) 6.37527 43
−0.8	(−1) −6.30239 72	(−1) 1.86164 63	(−1) 4.58166 34	(−1) 5.94088 89	(−1) 6.75592 38
−0.7	(−1) −4.39817 97	(−1) 2.81785 03	(−1) 5.22143 72	(−1) 6.42219 72	(−1) 7.14199 30
−0.6	(−1) −2.45653 39	(−1) 3.79118 64	(−1) 5.87174 11	(−1) 6.91083 10	(−1) 7.53352 62
−0.5	(−2) −4.77093 96	(−1) 4.78181 44	(−1) 6.53266 92	(−1) 7.40685 28	(−1) 7.93056 84
−0.4	(−1) +1.54050 87	(−1) 5.78989 52	(−1) 7.20431 59	(−1) 7.91032 56	(−1) 8.33316 46
−0.3	(−1) 3.59664 50	(−1) 6.81559 07	(−1) 7.88677 63	(−1) 8.42131 28	(−1) 8.74136 01
−0.2	(−1) 5.69168 81	(−1) 7.85906 39	(−1) 8.58014 62	(−1) 8.93987 82	(−1) 9.15520 06
−0.1	(−1) 7.82601 37	(−1) 8.92047 86	(−1) 9.28452 18	(−1) 9.46608 57	(−1) 9.57473 18
0.0	(0) 1.00000 00	(0) 1.00000 00	(0) 1.00000 00	(0) 1.00000 00	(0) 1.00000 00
0.1	(0) 1.22140 28	(0) 1.10977 94	(0) 1.07266 78	(0) 1.05416 86	(0) 1.04310 51
0.2	(0) 1.44684 80	(0) 1.22140 28	(0) 1.14646 55	(0) 1.10912 09	(0) 1.08679 33
0.3	(0) 1.67637 41	(0) 1.33488 69	(0) 1.22140 28	(0) 1.16486 34	(0) 1.13106 91
0.4	(0) 1.91002 01	(0) 1.45024 87	(0) 1.29748 97	(0) 1.22140 28	(0) 1.17593 74
0.5	(0) 2.14782 49	(0) 1.56750 53	(0) 1.37473 61	(0) 1.27874 56	(0) 1.22140 28
0.6	(0) 2.38982 79	(0) 1.68667 37	(0) 1.45315 23	(0) 1.33689 87	(0) 1.26747 01
0.7	(0) 2.63606 85	(0) 1.80777 12	(0) 1.53274 81	(0) 1.39586 86	(0) 1.31414 41
0.8	(0) 2.88658 67	(0) 1.93081 51	(0) 1.61353 39	(0) 1.45566 22	(0) 1.36142 97
0.9	(0) 3.14142 25	(0) 2.05582 28	(0) 1.69551 97	(0) 1.51628 63	(0) 1.40933 17
1.0	(0) 3.40061 61	(0) 2.18281 20	(0) 1.77871 60	(0) 1.57774 76	(0) 1.45785 51

$a\backslash b$	0.6	0.7	0.8	0.9	1.0
−1.0	(−1) 6.66666 67	(−1) 7.14285 71	(−1) 7.50000 00	(−1) 7.77777 78	(−1) 8.00000 00
−0.9	(−1) 6.98070 53	(−1) 7.41302 26	(−1) 7.73716 33	(−1) 7.98920 01	(−1) 8.19077 41
−0.8	(−1) 7.29894 21	(−1) 7.68657 38	(−1) 7.97712 40	(−1) 8.20297 76	(−1) 8.38356 13
−0.7	(−1) 7.62141 04	(−1) 7.96353 68	(−1) 8.21990 25	(−1) 8.41912 68	(−1) 8.57837 54
−0.6	(−1) 7.94814 35	(−1) 8.24393 73	(−1) 8.46551 94	(−1) 8.63766 45	(−1) 8.77523 03
−0.5	(−1) 8.27917 51	(−1) 8.52780 14	(−1) 8.71399 57	(−1) 8.85860 76	(−1) 8.97413 99
−0.4	(−1) 8.61453 89	(−1) 8.81515 54	(−1) 8.96535 20	(−1) 9.08197 30	(−1) 9.17511 81
−0.3	(−1) 8.95426 91	(−1) 9.10602 57	(−1) 9.21960 95	(−1) 9.30777 78	(−1) 9.37817 91
−0.2	(−1) 9.29839 97	(−1) 9.40043 88	(−1) 9.47678 92	(−1) 9.53603 91	(−1) 9.58333 69
−0.1	(−1) 9.64696 51	(−1) 9.69842 13	(−1) 9.73691 22	(−1) 9.76677 40	(−1) 9.79060 58
0.0	(0) 1.00000 00	(0) 1.00000 00	(0) 1.00000 00	(0) 1.00000 00	(0) 1.00000 00
0.1	(0) 1.03575 39	(0) 1.03052 02	(0) 1.02660 74	(0) 1.02357 34	(0) 1.02115 34
0.2	(0) 1.07196 17	(0) 1.06140 54	(0) 1.05351 56	(0) 1.04739 95	(0) 1.04252 22
0.3	(0) 1.10862 70	(0) 1.09265 84	(0) 1.08072 66	(0) 1.07147 98	(0) 1.06410 78
0.4	(0) 1.14575 32	(0) 1.12428 18	(0) 1.10824 29	(0) 1.09581 63	(0) 1.08591 18
0.5	(0) 1.18334 39	(0) 1.15627 85	(0) 1.13606 64	(0) 1.12041 07	(0) 1.10793 56
0.6	(0) 1.22140 28	(0) 1.18865 12	(0) 1.16419 94	(0) 1.14526 47	(0) 1.13018 06
0.7	(0) 1.25993 33	(0) 1.22140 28	(0) 1.19264 41	(0) 1.17038 02	(0) 1.15264 83
0.8	(0) 1.29893 91	(0) 1.25453 59	(0) 1.22140 28	(0) 1.19575 89	(0) 1.17534 02
0.9	(0) 1.33842 39	(0) 1.28805 34	(0) 1.25047 76	(0) 1.22140 28	(0) 1.19825 79
1.0	(0) 1.37839 12	(0) 1.32195 81	(0) 1.27987 08	(0) 1.24731 35	(0) 1.22140 28

Table 13.1　　　　　　　　**CONFLUENT HYPERGEOMETRIC FUNCTION** $M(a, b, x)$

$$x = 0.3$$

$a\backslash b$	0.1	0.2	0.3	0.4	0.5
−1.0	(0) −2.00000 00	(−1) −5.00000 00	0.00000 00	(−1) 2.50000 00	(−1) 4.00000 00
−0.9	(0) −1.73884 94	(−1) −3.67762 19	(−2) 8.90939 59	(−1) 3.17420 35	(−1) 4.54351 25
−0.8	(0) −1.46940 36	(−1) −2.31724 76	(−1) 1.80524 85	(−1) 3.86467 39	(−1) 5.09916 51
−0.7	(0) −1.19153 81	(−2) −9.18332 95	(−1) 2.74324 64	(−1) 4.57162 39	(−1) 5.66711 03
−0.6	(−1) −9.05127 09	(−2) +5.19671 16	(−1) 3.70525 58	(−1) 5.29526 85	(−1) 6.24750 17
−0.5	(−1) −6.10043 44	(−1) 1.99731 93	(−1) 4.69160 23	(−1) 6.03582 44	(−1) 6.84049 44
−0.4	(−1) −3.06158 84	(−1) 3.51517 11	(−1) 5.70261 46	(−1) 6.79351 05	(−1) 7.44624 48
−0.3	(−3) +6.65629 62	(−1) 5.07379 19	(−1) 6.73862 42	(−1) 7.56854 74	(−1) 8.06491 07
−0.2	(−1) 3.28532 83	(−1) 6.67375 21	(−1) 7.79996 60	(−1) 8.36115 78	(−1) 8.69665 13
−0.1	(−1) 6.59602 92	(−1) 8.31562 77	(−1) 8.88697 76	(−1) 9.17156 65	(−1) 9.34162 71
0.0	(0) 1.00000 00	(0) 1.00000 00	(0) 1.00000 00	(0) 1.00000 00	(0) 1.00000 00
0.1	(0) 1.34985 88	(0) 1.17274 56	(0) 1.11393 77	(0) 1.08466 87	(0) 1.06719 33
0.2	(0) 1.70931 54	(0) 1.34985 88	(0) 1.23054 56	(0) 1.17118 59	(0) 1.13575 92
0.3	(0) 2.07850 71	(0) 1.53139 94	(0) 1.34985 88	(0) 1.25957 47	(0) 1.20571 42
0.4	(0) 2.45757 28	(0) 1.71742 78	(0) 1.47191 26	(0) 1.34985 88	(0) 1.27707 51
0.5	(0) 2.84665 23	(0) 1.90800 49	(0) 1.59674 26	(0) 1.44206 18	(0) 1.34985 88
0.6	(0) 3.24588 71	(0) 2.10319 22	(0) 1.72438 49	(0) 1.53620 75	(0) 1.42408 24
0.7	(0) 3.65541 99	(0) 2.30305 18	(0) 1.85487 58	(0) 1.63232 02	(0) 1.49976 30
0.8	(0) 4.07539 50	(0) 2.50764 63	(0) 1.98825 19	(0) 1.73042 41	(0) 1.57691 80
0.9	(0) 4.50595 77	(0) 2.71703 89	(0) 2.12455 03	(0) 1.83054 38	(0) 1.65556 49
1.0	(0) 4.94725 50	(0) 2.93129 36	(0) 2.26380 82	(0) 1.93270 41	(0) 1.73572 13

$a\backslash b$	0.6	0.7	0.8	0.9	1.0
−1.0	(−1) 5.00000 00	(−1) 5.71428 57	(−1) 6.25000 00	(−1) 6.66666 67	(−1) 7.00000 00
−0.9	(−1) 5.45594 63	(−1) 6.10737 55	(−1) 6.59572 25	(−1) 6.97537 97	(−1) 7.27897 71
−0.8	(−1) 5.92137 29	(−1) 6.50811 03	(−1) 6.94776 02	(−1) 7.28940 91	(−1) 7.56249 82
−0.7	(−1) 6.39639 42	(−1) 6.91657 86	(−1) 7.30618 39	(−1) 7.60881 20	(−1) 7.85061 06
−0.6	(−1) 6.88112 54	(−1) 7.33287 00	(−1) 7.67106 45	(−1) 7.93364 63	(−1) 8.14336 18
−0.5	(−1) 7.37568 28	(−1) 7.75707 44	(−1) 8.04247 38	(−1) 8.26397 01	(−1) 8.44079 99
−0.4	(−1) 7.88018 36	(−1) 8.18928 28	(−1) 8.42048 41	(−1) 8.59984 20	(−1) 8.74297 33
−0.3	(−1) 8.39474 59	(−1) 8.62958 68	(−1) 8.80516 81	(−1) 8.94132 11	(−1) 9.04993 07
−0.2	(−1) 8.91948 91	(−1) 9.07807 88	(−1) 9.19659 93	(−1) 9.28846 71	(−1) 9.36172 12
−0.1	(−1) 9.45453 34	(−1) 9.53485 19	(−1) 9.59485 17	(−1) 9.64133 99	(−1) 9.67839 44
0.0	(0) 1.00000 00	(0) 1.00000 00	(0) 1.00000 00	(0) 1.00000 00	(0) 1.00000 00
0.1	(0) 1.05560 11	(0) 1.04736 18	(0) 1.04121 19	(0) 1.03645 08	(0) 1.03265 88
0.2	(0) 1.11226 90	(0) 1.09558 01	(0) 1.08312 85	(0) 1.07349 27	(0) 1.06582 10
0.3	(0) 1.17001 62	(0) 1.14466 45	(0) 1.12575 75	(0) 1.11113 16	(0) 1.09949 16
0.4	(0) 1.22885 51	(0) 1.19462 48	(0) 1.16910 65	(0) 1.14937 40	(0) 1.13367 58
0.5	(0) 1.28879 84	(0) 1.24547 07	(0) 1.21318 32	(0) 1.18822 61	(0) 1.16837 88
0.6	(0) 1.34985 88	(0) 1.29721 20	(0) 1.25799 56	(0) 1.22769 42	(0) 1.20360 57
0.7	(0) 1.41204 93	(0) 1.34985 88	(0) 1.30355 15	(0) 1.26778 47	(0) 1.23936 18
0.8	(0) 1.47538 27	(0) 1.40342 10	(0) 1.34985 88	(0) 1.30850 41	(0) 1.27565 25
0.9	(0) 1.53987 22	(0) 1.45790 88	(0) 1.39692 56	(0) 1.34985 88	(0) 1.31248 30
1.0	(0) 1.60553 08	(0) 1.51333 23	(0) 1.44475 99	(0) 1.39185 54	(0) 1.34985 88

CONFLUENT HYPERGEOMETRIC FUNCTION $M(a, b, x)$ Table 13.1

$x = 0.4$

$a \backslash b$	0.1	0.2	0.3	0.4	0.5
-1.0	(0)-3.00000 00	(0)-1.00000 00	(-1)-3.33333 33	0.00000 00	(-1) 2.00000 00
-0.9	(0)-2.67035 54	(-1)-8.32139 43	(-1)-2.19718 27	(-2) 8.63057 33	(-1) 2.69801 05
-0.8	(0)-2.32590 02	(-1)-6.57495 96	(-1)-1.01932 12	(-1) 1.75514 40	(-1) 3.41768 30
-0.7	(0)-1.96633 24	(-1)-4.75937 91	(-2)+2.01024 24	(-1) 2.67677 48	(-1) 4.15938 56
-0.6	(0)-1.59134 63	(-1)-2.87331 90	(-1) 1.46463 65	(-1) 3.62847 08	(-1) 4.92349 10
-0.5	(0)-1.20063 19	(-2)-9.15428 01	(-1) 2.77230 84	(-1) 4.61075 95	(-1) 5.71037 59
-0.4	(-1)-7.93875 31	(-1)+1.11566 21	(-1) 4.12484 23	(-1) 5.62417 45	(-1) 6.52042 19
-0.3	(-1)-3.70758 28	(-1) 3.22133 74	(-1) 5.52305 08	(-1) 6.66925 61	(-1) 7.35401 47
-0.2	(-2)+6.90415 20	(-1) 5.40300 15	(-1) 6.96775 63	(-1) 7.74655 09	(-1) 8.21154 46
-0.1	(-1) 5.25850 66	(-1) 7.66207 59	(-1) 8.45979 18	(-1) 8.85661 23	(-1) 9.09340 66
0.0	(0) 1.00000 00	(0) 1.00000 00	(0) 1.00000 00	(0) 1.00000 00	(0) 1.00000 00
0.1	(0) 1.49182 47	(0) 1.24182 32	(0) 1.15892 34	(0) 1.11772 81	(0) 1.09317 29
0.2	(0) 2.00166 43	(0) 1.49182 47	(0) 1.32283 59	(0) 1.23890 28	(0) 1.18890 02
0.3	(0) 2.52986 27	(0) 1.75015 41	(0) 1.49182 47	(0) 1.36358 21	(0) 1.28722 33
0.4	(0) 3.07676 82	(0) 2.01696 26	(0) 1.66597 84	(0) 1.49182 47	(0) 1.38818 41
0.5	(0) 3.64273 38	(0) 2.29240 35	(0) 1.84538 67	(0) 1.62369 00	(0) 1.49182 47
0.6	(0) 4.22811 68	(0) 2.57663 20	(0) 2.03014 00	(0) 1.75923 82	(0) 1.59818 80
0.7	(0) 4.83327 91	(0) 2.86980 51	(0) 2.22033 03	(0) 1.89852 99	(0) 1.70731 73
0.8	(0) 5.45858 73	(0) 3.17208 18	(0) 2.41605 02	(0) 2.04162 67	(0) 1.81925 64
0.9	(0) 6.10441 27	(0) 3.48362 30	(0) 2.61739 39	(0) 2.18859 08	(0) 1.93404 94
1.0	(0) 6.77113 12	(0) 3.80459 19	(0) 2.82445 63	(0) 2.33948 51	(0) 2.05174 12

$a \backslash b$	0.6	0.7	0.8	0.9	1.0
-1.0	(-1) 3.33333 33	(-1) 4.28571 43	(-1) 5.00000 00	(-1) 5.55555 56	(-1) 6.00000 00
-0.9	(-1) 3.92050 85	(-1) 4.79315 51	(-1) 5.44722 84	(-1) 5.95564 45	(-1) 6.36214 28
-0.8	(-1) 4.52459 74	(-1) 5.31423 36	(-1) 5.90572 12	(-1) 6.36521 50	(-1) 6.73238 89
-0.7	(-1) 5.14587 62	(-1) 5.84916 36	(-1) 6.37564 87	(-1) 6.78440 52	(-1) 7.11085 21
-0.6	(-1) 5.78462 40	(-1) 6.39816 17	(-1) 6.85718 29	(-1) 7.21335 46	(-1) 7.49764 78
-0.5	(-1) 6.44112 32	(-1) 6.96144 64	(-1) 7.35049 77	(-1) 7.65220 44	(-1) 7.89289 21
-0.4	(-1) 7.11565 94	(-1) 7.53923 92	(-1) 7.85576 88	(-1) 8.10109 70	(-1) 8.29670 27
-0.3	(-1) 7.80852 14	(-1) 8.13176 35	(-1) 8.37317 41	(-1) 8.56017 66	(-1) 8.70919 82
-0.2	(-1) 8.52000 13	(-1) 8.73924 56	(-1) 8.90289 30	(-1) 9.02958 86	(-1) 9.13049 86
-0.1	(-1) 9.25039 46	(-1) 9.36191 40	(-1) 9.44510 72	(-1) 9.50948 02	(-1) 9.56072 51
0.0	(0) 1.00000 00	(0) 1.00000 00	(0) 1.00000 00	(0) 1.00000 00	(0) 1.00000 00
0.1	(0) 1.07691 20	(0) 1.06537 37	(0) 1.05677 57	(0) 1.05012 98	(0) 1.04484 47
0.2	(0) 1.15580 59	(0) 1.13233 62	(0) 1.11485 65	(0) 1.10135 26	(0) 1.09061 91
0.3	(0) 1.23671 28	(0) 1.20091 13	(0) 1.17426 15	(0) 1.15368 38	(0) 1.13733 58
0.4	(0) 1.31966 37	(0) 1.27112 31	(0) 1.23500 97	(0) 1.20713 88	(0) 1.18500 76
0.5	(0) 1.40469 04	(0) 1.34299 62	(0) 1.29712 04	(0) 1.26173 33	(0) 1.23364 74
0.6	(0) 1.49182 47	(0) 1.41655 50	(0) 1.36061 33	(0) 1.31748 31	(0) 1.28326 80
0.7	(0) 1.58109 90	(0) 1.49182 47	(0) 1.42550 81	(0) 1.37440 41	(0) 1.33388 28
0.8	(0) 1.67254 59	(0) 1.56883 03	(0) 1.49182 47	(0) 1.43251 25	(0) 1.38550 48
0.9	(0) 1.76619 84	(0) 1.64759 75	(0) 1.55958 33	(0) 1.49182 47	(0) 1.43814 76
1.0	(0) 1.86208 99	(0) 1.72815 18	(0) 1.62880 44	(0) 1.55235 70	(0) 1.49182 47

Table 13.1   CONFLUENT HYPERGEOMETRIC FUNCTION $M(a, b, x)$

$$x = 0.5$$

$a \backslash b$	0.1	0.2	0.3	0.4	0.5
-1.0	(0)-4.00000 00	(0)-1.50000 00	(-1)-6.66666 67	(-1)-2.50000 00	0.00000 00
-0.9	(0)-3.61201 86	(0)-1.30112 70	(-1)-5.31342 47	(-1)-1.46751 27	(-2) 8.38114 43
-0.8	(0)-3.20079 89	(0)-1.09161 33	(-1)-3.89475 90	(-2)-3.89499 09	(-1) 1.71019 66
-0.7	(0)-2.76573 85	(-1)-8.71196 18	(-1)-2.40912 78	(-2)+7.35066 66	(-1) 2.61697 96
-0.6	(0)-2.30622 47	(-1)-6.39608 65	(-2)-8.54965 30	(-1) 1.90722 60	(-1) 3.55920 78
-0.5	(0)-1.82163 45	(-1)-3.96579 38	(-2)+7.69319 06	(-1) 3.12803 64	(-1) 4.53763 61
-0.4	(0)-1.31133 45	(-1)-1.41832 63	(-1) 2.46534 08	(-1) 4.39857 14	(-1) 5.55303 09
-0.3	(-1)-7.74681 00	(-1)+1.24911 75	(-1) 4.23474 05	(-1) 5.71992 06	(-1) 6.60617 00
-0.2	(-1)-2.11019 41	(-1) 4.03938 42	(-1) 6.07918 46	(-1) 7.09319 04	(-1) 7.69784 21
-0.1	(-1)+3.80315 52	(-1) 6.95536 57	(-1) 8.00036 50	(-1) 8.51950 36	(-1) 8.82884 81
0.0	(0) 1.00000 00	(0) 1.00000 00	(0) 1.00000 00	(0) 1.00000 00	(0) 1.00000 00
0.1	(0) 1.64872 13	(0) 1.31762 72	(0) 1.20798 34	(0) 1.15358 36	(0) 1.12121 22
0.2	(0) 2.32717 78	(0) 1.64872 13	(0) 1.42416 39	(0) 1.31281 87	(0) 1.24660 50
0.3	(0) 3.03607 92	(0) 1.99359 02	(0) 1.64872 13	(0) 1.47782 42	(0) 1.37626 32
0.4	(0) 3.77614 69	(0) 2.35254 68	(0) 1.88183 81	(0) 1.64872 13	(0) 1.51027 29
0.5	(0) 4.54811 35	(0) 2.72590 86	(0) 2.12369 98	(0) 1.82563 24	(0) 1.64872 13
0.6	(0) 5.35272 38	(0) 3.11399 83	(0) 2.37449 45	(0) 2.00868 23	(0) 1.79169 69
0.7	(0) 6.19073 40	(0) 3.51714 35	(0) 2.63441 32	(0) 2.19799 70	(0) 1.93928 94
0.8	(0) 7.06291 26	(0) 3.93567 68	(0) 2.90364 98	(0) 2.39370 49	(0) 2.09159 01
0.9	(0) 7.97004 04	(0) 4.36993 59	(0) 3.18240 09	(0) 2.59593 60	(0) 2.24869 11
1.0	(0) 8.91291 03	(0) 4.82026 39	(0) 3.47086 63	(0) 2.80482 21	(0) 2.41068 61

$a \backslash b$	0.6	0.7	0.8	0.9	1.0
-1.0	(-1) 1.66666 67	(-1) 2.85714 29	(-1) 3.75000 00	(-1) 4.44444 44	(-1) 5.00000 00
-0.9	(-1) 2.37390 35	(-1) 3.46998 42	(-1) 4.29138 21	(-1) 4.92975 27	(-1) 5.44007 21
-0.8	(-1) 3.10765 94	(-1) 4.10420 52	(-1) 4.85042 16	(-1) 5.42992 21	(-1) 5.89284 39
-0.7	(-1) 3.86848 36	(-1) 4.76023 18	(-1) 5.42745 70	(-1) 5.94522 72	(-1) 6.35854 17
-0.6	(-1) 4.65693 33	(-1) 5.43849 54	(-1) 6.02283 14	(-1) 6.47594 62	(-1) 6.83739 50
-0.5	(-1) 5.47357 40	(-1) 6.13943 38	(-1) 6.63689 23	(-1) 7.02236 09	(-1) 7.32963 60
-0.4	(-1) 6.31897 89	(-1) 6.86349 09	(-1) 7.26999 22	(-1) 7.58475 70	(-1) 7.83550 00
-0.3	(-1) 7.19372 99	(-1) 7.61111 66	(-1) 7.92248 85	(-1) 8.16342 38	(-1) 8.35522 55
-0.2	(-1) 8.09841 67	(-1) 8.38276 72	(-1) 8.59474 31	(-1) 8.75865 45	(-1) 8.88905 38
-0.1	(-1) 9.03363 78	(-1) 9.17890 54	(-1) 9.28712 29	(-1) 9.37074 63	(-1) 9.43722 94
0.0	(0) 1.00000 00	(0) 1.00000 00	(0) 1.00000 00	(0) 1.00000 00	(0) 1.00000 00
0.1	(0) 1.09981 19	(0) 1.08465 27	(0) 1.07337 51	(0) 1.06467 21	(0) 1.05776 16
0.2	(0) 1.20286 18	(0) 1.17189 67	(0) 1.14887 58	(0) 1.13112 17	(0) 1.11703 33
0.3	(0) 1.30921 31	(0) 1.26178 10	(0) 1.22654 08	(0) 1.19938 02	(0) 1.17784 06
0.4	(0) 1.41892 99	(0) 1.35435 51	(0) 1.30640 94	(0) 1.26947 93	(0) 1.24020 96
0.5	(0) 1.53207 73	(0) 1.44966 91	(0) 1.38852 11	(0) 1.34145 10	(0) 1.30416 68
0.6	(0) 1.64872 13	(0) 1.54777 40	(0) 1.47291 64	(0) 1.41532 79	(0) 1.36973 88
0.7	(0) 1.76892 87	(0) 1.64872 13	(0) 1.55963 60	(0) 1.49114 29	(0) 1.43695 27
0.8	(0) 1.89276 74	(0) 1.75256 32	(0) 1.64872 13	(0) 1.56892 95	(0) 1.50583 59
0.9	(0) 2.02030 62	(0) 1.85935 29	(0) 1.74021 40	(0) 1.64872 13	(0) 1.57641 61
1.0	(0) 2.15161 47	(0) 1.96914 38	(0) 1.83415 67	(0) 1.73055 26	(0) 1.64872 13

CONFLUENT HYPERGEOMETRIC FUNCTION $M(a, b, x)$ Table 13.1

$x = 0.6$

$a \backslash b$	0.1	0.2	0.3	0.4	0.5
−1.0	(0) −5.00000 00	(0) −2.00000 00	(0) −1.00000 00	(−1) −5.00000 00	(−1) −2.00000 00
−0.9	(0) −4.56442 36	(0) −1.77497 83	(−1) −8.45926 51	(−1) −3.81848 50	(−1) −1.03687 14
−0.8	(0) −4.09525 03	(0) −1.53457 51	(−1) −6.82397 09	(−1) −2.57117 79	(−3) −2.46606 50
−0.7	(0) −3.59141 57	(0) −1.27832 65	(−1) −5.09139 76	(−1) −1.25627 00	(−1) +1.03792 44
−0.6	(0) −3.05183 34	(0) −1.00575 96	(−1) −3.25877 35	(−2) +1.28080 81	(−1) 2.15219 91
−0.5	(0) −2.47539 54	(−1) −7.16392 12	(−1) −1.32327 40	(−1) 1.58375 09	(−1) 3.31950 22
−0.4	(0) −1.86097 11	(−1) −4.09732 38	(−2) +7.17978 94	(−1) 3.11265 10	(−1) 4.54119 67
−0.3	(0) −1.20740 73	(−2) −8.52791 51	(−1) 2.86791 75	(−1) 4.71672 67	(−1) 5.81866 96
−0.2	(−1) −5.13527 80	(−1) +2.57478 49	(−1) 5.12952 90	(−1) 6.39795 93	(−1) 7.15333 26
−0.1	(−1) +2.21866 89	(−1) 6.19061 29	(−1) 7.50585 66	(−1) 8.15836 59	(−1) 8.54662 21
0.0	(0) 1.00000 00	(0) 1.00000 00	(0) 1.00000 00	(0) 1.00000 00	(0) 1.00000 00
0.1	(0) 1.82211 88	(0) 1.40083 55	(0) 1.26151 16	(0) 1.19249 52	(0) 1.15149 54
0.2	(0) 2.68949 50	(0) 1.82211 88	(0) 1.53544 21	(0) 1.39353 51	(0) 1.30929 96
0.3	(0) 3.60342 49	(0) 2.26441 16	(0) 1.82211 88	(0) 1.60333 61	(0) 1.47356 68
0.4	(0) 4.56523 01	(0) 2.72828 58	(0) 2.12187 52	(0) 1.82211 88	(0) 1.64445 34
0.5	(0) 5.57625 77	(0) 3.21432 45	(0) 2.43505 08	(0) 2.05010 75	(0) 1.82211 88
0.6	(0) 6.63788 04	(0) 3.72312 11	(0) 2.76199 12	(0) 2.28753 06	(0) 2.00672 51
0.7	(0) 7.75149 76	(0) 4.25528 05	(0) 3.10304 83	(0) 2.53462 03	(0) 2.19843 71
0.8	(0) 8.91853 48	(0) 4.81141 85	(0) 3.45858 04	(0) 2.79161 30	(0) 2.39742 24
0.9	(1) 1.01404 45	(0) 5.39216 24	(0) 3.82895 20	(0) 3.05874 93	(0) 2.60385 15
1.0	(1) 1.14187 08	(0) 5.99815 10	(0) 4.21453 44	(0) 3.33627 37	(0) 2.81789 78

$a \backslash b$	0.6	0.7	0.8	0.9	1.0
−1.0	0.00000 00	(−1) 1.42857 14	(−1) 2.50000 00	(−1) 3.33333 33	(−1) 4.00000 00
−0.9	(−2) 8.15612 80	(−1) 2.13746 25	(−1) 3.12786 69	(−1) 3.89744 84	(−1) 4.51255 49
−0.8	(−1) 1.66954 03	(−1) 2.87723 99	(−1) 3.78124 01	(−1) 4.48302 85	(−1) 5.04345 12
−0.7	(−1) 2.56274 99	(−1) 3.64865 28	(−1) 4.46071 49	(−1) 5.09055 63	(−1) 5.59308 68
−0.6	(−1) 3.49622 62	(−1) 4.45246 33	(−1) 5.16689 67	(−1) 5.72052 24	(−1) 6.16186 59
−0.5	(−1) 4.47097 05	(−1) 5.28944 63	(−1) 5.90040 05	(−1) 6.37342 52	(−1) 6.75019 92
−0.4	(−1) 5.48800 20	(−1) 6.16039 00	(−1) 6.66185 18	(−1) 7.04977 12	(−1) 7.35850 35
−0.3	(−1) 6.54835 72	(−1) 7.06609 56	(−1) 7.45188 61	(−1) 7.75007 48	(−1) 7.98720 24
−0.2	(−1) 7.65309 05	(−1) 8.00737 79	(−1) 8.27114 95	(−1) 8.47485 87	(−1) 8.63672 59
−0.1	(−1) 8.80327 45	(−1) 8.98506 53	(−1) 9.12029 84	(−1) 9.22465 40	(−1) 9.30751 06
0.0	(0) 1.00000 00	(0) 1.00000 00	(0) 1.00000 00	(0) 1.00000 00	(0) 1.00000 00
0.1	(0) 1.12443 77	(0) 1.10530 38	(0) 1.09109 32	(0) 1.08014 45	(0) 1.07146 44
0.2	(0) 1.25375 32	(0) 1.21450 50	(0) 1.18537 84	(0) 1.16295 44	(0) 1.14519 01
0.3	(0) 1.38806 15	(0) 1.32769 20	(0) 1.28292 55	(0) 1.24848 64	(0) 1.22122 33
0.4	(0) 1.52747 91	(0) 1.44495 47	(0) 1.38380 56	(0) 1.33679 79	(0) 1.29961 13
0.5	(0) 1.67212 47	(0) 1.56638 46	(0) 1.48809 10	(0) 1.42794 70	(0) 1.38040 19
0.6	(0) 1.82211 88	(0) 1.69207 45	(0) 1.59585 51	(0) 1.52199 31	(0) 1.46364 36
0.7	(0) 1.97758 41	(0) 1.82211 88	(0) 1.70717 25	(0) 1.61899 63	(0) 1.54938 57
0.8	(0) 2.13864 53	(0) 1.95661 34	(0) 1.82211 88	(0) 1.71901 75	(0) 1.63767 83
0.9	(0) 2.30542 91	(0) 2.09565 57	(0) 1.94077 10	(0) 1.82211 88	(0) 1.72857 22
1.0	(0) 2.47806 43	(0) 2.23934 48	(0) 2.06320 72	(0) 1.92836 31	(0) 1.82211 88

Table 13.1   **CONFLUENT HYPERGEOMETRIC FUNCTION $M(a, b, x)$**

$$x = 0.7$$

$a \backslash b$	0.1	0.2	0.3	0.4	0.5
-1.0	(0)-6.00000 00	(0)-2.50000 00	(0)-1.33333 33	(-1)-7.50000 00	(-1)-4.00000 00
-0.9	(0)-5.52819 79	(0)-2.25396 47	(0)-1.16362 83	(-1)-6.19090 30	(-1)-2.92768 78
-0.8	(0)-5.01049 23	(0)-1.98691 64	(-1)-9.81007 11	(-1)-4.79194 87	(-1)-1.78834 77
-0.7	(0)-4.44515 47	(0)-1.69810 26	(-1)-7.85028 60	(-1)-3.30020 58	(-2)-5.79886 90
-0.6	(0)-3.83041 49	(0)-1.38675 31	(-1)-5.75241 82	(-1)-1.71267 91	(-2)+6.99831 62
-0.5	(0)-3.16446 06	(0)-1.05207 99	(-1)-3.51185 70	(-3)-2.63083 59	(-1) 2.05299 00
-0.4	(0)-2.44543 68	(-1)-6.93277 09	(-1)-1.12388 92	(-1)+1.76203 27	(-1) 3.48181 61
-0.3	(0)-1.67144 46	(-1)-3.09520 29	(-1)+1.41630 28	(-1) 3.65553 75	(-1) 4.98858 44
-0.2	(-1)-8.40541 00	(-1)+1.00033 57	(-1) 4.11364 25	(-1) 5.65746 78	(-1) 6.57561 66
-0.1	(-2)+4.92624 47	(-1) 5.36246 53	(-1) 6.97316 13	(-1) 7.77115 48	(-1) 8.24528 23
0.0	(0) 1.00000 00	(0) 1.00000 00	(0) 1.00000 00	(0) 1.00000 00	(0) 1.00000 00
0.1	(0) 2.01375 27	(0) 1.49219 50	(0) 1.31994 11	(0) 1.23474 77	(0) 1.18422 38
0.2	(0) 3.09264 92	(0) 2.01375 27	(0) 1.65767 60	(0) 1.48171 31	(0) 1.37745 14
0.3	(0) 4.23886 64	(0) 2.56561 44	(0) 2.01375 27	(0) 1.74125 83	(0) 1.57993 98
0.4	(0) 5.45463 06	(0) 3.14874 21	(0) 2.38873 10	(0) 2.01375 27	(0) 1.79195 11
0.5	(0) 6.74221 79	(0) 3.76411 90	(0) 2.78318 26	(0) 2.29957 36	(0) 2.01375 27
0.6	(0) 8.10395 56	(0) 4.41274 94	(0) 3.19769 12	(0) 2.59910 58	(0) 2.24561 74
0.7	(0) 9.54222 25	(0) 5.09565 95	(0) 3.63285 27	(0) 2.91274 21	(0) 2.48782 35
0.8	(1) 1.10594 50	(0) 5.81389 76	(0) 4.08927 57	(0) 3.24088 34	(0) 2.74065 46
0.9	(1) 1.26581 24	(0) 6.56853 43	(0) 4.56758 14	(0) 3.58393 85	(0) 3.00440 00
1.0	(1) 1.43407 83	(0) 7.36066 31	(0) 5.06840 38	(0) 3.94232 46	(0) 3.27935 49

$a \backslash b$	0.6	0.7	0.8	0.9	1.0
-1.0	(-1)-1.66666 67	0.00000 00	(-1) 1.25000 00	(-1) 2.22222 22	(-1) 3.00000 00
-0.9	(-2)-7.54915 03	(-2) 7.95165 75	(-1) 1.95634 74	(-1) 2.85846 10	(-1) 3.57936 92
-0.8	(-2)+2.09154 67	(-1) 1.63250 20	(-1) 2.69751 66	(-1) 3.52400 18	(-1) 4.18377 43
-0.7	(-1) 1.22710 86	(-1) 2.51322 11	(-1) 3.47447 03	(-1) 4.21962 49	(-1) 4.81385 81
-0.6	(-1) 2.30054 51	(-1) 3.43855 96	(-1) 4.28819 01	(-1) 4.94612 53	(-1) 5.47027 56
-0.5	(-1) 3.43109 52	(-1) 4.40977 87	(-1) 5.13967 66	(-1) 5.70431 32	(-1) 6.15369 36
-0.4	(-1) 4.62042 36	(-1) 5.42816 47	(-1) 6.02994 98	(-1) 6.49501 40	(-1) 6.86479 13
-0.3	(-1) 5.87022 82	(-1) 6.49502 91	(-1) 6.96004 90	(-1) 7.31906 85	(-1) 7.60426 03
-0.2	(-1) 7.18224 16	(-1) 7.61170 97	(-1) 7.93103 40	(-1) 8.17733 33	(-1) 8.37280 46
-0.1	(-1) 8.55823 13	(-1) 8.77956 99	(-1) 8.94398 42	(-1) 9.07068 09	(-1) 9.17114 12
0.0	(0) 1.00000 00	(0) 1.00000 00	(0) 1.00000 00	(0) 1.00000 00	(0) 1.00000 00
0.1	(0) 1.15093 86	(0) 1.12744 17	(0) 1.11002 02	(0) 1.09661 96	(0) 1.08601 24
0.2	(0) 1.30882 66	(0) 1.26042 67	(0) 1.22457 33	(0) 1.19701 89	(0) 1.17522 70
0.3	(0) 1.47385 50	(0) 1.39910 20	(0) 1.34377 57	(0) 1.30129 20	(0) 1.26772 07
0.4	(0) 1.64621 90	(0) 1.54361 79	(0) 1.46774 58	(0) 1.40953 43	(0) 1.36357 19
0.5	(0) 1.82611 74	(0) 1.69412 73	(0) 1.59660 44	(0) 1.52184 32	(0) 1.46286 04
0.6	(0) 2.01375 27	(0) 1.85078 59	(0) 1.73047 46	(0) 1.63831 77	(0) 1.56566 72
0.7	(0) 2.20933 17	(0) 2.01375 27	(0) 1.86948 15	(0) 1.75905 87	(0) 1.67207 52
0.8	(0) 2.41306 50	(0) 2.18318 94	(0) 2.01375 27	(0) 1.88416 89	(0) 1.78216 81
0.9	(0) 2.62516 74	(0) 2.35926 09	(0) 2.16341 82	(0) 2.01375 27	(0) 1.89603 16
1.0	(0) 2.84585 75	(0) 2.54213 50	(0) 2.31861 02	(0) 2.14791 66	(0) 2.01375 27

CONFLUENT HYPERGEOMETRIC FUNCTION $M(a, b, x)$ Table 13.1

$x = 0.8$

$a \backslash b$	0.1	0.2	0.3	0.4	0.5
−1.0	(0) −7.00000 00	(0) −3.00000 00	(0) −1.66666 67	(0) −1.00000 00	(−1) −6.00000 00
−0.9	(0) −6.50401 48	(0) −2.73837 67	(0) −1.48461 68	(−1) −8.58588 03	(−1) −4.83512 37
−0.8	(0) −5.94785 78	(0) −2.44921 23	(0) −1.28563 99	(−1) −7.05401 18	(−1) −3.58242 29
−0.7	(0) −5.32888 96	(0) −2.13135 83	(0) −1.06906 32	(−1) −5.39992 81	(−1) −2.23871 07
−0.6	(0) −4.64439 77	(0) −1.78363 55	(−1) −8.34197 05	(−1) −3.61905 04	(−2) −8.00722 55
−0.5	(0) −3.89159 56	(0) −1.40483 36	(−1) −5.80333 58	(−1) −1.70668 54	(−2) +7.34885 63
−0.4	(0) −3.06762 06	(−1) −9.93710 17	(−1) −3.06747 02	(−2) +3.41976 74	(−1) 2.37153 85
−0.3	(0) −2.16953 29	(−1) −5.48990 22	(−2) −1.26930 95	(−1) 2.53186 47	(−1) 4.11274 30
−0.2	(0) −1.19431 35	(−2) −6.93656 36	(−1) +3.02591 28	(−1) 4.86802 83	(−1) 5.96208 97
−0.1	(−1) −1.38863 05	(−1) +4.46505 60	(−1) 6.39888 38	(−1) 7.35564 06	(−1) 7.92325 45
0.0	(0) +1.00000 00	(0) 1.00000 00	(0) 1.00000 00	(0) 1.00000 00	(0) 1.00000 00
0.1	(0) 2.22554 09	(0) 1.59252 93	(0) 1.38374 79	(0) 1.28065 33	(0) 1.21961 77
0.2	(0) 3.54111 04	(0) 2.22554 09	(0) 1.79197 39	(0) 1.57807 97	(0) 1.45157 28
0.3	(0) 4.95014 63	(0) 2.90051 91	(0) 2.22554 09	(0) 1.89284 81	(0) 1.69626 83
0.4	(0) 6.45617 50	(0) 3.61898 52	(0) 2.68533 25	(0) 2.22554 09	(0) 1.95411 70
0.5	(0) 8.06281 37	(0) 4.38249 84	(0) 3.17225 39	(0) 2.57675 45	(0) 2.22554 09
0.6	(0) 9.77377 18	(0) 5.19265 68	(0) 3.68723 21	(0) 2.94709 89	(0) 2.51097 18
0.7	(1) 1.15928 53	(0) 6.05109 78	(0) 4.23121 63	(0) 3.33719 88	(0) 2.81085 12
0.8	(1) 1.35239 56	(0) 6.95949 89	(0) 4.80517 86	(0) 3.74769 30	(0) 3.12563 06
0.9	(1) 1.55710 78	(0) 7.91957 87	(0) 5.41011 38	(0) 4.17923 55	(0) 3.45577 20
1.0	(1) 1.77383 16	(0) 8.93309 73	(0) 6.04704 06	(0) 4.63249 51	(0) 3.80174 73

$a \backslash b$	0.6	0.7	0.8	0.9	1.0
−1.0	(−1) −3.33333 33	(−1) −1.42857 14	0.00000 00	(−1) 1.11111 11	(−1) 2.00000 00
−0.9	(−1) −2.33826 62	(−2) −5.57356 94	(−2) 7.76467 88	(−1) 1.81250 42	(−1) 2.64028 04
−0.8	(−1) −1.27465 48	(−2) +3.69102 15	(−1) 1.59854 95	(−1) 2.55227 74	(−1) 3.31335 07
−0.7	(−2) −1.40115 64	(−1) 1.35264 99	(−1) 2.46770 86	(−1) 3.33161 66	(−1) 4.02018 75
−0.6	(−1) +1.06779 15	(−1) 2.39517 31	(−1) 3.38544 19	(−1) 4.15173 34	(−1) 4.76178 82
−0.5	(−1) 2.35156 45	(−1) 3.49860 15	(−1) 4.35327 95	(−1) 5.01386 60	(−1) 5.53917 14
−0.4	(−1) 3.71375 95	(−1) 4.66490 92	(−1) 5.37278 55	(−1) 5.91927 92	(−1) 6.35337 71
−0.3	(−1) 5.15699 27	(−1) 5.89611 50	(−1) 6.44555 87	(−1) 6.86926 51	(−1) 7.20546 73
−0.2	(−1) 6.68394 10	(−1) 7.19428 36	(−1) 7.57323 29	(−1) 7.86514 37	(−1) 8.09652 62
−0.1	(−1) 8.29734 28	(−1) 8.56152 59	(−1) 8.75747 79	(−1) 8.90826 31	(−1) 9.02766 05
0.0	(0) 1.00000 00	(0) 1.00000 00	(0) 1.00000 00	(0) 1.00000 00	(0) 1.00000 00
0.1	(0) 1.17947 78	(0) 1.15119 12	(0) 1.13025 42	(0) 1.11417 60	(0) 1.10146 98
0.2	(0) 1.36846 08	(0) 1.30995 18	(0) 1.26668 86	(0) 1.23349 80	(0) 1.20729 30
0.3	(0) 1.56724 87	(0) 1.47651 22	(0) 1.40948 49	(0) 1.35811 24	(0) 1.31758 99
0.4	(0) 1.77614 79	(0) 1.65110 80	(0) 1.55882 92	(0) 1.48816 89	(0) 1.43248 29
0.5	(0) 1.99547 19	(0) 1.83397 98	(0) 1.71491 10	(0) 1.62382 02	(0) 1.55209 71
0.6	(0) 2.22554 09	(0) 2.02537 37	(0) 1.87792 43	(0) 1.76522 23	(0) 1.67656 00
0.7	(0) 2.46668 24	(0) 2.22554 09	(0) 2.04806 69	(0) 1.91253 43	(0) 1.80600 17
0.8	(0) 2.71923 11	(0) 2.43473 81	(0) 2.22554 09	(0) 2.06591 86	(0) 1.94055 51
0.9	(0) 2.98352 90	(0) 2.65322 74	(0) 2.41055 26	(0) 2.22554 09	(0) 2.08035 55
1.0	(0) 3.25992 56	(0) 2.88127 68	(0) 2.60331 27	(0) 2.39157 03	(0) 2.22554 09

Table 13.1   CONFLUENT HYPERGEOMETRIC FUNCTION $M(a, b, x)$

$x = 0.9$

$a\backslash b$	0.1	0.2	0.3	0.4	0.5
−1.0	(0) −8.00000 00	(0) −3.50000 00	(0) −2.00000 00	(0) −1.25000 00	(−1) −8.00000 00
−0.9	(0) −7.49259 77	(0) −3.22852 60	(0) −1.80907 26	(0) −1.10046 05	(−1) −6.76001 98
−0.8	(0) −6.90878 25	(0) −2.92208 06	(0) −1.59665 35	(−1) −9.35972 27	(−1) −5.40855 15
−0.7	(0) −6.24470 96	(0) −2.57899 21	(0) −1.36176 43	(−1) −7.55885 89	(−1) −3.94096 49
−0.6	(0) −5.49641 35	(0) −2.19753 81	(0) −1.10339 79	(−1) −5.59533 56	(−1) −2.35250 18
−0.5	(0) −4.65980 55	(0) −1.77594 43	(−1) −8.20518 02	(−1) −3.46228 53	(−2) −6.38272 88
−0.4	(0) −3.73067 11	(0) −1.31238 34	(−1) −5.12058 10	(−1) −1.15264 70	(−1) +1.20674 49
−0.3	(0) −2.70466 65	(−1) −8.04973 88	(−1) −1.76920 97	(−1) +1.34083 75	(−1) 3.18771 09
−0.2	(0) −1.57731 62	(−1) −2.51778 79	(−1) +1.86021 91	(−1) 4.02562 81	(−1) 5.30992 39
−0.1	(−1) −3.44010 11	(−1) +3.49195 37	(−1) 5.77931 14	(−1) 6.90939 03	(−1) 7.57882 50
0.0	(0) +1.00000 00	(0) 1.00000 00	(0) 1.00000 00	(0) 1.00000 00	(0) 1.00000 00
0.1	(0) 2.45960 31	(0) 1.70274 56	(0) 1.45345 52	(0) 1.33055 47	(0) 1.25791 83
0.2	(0) 4.03983 23	(0) 2.45960 31	(0) 1.93955 77	(0) 1.68343 42	(0) 1.53222 60
0.3	(0) 5.74586 78	(0) 3.27280 52	(0) 2.45960 31	(0) 2.05949 16	(0) 1.82352 69
0.4	(0) 7.58304 06	(0) 4.14464 74	(0) 3.01492 28	(0) 2.45960 31	(0) 2.13244 07
0.5	(0) 9.55683 50	(0) 5.07749 00	(0) 3.60688 44	(0) 2.88466 81	(0) 2.45960 31
0.6	(1) 1.16728 93	(0) 6.07375 88	(0) 4.23689 27	(0) 3.33560 96	(0) 2.80566 62
0.7	(1) 1.39370 17	(0) 7.13594 69	(0) 4.90639 03	(0) 3.81337 52	(0) 3.17129 88
0.8	(1) 1.63551 72	(0) 8.26661 58	(0) 5.61685 85	(0) 4.31893 69	(0) 3.55718 66
0.9	(1) 1.89334 94	(0) 9.46839 74	(0) 6.36981 80	(0) 4.85329 20	(0) 3.96403 28
1.0	(1) 2.16782 87	(1) 1.07439 95	(0) 7.16683 00	(0) 5.41746 38	(0) 4.39255 83

$a\backslash b$	0.6	0.7	0.8	0.9	1.0
−1.0	(−1) −5.00000 00	(−1) −2.85714 29	(−1) −1.25000 00	0.00000 00	(−1) 1.00000 00
−0.9	(−1) −3.93506 44	(−1) −1.92058 43	(−2) −4.12148 81	(−2) 7.59274 35	(−1) 1.69504 02
−0.8	(−1) −2.78312 29	(−2) −9.13906 92	(−2) +4.83592 97	(−1) 1.56725 54	(−1) 2.43169 00
−0.7	(−1) −1.54071 44	(−2) +1.65565 38	(−1) 1.43934 85	(−1) 2.42566 24	(−1) 3.21136 46
−0.6	(−2) −2.04284 74	(−1) 1.32057 89	(−1) 2.45729 51	(−1) 3.33625 68	(−1) 4.03551 32
−0.5	(−1) +1.22981 53	(−1) 2.55395 12	(−1) 3.53966 52	(−1) 4.30084 39	(−1) 4.90562 01
−0.4	(−1) 2.76533 21	(−1) 3.86857 31	(−1) 4.68874 74	(−1) 5.32127 33	(−1) 5.82320 50
−0.3	(−1) 4.40611 09	(−1) 5.26740 93	(−1) 5.90688 76	(−1) 6.39943 94	(−1) 6.78982 39
−0.2	(−1) 6.15609 81	(−1) 6.75350 07	(−1) 7.19649 04	(−1) 7.53728 29	(−1) 7.80706 95
−0.1	(−1) 8.01934 30	(−1) 8.32996 53	(−1) 8.56001 96	(−1) 8.73679 14	(−1) 8.87657 20
0.0	(0) 1.00000 00	(0) 1.00000 00	(0) 1.00000 00	(0) 1.00000 00	(0) 1.00000 00
0.1	(0) 1.21023 31	(0) 1.17668 82	(0) 1.15190 18	(0) 1.13289 93	(0) 1.11790 61
0.2	(0) 1.43307 07	(0) 1.36339 71	(0) 1.31197 24	(0) 1.27259 03	(0) 1.24155 02
0.3	(0) 1.66896 10	(0) 1.56047 09	(0) 1.48048 31	(0) 1.41929 15	(0) 1.37111 10
0.4	(0) 1.91836 37	(0) 1.76826 25	(0) 1.65771 19	(0) 1.57322 64	(0) 1.50677 14
0.5	(0) 2.18175 01	(0) 1.98713 34	(0) 1.84394 34	(0) 1.73462 38	(0) 1.64871 85
0.6	(0) 2.45960 31	(0) 2.21745 38	(0) 2.03946 90	(0) 1.90371 79	(0) 1.79714 36
0.7	(0) 2.75241 80	(0) 2.45960 31	(0) 2.24458 71	(0) 2.08074 81	(0) 1.95224 22
0.8	(0) 3.06070 20	(0) 2.71396 99	(0) 2.45960 31	(0) 2.26595 96	(0) 2.11421 45
0.9	(0) 3.38497 53	(0) 2.98095 21	(0) 2.68482 96	(0) 2.45960 31	(0) 2.28326 51
1.0	(0) 3.72577 04	(0) 3.26095 72	(0) 2.92058 65	(0) 2.66193 52	(0) 2.45960 31

CONFLUENT HYPERGEOMETRIC FUNCTION $M(a, b, x)$ Table 13.1

$x = 1.0$

$a\backslash b$	0.1	0.2	0.3	0.4	0.5
−1.0	(0)−9.00000 00	(0)−4.00000 00	(0)−2.33333 33	(0)−1.50000 00	(0)−1.00000 00
−0.9	(0)−8.49472 34	(0)−3.72474 63	(0)−2.13718 91	(0)−1.34483 48	(−1)−8.70327 28
−0.8	(0)−7.89481 34	(0)−3.40618 57	(0)−1.91443 23	(0)−1.17116 05	(−1)−7.26851 39
−0.7	(0)−7.19487 27	(0)−3.04197 32	(0)−1.66369 18	(−1)−9.78067 35	(−1)−5.68924 14
−0.6	(0)−6.38931 44	(0)−2.62968 42	(0)−1.38355 11	(−1)−7.64616 83	(−1)−3.95877 20
−0.5	(0)−5.47235 71	(0)−2.16681 22	(0)−1.07254 74	(−1)−5.29840 46	(−1)−2.07021 66
−0.4	(0)−4.43802 02	(0)−1.65076 69	(−1)−7.29170 37	(−1)−2.72739 30	(−3)−1.64753 21
−0.3	(0)−3.28011 86	(0)−1.07887 24	(−1)−3.51861 30	(−3)+7.71680 36	(−1)+2.20976 75
−0.2	(0)−1.99225 77	(−1)−4.48364 63	(−2)+6.09884 13	(−1) 3.12589 94	(−1) 4.61604 79
−0.1	(−1)−5.67828 07	(−1)+2.43610 69	(−1) 5.11038 28	(−1) 6.42974 92	(−1) 7.21012 79
0.0	(0)+1.00000 00	(0) 1.00000 00	(0) 1.00000 00	(0) 1.00000 00	(0) 1.00000 00
0.1	(0) 2.71828 18	(0) 1.82384 44	(0) 1.52963 87	(0) 1.38482 77	(0) 1.29938 93
0.2	(0) 4.59430 40	(0) 2.71828 18	(0) 2.10177 40	(0) 1.79865 55	(0) 1.62002 78
0.3	(0) 6.63559 00	(0) 3.68654 94	(0) 2.71828 18	(0) 2.24271 69	(0) 1.96278 70
0.4	(0) 8.84990 62	(0) 4.73198 60	(0) 3.38109 51	(0) 2.71828 18	(0) 2.32856 41
0.5	(1) 1.12452 68	(0) 5.85803 42	(0) 4.09220 54	(0) 3.22665 79	(0) 2.71828 18
0.6	(1) 1.38299 44	(0) 7.06824 32	(0) 4.85366 43	(0) 3.76919 11	(0) 3.13288 93
0.7	(1) 1.66124 65	(0) 8.36627 13	(0) 5.66758 48	(0) 4.34726 65	(0) 3.57336 26
0.8	(1) 1.96016 30	(0) 9.75588 81	(0) 6.53614 27	(0) 4.96230 95	(0) 4.04070 56
0.9	(1) 2.28065 08	(1) 1.12409 78	(0) 7.46157 79	(0) 5.61578 62	(0) 4.53595 02
1.0	(1) 2.62364 52	(1) 1.28255 41	(0) 8.44619 60	(0) 6.30920 50	(0) 5.06015 69

$a\backslash b$	0.6	0.7	0.8	0.9	1.0
−1.0	(−1)−6.66666 67	(−1)−4.28571 43	(−1)−2.50000 00	(−1)−1.11111 11	0.00000 00
−0.9	(−1)−5.54597 35	(−1)−3.29502 50	(−1)−1.60990 29	(−2)−3.01549 81	(−2) 7.43386 23
−0.8	(−1)−4.31756 71	(−1)−2.21753 45	(−2)−6.48146 54	(−2)+5.68299 01	(−1) 1.53827 23
−0.7	(−1)−2.97660 48	(−1)−1.04950 02	(−2)+3.88236 65	(−1) 1.50083 68	(−1) 2.38663 42
−0.6	(−1)−1.51809 81	(−2)+2.12929 76	(−1) 1.50229 88	(−1) 2.49853 18	(−1) 3.29050 15
−0.5	(−3)+6.30910 70	(−1) 1.57371 99	(−1) 2.69717 87	(−1) 3.56392 05	(−1) 4.25195 83
−0.4	(−1) 1.77225 36	(−1) 3.03694 92	(−1) 3.97610 35	(−1) 4.69960 88	(−1) 5.27314 45
−0.3	(−1) 3.61483 67	(−1) 4.60681 41	(−1) 5.34239 08	(−1) 5.90827 38	(−1) 6.35625 70
−0.2	(−1) 5.59644 73	(−1) 6.28763 08	(−1) 6.79945 04	(−1) 7.19266 55	(−1) 7.50355 07
−0.1	(−1) 7.72285 59	(−1) 8.08383 81	(−1) 8.35078 67	(−1) 8.55560 76	(−1) 8.71734 01
0.0	(0) 1.00000 00	(0) 1.00000 00	(0) 1.00000 00	(0) 1.00000 00	(0) 1.00000 00
0.1	(0) 1.24339 88	(0) 1.20408 08	(0) 1.17507 89	(0) 1.15288 20	(0) 1.13539 67
0.2	(0) 1.50311 03	(0) 1.42110 86	(0) 1.36069 55	(0) 1.31451 22	(0) 1.27817 41
0.3	(0) 1.77978 05	(0) 1.65157 89	(0) 1.55723 97	(0) 1.48520 44	(0) 1.42858 86
0.4	(0) 2.07407 40	(0) 1.89600 10	(0) 1.76511 25	(0) 1.66528 05	(0) 1.58690 33
0.5	(0) 2.38667 38	(0) 2.15489 81	(0) 1.98472 52	(0) 1.85507 07	(0) 1.75338 77
0.6	(0) 2.71828 18	(0) 2.42880 78	(0) 2.21650 01	(0) 2.05491 39	(0) 1.92831 84
0.7	(0) 3.06961 97	(0) 2.71828 18	(0) 2.46087 06	(0) 2.26515 76	(0) 2.11197 89
0.8	(0) 3.44142 89	(0) 3.02388 72	(0) 2.71828 18	(0) 2.48615 84	(0) 2.30465 98
0.9	(0) 3.83447 12	(0) 3.34620 59	(0) 2.98919 01	(0) 2.71828 18	(0) 2.50665 90
1.0	(0) 4.24952 89	(0) 3.68583 55	(0) 3.27406 39	(0) 2.96190 29	(0) 2.71828 18

Table 13.1 **CONFLUENT HYPERGEOMETRIC FUNCTION $M(a, b, x)$**

$x = 2.0$

$a\backslash b$	0.1	0.2	0.3	0.4	0.5
-1.0	(1)-1.90000 00	(0)-9.00000 00	(0)-5.66666 67	(0)-4.00000 00	(0)-3.00000 00
-0.9	(1)-1.94803 05	(0)-9.11450 17	(0)-5.67351 46	(0)-3.96130 19	(0)-2.93919 07
-0.8	(1)-1.95774 57	(0)-9.05346 68	(0)-5.57239 85	(0)-3.84746 13	(0)-2.82231 32
-0.7	(1)-1.92363 39	(0)-8.79313 67	(0)-5.34952 69	(0)-3.64939 40	(0)-2.64293 64
-0.6	(1)-1.83976 09	(0)-8.30798 80	(0)-4.99011 57	(0)-3.35738 15	(0)-2.39419 32
-0.5	(1)-1.69974 68	(0)-7.57063 96	(0)-4.47833 69	(0)-2.96103 91	(0)-2.06875 95
-0.4	(1)-1.49674 24	(0)-6.55175 56	(0)-3.79726 52	(0)-2.44928 29	(0)-1.65883 14
-0.3	(1)-1.22340 44	(0)-5.21994 53	(0)-2.92882 34	(0)-1.81029 53	(0)-1.15610 27
-0.2	(0)-8.71869 85	(0)-3.54165 86	(0)-1.85372 46	(0)-1.03148 90	(-1)-5.51740 45
-0.1	(0)-4.33729 58	(0)-1.48107 68	(-1)-5.51412 64	(-2)-9.94703 39	(-1)+1.63639 81
0.0	(0)+1.00000 00	(0)+1.00000 00	(0)+1.00000 00	(0)+1.00000 00	(0) 1.00000 00
0.1	(0) 7.38905 61	(0) 3.94227 09	(0) 2.82379 65	(0) 2.28204 66	(0) 1.96790 63
0.2	(1) 1.49320 73	(0) 7.38905 61	(0) 4.94472 25	(0) 3.76272 10	(0) 3.07855 71
0.3	(1) 2.37378 96	(1) 1.13864 24	(0) 7.38905 61	(0) 5.45904 52	(0) 4.34381 17
0.4	(1) 3.39223 44	(1) 1.59833 25	(1) 1.01846 79	(0) 7.38905 61	(0) 5.77622 05
0.5	(1) 4.56085 43	(1) 2.12317 23	(1) 1.33611 54	(0) 9.57185 22	(0) 7.38905 61
0.6	(1) 5.89272 84	(1) 2.71867 46	(1) 1.69497 98	(1) 1.20276 42	(0) 9.19634 52
0.7	(1) 7.40173 79	(1) 3.39068 27	(1) 2.09837 67	(1) 1.47777 93	(1) 1.12129 02
0.8	(1) 9.10260 50	(1) 4.14538 60	(1) 2.54981 38	(1) 1.78448 86	(1) 1.34543 65
0.9	(2) 1.10109 32	(1) 4.98933 60	(1) 3.05299 98	(1) 2.12527 66	(1) 1.59372 26
1.0	(2) 1.31432 41	(1) 5.92946 26	(1) 3.61185 28	(1) 2.50266 00	(1) 1.86788 78

$a\backslash b$	0.6	0.7	0.8	0.9	1.0
-1.0	(0)-2.33333 33	(0)-1.85714 29	(0)-1.50000 00	(0)-1.22222 22	(0)-1.00000 00
-0.9	(0)-2.26126 09	(0)-1.77944 34	(0)-1.41981 77	(0)-1.14139 10	(-1)-9.19616 98
-0.8	(0)-2.14541 69	(0)-1.66645 90	(0)-1.31049 88	(0)-1.03604 27	(-1)-8.18288 30
-0.7	(0)-1.98102 67	(0)-1.51452 14	(0)-1.16915 08	(-1)-9.03849 17	(-1)-6.94107 82
-0.6	(0)-1.76300 12	(0)-1.31972 79	(-1)-9.92701 33	(-1)-7.42341 04	(-1)-5.45057 11
-0.5	(0)-1.48592 22	(0)-1.07793 00	(-1)-7.77889 97	(-1)-5.48901 84	(-1)-3.69000 42
-0.4	(0)-1.14402 63	(-1)-7.84722 05	(-1)-5.21259 33	(-1)-3.20761 19	(-1)-1.63679 56
-0.3	(-1)-7.31188 76	(-1)-4.35429 49	(-1)-2.19146 36	(-2)-5.49879 73	(-2)+7.32914 71
-0.2	(-1)-2.40906 72	(-2)-2.50963 14	(-1)+1.32327 01	(-1)+2.51516 76	(-1) 3.44431 99
-0.1	(-1)+3.33718 60	(-1)+4.51527 65	(-1) 5.37263 41	(-1) 6.02027 13	(-1) 6.52400 38
0.0	(0) 1.00000 00	(0) 1.00000 00	(0) 1.00000 00	(0) 1.00000 00	(0) 1.00000 00
0.1	(0) 1.76568 32	(0) 1.62619 96	(0) 1.52511 88	(0) 1.44908 29	(0) 1.39018 53
0.2	(0) 2.63896 63	(0) 2.33634 06	(0) 2.11745 72	(0) 1.95312 22	(0) 1.82606 83
0.3	(0) 3.62852 02	(0) 3.13698 76	(0) 2.78211 92	(0) 2.51617 15	(0) 2.31092 49
0.4	(0) 4.74350 99	(0) 4.03507 07	(0) 3.52448 69	(0) 3.14250 04	(0) 2.84820 19
0.5	(0) 5.99361 56	(0) 5.03790 12	(0) 4.35023 19	(0) 3.83660 34	(0) 3.44152 39
0.6	(0) 7.38905 61	(0) 6.15318 83	(0) 5.26532 81	(0) 4.60320 94	(0) 4.09470 06
0.7	(0) 8.94061 15	(0) 7.38905 61	(0) 6.27606 41	(0) 5.44729 15	(0) 4.81173 45
0.8	(1) 1.06596 48	(0) 8.75406 09	(0) 7.38905 61	(0) 6.37407 66	(0) 5.59682 82
0.9	(1) 1.25581 43	(1) 1.02572 10	(0) 8.61126 21	(0) 7.38905 61	(0) 6.45439 28
1.0	(1) 1.46487 09	(1) 1.19079 79	(0) 9.94999 53	(0) 8.49799 64	(0) 7.38905 61

CONFLUENT HYPERGEOMETRIC FUNCTION $M(a, b, x)$ Table 13.1

$x = 3.0$

$a\backslash b$	0.1	0.2	0.3	0.4	0.5
−1.0	(1)−2.90000 00	(1)−1.40000 00	(0)−9.00000 00	(0)−6.50000 00	(0)−5.00000 00
−0.9	(1)−3.33062 11	(1)−1.57397 85	(0)−9.93407 08	(0)−7.05978 63	(0)−5.35304 11
−0.8	(1)−3.67972 78	(1)−1.71028 23	(1)−1.06346 98	(0)−7.45607 06	(0)−5.58342 63
−0.7	(1)−3.92295 55	(1)−1.79849 94	(1)−1.10419 34	(0)−7.64967 21	(0)−5.66362 13
−0.6	(1)−4.03286 65	(1)−1.82694 57	(1)−1.10887 39	(0)−7.59691 35	(0)−5.56302 55
−0.5	(1)−3.97869 07	(1)−1.78256 05	(1)−1.07004 00	(0)−7.24926 51	(0)−5.24773 50
−0.4	(1)−3.72604 95	(1)−1.65079 47	(0)−9.79393 09	(0)−6.55296 82	(0)−4.68029 11
−0.3	(1)−3.23666 24	(1)−1.41549 22	(0)−8.27742 10	(0)−5.44863 43	(0)−3.81941 32
−0.2	(1)−2.46803 49	(1)−1.05876 41	(0)−6.04935 06	(0)−3.87082 13	(0)−2.61971 67
−0.1	(1)−1.37312 67	(0)−5.60854 66	(0)−2.99786 41	(0)−1.74758 43	(0)−1.03141 44
0.0	(0)+1.00000 00	(0)+1.00000 00	(0)+1.0000 00	(0)+1.00000 00	(0)+1.00000 00
0.1	(1) 2.00855 37	(0) 9.47722 60	(0) 6.07912 54	(0) 4.45833 69	(0) 3.53408 59
0.2	(1) 4.41540 99	(1) 2.00855 37	(1) 1.23871 81	(0) 8.72184 59	(0) 6.63580 90
0.3	(1) 7.38953 06	(1) 3.31122 04	(1) 2.00855 37	(1) 1.38935 23	(1) 1.03759 15
0.4	(2) 1.10064 09	(1) 4.88711 46	(1) 2.93502 26	(1) 2.00855 37	(1) 1.48313 21
0.5	(2) 1.53485 39	(1) 6.77048 23	(1) 4.03729 70	(1) 2.74198 55	(1) 2.00855 37
0.6	(2) 2.05059 14	(1) 8.99862 23	(1) 5.33622 57	(1) 3.60289 07	(1) 2.62290 97
0.7	(2) 2.65765 56	(2) 1.16120 98	(1) 6.85444 79	(1) 4.60562 86	(1) 3.33600 27
0.8	(2) 3.36670 66	(2) 1.46549 60	(1) 8.61651 37	(1) 5.76574 86	(1) 4.15843 31
0.9	(2) 4.18932 19	(2) 1.81749 79	(2) 1.06490 11	(1) 7.10006 77	(1) 5.10165 02
1.0	(2) 5.13805 80	(2) 2.22239 01	(2) 1.29806 99	(1) 8.62675 30	(1) 6.17800 67

$a\backslash b$	0.6	0.7	0.8	0.9	1.0
−1.0	(0)−4.00000 00	(0)−3.28571 43	(0)−2.75000 00	(0)−2.33333 33	(0)−2.00000 00
−0.9	(0)−4.22698 22	(0)−3.43076 30	(0)−2.83937 20	(0)−2.38362 40	(0)−2.02218 41
−0.8	(0)−4.35776 62	(0)−3.49795 59	(0)−2.86423 28	(0)−2.37946 93	(0)−1.99773 27
−0.7	(0)−4.37205 21	(0)−3.47180 10	(0)−2.81244 38	(0)−2.31115 68	(0)−1.91873 96
−0.6	(0)−4.24734 55	(0)−3.33517 91	(0)−2.67062 69	(0)−2.16800 92	(0)−1.77653 50
−0.5	(0)−3.95879 09	(0)−3.06922 34	(0)−2.42407 50	(0)−1.93831 65	(0)−1.56163 15
−0.4	(0)−3.47899 58	(0)−2.65319 12	(0)−2.05665 59	(0)−1.60926 29	(0)−1.26366 85
−0.3	(0)−2.77784 38	(0)−2.06432 89	(0)−1.55071 23	(0)−1.16684 98	(−1)−8.71351 71
−0.2	(0)−1.82229 72	(0)−1.27772 88	(−1)−8.86954 74	(−1)−5.95815 42	(−1)−3.72391 35
−0.1	(−1)−5.76188 60	(−1)−2.66178 30	(−2)−4.43495 10	(−1)+1.20451 21	(−1)+2.46564 64
0.0	(0)+1.00000 00	(0)+1.00000 00	(0)+1.00000 00	(0) 1.00000 00	(0) 1.00000 00
0.1	(0) 2.94937 02	(0) 2.55311 64	(0) 2.27097 84	(0) 2.06241 49	(0) 1.90360 36
0.2	(0) 5.31885 34	(0) 4.42829 20	(0) 3.79559 01	(0) 3.32891 38	(0) 2.97434 69
0.3	(0) 8.15947 04	(0) 6.66364 61	(0) 5.60309 84	(0) 4.82245 42	(0) 4.23056 48
0.4	(1) 1.15266 06	(0) 9.30049 38	(0) 7.72517 18	(0) 6.56784 35	(0) 5.69204 18
0.5	(1) 1.54802 96	(1) 1.23835 54	(1) 1.01960 38	(0) 8.59185 66	(0) 7.38010 13
0.6	(1) 2.00855 37	(1) 1.59611 70	(1) 1.30526 48	(1) 1.09233 58	(0) 9.31770 09
0.7	(1) 2.54126 00	(1) 2.00855 37	(1) 1.63348 43	(1) 1.35934 30	(1) 1.15295 31
0.8	(1) 3.15373 75	(1) 2.48129 50	(1) 2.00855 37	(1) 1.66355 12	(1) 1.40421 20
0.9	(1) 3.85417 22	(1) 3.02040 57	(1) 2.43509 06	(1) 2.00855 37	(1) 1.68839 43
1.0	(1) 4.65138 52	(1) 3.63241 26	(1) 2.91805 85	(1) 2.39820 88	(1) 2.00855 37

Table 13.1   CONFLUENT HYPERGEOMETRIC FUNCTION $M(a, b, x)$

$x = 4.0$

$a\backslash b$	0.1	0.2	0.3	0.4	0.5
-1.0	(1)-3.90000 00	(1)-1.90000 00	(1)-1.23333 33	(0)-9.00000 00	(0)-7.00000 00
-0.9	(1)-5.28985 40	(1)-2.48147 20	(1)-1.55982 88	(1)-1.10723 65	(0)-8.40761 69
-0.8	(1)-6.56662 17	(1)-3.00867 57	(1)-1.85166 07	(1)-1.28958 24	(0)-9.62460 70
-0.7	(1)-7.65252 34	(1)-3.44868 41	(1)-2.09004 11	(1)-1.43486 25	(1)-1.05661 02
-0.6	(1)-8.45540 43	(1)-3.76267 54	(1)-2.25292 22	(1)-1.52885 30	(1)-1.11333 79
-0.5	(1)-8.86704 80	(1)-3.90525 49	(1)-2.31462 88	(1)-1.55505 56	(1)-1.12123 61
-0.4	(1)-8.76134 25	(1)-3.82372 05	(1)-2.24546 12	(1)-1.49445 23	(1)-1.06719 99
-0.3	(1)-7.99228 75	(1)-3.45726 34	(1)-2.01126 30	(1)-1.32524 14	(0)-9.36252 11
-0.2	(1)-6.39183 19	(1)-2.73610 36	(1)-1.57295 45	(1)-1.02255 01	(0)-7.11353 67
-0.1	(1)-3.76752 93	(1)-1.58055 26	(0)-8.86027 55	(0)-5.58125 37	(0)-3.73199 87
0.0	(0)+1.00000 00	(0)+1.00000 00	(0)+1.00000 00	(0)+1.00000 00	(0)+1.00000 00
0.1	(1) 5.45981 50	(1) 2.40818 08	(1) 1.44217 35	(0) 9.87867 71	(0) 7.32759 68
0.2	(2) 1.25936 21	(1) 5.45981 50	(1) 3.20473 65	(1) 2.14598 18	(1) 1.55257 11
0.3	(2) 2.18189 72	(1) 9.38520 09	(1) 5.45981 50	(1) 3.61972 65	(1) 2.59017 89
0.4	(2) 3.34927 25	(2) 1.43304 83	(1) 8.28815 42	(1) 5.45981 50	(1) 3.87987 49
0.5	(2) 4.80147 67	(2) 2.04591 31	(2) 1.17799 11	(1) 7.72277 23	(1) 5.45981 50
0.6	(2) 6.58320 17	(2) 2.79535 32	(2) 1.60355 04	(2) 1.04714 53	(1) 7.37235 87
0.7	(2) 8.74427 45	(2) 3.70166 95	(2) 2.11665 31	(2) 1.37755 99	(1) 9.66443 28
0.8	(3) 1.13401 20	(2) 4.78740 93	(2) 2.72967 48	(2) 1.77124 33	(2) 1.23879 22
0.9	(3) 1.44322 61	(2) 6.07756 33	(2) 3.45631 21	(2) 2.23672 99	(2) 1.56000 85
1.0	(3) 1.80888 49	(2) 7.59977 67	(2) 4.31169 57	(2) 2.78343 47	(2) 1.93640 05

$a\backslash b$	0.6	0.7	0.8	0.9	1.0
-1.0	(0)-5.66666 67	(0)-4.71428 57	(0)-4.00000 00	(0)-3.44444 44	(0)-3.00000 00
-0.9	(0)-6.66432 27	(0)-5.44175 41	(0)-4.54078 84	(0)-3.85159 75	(0)-3.30880 92
-0.8	(0)-7.50985 56	(0)-6.04428 51	(0)-4.97675 07	(0)-4.16932 54	(0)-3.54030 67
-0.7	(0)-8.14117 89	(0)-6.47484 53	(0)-5.27129 22	(0)-4.36854 34	(0)-3.67096 90
-0.6	(0)-8.48636 64	(0)-6.67916 15	(0)-5.38234 50	(0)-4.41593 73	(0)-3.67394 51
-0.5	(0)-8.46261 04	(0)-6.59496 95	(0)-5.26181 06	(0)-4.27354 17	(0)-3.51873 12
-0.4	(0)-7.97509 54	(0)-6.15120 28	(0)-4.85495 90	(0)-3.89828 45	(0)-3.17081 98
-0.3	(0)-6.91578 17	(0)-5.26711 67	(0)-4.09978 13	(0)-3.24149 77	(0)-2.59132 26
-0.2	(0)-5.16209 26	(0)-3.85134 51	(0)-2.92629 19	(0)-2.24839 06	(0)-1.73656 51
-0.1	(0)-2.57549 99	(0)-1.80088 43	(0)-1.25577 95	(-1)-8.57483 35	(-1)-5.57651 91
0.0	(0)+1.00000 00	(0)+1.00000 00	(0)+1.00000 00	(0)+1.00000 00	(0)+1.00000 00
0.1	(0) 5.73952 56	(0) 4.68094 79	(0) 3.93968 87	(0) 3.40078 42	(0) 2.99716 17
0.2	(1) 1.18390 73	(0) 9.38676 76	(0) 7.67325 59	(0) 6.43024 18	(0) 5.50132 78
0.3	(1) 1.95174 11	(1) 1.52787 90	(1) 1.23229 94	(1) 1.01831 42	(0) 8.58729 05
0.4	(1) 2.90181 11	(1) 2.25363 21	(1) 1.80245 87	(1) 1.47644 52	(1) 1.23377 53
0.5	(1) 4.06117 30	(1) 3.13582 01	(1) 2.49282 52	(1) 2.02901 97	(1) 1.68439 84
0.6	(1) 5.45981 50	(1) 4.19644 69	(1) 3.31999 64	(1) 2.68883 75	(1) 2.22065 21
0.7	(1) 7.13090 76	(1) 5.45981 50	(1) 4.30227 62	(1) 3.46999 38	(1) 2.85359 16
0.8	(1) 9.11107 21	(1) 6.95271 64	(1) 5.45981 50	(1) 4.38798 40	(1) 3.59535 37
0.9	(2) 1.14406 67	(1) 8.70463 66	(1) 6.81475 87	(1) 5.45981 50	(1) 4.45924 13
1.0	(2) 1.41640 95	(2) 1.07479 72	(1) 8.39140 83	(1) 6.70412 50	(1) 5.45981 50

CONFLUENT HYPERGEOMETRIC FUNCTION $M(a, b, x)$ — Table 13.1

$x = 5.0$

$a\backslash b$	0.1	0.2	0.3	0.4	0.5
−1.0	(1)−4.90000 00	(1)−2.40000 00	(1)−1.56666 67	(1)−1.15000 00	(0)−9.00000 00
−0.9	(1)−8.48135 46	(1)−3.90138 34	(1)−2.41382 36	(1)−1.69201 76	(1)−1.27235 43
−0.8	(2)−1.20177 53	(1)−5.37054 86	(1)−3.23511 34	(1)−2.21244 58	(1)−1.62630 91
−0.7	(2)−1.52985 90	(1)−6.71922 90	(1)−3.98065 33	(1)−2.67925 47	(1)−1.93973 31
−0.6	(2)−1.80596 42	(1)−7.83737 80	(1)−4.58862 62	(1)−3.05298 12	(1)−2.18551 10
−0.5	(2)−1.99749 08	(1)−8.58991 93	(1)−4.98353 39	(1)−3.28566 20	(1)−2.33084 19
−0.4	(2)−2.06475 40	(1)−8.81313 79	(1)−5.07426 08	(1)−3.31965 25	(1)−2.33646 31
−0.3	(2)−1.95997 71	(1)−8.31068 13	(1)−4.75193 11	(1)−3.08632 11	(1)−2.15579 45
−0.2	(2)−1.62617 59	(1)−6.84913 57	(1)−3.88754 12	(1)−2.50460 94	(1)−1.73399 46
−0.1	(1) 9.95925 89	(1)−4.15313 99	(1)−2.32934 93	(1)−1.47944 56	(1)−1.00692 28
0.0	(0)+1.00000 00	(0)+1.00000 00	(0)+1.00000 00	(0)+1.00000 00	(0)+1.00000 00
0.1	(2) 1.48413 16	(1) 6.28624 01	(1) 3.60663 62	(1) 2.36223 07	(1) 1.67304 26
0.2	(2) 3.53395 30	(2) 1.48413 16	(1) 8.42893 34	(1) 5.45552 50	(1) 3.81153 30
0.3	(2) 6.28371 74	(2) 2.62678 96	(2) 1.48413 16	(1) 9.55023 72	(1) 6.62935 70
0.4	(2) 9.87643 86	(2) 4.11434 26	(2) 2.31584 25	(2) 1.48413 16	(2) 1.02565 96
0.5	(3) 1.44760 74	(2) 6.01287 11	(2) 3.37396 77	(2) 2.15510 54	(2) 1.48413 16
0.6	(3) 2.02699 13	(2) 8.39773 11	(2) 4.69942 40	(2) 2.99320 90	(2) 2.05515 14
0.7	(3) 2.74711 92	(3) 1.13545 79	(2) 6.33864 72	(2) 4.02706 82	(2) 2.75772 43
0.8	(3) 3.63219 45	(3) 1.49804 92	(2) 8.34418 40	(2) 5.28902 72	(2) 3.61329 22
0.9	(3) 4.70961 17	(3) 1.93851 85	(3) 1.07753 37	(2) 6.81553 64	(2) 4.64598 46
1.0	(3) 6.01029 56	(3) 2.46923 43	(3) 1.36988 66	(2) 8.64757 36	(2) 5.88289 14

$a\backslash b$	0.6	0.7	0.8	0.9	1.0
−1.0	(0)−7.33333 33	(0)−6.14285 71	(0)−5.25000 00	(0)−4.55555 56	(0)−4.00000 00
−0.9	(1)−1.00125 62	(0)−8.13469 15	(0)−6.76712 82	(0)−5.73274 31	(0)−4.92670 46
−0.8	(1)−1.25327 68	(0)−9.98761 99	(0)−8.16187 54	(0)−6.80132 29	(0)−5.75641 51
−0.7	(1)−1.47334 02	(1)−1.15809 94	(0)−9.34109 21	(0)−7.68780 55	(0)−6.43011 23
−0.6	(1)−1.64188 17	(1)−1.27685 52	(1)−1.01924 14	(0)−8.30396 66	(0)−6.87726 99
−0.5	(1)−1.73534 19	(1)−1.33749 40	(1)−1.05817 04	(0)−8.54492 28	(0)−7.01437 97
−0.4	(1)−1.72563 11	(1)−1.31918 93	(1)−1.03502 42	(0)−8.28701 58	(0)−6.74333 16
−0.3	(1)−1.57953 99	(1)−1.19740 11	(0)−9.31162 41	(0)−7.38548 98	(0)−5.94963 73
−0.2	(1)−1.25808 94	(0)−9.43413 73	(0)−7.24837 36	(0)−5.67194 55	(0)−4.50048 61
−0.1	(0)−7.15818 24	(0)−5.23827 09	(0)−3.90821 47	(0)−2.95155 22	(0)−2.24261 78
0.0	(0)+1.00000 00	(0)+1.00000 00	(0)+1.00000 00	(0)+1.00000 00	(0)+1.00000 00
0.1	(1) 1.25021 43	(0) 9.72559 33	(0) 7.81074 40	(0) 6.43982 88	(0) 5.42870 50
0.2	(1) 2.80473 44	(1) 2.14485 95	(1) 1.69066 81	(1) 1.36614 90	(1) 1.12729 02
0.3	(1) 4.84355 66	(1) 3.67515 33	(1) 2.87239 67	(1) 2.29989 34	(1) 1.87930 66
0.4	(1) 7.45788 26	(1) 5.62973 09	(1) 4.37580 33	(1) 3.48308 09	(1) 2.82840 13
0.5	(2) 1.07513 41	(1) 8.08378 40	(1) 6.25698 73	(1) 4.95851 46	(1) 4.00784 46
0.6	(2) 1.48413 16	(2) 1.11223 46	(1) 8.57928 78	(1) 6.77444 40	(1) 5.45508 08
0.7	(2) 1.98603 96	(2) 1.48413 16	(2) 1.14140 27	(1) 8.98511 69	(1) 7.21214 61
0.8	(2) 2.59579 43	(2) 1.93485 65	(2) 1.48413 16	(2) 1.16513 78	(1) 9.32612 06
0.9	(2) 3.33018 07	(2) 2.47651 46	(2) 1.89509 28	(2) 1.48413 16	(2) 1.18496 18
1.0	(2) 4.20801 74	(2) 3.12265 96	(2) 2.38432 45	(2) 1.86309 66	(2) 1.48413 16

Table 13.1   CONFLUENT HYPERGEOMETRIC FUNCTION $M(a, b, x)$

$$x = 6.0$$

$a\backslash b$	0.1	0.2	0.3	0.4	0.5
-1.0	(1) -5.90000 00	(1) -2.90000 00	(1) -1.90000 00	(1) -1.40000 00	(1) -1.10000 00
-0.9	(2) -1.44132 92	(1) -6.43961 14	(1) -3.88390 81	(1) -2.66287 93	(1) -1.96459 57
-0.8	(2) -2.33128 14	(2) -1.01116 95	(1) -5.92627 62	(1) -3.95288 49	(1) -2.84081 83
-0.7	(2) -3.20791 31	(2) -1.37008 05	(1) -7.90656 11	(1) -5.19335 87	(1) -3.67618 94
-0.6	(2) -4.00174 16	(2) -1.69209 38	(1) -9.66592 36	(1) -6.28400 93	(1) -4.40252 67
-0.5	(2) -4.62243 63	(2) -1.94024 69	(2) -1.10002 61	(1) -7.09668 98	(1) -4.93318 77
-0.4	(2) -4.95505 80	(2) -2.06773 13	(2) -1.16523 15	(1) -7.47062 14	(1) -5.15995 73
-0.3	(2) -4.85579 61	(2) -2.01621 45	(2) -1.13027 51	(1) -7.20700 55	(1) -4.94954 27
-0.2	(2) -4.14715 07	(2) -1.71394 56	(1) -9.56011 20	(1) -6.06296 12	(1) -4.13963 47
-0.1	(2) -2.61250 17	(2) -1.07362 31	(1) -5.94951 89	(1) -3.74471 97	(1) -2.53449 16
0.0	(0) +1.00000 00	(0) +1.00000 00	(0) +1.00000 00	(0) +1.00000 00	(0) +1.00000 00
0.1	(2) 4.03428 79	(2) 1.66280 07	(1) 9.26969 34	(1) 5.89051 37	(1) 4.04184 10
0.2	(2) 9.83405 67	(2) 4.03428 79	(2) 2.23669 33	(2) 1.41226 82	(1) 9.61906 66
0.3	(3) 1.78513 43	(2) 7.30095 48	(2) 4.03428 79	(2) 2.53795 01	(2) 1.72165 84
0.4	(3) 2.86060 97	(3) 1.16700 13	(2) 6.43121 54	(2) 4.03428 79	(2) 2.72837 67
0.5	(3) 4.27068 45	(3) 1.73835 48	(2) 9.55746 91	(2) 5.98067 12	(2) 4.03428 79
0.6	(3) 6.08625 44	(3) 2.47231 35	(3) 1.35639 99	(2) 8.46913 69	(2) 5.69983 97
0.7	(3) 8.38957 36	(3) 3.40149 55	(3) 1.86253 97	(3) 1.16059 73	(2) 7.79473 21
0.8	(4) 1.12757 14	(3) 4.56354 65	(3) 2.49428 70	(3) 1.55134 92	(3) 1.03990 56
0.9	(4) 1.48541 80	(3) 6.00176 64	(3) 3.27475 26	(3) 2.03319 84	(3) 1.36045 49
1.0	(4) 1.92506 91	(3) 7.76580 14	(3) 4.23039 92	(3) 2.62218 79	(3) 1.75159 77

$a\backslash b$	0.6	0.7	0.8	0.9	1.0
-1.0	(0) -9.00000 00	(0) -7.57142 86	(0) -6.50000 00	(0) -5.66666 67	(0) -5.00000 00
-0.9	(1) -1.52103 70	(1) -1.21887 04	(1) -1.00236 52	(0) -8.41150 68	(0) -7.17389 32
-0.8	(1) -2.14539 69	(1) -1.67928 88	(1) -1.35080 52	(1) -1.11025 64	(0) -9.28639 79
-0.7	(1) -2.73534 89	(1) -2.11028 68	(1) -1.67379 50	(1) -1.35713 62	(1) -1.12032 42
-0.6	(1) -3.24219 87	(1) -2.47582 00	(1) -1.94390 70	(1) -1.56045 26	(1) -1.27553 63
-0.5	(1) -3.60439 87	(1) -2.73056 65	(1) -2.12682 93	(1) -1.69364 40	(1) -1.37333 18
-0.4	(1) -3.74541 77	(1) -2.81841 55	(1) -2.18026 23	(1) -1.72410 15	(1) -1.38810 25
-0.3	(1) -3.57134 39	(1) -2.67076 84	(1) -2.05268 12	(1) -1.61224 68	(1) -1.28887 64
-0.2	(1) -2.96819 67	(1) -2.20463 65	(1) -1.68195 09	(1) -1.31050 12	(1) -1.03853 60
-0.1	(1) -1.79891 61	(1) -1.32051 32	(0) -9.93780 50	(0) -7.62137 49	(0) -5.92948 86
0.0	(0) +1.00000 00	(0) +1.00000 00	(0) +1.00000 00	(0) +1.00000 00	(0) +1.00000 00
0.1	(1) 2.92224 67	(1) 2.19683 71	(1) 1.70335 65	(1) 1.35491 58	(1) 1.10148 13
0.2	(1) 6.89588 66	(1) 5.13440 78	(1) 3.93817 92	(1) 3.09503 99	(1) 2.48291 09
0.3	(2) 1.22879 89	(1) 9.10486 02	(1) 6.94664 31	(1) 5.42797 37	(1) 4.32726 56
0.4	(2) 1.94097 77	(2) 1.43316 97	(2) 1.08938 21	(1) 8.47842 06	(1) 6.73053 68
0.5	(2) 2.86223 27	(2) 2.10737 78	(2) 1.59705 69	(2) 1.23903 18	(1) 9.80333 40
0.6	(2) 4.03428 79	(2) 2.96297 41	(2) 2.23967 22	(2) 1.73291 89	(2) 1.36726 52
0.7	(2) 5.50517 98	(2) 4.03428 79	(2) 3.04245 98	(2) 2.34847 33	(2) 1.84838 13
0.8	(2) 7.33002 58	(2) 5.36065 25	(2) 4.03428 79	(2) 3.10736 70	(2) 2.44026 08
0.9	(2) 9.57187 15	(2) 6.98699 63	(2) 5.24808 61	(2) 4.03428 79	(2) 3.16176 35
1.0	(3) 1.23026 21	(2) 8.96449 42	(2) 6.72131 30	(2) 5.15728 26	(2) 4.03428 79

CONFLUENT HYPERGEOMETRIC FUNCTION $M(a, b, x)$ Table 13.1

$x = 7.0$

$a \backslash b$	0.1	0.2	0.3	0.4	0.5
-1.0	(1)-6.90000 00	(1)-3.40000 00	(1)-2.23333 33	(1)-1.65000 00	(1)-1.30000 00
-0.9	(2)-2.66288 80	(2)-1.15002 17	(1)-6.72111 28	(1)-4.47674 11	(1)-3.21693 87
-0.8	(2)-4.82834 55	(2)-2.03315 80	(1)-1.15809 32	(1)-7.51697 57	(1)-5.26450 27
-0.7	(2)-7.06530 95	(2)-2.93971 82	(2)-1.65375 76	(2)-1.05973 99	(1)-7.32517 82
-0.6	(2)-9.19980 13	(2)-3.79893 33	(2)-2.12025 19	(2)-1.34754 31	(1)-9.23583 79
-0.5	(3)-1.09929 51	(2)-4.51426 47	(2)-2.50491 09	(2)-1.58243 03	(2)-1.07780 84
-0.4	(3)-1.21270 91	(2)-4.95796 49	(2)-2.73838 73	(2)-1.72158 27	(2)-1.16671 10
-0.3	(3)-1.21896 61	(2)-4.96479 64	(2)-2.73134 11	(2)-1.71005 68	(2)-1.15389 05
-0.2	(3)-1.06546 71	(2)-4.32480 32	(2)-2.37063 77	(2)-1.47850 91	(1)-9.93558 67
-0.1	(2)-6.86139 84	(2)-2.77502 15	(2)-1.51499 28	(1)-9.40594 48	(1)-6.28867 03
0.0	(0)+1.00000 00	(0)+1.00000 00	(0)+1.00000 00	(0)+1.00000 00	(0)+1.00000 00
0.1	(3) 1.09663 32	(2) 4.42900 71	(2) 2.41753 11	(2) 1.50292 87	(2) 1.00798 98
0.2	(3) 2.72330 73	(3) 1.09663 32	(2) 5.96600 60	(2) 3.69501 44	(2) 2.46763 45
0.3	(3) 5.02903 83	(3) 2.02058 34	(3) 1.09663 32	(2) 6.77457 83	(2) 4.51182 31
0.4	(3) 8.19139 01	(3) 3.28466 83	(3) 1.77901 54	(3) 1.09663 32	(2) 7.28692 93
0.5	(4) 1.24220 89	(3) 4.97211 80	(3) 2.68791 51	(3) 1.65368 85	(3) 1.09663 32
0.6	(4) 1.79722 28	(3) 7.18148 47	(3) 3.87554 96	(3) 2.38009 49	(3) 1.57543 68
0.7	(4) 2.51381 30	(4) 1.00289 02	(3) 5.40336 15	(3) 3.31282 90	(3) 2.18907 73
0.8	(4) 3.42679 34	(4) 1.36506 23	(3) 7.34333 78	(3) 4.49515 29	(3) 2.96556 40
0.9	(4) 4.57689 88	(4) 1.82058 62	(3) 9.77948 66	(3) 5.97748 66	(3) 3.93749 79
1.0	(4) 6.01161 32	(4) 2.38799 82	(4) 1.28094 89	(3) 7.81838 27	(3) 5.14269 05

$a \backslash b$	0.6	0.7	0.8	0.9	1.0
-1.0	(1)-1.06666 67	(0)-9.00000 00	(0)-7.75000 00	(0)-6.77777 78	(0)-6.00000 00
-0.9	(1)-2.43203 85	(1)-1.90770 95	(1)-1.53927 06	(1)-1.27012 46	(1)-1.06732 11
-0.8	(1)-3.88035 55	(1)-2.96917 41	(1)-2.33863 78	(1)-1.88526 21	(1)-1.54912 65
-0.7	(1)-5.32790 43	(1)-4.02257 88	(1)-3.12617 60	(1)-2.48676 78	(1)-2.01662 21
-0.6	(1)-6.65941 15	(1)-4.98346 93	(1)-3.83826 01	(1)-3.02562 11	(1)-2.43133 06
-0.5	(1)-7.72147 28	(1)-5.74011 58	(1)-4.39120 14	(1)-3.43770 69	(1)-2.74320 50
-0.4	(1)-8.31498 75	(1)-6.14818 51	(1)-4.67738 87	(1)-3.64095 75	(1)-2.88847 09
-0.3	(1)-8.18647 83	(1)-6.02463 60	(1)-4.56087 46	(1)-3.53208 76	(1)-2.78716 65
-0.2	(1)-7.01816 36	(1)-5.14074 94	(1)-3.87234 20	(1)-2.98287 74	(1)-2.34034 55
-0.1	(1)-4.41663 81	(1)-3.21419 15	(1)-2.40338 13	(1)-1.83595 18	(1)-1.42690 55
0.0	(0)+1.00000 00	(0)+1.00000 00	(0)+1.00000 00	(0)+1.00000 00	(0)+1.00000 00
0.1	(1) 7.11674 98	(1) 5.21962 63	(1) 3.94472 08	(1) 3.05562 65	(1) 2.41701 00
0.2	(2) 1.73382 30	(2) 1.26468 67	(1) 9.49891 56	(1) 7.30700 42	(1) 5.73511 61
0.3	(2) 3.16073 31	(2) 2.29812 96	(2) 1.72012 72	(2) 1.31824 90	(2) 1.03047 87
0.4	(2) 5.09262 36	(2) 3.69345 22	(2) 2.75715 27	(2) 2.10704 18	(2) 1.64217 15
0.5	(2) 7.64800 47	(2) 5.53466 48	(2) 4.12222 44	(2) 3.14277 19	(2) 2.44332 54
0.6	(3) 1.09663 32	(2) 7.92047 08	(2) 5.88720 07	(2) 4.47895 79	(2) 3.47456 13
0.7	(3) 1.52109 75	(3) 1.09663 32	(2) 8.13601 69	(2) 6.17802 12	(2) 4.78318 84
0.8	(3) 2.05725 48	(3) 1.48067 73	(3) 1.09663 32	(2) 8.31248 87	(2) 6.42409 85
0.9	(3) 2.72726 12	(3) 1.95979 60	(3) 1.44913 63	(3) 1.09663 32	(2) 8.46076 16
1.0	(3) 3.55678 22	(3) 2.55205 62	(3) 1.88419 29	(3) 1.42364 54	(3) 1.09663 32

Table 13.1   **CONFLUENT HYPERGEOMETRIC FUNCTION** $M(a, b, x)$

$$x = 8.0$$

$a\backslash b$	0.1	0.2	0.3	0.4	0.5
−1.0	(1)−7.90000 00	(1)−3.90000 00	(1)−2.56666 67	(1)−1.90000 00	(1)−1.50000 00
−0.9	(2)−5.35947 58	(2)−2.23970 82	(2)−1.26764 73	(1)−8.18608 14	(1)−5.71092 02
−0.8	(3)−1.05913 37	(2)−4.34517 66	(2)−2.41159 61	(2)−1.52562 18	(2)−1.04182 83
−0.7	(3)−1.62135 82	(2)−6.59589 37	(2)−3.62791 31	(2)−2.27325 01	(2)−1.53682 58
−0.6	(3)−2.18025 86	(2)−8.82153 60	(2)−4.82414 97	(2)−3.00441 34	(2)−2.01811 79
−0.5	(3)−2.67429 61	(3)−1.07763 74	(2)−5.86783 06	(2)−3.63786 60	(2)−2.43202 00
−0.4	(3)−3.01799 53	(3)−1.21208 08	(2)−6.57678 93	(2)−4.06244 15	(2)−2.70544 00
−0.3	(3)−3.09632 67	(3)−1.23996 24	(2)−6.70780 36	(2)−4.13029 89	(2)−2.74155 31
−0.2	(3)−2.75810 97	(3)−1.10164 91	(2)−5.94329 13	(2)−3.64902 75	(2)−2.41475 59
−0.1	(3)−1.80829 89	(2)−7.20419 31	(2)−3.87580 16	(2)−2.37245 74	(2)−1.56480 05
0.0	(0)+1.00000 00	(0)+1.00000 00	(0)+1.00000 00	(0)+1.00000 00	(0)+1.00000 00
0.1	(3) 2.98095 80	(3) 1.18444 63	(2) 6.35818 11	(2) 3.88567 25	(2) 2.56061 41
0.2	(3) 7.51808 32	(3) 2.98095 80	(3) 1.59656 00	(2) 9.73282 54	(2) 6.39631 86
0.3	(4) 1.40881 29	(3) 5.57611 41	(3) 2.98095 80	(3) 1.81369 75	(3) 1.18950 58
0.4	(4) 2.32720 88	(3) 9.19616 72	(3) 4.90796 57	(3) 2.98095 80	(3) 1.95153 01
0.5	(4) 3.57745 28	(4) 1.41150 69	(3) 7.52139 08	(3) 4.56094 12	(3) 2.98095 80
0.6	(4) 5.24445 76	(4) 2.06625 00	(4) 1.09940 42	(3) 6.65669 18	(3) 4.34399 08
0.7	(4) 7.42998 57	(4) 2.92330 17	(4) 1.55324 53	(3) 9.39119 38	(3) 6.11953 13
0.8	(5) 1.02553 76	(4) 4.02964 70	(4) 2.13822 46	(4) 1.29105 19	(3) 8.40117 14
0.9	(5) 1.38646 40	(4) 5.44098 22	(4) 2.88342 27	(4) 1.73873 91	(4) 1.12994 43
1.0	(5) 1.84279 80	(4) 7.22305 38	(4) 3.82312 68	(4) 2.30252 22	(4) 1.49443 61

$a\backslash b$	0.6	0.7	0.8	0.9	1.0
−1.0	(1)−1.23333 33	(1)−1.04285 71	(0)−9.00000 00	(0)−7.88888 89	(0)−7.00000 00
−0.9	(1)−4.19816 11	(1)−3.20746 94	(1)−2.52522 99	(1)−2.03685 45	(1)−1.67621 46
−0.8	(1)−7.49216 65	(1)−5.59749 62	(1)−4.30847 38	(1)−3.39751 08	(1)−2.73380 70
−0.7	(2)−1.09361 95	(1)−8.08183 59	(1)−6.15107 90	(1)−4.79493 78	(1)−3.81325 44
−0.6	(2)−1.42648 08	(2)−1.04680 37	(1)−7.90952 94	(1)−6.11965 64	(1)−4.82945 42
−0.5	(2)−1.71051 24	(2)−1.24874 83	(1)−9.38477 69	(1)−7.22077 10	(1)−5.66582 71
−0.4	(2)−1.89519 44	(2)−1.37780 10	(2)−1.03097 46	(1)−7.89678 13	(1)−6.16743 32
−0.3	(2)−1.91386 58	(2)−1.38635 99	(2)−1.03347 63	(1)−7.88488 72	(1)−6.13297 12
−0.2	(2)−1.68033 35	(2)−1.21307 63	(1)−9.01063 22	(1)−6.84858 28	(1)−5.30551 30
−0.1	(2)−1.08493 76	(1)−7.80116 43	(1)−5.76904 74	(1)−4.36332 11	(1)−3.36181 13
0.0	(0)+1.00000 00	(0)+1.00000 00	(0)+1.00000 00	(0)+1.00000 00	(0)+1.00000 00
0.1	(2) 1.77542 34	(2) 1.27804 07	(1) 9.47420 10	(1) 7.19400 22	(1) 5.57451 38
0.2	(2) 4.42157 41	(2) 3.17224 03	(2) 2.34287 19	(2) 1.77165 46	(2) 1.36651 86
0.3	(2) 8.20490 47	(2) 5.87308 59	(2) 4.32702 55	(2) 3.26355 40	(2) 2.51027 48
0.4	(3) 1.34359 84	(2) 9.59878 19	(2) 7.05759 09	(2) 5.31172 06	(2) 4.07661 58
0.5	(3) 2.04885 12	(3) 1.46114 76	(3) 1.07237 41	(2) 8.05582 19	(2) 6.17064 03
0.6	(3) 2.98095 80	(3) 2.12243 36	(3) 1.55511 32	(3) 1.16622 16	(2) 8.91734 62
0.7	(3) 4.19313 16	(3) 2.98095 80	(3) 2.18075 96	(3) 1.63280 79	(3) 1.24646 81
0.8	(3) 5.74840 89	(3) 4.08075 63	(3) 2.98095 80	(3) 2.22860 68	(3) 1.69869 84
0.9	(3) 7.72114 36	(3) 5.47370 48	(3) 3.99294 06	(3) 2.98095 80	(3) 2.26888 68
1.0	(4) 1.01986 91	(3) 7.22067 87	(3) 5.26034 65	(3) 3.92186 75	(3) 2.98095 80

CONFLUENT HYPERGEOMETRIC FUNCTION $M(a, b, x)$ Table 13.1

$x = 9.0$

$a \backslash b$	0.1	0.2	0.3	0.4	0.5
-1.0	(1)-8.90000 00	(1)-4.40000 00	(1)-2.90000 00	(1)-2.15000 00	(1)-1.70000 00
-0.9	(3)-1.15822 92	(2)-4.70696 01	(2)-2.58988 67	(2)-1.62573 25	(2)-1.10263 21
-0.8	(3)-2.42781 38	(2)-9.74816 44	(2)-5.29323 09	(2)-3.27532 02	(2)-2.18739 83
-0.7	(3)-3.83823 48	(3)-1.53240 98	(2)-8.26992 61	(2)-5.08337 71	(2)-3.37079 66
-0.6	(3)-5.28795 76	(3)-2.10310 78	(3)-1.13032 66	(2)-6.91755 27	(2)-4.56573 11
-0.5	(3)-6.62068 16	(3)-2.62521 11	(3)-1.40643 82	(2)-8.57840 43	(2)-5.64186 81
-0.4	(3)-7.60990 61	(3)-3.00975 26	(3)-1.60814 10	(2)-9.78118 66	(2)-6.41404 87
-0.3	(3)-7.94036 79	(3)-3.13336 92	(3)-1.67025 41	(3)-1.01340 64	(2)-6.62844 84
-0.2	(3)-7.18584 92	(3)-2.82979 30	(3)-1.50519 87	(2)-9.11218 60	(2)-5.94613 42
-0.1	(3)-4.78270 15	(3)-1.87974 72	(2)-9.97775 31	(2)-6.02698 67	(2)-3.92362 38
0.0	(0)+1.00000 00	(0)+1.00000 00	(0)+1.00000 00	(0)+1.00000 00	(0)+1.00000 00
0.1	(3) 8.10308 39	(3) 3.17569 47	(3) 1.68114 27	(3) 1.01296 25	(2) 6.57992 17
0.2	(4) 2.07097 19	(3) 8.10308 39	(3) 4.28218 60	(3) 2.57548 14	(3) 1.66969 38
0.3	(4) 3.93063 86	(4) 1.53566 77	(3) 8.10308 39	(3) 4.86584 85	(3) 3.14939 49
0.4	(4) 6.57367 60	(4) 2.56471 76	(4) 1.35137 30	(3) 8.10308 39	(3) 5.23683 11
0.5	(5) 1.02271 23	(4) 3.98485 11	(4) 2.09683 16	(4) 1.25557 31	(3) 8.10308 39
0.6	(5) 1.51686 28	(4) 5.90279 86	(4) 3.10207 78	(4) 1.85508 62	(4) 1.19562 36
0.7	(5) 2.17356 27	(4) 8.44810 69	(4) 4.43426 09	(4) 2.64844 50	(4) 1.70478 81
0.8	(5) 3.03359 16	(5) 1.17771 47	(4) 6.17433 59	(4) 3.68332 96	(4) 2.36805 96
0.9	(5) 4.14598 16	(5) 1.60777 16	(4) 8.41941 52	(4) 5.01687 01	(4) 3.22165 07
1.0	(5) 5.56941 19	(5) 2.15743 14	(5) 1.12854 63	(4) 6.71721 10	(4) 4.30870 75

$a \backslash b$	0.6	0.7	0.8	0.9	1.0
-1.0	(1)-1.40000 00	(1)-1.18571 43	(1)-1.02500 00	(0)-9.00000 00	(0)-8.00000 00
-0.9	(1)-7.88310 88	(1)-5.86101 35	(1)-4.49394 10	(1)-3.53363 88	(1)-2.83797 81
-0.8	(2)-1.53831 87	(2)-1.12401 55	(1)-8.46300 77	(1)-6.53007 44	(1)-5.14354 17
-0.7	(2)-2.35259 85	(2)-1.70516 69	(1)-1.27296 76	(1)-9.73476 07	(1)-7.59652 04
-0.6	(2)-3.17089 67	(2)-2.28631 95	(2)-1.69747 84	(2)-1.29066 47	(2)-1.00113 60
-0.5	(2)-3.90366 91	(2)-2.80365 84	(2)-2.07304 42	(2)-1.56947 14	(2)-1.21196 37
-0.4	(2)-4.42433 15	(2)-3.16741 38	(2)-2.33416 78	(2)-1.76099 80	(2)-1.35492 40
-0.3	(2)-4.56001 78	(2)-3.25546 25	(2)-2.39208 63	(2)-1.79922 96	(2)-1.37997 11
-0.2	(2)-4.08061 95	(2)-2.90574 94	(2)-2.12938 18	(2)-1.59711 34	(2)-1.22131 75
-0.1	(2)-2.68584 35	(2)-1.90735 35	(2)-1.39363 74	(2)-1.04195 05	(1)-7.94021 75
0.0	(0)+1.00000 00	(0)+1.00000 00	(0)+1.00000 00	(0)+1.00000 00	(0)+1.00000 00
0.1	(2) 4.49581 13	(2) 3.18820 43	(2) 2.32750 60	(2) 1.73981 39	(2) 1.32662 16
0.2	(3) 1.13844 85	(2) 8.05506 28	(2) 5.86608 76	(2) 4.37321 78	(2) 3.32490 16
0.3	(3) 2.14370 76	(3) 1.51408 89	(3) 1.10059 12	(2) 8.18906 59	(2) 6.21332 82
0.4	(3) 3.55908 19	(3) 2.50977 29	(3) 1.82136 70	(3) 1.35291 34	(3) 1.02470 26
0.5	(3) 5.49915 09	(3) 3.87215 54	(3) 2.80582 25	(3) 2.08094 05	(3) 1.57360 49
0.6	(3) 8.10308 39	(3) 5.69778 22	(3) 4.12286 14	(3) 3.05330 38	(3) 2.30549 09
0.7	(4) 1.15389 32	(3) 8.10308 39	(3) 5.85547 03	(3) 4.33052 37	(3) 3.26534 78
0.8	(4) 1.60085 54	(4) 1.12277 41	(3) 8.10308 39	(3) 5.98502 62	(3) 4.50694 55
0.9	(4) 2.17532 51	(4) 1.52385 32	(4) 1.09842 88	(3) 8.10308 39	(3) 6.09425 86
1.0	(4) 2.90602 06	(4) 2.03337 24	(4) 1.46399 00	(4) 1.07870 28	(3) 8.10308 39

Table 13.1   **CONFLUENT HYPERGEOMETRIC FUNCTION $M(a, b, x)$**

$$x = 10.0$$

$a\backslash b$	0.1	0.2	0.3	0.4	0.5
-1.0	(1)-9.90000 00	(1)-4.90000 00	(1)-3.23333 33	(1)-2.40000 00	(1)-1.90000 00
-0.9	(3)-2.63572 95	(3)-1.04774 98	(2)-5.63504 48	(2)-3.45535 97	(2)-2.28812 39
-0.8	(3)-5.74321 45	(3)-2.26606 51	(2)-1.20865 20	(2)-7.34339 26	(2)-4.81371 33
-0.7	(3)-9.29414 29	(3)-3.65315 21	(3)-1.94041 89	(2)-1.17365 02	(2)-7.65615 62
-0.6	(4)-1.30473 07	(3)-5.11412 18	(3)-2.70839 91	(3)-1.63300 24	(3)-1.06170 13
-0.5	(4)-1.66086 19	(3)-6.49508 42	(3)-3.43144 26	(3)-2.06370 40	(3)-1.33814 35
-0.4	(4)-1.93829 90	(3)-7.56478 22	(3)-3.98819 28	(3)-2.39329 23	(3)-1.54831 36
-0.3	(4)-2.05153 93	(3)-7.99213 74	(3)-4.20553 66	(3)-2.51877 45	(3)-1.62617 94
-0.2	(4)-1.88191 87	(3)-7.31898 36	(3)-3.84460 18	(3)-2.29844 83	(3)-1.48115 57
-0.1	(4)-1.26894 82	(3)-4.92715 82	(3)-2.58388 05	(3)-1.54205 59	(2)-9.91916 94
0.0	(0)+1.00000 00	(0)+1.00000 00	(0)+1.00000 00	(0)+1.00000 00	(0)+1.00000 00
0.1	(4) 2.20264 66	(3) 8.52983 30	(3) 4.46140 89	(3) 2.65569 71	(3) 1.70399 66
0.2	(4) 5.69563 19	(4) 2.20264 66	(4) 1.15043 71	(3) 6.83804 74	(3) 4.38084 00
0.3	(5) 1.09330 93	(4) 4.22272 41	(4) 2.20264 66	(4) 1.30747 73	(3) 8.36496 74
0.4	(5) 1.84869 24	(4) 7.13160 87	(4) 3.71537 68	(4) 2.20264 66	(4) 1.40739 54
0.5	(5) 2.90713 00	(5) 1.12016 64	(4) 5.82887 58	(4) 3.45147 55	(4) 2.20264 66
0.6	(5) 4.35713 28	(5) 1.67700 20	(4) 8.71652 20	(4) 5.15540 77	(4) 3.28620 65
0.7	(5) 6.30765 47	(5) 2.42511 79	(5) 1.25912 31	(4) 7.43887 06	(4) 4.73642 75
0.8	(5) 8.89199 75	(5) 3.41517 02	(5) 1.77129 13	(5) 1.04535 82	(4) 6.64873 73
0.9	(6) 1.22723 53	(5) 4.70872 70	(5) 2.43971 24	(5) 1.43835 42	(4) 9.13874 32
1.0	(6) 1.66450 66	(5) 6.38024 53	(5) 3.30250 83	(5) 1.94508 11	(5) 1.23458 19

$a\backslash b$	0.6	0.7	0.8	0.9	1.0
-1.0	(1)-1.56666 67	(1)-1.32857 14	(1)-1.15000 00	(1)-1.01111 11	(0)-9.00000 00
-0.9	(2)-1.59656 19	(2)-1.15824 17	(1)-8.66482 26	(1)-6.64811 79	(1)-5.21121 29
-0.8	(2)-3.32180 59	(2)-2.38103 41	(2)-1.75833 05	(2)-1.33052 77	(1)-1.02772 90
-0.7	(2)-5.25566 60	(2)-3.74603 08	(2)-2.74969 50	(2)-2.06733 55	(2)-1.58596 75
-0.6	(2)-7.26224 96	(2)-5.15669 48	(2)-3.77001 68	(2)-2.82246 37	(2)-2.15560 45
-0.5	(2)-9.12749 57	(2)-6.46204 50	(2)-4.70972 63	(2)-3.51454 04	(2)-2.67503 59
-0.4	(3)-1.05359 27	(2)-7.44065 06	(2)-5.40890 80	(2)-4.02538 09	(2)-3.05522 11
-0.3	(3)-1.10424 16	(2)-7.78122 74	(2)-5.64358 20	(2)-4.19006 43	(2)-3.17236 75
-0.2	(3)-1.00381 19	(2)-7.05925 89	(2)-5.10920 02	(2)-3.78501 43	(2)-2.85915 68
-0.1	(2)-6.70959 43	(2)-4.70898 38	(2)-3.40090 10	(2)-2.51375 92	(2)-1.89427 82
0.0	(0)+1.00000 00	(0)+1.00000 00	(0)+1.00000 00	(0)+1.00000 00	(0)+1.00000 00
0.1	(3) 1.14989 01	(2) 8.05237 11	(2) 5.80387 50	(2) 4.28243 19	(2) 3.22252 43
0.2	(3) 2.95153 65	(3) 2.06339 28	(3) 1.48456 77	(3) 1.09332 07	(2) 8.21055 88
0.3	(3) 5.62785 57	(3) 3.92867 40	(3) 2.82236 24	(3) 2.07532 55	(3) 1.55600 88
0.4	(3) 9.45635 54	(3) 6.59238 53	(3) 4.72945 31	(3) 3.47272 61	(3) 2.59995 59
0.5	(4) 1.47812 55	(4) 1.02914 95	(3) 7.37367 65	(3) 5.40715 90	(3) 4.04275 54
0.6	(4) 2.20264 66	(4) 1.53174 58	(4) 1.09611 92	(3) 8.02783 98	(3) 5.99449 62
0.7	(4) 3.17106 89	(4) 2.20264 66	(4) 1.57436 46	(4) 1.15166 83	(3) 8.58922 62
0.8	(4) 4.44649 42	(4) 3.08513 39	(4) 2.20264 66	(4) 1.60942 26	(4) 1.19892 63
0.9	(4) 6.10528 43	(4) 4.23152 76	(4) 3.01784 47	(4) 2.20264 66	(4) 1.63901 69
1.0	(4) 8.23940 35	(4) 5.70477 12	(4) 4.06428 07	(4) 2.96327 38	(4) 2.20264 66

ZEROS OF $M(a, b, x)$ Table 13.2

$a \backslash b$	0.1	0.2	0.3	0.4	0.5
−1.0	0.10000 00	0.20000 00	0.30000 00	0.40000 00	0.50000 00
−0.9	0.11054 47	0.22012 64	0.32894 15	0.43713 15	0.54480 16
−0.8	0.12357 83	0.24477 52	0.36411 44	0.48196 35	0.59858 98
−0.7	0.14010 11	0.27567 24	0.40779 72	0.53721 21	0.66443 91
−0.6	0.16173 42	0.31555 72	0.46354 99	0.60707 04	0.74705 02
−0.5	0.19128 98	0.36906 09	0.53728 03	0.69839 96	0.85403 26
−0.4	0.23411 73	0.44470 78	0.63961 58	0.82334 00	0.99868 55
−0.3	0.30182 31	0.56019 88	0.79200 44	1.00591 69	1.20695 84
−0.2	0.42537 31	0.75993 80	1.04632 32	1.30289 37	1.53918 36
−0.1	0.72703 16	1.20342 40	1.58016 05	1.90320 51	2.19258 90

$a \backslash b$	0.6	0.7	0.8	0.9	1.0
−1.0	0.60000 00	0.70000 00	0.80000 00	0.90000 00	1.00000 00
−0.9	0.65203 19	0.75888 50	0.86541 05	0.97164 85	1.07763 19
−0.8	0.71419 38	0.82892 89	0.94291 59	1.05625 10	1.16901 22
−0.7	0.78986 07	0.91376 55	1.03637 62	1.15786 85	1.27838 33
−0.6	0.88415 45	1.01887 44	1.15158 21	1.28256 70	1.41205 79
−0.5	1.00529 53	1.15298 99	1.29771 21	1.43991 63	1.57995 68
−0.4	1.16751 37	1.33112 03	1.49044 27	1.64618 10	1.79887 13
−0.3	1.39828 59	1.58200 88	1.75960 56	1.93215 19	2.10045 49
−0.2	1.76075 91	1.97114 63	2.17271 84	2.36714 89	2.55566 24
−0.1	2.45881 88	2.70808 56	2.94434 51	3.17028 02	3.38779 57

Table 13.2 gives the smallest zeros in x of $M(a, b, x)$, near $a=b=0$, that is, the smallest positive roots in x of the equation $M(a, b, x) = 0$. Linear interpolation gives 3–4S. Interpolation by the Lagrange six-point formula in two dimensions gives 7S.

14. Coulomb Wave Functions

Milton Abramowitz [1]

Contents

$$F_0(\eta,\rho), \quad \frac{d}{d\rho}F_0(\eta,\rho), \qquad G_0(\eta,\rho), \frac{d}{d\rho}G_0(\eta,\rho)$$

$$\eta=.5(.5)20, \quad \rho=1(1)20, \quad 5S$$

$$\eta=0(.05)3, \quad 6S$$

The author wishes to acknowledge the assistance of David S. Liepman in checking the formulas and tables.

---

[1] National Bureau of Standards (deceased).

14. Coulomb Wave Functions

Mathematical Properties

14.1. Differential Equation, Series Expansions

Differential Equation

14.1.1

$$\frac{d^2w}{d\rho^2}+[1-\frac{2\eta}{\rho}-\frac{L(L+1)}{\rho^2}]w=0$$

$(\rho>0, -\infty<\eta<\infty, L$ a non-negative integer)

The Coulomb wave equation has a regular singularity at $\rho=0$ with indices $L+1$ and $-L$; it has an irregular singularity at $\rho=\infty$.

General Solution

14.1.2
$$w=C_1F_L(\eta, \rho)+C_2G_L(\eta, \rho) \qquad (C_1, C_2 \text{ constants})$$

where $F_L(\eta, \rho)$ is the regular Coulomb wave function and $G_L(\eta, \rho)$ is the irregular (logarithmic) Coulomb wave function.

Regular Coulomb Wave Function $F_L(\eta, \rho)$

14.1.3

$$F_L(\eta, \rho)=C_L(\eta)\rho^{L+1}e^{-i\rho}M(L+1-i\eta, 2L+2, 2i\rho)$$

14.1.4

$$=C_L(\eta)\rho^{L+1}\Phi_L(\eta, \rho)$$

14.1.5

$$\Phi_L(\eta, \rho)=\sum_{k=L+1}^{\infty} A_k^L(\eta)\rho^{k-L-1}$$

14.1.6

$$A_{L+1}^L=1, \quad A_{L+2}^L=\frac{\eta}{L+1},$$

$$(k+L)(k-L-1)A_k^L=2\eta A_{k-1}^L-A_{k-2}^L \qquad (k>L+2)$$

14.1.7 $\qquad C_L(\eta)=\dfrac{2^Le^{-\frac{\pi\eta}{2}}|\Gamma(L+1+i\eta)|}{\Gamma(2L+2)}$

(See chapter **6**.)

14.1.8 $\qquad C_0^2(\eta)=2\pi\eta(e^{2\pi\eta}-1)^{-1}$

14.1.9 $\qquad C_L^2(\eta)=\dfrac{p_L(\eta)C_0^2(\eta)}{2\eta(2L+1)}$

14.1.10 $\qquad C_L(\eta)=\dfrac{(L^2+\eta^2)^{\frac{1}{2}}}{L(2L+1)}C_{L-1}(\eta)$

14.1.11 $\quad \dfrac{p_L(\eta)}{2\eta}=\dfrac{(1+\eta^2)(4+\eta^2)\ldots(L^2+\eta^2)2^{2L}}{(2L+1)[(2L)!]^2}$

14.1.12 $\quad F_L'=\dfrac{d}{d\rho}F_L(\eta, \rho)=C_L(\eta)\rho^L\Phi_L^*(\eta, \rho)$

14.1.13 $\qquad \Phi_L^*(\eta, \rho)=\sum_{k=L+1}^{\infty} kA_k^L(\eta)\rho^{k-L-1}$

Irregular Coulomb Wave Function $G_L(\eta, \rho)$

14.1.14

$$G_L(\eta, \rho)=\frac{2\eta}{C_0^2(\eta)}F_L(\eta, \rho)[\ln 2\rho+\frac{q_L(\eta)}{p_L(\eta)}]+\theta_L(\eta, \rho)$$

14.1.15 $\qquad \theta_L(\eta, \rho)=D_L(\eta)\rho^{-L}\psi_L(\eta, \rho)$

14.1.16 $\qquad D_L(\eta)C_L(\eta)=\dfrac{1}{2L+1}$

14.1.17 $\qquad \psi_L(\eta, \rho)=\sum_{k=-L}^{\infty} a_k^L(\eta)\rho^{k+L}$

14.1.18

$$a_{-L}^L=1, \quad a_{L+1}^L=0,$$

$$(k-L-1)(k+L)a_k^L=2\eta a_{k-1}^L-a_{k-2}^L-(2k-1)p_L(\eta)A_k^L$$

14.1.19

$$\frac{q_L(\eta)}{p_L(\eta)}=\sum_{s=1}^{L}\frac{s}{s^2+\eta^2}-\sum_{s=1}^{2L+1}\frac{1}{s}$$

$$+\mathcal{R}\{\frac{\Gamma'(1+i\eta)}{\Gamma(1+i\eta)}\}+2\gamma+\frac{r_L(\eta)}{p_L(\eta)}$$

(See **Table 6.8.**)

14.1.20

$$r_L(\eta)=\frac{(-1)^{L+1}}{(2L)!}\mathscr{I}\{\frac{1}{2L+1}+\frac{2(i\eta-L)}{2L(1!)}$$

$$+\frac{2^2(i\eta-L)(i\eta-L+1)}{(2L-1)(2!)}+\ldots$$

$$+\frac{2^{2L}(i\eta-L)(i\eta-L+1)\ldots(i\eta+L-1)}{(2L)!}\}$$

14.1.21

$$G_L'=\frac{dG_L}{d\rho}=\frac{2\eta}{C_0^2(\eta)}\{F_L'[\ln 2\rho+\frac{q_L(\eta)}{p_L(\eta)}]+\rho^{-1}F_L(\eta, \rho)\}$$

$$+\theta_L'(\eta, \rho)$$

14.1.22 $\quad \theta_L' = \dfrac{d}{d\rho}\,\theta_L(\eta,\rho) = D_L(\eta)\,\rho^{-L-1}\psi_L^*(\eta,\rho)$

14.1.23 $\quad \psi_L^*(\eta,\rho) = \sum\limits_{k=-L}^{\infty} ka_k^L(\eta)\,\rho^{k+L}$

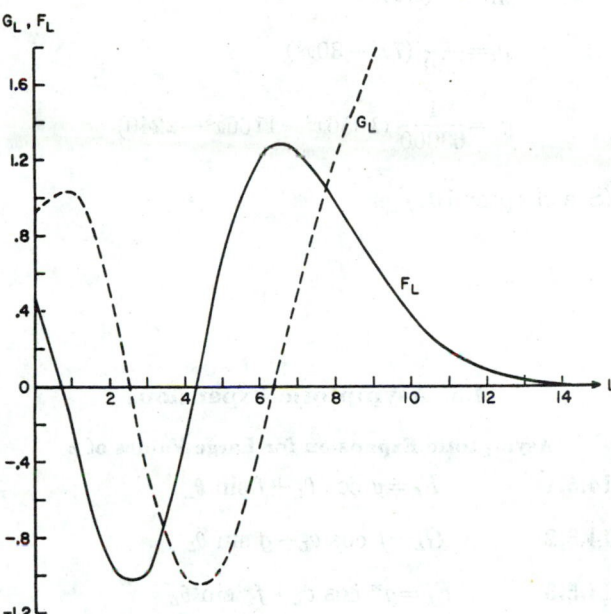

FIGURE 14.1. $\quad F_L(\eta,\rho),\ G_L(\eta,\rho).$

$$\eta=1,\ \rho=10$$

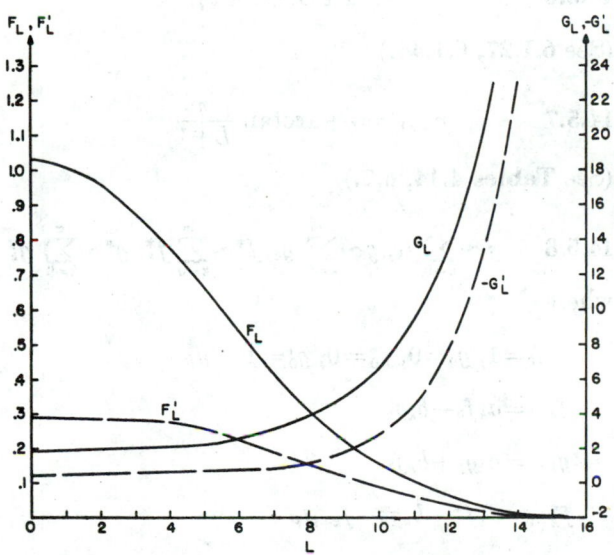

FIGURE 14.2. $\quad F_L,\ F_L',\ G_L$ and $G_L'.$

$$\eta=10,\ \rho=20$$

14.2. Recurrence and Wronskian Relations

Recurrence Relations

If $u_L = F_L(\eta,\rho)$ or $G_L(\eta,\rho)$,

14.2.1 $\quad L\,\dfrac{du_L}{d\rho} = (L^2+\eta^2)^{\frac{1}{2}}u_{L-1} - \left(\dfrac{L^2}{\rho}+\eta\right)u_L$

14.2.2

$$(L+1)\,\dfrac{du_L}{d\rho} = \left[\dfrac{(L+1)^2}{\rho}+\eta\right]u_L - [(L+1)^2+\eta^2]^{\frac{1}{2}}u_{L+1}$$

14.2.3

$$L[(L+1)^2+\eta^2]^{\frac{1}{2}}u_{L+1} = (2L+1)\left[\eta+\dfrac{L(L+1)}{\rho}\right]u_L$$
$$-(L+1)[L^2+\eta^2]^{\frac{1}{2}}u_{L-1}$$

Wronskian Relations

14.2.4 $\quad F_L'G_L - F_L G_L' = 1$

14.2.5 $\quad F_{L-1}G_L - F_L G_{L-1} = L(L^2+\eta^2)^{-\frac{1}{2}}$

14.3. Integral Representations

14.3.1

$$F_L + iG_L = \dfrac{ie^{-i\rho}\rho^{-L}}{(2L+1)!\,C_L(\eta)}\int_0^\infty e^{-t}t^{L-i\eta}(t+2i\rho)^{L+i\eta}dt$$

14.3.2

$$F_L - iG_L =$$
$$\dfrac{e^{-\pi\eta}\rho^{L+1}}{(2L+1)!\,C_L(\eta)}\int_{-1}^{-i\infty} e^{-i\rho t}(1-t)^{L-i\eta}(1+t)^{L+i\eta}dt$$

14.3.3

$$F_L + iG_L = \dfrac{e^{-\pi\eta}\rho^{L+1}}{(2L+1)!\,C_L(\eta)}$$
$$\int_0^\infty \{(1-\tanh^2 t)^{L+1}\exp[-i(\rho\tanh t-2\eta t)]$$
$$+i(1+t^2)^L\exp[-\rho t+2\eta\arctan t]\}dt$$

14.4. Bessel Function Expansions

Expansion in Terms of Bessel-Clifford Functions

14.4.1

$$F_L(\eta,\rho) = C_L(\eta)\,\dfrac{(2L+1)!}{(2\eta)^{2L+1}}\,\rho^{-L}\sum\limits_{k=2L+1}^{\infty} b_k t^{k/2}I_k(2\sqrt{t})$$
$$(t=2\eta\rho,\ \eta>0)$$

14.4.2

$$G_L(\eta,\rho) \sim D_L(\eta)\lambda_L(\eta)\,\rho^{-L}\sum\limits_{k=2L+1}^{\infty}(-1)^k b_k t^{k/2}K_k(2\sqrt{t})$$

14.4.3

$$b_{2L+1}=1, \quad b_{2L+2}=0,$$

$$4\eta^2(k-2L)b_{k+1}+kb_{k-1}+b_{k-2}=0 \qquad (k>2L+2)$$

14.4.4

$$\lambda_L(\eta)\sum_{k=2L+1}^{\infty}(-1)^k(k-1)!b_k=2$$

(See chapter **9**.)

Expansion in Terms of Spherical Bessel Functions

14.4.5

$$F_L(\eta,\rho)=$$
$$1\cdot3\cdot5\ldots(2L+1)\rho C_L(\eta)\sum_{k=L}^{\infty}b_k\sqrt{\frac{\pi}{2\rho}}J_{k+\frac{1}{2}}(\rho)$$

14.4.6

$$b_L=1, \, b_{L+1}=\frac{2L+3}{L+1}\eta$$

$$b_k=\frac{(2k+1)}{k(k+1)-L(L+1)}$$
$$\left\{2\eta b_{k-1}-\frac{(k-1)(k-2)-L(L+1)}{2k-3}b_{k-2}\right\}$$
$$(k>L+1)$$

14.4.7

$$F'_L(\eta,\rho)=1\cdot3\cdot5\ldots(2L+1)\rho C_L(\eta)$$
$$\left\{\frac{(L+1)}{(2L+1)}b_L\sqrt{\frac{\pi}{2\rho}}J_{L-\frac{1}{2}}(\rho)+\frac{(L+2)}{(2L+3)}\cdot b_{L+1}\right.$$
$$\left.\cdot\sqrt{\frac{\pi}{2\rho}}J_{L+\frac{1}{2}}(\rho)+\sum_{k=L+1}^{\infty}b'_k\sqrt{\frac{\pi}{2\rho}}J_{k+\frac{1}{2}}(\rho)\right\}$$

14.4.8 $\quad b'_k=\dfrac{(k+2)}{(2k+3)}b_{k+1}-\dfrac{(k-1)}{(2k-1)}b_{k-1}$

Expansion in Terms of Airy Functions

$$x=(2\eta-\rho)/(2\eta)^{1/3} \quad \mu=(2\eta)^{2/3}, \, \eta>>0$$
$$|\rho-2\eta|<2\eta$$

14.4.9

$$\genfrac{}{}{0pt}{}{F_0(\eta,\rho)}{G_0(\eta,\rho)}=\pi^{\frac{1}{2}}(2\eta)^{\frac{1}{6}}\left\{\genfrac{}{}{0pt}{}{\mathrm{Ai}(x)}{\mathrm{Bi}(x)}\left[1+\frac{g_1}{\mu}+\frac{g_2}{\mu^2}+\cdots\right]\right.$$
$$\left.+\genfrac{}{}{0pt}{}{\mathrm{Ai}'(x)}{\mathrm{Bi}'(x)}\left[\frac{f_1}{\mu}+\frac{f_2}{\mu^2}+\cdots\right]\right\}$$

14.4.10

$$\genfrac{}{}{0pt}{}{F'_0(\eta,\rho)}{G'_0(\eta,\rho)}=-\pi^{\frac{1}{2}}(2\eta)^{-\frac{1}{6}}\left\{\genfrac{}{}{0pt}{}{\mathrm{Ai}(x)}{\mathrm{Bi}(x)}\left[\frac{g'_1+xf_1}{\mu}\right.\right.$$
$$\left.+\frac{g'_2+xf_2}{\mu^2}+\cdots\right]+\genfrac{}{}{0pt}{}{\mathrm{Ai}'(x)}{\mathrm{Bi}'(x)}\left[1+\frac{(g_1+f'_1)}{\mu}\right.$$
$$\left.\left.+\frac{(g_2+f'_2)}{\mu^2}+\cdots\right]\right\}$$

$$f_1=(1/5)x^2$$

$$f_2=\frac{1}{35}(2x^3+6)$$

$$f_3=\frac{1}{63000}(84x^7+1480x^4+2320x)$$

$$g_1=-(1/5)x$$

$$g_2=\frac{1}{350}(7x^5-30x^2)$$

$$g_3=\frac{1}{63000}(1056x^6-1160x^3-2240)$$

(See chapter **10**.)

14.5. Asymptotic Expansions

Asymptotic Expansion for Large Values of ρ

14.5.1 $\qquad F_L=g\cos\theta_L+f\sin\theta_L$

14.5.2 $\qquad G_L=f\cos\theta_L-g\sin\theta_L$

14.5.3 $\qquad F'_L=g^*\cos\theta_L+f^*\sin\theta_L$

14.5.4 $\quad G'_L=f^*\cos\theta_L-g^*\sin\theta_L, \, gf^*-fg^*=1$

14.5.5 $\qquad \theta_L=\rho-\eta\ln2\rho-L\frac{\pi}{2}+\sigma_L$

14.5.6 $\qquad \sigma_L=\arg\Gamma(L+1+i\eta)$

(See **6.1.27, 6.1.44**.)

14.5.7 $\qquad \sigma_{L+1}=\sigma_L+\arctan\frac{\eta}{L+1}$

(See **Tables 4.14, 6.7**.)

14.5.8 $\quad f\sim\sum_{k=0}^{\infty}f_k, \, g\sim\sum_{\kappa=0}^{\infty}g_k, f^*\sim\sum_{k=0}^{\infty}f^*_k, \, g^*\sim\sum_{k=0}^{\infty}g^*_k$

where

$$f_0=1, \, g_0=0, f^*_0=0, \, g^*_0=1-\eta/\rho$$

$$f_{k+1}=a_kf_k-b_kg_k$$

$$g_{k+1}=a_kg_k+b_kf_k$$

$$f^*_{k+1}=a_kf^*_k-b_kg^*_k-f_{k+1}/\rho$$

$$g^*_{k+1}=a_kg^*_k+b_kf^*_k-g_{k+1}/\rho$$

$$a_k=\frac{(2k+1)\eta}{(2k+2)\rho}, \quad b_k=\frac{L(L+1)-k(k+1)+\eta^2}{(2k+2)\rho}$$

14.5.9

$$f+ig \sim 1 + \frac{(i\eta-L)(i\eta+L+1)}{1!(2i\rho)} + \frac{(i\eta-L)(i\eta-L+1)(i\eta+L+1)(i\eta+L+2)}{2!(2i\rho)^2}$$

$$+ \frac{(i\eta-L)(i\eta-L+1)(i\eta-L+2)(i\eta+L+1)(i\eta+L+2)(i\eta+L+3)}{3!(2i\rho)^3} + \cdots$$

Asymptotic Expansion for $L=0$, $\rho=2\eta \gg 0$

14.5.10

$$\begin{matrix} F_0(2\eta) \\ G_0(2\eta)/\sqrt{3} \end{matrix} \sim \frac{\Gamma(1/3)\beta^{1/2}}{2\sqrt{\pi}} \{1 \mp \frac{2}{35} \frac{\Gamma(2/3)}{\Gamma(1/3)} \frac{1}{\beta^4} - \frac{32}{8100} \frac{1}{\beta^6} \mp \frac{92672}{7371\cdot10^4} \frac{\Gamma(2/3)}{\Gamma(1/3)} \frac{1}{\beta^{10}} - \cdots\}$$

14.5.11

$$\begin{matrix} F_0'(2\eta) \\ G_0'(2\eta)/\sqrt{3} \end{matrix} \sim \frac{\Gamma(2/3)}{2\sqrt{\pi}\beta^{1/2}} \{\pm 1 + \frac{1}{15} \frac{\Gamma(1/3)}{\Gamma(2/3)} \frac{1}{\beta^2} \pm \frac{8}{56700} \frac{1}{\beta^6} + \frac{11488}{18711\cdot10^3} \frac{\Gamma(1/3)}{\Gamma(2/3)} \frac{1}{\beta^8} \pm \cdots\}$$

$$\beta=(2\eta/3)^{1/6}, \quad \Gamma(1/3)=2.6789\ 38534\ldots, \quad \Gamma(2/3)=1.3541\ 17939\ldots$$

14.5.12

$$\begin{matrix} F_0(2\eta) \\ G_0(2\eta) \end{matrix} \sim \begin{Bmatrix} .70633\ 26373 \\ 1.22340\ 4016 \end{Bmatrix} \eta^{1/6} \{1 \mp \frac{.04959\ 570165}{\eta^{2/3}} - \frac{.00888\ 88888\ 89}{\eta^2}$$

$$\mp \frac{.00245\ 51991\ 81}{\eta^{10/3}} - \frac{.00091\ 08958\ 061}{\eta^4} \mp \frac{.00025\ 34684\ 115}{\eta^{16/3}} - \cdots\}$$

14.5.13

$$\begin{matrix} F_0'(2\eta) \\ G_0'(2\eta) \end{matrix} \sim \begin{Bmatrix} .40869\ 57323 \\ -.70788\ 17734 \end{Bmatrix} \eta^{-1/6} \{1 \pm \frac{.17282\ 60369}{\eta^{2/3}} + \frac{.00031\ 74603\ 174}{\eta^2}$$

$$\pm \frac{.00358\ 12148\ 50}{\eta^{4/3}} + \frac{.00031\ 17824\ 680}{\eta^4} \pm \frac{.00090\ 73966\ 427}{\eta^{14/3}} + \cdots\}$$

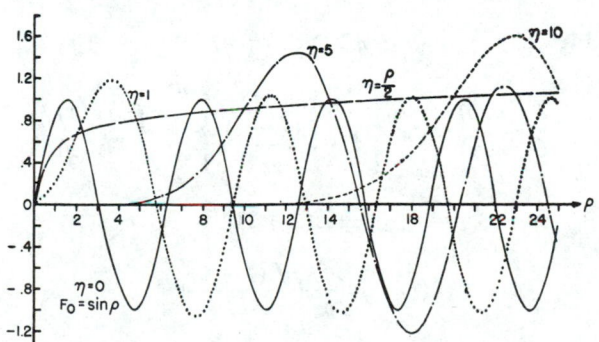

FIGURE 14.3. $F_0(\eta, \rho)$.

$\eta=0, 1, 5, 10, \rho/2$

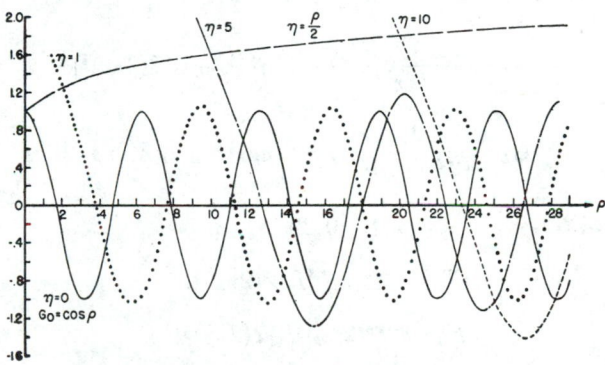

FIGURE 14.5. $G_0(\eta, \rho)$.

$\eta=0, 1, 5, 10, \rho/2$

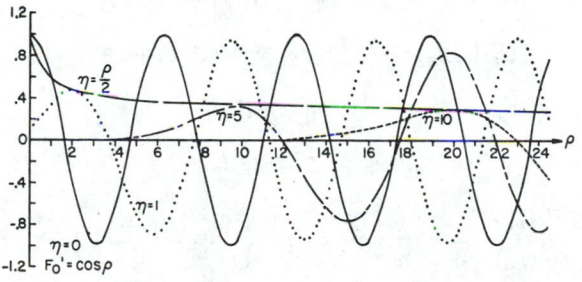

FIGURE 14.4. $F_0'(\eta, \rho)$.

$\eta=0, 1, 5, 10, \rho/2$

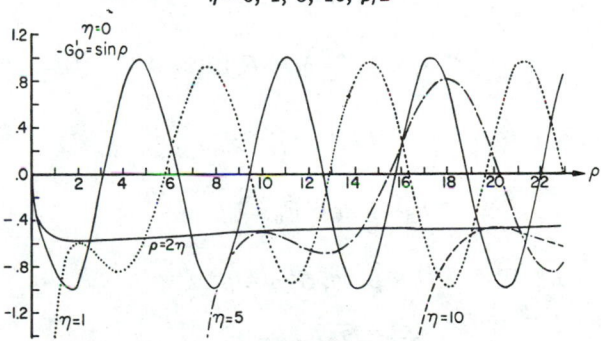

FIGURE 14.6. $G_0'(\eta, \rho)$.

$\eta=0, 1, 5, 10, \rho/2$

14.6. Special Values and Asymptotic Behavior

14.6.1
$$L>0,\ \rho=0$$
$$F_L=0,\ F_L'=0$$
$$G_L=\infty,\ G_L'=-\infty$$

14.6.2
$$L=0,\ \rho=0$$
$$F_0=0,\ F_0'=C_0(\eta)$$
$$G_0=1/C_0(\eta),\ G_0'=-\infty$$

14.6.3
$$L\to\infty$$
$$F_L\sim C_L(\eta)\rho^{L+1},\ G_L\sim D_L(\eta)\rho^{-L}$$

14.6.4
$$L=0,\ \eta=0$$
$$F_0=\sin\rho,\ F_0'=\cos\rho$$
$$G_0=\cos\rho,\ G_0'=-\sin\rho$$

14.6.5
$$\rho\to\infty$$
$$G_L+iF_L\sim\exp i[\rho-\eta\ln 2\rho-\frac{L\pi}{2}+\sigma_L]$$

14.6.6
$$L\geq 0,\ \eta=0$$
$$F_L=(\tfrac{1}{2}\pi\rho)^{\frac{1}{2}}J_{L+\frac{1}{2}}(\rho)$$
$$G_L=(-1)^L(\tfrac{1}{2}\pi\rho)^{\frac{1}{2}}J_{-(L+\frac{1}{2})}(\rho)$$

14.6.7
$$L\geq 0,\ 2\eta>>\rho$$
$$F_L\sim\frac{(2L+1)!\,C_L(\eta)}{(2\eta)^{L+1}}\,(2\eta\rho)^{\frac{1}{2}}I_{2L+1}[2(2\eta\rho)^{\frac{1}{2}}]$$
$$G_L\sim\frac{2(2\eta)^L}{(2L+1)!\,C_L(\eta)}\,(2\eta\rho)^{\frac{1}{2}}K_{2L+1}[2(2\eta\rho)^{\frac{1}{2}}]$$

14.6.8
$$L=0,\ 2\eta>>\rho$$
$$F_0\sim e^{-\pi\eta}(\pi\rho)^{\frac{1}{2}}I_1[2(2\eta\rho)^{\frac{1}{2}}]$$
$$F_0'\sim e^{-\pi\eta}(2\pi\eta)^{\frac{1}{2}}I_0[2(2\eta\rho)^{\frac{1}{2}}]$$
$$G_0\sim 2e^{\pi\eta}\left(\frac{\rho}{\pi}\right)^{\frac{1}{2}}K_1[2(2\eta\rho)^{\frac{1}{2}}]$$
$$G_0'\sim-2\left(\frac{2\eta}{\pi}\right)^{\frac{1}{2}}e^{\pi\eta}K_0[2(2\eta\rho)^{\frac{1}{2}}]$$

14.6.9
$$L=0,\ 2\eta>>\rho$$
$$F_0\sim\frac{1}{2}\beta e^{\alpha};\ F_0'\sim\frac{1}{2}\beta^{-1}e^{\alpha}$$
$$G_0\sim\beta e^{-\alpha};\ G_0'\sim-\beta^{-1}e^{-\alpha}$$
$$\alpha=2\sqrt{2\eta\rho}-\pi\eta$$
$$\beta=(\rho/2\eta)^{\frac{1}{2}}$$

14.6.10
$$L=0,\ 2\eta>>\rho$$
$$F_0\sim\frac{1}{2}\beta e^{\alpha};\ F_0'\sim\{\beta^{-2}+\frac{1}{8\eta}\,t^{-2}\beta^4\}F_0$$
$$G_0\sim\beta e^{-\alpha};\ G_0'\sim\{-\beta^{-2}+\frac{1}{8\eta}\,t^{-2}\beta^4\}G_0$$
$$t=\rho/2\eta$$
$$\alpha=2\eta\{[t(1-t)]^{\frac{1}{2}}+\arcsin t^{\frac{1}{2}}-\tfrac{1}{2}\pi\}$$
$$\beta=\{t/(1-t)\}^{\frac{1}{4}}$$

14.6.11
$$L=0,\ \rho>>2\eta$$
$$F_0=\alpha\sin\beta;\ F_0'=-t^2(bF_0-aG_0)$$
$$G_0=\alpha\cos\beta;\ G_0'=-t^2(aF_0+bG_0)$$
$$t=\frac{2\eta}{\rho}$$
$$\alpha=\left(\frac{1}{1-t}\right)^{\frac{1}{4}}\exp\left[-\frac{8t^3-3t^4}{64(2\eta)^2(1-t)^3}\right]$$
$$\beta=\frac{\pi}{4}+2\eta\left\{\frac{(1-t)^{\frac{1}{2}}}{t}+\frac{1}{2}\ln\left[\frac{1-(1-t)^{\frac{1}{2}}}{1+(1-t)^{\frac{1}{2}}}\right]\right\}$$
$$a=t^{-2}(1-t)^{\frac{1}{2}},\ b=[8\eta(1-t)]^{-1}$$

14.6.12
$$\eta>>0,\ 2\eta\sim\rho$$
$$\left.\begin{matrix}F_L(\eta,\rho)\\G_L(\eta,\rho)\end{matrix}\right\}\sim\sqrt{\pi}\left\{\frac{\rho_L}{1+\dfrac{L(L+1)}{\rho_L^2}}\right\}^{1/6}\left\{\begin{matrix}\text{Ai}\,(x)\\\text{Bi}\,(x)\end{matrix}\right\}$$
$$\rho_L=\eta+[\eta^2+L(L+1)]^{1/2}$$
$$x=(\rho_L-\rho)[\frac{1}{\rho_L}+\frac{L(L+1)}{\rho_L^3}]^{1/3}$$

14.6.13
$$\eta>>0,\ 2\eta\sim\rho$$
$$x=(2\eta-\rho)(2\eta)^{-1/3}$$
$$[G_0+iF_0]\sim\pi^{1/2}(2\eta)^{1/6}[\text{Bi}(x)+i\text{Ai}(x)]$$
$$[G_0'+iF_0']\sim-\pi^{1/2}(2\eta)^{-1/6}[\text{Bi}'(x)+i\text{Ai}'(x)]$$

14.6.14
$$\eta>>0$$
$$\rho_L=\eta+[\eta^2+L(L+1)]^{1/2}$$
$$\left.\begin{matrix}F_L(\rho_L)\\G_L(\rho_L)/\sqrt{3}\end{matrix}\right\}\sim\frac{\Gamma(1/3)}{2\sqrt{\pi}}\left(\frac{\rho_L}{3}\right)^{1/6}\left\{1+\frac{L(L+1)}{\rho_L^2}\right\}^{-1/6}$$
$$\left.\begin{matrix}F_L'(\rho_L)\\G_L'(\rho_L)/\sqrt{3}\end{matrix}\right\}\sim\pm\frac{\Gamma(2/3)}{2\sqrt{\pi}}\left(\frac{\rho_L}{3}\right)^{-1/6}\left\{1+\frac{L(L+1)}{\rho_L^2}\right\}^{1/6}$$

14.6.15
$$\rho = 2\eta \gg 0$$

$$\frac{F_0}{G_0/\sqrt{3}} \sim \frac{\Gamma(1/3)}{2\sqrt{\pi}} \left(\frac{2\eta}{3}\right)^{1/6}$$

$$\frac{F_0'}{-G_0'/\sqrt{3}} \sim \frac{\Gamma(2/3)}{2\sqrt{\pi}(2\eta/3)^{1/6}}$$

14.6.16
$$\eta \to \infty$$

$$\sigma_0(\eta) \sim [\frac{\pi}{4} + \eta(\ln \eta - 1)]$$

$$C_0(\eta) \sim (2\pi\eta)^{1/2} e^{-\pi\eta},$$

(Equality to 8S for $\eta > 3$.)

14.6.17
$$\eta \to 0$$

$$\sigma_0(\eta) \sim -\gamma\eta \qquad (\gamma = \text{Euler's constant})$$

$$C_L(\eta) \sim \frac{2^L L!}{(2L+1)!}$$

14.6.18
$$L \to \infty$$

$$C_L(\eta) \sim \frac{2^L L!}{(2L+1)!} e^{-\pi\eta/2}$$

Numerical Methods

14.7. Use and Extension of the Tables

In general the tables as presented are not simply interpolable. However, values for $L > 0$ may be obtained with the help of the recurrence relations. The values of $G_L(\eta, \rho)$ may be obtained by applying the recurrence relations in increasing order of L. Forward recurrence may be used for $F_L(\eta, \rho)$ as long as the instability does not produce errors in excess of the accuracy needed. In this case the backwards recurrence scheme (see **Example 1**) should be used.

Example 1. Compute $F_L(\eta, \rho)$ and $F_L'(\eta, \rho)$ for $\eta = 2$, $\rho = 5$, $L = 0(1)5$. Starting with $F_{10}^* = 1$, $F_{11}^* = 0$, where $F_L^* = cF_L$, we compute from **14.2.3** in decreasing order of L:

L	(1) F_L^*	(2) F_L	(3) F_L	(4) F_L'
11	0.			
10	1.			
9	4.49284			
8	17.5225			
7	61.3603			
6	191.238			
5	523.472	.090791	.091	.1043
4	1238.53	.21481	.215	.2030
3	2486.72	.43130	.4313	.3205
2	4158.46	.72124	.72125	.3952
1	5727.97	.99346	.99347	.3709
0	6591.81	1.1433	1.1433	.29380

$F_0/F_0^* = 1.7344 \times 10^{-4} = c^{-1}$.

The values in the second column are obtained from those in the first by multiplying by the normalization constant, F_0/F_0^* where F_0 is the known value obtained from **Table 14.1**.

Repetition starting with $F_{15}^* = 1$ and $F_{16}^* = 0$ yields the same results.

In column 3, the results have been given when **14.2.3** is used in increasing order of L.

F_L' (column 4) follows from **14.2.2**.

Example 2. Compute $G_L(\eta, \rho)$ and $G_L'(\eta, \rho)$ for $\eta = 2$, $\rho = 5$, $L = 1(1)5$.

Using **14.2.2** and $G_0(2, 5) = .79445$, $G_0' = -.67049$ from **Table 14.1** we find $G_1(2, 5) = 1.0815$. Then by forward recurrence using **14.2.3** we find:

L	G_L	$-G_L'$	*
1	1.0815	.60286	
2	1.4969	.56619	
3	2.0487	.79597	
4	3.0941	1.7318	
5	5.6298	4.5493	

The values of G_L' are obtained with **14.2.1**.

Example 3. Compute $G_0(\eta, \rho)$ for $\eta = 2$, $\rho = 2.5$.

From **Table 14.1**, $G_0(2, 2) = 3.5124$, $G_0'(2, 2) = -2.5554$. Successive differentiation of **14.1.1** for $L = 0$ gives

$$\rho \frac{d^{k+2}w}{d\rho^{k+2}} = (2\eta - \rho) \frac{d^k w}{d\rho^k} - k\left\{\frac{d^{k+1}w}{d\rho^{k+1}} + \frac{d^{k-1}w}{d\rho^{k-1}}\right\}$$

Taylor's expansion is $w(\rho + \Delta\rho) = w(\rho) + (\Delta\rho)w' + \frac{(\Delta\rho)^2}{2!} w'' + \ldots$ With $w = G_0(\eta, \rho)$ and $\Delta\rho = .5$ we get:

k	$\dfrac{d^k G_0}{d\rho^k}$	$\dfrac{(\Delta\rho)^k}{k!} \dfrac{d^k G_0}{d\rho^k}$
0	3.5124	3.5124
1	-2.5554	-1.2777
2	3.5124	.43905
3	-6.0678	-.12641
4	12.136	.03160
5	-29.540	-.00769
6	83.352	.00181
7	-268.26	-.00042

$$G_0(2, 2.5) = 2.5726$$

As a check the result is obtained with $\eta = 2$, $\rho = 3$, $\Delta\rho = -.5$. The derivative $G_0'(\eta, \rho)$ may be obtained using Taylor's formula with $w = G_0'(\eta, \rho)$.

*See page II.

References

Texts

[14.1] M. Abramowitz and H. A. Antosiewicz, Coulomb wave functions in the transition region. Phys. Rev. **96**, 75–77 (1954).

[14.2] Biedenharn, Gluckstern, Hull, and Breit, Coulomb functions for large charges and small velocities. Phys. Rev. **97**, 542 (1955).

[14.3] I. Bloch et al., Coulomb functions for reactions of protons and alpha-particles with the lighter nuclei. Rev. Mod. Phys. **23**, 147–182 (1951).

[14.4] Carl-Erik Fröberg, Numerical treatment of Coulomb wave functions. Rev. Mod. Phys. **27**, 399–411 (1955).

[14.5] I. A. Stegun and M. Abramowitz, Generation of Coulomb wave functions from their recurrence relations. Phys. Rev. **98**, 1851 (1955).

Tables

[14.6] M. Abramowitz and P. Rabinowitz, Evaluation of Coulomb wave functions along the transition line. Phys. Rev. **96**, 77–79 (1954). Tabulates F_0, F_0', G_0, G_0' for $\rho = 2\eta = 0(.5)20(2)50$, 8S.

[14.7] National Bureau of Standards, Tables of Coulomb wave functions, vol. I, Applied Math. Series 17 (U.S. Government Printing Office, Washington, D.C., 1952). Tabulates

$$\Phi_L(\eta, \rho) \text{ and } \frac{1}{k!}\frac{d^k \Phi_k(\eta, \rho)}{d\eta^k} \text{ for } \rho = 0(.2)5,$$

$$\eta = -5(1)5, \ L = 0(1)5, 10, 11, 20, 21, \quad \textbf{7D}.$$

[14.8] Numerical Computation Bureau, Tables of Whittaker functions (Wave functions in a Coulomb field). Report No. 9, Japan (1956).

[14.9] A. Tubis, Tables of non-relativistic Coulomb wave functions, Los Alamos Scientific Laboratory La–2150, Los Alamos, N. Mex. (1958). Values of F_0, F_0', G_0, G_0', $\rho = 0(.2)40$; $\eta = 0(.05)12$, 5S.

Table 14.1   **COULOMB WAVE FUNCTIONS OF ORDER ZERO**

$$F_0(\eta,\rho)$$

$\eta \backslash \rho$	1	2	3	4	5
0.5	(− 1) 5.1660	(0) 1.0211	(0) 1.0432	(− 1) 4.1924	(− 1) −4.9046
1.0	(− 1) 2.2753	(− 1) 6.6178	(0) 1.0841	(0) 1.1571	(− 1) +6.8494
1.5	(− 2) 8.4815	(− 1) 3.3159	(− 1) 7.3013	(0) 1.1186	(0) 1.2327
2.0	(− 2) 2.8898	(− 1) 1.4445	(− 1) 3.9861	(− 1) 7.7520	(0) 1.1433
2.5	(− 3) 9.3008	(− 2) 5.7560	(− 1) 1.9162	(− 1) 4.4865	(− 1) 8.0955
3.0	(− 3) 2.8751	(− 2) 2.1538	(− 2) 8.4417	(− 1) 2.3093	(− 1) 4.8882
3.5	(− 4) 8.6200	(− 3) 7.6857	(− 2) 3.4863	(− 1) 1.0927	(− 1) 2.6473
4.0	(− 4) 2.5224	(− 3) 2.6417	(− 2) 1.3692	(− 2) 4.8493	(− 1) 1.3227
4.5	(− 5) 7.2358	(− 4) 8.8072	(− 3) 5.1636	(− 2) 2.0448	(− 2) 6.2060
5.0	(− 5) 2.0413	(− 4) 2.8622	(− 3) 1.8829	(− 3) 8.2690	(− 2) 2.7673
5.5	(− 6) 5.6770	(− 5) 9.1017	(− 4) 6.6735	(− 3) 3.2283	(− 2) 1.1829
6.0	(− 6) 1.5593	(− 5) 2.8403	(− 4) 2.3080	(− 3) 1.2230	(− 3) 4.8778
6.5	(− 7) 4.2367	(− 6) 8.7187	(− 5) 7.8131	(− 4) 4.5136	(− 3) 1.9502
7.0	(− 7) 1.1400	(− 6) 2.6375	(− 5) 2.5954	(− 4) 1.6280	(− 4) 7.5886
7.5	(− 8) 3.0407	(− 7) 7.8750	(− 6) 8.4780	(− 5) 5.7536	(− 4) 2.8831
8.0	(− 9) 8.0474	(− 7) 2.3238	(− 6) 2.7278	(− 5) 1.9966	(− 4) 1.0722
8.5	(− 9) 2.1146	(− 8) 6.7842	(− 7) 8.6573	(− 6) 6.8154	(− 5) 3.9115
9.0	(−10) 5.5203	(− 8) 1.9614	(− 7) 2.7136	(− 6) 2.2918	(− 5) 1.4023
9.5	(−10) 1.4325	(− 9) 5.6202	(− 8) 8.4089	(− 7) 7.6019	(− 6) 4.9481
10.0	(−11) 3.6966	(− 9) 1.5971	(− 8) 2.5785	(− 7) 2.4900	(− 6) 1.7207
10.5	(−12) 9.4903	(−10) 4.5043	(− 9) 7.8306	(− 8) 8.0621	(− 7) 5.9043
11.0	(−12) 2.4248	(−10) 1.2613	(− 9) 2.3567	(− 8) 2.5824	(− 7) 2.0009
11.5	(−13) 6.1679	(−11) 3.5086	(−10) 7.0332	(− 9) 8.1895	(− 8) 6.7032
12.0	(−13) 1.5623	(−12) 9.6998	(−10) 2.0826	(− 9) 2.5730	(− 8) 2.2216
12.5	(−14) 3.9419	(−12) 2.6660	(−11) 6.1216	(−10) 8.0134	(− 9) 7.2896
13.0	(−15) 9.9089	(−13) 7.2878	(−11) 1.7870	(−10) 2.4754	(− 9) 2.3694
13.5	(−15) 2.4822	(−13) 1.9819	(−12) 5.1827	(−11) 7.5877	(−10) 7.6337
14.0	(−16) 6.1972	(−14) 5.3636	(−12) 1.4939	(−11) 2.3090	(−10) 2.4390
14.5	(−16) 1.5424	(−14) 1.4449	(−13) 4.2812	(−12) 6.9781	(−11) 7.7314
15.0	(−17) 3.8274	(−15) 3.8752	(−13) 1.2201	(−12) 2.0952	(−11) 2.4326
15.5	(−18) 9.4708	(−15) 1.0350	(−14) 3.4592	(−13) 6.2521	(−12) 7.5998
16.0	(−18) 2.3372	(−16) 2.7536	(−15) 9.7586	(−13) 1.8547	(−12) 2.3584
16.5	(−19) 5.7529	(−17) 7.2980	(−15) 2.7399	(−14) 5.4712	(−13) 7.2719
17.0	(−19) 1.4126	(−17) 1.9272	(−16) 7.6580	(−14) 1.6053	(−13) 2.2286
17.5	(−20) 3.4602	(−18) 5.0719	(−16) 2.1311	(−15) 4.6864	(−14) 6.7904
18.0	(−21) 8.4571	(−18) 1.3304	(−17) 5.9063	(−15) 1.3614	(−14) 2.0575
18.5	(−21) 2.0625	(−19) 3.4785	(−17) 1.6304	(−16) 3.9364	(−15) 6.2009
19.0	(−22) 5.0197	(−20) 9.0677	(−18) 4.4834	(−16) 1.1331	(−15) 1.8594
19.5	(−22) 1.2192	(−20) 2.3568	(−18) 1.2284	(−17) 3.2476	(−16) 5.5480
20.0	(−23) 2.9556	(−21) 6.1087	(−19) 3.3538	(−18) 9.2696	(−16) 1.6477

$$\frac{d}{d\rho}F_0(\eta,\rho)$$

$\eta \backslash \rho$	1	2	3	4	5
0.5	(− 1) 5.9292	(− 1) 3.2960	(− 1) −3.1699	(− 1) −8.6672	(− 1) −8.3314
1.0	(− 1) 3.4873	(− 1) 4.8156	(− 1) +3.0192	(− 1) −1.9273	(− 1) −7.2364
1.5	(− 1) 1.5684	(− 1) 3.3631	(− 1) 4.3300	(− 1) +2.9671	(− 1) −1.0456
2.0	(− 2) 6.1308	(− 1) 1.7962	(− 1) 3.2695	(− 1) 4.0401	(− 1) +2.9380
2.5	(− 2) 2.1980	(− 2) 8.2804	(− 1) 1.9237	(− 1) 3.1922	(− 1) 3.8386
3.0	(− 3) 7.4239	(− 2) 3.4693	(− 2) 9.8019	(− 1) 2.0030	(− 1) 3.1264
3.5	(− 3) 2.3993	(− 2) 1.3575	(− 2) 4.5336	(− 1) 1.0945	(− 1) 2.0555
4.0	(− 4) 7.4933	(− 3) 5.0436	(− 2) 1.9532	(− 2) 5.4362	(− 1) 1.1839
4.5	(− 4) 2.2767	(− 3) 1.7984	(− 3) 7.9650	(− 2) 2.5140	(− 2) 6.2113
5.0	(− 5) 6.7615	(− 4) 6.2008	(− 3) 3.1077	(− 2) 1.0992	(− 2) 3.0360
5.5	(− 5) 1.9700	(− 4) 2.0789	(− 3) 1.1690	(− 3) 4.5914	(− 2) 1.4028
6.0	(− 6) 5.6457	(− 5) 6.8046	(− 4) 4.2638	(− 3) 1.8462	(− 3) 6.1885
6.5	(− 6) 1.5950	(− 5) 2.1817	(− 4) 1.5145	(− 4) 7.1867	(− 3) 2.6259
7.0	(− 7) 4.4497	(− 6) 6.8691	(− 5) 5.2563	(− 4) 2.7200	(− 3) 1.0777
7.5	(− 7) 1.2276	(− 6) 2.1283	(− 5) 1.7875	(− 4) 1.0045	(− 4) 4.2964
8.0	(− 8) 3.3527	(− 7) 6.5001	(− 6) 5.9696	(− 5) 5.6292	(− 4) 1.6695
8.5	(− 9) 9.0744	(− 7) 1.9597	(− 6) 1.9614	(− 5) 1.2859	(− 5) 6.3417
9.0	(− 9) 2.4359	(− 8) 5.8395	(− 7) 6.3501	(− 6) 4.4771	(− 5) 2.3601
9.5	(−10) 6.4900	(− 8) 1.7215	(− 7) 2.0285	(− 6) 1.5341	(− 6) 8.6225
10.0	(−10) 1.7173	(− 9) 5.0256	(− 8) 6.4011	(− 7) 5.1804	(− 6) 3.0976
10.5	(−11) 4.5150	(− 9) 1.4539	(− 8) 1.9973	(− 7) 1.7262	(− 6) 1.0958
11.0	(−11) 1.1801	(−10) 4.1713	(− 9) 6.1672	(− 8) 5.6813	(− 7) 3.8219
11.5	(−12) 3.0676	(−10) 1.1875	(− 9) 1.8860	(− 8) 1.8487	(− 7) 1.3157
12.0	(−13) 7.9334	(−11) 3.3562	(−10) 5.7160	(− 9) 5.9521	(− 8) 4.4743
12.5	(−13) 2.0420	(−12) 9.4217	(−10) 1.7179	(− 9) 1.8975	(− 8) 1.5045
13.0	(−14) 5.2322	(−12) 2.6282	(−11) 5.1227	(−10) 5.9935	(− 9) 5.0060
13.5	(−14) 1.3350	(−13) 7.2879	(−11) 1.5163	(−10) 1.8768	(− 9) 1.6492
14.0	(−15) 3.3929	(−13) 2.0096	(−12) 4.4571	(−11) 5.8291	(−10) 5.3830
14.5	(−16) 8.5905	(−14) 5.5121	(−12) 1.3016	(−11) 1.7966	(−10) 1.7417
15.0	(−16) 2.1673	(−14) 1.5043	(−13) 3.7774	(−12) 5.4972	(−11) 5.5888
15.5	(−17) 5.4495	(−15) 4.0861	(−13) 1.0899	(−12) 1.6705	(−11) 1.7794
16.0	(−17) 1.3659	(−15) 1.1049	(−14) 3.1270	(−13) 5.0433	(−12) 5.6234
16.5	(−18) 3.4129	(−16) 2.9747	(−15) 8.9243	(−13) 1.5132	(−12) 1.7647
17.0	(−19) 8.5032	(−17) 7.9764	(−15) 2.5341	(−14) 4.5133	(−13) 5.5009
17.5	(−19) 2.1127	(−17) 2.1304	(−16) 7.1612	(−14) 1.3386	(−13) 1.7038
18.0	(−20) 5.2352	(−18) 5.6690	(−16) 2.0144	(−15) 3.9490	(−14) 5.2453
18.5	(−20) 1.2940	(−18) 1.5031	(−17) 5.6414	(−15) 1.1590	(−14) 1.6054
19.0	(−21) 3.1905	(−19) 3.9718	(−17) 1.5733	(−16) 3.3848	(−15) 4.8863
19.5	(−22) 7.8484	(−19) 1.0461	(−18) 4.3698	(−17) 9.8388	(−15) 1.4793
20.0	(−22) 1.9263	(−20) 2.7464	(−18) 1.2090	(−17) 2.8470	(−16) 4.4556

For use of this table see **Examples 1–3**.

COULOMB WAVE FUNCTIONS OF ORDER ZERO   Table 14.1

$$G_0(\eta,\rho)$$

$\eta\backslash\rho$	1	2	3	4	5
0.5	(0) 1.1975	(− 1) 5.3221	(− 1) −3.4105	(− 1) −9.8570	(− 1) −9.3493
1.0	(0) 2.0431	(0) 1.2758	(− 1) +6.2704	(− 1) −1.8901	(− 1) −8.9841
1.5	(0) 4.0886	(0) 2.0276	(0) 1.3423	(− 1) +7.1836	(− 2) −5.3716
2.0	(0) 9.8003	(0) 3.5124	(0) 2.0405	(0) 1.3975	(− 1) +7.9445
2.5	(1) 2.6401	(0) 7.1318	(0) 3.2733	(0) 2.0592	(0) 1.4442
3.0	(1) 7.6551	(1) 1.6390	(0) 6.0195	(0) 3.1445	(0) 2.0788
3.5	(2) 2.3355	(1) 4.0982	(1) 1.2493	(0) 5.4049	(0) 3.0657
4.0	(2) 7.4015	(2) 1.0878	(1) 2.8313	(1) 1.0423	(0) 5.0146
4.5	(3) 2.4167	(2) 3.0209	(1) 6.8403	(1) 2.1964	(0) 9.1424
5.0	(3) 8.0855	(2) 8.6969	(2) 1.7354	(1) 4.9434	(1) 1.8193
5.5	(4) 2.7606	(3) 2.5792	(2) 4.5790	(2) 1.1708	(1) 3.8704
6.0	(4) 9.5899	(3) 7.8428	(3) 1.2482	(2) 2.8891	(1) 8.6736
6.5	(5) 3.3815	(4) 2.4367	(3) 3.3782	(2) 7.3782	(2) 2.0275
7.0	(6) 1.2081	(4) 7.7137	(4) 1.0041	(3) 1.9403	(2) 4.9101
7.5	(6) 4.3664	(5) 2.4826	(4) 2.9432	(3) 5.2344	(3) 1.2258
8.0	(7) 1.5946	(5) 8.1086	(4) 8.7893	(4) 1.4441	(3) 3.1422
8.5	(7) 5.8778	(6) 2.6837	(5) 2.6689	(4) 4.0648	(3) 8.2458
9.0	(8) 2.1850	(6) 8.9891	(5) 8.2266	(5) 1.1648	(4) 2.2097
9.5	(8) 8.1855	(7) 3.0439	(6) 2.5706	(5) 3.3928	(4) 6.0344
10.0	(9) 3.0882	(8) 1.0411	(6) 8.1333	(6) 1.0029	(5) 1.6764
10.5	(10) 1.1727	(8) 3.5934	(7) 2.6029	(6) 3.0052	(5) 4.7305
11.0	(10) 4.4801	(9) 1.2509	(7) 8.4187	(6) 9.1181	(6) 1.3542
11.5	(11) 1.7211	(9) 4.3888	(8) 2.7496	(7) 2.7986	(6) 3.9285
12.0	(11) 6.6465	(10) 1.5511	(8) 9.0625	(7) 8.6825	(7) 1.1537
12.5	(12) 2.5793	(10) 5.5199	(9) 3.0124	(8) 2.7207	(7) 3.4272
13.0	(13) 1.0055	(11) 1.9769	(10) 1.0093	(8) 8.6053	(8) 1.0290
13.5	(13) 3.9366	(11) 7.1230	(10) 3.4069	(9) 2.7457	(8) 3.1205
14.0	(14) 1.5474	(12) 2.5811	(11) 1.1581	(9) 8.8331	(8) 9.5523
14.5	(14) 6.1061	(12) 9.4029	(11) 3.9629	(10) 2.8638	(9) 2.9500
15.0	(15) 2.4181	(13) 3.4429	(12) 1.3645	(10) 9.3530	(9) 9.1867
15.5	(15) 9.6091	(14) 1.2667	(12) 4.7264	(11) 3.0758	(10) 2.8835
16.0	(16) 3.8309	(14) 4.6814	(13) 1.6463	(12) 1.0182	(10) 9.1182
16.5	(17) 1.5320	(15) 1.7377	(13) 5.7652	(12) 3.3917	(11) 2.9039
17.0	(17) 6.1445	(15) 6.4769	(14) 2.0292	(13) 1.1365	(11) 9.3107
17.5	(18) 2.4714	(16) 2.4236	(14) 7.1771	(13) 3.8299	(12) 3.0045
18.0	(18) 9.9670	(16) 9.1034	(15) 2.5502	(14) 1.2976	(12) 9.7548
18.5	(19) 4.0300	(17) 3.4316	(15) 9.1019	(14) 4.4194	(13) 3.1857
19.0	(20) 1.6335	(18) 1.2981	(16) 3.2623	(15) 1.5126	(14) 1.0462
19.5	(20) 6.6365	(18) 4.9263	(17) 1.1741	(15) 5.2016	(14) 3.4544
20.0	(21) 2.7024	(19) 1.8756	(17) 4.2418	(16) 1.7969	(15) 1.1464

$$\frac{d}{d\rho}G_0(\eta,\rho)$$

$\eta\backslash\rho$	1	2	3	4	5
0.5	(− 1) −5.6132	(− 1) −8.0753	(− 1) −8.5494	(− 1) −3.4747	(− 1) +4.5076
1.0	(0) −1.2636	(− 1) −5.8273	(− 1) −7.4783	(− 1) −8.3273	(− 1) −5.1080
1.5	(0) −4.2300	(− 1) −9.5930	(− 1) −5.7358	(− 1) −7.0346	(− 1) −8.0665
2.0	(1) −1.3813	(0) −2.5554	(− 1) −8.3499	(− 1) −5.6167	(− 1) −6.7049
2.5	(1) −4.5128	(0) −7.1137	(0) −1.9326	(− 1) −7.6379	(− 1) −5.5046
3.0	(2) −1.5015	(1) −2.0029	(0) −4.8566	(0) −1.6029	(− 1) −7.1618
3.5	(2) −5.1001	(1) −5.7725	(1) −1.2438	(0) −3.7375	(0) −1.3970
4.0	(3) −1.7657	(2) −1.7086	(1) −3.2646	(0) −8.9366	(0) −3.0719
4.5	(3) −6.2161	(2) −5.1859	(1) −8.8150	(1) −2.1901	(0) −6.9633
5.0	(4) −2.2206	(3) −1.6097	(2) −2.4467	(1) −5.5222	(1) −1.6176
5.5	(4) −8.0354	(3) −5.0961	(2) −6.9635	(2) −1.4325	(1) −3.8641
6.0	(5) −2.9409	(4) −1.6418	(3) −2.0268	(2) −3.8154	(1) −9.4968
6.5	(6) −1.0873	(4) −5.3723	(3) −6.0185	(3) −1.0408	(2) −2.3977
7.0	(6) −4.0566	(5) −1.7825	(4) −1.8195	(3) −2.9006	(2) −6.2044
7.5	(7) −1.5259	(5) −5.9890	(4) −5.5897	(3) −8.2422	(3) −1.6419
8.0	(7) −5.7831	(6) −2.0352	(5) −1.7425	(4) −2.3835	(3) −4.4339
8.5	(8) −2.2067	(6) −6.9879	(5) −5.5045	(4) −7.0031	(4) −1.2197
9.0	(8) −8.4732	(7) −2.4222	(6) −1.7601	(5) −2.0878	(4) −3.4122
9.5	(9) −3.2724	(7) −8.4693	(6) −5.6909	(5) −6.3080	(4) −9.6943
10.0	(10) −1.2706	(8) −2.9853	(7) −1.8591	(6) −1.9295	(5) −2.7937
10.5	(10) −4.9580	(9) −1.0602	(7) −6.1315	(6) −5.9693	(5) −8.1574
11.0	(11) −1.9437	(9) −3.7915	(8) −2.0402	(7) −1.8664	(6) −2.4111
11.5	(11) −7.6530	(10) −1.3647	(8) −6.8449	(7) −5.8932	(6) −7.2077
12.0	(12) −3.0256	(10) −4.9424	(9) −2.3143	(8) −1.8780	(7) −2.1776
12.5	(13) −1.2008	(11) −1.8002	(9) −7.8819	(8) −6.0367	(7) −6.6446
13.0	(13) −4.7827	(11) −6.5922	(10) −2.7027	(9) −1.9562	(8) −2.0464
13.5	(14) −1.9115	(12) −2.4263	(10) −9.3274	(9) −6.3878	(8) −6.3581
14.0	(14) −7.6643	(12) −8.9735	(11) −3.2386	(10) −2.1009	(9) −1.9918
14.5	(15) −3.0826	(13) −3.3339	(12) −1.1310	(10) −6.9573	(9) −6.2887
15.0	(16) −1.2434	(14) −1.2440	(12) −3.9713	(11) −2.3188	(10) −2.0003
15.5	(16) −5.0296	(14) −4.6610	(13) −1.4017	(11) −7.7763	(10) −6.4071
16.0	(17) −2.0399	(15) −1.7532	(13) −4.9720	(12) −2.6230	(11) −2.0660
16.5	(17) −8.2941	(15) −6.6194	(14) −1.7719	(12) −8.8973	(11) −6.7044
17.0	(18) −3.3805	(16) −2.5081	(14) −6.3433	(13) −3.0340	(12) −2.1889
17.5	(19) −1.3810	(16) −9.5361	(15) −2.2806	(14) −1.0399	(12) −7.1879
18.0	(19) −5.6545	(17) −3.6376	(15) −8.2334	(14) −3.5813	(13) −2.3735
18.5	(20) −2.3201	(18) −1.3919	(16) −2.9841	(15) −1.2392	(13) −7.8789
19.0	(20) −9.5394	(18) −5.3424	(17) −1.0857	(15) −4.3069	(14) −2.6288
19.5	(21) −3.9299	(19) −2.0564	(17) −3.9642	(16) −1.5033	(14) −8.8139
20.0	(22) −1.6221	(19) −7.9378	(18) −1.4526	(16) −5.2691	(15) −2.9690

*See page II.

Table 14.1 COULOMB WAVE FUNCTIONS OF ORDER ZERO

$$F_0(\eta,\rho)$$

$\eta\backslash\rho$	6	7	8	9	10
0.5	(0)-1.0286	(-1)-7.6744	(-1)+1.0351	(-1)+8.8802	(-1)+9.3919
1.0	(-1)-1.6718	(-1)-9.0632	0)-1.0333	(-1)-4.3441	(-1)+4.7756
1.5	(-1)+8.7682	(-1)+1.1034	(-1)-7.0763	0)-1.1015	(-1)-8.0125
2.0	(0) 1.2850	0) 1.0148	(-1)+3.3340	0)-4.9930	0)-1.0616
2.5	(0) 1.1633	0) 1.3237	0) 1.1181	(-1)+5.1312	(-1)-3.0351
3.0	(-1) 8.3763	0) 1.1803	0) 1.3540	0) 1.1984	(-1)+6.6010
3.5	(-1) 5.2251	(-1) 8.6154	0) 1.1952	0) 1.3786	0) 1.2627
4.0	(-1) 2.9445	(-1) 5.5158	(-1) 8.8245	0) 1.2085	0) 1.3992
4.5	(-1) 1.5362	(-1) 3.2100	(-1) 5.7720	(-1) 9.0109	0) 1.2207
5.0	(-2) 7.5384	(-1) 1.7351	(-1) 3.4502	(-1) 6.0014	(-1) 9.1794
5.5	(-2) 3.5181	(-2) 8.8379	(-1) 1.9214	(-1) 3.6697	(-1) 6.2092
6.0	(-2) 1.5740	(-2) 4.2849	(-1) 1.0100	(-1) 2.0964	(-1) 3.8720
6.5	(-3) 6.7927	(-2) 1.9924	(-2) 5.0593	(-1) 1.1325	(-1) 2.2615
7.0	(-3) 2.8407	(-3) 8.9366	(-2) 2.4318	(-1) 5.8352	(-1) 1.2511
7.5	(-3) 1.1557	(-3) 3.8839	(-2) 1.1277	(-2) 2.8870	(-2) 6.6087
8.0	(-4) 4.5875	(-3) 1.6415	(-3) 5.0678	(-2) 1.3786	(-2) 3.3543
8.5	(-4) 1.7814	(-4) 6.7674	(-3) 2.2145	(-3) 6.3805	(-2) 1.6440
9.0	(-5) 6.7813	(-4) 2.7281	(-4) 9.4374	(-3) 2.8716	(-3) 7.8106
9.5	(-5) 2.5352	(-4) 1.0776	(-4) 3.9317	(-3) 1.2603	(-3) 3.6091
10.0	(-6) 9.3224	(-5) 4.1786	(-4) 1.6046	(-4) 5.4065	(-3) 1.6263
10.5	(-6) 3.3763	(-5) 1.5930	(-5) 6.4260	(-4) 2.2716	(-4) 7.1627
11.0	(-6) 1.2058	(-6) 5.9782	(-5) 2.5293	(-5) 9.3643	(-4) 3.0895
11.5	(-7) 4.2504	(-6) 2.2113	(-6) 9.7972	(-5) 3.7930	(-4) 1.3072
12.0	(-7) 1.4802	(-7) 8.0697	(-6) 3.7389	(-5) 1.5115	(-5) 5.4341
12.5	(-8) 5.0971	(-7) 2.9081	(-6) 1.4073	(-6) 5.9333	(-5) 2.2220
13.0	(-8) 1.7367	(-7) 1.0358	(-7) 5.2291	(-6) 2.2964	(-6) 8.9480
13.5	(-9) 5.8586	(-8) 3.6487	(-7) 1.9195	(-6) 8.7713	(-6) 3.5521
14.0	(-9) 1.9579	(-8) 1.2720	(-8) 6.9669	(-7) 3.3091	(-6) 1.3913
14.5	(-10) 6.4858	(-9) 4.3915	(-8) 2.5016	(-7) 1.2340	(-7) 5.3814
15.0	(-10) 2.1306	(-9) 1.5022	(-9) 8.8925	(-8) 4.5511	(-7) 2.0569
15.5	(-11) 6.9438	(-10) 5.0935	(-9) 3.1309	(-8) 1.6612	(-8) 7.7746
16.0	(-11) 2.2461	(-10) 1.7129	(-9) 1.0924	(-9) 6.0045	(-8) 2.9076
16.5	(-12) 7.2135	(-11) 5.7147	(-10) 3.7787	(-9) 2.1502	(-8) 1.0765
17.0	(-12) 2.3009	(-11) 1.8924	(-10) 1.2965	(-10) 7.6316	(-9) 3.9479
17.5	(-13) 7.2918	(-12) 6.2217	(-11) 4.4135	(-10) 2.6859	(-9) 1.4347
18.0	(-13) 2.2965	(-12) 2.0316	(-11) 1.4913	(-11) 9.3772	(-10) 5.1691
18.5	(-14) 7.1900	(-13) 6.5907	(-12) 5.0033	(-11) 3.2487	(-10) 1.8470
19.0	(-14) 2.2382	(-13) 2.1247	(-12) 1.6672	(-11) 1.1173	(-11) 6.5478
19.5	(-15) 6.9296	(-14) 6.8088	(-13) 5.5194	(-12) 3.8154	(-11) 2.3038
20.0	(-15) 2.1342	(-14) 2.1694	(-13) 1.8158	(-12) 1.2942	(-12) 8.0470

$$\frac{d}{d\rho}F_0(\eta,\rho)$$

$\eta\backslash\rho$	6	7	8	9	10
0.5	(-1)-1.6439	(-1)+6.5317	(-1)+9.6217	(-1)+4.8856	(-1)-3.9577
1.0	(-1)-8.9251	(-1)-4.9515	(-1)+2.6293	(-1)+8.6117	(-1)+8.4114
1.5	(-1)-5.9833	(-1)-8.7151	(-1)-6.7918	(-2)-5.9095	(-1)+6.3051
2.0	(-2)-4.4197	(-1)-4.9758	(-1)-8.2026	(-1)-7.7036	(-1)-2.9353
2.5	(-1)+2.9104	(-3)-1.2700	(-1)-4.1714	(-1)-7.6083	(-1)-8.0858
3.0	(-1) 3.6867	(-1)+2.8830	(-2)+3.0507	(-1)-3.5216	(-1)-7.0180
3.5	(-1) 3.0694	(-1) 3.5660	(-1) 2.8559	(-2)+5.4822	(-1)-2.9887
4.0	(-1) 2.0917	(-1) 3.0193	(-1) 3.4667	(-1) 2.8296	(-2)+7.3929
4.5	(-1) 1.2557	(-1) 2.1173	(-1) 2.9748	(-1) 3.3827	(-1) 2.8044
5.0	(-2) 6.8842	(-1) 1.3148	(-1) 2.1357	(-1) 2.9346	(-1) 3.3103
5.5	(-2) 3.5199	(-2) 7.4742	(-1) 1.3640	(-1) 2.1489	(-1) 2.8982
6.0	(-2) 1.7018	(-2) 3.9680	(-2) 7.9960	(-1) 1.4058	(-1) 2.1583
6.5	(-3) 7.8549	(-2) 1.9931	(-2) 4.3832	(-2) 8.4608	(-1) 1.4416
7.0	(-3) 3.4861	(-3) 9.5595	(-2) 2.2750	(-2) 4.7685	(-2) 8.8777
7.5	(-3) 1.4956	(-3) 4.4083	(-2) 1.1280	(-2) 2.5468	(-2) 5.1268
8.0	(-4) 6.2296	(-3) 1.9647	(-3) 5.3775	(-2) 1.2999	(-2) 2.8081
8.5	(-4) 2.5276	(-4) 8.4983	(-3) 2.4777	(-3) 6.3815	(-2) 1.4707
9.0	(-4) 1.0018	(-4) 3.5795	(-3) 1.1077	(-3) 3.0279	(-3) 7.4103
9.5	(-5) 3.8880	(-4) 1.4721	(-4) 4.8216	(-3) 1.3940	(-3) 3.6095
10.0	(-5) 1.4803	(-5) 5.9256	(-4) 2.0487	(-4) 6.2477	(-3) 1.7060
10.5	(-6) 5.5384	(-5) 2.3388	(-5) 8.5166	(-4) 2.7329	(-4) 7.8494
11.0	(-6) 2.0392	(-6) 9.0675	(-5) 3.4707	(-4) 1.1694	(-4) 3.5246
11.5	(-7) 7.3981	(-6) 3.4579	(-5) 1.3887	(-5) 4.9038	(-4) 1.5479
12.0	(-7) 2.6475	(-6) 1.2988	(-6) 5.4642	(-5) 2.0187	(-5) 6.6617
12.5	(-8) 9.3549	(-7) 4.8095	(-6) 2.1167	(-5) 8.1695	(-5) 2.8139
13.0	(-8) 3.2665	(-7) 1.7578	(-7) 8.0818	(-6) 3.2541	(-5) 1.1682
13.5	(-8) 1.1280	(-8) 6.3458	(-7) 3.0443	(-6) 1.2772	(-6) 4.7727
14.0	(-9) 3.8550	(-8) 2.2647	(-7) 1.1324	(-7) 4.9445	(-6) 1.9209
14.5	(-9) 1.3046	(-9) 7.9952	(-8) 4.1623	(-7) 1.8896	(-7) 7.6241
15.0	(-10) 4.3743	(-9) 2.7940	(-8) 1.5130	(-8) 7.1342	(-7) 2.9865
15.5	(-10) 1.4540	(-10) 9.6701	(-9) 5.4422	(-8) 2.6629	(-7) 1.1555
16.0	(-11) 4.7930	(-10) 3.3165	(-9) 1.9382	(-9) 9.8333	(-8) 4.4191
16.5	(-11) 1.5677	(-10) 1.1277	(-10) 6.8378	(-9) 3.5942	(-8) 1.6715
17.0	(-12) 5.0893	(-11) 3.8030	(-10) 2.3909	(-9) 1.3011	(-9) 6.2571
17.5	(-12) 1.6405	(-11) 1.2726	(-11) 8.2893	(-10) 4.6667	(-9) 2.3192
18.0	(-13) 5.2523	(-12) 4.2267	(-11) 2.8507	(-10) 1.6593	(-10) 8.5155
18.5	(-13) 1.6708	(-12) 1.3939	(-12) 9.7283	(-11) 5.8508	(-10) 3.0988
19.0	(-14) 5.2819	(-13) 4.5659	(-12) 3.2955	(-11) 2.0467	(-10) 1.1181
19.5	(-14) 1.6599	(-13) 1.4859	(-12) 1.1085	(-12) 7.1053	(-11) 4.0014
20.0	(-15) 5.1871	(-14) 4.8057	(-13) 3.7036	(-12) 2.4488	(-11) 1.4209

COULOMB WAVE FUNCTIONS OF ORDER ZERO Table 14.1

$$G_0(\eta,\rho)$$

$\eta\backslash\rho$	6	7	8	9	10
0.5	(− 1) −1.8864	(− 1) +7.0005	0) +1.0284	(− 1) +5.2116	(− 1) −4.1435
1.0	0) −1.0908	(− 1) −5.9842	(− 1) +2.9114	(− 1) +9.7148	(− 1) +9.4287
1.5	(− 1) −7.8946	0) −1.1403	(− 1) −8.7095	(− 2) −9.0032	(− 1) +7.4235
2.0	(− 2) +5.7313	(− 1) −6.8409	0) −1.1353	(− 1) −0.0415	(− 1) −3.9931
2.5	(− 1) 8.5834	(− 1) +1.4966	(− 1) −5.8782	0) −1.1041	0) −1.1456
3.0	0) 1.4847	(− 1) 9.1321	(− 1) +2.2822	(− 1) −5.0095	0) −1.0601
3.5	0) 2.0980	0) 1.5205	(− 1) 9.6127	(− 1) +2.9641	(− 1) −4.2253
4.0	0) 3.0138	0) 2.1165	0) 1.5526	0) 1.0040	(− 1) +3.5656
4.5	0) 4.7449	0) 2.9779	0) 2.1340	0) 1.5818	0) 1.0426
5.0	0) 8.2720	0) 4.5475	0) 2.9524	0) 2.1507	0) 1.6085
5.5	1) 1.5713	0) 7.6426	0) 4.3971	0) 2.9338	0) 2.1665
6.0	1) 3.1910	1) 1.3964	0) 7.1665	0) 4.2789	0) 2.9202
6.5	1) 6.8300	1) 2.7266	1) 1.2667	0) 6.7939	0) 4.1837
7.0	2) 1.5259	1) 5.6125	1) 2.3913	1) 1.1669	0) 6.4944
7.5	2) 3.5340	2) 1.2063	1) 4.7587	1) 2.1389	1) 1.0879
8.0	2) 8.4429	2) 2.6887	1) 9.8888	1) 4.1320	1) 1.9428
8.5	3) 2.0726	2) 6.1843	2) 2.1316	1) 8.3352	1) 3.6553
9.0	3) 5.2121	3) 1.4623	2) 4.7425	2) 1.7442	1) 7.1811
9.5	4) 1.3393	3) 3.5436	3) 1.0850	2) 3.7678	2) 1.4634
10.0	4) 3.5096	3) 8.7792	3) 2.5448	2) 8.3709	2) 3.0787
10.5	4) 9.3615	4) 2.2190	3) 6.1041	3) 1.9070	2) 6.6618
11.0	5) 2.5381	4) 5.7119	4) 1.4943	3) 4.4437	3) 1.4783
11.5	5) 6.9851	5) 1.4951	4) 3.7266	4) 1.0570	3) 3.3559
12.0	6) 1.9492	5) 3.9745	4) 9.4543	4) 2.5623	3) 7.7783
12.5	6) 5.5096	6) 1.0718	5) 2.4367	4) 6.3199	4) 1.8375
13.0	7) 1.5761	6) 2.9290	5) 6.3731	5) 1.5841	4) 4.4178
13.5	7) 4.5596	6) 8.1041	6) 1.6898	5) 4.0302	5) 1.0796
14.0	8) 1.3330	7) 2.2686	6) 4.5378	6) 1.0398	5) 2.6784
14.5	8) 3.9356	7) 6.4200	7) 1.2333	6) 2.7177	5) 6.7399
15.0	9) 1.1728	8) 1.8356	7) 3.3897	6) 7.1908	6) 1.7186
15.5	9) 3.5260	8) 5.2995	7) 9.4158	7) 1.9247	6) 4.4374
16.0	10) 1.0689	9) 1.5441	8) 2.6418	7) 5.2078	7) 1.1592
16.5	10) 3.2661	9) 4.5382	8) 7.4830	8) 1.4237	7) 3.0621
17.0	11) 1.0055	10) 1.3449	9) 2.1387	8) 3.9301	7) 8.1738
17.5	11) 3.1176	10) 4.0168	9) 6.1650	9) 1.0950	8) 2.2037
18.0	11) 9.7326	11) 1.2087	10) 1.7916	9) 3.0778	8) 5.9978
18.5	12) 3.0582	11) 3.6634	10) 5.2473	9) 8.7237	9) 1.6472
19.0	12) 9.6692	12) 1.1179	11) 1.5483	10) 2.4925	9) 4.5626
19.5	13) 3.0754	12) 3.4335	11) 4.6007	10) 7.1762	10) 1.2742
20.0	13) 9.8379	13) 1.0612	12) 1.3764	11) 2.0813	10) 3.5867

$$\frac{d}{d\rho}G_0(\eta,\rho)$$

$\eta\backslash\rho$	6	7	8	9	10
0.5	(− 1) +9.4204	(− 1) +7.0722	(− 1) −1.0134	(− 1) −8.3938	(− 1) −8.9014
1.0	(− 1) +1.5804	(− 1) +7.7643	(− 1) +8.9368	(− 1) +3.7613	(− 1) −4.3326
1.5	(− 1) −6.0177	(− 2) −5.6347	(− 1) +5.7724	(− 1) +9.0303	(− 1) +6.6389
2.0	(− 1) −7.8017	(− 1) −6.4998	(− 1) −2.0611	(− 1) +3.9589	(− 1) +8.3156
2.5	(− 1) −6.4488	(− 1) −7.5558	(− 1) −6.7507	(− 1) −3.1180	(− 1) +2.4273
3.0	(− 1) −5.4037	(− 1) −6.2420	(− 1) −7.3342	(− 1) −6.8725	(− 1) −3.8780
3.5	(− 1) −6.8137	(− 1) −5.3136	(− 1) −6.0700	(− 1) −7.1359	(− 1) −6.9193
4.0	0) −1.2552	(− 1) −6.5441	(− 1) −5.2327	(− 1) −5.9237	(− 1) −6.9585
4.5	0) −2.6310	0) −1.1510	(− 1) −6.3266	(− 1) −5.1597	(− 1) −5.7969
5.0	0) −5.7112	0) −2.3175	0) −1.0709	(− 1) −6.1460	(− 1) −5.0932
5.5	1) −1.2704	0) −4.8515	0) −2.0829	0) −1.0071	(− 1) −5.9925
6.0	1) −2.9032	1) −1.0407	0) −4.2272	0) −1.9007	(− 1) −9.5489
6.5	1) −6.8237	1) −2.2915	0) −8.7913	0) −3.7545	0) −1.7550
7.0	2) −1.6477	1) −5.1862	1) −1.8751	0) −7.6010	0) −3.3846
7.5	2) −4.0793	2) −1.2056	1) −4.1077	1) −1.5769	0) −6.6920
8.0	3) −1.0333	2) −2.8738	1) −9.2394	1) −3.3574	1) −1.3548
8.5	3) −2.6728	2) −7.0107	2) −2.1308	1) −7.3362	1) −2.8128
9.0	3) −7.0464	3) −1.7469	2) −5.0295	2) −1.6432	1) −5.9900
9.5	4) −1.8904	3) −4.4387	3) −1.2129	2) −3.7670	2) −1.3072
10.0	4) −5.1540	4) −1.1482	3) −2.9831	2) −8.8229	2) −2.9193
10.5	5) −1.4262	4) −3.0197	3) −7.4717	3) −2.1080	2) −6.6607
11.0	5) −4.0011	4) −8.0639	4) −1.9033	3) −5.1298	3) −1.5503
11.5	6) −1.1369	5) −2.1843	4) −4.9246	4) −1.2698	3) −3.6759
12.0	6) −3.2694	5) −5.9953	5) −1.2929	4) −3.1937	3) −8.8669
12.5	6) −9.5069	6) −1.6661	5) −3.4407	4) −8.1522	4) −2.1734
13.0	7) −2.7936	6) −4.6839	5) −9.2739	5) −2.1099	4) −5.4080
13.5	7) −8.2899	7) −1.3312	6) −2.5296	5) −5.5322	5) −1.3647
14.0	8) −2.4829	7) −3.8226	6) −6.9781	6) −1.4684	5) −3.4894
14.5	8) −7.5021	8) −1.1083	7) −1.9454	6) −3.9424	5) −9.0337
15.0	9) −2.2856	8) −3.2430	7) −5.4781	7) −1.0701	6) −2.3663
15.5	9) −7.0183	8) −9.5716	8) −1.5573	7) −2.9344	6) −6.2673
16.0	10) −2.1712	9) −2.8485	8) −4.4670	7) −8.1256	7) −1.6775
16.5	10) −6.7650	9) −8.5435	9) −1.2923	8) −2.2710	7) −4.5347
17.0	11) −2.1221	10) −2.5817	9) −3.7692	8) −6.4031	8) −1.2375
17.5	11) −6.7001	10) −7.8569	10) −1.1079	9) −1.8206	8) −3.4078
18.0	12) −2.1285	11) −2.4075	10) −3.2807	9) −5.2180	8) −9.4651
18.5	12) −6.8019	11) −7.4250	10) −9.7840	10) −1.5070	9) −2.6506
19.0	13) −2.1860	12) −2.3043	11) −2.9377	10) −4.3845	9) −7.4812
19.5	13) −7.0638	12) −7.1939	11) −8.8779	11) −1.2846	10) −2.1275
20.0	14) −2.2945	13) −2.2589	12) −2.6998	11) −3.7889	10) −6.0938

Table 14.1   **COULOMB WAVE FUNCTIONS OF ORDER ZERO**

$$F_0(\eta,\rho)$$

$\eta \backslash \rho$	11	12	13	14	15
0.5	(−1)+2.0734	(−1)−6.9792	(0)−1.0101	(−1)−4.5964	(−1)+4.8492
1.0	(0)+1.0298	(−1)+7.9515	(−2)−5.5932	(−1)−8.6120	(−1)−9.7879
1.5	(−2)+2.4612	(−1)+8.3008	(0)+1.0493	(−1)+5.1243	(−1)−3.9930
2.0	(0)−1.0170	(−1)−3.6119	(−1)+5.1844	(0)+1.0566	(−1)−8.8343
2.5	(−1)−9.6841	(0)−1.1262	(−1)−6.5977	(−1)+1.8869	(−1)+9.1875
3.0	(−1)−1.2613	(−1)−8.5079	(0)−1.1642	(−1)−8.7866	(−1)−1.1758
3.5	(−1)+7.8227	(−2)+3.2549	(−1)−7.2395	(−1)−1.1551	(0)−1.0318
4.0	(0) 1.3156	(−1) 8.8532	(−1)+1.7404	(−1)−5.9595	(0)−1.1153
4.5	(0) 1.4169	(0) 1.3600	(−1) 9.7341	(−1)+3.0035	(−1)−4.7101
5.0	(0) 1.2318	(0) 1.4324	(0) 1.3978	(0) 1.0496	(−1)+4.1342
5.5	(−1) 9.3335	(0) 1.2422	(0) 1.4462	(0) 1.4305	(0) 1.1161
6.0	(−1) 6.3994	(−1) 9.4757	(0) 1.2519	(0) 1.4586	(0) 1.4592
6.5	(−1) 4.0596	(−1) 6.5749	(−1) 9.6077	(0) 1.2610	(0) 1.4698
7.0	(−1) 2.4178	(−1) 4.2347	(−1) 6.7378	(−1) 9.7312	(0) 1.2697
7.5	(−1) 1.3660	(−1) 2.5662	(−1) 4.3989	(−1) 6.8900	(−1) 9.8472
8.0	(−2) 7.3768	(−1) 1.4773	(−1) 2.7074	(−1) 4.5535	(−1) 7.0328
8.5	(−2) 3.8306	(−2) 8.1375	(−1) 1.5852	(−1) 2.8422	(−1) 4.6997
9.0	(−2) 1.9215	(−2) 4.3132	(−2) 8.8895	(−1) 1.6898	(−1) 2.9711
9.5	(−3) 9.3472	(−2) 2.2096	(−2) 4.8001	(−2) 9.6316	(−1) 1.7913
10.0	(−3) 4.4228	(−2) 1.0980	(−2) 2.5064	(−2) 5.2898	(−1) 1.0363
10.5	(−3) 2.0410	(−3) 5.3087	(−2) 1.2700	(−2) 2.8108	(−2) 5.7809
11.0	(−4) 9.2064	(−3) 2.5036	(−3) 6.2624	(−2) 1.4498	(−2) 3.1214
11.5	(−4) 4.0667	(−3) 1.1541	(−3) 3.0126	(−3) 7.2798	(−2) 1.6367
12.0	(−4) 1.7621	(−4) 5.2102	(−3) 1.4168	(−3) 3.5666	(−3) 8.3567
12.5	(−5) 7.5001	(−4) 2.3072	(−4) 6.5253	(−3) 1.7085	(−3) 4.1640
13.0	(−5) 3.1398	(−4) 1.0036	(−4) 2.9480	(−4) 8.0157	(−3) 2.0290
13.5	(−5) 1.2943	(−5) 4.2931	(−4) 1.3082	(−4) 3.6890	(−4) 9.6841
14.0	(−6) 5.2587	(−5) 1.8082	(−5) 5.7090	(−4) 1.6677	(−4) 4.5343
14.5	(−6) 2.1078	(−6) 7.5055	(−5) 2.4529	(−5) 7.4139	(−4) 2.0854
15.0	(−7) 8.3417	(−6) 3.0731	(−5) 1.0386	(−5) 3.2448	(−5) 9.4326
15.5	(−7) 3.2617	(−6) 1.2422	(−6) 4.3371	(−5) 1.3994	(−5) 4.2002
16.0	(−7) 1.2609	(−7) 4.9601	(−6) 1.7878	(−6) 5.9525	(−5) 1.8429
16.5	(−8) 4.8223	(−7) 1.9580	(−7) 7.2797	(−6) 2.4990	(−6) 7.9746
17.0	(−8) 1.8255	(−8) 7.6449	(−7) 2.9299	(−6) 1.0363	(−6) 3.4058
17.5	(−9) 6.8436	(−8) 2.9542	(−7) 1.1663	(−7) 4.2471	(−6) 1.4366
18.0	(−9) 2.5420	(−8) 1.1303	(−8) 4.5940	(−7) 1.7213	(−7) 5.9886
18.5	(−10) 9.3587	(−9) 4.2845	(−8) 1.7916	(−8) 6.9031	(−7) 2.4686
19.0	(−10) 3.4166	(−9) 1.6095	(−9) 6.9206	(−8) 2.7406	(−7) 1.0068
19.5	(−10) 1.2373	(−10) 5.9943	(−9) 2.6491	(−8) 1.0776	(−8) 4.0646
20.0	(−11) 4.4462	(−10) 2.2143	(−9) 1.0052	(−9) 4.1981	(−8) 1.6250

$$\frac{d}{d\rho}F_0(\eta,\rho)$$

$\eta \backslash \rho$	11	12	13	14	15
0.5	(−1)−9.5680	(−1)−7.1349	(−1)+1.3869	(−1)+8.7670	(−1)+8.6352
1.0	(−1)+1.8546	(−1)−6.2449	(−1)−9.5769	(−1)−5.3599	(−1)+3.1951
1.5	(−1)+9.2360	(−1)+5.8520	(−1)−1.7814	(−1)−8.2728	(−1)−8.7421
2.0	(−1)+3.8476	(−1)+8.5839	(−1)+7.9972	(−1)+2.0967	(−1)−5.3804
2.5	(−1)−4.5774	(−1)+1.6399	(−1)+7.2679	(−1)+8.8132	(−1)+4.9591
3.0	(−1)−8.1670	(−1)−5.7064	(−2)−2.2037	(−1)+5.7220	(−1)+8.7738
3.5	(−1)−6.4636	(−1)−8.0763	(−1)−6.4688	(−1)−1.7427	(−1)+4.1643
4.0	(−1)−2.5453	(−1)−5.9550	(−1)−7.8882	(−1)−6.9700	(−1)−2.9695
4.5	(−2)+8.9270	(−1)−2.1713	(−1)−5.4930	(−1)−7.6466	(−1)−7.2842
5.0	(−1) 2.7803	(−1)+1.0181	(−1)−1.8523	(−1)−5.0747	(−1)−7.3777
5.5	(−1) 3.2469	(−1) 2.7572	(−1)+1.1221	(−1)−1.5772	(−1)−4.6963
6.0	(−1) 2.8649	(−1) 3.1907	(−1) 2.7353	(−1)+1.2094	(−1)−1.3378
6.5	(−1) 2.1649	(−1) 2.8342	(−1) 3.1402	(−1) 2.7144	(−1)+1.2836
7.0	(−1) 1.4725	(−1) 2.1694	(−1) 2.8059	(−1) 3.0946	(−1) 2.6945
7.5	(−2) 9.2538	(−1) 1.4994	(−1) 2.1722	(−1) 2.7794	(−1) 3.0530
8.0	(−2) 5.4607	(−2) 9.5947	(−1) 1.5231	(−1) 2.1737	(−1) 2.7548
8.5	(−2) 3.0589	(−2) 5.7724	(−2) 9.9053	(−1) 1.5440	(−1) 2.1743
9.0	(−2) 1.6394	(−2) 3.2995	(−2) 6.0640	(−1) 1.0189	(−1) 1.5625
9.5	(−3) 8.4560	(−2) 1.8054	(−2) 3.5301	(−2) 6.3375	(−1) 1.0450
10.0	(−3) 4.2172	(−3) 9.5118	(−2) 1.9665	(−2) 3.7513	(−2) 6.5943
10.5	(−3) 2.0412	(−3) 4.8467	(−2) 1.0573	(−2) 2.1282	(−2) 3.9633
11.0	(−4) 9.6175	(−3) 2.3971	(−3) 5.4937	(−2) 1.1634	(−2) 2.2844
11.5	(−4) 4.4224	(−3) 1.1542	(−3) 2.7714	(−3) 6.1551	(−2) 1.2693
12.0	(−4) 1.9888	(−4) 5.4237	(−3) 1.3612	(−3) 3.1620	(−3) 6.8276
12.5	(−5) 8.7636	(−4) 2.4927	(−4) 6.5256	(−3) 1.5818	(−3) 3.5670
13.0	(−5) 3.7897	(−4) 1.1224	(−4) 3.0596	(−4) 7.7243	(−3) 1.8150
13.5	(−5) 1.6105	(−5) 4.9597	(−4) 1.4055	(−4) 3.6892	(−4) 9.0158
14.0	(−6) 6.7342	(−5) 2.1535	(−5) 6.3355	(−4) 1.7264	(−4) 4.3806
14.5	(−6) 2.7736	(−6) 9.1993	(−5) 2.8061	(−5) 7.9271	(−4) 2.0855
15.0	(−6) 1.1263	(−6) 3.8704	(−5) 1.2227	(−5) 3.5765	(−5) 9.7427
15.5	(−7) 4.5133	(−6) 1.6053	(−6) 5.2466	(−5) 1.5873	(−5) 4.4720
16.0	(−7) 1.7861	(−7) 6.5690	(−6) 2.2191	(−6) 6.9375	(−5) 2.0192
16.5	(−8) 6.9850	(−7) 2.6544	(−7) 9.2602	(−6) 2.9885	(−6) 8.9777
17.0	(−8) 2.7014	(−7) 1.0598	(−7) 3.8151	(−6) 1.2700	(−6) 3.9341
17.5	(−8) 1.0337	(−8) 4.1839	(−7) 1.5529	(−7) 5.3278	(−6) 1.7006
18.0	(−9) 3.9159	(−8) 1.6340	(−8) 6.2491	(−7) 2.2081	(−7) 7.2565
18.5	(−9) 1.4693	(−9) 6.3169	(−8) 2.4875	(−8) 9.0465	(−7) 3.0587
19.0	(−10) 5.4629	(−9) 2.4184	(−9) 9.8001	(−8) 3.6658	(−7) 1.2744
19.5	(−10) 2.0135	(−10) 9.1730	(−9) 3.8231	(−8) 1.4700	(−8) 5.2514
20.0	(−11) 7.3598	(−10) 3.4487	(−9) 1.4774	(−9) 5.8367	(−8) 2.1413

COULOMB WAVE FUNCTIONS OF ORDER ZERO — Table 14.1

$$G_0(\eta,\rho)$$

$\eta \backslash \rho$	11	12	13	14	15
0.5	(0)-1.0028	(- 1)-7.4645	(- 1)+1.4266	(- 1)+9.0905	(- 1)+8.9435
1.0	(- 1)+2.1054	(- 1)-6.8021	(0)-1.0410	(- 1)-5.8152	(- 1)+3.4046
1.5	(0)+1.0819	(- 1)+6.8165	(- 1)-1.9619	(- 1)-9.3005	(- 1)-9.7885
2.0	(- 1)+4.6526	(0)+1.0451	(- 1)+9.6524	(- 1)+2.5664	(- 1)-6.2172
2.5	(- 1)-6.4066	(- 1)+1.9303	(- 1)+9.1486	(0)+1.0999	(- 1)+6.1593
3.0	(0)-1.2065	(- 1)-8.2667	(- 2)-5.4999	(- 1)+7.4014	(0)+1.1292
3.5	(0)-1.0105	(0)-1.2387	(- 1)-9.6933	(- 1)-2.7342	(- 1)+5.4881
4.0	(- 1)-3.5145	(- 1)-9.5867	(0)-1.2515	(- 1)-1.0783	(- 1)-4.6254
4.5	(- 1)+4.1032	(- 1)-2.8667	(- 1)-9.0670	(0)-1.2510	(0)-1.1612
5.0	(0) 1.0777	(- 1)+4.5891	(- 1)-2.2730	(- 1)-8.5560	(0)-1.2413
5.5	(0) 1.6333	(0) 1.1100	(- 1)+5.0322	(- 1)-1.7259	(- 1)-8.0595
6.0	(0) 2.1816	(0) 1.6563	(0) 1.1399	(- 1)+5.4393	(- 1)-1.2194
6.5	(0) 2.9102	(0) 2.1960	(0) 1.6778	(0) 1.1677	(- 1)+5.8159
7.0	(0) 4.1056	(0) 2.9029	(0) 2.2097	(0) 1.6980	(0) 1.1937
7.5	(0) 6.2486	(0) 4.0404	(0) 2.8977	(0) 2.2229	(0) 1.7172
8.0	(1) 1.0238	(0) 6.0432	(0) 3.9853	(0) 2.8940	(0) 2.2355
8.5	(1) 1.7863	(0) 9.7072	(0) 5.8691	(0) 3.9383	(0) 2.8916
9.0	(1) 3.2824	(1) 1.6587	(0) 9.2614	(0) 5.7197	(0) 3.8977
9.5	(1) 6.2966	(1) 2.9836	(1) 1.5529	(0) 8.8817	(0) 5.5902
10.0	(2) 1.2529	(1) 5.6013	(1) 2.7395	(1) 1.4638	(0) 8.5544
10.5	(2) 2.5735	(2) 1.0906	(1) 5.0429	(1) 2.5369	(1) 1.3878
11.0	(2) 5.4370	(2) 2.1919	(1) 9.6258	(1) 4.5863	(1) 2.3662
11.5	(3) 1.1780	(2) 4.5309	(2) 1.8964	(1) 8.5960	(1) 4.2071
12.0	(3) 2.6115	(2) 9.6054	(2) 3.8424	(2) 1.6627	(1) 7.7536
12.5	(3) 5.9114	(3) 2.0835	(2) 7.9840	(2) 3.3072	(2) 1.4744
13.0	(4) 1.3640	(3) 4.6148	(3) 1.6974	(2) 6.7457	(2) 2.8830
13.5	(4) 3.2036	(4) 1.0421	(3) 3.6852	(3) 1.4078	(2) 5.7803
14.0	(4) 7.6488	(4) 2.3953	(3) 8.1567	(3) 3.0002	(3) 1.1857
14.5	(5) 1.8544	(4) 5.5978	(4) 1.8380	(3) 6.5186	(3) 2.4836
15.0	(5) 4.5606	(5) 1.3286	(4) 4.2110	(4) 1.4419	(3) 5.3038
15.5	(6) 1.1368	(5) 3.1990	(4) 9.7988	(4) 3.2432	(4) 1.1531
16.0	(6) 2.8697	(5) 7.8082	(5) 2.3136	(4) 7.4095	(4) 2.5494
16.5	(6) 7.3309	(6) 1.9303	(5) 5.5378	(5) 1.7177	(4) 5.7251
17.0	(7) 1.8940	(6) 4.8301	(6) 1.3427	(5) 4.0372	(5) 1.3047
17.5	(7) 4.9456	(7) 1.2225	(6) 3.2955	(5) 9.6130	(5) 3.0146
18.0	(8) 1.3046	(7) 3.1276	(6) 8.1823	(6) 2.3172	(5) 7.0570
18.5	(8) 3.4746	(7) 8.0845	(7) 2.0539	(6) 5.6510	(6) 1.6726
19.0	(8) 9.3396	(8) 2.1103	(7) 5.2096	(7) 1.3934	(6) 4.0107
19.5	(9) 2.5325	(8) 5.5602	(8) 1.3345	(7) 3.4722	(6) 9.7253
20.0	(9) 6.9249	(9) 1.4781	(8) 3.4512	(7) 8.7394	(7) 2.3833

$$\frac{d}{d\rho}G_0(\eta,\rho)$$

$\eta \backslash \rho$	11	12	13	14	15
0.5	(- 1)-1.9549	(- 1)+6.6972	(- 1)+9.7040	(- 1)+4.4173	(- 1)-4.6958
1.0	(- 1)-9.3312	(- 1)-7.2341	(- 2)+5.5060	(- 1)+7.9924	(- 1)+9.1053
1.5	(- 2)-3.0001	(- 1)-7.2415	(- 1)-9.1975	(- 1)-4.4998	(- 1)+3.6132
2.0	(- 1)+8.0730	(- 1)+2.8479	(- 1)-4.3994	(- 1)-8.9553	(- 1)-7.5330
2.5	(- 1)+7.2980	(- 1)+8.5982	(- 1)+5.0789	(- 1)-1.6218	(- 1)-7.5598
3.0	(- 1)+1.1621	(- 1)+6.2091	(- 1)+8.5795	(- 1)+6.5611	(- 2)+7.8968
3.5	(- 1)-4.4342	(- 2)+1.2156	(- 1)+5.1517	(- 1)+8.2450	(- 1)+7.4771
4.0	(- 1)-6.9211	(- 1)-4.8470	(- 2)-7.3596	(- 1)+4.1682	(- 1)+7.7350
4.5	(- 1)-6.7991	(- 1)-6.8955	(- 1)-5.1566	(- 1)-1.4460	(- 1)+3.2728
5.0	(- 1)-5.6855	(- 1)-6.6551	(- 1)-6.8530	(- 1)-5.3907	(- 1)-2.0374
5.5	(- 1)-5.0324	(- 1)-5.5863	(- 1)-6.5243	(- 1)-6.8002	(- 1)-5.5683
6.0	(- 1)-5.8597	(- 1)-4.9764	(- 1)-5.4972	(- 1)-6.4050	(- 1)-6.7414
6.5	(- 1)-9.1132	(- 1)-5.7431	(- 1)-4.9245	(- 1)-5.4165	(- 1)-6.2956
7.0	(0)-1.6356	(- 1)-8.7431	(- 1)-5.6396	(- 1)-4.8763	(- 1)-5.3428
7.5	(0)-3.0877	(0)-1.5360	(- 1)-8.4240	(- 1)-5.5466	(- 1)-4.8313
8.0	(0)-5.9776	(0)-2.8442	(0)-1.4516	(- 1)-8.1456	(- 1)-5.4626
8.5	(1)-1.1842	(0)-5.4029	(0)-2.6410	(0)-1.3790	(- 1)-7.9001
9.0	(1)-2.4038	(1)-1.0496	(0)-4.9315	(0)-2.4689	(0)-1.3159
9.5	(1)-5.0022	(1)-2.0879	(0)-9.4124	(0)-4.5385	(0)-2.3213
10.0	(2)-1.0663	(1)-4.2551	(1)-1.8382	(0)-8.5238	(0)-4.2061
10.5	(2)-2.3257	(1)-8.8802	(1)-3.6758	(1)-1.6369	(0)-7.7837
11.0	(2)-5.1822	(2)-1.8956	(1)-7.5239	(1)-3.2170	(1)-1.4720
11.5	(3)-1.1779	(2)-4.1335	(2)-1.5749	(1)-6.4688	(1)-2.8470
12.0	(3)-2.7275	(2)-9.1940	(2)-3.3666	(2)-1.3297	(1)-5.6316
12.5	(3)-6.4259	(3)-2.0833	(2)-7.3407	(2)-2.7912	(2)-1.1385
13.0	(4)-1.5386	(3)-4.8031	(3)-1.6305	(2)-5.9750	(2)-2.3496
13.5	(4)-3.7400	(4)-1.1255	(3)-3.6849	(3)-1.3029	(2)-4.9448
14.0	(4)-9.2211	(4)-2.6777	(3)-8.4644	(3)-2.8906	(3)-1.0599
14.5	(5)-2.3041	(4)-6.4624	(4)-1.9742	(3)-6.5183	(3)-2.3115
15.0	(5)-5.8301	(5)-1.5808	(4)-4.6712	(4)-1.4925	(3)-5.1233
15.5	(6)-1.4929	(5)-3.9163	(5)-1.1203	(4)-3.4670	(4)-1.1531
16.0	(6)-3.8658	(5)-9.8198	(5)-2.7217	(4)-8.1642	(4)-2.6329
16.5	(7)-1.0118	(6)-2.4904	(5)-6.6925	(5)-1.9474	(4)-6.0946
17.0	(7)-2.6753	(6)-6.3846	(6)-1.6647	(5)-4.7022	(5)-1.4291
17.5	(7)-7.1420	(7)-1.6537	(6)-4.1862	(6)-1.1486	(5)-3.3924
18.0	(8)-1.9243	(7)-4.3256	(7)-1.0637	(6)-2.8369	(5)-8.1473
18.5	(8)-5.2302	(8)-1.1421	(7)-2.7299	(6)-7.0806	(6)-1.9785
19.0	(9)-1.4335	(8)-3.0423	(7)-7.0724	(7)-1.7850	(6)-4.8557
19.5	(9)-3.9609	(8)-8.1738	(7)-1.8489	(7)-4.5433	(7)-1.2038
20.0	(10)-1.1028	(9)-2.2141	(8)-4.8757	(8)-1.1670	(7)-3.0133

Table 14.1   COULOMB WAVE FUNCTIONS OF ORDER ZERO

$$F_0(\eta,\rho)$$

η\ρ	16	17	18	19	20
0.5	(0)+1.0105	(−1)+6.6039	(−1)−2.6356	(−1)−9.5714	(−1)−8.1320
1.0	(−1)−3.0813	(−1)+6.1193	(0)+1.0298	(−1)+5.9819	(−1)−3.2923
1.5	(0)−1.0106	(−1)−8.5450	(−2)−4.2659	(−1)+8.0098	(0)+1.0154
2.0	(−1)+1.0271	(−1)−7.4809	(0)−1.0610	(−1)−6.0110	(−1)+3.0159
2.5	(0)+1.0681	(−1)+5.2505	(−1)−3.6504	(0)−1.0050	(−1)−9.4813
3.0	(−1)+7.0689	(0)+1.1097	(−1)+8.3235	(−2)+3.2093	(−1)−7.8654
3.5	(−1)−3.8460	(−1)+4.6531	(−1)+1.0517	(−1)+1.0266	(−1)+3.8780
4.0	(0)−1.1328	(−1)−6.0877	(−1)+2.2016	(−1)+9.2908	(0)+1.1240
4.5	(0)−1.0557	(0)−1.1932	(−1)−7.9196	(−2)−1.3928	(−1)+7.6776
5.0	(−1)−3.5128	(−1)−9.8377	(0)−1.2226	(−1)−9.3827	(−1)−2.2935
5.5	(−1)+5.1503	(−1)−2.3772	(−1)−9.0447	(0)−1.2281	(0)−1.0524
6.0	(0) 1.1748	(−1)+6.0673	(−1)−1.3066	(−1)−8.2121	(0)−1.2155
6.5	(0) 1.4845	(0) 1.2270	(−1)+6.8982	(−2)−3.0049	(−1)−7.3630
7.0	(0) 1.4802	(0) 1.5072	(0) 1.2736	(−1)+7.6541	(−2)+6.4345
7.5	(0) 1.2778	(0) 1.4897	(0) 1.5276	(0) 1.3157	(−1) 8.3446
8.0	(−1) 9.9567	(0) 1.2856	(0) 1.4986	(0) 1.5461	(0) 1.3538
8.5	(−1) 7.1674	(0) 1.0060	(0) 1.2930	(0) 1.5069	(0) 1.5630
9.0	(−1) 4.8384	(−1) 7.2948	(0) 1.0159	(0) 1.3001	(0) 1.5147
9.5	(−1) 3.0947	(−1) 4.9703	(−1) 7.4157	(0) 1.0253	(0) 1.3070
10.0	(−1) 1.8899	(−1) 3.2134	(−1) 5.0960	(−1) 7.5308	(0) 1.0343
10.5	(−1) 1.1084	(−1) 1.9857	(−1) 3.3276	(−1) 5.2163	(−1) 7.6406
11.0	(−2) 6.2723	(−1) 1.1794	(−1) 2.0789	(−1) 3.4376	(−1) 5.3315
11.5	(−2) 3.4374	(−2) 6.7632	(−1) 1.2493	(−1) 2.1696	(−1) 3.5437
12.0	(−2) 1.8300	(−2) 3.7577	(−2) 7.2527	(−1) 1.3181	(−1) 2.2578
12.5	(−3) 9.4892	(−2) 2.0290	(−2) 4.0816	(−2) 7.7405	(−1) 1.3858
13.0	(−3) 4.8032	(−2) 1.0674	(−2) 2.2331	(−2) 4.4084	(−2) 8.2258
13.5	(−3) 2.3779	(−3) 5.4824	(−2) 1.1907	(−2) 2.4418	(−2) 4.7375
14.0	(−3) 1.1532	(−3) 2.7546	(−3) 6.2000	(−2) 1.3185	(−2) 2.6546
14.5	(−4) 5.4870	(−3) 1.3560	(−3) 3.1586	(−3) 6.9542	(−2) 1.4504
15.0	(−4) 2.5646	(−4) 6.5497	(−3) 1.5768	(−3) 3.5893	(−2) 7.7433
15.5	(−4) 1.1789	(−4) 3.1079	(−4) 7.7245	(−3) 1.8156	(−3) 4.0459
16.0	(−5) 5.3346	(−4) 1.4504	(−4) 3.7177	(−4) 9.0130	(−3) 2.0721
16.5	(−5) 2.3787	(−5) 6.6636	(−4) 1.7598	(−4) 4.3962	(−3) 1.0416
17.0	(−5) 1.0460	(−5) 3.0167	(−5) 8.2016	(−4) 2.1092	(−4) 5.1452
17.5	(−6) 4.5399	(−5) 1.3469	(−5) 3.7665	(−5) 9.9629	(−4) 2.5000
18.0	(−6) 1.9459	(−5) 5.9345	(−5) 1.7058	(−5) 4.6375	(−4) 1.1961
18.5	(−7) 8.2424	(−6) 2.5824	(−6) 7.6243	(−5) 2.1289	(−5) 6.5392
19.0	(−7) 3.4522	(−6) 1.1105	(−6) 3.3654	(−6) 9.6448	(−5) 2.6221
19.5	(−7) 1.4304	(−7) 4.7213	(−6) 1.4679	(−6) 4.3152	(−5) 1.2032
20.0	(−8) 5.8668	(−7) 1.9859	(−7) 6.3305	(−6) 1.9078	(−6) 5.4529

$$\frac{d}{d\rho}F_0(\eta,\rho)$$

η\ρ	16	17	18	19	20
0.5	(−1)+1.0374	(−1)−7.4873	(−1)−9.5176	(−1)−3.2396	(−1)+5.8913
1.0	(−1)+9.2398	(−1)+7.7918	(−3)−6.9768	(−1)−7.9198	(−1)−9.2215
1.5	(−1)−2.6352	(−1)+5.5592	(−1)+9.5486	(−1)+6.1234	(−1)−2.1544
2.0	(−1)−9.2711	(−1)−6.6487	(−2)+8.1839	(−1)+7.7886	(−1)+9.0561
2.5	(−1)−2.1794	(−1)−8.0683	(−1)−8.6636	(−1)−3.3293	(−1)+4.4171
3.0	(−1)+6.8521	(−2)+7.3796	(−1)−6.0115	(−1)−9.0956	(−1)−6.3111
3.5	(−1)+8.2181	(−1)+7.9551	(−1)+3.1511	(−1)−3.6640	(−1)−8.4454
4.0	(−1)+2.6981	(−1)+7.3722	(−1)+8.4585	(−1)+5.0199	(−1)−1.3528
4.5	(−1)−3.9491	(−1)+1.3669	(−1)+6.3816	(−1)+8.5260	(−1)+6.3846
5.0	(−1)−7.4641	(−1)−4.7259	(−2)+1.8327	(−1)+5.3380	(−1)+8.2868
5.5	(−1)−7.0977	(−1)−7.5469	(−1)−5.3380	(−2)−8.5571	(−1)+4.2976
6.0	(−1)−4.3534	(−1)−6.8162	(−1)−7.5595	(−1)−5.8167	(−1)−1.7601
6.5	(−1)−1.1279	(−1)−4.0420	(−1)−6.5393	(−1)−7.5212	(−1)−6.1873
7.0	(−1)+1.3471	(−2)−9.4232	(−1)−3.7584	(−1)−6.2703	(−1)−7.4462
7.5	(−1) 2.6755	(−1)+1.4020	(−2)−7.7728	(−1)−3.4994	(−1)−6.0113
8.0	(−1) 3.0148	(−1) 2.6574	(−1)+1.4497	(−2)−6.2964	(−1)−3.2623
8.5	(−1) 2.7316	(−1) 2.9796	(−1) 2.6401	(−1)+1.4915	(−2)−4.9686
9.0	(−1) 2.1740	(−1) 2.7098	(−1) 2.9470	(−1) 2.6235	(−1)+1.5282
9.5	(−1) 1.5790	(−1) 2.1730	(−1) 2.6893	(−1) 2.9166	(−1) 2.6076
10.0	(−1) 1.0690	(−1) 1.5938	(−1) 2.1715	(−1) 2.6698	(−1) 2.8881
10.5	(−2) 6.8361	(−1) 1.0912	(−1) 1.6072	(−1) 2.1696	(−1) 2.6513
11.0	(−2) 4.1667	(−2) 7.0640	(−1) 1.1118	(−1) 1.6191	(−1) 2.1673
11.5	(−2) 2.4370	(−2) 4.3620	(−2) 7.2792	(−1) 1.1309	(−1) 1.6300
12.0	(−2) 1.3747	(−2) 2.5860	(−2) 4.5494	(−2) 7.4828	(−1) 1.1487
12.5	(−3) 7.5088	(−2) 1.4792	(−2) 2.7313	(−2) 4.7295	(−2) 7.6757
13.0	(−3) 3.9846	(−3) 8.1964	(−2) 1.5829	(−2) 2.8730	(−2) 4.9026
13.5	(−3) 2.0598	(−3) 4.4133	(−3) 8.8884	(−2) 1.6854	(−2) 3.0112
14.0	(−3) 1.0396	(−3) 2.3153	(−3) 4.8514	(−3) 9.5832	(−2) 1.7867
14.5	(−4) 5.1328	(−3) 1.1861	(−3) 2.5805	(−3) 5.2978	(−2) 1.0279
15.0	(−4) 2.4832	(−4) 5.9443	(−3) 1.3405	(−3) 2.8547	(−3) 5.7512
15.5	(−4) 1.1789	(−4) 2.9194	(−4) 6.8135	(−3) 1.5025	(−3) 3.1370
16.0	(−5) 5.4992	(−4) 1.4071	(−4) 3.3940	(−4) 7.7388	(−3) 1.6717
16.5	(−5) 2.5233	(−5) 6.6637	(−4) 1.6592	(−4) 3.9067	(−4) 8.7182
17.0	(−5) 1.1401	(−5) 3.1043	(−5) 7.9706	(−4) 1.9356	(−4) 4.4568
17.5	(−6) 5.0769	(−5) 1.4240	(−5) 3.7665	(−5) 9.4242	(−4) 2.2364
18.0	(−6) 2.2300	(−6) 6.4378	(−5) 1.7526	(−5) 4.5139	(−4) 1.1028
18.5	(−7) 9.6688	(−6) 2.8708	(−6) 8.0374	(−5) 2.1289	(−5) 5.3499
19.0	(−7) 4.1409	(−6) 1.2636	(−6) 3.6355	(−6) 9.8957	(−5) 2.5557
19.5	(−7) 1.7529	(−7) 5.4935	(−6) 1.6231	(−6) 4.5369	(−5) 1.2033
20.0	(−8) 7.3379	(−7) 2.3605	(−7) 7.1576	(−6) 2.0531	(−6) 5.5878

COULOMB WAVE FUNCTIONS OF ORDER ZERO Table 14.1

$$G_0(\eta,\rho)$$

$\eta \backslash \rho$	16	17	18	19	20
0.5	(− 1) +1.0821	(− 1) −7.7111	(− 1) −9.7953	(− 1) −3.3354	(− 1) +6.0387
1.0	(− 1) +9.8687	(− 1) +8.3065	(− 3) −5.5146	(− 1) −8.3622	(− 1) −9.7243
1.5	(− 1) −2.9626	(− 1) +6.0950	(0) +1.0457	(− 1) +6.6931	(− 1) −2.3123
2.0	(0) −1.0694	(− 1) −7.6383	(− 2) +8.8035	(− 1) +8.7398	(0) +1.0133
2.5	(− 1) −2.5363	(− 1) −9.5594	(0) −1.0212	(− 1) −3.9315	(− 1) +5.0534
3.0	(− 1) +8.7388	(− 1) +1.0254	(− 1) −7.2872	(0) −1.0987	(− 1) −7.5896
3.5	(0) +1.0876	(0) +1.0419	(− 1) +4.1434	(− 1) −4.5088	(0) −1.0436
4.0	(− 1) +3.5629	(0) +1.0004	(− 1) +1.1362	(− 1) +6.7042	(− 1) −1.6256
4.5	(− 1) −6.2482	(− 1) +1.7088	(− 1) +8.8526	(0) +1.1729	(− 1) +8.7013
5.0	(0) −1.2237	(− 1) −7.6338	(− 3) −3.2476	(− 1) +7.5425	(0) +1.1657
5.5	(0) −1.2251	(0) −1.2701	(− 1) −8.8135	(− 1) −1.6427	(− 1) +6.1562
6.0	(− 1) −7.5801	(0) −1.2045	(0) −1.3038	(− 1) −9.8158	(− 1) −3.1172
6.5	(− 2) −7.4816	(− 1) −7.1189	(0) −1.1808	(0) −1.3275	(0) −1.0666
7.0	(− 1) +6.1662	(− 2) −3.0805	(− 1) −6.6763	(0) −1.1549	(0) −1.3430
7.5	(0) 1.2182	(− 1) +6.4936	(− 2) +1.0458	(− 1) −6.2518	(0) −1.1277
8.0	(0) 1.7353	(0) 1.2413	(− 1) 6.8010	(− 2) +4.9276	(− 1) −5.8448
8.5	(0) 2.2476	(0) 1.7525	(0) 1.2631	(− 1) 7.0906	(− 2) +8.5910
9.0	(0) 2.8903	(0) 2.2593	(0) 1.7689	(0) 1.2839	(− 1) 7.3645
9.5	(0) 3.8625	(0) 2.8897	(0) 2.2705	(0) 1.7846	(0) 1.3037
10.0	(0) 5.4768	(0) 3.8316	(0) 2.8898	(0) 2.2814	(0) 1.7997
10.5	(0) 8.2695	(0) 5.3768	(0) 3.8044	(0) 2.8904	(0) 2.2919
11.0	(1) 1.3223	(0) 8.0193	(0) 5.2879	(0) 3.7803	(0) 2.8915
11.5	(1) 2.2207	(1) 1.2652	(0) 7.7978	(0) 5.2085	(0) 3.7589
12.0	(1) 3.8880	(1) 2.0953	(1) 1.2151	(0) 7.6004	(0) 5.1370
12.5	(1) 7.0544	(1) 3.6163	(1) 1.9863	(1) 1.1707	(0) 7.4234
13.0	(2) 1.3205	(1) 6.4666	(1) 3.3826	(1) 1.8906	(1) 1.1312
13.5	(2) 2.5411	(2) 1.1927	(1) 5.9669	(1) 3.1797	(1) 1.8061
14.0	(2) 5.0139	(2) 2.2615	(2) 1.0855	(1) 5.5380	(1) 3.0021
14.5	(3) 1.0121	(2) 4.3958	(2) 2.0297	(1) 9.9453	(1) 5.1664
15.0	(3) 2.0860	(2) 8.7404	(2) 3.8903	(2) 1.8354	(1) 9.1659
15.5	(3) 4.3833	(3) 1.7745	(2) 7.6267	(2) 3.4717	(2) 1.6708
16.0	(3) 9.3774	(3) 3.6727	(3) 1.5265	(2) 6.7162	(2) 3.1213
16.5	(4) 2.0400	(3) 7.7388	(3) 3.1148	(3) 1.3264	(2) 5.9630
17.0	(4) 4.5079	(4) 1.6582	(3) 6.4702	(3) 2.6703	(3) 1.1629
17.5	(5) 1.0109	(4) 3.6090	(4) 1.3667	(3) 5.4726	(3) 2.3115
18.0	(5) 2.2987	(4) 7.9717	(4) 2.9323	(4) 1.1404	(3) 4.6772
18.5	(5) 5.2957	(5) 1.7855	(4) 6.3851	(4) 2.4141	(3) 9.6229
19.0	(6) 1.2353	(5) 4.0519	(5) 1.4098	(4) 5.1860	(4) 2.0110
19.5	(6) 2.9156	(5) 9.3105	(5) 3.1542	(5) 1.1297	(4) 4.2650
20.0	(6) 6.9590	(6) 2.1648	(5) 7.1454	(5) 2.4935	(4) 9.1723

$$\frac{d}{d\rho}G_0(\eta,\rho)$$

$\eta \backslash \rho$	16	17	18	19	20
0.5	(− 1) −9.7855	(− 1) −6.4000	(− 1) +2.5695	(− 1) +9.3189	(− 1) +7.9224
1.0	(− 1) +2.8609	(− 1) −5.7650	(− 1) −9.7102	(− 1) −5.6460	(− 1) +3.1370
1.5	(− 1) +9.1227	(− 1) +7.7374	(− 2) +3.6067	(− 1) −7.3679	(− 1) −9.3578
2.0	(− 2) −8.3491	(− 1) +6.5787	(− 1) +9.3570	(− 1) +5.3119	(− 1) −2.7296
2.5	(− 1) −8.8452	(− 1) −4.3562	(− 1) +3.1578	(− 1) +8.6483	(− 1) +8.1928
3.0	(− 1) −5.6757	(− 1) −8.9431	(− 1) −6.7512	(− 2) −1.9960	(− 1) +6.6241
3.5	(− 1) +2.7609	(− 1) −3.6790	(− 1) −8.2667	(− 1) −7.1410	(− 1) −3.0592
4.0	(− 1) +7.9794	(− 1) +4.3113	(− 1) −1.7673	(− 1) −7.1410	(− 1) −8.7013
4.5	(− 1) +7.1352	(− 1) +8.1848	(− 1) +5.4934	(− 3) −3.4829	(− 1) −5.7890
5.0	(− 1) +2.4665	(− 1) +6.4978	(− 1) +8.1799	(− 1) +6.3669	(− 1) +1.4822
5.5	(− 1) −2.5327	(− 1) +1.7444	(− 1) +5.8546	(− 1) +8.0282	(− 1) +6.9880
6.0	(− 1) −5.7031	(− 1) −2.9499	(− 1) +1.0993	(− 1) +5.2246	(− 1) +7.7756
6.5	(− 1) −6.6792	(− 1) −5.8050	(− 1) −3.3031	(− 2) +5.2317	(− 1) +4.6186
7.0	(− 1) −6.1949	(− 1) −6.6155	(− 1) −5.8814	(− 1) −3.6035	(− 4) +8.3738
7.5	(− 1) −5.2752	(− 1) −6.1017	(− 1) −6.5515	(− 1) −5.9378	(− 1) −3.8601
8.0	(− 1) −4.7892	(− 1) −5.2127	(− 1) −6.0151	(− 1) −6.4880	(− 1) −5.9783
8.5	(− 1) −5.3860	(− 1) −4.7495	(− 1) −5.1547	(− 1) −5.9344	(− 1) −6.4254
9.0	(− 1) −7.6818	(− 1) −5.3157	(− 1) −4.7121	(− 1) −5.1007	(− 1) −5.8590
9.5	(0) −1.2605	(− 1) −7.4860	(− 1) −5.2509	(− 1) −4.6767	(− 1) −5.0502
10.0	(0) −2.1932	(0) −1.2115	(− 1) −7.3093	(− 1) −5.1908	(− 1) −4.6431
10.5	(0) −3.9217	(0) −2.0812	(0) −1.1677	(− 1) −7.1488	(− 1) −5.1349
11.0	(0) −7.1592	(0) −3.6757	(0) −1.9822	(0) −1.1284	(− 1) −7.0023
11.5	(1) −1.3348	(0) −6.6261	(0) −3.4609	(0) −1.8942	(0) −1.0929
12.0	(1) −2.5439	(1) −1.2193	(0) −6.1663	(0) −3.2719	(0) −1.8154
12.5	(1) −4.9562	(1) −2.2921	(1) −1.1209	(0) −5.7662	(0) −3.1044
13.0	(1) −9.8652	(1) −4.4031	(1) −2.0805	(1) −1.0363	(0) −5.4152
13.5	(2) −2.0042	(1) −8.6387	(1) −3.9443	(1) −1.9007	(0) −9.6285
14.0	(2) −4.1515	(2) −1.7295	(1) −7.6350	(1) −3.5594	(1) −1.7465
14.5	(2) −8.7576	(2) −3.5297	(2) −1.5077	(1) −6.8033	(1) −3.2330
15.0	(3) −1.8795	(2) −7.3354	(2) −3.0346	(2) −1.3263	(1) −6.1066
15.5	(3) −4.0993	(3) −1.5507	(2) −6.2186	(2) −2.6348	(2) −1.1761
16.0	(3) −9.0788	(3) −3.3317	(3) −1.2962	(2) −5.3284	(2) −2.3079
16.5	(4) −2.0399	(3) −7.2680	(3) −2.7456	(3) −1.0960	(2) −4.6095
17.0	(4) −4.6466	(4) −1.6085	(3) −5.9047	(3) −2.2906	(2) −9.3627
17.5	(5) −1.0722	(4) −3.6089	(4) −1.2883	(3) −4.8605	(3) −1.9322
18.0	(5) −2.5048	(4) −8.2028	(4) −2.8495	(4) −1.0463	(3) −4.0483
18.5	(5) −5.9202	(5) −1.8875	(4) −6.3850	(4) −2.2832	(3) −8.6039
19.0	(6) −1.4150	(5) −4.3947	(5) −1.4484	(4) −5.0474	(4) −1.8537
19.5	(6) −3.4181	(6) −1.0347	(5) −3.3247	(5) −1.1297	(4) −4.0457
20.0	(6) −8.3412	(6) −2.4624	(5) −7.7176	(5) −2.5583	(4) −8.9396

Table 14.2 $C_0(\eta) = e^{-\frac{1}{2}\pi\eta}|\Gamma(1+i\eta)|$

η	$C_0(\eta)$	η	$C_0(\eta)$	η	$C_0(\eta)$
0.00	1.000000	1.00	(−1)1.08423	2.00	(−3)6.61992
0.05	0.922568	1.05	(−2)9.49261	2.05	(−3)5.72791
0.10	0.847659	1.10	(−2)8.30211	2.10	(−3)4.95461
0.15	0.775700	1.15	(−2)7.25378	2.15	(−3)4.28450
0.20	0.707063	1.20	(−2)6.33205	2.20	(−3)3.70402
0.25	0.642052	1.25	(−2)5.52279	2.25	(−3)3.20136
0.30	0.580895	1.30	(−2)4.81320	2.30	(−3)2.76623
0.35	0.523742	1.35	(−2)4.19173	2.35	(−3)2.38968
0.40	0.470665	1.40	(−2)3.64804	2.40	(−3)2.06392
0.45	0.421667	1.45	(−2)3.17287	2.45	(−3)1.78218
0.50	0.376686	1.50	(−2)2.75796	2.50	(−3)1.53858
0.55	0.335605	1.55	(−2)2.39599	2.55	(−3)1.32801
0.60	0.298267	1.60	(−2)2.08045	2.60	(−3)1.14604
0.65	0.264478	1.65	(−2)1.80558	2.65	(−4)9.88816
0.70	0.234025	1.70	(−2)1.56632	2.70	(−4)8.53013
0.75	0.206680	1.75	(−2)1.35817	2.75	(−4)7.35735
0.80	0.182206	1.80	(−2)1.17720	2.80	(−4)6.34476
0.85	0.160370	1.85	(−2)1.01996	2.85	(−4)5.47066
0.90	0.140940	1.90	(−3)8.83391	2.90	(−4)4.71626
0.95	0.123694	1.95	(−3)7.64847	2.95	(−4)4.06528
1.00	0.108423	2.00	(−3)6.61992	3.00	(−4)3.50366

$$\begin{bmatrix} (-4)5 \\ 5 \end{bmatrix}$$

For $\ln \Gamma(1+iy)$, see **Table 6.7.**

15. Hypergeometric Functions

FRITZ OBERHETTINGER [1]

Contents

[1] National Bureau of Standards. (Presently, Oregon State University, Corvallis, Oregon.)

15. Hypergeometric Functions

Mathematical Properties

15.1. Gauss Series, Special Elementary Cases, Special Values of the Argument

Gauss Series

The circle of convergence of the Gauss hypergeometric series

15.1.1

$$F(a, b; c; z) = {}_2F_1(a, b; c; z)$$

$$= F(b, a; c; z) = \sum_{n=0}^{\infty} \frac{(a)_n (b)_n}{(c)_n} \frac{z^n}{n!}$$

$$= \frac{\Gamma(c)}{\Gamma(a)\Gamma(b)} \sum_{n=0}^{\infty} \frac{\Gamma(a+n)\Gamma(b+n)}{\Gamma(c+n)} \frac{z^n}{n!}$$

is the unit circle $|z|=1$. The behavior of this series on its circle of convergence is:

(a) Divergence when $\mathscr{R}(c-a-b) \leq -1$.

(b) Absolute convergence when $\mathscr{R}(c-a-b) > 0$.

(c) Conditional convergence when $-1 < \mathscr{R}(c-a-b) \leq 0$; the point $z=1$ is excluded. The Gauss series reduces to a polynomial of degree n in z when a or b is equal to $-n$, $(n=0, 1, 2, \ldots)$. (For these cases see also **15.4**.) The series **15.1.1** is not defined when c is equal to $-m$, $(m=0, 1, 2, \ldots)$, provided a or b is not a negative integer n with $n < m$. For $c = -m$

15.1.2

$$\lim_{c \to -m} \frac{1}{\Gamma(c)} F(a, b; c; z) =$$

$$\frac{(a)_{m+1}(b)_{m+1}}{(m+1)!} z^{m+1} F(a+m+1, b+m+1; m+2; z)$$

Special Elementary Cases of Gauss Series

(For cases involving higher functions see **15.4**.)

15.1.3 $\quad F(1, 1; 2; z) = -z^{-1} \ln (1-z) \quad *$

15.1.4 $\quad F(\tfrac{1}{2}, 1; \tfrac{3}{2}; z^2) = \tfrac{1}{2} z^{-1} \ln \left(\frac{1+z}{1-z}\right)$

15.1.5 $\quad F(\tfrac{1}{2}, 1; \tfrac{3}{2}; -z^2) = z^{-1} \arctan z$

15.1.6

$$F(\tfrac{1}{2}, \tfrac{1}{2}; \tfrac{3}{2}; z^2) = (1-z^2)^{\frac{1}{2}} F(1, 1; \tfrac{3}{2}; z^2) = z^{-1} \arcsin z$$

15.1.7

$$F(\tfrac{1}{2}, \tfrac{1}{2}; \tfrac{3}{2}; -z^2) = (1+z^2)^{\frac{1}{2}} F(1, 1; \tfrac{3}{2}; -z^2)$$

$$= z^{-1} \ln [z + (1+z^2)^{\frac{1}{2}}]$$

15.1.8 $\qquad F(a, b; b; z) = (1-z)^{-a}$

15.1.9 $\quad F(a, \tfrac{1}{2}+a; \tfrac{1}{2}; z^2) = \tfrac{1}{2}[(1+z)^{-2a} + (1-z)^{-2a}]$

15.1.10

$$F(a, \tfrac{1}{2}+a; \tfrac{3}{2}; z^2) =$$

$$\tfrac{1}{2} z^{-1} (1-2a)^{-1} [(1+z)^{1-2a} - (1-z)^{1-2a}]$$

15.1.11

$$F(-a, a; \tfrac{1}{2}; -z^2) = \tfrac{1}{2} \{ [(1+z^2)^{\frac{1}{2}} + z]^{2a} + [(1+z^2)^{\frac{1}{2}} - z]^{2a} \}$$

15.1.12

$$F(a, 1-a; \tfrac{1}{2}; -z^2) =$$

$$\tfrac{1}{2}(1+z^2)^{-\frac{1}{2}} \{ [(1+z^2)^{\frac{1}{2}} + z]^{2a-1} + [(1+z^2)^{\frac{1}{2}} - z]^{2a-1} \}$$

15.1.13

$$F(a, \tfrac{1}{2}+a; 1+2a; z) = 2^{2a} [1+(1-z)^{\frac{1}{2}}]^{-2a}$$

$$= (1-z)^{\frac{1}{2}} F(1+a, \tfrac{1}{2}+a; 1+2a; z)$$

15.1.14

$$F(a, \tfrac{1}{2}+a; 2a; z) = 2^{2a-1}(1-z)^{-\frac{1}{2}} [1+(1-z)^{\frac{1}{2}}]^{1-2a}$$

15.1.15 $\quad F(a, 1-a; \tfrac{3}{2}; \sin^2 z) = \frac{\sin [(2a-1)z]}{(2a-1) \sin z}$

15.1.16 $\quad F(a, 2-a; \tfrac{3}{2}; \sin^2 z) = \frac{\sin [(2a-2)z]}{(a-1) \sin (2z)}$

15.1.17 $\qquad F(-a, a; \tfrac{1}{2}; \sin^2 z) = \cos (2az)$

15.1.18 $\quad F(a, 1-a; \tfrac{1}{2}; \sin^2 z) = \frac{\cos [(2a-1)z]}{\cos z}$

15.1.19 $\quad F(a, \tfrac{1}{2}+a; \tfrac{1}{2}; -\tan^2 z) = \cos^{2a} z \cos (2az)$

Special Values of the Argument

15.1.20

$$F(a, b; c; 1) = \frac{\Gamma(c)\Gamma(c-a-b)}{\Gamma(c-a)\Gamma(c-b)}$$

$$(c \neq 0, -1, -2, \ldots, \mathscr{R}(c-a-b) > 0)$$

*See page II.

556

15.1.21

$$F(a, b; a-b+1; -1)=2^{-a}\pi^{\frac{1}{2}}\frac{\Gamma(1+a-b)}{\Gamma(1+\frac{1}{2}a-b)\Gamma(\frac{1}{2}+\frac{1}{2}a)}$$

$$(1+a-b\neq 0, -1, -2, \ldots)$$

15.1.22

$$F(a, b; a-b+2; -1)=2^{-a}\pi^{1/2}(b-1)^{-1}\Gamma(a-b+2)$$

$$\left[\frac{1}{\Gamma(\frac{1}{2}a)\Gamma(\frac{3}{2}+\frac{1}{2}a-b)}-\frac{1}{\Gamma(\frac{1}{2}+\frac{1}{2}a)\Gamma(1+\frac{1}{2}a-b)}\right]$$

$$(a-b+2\neq 0, -1, -2, \ldots)$$

15.1.23 $\quad F(1, a; a+1; -1)=\frac{1}{2}a[\psi(\frac{1}{2}+\frac{1}{2}a)-\psi(\frac{1}{2}a)]$

15.1.24

$$F(a, b; \frac{1}{2}a+\frac{1}{2}b+\frac{1}{2};\frac{1}{2})=\pi^{\frac{1}{2}}\frac{\Gamma(\frac{1}{2}+\frac{1}{2}a+\frac{1}{2}b)}{\Gamma(\frac{1}{2}+\frac{1}{2}a)\Gamma(\frac{1}{2}+\frac{1}{2}b)}$$

$$(\frac{1}{2}a+\frac{1}{2}b+\frac{1}{2}\neq 0, -1, -2, \ldots)$$

15.1.25

$$F(a, b; \frac{1}{2}a+\frac{1}{2}b+1; \frac{1}{2})=2\pi^{\frac{1}{2}}(a-b)^{-1}\Gamma(1+\frac{1}{2}a+\frac{1}{2}b)$$

$$\{[\Gamma(\frac{1}{2}a)\Gamma(\frac{1}{2}+\frac{1}{2}b)]^{-1}-[\Gamma(\frac{1}{2}+\frac{1}{2}a)\Gamma(\frac{1}{2}b)]^{-1}\}$$

$$(\frac{1}{2}(a+b)+1\neq 0, -1, -2, \ldots)$$

15.1.26

$$F(a, 1-a; b; \frac{1}{2})=$$

$$2^{1-b}\pi^{\frac{1}{2}}\Gamma(b)\left[\Gamma(\frac{1}{2}a+\frac{1}{2}b)\Gamma(\frac{1}{2}+\frac{1}{2}b-\frac{1}{2}a)\right]^{-1}$$

$$(b\neq 0, -1, -2, \ldots)$$

15.1.27

$$F(1, 1; a+1; \frac{1}{2})=a[\psi(\frac{1}{2}+\frac{1}{2}a)-\psi(\frac{1}{2}a)]$$

$$(a\neq -1, -2, -3, \ldots)$$

15.1.28

$$F(a, a; a+1; \frac{1}{2})=2^{a-1}a[\psi(\frac{1}{2}+\frac{1}{2}a)-\psi(\frac{1}{2}a)]$$

$$(a\neq -1, -2, -3, \ldots)$$

15.1.29

$$F(a, \frac{1}{2}+a; \frac{3}{2}-2a;-\frac{1}{3})=(\frac{8}{9})^{-2a}\frac{\Gamma(\frac{4}{3})\Gamma(\frac{3}{2}-2a)}{\Gamma(\frac{3}{2})\Gamma(\frac{4}{3}-2a)}$$

$$(\frac{3}{2}-2a\neq 0, -1, -2, \ldots)$$

15.1.30

$$F(a, \frac{1}{2}+a; \frac{5}{6}+a; \frac{1}{9})=(\frac{3}{4})^a\pi^{\frac{1}{2}}\frac{\Gamma(\frac{5}{6}+\frac{2}{3}a)}{\Gamma(\frac{1}{2}+\frac{1}{3}a)\Gamma(\frac{5}{6}+\frac{1}{3}a)}$$

$$(\frac{5}{6}+\frac{2}{3}a\neq 0, -1, -2, \ldots)$$

15.1.31

$$F(a, \frac{1}{3}a+\frac{1}{3}; \frac{2}{3}a+\frac{2}{3}; e^{i\pi/3})$$

$$=2^{\frac{2}{3}a+\frac{2}{3}}\pi^{\frac{1}{2}}3^{-\frac{1}{2}(a+1)}e^{i\pi a/6}\frac{\Gamma(\frac{1}{3}a+\frac{5}{6})}{\Gamma(\frac{1}{3}a+\frac{2}{3})\Gamma(\frac{2}{3})}$$

$$(\frac{1}{3}a\neq -\frac{5}{6}, -\frac{11}{6}, -\frac{17}{6}, \ldots)$$

15.2. Differentiation Formulas and Gauss' Relations for Contiguous Functions

Differentiation Formulas

15.2.1 $\quad \dfrac{d}{dz}F(a, b; c; z)=\dfrac{ab}{c}F(a+1, b+1; c+1; z)$

15.2.2

$$\frac{d^n}{dz^n}F(a, b; c; z)=\frac{(a)_n(b)_n}{(c)_n}F(a+n, b+n; c+n; z)$$

15.2.3

$$\frac{d^n}{dz^n}[z^{a+n-1}F(a, b; c; z)]=(a)_n z^{a-1}F(a+n, b; c; z)$$

15.2.4

$$\frac{d^n}{dz^n}[z^{c-1}F(a, b; c; z)]=(c-n)_n z^{c-n-1}F(a, b; c-n; z)$$

15.2.5

$$\frac{d^n}{dz^n}[z^{c-a+n-1}(1-z)^{a+b-c}F(a, b; c; z)]$$

$$=(c-a)_n z^{c-a-1}(1-z)^{a+b-c-n}F(a-n, b; c; z)$$

15.2.6

$$\frac{d^n}{dz^n}[(1-z)^{a+b-c}F(a, b; c; z)]$$

$$=\frac{(c-a)_n(c-b)_n}{(c)_n}(1-z)^{a+b-c-n}F(a, b; c+n; z)$$

15.2.7

$$\frac{d^n}{dz^n}[(1-z)^{a+n-1}F(a, b; c; z)]$$

$$=\frac{(-1)^n(a)_n(c-b)_n}{(c)_n}(1-z)^{a-1}F(a+n, b; c+n; z)$$

15.2.8

$$\frac{d^n}{dz^n}[z^{c-1}(1-z)^{b-c+n}F(a, b; c; z)]$$

$$=(c-n)_n z^{c-n-1}(1-z)^{b-c}F(a-n, b; c-n; z)$$

15.2.9

$$\frac{d^n}{dz^n}[z^{c-1}(1-z)^{a+b-c}F(a, b; c; z)]$$

$$=(c-n)_n z^{c-n-1}(1-z)^{a+b-c-n}F(a-n, b-n; c-n; z)$$

Gauss' Relations for Contiguous Functions

The six functions $F(a\pm 1, b; c; z)$, $F(a, b\pm 1; c; z)$, $F(a, b; c\pm 1; z)$ are called contiguous to $F(a, b; c; z)$. Relations between $F(a, b; c; z)$ and any two contiguous functions have been given by Gauss. By repeated application of these relations the function $F(a+m, b+n; c+l; z)$ with integral $m, n, l(c+l \neq 0, -1, -2, \ldots)$ can be expressed as a linear combination of $F(a, b; c; z)$ and one of its contiguous functions with coefficients which are rational functions of a, b, c, z.

15.2.10

$$(c-a)F(a-1, b; c; z)+(2a-c-az+bz)F(a,b;c;z)$$
$$+a(z-1)F(a+1, b; c; z)=0$$

15.2.11

$$(c-b)F(a, b-1; c; z)+(2b-c-bz+az)F(a,b;c;z)$$
$$+b(z-1)F(a, b+1; c; z)=0$$

15.2.12

$$c(c-1)(z-1)F(a, b; c-1; z)$$
$$+c[c-1-(2c-a-b-1)z]F(a, b; c; z)$$
$$+(c-a)(c-b)zF(a, b; c+1; z)=0$$

15.2.13

$$[c-2a-(b-a)z]F(a, b; c; z)$$
$$+a(1-z)F(a+1, b; c; z)$$
$$-(c-a)F(a-1, b; c; z)=0$$

15.2.14

$$(b-a)F(a, b; c; z)+aF(a+1, b; c; z)$$
$$-bF(a, b+1; c; z)=0$$

15.2.15

$$(c-a-b)F(a, b; c; z)+a(1-z)F(a+1, b; c; z)$$
$$-(c-b)F(a, b-1; c; z)=0$$

15.2.16

$$c[a-(c-b)z]F(a, b; c; z)-ac(1-z)F(a+1, b; c; z)$$
$$+(c-a)(c-b)zF(a, b; c+1; z)=0$$

15.2.17

$$(c-a-1)F(a, b; c; z)+aF(a+1, b; c; z)$$
$$-(c-1)F(a, b; c-1; z)=0$$

15.2.18

$$(c-a-b)F(a, b; c; z)-(c-a)F(a-1, b; c; z)$$
$$+b(1-z)F(a, b+1; c; z)=0$$

15.2.19

$$(b-a)(1-z)F(a, b; c; z)-(c-a)F(a-1, b; c; z)$$
$$+(c-b)F(a, b-1; c; z)=0$$

15.2.20

$$c(1-z)F(a, b; c; z)-cF(a-1, b; c; z)$$
$$+(c-b)zF(a, b; c+1; z)=0$$

15.2.21

$$[a-1-(c-b-1)z]F(a, b; c; z)$$
$$+(c-a)F(a-1, b; c; z)$$
$$-(c-1)(1-z)F(a, b; c-1; z)=0$$

15.2.22

$$[c-2b+(b-a)z]F(a, b; c; z)$$
$$+b(1-z)F(a, b+1; c; z)$$
$$-(c-b)F(a, b-1; c; z)=0$$

15.2.23

$$c[b-(c-a)z]F(a,b;c;z)-bc(1-z)F(a,b+1;c;z)$$
$$+(c-a)(c-b)zF(a, b; c+1; z)=0$$

15.2.24

$$(c-b-1)F(a, b; c; z)+bF(a, b+1; c; z)$$
$$-(c-1)F(a, b; c-1; z)=0$$

15.2.25

$$c(1-z)F(a, b; c; z)-cF(a, b-1; c; z)$$
$$* \quad +(c-a)z\,F(a, b; c+1; z)=0$$

15.2.26

$$[b-1-(c-a-1)z]F(a, b; c; z)$$
$$+(c-b)F(a, b-1; c; z)$$
$$-(c-1)(1-z)F(a, b; c-1; z)=0$$

15.2.27

$$c[c-1-(2c-a-b-1)z]F(a, b; c; z)$$
$$+(c-a)(c-b)zF(a, b; c+1; z)$$
$$-c(c-1)(1-z)F(a, b; c-1; z)=0$$

15.3. Integral Representations and Transformation Formulas

Integral Representations

15.3.1

$$F(a, b; c; z)=$$

$$\frac{\Gamma(c)}{\Gamma(b)\Gamma(c-b)}\int_0^1 t^{b-1}(1-t)^{c-b-1}(1-tz)^{-a}dt$$

$$(\mathscr{R}c > \mathscr{R}b > 0)$$

The integral represents a one valued analytic function in the z-plane cut along the real axis from 1 to ∞ and hence **15.3.1** gives the analytic continuation of **15.1.1**, $F(a, b; c; z)$. Another integral representation is in the form of a Mellin-Barnes integral

15.3.2 $\quad F(a, b; c; z) = \dfrac{\Gamma(c)}{2\pi i \Gamma(a)\Gamma(b)} \displaystyle\int_{-i\infty}^{i\infty} \dfrac{\Gamma(a+s)\Gamma(b+s)\Gamma(-s)}{\Gamma(c+s)} (-z)^s ds$

$$= \tfrac{1}{2}i \dfrac{\Gamma(c)}{\Gamma(a)\Gamma(b)} \int_{-i\infty}^{i\infty} \dfrac{\Gamma(a+s)\Gamma(b+s)}{\Gamma(1+s)\Gamma(c+s)} \csc(\pi s)(-z)^s ds$$

Here $-\pi < \arg(-z) < \pi$ and the path of integration is chosen such that the poles of $\Gamma(a+s)$ and $\Gamma(b+s)$ i.e. the points $s=-a-n$ and $s=-b-m (n, m=0, 1, 2, \ldots)$ respectively, are at its left side and the poles of $\csc(\pi s)$ or $\Gamma(-s)$ i.e. $s=0, 1, 2,$ are at its right side. The cases in which $-a, -b$ or $-c$ are non-negative integers or $a-b$ equal to an integer are excluded.

Linear Transformation Formulas

From **15.3.1** and **15.3.2** a number of transformation formulas for $F(a, b; c; z)$ can be derived.

15.3.3 $\quad F(a, b; c; z) = (1-z)^{c-a-b} F(c-a, c-b; c; z)$

15.3.4 $\qquad\qquad = (1-z)^{-a} F\left(a, c-b; c; \dfrac{z}{z-1}\right)$

15.3.5 $\qquad\qquad = (1-z)^{-b} F\left(b, c-a; c; \dfrac{z}{z-1}\right)$

15.3.6 $\qquad\qquad = \dfrac{\Gamma(c)\Gamma(c-a-b)}{\Gamma(c-a)\Gamma(c-b)} F(a, b; a+b-c+1; 1-z)$

$$+ (1-z)^{c-a-b} \dfrac{\Gamma(c)\Gamma(a+b-c)}{\Gamma(a)\Gamma(b)} F(c-a, c-b; c-a-b+1; 1-z)$$

$$(|\arg(1-z)| < \pi)$$

15.3.7 $\qquad\qquad = \dfrac{\Gamma(c)\Gamma(b-a)}{\Gamma(b)\Gamma(c-a)} (-z)^{-a} F\left(a, 1-c+a; 1-b+a; \dfrac{1}{z}\right)$

$$+ \dfrac{\Gamma(c)\Gamma(a-b)}{\Gamma(a)\Gamma(c-b)} (-z)^{-b} F\left(b, 1-c+b; 1-a+b; \dfrac{1}{z}\right) \qquad (|\arg(-z)| < \pi)$$

15.3.8 $\qquad\qquad = (1-z)^{-a} \dfrac{\Gamma(c)\Gamma(b-a)}{\Gamma(b)\Gamma(c-a)} F\left(a, c-b; a-b+1; \dfrac{1}{1-z}\right)$

$$+ (1-z)^{-b} \dfrac{\Gamma(c)\Gamma(a-b)}{\Gamma(a)\Gamma(c-b)} F\left(b, c-a; b-a+1; \dfrac{1}{1-z}\right) \qquad (|\arg(1-z)| < \pi)$$

15.3.9 $\qquad\qquad = \dfrac{\Gamma(c)\Gamma(c-a-b)}{\Gamma(c-a)\Gamma(c-b)} z^{-a} F\left(a, a-c+1; a+b-c+1; 1-\dfrac{1}{z}\right)$

$$+ \dfrac{\Gamma(c)\Gamma(a+b-c)}{\Gamma(a)\Gamma(b)} (1-z)^{c-a-b} z^{a-c} F\left(c-a, 1-a; c-a-b+1; 1-\dfrac{1}{z}\right)$$

$$(|\arg z| < \pi, |\arg(1-z)| < \pi)$$

Each term of **15.3.6** has a pole when $c=a+b\pm m, (m=0, 1, 2, \ldots)$; this case is covered by

15.3.10 $\quad F(a, b; a+b; z) = \dfrac{\Gamma(a+b)}{\Gamma(a)\Gamma(b)} \displaystyle\sum_{n=0}^{\infty} \dfrac{(a)_n(b)_n}{(n!)^2} [2\psi(n+1)-\psi(a+n)-\psi(b+n)-\ln(1-z)](1-z)^n$

$$(|\arg(1-z)| < \pi, |1-z| < 1)$$

Furthermore for $m=1, 2, 3, \ldots$

15.3.11 $\quad F(a, b; a+b+m; z) = \dfrac{\Gamma(m)\Gamma(a+b+m)}{\Gamma(a+m)\Gamma(b+m)} \displaystyle\sum_{n=0}^{m-1} \dfrac{(a)_n(b)_n}{n!(1-m)_n} (1-z)^n$

$$- \dfrac{\Gamma(a+b+m)}{\Gamma(a)\Gamma(b)} (z-1)^m \sum_{n=0}^{\infty} \dfrac{(a+m)_n(b+m)_n}{n!(n+m)!} (1-z)^n [\ln(1-z)-\psi(n+1)$$

$$-\psi(n+m+1)+\psi(a+n+m)+\psi(b+n+m)] \qquad (|\arg(1-z)| < \pi, |1-z| < 1)$$

15.3.12 $F(a, b; a+b-m; z)=\dfrac{\Gamma(m)\Gamma(a+b-m)}{\Gamma(a)\Gamma(b)}(1-z)^{-m}\displaystyle\sum_{n=0}^{m-1}\dfrac{(a-m)_n(b-m)_n}{n!(1-m)_n}(1-z)^n$

$$-\dfrac{(-1)^m\Gamma(a+b-m)}{\Gamma(a-m)\Gamma(b-m)}\sum_{n=0}^{\infty}\dfrac{(a)_n(b)_n}{n!(n+m)!}(1-z)^n[\ln(1-z)-\psi(n+1)$$

$$-\psi(n+m+1)+\psi(a+n)+\psi(b+n)]$$

$$(|\arg(1-z)|<\pi,|1-z|<1)$$

Similarly each term of **15.3.7** has a pole when $b=a\pm m$ or $b-a=\pm m$ and the case is covered by

15.3.13 $F(a,a;c;z)=\dfrac{\Gamma(c)}{\Gamma(a)\Gamma(c-a)}(-z)^{-a}\displaystyle\sum_{n=0}^{\infty}\dfrac{(a)_n(1-c+a)_n}{(n!)^2}z^{-n}[\ln(-z)+2\psi(n+1)-\psi(a+n)-\psi(c-a-n)]$

$$(|\arg(-z)|<\pi,|z|>1,(c-a)\neq0,\pm1,\pm2,\ldots)$$

The case $b-a=m$, $(m=1, 2, 3, \ldots)$ is covered by

15.3.14 $F(a,a+m;c;z)=F(a+m,a;c;z)$

$$=\dfrac{\Gamma(c)(-z)^{-a-m}}{\Gamma(a+m)\Gamma(c-a)}\sum_{n=0}^{\infty}\dfrac{(a)_{n+m}(1-c+a)_{n+m}}{n!(n+m)!}z^{-n}[\ln(-z)+\psi(1+m+n)+\psi(1+n)$$

$$-\psi(a+m+n)-\psi(c-a-m-n)]+(-z)^{-a}\dfrac{\Gamma(c)}{\Gamma(a+m)}\sum_{n=0}^{m-1}\dfrac{\Gamma(m-n)(a)_n}{n!\Gamma(c-a-n)}z^{-n}$$

$$(|\arg(-z)|<\pi,|z|>1,(c-a)\neq0,\pm1,\pm2,\ldots)$$

The case $c-a=0, -1, -2, \ldots$ becomes elementary, **15.3.3**, and the case $c-a=1, 2, 3, \ldots$ can be obtained from **15.3.14**, by a limiting process (see [15.2]).

Quadratic Transformation Formulas

If, and only if the numbers $\pm(1-c)$, $\pm(a-b)$, $\pm(a+b-c)$ are such, that two of them are equal or one of them is equal to $\frac{1}{2}$, then there exists a quadratic transformation. The basic formulas are due to Kummer [15.7] and a complete list is due to Goursat [15.3]. See also [15.2].

15.3.15 $F(a,b;2b;z)=(1-z)^{-\frac{1}{2}a}F\left(\frac{1}{2}a,b-\frac{1}{2}a;b+\frac{1}{2};\dfrac{z^2}{4z-4}\right)$

15.3.16 $=(1-\frac{1}{2}z)^{-a}F(\frac{1}{2}a,\frac{1}{2}+\frac{1}{2}a;b+\frac{1}{2};z^2(2-z)^{-2})$

15.3.17 $=(\frac{1}{2}+\frac{1}{2}\sqrt{1-z})^{-2a}F\left[a,a-b+\frac{1}{2};b+\frac{1}{2};\left(\dfrac{1-\sqrt{1-z}}{1+\sqrt{1-z}}\right)^2\right]$

15.3.18 $=(1-z)^{-\frac{1}{2}a}F\left(a,2b-a;b+\frac{1}{2};-\dfrac{(1-\sqrt{1-z})^2}{4\sqrt{1-z}}\right)$

15.3.19 $F(a,a+\frac{1}{2};c;z)=(\frac{1}{2}+\frac{1}{2}\sqrt{1-z})^{-2a}F\left(2a,2a-c+1;c;\dfrac{1-\sqrt{1-z}}{1+\sqrt{1-z}}\right)$

15.3.20 $=(1\pm\sqrt{z})^{-2a}F\left(2a,c-\frac{1}{2};2c-1;\pm\dfrac{2\sqrt{z}}{1\pm\sqrt{z}}\right)$

15.3.21 $=(1-z)^{-a}F\left(2a,2c-2a-1;c;\dfrac{\sqrt{1-z}-1}{2\sqrt{1-z}}\right)$

15.3.22 $F(a,b;a+b+\frac{1}{2};z)=F(2a,2b;a+b+\frac{1}{2};\frac{1}{2}-\frac{1}{2}\sqrt{1-z})$

15.3.23 $=(\frac{1}{2}+\frac{1}{2}\sqrt{1-z})^{-2a}F\left(2a,a-b+\frac{1}{2};a+b+\frac{1}{2};\dfrac{\sqrt{1-z}-1}{\sqrt{1-z}+1}\right)$

15.3.24 $\quad F(a, b; a+b-\tfrac{1}{2}; z)=(1-z)^{-\frac{1}{2}}F(2a-1,2b-1;a+b-\tfrac{1}{2};\tfrac{1}{2}-\tfrac{1}{2}\sqrt{1-z})$

15.3.25 $\quad\quad\quad\quad =(1-z)^{-\frac{1}{2}}(\tfrac{1}{2}+\tfrac{1}{2}\sqrt{1-z})^{1-2a}\,F\left(2a-1,\,a-b+\tfrac{1}{2};\,a+b-\tfrac{1}{2};\,\dfrac{\sqrt{1-z}-1}{\sqrt{1-z}+1}\right)$

15.3.26 $\quad F(a, b; a-b+1; z)=(1+z)^{-a}F\left(\tfrac{1}{2}a,\,\tfrac{1}{2}a+\tfrac{1}{2};\,a-b+1;\,4z(1+z)^{-2}\right)$

15.3.27 $\quad\quad\quad\quad =(1\pm\sqrt{z})^{-2a}F(a,\,a-b+\tfrac{1}{2};\,2a-2b+1;\,\pm4\sqrt{z}(1\pm\sqrt{z})^{-2})$

15.3.28 $\quad\quad\quad\quad =(1-z)^{-a}F(\tfrac{1}{2}a,\,\tfrac{1}{2}a-b+\tfrac{1}{2};\,a-b+1;\,-4z(1-z)^{-2})$

15.3.29 $\quad F(a, b; \tfrac{1}{2}a+\tfrac{1}{2}b+\tfrac{1}{2}; z)=(1-2z)^{-a}F\left(\tfrac{1}{2}a,\,\tfrac{1}{2}a+\tfrac{1}{2};\,\tfrac{1}{2}a+\tfrac{1}{2}b+\tfrac{1}{2};\,\dfrac{4z^2-4z}{(1-2z)^2}\right)$

15.3.30 $\quad\quad\quad\quad =F(\tfrac{1}{2}a,\,\tfrac{1}{2}b;\,\tfrac{1}{2}a+\tfrac{1}{2}b+\tfrac{1}{2};\,4z-4z^2)$

15.3.31 $\quad F(a, 1-a; c; z)=(1-z)^{c-1}F(\tfrac{1}{2}c-\tfrac{1}{2}a,\,\tfrac{1}{2}c+\tfrac{1}{2}a-\tfrac{1}{2};\,c;\,4z-4z^2)$

15.3.32 $\quad\quad\quad\quad =(1-z)^{c-1}(1-2z)^{a-c}F\left(\tfrac{1}{2}c-\tfrac{1}{2}a,\,\tfrac{1}{2}c-\tfrac{1}{2}a+\tfrac{1}{2};\,c;\,(4z^2-4z)(1-2z)^{-2}\right)$

Cubic transformations are listed in [15.2] and [15.3].

In the formulas above, the square roots are defined so that their value is real and positive when $0 \leqq z < 1$. All formulas are valid in the neighborhood of $z=0$.

15.4. Special Cases of $F(a, b; c; z)$

Polynomials

When a or b is equal to a negative integer, then

15.4.1 $\quad F(-m, b; c; z)=\displaystyle\sum_{n=0}^{m}\frac{(-m)_n(b)_n}{(c)_n}\frac{z^n}{n!}$

This formula is also valid when $c=-m-l;\ m,\ l=0, 1, 2, \ldots$

15.4.2 $\quad F(-m, b;-m-l; z)=\displaystyle\sum_{n=0}^{m}\frac{(-m)_n(b)_n}{(-m-l)_n}\frac{z^n}{n!}$

Some particular cases are

15.4.3 $\quad F(-n, n; \tfrac{1}{2}; x)=T_n(1-2x)$

15.4.4 $\quad F(-n, n+1; 1; x)=P_n(1-2x)$

15.4.5 $\quad F\left(-n, n+2\alpha; \alpha+\dfrac{1}{2}; x\right)=\dfrac{n!}{(2\alpha)_n}\,C_n^{(\alpha)}(1-2x)$

15.4.6 $\quad F(-n, \alpha+1+\beta+n; \alpha+1; x)=\dfrac{n!}{(\alpha+1)_n}\,P_n^{(\alpha,\beta)}(1-2x)$

Here T_n, P_n, $C_n^{(\alpha)}$, $P_n^{(\alpha,\beta)}$ denote Chebyshev, Legendre's, Gegenbauer's and Jacobi's polynomials respectively (see chapter **22**).

Legendre Functions

Legendre functions are connected with those special cases of the hypergeometric function for which a quadratic transformation exists (see **15.3**).

15.4.7 $\quad F(a, b; 2b; z)=2^{2b-1}\Gamma(\tfrac{1}{2}+b)z^{\frac{1}{2}-b}(1-z)^{\frac{1}{2}(b-a-\frac{1}{2})}P_{a-b-\frac{1}{2}}^{\frac{1}{2}-b}\left[\left(1-\dfrac{z}{2}\right)(1-z)^{-\frac{1}{2}}\right]$

15.4.8 $\quad\quad\quad\quad =2^{2b}\pi^{-\frac{1}{2}}\dfrac{\Gamma(\tfrac{1}{2}+b)}{\Gamma(2b-a)}\,z^{-b}(1-z)^{\frac{1}{2}(b-a)}e^{i\pi(a-b)}P_{b-1}^{b-a}\left(\dfrac{2}{z}-1\right)$

15.4.9 $\quad F(a, b; 2b;-z)=2^{2b}\pi^{-\frac{1}{2}}\dfrac{\Gamma(\tfrac{1}{2}+b)}{\Gamma(a)}\,z^{-b}(1+z)^{\frac{1}{2}(b-a)}e^{-i\pi(a-b)}P_{b-1}^{a-b}\left(1+\dfrac{2}{z}\right)\ \ (|\arg z|<\pi,\,|\arg(1\pm z)|<\pi)$

15.4.10 $\quad F(a, a+\tfrac{1}{2}; c; z)=2^{c-1}\Gamma(c)z^{\frac{1}{2}-\frac{1}{2}c}(1-z)^{\frac{1}{2}c-a-\frac{1}{2}}P_{2a-c}^{1-c}[(1-z)^{-\frac{1}{2}}]$

$(|\arg z|<\pi, |\arg(1-z)|<\pi, z \text{ not between } 0 \text{ and } -\infty)$

15.4.11 $\quad F(a, a+\tfrac{1}{2}; c; x)=2^{c-1}\Gamma(c)(-x)^{\frac{1}{2}-\frac{1}{2}c}(1-x)^{\frac{1}{2}c-a-\frac{1}{2}}P_{2a-c}^{1-c}[(1-x)^{-\frac{1}{2}}]$ $\qquad(-\infty<x<0)$

15.4.12 $\quad F(a, b; a+b+\tfrac{1}{2}; z)=2^{a+b-\frac{1}{2}}\Gamma(\tfrac{1}{2}+a+b)(-z)^{\frac{1}{2}(\frac{1}{2}-a-b)}P_{a-b-\frac{1}{2}}^{\frac{1}{2}-a-b}[(1-z)^{\frac{1}{2}}]$

$(|\arg(-z)|<\pi, z \text{ not between } 0 \text{ and } 1)$

15.4.13 $\quad F(a, b; a+b+\tfrac{1}{2}; x)=2^{a+b-\frac{1}{2}}\Gamma(\tfrac{1}{2}+a+b)x^{\frac{1}{2}(\frac{1}{2}-a-b)}P_{a-b-\frac{1}{2}}^{\frac{1}{2}-a-b}[(1-x)^{\frac{1}{2}}]$ $\qquad(0<x<1)$

15.4.14 $\quad F(a, b; a-b+1; z)=\Gamma(a-b+1)z^{\frac{1}{2}b-\frac{1}{2}a}(1-z)^{-b}P_{-b}^{b-a}\left(\dfrac{1+z}{1-z}\right)$

$(|\arg(1-z)|<\pi, z \text{ not between } 0 \text{ and } -\infty)$

15.4.15 $\quad F(a, b; a-b+1; x)=\Gamma(a-b+1)(1-x)^{-b}(-x)^{\frac{1}{2}b-\frac{1}{2}a}P_{-b}^{b-a}\left(\dfrac{1+x}{1-x}\right)$ $\qquad(-\infty<x<0)$

15.4.16 $\quad F(a, 1-a; c; z)=\Gamma(c)(-z)^{\frac{1}{2}-\frac{1}{2}c}(1-z)^{\frac{1}{2}c-\frac{1}{2}}P_{-a}^{1-c}(1-2z)$

$(|\arg(-z)|<\pi, |\arg(1-z)|<\pi, z \text{ not between } 0 \text{ and } 1)$

15.4.17 $\quad F(a, 1-a; c; x)=\Gamma(c)x^{\frac{1}{2}-\frac{1}{2}c}(1-x)^{\frac{1}{2}c-\frac{1}{2}}P_{-a}^{1-c}(1-2x)$ $\qquad(0<x<1)$

15.4.18 $\quad F(a, b; \tfrac{1}{2}a+\tfrac{1}{2}b+\tfrac{1}{2}; z)=\Gamma(\tfrac{1}{2}+\tfrac{1}{2}a+\tfrac{1}{2}b)[z(z-1)]^{\frac{1}{4}(1-a-b)}P_{\frac{1}{2}(a-b-1)}^{\frac{1}{2}(1-a-b)}(1-2z)$

$(|\arg z|<\pi, |\arg(z-1)|<\pi, z \text{ not between } 0 \text{ and } 1)$

15.4.19 $\quad F(a, b; \tfrac{1}{2}a+\tfrac{1}{2}b+\tfrac{1}{2}; x)=\Gamma(\tfrac{1}{2}+\tfrac{1}{2}a+\tfrac{1}{2}b)(x-x^2)^{\frac{1}{4}(1-a-b)}P_{\frac{1}{2}(a-b-1)}^{\frac{1}{2}(1-a-b)}(1-2x)$ $\qquad(0<x<1)$

15.4.20 $\quad F(a, b; a+b-\tfrac{1}{2}; z)=2^{a+b-\frac{3}{2}}\Gamma(a+b-\tfrac{1}{2})(-z)^{\frac{1}{2}(\frac{1}{2}-a-b)}(1-z)^{-\frac{1}{2}}P_{b-a-\frac{1}{2}}^{\frac{3}{2}-a-b}[(1-z)^{\frac{1}{2}}]$

$(|\arg(-z)|<\pi, |\arg(1-z)|<\pi, \mathscr{R}[(1-z)^{\frac{1}{2}}]>0, z \text{ not between } 0 \text{ and } 1)$

15.4.21 $\quad F(a, b; a+b-\tfrac{1}{2}; x)=2^{a+b-\frac{3}{2}}\Gamma(a+b-\tfrac{1}{2})x^{\frac{1}{2}(\frac{1}{2}-a-b)}(1-x)^{-\frac{1}{2}}P_{b-a-\frac{1}{2}}^{\frac{3}{2}-a-b}[(1-x)^{\frac{1}{2}}]$ $\qquad(0<x<1)$

15.4.22 $\quad F(a, b; \tfrac{1}{2}; z)=\pi^{-\frac{1}{2}}2^{a+b-\frac{3}{2}}\Gamma(\tfrac{1}{2}+a)\Gamma(\tfrac{1}{2}+b)(z-1)^{\frac{1}{2}(\frac{1}{2}-a-b)}[P_{a-b-\frac{1}{2}}^{\frac{1}{2}-a-b}(z^{\frac{1}{2}})+P_{a-b-\frac{1}{2}}^{\frac{1}{2}-a-b}(-z^{\frac{1}{2}})]$

$(|\arg z|<\pi, |\arg(z-1)|<\pi, z \text{ not between } 0 \text{ and } 1)$

15.4.23 $\quad F(a, b; \tfrac{1}{2}; x)=\pi^{-\frac{1}{2}}2^{a+b-\frac{3}{2}}\Gamma(\tfrac{1}{2}+a)\Gamma(\tfrac{1}{2}+b)(1-x)^{\frac{1}{2}(\frac{1}{2}-a-b)}[P_{a-b-\frac{1}{2}}^{\frac{1}{2}-a-b}(x^{\frac{1}{2}})+P_{a-b-\frac{1}{2}}^{\frac{1}{2}-a-b}(-x^{\frac{1}{2}})]$ $\qquad(0<x<1)$

15.4.24 $\quad F(a, b; \tfrac{1}{2}; -z)=\pi^{-\frac{1}{2}}2^{a-b-1}\Gamma(\tfrac{1}{2}+a)\Gamma(1-b)(z+1)^{-\frac{1}{2}a-\frac{1}{2}b}e^{\pm i\frac{\pi}{2}(b-a)}\{P_{a+b-1}^{b-a}[z^{\frac{1}{2}}(1+z)^{-\frac{1}{2}}]$

$+P_{a+b-1}^{b-a}[-z^{\frac{1}{2}}(1+z)^{-\frac{1}{2}}]\}$

$(\pm \text{ according as } \mathscr{I}z \gtrless 0, z \text{ not between } 0 \text{ and } \infty)$

*15.4.25 $\quad F(a, b; \tfrac{1}{2}; -x)=\pi^{-\frac{1}{2}}2^{a-b-1}\Gamma(\tfrac{1}{2}+a)\Gamma(1-b)(1+x)^{-\frac{1}{2}a-\frac{1}{2}b}\{P_{a+b-1}^{b-a}[x^{\frac{1}{2}}(1+x)^{-\frac{1}{2}}]+P_{a+b-1}^{b-a}[-x^{1/2}(1+x)^{-\frac{1}{2}}]\}$

$(0<x<\infty)$

15.4.26 $\quad F(a, b; \tfrac{3}{2}; x)=-\pi^{-\frac{1}{2}}2^{a+b-\frac{3}{2}}\Gamma(a-\tfrac{1}{2})\Gamma(b-\tfrac{1}{2})x^{-\frac{1}{2}}(1-x)^{\frac{1}{2}(\frac{1}{2}-a-b)}\{P_{a-b-\frac{1}{2}}^{\frac{1}{2}-a-b}(x^{\frac{1}{2}})-P_{a-b-\frac{1}{2}}^{\frac{1}{2}-a-b}(-x^{\frac{1}{2}})\}$ $\qquad(0<x<1)$

15.5. The Hypergeometric Differential Equation

The hypergeometric differential equation

15.5.1 $\quad z(1-z)\dfrac{d^2w}{dz^2}+[c-(a+b+1)z]\dfrac{dw}{dz}-abw=0$

---

*See page II.

has three (regular) singular points $z=0, 1, \infty$. The pairs of exponents at these points are

15.5.2 $\qquad \rho_{1,2}^{(0)}=0, 1-c, \qquad \rho_{1,2}^{(1)}=0, c-a-b, \qquad \rho_{1,2}^{(\infty)}=a, b$

respectively. The general theory of differential equations of the Fuchsian type distinguishes between the following cases.

A. None of the numbers c, $c-a-b$; $a-b$ is equal to an integer. Then two linearly independent solutions of **15.5.1** in the neighborhood of the singular points $0, 1, \infty$ are respectively

15.5.3 $\quad w_{1(0)}=F(a, b; c; z)=(1-z)^{c-a-b}F(c-a, c-b; c; z)$

15.5.4 $\quad w_{2(0)}=z^{1-c}F(a-c+1, b-c+1; 2-c; z)=z^{1-c}(1-z)^{c-a-b}F(1-a, 1-b; 2-c; z)$

15.5.5 $\quad w_{1(1)}=F(a, b; a+b+1-c; 1-z)=z^{1-c}F(1+b-c, 1+a-c; a+b+1-c; 1-z)$

15.5.6 $\quad w_{2(1)}=(1-z)^{c-a-b}F(c-b, c-a; c-a-b+1; 1-z)=z^{1-c}(1-z)^{c-a-b}F(1-a, 1-b; c-a-b+1; 1-z)$

15.5.7 $\quad w_{1(\infty)}=z^{-a}F(a, a-c+1; a-b+1; z^{-1})=z^{b-c}(z-1)^{c-a-b}F(1-b, c-b; a-b+1; z^{-1})$

15.5.8 $\quad w_{2(\infty)}=z^{-b}F(b, b-c+1; b-a+1; z^{-1})=z^{a-c}(z-1)^{c-a-b}F(1-a, c-a; b-a+1; z^{-1})$

The second set of the above expressions is obtained by applying **15.3.3** to the first set.

Another set of representations is obtained by applying **15.3.4** to **15.5.3** through **15.5.8**. This gives **15.5.9–15.5.14**.

15.5.9 $\quad w_{1(0)}=(1-z)^{-a}F\left(a, c-b; c; \dfrac{z}{z-1}\right)=(1-z)^{-b}F\left(b, c-a; c; \dfrac{z}{z-1}\right)$

15.5.10 $\quad w_{2(0)}=z^{1-c}(1-z)^{c-a-1}F\left(a-c+1, 1-b; 2-c; \dfrac{z}{z-1}\right)=z^{1-c}(1-z)^{c-b-1}F\left(b-c+1, 1-a; 2-c; \dfrac{z}{z-1}\right)$

15.5.11 $\quad w_{1(1)}=z^{-a}F(a, a-c+1; a+b-c+1; 1-z^{-1})=z^{-b}F(b, b-c+1; a+b-c+1; 1-z^{-1})$

15.5.12
$w_{2(1)}=z^{a-c}(1-z)^{c-a-b}F(c-a, 1-a; c-a-b+1; 1-z^{-1})=z^{b-c}(1-z)^{c-a-b}F(c-b, 1-b; c-a-b+1; 1-z^{-1})$

15.5.13 $\quad w_{1(\infty)}=(z-1)^{-a}F\left(a, c-b; a-b+1; \dfrac{1}{1-z}\right)=(z-1)^{-b}F\left(b, c-a; b-a+1; \dfrac{1}{1-z}\right)$

15.5.14
$w_{2(\infty)}=z^{1-c}(z-1)^{c-a-1}F\left(a-c+1, 1-b; a-b+1; \dfrac{1}{1-z}\right)=z^{1-c}(z-1)^{c-b-1}F\left(b-c+1, 1-a; b-a+1; \dfrac{1}{1-z}\right)$

15.5.3 to **15.5.14** constitute Kummer's 24 solutions of the hypergeometric equation. The analytic continuation of $w_{1,2(0)}(z)$ can then be obtained by means of **15.3.3** to **15.3.9**.

B. One of the numbers a, b, $c-a$, $c-b$ is an integer. Then one of the hypergeometric series for instance $w_{1,2(0)}$, **15.5.3**, **15.5.4** terminates and the corresponding solution is of the form

15.5.15 $\quad w=z^{\alpha}(1-z)^{\beta}p_n(z)$

where $p_n(z)$ is a polynomial in z of degree n. This case is referred to as the degenerate case of the hypergeometric differential equation and its solutions are listed and discussed in great detail in [15.2].

C. The number $c-a-b$ is an integer, c nonintegral. Then **15.3.10** to **15.3.12** give the analytic continuation of $w_{1,2(0)}$ into the neighborhood of $z=1$. Similarly **15.3.13** and **15.3.14** give the analytic continuation of $w_{1,2(0)}$ into the neighborhood of $z=\infty$ in case $a-b$ is an integer but not c, subject of course to the further restrictions $c-a=0, \pm1, \pm2 \ldots$ (For a detailed discussion of all possible cases, see [15.2]).

D. The number $c=1$. Then **15.5.3**, **15.5.4** are replaced by

15.5.16 $\quad w_{1(0)}=F(a, b; 1; z)$

15.5.17 $\quad w_{2(0)} = F(a,b;1;z) \ln z + \sum_{n=1}^{\infty} \frac{(a)_n (b)_n}{(n!)^2} z^n [\psi(a+n) - \psi(a) + \psi(b+n) - \psi(b) - 2\psi(n+1) + 2\psi(1)]$ $\quad (|z|<1)$

E. The number $c = m+1$, $m = 1, 2, 3, \ldots$.   A fundamental system is

15.5.18 $\quad w_{1(0)} = F(a,b;m+1;z)$

15.5.19 $\quad w_{2(0)} = F(a,b;m+1;z) \ln z + \sum_{n=1}^{\infty} \frac{(a)_n (b)_n}{(1+m)_n n!} z^n [\psi(a+n) - \psi(a) + \psi(b+n) - \psi(b) - \psi(m+1+n)$

$$+ \psi(m+1) - \psi(n+1) + \psi(1)] - \sum_{n=1}^{m} \frac{(n-1)!(-m)_n}{(1-a)_n (1-b)_n} z^{-n} \quad (|z| < 1 \text{ and } a, b \neq 0, 1, 2, \ldots (m-1))$$

F. The number $c = 1-m$, $m = 1, 2, 3, \ldots$.   A fundamental system is

15.5.20 $\quad w_{1(0)} = z^m F(a+m, b+m; 1+m; z)$

15.5.21

$$w_{2(0)} = z^m F(a+m, b+m; 1+m; z) \ln z + z^m \sum_{n=1}^{\infty} z^n \frac{(a+m)_n (b+m)_n}{(1+m)_n n!} [\psi(a+m+n) - \psi(a+m) + \psi(b+m+n)$$

$$- \psi(b+m) - \psi(m+1+n) + \psi(m+1) - \psi(n+1) + \psi(1)] - \sum_{n=1}^{m} \frac{(n-1)!(-m)_n}{(1-a-m)_n (1-b-m)_n} z^{m-n}$$

$$(|z|<1 \text{ and } a, b \neq 0, -1, -2, \ldots -(m-1))$$

15.6. Riemann's Differential Equation

The hypergeometric differential equation **15.5.1** with the (regular) singular points 0, 1, ∞ is a special case of Riemann's differential equation with three (regular) singular points a, b, c

15.6.1

$$\frac{d^2 w}{dz^2} + \left[\frac{1-\alpha-\alpha'}{z-a} + \frac{1-\beta-\beta'}{z-b} + \frac{1-\gamma-\gamma'}{z-c} \right] \frac{dw}{dz}$$

$$+ \left[\frac{\alpha\alpha'(a-b)(a-c)}{z-a} + \frac{\beta\beta'(b-c)(b-a)}{z-b} \right.$$

$$\left. + \frac{\gamma\gamma'(c-a)(c-b)}{z-c} \right] \frac{w}{(z-a)(z-b)(z-c)} = 0$$

The pairs of the exponents with respect to the singular points a; b; c are α, α'; β, β'; γ, γ' respectively subject to the condition

15.6.2 $\quad\quad \alpha + \alpha' + \beta + \beta' + \gamma + \gamma' = 1$

The complete set of solutions of **15.6.1** is denoted by the symbol

15.6.3

$$w = P \left\{ \begin{matrix} a & b & c & \\ \alpha & \beta & \gamma & z \\ \alpha' & \beta' & \gamma' & \end{matrix} \right\}$$

Special Cases of Riemann's P Function

(a) The generalized hypergeometric function

15.6.4

$$w = P \left\{ \begin{matrix} 0 & \infty & 1 & \\ \alpha & \beta & \gamma & z \\ \alpha' & \beta' & \gamma' & \end{matrix} \right\}$$

(b) The hypergeometric function $F(a, b; c; z)$

15.6.5

$$w = P \left\{ \begin{matrix} 0 & \infty & 1 & \\ 0 & a & 0 & z \\ 1-c & b & c-a-b & \end{matrix} \right\}$$

(c) The Legendre functions $P_\nu^\mu(z)$, $Q_\nu^\mu(z)$

15.6.6

$$w = P \left\{ \begin{matrix} 0 & \infty & 1 & \\ -\frac{1}{2}\nu & \frac{1}{2}\mu & 0 & (1-z^2)^{-1} \\ \frac{1}{2} + \frac{1}{2}\nu & -\frac{1}{2}\mu & \frac{1}{2} & \end{matrix} \right\}$$

(d) The confluent hypergeometric function

15.6.7

$$w = P \left\{ \begin{matrix} 0 & \infty & c & \\ \frac{1}{2}+u & -c & c-k & z \\ \frac{1}{2}-u & 0 & k & \end{matrix} \right\}$$

provided $\lim c \to \infty$.

Transformation Formulas for Riemann's P Function

15.6.8 $\left(\dfrac{z-a}{z-b}\right)^k \left(\dfrac{z-c}{z-b}\right)^l P \left\{ \begin{array}{cccc} a & b & c & \\ \alpha & \beta & \gamma & z \\ \alpha' & \beta' & \gamma' & \end{array} \right\} = P \left\{ \begin{array}{cccc} a & b & c & \\ \alpha+k & \beta-k-l & \gamma+l & z \\ \alpha'+k & \beta'-k-l & \gamma'+l & \end{array} \right\}$

15.6.9 $P \left\{ \begin{array}{cccc} a & b & c & \\ \alpha & \beta & \gamma & z \\ \alpha' & \beta' & \gamma' & \end{array} \right\} = P \left\{ \begin{array}{cccc} a_1 & b_1 & c_1 & \\ \alpha & \beta & \gamma & z_1 \\ \alpha' & \beta' & \gamma' & \end{array} \right\}$

where

15.6.10 $z = \dfrac{Az_1+B}{Cz_1+D},\ a = \dfrac{Aa_1+B}{Ca_1+D},\ b = \dfrac{Ab_1+B}{Cb_1+D},\ c = \dfrac{Ac_1+B}{Cc_1+D}$

and A, B, C, D are arbitrary constants such that $AD-BC \neq 0$.

Riemann's P function reduced to the hypergeometric function is

15.6.11 $P \left\{ \begin{array}{cccc} a & b & c & \\ \alpha & \beta & \gamma & z \\ \alpha' & \beta' & \gamma' & \end{array} \right\} = \left(\dfrac{z-a}{z-b}\right)^\alpha \left(\dfrac{z-c}{z-b}\right)^\gamma P \left\{ \begin{array}{cccc} 0 & \infty & 1 & \\ 0 & \alpha+\beta+\gamma & 0 & \dfrac{(z-a)(c-b)}{(z-b)(c-a)} \\ \alpha'-\alpha & \alpha+\beta'+\gamma & \gamma'-\gamma & \end{array} \right\}$

The P function on the right hand side is Gauss' hypergeometric function (see **15.6.5**). If it is replaced by Kummer's 24 solutions **15.5.3** to **15.5.14** the complete set of 24 solutions for Riemann's differential equation **15.6.1** is obtained. The first of these solutions is for instance by **15.5.3** and **15.6.5**

15.6.12 $w = \left(\dfrac{z-a}{z-b}\right)^\alpha \left(\dfrac{z-c}{z-b}\right)^\gamma F\left[\alpha+\beta+\gamma, \alpha+\beta'+\gamma; 1+\alpha-\alpha'; \dfrac{(z-a)(c-b)}{(z-b)(c-a)}\right]$

15.7. Asymptotic Expansions

The behavior of $F(a, b; c; z)$ for large $|z|$ is described by the transformation formulas of **15.3.**

For fixed a, b, z and large $|c|$ one has [15.8]

15.7.1

$$F(a, b; c; z) = \sum_{n=0}^{m} \frac{(a)_n (b)_n}{(c)_n} \frac{z^n}{n!} + O(|c|^{-m-1})$$

For fixed a, c, z, ($c \neq 0$, -1, -2, . . . , $0 < |z| < 1$) and large $|b|$ one has [15.2]

15.7.2

$$F(a, b; c; z) = e^{-i\pi a}[\Gamma(c)/\Gamma(c-a)](bz)^{-a}[1+O(|bz|^{-1})]$$
$$+ [\Gamma(c)/\Gamma(a)] e^{bz}(bz)^{a-c}[1+O(|bz|^{-1})]$$
$$\left(-\frac{3\pi}{2} < \arg(bz) < \frac{1}{2}\pi\right)$$

15.7.3

$$F(a, b; c; z) = e^{i\pi a}[\Gamma(c)/\Gamma(c-a)](bz)^{-a}[1+O(|bz|^{-1})]$$
$$+ [\Gamma(c)/\Gamma(a)] e^{bz}(bz)^{a-c}[1+O(|bz|^{-1})]$$
$$\left(-\frac{1}{2}\pi < \arg(bz) < \frac{3}{2}\pi\right)$$

For the case when more than one of the parameters are large consult [15.2].

References

[15.1] P. Appell and J. Kampé de Fériet, Fonctions hypergéométriques et hypersphériques (Gauthiers-Villars, Paris, France, 1926).

[15.2] A. Erdélyi et al., Higher transcendental functions, vol. 1 (McGraw-Hill Book Co., Inc., New York, N.Y., 1953).

[15.3] E. Goursat, Ann. Sci. École Norm. Sup(2)**10**, 3–142(1881).

[15.4] E. Goursat, Propriétés genérales de l'équation d'Euler et de Gauss (Actualités scientifiques et industriélles 333, Paris, France, 1936).

[15.5] J. Kampé de Fériet, La fonction hypergéométrique (Gauthiers-Villars, Paris, France, 1937).

[15.6] F. Klein, Vorlesungen über die hypergeometrische Funktion (B. G. Teubner, Berlin, Germany, 1933).

[15.7] E. E. Kummer, Über die hypergeometrische Reihe, J. Reine Angew. Math. 15, 39–83, 127–172(1836).

[15.8] T. M. MacRobert, Proc. Edinburgh Math. Soc. 42, 84–88(1923).

[15.9] T. M. MacRobert, Functions of a complex variable, 4th ed. (Macmillan and Co., Ltd., London, England, 1954).

[15.10] E. G. C. Poole, Introduction to the theory of linear differential equations (Clarendon Press, Oxford, England, 1936).

[15.11] C. Snow, The hypergeometric and Legendre functions with applications to integral equations of potential theory, Applied Math. Series 19 (U.S. Government Printing Office, Washington, D.C., 1952).

[15.12] E. T. Whittaker and G. N. Watson, A course of modern analysis, 4th ed. (Cambridge Univ. Press, Cambridge, England, 1952).

16. Jacobian Elliptic Functions and Theta Functions

L. M. Milne-Thomson [1]

Contents

[1] University of Arizona. (Prepared under contract with the National Bureau of Standards.)

$\vartheta_s(\epsilon^\circ \backslash \alpha^\circ), \sqrt{\sec \alpha}\, \vartheta_c(\epsilon_1^\circ \backslash \alpha^\circ)$

$\vartheta_n(\epsilon^\circ \backslash \alpha^\circ), \sqrt{\sec \alpha}\, \vartheta_d(\epsilon_1^\circ \backslash \alpha^\circ)$

$\alpha = 0^\circ(5^\circ)85^\circ, \epsilon, \epsilon_1 = 0^\circ(5^\circ)90^\circ, \quad 9\text{–}10D$

$\dfrac{d}{du} \ln \vartheta_s(u) = f(\epsilon^\circ \backslash \alpha^\circ)$

$\dfrac{d}{du} \ln \vartheta_c(u) = -f(\epsilon_1^\circ \backslash \alpha^\circ)$

$\dfrac{d}{du} \ln \vartheta_n(u) = g(\epsilon^\circ \backslash \alpha^\circ)$

$\dfrac{d}{du} \ln \vartheta_d(u) = -g(\epsilon_1^\circ \backslash \alpha^\circ)$

$\alpha = 0^\circ(5^\circ)85^\circ, \epsilon, \epsilon_1 = 0^\circ(5^\circ)90^\circ, \quad 5\text{–}6D$

The author wishes to acknowledge his great indebtedness to his friend, the late Professor E. H. Neville, for invaluable assistance in reading and criticizing the manuscript. Professor Neville generously supplied material from his own work and was responsible for many improvements in matter and arrangement.

The author's best thanks are also due to David S. Liepman and Ruth Zucker for the preparation and checking of the tables and graphs.

16. Jacobian Elliptic Functions and Theta Functions

Mathematical Properties

Jacobian Elliptic Functions

16.1. Introduction

A doubly periodic meromorphic function is called an *elliptic function*.

Let m, m_1 be numbers such that

$$m + m_1 = 1.$$

We call m the *parameter*, m_1 the *complementary parameter*.

In what follows we shall assume that the parameter m is a real number. Without loss of generality we can then suppose that $0 \leq m \leq 1$ (see **16.10, 16.11**).

We define *quarter-periods* K and iK' by

16.1.1

$$K(m) = K = \int_0^{\pi/2} \frac{d\theta}{(1 - m \sin^2 \theta)^{1/2}},$$

$$iK'(m) = iK' = i \int_0^{\pi/2} \frac{d\theta}{(1 - m_1 \sin^2 \theta)^{1/2}}$$

so that K and K' are real numbers. K is called the real, iK' the imaginary quarter-period.

We note that

16.1.2 $\qquad K(m) = K'(m_1) = K'(1 - m).$

We also note that if any *one* of the numbers m, m_1, $K(m)$, $K'(m)$, $K'(m)/K(m)$ is given, all the rest are determined. Thus K and K' can not both be chosen arbitrarily.

In the Argand diagram denote the points 0, K, $K + iK'$, iK' by s, c, d, n respectively. These points are at the vertices of a rectangle. The translations of this rectangle by λK, $\mu iK'$, where λ, μ are given all integral values positive or negative, will lead to the lattice

.s	.c	.s	.c
.n	.d	.n	.d
.s	.c	.s	.c
.n	.d	.n	.d

the pattern being repeated indefinitely on all sides.

Let p, q be any two of the letters s, c, d, n. Then p, q determine in the lattice a minimum rectangle whose sides are of length K and K' and whose vertices s, c, d, n are in counterclockwise order.

Definition

The Jacobian elliptic function pq u is defined by the following three properties.

(i) pq u has a simple zero at p and a simple pole at q.

(ii) The step from p to q is a half-period of pq u. Those of the numbers $K, iK', K+iK'$ which differ from this step are only quarter-periods.

(iii) The coefficient of the leading term in the expansion of pq u in ascending powers of u about $u = 0$ is unity. With regard to (iii) the leading term is u, $1/u$, 1 according as $u = 0$ is a zero, a pole, or an ordinary point.

Thus the functions with a pole or zero at the origin (i.e., the functions in which one letter is s) are odd, and the others are even.

Should we wish to call explicit attention to the value of the parameter, we write pq $(u|m)$ instead of pq u.

The Jacobian elliptic functions can also be defined with respect to certain integrals. Thus if

16.1.3 $\qquad u = \int_0^{\varphi} \frac{d\theta}{(1 - m \sin^2 \theta)^{1/2}},$

the angle φ is called the *amplitude*

16.1.4 $\quad \varphi = \text{am } u$

and we define

16.1.5

$$\text{sn } u = \sin \varphi, \quad \text{cn } u = \cos \varphi,$$

$$\text{dn } u = (1 - m \sin^2 \varphi)^{1/2} = \Delta(\varphi).$$

Similarly all the functions pq u can be expressed in terms of φ. This second set of definitions, although seemingly different, is mathematically equivalent to the definition previously given in terms of a lattice. For further explanation of notations, including the interpretation, of such expressions as sn $(\varphi \backslash \alpha)$, cn $(u|m)$, dn (u, k), see **17.2**.

16.2. Classification of the Twelve Jacobian Elliptic Functions

According to Poles and Half-Periods

	Pole iK'	Pole $K+iK'$	Pole K	Pole 0	
Half period iK'	sn u	cd u	dc u	ns u	Periods $2iK'$, $4K+4iK'$, $4K$
Half period $K+iK'$	cn u	sd u	nc u	ds u	Periods $4iK'$, $2K+2iK'$, $4K$
Half period K	dn u	nd u	sc u	cs u	Periods $4iK'$, $4K+4iK'$, $2K$

The three functions in a vertical column are *copolar*.

The four functions in a horizontal line are *coperiodic*. Of the periods quoted in the last line of each row only two are independent.

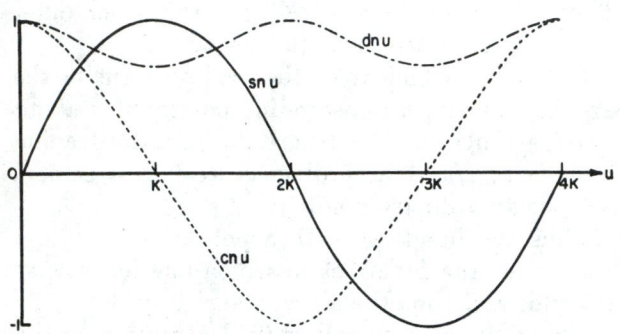

FIGURE 16.1.   *Jacobian elliptic functions*

sn u, cn u, dn u

$$m=\tfrac{1}{2}$$

The curve for cn $(u|\tfrac{1}{2})$ is the boundary between those which have an inflexion and those which have not.

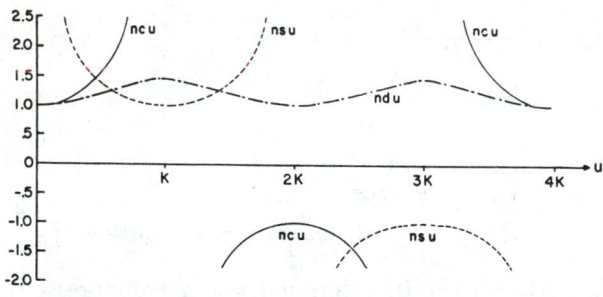

FIGURE 16.2.   *Jacobian elliptic functions*

ns u, nc u, nd u

$$m=\tfrac{1}{2}$$

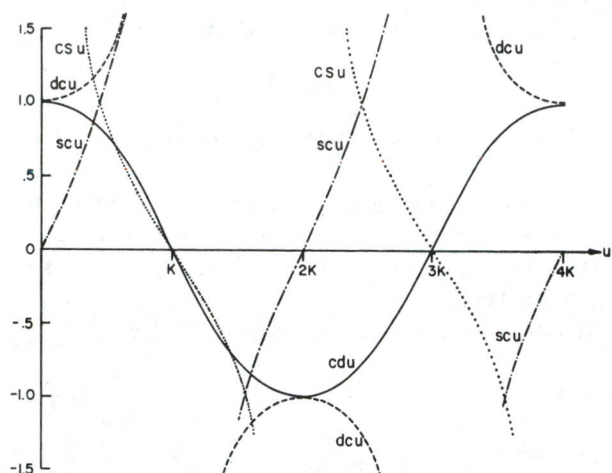

FIGURE 16.3.   *Jacobian elliptic functions*

sc u, cs u, cd u, dc u

$$m=\tfrac{1}{2}$$

16.3. Relation of the Jacobian Functions to the Copolar Trio sn u, cn u, dn u

16.3.1 $\text{cd } u=\dfrac{\text{cn } u}{\text{dn } u}$ $\text{dc } u=\dfrac{\text{dn } u}{\text{cn } u}$ $\text{ns } u=\dfrac{1}{\text{sn } u}$

16.3.2 $\text{sd } u=\dfrac{\text{sn } u}{\text{dn } u}$ $\text{nc } u=\dfrac{1}{\text{cn } u}$ $\text{ds } u=\dfrac{\text{dn } u}{\text{sn } u}$

16.3.3 $\text{nd } u=\dfrac{1}{\text{dn } u}$ $\text{sc } u=\dfrac{\text{sn } u}{\text{cn } u}$ $\text{cs } u=\dfrac{\text{cn } u}{\text{sn } u}$

And generally if p, q, r are any three of the letters s, c, d, n,

16.3.4 $$\text{pq } u=\dfrac{\text{pr } u}{\text{qr } u}$$

provided that when two letters are the same, e.g., pp u, the corresponding function is put equal to unity.

16.4. Calculation of the Jacobian Functions by Use of the Arithmetic-Geometric Mean (A.G.M.)

For the A.G.M. scale see 17.6.

To calculate sn $(u|m)$, cn $(u|m)$, and dn $(u|m)$ form the A.G.M. scale starting with

16.4.1 $$a_0=1,\ b_0=\sqrt{m_1},\ c_0=\sqrt{m},$$

terminating at the step N when c_N is negligible to the accuracy required. Find φ_N in degrees where

16.4.2 $$\varphi_N=2^N a_N u\,\frac{180°}{\pi}$$

and then compute successively $\varphi_{N-1},\ \varphi_{N-2},\ \ldots,$ $\varphi_1,\ \varphi_0$ from the recurrence relation

16.4.3 $$\sin\,(2\varphi_{n-1}-\varphi_n)=\frac{c_n}{a_n}\sin\,\varphi_n.$$

Then

16.4.4
$$\text{sn }(u|m)=\sin\,\varphi_0,\quad \text{cn }(u|m)=\cos\,\varphi_0$$
$$\text{dn }(u|m)=\frac{\cos\,\varphi_0}{\cos\,(\varphi_1-\varphi_0)}.$$

From these all the other functions can be determined.

16.5. Special Arguments

	u	sn u	cn u	dn u
16.5.1	0	0	1	1
16.5.2	$\frac{1}{2}K$	$\dfrac{1}{(1+m_1^{1/2})^{1/2}}$	$\dfrac{m_1^{1/4}}{(1+m_1^{1/2})^{1/2}}$	$m_1^{1/4}$
16.5.3	K	1	0	$m_1^{1/2}$
16.5.4	$\frac{1}{2}(iK')$	$im^{-1/4}$	$\dfrac{(1+m^{1/2})^{1/2}}{m^{1/4}}$	$(1+m^{1/2})^{1/2}$
16.5.5	$\frac{1}{2}(K+iK')$	$2^{-1/2}m^{-1/4}[(1+m^{1/2})^{1/2}+i(1-m^{1/2})^{1/2}]$	$\left(\dfrac{m_1}{4m}\right)^{1/4}(1-i)$	$\left(\dfrac{m_1}{4}\right)^{1/4}[(1+m_1^{1/2})^{1/2}-i(1-m_1^{1/2})^{1/2}]$
16.5.6	$K+\frac{1}{2}(iK')$	$m^{-1/4}$	$-i\left(\dfrac{1-m^{1/2}}{m^{1/2}}\right)^{1/2}$	$(1-m^{1/2})^{1/2}$
16.5.7	iK'	∞	∞	∞
16.5.8	$\frac{1}{2}K+iK'$	$(1-m_1^{1/2})^{-1/2}$	$-i\left(\dfrac{m_1^{1/2}}{1-m_1^{1/2}}\right)^{1/2}$	$-im_1^{1/4}$
16.5.9	$K+iK'$	$m^{-1/2}$	$-i(m_1/m)^{1/2}$	0

16.6. Jacobian Functions when $m=0$ or 1

		$m=0$	$m=1$
16.6.1	sn $(u|m)$	$\sin u$	$\tanh u$
16.6.2	cn $(u|m)$	$\cos u$	$\text{sech } u$
16.6.3	dn $(u|m)$	1	$\text{sech } u$
16.6.4	cd $(u|m)$	$\cos u$	1
16.6.5	sd $(u|m)$	$\sin u$	$\sinh u$
16.6.6	nd $(u|m)$	1	$\cosh u$
16.6.7	dc $(u|m)$	$\sec u$	1
16.6.8	nc $(u|m)$	$\sec u$	$\cosh u$
16.6.9	sc $(u|m)$	$\tan u$	$\sinh u$
16.6.10	ns $(u|m)$	$\csc u$	$\coth u$
16.6.11	ds $(u|m)$	$\csc u$	$\text{csch } u$
16.6.12	cs $(u|m)$	$\cot u$	$\text{csch } u$
16.6.13	am $(u|m)$	u	$\text{gd } u$

*See page II.

16.7. Principal Terms

When the elliptic function pq u is expanded in ascending powers of $(u-K_r)$, where K_r is one of 0, K, iK', $K+iK'$, the first term of the expansion is called the principal term and has one of the forms A, $B\times(u-K_r)$, $C\div(u-K_r)$ according as K_r is an ordinary point, a zero, or a pole of pq u. The following list gives these forms, where $\times$ means that the factor $(u-K_r)$ has to be supplied and $\div$ means that the divisor $(u-K_r)$ has to be supplied.

	$K_r=$	0	K	iK'	$K+iK'$
16.7.1	sn u	$1\times$	1	$m^{-1/2}\div$	$m^{-1/2}$
16.7.2	cn u	1	$-m_1^{1/2}\times$	$-im^{-1/2}\div$	$-i\left(\dfrac{m_1}{m}\right)^{1/2}$
16.7.3	dn u	1	$m_1^{1/2}$	$-i\div$	$im_1^{1/2}\times$
16.7.4	cd u	1	$-1\times$	$m^{-1/2}$	$-m^{-1/2}\div$
16.7.5	sd u	$1\times$	$m_1^{-1/2}$	$im^{-1/2}$	$-i\dfrac{1}{(mm_1)^{1/2}}\div$
16.7.6	nd u	1	$m_1^{-1/2}$	$i\times$	$-im_1^{-1/2}\div$
16.7.7	dc u	1	$-1\div$	$m^{1/2}$	$-m^{1/2}\times$
16.7.8	nc u	1	$-m_1^{-1/2}\div$	$im^{1/2}\times$	$i\left(\dfrac{m}{m_1}\right)^{1/2}$
16.7.9	sc u	$1\times$	$-m_1^{-1/2}\div$	i	$im_1^{-1/2}$
16.7.10	ns u	$1\div$	1	$m^{1/2}\times$	$m^{1/2}$
16.7.11	ds u	$1\div$	$m_1^{1/2}$	$-im^{1/2}$	$i(mm_1)^{1/2}\times$
16.7.12	cs u	$1\div$	$-m_1^{1/2}\times$	$-i$	$-im_1^{1/2}$

16.8. Change of Argument

		u	$-u$	$u+K$	$u-K$	$K-u$	$u+2K$	$u-2K$	$2K-u$	$u+iK'$	$u+2iK'$	$u+K+iK'$	$u+2K$ $+2iK'$
16.8.1	sn	sn u	$-$ sn u	cd u	$-$ cd u	cd u	$-$sn u	$-$sn u	sn u	$m^{-1/2}$ns u	sn u	$m^{-1/2}$dc u	$-$sn u
16.8.2	cn	cn u	cn u	$-m_1^{1/2}$sd u	$m_1^{1/2}$sd u	$m_1^{1/2}$sd u	$-$cn u	$-$cn u	$-$cn u	$-im^{-1/2}$ds u	$-$cn u	$-im_1^{1/2}m^{-1/2}$nc u	cn u
16.8.3	dn	dn u	dn u	$m_1^{1/2}$nd u	$m_1^{1/2}$nd u	$m_1^{1/2}$nd u	dn u	dn u	dn u	$-i$cs u	$-$dn u	$im_1^{1/2}$sc u	$-$dn u
16.8.4	cd	cd u	cd u	$-$sn u	sn u	sn u	$-$cd u	$-$cd u	$-$cd u	$m^{-1/2}$dc u	cd u	$-m^{-1/2}$ns u	$-$cd u
16.8.5	sd	sd u	$-$sd u	$m_1^{-1/2}$cn u	$-m_1^{-1/2}$cn u	$m_1^{-1/2}$cn u	$-$sd u	$-$sd u	sd u	$im^{-1/2}$nc u	$-$sd u	$-im_1^{-1/2}m^{-1/2}$ds u	sd u
16.8.6	nd	nd u	nd u	$m_1^{-1/2}$dn u	$m_1^{-1/2}$dn u	$m_1^{-1/2}$dn u	nd u	nd u	nd u	$i$sc u	$-$nd u	$-im_1^{-1/2}$cs u	$-$nd u
16.8.7	dc	dc u	dc u	$-$ns u	ns u	ns u	$-$dc u	$-$dc u	$-$dc u	$m^{1/2}$cd u	dc u	$-m^{1/2}$sn u	$-$dc u
16.8.8	nc	nc u	nc u	$-m_1^{-1/2}$ds u	$m_1^{-1/2}$ds u	$m_1^{-1/2}$ds u	$-$nc u	$-$nc u	$-$nc u	$im^{1/2}$sd u	$-$nc u	$im_1^{-1/2}m^{1/2}$cn u	nc u
16.8.9	sc	sc u	$-$sc u	$-m_1^{-1/2}$cs u	$-m_1^{-1/2}$cs u	$m_1^{-1/2}$cs u	sc u	sc u	$-$sc u	ind u	$-$sc u	$im_1^{-1/2}$dn u	$-$sc u
16.8.10	ns	ns u	$-$ns u	dc u	$-$dc u	dc u	$-$ns u	$-$ns u	ns u	$m^{1/2}$sn u	ns u	$m^{1/2}$cd u	$-$ns u
16.8.11	ds	ds u	$-$ds u	$m_1^{1/2}$nc u	$-m_1^{1/2}$nc u	$m_1^{1/2}$nc u	$-$ds u	$-$ds u	ds u	$-im^{1/2}$cn u	$-$ds u	$im_1^{1/2}m^{1/2}$sd u	ds u
16.8.12	cs	cs u	$-$cs u	$-m_1^{1/2}$sc u	$-m_1^{1/2}$sc u	$m_1^{1/2}$sc u	cs u	cs u	$-$cs u	$-i$dn u	$-$cs u	$-im_1^{1/2}$nd u	$-$cs u

16.9. Relations Between the Squares of the Functions

16.9.1 $\quad -\operatorname{dn}^2u+m_1=-m\ \operatorname{cn}^2u=m\ \operatorname{sn}^2u-m$

16.9.2 $\quad -m_1\operatorname{nd}^2u+m_1=-mm_1\operatorname{sd}^2u=m\ \operatorname{cd}^2u-m$

16.9.3 $\quad m_1\operatorname{sc}^2u+m_1=m_1\operatorname{nc}^2u=\operatorname{dc}^2u-m$

16.9.4 $\quad \operatorname{cs}^2u+m_1=\operatorname{ds}^2u=\operatorname{ns}^2u-m$

In using the above results remember that $m+m_1=1$.

If pq u, rt u are any two of the twelve functions, one entry expresses tq^2u in terms of pq^2u and another expresses qt^2u in terms of rt^2u. Since tq$^2u\cdot$ qt$^2u=1$, we can obtain from the table the bilinear relation between pq^2u and rt^2u. Thus for the functions cd u, sn u we have

16.9.5 $\quad \operatorname{nd}^2u=\dfrac{1-m\ \operatorname{cd}^2u}{m_1},\ \operatorname{dn}^2u=1-m\ \operatorname{sn}^2u$

and therefore

16.9.6 $\quad (1-m\ \operatorname{cd}^2u)(1-m\ \operatorname{sn}^2u)=m_1.$

16.10. Change of Parameter

Negative Parameter

If m is a positive number, let

16.10.1 $\quad \mu=\dfrac{m}{1+m},\ \mu_1=\dfrac{1}{1+m},\ v=\dfrac{u}{\mu_1^{\frac{1}{2}}}\qquad (0<\mu<1)$

16.10.2 $\quad \operatorname{sn}\ (u|-m)=\mu_1^{\frac{1}{2}}\operatorname{sd}\ (v|\mu)$

16.10.3 $\quad \operatorname{cn}\ (u|-m)=\operatorname{cd}\ (v|\mu)$

16.10.4 $\quad \operatorname{dn}\ (u|-m)=\operatorname{nd}\ (v|\mu).$

16.11. Reciprocal Parameter (Jacobi's Real Transformation)

16.11.1 $\quad m>0,\ \mu=m^{-1},\ v=um^{1/2}$

16.11.2 $\quad \operatorname{sn}\ (u|m)=\mu^{1/2}\operatorname{sn}\ (v|\mu)$

16.11.3 $\quad \operatorname{cn}\ (u|m)=\operatorname{dn}\ (v|\mu)$

16.11.4 $\quad \operatorname{dn}\ (u|m)=\operatorname{cn}\ (v|\mu)$

Here if $m>1$ then $m^{-1}=\mu<1$.

Thus elliptic functions whose parameter is real can be made to depend on elliptic functions whose parameter lies between 0 and 1.

16.12. Descending Landen Transformation (Gauss' Transformation)

To decrease the parameter, let

16.12.1 $\quad \mu=\left(\dfrac{1-m_1^{1/2}}{1+m_1^{1/2}}\right)^2,\ v=\dfrac{u}{1+\mu^{1/2}},$

then

16.12.2 $\quad \operatorname{sn}\ (u|\,m)=\dfrac{(1+\mu^{1/2})\ \operatorname{sn}\ (v|\mu)}{1+\mu^{1/2}\operatorname{sn}^2\ (v|\mu)}$

16.12.3 $\quad \operatorname{cn}\ (u|\,m)=\dfrac{\operatorname{cn}\ (v|\mu)\ \operatorname{dn}\ (v|\mu)}{1+\mu^{1/2}\operatorname{sn}^2\ (v|\mu)}$

16.12.4 $\quad \operatorname{dn}\ (u|\,m)=\dfrac{\operatorname{dn}^2(v|\mu)-(1-\mu^{1/2})}{(1+\mu^{1/2})-\operatorname{dn}^2(v|\mu)}.$

Note that successive applications can be made conveniently to find sn $(u|m)$ in terms of sn $(v|\mu)$ and dn $(u|m)$ in terms of dn $(v|\mu)$, but that the calculation of cn $(u|m)$ requires all three functions.

16.13. Approximation in Terms of Circular Functions

When the parameter m is so small that we may neglect m^2 and higher powers, we have the approximations

16.13.1
$$\operatorname{sn}\ (u|m)\approx\sin u-\tfrac{1}{4}\ m(u-\sin u\cos u)\cos u$$

16.13.2
$$\operatorname{cn}\ (u|m)\approx\cos u+\tfrac{1}{4}\ m(u-\sin u\cos u)\sin u$$

16.13.3 $\quad \operatorname{dn}\ (u|m)\approx 1-\tfrac{1}{2}\ m\ \sin^2 u$

16.13.4 $\quad \operatorname{am}\ (u|m)\approx u-\tfrac{1}{4}\ m(u-\sin u\cos u).$

One way of calculating the Jacobian functions is to use Landen's descending transformation to reduce the parameter sufficiently for the above formulae to become applicable. See also **16.14**.

16.14. Ascending Landen Transformation

To increase the parameter, let

16.14.1 $\quad \mu=\dfrac{4m^{1/2}}{(1+m^{1/2})^2},\ \mu_1=\left(\dfrac{1-m^{1/2}}{1+m^{1/2}}\right)^2,\ v=\dfrac{u}{1+\mu_1^{1/2}}$

16.14.2 $\quad \operatorname{sn}\ (u|\,m)=(1+\mu_1^{1/2})\dfrac{\operatorname{sn}\ (v|\mu)\ \operatorname{cn}\ (v|\mu)}{\operatorname{dn}\ (v|\mu)}$

16.14.3 $\quad \operatorname{cn}\ (u|\,m)=\dfrac{1+\mu_1^{1/2}}{\mu}\dfrac{\operatorname{dn}^2(v|\mu)-\mu_1^{1/2}}{\operatorname{dn}\ (v|\mu)}$

16.14.4 $\quad \operatorname{dn}\ (u|\,m)=\dfrac{1-\mu_1^{1/2}}{\mu}\dfrac{\operatorname{dn}^2\ (v|\mu)+\mu_1^{1/2}}{\operatorname{dn}\ (v|\mu)}$

Note that, when successive applications are to be made, it is simplest to calculate dn $(u|m)$ since this is expressed always in terms of the same function. The calculation of cn $(u|m)$ leads to that of dn $(v|\mu)$.

The calculation of sn $(u|m)$ necessitates the evaluation of all three functions.

16.15. Approximation in Terms of Hyperbolic Functions

When the parameter m is so close to unity that m_1^2 and higher powers of m_1 can be neglected we have the approximations

16.15.1

$$\operatorname{sn}(u|m) \approx \tanh u + \tfrac{1}{4} m_1 (\sinh u \cosh u - u) \operatorname{sech}^2 u$$

16.15.2

$$\operatorname{cn}(u|m) \approx \operatorname{sech} u$$
$$- \tfrac{1}{4} m_1 (\sinh u \cosh u - u) \tanh u \operatorname{sech} u$$

16.15.3

$$\operatorname{dn}(u|m) \approx \operatorname{sech} u$$
$$+ \tfrac{1}{4} m_1 (\sinh u \cosh u + u) \tanh u \operatorname{sech} u$$

16.15.4

$$\operatorname{am}(u|m) \approx \operatorname{gd} u + \tfrac{1}{4} m_1 (\sinh u \cosh u - u) \operatorname{sech} u.$$

Another way of calculating the Jacobian functions is to use Landen's ascending transformation to increase the parameter sufficiently for the above formulae to become applicable. See also **16.13**.

16.16. Derivatives

	Func-tion	Derivative	
16.16.1	sn u	cn u dn u	
16.16.2	cn u	$-$sn u dn u	Pole n
16.16.3	dn u	$-m$ sn u cn u	
16.16.4	cd u	$-m_1$ sd u nd u	
16.16.5	sd u	cd u nd u	Pole d
16.16.6	nd u	m sd u cd u	
16.16.7	dc u	m_1 sc u nc u	
16.16.8	nc u	sc u dc u	Pole c
16.16.9	sc u	dc u nc u	
16.16.10	ns u	$-$ds u cs u	
16.16.11	ds u	$-$cs u ns u	Pole s
16.16.12	cs u	$-$ns u ds u	

Note that the derivative is proportional to the product of the two copolar functions.

16.17. Addition Theorems

16.17.1 sn$(u+v)$

$$= \frac{\operatorname{sn} u \cdot \operatorname{cn} v \cdot \operatorname{dn} v + \operatorname{sn} v \cdot \operatorname{cn} u \cdot \operatorname{dn} u}{1 - m \operatorname{sn}^2 u \cdot \operatorname{sn}^2 v}$$

16.17.2 cn$(u+v)$

$$= \frac{\operatorname{cn} u \cdot \operatorname{cn} v - \operatorname{sn} u \cdot \operatorname{dn} u \cdot \operatorname{sn} v \cdot \operatorname{dn} v}{1 - m \operatorname{sn}^2 u \cdot \operatorname{sn}^2 v}$$

16.17.3 dn$(u+v) = \dfrac{\operatorname{dn} u \cdot \operatorname{dn} v - m \operatorname{sn} u \cdot \operatorname{cn} u \cdot \operatorname{sn} v \cdot \operatorname{cn} v}{1 - m \operatorname{sn}^2 u \cdot \operatorname{sn}^2 v}$

Addition theorems are derivable one from another and are expressible in a great variety of forms. Thus $\operatorname{ns}(u+v)$ comes from $1/\operatorname{sn}(u+v)$ in the form $(1 - m \operatorname{sn}^2 u \operatorname{sn}^2 v)/(\operatorname{sn} u \operatorname{cn} v \operatorname{dn} v + \operatorname{sn} v \operatorname{cn} u \operatorname{dn} u)$ from **16.17.1**.

Alternatively $\operatorname{ns}(u+v) = m^{1/2} \operatorname{sn}\{(iK'-u)-v\}$ which again from **16.17.1** yields the form $(\operatorname{ns} u \operatorname{cs} v \operatorname{ds} u - \operatorname{ns} v \operatorname{cs} u \operatorname{ds} v)/(\operatorname{ns}^2 u - \operatorname{ns}^2 v)$.

The function pq$(u+v)$ is a rational function of the four functions pq u, pq v, pq$'u$, pq$'v$.

16.18. Double Arguments

16.18.1 sn $2u$

$$= \frac{2\operatorname{sn} u \cdot \operatorname{cn} u \cdot \operatorname{dn} u}{1 - m\operatorname{sn}^4 u} = \frac{2\operatorname{sn} u \cdot \operatorname{cn} u \cdot \operatorname{dn} u}{\operatorname{cn}^2 u + \operatorname{sn}^2 u \cdot \operatorname{dn}^2 u}$$

16.18.2 cn $2u$

$$= \frac{\operatorname{cn}^2 u - \operatorname{sn}^2 u \cdot \operatorname{dn}^2 u}{1 - m\operatorname{sn}^4 u} = \frac{\operatorname{cn}^2 u - \operatorname{sn}^2 u \cdot \operatorname{dn}^2 u}{\operatorname{cn}^2 u + \operatorname{sn}^2 u \cdot \operatorname{dn}^2 u}$$

16.18.3 dn $2u$

$$= \frac{\operatorname{dn}^2 u - m\operatorname{sn}^2 u \cdot \operatorname{cn}^2 u}{1 - m\operatorname{sn}^4 u} = \frac{\operatorname{dn}^2 u + \operatorname{cn}^2 u(\operatorname{dn}^2 u - 1)}{\operatorname{dn}^2 u - \operatorname{cn}^2 u(\operatorname{dn}^2 u - 1)}$$

16.18.4 $\quad \dfrac{1 - \operatorname{cn} 2u}{1 + \operatorname{cn} 2u} = \dfrac{\operatorname{sn}^2 u \cdot \operatorname{dn}^2 u}{\operatorname{cn}^2 u}$

16.18.5 $\quad \dfrac{1 - \operatorname{dn} 2u}{1 + \operatorname{dn} 2u} = \dfrac{m\operatorname{sn}^2 u \cdot \operatorname{cn}^2 u}{\operatorname{dn}^2 u}$

16.19. Half Arguments

16.19.1 $\quad \operatorname{sn}^2 \tfrac{1}{2}u = \dfrac{1 - \operatorname{cn} u}{1 + \operatorname{dn} u}$

16.19.2 $\quad \operatorname{cn}^2 \tfrac{1}{2}u = \dfrac{\operatorname{dn} u + \operatorname{cn} u}{1 + \operatorname{dn} u}$

16.19.3 $\quad \operatorname{dn}^2 \tfrac{1}{2} u = \dfrac{m_1 + \operatorname{dn} u + m\operatorname{cn} u}{1 + \operatorname{dn} u}$

16.20. Jacobi's Imaginary Transformation

16.20.1 $\quad \operatorname{sn}(iu|m) = i\operatorname{sc}(u|m_1)$

16.20.2 $\quad \operatorname{cn}(iu|m) = \operatorname{nc}(u|m_1)$

16.20.3 $\quad \operatorname{dn}(iu|m) = \operatorname{dc}(u|m_1)$

16.21. Complex Arguments

With the abbreviations

16.21.1

$$s=\mathrm{sn}(x|m),\ c=\mathrm{cn}(x|m),\ d=\mathrm{dn}(x|m),\ s_1=\mathrm{sn}(y|m_1),$$
$$c_1=\mathrm{cn}(y|m_1),\ d_1=\mathrm{dn}(y|m_1)$$

16.21.2 $\quad \mathrm{sn}(x+iy|m)=\dfrac{s\cdot d_1+ic\cdot d\cdot s_1\cdot c_1}{c_1^2+ms^2\cdot s_1^2}$

16.21.3 $\quad \mathrm{cn}(x+iy|m)=\dfrac{c\cdot c_1-is\cdot d\cdot s_1\cdot d_1}{c_1^2+ms^2\cdot s_1^2}$

16.21.4 $\quad \mathrm{dn}(x+iy|m)=\dfrac{d\cdot c_1\cdot d_1-ims\cdot c\cdot s_1}{c_1^2+ms^2\cdot s_1^2}$

16.22. Leading Terms of the Series in Ascending Powers of u

16.22.1

$$\mathrm{sn}(u|m)=u-(1+m)\frac{u^3}{3!}+(1+14m+m^2)\frac{u^5}{5!}$$
$$-(1+135m+135m^2+m^3)\frac{u^7}{7!}+\dots$$

16.22.2

$$\mathrm{cn}(u|m)=1-\frac{u^2}{2!}+(1+4m)\frac{u^4}{4!}$$
$$-(1+44m+16m^2)\frac{u^6}{6!}+\dots$$

16.22.3

$$\mathrm{dn}(u|m)=1-m\frac{u^2}{2!}+m(4+m)\frac{u^4}{4!}$$
$$-m(16+44m+m^2)\frac{u^6}{6!}+\dots$$

No formulae are known for the general coefficients in these series.

16.23. Series Expansions in Terms of the Nome $q=e^{-\pi K'/K}$ and the Argument $v=\pi u/(2K)$

16.23.1 $\quad \mathrm{sn}\,(u|m)=\dfrac{2\pi}{m^{1/2}K}\displaystyle\sum_{n=0}^{\infty}\dfrac{q^{n+1/2}}{1-q^{2n+1}}\sin\,(2n+1)v$

16.23.2 $\quad \mathrm{cn}\,(u|m)=\dfrac{2\pi}{m^{1/2}K}\displaystyle\sum_{n=0}^{\infty}\dfrac{q^{n+1/2}}{1+q^{2n+1}}\cos\,(2n+1)v$

16.23.3 $\quad \mathrm{dn}\,(u|m)=\dfrac{\pi}{2K}+\dfrac{2\pi}{K}\displaystyle\sum_{n=1}^{\infty}\dfrac{q^n}{1+q^{2n}}\cos 2nv$

16.23.4

$$\mathrm{cd}\,(u|m)=\frac{2\pi}{m^{1/2}K}\sum_{n=0}^{\infty}\frac{(-1)^n q^{n+1/2}}{1-q^{2n+1}}\cos\,(2n+1)v$$

16.23.5

$$\mathrm{sd}\,(u|m)=\frac{2\pi}{(mm_1)^{1/2}K}\sum_{n=0}^{\infty}(-1)^n\frac{q^{n+1/2}}{1+q^{2n+1}}\sin\,(2n+1)v$$

16.23.6

$$\mathrm{nd}\,(u|m)=\frac{\pi}{2m_1^{1/2}K}+\frac{2\pi}{m_1^{1/2}K}\sum_{n=1}^{\infty}(-1)^n\frac{q^n}{1+q^{2n}}\cos 2nv$$

16.23.7

$$\mathrm{dc}\,(u|m)=\frac{\pi}{2K}\sec v$$
$$+\frac{2\pi}{K}\sum_{n=0}^{\infty}(-1)^n\frac{q^{2n+1}}{1-q^{2n+1}}\cos\,(2n+1)v$$

16.23.8

$$\mathrm{nc}\,(u|m)=\frac{\pi}{2m_1^{1/2}K}\sec v$$
$$-\frac{2\pi}{m_1^{1/2}K}\sum_{n=0}^{\infty}(-1)^n\frac{q^{2n+1}}{1+q^{2n+1}}\cos\,(2n+1)v$$

16.23.9

$$\mathrm{sc}\,(u|m)=\frac{\pi}{2m_1^{1/2}K}\tan v$$
$$+\frac{2\pi}{m_1^{1/2}K}\sum_{n=1}^{\infty}(-1)^n\frac{q^{2n}}{1+q^{2n}}\sin 2nv$$

16.23.10

$$\mathrm{ns}\,(u|m)=\frac{\pi}{2K}\csc v-\frac{2\pi}{K}\sum_{n=0}^{\infty}\frac{q^{2n+1}}{1-q^{2n+1}}\sin\,(2n+1)v$$

16.23.11

$$\mathrm{ds}\,(u|m)=\frac{\pi}{2K}\csc v-\frac{2\pi}{K}\sum_{n=0}^{\infty}\frac{q^{2n+1}}{1+q^{2n+1}}\sin\,(2n+1)v$$

16.23.12

$$\mathrm{cs}\,(u|m)=\frac{\pi}{2K}\cot v-\frac{2\pi}{K}\sum_{n=1}^{\infty}\frac{q^{2n}}{1+q^{2n}}\sin 2nv$$

16.24. Integrals of the Twelve Jacobian Elliptic Functions

16.24.1 $\quad \int \mathrm{sn}\,u\,du=m^{-1/2}\ln\,(\mathrm{dn}\,u-m^{1/2}\mathrm{cn}\,u)$

16.24.2 $\quad \int \mathrm{cn}\,u\,du=m^{-1/2}\arccos\,(\mathrm{dn}\,u)$

16.24.3 $\quad \int \mathrm{dn}\,u\,du=\arcsin\,(\mathrm{sn}\,u)$

16.24.4 $\quad \int \mathrm{cd}\,u\,du=m^{-1/2}\ln\,(\mathrm{nd}\,u+m^{1/2}\mathrm{sd}\,u)$

16.24.5 $\quad \int \mathrm{sd}\,u\,du=(mm_1)^{-1/2}\arcsin\,(-m^{1/2}\mathrm{cd}\,u)$

16.24.6 $\quad \int \mathrm{nd}\,u\,du=m_1^{-1/2}\arccos\,(\mathrm{cd}\,u)$

16.24.7 $\quad \int \mathrm{dc}\,u\,du=\ln\,(\mathrm{nc}\,u+\mathrm{sc}\,u)$

16.24.8 $\quad \int \mathrm{nc}\,u\,du=m_1^{-1/2}\ln\,(\mathrm{dc}\,u+m_1^{1/2}\mathrm{sc}\,u)$

16.24.9 $\quad \int \mathrm{sc}\,u\,du=m_1^{-1/2}\ln\,(\mathrm{dc}\,u+m_1^{1/2}\mathrm{nc}\,u)$

16.24.10 $\quad \int \mathrm{ns}\,u\,du=\ln\,(\mathrm{ds}\,u-\mathrm{cs}\,u)$

16.24.11 $\quad \int \mathrm{ds}\,u\,du=\ln\,(\mathrm{ns}\,u-\mathrm{cs}\,u)$

16.24.12 $\quad \int \mathrm{cs}\,u\,du=\ln\,(\mathrm{ns}\,u-\mathrm{ds}\,u)$

In numerical use of the above table certain restrictions must be put on u in order to keep the arguments of the logarithms positive and to avoid trouble with many-valued inverse circular functions.

16.25. Notation for the Integrals of the Squares of the Twelve Jacobian Elliptic Functions

16.25.1　　$\mathrm{Pq}\,u=\displaystyle\int_0^u \mathrm{pq}^2 t\,dt$ when $\mathrm{q}\neq \mathrm{s}$

16.25.2　　$\mathrm{Ps}\,u=\displaystyle\int_0^u\left(\mathrm{pq}^2 t-\frac{1}{t^2}\right)dt-\frac{1}{u}$

Examples

$$\mathrm{Cd}\,u=\int_0^u \mathrm{cd}^2 t\,dt,\ \mathrm{Ns}\,u=\int_0^u\left(\mathrm{ns}^2 t-\frac{1}{t^2}\right)dt-\frac{1}{u}$$

16.26. Integrals in Terms of the Elliptic Integral of the Second Kind (see 17.4)

16.26.1　　　$m\mathrm{Sn}\,u=-E(u)+u$

16.26.2　　$m\mathrm{Cn}\,u=E(u)-m_1 u$　　　　　Pole n

16.26.3　　　　$\mathrm{Dn}\,u=E(u)$

16.26.4　$m\mathrm{Cd}\,u=-E(u)+u+m\mathrm{sn}\,u\,\mathrm{cd}\,u$

16.26.5

$mm_1\mathrm{Sd}\,u=E(u)-m_1 u-m\mathrm{sn}\,u\,\mathrm{cd}\,u$　　　Pole d

16.26.6　$m_1\mathrm{Nd}\,u=E(u)-m\mathrm{sn}\,u\,\mathrm{cd}\,u$

16.26.7　$\mathrm{Dc}\,u=-E(u)+u+\mathrm{sn}\,u\,\mathrm{dc}\,u$

16.26.8

$m_1\mathrm{Nc}\,u=-E(u)+m_1 u+\mathrm{sn}\,u\,\mathrm{dc}\,u$　　　Pole c

16.26.9　$m_1\mathrm{Sc}\,u=-E(u)+\mathrm{sn}\,u\,\mathrm{dc}\,u$

16.26.10　$\mathrm{Ns}\,u=-E(u)+u-\mathrm{cn}\,u\,\mathrm{ds}\,u$

16.26.11

$\mathrm{Ds}\,u=-E(u)+m_1 u-\mathrm{cn}\,u\,\mathrm{ds}\,u$　　　Pole s

16.26.12　$\mathrm{Cs}\,u=-E(u)-\mathrm{cn}\,u\,\mathrm{ds}\,u$

All the above may be expressed in terms of Jacobi's zeta function (see **17.4.27**).

$$Z(u)=E(u)-\frac{E}{K}u,\text{ where } E=E(K)$$

16.27. Theta Functions; Expansions in Terms of the Nome q

16.27.1

$$\vartheta_1(z,\,q)=\vartheta_1(z)=2q^{1/4}\sum_{n=0}^{\infty}(-1)^n q^{n(n+1)}\sin(2n+1)z$$

16.27.2

$$\vartheta_2(z,q)=\vartheta_2(z)=2q^{1/4}\sum_{n=0}^{\infty}q^{n(n+1)}\cos(2n+1)z$$

16.27.3　$\vartheta_3(z,\,q)=\vartheta_3(z)=1+2\displaystyle\sum_{n=1}^{\infty}q^{n^2}\cos 2nz$

16.27.4

$$\vartheta_4(z,\,q)=\vartheta_4(z)=1+2\sum_{n=1}^{\infty}(-1)^n q^{n^2}\cos 2nz$$

Theta functions are important because every one of the Jacobian elliptic functions can be expressed as the ratio of two theta functions. See **16.36**.

The notation shows these functions as depending on the variable z and the nome q, $|q|<1$. In this case, here and elsewhere, the convergence is not dependent on the trigonometrical terms. In their relation to the Jacobian elliptic functions, we note that the nome q is given by

$$q=e^{-\pi K'/K},$$

where K and iK' are the quarter periods. Since $q=q(m)$ is determined when the parameter m is given, we can also regard the theta functions as dependent upon m and then we write

$$\vartheta_a(z,\,q)=\vartheta_a(z|m),\ a=1,\,2,\,3,\,4$$

but when no ambiguity is to be feared, we write $\vartheta_a(z)$ simply.

The above notations are those given in Modern Analysis [16.6].

There is a bewildering variety of notations, for example the function $\vartheta_4(z)$ above is sometimes denoted by $\vartheta_0(z)$ or $\vartheta(z)$; see the table given in Modern Analysis [16.6]. Further the argument $u=2Kz/\pi$ is frequently used so that in consulting books caution should be exercised.

16.28. Relations Between the Squares of the Theta Functions

16.28.1　$\vartheta_1^2(z)\vartheta_4^2(0)=\vartheta_3^2(z)\vartheta_2^2(0)-\vartheta_2^2(z)\vartheta_3^2(0)$

16.28.2　$\vartheta_2^2(z)\vartheta_4^2(0)=\vartheta_4^2(z)\vartheta_2^2(0)-\vartheta_1^2(z)\vartheta_3^2(0)$

16.28.3　$\vartheta_3^2(z)\vartheta_4^2(0)=\vartheta_4^2(z)\vartheta_3^2(0)-\vartheta_1^2(z)\vartheta_2^2(0)$

16.28.4　$\vartheta_4^2(z)\vartheta_4^2(0)=\vartheta_3^2(z)\vartheta_3^2(0)-\vartheta_2^2(z)\vartheta_2^2(0)$

16.28.5　　　$\vartheta_2^4(0)+\vartheta_4^4(0)=\vartheta_3^4(0)$

Note also the important relation

16.28.6　$\vartheta_1'(0)=\vartheta_2(0)\vartheta_3(0)\vartheta_4(0)$ or $\vartheta_1'=\vartheta_2\vartheta_3\vartheta_4$

16.29. Logarithmic Derivatives of the Theta Functions

16.29.1　$\dfrac{\vartheta_1'(u)}{\vartheta_1(u)}=\cot u+4\displaystyle\sum_{n=1}^{\infty}\frac{q^{2n}}{1-q^{2n}}\sin 2nu$

16.29.2

$$\frac{\vartheta_2'(u)}{\vartheta_2(u)} = -\tan u + 4 \sum_{n=1}^{\infty} (-1)^n \frac{q^{2n}}{1-q^{2n}} \sin 2nu$$

16.29.3 $\quad \dfrac{\vartheta_3'(u)}{\vartheta_3(u)} = 4 \sum_{n=1}^{\infty} (-1)^n \dfrac{q^n}{1-q^{2n}} \sin 2nu$

16.29.4 $\quad \dfrac{\vartheta_4'(u)}{\vartheta_4(u)} = 4 \sum_{n=1}^{\infty} \dfrac{q^n}{1-q^{2n}} \sin 2nu$

16.30. Logarithms of Theta Functions of Sum and Difference

16.30.1

$$\ln \frac{\vartheta_1(\alpha+\beta)}{\vartheta_1(\alpha-\beta)} = \ln \frac{\sin(\alpha+\beta)}{\sin(\alpha-\beta)}$$
$$+ 4 \sum_{n=1}^{\infty} \frac{1}{n} \frac{q^{2n}}{1-q^{2n}} \sin 2n\alpha \sin 2n\beta$$

16.30.2

$$\ln \frac{\vartheta_2(\alpha+\beta)}{\vartheta_2(\alpha-\beta)} = \ln \frac{\cos(\alpha+\beta)}{\cos(\alpha-\beta)}$$
$$+ 4 \sum_{n=1}^{\infty} \frac{(-1)^n}{n} \frac{q^{2n}}{1-q^{2n}} \sin 2n\alpha \sin 2n\beta$$

16.30.3

$$\ln \frac{\vartheta_3(\alpha+\beta)}{\vartheta_3(\alpha-\beta)} = 4 \sum_{n=1}^{\infty} \frac{(-1)^n}{n} \frac{q^n}{1-q^{2n}} \sin 2n\alpha \sin 2n\beta$$

16.30.4

$$\ln \frac{\vartheta_4(\alpha+\beta)}{\vartheta_4(\alpha-\beta)} = 4 \sum_{n=1}^{\infty} \frac{1}{n} \frac{q^n}{1-q^{2n}} \sin 2n\alpha \sin 2n\beta$$

The corresponding expressions when $\beta = i\gamma$ are easily deduced by use of the formulae 4.3.55 and 4.3.56.

16.31. Jacobi's Notation for Theta Functions

16.31.1 $\quad \Theta(u|m) = \Theta(u) = \vartheta_4(v), \qquad v = \dfrac{\pi u}{2K}$

16.31.2 $\quad \Theta_1(u|m) = \Theta_1(u) = \vartheta_3(v) = \Theta(u+K)$

16.31.3 $\quad \mathrm{H}(u|m) = \mathrm{H}(u) = \vartheta_1(v)$

16.31.4 $\quad \mathrm{H}_1(u|m) = \mathrm{H}_1(u) = \vartheta_2(v) = \mathrm{H}(u+K)$

16.32. Calculation of Jacobi's Theta Function $\Theta(u|m)$ by Use of the Arithmetic-Geometric Mean

Form the A.G.M. scale starting with

16.32.1 $\quad a_0 = 1, \; b_0 = \sqrt{m_1}, \; c_0 = \sqrt{m}$

terminating with the Nth step when c_N is negligible to the accuracy required. Find φ_N in degrees, where

16.32.2 $\quad \varphi_N = 2^N a_N u \dfrac{180°}{\pi}$

and then compute successively $\varphi_{N-1}, \; \varphi_{N-2}, \; \cdots, \varphi_1, \; \varphi_0$ from the recurrence relation

16.32.3 $\quad \sin(2\varphi_{n-1} - \varphi_n) = \dfrac{c_n}{a_n} \sin \varphi_n.$

Then

16.32.4

$$\ln \Theta(u|m) = \frac{1}{2} \ln \frac{2 m_1^{1/2} K(m)}{\pi} + \frac{1}{2} \ln \frac{\cos(\varphi_1 - \varphi_0)}{\cos \varphi_0}$$
$$+ \frac{1}{4} \ln \sec(2\varphi_0 - \varphi_1) + \frac{1}{8} \ln \sec(2\varphi_1 - \varphi_2) + \cdots$$
$$+ \frac{1}{2^{N+1}} \ln \sec(2\varphi_{N-1} - \varphi_N)$$

16.33. Addition of Quarter-Periods to Jacobi's Eta and Theta Functions

u	$-u$	$u+K$	$u+2K$	$u+iK'$	$u+2iK'$	$u+K+iK'$	$u+2K+2iK'$
16.33.1 $\mathrm{H}(u)$	$-\mathrm{H}(u)$	$\mathrm{H}_1(u)$	$-\mathrm{H}(u)$	$iM(u)\Theta(u)$	$-N(u)\mathrm{H}(u)$	$M(u)\Theta_1(u)$	$N(u)\mathrm{H}(u)$
16.33.2 $\mathrm{H}_1(u)$	$\mathrm{H}_1(u)$	$-\mathrm{H}(u)$	$-\mathrm{H}_1(u)$	$M(u)\Theta_1(u)$	$N(u)\mathrm{H}_1(u)$	$-iM(u)\Theta(u)$	$-N(u)\mathrm{H}_1(u)$
16.33.3 $\Theta_1(u)$	$\Theta_1(u)$	$\Theta(u)$	$\Theta_1(u)$	$M(u)\mathrm{H}_1(u)$	$N(u)\Theta_1(u)$	$iM(u)\mathrm{H}(u)$	$N(u)\Theta_1(u)$
16.33.4 $\Theta(u)$	$\Theta(u)$	$\Theta_1(u)$	$\Theta(u)$	$iM(u)\mathrm{H}(u)$	$-N(u)\Theta(u)$	$M(u)\mathrm{H}_1(u)$	$-N(u)\Theta(u)$

where

$$M(u) = \left[\exp\left(-\frac{\pi i u}{2K} \right) \right] q^{-\frac{1}{4}},$$

$$N(u) = \left[\exp\left(-\frac{\pi i u}{K} \right) \right] q^{-1}$$

$\mathrm{H}(u)$ and $\mathrm{H}_1(u)$ have the period $4K$. $\Theta(u)$ and $\Theta_1(u)$ have the period $2K$.

$2iK'$ is a quasi-period for all four functions, that is to say, increase of the argument by $2iK'$ multiplies the function by a factor.

16.34. Relation of Jacobi's Zeta Function to the Theta Functions

$$Z(u) = \frac{\partial}{\partial u} \ln \Theta(u)$$

16.34.1 $\quad Z(u) = \frac{\pi}{2K} \frac{\vartheta_1'\left(\frac{\pi u}{2K}\right)}{\vartheta_1\left(\frac{\pi u}{2K}\right)} - \frac{\text{cn } u \text{ dn } u}{\text{sn } u}$

16.34.2 $\quad = \frac{\pi}{2K} \frac{\vartheta_2'\left(\frac{\pi u}{2K}\right)}{\vartheta_2\left(\frac{\pi u}{2K}\right)} + \frac{\text{dn } u \text{ sn } u}{\text{cn } u}$

16.34.3 $\quad = \frac{\pi}{2K} \frac{\vartheta_3'\left(\frac{\pi u}{2K}\right)}{\vartheta_3\left(\frac{\pi u}{2K}\right)} - m \frac{\text{sn } u \text{ cn } u}{\text{dn } u}$

16.34.4 $\quad = \frac{\pi}{2K} \frac{\vartheta_4'\left(\frac{\pi u}{2K}\right)}{\vartheta_4\left(\frac{\pi u}{2K}\right)}$

16.35. Calculation of Jacobi's Zeta Function $Z(u|m)$ by Use of the Arithmetic-Geometric Mean

Form the A.G.M. scale **17.6** starting with

16.35.1 $\qquad a_0 = 1, \ b_0 = \sqrt{m_1}, \ c_0 = \sqrt{m}$

terminating at the Nth step when c_N is negligible to the accuracy required. Find φ_N in degrees where

16.35.2 $\qquad \varphi_N = 2^N a_N u \frac{180°}{\pi}$

and then compute successively $\varphi_{N-1}, \ \varphi_{N-2}, \ . . .,$ $\varphi_1, \ \varphi_0$ from the recurrence relation

16.35.3 $\qquad \sin (2\varphi_{n-1} - \varphi_n) = \frac{c_n}{a_n} \sin \varphi_n.$

Then

16.35.4

$$Z(u|m) = c_1 \sin \varphi_1 + c_2 \sin \varphi_2 + . . . + c_N \sin \varphi_N.$$

16.36. Neville's Notation for Theta Functions

These functions are defined in terms of Jacobi's theta functions of **16.31** by

16.36.1 $\qquad \vartheta_s(u) = \frac{H(u)}{H'(0)}, \ \vartheta_c(u) = \frac{H(u+K)}{H(K)}$

16.36.2 $\qquad \vartheta_d(u) = \frac{\Theta(u+K)}{\Theta(K)}, \ \vartheta_n(u) = \frac{\Theta(u)}{\Theta(0)}.$

If λ, μ are any integers positive, negative, or zero the points $u_0 + 2\lambda K + 2\mu iK'$ are said to be *congruent to* u_0.

$\vartheta_s(u)$ has zeros at the points congruent to 0
$\vartheta_c(u)$ has zeros at the points congruent to K
$\vartheta_n(u)$ has zeros at the points congruent to iK'
$\vartheta_d(u)$ has zeros at the points congruent to $K+iK'$

Thus the suffix secures that the function $\vartheta_p(u)$ has zeros at the points marked p in the introductory diagram in **16.1.2**, and the constant by which Jacobi's function is divided secures that the leading coefficient of $\vartheta_p(u)$ at the origin is unity. Therefore the functions have the fundamentally important property that if p, q are any two of the letters s, c, n, d, the Jacobian elliptic function pq u is given by

16.36.3 $\qquad \text{pq } u = \frac{\vartheta_p(u)}{\vartheta_q(u)}.$

These functions also have the property

16.36.4 $\qquad m_1^{-1/4} \vartheta_c(K-u) = \vartheta_s(u)$

16.36.5 $\qquad m_1^{-1/4} \vartheta_d(K-u) = \vartheta_n(u),$

for complementary arguments u and $K-u$.

In terms of the theta functions defined in **16.27**, let $v = \pi u/(2K)$, then

16.36.6 $\qquad \vartheta_s(u) = \frac{2K\vartheta_1(v)}{\vartheta_1'(0)}, \ \vartheta_c(u) = \frac{\vartheta_2(v)}{\vartheta_2(0)}$

16.36.7 $\qquad \vartheta_d(u) = \frac{\vartheta_3(v)}{\vartheta_3(0)}, \ \vartheta_n(u) = \frac{\vartheta_4(v)}{\vartheta_4(0)}.$

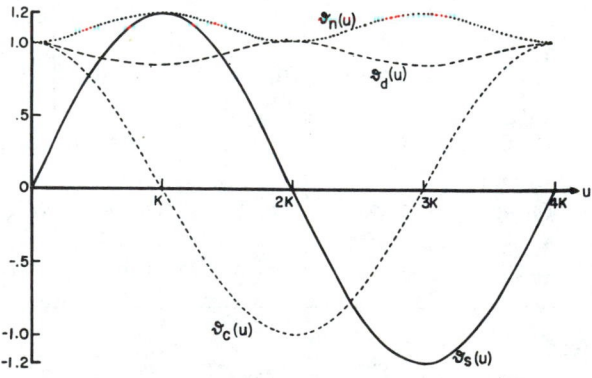

FIGURE 16.4. *Neville's theta functions*
$\vartheta_s(u), \ \vartheta_c(u), \ \vartheta_d(u), \ \vartheta_n(u)$
$m = \frac{1}{2}$

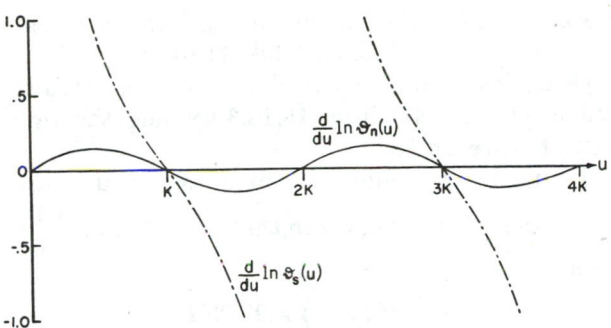

FIGURE 16.5. *Logarithmic derivatives of theta functions*

$$\frac{d}{du}\ln\vartheta_s(u), \frac{d}{du}\ln\vartheta_n(u)$$

$$m=\frac{1}{2}$$

16.37. Expression as Infinite Products

$$q=q(m),\ v=\pi u/(2K)$$

16.37.1
$$\vartheta_s(u)=\left(\frac{16q}{mm_1}\right)^{1/6}\sin v\prod_{n=1}^{\infty}(1-2q^{2n}\cos 2v+q^{4n})$$

16.37.2
$$\vartheta_c(u)=\left(\frac{16qm_1^{1/2}}{m}\right)^{1/6}\cos v\prod_{n=1}^{\infty}(1+2q^{2n}\cos 2v+q^{4n})$$

16.37.3
$$\vartheta_d(u)=\left(\frac{mm_1}{16q}\right)^{1/12}\prod_{n=1}^{\infty}(1+2q^{2n-1}\cos 2v+q^{4n-2})$$

16.37.4
$$\vartheta_n(u)=\left(\frac{m}{16qm_1^2}\right)^{1/12}\prod_{n=1}^{\infty}(1-2q^{2n-1}\cos 2v+q^{4n-2})$$

16.38. Expression as Infinite Series

Let $v=\pi u/(2K)$

16.38.1
$$\vartheta_s(u)=\left[\frac{2\pi q^{1/2}}{m^{1/2}m_1^{1/2}K}\right]^{1/2}\sum_{n=0}^{\infty}(-1)^n q^{n(n+1)}\sin(2n+1)v$$

16.38.2
$$\vartheta_c(u)=\left[\frac{2\pi q^{1/2}}{m^{1/2}K}\right]^{1/2}\sum_{n=0}^{\infty}q^{n(n+1)}\cos(2n+1)v$$

16.38.3
$$\vartheta_d(u)=\left[\frac{\pi}{2K}\right]^{1/2}\{1+2\sum_{n=1}^{\infty}q^{n^2}\cos 2nv\}$$

16.38.4
$$\vartheta_n(u)=\left[\frac{\pi}{2m_1^{1/2}K}\right]^{1/2}\{1+2\sum_{n=1}^{\infty}(-1)^n q^{n^2}\cos 2nv\}$$

16.38.5 $\quad(2K/\pi)^{1/2}=1+2q+2q^4+2q^9+\ldots=\vartheta_3(0,q)$

16.38.6
$$(2K'/\pi)^{1/2}=1+2q_1+2q_1^4+2q_1^9+\ldots=\vartheta_3(0,q_1)$$

16.38.7
$$(2m^{1/2}K/\pi)^{1/2}=2q^{1/4}(1+q^2+q^6+q^{12}+q^{20}+\ldots)$$
$$=\vartheta_2(0,q)$$

16.38.8
$$(2m_1^{1/2}K/\pi)^{1/2}=1-2q+2q^4-2q^9+\ldots=\vartheta_4(0,q).$$

Numerical Methods

16.39. Use and Extension of the Tables

Example 1. Calculate nc $(1.99650|.64)$ to 4S. From **Table 17.1**, $1.99650=K+.001$. From the table of principal terms

$$\text{nc } u=-m_1^{-1/2}/(u-K)+\ldots$$

$$\text{nc }(K+.001|.64)=\frac{-(.36)^{-1/2}}{.001}+\ldots$$
$$=-\frac{10000}{6}+\ldots$$
$$=-1667+\ldots$$

and since the next term is of order .001 this value -1667 is correct to at least 4S.

Example 2. Use the descending Landen transformation to calculate dn $(.20|.19)$ to 6D.
Here $m=.19$, $m_1^{1/2}=.9$ and so from **16.12.1**

$$\mu=\left(\frac{1}{19}\right)^2,\ 1+\mu^{1/2}=\frac{20}{19},\ v=.19.$$

Also

$$\mu^2=\left(\frac{1}{19}\right)^4=10^{-6}\times 7.67$$

which is negligible.
From **16.12.4**

$$\text{dn}(.20|.19)=\frac{\text{dn}^2\left[.19\left|\left(\frac{1}{19}\right)^2\right.\right]-\left(1-\frac{1}{19}\right)}{\left(1+\frac{1}{19}\right)-\text{dn}^2\left[.19\left|\left(\frac{1}{19}\right)^2\right.\right]}.$$

Now from **16.13.3**

$$\text{dn}\left[.19\left|\left(\frac{1}{19}\right)^2\right.\right]=.999951$$

whence dn $(.20|.19)=.996253$.

Example 3. Use the ascending Landen transformation to calculate dn $(.20|.81)$ to 5D.
From **16.14.1**

$$\mu=\frac{4(.9)}{(1.9)^2}=\frac{360}{361},\ \mu_1=\left(\frac{1}{19}\right)^2$$

$$1+\mu_1^{1/2}=\frac{20}{19},\quad v=\frac{19}{20}\times.20=.19,$$

μ_1^2 is negligible to 4D. **Thus**

$$\mathrm{dn}\,(.20|.81)=\frac{19}{20}\times\frac{\mathrm{dn}^2\left(.19\left|\frac{360}{361}\right.\right)+\frac{1}{19}}{\mathrm{dn}\left(.19\left|\frac{360}{361}\right.\right)}.$$

From **16.15.3**

$$\mathrm{dn}\left(.19\left|\frac{360}{361}\right.\right)=\mathrm{sech}\,(.19)+\frac{1}{4}\times\frac{1}{361}\,\tanh .19\,\mathrm{sech}\,.19$$

$$[\sinh .19\cosh .19+.19]$$

$$=.982218+\frac{1}{4}\times\frac{1}{361}\,(.187746)(.982218)$$

$$[(.191145)(1.01810)+.19]$$

$$=.982218+\frac{1}{4}\times\frac{1}{361}\,(.184408)[.384605]$$

$$=.982218+.000049=.982267.$$

Thus dn $(.20|.81)=.98406.$

Example 4. Use the ascending Landen transformation to calculate cn $(.20|.81)$ to 6D.

Using **16.14.4**, we calculate dn $(.20|.81)$ and deduce cn $(.20|.81)$ from **16.14.3** settling the sign from **Figure 16.1**.

As in the preceding example, we reduce the calculation of dn $(.20|.81)$ to that of dn $\left(.19\left|\frac{360}{361}\right.\right)$, when

$$\mathrm{dn}\left(.19\left|\frac{360}{361}\right.\right)=.982267$$

$$\mathrm{dn}\,(.20|.81)=.984056$$

$$\mathrm{cn}\,(.20|.81)=.980278.$$

Example 5. Use the A.G.M. scale to compute dc $(.672|.36)$ to 4D.

From **16.9.6** we have $\mathrm{dc}^2(.672|.36)=.36+\dfrac{.64}{1-\mathrm{sn}^2(.672|.36)}$. We now calculate $\mathrm{sn}(.672|.36)$ by the method given in **16.4**. Form the A.G.M. scale

n	a_n	b_n	c_n	$\dfrac{c_n}{a_n}$	φ_n	$\sin \varphi_n$	$\sin (2\varphi_{n-1}-\varphi_n)$	$2\varphi_{n-1}-\varphi_n$
0	1	.8	.6	.6	.65546	.60952		
1	.9	.89443	.1	.11111	1.2069	.93452	.10383	.10402
2	.89721	.89721	.00279	.00311	2.4117	.66679	.00207	.00207
3	.89721	.89721	0	0	4.8234	−.99384	0	0

$$\varphi_n=2^n a_n u \qquad \varphi_3=2^3(.89721)(.672)=4.8234$$

continuing until $c_n=0$ to 5D.

Then complete as indicated in **16.4** to find φ_0 and so sn u and hence dc u,

$$\varphi_0=.65546 \qquad \mathrm{sn}\,u=.60952 \qquad \mathrm{dc}\,u=1.1740.$$

Example 6. Use the A.G.M. scale to compute $\Theta(.6|.36)$ to 5D.

We use the method explained in **16.32** with $a_0=1$, $b_0=.8$, $c_0=.6$.

Computing the A.G.M. as explained in **17.6**, we find

(For values of a_n, b_n, c_n, see **Example 5**.)

n	φ_n	$\sin \varphi_n$	$\sin (2\varphi_{n-1}-\varphi_n)$	$2\varphi_{n-1}-\varphi_n$	$\sec (2\varphi_{n-1}-\varphi_n)$	$\dfrac{1}{2^{n+1}}\ln \sec (2\varphi_{n-1}-\varphi_n)$
0	.58803	.55472				
1	1.0780	.88101	.09789	.09805	1.0048	.00120
2	2.1533	.83509	.00260	.00260	1.	0
3	4.3066	−.91879	0	0	1.	0

and then complete the calculation outlined in **16.32** to give

$$\ln \Theta(u|m)=-.05734+.02935+.00120$$
$$=-.02679$$
$$\Theta(u|m)=.97357.$$

The series expansion for Θ is preferable.

Example 7. Use the q-series to compute cs $(.53601\ 62|.09)$.

Here we use the series **16.23.12**, $K=1.60804\ 862$, $q=.00589\ 414$, $v=\frac{\pi u}{2K}=\frac{\pi}{6}$ radians or $30°$.

Since q^4 is negligible to 8D, we have to 7D

$$
\begin{aligned}
\text{cs }(.53601\ 62|.09) \\
&=\frac{\pi}{2K}\cot 30°-\frac{2\pi}{K}\left\{\frac{q^2}{1+q^2}\sin 60°\right\} \\
&=(.97683\ 3852)(1.73205\ 081) \\
&\quad-3.90733\ 541[(.00003\ 4740)(.86602\ 5404)] \\
&=1.69180\ 83.
\end{aligned}
$$

Example 8. Use theta functions to compute sn $(.61802|.5)$ to 5D.

Here $K(\tfrac{1}{2})=1.85407$

$$
\epsilon°=\frac{.61802}{1.85407}\times 90°=30°
$$

$$
\sin^2 \alpha=1/2,\ \alpha=45°.
$$

Thus

$$
\begin{aligned}
\text{sn }(.61802|.5)&=\frac{\vartheta_s(30°\backslash 45°)}{\vartheta_n(30°\backslash 45°)} \\
&=\frac{.59128}{1.04729}=.56458
\end{aligned}
$$

from **Table 16.1.**

Example 9. Use theta functions to compute sc $(.61802|.5)$ to 5D.

As in the preceding example

$$
\epsilon°=30°,\ \alpha°=45°
$$

so that

$$
\text{sc }(.61802|.5)=\frac{\vartheta_s(30°\backslash 45°)}{\vartheta_c(30°\backslash 45°)}.
$$

We use **Table 16.1** to give

$$
\vartheta_s(30°\backslash 45°)=.59128
$$

$$
(\sec 45°)^{\frac{1}{2}}\vartheta_c(30°\backslash 45°)=1.02796.
$$

Therefore

$$
\text{sc }(.61802|.5)=\frac{.59128}{1.02796}\ (\sec 45°)^{\frac{1}{2}}
$$

$$
=.68402.
$$

Example 10. Find sn $(.75342|.7)$ by inverse interpolation in **Table 17.5.**

This method is explained in chapter **17, Example 7.**

Example 11. Find u, given that cs $(u|.5)=.75$. From **16.9.4** we have

$$
\text{sn}^2\ u=\frac{1}{1+\text{cs }^2 u}.
$$

Thus

$$
\text{sn}^2\ (u|.5)=.64
$$

and

$$
\text{sn }(u|.5)=.8.
$$

We have therefore replaced the problem by that of finding u given sn $(u|m)$, where m is known. If $\varphi=\text{am }u$

$$
\sin \varphi=\text{sn }u \text{ and so}
$$

$$
\varphi=.9272952 \text{ radians or } 53.13010°.
$$

From **Table 17.5,**

$$
u=F(53.13010°\backslash 45°)=.99391.
$$

Alternatively, starting with the above value of φ we can use the A.G.M. scale to calculate $F(\varphi\backslash\alpha)$ as explained in **17.6.** This method is to be preferred if more figures are required, or if α differs from a tabular value in **Table 17.5.**

References

Texts

[16.1] A. Erdélyi et al., Higher transcendental functions, vol. 2 (McGraw-Hill Book Co., Inc., New York, N.Y., 1953).

[16.2] L. V. King, On the direct numerical calculation of elliptic functions and integrals (Cambridge Univ. Press, Cambridge, England, 1924).

[16.3] W. Magnus and F. Oberhettinger, Formulas and theorems for the special functions of mathematical physics (Chelsea Publishing Co., New York, N.Y., 1949).

[16.4] E. H. Neville, Jacobian elliptic functions, 2d ed. (Oxford Univ. Press, London, England, 1951).

[16.5] F. Tricomi, Elliptische Funktionen (Akademische Verlagsgesellschaft, Leipzig, Germany, 1948).

[16.6] E. T. Whittaker and G. N. Watson, A course of modern analysis, chs. 20, 21, 22, 4th ed. (Cambridge Univ. Press, Cambridge, England, 1952).

Tables

[16.7] E. P. Adams and R. L. Hippisley, Smithsonian mathematical formulae and tables of elliptic functions, 3d reprint (The Smithsonian Institution, Washington, D.C., 1957).

[16.8] J. Hoüel, Recueil de formules et de tables numériques (Gauthier-Villars, Paris, France, 1901).

[16.9] E. Jahnke and F. Emde, Tables of functions, 4th ed. (Dover Publications, Inc., New York, N.Y., 1945).

[16.10] L. M. Milne-Thomson, Die elliptischen Funktionen von Jacobi (Julius Springer, Berlin, Germany, 1931).

[16.11] L. M. Milne-Thomson, Jacobian elliptic function tables (Dover Publications, Inc., New York, N.Y., 1956).

[16.12] G. W. and R. M. Spenceley, Smithsonian elliptic function tables, Smithsonian Miscellaneous Collection, vol. 109 (Washington, D.C., 1947).

Table 16.1   THETA FUNCTIONS

$$\vartheta_s\left(\epsilon\backslash\alpha\right)$$

$\epsilon\backslash\alpha$	0°	5°	10°	15°	20°	25°	α/ϵ_1
0°	0.00000 0000	0.00000 0000	0.00000 0000	0.00000 0000	0.00000 0000	0.00000 0000	90°
5	0.08715 5743	0.08732 1966	0.08782 4152	0.08867 3070	0.08988 7414	0.09149 5034	85
10	0.17364 8178	0.17397 9362	0.17497 9362	0.17667 1584	0.17909 1708	0.18229 6223	80
15	0.25881 9045	0.25931 2677	0.26080 4191	0.26332 6099	0.26693 4892	0.27171 4833	75
20	0.34202 0143	0.34267 2476	0.34464 3695	0.34797 7361	0.35274 9211	0.35907 2325	70
25	0.42261 8262	0.42342 4343	0.42586 0446	0.42998 1306	0.43588 2163	0.44370 5382	65
30	0.50000 0000	0.50095 3708	0.50383 6358	0.50871 3952	0.51570 1435	0.52497 0857	60
35	0.57357 6436	0.57467 0526	0.57797 7994	0.58357 6134	0.59159 9683	0.60225 0597	55
40	0.64278 7610	0.64401 3768	0.64772 1085	0.65399 8067	0.66299 9145	0.67495 6130	50
45	0.70710 6781	0.70845 5688	0.71253 4820	0.71944 3681	0.72935 6053	0.74253 3161	45
50	0.76604 4443	0.76750 5843	0.77192 5893	0.77941 4712	0.79016 4790	0.80446 5863	40
55	0.81915 2044	0.82071 4821	0.82544 2256	0.83345 4505	0.84496 1783	0.86028 0899	35
60	0.86602 5404	0.86767 7668	0.87267 6562	0.88115 1505	0.89332 9083	0.90955 1166	30
65	0.90630 7787	0.90803 6964	0.91326 9273	0.92214 2410	0.93489 7610	0.95189 9199	25
70	0.93969 2621	0.94148 5546	0.94691 1395	0.95611 4956	0.96935 0025	0.98700 0216	20
75	0.96592 5826	0.96776 8848	0.97334 6839	0.98281 0311	0.99642 3213	1.01458 4761	15
80	0.98480 7753	0.98668 6836	0.99237 4367	1.00202 5068	1.01591 0350	1.03444 0908	10
85	0.99619 4698	0.99809 5528	1.00384 9133	1.01361 2807	1.02766 2527	1.04641 6011	5
90	1.00000 0000	1.00190 8098	1.00768 3786	1.01748 5224	1.03158 9925	1.05041 7974	0

$\epsilon\backslash\alpha$	30°	35°	40°	45°	50°	55°	α/ϵ_1
0°	0.00000 0000	0.00000 0000	0.00000 0000	0.00000 0000	0.00000 0000	0.00000 0000	90°
5	0.09353 4894	0.09606 0073	0.09914 2353	0.10287 9331	0.10740 5819	0.11291 2907	85
10	0.18636 3367	0.19139 9811	0.19754 9961	0.20501 0420	0.21405 3194	0.22506 4618	80
15	0.27778 4006	0.28530 3629	0.29449 2321	0.30564 8349	0.31918 5434	0.33569 3043	75
20	0.36710 5393	0.37706 5455	0.38924 7478	0.40405 4995	0.42204 9614	0.44403 4769	70
25	0.45365 1078	0.46599 3521	0.48110 6437	0.49950 2749	0.52189 9092	0.54932 5515	65
30	0.53676 4494	0.55141 5176	0.56937 7735	0.59127 8602	0.61799 6720	0.65080 1843	60
35	0.61581 3814	0.63268 1725	0.65339 2178	0.67868 8658	0.70961 8904	0.74770 4387	55
40	0.69019 6708	0.70917 3264	0.73250 7761	0.76106 3101	0.79606 0581	0.83928 2749	50
45	0.75934 4980	0.78030 3503	0.80611 4729	0.83776 1607	0.87664 1114	0.92480 2089	45
50	0.82272 9031	0.84552 4503	0.87364 0739	0.90817 9128	0.95071 1025	1.00355 1297	40
55	0.87986 2121	0.90433 1298	0.93455 6042	0.97175 1955	1.01765 9399	1.07485 2509	35
60	0.93030 4365	0.95626 6326	0.98837 8598	1.02796 3895	1.07692 1759	1.13807 1621	30
65	0.97366 6431	1.00092 3589	1.03467 8996	1.07635 2410	1.12798 8100	1.19262 9342	25
70	1.00961 2870	1.03795 2481	1.07308 5074	1.11651 4503	1.17041 0792	1.23801 2299	20
75	1.03786 5044	1.06706 1179	1.10328 6100	1.14811 2152	1.20381 2008	1.27378 3626	15
80	1.05820 3585	1.08801 9556	1.12503 6391	1.17087 7087	1.22789 0346	1.29959 2533	10
85	1.07047 0366	1.10066 1511	1.13815 8265	1.18461 4727	1.24242 6337	1.31518 2322	5
90	1.07456 9932	1.10488 6686	1.14254 4218	1.18920 7115	1.24728 6586	1.32039 6454	0

$\epsilon\backslash\alpha$	60°	65°	70°	75°	80°	85°	α/ϵ_1
0°	0.00000 0000	0.00000 0000	0.00000 0000	0.00000 0000	0.00000 0000	0.00000 0000	90°
5	0.11968 1778	0.12814 8474	0.13904 1489	0.15372 0475	0.17522 3596	0.21321 7690	85
10	0.23861 4577	0.25558 9564	0.27747 6571	0.30706 5715	0.35063 9262	0.42844 3440	80
15	0.35604 4091	0.38160 3032	0.41467 2740	0.45960 9511	0.52633 5260	0.64743 4941	75
20	0.47120 6153	0.50544 4270	0.54994 7578	0.61082 7702	0.70219 9693	0.87146 4767	70
25	0.58332 3727	0.62633 5361	0.68254 9331	0.76005 8920	0.87783 8622	1.10111 6239	65
30	0.69160 6043	0.74345 9784	0.81164 3704	0.90647 6281	1.05251 4778	1.33612 3616	60
35	0.79525 0355	0.85596 1570	0.93630 8263	1.04907 2506	1.22511 1680	1.57526 8297	55
40	0.89344 6594	0.96294 9380	1.05553 5305	1.18666 0037	1.39412 6403	1.81633 9939	50
45	0.98538 4972	1.06350 5669	1.16824 3466	1.31788 6740	1.55769 2334	2.05616 7815	45
50	1.07026 6403	1.15670 0687	1.27329 7730	1.44126 6644	1.71363 1283	2.29072 3417	40
55	1.14731 5349	1.24161 0747	1.36953 6895	1.55522 4175	1.85953 2258	2.51529 0558	35
60	1.21579 4546	1.31733 9855	1.45580 7011	1.65814 9352	1.99285 2358	2.72469 4161	30
65	1.27502 0900	1.38304 3549	1.53099 8883	1.74846 0610	2.11103 3523	2.91357 4159	25
70	1.32438 1718	1.43795 3601	1.59408 7380	1.82467 1332	2.21162 7685	3.07668 6743	20
75	1.36335 0417	1.48140 2159	1.64417 0149	1.88545 5864	2.29242 2061	3.20921 2227	15
80	1.39150 0813	1.51284 3876	1.68050 3336	1.92971 0721	2.35155 6149	3.30704 7313	10
85	1.40851 9209	1.53187 4716	1.70253 2036	1.95660 6998	2.38762 2438	3.36705 9918	5
90	1.41421 3562	1.53824 6269	1.70991 3565	1.96563 0511	2.39974 3837	3.38728 7004	0

$$\sqrt{\sec\alpha}\;\vartheta_c(\epsilon_1\backslash\alpha)$$

$$\epsilon°=\frac{u}{K}\cdot 90° \qquad \epsilon_1°=90°-\epsilon° \qquad \alpha=\arcsin\sqrt{m} \qquad \vartheta_s(u|m)=\vartheta_s(\epsilon°\backslash\alpha°)$$

In calculating elliptic functions from theta functions, when the modular angle exceeds about 60°, use the descending Landen transformation 16.12 to induce dependence on a smaller modular angle.

Compiled from E. P. Adams and R. L. Hippisley, Smithsonian mathematical formulae and tables of elliptic functions, 3d reprint (The Smithsonian Institution, Washington, D.C., 1957) (with permission).

THETA FUNCTIONS

$$\vartheta_n(\epsilon\backslash\alpha)$$

Table 16.1

$\epsilon\backslash\alpha$	$0°$	$5°$	$10°$	$15°$	$20°$	$25°$	α/ϵ_1
$0°$	1	1.00000 00000	1.00000 00000	1.00000 00000	1.00000 00000	1.00000 00000	$90°$
5	1	1.00001 44942	1.00005 83670	1.00013 28199	1.00023 99605	1.00038 29783	85
10	1	1.00005 75362	1.00023 16945	1.00052 72438	1.00095 25510	1.00152 02770	80
15	1	1.00012 78184	1.00051 47160	1.00117 12875	1.00211 61200	1.00337 73404	75
20	1	1.00022 32051	1.00089 88322	1.00204 53820	1.00369 53131	1.00589 77438	70
25	1	1.00034 07982	1.00137 23717	1.00312 29684	1.00564 21475	1.00900 49074	65
30	1	1.00047 70246	1.00192 09464	1.00437 13049	1.00789 74700	1.01260 44231	60
35	1	1.00062 77451	1.00252 78880	1.00575 24612	1.01039 27539	1.01658 69227	55
40	1	1.00078 83803	1.00317 47551	1.00722 44718	1.01305 21815	1.02083 14013	50
45	1	1.00095 40492	1.00384 18928	1.00874 26104	1.01579 49474	1.02520 88930	45
50	1	1.00111 97181	1.00450 90305	1.01026 07491	1.01853 77143	1.02958 63905	40
55	1	1.00128 03532	1.00515 58975	1.01173 27599	1.02119 71444	1.03383 08852	35
60	1	1.00143 10738	1.00576 28392	1.01311 39167	1.02369 24323	1.03781 34098	30
65	1	1.00156 73002	1.00631 14139	1.01436 22536	1.02594 77596	1.04141 29561	25
70	1	1.00168 48932	1.00678 49535	1.01543 98405	1.02789 45992	1.04452 01522	20
75	1	1.00178 02800	1.00716 90696	1.01631 39354	1.02947 37972	1.04704 05862	15
80	1	1.00185 05621	1.00745 20912	1.01695 79795	1.03063 73701	1.04889 76746	10
85	1	1.00189 36042	1.00762 54187	1.01735 24037	1.03134 99632	1.05003 49895	5
90	1	1.00190 80984	1.00768 37857	1.01748 52237	1.03158 99246	1.05041 79735	0

$\epsilon\backslash\alpha$	$30°$	$35°$	$40°$	$45°$	$50°$	$55°$	α/ϵ_1
$0°$	1.00000 00000	1.00000 00000	1.00000 00000	1.00000 00000	1.00000 00000	1.00000 00000	$90°$
5	1.00056 64294	1.00079 66833	1.00108 26253	1.00143 67802	1.00187 71775	1.00243 05914	85
10	1.00224 85079	1.00316 25308	1.00429 76203	1.00570 35065	1.00745 17850	1.00964 88003	80
15	1.00499 51300	1.00702 56701	1.00954 73402	1.01267 06562	1.01655 47635	1.02143 61311	75
20	1.00872 28461	1.01226 87413	1.01667 23379	1.02212 67193	1.02891 00179	1.03743 56974	70
25	1.01331 83978	1.01873 24599	1.02545 62012	1.03378 46028	1.04414 27466	1.05716 29130	65
30	1.01864 21583	1.02622 04548	1.03563 21191	1.04729 03271	1.06179 07561	1.08002 00285	60
35	1.02453 23743	1.03450 52308	1.04689 09786	1.06223 37524	1.08131 84270	1.10531 40947	55
40	1.03081 00797	1.04333 50787	1.05889 07481	1.07816 10137	1.10213 29153	1.13227 78297	50
45	1.03728 45330	1.05244 17208	1.07126 68617	1.09458 82886	1.12360 21058	1.16009 27802	45
50	1.04375 90125	1.06154 84606	1.08364 32917	1.11101 64844	1.14507 37802	1.18791 40899	40
55	1.05003 67930	1.07037 85902	1.09564 39724	1.12694 63970	1.16589 54205	1.21489 61356	35
60	1.05592 71242	1.07866 37978	1.10690 42279	1.14189 38846	1.18543 40490	1.24021 82552	30
65	1.06125 10260	1.08615 23221	1.11708 18582	1.15540 45920	1.20309 54999	1.26310 97835	25
70	1.06584 67280	1.09261 66042	1.12586 75438	1.16706 77783	1.21834 25328	1.28287 36204	20
75	1.06957 45853	1.09786 02047	1.13299 42539	1.17652 88244	1.23071 12287	1.29890 75994	15
80	1.07232 13226	1.10172 37756	1.13824 53698	1.18350 00363	1.23982 51648	1.31072 29838	10
85	1.07400 34764	1.10408 99048	1.14146 12760	1.18776 94140	1.24540 69243	1.31795 95033	5
90	1.07456 99318	1.10488 66859	1.14254 42177	1.18920 71150	1.24728 65857	1.32039 64540	0

$\epsilon\backslash\alpha$	$60°$	$65°$	$70°$	$75°$	$80°$	$85°$	α/ϵ_1
$0°$	1.00000 00000	1.00000 00000	1.00000 00000	1.00000 00000	1.00000 00000	1.00000 00000	$90°$
5	1.00313 85295	1.00406 92257	1.00534 44028	1.00720 88997	1.01026 06485	1.01663 88247	85
10	1.01245 94672	1.01615 50083	1.02121 95717	1.02862 79374	1.04076 43440	1.06618 38299	80
15	1.02768 16504	1.03589 51569	1.04715 56657	1.06363 90673	1.09068 07598	1.14751 59063	75
20	1.04834 57003	1.06269 75825	1.08238 38086	1.11122 86903	1.15864 11101	1.25875 62174	70
25	1.07382 76019	1.09575 73598	1.12585 71388	1.17001 24008	1.24276 19421	1.39725 25218	65
30	1.10335 71989	1.13408 00433	1.17627 97795	1.23826 96285	1.34068 05139	1.55957 26706	60
35	1.13604 11010	1.17651 06705	1.23214 31946	1.31398 80140	1.44960 33094	1.74151 57980	55
40	1.17088 93642	1.22176 77148	1.29176 91861	1.39491 71251	1.56636 90138	1.93815 19599	50
45	1.20684 51910	1.26848 10938	1.35335 85717	1.47863 07744	1.68752 66770	2.14389 95792	45
50	1.24281 67937	1.31523 31927	1.41504 43413	1.56259 67789	1.80942 88493	2.35264 71220	40
55	1.27771 04815	1.36060 17261	1.47494 78852	1.64425 25175	1.92833 82823	2.55792 12198	35
60	1.31046 39783	1.40320 31647	1.53123 64694	1.72108 41609	2.04054 54606	2.75309 84351	30
65	1.34007 89457	1.44173 53793	1.58218 06891	1.79070 70015	2.14249 29245	2.93165 25995	25
70	1.36565 16965	1.47501 81348	1.62620 90720	1.85094 39670	2.23090 12139	3.08742 47870	20
75	1.38640 11169	1.50203 00916	1.66195 87940	1.89989 92030	2.30289 04563	3.21489 91220	15
80	1.40169 28947	1.52194 10514	1.68832 00831	1.93602 35909	2.35609 12550	3.30946 52989	10
85	1.41105 92570	1.53413 83232	1.70447 27784	1.95816 92561	2.38873 86793	3.36764 82512	5
90	1.41421 35624	1.53824 62687	1.70991 35651	1.96563 05108	2.39974 38370	3.38728 70037	0

$$\sqrt{\sec\alpha}\;\vartheta_d(\epsilon_1\backslash\alpha)$$

$$\epsilon° = \frac{u}{K}90° \qquad \epsilon_1° = 90° - \epsilon° \qquad \alpha = \arcsin\sqrt{m} \qquad \vartheta_n(u\,|\,m) = \vartheta_n(\epsilon\backslash\alpha)$$

In calculating elliptic functions from theta functions, when the modular angle exceeds about 60°, use the descending Landen transformation **16.12** to induce dependence on a smaller modular angle.

Table 16.2   **LOGARITHMIC DERIVATIVES OF THETA FUNCTIONS**

$$\frac{d}{du}\ln\vartheta_s(u)=f(\epsilon\backslash\alpha)$$

$\epsilon\backslash\alpha$	0°	5°	10°	15°	20°	25°	α/ϵ_1
0°	∞	∞	∞	∞	∞	∞	90°
5	11.43005	11.40829	11.34306	11.23449	11.08275	10.88811	85
10	5.67128	5.66049	5.62812	5.57427	5.49902	5.40253	80
15	3.73205	3.72495	3.70365	3.66823	3.61876	3.55536	75
20	2.74748	2.74225	2.72658	2.70051	2.66414	2.61756	70
25	2.14451	2.14043	2.12820	2.10787	2.07952	2.04325	65
30	1.73205	1.72875	1.71888	1.70248	1.67962	1.65041	60
35	1.42815	1.42543	1.41729	1.40378	1.38497	1.36096	55
40	1.19175	1.18949	1.18270	1.17143	1.15577	1.13581	50
45	1.00000	0.99810	0.99240	0.98296	0.96985	0.95315	45
50	0.83910	0.83750	0.83273	0.82481	0.81383	0.79987	40
55	0.70021	0.69888	0.69489	0.68830	0.67915	0.66754	35
60	0.57735	0.57625	0.57297	0.56754	0.56001	0.55047	30
65	0.46631	0.46542	0.46277	0.45839	0.45232	0.44464	25
70	0.36397	0.36328	0.36121	0.35779	0.35306	0.34708	20
75	0.26795	0.26744	0.26592	0.26340	0.25992	0.25553	15
80	0.17633	0.17599	0.17499	0.17334	0.17105	0.16816	10
85	0.08749	0.08732	0.08683	0.08600	0.08487	0.08344	5
90	0.00000	0.00000	0.00000	0.00000	0.00000	0.00000	0

$\epsilon\backslash\alpha$	30°	35°	40°	45°	50°	55°	α/ϵ_1
0°	∞	∞	∞	∞	∞	∞	90°
5	10.65083	10.37113	10.04914	9.68479	9.27764	8.82657	85
10	5.28496	5.14645	4.98711	4.80696	4.60585	4.38332	80
15	3.47816	3.38730	3.28290	3.16502	3.03365	2.88859	75
20	2.56090	2.49430	2.41789	2.33179	2.23605	2.13062	70
25	1.99919	1.94749	1.88828	1.82172	1.74793	1.66695	65
30	1.61498	1.57348	1.52607	1.47292	1.41419	1.35001	60
35	1.33189	1.29791	1.25919	1.21591	1.16828	1.11647	55
40	1.11167	1.08352	1.05154	1.01592	0.97687	0.93462	50
45	0.93301	0.90958	0.88302	0.85355	0.82139	0.78679	45
50	0.78307	0.76355	0.74151	0.71714	0.69066	0.66232	40
55	0.65359	0.63743	0.61923	0.59918	0.57749	0.55441	35
60	0.53902	0.52579	0.51093	0.49462	0.47705	0.45846	30
65	0.43543	0.42482	0.41292	0.39991	0.38595	0.37125	25
70	0.33992	0.33169	0.32248	0.31242	0.30168	0.29042	20
75	0.25028	0.24424	0.23751	0.23017	0.22235	0.21419	15
80	0.16471	0.16076	0.15634	0.15155	0.14645	0.14114	10
85	0.08173	0.07977	0.07759	0.07522	0.07270	0.07009	5
90	0.00000	0.00000	0.00000	0.00000	0.00000	0.00000	0

$\epsilon\backslash\alpha$	60°	65°	70°	75°	80°	85°	α/ϵ_1
0°	∞	∞	∞	∞	∞	∞	90°
5	8.32941	7.78200	7.17654	6.49756	5.71041	4.71263	85
10	4.13843	3.86930	3.57238	3.24056	2.85790	2.37760	80
15	2.72935	2.55490	2.36323	2.15026	1.90678	1.60605	75
20	2.01530	1.88950	1.75208	1.60057	1.42943	1.22261	70
25	1.57876	1.48308	1.37931	1.26603	1.13996	0.99169	65
30	1.28047	1.20552	1.12492	1.03795	0.94288	0.83453	60
35	1.06066	1.00096	0.93737	0.86969	0.79715	0.71737	55
40	0.88940	0.84142	0.79086	0.73784	0.68225	0.62344	50
45	0.75000	0.71131	0.67101	0.62941	0.58682	0.54358	45
50	0.63242	0.60125	0.56918	0.53662	0.50411	0.47247	40
55	0.53023	0.50526	0.47987	0.45454	0.42988	0.40690	35
60	0.43911	0.41932	0.39943	0.37992	0.36140	0.34488	30
65	0.35605	0.34063	0.32532	0.31054	0.29684	0.28513	25
70	0.27885	0.26719	0.25574	0.24484	0.23497	0.22685	20
75	0.20584	0.19749	0.18935	0.18170	0.17490	0.16949	15
80	0.13572	0.13034	0.12512	0.12026	0.11601	0.11272	10
85	0.06742	0.06478	0.06224	0.05988	0.05784	0.05628	5
90	0.00000	0.00000	0.00000	0.00000	0.00000	0.00000	0

$$\frac{d}{du}\ln\vartheta_c(u)=-f(\epsilon_1\backslash\alpha)$$

In calculating elliptic functions from theta functions, when the modular angle exceeds about 60°, use the descending Landen transformation **16.12** to induce dependence on a smaller modular angle.

LOGARITHMIC DERIVATIVES OF THETA FUNCTIONS Table 16.2

$$\frac{d}{du}\ln\vartheta_n(u)=g(\epsilon\backslash\alpha)$$

$\epsilon\backslash\alpha$	0°	5°	10°	15°	20°	25°	α/ϵ_1
0°	0	0.000000	0.000000	0.000000	0.000000	0.000000	90°
5	0	0.000331	0.001324	0.002984	0.005318	0.008337	85
10	0	0.000651	0.002607	0.005875	0.010466	0.016401	80
15	0	0.000952	0.003811	0.008583	0.015283	0.023933	75
20	0	0.001224	0.004897	0.011024	0.019616	0.030690	70
25	0	0.001458	0.005833	0.013124	0.023332	0.036462	65
30	0	0.001649	0.006591	0.014819	0.026318	0.041075	60
35	0	0.001788	0.007147	0.016057	0.028487	0.044394	55
40	0	0.001874	0.007486	0.016804	0.029776	0.046332	50
45	0	0.001903	0.007596	0.017037	0.030154	0.046846	45
50	0	0.001873	0.007476	0.016753	0.029616	0.045938	40
55	0	0.001787	0.007129	0.015962	0.028185	0.043654	35
60	0	0.001647	0.006566	0.014691	0.025912	0.040077	30
65	0	0.001457	0.005805	0.012979	0.022871	0.035328	25
70	0	0.001222	0.004868	0.010879	0.019154	0.029556	20
75	0	0.000951	0.003786	0.008455	0.014877	0.022935	15
80	0	0.000650	0.002589	0.005780	0.010165	0.015661	10
85	0	0.000330	0.001314	0.002933	0.005157	0.007942	5
90	0	0.000000	0.000000	0.000000	0.000000	0.000000	0

$\epsilon\backslash\alpha$	30°	35°	40°	45°	50°	55°	α/ϵ_1
0°	0.000000	0.000000	0.000000	0.000000	0.000000	0.000000	90°
5	0.012059	0.016511	0.021734	0.027787	0.034760	0.042791	85
10	0.023711	0.032444	0.042671	0.054498	0.068087	0.083685	80
15	0.034569	0.047248	0.062057	0.079124	0.098650	0.120939	75
20	0.044277	0.060427	0.079221	0.100783	0.125308	0.153099	70
25	0.052528	0.071558	0.093605	0.118758	0.147169	0.179081	65
30	0.059074	0.080308	0.104784	0.132533	0.163627	0.198206	60
35	0.063730	0.086442	0.112477	0.141791	0.174358	0.210188	55
40	0.066384	0.089827	0.116544	0.146411	0.179298	0.215082	50
45	0.066987	0.090424	0.116978	0.146447	0.178606	0.213212	45
50	0.065561	0.088287	0.113888	0.142097	0.172615	0.205102	40
55	0.062183	0.083549	0.107483	0.133678	0.161784	0.191402	35
60	0.056989	0.076408	0.098051	0.121592	0.146658	0.172831	30
65	0.050157	0.067122	0.085943	0.106302	0.127835	0.150136	25
70	0.041905	0.055989	0.071553	0.088310	0.105932	0.124058	20
75	0.032483	0.043344	0.055309	0.068143	0.081578	0.095321	15
80	0.022163	0.029545	0.037660	0.046339	0.055395	0.064622	10
85	0.011235	0.014968	0.019067	0.023443	0.028000	0.032631	5
90	0.000000	0.000000	0.000000	0.000000	0.000000	0.000000	0

$\epsilon\backslash\alpha$	60°	65°	70°	75°	80°	85°	α/ϵ_1
0°	0.000000	0.000000	0.000000	0.000000	0.000000	0.000000	90°
5	0.052098	0.063034	0.076222	0.092860	0.115687	0.153481	85
10	0.101680	0.122704	0.147856	0.179293	0.221544	0.289421	80
15	0.146471	0.176024	0.210938	0.253725	0.309882	0.395712	75
20	0.184635	0.220691	0.262588	0.312762	0.376371	0.467893	70
25	0.214885	0.255225	0.301193	0.354775	0.420046	0.507818	65
30	0.236514	0.278976	0.326329	0.379918	0.442452	0.520777	60
35	0.249349	0.292010	0.338517	0.389553	0.446532	0.512966	55
40	0.253651	0.294931	0.338908	0.385698	0.435687	0.490013	50
45	0.250000	0.288691	0.328990	0.370590	0.413176	0.456422	45
50	0.239181	0.274426	0.310353	0.346389	0.381811	0.415539	40
55	0.222085	0.253326	0.284538	0.315020	0.343874	0.369741	35
60	0.199639	0.226549	0.252950	0.278119	0.301140	0.320668	30
65	0.172751	0.195171	0.216820	0.237026	0.254956	0.269431	25
70	0.142285	0.160167	0.177204	0.192823	0.206331	0.216780	20
75	0.109049	0.122405	0.134996	0.146375	0.156015	0.163217	15
80	0.073794	0.082664	0.090960	0.098382	0.104574	0.109083	10
85	0.037222	0.041645	0.045763	0.049423	0.052449	0.054618	5
90	0.000000	0.000000	0.000000	0.000000	0.000000	0.000000	0

$$\frac{d}{du}\ln\vartheta_d(u)=-g(\epsilon_1\backslash\alpha)$$

In calculating elliptic functions from theta functions, when the modular angle exceeds about 60°, use the descending Landen transformation **16.12** to induce dependence on a smaller modular angle.

17. Elliptic Integrals

L. M. Milne-Thomson [1]

Contents

---

[1] University of Arizona. (Prepared under contract with the National Bureau of Standards.)

The author acknowledges with thanks the assistance of Ruth Zucker in the computation of the examples, Ruth E. Capuano for **Table 17.3,** David S. Liepman for **Table 17.4,** and Andreas Schopf for **Table 17.9.**

17. Elliptic Integrals

Mathematical Properties

17.1. Definition of Elliptic Integrals

If $R(x, y)$ is a rational function of x and y, where y^2 is equal to a cubic or quartic polynomial in x, the integral

17.1.1
$$\int R(x,y)dx$$

is called an *elliptic integral*.

The elliptic integral just defined can not, in general, be expressed in terms of elementary functions.

Exceptions to this are

(i) when $R(x, y)$ contains no odd powers of y.
(ii) when the polynomial y^2 has a repeated factor.

We therefore exclude these cases.

By substituting for y^2 and denoting by $p_s(x)$ a polynomial in x we get [2]

$$R(x,y) = \frac{p_1(x) + yp_2(x)}{p_3(x) + yp_4(x)}$$

$$= \frac{[p_1(x) + yp_2(x)][p_3(x) - yp_4(x)]y}{\{[p_3(x)]^2 - y^2[p_4(x)]^2\}y}$$

$$= \frac{p_5(x) + yp_6(x)}{yp_7(x)} = R_1(x) + \frac{R_2(x)}{y}$$

where $R_1(x)$ and $R_2(x)$ are rational functions of x. Hence, by expressing $R_2(x)$ as the sum of a polynomial and partial fractions

$$\int R(x,y)dx = \int R_1(x)dx + \Sigma_s A_s \int x^s y^{-1}dx$$
$$+ \Sigma_s B_s \int [(x-c)^s y]^{-1}dx$$

Reduction Formulae

Let

17.1.2

$$y^2 = a_0 x^4 + a_1 x^3 + a_2 x^2 + a_3 x + a_4 \qquad (|a_0| + |a_1| \neq 0)$$
$$= b_0(x-c)^4 + b_1(x-c)^3 + b_2(x-c)^2 + b_3(x-c) + b_4$$
$$(|b_0| + |b_1| \neq 0)$$

17.1.3 $\quad I_s = \int x^s y^{-1} dx, \; J_s = \int [y(x-c)^s]^{-1} dx$

By integrating the derivatives of yx^s and $y(x-c)^{-s}$ we get the reduction formulae

17.1.4

$$(s+2)a_0 I_{s+3} + \tfrac{1}{2}a_1(2s+3)I_{s+2} + a_2(s+1)I_{s+1}$$
$$+ \tfrac{1}{2}a_3(2s+1)I_s + sa_4 I_{s-1} = x^s y \quad (s=0, 1, 2, \ldots)$$

[2] See [17.7] **22.72.**

17.1.5

$$(2-s)b_0 J_{s-3} + \tfrac{1}{2}b_1(3-2s)J_{s-2} + b_2(1-s)J_{s-1}$$
$$+ \tfrac{1}{2}b_3(1-2s)J_s - sb_4 J_{s+1} = y(x-c)^{-s}$$
$$(s=1, 2, 3, \ldots)$$

By means of these reduction formulae and certain transformations (see **Examples 1** and **2**) every elliptic integral can be brought to depend on the integral of a rational function and on three canonical forms for elliptic integrals.

17.2. Canonical Forms

Definitions

17.2.1

$m = \sin^2 \alpha$; m is the parameter,
$\qquad\qquad \alpha$ is the modular angle

17.2.2 $\qquad\qquad x = \sin \varphi = \operatorname{sn} u$

17.2.3 $\qquad\qquad \cos \varphi = \operatorname{cn} u$

17.2.4

$(1 - m \sin^2 \varphi)^{\frac{1}{2}} = \operatorname{dn} u = \Delta(\varphi)$, the delta amplitude

17.2.5 $\quad \varphi = \arcsin (\operatorname{sn} u) = \operatorname{am} u$, the amplitude

Elliptic Integral of the First Kind

17.2.6 $\quad F(\varphi \backslash \alpha) = F(\varphi|m) = \int_0^\varphi (1 - \sin^2 \alpha \sin^2 \theta)^{-\frac{1}{2}} d\theta$

17.2.7 $\qquad = \int_0^x [(1-t^2)(1-mt^2)]^{-\frac{1}{2}} dt$

$\qquad\qquad = \int_0^u dw = u$

Elliptic Integral of the Second Kind

17.2.8 $\quad E(\varphi \backslash \alpha) = E(u|m) = \int_0^x (1-t^2)^{-\frac{1}{2}}(1-mt^2)^{\frac{1}{2}} dt$

17.2.9 $\qquad = \int_0^\varphi (1 - \sin^2 \alpha \sin^2 \theta)^{\frac{1}{2}} d\theta$

17.2.10 $\qquad = \int_0^u \operatorname{dn}^2 w \, dw$

17.2.11 $\qquad = m_1 u + m \int_0^u \operatorname{cn}^2 w \, dw$

17.2.12 $\quad E(\varphi\backslash\alpha)=u-m\int_0^u \operatorname{sn}^2 w\,dw$

17.2.13 $\quad\quad =\dfrac{\pi}{2K(m)}\dfrac{\vartheta_4'(\pi u/2K)}{\vartheta_4(\pi u/2K)}+\dfrac{E(m)u}{K(m)}$

(For theta functions, see chapter **16**.)

Elliptic Integral of the Third Kind

17.2.14
$$\Pi(n;\varphi\backslash\alpha)=\int_0^\varphi (1-n\sin^2\theta)^{-1}[1-\sin^2\alpha\sin^2\theta]^{-1/2}d\theta$$

If $x=\operatorname{sn}(u|m)$,

17.2.15
$$\Pi(n;u|m)=\int_0^x (1-nt^2)^{-1}[(1-t^2)(1-mt^2)]^{-1/2}dt$$

17.2.16 $\quad =\int_0^u (1-n\operatorname{sn}^2(w|m))^{-1}dw$

The Amplitude φ

17.2.17 $\quad \varphi=\operatorname{am}u=\arcsin(\operatorname{sn}u)=\arcsin x$

can be calculated from **Tables 17.5** and **4.14**.

The Parameter m

Dependence on the *parameter m* is denoted by a vertical stroke preceding the parameter, e.g., $F(\varphi|m)$.

Together with the parameter we define the *complementary parameter m_1* by

17.2.18 $\quad\quad\quad m+m_1=1$

When the parameter is real, it can always be arranged, see **17.4**, that $0\le m\le 1$.

The Modular Angle α

Dependence on the modular angle α, defined in terms of the parameter by **17.2.1**, is denoted by a backward stroke $\backslash$ preceding the modular angle, thus $E(\varphi\backslash\alpha)$. The *complementary modular* angle is $\pi/2-\alpha$ or $90°-\alpha$ according to the unit and thus $m_1=\sin^2(90°-\alpha)=\cos^2\alpha$.

The Modulus k

In terms of Jacobian elliptic functions (chapter **16**), the modulus k and the complementary modulus are defined by

17.2.19 $\quad k=\operatorname{ns}(K+iK'),\ k'=\operatorname{dn}K$.

They are related to the parameter by $k^2=m$, $k'^2=m_1$.

Dependence on the modulus is denoted by a comma preceding it, thus $\Pi(n;u,k)$.

In computation the modulus is of minimal importance, since it is the parameter and its complement which arise naturally. The parameter and the modular angle will be employed in this chapter to the exclusion of the modulus.

The Characteristic n

The elliptic integral of the third kind depends on three variables namely (i) the parameter, (ii) the amplitude, (iii) the characteristic n. When real, the characteristic may be any number in the interval $(-\infty,\infty)$. The properties of the integral depend upon the location of the characteristic in this interval, see **17.7**.

17.3. Complete Elliptic Integrals of the First and Second Kinds

Referred to the canonical forms of **17.2**, the elliptic integrals are said to be *complete* when the amplitude is $\frac12\pi$ and so $x=1$. These complete integrals are designated as follows

17.3.1
$$[K(m)]=K=\int_0^1 [(1-t^2)(1-mt^2)]^{-1/2}dt$$
$$=\int_0^{\pi/2}(1-m\sin^2\theta)^{-1/2}d\theta$$

17.3.2 $\quad\quad K=F(\tfrac12\pi|m)=F(\tfrac12\pi\backslash\alpha)$

17.3.3
$$E[K(m)]=E=\int_0^1 (1-t^2)^{-1/2}(1-mt^2)^{1/2}dt$$
$$=\int_0^{\pi/2}(1-m\sin^2\theta)^{1/2}d\theta$$

17.3.4 $\quad\quad E=E[K(m)]=E(m)=E(\tfrac12\pi\backslash\alpha)$

We also define

17.3.5
$$K'=K(m_1)=K(1-m)=\int_0^{\pi/2}(1-m_1\sin^2\theta)^{-1/2}d\theta$$

17.3.6 $\quad\quad K'=F(\tfrac12\pi|m_1)=F(\tfrac12\pi\backslash\tfrac12\pi-\alpha)$

17.3.7
$$E'=E(m_1)=E(1-m)=\int_0^{\pi/2}(1-m_1\sin^2\theta)^{1/2}d\theta$$

17.3.8 $\quad E'=E[K(m_1)]=E(m_1)=E(\tfrac12\pi\backslash\tfrac12\pi-\alpha)$

K and iK' are the "real" and "imaginary" *quarter-periods* of the corresponding Jacobian elliptic functions (see chapter **16**).

Relation to the Hypergeometric Function
(see chapter 15)

17.3.9 $\qquad K = \tfrac{1}{2}\pi F(\tfrac{1}{2}, \tfrac{1}{2}; 1; m)$

17.3.10 $\qquad E = \tfrac{1}{2}\pi F(-\tfrac{1}{2}, \tfrac{1}{2}; 1; m)$

Infinite Series

17.3.11

$$K(m) = \frac{1}{2}\pi\left[1 + \left(\frac{1}{2}\right)^2 m + \left(\frac{1\cdot 3}{2\cdot 4}\right)^2 m^2 + \left(\frac{1\cdot 3\cdot 5}{2\cdot 4\cdot 6}\right)^2 m^3 + \ldots\right]\qquad (|m|<1)$$

17.3.12

$$E(m) = \frac{1}{2}\pi\left[1 - \left(\frac{1}{2}\right)^2 \frac{m}{1} - \left(\frac{1\cdot 3}{2\cdot 4}\right)^2 \frac{m^2}{3} - \left(\frac{1\cdot 3\cdot 5}{2\cdot 4\cdot 6}\right)^2 \frac{m^3}{5} - \ldots\right]\qquad (|m|<1)$$

Legendre's Relation

17.3.13 $\qquad EK' + E'K - KK' = \tfrac{1}{2}\pi$

Auxiliary Function

17.3.14 $\qquad L(m) = \dfrac{K'(m)}{\pi}\ln\dfrac{16}{m_1} - K(m)$

17.3.15 $\quad m = 1 - 16\exp\left[-\pi(K(m)+L(m))/K'(m)\right]$

17.3.16 $\quad m = 16\exp\left[-\pi(K'(m)+L(m_1))/K(m)\right]$

The function $L(m)$ is tabulated in **Table 17.4.**

q-Series

The Nome q and the Complementary Nome q_1

17.3.17 $\qquad q = q(m) = \exp[-\pi K'/K]$

17.3.18 $\qquad q_1 = q(m_1) = \exp[-\pi K/K']$

17.3.19 $\qquad \ln\dfrac{1}{q}\ln\dfrac{1}{q_1} = \pi^2$

17.3.20

$$\log_{10}\frac{1}{q}\log_{10}\frac{1}{q_1} = (\pi\log_{10}e)^2 = 1.86152\ 28349 \text{ to 10D}$$

17.3.21

$$q = \exp[-\pi K'/K] = \frac{m}{16} + 8\left(\frac{m}{16}\right)^2 + 84\left(\frac{m}{16}\right)^3 + 992\left(\frac{m}{16}\right)^4 + \ldots \qquad (|m|<1)$$

17.3.22 $\qquad K = \dfrac{1}{2}\pi + 2\pi\displaystyle\sum_{s=1}^{\infty}\frac{q^s}{1+q^{2s}}$

17.3.23

$$\frac{E}{K} = \frac{1}{3}(1+m_1) + (\pi/K)^2\left[1/12 - 2\sum_{s=1}^{\infty}q^{2s}(1-q^{2s})^{-2}\right]$$

17.3.24 $\quad \text{am } u = v + \displaystyle\sum_{s=1}^{\infty}\frac{2q^s\sin 2sv}{s(1+q^{2s})}$ where $v = \pi u/(2K)$

Limiting Values

17.3.25 $\qquad \displaystyle\lim_{m\to 0} K'(E-K) = 0$

17.3.26 $\qquad \displaystyle\lim_{m\to 1}[K - \tfrac{1}{2}\ln(16/m_1)] = 0$

17.3.27 $\qquad \displaystyle\lim_{m\to 0} m^{-1}(K-E) = \lim_{m\to 0} m^{-1}(E-m_1 K) = \pi/4$

17.3.28 $\qquad \displaystyle\lim_{m\to 0} q/m = \lim_{m_1\to 1} q_1/m_1 = 1/16$

Alternative Evaluations of K and E (see also 17.5)

17.3.29

$$K(m) = 2[1+m_1^{1/2}]^{-1}K[(1-m_1^{1/2})/(1+m_1^{1/2})]^2$$

17.3.30

$$E(m) = (1+m_1^{1/2})E([(1-m_1^{1/2})/(1+m_1^{1/2})]^2) - 2m_1^{1/2}(1+m_1^{1/2})^{-1}K([(1-m_1^{1/2})/(1+m_1^{1/2})]^2)$$

17.3.31 $\qquad K(\alpha) = 2F(\arctan(\sec^{1/2}\alpha)\backslash\alpha)$

17.3.32 $\qquad E(\alpha) = 2E(\arctan(\sec^{1/2}\alpha)\backslash\alpha) - 1 + \cos\alpha$

Polynomial Approximations [3] ($0 \le m < 1$)

17.3.33

$$K(m) = [a_0 + a_1 m_1 + a_2 m_1^2] + [b_0 + b_1 m_1 + b_2 m_1^2]\ln(1/m_1) + \epsilon(m)$$
$$|\epsilon(m)| \le 3\times 10^{-5}$$

$$\begin{aligned}a_0 &= 1.38629\ 44 & b_0 &= .5\\ a_1 &= .11197\ 23 & b_1 &= .12134\ 78\\ a_2 &= .07252\ 96 & b_2 &= .02887\ 29\end{aligned}$$

17.3.34

$$K(m) = [a_0 + a_1 m_1 + \ldots + a_4 m_1^4] + [b_0 + b_1 m_1 + \ldots + b_4 m_1^4]\ln(1/m_1) + \epsilon(m)$$
$$|\epsilon(m)| \le 2\times 10^{-8}$$

$$\begin{aligned}a_0 &= 1.38629\ 436112 & b_0 &= .5\\ a_1 &= .09666\ 344259 & b_1 &= .12498\ 593597\\ a_2 &= .03590\ 092383 & b_2 &= .06880\ 248576\\ a_3 &= .03742\ 563713 & b_3 &= .03328\ 355346\\ a_4 &= .01451\ 196212 & b_4 &= .00441\ 787012\end{aligned}$$

---

[3] The approximations **17.3.33–17.3.36** are from C. Hastings, Jr., Approximations for Digital Computers, Princeton Univ. Press, Princeton, N. J. (with permission).

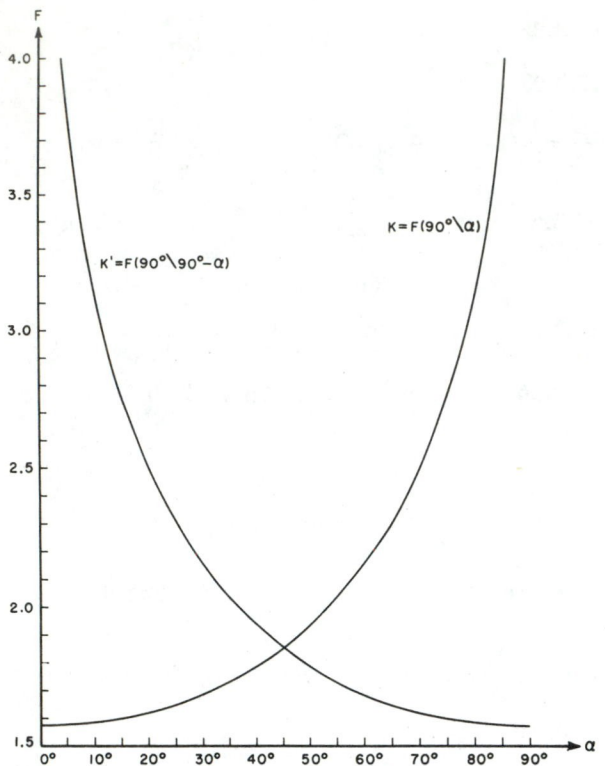

FIGURE 17.1. *Complete elliptic integral of the first kind.*

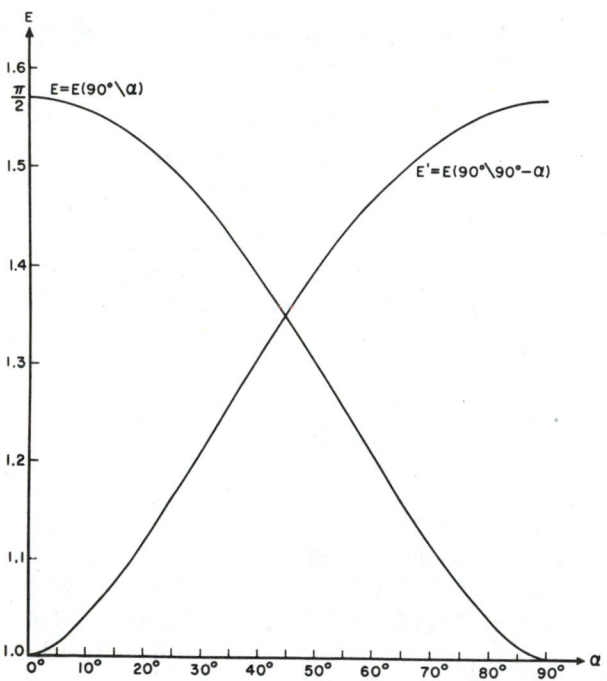

FIGURE 17.2. *Complete elliptic integral of the second kind.*

17.3.35

$$E(m)=[1+a_1m_1+a_2m_1^2]+[b_1m_1+b_2m_1^2]\ln(1/m_1)$$
$$+\epsilon(m)$$
$$|\epsilon(m)|<4\times10^{-5}$$

$$a_1=.46301\ 51 \qquad b_1=.24527\ 27$$
$$a_2=.10778\ 12 \qquad b_2=.04124\ 96$$

17.3.36

$$E(m)=[1+a_1m_1+\ldots+a_4m_1^4]+[b_1m_1+\ldots$$
$$+b_4m_1^4]\ln(1/m_1)+\epsilon(m)$$
$$|\epsilon(m)|<2\times10^{-8}$$

$$a_1=.44325\ 141463 \qquad b_1=.24998\ 368310$$
$$a_2=.06260\ 601220 \qquad b_2=.09200\ 180037$$
$$a_3=.04757\ 383546 \qquad b_3=.04069\ 697526$$
$$a_4=.01736\ 506451 \qquad b_4=.00526\ 449639$$

17.4. Incomplete Elliptic Integrals of the First and Second Kinds

Extension of the Tables

Negative Amplitude

17.4.1 $$F(-\varphi|m)=-F(\varphi|m)$$

17.4.2 $$E(-\varphi|m)=-E(\varphi|m)$$

Amplitude of Any Magnitude

17.4.3 $$F(s\pi\pm\varphi|m)=2sK\pm F(\varphi|m)$$

17.4.4 $$E(u+2K)=E(u)+2E$$

17.4.5 $$E(u+2iK')=E(u)+2i(K'-E')$$

17.4.6

$$E(u+2mK+2niK')=E(u)+2mE+2ni(K'-E')$$

17.4.7 $$E(K-u)=E-E(u)+m\,\mathrm{sn}\,u\,\mathrm{cd}\,u$$

Imaginary Amplitude

If $\tan\theta=\sinh\varphi$

17.4.8 $$F(i\varphi\backslash\alpha)=iF(\theta\backslash\tfrac{1}{2}\pi-\alpha)$$

17.4.9

$$E(i\varphi\backslash\alpha)=-iE(\theta\backslash\tfrac{1}{2}\pi-\alpha)+iF(\theta\backslash\tfrac{1}{2}\pi-\alpha)$$
$$+i\tan\theta(1-\cos^2\alpha\sin^2\theta)^{\frac{1}{2}}$$

Jacobi's Imaginary Transformation

17.4.10

$$E(iu|m)=i[u+\mathrm{dn}(u|m_1)\mathrm{sc}(u|m_1)-E(u|m_1)]$$

Complex Amplitude

17.4.11 $$F(\varphi+i\psi|m)=F(\lambda|m)+iF(\mu|m_1)$$

where $\cot^2 \lambda$ is the positive root of the equation $x^2 - [\cot^2 \varphi + m \sinh^2\psi \csc^2\varphi - m_1]x - m_1 \cot^2\varphi = 0$ and $m \tan^2 \mu = \tan^2\varphi \cot^2\lambda - 1$.

17.4.12

$$E(\varphi + i\psi \backslash \alpha) = E(\lambda \backslash \alpha) - iE(\mu \backslash 90° - \alpha) + iF(\mu \backslash 90° - \alpha) + \frac{b_1 + ib_2}{b_3}$$

where

$$b_1 = \sin^2 \alpha \sin \lambda \cos \lambda \sin^2 \mu (1 - \sin^2 \alpha \sin^2 \lambda)^{\frac{1}{2}}$$

$$b_2 = (1 - \sin^2 \alpha \sin^2 \lambda)(1 - \cos^2 \alpha \sin^2 \mu)^{\frac{1}{2}} \sin \mu \cos \mu$$

$$b_3 = \cos^2 \mu + \sin^2 \alpha \sin^2 \lambda \sin^2 \mu$$

Amplitude Near to $\pi/2$ (see also 17.5)

If $\cos \alpha \tan \varphi \tan \psi = 1$

17.4.13 $\quad F(\varphi \backslash \alpha) + F(\psi \backslash \alpha) = F(\pi/2 \backslash \alpha) = K$

17.4.14

$$E(\varphi \backslash \alpha) + E(\psi \backslash \alpha) = E(\pi/2 \backslash \alpha) + \sin^2\alpha \sin\varphi \sin\psi$$

Values when φ is near to $\pi/2$ and m is near to unity can be calculated by these formulae.

Parameter Greater Than Unity

17.4.15 $\quad F(\varphi|m) = m^{-\frac{1}{2}}F(\theta|m^{-1}), \sin \theta = m^{\frac{1}{2}} \sin \varphi$

17.4.16 $\quad E(u|m) = m^{\frac{1}{2}}E(um^{\frac{1}{2}}|m^{-1}) - (m-1)u$

by which a parameter greater than unity can be replaced by a parameter less than unity.

Negative Parameter

17.4.17

$$F(\varphi|-m) = (1+m)^{-\frac{1}{2}}K(m(1+m)^{-1})$$
$$- (1+m)^{-\frac{1}{2}}F\left(\frac{\pi}{2} - \varphi \,\middle|\, m(1+m)^{-1}\right)$$

17.4.18

$$E(u|-m) = (1+m)^{\frac{1}{2}}\{E(u(1+m)^{\frac{1}{2}}|m(m+1)^{-1})$$
$$- m(1+m)^{-\frac{1}{2}}\operatorname{sn}(u(1+m)^{\frac{1}{2}}|m(1+m)^{-1})$$
$$\operatorname{cd}(u(1+m)^{\frac{1}{2}}|m(1+m)^{-1})\}$$

whereby computations can be made for negative parameters, and therefore for pure imaginary modulus.

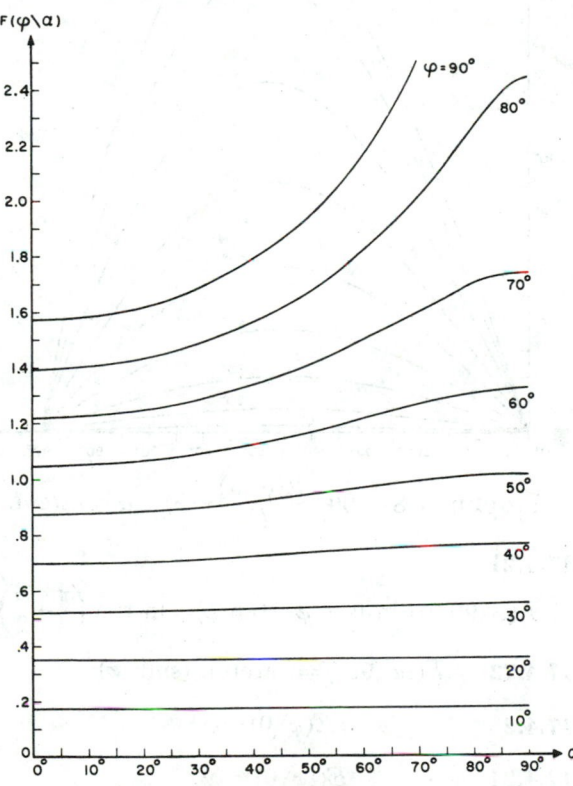

FIGURE 17.3. *Incomplete elliptic integral of the first kind.*

$F(\varphi \backslash \alpha), \quad \varphi$ constant

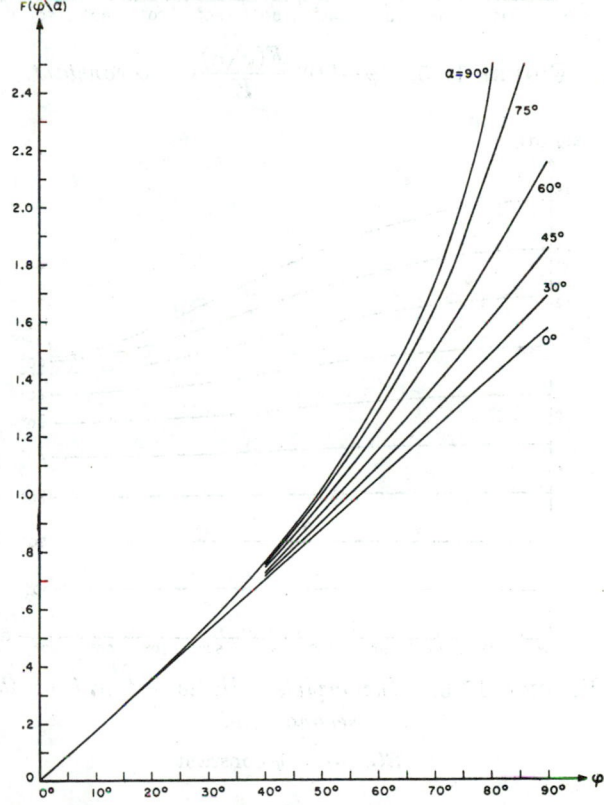

FIGURE 17.4. *Incomplete elliptic integral of the first kind.*

$F(\varphi \backslash \alpha), \quad \alpha$ constant

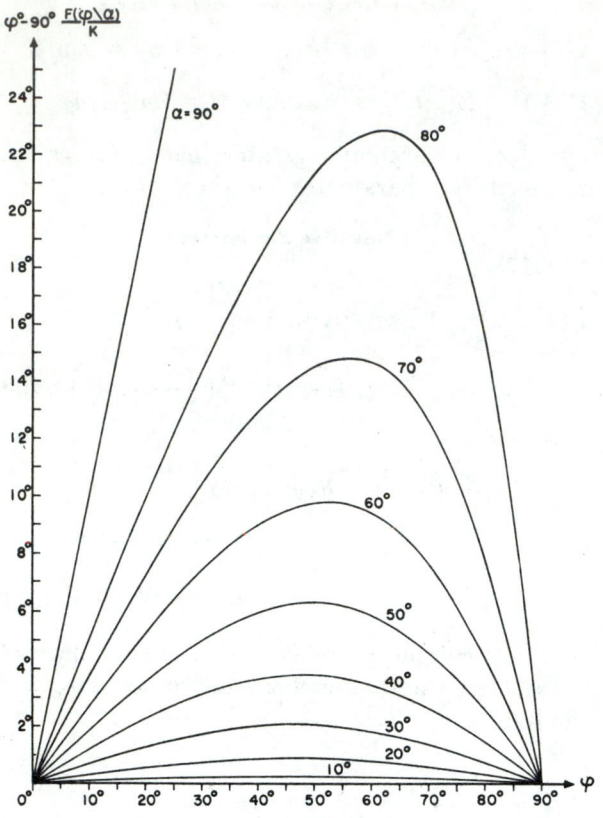

FIGURE 17.5. $\varphi - 90° \dfrac{F(\varphi \backslash \alpha)}{K}$, α constant.

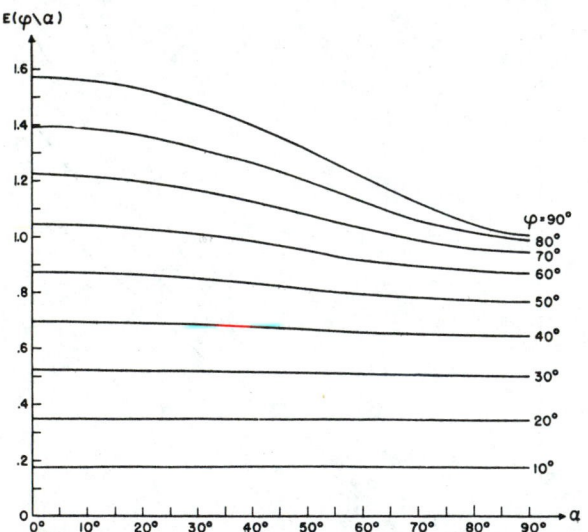

FIGURE 17.6. Incomplete elliptic integral of the second kind.

$E(\varphi \backslash \alpha)$, φ constant

Special Cases

17.4.19 $\qquad F(\varphi \backslash 0) = \varphi$

17.4.20 $\qquad F(i\varphi \backslash 0) = i\varphi$

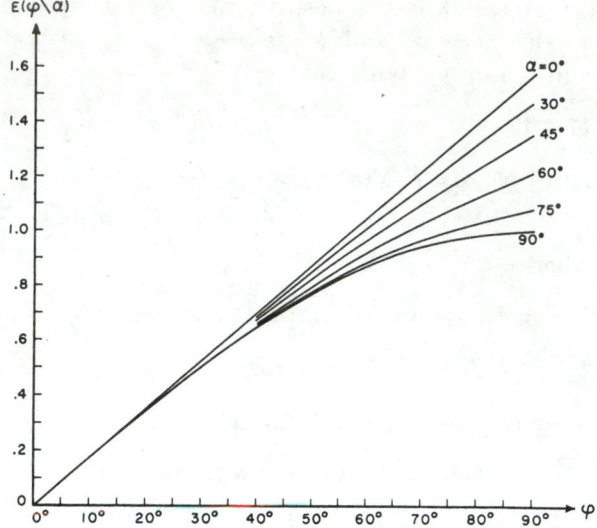

FIGURE 17.7. Incomplete elliptic integral of the second kind.

$E(\varphi \backslash \alpha)$, α constant

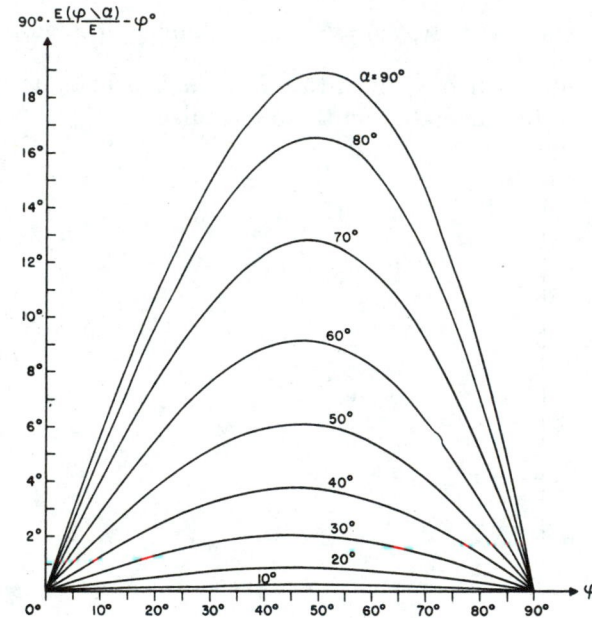

FIGURE 17.8. $90° \dfrac{E(\varphi \backslash \alpha)}{E} - \varphi$, α constant.

17.4.21
$$F(\varphi \backslash 90°) = \ln(\sec \varphi + \tan \varphi) = \ln \tan \left(\frac{\pi}{4} + \frac{\varphi}{2} \right)$$

17.4.22 $\qquad F(i\varphi \backslash 90°) = i \arctan (\sinh \varphi)$

17.4.23 $\qquad E(\varphi \backslash 0) = \varphi$

17.4.24 $\qquad E(i\varphi \backslash 0) = i\varphi$

17.4.25 $\qquad E(\varphi \backslash 90°) = \sin \varphi$

17.4.26 $\qquad E(i\varphi \backslash 90°) = i \sinh \varphi$

Jacobi's Zeta Function

17.4.27 $\quad Z(\varphi\backslash\alpha)=E(\varphi\backslash\alpha)-E(\alpha)F(\varphi\backslash\alpha)/K(\alpha)$

17.4.28 $\quad Z(u|m)=Z(u)=E(u)-uE(m)/K(m)$

17.4.29 $\quad Z(-u)=-Z(u)$

17.4.30 $\quad Z(u+2K)=Z(u)$

17.4.31 $\quad Z(K-u)=-Z(K+u)$

17.4.32 $\quad Z(u)=Z(u-K)-m\operatorname{sn}(u-K)\operatorname{cd}(u-K)$

Special Values

17.4.33 $\quad Z(u|0)=0$

17.4.34 $\quad Z(u|1)=\tanh u$

Addition Theorem

17.4.35

$$Z(u+v)=Z(u)+Z(v)-m\operatorname{sn}u\,\operatorname{sn}v\,\operatorname{sn}(u+v)$$

Jacobi's Imaginary Transformation

17.4.36

$$iZ(iu|m)=Z(u|m_1)+\frac{\pi u}{2KK'},-\operatorname{dn}(u|m_1)\operatorname{sc}(u|m_1)$$

Relation to Jacobi's Theta Function

17.4.37 $\quad Z(u)=\Theta'(u)/\Theta(u)=\dfrac{d}{du}\ln\,\Theta(u)$

q-Series

17.4.38 $\quad Z(u)=\dfrac{2\pi}{K}\displaystyle\sum_{s=1}^{\infty}q^{s}(1-q^{2s})^{-1}\sin\,(\pi su/K)$

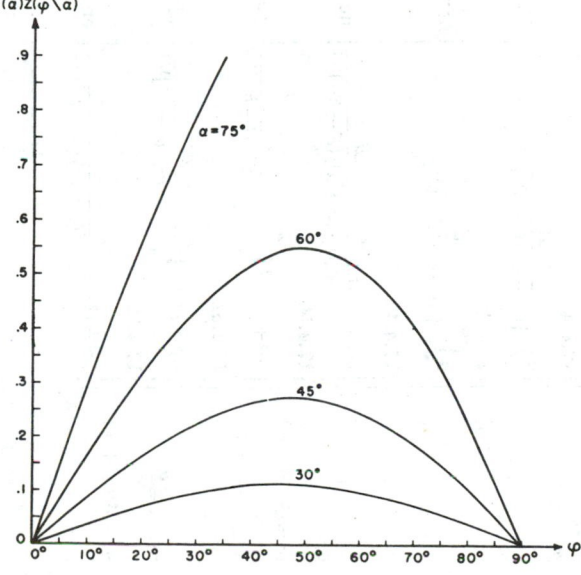

FIGURE 17.9. *Jacobian zeta function* $K(\alpha)Z(\varphi\backslash\alpha)$.

Heuman's Lambda Function

17.4.39

$$\Lambda_0(\varphi\backslash\alpha)=\frac{F(\varphi\backslash90°-\alpha)}{K'(\alpha)}+\frac{2}{\pi}K(\alpha)Z(\varphi\backslash90°-\alpha)$$

17.4.40

$$=\frac{2}{\pi}\{K(\alpha)E(\varphi\backslash90°-\alpha)$$

$$-[K(\alpha)-E(\alpha)]F(\varphi\backslash90°-\alpha)\}$$

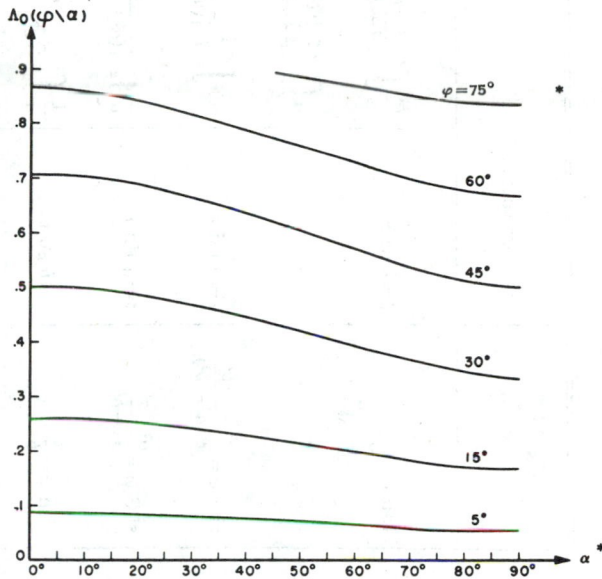

FIGURE 17.10. *Heuman's lambda function* $\Lambda_0(\varphi\backslash\alpha)$.

Numerical Evaluation of Incomplete Integrals of the First and Second Kinds

For the numerical evaluation of an elliptic integral the quartic (or cubic [4]) under the radical should first be expressed in terms of t^2, see **Examples 1** and **2**. In the resulting quartic there are only six possible sign patterns or combinations of the factors namely

$$(t^2+a^2)(t^2+b^2),\quad(a^2-t^2)(t^2-b^2),$$
$$(a^2-t^2)(b^2-t^2),\quad(t^2-a^2)(t^2-b^2),\quad(t^2+a^2)(t^2-b^2),$$
$$(t^2+a^2)(b^2-t^2).$$

The list which follows is then exhaustive for integrals which reduce to $F(\varphi\backslash\alpha)$ or $E(\varphi\backslash\alpha)$.

The value of the elliptic integral of the first kind is also expressed as an *inverse* Jacobian elliptic function. Here, for example, the notation $u=\operatorname{sn}^{-1}x$ means that $x=\operatorname{sn}u$.

The column headed "t substitution" gives the Jacobian elliptic function substitution which is appropriate to reduce every elliptic integral which contains the given quartic.

---

[4] For an alternate treatment of cubics see **17.4.61** and **17.4.70**.

Conditions	$F(\varphi\backslash\alpha)$	Equivalent Inverse Jacobian Elliptic Function	φ	t Substitution	$E(\varphi\backslash\alpha)$
$\cos\alpha=b/a$ $a>b$ $m=(a^2-b^2)/a^2$	**17.4.41** $a\displaystyle\int_0^x \frac{dt}{[(t^2+a^2)(t^2+b^2)]^{1/2}}$	$\mathrm{sc}^{-1}\left(\dfrac{x}{b}\Big|\dfrac{a^2-b^2}{a^2}\right)$	$\tan\varphi=\dfrac{x}{b}$	$t=b\,\mathrm{sc}\,v$	$\dfrac{b^2}{a}\displaystyle\int_0^x \dfrac{t^2+a^2}{t^2+b^2}\,\dfrac{dt}{[(t^2+a^2)(t^2+b^2)]^{1/2}}$
	17.4.42 $a\displaystyle\int_x^\infty \frac{dt}{[(t^2+a^2)(t^2+b^2)]^{1/2}}$	$\mathrm{cs}^{-1}\left(\dfrac{x}{a}\Big|\dfrac{a^2-b^2}{a^2}\right)$	$\tan\varphi=\dfrac{a}{x}$	$t=a\,\mathrm{cs}\,v$	$a\displaystyle\int_x^\infty \dfrac{t^2+b^2}{t^2+a^2}\,\dfrac{dt}{[(t^2+a^2)(t^2+b^2)]^{1/2}}$
	17.4.43 $a\displaystyle\int_b^x \frac{dt}{[(a^2-t^2)(t^2-b^2)]^{1/2}}$	$\mathrm{nd}^{-1}\left(\dfrac{x}{b}\Big|\dfrac{a^2-b^2}{a^2}\right)$	$\sin^2\varphi=\dfrac{a^2(x^2-b^2)}{x^2(a^2-b^2)}$	$t=b\,\mathrm{nd}\,v$	$ab^2\displaystyle\int_b^x \dfrac{1}{t^2}\,\dfrac{dt}{[(a^2-t^2)(t^2-b^2)]^{1/2}}$
	17.4.44 $a\displaystyle\int_x^a \frac{dt}{[(a^2-t^2)(t^2-b^2)]^{1/2}}$	$\mathrm{dn}^{-1}\left(\dfrac{x}{b}\Big|\dfrac{a^2-b^2}{a^2}\right)$	$\sin^2\varphi=\dfrac{a^2-x^2}{a^2-b^2}$	$t=a\,\mathrm{dn}\,v$	$\dfrac{1}{a}\displaystyle\int_x^a \dfrac{t^2\,dt}{[(a^2-t^2)(t^2-b^2)]^{1/2}}$
$\sin\alpha=b/a$ $a>b$ $m=b^2/a^2$	**17.4.45** $a\displaystyle\int_0^x \frac{dt}{[(a^2-t^2)(b^2-t^2)]^{1/2}}$	$\mathrm{sn}^{-1}\left(\dfrac{x}{b}\Big|\dfrac{b^2}{a^2}\right)$	$\sin\varphi=\dfrac{x}{b}$	$t=b\,\mathrm{sn}\,v$	$\dfrac{1}{a}\displaystyle\int_0^x \dfrac{(a^2-t^2)\,dt}{[(a^2-t^2)(b^2-t^2)]^{1/2}}$
	17.4.46 $a\displaystyle\int_x^b \frac{dt}{[(a^2-t^2)(b^2-t^2)]^{1/2}}$	$\mathrm{cd}^{-1}\left(\dfrac{x}{b}\Big|\dfrac{b^2}{a^2}\right)$	$\sin^2\varphi=\dfrac{a^2(b^2-x^2)}{b^2(a^2-x^2)}$	$t=b\,\mathrm{cd}\,v$	$a(a^2-b^2)\displaystyle\int_x^b \left(\dfrac{1}{a^2-t^2}\right)\dfrac{dt}{[(a^2-t^2)(b^2-t^2)]^{1/2}}$
	17.4.47 $a\displaystyle\int_a^x \frac{dt}{[(t^2-a^2)(t^2-b^2)]^{1/2}}$	$\mathrm{dc}^{-1}\left(\dfrac{x}{b}\Big|\dfrac{b^2}{a^2}\right)$	$\sin^2\varphi=\dfrac{x^2-a^2}{x^2-b^2}$	$t=a\,\mathrm{dc}\,v$	$\dfrac{a^2-b^2}{a}\displaystyle\int_a^x \left(\dfrac{t^2}{t^2-b^2}\right)\dfrac{dt}{[(t^2-a^2)(t^2-b^2)]^{1/2}}$
	17.4.48 $a\displaystyle\int_x^\infty \frac{dt}{[(t^2-a^2)(t^2-b^2)]^{1/2}}$	$\mathrm{ns}^{-1}\left(\dfrac{x}{a}\Big|\dfrac{b^2}{a^2}\right)$	$\sin\varphi=\dfrac{a}{x}$	$t=a\,\mathrm{ns}\,v$	$a\displaystyle\int_x^\infty \left(\dfrac{t^2-b^2}{t^2}\right)\dfrac{dt}{[(t^2-a^2)(t^2-b^2)]^{1/2}}$
$\cot\alpha=\dfrac{b}{a}$ $m=a^2/(a^2+b^2)$	**17.4.49** $(a^2+b^2)^{1/2}\displaystyle\int_b^x \frac{dt}{[(t^2+a^2)(t^2-b^2)]^{1/2}}$	$\mathrm{nc}^{-1}\left(\dfrac{x}{b}\Big|\dfrac{a^2}{a^2+b^2}\right)$	$\cos\varphi=\dfrac{b}{x}$	$t=b\,\mathrm{nc}\,v$	$\dfrac{b^2}{(a^2+b^2)^{1/2}}\displaystyle\int_b^x \dfrac{t^2+a^2}{t^2}\,\dfrac{dt}{[(t^2+a^2)(t^2-b^2)]^{1/2}}$
	17.4.50 $(a^2+b^2)^{1/2}\displaystyle\int_x^\infty \frac{dt}{[(t^2+a^2)(t^2-b^2)]^{1/2}}$	$\mathrm{ds}^{-1}\left(\dfrac{x}{(a^2+b^2)^{1/2}}\Big|\dfrac{a^2}{a^2+b^2}\right)$	$\sin^2\varphi=\dfrac{a^2+b^2}{a^2+x^2}$	$t=(a^2+b^2)^{1/2}\,\mathrm{ds}\,v$	$(a^2+b^2)^{1/2}\displaystyle\int_x^\infty \dfrac{t^2}{t^2+a^2}\,\dfrac{dt}{[(t^2+a^2)(t^2-b^2)]^{1/2}}$
$\tan\alpha=\dfrac{b}{a}$ $m=b^2/(a^2+b^2)$	**17.4.51** $(a^2+b^2)^{1/2}\displaystyle\int_0^x \frac{dt}{[(t^2+a^2)(b^2-t^2)]^{1/2}}$	$\mathrm{sd}^{-1}\left(\dfrac{x(a^2+b^2)^{1/2}}{ab}\Big|\dfrac{b^2}{a^2+b^2}\right)$	$\sin^2\varphi=\dfrac{x^2(a^2+b^2)}{b^2(a^2+x^2)}$	$t=\dfrac{ab}{(a^2+b^2)^{1/2}}\,\mathrm{sd}\,v$	$a^2(a^2+b^2)^{1/2}\displaystyle\int_0^x \dfrac{1}{(a^2+t^2)}\,\dfrac{dt}{[(t^2+a^2)(b^2-t^2)]^{1/2}}$
	17.4.52 $(a^2+b^2)^{1/2}\displaystyle\int_x^b \frac{dt}{[(t^2+a^2)(b^2-t^2)]^{1/2}}$	$\mathrm{cn}^{-1}\left(\dfrac{x}{b}\Big|\dfrac{b^2}{a^2+b^2}\right)$	$\cos\varphi=\dfrac{x}{b}$	$t=b\,\mathrm{cn}\,v$	$\dfrac{1}{(a^2+b^2)^{1/2}}\displaystyle\int_x^b \dfrac{(t^2+a^2)\,dt}{[(t^2+a^2)(b^2-t^2)]^{1/2}}$

Some Important Special Cases

$\tfrac{1}{2}F(\varphi\backslash\alpha)$	$\cos\varphi$	α	$\dfrac{1}{3^{1/4}}F(\varphi\backslash\alpha)$	$\cos\varphi$	α
17.4.53 $\displaystyle\int_x^\infty \frac{dt}{(1+t^4)^{\frac12}}$	$\dfrac{x^2-1}{x^2+1}$	$45°$	**17.4.57** $\displaystyle\int_x^\infty \frac{dt}{(t^3-1)^{\frac12}}$	$\dfrac{x-1-\sqrt3}{x-1+\sqrt3}$	$15°$
17.4.54 $\displaystyle\int_0^x \frac{dt}{(1+t^4)^{\frac12}}$	$\dfrac{1-x^2}{1+x^2}$	$45°$	**17.4.58** $\displaystyle\int_1^x \frac{dt}{(t^3-1)^{\frac12}}$	$\dfrac{\sqrt3+1-x}{\sqrt3-1+x}$	$15°$
17.4.55 $\dfrac{1}{2^{1/2}}\displaystyle\int_1^x \frac{dt}{(t^4-1)^{\frac12}}$	$\dfrac{1}{x}$	$45°$	**17.4.59** $\displaystyle\int_x^1 \frac{dt}{(1-t^3)^{\frac12}}$	$\dfrac{\sqrt3-1+x}{\sqrt3+1-x}$	$75°$
17.4.56 $\dfrac{1}{2^{1/2}}\displaystyle\int_x^1 \frac{dt}{(1-t^4)^{\frac12}}$	x	$45°$	**17.4.60** $\displaystyle\int_{-\infty}^x \frac{dt}{(1-t^3)^{\frac12}}$	$\dfrac{1-\sqrt3-x}{1+\sqrt3-x}$	$75°$

Reduction of $\int dt/\sqrt{P}$ where $P=P(t)$ is a cubic polynomial with three real factors $P=(t-\beta_1)(t-\beta_2)(t-\beta_3)$ where $\beta_1>\beta_2>\beta_3$. Write

17.4.61

$$\lambda=\frac{1}{2}(\beta_1-\beta_3)^{1/2},\quad m=\sin^2\alpha=\frac{\beta_2-\beta_3}{\beta_1-\beta_3},$$
$$m_1=\cos^2\alpha=\frac{\beta_1-\beta_2}{\beta_1-\beta_3}$$

17.4.62 $\lambda\displaystyle\int_{\beta_3}^x \frac{dt}{\sqrt{P}}$	$F(\varphi\backslash\alpha)$	$\sin^2\varphi=\dfrac{x-\beta_3}{\beta_2-\beta_3}$
17.4.63 $\lambda\displaystyle\int_x^{\beta_2} \frac{dt}{\sqrt{P}}$	$F(\varphi\backslash\alpha)$	$\cos^2\varphi=\dfrac{(\beta_1-\beta_2)(x-\beta_3)}{(\beta_2-\beta_3)(\beta_1-x)}$
17.4.64 $\lambda\displaystyle\int_{\beta_1}^x \frac{dt}{\sqrt{P}}$	$F(\varphi\backslash\alpha)$	$\sin^2\varphi=\dfrac{x-\beta_1}{x-\beta_2}$
17.4.65 $\lambda\displaystyle\int_x^\infty \frac{dt}{\sqrt{P}}$	$F(\varphi\backslash\alpha)$	$\cos^2\varphi=\dfrac{x-\beta_1}{x-\beta_3}$
17.4.66 $\lambda\displaystyle\int_{-\infty}^x \frac{dt}{\sqrt{-P}}$	$F(\varphi\backslash(90°-\alpha°))$	$\sin^2\varphi=\dfrac{\beta_1-\beta_3}{\beta_1-x}$
17.4.67 $\lambda\displaystyle\int_x^{\beta_3} \frac{dt}{\sqrt{-P}}$	$F(\varphi\backslash(90°-\alpha°))$	$\cos^2\varphi=\dfrac{\beta_2-\beta_3}{\beta_2-x}$
17.4.68 $\lambda\displaystyle\int_{\beta_2}^x \frac{dt}{\sqrt{-P}}$	$F(\varphi\backslash(90°-\alpha°))$	$\sin^2\varphi=\dfrac{(\beta_1-\beta_3)(x-\beta_2)}{(\beta_1-\beta_2)(x-\beta_3)}$
17.4.69 $\lambda\displaystyle\int_x^{\beta_1} \frac{dt}{\sqrt{-P}}$	$F(\varphi\backslash(90°-\alpha°))$	$\cos^2\varphi=\dfrac{x-\beta_2}{\beta_1-\beta_2}$

Reduction of $\int dt/\sqrt{P}$ when $P=P(t)=t^3+a_1t^2+a_2t+a_3$ is a cubic polynomial with only one real root $t=\beta$. We form the first and second derivatives $P'(t)$, $P''(t)$ with respect to t and then write

17.4.70 $\quad \lambda^2=[P'(\beta)]^{1/2},\quad m=\sin^2\alpha=\dfrac{1}{2}-\dfrac{1}{8}\dfrac{P''(\beta)}{[P'(\beta)]^{1/2}}$

17.4.71 $\lambda\displaystyle\int_\beta^x \frac{dt}{\sqrt{P}}$	$F(\varphi\backslash\alpha)$	$\cos\varphi=\dfrac{\lambda^2-(x-\beta)}{\lambda^2+(x-\beta)}$
17.4.72 $\lambda\displaystyle\int_x^\infty \frac{dt}{\sqrt{P}}$	$F(\varphi\backslash\alpha)$	$\cos\varphi=\dfrac{(x-\beta)-\lambda^2}{(x-\beta)+\lambda^2}$
17.4.73 $\lambda\displaystyle\int_{-\infty}^x \frac{dt}{\sqrt{(-P)}}$	$F(\varphi\backslash(90°-\alpha°))$	$\cos\varphi=\dfrac{(\beta-x)-\lambda^2}{(\beta-x)+\lambda^2}$
17.4.74 $\lambda\displaystyle\int_x^\beta \frac{dt}{\sqrt{(-P)}}$	$F(\varphi\backslash(90°-\alpha°))$	$\cos\varphi=\dfrac{\lambda^2-(\beta-x)}{\lambda^2+(\beta-x)}$

17.5. Landen's Transformation

Descending Landen Transformation [5]

Let α_n, α_{n+1} be two modular angles such that

17.5.1 $\quad (1+\sin\alpha_{n+1})(1+\cos\alpha_n)=2 \qquad (\alpha_{n+1}<\alpha_n)$

and let φ_n, φ_{n+1} be two corresponding amplitudes such that

17.5.2 $\quad \tan(\varphi_{n+1}-\varphi_n)=\cos\alpha_n\tan\varphi_n \quad (\varphi_{n+1}>\varphi_n)$

[5] The emphasis here is on the modular angle since this is an argument of the Tables. All formulae concerning Landen's transformation may also be expressed in terms of the modulus $k=m^{\frac12}=\sin\alpha$ and its complement $k'=m_1^{\frac12}=\cos\alpha$.

Thus the step from n to $n+1$ decreases the modular angle but increases the amplitude. By iterating the process we can descend from a given modular angle to one whose magnitude is negligible, when **17.4.19** becomes applicable.

With $\alpha_0 = \alpha$ we have

17.5.3

$$F(\varphi \setminus \alpha) = (1 + \cos \alpha)^{-1} F(\varphi_1 \setminus \alpha_1)$$
$$= \tfrac{1}{2}(1 + \sin \alpha_1) F(\varphi_1 \setminus \alpha_1)$$

17.5.4 $\quad F(\varphi \setminus \alpha) = 2^{-n} \prod_{s=1}^{n} (1 + \sin \alpha_s) \dot{F}(\varphi_n \setminus \alpha_n)$

17.5.5 $\quad F(\varphi \setminus \alpha) = \Phi \prod_{s=1}^{\infty} (1 + \sin \alpha_s)$

17.5.6 $\quad \Phi = \lim_{n \to \infty} \dfrac{1}{2^n} F(\varphi_n \setminus \alpha_n) = \lim_{n \to \infty} \dfrac{\varphi_n}{2^n}$

17.5.7 $\quad K = F(\tfrac{1}{2}\pi \setminus \alpha) = \tfrac{1}{2}\pi \prod_{s=1}^{\infty} (1 + \sin \alpha_s)$

17.5.8 $\quad F(\varphi \setminus \alpha) = 2\pi^{-1} K \Phi$

17.5.9

$$E(\varphi \setminus \alpha) = F(\varphi \setminus \alpha)\left[1 - \frac{1}{2}\sin^2\alpha\left(1 + \frac{1}{2}\sin\alpha_1\right.\right.$$
$$\left.\left. + \frac{1}{2^2}\sin\alpha_1\sin\alpha_2 + \dots\right)\right] + \sin\alpha\left[\frac{1}{2}(\sin\alpha_1)^{1/2}\sin\varphi_1\right.$$
$$\left. + \frac{1}{2^2}(\sin\alpha_1\sin\alpha_2)^{1/2}\sin\varphi_2 + \dots\right]$$

17.5.10

$$E = K\left[1 - \frac{1}{2}\sin^2\alpha\left(1 + \frac{1}{2}\sin\alpha_1 + \frac{1}{2^2}\sin\alpha_1\sin\alpha_2\right.\right.$$
$$\left.\left. + \frac{1}{2^3}\sin\alpha_1\sin\alpha_2\sin\alpha_3 + \dots\right)\right]$$

Ascending Landen Transformation

Let α_n, α_{n+1} be two modular angles such that

17.5.11 $\quad (1 + \sin \alpha_n)(1 + \cos \alpha_{n+1}) = 2 \qquad (\alpha_{n+1} > \alpha_n)$

and let φ_n, φ_{n+1} be two corresponding amplitudes such that

17.5.12 $\quad \sin(2\varphi_{n+1} - \varphi_n) = \sin\alpha_n \sin\varphi_n \qquad (\varphi_{n+1} < \varphi_n)$

Thus the step from n to $n+1$ increases the modular angle but decreases the amplitude. By iterating the process we can ascend from a given modular angle to one whose difference from a right angle is so small that **17.4.21** becomes applicable.

With $\alpha_0 = \alpha$ we have

17.5.13 $\quad F(\varphi \setminus \alpha) = 2(1 + \sin \alpha)^{-1} F(\varphi_1 \setminus \alpha_1)$

17.5.14 $\quad F(\varphi \setminus \alpha) = 2^n \prod_{s=0}^{n=1} (1 + \sin \alpha_s)^{-1} F(\varphi_n \setminus \alpha_n)$

17.5.15 $\quad F(\varphi \setminus \alpha) = \prod_{s=1}^{n} (1 + \cos \alpha_s) F(\varphi_n \setminus \alpha_n)$

17.5.16 $\quad F(\varphi \setminus \alpha) = [\csc \alpha \prod_{s=1}^{\infty} \sin \alpha_s]^{\frac{1}{2}} \ln \tan (\tfrac{1}{4}\pi + \tfrac{1}{2}\Phi)$

17.5.17 $\qquad \Phi = \lim_{n \to \infty} \varphi_n$

Neighborhood of a Right Angle (see also **17.4.13**)

When both φ and α are near to a right angle, interpolation in the table $F(\varphi \setminus \alpha)$ is difficult. Either Landen's transformation can then be used with advantage to increase the modular angle and decrease the amplitude or vice-versa.

17.6. The Process of the Arithmetic-Geometric Mean

Starting with a given number triple (a_0, b_0, c_0) we proceed to determine number triples (a_1, b_1, c_1), (a_2, b_2, c_2), ..., (a_N, b_N, c_N) according to the following scheme of arithmetic and geometric means

17.6.1

$$
\begin{array}{lll}
a_0 & b_0 & \\
a_1 = \tfrac{1}{2}(a_0 + b_0) & b_1 = (a_0 b_0)^{\frac{1}{2}} & c_0 \\
a_2 = \tfrac{1}{2}(a_1 + b_1) & b_2 = (a_1 b_1)^{\frac{1}{2}} & c_1 = \tfrac{1}{2}(a_0 - b_0) \\
\vdots & \vdots & c_2 = \tfrac{1}{2}(a_1 - b_1) \\
& & \vdots \\
a_N = \tfrac{1}{2}(a_{N-1} + b_{N-1}) & b_N = (a_{N-1} b_{N-1})^{\frac{1}{2}} & c_N = \tfrac{1}{2}(a_{N-1} - b_{N-1}).
\end{array}
$$

We stop at the Nth step when $a_N = b_N$, i.e., when $c_N = 0$ to the degree of accuracy to which the numbers are required.

To determine the complete elliptic integrals $K(\alpha)$, $E(\alpha)$ we start with

17.6.2 $\qquad a_0 = 1, \; b_0 = \cos \alpha, \; c_0 = \sin \alpha$

whence

17.6.3 $\qquad\qquad K(\alpha) = \dfrac{\pi}{2a_N}$

17.6.4 $\dfrac{K(\alpha)-E(\alpha)}{K(\alpha)}=\dfrac{1}{2}\,[c_0^2+2c_1^2+2^2c_2^2+\,\ldots\,+2^Nc_N^2]$

To determine $K'(\alpha)$, $E'(\alpha)$ we start with

17.6.5 $\qquad a_0'=1,\; b_0'=\sin\alpha,\; c_0'=\cos\alpha$

whence

17.6.6 $\qquad\qquad K'(\alpha)=\dfrac{\pi}{2a_N'}$

17.6.7

$$\dfrac{K'(\alpha)-E'(\alpha)}{K'(\alpha)}=\dfrac{1}{2}\,[c_0'^2+2c_1'^2+2^2c_2'^2+\,\ldots\,+2^Nc_N'^2]$$

To calculate $F(\varphi\backslash\alpha)$, $E(\varphi\backslash\alpha)$ start from **17.5.2** which corresponds to the descending Landen transformation and determine φ_1, φ_2, $\ldots$, φ_N successively from the relation

17.6.8 $\quad\tan\,(\varphi_{n+1}-\varphi_n)=(b_n/a_n)\,\tan\varphi_n,\;\varphi_0=\varphi$

Then to the prescribed accuracy

17.6.9 $\qquad\quad F(\varphi\backslash\alpha)=\varphi_N/(2^Na_N)\qquad$ *

17.6.10

$Z(\varphi\backslash\alpha)=E(\varphi\backslash\alpha)-(E/K)F(\varphi\backslash\alpha)$

* $\qquad\quad =c_1\sin\varphi_1+c_2\sin\varphi_2+\,\ldots\,+c_N\sin\varphi_N$

17.7. Elliptic Integrals of the Third Kind

17.7.1

$\Pi(n;\varphi\backslash\alpha)=\displaystyle\int_0^\varphi\,(1-n\sin^2\theta)^{-1}(1-\sin^2\alpha\sin^2\theta)^{-\frac{1}{2}}d\theta$

17.7.2 $\qquad\qquad \Pi\,(n;\tfrac{1}{2}\pi\backslash\alpha)=\Pi\,(n\backslash\alpha)$

Case (i) Hyperbolic Case $0<n<\sin^2\alpha$

$\epsilon=\arcsin\,(n/\sin^2\alpha)^{\frac{1}{2}},\qquad 0\leq\epsilon\leq\tfrac{1}{2}\pi$

$\beta=\tfrac{1}{2}\pi F(\epsilon\backslash\alpha)/K(\alpha)$

$q=q(\alpha)$

$v=\tfrac{1}{2}\pi F(\varphi\backslash\alpha)/K(\alpha),$

$\delta_1=[n(1-n)^{-1}(\sin^2\alpha-n)^{-1}]^{\frac{1}{2}}$

17.7.3

$\Pi(n;\varphi\backslash\alpha)=\delta_1\,[-\tfrac{1}{2}\ln\,[\vartheta_4(v+\beta)/\vartheta_4(v-\beta)]$

$\qquad\qquad\qquad\qquad +v\vartheta_1'(\beta)/\vartheta_1(\beta)]$

17.7.4

$\tfrac{1}{2}\ln\dfrac{\vartheta_4(v+\beta)}{\vartheta_4(v-\beta)}=2\,\displaystyle\sum_{s=1}^\infty\,s^{-1}q^s(1-q^{2s})^{-1}\sin 2sv\sin 2s\beta$

17.7.5

$\dfrac{\vartheta_1'(\beta)}{\vartheta_1(\beta)}=\cot\,\beta+4\,\displaystyle\sum_{s=1}^\infty\,q^{2s}(1-2q^{2s}\cos 2\beta+q^{4s})^{-1}\sin 2\beta$

In the above we can also use Neville's theta functions **16.36.**

17.7.6 $\qquad \Pi(n\backslash\alpha)=K(\alpha)+\delta_1K(\alpha)Z(\epsilon\backslash\alpha)$

Case (ii) Hyperbolic Case $n>1$

The case $n>1$ can be reduced to the case $0<N<\sin^2\alpha$ by writing

17.7.7 $\quad N=n^{-1}\sin^2\alpha,\;p_1=[(n-1)(1-n^{-1}\sin^2\alpha)]^{\frac{1}{2}}$

17.7.8

$\Pi(n;\,\varphi\backslash\alpha)=-\Pi(N;\,\varphi\backslash\alpha)+F(\varphi\backslash\alpha)$

$\qquad\qquad +\dfrac{1}{2p_1}\ln\,[(\Delta(\varphi)+p_1\tan\varphi)(\Delta(\varphi)-p_1\tan\varphi)^{-1}]$

where $\Delta(\varphi)$ is the delta amplitude, **17.2.4.**

17.7.9 $\qquad\qquad \Pi(n\backslash\alpha)=K(\alpha)-\Pi(N\backslash\alpha)$

Case (iii) Circular Case $\sin^2\alpha<n<1$

$\epsilon=\arcsin\,[(1-n)/\cos^2\alpha]^{\frac{1}{2}}\qquad 0\leq\epsilon\leq\tfrac{1}{2}\pi$

$\beta=\tfrac{1}{2}\pi F(\epsilon\backslash 90°-\alpha)/K(\alpha)$

$q=q(\alpha)$

17.7.10

$v=\tfrac{1}{2}\pi F(\varphi\backslash\alpha)/K(\alpha),\;\delta_2=[n(1-n)^{-1}(n-\sin^2\alpha)^{-1}]^{\frac{1}{2}}$

17.7.11 $\qquad\qquad \Pi(n;\,\varphi\backslash\alpha)=\delta_2(\lambda-4\mu v)$

17.7.12

$\lambda=\arctan\,(\tanh\beta\tan v)$

$\qquad +2\,\displaystyle\sum_{s=1}^\infty\,(-1)^{s-1}s^{-1}q^{2s}(1-q^{2s})^{-1}\sin 2sv\sinh 2s\beta$

17.7.13

$\mu=\Bigg[\displaystyle\sum_{s=1}^\infty\,sq^{s^2}\sinh 2s\beta\Bigg]\Bigg[1+2\displaystyle\sum_{s=1}^\infty\,q^{s^2}\cosh 2s\beta\Bigg]^{-1}$

17.7.14 $\quad \Pi(n\backslash\alpha)=K(\alpha)+\tfrac{1}{2}\pi\delta_2[1-\Lambda_0(\epsilon\backslash\alpha)]$

where Λ_0 is Heuman's Lambda function, **17.4.39.**

---

*See page II.

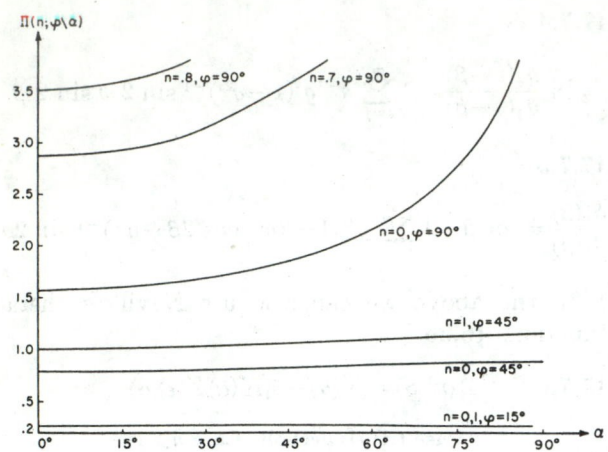

FIGURE 17.11. *Elliptic integral of the third kind* $\Pi(n; \varphi\backslash\alpha)$.

Case (iv) Circular Case $n<0$

The case $n<0$ can be reduced to the case $\sin^2\alpha<N<1$ by writing

17.7.15

$$N=(\sin^2\alpha-n)(1-n)^{-1}$$

$$p_2=[-n(1-n)^{-1}(\sin^2\alpha-n)]^{\frac{1}{2}}$$

17.7.16

$$[(1-n)(1-n^{-1}\sin^2\alpha)]^{\frac{1}{2}}\Pi(n;\varphi\backslash\alpha)$$
$$=[(1-N)(1-N^{-1}\sin^2\alpha)]^{\frac{1}{2}}\Pi(N;\varphi\backslash\alpha)$$
$$+p_2^{-1}\sin^2\alpha F(\varphi\backslash\alpha)+\arctan[\tfrac{1}{2}p_2\sin 2\varphi/\Delta(\varphi)]$$

17.7.17

$$\Pi(n\backslash\alpha)=(-n\cos^2\alpha)(1-n)^{-1}(\sin^2\alpha-n)^{-1}\Pi(N\backslash\alpha)$$
$$+\sin^2\alpha(\sin^2\alpha-n)^{-1}K(\alpha)$$

Special Cases

17.7.18
$$n=0$$
$$\Pi(0;\varphi\backslash\alpha)=F(\varphi\backslash\alpha)$$

17.7.19
$$n=0,\ \alpha=0$$
$$\Pi(0;\varphi\backslash0)=\varphi$$

17.7.20
$$\alpha=0$$
$$\Pi(n;\varphi\backslash0)=(1-n)^{-\frac{1}{2}}\arctan[(1-n)^{\frac{1}{2}}\tan\varphi], \qquad *$$
$$n<1$$
$$=(n-1)^{-\frac{1}{2}}\operatorname{arctanh}[(n-1)^{\frac{1}{2}}\tan\varphi],$$
$$n>1$$
$$=\tan\varphi \qquad n=1$$

17.7.21
$$\alpha=\pi/2$$
$$\Pi(n;\varphi\backslash\pi/2)=(1-n)^{-1}[\ln(\tan\varphi+\sec\varphi)$$
$$-\tfrac{1}{2}n^{\frac{1}{2}}\ln(1+n^{\frac{1}{2}}\sin\varphi)(1-n^{\frac{1}{2}}\sin\varphi)^{-1}] \qquad n\neq1$$

17.7.22
$$n=\pm\sin\alpha$$
$$(1\mp\sin\alpha)\{2\Pi(\pm\sin\alpha;\varphi\backslash\alpha)-F(\varphi\backslash\alpha)\}$$
$$=\arctan[(1\mp\sin\alpha)\tan\varphi/\Delta(\varphi)]$$

17.7.23
$$n=1\pm\cos\alpha$$
$$2\cos\alpha\Pi(1\pm\cos\alpha;\varphi\backslash\alpha)=\pm\tfrac{1}{2}\ln[(1+\tan\varphi$$
$$\cdot\Delta(\varphi))(1-\tan\varphi\cdot\Delta(\varphi))^{-1}]+\tfrac{1}{2}\ln[(\Delta(\varphi)$$
$$+\cos\alpha\cdot\tan\varphi)(\Delta(\varphi)-\cos\alpha\tan\varphi)^{-1}]$$
$$\mp(1\mp\cos\alpha)F(\varphi\backslash\alpha)$$

17.7.24
$$n=\sin^2\alpha$$
$$\Pi(\sin^2\alpha;\varphi\backslash\alpha)=\sec^2\alpha E(\varphi\backslash\alpha)-(\tan^2\alpha\sin 2\varphi)/(2\Delta(\varphi))$$

17.7.25
$$n=1$$
$$\Pi(1;\varphi\backslash\alpha)=F(\varphi\backslash\alpha)-\sec^2\alpha E(\varphi\backslash\alpha)+\sec^2\alpha\tan\varphi\Delta(\varphi)$$

Numerical Methods

17.8. Use and Extension of the Tables

Example 1. Reduce to canonical form $\int y^{-1}dx$, where

$$y^2=-3x^4+34x^3-119x^2+172x-90$$

By inspection or by solving an equation of the fourth degree we find that

$$y^2=Q_1Q_2 \text{ where } Q_1=3x^2-10x+9,\ Q_2=-x^2+8x-10$$

First Method

$Q_1-\lambda Q_2=(3+\lambda)x^2-(10+8\lambda)x+9+10\lambda$ is a perfect square if the discriminant

$$(10+8\lambda)^2-4(3+\lambda)(9+10\lambda)=0;\text{ i.e., if }\lambda=-\frac{2}{3}\text{ or }\frac{1}{2}$$

and then

$$Q_1+\frac{2}{3}Q_2=\frac{7}{3}(x-1)^2,\ Q_1-\frac{1}{2}Q_2=\frac{7}{2}(x-2)^2$$

Solving for Q_1 and Q_2 we get

$$Q_1=(x-1)^2+2(x-2)^2,\ Q_2=2(x-1)^2-3(x-2)^2$$

The substitution $t=(x-1)/(x-2)$ then gives

$$\int y^{-1}dx=\pm\int[(t^2+2)(2t^2-3)]^{-\frac{1}{2}}dt$$

*See page II.

If the quartic $y^2=0$ has four real roots in x (or in the case of a cubic all three roots are real), we must so combine the factors that no root of $Q_1=0$ lies between the roots of $Q_2=0$ and no root of $Q_2=0$ lies between the roots of $Q_1=0$. Provided this condition is observed the method just described will always lead to real values of λ. These values may, however, be irrational.

Second Method

Write

$$t^2=\frac{Q_1}{Q_2}=\frac{3x^2-10x+9}{-x^2+8x-10}$$

and let the discriminant of $Q_2t^2-Q_1$ be

$$4T^2=(8t^2+10)^2-4(t^2+3)(10t^2+9)$$
$$=4(3t^2+2)(2t^2-1)$$

Then

$$\int y^{-1}dx=\pm\int T^{-1}dt=\pm\int[(3t^2+2)(2t^2-1)]^{-\frac{1}{2}}dt$$

This method will succeed if, as here, T^2 as a function of t^2 has real factors. If the coefficients of the given quartic are rational numbers, the factors of T^2 will likewise be rational.

Third Method

Write

$$w=\frac{Q_1}{Q_2}=\frac{3x^2-10x+9}{-x^2+8x-10}$$

and let the discriminant of Q_2w-Q_1 be

$$4W=4(3w+2)(2w-1)=4(Aw^2+Bw+C)$$

Then if

$$z^2=W/w \text{ and } Z^2=(B-z^2)^2-4AC=(z^2-1)^2+48$$

$$\int y^{-1}dx=\pm\int Z^{-1}dz$$

However, in this case the factors of Z are complex and the method fails.

Of the second and third methods one will always succeed where the other fails, and if the coefficients of the given quartic are rational numbers, the factors of T^2 or Z^2, as the case may be, will be rational.

Example 2. Reduce to canonical form $\int y^{-1}dx$ where $y^2=x(x-1)(x-2)$.

We use the third method of **Example 1** taking $Q_1=(x-1)$, $Q_2=x(x-2)$ and writing

$$w=\frac{Q_1}{Q_2}=\frac{x-1}{x^2-2x}$$

The discriminant of $Q_2w-Q_1=x^2w-(2w+1)x+1$ is

$$4W=(2w+1)^2-4w=4w^2+1$$

so that

$$W=Aw^2+Bw+C \text{ where } A=1,\ B=0,\ C=\frac{1}{4}$$

and if we write $z^2=W/w$ and

$$Z^2=(B-z^2)^2-4AC=(z^2)^2-1=(z^2-1)(z^2+1),$$

$$\int y^{-1}dx=\pm\int[(z^2-1)(z^2+1)]^{-1/2}\,dz$$

The first method of **Example 1** fails with the above values of Q_1 and Q_2 since the root of $Q_1=0$ lies between the roots of $Q_2=0$, and we get imaginary values of λ. The method succeeds, however, if we take $Q_1=x$, $Q_2=(x-1)(x-2)$, for then the roots of $Q_1=0$ do not lie between those of $Q_2=0$.

Example 3. Find $K(80/81)$.

First Method

Use **17.3.29** with $m=80/81$, $m_1=1/81$, $m_1^{1/2}=1/9$. Since $[(1-m_1^{1/2})(1+m_1^{1/2})^{-1}]^2=.64$, $K(80/81)=1.8\ K(.64)=3.59154\ 500$ to 8D, taking $K(.64)$ from **Table 17.1**.

Second Method

Table 17.4 giving $L(m)$ is useful for computing $K(m)$ when m is near unity or $K'(m)$ when m is near zero.

$$K(80/81)=\frac{1}{\pi}\ K'(80/81)\ \ln\ (16\times81)-L(80/81).$$

By interpolation in **Tables 17.1** and **17.4**, since $80/81=.98765\ 43210$,

$$K'(80/81)=1.57567\ 8423$$
$$L(80/81)=.00311\ 16543$$
$$K(80/81)=\pi^{-1}(1.57567\ 8423)(7.16703\ 7877)$$
$$-.00311\ 16543$$
$$=3.59154\ 5000 \text{ to 9D.}$$

Third Method

The polynomial approximation **17.3.34** gives to 8D

$$K(80/81)=3.59154\ 501$$

Fourth Method, Arithmetic-Geometric Mean

Here $\sin^2\alpha=80/81$ and we start with

$$a_0=1,\ b_0=\frac{1}{9},\ c_0=\sqrt{80/81}=.99380\ 79900$$

giving

n	a_n	b_n	c_n
0	1.00000 00000	.11111 11111	.99380 79900
1	.55555 55555	.33333 33333	.44444 44444
2	.44444 44444	.43033 14829	.11111 11111
3	.43738 79636	.43733 10380	.00705 64808
4	.43735 95008	.43735 94999	.00002 84628
5	.43735 95003	.43735 95003	0

Thus $K(80/81) = \frac{1}{2} \pi a_5^{-1} = 3.59154\ 5001$.

Example 4. Find $E(80/81)$.

First Method

Use **17.3.30** which gives, with $m = 80/81$

$$E(80/81) = \frac{10}{9} E(.64) - \frac{1}{5} K(.64)$$
$$= 1.01910\ 6047$$

taking $E(.64)$ and $K(.64)$ from **Table 17.1.**

Second Method

Polynomial approximation, **17.3.36** gives $E(80/81) = 1.01910\ 6060$. The last two figures must be dropped to $\overline{\text{keep}}$ within the limit of accuracy of the method.

Third Method

Arithmetic-geometric mean, **17.6.** The numbers were calculated in **Example 3,** fourth method, and we have

$$\frac{K(80/81) - E(80/81)}{K(80/81)} = \frac{1}{2} [c_0^2 + 2c_1^2 + 2^2 c_2^2 + \ldots + 2^5 c_5^2]$$

$$= \frac{1}{2} [1.43249\ 71298]$$

$$= .71624\ 85649.$$

Using the value of $K(80/81)$ found in **Example 3,** fourth method, we have

$$E(80/81) = 1.01910\ 6048 \text{ to } 9D.$$

Example 5. Find q when $m = .9995$.
Here $m_1 = .0005$ and so from **Table 17.4**

$$Q(m) = .06251\ 563013$$
$$q_1 = m_1 Q(m) = .00003\ 12578\ 15.$$

From **17.3.19**

$$\ln\left(\frac{1}{q}\right) = \pi^2 / \ln\left(\frac{1}{q_1}\right) = \pi^2 / 10.37324\ 1132$$

$$= .95144\ 84701$$

$$q = .38618\ 125.$$

The computation could also be made using common logarithms with the aid of **17.3.20**. The point of this procedure is that it enables us to calculate q_1 without the loss of significant figures which would result from direct interpolation in **Table 17.1**. By this means $\ln (1/q_1)$ can be found without loss of accuracy.

Example 6. Find m to 10D when $K'/K = .25$ and when $K'/K = 3.5$.

From **17.3.15** with $K'/K = .25$ we can write the iteration formula

$$m^{(n+1)} = 1 - 16e^{-4\pi} \exp\left[-\pi L(m^{(n)}) / K'(m^{(n)})\right].$$

Then by iteration using **Tables 17.1** and **17.4**

n	$m^{(n)}$
0	1.
1	.99994 42025
2	.99994 42041
3	.99994 42041

Thus $m = .99994\ 42041$.

From **17.3.16** with $K'/K = 3.5$ we can write the iteration formula,

$$m^{(n+1)} = 16e^{-3.5\pi} \exp\left[-\pi L(m_1^{(n)}) / K(m^{(n)})\right]$$

n	$m^{(n)}$
0	0
1	.(3)26841 25043
2	.(3)26837 65
3	.(3)26837 65

Thus $m = .00026\ 83765$.

The above methods in conjunction with the auxiliary **Table 17.4** of $L(m)$ enable us to extend **Table 17.3** for $K'/K > 3$, and for $K'/K < .3$.

Example 7. Calculate to 5D the Jacobian elliptic function sn $(.75342|.7)$ using **Table 17.5.**
Here

$$m = \sin^2 \alpha = .7, \quad \alpha = 56.789089°.$$

Thus, sn $(.75342|.7) = \sin \varphi$ where φ is determined from

$$F(\varphi \backslash 56.789089°) = .75342.$$

Inspection of **Table 17.5** shows that φ lies between 40° and 45°. We have from the table of $F(\varphi \backslash \alpha)$

α φ	56°	58°	60°
35°	.63803	.63945	.64085
40°	.73914	.74138	.74358
45°	.84450	.84788	.85122
50°	.95479	.95974	.96465

From this we form the table of $F(\varphi \backslash 56.789089°)$

φ	F	Δ	Δ_2	Δ_3
35°	.63859			
		10144		
40°	.74003		437	
		10581		72
45°	.84584		509	
		11090		
50°	.95674			

A rough estimate now shows that φ lies between 40° and 41°. We therefore form the following table of $F(\varphi \backslash 56.789089°)$ by direct interpolation in the foregoing table

φ	F
40.0°	.74003
40.5°	.75040
41.0°	.76082

whence by linear inverse interpolation

$$\varphi = 40.5° + .5° \left[\frac{.75342 - .75040}{.76082 - .75040} \right] = 40.6449°$$

and so $\sin \varphi = .65137 = \text{sn } (.75342|.7)$.

This method of bivariate interpolation is given merely as an illustration. Other more direct methods such as that of the arithmetic-geometric mean described in **17.6** and illustrated for the Jacobian functions in chapter **16** are less laborious.

Example 8. Evaluate

$$\int_2^3 [(2t^2 + 1)(t^2 - 2)]^{-1/2} dt.$$

First Method, Bivariate Interpolation

From **17.4.50** we have

$$\sqrt{5} \int_2^3 [(2t^2 + 1)(t^2 - 2)]^{-1/2} dt = F(\varphi_1 \backslash \alpha) - F(\varphi_2 \backslash \alpha)$$

where

$$\sin^2 \alpha = \frac{1}{5}, \cos \varphi_1 = \frac{\sqrt{2}}{3}, \cos \varphi_2 = \frac{\sqrt{2}}{2}$$

Thus $\alpha = 26.56505 \ 12°$, $\varphi_1 = 61.87449 \ 43°$, $\varphi_2 = 45°$, $F(\varphi_1 \backslash \alpha) = 1.115921$ and $F(\varphi_2 \backslash \alpha) = .800380$ and **therefore** the integral is equal to .141114.

Second Method, Numerical Quadrature

Simpson's formula with 11 ordinates and interval .1 gives .141117.

Example 9. Evaluate

$$\int_2^4 [(t^2 - 2)(t^2 - 4)]^{-\frac{1}{2}} dt.$$

First Method, Reduction to Standard Form and Bivariate Interpolation

Here we can use **17.4.48** noting that $a^2 = 4$, $b^2 = 2$, and that

$$\int_2^4 [(t^2 - 2)(t^2 - 4)]^{-\frac{1}{2}} dt = \int_2^\infty - \int_4^\infty$$

$$= \frac{1}{2} [F(\varphi_1 \backslash 45°) - F(\varphi_2 \backslash 45°)]$$

$$= \frac{1}{2} [1.854075 - .535623] = .659226$$

where

$$\sin \varphi_1 = \frac{2}{2}, \sin \varphi_2 = \frac{2}{4}, \sin^2 \alpha = \frac{2}{4}.$$

Thus

$$\alpha = 45°, \varphi_1 = 90°, \varphi_2 = 30°.$$

Second Method, Numerical Integration

If we wish to use numerical integration we must observe that the integrand has a singularity at $t = 2$ where it behaves like $[8(t - 2)]^{-\frac{1}{2}}$.

We remove the singularity at $t = 2$, by writing

$$\int_2^4 [(t^2 - 2)(t^2 - 4)]^{-\frac{1}{2}} dt = \int_2^4 f(t) dt + \int_2^4 [8(t - 2)]^{-\frac{1}{2}} dt$$

where

$$f(t) = [(t^2 - 2)(t^2 - 4)]^{-\frac{1}{2}} - [8(t - 2)]^{-\frac{1}{2}}.$$

If we define $f(2) = 0$,

$$\int_2^4 f(t) dt$$

can be calculated by numerical quadrature. Also

$$\int_2^4 [8(t - 2)]^{-\frac{1}{2}} dt = \left[\frac{1}{\sqrt{2}} (t - 2)^{\frac{1}{2}} \right]_2^4 = 1$$

and thus we calculate the integral as

$$1 + \int_2^4 f(t) dt = 1 - .340773 = .659227.$$

Example 10. Evaluate

$$u = \int_{17}^\infty (x^3 - 7x + 6)^{-\frac{1}{2}} dx.$$

$x^3 - 7x + 6 = (x - 1)(x - 2)(x + 3)$ and we use **17.4.65** with $\beta_1 = 2$, $\beta_2 = 1$, $\beta_3 = -3$,

$$m=\sin^2\alpha=4/5,\ \lambda=\sqrt{5}/2,\ \cos^2\varphi=3/4.$$

Thus $\alpha=63.434949°$, $\varphi=30°$ and

$$u=2(5)^{-\frac{1}{2}}F(30°\backslash63.434949°)$$
$$=2(5)^{-\frac{1}{2}}(.543604)=.486214 \text{ from } \textbf{Table 17.5}.$$

The above integral is of the Weierstrass type and in fact $17=\mathcal{P}(\tfrac{1}{2}u;\ 28,\ -24)$ (see chapter **18**).

Example 11. Evaluate

$$\int_0^{2/3}(24-12t+2t^2-t^3)^{-1/2}\,dt.$$

We have

$$24-12t+2t^2-t^3=-(t-2)(t^2+12)=-P(t).$$

There is only one real zero and we therefore use **17.4.74** with $P(t)=t^3-2t^2+12t-24$, $\beta=2$ so that $P'(2)=16$, $P''(2)=8$, $\lambda=2$ and therefore

$$m=\sin^2\alpha=\frac{1}{4},\quad\alpha=30°$$

Therefore the given integral is

$$\int_0^2-\int_{2/3}^2=\frac{1}{2}\,[F(\varphi_1\backslash60°)-F(\varphi_2\backslash60°)]$$

where

$$\cos\varphi_1=\frac{1}{3},\qquad\varphi_1=70.52877\ 93°$$

$$\cos\varphi_2=\frac{1}{2},\qquad\varphi_2=60°$$

and the integral $=\frac{1}{2}[1.510344-1.212597]=.148874$.

Example 12. Use Landen's transformation to evaluate

$$\int_0^{\pi/2}\left(1-\frac{1}{4}\sin^2\theta\right)^{-1/2}d\theta \text{ to 5D}.$$

First Method, Descending Transformation

We use **17.5.1** to give

$$1+\sin\alpha_1=\frac{2}{1+\cos30°}=1.071797$$

$$\cos\alpha_1=[(1-\sin\alpha_1)(1+\sin\alpha_1)]^{1/2}=.997419$$

$$1+\sin\alpha_2=\frac{2}{1+\cos\alpha_1}=1.001292;\ \cos\alpha_2=.999999$$

$$1+\sin\alpha_3=\frac{2}{1+\cos\alpha_2}=1.000000$$

Thus from **17.5.7**,

the integral $=F(90°\backslash30°)=\dfrac{\pi}{2}\,(1.071797)(1.001292)$

$$=1.68575 \text{ to 5D}.$$

Second Method, Ascending Transformation

We use **17.5.11** to give

$$1+\cos\alpha_{n+1}=2/(1+\sin\alpha_n)$$

n	$\cos\alpha_n$	$\sin\alpha_n$
1	.33333 333	.94280 904
2	.02943 725	.99956 663
3	.00021 673	.99999 998

$$\sin(2\varphi_1-90°)=\sin30°,\qquad\varphi_1=60°$$
$$\sin(2\varphi_2-\varphi_1)=\sin\alpha_1\sin\varphi_1,\qquad\varphi_2=57.367805°$$
$$\sin(2\varphi_3-\varphi_2)=\sin\alpha_2\sin\varphi_2,\qquad\varphi_3=57.348426°$$
$$\sin(2\varphi_4-\varphi_3)=\sin\alpha_3\sin\varphi_3,\qquad\varphi_4=57.348425°=\Phi.$$

From **17.5.16**

$$F(90°\backslash30°)=\frac{2}{1.5}\ \frac{2}{1.94280\ 904}\ \frac{2}{1.99956\ 663}$$

$$\frac{2}{1.99999\ 998}\ \ln\tan\left(45°+\frac{1}{2}\,\Phi\right)$$

$$=1.37288\ 050\ \ln\tan73.674213°$$

$$=1.37288\ 050(1.22789\ 30)$$

$F(90°\backslash30°)=1.68575$ to 5D.

Example 13. Find the value of $F(89.5°\backslash89.5°)$.

First Method

This is a case where interpolation in **Table 17.5** is not possible. We use **17.4.13** which gives

$$F(89.5°\backslash89.5°)=F(90°\backslash89.5°)-F(\psi\backslash89.5°)$$

where

$$\cot\psi=\sin(.5°)\cot(.5°)=\cos(.5°)$$
$$\psi=45.00109\ 084°$$

and $F(\psi\backslash89.5°)=.881390$ from **Table 17.5**.

$$F(90°\backslash89.5°)=K(\sin^2 89.5°)=K(.99992\ 38476)$$
$$=6.12777\ 88$$

Thus $F(89.5°\backslash89.5°)=5.246389$.

Second Method

Landen's ascending transformation, **17.5.11**, gives

$\cos \alpha_1 = (1-\sin 89.5°)/(1+\sin 89.5°)$

$\sin \alpha_1 = [(1-\cos \alpha_1)(1+\cos \alpha_1)]^{\frac{1}{2}} = .99999\ 99997$

$\cos \alpha_2 = 0$

$\sin \alpha_2 = 1.$

17.5.12 then gives

$$\sin (2\varphi_1 - 89.5°) = \sin 89.5° \sin 89.5°$$

$$= .99992\ 38476$$

$$2\varphi_1 - 89.5° = 89.2929049°, \quad \varphi_1 = 89.39645\ 245°$$

$\sin (2\varphi_2 - \varphi_1) = \sin \alpha_1 \sin \varphi_1, \quad \varphi_2 = 89.39645\ 602°$

$\sin (2\varphi_3 - \varphi_2) = \sin \varphi_2, \quad \varphi_3 = \varphi_2 = \Phi.$

Thus **17.5.16** gives

$$F(89.5°\backslash 89.5°) =$$

$$\left(\frac{1}{.99996\ 19231}\right)^{\frac{1}{2}} \ln (\tan 89.69822\ 801°) = 5.24640.$$

Example 14. Evaluate

$$\int_1^2 [(9-t^2)(16+t^2)^3]^{-\frac{1}{2}} dt \text{ to } 5\mathbf{D}.$$

From **17.4.51** the given integral

$$= \int_0^2 - \int_0^1 = \frac{1}{80} [E(\varphi_1\backslash\alpha) - E(\varphi_2\backslash\alpha)]$$

where

$$\sin \alpha = \tfrac{3}{5}, \quad \alpha = 36.86990°$$

$$\sin \varphi_1 = \tfrac{1}{3}\sqrt{5}, \quad \varphi_1 = 48.18968°$$

$$\sin \varphi_2 = \frac{5}{3\sqrt{17}}, \quad \varphi_2 = 23.84264°.$$

By bivariate interpolation in **Table 17.6** we find that the given integral

$$= \frac{1}{80}[.80904 - .41192] = .00496.$$

Simpson's rule with 3 ordinates gives

$$\tfrac{1}{6}[.00504 + .01975 + .005] = .00496.$$

Example 15. Evaluate

$$\Pi(\tfrac{1}{16}; 45°\backslash 30°) =$$

$$\int_0^{\pi/4} (1-\tfrac{1}{16}\sin^2\theta)^{-1}(1-\tfrac{1}{4}\sin^2\theta)^{-\frac{1}{2}} d\theta \text{ to } 6\mathbf{D}.$$

This is case (i) of integrals of the third kind, $0 < n < \sin^2 \alpha$, **17.7.3**

$$n = \tfrac{1}{16}, \quad \varphi = 45°, \quad \alpha = 30°,$$

$$\epsilon = \arcsin (n/\sin^2 \alpha)^{\frac{1}{2}} = 30°,$$

$$\beta = \tfrac{1}{2}\pi F(30°\backslash 30°)/K(30°) = .49332\ 60$$

$$v = \tfrac{1}{2}\pi F(45°\backslash 30°)/K(30°) = .74951\ 51,$$

$$\delta_1 = (16/45)^{\frac{1}{2}}$$

and so from **17.7.3**

$$\Pi(\tfrac{1}{16}; 45°\backslash 30°) =$$

$$(16/45)^{\frac{1}{2}} \left\{ -\tfrac{1}{2}\ln \frac{\vartheta_4(v+\beta)}{\vartheta_4(v-\beta)} + \frac{\vartheta_1'(\beta)}{\vartheta_1(\beta)} v \right\}$$

$$q = .01797\ 24.$$

Using the q-series, **16.27**, for the ϑ functions we get

$$\Pi(\tfrac{1}{16}; 45°\backslash 30°) = (16/45)^{\frac{1}{2}}\{ -.02995\ 89$$

$$+ (1.86096\ 21)(.74951\ 51) \} = .813845.$$

Table 17.9 gives .81385 with 4 point Lagrangian interpolation.

Example 16. Evaluate the complete elliptic integral

$$\Pi(\tfrac{1}{16}\backslash 30°) \text{ to } 6\mathbf{D}.$$

From **17.7.6** we have

$$\Pi(\tfrac{1}{16}\backslash 30°) = K(30°) + (16/45)^{1/2}K(\alpha)Z(\epsilon\backslash 30°)$$

where $\quad \epsilon = \arcsin(n/\sin^2\alpha)^{\frac{1}{2}} = 30°.$ Thus using **Table 17.7**

$$\Pi(\tfrac{1}{16}\backslash 30°) = 1.743055.$$

Table 17.9 gives 1.74302 with 5 point Lagrangian interpolation.

Example 17. Evaluate

$$\Pi(\tfrac{5}{8}; 45°\backslash 30°)$$

$$= \int_0^{\pi/4} (1-\tfrac{5}{8}\sin^2\theta)^{-1}(1-\tfrac{1}{4}\sin^2\theta)^{-1/2} d\theta$$

to $6\mathbf{D}.$

This is case (iii) of integrals of the third kind, $\sin^2 \alpha < n < 1,$

$$n = \tfrac{5}{8}, \quad \varphi = 45°, \quad \alpha = 30°$$

$$\epsilon = \arcsin \ [(1-n)/\cos^2 \alpha]^{\frac{1}{2}} = 45°$$

$$\beta = \tfrac{1}{2}\pi F(45° \backslash 60°)/K(30°) = .79317\ 74$$

$$v = \tfrac{1}{2}\pi F(45° \backslash 30°)/K(30°) = .74951\ 51$$

$$\delta_2 = (40/9)^{\frac{1}{2}}$$

$$q = .01797\ 24$$

and so from **17.7.11**

$$\Pi\ (\tfrac{5}{8};\ 45° \backslash 30°) = (40/9)^{1/2}(\lambda - 4\mu v)$$

$$= 2.10818\ 51 \{.55248\ 32 - 4(.03854\ 26)$$

$$(.74951\ 51)\} = .921129.$$

Table 17.9 gives .92113 with 4 point Lagrangian interpolation.

Example 18. Evaluate the complete elliptic integral

$$\Pi\ (\tfrac{5}{8} \backslash 30°) \ \text{to 5D.}$$

From **17.7.14** we have

$$\Pi\ (\tfrac{5}{8} \backslash 30°) = K(30°) + \frac{\pi}{2}\sqrt{\frac{40}{9}}\ [1 - \Lambda_0(\epsilon \backslash 30°)]$$

where $\epsilon = \arcsin\ [(1-n)/\cos^2 \alpha]^{1/2} = 45°$. Thus using **Table 17.8**

$$\Pi\ (\tfrac{5}{8} \backslash 30°) = 2.80099.$$

Table 17.9 gives 2.80126 by 6 point Lagrangian interpolation. The discrepancy results from interpolation with respect to n for $\varphi = 90°$ in **Table 17.9**.

Example 19. Evaluate

$$\Pi\ (\tfrac{5}{4};\ 45° \backslash 30°)$$

$$= \int_0^{\pi/4} (1 - \tfrac{5}{4}\sin^2 \theta)^{-1}(1 - \tfrac{1}{4}\sin^2 \theta)^{-1/2}\ d\theta$$

to 5D.

Here $n = \dfrac{5}{4}$, $\varphi = 45°$, $\alpha = 30°$ and since the characteristic is greater than unity we use **17.7.7**

$$N = n^{-1}\sin^2 \alpha = .2, \ p_1 = (1/5)^{\frac{1}{2}}$$

$$\Pi\ (\tfrac{5}{4};\ 45° \backslash 30°) = -\Pi(\ 2;\ 45° \backslash 30°) + F(45° \backslash 30°)$$

$$+ (\tfrac{1}{2}\sqrt{5})\ \ln\frac{(7/8)^{\frac{1}{2}} + (1/5)^{\frac{1}{2}}}{(7/8)^{\frac{1}{2}} - (1/5)^{\frac{1}{2}}}$$

$$= -.83612 + .80437$$

$$+ \tfrac{1}{2}\sqrt{5}\ \ln\frac{\sqrt{35} + \sqrt{8}}{\sqrt{35} - \sqrt{8}}$$

$$= 1.13214.$$

Numerical quadrature gives the same result.

Example 20. Evaluate

$$\Pi\ (-\tfrac{1}{4};\ 45° \backslash 30°)$$

$$= \int_0^{\pi/4} (1 + \tfrac{1}{4}\sin^2 \theta)^{-1}(1 - \tfrac{1}{4}\sin^2 \theta)^{-\frac{1}{2}}d\theta$$

to 5D.

Here the characteristic is negative and we therefore use **17.7.15** with $n = -\dfrac{1}{4}$, $\sin^2 \alpha = \dfrac{1}{4}$

$$N = (1-n)^{-1}(\sin^2 \alpha - n) = .4, \ p_2 = \sqrt{.1}$$

and therefore

$$(5/2)^{\frac{1}{2}}\ \Pi\ (-\tfrac{1}{4};\ 45° \backslash 30°) = (9/40)^{\frac{1}{2}}\Pi\ (\tfrac{2}{5};\ 45° \backslash 30°)$$

$$+ \frac{1}{2}(5/2)^{\frac{1}{2}}F(45° \backslash 30°) + \arctan\ (35)^{-\frac{1}{2}}$$

Using **Tables 4.14, 17.5,** and **17.9** we get

$$\Pi\ (-\tfrac{1}{4};\ 45° \backslash 30°) = .76987$$

References

Texts

[17.1] A. Cayley, An elementary treatise on elliptic functions (Dover Publications, Inc., New York, N.Y., 1956).

[17.2] A. Erdélyi et al., Higher transcendental functions, vol. 2, ch. 23 (McGraw-Hill Book Co., Inc., New York, N.Y., 1953).

[17.3] L. V. King, On the direct numerical calculation of elliptic functions and integrals (Cambridge Univ. Press, Cambridge, England, 1924).

[17.4] E. H. Neville, Jacobian elliptic functions, 2d ed. (Oxford Univ. Press, London, England, 1951).

[17.5] F. Oberhettinger and W. Magnus, Anwendung der elliptischen Funktionen in Physik und Technik (Springer-Verlag, Berlin, Germany, 1949).

[17.6] F. Tricomi, Elliptische Funktionen (Akademische Verlagsgesellschaft, Leipzig, Germany, 1948).

[17.7] E. T. Whittaker and G. N. Watson, A course of modern analysis, chs. 20, 21, 22, 4th ed. (Cambridge Univ. Press, Cambridge, England, 1952).

Tables

[17.8] P. F. Byrd and M. D. Friedman, Handbook of elliptic integrals for engineers and physicists (Springer-Verlag, Berlin, Germany, 1954).

[17.9] C. Heuman, Tables of complete elliptic integrals, J. Math. Phys. **20**, 127–206 (1941).

[17.10] J. Hoüel, Recueil de formules et de tables numériques (Gauthier-Villars, Paris, France, 1901).

[17.11] E. Jahnke and F. Emde, Tables of functions, 4th ed. (Dover Publications, Inc., New York, N.Y., 1945).

[17.12] L. M. Milne-Thomson, Jacobian elliptic function tables (Dover Publications, Inc., New York, N.Y., 1956).

[17.13] L. M. Milne-Thomson, Ten-figure table of the complete elliptic integrals K, K', E, E' and a table of $\dfrac{1}{\vartheta_3^2(0|\tau)}$, $\dfrac{1}{\vartheta_3^{2\prime\prime}(0|\tau)}$, Proc. London Math. Soc. 2, 33 (1931).

[17.14] L. M. Milne-Thomson, The Zeta function of Jacobi, Proc. Roy. Soc. Edinburgh **52** (1931).

[17.15] L. M. Milne-Thomson, Die elliptischen Funktionen von Jacobi (Julius Springer, Berlin, Germany, 1931).

[17.16] K. Pearson, Tables of the complete and incomplete elliptic integrals (Cambridge Univ. Press, Cambridge, England, 1934).

[17.17] G. W. and R. M. Spenceley, Smithsonian elliptic function tables, Smithsonian Miscellaneous Collection, vol. 109 (Washington, D.C., 1947).

Table 17.1 COMPLETE ELLIPTIC INTEGRALS OF THE FIRST AND SECOND KINDS AND THE NOME q WITH ARGUMENT THE PARAMETER m

$$K(m)=\int_0^{\frac{\pi}{2}} (1-m\sin^2\theta)^{-\frac{1}{2}}d\theta \qquad K'(m)=K(m_1)$$

$$E(m)=\int_0^{\frac{\pi}{2}} (1-m\sin^2\theta)^{\frac{1}{2}}d\theta \qquad E'(m)=E(m_1)$$

$$q(m)=\exp\left[-\pi K'(m)/K(m)\right] \qquad q_1(m)=q(m_1)$$

m	$K(m)$	$K'(m)$	$q(m)$	m_1
0.00	1.57079 63267 94897	∞	0.00000 00000 00000	1.00
0.01	1.57474 55615 17356	3.69563 73629 89875	0.00062 81456 60383	0.99
0.02	1.57873 99120 07773	3.35414 14456 99160	0.00126 26665 23204	0.98
0.03	1.58278 03424 06373	3.15587 49478 91841	0.00190 36912 69025	0.97
0.04	1.58686 78474 54166	3.01611 24924 77648	0.00255 13525 13689	0.96
0.05	1.59100 34537 90792	2.90833 72484 44552	0.00320 57869 70686	0.95
0.06	1.59518 82213 21610	2.82075 24967 55872	0.00386 71356 22010	0.94
0.07	1.59942 32446 58510	2.74707 30040 24667	0.00453 55438 98018	0.93
0.08	1.60370 96546 39253	2.68355 14063 15229	0.00521 11618 66885	0.92
0.09	1.60804 86199 30513	2.62777 33320 84344	0.00589 41444 34269	0.91
0.10	1.61244 13487 20219	2.57809 21133 48173	0.00658 46515 53858	0.90
0.11	1.61688 90905 05203	2.53333 45460 02200	0.00728 28484 49518	0.89
0.12	1.62139 31379 80658	2.49263 53232 39716	0.00798 89058 49815	0.88
0.13	1.62595 48290 38433	2.45533 80283 21380	0.00870 30002 35762	0.87
0.14	1.63057 55488 81754	2.42093 29603 44303	0.00942 53141 02678	0.86
0.15	1.63525 67322 64580	2.38901 64863 25580	0.01015 60362 37153	0.85
0.16	1.63999 98658 64511	2.35926 35547 45007	0.01089 53620 10173	0.84
0.17	1.64480 64907 98881	2.33140 85677 50251	0.01164 34936 87540	0.83
0.18	1.64967 82052 94514	2.30523 17368 77189	0.01240 06407 58856	0.82
0.19	1.65461 66675 22527	2.28054 91384 22770	0.01316 70202 86392	0.81
0.20	1.65962 35986 10528	2.25722 53268 20854	0.01394 28572 75318	0.80
0.21	1.66470 07858 45692	2.23506 77552 60349	0.01472 83850 66891	0.79
0.22	1.66985 00860 83368	2.21402 24978 46332	0.01552 38457 56320	0.78
0.23	1.67507 34293 77219	2.19397 09253 19189	0.01632 94906 37206	0.77
0.24	1.68037 28228 48361	2.17482 70902 46414	0.01714 55806 74605	0.76
0.25	1.68575 03548 12596	2.15651 56474 99643	0.01797 23870 08967	0.75
0.26	1.69120 81991 86631	2.13897 01837 52114	0.01881 01914 93399	0.74
0.27	1.69674 86201 96168	2.12213 18631 57396	0.01965 92872 66940	0.73
0.28	1.70237 39774 10990	2.10594 83200 52758	0.02051 99793 66788	0.72
0.29	1.70808 67311 34606	2.09037 27465 52360	0.02139 25853 82708	0.71
0.30	1.71388 94481 78791	2.07536 31352 92469	0.02227 74361 57154	0.70
0.31	1.71978 48080 56405	2.06088 16467 30131	0.02317 48765 35013	0.69
0.32	1.72577 56096 29320	2.04689 40772 10577	0.02408 52661 67250	0.68
0.33	1.73186 47782 52098	2.03336 94091 52233	0.02500 89803 73177	0.67
0.34	1.73805 53734 56358	2.02027 94286 03592	0.02594 64110 66576	0.66
0.35	1.74435 05972 25613	2.00759 83984 24376	0.02689 79677 51443	0.65
0.36	1.75075 38029 15753	1.99530 27776 64729	0.02786 40785 93729	0.64
0.37	1.75726 85048 82456	1.98337 09795 27821	0.02884 51915 76181	0.63
0.38	1.76389 83888 83731	1.97178 31617 25656	0.02984 17757 44138	0.62
0.39	1.77064 73233 .33534	1.96052 10441 65830	0.03085 43225 51033	0.61
0.40	1.77751 93714 91253	1.94956 77498 06026	0.03188 33473 13363	0.60
0.41	1.78451 88046 81873	1.93890 76652 34220	0.03292 93907 86003	0.59
0.42	1.79165 01166 52966	1.92852 63181 14418	0.03399 30208 70043	0.58
0.43	1.79891 80391 87685	1.91841 02691 09912	0.03507 48344 66773	0.57
0.44	1.80632 75591 07699	1.90854 70162 81211	0.03617 54594 93133	0.56
0.45	1.81388 39368 16983	1.89892 49102 71554	0.03729 55570 75822	0.55
0.46	1.82159 27265 56821	1.88953 30788 53096	0.03843 58239 43468	0.54
0.47	1.82945 97985 64730	1.88036 13596 22178	0.03959 69950 38753	0.53
0.48	1.83749 13633 55796	1.87140 02398 11034	0.04077 98463 75263	0.52
0.49	1.84569 39983 74724	1.86264 08023 32739	0.04198 51981 67183	0.51
0.50	1.85407 46773 01372	1.85407 46773 01372	0.04321 39182 63772	0.50
m_1	$K'(m)$	$K(m)$	$q_1(m)$	m

$$\begin{bmatrix} (-5)2 \\ 11 \end{bmatrix} \qquad\qquad\qquad\qquad \begin{bmatrix} (-6)3 \\ 9 \end{bmatrix}$$

See Examples 3–4.

$E(m)$ and $E'(m)$ from L. M. Milne-Thomson, Ten-figure table of the complete elliptic integrals K, K', E, E' and a table of $\dfrac{1}{\vartheta_3^2(0|\tau)}$, $\dfrac{1}{\vartheta_3''^2(0|\tau)}$, Proc. London Math. Soc.(2)**33**, 1931(with permission).

COMPLETE ELLIPTIC INTEGRALS OF THE FIRST AND SECOND KINDS   Table 17.1
AND THE NOME q WITH ARGUMENT THE PARAMETER m

$$K(m)=\int_0^{\frac{\pi}{2}}(1-m\sin^2\theta)^{-\frac{1}{2}}d\theta \qquad K'(m)=K(m_1)$$

$$E(m)=\int_0^{\frac{\pi}{2}}(1-m\sin^2\theta)^{\frac{1}{2}}d\theta \qquad E'(m)=E(m_1)$$

$$q(m)=\exp\left[-\pi K'(m)/K(m)\right] \qquad q_1(m)=q(m_1)$$

m	$q_1(m)$	$E(m)$	$E'(m)$	m_1
0.00	1.00000 00000 00000	1.57079 6327	1.00000 0000	1.00
0.01	0.26219 62679 17709	1.56686 1942	1.01599 3546	0.99
0.02	0.22793 45740 67492	1.56291 2645	1.02859 4520	0.98
0.03	0.20687 98108 47842	1.55894 8244	1.03994 6861	0.97
0.04	0.19149 63082 09940	1.55496 8546	1.05050 2227	0.96
0.05	0.17931 60069 55723	1.55097 3352	1.06047 3728	0.95
0.06	0.16920 75311 46133	1.54696 2456	1.06998 6130	0.94
0.07	0.16055 42010 73011	1.54293 5653	1.07912 1407	0.93
0.08	0.15298 14810 09741	1.53889 2730	1.08793 7503	0.92
0.09	0.14624 42694 73236	1.53483 3465	1.09647 7517	0.91
0.10	0.14017 31269 54262	1.53075 7637	1.10477 4733	0.90
0.11	0.13464 58847 92091	1.52666 5017	1.11285 5607	0.89
0.12	0.12957 14695 20553	1.52255 5369	1.12074 1661	0.88
0.13	0.12488 01223 52049	1.51842 8454	1.12845 0735	0.87
0.14	0.12051 71957 28729	1.51428 4027	1.13599 7843	0.86
0.15	0.11643 90607 17472	1.51012 1831	1.14339 5792	0.85
0.16	0.11261 03164 23363	1.50594 1612	1.15065 5629	0.84
0.17	0.10900 18330 23834	1.50174 3101	1.15778 6979	0.83
0.18	0.10558 93457 98477	1.49752 6026	1.16479 8293	0.82
0.19	0.10235 24235 13544	1.49329 0109	1.17169 7053	0.81
0.20	0.09927 36973 38825	1.48903 5058	1.17848 9924	0.80
0.21	0.09633 82749 65990	1.48476 0581	1.18518 2883	0.79
0.22	0.09353 32888 80648	1.48046 6375	1.19178 1311	0.78
0.23	0.09084 75434 60707	1.47615 2126	1.19829 0087	0.77
0.24	0.08827 12359 87862	1.47181 7514	1.20471 3641	0.76
0.25	0.08579 57337 02195	1.46746 2209	1.21105 6028	0.75
0.26	0.08341 33938 83117	1.46308 5873	1.21732 0955	0.74
0.27	0.08111 74173 41165	1.45868 8155	1.22351 1839	0.73
0.28	0.07890 17281 26084	1.45426 8698	1.22963 1828	0.72
0.29	0.07676 08740 04317	1.44982 7128	1.23568 3836	0.71
0.30	0.07468 99435 37179	1.44536 3064	1.24167 0567	0.70
0.31	0.07268 44965 37110	1.44087 6115	1.24759 4538	0.69
0.32	0.07074 05053 87511	1.43636 5871	1.25345 8093	0.68
0.33	0.06885 43052 47167	1.43183 1919	1.25926 3421	0.67
0.34	0.06702 25515 69108	1.42727 3821	1.26501 2576	0.66
0.35	0.06524 21836 78738	1.42269 1133	1.27070 7480	0.65
0.36	0.06351 03934 00746	1.41808 3394	1.27634 9943	0.64
0.37	0.06182 45979 15898	1.41345 0127	1.28194 1668	0.63
0.38	0.06018 24161 79938	1.40879 0839	1.28748 4262	0.62
0.39	0.05858 16483 56838	1.40410 5019	1.29297 9239	0.61
0.40	0.05702 02578 14610	1.39939 2139	1.29842 8034	0.60
0.41	0.05549 63553 09081	1.39465 1652	1.30383 2008	0.59
0.42	0.05400 81850 43499	1.38988 2992	1.30919 2448	0.58
0.43	0.05255 41123 42653	1.38508 5568	1.31451 0576	0.57
0.44	0.05113 26127 21764	1.38025 8774	1.31978 7557	0.56
0.45	0.04974 22621 64574	1.37540 1972	1.32502 4498	0.55
0.46	0.04838 17284 53289	1.37051 4505	1.33022 2453	0.54
0.47	0.04704 97634 16424	1.36559 5691	1.33538 2430	0.53
0.48	0.04574 51959 80149	1.36064 4814	1.34050 5388	0.52
0.49	0.04446 69259 25028	1.35566 1135	1.34559 2245	0.51
0.50	0.04321 39182 63772	1.35064 3881	1.35064 3881	0.50
m_1	$q(m)$	$E'(m)$	$E(m)$	m

$$\begin{bmatrix} (-6)4 \\ 6 \end{bmatrix}$$

Table 17.2   COMPLETE ELLIPTIC INTEGRALS OF THE FIRST AND SECOND KINDS AND THE NOME q WITH ARGUMENT THE MODULAR ANGLE α

$$K(\alpha)=\int_0^{\frac{\pi}{2}}(1-\sin^2\alpha\,\sin^2\theta)^{-\frac{1}{2}}d\theta \qquad K'(\alpha)=K(90°-\alpha)$$

$$E(\alpha)=\int_0^{\frac{\pi}{2}}(1-\sin^2\alpha\,\sin^2\theta)^{\frac{1}{2}}d\theta \qquad E'(\alpha)=E(90°-\alpha)$$

$$q(\alpha)=\exp\left[-\pi K'(\alpha)/K(\alpha)\right] \qquad q_1(\alpha)=q(90°-\alpha)$$

α	$K(\alpha)$	$K'(\alpha)$	$q(\alpha)$	$90°-\alpha$
0°	1.57079 63267 94897	∞	0.00000 00000 00000	90°
1	1.57091 59581 27243	5.43490 98296 25564	0.00001 90395 55387	89
2	1.57127 49523 72225	4.74271 72652 78886	0.00007 61698 24680	88
3	1.57187 36105 14009	4.33865 39759 99725	0.00017 14256 42257	87
4	1.57271 24349 95227	4.05275 81695 49437	0.00030 48651 48814	86
5	1.57379 21309 24768	3.83174 19997 84146	0.00047 65699 16867	85
6	1.57511 36077 77251	3.65185 59694 78752	0.00068 66451 27305	84
7	1.57667 79815 92838	3.50042 24991 71838	0.00093 52197 97816	83
8	1.57848 65776 88648	3.36986 80266 68445	0.00122 24470 64294	82
9	1.58054 09338 95721	3.25530 29421 43555	0.00154 85045 16579	81
10	1.58284 28043 38351	3.15338 52518 87839	0.00191 35945 90170	80
11	1.58539 41637 75538	3.06172 86120 38789	0.00231 79450 15821	79
12	1.58819 72125 27520	2.97856 89511 81384	0.00276 18093 29252	78
13	1.59125 43820 13687	2.90256 49406 70027	0.00324 54674 43525	77
14	1.59456 83409 31825	2.83267 25829 18100	0.00376 92262 86978	76
15	1.59814 20021 12540	2.76806 31453 68768	0.00433 34205 09983	75
16	1.60197 85300 86952	2.70806 76145 90486	0.00493 84132 64213	74
17	1.60608 13494 10364	2.65213 80046 30204	0.00558 45970 58517	73
18	1.61045 41537 89663	2.59981 97300 61099	0.00627 23946 95994	72
19	1.61510 09160 67722	2.55073 14496 27254	0.00700 22602 97383	71
20	1.62002 58991 24204	2.50455 00790 01634	0.00777 46804 16442	70
21	1.62523 36677 58843	2.46099 94583 04126	0.00859 01752 53626	69
22	1.63072 91016 30788	2.41984 16537 39137	0.00944 92999 75082	68
23	1.63651 74093 35819	2.38087 01906 04429	0.01035 26461 44729	67
24	1.64260 41437 12491	2.34390 47244 46913	0.01130 08432 78049	66
25	1.64899 52184 78530	2.30878 97981 67196	0.01229 45605 27181	65
26	1.65569 69263 10344	2.27537 64296 11676	0.01333 45085 07947	64
27	1.66271 59584 91370	2.24354 93416 98626	0.01442 14412 80638	63
28	1.67005 94262 69580	2.21319 46949 79374	0.01555 61584 97708	62
29	1.67773 48840 80745	2.18421 32169 49248	0.01673 95077 33023	61
30	1.68575 03548 12596	2.15651 56474 99643	0.01797 23870 08967	60
31	1.69411 43573 05914	2.13002 14383 99325	0.01925 57475 39635	59
32	1.70283 59363 12341	2.10465 76584 91159	0.02059 05967 10437	58
33	1.71192 46951 55678	2.08035 80666 91578	0.02197 80013 16901	57
34	1.72139 08313 74249	2.05706 23227 97365	0.02341 90910 88188	56
35	1.73124 51756 57058	2.03471 53121 85791	0.02491 50625 23981	55
36	1.74149 92344 26774	2.01326 65652 05468	0.02646 71830 76961	54
37	1.75216 52364 68845	1.99266 97557 34209	0.02807 67957 17219	53
38	1.76325 61840 59342	1.97288 22662 74650	0.02974 53239 19583	52
39	1.77478 59091 05608	1.95386 48092 51663	0.03147 42771 20286	51
40	1.78676 91348 85021	1.93558 10960 04722	0.03326 52566 95577	50
41	1.79922 15440 49811	1.91799 75464 36423	0.03511 99625 22096	49
42	1.81215 98536 62126	1.90108 30334 63664	0.03704 02001 87133	48
43	1.82560 18981 35889	1.88480 86573 80404	0.03902 78889 26607	47
44	1.83956 67210 93652	1.86914 75460 26462	0.04108 50703 79885	46
45	1.85407 46773 01372	1.85407 46773 01372	0.04321 39182 63772	45
$90°-\alpha$	$K'(\alpha)$	$K(\alpha)$	$q_1(\alpha)$	α

$$\left[\begin{matrix}(-5)7\\11\end{matrix}\right] \qquad\qquad\qquad \left[\begin{matrix}(-6)9\\9\end{matrix}\right]$$

Compiled from G. W. and R. M. Spenceley, Smithsonian elliptic function tables, Smithsonian Miscellaneous Collection, vol. 109, Washington, D.C., 1947 (with permission).

COMPLETE ELLIPTIC INTEGRALS OF THE FIRST AND SECOND KINDS   Table 17.2
AND THE NOME q WITH ARGUMENT THE MODULAR ANGLE α

$$K(\alpha)=\int_0^{\frac{\pi}{2}} (1-\sin^2\alpha\,\sin^2\theta)^{-\frac{1}{2}}d\theta \qquad K'(\alpha)=K(90°-\alpha)$$

$$E(\alpha)=\int_0^{\frac{\pi}{2}} (1-\sin^2\alpha\,\sin^2\theta)^{\frac{1}{2}}d\theta \qquad E'(\alpha)=E(90°-\alpha)$$

$$q(\alpha)=\exp\left[-\pi K'(\alpha)/K(\alpha)\right] \qquad q_1(\alpha)=q(90°-\alpha)$$

α	$q_1(\alpha)$	$E(\alpha)$	$E'(\alpha)$	$90°-\alpha$
0°	1.00000 00000 00000	1.57079 63267 94897	1.00000 00000 00000	90°
1	0.40330 93063 38378	1.57067 67091 27960	1.00075 15777 01834	89
2	0.35316 56482 96037	1.57031 79198 97448	1.00258 40855 27552	88
3	0.32040 03371 34866	1.56972 01504 23979	1.00525 85872 09152	87
4	0.29548 83855 58691	1.56888 37196 07763	1.00864 79569 07096	86
5	0.27517 98048 73563	1.56780 90739 77622	1.01266 35062 34396	85
6	0.25794 01957 66337	1.56649 67877 60132	1.01723 69183 41019	84
7	0.24291 29743 06665	1.56494 75629 69419	1.02231 25881 67584	83
8	0.22956 71598 81194	1.56316 22295 18261	1.02784 36197 40833	82
9	0.21754 89496 99726	1.56114 17453 51334	1.03378 94623 90754	81
10	0.20660 97552 00965	1.55888 71966 01596	1.04011 43957 06010	80
11	0.19656 76611 43642	1.55639 97977 70947	1.04678 64993 44049	79
12	0.18728 51836 10217	1.55368 08919 36509	1.05377 69204 07046	78
13	0.17865 56628 04653	1.55073 19509 84013	1.06105 93337 53857	77
14	0.17059 45383 49477	1.54755 45758 69993	1.06860 95329 78401	76
15	0.16303 35348 21581	1.54415 04969 14673	1.07640 51130 76403	75
16	0.15591 66592 65792	1.54052 15741 27631	1.08442 52193 72543	74
17	0.14919 73690 67429	1.53666 97975 68556	1.09265 03455 37715	73
18	0.14283 65198 36280	1.53259 72877 45636	1.10106 21687 57941	72
19	0.13680 08474 28619	1.52830 62960 54359	1.10964 34135 42761	71
20	0.13106 18244 99858	1.52379 92052 59774	1.11837 77379 69864	70
21	0.12559 47852 09819	1.51907 85300 25531	1.12724 96377 57702	69
22	0.12037 82455 07894	1.51414 69174 93342	1.13624 43646 84239	68
23	0.11539 33684 49987	1.50900 71479 16775	1.14534 78566 80849	67
24	0.11062 35386 78854	1.50366 21353 53715	1.15454 66775 24465	66
25	0.10605 40201 85996	1.49811 49284 22116	1.16382 79644 93139	65
26	0.10167 16783 93444	1.49236 87111 24151	1.17317 93826 83722	64
27	0.09746 47524 70352	1.48642 68037 44253	1.18258 90849 45384	63
28	0.09342 26672 88483	1.48029 26638 27039	1.19204 56765 79886	62
29	0.08953 58769 52553	1.47396 98872 41625	1.20153 81841 13662	61
30	0.08579 57337 02195	1.46746 22093 39427	1.21105 60275 68459	60
31	0.08219 43773 66408	1.46077 35062 13127	1.22058 89957 54247	59
32	0.07872 46415 92073	1.45390 77960 65210	1.23012 72241 85949	58
33	0.07537 99738 58803	1.44686 92406 95183	1.23966 11752 88672	57
34	0.07215 43668 98737	1.43966 21471 15459	1.24918 16206 07472	56
35	0.06904 22996 09032	1.43229 09693 06756	1.25867 96247 79997	55
36	0.06603 86859 10861	1.42476 03101 24890	1.26814 65310 65206	54
37	0.06313 88302 96461	1.41707 49233 71952	1.27757 39482 50391	53
38	0.06033 83890 33716	1.40923 97160 46096	1.28695 37387 83001	52
39	0.05763 33361 79494	1.40125 97507 85523	1.29627 80079 94134	51
40	0.05501 99336 98829	1.39314 02485 23812	1.30553 90942 97794	50
41	0.05249 47051 04844	1.38488 65913 75413	1.31472 95602 64623	49
42	0.05005 44121 29953	1.37650 43257 72082	1.32384 21844 81263	48
43	0.04769 60340 17056	1.36799 91658 73159	1.33286 99541 17179	47
44	0.04541 67490 83529	1.35937 69972 75008	1.34180 60581 29911	46
45	0.04321 39182 63772	1.35064 38810 47676	1.35064 38810 47676	45
$90°-\alpha$	$q(\alpha)$	$E'(\alpha)$	$E(\alpha)$	α

$$\left[\begin{array}{c}(-5)3\\9\end{array}\right]$$

Table 17.3   PARAMETER m WITH ARGUMENT $K'(m)/K(m)$

$\dfrac{K'}{K}$	m	$\dfrac{K'}{K}$	m	$\dfrac{K'}{K}$	m
0.30	0.99954 69976	1.20	0.30866 25998	2.10	0.02158 74007
0.32	0.99912 85258	1.22	0.29292 52811	2.12	0.02028 61803
0.34	0.99844 79307	1.24	0.27782 39170	2.14	0.01906 26278
0.36	0.99740 80762	1.26	0.26335 17107	2.16	0.01791 21974
0.38	0.99590 01861	1.28	0.24949 94512	2.18	0.01683 05990
0.40	0.99380 79974	1.30	0.23625 58558	2.20	0.01581 37845
0.42	0.99101 23521	1.32	0.22360 78874	2.22	0.01485 79356
0.44	0.98739 58502	1.34	0.21154 10467	2.24	0.01395 94517
0.46	0.98284 72586	1.36	0.20003 96393	2.26	0.01311 49385
0.48	0.97726 54540	1.38	0.18908 70181	2.28	0.01232 11967
0.50	0.97056 27485	1.40	0.17866 58032	2.30	0.01157 52117
0.52	0.96266 75125	1.42	0.16875 80773	2.32	0.01087 41433
0.54	0.95352 60602	1.44	0.15934 55603	2.34	0.01021 53165
0.56	0.94310 38029	1.46	0.15040 97635	2.36	0.00959 62118
0.58	0.93138 57063	1.48	0.14193 21249	2.38	0.00901 44574
0.60	0.91837 61134	1.50	0.13389 41273	2.40	0.00846 78199
0.62	0.90409 80105	1.52	0.12627 73987	2.42	0.00795 41974
0.64	0.88859 18214	1.54	0.11906 38004	2.44	0.00747 16117
0.66	0.87191 38254	1.56	0.11223 54993	2.46	0.00701 82011
0.68	0.85413 42916	1.58	0.10577 50300	2.48	0.00659 22140
0.70	0.83533 54217	1.60	0.09966 53447	2.50	0.00619 20026
0.72	0.81560 91841	1.62	0.09388 98538	2.52	0.00581 60167
0.74	0.79505 51193	1.64	0.08843 24583	2.54	0.00546 27984
0.76	0.77377 81814	1.66	0.08327 75739	2.56	0.00513 09763
0.78	0.75188 66711	1.68	0.07841 01486	2.58	0.00481 92610
0.80	0.72949 03078	1.70	0.07381 56747	2.60	0.00452 64398
0.82	0.70669 84707	1.72	0.06948 01950	2.62	0.00425 13725
0.84	0.68361 86358	1.74	0.06539 03054	2.64	0.00399 29873
0.86	0.66035 50204	1.76	0.06153 31533	2.66	0.00375 02764
0.88	0.63700 74395	1.78	0.05789 64327	2.68	0.00352 22924
0.90	0.61367 03730	1.80	0.05446 83767	2.70	0.00330 81448
0.92	0.59043 22404	1.82	0.05123 77481	2.72	0.00310 69966
0.94	0.56737 48621	1.84	0.04819 38272	2.74	0.00291 80610
0.96	0.54457 30994	1.86	0.04532 63995	2.76	0.00274 05988
0.98	0.52209 46531	1.88	0.04262 57408	2.78	0.00257 39151
1.00	0.50000 00000	1.90	0.04008 26022	2.80	0.00241 73568
1.02	0.47834 24497	1.92	0.03768 81947	2.82	0.00227 03103
1.04	0.45716 83054	1.94	0.03543 41720	2.84	0.00213 21990
1.06	0.43651 71048	1.96	0.03331 26147	2.86	0.00200 24811
1.08	0.41642 19278	1.98	0.03131 60134	2.88	0.00188 06475
1.10	0.39690 97552	2.00	0.02943 72515	2.90	0.00176 62198
1.12	0.37800 18621	2.02	0.02766 95892	2.92	0.00165 87487
1.14	0.35971 42366	2.04	0.02600 66464	2.94	0.00155 78119
1.16	0.34205 80100	2.06	0.02444 23873	2.96	0.00146 30127
1.18	0.32503 98919	2.08	0.02297 11038	2.98	0.00137 39785
1.20	0.30866 25998	2.10	0.02158 74007	3.00	0.00129 03591
	$\begin{bmatrix}(-4)2\\9\end{bmatrix}$		$\begin{bmatrix}(-5)8\\7\end{bmatrix}$		$\begin{bmatrix}(-5)1\\6\end{bmatrix}$

For $\dfrac{K'}{K} > 3.0$, $\dfrac{K'}{K} < 0.3$, see **Example 6.**

Table 17.4

AUXILIARY FUNCTIONS FOR COMPUTATION OF THE NOME q AND THE PARAMETER m

$$Q(m) = \frac{q_1(m)}{m_1} \qquad\qquad L(m) = -K(m) + \frac{K'(m)}{\pi}\ln\frac{16}{m_1}$$

m_1	$Q(m)$	$L(m)$	m_1	$Q(m)$	$L(m)$
0.00	0.06250 00000 00000	0.00000 00000	0.08	0.06513 95233 36060	0.02111 58281
0.01	0.06281 45660 38302	0.00251 65276	0.09	0.06549 04937 14101	0.02392 34345
0.02	0.06313 33261 60188	0.00506 66040	0.10	0.06584 65155 38584	0.02677 14110
0.03	0.06345 63756 34180	0.00765 09870	0.11	0.06620 77131 77434	0.02966 07472
0.04	0.06378 38128 42217	0.01027 04595	0.12	0.06657 42154 15123	0.03259 24678
0.05	0.06411 57394 13714	0.01292 58301	0.13	0.06694 61556 59704	0.03556 76342
0.06	0.06445 22603 66828	0.01561 79344	0.14	0.06732 36721 61983	0.03858 73466
0.07	0.06479 34842 57396	0.01834 76360	0.15	0.06770 69082 47689	0.04165 27452
	$\begin{bmatrix}(-7)6\\8\end{bmatrix}$	$\begin{bmatrix}(-6)5\\5\end{bmatrix}$		$\begin{bmatrix}(-7)7\\8\end{bmatrix}$	$\begin{bmatrix}(-6)6\\5\end{bmatrix}$

See Examples 3, 5 and 6.

ELLIPTIC INTEGRAL OF THE FIRST KIND $F(\varphi \backslash \alpha)$

Table 17.5

$$F(\varphi \backslash \alpha) = \int_0^\varphi (1 - \sin^2 \alpha \sin^2 \theta)^{-\frac{1}{2}} d\theta$$

$\alpha \backslash \varphi$	0°	5°	10°	15°	20°	25°	30°
0°	0	0.08726 646	0.17453 293	0.26179 939	0.34906 585	0.43633 231	0.52359 878
2	0	0.08726 660	0.17453 400	0.26180 298	0.34907 428	0.43634 855	0.52362 636
4	0	0.08726 700	0.17453 721	0.26181 374	0.34909 952	0.43639 719	0.52370 903
6	0	0.08726 767	0.17454 255	0.26183 163	0.34914 148	0.43647 806	0.52384 653
8	0	0.08726 860	0.17454 999	0.26185 656	0.34919 998	0.43659 086	0.52403 839
10	0	0.08726 980	0.17455 949	0.26188 842	0.34927 479	0.43673 518	0.52428 402
12	0	0.08727 124	0.17457 102	0.26192 707	0.34936 558	0.43691 046	0.52458 259
14	0	0.08727 294	0.17458 451	0.26197 234	0.34947 200	0.43711 606	0.52493 314
16	0	0.08727 487	0.17459 991	0.26202 402	0.34959 358	0.43735 119	0.52533 449
18	0	0.08727 703	0.17461 714	0.26208 189	0.34972 983	0.43761 496	0.52578 529
20	0	0.08727 940	0.17463 611	0.26214 568	0.34988 016	0.43790 635	0.52628 399
22	0	0.08728 199	0.17465 675	0.26221 511	0.35004 395	0.43822 422	0.52682 887
24	0	0.08728 477	0.17467 895	0.26228 985	0.35022 048	0.43856 733	0.52741 799
26	0	0.08728 773	0.17470 261	0.26236 958	0.35040 901	0.43893 430	0.52804 924
28	0	0.08729 086	0.17472 762	0.26245 392	0.35060 870	0.43932 365	0.52872 029
30	0	0.08729 413	0.17475 386	0.26254 249	0.35081 868	0.43973 377	0.52942 863
32	0	0.08729 755	0.17478 119	0.26263 487	0.35103 803	0.44016 296	0.53017 153
34	0	0.08730 108	0.17480 950	0.26273 064	0.35126 576	0.44060 939	0.53094 608
36	0	0.08730 472	0.17483 864	0.26282 934	0.35150 083	0.44107 115	0.53174 916
38	0	0.08730 844	0.17486 848	0.26293 052	0.35174 218	0.44154 622	0.53257 745
40	0	0.08731 222	0.17489 887	0.26303 369	0.35198 869	0.44203 247	0.53342 745
42	0	0.08731 606	0.17492 967	0.26313 836	0.35223 920	0.44252 769	0.53429 546
44	0	0.08731 992	0.17496 073	0.26324 404	0.35249 254	0.44302 960	0.53517 761
46	0	0.08732 379	0.17499 189	0.26335 019	0.35274 748	0.44353 584	0.53606 986
48	0	0.08732 765	0.17502 300	0.26345 633	0.35300 280	0.44404 397	0.53696 798
50	0	0.08733 149	0.17505 392	0.26356 191	0.35325 724	0.44455 151	0.53786 765
52	0	0.08733 528	0.17508 448	0.26366 643	0.35350 955	0.44505 593	0.53876 438
54	0	0.08733 901	0.17511 455	0.26376 936	0.35375 845	0.44555 469	0.53965 358
56	0	0.08734 265	0.17514 397	0.26387 020	0.35400 269	0.44604 519	0.54053 059
58	0	0.08734 620	0.17517 260	0.26396 842	0.35424 101	0.44652 487	0.54139 069
60	0	0.08734 962	0.17520 029	0.26406 355	0.35447 217	0.44699 117	0.54222 911
62	0	0.08735 291	0.17522 690	0.26415 509	0.35469 497	0.44744 153	0.54304 111
64	0	0.08735 605	0.17525 232	0.26424 258	0.35490 823	0.44787 348	0.54382 197
66	0	0.08735 902	0.17527 640	0.26432 556	0.35511 081	0.44828 459	0.54456 704
68	0	0.08736 182	0.17529 903	0.26440 362	0.35530 160	0.44867 252	0.54527 182
70	0	0.08736 442	0.17532 010	0.26447 634	0.35547 959	0.44903 502	0.54593 192
72	0	0.08736 681	0.17533 949	0.26454 334	0.35564 377	0.44936 997	0.54654 316
74	0	0.08736 898	0.17535 712	0.26460 428	0.35579 326	0.44967 538	0.54710 162
76	0	0.08737 092	0.17537 289	0.26465 883	0.35592 721	0.44994 944	0.54760 364
78	0	0.08737 262	0.17538 672	0.26470 671	0.35604 488	0.45019 046	0.54804 587
80	0	0.08737 408	0.17539 854	0.26474 766	0.35614 560	0.45039 699	0.54842 535
82	0	0.08737 528	0.17540 830	0.26478 147	0.35622 881	0.45056 775	0.54873 947
84	0	0.08737 622	0.17541 594	0.26480 795	0.35629 402	0.45070 168	0.54898 608
86	0	0.08737 689	0.17542 143	0.26482 697	0.35634 086	0.45079 795	0.54916 348
88	0	0.08737 730	0.17542 473	0.26483 842	0.35636 908	0.45085 596	0.54927 042
90	0	0.08737 744	0.17542 583	0.26484 225	0.35637 851	0.45087 533	0.54930 614
		$\begin{bmatrix} (-8)3 \\ 3 \end{bmatrix}$	$\begin{bmatrix} (-7)3 \\ 4 \end{bmatrix}$	$\begin{bmatrix} (-6)1 \\ 4 \end{bmatrix}$	$\begin{bmatrix} (-6)2 \\ 5 \end{bmatrix}$	$\begin{bmatrix} (-6)5 \\ 5 \end{bmatrix}$	$\begin{bmatrix} (-6)9 \\ 5 \end{bmatrix}$
5	0	0.08726 730	0.17453 962	0.26182 180	0.34911 842	0.43643 361	0.52377 095
15	0	0.08727 387	0.17459 198	0.26199 739	0.34953 092	0.43722 998	0.52512 754
25	0	0.08728 623	0.17469 061	0.26232 912	0.35031 330	0.43874 792	0.52772 849
35	0	0.08730 289	0.17482 397	0.26277 965	0.35138 244	0.44083 848	0.53134 425
45	0	0.08732 185	0.17497 630	0.26329 709	0.35261 989	0.44328 233	0.53562 273
55	0	0.08734 084	0.17512 935	0.26382 007	0.35388 123	0.44580 113	0.54009 391
65	0	0.08735 756	0.17526 454	0.26428 466	0.35501 092	0.44808 179	0.54419 926
75	0	0.08736 998	0.17536 525	0.26463 238	0.35586 223	0.44981 645	0.54735 991
85	0	0.08737 659	0.17541 895	0.26481 840	0.35631 976	0.45075 457	0.54908 352

The table can also be used inversely to find $\varphi = \text{am } u$ where $u = F(\varphi \backslash \alpha)$ and so the Jacobian elliptic functions, for example $\text{sn } u = \sin \varphi$, $\text{cn } u = \cos \varphi$, $\text{dn } u = (1 - \sin^2 \alpha \sin^2 \varphi)^{1/2}$. See **Examples 7–11.** Compiled from K. Pearson, Tables of the complete and incomplete elliptic integrals, Cambridge Univ. Press, Cambridge, England, 1934 (with permission). Known errors have been corrected.

Table 17.5   **ELLIPTIC INTEGRAL OF THE FIRST KIND** $F(\varphi\backslash\alpha)$

$$F(\varphi\backslash\alpha)=\int_0^\varphi (1-\sin^2\alpha\sin^2\theta)^{-\frac{1}{2}}d\theta$$

$\alpha\backslash\varphi$	35°	40°	45°	50°	55°	60°
0°	0.61086 524	0.69813 170	0.78539 816	0.87266 463	0.95993 109	1.04719 755
2	0.61090 819	0.69819 436	0.78548 509	0.87278 045	0.96008 037	1.04738 465
4	0.61103 691	0.69838 220	0.78574 574	0.87312 784	0.96052 821	1.04794 603
•6	0.61125 108	0.69869 484	0.78617 974	0.87370 649	0.96127 450	1.04888 194
8	0.61155 010	0.69913 161	0.78678 644	0.87451 593	0.96231 911	1.05019 278
10	0.61193 318	0.69969 159	0.78756 494	0.87555 545	0.96366 180	1.05187 911
12	0.61239 927	0.70037 358	0.78851 403	0.87682 412	0.96530 224	1.05394 160
14	0.61294 707	0.70117 608	0.78963 221	0.87832 076	0.96723 998	1.05638 099
16	0.61357 504	0.70209 730	0.79091 768	0.88004 389	0.96947 438	1.05919 813
18	0.61428 140	0.70313 511	0.79236 827	0.88199 174	0.97200 462	1.06239 384
20	0.61506 406	0.70428 706	0.79398 143	0.88416 214	0.97482 960	1.06596 891
22	0.61592 071	0.70555 037	0.79575 422	0.88655 254	0.97794 790	1.06992 405
24	0.61684 871	0.70692 183	0.79768 324	0.88915 992	0.98135 773	1.07425 976
26	0.61784 515	0.70839 788	0.79976 461	0.89198 071	0.98505 681	1.07897 628
28	0.61890 682	0.70997 451	0.80199 389	0.89501 076	0.98904 227	1.08407 347
30	0.62003 018	0.71164 728	0.80436 610	0.89824 524	0.99331 059	1.08955 067
32	0.62121 138	0.71341 124	0.80687 558	0.90167 852	0.99785 743	1.09540 656
34	0.62244 622	0.71526 098	0.80951 599	0.90530 415	1.00267 749	1.10163 899
36	0.62373 019	0.71719 052	0.81228 024	0.90911 465	1.00776 438	1.10824 474
38	0.62505 840	0.71919 335	0.81516 039	0.91310 148	1.01311 039	1.11521 933
40	0.62642 563	0.72126 235	0.81814 765	0.91725 487	1.01870 633	1.12255 667
42	0.62782 630	0.72338 982	0.82123 227	0.92156 370	1.02454 127	1.13024 880
44	0.62925 446	0.72556 741	0.82440 346	0.92601 535	1.03060 230	1.13828 546
46	0.63070 385	0.72778 615	0.82764 941	0.93059 558	1.03687 427	1.14665 369
48	0.63216 783	0.73003 640	0.83095 712	0.93528 835	1.04333 948	1.15533 731
50	0.63363 947	0.73230 789	0.83431 247	0.94007 568	1.04997 735	1.16431 637
52	0.63511 150	0.73458 970	0.83770 010	0.94493 756	1.05676 412	1.17356 652
54	0.63657 639	0.73687 028	0.84110 344	0.94985 177	1.06367 248	1.18305 833
56	0.63802 636	0.73913 751	0.84450 468	0.95479 381	1.07067 128	1.19275 650
58	0.63945 343	0.74137 870	0.84788 483	0.95973 682	1.07772 516	1.20261 907
60	0.64084 944	0.74358 071	0.85122 375	0.96465 156	1.08479 434	1.21259 661
62	0.64220 613	0.74572 998	0.85450 024	0.96950 647	1.09183 436	1.22263 139
64	0.64351 521	0.74781 266	0.85769 220	0.97426 773	1.09879 601	1.23265 660
66	0.64476 839	0.74981 471	0.86077 677	0.97889 946	1.10562 535	1.24259 576
68	0.64595 751	0.75172 208	0.86373 057	0.98336 406	1.11226 392	1.25236 238
70	0.64707 458	0.75352 078	0.86652 996	0.98762 253	1.11864 920	1.26185 988
72	0.64811 189	0.75519 716	0.86915 135	0.99163 507	1.12471 530	1.27098 218
74	0.64906 209	0.75673 800	0.87157 159	0.99536 166	1.13039 401	1.27961 482
76	0.64991 829	0.75813 076	0.87376 830	0.99876 287	1.13561 610	1.28763 696
78	0.65067 415	0.75936 376	0.87572 037	1.00180 067	1.14031 304	1.29492 436
80	0.65132 394	0.76042 640	0.87740 833	1.00443 942	1.14441 892	1.30135 321
82	0.65186 270	0.76130 931	0.87881 481	1.00664, 678	1.14787 262	1.30680 495
84	0.65228 622	0.76200 457	0.87992 495	1.00839 470	1.15062 010	1.31117 166
86	0.65259 116	0.76250 582	0.88072 675	1.00966 028	1.15261 652	1.31436 170
88	0.65277 510	0.76280 846	0.88121 143	1.01042 658	1.15382 828	1.31630 510
90	0.65283 658	0.76290 965	0.88137 359	1.01068 319	1.15423 455	1.31695 790
	$\left[\begin{smallmatrix}(-5)2\\5\end{smallmatrix}\right]$	$\left[\begin{smallmatrix}(-5)3\\6\end{smallmatrix}\right]$	$\left[\begin{smallmatrix}(-5)4\\6\end{smallmatrix}\right]$	$\left[\begin{smallmatrix}(-5)6\\6\end{smallmatrix}\right]$	$\left[\begin{smallmatrix}(-4)1\\7\end{smallmatrix}\right]$	$\left[\begin{smallmatrix}(-4)2\\7\end{smallmatrix}\right]$
5	0.61113 335	0.69852 295	0.78594 111	0.87338 828	0.96086 405	1.04836 715
15	0.61325 114	0.70162 198	0.79025 416	0.87915 412	0.96832 014	1.05774 229
25	0.61733 857	0.70764 702	0.79870 514	0.89054 388	0.98317 128	1.07657 042
35	0.62308 236	0.71621 617	0.81088 311	0.90718 679	1.00518 803	1.10489 545
45	0.62997 691	0.72667 222	0.82601 788	0.92829 036	1.03371 296	1.14242 906
55	0.63730 374	0.73800 634	0.84280 548	0.95232 094	1.06716 268	1.18788 407
65	0.64414 930	0.74882 464	0.85924 936	0.97660 210	1.10223 077	1.23764 210
75	0.64950 235	0.75745 364	0.87269 924	0.99710 535	1.13306 645	1.28370 993
85	0.65245 368	0.76227 978	0.88036 502	1.00908 899	1.15171 457	1.31291 870

ELLIPTIC INTEGRAL OF THE FIRST KIND $F(\varphi\backslash\alpha)$ Table 17.5

$$F(\varphi\backslash\alpha)=\int_0^\varphi (1-\sin^2\alpha\,\sin^2\theta)^{-\frac{1}{2}}\,d\theta$$

$\alpha\backslash\varphi$	65°	70°	75°	80°	85°	90°
0°	1.13446 401	1.22173 048	1.30899 694	1.39626 340	1.48352 986	1.57079 633
2	1.13469 294	1.22200 477	1.30931 959	1.39663 672	1.48395 543	1.57127 495
4	1.13537 994	1.22282 810	1.31028 822	1.39775 763	1.48523 342	1.57271 244
6	1.13652 576	1.22420 180	1.31190 491	1.39962 909	1.48736 769	1.57511 361
8	1.13813 158	1.22612 810	1.31417 314	1.40225 598	1.49036 470	1.57848 658
10	1.14019 906	1.22861 010	1.31709 778	1.40564 522	1.49423 361	1.58284 280
12	1.14273 032	1.23165 180	1.32068 514	1.40980 577	1.49898 627	1.58819 721
14	1.14572 789	1.23525 808	1.32494 296	1.41474 871	1.50463 742	1.59456 834
16	1.14919 471	1.23943 470	1.32988 047	1.42048 728	1.51120 474	1.60197 853
18	1.15313 409	1.24418 827	1.33550 840	1.42703 700	1.51870 904	1.61045 415
20	1.15754 967	1.24952 627	1.34183 901	1.43441 578	1.52717 445	1.62002 590
22	1.16244 535	1.25545 700	1.34888 616	1.44264 399	1.53662 865	1.63072 910
24	1.16782 525	1.26198 957	1.35666 531	1.45174 466	1.54710 309	1.64260 414
26	1.17369 362	1.26913 385	1.36519 359	1.46174 360	1.55863 334	1.65569 693
28	1.18005 472	1.27690 045	1.37448 981	1.47266 958	1.57125 942	1.67005 943
30	1.18691 274	1.28530 059	1.38457 455	1.48455 455	1.58502 624	1.68575 035
32	1.19427 162	1.29434 605	1.39547 013	1.49743 384	1.59998 406	1.70283 594
34	1.20213 489	1.30404 906	1.40720 064	1.51134 644	1.61618 906	1.72139 083
36	1.21050 542	1.31442 210	1.41979 198	1.52633 523	1.63370 398	1.74149 923
38	1.21938 520	1.32547 772	1.43327 179	1.54244 734	1.65259 894	1.76325 618
40	1.22877 499	1.33722 824	1.44766 938	1.55973 441	1.67295 226	1.78676 913
42	1.23867 392	1.34968 545	1.46301 565	1.57825 301	1.69485 156	1.81215 985
44	1.24907 904	1.36286 013	1.47934 287	1.59806 493	1.71839 498	1.83956 672
46	1.25998 475	1.37676 148	1.49668 437	1.61923 762	1.74369 264	1.86914 755
48	1.27138 210	1.39139 640	1.51507 416	1.64184 453	1.77086 836	1.90108 303
50	1.28325 798	1.40676 855	1.53454 619	1.66596 542	1.80006 176	1.93558 110
52	1.29559 414	1.42287 717	1.55513 354	1.69168 665	1.83143 068	1.97288 227
54	1.30836 604	1.43971 560	1.57686 709	1.71910 125	1.86515 414	2.01326 657
56	1.32154 149	1.45726 935	1.59977 378	1.74830 880	1.90143 591	2.05706 232
58	1.33507 910	1.47551 372	1.62387 409	1.77941 482	1.94050 873	2.10465 766
60	1.34892 643	1.49441 087	1.64917 867	1.81252 953	1.98263 957	2.15651 565
62	1.36301 803	1.51390 609	1.67568 359	1.84776 547	2.02813 570	2.21319 470
64	1.37727 323	1.53392 332	1.70336 398	1.88523 335	2.07735 219	2.27537 643
66	1.39159 384	1.55435 972	1.73216 516	1.92503 509	2.13070 052	2.34390 472
68	1.40586 195	1.57507 940	1.76199 085	1.96725 237	2.18865 839	2.41984 165
70	1.41993 796	1.59590 624	1.79268 736	2.01192 798	2.25177 995	2.50455 008
72	1.43365 925	1.61661 644	1.82402 292	2.05903 582	2.32070 416	2.59981 973
74	1.44684 001	1.63693 134	1.85566 175	2.10843 282	2.39615 610	2.70806 762
76	1.45927 266	1.65651 218	1.88713 308	2.15978 295	2.47892 739	2.83267 258
78	1.47073 163	1.67495 873	1.91779 814	2.21243 977	2.56980 281	2.97856 895
80	1.48098 006	1.69181 489	1.94682 231	2.26527 326	2.66935 045	3.15338 525
82	1.48977 975	1.70658 456	1.97316 666	2.31643 897	2.77736 748	3.36986 803
84	1.49690 410	1.71876 033	1.99562 118	2.36313 736	2.89146 664	3.65185 597
86	1.50215 336	1.72786 543	2.01290 452	2.40153 358	3.00370 926	4.05275 817
88	1.50537 033	1.73350 464	2.02384 126	2.42718 003	3.09448 898	4.74271 727
90	1.50645 424	1.73541 516	2.02758 942	2.43624 605	3.13130 133	∞
	$\left[\dfrac{(-4)3}{8}\right]$	$\left[\dfrac{(-4)5}{8}\right]$	$\left[\dfrac{(-4)9}{10}\right]$	$\left[\dfrac{(-3)2}{10}\right]$	$\left[(-3)7\right]$	
5	1.13589 544	1.22344 604	1.31101 537	1.39859 928	1.48619 317	1.57379 213
15	1.14740 244	1.23727 471	1.32732 612	1.41751 762	1.50780 533	1.59814 200
25	1.17069 811	1.26548 460	1.36083 467	1.45663 012	1.55273 384	1.64899 522
35	1.20625 660	1.30915 104	1.41338 702	1.51870 347	1.62477 858	1.73124 518
45	1.25446 980	1.36971 948	1.48788 472	1.60847 673	1.73081 713	1.85407 468
55	1.31490 567	1.44840 433	1.58817 233	1.73347 444	1.88296 142	2.03471 531
65	1.38443 225	1.54409 676	1.71762 935	1.90483 674	2.10348 169	2.30878 680
75	1.45316 359	1.64683 711	1.87145 396	2.13389 514	2.43657 614	2.76806 315
85	1.49977 412	1.72372 395	2.00498 776	2.38364 709	2.94868 876	3.83174 200

Table 17.6

ELLIPTIC INTEGRAL OF THE SECOND KIND $E(\varphi\backslash\alpha)$

$$E(\varphi\backslash\alpha) = \int_0^\varphi (1-\sin^2\alpha\,\sin^2\theta)^{\frac{1}{2}}\,d\theta$$

$\alpha\backslash\varphi$	0°	5°	10°	15°	20°	25°	30°
0°	0	0.08726 646	0.17453 293	0.26179 939	0.34906 585	0.43633 231	0.52359 878
2	0	0.08726 633	0.17453 185	0.26179 579	0.34905 742	0.43631 608	0.52357 119
4	0	0.08726 592	0.17452 864	0.26178 503	0.34903 218	0.43626 745	0.52348 856
6	0	0.08726 525	0.17452 330	0.26176 715	0.34899 025	0.43618 665	0.52335 123
8	0	0.08726 432	0.17451 587	0.26174 224	0.34893 181	0.43607 403	0.52315 981
10	0	0.08726 313	0.17450 636	0.26171 041	0.34885 714	0.43593 011	0.52291 511
12	0	0.08726 168	0.17449 485	0.26167 182	0.34876 657	0.43575 552	0.52261 821
14	0	0.08725 999	0.17448 137	0.26162 664	0.34866 055	0.43555 106	0.52227 039
16	0	0.08725 806	0.17446 599	0.26157 510	0.34853 954	0.43531 765	0.52187 317
18	0	0.08725 590	0.17444 879	0.26151 743	0.34840 412	0.43505 633	0.52142 828
20	0	0.08725 352	0.17442 985	0.26145 391	0.34825 492	0.43476 831	0.52093 770
22	0	0.08725 094	0.17440 926	0.26138 485	0.34809 262	0.43445 488	0.52040 357
24	0	0.08724 816	0.17438 712	0.26131 056	0.34791 800	0.43411 749	0.51982 827
26	0	0.08724 521	0.17436 353	0.26123 141	0.34773 187	0.43375 767	0.51921 436
28	0	0.08724 208	0.17433 862	0.26114 778	0.34753 510	0.43337 709	0.51856 461
30	0	0.08723 881	0.17431 250	0.26106 005	0.34732 863	0.43297 749	0.51788 193
32	0	0.08723 540	0.17428 529	0.26096 867	0.34711 342	0.43256 075	0.51716 944
34	0	0.08723 187	0.17425 714	0.26087 405	0.34689 050	0.43212 880	0.51643 040
36	0	0.08722 824	0.17422 817	0.26077 666	0.34666 093	0.43168 368	0.51566 820
38	0	0.08722 453	0.17419 852	0.26067 697	0.34642 580	0.43122 748	0.51488 638
40	0	0.08722 075	0.17416 835	0.26057 545	0.34618 625	0.43076 236	0.51408 862
42	0	0.08721 692	0.17413 779	0.26047 261	0.34594 343	0.43029 055	0.51327 866
44	0	0.08721 307	0.17410 700	0.26036 893	0.34569 850	0.42981 431	0.51246 037
46	0	0.08720 920	0.17407 613	0.26026 492	0.34545 266	0.42933 594	0.51163 767
48	0	0.08720 535	0.17404 531	0.26016 110	0.34520 710	0.42885 776	0.51081 454
50	0	0.08720 152	0.17401 472	0.26005 795	0.34496 302	0.42838 212	0.50999 501
52	0	0.08719 774	0.17398 449	0.25995 600	0.34472 162	0.42791 134	0.50918 310
54	0	0.08719 402	0.17395 477	0.25985 574	0.34448 409	0.42744 775	·0.50838 287
56	0	0.08719 039	0.17392 571	0.25975 765	0.34425 159	0.42699 368	0.50759 831
58	0	0.08718 686	0.17389 745	0.25966 224	0.34402 529	0.42655 138	0.50683 341
60	0	0.08718 345	0.17387 013	0.25956 996	0.34380 631	0.42612 308	0.50609 207
62	0	0.08718 017	0.17384 388	0.25948 126	0.34359 575	0.42571 097	0.50537 811
64	0	0.08717 704	0.17381 883	0.25939 660	0.34339 465	0.42531 712	0.50469 523
66	0	0.08717 408	0.17379 511	0.25931 640	0.34320 404	0.42494 358	0.50404 700
68	0	0.08717 130	0.17377 283	0.25924 104	0.34302 487	0.42459 224	0.50343 686
70	0	0.08716 871	0.17375 210	0.25917 090	0.34285 805	0.42426 495	0.50286 804
72	0	0.08716 633	0.17373 302	0.25910 634	0.34270 443	0.42396 339	0.50234 359
74	0	0.08716 416	0.17371 568	0.25904 767	0.34256 478	0.42368 913	0.50186 633
76	0	0.08716 223	0.17370 018	0.25899 519	0.34243 984	0.42344 363	0.50143 886
78	0	0.08716 053	0.17368 659	0.25894 917	0.34233 022	0.42322 817	0.50106 351
80	0	0.08715 909	0.17367 498	0.25890 983	0.34223 650	0.42304 389	0.50074 232
82	0	0.08715 789	0.17366 539	0.25887 737	0.34215 915	0.42289 175	0.50047 707
84	0	0.08715 695	0.17365 789	0.25885 195	0.34209 857	0.42277 258	0.50026 923
86	0	0.08715 628	0.17365 250	0.25883 370	0.34205 507	0.42268 700	0.50011 993
88	0	0.08715 588	0.17364 926	0.25882 271	0.34202 889	0.42263 547	0.50003 003
90	0	0.08715 574	0.17364 818	0.25881 905	0.34202 014	0.42261 826	0.50000 000
		$\begin{bmatrix}(-8)4\\3\end{bmatrix}$	$\begin{bmatrix}(-7)3\\4\end{bmatrix}$	$\begin{bmatrix}(-7)9\\4\end{bmatrix}$	$\begin{bmatrix}(-6)2\\5\end{bmatrix}$	$\begin{bmatrix}(-6)4\\5\end{bmatrix}$	$\begin{bmatrix}(-6)7\\5\end{bmatrix}$
5	0	0.08726 562	0.17452 624	0.26177 698	0.34901 329	0.43623 105	0.52342 670
15	0	0.08725 905	0.17447 391	0.26160 165	0.34860 188	0.43543 791	0.52207 785
25	0	0.08724 671	0.17437 550	0.26127 157	0.34782 632	0.43394 028	0.51952 597
35	0	0.08723 006	0.17424 275	0.26082 567	0.34677 648	0.43190 776	0.51605 197
45	0	0.08721 113	0.17409 157	0.26031 693	0.34557 562	0.42957 525	0.51204 932
55	0	0.08719 220	0.17394 015	0.25980 639	0.34436 714	0.42721 938	0.50798 838
65	0	0.08717 554	0.17380 680	0.25935 592	0.34329 797	0.42512 769	0.50436 656
75	0	0.08716 317	0.17370 770	0.25902 064	0.34250 043	0.42356 271	0.50164 622
85	0	0.08715 659	0.17365 493	0.25884 192	0.34207 467	0.42272 556	0.50018 720

See Example 14.

Compiled from K. Pearson, Tables of the complete and incomplete elliptic integrals, Cambridge Univ. Press, Cambridge, England, 1934 (with permission). Known errors have been corrected.

ELLIPTIC INTEGRAL OF THE SECOND KIND $E(\varphi\backslash\alpha)$ Table 17.6

$$E(\varphi\backslash\alpha)=\int_0^\varphi (1-\sin^2\alpha\,\sin^2\theta)^{\frac{1}{2}}d\theta$$

$\alpha\backslash\varphi$	35°	40°	45°	50°	55°	60°
0°	0.61086 524	0.69813 170	0.78539 816	0.87266 463	0.95993 109	1.04719 755
2	0.61082 230	0.69806 905	0.78531 125	0.87254 883	0.95978 184	1.04701 051
4	0.61069 365	0.69788 136	0.78505 085	0.87220 183	0.95933 459	1.04644 996
6	0.61047 983	0.69756 935	0.78461 792	0.87162 487	0.95859 083	1.04551 764
8	0.61018 171	0.69713 427	0.78401 409	0.87081 998	0.95755 301	1.04421 646
10	0.60980 055	0.69657 784	0.78324 162	0.86979 001	0.95622 460	1.04255 047
12	0.60933 793	0.69590 226	0.78230 343	0.86853 863	0.95461 005	1.04052 491
14	0.60879 577	0.69511 023	0.78120 308	0.86707 031	0.95271 478	1.03814 615
16	0.60817 636	0.69420 492	0.77994 473	0.86539 034	0.95054 522	1.03542 177
18	0.60748 229	0.69318 999	0.77853 323	0.86350 481	0.94810 878	1.03236 049
20	0.60671 652	0.69206 954	0.77697 402	0.86142 062	0.94541 386	1.02897 221
22	0.60588 229	0.69084 814	0.77527 316	0.85914 545	0.94246 984	1.02526 804
24	0.60498 319	0.68953 083	0.77343 735	0.85668 781	0.93928 709	1.02126 023
26	0.60402 308	0.68812 308	0.77147 387	0.85405 695	0.93587 699	1.01696 224
28	0.60300 616	0.68663 077	0.76939 059	0.85126 295	0.93225 186	1.01238 873
30	0.60193 687	0.68506 023	0.76719 599	0.84831 663	0.92842 504	1.00755 556
32	0.60081 994	0.68341 817	0.76489 908	0.84522 958	0.92441 083	1.00247 977
34	0.59966 035	0.68171 170	0.76250 947	0.84201 414	0.92022 452	0.99717 966
36	0.59846 332	0.67994 830	0.76003 726	0.83868 340	0.91588 234	0.99167 469
38	0.59723 431	0.67813 578	0.75749 309	0.83525 115	0.91140 150	0.98598 560
40	0.59597 897	0.67628 229	0.75488 809	0.83173 189	0.90680 017	0.98012 430
42	0.59470 312	0.67439 630	0.75223 383	0.82814 080	0.90209 742	0.97414 397
44	0.59341 278	0.67248 651	0.74954 234	0.82449 369	0.89731 325	0.96803 899
46	0.59211 406	0.67056 191	0.74682 605	0.82080 700	0.89246 858	0.96184 497
48	0.59081 324	0.66863 167	0.74409 773	0.81709 775	0.88758 513	0.95558 873
50	0.58951 664	0.66670 515	0.74137 047	0.81338 346	0.88268 551	0.94929 830
52	0.58823 065	0.66479 183	0.73865 766	0.80968 217	0.87779 305	0.94300 285
54	0.58696 171	0.66290 130	0.73597 286	0.80601 230	0.87293 184	0.93673 272
56	0.58571 622	0.66104 317	0.73332 979	0.80239 262	0.86812 660	0.93051 931
58	0.58450 056	0.65922 707	0.73074 229	0.79884 217	0.86340 261	0.92439 505
60	0.58332 103	0.65746 255	0.72822 416	0.79538 015	0.85878 561	0.91839 329
62	0.58218 382	0.65575 905	0.72578 915	0.79202 582	0.85430 169	0.91254 821
64	0.58109 497	0.65412 585	0.72345 085	0.78879 839	0.84997 726	0.90689 460
66	0.58006 032	0.65257 197	0.72122 260	0.78571 685	0.84583 811	0.90146 778
68	0.57908 549	0.65110 612	0.71911 737	0.78279 987	0.84191 082	0.89630 323
70	0.57817 584	0.64973 667	0.71714 767	0.78006 562	0.83822 090	0.89143 642
72	0.57733 641	0.64847 154	0.71532 545	0.77753 157	0.83479 335	0.88690 237
74	0.57657 189	0.64731 812	0.71366 196	0.77521 434	0.83165 223	0.88273 530
76	0.57588 663	0.64628 328	0.71216 766	0.77312 952	0.82882 031	0.87896 810
78	0.57528 450	0.64537 322	0.71085 210	0.77129 143	0.82631 879	0.87563 185
80	0.57476 897	0.64459 347	0.70972 381	0.76971 298	0.82416 694	0.87275 520
82	0.57434 302	0.64394 879	0.70877 019	0.76840 544	0.82238 177	0.87036 381
84	0.57400 912	0.64344 316	0.70805 745	0.76737 830	0.82097 770	0.86847 970
86	0.57376 921	0.64307 973	0.70753 050	0.76663 912	0.81996 631	0.86712 068
88	0.57362 470	0.64286 075	0.70721 289	0.76619 339	0.81935 604	0.86629 990
90	0.57357 644	0.64278 761	0.70710 678	0.76604 444	0.81915 204	0.86602 540
	$\begin{bmatrix}(-5)1\\5\end{bmatrix}$	$\begin{bmatrix}(-5)2\\5\end{bmatrix}$	$\begin{bmatrix}(-5)3\\5\end{bmatrix}$	$\begin{bmatrix}(-5)4\\6\end{bmatrix}$	$\begin{bmatrix}(-5)5\\6\end{bmatrix}$	$\begin{bmatrix}(-5)7\\6\end{bmatrix}$
5	0.61059 734	0.69774 083	0.78485 586	0.87194 199	0.95899 964	1.04603 012
15	0.60849 557	0.69467 152	0.78059 337	0.86625 642	0.95166 385	1.03682 664
25	0.60451 051	0.68883 790	0.77247 109	0.85539 342	0.93760 971	1.01914 662
35	0.59906 618	0.68083 664	0.76128 304	0.84036 234	0.91807 186	0.99445 152
45	0.59276 408	0.67152 549	0.74818 650	0.82265 424	0.89489 714	0.96495 146
55	0.58633 563	0.66196 758	0.73464 525	0.80419 500	0.87052 066	0.93361 692
65	0.58057 051	0.65333 844	0.72232 215	0.78723 820	0.84788 276	0.90415 063
75	0.57621 910	0.64678 548	0.71289 304	0.77414 195	0.83019 625	0.88079 972
85	0.57387 732	0.64324 351	0.70776 799	0.76697 232	0.82042 232	0.86773 361

Table 17.6   **ELLIPTIC INTEGRAL OF THE SECOND KIND** $E(\varphi\backslash\alpha)$

$$E(\varphi\backslash\alpha)=\int_0^\varphi (1-\sin^2\alpha\sin^2\theta)^{\frac{1}{2}} d\theta$$

$\alpha\backslash\varphi$	65°	70°	75°	80°	85°	90°
0°	1.13446 401	1.22173 048	1.30899 694	1.39626 340	1.48352 986	1.57079 633
2	1.13423 517	1.22145 628	1.30867 442	1.39589 024	1.48310 448	1.57031 792
4	1.13354 929	1.22063 443	1.30770 767	1.39477 165	1.48182 929	1.56888 372
6	1.13240 837	1.21926 717	1.30609 916	1.39291 030	1.47970 717	1.56649 679
8	1.13081 573	1.21735 820	1.30385 297	1.39031 062	1.47674 288	1.56316 223
10	1.12877 602	1.21491 274	1.30097 484	1.38697 886	1.47294 312	1.55888 720
12	1.12629 522	1.21193 748	1.29747 215	1.38292 302	1.46831 652	1.55368 089
14	1.12338 066	1.20844 065	1.29335 393	1.37815 292	1.46287 363	1.54755 458
16	1.12004 099	1.20443 195	1.28863 089	1.37268 017	1.45662 693	1.54052 157
18	1.11628 624	1.19992 262	1.28331 541	1.36651 823	1.44959 085	1.53259 729
20	1.11212 778	1.19492 542	1.27742 153	1.35968 233	1.44178 179	1.52379 921
22	1.10757 834	1.18945 465	1.27096 502	1.35218 961	1.43321 813	1.51414 692
24	1.10265 204	1.18352 618	1.26396 337	1.34405 903	1.42392 023	1.50366 214
26	1.09736 439	1.17715 743	1.25643 578	1.33531 146	1.41391 049	1.49236 871
28	1.09173 228	1.17036 745	1.24840 326	1.32596 967	1.40321 335	1.48029 266
30	1.08577 404	1.16317 686	1.23988 858	1.31605 841	1.39185 532	1.46746 221
32	1.07950 942	1.15560 796	1.23091 635	1.30560 436	1.37986 503	1.45390 780
34	1.07295 961	1.14768 469	1.22151 305	1.29463 629	1.36727 328	1.43966 215
36	1.06614 728	1.13943 273	1.21170 705	1.28318 499	1.35411 306	1.42476 031
38	1.05909 660	1.13087 946	1.20152 870	1.27128 343	1.34041 965	1.40923 972
40	1.05183 322	1.12205 408	1.19101 036	1.25896 675	1.32623 066	1.39314 025
42	1.04438 435	1.11298 760	1.18018 648	1.24627 240	1.31158 614	1.37650 433
44	1.03677 875	1.10371 291	1.16909 366	1.23324 019	1.29652 865	1.35937 700
46	1.02904 677	1.09426 484	1.15777 077	1.21991 241	1.28110 340	1.34180 606
48	1.02122 034	1.08468 023	1.14625 899	1.20633 398	1.26535 837	1.32384 218
50	1.01333 305	1.07499 796	1.13460 200	1.19255 255	1.24934 449	1.30553 909
52	1.00542 010	1.06525 908	1.12284 604	1.17861 873	1.23311 580	1.28695 374
54	0.99751 835	1.05550 682	1.11104 010	1.16458 621	1.21672 971	1.26814 653
56	0.98966 632	1.04578 671	1.09923 604	1.15051 210	1.20024 724	1.24918 162
58	0.98190 414	1.03614 663	1.08748 883	1.13645 710	1.18373 339	1.23012 722
60	0.97427 354	1.02663 689	1.07585 669	1.12248 590	1.16725 747	1.21105 603
62	0.96681 780	1.01731 023	1.06440 132	1.10866 752	1.15089 364	1.19204 568
64	0.95958 158	1.00822 192	1.05318 814	1.09507 580	1.13472 145	1.17317 938
66	0.95261 084	0.99942 966	1.04228 653	1.08178 986	1.11882 658	1.15454 668
68	0.94595 256	0.99099 354	1.03176 998	1.06889 476	1.10330 172	1.13624 437
70	0.93965 447	0.98297 583	1.02171 634	1.05648 221	1.08824 773	1.11837 774
72	0.93376 462	0.97544 068	1.01220 781	1.04465 133	1.07377 505	1.10106 217
74	0.92833 088	0.96845 360	1.00333 091	1.03350 951	1.06000 556	1.08442 522
76	0.92340 024	0.96208 074	0.99517 606	1.02317 331	1.04707 504	1.06860 953
78	0.91901 802	0.95638 776	0.98783 670	1.01376 904	1.03513 640	1.05377 692
80	0.91522 691	0.95143 847	0.98140 781	1.00543 295	1.02436 393	1.04011 440
82	0.91206 588	0.94729 297	0.97598 331	0.99831 000	1.01495 896	1.02784 362
84	0.90956 905	0.94400 544	0.97165 228	0.99255 019	1.00715 650	1.01723 692
86	0.90776 445	0.94162 171	0.96849 392	0.98830 025	1.00123 026	1.00864 796
88	0.90667 305	0.94017 677	0.96657 142	0.98568 915	0.99748 392	1.00258 409
90	0.90630 779	0.93969 262	0.96592 583	0.98480 775	0.99619 470	1.00000 000
	$\left[\begin{smallmatrix}(-5)9\\6\end{smallmatrix}\right]$	$\left[\begin{smallmatrix}(-4)1\\7\end{smallmatrix}\right]$	$\left[\begin{smallmatrix}(-4)2\\7\end{smallmatrix}\right]$	$\left[\begin{smallmatrix}(-4)2\\9\end{smallmatrix}\right]$	$\left[\begin{smallmatrix}(-4)3\\9\end{smallmatrix}\right]$	$\left[\begin{smallmatrix}(-4)4\\10\end{smallmatrix}\right]$
5	1.13303 553	1.22001 878	1.30698 342	1.39393 358	1.48087 384	1.56780 907
15	1.12176 337	1.20649 962	1.29106 728	1.37550 358	1.45984 990	1.54415 050
25	1.10005 236	1.18039 569	1.26026 405	1.33976 099	1.41900 286	1.49811 493
35	1.06958 479	1.14359 813	1.21665 853	1.28896 903	1.36076 208	1.43229 097
45	1.03292 660	1.09900 829	1.16345 846	1.22661 050	1.28885 906	1.35064 388
55	0.99358 365	1.05063 981	1.10513 448	1.15755 065	1.20849 656	1.25867 963
65	0.95606 011	1.00378 508	1.04769 389	1.08838 943	1.12673 373	1.16382 796
75	0.92579 978	0.96518 626	0.99915 744	1.02823 305	1.05342 632	1.07640 511
85	0.90857 873	0.94269 813	0.96992 212	0.99022 779	1.00394 027	1.01266 351

JACOBIAN ZETA FUNCTION $Z(\varphi\backslash\alpha)$

Table 17.7

$$K(\alpha)Z(\varphi\backslash\alpha) = K(\alpha)E(\varphi\backslash\alpha) - E(\alpha)F(\varphi\backslash\alpha)$$

$$K(90°)Z(\varphi\backslash\alpha) = K(90°)Z(u|1) = K(90°)\tanh u = \infty \text{ for all } u$$

$\alpha\backslash\varphi$	0°	5°	10°	15°	20°	25°	30°
0°	0	0.000000	0.000000	0.000000	0.000000	0.000000	0.000000
2	0	0.000083	0.000164	0.000239	0.000308	0.000367	0.000414
4	0	0.000332	0.000655	0.000957	0.001231	0.001467	0.001658
6	0	0.000748	0.001474	0.002155	0.002770	0.003302	0.003734
8	0	0.001331	0.002621	0.003832	0.004928	0.005875	0.006644
10	0	0.002080	0.004098	0.005992	0.007706	0.009188	0.010393
12	0	0.002997	0.005905	0.008635	0.011107	0.013246	0.014987
14	0	0.004082	0.008043	0.011765	0.015136	0.018055	0.020433
16	0	0.005337	0.010516	0.015384	0.019796	0.023621	0.026740
18	0	0.006761	0.013324	0.019496	0.025094	0.029951	0.033919
20	0	0.008357	0.016470	0.024105	0.031035	0.037055	0.041981
22	0	0.010125	0.019958	0.029216	0.037627	0.044942	0.050941
24	0	0.012067	0.023791	0.034834	0.044878	0.053626	0.060814
26	0	0.014186	0.027972	0.040968	0.052799	0.063119	0.071617
28	0	0.016483	0.032508	0.047624	0.061401	0.073438	0.083373
30	0	0.018962	0.037403	0.054811	0.070696	0.084599	0.096103
32	0	0.021625	0.042664	0.062540	0.080700	0.096624	0.109834
34	0	0.024476	0.048298	0.070823	0.091430	0.109534	0.124596
36	0	0.027520	0.054315	0.079674	0.102905	0.123356	0.140421
38	0	0.030761	0.060725	0.089108	0.115148	0.138120	0.157347
40	0	0.034205	0.067540	0.099145	0.128185	0.153860	0.175418
42	0	0.037860	0.074774	0.109807	0.142046	0.170614	0.194683
44	0	0.041734	0.082444	0.121118	0.156765	0.188428	0.215197
46	0	0.045835	0.090569	0.133109	0.172383	0.207353	0.237025
48	0	0.050177	0.099172	0.145813	0.188947	0.227450	0.260240
50	0	0.054771	0.108280	0.159273	0.206513	0.248789	0.284929
52	0	0.059634	0.117925	0.173536	0.225145	0.271452	0.311193
54	0	0.064786	0.128146	0.188661	0.244921	0.295538	0.339150
56	0	0.070249	0.138989	0.204716	0.265933	0.321161	0.368940
58	0	0.076052	0.150510	0.221785	0.288294	0.348462	0.400731
60	0	0.082227	0.162776	0.239971	0.312138	0.377610	0.434726
62	0	0.088818	0.175872	0.259398	0.337632	0.408811	0.471170
64	0	0.095876	0.189901	0.280221	0.364981	0.442321	0.510371
66	0	0.103468	0.204994	0.302637	0.394446	0.478462	0.552710
68	0	0.111676	0.221320	0.326895	0.426356	0.517644	0.598675
70	0	0.120612	0.239097	0.353322	0.461145	0.560402	0.648900
72	0	0.130420	0.258615	0.382351	0.499384	0.607444	0.704225
74	0	0.141301	0.280272	0.414575	0.541857	0.659739	0.765797
76	0	0.153537	0.304631	0.450832	0.589673	0.718657	0.835238
78	0	0.167542	0.332519	0.492356	0.644462	0.786214	0.914934
80	0	0.183967	0.365230	0.541075	0.708771	0.865556	1.008608
82	0	0.203902	0.404937	0.600229	0.786884	0.961976	1.122523
84	0	0.229402	0.455734	0.675918	0.886859	1.085434	1.268462
86	0	0.265091	0.526833	0.781873	1.026844	1.258352	1.472953
88	0	0.325753	0.647691	0.962000	1.264856	1.552420	1.820811
90	∞	∞	∞	∞	∞	∞	∞
5	0	0.000519	0.001023	0.001496	0.001923	0.002292	0.002592
15	0	0.004688	0.009238	0.013513	0.017387	0.020743	0.023479
25	0	0.013105	0.025838	0.037836	0.048754	0.058271	0.066098
35	0	0.025973	0.051258	0.075176	0.097073	0.116329	0.132373
45	0	0.043755	0.086448	0.127026	0.164459	0.197748	0.225942
55	0	0.067477	0.133487	0.196567	0.255266	0.308149	0.353807
65	0	0.099601	0.197305	0.291216	0.379430	0.460039	0.531121
75	0	0.147228	0.292070	0.432134	0.565011	0.688264	0.799407
85	0	0.245478	0.487761	0.723644	0.949910	1.163313	1.360551

See Example 16.

Compiled from P. F. Byrd and M.D. Friedman, Handbook of elliptic integrals for engineers and physicists, Springer-Verlag, Berlin, Germany, 1954 (with permission).

Table 17.7   **JACOBIAN ZETA FUNCTION** $Z(\varphi\backslash\alpha)$

$$K(\alpha)Z(\varphi\backslash\alpha)=K(\alpha)E(\varphi\backslash\alpha)-E(\alpha)F(\varphi\backslash\alpha)$$

$$K(90°)Z(\varphi\backslash\alpha)=K(90°)Z(u|1)=K(90°)\tanh u=\infty \text{ for all } u$$

$\alpha\backslash\varphi$	35°	40°	45°	50°	55°	60°
0°	0.000000	0.000000	0.000000	0.000000	0.000000	0.000000
2	0.000450	0.000471	0.000479	0.000471	0.000450	0.000415
4	0.001800	0.001886	0.001916	0.001887	0.001800	0.001659
6	0.004052	0.004248	0.004314	0.004250	0.004056	0.003739
8	0.007212	0.007561	0.007681	0.007567	0.007224	0.006660
10	0.011284	0.011833	0.012023	0.011849	0.011313	0.010433
12	0.016276	0.017073	0.017353	0.017106	0.016337	0.015070
14	0.022197	0.023293	0.023683	0.023354	0.022312	0.020588
16	0.029060	0.030505	0.031029	0.030610	0.029257	0.027006
18	0.036876	0.038728	0.039411	0.038897	0.037194	0.034347
20	0.045662	0.047979	0.048850	0.048238	0.046150	0.042639
22	0.055435	0.058279	0.059372	0.058663	0.056156	0.051912
24	0.066216	0.069655	0.071005	0.070203	0.067246	0.062203
26	0.078026	0.082132	0.083783	0.082895	0.079461	0.073551
28	0.090893	0.095744	0.097742	0.096782	0.092844	0.086003
30	0.104844	0.110525	0.112924	0.111909	0.107447	0.099613
32	0.119914	0.126515	0.129375	0.128330	0.123327	0.114438
34	0.136138	0.143758	0.147147	0.146103	0.140549	0.130548
36	0.153557	0.162305	0.166300	0.165296	0.159186	0.148018
38	0.172220	0.182211	0.186898	0.185983	0.179319	0.166934
40	0.192178	0.203541	0.209016	0.208248	0.201042	0.187395
42	0.213492	0.226365	0.232738	0.232187	0.224459	0.209512
44	0.236228	0.250764	0.258158	0.257907	0.249691	0.233413
46	0.260466	0.276831	0.285383	0.285531	0.276871	0.259243
48	0.286295	0.304671	0.314535	0.315196	0.306156	0.287169
50	0.313816	0.334405	0.345755	0.347064	0.337723	0.317383
52	0.343151	0.366173	0.379203	0.381317	0.371776	0.350108
54	0.374438	0.400138	0.415067	0.418166	0.408552	0.385601
56	0.407844	0.436490	0.453565	0.457861	0.448328	0.424167
58	0.443565	0.475457	0.494956	0.500691	0.491428	0.466161
60	0.481836	0.517310	0.539547	0.547003	0.538238	0.512007
62	0.522947	0.562378	0.587709	0.597211	0.589220	0.562214
64	0.567251	0.611064	0.639896	0.651822	0.644933	0.617399
66	0.615191	0.663870	0.696670	0.711460	0.706068	0.678320
68	0.667330	0.721434	0.758741	0.776910	0.773487	0.745922
70	0.724397	0.784577	0.827024	0.849178	0.848294	0.821411
72	0.787359	0.854390	0.902728	0.929590	0.931931	0.906356
74	0.857536	0.932355	0.987491	1.019938	1.026343	1.002860
76	0.936789	1.020563	1.083621	1.122735	1.134246	1.113848
78	1.027859	1.122089	1.194508	1.241670	1.259612	1.243568
80	1.135017	1.241721	1.325428	1.382470	1.408589	1.398577
82	1.265447	1.387516	1.485245	1.554749	1.591484	1.589820
84	1.432669	1.574623	1.690632	1.776579	1.827639	1.837791
86	1.667113	1.837147	1.979107	2.088611	2.160541	2.188502
88	2.066078	2.284127	2.470622	2.620801	2.729164	2.788909
90	∞	∞	∞	∞	∞	∞
5	0.002813	0.002948	0.002994	0.002949	0.002815	0.002594
15	0.025510	0.026774	0.027228	0.026855	0.025662	0.023683
25	0.071991	0.075754	0.077249	0.076403	0.073210	0.067742
35	0.144695	0.152865	0.156547	0.155518	0.149686	0.139108
45	0.248154	0.263583	0.271538	0.271473	0.263028	0.246077
55	0.390865	0.418002	0.433972	0.437641	0.428046	0.404479
65	0.590735	0.636916	0.667669	0.680968	0.674774	0.647089
75	0.895883	0.975016	1.033955	1.069585	1.078397	1.056317
85	1.538234	1.692810	1.820471	1.916972	1.977347	1.995386

JACOBIAN ZETA FUNCTION $Z(\varphi\backslash\alpha)$　　　Table 17.7

$$K(\alpha)Z(\varphi\backslash\alpha)=K(\alpha)E(\varphi\backslash\alpha)-E(\alpha)F(\varphi\backslash\alpha)$$

$$K(90°)Z(\varphi\backslash\alpha)=K(90°)Z(u|1)=K(90°)\tanh u=\infty \text{ for all } u$$

$\alpha\backslash\varphi$	65°	70°	75°	80°	85°	90°
0°	0.000000	0.000000	0.000000	0.000000	0.000000	0
2	0.000367	0.000308	0.000239	0.000164	0.000083	0
4	0.001468	0.001232	0.000958	0.000656	0.000333	0
6	0.003308	0.002776	0.002160	0.001477	0.000750	0
8	0.005893	0.004946	0.003849	0.002633	0.001337	0
10	0.009233	0.007751	0.006032	0.004127	0.002096	0
12	0.013341	0.011202	0.008718	0.005966	0.003030	0
14	0.018231	0.015312	0.011920	0.008158	0.004143	0
16	0.023922	0.020098	0.015649	0.010713	0.005442	0
18	0.030438	0.025581	0.019924	0.013642	0.006930	0
20	0.037803	0.031783	0.024763	0.016959	0.008617	0
22	0.046047	0.038732	0.030188	0.020680	0.010509	0
24	0.055206	0.046459	0.036225	0.024823	0.012617	0
26	0.065319	0.055000	0.042905	0.029411	0.014952	0
28	0.076431	0.064397	0.050260	0.034466	0.017526	0
30	0.088594	0.074696	0.058332	0.040018	0.020354	0
32	0.101867	0.085951	0.067164	0.046099	0.023454	0
34	0.116315	0.098224	0.076808	0.052747	0.026845	0
36	0.132015	0.111585	0.087324	0.060004	0.030550	0
38	0.149053	0.126114	0.098779	0.067920	0.034595	0
40	0.167527	0.141905	0.111254	0.076554	0.039011	0
42	0.187551	0.159064	0.124839	0.085973	0.043833	0
44	0.209254	0.177713	0.139641	0.096255	0.049104	0
46	0.232785	0.197996	0.155784	0.107493	0.054874	0
48	0.258315	0.220078	0.173414	0.119798	0.061201	0
50	0.286045	0.244154	0.192704	0.133299	0.068157	0
52	0.316206	0.270454	0.213858	0.148154	0.075826	0
54	0.349070	0.299246	0.237121	0.164550	0.084312	0
56	0.384960	0.330854	0.262789	0.182720	0.093745	0
58	0.424255	0.365664	0.291220	0.202947	0.104281	0
60	0.467411	0.404143	0.322854	0.225584	0.116121	0
62	0.514976	0.446860	0.358236	0.251076	0.129521	0
64	0.567621	0.494517	0.398048	0.279993	0.144812	0
66	0.626169	0.547987	0.443155	0.313069	0.162430	0
68	0.691653	0.608372	0.494668	0.351277	0.182965	0
70	0.765385	0.677086	0.554038	0.395917	0.207230	0
72	0.849072	0.755975	0.623195	0.448779	0.236382	0
74	0.944993	0.847508	0.704762	0.512376	0.272114	0
76	1.056298	0.955095	0.802400	0.590350	0.317015	0
78	1.187535	1.083634	0.921408	0.688163	0.375226	0
80	1.345674	1.240571	1.069839	0.814374	0.453784	0
82	1.542281	1.438150	1.260828	0.983236	0.565578	0
84	1.798909	1.698985	1.518315	1.220780	0.736684	0
86	2.163806	2.073357	1.894760	1.583040	1.028059	0
88	2.790834	2.721008	2.555104	2.241393	1.628299	0
90	∞	∞	∞	∞	∞	∞
5	0.002295	0.001926	0.001498	0.001025	0.000520	0
15	0.020975	0.017619	0.013718	0.009390	0.004769	0
25	0.060141	0.050625	0.039483	0.027060	0.013755	0
35	0.124003	0.104764	0.081953	0.056296	0.028657	0
45	0.220781	0.187640	0.147536	0.101748	0.051923	0
55	0.366615	0.314676	0.249634	0.173397	0.088901	0
65	0.596098	0.520463	0.419877	0.295957	0.153297	0
75	0.998480	0.899033	0.751288	0.549278	0.293208	0
85	1.962673	1.866624	1.686113	1.380465	0.860811	0

Table 17.8   **HEUMAN'S LAMBDA FUNCTION** $\Lambda_0(\varphi\backslash\alpha)$

$$\Lambda_0(\varphi\backslash\alpha)=\frac{F(\varphi\backslash90^\circ-\alpha)}{K'(\alpha)}+\frac{2}{\pi}K(\alpha)Z(\varphi\backslash90^\circ-\alpha)=\frac{2}{\pi}\{K(\alpha)E(\varphi\backslash90^\circ-\alpha)-[K(\alpha)-E(\alpha)]F(\varphi\backslash90^\circ-\alpha)\}$$

$\alpha\backslash\varphi$	0°	5°	10°	15°	20°	25°	30°
0°	0	0.087156	0.173648	0.258819	0.342020	0.422618	0.500000
2	0	0.087129	0.173595	0.258740	0.341916	0.422490	0.499848
4	0	0.087050	0.173437	0.258504	0.341604	0.422104	0.499391
6	0	0.086917	0.173173	0.258111	0.341084	0.421462	0.498633
8	0	0.086732	0.172804	0.257562	0.340359	0.420566	0.497574
10	0	0.086495	0.172332	0.256858	0.339430	0.419419	0.496219
12	0	0.086206	0.171757	0.256001	0.338299	0.418024	0.494572
14	0	0.085866	0.171080	0.254994	0.336969	0.416385	0.492638
16	0	0.085476	0.170303	0.253838	0.335445	0.414506	0.490424
18	0	0.085037	0.169429	0.252536	0.333729	0.412394	0.487937
20	0	0.084549	0.168458	0.251092	0.331827	0.410054	0.485184
22	0	0.084013	0.167393	0.249509	0.329743	0.407492	0.482176
24	0	0.083432	0.166236	0.247790	0.327483	0.404717	0.478920
26	0	0.082806	0.164991	0.245941	0.325052	0.401736	0.475428
28	0	0.082136	0.163661	0.243966	0.322458	0.398558	0.471710
30	0	0.081425	0.162247	0.241870	0.319707	0.395191	0.467777
32	0	0.080674	0.160755	0.239657	0.316806	0.391645	0.463642
34	0	0.079884	0.159187	0.237335	0.313764	0.387930	0.459316
36	0	0.079058	0.157548	0.234908	0.310587	0.384057	0.454813
38	0	0.078198	0.155842	0.232383	0.307286	0.380037	0.450147
40	0	0.077307	0.154073	0.229767	0.303869	0.375880	0.445330
42	0	0.076385	0.152246	0.227068	0.300346	0.371600	0.440378
44	0	0.075436	0.150367	0.224292	0.296727	0.367209	0.435306
46	0	0.074463	0.148439	0.221447	0.293022	0.362720	0.430127
48	0	0.073469	0.146470	0.218543	0.289242	0.358145	0.424860
50	0	0.072455	0.144464	0.215587	0.285399	0.353500	0.419519
52	0	0.071426	0.142428	0.212589	0.281505	0.348799	0.414121
54	0	0.070385	0.140370	0.209558	0.277573	0.344057	0.408685
56	0	0.069336	0.138295	0.206506	0.273616	0.339290	0.403228
58	0	0.068281	0.136211	0.203443	0.269648	0.334516	0.397769
60	0	0.067226	0.134126	0.200380	0.265684	0.329751	0.392328
62	0	0.066175	0.132049	0.197331	0.261739	0.325015	0.386926
64	0	0.065131	0.129989	0.194307	0.257832	0.320328	0.381586
66	0	0.064100	0.127955	0.191324	0.253979	0.315710	0.376331
68	0	0.063088	0.125958	0.188396	0.250200	0.311185	0.371186
70	0	0.062100	0.124009	0.185540	0.246517	0.306778	0.366180
72	0	0.061143	0.122121	0.182774	0.242952	0.302515	0.361342
74	0	0.060223	0.120307	0.180119	0.239531	0.298427	0.356706
76	0	0.059348	0.118583	0.177596	0.236282	0.294547	0.352309
78	0	0.058528	0.116967	0.175231	0.233238	0.290914	0.348194
80	0	0.057773	0.115479	0.173054	0.230436	0.287571	0.344410
82	0	0.057095	0.114143	0.171099	0.227922	0.284573	0.341017
84	0	0.056508	0.112988	0.169410	0.225750	0.281983	0.338088
86	0	0.056034	0.112053	0.168043	0.223992	0.279887	0.335718
88	0	0.055698	0.111392	0.167078	0.222751	0.278408	0.334046
90	0	0.055556	0.111111	0.166667	0.222222	0.277778	0.333333
		$\left[\begin{smallmatrix}(-5)2\\5\end{smallmatrix}\right]$	$\left[\begin{smallmatrix}(-5)5\\5\end{smallmatrix}\right]$	$\left[\begin{smallmatrix}(-5)7\\6\end{smallmatrix}\right]$	$\left[\begin{smallmatrix}(-5)9\\6\end{smallmatrix}\right]$	$\left[\begin{smallmatrix}(-4)1\\6\end{smallmatrix}\right]$	$\left[\begin{smallmatrix}(-4)1\\6\end{smallmatrix}\right]$
5	0	0.086990	0.173318	0.258327	0.341370	0.421815	0.499050
15	0	0.085677	0.170704	0.254434	0.336231	0.415475	0.491565
25	0	0.083124	0.165625	0.246882	0.326288	0.403252	0.477203
35	0	0.079476	0.158377	0.236134	0.312192	0.386013	0.457086
45	0	0.074953	0.149408	0.222878	0.294884	0.364976	0.432729
55	0	0.069861	0.139334	0.208034	0.275597	0.341676	0.405958
65	0	0.064614	0.128968	0.192809	0.255897	0.318009	0.378946
75	0	0.059779	0.119433	0.178839	0.237883	0.296459	0.354475
85	0	0.056256	0.112490	0.168682	0.224814	0.280867	0.336826

Compiled from C. Heuman, Tables of complete elliptic integrals, J. Math. Phys. **20**, 127–206, 1941 (with permission).

HEUMAN'S LAMBDA FUNCTION $\Lambda_0(\varphi\backslash\alpha)$ Table 17.8

$$\Lambda_0(\varphi\backslash\alpha)=\frac{F(\varphi\backslash 90°-\alpha)}{K'(\alpha)}+\frac{2}{\pi}K(\alpha)Z(\varphi\backslash 90°-\alpha)=\frac{2}{\pi}\{K(\alpha)E(\varphi\backslash 90°-\alpha)-[K(\alpha)-E(\alpha)]F(\varphi\backslash 90°-\alpha)\}$$

$\alpha\backslash\varphi$	35°	40°	45°	50°	55°	60°
0°	0.573576	0.642788	0.707107	0.766044	0.819152	0.866025
2	0.573402	0.642592	0.706891	0.765811	0.818903	0.865762
4	0.572878	0.642006	0.706247	0.765113	0.818157	0.864975
6	0.572009	0.641032	0.705177	0.763956	0.816922	0.863674
8	0.570795	0.639674	0.703687	0.762347	0.815210	0.861876
10	0.569244	0.637940	0.701786	0.760298	0.813034	0.859602
12	0.567360	0.635836	0.699484	0.757822	0.810416	0.856877
14	0.565150	0.633373	0.696794	0.754937	0.807375	0.853731
16	0.562623	0.630561	0.693729	0.751660	0.803935	0.850194
18	0.559789	0.627412	0.690306	0.748011	0.800123	0.846297
20	0.556657	0.623939	0.686540	0.744012	0.795963	0.842073
22	0.553238	0.620157	0.682450	0.739683	0.791483	0.837553
24	0.549546	0.616080	0.678054	0.735049	0.786709	0.832766
26	0.545591	0.611725	0.673372	0.730130	0.781667	0.827743
28	0.541389	0.607107	0.668422	0.724951	0.776384	0.822510
30	0.536953	0.602244	0.663225	0.719533	0.770883	0.817093
32	0.532297	0.597153	0.657801	0.713900	0.765190	0.811517
34	0.527437	0.591851	0.652170	0.708073	0.759326	0.805804
36	0.522388	0.586356	0.646351	0.702074	0.753314	0.799976
38	0.517165	0.580687	0.640365	0.695923	0.747177	0.794052
40	0.511786	0.574862	0.634231	0.689642	0.740932	0.788051
42	0.506266	0.568898	0.627970	0.683251	0.734602	0.781992
44	0.500622	0.562815	0.621600	0.676769	0.728203	0.775891
46	0.494873	0.556632	0.615142	0.670217	0.721756	0.769764
48	0.489034	0.550366	0.608615	0.663613	0.715277	0.763627
50	0.483125	0.544038	0.602038	0.656976	0.708785	0.757496
52	0.477164	0.537668	0.595432	0.650326	0.702298	0.751385
54	0.471170	0.531275	0.588817	0.643682	0.695832	0.745310
56	0.465163	0.524879	0.582212	0.637064	0.689405	0.739286
58	0.459163	0.518502	0.575640	0.630491	0.683037	0.733329
60	0.453192	0.512167	0.569122	0.623985	0.676745	0.727455
62	0.447272	0.505895	0.562680	0.617567	0.670549	0.721680
64	0.441428	0.499711	0.556339	0.611258	0.664469	0.716024
66	0.435683	0.493642	0.550124	0.605085	0.658528	0.710504
68	0.430065	0.487715	0.544062	0.599072	0.652749	0.705142
70	0.424604	0.481959	0.538183	0.593247	0.647159	0.699961
72	0.419332	0.476408	0.532519	0.587641	0.641784	0.694985
74	0.414284	0.471098	0.527106	0.582290	0.636659	0.690244
76	0.409500	0.466070	0.521985	0.577231	0.631818	0.685770
78	0.405026	0.461371	0.517202	0.572511	0.627303	0.681601
80	0.400915	0.457055	0.512813	0.568181	0.623166	0.677782
82	0.397229	0.453189	0.508883	0.564307	0.619464	0.674368
84	0.394049	0.449853	0.505494	0.560967	0.616276	0.671427
86	0.391477	0.447157	0.502754	0.558268	0.613700	0.669053
88	0.389662	0.445255	0.500823	0.556366	0.611884	0.667379
90	0.388889	0.444444	0.500000	0.555556	0.611111	0.666667
	$\left[\begin{smallmatrix}(-4)1\\6\end{smallmatrix}\right]$	$\left[\begin{smallmatrix}(-4)1\\6\end{smallmatrix}\right]$	$\left[\begin{smallmatrix}(-4)1\\6\end{smallmatrix}\right]$	$\left[\begin{smallmatrix}(-4)1\\6\end{smallmatrix}\right]$	$\left[\begin{smallmatrix}(-4)1\\6\end{smallmatrix}\right]$	$\left[\begin{smallmatrix}(-4)1\\6\end{smallmatrix}\right]$
5	0.572487	0.641567	0.705765	0.764592	0.817600	0.864388
15	0.563926	0.632010	0.695307	0.753346	0.805703	0.852010
25	0.547600	0.613936	0.675748	0.732623	0.784220	0.830282
35	0.524935	0.589127	0.649283	0.705094	0.756337	0.802903
45	0.497760	0.559735	0.618381	0.673501	0.724985	0.772830
55	0.468167	0.528076	0.585512	0.640369	0.692612	0.742291
65	0.438541	0.496661	0.553214	0.608153	0.661480	0.713246
75	0.411857	0.468546	0.524506	0.579721	0.634200	0.687972
85	0.392679	0.448417	0.504034	0.559529	0.614903	0.670162

Table 17.8   **HEUMAN'S LAMBDA FUNCTION** $\Lambda_0(\varphi\backslash\alpha)$

$$\Lambda_0(\varphi\backslash\alpha) = \frac{F(\varphi\backslash90°-\alpha)}{K'(\alpha)} + \frac{2}{\pi} K(\alpha) Z(\varphi\backslash90°-\alpha)$$

$$= \frac{2}{\pi}\{K(\alpha)E(\varphi\backslash90°-\alpha) - [K(\alpha)-E(\alpha)]F(\varphi\backslash90°-\alpha)\}$$

$\alpha\backslash\varphi$	65°	70°	75°	80°	85°	90°
0°	0.906308	0.939693	0.965926	0.984808	0.996195	1
2	0.906032	0.939407	0.965633	0.984511	0.995903	1
4	0.905210	0.938559	0.964769	0.983652	0.995130	1
6	0.903857	0.937172	0.963376	0.982315	0.994063	1
8	0.901997	0.935282	0.961512	0.980599	0.992833	1
10	0.899660	0.932934	0.959244	0.978597	0.991511	1
12	0.896881	0.930177	0.956638	0.976384	0.990135	1
14	0.893699	0.927061	0.953755	0.974016	0.988727	1
16	0.890152	0.923634	0.950646	0.971534	0.987299	1
18	0.886280	0.919940	0.947355	0.968969	0.985858	1
20	0.882119	0.916018	0.943918	0.966343	0.984410	1
22	0.877704	0.911904	0.940364	0.963671	0.982958	1
24	0.873068	0.907630	0.936718	0.960968	0.981506	1
26	0.868240	0.903221	0.933000	0.958241	0.980054	1
28	0.863249	0.898703	0.929226	0.955500	0.978604	1
30	0.858117	0.894095	0.925409	0.952751	0.977159	1
32	0.852869	0.889416	0.921563	0.949998	0.975719	1
34	0.847523	0.884681	0.917695	0.947247	0.974286	1
36	0.842100	0.879904	0.913817	0.944502	0.972861	1
38	0.836615	0.875099	0.909935	0.941766	0.971445	1
40	0.831085	0.870277	0.906056	0.939042	0.970039	1
42	0.825524	0.865449	0.902188	0.936335	0.968644	1
44	0.819946	0.860625	0.898337	0.933647	0.967262	1
46	0.814365	0.855814	0.894508	0.930981	0.965894	1
48	0.808792	0.851026	0.890708	0.928341	0.964540	1
50	0.803241	0.846269	0.886942	0.925731	0.963204	1
52	0.797724	0.841553	0.883216	0.923152	0.961885	1
54	0.792252	0.836887	0.879537	0.920610	0.960586	1
56	0.786839	0.832280	0.875911	0.918108	0.959309	1
58	0.781496	0.827742	0.872345	0.915649	0.958055	1
60	0.776237	0.823283	0.868846	0.913240	0.956826	1
62	0.771077	0.818913	0.865421	0.910884	0.955626	1
64	0.766029	0.814645	0.862080	0.908588	0.954457	1
66	0.761110	0.810490	0.858831	0.906357	0.953321	1
68	0.756338	0.806464	0.855685	0.904198	0.952223	1
70	0.751731	0.802581	0.852654	0.902119	0.951166	1
72	0.747312	0.798860	0.849751	0.900129	0.950154	1
74	0.743104	0.795319	0.846990	0.898237	0.949193	1
76	0.739137	0.791983	0.844390	0.896456	0.948288	1
78	0.735442	0.788877	0.841972	0.894800	0.947446	1
80	0.732059	0.786036	0.839759	0.893286	0.946677	1
82	0.729036	0.783497	0.837783	0.891933	0.945990	1
84	0.726434	0.781312	0.836083	0.890770	0.945400	1
86	0.724333	0.779549	0.834711	0.889831	0.944923	1
88	0.722852	0.778307	0.833745	0.889170	0.944587	1
90	0.722222	0.777778	0.833333	0.888889	0.944444	1
	$\begin{bmatrix}(-4)1\\6\end{bmatrix}$	$\begin{bmatrix}(-5)9\\6\end{bmatrix}$	$\begin{bmatrix}(-5)7\\6\end{bmatrix}$	$\begin{bmatrix}(-5)5\\5\end{bmatrix}$	$\begin{bmatrix}(-5)2\\5\end{bmatrix}$	
5	0.904599	0.937930	0.964135	0.983037	0.994624	1
15	0.891969	0.925384	0.952226	0.972787	0.988015	1
25	0.870676	0.905441	0.934867	0.959607	0.980779	1
35	0.844820	0.882297	0.915757	0.945873	0.973573	1
45	0.817155	0.858217	0.896419	0.932311	0.966576	1
55	0.789537	0.834576	0.877717	0.919353	0.959944	1
65	0.763552	0.812552	0.860443	0.907464	0.953885	1
75	0.741089	0.793624	0.845669	0.897332	0.948733	1
85	0.725315	0.780373	0.835352	0.890270	0.945145	1

ELLIPTIC INTEGRAL OF THE THIRD KIND $\Pi(n; \varphi \backslash \alpha)$ Table 17.9

$$\Pi(n; \varphi \backslash \alpha) = \int_0^\varphi (1-n \sin^2 \theta)^{-1}[1-\sin^2 \alpha \sin^2 \theta]^{-\frac{1}{2}} d\theta$$

n	$\alpha \backslash \varphi$	$0°$	$15°$	$30°$	$45°$	$60°$	$75°$	$90°$
0.0	0°	0	0.26180	0.52360	0.78540	1.04720	1.30900	1.57080
0.0	15	0	0.26200	0.52513	0.79025	1.05774	1.32733	1.59814
0.0	30	0	0.26254	0.52943	0.80437	1.08955	1.38457	1.68575
0.0	45	0	0.26330	0.53562	0.82602	1.14243	1.48788	1.85407
0.0	60	0	0.26406	0.54223	0.85122	1.21260	1.64918	2.15651
0.0	75	0	0.26463	0.54736	0.87270	1.28371	1.87145	2.76806
0.0	90	0	0.26484	0.54931	0.88137	1.31696	2.02759	∞
0.1	0	0	0.26239	0.52820	0.80013	1.07949	1.36560	1.65576
0.1	15	0	0.26259	0.52975	0.80514	1.09058	1.38520	1.68536
0.1	30	0	0.26314	0.53412	0.81972	1.12405	1.44649	1.78030
0.1	45	0	0.26390	0.54041	0.84210	1.17980	1.55739	1.96326
0.1	60	0	0.26467	0.54712	0.86817	1.25393	1.73121	2.29355
0.1	75	0	0.26524	0.55234	0.89040	1.32926	1.97204	2.96601
0.1	90	0	0.26545	0.55431	0.89939	1.36454	2.14201	∞
0.2	0	0	0.26299	0.53294	0.81586	1.11534	1.43078	1.75620
0.2	15	0	0.26319	0.53452	0.82104	1.12705	1.45187	1.78850
0.2	30	0	0.26374	0.53896	0.83612	1.16241	1.51792	1.89229
0.2	45	0	0.26450	0.54535	0.85928	1.22139	1.63775	2.09296
0.2	60	0	0.26527	0.55217	0.88629	1.30003	1.82643	2.45715
0.2	75	0	0.26585	0.55747	0.90934	1.38016	2.08942	3.20448
0.2	90	0	0.26606	0.55948	0.91867	1.41777	2.27604	∞
0.3	0	0	0.26359	0.53784	0.83271	1.15551	1.50701	1.87746
0.3	15	0	0.26379	0.53945	0.83808	1.16791	1.52988	1.91309
0.3	30	0	0.26434	0.54396	0.85370	1.20543	1.60161	2.02779
0.3	45	0	0.26511	0.55046	0.87771	1.26812	1.73217	2.25038
0.3	60	0	0.26588	0.55739	0.90574	1.35193	1.93879	2.65684
0.3	75	0	0.26646	0.56278	0.92969	1.43759	2.22876	3.49853
0.3	90	0	0.26667	0.56483	0.93938	1.47789	2.43581	∞
0.4	0	0	0.26420	0.54291	0.85084	1.20098	1.59794	2.02789
0.4	15	0	0.26440	0.54454	0.85641	1.21419	1.62298	2.06774
0.4	30	0	0.26495	0.54912	0.87262	1.25419	1.70165	2.19629
0.4	45	0	0.26572	0.55573	0.89756	1.32117	1.84537	2.44683
0.4	60	0	0.26650	0.56278	0.92670	1.41098	2.07413	2.90761
0.4	75	0	0.26708	0.56827	0.95162	1.50309	2.39775	3.87214
0.4	90	0	0.26729	0.57035	0.96171	1.54653	2.63052	∞
0.5	0	0	0.26481	0.54814	0.87042	1.25310	1.70919	2.22144
0.5	15	0	0.26501	0.54980	0.87621	1.26726	1.73695	2.26685
0.5	30	0	0.26557	0.55447	0.89307	1.31017	1.82433	2.41367
0.5	45	0	0.26634	0.56119	0.91902	1.38218	1.98464	2.70129
0.5	60	0	0.26712	0.56837	0.94939	1.47906	2.24155	3.23477
0.5	75	0	0.26770	0.57394	0.97538	1.57881	2.60846	4.36620
0.5	90	0	0.26792	0.57606	0.98591	1.62599	2.87468	∞
0.6	0	0	0.26543	0.55357	0.89167	1.31379	1.85002	2.48365
0.6	15	0	0.26563	0.55525	0.89770	1.32907	1.88131	2.53677
0.6	30	0	0.26619	0.56000	0.91527	1.37544	1.98005	2.70905
0.6	45	0	0.26696	0.56684	0.94235	1.45347	2.16210	3.04862
0.6	60	0	0.26775	0.57414	0.97406	1.55884	2.45623	3.68509
0.6	75	0	0.26833	0.57982	1.00123	1.66780	2.88113	5.05734
0.6	90	0	0.26855	0.58198	1.01225	1.71951	3.19278	∞
			$\begin{bmatrix} (-5)5 \\ 4 \end{bmatrix}$	$\begin{bmatrix} (-4)4 \\ 6 \end{bmatrix}$	$\begin{bmatrix} (-3)2 \\ 7 \end{bmatrix}$	$\begin{bmatrix} (-3)7 \\ 7 \end{bmatrix}$		

See Examples 15–20.

Table 17.9   **ELLIPTIC INTEGRAL OF THE THIRD KIND $\Pi(n;\varphi\backslash\alpha)$**

$$\Pi(n;\varphi\backslash\alpha)=\int_0^\varphi (1-n\sin^2\theta)^{-1}[1-\sin^2\alpha\sin^2\theta]^{-\frac{1}{2}}d\theta$$

n	$\alpha\backslash\varphi$	0°	15°	30°	45°	60°	75°	90°
0.7	0°	0	0.26605	0.55918	0.91487	1.38587	2.03720	2.86787
0.7	15	0	0.26625	0.56090	0.92116	1.40251	2.07333	2.93263
0.7	30	0	0.26681	0.56573	0.93952	1.45309	2.18765	3.14339
0.7	45	0	0.26759	0.57270	0.96784	1.53846	2.39973	3.56210
0.7	60	0	0.26838	0.58014	1.00104	1.65425	2.74586	4.35751
0.7	75	0	0.26897	0.58592	1.02954	1.77459	3.25315	6.11030
0.7	90	0	0.26918	0.58812	1.04110	1.83192	3.63042	∞
0.8	0	0	0.26668	0.56501	0.94034	1.47370	2.30538	3.51240
0.8	15	0	0.26688	0.56676	0.94694	1.49205	2.34868	3.59733
0.8	30	0	0.26745	0.57168	0.96618	1.54790	2.48618	3.87507
0.8	45	0	0.26823	0.57877	0.99588	1.64250	2.74328	4.43274
0.8	60	0	0.26902	0.58635	1.03076	1.77145	3.16844	5.51206
0.8	75	0	0.26961	0.59225	1.06073	1.90629	3.80370	7.96669
0.8	90	0	0.26982	0.59449	1.07290	1.97080	4.28518	∞
0.9	0	0	0.26731	0.57106	0.96853	1.58459	2.74439	4.96729
0.9	15	0	0.26752	0.57284	0.97547	1.60515	2.79990	5.09958
0.9	30	0	0.26808	0.57785	0.99569	1.66788	2.97710	5.53551
0.9	45	0	0.26887	0.58508	1.02695	1.77453	3.31210	6.42557
0.9	60	0	0.26966	0.59281	1.06372	1.92081	3.87661	8.20086
0.9	75	0	0.27025	0.59882	1.09535	2.07487	4.74432	12.46407
0.9	90	0	0.27047	0.60110	1.10821	2.14899	5.42125	∞
1.0	0	0	0.26795	0.57735	1.00000	1.73205	3.73205	∞
1.0	15	0	0.26816	0.57916	1.00731	1.75565	3.81655	∞
1.0	30	0	0.26872	0.58428	1.02866	1.82781	4.08864	∞
1.0	45	0	0.26951	0.59165	1.06170	1.95114	4.61280	∞
1.0	60	0	0.27031	0.59953	1.10060	2.12160	5.52554	∞
1.0	75	0	0.27090	0.60566	1.13414	2.30276	7.00372	∞
1.0	90	0	0.27112	0.60799	1.14779	2.39053	8.22356	∞
			$\begin{bmatrix}(-5)5\\4\end{bmatrix}$	$\begin{bmatrix}(-4)5\\6\end{bmatrix}$	$\begin{bmatrix}(-3)2\\7\end{bmatrix}$	$\begin{bmatrix}(-2)1\\7\end{bmatrix}$		

18. Weierstrass Elliptic and Related Functions

Thomas H. Southard [1]

Contents

[1] University of California, National Bureau of Standards, and California State College at Hayward.

Table 18.1. Table for Obtaining Periods for Invariants g_2 and g_3 Page
$(\overline{g}_2 = g_2 g_3^{-2/3})$.   673

Non-Negative Discriminant $(3 \leq \overline{g}_2 \leq \infty)$

$$\omega g_3^{1/6}, \; \frac{\omega' g_3^{1/6}}{i} + \frac{\sqrt{6}}{12} \ln(\overline{g}_2 - 3); \; \overline{g}_2 = 3(.05)3.4, \, 7D$$

$$\omega g_3^{1/6}, \; \omega' g_3^{1/6}/i; \; \overline{g}_2 = 3.4(.1)5(.2)10, \, 7D$$

$$\omega g_3^{1/6} \overline{g}_2^{1/4}, \; \omega' g_3^{1/6} \overline{g}_2^{1/4}/i; \; \overline{g}_2^{-1} = .1(-.01)0, \, 7D$$

Non-Positive Discriminant $(-\infty \leq \overline{g}_2 \leq 3)$

$$\omega_2 g_3^{1/6} |\overline{g}_2|^{1/4}, \; \omega'_2 g_3^{1/6} |\overline{g}_2|^{1/4}/i; \; \overline{g}_2^{-1} = 0(-.01)-.2, \, 7D$$

$$\omega_2 g_3^{1/6}, \; \omega'_2 g_3^{1/6}/i; \; \overline{g}_2^{-1} = -.2(-.05)-1, \, 7D$$

$$\omega_2 g_3^{1/6}, \; \frac{\omega'_2}{i} \, g_3^{1/6} + \frac{\sqrt{6}}{6} \ln(3 - \overline{g}_2); \; \overline{g}_2 = -1(.2)3, \, 7D$$

Table 18.2. Table for Obtaining $\wp$, $\wp'$ and ζ on $0x$ and $0y$ (Unit
Real Half-Period—Period Ratio a).   674

Positive Discriminant $(0 \leq x \leq 1, \, 0 \leq y \leq a)$

$$z^2 \, \wp(z), \; z^3 \, \wp'(z), \; z\zeta(z), \; a = 1, \, 1.05, \, 1.1, \, 1.2, \, 1.4, \, 2, \, 4$$
$$x = 0(.05)1, \; y = 0(.05) \, 1.1, \, 1.2 \, (.2) \, a, \, 6-8D$$

Negative Discriminant $(0 \leq x \leq 1, \, 0 \leq y \leq a/2)$

$$z^2 \, \wp(z), \; z^3 \, \wp'(z), \; z\zeta(z), \; a = 1, \, 1.05, \, 1.15, \, 1.3, \, 1.5, \, 2, \, 4$$
$$x = 0(.05)1, \; y = 0(.05)1 \; (.1)b(b \geq a/2), \, 7D$$

Table 18.3. Invariants and Values at Half-Periods $(1 \leq a \leq \infty)$ (Unit
Real Half-Period).   680
$$a = 1(.02)1.6(.05)2.3(.1) \, 4, \, \infty, \, 6-8D$$
Non-Negative Discriminant

$$g_2, \, g_3, \, e_1 = \wp(1), \, e_3 = \wp(\omega'), \, \eta = \zeta(1), \, \eta'/i = \zeta(\omega')/i, \, \sigma(1), \, \sigma(\omega')/i, \, \sigma(\omega_2)$$

Non-Positive Discriminant

$$g_2, \, g_3, \, e_1, \, \eta_2 = \zeta(1), \, \eta'_2/i = \zeta(\omega'_2)/i, \, \sigma(1), \, \sigma(\omega'_2)/i, \, \sigma(\omega')$$

The author gratefully acknowledges the assistance and encouragement of the personnel of Numerical Analysis Research, UCLA (especially Dr. C. B. Tompkins for generating the author's interest in the project, and Mrs. H. O. Rosay for programming and computing, hand calculation and formula checking) and the personnel of the Computation Laboratory (especially R. Capuano, E. Godefroy, D. Liepman, B. Walter and R. Zucker for the preparation and checking of the tables and maps).

18. Weierstrass Elliptic and Related Functions

Mathematical Properties

18.1. Definitions, Symbolism, Restrictions and Conventions

An elliptic function is a single-valued doubly periodic function of a single complex variable which is analytic except at poles and whose only singularities in the finite plane are poles. If ω and ω' are a pair of (primitive) half-periods of such a function $f(z)$, then $f(z+2M\omega+2N\omega')=f(z)$, M and N being integers. Thus the study of any such function can be reduced to consideration of its behavior in a *fundamental period parallelogram* (FPP). An elliptic function has a finite number of poles (and the same number of zeros) in a FPP; the number of such poles (zeros) (an irreducible set) is the *order* of the function (poles and zeros are counted according to their multiplicity). All other poles (zeros) are called *congruent* to the irreducible set. The simplest (nontrivial) elliptic functions are of order two. One may choose as the standard function of order two either a function with two simple poles (Jacobi's choice) or one double pole (Weierstrass' choice) in a FPP.

Weierstrass' $\wp$-Function. Let ω, ω' denote a pair of complex numbers with $\mathscr{I}(\omega'/\omega)>0$. Then $\wp(z)=\wp(z|\omega, \omega')$ is an elliptic function of order two with periods 2ω, $2\omega'$ and having a double pole at $z=0$, whose principal part is z^{-2}; $\wp(z)-z^{-2}$ is analytic in a neighborhood of the origin and vanishes at $z=0$.

Weierstrass' ζ-Function $\zeta(z)=\zeta(z|\omega, \omega')$ satisfies the condition $\zeta'(z)=-\wp(z)$; further, $\zeta(z)$ has a simple pole at $z=0$ whose principal part is z^{-1}; $\zeta(z)-z^{-1}$ vanishes at $z=0$ and is analytic in a neighborhood of the origin. $\zeta(z)$ is *NOT* an elliptic function, since it is not periodic. However, it is quasi-periodic (see "period" relations), so reduction to FPP is possible.

Weierstrass' σ-Function $\sigma(z)=\sigma(z|\omega, \omega')$ satisfies the condition $\sigma'(z)/\sigma(z)=\zeta(z)$; further, $\sigma(z)$ is an entire function which vanishes at the origin. Like ζ, it is *NOT* an elliptic function, since it is not periodic. However, it is quasi-periodic (see "period" relations), so reduction to FPP is possible.

Invariants g_2 and g_3

Let $W=2M\omega+2N\omega'$, M and N being integers. Then

18.1.1 $\qquad g_2=60\Sigma'W^{-4}$ and $g_3=140\Sigma'W^{-6}$

are the INVARIANTS, summation being over all pairs M, N except $M=N=0$.

Alternate Symbolism Emphasizing Invariants

18.1.2 $\qquad\qquad \wp(z)=\wp(z; g_2, g_3)$
18.1.3 $\qquad\qquad \wp'(z)=\wp'(z; g_2, g_3)$
18.1.4 $\qquad\qquad \zeta(z)=\zeta(z; g_2, g_3)$
18.1.5 $\qquad\qquad \sigma(z)=\sigma(z; g_2, g_3)$

Fundamental Differential Equation, Discriminant and Related Quantities

18.1.6 $\qquad \wp'^2(z)=4\wp^3(z)-g_2\wp(z)-g_3$

18.1.7
$$=4(\wp(z)-e_1)(\wp(z)-e_2)(\wp(z)-e_3)$$

18.1.8
$$\Delta=g_2^3-27g_3^2=16(e_2-e_3)^2(e_3-e_1)^2(e_1-e_2)^2$$

18.1.9
$$g_2=-4(e_1e_2+e_1e_3+e_2e_3)=2(e_1^2+e_2^2+e_3^2)$$

18.1.10 $\qquad g_3=4e_1e_2e_3=\tfrac{4}{3}(e_1^3+e_2^3+e_3^3)$
18.1.11 $\qquad\qquad e_1+e_2+e_3=0$
18.1.12 $\qquad\qquad e_1^4+e_2^4+e_3^4=g_2^2/8$
18.1.13 $\qquad 4e_i^3-g_2e_i-g_3=0 (i=1, 2, 3)$

Agreement about Values of Invariants (and Discriminant)

We shall consider, in this chapter, only *real* g_2 and g_3 (this seems to cover most applications)—hence Δ is real. We shall dichotomize most of what follows (either $\Delta>0$ or $\Delta<0$). Homogeneity relations **18.2.1–18.2.15** enable a further restriction to non-negative g_3 (except for one case when $\Delta=0$).

Note on Symbolism for Roots of Complex Numbers and for Conjugate Complex Numbers

In this chapter, $z^{1/n}$ (n a positive integer) is used to denote the principal nth root of z, as in chapter 3; $\bar{z}$ is used to denote the complex conjugate of z.

FPP's, Symbols for Periods, etc.

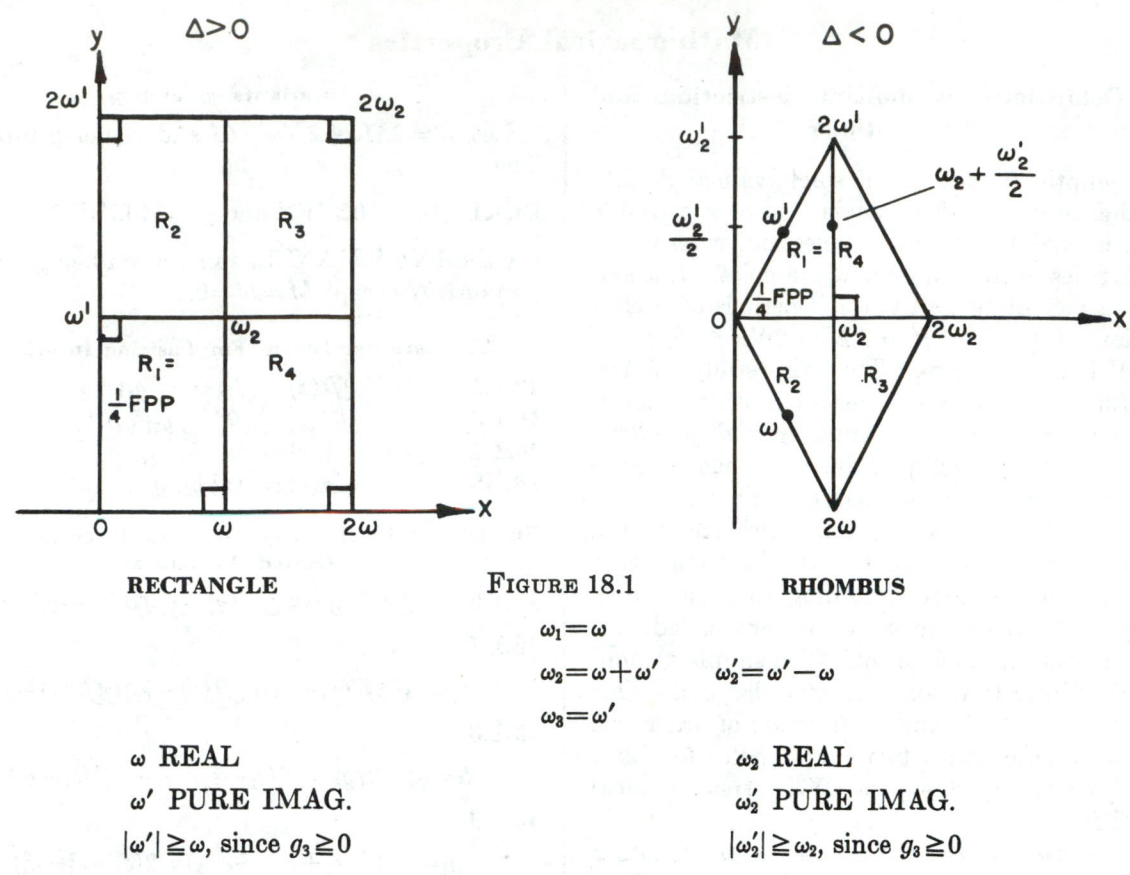

FIGURE 18.1

RECTANGLE   RHOMBUS

$$\omega_1 = \omega$$

$$\omega_2 = \omega + \omega' \qquad \omega_2' = \omega' - \omega$$

$$\omega_3 = \omega'$$

ω REAL ω_2 REAL

ω' PURE IMAG. ω_2' PURE IMAG.

$|\omega'| \geqq \omega$, since $g_3 \geqq 0$ $|\omega_2'| \geqq \omega_2$, since $g_3 \geqq 0$

Fundamental Rectangles

Study of all four functions ($\wp$, $\wp'$, ζ, σ) can be reduced to consideration of their values in a Fundamental Rectangle including the origin (see **18.2** on homogeneity relations, reduction formulas and processes).

$\Delta > 0$ $\Delta < 0$

Fundamental Rectangle is $\frac{1}{4}$ FPP, which has vertices 0, ω, ω_2 and ω' Fundamental Rectangle has vertices 0, ω_2, $\omega_2 + \frac{\omega_2'}{2}$, $\frac{\omega_2'}{2}$.

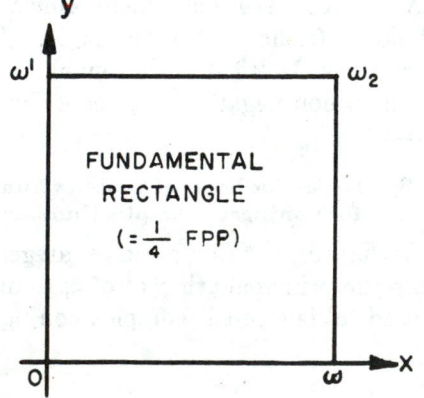 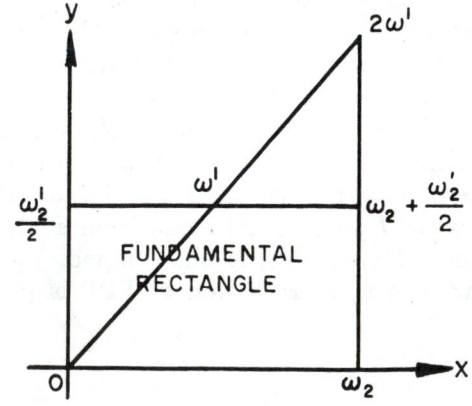

FIGURE 18.2

There is a point on the right boundary of Fundamental Rectangle where $\wp = 0$. Denote it by z_0.

18.2. Homogeneity Relations, Reduction Formulas and Processes

Homogeneity Relations (Suppose $t \neq 0$)

Note that Period Ratio is preserved.

18.2.1 $\quad \mathcal{P}'(tz|t\omega, t\omega') = t^{-3}\mathcal{P}'(z|\omega, \omega')$

18.2.2 $\quad \mathcal{P}(tz|t\omega, t\omega') = t^{-2}\mathcal{P}(z|\omega, \omega')$

18.2.3 $\quad \zeta(tz|t\omega, t\omega') = t^{-1}\zeta(z|\omega, \omega')$

18.2.4 $\quad \sigma(tz|t\omega, t\omega') = t\sigma(z|\omega, \omega')$

18.2.5 $\quad g_2(t\omega, t\omega') = t^{-4}g_2(\omega, \omega')$

18.2.6 $\quad g_3(t\omega, t\omega') = t^{-6}g_3(\omega, \omega')$

18.2.7 $\quad e_i(t\omega, t\omega') = t^{-2}e_i(\omega, \omega'), \ i=1, 2, 3$

18.2.8 $\quad \Delta(t\omega, t\omega') = t^{-12}\Delta(\omega, \omega')$

18.2.9 $\quad H_i(t\omega, t\omega') = t^{-2}H_i(\omega, \omega'), \ i=1, 2, 3$

$$\text{(See } \mathbf{18.3})$$

18.2.10 $\quad q(t\omega, t\omega') = q(\omega, \omega') \quad \text{(See } \mathbf{18.10})$

18.2.11 $\quad m(t\omega, t\omega') = m(\omega, \omega') \quad \text{(See } \mathbf{18.9})$

18.2.12 $\quad \mathcal{P}'(tz; t^{-4}g_2, t^{-6}g_3) = t^{-3}\mathcal{P}'(z; g_2, g_3)$

18.2.13 $\quad \mathcal{P}(tz; t^{-4}g_2, t^{-6}g_3) = t^{-2}\mathcal{P}(z; g_2, g_3)$

18.2.14 $\quad \zeta(tz; t^{-4}g_2, t^{-6}g_3) = t^{-1}\zeta(z; g_2, g_3)$

18.2.15 $\quad \sigma(tz; t^{-4}g_2, t^{-6}g_3) = t\sigma(z; g_2, g_3)$

The Case $g_3 < 0$

Put $t = i$ and obtain, e.g.,

18.2.16 $\quad \mathcal{P}(z; g_2, g_3) = -\mathcal{P}(iz; g_2, -g_3)$

Thus the case $g_3 < 0$ can be reduced to one where $g_3 > 0$.

"Period" Relations and Reduction to the FPP (M, N integers)

18.2.17 $\quad \mathcal{P}'(z + 2M\omega + 2N\omega') = \mathcal{P}'(z)$

18.2.18 $\quad \mathcal{P}(z + 2M\omega + 2N\omega') = \mathcal{P}(z)$

18.2.19
$$\zeta(z + 2M\omega + 2N\omega') = \zeta(z) + 2M\eta + 2N\eta'$$

18.2.20
$$\sigma(z + 2M\omega + 2N\omega')$$
$$= (-1)^{M+N+MN}\sigma(z) \exp[(z + M\omega + N\omega')(2M\eta + 2N\eta')]$$

18.2.21 $\quad$ where $\eta = \zeta(\omega), \ \eta' = \zeta(\omega')$

"Conjugate" Values

$f(\bar{z}) = \bar{f}(z)$, where f is any one of the functions $\mathcal{P}, \mathcal{P}', \zeta, \sigma$.

Reduction to ¼ FPP (See Figure 18.1)

$\Delta > 0$ $\qquad\qquad$ ($\bar{s}$ denotes conjugate of s) $\qquad\qquad$ $\Delta < 0$

Point z_4 in R_4

18.2.22 $\quad \mathcal{P}'(z_4) = -\overline{\mathcal{P}'(2\omega - z_4)} \qquad\qquad \mathcal{P}'(\bar{z}_4) = -\overline{\mathcal{P}'(2\omega_2 - z_4)}$

18.2.23 $\quad \mathcal{P}(z_4) = \overline{\mathcal{P}(2\omega - z_4)} \qquad\qquad \mathcal{P}(z_4) = \overline{\mathcal{P}(2\omega_2 - z_4)}$

18.2.24 $\quad \zeta(z_4) = -\overline{\zeta(2\omega - z_4)} + 2\eta \qquad\qquad \zeta(z_4) = -\overline{\zeta(2\omega_2 - z_4)} + 2(\eta + \eta')$

18.2.25 $\quad \sigma(z_4) = \overline{\sigma(2\omega - z_4)} \exp[2\eta(z_4 - \omega)] \qquad \sigma(z_4) = \overline{\sigma(2\omega_2 - z_4)} \exp[2(\eta + \eta')(z_4 - \omega_2)]$

Point z_3 in R_3

18.2.26 $\quad \mathcal{P}'(z_3) = -\mathcal{P}'(2\omega_2 - z_3) \qquad\qquad \mathcal{P}'(z_3) = -\mathcal{P}'(2\omega_2 - z_3)$

18.2.27 $\quad \mathcal{P}(z_3) = \mathcal{P}(2\omega_2 - z_3) \qquad\qquad \mathcal{P}(z_3) = \mathcal{P}(2\omega_2 - z_3)$

18.2.28 $\quad \zeta(z_3) = -\zeta(2\omega_2 - z_3) + 2(\eta + \eta') \qquad\qquad \zeta(z_3) = -\zeta(2\omega_2 - z_3) + 2(\eta + \eta')$

18.2.29 $\quad \sigma(z_3) = \sigma(2\omega_2 - z_3) \exp[2(\eta + \eta')(z_3 - \omega_2)] \qquad \sigma(z_3) = \sigma(2\omega_2 - z_3) \exp[2(\eta + \eta')(z_3 - \omega_2)]$

Point z_2 in R_2

18.2.30 $\quad \mathcal{P}'(z_2) = \overline{\mathcal{P}'(z_2 - 2\omega')} \qquad\qquad \mathcal{P}'(z_2) = \overline{\mathcal{P}'(\bar{z}_2)}$

18.2.31 $\quad \mathcal{P}(z_2) = \overline{\mathcal{P}(z_2 - 2\omega')} \qquad\qquad \mathcal{P}(z_2) = \overline{\mathcal{P}(\bar{z}_2)}$

18.2.32 $\quad \zeta(z_2) = \overline{\zeta(z_2 - 2\omega')} + 2\eta' \qquad\qquad \zeta(z_2) = \overline{\zeta(\bar{z}_2)}$

18.2.33 $\quad \sigma(z_2) = -\overline{\sigma(z_2 - 2\omega')} \exp[2\eta'(z_2 - \omega')] \qquad \sigma(z_2) = \overline{\sigma(\bar{z}_2)}$

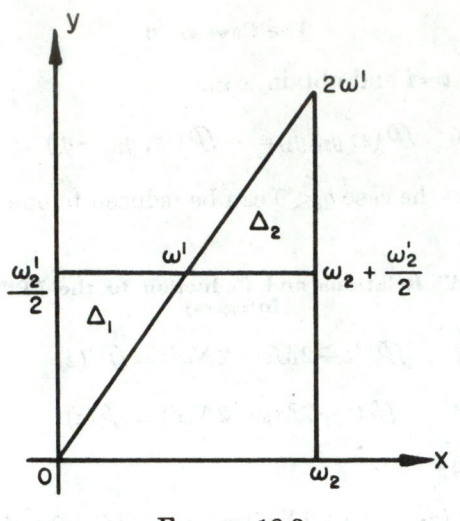

FIGURE 18.3

Reduction from ¼ FPP to Fundamental Rectangle in Case $\Delta < 0$

We need only be concerned with the case when z is in triangle Δ_2 (therefore $2\omega' - z$ is in triangle Δ_1).

18.2.34 $$\mathcal{P}(z) = \mathcal{P}(2\omega' - z)$$

18.2.35 $$\mathcal{P}'(z) = -\mathcal{P}'(2\omega' - z)$$

18.2.36 $$\zeta(z) = 2\eta' - \zeta(2\omega' - z)$$

18.2.37 $$\sigma(z) = \sigma(2\omega' - z) \exp[2\eta'(z - \omega')]$$

Reduction to Case where Real Half-Period is Unity

(preserving period ratio)

	$\Delta > 0$	$\Delta < 0$
		$(\omega_2 = \omega + \omega')$

18.2.38 $$\mathcal{P}'(z|\omega, \omega') = \omega^{-3} \mathcal{P}'\left(z\omega^{-1}\middle|1, \frac{\omega'}{\omega}\right) \qquad \mathcal{P}'(z|\omega, \omega') = \omega_2^{-3} \mathcal{P}'\left(z\omega_2^{-1}\middle|\frac{\omega}{\omega_2}, \frac{\omega'}{\omega_2}\right)$$

18.2.39 $$\mathcal{P}(z|\omega, \omega') = \omega^{-2} \mathcal{P}\left(z\omega^{-1}\middle|1, \frac{\omega'}{\omega}\right) \qquad \mathcal{P}(z|\omega, \omega') = \omega_2^{-2} \mathcal{P}\left(z\omega_2^{-1}\middle|\frac{\omega}{\omega_2}, \frac{\omega'}{\omega_2}\right)$$

18.2.40 $$\zeta(z|\omega, \omega') = \omega^{-1} \zeta\left(z\omega^{-1}\middle|1, \frac{\omega'}{\omega}\right) \qquad \zeta(z|\omega, \omega') = \omega_2^{-1} \zeta\left(z\omega_2^{-1}\middle|\frac{\omega}{\omega_2}, \frac{\omega'}{\omega_2}\right)$$

18.2.41 $$\sigma(z|\omega, \omega') = \omega \sigma\left(z\omega^{-1}\middle|1, \frac{\omega'}{\omega}\right) \qquad \sigma(z|\omega, \omega') = \omega_2 \sigma\left(z\omega_2^{-1}\middle|\frac{\omega}{\omega_2}, \frac{\omega'}{\omega_2}\right)$$

18.2.42 $$g_2(\omega, \omega') = \omega^{-4} g_2\left(1, \frac{\omega'}{\omega}\right) \qquad g_2(\omega, \omega') = \omega_2^{-4} g_2\left(\frac{\omega}{\omega_2}, \frac{\omega'}{\omega_2}\right)$$

18.2.43 $$g_3(\omega, \omega') = \omega^{-6} g_3\left(1, \frac{\omega'}{\omega}\right) \qquad g_3(\omega, \omega') = \omega_2^{-6} g_3\left(\frac{\omega}{\omega_2}, \frac{\omega'}{\omega_2}\right)$$

18.2.44 $$e_i(\omega, \omega') = \omega^{-2} e_i\left(1, \frac{\omega'}{\omega}\right) \qquad e_i(\omega, \omega') = \omega_2^{-2} e_i\left(\frac{\omega}{\omega_2}, \frac{\omega'}{\omega_2}\right)$$

$$(i = 1, 2, 3) \qquad\qquad (i = 1, 2, 3)$$

18.2.45 $$\Delta(\omega, \omega') = \omega^{-12} \Delta\left(1, \frac{\omega'}{\omega}\right) \qquad \Delta(\omega, \omega') = \omega_2^{-12} \Delta\left(\frac{\omega}{\omega_2}, \frac{\omega'}{\omega_2}\right)$$

NOTE: New real half-period is

$$\frac{\omega}{\omega_2} + \frac{\omega'}{\omega_2} = \frac{\omega + \omega'}{\omega_2} = 1$$

18.3. Special Values and Relations

Values at Periods

$\mathscr{P}$, $\mathscr{P}'$, and ζ are infinite, σ is zero at $z=2\omega_1$, $i=1,2,3$ and at $2\omega_2'(\Delta<0)$.

$$\Delta>0 \qquad\qquad\qquad\qquad \Delta<0$$

Half-Periods

18.3.1 $\qquad\qquad \mathscr{P}(\omega_i)=e_i(i=1,2,3)$

18.3.2 $\qquad\qquad \mathscr{P}'(\omega_i)=0(i=1,2,3)$

18.3.3 $\qquad\qquad \eta_i=\zeta(\omega_i)(i=1,2,3)$

18.3.4 $\qquad\qquad \eta_1=\eta,\ \eta_2=\eta+\eta',\ \eta_3=\eta'$

18.3.5 $\qquad\qquad H_i^2=2e_i^2+e_je_k\ (i,j,k=1,2,3;\ i\neq j,i\neq k,j\neq k)$

18.3.6 $\qquad\qquad =(e_i-e_j)(e_i-e_k)=2e_i^2+\dfrac{g_3}{4e_i}=3e_i^2-\dfrac{g_2}{4}$

18.3.7 $\qquad e_i$ real $\qquad\qquad\qquad\qquad e_2$ real and non-negative

18.3.8 $\qquad e_1>0\geq e_2>e_3 \qquad\qquad\qquad (e_2=0$ when $g_3=0)$

$\qquad\qquad$ (equality when $g_3=0$) $\qquad\qquad\qquad e_1=-\alpha+i\beta,\ e_3=\overline{e}_1$

$\qquad\qquad\qquad\qquad\qquad\qquad\qquad\qquad\qquad$ where $\alpha\geq 0,\beta>0$

$\qquad\qquad\qquad\qquad\qquad\qquad\qquad\qquad\qquad$ (equality when $g_3=0$)

18.3.9 $\qquad \eta>0 \qquad\qquad\qquad\qquad \eta_2'=\zeta(\omega_2')=\eta'-\eta$

18.3.10 $\qquad \eta'/i\lesseqgtr 0$ if $\qquad\qquad\qquad \eta_2>0$

18.3.11 $\qquad |\omega'|/\omega\lesseqgtr 1.91014\ 050$ (approx.) $\qquad \eta_2'/i\lesseqgtr 0$ if $|\omega_2'|/\omega_2\gtrless 3.81915\ 447$ (approx.)

18.3.12 $\qquad H_1>0,\ H_3>0 \qquad\qquad\qquad H_2>0$

18.3.13 $\qquad H_2\equiv i\sqrt{-H_2^2} \qquad\qquad\qquad \pi/4<\arg(H_3)\leq \pi/2$ (equality if $g_3=0$); $H_1=\overline{H}_3$

18.3.14 $\qquad \sigma(\omega)=e^{\eta\omega/2}/H_1^{1/2} \qquad\qquad \sigma(\omega_2)=e^{\eta_2\omega_2/2}/H_2^{1/2}$

18.3.15 $\qquad \sigma(\omega')=ie^{\eta'\omega'/2}/H_3^{1/2} \qquad\qquad \sigma(\omega_2')=ie^{\eta_2'\omega_2'/2}/H_2^{1/2}$

18.3.16 $\qquad \sigma^2(\omega_2)=e^{\eta_2\omega_2}/(-H_2) \qquad\qquad \sigma^2(\omega')=e^{\eta'\omega'}/(-H_3)$

18.3.17 $\qquad \arg[\sigma(\omega_2)]=\dfrac{\eta'\omega}{i}+\dfrac{\pi}{2} \qquad\qquad \arg[\sigma(\omega')]=\dfrac{\eta_2\omega_2}{4i}+\dfrac{\pi}{2}-\dfrac{1}{2}\arg(e_2+H_2-e_i)$

Quarter Periods

18.3.18 $\qquad \mathscr{P}(\omega/2)=e_1+H_1>e_1 \qquad\qquad \mathscr{P}(\omega_2/2)=e_2+H_2>e_2$

18.3.19 $\qquad \mathscr{P}'(\omega/2)=-2H_1\sqrt{2H_1+3e_1} \qquad\qquad \mathscr{P}'(\omega_2/2)=-2H_2\sqrt{2H_2+3e_2}$

18.3.20 $\qquad \zeta(\omega/2)=\tfrac{1}{2}[\eta+\sqrt{2H_1+3e_1}] \qquad\qquad \zeta(\omega_2/2)=\tfrac{1}{2}[\eta_2+\sqrt{2H_2+3e_2}]$

$$\Delta > 0 \qquad\qquad\qquad \Delta < 0$$

18.3.21 $\qquad \sigma(\omega/2) = \dfrac{e^{\eta\omega/8}}{2^{1/4}H_1^{3/8}(2H_1+3e_1)^{1/8}} \qquad \sigma(\omega_2/2) = \dfrac{e^{\eta_2\omega_2/8}}{2^{1/4}H_2^{3/8}(2H_2+3e_2)^{1/8}}$

18.3.22 $\qquad \wp(\omega'/2) = e_3 - H_3 < e_3 < 0 \qquad\qquad \wp(\omega_2'/2) = e_2 - H_2 = \wp(\omega_2+\omega_2'/2) < e_2 < 0$

18.3.23 $\qquad \wp'(\omega'/2) = -2H_3 i\sqrt{2H_3-3e_3} \qquad\qquad \wp'(\omega_2'/2) = -2H_2 i\sqrt{2H_2-3e_2} = \overline{\wp}'(\omega_2+\omega_2'/2)$

18.3.24 $\qquad \zeta(\omega'/2) = \tfrac{1}{2}[\eta' - i\sqrt{2H_3-3e_3}] \qquad\qquad \zeta(\omega_2'/2) = \tfrac{1}{2}[\eta_2' - i\sqrt{2H_2-3e_2}] = -\zeta(\omega_2+\omega_2'/2)+2\eta'$

18.3.25 $\qquad \sigma(\omega'/2) = \dfrac{ie^{\eta'\omega'/8}}{2^{1/4}H_3^{3/8}(2H_3-3e_3)^{1/8}} \qquad \sigma(\omega_2'/2) = \dfrac{ie^{\eta_2'\omega_2'/8}}{2^{1/4}H_2^{3/8}(2H_2-3e_2)^{1/8}}$

$$= \sigma(\omega_2+\omega_2'/2)\exp[-\eta'\omega_2]$$

18.3.26 $\qquad \wp(\omega_2/2) = e_2 - H_2 \qquad\qquad\qquad \wp(\omega'/2) = e_3 - H_3$

18.3.27 $\qquad \wp'(\omega_2/2) = -2H_2 i(2H_2-3e_2)^{\frac{1}{2}} \qquad\qquad \wp'(\omega'/2) = -2iH_3(2H_3-3e_3)^{\frac{1}{2}}$

18.3.28 $\qquad \zeta(\omega_2/2) = \tfrac{1}{2}[\eta_2 - i(2H_2-3e_2)^{\frac{1}{2}}] \qquad\qquad \zeta(\omega'/2) = \tfrac{1}{2}[\eta' - i(2H_3-3e_3)^{\frac{1}{2}}]$

18.3.29 $\qquad \sigma(\omega_2/2) = \dfrac{e^{\eta_2\omega_2/8}e^{i\pi/4}}{[4H_2^3(2H_2-3e_2)]^{1/8}} \qquad \sigma(\omega'/2) = \dfrac{e^{\eta'\omega'/8}e^{i\pi/4}}{[4H_3^3(2H_3-3e_3)]^{1/8}}$

One-Third Period Relations

$$\text{At } z = 2\omega_i/3 \,(i=1, 2, 3) \text{ or } 2\omega_2'/3,\ \wp''^2 = 12\wp\wp'^2;$$

equivalently:

18.3.30 $$48\wp^4 - 24g_2\wp^2 - 48g_3\wp - g_2^2 = 0$$

$$\Delta > 0 \qquad\qquad\qquad \Delta < 0$$

18.3.31 $\qquad \zeta(2\omega/3) = \dfrac{2\eta}{3} + \left[\dfrac{\wp(2\omega/3)}{3}\right]^{\frac{1}{2}} \qquad\qquad \zeta(2\omega_2/3) = \dfrac{2\eta_2}{3} + \left[\dfrac{\wp(2\omega_2/3)}{3}\right]^{\frac{1}{2}}$

18.3.32 $\qquad \zeta(2\omega'/3) = \dfrac{2\eta'}{3} - \left[\dfrac{\wp(2\omega'/3)}{3}\right]^{\frac{1}{2}} \qquad\qquad \zeta(2\omega_2'/3) = \dfrac{2\eta_2'}{3} - \left[\dfrac{\wp(2\omega_2'/3)}{3}\right]^{\frac{1}{2}}$

18.3.33 $\qquad \zeta(2\omega_2/3) = \dfrac{2\eta_2}{3} + \left[\dfrac{\wp(2\omega_2/3)}{3}\right]^{\frac{1}{2}} \qquad\qquad \zeta(2\omega'/3) = \dfrac{2\eta'}{3} + \left[\dfrac{\wp(2\omega'/3)}{3}\right]^{\frac{1}{2}}$

18.3.34 $\qquad \sigma(2\omega/3) = \dfrac{-\exp[2\eta\omega/9]}{\sqrt[3]{\wp'(2\omega/3)}} \qquad\qquad \sigma(2\omega_2/3) = \dfrac{-\exp[2\eta_2\omega_2/9]}{\sqrt[3]{\wp'(2\omega_2/3)}}$

18.3.35 $\qquad \sigma(2\omega'/3) = \dfrac{-\exp[2\eta'\omega'/9]}{[\wp'(2\omega'/3)]^{1/3}e^{2\pi i/3}} \qquad\qquad \sigma(2\omega_2'/3) = \dfrac{-\exp[2\eta_2'\omega_2'/9]}{[\wp'(2\omega_2'/3)]^{1/3}e^{2\pi i/3}}$

18.3.36 $\qquad \sigma(2\omega_2/3) = \dfrac{-\exp[2\eta_2\omega_2/9]}{[\wp'(2\omega_2/3)]^{1/3}e^{2\pi i/3}} \qquad\qquad \sigma(2\omega'/3) = \dfrac{-\exp[2\eta'\omega'/9]}{[\wp'(2\omega'/3)]^{1/3}e^{2\pi i/3}}$

Legendre's Relation

18.3.37 $\qquad \eta\omega' - \eta'\omega = \pi i/2 \qquad\qquad\qquad \eta_2\omega_2' - \eta_2'\omega_2 = \pi i$

(also valid for $\Delta < 0$)

Relations Among the H_i

18.3.38 $$H_1^2 + H_2^2 + H_3^2 = 3g_2/4$$

18.3.39 $$H_1^2 H_2^2 + H_2^2 H_3^2 + H_3^2 H_1^2 = 0$$

18.3.40
$$H_1^2 H_2^2 H_3^2 = -\Delta/16$$

18.3.41
$$16 H_i^6 - 12 g_2 H_i^4 + \Delta = 0 (i = 1, 2, 3)$$

18.4. Addition and Multiplication Formulas

Addition Formulas[2] $(z_1 \neq z_2)$

18.4.1
$$\wp(z_1 + z_2) = \frac{1}{4} \left[\frac{\wp'(z_1) - \wp'(z_2)}{\wp(z_1) - \wp(z_2)} \right]^2 - \wp(z_1) - \wp(z_2)$$

18.4.2
$$\wp'(z_1 + z_2) = \frac{\wp(z_1 + z_2)[\wp'(z_1) - \wp'(z_2)] + \wp(z_1)\wp'(z_2) - \wp'(z_1)\wp(z_2)}{\wp(z_2) - \wp(z_1)}$$

18.4.3
$$\zeta(z_1 + z_2) = \zeta(z_1) + \zeta(z_2) + \frac{1}{2} \frac{\wp'(z_1) - \wp'(z_2)}{\wp(z_1) - \wp(z_2)}$$

18.4.4
$$\sigma(z_1 + z_2)\sigma(z_1 - z_2) = -\sigma^2(z_1)\sigma^2(z_2)[\wp(z_1) - \wp(z_2)]$$

Duplication and Triplication Formulas

$$\left[\text{Note that } \wp'' = 6\wp^2(z) - \frac{g_2}{2}, \; \wp'^2(z) = 4\wp^3(z) - g_2\wp(z) - g_3 \text{ and } \wp'''(z) = 12\wp(z)\wp'(z) \right]$$

18.4.5
$$\wp(2z) = -2\wp(z) + \left[\frac{\wp''(z)}{2\wp'(z)} \right]^2$$

18.4.6
$$\wp'(2z) = \frac{-4\wp'^4(z) + 12\wp(z)\wp'^2(z)\wp''(z) - \wp''^3(z)}{4\wp'^3(z)}$$

18.4.7
$$\zeta(2z) = 2\zeta(z) + \wp''(z)/2\wp'(z)$$

18.4.8
$$\sigma(2z) = -\wp'(z)\sigma^4(z)$$

18.4.9
$$\zeta(3z) = 3\zeta(z) + \frac{4\wp'^3(z)}{\wp'(z)\wp'''(z) - \wp''^2(z)}$$

18.4.10
$$\sigma(3z) = -\wp'^2(z)\sigma^9(z)[\wp(2z) - \wp(z)]$$

18.5. Series Expansions

Laurent Series

18.5.1
$$\wp(z) = z^{-2} + \sum_{k=2}^{\infty} c_k z^{2k-2}$$

18.5.2 where
$$c_2 = g_2/20, \; c_3 = g_3/28$$

and

18.5.3
$$c_k = \frac{3}{(2k+1)(k-3)} \sum_{m=2}^{k-2} c_m c_{k-m}, \; k \geq 4$$

18.5.4
$$\wp'(z) = -2z^{-3} + \sum_{k=2}^{\infty} (2k-2)c_k z^{2k-3}$$

18.5.5
$$\zeta(z) = z^{-1} - \sum_{k=2}^{\infty} c_k z^{2k-1}/(2k-1)$$

18.5.6
$$\sigma(z) = \sum_{m,n=0}^{\infty} a_{m,n} (\tfrac{1}{2}g_2)^m (2g_3)^n \cdot \frac{z^{4m+6n+1}}{(4m+6n+1)!}$$

---

[2] Formulas for ζ and σ are *not* true algebraic addition formulas.

18.5.7 $$\text{where } a_{0,0}=1 \text{ and}$$

18.5.8 $$a_{m,n}=3(m+1)a_{m+1,n-1}+\frac{16}{3}(n+1)a_{m-2,n+1}-\frac{1}{3}(2m+3n-1)(4m+6n-1)a_{m-1,n},$$

it being understood that $a_{m,n}=0$ if either subscript is negative.

(The radius of convergence of the above series for $\wp-z^{-2}$, $\wp'+2z^{-3}$ and $\zeta-z^{-1}$ is equal to the smallest of $|2\omega|$, $|2\omega'|$ and $|2\omega\pm2\omega'|$; series for σ converges for all z.)

Values of Coefficients[3] c_k in Terms of c_2 and c_3

18.5.9 $$c_4=c_2^2/3$$

18.5.10 $$c_5=3c_2c_3/11$$

18.5.11 $$c_6=[2c_2^3+3c_3^2]/39$$

18.5.12 $$c_7=2c_2^2c_3/33$$

18.5.13 $$c_8=5c_2(11c_2^3+36c_3^2)/7293$$

18.5.14 $$c_9=c_3(29c_2^3+11c_3^2)/2717$$

18.5.15 $$c_{10}=(242c_2^5+1455c_2^2c_3^2)/240669$$

18.5.16 $$c_{11}=14c_2c_3(389c_2^3+369c_3^2)/3187041$$

18.5.17 $$c_{12}=(114950c_2^6+1080000c_2^3c_3^2+166617c_3^4)/891678645$$

18.5.18 $$c_{13}=10c_2^2c_3(297c_2^3+530c_3^2)/11685817$$

18.5.19 $$c_{14}=\frac{2c_2(528770c_2^6+7164675c_2^3c_3^2+2989602c_3^4)}{(306735)(215441)}$$

18.5.20 $$c_{15}=\frac{4c_3(62921815c_2^6+179865450c_2^3c_3^2+14051367c_3^4)}{(179685)(38920531)}$$

18.5.21 $$c_{16}=\frac{c_2^2(58957855c_2^6+1086511320c_2^3c_3^2+875341836c_3^4)}{(5909761)(5132565)}$$

18.5.22 $$c_{17}=\frac{c_2c_3(30171955c_2^6+126138075c_2^3c_3^2+28151739c_3^4)}{(920205)(6678671)}$$

18.5.23 $$c_{18}=\frac{1541470\cdot949003c_2^9+30458088737\cdot1155c_2^6c_3^2+122378650673\cdot378c_2^3c_3^4+2348703\cdot887777c_3^6}{(1342211013)(4695105713)}$$

18.5.24 $$c_{19}=\frac{2c_2^2c_3(3365544215c_2^6+429852433\cdot45c_2^3c_3^2+8527743477c_3^4)}{(91100295)(113537407)}$$

[3] *NOTES:*

1. c_4-c_{16} were computed and checked independently by D. H. Lehmer; these were double-checked by substituting $g_2=20\ c_2$, $g_3=28\ c_3$ in values given in [18.10].

2. $c_{17}-c_{18}$ were derived from values in [18.10] by the same substitution. These were checked (numerically) for particular values of g_2, g_3.

3. c_{19} is given incorrectly in [18.12] (factor 13 is missing in denominator of third term of bracket); this value was computed independently.

4. No factors of any of the above integers with more than ten digits are known to the author. This is not necessarily true of smaller integers, which have, in many instances, been arranged for convenient use with a desk calculator.

Value [4] of Coefficients $a_{m,n}$

$n \uparrow$

n	m=0	1	2	3	4	5	6	7	8	9	10	11	12
8	$-2 \cdot 3^7 \cdot 5 \cdot 59$ ·107895773												
7	$-2 \cdot 3^7 \cdot 5 \cdot 23^{10}$ ·257·18049	$-2 \cdot 3^7 \cdot 5 \cdot 59^{11}$ ·107895773											
6	$-2 \cdot 3^6 \cdot 5^9$ ·229·2683 ·18049	$-2 \cdot 3^7 \cdot 5 \cdot 23^9$ ·257	$-2 \cdot 3^6 \cdot 5 \cdot 7^{11}$ ·181·1699 ·2803	$-2 \cdot 3^8 \cdot 5 \cdot 7^{10}$ ·41·6047 ·4922497									
5	$2 \cdot 3^8 \cdot 5^8$ ·9103	$-2 \cdot 3^6 \cdot 5 \cdot 229^8$ ·2683	$-2 \cdot 3^5 \cdot 5^{10}$ ·40570423	$-2 \cdot 3^6 \cdot 5 \cdot 7^9$ ·59·179 ·142231	$-2 \cdot 3^4 \cdot 5 \cdot 7^{10}$ ·1321 ·1415535763								
4	$2 \cdot 3^8 \cdot 5^{8\cdot 2}$ ·31	$2 \cdot 3 \cdot 5 \cdot 9103^7$	$-2 \cdot 3^5 \cdot 5 \cdot 7^8$ ·13·37·41	$-2 \cdot 3^5 \cdot 5 \cdot 691^8$ ·83609	$-2 \cdot 3 \cdot 5 \cdot 11 \cdot 31^8$ ·313·190387	$-2 \cdot 3^4 \cdot 5 \cdot 7 \cdot 23^{10,2}$ ·263·4848953	$-2 \cdot 3 \cdot 5 \cdot 7 \cdot 19^9$ ·1752686144977						
3	$2 \cdot 3 \cdot 23^{8\cdot 4}$	$2 \cdot 3 \cdot 5 \cdot 31^{3\cdot 4\cdot 2}$	$2 \cdot 3 \cdot 5 \cdot 17^{3\cdot 7}$ ·109	$-2 \cdot 3 \cdot 5 \cdot 83^{5}$ ·3911	$-2 \cdot 3 \cdot 5 \cdot 503^{4\cdot 6}$ ·156217	$-2 \cdot 3 \cdot 31^{4\cdot 8}$ ·315989669	$-2 \cdot 3 \cdot 5 \cdot 7 \cdot 11 \cdot 29^{4\cdot 7}$ ·83·1129·9551	$-2 \cdot 3 \cdot 5 \cdot 7 \cdot 613^{4\cdot 8}$ ·17605225081					
2	$-2 \cdot 3^3$	$2 \cdot 3 \cdot 23^{3\cdot 3}$	$2 \cdot 3 \cdot 5 \cdot 53^{3\cdot 5}$	$2 \cdot 3 \cdot 5 \cdot 37^{3\cdot 4}$ ·167	$-2 \cdot 3 \cdot 5 \cdot 17^6$ ·3037	$-2 \cdot 3 \cdot 61 \cdot 151^{4\cdot 7}$ ·653	$-2 \cdot 3^{3\cdot 7}$ ·2387260103	$-2 \cdot 3 \cdot 5 \cdot 7 \cdot 17 \cdot 53^{4\cdot 7}$ ·2957·41189	$-2 \cdot 3 \cdot 5 \cdot 7 \cdot 17^{9\cdot 2}$ ·67·195651059	$2 \cdot 3 \cdot 5 \cdot 7^{3\cdot 7\cdot 2}$ ·35866647631901			
1	-3	$-2 \cdot 3^2$	$3^3 \cdot 19$	$2 \cdot 3 \cdot 311^{2\cdot 3}$	$3 \cdot 5 \cdot 20807^3$	$-2 \cdot 3 \cdot 11^6$ ·2609	$-3 \cdot 17^5$ ·1578257	$-2 \cdot 3 \cdot 7 \cdot 13^{3\cdot 5}$ ·2742587	-3^7 ·248882935409	$-2 \cdot 3 \cdot 5 \cdot 7 \cdot 193^{5\cdot 2\cdot 7}$ ·13679·274973	$3 \cdot 7^6$ ·89555603641079		
0	1	-1	-3^3	$3 \cdot 23$	$3 \cdot 107$	$3 \cdot 7 \cdot 23 \cdot 37^3$	$3 \cdot 313 \cdot 503^3$	$-3 \cdot 7^4$ ·685973	$3 \cdot 11 \cdot 37^5$ ·257981	$-3 \cdot 7 \cdot 23 \cdot 14387^4$ ·40763	$-3 \cdot 71^4$ ·176302760639	$-3 \cdot 7 \cdot 11 \cdot 23^{6\cdot 2}$ ·383·739·18539	$3 \cdot 7 \cdot 24733^{5\cdot 3}$ ·198922785511
	0	1	2	3	4	5	6	7	8	9	10	11	12

$\rightarrow m$

[4] Values of $a_{m,n}$ in unfactored form for $4m+6n+1\le35$ are given in [18.25], p. 7; of $(a_{m,n})3^{-n}$ in factored form in [18.15], Vol. 4, p. 89 for $4m+6n+1\le25$. Additional values were computed and checked on desk calculators; primality of large factors was established with the aid of SWAC (National Bureau of Standards Western Automatic Computer).

Reversed Series[5] for Large $|\mathscr{P}|$

18.5.25

$$z=\frac{1}{2}\left[2u+c_2u^5+c_3u^7+\frac{\alpha_2^2}{3}u^9+\frac{6\alpha_2\alpha_3}{11}u^{11}\right.$$

$$+\frac{1}{13}(3\alpha_3^2+5\alpha_2^3)u^{13}+\alpha_2^2\alpha_3u^{15}+\frac{5\alpha_2}{68}(12\alpha_3^2+7\alpha_2^3)u^{17}$$

$$+\frac{5\alpha_3}{19}(\alpha_3^2+7\alpha_2^3)u^{19}+\frac{\alpha_2^2}{4}(3\alpha_2^3+10\alpha_3^2)u^{21}$$

$$+\frac{35\alpha_2\alpha_3}{92}(9\alpha_2^3+4\alpha_3^2)u^{23}$$

$$+\frac{7}{200}(33\alpha_2^6+180\alpha_2^3\alpha_3^2+10\alpha_3^4)u^{25}$$

$$+\frac{7\alpha_2^2\alpha_3}{12}(11\alpha_2^3+10\alpha_3^2)u^{27}$$

$$+\frac{3\alpha_2}{2^3\cdot29}(143\alpha_2^6+1155\alpha_2^3\alpha_3^2+210\alpha_3^4)u^{29}$$

$$+\frac{21\alpha_3}{2^3\cdot31}(143\alpha_2^6+220\alpha_2^3\alpha_3^2+6\alpha_3^4)u^{31}$$

$$+\frac{3\alpha_2^2}{2^6}(65\alpha_2^6+728\alpha_2^3\alpha_3^2+280\alpha_3^4)u^{33}$$

$$+\frac{33\alpha_2\alpha_3}{2^3\cdot5\cdot7}(195\alpha_2^6+455\alpha_2^3\alpha_3^2+42\alpha_3^4)u^{35}$$

$$+\frac{11}{2^6\cdot37}(1105\alpha_2^9+16380\alpha_2^6\alpha_3^2+10920\alpha_2^3\alpha_3^4$$

$$+168\alpha_3^6)u^{37}+\frac{33\alpha_2^2\alpha_3}{2^6}(85\alpha_2^6+280\alpha_2^3\alpha_3^2+56\alpha_3^4)u^{39}$$

$$+\frac{143\alpha_2}{2^7\cdot41}(323\alpha_2^9+6120\alpha_2^6\alpha_3^2+6300\alpha_2^3\alpha_3^4+336\alpha_3^6)u^{41}$$

$$+\frac{143\alpha_3}{2^6\cdot43}(1615\alpha_2^9+7140\alpha_2^6\alpha_3^2+2520\alpha_2^3\alpha_3^4+24\alpha_3^6)u^{43}$$

$$\left.+O(u^{45})\right],$$

18.5.26 where $\alpha_2=g_2/8$

18.5.27 $\qquad\alpha_3=g_3/8$

18.5.28 $\qquad u=(\mathscr{P}^{-1})^{\frac{1}{2}}$

Reversed Series for Large $|\mathscr{P}'|$

18.5.29 $z=A_1u+A_5u^5+A_7u^7+A_9u^9+\ldots$

18.5.30 where $u=(\mathscr{P}'^{1/3})^{-1}e^{i\pi/3}$

18.5.31 $\qquad A_1=2^{1/3}$

18.5.32 $\qquad A_5=-\frac{a_2}{5}A_1^2$

18.5.33 $\qquad A_7=\frac{-4a_3A_1}{7}$

18.5.34 $\qquad A_9=0$

18.5.35 $\qquad A_{11}=8a_2a_3A_1^2/11$

18.5.36 $\qquad A_{13}=\frac{10A_1}{39}(a_2^3+6a_3^2)$

18.5.37 $\qquad A_{15}=-96a_2^2a_3/175$

18.5.38 $\qquad A_{17}=-\frac{14a_2A_1^2}{51}(a_2^3+12a_3^2)$

18.5.39 where $a_2=g_2/6,\ a_3=g_3/6$

Reversed Series for Large $|\zeta|$

18.5.40 $z=u+A_5u^5+A_7u^7+A_9u^9+\ldots$

18.5.41 where $u=\zeta^{-1}$

18.5.42 $\qquad A_5=-\delta_2/5$

18.5.43 $\qquad A_7=-\delta_3/7$

18.5.44 $\qquad A_9=\delta_2^2/7$

18.5.45 $\qquad A_{11}=3\delta_2\delta_3/11$

18.5.46 $\qquad A_{13}=\frac{17}{1001}(-8\delta_2^3+7\delta_3^2)$

18.5.47 $\qquad A_{15}=-41\delta_2^2\delta_3/91$

18.5.48 $\qquad A_{17}=\frac{\delta_2}{9163}(1349\delta_2^3-4116\delta_3^2)$

18.5.49 $A_{19}=\frac{2\delta_3}{323323}(115431\delta_2^3-22568\delta_3^2)$

18.5.50 where $\delta_2=g_2/12$

18.5.51 $\qquad\delta_3=g_3/20$

---

[5] In this and other series a choice of the value of the root has been made so that z will be in the Fundamental Rectangle **(Figure 18.2)**, whenever the value of the given function is appropriate.

Other Series Involving $\mathscr{P}$

Series near z_0 $[\mathscr{P}(z_0)=0]$

18.5.52

$$\mathscr{P}=\mathscr{P}_0'u\left[1-3c_2u^4-4c_3u^6+\frac{10c_2^2}{3}u^8+\frac{114c_2c_3}{11}u^{10}\right.$$

$$+\frac{7(12c_3^2-5c_2^3)}{13}u^{12}-\frac{488c_2^2c_3}{33}u^{14}\right]+u^2\left[-5c_2-14c_3u^2\right.$$

$$+5c_2^2u^4+33c_2c_3u^6+\frac{84c_3^2-10c_2^3}{3}u^8-\frac{1363c_2^2c_3u^{10}}{33}$$

$$+\frac{5c_2(55c_2^3-2316c_0^2)u^{12}}{143}\right]+\cdots$$

18.5.53

where $u=(z-z_0),\mathscr{P}_0'\equiv\mathscr{P}'(z_0)=i\sqrt{g_3}$

18.5.54

$$u=\mathscr{P}_0'[v+av^2+2a^2v^3+\left(\frac{g_3\mathscr{P}_0'^2}{2}+5a^3\right)v^4+\frac{a}{5}(3\mathscr{P}_0'^4$$

$$+15g_3\mathscr{P}_0'^2+70a^3)v^5+2a^2(2\mathscr{P}_0'^4+7g_3\mathscr{P}_0'^2+21a^3)v^6$$

$$+\left(\frac{g_3\mathscr{P}_0'^6}{7}+\{g_3^2+20a^3\}\mathscr{P}_0'^4+15a^2g_3\mathscr{P}_0'^2+132a^6\right)v^7$$

$$+15a\left(\frac{g_3\mathscr{P}_0'^6}{4}+\left\{\frac{3g_3^2}{4}+6a^3\right\}\mathscr{P}_0'^4+\frac{33ag_3}{2}\mathscr{P}_0'^2\right.$$

$$+\frac{143a^6}{5}\right)v^8+\frac{5a^2}{2}\left(\frac{2}{3}\mathscr{P}_0'^8+15g_3\mathscr{P}_0'^6\right.$$

$$+\{154a^3+33g_3^2\}\mathscr{P}_0'^4+\frac{2002a^3g_3\mathscr{P}_0'^2}{5}+572a^6\right)v^9$$

$$+\frac{1}{4}\left(3\{28a^3+g_3^2\}\mathscr{P}_0'^8+11g_3\{98a^3+g_3^2\}\mathscr{P}_0'^6\right.$$

$$+2002a^3\left\{\frac{16}{5}a^3+g_3^2\right\}\mathscr{P}_0'^4$$

$$+16016\,a^6g_3\mathscr{P}_0'^2+19448\,a^9)v^{10}]+\cdots$$

18.5.55   where $v=\mathscr{P}/(\mathscr{P}_0')^2$ and $a=g_2/4$

Series near ω_i

18.5.56

$$(\mathscr{P}-e_i)=(3e_i^2-5c_2)u+(10c_2e_i+21c_3)u^2+(7c_2e_i^2$$

$$+21c_3e_i+5c_2^2)u^3+(18c_3e_i^2+30c_2^2e_i$$

$$+33c_2c_3)u^4+\left(22c_2^2e_i^2+92c_2c_3e_i+105c_3^2\right.$$

$$-\frac{10c_2^3}{3}\right)u^5+\left(\frac{728}{11}c_2c_3e_i^2+\frac{220}{3}c_2^3e_i+84c_3^2e_i\right.$$

$$+\frac{1214}{11}c_2^2c_3\right)u^6+\left(\frac{635}{13}c_2^3e_i^2+\frac{855}{13}c_3^2e_i^2\right.$$

$$+\frac{3405}{11}c_2^2c_3e_i+\frac{45750}{143}c_2c_3^2+\frac{25}{13}c_2^4\right)u^7+\cdots,$$

18.5.57   where $u=(z-\omega_i)^2$

Other Series Involving $\mathscr{P}'$

Series near z_0

18.5.58

$$(\mathscr{P}'-\mathscr{P}_0')=\left[-10c_2u-56c_3u^3+30c_2^2u^5+264c_2c_3u^7\right.$$

$$+\frac{(840c_3^2-100c_2^3)}{3}u^9-\frac{5452c_2^2c_3}{11}u^{11}$$

$$+\frac{70c_2(55c_3^2-2316c_3^2)}{143}u^{13}\right]$$

$$+\mathscr{P}_0'\left[-15c_2u^4-28c_3u^6+30c_2^2u^8+114c_2c_3u^{10}\right.$$

$$+7(12c_3^2-5c_2^3)u^{12}-\frac{2440c_2^2c_3}{11}u^{14}\right]+\cdots$$

18.5.59   where $u=(z-z_0)$

18.5.60

$$(z-z_0)=A-bA^3-\frac{3\mathscr{P}_0'}{2}A^4+3(c_2+b^2)A^5$$

$$+10b\mathscr{P}_0'A^6-3[36c_3-3\mathscr{P}_0'+4b^3]A^7$$

$$-3\mathscr{P}_0'\left(\frac{25}{2}c_2+21b^2\right)A^8+\frac{5}{12}\left(285b^2c_2\right.$$

$$+100c_2^2-279\mathscr{P}_0'^2b+132b^4\right)A^9+\cdots$$

18.5.61   where $A=(\mathscr{P}'-\mathscr{P}_0')/(-10c_2)$

18.5.62   and $b=4g_3/g_2$

Series near ω_i

18.5.63

$$\mathscr{P}'=2(3e_i^2-5c_2)\alpha+4(10c_2e_i+21c_3)\alpha^3+6(7c_2e_i^2$$

$$+21c_3e_i+5c_2^2)\alpha^5+24(6c_3e_i^2+10c_2^2e_i$$

$$+11c_2c_3)\alpha^7+10\left(22c_2^2e_i^2+92c_2c_3e_i+105c_3^2\right.$$

$$-\frac{10c_2^3}{3}\right)\alpha^9+24\left(\frac{364}{11}c_2c_3e_i^2+\frac{110}{3}c_2^3e_i\right.$$

$$+42c_3^2e_i+\frac{607}{11}c_2^2c_3\right)\alpha^{11}+70\left(\frac{127}{13}c_2^3e_i^2\right.$$

$$+\frac{171}{13}c_3^2e_i^2+\frac{681}{11}c_2^2c_3e_i+\frac{9150}{143}c_2c_3^2+\frac{5}{13}c_2^4\right)\alpha^{13}$$

$$+\cdots,$$

18.5.64   where $\alpha=(z-\omega_i)$.

Other Series Involving ζ

Series near z_0 $[\wp(z_0)=0]$

18.5.65

$$\zeta-\zeta_0=\wp_0'\left[-\frac{u^2}{2}+\frac{c_2u^6}{2}+\frac{c_3u^8}{2}-\frac{c_2^2u^{10}}{3}-\frac{19c_2c_3u^{12}}{22}\right.$$

$$\left.+\frac{(5c_2^3-12c_3^2)}{26}u^{14}+\frac{61c_2^2c_3u^{16}}{66}\right]+\left[\frac{5c_2u^3}{3}\right.$$

$$+\frac{7c_3u^5}{2}-\frac{5c_2^2u^7}{7}-\frac{11c_2c_3u^9}{3}+\frac{(10c_3^3-84c_3^2)}{33}u^{11}$$

$$\left.+\frac{1363c_2^2c_3}{429}u^{13}+\frac{c_2(2316c_3^2-55c_2^3)}{429}u^{15}\right]+\ldots,$$

18.5.66   where $u=(z-z_0)$,

18.5.67 $\zeta_0\equiv\zeta(z_0)$

Series near ω_i

18.5.68

$$(\zeta-\eta_i)=-e_i\alpha-\frac{(3e_i^2-5c_2)}{3}\alpha^3-\frac{(10c_2e_i+21c_3)\alpha^5}{5}$$

$$-\frac{(7c_2e_i^2+21c_3e_i+5c_2^2)\alpha^7}{7}$$

$$-\frac{(6c_3e_i^2+10c_2^2e_i+11c_2c_3)\alpha^9}{3}$$

$$-\frac{\left(22c_2^2e_i^2+92c_2c_3e_i+105c_3^2-\frac{10}{3}c_2^3\right)\alpha^{11}}{11}$$

$$-\frac{2}{13}\left(\frac{364}{11}c_2c_3e_i^2+\frac{110}{3}c_2^3e_i+42c_3^2e_i\right.$$

$$\left.+\frac{607}{11}c_2^2c_3\right)\alpha^{13}-\frac{1}{3}\left(\frac{127}{13}c_2^3e_i^2+\frac{171}{13}c_3^2e_i^2\right.$$

$$\left.+\frac{681}{11}c_2^2c_3e_i+\frac{9150}{143}c_2c_3^2+\frac{5}{13}c_2^4\right)\alpha^{15}-\ldots,$$

18.5.69

where $\alpha=(z-\omega_i)$

Reversed Series for Small $|\sigma|$

18.5.70

$$z=\sigma+\frac{\gamma_2}{5}\sigma^5+\frac{\gamma_3}{7}\sigma^7+\frac{3\gamma_2^2}{14}\sigma^9$$

$$+\frac{19\gamma_2\gamma_3}{55}\sigma^{11}+\frac{3842\gamma_2^3+861\gamma_3^2}{6006}\sigma^{13}+\ldots,$$

18.5.71   where $\gamma_2=g_2/48$

18.5.72 $\gamma_3=g_3/120$

For reversion of Maclaurin series, see **3.6.25** and [**18.18**].

18.6. Derivatives and Differential Equations

Ordinary $(c_2=g_2/20,\ c_3=g_3/28)$

18.6.1 $\zeta'(z)=-\wp(z)$

18.6.2 $\sigma'(z)/\sigma(z)=\zeta(z)$

18.6.3

$$\wp'^2(z)=4\wp^3(z)-g_2\wp(z)-g_3=4(\wp^3-5c_2\wp-7c_3)$$

18.6.4 $\wp''(z)=6\wp^2(z)-\tfrac{1}{2}g_2=6\wp^2-10c_2$

18.6.5 $\wp'''(z)=12\wp\wp'$

18.6.6

$$\wp^{(4)}(z)=12(\wp\wp''+\wp'\wp')$$

$$=5!\left[\wp^3-3c_2\wp-\frac{14c_3}{5}\right]$$

18.6.7

$$\wp^{(5)}(z)=12(\wp\wp'''+2\wp'\wp''+\wp''\wp')$$

$$=3\cdot5!\,\wp'[\wp^2-c_2]$$

18.6.8

$$\wp^{(6)}(z)=12(\wp\wp^{(4)}+3\wp'\wp'''+3\wp''\wp''$$

$$+\wp'''\wp')$$

18.6.9 $=7![\wp^4-4c_2\wp^2-4c_3\wp+5c_2^2/7]$

18.6.10 $\wp^{(7)}(z)=4\cdot7!\,\wp'[\wp^3-2c_2\wp-c_3]$

18.6.11

$$\wp^{(8)}(z)=9![\wp^5-5c_2\wp^3-5c_3\wp^2$$

$$+(10c_2^2\wp+11c_2c_3)/3]$$

18.6.12

$$\wp^{(9)}(z)=5\cdot9!\,\wp'[\wp^4-3c_2\wp^2-2c_3\wp+2c_2^2/3]$$

18.6.13

$$\wp^{(10)}(z)=11![\wp^6-6c_2\wp^4-6c_3\wp^3+7c_2^2\wp^2$$

$$+(342c_2c_3\wp+84c_3^2-10c_2^3)/33]$$

18.6.14

$$\wp^{(11)}(z)=6\cdot11!\,\wp'[\wp^5-4c_2\wp^3-3c_3\wp^2$$

$$+(77c_2^2\wp+57c_2c_3)/33]$$

18.6.15

$$\wp^{(12)}(z)=13![\wp^7-7c_2\wp^5-7c_3\wp^4+35c_2^2\wp^3/3$$

$$+210c_2c_3\wp^2/11+(84c_3^2-35c_2^3)\wp/13-1363c_2^2c_3/429]$$

18.6.16

$$\wp^{(13)}(z)=7\cdot13!\,\wp'[\wp^6-5c_2\wp^4-4c_3\wp^3$$

$$+5c_2^2\wp^2+60c_2c_3\wp/11+(12c_3^2-5c_2^3)/13]$$

18.6.17

$$\wp^{(14)}(z)=15![\wp^8-8c_2\wp^6-8c_3\wp^5+52c_2^2\wp^4/3$$

$$+328c_2c_3\wp^3/11+(444c_3^2-328c_2^3)\wp^2/39$$

$$-488c_2^2c_3\wp/33+c_2(55c_2^3-2316c_3^2)/429]$$

18.6.18

$$\wp^{(15)}(z)=8\cdot15!\,\wp'[\wp^7-6c_2\wp^5-5c_3\wp^4+26c_2^2\wp^3/3$$

$$+123c_2c_3\wp^2/11+(111c_3^2-82c_2^3)\wp/39-61c_2^2c_3/33]$$

Partial Derivatives with Respect to Invariants

18.6.19

$$\Delta \frac{\partial \mathscr{P}}{\partial g_3} = \mathscr{P}'\left(3g_2\zeta - \frac{9}{2}g_3 z\right) + 6g_2\mathscr{P}^2 - 9g_3\mathscr{P} - g_2^2$$

18.6.20

$$\Delta \frac{\partial \mathscr{P}}{\partial g_2} = \mathscr{P}'\left(-\frac{9}{2}g_3\zeta + \frac{g_2^2 z}{4}\right) - 9g_3\mathscr{P}^2 + \frac{g_2^2}{2}\mathscr{P} + \frac{3}{2}g_2 g_3$$

18.6.21

$$\Delta \frac{\partial \zeta}{\partial g_3} = -3\zeta\left(g_2\mathscr{P} + \frac{3}{2}g_3\right)$$
$$+ \frac{1}{2}z\left(9g_3\mathscr{P} + \frac{1}{2}g_2^2\right) - \frac{3}{2}g_2\mathscr{P}'$$

18.6.22

$$\Delta \frac{\partial \zeta}{\partial g_2} = \frac{1}{2}\zeta\left(9g_3\mathscr{P} + \frac{1}{2}g_2^2\right)$$
$$- \frac{1}{2}g_2 z\left(\frac{1}{2}g_2\mathscr{P} + \frac{3}{4}g_3\right) + \frac{9}{4}g_3\mathscr{P}'$$

18.6.23 $\Delta \dfrac{\partial \sigma}{\partial g_3} = \dfrac{3}{2}g_2\sigma'' + \dfrac{9}{2}g_3\sigma + \dfrac{1}{8}g_2^2 z^2\sigma - \dfrac{9}{2}g_2 z\sigma'$

18.6.24

$$\Delta \frac{\partial \sigma}{\partial g_2} = -\frac{9}{4}g_3\sigma'' - \frac{1}{4}g_2^2\sigma - \frac{3}{16}g_2 g_3 z^2\sigma + \frac{1}{4}g_2^2 z\sigma'$$

$$\left(\text{here } ' \text{ denotes } \frac{\partial}{\partial z}\right)$$

Differential Equations

18.6.25

Equation	Solution
$y'^3 = y^2(y-a)^2$	$y = \dfrac{a}{2} + \dfrac{27}{16}\mathscr{P}'\left(\dfrac{z}{2}; 0, -\dfrac{64a^2}{729}\right)$

18.6.26

$y'^3 = (y^3 - 3ay^2 + 3y)^2$ $y = \dfrac{2}{a - 3\mathscr{P}'(z; 0, g_3)}$,

$$g_3 = \frac{4 - 3a^2}{27}$$

18.6.27

$y'^4 = \dfrac{128}{3}(y+a)^2(y+b)^3$ $y = 6\mathscr{P}^2(z; g_2, 0) - b$,

$$g_2 = -\frac{2}{3}(a-b)$$

$y'' = [a\mathscr{P}(z) + b]y$ (Lamé's equation)—see [18.8], 2.26

For other (more specialized) equations (of orders 1–3) involving $\mathscr{P}(z)$, see [18.8], nos. 1.49, 2.28, 2.72–3, 2.439–440, 3.9–12.

For the use of $\mathscr{P}(z)$ in solving differential equations of the form $y'^m + A(z,y) = 0$, where $A(z,y)$ is a polynomial in y of degree $2m$, with coefficients which are analytic functions of z, see [18.7], p. 312ff.

18.7. Integrals

Indefinite

18.7.1 $\displaystyle\int \mathscr{P}^2(z)dz = \frac{1}{6}\mathscr{P}'(z) + \frac{1}{12}g_2 z$

18.7.2 $\displaystyle\int \mathscr{P}^3(z)dz = \frac{1}{120}\mathscr{P}'''(z) - \frac{3}{20}g_2\zeta(z) + \frac{1}{10}g_3 z$

(formulas for higher powers may be derived by integration of formulas for $\mathscr{P}^{(2k)}(z)$)

For $\int \mathscr{P}^n(z)dz$, n any positive integer, see [18.15] vol. 4, pp. 108–9.

If $\mathscr{P}'(a) \neq 0$

18.7.3

$$\mathscr{P}'(a)\int \frac{dz}{\mathscr{P}(z) - \mathscr{P}(a)}$$
$$= 2z\zeta(a) + \ln\sigma(z-a) - \ln\sigma(z+a)$$

For $\int dz/[\mathscr{P}(z) - \mathscr{P}(a)]^n$, $(\mathscr{P}'(a) \neq 0)$ n any positive integer, see [18.15], vol. 4, pp. 109–110.

Definite

	$\Delta > 0$	$\Delta < 0$

18.7.4 $\displaystyle\omega = \int_{e_1}^{\infty} \frac{dt}{\sqrt{s(t)}}$ $\displaystyle\omega_2 = \int_{e_2}^{\infty} \frac{dt}{\sqrt{s(t)}}$

18.7.5 $\displaystyle\omega' = i\int_{-\infty}^{e_3} \frac{dt}{\sqrt{|s(t)|}}$ $\displaystyle\omega_2' = i\int_{-\infty}^{e_2} \frac{dt}{\sqrt{|s(t)|}}$

18.7.6   where t is real and

18.7.7 $s(t) = 4t^3 - g_2 t - g_3$

18.8 Conformal Mapping

$$w = u + iv$$

$\Delta > 0$

$w = \wp(z)$ maps the Fundamental Rectangle onto the half-plane $v \leq 0$; if $|\omega'| = \omega(g_3 = 0)$, the isosceles triangle $0\omega\omega_2$ is mapped onto $u \geq 0$, $v \leq 0$.

$w = \wp'(z)$ maps the Fundamental Rectangle onto the w-plane less quadrant III; if $|\omega'| = \omega$, the triangle $0\omega\omega_2$ is mapped onto $v \geq 0$, $v \geq u$.

$\Delta < 0$

$w = \wp(z)$ maps the Fundamental Rectangle onto the half-plane $v \leq 0$; if $|\omega_2'| = \omega_2(g_3 = 0)$, the isosceles triangle $0\omega_2\omega'$ is mapped onto $u \geq 0$, $v \leq 0$.

$w = \wp'(z)$ maps the Fundamental Rectangle onto most of the w-plane less quadrant III; if $|\omega_2'| = \omega_2$, the triangle $0\omega_2\omega'$ is mapped onto $v \geq 0$, $v \geq u$.

$(a = \text{period ratio})$

$w = \zeta(z)$ maps the Fundamental Rectangle onto the half-plane $u \geq 0$. If $a \leq 1.9$ (approx.), $v \leq 0$; otherwise the image extends into quadrant I. For very large a, the image has a large area in quadrant I.

$w = \sigma(z)$ maps the Fundamental Rectangle onto quadrant I if $a < 1.9$ (approx.), onto quadrants I and II if $1.9 \leq a < 3.8$ (approx.). For large a, $\arg[\sigma(\omega_2)] \approx \dfrac{\pi^2 a}{12}$; consequently the image winds around the origin for large a.

Other maps are described in [18.23] arts. 13.7 (square on circle), 13.11 (ring on plane with 2 slits in line) and in [18.24], p. 35 (double half equilateral triangle on half-plane).

$w = \zeta(z)$ maps the Fundamental Rectangle onto the half-plane $u \geq 0$. The image is mostly in quadrant IV for small a, entirely so for (approx.) $1.3 \leq a \leq 3.8$. For very large a, the image has a large area in quadrant I.

$w = \sigma(z)$ maps the Fundamental Rectangle onto quadrant I if $a < 3.8$ (approx.), onto quadrants I and II if $3.8 \leq a < 7.6$ (approx.). For large a, $\arg\left[\sigma\left(\omega_2 + \dfrac{\omega_2'}{2}\right)\right] \approx \dfrac{\pi^2 a}{24}$; consequently the image winds around the origin for large a.

Other maps are described in [18.23] arts. 13.8 (equilateral triangle on half-plane) and 13.9 (isosceles triangle on half-plane).

Obtaining $\wp'$ from $\wp'^2$

Fundamental Rectangle

$\Delta > 0$

FUNDAMENTAL RECTANGLE

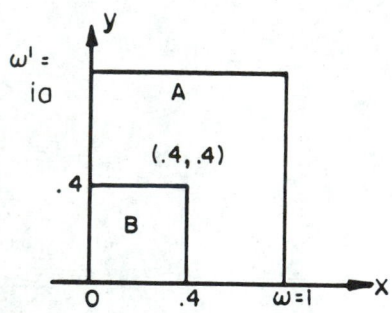

Fundamental Rectangle

$\Delta < 0$

FUNDAMENTAL RECTANGLE

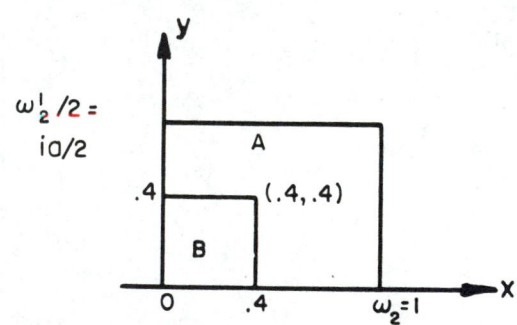

FIGURE 18.4

In region A

$\mathscr{R}(\wp') \geq 0$ if $y \geq .4$ and $x \leq .5$; $\mathscr{I}(\wp') \geq 0$ elsewhere

In region A

(1) If $a \geq 1.05$, use criterion for region A for $\Delta > 0$.

(2) If $1 \leq a < 1.05$: $\mathscr{R}(\wp') \geq 0$ if $y \geq .4$ and $x \leq .4$, $-\pi/4 < \arg(\wp') < 3\pi/4$ if $.4 < y \leq .5$ and $.4 < x \leq .5$. $\mathscr{I}(\wp') \geq 0$ elsewhere

In region B

The sign (indeed, perhaps one or more significant digits) of $\wp'$ is obtainable from the first term, $-2/z^3$, of the Laurent series for $\wp'$.

In region B

Use the criterion for region B for $\Delta>0$.

(Precisely similar criteria apply when the real half-period $\neq 1$)

$$\Delta>0 \qquad \omega=1$$

Map: $\wp(z)=u+iv$

Near zero: $\wp(z)=\dfrac{1}{z^2}+\epsilon_1$

$$\wp(z)=\dfrac{1}{z^2}+c_2z^2+\epsilon_2$$

$$\omega'=i$$

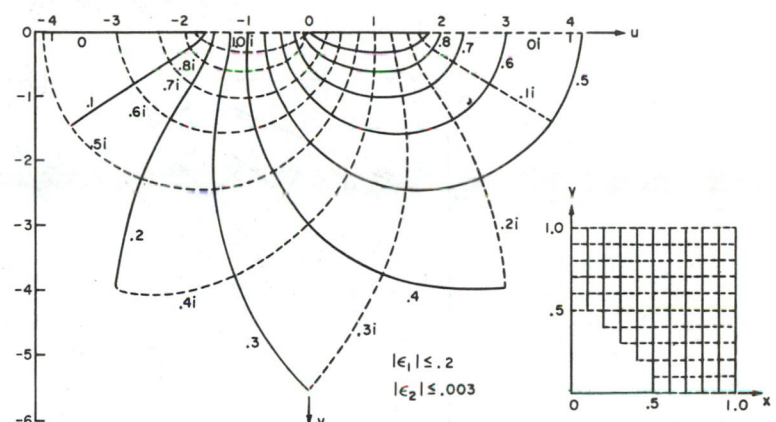

$$\omega'=1.4i$$

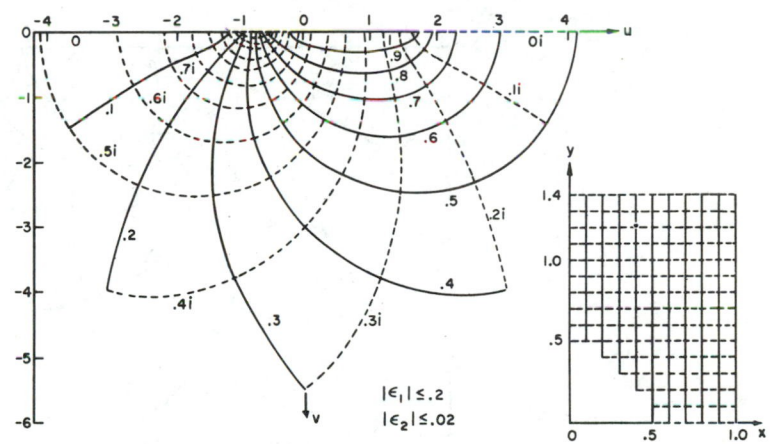

$$\omega'=2.0i$$

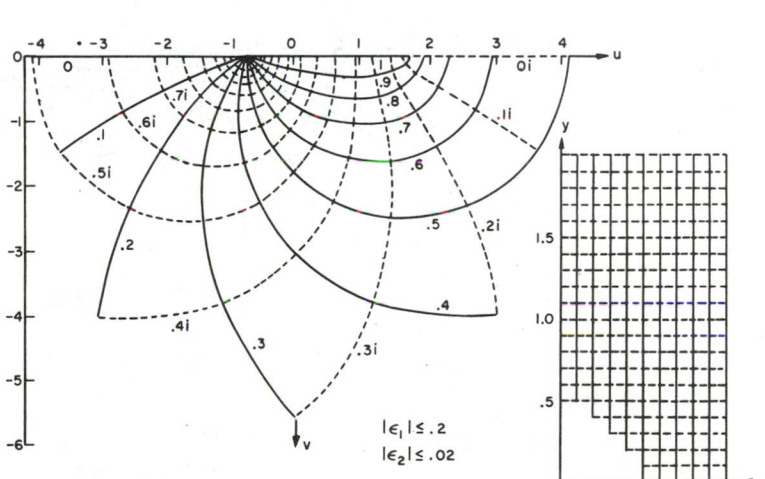

FIGURE 18.5

$$\Delta < 0 \qquad \omega_2 = 1$$

$$\text{Map: } \wp(z) = u + iv$$

$$\text{Near zero: } \wp(z) = \frac{1}{z^2} + \epsilon_1$$

$$\wp(z) = \frac{1}{z^2} + c_2 z^2 + \epsilon_2$$

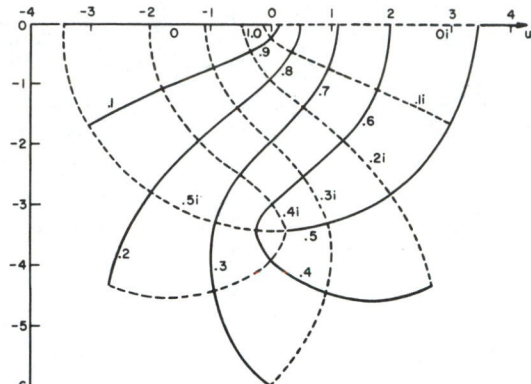
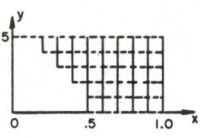

$\omega_2' = i$

$|\epsilon_1| \leq .7$
$|\epsilon_2| \leq .05$

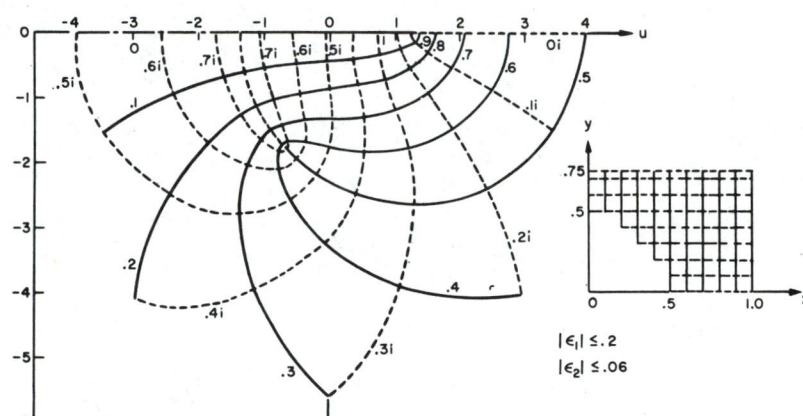

$\omega_2' = 1.5i$

$|\epsilon_1| \leq .2$
$|\epsilon_2| \leq .06$

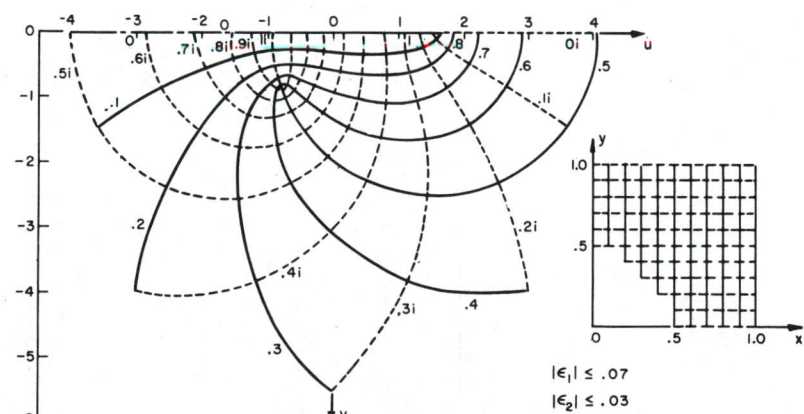

$\omega_2' = 2.0i$

$|\epsilon_1| \leq .07$
$|\epsilon_2| \leq .03$

FIGURE 18.6

$$\Delta > 0 \qquad \omega = 1$$

Map: $\zeta(z) = u + iv$

Near zero: $\zeta(z) = \dfrac{1}{z} + \epsilon_1$

$$\zeta(z) = \dfrac{1}{z} - \dfrac{c_2 z^3}{3} + \epsilon_2$$

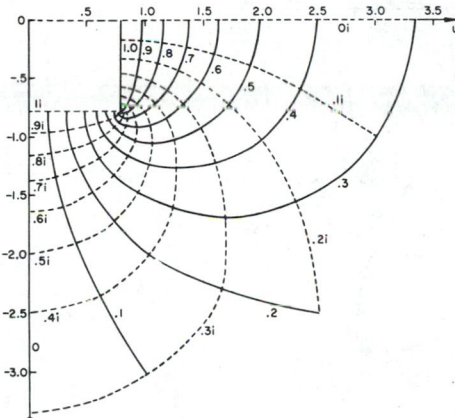

$\omega' = i$

$|\epsilon_1| \leq .01$
$|\epsilon_2| \leq 2 \times 10^{-5}$

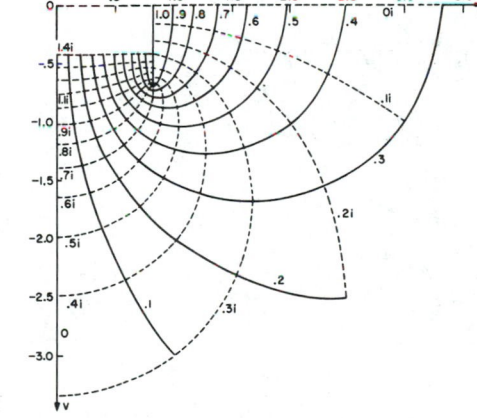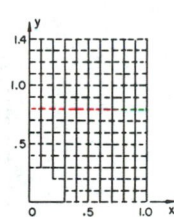

$\omega' = 1.4i$

$|\epsilon_1| \leq .007$
$|\epsilon_2| \leq .0002$

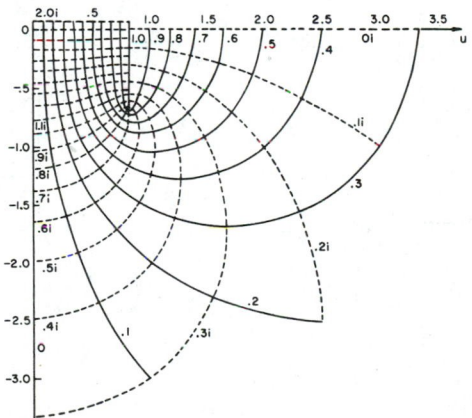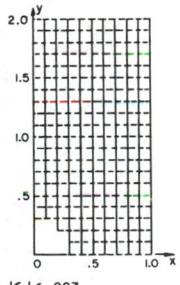

$\omega' = 2.0i$

$|\epsilon_1| \leq .007$
$|\epsilon_2| \leq .0002$

FIGURE 18.7

$$\Delta < 0 \qquad \omega_2 = 1$$

$$\text{Map: } \zeta(z) = u + iv$$

$$\text{Near zero: } \zeta(z) = \frac{1}{z} + \epsilon_1$$

$$\zeta(z) = \frac{1}{z} - \frac{c_2 z^3}{3} + \epsilon_2$$

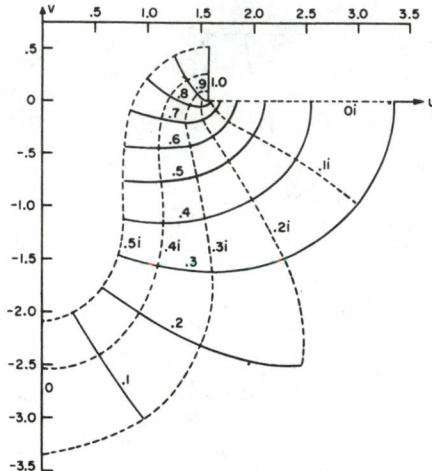

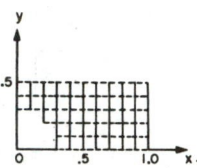

$$|\epsilon_1| \leq .04$$
$$|\epsilon_2| \leq .0002$$

$\omega_2' = i$

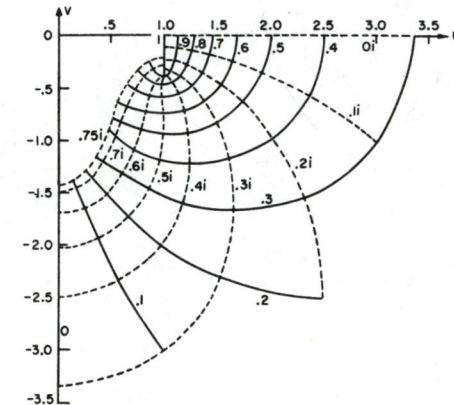

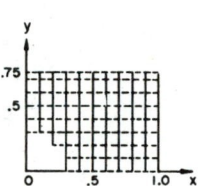

$$|\epsilon_1| \leq .007$$
$$|\epsilon_2| \leq .0009$$

$\omega_2' = 1.5i$

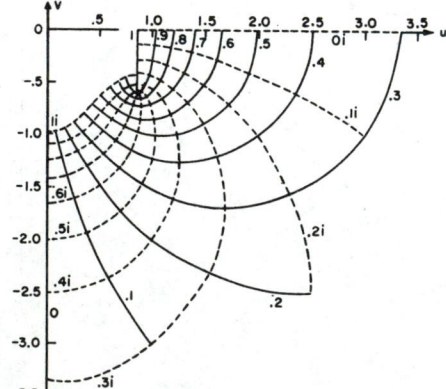

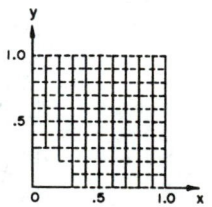

$$|\epsilon_1| \leq .004$$
$$|\epsilon_2| \leq .0004$$

$\omega_2' = 2.0i$

FIGURE 18.8

$$\Delta > 0 \qquad \omega = 1$$

$$\text{Map: } \sigma(z) = u + iv$$

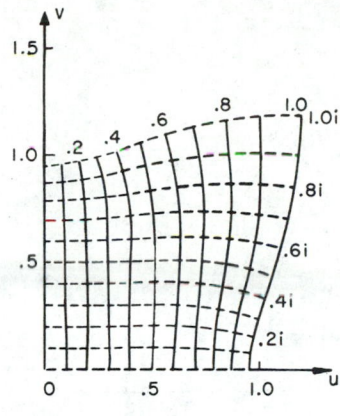
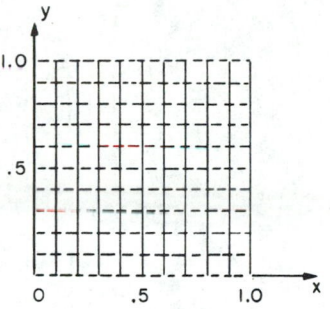

$\omega' = i$

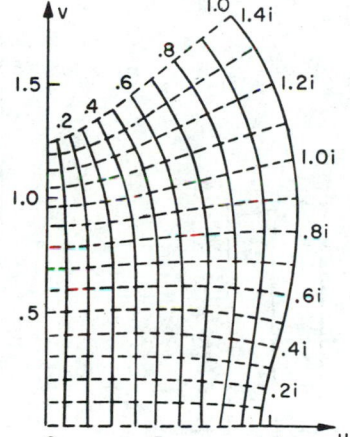
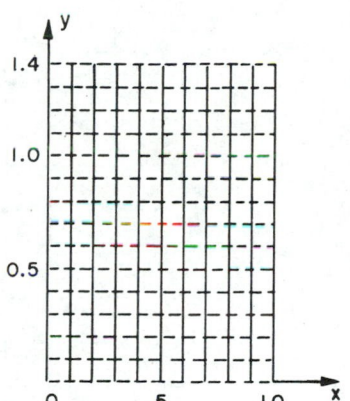

$\omega' = 1.4i$

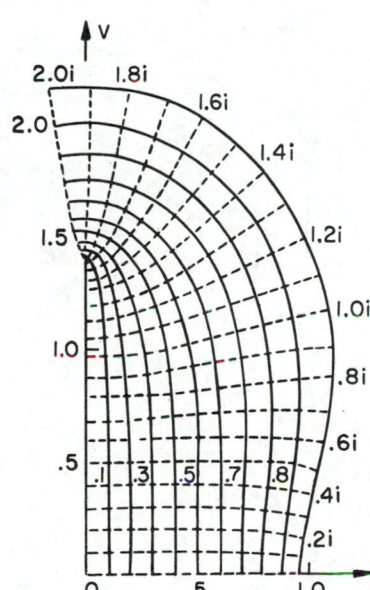
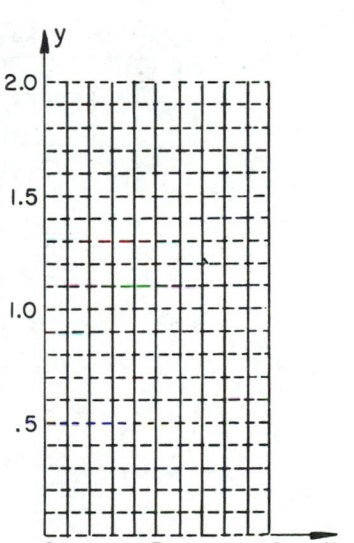

$\omega' = 2.0i$

FIGURE 18.9

$$\Delta < 0 \qquad \omega_2 = 1$$

$$\text{Map: } \sigma(z) = u + iv$$

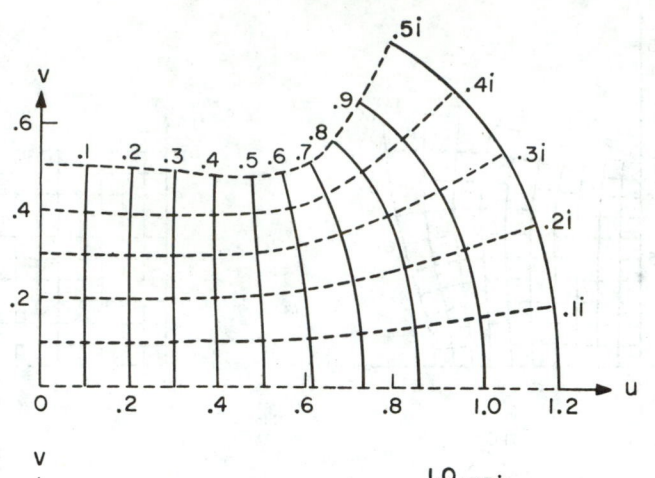
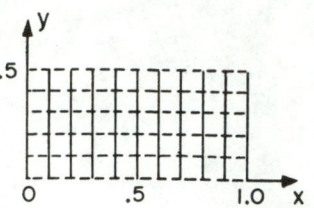

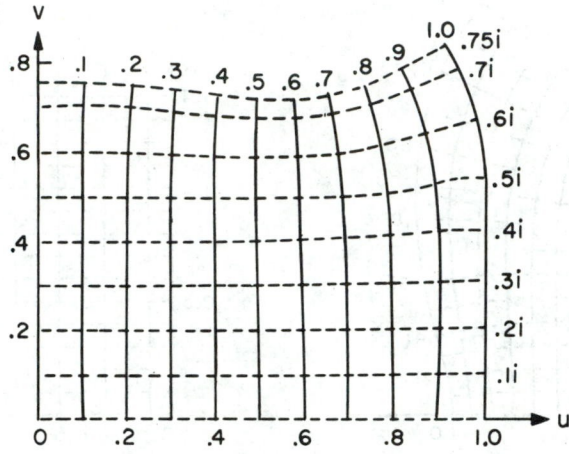
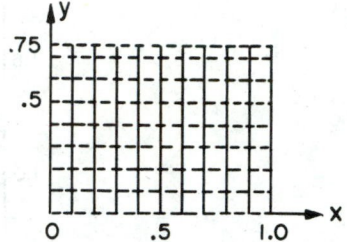
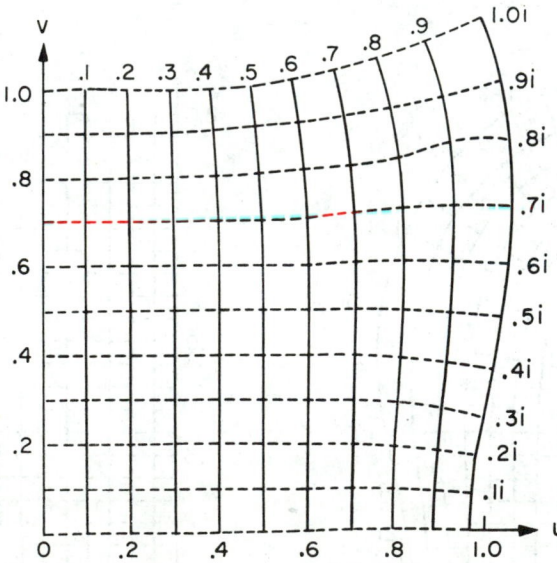
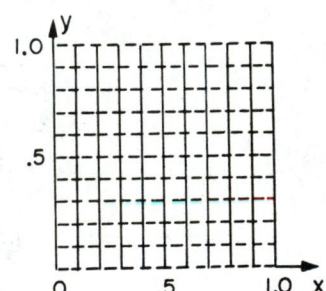

$$\omega_2' = i$$

$$\omega_2' = 1.5i$$

$$\omega_2' = 2.0i$$

FIGURE 18.10

18.9. Relations with Complete Elliptic Integrals K and K' and Their Parameter m

and with Jacobi's Elliptic Functions (see chapter 16)

(Here $K(m)$ and $K'(m) = K(1-m)$ are complete elliptic integrals of the 1st kind; see chapter 17.)

	$\Delta > 0$	$\Delta < 0$

18.9.1
$$e_1 = \frac{(2-m)K^2(m)}{3\omega^2} \qquad e_1 = \frac{(2m-1)+6i\sqrt{m-m^2}}{3\omega_2^2} \cdot K^2(m)$$

18.9.2
$$e_2 = \frac{(2m-1)K^2(m)}{3\omega^2} \qquad e_2 = \frac{2(1-2m)K^2(m)}{3\omega_2^2}$$

18.9.3
$$e_3 = \frac{-(m+1)K^2(m)}{3\omega^2} \qquad e_3 = \frac{(2m-1)-6i\sqrt{m-m^2}}{3\omega_2^2} \cdot K^2(m)$$

18.9.4
$$g_2 = \frac{4(m^2-m+1)K^4(m)}{3\omega^4} \qquad g_2 = \frac{4(16m^2-16m+1)K^4(m)}{3\omega_2^4}$$

18.9.5
$$g_3 = \frac{4(m-2)(2m-1)(m+1)K^6(m)}{27\omega^6} \qquad g_3 = \frac{8(2m-1)(32m^2-32m-1)K^6(m)}{27\omega_2^6}$$

18.9.6
$$\Delta = \frac{16m^2(m-1)^2 K^{12}(m)}{\omega^{12}} \qquad \Delta = \frac{-256(m-m^2)K^{12}(m)}{\omega_2^{12}}$$

18.9.7
$$\omega' = \frac{iK'(m)\omega}{K(m)} \qquad \omega_2' = \frac{iK'(m)\omega_2}{K(m)}$$

18.9.8
$$\omega = K(m)/(e_1-e_3)^{1/2} \qquad \omega_2 = K(m)/H_2^{1/2}$$

18.9.9
$$m = (e_2-e_3)/(e_1-e_3) \qquad m = \frac{1}{2} - \frac{3e_2}{4H_2}$$

18.9.10 $\qquad\qquad\qquad [0 < m \le \tfrac{1}{2},\ \text{since}\ g_3 \ge 0]$

18.9.11
$$\mathscr{P}(z) = e_3 + (e_1-e_3)/\mathrm{sn}^2(z^*|m) \qquad \mathscr{P}(z) = e_2 + H_2 \frac{1+\mathrm{cn}(z'|m)}{1-\mathrm{cn}(z'|m)}$$

18.9.12
$$\mathscr{P}'(z) = -2(e_1-e_3)^{3/2} \cdot \mathrm{cn}(z^*|m)\mathrm{dn}(z^*|m)/\mathrm{sn}^3(z^*|m)$$
where
$$z^* = (e_1-e_3)^{\frac{1}{2}} z$$

$$\mathscr{P}'(z) = \frac{-4H_2^{3/2}\mathrm{sn}(z'|m)\mathrm{dn}(z'|m)}{[1-\mathrm{cn}(z'|m)]^2}$$
where
$$z' = 2zH_2^{1/2}$$

18.9.13
$$\eta = \zeta(\omega) = \frac{K(m)}{3\omega}[3E(m)+(m-2)K(m)] \qquad \eta_2 = \zeta(\omega_2) = \frac{K(m)}{3\omega_2}[6E(m)+(4m-5)K(m)]$$

18.9.14
$$\eta' = \zeta(\omega') = \frac{\eta\omega'-\frac{1}{2}\pi i}{\omega} \qquad \eta_2' = \zeta(\omega_2') = \frac{\eta_2\omega_2'-\pi i}{\omega_2}$$

$[E(m)$ is a complete elliptic integral of the 2d kind (see chapter 17).]

18.10. Relations with Theta Functions (chapter 16)

The formal definitions of the four ϑ functions are given by the series **16.27.1–16.27.4** which converge for all complex z and all q defined below. (Some authors use πz, instead of z, as the independent variable.) These functions depend on z and on a parameter q, which is usually suppressed. Note that

$$\vartheta_1'(0)=\vartheta_2(0)\vartheta_3(0)\vartheta_4(0), \text{ where } \vartheta_i(0)=\vartheta_i(0, q).$$

	$\Delta>0$	$\Delta<0$

18.10.1 $\tau=\omega'/\omega$ $\tau_2=\omega_2'/2\omega_2$

18.10.2 $q=e^{i\pi\tau}=e^{-\pi K'/K}$ $q=iq_2=ie^{i\pi\tau_2}=ie^{-\pi|\omega_2'|/2\omega_2}$

18.10.3

q is real and since $g_3\geq0(|\omega'|\geq\omega)$, $0<q\leq e^{-\pi}$ q is pure imaginary and since $g_3\geq0(|\omega_2'|\geq\omega_2)$, $0<|q|\leq e^{-\pi/2}$

18.10.4 $(v=\pi z/2\omega)$ $(v=\pi z/2\omega_2)$

18.10.5 $\wp(z)=e_j+\dfrac{\pi^2}{4\omega^2}\left[\dfrac{\vartheta_1'(0)\vartheta_{j+1}(v)}{\vartheta_{j+1}(0)\vartheta_1(v)}\right]^2$ $\wp(z)=e_2+\dfrac{\pi^2}{4\omega_2^2}\left[\dfrac{\vartheta_1'(0)\vartheta_2(v)}{\vartheta_2(0)\vartheta_1(v)}\right]^2$

$$j=1, 2, 3$$

18.10.6 $\wp'(z)=-\dfrac{\pi^3}{4\omega^3}\dfrac{\vartheta_2(v)\vartheta_3(v)\vartheta_4(v)\vartheta_1'^3(0)}{\vartheta_2(0)\vartheta_3(0)\vartheta_4(0)\vartheta_1^3(v)}$ $\wp'(z)=-\dfrac{\pi^3}{4\omega_2^3}\dfrac{\vartheta_2(v)\vartheta_3(v)\vartheta_4(v)\vartheta_1'^3(0)}{\vartheta_2(0)\vartheta_3(0)\vartheta_4(0)\vartheta_1^3(v)}$

18.10.7 $\zeta(z)=\dfrac{\eta z}{\omega}+\dfrac{\pi\vartheta_1'(v)}{2\omega\vartheta_1(v)}$ $\zeta(z)=\dfrac{\eta_2 z}{\omega_2}+\dfrac{\pi\vartheta_1'(v)}{2\omega_2\vartheta_1(v)}$

18.10.8 $\sigma(z)=\dfrac{2\omega}{\pi}\exp\left(\dfrac{\eta z^2}{2\omega}\right)\dfrac{\vartheta_1(v)}{\vartheta_1'(0)}$ $\sigma(z)=\dfrac{2\omega_2}{\pi}\exp\left(\dfrac{\eta_2 z^2}{2\omega_2}\right)\dfrac{\vartheta_1(v)}{\vartheta_1'(0)}$

18.10.9 $12\omega^2 e_1=\pi^2[\vartheta_3^4(0)+\vartheta_4^4(0)]$ $12\omega_2^2 e_1=\pi^2[\vartheta_2^4(0)-\vartheta_4^4(0)]$

18.10.10 $12\omega^2 e_2=\pi^2[\vartheta_2^4(0)-\vartheta_4^4(0)]$ $12\omega_2^2 e_2=\pi^2[\vartheta_3^4(0)+\vartheta_4^4(0)]$

18.10.11 $12\omega^2 e_3=-\pi^2[\vartheta_2^4(0)+\vartheta_3^4(0)]$ $12\omega_2^2 e_3=-\pi^2[\vartheta_2^4(0)+\vartheta_3^4(0)]$

18.10.12 $(e_2-e_3)^{\frac{1}{2}}=-i(e_3-e_2)^{\frac{1}{2}}=\dfrac{\pi}{2\omega}\vartheta_2^2(0)$ $(e_2-e_3)^{\frac{1}{2}}=i(e_3-e_2)^{\frac{1}{2}}=\dfrac{\pi}{2\omega_2}\vartheta_3^2(0)$

18.10.13 $(e_1-e_3)^{\frac{1}{2}}=-i(e_3-e_1)^{\frac{1}{2}}=\dfrac{\pi}{2\omega}\vartheta_3^2(0)$ $(e_1-e_3)^{\frac{1}{2}}=i(e_3-e_1)^{\frac{1}{2}}=\dfrac{\pi}{2\omega_2}\vartheta_2^2(0)$

18.10.14 $(e_1-e_2)^{\frac{1}{2}}=-i(e_2-e_1)^{\frac{1}{2}}=\dfrac{\pi}{2\omega}\vartheta_4^2(0)$ $(e_2-e_1)^{\frac{1}{2}}=-i(e_1-e_2)^{\frac{1}{2}}=\dfrac{\pi}{2\omega_2}\vartheta_4^2(0)$

18.10.15 $g_2=\dfrac{2}{3}\left(\dfrac{\pi}{2\omega}\right)^4[\vartheta_2^8(0)+\vartheta_3^8(0)+\vartheta_4^8(0)]$ $g_2=\dfrac{2}{3}\left(\dfrac{\pi}{2\omega_2}\right)^4[\vartheta_2^8(0)+\vartheta_3^8(0)+\vartheta_4^8(0)]$

18.10.16 $g_3=4e_1e_2e_3$ $g_3=4e_1e_2e_3$

18.10.17 $\Delta^{\frac{1}{2}}=\dfrac{\pi^3}{4\omega^3}\vartheta_1'^2(0)$ $(-\Delta)^{\frac{1}{2}}=\dfrac{\pi^3}{4\omega_2^3}\vartheta_1'^2(0)\,e^{-i\pi/4}$

18.10.18 $\eta\equiv\zeta(\omega)=-\dfrac{\pi^2\vartheta_1'''(0)}{12\omega\vartheta_1'(0)}$ $\eta_2\equiv\zeta(\omega_2)=-\dfrac{\pi^2\vartheta_1'''(0)}{12\omega_2\vartheta_1'(0)}$

18.10.19 $\eta'\equiv\zeta(\omega')=\dfrac{\eta\omega'-\frac{1}{2}\pi i}{\omega}$ $\eta_2'\equiv\zeta(\omega_2')=\dfrac{\eta_2\omega_2'-\pi i}{\omega_2}$

Series

18.10.20
$$\vartheta_1(0) = 0$$

18.10.21
$$\vartheta_2(0) = 2q^{\frac{1}{4}}[1 + q^{1\cdot2} + q^{2\cdot3} + q^{3\cdot4} + \ldots + q^{n(n+1)} + \ldots]$$

18.10.22
$$\vartheta_3(0) = 1 + 2[q + q^4 + q^9 + \ldots + q^{n^2} + \ldots]$$

18.10.23
$$\vartheta_4(0) = 1 + 2[-q + q^4 - q^9 + \ldots + (-1)^n q^{n^2} + \ldots]$$

Attainable Accuracy

$\Delta > 0$	$\Delta < 0$
Note: $\vartheta_j(0) > 0$, $j = 2, 3, 4$	Note: $\vartheta_2(0) = Ae^{i\pi/8}$, $A > 0$; $\mathscr{R}\,\vartheta_3(0) > 0$; $\vartheta_4(0) = \overline{\vartheta_3(0)}$

$\vartheta_j(0)$:		
	2 terms give at least 5S	2 terms give at least 3S
$j = 2, 3, 4$	3 terms give at least 11S	3 terms give at least 5S
	4 terms give at least 21S	4 terms give at least 10S

18.11 Expressing any Elliptic Function in Terms of $\mathscr{P}$ and $\mathscr{P}'$

If $f(z)$ is any elliptic function and $\mathscr{P}(z)$ has same periods, write

18.11.1
$$f(z) = \tfrac{1}{2}[f(z) + f(-z)] + \tfrac{1}{2}[\{f(z) - f(-z)\}\{\mathscr{P}'(z)\}^{-1}]\mathscr{P}'(z).$$

Since both brackets represent even elliptic functions, we ask how to express an even elliptic function $g(z)$ (of order $2k$) in terms of $\mathscr{P}(z)$. Because of the evenness, an irreducible set of zeros can be denoted by a_i ($i = 1, 2, \ldots, k$) and the set of points congruent to $-a_i$ ($i = 1, 2, \ldots, k$); correspondingly in connection with the poles we consider the points $\pm b_i$, $i = 1, 2, \ldots, k$. Then

18.11.2
$$g(z) = A \prod_{i=1}^{k}\left\{\frac{\mathscr{P}(z) - \mathscr{P}(a_i)}{\mathscr{P}(z) - \mathscr{P}(b_i)}\right\}, \quad \text{where } A \text{ is}$$

a constant. If any a_i or b_i is congruent to the origin, the corresponding factor is omitted from the product. Factors corresponding to multiple poles (zeros) are repeated according to the multiplicity.

18.12. Case $\Delta = 0\,(c > 0)$

Subcase I

18.12.1 $g_2 > 0$, $g_3 < 0$: $(e_1 = e_2 = c,\ e_3 = -2c)$

18.12.2 $H_1 = H_2 = 0$, $H_3 = 3c$

18.12.3
$$\mathscr{P}(z; 12c^2, -8c^3) = c + 3c\{\sinh[(3c)^{\frac{1}{2}}z]\}^{-2}$$

18.12.4
$$\zeta(z; 12c^2, -8c^3) = -cz + (3c)^{\frac{1}{2}}\coth[(3c)^{\frac{1}{2}}z]$$

18.12.5
$$\sigma(z; 12c^2, -8c^3) = (3c)^{-\frac{1}{2}}\sinh[(3c)^{\frac{1}{2}}z]e^{-cz^2/2}$$

18.12.6 $\omega = \infty$, $\omega' = (12c)^{-\frac{1}{2}}\pi i$

18.12.7 $\eta = \zeta(\omega) = -\infty$

18.12.8 $\eta' = \zeta(\omega') = -c\omega'$

18.12.9 $q = 1$, $m = 1$

18.12.10 $\sigma(\omega) = 0$

18.12.11
$$\sigma(\omega') = \frac{2\omega' e^{\pi^2/24}}{\pi}$$

18.12.12 $\sigma(\omega_2) = 0$

18.12.13 $\mathscr{P}(\omega/2) = c$

18.12.14 $\mathscr{P}'(\omega/2) = 0$

18.12.15 $\zeta(\omega/2) = -\infty$

18.12.16 $\sigma(\omega/2) = 0$

18.12.17 $\mathscr{P}(\omega'/2) = -5c$

18.12.18
$$\mathscr{P}'(\omega'/2) = \frac{-\pi^3}{2\omega'^3}$$

18.12.19 $\zeta(\omega'/2) = \tfrac{1}{2}(-c\omega' + \pi/\omega')$

18.12.20
$$\sigma(\omega'/2) = \frac{\omega'\, e^{\pi^2/96}\sqrt{2}}{\pi}$$

18.12.21
$$\mathcal{P}(\omega_2/2) = c$$

18.12.22
$$\mathcal{P}'(\omega_2/2) = 0$$

18.12.23
$$\zeta(\omega_2/2) = -\infty - \frac{c\omega'}{2}$$

18.12.24
$$\sigma(\omega_2/2) = 0$$

Subcase II

18.12.25
$$g_2 > 0,\ g_3 > 0:\ (e_1 = 2c,\ e_2 = e_3 = -c)$$

18.12.26
$$H_1 = 3c,\ H_2 = H_3 = 0$$

18.12.27
$$\mathcal{P}(z; 12c^2, 8c^3) = -c + 3c\{\sin[(3c)^{\frac{1}{2}}z]\}^{-2}$$

18.12.28
$$\zeta(z; 12c^2, 8c^3) = cz + (3c)^{\frac{1}{2}}\cot[(3c)^{\frac{1}{2}}z]$$

18.12.29
$$\sigma(z; 12c^2, 8c^3) = (3c)^{-\frac{1}{2}}\sin[(3c)^{\frac{1}{2}}z]e^{cz^2/2}$$

18.12.30
$$\omega = (12c)^{-\frac{1}{2}}\pi,\ \omega' = i\infty$$

18.12.31
$$\eta = \zeta(\omega) = c\omega$$

18.12.32
$$\eta' = \zeta(\omega') = i\infty$$

18.12.33
$$q = 0,\qquad m = 0$$

18.12.34
$$\sigma(\omega) = \frac{2\omega e^{\pi^2/24}}{\pi}$$

18.12.35
$$\sigma(\omega') = 0$$

18.12.36
$$\sigma(\omega_2) = 0$$

18.12.37
$$\mathcal{P}(\omega/2) = 5c$$

18.12.38
$$\mathcal{P}'(\omega/2) = \frac{-\pi^3}{2\omega^3}$$

18.12.39
$$\zeta(\omega/2) = \tfrac{1}{2}(c\omega + \pi/\omega)$$

18.12.40
$$\sigma(\omega/2) = \frac{e^{\pi^2/96}\omega\sqrt{2}}{\pi}$$

18.12.41
$$\mathcal{P}(\omega'/2) = -c$$

18.12.42
$$\mathcal{P}'(\omega'/2) = 0$$

18.12.43
$$\zeta(\omega'/2) = +i\infty$$

18.12.44
$$\sigma(\omega'/2) = 0$$

18.12.45
$$\mathcal{P}(\omega_2/2) = -c$$

18.12.46
$$\mathcal{P}'(\omega_2/2) = 0$$

18.12.47
$$\zeta(\omega_2/2) = \frac{c\omega}{2} + i\infty$$

18.12.48
$$\sigma(\omega_2/2) = 0$$

Subcase III

18.12.49
$$g_2 = 0,\ g_3 = 0\ (e_1 = e_2 = e_3 = 0)$$

18.12.50
$$\mathcal{P}(z; 0, 0) = z^{-2}$$

18.12.51
$$\zeta(z; 0, 0) = z^{-1}$$

18.12.52
$$\sigma(z; 0, 0) = z$$

18.12.53
$$\omega = -i\omega' = \infty$$

18.13. Equianharmonic Case ($g_2 = 0$, $g_3 = 1$)

If $g_2 = 0$ and $g_3 > 0$, homogeneity relations allow us to reduce our considerations of $\mathcal{P}$ to $\mathcal{P}(z; 0, 1)$ ($\mathcal{P}'$, ζ and σ are handled similarly). Thus $\mathcal{P}(z; 0, g_3) = g_3^{1/3}\mathcal{P}(zg_3^{1/6}; 0, 1)$. The case $g_2 = 0$, $g_3 = 1$ is called the EQUIANHARMONIC case.

$\frac{1}{4}$ FPP; *Reduction to Fundamental Triangle*

$\Delta_1 \equiv \Delta 0\omega_2 z_0$ is the Fundamental Triangle

Let ϵ denote $e^{i\pi/3}$ throughout **18.13**.

$$\omega_2 \approx 1.5299\ 54037\ 05719\ 28749\ 13194\ 17231^6$$

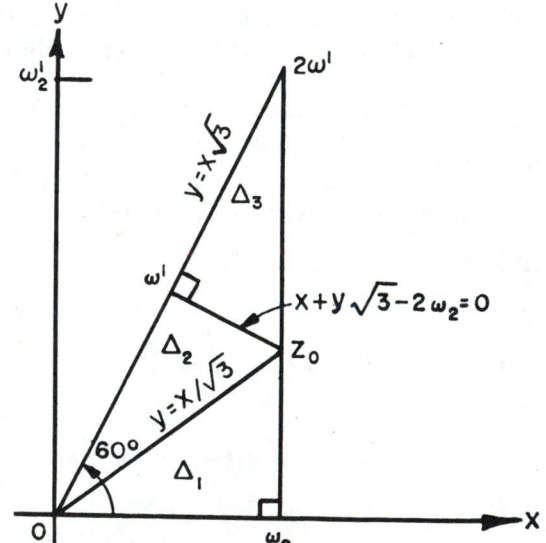

FIGURE 18.11

[6] This value was computed and checked by multiple precision on a desk calculator and is believed correct to 30S.

Reduction for z_2 in Δ_2: $z_1 = \epsilon \bar{z}_2$ is in Δ_1.

18.13.1 $\qquad \wp(z_2) = \epsilon^{-2} \overline{\wp}(z_1)$

18.13.2 $\qquad \wp'(z_2) = -\overline{\wp}'(z_1)$

18.13.3 $\qquad \zeta(z_2) = \epsilon^{-1} \bar{\zeta}(z_1)$

18.13.4 $\qquad \sigma(z_2) = \epsilon \bar{\sigma}(z_1)$

Reduction for z_3 in Δ_3: $z_1 = \epsilon^{-1}(2\omega' - z_3)$ is in Δ_1

18.13.5 $\qquad \wp(z_3) = \epsilon^{-2} \wp(z_1)$

18.13.6 $\qquad \wp'(z_3) = \wp'(z_1)$

18.13.7 $\quad \zeta(z_3) = -\epsilon^{-1}\zeta(z_1) + 2\eta', \qquad \eta' = \zeta(\omega')$

18.13.8 $\quad \sigma(z_3) = \epsilon \sigma(z_1) \exp\left[(z_3 - \omega')(2\eta')\right]$

Special Values and Formulas

18.13.9 $\quad \Delta = -27, \qquad H_1 = \sqrt{3}(4^{-1/3})\bar{\epsilon},$
$$H_2 = \sqrt{3}(4^{-1/3}), \qquad H_3 = \sqrt{3}(4^{-1/3})\epsilon$$

18.13.10 $\quad m = \sin^2 15° = \dfrac{2-\sqrt{3}}{4}, \qquad q = ie^{-\pi\sqrt{3}/2}$

18.13.11 $\qquad \vartheta_2(0) = Ae^{i\pi/8}$

18.13.12 $\qquad \vartheta_3(0) = Ae^{i\pi/24}$

18.13.13 $\qquad \vartheta_4(0) = Ae^{-i\pi/24}$

18.13.14

$\qquad$ where $A = (\omega_2/\pi)^{1/2} 2^{1/3} 3^{1/8} \approx 1.0086\ 67$

18.13.15 $\qquad \omega_2 = \dfrac{K(m)2^{1/3}}{3^{1/4}} = \dfrac{\Gamma^3(1/3)}{4\pi}$

Values at Half-periods

	$\wp$	$\wp'$	ζ	σ
18.13.16 $\omega \equiv \omega_1$	$e_1 = 4^{-1/3}\epsilon^2$	0	$\eta = \epsilon\pi/2\omega_2\sqrt{3}$	$\epsilon^{-1}\sigma(\omega_2)$
18.13.17 ω_2	$e_2 = 4^{-1/3}$	0	$\eta_2 = \eta + \eta' = \pi/2\omega_2\sqrt{3}$	$\dfrac{e^{\pi/4\sqrt{3}}(2^{1/3})}{3^{\frac{1}{4}}}$
18.13.18 $\omega' \equiv \omega_3$	$e_3 = 4^{-1/3}\epsilon^{-2}$	0	$\eta' = \epsilon^{-1}\pi/2\omega_2\sqrt{3}$	$\epsilon\sigma(w_2)$
18.13.19 ω_2'	$e_2 = 4^{-1/3}$	0	$\eta_2' = -\pi i/2\omega_2 = \eta' - \eta$	$\dfrac{ie^{3\pi/4\sqrt{3}}(2^{1/3})}{3^{\frac{1}{4}}}$

Values [7] along $(0, \omega_2)$

	$\wp$	$\wp'$	ζ	σ
18.13.20 $2\omega_2/9$	$\dfrac{\sqrt[3]{\cos 80°}}{\sqrt[3]{\cos 20°} - \sqrt[3]{\cos 40°}}$	$\dfrac{-\sqrt{3}[\sqrt[3]{\cos 20°} + \sqrt[3]{\cos 40°}]}{\sqrt[3]{\cos 20°} - \sqrt[3]{\cos 40°}}$		
18.13.21 $\omega_2/3$	$1/(2^{1/3}-1)$	$-\sqrt{3}(2^{1/3}+1)/(2^{1/3}-1)$	$\dfrac{\eta_2}{3} + \dfrac{\sqrt{3}(2^{2/3}+2+2^{4/3})}{6}$	$\dfrac{e^{\pi/36\sqrt{3}}}{3^{1/6}}\sqrt[4]{\dfrac{2^{1/3}-1}{2^{1/3}+1}}$
18.13.22 $4\omega_2/9$	$\dfrac{\sqrt[3]{\cos 40°}}{\sqrt[3]{\cos 20°} - \sqrt[3]{\cos 80°}}$	$\dfrac{-\sqrt{3}[\sqrt[3]{\cos 20°} + \sqrt[3]{\cos 80°}]}{\sqrt[3]{\cos 20°} - \sqrt[3]{\cos 80°}}$		
18.13.23 $\omega_2/2$	$e_2 + H_2$	$-3^{3/4}\sqrt{2+\sqrt{3}}$	$(\pi/4\omega_2\sqrt{3}) + (3^{1/4}\sqrt{2+\sqrt{3}}/2^{4/3})$	$\dfrac{e^{\pi/16\sqrt{3}}(2^{1/12})}{3^{1/4}\sqrt[8]{2+\sqrt{3}}}$
18.13.24 $2\omega_2/3$	1	$-\sqrt{3}$	$\tfrac{2}{3}(\eta_2) + 3^{-1/2}$	$e^{\pi/9\sqrt{3}}/3^{1/6}$
18.13.25 $8\omega_2/9$	$\dfrac{\sqrt[3]{\cos 20°}}{\sqrt[3]{\cos 40°} + \sqrt[3]{\cos 80°}}$	$\dfrac{-\sqrt{3}[\sqrt[3]{\cos 40°} - \sqrt[3]{\cos 80°}]}{\sqrt[3]{\cos 40°} + \sqrt[3]{\cos 80°}}$		

[7] Values at $2\omega_2/9$, $4\omega_2/9$ and $8\omega_2/9$ from [18.14].

Values along $(0, z_0)$

	$\mathscr{P}$	$\mathscr{P}'$	ζ	σ
18.13.26 $z_0/2$	$-2^{1/3}\epsilon^2$	$3i$	$\left[\dfrac{\eta_2}{\sqrt{3}}+2^{-1/3}\right]e^{-i\pi/6}$	$\dfrac{e^{\pi/12\sqrt{3}}e^{i\pi/6}}{3^{1/4}}$
18.13.27 $3z_0/4$	$\epsilon^2(e_2-H_2)$	$i(3^{3/4})\sqrt{2-\sqrt{3}}$	$\left[\dfrac{\pi}{4\omega_2}+\dfrac{3^{1/4}\sqrt{2-\sqrt{3}}}{2^{4/3}}\right]e^{-i\pi/6}$	$\dfrac{e^{3\pi/16\sqrt{3}}(2^{1/12})e^{i\pi/6}}{3^{1/4}\sqrt[8]{2-\sqrt{3}}}$
18.13.28 z_0	0	i	$\dfrac{2\eta_2}{\sqrt{3}}e^{-i\pi/6}$	$e^{\pi/3\sqrt{3}}\cdot e^{i\pi/6}$

Duplication Formulas

18.13.29 $\quad \mathscr{P}(2z)=\dfrac{\mathscr{P}(z)[\mathscr{P}^3(z)+2]}{4\mathscr{P}^3(z)-1}$

18.13.30 $\quad \mathscr{P}'(2z)=\dfrac{2\mathscr{P}^6(z)-10\mathscr{P}^3(z)-1}{[\mathscr{P}'(z)]^3}$

18.13.31 $\quad \zeta(2z)=2\zeta(z)+\dfrac{3\mathscr{P}^2(z)}{\mathscr{P}'(z)}$

18.13.32 $\quad \sigma(2z)=-\mathscr{P}'(z)\sigma^4(z)$

Trisection Formulas (x real)

18.13.33 $\quad \mathscr{P}\left(\dfrac{x}{3}\right)=\dfrac{\sqrt[3]{\cos\dfrac{\phi-\pi}{3}}}{\sqrt[3]{\cos\dfrac{\phi}{3}}-\sqrt[3]{\cos\dfrac{\phi+\pi}{3}}}$

18.13.34 $\quad \mathscr{P}'\left(\dfrac{x}{3}\right)=-\sqrt{3}\,\dfrac{\sqrt[3]{\cos\dfrac{\phi}{3}}+\sqrt[3]{\cos\dfrac{\phi+\pi}{3}}}{\sqrt[3]{\cos\dfrac{\phi}{3}}-\sqrt[3]{\cos\dfrac{\phi+\pi}{3}}}$

where $\tan\ \phi=\mathscr{P}'(x)$, $0<x<2\omega_2$ and we must choose ϕ in intervals

$$\left(-\frac{\pi}{2},\frac{\pi}{2}\right),\ \left(\frac{\pi}{2},\frac{3\pi}{2}\right),\ \left(\frac{3\pi}{2},\frac{5\pi}{2}\right) \text{ to get}$$

$$\mathscr{P}\left(\frac{x}{3}\right),\ \mathscr{P}\left(\frac{x}{3}+\frac{2\omega_2}{3}\right),\ \mathscr{P}\left(\frac{x}{3}+\frac{4\omega_2}{3}\right),\ \text{respectively.}$$

Complex Multiplication

18.13.35 $\quad \mathscr{P}(\epsilon z)=\epsilon^{-2}\mathscr{P}(z)$

18.13.36 $\quad \mathscr{P}'(\epsilon z)=-\mathscr{P}'(z)$

18.13.37 $\quad \zeta(\epsilon z)=\epsilon^{-1}\zeta(z)$

18.13.38 $\quad \sigma(\epsilon z)=\epsilon\sigma(z)$

In the above, ϵ denotes (as it does throughout section **18.13**), $e^{i\pi/3}$. The above equations are useful as follows, e.g.:

If z is real, ϵz is on $0\omega'$ (**Figure 18.11**); if ϵz were purely imaginary, z would be on $0z_0$ (**Figure 18.11**).

Conformal Maps

Equianharmonic Case

Map: $f(z)=u+iv$

$$\mathscr{P}(z)$$

Near zero: $\mathscr{P}(z)=\dfrac{1}{z^2}+\epsilon_1$

$$\mathscr{P}(z)=\dfrac{1}{z^2}+\dfrac{z^4}{28}+\epsilon_2$$

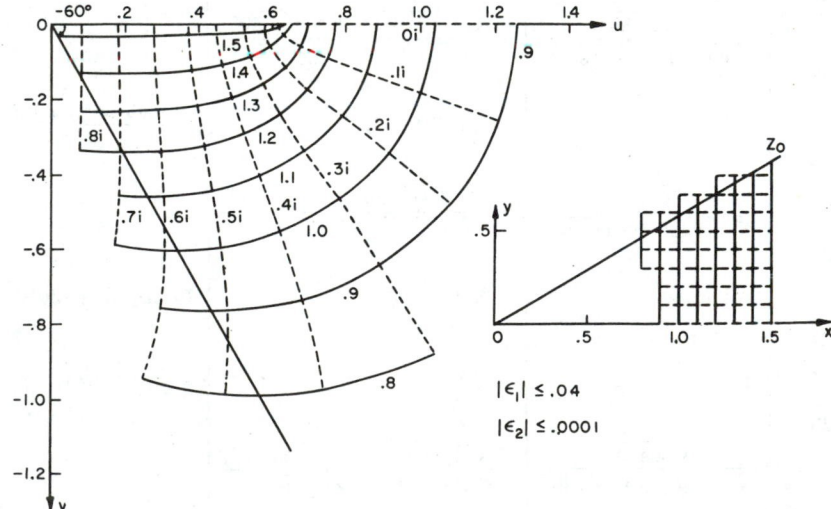

$|\epsilon_1|\leq .04$

$|\epsilon_2|\leq .0001$

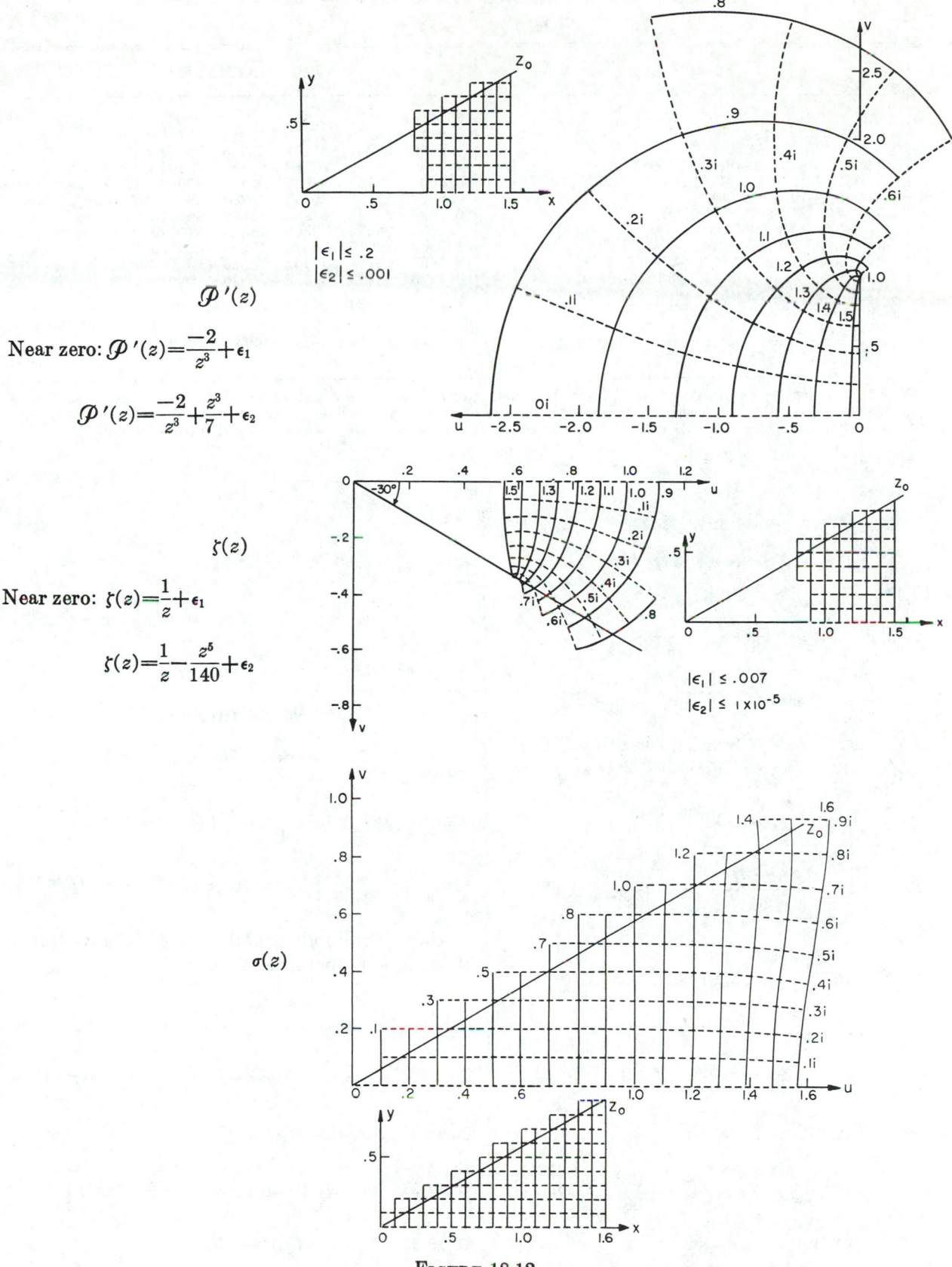

$\mathscr{P}'(z)$

Near zero: $\mathscr{P}'(z) = \dfrac{-2}{z^3} + \epsilon_1$

$$\mathscr{P}'(z) = \dfrac{-2}{z^3} + \dfrac{z^3}{7} + \epsilon_2$$

$|\epsilon_1| \leq .2$
$|\epsilon_2| \leq .001$

$\zeta(z)$

Near zero: $\zeta(z) = \dfrac{1}{z} + \epsilon_1$

$$\zeta(z) = \dfrac{1}{z} - \dfrac{z^5}{140} + \epsilon_2$$

$|\epsilon_1| \leq .007$
$|\epsilon_2| \leq 1 \times 10^{-5}$

$\sigma(z)$

FIGURE 18.12

Coefficients for Laurent Series for $\wp$, $\wp'$ and ζ

$$(c_m = 0 \text{ for } m \neq 3k)$$

k	EXACT c_{3k}	APPROXIMATE c_{3k}
1	$1/28$	3. 5714 28571 42857 . . . $\times 10^{-2}$
2	$1/(13\cdot 28^2) = 1/10192$	9. 8116 16954 47409 73312 40188 $\times 10^{-5}$
3	$1/(13\cdot 19\cdot 28^3) = 1/5422144$	1. 8442 88901 21693 55885 78983 $\times 10^{-7}$
4	$3/(5\cdot 13^2\cdot 19\cdot 28^4) = 234375/(7709611 \times 10^8)$	3. 0400 36650 35758 61350 20301 $\times 10^{-10}$
5	$4/(5\cdot 13^2\cdot 19\cdot 31\cdot 28^5) = 78125/(16729\ 85587 \times 10^8)$	4. 6697 95161 83961 00384 33643 $\times 10^{-13}$
6	$(7\cdot 43)/(13^3\cdot 19^2\cdot 31\cdot 37\cdot 28^6)$	6. 8662 18676 79393 36788 98 $\times 10^{-16}$
7	$(6\cdot 431)/(5\cdot 13^3\cdot 19^2\cdot 31\cdot 37\cdot 43\cdot 28^7)$	9. 7990 31742 57961 41839 66 $\times 10^{-19}$
8	$(3\cdot 7\cdot 313)/(5^2\cdot 13^4\cdot 19^2\cdot 31\cdot 37\cdot 43\cdot 28^8)$	1. 3685 06574 79360 13026 87 $\times 10^{-21}$
9	$(4\cdot 1201)/(5^2\cdot 13^4\cdot 19^3\cdot 31\cdot 37\cdot 43\cdot 28^9)$	1. 8800 72610 01329 79236 40 $\times 10^{-24}$
10	$(2^2\cdot 3\cdot 41\cdot 1823)/(5\cdot 13^5\cdot 19^3\cdot 31^2\cdot 37\cdot 43\cdot 61\cdot 28^{10})$	2. 5497 66946 68202 63683 $\times 10^{-27}$
11	$(3\cdot 79\cdot 733)/(5\cdot 13^4\cdot 19^3\cdot 31^2\cdot 37\cdot 43\cdot 61\cdot 67\cdot 28^{11})$	3. 4222 48599 51463 05316 $\times 10^{-30}$
12	$\dfrac{3\cdot 1153\cdot 13963\cdot 29059}{5^3\cdot 13^6\cdot 19^4\cdot 31^2\cdot 37^2\cdot 43\cdot 61\cdot 67\cdot 73\cdot 28^{12}}$	4. 5541 38864 99184 30391 $\times 10^{-33}$
13	$\dfrac{2^2\cdot 3^2\cdot 7\cdot 11\cdot 2647111}{5^3\cdot 13^5\cdot 19^4\cdot 31^2\cdot 37^2\cdot 61\cdot 67\cdot 73\cdot 79\cdot 28^{13}}$	6. 0171 15776 98241 99591 $\times 10^{-36}$

First 5 approximate values determined from exact values of c_{3k}; subsequent values determined by using exact ratios c_{3k}/c_{3k-3}, using at least double precision arithmetic with a desk calculator. All approximate c's were checked with the use of the recursion relation; $c_3 - c_{27}$ are believed correct to at least 21S; $c_{30} - c_{39}$ are believed correct to 20S.

$$c_{3k} \leq \frac{c_3}{13^{k-1}\cdot 28^{k-1}}, \quad k = 2, 3, 4, \ldots$$

Other Series Involving $\wp$

Reversed Series for Large $|\wp|$

18.13.39

$$z = (\wp^{-1})^{1/2}\left[1 + \frac{u}{7} + \frac{3u^2}{26} + \frac{5u^3}{38} + \frac{7u^4}{40} + \frac{63u^5}{248}\right.$$
$$\left. + \frac{231u^6}{592} + \frac{429u^7}{688} + O(u^8)\right],$$

18.13.40 where $u = \wp^{-3}/8$ and z is in the Fundamental Triangle (**Figure 18.11**) if $\wp$ has an appropriate value.

Series near z_0

18.13.41

$$\wp = iu\left[1 - \frac{u^6}{7} + \frac{3u^{12}}{364}\right] + u^4\left[-\frac{1}{2} + \frac{u^6}{28}\right] + O(u^{16})$$

18.13.42

$$u = -i\wp\left[1 + \frac{\wp^3}{2} + \frac{6\wp^6}{7} + 2\wp^9 + \frac{70\wp^{12}}{13} + O(\wp^{15})\right],$$

18.13.43 where $u = (z - z_0)$

Series near ω_2

18.13.44

$$(\wp - e_2) = 3e_2^2 u\left[1 + x + x^2 + \frac{6}{7}x^3\right.$$
$$\left. + \frac{5}{7}x^4 + \frac{4}{7}x^5 + \frac{285}{637}x^6 + O(x^7)\right],$$

18.13.45 where $u = (z - \omega_2)^2$, $x = e_2 u$

18.13.46

$$u = e_2^{-1}\left[w - w^2 + w^3 - \frac{6}{7}w^4 + \frac{3}{7}w^5\right.$$
$$\left. + \frac{3}{7}w^6 - \frac{1143}{637}w^7 + O(w^8)\right],$$

18.13.47 where $w = (\wp - e_2)/3e_2$

Other Series Involving $\wp'$

Reversed Series for Large $|\wp'|$

18.13.48

$$z = 2^{1/3}(\wp'^{1/3})^{-1}e^{i\pi/3}\left[1 - \frac{2}{21}(\wp')^{-2}\right.$$
$$\left. + \frac{5}{117}(\wp')^{-4} + O(\wp'^{-6})\right],$$

z being in the Fundamental Triangle (**Figure 18.11**) if $\wp'$ has an appropriate value.

Series near z_0

18.13.49

$$(\wp' - i) = x\left[-2 - ix + \frac{5}{14}x^2 + \frac{3i}{28}x^3 + O(x^4)\right]$$

18.13.50 where $x = (z - z_0)^3$

18.13.51 $x = 2\alpha\left[1 - i\alpha - \frac{9}{7}\alpha^2 + \frac{13i\alpha^3}{7} + O(\alpha^4)\right],$

18.13.52 where $\alpha = (\wp' - i)/(-4)$

Series near ω_2

18.13.53

$$\mathscr{P}'=6e_2^2(z-\omega_2)\left[1+2v+3v^2+\frac{24}{7}v^3\right.$$

$$\left.+\frac{25}{7}v^4+\frac{24}{7}v^5+\frac{285}{91}v^6+O(v^7)\right],$$

18.13.54 where $v=e_2(z-\omega_2)^2$

18.13.55

$$(z-\omega_2)=(\mathscr{P}'/6e_2^2)\left[1-2w+9w^2-\frac{360}{7}w^3\right.$$

$$\left.+330w^4-2268w^5+\frac{212058}{13}w^6+O(w^7)\right],$$

18.13.56 where $w=\mathscr{P}'^2/9$

Other Series Involving ζ

Reversed Series for Large $|\zeta|$

18.13.57

$$z=\zeta^{-1}\left[1-\frac{\gamma}{7}+\frac{17\gamma^2}{143}-\frac{496\gamma^3}{3553}+O(\gamma^4)\right],$$

18.13.58

$$\gamma=\zeta^{-6}/20$$

Series near z_0

18.13.59

$$(\zeta-\zeta_0)=i\left[-\frac{u^2}{2}+\frac{u^8}{56}-\frac{3u^{14}}{5096}\right]+\left[\frac{u^5}{8}-\frac{u^{11}}{308}\right]+O(u^{17}),$$

18.13.60 where $u=(z-z_0)$

Series near ω_2

18.13.61

$$(\zeta-\eta_2)=-e_2(z-\omega_2)\left[1+v+\frac{3}{5}v^2+\frac{3}{7}v^3+\frac{2}{7}v^4\right.$$

$$\left.+\frac{15}{77}v^5+\frac{12}{91}v^6+\frac{57}{637}v^7+O(v^8)\right],$$

18.13.62

$$v=e_2(z-\omega_2)^2$$

18.13.63

$$(z-\omega_2)=\frac{(\zeta-\eta_2)}{-e_2}\left[1-w+\frac{12w^2}{5}-\frac{267w^3}{35}+\frac{139w^4}{5}\right.$$

$$\left.-\frac{30192w^5}{275}+\frac{1634208}{3575}w^6+O(w^7)\right],$$

18.13.64

$$w=(\zeta-\eta_2)^2/e_2$$

Series Involving σ

18.13.65

$$\sigma=z-\frac{2\cdot3}{7!}z^7-\frac{2^3\cdot3^3}{13!}z^{13}+\frac{2^6\cdot3^4\cdot23}{19!}z^{19}$$

$$+\frac{2^7\cdot3^5\cdot5^2\cdot31}{25!}z^{25}+\frac{2^8\cdot3^8\cdot5\cdot9103}{31!}z^{31}$$

$$-\frac{2^{12}\cdot3^9\cdot5\cdot229\cdot2683}{37!}z^{37}$$

$$-\frac{2^{14}\cdot3^{10}\cdot5\cdot23\cdot257\cdot18049}{43!}z^{43}$$

$$-\frac{2^{15}\cdot3^{12}\cdot5\cdot59\cdot107895773}{49!}z^{49}+O(z^{55})$$

18.13.66

$$z=\sigma+\frac{\sigma^7}{2^3\cdot3\cdot5\cdot7}+\frac{41\sigma^{13}}{2^7\cdot3^2\cdot5^2\cdot11\cdot13}+\frac{13\cdot337\sigma^{19}}{2^{10}\cdot3^4\cdot5^3\cdot11\cdot17\cdot19}$$

$$+\frac{31\cdot101\sigma^{25}}{2^{15}\cdot3^5\cdot5\cdot11^2\cdot17\cdot23}+O(\sigma^{31})$$

Economized Polynomials $(0\leq x\leq1.53)$

18.13.67 $\quad x^2\mathscr{P}(x)=\sum_0^6 a_n x^{6n}+\epsilon(x)$

$$|\epsilon(x)|<2\times10^{-7}$$

$a_0=(-1)9.99999\ 96 \qquad a_4=-(-9)2.20892\ 47$

$a_1=(-2)3.57143\ 20 \qquad a_5=(-10)1.74915\ 35$

$a_2=(-5)9.80689\ 93 \qquad a_6=-(-12)4.46863\ 93$

$a_3=(-7)2.00835\ 02$

18.13.68 $\quad x^3\mathscr{P}'(x)=\sum_0^6 a_n x^{6n}+\epsilon(x)$

$$|\epsilon(x)|<4\times10^{-7}$$

$a_0=-2.00000\ 00 \qquad a_4=-(-9)2.12719\ 66$

$a_1=(-1)1.42857\ 22 \qquad a_5=(-10)6.53654\ 67$

$a_2=(-4)9.81018\ 03 \qquad a_6=-(-11)1.70510\ 78$

$a_3=(-6)3.00511\ 93$

18.13.69 $\quad x\zeta(x)=\sum_0^6 a_n x^{6n}+\epsilon(x)$

$$|\epsilon(x)|<3\times10^{-8}$$

$a_0=(-1)9.99999\ 98 \qquad a_4=(-10)6.12486\ 14$

$a_1=-(-3)7.14285\ 86 \qquad a_5=(-11)4.66919\ 85$

$a_2=-(-6)8.91165\ 65 \qquad a_6=(-12)1.25014\ 65$

$a_3=-(-8)1.44381\ 84$

18.14. Lemniscatic Case
$$(g_2=1, \ g_3=0)$$

If $g_2>0$ and $g_3=0$, homogeneity relations allow us to reduce our consideration of $\wp$ to $\wp\,(z; 1, 0)$ ($\wp'$, ζ and σ are handled similarly). Thus $\wp\,(z; g_2, 0)=g_2^{\frac{1}{2}}\wp\,(zg_2^{\frac{1}{4}}; 1, 0)$. The case $g_2=1, \ g_3=0$ is called the LEMNISCATIC case.

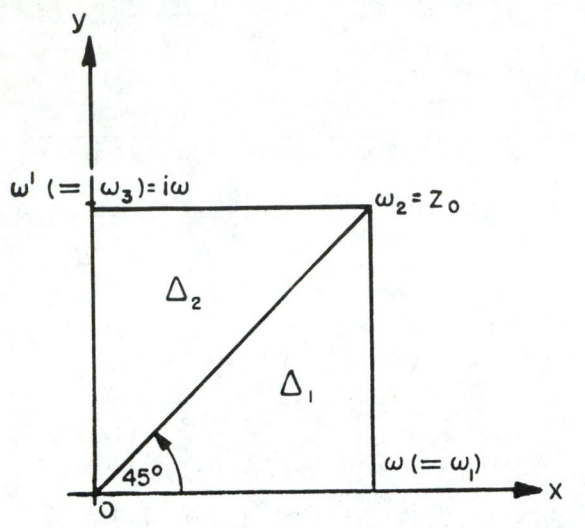

FIGURE 18.13

$\frac{1}{4}FPP$; *Reduction to Fundamental Triangle*

$\Delta_1\equiv\Delta 0\omega\omega_2$ is the Fundamental Triangle

$\omega\approx 1.8540\ 74677\ 30137\ 192$ [8]

Reduction for z_2 in Δ_2: $z_1=i\bar{z}_2$ is in Δ_1

18.14.1 $$\wp\,(z_2)=-\overline{\wp}\,(z_1)$$

18.14.2 $$\wp\,'(z_2)=i\,\overline{\wp'}\,(z_1)$$

18.14.3 $$\zeta(z_2)=-i\overline{\zeta}(z_1)$$

18.14.4 $$\sigma(z_2)=i\overline{\sigma}(z_1)$$

Special Values and Formulas

18.14.5

$$\Delta=1, H_1=H_3=2^{-\frac{1}{2}}, H_2=i/2,$$
$$m=\sin^2 45°=\tfrac{1}{2}, q=e^{-\pi}$$

18.14.6 $$\vartheta_2(0)=\vartheta_4(0)=(\omega\sqrt{2}/\pi)^{\frac{1}{2}}; \ \vartheta_3(0)=(2\omega/\pi)^{\frac{1}{2}}$$

18.14.7 $$\omega=K(\sin^2 45°)=\frac{\Gamma^2(\tfrac{1}{4})}{4\sqrt{\pi}}=\frac{\tilde{\omega}}{\sqrt{2}} \text{ where}$$

$\tilde{\omega}\approx 2.62205\ 75542\ 92119\ 81046\ 48395\ 89891\ 11941\ 36827\ 54951\ 43162$ is the Lemniscate constant [18.9]

[8] This value was computed and checked by double precision methods on a desk calculator and is believed correct to 18S.

Values at Half-periods

	$\wp$	$\wp'$	ζ	σ
18.14.8 $\omega=\omega_1$	$e_1=\tfrac{1}{2}$	0	$\eta=\pi/4\omega$	$e^{\pi/8}(2^{1/4})$
18.14.9 $\omega_2=z_0$	$e_2=0$	0	$\eta+\eta'$	$e^{\pi/4}(\sqrt{2})e^{i\pi/4}$
18.14.10 $\omega'=\omega_3$	$e_3=-\tfrac{1}{2}$	0	$\eta'=-\pi i/4\omega$	$ie^{\pi/8}(2^{1/4})$

Values along $(0, \omega)$

	$\wp$	$\wp'$	ζ	σ
18.14.11 $\omega/4$	$\frac{\sqrt{\alpha}}{2}(\sqrt{\alpha}+2^{1/4})(1+2^{1/4})$			
18.14.12 $\omega/2$	$\alpha/2$	$-\alpha$	$\frac{\pi}{8\omega}+\frac{\alpha}{2\sqrt{2}}$	$\frac{e^{\pi/32}(2^{1/16})}{\alpha^{\frac{1}{4}}}$
18.14.13 $2\omega/3$	$\tfrac{1}{2}\sqrt{1+\sec 30°}$	$-\frac{\sqrt[4]{2\sqrt{3}+3}}{\sqrt{3}}$	$\frac{2\eta}{3}+\sqrt{\frac{\wp\,(2\omega/3)}{3}}$	$\frac{e^{\pi/18}(3^{1/8})}{(2+\sqrt{3})^{1/12}}$
18.14.14 $3\omega/4$	$\frac{\sqrt{\alpha}}{2}(\sqrt{\alpha}-2^{\frac{1}{4}})(1+2^{\frac{1}{4}})$			

$$\alpha=1+\sqrt{2}$$

Values along $(0, z_0)$

	$\mathscr{P}$	$\mathscr{P}'$	ζ	σ
18.14.15 $z_0/4$	$-\dfrac{i}{2}(\alpha+\sqrt{2\alpha})$	$\alpha(\sqrt{\alpha}+\sqrt{2})e^{i\pi/4}$		$\dfrac{e^{\pi/64}(2^{1/32})}{\alpha^{1/4}(\sqrt{\alpha}+\sqrt{2})^{1/4}}\,e^{i\pi/4}$
18.14.16 $z_0/2$	$-i/2$	$e^{i\pi/4}$	$\left[\dfrac{\pi}{4\omega\sqrt{2}}+\dfrac{1}{2}\right]e^{-i\pi/4}$	$e^{\pi/16}(2^{1/8})e^{i\pi/4}$
18.14.17 $2z_0/3$	$\dfrac{-i}{2}\sqrt{\sec 30°-1}$	$\dfrac{e^{i\pi/4}\sqrt[4]{2\sqrt{3}-3}}{\sqrt{3}}$	$\dfrac{2\eta_2}{3}+\left[\dfrac{\mathscr{P}\,(2z_0/3)}{3}\right]^{1/2}$	$\dfrac{e^{\pi/9}e^{i\pi/4}(3^{1/6})}{\sqrt[12]{2\sqrt{3}-3}}$
18.14.18 $3z_0/4$	$-\dfrac{i}{2}\,(\alpha-\sqrt{2\alpha})$	$\alpha(\sqrt{\alpha}-\sqrt{2})e^{i\pi/4}$		$\dfrac{e^{9\pi/64}(2^{1/32})}{\alpha^{1/4}(\sqrt{\alpha}-\sqrt{2})^{1/4}}\,e^{i\pi/4}$

$$\alpha=1+\sqrt{2}$$

Duplication Formulas

18.14.19
$$\mathscr{P}(2z)$$
$$=[\,\mathscr{P}^2(z)+\tfrac{1}{4}]^2/\{\,\mathscr{P}(z)[4\,\mathscr{P}^2(z)-1]\}$$

18.14.20
$$\mathscr{P}'(2z)=(\beta+1)(\beta^2-6\beta+1)/[32\,\mathscr{P}'^3(z)],\ \beta=4\,\mathscr{P}^2(z)$$

18.14.21
$$\zeta(2z)=2\zeta(z)+\frac{6\,\mathscr{P}^2(z)-\tfrac{1}{2}}{2\,\mathscr{P}'(z)}$$

18.14.22
$$\sigma(2z)=-\mathscr{P}'(z)\sigma^4(z)$$

Bisection Formulas $(0<x<2\omega)$

18.14.23
$$\mathscr{P}\left(\frac{x}{2}\right)$$
$$=[\mathscr{P}^{\frac{1}{2}}(x)+\{\mathscr{P}(x)+\tfrac{1}{2}\}^{\frac{1}{2}}]\,[\mathscr{P}^{\frac{1}{2}}(x)\pm\{\mathscr{P}(x)-\tfrac{1}{2}\}^{\frac{1}{2}}]$$

[Use $+$ on $0<x\leq\omega$, $-$ on $\omega\leq x<2\omega$]

18.14.24
$$\tfrac{1}{2}\mathscr{P}'\left(\frac{x}{2}\right)=\mathscr{P}'(x)\mp[2\mathscr{P}(x)+\tfrac{1}{2}]\sqrt{\mathscr{P}(x)-\tfrac{1}{2}}$$
$$-[2\mathscr{P}(x)-\tfrac{1}{2}]\sqrt{\mathscr{P}(x)+\tfrac{1}{2}}$$
$$-2\mathscr{P}^{3/2}(x)\ \text{(See [18.13].)}$$

[Use $-$ on $0<x\leq\omega$, $+$ on $\omega\leq x<2\omega$]

Complex Multiplication

18.14.25
$$\mathscr{P}(iz)=-\mathscr{P}(z)$$

18.14.26
$$\mathscr{P}'(iz)=i\mathscr{P}'(z)$$

18.14.27
$$\zeta(iz)=-i\zeta(z)$$

18.14.28
$$\sigma(iz)=i\sigma(z)$$

The above equations could be used as follows, e.g.: If z were real, iz would be purely imaginary.

Conformal Maps

Lemniscatic Case
Map: $f(z)=u+iv$
$$\mathscr{P}(z)$$

Near zero: $\mathscr{P}(z)=\dfrac{1}{z^2}+\epsilon_1$

$$\mathscr{P}(z)=\frac{1}{z^2}+\frac{z^2}{20}+\epsilon_2,\ |z|<1$$

Near z_0: $\mathscr{P}(z)=\dfrac{-(z-z_0)^2}{4}+\epsilon_3,$
$$|z-z_0|<\sqrt{2}$$

$$\mathscr{P}(z)=\frac{-(z-z_0)^2}{4}+\frac{(z-z_0)^6}{80}+\epsilon_4$$

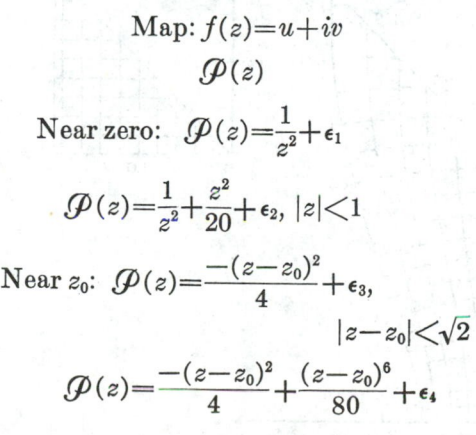

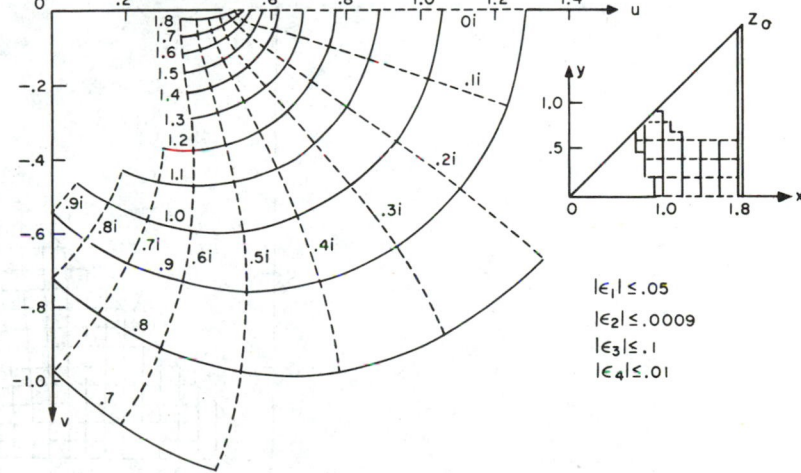

$|\epsilon_1|\leq.05$
$|\epsilon_2|\leq.0009$
$|\epsilon_3|\leq.1$
$|\epsilon_4|\leq.01$

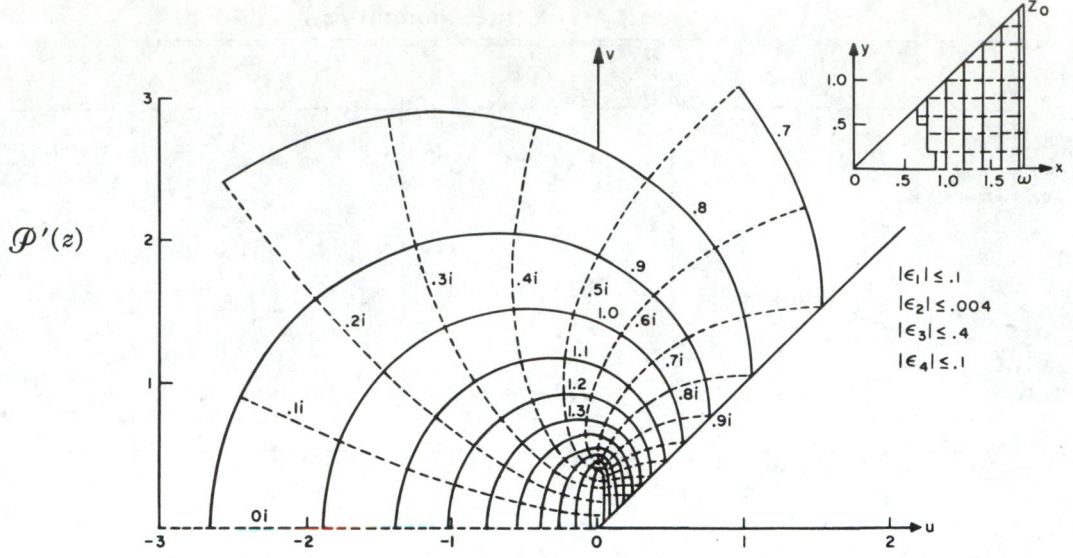

Near zero: $\mathscr{P}'(z) = \dfrac{-2}{z^3} + \epsilon_1$

$\mathscr{P}'(z) = \dfrac{-2}{z^3} + \dfrac{z}{10} + \epsilon_2$

Near z_0: $\mathscr{P}'(z) = \dfrac{-(z-z_0)}{2} + \epsilon_3$

$\mathscr{P}'(z) = \dfrac{-(z-z_0)}{2} + \dfrac{3(z-z_0)^5}{40} + \epsilon_4$

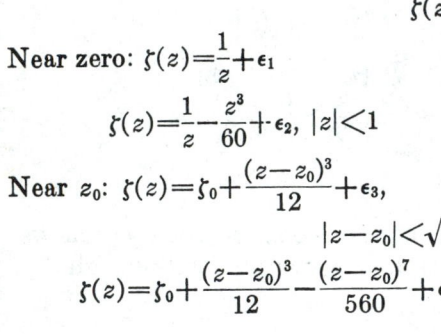

Near zero: $\zeta(z) = \dfrac{1}{z} + \epsilon_1$

$\zeta(z) = \dfrac{1}{z} - \dfrac{z^3}{60} + \epsilon_2$, $|z| < 1$

Near z_0: $\zeta(z) = \zeta_0 + \dfrac{(z-z_0)^3}{12} + \epsilon_3$,

$|z - z_0| < \sqrt{2}$

$\zeta(z) = \zeta_0 + \dfrac{(z-z_0)^3}{12} - \dfrac{(z-z_0)^7}{560} + \epsilon_4$

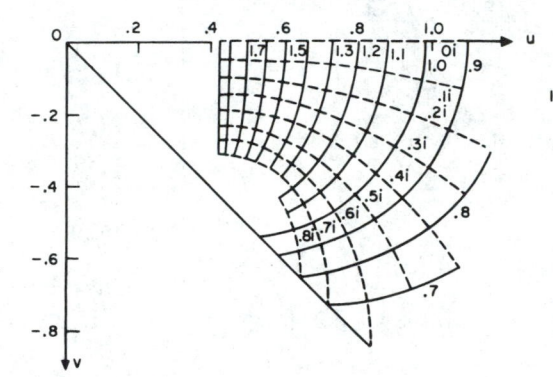

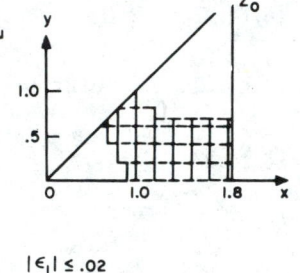

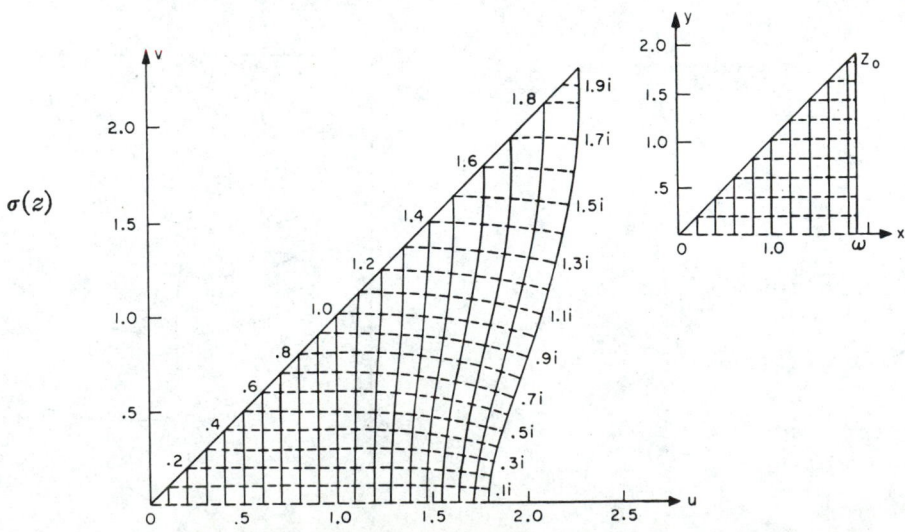

FIGURE 18.14

Coefficients for Laurent Series for $\wp$, $\wp'$, and ζ

$(c_m = 0$ for m odd$)$

k	EXACT c_{2k}	APPROXIMATE c_{2k}
1	$1/20$	$.05$
2	$1/(3 \cdot 20^2) = 1/1200$	$.8333 \ldots \times 10^{-3}$
3	$2/(3 \cdot 13 \cdot 20^3) = 1/156000$	$.641025 \quad 641025 \ldots \times 10^{-5}$
4	$5/(3 \cdot 13 \cdot 17 \cdot 20^4) = 1/21216000$	$.47134 \quad 23831 \quad 07088 \quad 98944 \times 10^{-7}$
5	$2/(3^2 \cdot 13 \cdot 17 \cdot 20^5) = 1/(31824 \times 10^5)$	$.31422 \quad 82554 \quad 04725 \quad 99296 \times 10^{-9}$
6	$10/(3^3 \cdot 13^2 \cdot 17 \cdot 20^6) = 1/(4964544 \times 10^5)$	$.20142 \quad 83688 \quad 49183 \quad 32882 \times 10^{-11}$
7	$4/(3 \cdot 13^2 \cdot 17 \cdot 29 \cdot 20^7) = 1/(7998432 \times 10^7)$	$.12502 \quad 45048 \quad 02941 \quad 37651 \times 10^{-13}$
8	$2453/(3^4 \cdot 11 \cdot 13^2 \cdot 17^2 \cdot 29 \cdot 20^8) = 958203125/(1262002599 \times 10^{16})$	$.75927 \quad 19109 \quad 76468 \quad 59917 \times 10^{-16}$
9	$2 \cdot 5 \cdot 7 \cdot 61/(3^3 \cdot 13^3 \cdot 17^2 \cdot 29 \cdot 37 \cdot 20^9) = 833984375/(18394643943 \times 10^{17})$	$.45338 \quad 43533 \quad 93461 \quad 06092 \times 10^{-18}$

$$c_{2k} \leq \frac{c_2^k}{3^{k-1}}, \quad k = 1, 2, \ldots$$

Other Series Involving $\wp$

Reversed Series for Large $|\wp|$

18.14.29

$$z = (\wp^{-1})^{1/2} \left[1 + \frac{w}{5} + \frac{w^2}{6} + \frac{5w^3}{26} + \frac{35w^4}{136} \right.$$
$$+ \frac{3w^5}{8} + \frac{231w^6}{400} + \frac{429w^7}{464} + \frac{195w^8}{128}$$
$$\left. + \frac{12155w^9}{4736} + \frac{46189w^{10}}{10496} + O(w^{11}) \right],$$

18.14.30 $w = \wp^{-2}/8$, and z is in the Fundamental Triangle **(Figure 18.13)** if $\wp$ has an appropriate value.

Series near z_0

18.14.31 $\quad 2\wp = -x + \dfrac{x^3}{5} - \dfrac{2x^5}{75} + \dfrac{x^7}{325} + O(x^9),$

18.14.32 $\quad x = (z - z_0)^2/2$

18.14.33 $\quad x = -\left[w + \dfrac{w^3}{5} + \dfrac{7w^5}{75} + \dfrac{11w^7}{195} + O(w^9) \right]$
$$w = 2\wp$$

Series near ω

18.14.34

$$(\wp - e_1) = v + v^2 + \frac{4v^3}{5} + \frac{3v^4}{5} + \frac{32v^5}{75} + \frac{22v^6}{75} + \frac{64v^7}{325} + O(v^8),$$

18.14.35 $\quad v = (z - \omega)^2/2$

18.14.36

$$v = y \left[1 - y + \frac{6y^2}{5} - \frac{8y^3}{5} + \frac{172y^4}{75} \right.$$
$$\left. - \frac{52y^5}{15} + \frac{1064y^6}{195} + O(y^7) \right],$$

18.14.37 $\quad y = (\wp - e_1)$

Other Series Involving $\wp'$

Reversed Series for Large $|\wp'|$

18.14.38

$$z = Au \left[1 - \frac{v}{5} + \frac{5v^3}{39} - \frac{7v^4}{51} + O(v^5) \right], \quad u = (\wp'^{1/3})^{-1} e^{i\pi/3},$$

18.14.39 $A = 2^{1/3}$, $v = Au^4/6$, and z is in the Fundamental Triangle **(Figure 18.13)** if $\wp'$ has an appropriate value.

Series near z_0

18.14.40

$$\wp' = \frac{1}{2}(z - z_0) \left[-1 + 3w - \frac{10w^2}{3} + \frac{35w^3}{13} + O(w^4) \right],$$

18.14.41 $\quad w = (z - z_0)^4/20$

18.14.42

$$(z - z_0) = 2\wp' \left[1 + \frac{3u}{5} + \frac{5u^2}{3} + \frac{84u^3}{13} + O(u^4) \right],$$

18.14.43 $\quad u = 4\wp'^4$

Series near ω

18.14.44

$$\wp' = x \left[1 + x^2 + \frac{3}{5}x^4 + \frac{3}{10}x^6 + \frac{2}{15}x^8 + \frac{11}{200}x^{10} + O(x^{12}) \right],$$

18.14.45 $\quad x = (z - \omega)$

18.14.46

$$x = \wp' - \wp'^3 + \frac{12\wp'^5}{5} - \frac{15\wp'^7}{2}$$
$$+ \frac{80\wp'^9}{3} - \frac{819\wp'^{11}}{8} + O(\wp'^{13})$$

Other Series Involving ζ

Reversed Series for Large $|\zeta|$

18.14.47 $\quad z = \zeta^{-1} \left[1 - \dfrac{v}{5} + \dfrac{v^2}{7} - \dfrac{136v^3}{1001} + \dfrac{1349v^4}{9163} + O(v^5) \right],$

18.14.48 $\quad v = \zeta^{-4}/12$

Series near z_0

18.14.49

$$(\zeta-\zeta_0)=\frac{1}{4}(z-z_0)^3\left[\frac{1}{3}-\frac{v}{7}+\frac{2v^2}{33}-\frac{v^3}{39}+O(v^4)\right],$$

18.14.50

$$v=(z-z_0)^4/20$$

Series near ω

18.14.51

$$(\zeta-\eta)=-\frac{x}{2}-\frac{x^3}{6}-\frac{x^5}{20}-\frac{x^7}{70}-\frac{x^9}{240}$$
$$-\frac{x^{11}}{825}-\frac{11x^{13}}{31200}-\frac{x^{15}}{9750}+O(x^{17}),$$

18.14.52

$$x=(z-\omega)$$

18.14.53

$$x=w-\frac{w^3}{3}+\frac{7w^5}{30}-\frac{13w^7}{63}+\frac{929w^9}{4536}-\frac{194w^{11}}{891}+\frac{942883w^{13}}{3891888}$$
$$+O(w^{15})$$

18.14.54 $w=-2(\zeta-\eta)$

Series Involving σ

18.14.55

$$\sigma=z-\frac{z^5}{2\cdot5!}-\frac{3^2z^9}{2^2\cdot9!}+\frac{3\cdot23z^{13}}{2^3\cdot13!}+\frac{3\cdot107z^{17}}{2^4\cdot17!}+\frac{3^3\cdot7\cdot23\cdot37z^{21}}{2^5\cdot21!}$$
$$+\frac{3^2\cdot313\cdot503z^{25}}{2^6\cdot25!}-\frac{3^4\cdot7\cdot685973z^{29}}{2^7\cdot29!}+O(z^{33})$$

18.14.56

$$z=\sigma+\frac{\sigma^5}{2^4\cdot3\cdot5}+\frac{\sigma^9}{2^9\cdot3\cdot7}+\frac{17\cdot113\sigma^{13}}{2^{13}\cdot3^4\cdot7\cdot11\cdot13}$$
$$+\frac{122051\sigma^{17}}{2^{19}\cdot3^5\cdot7^2\cdot11\cdot17}+\frac{5\cdot13\sigma^{21}}{2^{23}\cdot3^2\cdot11\cdot19}+O(\sigma^{25})$$

Economized Polynomials ($0\le x\le1.86$)

18.14.57 $$x^2\mathscr{P}(x)=\sum_0^6a_nx^{4n}+\epsilon(x)$$

$$|\epsilon(x)|<2\times10^{-7}$$

$a_0=(-1)9.99999\ 98$ $a_4=(-8)4.81438\ 20$

$a_1=(-2)4.99999\ 62$ $a_5=(-10)2.29729\ 21$

$a_2=(-4)8.33352\ 77$ $a_6=(-12)4.94511\ 45$

$a_3=(-6)6.40412\ 86$

18.14.58 $$x^3\mathscr{P}'(x)=\sum_0^6a_nx^{4n}+\epsilon(x)$$

$$|\epsilon(x)|<4\times10^{-7}$$

$a_0=-2.00000\ 00$ $a_4=(-7)6.58947\ 52$

$a_1=(-1)1.00000\ 02$ $a_5=(-9)5.59262\ 49$

$a_2=(-3)4.99995\ 38$ $a_6=(-11)5.54177\ 69$

$a_3=(-5)6.41145\ 59$

18.14.59 $$x\zeta(x)=\sum_0^6a_nx^{4n}+\epsilon(x)$$

$$|\epsilon(x)|<3\times10^{-8}$$

$a_0=(-1)9.99999\ 99$ $a_4=-(-9)2.57492\ 62$

$a_1=-(-2)1.66666\ 74$ $a_5=-(-11)5.67008\ 00$

$a_2=-(-4)1.19036\ 70$ $a_6=-(-13)9.70015\ 80$

$a_3=-(-7)5.86451\ 63$

18.15. Pseudo-Lemniscatic Case

$$(g_2=-1,\ g_3=0)$$

If $g_2<0$ and $g_3=0$, homogeneity relations allow us to reduce our consideration of $\mathscr{P}$ to $\mathscr{P}(z;-1,0)$. Thus

18.15.1 $$\mathscr{P}(z;g_2,0)=|g_2|^{1/2}\mathscr{P}(z|g_2|^{1/4};-1,0)$$

[$\mathscr{P}'$, ζ and σ are handled similarly]. Because of its similarity to the lemniscatic case, we refer to the case $g_2=-1$, $g_3=0$ as the pseudo-lemniscatic case. It plays the same role (period ratio unity) for $\Delta<0$ as does the lemniscatic case for $\Delta>0$.

$$\omega_2=\sqrt{2}\times\text{(real half-period for lemniscatic case)}$$
$$=\tilde{\omega}\ \text{(the Lemniscate Constant—see \textbf{18.14.7})}$$

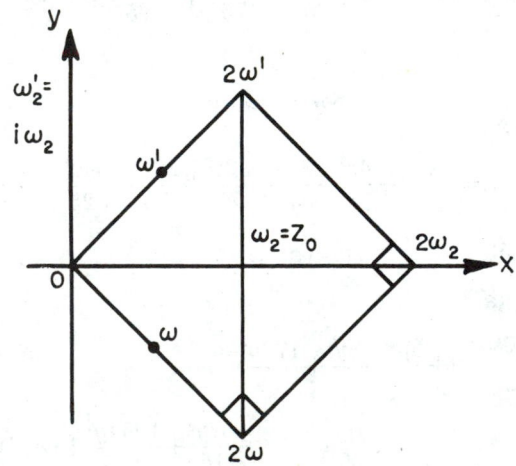

FIGURE 18.15

Special Values and Relations

18.15.2 $\qquad \Delta = -1, \; g_2 = -1, \; g_3 = 0$

18.15.4

18.15.3

$$\vartheta_2(0) = R 2^{1/4} e^{i\pi/8}, \; \vartheta_3(0) = R e^{i\pi/8}, \; \vartheta_4(0) = R e^{-i\pi/8},$$

$H_1 = -i/\sqrt{2}, \; H_2 = \tfrac{1}{2}, \; H_3 = i/\sqrt{2}, \qquad m = \tfrac{1}{2}, \; q = i e^{-\pi/2}$

18.15.5 $\qquad\qquad$ where $R = \sqrt{\omega_2 \sqrt{2}/\pi}$

Values at Half-Periods

		$\wp$	$\wp'$	ζ	σ
18.15.6	$\omega \equiv \omega_1$	$i/2$		$\tfrac{1}{2}(\eta_2 - \eta_2')$	$e^{-i\pi/4} e^{\pi/8}(2^{1/4})$
18.15.7	ω_2		0	$\eta_2 = \pi/2\omega_2$	$e^{\pi/4}\sqrt{2}$
18.15.8	$\omega' = \omega_3$	$-i/2$		$\tfrac{1}{2}(\eta_2 + \eta_2')$	$e^{i\pi/4} e^{\pi/8}(2^{1/4})$
18.15.9	ω_2'		0	$\eta_2' = -i\eta_2$	$i\sigma(\omega_2)$

Relations with Lemniscatic Values

18.15.10 $\quad \wp(z; -1, 0) = i\wp(z e^{i\pi/4}; 1, 0)$

18.15.12 $\quad \zeta(z; -1, 0) = e^{i\pi/4} \zeta(z e^{i\pi/4}; 1, 0)$

18.15.11 $\quad \wp'(z; -1, 0) = e^{3\pi i/4} \wp'(z e^{i\pi/4}; 1, 0)$

18.15.13 $\quad \sigma(z; -1, 0) = e^{-i\pi/4} \sigma(z e^{i\pi/4}; 1, 0)$

Numerical Methods

18.16. Use and Extension of the Tables

Example 1. Lemniscatic Case

(a) Given $z = x + iy$ in the Fundamental Triangle, find $\wp(\wp', \zeta, \sigma)$ more accurately than can be done with the maps.

σ—Use Maclaurin series throughout the Fundamental Triangle. Five terms give at least six significant figures, six terms at least ten. $\wp, \zeta$—Use Laurent's series directly "near" 0, (if $|z| < 1$, four terms give at least eight significant figures for $\wp$, nine for ζ; five terms at least ten significant figures for $\wp$, eleven for ζ). Use Taylor's series directly "near" z_0. Elsewhere (unless approximately seven or eight significant figures are insufficient) use economized polynomials to obtain $\wp(x)$, $\wp'(x)$ and/or $\zeta(x)$ as appropriate. To get $\wp(iy)$, $\wp'(iy)$ and/or $\zeta(iy)$, use Laurent's series for "small" y, otherwise use economized polynomials to compute $\wp(y)$, $\wp'(y)$ and/or $\zeta(y)$, then use complex multiplication to obtain $\wp(iy)$, $\wp'(iy)$ and/or $\zeta(iy)$. Finally, use appropriate addition formula to get $\wp(z)$ and/or $\zeta(z)$.

$\wp'$—Use Laurent's series directly "near" 0 (if $|z| < 1$, four terms give at least six significant figures, five terms at least eight significant figures). Elsewhere, either use economized polynomials and addition formula as for $\wp$ and ζ, or get $\wp'^2 = 4\wp^3 - \wp$ and extract appropriate square root ($\mathcal{I}\wp' \geq 0$).

(b) Given $\wp(\wp', \zeta, \sigma)$ corresponding to a point in the Fundamental Triangle, compute z more accurately than can be done with the maps. Only a few significant figures are obtainable from the use of any of the given (truncated) reversed series, except in a small neighborhood of the center of the series. For greater accuracy, use inverse interpolation procedures.

Example 2. Equianharmonic Case

(a) Given $z = x + iy$ in the Fundamental Triangle, find $\wp(\wp', \zeta, \sigma)$ more accurately than can be done with the maps.

σ—Use Maclaurin series throughout the Fundamental Triangle. Four terms give at least eleven significant figures, five terms at least twenty one.

$\wp, \zeta$—Use Laurent's series directly "near" 0 (if $|z| < 1$, four terms give at least 10S for $\wp$, 11S for ζ; five terms at least 13S for $\wp$, 14S for ζ). Elsewhere (unless approximately seven or eight significant figures are insufficient) use economized polynomials to obtain $\wp(x)$, $\wp'(x)$ and/or $\zeta(x)$, as appropriate. To get $\wp(iy)$, $\wp'(iy)$ and/or $\zeta(iy)$, use Laurent's series. Then use appropriate addition formula to get $\wp(z)$ and/or $\zeta(z)$.

$\mathscr{P}'$—Use Laurent's series directly "near" 0 (if $|z|<1$, four terms give at least 8S, five terms at least 11S). Elsewhere, either proceed as for $\mathscr{P}$ and ζ, or get $\mathscr{P}'^2=4\mathscr{P}^3-1$ and extract appropriate square root ($\mathscr{I}\mathscr{P}'\geqq 0$).

(b) Given $\mathscr{P}(\mathscr{P}',\zeta,\sigma)$ corresponding to a point in the Fundamental Triangle, compute z more accurately than can be done with the maps. Only a few significant figures are obtainable from the use of any of the given (truncated) reversed series, except in a small neighborhood of the center of the series. For greater accuracy, use inverse interpolation procedures.

Example 3. Given period ratio a, find parameters m (of elliptic integrals and Jacobi's functions of chapter **16**) and q (of ϑ functions).

m—In both the cases $\Delta>0$ and $\Delta<0$, the period ratio is equal to $K'(m)/K(m)$ (see **18.9**). Knowing K'/K, if $1<K'/K\leqq 3$, use **Table 17.3** to find m; if $K'/K>3$, use the method of **Example 6** in chapter **17**. An alternative method is to use **Table 18.3** to obtain the necessary entries, thence use

$$m=(e_2-e_3)/(e_1-e_3) \text{ in case } \Delta>0,$$

$$m=\tfrac{1}{2}-3e_2/4H_2 \text{ in case } \Delta<0.$$

q—In both the cases $\Delta>0$ and $\Delta<0$, the period ratio determines the exponent for $q[q=e^{-\pi a}$ if $\Delta>0$, $q=ie^{-\pi a/2}$ if $\Delta<0]$. Hence enter **Table 4.16** $[e^{-\pi x}, x=0(.01)1]$ and multiply the results as appropriate $[\text{e.g.}, e^{-4.72\pi}=(e^{-\pi})^4(e^{-.72\pi})]$.

Determination of Values at Half-Periods, Invariants and Related Quantities from Given Periods (Table 18.3)

$\Delta>0$	$\Delta<0$
Given ω and ω', form $\omega'/i\omega$ and enter **Table 18.3.** Multiply the results obtained by the appropriate power of ω (see footnotes of **Table 18.3**) to obtain value desired.	Given ω_2 and ω_2', form $\omega_2'/i\omega_2$ and enter **Table 18.3.** Multiply the results obtained by the appropriate power of ω_2 (see footnotes of **Table 18.3**) to obtain value desired.

Example 4.

Given $\omega=10$, $\omega'=11i$, find e_i, g_i, and Δ.

Here $\omega'/i\omega=1.1$, so that direct reading of **Table 18.3** gives

$$e_1(1)=1.6843\ 041$$
$$e_2(1)=-.2166\ 258\ (=-e_1-e_3)$$
$$e_3(1)=-1.4676\ 783$$
$$g_2(1)=10.0757\ 7364$$
$$g_3(1)=2.1420\ 1000.$$

Multiplying by appropriate powers of $\omega=10$ we obtain

$$e_1=.01684\ 3041$$
$$e_2=-.00216\ 6258$$
$$e_3=-.01467\ 6783$$
$$g_2=1.0075\ 77364 \times 10^{-3}$$
$$g_3=2.1420\ 1000 \times 10^{-6}$$

whence
$$\Delta=8.9902\ 3191 \times 10^{-10}$$

Example 4.

Given $\omega_2=10$, $\omega_2'=11i$, find e_i, g_i, and Δ.

Here $\omega_2'/i\omega_2=1.1$, so that direct reading of **Table 18.3** gives

$$e_1(1)=-.2166\ 2576+3.0842\ 589i$$
$$e_2(1)=.4332\ 5152=-2\mathscr{R}(e_1)$$
$$e_3(1)=\bar{e}_1(1)$$
$$g_2(1)=-37.4874\ 912$$
$$g_3(1)=16.5668\ 099.$$

Multiplying by appropriate powers of $\omega_2=10$ we obtain

$$e_1=-.00216\ 62576+.03084\ 2589i$$
$$e_2=.00433\ 25152$$
$$e_3=\bar{e}_1$$
$$g_2=-3.7487\ 4912 \times 10^{-3}$$
$$g_3=1.6566\ 8099 \times 10^{-5}$$

whence
$$\Delta=-6.0092\ 019 \times 10^{-8}$$

Example 5. ($\Delta > 0$)

Given $\omega = 10$, $\omega' = 55i$, find η, η', $\sigma(\omega)$, $\sigma(\omega')$ and $\sigma(\omega_2)$.

Forming $\omega'/i\omega = 5.5$ and entering **Table 18.3** we obtain $\eta = .82246704$, $\sigma(\omega) = .96045\ 40$. Using Legendre's relation we find $\eta' = \eta\omega' - \pi i/2 = 2.9527\ 723i$. Since interpolation for $\sigma(\omega')$ and $\sigma(\omega + \omega')$ is difficult, use is made of **18.3.15–18.3.17** together with **18.3.4** and **18.3.6**. Values of g_2, g_3 and e_1 can be read directly to eight significant figures and e_3 to about five significant figures giving $g_2 = 8.1174\ 243$, $g_3 = 4.4508\ 759$, $e_1 = 1.6449\ 341$, and $e_3 = -.82247$. Use of **18.3.6** yields $H_3 = .00174\ 69$ and $H_2 = .00174\ 69i$. Application of **18.3.15–18.3.17** yields $\sigma(\omega')/i = .0071177$ and $\sigma(\omega_2) = -.002016 - .01055i$. Multiplying the results obtained by the appropriate powers of ω we obtain $\eta = .08224\ 6704$, $\eta' = .29527\ 723i$, $\sigma(\omega) = 9.6045\ 40$, $\sigma(\omega') = .071177i$ and $\sigma(\omega_2) = -.02016 - .1055i$.

Example 5. ($\Delta < 0$)

Given $\omega_2 = 1000$, $\omega_2' = 1004i$, find η_2, η_2', $\sigma(\omega_2)$, $\sigma(\omega_2')$ and $\sigma(\omega')$.

With $\omega_2'/i\omega_2 = 1.004$, four point interpolation in **Table 18.3** gives $\eta_2 = 1.5626\ 756$, $\eta_2' = -1.5726\ 664i$, $\sigma(\omega_2) = 1.1805\ 028$, $\sigma(\omega_2') = 1.1901\ 52i$ and $\sigma(\omega') = .475084 + .476717i$.

Multiplying the results obtained by the appropriate powers of ω_2 gives $\eta_2 = .00156\ 26756$, $\eta_2' = -.00157\ 26664i$, $\sigma(\omega_2) = 1180.5028$, $\sigma(\omega_2') = 1190.152i$ and $\sigma(\omega') = 475.084 + 476.717i$.

Determination of Periods from Given Invariants (Table 18.1.)

$\Delta > 0$

Given $g_2 > 0$ and $g_3 > 0$ such that $\Delta = g_2^3 - 27g_3^2 > 0$ (if $g_3 = 0$, $|\omega'| = \omega$; see lemniscatic case), compute $\bar{g}_2 = g_2 g_3^{-2/3}$. From **Table 18.1**, determine $\omega g_3^{1/6}$ and $\omega' g_3^{1/6}$, thence ω and ω'.

$\Delta < 0$

Given g_2 and $g_3 > 0$ such that $\Delta = g_2^3 - 27g_3^2 < 0$ (if $g_3 = 0$, $|\omega_2'| = \omega_2$; see pseudo-lemniscatic case), compute $\bar{g}_2 = g_2 g_3^{-2/3}$. From **Table 18.1**, determine $\omega_2 g_3^{1/6}$ and $\omega_2' g_3^{1/6}$, thence ω_2 and ω_2'.

Example 6.

Given $g_2 = 10$, $g_3 = 2$, find ω and ω'. With $\bar{g}_2 = g_2 g_3^{-2/3} = 6.2996\ 05249$, from **Table 18.1** $\omega g_3^{1/6} = 1.1267\ 806$ and $\omega' g_3^{1/6} = 1.2324\ 295i$ whence $\omega = 1.003847$ and $\omega' = 1.097970i$.

Example 6.

Given $g_2 = -10$, $g_3 = 2$, find ω_2 and ω_2'. With $\bar{g}_2 = g_2 g_3^{-2/3} = -10/1.5874\ 0105 = -6.2996\ 053$, from **Table 18.1** $\omega_2 g_3^{1/6} = 1.5741\ 349$ and $\omega_2' g_3^{1/6} = 1.7124\ 396i$ whence $\omega_2 = 1.40239\ 48$ and $\omega_2' = 1.52561\ 02i$.

Example 7.

Given $g_2 = 8$, $g_3 = 4$, find ω and ω'. With $\bar{g}_2 = g_2 g_3^{-2/3} = 3.1748\ 02104$, from **Table 18.1** $\omega g_3^{1/6} = 1.2718\ 310$ and $\omega' g_3^{1/6} = 1.8702\ 425i$ whence $\omega = 1.009453$ and $\omega' = 1.484413i$.

Example 7.

Given $g_2 = 7$, $g_3 = 6$, find ω_2 and ω_2'. With $\bar{g}_2 = g_2 g_3^{-2/3} = 7/3.3019\ 2725 = 2.119974$, from **Table 18.1** $\omega_2 g_3^{1/6} = 1.3423\ 442$ and $\omega_2' g_3^{1/6} = 3.1441\ 141i$ whence $\omega_2 = .99579\ 976$ and $\omega_2' = 2.33241\ 83i$.

Computation of $\wp$, $\wp'$, or ζ for Given z and Arbitrary g_2, g_3

(or arbitrary periods from which g_2 and g_3 can be computed— in any case, periods must be known, at least approximately)

First reduce the problem (if necessary) to computation for a point z in the Fundamental Rectangle by use of appropriate results from **18.2**.

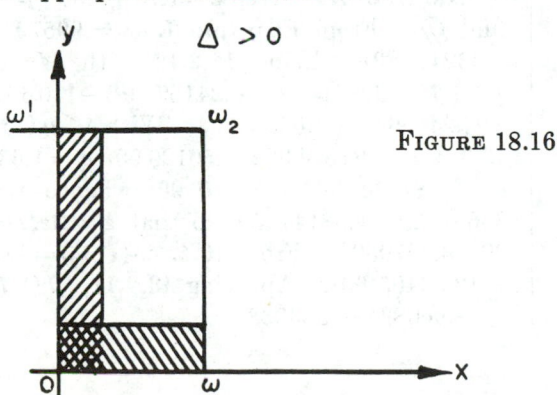
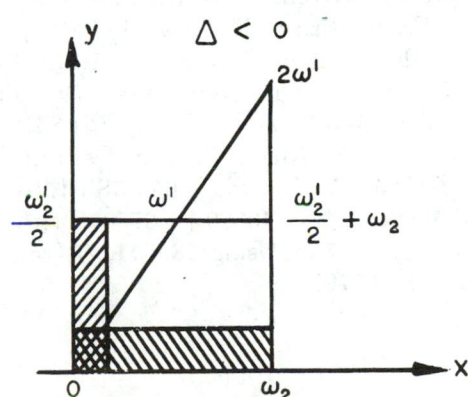

FIGURE 18.16

Method 1 (as accurate as desired)

If both x and y are "small," (point in double-cross hatched region) use Laurent's series in z directly. If either x or y is "large," use Laurent's series on $0x$, then on $0y$ and finally use an addition formula. (For $\wp'$ an alternative is to get $\wp$, then compute the appropriate root of $\wp'^2 = 4\wp^3 - g_2\wp - g_3$; see **18.8**.)

$\Delta > 0$

Method 2 (for $\wp$ or $\wp'$ only)

Compute $e_i(i=1,2,3)$ (if only g_2, g_3 are given use **Table 18.1** to get the periods, then get e_i in **Table 18.3**; if periods are also given, use **Table 18.3** directly). In any case, obtain $m(=[e_2-e_3]/[e_1-e_3])$, thence Jacobi's functions $\mathrm{sn}(z^*|m)$, $\mathrm{cn}(z^*|m)$, $\mathrm{dn}(z^*|m)$, from **16.4** and **16.21** and $\wp$ or $\wp'$ from **18.9.11–18.9.12**.

Method 3 (accuracy limited by **Table 4.16** of $e^{-\pi z}$ and by the method of getting periods).

Obtain periods, their ratio a, then $q=e^{-\pi a}$ from **Table 4.16**. Hence get $\vartheta_i(0)$, $i=2,3,4$ from truncated series **18.10.21–.23**. Compute appropriate ϑ functions for $z=x$ and for $z=iy$, whence get $\wp(x)$, $\wp'(x)$ and/or $\zeta(x)$, $\wp(iy)$, $\wp'(iy)$ and/or $\zeta(iy)$, then use an addition formula (if either x or y is "small", it is probably easier to use Laurent's series).

Example 8. Given $z=.07+.1i$, $g_2=10$, $g_3=2$, find $\wp$.

Using Laurent's series directly with
$$c_2 = .5$$
$$c_3 = .07142\ 85714$$
$$c_4 = .08333\ 33333$$
$$c_5 = .00974\ 02597$$
$$\begin{aligned} z^{-2} &= -22.97193\ 820 - 63.06022\ 25i \\ +c_2 z^2 &= -\quad .00255\ 000 + \quad .00700\ 00i \\ +c_3 z^4 &= -\quad .00001\ 214 - \quad .00001\ 02i \\ +c_4 z^6 &= +\quad .00000\ 024 - \quad .00000\ 01i \end{aligned}$$

$$\wp(z) = -22.97450\ 010 - 63.05323\ 28i.$$

Example 9. Given $z=15+73i$, $g_2=8$, $g_3=4$, find $\wp$. From **Example 7**, $\omega=1.009453$, $\omega'=1.484413i$. From **Table 18.3**, $e_1=1.61803\ 37$, $e_3=-.99999\ 96$, whence $m=.14589\ 79$. From **18.2.18** with $M=7$ and $N=24$, $\wp(.867658+1.748176i)=\wp(15+73i)$. Since z lies in R_2, by **18.2.31** $\wp(15+73i)=\overline{\wp}(.867658+1.22065i)$. From **16.4** with $z^*=1.40390+1.97505i$, $\mathrm{sn}(z^*|m)=2.46550+1.96527i$. Using **18.9.11**, $\wp(15+73i)=-.57743+.067797i$.

$\Delta < 0$

Method 2 (for $\wp$ or $\wp'$ only)

Compute e_2 and H_2 (if only g_2,g_3 are given, use **Table 18.1** to get the periods, then get e_i in **Table 18.3**; if periods are also given use **Table 18.3** directly). In any case, obtain $m(=\frac{1}{2}-3e_2/4H_2)$, thence Jacobi's functions $\mathrm{sn}(z'|m)$, $\mathrm{cn}(z'|m)$, $\mathrm{dn}(z'|m)$, from **16.4** and **16.21** and $\wp$ or $\wp'$ from **18.9.11–18.9.12**.

Method 3 (accuracy limited as in the case $\Delta > 0$).

Obtain periods, their ratio a, thence $q_2=e^{-\pi a/2}$ from **Table 4.16**. Then proceed as in the case $\Delta > 0$, using corresponding formulas.

Example 8. Given $z=.1+.03i$, $g_2=-10$, $g_3=2$, find $\wp$.

Using Laurent's series directly with
$$c_2 = -.5$$
$$c_3 = .07142\ 85714$$
$$c_4 = .80333\ 33333$$
$$z^{-2} = 76.59287\ 938 - 50.50079\ 960i$$
$$c_2 z^2 = -.00455\ 000 - \quad .00300\ 000i$$
$$c_3 z^4 = +.00000\ 334 + \quad .00000\ 780i$$
$$c_4 z^6 = -.00000\ 002 + \quad .00000\ 011i$$

$$\wp(z) = 76.58833\ 270 - 50.50379\ 169i.$$

Example 9. Given $z=1.75+3.6i$, $g_2=7$, $g_3=6$, find $\wp$. From **Example 7**, $\omega_2=.99579\ 98$, $\omega_2'=2.33241\ 83i$. Using **18.2.18** with $M=1$, $N=1$, $\wp(1.75+3.6i)=\wp(-.24159\ 96-1.064836i)=\wp(.24159\ 96+1.0648\ 36i)$. With $\Delta < 0$ from **Table 18.3**, $e_1=-.81674\ 362+.50120\ 90i$, $e_2=1.63348\ 724$, $e_3=-.81674\ 362-.50120\ 90i$ whence $m=.01014\ 3566$, $H_2^{\frac{1}{2}}=1.58144\ 50$, so that $z'=2zH_2^{\frac{1}{2}}=.76415\ 29+3.367959i$. From **16.4**, $\mathrm{cn}(z'|m)=4.00543\ 66-12.32465\ 69i$. Applying **18.9.11**, $\wp(1.75+3.6i)=-.960894-.383068i$.

$\Delta>0$

Example 10. Given $\omega=10$, $\omega'=20i$, find $\zeta(9+19i)$ by use of theta functions, **18.10** and addition formulas.

For the period ratio $a=\omega'/\omega i=2$ with the aid of **Table 4.16,** $q=e^{-2\pi}=.00186\ 74427.$

Using the truncated approximations **18.10.21–18.10.23** we compute the theta functions for argument zero. Using **16.27.1–16.27.4** we compute the theta functions for arguments v where $z=x$ and $z=iy$. Then, with **18.10.5–18.10.7** together with **18.10.9** and **18.10.18** we obtain $\zeta(9)=.09889\ 5484$, $\zeta(19i)=-.00120\ 0155i$, $\wp(9)=.01706\ 9647$, $\wp'(9)=-.00125\ 8460$, $\wp(19i)=-.00861\ 2615$, $\wp'(19i)=-.00003\ 757i$. Using the addition formula **18.4.3,** we obtain $\zeta(9+19i)=.07439\ 49-.00046\ 88i.$

$\Delta<0$

Example 10. Given $\omega_2=5$, $\omega_2'=7i$ find $\wp'(3+2i)$ by use of theta functions, **18.10** and addition formulas.

With the use of **Table 4.16** and **18.10.2,** $q=ie^{-.7\pi}=.11090\ 12784i.$

The theta functions are computed for argument zero using **18.10.21–18.10.23** and the theta functions for arguments v_1 and v_2 corresponding to $z=z_1+z_2$ using **16.27.1–16.27.4.** Using **18.10.5–18.10.6** together with **18 10.10,** we find $\wp(3)=.10576\ 946$, $\wp(2i)=-.24497\ 773$, $\wp'(3)=.07474140$, $\wp'(2i)=-.25576\ 007i$. The addition formula **18.4.1** yields $\wp(3+2i)=.01763\ 210-.07769\ 187i$, and **18.4.2** yields $\wp'(3+2i)=-.00069\ 182+.04771\ 305i.$

Use of Table 18.2 in Computing $\wp$, $\wp'$, ζ for Special Period Ratios

If the problem is reduced to computing $\wp$, $\wp'$, ζ in the Fundamental Rectangle for the case when the real half-period is unity and pure imaginary half-period is ia, for certain values of a **Table 18.2** may be used. Consider $\wp$ as an example. If $|z|$ is "small", then use Laurent's series directly for $\wp(z)$ [invariants for use in the series are given in **Table 18.3**].

If x is "large" and y "small" use **Table 18.2** to obtain $x^2\wp(x)$ and $x^3\wp'(x)$, thence $\wp(x)$ and $\wp'(x)$; use Laurent's series to obtain $\wp(iy)$ and $\wp'(iy)$; finally, use addition formula **18.4.1.**

For x "small" and y "large", reverse the procedure. For both x and y "large," use **Table 18.2** to obtain $\wp(x)$, $\wp'(x)$, $\wp(iy)$ and $\wp'(iy)$, thence use addition formula **18.4.1.**

Similar procedures apply to $\wp'$ or ζ. For $\wp'$, one can also first obtain $\wp$, then compute $\wp'^2=4\wp^3-g_2\wp-g_3$ and extract the appropriate square root (see **18.8** re choice of sign for $\wp'$).

$\Delta>0$

Example 11. Compute $\wp(.8+i)$ when $a=1.2$. Using **Table 18.2** or Laurent's series **18.5.1–4** with $g_2=9.15782\ 851$ and $g_3=3.23761\ 717$ from **Table 18.3,** $\wp(.8)=1.92442\ 11$, $\wp'(.8)=-2.76522\ 05$, $\wp(i)=-1.40258\ 06$ and $\wp'(i)=-1.19575\ 58i$. Using the addition formula **18.4.1** $\wp(.8+i)=-.381433-.149361i.$

Example 12. Compute $\zeta(.02+3i)$ for $a=4$. Using **Table 18.2** or Laurent's series **18.5.1–5** with
$$g_2=8.11742\ 426$$
$$g_3=4.45087\ 587$$
from **Table 18.3,**
$$\zeta(.02)=49.99999\ 89,$$
$$\wp(.02)=2500.00016,$$
$$\wp'(.02)=-249999.98376,$$
$$\zeta(3i)=.89635\ 173i,$$
$$\wp(3i)=-.82326\ 511,$$
$$\wp'(3i)=-.00249\ 829i.$$
Applying the addition formula **18.4.3,**
$$\zeta(.02+3i)=.016465+.89635i.$$

$\Delta<0$

Example 11. Compute $\wp(.9+.1i)$ for $a=1.05$. Using **Table 18.2** or Laurent's series **18.5.1–4** with $g_2=-42.41653\ 54$ and $g_3=9.92766\ 62$ from **Table 18.3,** $\wp(.9)=.34080\ 33$, $\wp'(.9)=-2.164801$, $\wp(.1i)=-99.97876$, $\wp'(.1i)=-2000.4255i$. With the addition formula **18.4.1** $\wp(.9+.1i)=.231859-.215149i.$

Example 12. Compute $\wp'(.4+.9i)$ for $a=2$. Using **Table 18.2** or Laurent's series **18.5.1–4,** with
$$g_2=4.54009\ 85,$$
$$g_3=8.38537\ 94$$
from **Table 18.3,**
$$\wp(.4)=6.29407\ 07,$$
$$\wp'(.4)=-30.99041,$$
$$\wp(.9i)=-1.225548,$$
$$\wp'(.9i)=-3.19127\ 03i.$$
Using the addition formulas **18.4.1–2,**
$$\wp'(.4+.9i)=1.10519\ 76-.56489\ 00i.$$

Computation of σ for Given z and Arbitrary g_2 and g_3

(or periods from which g_2 and g_3 can be computed—in any case, periods must be known, at least approximately)

First reduce the problem (if necessary) to computation for a point z in the Fundamental Rectangle (see **18.2**). After final reduction let z denote the point obtained.

$\Delta > 0$

If $\mathscr{R}z > \omega/2$ or,

$\mathscr{I}z > \omega'/2$, use duplication formula

$$\sigma(z) = -\mathscr{P}'(z/2)\sigma^4(z/2),$$

obtaining $\sigma(z/2)$ by use of Maclaurin series for σ and $\mathscr{P}'(z/2)$ by method explained above. Otherwise, simply use Maclaurin series for σ directly.

$\Delta < 0$

If $\mathscr{R}z > \omega_2/2$ or
$\mathscr{I}z > \omega_2'/4$, use duplication formula as in case $\Delta > 0$. Otherwise, use Maclaurin series for σ directly.

An alternate method is to use theta functions **18.10** first computing q and $\vartheta_i(0)$, $i = 2, 3, 4$.

$\Delta > 0$

Example 13. Compute $\sigma(.4 + 1.3i)$ for $g_2 = 8$, $g_3 = 4$. From **Example 7**, $\omega = 1.009453$ and $\omega' = 1.484413i$. Since $\mathscr{I}z > \omega'/2$, the Maclaurin series **18.5.6** is used to obtain $\sigma(z/2) = \sigma(.2 + .65i) = .19543\ 86 + .64947\ 28i$, the Laurent series **18.5.4** to obtain $\mathscr{P}'(.2 + .65i) = 5.02253\ 80 - 3.56066\ 93i$. The duplication formula **18.4.8** gives $\sigma(.4 + 1.3i) = .278080 + 1.272785i$.

$\Delta < 0$

Example 13. Compute $\sigma(.8 + .4i)$ for $g_2 = 7$, $g_3 = 6$. From **Example 7**, $\omega_2 = .99579\ 976$, $\omega_2' = 2.33241\ 83i$. Since $\mathscr{R}z > \omega_2/2$, the Maclaurin series **18.5.6** is used to obtain $\sigma(z/2) = \sigma(.4 + .2i) = .40038\ 019 + .19962\ 017i$, the Laurent series **18.5.4** to obtain $\mathscr{P}'(.4 + .2i) = -3.70986\ 70 + 22.218544i$. The duplication formula **18.4.8** gives $\sigma(.8 + .4i) = .81465\ 765 + .38819\ 473i$.

Given $\sigma(\mathscr{P}, \mathscr{P}', \zeta)$ corresponding to a point in the Fundamental Rectangle, as well as g_2 and g_3 or the equivalent, find z.

Only a few significant figures are obtainable from the use of any of the given (truncated) reversed series, except in a small neighborhood of the center of the series. For greater accuracy, use inverse interpolation procedures.

If the given function does not correspond to a value of z in the Fundamental Rectangle (see Conformal Maps) the problem can always be reduced to this case by the use of appropriate reduction formulas in **18.2**. This process is relatively simple for $\mathscr{P}(z)$, more difficult for the other functions (e.g. if $\Delta > 0$ and $\mathscr{P} = a + ib$, where $b > 0$, simply consider $\overline{\mathscr{P}} = a - ib$ and find z_1 in R_1 [**Figure 18.1**]; then compute $z_2 = \bar{z}_1 + 2\omega'$, the point in R_2 corresponding to the given $\mathscr{P}$).

$\Delta > 0$

Example 14. Given $\mathscr{P} = 1 - i$, $g_2 = 10$, $g_3 = 2$, find z. Using the first three terms of the reversed series **18.5.25** $z_1 \approx .727 + .423i$. The Laurent series **18.5.1** gives

$$\mathscr{P}(z_1) = \mathscr{P}(.727 + .423i) = .825 - .895i$$

and

$$\mathscr{P}(z_2) = \mathscr{P}(.697 + .393i) = .938 - 1.038i.$$

Inverse interpolation gives $z_1^{(1)} = .707 + .380i$. Repeated applications of the above procedure yield $z = .706231 + .379893i$.

$\Delta < 0$

Example 14. Given $\mathscr{P} = 1 + i$, $g_2 = -10$, $g_3 = 2$, find z. From **Example 6**, $\omega_2 = 1.40239\ 48$ and $\omega_2' = 1.52561\ 02i$. Since $b > 0$, z exists in R_2 and z is computed with $\overline{\mathscr{P}}$. Using **18.5.25** with $\alpha_2 = -1.25$, $\alpha_3 = .25$, $u = [(\overline{\mathscr{P}})^{-1}]^{1/2}$ and the coefficients c_n from **Example 8**

$$2u = 1.55377\ 3973 + .64359\ 42493i$$

$$c_2 u^5 = .08044\ 9281 - .19422\ 17466i$$

$$c_3 u^7 = -.01961\ 9359 + .00812\ 66047i$$

$$\frac{\alpha_2^2 u^9}{3} = -.10115\ 7160 - .04190\ 06673i$$

$\Delta>0$

$\Delta<0$

Stopping with the term in u^7, $z_1 \approx .81+.23i$. Assuming $\Delta z = -.03-.01i$, using **18.5.1**, $\mathcal{P}(.81+.23i)=.91410\ 95-.86824\ 37i$, $\mathcal{P}(.78+.22i)=1.03191\ 60-.91795\ 22i$; with inverse interpolation $z_1^{(1)}=.7725+.2404i$. Repeated applications of inverse interpolation yield $z=.772247-.239258i$.

Example 15. Given $\zeta=10-15i$, $g_2=8$, $g_3=4$, find z. Using the reversed series **18.5.40** with

$$A_5=-.13333\ 333,$$

$$A_7=-.02857\ 14286,$$

$$u=\ .03076\ 923076+.04615\ 384615i$$

$$A_5u^5=-.00000\ 001402+.00000\ 006860i$$

$$A_7u^7=-.00000\ 000004-.00000\ 000003i$$

$$z=\ .03076\ 921670+.04615\ 391472i.$$

Example 15. Given $\sigma=.4+.1i$, $g_2=7$, $g_3=6$, find z. Using the reversed series **18.5.70** with $\gamma_2=.14583$, $\gamma_3=.05$

$$\sigma=+.40000\ 000+.10000\ 000i$$

$$\frac{\gamma_2\sigma^5}{5}=+.00011\ 783+.00032\ 696i$$

$$\frac{\gamma_3\sigma^7}{7}=-.00000\ 208+.00001\ 432i$$

$$\frac{3\gamma_2^2\sigma^9}{14}=-.00000\ 093+.00000\ 126i$$

$$\frac{19\gamma_2\gamma_3\sigma^{11}}{55}=-.00000\ 013+.00000\ 006i$$

$$z=.40011\ 469+.10034\ 260i$$

Methods of Computation of $\mathcal{P}$ ($\mathcal{P}'$, ζ or σ) for Given z and Given g_2, g_3 (or the equivalent), with the Use of Automatic Digital Computing Machinery

(a) Integration of Differential Equation

$\mathcal{P}$ and $\mathcal{P}'$ may be generated for any z close enough to a "known point" z^* ($\mathcal{P}(z^*)$ and $\mathcal{P}'(z^*)$ being given) by integrating $\mathcal{P}''=6\mathcal{P}^2-g_2/2$. A program to do this on SWAC, via a modification of the Hammer-Hollingsworth method (MTAC, July 1955, pp. 92–96) due to Dr. P. Henrici, exists at Numerical Analysis Research, UCLA (code number 00600, written by W. L. Wilson, Jr.). The program has been tested numerically in the equianharmonic case, using integration steps of various sizes. For example, if one starts with $z^*=\omega_2$, using an "integration step" (h,k), where h and k are respectively the horizontal and vertical components of a step, with (h,k) having one of the six values $(\pm2h_0,0)$, $(\pm h_0,\pm k_0)$, $h_0=\omega_2/2000$, $k_0=|\omega_2'|/2000$, one can expect almost 8S in $\mathcal{P}$ and 7S in $\mathcal{P}'$ after 1000 steps, unless z is too near a pole.

(b) Use of Series

The process of reducing the computation problem to one in which z is in the Fundamental Rectangle can obviously be mechanized. Inside the Fundamental Rectangle the direct use of Laurent's series is appropriate when the period

ratio a is not too large. However, if $a\geqq\sqrt{3}(\Delta>0)$ or $a\geqq2\sqrt{3}(\Delta<0)$, the series will diverge at the far corner of the Fundamental Rectangle, so that use may be made of an appropriate duplication formula. Alternatively, one may compute the functions on $0x$ and $0y$, then use an addition formula. Even so, the series will diverge at $z=ia$ if $a\geqq2(\Delta>0)$ and at $z=ia/2$ if $a\geqq4(\Delta<0)$.

For great accuracy, multiple precision operations might be necessary. Double precision floating point mode has been used in a program, written for SWAC, to compute $\mathcal{P}$, $\mathcal{P}'$ and ζ.

For computation of σ, use of the Maclaurin series throughout the Fundamental Rectangle is probably simplest (series converges for all z).

Mention should be made of the possible use of the series defining the ϑ functions. These series converge for all complex v, and the computation of $\mathcal{P}$, $\mathcal{P}'$, ζ and σ by **18.10.5–18.10.8** could easily be mechanized. The series involved have the advantage of converging very fast, even in case $\Delta<0$, where $|q|\leqq e^{-\pi/2}(q\leqq e^{-\pi}$ if $\Delta>0)$.

Use of Maps

If the problem (of computing $\mathcal{P}$, $\mathcal{P}'$, ζ or σ for given z) is reduced to the case where the real half-period is unity and imaginary half-period is one of those used in the maps in **18.8** inspection of the appropriate figure will give the value of $\wp(z)$ [$\zeta(z)$ or $\sigma(z)$] to 2–3S. If $\wp'$ is wanted instead, get $\wp$, use **18.6.3** to obtain $\wp'^2$ and select sign (s) of $\wp'$ appropriately. (See Conformal Mapping **(18.8)** for choice of sign of square root of $\wp'^2$).

Computation of z_0

Given g_2, g_3 (or equivalent)

Since $z_0^2\wp(z_0)=0$, the Laurent's series gives

$$0=1+c_2u^2+c_3u^3+c_4u^4+\;.\;.\;.$$

where $u=z_0^2$. We may solve this equation [by Graeffe's (root-squaring) process or otherwise] for its absolutely smallest root [having found an

approximation to $|z_0|$ by Graeffe's process, we may use the fact that $z_0=\omega+iy_0(\Delta>0)$, $z_0=\omega_2 +iy_0(\Delta<0)$ to obtain an approximation to z_0].

It is noted that y_0/ω is a monotonic decreasing function of (period ratio) $a\geq 1$ for $\Delta>0$ and

$$[1\geq y_0/\omega>\frac{2}{\pi}\;\text{arccosh}\;\sqrt{3}(\approx .7297)].$$

y_0/ω_2 is a monotonic increasing function of a for $\Delta<0$ and

$$[0\leq y_0/\omega_2<\frac{2}{\pi}\;\text{arccosh}\;\sqrt{3}]$$

Further data is available from **Table 18.2** or from Conformal Maps defined by $\wp(z)$.

References

Texts and Articles

[18.1] P. Appell and E. Lacour, Principes de la théorie des fonctions elliptiques et applications (Gauthier-Villars, Paris, France, 1897).

[18.2] A. Erdélyi et al., Higher transcendental functions, vol. 2, ch. 13 (McGraw-Hill Book Co., Inc., New York, N.Y., 1953).

[18.3] E. Graeser, Einführung in die Theorie der elliptischen Funktionen und deren Anwendungen (R. Oldenbourg, Munich, Germany, 1950).

[18.4] G. H. Halphen, Traité des fonctions elliptiques et de leurs applications, 1 (Gauthier-Villars, Paris, France, 1886).

[18.5] H. Hancock, Lectures on the theory of elliptic functions, vol. 1 (John Wiley & Sons, Inc., New York, N.Y., 1910, reprinted, Dover Publications, Inc., New York, N.Y., 1958).

[18.6] A. Hurwitz and R. Courant, Vorlesungen über allgemeine Funktionentheorie und elliptische Funktionen, 3d ed. (Springer, Berlin, Germany, 1929).

[18.7] E. L. Ince, Ordinary differential equations (Dover Publications, Inc., New York, N.Y., 1944).

[18.8] E. Kamke, Differentialgleichungen, Lösungsmethoden und Lösungen, vol. 1, 2d ed. (Akademische Verlagsgesellschaft, Leipzig, Germany, 1943).

[18.9] D. H. Lehmer, The lemniscate constant, Math. Tables Aids Comp. **3**, 550–551 (1948–49).

[18.10] S. C. Mitra, On the expansion of the Weierstrassian and Jacobian elliptic functions in powers of the argument, Bull. Calcutta Math. Soc. 17, 159–172 (1926).

[18.11] F. Oberhettinger and W. Magnus, Anwendung der elliptischen Funktionen in Physik und Technik (Springer, Berlin, Germany, 1949).

[18.12] G. Prasad, An introduction to the theory of elliptic functions and higher transcendentals (Univ. of Calcutta, India, 1928).

[18.13] U. Richard, Osservazioni sulla bisezione delle funzioni ellittiche di Weierstrass, Boll. Un. Mat. Ital. **3**, **4**, 395–397 (1949).

[18.14] E. S. Selmer, A simple trisection formula for the elliptic function of Weierstrass in the equianharmonic case, Norske Vid. Selsk. Forh. Trondheim **19**, 29, 116–119 (1947).

[18.15] J. Tannery and J. Molk, Éléments de la théorie des fonctions elliptiques, 4 vols. (Gauthier-Villars, Paris, France, 1893–1902).

[18.16] F. Tricomi, Elliptische Funktionen (Akademische Verlagsgesellschaft, Leipzig, Germany, 1948).

[18.17] F. Tricomi, Funzioni ellittiche, 2d ed. (Bologna, Italy, 1951).

[18.18] C. E. Van Orstrand, Reversion of power series, Phil. Mag. (6) 19, 366–376 (Jan.–June 1910).

[18.19] E. T. Whittaker and G. N. Watson, A course of modern analysis, ch. 20, 4th ed. (Cambridge Univ. Press, Cambridge, England, 1952).

Guides, Collections of Formulas, etc.

[18.20] P. F. Byrd and M. D. Friedman, Handbook of elliptic integrals for engineers and physicists, Appendix, sec. 1030 (Springer-Verlag, Berlin, Germany, 1954).

[18.21] A. Fletcher, Guide to tables of elliptic functions, Math. Tables Aids Comp. **3**, 247–249 (1948–49).

[18.22] S. Flügge, Handbuch der Physik, vol. 1, pp. 120-146 (Springer-Verlag, Berlin, Germany, 1956).

[18.23] H. Kober, Dictionary of conformal representations (Dover Publications, Inc., New York, N.Y., 1952).

[18.24] L. M. Milne-Thomson, Jacobian elliptic function tables (Dover Publications, Inc., New York, N.Y., 1950).

[18.25] K. Weierstrass and H. A. Schwarz, Formeln und Lehrsätze zum Gebrauche der elliptischen Funktionen. Nach Vorlesungen und Aufzeichnungen des Herrn K. Weierstrass bearbeitet und herausgegeben von H. A. Schwarz, 2d ed. (Springer, Berlin, Germany, 1893).

Tables

[18.26] Chih-Bing Ling, Evaluation at half-periods of Weierstrass' elliptic function with rectangular primitive period-parallelogram, Math. Comp. **14**, 69, 67–70 (1960). Values of e_i ($i=1, 2, 3$) to 15D for various period ratios in case $\Delta > 0$.

[18.27] E. Jahnke and F. Emde, Tables of functions, 4th ed., pp. 100–106 (Dover Publications, Inc., New York, N.Y., 1945). Equianharmonic case, real argument, $\wp(u), \wp'(u), \zeta(u), \sigma(u), u=0 \left(\dfrac{\omega_2}{180} \right) \dfrac{4\omega_2}{3}$, mostly 4D.

[18.28] T. H. Southard, Approximation and table of the Weierstrass $\wp$ function in the equianharmonic case for real argument, Math. Tables Aids Comp. 11, 58, 99–100 (Apr. 1957). $f(u) = \wp(u) - \dfrac{1}{u^2}$ to 7D, with modified central differences, $u=0(.1).8(.05)1.55$.

[18.29] D. A. Strayhorne, A study of an elliptic function (Thesis, Chicago, Ill., 1946). Air Documents Division T–2, AMC, Wright Field, Microfilm No. Rc–734F15000. $\wp(z; 37, -42)$, 4D, $z=.04i(.04i)1.36i$.

Table 18.1

TABLE FOR OBTAINING PERIODS FOR INVARIANTS g_2 AND g_3

$$(\bar{g}_2 = g_2 g_3^{-\frac{2}{3}})$$

Non-Negative Discriminant

$\bar{g}_2$	$\omega g_3^{\frac{1}{6}}$	$\dfrac{\omega' g_3^{\frac{1}{6}}}{i} + \dfrac{\sqrt{6}}{12}\ln(\bar{g}_2-3)$	
3.00	1.28254 98	1.52168 83	$\Delta = 0$
3.05	1.27944 73	1.51892 22	
3.10	1.27637 43	1.51685 48	
3.15	1.27333 03	1.51505 45	
3.20	1.27031 49	1.51342 84	
3.25	1.26732 80	1.51193 18	
3.30	1.26436 90	1.51053 84	
3.35	1.26143 77	1.50923 08	
3.40	1.25853 38	1.50799 63	

$\bar{g}_2$	$\omega g_3^{\frac{1}{6}}$	$\omega' g_3^{\frac{1}{6}}/i$
3.4	1.25853 38	1.69503 33
3.5	1.25280 64	1.64719 87
3.6	1.24718 42	1.60789 93
3.7	1.24166 45	1.57451 65
3.8	1.23624 47	1.54548 31
3.9	1.23092 23	1.51978 54
4.0	1.22569 47	1.49672 94
4.1	1.22055 95	1.47581 86
4.2	1.21551 44	1.45668 57
4.3	1.21055 69	1.43905 10
4.4	1.20568 50	1.42269 63
4.5	1.20089 62	1.40744 84
4.6	1.19618 86	1.39316 72
4.7	1.19156 00	1.37973 79
4.8	1.18700 83	1.36706 51
4.9	1.18253 18	1.35506 88
5.0	1.17812 83	1.34368 10
5.2	1.16953 35	1.32250 70
5.4	1.16120 96	1.30316 60
5.6	1.15314 34	1.28537 08
5.8	1.14532 23	1.26889 69
6.0	1.13773 46	1.25356 57
6.2	1.13036 91	1.23923 29
6.4	1.12321 55	1.22577 98
6.6	1.11626 38	1.21310 78
6.8	1.10950 49	1.20113 41
7.0	1.10293 00	1.18978 83
7.2	1.09653 11	1.17901 03
7.4	1.09030 03	1.16874 82
7.6	1.08423 04	1.15895 67
7.8	1.07831 46	1.14959 65
8.0	1.07254 63	1.14063 29
8.2	1.06691 95	1.13203 51
8.4	1.06142 83	1.12377 59
8.6	1.05606 74	1.11583 09
8.8	1.05083 15	1.10817 84
9.0	1.04571 58	1.10079 87
9.2	1.04071 56	1.09367 40
9.4	1.03582 65	1.08678 83
9.6	1.03104 44	1.08012 69
9.8	1.02636 52	1.07367 66
10.0	1.02178 54	1.06742 51

$\bar{g}_2^{-1}$	$\omega g_3^{\frac{1}{6}} \bar{g}_2^{\frac{1}{4}}$	$\omega' g_3^{\frac{1}{6}} \bar{g}_2^{\frac{1}{4}}/i$	$\langle \bar{g}_2 \rangle$
0.10	1.81701 99	1.89818 01	10
0.09	1.82207 90	1.89119 06	11
0.08	1.82696 90	1.88476 56	13
0.07	1.83165 87	1.87888 68	14
0.06	1.83611 17	1.87354 40	17
0.05	1.84028 47	1.86873 53	20
0.04	1.84412 45	1.86447 02	25
0.03	1.84756 35	1.86077 37	33
0.02	1.85050 78	1.85769 72	50
0.01	1.85280 73	1.85534 90	100
0.00	1.85407 47	1.85407 47	∞

$$\begin{bmatrix}(-4)1\\10\end{bmatrix} \qquad \begin{bmatrix}(-4)1\\10\end{bmatrix}$$

$$\frac{\sqrt{6}}{12} = 0.20412\ 4145$$

Non-Positive Discriminant

$\bar{g}_2^{-1}$	$\omega_2 g_3^{\frac{1}{6}}|\bar{g}_2|^{\frac{1}{4}}$	$\omega'_2 g_3^{\frac{1}{6}}|\bar{g}_2|^{\frac{1}{4}}/i$	$\langle \bar{g}_2 \rangle$
-0.00	2.62205 76	2.62205 76	- ∞
-0.01	2.62025 54	2.62384 98	-100
-0.02	2.61693 53	2.62710 11	- 50
-0.03	2.61258 87	2.63126 10	- 33
-0.04	2.60737 43	2.63611 20	- 25
-0.05	2.60137 48	2.64151 34	- 20
-0.06	2.59464 00	2.64735 75	- 17
-0.07	2.58720 37	2.65355 47	- 14
-0.08	2.57909 05	2.66002 55	- 13
-0.09	2.57032 09	2.66669 74	- 11
-0.10	2.56091 33	2.67350 25	- 10
-0.11	2.55088 61	2.68037 66	- 9
-0.12	2.54025 86	2.68725 88	- 8
-0.13	2.52905 23	2.69409 09	- 8
-0.14	2.51729 09	2.70081 77	- 7
-0.15	2.50500 11	2.70738 70	- 7
-0.16	2.49221 23	2.71375 03	- 6
-0.17	2.47895 70	2.71986 26	- 6
-0.18	2.46527 01	2.72568 31	- 6
-0.19	2.45118 90	2.73117 52	- 5
-0.20	2.43675 29	2.73630 70	- 5

$\bar{g}_2^{-1}$	$\omega_2 g_3^{\frac{1}{6}}$	$\omega'_2 g_3^{\frac{1}{6}}/i$	$\langle \bar{g}_2 \rangle$
-0.20	1.62955 49	1.82987 88	- 5
-0.25	1.66926 74	1.94863 05	- 4
-0.30	1.68880 94	2.04569 84	- 3
-0.35	1.69574 71	2.12452 34	- 3
-0.40	1.69529 14	2.18836 87	- 3
-0.45	1.69080 53	2.24023 31	- 2
-0.50	1.68433 20	2.28267 03	- 2
-0.55	1.67705 44	2.31773 31	- 2
-0.60	1.66962 98	2.34701 74	- 2
-0.65	1.66240 65	2.37174 42	- 2
-0.70	1.65555 57	2.39284 34	- 1
-0.75	1.64914 98	2.41102 56	- 1
-0.80	1.64320 64	2.42683 68	- 1
-0.85	1.63771 44	2.44070 05	- 1
-0.90	1.63264 84	2.45294 88	- 1
-0.95	1.62797 70	2.46384 40	- 1
-1.00	1.62366 67	2.47359 62	- 1

$\bar{g}_2$	$\omega_2 g_3^{\frac{1}{6}}$	$\dfrac{\omega'_2 g_3^{\frac{1}{6}}}{i} + \dfrac{\sqrt{6}}{6}\ln(3-\bar{g}_2)$	
-1.0	1.62366 67	3.03954 85	
-0.8	1.60646 93	3.05518 40	
-0.6	1.58820 63	3.06892 24	
-0.4	1.56918 06	3.08070 50	
-0.2	1.54967 81	3.09053 50	
0.0	1.52995 40	3.09846 47	
0.2	1.51022 67	3.10458 18	
0.4	1.49067 44	3.10899 55	
0.6	1.47143 75	3.11182 48	
0.8	1.45262 13	3.11318 95	
1.0	1.43430 15	3.11320 22	
1.2	1.41652 88	3.11196 36	
1.4	1.39933 41	3.10955 78	
1.6	1.38273 24	3.10604 84	
1.8	1.36672 71	3.10147 38	
2.0	1.35131 24	3.09584 00	
2.2	1.33647 63	3.08910 74	
2.4	1.32220 24	3.08116 35	
2.6	1.30847 11	3.07175 37	
2.8	1.29526 10	3.06025 10	
3.0	1.28254 98	3.04337 67	$\Delta = 0$

$$\begin{bmatrix}(-3)3\\10\end{bmatrix} \qquad \begin{bmatrix}(-3)3\\11\end{bmatrix}$$

$$\frac{\sqrt{6}}{6} = 0.40824\ 829$$

Table 18.2　　　　TABLE FOR OBTAINING $\mathscr{P}$, $\mathscr{P}'$ AND ζ ON $0x$ AND $0y$

(Positive Discriminant—Unit Real Half-Period)

$$z^2\mathscr{P}(z)$$

$z=x\backslash a$	1.00	1.05	1.1	1.2	1.4	2.0	4.0
0.00	1.00000 00	1.00000 00	1.00000 00	1.00000 00	1.00000 00	1.00000 00	1.00000 00
0.05	1.00000 37	1.00000 34	1.00000 32	1.00000 29	1.00000 26	1.00000 25	1.00000 25
0.10	1.00005 91	1.00005 41	1.00005 05	1.00004 59	1.00004 22	1.00004 08	1.00004 07
0.15	1.00029 91	1.00027 41	1.00025 59	1.00023 31	1.00021 46	1.00020 75	1.00020 73
0.20	1.00094 57	1.00086 77	1.00081 12	1.00074 02	1.00068 25	1.00066 02	1.00065 97
0.25	1.00230 98	1.00212 32	1.00198 79	1.00181 79	1.00167 98	1.00162 64	1.00162 51
0.30	1.00479 35	1.00441 61	1.00414 21	1.00379 79	1.00351 80	1.00340 97	1.00340 71
0.35	1.00889 27	1.00821 33	1.00772 00	1.00709 99	1.00659 56	1.00640 03	1.00639 57
0.40	1.01520 23	1.01408 14	1.01326 70	1.01224 31	1.01140 98	1.01108 69	1.01107 93
0.45	1.02442 50	1.02269 65	1.02144 05	1.01985 94	1.01857 24	1.01807 36	1.01806 19
0.50	1.03738 54	1.03486 08	1.03302 47	1.03071 36	1.02883 08	1.02810 10	1.02808 38
0.55	1.05504 92	1.05152 36	1.04895 81	1.04572 73	1.04309 40	1.04207 28	1.04204 87
0.60	1.07855 23	1.07381 21	1.07036 11	1.06601 29	1.06246 70	1.06109 15	1.06105 91
0.65	1.10923 99	1.10307 22	1.09857 95	1.09291 64	1.08829 58	1.08650 29	1.08646 07
0.70	1.14872 15	1.14092 35	1.13524 09	1.12807 45	1.12222 46	1.11995 41	1.11990 05
0.75	1.19894 38	1.18933 40	1.18232 81	1.17348 94	1.16627 18	1.16346 98	1.16340 37
0.80	1.26229 01	1.25071 86	1.24227 98	1.23162 95	1.22292 96	1.21955 14	1.21947 17
0.85	1.34171 37	1.32807 28	1.31812 18	1.30556 03	1.29529 60	1.29130 97	1.29121 57
0.90	1.44091 81	1.42515 17	1.41364 80	1.39912 31	1.38725 23	1.38264 14	1.38253 27
0.95	1.56460 22	1.54671 40	1.53366 04	1.51717 65	1.50370 31	1.49846 94	1.49834 59
1.00	1.71879 62	1.69885 59	1.68430 41	1.66592 77	1.65090 68	1.64507 17	1.64493 41
	$\begin{bmatrix}(-3)4\\8\end{bmatrix}$	$\begin{bmatrix}(-3)4\\8\end{bmatrix}$	$\begin{bmatrix}(-3)4\\8\end{bmatrix}$	$\begin{bmatrix}(-3)4\\8\end{bmatrix}$	$\begin{bmatrix}(-3)4\\8\end{bmatrix}$	$\begin{bmatrix}(-3)4\\8\end{bmatrix}$	$\begin{bmatrix}(-3)4\\8\end{bmatrix}$

$z/i=y\backslash a$	1.00	1.05	1.1	1.2	1.4	2.0	4.0
0.00	1.00000 00	1.00000 00	1.00000 00	1.00000 00	1.00000 00	1.00000 00	1.00000 00
0.05	1.00000 37	1.00000 34	1.00000 31	1.00000 29	1.00000 26	1.00000 25	1.00000 25
0.10	1.00005 91	1.00005 03	1.00005 03	1.00004 57	1.00004 19	1.00004 05	1.00004 04
0.15	1.00029 91	1.00027 31	1.00025 42	1.00023 05	1.00021 13	1.00020 39	1.00020 37
0.20	1.00094 57	1.00086 20	1.00080 14	1.00072 54	1.00066 38	1.00063 99	1.00063 94
0.25	1.00230 98	1.00210 14	1.00195 05	1.00176 15	1.00160 81	1.00154 88	1.00154 75
0.30	1.00479 35	1.00435 08	1.00403 04	1.00362 91	1.00330 38	1.00317 81	1.00317 52
0.35	1.00889 27	1.00804 86	1.00743 81	1.00667 40	1.00605 50	1.00581 59	1.00581 03
0.40	1.01520 23	1.01371 37	1.01263 81	1.01129 28	1.01020 38	1.00978 33	1.00977 34
0.45	1.02442 50	1.02194 93	1.02016 25	1.01792 92	1.01612 33	1.01542 64	1.01540 99
0.50	1.03738 54	1.03345 04	1.03061 34	1.02707 18	1.02421 09	1.02310 77	1.02308 17
0.55	1.05504 92	1.04901 44	1.04466 92	1.03925 21	1.03488 20	1.03319 83	1.03315 85
0.60	1.07855 23	1.06955 87	1.06309 37	1.05504 64	1.04856 45	1.04606 96	1.04601 09
0.65	1.10923 99	1.09614 60	1.08675 16	1.07507 92	1.06569 47	1.06208 70	1.06200 18
0.70	1.14872 15	1.13001 89	1.11663 04	1.10003 09	1.08671 44	1.08160 18	1.08148 16
0.75	1.19894 38	1.17264 63	1.15387 03	1.13065 03	1.11207 03	1.10494 84	1.10478 09
0.80	1.26229 01	1.22578 78	1.19980 68	1.16777 18	1.14221 52	1.13243 76	1.13220 79
0.85	1.34171 37	1.29157 86	1.25602 53	1.21233 97	1.17761 18	1.16435 46	1.16404 34
0.90	1.44091 81	1.37264 39	1.32443 52	1.26544 15	1.21873 89	1.20095 66	1.20053 95
0.95	1.56460 22	1.47224 79	1.40736 61	1.32835 02	1.26610 10	1.24247 14	1.24191 74
1.00	1.71879 62	1.59449 89	1.50769 66	1.40258 06	1.32024 17	1.28909 73	1.28836 81
1.05		1.74462 36	1.62902 39				
1.10			1.77589 10				
	$\begin{bmatrix}(-3)4\\8\end{bmatrix}$	$\begin{bmatrix}(-3)3\\8\end{bmatrix}$	$\begin{bmatrix}(-3)3\\7\end{bmatrix}$	$\begin{bmatrix}(-3)1\\7\end{bmatrix}$	$\begin{bmatrix}(-4)8\\6\end{bmatrix}$	$\begin{bmatrix}(-4)6\\6\end{bmatrix}$	$\begin{bmatrix}(-4)6\\6\end{bmatrix}$

$z/i=y\backslash a$	1.00	1.05	1.1	1.2	1.4	2.0	4.0
1.0	1.71879 62	1.59449 89	1.50769 66	1.40258 06	1.32024 17	1.28909 73	1.288368
1.2				1.85616 29	1.61789 95	1.52970 17	1.527649
1.4					2.09401 44	1.86127 05	1.855916
1.6						2.28676 23	2.273495
1.8						2.80921 52	2.777516
2.0						3.43759 29	3.363868
2.2							4.028426
2.4							4.767658
2.6							5.578809
2.8							6.459856
3.0							7.409386
3.2							8.426442
3.4							9.510400
3.6							10.660867
3.8							11.877621
4.0							13.160574

If the real half-period $\neq 1$, see **18.2** Homogeneity Relations. Interpolation with respect to a will, in general, be difficult because of the non-uniform subintervals involved. Aitken's interpolation may be used in this case. As few as 3S may be obtained. For the computation of $\mathscr{P}$, $\mathscr{P}'$ or ζ at $z=x+iy$, an addition formula may be used (**18.4** and **Examples 11–12**).

TABLE FOR OBTAINING $\wp, \wp'$ AND ζ ON $0x$ AND $0y$ — Table 18.2

(Positive Discriminant—Unit Real Half-Period)

$$z^3 \wp'(z)$$

$z=x \backslash a$	1.00	1.05	1.1	1.2	1.4	2.0	4.0
0.00	-2.00000 00	-2.00000 00	-2.00000 00	-2.00000 00	-2.00000 00	-2.00000 00	-2.00000 00
0.05	-1.99999 26	-1.99999 32	-1.99999 37	-1.99999 43	-1.99999 47	-1.99999 49	-1.99999 49
0.10	-1.99988 18	-1.99989 17	-1.99989 89	-1.99990 80	-1.99991 53	-1.99991 81	-1.99991 82
0.15	-1.99940 16	-1.99945 07	-1.99948 63	-1.99953 10	-1.99956 73	-1.99958 14	-1.99958 17
0.20	-1.99810 75	-1.99825 79	-1.99836 70	-1.99850 41	-1.99861 55	-1.99865 86	-1.99865 97
0.25	-1.99537 33	-1.99572 57	-1.99598 17	-1.99630 33	-1.99656 50	-1.99666 63	-1.99666 88
0.30	-1.99038 23	-1.99107 69	-1.99158 17	-1.99221 67	-1.99273 38	-1.99293 42	-1.99293 89
0.35	-1.98210 95	-1.98332 00	-1.98420 07	-1.98530 95	-1.98621 31	-1.98656 35	-1.98657 17
0.40	-1.96928 90	-1.97121 06	-1.97260 49	-1.97437 35	-1.97581 22	-1.97637 02	-1.97638 34
0.45	-1.95036 13	-1.95319 16	-1.95525 47	-1.95785 77	-1.95998 33	-1.96080 82	-1.96082 78
0.50	-1.92339 01	-1.92730 50	-1.93016 21	-1.93377 03	-1.93671 95	-1.93786 53	-1.93789 23
0.55	-1.88593 83	-1.89106 43	-1.89480 97	-1.89954 33	-1.90341 73	-1.90492 32	-1.90495 86
0.60	-1.83488 99	-1.84127 27	-1.84594 09	-1.85184 82	-1.85668 71	-1.85856 93	-1.85861 37
0.65	-1.76619 53	-1.77376 97	-1.77931 45	-1.78633 89	-1.79209 80	-1.79433 95	-1.79439 25
0.70	-1.67451 43	-1.68307 45	-1.60934 72	-1.69729 96	-1.70382 60	-1.70636 76	-1.70642 75
0.75	-1.55271 74	-1.56189 13	-1.56861 96	-1.57715 61	-1.58416 75	-1.58689 93	-1.58696 39
0.80	-1.39118 65	-1.40041 70	-1.40719 15	-1.41579 29	-1.42286 23	-1.42561 79	-1.42568 30
0.85	-1.17683 20	-1.18536 53	-1.19163 25	-1.19959 24	-1.20613 88	-1.20869 13	-1.20875 17
0.90	-0.89169 81	-0.89858 18	-0.90364 00	-0.91006 69	-0.91535 50	-0.91741 70	-0.91746 57
0.95	-0.51095 87	-0.51505 33	-0.51806 28	-0.52188 70	-0.52503 45	-0.52626 26	-0.52629 14
1.00	0.00000 00	0.00000 00	0.00000 00	0.00000 00	0.00000 00	0.00000 00	0.00000 00
	$\begin{bmatrix}(-2)2\\9\end{bmatrix}$	$\begin{bmatrix}(-2)2\\9\end{bmatrix}$	$\begin{bmatrix}(-2)2\\9\end{bmatrix}$	$\begin{bmatrix}(-2)2\\9\end{bmatrix}$	$\begin{bmatrix}(-2)2\\9\end{bmatrix}$	$\begin{bmatrix}(-2)2\\9\end{bmatrix}$	$\begin{bmatrix}(-2)2\\9\end{bmatrix}$

$z/i=y \backslash a$	1.00	1.05	1.1	1.2	1.4	2.0	4.0
0.00	-2.00000 00	-2.00000 00	-2.00000 00	-2.00000 00	-2.00000 00	-2.00000 00	-2.00000 00
0.05	-1.99999 26	-1.99999 32	-1.99999 37	-1.99999 43	-1.99999 48	-1.99999 49	-1.99999 49
0.10	-1.99988 18	-1.99989 21	-1.99989 95	-1.99990 89	-1.99991 65	-1.99991 94	-1.99991 95
0.15	-1.99940 16	-1.99945 48	-1.99949 33	-1.99954 15	-1.99958 07	-1.99959 59	-1.99959 62
0.20	-1.99810 75	-1.99828 08	-1.99840 62	-1.99856 33	-1.99869 07	-1.99873 99	-1.99874 11
0.25	-1.99537 33	-1.99581 31	-1.99613 14	-1.99652 94	-1.99685 19	-1.99697 66	-1.99697 95
0.30	-1.99038 23	-1.99133 82	-1.99202 89	-1.99289 25	-1.99359 12	-1.99386 12	-1.99386 76
0.35	-1.98210 95	-1.98398 06	-1.98533 03	-1.98701 63	-1.98837 91	-1.98890 48	-1.98891 71
0.40	-1.96928 90	-1.97268 69	-1.97513 44	-1.97818 68	-1.98065 01	-1.98159 94	-1.98162 18
0.45	-1.95036 13	-1.95619 80	-1.96039 48	-1.96561 82	-1.96982 60	-1.97144 57	-1.97148 38
0.50	-1.92339 01	-1.93299 84	-1.93989 10	-1.94845 17	-1.95533 26	-1.95797 74	-1.95803 95
0.55	-1.88593 83	-1.90123 75	-1.91218 25	-1.92574 23	-1.93661 23	-1.94078 35	-1.94088 17
0.60	-1.83488 99	-1.85861 50	-1.87553 39	-1.89643 16	-1.91313 16	-1.91952 74	-1.91967 77
0.65	-1.76619 53	-1.80221 44	-1.82780 48	-1.85930 08	-1.88437 77	-1.89395 96	-1.89418 46
0.70	-1.67451 43	-1.72827 05	-1.76629 64	-1.81290 09	-1.84984 78	-1.86392 68	-1.86425 71
0.75	-1.55271 74	-1.63184 71	-1.68753 62	-1.75545 41	-1.80902 61	-1.82937 52	-1.82985 21
0.80	-1.39118 65	-1.50639 22	-1.58698 80	-1.68471 79	-1.76134 96	-1.79034 89	-1.79102 80
0.85	-1.17683 20	-1.34312 50	-1.45865 26	-1.59780 32	-1.70615 96	-1.74698 46	-1.74793 96
0.90	-0.89169 81	-1.13018 63	-1.29452 95	-1.49093 18	-1.64263 75	-1.69950 14	-1.70082 95
0.95	-0.51095 87	-0.85145 23	-1.08387 84	-1.35912 08	-1.56972 20	-1.64818 82	-1.65001 75
1.00	0.00000 00	-0.48485 79	-0.81220 52	-1.19575 58	-1.48600 58	-1.59338 85	-1.59588 68
1.05		0.00000 00	-0.45984 59				
1.10			0.00000 00				
	$\begin{bmatrix}(-2)2\\9\end{bmatrix}$	$\begin{bmatrix}(-2)1\\9\end{bmatrix}$	$\begin{bmatrix}(-2)1\\9\end{bmatrix}$	$\begin{bmatrix}(-3)4\\9\end{bmatrix}$	$\begin{bmatrix}(-3)1\\7\end{bmatrix}$	$\begin{bmatrix}(-4)4\\6\end{bmatrix}$	$\begin{bmatrix}(-4)6\\6\end{bmatrix}$

$z/i=y \backslash a$	1.00	1.05	1.1	1.2	1.4	2.0	4.0
1.0	0.00000 00	-0.48485 79	-0.81220 52	-1.19575 58	-1.48600 58	-1.59338 85	-1.59588 68
1.2				0.00000 00	-0.99449 51	-1.34717 40	-1.35527 93
1.4					0.00000 00	-1.07521 03	-1.09935 83
1.6						-0.78786 76	-0.85550 88
1.8						-0.46104 27	-0.64191 20
2.0						0.00000 00	-0.46669 27
2.2							-0.33022 92
2.4							-0.22828 89
2.6							-0.15467 43
2.8							-0.10296 79
3.0							-0.06745 48
3.2							-0.04346 22
3.4							-0.02734 75
3.6							-0.01629 07
3.8							-0.00795 66
4.0							0.00000 00

Table 18.2   **TABLE FOR OBTAINING $\mathcal{P}$, $\mathcal{P}'$ AND ζ ON $0x$ AND $0y$**

(Positive Discriminant—Unit Real Half-Period)

$z\zeta(z)$

$z=x\backslash a$	1.00	1.05	1.1	1.2	1.4	2.0	4.0
0.00	1.00000 000	1.00000 000	1.00000 000	1.00000 000	1.00000 000	1.00000 000	1.00000 000
0.05	0.99999 876	0.99999 887	0.99999 895	0.99999 905	0.99999 912	0.99999 915	0.99999 915
0.10	0.99998 031	0.99998 198	0.99998 319	0.99998 471	0.99998 595	0.99998 643	0.99998 644
0.15	0.99990 029	0.99990 871	0.99991 481	0.99992 246	0.99992 868	0.99993 109	0.99993 115
0.20	0.99968 483	0.99971 119	0.99973 030	0.99975 429	0.99977 377	0.99978 130	0.99978 148
0.25	0.99923 041	0.99929 399	0.99934 010	0.99939 799	0.99944 501	0.99946 321	0.99946 364
0.30	0.99840 360	0.99853 355	0.99862 782	0.99874 617	0.99884 235	0.99887 957	0.99888 045
0.35	0.99704 076	0.99727 741	0.99744 912	0.99766 478	0.99784 008	0.99790 793	0.99790 954
0.40	0.99494 715	0.99534 298	0.99563 028	0.99599 122	0.99628 469	0.99639 831	0.99640 099
0.45	0.99189 577	0.99251 583	0.99296 602	0.99353 179	0.99399 196	0.99417 016	0.99417 438
0.50	0.98762 541	0.98854 726	0.98921 683	0.99005 855	0.99074 340	0.99100 867	0.99101 490
0.55	0.98183 783	0.98315 105	0.98410 521	0.98530 511	0.98628 174	0.98666 012	0.98666 904
0.60	0.97419 386	0.97599 894	0.97731 096	0.97896 146	0.98030 531	0.98082 605	0.98083 833
0.65	0.96430 782	0.96671 478	0.96846 489	0.97066 726	0.97246 106	0.97315 633	0.97317 272
0.70	0.95174 028	0.95486 674	0.95714 079	0.96000 343	0.96233 582	0.96324 002	0.96326 132
0.75	0.93598 819	0.93995 720	0.94284 503	0.94648 146	0.94944 525	0.95059 446	0.95062 155
0.80	0.91647 208	0.92140 960	0.92500 321	0.92952 973	0.93322 007	0.93465 128	0.93468 503
0.85	0.89251 910	0.89855 136	0.90294 299	0.90847 617	0.91298 848	0.91473 876	0.91478 003
0.90	0.86334 108	0.87059 177	0.87587 177	0.88252 588	0.88795 364	0.89005 936	0.89010 902
0.95	0.82800 562	0.83659 307	0.84284 790	0.85073 222	0.85716 486	0.85966 076	0.85971 964
1.00	0.78539 822	0.79543 267	0.80274 283	0.81195 906	0.81947 977	0.82239 820	0.82246 703
	$\begin{bmatrix}(-4)9\\7\end{bmatrix}$	$\begin{bmatrix}(-4)9\\8\end{bmatrix}$	$\begin{bmatrix}(-4)9\\8\end{bmatrix}$	$\begin{bmatrix}(-4)9\\8\end{bmatrix}$	$\begin{bmatrix}(-4)9\\8\end{bmatrix}$	$\begin{bmatrix}(-4)9\\8\end{bmatrix}$	$\begin{bmatrix}(-4)9\\8\end{bmatrix}$

$z/i=y\backslash a$	1.00	1.05	1.1	1.2	1.4	2.0	4.0
0.00	1.00000 000	1.00000 000	1.00000 000	1.00000 000	1.00000 000	1.00000 000	1.00000 000
0.05	0.99999 876	0.99999 887	0.99999 895	0.99999 905	0.99999 912	0.99999 915	0.99999 916
0.10	0.99998 031	0.99998 200	0.99998 322	0.99998 476	0.99998 601	0.99998 649	0.99998 650
0.15	0.99990 029	0.99990 891	0.99991 516	0.99992 299	0.99992 935	0.99993 181	0.99993 187
0.20	0.99968 483	0.99971 234	0.99973 226	0.99975 725	0.99977 752	0.99978 537	0.99978 555
0.25	0.99923 041	0.99929 836	0.99934 758	0.99940 928	0.99945 935	0.99947 871	0.99947 917
0.30	0.99840 360	0.99854 660	0.99865 014	0.99877 991	0.99888 517	0.99892 586	0.99892 682
0.35	0.99704 076	0.99731 033	0.99750 544	0.99774 989	0.99794 811	0.99802 472	0.99802 653
0.40	0.99494 715	0.99541 639	0.99575 586	0.99618 100	0.99652 557	0.99665 871	0.99666 184
0.45	0.99189 577	0.99266 485	0.99322 092	0.99391 695	0.99448 077	0.99469 855	0.99470 368
0.50	0.98762 541	0.98882 817	0.98969 725	0.99078 438	0.99166 445	0.99200 425	0.99201 225
0.55	0.98183 783	0.98364 988	0.98495 820	0.98659 357	0.98791 646	0.98842 700	0.98843 902
0.60	0.97419 386	0.97684 238	0.97875 291	0.98113 896	0.98306 740	0.98381 123	0.98382 874
0.65	0.96430 782	0.96808 373	0.97080 464	0.97419 926	0.97694 003	0.97799 651	0.97802 138
0.70	0.95174 028	0.95701 320	0.96080 810	0.96553 710	0.96935 061	0.97081 949	0.97085 406
0.75	0.93598 819	0.94322 518	0.94842 600	0.95489 807	0.96010 986	0.96211 557	0.96216 276
0.80	0.91647 208	0.92626 102	0.93328 385	0.94200 908	0.94902 381	0.95172 061	0.95178 405
0.85	0.89251 910	0.90559 833	0.91496 295	0.92657 574	0.93589 412	0.93947 230	0.93955 644
0.90	0.86334 108	0.88063 688	0.89299 175	0.90827 878	0.92051 815	0.92521 144	0.92532 176
0.95	0.82800 562	0.85068 069	0.86683 386	0.88676 908	0.90268 849	0.90878 307	0.90892 628
1.00	0.78539 822	0.81491 420	0.83587 315	0.86166 128	0.88219 209	0.89003 731	0.89022 154
1.05		0.77237 164	0.79939 419				
1.10			0.75655 714				
	$\begin{bmatrix}(-4)9\\7\end{bmatrix}$	$\begin{bmatrix}(-4)8\\8\end{bmatrix}$	$\begin{bmatrix}(-4)8\\7\end{bmatrix}$	$\begin{bmatrix}(-4)4\\7\end{bmatrix}$	$\begin{bmatrix}(-4)3\\6\end{bmatrix}$	$\begin{bmatrix}(-4)3\\6\end{bmatrix}$	$\begin{bmatrix}(-4)3\\6\end{bmatrix}$

$z/i=y\backslash a$	1.00	1.05	1.1	1.2	1.4	2.0	4.0
1.0	0.78539 822	0.81491 420	0.83587 315	0.86166 128	0.88219 209	0.89003 731	0.89022 15
1.2				0.71573 454	0.76897 769	0.78909 505	0.78956 60
1.4					0.59293 450	0.64073 496	0.64184 73
1.6						0.43846 099	0.44095 77
1.8						+0.17708 802	+0.18250 43
2.0						−0.14800 012	−0.13652 01
2.2							−0.51809 61
2.4							−0.96348 97
2.6							−1.47349 03
2.8							−2.04858 16
3.0							−2.68905 52
3.2							−3.39508 38
3.4							−4.16677 17
3.6							−5.00417 86
3.8							−5.90734 21
4.0							−6.87630 32
						$\begin{bmatrix}(-3)8\end{bmatrix}$	$\begin{bmatrix}(-3)8\\10\end{bmatrix}$

TABLE FOR OBTAINING $\mathscr{P}, \mathscr{P}'$ AND ζ ON $0x$ AND $0y$ Table 18.2

(Negative Discriminant—Unit Real Half-Period)

$z^2\mathscr{P}(z)$

$z=x\backslash a$	1.00	1.05	1.15	1.3	1.5	2.0	4.0
0.00	1.00000 00	1.00000 00	1.00000 00	1.00000 00	1.00000 00	1.00000 00	1.00000 00
0.05	0.99998 52	0.99998 68	0.99998 98	0.99999 38	0.99999 75	1.00000 14	1.00000 25
0.10	0.99976 37	0.99978 83	0.99983 74	0.99990 10	0.99996 06	1.00002 30	1.00004 07
0.15	0.99880 40	0.99893 08	0.99918 15	0.99950 43	0.99980 51	1.00011 83	1.00020 71
0.20	0.99622 33	0.99663 32	0.99743 55	0.99845 77	0.99940 30	1.00038 24	1.00065 92
0.25	0.99079 63	0.99182 47	0.99381 16	0.99631 17	0.99860 26	1.00096 01	1.00162 38
0.30	0.98097 82	0.98317 67	0.98736 11	0.99255 06	0.99725 51	1.00205 83	1.00340 46
0.35	0.96495 11	0.96915 65	0.97703 14	0.98664 20	0.99525 02	1.00396 16	1.00639 11
0.40	0.94070 57	0.94811 25	0.96174 61	0.97810 01	0.99255 94	1.00705 13	1.01107 17
0.45	0.90617 03	0.91839 70	0.94051 05	0.96656 45	0.98928 71	1.01183 11	1.01805 02
0.50	0.85939 83	0.87853 56	0.91254 55	0.95189 16	0.98573 01	1.01895 42	1.02806 66
0.55	0.79882 11	0.82744 45	0.87744 80	0.93426 12	0.98244 30	1.02925 89	1.04202 47
0.60	0.72356 52	0.76469 39	0.83537 63	0.91429 23	0.98031 24	1.04381 01	1.06102 67
0.65	0.63382 07	0.69080 48	0.78725 05	0.89316 80	0.98063 64	1.06395 05	1.08641 83
0.70	0.53123 69	0.60756 14	0.73495 90	0.87276 38	0.98521 20	1.09136 32	1.11984 70
0.75	0.41930 23	0.51830 84	0.68155 50	0.85577 68	0.99643 13	1.12815 05	1.16333 76
0.80	0.30366 33	0.42820 16	0.63143 16	0.84585 35	1.01739 07	1.17693 44	1.21939 20
0.85	0.19233 10	0.34438 12	0.59046 32	0.84771 96	1.05201 81	1.24098 76	1.29112 16
0.90	0.09574 08	0.27605 07	0.56611 51	0.86731 78	1.10523 21	1.32440 72	1.38242 38
0.95	0.02666 27	0.23446 42	0.56753 12	0.91197 25	1.18314 77	1.43234 85	1.49822 24
1.00	0.00000 00	0.23286 11	0.60563 48	0.99060 83	1.29335 96	1.57134 70	1.64479 64
	$\begin{bmatrix} (-3)5 \\ 8 \end{bmatrix}$	$\begin{bmatrix} (-3)5 \\ 8 \end{bmatrix}$	$\begin{bmatrix} (-3)5 \\ 8 \end{bmatrix}$	$\begin{bmatrix} (-3)4 \\ 8 \end{bmatrix}$	$\begin{bmatrix} (-3)4 \\ 8 \end{bmatrix}$	$\begin{bmatrix} (-3)4 \\ 8 \end{bmatrix}$	$\begin{bmatrix} (-3)4 \\ 8 \end{bmatrix}$

$z/i=y\backslash a$	1.00	1.05	1.15	1.3	1.5	2.0	4.0
0.00	1.00000 00	1.00000 00	1.00000 00	1.00000 00	1.00000 00	1.00000 00	1.00000 00
0.05	0.99998 52	0.99998 67	0.99998 98	0.99999 37	0.99999 75	1.00000 14	1.00000 32
0.10	0.99976 37	0.99978 76	0.99983 59	0.99989 93	0.99995 93	1.00002 24	1.00004 04
0.15	0.99880 40	0.99892 27	0.99916 47	0.99948 51	0.99978 96	1.00011 15	1.00020 35
0.20	0.99622 33	0.99658 78	0.99734 10	0.99834 96	0.99931 61	1.00034 41	1.00063 88
0.25	0.99079 63	0.99165 20	0.99345 16	0.99589 95	0.99827 12	1.00081 39	1.00154 61
0.30	0.98097 82	0.98266 22	0.98628 83	0.99132 10	0.99626 60	1.00162 14	1.00317 22
0.35	0.96495 11	0.96786 42	0.97433 43	0.98354 71	0.99275 81	1.00285 94	1.00580 47
0.40	0.94070 57	0.94525 04	0.95576 47	0.97122 41	0.98701 30	1.00459 41	1.00976 35
0.45	0.90617 03	0.91264 56	0.92846 67	0.95268 27	0.97806 19	1.00684 49	1.01539 36
0.50	0.85939 83	0.86784 46	0.89009 57	0.92592 17	0.96465 71	1.00955 92	1.02305 58
0.55		0.80881 13	0.83817 66	0.88861 10	0.94522 83	1.01258 51	1.03311 90
0.60			0.77024 24	0.83812 71	0.91784 50	1.01563 95	1.04595 22
0.65				0.77163 28	0.88019 00	1.01827 41	1.06191 71
0.70					0.82955 45	1.01983 61	1.08136 14
0.75					0.76286 31	1.01942 61	1.10461 36
0.80						1.01585 25	1.13197 83
0.85						1.00758 28	1.16373 23
0.90						0.99269 39	1.20012 24
0.95						0.96882 29	1.24136 39
1.00						0.93312 29	1.28763 91
	$\begin{bmatrix} (-3)2 \\ 7 \end{bmatrix}$	$\begin{bmatrix} (-3)2 \\ 7 \end{bmatrix}$	$\begin{bmatrix} (-3)2 \\ 8 \end{bmatrix}$	$\begin{bmatrix} (-3)2 \\ 7 \end{bmatrix}$	$\begin{bmatrix} (-3)2 \\ 7 \end{bmatrix}$	$\begin{bmatrix} (-3)1 \\ 6 \end{bmatrix}$	$\begin{bmatrix} (-4)6 \\ 6 \end{bmatrix}$

$z/i=y\backslash a$	4.0
1.1	1.39585 80
1.2	1.52559 80
1.3	1.67719 97
1.4	1.85056 87
1.5	2.04521 26
1.6	2.26025 62
1.7	2.49441 96
1.8	2.74594 50
1.9	3.01245 16
2.0	3.29069 52
	$\begin{bmatrix} (-3)3 \\ 7 \end{bmatrix}$

If the real half-period $\neq 1$, see **18.2** Homogeneity Relations. Interpolation with respect to a will, in general, be difficult because of the non-uniform subintervals involved. Aitken's interpolation may be used in this case. As few as 3S may be obtained. For the computation of $\mathscr{P}, \mathscr{P}'$ or ζ at $z=x+iy$, an addition formula may be used (**18.4** and **Examples 11–12**).

Table 18.2   **TABLE FOR OBTAINING $\mathcal{P}$, $\mathcal{P}'$ AND ζ ON $0x$ AND $0y$**
(Negative Discriminant—Unit Real Half-Period)
$$z^3\mathcal{P}'(z)$$

$z=x\backslash a$	1.00	1.05	1.15	1.3	1.5	2.0	4.0
0.00	-2.00000 00	-2.00000 00	-2.00000 00	-2.00000 00	-2.00000 00	-2.00000 00	-2.00000 00
0.05	-2.00002 95	-2.00002 65	-2.00002 04	-2.00001 24	-2.00000 50	-1.99999 71	-1.99999 49
0.10	-2.00047 25	-2.00042 27	-2.00032 37	-2.00019 63	-2.00007 74	-1.99995 34	-1.99991 83
0.15	-2.00239 01	-2.00212 89	-2.00161 92	-2.00097 17	-2.00037 44	-1.99975 65	-1.99958 21
0.20	-2.00753 43	-2.00667 30	-2.00502 56	-2.00297 32	-2.00110 66	-1.99919 66	-1.99866 07
0.25	-2.01829 41	-2.01608 73	-2.01196 38	-2.00694 49	-2.00246 05	-1.99793 23	-1.99667 11
0.30	-2.03755 78	-2.03274 55	-2.02397 99	-2.01358 73	-2.00448 84	-1.99544 16	-1.99294 36
0.35	-2.06843 88	-2.05907 94	-2.04247 95	-2.02334 71	-2.00696 68	-1.99095 58	-1.98657 99
0.40	-2.11379 74	-2.09713 03	-2.06835 37	-2.03614 78	-2.00922 15	-1.98338 63	-1.97639 65
0.45	-2.17550 18	-2.14789 87	-2.10148 48	-2.05106 10	-2.00992 37	-1.97120 64	-1.96084 72
0.50	-2.25339 16	-2.21047 72	-2.14013 46	-2.06592 49	-2.00685 64	-1.95234 05	-1.93791 93
0.55	-2.34395 53	-2.28098 85	-2.18023 97	-2.07692 41	-1.99665 49	-1.92399 70	-1.90499 42
0.60	-2.43881 27	-2.35140 73	-2.21466 43	-2.07815 03	-1.97452 31	-1.88246 83	-1.85865 81
0.65	-2.52318 49	-2.40840 49	-2.23248 50	-2.06116 83	-1.93392 01	-1.82286 83	-1.79444 54
0.70	-2.57463 40	-2.43241 27	-2.21839 89	-2.01460 73	-1.86620 81	-1.73878 53	-1.70648 76
0.75	-2.56240 86	-2.39712 18	-2.15233 79	-1.92378 08	-1.76023 25	-1.62181 13	-1.58702 84
0.80	-2.44770 16	-2.26959 69	-2.00933 39	-1.77031 11	-1.60178 75	-1.46089 21	-1.42574 81
0.85	-2.18496 84	-2.01105 50	-1.75959 77	-1.53168 32	-1.37288 13	-1.24141 08	-1.20881 20
0.90	-1.72414 78	-1.57813 99	-1.36864 82	-1.18057 88	-1.05066 42	-0.94387 76	-0.91751 44
0.95	-1.01321 01	-0.92423 16	-0.79716 03	-0.68374 39	-0.60580 78	-0.54202 52	-0.52632 04
1.00	0.00000 00	0.00000 00	0.00000 00	0.00000 00	0.00000 00	0.00000 00	0.00000 00
	$\begin{bmatrix}(-2)4\\10\end{bmatrix}$	$\begin{bmatrix}(-2)3\\9\end{bmatrix}$	$\begin{bmatrix}(-2)3\\9\end{bmatrix}$	$\begin{bmatrix}(-2)2\\9\end{bmatrix}$	$\begin{bmatrix}(-2)2\\9\end{bmatrix}$	$\begin{bmatrix}(-2)2\\9\end{bmatrix}$	$\begin{bmatrix}(-2)2\\9\end{bmatrix}$

$z/i=y\backslash a$	1.00	1.05	1.15	1.3	1.5	2.0	4.0
0.00	-2.00000 00	-2.00000 00	-2.00000 00	-2.00000 00	-2.00000 00	-2.00000 00	-2.00000 00
0.05	-2.00002 95	-2.00002 65	-2.00002 05	-2.00001 25	-2.00000 50	-1.99999 72	-1.99999 49
0.10	-2.00047 25	-2.00042 55	-2.00032 97	-2.00020 30	-2.00008 28	-1.99995 74	-1.99991 95
0.15	-2.00239 01	-2.00216 12	-2.00168 65	-2.00104 87	-2.00043 62	-1.99978 38	-1.99959 66
0.20	-2.00753 43	-2.00685 42	-2.00540 32	-2.00340 55	-2.00145 41	-1.99935 00	-1.99874 22
0.25	-2.01829 41	-2.01677 67	-2.01340 12	-2.00859 22	-2.00378 54	-1.99851 75	-1.99698 24
0.30	-2.03755 78	-2.03479 40	-2.02825 59	-2.01849 50	-2.00844 10	-1.99718 99	-1.99387 40
0.35	-2.06843 88	-2.06420 40	-2.05319 59	-2.03567 60	-2.01691 87	-1.99536 97	-1.98892 95
0.40	-2.11379 74	-2.10841 06	-2.09200 85	-2.06346 12	-2.03134 51	-1.99323 08	-1.98164 41
0.45	-2.17550 18	-2.17036 66	-2.14879 02	-2.10597 25	-2.05462 43	-1.99120 21	-1.97152 19
0.50	-2.25339 16	-2.25173 01	-2.22747 67	-2.16805 61	-2.09057 56	-1.99006 63	-1.95810 18
0.55		-2.35170 68	-2.33108 42	-2.25504 79	-2.14403 61	-1.99107 16	-1.94097 97
0.60			-2.46061 76	-2.37230 39	-2.22089 13	-1.99605 96	-1.91982 80
0.65				-2.52442 19	-2.32798 29	-2.00760 83	-1.89440 95
0.70					-2.47283 02	-2.02919 12	-1.86458 73
0.75					-2.66308 69	-2.06534 90	-1.83032 90
0.80						-2.12187 04	-1.79170 68
0.85						-2.20596 83	-1.74889 39
0.90						-2.32643 60	-1.70215 68
0.95						-2.49375 12	-1.65184 57
1.00						-2.72008 43	-1.59838 35
	$\begin{bmatrix}(-3)2\\8\end{bmatrix}$	$\begin{bmatrix}(-3)2\\8\end{bmatrix}$	$\begin{bmatrix}(-3)3\\8\end{bmatrix}$	$\begin{bmatrix}(-3)4\\8\end{bmatrix}$	$\begin{bmatrix}(-3)6\\8\end{bmatrix}$	$\begin{bmatrix}(-3)7\\8\end{bmatrix}$	$\begin{bmatrix}(-4)6\\6\end{bmatrix}$

$z/i=y\backslash a$	4.0
1.1	-1.48398 95
1.2	-1.36337 47
1.3	-1.24144 17
1.4	-1.12345 13
1.5	-1.01509 75
1.6	-0.92286 21
1.7	-0.85472 55
1.8	-0.82134 27
1.9	-0.83783 54
2.0	-0.92645 86
	$\begin{bmatrix}(-3)9\\9\end{bmatrix}$

TABLE FOR OBTAINING $\wp$, $\wp'$ AND ζ ON $0x$ AND $0y$ Table 18.2

(Negative Discriminant—Unit Real Half-Period)

$z\zeta(z)$

$z=x\backslash a$	1.00	1.05	1.15	1.3	1.5	2.0	4.0
0.00	1.00000 00	1.00000 00	1.00000 00	1.00000 00	1.00000 00	1.00000 00	1.00000 00
0.05	1.00000 49	1.00000 44	1.00000 34	1.00000 21	1.00000 08	0.99999 95	0.99999 92
0.10	1.00007 88	1.00007 06	1.00005 43	1.00003 31	1.00001 32	0.99999 24	0.99998 65
0.15	1.00039 88	1.00035 70	1.00027 40	1.00016 65	1.00006 60	0.99996 16	0.99993 12
0.20	1.00125 98	1.00112 60	1.00086 16	1.00052 15	1.00020 48	0.99987 51	0.99978 17
0.25	1.00307 33	1.00274 09	1.00208 94	1.00125 79	1.00048 81	0.99968 98	0.99946 41
0.30	1.00636 38	1.00566 06	1.00429 54	1.00256 91	1.00098 15	0.99934 32	0.99888 13
0.35	1.01176 23	1.01043 07	1.00787 32	1.00467 27	1.00175 16	0.99875 38	0.99791 11
0.40	1.01999 45	1.01767 00	1.01325 74	1.00779 77	1.00285 61	0.99781 57	0.99640 37
0.45	1.03186 18	1.02805 07	1.02090 50	1.01217 02	1.00433 47	0.99639 49	0.99417 86
0.50	1.04821 35	1.04227 15	1.03127 19	1.01799 52	1.00619 68	0.99432 31	0.99102 12
0.55	1.06990 78	1.06102 21	1.04478 39	1.02543 63	1.00840 79	0.99139 16	0.98667 79
0.60	1.09776 14	1.08493 81	1.06180 26	1.03459 22	1.01087 54	0.98734 37	0.98085 06
0.65	1.13248 70	1.11454 88	1.08258 64	1.04547 13	1.01343 17	0.98186 55	0.97318 91
0.70	1.17462 06	1.15021 58	1.10724 76	1.05796 45	1.01581 69	0.97457 57	0.96328 27
0.75	1.22444 09	1.19206 86	1.13570 79	1.07181 59	1.01765 94	0.96501 30	0.95064 87
0.80	1.28188 76	1.23993 78	1.16765 25	1.08659 33	1.01845 50	0.95262 09	0.93471 88
0.85	1.34648 26	1.29329 24	1.20248 62	1.10165 80	1.01754 41	0.93672 94	0.91482 13
0.90	1.41726 20	1.35118 37	1.23929 22	1.11613 35	1.01408 58	0.91653 15	0.89015 86
0.95	1.49272 42	1.41220 03	1.27679 52	1.12887 36	1.00702 73	0.89105 46	0.85977 85
1.00	1.57079 62	1.47443 48	1.31332 66	1.13842 65	0.99506 76	0.85912 29	0.82253 59
	$\begin{bmatrix}(-3)1\\7\end{bmatrix}$	$\begin{bmatrix}(-4)8\\7\end{bmatrix}$	$\begin{bmatrix}(-4)5\\6\end{bmatrix}$	$\begin{bmatrix}(-4)4\\6\end{bmatrix}$	$\begin{bmatrix}(-4)6\\6\end{bmatrix}$	$\begin{bmatrix}(-4)8\\6\end{bmatrix}$	$\begin{bmatrix}(-4)9\\6\end{bmatrix}$

$z/i=y\backslash a$	1.00	1.05	1.15	1.3	1.5	2.0	4.0
0.00	1.00000 00	1.00000 00	1.00000 00	1.00000 00	1.00000 00	1.00000 00	1.00000 00
0.05	1.00000 49	1.00000 44	1.00000 34	1.00000 21	1.00000 08	0.99999 95	0.99999 92
0.10	1.00007 88	1.00007 08	1.00005 46	1.00003 34	1.00001 35	0.99999 25	0.99998 65
0.15	1.00039 88	1.00035 86	1.00027 81	1.00017 04	1.00006 91	0.99996 24	0.99993 19
0.20	1.00125 98	1.00113 51	1.00088 05	1.00054 31	1.00022 22	0.99988 28	0.99978 57
0.25	1.00307 33	1.00277 55	1.00216 14	1.00134 04	1.00055 43	0.99971 90	0.99947 96
0.30	1.00636 38	1.00576 38	1.00451 03	1.00281 53	1.00117 94	0.99943 06	0.99892 78
0.35	1.01176 23	1.01069 02	1.00841 42	1.00529 28	1.00225 03	0.99897 41	0.99802 83
0.40	1.01999 45	1.01824 62	1.01445 97	1.00917 72	1.00396 67	0.99830 68	0.99666 50
0.45	1.03186 18	1.02921 31	1.02333 32	1.01496 03	1.00658 42	0.99739 10	0.99470 88
0.50	1.04821 35	1.04444 39	1.03581 72	1.02322 84	1.01042 41	0.99619 89	0.99202 03
0.55		1.06483 58	1.05277 97	1.03466 71	1.01588 39	0.99471 80	0.98845 10
0.60			1.07515 67	1.05006 29	1.02344 73	0.99295 77	0.98384 63
0.65				1.07029 97	1.03369 45	0.99095 58	0.97804 63
0.70					1.04730 93	0.98878 64	0.97088 86
0.75				1.06508 51		0.98656 79	0.96221 00
0.80						0.98447 25	0.95184 75
0.85						0.98273 54	0.93964 06
0.90						0.98166 56	0.92543 21
0.95						0.98165 63	0.90906 94
1.00						0.98319 64	0.89040 57
	$\begin{bmatrix}(-4)6\\6\end{bmatrix}$	$\begin{bmatrix}(-4)6\\6\end{bmatrix}$	$\begin{bmatrix}(-4)7\\6\end{bmatrix}$	$\begin{bmatrix}(-4)6\\6\end{bmatrix}$	$\begin{bmatrix}(-4)5\\6\end{bmatrix}$	$\begin{bmatrix}(-4)2\\6\end{bmatrix}$	$\begin{bmatrix}(-4)3\\5\end{bmatrix}$

$z/i=y\backslash a$	4.0
1.1	0.84561 98
1.2	0.79003 67
1.3	0.72274 36
1.4	0.64295 89
1.5	0.55003 38
1.6	0.44345 14
1.7	0.32282 70
1.8	0.18790 92
1.9	+0.03858 90
2.0	−0.12508 40
	$\begin{bmatrix}(-3)2\\6\end{bmatrix}$

Table 18.3   **INVARIANTS AND VALUES AT HALF-PERIODS**
(Non-Negative Discriminant—Unit Real Half-Period)

$a=\omega'/i$	g_2	g_3	$e_1=\wp(1)$	$e_3=\wp(\omega')$	$\eta=\zeta(1)$	$\eta'/i=\zeta(\omega')/i$
1.00	11.81704 500	0.00000 000	1.71879 64	−1.71879 64	0.78539 816	−0.78539 82
1.02	11.37372 384	0.55318 992	1.71005 96	−1.66138 15	0.78979 718	−0.76520 32
1.04	10.98419 107	1.03485 699	1.70235 77	−1.60783 69	0.79367 192	−0.74537 75
1.06	10.64177 347	1.45484 521	1.69556 79	−1.55787 59	0.79708 535	−0.72588 58
1.08	10.34065 794	1.82151 890	1.68958 18	−1.51123 63	0.80009 279	−0.70669 61
1.10	10.07577 364	2.14201 000	1.68430 41	−1.46767 83	0.80274 283	−0.68777 92
1.12	9.84269 185	2.42241 937	1.67965 08	−1.42698 19	0.80507 817	−0.66910 88
1.14	9.63754 049	2.66798 153	1.67554 80	−1.38894 48	0.80713 637	−0.65066 09
1.16	9.45693 072	2.88320 000	1.67193 04	−1.35338 12	0.80895 045	−0.63241 38
1.18	9.29789 413	3.07195 918	1.66874 05	−1.32011 96	0.81054 949	−0.61434 79
1.20	9.15782 851	3.23761 717	1.66592 77	−1.28900 20	0.81195 906	−0.59644 54
1.22	9.03445 117	3.38308 317	1.66344 74	−1.25988 23	0.81320 168	−0.57869 03
1.24	8.92575 843	3.51088 223	1.66126 03	−1.23262 55	0.81429 717	−0.56106 78
1.26	8.82999 055	3.62320 977	1.65933 17	−1.20710 65	0.81526 299	−0.54356 50
1.28	8.74560 138	3.72197 756	1.65763 09	−1.18320 95	0.81611 453	−0.52616 97
1.30	8.67123 169	3.80885 265	1.65613 11	−1.16082 70	0.81686 533	−0.50887 14
1.32	8.60568 628	3.88529 056	1.65480 86	−1.13985 91	0.81752 732	−0.49166 03
1.34	8.54791 374	3.95256 351	1.65364 22	−1.12021 33	0.81811 103	−0.47452 75
1.36	8.49698 890	4.01178 462	1.65261 37	−1.10180 31	0.81862 572	−0.45746 53
1.38	8.45209 746	4.06392 870	1.65170 67	−1.08454 85	0.81907 958	−0.44046 65
1.40	8.41252 263	4.10985 014	1.65090 68	−1.06837 47	0.81947 977	−0.42352 46
1.42	8.37763 305	4.15029 819	1.65020 13	−1.05321 20	0.81983 269	−0.40663 39
1.44	8.34687 283	4.18593 045	1.64957 92	−1.03899 58	0.82014 389	−0.38978 91
1.46	8.31975 228	4.21732 438	1.64903 06	−1.02566 55	0.82041 831	−0.37298 56
1.48	8.29583 997	4.24498 728	1.64854 68	−1.01316 45	0.82066 031	−0.35621 91
1.50	8.27475 580	4.26936 502	1.64812 02	−1.00144 04	0.82087 370	−0.33948 58
1.52	8.25616 484	4.29084 965	1.64774 39	−0.99044 37	0.82106 191	−0.32278 22
1.54	8.23977 191	4.30978 602	1.64741 20	−0.98012 84	0.82122 787	−0.30610 54
1.56	8.22531 684	4.32647 752	1.64711 94	−0.97045 19	0.82137 423	−0.28945 25
1.58	8.21257 036	4.34119 120	1.64686 13	−0.96137 37	0.82150 329	−0.27282 11
1.60	8.20133 033	4.35416 210	1.64663 38	−0.95285 64	0.82161 711	−0.25620 90
1.65	8.17870 308	4.38026 291	1.64617 54	−0.93379 17	0.82184 628	−0.21475 00
1.70	8.16217 907	4.39931 441	1.64584 08	−0.91752 88	0.82201 364	−0.17337 32
1.75	8.15011 147	4.41322 294	1.64559 63	−0.90365 18	0.82213 589	−0.13205 85
1.80	8.14129 812	4.42337 818	1.64541 78	−0.89180·82	0.82222 516	−0.09079 10
1.85	8.13486 127	4.43079 368	1.64528 73	−0.88169 76	0.82229 038	−0.04955 91
1.90	8.13016 001	4.43620 896	1.64519 21	−0.87306 52	0.82233 800	−0.00835 41
1.95	8.12672 634	4.44016 375	1.64512 25	−0.86569 37	0.82237 281	+0.03283 07
2.00	8.12421 844	4.44305 205	1.64507 17	−0.85939 82	0.82239 820	0.07400 01
2.05	8.12238 671	4.44516 152	1.64503 45	−0.85402 10	0.82241 676	0.11515 80
2.10	8.12104 883	4.44670 219	1.64500 74	−0.84942 78	0.82243 032	0.15630 73
2.15	8.12007 164	4.44782 746	1.64498 76	−0.84550 41	0.82244 022	0.19745 01
2.20	8.11935 791	4.44864 934	1.64497 32	−0.84215 20	0.82244 745	0.23858 81
2.25	8.11883 660	4.44924 963	1.64496 26	−0.83928 80	0.82245 274	0.27972 23
2.30	8.11845 583	4.44968 808	1.64495 49	−0.83684 11	0.82245 659	0.32085 38
2.4	8.11797 459	4.45024 222	1.64494 51	−0.83296 37	0.82246 146	0.40311 12
2.5	8.11771 785	4.45053 785	1.64494 00	−0.83013 28	0.82246 406	0.48536 38
2.6	8.11758 087	4.45069 555	1.64493 71	−0.82806 54	0.82246 546	0.56761 39
2.7	8.11750 782	4.45077 969	1.64493 57	−0.82655 58	0.82246 619	0.64986 24
2.8	8.11746 884	4.45082 457	1.64493 49	−0.82545 33	0.82246 659	0.73211 01
2.9	8.11744 804	4.45084 852	1.64493 45	−0.82464 81	0.82246 680	0.81435 74
3.0	8.11743 694	4.45086 130	1.64493 43	−0.82406 01	0.82246 691	0.89660 44
3.1	8.11743 103	4.45086 811	1.64493 42	−0.82363 06	0.82246 698	0.97885 13
3.2	8.11742 787	4.45087 174	1.64493 41	−0.82331 68	0.82246 701	1.06109 81
3.3	8.11742 619	4.45087 368	1.64493 41	−0.82308 78	0.82246 702	1.14334 48
3.4	8.11742 529	4.45087 472	1.64493 41	−0.82292 04	0.82246 703	1.22559 16
3.5	8.11742 481	4.45087 528	1.64493 41	−0.82279 82	0.82246 703	1.30783 83
3.6	8.11742 455	4.45087 556	1.64493 41	−0.82270 89	0.82246 703	1.39008 50
3.7	8.11742 441	4.45087 572	1.64493 41	−0.82264 37	0.82246 704	1.47233 17
3.8	8.11742 434	4.45087 581	1.64493 41	−0.82259 61	0.82246 704	1.55457 84
3.9	8.11742 430	4.45087 585	1.64493 41	−0.82256 13	0.82246 704	1.63682 51
4.0	8.11742 426	4.45087 587	1.64493 41	−0.82253 59	0.82246 704	1.71907 18
∞ $\Delta=0$	8.11742 426 $\begin{bmatrix}(-3)7\\8\end{bmatrix}$	4.45087 590 $\begin{bmatrix}(-3)9\\8\end{bmatrix}$	1.64493 41 $\begin{bmatrix}(-4)1\\5\end{bmatrix}$	−0.82246 70 $\begin{bmatrix}(-4)5\\6\end{bmatrix}$	0.82246 704 $\begin{bmatrix}(-5)7\\6\end{bmatrix}$	$\begin{bmatrix}\infty\\(-5)5\\5\end{bmatrix}$

For $a=1$: $g_2=\omega^4$, $g_3=0$, $e_1=\omega^2/2$, $e_3=-\omega^2/2$, $\eta=\pi/4$, $\eta'/i=-\pi/4$.

For $a=\infty$: $g_2=\pi^4/12$, $g_3=\pi^6/216$, $e_1=\pi^2/6$, $e_3=-\pi^2/12$, $\eta=\pi^2/12$, $\eta'/i=\infty$.

($\omega=1.85407\ 4677$ is the real half-period in the Lemniscatic case **18.14**.)

For $4<a<\infty$, to obtain η' use Legendre's relation $\eta'=\eta\omega'-\pi i/2$.

To obtain the corresponding values of tabulated quantities when the real half-period $\omega\neq1$, multiply g_2 by ω^{-4}, g_3 by ω^{-6}, e_i by ω^{-2} and η by ω^{-1}.

INVARIANTS AND VALUES AT HALF-PERIODS Table 18.3

(Non-Negative Discriminant—Unit Real Half-Period)

$a=\omega'/i$	$\sigma(1)$	$\sigma(\omega')/i$	$\mathscr{R}\sigma(\omega_2)$	$\mathscr{I}\sigma(\omega_2)$
1.00	0.94989 88	0.949899	1.182951	1.182951
1.02	0.95114 80	0.967481	1.170397	1.218650
1.04	0.95224 92	0.984884	1.157316	1.253864
1.06	0.95321 98	1.002097	1.143695	1.288619
1.08	0.95407 54	1.019107	1.129522	1.322935
1.10	0.95482 97	1.035904	1.114782	1.356827
1.12	0.95549 47	1.052476	1.099457	1.390301
1.14	0.95608 10	1.068811	1.083531	1.423362
1.16	0.95659 79	1.084899	1.066989	1.456007
1.18	0.95705 36	1.100727	1.049814	1.488231
1.20	0.95745 55	1.116285	1.031991	1.520022
1.22	0.95780 98	1.131562	1.013507	1.551369
1.24	0.95812 22	1.146546	0.994349	1.582254
1.26	0.95839 77	1.161227	0.974506	1.612657
1.28	0.95864 07	1.175594	0.953970	1.642557
1.30	0.95885 49	1.189636	0.932733	1.671930
1.32	0.95904 38	1.203344	0.910790	1.700750
1.34	0.95921 04	1.216707	0.888138	1.728989
1.36	0.95935 73	1.229716	0.864776	1.756618
1.38	0.95948 68	1.242361	0.840704	1.783607
1.40	0.95960 10	1.254633	0.815927	1.809925
1.42	0.95970 18	1.266522	0.790449	1.835542
1.44	0.95979 06	1.278021	0.764278	1.860425
1.46	0.95986 89	1.289120	0.737425	1.884541
1.48	0.95993 80	1.299811	0.709900	1.907860
1.50	0.95999 90	1.310087	0.681719	1.930348
1.52	0.96005 27	1.319941	0.652896	1.951974
1.54	0.96010 01	1.329364	0.623452	1.972707
1.56	0.96014 19	1.338351	0.593404	1.992515
1.58	0.96017 87	1.346895	0.562777	2.011370
1.60	0.96021 13	1.354990	0.531593	2.029242
1.65	0.96027 67	1.373224	0.451372	2.069439
1.70	0.96032 45	1.388539	0.368286	2.102914
1.75	0.96035 94	1.400869	0.282840	2.129313
1.80	0.96038 49	1.410170	0.195588	2.148344
1.85	0.96040 35	1.416408	0.107125	2.159783
1.90	0.96041 71	1.419573	+0.018074	2.163478
1.95	0.96042 70	1.419665	−0.070918	2.159353
2.00	0.96043 43	1.416707	−0.159199	2.147412
2.05	0.96043 96	1.410733	−0.246114	2.127732
2.10	0.96044 35	1.401800	−0.331019	2.100473
2.15	0.96044 63	1.389977	−0.413290	2.065864
2.20	0.96044 84	1.375349	−0.492330	2.024211
2.25	0.96044 99	1.358018	−0.567579	1.975882
2.30	0.96045 10	1.338098	−0.638522	1.921308
2.4	0.96045 24	1.291016	−0.765682	1.795415
2.5	0.96045 31	1.235264	−0.870782	1.650936
2.6	0.96045 35	1.172151	−0.951807	1.492779
2.7	0.96045 37	1.103091	−1.007808	1.326086
2.8	0.96045 38	1.029557	−1.038896	1.155967
2.9	0.96045 39	0.953025	−1.046157	0.987255
3.0	0.96045 40	0.874937	−1.031530	0.824296
3.1	0.96045 40	0.796655	−0.997636	0.670787
3.2	0.96045 40	0.719428	−0.947586	0.529666
3.3	0.96045 40	0.644360	−0.884775	0.403050
3.4	0.96045 40	0.572395	−0.812687	0.292246
3.5	0.96045 40	0.504299	−0.734720	0.197780
3.6	0.96045 40	0.440663	−0.654024	0.119493
3.7	0.96045 40	0.381903	−0.573398	0.056643
3.8	0.96045 40	0.328268	−0.495196	+0.008033
3.9	0.96045 40	0.279851	−0.421291	−0.027857
4.0	0.96045 40	0.236623	−0.353075	−0.052740
∞	0.96045 40	0.000000	0.000000	0.000000
$\Delta=0$	$\begin{bmatrix}(-5)2\\4\end{bmatrix}$	$\begin{bmatrix}(-4)9\\5\end{bmatrix}$	$\begin{bmatrix}(-3)3\\6\end{bmatrix}$	$\begin{bmatrix}(-3)2\\6\end{bmatrix}$

$\omega_2=1+\omega'$, $e_2=\mathscr{P}(1+\omega')=-(e_1+e_3)$, $\eta_2=\zeta(1+\omega')=\eta+\eta'$.

For $a=1$: $\sigma(1)=e^{\pi/8}2^{1/4}/\omega$, $\sigma(\omega')=i\sigma(1)$, $\sigma(\omega_2)=\sqrt{2}e^{\pi/4}e^{i\pi/4}/\omega$.

For $a=\infty$: $\sigma(1)=2e^{\pi^2/24}/\pi$, $\sigma(\omega')=0$, $\sigma(\omega_2)=0$.

($\omega=1.85407\ 4677$ is the real half-period in the Lemniscatic case **18.14**.)

To obtain the corresponding values of tabulated quantities when the real half-period $\omega\neq1$, multiply σ by ω.

Table 18.3   INVARIANTS AND VALUES AT HALF-PERIODS

(Non-Positive Discriminant—Unit Real Half-Period)

$a=\omega_2'/i$	g_2	g_3	$\mathcal{R}e_1=$ $\mathcal{R}\wp\left(\frac{1}{2}-\frac{\omega_2'}{2}\right)$	$\mathcal{I}e_1=$ $\mathcal{I}\wp\left(\frac{1}{2}-\frac{\omega_2'}{2}\right)$	$\eta_2=\zeta(1)$	$\eta_2'/i=\zeta(\omega_2')/i$
1.00	−47.26818 00	0.00000 00	0.00000 000	3.43759 29	1.57079 63	−1.57079 63
1.02	−45.35272 19	4.41906 00	−0.04867 810	3.36827 69	1.53091 63	−1.58005 81
1.04	−43.40071 30	8.23156 58	−0.09452 083	3.29802 68	1.49282 30	−1.58905 67
1.06	−41.42954 84	11.49257 28	−0.13769 202	3.22711 39	1.45647 87	−1.59772 52
1.08	−39.45420 53	14.25448 26	−0.17834 547	3.15578 40	1.42184 01	−1.60600 53
1.10	−37.48749 12	16.56680 99	−0.21662 576	3.08425 89	1.38885 99	−1.61384 68
1.12	−35.54027 17	18.47603 08	−0.25266 894	3.01273 84	1.35748 74	−1.62120 68
1.14	−33.62168 02	20.02550 17	−0.28660 315	2.94140 17	1.32766 96	−1.62804 93
1.16	−31.73930 91	21.25543 82	−0.31854 915	2.87040 90	1.29935 18	−1.63434 46
1.18	−29.89938 64	22.20294 45	−0.34862 086	2.79990 29	1.27247 81	−1.64006 85
1.20	−28.10693 45	22.90208 34	−0.37692 571	2.73000 96	1.24699 24	−1.64520 18
1.22	−26.36591 62	23.38397 82	−0.40356 512	2.66084 07	1.22283 82	−1.64973 00
1.24	−24.67936 58	23.67693 85	−0.42863 481	2.59249 39	1.19995 95	−1.65364 28
1.26	−23.04950 83	23.80660 45	−0.45222 513	2.52505 44	1.17830 09	−1.65693 36
1.28	−21.47786 60	23.79610 09	−0.47442 139	2.45859 58	1.15780 77	−1.65959 88
1.30	−19.96535 52	23.66620 08	−0.49530 414	2.39318 14	1.13842 65	−1.66163 82
1.32	−18.51237 16	23.43548 95	−0.51494 941	2.32886 49	1.12010 02	−1.66305 38
1.34	−17.11886 71	23.12052 98	−0.53342 897	2.26569 11	1.10279 31	−1.66384 99
1.36	−15.78441 82	22.73602 29	−0.55081 058	2.20369 72	1.08644 09	−1.66403 31
1.38	−14.50828 67	22.29496 60	−0.56715 817	2.14291 32	1.07100 10	−1.66361 13
1.40	−13.28947 27	21.80880 22	−0.58253 209	2.08336 24	1.05642 75	−1.66259 42
1.42	−12.12676 19	21.28756 31	−0.59698 926	2.02506 27	1.04267 61	−1.66099 26
1.44	−11.01876 70	20.74000 36	−0.61058 339	1.96802 64	1.02970 43	−1.65881 85
1.46	− 9.96396 40	20.17372 81	−0.62336 513	1.91226 13	1.01747 14	−1.65608 44
1.48	− 8.96072 32	19.59530 70	−0.63538 226	1.85777 09	1.00593 83	−1.65280 40
1.50	− 8.00733 71	19.01038 59	−0.64667 980	1.80455 50	0.99506 76	−1.64899 13
1.52	− 7.10204 36	18.42378 52	−0.65730 023	1.75261 00	0.98482 36	−1.64466 08
1.54	− 6.24304 63	17.83959 12	−0.66728 357	1.70192 94	0.97517 21	−1.63982 76
1.56	− 5.42853 20	17.26123 98	−0.67666 751	1.65250 41	0.96608 09	−1.63450 65
1.58	− 4.65668 53	16.69159 27	−0.68548 761	1.60432 26	0.95751 90	−1.62871 26
1.60	− 3.92570 12	16.13300 57	−0.69377 734	1.55737 16	0.94945 69	−1.62246 17
1.65	− 2.26537 64	14.79653 23	−0.71238 375	1.44527 36	0.93130 88	−1.60493 31
1.70	− 0.82241 58	13.56033 77	−0.72831 198	1.34049 21	0.91571 53	−1.58487 67
1.75	+ 0.42844 48	12.43388 94	−0.74194 441	1.24271 21	0.90232 74	−1.56251 97
1.80	1.51045 44	11.41927 28	−0.75360 961	1.15159 40	0.89084 07	−1.53807 94
1.85	2.44471 18	10.51370 92	−0.76358 973	1.06678 48	0.88099 10	−1.51175 93
1.90	3.25015 81	9.71138 21	−0.77212 691	0.98792 73	0.87254 91	−1.48374 94
1.95	3.94365 25	9.00473 54	−0.77942 883	0.91466 65	0.86531 67	−1.45422 51
2.00	4.54009 85	8.38537 94	−0.78567 351	0.84665 46	0.85912 29	−1.42334 69
2.05	5.05259 79	7.84470 38	−0.79101 353	0.78355 46	0.85382 00	−1.39126 17
2.10	5.49261 57	7.37428 09	−0.79557 957	0.72504 25	0.84928 11	−1.35810 23
2.15	5.87014 76	6.96611 56	−0.79948 352	0.67080 91	0.84539 69	−1.32398 93
2.20	6.19388 05	6.61278 90	−0.80282 119	0.62056 06	0.84207 37	−1.28903 05
2.25	6.47134 49	6.30752 86	−0.80567 458	0.57401 95	0.83923 09	−1.25332 31
2.30	6.70905 42	6.04422 78	−0.80811 383	0.53092 40	0.83679 93	−1.21695 43
2.4	7.08692 59	5.62231 14	−0.81198 137	0.45410 32	0.83294 16	−1.14253 28
2.5	7.36377 30	5.31058 54	−0.81480 718	0.38831 56	0.83012 09	−1.06629 03
2.6	7.56643 61	5.08099 59	−0.81687 167	0.33200 75	0.82805 92	−0.98863 87
2.7	7.71470 39	4.91228 49	−0.81837 985	0.28383 23	0.82655 25	−0.90990 09
2.8	7.82312 83	4.78851 39	−0.81948 158	0.24262 75	0.82545 16	−0.83032 82
2.9	7.90239 07	4.69782 05	−0.82028 636	0.20739 21	0.82464 72	−0.75011 58
3.0	7.96032 11	4.63142 26	−0.82087 422	0.17726 58	0.82405 96	−0.66941 39
3.1	8.00265 32	4.58284 25	−0.82130 361	0.15151 09	0.82363 03	−0.58833 87
3.2	8.03358 32	4.54731 53	−0.82161 725	0.12949 50	0.82331 67	−0.50697 92
3.3	8.05618 01	4.52134 25	−0.82184 634	0.11067 62	0.82308 77	−0.42540 32
3.4	8.07268 80	4.50235 93	−0.82201 368	0.09459 10	0.82292 04	−0.34366 33
3.5	8.08474 69	4.48848 72	−0.82213 590	0.08084 29	0.82279 82	−0.26179 91
3.6	8.09355 57	4.47835 14	−0.82222 517	0.06909 25	0.82270 89	−0.17984 06
3.7	8.09999 01	4.47094 62	−0.82229 038	0.05904 97	0.82264 37	−0.09781 10
3.8	8.10469 00	4.46553 65	−0.82233 800	0.05046 65	0.82259 61	−0.01572 75
3.9	8.10812 30	4.46158 47	−0.82237 279	0.04313 08	0.82256 13	+0.06639 64
4.0	8.11063 05	4.45869 80	−0.82239 820	0.03686 13	0.82253 59	+0.14855 08
∞	8.11742 43	4.45087 59	−0.82246 703	0.00000 00	0.82246 70	∞
Δ=0	$\left[\begin{matrix}(-2)3\\9\end{matrix}\right]$	$\left[\begin{matrix}(-2)8\\8\end{matrix}\right]$	$\left[\begin{matrix}(-4)4\\7\end{matrix}\right]$	$\left[\begin{matrix}(-3)1\\6\end{matrix}\right]$	$\left[\begin{matrix}(-4)3\\6\end{matrix}\right]$	$\left[\begin{matrix}(-4)3\\6\end{matrix}\right]$

For $a=1$: $g_2=-4\omega^4$, $g_3=0$, $\mathcal{R}e_1=0$, $\mathcal{I}e_1=\omega^2$, $\eta_2=\pi/2$, $\eta_2'/i=-\pi/2$.

For $a=\infty$: $g_2=\pi^4/12$, $g_3=\pi^6/216$, $\mathcal{R}e_1=-\pi^2/12$, $\mathcal{I}e_1=0$, $\eta_2=\pi^2/12$, $\eta_2'/i=\infty$.

($\omega=1.85407\ 4677$ is the real half-period in the Lemniscatic case 18.14.)

For $4<a<\infty$, to obtain η_2' use Legendre's relation $\eta_2'=\eta_2\omega_2'-\pi i$.

To obtain the corresponding values of tabulated quantities when the real half-period $\omega_2\neq1$,

multiply g_2 by ω_2^{-4}, g_3 by ω_2^{-6}, e_i by ω_2^{-2} and η by ω_2^{-1}.

INVARIANTS AND VALUES AT HALF-PERIODS Table 18.3

(Non-Positive Discriminant—Unit Real Half-Period)

$a=\omega_2'/i$	$\sigma(1)$	$\sigma(\omega_2')/i$	$\mathcal{R}\sigma(\omega')$	$\mathcal{I}\sigma(\omega')$
1.00	1.18295 13	1.182951	0.474949	0.474949
1.02	1.17091 79	1.219157	0.475654	0.483826
1.04	1.15940 62	1.255842	0.476433	0.492792
1.06	1.14841 45	1.292964	0.477275	0.501851
1.08	1.13793 68	1.330480	0.478169	0.511006
1.10	1.12796 39	1.368342	0.479107	0.520259
1.12	1.11848 38	1.406502	0.480078	0.529611
1.14	1.10948 26	1.444910	0.481074	0.539064
1.16	1.10094 49	1.483513	0.482085	0.548616
1.18	1.09285 44	1.522257	0.483104	0.558268
1.20	1.08519 40	1.561089	0.484122	0.568019
1.22	1.07794 61	1.599952	0.485132	0.577866
1.24	1.07109 31	1.638790	0.486126	0.587809
1.26	1.06461 72	1.677548	0.487098	0.597843
1.28	1.05050 11	1.716167	0.488041	0.607968
1.30	1.05272 75	1.754591	0.488949	0.618179
1.32	1.04727 97	1.792765	0.489817	0.628474
1.34	1.04214 12	1.830630	0.490639	0.638850
1.36	1.03729 63	1.868133	0.491410	0.649302
1.38	1.03272 96	1.905218	0.492126	0.659828
1.40	1.02842 64	1.941832	0.492783	0.670422
1.42	1.02437 26	1.977922	0.493376	0.681082
1.44	1.02055 48	2.013437	0.493902	0.691804
1.46	1.01696 00	2.048327	0.494357	0.702582
1.48	1.01357 57	2.082544	0.494739	0.713414
1.50	1.01039 05	2.116040	0.495045	0.724295
1.52	1.00739 28	2.148771	0.495272	0.735221
1.54	1.00457 23	2.180693	0.495418	0.746189
1.56	1.00191 88	2.211766	0.495480	0.757192
1.58	0.99942 27	2.241950	0.495458	0.768229
1.60	0.99707 51	2.271208	0.495348	0.779295
1.65	0.99179 98	2.340071	0.494687	0.807059
1.70	0.98727 79	2.402437	0.493456	0.834917
1.75	0.98340 36	2.457895	0.491645	0.862812
1.80	0.98008 56	2.506120	0.489246	0.890687
1.85	0.97724 49	2.546866	0.486255	0.918490
1.90	0.97481 36	2.579972	0.482673	0.946170
1.95	0.97273 30	2.605345	0.478503	0.973680
2.00	0.97095 31	2.622973	0.473748	1.000975
2.05	0.96943 05	2.632902	0.468417	1.028011
2.10	0.96812 82	2.635245	0.462516	1.054750
2.15	0.96701 46	2.630169	0.456054	1.081151
2.20	0.96606 23	2.617892	0.449041	1.107179
2.25	0.96524 80	2.598678	0.441488	1.132799
2.30	0.96455 19	2.572828	0.433405	1.157978
2.4	0.96344 79	2.502604	0.415693	1.206881
2.5	0.96264 13	2.410244	0.395997	1.253647
2.6	0.96205 18	2.299090	0.374417	1.298044
2.7	0.96162 12	2.172666	0.351055	1.339858
2.8	0.96130 65	2.034544	0.326022	1.378884
2.9	0.96107 67	1.888235	0.299435	1.414929
3.0	0.96090 89	1.737097	0.271420	1.447812
3.1	0.96078 62	1.584242	0.242114	1.477367
3.2	0.96069 67	1.432486	0.211664	1.503441
3.3	0.96063 12	1.284291	0.180224	1.525899
3.4	0.96058 34	1.141740	0.147962	1.544621
3.5	0.96054 86	1.006520	0.115052	1.559512
3.6	0.96052 31	0.879924	0.081678	1.570495
3.7	0.96050 44	0.762869	0.048028	1.577518
3.8	0.96049 08	0.655914	+0.014297	1.580552
3.9	0.96048 09	0.559298	−0.019318	1.579595
4.0	0.96047 37	0.472982	−0.052618	1.574671
∞	0.96045 40	0.000000	0.000000	0.000000
$\Delta=0$	$\begin{bmatrix}(-5)9\\6\end{bmatrix}$	$\begin{bmatrix}(-3)3\\6\end{bmatrix}$	$\begin{bmatrix}(-4)2\\4\end{bmatrix}$	$\begin{bmatrix}(-4)5\\5\end{bmatrix}$

$\omega'=\frac{1}{2}+\frac{\omega_2'}{2}$, $e_3=\mathcal{P}\left(\frac{1}{2}+\frac{\omega_2'}{2}\right)=\bar{e}_1$, $e_2=\mathcal{P}(1)=-2\mathcal{R}e_1$, $\eta'=\zeta\left(\frac{1}{2}+\frac{\omega_2'}{2}\right)=\frac{1}{2}(\eta_2+\eta_2')$.

For $a=1$: $\sigma(1)=e^{\pi/4}/\omega$, $\sigma(\omega_2')=i\sigma(1)$, $\sigma(\omega')=e^{\pi/8}e^{i\pi/4}/2^{1/4}\omega$.

For $a=\infty$: $\sigma(1)=2e^{\pi^2/24}/\pi$, $\sigma(\omega_2')=0$, $\sigma(\omega')=0$.

($\omega=1.85407\ 4677$ is the real half-period in the Lemniscatic case 18.14.)

To obtain the corresponding values of tabulated quantities when the real half-period $\omega_2\neq1$, multiply σ by ω_2.

19. Parabolic Cylinder Functions

J. C. P. MILLER [1]

Contents

The Equation $\dfrac{d^2y}{dx^2}-(\tfrac{1}{4}x^2 \mid a)y=0$

The Equation $\dfrac{d^2y}{dx^2}+(\tfrac{1}{4}x^2-a)y=0$

The author acknowledges permission from H.M. Stationery Office to draw freely from [19.11] the material in the introduction, and the tabular values of $W(a, x)$ for $a = -5(1)5$, $\pm x = 0(.1)5$.   Other tables of $W(a, x)$ and the tables of $U(a, x)$ and $V(a, x)$ were prepared on EDSAC 2 at the University Mathematical Laboratory, Cambridge, England, using a program prepared by Miss Joan Walsh for solution of general second order linear homogeneous differential equations with quadratic polynomial coefficients.   The auxiliary tables were prepared at the Computation Laboratory of the National Bureau of Standards.

[1] The University Mathematical Laboratory, Cambridge, England.   (Prepared under contract with the National Bureau of Standards.)

19. Parabolic Cylinder Functions

Mathematical Properties

19.1. The Parabolic Cylinder Functions

Introductory

These are solutions of the differential equation

19.1.1
$$\frac{d^2y}{dx^2} + (ax^2 + bx + c)y = 0$$

with two real and distinct standard forms

19.1.2
$$\frac{d^2y}{dx^2} - (\tfrac{1}{4}x^2 + a)y = 0$$

19.1.3
$$\frac{d^2y}{dx^2} + (\tfrac{1}{4}x^2 - a)y = 0$$

The functions

19.1.4
$$y(a, x) \qquad y(a, -x) \qquad y(-a, ix) \qquad y(-a, -ix)$$

are all solutions either of **19.1.2** or of **19.1.3** if any one is such a solution.

Replacement of a by $-ia$ and x by $xe^{\frac{1}{4}i\pi}$ converts **19.1.2** into **19.1.3**. If $y(a, x)$ is a solution of **19.1.2**, then **19.1.3** has solutions:

19.1.5
$$y(-ia, xe^{\frac{1}{4}i\pi}) \qquad y(-ia, -xe^{\frac{1}{4}i\pi})$$
$$y(ia, -xe^{-\frac{1}{4}i\pi}) \qquad y(ia, xe^{-\frac{1}{4}i\pi})$$

Both variable x and the parameter a may take on general complex values in this section and in many subsequent sections. Practical applications appear to be confined to real solutions of real equations; therefore attention is confined to such solutions, and, in general, formulas are given for the two equations **19.1.2** and **19.1.3** independently. The principal computational consequence of the remarks above is that reflection in the y-axis produces an independent solution in almost all cases (Hermite functions provide an exception), so that tables may be confined either to positive x or to a single solution of **19.1.2** or **19.1.3**.

The Equation $\frac{d^2y}{dx^2} - \left(\frac{1}{4}x^2 + a\right)y = 0$

19.2. Power Series in x

Even and odd solutions of **19.1.2** are given by

19.2.1
$$y_1 = e^{-\frac{1}{4}x^2} M(\tfrac{1}{2}a + \tfrac{1}{4}, \tfrac{1}{2}, \tfrac{1}{2}x^2)$$
$$= e^{-\frac{1}{4}x^2}\left\{1 + (a + \tfrac{1}{2})\frac{x^2}{2!} + (a + \tfrac{1}{2})(a + \tfrac{5}{2})\frac{x^4}{4!} + \ldots\right\}$$
$$= e^{-\frac{1}{4}x^2} {}_1F_1(\tfrac{1}{2}a + \tfrac{1}{4}; \tfrac{1}{2}; \tfrac{1}{2}x^2)$$

19.2.2
$$= e^{\frac{1}{4}x^2} M(-\tfrac{1}{2}a + \tfrac{1}{4}, \tfrac{1}{2}, -\tfrac{1}{2}x^2)$$
$$= e^{\frac{1}{4}x^2}\left\{1 + (a - \tfrac{1}{2})\frac{x^2}{2!} + (a - \tfrac{1}{2})(a - \tfrac{5}{2})\frac{x^4}{4!} + \ldots\right\}$$

19.2.3
$$y_2 = xe^{-\frac{1}{4}x^2} M(\tfrac{1}{2}a + \tfrac{3}{4}, \tfrac{3}{2}, \tfrac{1}{2}x^2)$$
$$= e^{-\frac{1}{4}x^2}\left\{x + (a + \tfrac{3}{2})\frac{x^3}{3!} + (a + \tfrac{3}{2})(a + \tfrac{7}{2})\frac{x^5}{5!} + \ldots\right\}$$

19.2.4
$$= xe^{\frac{1}{4}x^2} M(-\tfrac{1}{2}a + \tfrac{3}{4}, \tfrac{3}{2}, -\tfrac{1}{2}x^2)$$
$$= e^{\frac{1}{4}x^2}\left\{x + (a - \tfrac{3}{2})\frac{x^3}{3!} + (a - \tfrac{3}{2})(a - \tfrac{7}{2})\frac{x^5}{5!} + \ldots\right\}$$

these series being convergent for all values of x (see chapter 13 for $M(a, c, z)$).

Alternatively,

19.2.5
$$y_1 = 1 + a\frac{x^2}{2!} + \left(a^2 + \tfrac{1}{2}\right)\frac{x^4}{4!} + \left(a^3 + \tfrac{7}{2}a\right)\frac{x^6}{6!}$$
$$+ \left(a^4 + 11a^2 + \tfrac{15}{4}\right)\frac{x^8}{8!} + \left(a^5 + 25a^3 + \tfrac{211}{4}a\right)\frac{x^{10}}{10!} + \ldots$$

19.2.6
$$y_2 = x + a\frac{x^3}{3!} + \left(a^2 + \tfrac{3}{2}\right)\frac{x^5}{5!} + \left(a^3 + \tfrac{13}{2}a\right)\frac{x^7}{7!}$$
$$+ \left(a^4 + 17a^2 + \tfrac{63}{4}\right)\frac{x^9}{9!} + \left(a^5 + 35a^3 + \tfrac{531}{4}a\right)\frac{x^{11}}{11!} + \ldots$$

in which non-zero coefficients a_n of $x^n/n!$ are connected by

19.2.7
$$a_{n+2} = a \cdot a_n + \tfrac{1}{4}n(n-1)a_{n-2}$$

19.3. Standard Solutions

These have been chosen to have the asymptotic behavior exhibited in **19.8**. The first is Whittaker's function [19.8, 19.9] in a more symmetrical notation.

19.3.1
$$U(a, x) = D_{-a-\frac{1}{2}}(x) = \cos \pi(\tfrac{1}{4} + \tfrac{1}{2}a) \cdot Y_1$$
$$- \sin \pi(\tfrac{1}{4} + \tfrac{1}{2}a) \cdot Y_2$$

19.3.2
$$V(a, x) = \frac{1}{\Gamma(\tfrac{1}{2}-a)} \{ \sin \pi(\tfrac{1}{4} + \tfrac{1}{2}a) \cdot Y_1$$
$$+ \cos \pi(\tfrac{1}{4} + \tfrac{1}{2}a) \cdot Y_2 \}$$

in which

19.3.3 $\quad Y_1 = \dfrac{1}{\sqrt{\pi}} \dfrac{\Gamma(\tfrac{1}{4} - \tfrac{1}{2}a)}{2^{\frac{1}{2}a+\frac{1}{4}}} y_1 = \sqrt{\pi} \dfrac{\sec \pi(\tfrac{1}{4} + \tfrac{1}{2}a)}{2^{\frac{1}{2}a+\frac{1}{4}} \Gamma(\tfrac{3}{4} + \tfrac{1}{2}a)} y_1$

19.3.4 $\quad Y_2 = \dfrac{1}{\sqrt{\pi}} \dfrac{\Gamma(\tfrac{3}{4} - \tfrac{1}{2}a)}{2^{\frac{1}{2}a-\frac{1}{4}}} y_2 = \sqrt{\pi} \dfrac{\csc \pi(\tfrac{1}{4} + \tfrac{1}{2}a)}{2^{\frac{1}{2}a-\frac{1}{4}} \Gamma(\tfrac{1}{4} + \tfrac{1}{2}a)} y_2$

19.3.5
$$U(a, 0) = \frac{\sqrt{\pi}}{2^{\frac{1}{2}a+\frac{1}{4}} \Gamma(\tfrac{3}{4} + \tfrac{1}{2}a)}$$
$$U'(a, 0) = -\frac{\sqrt{\pi}}{2^{\frac{1}{2}a-\frac{1}{4}} \Gamma(\tfrac{1}{4} + \tfrac{1}{2}a)}$$

19.3.6
$$V(a, 0) = \frac{2^{\frac{1}{2}a+\frac{1}{4}} \sin \pi(\tfrac{3}{4} - \tfrac{1}{2}a)}{\Gamma(\tfrac{3}{4} - \tfrac{1}{2}a)}$$
$$V'(a, 0) = \frac{2^{\frac{1}{2}a+\frac{3}{4}} \sin \pi(\tfrac{1}{4} - \tfrac{1}{2}a)}{\Gamma(\tfrac{1}{4} - \tfrac{1}{2}a)}$$

In terms of the more familiar $D_n(x)$ of Whittaker,

19.3.7 $\qquad U(a, x) = D_{-a-\frac{1}{2}}(x)$

19.3.8
$$V(a, x) = \frac{1}{\pi} \Gamma(\tfrac{1}{2} + a) \{ \sin \pi a \cdot D_{-a-\frac{1}{2}}(x) + D_{-a-\frac{1}{2}}(-x) \}$$

19.4. Wronskian and Other Relations

19.4.1 $\qquad W\{U, V\} = \sqrt{2/\pi}$

19.4.2
$$\pi V(a, x) = \Gamma(\tfrac{1}{2} + a) \{ \sin \pi a \cdot U(a, x) + U(a, -x) \}$$

19.4.3
$$\Gamma(\tfrac{1}{2} + a) U(a, x) = \pi \sec^2 \pi a \{ V(a, -x)$$
$$- \sin \pi a \cdot V(a, x) \}$$

19.4.4
$$\frac{\Gamma(\tfrac{1}{4} - \tfrac{1}{2}a) \cos \pi(\tfrac{1}{4} + \tfrac{1}{2}a)}{\sqrt{\pi} 2^{\frac{1}{2}a-\frac{1}{4}}} y_1 = 2 \sin \pi(\tfrac{3}{4} + \tfrac{1}{2}a) \cdot Y_1$$
$$= U(a, x) + U(a, -x)$$

19.4.5
$$-\frac{\Gamma(\tfrac{3}{4} - \tfrac{1}{2}a) \sin \pi(\tfrac{1}{4} + \tfrac{1}{2}a)}{\sqrt{\pi} 2^{\frac{1}{2}a-\frac{1}{4}}} y_2 = 2 \cos \pi(\tfrac{3}{4} + \tfrac{1}{2}a) \cdot Y_2$$
$$= U(a, x) - U(a, -x)$$

19.4.6
$$\sqrt{2\pi} U(-a, \pm ix) =$$
$$\Gamma(\tfrac{1}{2} + a) \{ e^{-i\pi(\frac{1}{2}a-\frac{1}{4})} U(a, \pm x) + e^{i\pi(\frac{1}{2}a-\frac{1}{4})} U(a, \mp x) \}$$

19.4.7
$$\sqrt{2\pi} U(a, \pm x) =$$
$$\Gamma(\tfrac{1}{2} - a) \{ e^{-i\pi(\frac{1}{2}a+\frac{1}{4})} U(-a, \pm ix) + e^{i\pi(\frac{1}{2}a+\frac{1}{4})} U(-a, \mp ix) \}$$

19.5. Integral Representations

A full treatment is given in [19.11] section 4. Representations are given here for $U(a, z)$ only; others may be derived by use of the relations given in **19.4**.

19.5.1 $\quad U(a, z) = \dfrac{\Gamma(\tfrac{1}{2} - a)}{2\pi i} e^{-\frac{1}{4}z^2} \displaystyle\int_\alpha e^{zs - \frac{1}{2}s^2} s^{a-\frac{1}{2}} ds$

19.5.2 $\qquad = \dfrac{\Gamma(\tfrac{1}{2} - a)}{2\pi i} e^{\frac{1}{4}z^2} \displaystyle\int_\beta e^{-\frac{1}{2}t^2} (z+t)^{a-\frac{1}{2}} dt$

where α and β are the contours shown in **Figures 19.1** and **19.2**.

When $a + \tfrac{1}{2}$ is a positive integer these integrals become indeterminate; in this case

19.5.3 $\quad U(a, z) = \dfrac{1}{\Gamma(\tfrac{1}{2} + a)} e^{-\frac{1}{4}z^2} \displaystyle\int_0^\infty e^{-zs - \frac{1}{2}s^2} s^{a-\frac{1}{2}} ds$

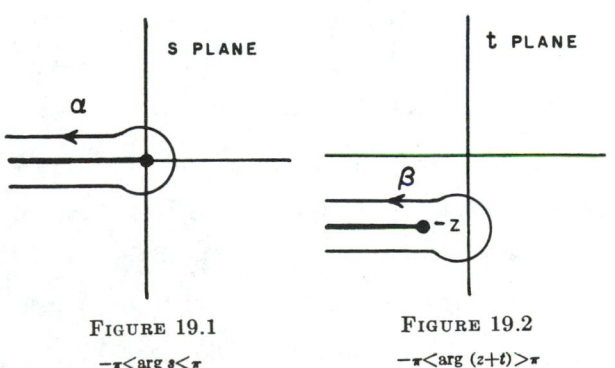

FIGURE 19.1
$-\pi < \arg s < \pi$

FIGURE 19.2
$-\pi < \arg (z+t) > \pi$

19.5.4 $\quad U(a,z)=\dfrac{1}{\sqrt{2\pi i}}\,e^{\frac{1}{4}z^2}\displaystyle\int_\epsilon e^{-zs+\frac{1}{2}s^2}s^{-a-\frac{1}{2}}ds$

19.5.5 $\quad =\dfrac{e^{(a-\frac{1}{2})\pi i}}{\sqrt{2\pi i}}\,e^{\frac{1}{4}z^2}\displaystyle\int_{\epsilon_3} e^{zs+\frac{1}{2}s^2}s^{-a-\frac{1}{2}}ds$

19.5.6 $\quad =\dfrac{e^{-(a-\frac{1}{2})\pi i}}{\sqrt{2\pi i}}\,e^{\frac{1}{4}z^2}\displaystyle\int_{\epsilon_4} e^{zs+\frac{1}{2}s^2}s^{-a-\frac{1}{2}}ds$

where ϵ, ϵ_3 and ϵ_4 are shown in **Figures 19.3** and **19.4**.

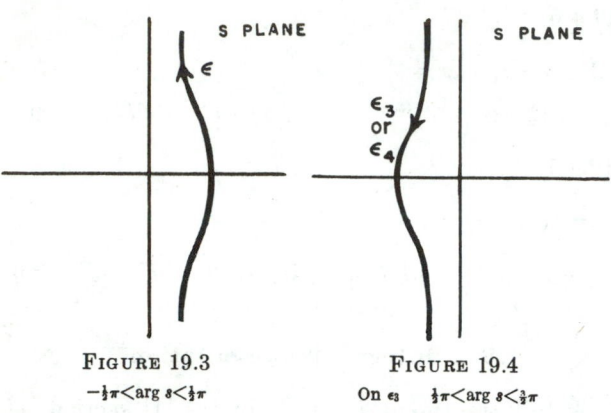

FIGURE 19.3
$-\frac{1}{2}\pi<\arg s<\frac{1}{2}\pi$

FIGURE 19.4
On ϵ_3 $\quad\frac{1}{2}\pi<\arg s<\frac{3}{2}\pi$
On ϵ_4 $\quad-\frac{3}{2}\pi<\arg s<-\frac{1}{2}\pi$

19.5.7

$$U(a,z)=\frac{\Gamma(\frac{3}{4}-\frac{1}{2}a)}{2^{\frac{1}{2}a+\frac{1}{4}}\pi i}\int_{(\zeta_1)}e^{\frac{1}{4}z^2t}(1+t)^{\frac{1}{2}a-\frac{1}{4}}(1-t)^{-\frac{1}{2}a-\frac{1}{4}}dt$$

19.5.8

$$=\frac{\Gamma(\frac{3}{4}-\frac{1}{2}a)}{2^{\frac{1}{2}a+\frac{1}{4}}\pi i}\int_{\zeta_1}\tfrac{1}{2}ze^v(\tfrac{1}{4}z^2+v)^{\frac{1}{2}a-\frac{1}{4}}(\tfrac{1}{4}z^2-v)^{-\frac{1}{2}a-\frac{1}{4}}dv$$

19.5.9

$$U(a,z)$$
$$=\frac{i\Gamma(\frac{1}{4}-\frac{1}{2}a)}{2^{\frac{1}{2}a+\frac{1}{4}}\pi}\int_{(\eta_1)}\tfrac{1}{2}ze^{-\frac{1}{4}z^2t}(1+t)^{-\frac{1}{2}a-\frac{1}{4}}(1-t)^{\frac{1}{2}a-\frac{1}{4}}dt$$

19.5.10

$$=\frac{i\Gamma(\frac{1}{4}-\frac{1}{2}a)}{2^{\frac{1}{2}a+\frac{1}{4}}\pi}\int_{\eta_1}e^{-v}(\tfrac{1}{4}z^2+v)^{-\frac{1}{2}a-\frac{1}{4}}(\tfrac{1}{4}z^2-v)^{\frac{1}{2}a-\frac{1}{4}}dv$$

The contour ζ_1 is such that $(\frac{1}{4}z^2+v)$ goes from $\infty e^{-i\pi}$ to $\infty e^{i\pi}$ while $v=\frac{1}{4}z^2$ is *not* encircled; $(\frac{1}{4}z^2-v)^{-\frac{1}{2}a-\frac{1}{4}}$ has its principal value except possibly in the immediate neighborhood of the branch-point when encirclement is being avoided. Likewise η_1 is such that $(\frac{1}{4}z^2-v)$ goes from $\infty e^{i\pi}$ to $\infty e^{-i\pi}$ while encirclement of $v=-\frac{1}{4}z^2$ is similarly avoided. The contours (ζ_1) and (η_1) may be obtained from ζ_1 and η_1 by use of the substitution $v=\frac{1}{4}z^2t$.

The expressions **19.5.7** and **19.5.8** become indeterminate when $a=\frac{3}{2},\frac{7}{2},\frac{11}{2},\ldots$; for these values

19.5.11

$$U(a,z)=\frac{1}{\Gamma(\frac{1}{4}+\frac{1}{2}a)}\,ze^{-\frac{1}{4}z^2}\int_0^\infty e^{-s}s^{\frac{1}{2}a-\frac{1}{4}}(z^2+2s)^{-\frac{1}{2}a-\frac{1}{4}}ds$$

Again **19.5.9** and **19.5.10** become indeterminate when $a=\frac{1}{2},\frac{5}{2},\frac{9}{2},\ldots$; for these values

19.5.12

$$U(a,z)=\frac{1}{\Gamma(\frac{3}{4}+\frac{1}{2}a)}\,e^{-\frac{1}{4}z^2}\int_0^\infty e^{-s}s^{\frac{1}{2}a-\frac{1}{4}}(z^2+2s)^{-\frac{1}{2}a-\frac{1}{4}}ds$$

Barnes-Type Integrals

19.5.13 $\quad U(a,z)=\dfrac{e^{-\frac{1}{4}z^2}}{2\pi i}\,z^{-a-\frac{1}{2}}\displaystyle\int_{-\infty i}^{+\infty i}\frac{\Gamma(s)\,\Gamma(\frac{1}{2}+a-2s)}{\Gamma(\frac{1}{2}+a)}\,(\sqrt{2}z)^{2s}ds$ $\qquad\qquad(|\arg z|<\frac{3}{4}\pi)$

where the contour separates the zeros of $\Gamma(s)$ from those of $\Gamma(a+\frac{1}{2}-2s)$. Similarly

19.5.14 $\quad V(a,z)=\sqrt{\dfrac{2}{\pi}}\,\dfrac{e^{\frac{1}{4}z^2}}{2\pi i}\,z^{a-\frac{1}{2}}\displaystyle\int_{-\infty i}^{+\infty i}\frac{\Gamma(s)\,\Gamma(\frac{1}{2}-a-2s)}{\Gamma(\frac{1}{2}-a)}\,(\sqrt{2}z)^{2s}\cos s\pi\,ds$ $\qquad\qquad(|\arg z|<\frac{1}{4}\pi)$

19.6. Recurrence Relations

19.6.1 $\quad U'(a,x)+\frac{1}{2}xU(a,x)+(a+\frac{1}{2})U(a+1,x)=0$

19.6.2 $\quad U'(a,x)-\frac{1}{2}xU(a,x)+U(a-1,x)=0$

19.6.3 $\quad 2U'(a,x)+U(a-1,x)+(a+\frac{1}{2})U(a+1,x)=0$

19.6.4 $\quad xU(a,x)-U(a-1,x)+(a+\frac{1}{2})U(a+1,x)=0$

These are also satisfied by $\Gamma(\frac{1}{2}-a)V(a,x)$.

19.6.5 $\quad V'(a,x)-\frac{1}{2}xV(a,x)-(a-\frac{1}{2})V(a-1,x)=0$

19.6.6 $\quad V'(a,x)+\frac{1}{2}xV(a,x)-V(a+1,x)=0$

19.6.7

$$2V'(a,x)-V(a+1,x)-(a-\tfrac{1}{2})V(a-1,x)=0$$

19.6.8

$$xV(a,x)-V(a+1,x)+(a-\tfrac{1}{2})V(a-1,x)=0$$

These are also satisfied by $U(a,x)/\Gamma(\frac{1}{2}-a)$

19.6.9 $\quad y_1'(a,x)+\frac{1}{2}xy_1(a,x)=(a+\frac{1}{2})y_2(a+1,x)$

19.6.10 $\quad y_1'(a,x)-\frac{1}{2}xy_1(a,x)=(a-\frac{1}{2})y_2(a-1,x)$

19.6.11 $\quad y_2'(a,x)+\frac{1}{2}xy_2(a,x)=y_1(a+1,x)$

19.6.12 $\quad y_2'(a,x)-\frac{1}{2}xy_2(a,x)=y_1(a-1,x)$

Asymptotic Expansions

19.7. Expressions in Terms of Airy Functions

When a is large and negative, write, for $0\leq x<\infty$

$$x=2\sqrt{|a|}\xi \qquad t=(4|a|)^{\frac{2}{3}}\tau$$

19.7.1

$$\tau=-(\tfrac{3}{2}\vartheta_3)^{\frac{2}{3}}$$

$$\vartheta_3=\tfrac{1}{2}\int_\xi^1\sqrt{1-s^2}ds=\tfrac{1}{4}\arccos\xi-\tfrac{1}{4}\xi\sqrt{1-\xi^2}\quad(\xi\leq1)$$

19.7.2

$$\tau=+(\tfrac{3}{2}\vartheta_2)^{\frac{2}{3}}$$

$$\vartheta_2=\tfrac{1}{2}\int_1^\xi\sqrt{s^2-1}ds=\tfrac{1}{4}\xi\sqrt{\xi^2-1}-\tfrac{1}{4}\operatorname{arccosh}\xi\quad(\xi\geq1)$$

Then for $x\geq0,\,a\to-\infty$

19.7.3

$$U(a,x)\sim2^{-\frac{1}{4}-\frac{1}{2}a}\Gamma\ (\tfrac{1}{4}-\tfrac{1}{2}a)\left(\frac{t}{\xi^2-1}\right)^{\frac{1}{4}}\operatorname{Ai}(t)$$

19.7.4

$$\Gamma\ (\tfrac{1}{2}-a)\,V(a,x)\sim2^{-\frac{1}{4}-\frac{1}{2}a}\Gamma\ (\tfrac{1}{4}-\tfrac{1}{2}a)\left(\frac{t}{\xi^2-1}\right)^{\frac{1}{4}}\operatorname{Bi}(t)$$

Table 19.3 gives τ as a function of ξ.

See [19.5] for further developments.

19.8. Expansions for x Large and a Moderate

When $x\gg|a|$

19.8.1

$$U(a,x)\sim e^{-\frac{1}{4}x^2}x^{-a-\frac{1}{2}}\left\{1-\frac{(a+\tfrac{1}{2})(a+\tfrac{3}{2})}{2x^2}\right.$$

$$\left.+\frac{(a+\tfrac{1}{2})(a+\tfrac{3}{2})(a+\tfrac{5}{2})(a+\tfrac{7}{2})}{2\cdot4x^4}-\cdots\right\}$$

$$(x\to+\infty)$$

19.8.2

$$V(a,x)\sim\sqrt{\frac{2}{\pi}}e^{\frac{1}{4}x^2}x^{a-\frac{1}{2}}\left\{1+\frac{(a-\tfrac{1}{2})(a-\tfrac{3}{2})}{2x^2}\right.$$

$$\left.+\frac{(a-\tfrac{1}{2})(a-\tfrac{3}{2})(a-\tfrac{5}{2})(a-\tfrac{7}{2})}{2\cdot4x^4}+\cdots\right\}$$

$$(x\to+\infty)$$

These expansions form the basis for the choice of standard solutions in **19.3**. The former is valid for complex x, with $|\arg x|<\tfrac{1}{2}\pi$, in the complete

sense of Watson [19.6], although valid for a wider range of $|\arg x|$ in Poincaré's sense; the second series is completely valid *only for x real and positive*.

19.9. Expansions for a Large With x Moderate

(i) a positive

When $a\gg x^2$, with $p=\sqrt{a}$, then

19.9.1 $\quad U(a,x)=\dfrac{\sqrt{\pi}}{2^{\frac{1}{2}a+\frac{1}{4}}\Gamma(\tfrac{3}{4}+\tfrac{1}{2}a)}\exp\ (-px+v_1)$

19.9.2 $\quad U(a,-x)=\dfrac{\sqrt{\pi}}{2^{\frac{1}{2}a+\frac{1}{4}}\Gamma(\tfrac{3}{4}+\tfrac{1}{2}a)}\exp\ (px+v_2)$

where

19.9.3

$$v_1,v_2\sim\mp\frac{\tfrac{2}{3}(\tfrac{1}{2}x)^3}{2p}-\frac{(\tfrac{1}{2}x)^2}{(2p)^2}\mp\frac{\tfrac{1}{2}x-\tfrac{2}{5}(\tfrac{1}{2}x)^5}{(2p)^3}$$

$$+\frac{2(\tfrac{1}{2}x)^4}{(2p)^4}\pm\frac{(\tfrac{16}{3}\tfrac{1}{2}x)^3-\tfrac{4}{7}(\tfrac{1}{2}x)^7}{(2p)^5}+\cdots$$

$$(a\to+\infty)$$

The upper sign gives the first function, and the lower sign the second function.

(ii) a negative

When $-a\gg x^2$, with $p=\sqrt{-a}$, then

19.9.4

$$U(a,x)+i\Gamma\ (\tfrac{1}{2}-a)\cdot V(a,x)$$

$$=\frac{e^{i\pi(\frac{1}{4}+\frac{1}{2}a)}\Gamma(\tfrac{1}{4}-\tfrac{1}{2}a)}{2^{\frac{1}{2}a+\frac{1}{4}}\sqrt{\pi}}\,e^{ipx}\exp\ (v_r+iv_i)$$

where

19.9.5

$$v_r\sim+\frac{(\tfrac{1}{2}x)^2}{(2p)^2}+\frac{2(\tfrac{1}{2}x)^4}{(2p)^4}-\frac{9(\tfrac{1}{2}x)^2-\tfrac{16}{3}(\tfrac{1}{2}x)^6}{(2p)^6}-\cdots$$

$$v_i\sim-\frac{\tfrac{2}{3}(\tfrac{1}{2}x)^3}{2p}+\frac{\tfrac{1}{2}x+\tfrac{2}{5}(\tfrac{1}{2}x)^5}{(2p)^3}+\frac{\tfrac{16}{3}(\tfrac{1}{2}x)^3-\tfrac{4}{7}(\tfrac{1}{2}x)^7}{(2p)^5}-\cdots$$

$$(a\to-\infty)$$

Further expansions of a similar type will be found in [19.11].

19.10. Darwin's Expansions

(i) a positive, x^2+4a large. Write

19.10.1 $\qquad X=\sqrt{x^2+4a}$

$$\theta=4a\vartheta_1(x/2\sqrt{a})=\tfrac{1}{2}\int_0^x Xdx=\tfrac{1}{4}xX+a\ln\frac{x+X}{2\sqrt{a}}$$

$$=\frac{x}{4}\sqrt{x^2+4a}+a\operatorname{arcsinh}\frac{x}{2\sqrt{a}}$$

(see **Table 19.3** for ϑ_1), then

19.10.2 $\quad U(a, x) = \dfrac{(2\pi)^{1/4}}{\sqrt{\Gamma(\frac{1}{2}+a)}} \exp\{-\theta+v(a, x)\}$

19.10.3 $\quad U(a, -x) = \dfrac{(2\pi)^{1/4}}{\sqrt{\Gamma(\frac{1}{2}+a)}} \exp\{\theta+v(a, -x)\}$

where

19.10.4

$$v(a, x) \sim -\tfrac{1}{2}\ln X + \sum_{s=1} (-1)^s d_{3s}/X^{3s}$$
$$(a>0, \; x^2+4a \to +\infty)$$

and d_{3s} is given by **19.10.13**.

(ii) a negative, x^2+4a large and positive. Write

19.10.5 $\qquad X = \sqrt{x^2-4|a|}$

$$\theta = 4|a|\vartheta_2(x/2\sqrt{|a|}) = \tfrac{1}{2}\int_{2\sqrt{|a|}}^{x} X\,dx = \tfrac{1}{4}xX + a\ln\frac{x+X}{2\sqrt{|a|}}$$
$$= \tfrac{1}{4}x\sqrt{x^2-4|a|} + a\operatorname{arccosh}\frac{x}{2\sqrt{|a|}}$$

(see **Table 19.3** for ϑ_2), then

19.10.6 $\quad U(a, x) = \dfrac{\sqrt{\Gamma(\frac{1}{2}-a)}}{(2\pi)^{1/4}} \exp\{-\theta+v(a, x)\}$

19.10.7

$$V(a, x) = \frac{2}{(2\pi)^{1/4}\sqrt{\Gamma(\frac{1}{2}-a)}} \exp\{\theta+v(a, -x)\}$$

where again

19.10.8

$$v(a, x) \sim -\tfrac{1}{2}\ln X + \sum_{s=1} (-1)^s d_{3s}/X^{3s}$$
$$(a<0, \; x^2+4a \to +\infty)$$

and d_{3s} is given by **19.10.13**.

(iii) a large and negative and x moderate. Write

19.10.9 $\qquad Y = \sqrt{4|a|-x^2}$

$$\theta = 4|a|\vartheta_4(x/2\sqrt{|a|})$$
$$= \tfrac{1}{2}\int_0^x Y\,dx = \tfrac{1}{4}xY + |a|\arcsin\frac{x}{2\sqrt{|a|}}$$

(see **Table 19.3** for $\vartheta_4 = \tfrac{1}{8}\pi - \vartheta_3$), then

19.10.10

$$U(a, x) = \frac{2\sqrt{\Gamma(\frac{1}{2}-a)}}{(2\pi)^{1/4}} e^{v_r} \cos\left\{\tfrac{1}{4}\pi+\tfrac{1}{2}\pi a+\theta+v_i\right\}$$

19.10.11

$$V(a, x) =$$
$$\frac{2}{(2\pi)^{\frac{1}{4}}\sqrt{\Gamma(\frac{1}{2}-a)}} e^{v_r} \sin\left\{\tfrac{1}{4}\pi+\tfrac{1}{2}\pi a+\theta+v_i\right\}$$

where

19.10.12 $\quad v_r \sim -\tfrac{1}{2}\ln Y - \dfrac{d_6}{Y^6} + \dfrac{d_{12}}{Y^{12}} - \cdots$

$$v_i \sim \frac{d_3}{Y^3} - \frac{d_9}{Y^9} + \cdots \qquad (x^2+4a \to -\infty)$$

In each case the coefficients d_{3r} are given by

19.10.13

$$d_3 = \frac{1}{a}\left(\frac{x^3}{48} + \tfrac{1}{2}ax\right)$$

$$d_6 = \tfrac{3}{4}x^2 - 2a$$

$$d_9 = \frac{1}{a^3}\left(-\frac{7}{5760}x^9 - \frac{7}{320}ax^7 - \frac{49}{320}a^2x^5\right.$$
$$\left. + \frac{31}{12}a^3x^3 - 19a^4x\right)$$

$$d_{12} = \frac{153}{8}x^4 - 186ax^2 + 80a^2$$

See [19.11] for $d_{15}, \ldots, d_{24}$, and [19.5] for an alternative form.

19.11. Modulus and Phase

When a is negative and $|x|<2\sqrt{|a|}$, the functions U and V are oscillatory and it is sometimes convenient to write

19.11.1 $\quad U(a, x)+i\Gamma(\tfrac{1}{2}-a)V(a, x) = F(a, x)e^{i\chi(a, x)}$

19.11.2 $\quad U'(a, x)+i\Gamma(\tfrac{1}{2}-a)V'(a, x) = -G(a, x)e^{i\psi(a, x)}$

Then, when $a<0$ and $|a| \gg x^2$,

19.11.3

$$F = \frac{\Gamma(\frac{1}{4}-\frac{1}{2}a)}{2^{\frac{1}{2}a+\frac{1}{4}}\sqrt{\pi}} e^{v_r}, \qquad \chi = (\tfrac{1}{2}a+\tfrac{1}{4})\pi + px + v_i$$

where v_r, v_i are given by **19.9.5** and $p = \sqrt{-a}$.

Alternatively, with $p = \sqrt{|a|}$, and again $-a \gg x^2$,

19.11.4

$$F \sim \frac{\Gamma(\frac{1}{4}-\frac{1}{2}a)}{2^{\frac{1}{2}a+\frac{1}{4}}\sqrt{\pi}}\left\{1 + \frac{x^2}{(4p)^2} + \frac{\frac{5}{2}x^4}{(4p)^4}\right.$$
$$\left. + \frac{\frac{15}{2}x^6-144x^2}{(4p)^6} + \cdots\right\}$$

19.11.5
$$\chi \sim (\tfrac{1}{2}a+\tfrac{1}{4})\pi+px\left\{1-\frac{\tfrac{2}{3}x^2}{(4p)^2}-\frac{\tfrac{2}{5}x^4-16}{(4p)^4}\right.$$
$$\left.-\frac{\tfrac{4}{7}x^6-\tfrac{256}{3}x^2}{(4p)^6}-\cdots\right\}$$

19.11.6
$$G \sim \frac{\Gamma(\tfrac{3}{4}-\tfrac{1}{2}a)}{2^{\frac{1}{2}a-\frac{1}{4}}\sqrt{\pi}}\left\{1-\frac{x^2}{(4p)^2}-\frac{\tfrac{3}{2}x^4}{(4p)^4}\right.$$
$$\left.-\frac{\tfrac{7}{2}x^6-176x^2}{(4p)^6}-\cdots\right\}$$

19.11.7
$$\psi \sim (\tfrac{1}{2}a-\tfrac{1}{4})\pi+px\left\{1-\frac{\tfrac{2}{3}x^2}{(4p)^2}-\frac{\tfrac{2}{5}x^4+16}{(4p)^4}\right.$$
$$\left.-\frac{\tfrac{4}{7}x^6+\tfrac{320}{3}x^2}{(4p)^6}-\cdots\right\}$$

Again, when x^2+4a is large and negative, with $Y=\sqrt{4|a|-x^2}$, then

19.11.8
$$F=\frac{2\sqrt{\Gamma(\tfrac{1}{2}-a)}}{(2\pi)^{\frac{1}{4}}}\,e^{v_r}\qquad \chi=\tfrac{1}{4}\pi+\tfrac{1}{2}\pi a+\theta+v_i$$

where θ, v_r and v_i are given by **19.10.9** and **19.10.12**.

Another form is

19.11.9
$$F\sim\frac{2\sqrt{\Gamma(\tfrac{1}{2}-a)}}{(2\pi)^{\frac{1}{4}}\sqrt{Y}}\left(1+\frac{3}{4Y^4}+\frac{5a}{Y^6}+\frac{621}{32Y^8}+\cdots\right)$$
$$(x^2+4a\to-\infty)$$

19.11.10
$$G\sim\frac{\sqrt{Y}\sqrt{\Gamma(\tfrac{1}{2}-a)}}{(2\pi)^{\frac{1}{4}}}\left(1-\frac{5}{4Y^4}-\frac{7a}{Y^6}-\frac{835}{32Y^8}-\cdots\right)$$
$$(x^2+4a\to-\infty)$$

while ψ and χ are connected by

19.11.11
$$\psi-\chi\sim-\tfrac{1}{2}\pi-\frac{x}{Y^3}\left(1+\frac{47}{6Y^4}+\frac{214a}{3Y^6}+\frac{14483}{40Y^8}+\cdots\right)$$
$$(x^2+4a\to-\infty)$$

Connections With Other Functions

19.12. Connection With Confluent Hypergeometric Functions (see chapter 13)

19.12.1
$$U(a,\pm x)=\frac{\sqrt{\pi}2^{-\frac{1}{2}a}x^{-\frac{1}{2}}}{\Gamma(\tfrac{3}{4}+\tfrac{1}{2}a)}\,M_{-\frac{1}{2}a,-\frac{1}{4}}(\tfrac{1}{2}x^2)$$
$$\mp\frac{\sqrt{\pi}2^{1-\frac{1}{2}a}x^{-\frac{1}{2}}}{\Gamma(\tfrac{1}{4}+\tfrac{1}{2}a)}\,M_{-\frac{1}{2}a,\frac{1}{4}}(\tfrac{1}{2}x^2)$$

19.12.2
$$U(a,x)=2^{-\frac{1}{4}a}x^{-\frac{1}{2}}W_{-\frac{1}{2}a,-\frac{1}{4}}(\tfrac{1}{2}x^2)$$

19.12.3
$$U(a,\pm x)=\frac{\sqrt{\pi}2^{-\frac{1}{4}-\frac{1}{2}a}e^{-\frac{1}{4}x^2}}{\Gamma(\tfrac{3}{4}+\tfrac{1}{2}a)}\,M(\tfrac{1}{2}a+\tfrac{1}{4},\tfrac{1}{2},\tfrac{1}{2}x^2)$$
$$\mp\frac{\sqrt{\pi}2^{\frac{1}{4}-\frac{1}{2}a}xe^{-\frac{1}{4}x^2}}{\Gamma(\tfrac{1}{4}+\tfrac{1}{2}a)}\,M(\tfrac{1}{2}a+\tfrac{3}{4},\tfrac{3}{2},\tfrac{1}{2}x^2)$$

19.12.4
$$U(a,x)=2^{-\frac{1}{4}-\frac{1}{2}a}e^{-\frac{1}{4}x^2}U(\tfrac{1}{2}a+\tfrac{1}{4},\tfrac{1}{2},\tfrac{1}{2}x^2)$$
$$=2^{-\frac{1}{4}-\frac{1}{2}a}xe^{-\frac{1}{4}x^2}U(\tfrac{1}{2}a+\tfrac{3}{4},\tfrac{3}{2},\tfrac{1}{2}x^2)$$

Expressions for $V(a,x)$ may be obtained from these by use of **19.4.2**.

19.13. Connection With Hermite Polynomials and Functions

When n is a non-negative integer

19.13.1
$$U(-n-\tfrac{1}{2},x)=e^{-\frac{1}{4}x^2}He_n(x)=2^{-\frac{1}{2}n}e^{-\frac{1}{4}x^2}H_n(x/\sqrt{2})$$

19.13.2
$$V(n+\tfrac{1}{2},x)=\sqrt{2/\pi}e^{\frac{1}{4}x^2}He_n^*(x)=2^{-\frac{1}{2}n}e^{\frac{1}{4}x^2}H_n^*(x/\sqrt{2})$$

in which $H_n(x)$ and $He_n(x)$ are Hermite polynomials (see chapter **22**) while

19.13.3 $\quad He_n^*(x)=e^{-\frac{1}{2}x^2}\dfrac{d^n}{dx^n}\,e^{\frac{1}{2}x^2}=(-i)^nHe_n(ix)$

19.13.4 $\quad H_n^*(x)=e^{-x^2}\dfrac{d^n}{dx^n}\,e^{x^2}=(-i)^nH_n(ix)$

This gives one elementary solution to **19.1.2** whenever $2a$ is an odd integer, positive or negative.

19.14. Connection With Probability Integrals and Dawson's Integral (see chapter 7)

If, as in [19.10]

19.14.1 $\quad Hh_{-1}(x)=e^{-\frac{1}{2}x^2}$

19.14.2
$$Hh_n(x)=\int_x^\infty Hh_{n-1}(t)dt=(1/n!)\int_x^\infty(t-x)^ne^{-\frac{1}{2}t^2}dt$$
$$(n\geq0)$$

then

19.14.3 $\quad U(n+\tfrac{1}{2},x)=e^{\frac{1}{4}x^2}Hh_n(x)\qquad (n\geq-1)$

Correspondingly

19.14.4
$$V(\tfrac{1}{2},x)=\sqrt{2/\pi}e^{\frac{1}{4}x^2}$$

and

19.14.5
$$V(-n-\tfrac{1}{2},x)=e^{-\frac{1}{4}x^2}\left\{\int_0^x e^{-\frac{1}{4}t^2}V(-n+\tfrac{1}{2},\,t)dt\right.$$
$$\left.-\frac{\sin\tfrac{1}{2}n\pi}{2^{\frac{1}{2}n}\Gamma(\tfrac{1}{2}n+1)}\right\}\ (n\geq0)$$

Here $V(-\tfrac{1}{2},\ x)$ is closely related to Dawson's integral

$$\int_0^x e^{t^2}dt$$

These relations give a second solution of **19.1.2** whenever $2a$ is an odd integer, and a second solution is unobtainable from $U(a,\ x)$ by reflection in the y-axis.

19.15. Explicit Formula in Terms of Bessel Functions When $2a$ Is an Integer

Write

19.15.1
$$I_{-n}-I_n=(2/\pi)\sin n\pi\cdot K_n$$

19.15.2
$$I_{-n}+I_n=\cos n\pi\cdot\mathscr{I}_n$$

where the argument of all modified Bessel functions is $\tfrac{1}{4}x^2$. Then

19.15.3
$$U(1,x)=2\pi^{-\frac{1}{2}}(\tfrac{1}{2}x)^{\frac{3}{2}}(-K_{\frac{1}{4}}+K_{\frac{3}{4}})$$

19.15.4
$$U(2,x)=2\cdot\tfrac{2}{3}\pi^{-\frac{1}{2}}(\tfrac{1}{2}x)^{\frac{5}{2}}(2K_{\frac{1}{4}}-3K_{\frac{3}{4}}+K_{\frac{5}{4}})$$

19.15.5
$$U(3,x)=2\cdot\tfrac{2}{3}\cdot\tfrac{2}{5}\pi^{-\frac{1}{2}}(\tfrac{1}{2}x)^{\frac{7}{2}}(-5K_{\frac{1}{4}}+9K_{\frac{3}{4}}-5K_{\frac{5}{4}}+K_{\frac{7}{4}})$$

19.15.6
$$V(1,x)=\tfrac{1}{2}(\tfrac{1}{2}x)^{\frac{3}{2}}(\mathscr{I}_{\frac{1}{4}}-\mathscr{I}_{\frac{3}{4}})$$

19.15.7
$$V(2,x)=\tfrac{1}{2}(\tfrac{1}{2}x)^{\frac{5}{2}}(2\mathscr{I}_{\frac{1}{4}}-3\mathscr{I}_{\frac{3}{4}}+\mathscr{I}_{\frac{5}{4}})$$

19.15.8
$$V(3,x)=\tfrac{1}{2}(\tfrac{1}{2}x)^{\frac{7}{2}}(5\mathscr{I}_{\frac{1}{4}}-9\mathscr{I}_{\frac{3}{4}}+5\mathscr{I}_{\frac{5}{4}}-\mathscr{I}_{\frac{7}{4}})$$

19.15.9
$$U(0,x)=\pi^{-\frac{1}{2}}(\tfrac{1}{2}x)^{\frac{1}{2}}K_{\frac{1}{4}}$$

19.15.10
$$U(-1,x)=\pi^{-\frac{1}{2}}(\tfrac{1}{2}x)^{\frac{3}{2}}(K_{\frac{1}{4}}+K_{\frac{3}{4}})$$

19.15.11
$$U(-2,x)=\pi^{-\frac{1}{2}}(\tfrac{1}{2}x)^{\frac{5}{2}}(2K_{\frac{1}{4}}+3K_{\frac{3}{4}}-K_{\frac{5}{4}})$$

19.15.12
$$U(-3,x)=\pi^{-\frac{1}{2}}(\tfrac{1}{2}x)^{\frac{7}{2}}(5K_{\frac{1}{4}}+9K_{\frac{3}{4}}-5K_{\frac{5}{4}}-K_{\frac{7}{4}})$$

19.15.13
$$V(0,x)=\tfrac{1}{2}(\tfrac{1}{2}x)^{\frac{1}{2}}\mathscr{I}_{\frac{1}{4}}$$

19.15.14
$$V(-1,x)=(\tfrac{1}{2}x)^{\frac{3}{2}}(\mathscr{I}_{\frac{1}{4}}+\mathscr{I}_{\frac{3}{4}})$$

19.15.15
$$V(-2,x)=\tfrac{2}{3}(\tfrac{1}{2}x)^{\frac{5}{2}}(2\mathscr{I}_{\frac{1}{4}}+3\mathscr{I}_{\frac{3}{4}}-\mathscr{I}_{\frac{5}{4}})$$

19.15.16
$$V(-3,x)=\tfrac{2}{3}\cdot\tfrac{2}{5}(\tfrac{1}{2}x)^{\frac{7}{2}}(5\mathscr{I}_{\frac{1}{4}}+9\mathscr{I}_{\frac{3}{4}}-5\mathscr{I}_{\frac{5}{4}}-\mathscr{I}_{\frac{7}{4}})$$

19.15.17
$$U(-\tfrac{1}{2},x)=\sqrt{2/\pi}(\tfrac{1}{2}x)K_{\frac{1}{2}}$$

19.15.18
$$U(-\tfrac{3}{2},x)=\sqrt{2/\pi}(\tfrac{1}{2}x)^2 2K_{\frac{1}{2}}$$

19.15.19
$$U(-\tfrac{5}{2},x)=\sqrt{2/\pi}(\tfrac{1}{2}x)^3(5K_{\frac{1}{2}}-K_{\frac{3}{2}})$$

19.15.20
$$V(\tfrac{1}{2},x)=(\tfrac{1}{2}x)(I_{\frac{1}{2}}+I_{-\frac{1}{2}})$$

19.15.21
$$V(\tfrac{3}{2},x)=(\tfrac{1}{2}x)^2(2I_{\frac{1}{2}}+2I_{-\frac{1}{2}})$$

19.15.22
$$V(\tfrac{5}{2},x)=(\tfrac{1}{2}x)^3(5I_{\frac{1}{2}}+5I_{-\frac{1}{2}}-I_{\frac{3}{2}}-I_{-\frac{3}{2}})$$

The Equation $\dfrac{d^2y}{dx^2}+\left(\dfrac{1}{4}x^2-a\right)y=0$

19.16. Power Series in x

Even and odd solutions are given by **19.2.1** to **19.2.4** with $-ia$ written for a and $xe^{\frac{1}{4}i\pi}$ for x; the series involves complex quantities in which the imaginary part of the sum vanishes identically.
Alternatively,

19.16.1
$$y_1=1+a\frac{x^2}{2!}+(a^2-\tfrac{1}{2})\frac{x^4}{4!}+(a^3-\tfrac{7}{2}a)\frac{x^6}{6!}$$
$$+(a^4-11a^2+\tfrac{15}{4})\frac{x^8}{8!}+(a^5-25a^3+\tfrac{211}{4}a)\frac{x^{10}}{10!}+\cdots$$

19.16.2
$$y_2=x+a\frac{x^3}{3!}+(a^2-\tfrac{3}{2})\frac{x^5}{5!}+(a^3-\tfrac{13}{2}a)\frac{x^7}{7!}$$
$$+(a^4-17a^2+\tfrac{63}{4})\frac{x^9}{9!}+(a^5-35a^3+\tfrac{531}{4}a)\frac{x^{11}}{11!}+\cdots$$

in which non-zero coefficients a_n of $x^n/n!$ are connected by

19.16.3
$$a_{n+2}=a\cdot a_n-\tfrac{1}{4}n(n-1)a_{n-2}$$

19.17. Standard Solutions (see [19.4])

19.17.1
$$W(a,\pm x)=\frac{(\cosh\pi a)^{\frac{1}{4}}}{2\sqrt{\pi}}(G_1y_1\mp\sqrt{2}G_3y_2)$$

19.17.2
$$=2^{-3/4}\left(\sqrt{\frac{G_1}{G_3}}\,y_1\mp\sqrt{\frac{2G_3}{G_1}}\,y_2\right)$$

where

19.17.3
$$G_1=|\Gamma(\tfrac{1}{4}+\tfrac{1}{2}ia)|\qquad G_3=|\Gamma(\tfrac{3}{4}+\tfrac{1}{2}ia)|$$

At $x=0$,

19.17.4 $W(a,0)=\dfrac{1}{2^{\frac{3}{4}}}\left|\dfrac{\Gamma(\frac{1}{4}+\frac{1}{2}ia)}{\Gamma(\frac{3}{4}+\frac{1}{2}ia)}\right|^{\frac{1}{2}}=\dfrac{1}{2^{\frac{3}{4}}}\sqrt{\dfrac{G_1}{G_3}}$

19.17.5

$$W'(a,0)=-\dfrac{1}{2^{\frac{1}{4}}}\left|\dfrac{\Gamma(\frac{3}{4}+\frac{1}{2}ia)}{\Gamma(\frac{1}{4}+\frac{1}{2}ia)}\right|^{\frac{1}{2}}=-\dfrac{1}{2^{\frac{1}{4}}}\sqrt{\dfrac{G_3}{G_1}}$$

Complex Solutions

19.17.6 $E(a,x)=k^{-\frac{1}{2}}W(a,x)+ik^{\frac{1}{2}}W(a,-x)$

19.17.7 $E^*(a,x)=k^{-\frac{1}{2}}W(a,x)-ik^{\frac{1}{2}}W(a,-x)$

where

19.17.8 $k=\sqrt{1+e^{2\pi a}}-e^{\pi a}$ $1/k=\sqrt{1+e^{2\pi a}}+e^{\pi a}$

In terms of $U(a,x)$ of **19.3**,

19.17.9 $E(a,x)=\sqrt{2}e^{\frac{1}{2}\pi a+\frac{1}{4}i\pi+\frac{1}{2}i\phi_2}U(ia,xe^{-\frac{1}{4}i\pi})$

with

19.17.10 $\phi_2=\arg\Gamma(\tfrac{1}{2}+ia)$

where the branch is defined by $\phi_2=0$ when $a=0$ and by continuity elsewhere.

Also

19.17.11

$$\sqrt{2\pi}U(ia,xe^{-\frac{1}{4}i\pi})=\Gamma(\tfrac{1}{2}-ia)\{e^{\frac{1}{2}\pi a-\frac{1}{4}i\pi}U(-ia,xe^{\frac{1}{4}i\pi})$$
$$+e^{-\frac{1}{2}\pi a+\frac{1}{4}i\pi}U(-ia,-xe^{\frac{1}{4}i\pi})\}$$

19.18. Wronskian and Other Relations

19.18.1 $W\{W(a,x),W(a,-x)\}=1$

19.18.2 $W\{E(a,x),E^*(a,x)\}=-2i$

19.18.3 $\sqrt{1+e^{2\pi a}}E(a,x)=e^{\pi a}E^*(a,x)+iE^*(a,-x)$

19.18.4 $E^*(a,x)=e^{-i(\phi_2+\frac{1}{2}\pi)}E(-a,ix)$

19.18.5

$$\sqrt{\Gamma(\tfrac{1}{2}+ia)}E^*(a,x)=e^{-\frac{1}{4}i\pi}\sqrt{\Gamma(\tfrac{1}{2}-ia)}E(-a,ix)$$

19.19. Integral Representations

These are covered for **19.1.3** as well as for **19.1.2** in **19.5** (general complex argument).

Asymptotic Expansions

19.20. Expressions in Terms of Airy Functions

When a is large and positive, write, for $0\le x<\infty$

$$x=2\sqrt{a}\,\xi\qquad t=(4a)^{\frac{1}{3}}\tau$$

19.20.1

$$\tau=-(\tfrac{3}{2}\vartheta_3)^{\frac{2}{3}}$$
$$\vartheta_3=\frac{1}{2}\int_\xi^1\sqrt{1-s^2}\,ds=\tfrac{1}{4}\arccos\xi-\tfrac{1}{4}\xi\sqrt{1-\xi^2}\qquad(\xi\le1)$$

19.20.2

$$\tau=+(\tfrac{3}{2}\vartheta_2)^{\frac{2}{3}}$$
$$\vartheta_2=\tfrac{1}{2}\int_1^\xi\sqrt{s^2-1}\,ds=\tfrac{1}{4}\xi\sqrt{\xi^2-1}-\tfrac{1}{4}\arccosh\xi\qquad(\xi\ge1)$$

Then for $x>0$, $a\to+\infty$

19.20.3

$$W(a,x)\sim\sqrt{\pi}(4a)^{-\frac{1}{4}}e^{-\frac{1}{2}\pi a}\left(\dfrac{t}{\xi^2-1}\right)^{\frac{1}{4}}\mathrm{Bi}(-t)$$

19.20.4

$$W(a,-x)\sim2\sqrt{\pi}(4a)^{-\frac{1}{4}}e^{\frac{1}{2}\pi a}\left(\dfrac{t}{\xi^2-1}\right)^{\frac{1}{4}}\mathrm{Ai}(-t)$$

Table 19.3 gives τ as a function of ξ. See [19.5] for further developments.

19.21. Expansions for x Large and a Moderate

When $x\gg|a|$,

19.21.1

$$E(a,x)=\sqrt{2/x}\,\exp\,\{i(\tfrac{1}{4}x^2-a\ln x+\tfrac{1}{2}\phi_2+\tfrac{1}{4}\pi)\}s(a,x)$$

19.21.2

$$W(a,x)=\sqrt{2k/x}\{s_1(a,x)\cos(\tfrac{1}{4}x^2-a\ln x+\tfrac{1}{4}\pi+\tfrac{1}{2}\phi_2)$$
$$-s_2(a,x)\sin(\tfrac{1}{4}x^2-a\ln x+\tfrac{1}{4}\pi+\tfrac{1}{2}\phi_2)\}$$

19.21.3

$$W(a,-x)=\sqrt{2/kx}\{s_1(a,x)\sin(\tfrac{1}{4}x^2-a\ln x+\tfrac{1}{4}\pi+\tfrac{1}{2}\phi_2)$$
$$+s_2(a,x)\cos(\tfrac{1}{4}x^2-a\ln x+\tfrac{1}{4}\pi+\tfrac{1}{2}\phi_2)\}$$

where ϕ_2 is defined by **19.17.10** and

19.21.4 $s(a,x)=s_1(a,x)+is_2(a,x)$

19.21.5

$$s_1(a,x)\sim1+\dfrac{v_2}{1!2x^2}-\dfrac{u_4}{2!2^2x^4}-\dfrac{v_6}{3!2^3x^6}+\dfrac{u_8}{4!2^4x^8}+\cdots$$

19.21.6

$$s_2(a,x)\sim-\dfrac{u_2}{1!2x^2}-\dfrac{v_4}{2!2^2x^4}+\dfrac{u_6}{3!2^3x^6}+\dfrac{v_8}{4!2^4x^8}-\cdots$$

with

$$(x\to+\infty)$$

19.21.7 $u_r + iv_r = \Gamma(r + \tfrac{1}{2} + ia)/\Gamma(\tfrac{1}{2} + ia)$

or

19.21.8 $s(a, x) \sim \sum_{r=0}^{\infty} (-i)^r \dfrac{\Gamma(2r + \tfrac{1}{2} + ia)}{\Gamma(\tfrac{1}{2} + ia)} \dfrac{1}{2^r r! x^{2r}}$

19.22. Expansions for a Large With x Moderate

(i) a positive

When $a \gg x^2$, with $p = \sqrt{a}$, then

19.22.1 $W(a, x) = W(a, 0) \exp(-px + v_1)$

19.22.2 $W(a, -x) = W(a, 0) \exp(px + v_2)$

where $W(a, 0)$ is given by **19.17.4,** and

19.22.3

$$v_1, v_2 \sim \pm \frac{\tfrac{2}{3}(\tfrac{1}{2}x)^3}{2p} + \frac{(\tfrac{1}{2}x)^2}{(2p)^2} \pm \frac{\tfrac{1}{2}x + \tfrac{2}{5}(\tfrac{1}{2}x)^5}{(2p)^3}$$

$$+ \frac{2(\tfrac{1}{2}x)^4}{(2p)^4} \pm \frac{\tfrac{16}{3}(\tfrac{1}{2}x)^3 + \tfrac{4}{7}(\tfrac{1}{2}x)^7}{(2p)^5} + \cdots$$

$$(a \to +\infty)$$

The upper sign gives the first function, and the lower sign the second function.

(ii) a negative

When $-a \gg x^2$, with $p = \sqrt{-a}$, then

19.22.4

$$W(a, x) + iW(a, -x)$$
$$= \sqrt{2} W(a, 0) \exp\{v_r + i(px + \tfrac{1}{4}\pi + v_i)\}$$

where $W(a, 0)$ is given by **19.17.4,** and

19.22.5

$$v_r \sim -\frac{(\tfrac{1}{2}x)^2}{(2p)^2} + \frac{2(\tfrac{1}{2}x)^4}{(2p)^4} - \frac{9(\tfrac{1}{2}x)^2 + \tfrac{16}{3}(\tfrac{1}{2}x)^6}{(2p)^6} + \cdots$$

$$v_i \sim \frac{\tfrac{2}{3}(\tfrac{1}{2}x)^3}{2p} - \frac{\tfrac{1}{2}x + \tfrac{2}{5}(\tfrac{1}{2}x)^5}{(2p)^3} + \frac{\tfrac{16}{3}(\tfrac{1}{2}x)^3 + \tfrac{4}{7}(\tfrac{1}{2}x)^7}{(2p)^5} - \cdots$$

$$(a \to -\infty)$$

Further expansions of a similar type will be found in [19.3].

19.23. Darwin's Expansions

(i) a positive, $x^2 - 4a \gg 0$

Write

19.23.1

$$X = \sqrt{x^2 - 4a} \qquad \theta = 4a\vartheta_2(x/2\sqrt{a}) = \tfrac{1}{2}\int_{2\sqrt{a}}^{x} X dx$$

$$= \tfrac{1}{4}xX - a \ln \frac{x + X}{2\sqrt{a}}$$

$$= \tfrac{1}{4}x\sqrt{x^2 - 4a} - a \operatorname{arccosh} \frac{x}{2\sqrt{a}}$$

(see **Table 19.3** for ϑ_2), then

19.23.2 $W(a, x) = \sqrt{2k}e^{v_r} \cos(\tfrac{1}{4}\pi + \theta + v_i)$

19.23.3 $W(a, -x) = \sqrt{2/k}e^{v_r} \sin(\tfrac{1}{4}\pi + \theta + v_i)$

where

19.23.4 $v_r \sim -\tfrac{1}{2}\ln X - \dfrac{d_6}{X^6} + \dfrac{d_{12}}{X^{12}} - \cdots$

$$v_i \sim -\frac{d_3}{X^3} + \frac{d_9}{X^9} - \frac{d_{15}}{X^{15}} + \cdots$$

$$(x^2 - 4a \to \infty)$$

and d_{3r} is given by **19.23.12**.

(ii) a positive, $4a - x^2 \gg 0$

Write

19.23.5

$$Y = \sqrt{4a - x^2} \qquad \theta = 4a\vartheta_4(x/2\sqrt{a})$$

$$= \tfrac{1}{2}\int_0^x Y dx = \tfrac{1}{4}xY + a \arcsin \frac{x}{2\sqrt{a}}$$

(see **Table 19.3** for $\vartheta_4 = \tfrac{1}{8}\pi - \vartheta_3$), then

19.23.6 $W(a, x) = \exp\{-\theta + v(a, x)\}$

19.23.7 $W(a, -x) = \exp\{\theta + v(a, -x)\}$

where

19.23.8

$$v(a, x) \sim -\tfrac{1}{2}\ln Y + \frac{d_3}{Y^3} + \frac{d_6}{Y^6} + \frac{d_9}{Y^9} + \cdots$$

$$(x^2 - 4a \to -\infty)$$

and d_{3r} is again given by **19.23.12**.

(iii) a negative, $x^2 - 4a \gg 0$

Write

19.23.9

$$X = \sqrt{x^2 + 4|a|} \qquad \theta = 4|a|\vartheta_1(x/2\sqrt{|a|}) = \tfrac{1}{2}\int_0^x X dx$$

$$= \tfrac{1}{4}xX - a \ln \frac{x + X}{2\sqrt{|a|}}$$

$$= \tfrac{1}{4}x\sqrt{x^2 + 4|a|} - a \operatorname{arcsinh} \frac{x}{2\sqrt{|a|}}$$

(see **Table 19.3** for ϑ_1) then

19.23.10 $W(a, x) = \sqrt{2k}e^{v_r} \cos(\tfrac{1}{4}\pi + \theta + v_i)$

19.23.11 $W(a, -x) = \sqrt{2/k}e^{v_r} \sin(\tfrac{1}{4}\pi + \theta + v_i)$

where v_r and v_i are again given by **19.23.4**.   In each case the coefficients d_{3r} are given by

19.23.12

$$d_3=-\frac{1}{a}\left(\frac{x^3}{48}-\frac{1}{2}ax\right)$$

$$d_6=\frac{3}{4}x^2+2a$$

$$d_9=\frac{1}{a^3}\left(\frac{7}{5760}x^9-\frac{7}{320}ax^7+\frac{49}{320}a^2x^5+\frac{31}{12}a^3x^3+19a^4x\right)$$

$$d_{12}=\frac{153}{8}x^4+186ax^2+80a^2$$

See [19.11] for $d_{15}, \ldots, d_{24}$, and [19.5] for an alternative form.

19.24. Modulus and Phase

When a is positive, the function $W(a, x)$ is oscillatory when $x<-2\sqrt{a}$ and when $x>2\sqrt{a}$; when a is negative, the function is oscillatory for all x. In such cases it is sometimes convenient to write

19.24.1

$$k^{-\frac{1}{2}}W(a, x)+ik^{\frac{1}{2}}W(a,-x)=E(a, x)=Fe^{ix} \qquad (x>0)$$

19.24.2

$$k^{-\frac{1}{2}}\frac{dW(a, x)}{dx}+ik^{\frac{1}{2}}\frac{dW(a,-x)}{dx}=E'(a, x)=-Ge^{i\psi}$$
$$(x>0)$$

Then, when $x^2>>|a|$,

19.24.3

$$F\sim\sqrt{\frac{2}{x}}\left(1+\frac{a}{x^2}+\frac{10a^2-3}{4x^4}+\frac{30a^3-47a}{4x^6}+\cdots\right)$$

19.24.4

$$\chi\sim\frac{1}{4}x^2-a\ln x+\frac{1}{2}\phi_2+\frac{1}{4}\pi+\frac{4a^2-3}{8x^2}+\frac{4a^3-19a}{8x^4}+\cdots$$

19.24.5

$$G\sim\sqrt{\frac{x}{2}}\left(1-\frac{a}{x^2}-\frac{6a^2-5}{4x^4}-\frac{14a^3-63a}{4x^6}-\cdots\right)$$

19.24.6

$$\psi\sim\frac{1}{4}x^2-a\ln x+\frac{1}{2}\phi_2-\frac{1}{4}\pi+\frac{4a^2+5}{8x^2}+\frac{4a^3+29a}{8x^4}+\cdots$$

where ϕ_2 is defined by **19.17.10**.

When $a<0$, $|a|>>x^2$

19.24.7 $$F\sim\sqrt{2}W(a, 0)e^{v_r}$$

where v_r is given by **19.22.5** with $p=\sqrt{-a}$. Also

19.24.8

$$F\sim\frac{1}{\sqrt{p}}\left(1-\frac{x^2}{(4p)^2}+\frac{\frac{5}{2}x^4+8}{(4p)^4}-\frac{\frac{15}{2}x^6+152x^2}{(4p)^6}+\cdots\right)$$

19.24.9

$$\chi\sim\frac{1}{4}\pi+px\left(1+\frac{\frac{2}{3}x^2}{(4p)^2}-\frac{\frac{2}{5}x^4+16}{(4p)^4}+\frac{\frac{4}{7}x^6+\frac{256}{3}x^2}{(4p)^6}-\cdots\right)$$

19.24.10

$$G\sim\sqrt{p}\left(1+\frac{x^2}{(4p)^2}-\frac{\frac{3}{2}x^4+8}{(4p)^4}+\frac{\frac{7}{2}x^6+168x^2}{(4p)^6}-\cdots\right)$$

19.24.11

$$\psi\sim-\frac{1}{4}\pi+px\left(1+\frac{\frac{2}{3}x^2}{(4p)^2}-\frac{\frac{2}{5}x^4-16}{(4p)^4}+\frac{\frac{4}{7}x^6-\frac{320}{3}x^2}{(4p)^6}-\cdots\right)$$

Again, when $a<0$, $x^2-4a\gg0$, with $X=\sqrt{x^2+4|a|}$, then

19.24.12 $$F\sim\sqrt{2}e^{v_r} \qquad \chi=\frac{1}{4}\pi+\theta+v_i$$

where θ, v_r and v_i are given by **19.23.4** and **19.23.9**.

Another form also when $a>0$, $x^2-4a\rightarrow\infty$ is

19.24.13

$$F\sim\sqrt{\frac{2}{X}}\left(1-\frac{3}{4X^4}-\frac{5a}{X^6}+\frac{621}{32X^8}+\frac{1371a}{4X^{10}}-\cdots\right)$$

19.24.14

$$G\sim\sqrt{\frac{X}{2}}\left(1+\frac{5}{4X^4}+\frac{7a}{X^6}-\frac{835}{32X^8}-\frac{1729a}{4X^{10}}+\cdots\right)$$

while ψ and χ are connected by

19.24.15

$$\psi-\chi\sim-\frac{1}{2}\pi+\frac{x}{X^3}\left(1-\frac{47}{6X^4}-\frac{214a}{3X^6}+\frac{14483}{40X^8}+\cdots\right)$$

19.25. Connections With Other Functions

Connection With Confluent Hypergeometric and Bessel Functions

19.25.1

$$W(a, \pm x)=2^{-\frac{3}{4}}\left\{\sqrt{\frac{G_1}{G_3}}H(-\frac{3}{4}, \frac{1}{2}a, \frac{1}{4}x^2)\right.$$
$$\left.\pm\sqrt{\frac{2G_3}{G_1}}xH(-\frac{1}{4}, \frac{1}{2}a, \frac{1}{4}x^2)\right\}$$

where

19.25.2

$$H(m, n, x)=e^{-ix}{}_1F_1(m+1-in; 2m+2; 2ix)$$

19.25.3 $$=e^{-ix}M(m+1-in, 2m+2, 2ix)$$

19.25.4

$$W(0, \pm x)=2^{-\frac{3}{4}}\sqrt{\pi x}\{J_{-\frac{1}{4}}(\frac{1}{4}x^2)\pm J_{\frac{1}{4}}(\frac{1}{4}x^2)\} \qquad (x\geq0)$$

19.25.5

$$\frac{d}{dx} W(0, \pm x) = -2^{-\frac{1}{2}}x\sqrt{\pi x}\{ J_{\frac{1}{4}}(\tfrac{1}{4}x^2) \pm J_{-\frac{1}{4}}(\tfrac{1}{4}x^2) \}$$

$$(x \geq 0)$$

19.26. Zeros

Zeros of solutions $U(a, x)$, $V(a, x)$ of **19.1.2** occur only for $|x| < 2\sqrt{-a}$ when a is negative. A single exceptional zero is possible, for any a, in the general solution; neither $U(a, x)$ nor $V(a, x)$ has such a zero for $x > 0$.

Approximations may be obtained by reverting the series for ψ (or χ for zeros of derivatives) in **19.11**, giving ψ (or χ) values that are multiples of $\frac{1}{2}\pi$, odd multiples for $U(a, x)$, even multiples for $V(a, x)$. Writing

$$\alpha = (\tfrac{1}{2}r - \tfrac{1}{2}a - \tfrac{1}{4})\pi$$

as an approximation to a zero of the function, or

$$\beta = (\tfrac{1}{2}r - \tfrac{1}{2}a + \tfrac{1}{4})\pi$$

as an approximation to a zero of the derivative, we obtain for the corresponding zero c or c', with $-a = p^2$ the expressions

19.26.1 $c \approx \dfrac{\alpha}{p} + \dfrac{2\alpha^3 - 3\alpha}{48p^5} + \dfrac{52\alpha^5 - 240\alpha^3 + 315\alpha}{7680 p^9} + \cdots$

19.26.2 $c' \approx \dfrac{\beta}{p} + \dfrac{2\beta^3 + 3\beta}{48p^5} + \dfrac{52\beta^5 + 280\beta^3 - 285\beta}{7680 p^9} + \cdots$

These expansions, however, are of little value in the neighborhood of the turning point $x = 2\sqrt{-a}$. Here first approximations may be obtained by use of the formulas of **19.7**. If a_n (negative) is a zero of $\mathrm{Ai}(t)$, the corresponding zero c of $U(a, x)$ is obtained approximately by solving

19.26.3

$$\vartheta_3 = \tfrac{1}{4}\{ \arccos \xi - \xi\sqrt{1-\xi^2} \} = \frac{(-a_n)^{\frac{3}{2}}}{6|a|}$$

$$c = 2\sqrt{|a|}\,\xi \qquad (a \ll 0)$$

This may be done by inverse use of **Table 19.3**. For a zero of $V(a, x)$, a_n must be replaced by b_n, a zero of $\mathrm{Bi}(t)$. For further developments see [19.5].

Zeros of solutions $W(a, x)$, $W(a, -x)$ of **19.1.3** occur for $|x| > 2\sqrt{a}$ when a is positive; the general solution may, however, have a single zero between $-2\sqrt{a}$ and $+2\sqrt{a}$. If a is negative, zeros are unrestricted in range.

Approximations may be obtained by reverting the series for ψ (or χ) in **19.24**. With $-a = p^2$, $\alpha = (\tfrac{1}{2}r - \tfrac{1}{4})\pi$, $\beta = (\tfrac{1}{2}r + \tfrac{1}{4})\pi$, $r \geq 0$ being an odd

integer for $W(a, x)$ or its derivative, or an even integer for $W(a, -x)$ or its derivative, the zeros $\pm c$, $\pm c'$ have expansions

19.26.4 $c \approx \dfrac{\alpha}{p} - \dfrac{2\alpha^3 - 3\alpha}{48p^5} + \dfrac{52\alpha^5 - 240\alpha^3 + 315\alpha}{7680 p^9} + \cdots$

19.26.5 $c' \approx \dfrac{\beta}{p} - \dfrac{2\beta^3 + 3\beta}{48p^5} + \dfrac{52\beta^5 + 280\beta^3 - 285\beta}{7680 p^9} + \cdots$

When x is large and a moderate, we may solve inversely the series **19.24.4** or **19.24.6** with $\alpha = \frac{1}{2}(r\pi - \frac{1}{2}\pi - \phi_2)$, $\beta = \frac{1}{2}(r\pi + \frac{1}{2}\pi - \phi_2)$, r odd or even as above; the presence of the logarithm makes it inconvenient to revert formally.

The expansions **19.26.4** and **19.26.5** fail when x is in the neighborhood of $2\sqrt{|a|}$. When a is positive, a zero c of $W(a, -x)$ is obtained approximately by solving

19.26.6

$$\vartheta_2 = \tfrac{1}{4}\{ \xi\sqrt{\xi^2-1} - \operatorname{arccosh} \xi \} = \frac{(-a_n)^{\frac{3}{2}}}{6a}$$

$$c = 2\sqrt{a}\,\xi \qquad (a \gg 0)$$

with the aid of **Table 19.3**. For a zero of $W(a, x)$ we replace a_n by b_n. When a is negative we solve, again with the aid of **Table 19.3**,

19.26.7

$$\vartheta_1 = \tfrac{1}{4}\{ \xi\sqrt{\xi^2+1} + \operatorname{arcsinh} \xi \} = \frac{(n-\frac{1}{4})\pi}{4|a|}$$

$$c = 2\sqrt{|a|}\,\xi \qquad (-a \gg 0)$$

where $n = 1, 2, 3, \ldots$ for an approximate zero of $W(a, -x)$, and $n = \frac{1}{2}, \frac{3}{2}, \frac{5}{2}, \ldots$ for an approximate zero of $W(a, x)$. Further developments are given in [19.5].

Any of the approximations to zeros obtained above may readily be improved as follows:

Let c be a zero of y, and c' a zero of y', where y is a solution of

19.26.8 $y'' - Iy = 0$

Here $I = a \pm \frac{1}{4}x^2$, $I' = \pm \frac{1}{2}x$, $I'' = \pm \frac{1}{2}$; the method is general and the following formulae may be used whenever $I''' = 0$. Then if γ, γ' are approximations to the zeros c, c' and

19.26.9 $u = y(\gamma)/y'(\gamma)$ $v = y'(\gamma')/I^2 y(\gamma')$

with $I \equiv I(\gamma)$ or $I \equiv I(\gamma')$ respectively, then

19.26.10

$$c \sim \gamma - u - \tfrac{1}{3} I u^3 + \tfrac{1}{12} I' u^4$$
$$- (\tfrac{1}{60} I'' + \tfrac{1}{5} I^2) u^5 + \tfrac{11}{90} II' u^6 + \cdots$$

19.26.11

$$y'(c) \sim y'(\gamma) \{1 - \tfrac{1}{2} I u^2 + \tfrac{1}{6} I' u^3$$
$$- (\tfrac{1}{24} I'' + \tfrac{1}{8} I^2) u^4 + \tfrac{7}{60} II' u^5 + \cdots \}$$

19.26.12

$$c' \sim \gamma' - Iv - \tfrac{1}{2} II' v^2 + (\tfrac{1}{6} I^2 I'' - \tfrac{1}{2} II'^2 - \tfrac{1}{3} I^4) v^3$$
$$+ (\tfrac{5}{12} I^2 I' I'' - \tfrac{5}{8} II'^3 - \tfrac{5}{12} I^4 I') v^4 + \cdots$$

19.26.13

$$y(c') \sim y(\gamma') \{1 - \tfrac{1}{2} I^3 v^2 - \tfrac{1}{6} I^3 I' v^3$$
$$- (\tfrac{1}{8} I^3 I'^2 - \tfrac{1}{24} I^4 I'' + \tfrac{1}{8} I^6) v^4 + \cdots \}$$

The process can be repeated, if necessary, using as many terms at any stage as seems convenient.

Note the relations, holding at zeros,

19.26.14 $\quad U'(a, c) = -\sqrt{2/\pi}/V(a, c)$

19.26.15 $\quad V'(a, c') = \sqrt{2/\pi}/U(a, c')$

19.26.16 $\quad W'(a, c) = -1/W(a, -c)$

19.26.17

$$W(a, c') = 1/\left\{ \frac{d}{dx} W(a, -x) \right\}_{x=c'} = -1/W'(a, -c')$$

19.27. Bessel Functions of Order $\pm\tfrac{1}{4}$, $\pm\tfrac{3}{4}$ as Parabolic Cylinder Functions

Most applications of these functions refer to cases where parabolic cylinder functions would be more appropriate. We have

19.27.1 $\quad J_{\pm\frac{1}{4}}(\tfrac{1}{4}x^2) = \dfrac{2^{\frac{1}{4}}}{\sqrt{\pi x}} \{ W(0, -x) \mp W(0, x) \}$

19.27.2 $\quad J_{\pm\frac{3}{4}}(\tfrac{1}{4}x^2) = \dfrac{-2^{\frac{1}{4}}}{x\sqrt{\pi x}} \{ W(0, x) \pm W(0, -x) \}$

Functions of other orders may be obtained by use of the recurrence relation **10.1.22,** which here becomes

19.27.3 $\quad \tfrac{1}{4}x^2 J_{\nu+1}(\tfrac{1}{4}x^2) - 2\nu J_\nu(\tfrac{1}{4}x^2) + \tfrac{1}{4}x^2 J_{\nu-1}(\tfrac{1}{4}x^2) = 0$

Again

19.27.4 $\quad I_{-\frac{1}{4}}(\tfrac{1}{4}x^2) + I_{\frac{1}{4}}(\tfrac{1}{4}x^2) = \dfrac{2}{\sqrt{x}} V(0, x)$

19.27.5

$$\frac{\sqrt{2}}{\pi} K_{\frac{1}{4}}(\tfrac{1}{4}x^2) = I_{-\frac{1}{4}}(\tfrac{1}{4}x^2) - I_{\frac{1}{4}}(\tfrac{1}{4}x^2) = \frac{2}{\sqrt{\pi x}} U(0, x)$$

19.27.6 $\quad I_{-\frac{3}{4}}(\tfrac{1}{4}x^2) + I_{\frac{3}{4}}(\tfrac{1}{4}x^2) = -\dfrac{4}{x\sqrt{x}} \dfrac{d}{dx} V(0, x)$

19.27.7

$$\frac{\sqrt{2}}{\pi} K_{\frac{3}{4}}(\tfrac{1}{4}x^2) = I_{-\frac{3}{4}}(\tfrac{1}{4}x^2) - I_{\frac{3}{4}}(\tfrac{1}{4}x^2)$$
$$= -\frac{4}{x\sqrt{\pi x}} \frac{d}{dx} U(0, x)$$

As before, Bessel functions of other orders may be obtained by use of the recurrence relation **10.2.23,** which here becomes

19.27.8 $\quad \tfrac{1}{4}x^2 I_{\nu+1}(\tfrac{1}{4}x^2) + 2\nu I_\nu(\tfrac{1}{4}x^2) - \tfrac{1}{4}x^2 I_{\nu-1}(\tfrac{1}{4}x^2) = 0$

19.27.9 $\quad \tfrac{1}{4}x^2 K_{\nu+1}(\tfrac{1}{4}x^2) - 2\nu K_\nu(\tfrac{1}{4}x^2) - \tfrac{1}{4}x^2 K_{\nu-1}(\tfrac{1}{4}x^2) = 0$

Numerical Methods

19.28. Use and Extension of the Tables

For $U(a, x)$, $V(a, x)$ and $W(a, x)$, interpolation x-wise may be carried out to 5-figure accuracy almost everywhere by using 5-point or 6-point Lagrangian interpolation. For $|a| \leq 1$, comparable accuracy a-wise may be obtained with 5- or 6-point interpolation.

For $|a| > 1$, $U(a, x)$ and $V(a, x)$ may be obtained by use of recurrence relations from two values, possibly obtained by interpolation, with $|a| \leq 1$; such a procedure is not available for $W(a, \pm x)$, $|a| > 1$.

In cases where straightforward use of the a-wise recurrence relation results in loss of accuracy by cancellation of leading digits, it may be worth while to remark that greater accuracy is usually attainable by use of the recurrence relation in the reverse direction, from arbitrary starting values (often 1 and 0) for two values of a somewhat beyond the last value desired. This is because the recurrence relation is a second order homogeneous linear difference equation, and has two independent solutions. Loss of accuracy by cancellation occurs when the solution desired is diminishing as a varies, while the companion solution is increasing. By reversing the direction of progress in a, the roles of the two solutions are interchanged, and the contribution of the desired solution now increases, while the unwanted solution diminishes to the point of negligibility. By starting sufficiently beyond the last value of a for which the function is desired, we can ensure that the unwanted solution is negligible but, because the starting values were arbitrary, we have an unknown multiple of the solution desired. The computation is then carried back until a value of a with $|a| \leq 1$ is reached, when the precise multiple that we have of the desired solution may be determined and hence removed throughout. Compare also **9.12, Example 1.**

Example 1. Evaluate $U(a, 5)$ for $a = 5, 6, 7,$ . . ., using **19.6.4.**

$$(a + \tfrac{1}{2}) U(a+1, x) + x U(a, x) - U(a-1, x) = 0$$

a	Forward Recurrence		Backward Recurrence		Final Values	
3	(-6)	5. 2847*	(12)	1. 59035	(-6)	5. 2847**
4	(-7)	9. 172*	(11)	2. 76028	(-7)	9. 1724
5	(-7)	1. 5527	(10)	4. 67131	(-7)	1. 55227
6	(-8)	2. 5609	(9)	7. 72041	(-8)	2. 5655
7	(-9)	4. 1885	(9)	1. 24785	(-9)	4. 1466
8	(-10)	6. 2220	(8)	1. 97488	(-10)	6. 5625
9	(-10)	$+1.$ 2676	(7)	3. 06369	(-10)	1. 01806
10	(-11)	$-0.$ 1221	(6)	4. 66352	(-11)	1. 5497
11	(-11)	$+1.$ 2654	(0)	697082	(-12)	2. 3164
12	(-12)	$-5.$ 6079		102444	(-13)	3. 404
13	(-12)	$+3.$ 2555		14789	(-14)	4. 91
14				2111	(-15)	7. 01
15				292	(-16)	9. 7
16				42		
17				5		
18				$1+$		
19				$0+$		

*From tables.   +Starting values.

**This value was used to obtain the constant multiplier $\dfrac{d}{k^*} = \dfrac{(-6)5.2847}{(12)1.59035} = (-18)3.32298$ for converting the previous column into this one.

The second column shows forward recurrence starting with values at $a = 3, 4$ from **Table 19.1.** Backward recurrence starts with values 0 and 1 at $a = 19$ and 18, containing a multiple $kU(a, 5)$ and a subsequently negligible multiple of the other solution $\Gamma(\tfrac{1}{2} - a) V(a, 5)$. Rounding errors convert $kU(a, x)$ into $k^*U(a, x)$ *without affecting the values in the last column.* The value of $1/k^*$ is identified from the known value of $U(3, 5)$, and used to obtain the final column by multiplying throughout by $1/k^*$. The improvement in $U(5, 5)$ is evident by comparison with **Table 19.1.**

Derivatives. These are not tabulated here. Since the functions $U(a, x)$, $V(a, x)$ and $W(a, x)$ satisfy differential equations, values of derivatives are often required.

For all these functions the equation is second order with first derivative absent, so that *second derivatives* may be readily obtained from function values by use of the differential equation.

First derivatives can be obtained for $U(a, x)$ and $V(a, x)$ by applying the appropriate recurrence

relations **19.6.1–2.** If less accuracy is needed they can be found by use of mean central differences of $U(a, x)$, $V(a, x)$ and also of $W(a, x)$ with the formula

$$hu' = h \frac{du}{dx} = \mu \delta u - \tfrac{1}{6}\mu \delta^3 u + \tfrac{1}{30}\mu \delta^5 u - \ldots$$

using $h = .1$; this usually gives a 3- or 4-figure value of du/dx.

If greater accuracy is needed for $dW(a, x)/dx$ it may be obtained by evaluating d^2W/dx^2 with the help of the differential equation satisfied by W and integrating this second derivative numerically. This requires one accurate value of dW/dx to start off the integration; we describe two methods for obtaining this, both making use of the difference between two fairly widely separated values of W, for example, separated by 5 or 10 tabular intervals.

(i) Write f_r, f_r', f_r'' for $W(a, x_0 + rh)$ and its first two derivatives, then f_0' may be found from

$$hf_0' = \frac{1}{2n}(f_n - f_{-n}) - \frac{h^2}{2n}\sum_1^{n-1}(n-r)(f_r'' - f_{-r}'')$$

$$-\frac{h^2}{2n}\{\tfrac{1}{12} - \tfrac{1}{240}\delta^2 + \tfrac{31}{60480}\delta^4 - \ldots\}(f_n'' - f_{-n}'')$$

$$-h^2\{\tfrac{1}{12}\mu\delta - \tfrac{11}{720}\mu\delta^3 + \tfrac{191}{60480}\mu\delta^5 - \ldots\}f_0''$$

(ii) Consider a solution y of the differential equation for $W(a, x)$, namely $y'' = (-\tfrac{1}{4}x^2 + a)y$. If we are given values y and y' at a particular $x = x_0$ and write $T_n = H^n y^{(n)}/n!$, $T_{-1} = T_{-2} = 0$, then we may compute $T_2, T_3, T_4, \ldots$ in succession by use of the recurrence relation obtained from the differential equation,

$$T_{n+2} = \frac{H^2}{(n+1)(n+2)}[(-\tfrac{1}{4}x_0^2 + a)T_n - \tfrac{1}{2}Hx_0 T_{n-1}$$

$$-\tfrac{1}{4}H^2 T_{n-2}]$$

These are computed, to a fixed number of decimals until they become negligible, thus giving

$$y(x_0 \pm H) = T_0 \pm T_1 + T_2 \pm T_3 + \ldots$$

This may be applied, with $H = rh$, h being the tabular interval, and r a small integer, say $r = 5$, to the solutions $y = y_1$, $y = y_2$ having

$y_1(x_0) = W(a, x_0)$	$y_1'(x_0) = W^{*\prime}(a, x_0)$
$y_2(x_0) = 0$	$y_2'(x_0) = 1$

in which $W^{*\prime}(a, x_0)$ is an approximation to $W'(a, x_0)$, not necessarily a good one; it may be obtained from differences, for example. We thus obtain $y_1(x_0 \pm H)$ and $y_2(x_0 \pm H)$.

Now suppose

$$W'(a, x_0) = W^{*'}(a, x_0) + \lambda$$

then, for all x

$$W(a, x) = y_1(x) + \lambda y_2(x)$$

and in particular

$$W(a, x_0 \pm H) = y_1(x_0 \pm H) + \lambda y_2(x_0 \pm H)$$

The values of $W(a, x_0 \pm H)$ may be read from the tables and two independent estimates of λ obtained, whence

$$W'(a, x_0) = W^{*'}(a, x_0) + \lambda$$

to a suitable accuracy.

Example 2. Evaluate $W'(-3, 1)$ using $r=5$. From **Table 19.2**

$$W(-3, .5) = -.05857 \qquad W(-3, 1) = -.61113$$
$$W(-3, 1.5) = -.69502$$

(i) Using the first method

x	$W(-3, x)$	$W''(-3, x)$	δ	δ^2	δ^3
0.4	$+0.07298$	-0.22186			
0.5	$-.05857$	$+.17937$		$+131$	
0.6	$-.18832$	$.58191$			
0.7	$-.31226$	$.97503$			
0.8	$-.42646$	1.34761			
			34081		
0.9	$-.52722$	1.68842			
			29775		-1095
1.0	$-.61113$	1.98617			
			24374		-1032
1.1	$-.67522$	2.22991			
			17941		
1.2	$-.71706$	2.40932			
1.3	$-.73488$	2.51513			
1.4	$-.72761$	2.53936			
1.5	$-.69502$	2.47601			-9129
1.6	$-.63774$	2.32137			

The fifth decimal in $W''(-3, x)$ is only a guard figure which is hardly needed. Only the differences needed have been computed.

Then

$$\tfrac{1}{10}W'(-3, 1)$$
$$= \tfrac{1}{10}(-.69502 + .05857) - \frac{1}{1000}(10.38874)$$
$$- \frac{1}{1000}\left\{\tfrac{1}{12}(2.29664) - \frac{1}{240}(-.09260)\right\}$$
$$- \frac{1}{100}\left\{\frac{1}{24}(.54149) - \frac{11}{1440}(-.02127)\right\}$$
$$= -.0636450 - .0103887 - .0001918 - .0002272$$
$$= -.0744527$$

Thus $W'(-3, 1) = -.74453$. This might have an error up to about $1\tfrac{1}{2}$ units in the last figure but is, in fact, correct to 5 decimals.

(ii) Using the second method, with

$$y_1(1) = W(-3, 1) = -.61113 \qquad \text{to 5 decimals}$$
$$y_1'(1) = -.745 \qquad \text{to about 3 decimals}$$

the following values result, with $H = .5$,

	y_1	y_2	$W(-3, x) = y_1 + \lambda y_2$
T_0	$-.61113$	$.0000$	At $x = 1.5$
T_1	$-.37250$	$+.5000$	$x - .695223 + .4323\lambda$
			$= -.69502$
T_2	$+.24827\ 2$	$.0000$	$\lambda = .000203/.4323$
T_3	$+\ 5680\ 9$	$-\ 677$	$= .000470$
T_4	$-\ 1407\ 4$	$-\ 26$	So $W'(-3, 1)$
			$= -.745 + \lambda$
			$= -.744530$
T_5	$-\ 279\ 3$	$+\ 24$	At $x = .5$
T_6	$+\ 13\ 4$	$+\ 2$	$-.058363 - .4371\lambda$
			$= -.05857$
T_7	$+\ 5\ 4$		$\lambda = .000207/.4371$
T_8	$+\ 5$		$= .000474$
$y(1.5)$	$-.695223$	$+.4323$	So $W'(-3, 1)$
$y(.5)$	$-.058363$	$-.4371$	$= -.745 + \lambda$
			$= -.744526$

Thus $W'(-3, 1) = -.74453$ which is correct to 5 decimals.

Example 3. Evaluate the positive zero of $U(-3, x)$.

We use **19.7.3** to obtain a first approximation, see **19.26.3**. The appropriate zero of $\text{Ai}(t)$ is at

$$t = (4|a|)^{\frac{1}{3}}\tau = -2.338$$

whence

$$\tau = -(2.338) \times (12)^{-\frac{1}{3}} = -.4461$$

Hence, from **Table 19.3**, $\xi = .3990$ and the approximate zero is $x = 2\sqrt{|a|}\xi = 1.382$.

We improve this by using **19.26.10**, but take, for convenience, $x = 1.4$ as an approximation, so that the value of U can be read directly from the tables. U' can be obtained as in the section following **Example 1.**

We find

$$U(-3, 1.4) = .02627 \qquad U'(-3, 1.4) = 2.0637$$

Then **19.26.9** gives

$$u = U/U' = .012730 \qquad I = -2.51$$
$$I' = .7 \qquad I'' = .5$$

and

$c = 1.4 - .012730 + .000002 = 1.38727$

which is correct to 5 decimals, while **19.26.11** gives

$y'(c) = 2.0637(1 + .000203) = 2.0641$

compared with the correct value 2.06416.

References

Texts

[19.1] H. Buchholz, Die konfluente hypergeometrische Funktion (Springer-Verlag, Berlin, Germany, 1953).

[19.2] C. G. Darwin, On Weber's function, Quart. J. Mech. Appl. Math. **2**, 311–320 (1949).

[19.3] A. Erdélyi et al., Higher transcendental functions, vol. 2 (McGraw-Hill Book Co., Inc., New York, N.Y., 1953).

[19.4] J. C. P. Miller, On the choice of standard solutions to Weber's equation, Proc. Cambridge Philos. Soc. **48**, 428–435 (1952).

[19.5] F. W. J. Olver, Uniform asymptotic expansions for Weber parabolic cylinder functions of large order, J. Research NBS **63B**, 2, 131–169 (1959), RP63B2–14.

[19.6] G. N. Watson, A theory of asymptotic series, Philos. Trans. Roy. Soc. London, **A 211**, 279–313 (1911).

[19.7] H. F. Weber, Ueber die Integration der partiellen Differential-gleichung: $\partial^2 u/\partial x^2 + \partial^2 u/\partial y^2 + k^2 u = 0$, Math. Ann. **1**, 1–36 (1869).

[19.8] E. T. Whittaker, On the functions associated with the parabolic cylinder in harmonic analysis, Proc. London Math. Soc. **35**, 417–427 (1903).

[19.9] E. T. Whittaker and G. N. Watson, A course of modern analysis, 4th ed. (Cambridge Univ. Press, Cambridge, England, 1952).

Tables

[19.10] British Association for the Advancement of Science, Mathematical Tables, vol. I, Circular and hyperbolic functions, exponential, sine and cosine integrals, factorial (gamma) and derived functions, integrals of probability integral, 1st ed. (British Association, London, England, 1931; Cambridge Univ. Press, Cambridge, England, 2d ed., 1946, 3d ed., 1951).

[19.11] National Physical Laboratory, Tables of Weber parabolic cylinder functions. Computed by Scientific Computing Service Ltd. Mathematical Introduction by J. C. P. Miller. (Her Majesty's Stationery Office, London, England, 1955).

[19.12] National Physical Laboratory Mathematical Tables, vol. 4, Tables of Weber parabolic cylinder functions and other functions for large arguments, by L. Fox (Her Majesty's Stationery Office, London, England, 1960).

Table 19.1

x	$U(-5.0, x)$	$U(-4.5, x)$	$U(-4.0, x)$	$U(-3.5, x)$	$U(-3.0, x)$	$U(-2.5, x)$	$U(-2.0, x)$	$U(-1.5, x)$
0.0	(0) 3.0522	(0) 3.0000	(0) 1.5204	0.0000	(0)-0.8721	(0)-1.0000	(-1)-6.0814	0.0000
0.1	(0) 3.6547	(0) 2.9328	(0) 1.1869	(-1)-2.9825	(0)-1.0103	(-1)-9.8753	(-1)-5.1516	(-1)0.9975
0.2	(0) 4.0753	(0) 2.7341	(-1) 8.0608	(-1)-5.8611	(0)-1.1183	(-1)-9.5045	(-1)-4.1190	(-1)1.9801
0.3	(0) 4.2934	(0) 2.4132	(-1)+3.9325	(-1)-8.5358	(0)-1.1930	(-1)-8.8975	(-1)-3.0046	(-1)2.9333
0.4	(0) 4.2988	(0) 1.9846	(-1)-0.3518	(0)-1.0915	(0)-1.2322	(-1)-8.0706	(-1)-1.8308	(-1)3.8432
0.5	(0) 4.0918	(0) 1.4678	(-1)-4.6224	(0)-1.2917	(0)-1.2351	(-1)-7.0456	(-1)-0.6213	(-1)4.6971
0.6	(0) 3.6836	(-1) 8.8615	(-1)-8.7118	(0)-1.4477	(0)-1.2018	(-1)-5.8492	(-1)+0.6004	(-1)5.4836
0.7	(0) 3.0953	(-1)+2.6550	(0)-1.2462	(0)-1.5544	(0)-1.1336	(-1)-4.5120	(-1) 1.8107	(-1)6.1929
0.8	(0) 2.3566	(-1)-3.6676	(0)-1.5731	(0)-1.6088	(0)-1.0329	(-1)-3.0677	(-1) 2.9871	(-1)6.8172
0.9	(0) 1.5042	(-1)-9.8321	(0)-1.8397	(0)-1.6097	(-1)-9.0285	(-1)-1.5517	(-1) 4.1087	(-1)7.3502
1.0	(0)+0.5799	0)-1.5576	(0)-2.0368	(0)-1.5576	(-1)-7.4764	0.0000	(-1) 5.1567	(-1)7.7880
1.1	(0)-0.3719	0)-2.0661	(0)-2.1578	(0)-1.4550	(-1)-5.7190	(-1) 1.5518	(-1) 6.1146	(-1)8.1287
1.2	(0)-1.3064	0)-2.4882	(0)-2.1992	(0)-1.3061	(-1)-3.8076	(-1) 3.0698	(-1) 6.9691	(-1)8.3721
1.3	(0)-2.1806	0)-2.8077	(0)-2.1608	(0)-1.1162	(-1)-1.7956	(-1) 4.5223	(-1) 7.7095	(-1)8.5203
1.4	(0)-2.9554	0)-3.0131	(0)-2.0454	(-1)-8.9198	(-1)+0.2627	(-1) 5.8812	(-1) 8.3285	(-1)8.5768
1.5	(0)-3.5976	0)-3.0982	(0)-1.8583	(-1)-6.4101	(-1) 2.3147	(-1) 7.1223	(-1) 8.8221	(-1)8.5467
1.6	(0)-4.0808	0)-3.0617	(0)-1.6076	(-1)-3.7121	(-1) 4.3106	(-1) 8.2258	(-1) 9.1890	(-1)8.4367
1.7	(0)-4.3868	0)-2.9073	(0)-1.3029	(-1)-0.9080	(-1) 6.2053	(-1) 9.1766	(-1) 9.4313	(-1)8.2541
1.8	(0)-4.5059	0)-2.6435	(-1)-9.5564	(-1)+1.9218	(-1) 7.9592	(-1) 9.9648	(-1) 9.5532	(-1)8.0074
1.9	(0)-4.4368	0)-2.2824	(-1)-5.7791	(-1) 4.7004	(-1) 9.5394	(0) 1.0585	(-1) 9.5616	(-1)7.7055
2.0	(0)-4.1866	0)-1.8394	(-1)-1.8226	(-1) 7.3576	(0) 1.0920	(0) 1.1036	(-1) 9.4652	(-1)7.3576
2.1	(0)-3.7694	0)-1.3321	(-1)+2.1890	(-1) 9.8317	(0) 1.2083	(0) 1.1323	(-1) 9.2742	(-1)6.9728
2.2	(0)-3.2057	(-1)-7.7961	(-1) 6.1381	(0) 1.2071	(0) 1.3017	(0) 1.1451	(-1) 9.0001	(-1)6.5603
2.3	(0)-2.5208	(-1)-2.0142	(-1) 9.9170	(0) 1.4035	(0) 1.3719	(0) 1.1431	(-1) 8.6549	(-1)6.1288
2.4	(0)-1.7434	(-1)+3.8325	(0) 1.3432	(0) 1.5694	(0) 1.4191	(0) 1.1278	(-1) 8.2510	(-1)5.6863
2.5	(0)-0.9039	(-1) 9.5635	(0) 1.6604	(0) 1.7031	(0) 1.4443	(0) 1.1005	(-1) 7.8009	(-1)5.2403
2.6	(0)-0.0332	0) 1.5015	(0) 1.9373	(0) 1.8039	(0) 1.4487	(0) 1.0628	(-1) 7.3167	(-1)4.7975
2.7	(0)+0.8387	0) 2.0048	(0) 2.1696	(0) 1.8721	(0) 1.4341	(0) 1.0166	(-1) 6.8097	(-1)4.3638
2.8	(0) 1.6842	0) 2.4545	(0) 2.3548	(0) 1.9089	(0) 1.4027	(-1) 9.6347	(-1) 6.2905	(-1)3.9440
2.9	(0) 2.4789	0) 2.8422	(0) 2.4921	(0) 1.9164	(0) 1.3567	(-1) 9.0514	(-1) 5.7687	(-1)3.5424
3.0	(0) 3.2021	0) 3.1620	(0) 2.5823	(0) 1.8972	(0) 1.2985	(-1) 8.4319	(-1) 5.2527	(-1)3.1620
3.1	(0) 3.8377	0) 3.4108	(0) 2.6273	(0) 1.8543	(0) 1.2306	(-1) 7.7913	(-1) 4.7497	(-1)2.8052
3.2	(0) 4.3739	0) 3.5883	(0) 2.6304	(0) 1.7910	(0) 1.1553	(-1) 7.1430	(-1) 4.2658	(-1)2.4738
3.3	(0) 4.8038	0) 3.6963	(0) 2.5957	(0) 1.7109	(0) 1.0749	(-1) 6.4987	(-1) 3.8056	(-1)2.1684
3.4	(0) 5.1246	0) 3.7388	(0) 2.5279	(0) 1.6175	(-1) 9.9150	(-1) 5.8688	(-1) 3.3729	(-1)1.8896
3.5	(0) 5.3376	0) 3.7212	(0) 2.4320	(0) 1.5142	(-1) 9.0701	(-1) 5.2617	(-1) 2.9700	(-1)1.6370
3.6	(0) 5.4473	0) 3.6501	(0) 2.3134	(0) 1.4043	(-1) 8.2306	(-1) 4.6840	(-1) 2.5987	(-1)1.4099
3.7	(0) 5.4614	0) 3.5331	(0) 2.1771	(0) 1.2906	(-1) 7.4107	(-1) 4.1408	(-1) 2.2595	(-1)1.2073
3.8	(0) 5.3895	0) 3.3781	(0) 2.0282	(0) 1.1760	(-1) 6.6219	(-1) 3.6358	(-1) 1.9525	(-1)1.0280
3.9	(0) 5.2427	0) 3.1929	(0) 1.8714	(0) 1.0626	(-1) 5.8733	(-1) 3.1709	(-1) 1.6768	(-2)8.7028
4.0	(0) 5.0332	0) 2.9854	(0) 1.7108	(-1) 9.5241	(-1) 5.1716	(-1) 2.7473	(-1) 1.4313	(-2)7.3263
4.1	(0) 4.7733	0) 2.7630	(0) 1.5502	(-1) 8.4694	(-1) 4.5215	(-1) 2.3649	(-1) 1.2144	(-2)6.1328
4.2	(0) 4.4753	0) 2.5323	(0) 1.3927	(-1) 7.4740	(-1) 3.9256	(-1) 2.0226	(-1) 1.0242	(-2)5.1052
4.3	(0) 4.1508	0) 2.2992	(0) 1.2408	(-1) 6.5463	(-1) 3.3849	(-1) 1.7190	(-2) 8.5874	(-2)4.2261
4.4	(0) 3.8106	0) 2.0689	(0) 1.0967	(-1) 5.6918	(-1) 2.8991	(-1) 1.4517	(-2) 7.1578	(-2)3.4791
4.5	(0) 3.4641	0) 1.8455	(-1) 9.6165	(-1) 4.9134	(-1) 2.4665	(-1) 1.2185	(-2) 5.9314	(-2)2.8484
4.6	(0) 3.1197	0) 1.6324	(-1) 8.3683	(-1) 4.2117	(-1) 2.0848	(-1) 1.0164	(-2) 4.8867	(-2)2.3192
4.7	(0) 2.7843	0) 1.4322	(-1) 7.2277	(-1) 3.5852	(-1) 1.7507	(-2) 8.4272	(-2) 4.0029	(-2)1.8780
4.8	(0) 2.4632	0) 1.2466	(-1) 6.1969	(-1) 3.0311	(-1) 1.4608	(-2) 6.9451	(-2) 3.2603	(-2)1.5125
4.9	(0) 2.1608	0) 1.0766	(-1) 5.2750	(-1) 2.5455	(-1) 1.2112	(-2) 5.6894	(-2) 2.6403	(-2)1.2116
5.0	(0) 1.8800	(-1) 9.2276	(-1) 4.4586	(-1) 2.1235	(-2) 9.9802	(-2) 4.6331	(-2) 2.1262	(-3)9.6523

For interpolation, see **19.28.**

Table 19.1

x	$V(-5.0, x)$	$V(-4.5, x)$	$V(-4.0, x)$	$V(-3.5, x)$	$V(-3.0, x)$	$V(-2.5, x)$	$V(-2.0, x)$	$V(-1.5, x)$
0.0	(−2)−5.8311	0.0000	(−1) 1.3071	(−1) 2.6596	(−1) 2.6240	0.0000	(−1)−4.5748	(−1)−7.9788
0.1	(−2)−4.3898	(−2) 2.6397	(−1) 1.5417	(−1) 2.6132	(−1) 2.1296	(−1)−0.7946	(−1)−5.1829	(−1)−7.9191
0.2	(−2)−2.7299	(−2) 5.1612	(−1) 1.7149	(−1) 2.4757	(−1) 1.5714	(−1)−1.5693	(−1)−5.6877	(−1)−7.7409
0.3	(−2)−0.9344	(−2) 7.4519	(−1) 1.8199	(−1) 2.2520	(−2) 9.6646	(−1)−2.3051	(−1)−6.0796	(−1)−7.4476
0.4	(−2)+0.9074	(−2) 9.4102	(−1) 1.8527	(−1) 1.9503	(−2)+3.3275	(−1)−2.9840	(−1)−6.3515	(−1)−7.0444
0.5	(−2) 2.7045	(−1) 1.0950	(−1) 1.8125	(−1) 1.5812	(−2)−3.1080	(−1)−3.5896	(−1)−6.4991	(−1)−6.5385
0.6	(−2) 4.3687	(−1) 1.2007	(−1) 1.7011	(−1) 1.1580	(−2)−9.4527	(−1)−4.1079	(−1)−6.5210	(−1)−5.9387
0.7	(−2) 5.8194	(−1) 1.2536	(−1) 1.5234	(−2) 6.9534	(−1)−1.5523	(−1)−4.5275	(−1)−6.4186	(−1)−5.2553
0.8	(−2) 6.9875	(−1) 1.2518	(−1) 1.2869	(−2)+2.0926	(−1)−2.1149	(−1)−4.8397	(−1)−6.1959	(−1)−4.4995
0.9	(−2) 7.8188	(−1) 1.1958	(−1) 1.0010	(−2)−2.8383	(−1)−2.6176	(−1)−5.0388	(−1)−5.8594	(−1)−3.6835
1.0	(−2) 8.2767	(−1) 1.0887	(−2) 6.7728	(−2)−7.6762	(−1)−3.0472	(−1)−5.1225	(−1)−5.4177	(−1)−2.8197
1.1	(−2) 8.3429	(−2) 9.3549	(−2)+3.2819	(−1)−1.2266	(−1)−3.3933	(−1)−5.0912	(−1)−4.8813	(−1)−1.9206
1.2	(−2) 8.0189	(−2) 7.4311	(−2)−0.3303	(−1)−1.6465	(−1)−3.6481	(−1)−4.9482	(−1)−4.2621	(−1)−0.9984
1.3	(−2) 7.3241	(−2) 5.2005	(−2)−3.9309	(−1)−2.0148	(−1)−3.8069	(−1)−4.6995	(−1)−3.5731	(−1)−0.0648
1.4	(−2) 6.2954	(−2) 2.7584	(−2)−7.3916	(−1)−2.3214	(−1)−3.8677	(−1)−4.3533	(−1)−2.8278	(−1)+0.8696
1.5	(−2) 4.9836	(−2)+0.2057	(−1)−1.0594	(−1)−2.5583	(−1)−3.8317	(−1)−3.9197	(−1)−2.0397	(−1) 1.7953
1.6	(−2) 3.4514	(−2)−2.3553	(−1)−1.3434	(−1)−2.7203	(−1)−3.7025	(−1)−3.4103	(−1)−1.2222	(−1) 2.7043
1.7	(−2) 1.7690	(−2)−4.8261	(−1)−1.5824	(−1)−2.8047	(−1)−3.4861	(−1)−2.8375	(−1)−0.3880	(−1) 3.5902
1.8	(−2)+0.0110	(−2)−7.1155	(−1)−1.7697	(−1)−2.8113	(−1)−3.1904	(−1)−2.2142	(−1)+0.4512	(−1) 4.4484
1.9	(−2)−1.7477	(−2)−9.1435	(−1)−1.9008	(−1)−2.7426	(−1)−2.8250	(−1)−1.5535	(−1) 1.2852	(−1) 5.2761
2.0	(−2)−3.4354	(−1)−1.0844	(−1)−1.9731	(−1)−2.6027	(−1)−2.4003	(−1)−0.8679	(−1) 2.1053	(−1) 6.0723
2.1	(−2)−4.9863	(−1)−1.2166	(−1)−1.9864	(−1)−2.3979	(−1)−1.9277	(−1)−0.1692	(−1) 2.9044	(−1) 6.8384
2.2	(−2)−6.3439	(−1)−1.3076	(−1)−1.9423	(−1)−2.1357	(−1)−1.4184	(−1)+0.5320	(−1) 3.6777	(−1) 7.5775
2.3	(−2)−7.4620	(−1)−1.3558	(−1)−1.8442	(−1)−1.8247	(−2)−8.8371	(−1) 1.2264	(−1) 4.4221	(−1) 8.2948
2.4	(−2)−8.3067	(−1)−1.3610	(−1)−1.6967	(−1)−1.4739	(−2)−3.3411	(−1) 1.9066	(−1) 5.1367	(−1) 8.9975
2.5	(−2)−8.8568	(−1)−1.3246	(−1)−1.5059	(−1)−1.0927	(−2)+2.2080	(−1) 2.5667	(−1) 5.8227	(−1) 9.6950
2.6	(−2)−9.1035	(−1)−1.2495	(−1)−1.2784	(−2)−6.9034	(−2) 7.7266	(−1) 3.2030	(−1) 6.4834	(0) 1.0399
2.7	(−2)−9.0496	(−1)−1.1392	(−1)−1.0214	(−2)−2.7540	(−1) 1.3145	(−1) 3.8134	(−1) 7.1242	(0) 1.1122
2.8	(−2)−8.7090	(−2)−9.9858	(−2)−7.4214	(−2)+1.4424	(−1) 1.8411	(−1) 4.3982	(−1) 7.7525	(0) 1.1882
2.9	(−2)−8.1043	(−2)−8.3257	(−2)−4.4770	(−2) 5.6176	(−1) 2.3486	(−1) 4.9594	(−1) 8.3779	(0) 1.2697
3.0	(−2)−7.2651	(−2)−6.4659	(−2)−1.4470	(−2) 9.7155	(−1) 2.8352	(−1) 5.5010	(−1) 9.0120	(0) 1.3588
3.1	(−2)−6.2264	(−2)−4.4605	(−2)+1.6090	(−1) 1.3693	(−1) 3.3007	(−1) 6.0291	(−1) 9.6689	(0) 1.4582
3.2	(−2)−5.0260	(−2)−2.3612	(−2) 4.6402	(−1) 1.7522	(−1) 3.7466	(−1) 6.5514	(0) 1.0365	(0) 1.5708
3.3	(−2)−3.7030	(−2)−0.2157	(−2) 7.6054	(−1) 2.1187	(−1) 4.1761	(−1) 7.0778	(0) 1.1119	(0) 1.7001
3.4	(−2)−2.2954	(−2)+1.9344	(−1) 1.0474	(−1) 2.4688	(−1) 4.5942	(−1) 7.6202	(0) 1.1954	(0) 1.8502
3.5	(−2)−0.8391	(−2) 4.0539	(−1) 1.3228	(−1) 2.8040	(−1) 5.0074	(−1) 8.1924	(0) 1.2896	(0) 2.0262
3.6	(−2)+0.6339	(−2) 6.1158	(−1) 1.5859	(−1) 3.1270	(−1) 5.4239	(−1) 8.8110	(0) 1.3975	(0) 2.2339
3.7	(−2) 2.0962	(−2) 8.1014	(−1) 1.8370	(−1) 3.4421	(−1) 5.8535	(−1) 9.4951	(0) 1.5228	(0) 2.4806
3.8	(−2) 3.5259	(−1) 1.0000	(−1) 2.0775	(−1) 3.7545	(−1) 6.3080	(0) 1.0267	(0) 1.6699	(0) 2.7751
3.9	(−2) 4.9072	(−1) 1.1811	(−1) 2.3101	(−1) 4.0712	(−1) 6.8012	(0) 1.1153	(0) 1.8439	(0) 3.1285
4.0	(−2) 6.2301	(−1) 1.3540	(−1) 2.5382	(−1) 4.4004	(−1) 7.3492	(0) 1.2186	(0) 2.0513	(0) 3.5541
4.1	(−2) 7.4913	(−1) 1.5202	(−1) 2.7664	(−1) 4.7517	(−1) 7.9710	(0) 1.3401	(0) 2.2999	(0) 4.0690
4.2	(−2) 8.6933	(−1) 1.6819	(−1) 3.0002	(−1) 5.1365	(−1) 8.6890	(0) 1.4846	(0) 2.5993	(0) 4.6942
4.3	(−2) 9.8444	(−1) 1.8422	(−1) 3.2465	(−1) 5.5683	(−1) 9.5300	(0) 1.6575	(0) 2.9616	(0) 5.4567
4.4	(−1) 1.0959	(−1) 2.0048	(−1) 3.5131	(−1) 6.0629	(0) 1.0526	(0) 1.8657	(0) 3.4019	(0) 6.3903
4.5	(−1) 1.2056	(−1) 2.1743	(−1) 3.8093	(−1) 6.6389	(0) 1.1717	(0) 2.1178	(0) 3.9393	(0) 7.5384
4.6	(−1) 1.3161	(−1) 2.3561	(−1) 4.1462	(−1) 7.3192	(0) 1.3150	(0) 2.4244	(0) 4.5978	(0) 8.9563
4.7	(−1) 1.4305	(−1) 2.5567	(−1) 4.5368	(−1) 8.1309	(0) 1.4885	(0) 2.7989	(0) 5.4083	(1) 1.0715
4.8	(−1) 1.5525	(−1) 2.7834	(−1) 4.9967	(−1) 9.1078	(0) 1.6998	(0) 3.2584	(0) 6.4102	(1) 1.2908
4.9	(−1) 1.6863	(−1) 3.0454	(−1) 5.5449	(0) 1.0291	(0) 1.9582	(0) 3.8246	(0) 7.6545	(1) 1.5653
5.0	(−1) 1.8370	(−1) 3.3533	(−1) 6.2047	(0) 1.1734	(0) 2.2757	(0) 4.5254	(0) 9.2067	(1) 1.9107

Table 19.1

x	$U(-1.0,x)$	$U(-0.9,x)$	$U(-0.8,x)$	$U(-0.7,x)$	$U(-0.6,x)$	$U(-0.5,x)$	$U(-0.4,x)$
0.0	(−1) 5.8137	(−1) 6.8058	(−1) 7.7241	(−1) 8.5642	(−1) 9.3233	(0) 1.0000	(0) 1.0594
0.1	(−1) 6.3918	(−1) 7.2692	(−1) 8.0677	(−1) 8.7853	(−1) 9.4211	(−1) 9.9750	(0) 1.0448
0.2	(−1) 6.9062	(−1) 7.6673	(−1) 8.3471	(−1) 8.9453	(−1) 9.4626	(−1) 9.9005	(0) 1.0261
0.3	(−1) 7.3523	(−1) 7.9973	(−1) 8.5606	(−1) 9.0436	(−1) 9.4483	(−1) 9.7775	(0) 1.0035
0.4	(−1) 7.7267	(−1) 8.2572	(−1) 8.7077	(−1) 9.0807	(−1) 9.3796	(−1) 9.6079	(−1) 9.7698
0.5	(−1) 8.0270	(−1) 8.4462	(−1) 8.7886	(−1) 9.0580	(−1) 9.2584	(−1) 9.3941	(−1) 9.4700
0.6	(−1) 8.2522	(−1) 8.5646	(−1) 8.8049	(−1) 8.9776	(−1) 9.0874	(−1) 9.1393	(−1) 9.1382
0.7	(−1) 8.4023	(−1) 8.6136	(−1) 8.7586	(−1) 8.8425	(−1) 8.8702	(−1) 8.8471	(−1) 8.7781
0.8	(−1) 8.4788	(−1) 8.5958	(−1) 8.6531	(−1) 8.6563	(−1) 8.6107	(−1) 8.5214	(−1) 8.3937
0.9	(−1) 8.4842	(−1) 8.5144	(−1) 8.4923	(−1) 8.4235	(−1) 8.3133	(−1) 8.1669	(−1) 7.9892
1.0	(−1) 8.4220	(−1) 8.3737	(−1) 8.2808	(−1) 8.1488	(−1) 7.9828	(−1) 7.7880	(−1) 7.5689
1.1	(−1) 8.2967	(−1) 8.1787	(−1) 8.0238	(−1) 7.8374	(−1) 7.6245	(−1) 7.3897	(−1) 7.1372
1.2	(−1) 8.1136	(−1) 7.9348	(−1) 7.7269	(−1) 7.4949	(−1) 7.2435	(−1) 6.9768	(−1) 6.6986
1.3	(−1) 7.8786	(−1) 7.6480	(−1) 7.3960	(−1) 7.1269	(−1) 6.8451	(−1) 6.5541	(−1) 6.2573
1.4	(−1) 7.5982	(−1) 7.3248	(−1) 7.0371	(−1) 6.7392	(−1) 6.4345	(−1) 6.1263	(−1) 5.8173
1.5	(−1) 7.2789	(−1) 6.9716	(−1) 6.6565	(−1) 6.3372	(−1) 6.0168	(−1) 5.6978	(−1) 5.3826
1.6	(−1) 6.9279	(−1) 6.5948	(−1) 6.2600	(−1) 5.9266	(−1) 5.5968	(−1) 5.2729	(−1) 4.9566
1.7	(−1) 6.5519	(−1) 6.2008	(−1) 5.8535	(−1) 5.5123	(−1) 5.1791	(−1) 4.8554	(−1) 4.5424
1.8	(−1) 6.1577	(−1) 5.7958	(−1) 5.4424	(−1) 5.0993	(−1) 4.7676	(−1) 4.4486	(−1) 4.1429
1.9	(−1) 5.7517	(−1) 5.3855	(−1) 5.0319	(−1) 4.6918	(−1) 4.3662	(−1) 4.0555	(−1) 3.7603
2.0	(−1) 5.3401	(−1) 4.9754	(−1) 4.6264	(−1) 4.2938	(−1) 3.9779	(−1) 3.6788	(−1) 3.3965
2.1	(−1) 4.9285	(−1) 4.5701	(−1) 4.2301	(−1) 3.9086	(−1) 3.6054	(−1) 3.3204	(−1) 3.0532
2.2	(−1) 4.5219	(−1) 4.1741	(−1) 3.8466	(−1) 3.5391	(−1) 3.2511	(−1) 2.9820	(−1) 2.7312
2.3	(−1) 4.1247	(−1) 3.7910	(−1) 3.4788	(−1) 3.1876	(−1) 2.9165	(−1) 2.6647	(−1) 2.4313
2.4	(−1) 3.7407	(−1) 3.4238	(−1) 3.1292	(−1) 2.8559	(−1) 2.6029	(−1) 2.3693	(−1) 2.1538
2.5	(−1) 3.3732	(−1) 3.0751	(−1) 2.7995	(−1) 2.5453	(−1) 2.3112	(−1) 2.0961	(−1) 1.8987
2.6	(−1) 3.0246	(−1) 2.7467	(−1) 2.4912	(−1) 2.2566	(−1) 2.0418	(−1) 1.8452	(−1) 1.6657
2.7	(−1) 2.6968	(−1) 2.4399	(−1) 2.2049	(−1) 1.9903	(−1) 1.7945	(−1) 1.6162	(−1) 1.4541
2.8	(−1) 2.3911	(−1) 2.1556	(−1) 1.9412	(−1) 1.7462	(−1) 1.5691	(−1) 1.4086	(−1) 1.2632
2.9	(−1) 2.1084	(−1) 1.8942	(−1) 1.7000	(−1) 1.5241	(−1) 1.3651	(−1) 1.2215	(−1) 1.0920
3.0	(−1) 1.8488	(−1) 1.6555	(−1) 1.4809	(−1) 1.3234	(−1) 1.1816	(−1) 1.0540	(−2) 9.3934
3.1	(−1) 1.6124	(−1) 1.4391	(−1) 1.2832	(−1) 1.1432	(−1) 1.0175	(−2) 9.0491	(−2) 8.0408
3.2	(−1) 1.3985	(−1) 1.2443	(−1) 1.1061	(−2) 9.8240	(−2) 8.7182	(−2) 7.7305	(−2) 6.8492
3.3	(−1) 1.2064	(−1) 1.0701	(−2) 9.4842	(−2) 8.3989	(−2) 7.4318	(−2) 6.5710	(−2) 5.8055
3.4	(−1) 1.0351	(−2) 9.1545	(−2) 8.0899	(−2) 7.1436	(−2) 6.3032	(−2) 5.5576	(−2) 4.8967
3.5	(−2) 8.8335	(−2) 7.7900	(−2) 6.8646	(−2) 6.0447	(−2) 5.3190	(−2) 4.6771	(−2) 4.1098
3.6	(−2) 7.4981	(−2) 6.5939	(−2) 5.7946	(−2) 5.0887	(−2) 4.4657	(−2) 3.9164	(−2) 3.4324
3.7	(−2) 6.3306	(−2) 5.5521	(−2) 4.8660	(−2) 4.2619	(−2) 3.7304	(−2) 3.2631	(−2) 2.8525
3.8	(−2) 5.3165	(−2) 4.6503	(−2) 4.0651	(−2) 3.5512	(−2) 3.1004	(−2) 2.7052	(−2) 2.3589
3.9	(−2) 4.4411	(−2) 3.8747	(−2) 3.3784	(−2) 2.9439	(−2) 2.5638	(−2) 2.2315	(−2) 1.9411
4.0	(−2) 3.6903	(−2) 3.2115	(−2) 2.7932	(−2) 2.4280	(−2) 2.1094	(−2) 1.8316	(−2) 1.5895
4.1	(−2) 3.0502	(−2) 2.6480	(−2) 2.2975	(−2) 1.9923	(−2) 1.7268	(−2) 1.4958	(−2) 1.2951
4.2	(−2) 2.5079	(−2) 2.1720	(−2) 1.8800	(−2) 1.6265	(−2) 1.4064	(−2) 1.2155	(−2) 1.0500
4.3	(−2) 2.0512	(−2) 1.7723	(−2) 1.5305	(−2) 1.3211	(−2) 1.1397	(−3) 9.8282	(−3) 8.4709
4.4	(−2) 1.6688	(−2) 1.4386	(−2) 1.2396	(−2) 1.0676	(−3) 9.1898	(−3) 7.9071	(−3) 6.8002
4.5	(−2) 1.3507	(−2) 1.1618	(−3) 9.9881	(−3) 8.5831	(−3) 7.3725	(−3) 6.3297	(−3) 5.4320
4.6	(−2) 1.0875	(−3) 9.3333	(−3) 8.0067	(−3) 6.8657	(−3) 5.8847	(−3) 5.0418	(−3) 4.3177
4.7	(−3) 8.7099	(−3) 7.4594	(−3) 6.3856	(−3) 5.4641	(−3) 4.6736	(−3) 3.9958	(−3) 3.4150
4.8	(−3) 6.9398	(−3) 5.9310	(−3) 5.0667	(−3) 4.3266	(−3) 3.6931	(−3) 3.1511	(−3) 2.6876
4.9	(−3) 5.5007	(−3) 4.6914	(−3) 3.9996	(−3) 3.4085	(−3) 2.9036	(−3) 2.4726	(−3) 2.1047
5.0	(−3) 4.3375	(−3) 3.6919	(−3) 3.1412	(−3) 2.6716	(−3) 2.2714	(−3) 1.9305	(−3) 1.6401

x	$V(-1.0,x)$	$V(-0.9,x)$	$V(-0.8,x)$	$V(-0.7,x)$	$V(-0.6,x)$	$V(-0.5,x)$	$V(-0.4,x)$
0.0	(−1)−6.5600	(−1)−5.5730	(−1)−4.3852	(−1)−3.0307	(−1)−1.5522	0.0000	(−1)1.5701
0.1	(−1)−5.8422	(−1)−4.7818	(−1)−3.5487	(−1)−2.1784	(−1)−0.7135	(−1)0.7972	(−1)2.3012
0.2	(−1)−5.0662	(−1)−3.9477	(−1)−2.6839	(−1)−1.3109	(−1)+0.1294	(−1)1.5905	(−1)3.0232
0.3	(−1)−4.2400	(−1)−3.0785	(−1)−1.7980	(−1)−0.4343	(−1) 0.9716	(−1)2.3760	(−1)3.7334
0.4	(−1)−3.3725	(−1)−2.1823	(−1)−0.8980	(−1)+0.4451	(−1) 1.8082	(−1)3.1502	(−1)4.4296
0.5	(−1)−2.4725	(−1)−1.2674	(−1)+0.0088	(−1) 1.3217	(−1) 2.6347	(−1)3.9099	(−1)5.1099
0.6	(−1)−1.5494	(−1)−0.3418	(−1) 0.9156	(−1) 2.1900	(−1) 3.4471	(−1)4.6526	(−1)5.7729
0.7	(−1)−0.6122	(−1)+0.5867	(−1) 1.8159	(−1) 3.0449	(−1) 4.2420	(−1)5.3763	(−1)6.4182
0.8	(−1)+0.3305	(−1) 1.5106	(−1) 2.7040	(−1) 3.8823	(−1) 5.0167	(−1)6.0797	(−1)7.0457
0.9	(−1) 1.2704	(−1) 2.4234	(−1) 3.5749	(−1) 4.6988	(−1) 5.7694	(−1)6.7626	(−1)7.6563
1.0	(−1) 2.2004	(−1) 3.3194	(−1) 4.4245	(−1) 5.4920	(−1) 6.4993	(−1)7.4254	(−1)8.2519
1.1	(−1) 3.1139	(−1) 4.1939	(−1) 5.2498	(−1) 6.2606	(−1) 7.2065	(−1)8.0697	(−1)8.8353
1.2	(−1) 4.0057	(−1) 5.0435	(−1) 6.0492	(−1) 7.0044	(−1) 7.8924	(−1)8.6982	(−1)9.4101
1.3	(−1) 4.8721	(−1) 5.8660	(−1) 6.8220	(−1) 7.7246	(−1) 8.5594	(−1)9.3147	(−1)9.9812
1.4	(−1) 5.7105	(−1) 6.6605	(−1) 7.5693	(−1) 8.4234	(−1) 9.2113	(−1)9.9240	(0)1.0555
1.5	(−1) 6.5198	(−1) 7.4279	(−1) 8.2931	(−1) 9.1046	(−1) 9.8533	(0)1.0532	(0)1.1138
1.6	(−1) 7.3008	(−1) 8.1704	(−1) 8.9974	(−1) 9.7734	(0) 1.0492	(0)1.1148	(0)1.1739
1.7	(−1) 8.0557	(−1) 8.8917	(−1) 9.6875	(0) 1.0437	(0) 1.1134	(0)1.1778	(0)1.2369
1.8	(−1) 8.7883	(−1) 9.5974	(0) 1.0370	(0) 1.1102	(0) 1.1791	(0)1.2436	(0)1.3038
1.9	(−1) 9.5044	(0) 1.0295	(0) 1.1054	(0) 1.1780	(0) 1.2472	(0)1.3132	(0)1.3762
2.0	(0) 1.0211	(0) 1.0992	(0) 1.1749	(0) 1.2482	(0) 1.3191	(0)1.3881	(0)1.4554
2.1	(0) 1.0918	(0) 1.1701	(0) 1.2468	(0) 1.3222	(0) 1.3964	(0)1.4699	(0)1.5435
2.2	(0) 1.1637	(0) 1.2434	(0) 1.3225	(0) 1.4015	(0) 1.4806	(0)1.5607	(0)1.6424
2.3	(0) 1.2380	(0) 1.3205	(0) 1.4037	(0) 1.4879	(0) 1.5740	(0)1.6625	(0)1.7546
2.4	(0) 1.3163	(0) 1.4032	(0) 1.4922	(0) 1.5837	(0) 1.6787	(0)1.7781	(0)1.8830
2.5	(0) 1.4005	(0) 1.4936	(0) 1.5902	(0) 1.6912	(0) 1.7975	(0)1.9104	(0)2.0311
2.6	(0) 1.4925	(0) 1.5939	(0) 1.7005	(0) 1.8134	(0) 1.9338	(0)2.0631	(0)2.2029
2.7	(0) 1.5949	(0) 1.7068	(0) 1.8259	(0) 1.9535	(0) 2.0911	(0)2.2404	(0)2.4032
2.8	(0) 1.7104	(0) 1.8355	(0) 1.9700	(0) 2.1157	(0) 2.2741	(0)2.4474	(0)2.6378
2.9	(0) 1.8424	(0) 1.9837	(0) 2.1371	(0) 2.3045	(0) 2.4881	(0)2.6902	(0)2.9136
3.0	(0) 1.9948	(0) 2.1558	(0) 2.3321	(0) 2.5258	(0) 2.7396	(0)2.9763	(0)3.2392
3.1	(0) 2.1722	(0) 2.3571	(0) 2.5609	(0) 2.7864	(0) 3.0365	(0)3.3147	(0)3.6249
3.2	(0) 2.3801	(0) 2.5940	(0) 2.8310	(0) 3.0945	(0) 3.3882	(0)3.7163	(0)4.0834
3.3	(0) 2.6253	(0) 2.8740	(0) 3.1511	(0) 3.4604	(0) 3.8066	(0)4.1947	(0)4.6305
3.4	(0) 2.9159	(0) 3.2066	(0) 3.5319	(0) 3.8966	(0) 4.3061	(0)4.7667	(0)5.2855
3.5	(0) 3.2618	(0) 3.6032	(0) 3.9868	(0) 4.4183	(0) 4.9045	(0)5.4531	(0)6.0726
3.6	(0) 3.6752	(0) 4.0781	(0) 4.5323	(0) 5.0449	(0) 5.6242	(0)6.2797	(0)7.0220
3.7	(0) 4.1712	(0) 4.6487	(0) 5.1887	(0) 5.8001	(0) 6.4930	(0)7.2790	(0)8.1716
3.8	(0) 4.7686	(0) 5.3371	(0) 5.9818	(0) 6.7138	(0) 7.5458	(0)8.4920	(0)9.5693
3.9	(0) 5.4910	(0) 6.1706	(0) 6.9437	(0) 7.8238	(0) 8.8266	(0)9.9703	(1)1.1276
4.0	(0) 6.3680	(0) 7.1841	(0) 8.1149	(0) 9.1775	(1) 1.0391	(1)1.1779	(1)1.3367
4.1	(0) 7.4368	(0) 8.4212	(0) 9.5470	(1) 1.0835	(1) 1.2311	(1)1.4002	(1)1.5942
4.2	(0) 8.7448	(0) 9.9377	(1) 1.1305	(1) 1.2875	(1) 1.4676	(1)1.6747	(1)1.9127
4.3	(1) 1.0352	(1) 1.1805	(1) 1.3474	(1) 1.5394	(1) 1.7604	(1)2.0149	(1)2.3082
4.4	(1) 1.2337	(1) 1.4113	(1) 1.6160	(1) 1.8520	(1) 2.1243	(1)2.4386	(1)2.8017
4.5	(1) 1.4797	(1) 1.6981	(1) 1.9502	(1) 2.2417	(1) 2.5787	(1)2.9687	(1)3.4202
4.6	(1) 1.7862	(1) 2.0559	(1) 2.3680	(1) 2.7297	(1) 3.1489	(1)3.6350	(1)4.1991
4.7	(1) 2.1698	(1) 2.5044	(1) 2.8928	(1) 3.3437	(1) 3.8676	(1)4.4765	(1)5.1846
4.8	(1) 2.6520	(1) 3.0694	(1) 3.5549	(1) 4.1199	(1) 4.7777	(1)5.5441	(1)6.4372
4.9	(1) 3.2611	(1) 3.7844	(1) 4.3944	(1) 5.1058	(1) 5.9359	(1)6.9051	(1)8.0370
5.0	(1) 4.0344	(1) 4.6937	(1) 5.4639	(1) 6.3641	(1) 7.4168	(1)8.6484	(2)1.0090

Table 19.1

x	$U(-0.3,x)$	$U(-0.2,x)$	$U(-0.1,x)$	$U(0,x)$	$U(0.1,x)$	$U(0.2,x)$	$U(0.3,x)$
0.0	(0)1.1105	(0)1.1535	(0)1.1887	(0)1.2163	(0)1.2366	(0)1.2500	(0)1.2570
0.1	(0)1.0843	(0)1.1161	(0)1.1406	(0)1.1581	(0)1.1691	(0)1.1740	(0)1.1732
0.2	(0)1.0548	(0)1.0764	(0)1.0914	(0)1.1000	(0)1.1029	(0)1.1004	(0)1.0930
0.3	(0)1.0223	(0)1.0347	(0)1.0412	(0)1.0421	(0)1.0379	(0)1.0291	(0)1.0161
0.4	(−1)9.8697	(−1)9.9120	(−1)9.9016	(−1)9.8431	(−1)9.7411	(−1)9.6004	(−1)9.4255
0.5	(−1)9.4906	(−1)9.4609	(−1)9.3856	(−1)9.2695	(−1)9.1173	(−1)8.9333	(−1)8.7218
0.6	(−1)9.0890	(−1)8.9968	(−1)8.8661	(−1)8.7018	(−1)8.5082	(−1)8.2895	(−1)8.0498
0.7	(−1)8.6684	(−1)8.5228	(−1)8.3458	(−1)8.1419	(−1)7.9153	(−1)7.6699	(−1)7.4093
0.8	(−1)8.2324	(−1)8.0421	(−1)7.8273	(−1)7.5920	(−1)7.3400	(−1)7.0750	(−1)6.8000
0.9	(−1)7.7849	(−1)7.5583	(−1)7.3135	(−1)7.0542	(−1)6.7838	(−1)6.5055	(−1)6.2220
1.0	(−1)7.3298	(−1)7.0747	(−1)6.8072	(−1)6.5307	(−1)6.2482	(−1)5.9622	(−1)5.6753
1.1	(−1)6.8710	(−1)6.5946	(−1)6.3111	(−1)6.0235	(−1)5.7343	(−1)5.4457	(−1)5.1597
1.2	(−1)6.4124	(−1)6.1212	(−1)5.8278	(−1)5.5346	(−1)5.2436	(−1)4.9566	(−1)4.6753
1.3	(−1)5.9576	(−1)5.6576	(−1)5.3596	(−1)5.0655	(−1)4.7769	(−1)4.4953	(−1)4.2217
1.4	(−1)5.5101	(−1)5.2066	(−1)4.9087	(−1)4.6178	(−1)4.3352	(−1)4.0619	(−1)3.7986
1.5	(−1)5.0730	(−1)4.7706	(−1)4.4769	(−1)4.1927	(−1)3.9191	(−1)3.6565	(−1)3.4055
1.6	(−1)4.6492	(−1)4.3519	(−1)4.0657	(−1)3.7912	(−1)3.5288	(−1)3.2790	(−1)3.0417
1.7	(−1)4.2412	(−1)3.9524	(−1)3.6765	(−1)3.4139	(−1)3.1647	(−1)2.9290	(−1)2.7065
1.8	(−1)3.8510	(−1)3.5734	(−1)3.3102	(−1)3.0613	(−1)2.8266	(−1)2.6060	(−1)2.3990
1.9	(−1)3.4805	(−1)3.2162	(−1)2.9673	(−1)2.7334	(−1)2.5142	(−1)2.3093	(−1)2.1181
2.0	(−1)3.1309	(−1)2.8816	(−1)2.6482	(−1)2.4302	(−1)2.2270	(−1)2.0381	(−1)1.8627
2.1	(−1)2.8032	(−1)2.5700	(−1)2.3529	(−1)2.1513	(−1)1.9643	(−1)1.7913	(−1)1.6315
2.2	(−1)2.4980	(−1)2.2816	(−1)2.0812	(−1)1.8960	(−1)1.7252	(−1)1.5678	(−1)1.4232
2.3	(−1)2.2155	(−1)2.0162	(−1)1.8326	(−1)1.6637	(−1)1.5086	(−1)1.3665	(−1)1.2363
2.4	(−1)1.9556	(−1)1.7734	(−1)1.6064	(−1)1.4534	(−1)1.3136	(−1)1.1859	(−1)1.0695
2.5	(−1)1.7179	(−1)1.5526	(−1)1.4017	(−1)1.2640	(−1)1.1387	(−1)1.0248	(−2)9.2134
2.6	(−1)1.5020	(−1)1.3529	(−1)1.2174	(−1)1.0944	(−2)9.8278	(−2)8.8173	(−2)7.9031
2.7	(−1)1.3069	(−1)1.1734	(−1)1.0525	(−2)9.4322	(−2)8.4445	(−2)7.5534	(−2)6.7502
2.8	(−1)1.1317	(−1)1.0129	(−2)9.0579	(−2)8.0925	(−2)7.2235	(−2)6.4422	(−2)5.7406
2.9	(−2)9.7528	(−2)8.7027	(−2)7.7589	(−2)6.9114	(−2)6.1513	(−2)5.4703	(−2)4.8608
3.0	(−2)8.3643	(−2)7.4416	(−2)6.6151	(−2)5.8757	(−2)5.2146	(−2)4.6244	(−2)4.0978
3.1	(−2)7.1389	(−2)6.3330	(−2)5.6137	(−2)4.9721	(−2)4.4006	(−2)3.8918	(−2)3.4393
3.2	(−2)6.0636	(−2)5.3640	(−2)4.7415	(−2)4.1881	(−2)3.6967	(−2)3.2606	(−2)2.8739
3.3	(−2)5.1253	(−2)4.5215	(−2)3.9860	(−2)3.5114	(−2)3.0912	(−2)2.7194	(−2)2.3907
3.4	(−2)4.3112	(−2)3.7932	(−2)3.3351	(−2)2.9303	(−2)2.5730	(−2)2.2577	(−2)1.9799
3.5	(−2)3.6089	(−2)3.1669	(−2)2.7772	(−2)2.4340	(−2)2.1318	(−2)1.8659	(−2)1.6322
3.6	(−2)3.0063	(−2)2.6314	(−2)2.3018	(−2)2.0122	(−2)1.7580	(−2)1.5351	(−2)1.3396
3.7	(−2)2.4921	(−2)2.1759	(−2)1.8986	(−2)1.6558	(−2)1.4431	(−2)1.2571	(−2)1.0944
3.8	(−2)2.0558	(−2)1.7906	(−2)1.5587	(−2)1.3560	(−2)1.1791	(−2)1.0247	(−3)8.9001
3.9	(−2)1.6876	(−2)1.4664	(−2)1.2735	(−2)1.1053	(−3)9.5887	(−3)8.3139	(−3)7.2048
4.0	(−2)1.3786	(−2)1.1951	(−2)1.0355	(−3)8.9669	(−3)7.7613	(−3)6.7143	(−3)5.8057
4.1	(−2)1.1207	(−3)9.6928	(−3)8.3792	(−3)7.2400	(−3)6.2526	(−3)5.3973	(−3)4.6568
4.2	(−3)9.0656	(−3)7.8234	(−3)6.7481	(−3)5.8179	(−3)5.0135	(−3)4.3184	(−3)3.7179
4.3	(−3)7.2976	(−3)6.2839	(−3)5.4085	(−3)4.6529	(−3)4.0011	(−3)3.4390	(−3)2.9546
4.4	(−3)5.8457	(−3)5.0228	(−3)4.3139	(−3)3.7034	(−3)3.1779	(−3)2.7259	(−3)2.3371
4.5	(−3)4.6596	(−3)3.9954	(−3)3.4243	(−3)2.9336	(−3)2.5122	(−3)2.1504	(−3)1.8400
4.6	(−3)3.6961	(−3)3.1626	(−3)2.7050	(−3)2.3127	(−3)1.9765	(−3)1.6885	(−3)1.4419
4.7	(−3)2.9173	(−3)2.4912	(−3)2.1265	(−3)1.8145	(−3)1.5477	(−3)1.3195	(−3)1.1246
4.8	(−3)2.2914	(−3)1.9528	(−3)1.6637	(−3)1.4168	(−3)1.2061	(−3)1.0263	(−4)8.7305
4.9	(−3)1.7909	(−3)1.5233	(−3)1.2952	(−3)1.1009	(−4)9.3540	(−4)7.9449	(−4)6.7457
5.0	(−3)1.3929	(−3)1.1825	(−3)1.0035	(−4)8.5136	(−4)7.2201	(−4)6.1210	(−4)5.1875

x	$V(-0.3,x)$	$V(-0.2,x)$	$V(-0.1,x)$	$V(0,x)$	$V(0.1,x)$	$V(0.2,x)$	$V(0.3,x)$
0.0	(−1) 3.0993	(−1) 4.5280	(−1) 5.7994	(−1) 6.8621	(−1) 7.6731	(−1) 8.2008	(−1) 8.4269
0.1	(−1) 3.7442	(−1) 5.0724	(−1) 6.2358	(−1) 7.1901	(−1) 7.9000	(−1) 8.3406	(−1) 8.5002
0.2	(−1) 4.3780	(−1) 5.6069	(−1) 6.6661	(−1) 7.5184	(−1) 8.1349	(−1) 8.4974	(−1) 8.5993
0.3	(−1) 4.9991	(−1) 6.1307	(−1) 7.0905	(−1) 7.8474	(−1) 8.3788	(−1) 8.6720	(−1) 8.7250
0.4	(−1) 5.6064	(−1) 6.6436	(−1) 7.5093	(−1) 8.1782	(−1) 8.6331	(−1) 8.8660	(−1) 8.8790
0.5	(−1) 6.1992	(−1) 7.1460	(−1) 7.9238	(−1) 8.5124	(−1) 8.8994	(−1) 9.0813	(−1) 9.0632
0.6	(−1) 6.7773	(−1) 7.6386	(−1) 8.3353	(−1) 8.8519	(−1) 9.1803	(−1) 9.3205	(−1) 9.2803
0.7	(−1) 7.3412	(−1) 8.1229	(−1) 8.7460	(−1) 9.1994	(−1) 9.4787	(−1) 9.5867	(−1) 9.5336
0.8	(−1) 7.8922	(−1) 8.6009	(−1) 9.1588	(−1) 9.5583	(−1) 9.7982	(−1) 9.8840	(−1) 9.8273
0.9	(−1) 8.4321	(−1) 9.0756	(−1) 9.5771	(−1) 9.9325	(0) 1.0143	(0) 1.0217	(0) 1.0166
1.0	(−1) 8.9640	(−1) 9.5505	(0) 1.0005	(0) 1.0327	(0) 1.0519	(0) 1.0591	(0) 1.0556
1.1	(−1) 9.4914	(0) 1.0030	(0) 1.0449	(0) 1.0747	(0) 1.0932	(0) 1.1013	(0) 1.1005
1.2	(0) 1.0019	(0) 1.0521	(0) 1.0913	(0) 1.1200	(0) 1.1389	(0) 1.1490	(0) 1.1520
1.3	(0) 1.0553	(0) 1.1028	(0) 1.1406	(0) 1.1693	(0) 1.1898	(0) 1.2032	(0) 1.2110
1.4	(0) 1.1100	(0) 1.1559	(0) 1.1936	(0) 1.2236	(0) 1.2470	(0) 1.2649	(0) 1.2789
1.5	(0) 1.1668	(0) 1.2125	(0) 1.2513	(0) 1.2839	(0) 1.3115	(0) 1.3353	(0) 1.3569
1.6	(0) 1.2267	(0) 1.2734	(0) 1.3147	(0) 1.3515	(0) 1.3848	(0) 1.4160	(0) 1.4466
1.7	(0) 1.2908	(0) 1.3400	(0) 1.3853	(0) 1.4277	(0) 1.4683	(0) 1.5085	(0) 1.5499
1.8	(0) 1.3603	(0) 1.4136	(0) 1.4645	(0) 1.5142	(0) 1.5639	(0) 1.6150	(0) 1.6692
1.9	(0) 1.4368	(0) 1.4958	(0) 1.5542	(0) 1.6130	(0) 1.6738	(0) 1.7379	(0) 1.8070
2.0	(0) 1.5220	(0) 1.5886	(0) 1.6563	(0) 1.7265	(0) 1.8005	(0) 1.8799	(0) 1.9665
2.1	(0) 1.6178	(0) 1.6941	(0) 1.7734	(0) 1.8572	(0) 1.9470	(0) 2.0446	(0) 2.1517
2.2	(0) 1.7267	(0) 1.8149	(0) 1.9083	(0) 2.0085	(0) 2.1171	(0) 2.2360	(0) 2.3672
2.3	(0) 1.8513	(0) 1.9541	(0) 2.0645	(0) 2.1841	(0) 2.3149	(0) 2.4589	(0) 2.6185
2.4	(0) 1.9950	(0) 2.1153	(0) 2.2459	(0) 2.3887	(0) 2.5457	(0) 2.7195	(0) 2.9124
2.5	(0) 2.1614	(0) 2.3028	(0) 2.4576	(0) 2.6278	(0) 2.8159	(0) 3.0247	(0) 3.2572
2.6	(0) 2.3551	(0) 2.5218	(0) 2.7053	(0) 2.9080	(0) 3.1330	(0) 3.3834	(0) 3.6627
2.7	(0) 2.5818	(0) 2.7785	(0) 2.9961	(0) 3.2376	(0) 3.5064	(0) 3.8063	(0) 4.1415
2.8	(0) 2.8478	(0) 3.0803	(0) 3.3387	(0) 3.6263	(0) 3.9474	(0) 4.3064	(0) 4.7084
2.9	(0) 3.1612	(0) 3.4366	(0) 3.7435	(0) 4.0864	(0) 4.4700	(0) 4.8998	(0) 5.3820
3.0	(0) 3.5318	(0) 3.8584	(0) 4.2236	(0) 4.6326	(0) 5.0914	(0) 5.6065	(0) 6.1855
3.1	(0) 3.9715	(0) 4.3596	(0) 4.7948	(0) 5.2835	(0) 5.8328	(0) 6.4510	(0) 7.1472
3.2	(0) 4.4950	(0) 4.9572	(0) 5.4768	(0) 6.0617	(0) 6.7208	(0) 7.4640	(0) 8.3029
3.3	(0) 5.1205	(0) 5.6722	(0) 6.2941	(0) 6.9957	(0) 7.7882	(0) 8.6838	(0) 9.6969
3.4	(0) 5.8704	(0) 6.5308	(0) 7.2770	(0) 8.1210	(0) 9.0763	(1) 1.0158	(1) 1.1385
3.5	(0) 6.7730	(0) 7.5658	(0) 8.4638	(0) 9.4818	(1) 1.0637	(1) 1.1948	(1) 1.3438
3.6	(0) 7.8635	(0) 8.8182	(0) 9.9023	(1) 1.1134	(1) 1.2535	(1) 1.4130	(1) 1.5945
3.7	(0) 9.1860	(1) 1.0340	(1) 1.1653	(1) 1.3149	(1) 1.4854	(1) 1.6799	(1) 1.9019
3.8	(1) 1.0797	(1) 1.2196	(1) 1.3793	(1) 1.5616	(1) 1.7699	(1) 2.0080	(1) 2.2804
3.9	(1) 1.2766	(1) 1.4470	(1) 1.6419	(1) 1.8649	(1) 2.1203	(1) 2.4130	(1) 2.7486
4.0	(1) 1.5185	(1) 1.7268	(1) 1.9656	(1) 2.2395	(1) 2.5539	(1) 2.9150	(1) 3.3300
4.1	(1) 1.8169	(1) 2.0725	(1) 2.3663	(1) 2.7041	(1) 3.0927	(1) 3.5401	(1) 4.0554
4.2	(1) 2.1864	(1) 2.5016	(1) 2.8646	(1) 3.2829	(1) 3.7653	(1) 4.3219	(1) 4.9644
4.3	(1) 2.6464	(1) 3.0366	(1) 3.4870	(1) 4.0073	(1) 4.6086	(1) 5.3040	(1) 6.1085
4.4	(1) 3.2213	(1) 3.7065	(1) 4.2680	(1) 4.9179	(1) 5.6708	(1) 6.5433	(1) 7.5550
4.5	(1) 3.9432	(1) 4.5494	(1) 5.2524	(1) 6.0680	(1) 7.0147	(1) 8.1143	(1) 9.3921
4.6	(1) 4.8541	(1) 5.6148	(1) 6.4990	(1) 7.5270	(1) 8.7230	(2) 1.0115	(2) 1.1736
4.7	(1) 6.0085	(1) 6.9677	(1) 8.0849	(1) 9.3866	(2) 1.0904	(2) 1.2674	(2) 1.4740
4.8	(1) 7.4787	(1) 8.6937	(2) 1.0112	(2) 1.1768	(2) 1.3703	(2) 1.5964	(2) 1.8608
4.9	(1) 9.3598	(2) 1.0906	(2) 1.2715	(2) 1.4831	(2) 1.7309	(2) 2.0211	(2) 2.3611
5.0	(2) 1.1778	(2) 1.3756	(2) 1.6073	(2) 1.8791	(2) 2.1979	(2) 2.5720	(2) 3.0112

Table 19.1

x	$U(0.4,x)$	$U(0.5,x)$	$U(0.6,x)$	$U(0.7,x)$	$U(0.8,x)$	$U(0.9,x)$	$U(1.0,x)$
0.0	(0)1.2579	(0)1.2533	(0)1.2436	(0)1.2292	(0)1.2106	(0)1.1883	(0)1.1627
0.1	(0)1.1672	(0)1.1564	(0)1.1413	(0)1.1223	(0)1.1000	(0)1.0746	(0)1.0467
0.2	(0)1.0811	(0)1.0652	(0)1.0458	(0)1.0233	(−1)9.9813	(−1)9.7063	(−1)9.4122
0.3	(−1)9.9946	(−1)9.7955	(−1)9.5680	(−1)9.3162	(−1)9.0440	(−1)8.7549	(−1)8.4523
0.4	(−1)9.2205	(−1)8.9898	(−1)8.7372	(−1)8.4665	(−1)8.1811	(−1)7.8843	(−1)7.5790
0.5	(−1)8.4870	(−1)8.2327	(−1)7.9624	(−1)7.6795	(−1)7.3870	(−1)7.0879	(−1)6.7845
0.6	(−1)7.7928	(−1)7.5219	(−1)7.2403	(−1)6.9511	(−1)6.6567	(−1)6.3597	(−1)6.0622
0.7	(−1)7.1368	(−1)6.8555	(−1)6.5683	(−1)6.2776	(−1)5.9857	(−1)5.6945	(−1)5.4060
0.8	(−1)6.5181	(−1)6.2318	(−1)5.9437	(−1)5.6558	(−1)5.3699	(−1)5.0877	(−1)4.8105
0.9	(−1)5.9358	(−1)5.6493	(−1)5.3643	(−1)5.0826	(−1)4.8057	(−1)4.5347	(−1)4.2709
1.0	(−1)5.3894	(−1)5.1064	(−1)4.8280	(−1)4.5553	(−1)4.2896	(−1)4.0318	(−1)3.7826
1.1	(−1)4.8780	(−1)4.6019	(−1)4.3327	(−1)4.0713	(−1)3.8187	(−1)3.5753	(−1)3.3417
1.2	(−1)4.4008	(−1)4.1343	(−1)3.8765	(−1)3.6282	(−1)3.3898	(−1)3.1618	(−1)2.9443
1.3	(−1)3.9571	(−1)3.7022	(−1)3.4575	(−1)3.2235	(−1)3.0003	(−1)2.7881	(−1)2.5870
1.4	(−1)3.5459	(−1)3.3042	(−1)3.0739	(−1)2.8550	(−1)2.6475	(−1)2.4514	(−1)2.2665
1.5	(−1)3.1663	(−1)2.9390	(−1)2.7238	(−1)2.5204	(−1)2.3288	(−1)2.1487	(−1)1.9797
1.6	(−1)2.8171	(−1)2.6050	(−1)2.4053	(−1)2.2177	(−1)2.0419	(−1)1.8774	(−1)1.7240
1.7	(−1)2.4972	(−1)2.3007	(−1)2.1167	(−1)1.9447	(−1)1.7844	(−1)1.6351	(−1)1.4965
1.8	(−1)2.2054	(−1)2.0246	(−1)1.8561	(−1)1.6994	(−1)1.5540	(−1)1.4193	(−1)1.2948
1.9	(−1)1.9402	(−1)1.7749	(−1)1.6216	(−1)1.4798	(−1)1.3487	(−1)1.2278	(−1)1.1165
2.0	(−1)1.7003	(−1)1.5501	(−1)1.4115	(−1)1.2838	(−1)1.1664	(−1)1.0585	(−2)9.5952
2.1	(−1)1.4842	(−1)1.3486	(−1)1.2240	(−1)1.1097	(−1)1.0050	(−2)9.0923	(−2)8.2173
2.2	(−1)1.2904	(−1)1.1687	(−1)1.0574	(−2)9.5563	(−2)8.6280	(−2)7.7820	(−2)7.0122
2.3	(−1)1.1174	(−1)1.0088	(−2)9.0985	(−2)8.1979	(−2)7.3793	(−2)6.6361	(−2)5.9622
2.4	(−2)9.6358	(−2)8.6728	(−2)7.7984	(−2)7.0055	(−2)6.2874	(−2)5.6377	(−2)5.0508
2.5	(−2)8.2754	(−2)7.4258	(−2)6.6573	(−2)5.9630	(−2)5.3363	(−2)4.7714	(−2)4.2627
2.6	(−2)7.0773	(−2)6.3320	(−2)5.6603	(−2)5.0555	(−2)4.5115	(−2)4.0227	(−2)3.5839
2.7	(−2)6.0272	(−2)5.3770	(−2)4.7930	(−2)4.2689	(−2)3.7990	(−2)3.3782	(−2)3.0017
2.8	(−2)5.1111	(−2)4.5470	(−2)4.0418	(−2)3.5900	(−2)3.1863	(−2)2.8258	(−2)2.5042
2.9	(−2)4.3157	(−2)3.8288	(−2)3.3942	(−2)3.0068	(−2)2.6615	(−2)2.3543	(−2)2.0810
3.0	(−2)3.6284	(−2)3.2104	(−2)2.8384	(−2)2.5078	(−2)2.2142	(−2)1.9535	(−2)1.7224
3.1	(−2)3.0372	(−2)2.6803	(−2)2.3636	(−2)2.0830	(−2)1.8344	(−2)1.6144	(−2)1.4199
3.2	(−2)2.5313	(−2)2.2281	(−2)1.9598	(−2)1.7228	(−2)1.5134	(−2)1.3287	(−2)1.1658
3.3	(−2)2.1004	(−2)1.8441	(−2)1.6181	(−2)1.4189	(−2)1.2434	(−2)1.0890	(−3)9.5318
3.4	(−2)1.7351	(−2)1.5196	(−2)1.3301	(−2)1.1636	(−2)1.0172	(−3)8.8881	(−3)7.7615
3.5	(−2)1.4270	(−2)1.2468	(−2)1.0887	(−3)9.5009	(−3)8.2868	(−3)7.2238	(−3)6.2937
3.6	(−2)1.1683	(−2)1.0184	(−3)8.8715	(−3)7.7243	(−3)6.7217	(−3)5.8462	(−3)5.0820
3.7	(−3)9.5224	(−3)8.2810	(−3)7.1975	(−3)6.2525	(−3)5.4288	(−3)4.7111	(−3)4.0863
3.8	(−3)7.7263	(−3)6.7038	(−3)5.8136	(−3)5.0391	(−3)4.3655	(−3)3.7801	(−3)3.2716
3.9	(−3)6.2406	(−3)5.4026	(−3)4.6749	(−3)4.0432	(−3)3.4952	(−3)3.0200	(−3)2.6082
4.0	(−3)5.0176	(−3)4.3344	(−3)3.7425	(−3)3.2298	(−3)2.7861	(−3)2.4023	(−3)2.0704
4.1	(−3)4.0160	(−3)3.4617	(−3)2.9826	(−3)2.5686	(−3)2.2111	(−3)1.9025	(−3)1.6363
4.2	(−3)3.1995	(−3)2.7521	(−3)2.3663	(−3)2.0336	(−3)1.7470	(−3)1.5001	(−3)1.2876
4.3	(−3)2.5373	(−3)2.1781	(−3)1.8689	(−3)1.6029	(−3)1.3742	(−3)1.1776	(−3)1.0088
4.4	(−3)2.0029	(−3)1.7158	(−3)1.4693	(−3)1.2577	(−3)1.0761	(−4)9.2036	(−4)7.8686
4.5	(−3)1.5738	(−3)1.3455	(−3)1.1499	(−4)9.8235	(−4)8.3889	(−4)7.1610	(−4)6.1105
4.6	(−3)1.2308	(−3)1.0503	(−4)8.9583	(−4)7.6382	(−4)6.5103	(−4)5.5468	(−4)4.7242
4.7	(−4)9.5815	(−4)8.1601	(−4)6.9470	(−4)5.9121	(−4)5.0295	(−4)4.2772	(−4)3.6361
4.8	(−4)7.4240	(−4)6.3107	(−4)5.3625	(−4)4.5551	(−4)3.8680	(−4)3.2833	(−4)2.7861
4.9	(−4)5.7255	(−4)4.8579	(−4)4.1203	(−4)3.4935	(−4)2.9611	(−4)2.5090	(−4)2.1252
5.0	(−4)4.3948	(−4)3.7221	(−4)3.1512	(−4)2.6671	(−4)2.2566	(−4)1.9086	(−4)1.6138

Table 19.1

x	$V(0.4,x)$	$V(0.5,x)$	$V(0.6,x)$	$V(0.7,x)$	$V(0.8,x)$	$V(0.9,x)$	$V(1.0,x)$
0.0	(−1) 8.3485	(−1) 7.9788	(−1) 7.3474	(−1) 6.4988	(−1) 5.4912	(−1) 4.3932	(−1) 3.2800
0.1	(−1) 8.3808	(−1) 7.9988	(−1) 7.3851	(−1) 6.5836	(−1) 5.6492	(−1) 4.6453	(−1) 3.6401
0.2	(−1) 8.4468	(−1) 8.0590	(−1) 7.4675	(−1) 6.7147	(−1) 5.8526	(−1) 4.9394	(−1) 4.0368
0.3	(−1) 8.5475	(−1) 8.1604	(−1) 7.5954	(−1) 6.8936	(−1) 6.1035	(−1) 5.2785	(−1) 4.4742
0.4	(−1) 8.6844	(−1) 8.3045	(−1) 7.7707	(−1) 7.1224	(−1) 6.4046	(−1) 5.6664	(−1) 4.9575
0.5	(−1) 8.8595	(−1) 8.4934	(−1) 7.9958	(−1) 7.4039	(−1) 6.7596	(−1) 6.1076	(−1) 5.4924
0.6	(−1) 9.0757	(−1) 8.7302	(−1) 8.2739	(−1) 7.7419	(−1) 7.1730	(−1) 6.6077	(−1) 6.0858
0.7	(−1) 9.3364	(−1) 9.0186	(−1) 8.6092	(−1) 8.1412	(−1) 7.6504	(−1) 7.1733	(−1) 6.7457
0.8	(−1) 9.6460	(−1) 9.3633	(−1) 9.0068	(−1) 8.6076	(−1) 8.1984	(−1) 7.8124	(−1) 7.4814
0.9	(0) 1.0010	(−1) 9.7698	(−1) 9.4730	(−1) 9.1481	(−1) 8.8253	(−1) 8.5344	(−1) 8.3040
1.0	(0) 1.0434	(0) 1.0245	(0) 1.0015	(−1) 9.7713	(−1) 9.5408	(−1) 9.3507	(−1) 9.2267
1.1	(0) 1.0926	(0) 1.0797	(0) 1.0643	(0) 1.0488	(0) 1.0357	(0) 1.0275	(0) 1.0265
1.2	(0) 1.1495	(0) 1.1436	(0) 1.1367	(0) 1.1309	(0) 1.1287	(0) 1.1323	(0) 1.1437
1.3	(0) 1.2151	(0) 1.2174	(0) 1.2200	(0) 1.2251	(0) 1.2348	(0) 1.2514	(0) 1.2765
1.4	(0) 1.2908	(0) 1.3024	(0) 1.3158	(0) 1.3330	(0) 1.3561	(0) 1.3870	(0) 1.4276
1.5	(0) 1.3779	(0) 1.4003	(0) 1.4260	(0) 1.4569	(0) 1.4949	(0) 1.5420	(0) 1.5999
1.6	(0) 1.4784	(0) 1.5132	(0) 1.5528	(0) 1.5992	(0) 1.6542	(0) 1.7196	(0) 1.7973
1.7	(0) 1.5943	(0) 1.6433	(0) 1.6989	(0) 1.7629	(0) 1.8373	(0) 1.9238	(0) 2.0243
1.8	(0) 1.7281	(0) 1.7936	(0) 1.8675	(0) 1.9518	(0) 2.0484	(0) 2.1592	(0) 2.2862
1.9	(0) 1.8829	(0) 1.9674	(0) 2.0625	(0) 2.1703	(0) 2.2926	(0) 2.4317	(0) 2.5896
2.0	(0) 2.0622	(0) 2.1689	(0) 2.2886	(0) 2.4236	(0) 2.5760	(0) 2.7481	(0) 2.9424
2.1	(0) 2.2705	(0) 2.4030	(0) 2.5514	(0) 2.7182	(0) 2.9058	(0) 3.1169	(0) 3.3542
2.2	(0) 2.5130	(0) 2.6757	(0) 2.8578	(0) 3.0620	(0) 3.2911	(0) 3.5483	(0) 3.8368
2.3	(0) 2.7961	(0) 2.9943	(0) 3.2160	(0) 3.4644	(0) 3.7428	(0) 4.0548	(0) 4.4044
2.4	(0) 3.1275	(0) 3.3676	(0) 3.6363	(0) 3.9371	(0) 4.2741	(0) 4.6517	(0) 5.0747
2.5	(0) 3.5166	(0) 3.8065	(0) 4.1310	(0) 4.4944	(0) 4.9015	(0) 5.3578	(0) 5.8692
2.6	(0) 3.9749	(0) 4.3241	(0) 4.7153	(0) 5.1536	(0) 5.6451	(0) 6.1963	(0) 6.8146
2.7	(0) 4.5165	(0) 4.9368	(0) 5.4079	(0) 5.9365	(0) 6.5297	(0) 7.1959	(0) 7.9440
2.8	(0) 5.1589	(0) 5.6644	(0) 6.2320	(0) 6.8696	(0) 7.5862	(0) 8.3921	(0) 9.2985
2.9	(0) 5.9235	(0) 6.5320	(0) 7.2162	(0) 7.9862	(0) 8.8529	(0) 9.8292	(1) 1.0929
3.0	(0) 6.8368	(0) 7.5701	(0) 8.3962	(0) 9.3274	(1) 1.0378	(1) 1.1563	(1) 1.2900
3.1	(0) 7.9320	(0) 8.8172	(0) 9.8164	(1) 1.0945	(1) 1.2220	(1) 1.3662	(1) 1.5293
3.2	(0) 9.2504	(1) 1.0321	(1) 1.1533	(1) 1.2903	(1) 1.4455	(1) 1.6214	(1) 1.8207
3.3	(1) 1.0844	(1) 1.2142	(1) 1.3615	(1) 1.5284	(1) 1.7178	(1) 1.9329	(1) 2.1773
3.4	(1) 1.2777	(1) 1.4357	(1) 1.6151	(1) 1.8190	(1) 2.0509	(1) 2.3148	(1) 2.6153
3.5	(1) 1.5132	(1) 1.7060	(1) 1.9253	(1) 2.1752	(1) 2.4601	(1) 2.7849	(1) 3.1555
3.6	(1) 1.8014	(1) 2.0373	(1) 2.3064	(1) 2.6137	(1) 2.9646	(1) 3.3658	(1) 3.8246
3.7	(1) 2.1555	(1) 2.4452	(1) 2.7765	(1) 3.1556	(1) 3.5896	(1) 4.0868	(1) 4.6566
3.8	(1) 2.5923	(1) 2.9495	(1) 3.3588	(1) 3.8282	(1) 4.3669	(1) 4.9853	(1) 5.6956
3.9	(1) 3.1336	(1) 3.5756	(1) 4.0833	(1) 4.6667	(1) 5.3377	(1) 6.1098	(1) 6.9986
4.0	(1) 3.8072	(1) 4.3563	(1) 4.9884	(1) 5.7165	(1) 6.5556	(1) 7.5232	(1) 8.6395
4.1	(1) 4.6493	(1) 5.3341	(1) 6.1242	(1) 7.0364	(1) 8.0899	(1) 9.3073	(2) 1.0715
4.2	(1) 5.7065	(1) 6.5642	(1) 7.5559	(1) 8.7031	(2) 1.0031	(2) 1.1569	(2) 1.3351
4.3	(1) 7.0397	(1) 8.1183	(1) 9.3682	(2) 1.0817	(2) 1.2498	(2) 1.4449	(2) 1.6714
4.4	(1) 8.7286	(2) 1.0091	(2) 1.1673	(2) 1.3511	(2) 1.5647	(2) 1.8131	(2) 2.1022
4.5	(2) 1.0878	(2) 1.2605	(2) 1.4616	(2) 1.6957	(2) 1.9684	(2) 2.2861	(2) 2.6566
4.6	(2) 1.3624	(2) 1.5826	(2) 1.8392	(2) 2.1387	(2) 2.4882	(2) 2.8963	(2) 3.3731
4.7	(2) 1.7151	(2) 1.9968	(2) 2.3259	(2) 2.7106	(2) 3.1606	(2) 3.6870	(2) 4.3032
4.8	(2) 2.1701	(2) 2.5321	(2) 2.9559	(2) 3.4524	(2) 4.0341	(2) 4.7161	(2) 5.5160
4.9	(2) 2.7596	(2) 3.2270	(2) 3.7752	(2) 4.4187	(2) 5.1742	(2) 6.0616	(2) 7.1043
5.0	(2) 3.5270	(2) 4.1331	(2) 4.8456	(2) 5.6833	(2) 6.6688	(2) 7.8285	(2) 9.1938

Table 19.1

x	$U(1.5,x)$	$U(2.0,x)$	$U(2.5,x)$	$U(3.0,x)$	$U(3.5,x)$	$U(4.0,x)$	$U(4.5,x)$	$U(5.0,x)$
0.0	(0)1.0000	(−1)8.1085	(−1)6.2666	(−1)4.6509	(−1)3.3333	(−1)2.3167	(−1)1.5666	(−1)1.0335
0.1	(−1)8.8187	(−1)7.0232	(−1)5.3409	(−1)3.9060	(−1)2.7615	(−1)1.8950	(−1)1.2662	(−2)8.2588
0.2	(−1)7.7700	(−1)6.0787	(−1)4.5492	(−1)3.2786	(−1)2.2867	(−1)1.5494	(−1)1.0230	(−2)6.5971
0.3	(−1)6.8389	(−1)5.2566	(−1)3.8719	(−1)2.7501	(−1)1.8924	(−1)1.2662	(−2)8.2604	(−2)5.2673
0.4	(−1)6.0120	(−1)4.5410	(−1)3.2925	(−1)2.3050	(−1)1.5650	(−1)1.0340	(−2)6.6663	(−2)4.2032
0.5	(−1)5.2778	(−1)3.9182	(−1)2.7969	(−1)1.9302	(−1)1.2931	(−2)8.4374	(−2)5.3758	(−2)3.3518
0.6	(−1)4.6262	(−1)3.3763	(−1)2.3731	(−1)1.6146	(−1)1.0674	(−2)6.8788	(−2)4.3316	(−2)2.6707
0.7	(−1)4.0482	(−1)2.9051	(−1)2.0109	(−1)1.3490	(−2)8.8019	(−2)5.6025	(−2)3.4869	(−2)2.1262
0.8	(−1)3.5360	(−1)2.4957	(−1)1.7015	(−1)1.1256	(−2)7.2491	(−2)4.5579	(−2)2.8040	(−2)1.6910
0.9	(−1)3.0825	(−1)2.1403	(−1)1.4375	(−2)9.3785	(−2)5.9624	(−2)3.7035	(−2)2.2523	(−2)1.3434
1.0	(−1)2.6816	(−1)1.8321	(−1)1.2124	(−2)7.8022	(−2)4.8971	(−2)3.0053	(−2)1.8068	(−2)1.0660
1.1	(−1)2.3276	(−1)1.5651	(−1)1.0208	(−2)6.4802	(−2)4.0160	(−2)2.4351	(−2)1.4475	(−3)8.4479
1.2	(−1)2.0157	(−1)1.3343	(−2)8.5773	(−2)5.3727	(−2)3.2880	(−2)1.9701	(−2)1.1579	(−3)6.6856
1.3	(−1)1.7412	(−1)1.1350	(−2)7.1928	(−2)4.4461	(−2)2.6872	(−2)1.5913	(−3)9.2486	(−3)5.2831
1.4	(−1)1.5003	(−2)9.6317	(−2)6.0190	(−2)3.6721	(−2)2.1922	(−2)1.2831	(−3)7.3749	(−3)4.1683
1.5	(−1)1.2893	(−2)8.1541	(−2)5.0255	(−2)3.0265	(−2)1.7849	(−2)1.0327	(−3)5.8705	(−3)3.2833
1.6	(−1)1.1049	(−2)6.8857	(−2)4.1862	(−2)2.4890	(−2)1.4503	(−3)8.2953	(−3)4.6645	(−3)2.5816
1.7	(−2)9.4412	(−2)5.7994	(−2)3.4786	(−2)2.0423	(−2)1.1759	(−3)6.6500	(−3)3.6991	(−3)2.0262
1.8	(−2)8.0438	(−2)4.8712	(−2)2.8833	(−2)1.6718	(−3)9.5127	(−3)5.3198	(−3)2.9276	(−3)1.5873
1.9	(−2)6.8324	(−2)4.0801	(−2)2.3837	(−2)1.3652	(−3)7.6780	(−3)4.2463	(−3)2.3122	(−3)1.2409
2.0	(−2)5.7853	(−2)3.4076	(−2)1.9653	(−2)1.1120	(−3)6.1823	(−3)3.3818	(−3)1.8222	(−4)9.6810
2.1	(−2)4.8830	(−2)2.8375	(−2)1.6159	(−3)9.0339	(−3)4.9656	(−3)2.6869	(−3)1.4328	(−4)7.5364
2.2	(−2)4.1080	(−2)2.3556	(−2)1.3248	(−3)7.3193	(−3)3.9782	(−3)2.1296	(−3)1.1240	(−4)5.8538
2.3	(−2)3.4444	(−2)1.9495	(−2)1.0829	(−3)5.9138	(−3)3.1787	(−3)1.6837	(−4)8.7960	(−4)4.5364
2.4	(−2)2.8782	(−2)1.6082	(−3)8.8260	(−3)4.7646	(−3)2.5331	(−3)1.3277	(−4)6.8665	(−4)3.5071
2.5	(−2)2.3966	(−2)1.3223	(−3)7.1710	(−3)3.8275	(−3)2.0129	(−3)1.0442	(−4)5.3467	(−4)2.7047
2.6	(−2)1.9886	(−2)1.0837	(−3)5.8081	(−3)3.0655	(−3)1.5951	(−4)8.1895	(−4)4.1523	(−4)2.0806
2.7	(−2)1.6441	(−3)8.8509	(−3)4.6891	(−3)2.4478	(−3)1.2603	(−4)6.4052	(−4)3.2161	(−4)1.5964
2.8	(−2)1.3544	(−3)7.2040	(−3)3.7734	(−3)1.9484	(−4)9.9277	(−4)4.9954	(−4)2.4841	(−4)1.2216
2.9	(−2)1.1116	(−3)5.8431	(−3)3.0264	(−3)1.5460	(−4)7.7967	(−4)3.8845	(−4)1.9134	(−5)9.3228
3.0	(−3)9.0885	(−3)4.7224	(−3)2.4191	(−3)1.2228	(−4)6.1042	(−4)3.0117	(−4)1.4695	(−5)7.0950
3.1	(−3)7.4028	(−3)3.8030	(−3)1.9270	(−4)9.6394	(−4)4.7641	(−4)2.3279	(−4)1.1253	(−5)5.3843
3.2	(−3)6.0067	(−3)3.0513	(−3)1.5296	(−4)7.5735	(−4)3.7062	(−4)1.7938	(−5)8.5914	(−5)4.0742
3.3	(−3)4.8549	(−3)2.4392	(−3)1.2099	(−4)5.9301	(−4)2.8738	(−4)1.3778	(−5)6.5394	(−5)3.0738
3.4	(−3)3.9086	(−3)1.9426	(−4)9.5361	(−4)4.6274	(−4)2.2210	(−4)1.0550	(−5)4.9621	(−5)2.3121
3.5	(−3)3.1342	(−3)1.5412	(−4)7.4887	(−4)3.5982	(−4)1.7107	(−5)8.0514	(−5)3.7534	(−5)1.7338
3.6	(−3)2.5032	(−3)1.2181	(−4)5.8592	(−4)2.7880	(−4)1.3131	(−5)6.1244	(−5)2.8300	(−5)1.2961
3.7	(−3)1.9912	(−4)9.5895	(−4)4.5672	(−4)2.1526	(−4)1.0045	(−5)4.6430	(−5)2.1269	(−6)9.6590
3.8	(−3)1.5775	(−4)7.5202	(−4)3.5468	(−4)1.6559	(−5)7.6567	(−5)3.5080	(−5)1.5932	(−6)7.1749
3.9	(−3)1.2446	(−4)5.8741	(−4)2.7439	(−4)1.2692	(−5)5.8157	(−5)2.6413	(−5)1.1894	(−6)5.3123
4.0	(−4)9.7788	(−4)4.5702	(−4)2.1146	(−5)9.6913	(−5)4.4015	(−5)1.9818	(−6)8.8495	(−6)3.9203
4.1	(−4)7.6513	(−4)3.5414	(−4)1.6233	(−5)7.3727	(−5)3.3191	(−5)1.4817	(−6)6.5617	(−6)2.8834
4.2	(−4)5.9616	(−4)2.7331	(−4)1.2413	(−5)5.5875	(−5)2.4937	(−5)1.1039	(−6)4.8485	(−6)2.1136
4.3	(−4)4.6255	(−4)2.1007	(−5)9.4547	(−5)4.2185	(−5)1.8667	(−6)8.1946	(−6)3.5701	(−6)1.5440
4.4	(−4)3.5736	(−4)1.6081	(−5)7.1727	(−5)3.1726	(−5)1.3920	(−6)6.0609	(−6)2.6194	(−6)1.1240
4.5	(−4)2.7491	(−4)1.2259	(−5)5.4198	(−5)2.3767	(−5)1.0342	(−6)4.4663	(−6)1.9150	(−7)8.1539
4.6	(−4)2.1058	(−5)9.3061	(−5)4.0787	(−5)1.7736	(−6)7.6538	(−6)3.2790	(−6)1.3949	(−7)5.8942
4.7	(−4)1.6061	(−5)7.0352	(−5)3.0571	(−5)1.3183	(−6)5.6428	(−6)2.3983	(−6)1.0124	(−7)4.2455
4.8	(−4)1.2197	(−5)5.2961	(−5)2.2819	(−6)9.7593	(−6)4.1440	(−6)1.7475	(−7)7.3205	(−7)3.0469
4.9	(−5)9.2216	(−5)3.9701	(−5)1.6964	(−6)7.1961	(−6)3.0315	(−6)1.2685	(−7)5.2737	(−7)2.1788
5.0	(−5)6.9418	(−5)2.9634	(−5)1.2558	(−6)5.2847	(−6)2.2089	(−7)9.1724	(−7)3.7849	(−7)1.5523

Table 19.1

x	$V(1.5,x)$	$V(2.0,x)$	$V(2.5,x)$	$V(3.0,x)$	$V(3.5,x)$	$V(4.0,x)$	$V(4.5,x)$	$V(5.0,x)$
0.0	0.0000	(−1)3.4311	(−1)7.9788	(−1)4.9200	0.0000	(0)0.8578	(0)2.3937	(0)1.7220
0.1	(−1)0.7999	(−1)3.9591	(−1)8.0788	(−1)5.8561	(−1)2.4076	(0)1.0483	(0)2.4477	(0)2.1545
0.2	(−1)1.6118	(−1)4.5665	(−1)8.3814	(−1)6.9684	(−1)4.8999	(0)1.2810	(0)2.6124	(0)2.6952
0.3	(−1)2.4481	(−1)5.2660	(−1)8.8948	(−1)8.2911	(−1)7.5647	(0)1.5652	(0)2.8954	(0)3.3715
0.4	(−1)3.3218	(−1)6.0721	(−1)9.6332	(−1)9.8651	(0)1.0497	(0)1.9126	(0)3.3098	(0)4.2178
0.5	(−1)4.2467	(−1)7.0024	(0)1.0617	(0)1.1740	(0)1.3802	(0)2.3376	(0)3.8751	(0)5.2778
0.6	(−1)5.2381	(−1)8.0774	(0)1.1873	(0)1.3975	(0)1.7600	(0)2.8579	(0)4.6180	(0)6.6060
0.7	(−1)6.3130	(−1)9.3217	(0)1.3438	(0)1.6644	(0)2.2033	(0)3.4955	(0)5.5736	(0)8.2721
0.8	(−1)7.4906	(0)1.0764	(0)1.5356	(0)1.9833	(0)2.7266	(0)4.2777	(0)6.7880	(1)1.0364
0.9	(−1)8.7928	(0)1.2440	(0)1.7683	(0)2.3652	(0)3.3501	(0)5.2386	(0)8.3200	(1)1.2993
1.0	(0)1.0245	(0)1.4390	(0)2.0490	(0)2.8230	(0)4.0980	(0)6.4206	(1)1.0245	(1)1.6301
1.1	(0)1.1877	(0)1.6665	(0)2.3862	(0)3.3729	(0)5.0002	(0)7.8765	(1)1.2659	(1)2.0469
1.2	(0)1.3724	(0)1.9325	(0)2.7905	(0)4.0346	(0)6.0933	(0)9.6727	(1)1.5683	(1)2.5728
1.3	(0)1.5826	(0)2.2442	(0)3.2748	(0)4.8322	(0)7.4224	(1)1.1892	(1)1.9473	(1)3.2373
1.4	(0)1.8234	(0)2.6104	(0)3.8551	(0)5.7959	(0)9.0439	(1)1.4640	(1)2.4227	(1)4.0782
1.5	(0)2.1005	(0)3.0418	(0)4.5511	(0)6.9626	(1)1.1028	(1)1.8048	(1)3.0195	(1)5.1442
1.6	(0)2.4211	(0)3.5514	(0)5.3869	(0)8.3782	(1)1.3461	(1)2.2284	(1)3.7699	(1)6.4978
1.7	(0)2.7936	(0)4.1551	(0)6.3925	(1)1.0100	(1)1.6454	(1)2.7558	(1)4.7150	(1)8.2198
1.8	(0)3.2284	(0)4.8722	(0)7.6047	(1)1.2199	(1)2.0145	(1)3.4139	(1)5.9076	(2)1.0415
1.9	(0)3.7380	(0)5.7267	(0)9.0697	(1)1.4765	(1)2.4708	(1)4.2370	(1)7.4155	(2)1.3218
2.0	(0)4.3378	(0)6.7480	(1)1.0844	(1)1.7910	(1)3.0364	(1)5.2689	(1)9.3262	(2)1.6806
2.1	(0)5.0463	(0)7.9725	(1)1.3000	(1)2.1774	(1)3.7393	(1)6.5656	(2)1.1753	(2)2.1408
2.2	(0)5.8865	(0)9.4452	(1)1.5626	(1)2.6535	(1)4.6150	(1)8.1989	(2)1.4841	(2)2.7325
2.3	(0)6.8869	(1)1.1222	(1)1.8834	(1)3.2418	(1)5.7092	(2)1.0262	(2)1.8781	(2)3.4948
2.4	(0)8.0823	(1)1.3374	(1)2.2765	(1)3.9709	(1)7.0801	(2)1.2873	(2)2.3822	(2)4.4794
2.5	(0)9.5162	(1)1.5987	(1)2.7597	(1)4.8771	(1)8.8025	(2)1.6189	(2)3.0285	(2)5.7544
2.6	(1)1.1243	(1)1.9172	(1)3.3555	(1)6.0069	(2)1.0973	(2)2.0411	(2)3.8596	(2)7.4093
2.7	(1)1.3329	(1)2.3068	(1)4.0926	(1)7.4199	(2)1.3716	(2)2.5801	(2)4.9310	(2)9.5631
2.8	(1)1.5860	(1)2.7849	(1)5.0074	(1)9.1925	(2)1.7193	(2)3.2701	(2)6.3162	(3)1.2374
2.9	(1)1.8943	(1)3.3738	(1)6.1466	(2)1.1423	(2)2.1614	(2)4.1562	(2)8.1119	(3)1.6051
3.0	(1)2.2710	(1)4.1018	(1)7.5701	(2)1.4240	(2)2.7252	(2)5.2976	(3)1.0447	(3)2.0877
3.1	(1)2.7333	(1)5.0049	(1)9.3551	(2)1.7809	(2)3.4467	(2)6.7721	(3)1.3491	(3)2.7227
3.2	(1)3.3028	(1)6.1295	(2)1.1601	(2)2.2345	(2)4.3729	(2)8.6829	(3)1.7474	(3)3.5606
3.3	(1)4.0070	(1)7.5350	(2)1.4437	(2)2.8131	(2)5.5657	(3)1.1167	(3)2.2698	(3)4.6697
3.4	(1)4.8812	(1)9.2982	(2)1.8032	(2)3.5537	(2)7.1071	(3)1.4407	(3)2.9574	(3)6.1422
3.5	(1)5.9708	(2)1.1519	(2)2.2604	(2)4.5048	(2)9.1055	(3)1.8646	(3)3.8650	(3)8.1029
3.6	(1)7.3343	(2)1.4325	(2)2.8441	(2)5.7308	(3)1.1705	(3)2.4212	(3)5.0672	(4)1.0722
3.7	(1)9.0472	(2)1.7887	(2)3.5920	(2)7.3166	(3)1.5100	(3)3.1543	(3)6.6645	(4)1.4232
3.8	(2)1.1208	(2)2.2424	(2)4.5540	(2)9.3755	(3)1.9547	(3)4.1233	(3)8.7939	(4)1.8950
3.9	(2)1.3945	(2)2.8227	(2)5.7960	(3)1.2058	(3)2.5393	(3)5.4084	(4)1.1642	(4)2.5313
4.0	(2)1.7425	(2)3.5678	(2)7.4057	(3)1.5567	(3)3.3108	(3)7.1188	(4)1.5465	(4)3.3924
4.1	(2)2.1870	(2)4.5283	(2)9.5001	(3)2.0173	(3)4.3324	(3)9.4032	(4)2.0613	(4)4.5614
4.2	(2)2.7569	(2)5.7716	(3)1.2236	(3)2.6243	(3)5.6903	(4)1.2465	(4)2.7570	(4)6.1538
4.3	(2)3.4909	(2)7.3873	(3)1.5823	(3)3.4272	(3)7.5019	(4)1.6584	(4)3.7005	(4)8.3306
4.4	(2)4.4399	(2)9.4956	(3)2.0545	(3)4.4934	(3)9.9277	(4)2.2145	(4)4.9845	(5)1.1316
4.5	(2)5.6724	(3)1.2258	(3)2.6786	(3)5.9146	(4)1.3188	(4)2.9680	(4)6.7384	(5)1.5426
4.6	(2)7.2797	(3)1.5893	(3)3.5069	(3)7.8166	(4)1.7588	(4)3.9929	(4)9.1425	(5)2.1103
4.7	(2)9.3849	(3)2.0695	(3)4.6106	(4)1.0372	(4)2.3547	(4)5.3922	(5)1.2450	(5)2.8973
4.8	(3)1.2154	(3)2.7065	(3)6.0871	(4)1.3819	(4)3.1649	(4)7.3096	(5)1.7018	(5)3.9923
4.9	(3)1.5812	(3)3.5553	(3)8.0706	(4)1.8487	(4)4.2708	(4)9.9472	(5)2.3348	(5)5.5212
5.0	(3)2.0666	(3)4.6909	(4)1.0746	(4)2.4833	(4)5.7864	(5)1.3589	(5)3.2156	(5)7.6639

Table 19.2

x	$W(-5.0,x)$	$W(-4.0,x)$	$W(-3.0,x)$	$W(-2.0,x)$	$W(-5.0,-x)$	$W(-4.0,-x)$	$W(-3.0,-x)$	$W(-2.0,-x)$
0.0	0.47348	0.50102	0.53933	0.60027	0.47348	0.50102	0.53933	0.60027
0.1	0.35697	0.39190	0.43901	0.51126	0.56641	0.59017	0.62350	0.67730
0.2	0.22267	0.26715	0.32555	0.41203	0.63113	0.65576	0.68900	0.74078
0.3	+0.07727	+0.13172	0.20231	0.30453	0.66435	0.69515	0.73381	0.78939
0.4	−0.07200	−0.00899	+0.07298	0.19088	0.66434	0.70666	0.75649	0.82206
0.5	−0.21764	−0.14933	−0.05857	+0.07334	0.63099	0.68972	0.75622	0.83798
0.6	−0.35231	−0.28362	−0.18832	−0.04569	0.56583	0.64485	0.73285	0.83665
0.7	−0.46911	−0.40634	−0.31226	−0.16377	0.47199	0.57370	0.68690	0.81785
0.8	−0.56198	−0.51236	−0.42646	−0.27838	0.35408	0.47898	0.61955	0.78173
0.9	−0.62597	−0.59713	−0.52722	−0.38697	0.21799	0.36441	0.53268	0.72875
1.0	−0.65752	−0.65688	−0.61113	−0.48704	+0.07061	0.23458	0.42880	0.65975
1.1	−0.65470	−0.68881	−0.67522	−0.57617	−0.08044	+0.09483	0.31103	0.57594
1.2	−0.61732	−0.69121	−0.71706	−0.65204	−0.22724	−0.04897	0.18303	0.47890
1.3	−0.54700	−0.66357	−0.73488	−0.71255	−0.36189	−0.19063	+0.04890	0.37059
1.4	−0.44716	−0.60670	−0.72761	−0.75583	−0.47700	−0.32388	−0.08688	0.25333
1.5	−0.32290	−0.52270	−0.69502	−0.78031	−0.56602	−0.44262	−0.21962	0.12978
1.6	−0.18077	−0.41495	−0.63774	−0.78484	−0.62369	−0.54122	−0.34454	+0.00294
1.7	−0.02851	−0.28803	−0.55733	−0.76869	−0.64634	−0.61480	−0.45694	−0.12397
1.8	+0.12535	−0.14758	−0.45625	−0.73166	−0.63218	−0.65945	−0.55237	−0.24749
1.9	0.27194	−0.00009	−0.33785	−0.67412	−0.58147	−0.67250	−0.62680	−0.36405
2.0	0.40253	+0.14739	−0.20633	−0.59707	−0.49661	−0.65271	−0.67684	−0.47006
2.1	0.50907	0.28751	−0.06661	−0.50217	−0.38212	−0.60042	−0.69989	−0.56198
2.2	0.58468	0.41299	+0.07581	−0.39174	−0.24445	−0.51764	−0.69432	−0.63649
2.3	0.62416	0.51702	0.21503	−0.26879	−0.09171	−0.40802	−0.65962	−0.69061
2.4	0.62438	0.59364	0.34495	−0.13696	+0.06678	−0.27680	−0.59652	−0.72184
2.5	0.58460	0.63810	0.45960	−0.00046	0.22095	−0.13062	−0.50704	−0.72830
2.6	0.50668	0.64722	0.55333	+0.13603	0.36067	+0.02276	−0.39454	−0.70889
2.7	0.39507	0.61968	0.62119	0.26749	0.47637	0.17482	−0.26363	−0.66340
2.8	0.25669	0.55625	0.65920	0.38872	0.55973	0.31672	−0.12008	−0.59265
2.9	+0.10057	0.45985	0.66463	0.49459	0.60434	0.43980	+0.02936	−0.49853
3.0	−0.06260	0.33555	0.63631	0.58021	0.60627	0.53615	0.17727	−0.38404
3.1	−0.22123	0.19042	0.57472	0.64123	0.56451	0.59915	0.31588	−0.25332
3.2	−0.36354	+0.03320	0.48225	0.67411	0.48124	0.62397	0.43747	−0.11153
3.3	−0.47850	−0.12614	0.36312	0.67637	0.36184	0.60808	0.53481	+0.03530
3.4	−0.55672	−0.27701	0.22333	0.64681	0.21471	0.55155	0.60167	0.18042
3.5	−0.59128	−0.40886	+0.07050	0.58576	+0.05079	0.45725	0.63325	0.31672
3.6	−0.57849	−0.51196	−0.08654	0.49519	−0.11714	0.33088	0.62663	0.43701
3.7	−0.51836	−0.57820	−0.23816	0.37883	−0.27544	0.18074	0.58111	0.53447
3.8	−0.41490	−0.60177	−0.37452	0.24205	−0.41066	+0.01731	0.49849	0.60305
3.9	−0.27601	−0.57982	−0.48622	+0.09180	−0.51073	−0.14737	0.38313	0.63793
4.0	−0.11306	−0.51295	−0.56500	−0.06370	−0.56615	−0.30058	0.24189	0.63597
4.1	+0.05995	−0.40534	−0.60443	−0.21535	−0.57098	−0.42985	+0.08387	0.59605
4.2	0.22741	−0.26474	−0.60059	−0.35365	−0.52367	−0.52406	−0.08010	0.51937
4.3	0.37359	−0.10210	−0.55252	−0.46937	−0.42750	−0.57448	−0.23812	0.40960
4.4	0.48406	+0.06923	−0.46263	−0.55413	−0.29056	−0.57571	−0.37804	0.27290
4.5	0.54726	0.23443	−0.33674	−0.60118	−0.12531	−0.52643	−0.48847	+0.11769
4.6	0.55583	0.37847	−0.18393	−0.60601	+0.05237	−0.42982	−0.55975	−0.04573
4.7	0.50770	0.48758	−0.01604	−0.56693	0.22465	−0.29363	−0.58492	−0.20576
4.8	0.40664	0.55059	+0.15314	−0.48549	0.37342	−0.12977	−0.56059	−0.35036
4.9	0.26226	0.56028	0.30893	−0.36666	0.48233	+0.04660	−0.48753	−0.46788
5.0	0.08936	0.51440	0.43707	−0.21874	0.53861	0.21827	−0.37095	−0.54818
	$\begin{bmatrix} (-3)7 \\ 6 \end{bmatrix}$	$\begin{bmatrix} (-3)7 \\ 6 \end{bmatrix}$	$\begin{bmatrix} (-3)6 \\ 6 \end{bmatrix}$	$\begin{bmatrix} (-3)5 \\ 6 \end{bmatrix}$	$\begin{bmatrix} (-3)7 \\ 6 \end{bmatrix}$	$\begin{bmatrix} (-3)6 \\ 6 \end{bmatrix}$	$\begin{bmatrix} (-3)6 \\ 6 \end{bmatrix}$	$\begin{bmatrix} (-3)5 \\ 6 \end{bmatrix}$

Values of $W(a,x)$ for integral values of a are from National Physical Laboratory, Tables of Weber parabolic cylinder functions. Computed by Scientific Computing Service Ltd. Mathematical Introduction by J. C. P. Miller. Her Majesty's Stationery Office, London, England, 1955 (with permission).

Table 19.2

x	$W(2.0,x)$	$W(3.0,x)$	$W(4.0,x)$	$W(5.0,x)$	$W(2.0,-x)$	$W(3.0,-x)$	$W(4.0,-x)$	$W(5.0,-x)$
0.0	(−1) 6.0027	(−1) 5.3933	(−1) 5.0102	(−1)4.7348	(−1) 6.0027	(−1)5.3933	(−1)5.0102	(−1)4.7348
0.1	(−1) 5.2271	(−1) 4.5427	(−1) 4.1061	(−1)3.7888	(−1) 6.8986	(−1)6.4061	(−1)6.1154	(−1)5.9185
0.2	(−1) 4.5561	(−1) 3.8285	(−1) 3.3667	(−1)3.0330	(−1) 7.9324	(−1)7.6114	(−1)7.4658	(−1)7.3991
0.3	(−1) 3.9758	(−1) 3.2292	(−1) 2.7621	(−1)2.4291	(−1) 9.1243	(−1)9.0448	(−1)9.1150	(−1)9.2505
0.4	(−1) 3.4744	(−1) 2.7262	(−1) 2.2677	(−1)1.9466	(0) 1.0497	(0)1.0748	(0)1.1128	(0)1.1564
0.5	(−1) 3.0411	(−1) 2.3041	(−1) 1.8634	(−1)1.5611	(0) 1.2075	(0)1.2770	(0)1.3583	(0)1.4454
0.6	(−1) 2.6668	(−1) 1.9499	(−1) 1.5327	(−1)1.2530	(0) 1.3888	(0)1.5168	(0)1.6574	(0)1.8059
0.7	(−1) 2.3436	(−1) 1.6525	(−1) 1.2621	(−1)1.0067	(0) 1.5967	(0)1.8008	(0)2.0215	(0)2.2555
0.8	(−1) 2.0644	(−1) 1.4028	(−1) 1.0407	(−2)8.0964	(0) 1.8345	(0)2.1368	(0)2.4643	(0)2.8155
0.9	(−1) 1.8233	(−1) 1.1931	(−2) 8.5930	(−2)6.5197	(0) 2.1061	(0)2.5335	(0)3.0019	(0)3.5123
1.0	(−1) 1.6151	(−1) 1.0168	(−2) 7.1069	(−2)5.2572	(0) 2.4156	(0)3.0013	(0)3.6538	(0)4.3782
1.1	(−1) 1.4351	(−2) 8.6859	(−2) 5.8882	(−2)4.2455	(0) 2.7674	(0)3.5517	(0)4.4431	(0)5.4528
1.2	(−1) 1.2795	(−2) 7.4385	(−2) 4.8880	(−2)3.4340	(0) 3.1662	(0)4.1980	(0)5.3970	(0)6.7844
1.3	(−1) 1.1450	(−2) 6.3880	(−2) 4.0663	(−2)2.7825	(0) 3.6169	(0)4.9554	(0)6.5479	(0)8.4318
1.4	(−1) 1.0286	(−2) 5.5025	(−2) 3.3906	(−2)2.2590	(0) 4.1247	(0)5.8406	(0)7.9336	(1)1.0466
1.5	(−2) 9.2770	(−2) 4.7556	(−2) 2.8343	(−2)1.8377	(0) 4.6948	(0)6.8726	(0)9.5984	(1)1.2975
1.6	(−2) 8.4018	(−2) 4.1248	(−2) 2.3757	(−2)1.4984	(0) 5.3324	(0)8.0723	(1)1.1594	(1)1.6060
1.7	(−2) 7.6411	(−2) 3.5917	(−2) 1.9973	(−2)1.2246	(0) 6.0424	(0)9.4626	(1)1.3979	(1)1.9848
1.8	(−2) 6.9782	(−2) 3.1406	(−2) 1.6845	(−2)1.0035	(0) 6.8296	(1)1.1069	(1)1.6824	(1)2.4487
1.9	(−2) 6.3984	(−2) 2.7584	(−2) 1.4256	(−3)8.2455	(0) 7.6980	(1)1.2917	(1)2.0206	(1)3.0155
2.0	(−2) 5.8890	(−2) 2.4342	(−2) 1.2111	(−3)6.7954	(0) 8.6507	(1)1.5037	(1)2.4216	(1)3.7062
2.1	(−2) 5.4386	(−2) 2.1588	(−2) 1.0330	(−3)5.6183	(0) 9.6899	(1)1.7457	(1)2.8952	(1)4.5455
2.2	(−2) 5.0372	(−2) 1.9245	(−3) 8.8491	(−3)4.6610	(1) 1.0816	(1)2.0209	(1)3.4529	(1)5.5623
2.3	(−2) 4.6755	(−2) 1.7247	(−3) 7.6160	(−3)3.8810	(1) 1.2027	(1)2.3322	(1)4.1069	(1)6.7904
2.4	(−2) 4.3456	(−2) 1.5540	(−3) 6.5875	(−3)3.2443	(1) 1.3319	(1)2.6827	(1)4.8711	(1)8.2686
2.5	(−2) 4.0402	(−2) 1.4075	(−3) 5.7281	(−3)2.7236	(1) 1.4686	(1)3.0749	(1)5.7600	(2)1.0042
2.6	(−2) 3.7524	(−2) 1.2813	(−3) 5.0088	(−3)2.2968	(1) 1.6117	(1)3.5113	(1)6.7894	(2)1.2161
2.7	(−2) 3.4763	(−2) 1.1719	(−3) 4.4055	(−3)1.9464	(1) 1.7597	(1)3.9937	(1)7.9756	(2)1.4683
2.8	(−2) 3.2064	(−2) 1.0764	(−3) 3.8984	(−3)1.6580	(1) 1.9108	(1)4.5230	(1)9.3355	(2)1.7672
2.9	(−2) 2.9379	(−3) 9.9205	(−3) 3.4711	(−3)1.4202	(1) 2.0626	(1)5.0992	(2)1.0886	(2)2.1198
3.0	(−2) 2.6664	(−3) 9.1665	(−3) 3.1099	(−3)1.2237	(1) 2.2123	(1)5.7210	(2)1.2643	(2)2.5340
3.1	(−2) 2.3883	(−3) 8.4815	(−3) 2.8032	(−3)1.0610	(1) 2.3564	(1)6.3856	(2)1.4620	(2)3.0179
3.2	(−2) 2.1007	(−3) 7.8473	(−3) 2.5414	(−4)9.2596	(1) 2.4910	(1)7.0882	(2)1.6831	(2)3.5801
3.3	(−2) 1.8013	(−3) 7.2477	(−3) 2.3163	(−4)8.1356	(1) 2.6116	(1)7.8218	(2)1.9284	(2)4.2298
3.4	(−2) 1.4891	(−3) 6.6685	(−3) 2.1209	(−4)7.1975	(1) 2.7132	(1)8.5768	(2)2.1983	(2)4.9757
3.5	(−2) 1.1637	(−3) 6.0967	(−3) 1.9491	(−4)6.4117	(1) 2.7908	(1)9.3410	(2)2.4925	(2)5.8266
3.6	(−3) 8.2597	(−3) 5.5212	(−3) 1.7956	(−4)5.7506	(1) 2.8386	(2)1.0099	(2)2.8101	(2)6.7902
3.7	(−3) 4.7816	(−3) 4.9326	(−3) 1.6558	(−4)5.1910	(1) 2.8513	(2)1.0833	(2)3.1488	(2)7.8732
3.8	(−3)+1.2365	(−3) 4.3233	(−3) 1.5256	(−4)4.7135	(1) 2.8234	(2)1.1520	(2)3.5057	(2)9.0802
3.9	(−3)−2.3273	(−3) 3.6879	(−3) 1.4014	(−4)4.3017	(1) 2.7502	(2)1.2137	(2)3.8760	(3)1.0413
4.0	(−3)−5.8480	(−3) 3.0231	(−3) 1.2800	(−4)3.9416	(1) 2.6275	(2)1.2657	(2)4.2539	(3)1.1870
4.1	(−3)−9.2508	(−3) 2.3283	(−3) 1.1586	(−4)3.6211	(1) 2.4523	(2)1.3050	(2)4.6317	(3)1.3446
4.2	(−2)−1.2449	(−3) 1.6058	(−3) 1.0349	(−4)3.3295	(1) 2.2234	(2)1.3286	(2)4.9999	(3)1.5128
4.3	(−2)−1.5347	(−3) 0.8609	(−4) 9.0706	(−4)3.0577	(1) 1.9410	(2)1.3334	(2)5.3475	(3)1.6899
4.4	(−2)−1.7842	(−3)+0.1023	(−4) 7.7357	(−4)2.7975	(1) 1.6079	(2)1.3167	(2)5.6617	(3)1.8733
4.5	(−2)−1.9831	(−3)−0.6579	(−4) 6.3364	(−4)2.5418	(1) 1.2294	(2)1.2758	(2)5.9283	(3)2.0596
4.6	(−2)−2.1213	(−3)−1.4043	(−4) 4.8704	(−4)2.2847	(0) 8.1345	(2)1.2086	(2)6.1317	(3)2.2445
4.7	(−2)−2.1898	(−3)−2.1182	(−4) 3.3422	(−4)2.0210	(0)+3.7101	(2)1.1138	(2)6.2561	(3)2.4229
4.8	(−2)−2.1815	(−3)−2.7786	(−4) 1.7637	(−4)1.7468	(0)−0.8430	(1)9.9105	(2)6.2853	(3)2.5885
4.9	(−2)−2.0914	(−3)−3.3622	(−4)+0.1548	(−4)1.4595	(0)−5.3626	(1)8.4104	(2)6.2040	(3)2.7344
5.0	(−2)−1.9179	(−3)−3.8449	(−4)−1.4564	(−4)1.1577	(0)−9.6664	(1)6.6590	(2)5.9987	(3)2.8528

For interpolation, see **19.28.**

Table 19.2

x	$W(-1.0,x)$	$W(-0.9,x)$	$W(-0.8,x)$	$W(-0.7,x)$	$W(-0.6,x)$	$W(-0.5,x)$	$W(-0.4,x)$
0.0	0.73148	0.75416	0.77982	0.80879	0.84130	0.87718	0.91553
0.1	0.65958	0.68457	0.71267	0.74421	0.77940	0.81803	0.85912
0.2	0.58108	0.60881	0.63980	0.67441	0.71281	0.75477	0.79925
0.3	0.49671	0.52750	0.56175	0.59981	0.64187	0.68766	0.73610
0.4	0.40726	0.44133	0.47908	0.52089	0.56693	0.61696	0.66984
0.5	0.31359	0.35102	0.39240	0.43811	0.48837	0.54293	0.60064
0.6	0.21659	0.25734	0.30233	0.35200	0.40658	0.46584	0.52866
0.7	0.11723	0.16111	0.20958	0.26311	0.32198	0.38601	0.45409
0.8	+0.01657	+0.06324	0.11490	0.17206	0.23506	0.30379	0.37715
0.9	−0.08429	−0.03529	+0.01912	+0.07954	0.14637	0.21956	0.29811
1.0	−0.18412	−0.13342	−0.07684	−0.01369	+0.05650	0.13380	0.21727
1.1	−0.28164	−0.23002	−0.17198	−0.10679	−0.03384	+0.04704	0.13503
1.2	−0.37549	−0.32384	−0.26523	−0.19880	−0.12386	−0.04009	+0.05185
1.3	−0.46422	−0.41357	−0.35538	−0.28870	−0.21269	−0.12687	−0.03172
1.4	−0.54635	−0.49783	−0.44119	−0.37536	−0.29933	−0.21246	−0.11502
1.5	−0.62034	−0.57517	−0.52130	−0.45753	−0.38270	−0.29594	−0.19728
1.6	−0.68464	−0.64409	−0.59431	−0.53393	−0.46162	−0.37627	−0.27764
1.7	−0.73771	−0.70310	−0.65875	−0.60317	−0.53480	−0.45231	−0.35510
1.8	−0.77808	−0.75070	−0.71317	−0.66382	−0.60091	−0.52280	−0.42857
1.9	−0.80439	−0.78547	−0.75611	−0.71446	−0.65854	−0.58645	−0.49684
2.0	−0.81541	−0.80610	−0.78618	−0.75365	−0.70628	−0.64186	−0.55864
2.1	−0.81014	−0.81144	−0.80212	−0.78003	−0.74273	−0.68765	−0.61261
2.2	−0.78787	−0.80054	−0.80282	−0.79238	−0.76654	−0.72243	−0.65738
2.3	−0.74822	−0.77279	−0.78741	−0.78960	−0.77649	−0.74486	−0.69156
2.4	−0.69124	−0.72790	−0.75531	−0.77089	−0.77153	−0.75373	−0.71385
2.5	−0.61743	−0.66601	−0.70633	−0.73570	−0.75086	−0.74799	−0.72301
2.6	−0.52785	−0.58777	−0.64071	−0.68391	−0.71398	−0.72686	−0.71801
2.7	−0.42412	−0.49436	−0.55918	−0.61582	−0.66079	−0.68984	−0.69802
2.8	−0.30847	−0.38753	−0.46303	−0.53224	−0.59164	−0.63684	−0.66256
2.9	−0.18374	−0.26968	−0.35416	−0.43455	−0.50739	−0.56821	−0.61149
3.0	−0.05335	−0.14378	−0.23506	−0.32474	−0.40948	−0.48485	−0.54517
3.1	+0.07873	−0.01339	−0.10884	−0.20540	−0.29995	−0.38820	−0.46444
3.2	0.20811	+0.11741	+0.02083	−0.07973	−0.18146	−0.28034	−0.37075
3.3	0.33006	0.24412	0.14977	+0.04850	−0.05729	−0.16395	−0.26614
3.4	0.43974	0.36198	0.27340	0.17504	+0.06875	−0.04232	−0.15327
3.5	0.53233	0.46613	0.38695	0.29527	0.19236	+0.08071	−0.03541
3.6	0.60334	0.55184	0.48557	0.40440	0.30891	0.20083	+0.08365
3.7	0.64885	0.61476	0.56460	0.49761	0.41360	0.31342	0.19963
3.8	0.66575	0.65118	0.61986	0.57035	0.50168	0.41373	0.30797
3.9	0.65207	0.65834	0.64786	0.61858	0.56868	0.49706	0.40397
4.0	0.60721	0.63466	0.64616	0.63904	0.61072	0.55906	0.48303
4.1	0.53214	0.58002	0.61356	0.62958	0.62476	0.59598	0.54088
4.2	0.42952	0.49593	0.55042	0.58939	0.60892	0.60496	0.57391
4.3	0.30382	0.38565	0.45874	0.51923	0.56270	0.58437	0.57944
4.4	0.16115	0.25422	0.34234	0.42158	0.48725	0.53398	0.55599
4.5	+0.00918	+0.10831	0.20677	0.30072	0.38544	0.45522	0.50355
4.6	−0.14329	−0.04397	+0.05918	0.16266	0.26194	0.35129	0.42375
4.7	−0.28674	−0.19348	−0.09193	+0.01497	+0.12315	0.22716	0.31998
4.8	−0.41153	−0.33057	−0.23720	−0.13360	−0.02310	+0.08947	0.19740
4.9	−0.50861	−0.44572	−0.36694	−0.27352	−0.16782	−0.05374	+0.06277
5.0	−0.57025	−0.53023	−0.47182	−0.39516	−0.30146	−0.19341	−0.07580
	$\left[\begin{smallmatrix}(-3)4\\5\end{smallmatrix}\right]$	$\left[\begin{smallmatrix}(-3)4\\5\end{smallmatrix}\right]$	$\left[\begin{smallmatrix}(-3)4\\5\end{smallmatrix}\right]$	$\left[\begin{smallmatrix}(-3)4\\5\end{smallmatrix}\right]$	$\left[\begin{smallmatrix}(-3)4\\5\end{smallmatrix}\right]$	$\left[\begin{smallmatrix}(-3)4\\5\end{smallmatrix}\right]$	$\left[\begin{smallmatrix}(-3)4\\5\end{smallmatrix}\right]$

Table 19.2

x	$W(-1.0,-x)$	$W(-0.9,-x)$	$W(-0.8,-x)$	$W(-0.7,-x)$	$W(-0.6,-x)$	$W(-0.5,-x)$	$W(-0.4,-x)$
0.0	0.73148	0.75416	0.77982	0.80879	0.84130	0.87718	0.91553
0.1	0.79607	0.81697	0.84073	0.86771	0.89814	0.93193	0.96827
0.2	0.85267	0.87241	0.89490	0.92053	0.94958	0.98201	1.01711
0.3	0.90067	0.91990	0.94182	0.96682	0.99522	1.02707	1.06178
0.4	0.93946	0.95892	0.98099	1.00612	1.03467	1.06677	1.10197
0.5	0.96849	0.98892	1.01192	1.03797	1.06749	1.10070	1.13729
0.6	0.98722	1.00940	1.03413	1.06191	1.09323	1.12843	1.16736
0.7	0.99521	1.01990	1.04713	1.07745	1.11143	1.14951	1.19170
0.8	0.99202	1.01997	1.05048	1.08414	1.12160	1.16343	1.20981
0.9	0.97734	1.00923	1.04374	1.08151	1.12325	1.16966	1.22114
1.0	0.95092	0.98738	1.02655	1.06912	1.11589	1.16769	1.22511
1.1	0.91262	0.95418	0.99859	1.04657	1.09904	1.15695	1.22112
1.2	0.86244	0.90952	0.95962	1.01355	1.07228	1.13693	1.20855
1.3	0.80055	0.85341	0.90954	0.96978	1.03523	1.10714	1.18680
1.4	0.72729	0.78603	0.84835	0.91515	0.98760	1.06714	1.15529
1.5	0.64322	0.70774	0.77623	0.84963	0.92923	1.01659	1.11351
1.6	0.54911	0.61912	0.69355	0.77341	0.86006	0.95525	1.06102
1.7	0.44603	0.52099	0.60091	0.68684	0.78025	0.88304	0.99750
1.8	0.33528	0.41443	0.49914	0.59053	0.69014	0.80004	0.92281
1.9	0.21849	0.30081	0.38936	0.48532	0.59032	0.70659	0.83697
2.0	+0.09757	0.18179	0.27298	0.37236	0.48166	0.60326	0.74025
2.1	−0.02528	+0.05934	0.15171	0.25309	0.36531	0.49090	0.63319
2.2	−0.14758	−0.06427	+0.02758	0.12930	0.24278	0.37070	0.51665
2.3	−0.26660	−0.18651	−0.09709	+0.00305	+0.11588	0.24419	0.39182
2.4	−0.37941	−0.30459	−0.21967	−0.12323	−0.01322	+0.11327	0.26028
2.5	−0.48297	−0.41552	−0.33731	−0.24685	−0.14203	−0.01983	+0.12398
2.6	−0.57415	−0.51623	−0.44698	−0.36487	−0.26774	−0.15248	−0.01472
2.7	−0.64990	−0.60356	−0.54551	−0.47416	−0.38730	−0.28178	−0.15309
2.8	−0.70733	−0.67449	−0.62975	−0.57149	−0.49748	−0.40451	−0.28802
2.9	−0.74387	−0.72615	−0.69663	−0.65363	−0.59492	−0.51729	−0.41615
3.0	−0.75737	−0.75605	−0.74331	−0.71748	−0.67629	−0.61660	−0.53384
3.1	−0.74633	−0.76219	−0.76738	−0.76019	−0.73841	−0.69897	−0.63739
3.2	−0.70996	−0.74323	−0.76692	−0.77937	−0.77841	−0.76108	−0.72310
3.3	−0.64841	−0.69863	−0.74077	−0.77320	−0.79386	−0.79994	−0.78743
3.4	−0.56281	−0.62881	−0.68862	−0.74065	−0.78300	−0.81309	−0.82721
3.5	−0.45542	−0.53525	−0.61114	−0.68160	−0.74490	−0.79874	−0.83985
3.6	−0.32961	−0.42059	−0.51016	−0.59701	−0.67961	−0.75603	−0.82349
3.7	−0.18992	−0.28860	−0.38867	−0.48899	−0.58833	−0.68515	−0.77725
3.8	−0.04191	−0.14423	−0.25086	−0.36092	−0.47349	−0.58750	−0.70141
3.9	+0.10799	+0.00657	−0.10208	−0.21739	−0.33883	−0.46582	−0.59756
4.0	0.25266	0.15702	+0.05134	−0.06416	−0.18934	−0.32421	−0.46872
4.1	0.38471	0.29976	0.20225	+0.09203	−0.03124	−0.16811	−0.31938
4.2	0.49679	0.42722	0.34303	0.24366	+0.12831	−0.00420	−0.15545
4.3	0.58208	0.53205	0.46597	0.38285	0.28140	+0.15987	+0.01587
4.4	0.63477	0.60759	0.56372	0.50171	0.41981	0.31572	0.18634
4.5	0.65055	0.64841	0.62979	0.59285	0.53543	0.45473	0.34702
4.6	0.62708	0.65075	0.65910	0.64997	0.62083	0.56851	0.48877
4.7	0.56440	0.61301	0.64846	0.66833	0.66982	0.64950	0.60280
4.8	0.46513	0.53614	0.59705	0.64531	0.67800	0.69154	0.68125
4.9	0.33464	0.42379	0.50672	0.58085	0.64328	0.69050	0.71794
5.0	0.18091	0.28240	0.38215	0.47771	0.56635	0.64481	0.70889
	$\begin{bmatrix} (-3)5 \\ 5 \end{bmatrix}$	$\begin{bmatrix} (-3)5 \\ 5 \end{bmatrix}$	$\begin{bmatrix} (-3)5 \\ 5 \end{bmatrix}$	$\begin{bmatrix} (-3)5 \\ 5 \end{bmatrix}$	$\begin{bmatrix} (-3)5 \\ 5 \end{bmatrix}$	$\begin{bmatrix} (-3)6 \\ 5 \end{bmatrix}$	$\begin{bmatrix} (-3)6 \\ 5 \end{bmatrix}$

Table 19.2

x	$W(-0.3,x)$	$W(-0.2,x)$	$W(-0.1,x)$	$W(0,x)$	$W(0.1,x)$	$W(0.2,x)$	$W(0.3,x)$
0.0	0.95411	0.98880	1.01364	1.02277	1.01364	0.98880	0.95411
0.1	0.90030	0.93725	0.96381	0.97388	0.96480	0.93920	0.90311
0.2	0.84377	0.88381	0.91299	0.92496	0.91691	0.89145	0.85480
0.3	0.78461	0.82851	0.86116	0.87595	0.86984	0.84540	0.80896
0.4	0.72293	0.77137	0.80828	0.82673	0.82344	0.80084	0.76536
0.5	0.65878	0.71237	0.75426	0.77719	0.77753	0.75757	0.72375
0.6	0.59225	0.65150	0.69902	0.72716	0.73192	0.71533	0.68386
0.7	0.52341	0.58875	0.64245	0.67647	0.68637	0.67388	0.64540
0.8	0.45236	0.52410	0.58445	0.62496	0.64067	0.63296	0.60809
0.9	0.37924	0.45756	0.52493	0.57244	0.59459	0.59228	0.57163
1.0	0.30421	0.38918	0.46383	0.51877	0.54790	0.55160	0.53573
1.1	0.22751	0.31906	0.40111	0.46381	0.50038	0.51063	0.50010
1.2	0.14946	0.24734	0.33677	0.40744	0.45186	0.46915	0.46446
1.3	+0.07042	0.17425	0.27090	0.34961	0.40217	0.42691	0.42854
1.4	−0.00912	0.10007	0.20361	0.29032	0.35118	0.38374	0.39209
1.5	−0.08857	+0.02522	0.13514	0.22960	0.29883	0.33945	0.35491
1.6	−0.16725	−0.04982	+0.06577	0.16760	0.24510	0.29393	0.31679
1.7	−0.24435	−0.12443	−0.00407	0.10454	0.19006	0.24713	0.27761
1.8	−0.31894	−0.19788	−0.07387	+0.04073	0.13384	0.19904	0.23725
1.9	−0.38999	−0.26933	−0.14299	−0.02340	0.07667	0.14975	0.19569
2.0	−0.45633	−0.33779	−0.21066	−0.08731	+0.01891	0.09941	0.15296
2.1	−0.51674	−0.40219	−0.27600	−0.15034	−0.03902	+0.04828	0.10917
2.2	−0.56989	−0.46135	−0.33802	−0.21170	−0.09655	−0.00327	0.06450
2.3	−0.61444	−0.51400	−0.39560	−0.27048	−0.15300	−0.05478	+0.01926
2.4	−0.64903	−0.55882	−0.44755	−0.32569	−0.20756	−0.10567	−0.02617
2.5	−0.67233	−0.59448	−0.49261	−0.37619	−0.25934	−0.15523	−0.07129
2.6	−0.68311	−0.61966	−0.52947	−0.42082	−0.30731	−0.20267	−0.11551
2.7	−0.68033	−0.63315	−0.55686	−0.45833	−0.35040	−0.24709	−0.15811
2.8	−0.66313	−0.63385	−0.57356	−0.48749	−0.38745	−0.28749	−0.19829
2.9	−0.63097	−0.62088	−0.57846	−0.50710	−0.41729	−0.32283	−0.23518
3.0	−0.58369	−0.59365	−0.57063	−0.51607	−0.43878	−0.35203	−0.26783
3.1	−0.52157	−0.55190	−0.54943	−0.51344	−0.45085	−0.37401	−0.29526
3.2	−0.44541	−0.49584	−0.51451	−0.49851	−0.45256	−0.38777	−0.31648
3.3	−0.35655	−0.42613	−0.46594	−0.47084	−0.44315	−0.39239	−0.33055
3.4	−0.25697	−0.34402	−0.40427	−0.43039	−0.42215	−0.38713	−0.33663
3.5	−0.14924	−0.25134	−0.33055	−0.37754	−0.38941	−0.37148	−0.33401
3.6	−0.03654	−0.15050	−0.24643	−0.31318	−0.34517	−0.34523	−0.32218
3.7	+0.07742	−0.04453	−0.15413	−0.23871	−0.29013	−0.30852	−0.30091
3.8	0.18846	+0.06302	−0.05645	−0.15612	−0.22549	−0.26190	−0.27027
3.9	0.29213	0.16814	+0.04330	−0.06794	−0.15299	−0.20639	−0.23072
4.0	0.38382	0.26651	0.14132	+0.02278	−0.07486	−0.14349	−0.18313
4.1	0.45904	0.35370	0.23354	0.11257	+0.00615	−0.07518	−0.12880
4.2	0.51364	0.42535	0.31572	0.19762	0.08689	−0.00389	−0.06948
4.3	0.54413	0.47744	0.38368	0.27395	0.16386	+0.06754	−0.00733
4.4	0.54793	0.50658	0.43357	0.33764	0.23342	0.13597	+0.05511
4.5	0.52370	0.51029	0.46212	0.38503	0.29194	0.19809	0.11504
4.6	0.47151	0.48726	0.46690	0.41300	0.33601	0.25059	0.16948
4.7	0.39312	0.43762	0.44663	0.41921	0.36270	0.29037	0.21549
4.8	0.29197	0.36308	0.40138	0.40237	0.36981	0.31476	0.25027
4.9	0.17327	0.26703	0.33274	0.36248	0.35608	0.32171	0.27144
5.0	0.04376	0.15455	0.24393	0.30095	0.32145	0.31009	0.27719
	$\left[\begin{matrix}(-3)4\\5\end{matrix}\right]$	$\left[\begin{matrix}(-3)3\\5\end{matrix}\right]$	$\left[\begin{matrix}(-3)3\\5\end{matrix}\right]$	$\left[\begin{matrix}(-3)3\\5\end{matrix}\right]$	$\left[\begin{matrix}(-3)3\\5\end{matrix}\right]$	$\left[\begin{matrix}(-3)2\\5\end{matrix}\right]$	$\left[\begin{matrix}(-3)2\\5\end{matrix}\right]$

Table 19.2

x	$W(-0.3,-x)$	$W(-0.2,-x)$	$W(-0.1,-x)$	$W(0,-x)$	$W(0.1,-x)$	$W(0.2,-x)$	$W(0.3,-x)$
0.0	0.95411	0.98880	1.01364	1.02277	1.01364	0.98880	0.95411
0.1	1.00506	1.03835	1.06245	1.07165	1.06348	1.04037	1.00797
0.2	1.05296	1.08581	1.11016	1.12050	1.11435	1.09399	1.06483
0.3	1.09759	1.13097	1.15665	1.16924	1.16622	1.14968	1.12477
0.4	1.13868	1.17362	1.20172	1.21771	1.21899	1.20741	1.18782
0.5	1.17589	1.21344	1.24510	1.26568	1.27248	1.26706	1.25396
0.6	1.20884	1.25007	1.28645	1.31285	1.32644	1.32845	1.32307
0.7	1.23706	1.28307	1.32534	1.35884	1.38053	1.39129	1.39494
0.8	1.26006	1.31193	1.36129	1.40315	1.43429	1.45520	1.46928
0.9	1.27725	1.33606	1.39368	1.44521	1.48719	1.51968	1.54567
1.0	1.28802	1.35480	1.42185	1.48433	1.53055	1.58412	1.62356
1.1	1.29171	1.36744	1.44504	1.51974	1.58760	1.64775	1.70224
1.2	1.28761	1.37321	1.46241	1.55054	1.63341	1.70967	1.78087
1.3	1.27501	1.37129	1.47304	1.57575	1.67498	1.76885	1.85841
1.4	1.25320	1.36083	1.47598	1.59429	1.71113	1.82408	1.93366
1.5	1.22150	1.34098	1.47020	1.60502	1.74059	1.87401	2.00522
1.6	1.17926	1.31091	1.45469	1.60672	1.76201	1.91713	2.07150
1.7	1.12596	1.26983	1.42841	1.59813	1.77390	1.95181	2.13072
1.8	1.06115	1.21705	1.39039	1.57800	1.77474	1.97628	2.18093
1.9	0.98458	1.15200	1.33973	1.54509	1.76299	1.98870	2.22000
2.0	0.89620	1.07426	1.27565	1.49825	1.73709	1.98714	2.24569
2.1	0.79618	0.98365	1.19757	1.43644	1.69557	1.96968	2.25565
2.2	0.68503	0.88026	1.10510	1.35882	1.63706	1.93446	2.24752
2.3	0.56357	0.76448	0.99819	1.26478	1.56041	1.87972	2.21894
2.4	0.43300	0.63710	0.87711	1.15405	1.46471	1.80390	2.16770
2.5	0.29492	0.49932	0.74256	1.02673	1.34942	1.70575	2.09177
2.6	0.15140	0.35277	0.59571	0.88342	1.21444	1.58440	1.98946
2.7	+0.00489	0.19959	0.43825	0.72523	1.06021	1.43949	1.85956
2.8	−0.14168	+0.04242	0.27241	0.55388	0.88776	1.27129	1.70140
2.9	−0.28503	−0.11563	+0.10100	0.37173	0.69887	1.08078	1.51507
3.0	−0.42150	−0.27098	−0.07258	+0.18182	0.49606	0.86979	1.30151
3.1	−0.54722	−0.41967	−0.24442	−0.01213	0.28264	0.64105	1.06267
3.2	−0.65815	−0.55742	−0.41011	−0.20574	+0.06279	0.39827	0.80159
3.3	−0.75027	−0.67978	−0.56487	−0.39404	−0.15855	+0.14618	0.52249
3.4	−0.81974	−0.78229	−0.70368	−0.57158	−0.37567	−0.10952	+0.23083
3.5	−0.86311	−0.86067	−0.82147	−0.73259	−0.58228	−0.36221	−0.06670
3.6	−0.87754	−0.91101	−0.91331	−0.87118	−0.77162	−0.60449	−0.36232
3.7	−0.86098	−0.93010	−0.97470	−0.98158	−0.93674	−0.82836	−0.64721
3.8	−0.81248	−0.91559	−1.00185	−1.05844	−1.07077	−1.02554	−0.91187
3.9	−0.73233	−0.86631	−0.99193	−1.09719	−1.16728	−1.18779	−1.14634
4.0	−0.62227	−0.78249	−0.94343	−1.09434	−1.22069	−1.30732	−1.34070
4.1	−0.48559	−0.66595	−0.85640	−1.04786	−1.22662	−1.37730	−1.48554
4.2	−0.32717	−0.52024	−0.73270	−0.95753	−1.18240	−1.39231	−1.57256
4.3	−0.15346	−0.35070	−0.57611	−0.82515	−1.08743	−1.34891	−1.59514
4.4	+0.02771	−0.16437	−0.39249	−0.65483	−0.94350	−1.24610	−1.54901
4.5	0.20739	+0.03014	−0.18962	−0.45301	−0.75508	−1.08573	−1.43285
4.6	0.37594	0.22299	+0.02291	−0.22843	−0.52942	−0.87285	−1.24877
4.7	0.52351	0.40359	0.23414	+0.00810	−0.27649	−0.61582	−1.00271
4.8	0.64069	0.56113	0.43218	0.24408	−0.00874	−0.32626	−0.70462
4.9	0.71919	0.68534	0.60494	0.46598	+0.25940	−0.01876	−0.36835
5.0	0.75259	0.76721	0.74090	0.65996	0.51219	+0.28970	−0.01132
	$\begin{bmatrix}(-3)6\\5\end{bmatrix}$	$\begin{bmatrix}(-3)5\\5\end{bmatrix}$	$\begin{bmatrix}(-3)5\\5\end{bmatrix}$	$\begin{bmatrix}(-3)5\\5\end{bmatrix}$	$\begin{bmatrix}(-3)6\\5\end{bmatrix}$	$\begin{bmatrix}(-3)7\\5\end{bmatrix}$	$\begin{bmatrix}(-3)9\\5\end{bmatrix}$

Table 19.2

x	$W(0.4,x)$	$W(0.5,x)$	$W(0.6,x)$	$W(0.7,x)$	$W(0.8,x)$	$W(0.9,x)$	$W(1.0,x)$
0.0	0.91553	0.87718	0.84130	0.80879	0.77982	0.75416	0.73148
0.1	0.86271	0.82232	0.78433	0.74973	0.71874	0.69116	0.66667
0.2	0.81331	0.77155	0.73205	0.69590	0.66339	0.63436	0.60852
0.3	0.76709	0.72456	0.68408	0.64687	0.61328	0.58321	0.55639
0.4	0.72376	0.68104	0.64007	0.60222	0.56794	0.53718	0.50970
0.5	0.68304	0.64064	0.59964	0.56155	0.52692	0.49578	0.46791
0.6	0.64462	0.60305	0.56244	0.52446	0.48979	0.45853	0.43051
0.7	0.60820	0.56793	0.52810	0.49058	0.45614	0.42499	0.39703
0.8	0.57347	0.53495	0.49629	0.45952	0.42558	0.39476	0.36704
0.9	0.54011	0.50380	0.46666	0.43095	0.39774	0.36745	0.34013
1.0	0.50782	0.47414	0.43889	0.40452	0.37228	0.34271	0.31594
1.1	0.47630	0.44567	0.41266	0.37992	0.34888	0.32020	0.29412
1.2	0.44523	0.41808	0.38765	0.35682	0.32720	0.29960	0.27435
1.3	0.41435	0.39108	0.36358	0.33494	0.30697	0.28063	0.25634
1.4	0.38338	0.36438	0.34015	0.31399	0.28790	0.26299	0.23981
1.5	0.35206	0.33771	0.31709	0.29370	0.26973	0.24643	0.22451
1.6	0.32018	0.31084	0.29416	0.27382	0.25219	0.23071	0.21019
1.7	0.28752	0.28354	0.27111	0.25410	0.23506	0.21559	0.19662
1.8	0.25395	0.25561	0.24773	0.23433	0.21812	0.20085	0.18361
1.9	0.21934	0.22689	0.22384	0.21430	0.20115	0.18629	0.17094
2.0	0.18363	0.19726	0.19927	0.19384	0.18398	0.17173	0.15845
2.1	0.14682	0.16665	0.17390	0.17280	0.16644	0.15700	0.14595
2.2	0.10899	0.13504	0.14767	0.15107	0.14841	0.14195	0.13331
2.3	0.07029	0.10248	0.12054	0.12857	0.12976	0.12647	0.12038
2.4	+0.03094	0.06908	0.09255	0.10528	0.11045	0.11045	0.10707
2.5	−0.00872	0.03504	0.06378	0.08121	0.09043	0.09385	0.09330
2.6	−0.04827	+0.00063	0.03440	0.05645	0.06972	0.07662	0.07900
2.7	−0.08719	−0.03378	+0.00466	0.03113	0.04840	0.05879	0.06416
2.8	−0.12486	−0.06773	−0.02513	+0.00547	0.02659	0.04042	0.04879
2.9	−0.16058	−0.10069	−0.05457	−0.02025	+0.00447	0.02163	0.03296
3.0	−0.19356	−0.13202	−0.08319	−0.04569	−0.01769	+0.00259	0.01677
3.1	−0.22295	−0.16105	−0.11043	−0.07041	−0.03960	−0.01649	+0.00038
3.2	−0.24788	−0.18700	−0.13568	−0.09392	−0.06087	−0.03531	−0.01602
3.3	−0.26746	−0.20910	−0.15826	−0.11569	−0.08106	−0.05355	−0.03216
3.4	−0.28083	−0.22656	−0.17749	−0.13511	−0.09969	−0.07080	−0.04774
3.5	−0.28722	−0.23861	−0.19265	−0.15158	−0.11623	−0.08664	−0.06242
3.6	−0.28598	−0.24455	−0.20307	−0.16446	−0.13014	−0.10061	−0.07581
3.7	−0.27664	−0.24381	−0.20814	−0.17317	−0.14088	−0.11222	−0.08750
3.8	−0.25895	−0.23596	−0.20735	−0.17718	−0.14793	−0.12101	−0.09707
3.9	−0.23299	−0.22079	−0.20033	−0.17604	−0.15084	−0.12652	−0.10411
4.0	−0.19913	−0.19835	−0.18692	−0.16946	−0.14922	−0.12836	−0.10824
4.1	−0.15813	−0.16901	−0.16717	−0.15730	−0.14284	−0.12624	−0.10912
4.2	−0.11115	−0.13343	−0.14143	−0.13965	−0.13162	−0.11996	−0.10653
4.3	−0.05975	−0.09266	−0.11032	−0.11684	−0.11566	−0.10948	−0.10030
4.4	−0.00585	−0.04811	−0.07481	−0.08947	−0.09531	−0.09494	−0.09046
4.5	+0.04828	−0.00149	−0.03614	−0.05843	−0.07112	−0.07669	−0.07716
4.6	0.10016	+0.04518	+0.00411	−0.02485	−0.04392	−0.05525	−0.06075
4.7	0.14714	0.08968	0.04416	+0.00985	−0.01477	−0.03141	−0.04174
4.8	0.18659	0.12967	0.08203	0.04406	+0.01506	−0.00614	−0.02086
4.9	0.21607	0.16286	0.11567	0.07604	0.04414	+0.01943	+0.00100
5.0	0.23350	0.18712	0.14307	0.10399	0.07092	0.04399	0.02281
	$\begin{bmatrix}(-3)2\\5\end{bmatrix}$	$\begin{bmatrix}(-3)1\\5\end{bmatrix}$	$\begin{bmatrix}(-4)8\\5\end{bmatrix}$	$\begin{bmatrix}(-4)7\\4\end{bmatrix}$	$\begin{bmatrix}(-4)7\\4\end{bmatrix}$	$\begin{bmatrix}(-4)8\\4\end{bmatrix}$	$\begin{bmatrix}(-4)8\\4\end{bmatrix}$

Table 19.2

x	$W(0.4,-x)$	$W(0.5,-x)$	$W(0.6,-x)$	$W(0.7,-x)$	$W(0.8,-x)$	$W(0.9,-x)$	$W(1.0,-x)$
0.0	0.91553	0.87718	0.84130	0.80879	0.77982	0.75416	0.73148
0.1	0.97201	0.93642	0.90331	0.87352	0.84714	0.82396	0.80361
0.2	1.03235	1.00031	0.97072	0.94433	0.92122	0.90115	0.88375
0.3	1.09671	1.06911	1.04386	1.02166	1.00258	0.98636	0.97265
0.4	1.16520	1.14300	1.12302	1.10591	1.09173	1.08022	1.07106
0.5	1.23789	1.22215	1.20846	1.19746	1.18917	1.18338	1.17975
0.6	1.31475	1.30664	1.30040	1.29663	1.29538	1.29644	1.29949
0.7	1.39567	1.39648	1.39896	1.40371	1.41079	1.42000	1.43106
0.8	1.48046	1.49158	1.50419	1.51888	1.53574	1.55459	1.57519
0.9	1.56879	1.59174	1.61602	1.64225	1.67051	1.70068	1.73254
1.0	1.6602	1.6966	1.7343	1.7738	1.8153	1.8586	1.9037
1.1	1.7541	1.8057	1.8586	1.9133	1.9700	2.0286	2.0891
1.2	1.8497	1.9184	1.9884	2.0603	2.1345	2.2107	2.2891
1.3	1.9460	2.0337	2.1230	2.2144	2.3083	2.4048	2.5037
1.4	2.0418	2.1506	2.2613	2.3746	2.4909	2.6102	2.7327
1.5	2.1358	2.2677	2.4020	2.5397	2.6811	2.8264	2.9756
1.6	2.2263	2.3833	2.5437	2.7083	2.8777	3.0520	3.2316
1.7	2.3115	2.4956	2.6843	2.8785	3.0788	3.2856	3.4991
1.8	2.3891	2.6023	2.8216	3.0480	3.2823	3.5249	3.7762
1.9	2.4570	2.7009	2.9529	3.2141	3.4854	3.7674	4.0605
2.0	2.5125	2.7886	3.0752	3.3737	3.6849	4.0097	4.3487
2.1	2.5529	2.8623	3.1853	3.5231	3.8770	4.2479	4.6368
2.2	2.5754	2.9188	3.2793	3.6583	4.0573	4.4775	4.9201
2.3	2.5770	2.9546	3.3532	3.7748	4.2209	4.6931	5.1930
2.4	2.5548	2.9660	3.4030	3.8678	4.3624	4.8889	5.4490
2.5	2.5061	2.9496	3.4241	3.9321	4.4760	5.0582	5.6811
2.6	2.4283	2.9018	3.4124	3.9626	4.5555	5.1940	5.8811
2.7	2.3192	2.8196	3.3634	3.9538	4.5944	5.2887	6.0405
2.8	2.1772	2.7001	3.2734	3.9007	4.5863	5.3346	6.1502
2.9	2.0013	2.5413	3.1389	3.7984	4.5251	5.3240	6.2008
3.0	1.7914	2.3419	2.9573	3.6430	4.4050	5.2495	6.1832
3.1	1.5484	2.1015	2.7270	3.4312	4.2211	5.1041	6.0883
3.2	1.2746	1.8213	2.4478	3.1612	3.9697	4.8822	5.9081
3.3	0.9733	1.5038	2.1206	2.8324	3.6486	4.5794	5.6359
3.4	0.6496	1.1529	1.7487	2.4466	3.2576	4.1934	5.2669
3.5	+0.3098	0.7746	1.3369	2.0074	2.7987	3.7241	4.7985
3.6	−0.0381	+0.3767	0.8923	1.5210	2.2767	3.1746	4.2315
3.7	−0.3848	−0.0314	+0.4244	0.9962	1.6994	2.5511	3.5700
3.8	−0.7198	−0.4385	−0.0553	+0.4445	1.0779	1.8636	2.8225
3.9	−1.0317	−0.8319	−0.5332	−0.1199	+0.4263	1.1259	2.0016
4.0	−1.3084	−1.1977	−0.9940	−0.6804	−0.2378	+0.3558	1.1251
4.1	−1.5382	−1.5216	−1.4209	−1.2184	−0.8941	−0.4249	+0.2152
4.2	−1.7095	−1.7893	−1.7966	−1.7136	−1.5199	−1.1915	−0.7013
4.3	−1.8124	−1.9871	−2.1039	−2.1453	−2.0907	−1.9160	−1.5936
4.4	−1.8391	−2.1032	−2.3268	−2.4930	−2.5817	−2.5692	−2.4280
4.5	−1.7844	−2.1283	−2.4513	−2.7376	−2.9685	−3.1213	−3.1692
4.6	−1.6469	−2.0567	−2.4668	−2.8632	−3.2291	−3.5437	−3.7818
4.7	−1.4292	−1.8870	−2.3670	−2.8579	−3.3452	−3.8110	−4.2326
4.8	−1.1387	−1.6231	−2.1513	−2.7153	−3.3040	−3.9027	−4.4924
4.9	−0.7876	−1.2742	−1.8252	−2.4359	−3.0995	−3.8054	−4.5392
5.0	−0.3927	−0.8557	−1.4010	−2.0281	−2.7346	−3.5149	−4.3599
	$\begin{bmatrix}(-2)1\\5\end{bmatrix}$	$\begin{bmatrix}(-2)1\\5\end{bmatrix}$	$\begin{bmatrix}(-2)1\\5\end{bmatrix}$	$\begin{bmatrix}(-2)2\\5\end{bmatrix}$	$\begin{bmatrix}(-2)2\\5\end{bmatrix}$	$\begin{bmatrix}(-2)2\\5\end{bmatrix}$	$\begin{bmatrix}(-2)3\\5\end{bmatrix}$

Table 19.3   **AUXILIARY FUNCTIONS**

The functions ϑ_1, ϑ_2, ϑ_3 of **19.10** and **19.23** are needed in Darwin's expansion and also the function τ of **19.7** and **19.20**.

ξ	ϑ_1	ϑ_3	τ	ξ	ϑ_1	ϑ_2	τ
0.0	0.00000	0.39270	-0.70270	5.0	6.9519	5.5506	4.1079
0.1	0.05008	0.34278	-0.64181	5.1	7.2093	5.7981	4.2291
0.2	0.10066	0.29337	-0.57855	5.2	7.4716	6.0507	4.3511
0.3	0.15222	0.24498	-0.51304	5.3	7.7388	6.3084	4.4738
0.4	0.20521	0.19817	-0.44540	5.4	8.0109	6.5712	4.5972
0.5	0.26006	0.15355	-0.37574	5.5	8.2880	6.8391	4.7213
0.6	0.31713	0.11182	-0.30415	5.6	8.5700	7.1120	4.8461
0.7	0.37678	0.07387	-0.23071	5.7	8.8569	7.3901	4.9716
0.8	0.43929	0.04088	-0.15549	5.8	9.1487	7.6732	5.0977
0.9	0.50492	0.01468	-0.07857	5.9	9.4454	7.9614	5.2246

ξ	ϑ_1	ϑ_2	τ	ξ	ϑ_1	ϑ_2	τ
1.0	0.57390	0.00000	0.00000	6.0	9.7471	8.2546	5.3521
1.1	0.64640	0.01513	0.08015	6.1	10.0537	8.5530	5.4803
1.2	0.72261	0.04341	0.16185	6.2	10.3652	8.8564	5.6092
1.3	0.80265	0.08086	0.24502	6.3	10.6817	9.1649	5.7387
1.4	0.88666	0.12617	0.32964	6.4	11.0031	9.4784	5.8688
1.5	0.97473	0.17866	0.41566	6.5	11.3295	9.7970	5.9996
1.6	1.06696	0.23786	0.50304	6.6	11.6608	10.1207	6.1310
1.7	1.16344	0.30347	0.59175	6.7	11.9970	10.4494	6.2631
1.8	1.26422	0.37527	0.68175	6.8	12.3382	10.7832	6.3958
1.9	1.36937	0.45309	0.77300	6.9	12.6843	11.1220	6.5290
2.0	1.47894	0.53679	0.86549	7.0	13.0354	11.4659	6.6629
2.1	1.59299	0.62626	0.95917	7.1	13.3914	11.8148	6.7974
2.2	1.71155	0.72142	1.05403	7.2	13.7524	12.1688	6.9325
2.3	1.83466	0.82220	1.15004	7.3	14.1183	12.5278	7.0682
2.4	1.96236	0.92853	1.24716	7.4	14.4892	12.8919	7.2045
2.5	2.09467	1.04036	1.34539	7.5	14.8651	13.2610	7.3414
2.6	2.23163	1.15764	1.44470	7.6	15.2459	13.6352	7.4789
2.7	2.37325	1.28034	1.54506	7.7	15.6316	14.0144	7.6169
2.8	2.51956	1.40843	1.64646	7.8	16.0223	14.3987	7.7555
2.9	2.67058	1.54187	1.74888	7.9	16.4180	14.7880	7.8947
3.0	2.82632	1.68063	1.85229	8.0	16.8186	15.1823	8.0344
3.1	2.98681	1.82470	1.95669	8.1	17.2242	15.5817	8.1747
3.2	3.15205	1.97406	2.06206	8.2	17.6348	15.9861	8.3155
3.3	3.32207	2.12867	2.16837	8.3	18.0503	16.3956	8.4569
3.4	3.49688	2.28853	2.27562	8.4	18.4708	16.8101	8.5989
3.5	3.67648	2.45363	2.38378	8.5	18.8962	17.2296	8.7413
3.6	3.86089	2.62394	2.49285	8.6	19.3266	17.6542	8.8844
3.7	4.05011	2.79946	2.60281	8.7	19.7620	18.0838	9.0279
3.8	4.24416	2.98017	2.71365	8.8	20.2024	18.5184	9.1720
3.9	4.44305	3.16606	2.82536	8.9	20.6477	18.9581	9.3166
4.0	4.64678	3.35712	2.93791	9.0	21.0980	19.4028	9.4617
4.1	4.85537	3.55335	3.05131	9.1	21.5532	19.8525	9.6074
4.2	5.06880	3.75474	3.16554	9.2	22.0135	20.3073	9.7535
4.3	5.28711	3.96127	3.28058	9.3	22.4787	20.7671	9.9002
4.4	5.51028	4.17295	3.39643	9.4	22.9488	21.2319	10.0474
4.5	5.73833	4.38976	3.51308	9.5	23.4240	21.7017	10.1951
4.6	5.97126	4.61169	3.63051	9.6	23.9041	22.1766	10.3433
4.7	6.20908	4.83875	3.74872	9.7	24.3892	22.6565	10.4920
4.8	6.45178	5.07093	3.86770	9.8	24.8792	23.1414	10.6411
4.9	6.69938	5.30822	3.98743	9.9	25.3742	23.6314	10.7908
5.0	6.95188	5.55062	4.10792	10.0	25.8742	24.1264	10.9410
	$\begin{bmatrix}(-4)6\\4\end{bmatrix}$	$\begin{bmatrix}(-3)2\\6\end{bmatrix}$	$\begin{bmatrix}(-4)3\\3\end{bmatrix}$		$\begin{bmatrix}(-4)6\\3\end{bmatrix}$	$\begin{bmatrix}(-4)7\\3\end{bmatrix}$	$\begin{bmatrix}(-4)1\\3\end{bmatrix}$

When interpolating for ϑ_2 and ϑ_3 for ξ near unity, it is better to interpolate for τ and then use

$$\vartheta_2 = \tfrac{2}{3}\tau^{3/2} \text{ or } \vartheta_3 = \tfrac{2}{3}(-\tau)^{3/2}.$$

20. Mathieu Functions

Gertrude Blanch [1]

Contents

Even Solutions

$$a_r,\ ce_r(0, q),\ ce_r\left(\frac{\pi}{2}, q\right),\ ce'_r\left(\frac{\pi}{2}, q\right),\ (4q)^{\frac{r}{2}} g_{e,r}(q),\ (4q)^r f_{e,r}(q)$$

Odd Solutions

$$b_r,\ se'_r(0, q),\ se_r\left(\frac{\pi}{2}, q\right),\ se'_r\left(\frac{\pi}{2}, q\right),\ (4q)^{\frac{r}{2}} g_{o,r}(q),\ (4q)^r f_{o,r}(q)$$

$$q = 0(5)25, \quad 8\text{D or S}$$

$$a_r + 2q - (4r+2)\sqrt{q},\ b_r + 2q - (4r-2)\sqrt{q}$$

$$q^{-\frac{1}{2}} = .16(-.04)0, \quad 8\text{D}$$

$$r = 0,\ 1,\ 2,\ 5,\ 10,\ 15$$

$$q = 5,\ 25;\ r = 0,\ 1,\ 2,\ 5,\ 10,\ 15, \quad 9\text{D}$$

---

[1] Aeronautical Research Laboratories, Wright-Patterson Air Force Base, Ohio.

20. Mathieu Functions

Mathematical Properties

20.1. Mathieu's Equation

Canonical Form of the Differential Equation

20.1.1
$$\frac{d^2y}{dv^2}+(a-2q\cos 2v)y=0$$

Mathieu's Modified Differential Equation

20.1.2 $\quad \frac{d^2f}{du^2}-(a-2q\cosh 2u)f=0 \qquad (v=iu,\, y=f)$

Relation Between Mathieu's Equation and the Wave Equation for the Elliptic Cylinder

The wave equation in Cartesian coordinates is

20.1.3
$$\frac{\partial^2W}{\partial x^2}+\frac{\partial^2W}{\partial y^2}+\frac{\partial^2W}{\partial z^2}+k^2W=0$$

A solution W is obtainable by separation of variables in elliptical coordinates. Thus, let

$$x=\rho\cosh u\cos v;\ y=\rho\sinh u\sin v;\ z=z;$$

ρ a positive constant; **20.1.3** becomes

20.1.4
$$*\frac{\partial^2W}{\partial z^2}+\frac{2}{\rho^2(\cosh 2u-\cos 2v)}\left(\frac{\partial^2W}{\partial u^2}+\frac{\partial^2W}{\partial v^2}\right)+k^2W=0$$

Assuming a solution of the form

$$W=\varphi(z)f(u)g(v)$$

and substituting the above into **20.1.4** one obtains, after dividing through by W,

$$\frac{1}{\varphi}\frac{d^2\varphi}{dz^2}+G=0$$

where

$$*G=\frac{2}{\rho^2(\cosh 2u-\cos 2v)}\left\{\frac{d^2f}{du^2}\frac{1}{f}+\frac{d^2g}{dv^2}\frac{1}{g}\right\}+k^2$$

Since $z,\,u,\,v$ are independent variables, it follows that

20.1.5
$$\frac{d^2\varphi}{dz^2}+c\varphi=0$$

where c is a constant.

Again, from the fact that $G=c$ and that $u,\,v$ are independent variables, one sets

20.1.6
$$*a=\frac{d^2f}{du^2}\frac{1}{f}+\frac{(k^2-c)}{2}\rho^2\cosh 2u$$

$$a=-\frac{d^2g}{dv^2}\frac{1}{g}+\frac{(k^2-c)}{2}\rho^2\cos 2v$$

where a is a constant. The above are equivalent to **20.1.1** and **20.1.2**. The constants c and a are often referred to as *separation constants*, due to the role they play in **20.1.5** and **20.1.6**.

For some physically important solutions, the function g must be periodic, of period π or 2π. It can be shown that there exists a countably infinite set of *characteristic values* $a_r(q)$ which yield even periodic solutions of **20.1.1**; there is another countably infinite sequence of *characteristic values* $b_r(q)$ which yield odd periodic solutions of **20.1.1**.

It is known that there exist periodic solutions of period $k\pi$, where k is any positive integer. In what follows, however, the term *characteristic value* will be reserved for a value associated with solutions of period π or 2π only. These characteristic values are of basic importance to the general theory of the differential equation for arbitrary parameters a and q.

An Algebraic Form of Mathieu's Equation

20.1.7
$$(1-t^2)\frac{d^2y}{dt^2}-t\frac{dy}{dt}+(a+2q-4qt^2)y=0 \qquad (\cos v=t)$$

Relation to Spheroidal Wave Equation

20.1.8 $\quad (1-t^2)\frac{d^2y}{dt^2}-2(b+1)t\frac{dy}{dt}+(c-4qt^2)y=0 \qquad *$

Thus, Mathieu's equation is a special case of **20.1.8**, with $b=-\frac{1}{2}$, $c=a+2q$.

20.2. Determination of Characteristic Values

A solution of **20.1.1** with v replaced by z, having period π or 2π is of the form

20.2.1
$$y=\sum_{m=0}^{\infty}(A_m\cos mz+B_m\sin mz)$$

where B_0 can be taken as zero. If the above is substituted into **20.1.1** one obtains

20.2.2
$$\sum_{m=-2}^{\infty}[(a-m^2)A_m-q(A_{m-2}+A_{m+2})]\cos mz$$
$$+\sum_{m=-1}^{\infty}[(a-m^2)B_m-q(B_{m-2}+B_{m+2})]\sin mz=0$$
$$A_{-m},\,B_{-m}=0 \qquad\qquad m>0$$

*See page II.

Equation **20.2.2** can be reduced to one of four simpler types, given in **20.2.3** and **20.2.4** below

20.2.3 $\quad y_0 = \sum_{m=0}^{\infty} A_{2m+p} \cos (2m+p)z, \qquad p=0$ or 1

20.2.4 $\quad y_1 = \sum_{m=0}^{\infty} B_{2m+p} \sin (2m+p)z, \qquad p=0$ or 1

If $p=0$, the solution is of period π; if $p=1$, the solution is of period 2π.

Recurrence Relations Among the Coefficients

Even solutions of period π:

20.2.5 $\qquad\qquad aA_0 - qA_2 = 0$

20.2.6 $\qquad (a-4)A_2 - q(2A_0 + A_4) = 0$

20.2.7 $\quad (a-m^2)A_m - q(A_{m-2} + A_{m+2}) = 0 \qquad (m \geq 3)$

Even solutions of period 2π:

20.2.8 $\qquad (a-1)A_1 - q(A_1 + A_3) = 0$,

along with **20.2.7** for $m \geq 3$.

Odd solutions of period π:

20.2.9 $\qquad\qquad (a-4)B_2 - qB_4 = 0$

* **20.2.10** $\quad (a-m^2)B_m - q(B_{m-2} + B_{m+2}) = 0 \qquad (m \geq 3)$

Odd solutions of period 2π:

20.2.11 $\qquad (a-1)B_1 + q(B_1 - B_3) = 0$,

along with **20.2.10** for $m \geq 3$.

Let

20.2.12 $\quad Ge_m = A_m/A_{m-2}, \quad Go_m = B_m/B_{m-2};$

$G_m = Ge_m$ or Go_m when the same operations apply to both, and no ambiguity is likely to arise. Further let

20.2.13 $\qquad\qquad V_m = (a-m^2)/q.$

Equations **20.2.5–20.2.7** are equivalent to

20.2.14 $\qquad Ge_2 = V_0; \quad Ge_4 = V_2 - \dfrac{2}{Ge_2}$

20.2.15 $\quad G_m = 1/(V_m - G_{m+2}) \qquad (m \geq 3),$

for even solutions of period π.
Similarly

20.2.16 $\quad V_1 - 1 = Ge_3$; for even solutions of period 2π, along with **20.2.15**

20.2.17 $\quad V_1 + 1 = Go_3$, for odd solutions of period 2π, along with **20.2.15**

*See page II.

20.2.18 $\quad V_2 = Go_4$, for odd solutions of period π, along with **20.2.15**

These three-term recurrence relations among the coefficients indicate that every G_m can be developed into two types of continued fractions. Thus **20.2.15** is equivalent to

20.2.19

$$G_m = \frac{1}{V_m - G_{m+2}} = \frac{1}{V_m -} \frac{1}{V_{m+2} -} \frac{1}{V_{m+4} -} \cdots \quad (m \geq 3)$$

20.2.20

$$G_{m+2} = V_m - 1/G_m$$
$$= V_m - \frac{1}{V_{m-2} -} \frac{1}{V_{m-4} -} \cdots \frac{\varphi_0}{V_{0+d} + \varphi_1} \qquad (m \geq 3)$$

where

$\varphi_1 = d = 0; \; \varphi_0 = 2$, if $G_{m+2} = A_{2s}/A_{2s-2}$

$\varphi_1 = d = \varphi_0 = 0$, if $G_{m+2} = B_{2s}/B_{2s-2}$

$\varphi_1 = -1; \; \varphi_0 = d = 1$, if $G_{m+2} = A_{2s+1}/A_{2s-1}$

$\varphi_1 = d = \phi_0 = 1$, if $G_{m+2} = B_{2s+1}/B_{2s-1}$

The four choices of the parameters φ_1, φ_0, d correspond to the four types of solutions **20.2.3–20.2.4**. Hereafter, it will be convenient to separate the characteristic values a into two major subsets:

$a = a_r$, associated with even periodic solutions

$a = b_r$, associated with odd periodic solutions

If **20.2.19** is suitably combined with **20.2.13–20.2.18** there result four types of continued fractions, the roots of which yield the required characteristic values

20.2.21 $\quad V_0 - \dfrac{2}{V_2 -} \dfrac{1}{V_4 -} \dfrac{1}{V_6 -} \cdots = 0 \quad$ Roots: a_{2r}

20.2.22

$$V_1 - 1 - \frac{1}{V_3 -} \frac{1}{V_5 -} \frac{1}{V_7 -} \cdots = 0 \quad \text{Roots: } a_{2r+1}$$

20.2.23 $\quad V_2 - \dfrac{1}{V_4 -} \dfrac{1}{V_6 -} \dfrac{1}{V_8 -} \cdots = 0 \quad$ Roots: b_{2r}

20.2.24

$$V_1 + 1 - \frac{1}{V_3 -} \frac{1}{V_5 -} \frac{1}{V_7 -} \cdots = 0 \quad \text{Roots: } b_{2r+1}$$

If a is a root of **20.2.21–20.2.24,** then the corresponding solution exists and is an entire function of z, for general complex values of q.

If q is real, then the Sturmian theory of second order linear differential equations yields the following:

(a) For a fixed real q, characteristic values a_r and b_r are real and distinct, if $q\neq0$; $a_0<b_1<a_1<b_2<a_2<\ldots$, $q>0$ and $a_r(q)$, $b_r(q)$ approach r^2 as q approaches zero.

(b) A solution of **20.1.1** associated with a_r or b_r has r zeros in the interval $0\leq z<\pi$, (q real).

(c) The form of **20.2.21** and **20.2.23** shows that if a_{2r} is a root of **20.2.21** and q is different from zero, then a_{2r} cannot be a root of **20.2.23**; similarly, no root of **20.2.22** can be a root of **20.2.24** if $q\neq0$. It may be shown from other considerations that for a given point (a, q) there can be at most one periodic solution of period π or 2π if $q\neq0$. This no longer holds for solutions of period $s\pi$, $s\geq3$; for these all solutions are periodic, if one is.

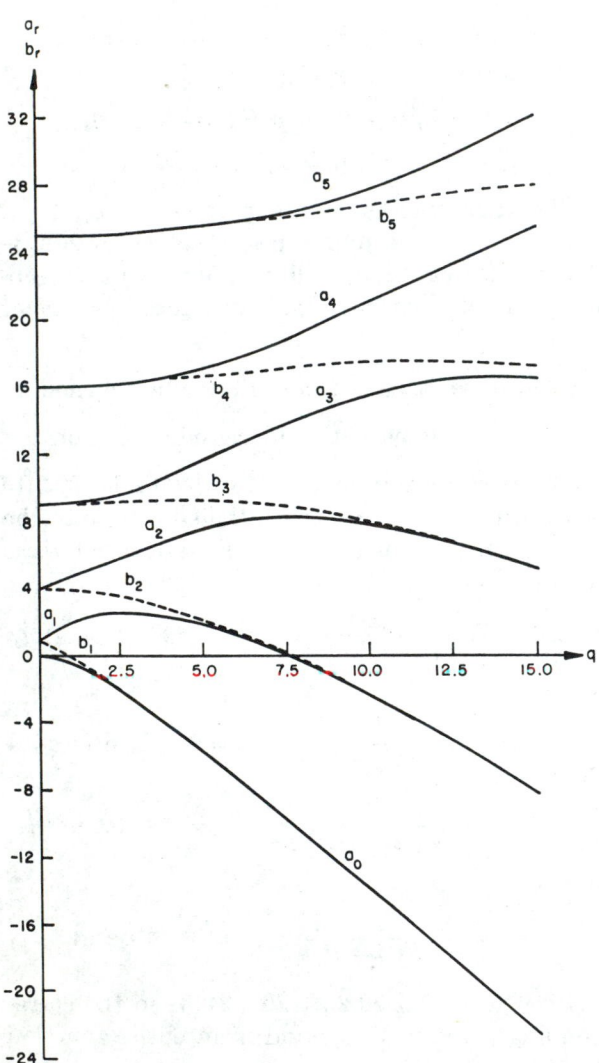

FIGURE 20.1. *Characteristic Values* a_r, b_r $r=0,1(1)5$

Power Series for Characteristic Values

20.2.25

$$a_0(q)=-\frac{q^2}{2}+\frac{7q^4}{128}-\frac{29q^6}{2304}+\frac{68687q^8}{18874368}+\cdots$$

$$\begin{aligned}a_1(-q)\\b_1(q)\end{aligned}=1-q-\frac{q^2}{8}+\frac{q^3}{64}-\frac{q^4}{1536}-\frac{11q^5}{36864}+\frac{49q^6}{589824}$$
$$-\frac{55q^7}{9437184}-\frac{83q^8}{35389440}+\cdots$$

$$b_2(q)=4-\frac{q^2}{12}+\frac{5q^4}{13824}-\frac{289q^6}{79626240}$$
$$+\frac{21391q^8}{458647142400}+\cdots$$

$$a_2(q)=4+\frac{5q^2}{12}-\frac{763q^4}{13824}+\frac{1002401q^6}{79626240}$$
$$-\frac{1669068401q^8}{458647142400}+\cdots$$

$$\begin{aligned}a_3(-q)\\b_3(q)\end{aligned}=9+\frac{q^2}{16}-\frac{q^3}{64}+\frac{13q^4}{20480}+\frac{5q^5}{16384}$$
$$-\frac{1961q^6}{23592960}+\frac{609q^7}{104857600}+\cdots$$

$$b_4(q)=16+\frac{q^2}{30}-\frac{317q^4}{864000}+\frac{10049q^6}{2721600000}+\cdots$$

$$a_4(q)=16+\frac{q^2}{30}+\frac{433q^4}{864000}-\frac{5701q^6}{2721600000}+\cdots$$

$$\begin{aligned}a_5(-q)\\b_5(q)\end{aligned}=25+\frac{q^2}{48}+\frac{11q^4}{774144}-\frac{q^5}{147456}$$
$$+\frac{37q^6}{891813888}+\cdots$$

$$b_6(q)=36+\frac{q^2}{70}+\frac{187q^4}{43904000}-\frac{5861633q^6}{92935987200000}+\cdots$$

$$a_6(q)=36+\frac{q^2}{70}+\frac{187q^4}{43904000}+\frac{6743617q^6}{92935987200000}+\cdots$$

For $r\geq7$, and $|q|$ not too large, a_r is approximately equal to b_r, and the following approximation may be used

20.2.26

$$\left.\begin{aligned}a_r\\b_r\end{aligned}\right\}=r^2+\frac{q^2}{2(r^2-1)}+\frac{(5r^2+7)q^4}{32(r^2-1)^3(r^2-4)}$$
$$+\frac{(9r^4+58r^2+29)q^6}{64(r^2-1)^5(r^2-4)(r^2-9)}+\cdots$$

The above expansion is not limited to integral values of r, and it is a very good approximation for r of the form $n+\frac{1}{2}$ where n is an integer. In case of integral values of $r=n$, the series holds only up to terms not involving r^2-n^2 in the denominator. Subsequent terms must be derived specially (as shown by Mathieu). Mulholland and Goldstein [20.38] have computed characteristic values for purely imaginary q and found that a_0 and a_2 have a common real value for $|q|$ in the neighborhood of 1.468; Bouwkamp [20.5] has computed this number as $q_0=\pm i\,1.46876852$ to 8 decimals. For values of $-iq>-iq_0$, a_0 and a_2 are conjugate complex numbers. From equation **20.2.25** it follows that the radius of convergence for the series defining a_0 is no greater than $|q_0|$. It is shown in [20.36], section **2.25** that the radius of convergence for $a_{2n}(q)$, $n\geq2$ is greater than 3. Furthermore

$$a_r-b_r=O(q^r/r^{r-1}),\ r\to\infty.$$

Power Series in q for the Periodic Functions (for sufficiently small $|q|$)

20.2.27

$$ce_0(z, q)=2^{-\frac{1}{2}}\left[1-\frac{q}{2}\cos 2z+q^2\left(\frac{\cos 4z}{32}-\frac{1}{16}\right)\right.$$
$$\left.-q^3\left(\frac{\cos 6z}{1152}-\frac{11\cos 2z}{128}\right)+\cdots\right]$$

$$ce_1(z, q)=\cos z-\frac{q}{8}\cos 3z$$
$$+q^2\left[\frac{\cos 5z}{192}-\frac{\cos 3z}{64}-\frac{\cos z}{128}\right]$$
$$-q^3\left[\frac{\cos 7z}{9216}-\frac{\cos 5z}{1152}-\frac{\cos 3z}{3072}+\frac{\cos z}{512}\right]+\cdots$$

$$se_1(z, q)=\sin z-\frac{q}{8}\sin 3z$$
$$+q^2\left[\frac{\sin 5z}{192}+\frac{\sin 3z}{64}-\frac{\sin z}{128}\right]$$
$$-q^3\left[\frac{\sin 7z}{9216}+\frac{\sin 5z}{1152}-\frac{\sin 3z}{3072}-\frac{\sin z}{512}\right]+\cdots$$

$$ce_2(z, q)=\cos 2z-q\left(\frac{\cos 4z}{12}-\frac{1}{4}\right)+q^2\left(\frac{\cos 6z}{384}-\frac{19\cos 2z}{288}\right)+\cdots$$

$$se_2(z, q)=\sin 2z-q\,\frac{\sin 4z}{12}+q^2\left(\frac{\sin 6z}{384}-\frac{\sin 2z}{288}\right)+\cdots$$

20.2.28

$$\begin{matrix}ce_r(z, q)\\se_r(z, q)\end{matrix}=\cos(rz-p(\pi/2))-q\left\{\frac{\cos\left[(r+2)z-p\dfrac{\pi}{2}\right]}{4(r+1)}\right.$$
$$\left.-\frac{\cos[(r-2)z-p(\pi/2)]}{4(r-1)}\right\}$$
$$+q^2\left\{\frac{\cos[(r+4)z-p(\pi/2)]}{32(r+1)(r+2)}+\frac{\cos[(r-4)z-p(\pi/2)]}{32(r-1)(r-2)}\right.$$
$$\left.-\frac{\cos[rz-p(\pi/2)]}{32}\left[\frac{2(r^2+1)}{(r^2-1)^2}\right]\right\}+\cdots$$

with $p=0$ for $ce_r(z, q)$, $p=1$ for $se_r(z, q)$, $r\geq3$.

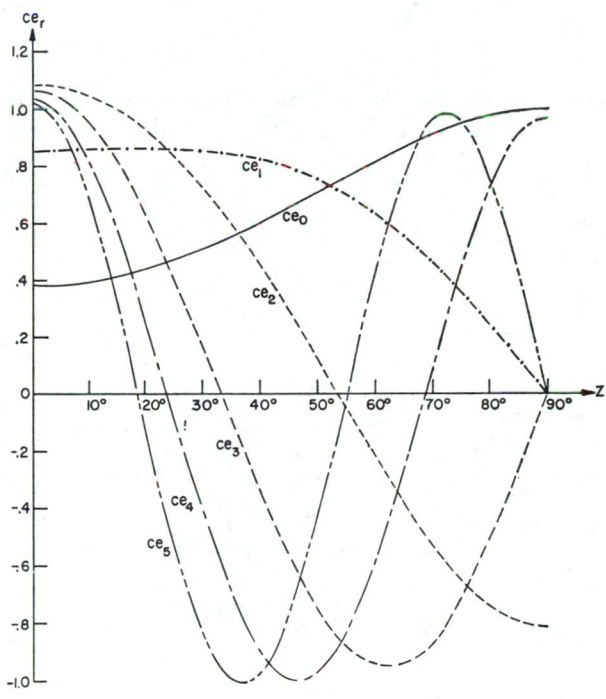

FIGURE 20.2. *Even Periodic Mathieu Functions, Orders 0–5 $q=1$.*

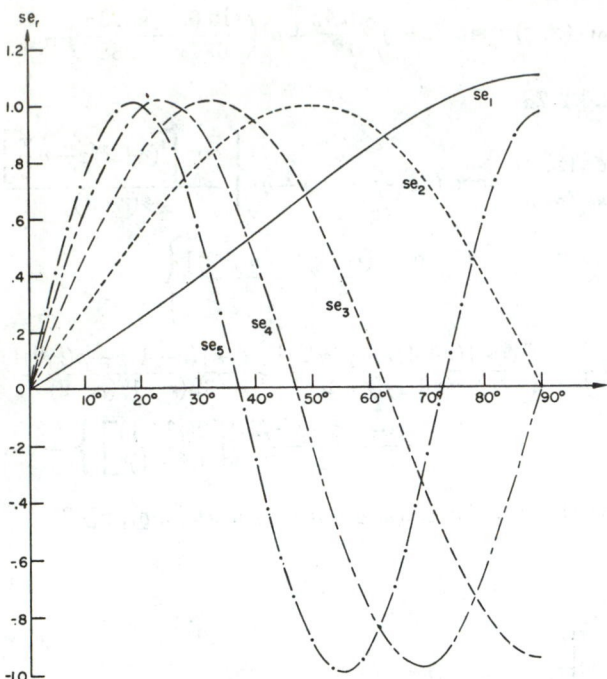

FIGURE 20.3. *Odd Periodic Mathieu Functions, Orders 1–5*
$q=1$.

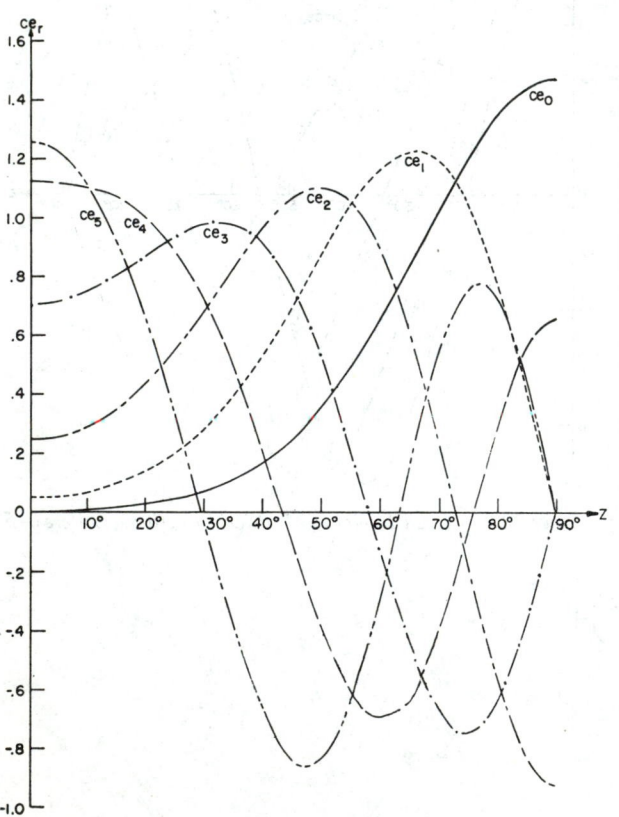

FIGURE 20.4. *Even Periodic Mathieu Functions, Orders 0–5*
$q=10$.

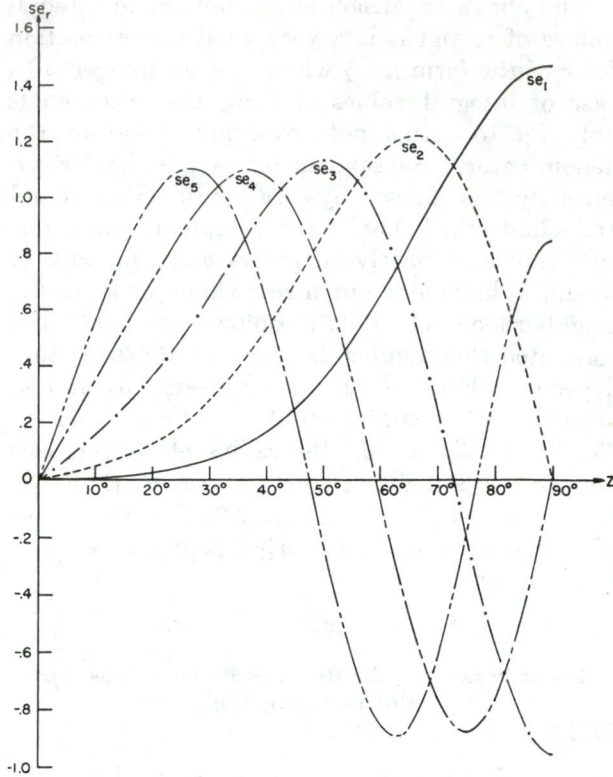

FIGURE 20.5. *Odd Periodic Mathieu Functions, Orders 1–5*
$q=10$.

For coefficients associated with above functions

20.2.29

$$A_0^0(0)=2^{-\frac{1}{2}};\ A_r^r(0)=B_r^r(0)=1,\ r>0$$

$$A_{2s}^0=[(-1)^s q^s/s!\ s!\ 2^{2s-1}]\ A_0^0+\ \ldots,\ s>0$$

$$\begin{aligned}A_{r+2s}^r\\B_{r+2s}^r\end{aligned}=[(-1)^s r!\ q^s/4^s(r+s)!\ s!]\ C_r^r+\ \ldots$$

$$rs>0,\ C_r^r=A_r^r\ \text{or}\ B_r^r$$

$$A_{r-2s}^r\ \text{or}\ B_{r-2s}^r=\frac{(r-s-1)!}{s!(r-1)!}\frac{q^s}{4^s}C_r^r+\ \ldots$$

Asymptotic Expansion for Characteristic Values, $q\gg1$

Let $w=2r+1$, $q=w^4\varphi$, φ real. Then

20.2.30

$$a\sim b_{r+1}\sim -2q+2w\sqrt{q}-\frac{w^2+1}{8}-\frac{\left(w+\dfrac{3}{w}\right)}{2^7\sqrt{\varphi}}$$

$$-\frac{d_1}{2^{12}\varphi}-\frac{d_2}{2^{17}\varphi^{3/2}}-\frac{d_3}{2^{20}\varphi^2}-\frac{d_4}{2^{25}\varphi^{5/2}}-\ \cdots$$

where

$$d_1=5+\frac{34}{w^2}+\frac{9}{w^4}$$

$$d_2=\frac{33}{w}+\frac{410}{w^3}+\frac{405}{w^5}$$

$$d_3 = \frac{63}{w^2} + \frac{1260}{w^4} + \frac{2943}{w^6} + \frac{486}{w^8}$$

$$d_4 = \frac{527}{w^3} + \frac{15617}{w^5} + \frac{69001}{w^7} + \frac{41607}{w^9}$$

20.2.31 $b_{r+1} - a_r \sim 2^{4r+5}\sqrt{2/\pi}\, q^{\frac{1}{2}r+\frac{3}{4}} e^{-4\sqrt{q}}/r!, \qquad q \to \infty$

(given in [20.36] without proof.)

20.3. Floquet's Theorem and Its Consequences

Since the coefficients of Mathieu's equation

20.3.1 $y'' + (a - 2q\cos 2z)y = 0$

are periodic functions of z, it follows from the known theory relating to such equations that there exists a solution of the form

20.3.2 $F_\nu(z) = e^{i\nu z}P(z),$

where ν depends on a and q, and $P(z)$ is a periodic function, of the same period as that of the coefficients in **20.3.1**, namely π. (Floquet's theorem; see [20.16] or [20.22] for its more general form.) The constant ν is called the *characteristic exponent*. Similarly

20.3.3 $F_\nu(-z) = e^{-i\nu z}P(-z)$

satisfies **20.3.1** whenever **20.3.2** does. Both $F_\nu(z)$ and $F_\nu(-z)$ have the property

20.3.4

$y(z+k\pi) = C^k y(z),\ y = F_\nu(z)$ or $F_\nu(-z),$
$\qquad C = e^{i\nu\pi}$ for $F_\nu(z),\ C = e^{-i\nu\pi}$ for $F_\nu(-z)$

Solutions having the property **20.3.4** will hereafter be termed *Floquet* solutions. Whenever $F_\nu(z)$ and $F_\nu(-z)$ are linearly independent, the general solution of **20.3.1** can be put into the form

20.3.5 $y = AF_\nu(z) + BF_\nu(-z)$

If $AB \neq 0$, the above solution will *not be a Floquet solution*. It will be seen later, from the method for determining ν when a and q are given, that there is some ambiguity in the definition of ν; namely, ν can be replaced by $\nu + 2k$, where k is an arbitrary integer. This is as it should be, since the addition of the factor $\exp(2ikz)$ in **20.3.2** still leaves a periodic function of period π for the coefficient of $\exp i\nu z$.

It turns out that when a belongs to the set of characteristic values a_r and b_r of **20.2**, then ν is zero or an integer. It is convenient to associate $\nu = r$ with $a_r(q)$, and $\nu = -r$ with $b_r(q)$; see [20.36]. In the special case when ν is an integer, $F_\nu(z)$ is

proportional to $F_\nu(-z)$; the second, independent solution of **20.3.1** then has the form

20.3.6 $y_2 = z ce_r(z, q) + \sum_{k=0}^{\infty} d_{2k+p} \sin(2k+p)z,$

associated with $ce_r(z, q)$

20.3.7 $y_2 = z se_r(z, q) + \sum_{k=0}^{\infty} f_{2k+p} \cos(2k+p)z,$

associated with $se_r(z, q)$

The coefficients d_{2k+p} and f_{2k+p} depend on the corresponding coefficients A_m and B_m, respectively, of **20.2**, as well as on a and q. See [20.30], section (7.50)–(7.51) and [20.58], section V, for details.

If ν is not an integer, then the Floquet solutions $F_\nu(z)$ and $F_\nu(-z)$ are linearly independent. It is clear that **20.3.2** can be written in the form

20.3.8 $F_\nu(z) = \sum_{k=-\infty}^{\infty} c_{2k}e^{i(\nu+2k)z}.$

From **20.3.8** it follows that if ν is a proper fraction m_1/m_2, then every solution of **20.3.1** is periodic, and of period at most $2\pi m_2$. This agrees with results already noted in **20.2**; i.e., both independent solutions are periodic, if one is, provided the period is different from π and 2π.

Method of Generating the Characteristic Exponent

Define two linearly independent solutions of **20.3.1**, for fixed a, q by

20.3.9

$y_1(0) = 1;\ y_1'(0) = 0.$

$y_2(0) = 0;\ y_2'(0) = 1.$

Then it can be shown that

20.3.10 $\cos \pi\nu - y_1(\pi) = 0$

20.3.11 $\cos \pi\nu - 1 - 2y_1'\left(\frac{\pi}{2}\right) y_2\left(\frac{\pi}{2}\right) = 0$

Thus ν may be obtained from a knowledge of $y_1(\pi)$ or from a knowledge of both $y_1'\left(\frac{\pi}{2}\right)$ and $y_2\left(\frac{\pi}{2}\right)$. For numerical purposes **20.3.11** may be more desirable because of the shorter range of integration, and hence the lesser accumulation of round-off errors. Either ν, $-\nu$, or $\pm\nu + 2k$ (k an arbitrary integer) can be taken as the solution of **20.3.11**. Once ν has been fixed, the coefficients of **20.3.8** can be determined, except for an arbitrary multiplier which is independent of z.

The characteristic exponent can also be computed from a continued fraction, in a manner analogous to developments in **20.2**, if a sufficiently close first approximation to ν is available. For systematic tabulation, this method is considerably faster than the method of numerical integration. Thus, when **20.3.8** is substituted into **20.3.1,** there result the following recurrence relations:

20.3.12 $$V_{2n}c_{2n}=c_{2n-2}+c_{2n+2}$$

where

20.3.13 $V_{2n}=[a-(2n+\nu)^2]/q, \quad -\infty < n < \infty.$

When ν is complex, the coefficients V_{2n} may also be complex. As in **20.2,** it is possible to generate the ratios

$$G_m=c_m/c_{m-2} \text{ and } H_{-m}=c_{-m-2}/c_{-m}$$

from the continued fractions

20.3.14

$$G_m=\frac{1}{V_m-}\ \frac{1}{V_{m+2}-}\ \cdots, \qquad m\geq 0$$

$$H_{-m}=\frac{1}{V_{-m-2}-}\ \frac{1}{V_{-m-4}-}\ \cdots, \qquad m\geq 0.$$

From the form of **20.3.13** and the known properties of continued fractions it is assured that for sufficiently large values of $|m|$ both $|G_m|$ and $|H_{-m}|$ converge. Once values of G_m and H_{-m} are available for some sufficiently large value of m, then the finite number of ratios $G_{m-2}, G_{m-4}, \ldots, G_0$ can be computed in turn, if they exist. Similarly for $H_{-m+2}, \ldots, H_0$. It is easy to show that ν is the correct characteristic exponent, appropriate for the point (a, q), if and only if $H_0G_0=1$. An iteration technique can be used to improve the value of ν, by the method suggested in [20.3]. One coefficient c_j can be assigned arbitrarily; the rest are then completely determined. After all the c_j become available, a multiplier (depending on q but not on z) can be found to satisfy a prescribed normalization.

It is well known that continued fractions can be converted to determinantal form. Equation **20.3.14** can in fact be written as a determinant with an infinite number of rows—a special case of Hill's determinant. See [20.19], [20.36], [20.15], or [20.30] for details. Although the determinant has actually been used in computations where high-speed computers were available, the direct use of the continued fraction seems much less laborious.

Special Cases (a, q Real)

Corresponding to $q=0$, $y_1=\cos \sqrt{a}z$, $y_2=\sin \sqrt{a}z$; the Floquet solutions are $\exp(iaz)$ and $\exp(-iaz)$. As a, q vary continuously in the $q-a$ plane, ν describes curves; ν is real when (q, a), $q\geq 0$ lies in the region between $a_r(q)$ and $b_{r+1}(q)$ and

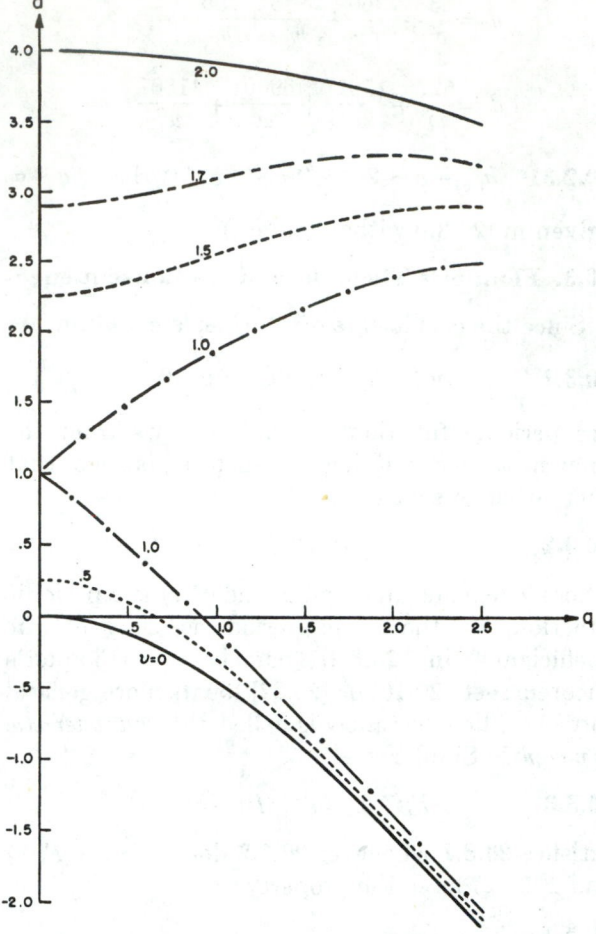

FIGURE 20.6. *Characteristic Exponent-First Two Stable Regions* $y=e^{i\nu z}P(x)$ *where* $P(x)$ *is a periodic function of period* π.

Definition of ν;
In first stable region, $0\leq\nu\leq 1$,
In second stable region, $1\leq\nu\leq 2$.

(Constructed from tabular values supplied by T. Tamir, Brooklyn Polytechnic Institute)

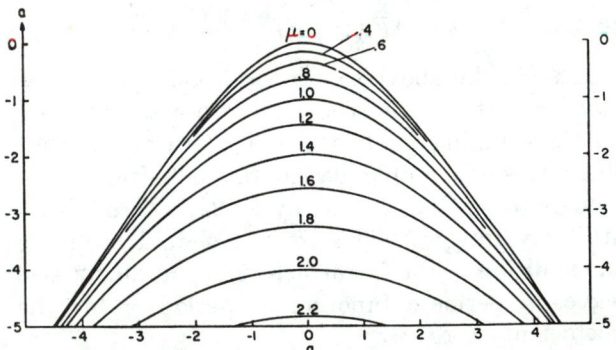

FIGURE 20.7. *Characteristic Exponent in First Unstable Region. Differential equation:* $y''+(a-2q\cos 2x)y=0$. *The Floquet solution* $y=e^{i\nu z}P(x)$, *where* $P(x)$ *is a periodic function of period* π. *In the first unstable region,* $\nu=i\mu$; μ *is given for* $a\geq -5$. (Constructed at NBS.)

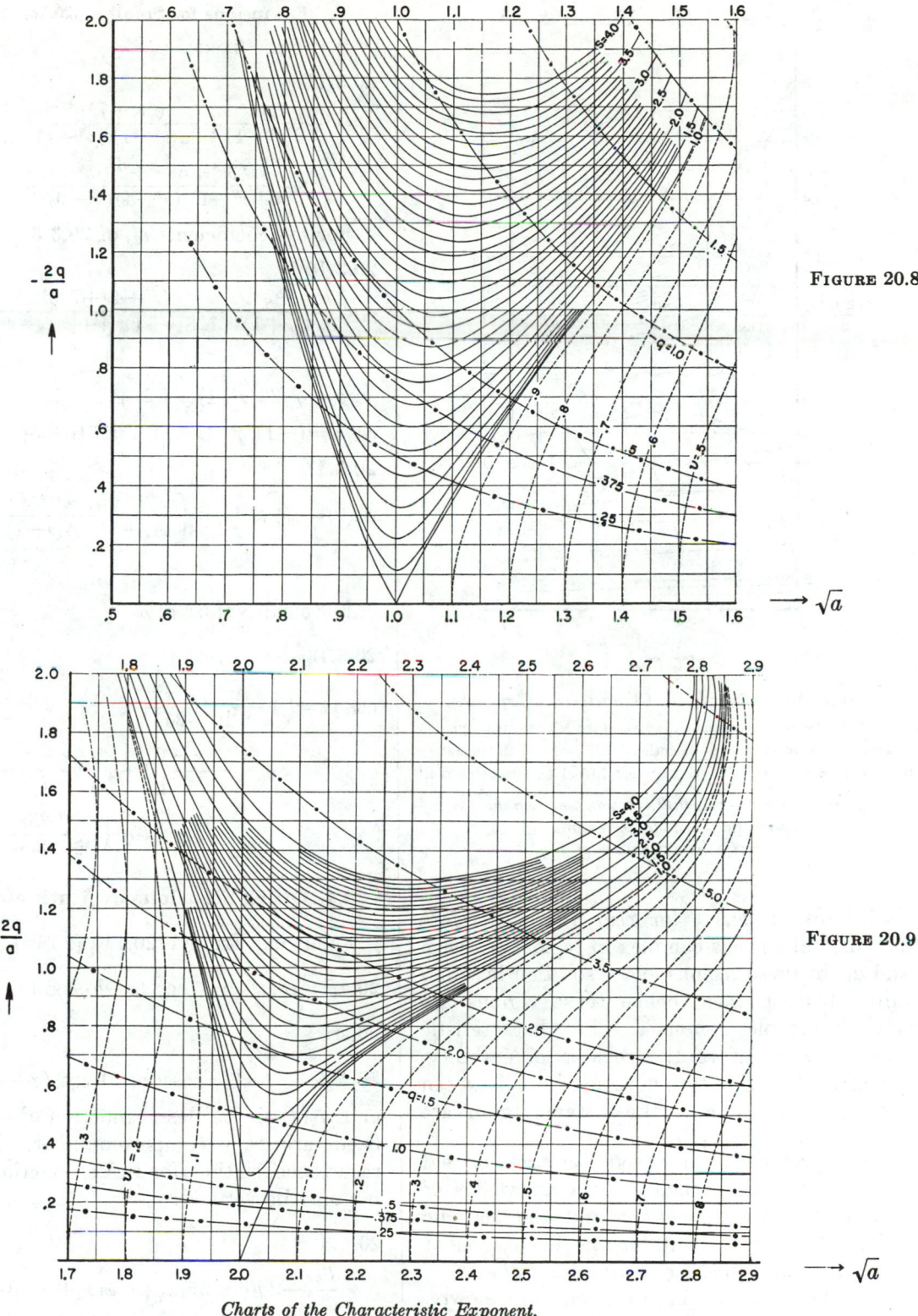

FIGURE 20.8

FIGURE 20.9

Charts of the Characteristic Exponent.

(From S. J. Zaroodny, An elementary review of the Mathieu-Hill equation of real variable based on numerical solutions, Ballistic Research Laboratory Memo. Rept. 878, Aberdeen Proving Ground, Md., 1955, with permission.)

——— $s = e^{i\nu\pi} = constant$; *in unstable regions*

– – – – $\nu = constant$; *in stable regions*

– . – . – *Lines of constant values of* $-q$.

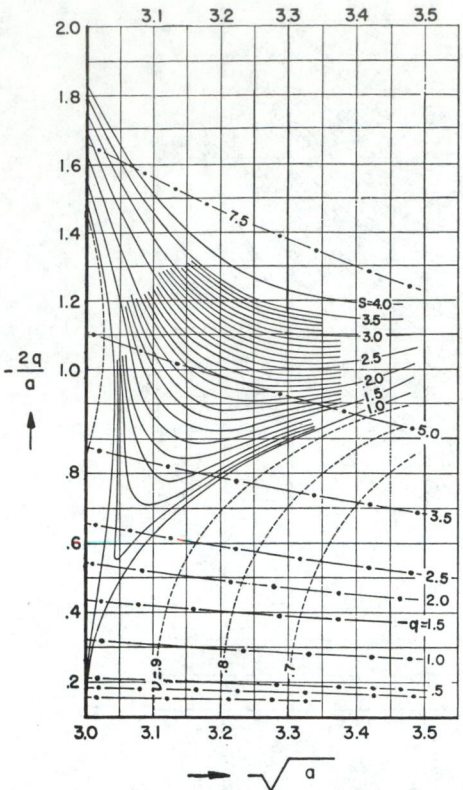

FIGURE 20.10. *Chart of the Characteristic Exponent.*

(From S. J. Zaroodny, An elementary review of the Mathieu-Hill equation of real variable based on numerical solutions, Ballistic Research Laboratory Memo. Rept. 878, Aberdeen Proving Ground, Md., 1955, with permission)

——— $s = e^{i\nu\pi} = constant$; *in unstable regions*

– – – – $\nu = constant$; *in stable regions*

–.–.– *Lines of constant values of* $-q$.

all solutions of **20.1.1** for real z are therefore bounded (stable); ν is complex in regions between b_r and a_r; in these regions every solution becomes infinite at least once; hence these regions are termed "unstable regions". The characteristic curves a_r, b_r separate the regions of stability. For negative q, the stable regions are between b_{2r+1} and b_{2r+2}, a_{2r} and a_{2r+1}; the unstable regions are between a_{2r+1} and b_{2r+1}, a_{2r} and b_{2r}.

In some problems solutions are required for real values of z only. In such cases a knowledge of the characteristic exponent ν and the periodic function $P(z)$ is sufficient for the evaluation of the required functions. For complex values of z, however, the series defining $P(z)$ converges slowly. Other solutions will be determined in the next section; they all have the remarkable property that they depend on the same coefficients c_m developed in connection with Floquet's theorem (except for an arbitrary normalization factor).

Expansions for Small q ([20.36] chapter 2)

If ν, q are fixed:

20.3.15

$$a = \nu^2 + \frac{q^2}{2(\nu^2-1)} + \frac{(5\nu^2+7)q^4}{32(\nu^2-1)^3(\nu^2-4)}$$
$$+ \frac{(9\nu^4+58\nu^2+29)q^6}{64(\nu^2-1)^5(\nu^2-4)(\nu^2-9)} + \dots \quad (\nu \neq 1, 2, 3).$$

For the coefficients c_{2j} of **20.3.8**

20.3.16

$$c_2/c_0 = \frac{-q}{4(\nu+1)} - \frac{(\nu^2+4\nu+7)q^3}{128(\nu+1)^3(\nu+2)(\nu-1)} + \dots$$
$$(\nu \neq 1, 2)$$

$$c_4/c_0 = q^2/32(\nu+1)(\nu+2) + \dots$$

$$c_{2s}/c_0 = (-1)^s q^s \Gamma(\nu+1)/2^{2s} s! \Gamma(\nu+s+1) + \dots$$

20.3.17

$$F_\nu(z) = c_0\left[e^{i\nu z} - q\left\{\frac{e^{i(\nu+2)z}}{4(\nu+1)} - \frac{e^{i(\nu-2)z}}{4(\nu-1)}\right\}\right] + \dots$$
$$(\nu \text{ not an integer})$$

For small values of a

20.3.18

$$\cos \nu\pi = \left(1 - \frac{a\pi^2}{2} + \frac{a^2\pi^4}{24} + \dots\right)$$
$$- \frac{q^2\pi^2}{4}\left[1 + a\left(1 - \frac{\pi^2}{6}\right) + \dots\right]$$
$$+ q^4\left(\frac{\pi^4}{96} - \frac{25\pi^2}{256} + \dots\right) + \dots$$

20.4. Other Solutions of Mathieu's Equation

Following Erdélyi [20.14], [20.15], define

20.4.1 $\varphi_k(z) = [e^{i\pi}\cos(z-b)/\cos(z+b)]^{\frac{1}{2}k}J_k(f)$

where

20.4.2 $f = 2[q\cos(z-b)\cos(z+b)]^{\frac{1}{2}}$,

and $J_k(f)$ is the Bessel function of order k; b is a fixed, arbitrary complex number. By using the recurrence relations for Bessel functions the following may be verified:

20.4.3

$$\frac{d^2\varphi_k}{dz^2} - 2q(\cos 2z)\varphi_k + q(\varphi_{k-2}+\varphi_{k+2}) + k^2\varphi_k = 0.$$

It follows that a formal solution of **20.1.1** is given by

20.4.4 $y = \sum\limits_{n=-\infty}^{\infty} c_{2n}\varphi_{2n+\nu}$

where the coefficients c_{2n} are those associated with Floquet's solution. In the above, ν may be complex. Except for the special case when ν is an integer, the following holds:

$$\frac{\varphi_{2n+\nu-2}}{\varphi_{2n+\nu}} \sim \frac{\varphi_{-2n+\nu}}{\varphi_{-2n+\nu+2}} \sim \frac{-4n^2}{q[\cos{(z-b)}]^2} \qquad (n \to \infty)$$

If ν and n are integers, $J_{-2n+\nu}(f) = (-1)^\nu J_{2n-\nu}(f)$.

$$[\varphi_{2n+\nu}/\varphi_{2n+\nu-2}] \sim -[\cos{(z-b)}]^2 q/4n^2$$

$$[\varphi_{-2n+\nu}/\varphi_{-2n+\nu+2}] \sim -4n^2/q[\cos{(z-b)}]^2$$

On the other hand

$$\frac{c_{2n}}{c_{2n-2}} \sim \frac{c_{-2n}}{c_{-2n+2}} \sim \frac{-q}{4n^2} \qquad (n \to \infty)$$

It follows that **20.4.4** converges absolutely and uniformly in every closed region where

$|\cos{(z-b)}| > d_1 > 1.$

There are two such disjoint regions:

(I) $\mathscr{I}(z-b) > d_2 > 0;$ $(|\cos{(z-b)}| > d_1 > 1)$

(II) $\mathscr{I}(z-b) < -d_2 < 0;$ $(|\cos{(z-b)}| > d_1 > 1)$

If ν is an integer **20.4.4** converges for all values of z. Various representations are found by specializing b.

20.4.5

If $b=0$, $y = e^{i\pi\nu/2} \sum_{n=-\infty}^{\infty} c_{2n}(-1)^n J_{2n+\nu}(2\sqrt{q}\cos z)$

$$(|\cos z| > 1, |\arg 2\sqrt{q}\cos z| \leq \pi)$$

20.4.6

If $b=\frac{\pi}{2}$, $y = \sum_{n=-\infty}^{\infty} c_{2n} J_{2n+\nu}(2i\sqrt{q}\sin z)$

$$(|\sin z| > 1, |\arg 2\sqrt{q}\sin z| \leq \pi)$$

If $b \to \infty i$, y reduces to a multiple of the solution **20.3.8**. The fact that **20.3.8**, **20.4.5**, and **20.4.6** are special cases of **20.4.4** explains why it is that these apparently dissimilar expansions involve the same set of coefficients c_{2n}.

Since **20.4.4** results from the recurrence properties of Bessel functions, $J_k(f)$ can be replaced by $H_k^{(j)}(f)$, $j=1, 2$, where $H_k^{(j)}$ is the Hankel function, at least formally. Thus let

$$\psi_k^j = [e^{i\pi}\cos{(z-b)}/\cos{(z+b)}]^{\frac{1}{2}k} H_k^{(j)}(f)$$

where f satisfies **20.4.2**. An examination of the ratios $\psi_{2n+\nu}/\psi_{2n+\nu-2}$ shows that

$$y = \sum_{n=-\infty}^{\infty} c_{2n}\psi_{2n+\nu}^{(j)}$$

will be a solution provided

$$|\cos{(z-b)}| > 1; |\cos{(z+b)}| > 1.$$

The above two conditions are necessary even when ν is an integer. Once b is fixed, the regions in which the solutions converge can be readily established.

Following [20.36] let

20.4.7

$$J_p(x) = Z_p^{(1)}(x); \quad Y_p(x) = Z_p^{(2)}(x);$$
$$H_p^{(1)}(x) = Z_p^{(3)}(x); \quad H_p^{(2)}(x) = Z_p^{(4)}(x)$$

If z is replaced by $-iz$ in **20.4.5** and **20.4.6** solutions of **20.1.2** are obtained. Thus

20.4.8

$$y_1^{(j)}(z) = \sum_{n=-\infty}^{\infty} c_{2n}(-1)^n Z_{2n+\nu}^{(j)}(2\sqrt{q}\cosh z)$$

$$(|\cosh z| > 1)$$

20.4.9

$$y_2^{(j)}(z) = \sum_{n=-\infty}^{\infty} c_{2n} Z_{2n+\nu}^{(j)}(2\sqrt{q}\sinh z)$$

$$(|\sinh z| > 1, j=1, 2, 3, 4)$$

The relation between $y_1^{(j)}(z)$ and $y_2^{(j)}(z)$ can be determined from the asymptotic properties of the Bessel functions for large values of argument. It can be shown that

20.4.10

$$y_1^{(j)}(z)/y_2^{(j)}(z) = [F_\nu(0)/F_\nu\left(\frac{\pi}{2}\right)]e^{i\nu\pi/2} \qquad (\mathscr{R} z > 0).$$

When ν is not an integer, the above solutions do not vanish identically. See **20.6** for integral values of ν.

Solutions Involving Products of Bessel Functions

20.4.11

$$y_3^{(j)}(z) = \frac{1}{c_{2s}} \sum_{n=-\infty}^{\infty} c_{2n}(-1)^n Z_{n+\nu+s}^{(j)}(\sqrt{q}e^{iz}) J_{n-s}(\sqrt{q}e^{-iz})$$

$$(j=1, 2, 3, 4)$$

satisfies **20.1.1**, where $Z_n^{(j)}(u)$ is defined in **20.4.7**, the coefficients c_{2n} belong to the Floquet solution, and s is an arbitrary integer, $c_{2s} \neq 0$. The solution converges over the entire complex z-plane if $q \neq 0$. Written with z replaced by $-iz$, one obtains solutions of **20.1.2**.

20.4.12

$$M_\nu^j(z, q) = \frac{1}{c_{2s}^\nu} \sum_{n=-\infty}^{\infty} c_{2n}^\nu (-1)^n Z_{n+\nu+s}^{(j)}(\sqrt{q}e^z) J_{n-s}(\sqrt{q}e^{-z})$$

It can be verified from **20.4.8** and **20.4.12** that

20.4.13 $\dfrac{y_1^{(j)}(z)}{M_j^j(z, q)} = F_\nu(0),$ $(\mathscr{R}z > 0)$

provided $c_{2s} \neq 0$. If $c_{2s} = 0$, the coefficient of $1/c_{2s}$ in **20.4.11** vanishes identically. For details see [20.43], [20.15], [20.36].

If s is chosen so that $|c_{2s}|$ is the largest coefficient of the set $|c_{2j}|$, then rapid convergence of **20.4.12** is obtained, when $\mathscr{R}z > 0$. Even then one must be on guard against the possible loss of significant figures in the process of summing the series, especially so when q is large, and $|z|$ small. (If $j \neq 1$, then the phase of the logarithmic terms occurring in **20.4.12** must be defined, to make the functions single-valued.)

20.5. Properties of Orthogonality and Normalization

If $a(\nu+2p, q)$, $a(\nu+2s, q)$ are simple roots of **20.3.10** then

20.5.1 $\displaystyle\int_0^\pi F_{\nu+2p}(z) F_{\nu+2s}(-z)dz = 0,$ if $p \neq s$.

Define

20.5.2 $ce_\nu(z, q) = \dfrac{1}{2}[F_\nu(z) + F_\nu(-z)];$

$se_\nu(z, q) = -i\dfrac{1}{2}[F_\nu(z) - F_\nu(-z)]$

$ce_\nu(z, q)$, $se_\nu(z, q)$ are thus even and odd functions of z, respectively, for all ν (when not identically zero).

If ν is an integer, then $ce_\nu(z, q)$, $se_\nu(z, q)$ are either Floquet solutions or identically zero. The solutions $ce_r(z, q)$ are associated with a_r; $se_r(z, q)$ are associated with b_r; r an integer.

Normalization for Integral Values of ν and Real q

20.5.3 $\displaystyle\int_0^{2\pi}[ce_r(z, q)]^2 dz = \int_0^{2\pi}[se_r(z, q)]^2 dz = \pi$

For integral values of ν the summation in **20.3.8** reduces to the simpler forms **20.2.3–20.2.4**; on account of **20.5.3**, the coefficients A_m and B_m (for all orders r) have the property

20.5.4

$$2A_0^2 + A_2^2 + \ldots = A_1^2 + A_3^2 + \ldots$$
$$= B_1^2 + B_3^2 + \ldots = B_2^2 + B_4^2 + \ldots = 1.$$

20.5.5

$$A_0^{2s} = \frac{1}{2\pi}\int_0^{2\pi} ce_{2s}(z, q)dz; \quad A_n^r = \frac{1}{\pi}\int_0^{2\pi} ce_r(z, q)\cos nz dz$$

$$B_n^r = \frac{1}{\pi}\int_0^{2\pi} se_r(z, q)\sin nz dz \qquad n \neq 0$$

For integral values of ν, the functions $ce_r(z, q)$ and $se_r(z, q)$ form a complete orthogonal set for the interval $0 \leq z \leq 2\pi$. Each of the four systems $ce_{2r}(z)$, $ce_{2r+1}(z)$, $se_{2r}(z)$, $se_{2r+1}(z)$ is complete in the smaller interval $0 \leq z \leq \frac{1}{2}\pi$, and each of the systems $ce_r(z)$, $se_r(z)$ is complete in $0 \leq z \leq \pi$.

If q is not real, there exist multiple roots of **20.3.10**; for such special values of $a(q)$, the integrals in **20.5.3** vanish, and the normalization is therefore impossible. In applications, the particular normalization adopted is of little importance, except possibly for obtaining quantitative relations between solutions of various types. For this reason the normalization of $F_\nu(z)$, for arbitrary complex values of a, q, will not be specified here. It is worth noting, however, that solutions

$$\alpha ce_r(z, q), \qquad \beta se_r(z, q)$$

defined so that

$$\alpha ce_r(0, q) = 1; \qquad \left[\frac{d}{dz}\beta se_r(z, q)\right]_{z=0} = 1$$

are always possible. This normalization has in fact been used in [20.59], and also in [20.58], where the most extensive tabular material is available. The tabulated entries in [20.58] supply the conversion factors $A = 1/\alpha$, $B = 1/\beta$, along with the coefficients. Thus conversion from one normalization to another is rather easy.

In a similar vein, no general normalization will be imposed on the functions defined in **20.4.8**.

20.6. Solutions of Mathieu's Modified Equation 20.1.2 for Integral ν (Radial Solutions)

Solutions of the first kind

20.6.1

$$Ce_{2r+p}(z, q) = ce_{2r+p}(iz, q)$$

$$= \sum_{k=0}^{\infty} A_{2k+p}^{2r+p}(q)\cosh(2k+p)z$$

associated with a_r

20.6.2 $\qquad Se_{2r+p}(z, q) = -ise_{2r+p}(iz, q) = \sum\limits_{k=0}^{\infty} B_{2k+p}^{2r+p}(q) \sinh(2k+p)z$, associated with b_r

writing $A_{2k+p}^{2r+p}(q) = A_{2k+p}$ for brevity; similarly for B_{2k+p}; $p = 0, 1,$

20.6.3 $\qquad Ce_{2r}(z, q) = \dfrac{ce_{2r}\left(\dfrac{\pi}{2}, q\right)}{A_0^{2r}} \sum\limits_{k=0}^{\infty} (-1)^k A_{2k} J_{2k}(2\sqrt{q}\,\cosh z) = \dfrac{ce_{2r}(0, q)}{A_0^{2r}} \sum\limits_{k=0}^{\infty} A_{2k} J_{2k}(2\sqrt{q}\,\sinh z)$

20.6.4 $\qquad Ce_{2r+1}(z, q) = \dfrac{ce_{2r+1}'\left(\dfrac{\pi}{2}, q\right)}{\sqrt{q}A_1^{2r+1}} \sum\limits_{k=0}^{\infty} (-1)^{k+1} A_{2k+1} J_{2k+1}(2\sqrt{q}\,\cosh z)$

$\qquad\qquad\qquad = \dfrac{ce_{2r+1}(0, q)}{\sqrt{q}A_1^{2r+1}} \coth z \sum\limits_{k=0}^{\infty} (2k+1) A_{2k+1} J_{2k+1}(2\sqrt{q}\,\sinh z)$

20.6.5 $\qquad Se_{2r}(z, q) = \dfrac{se_{2r}'\left(\dfrac{\pi}{2}, q\right)\tanh z}{qB_2^{2r}} \sum\limits_{k=1}^{\infty} (-1)^k 2k B_{2k} J_{2k}(2\sqrt{q}\,\cosh z)$

$\qquad\qquad\qquad = \dfrac{se_{2r}'(0, q)}{qB_2^{2r}} \coth z \sum\limits_{k=1}^{\infty} 2k B_{2k} J_{2k}(2\sqrt{q}\,\sinh z)$

20.6.6 $\qquad Se_{2r+1}(z, q) = \dfrac{se_{2r+1}\left(\dfrac{\pi}{2}, q\right)}{\sqrt{q}B_1^{2r+1}} \tanh z \sum\limits_{k=0}^{\infty} (-1)^k (2k+1) B_{2k+1} J_{2k+1}(2\sqrt{q}\,\cosh z)$

$\qquad\qquad\qquad = \dfrac{se_{2r+1}'(0,q)}{\sqrt{q}B_1^{2r+1}} \sum\limits_{k=0}^{\infty} B_{2k+1} J_{2k+1}(2\sqrt{q}\,\sinh z)$

See [20.30] for still other forms.

Solutions of the second kind, as well as solutions of the third and fourth kind (analogous to Hankel functions) are obtainable from **20.4.12.**

20.6.7 $\qquad Mc_{2r}^{(j)}(z, q) = \sum\limits_{k=0}^{\infty} (-1)^{r+k} A_{2k}^{2r}(q)[J_{k-s}(u_1)Z_{k+s}^{(j)}(u_2) + J_{k+s}(u_1)Z_{k-s}^{(j)}(u_2)]/\epsilon_s A_{2s}^{2r}$

$\qquad\qquad$ where $\epsilon_0 = 2$, $\epsilon_s = 1$, for $s = 1, 2, \ldots$; s arbitrary, associated with a_{2r}

20.6.8 $\qquad Mc_{2r+1}^{(j)}(z, q) = \sum\limits_{k=0}^{\infty} (-1)^{r+k} A_{2k+1}^{2r+1}(q)[J_{k-s}(u_1)Z_{k+s+1}^{(j)}(u_2) + J_{k+s+1}(u_1)Z_{k-s}^{(j)}(u_2)]/A_{2s+1}^{2r+1}$

$\qquad\qquad\qquad\qquad\qquad\qquad\qquad\qquad\qquad\qquad\qquad\qquad$ associated with a_{2r+1}

20.6.9 $\qquad Ms_{2r}^{(j)}(z, q) = \sum\limits_{k=1}^{\infty} (-1)^{k+r} B_{2k}^{2r}(q)[J_{k-s}(u_1)Z_{k+s}^{(j)}(u_2) - J_{k+s}(u_1)Z_{k-s}^{(j)}(u_2)]/B_{2s}^{2r}$, associated with b_{2r}

20.6.10 $\qquad Ms_{2r+1}^{(j)}(z, q) = \sum\limits_{k=0}^{\infty} (-1)^{k+r} B_{2k+1}^{2r+1}(q)[J_{k-s}(u_1)Z_{k+s+1}^{(j)}(u_2) - J_{k+s+1}(u_1)Z_{k-s}^{(j)}(u_2)]/B_{2s+1}^{2r+1}$

$\qquad\qquad\qquad\qquad\qquad\qquad\qquad\qquad\qquad\qquad\qquad\qquad$ associated with b_{2r+1}

where

$$u_1 = \sqrt{q}e^{-z}, \quad u_2 = \sqrt{q}e^z, \quad B_{2s+p}^{2r+p}, \quad A_{2s+p}^{2r+p} \neq 0, \, p = 0, 1.$$

See **20.4.7** for definition of $Z_m^{(j)}(x)$.

Solutions **20.6.7–20.6.10** converge for all values of z, when $q \neq 0$. If $j = 2, 3, 4$ the logarithmic terms entering into the Bessel functions $Y_m(u_2)$ must be defined, to make the functions single-valued. This can be accomplished as follows:

Define (as in [20.58])

20.6.11 $\qquad\qquad\qquad\qquad\qquad \ln(\sqrt{q}e^z) = \ln(\sqrt{q}) + z$

See [20.15] and [20.36], section **2.75** for derivation.

Other Expressions for the Radial Functions (Valid Over More Limited Regions)

20.6.12
$$Mc_{2r}^{(j)}(z,q)=[ce_{2r}(0,q)]^{-1}\sum_{k=0}^{\infty}(-1)^{k+r}A_{2k}^{2r}(q)Z_{2k}^{(j)}(2\sqrt{q}\cosh z)$$

$$Mc_{2r+1}^{(j)}(z,q)=[ce_{2r+1}(0,q)]^{-1}\sum_{k=0}^{\infty}(-1)^{k+r}A_{2k+1}^{2r+1}(q)Z_{2k+1}^{(j)}(2\sqrt{q}\cosh z)$$

20.6.13
$$Ms_{2r}^{(j)}(z,q)=[se_{2r}'(0,q)]^{-1}\tanh z\sum_{k=1}^{\infty}(-1)^{k+r}2kB_{2k}^{2r}(q)Z_{2k}^{(j)}(2\sqrt{q}\cosh z)$$

$$Ms_{2r+1}^{(j)}(z,q)=[se_{2r+1}'(0,q)]^{-1}\tanh z\sum_{k=0}^{\infty}(-1)^{k+r}(2k+1)B_{2k+1}^{2r+1}(q)Z_{2k+1}^{(j)}(2\sqrt{q}\cosh z)$$

Valid for $\mathscr{R}z>0$, $|\cosh z|>1$; if $j=1$, valid for all z. They agree with **20.6.7–20.6.10** if the Bessel functions $Y_m(2q^{\frac{1}{2}}\cosh z)$ are made single-valued in a suitable way. For example, let

$$Y_m(u)=\frac{2}{\pi}(\ln u)J_m(u)+\phi(u)$$

where $\phi(u)$ is single-valued for all finite values of u. With $u=2q^{\frac{1}{2}}\cosh z$, define

20.6.14
$$\ln(2q^{\frac{1}{2}}\cosh z)=\ln 2q^{\frac{1}{2}}+z+\ln\tfrac{1}{2}(1+e^{-2z})\qquad\qquad -\frac{\pi}{2}\leq\arg\tfrac{1}{2}(1+e^{-2z})\leq\frac{\pi}{2}.$$

(If q is not positive, the phase of $\ln 2q^{\frac{1}{2}}$ must also be specified, although this specification will not affect continuity with respect to z. If $Y_m(u)$ is defined from some other expression, the definition must be compatible with **20.6.14**.)

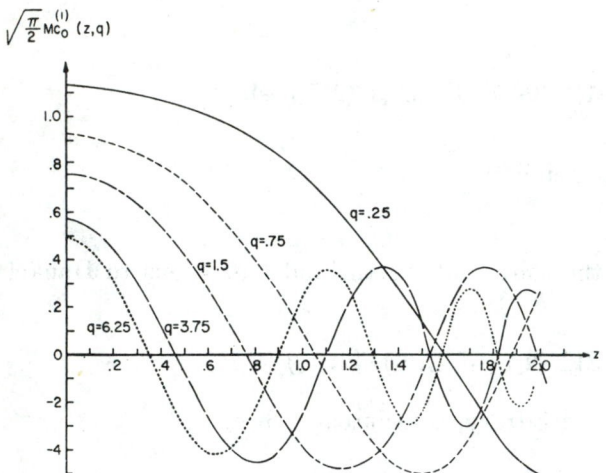

FIGURE 20.11. *Radial Mathieu Function of the First Kind.*

(From J. C. Wiltse and M. J. King, Values of the Mathieu functions, The Johns Hopkins Univ. Radiation Laboratory Tech. Rept. AF-53, 1958, with permission)

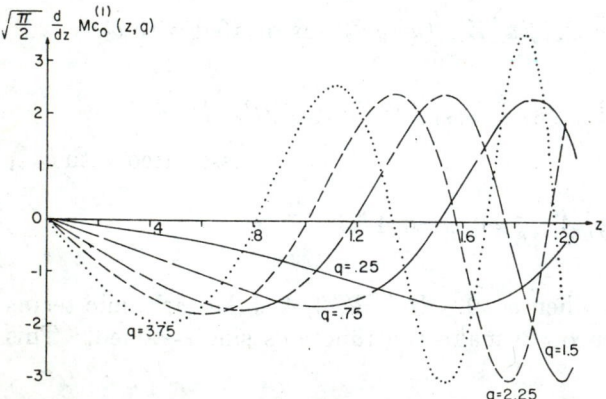

FIGURE 20.12. *Derivative of the Radial Mathieu Function of the First Kind.*

(From J. C. Wiltse and M. J. King, Derivatives, zeros, and other data pertaining to Mathieu functions, The Johns Hopkins Univ. Radiation Laboratory Tech. Rept. AF-57, 1958, with permission)

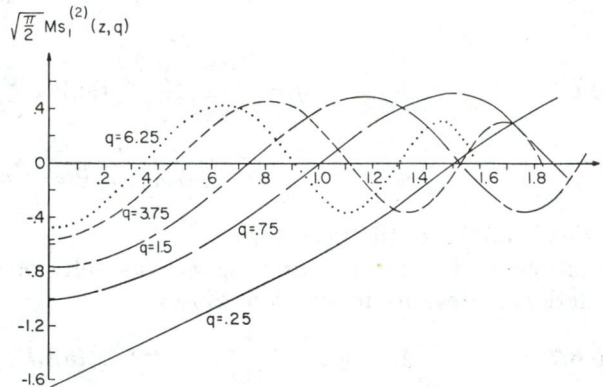

FIGURE 20.13. *Radial Mathieu Function of the Second Kind.*

(From J. C. Wiltse and M. J. King, Values of the Mathieu functions, The Johns Hopkins Univ. Radiation Laboratory Tech. Rept. AF-53, 1958, with permission)

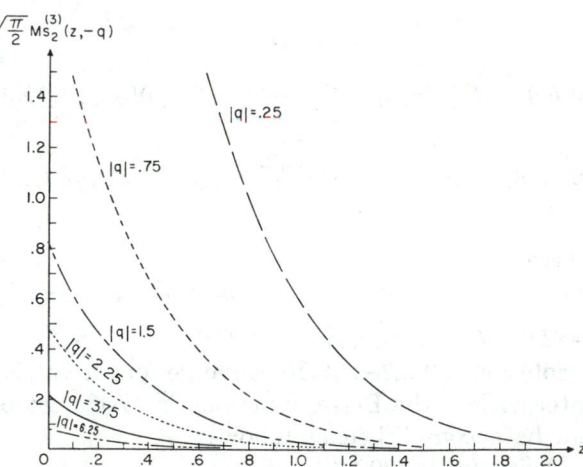

FIGURE 20.14. *Radial Mathieu Function of the Third Kind.*

(From J. C. Wiltse and M. J. King, Values of the Mathieu functions, The Johns Hopkins Univ. Radiation Laboratory Tech. Rept. AF-53, 1958, with permission)

If $j=1$, $Mc_{2r+p}^{(1)}$ and $Ms_{2r+p}^{(1)}$, $p=0, 1$ are solutions of the first kind, proportional to Ce_{2r+p} and Se_{2r+p}, respectively.

Thus

20.6.15

$$Ce_{2r}(z, q) = \frac{ce_{2r}\left(\frac{\pi}{2}, q\right) ce_{2r}(0, q)}{(-1)^r A_0^{2r}} Mc_{2r}^{(1)}(z, q)$$

$$Ce_{2r+1}(z, q) = \frac{ce_{2r+1}'\left(\frac{\pi}{2}, q\right) ce_{2r+1}(0, q)}{(-1)^{r+1}\sqrt{q} A_1^{2r+1}} Mc_{2r+1}^{(1)}(z, q)$$

$$Se_{2r}(z, q) = \frac{se_{2r}'(0, q) se_{2r}'\left(\frac{\pi}{2}, q\right)}{(-1)^r q B_2^{2r}} Ms_{2r}^{(1)}(z, q)$$

$$Se_{2r+1}(z, q) = \frac{se_{2r+1}'(0, q) se_{2r+1}\left(\frac{\pi}{2}, q\right)}{(-1)^r \sqrt{q} B_1^{2r+1}} Ms_{2r+1}^{(1)}(z, q)$$

The Mathieu-Hankel functions are

20.6.16

$$M_r^{(3)}(z, q) = M_r^{(1)}(z, q) + i M_r^{(2)}(z, q)$$

$$M_r^{(4)}(z, q) = M_r^{(1)}(z, q) - i M_r^{(2)}(z, q)$$

$$M_r^{(j)} = Mc_r^{(j)} \text{ or } Ms_r^{(j)}.$$

From **20.6.7–20.6.11** and the known properties of Bessel functions one obtains

20.6.17

$$M_{2r+p}^{(2)}(z+in\pi, q)$$
$$= (-1)^{np}[M_{2r+p}^{(2)}(z, q) + 2ni M_{2r+p}^{(1)}(z, q)]$$

$$M_{2r+p}^{(3)}(z+in\pi, q)$$
$$= (-1)^{np}[M_{2r+p}^{(3)}(z, q) - 2n M_{2r+p}^{(1)}(z, q)]$$

$$M_{2r+p}^{(4)}(z+in\pi, q)$$
$$= (-1)^{np}[M_{2r+p}^{(4)}(z, q) + 2n M_{2r+p}^{(1)}(z, q)]$$

where $M=Mc$ or Ms throughout any of the above equations.

Other Properties of Characteristic Functions, q Real (Associated With a_r and b_r)

Consider

20.6.18

$$X_1 = Mc_r^{(2)}(z, q) + Mc_r^{(2)}(-z, q);$$
$$X_2 = Ms_r^{(2)}(z, q) - Ms_r^{(2)}(-z, q)$$

Since X_1 is an even solution it must be proportional to $Mc_r^{(1)}(z, q)$; for **20.1.2** admits of only one even solution (aside from an arbitrary constant factor). Similarly, X_2 is proportional to $Ms_r^{(1)}(z, q)$. The proportionality factors can be found by considering values of the functions at $z=0$. Define, therefore,

20.6.19

$$Mc_r^{(2)}(-z, q) = -Mc_r^{(2)}(z, q) - 2f_{e, r} Mc_r^{(1)}(z, q)$$

20.6.20

$$Ms_r^{(2)}(-z, q) = Ms_r^{(2)}(z, q) - 2f_{o, r} Ms_r^{(1)}(z, a)$$

where

20.6.21

$$f_{e, r} = -Mc_r^{(2)}(0, q)/Mc_r^{(1)}(0, q)$$

$$f_{o, r} = \left[\frac{d}{dz} Ms_r^{(2)}(z, q) / \frac{d}{dz} Ms_r^{(1)}(z, q)\right]_{z=0}$$

See [20.58].

In particular the above equations can be used to extend solutions of **20.6.12–20.6.13** when $\mathscr{R}z<0$. For although the latter converge for $\mathscr{R}z<0$, provided only $|\cosh z|>1$, they do not represent the same functions as **20.6.9–20.6.10**.

20.7. Representations by Integrals and Some Integral Equations

Let

20.7.1
$$G(u) = \oint_C K(u, t) V(t) dt$$

be defined for u in a domain U and let the contour C belong to the region T of the complex t-plane, with $t=\gamma_0$ as the starting point of the contour and $t=\gamma_1$ as its end-point. The kernel $K(u, t)$ and the function $V(t)$ satisfy **20.7.3** and the hypotheses in **20.7.2**.

20.7.2 $K(u, t)$ and its first two partial derivatives with respect to u and t are continuous for t on C and u in U; V and $\frac{dV}{dt}$ are continuous in t.

20.7.3
$$\left[\frac{\partial K}{\partial t} V - \frac{dV}{dt} K\right]_{\gamma_0}^{\gamma_1} = 0; \quad \frac{d^2V}{dt^2} + (a - 2q \cos 2t) V = 0.$$

If K satisfies

20.7.4 $\quad \frac{\partial^2 K}{\partial u^2} + \frac{\partial^2 K}{\partial t^2} + 2q(\cosh 2u - \cos 2t) K = 0$

then $G(u)$ is a solution of Mathieu's modified equation **20.1.2**.

If $K(u, t)$ satisfies

20.7.5 $\quad \frac{\partial^2 K}{\partial u^2} + \frac{\partial^2 K}{\partial t^2} + 2q(\cos 2u - \cos 2t) K = 0$

then $G(u)$ is a solution of Mathieu's equation **20.1.1**, with u replacing v.

Kernels $K_1(z, t)$ and $K_2(z, t)$

20.7.6 $K_1(z, t) = Z_\nu^{(j)}(u)[M(z, t)]^{-\nu/2}$, $(\mathscr{R}z > 0)$

where

20.7.7 $u = \sqrt{2q(\cosh 2z + \cos 2t)}$

20.7.8 $M(z, t) = \cosh (z + it)/\cosh (z - it)$

To make $M^{-\frac{1}{2}\nu}$ single-valued, define

20.7.9

$$\cosh (z + i\pi) = e^{i\pi} \cosh z$$
$$\cosh (z - i\pi) = e^{-i\pi} \cosh z$$
$$M(z, 0) = 1$$
$$[M(z, \pi)]^{-\frac{1}{2}\nu} = e^{-i\nu\pi}M(z, 0)$$

Let

20.7.10 $G(z, q) = \dfrac{1}{\pi} \displaystyle\int_0^\pi K_1(u, t)F_\nu(t)dt$, $(\mathscr{R}z > 0)$

where $F_\nu(t)$ is defined in **20.3.8**. It may be verified that K_1F_ν satisfies **20.7.3**, K satisfies **20.7.2** and **20.7.4**. Hence G is a solution of **20.1.2** (with z replacing u). It can be shown that K_1 may be replaced by the more general function

20.7.11

$K_2(z, t) = Z_{\nu+2s}^{(j)}(u)[M(z, t)]^{-\frac{1}{2}\nu+s}$, s any integer.

See **20.4.7** for definition of $Z_{\nu+2s}^{(j)}(u)$.

From the known expansions for $Z_{\nu+2s}^{(j)}(u)$ when $\mathscr{R}z$ is large and positive it may be verified that

20.7.12

$M_\nu^{(j)}(z, q) =$

$$\frac{(-1)^s}{\pi c_{2s}} \int_0^\pi Z_{\nu+2s}^{(j)}(u) \left[\frac{\cosh z + it}{\cosh z - it}\right]^{-\frac{1}{2}\nu - s} F_\nu(t)dt$$
$$(\mathscr{R}z > 0, \ \mathscr{R}(\nu + \tfrac{1}{2}) > 0)$$

where $M_\nu^{(j)}(z, q)$ is given by **20.4.12**, $s = 0, 1, \ldots,$ $c_{2s} \neq 0$, and $F_\nu(t)$ is the Floquet solution, **20.3.8**.

Kernel $K_3(z, t, a)$

20.7.13 $K_3(z, t, a) = e^{2i\sqrt{q}\,w}$

where

20.7.14 $w = \cosh z \cos a \cos t + \sinh z \sin a \sin t$

20.7.15 $G(z, q, a) = \dfrac{1}{\pi} \displaystyle\oint_C e^{2i\sqrt{q}\,w}F_\nu(t)dt$

where $F_\nu(t)$ is the Floquet solution **20.3.8**. The path C is chosen so that $G(z, t, a)$ exists, and **20.7.2**, **20.7.3** are satisfied. Then it may be verified that $K_3(z, t, a)$, considered as a function of z and t, satisfies **20.7.4**; also, considered as a function of a and t, K_3 satisfies **20.7.5**. Consequently $G(z, q, a) = Y(z, q)y(a, q)$, where Y and y satisfy **20.1.2** and **20.1.1**, respectively.

Choice of Path C. Three paths will be defined:

20.7.16

Path C_3: from $-d_1 + i\infty$ to $d_2 - i\infty$, d_1, d_2 real

$$-d_1 < \arg [\sqrt{q}\{\cosh (z + ia) \pm 1\}] < \pi - d_1$$
$$-d_2 < \arg [\sqrt{q}\{\cosh (z - ia) \pm 1\}] < \pi - d_2$$

20.7.17

Path C_4: from $d_2 - i\infty$ to $2\pi + i\infty - d_1$

(same d_1, d_2 as in **20.7.16**)

20.7.18

$$F_\nu(a)M_\nu^j(z, q) = \frac{e^{-i\nu\frac{\pi}{2}}}{\pi} \oint_{C_j} e^{2i\sqrt{q}\,w}F_\nu(t)dt \qquad j = 3, 4$$

where $M_\nu^j(z, q)$ is also given by **20.4.12**.

20.7.19   Path C_1: from $-d_1 + i\infty$ to $2\pi - d_1 + i\infty$

$$F_\nu(a)M_\nu^{(1)}(z, q) = \frac{e^{-i\nu\frac{\pi}{2}}}{2\pi} \oint_{C_1} e^{2i\sqrt{q}\,w}F_\nu(t)dt$$

See [20.36], section **2.68**.

If ν is an integer the paths can be simplified; for in that case $F_\nu(t)$ is periodic and the integrals exist when the path is taken from 0 to 2π. Still further simplifications are possible, if z is also real.

The following are among the more important integral representations for the periodic functions $ce_r(z, q)$, $se_r(z, q)$ and for the associated radial solutions.

Let $r = 2s + p$, $p = 0$ or 1

20.7.20

$$ce_r(z, q) = \rho_r \int_0^{\pi/2} \cos\left(2\sqrt{q} \cos z \cos t - p\frac{\pi}{2}\right) ce_r(t, q)dt$$

20.7.21
$$ce_r(z, q) = \sigma_r \int_0^{\pi/2} \cosh(2\sqrt{q}\,\sin z \sin t)[(1-p) + p \cos z \cos t]ce_r(t, q)dt$$

20.7.22
$$se_r(z, q) = \rho_r \int_0^{\pi/2} \sin\left(2\sqrt{q}\cos z \cos t + p\frac{\pi}{2}\right)\sin z \sin t\, se_r(t, q)dt$$

20.7.23
$$se_r(z, q) = \sigma_r \int_0^{\pi/2} \sinh(2\sqrt{q}\,\sin z \sin t)[(1-p)\cos z \cos t + p]se_r(t, q)dt$$

where

20.7.24 $\quad \rho_r = \frac{2}{\pi}ce_{2s}\left(\frac{\pi}{2}, q\right)/A_0^{2s}(q); p=0\; \rho_r = \frac{-2}{\pi}ce_{2s+1}'\left(\frac{\pi}{2}, q\right)/\sqrt{q}A_1^{2s+1}(q)$ if $p=1$, for functions $ce_r(z, q)$

$$\rho_r = \frac{-4}{\pi}se_{2s}'\left(\frac{\pi}{2}, q\right)/\sqrt{q}B_2^{2s}(q); \; \rho_r = \frac{4}{\pi}se_{2s+1}\left(\frac{\pi}{2}, q\right)/B_1^{2s+1}(q), \text{ for functions } se_r(z, q)$$

$$\sigma_r = \frac{2}{\pi}ce_{2s}(0, q)/A_0^{2s}(q) \text{ if } p=0; \qquad \sigma_r = \frac{4}{\pi}ce_{2s+1}(0, q)/A_1^{2s+1}(q), \text{ if } p=1; \text{ associated with functions } ce_r(z, q)$$

$$\sigma_r = \frac{4}{\pi}se_{2s}'(0, q)/\sqrt{q}B_2^{2s}(q), \text{ if } p=0; \qquad \sigma_r = \frac{2}{\pi}se_{2s+1}'(0, q)/\sqrt{q}B_1^{2s+1}(q), \text{ if } p=1; \text{ associated with } se_r(z, q)$$

Integrals Involving Bessel Function Kernels

Let

20.7.25 $\qquad u = \sqrt{2q(\cosh 2z + \cos 2t)}, \; (\mathscr{R}\cosh 2z > 1; \text{ if } j=1, \text{ valid also when } z=0)$

20.7.26

$$Mc_{2r}^{(j)}(z, q) = \frac{(-1)^r 2}{\pi A_0^{2r}}\int_0^{\frac{\pi}{2}} Z_0^{(j)}(u)ce_{2r}(t, q)dt; \; Mc_{2r+1}^{(j)}(z, q) = \frac{(-1)^r 8\sqrt{q}\cosh z}{\pi A_1^{2r+1}}\int_0^{\frac{\pi}{2}}\frac{Z_1^{(j)}(u)\cos t}{u}ce_{2r+1}(t, q)dt$$

20.7.27
$$Ms_{2r}^{(j)}(z, q) = \frac{(-1)^{r+1}8q\sinh 2z}{\pi B_2^{2r}}\int_0^{\frac{\pi}{2}}\frac{Z_2^{(j)}(u)\sin 2t\, se_{2r}(t, q)dt}{u^2}$$

$$Ms_{2r+1}^{(j)}(z, q) = \frac{(-1)^r 8\sqrt{q}\sinh z}{\pi B_1^{2r+1}}\int_0^{\frac{\pi}{2}}\frac{Z_1^{(j)}(u)\sin t\, se_{2r+1}(t, q)dt}{u}$$

In the above the j-convention of **20.4.7** applies and the functions Mc, Ms are defined in **20.5.1–20.5.4**. (These solutions are normalized so that they approach the corresponding Bessel-Hankel functions as $\mathscr{R}z \to \infty$.)

Other Integrals for $Mc_r^{(1)}(z, q)$ and $Ms_r^{(1)}(z, q)$

20.7.28
$$Mc_r^{(1)}(z, q) = \frac{(-1)^s 2}{\pi ce_r(0, q)}\int_0^{\frac{\pi}{2}}\cos\left(2\sqrt{q}\cosh z \cos t - p\frac{\pi}{2}\right)ce_r(t, q)dt$$

20.7.29
$$Mc_r^{(1)}(z, q) = \tau_r \int_0^{\frac{\pi}{2}}[(1-p) + p\cosh z \cos t]\cos(2\sqrt{q}\sinh z \sin t)ce_r(t, q)dt$$

$$r = 2s+p, p=0,1; \; \tau_r = \frac{2}{\pi}(-1)^s/ce_{2s}\left(\frac{\pi}{2}, q\right), \text{ if } p=0; \; \tau_r = \frac{2}{\pi}(-1)^{s+1}2\sqrt{q}/ce_{2s+1}'\left(\frac{\pi}{2}, q\right)$$

20.7.30
$$Ms_{2r+1}^{(1)}(z, q) = \frac{2}{\pi}\frac{(-1)^r}{se_{2r+1}\left(\frac{\pi}{2}, q\right)}\int_0^{\frac{\pi}{2}}\sin(2\sqrt{q}\sinh z \sin t)se_{2r+1}(t, q)dt$$

20.7.31
$$Ms_{2r+1}^{(1)}(z, q) = \frac{4}{\pi}\frac{\sqrt{q}(-1)^r}{se_{2r+1}'(0, q)}\int_0^{\frac{\pi}{2}}\sinh z \sin t \cos(2\sqrt{q}\cosh z \cos t)se_{2r+1}(t, q)dt$$

20.7.32
$$Ms_{2r}^{(1)}(z, q) = \frac{4}{\pi}\sqrt{q}\frac{(-1)^{r+1}}{se_{2r}'(0, q)}\int_0^{\frac{\pi}{2}}\sin(2\sqrt{q}\cosh z \cos t)[\sinh z \sin t\, se_{2r}(t, q)]dt$$

20.7.33
$$Ms_{2r}^{(1)}(z, q) = \frac{4}{\pi}\frac{(-1)^r\sqrt{q}}{se_{2r}'\left(\frac{\pi}{2}, q\right)}\int_0^{\frac{\pi}{2}}\sin(2\sqrt{q}\sinh z \sin t)[\cosh z \cos t\, se_{2r}(t, q)]dt$$

Further with $w = \cosh z \cos \alpha \cos t + \sinh z \sin \alpha \sin t$

20.7.34
$$ce_r(\alpha, q)Mc_r^{(1)}(z, q) = \frac{(-1)^s(i)^{-p}}{2\pi} \int_0^{2\pi} e^{2i\sqrt{q}\,w} ce_r(t, q)dt$$

20.7.35
$$se_r(\alpha, q)Ms_r^{(1)}(z, q) = \frac{(-1)^s(-i)^p}{2\pi} \int_0^{2\pi} e^{2i\sqrt{q}\,w} se_r(t, q)dt.$$

The above can be differentiated with respect to α, and we obtain

20.7.36
$$ce_r'(\alpha, q)Mc_r^{(1)}(z, q) = \frac{(-1)^s(i)^{-p+1}\sqrt{q}}{\pi} \int_0^{2\pi} e^{2i\sqrt{q}\,w} \frac{\partial w}{\partial \alpha} ce_r(t, q)dt$$

20.7.37
$$se_r'(\alpha, q)Ms_r^{(1)}(z, q) = \frac{(-1)^{s+p}(i)^{-p+1}\sqrt{q}}{\pi} \int_0^{2\pi} e^{2i\sqrt{q}\,w} \frac{\partial w}{\partial \alpha} se_r(t, q)dt$$

Integrals With Infinite Limits

$$r = 2s + p$$

In **20.7.38–20.7.41** below, z and q are positive.

20.7.38
$$Mc_r^{(1)}(z, q) = \gamma_r \int_0^\infty \sin\left(2\sqrt{q}\cosh z \cosh t + p\frac{\pi}{2}\right) Mc_r^{(1)}(t, q)dt$$

$$\gamma_r = 2ce_{2s}\left(\frac{\pi}{2}, q\right)/\pi A_0^{2s}, \text{ if } p=0 \qquad \gamma_r = 2ce_{2s+1}'\left(\frac{\pi}{2}, q\right)/\sqrt{q}\,\pi A_1^{2s+1}, \text{ if } p=1$$

20.7.39
$$Ms_r^{(1)}(z, q) = \gamma_r \int_0^\infty \sinh z \sinh t \left[\cos\left(2\sqrt{q}\cosh z \cosh t - p\frac{\pi}{2}\right)\right] Ms_r^{(1)}(t, q)dt$$

$$\gamma_r = -4se_{2s}'\left(\frac{\pi}{2}, q\right)/\sqrt{q}\pi B_2^{2s}, \text{ if } p=0 \qquad \gamma_r = -4se_{2s+1}\left(\frac{\pi}{2}, q\right)/\pi B_1^{2s+1}, \text{ if } p=1$$

20.7.40
$$Mc_r^{(2)}(z, q) = \gamma_r \int_0^\infty \cos\left(2\sqrt{q}\cosh z \cosh t - p\frac{\pi}{2}\right) Mc_r^{(1)}(t, q)dt$$

$$\gamma_r = -2ce_{2s}(\tfrac{1}{2}\pi, q)/\pi A_0^{2s}, \text{ if } p=0 \qquad \gamma_r = 2ce_{2s+1}'(\tfrac{1}{2}\pi, q)/\pi\sqrt{q}A_1^{2s+1}, \text{ if } p=1$$

20.7.41
$$Ms_r^{(2)}(z, q) = \gamma_r \int_0^\infty \sin\left(2\sqrt{q}\cosh z \cosh t + p\frac{\pi}{2}\right) \sinh z \sinh t\, Ms_r^{(1)}(t, q)dt$$

$$\gamma_r = -4se_{2s}'(\tfrac{1}{2}\pi, q)/\sqrt{q}\,\pi B_2^{2s}, \text{ if } p=0 \qquad \gamma_r = 4se_{2s+1}(\tfrac{1}{2}\pi, q)/\pi B_1^{2s+1}, \text{ if } p=1$$

Additional forms in [20.30], [20.36], [20.15].

20.8. Other Properties

Relations Between Solutions for Parameters q and $-q$

Replacing z by $\frac{1}{2}\pi - z$ in **20.1.1** one obtains

20.8.1 $\qquad y'' + (a + 2q\cos 2z)y = 0$

Hence if $u(z)$ is a solution of **20.1.1** then $u(\frac{1}{2}\pi - z)$ satisfies **20.8.1**. It can be shown that

20.8.2

$$a(-\nu, q) = a(\nu, -q) = a(\nu, q), \nu \text{ not an integer}$$

$$c_{2m}^\nu(-q) = \rho(-1)^m c_{2m}^\nu(q), \nu \text{ not an integer}$$

(c_{2m} defined in **20.3.8**) and ρ depending on the normalization;

$$F_\nu(z, -q) = \rho e^{-i\nu\pi/2}F_\nu\left(z + \frac{\pi}{2}, q\right) = \rho e^{i\nu\pi/2}F_\nu\left(z - \frac{\pi}{2}, q\right)$$

20.8.3

$$a_{2r}(-q) = a_{2r}(q) \; ; \; b_{2r}(-q) = b_{2r}(q), \text{ for integral } \nu$$

$$a_{2r+1}(-q) = b_{2r+1}(q), \; b_{2r+1}(-q) = a_{2r+1}(q)$$

20.8.4

$$ce_{2r}(z, -q) = (-1)^r ce_{2r}(\tfrac{1}{2}\pi - z, q)$$

$$ce_{2r+1}(z, -q) = (-1)^r se_{2r+1}(\tfrac{1}{2}\pi - z, q)$$

$$se_{2r+1}(z, -q) = (-1)^r ce_{2r+1}(\tfrac{1}{2}\pi - z, q)$$

$$se_{2r}(z, -q) = (-1)^{r-1} se_{2r}(\tfrac{1}{2}\pi - z, q)$$

For the coefficients associated with the above solutions for integral ν:

20.8.5

$$A_{2m}^{2r}(-q) = (-1)^{m-r} A_{2m}^{2r}(q);$$
$$B_{2m}^{2r}(-q) = (-1)^{m-r} B_{2m}^{2r}(q)$$

$$A_{2m+1}^{2r+1}(-q) = (-1)^{m-r} B_{2m+1}^{2r+1}(q);$$
$$B_{2m}^{2r+1}(-q) = (-1)^{m-r} A_{2m+1}^{2r+1}(q).$$

For the corresponding modified equation

20.8.6 $\qquad y'' - (a + 2q \cosh 2z)y = 0$

20.8.7

$$M_\nu^{(j)}(z, -q) = M_\nu^{(j)}\left(z + i\frac{\pi}{2}, q\right),$$

$\qquad\qquad M_r^{(j)}(z, q)$ defined in **20.4.12**.

For integral values of ν let

20.8.8

$$Ie_{2r}(z, q) = \sum_{k=0}^{\infty} (-1)^{k+s} A_{2k}[I_{k-s}(u_1)I_{k+s}(u_2) + I_{k+s}(u_1)I_{k-s}(u_2)]/A_{2s}\epsilon_s$$

$$Io_{2r}(z, q) = \sum_{k=1}^{\infty} (-1)^{k+s} B_{2k}[I_{k-s}(u_1)I_{k+s}(u_2) - I_{k+s}(u_1)I_{k-s}(u_2)]/B_{2s}$$

$$Ie_{2r+1}(z, q) = \sum_{k=0}^{\infty} (-1)^{k+s} B_{2k+1}[I_{k-s}(u_1)I_{k+s+1}(u_2) + I_{k+s+1}(u_1)I_{k-s}(u_2)]/B_{2s+1}$$

$$Io_{2r+1}(z, q) = \sum_{k=0}^{\infty} (-1)^{k+s} A_{2k+1}[I_{k-s}(u_1)I_{k+s+1}(u_2) - I_{k+s+1}(u_1)I_{k-s}(u_2)]/A_{2s+1}$$

20.8.9

$$Ke_{2r}(z, q) = \sum_{k=0}^{\infty} A_{2k}[I_{k-s}(u_1)K_{k+s}(u_2) + I_{k+s}(u_1)K_{k-s}(u_2)]/A_{2s}\epsilon_s$$

* $$Ko_{2r}(z, q) = \sum_{k=0}^{\infty} B_{2k}[I_{k-s}(u_1)K_{k+s}(u_2)$$
* $$\qquad\qquad - I_{k+s}(u_1)K_{k-s}(u_2)]/B_{2s}$$

$$Ke_{2r+1}(z, q) = \sum_{k=0}^{\infty} B_{2k+1}[I_{k-s}(u_1)K_{k+s+1}(u_2) - I_{k+s+1}(u_1)K_{k-s}(u_2)]/B_{2s+1}$$

---

*See page II.

$$Ko_{2r+1}(z, q) = \sum_{k=0}^{\infty} A_{2k+1}[I_{k-s}(u_1)K_{k+s+1}(u_2) + I_{k+s+1}(u_1)K_{k-s}(u_2)]/A_{2s+1}$$

where $I_m(x)$, $K_m(x)$ are the modified Bessel functions, u_1, u_2 are defined below **20.6.10**. Superscripts are omitted, $\epsilon_s = 2$, if $s = 0$, $\epsilon_s = 1$ if $s \neq 0$.

Then for functions of first kind:

20.8.10

$$Mc_{2r}^{(1)}(z, -q) = (-1)^r Ie_{2r}(z, q)$$

$$Ms_{2r}^{(1)}(z, -q) = (-1)^r Io_{2r}(z, q)$$

$$Mc_{2r+1}^{(1)}(z, -q) = (-1)^r i Ie_{2r+1}(z, q)$$

$$Ms_{2r+1}^{(1)}(z, -q) = (-1)^r i Io_{2r+1}(z, q)$$

For the Mathieu-Hankel function of first kind:

20.8.11

$$Mc_{2r}^{(3)}(z, -q) = (-1)^{r+1} i \frac{2}{\pi} Ke_{2r}(z, q)$$

$$Ms_{2r}^{(3)}(z, -q) = (-1)^{r+1} i \frac{2}{\pi} Ko_{2r}(z, q)$$

$$Mc_{2r+1}^{(3)}(z, -q) = (-1)^{r+1} \frac{2}{\pi} Ke_{2r+1}(z, q)$$

$$Ms_{2r+1}^{(3)}(z, -q) = (-1)^{r+1} \frac{2}{\pi} Ko_{2r+1}(z, q)$$

For $M_r^{(j)}(z, -q)$, $j = 2, 4$, one may use the definitions

$$M_r^{(2)} = -i(M_r^{(3)} - M_r^{(1)}); \; M_r = Mc_r \text{ or } Ms_r$$

also

$$M_r^{(4)}(z, -q) = 2M_r^{(1)}(z, -q) - M_r^{(3)}(z, -q)$$

$$M = Mc \text{ or } Ms; \text{ for real } z, q, M_r^{(j)}(z, -q)$$

are in general complex if $j = 2, 4$.

Zeros of the Functions for Real Values of q.

See [20.36], section **2.8** for further results.

Zeros of $ce_r(z, q)$ and $se_r(z, q)$, $Mc_r^{(1)}(z, q)$, $Ms_r^{(1)}(z, q)$.

In $0 \leq z < \pi$, $ce_r(z, q)$ and $se_r(z, q)$ have r real * zeros.

There are complex zeros if $q > 0$.

If $z_0 = x_0 + iy_0$ is any zero of $ce_r(z, q)$, $se_r(z, q)$ in

$$-\frac{\pi}{2} < x_0 < \frac{\pi}{2}, \text{ then } k\pi \pm z_0, \; k\pi \pm \bar{z}_0$$

are also zeros, k an integer.

In the strip $-\frac{\pi}{2}<x_0<\frac{\pi}{2}$, the imaginary zeros of $ce_r(z, q)$, $se_r(z, q)$ are the real zeros of $Ce_r(z, q)$, $Se_r(z, q)$, hence also the real zeros of $Mc_r^{(1)}(z, q)$ and $Ms_r^{(1)}(z, q)$, respectively.

For small q, the large zeros of $Ce_r(z, q)$, $Se_r(z, q)$ approach the zeros of $J_r(2\sqrt{q}\cosh z)$.

Tabulation of Zeros

Ince [20.56] tabulates the first "non-trivial" zero $\left(\text{i.e. different from } 0, \frac{\pi}{2}, \pi\right)$ for $ce_r(z)$, $se_r(z)$, $r=2(1)5$ and for $se_6(z)$ to within °10^{-4}, for $q=0(1)$ $10(2)40$. He also gives the "turning" points (zeros of the derivative) and also expansions for them for small q. Wiltse and King [20.61,2] tabulate the first two (non-trivial) zeros of $Mc_r^{(1)}(z, q)$ and $Ms_r^{(1)}(z, q)$ and of their derivatives $r=0, 1, 2$ for 6 or 7 values of q between .25 and 10. The graphs reproduced here indicate their location.

Between two real zeros of $Mc_r^{(1)}(z, q)$, $Ms_r^{(1)}(z, q)$ there is a zero of $Mc_r^{(2)}(z, q)$, $Ms_r^{(2)}(z, q)$, respectively. No tabulation of such zeros exists yet.

Available tables are described in the References. The most comprehensive tabulation of the characteristic values a_r, b_r (in a somewhat different notation) and of the coefficients proportional to A_m and B_m as defined in **20.5.4** and **20.5.5** can be found in [20.58]. In addition, the table contains certain important "joining factors", with the aid of which it is possible to obtain values of $Mc_r^{(j)}(z, q)$ and $Ms_r^{(j)}(z, q)$ as well as their derivatives, at $x=0$. Values of the functions $ce_r(x, q)$ and $se_r(x, q)$ for orders up to five or six can be found in [20.56]. Tabulations of less extensive character, but important in some aspects, are outlined in the other references cited. In this chapter only representative values of the various functions are given, along with several graphs.

Special Values for Arguments 0 and $\frac{\pi}{2}$

20.8.12

$$ce_{2r}\left(\frac{\pi}{2}, q\right)=(-1)^r g_{e,2r}(q)A_0^{2r}(q)\sqrt{\frac{\pi}{2}}$$

$$ce'_{2r+1}\left(\frac{\pi}{2}, q\right)=(-1)^{r+1}g_{e,2r+1}(q)A_1^{2r+1}(q)\sqrt{\frac{\pi}{2}q}$$

$$se'_{2r}\left(\frac{\pi}{2}, q\right)=(-1)^r g_{0,2r}(q)B_2^{2r}(q)\cdot q\sqrt{\frac{\pi}{2}}$$

$$se_{2r+1}\left(\frac{\pi}{2}, q\right)=(-1)^r g_{0,2r+1}(q)B_1^{2r+1}(q)\sqrt{\frac{\pi}{2}q}$$

$$Mc_r^{(1)}(0, q)=\sqrt{\frac{2}{\pi}}\frac{1}{g_{e,r}(q)}$$

$$Mc_r^{(2)}(0, q)=-\sqrt{\frac{2}{\pi}}f_{e,r}(q)/g_{e,r}(q)$$

$$\frac{d}{dz}[Mc_r^{(2)}(z, q)]_{z=0}=\sqrt{\frac{2}{\pi}}g_{e,r}(q)$$

$$\frac{d}{dz}[Ms_r^{(1)}(z, q)]_{z=0}=\sqrt{\frac{2}{\pi}}\frac{1}{g_{0,r}(q)}$$

$$\frac{d}{dz}\left[Ms_r^{(2)}(z, q)\right]_{z=0}=\sqrt{\frac{2}{\pi}}f_{0,r}(q)/g_{0,r}(q)$$

$$Ms_r^{(2)}(z, q)=-g_{0,r}(q)\sqrt{\frac{2}{\pi}}$$

The functions $f_{0,r}, g_{0,r}, f_{e,r}, g_{e,r}$ are tabulated in [20.58] for $q\leq25$.

20.9. Asymptotic Representations

The representations given below are applicable to the *characteristic solutions*, for real values of q, unless otherwise noted. The Floquet exponent ν is defined below, as in [20.36] to be as follows:

In solutions associated with a_r: $\nu=r$
In solutions associated with b_r: $\nu=-r$.

For the functions defined in **20.6.7–20.6.10**:

20.9.1

$$Mc_r^{(3)}(z, q)$$
$$(-1)^r Ms_r^{(3)}(z, q)$$

$$\sim\frac{e^{i\left(2\sqrt{q}\cosh z-\frac{\nu\pi}{2}-\frac{\pi}{4}\right)}}{\pi^{\frac{1}{2}}q^{1/4}(\cosh z-\sigma)^{\frac{1}{2}}}\sum_{m=0}^{\infty}\frac{D_m}{[-4i\sqrt{q}(\cosh z-\sigma)]^m}$$

where $D_{-1}=D_{-2}=0$; $D_0=1$, and the coefficients D_m are obtainable from the following recurrence formula:

20.9.2

$$(m+1)D_{m+1}+\left[\left(m+\frac{1}{2}\right)^2-\left(m+\frac{1}{4}\right)8i\sqrt{q}\,\sigma\right.$$
$$\left.+2q-a\right]D_m+\left(m-\frac{1}{2}\right)[16q(1-\sigma^2)-8i\sqrt{q}\,\sigma m]D_{m-1}$$
$$+4q(2m-3)(2m-1)(1-\sigma^2)D_{m-2}=0$$

20.9.3

$$Mc_r^{(4)}(z, q)$$
$$(-1)^r Ms_r^{(4)}(z, q)$$

$$\sim\frac{e^{-i\left[2\sqrt{q}\cosh z-\frac{1}{2}\nu\pi-\frac{1}{4}\pi\right]}}{\pi^{\frac{1}{2}}q^{1/4}(\cosh z-\sigma)^{\frac{1}{2}}}\sum_{m=0}^{\infty}\frac{d_m}{[4i\sqrt{q}(\cosh z-\sigma)]^m}$$

$d_{-1}=d_{-2}=0$; $d_0=1$, and

20.9.4

$$(m+1)d_{m+1}+\left[\left(m+\tfrac{1}{2}\right)^2+\left(m+\tfrac{1}{4}\right)8i\sqrt{q}\,\sigma\right.$$
$$\left.+2q-a\right]d_m+\left(m-\tfrac{1}{2}\right)[16q(1-\sigma^2)+8i\sqrt{q}\,\sigma m]d_{m-1}$$
$$+4q(2m-3)(2m-1)(1-\sigma^2)d_{m-2}=0.$$

In the above

$$-2\pi<\arg\sqrt{q}\cosh z<\pi$$
$$|\cosh z-\sigma|>|\sigma\pm1|,\ \mathscr{R}z>0,$$

but σ is otherwise arbitrary. If $\sigma^2=1$, **20.9.2** and **20.9.4** become three-term recurrence relations.

Formulas **20.9.1** and **20.9.3** are valid for arbitrary a, q, provided ν is also known; they give multiples of **20.4.12**, normalized so as to approach the corresponding Hankel functions $H_\nu^{(1)}(\sqrt{q}e^z)$, $H_\nu^{(2)}(\sqrt{q}e^z)$, as $z\rightarrow\infty$. See [20.36], section **2.63**. The formula is especially useful if $|\cosh z|$ is large and q is not too large; thus if $\sigma=-1$, the absolute ratio of two successive terms in the expansion is essentially

$$\left|\left(\frac{\sqrt{q}}{m}+\frac{m}{4\sqrt{q}}+2\right)/(\cosh z+1)\right|.$$

If a, q, z, ν are real, the real and imaginary components of $Mc_r^{(3)}(z,q)$ are $Mc_r^{(1)}(z,q)$ and $Mc_r^{(2)}(z,q)$, respectively; similarly for the components of $Ms_r^{(3)}(z,q)$. If the parameters are complex

20.9.5 $\quad Mc_r^{(1)}(z,q)=\tfrac{1}{2}[Mc_r^{(3)}(z,q)+Mc_r^{(4)}(z,q)]$

20.9.6 $\quad Mc_r^{(2)}(z,q)=-\dfrac{i}{2}[Mc_r^{(3)}(z,q)-Mc_r^{(4)}(z,q)]$

Replacing c by s in the above will yield corresponding relations among $Ms_r^{(j)}(z,q)$.

Formulas in which the parameter a does not enter explicitly:

Goldstein's Expansions

20.9.7
$$Mc_r^{(3)}(z,q)\sim iMs_{r+1}^{(3)}(z,q)$$
$$\approx[F_0(z)-iF_1(z)]e^{i\phi}/\pi^{\frac{1}{2}}q^{\frac{1}{4}}(\cosh z)^{\frac{1}{2}}$$

where

20.9.8

$$\phi=2\sqrt{q}\sinh z-\tfrac{1}{2}(2r+1)\arctan\sinh z,$$
$$\mathscr{R}z>0,\ q\gg1,\ w=2r+1$$

20.9.9

$$F_0(z)\sim1+\frac{w}{8\sqrt{q}\cosh^2 z}$$
$$+\frac{1}{2048q}\left[\frac{w^4+86w^2+105}{\cosh^4 z}-\frac{w^4+22w^2+57}{\cosh^2 z}\right]$$
$$+\frac{1}{16384q^{3/2}}\left[\frac{-(w^5+14w^3+33w)}{\cosh^2 z}\right.$$
$$\left.-\frac{(2w^5+124w^3+1122w)}{\cosh^4 z}+\frac{3w^5+290w^3+1627w}{\cosh^6 z}\right]+\cdots$$

20.9.10

$$F_1(z)\sim\frac{\sinh z}{\cosh^2 z}\left[\frac{w^2+3}{32\sqrt{q}}+\frac{1}{512q}\left(w^3+3w+\frac{4w^3+44w}{\cosh^2 z}\right)\right.$$
$$+\frac{1}{16384q^{\frac{3}{2}}}\left\{5w^4+34w^2+9\right.$$
$$-\frac{(w^6-47w^4+667w^2+2835)}{12\cosh^2 z}$$
$$\left.\left.+\frac{(w^6+505w^4+12139w^2+10395)}{12\cosh^4 z}\right\}\right]+\cdots$$

See [20.18] for details and an added term in $q^{-5/2}$; a correction to the latter is noted in [20.58].

The expansions **20.9.7** are especially useful when q is large and z is bounded away from zero. The order of magnitude of $Mc_r^2(0,q)$ cannot be obtained from the expansion. The expansion can also be used, with some success, for $z=ix$, when q is large, if $|\cos x|\gg0$; they fail at $x=\tfrac{1}{2}\pi$. Thus, if q, x are real, one obtains

20.9.11

$$ce_r(x,q)\sim\frac{ce_r(0,q)2^{r-\frac{1}{2}}}{F_0(0)}\{W_1[P_0(x)-P_1(x)]$$
$$+W_2[P_0(x)+P_1(x)]\}$$

20.9.12
$$se_{r+1}(x,q)\sim se'_{r+1}(0,q)\tau_{r+1}\{W_1[P_0(x)-P_1(x)]$$
$$-W_2[P_0(x)+P_1(x)]\}$$

In the above, $P_0(x)$ and $P_1(x)$ are obtainable from $F_0(z)$, $F_1(x)$ in **20.9.9–20.9.10** by replacing $\cosh z$ with $\cos x$ and $\sinh z$ with $\sin x$. Thus $P_0(x)=F_0(ix)$; $P_1(x)=-iF_1(ix)$:

20.9.13

$$W_1=e^{2\sqrt{q}\sin x}[\cos(\tfrac{1}{2}x+\tfrac{1}{4}\pi)]^{2r+1}/(\cos x)^{r+1}$$

$$W_2=e^{-2\sqrt{q}\sin x}[\sin(\tfrac{1}{2}x+\tfrac{1}{4}\pi)]^{2r+1}/(\cos x)^{r+1}$$

20.9.14

$$\tau_{r+1} \sim 2^{r-\frac{1}{2}} \bigg/ \left[2\sqrt{q} - \tfrac{1}{4}w - \frac{(2w^2+3)}{64\sqrt{q}} - \frac{(7w^3+47w)}{1024q} - \cdots\right]$$

See **20.9.23–20.9.24** for expressions relating to $ce_r(0, q)$ and $se_r'(0, q)$. When $|\cos x| > \sqrt{4r+2}/q^{\frac{1}{4}}$, **20.9.11–20.9.12** are useful. The approximations become poorer as r increases.

Expansions in Terms of Parabolic Cylinder Functions

(Good for angles close to $\tfrac{1}{2}\pi$, for large values of q, especially when $|\cos x| < 2^{\frac{1}{2}}/q^{\frac{1}{4}}$.) Due to Sips [20.44–20.46].

20.9.15 $ce_r(x, q) \sim C_r[Z_0(\alpha) + Z_1(\alpha)]$

20.9.16

$$se_{r+1}(x, q) \sim S_r[Z_0(\alpha) - Z_1(\alpha)] \sin x, \qquad \alpha = 2q^{\frac{1}{4}} \cos x.$$

Let $D_k = D_k(\alpha) = (-1)^k e^{\frac{1}{4}\alpha^2} \dfrac{d^k}{d\alpha^k} e^{-\frac{1}{2}\alpha^2}$.

20.9.17

$$Z_0(\alpha) \sim D_r + \frac{1}{4q^{\frac{1}{2}}} \left[-\frac{D_{r+4}}{16} + \frac{3}{2}\binom{r}{4} D_{r-4}\right]$$

$$+ \frac{1}{16q} \left[\frac{D_{r+8}}{512} - \frac{(r+2)D_{r+4}}{16} + \frac{3}{2}(r-1)\binom{r}{4} D_{r-4}\right.$$

$$\left. + \frac{315}{4}\binom{r}{8} D_{r-8}\right] + \cdots$$

20.9.18

$$Z_1(\alpha) \sim \frac{1}{4q^{\frac{1}{4}}} \left[-\frac{1}{4} D_{r+2} - \frac{r(r-1)}{4} D_{r-2}\right]$$

$$+ \frac{1}{16q} \left[\frac{D_{r+6}}{64} + \frac{(r^2-25r-36)}{64} D_{r+2}\right.$$

$$\left. + \frac{r(r-1)(-r^2-27r+10)}{64} D_{r-2} - \frac{45}{4}\binom{r}{6} D_{r-6} + \cdots\right]$$

20.9.19

$$C_r \sim \left(\frac{\pi}{2}\right)^{\frac{1}{4}} q^{\frac{1}{8}}/(r!)^{\frac{1}{2}} \left[1 + \frac{2r+1}{8q^{\frac{1}{2}}}\right.$$

$$\left. + \frac{r^4+2r^3+263r^2+262r+108}{2048q} + \frac{f_1}{16384q^{\frac{3}{2}}} + \cdots\right]^{-\frac{1}{2}}$$

$$f_1 = 6r^5 + 15r^4 + 1280r^3 + 1905r^2 + 1778r + 572$$

*See page II.

20.9.20

$$S_r \sim \left(\frac{\pi}{2}\right)^{\frac{1}{4}} q^{\frac{1}{8}}/(r!)^{\frac{1}{2}} \left[1 - \frac{2r+1}{8q^{\frac{1}{2}}}\right.$$

$$\left. + \frac{r^4+2r^3-121r^2-122r-84}{2048q} + \frac{f_2}{16384q^{\frac{3}{2}}} + \cdots\right]^{-\frac{1}{2}}$$

$$f_2 = 2r^5 + 5r^4 - 416r^3 - 629r^2 - 1162r - 476$$

It should be noted that **20.9.15** is also valid as an approximation for $se_{r+1}(x, q)$, but **20.9.16** may give slightly better results. See [20.4.]

Explicit Expansions for Orders 0, 1, to Terms in $q^{-3/2}$ (q Large)

20.9.21 For $r=0$:

$$Z_0 \sim D_0 - \frac{D_4}{64\sqrt{q}} + \frac{1}{16q}\left(-\frac{D_4}{8} + \frac{D_8}{512}\right) \quad *$$

$$+ \frac{1}{64q^{3/2}}\left(-\frac{99D_4}{256} + \frac{3D_8}{256} - \frac{D_{12}}{24576}\right) + \cdots$$

$$Z_1 \sim \frac{-D_2}{16\sqrt{q}} + \frac{1}{16q}\left(-\frac{9D_2}{16} + \frac{D_6}{64}\right)$$

$$+ \frac{1}{64q^{3/2}}\left(-\frac{61D_2}{32} + \frac{25D_6}{256} - \frac{5D_{10}}{10240}\right) + \cdots$$

20.9.22 For $r=1$:

$$Z_0 \sim D_1 - \frac{D_5}{64\sqrt{q}} + \frac{1}{16q}\left(-\frac{3D_5}{16} + \frac{D_9}{512}\right)$$

$$+ \frac{1}{64q^{3/2}}\left(-\frac{207D_5}{256} + \frac{D_9}{64} - \frac{D_{13}}{24576}\right) + \cdots$$

$$Z_1 \sim \frac{-D_3}{16\sqrt{q}} + \frac{1}{16q}\left(-\frac{15D_3}{16} + \frac{D_7}{64}\right)$$

$$+ \frac{1}{64q^{3/2}}\left(-\frac{153D_3}{32} + \frac{35D_7}{256} - \frac{D_{11}}{2048}\right) + \cdots$$

Formulas Involving $ce_r(0, q)$ and $se_r(0, q)$

20.9.23

$$\frac{ce_0(0, q)}{ce_0(\tfrac{1}{2}\pi, q)} \sim 2\sqrt{2}\, e^{-2\sqrt{q}} \left(1 + \frac{1}{16\sqrt{q}} + \frac{9}{256q} + \cdots\right)$$

$$\frac{ce_2(0, q)}{ce_2(\tfrac{1}{2}\pi, q)} \sim -32q\sqrt{2}\, e^{-2\sqrt{q}} \left(1 - \frac{1}{16\sqrt{q}} + \frac{29}{128q} + \cdots\right)$$

$$\frac{ce_1(0, q)}{ce_1'(\frac{1}{2}\pi, q)} \sim -4\sqrt{2}\,e^{-2\sqrt{q}}\left(1 + \frac{3}{16\sqrt{q}} + \frac{45}{256q} + \cdots\right)$$

$$\frac{ce_3(0, q)}{ce_3'(\frac{1}{2}\pi, q)} \sim \frac{64}{3}\,q\sqrt{2}\,e^{-2\sqrt{q}}\left(1 - \frac{3}{16\sqrt{q}} + \frac{47}{128q} + \cdots\right)$$

20.9.24

$$\frac{se_1'(0, q)}{se_1(\frac{1}{2}\pi, q)} \sim 4\,q\sqrt{2}\,e^{-2\sqrt{q}}\left(1 - \frac{3}{16\sqrt{q}} - \frac{11}{256q} + \cdots\right)$$

$$\frac{se_3'(0, q)}{se_3(\frac{1}{2}\pi, q)} \sim -64\,q\sqrt{2}\,e^{-2\sqrt{q}}\left(1 - \frac{21}{16\sqrt{q}} - \frac{17}{128q} + \cdots\right)$$

$$\frac{se_2'(0, q)}{se_2'(\frac{1}{2}\pi, q)} \sim -8\,q\sqrt{2}\,e^{-2\sqrt{q}}\left(1 - \frac{9}{16\sqrt{q}} - \frac{39}{256q} + \cdots\right)$$

$$\frac{se_4'(0, q)}{se_4'(\frac{1}{2}\pi, q)} \sim \frac{128}{3}\,q\sqrt{2}\,e^{-2\sqrt{q}}\left(1 - \frac{31}{16\sqrt{q}} - \frac{15}{128q} + \cdots\right)$$

For higher orders, these ratios are increasingly more difficult to obtain. One method of estimating values at the origin is to evaluate both **20.9.11** and **20.9.15** for some x where both expansions are satisfactory, and so to use **20.9.11** as a means to solve for $ce_r(0, q)$; similarly for $se_r'(0, q)$.

Other asymptotic expansions, valid over various regions of the complex z-plane, for real values of a, q, have been given by Langer [20.25]. It is not always easy, however, to determine the linear combinations of Langer's solutions which coincide with those defined here.

20.10. Comparative Notations

	This Volume	[20.58] NBS	[20.59] Stratton-Morse, etc.	[20.36] Meixner and Schäfke	[20.30] McLachlan	[20.15] Bateman Manuscript	Comments
Parameters in 20.1.1	a	$b=a+2q$	b	λ	a	h	
	q	$s=4q$	$c=2\sqrt{q}$	h^2	q	θ	
	a_r	$be_r=a_r+2q$	$b_r=a_r+2q$	a_r	a_r	a_r	
	b_r	$bo_r=b_r+2q$	$b'_r=b_r+2q$	b_r	b_r	b_r	
Periodic Solutions, of 20.1.1:							
Even	$ce_r(z,q)$	$A^r Se_r(s,x)$ *	$A^r Se_r^{(1)}(c,\cos x)$ *	$ce_r(z,h^2)$ *	$ce_r(z,q)$	$ce_r(z,\theta)$	See Note 1.
Odd	$se_r(z,q)$	$B^r So_r(s,x)$ *	$A^r So_r^{(1)}(c,\cos x)$ *	$se_r(z,h^2)$ *	$se_r(z,q)$	$se_r(z,\theta)$	
Coefficients in Periodic Solutions:							
Even	$A_m^r(q)$	$A^r De_m^r(s)$ *	$A^r D_m^r$ *	A_m^r	A_m^r	A_m^r	
Odd	$B_m^r(q)$	$B^r Do_m^r(s)$ *	$B^r F_m^r$ *	B_m^r	B_m^r	B_m^r	
$\frac{1}{\pi}\int_0^{2\pi} y^2\,dz$, y is the Standard Solution of 20.1.1.	1	$(A^r)^{-2}$ or $(B^r)^{-2}$	$(A^r)^{-2}$ or $(B^r)^{-2}$	1	1	1	See Note 1.
Floquet's Solutions 20.3.8	$F_r(z)$			$me_r(z,h^2)$	$\phi(z)$		
Characteristic Exponent	ν	$\mu=i\nu$		ν	$\mu=i\nu$	$\mu=i\nu$	
Normalizations of Floquet's Solutions.	Unspecified			$\frac{1}{\pi}\int_0^\pi (me_r(z,h^2)me_{-r}(z,h^2)=1$			
Solutions of Modified Equation 20.1.2.	$Ce_r(z,q)$	$Age_{,r}(s)Je_r(s,q)$	$Age_{,r}(s)Je_r(c,\cosh x)$	$Ce_r(z,q)$	$Ce_r(z,q)$	$Ce_r(z,\theta)$	
	$Se_r(z,q)$	$Bgo_{,r}(s)Jo_r(s,q)$	$Bgo_{,r}(s)Jo_r(c,\cosh x)$	$Se_r(z,q)$	$Se_r(z,q)$	$Se_r(z,\theta)$	
	$Mc_r^{(1)}(z,q)$	$\sqrt{\frac{2}{\pi}}Je_r(s,z)$	$\sqrt{\frac{2}{\pi}}Je_r(c,\cosh z)$	$Mc_r^{(1)}(z,h)$	$\sqrt{\frac{2}{\pi}}Ce_r(z,q)/Age_{,r}(q)$	$\sqrt{\frac{2}{\pi}}Ce_r(z,\theta)/Age_{,r}(q)$	
	$Ms_r^{(1)}(z,q)$	$\sqrt{\frac{2}{\pi}}Jo_r(s,z)$	$\sqrt{\frac{2}{\pi}}Jo_r(c,\cosh z)$	$Ms_r^{(1)}(z,h)$	$\sqrt{\frac{2}{\pi}}Se_r(z,q)/Bgo_{,r}(q)$	$\sqrt{\frac{2}{\pi}}Se_r(z,\theta)/Bgo_{,r}(q)$	
	$Mc_r^{(2)}(z,q)$	$\sqrt{\frac{2}{\pi}}Ne_r(s,z)$	$\sqrt{\frac{2}{\pi}}Ne_r(c,\cosh z)$	$Mc_r^{(2)}(z,h)$	$\sqrt{\frac{2}{\pi}}Fey_r(z,q)/Age_{,r}(q)$	$\sqrt{\frac{2}{\pi}}Fey_r(z,\theta)/Age_{,r}(q)$	
	$Ms_r^{(2)}(z,q)$	$\sqrt{\frac{2}{\pi}}No_r(s,z)$	$\sqrt{\frac{2}{\pi}}No_r(c,\cosh z)$	$Ms_r^{(2)}(z,h)$	$\sqrt{\frac{2}{\pi}}Gey_r(z,q)/Bgo_{,r}(q)$	$\sqrt{\frac{2}{\pi}}Gey_r(z,\theta)/Bgo_{,r}(q)$	
Joining Factors	$\sqrt{2/\pi}/Mc_r^{(1)}(0,q)$	$ge_{,r}(s)$	$\sqrt{2\pi}\,\lambda_r^{(e)}$	$\sqrt{2/\pi}/Mc_r^{(1)}(0,h)$	$(-1)^r p_r\sqrt{\frac{2}{\pi}}/A$	Same as [20.30]	See Note 2.
	$\sqrt{2/\pi}/\frac{d}{dz}[Ms_r^{(1)}(z,q)]_{z=0}$	$go_{,r}(s)$	$\sqrt{2\pi}\,\lambda_r^{(0)}$	$\sqrt{2/\pi}/\frac{d}{dz}[Ms_r^{(1)}(z,h)]_{z=0}$	$(-1)^r s_r\sqrt{\frac{2}{\pi}}/B$		
	$-Mc_r^{(2)}(0,q)/Mc_r^{(1)}(0,q)$	$fe_{,r}(s)$	$-\frac{2}{\pi}\frac{K_1'}{K_1}$	$-Mc_r^{(2)}(0,h)/Mc_r^{(1)}(0,h)$	$\frac{-Fey_r(0,q)}{Ce_r(0,q)}$	Same as [20.30]	See Note 3.
	$\left[\dfrac{\frac{d}{dz}Ms_r^{(2)}(z,q)}{\frac{d}{dz}Ms_r^{(1)}(z,q)}\right]_{z=0}$	$fo_{,r}(s)$	$\frac{2}{\pi}\frac{K_3'}{K_3}$	Same as this volume	$\left[\dfrac{\frac{d}{dz}Gey_r(z,q)}{\frac{d}{dz}Se_r(z,q)}\right]_{z=0}$	Same as [20.30]	

NOTE: 1. The conversion factors A^r and B^r are tabulated in [20.58] along with the coefficients.
2. The multipliers p_r and s_r are defined in [20.30], Appendix 1, section 3, equations 3, 4, 5, 6.
3. See [20.59], sections (5.3) and (5.5). In eq. (316) of (5.5), the first term should have a minus sign.

*See page II.

References

Texts

[20.1] W. G. Bickley, The tabulation of Mathieu functions, Math. Tables Aids Comp. **1**, 409–419 (1945).

[20.2] W. G. Bickley and N. W. McLachlan, Mathieu functions of integral order and their tabulation, Math. Tables Aids Comp. **2**, 1–11 (1946).

[20.3] G. Blanch, On the computation of Mathieu functions, J. Math. Phys. **25**, 1–20 (1946).

[20.4] G. Blanch, The asymptotic expansions for the odd periodic Mathieu functions, Trans. Amer. Math. Soc. **97**, 2, 357–366 (1960).

[20.5] C. J. Bouwkamp, A note on Mathieu functions, Kon. Nederl. Akad. Wetensch. Proc. **51**, 891–893 (1948).

[20.6] C. J. Bouwkamp, On spheroidal wave functions of order zero. J. Math. Phys. **26**, 79–92 (1947).

[20.7] M. R. Campbell, Sur les solutions de période 2 $s\pi$ de l'équation de Mathieu associée, C.R. Acad. Sci., Paris, **223**, 123–125 (1946).

[20.8] M. R. Campbell, Sur une catégorie remarquable de solutions de l'équation de Mathieu associée, C.R. Acad. Sci., Paris, **226**, 2114–2116 (1948).

[20.9] T. M. Cherry, Uniform asymptotic formulae for functions with transition points, Trans. Amer. Math. Soc. **68**, 224–257 (1950).

[20.10] S. C. Dhar, Mathieu functions (Calcutta Univ. Press, Calcutta, India, 1928).

[20.11] J. Dougall, The solution of Mathieu's differential equation, Proc. Edinburgh Math. Soc. **34**, 176–196 (1916).

[20.12] J. Dougall, On the solutions of Mathieu's differential equation, and their asymptotic expansions, Proc. Edinburgh Math. Soc. **41**, 26–48 (1923).

[20.13] A. Erdélyi, Über die Integration der Mathieuschen Differentialgleichung durch Laplacesche Integrale, Math. Z. **41**, 653–664 (1936).

[20.14] A. Erdélyi, On certain expansions of the solutions of Mathieu's differential equation, Proc. Cambridge Philos. Soc. **38**, 28–33 (1942).

[20.15] A. Erdélyi et al., Higher transcendental functions, vol. 3 (McGraw-Hill Book Co., Inc., New York, N.Y., 1955).

[20.16] G. Floquet, Sur les équations différentielles linéaires à coefficients périodiques, Ann. École Norm. Sup. **12**, 47 (1883).

[20.17] S. Goldstein, The second solution of Mathieu's differential equation, Proc. Cambridge Philos. Soc. **24**, 223–230 (1928).

[20.18] S. Goldstein, Mathieu functions, Trans. Cambridge Philos. Soc. **23**, 303–336 (1927).

[20.19] G. W. Hill, On the path of motion of the lunar perigee, Acta Math. **8**, 1 (1886).

[20.20] E. Hille, On the zeros of the Mathieu functions, Proc. London Math. Soc. **23**, 185–237 (1924).

[20.21] E. L. Ince, A proof of the impossibility of the coexistence of two Mathieu functions, Proc. Cambridge Philos. Soc. **21**, 117–120 (1922).

[20.22] E. L. Ince, Ordinary differential equations (Longmans, Green & Co., 1927, reprinted by Dover Publications, Inc., New York, N.Y., 1944).

[20.23] H. Jeffreys, On the modified Mathieu's equation, Proc. London Math. Soc. **23**, 449–454 (1924).

[20.24] V. D. Kupradze, Fundamental problems in the mathematical theory of diffraction (1935). Translated from the Russian by Curtis D. Benster, NBS Report 200 (Oct. 1952).

[20.25] R. E. Langer, The solutions of the Mathieu equation with a complex variable and at least one parameter large, Trans. Amer. Math. Soc. **36**, 637–695 (1934).

[20.26] S. Lubkin and J. J. Stoker, Stability of columns and strings under periodically varying forces, Quart. Appl. Math. **1**, 215–236 (1943).

[20.27] É. Mathieu, Mémoire sur le mouvement vibratoire d'une membrane de forme elliptique, J. Math. Pures Appl. **13**, 137–203 (1868).

[20.28] N. W. McLachlan, Mathieu functions and their classification, J. Math. Phys. **25**, 209–240 (1946).

[20.29] N. W. McLachlan, Mathieu functions of fractional order, J. Math. Phys. **26**, 29–41 (1947).

[20.30] N. W. McLachlan, Theory and application of Mathieu functions (Clarendon Press, Oxford, England, 1947).

[20.31] N. W. McLachlan, Application of Mathieu's equation to stability of non-linear oscillator, Math. Gaz. **35**, 105–107 (1951).

[20.32] J. Meixner, Über das asymptotische Verhalten von Funktionen, die durch Reihen nach Zylinderfunktionen dargestellt werden können, Math. Nachr. **3**, 9–13, Reihenentwicklungen von Produkten zweier Mathieuschen Funktionen nach Produkten von Zylinder und Exponentialfunktionen, 14–19 (1949).

[20.33] J. Meixner, Integralbeziehungen zwischen Mathieuschen Funktionen, Math. Nachr. **5**, 371–378 (1951).

[20.34] J. Meixner, Reihenentwicklungen vom Siegerschen Typus für die Spheroid Funktionen, Arch. Math. Oberwolfach **1**, 432–440 (1949).

[20.35] J. Meixner, Asymptotische Entwicklung der Eigenwerte und Eigenfunktionen der Differentialgleichungen der Sphäroidfunktionen und der Mathieuschen Funktionen, Z. Angew. Math. Mech. **28**, 304–310 (1948).

[20.36] J. Meixner and F. W. Schäfke, Mathieusche Funktionen und Sphäroidfunktionen (Springer-Verlag, Berlin, Germany, 1954).

[20.37] P. M. Morse and P. J. Rubinstein, The diffraction of waves by ribbons and by slits, Phys. Rev. **54**, 895–898 (1938).

[20.38] H. P. Mulholland and S. Goldstein, The characteristic numbers of the Mathieu equation with purely imaginary parameters, Phil. Mag. **8**, 834–840 (1929).

[20.39] L. Onsager, Solutions of the Mathieu equation of period 4π and certain related functions (Yale Univ. Dissertation, New Haven, Conn., 1935).

[20.40] F. W. Schäfke, Über die Stabilitätskarte der. Mathieuschen Differentialgleichung, Math. Nachr. **4**, 175–183 (1950).

[20.41] F. W. Schäfke, Das Additions theorem der Mathieuschen Funktionen, Math. Z. **58**, 436–447 (1953).

[20.42] F. W. Schäfke, Eine Methode zur Berechnung des charakteristischen Exponenten einer Hillschen Differentialgleichung, Z. Angew. Math. Mech. **33**, 279–280 (1953).

[20.43] B. Sieger, Die Beugung einer ebenen elektrischen Welle an einem Schirm von elliptischem Querschnitt, Ann. Physik. **4**, 27, 626–664 (1908)·

[20.44] R. Sips, Représentation asymptotique des fonctions de Mathieu et des fonctions d'onde sphéroidales, Trans. Amer. Math. Soc. **66**, 93–134 (1949).

[20.45] R. Sips, Représentation asymptotique des fonctions de Mathieu et des fonctions sphéroidales II, Trans. Amer. Math. Soc. **90**, 2, 340–368 (1959).

[20.46] R. Sips, Recherches sur les fonctions de Mathieu, Bull. Soc. Roy. Sci. Liège **22**, 341–355, 374–387, 444–455, 530–540 (1953); **23**, 37–47, 90–103 (1954).

[20.47] M. J. O. Strutt, Die Hillsche Differentialgleichung im komplexen Gebiet, Nieuw. Arch. Wisk. **18** 31–55 (1935).

[20.48] M. J. O. Strutt, Lamésche, Mathieusche und verwandte Funktionen in Physik und Technik, Ergeb. Math. Grenzgeb. **1**, 199–323 (1932).

[20.49] M. J. O. Strutt, On Hill's problems with complex parameters and a real periodic function, Proc. Roy. Soc. Edinburgh Sect. A **62**, 278–296 (1948).

[20.50] E. T. Whittaker, On functions associated with elliptic cylinders in harmonic analysis, Proc. Intl. Congr. Math. Cambr. **1**, 366 (1912).

[20.51] E. T. Whittaker, On the general solution of Mathieu's equation, Proc. Edinburgh Math. Soc. **32**, 75–80 (1914).

[20.52] E. T. Whittaker and G. N. Watson, A course of modern analysis, 4th ed. (Cambridge Univ. Press, Cambridge, England, 1952).

Tables

[20.53] G. Blanch and I. Rhodes, Table of characteristic values of Mathieu's equation for large values of the parameter, J. Washington Acad. Sci. **45**, 6, 166–196 (1955). $Be_r(t) = a_r(q) + 2q - 2(2r+1)\sqrt{q}$, $Bo_r(t) = b_r(q) + 2q - 2(2r-1)\sqrt{q}, t = 1/2\sqrt{q}$, $r = 0(1)15$, $0 \leq t \leq .1$, with δ^2, δ^4*; 8D (about); interpolable.

[20.54] J. G. Brainerd, H. J. Gray, and R. Merwin, Solu tion of the Mathieu equation, Am. Inst. Elec. Engrs. **67** (1948). Characteristic exponent over a wide range. μ, M for $\epsilon = 1(1)10$; $k = .1(.1)1$, 5D; $g(t)$, $h(t)$ for $t = 0(.1)3.1$, π, 5D; $\epsilon = 1(1)10$, $k = .1(.1)1$, where $g(t)$, $h(t)$ are solutions of $y'' + \epsilon(1 + k \cos t)y = 0$, with $g(0) = h'(0) = 1$, $g'(0) = h(0) = 0$, $\cos 2\pi\mu = 2g(\pi)h'(\pi) - 1$, $M = [-g(\pi)g'(\pi)/h(\pi)h'(\pi)]^{1/2}$.

[20.55] J. G. Brainerd and C. N. Weygandt, Solutions of Mathieu's equation, Phil. Mag. **30**, 458–477 (1940).

[20.56] E. L. Ince, Tables of the elliptic cylinder functions, Proc. Roy. Soc. Edinburgh **52**, 355–423; Zeros and turning points, **52**, 424–433 (1932). Characteristic values a_0, a_1, . . ., a_5, b_1, b_2, . . ., b_6, and coefficients for $\theta = 0(1)10(2)20(4)40$; 7D. Also $ce_r(x, \theta)$, $se_r(x, \theta)$, $\theta = 0(1)10$, $x = 0°(1°)90°$; 5D, corresponding to characteristic values in the tables. $a_r = be_r - 2q$; $b_r = bo_r - 2q$; $\theta = q$.

[20.57] E. T. Kirkpatrick, Tables of values of the modified Mathieu function, Math. Comp. **14**, 70 (1960). $Ce_r(u, q)$, $r = 0(1)5$, $Se_r(u, q)$, $r = 1(1)6$; $u = .1(.1)1$; $q = 1(1)20$.

[20.58] National Bureau of Standards, Tables relating to Mathieu functions (Columbia Univ. Press, New York, N.Y., 1951). Characteristic values $be_r(s)$, $bo_r(s)$ for $0 \leq s \leq 100$, along with δ^2*, interpolable to 8D; coefficients $De_k(s)$, $Do_k(s)$ and conversion factors for $ce_r(q)$, $se_r(q)$, same range, without differences but interpolable to 9D with Lagrangian formulas of order 7. "Joining factors" $s^{\frac{1}{2}r}g_{e,r}$, $s^{\frac{1}{2}r}g_{o,r}$, $s^r f_{e,r}$, $s^r f_{o,r}$ along with δ^2*; interpolable to 8S.

[20.59] J. A. Stratton, P. M. Morse, L. J. Chu and R. A. Hutner, Elliptic cylinder and spheroidal wave functions (John Wiley & Sons, Inc., New York, N.Y., 1941). Theory and tables for b_0, b_1, b_2, b_3, b_4, b_1', b_2', b_3', b_4', and coefficients for $Se_r(s, x)$ and $So_r(s, x)$ for $c = 0(.2)4.4$ and $.5(1)4.5$; mostly 5S; $c = 2q^{\frac{1}{2}}$, $b_r = a_r + 2q$, $b_r' = b_r + 2q$.

[20.60] T. Tamir, Characteristic exponents of Mathieu * equations, Math. Comp. **16**, 77 (1962). The Floquet exponent ν_r of the first three stable regions; namely $r = 0$, 1, 2; $q = .1(.1)2.5$; $a = r(.1)r + 1$, 5D.

[20.61] J. C. Wiltse and M. J. King, Values of the Mathieu functions, The Johns Hopkins Univ. Radiation Laboratory Technical Report AF–53, Baltimore, Md. (1958). (Notation of [20.58] used: $ce_n(v, q)/A$, $se_n(v, q)/B$ for 12 values of q between .25 and 10 and from 8 to 14 values of v; $\sqrt{\pi/2} Mc_r^j(u, q)$, $\sqrt{\pi/2} Ms_r^j(u, q)$, $j = 1$, 2 for 6 to 8 values of q between .25 and 10 and about 20 values of u, $r = 0$, 1, 2; $\sqrt{\pi/2} Mc_r^{(3)}(-|u|, q)$, $\sqrt{\pi/2} Ms_r^{(3)}(-|u|, q)$, $r = 0$, 1, 2 for about 9 values of u and q, 2 to 4 D in all.

[20.62] J. C. Wiltse and M. J. King, Derivatives, zeros, and other data pertaining to Mathieu functions, The Johns Hopkins Univ. Radiation Laboratory Technical Report AF–57, Baltimore, Md. (1958).

[20.63] S. J. Zaroodny, An elementary review of the Mathieu-Hill equation of real variable based on numerical solutions, Ballistic Research Laboratory Memorandum Report 878, Aberdeen Proving Ground, Md. (1955). Chart of the characteristic exponent.

See also [20.18]. It contains, among other tabulations, values of a_r, b_r and coefficients for $ce_r(x, q)$, $se_r(x, q)$, $q = 40(20)100(50)200$; 5D, $r \leq 2$.

---

*See page II.

Table 20.1   CHARACTERISTIC VALUES, JOINING FACTORS, SOME CRITICAL VALUES

EVEN SOLUTIONS

r	q	a_r	$ce_r(0, q)$	$ce_r(\tfrac{1}{2}\pi, q)$	$(4q)^{\frac{1}{4}r}g_{e,r}(q)$	$(4q)^r f_{e,r}(q)$
0	0	0.00000 000	(−1)7.07106 781	(−1) 7.07106 78	(−1)7.97884 56	∞
	5	− 5.80004 602	(−2)4.48001 817	1.33484 87	1.97009 00	(− 3)1.86132 97
	10	− 13.93697 996	(−3)7.62651 757	1.46866 05	2.40237 95	(− 5)5.54257 96
	15	− 22.51303 776	(−3)1.93250 832	1.55010 82	2.68433 53	(− 6)3.59660 89
	20	− 31.31339 007	(−4)6.03743 829	1.60989 09	2.90011 25	(− 7)3.53093 01
	25	− 40.25677 955	(−4)2.15863 018	1.65751 03	3.07743 91	(− 8)4.53098 68
2	0	4.00000 000	1.00000 000	−1.00000 00	(1)1.27661 53	(1)8.14873 31
	5	7.44910 974	(−1)7.35294 308	(−1)−7.24488 15	(1)2.63509 89	(2)1.68665 79
	10	7.71736 985	(−1)2.45888 349	(−1)−9.26759 26	(1)7.22275 58	(1)6.89192 56
	15	5.07798 320	(−2)7.87928 278	−1.01996 62	(2)1.32067 71	(1)1.73770 48
	20	+ 1.15428 288	(−2)2.86489 431	−1.07529 32	(2)1.98201 14	4.29953 32
	25	− 3.52216 473	(−2)1.15128 663	−1.11627 90	(2)2.69191 26	1.11858 69
10	0	100.00000 000	1.00000 000	−1.00000 00	(12)1.51800 43	(23)2.30433 72
	5	100.12636 922	1.02599 503	(−1)−9.75347 49	(12)1.48332 54	(23)2.31909 77
	10	100.50677 002	1.05381 599	(−1)−9.51645 32	(12)1.45530 39	(23)2.36418 54
	15	101.14520 345	1.08410 631	(−1)−9.28548 06	(12)1.43299 34	(23)2.44213 04
	20	102.04891 602	1.11778 863	(−1)−9.05710 78	(12)1.41537 24	(23)2.55760 55
	25	103.23020 480	1.15623 992	(−1)−8.82691 92	(12)1.40118 52	(23)2.71854 15

r	q	a_r	$ce_r(0, q)$	$ce'_r(\tfrac{1}{2}\pi, q)$	$(4q)^{\frac{1}{4}r}g_{e,r}(q)$	$(4q)^r f_{e,r}(q)$
1	0	1.00000 000	1.00000 000	−1.00000 00	1.59576 91	2.54647 91
	5	+ 1.85818 754	(−1)2.56542 879	−3.46904 21	7.26039 84	1.02263 46
	10	− 2.39914 240	(−2)5.35987 478	−4.85043 83	(1)1.35943 49	(− 2)9.72660 12
	15	− 8.10110 513	(−2)1.50400 665	−5.76420 64	(1)1.91348 51	(− 2)1.19739 95
	20	− 14.49130 142	(÷3)5.05181 376	−6.49056 58	(1)2.42144 01	(− 3)1.84066 20
	25	− 21.31489 969	(−3)1.91105 151	−7.10674 15	(1)2.89856 94	(− 4)3.33747 55
5	0	25.00000 000	1.00000 000	−5.00000 00	(4)4.90220 27	(8)4.80631 83
	5	25.54997 175	1.12480 725	−5.39248 61	(4)4.43075 22	(8)5.11270 71
	10	27.70376 873	1.25801 994	−5.32127 65	(4)4.19827 66	(8)6.83327 77
	15	31.95782 125	1.19343 223	−5.11914 99	(4)5.25017 04	(9)1.18373 72
	20	36.64498 973	(−1)9.36575 531	−5.77867 52	(4)8.96243 97	(9)1.85341 57
	25	40.05019 099	(−1)6.10694 310	−7.05988 45	(5)1.71582 55	(9)2.09679 12
15	0	225.00000 000	1.00000 000	(1) 1.50000 00	(20)5.60156 72	(40)2.09183 70
	5	225.05581 248	1.01129 373	(1) 1.51636 57	(20)5.54349 84	(40)2.09575 00
	10	225.22335 698	1.02287 828	(1) 1.53198 84	(20)5.49405 67	(40)2.10754 45
	15	225.50295 624	1.03479 365	(1) 1.54687 43	(20)5.45287 72	(40)2.12738 84
	20	225.89515 341	1.04708 434	(1) 1.56102 79	(20)5.41964 26	(40)2.15556 69
	25	226.40072 004	1.05980 044	(1) 1.57444 72	(20)5.39407 68	(40)2.19249 18

Compiled from National Bureau of Standards, Tables relating to Mathieu functions, Columbia Univ. Press, New York, N.Y., 1951 (with permission).

$$a_r + 2q - (4r+2)\sqrt{q}$$

$q^{-\frac{1}{2}}\backslash r$	0	1	2	5	10	15	$<q>$
0.16	−0.25532 994	−1.30027 212	−3.45639 483	−17.84809 551	−76.04295 314	− 80.93485 048	39
0.12	−0.25393 098	−1.28658 972	−3.39777 782	−16.92019 225	−76.84607 855	−141.64507 841	69
0.08	−0.25257 851	−1.27371 191	−3.34441 938	−16.25305 645	−63.58155 264	−162.30500 052	156
0.04	−0.25126 918	−1.26154 161	−3.29538 745	−15.70968 373	−58.63500 546	−132.08298 271	625
0.00	−0.25000 000	−1.25000 000	−3.25000 000	−15.25000 000	−55.25000 000	−120.25000 000	∞

For $g_{e,r}$ and $f_{e,r}$ see **20.8.12**.

$<q>$ = nearest integer to q.

Compiled from G. Blanch and I. Rhodes, Table of characteristic values of Mathieu's equation for large values of the parameter, Jour. Wash. Acad. Sci., **45**, 6, 1955 (with permission).

CHARACTERISTIC VALUES, JOINING FACTORS, SOME CRITICAL VALUES   Table 20.1

ODD SOLUTIONS

r	q	b_r	$se_r'(0, q)$	$se_r'(\tfrac{1}{2}\pi, q)$	$(4q)^{\frac{1}{2}r}g_{o,r}(q)$	$(4q)^r f_{o,r}(q)$
2	0	4.00000 000	2.00000 00	−2.00000 00	6.38307 65	(1)8.14873 31
	5	+ 2.09946 045	(−1)7.33166 22	−3.64051 79	(1)1.24474 88	(1)2.24948 08
	10	− 2.38215 824	(−1)2.48822 84	−4.86342 21	(1)1.86133 36	3.91049 85
	15	− 8.09934 680	(−2)9.18197 14	−5.76557 38	(1)2.42888 57	(− 1)7.18762 28
	20	− 14.49106 325	(−2)3.70277 78	−6.49075 22	(1)2.95502 89	(− 1)1.47260 95
	25	− 21.31486 062	(−2)1.60562 17	−7.10677 19	(1)3.44997 83	(− 2)3.33750 27
10	0	100.00000 000	(1)1.00000 00	(1)−1.00000 00	(11)1.51800 43	(23)2.30433 72
	5	100.12636 922	9.73417 32	(1)−1.02396 46	(11)1.56344 50	(23)2.31909 77
	10	100.50676 946	9.44040 54	(1)−1.04539 48	(11)1.62453 03	(23)2.36418 52
	15	101.14517 229	9.11575 13	(1)−1.06429 00	(11)1.70421 18	(23)2.44211 78
	20	102.04839 286	8.75554 51	(1)−1.08057 24	(11)1.80695 19	(23)2.55740 30
	25	103.22568 004	8.35267 84	(1)−1.09413 54	(11)1.93959 86	(23)2.71681 11

r	q	b_r	$se_r'(0, q)$	$se_r(\tfrac{1}{2}\pi, q)$	$(4q)^{\frac{1}{2}r}g_{o,r}(q)$	$(4q)^r f_{o,r}(q)$
1	0	+ 1.00000 000	1.00000 00	1.00000 00	1.59576 91	2.54647 91
	5	− 5.79008 060	(−1)1.74675 40	1.33743 39	2.27041 76	(− 2)3.74062 82
	10	− 13.93655 248	(−2)4.40225 66	1.46875 57	2.63262 99	(− 3)2.21737 88
	15	− 22.51300 350	(−2)1.39251 35	1.55011 51	2.88561 87	(− 4)2.15798 83
	20	− 31.31338 617	(−3)5.07788 49	1.60989 16	3.08411 21	(− 4)2.82474 71
	25	− 40.25677 898	(−3)2.04435 94	1.65751 04	3.24945 50	(− 6)4.53098 74
5	0	25.00000 000	5.00000 00	1.00000 00	(3)9.80440 55	(8)4.80631 83
	5	25.51081 605	4.33957 00	(−1) 9.06077 93	(4)1.14793 21	(8)5.05257 20
	10	26.76642 636	3.40722 68	(−1) 8.46038 43	(4)1.52179 77	(8)5.46799 57
	15	27.96788 060	2.41166 65	(−1) 8.37049 34	(4)2.20680 20	(8)5.27524 17
	20	28.46822 133	1.56889 69	(−1) 8.63543 12	(4)3.27551 12	(8)4.26215 66
	25	28.06276 590	(−1)9.64071 62	(−1) 8.99268 33	(4)4.76476 62	(8)2.94147 89
15	0	225.00000 000	(1)1.50000 00	−1.00000 00	(19)3.73437 81	(40)2.09183 70
	5	225.05581 248	(1)1.48287 89	(−1)−9.88960 70	(19)3.78055 49	(40)2.09575 00
	10	225.22335 698	(1)1.46498 60	(−1)−9.78142 35	(19)3.83604 43	(40)2.10754 45
	15	225.50295 624	(1)1.44630 01	(−1)−9.67513 70	(19)3.90140 52	(40)2.12738 84
	20	225.89515 341	(1)1.42679 46	(−1)−9.57045 25	(19)3.97732 29	(40)2.15556 69
	25	226.40072 004	(1)1.40643 73	(−1)−9.46708 70	(19)4.06462 83	(40)2.19249 18

$$b_r + 2q - (4r-2)\sqrt{q}$$

$q^{-\frac{1}{2}}\backslash r$	1	2	5	10	15	$<q>$
0.16	−0.25532 994	−1.30027 164	−11.53046 855	−51.32546 875	− 55.93485 112	39
0.12	−0.25393 098	−1.28658 971	−11.12574 983	−56.10964 961	−108.31442 060	69
0.08	−0.25257 851	−1.27371 191	−10.78895 146	−51.15347 975	−132.59692 424	156
0.04	−0.25126 918	−1.26154 161	−10.50135 748	−47.72149 533	−114.76358 461	625
0.00	−0.25000 000	−1.25000 000	−10.25000 000	−45.25000 000	−105.25000 000	∞

For $g_{o,r}$ and $f_{o,r}$ see **20.8.12**.

$<q>$ = nearest integer to q.

Table 20.2 COEFFICIENTS A_m AND B_m

$$A_m$$

$q=5$

$m\backslash r$	0	2	10
0	+0.54061 2446	+0.43873 7166	+0.00000 1679
2	-0.62711 5414	+0.65364 0260	+0.00003 3619
4	+0.14792 7090	-0.42657 8935	+0.00064 2987
6	-0.01784 8061	+0.07588 5673	+0.01078 4807
8	+0.00128 2863	-0.00674 1769	+0.13767 5121
10	-0.00006 0723	+0.00036 4942	+0.98395 5640
12	+0.00000 2028	-0.00001 3376	-0.11280 6780
14	-0.00000 0050	+0.00000 0355	+0.00589 2962
16	+0.00000 0001	-0.00000 0007	-0.00018 9166
18			+0.00000 4226
20			-0.00000 0071
22			+0.00000 0001

$m\backslash r$	1	5	15
1	+0.76246 3686	+0.07768 5798	0.00000 0000
3	-0.63159 6319	+0.30375 1030	+0.00000 0002
5	+0.13968 4806	+0.92772 8396	+0.00000 0106
7	-0.01491 5596	-0.20170 6148	+0.00000 4227
9	+0.00094 4842	+0.01827 4579	+0.00014 8749
11	-0.00003 9702	-0.00095 9038	+0.00428 1393
13	+0.00000 1189	+0.00003 3457	+0.08895 2014
15	-0.00000 0027	-0.00000 0839	+0.99297 4092
17	+0.00000 0001	+0.00000 0016	-0.07786 7946
19			-0.00286 6409
21			-0.00006 6394
23			+0.00000 1092
25			-0.00000 0014

$q=25$

$m\backslash r$	0	2	10
0	+0.42974 1038	+0.33086 5777	+0.00502 6361
2	-0.69199 9610	-0.04661 4551	+0.02075 4891
4	+0.36554 4890	-0.64770 5862	+0.07232 7761
6	-0.13057 5523	+0.55239 9372	+0.23161 1726
8	+0.03274 5863	-0.22557 4897	+0.55052 4391
10	-0.00598 3606	+0.05685 2843	+0.63227 5658
12	+0.00082 3792	-0.00984 6277	-0.46882 9197
14	-0.00008 7961	+0.00124 8919	+0.13228 7155
16	+0.00000 7466	-0.00012 1205	-0.02206 0893
18	-0.00000 0514	+0.00000 9296	+0.00252 2374
20	+0.00000 0029	-0.00000 0578	-0.00021 3672
22	-0.00000 0001	+0.00000 0030	+0.00001 4078
24		-0.00000 0001	-0.00000 0746
26			+0.00000 0032
28			-0.00000 0001

$m\backslash r$	1	5	15
1	+0.39125 2265	+0.65659 0398	+0.00000 4658
3	-0.74048 2467	+0.36900 8820	+0.00003 7337
5	+0.50665 3803	-0.19827 8625	+0.00032 0026
7	-0.19814 2336	-0.48837 4067	+0.00254 0806
9	+0.05064 0536	+0.37311 2810	+0.01770 9603
11	-0.00910 8920	-0.12278 1866	+0.10045 8755
13	+0.00121 2864	+0.02445 3933	+0.40582 7402
15	-0.00012 4121	-0.00335 1335	+0.83133 2650
17	+0.00001 0053	+0.00033 9214	-0.35924 8831
19	-0.00000 0660	-0.00002 6552	+0.06821 6074
21	+0.00000 0036	+0.00000 1661	-0.00802 4550
23	-0.00000 0002	-0.00000 0085	+0.00066 6432
25		+0.00000 0004	-0.00004 1930
27			+0.00000 2090
29			-0.00000 0085
31			+0.00000 0003

$$B_m$$

$q=5$

$m\backslash r$	2	10
2	+0.93342 9442	+0.00003 3444
4	-0.35480 3915	+0.00064 2976
6	+0.05296 3730	+0.01078 4807
8	-0.00429 5885	+0.13767 5120
10	+0.00021 9797	+0.98395 5640
12	-0.00000 7752	-0.11280 6780
14	+0.00000 0200	+0.00589 2962
16	-0.00000 0004	-0.00018 9166
18		+0.00000 4227
20		-0.00000 0070
22		+0.00000 0001

$m\backslash r$	1	5	15
1	+0.94001 9024	+0.05038 2462	0.00000 0000
3	-0.33654 1963	+0.29736 5513	+0.00000 0002
5	+0.05547 7529	+0.93156 6997	+0.00000 0106
7	-0.00508 9553	-0.20219 3638	+0.00000 4227
9	+0.00029 3879	+0.01830 5721	+0.00014 8749
11	-0.00001 1602	-0.00096 0277	+0.00428 1392
13	+0.00000 0332	+0.00003 3493	+0.08895 2014
15	-0.00000 0007	-0.00000 0842	+0.99297 4092
17		+0.00000 0017	-0.07786 7946
19			+0.00286 6409
21			-0.00006 6394
23			+0.00000 1093
25			-0.00000 0013

$q=25$

$m\backslash r$	2	10
2	+0.65743 9912	+0.01800 3596
4	-0.66571 9990	+0.07145 6762
6	+0.33621 0033	+0.23131 0990
8	-0.10507 3258	+0.55054 4783
10	+0.02236 2380	+0.63250 8750
12	-0.00344 2304	-0.46893 3949
14	+0.00040 0182	+0.13230 9765
16	-0.00003 6315	-0.02206 3990
18	+0.00000 2640	+0.00252 2676
20	-0.00000 0157	-0.00021 3694
22	+0.00000 0008	+0.00001 4079
24		-0.00000 0746
26		+0.00000 0033

$m\backslash r$	1	5	15
1	+0.81398 3846	+0.30117 4196	+0.00000 3717
3	-0.52931 0219	+0.62719 8468	+0.00003 7227
5	+0.22890 0813	+0.17707 1306	+0.00032 0013
7	-0.06818 2972	-0.60550 5349	+0.00254 0804
9	+0.01453 0886	+0.33003 2984	+0.01770 9603
11	-0.00229 5765	-0.09333 5984	+0.10045 8755
13	+0.00027 7422	+0.01694 2545	+0.40582 7403
15	-0.00002 6336	-0.00217 7430	+0.83133 2650
17	+0.00000 2009	+0.00021 0135	-0.35924 8830
19	-0.00000 0126	-0.00001 5851	+0.06821 6074
21	+0.00000 0007	+0.00000 0962	-0.00802 4551
23		-0.00000 0048	+0.00066 6432
25		+0.00000 0002	-0.00004 1930
27			+0.00000 2090
29			-0.00000 0086
31			+0.00000 0003

For A_m and B_m see **20.2.3–20.2.11**

Compiled from National Bureau of Standards, Tables relating to Mathieu functions, Columbia Univ. Press, New York, N.Y., 1951 (with permission).

21. Spheroidal Wave Functions

Arnold N. Lowan [1]

Contents

[1] Yeshiva University. (Prepared under contract with the National Bureau of Standards.) (Deceased.)

751

21. Spheroidal Wave Functions

Mathematical Properties

21.1. Definition of Elliptical Coordinates

21.1.1
$$\xi=\frac{r_1+r_2}{2f}; \quad \eta=\frac{r_1-r_2}{2f}$$

r_1 and r_2 are the distances to the foci of a family of confocal ellipses and hyperbolas; $2f$ is the distance between foci.

21.1.2
$$a=f\xi,\; b=f\sqrt{\xi^2-1}, \qquad e=\frac{f}{a}$$

$a=$semi-major axis; $b=$semi-minor axis; $e=$eccentricity.

Equation of Family of Confocal Ellipses

21.1.3
$$\frac{x^2}{\xi^2}+\frac{y^2}{\xi^2-1}=f^2 \qquad (1<\xi<\infty)$$

Equation of Family of Confocal Hyperbolas

21.1.4
$$\frac{x^2}{\eta^2}-\frac{y^2}{1-\eta^2}=f^2 \qquad (-1<\eta<1)$$

Relations Between Cartesian and Elliptical Coordinates

21.1.5
$$x=f\xi\eta; \; y=f\sqrt{(\xi^2-1)(1-\eta^2)}$$

21.2. Definition of Prolate Spheroidal Coordinates

If the system of confocal ellipses and hyberbolas referred to in **21.1.3** and **21.1.4** revolves around the major axis, then

21.2.1
$$\frac{x^2}{\xi^2}+\frac{r^2}{\xi^2-1}=f^2; \qquad \frac{x^2}{\eta^2}-\frac{r^2}{1-\eta^2}=f^2$$

$$y=r\cos\phi; \;\; z=r\sin\phi; \; 0\le\phi\le2\pi$$

where ξ, η and ϕ are prolate spheroidal coordinates.

Relations Between Cartesian and Prolate Spheroidal Coordinates

21.2.2
$$x=f\xi\eta; y=f\sqrt{(\xi^2-1)(1-\eta^2)}\cos\phi;$$
$$z=f\sqrt{(\xi^2-1)(1-\eta^2)}\sin\phi$$

21.3. Definition of Oblate Spheroidal Coordinates

If the system of confocal ellipses and hyperbolas referred to in **21.1.3** and **21.1.4** revolves around the minor axis, then

21.3.1
$$\frac{r^2}{\xi^2}+\frac{y^2}{\xi^2-1}=f^2; \qquad \frac{r^2}{\eta^2}-\frac{y^2}{1-\eta^2}=f^2$$

$$z=r\cos\phi; \; x=r\sin\phi; \; 0\le\phi\le2\pi$$

where ξ, η and ϕ are oblate spheroidal coordinates.

Relations Between Cartesian and Oblate Spheroidal Coordinates

21.3.2
$$x=f\xi\eta\sin\phi; y=f\sqrt{(\xi^2-1)(1-\eta^2)}; z=f\xi\eta\cos\dot\phi$$

21.4. Laplacian in Spheroidal Coordinates

21.4.1
$$\nabla^2=\frac{1}{h_\xi h_\eta h_\phi}\left[\frac{\partial}{\partial\xi}\left(\frac{h_\eta h_\phi}{h_\xi}\frac{\partial}{\partial\xi}\right)+\frac{\partial}{\partial\eta}\left(\frac{h_\xi h_\phi}{h_\eta}\frac{\partial}{\partial\eta}\right)+\frac{\partial}{\partial\phi}\left(\frac{h_\xi h_\eta}{h_\phi}\frac{\partial}{\partial\phi}\right)\right]$$

$$h_\xi^2=\left(\frac{\partial x}{\partial\xi}\right)^2+\left(\frac{\partial y}{\partial\xi}\right)^2+\left(\frac{\partial z}{\partial\xi}\right)^2$$

$$h_\eta^2=\left(\frac{\partial x}{\partial\eta}\right)^2+\left(\frac{\partial y}{\partial\eta}\right)^2+\left(\frac{\partial z}{\partial\eta}\right)^2$$

$$h_\phi^2=\left(\frac{\partial x}{\partial\phi}\right)^2+\left(\frac{\partial y}{\partial\phi}\right)^2+\left(\frac{\partial z}{\partial\phi}\right)^2$$

Metric Coefficients for Prolate Spheroidal Coordinates

21.4.2
$$h_\xi=f\sqrt{\frac{\xi^2-\eta^2}{\xi^2-1}}; \; h_\eta=f\sqrt{\frac{\xi^2-\eta^2}{1-\eta^2}}; \; h_\phi=f\sqrt{(\xi^2-1)(1-\eta^2)} \qquad *$$

Metric Coefficients for Oblate Spheroidal Coordinates

21.4.3
$$h_\xi=f\sqrt{\frac{\xi^2-\eta^2}{\xi^2-1}}; \; h_\eta=f\sqrt{\frac{\xi^2-\eta^2}{1-\eta^2}}; \; h_\phi=f\xi\eta \qquad *$$

21.5. Wave Equation in Prolate and Oblate Spheroidal Coordinates

Wave Equation in Prolate Spheroidal Coordinates

21.5.1
$$\nabla^2\Phi+k^2\Phi=\frac{\partial}{\partial\xi}\left[(\xi^2-1)\frac{\partial\Phi}{\partial\xi}\right]+\frac{\partial}{\partial\eta}\left[(1-\eta^2)\frac{\partial\Phi}{\partial\eta}\right]$$
$$+\frac{\xi^2-\eta^2}{(\xi^2-1)(1-\eta^2)}\frac{\partial^2\Phi}{\partial\phi^2}+c^2(\xi^2-\eta^2)\Phi=0$$
$$\left(c=\frac{1}{2}fk\right)$$

*See page II.

752

Wave Equation in Oblate Spheroidal Coordinates

21.5.2

$$\nabla^2 \Phi + k^2 \Phi = \frac{\partial}{\partial \xi}\left[(\xi^2+1)\frac{\partial \Phi}{\partial \xi}\right] + \frac{\partial}{\partial \eta}\left[(1-\eta^2)\frac{\partial \Phi}{\partial \eta}\right]$$

$$+ \frac{\xi^2+\eta^2}{(\xi^2+1)(1-\eta^2)}\frac{\partial^2 \Phi}{\partial \phi^2} + c^2(\xi^2+\eta^2)\Phi = 0$$

$$\left(c = \frac{1}{2}fk\right)$$

21.5.2 may be obtained from **21.5.1** by the transformations

$$\xi \to \pm i\xi, \quad c \to \mp ic.$$

21.6. Differential Equations for Radial and Angular Prolate Spheroidal Wave Functions

If in **21.5.1** we put

$$\Phi = R_{mn}(c,\xi)S_{mn}(c,\eta)\frac{\cos}{\sin}m\phi$$

then the "radial solution" $R_{mn}(c,\xi)$ and the "angular solution" $S_{mn}(c,\eta)$ satisfy the differential equations

21.6.1

$$\frac{d}{d\xi}\left[(\xi^2-1)\frac{d}{d\xi}R_{mn}(c,\xi)\right]$$

$$-\left(\lambda_{mn}-c^2\xi^2+\frac{m^2}{\xi^2-1}\right)R_{mn}(c,\xi)=0$$

21.6.2

$$\frac{d}{d\eta}\left[(1-\eta^2)\frac{d}{d\eta}S_{mn}(c,\eta)\right]$$

$$+\left(\lambda_{mn}-c^2\eta^2-\frac{m^2}{1-\eta^2}\right)S_{mn}(c,\eta)=0$$

where the separation constants (or eigenvalues) λ_{mn} are to be determined so that $R_{mn}(c,\xi)$ and $S_{mn}(c,\eta)$ are finite at $\xi=\pm 1$ and $\eta=\pm 1$ respectively.

(**21.6.1** and **21.6.2** are identical. Radial and angular prolate spheroidal functions satisfy the same differential equation over different ranges of the variable.)

Differential Equations for Radial and Angular Oblate Spheroidal Functions

21.6.3

$$\frac{d}{d\xi}\left[(\xi^2+1)\frac{d}{d\xi}R_{mn}(c,\xi)\right]$$

$$-\left(\lambda_{mn}-c^2\xi^2-\frac{m^2}{\xi^2+1}\right)R_{mn}(c,\xi)=0$$

21.6.4

$$\frac{d}{d\eta}\left[(1-\eta^2)\frac{d}{d\eta}S_{mn}(c,\eta)\right]$$

$$+\left(\lambda_{mn}+c^2\eta^2-\frac{m^2}{1-\eta^2}\right)S_{mn}(c,\eta)=0$$

(**21.6.3** may be obtained from **21.6.1** by the transformations $\xi \to \pm i\xi$, $c \to \mp ic$; **21.6.4** may be obtained from **21.6.2** by the transformation $c \to \mp ic$.)

21.7. Prolate Angular Functions

21.7.1

$$S_{mn}^{(1)}(c,\eta) = \sideset{}{'}\sum_{r=0,1}^{\infty} d_r^{mn}(c)P_{m+r}^m(\eta)$$

= Prolate angular function of the first kind

21.7.2

$$S_{mn}^{(2)}(c,\eta) = \sideset{}{'}\sum_{r=-\infty}^{\infty} d_r^{mn}(c)Q_{m+r}^m(\eta)$$

= Prolate angular function of the second kind

($P_n^m(\eta)$ and $Q_n^m(\eta)$ are associated Legendre functions of the first and second kinds respectively. However, for $-1 \le z \le 1$, $P_n^m(z) = (1-z^2)^{m/2}d^mP_n(z)/dz^m$ (see **8.6.6**). The summation is extended over even values or odd values of r.)

Recurrence Relations Between the Coefficients

21.7.3

$$\alpha_k d_{k+2} + (\beta_k - \lambda_{mn})d_k + \gamma_k d_{k-2} = 0$$

$$\alpha_k = \frac{(2m+k+2)(2m+k+1)c^2}{(2m+2k+3)(2m+2k+5)}$$

$$\beta_k = (m+k)(m+k+1)$$

$$+\frac{2(m+k)(m+k+1)-2m^2-1}{(2m+2k-1)(2m+2k+3)}c^2$$

$$\gamma_k = \frac{k(k-1)c^2}{(2m+2k-3)(2m+2k-1)}$$

Transcendental Equation for λ_{mn}

21.7.4

$$U(\lambda_{mn}) = U_1(\lambda_{mn}) + U_2(\lambda_{mn}) = 0$$

$$U_1(\lambda_{mn}) = \gamma_r^m - \lambda_{mn} - \frac{\beta_r^m}{\gamma_{r-2}^m - \lambda_{mn} -}\frac{\beta_{r-2}^m}{\gamma_{r-4}^m - \lambda_{mn} -}\cdots$$

$$U_2(\lambda_{mn}) = -\frac{\beta_{r+2}^m}{\gamma_{r+2}^m - \lambda_{mn} -}\frac{\beta_{r+4}^m}{\gamma_{r+4}^m - \lambda_{mn} -}\cdots$$

$$\beta_k^m = \frac{k(k-1)(2m+k)(2m+k-1)c^4}{(2m+2k-1)^2(2m+2k+1)(2m+2k-3)}$$

$$(k \ge 2)$$

$$\gamma_k^m = (m+k)(m+k+1)$$

$$+\frac{1}{2}c^2\left[1-\frac{4m^2-1}{(2m+2k-1)(2m+2k+3)}\right] \quad (k \ge 0)$$

(The choice of r in **21.7.4** is arbitrary.)

Power Series Expansion for λ_{mn}

21.7.5

$$\lambda_{mn}=\sum_{k=0}^{\infty} l_{2k}c^{2k}$$

$$l_0=n(n+1)$$

$$l_2=\tfrac{1}{2}\left[1-\frac{(2m-1)(2m+1)}{(2n-1)(2n+3)}\right]$$

$$l_4=\frac{-(n-m+1)(n-m+2)(n+m+1)(n+m+2)}{2(2n+1)(2n+3)^3(2n+5)}+\frac{(n-m-1)(n-m)(n+m-1)(n+m)}{2(2n-3)(2n-1)^3(2n+1)}$$

$$l_6=(4m^2-1)\left[\frac{(n-m+1)(n-m+2)(n+m+1)(n+m+2)}{(2n-1)(2n+1)(2n+3)^5(2n+5)(2n+7)}-\frac{(n-m-1)(n-m)(n+m-1)(n+m)}{(2n-5)(2n-3)(2n-1)^5(2n+1)(2n+3)}\right]$$

$$l_8=2(4m^2-1)^2A+\frac{1}{16}B+\frac{1}{8}C+\frac{1}{2}D$$

$$A=\frac{(n-m-1)(n-m)(n+m-1)(n+m)}{(2n-5)^2(2n-3)(2n-1)^7(2n+1)(2n+3)^2}-\frac{(n-m+1)(n-m+2)(n+m+1)(n+m+2)}{(2n-1)^2(2n+1)(2n+3)^7(2n+5)(2n+7)^2}$$

$$B=\frac{(n-m-3)(n-m-2)(n-m-1)(n-m)(n+m-3)(n+m-2)(n+m-1)(n+m)}{(2n-7)(2n-5)^2(2n-3)^3(2n-1)^4(2n+1)}$$

$$-\frac{(n-m+1)(n-m+2)(n-m+3)(n-m+4)(n+m+1)(n+m+2)(n+m+3)(n+m+4)}{(2n+1)(2n+3)^4(2n+5)^3(2n+7)^2(2n+9)}$$

$$C=\frac{(n-m+1)^2(n-m+2)^2(n+m+1)^2(n+m+2)^2}{(2n+1)^2(2n+3)^7(2n+5)^2}-\frac{(n-m-1)^2(n-m)^2(n+m-1)^2(n+m)^2}{(2n-3)^2(2n-1)^7(2n+1)^2}$$

$$D=\frac{(n-m-1)(n-m)(n-m+1)(n-m+2)(n+m-1)(n+m)(n+m+1)(n+m+2)}{(2n-3)(2n-1)^4(2n+1)^2(2n+3)^4(2n+5)}$$

Asymptotic Expansion for λ_{mn}

21.7.6

$$\lambda_{mn}(c)=cq+m^2-\frac{1}{8}(q^2+5)-\frac{q}{64c}(q^2+11-32m^2)$$

$$-\frac{1}{1024c^2}[5(q^4+26q^2+21)-384m^2(q^2+1)]$$

$$-\frac{1}{c^3}\left[\frac{1}{128^2}(33q^5+1594q^3+5621q)\right.$$

$$\left.-\frac{m^2}{128}(37q^3+167q)+\frac{m^4}{8}q\right]$$

$$-\frac{1}{c^4}\left[\frac{1}{256^2}(63q^6+4940q^4+43327q^2+22470)\right.$$

$$\left.-\frac{m^2}{512}(115q^4+1310q^2+735)+\frac{3m^4}{8}(q^2+1)\right]$$

$$-\frac{1}{c^5}\left[\frac{1}{1024^2}(527q^7+61529q^5+1043961q^3\right.$$

$$+2241599q)-\frac{m^2}{32\cdot1024}(5739q^5+127550q^3$$

$$\left.+298951q)+\frac{m^4}{512}(355q^3+1505q)-\frac{m^6q}{16}\right]+O(c^{-6})$$

$$q=2(n-m)+1$$

Refinement of Approximate Values of λ_{mn}

If $\lambda_{mn}^{(1)}$ is an approximation to λ_{mn} obtained either from **21.7.5** or **21.7.6** then

21.7.7

$$\lambda_{mn}=\lambda_{mn}^{(1)}+\delta\lambda_{mn}$$

$$\delta\lambda_{mn}=\frac{U_1(\lambda_{mn}^{(1)})+U_2(\lambda_{mn}^{(1)})}{\Delta_1+\Delta_2}$$

$$\Delta_1=1+\frac{\beta_r^m}{(N_r^m)^2}+\frac{\beta_r^m\beta_{r-2}^m}{(N_r^mN_{r-2}^m)^2}+\frac{\beta_r^m\beta_{r-2}^m\beta_{r-4}^m}{(N_r^mN_{r-2}^mN_{r-4}^m)^2}+\cdots$$

$$\Delta_2=\frac{(N_{r+2}^m)^2}{\beta_{r+2}^m}+\frac{(N_{r+2}^mN_{r+4}^m)^2}{\beta_{r+2}^m\beta_{r+4}^m}+\frac{(N_{r+2}^mN_{r+4}^mN_{r+6}^m)^2}{\beta_{r+2}^m\beta_{r+4}^m\beta_{r+6}^m}$$
$$+\cdots$$

$$N_r^m=\frac{(2m+r)(2m+r-1)c^2}{(2m+2r-1)(2m+2r+1)}\frac{d_r}{d_{r-2}}\qquad(r\geq2)$$

$$\beta_r^m=\frac{r(r-1)(2m+r)(2m+r-1)c^4}{(2m+2r-1)^2(2m+2r+1)(2m+2r-3)}$$

$$(r\geq2)$$

Evaluation of Coefficients

Step 1. Calculate N_r^m's from

21.7.8

$$N_{r+2}^m = \gamma_r^m - \lambda_{mn} - \frac{\beta_r^m}{N_r^m} \qquad (r \geq 2)$$

$$N_2^m = \gamma_0^m - \lambda_{mn}; \quad N_3^m = \gamma_1^m - \lambda_{mn}$$

$$\gamma_r^m = (m+r)(m+r+1)$$

$$+ \frac{1}{2} c^2 \left[1 - \frac{4m^2-1}{(2m+2r-1)(2m+2r+3)} \right] \quad (r \geq 0)$$

Step 2. Calculate ratios $\frac{d_0}{d_{2r}}$ and $\frac{d_1}{d_{2p+1}}$ from

21.7.9
$$\frac{d_0}{d_{2r}} = \left(\frac{d_0}{d_2}\right)\left(\frac{d_2}{d_4}\right) \cdots \left(\frac{d_{2r-2}}{d_{2r}}\right)$$

21.7.10
$$\frac{d_1}{d_{2p+1}} = \left(\frac{d_1}{d_3}\right)\left(\frac{d_3}{d_5}\right) \cdots \left(\frac{d_{2p-1}}{d_{2p+1}}\right)$$

and the formula for N_r^m in **21.7.7**.

The coefficients d_r^{mn} are determined to within the arbitrary factor d_0 for r even and d_1 for r odd. The choice of these factors depends on the normalization scheme adopted.

Normalization of Angular Functions
Meixner-Schäfke Scheme

21.7.11 $\displaystyle\int_{-1}^{1} [S_{mn}(c, \eta)]^2 d\eta = \frac{2}{2n+1} \frac{(n+m)!}{(n-m)!}$

Stratton-Morse-Chu-Little-Corbató Scheme

21.7.12 $\displaystyle\sideset{}{'}\sum_{r=0,1} \frac{(r+2m)!}{r!} d_r = \frac{(n+m)!}{(n-m)!}$

(This normalization has the effect that $S_{mn}(c, \eta) \to P_n^m(\eta)$ as $\eta \to 1$.)

Flammer Scheme [21.4]

21.7.13
$$S_{mn}(c, 0) = P_n^m(0) = \frac{(-1)^{\frac{n-m}{2}}(n+m)!}{2^n \left(\frac{n-m}{2}\right)!\left(\frac{n+m}{2}\right)!}$$
$$(n-m) \text{ even}$$

21.7.14
$$S_{mn}'(c, 0) = P_n^{m'}(0) = \frac{(-1)^{\frac{n-m-1}{2}}(n+m+1)!}{2^n \left(\frac{n-m-1}{2}\right)!\left(\frac{n+m+1}{2}\right)!}$$
$$(n-m) \text{ odd}$$

The above lead to the following conditions for d_r^{mn}

21.7.15
$$\sideset{}{'}\sum_{r=0}^{\infty} \frac{(-1)^{r/2}(r+2m)!}{2^r \left(\frac{r}{2}\right)!\left(\frac{r+2m}{2}\right)!} d_r^{mn} = \frac{(-1)^{\frac{n-m}{2}}(n+m)!}{2^{n-m}\left(\frac{n-m}{2}\right)!\left(\frac{n+m}{2}\right)!}$$
$$(n-m) \text{ even}$$

21.7.16
$$\sideset{}{'}\sum_{r=1}^{\infty} \frac{(-1)^{\frac{r-1}{2}}(r+2m+1)!}{2^r \left(\frac{r-1}{2}\right)!\left(\frac{r+2m+1}{2}\right)!} d_r^{mn}$$
$$= \frac{(-1)^{\frac{n-m-1}{2}}(n+m+1)!}{2^{n-m}\left(\frac{n-m-1}{2}\right)!\left(\frac{n+m+1}{2}\right)!} \quad (n-m) \text{ odd}$$

(The normalization scheme **21.7.13** and **21.7.14** is also used in [21.10].)

Asymptotic Expansions for $S_{mn}(c, \eta)$

21.7.17
$$S_{mn}(c, \eta) = (1-\eta^2)^{\frac{1}{2}} U_{mn}(c, \eta) \qquad (c \to \infty)$$

$$U_{mn}(x) = \sum_{r=-\infty}^{\infty} h_r^l D_{l+r}(x) \qquad l = n-m$$

where the $D_r(x)$'s are the parabolic cylinder functions (see chapter **19**).

$$D_r(x) = (-1)^r e^{x^2/4} \frac{d^2}{dx^2} e^{-x^2/2} = 2^{-r/2} e^{-x^2/4} H_r\left(\frac{x}{\sqrt{2}}\right)$$

and the $H_r(x)$ are the Hermite polynomials (see chapter **22**). (For tables of $h_{\pm r}^l/h_0^l$ see [21.4].)

Expansion of $S_{mn}(c, \eta)$ in Powers of η

21.7.18
$$S_{mn}(c, \eta) = (1-\eta^2)^{m/2} \sideset{}{'}\sum_{r=0,1}^{\infty} p_r^{mn}(c) \eta^r$$

$$(r+1)(r+2)p_{r+2}^{mn}(c) - [r(r+2m+1)+m(m+1)$$
$$- \lambda_{mn}(c)]p_r^{mn}(c) - c^2 p_{r-2}^{mn}(c) = 0$$

(The derivation of the transcendental equation for λ_{mn} is similar to the derivation of **21.7.4** from **21.7.3**.)

Expansion of $S_{mn}(c, \eta)$ in Powers of $(1-\eta^2)$

21.7.19
$$S_{mn}(c, \eta) = (1-\eta^2)^{m/2} \sum_{k=0}^{\infty} c_{2k}^{mn}(1-\eta^2)^k \qquad (n-m) \text{ even}$$

21.7.20

$$S_{mn}(c,\eta)=\eta(1-\eta^2)^{m/2}\sum_{k=0}^{\infty}c_{2k}^{mn}(1-\eta^2)^k \qquad (n-m)\text{ odd}$$

$$c_{2k}^{mn}=\frac{1}{2^m k!(m+k)!}\sum_{r=k}^{\infty}\frac{(2m+2r)!}{(2r)!}(-r)_k\left(m+r+\frac{1}{2}\right)_k d_{2r}^{mn}$$
$$(n-m)\text{ even}$$

$$c_{2k}^{mn}=$$
$$\frac{1}{2^m k!(m+k)!}\sum_{r=k}^{\infty}\frac{(2m+2r+1)!}{(2r+1)!}(-r)_k\left(m+r+\frac{3}{2}\right)_k d_{2r+1}^{mn}$$
$$(n-m)\text{ odd}$$

$$(\alpha)_k=\alpha(\alpha+1)(\alpha+2)\ldots(\alpha+k+1)$$

(The d_r^{mn}'s are the coefficients in **21.7.1**.)

Prolate Angular Functions—Second Kind

Expansion **21.7.2** ultimately leads to

21.7.21

$$S_{mn}^{(2)}(c,\eta)=\sum_{r=-2m,\,-2m+1}^{\infty}{}' d_r^{mn}Q_{m+r}^m(\eta)$$
$$+\sum_{r=2m+2,\,2m+1}^{\infty}{}' d_{\rho|r}^{mn}P_{r-m-1}^m(\eta)$$

(The coefficients d_r^{mn} are the same as in **21.7.1**; the coefficients $d_{\rho|r}^{mn}$ are tabulated in [21.4].)

21.8. Oblate Angular Functions

Power Series Expansion for Eigenvalues

21.8.1 $\lambda_{mn}=\sum_{k=0}^{\infty}(-1)^k l_{2k}c^{2k}$

where the l_k's are the same as in **21.7.5**.

Asymptotic Expansion for Eigenvalues [21.4]

21.8.2

$$\lambda_{mn}=-c^2+2c(2\nu+m+1)-2\nu(\nu+m+1)$$
$$-(m+1)+\Lambda_{mn}$$

$$\nu=\frac{1}{2}(n-m)\text{ for }(n-m)\text{ even};$$

$$\nu=\frac{1}{2}(n-m-1)\text{ for }(n-m)\text{ odd}$$

$$\Lambda_{mn}=\sum_{k=1}^{\infty}\beta_k^{mn}c^{-k}$$

$$\beta_1^{mn}=-2^{-3}q(q^2+1-m^2)$$

$$\beta_2^{mn}=-2^{-6}[5q^4+10q^2+1-2m^2(3q^2+1)+m^4]$$

$$\beta_3^{mn}=-2^{-9}q[33q^4+114q^2+37-2m^2(23q^2+25)$$
$$+13m^4]$$

$$\beta_4^{mn}=-2^{-10}[63q^6+340q^4+239q^2+14$$
$$-10m^2(10q^4+23q^2+3)+m^4(39q^2-18)-2m^6]$$

$$\beta_k^{mn}=\nu(\nu+m)a_k^{-1}+(\nu+1)(\nu+m+1)a_k^{+1}$$

$$q=n+1\text{ for }(n-m)\text{ even};\ q=n\text{ for }(n-m)\text{ odd}$$

(For the definition of $a_k^{\pm r}$ see **21.8.3**.)

Asymptotic Expansion for Oblate Angular Functions

21.8.3

$$S_{mn}(-ic,\eta)\sim(1-\eta^2)^{m/2}\sum_{s=-\nu}^{\infty}A_s^{mn}\{e^{-c(1-\eta)}L_{\nu+s}^{(m)}[2c(1-\eta)]$$
$$+(-1)^{n-m}e^{-c(1+\eta)}L_{\nu+s}^{(m)}[2c(1+\eta)]\}$$

where the $L_\nu^{(m)}(x)$ are Laguerre polynomials (see chapter **22**) and

$$\frac{A_{\pm r}^{mn}}{A_0^{mn}}=\sum_{k=r}^{\infty}a_k^{\pm r}(m,n)c^{-k}$$

(Expressions of $a_k^{\pm r}$ are given in [21.4].)

21.9. Radial Spheroidal Wave Functions

21.9.1

$$R_{mn}^{(p)}(c,\xi)=\left\{\sum_{r=0,\,1}^{\infty}{}'\frac{(2m+r)!}{r!}d_r^{mn}\right\}^{-1}\left(\frac{\xi^2-1}{\xi^2}\right)^{m/2}$$
$$\cdot\sum_{r=0,\,1}^{\infty}{}' i^{r+m-n}\frac{(2m+r)!}{r!}d_r^{mn}Z_{m+r}^{(p)}(c\xi)^{*}$$

$$Z_n^{(p)}(z)=\sqrt{\frac{\pi}{2z}}\,J_{n+\frac{1}{2}}(z) \qquad (p=1)$$
$$=\sqrt{\frac{\pi}{2z}}\,Y_{n+\frac{1}{2}}(z) \qquad (p=2)$$

($J_{n+\frac{1}{2}}(z)$ and $Y_{n+\frac{1}{2}}(z)$ are Bessel functions, order $n+\frac{1}{2}$, of the first and second kind respectively (see chapter **10**).)

21.9.2 $R_{mn}^{(3)}(c,\xi)=R_{mn}^{(1)}(c,\xi)+iR_{mn}^{(2)}(c,\xi)$

21.9.3 $R_{mn}^{(4)}(c,\xi)=R_{mn}^{(1)}(c,\xi)-iR_{mn}^{(2)}(c,\xi)$

Asymptotic Behavior of $R_{mn}^{(1)}(c,\xi)$ and $R_{mn}^{(2)}(c,\xi)$

21.9.4 $R_{mn}^{(1)}(c,\xi)\xrightarrow[c\xi\to\infty]{}\frac{1}{c\xi}\cos\left[c\xi-\frac{1}{2}(n+1)\pi\right]$

21.9.5 $R_{mn}^{(2)}(c,\xi)\xrightarrow[c\xi\to\infty]{}\frac{1}{c\xi}\sin\left[c\xi-\frac{1}{2}(n+1)\pi\right]$

*See page II.

21.10. Joining Factors for Prolate Spheroidal Wave Functions

21.10.1

$$S_{mn}^{(1)}(c,\xi)=\kappa_{mn}^{(1)}(c)R_{mn}^{(1)}(c,\xi)$$

$$\kappa_{mn}^{(1)}(c)=\frac{(2m+1)(n+m)!\sum\limits_{r=0}^{\infty}{}' d_r^{mn}(2m+r)!/r!}{2^{n+m}d_0^{mn}(c)c^m m!\left(\dfrac{n-m}{2}\right)!\left(\dfrac{n+m}{2}\right)!}$$

$$(n-m)\text{ even}$$

$$=\frac{(2m+3)(n+m+1)!\sum\limits_{r=1}^{\infty}{}' d_r^{mn}(2m+r)!/r!}{2^{n+m}d_1^{mn}(c)c^{m+1}m!\left(\dfrac{n-m-1}{2}\right)!\left(\dfrac{n+m+1}{2}\right)!}$$

$$(n-m)\text{ odd}$$

21.10.2

$$S_{mn}^{(2)}(c,\xi)=\kappa_{mn}^{(2)}(c)R_{mn}^{(2)}(c,\xi)$$

$$\kappa_{mn}^{(2)}(c)=\frac{2^{n-m}(2m)!\left(\dfrac{n-m}{2}\right)!\left(\dfrac{n+m}{2}\right)! d_{-2m}^{mn}(c)}{(2m-1)m!(n+m)!c^{m-1}}\sum\limits_{r=0}^{\infty}{}'\frac{(2m+r)!}{r!}d_r^{mn}(c) \qquad (n-m)\text{ even}$$

$$=-\frac{2^{n-m}(2m)!\left(\dfrac{n-m-1}{2}\right)!\left(\dfrac{n+m+1}{2}\right)! d_{-2m+1}^{mn}(c)}{(2m-3)(2m-1)m!(n+m+1)!c^{m-2}}\sum\limits_{r=1}^{\infty}{}'\frac{(2m+r)!}{r!}d_r^{mn}(c) \qquad (n-m)\text{ odd}$$

(The expression for joining factors appropriate to the oblate case may be obtained from the above formulas by the transformation $c\to-ic$.)

21.11. Notation

Notation for Prolate Spheroidal Wave Functions

	Ang. coord.	Rad. coord.	Independent variable	Ang. wave function	Rad. wave function	Eigenvalue	Normalization of angular functions	Remarks
Stratton, Morse, Chu, Little and Corbató	η	ξ	h	$S_{ml}(h, \eta)$	$je_{ml}(h, \xi)$ $ne_{ml}(h, \xi)$ $he_{ml}(h, \xi)$	$A_{ml}(h)$	$S_{ml}(h, 1) = P_l^m(1)$	$l=$Flammer's n $A_{ml}=\lambda_{mn}$
Flammer and this chapter	η	ξ	c	$S_{mn}(c, \eta)$	$R_{mn}^{(i)}(c, \xi)$	$\lambda_{mn}(c)$	$S_{mn}(c, 0) = P_n^m(0)\quad (n-m)$ even $S_{mn}'(c, 0) = P_n^{m'}(0)\quad (n-m)$ odd	
Chu and Stratton	η	ξ	c	$S_{ml}^{(1)}(c, \eta)$	$R_{ml}^{(i)}(c, \xi)$	A_{ml}	$S_{ml}^{(1)}(c, 0) = P_{m+l}^m(0)\quad (l$ even$)$ $S_{ml}^{(1)'}(c, 0) = P_{m+l}^m(0)\quad (l$ odd$)$	$l=$Flammer's $n-m$ $A_{ml}=\lambda_{m, n-m}$
Meixner and Schäfke	η	ξ	γ	$PS_n^m(\eta, \gamma^2)$	$S_n^{m(i)}(\xi, \gamma^2)$	$\lambda_n^m(\gamma^2)$	$\int_{-1}^1 [PS_n^m(\eta, \gamma^2)]^2 d\eta = \frac{2}{2n+1}\frac{(n+m)!}{(n-m)!}$	$\lambda_n^m(\gamma^2) = \lambda_{mn}(c) - c^2$
Morse and Feshbach	$\eta=\cos\vartheta$	$\xi=\cosh\mu$	h	$S_{ml}(h, \eta)$	$je_{ml}(h, \xi)$ $ne_{ml}(h, \xi)$ $he_{ml}(h, \xi)$	A_{ml}	$[(1-\eta^2)^{-m/2}S_{ml}(h, \eta)]_{\eta=1} = [(1-\eta^2)^{-m/2}P_n^m(\eta)]_{\eta=1}$	$l=$Flammer's n $A_{ml}=\lambda_{mn}$
Page	ξ	η	ϵ	$U_{lm}(\xi)$	$v_{lm}(\eta)$ $p_{lm}(\eta)$ $q_{lm}(\eta)$	α_{lm}	$[(1-\xi^2)^{-m/2}U_{lm}(\xi)]_{\xi=1} = 1$	$l=$Flammer's n $\alpha_{lm}=\lambda_{mn}-c^2$

Notation for Oblate Spheroidal Wave Functions

	Ang. coord.	Rad. coord.	Independent variable	Ang. wave function	Rad. wave function	Eigenvalue	Normalization of angular functions	Remarks
Stratton, Morse, Chu, Little and Corbató	η	ξ	g	$S_{ml}(ig, \eta)$	$je_{ml}(ig, -i\xi)$	A_{ml}	$S_{ml}(ig, 1) = P_l^m(1)$	$l=$Flammer's n $A_{ml}=\lambda_{mn}$
Flammer and this chapter	η	ξ	c	$S_{mn}(-ic, \eta)$	$R_{mn}^{(i)}(-ic, i\xi)$	$\lambda_{mn}(-ic)$	$S_{mn}(-ic, 0) = P_n^m(0)\quad (n-m)$ even $S_{mn}'(-ic, 0) = P_n^{m'}(0)\quad (n-m)$odd	
Chu and Stratton	η	ξ	c	$S_{ml}^{(1)}(-ic, \eta)$	$R_{ml}^{(1)}(-ic, i\xi)$	B_{ml}	$S_{ml}^{(1)}(-ic, 0) = P_{m+l}^m(0)\quad (l$ even$)$ $S_{ml}^{(1)'}(-ic, 0) = P_{m+l}^{m'}(0)\quad (l$ odd$)$	$l=$Flammer's $n-m$ $B_{lm}=-\lambda_{m, n-m}$
Meixner and Schäfke	η	ξ	γ	$ps_n^m(\eta, -\gamma^2)$	$S_n^{m(i)}(-i\xi, i\gamma^2)$	$\lambda_n^m(-\gamma^2)$	$\int_{-1}^1 [ps_n^m(\eta, -\gamma^2)]^2 d\eta = \frac{2}{2n+1}\frac{(n+m)!}{(n-m)!}$	$\lambda_n^m(-\gamma^2) = \lambda_{mn}(-ic) + c^2$
Morse and Feshbach	$\eta=\cos\vartheta$	$\xi=\sinh\mu$	g	$S_{ml}(ig, \eta)$	$je_{ml}(ig, -i\xi)$ $ne_{ml}(ig, -i\xi)$ $he_{ml}(ig, -i\xi)$	A_{ml}	$[(1-\eta^2)^{-m/2}S_{ml}(ig, \eta)]_{\eta=1} = [(1-\eta^2)^{-m/2}P_n^m(\eta)]_{\eta=1}$	$l=$Flammer's n $A_{ml}=\lambda_{mn}$
Leitner and Spence	η	ξ	ϵ	$U_{lm}(\eta)$	(i) $v_{lm}(\xi)$	α_{lm}	$[(1-\eta^2)^{-m/2}U_{lm}(\eta)]_{\eta=1} = 1$	$l=$Flammer's n $\alpha_{lm}=\lambda_{mn}+c^3$

The notation in this chapter closely follows the notation in [21.4].

References

[21.1] M. Abramowitz, Asymptotic expansion of spheroidal wave functions, J. Math. Phys. **28**, 195–199 (1949).

[21.2] G. Blanch, On the computation of Mathieu functions, J. Math. Phys. **25**, 1–20 (1946).

[21.3] C. J. Bouwkamp, Theoretische en numerieke behandeling van de buiging door en ronde opening, Diss. Groningen, Groningen-Batavia, (1941).

[21.4] C. Flammer, Spheroidal wave functions (Stanford Univ. Press, Stanford, Calif., 1957).

[21.5] A. Leitner and R. D. Spence, The oblate spheroidal wave functions, J. Franklin Inst. **249**, 299–321 (1950).

[21.6] J. Meixner and F. W. Schäfke, Mathieusche Funktionen und Sphäroidfunktionen (Springer-Verlag, Berlin, Göttingen, Heidelberg, Germany, 1954).

[21.7] P. M. Morse and H. Feshbach, Methods of theoretical physics (McGraw-Hill Book Co., Inc., New York, N.Y., 1953).

[21.8] L. Page, The electrical oscillations of a prolate spheroid, Phys. Rev. **65**, 98–117 (1944).

[21.9] J. A. Stratton, P. M. Morse, L. J. Chu and R. A. Hutner, Elliptic cylinder and spheroidal wave functions (John Wiley & Sons, Inc., New York, N.Y., 1941).

[21.10] J. A. Stratton, P. M. Morse, L. J. Chu, J. D. C. Little, and F. J. Corbató, Spheroidal wave functions (John Wiley & Sons, Inc., New York, N.Y., 1956).

Table 21.1   **EIGENVALUES—PROLATE AND OBLATE**

PROLATE

$$\lambda_{mn}(c) - m(m+1) \quad *$$

$$\lambda_{0n}(c)$$

$c^2 \backslash n$	0	1	2	3	4
0	0.000000	2.000000	6.000000	12.000000	20.000000
1	0.319000	2.593084	6.533471	12.514462	20.508274
2	0.611314	3.172127	7.084258	13.035830	21.020137
3	0.879933	3.736869	7.649317	13.564354	21.535636
4	1.127734	4.287128	8.225713	14.100203	22.054829
5	1.357356	4.822809	8.810735	14.643458	22.577779
6	1.571155	5.343903	9.401958	15.194110	23.104553
7	1.771183	5.850492	9.997251	15.752059	23.635223
8	1.959206	6.342739	10.594773	16.317122	24.169860
9	2.136732	6.820888	11.192938	16.889030	24.708534
10	2.305040	7.285254	11.790394	17.467444	25.251312
11	2.465217	7.736212	12.385986	18.051962	25.798254
12	2.618185	8.174189	12.978730	18.642128	26.349411
13	2.764731	8.599648	13.567791	19.237446	26.904827
14	2.905523	9.013085	14.152458	19.837389	27.464530
15	3.041137	9.415010	14.732130	20.441413	28.028539
16	3.172067	9.805943	15.306299	21.048960	28.596854
	$\begin{bmatrix}(-3)3 \\ 6\end{bmatrix}$	$\begin{bmatrix}(-3)2 \\ 5\end{bmatrix}$	$\begin{bmatrix}(-3)2 \\ 6\end{bmatrix}$	$\begin{bmatrix}(-4)9 \\ 5\end{bmatrix}$	$\begin{bmatrix}(-4)5 \\ 5\end{bmatrix}$

$$c^{-1}[\lambda_{0n}(c)]$$

$c^{-1} \backslash n$	0	1	2	3	4
0.25	0.793016	2.451485	3.826574	5.26224	7.14921
0.24	0.802442	2.477117	3.858771	5.25133	7.05054
0.23	0.811763	2.503218	3.895890	5.25040	6.96237
0.22	0.820971	2.529593	3.937869	5.26046	6.88638
0.21	0.830059	2.556036	3.984499	5.28251	6.82460
0.20	0.839025	2.582340	4.035382	5.31747	6.77941
0.19	0.847869	2.608310	4.089903	5.36610	6.75360
0.18	0.856592	2.633778	4.147207	5.42883	6.75030
0.17	0.865200	2.658616	4.206229	5.50551	6.77286
0.16	0.873698	2.682743	4.265772	5.59516	6.82451
0.15	0.882095	2.706127	4.324653	5.69566	6.90779
0.14	0.890399	2.728784	4.381878	5.80359	7.02356
0.13	0.898617	2.750762	4.436798	5.91452	7.16962
0.12	0.906758	2.772133	4.489168	6.02383	7.33916
0.11	0.914827	2.792971	4.539096	6.12806	7.52035
0.10	0.922830	2.813346	4.586895	6.22577	7.69932
0.09	0.930772	2.833316	4.632927	6.31730	7.86638
0.08	0.938657	2.852927	4.677506	6.40385	8.01951
0.07	0.946487	2.872213	4.720863	6.48655	8.16148
0.06	0.954267	2.891203	4.763160	6.56618	8.29538
0.05	0.961998	2.909920	4.804519	6.64326	8.42315
0.04	0.969683	2.928382	4.845033	6.71812	8.54594
0.03	0.977324	2.946608	4.884779	6.79104	8.66452
0.02	0.984923	2.964611	4.923820	6.86221	8.77945
0.01	0.992481	2.982404	4.962212	6.93182	8.89116
0.00	1.000000	3.000000	5.000000	7.00000	9.00000
	$\begin{bmatrix}(-5)2 \\ 4\end{bmatrix}$	$\begin{bmatrix}(-5)9 \\ 5\end{bmatrix}$	$\begin{bmatrix}(-4)6 \\ 6\end{bmatrix}$	$\begin{bmatrix}(-3)2 \\ 6\end{bmatrix}$	$\begin{bmatrix}(-3)4 \\ 9\end{bmatrix}$

*See page II.

EIGENVALUES—PROLATE AND OBLATE Table 21.1
OBLATE

$$\lambda_{mn}(-ic) - m(m+1) \qquad *$$
$$\lambda_{0n}(-ic)$$

$c^2\backslash n$	0	1	2	3	4
0	0.000000	2.000000	6.000000	12.000000	20.000000
1	−0.348602	1.393206	5.486800	11.492120	19.495276
2	−0.729391	0.773097	4.996484	10.990438	18.994079
3	−1.144328	+0.140119	4.531027	10.494512	18.496395
4	−1.594493	−0.505243	4.091509	10.003863	18.002228
5	−2.079934	−1.162477	3.677958	9.517982	17.511597
6	−2.599668	−1.831050	3.289357	9.036338	17.024540
7	−3.151841	−2.510421	2.923796	8.558395	16.541110
8	−3.733981	−3.200049	2.578730	8.083615	16.061382
9	−4.343292	−3.899400	2.251269	7.611465	15.585448
10	−4.976895	−4.607952	1.938419	7.141427	15.113424
11	−5.632021	−5.325200	1.637277	6.673001	14.645441
12	−6.306116	−6.050659	1.345136	6.205705	14.181652
13	−6.996903	−6.783867	1.059541	5.739084	13.722230
14	−7.702385	−7.524384	0.778305	5.272706	13.267364
15	−8.420841	−8.271795	0.499495	4.806165	12.817261
16	−9.150793	−9.025710	0.221407	4.339082	12.372144
	$\begin{bmatrix}(-3)4\\7\end{bmatrix}$	$\begin{bmatrix}(-3)2\\5\end{bmatrix}$	$\begin{bmatrix}(-3)3\\7\end{bmatrix}$	$\begin{bmatrix}(-4)8\\5\end{bmatrix}$	$\begin{bmatrix}(-4)6\\5\end{bmatrix}$

$$c^{-2}[\lambda_{0n}(-ic)]$$

$c^{-1}\backslash n$	0	1	2	3	4
0.25	−0.571924	−0.564106	+0.013837	0.271192	0.77325
0.24	−0.585248	−0.579552	−0.009136	0.213225	0.67822
0.23	−0.599067	−0.595037	−0.031481	0.157464	0.58772
0.22	−0.613349	−0.610591	−0.053477	0.103825	0.50191
0.21	−0.628058	−0.626242	−0.075480	0.052196	0.42099
0.20	−0.643161	−0.642016	−0.097943	+0.002437	0.34521
0.19	−0.658625	−0.657938	−0.121428	−0.045635	0.27490
0.18	−0.674418	−0.674031	−0.146603	−0.092251	0.21043
0.17	−0.690515	−0.690310	−0.174201	−0.137692	0.15215
0.16	−0.706891	−0.706792	−0.204894	−0.182301	0.10020
0.15	−0.723530	−0.723486	−0.239109	−0.226469	0.05428
0.14	−0.740416	−0.740399	−0.276886	−0.270627	+0.01332
0.13	−0.757541	−0.757535	−0.317881	−0.315206	−0.02476
0.12	−0.774896	−0.774894	−0.361548	−0.360594	−0.06337
0.11	−0.792476	−0.792476	−0.407352	−0.407081	−0.10723
0.10	−0.810279	−0.810279	−0.454896	−0.454839	−0.16065
0.09	−0.828301	−0.828301	−0.503937	−0.503928	−0.22419
0.08	−0.846539	−0.846539	−0.554337	−0.554337	−0.29513
0.07	−0.864992	−0.864992	−0.606021	−0.606021	−0.37117
0.06	−0.883657	−0.883657	−0.658931	−0.658931	−0.45125
0.05	−0.902532	−0.902532	−0.713025	−0.713025	−0.53495
0.04	−0.921616	−0.921616	−0.768262	−0.768262	−0.62200
0.03	−0.940906	−0.940906	−0.824608	−0.824608	−0.71218
0.02	−0.960402	−0.960402	−0.882031	−0.882031	−0.80533
0.01	−0.980100	−0.980100	−0.940503	−0.940503	−0.90131
0.00	−1.000000	−1.000000	−1.000000	−1.000000	−1.00000
	$\begin{bmatrix}(-5)6\\4\end{bmatrix}$	$\begin{bmatrix}(-5)3\\4\end{bmatrix}$	$\begin{bmatrix}(-4)4\\7\end{bmatrix}$	$\begin{bmatrix}(-4)3\\6\end{bmatrix}$	$\begin{bmatrix}(-3)1\\8\end{bmatrix}$

*See page II.

Table 21.1   **EIGENVALUES—PROLATE AND OBLATE**

PROLATE

$$\lambda_{mn}(c) - m(m+1) \qquad *$$

$$\lambda_{1n}(c) - 2 \qquad *$$

$c^2\backslash n$	1	2	3	4	5
0	0.000000	4.000000	10.000000	18.000000	28.000000
1	0.195548	4.424699	10.467915	18.481696	28.488065
2	0.382655	4.841718	10.937881	18.965685	28.977891
3	0.561975	5.251162	11.409266	19.451871	29.469456
4	0.734111	5.653149	11.881493	19.940143	29.962738
5	0.899615	6.047807	12.354034	20.430382	30.457716
6	1.058995	6.435272	12.826413	20.922458	30.954363
7	1.212711	6.815691	13.298196	21.416235	31.452653
8	1.361183	7.189213	13.768997	21.911569	31.952557
9	1.504795	7.555998	14.238466	22.408312	32.454044
10	1.643895	7.916206	14.706292	22.906311	32.957080
11	1.778798	8.270004	15.172199	23.405410	33.461629
12	1.909792	8.617558	15.635940	23.905451	33.967652
13	2.037141	8.959038	16.097297	24.406277	34.475109
14	2.161081	9.294612	16.556078	24.907729	34.983956
15	2.281832	9.624450	17.012115	25.409649	35.494147
16	2.399593	9.948719	17.465260	25.911881	36.005634
	$\begin{bmatrix}(-3)1\\5\end{bmatrix}$	$\begin{bmatrix}(-3)1\\5\end{bmatrix}$	$\begin{bmatrix}(-4)4\\5\end{bmatrix}$	$\begin{bmatrix}(-4)3\\4\end{bmatrix}$	$\begin{bmatrix}(-4)2\\4\end{bmatrix}$

$$c^{-1}[\lambda_{1n}(c) - 2] \qquad *$$

$c^{-1}\backslash n$	1	2	3	4	5
0.25	0.599898	2.487179	4.366315	6.47797	9.00140
0.24	0.613295	2.491544	4.338520	6.38296	8.80891
0.23	0.627023	2.497852	4.315609	6.29522	8.62445
0.22	0.641073	2.506130	4.297923	6.21556	8.44916
0.21	0.655431	2.516383	4.285792	6.14494	8.28436
0.20	0.670084	2.528591	4.279522	6.08438	8.13163
0.19	0.685014	2.542705	4.279366	6.03498	7.99282
0.18	0.700204	2.558644	4.285495	5.99788	7.87010
0.17	0.715632	2.576296	4.297965	5.97420	7.76598
0.16	0.731281	2.595516	4.316672	5.96496	7.68328
0.15	0.747129	2.616135	4.341320	5.97090	7.62508
0.14	0.763159	2.637968	4.371397	5.99230	7.59446
0.13	0.779353	2.660829	4.406191	6.02874	7.59407
0.12	0.795696	2.684536	4.444844	6.07889	7.62539
0.11	0.812174	2.708934	4.486445	6.14051	7.68773
0.10	0.828776	2.733891	4.530151	6.21063	7.77728
0.09	0.845493	2.759305	4.575277	6.28624	7.88714
0.08	0.862316	2.785099	4.621329	6.36482	8.00897
0.07	0.879237	2.811212	4.667984	6.44473	8.13579
0.06	0.896251	2.837600	4.715031	6.52505	8.26355
0.05	0.913352	2.864224	4.762333	6.60532	8.39048
0.04	0.930535	2.891056	4.809790	6.68528	8.51592
0.03	0.947796	2.918069	4.857332	6.76480	8.63963
0.02	0.965129	2.945243	4.904906	6.84378	8.76153
0.01	0.982531	2.972558	4.952472	6.92219	8.88164
0.00	1.000000	3.000000	5.000000	7.00000	9.00000
	$\begin{bmatrix}(-5)4\\4\end{bmatrix}$	$\begin{bmatrix}(-4)2\\5\end{bmatrix}$	$\begin{bmatrix}(-4)8\\6\end{bmatrix}$	$\begin{bmatrix}(-3)2\\6\end{bmatrix}$	$\begin{bmatrix}(-3)4\\7\end{bmatrix}$

*See page II.

EIGENVALUES—PROLATE AND OBLATE Table 21.1

OBLATE

$$\lambda_{mn}(-ic)-m(m+1) \qquad *$$

$$\lambda_{1n}(-ic)-2 \qquad *$$

$c^2 \backslash n$	1	2	3	4	5
0	0.000000	4.000000	10.000000	18.000000	28.000000
1	−0.204695	3.567527	9.534818	17.520683	27.513713
2	−0.419293	3.127202	9.073104	17.043817	27.029223
3	−0.644596	2.678958	8.615640	16.569461	26.546548
4	−0.881446	2.222747	8.163245	16.097655	26.065706
5	−1.130712	1.758534	7.716768	15.628426	25.586715
6	−1.393280	1.286300	7.277072	15.161786	25.109592
7	−1.670028	0.806045	6.845015	14.697727	24.634357
8	−1.961809	+0.317782	6.421425	14.236229	24.161031
9	−2.269420	−0.178458	6.007074	13.777252	23.689634
10	−2.593577	−0.682630	5.602649	13.320743	23.220190
11	−2.934882	−1.194673	5.208724	12.866634	22.752726
12	−3.293803	−1.714511	4.825732	12.414840	22.287271
13	−3.670646	−2.242055	4.453947	11.965266	21.823856
14	−4.065548	−2.777205	4.093464	11.517803	21.362516
15	−4.478470	−3.319848	3.744202	11.072331	20.903290
16	−4.909200	−3.869861	3.405903	10.628718	20.446222
	$\left[\begin{smallmatrix}(-3)2\\5\end{smallmatrix}\right]$	$\left[\begin{smallmatrix}(-3)1\\5\end{smallmatrix}\right]$	$\left[\begin{smallmatrix}(-3)1\\5\end{smallmatrix}\right]$	$\left[\begin{smallmatrix}(-4)3\\5\end{smallmatrix}\right]$	$\left[\begin{smallmatrix}(-4)3\\4\end{smallmatrix}\right]$

$$c^{-2}[\lambda_{1n}(-ic)-2] \qquad *$$

$c^{-1} \backslash n$	1	2	3	4	5
0.25	−0.306825	−0.241866	0.21286	0.66429	1.2778
0.24	−0.318148	−0.266693	0.17062	0.57759	1.1420
0.23	−0.330984	−0.291340	0.13125	0.49460	1.0120
0.22	−0.345469	−0.315894	0.09476	0.41533	0.8879
0.21	−0.361702	−0.340450	0.06107	0.33974	0.7697
0.20	−0.379735	−0.365113	0.03001	0.26779	0.6575
0.19	−0.399564	−0.389998	+0.00127	0.19942	0.5515
0.18	−0.421125	−0.415222	−0.02563	0.13449	0.4520
0.17	−0.444308	−0.440907	−0.05142	0.07282	0.3591
0.16	−0.468974	−0.467166	−0.07710	+0.01411	0.2735
0.15	−0.494976	−0.494104	−0.10406	−0.04205	0.1958
0.14	−0.522180	−0.521805	−0.13412	−0.09625	0.1271
0.13	−0.550474	−0.550335	−0.16924	−0.14929	0.0680
0.12	−0.579775	−0.579732	−0.21076	−0.20210	+0.0183
0.11	−0.610027	−0.610016	−0.25868	−0.25572	−0.0250
0.10	−0.641193	−0.641191	−0.31185	−0.31111	−0.0685
0.09	−0.673251	−0.673251	−0.36901	−0.36888	−0.1219
0.08	−0.706186	−0.706186	−0.42934	−0.42932	−0.1907
0.07	−0.739985	−0.739985	−0.49242	−0.49242	−0.2714
0.06	−0.774638	−0.774638	−0.55807	−0.55807	−0.3598
0.05	−0.810135	−0.810135	−0.62616	−0.62616	−0.4542
0.04	−0.846468	−0.846468	−0.69657	−0.69657	−0.5540
0.03	−0.883628	−0.883628	−0.76923	−0.76923	−0.6588
0.02	−0.921608	−0.921608	−0.84406	−0.84406	−0.7682
0.01	−0.960401	−0.960401	−0.92100	−0.92100	−0.8820
0.00	−1.000000	−1.000000	−1.00000	−1.00000	−1.0000
	$\left[\begin{smallmatrix}(-4)2\\5\end{smallmatrix}\right]$	$\left[\begin{smallmatrix}(-4)1\\5\end{smallmatrix}\right]$	$\left[\begin{smallmatrix}(-4)8\\6\end{smallmatrix}\right]$	$\left[\begin{smallmatrix}(-4)5\\5\end{smallmatrix}\right]$	$\left[\begin{smallmatrix}(-3)2\\7\end{smallmatrix}\right]$

*See page II.

Table 21.1   **EIGENVALUES—PROLATE AND OBLATE**

PROLATE

$$\lambda_{mn}(c) - m(m+1) \qquad *$$

$$\lambda_{2n}(c) - 6 \qquad *$$

$c^2 \backslash n$	2	3	4	5	6
0	0.000000	6.000000	14.000000	24.000000	36.000000
1	0.140948	6.331101	14.402353	24.436145	36.454889
2	0.278219	6.657791	14.804100	24.872744	36.910449
3	0.412006	6.980147	15.205077	25.309731	37.366657
4	0.542495	7.298250	15.605133	25.747043	37.823486
5	0.669857	7.612179	16.004126	26.184612	38.280913
6	0.794252	7.922016	16.401931	26.622373	38.738910
7	0.915832	8.227840	16.798429	27.060261	39.197451
8	1.034738	8.529734	17.193516	27.498208	39.656510
9	1.151100	8.827778	17.587093	27.936151	40.116059
10	1.265042	9.122052	17.979073	28.374023	40.576070
11	1.376681	9.412636	18.369377	28.811761	41.036514
12	1.486122	9.699610	18.757932	29.249302	41.497364
13	1.593469	9.983052	19.144675	29.686584	41.958589
14	1.698816	10.263039	19.529549	30.123544	42.420160
15	1.802252	10.539650	19.912501	30.560125	42.882048
16	1.903860	10.812958	20.293486	30.996267	43.344222
	$\begin{bmatrix} (-4)5 \\ 4 \end{bmatrix}$	$\begin{bmatrix} (-4)6 \\ 4 \end{bmatrix}$	$\begin{bmatrix} (-4)2 \\ 4 \end{bmatrix}$	$\begin{bmatrix} (-5)6 \\ 4 \end{bmatrix}$	$\begin{bmatrix} (-5)8 \\ 4 \end{bmatrix}$

$$c^{-1}[\lambda_{2n}(c) - 6] \qquad *$$

$c^{-1} \backslash n$	2	3	4	5	6
0.25	0.475965	2.703239	5.073371	7.74906	10.8360
0.24	0.489447	2.683149	4.994116	7.58138	10.5536
0.23	0.503526	2.665356	4.919290	7.41971	10.2781
0.22	0.518220	2.650003	4.849313	7.26479	10.0103
0.21	0.533551	2.637236	4.784640	7.11743	9.7512
0.20	0.549534	2.627196	4.725757	6.97858	9.5023
0.19	0.566185	2.620017	4.673177	6.84931	9.2649
0.18	0.583513	2.615819	4.627427	6.73081	9.0409
0.17	0.601526	2.614701	4.589031	6.62442	8.8323
0.16	0.620224	2.616735	4.558480	6.53155	8.6417
0.15	0.639604	2.621954	4.536196	6.45371	8.4718
0.14	0.659659	2.630349	4.522485	6.39236	8.3260
0.13	0.680376	2.641862	4.517479	6.34878	8.2078
0.12	0.701737	2.656384	4.521086	6.32389	8.1208
0.11	0.723722	2.673764	4.532956	6.31794	8.0678
0.10	0.746308	2.693817	4.552484	6.33030	8.0507
0.09	0.769471	2.716339	4.578871	6.35935	8.0688
0.08	0.793186	2.741120	4.611219	6.40263	8.1184
0.07	0.817429	2.767960	4.648642	6.45738	8.1932
0.06	0.842175	2.796673	4.690346	6.52096	8.2864
0.05	0.867402	2.827089	4.735658	6.59127	8.3919
0.04	0.893087	2.859059	4.784022	6.66670	8.5057
0.03	0.919209	2.892449	4.834980	6.74607	8.6249
0.02	0.945747	2.927138	4.888160	6.82849	8.7477
0.01	0.972684	2.963019	4.943252	6.91330	8.8730
0.00	1.000000	3.000000	5.000000	7.00000	9.0000
	$\begin{bmatrix} (-5)9 \\ 4 \end{bmatrix}$	$\begin{bmatrix} (-4)4 \\ 5 \end{bmatrix}$	$\begin{bmatrix} (-3)1 \\ 6 \end{bmatrix}$	$\begin{bmatrix} (-3)2 \\ 6 \end{bmatrix}$	$\begin{bmatrix} (-3)4 \\ 5 \end{bmatrix}$

*See page II.

EIGENVALUES—PROLATE AND OBLATE Table 21.1

OBLATE

$$\lambda_{mn}(-ic)-m(m+1) \qquad *$$

$$\lambda_{2n}(-ic)-6 \qquad *$$

$c^2\backslash n$	2	3	4	5	6
0	0.000000	6.000000	14.000000	24.000000	36.000000
1	−0.144837	5.664409	13.597220	23.564371	35.545806
2	−0.293786	5.324253	13.194206	23.129322	35.092330
3	−0.447086	4.979458	12.791168	22.694912	34.639597
4	−0.604989	4.629951	12.388328	22.261201	34.187627
5	−0.767764	4.275662	11.985928	21.828245	33.736444
6	−0.935698	3.916525	11.584224	21.396098	33.286069
7	−1.109090	3.552475	11.183489	20.964812	32.836522
8	−1.288259	3.183450	10.784014	20.534436	32.387826
9	−1.473539	2.809393	10.386106	20.105013	31.940000
10	−1.665278	2.430250	9.990084	19.676587	31.493066
11	−1.863838	2.045970	9.596286	19.249195	31.047043
12	−2.069595	1.656508	9.205059	18.822869	30.601952
13	−2.282933	1.261822	8.816762	18.397640	30.157814
14	−2.504245	0.861875	8.431761	17.973532	29.714648
15	−2.733927	0.456635	8.050424	17.550565	29.272476
16	−2.972375	0.046076	7.673121	17.128753	28.831317
	$\begin{bmatrix}(-3)1\\4\end{bmatrix}$	$\begin{bmatrix}(-4)7\\4\end{bmatrix}$	$\begin{bmatrix}(-4)5\\4\end{bmatrix}$	$\begin{bmatrix}(-4)1\\4\end{bmatrix}$	$\begin{bmatrix}(-4)1\\4\end{bmatrix}$

$$c^{-2}[\lambda_{2n}(-ic)-6] \qquad *$$

$c^{-1}\backslash n$	2	3	4	5	6
0.25	−0.185773	+0.002879	0.47957	1.07054	1.8019
0.24	−0.190754	−0.030028	0.41280	0.95365	1.6261
0.23	−0.196680	−0.062228	0.34933	0.84167	1.4577
0.22	−0.203790	−0.093813	0.28933	0.73461	1.2965
0.21	−0.212386	−0.124893	0.23297	0.63251	1.1428
0.20	−0.222841	−0.155607	0.18049	0.53537	0.9964
0.19	−0.235596	−0.186120	0.13215	0.44322	0.8574
0.18	−0.251126	−0.216631	0.08816	0.35607	0.7260
0.17	−0.269873	−0.247375	0.04864	0.27389	0.6022
0.16	−0.292149	−0.278624	+0.01342	0.19662	0.4863
0.15	−0.318047	−0.310677	−0.01813	0.12409	0.3785
0.14	−0.347414	−0.343847	−0.04727	+0.05600	0.2795
0.13	−0.379928	−0.378432	−0.07609	−0.00822	0.1901
0.12	−0.415213	−0.414688	−0.10778	−0.06954	0.1120
0.11	−0.452947	−0.452800	−0.14643	−0.12937	+0.0470
0.10	−0.492902	−0.492871	−0.19508	−0.18959	−0.0051
0.09	−0.534942	−0.534937	−0.25333	−0.25217	−0.0517
0.08	−0.578991	−0.578991	−0.31876	−0.31861	−0.1076
0.07	−0.625006	−0.625006	−0.38955	−0.38955	−0.1844
0.06	−0.672956	−0.672956	−0.46494	−0.46494	−0.2768
0.05	−0.722813	−0.722813	−0.54456	−0.54456	−0.3791
0.04	−0.774556	−0.774556	−0.62821	−0.62821	−0.4895
0.03	−0.828164	−0.828164	−0.71571	−0.71571	−0.6073
0.02	−0.883618	−0.883618	−0.80691	−0.80691	−0.7319
0.01	−0.940902	−0.940902	−0.90171	−0.90171	−0.8629
0.00	−1.000000	−1.000000	−1.00000	−1.00000	−1.0000
	$\begin{bmatrix}(-4)5\\6\end{bmatrix}$	$\begin{bmatrix}(-4)2\\5\end{bmatrix}$	$\begin{bmatrix}(-3)1\\8\end{bmatrix}$	$\begin{bmatrix}(-4)6\\6\end{bmatrix}$	$\begin{bmatrix}(-3)3\\8\end{bmatrix}$

*See page II.

Table 21.2 ANGULAR FUNCTIONS—PROLATE AND OBLATE
PROLATE

$$S_{mn}(c, \cos\theta)$$

m	n	c\θ	0°	10°	20°	30°	40°	50°	60°	70°	80°	90°
0	0	1	0.8481	0.8525	0.8651	0.8847	0.9091	0.9354	0.9606	0.9815	0.9952	1.000
		2	0.5315	0.5431	0.5772	0.6320	0.7032	0.7842	0.8654	0.9355	0.9831	1.000
		3	0.2675	0.2815	0.3242	0.3967	0.4980	0.6226	0.7571	0.8805	0.9682	1.000
		4	0.1194	0.1312	0.1689	0.2379	0.3442	0.4885	0.6589	0.8271	0.9530	1.000
		5	0.0502	0.0585	0.0861	0.1419	0.2380	0.3839	0.5742	0.7776	0.9383	1.000
0	1	1	0.9046	0.8936	0.8602	0.8035	0.7225	0.6169	0.4878	0.3381	0.1731	0
		2	0.6681	0.6665	0.6598	0.6429	0.6081	0.5472	0.4540	0.3270	0.1717	0
		3	0.4034	0.4099	0.4273	0.4489	0.4630	0.4543	0.4068	0.3110	0.1695	0
		4	0.2042	0.2138	0.2415	0.2833	0.3294	0.3618	0.3566	0.2929	0.1669	0
		5	0.0916	0.1001	0.1262	0.1703	0.2279	0.2840	0.3104	0.2752	0.1643	0
0	2	1	1.022	0.9795	0.8553	0.6621	0.4198	0.1556	−0.0988	−0.3105	−0.4509	−0.5000
		2	1.064	1.030	0.9271	0.7579	0.5296	0.2602	−0.0192	−0.2668	−0.4385	−0.5000
		3	1.041	1.023	0.9640	0.8497	0.6660	0.4104	+0.1061	−0.1938	−0.4171	−0.5000
		4	0.8730	0.8768	0.8787	0.8513	0.7549	0.5553	0.2512	−0.0998	−0.3879	−0.5000
		5	0.6018	0.6233	0.6792	0.7407	0.7537	0.6494	0.3844	+0.0008	−0.3542	−0.5000
0	3	1	0.9892	0.9042	0.6692	0.3400	−0.0045	−0.2816	−0.4259	−0.4085	−0.2467	0
		2	0.9590	0.8864	0.6816	0.3840	+0.0560	−0.2261	−0.3907	−0.3949	−0.2447	0
		3	0.9090	0.8546	0.6957	0.4485	0.1501	−0.1364	−0.3319	−0.3714	−0.2412	0
		4	0.8197	0.7877	0.6868	0.5087	0.2591	−0.0215	−0.2514	−0.3376	−0.2361	0
		5	0.6650	0.6560	0.6183	0.5245	0.3482	+0.0971	−0.1575	−0.2952	−0.2293	0
1	1	1	0	0.1578	0.3134	0.4643	0.6067	0.7355	0.8450	0.9290	0.9819	1.000
		2	0	0.1194	0.2437	0.3757	0.5149	0.6562	0.7892	0.9000	0.9740	1.000
		3	0	0.0776	0.1654	0.2724	0.4030	0.5546	0.7144	0.8597	0.9627	1.000
		4	0	0.0449	0.1018	0.1832	0.2994	0.4537	0.6353	0.8150	0.9497	1.000
		5	0	0.0239	0.0588	0.1179	0.2162	0.3650	0.5602	0.7698	0.9361	1.000
1	2	1	0	0.4788	0.9054	1.232	1.417	1.435	1.276	0.9562	0.5119	0
		2	0	0.3896	0.7509	1.052	1.253	1.316	1.212	0.9335	0.5088	0
		3	0	0.2780	0.5538	0.8148	1.030	1.149	1.118	0.8992	0.5039	0
		4	0	0.1762	0.3683	0.5813	0.7968	0.9643	1.008	0.8575	0.4979	0
		5	0	0.1011	0.2254	0.3896	0.5906	0.7879	0.8957	0.8127	0.4911	0
1	3	1	0	0.9928	1.745	2.075	1.903	1.280	0.3775	−0.5521	−1.244	−1.500
		2	0	0.9559	1.710	2.092	1.998	1.432	0.5298	−0.4541	−1.214	−1.500
		3	0	0.8745	1.611	2.063	2.097	1.640	0.7606	−0.2972	−1.174	−1.500
		4	0	0.7393	1.418	1.934	2.128	1.841	1.032	−0.0951	−1.097	−1.500
		5	0	0.5662	1.146	1.691	2.047	1.975	1.299	+0.1319	−1.017	−1.500
2	2	1	0	0.0844	0.3295	0.7111	1.189	1.710	2.211	2.627	2.903	3.000
		2	0	0.0690	0.2744	0.6092	1.054	1.572	2.101	2.566	2.886	3.000
		3	0	0.0500	0.2051	0.4773	0.8738	1.380	1.944	2.475	2.859	3.000
		4	0	0.0328	0.1405	0.3487	0.6876	1.171	1.764	2.367	2.827	3.000
		5	0	0.0198	0.0898	0.2414	0.5212	0.9701	1.580	2.251	2.791	3.000
2	3	1	0	0.4222	1.570	3.116	4.596	5.530	5.548	4.501	2.522	0
		2	0	0.3597	1.358	2.755	4.175	5.170	5.327	4.417	2.510	0
		3	0	0.2765	1.070	2.255	3.576	4.641	4.994	4.286	2.491	0
		4	0	0.1934	0.7758	1.723	2.909	4.025	4.588	4.122	2.466	0
		5	0	0.1244	0.5226	1.243	2.269	3.395	4.150	3.936	2.437	0

From C. Flammer, Spheroidal wave functions. Stanford Univ. Press, Stanford, Calif., 1957 (with permission).

ANGULAR FUNCTIONS—PROLATE AND OBLATE

Table 21.2

OBLATE

$$S_{mn}(-ic, \eta)$$

m	n	$c\backslash\eta$	0	0.1	0.2	0.3	0.4	0.5	0.6	0.7	0.8	0.9	1.0
0	0	1	1.000	1.002	1.007	1.016	1.028	1.044	1.064	1.088	1.115	1.147	1.183
		2	1.000	1.008	1.032	1.073	1.132	1.210	1.310	1.434	1.585	1.767	1.986
		3	1.000	1.022	1.089	1.205	1.377	1.617	1.940	2.366	2.923	3.648	4.589
		4	1.000	1.047	1.191	1.449	1.854	2.452	3.319	4.557	6.323	8.837	12.42
		5	1.000	1.083	1.341	1.835	2.648	3.952	6.000	9.211	14.23	22.11	34.48
0	1	1	0	0.1001	0.2009	0.3027	0.4065	0.5128	0.6222	0.7353	0.8530	0.9760	1.105
		2	0	0.1004	0.2034	0.3114	0.4274	0.5542	0.6952	0.8539	1.035	1.243	1.484
		3	0	0.1011	0.2079	0.3273	0.4664	0.6338	0.8398	1.098	1.425	1.842	2.378
		4	0	0.1016	0.2150	0.3526	0.5298	0.7681	1.096	1.552	2.195	3.105	4.396
		5	0	0.1032	0.2252	0.3884	0.6252	0.9804	1.525	2.369	3.684	5.741	8.970
0	2	1	−0.5000	−0.4863	−0.4450	−0.3757	−0.2779	−0.1507	+0.0070	0.1965	0.4197	0.6784	0.9749
		2	−0.5000	−0.4897	−0.4585	−0.4052	−0.3277	−0.2231	−0.0872	+0.0849	0.2999	0.5660	0.8930
		3	−0.5000	−0.4943	−0.4766	−0.4448	−0.3952	−0.3223	−0.2183	−0.0721	+0.1311	0.3845	0.7958
		4	−0.5000	−0.4994	−0.4966	−0.4891	−0.4716	−0.4356	−0.3681	−0.2485	−0.0458	0.2868	0.8201
		5	−0.5000	−0.5061	−0.5234	−0.5495	−0.5780	−0.5977	−0.5869	−0.5067	−0.2880	0.1892	1.132
0	3	1	0	−0.1477	−0.2810	−0.3855	−0.4466	−0.4491	−0.3768	−0.2130	+0.0600	0.4613	1.011
		2	0	−0.1480	−0.2839	−0.3947	−0.4668	−0.4839	−0.4275	−0.2757	−0.0015	0.4274	1.051
		3	0	−0.1486	−0.2885	−0.4097	−0.4998	−0.5421	−0.5140	−0.3841	−0.1091	0.3711	1.138
		4	0	−0.1495	−0.2949	−0.4306	−0.5415	−0.6270	−0.6432	−0.5540	−0.2765	0.2912	1.327
		5	0	−0.1504	−0.3033	−0.4589	−0.6123	−0.7489	−0.8356	−0.8080	−0.5447	0.1715	1.723
1	1	1	1.000	0.9961	0.9838	0.9628	0.9316	0.8884	0.8299	0.7506	0.6402	0.4731	0
		2	1.000	0.9994	0.9973	0.9923	0.9827	0.9652	0.9340	0.8802	0.7864	0.6118	0
		3	1.000	1.006	1.025	1.055	1.093	1.135	1.172	1.188	1.149	0.9724	0
		4	1.000	1.020	1.079	1.178	1.319	1.498	1.708	1.920	2.067	1.950	0
		5	1.000	1.041	1.174	1.406	1.776	2.242	2.878	3.642	4.400	4.651	0
1	2	1	0	0.2987	0.5897	0.8643	1.113	1.322	1.478	1.554	1.508	1.247	0
		2	0	0.2985	0.5950	0.8815	1.153	1.398	1.600	1.730	1.734	1.487	0
		3	0	0.3005	0.6043	0.9140	1.228	1.541	1.837	2.082	2.200	2.000	0
		4	0	0.3022	0.6213	0.9640	1.349	1.780	2.250	2.723	3.092	3.033	0
		5	0	0.2990	0.6400	1.040	1.537	2.165	2.947	3.868	4.786	5.138	0
1	3	1	−1.500	−1.421	−1.189	−0.8136	−0.3165	0.2710	0.9015	1.501	1.946	1.988	0
		2	−1.500	−1.431	−1.228	−0.8941	−0.4427	+0.1060	0.7174	1.329	1.826	1.951	0
		3	−1.500	−1.447	−1.289	−1.024	−0.6502	−0.1738	+0.3916	1.006	1.572	1.834	0
		4	−1.500	−1.467	−1.364	−1.184	−0.9148	−0.5415	−0.0538	0.5403	1.177	1.619	0
		5	−1.500	−1.486	−1.442	−1.353	−1.198	−0.9435	−0.5506	0.0161	0.7471	1.439	0
2	2	1	3.000	2.972	2.889	2.748	2.549	2.291	1.970	1.585	1.131	0.6041	0
		2	3.000	2.979	2.915	2.805	2.644	2.425	2.138	1.770	1.305	0.7234	0
		3	3.000	2.992	2.965	2.915	2.830	2.693	2.481	2.161	1.687	0.9944	0
		4	3.000	3.013	3.052	3.111	3.170	3.200	3.157	2.966	2.512	1.615	0
		5	3.000	3.052	3.211	3.469	3.813	4.202	4.564	4.746	4.460	3.188	0
2	3	1	0	1.486	2.886	4.115	5.086	5.704	5.877	5.503	4.477	2.683	0
		2	0	1.488	2.906	4.180	5.226	5.954	6.251	5.982	4.990	3.077	0
		3	0	1.494	2.943	4.295	5.482	6.413	6.951	6.904	6.008	3.879	0
		4	0	1.498	2.996	4.475	5.891	7.166	8.132	8.515	7.857	5.408	0
		5	0	1.509	3.073	4.738	6.515	8.347	10.07	11.28	11.21	8.354	0

Table 21.3   PROLATE RADIAL FUNCTIONS—FIRST AND SECOND KINDS

$$R_{mn}^{(1)}(c, \xi) \qquad\qquad\qquad R_{mn}^{(2)}(c, \xi)$$

m	n	$c\backslash\xi$	1.005	1.020	1.044	1.077	1.005	1.020	1.044	1.077
0	0	1	(−1) 9.468	(−1) 9.419	(−1) 9.339	(−1) 9.228	(0) −2.838	(0) −2.096	(0) −1.666	(0) −1.356
		2	(−1) 8.257	(−1) 8.077	(−1) 7.789	(−1) 7.392	(0) −1.244	(−1) −8.020	(−1) −5.341	(−1) −3.333
		3	(−1) 7.026	(−1) 6.662	(−1) 6.091	(−1) 5.330	(−1) −7.104	(−1) −3.422	(−1) −1.281	(−2) 3.51
		4	(−1) 6.054	(−1) 5.471	(−1) 4.585	(−1) 3.463	(−1) −4.508	(−1) −1.287	(−2) 6.61	(−1) 1.952
		5	(−1) 5.313	(−1) 4.488	(−1) 3.287	(−1) 1.869	(−1) −3.052	(−2) −1.02	(−1) 1.537	(−1) 2.291
0	1	1	(−1) 3.153	(−1) 3.190	(−1) 3.249	(−1) 3.328	(0) −6.912	(0) −4.801	(0) −3.669	(0) −2.920
		2	(−1) 5.289	(−1) 5.298	(−1) 5.308	(−1) 5.311	(0) −2.189	(0) −1.540	(0) −1.177	(−1) −9.216
		3	(−1) 6.064	(−1) 5.960	(−1) 5.786	(−1) 5.529	(0) −1.133	(−1) −7.365	(−1) −4.987	(−1) −3.207
		4	(−1) 5.892	(−1) 5.612	(−1) 5.162	(−1) 4.542	(−1) −6.741	(−1) −3.528	(−1) −1.534	(−3) −4.9
		5	(−1) 5.381	(−1) 4.888	(−1) 4.125	(−1) 3.137	(−1) −4.293	(−1) −1.390	(−2) 3.87	(−1) 1.594
0	2	1	(−2) 4.470	(−2) 4.655	(−2) 4.954	(−2) 5.373	(1) −3.593	(1) −2.185	(1) −1.484	(1) −1.056
		2	(−1) 1.696	(−1) 1.749	(−1) 1.833	(−1) 1.947	(0) −5.241	(0) −3.358	(0) −2.403	(0) −1.807
		3	(−1) 3.295	(−1) 3.346	(−1) 3.421	(−1) 3.509	(0) −2.031	(0) −1.364	(0) −1.007	(−1) −7.694
		4	(−1) 4.507	(−1) 4.477	(−1) 4.413	(−1) 4.293	(0) −1.095	(−1) −7.053	(−1) −4.783	(−1) −3.115
		5	(−1) 4.952	(−1) 4.763	(−1) 4.444	(−1) 3.976	(−1) −7.388	(−1) −4.417	(−1) −2.630	(−1) −1.340
0	3	1	(−3) 3.912	(−3) 4.249	(−3) 4.814	(−3) 5.638	(−2) −3.288	(2) −1.659	(2) −1.082	(1) −6.916
		2	(−2) 3.085	(−2) 3.317	(−2) 3.700	(−2) 4.249	(−1) −2.194	(1) −1.223	(0) −7.705	(0) −5.123
		3	(−2) 9.956	(−1) 1.054	(−1) 1.147	(−1) 1.275	(0) −5.020	(0) −2.966	(0) −1.985	(0) −1.408
		4	(−1) 2.107	(−1) 2.183	(−1) 2.298	(−1) 2.443	(0) −2.043	(0) −1.293	(−1) −9.141	(−1) −6.749
		5	(−1) 3.298	(−1) 3.329	(−1) 3.360	(−1) 3.362	(0) −1.149	(−1) −7.422	(−1) −5.182	(−1) −3.612
1	1	1	(−2) 3.270	(−2) 6.544	(−2) 9.716	(−1) 1.287	(1) −1.506	(0) −7.294	(0) −4.734	(0) −3.432
		2	(−2) 6.187	(−1) 1.227	(−1) 1.793	(−1) 2.323	(0) −4.079	(0) −2.077	(0) −1.417	(0) −1.071
		3	(−2) 8.596	(−1) 1.677	(−1) 2.386	(−1) 2.973	(0) −2.019	(0) −1.075	(−1) −7.453	(−1) −5.480
		4	(−1) 1.053	(−1) 2.007	(−1) 2.744	(−1) 3.221	(0) −1.273	(−1) −6.911	(−1) −4.585	(−1) −2.924
		5	(−1) 1.211	(−1) 2.235	(−1) 2.894	(−1) 3.118	(−1) −9.101	(−1) −4.885	(−1) −2.874	(−1) −1.248
1	2	1	(−3) 6.503	(−2) 1.322	(−2) 2.012	(−2) 2.754	(1) −7.295	(1) −3.269	(1) −1.939	(1) −1.275
		2	(−2) 2.378	(−2) 4.802	(−2) 7.227	(−2) 9.738	(1) −1.014	(0) −4.717	(0) −2.932	(0) −2.038
		3	(−2) 4.658	(−2) 9.296	(−1) 1.372	(−1) 1.798	(0) −3.552	(0) −1.751	(0) −1.156	(−1) −8.473
		4	(−2) 6.975	(−1) 1.367	(−1) 1.960	(−1) 2.460	(0) −1.842	(−1) −9.597	(−1) −6.533	(−1) −4.718
		5	(−2) 9.035	(−1) 1.739	(−1) 2.376	(−1) 2.803	(0) −1.778	(−1) −6.362	(−1) −4.170	(−1) −2.651
1	3	1	(−4) 7.586	(−3) 1.577	(−3) 2.483	(−3) 3.556	(2) −6.014	(2) −2.491	(2) −1.354	(1) −8.127
		2	(−3) 5.725	(−2) 1.183	(−2) 1.845	(−2) 2.607	(1) −4.027	(1) −1.707	(0) −9.553	(0) −5.934
		3	(−2) 1.737	(−2) 3.553	(−2) 5.453	(−2) 7.529	(0) −9.025	(0) −3.994	(0) −2.354	(0) −1.552
		4	(−2) 3.516	(−2) 7.089	(−1) 1.063	(−1) 1.418	(0) −3.449	(0) −1.629	(0) −1.032	(−1) −7.288
		5	(−2) 5.604	(−1) 1.108	(−1) 1.608	(−1) 2.048	(0) −1.692	(−1) −8.600	(−1) −5.214	(−1) −3.006
2	2	1	(−4) 6.612	(−3) 2.659	(−3) 5.898	(−2) 1.044	(2) −3.750	(1) −9.112	(1) −3.973	(1) −2.156
		2	(−3) 2.566	(−2) 1.025	(−2) 2.249	(−2) 3.920	(1) −4.852	(1) −1.203	(0) −5.417	(0) −3.077
		3	(−3) 5.520	(−2) 2.181	(−2) 4.698	(−2) 7.974	(1) −1.515	(0) −3.889	(0) −1.852	(0) −1.126
		4	(−3) 9.302	(−2) 3.616	(−2) 7.587	(−1) 1.239	(0) −6.821	(0) −1.843	(−1) −9.431	(−1) −6.132
		5	(−2) 1.372	(−2) 5.223	(−1) 1.058	(−1) 1.639	(0) −3.755	(0) −1.081	(−1) −5.907	(−1) −3.910
2	3	1	(−5) 9.415	(−4) 3.845	(−4) 8.736	(−3) 1.596	(3) −2.609	(2) −6.096	(2) −2.517	(2) −1.279
		2	(−4) 7.128	(−3) 2.896	(−3) 6.525	(−2) 1.178	(2) −1.728	(1) −4.095	(1) −1.727	(0) −9.031
		3	(−3) 2.208	(−3) 8.889	(−2) 1.974	(−2) 3.492	(1) −3.745	(0) −9.098	(0) −3.994	(0) −2.208
		4	(−3) 4.683	(−2) 1.862	(−2) 4.048	(−2) 6.946	(1) −1.334	(0) −3.370	(0) −1.573	(−1) −9.397
		5	(−3) 8.060	(−2) 3.150	(−2) 6.657	(−1) 1.096	(0) −6.274	(0) −1.671	(−1) −8.409	(−1) −5.379

From C. Flammer, Spheroidal wave functions. Stanford Univ. Press, Stanford, Calif., 1957 (with permission).

OBLATE RADIAL FUNCTIONS—FIRST AND SECOND KINDS   Table 21.4

			$R_{mn}^{(1)}(-ic, i\xi)$		$R_{mn}^{(2)}(-ic, i\xi)$	
m	n	$c\backslash\xi$	0	0.75	0	0.75
0	0	0.2	(−1)9.9557	(−1)9.9183	(0)−7.7864	(0)−4.5290
		0.5	(−1)9.7265	(−1)9.4976	(0)−2.9707	(0)−1.5906
		0.8	(−1)9.3168	(−1)8.7520	(0)−1.7002	(−1)−7.5527
		1.0	(−1)8.9565	(−1)8.1032	(0)−1.2524	(−1)−4.4277
		1.5	(−1)7.8320	(−1)6.1209	(−1)−6.2189	(−2)+1.2204
		2.0	(−1)6.5571	(−1)3.9526	(−1)−3.0356	(−1) 2.2634
		2.5	(−1)5.3430	(−1)1.9680	(−1)−1.3758	(−1) 3.0225
0	1	0.2	0	(−2)4.9808	(1)−7.5120	(1)−2.3239
		0.5	0	(−1)1.2202	(1)−1.2120	(0)−4.0338
		0.8	0	(−1)1.8802	(0)−4.8077	(0)−1.7744
		1.0	0	(−1)2.2696	(0)−3.1202	(0)−1.2314
		1.5	0	(−1)3.0132	(0)−1.4537	(−1)−6.3156
		2.0	0	(−1)3.3765	(−1)−8.7035	(−1)−3.4641
		2.5	0	(−1)3.3530	(−1)−6.0006	(−1)−1.5694
0	2	0.2	(−4)8.8992	(−3)2.3840	(3)−2.2106	(2)−3.4260
		0.5	(−3)5.5964	(−2)1.4744	(2)−1.4205	(1)−2.2700
		0.8	(−2)1.4489	(−2)3.6993	(1)−3.5130	(0)−5.9376
		1.0	(−2)2.2868	(−2)5.6728	(1)−1.8068	(0)−3.2496
		1.5	(−2)5.3150	(−1)1.1932	(0)−5.5629	(0)−1.2084
		2.0	(−2)9.7914	(−1)1.9147	(0)−2.5149	(−1)−6.5653
		2.5	(−1)1.5649	(−1)2.5730	(0)−1.4263	(−1)−3.9702
1	1	0.2	(−2)6.6454	(−2)8.2880	(1)−5.9560	(1)−2.1507
		0.5	(−1)1.6336	(−1)2.0133	(1)−1.0060	(0)−3.8583
		0.8	(−1)2.5333	(−1)3.0524	(0)−4.2765	(0)−1.7483
		1.0	(−1)3.0762	(−1)3.6283	(0)−2.9165	(0)−1.2196
		1.5	(−1)4.1708	(−1)4.5492	(0)−1.4980	(−1)−5.8081
		2.0	(−1)4.8229	(−1)4.6553	(−1)−9.1106	(−1)−2.3210
		2.5	(−1)5.0170	(−1)4.0221	(−1)−5.7028	(−3)+3.168
1	2	0.2	0	(−3)2.4923	(3)−1.8781	(2)−3.2287
		0.5	0	(−2)1.5314	(2)−1.2123	(1)−2.1474
		0.8	0	(−2)3.7974	(1)−3.0070	(0)−5.6543
		1.0	0	(−2)5.7617	(1)−1.5622	(0)−3.1109
		1.5	0	(−1)1.1699	(0)−4.8667	(0)−1.1709
		2.0	0	(−1)1.7976	(0)−2.1999	(−1)−6.4134
		2.5	0	(−1)2.3200	(0)−1.2282	(−1)−3.9677
1	3	0.2	(−5)1.5236	(−5)7.2462	(4)−9.6745	(3)−8.1316
		0.5	(−4)2.3850	(−3)1.1206	(3)−2.4841	(2)−2.1259
		0.8	(−4)9.7909	(−3)4.4965	(2)−3.8151	(1)−3.3786
		1.0	(−3)1.9166	(−3)8.6200	(2)−1.5721	(1)−1.4390
		1.5	(−3)6.5244	(−2)2.7259	(1)−3.1742	(0)−3.2838
		2.0	(−2)1.5669	(−2)5.8920	(1)−1.0386	(0)−1.2924
		2.5	(−2)3.1147	(−1)1.0193	(0)−4.4705	(−1)−6.9734
2	2	0.2	(−3)2.6602	(−3)4.1496	(3)−1.1093	(2)−2.6888
		0.5	(−2)1.6413	(−2)2.5393	(1)−7.2682	(1)−1.8121
		0.8	(−2)4.1024	(−2)6.2453	(1)−1.8724	(0)−4.9121
		1.0	(−2)6.2694	(−2)9.4031	(0)−9.9297	(0)−2.7508
		1.5	(−1)1.3055	(−1)1.8562	(0)−3.4267	(0)−1.0939
		2.0	(−1)2.0801	(−1)2.7317	(0)−1.7581	(−1)−6.0206
		2.5	(−1)2.8190	(−1)3.3111	(0)−1.0954	(−1)−3.3594

PROLATE JOINING FACTORS—FIRST KIND $\kappa_{mn}^{(1)}(c)$ Table 21.5

c	$\kappa_{00}^{(1)}$	$\kappa_{01}^{(1)}$	$\kappa_{02}^{(1)}$	$\kappa_{11}^{(1)}$	$\kappa_{12}^{(1)}$	$\kappa_{13}^{(1)}$	$\kappa_{22}^{(1)}$
1	(−1)8.943	(−1)9.422	(1)4.637	(0)2.770	(1)4.319	(2)7.919	(1)4.234
2	(−1)6.391	(0)1.586	(1)1.268	(0)1.095	(0)9.527	(2)1.002	(0)8.838
3	(−1)3.742	(0)1.829	(0)6.352	(−1)5.011	(0)3.417	(1)2.982	(0)2.935
4	(−1)1.909	(0)1.795	(0)3.867	(−1)2.294	(0)1.413	(1)1.222	(0)1.118
5	(−2)8.97	(0)1.665	(0)2.401	(−1)1.023	(−1)6.067	(0)5.725	(−1)4.455

From C. Flammer, Spheroidal wave functions. Stanford Univ. Press, Stanford, Calif., 1957 (with permission).

22. Orthogonal Polynomials

Urs W. Hochstrasser [1]

Contents

---

[1] Guest Worker, National Bureau of Standards, from The American University. (Presently, Atomic Energy Commission, Switzerland.)

22. Orthogonal Polynomials

Mathematical Properties

22.1. Definition of Orthogonal Polynomials

A system of polynomials $f_n(x)$, degree $[f_n(x)]=n$, is called orthogonal on the interval $a \leq x \leq b$, with respect to the weight function $w(x)$, if

22.1.1

$$\int_a^b w(x) f_n(x) f_m(x) dx = 0$$
$$(n \neq m; n, m=0, 1, 2, \ldots)$$

The weight function $w(x)[w(x) \geq 0]$ determines the system $f_n(x)$ up to a constant factor in each polynomial. The specification of these factors is referred to as standardization. For suitably standardized orthogonal polynomials we set

22.1.2

$$\int_a^b w(x) f_n^2(x) dx = h_n, f_n(x) = k_n x^n + k_n' x^{n-1} + \ldots$$
$$(n=0, 1, 2, \ldots)$$

These polynomials satisfy a number of relationships of the same general form. The most important ones are:

Differential Equation

22.1.3 $$g_2(x) f_n'' + g_1(x) f_n' + a_n f_n = 0$$

where $g_2(x)$, $g_1(x)$ are independent of n and a_n a constant depending only on n.

Recurrence Relation

22.1.4 $$f_{n+1} = (a_n + x b_n) f_n - c_n f_{n-1}$$

where

22.1.5

$$b_n = \frac{k_{n+1}}{k_n}, \quad a_n = b_n \left(\frac{k_{n+1}'}{k_{n+1}} - \frac{k_n'}{k_n}\right), \quad c_n = \frac{k_{n+1} k_{n-1} h_n}{k_n^2 h_{n-1}}$$

Rodrigues' Formula

22.1.6 $$f_n = \frac{1}{e_n w(x)} \frac{d^n}{dx_n} \{w(x)[g(x)]^n\}$$

where $g(x)$ is a polynomial in x independent of n. The system $\left\{\dfrac{df_n}{dx}\right\}$ consists again of orthogonal polynomials.

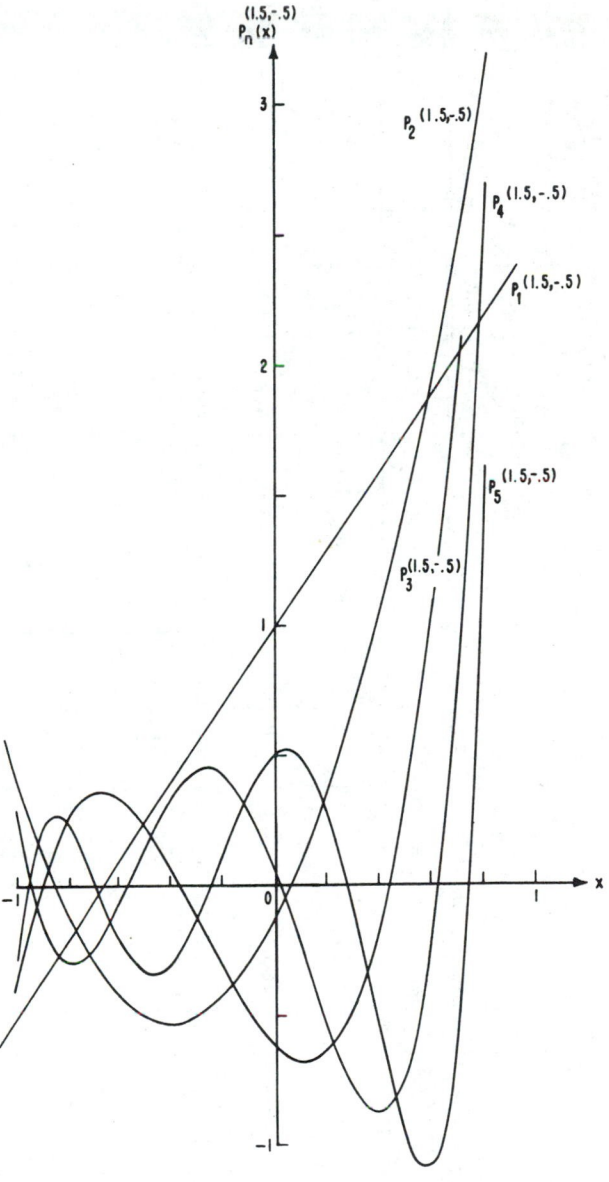

FIGURE 22.1. *Jacobi Polynomials* $P_n^{(\alpha, \beta)}(x)$, $\alpha = 1.5$, $\beta = -.5$, $n = 1(1)5$.

22.2. Orthogonality Relations

	$f_n(x)$	Name of Polynomial	a	b	$w(x)$	Standardization	h_n	Remarks
22.2.1	$P_n^{(\alpha,\beta)}(x)$	Jacobi	-1	1	$(1-x)^\alpha(1+x)^\beta$	$P_n^{(\alpha,\beta)}(1)=\binom{n+\alpha}{n}$	$\dfrac{2^{\alpha+\beta+1}}{2n+\alpha+\beta+1}\dfrac{\Gamma(n+\alpha+1)\Gamma(n+\beta+1)}{n!\Gamma(n+\alpha+\beta+1)}$	$\alpha>-1,\,\beta>-1$
22.2.2	$G_n(p,q,x)$	Jacobi	0	1	$(1-x)^{p-q}x^{q-1}$	$k_n=1$	$\dfrac{n!\,\Gamma(n+q)\,\Gamma(n+p)\,\Gamma(n+p-q+1)}{(2n+p)\,\Gamma^2(2n+p)}$	$p-q>-1,\,q>0$
22.2.3	$C_n^{(\alpha)}(x)$	Ultraspherical (Gegenbauer)	-1	1	$(1-x^2)^{\alpha-\frac{1}{2}}$	$C_n^{(\alpha)}(1)$ $=\binom{n+2\alpha-1}{n}$ $(\alpha\neq0)$	$\dfrac{\pi 2^{1-2\alpha}\Gamma(n+2\alpha)}{n!(n+\alpha)[\Gamma(\alpha)]^2}\quad \alpha\neq0$	$\alpha>-\frac{1}{2}$
						$C_n^{(0)}(1)=\dfrac{2}{n},$ $C_0^{(0)}(1)=1$	$\dfrac{2\pi}{n^2}\quad \alpha=0$	
22.2.4	$T_n(x)$	Chebyshev of the first kind	-1	1	$(1-x^2)^{-\frac{1}{2}}$	$T_n(1)=1$	$\begin{cases}\dfrac{\pi}{2} & n\neq0 \\ \pi & n=0\end{cases}$	
22.2.5	$U_n(x)$	Chebyshev of the second kind	-1	1	$(1-x^2)^{\frac{1}{2}}$	$U_n(1)=n+1$	$\dfrac{\pi}{2}$	
22.2.6	$C_n(x)$	Chebyshev of the first kind	-2	2	$\left(1-\dfrac{x^2}{4}\right)^{-\frac{1}{2}}$	$C_n(2)=2$	$\begin{cases}4\pi & n\neq0 \\ 8\pi & n=0\end{cases}$	
22.2.7	$S_n(x)$	Chebyshev of the second kind	-2	2	$\left(1-\dfrac{x^2}{4}\right)^{\frac{1}{2}}$	$S_n(2)=n+1$	π	
22.2.8	$T_n^*(x)$	Shifted Chebyshev of the first kind	0	1	$(x-x^2)^{-\frac{1}{2}}$	$T_n^*(1)=1$	$\begin{cases}\dfrac{\pi}{2} & n\neq0 \\ \pi & n=0\end{cases}$	
22.2.9	$U_n^*(x)$	Shifted Chebyshev of the second kind	0	1	$(x-x^2)^{\frac{1}{2}}$	$U_n^*(1)=n+1$	$\dfrac{\pi}{8}\quad *$	
22.2.10	$P_n(x)$	Legendre (Spherical)	-1	1	1	$P_n(1)=1$	$\dfrac{2}{2n+1}$	
22.2.11	$P_n^*(x)$	Shifted Legendre	0	1	1	$P_n^*(1)=1$	$\dfrac{1}{2n+1}$	

*See page II.

22.2. Orthogonality Relations—Continued

22.2.12	$L_n^{(\alpha)}(x)$	Generalized Laguerre	0	∞	$e^{-x}x^\alpha$	$k_n = \dfrac{(-1)^n}{n!}$	$\dfrac{\Gamma(\alpha+n+1)}{n!}$	$\alpha > -1$
22.2.13	$L_n(x)$	Laguerre	0	∞	e^{-x}	$k_n = \dfrac{(-1)^n}{n!}$	1	
22.2.14	$H_n(x)$	Hermite	$-\infty$	∞	e^{-x^2}	$e_n = (-1)^n$	$\sqrt{\pi}\,2^n n!$	
22.2.15	$He_n(x)$	Hermite	$-\infty$	∞	$e^{-\frac{x^2}{2}}$	$e_n = (-1)^n$	$\sqrt{2\pi}\,n!$	

*See page II.

22.3. Explicit Expressions

$$f_n(x) = d_n \sum_{m=0}^{N} c_m g_m(x)$$

	$f_n(x)$	N	d_n	c_m	$g_m(x)$	k_n	Remarks
22.3.1	$P_n^{(\alpha,\beta)}(x)$	n	$\dfrac{1}{2^n}$	$\dbinom{n+\alpha}{m}\dbinom{n+\beta}{n-m}$	$(x-1)^{n-m}(x+1)^m$	$\dfrac{1}{2^n}\dbinom{2n+\alpha+\beta}{n}$	$\alpha>-1,\ \beta>-1$
22.3.2	$P_n^{(\alpha,\beta)}(x)$	n	$\dfrac{\Gamma(\alpha+n+1)}{n!\Gamma(\alpha+\beta+n+1)}$	$\dbinom{n}{m}\dfrac{\Gamma(\alpha+\beta+n+m+1)}{2^m\Gamma(\alpha+m+1)}$	$(x-1)^m$	$\dfrac{1}{2^n}\dbinom{2n+\alpha+\beta}{n}$	$\alpha>-1,\ \beta>-1$
22.3.3	$G_n(p,q,x)$	n	$\dfrac{\Gamma(q+n)}{\Gamma(p+2n)}$	$(-1)^m\dbinom{n}{m}\dfrac{\Gamma(p+2n-m)}{\Gamma(q+n-m)}$	x^{n-m}	1	$p-q>-1,\ q>0$
22.3.4	$C_n^{(\alpha)}(x)$	$\left[\dfrac{n}{2}\right]$	$\dfrac{1}{\Gamma(\alpha)}$	$(-1)^m\dfrac{\Gamma(\alpha+n-m)}{m!(n-2m)!}$	$(2x)^{n-2m}$	$\dfrac{2^n}{n!}\dfrac{\Gamma(\alpha+n)}{\Gamma(\alpha)}$	$\alpha>-\frac{1}{2},\ \alpha\neq 0$
22.3.5	$C_n^{(0)}(x)$	$\left[\dfrac{n}{2}\right]$	1	$(-1)^m\dfrac{(n-m-1)!}{m!(n-2m)!}$	$(2x)^{n-2m}$	$\dfrac{2^n}{n}\quad n\neq 0$	$n\neq 0,\ C_0^{(0)}(1)=1$
22.3.6	$T_n(x)$	$\left[\dfrac{n}{2}\right]$	$\dfrac{n}{2}$	$(-1)^m\dfrac{(n-m-1)!}{m!(n-2m)!}$	$(2x)^{n-2m}$	2^{n-1}	
22.3.7	$U_n(x)$	$\left[\dfrac{n}{2}\right]$	1	$(-1)^m\dfrac{(n-m)!}{m!(n-2m)!}$	$(2x)^{n-2m}$	2^n	
22.3.8	$P_n(x)$	$\left[\dfrac{n}{2}\right]$	$\dfrac{1}{2^n}$	$(-1)^m\dbinom{n}{m}\dbinom{2n-2m}{n}$	x^{n-2m}	$\dfrac{(2n)!}{2^n(n!)^2}$	
22.3.9	$L_n^{(\alpha)}(x)$	n	1	$(-1)^m\dbinom{n+\alpha}{n-m}\dfrac{1}{m!}$	x^m	$\dfrac{(-1)^n}{n!}$	$\alpha>-1$
22.3.10	$H_n(x)$	$\left[\dfrac{n}{2}\right]$	$n!$	$(-1)^m\dfrac{1}{m!(n-2m)!}$	$(2x)^{n-2m}$	2^n	see 22.11
22.3.11	$He_n(x)$	$\left[\dfrac{n}{2}\right]$	$n!$	$(-1)^m\dfrac{1}{m!2^m(n-2m)!}$	x^{n-2m}	1	

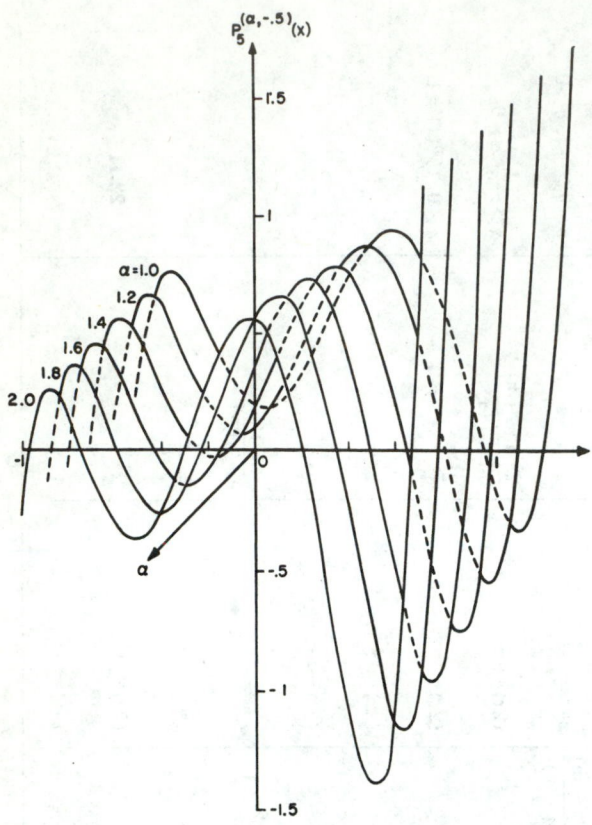

FIGURE 22.2. *Jacobi Polynomials* $P_n^{(\alpha,\beta)}(x)$, $\alpha=1(.2)2$, $\beta=-.5$, $n=5$.

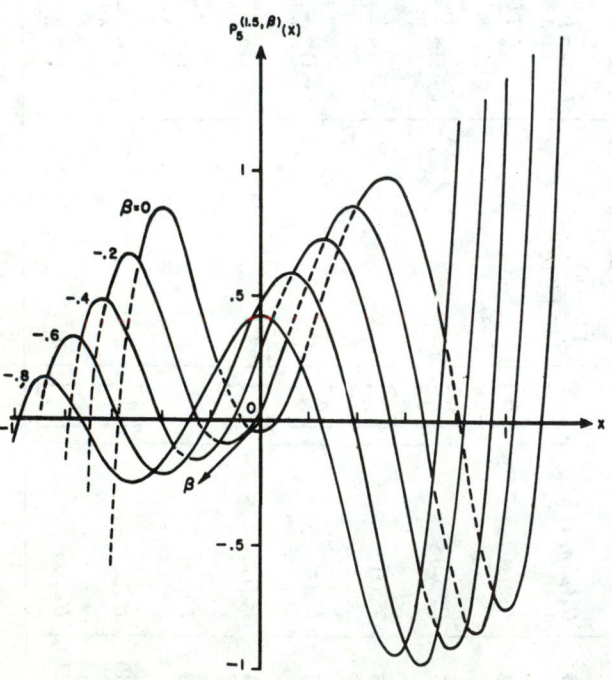

FIGURE 22.3. *Jacobi Polynomials* $P_n^{(\alpha,\beta)}(x)$, $\alpha=1.5$, $\beta=-.8(.2)0$, $n=5$.

Explicit Expressions Involving Trigonometric Functions

$$f_n(\cos\theta)=\sum_{m=0}^{n} a_m \cos (n-2m)\theta$$

	$f_n(\cos\theta)$	a_m	Remarks
22.3.12	$C_n^{(\alpha)}(\cos\theta)$	$\dfrac{\Gamma(\alpha+m)\,\Gamma(\alpha+n-m)}{m!\,(n-m)!\,[\Gamma(\alpha)]^2}$	$\alpha\neq 0$
22.3.13	$P_n(\cos\theta)$	$\dfrac{1}{4^n}\dbinom{2m}{m}\dbinom{2n-2m}{n-m}$	

22.3.14 $$C_n^{(0)}(\cos\theta)=\frac{2}{n}\cos n\theta$$

22.3.15 $$T_n(\cos\theta)=\cos n\theta$$

22.3.16 $$U_n(\cos\theta)=\frac{\sin (n+1)\theta}{\sin\theta}$$

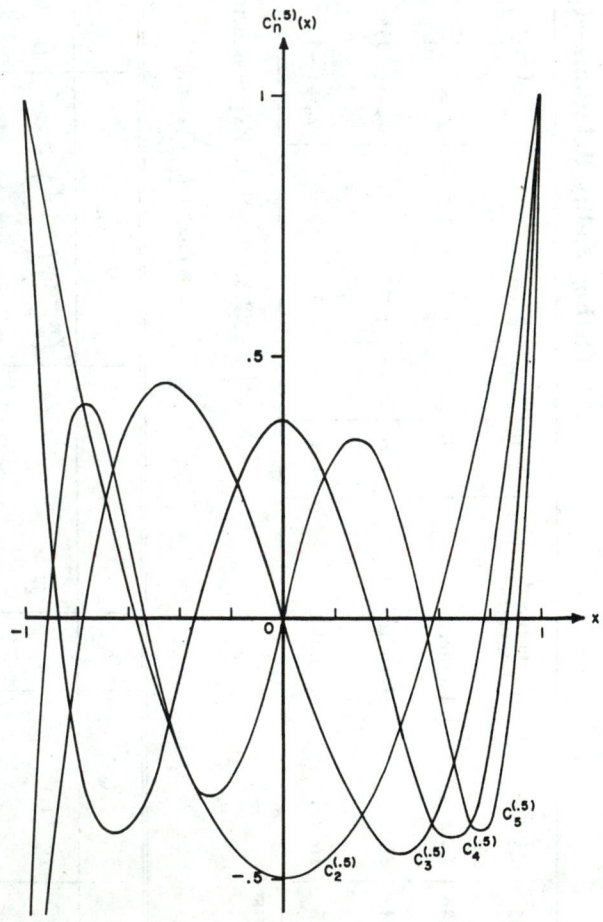

FIGURE 22.4. *Gegenbauer (Ultraspherical) Polynomials* $C_n^{(\alpha)}(x)$, $\alpha=.5$, $n=2(1)5$.

22.4. Special Values

	$f_n(x)$	$f_n(-x)$	$f_n(1)$	$f_n(0)$	$f_0(x)$	$f_1(x)$
22.4.1	$P_n^{(\alpha,\beta)}(x)$	$(-1)^n P_n^{(\beta,\alpha)}(x)$	$\binom{n+\alpha}{n}$ *		1	$\frac{1}{2}[\alpha-\beta+(\alpha+\beta+2)x]$
22.4.2	$C_n^{(\alpha)}(x)$, $\alpha\neq 0$	$(-1)^n C_n^{(\alpha)}(x)$	$\binom{n+2\alpha-1}{n}$	$\begin{cases}0,\ n=2m+1\\ (-1)^{n/2}\frac{\Gamma(\alpha+n/2)}{\Gamma(\alpha)(n/2)!},\ n=2m\end{cases}$	1	$2\alpha x$
22.4.3	$C_n^{(0)}(x)$	$(-1)^n C_n^{(0)}(x)$	$\frac{2}{n},\ n\neq 0$	$\begin{cases}\frac{(-1)^m}{m},\ n=2m\neq 0\\ 0,\ n=2m+1\end{cases}$	1	$2x$
22.4.4	$T_n(x)$	$(-1)^n T_n(x)$	1	$\begin{cases}(-1)^m,\ n=2m\\ 0,\ n=2m+1\end{cases}$	1	x
22.4.5	$U_n(x)$	$(-1)^n U_n(x)$	$n+1$	$\begin{cases}(-1)^m,\ n=2m\\ 0,\ n=2m+1\end{cases}$	1	$2x$
22.4.6	$P_n(x)$	$(-1)^n P_n(x)$	1	$\begin{cases}\frac{(-1)^m}{4^m}\binom{2m}{m},\ n=2m\ *\\ 0,\ n=2m+1\end{cases}$	1	x
22.4.7	$L_n^{(\alpha)}(x)$			$\binom{n+\alpha}{n}$	1	$-x+\alpha+1$
22.4.8	$H_n(x)$	$(-1)^n H_n(x)$		$\begin{cases}(-1)^m\frac{(2m)!}{m!},\ n=2m\\ 0,\ n=2m+1\end{cases}$	1	$2x$

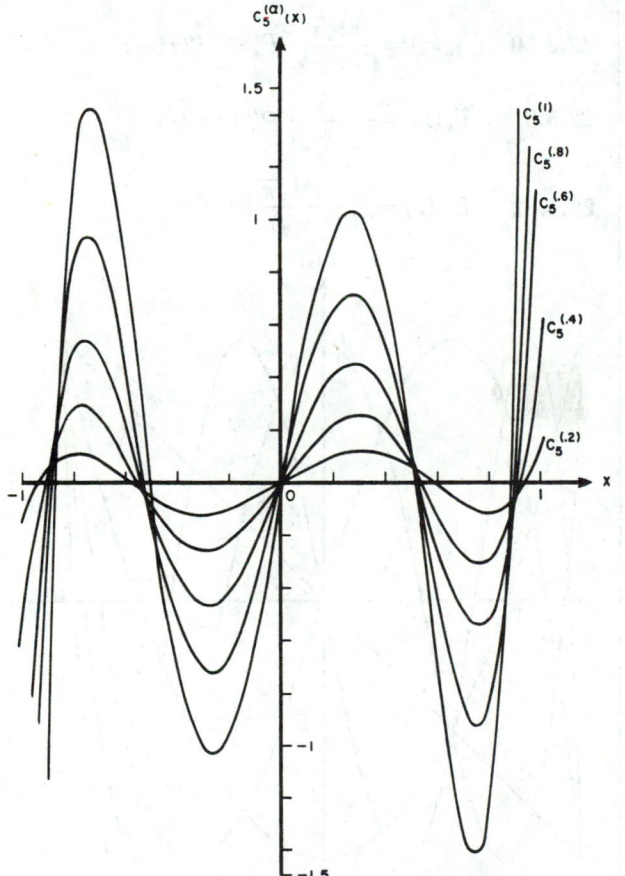

FIGURE 22.5. *Gegenbauer (Ultraspherical) Polynomials* $C_n^{(\alpha)}(x)$, $\alpha=.2(.2)1$, $n=5$.

22.5. Interrelations

Interrelations Between Orthogonal Polynomials of the Same Family

Jacobi Polynomials

22.5.1
$$P_n^{(\alpha,\beta)}(x)=\frac{\Gamma(2n+\alpha+\beta+1)}{n!\,\Gamma(n+\alpha+\beta+1)}\,G_n\left(\alpha+\beta+1,\beta+1,\frac{x+1}{2}\right)$$

22.5.2
$$G_n(p,q,x)=\frac{n!\,\Gamma(n+p)}{\Gamma(2n+p)}\,P_n^{(p-q,\,q-1)}(2x-1)$$
(see [22.21]).

22.5.3
$$F_n(p,q,x)=(-1)^n n!\,\frac{\Gamma(q)}{\Gamma(q+n)}\,P_n^{(p-q,\,q-1)}(2x-1)$$
(see [22.13]).

Ultraspherical Polynomials

22.5.4
$$C_n^{(0)}(x)=\lim_{\alpha\to 0}\frac{1}{\alpha}\,C_n^{(\alpha)}(x)$$

Chebyshev Polynomials

22.5.5
$$T_n(x)=\tfrac{1}{2}C_n(2x)=T_n^*\left(\frac{1+x}{2}\right)$$

22.5.6
$$T_n(x)=U_n(x)-xU_{n-1}(x)$$

22.5.7 $\quad T_n(x) = xU_{n-1}(x) - U_{n-2}(x)$

22.5.8 $\quad T_n(x) = \frac{1}{2}[U_n(x) - U_{n-2}(x)]$

22.5.9 $\quad U_n(x) = S_n(2x) = U_n^*\left(\frac{1+x}{2}\right)$

22.5.10 $\quad U_{n-1}(x) = \frac{1}{1-x^2}[xT_n(x) - T_{n+1}(x)]$

22.5.11 $\quad C_n(x) = 2T_n\left(\frac{x}{2}\right) = 2T_n^*\left(\frac{x+2}{4}\right)$

22.5.12 $\quad C_n(x) = S_n(x) - S_{n-2}(x)$

22.5.13 $\quad S_n(x) = U_n\left(\frac{x}{2}\right) = U_n^*\left(\frac{x+2}{4}\right)$

22.5.14 $\quad T_n^*(x) = T_n(2x-1) = \frac{1}{2}C_n(4x-2)$

(see [22.22]).

22.5.15 $\quad U_n^*(x) = S_n(4x-2) = U_n(2x-1)$

(see [22.22]).

Generalized Laguerre Polynomials

22.5.16 $\quad L_n^{(0)}(x) = L_n(x)$

22.5.17 $\quad L_n^{(m)}(x) = (-1)^m \frac{d^m}{dx^m}[L_{n+m}(x)]$

Hermite Polynomials

22.5.18 $\quad He_n(x) = 2^{-n/2} H_n\left(\frac{x}{\sqrt{2}}\right)$

(see [22.20]).

22.5.19 $\quad H_n(x) = 2^{n/2} He_n(x\sqrt{2})$

(see [22.13], [22.20]).

Interrelations Between Orthogonal Polynomials of Different Families

Jacobi Polynomials

22.5.20
$$P_n^{(\alpha-\frac{1}{2},\,\alpha-\frac{1}{2})}(x) = \frac{\Gamma(2\alpha)\,\Gamma(\alpha+n+\frac{1}{2})}{\Gamma(2\alpha+n)\,\Gamma(\alpha+\frac{1}{2})}\,C_n^{(\alpha)}(x)$$

22.5.21
$$P_n^{(\alpha,\,\frac{1}{2})}(x) = \frac{(\frac{1}{2})_{n+1}}{\sqrt{\frac{x+1}{2}}\,(\alpha+\frac{1}{2})_{n+1}}\,C_{2n+1}^{(\alpha+\frac{1}{2})}\left(\sqrt{\frac{x+1}{2}}\right)$$

22.5.22 $\quad P_n^{(\alpha,\,-\frac{1}{2})}(x) = \frac{(\frac{1}{2})_n}{(\alpha+\frac{1}{2})_n}\,C_{2n}^{(\alpha+\frac{1}{2})}\left(\sqrt{\frac{x+1}{2}}\right)$

22.5.23 $\quad P_n^{(-\frac{1}{2},\,-\frac{1}{2})}(x) = \frac{1}{4^n}\binom{2n}{n}T_n(x)$

22.5.24 $\quad P_n^{(0,\,0)}(x) = P_n(x)$

Ultraspherical Polynomials

22.5.25
$$C_{2n}^{(\alpha)}(x) = \frac{\Gamma(\alpha+n)n!2^{2n}}{\Gamma(\alpha)(2n)!}\,P_n^{(\alpha-\frac{1}{2},\,-\frac{1}{2})}(2x^2-1)$$
$$(\alpha \neq 0)$$

22.5.26
$$C_{2n+1}^{(\alpha)}(x) = \frac{\Gamma(\alpha+n+1)n!2^{2n+1}}{\Gamma(\alpha)(2n+1)!}\,xP_n^{(\alpha-\frac{1}{2},\,\frac{1}{2})}(2x^2-1)$$
$$(\alpha \neq 0)$$

22.5.27
$$C_n^{(\alpha)}(x) = \frac{\Gamma(\alpha+\frac{1}{2})\,\Gamma(2\alpha+n)}{\Gamma(2\alpha)\,\Gamma(\alpha+n+\frac{1}{2})}\,P_n^{(\alpha-\frac{1}{2},\,\alpha-\frac{1}{2})}(x)$$
$$(\alpha \neq 0)$$

22.5.28
$$C_n^{(0)}(x) = \frac{2}{n}\,T_n(x) = 2\,\frac{(n-1)!}{\Gamma(n+\frac{1}{2})}\,\sqrt{\pi}P_n^{(-\frac{1}{2},\,-\frac{1}{2})}(x) \qquad *$$

Chebyshev Polynomials

22.5.29 $\quad T_{2n+1}(x) = \frac{n!\sqrt{\pi}}{\Gamma(n+\frac{1}{2})}\,xP_n^{(-\frac{1}{2},\,\frac{1}{2})}(2x^2-1)$

22.5.30 $\quad U_{2n}(x) = \frac{n!\sqrt{\pi}}{\Gamma(n+\frac{1}{2})}\,P_n^{(\frac{1}{2},\,-\frac{1}{2})}(2x^2-1)$

22.5.31 $\quad T_n(x) = \frac{n!\sqrt{\pi}}{\Gamma(n+\frac{1}{2})}\,P_n^{(-\frac{1}{2},\,-\frac{1}{2})}(x)$

22.5.32 $\quad U_n(x) = \frac{(n+1)!\sqrt{\pi}}{2\Gamma(n+\frac{3}{2})}\,P_n^{(\frac{1}{2},\,\frac{1}{2})}(x)$

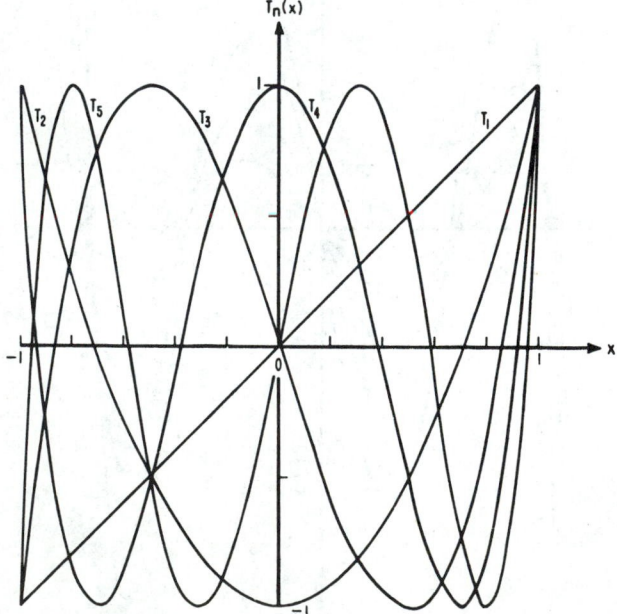

FIGURE 22.6. *Chebyshev Polynomials* $T_n(x)$, $n=1(1)5$.

*See page II.

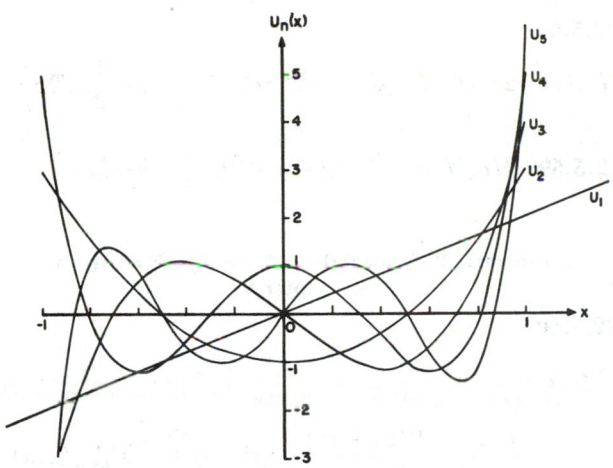

FIGURE 22.7. *Chebyshev Polynomials* $U_n(x)$, $n=1(1)5$.

22.5.33 $$T_n(x)=\frac{n}{2}\,C_n^{(0)}(x)$$

22.5.34 $$U_n(x)=C_n^{(1)}(x)$$

Legendre Polynomials

22.5.35 $$P_n(x)=P_n^{(0,0)}(x)$$

22.5.36 $$P_n(x)=C_n^{(1/2)}(x)$$

22.5.37
$$\frac{d^m}{dx^m}[P_n(x)]=1\cdot3\,\ldots\,(2m-1)C_{n-m}^{(m+\frac12)}(x)\qquad(m\leq n)$$

Generalized Laguerre Polynomials

22.5.38 $$L_n^{(-1/2)}(x)=\frac{(-1)^n}{n!2^{2n}}\,H_{2n}(\sqrt{x})$$

22.5.39 $$L_n^{(1/2)}(x)=\frac{(-1)^n}{n!2^{2n+1}\sqrt{x}}\,H_{2n+1}(\sqrt{x})$$

Hermite Polynomials

22.5.40 $$H_{2m}(x)=(-1)^m2^{2m}m!L_m^{(-1/2)}(x^2)$$

22.5.41 $$H_{2m+1}(x)=(-1)^m2^{2m+1}m!xL_m^{(1/2)}(x^2)$$

Orthogonal Polynomials as Hypergeometric Functions (see chapter 15)
$$f_n(x)=dF(a,\,b;\,c;\,g(x))$$

For each of the listed polynomials there are numerous other representations in terms of hypergeometric functions.

	$f_n(x)$	d	a	b	c	$g(x)$
22.5.42	$P_n^{(\alpha,\beta)}(x)$	$\binom{n+\alpha}{n}$	$-n$	$n+\alpha+\beta+1$	$\alpha+1$	$\dfrac{1-x}{2}$
22.5.43	$P_n^{(\alpha,\beta)}(x)$	$\binom{2n+\alpha+\beta}{n}\left(\dfrac{x-1}{2}\right)^n$	$-n$	$-n-\alpha$	$-2n-\alpha-\beta$	$\dfrac{2}{1-x}$
22.5.44	$P_n^{(\alpha,\beta)}(x)$	$\binom{n+\alpha}{n}\left(\dfrac{1+x}{2}\right)^n$	$-n$	$-n-\beta$	$\alpha+1$	$\dfrac{x-1}{x+1}$
22.5.45	$P_n^{(\alpha,\beta)}(x)$	$\binom{n+\beta}{n}\left(\dfrac{x-1}{2}\right)^n$	$-n$	$-n-\alpha$	$\beta+1$	$\dfrac{x+1}{x-1}$
22.5.46	$C_n^{(\alpha)}(x)$	$\dfrac{\Gamma(n+2\alpha)}{n!\Gamma(2\alpha)}$	$-n$	$n+2\alpha$	$\alpha+\tfrac12$	$\dfrac{1-x}{2}$
22.5.47	$T_n(x)$	1	$-n$	n	$\tfrac12$	$\dfrac{1-x}{2}$
22.5.48	$U_n(x)$	$n+1$	$-n$	$n+2$ *	$\tfrac32$	$\dfrac{1-x}{2}$
22.5.49	$P_n(x)$	1	$-n$	$n+1$	1	$\dfrac{1-x}{2}$
22.5.50	$P_n(x)$	$\binom{2n}{n}\left(\dfrac{x-1}{2}\right)^n$	$-n$	$-n$	$-2n$	$\dfrac{2}{1-x}$
22.5.51	$P_n(x)$	$\binom{2n}{n}\left(\dfrac{x}{2}\right)^n$	$-\dfrac{n}{2}$	$\dfrac{1-n}{2}$	$\tfrac12-n$	$\dfrac{1}{x^2}$
22.5.52	$P_{2n}(x)$	$(-1)^n\dfrac{(2n)!}{2^{2n}(n!)^2}$	$-n$	$n+\tfrac12$	$\tfrac12$	x^2
22.5.53	$P_{2n+1}(x)$	$(-1)^n\dfrac{(2n+1)!}{2^{2n}(n!)^2}x$	$-n$	$n+\tfrac32$	$\tfrac32$	x^2

*See page II.

Orthogonal Polynomials as Confluent Hypergeometric Functions (see chapter 13)

22.5.54 $\quad L_n^{(\alpha)}(x) = \binom{n+\alpha}{n} M(-n, \alpha+1, x)$

Orthogonal Polynomials as Parabolic Cylinder Functions (see chapter 19)

22.5.55 $\quad H_n(x) = 2^n U\left(\frac{1}{2} - \frac{1}{2}n, \frac{3}{2}, x^2\right)$

22.5.56 $\quad H_{2m}(x) = (-1)^m \frac{(2m)!}{m!} M\left(-m, \frac{1}{2}, x^2\right)$

22.5.57

* $H_{2m+1}(x) = (-1)^m \frac{(2m+1)!}{m!} 2x M\left(-m, \frac{3}{2}, x^2\right)$

22.5.58

$$H_n(x) = 2^{n/2} e^{x^2/2} D_n(\sqrt{2}x) = 2^{n/2} e^{x^2/2} U\left(-n-\frac{1}{2}, \sqrt{2}x\right)$$

22.5.59 $\quad He_n(x) = e^{x^2/4} D_n(x) = e^{x^2/4} U\left(-n-\frac{1}{2}, x\right)$

Orthogonal Polynomials as Legendre Functions (see chapter 8)

22.5.60

$$C_n^{(\alpha)}(x) =$$

$$\frac{\Gamma(\alpha+\frac{1}{2})\Gamma(2\alpha+n)}{n!\,\Gamma(2\alpha)} \left[\frac{1}{4}(x^2-1)\right]^{\frac{1}{4}-\frac{\alpha}{2}} P_{n+\alpha-\frac{1}{2}}^{(\frac{1}{2}-\alpha)}(x)$$
$$(\alpha \neq 0)$$

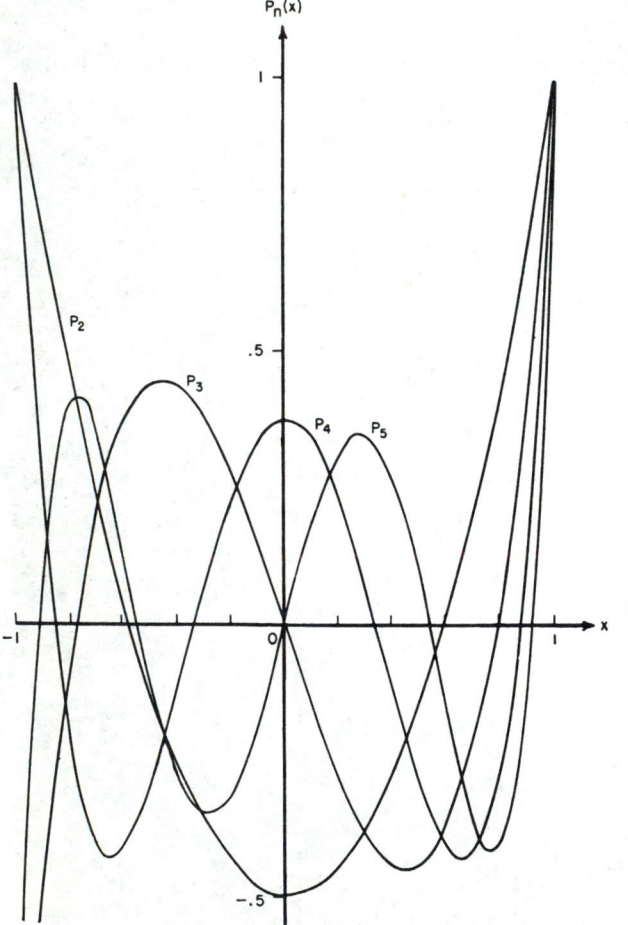

FIGURE 22.8. *Legendre Polynomials $P_n(x)$,*
n=2(1)5.

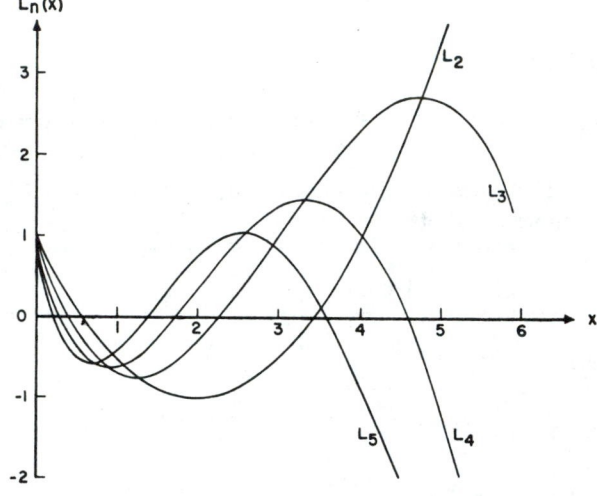

FIGURE 22.9. *Laguerre Polynomials $L_n(x)$,*
n=2(1)5.

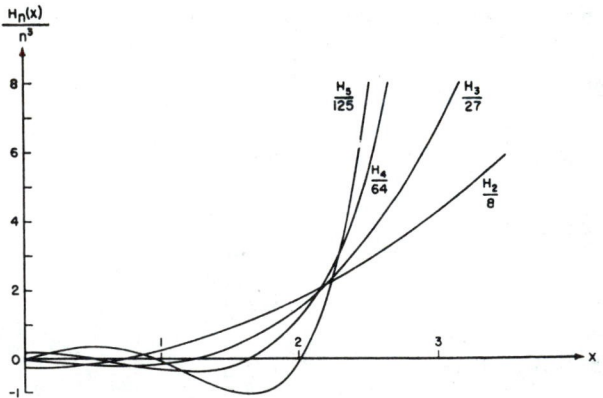

FIGURE 22.10. *Hermite Polynomials $\frac{H_n(x)}{n^3}$,*
n=2(1)5.

*See page II.

22.6. Differential Equations

$$g_2(x)\,y'' + g_1(x)\,y' + g_0(x)\,y = 0$$

	y	$g_2(x)$	$g_1(x)$	$g_0(x)$
22.6.1	$P_n^{(\alpha,\beta)}(x)$	$1-x^2$	$\beta-\alpha-(\alpha+\beta+2)x$	$n(n+\alpha+\beta+1)$
22.6.2	$(1-x)^\alpha(1+x)^\beta P_n^{(\alpha,\beta)}(x)$	$1-x^2$	$\alpha-\beta+(\alpha+\beta-2)x$	$(n+1)(n+\alpha+\beta)$
22.6.3	$(1-x)^{\frac{\alpha+1}{2}}(1+x)^{\frac{\beta+1}{2}}P_n^{(\alpha,\beta)}(x)$	1	0	$\dfrac{1}{4}\dfrac{1-\alpha^2}{(1-x)^2}+\dfrac{1}{4}\dfrac{1-\beta^2}{(1+x)^2}$ $\quad+\dfrac{2n(n+\alpha+\beta+1)+(\alpha+1)(\beta+1)}{2(1-x^2)}$
22.6.4	$\left(\sin\dfrac{x}{2}\right)^{\alpha+\frac{1}{2}}\left(\cos\dfrac{x}{2}\right)^{\beta+\frac{1}{2}}P_n^{(\alpha,\beta)}(\cos x)$	1	0	$\dfrac{1-4\alpha^2}{16\sin^2\dfrac{x}{2}}+\dfrac{1-4\beta^2}{16\cos^2\dfrac{x}{2}}$ $\quad+\left(n+\dfrac{\alpha+\beta+1}{2}\right)^2$
22.6.5	$C_n^{(\alpha)}(x)$	$1-x^2$	$-(2\alpha+1)x$	$n(n+2\alpha)$
22.6.6	$(1-x^2)^{\alpha-\frac{1}{2}}C_n^{(\alpha)}(x)$	$1-x^2$	$(2\alpha-3)x$	$(n+1)(n+2\alpha-1)$
22.6.7	$(1-x^2)^{\frac{\alpha}{2}+\frac{1}{4}}C_n^{(\alpha)}(x)$	1	0	$\dfrac{(n+\alpha)^2}{1-x^2}+\dfrac{2+4\alpha-4\alpha^2+x^2}{4(1-x^2)^2}$
22.6.8	$(\sin x)^\alpha C_n^{(\alpha)}(\cos x)$	1	0	$(n+\alpha)^2+\dfrac{\alpha(1-\alpha)}{\sin^2 x}$
22.6.9	$T_n(x)$	$1-x^2$	$-x$	n^2
22.6.10	$T_n(\cos x)$	1	0	n^2
22.6.11	$\dfrac{1}{\sqrt{1-x^2}}T_n(x);\; U_{n-1}(x)$ *	$1-x^2$	$-3x$	n^2-1
22.6.12	$U_n(x)$	$1-x^2$	$-3x$	$n(n+2)$
22.6.13	$P_n(x)$	$1-x^2$	$-2x$	$n(n+1)$
22.6.14	$\sqrt{1-x^2}\,P_n(x)$	1	0	$\dfrac{n(n+1)}{1-x^2}+\dfrac{1}{(1-x^2)^2}$
22.6.15	$L_n^{(\alpha)}(x)$	x	$\alpha+1-x$	n
22.6.16	$e^{-x}x^{\alpha/2}L_n^{(\alpha)}(x)$ *	x	$x+1$	$n+\dfrac{\alpha}{2}+1-\dfrac{\alpha^2}{4x}$
22.6.17	$e^{-x/2}x^{(\alpha+1)/2}L_n^{(\alpha)}(x)$	1	0	$\dfrac{2n+\alpha+1}{2x}+\dfrac{1-\alpha^2}{4x^2}-\dfrac{1}{4}$
22.6.18	$e^{-x^2/2}x^{\alpha+\frac{1}{2}}L_n^{(\alpha)}(x^2)$	1	0	$4n+2\alpha+2-x^2+\dfrac{1-4\alpha^2}{4x^2}$
22.6.19	$H_n(x)$	1	$-2x$	$2n$
22.6.20	$e^{-\frac{x^2}{2}}H_n(x)$	1	0	$2n+1-x^2$
22.6.21	$He_n(x)$	1	$-x$	n

*See page II.

22.7. Recurrence Relations

Recurrence Relations With Respect to the Degree n

$$a_{1n}f_{n+1}(x)=(a_{2n}+a_{3n}x)f_n(x)-a_{4n}f_{n-1}(x)$$

	f_n	a_{1n}	a_{2n}	a_{3n}	a_{4n}
22.7.1	$P_n^{(\alpha,\beta)}(x)$	$2(n+1)(n+\alpha+\beta+1)$ $(2n+\alpha+\beta)$	$(2n+\alpha+\beta+1)(\alpha^2-\beta^2)$	$(2n+\alpha+\beta)_3$	$2(n+\alpha)(n+\beta)$ $(2n+\alpha+\beta+2)$
22.7.2	$G_n(p,q,x)$	$(2n+p-2)_4(2n+p-1)$	$-[2n(n+p)+q(p-1)]$ $(2n+p-2)_3$	$(2n+p-2)_4$ $(2n+p-1)$	$n(n+q-1)(n+p-1)$ $(n+p-q)(2n+p+1)$
22.7.3	$C_n^{(\alpha)}(x)$	$n+1$	0	$2(n+\alpha)$	$n+2\alpha-1$
22.7.4	$T_n(x)$	1	0	2	1
22.7.5	$U_n(x)$	1	0	2	1
22.7.6	$S_n(x)$	1	0	1	1
22.7.7	$C_n(x)$	1	0	1	1
22.7.8	$T_n^*(x)$	1	-2	4	1
22.7.9	$U_n^*(x)$	1	-2	4	1
22.7.10	$P_n(x)$	$n+1$	0	$2n+1$	n
22.7.11	$P_n^*(x)$	$n+1$	$-2n-1$	$4n+2$	n
22.7.12	$L_n^{(\alpha)}(x)$	$n+1$	$2n+\alpha+1$	-1	$n+\alpha$
22.7.13	$H_n(x)$	1	0	2	$2n$
22.7.14	$He_n(x)$	1	0	1	n

Miscellaneous Recurrence Relations

Jacobi Polynomials

22.7.15

$$\left(n+\frac{\alpha}{2}+\frac{\beta}{2}+1\right)(1-x)P_n^{(\alpha+1,\beta)}(x)$$
$$=(n+\alpha+1)P_n^{(\alpha,\beta)}(x)-(n+1)P_{n+1}^{(\alpha,\beta)}(x)$$

22.7.16

$$\left(n+\frac{\alpha}{2}+\frac{\beta}{2}+1\right)(1+x)P_n^{(\alpha,\beta+1)}(x)$$
$$=(n+\beta+1)P_n^{(\alpha,\beta)}(x)+(n+1)P_{n+1}^{(\alpha,\beta)}(x)$$

22.7.17

$$(1-x)P_n^{(\alpha+1,\beta)}(x)+(1+x)P_n^{(\alpha,\beta+1)}(x)=2P_n^{(\alpha,\beta)}(x)$$

22.7.18

$$(2n+\alpha+\beta)P_n^{(\alpha-1,\beta)}(x)=(n+\alpha+\beta)P_n^{(\alpha,\beta)}(x)$$
$$-(n+\beta)P_{n-1}^{(\alpha,\beta)}(x)$$

22.7.19

$$(2n+\alpha+\beta)P_n^{(\alpha,\beta-1)}(x)=(n+\alpha+\beta)P_n^{(\alpha,\beta)}(x)$$
$$+(n+\alpha)P_{n-1}^{(\alpha,\beta)}(x)$$

22.7.20　$P_n^{(\alpha,\beta-1)}(x)-P_n^{(\alpha-1,\beta)}(x)=P_{n-1}^{(\alpha,\beta)}(x)$

Ultraspherical Polynomials

22.7.21

$$2\alpha(1-x^2)C_{n-1}^{(\alpha+1)}(x)=(2\alpha+n-1)C_{n-1}^{(\alpha)}(x)-nxC_n^{(\alpha)}(x)$$

22.7.22

$$=(n+2\alpha)xC_n^{(\alpha)}(x)$$
$$-(n+1)C_{n+1}^{(\alpha)}(x)$$

22.7.23　$(n+\alpha)C_{n+1}^{(\alpha-1)}(x)=(\alpha-1)[C_{n+1}^{(\alpha)}(x)-C_{n-1}^{(\alpha)}(x)]$

Chebyshev Polynomials

22.7.24

$$2T_m(x)T_n(x)=T_{n+m}(x)+T_{n-m}(x)\qquad(n\geq m)\qquad *$$

22.7.25

$$2(x^2-1)U_{m-1}(x)U_{n-1}(x)=T_{n+m}(x)-T_{n-m}(x)$$
$$(n\geq m)$$

22.7.26

$$2T_m(x)U_{n-1}(x)=U_{n+m-1}(x)+U_{n-m-1}(x)\qquad(n>m)$$

22.7.27

$$2T_n(x)U_{m-1}(x)=U_{n+m-1}(x)-U_{n-m-1}(x)\qquad(n>m)$$

22.7.28　　　$2T_n(x)U_{n-1}(x)=U_{2n-1}(x)$

*See page II.

Generalized Laguerre Polynomials

22.7.29

$$L_n^{(\alpha+1)}(x) = \frac{1}{x}\left[(x-n)L_n^{(\alpha)}(x) + (\alpha+n)L_{n-1}^{(\alpha)}(x)\right]$$

22.7.30 $\quad L_n^{(\alpha-1)}(x) = L_n^{(\alpha)}(x) - L_{n-1}^{(\alpha)}(x)$

22.7.31

$$L_n^{(\alpha+1)}(x) = \frac{1}{x}\left[(n+\alpha+1)L_n^{(\alpha)}(x) - (n+1)L_{n+1}^{(\alpha)}(x)\right]$$

22.7.32

$$L_n^{(\alpha-1)}(x) = \frac{1}{n+\alpha}\left[(n+1)L_{n+1}^{(\alpha)}(x) - (n+1-x)L_n^{(\alpha)}(x)\right]$$

22.8. Differential Relations

$$g_2(x)\frac{d}{dx}f_n(x) = g_1(x)f_n(x) + g_0(x)f_{n-1}(x)$$

	f_n	g_2	g_1	g_0
22.8.1	$P_n^{(\alpha,\beta)}(x)$	$(2n+\alpha+\beta)(1-x^2)$	$n[\alpha-\beta-(2n+\alpha+\beta)x]$	$2(n+\alpha)(n+\beta)$
22.8.2	$C_n^{(\alpha)}(x)$	$1-x^2$	$-nx$	$n+2\alpha-1$
22.8.3	$T_n(x)$	$1-x^2$	$-nx$	n
22.8.4	$U_n(x)$	$1-x^2$	$-nx$	$n+1$
22.8.5	$P_n(x)$	$1-x^2$	$-nx$	n
22.8.6	$L_n^{(\alpha)}(x)$	x	n	$-(n+\alpha)$
22.8.7	$H_n(x)$	1	0	$2n$
22.8.8	$He_n(x)$	1	0	n

22.9. Generating Functions

$$g(x,z) = \sum_{n=0}^{\infty} a_n f_n(x) z^n \qquad R = \sqrt{1-2xz+z^2}$$

	$f_n(x)$	a_n	$g(x,z)$	Remarks
22.9.1	$P_n^{(\alpha,\beta)}(x)$	$2^{-\alpha-\beta}$	$R^{-1}(1-z+R)^{-\alpha}(1+z+R)^{-\beta}$	$|z|<1$
22.9.2	$C_n^{(\alpha)}(x)$	$\dfrac{2^{\frac{1}{2}-\alpha}\Gamma(\alpha+\frac{1}{2}+n)\Gamma(2\alpha)}{\Gamma(\alpha+\frac{1}{2})\Gamma(2\alpha+n)}$	$R^{-1}(1-xz+R)^{\frac{1}{2}-\alpha}$	$|z|<1, \alpha\neq0$
22.9.3	$C_n^{(\alpha)}(x)$	1	$R^{-2\alpha}$	$|z|<1, \alpha\neq0$
22.9.4	$C_n^{(0)}(x)$	1	$-\ln R^2$	$|z|<1$
22.9.5	$C_n^{(\alpha)}(x)$	$\dfrac{\Gamma(2\alpha)}{\Gamma(\alpha+\frac{1}{2})\Gamma(2\alpha+n)}$	$e^{z\cos\theta}\left(\dfrac{z}{2}\sin\theta\right)^{\frac{1}{2}-\alpha}J_{\alpha-\frac{1}{2}}(z\sin\theta)$	$x=\cos\theta$
22.9.6	$T_n(x)$	2	$\left(\dfrac{1-z^2}{R^2}+1\right)$	$\begin{array}{c}-1<x<1\\|z|<1\end{array}$
22.9.7	$T_n(x)$	$\dfrac{\sqrt{2}}{4^n}\binom{2n}{n}$	$R^{-1}(1-xz+R)^{1/2}$	$\begin{array}{c}-1<x<1\\|z|<1\end{array}$
22.9.8	$T_n(x)$	$\dfrac{1}{n}$	$1-\frac{1}{2}\ln R^2$	$\begin{array}{c}a_0=1\\-1<x<1\\|z|<1\end{array}$
22.9.9	$T_n(x)$	1	$\dfrac{1-xz}{R^2}$	$\begin{array}{c}-1<x<1\\|z|<1\end{array}$
22.9.10	$U_n(x)$	1	R^{-2}	$\begin{array}{c}-1<x<1\\|z|<1\end{array}$
22.9.11	$U_n(x)$	$\dfrac{\sqrt{2}}{4^{n+1}}\binom{2n+2}{n+1}$	$\dfrac{1}{R}(1-xz+R)^{-1/2}$ *	$\begin{array}{c}-1<x<1\\|z|<1\end{array}$

*See page II.

22.9. Generating Functions—Continued

$$g(x, z) = \sum_{n=0}^{\infty} a_n f_n(x) z^n \qquad\qquad R = \sqrt{1 - 2xz + z^2}$$

	$f_n(x)$	a_n	$g(x, z)$	Remarks
22.9.12	$P_n(x)$	1	R^{-1}	$-1 < x < 1$ $\quad\;\; \lvert z \rvert < 1$
22.9.13	$P_n(x)$	$\dfrac{1}{n!}$	$e^z \cos\theta\, J_0(z \sin\theta)$	$x = \cos\theta$
22.9.14	$S_n(x)$	1	$(1 - xz + z^2)^{-1}$	$-2 < x < 2$ $\quad\;\; \lvert z \rvert < 1$
22.9.15	$L_n^{(\alpha)}(x)$	1	$(1 - z)^{-\alpha - 1} \exp\left(\dfrac{xz}{z-1}\right)$	$\lvert z \rvert < 1$
22.9.16	$L_n^{(\alpha)}(x)$	$\dfrac{1}{\Gamma(n+\alpha+1)}$	$(xz)^{-\frac{1}{2}\alpha} e^z J_\alpha[2(xz)^{1/2}]$	
22.9.17	$H_n(x)$	$\dfrac{1}{n!}$	$e^{2xz - z^2}$	
22.9.18	$H_{2n}(x)$	$\dfrac{(-1)^n}{(2n)!}$	$e^z \cos(2x\sqrt{z})$ $\quad$ *	
22.9.19	$H_{2n+1}(x)$	$\dfrac{(-1)^n}{(2n+1)!}$	$z^{-1/2} e^z \sin(2x\sqrt{z})$ $\quad$ *	

22.10. Integral Representations
Contour Integral Representations

$$f_n(x) = \frac{g_0(x)}{2\pi i} \int_C [g_1(z, x)]^n g_2(z, x)\, dz \quad \text{where } C \text{ is a closed contour taken around } z = a \text{ in the positive sense}$$

	$f_n(x)$	$g_0(x)$	$g_1(z,x)$	$g_2(z,x)$	a	Remarks
22.10.1	$P_n^{(\alpha,\beta)}(x)$	$\dfrac{1}{(1-x)^\alpha (1+x)^\beta}$	$\dfrac{z^2-1}{2(z-x)}$	$\dfrac{(1-z)^\alpha (1+z)^\beta}{z-x}$	x	± 1 outside C
22.10.2	$C_n^{(\alpha)}(x)$	1	$1/z$	$(1 - 2xz + z^2)^{-\alpha} z^{-1}$	0	Both zeros of $1-2xz+z^2$ outside C, $\alpha > 0$
22.10.3	$T_n(x)$	$1/2$	$1/z$	$\dfrac{1 - z^2}{z(1 - 2xz + z^2)}$	0	Both zeros of $1-2xz+z^2$ outside C
22.10.4	$U_n(x)$	1	$1/z$	$\dfrac{1}{z(1 - 2xz + z^2)}$	0	Both zeros of $1-2xz+z^2$ outside C
22.10.5	$P_n(x)$	1	$1/z$	$\dfrac{1}{z}(1 - 2xz + z^2)^{-1/2}$	0	Both zeros of $1-2xz+z^2$ outside C
22.10.6	$P_n(x)$	$\dfrac{1}{2^n}$	$\dfrac{z^2-1}{z-x}$	$\dfrac{1}{z-x}$	x	
22.10.7	$L_n^{(\alpha)}(x)$	$e^z x^{-\alpha}$	$\dfrac{z}{z-x}$	$\dfrac{z^\alpha}{z-x} e^{-z}$	x	Zero outside C
22.10.8	$L_n^{(\alpha)}(x)$	1	$1 + \dfrac{x}{z}$	$e^{-z}\left(1 + \dfrac{z}{x}\right)^\alpha 1/z$	0	$z = -x$ outside C
22.10.9	$H_n(x)$	$n!$	$1/z$	$\dfrac{e^{2xz - z^2}}{z}$	0	

Miscellaneous Integral Representations

$$\textbf{22.10.10} \quad C_n^{(\alpha)}(x) = \frac{2^{(1-2\alpha)} \Gamma(n+2\alpha)}{n! [\Gamma(\alpha)]^2} \int_0^\pi [x + \sqrt{x^2-1}\, \cos\phi]^n (\sin\phi)^{2\alpha-1} d\phi \qquad (\alpha > 0)$$

$$\textbf{22.10.11} \quad C_n^{(\alpha)}(\cos\theta) = \frac{2^{1-\alpha} \Gamma(n+2\alpha)}{n! [\Gamma(\alpha)]^2} (\sin\theta)^{1-2\alpha} \int_0^\theta \frac{\cos(n+\alpha)\phi}{(\cos\phi - \cos\theta)^{1-\alpha}} d\phi \qquad (\alpha > 0)$$

*See page II.

22.10.12 $\quad P_n(\cos\theta)=\dfrac{1}{\pi}\displaystyle\int_0^\pi (\cos\theta+i\sin\theta\cos\phi)^n d\phi$

22.10.13 $\quad P_n(\cos\theta)=\dfrac{\sqrt{2}}{\pi}\displaystyle\int_\theta^\pi \dfrac{\sin(n+\frac{1}{2})\phi d\phi}{(\cos\theta-\cos\phi)^{\frac{1}{2}}}$

22.10.14 $\quad L_n^{(\alpha)}(x)=\dfrac{e^x x^{-\frac{\alpha}{2}}}{n!}\displaystyle\int_0^\infty e^{-t}t^{n+\frac{\alpha}{2}}\,J_\alpha(2\sqrt{tx})dt$

22.10.15
$$H_n(x)=e^{x^2}\frac{2^{n+1}}{\sqrt{\pi}}\int_0^\infty e^{-t^2}t^n\cos\left(2xt-\frac{n}{2}\pi\right)dt$$

22.11. Rodrigues' Formula

$$f_n(x)=\frac{1}{a_n\rho(x)}\frac{d^n}{dx^n}\{\rho(x)(g(x))^n\}$$

The polynomials given in the following table are the only orthogonal polynomials which satisfy this formula.

	$f_n(x)$	a_n	$\rho(x)$	$g(x)$
22.11.1	$P_n^{(\alpha,\beta)}(x)$	$(-1)^n 2^n n!$	$(1-x)^\alpha(1+x)^\beta$	$1-x^2$
22.11.2	$C_n^{(\alpha)}(x)$	$(-1)^n 2^n n!\,\dfrac{\Gamma(2\alpha)\Gamma(\alpha+n+\frac{1}{2})}{\Gamma(\alpha+\frac{1}{2})\Gamma(n+2\alpha)}$	$(1-x^2)^{\alpha-\frac{1}{2}}$	$1-x^2$
22.11.3	$T_n(x)$	$(-1)^n 2^n\,\dfrac{\Gamma(n+\frac{1}{2})}{\sqrt{\pi}}$ $\quad*$	$(1-x^2)^{-\frac{1}{2}}$	$1-x^2$
22.11.4	$U_n(x)$	$(-1)^n 2^{n+1}\,\dfrac{\Gamma(n+\frac{3}{2})}{(n+1)\sqrt{\pi}}$	$(1-x^2)^{\frac{1}{2}}$	$1-x^2$
22.11.5	$P_n(x)$	$(-1)^n 2^n n!$	1	$1-x^2$
22.11.6	$L_n^{(\alpha)}(x)$	$n!$	$e^{-x}x^\alpha$	x
22.11.7	$H_n(x)$	$(-1)^n$	e^{-x^2}	1
22.11.8	$He_n(x)$	$(-1)^n$	$e^{-x^2/2}$	1

22.12. Sum Formulas

Christoffel-Darboux Formula

22.12.1
$$\sum_{m=0}^n \frac{1}{h_m}f_m(x)f_m(y)=\frac{k_n}{k_{n+1}h_n}\frac{f_{n+1}(x)f_n(y)-f_n(x)f_{n+1}(y)}{x-y}$$

Miscellaneous Sum Formulas (Only a Limited Selection Is Given Here.)

22.12.2 $\quad\displaystyle\sum_{m=0}^n T_{2m}(x)=\frac{1}{2}[1+U_{2n}(x)]$

22.12.3 $\quad\displaystyle\sum_{m=0}^{n-1} T_{2m+1}(x)=\frac{1}{2}U_{2n-1}(x)$

22.12.4 $\quad\displaystyle\sum_{m=0}^n U_{2m}(x)=\frac{1-T_{2n+2}(x)}{2(1-x^2)}$

22.12.5 $\quad\displaystyle\sum_{m=0}^{n-1} U_{2m+1}(x)=\frac{x-T_{2n+1}(x)}{2(1-x^2)}$

22.12.6 $\quad\displaystyle\sum_{m=0}^n L_m^{(\alpha)}(x)L_{n-m}^{(\beta)}(y)=L_n^{(\alpha+\beta+')}(x+y)$

22.12.7 $\quad\displaystyle\sum_{m=0}^n\binom{n+\alpha}{m}\mu^{n-m}(1-\mu)^m L_{n-m}^{(\alpha)}(x)=L_n^{(\alpha)}(\mu x)$

22.12.8
$$H_n(x+y)=\frac{1}{2^{n/2}}\sum_{k=0}^n\binom{n}{k}H_k(\sqrt{2}x)H_{n-k}(\sqrt{2}y)$$

22.13. Integrals Involving Orthogonal Polynomials

22.13.1
$$2n\int_0^x(1-y)^\alpha(1+y)^\beta P_n^{(\alpha,\beta)}(y)dy$$
$$=P_{n-1}^{(\alpha+1,\beta+1)}(0)-(1-x)^{\alpha+1}(1+x)^{\beta+1}P_{n-1}^{(\alpha+1,\beta+1)}(x)$$

22.13.2
$$\frac{n(2\alpha+n)}{2\alpha}\int_0^x(1-y^2)^{\alpha-\frac{1}{2}}C_n^{(\alpha)}(y)dy$$
$$=C_{n-1}^{(\alpha+1)}(0)-(1-x^2)^{\alpha+\frac{1}{2}}C_{n-1}^{(\alpha+1)}(x)$$

22.13.3 $\quad\displaystyle\fint_{-1}^1\frac{T_n(y)dy}{(y-x)\sqrt{1-y^2}}=\pi U_{n-1}(x)$

22.13.4 $\quad\displaystyle\fint_{-1}^1\frac{\sqrt{1-y^2}\,U_{n-1}(y)dy}{(y-x)}=-\pi T_n(x)\quad*$

22.13.5 $\quad\displaystyle\int_{-1}^1(1-x)^{-1/2}P_n(x)dx=\frac{2^{3/2}}{2n+1}\quad*$

22.13.6 $\quad\displaystyle\int_0^\pi P_{2n}(\cos\theta)d\theta=\frac{\pi}{16^n}\binom{2n}{n}^2$

22.13.7 $\quad\displaystyle\int_0^\pi P_{2n+1}(\cos\theta)\cos\theta d\theta=\frac{\pi}{4^{2n+1}}\binom{2n}{n}\binom{2n+2}{n+1}$

*See page II.

22.13.8

$$\int_0^1 x^\lambda P_{2n}(x)dx = \frac{(-1)^n \Gamma\left(n-\frac{\lambda}{2}\right)\Gamma\left(\frac{1}{2}+\frac{\lambda}{2}\right)}{2\Gamma\left(-\frac{\lambda}{2}\right)\Gamma\left(n+\frac{3}{2}+\frac{\lambda}{2}\right)} \quad (\lambda > -1)$$

22.13.9

$$\int_0^1 x^\lambda P_{2n+1}(x)dx = \frac{(-1)^n \Gamma\left(n+\frac{1}{2}-\frac{\lambda}{2}\right)\Gamma\left(1+\frac{\lambda}{2}\right)}{2\Gamma\left(n+2+\frac{\lambda}{2}\right)\Gamma\left(\frac{1}{2}-\frac{\lambda}{2}\right)}$$

$$(\lambda > -2)$$

22.13.10

$$\int_{-1}^x \frac{P_n(t)dt}{\sqrt{x-t}} = \frac{1}{(n+\frac{1}{2})\sqrt{1+x}}[T_n(x)+T_{n+1}(x)]$$

22.13.11

$$\int_x^1 \frac{P_n(t)dt}{\sqrt{t-x}} = \frac{1}{(n+\frac{1}{2})\sqrt{1-x}}[T_n(x)-T_{n+1}(x)]$$

22.13.12 $\displaystyle\int_x^\infty e^{-t}L_n^{(\alpha)}(t)dt = e^{-x}[L_n^{(\alpha)}(x)-L_{n-1}^{(\alpha)}(x)]$

22.13.13

$$\Gamma(\alpha+\beta+n+1)\int_0^x (x-t)^{\beta-1}t^\alpha L_n^{(\alpha)}(t)dt$$
$$= \Gamma(\alpha+n+1)\Gamma(\beta)x^{\alpha+\beta}L_n^{(\alpha+\beta)}(x)$$
$$(\mathscr{R}\alpha > -1,\ \mathscr{R}\beta > 0)$$

22.13.14

$$\int_0^x L_m(t)L_n(x-t)dt$$
$$= \int_0^x L_{m+n}(t)dt = L_{m+n}(x)-L_{m+n+1}(x)$$

22.13.15 $\displaystyle\int_0^x e^{-t^2}H_n(t)dt = H_{n-1}(0)-e^{-x^2}H_{n-1}(x)$

22.13.16 $\displaystyle\int_0^x H_n(t)dt = \frac{1}{2(n+1)}[H_{n+1}(x)-H_{n+1}(0)]$

22.13.17 $\displaystyle\int_{-\infty}^\infty e^{-t^2}H_{2m}(tx)dt = \sqrt{\pi}\,\frac{(2m)!}{m!}(x^2-1)^m$

22.13.18

$$\int_{-\infty}^\infty e^{-t^2}tH_{2m+1}(tx)dt = \sqrt{\pi}\,\frac{(2m+1)!}{m!}x(x^2-1)^m$$

22.13.19 $\displaystyle\int_{-\infty}^\infty e^{-t^2}t^n H_n(xt)dt = \sqrt{\pi}n!P_n(x)$

22.13.20

$$\int_0^\infty e^{-t^2}[H_n(t)]^2\cos(xt)dt = \sqrt{\pi}2^{n-1}n!e^{-\frac{1}{2}x^2}L_n\left(\frac{x^2}{2}\right)$$

22.14. Inequalities

22.14.1

$$|P_n^{(\alpha,\beta)}(x)| \lesssim \begin{cases} \binom{n+q}{n} \approx n^q,\ \text{if } q=\max(\alpha,\beta) \geq -1/2 \\ \qquad\qquad (\alpha > -1,\ \beta > -1) \\ |P_n^{(\alpha,\beta)}(x')| \approx \sqrt{\dfrac{1}{n}},\ \text{if } q < -\dfrac{1}{2} \end{cases}$$

x' maximum point nearest to $\dfrac{\beta-\alpha}{\alpha+\beta+1}$

22.14.2

$$|C_n^{(\alpha)}(x)| \leq \begin{cases} \binom{n+2\alpha-1}{n} & (\alpha > 0) \\ |C_n^{(\alpha)}(x')| & \left(-\dfrac{1}{2} < \alpha < 0\right) \end{cases}$$

$x'=0$ if $n=2m$; $x'=$ maximum point nearest zero if $n=2m+1$

22.14.3

$$|C_n^{(\alpha)}(\cos\theta)| < 2^{1-\alpha}\frac{n^{\alpha-1}}{(\sin\theta)^\alpha\Gamma(\alpha)}\ (0 < \alpha < 1, 0 < \theta < \pi)$$

22.14.4 $\quad |T_n(x)| \leq 1 \qquad (-1 \leq x \leq 1)$

22.14.5 $\quad \left|\dfrac{dT_n(x)}{dx}\right| \leq n^2 \qquad (-1 \leq x \leq 1)$

22.14.6 $\quad |U_n(x)| \leq n+1 \qquad (-1 \leq x \leq 1)$

22.14.7 $\quad |P_n(x)| \leq 1 \qquad (-1 \leq x \leq 1)$

22.14.8 $\quad \left|\dfrac{dP_n(x)}{dx}\right| \leq \dfrac{1}{2}n(n+1) \qquad (-1 \leq x \leq 1)$

22.14.9 $\quad |P_n(x)| \leq \sqrt{\dfrac{2}{\pi n}}\dfrac{1}{\sqrt[4]{1-x^2}} \quad (-1 < x \leq 1)$

22.14.10

$$P_n^2(x)-P_{n-1}(x)P_{n+1}(x) < \frac{2n+1}{3n(n+1)} \qquad (-1 \leq x \leq 1)$$

22.14.11

$$P_n^2(x)-P_{n-1}(x)P_{n+1}(x) \geq \frac{1-P_n^2(x)}{(2n-1)(n+1)}$$
$$(-1 \leq x \leq 1)$$

22.14.12 $\quad |L_n(x)| \leq e^{x/2} \qquad (x \geq 0)$

22.14.13 $\quad |L_n^{(\alpha)}(x)| \leq \dfrac{\Gamma(\alpha+n+1)}{n!\Gamma(\alpha+1)}e^{x/2} \quad (\alpha \geq 0, x \geq 0)$

22.14.14

$$|L_n^{(\alpha)}(x)| \leq \left[2-\frac{\Gamma(\alpha+n+1)}{n!\Gamma(\alpha+1)}\right]e^{x/2} \quad (-1 < \alpha < 0, x \geq 0)$$

22.14.15 $\quad |H_{2m}(x)| \leq e^{x^2/2} 2^{2m} m! \left[2 - \frac{1}{2^{2m}} \binom{2m}{m} \right]$

22.14.16 $\quad |H_{2m+1}(x)| \leq x e^{x^2/2} \frac{(2m+2)!}{(m+1)!} \quad (x \geq 0)$

22.14.17 $\quad |H_n(x)| < e^{x^2/2} k 2^{n/2} \sqrt{n!} \quad k \approx 1.086435$

22.15. Limit Relations

22.15.1

$$\lim_{n \to \infty} \left[\frac{1}{n^{\alpha}} P_n^{(\alpha, \beta)} \left(\cos \frac{x}{n} \right) \right]$$

$$= \lim_{n \to \infty} \frac{1}{n^{\alpha}} P_n^{(\alpha, \beta)} \left(1 - \frac{x^2}{2n^2} \right) = \left(\frac{2}{x} \right)^{\alpha} J_{\alpha}(x)$$

22.15.2 $\quad \lim_{n \to \infty} \left[\frac{1}{n^{\alpha}} L_n^{(\alpha)} \left(\frac{x}{n} \right) \right] = x^{-\alpha/2} J_{\alpha}(2\sqrt{x})$

22.15.3 $\quad \lim_{n \to \infty} \left[\frac{(-1)^n \sqrt{n}}{4^n n!} H_{2n} \left(\frac{x}{2\sqrt{n}} \right) \right] = \frac{1}{\sqrt{\pi}} \cos x$

22.15.4 $\quad \lim_{n \to \infty} \left[\frac{(-1)^n}{4^n n!} H_{2n+1} \left(\frac{x}{2\sqrt{n}} \right) \right] = \frac{2}{\sqrt{\pi}} \sin x$

22.15.5 $\quad \lim_{\beta \to \infty} P_n^{(\alpha, \beta)} \left(1 - \frac{2x}{\beta} \right) = L_n^{(\alpha)}(x)$

22.15.6 $\quad \lim_{\alpha \to \infty} \frac{1}{\alpha^{n/2}} C_n^{(\alpha)} \left(\frac{x}{\sqrt{\alpha}} \right) = \frac{1}{n!} H_n(x)$

For asymptotic expansions, see [22.5] and [22.17].

22.16. Zeros

For tables of the zeros and associated weight factors necessary for the Gaussian-type quadrature formulas see chapter **25**. All the zeros of the orthogonal polynomials are real, simple and located in the interior of the interval of orthogonality.

Explicit and Asymptotic Formulas and Inequalities

Notations:

$x_m^{(n)}$ mth zero of $f_n(x)$ $(x_1^{(n)} < x_2^{(n)} < \ldots < x_n^{(n)})$

$\theta_m^{(n)} = \arccos x_{n-m+1}^{(n)} (0 < \theta_1^{(n)} < \theta_2^{(n)} < \ldots < \theta_n^{(n)} < \pi)$

$j_{\alpha, m}$, mth positive zero of the Bessel function $J_{\alpha}(x)$

$0 < j_{\alpha, 1} < j_{\alpha, 2} < \ldots$

	$f_n(x)$	Relation
22.16.1	$P_n^{(\alpha, \beta)}(\cos \theta)$	$\lim_{n \to \infty} n\theta_m^{(n)} = j_{\alpha, m} \quad (\alpha > -1, \beta > -1)$
22.16.2	$C_n^{(\alpha)}(x)$	$x_m^{(n)} = 1 - \frac{j_{\alpha - \frac{1}{2}, m}^2}{2n^2} \left[1 - \frac{2\alpha}{n} + O\left(\frac{1}{n^2}\right) \right]$
22.16.3	$C_n^{(\alpha)}(\cos \theta)$	$\frac{(m+\alpha-1)\pi}{n+\alpha} \leq \theta_m^{(n)} \leq \frac{m\pi}{n+\alpha} \quad (0 \leq \alpha \leq 1)$
22.16.4	$T_n(x)$	$x_m^{(n)} = \cos \frac{2m-1}{2n} \pi$
22.16.5	$U_n(x)$	$x_m^{(n)} = \cos \frac{m}{n+1} \pi$
22.16.6	$P_n(\cos \theta)$	$\begin{cases} \frac{2m-1}{2n+1} \pi \leq \theta_m^{(n)} \leq \frac{2m}{2n+1} \pi \\ \theta_m^{(n)} = \frac{4m-1}{4n+2} \pi + \frac{1}{8n^2} \cot \frac{4m-1}{4n+2} \pi + O(n^{-3}) \end{cases}$
22.16.7	$P_n(x)$	$\begin{cases} x_m^{(n)} = 1 - \frac{j_{0, m}^2}{2n^2} [1 - \frac{1}{n} + O(n^{-2})] \\ x_m^{(n)} = 1 - \frac{4\xi_m^{(n)}}{2n+1+\xi_m^{(n)}}; \ \xi_m^{(n)} = \frac{j_{0, m}^2}{4n+2} \left[1 + \frac{j_{0, m-2}^2}{12(2n+1)^2} \right] + O\left(\frac{1}{n^5}\right) \end{cases}$
22.16.8	$L_n^{(\alpha)}(x)$	$\begin{cases} x_m^{(n)} > \frac{j_{\alpha, m}^2}{4k_n} \\ x_m^{(n)} < \frac{k_m}{k_n} \left(2k_m + \sqrt{4k_m^2 + \frac{1}{4} - \alpha^2} \right) \\ x_m^{(n)} = \frac{j_{\alpha, m}^2}{4k_n} \left(1 + \frac{2(\alpha^2-1) + j_{\alpha, m}^2}{48k_n^2} \right) + O(n^{-5}) \end{cases} \quad \left. \right\} k_r = r + \frac{\alpha+1}{2}$

For error estimates see [22.6].

22.17. Orthogonal Polynomials of a Discrete Variable

In this section some polynomials $f_n(x)$ are listed which are orthogonal with respect to the scalar product

22.17.1 $(f_n, f_m) = \sum_i w^*(x_i) f_n(x_i) f_m(x_i).$

The x_i are the integers in the interval $a \leq x_i \leq b$ and $w^*(x_i)$ is a positive function such that $\sum_i w^*(x_i)$ is finite. The constant factor which is still free in each polynomial when only the orthogonality condition is given is defined here by the explicit representation (which corresponds to the Rodrigues' formula)

22.17.2 $f_n(x) = \dfrac{1}{r_n w^*(x)} \Delta^n[w^*(x) g(x, n)]$

where $g(x, n) = g(x) g(x-1) \ldots g(x-n+1)$ and $g(x)$ is a polynomial in x independent of n.

Name	a	b	$w^*(x)$	r_n	$g(x, n)$	Remarks
Chebyshev	0	$N-1$	1	$1/n!$	$\binom{x}{n}\binom{x-N}{n}$	
Krawtchouk	0	N	$p^x q^{N-x} \binom{N}{x}$	$(-1)^n n!$	$\dfrac{q^n x!}{(x-n)!}$	$p, q > 0;$ $p+q=1$
Charlier	0	∞	$\dfrac{e^{-a} a^x}{x!}$	$(-1)^n \sqrt{a^n n!}$	$\dfrac{x!}{(x-n)!}$	$a > 0$
Meixner	0	∞	$\dfrac{c^x \Gamma(b+x)}{\Gamma(b) x!}$	c^n	$\dfrac{x!}{(x-n)!}$	$b > 0,\ 0 < c < 1$
Hahn	0	∞	$\dfrac{\Gamma(b)\Gamma(c+x)\Gamma(d+x)}{x!\Gamma(b+x)\Gamma(c)\Gamma(d)}$	$n!$	$\dfrac{x!\Gamma(b+x)}{(x-n)!\Gamma(b+x-n)}$	

For a more complete list of the properties of these polynomials see [22.5] and [22.17].

Numerical Methods

22.18. Use and Extension of the Tables

Evaluation of an orthogonal polynomial for which the coefficients are given numerically.

Example 1. Evaluate $L_6(1.5)$ and its first and second derivative using **Table 22.10** and the Horner scheme.

	1	-36	450	-2400	5400	-4320	720
$x = 1.5$		1.5	-51.75	597.375	-2703.9375	4044.09375	-413.859375
	1	-34.5	398.25	-1802.625	2696.0625	-275.90625	306.140625
1.5		1.5	-49.5	523.125	-1919.25	1165.21875	$L_6 = \dfrac{306.140625}{720}$ $= .42519\ 53$
	1	-33.0	348.75	-1279.500	776.8125	889.3125	
1.5		1.5	-47.25	452.250	-1240.875		$L_6' = \dfrac{889.3125}{720}$ $= 1.23515\ 625$
	1	-31.5	301.50	-827.250	-464.0625		$L_6'' = 2\dfrac{[-464.0625]}{720}$ $= -1.28906\ 25$

Evaluation of an orthogonal polynomial using the explicit representation when the coefficients are not given numerically.

If an isolated value of the orthogonal polynomial $f_n(x)$ is to be computed, use the proper explicit expression rewritten in the form

$$f_n(x) = d_n(x) a_0(x)$$

and generate $a_0(x)$ recursively, where

$$a_{m-1}(x) = 1 - \frac{b_m}{c_m} f(x) a_m(x) \qquad (m = n,\ n-1,\ \ldots,\ 2,\ 1,\ a_n(x) = 1).$$

The $d_n(x),\ b_m,\ c_m,\ f(x)$ for the polynomials of this chapter are listed in the following table:

$f_n(x)$	$d_n(x)$	b_m	c_m	$f(x)$
$P_n^{(\alpha,\beta)}$	$\binom{n+\alpha}{n}$	$(n-m+1)(\alpha+\beta+n+m)$	$2m(\alpha+m)$	$1-x$
$C_{2n}^{(\alpha)}$	$(-1)^n \dfrac{(\alpha)_n}{n!}$	$2(n-m+1)(\alpha+n+m-1)$	$m(2m-1)$	x^2
$C_{2n+1}^{(\alpha)}$	$(-1)^n \dfrac{(\alpha)_{n+1}}{n!} 2x$	$2(n-m+1)(\alpha+n+m)$	$m(2m+1)$	x^2
T_{2n}	$(-1)^n$	$2(n-m+1)(n+m-1)$	$m(2m-1)$	x^2
T_{2n+1}	$(-1)^n(2n+1)x$	$2(n-m+1)(n+m)$	$m(2m+1)$	x^2
U_{2n}	$(-1)^n$	$2(n-m+1)(n+m)$	$m(2m-1)$	x^2
U_{2n+1}	$(-1)^n 2(n+1)x$	$2(n-m+1)(n+m+1)$	$m(2m+1)$	x^2
P_{2n}	$\dfrac{(-1)^n}{4^n}\binom{2n}{n}$	$(n-m+1)(2n+2m-1)$	$m(2m-1)$	x^2
P_{2n+1}	$\dfrac{(-1)^n}{4^n}\binom{2n+1}{n}(n+1)x$	$(n-m+1)(2n+2m+1)$	$m(2m+1)$	x^2
$L_n^{(\alpha)}$	$\binom{n+\alpha}{n}$	$n-m+1$	$m(\alpha+m)$	x
H_{2n}	$(-1)^n \dfrac{(2n)!}{n!}$	$2(n-m+1)$	$m(2m-1)$	x^2
H_{2n+1}	$(-1)^n \dfrac{(2n+1)!}{n!} 2x$	$2(n-m+1)$	$m(2m+1)$	x^2

Example 2. Compute $P_8^{(1/2,\,3/2)}(2)$. Here $d_8 = \binom{8.5}{8} = 3.33847$, $f(2) = -1$.

m	8	7	6	5	4	3	2	1	0
a_m	1	1.132353	1.366667	1.841026	3.008392	6.849651	26.44156	223.1091	6545.533
b_m	18	34	48	60	70	78	84	88	90
c_m	136	105	78	55	36	21	10	3	0

$$P_8^{(1/2,\,3/2)}(2) = d_8 a_0(2) = (3.33847)(6545.533) = 21852.07$$

Evaluation of orthogonal polynomials by means of their recurrence relations

Example 3. Compute $C_n^{(\frac{1}{4})}(2.5)$ for $n = 2, 3, 4, 5, 6$.

From **Table 22.2** $C_0^{(\frac{1}{4})} = 1$, $C_1^{(\frac{1}{4})} = 1.25$ and from **22.7** the recurrence relation is

$$C_{n+1}^{(\frac{1}{4})}(2.5) = \left[5(n+\tfrac{1}{4}) C_n^{(\frac{1}{4})}(2.5) - (n-\tfrac{1}{2}) C_{n-1}^{(\frac{1}{4})}(2.5) \right] \frac{1}{n+1}.$$

n	2	3	4	5	6
$C_n^{(\frac{1}{4})}(2.5)$	3.65625	13.08594	50.87648	207.0649	867.7516

Check: Compute $C_6^{(\frac{1}{4})}(2.5)$ by the method of **Example 2.**

Change of Interval of Orthogonality

In some applications it is more convenient to use polynomials orthogonal on the interval $[0, 1]$. One can obtain the new polynomials from the ones given in this chapter by the substitution $x = 2\bar{x} - 1$. The coefficients of the new polynomial can be computed from the old by the following recursive scheme, provided the standardization is not changed. If

$$f_n(x) = \sum_{m=0}^{n} a_m x^m, \quad f_n^*(x) = f_n(2x-1) = \sum_{m=0}^{n} a_m^* x^m$$

then the a_m^* are given recursively by the a_m through the relations

$$a_m^{(j)} = 2a_m^{(j-1)} - a_{m+1}^{(j)}; \quad m = n-1, \ n-2, \ \ldots, \ j; \ j = 0, 1, 2, \ldots, n$$
$$a_m^{(-1)} = a_m/2, \ m = 0, 1, 2, \ldots, n$$
$$a_n^{(j)} = 2^j a_n, \ j = 0, 1, 2, \ldots, n \text{ and } a_m^{(m)} = a_m^*; \ m = 0, 1, 2, \ldots, n.$$

Example 4. Given $T_5(x) = 5x - 20x^3 + 16x^5$, find $T_5^*(x)$.

m / j	5	4	3	2	1	0
-1	$8 = a_5^{(-1)}$	0	$-10 = a_3^{(-1)}$	0	$2.5 = a_1^{(-1)}$	0
0	16	-16	-4	4	1	$-1 = a_0^*$
1	32	-64	56	-48	$50 = a_1^*$	
2	64	-192	304	$-400 = a_2^*$		
3	128	-512	$1120 = a_3^*$			
4	256	$-1280 = a_4^*$				
5	$512 = a_5^*$					

Hence, $T_5^*(x) = 512x^5 - 1280x^4 + 1120x^3 - 400x^2 + 50x - 1$.

22.19. Least Square Approximations

Problem: Given a function $f(x)$ (analytically or in form of a table) in a domain D (which may be a continuous interval or a set of discrete points).[2] Approximate $f(x)$ by a polynomial $F_n(x)$ of given degree n such that a weighted sum of the squares of the errors in D is least.

Solution: Let $w(x) \geq 0$ be the weight function chosen according to the relative importance of the errors in different parts of D. Let $f_m(x)$ be orthogonal polynomials in D relative to $w(x)$, i.e. $(f_m, f_n) = 0$ for $m \neq n$, where

$$(f, g) = \begin{cases} \displaystyle\int_D w(x) f(x) g(x) dx \\ \qquad\qquad \text{if } D \text{ is a continuous interval} \\ \displaystyle\sum_{m=1}^{N} w(x_m) f(x_m) g(x_m) \\ \qquad\qquad \text{if } D \text{ is a set of } N \text{ discrete points } x_m. \end{cases}$$

Then

$$F_n(x) = \sum_{m=0}^{n} a_m f_m(x)$$

where

*
$$a_m = (f, f_m)/(f_m, f_m).$$

[2] $f(x)$ has to be square integrable, see e.g. [22.17].

*See page II.

D a Continuous Interval

Example 5. Find a least square polynomial of degree 5 for $f(x) = \dfrac{1}{1+x}$, in the interval $2 \leq x \leq 5$, using the weight function

$$w(x) = \frac{1}{\sqrt{(x-2)(5-x)}}$$

which stresses the importance of the errors at the ends of the interval.

Reduction to interval $[-1, 1]$, $\quad t = \dfrac{2x-7}{3}$

$$w(x(t)) = \frac{2}{3} \frac{1}{\sqrt{1-t^2}}$$

From **22.2**, $f_m(t) = T_m(t)$ and

$$a_m = \frac{4}{3\pi} \int_{-1}^{1} \frac{1}{\sqrt{1-t^2}} \frac{1}{t+3} T_m(t) dt \qquad (m \neq 0)$$

$$a_0 = \frac{2}{3\pi} \int_{-1}^{1} \frac{1}{\sqrt{1-t^2}} \frac{dt}{t+3}$$

Evaluating the integrals numerically we get

$$\frac{1}{1+x} \sim .235703 - .080880 T_1 \left(\frac{2x-7}{3}\right) + .013876 T_2 \left(\frac{2x-7}{3}\right) - .002380 T_3 \left(\frac{2x-7}{3}\right)$$

$$+ .000408 T_4 \left(\frac{2x-7}{3}\right) - .000070 T_5 \left(\frac{2x-7}{3}\right)$$

D a Set of Discrete Points

If $x_m = m (m = 0, 1, 2, \ldots, N)$ and $w(x) = 1$, use the Chebyshev polynomials in the discrete range 22.17. It is convenient to introduce here a slightly different standardization such that

$$f_n(x) = \sum_{m=0}^{n} (-1)^m \binom{n}{m} \binom{n+m}{m} \frac{x!(N-m)!}{(x-m)!N!}$$

$$(f_n, f_n) = \frac{(N+n+1)!(N-n)!}{(2n+1)(N!)^2}$$

Recurrence relation: $f_0(x) = 1, f_1(x) = 1 - \dfrac{2x}{N}$

$$(n+1)(N-n)f_{n+1}(x) = (2n+1)(N-2x)f_n(x) - n(N+n+1)f_{n-1}(x)$$

Example 6. Approximate in the least square sense the function $f(x)$ given in the following table by a third degree polynomial.

x	$f(x)$	$\bar{x}=\dfrac{x-10}{2}$	$f_0(\bar{x})$	$f_1(\bar{x})$	$f_2(\bar{x})$	$f_3(\bar{x})$
10	.3162	0	1	1	1	1
12	.2887	1	1	1/2	$-1/2$	-2
14	.2673	2	1	0	-1	0
16	.2500	3	1	$-1/2$	$-1/2$	2
18	.2357	4	1	-1	1	-1

	$f_0(\bar{x})$	$f_1(\bar{x})$	$f_2(\bar{x})$	$f_3(\bar{x})$
$(f_n, f_n) = \sum_{\bar{x}=0}^{4} f_n^2(\bar{x})$	5	2.5	3.5	10
$(f, f_n) = \sum_{\bar{x}=0}^{4} f_n(\bar{x}) f(2\bar{x}+10)$	1.3579	.09985	.01525	.0031
$a_n = \dfrac{(f, f_n)}{(f_n, f_n)}$	.271580	.039940	.0043571	.000310

$$f(x) \sim .27158 + .03994(3.5 - .25x) + .0043571(23.5 - 3.5x + .125x^2) + .00031(266 - 59.8333x$$

$$+ 4.375x^2 - .10417x^3)$$

$$f(x) \sim .59447 - .043658x + .0019009x^2 - .000032292x^3$$

22.20. Economization of Series

Problem: Given $f(x) = \sum_{m=0}^{n} a_m x^m$ in the interval $-1 \le x \le 1$ and $R > 0$. Find $\bar{f}(x) = \sum_{m=0}^{N} b_m x^m$ with N as small as possible, such that $|\bar{f}(x) - f(x)| < R$.

Solution: Express $f(x)$ in terms of Chebyshev polynomials using **Table 22.3,**

$$f(x) = \sum_{m=0}^{n} b_m T_m(x)$$

Then, since $|T_m(x)| \le 1 (-1 \le x \le 1)$

$$\bar{f}(x) = \sum_{m=0}^{N} b_m T_m(x)$$

within the desired accuracy if

$$\sum_{m=N+1}^{n} |b_m| < R$$

$\bar{f}(x)$ is evaluated most conveniently by using the recurrence relation (see 22.7).

Example 7. Economize $f(x) = 1 + x/2 + x^2/3 + x^3/4 + x^4/5 + x^5/6$ with $R = .05$.

From **Table 22.3**

$$f(x) = \frac{1}{120}[149T_0(x) + 32T_2(x) + 3T_4(x)] + \frac{1}{96}[76T_1(x) + 11T_3(x) + T_5(x)]$$

so

$$\bar{f}(x) = \frac{1}{120}[149T_0(x) + 32T_2(x)] + \frac{1}{96}[76T_1(x) + 11T_3(x)]$$

since

$$|\bar{f}(x) - f(x)| \leq \frac{1}{40} + \frac{1}{96} < .05$$

References

Texts

[22.1] Bibliography on orthogonal polynomials, Bull. of the National Research Council No. 103, Washington, D.C. (1940).

[22.2] P. L. Chebyshev, Sur l'interpolation. Oeuvres, vol. 2, pp. 59–68.

[22.3] R. Courant and D. Hilbert, Methods of mathematical physics, vol. 1, ch. 7 (Interscience Publishers, New York, N.Y., 1953).

[22.4] G. Doetsch, Die in der Statistik seltener Ereignisse auftretenden Charlierschen Polynome und eine damit zusammenhängende Differential-differenzengleichung, Math. Ann. **109**, 257–266 (1934).

[22.5] A. Erdélyi et al., Higher transcendental functions, vol. 2, ch. 10 (McGraw-Hill Book Co., Inc., New York, N.Y., 1953).

[22.6] L. Gatteschi, Limitazione degli errori nelle formule asintotiche per le funzioni speciali, Rend. Sem. Mat. Univ. Torina **16**, 83–94 (1956–57).

[22.7] T. L. Geronimus, Teoria ortogonalnikh mnogochlenov (Moscow, U.S.S.R., 1950).

[22.8] W. Hahn, Über Orthogonalpolynome, die q-Differenzengleichungen genügen, Math. Nachr. **2**, 4–34 (1949).

[22.9] St. Kaczmarz and H. Steinhaus, Theorie der Orthogonalreihen, ch. 4 (Chelsea Publishing Co., New York, N.Y., 1951).

[22.10] M. Krawtchouk, Sur une généralisation des polynomes d'Hermite, C.R. Acad. Sci. Paris **187**, 620–622 (1929).

[22.11] C. Lanczos, Trigonometric interpolation of empirical and analytical functions, J. Math. Phys. **17**, 123–199 (1938).

[22.12] C. Lanczos, Applied analysis (Prentice-Hall, Inc., Englewood Cliffs, N.J., 1956).

[22.13] W. Magnus and F. Oberhettinger, Formeln und Sätze für die speziellen Funktionen der mathematischen Physik, ch. 5, 2d ed. (Springer-Verlag, Berlin, Germany, 1948).

[22.14] J. Meixner, Orthogonale Polynomsysteme mit einer besonderen Gestalt der erzeugenden Funktion, J. London Math. Soc. **9**, 6–13 (1934).

[22.15] G. Sansone, Orthogonal functions, Pure and Applied Mathematics, vol. IX (Interscience Publishers, New York, N.Y., 1959).

[22.16] J. Shohat, Théorie générale des polynomes orthogonaux de Tchebichef, Mém. Soc. Math. **66** (Gauthier-Villars, Paris, France, 1934).

[22.17] G. Szegö, Orthogonal polynomials, Amer. Math. Soc. Colloquium Publications 23, rev. ed. (1959).

[22.18] F. G. Tricomi, Vorlesungen über Orthogonalreihen, chs. 4, 5, 6 (Springer-Verlag, Berlin, Germany, 1955).

Tables

[22.19] British Association for the Advancement of Science, Legendre Polynomials, Mathematical Tables, Part vol. A (Cambridge Univ. Press, Cambridge, England, 1946). $P_n(x)$, $x = 0(.01)6$, $n = 1(1)12$, 7–8D.

[22.20] N. R. Jörgensen, Undersögelser over frekvensflader og korrelation (Busck, Copenhagen, Denmark, 1916). $He_n(x)$, $x = 0(.01)4$, $n = 1(1)6$, exact.

[22.21] L. N. Karmazina, Tablitsy polinomov Jacobi (Izdat. Akad. Nauk SSSR., Moscow, U.S.S.R., 1954). $G_n(p, q, x)$, $x = 0(.01)1$, $q = .1(.1)1$, $p = 1.1(.1)3$, $n = 1(1)5$, 7D.

[22.22] National Bureau of Standards, Tables of Chebyshev polynomials $S_n(x)$ and $C_n(x)$, Applied Math. Series 9 (U.S. Government Printing Office, Washington, D.C., 1952). $x = 0(.001)2$, $n = 2(1)12$, 12D; Coefficients for $T_n(x)$, $U_n(x)$, $C_n(x)$, $S_n(x)$ for $n = 0(1)12$.

[22.23] J. B. Russel, A table of Hermite functions, J. Math. Phys. **12**, 291–297 (1933). $e^{-x^2/2}H_n(x)$, $x = 0(.04)1(.1)4(.2)7(.5)8$, $n = 0(1)11$, 5D.

[22.24] N. Wiener, Extrapolation, interpolation and smoothing of stationary time series (John Wiley & Sons, Inc., New York, N.Y., 1949). $L_n(x)$, $n = 0(1)5$, $x = 0(.01).1(.1)18(.2)20(.5)21(1)26(2)30$, 3–5D.

Coefficients for the Jacobi Polynomials $P_n^{(\alpha,\beta)}(x)=a_n^{-1}\sum_{m=0}^{n}c_m(x-1)^m$ **Table 22.1**

	a_n	$(x-1)^0$	$(x-1)^1$	$(x-1)^2$	$(x-1)^3$	$(x-1)^4$	$(x-1)^5$	$(x-1)^6$
$P_0^{(\alpha,\beta)}$	1	1						
$P_1^{(\alpha,\beta)}$	2	$2(\alpha+1)$	$\alpha+\beta+2$					
$P_2^{(\alpha,\beta)}$	8	$4(\alpha+1)_2$	$4(\alpha+\beta+3)(\alpha+2)$	$(\alpha+\beta+3)_2$				
$P_3^{(\alpha,\beta)}$	48	$8(\alpha+1)_3$	$12(\alpha+\beta+4)(\alpha+2)_2$	$6(\alpha+\beta+4)_2(\alpha+3)$	$(\alpha+\beta+4)_3$			
$P_4^{(\alpha,\beta)}$	384	$16(\alpha+1)_4$	$32(\alpha+\beta+5)(\alpha+2)_3$	$24(\alpha+\beta+5)_2(\alpha+3)_2$	$8(\alpha+\beta+5)_3(\alpha+4)$	$(\alpha+\beta+5)_4$		
$P_5^{(\alpha,\beta)}$	3840	$32(\alpha+1)_5$	$80(\alpha+\beta+6)(\alpha+2)_4$	$80(\alpha+\beta+6)_2(\alpha+3)_3$	$40(\alpha+\beta+6)_3(\alpha+4)_2$	$10(\alpha+\beta+6)_4(\alpha+5)$	$(\alpha+\beta+6)_5$	
$P_6^{(\alpha,\beta)}$	46080	$64(\alpha+1)_6$	$192(\alpha+\beta+7)(\alpha+2)_5$	$240(\alpha+\beta+7)_2(\alpha+3)_4$	$160(\alpha+\beta+7)_3(\alpha+4)_3$	$60(\alpha+\beta+7)_4(\alpha+5)_2$	$12(\alpha+\beta+7)_5(\alpha+6)$	$(\alpha+\beta+7)_6$

$$(m)_n=m(m+1)(m+2)\ \ldots\ (m+n-1)$$

$$P_5^{(1,1)}(x)=\frac{1}{3840}\left[(8)_5(x-1)^5+10(8)_4(6)(x-1)^4+40(8)_3(5)_2(x-1)^3+80(8)_2(4)_3(x-1)^2+80(8)(3)_4(x-1)+32(2)_5\right]$$

$$P_5^{(1,1)}(x)=\frac{1}{3840}\left[95040(x-1)^5+475200(x-1)^4+864000(x-1)^3+691200(x-1)^2+230400(x-1)+23040\right]$$

Table 22.2   **Coefficients for the Ultraspherical Polynomials $C_n^{(\alpha)}(x)$ and for x^n in terms of $C_m^{(\alpha)}(x)$**

$$C_n^{(\alpha)}(x)=a_n^{-1}\sum_{m=0}^{n}c_m x^m \quad\text{and}\quad x^n=b_n^{-1}\sum_{m=0}^{n}d_m C_m^{(\alpha)}(x) \qquad (\alpha\neq 0)$$

		x^0	x^1	x^2	x^3	x^4	x^5	x^6	
	b_n	1	2α	$2(\alpha)_2$	$4(\alpha)_3$	$4(\alpha)_4$	$8(\alpha)_5$	$8(\alpha)_6$	
$C_0^{(\alpha)}$	a_n 1	1 1		α		$3\alpha(\alpha+3)$		$15\alpha(\alpha+4)(\alpha+5)$	$C_0^{(\alpha)}$
$C_1^{(\alpha)}$	1		2α 1		$3(\alpha+1)$		$15(\alpha+1)(\alpha+4)$		$C_1^{(\alpha)}$
$C_2^{(\alpha)}$	1	$-\alpha$		$2(\alpha)_2$ 1		$6(\alpha+2)$		$45(\alpha+2)(\alpha+5)$	$C_2^{(\alpha)}$
$C_3^{(\alpha)}$	3		$-6(\alpha)_2$		$4(\alpha)_3$ 3		$30(\alpha+3)$		$C_3^{(\alpha)}$
$C_4^{(\alpha)}$	6	$3(\alpha)_2$		$-12(\alpha)_3$		$4(\alpha)_4$ 6		$90(\alpha+4)$	$C_4^{(\alpha)}$
$C_5^{(\alpha)}$	15		$15(\alpha)_3$		$-20(\alpha)_4$		$4(\alpha)_5$ 30		$C_5^{(\alpha)}$
$C_6^{(\alpha)}$	90	$-15(\alpha)_3$		$90(\alpha)_4$		$-60(\alpha)_5$		$8(\alpha)_6$ 90	$C_6^{(\alpha)}$
		x^0	x^1	x^2	x^3	x^4	x^5	x^6	

$$(\alpha)_n=\alpha(\alpha+1)(\alpha+2)\;.\,.\,.\;(\alpha+n-1)$$

$$C_3^{(2)}(x)=\frac{1}{3}\left[4(2)_3 x^3-6(2)_2 x\right] \qquad x^3=\frac{1}{4(2)_3}\left[3(3)\,C_1^{(2)}(x)+3C_3^{(2)}(x)\right]$$

$$C_3^{(2)}(x)=\frac{1}{3}\left[96x^3-36x\right] \qquad x^3=\frac{1}{96}\left[9C_1^{(2)}(x)+3C_3^{(2)}(x)\right]$$

Table 22.3

Coefficients for the Chebyshev Polynomials $T_n(x)$ and for x^n in terms of $T_m(x)$

$$T_n(x) = \sum_{m=0}^{n} c_m x^m \qquad x^n = b_n^{-1} \sum_{m=0}^{n} d_m T_m(x)$$

	x^0	x^1	x^2	x^3	x^4	x^5	x^6	x^7	x^8	x^9	x^{10}	x^{11}	x^{12}	
b_n	1	1	2	4	8	16	32	64	128	256	512	1024	2048	
T_0	1 1		1		3		10		35		126		462	T_0
T_1		1 1		3		10		35		126		462		T_1
T_2	-1		2 1		4		15		56		210		792	T_2
T_3		-3		4 1		5		21		84		330		T_3
T_4	1		-8		8 1		6		28		120		495	T_4
T_5		5		-20		16 1		7		36		165		T_5
T_6	-1		18		-48		32 1		8		45		220	T_6
T_7		-7		56		-112		64 1		9		55		T_7
T_8	1		-32		160		-256		128 1		10		66	T_8
T_9		9		-120		432		-576		256 1		11		T_9
T_{10}	-1		50		-400		1120		-1280		512 1		12	T_{10}
T_{11}		-11		220		-1232		2816		-2816		1024 1		T_{11}
T_{12}	1		-72		840		-3584		6912		-6144		2048 1	T_{12}
	x^0	x^1	x^2	x^3	x^4	x^5	x^6	x^7	x^8	x^9	x^{10}	x^{11}	x^{12}	

$$T_6(x) = 32x^6 - 48x^4 + 18x^2 - 1 \qquad x^6 = \frac{1}{32}[10T_0 + 15T_2 + 6T_4 + T_6]$$

Chebyshev Polynomials $T_n(x)$
Table 22.4

$n\backslash x$	0.2	0.4	0.6	0.8	1.0
0	$+1.00000\ 00000$	$+1.00000\ 00000$	$+1.00000\ 00000$	$+1.00000\ 00000$	1
1	$+0.20000\ 00000$	$+0.40000\ 00000$	$+0.60000\ 00000$	$+0.80000\ 00000$	1
2	$-0.92000\ 00000$	$-0.68000\ 00000$	$-0.28000\ 00000$	$+0.28000\ 00000$	1
3	$-0.56800\ 00000$	$-0.94400\ 00000$	$-0.93600\ 00000$	$-0.35200\ 00000$	1
4	$+0.69280\ 00000$	$-0.07520\ 00000$	$-0.84320\ 00000$	$-0.84320\ 00000$	1
5	$+0.84512\ 00000$	$+0.88384\ 00000$	$-0.07584\ 00000$	$-0.99712\ 00000$	1
6	$-0.35475\ 20000$	$+0.78227\ 20000$	$+0.75219\ 20000$	$-0.75219\ 20000$	1
7	$-0.98702\ 08000$	$-0.25802\ 24000$	$+0.97847\ 04000$	$-0.20638\ 72000$	1
8	$-0.04005\ 63200$	$-0.98868\ 99200$	$+0.42197\ 24800$	$+0.42197\ 24800$	1
9	$+0.97099\ 82720$	$-0.53292\ 95360$	$-0.47210\ 34240$	$-0.88154\ 31680$	1
10	$+0.42845\ 56288$	$+0.56234\ 62912$	$-0.98849\ 65888$	$+0.98849\ 65888$	1
11	$-0.79961\ 60205$	$+0.98280\ 65690$	$-0.71409\ 24826$	$+0.70005\ 13741$	1
12	$-0.74830\ 20370$	$+0.22389\ 89640$	$+0.13158\ 56097$	$+0.13158\ 56097$	1

Table 22.5

Coefficients for the Chebyshev Polynomials $U_n(x)$ and for x^n in terms of $U_m(x)$

$$U_n(x) = \sum_{m=0}^{n} c_m x^m \qquad x^n = b_n^{-1} \sum_{m=0}^{n} d_m U_m(x)$$

	x^0	x^1	x^2	x^3	x^4	x^5	x^6	x^7	x^8	x^9	x^{10}	x^{11}	x^{12}	
b_n	1	2	4	8	16	32	64	128	256	512	1024	2048	4096	
U_0	1 1		1		2		5		14		42		132	U_0
U_1		2 1		2		5		14		42		132		U_1
U_2	−1		4 1		3		9		28		90		297	U_2
U_3		−4		8 1		4		14		48		165		U_3
U_4	1		−12		16 1		5		20		75		275	U_4
U_5		6		−32		32 1		6		27		110		U_5
U_6	−1		24		−80		64 1		7		35		154	U_6
U_7		−8		80		−192		128 1		8		44		U_7
U_8	1		−40		240		−448		256 1		9		54	U_8
U_9		10		−160		672		−1024		512 1		10		U_9
U_{10}	−1		60		−560		1792		−2304		1024 1		11	U_{10}
U_{11}		−12		280		−1792		4608		−5120		2048 1		U_{11}
U_{12}	1		−84		1120		−5376		11520		−11264		4096 1	U_{12}
	x^0	x^1	x^2	x^3	x^4	x^5	x^6	x^7	x^8	x^9	x^{10}	x^{11}	x^{12}	

$$U_6(x) = 64x^6 - 80x^4 + 24x^2 - 1 \qquad x^6 = \frac{1}{64}[5U_0 + 9U_2 + 5U_4 + U_6]$$

Table 22.6   ## Chebyshev Polynomials $U_n(x)$

$n \setminus x$	0.2	0.4	0.6	0.8	1.0
0	+1.00000 00000	+1.00000 00000	+1.00000 00000	+1.00000 00000	1
1	+0.40000 00000	+0.80000 00000	+1.20000 00000	+1.60000 00000	2
2	−0.84000 00000	−0.36000 00000	+0.44000 00000	+1.56000 00000	3
3	−0.73600 00000	−1.08800 00000	−0.67200 00000	+0.89600 00000	4
4	+0.54560 00000	−0.51040 00000	−1.24640 00000	−0.12640 00000	5
5	+0.95424 00000	+0.67968 00000	−0.82368 00000	−1.09824 00000	6
6	−0.16390 40000	+1.05414 40000	+0.25798 40000	−1.63078 40000	7
7	−1.01980 16000	+0.16363 52000	+1.13326 08000	−1.51101 44000	8
8	−0.24401 66400	−0.92323 58400	+1.10192 89600	−0.78683 90400	9
9	+0.92219 49440	−0.90222 38720	+0.18905 39520	+0.25207 19360	10
10	+0.61289 46176	+0.20145 67424	−0.87506 42176	+1.19015 41376	11
11	−0.67703 70970	+1.06338 92659	−1.23913 10131	+1.65217 46842	12
12	−0.88370 94564	+0.64925 46703	−0.61189 29981	+1.45332 53571	13

Table 22.7

Coefficients for the Chebyshev Polynomials $C_n(x)$ and for x^n in terms of $C_m(x)$

$$C_n(x) = \sum_{m=0}^{n} c_m x^m \qquad x^n = b_n^{-1}\sum_{m=0}^{n} d_m C_m(x)$$

	x^0	x^1	x^2	x^3	x^4	x^5	x^6	x^7	x^8	x^9	x^{10}	x^{11}	x^{12}	
b_n	2	1	1	1	1	1	1	1	1	1	1	1	1	
C_0	2 1		1		3		10		35		126		462	C_0
C_1		1 1		3		10		35		126		462		C_1
C_2	-2		1 1		4		15		56		210		792	C_2
C_3		-3		1 1		5		21		84		330		C_3
C_4	2		-4		1 1		6		28		120		495	C_4
C_5*		5		-5		1 1		7		36		165		C_5
C_6	-2		9		-6		1 1		8		45		220	C_6
C_7		-7		14		-7		1 1		9		55		C_7
C_8	2		-16		20		-8		1 1		10		66	C_8
C_9		9		-30		27		-9		1 1		11		C_9
C_{10}	-2		25		-50		35		-10		1 1		12	C_{10}
C_{11}		-11		55		-77		44		-11		1 1		C_{11}
C_{12}	2		-36		105		-112		54		-12		1 1	C_{12}
	x^0	x^1	x^2	x^3	x^4	x^5	x^6	x^7	x^8	x^9	x^{10}	x^{11}	x^{12}	

*See page II.

$$C_6(x) = x^6 - 6x^4 + 9x^2 - 2 \qquad x^6 = 10C_0 + 15C_2 + 6C_4 + C_6$$

Table 22.8

Coefficients for the Chebyshev Polynomials $S_n(x)$ and for x^n in terms of $S_m(x)$

$$S_n(x) = \sum_{m=0}^{n} c_m x^m \qquad x^n = \sum_{m=0}^{n} d_m S_m(x)$$

	x^0	x^1	x^2	x^3	x^4	x^5	x^6	x^7	x^8	x^9	x^{10}	x^{11}	x^{12}	
S_0	1 1		1		2		5		14		42		132	S_0
S_1		1 1		2		5		14		42		132		S_1
S_2	-1		1 1		3		9		28		90		297	S_2
S_3		-2		1 1		4		14		48		165		S_3
S_4	1		-3		1 1		5		20		75		275	S_4
S_5		3		-4		1 1		6		27		110		S_5
S_6	-1		6		-5		1 1		7		35		154	S_6
S_7		-4		10		-6		1 1		8		44		S_7
S_8	1		-10		15		-7		1 1		9		54	S_8
S_9		5		-20		21		-8		1 1		10		S_9
S_{10}	-1		15		-35		28		-9		1 1		11	S_{10}
S_{11}		-6		35		-56		36		-10		1 1		S_{11}
S_{12}	1		-21		70		-84		45		-11		1 1	S_{12}
	x^0	x^1	x^2	x^3	x^4	x^5	x^6	x^7	x^8	x^9	x^{10}	x^{11}	x^{12}	

$$S_6(x) = x^6 - 5x^4 + 6x^2 - 1 \qquad x^6 = 5S_0 + 9S_2 + 5S_4 + S_6$$

*See page II.

Table 22.9 **Coefficients for the Legendre Polynomials $P_n(x)$ and for x^n in terms of $P_m(x)$**

$$P_n(x)=a_n^{-1}\sum_{m=0}^{n}c_m x^m \qquad x^n=b_n^{-1}\sum_{m=0}^{n}d_m P_m(x)$$

	a_n b_n	x^0	x^1	x^2	x^3	x^4	x^5	x^6	x^7	x^8	x^9	x^{10}	x^{11}	x^{12}	
		1	1	3	5	35	63	231	429	6435	12155	46189	88179	676039	b_n
P_0	1	1 1		1		7		33		715		4199		52003	P_0
P_1	1		1 1		3		27		143		3315		20349		P_1
P_2	2	−1		3 2		20		110		2600		16150		208012	P_2
P_3	2		−3		5 2		28		182		4760		31654		P_3
P_4	8	3		−30		35 8		72		2160		15504		220248	P_4
P_5	8		15		−70		63 8		88		2992		23408		P_5
P_6	16	−5		105		−315		231 16		832		7904		133952	P_6
P_7	16		−35		315		−693		429 16		960		10080		P_7
P_8	128	35		−1260		6930		−12012		6435 128		2176		50048	P_8
P_9	128		315		−4620		18018		−25740		12155 128		2432		P_9
P_{10}	256	−63		3465		−30030		90090		−109395		46189 256		10752	P_{10}
P_{11}	256		−693		15015		−90090		218790		−230945		88179 256		P_{11}
P_{12}	1024	231		−18018		225225		−1021020		2078505		−1939938		676039 1024	P_{12}
		x^0	x^1	x^2	x^3	x^4	x^5	x^6	x^7	x^8	x^9	x^{10}	x^{11}	x^{12}	

$$P_6(x)=\frac{1}{16}[231x^6-315x^4+105x^2-5] \qquad x^6=\frac{1}{231}[33P_0+110P_2+72P_4+16P_6]$$

For values of $P_n(x)$, see chapter 8.

Coefficients for the Laguerre Polynomials $L_n(x)$ and for x^n in terms of $L_m(x)$

Table 22.10

$$L_n(x) = a_n^{-1} \sum_{m=0}^{n} c_m x^m \qquad x^n = \sum_{m=0}^{n} d_m L_m(x)$$

	a_n	x^0	x^1		x^2		x^3		x^4		x^5		x^6		x^7		x^8		x^9		x^{10}	x^{11}	x^{12}	
L_0	1	1		1		1		2		6		24		120		720		5040		40320	362880	3628800	39916800	479001600	L_0
L_1	1	1	−1	−1		−4		−18		−96		−600		−4320		−35280		−322560		−3265920	−36288000	−439084800	−5748019200	L_1
L_2	2	2		−4	1	2		18		144		1200		10800		105840		1128960		13063680	163296000	2195424000	31614105600	L_2
L_3	6	6		−18		9	−1	−6		−96		−1200		−14400		−176400		−2257920		−30481920	−435456000	−6586272000	−105380352000	L_3
L_4	24	24		−96		72		−16	1	24		600		10800		176400		2822400		45722880	762048000	13172544000	237105792000	L_4
L_5	120	120		−600		600		−200		25	−1	−120		−4320		−105840		−2257920		−45722880	−914457600	−18441561600	−379369267200	L_5
L_6	720	720		−4320		5400		−2400		450		−36	1	720		35280		1128960		30481920	762048000	18441561600	442597478400	L_6
L_7	5040	5040		−35280		52920		−29400		7350		−882		49	−1	−5040		−322560		−13063680	−435456000	−13172544000	−379369267200	L_7
L_8	40320	40320		−322560		564480		−376320		117600		−18816		1568		−64	1	40320		3269520	163296000	6586272000	237105792000	L_8
L_9	362880	362880		−3265920		6531840		−5080320		1905120		−381024		42336		−2592		81	−1	−362880	−36288000	−2195424000	−105380352000	L_9
L_{10}	3628800	3628800		−36288000		81648000		−72576000		31752000		−7620480		1058400		−86400		4050		−100	1	3628800	439084800	31614105600	L_{10}
L_{11}	39916800	39916800		−439084800		1097712000		−1097712000		548856000		−153679680		25613280		−2613600		163350		−6050	121	−1	−39916800	−5748019200	L_{11}
L_{12}	479001600	479001600		−5748019200		15807052800		−17563392000		9879408000		−3161410560		614718720		−75271680		5880600		−290400	8712	−144	1	479001600	L_{12}
	a_n	x^0	x^1		x^2		x^3		x^4		x^5		x^6		x^7		x^8		x^9		x^{10}	x^{11}	x^{12}	

$$L_6(x) = \frac{1}{720}\,[x^6 - 36x^5 + 450x^4 - 2400x^3 + 5400x^2 - 4320x + 720] \qquad x^6 = 720L_0 - 4320L_1 + 10800L_2 - 14400L_3 + 10800L_4 - 4320L_5 + 720L_6$$

Table 22.11 **Laguerre Polynomials $L_n(x)$**

$n\backslash x$	0.5	1.0	3.0	5.0	10.0
0	+1.00000 00000	+1.00000. 00000	+1.00000 00000	+1.00000 00000	+1.00000 00000
1	+0.50000 00000	0.00000 00000	−2.00000 00000	−4.00000 00000	−9.00000 00000
2	+0.12500 00000	−0.50000 00000	−0.50000 00000	+3.50000 00000	+31.00000 00000
3	−0.14583 33333	−0.66666 66667	+1.00000 00000	+2.66666 66667	−45.66666 66667
4	−0.33072 91667	−0.62500 00000	+1.37500 00000	−1.29166 66667	+11.00000 00000
5	−0.44557 29167	−0.46666 66667	+0.85000 00000	−3.16666 66667	+34.33333 33333
6	−0.50414 49653	−0.25694 44444	−0.01250 00000	−2.09027 77778	−3.44444 44444
7	−0.51833 92237	−0.04047 61905	−0.74642 85714	+0.32539 68254	−30.90476 19048
8	−0.49836 29984	+0.15399 30556	−1.10870 53571	+2.23573 90873	−16.30158 73016
9	−0.45291 95204	+0.30974 42681	−1.06116 07143	+2.69174 38272	+14.79188 71252
10	−0.38937 44141	+0.41894 59325	−0.70002 23214	+1.75627 61795	+27.98412 69841
11	−0.31390 72988	+0.48013 41791	−0.18079 95130	+0.10754 36909	+14.53695 68703
12	−0.23164 96389	+0.49621 22235	+0.34035 46063	−1.44860 42948	−9.90374 64593

Coefficients for the Hermite Polynomials $H_n(x)$ and for x^n in terms of $H_m(x)$

Table 22.12

$$H_n(x) = \sum_{m=0}^{n} c_m x^m \qquad x^n = b_n^{-1} \sum_{m=0}^{n} d_m H_m(x)$$

	x^0	x^1	x^2	x^3	x^4	x^5	x^6	x^7	x^8	x^9	x^{10}	x^{11}	x^{12}	
b_n	1	2	4	8	16	32	64	128	256	512	1024	2048	4096	b_n
H_0	1 1		2		12		120		1680		* 30240		665280	H_0
H_1		2 1		6		60		840		15120		332640		H_1
H_2	−2		4 1		12		180		3360		75600		1995840	H_2
H_3		−12		8 1		20		420		10080		277200		H_3
H_4	12		−48		16 1		30		840		25200		831600	H_4
H_5		120		−160		32 1		42		1512		55440		H_5
H_6	−120		720		−480		64 1		56		2520		110880	H_6
H_7		−1680		3360		−1344		128 1		72		3960		H_7
H_8	1680		−13440		13440		−3584		256 1		90		5940	H_8
H_9		30240		−80640		48384		−9216		512 1		110		H_9
H_{10}	−30240		302400		−403200		161280		−23040		1024 1		132	H_{10}
H_{11}		−665280		2217600		−1774080		506880		−56320		2048 1		H_{11}
H_{12}	665280		−7983360		13305600		−7096320		1520640		−135168		4096 1	H_{12}
	x^0	x^1	x^2	x^3	x^4	x^5	x^6	x^7	x^8	x^9	x^{10}	x^{11}	x^{12}	

$$H_6(x) = 64x^6 - 480x^4 + 720x^2 - 120 \qquad x^6 = \frac{1}{64}[120H_0 + 180H_2 + 30H_4 + H_6]$$

*See page II.

Table 22.13   **Hermite Polynomials $H_n(x)$**

$n\backslash x$	0.5	1.0	3.0	5.0	10.0
0	$+1.00000$	$+1.00000$	$+1.00000\ 00$	$1.00000\ 00000$	$1.00000\ 00000$
1	$+1.00000$	$+2.00000$	$+6.00000\ 00$	$(1)1.00000\ 00000$	$(1)2.00000\ 00000$
2	-1.00000	$+2.00000$	$(1)+3.40000\ 00$	$(1)9.80000\ 00000$	$(2)3.98000\ 00000$
3	-5.00000	-4.00000	$(2)+1.80000\ 00$	$(2)9.40000\ 00000$	$(3)7.88000\ 00000$
4	$+1.00000$	$(1)-2.00000$	$(2)+8.76000\ 00$	$(3)8.81200\ 00000$	$(5)1.55212\ 00000$
5	$(1)+4.10000$	$(0)-8.00000$	$(3)+3.81600\ 00$	$(4)8.06000\ 00000$	$(6)3.04120\ 00000$
6	$(1)+3.10000$	$(2)+1.84000$	$(4)+1.41360\ 00$	$(5)7.17880\ 00000$	$(7)5.92718\ 80000$
7	$(2)-4.61000$	$(2)+4.64000$	$(4)+3.90240\ 00$	$(6)6.21160\ 00000$	$(9)1.14894\ 32000$
8	$(2)-8.95000$	$(3)-1.64800$	$(4)+3.62400\ 00$	$(7)5.20656\ 80000$	$(10)2.21490\ 57680$
9	$(3)+6.48100$	$(4)-1.07200$	$(5)-4.06944\ 00$	$(8)4.21271\ 20000$	$(11)4.24598\ 06240$
10	$(4)+2.25910$	$(3)+8.22400$	$(6)-3.09398\ 40$	$(9)3.27552\ 97600$	$(12)8.09327\ 82098$
11	$(5)-1.07029$	$(5)+2.30848$	$(7)-1.04250\ 24$	$(10)2.43298\ 73600$	$(14)1.53373\ 60295$
12	$(5)-6.04031$	$(5)+2.80768$	$(6)+5.51750\ 40$	$(11)1.71237\ 08128$	$(15)2.88941\ 99383$

23. Bernoulli and Euler Polynomials— Riemann Zeta Function

Emilie V. Haynsworth [1] and Karl Goldberg [2]

Contents

$$\zeta(n)=\sum_{k=1}^{\infty}\frac{1}{k^n}, \quad 20\mathrm{D}$$

$$\eta(n)=\sum_{k=1}^{\infty}\frac{(-1)^{k-1}}{k^n}, \quad 20\mathrm{D}$$

$$\lambda(n)=\sum_{k=0}^{\infty}\frac{1}{(2k+1)^n}, \quad 20\mathrm{D}$$

$$\beta(n)=\sum_{k=0}^{\infty}\frac{(-1)^k}{(2k+1)^n}, \quad 18\mathrm{D}$$

$$n=1(1)42$$

$$\sum_{k=1}^{m} k^n, n=1(1)10, m=1(1)100$$

The authors acknowledge the assistance of Ruth E. Capuano in the preparation and checking of the tables.

---

[1] National Bureau of Standards. (Presently, Auburn University.)
[2] National Bureau of Standards.

23. Bernoulli and Euler Polynomials—Riemann Zeta Function

Mathematical Properties

23.1. Bernoulli and Euler Polynomials and the Euler-Maclaurin Formula

Generating Functions

23.1.1 $\dfrac{te^{xt}}{e^t-1}=\sum\limits_{n=0}^{\infty} B_n(x)\,\dfrac{t^n}{n!}$ $\qquad |t|<2\pi$ $\qquad \dfrac{2e^{xt}}{e^t+1}=\sum\limits_{n=0}^{\infty} E_n(x)\,\dfrac{t^n}{n!}$ $\qquad |t|<\pi$

Bernoulli and Euler Numbers

23.1.2 $B_n=B_n(0)$ $\qquad n=0,1,\ldots$ $\qquad E_n=2^n E_n\left(\dfrac{1}{2}\right)=$ integer $\qquad n=0,1,\ldots$

23.1.3 $B_0=1, B_1=-\dfrac{1}{2}, B_2=\dfrac{1}{6}, B_4=-\dfrac{1}{30}$ $\qquad E_0=1, E_2=-1, E_4=5$

(For occurrence of B_n and E_n in series expansions of circular functions, see chapter 4.)

Sums of Powers

23.1.4 $\sum\limits_{k=1}^{m} k^n=\dfrac{B_{n+1}(m+1)-B_{n+1}}{n+1}$ $\qquad \sum\limits_{k=1}^{m}(-1)^{m-k}k^n=\dfrac{E_n(m+1)+(-1)^m E_n(0)}{2}$

$\qquad m,n=1,2,\ldots$ $\qquad m,n=1,2,\ldots$

Derivatives and Differences

23.1.5 $B'_n(x)=nB_{n-1}(x)$ $\qquad n=1,2,\ldots$ $\qquad E'_n(x)=nE_{n-1}(x)$ $\qquad n=1,2,\ldots$

23.1.6 $B_n(x+1)-B_n(x)=nx^{n-1}$ $\qquad n=0,1,\ldots$ $\qquad E_n(x+1)+E_n(x)=2x^n$ $\qquad n=0,1,\ldots$

Expansions

23.1.7

$B_n(x+h)=\sum\limits_{k=0}^{n}\binom{n}{k}B_k(x)h^{n-k}$ $\qquad n=0,1,\ldots$ $\qquad E_n(x+h)=\sum\limits_{k=0}^{n}\binom{n}{k}E_k(x)h^{n-k}$ $\qquad n=0,1,\ldots$

$E_n(x)=\sum\limits_{k=0}^{n}\binom{n}{k}\dfrac{E_k}{2^k}\left(x-\dfrac{1}{2}\right)^{n-k}$ $\qquad n=0,1,\ldots$

Symmetry

23.1.8 $B_n(1-x)=(-1)^n B_n(x)$ $\qquad n=0,1,\ldots$ $\qquad E_n(1-x)=(-1)^n E_n(x)$ $\qquad n=0,1,\ldots$

23.1.9 $(-1)^n B_n(-x)=B_n(x)+nx^{n-1}$ $\qquad n=0,1,\ldots$ $\qquad (-1)^{n+1}E_n(-x)=E_n(x)-2x^n$ $\qquad n=0,1,\ldots$

Multiplication Theorem

23.1.10

$B_n(mx)=m^{n-1}\sum\limits_{k=0}^{m-1}B_n\left(x+\dfrac{k}{m}\right)$ $\qquad \begin{array}{l}n=0,1,\ldots\\ m=1,2,\ldots\end{array}$ $\qquad E_n(mx)=m^n\sum\limits_{k=0}^{m-1}(-1)^k E_n\left(x+\dfrac{k}{m}\right)$ $\qquad \begin{array}{l}n=0,1,\ldots\\ m=1,3,\ldots\end{array}$

$E_n(mx)=-\dfrac{2}{n+1}\,m^n\sum\limits_{k=0}^{m-1}(-1)^k B_{n+1}\left(x+\dfrac{k}{m}\right)$

$\qquad \begin{array}{l}n=0,1,\ldots\\ m=2,4,\ldots\end{array}$

Integrals

23.1.11 $\qquad \int_a^x B_n(t)dt = \dfrac{B_{n+1}(x) - B_{n+1}(a)}{n+1}$ $\qquad\qquad \int_a^x E_n(t)dt = \dfrac{E_{n+1}(x) - E_{n+1}(a)}{n+1}$

23.1.12 $\qquad \int_0^1 B_n(t)B_m(t)dt = (-1)^{n-1}\dfrac{m!n!}{(m+n)!}B_{m+n}$ $\qquad\qquad \int_0^1 E_n(t)E_m(t)dt$

$$m,n=1,2,\ldots$$

$$= (-1)^n 4(2^{m+n+2}-1)\dfrac{m!n!}{(m+n+2)!}B_{m+n+2}$$

$$m,n=0,1,\ldots$$

(The polynomials are orthogonal for $m+n$ odd.)

Inequalities

23.1.13 $\qquad |B_{2n}| > |B_{2n}(x)| \qquad n=1,2,\ldots, \qquad 1>x>0$ $\qquad 4^{-n}|E_{2n}| > (-1)^n E_{2n}(x) > 0 \qquad n=1,2,\ldots, \qquad \tfrac{1}{2}>x>0$

23.1.14

$$\dfrac{2(2n+1)!}{(2\pi)^{2n+1}}\left(\dfrac{1}{1-2^{-2n}}\right) > (-1)^{n+1}B_{2n+1}(x) > 0$$

$$n=1,2,\ldots, \qquad \tfrac{1}{2}>x>0$$

$$\dfrac{4(2n-1)!}{\pi^{2n}}\left(1+\dfrac{1}{2^{2n}-2}\right) > (-1)^n E_{2n-1}(x) > 0$$

$$n=1,2,\ldots, \qquad \tfrac{1}{2}>x>0$$

23.1.15

$$\dfrac{2(2n)!}{(2\pi)^{2n}}\left(\dfrac{1}{1-2^{1-2n}}\right) > (-1)^{n+1}B_{2n} > \dfrac{2(2n)!}{(2\pi)^{2n}}$$

$$n=1,2,\ldots$$

$$\dfrac{4^{n+1}(2n)!}{\pi^{2n+1}} > (-1)^n E_{2n} > \dfrac{4^{n+1}(2n)!}{\pi^{2n+1}}\left(\dfrac{1}{1+3^{-1-2n}}\right)$$

$$n=0,1,\ldots$$

Fourier Expansions

23.1.16

$$B_n(x) = -2\dfrac{n!}{(2\pi)^n}\sum_{k=1}^{\infty}\dfrac{\cos(2\pi kx - \tfrac{1}{2}\pi n)}{k^n}$$

$$n>1, 1\geq x\geq 0$$
$$n=1, 1>x>0$$

$$E_n(x) = 4\dfrac{n!}{\pi^{n+1}}\sum_{k=0}^{\infty}\dfrac{\sin((2k+1)\pi x - \tfrac{1}{2}\pi n)}{(2k+1)^{n+1}}$$

$$n>0, 1\geq x\geq 0$$
$$n=0, 1>x>0$$

23.1.17

$$B_{2n-1}(x) = \dfrac{(-1)^n 2(2n-1)!}{(2\pi)^{2n-1}}\sum_{k=1}^{\infty}\dfrac{\sin 2k\pi x}{k^{2n-1}}$$

$$n>1, 1\geq x\geq 0$$
$$n=1, 1>x>0$$

$$E_{2n-1}(x) = \dfrac{(-1)^n 4(2n-1)!}{\pi^{2n}}\sum_{k=0}^{\infty}\dfrac{\cos(2k+1)\pi x}{(2k+1)^{2n}}$$

$$n=1,2,\ldots, \qquad 1\geq x\geq 0$$

23.1.18

$$B_{2n}(x) = \dfrac{(-1)^{n-1}2(2n)!}{(2\pi)^{2n}}\sum_{k=1}^{\infty}\dfrac{\cos 2k\pi x}{k^{2n}}$$

$$n=1,2,\ldots, \qquad 1\geq x\geq 0$$

$$E_{2n}(x) = \dfrac{(-1)^n 4(2n)!}{\pi^{2n+1}}\sum_{k=0}^{\infty}\dfrac{\sin(2k+1)\pi x}{(2k+1)^{2n+1}}$$

$$n>0, 1\geq x\geq 0$$
$$n=0, 1>x>0$$

Special Values

23.1.19 $\qquad B_{2n+1}=0 \qquad\qquad n=1,2,\ldots$ $\qquad E_{2n+1}=0 \qquad\qquad n=0,1,\ldots$

23.1.20 $\qquad B_n(0) = (-1)^n B_n(1)$ $\qquad\qquad E_n(0) = -E_n(1)$

$$= B_n \qquad\qquad n=0,1,\ldots$$

$$= -2(n+1)^{-1}(2^{n+1}-1)B_{n+1} \qquad n=1,2,\ldots$$

23.1.21 $\qquad B_n(\tfrac{1}{2}) = -(1-2^{1-n})B_n \qquad n=0,1,\ldots$ $\qquad E_n(\tfrac{1}{2}) = 2^{-n}E_n \qquad\qquad n=0,1,\ldots$

23.1.22 $\quad B_n(\tfrac{1}{4})=(-1)^n B_n(\tfrac{3}{4})$

$$=-2^{-n}(1-2^{1-n})B_n-n4^{-n}E_{n-1}$$

$$n=1,2,\ldots$$

23.1.23 $\quad B_{2n}(\tfrac{1}{3})=B_{2n}(\tfrac{2}{3})$

$$=-2^{-1}(1-3^{1-2n})B_{2n} \qquad n=0,1,\ldots$$

23.1.24 $\quad B_{2n}(\tfrac{1}{6})=B_{2n}(\tfrac{5}{6})$

$$=2^{-1}(1-2^{1-2n})(1-3^{1-2n})B_{2n}$$

$$n=0,1,\ldots$$

$E_{2n-1}(\tfrac{1}{3})=-E_{2n-1}(\tfrac{2}{3})$

$$=-(2n)^{-1}(1-3^{1-2n})(2^{2n}-1)B_{2n}$$

$$n=1,2,\ldots$$

Symbolic Operations

23.1.25 $\quad p(B(x)+1)-p(B(x))=p'(x)$

23.1.26 $\quad B_n(x+h)=(B(x)+h)^n \qquad n=0,1,\ldots$

$p(E(x)+1)+p(E(x))=2p(x)$

$E_n(x+h)=(E(x)+h)^n \qquad n=0,1,\ldots$

Here $p(x)$ denotes a polynomial in x and after expanding we set $\{B(x)\}^n=B_n(x)$ and $\{E(x)\}^n=E_n(x)$.

Relations Between the Polynomials

23.1.27

$$E_{n-1}(x)=\frac{2^n}{n}\left\{B_n\left(\frac{x+1}{2}\right)-B_n\left(\frac{x}{2}\right)\right\}$$

$$=\frac{2}{n}\left\{B_n(x)-2^n B_n\left(\frac{x}{2}\right)\right\} \qquad n=1,2,\ldots$$

23.1.28

$$E_{n-2}(x)=2\binom{n}{2}^{-1}\sum_{k=0}^{n-2}\binom{n}{k}(2^{n-k}-1)B_{n-k}B_k(x)$$

$$n=2,3,\ldots$$

23.1.29

$$B_n(x)=2^{-n}\sum_{k=0}^{n}\binom{n}{k}B_{n-k}E_k(2x) \qquad n=0,1,\ldots$$

Euler-Maclaurin Formulas

Let $F(x)$ have its first $2n$ derivatives continuous on an interval (a, b). Divide the interval into m equal parts and let $h=(b-a)/m$. Then for some θ, $1>\theta>0$, depending on $F^{(2n)}(x)$ on (a, b), we have

23.1.30

$$\sum_{k=0}^{m}F(a+kh)=\frac{1}{h}\int_a^b F(t)dt+\frac{1}{2}\{F(b)+F(a)\}$$

$$+\sum_{k=1}^{n-1}\frac{h^{2k-1}}{(2k)!}B_{2k}\{F^{(2k-1)}(b)-F^{(2k-1)}(a)\}$$

$$+\frac{h^{2n}}{(2n)!}B_{2n}\sum_{k=0}^{m-1}F^{(2n)}(a+kh+\theta h)$$

Equivalent to this is

23.1.31

$$\frac{1}{h}\int_x^{x+h}F(t)dt=\frac{1}{2}\{F(x+h)+F(x)\}$$

$$-\sum_{k=1}^{n-1}\frac{h^{2k-1}}{(2k)!}B_{2k}\{F^{(2k-1)}(x+h)-F^{(2k-1)}(x)\}$$

$$-\frac{h^{2n}}{(2n)!}B_{2n}F^{(2n)}(x+\theta h) \qquad b-h\geq x\geq a$$

Let $\hat{B}_n(x)=B_n(x-[x])$. The Euler Summation Formula is

23.1.32

$$\sum_{k=0}^{m-1}F(a+kh+\omega h)=\frac{1}{h}\int_a^b F(t)dt$$

$$+\sum_{k=1}^{p}\frac{h^{k-1}}{k!}B_k(\omega)\{F^{(k-1)}(b)-F^{(k-1)}(a)\}$$

$$-\frac{h^p}{p!}\int_0^1 \hat{B}_p(\omega-t)\left\{\sum_{k=0}^{m-1}F^{(p)}(a+kh+th)\right\}dt$$

$$p\leq 2n,\ 1\geq\omega\geq 0$$

23.2. Riemann Zeta Function and Other Sums of Reciprocal Powers

23.2.1 $\quad \zeta(s) = \sum_{k=1}^{\infty} k^{-s} \qquad \mathscr{R}s > 1$

23.2.2 $\quad = \prod_{p} (1-p^{-s})^{-1} \qquad \mathscr{R}s > 1$

(product over all primes p).

23.2.3 $\quad = \dfrac{1}{s-1} + \dfrac{1}{2} + \sum_{k=1}^{n} \dfrac{B_{2k}}{2k} \binom{s+2k-2}{2k-1}$
$$- \binom{s+2n}{2n+1} \int_{1}^{\infty} \dfrac{B_{2n+1}(x-[x])}{x^{s+2n+1}} dx$$
$$s \neq 1, n=1, 2, \ldots, \quad \mathscr{R}s > -2n$$

* 23.2.4 $\quad = -\dfrac{\Gamma(1-s)}{2\pi i} \int_{C} \dfrac{(-z)^{s-1}}{e^{z}-1} dz$

23.2.5 $\quad = \dfrac{1}{s-1} + \sum_{n=0}^{\infty} \dfrac{(-1)^{n}}{n!} \gamma_{n}(s-1)^{n}$

where

$$\gamma_{n} = \lim_{m\to\infty} \left\{ \sum_{k=1}^{m} \dfrac{(\ln k)^{n}}{k} - \dfrac{(\ln m)^{n+1}}{n+1} \right\}$$
$$\mathscr{R}s > 0$$

23.2.6 $\quad = 2^{s} \pi^{s-1} \sin\left(\tfrac{1}{2}\pi s\right) \Gamma(1-s) \zeta(1-s)$

23.2.7 $\quad = \dfrac{1}{\Gamma(s)} \int_{0}^{\infty} \dfrac{x^{s-1}}{e^{x}-1} dx \qquad \mathscr{R}s > 1$

23.2.8 $\quad = \dfrac{1}{(1-2^{1-s})\Gamma(s)} \int_{0}^{\infty} \dfrac{x^{s-1}}{e^{x}+1} dx$

23.2.9 $\quad = \sum_{k=1}^{n} k^{-s} + (s-1)^{-1} n^{1-s} - s \int_{n}^{\infty} \dfrac{x-[x]}{x^{s+1}} dx$
$$n=1, 2, \ldots, \mathscr{R}s > 0$$

23.2.10 $\quad = \dfrac{\exp\left(\ln 2\pi - 1 - \tfrac{1}{2}\gamma\right)s}{2(s-1)\Gamma(\tfrac{1}{2}s+1)} \prod_{\rho} \left(1-\dfrac{s}{\rho}\right) e^{\frac{s}{\rho}}$

product over all zeros ρ of $\zeta(s)$ with $\mathscr{R}\rho > 0$.

The contour C in the fourth formula starts at infinity on the positive real axis, circles the origin once in the positive direction excluding the points $\pm 2ni\pi$ for $n=1, 2, \ldots$, and returns to the starting point. Therefore $\zeta(s)$ is regular for all values of s except for a simple pole at $s=1$ with residue 1.

Special Values

23.2.11 $\qquad \zeta(0) = -\tfrac{1}{2}$

23.2.12 $\qquad \zeta(1) = \infty$

23.2.13 $\qquad \zeta'(0) = -\tfrac{1}{2} \ln 2\pi$

23.2.14 $\quad \zeta(-2n) = 0 \qquad n=1, 2, \ldots$

23.2.15 $\quad \zeta(1-2n) = -\dfrac{B_{2n}}{2n} \qquad n=1, 2, \ldots$

23.2.16 $\quad \zeta(2n) = \dfrac{(2\pi)^{2n}}{2(2n)!} |B_{2n}| \qquad n=1, 2, \ldots$

23.2.17
$$\zeta(2n+1) = \dfrac{(-1)^{n+1}(2\pi)^{2n+1}}{2(2n+1)!} \int_{0}^{1} B_{2n+1}(x) \cot(\pi x) dx$$
$$n=1, 2, \ldots$$

Sums of Reciprocal Powers

The sums referred to are

23.2.18 $\quad \zeta(n) = \sum_{k=1}^{\infty} k^{-n} \qquad n=2, 3, \ldots$

23.2.19
$$\eta(n) = \sum_{k=1}^{\infty} (-1)^{k-1} k^{-n} = (1-2^{1-n}) \zeta(n) \qquad n=1, 2, \ldots$$

23.2.20
$$\lambda(n) = \sum_{k=0}^{\infty} (2k+1)^{-n} = (1-2^{-n}) \zeta(n) \qquad n=2, 3, \ldots$$

23.2.21
$$\beta(n) = \sum_{k=0}^{\infty} (-1)^{k} (2k+1)^{-n} \qquad n=1, 2, \ldots$$

These sums can be calculated from the Bernoulli and Euler polynomials by means of the last two formulas for special values of the zeta function (note that $\eta(1) = \ln 2$), and

23.2.22 $\quad \beta(2n+1) = \dfrac{(\pi/2)^{2n+1}}{2(2n)!} |E_{2n}| \qquad n=0, 1, \ldots$

23.2.23
$$\beta(2n) = \dfrac{(-1)^{n}\pi^{2n}}{4(2n-1)!} \int_{0}^{1} E_{2n-1}(x) \sec(\pi x) dx$$
$$n=1, 2, \ldots$$

$\beta(2)$ is known as Catalan's constant. Some other special values are

23.2.24 $\quad \zeta(2) = 1 + \dfrac{1}{2^{2}} + \dfrac{1}{3^{2}} + \ldots = \dfrac{\pi^{2}}{6}$

23.2.25 $\quad \zeta(4) = 1 + \dfrac{1}{2^{4}} + \dfrac{1}{3^{4}} + \ldots = \dfrac{\pi^{4}}{90}$

23.2.26 $\eta(2)=1-\dfrac{1}{2^2}+\dfrac{1}{3^2}-\ \ldots\ =\dfrac{\pi^2}{12}$

23.2.27 $\eta(4)=1-\dfrac{1}{2^4}+\dfrac{1}{3^4}-\ \ldots\ =\dfrac{7\pi^4}{720}$

23.2.28 $\lambda(2)=1+\dfrac{1}{3^2}+\dfrac{1}{5^2}+\ \ldots\ =\dfrac{\pi^2}{8}$

23.2.29 $\lambda(4)=1+\dfrac{1}{3^4}+\dfrac{1}{5^4}+\ \ldots\ =\dfrac{\pi^4}{96}$

23.2.30 $\beta(1)=1-\dfrac{1}{3}+\dfrac{1}{5}-\ \ldots\ =\dfrac{\pi}{4}$

23.2.31 $\beta(3)=1-\dfrac{1}{3^3}+\dfrac{1}{5^3}-\ \ldots\ =\dfrac{\pi^3}{32}$

References

Texts

[23.1] G. Boole, The calculus of finite differences, 3d ed. (Hafner Publishing Co., New York, N.Y., 1932).

[23.2] W. E. Briggs and S. Chowla, The power series coefficients of $\zeta(s)$, Amer. Math. Monthly **62**, 323–325 (1955).

[23.3] T. Fort, Finite differences (Clarendon Press, Oxford, England, 1948).

[23.4] C. Jordan, Calculus of finite differences, 2d ed. (Chelsea Publishing Co., New York, N.Y., 1960).

[23.5] K. Knopp, Theory and application of infinite series (Blackie and Son, Ltd., London, England, 1951).

[23.6] L. M. Milne-Thomson, Calculus of finite differences (Macmillan and Co., Ltd., London, England, 1951).

[23.7] N. E. Nörlund, Vorlesungen über Differenzenrechnung (Edwards Bros., Ann Arbor, Mich., 1945).

[23.8] C. H. Richardson, An introduction to the calculus of finite differences (D. Van Nostrand Co., Inc., New York, N.Y., 1954).

[23.9] J. F. Steffensen, Interpolation (Chelsea Publishing Co., New York, N.Y., 1950).

[23.10] E. C. Titchmarsh, The zeta-function of Riemann (Cambridge Univ. Press, Cambridge, England, 1930).

[23.11] A. D. Wheelon, A short table of summable series, Report No. SM-14642, Douglas Aircraft Co., Inc., Santa Monica, Calif. (1953).

Tables

[23.12] G. Blanch and R. Siegel, Table of modified Bernoulli polynomials, J. Research NBS **44**, 103–107 (1950) RP2060.

[23.13] H. T. Davis, Tables of the higher mathematical functions, vol. II (Principia Press, Bloomington, Ind., 1935).

[23.14] R. Hensman, Tables of the generalized Riemann Zeta function, Report No. T2111, Telecommunications Research Establishment, Ministry of Supply, Great Malvern, Worcestershire, England (1948). $\zeta(s,a)$, $s=-10(.1)0$, $a=0(.1)2$, 5D; $(s-1)\zeta(s,a)$, $s=0(.1)1$, $a=0(.1)2$, 5D.

[23.15] D. H. Lehmer, On the maxima and minima of Bernoulli polynomials, Amer. Math. Monthly **47**, 533–538 (1940).

[23.16] E. O. Powell, A table of the generalized Riemann Zeta function in a particular case, Quart. J. Mech. Appl. Math. **5**, 116–123 (1952). $\zeta(\tfrac{1}{2},a)$, $a=1(.01)2(.02)5(.05)10$, 10D.

COEFFICIENTS b_k OF THE BERNOULLI POLYNOMIALS $B_n(x)=\sum\limits_{k=0}^{n} b_k x^k$ Table 23.1

$n\backslash k$	0	1	2	3	4	5	6	7	8	9	10	11	12	13	14	15
0	1															
1	$-\frac{1}{2}$	1														
2	$\frac{1}{6}$	-1	1													
3	0	$\frac{1}{2}$	$-\frac{3}{2}$	1												
4	$-\frac{1}{30}$	0	1	-2	1											
5	0	$-\frac{1}{6}$	0	$\frac{5}{3}$	$-\frac{5}{2}$	1										
6	$\frac{1}{42}$	0	$-\frac{1}{2}$	0	$\frac{5}{2}$	-3	1									
7	0	$\frac{1}{6}$	0	$-\frac{7}{6}$	0	$\frac{7}{2}$	$-\frac{7}{2}$	1								
8	$-\frac{1}{30}$	0	$\frac{2}{3}$	0	$-\frac{7}{3}$	0	$\frac{14}{3}$	-4	1							
9	0	$-\frac{3}{10}$	0	2	0	$-\frac{21}{5}$	0	6	$-\frac{9}{2}$	1						
10	$\frac{5}{66}$	0	$-\frac{3}{2}$	0	5	0	-7	0	$\frac{15}{2}$	-5	1					
11	0	$\frac{5}{6}$	0	$-\frac{11}{2}$	0	11	0	-11	0	$\frac{55}{6}$	$-\frac{11}{2}$	1				
12	$-\frac{691}{2730}$	0	5	0	$-\frac{33}{2}$	0	22	0	$-\frac{33}{2}$	0	11	-6	1			
13	0	$-\frac{691}{210}$	0	$\frac{65}{3}$	0	$-\frac{429}{10}$	0	$\frac{286}{7}$	0	$-\frac{143}{6}$	0	13	$-\frac{13}{2}$	1		
14	$\frac{7}{6}$	0	$-\frac{691}{30}$	0	$\frac{455}{6}$	0	$-\frac{1001}{10}$	0	$\frac{143}{2}$	0	$-\frac{1001}{30}$	0	$\frac{91}{6}$	-7	1	
15	0	$\frac{35}{2}$	0	$-\frac{691}{6}$	0	$\frac{455}{2}$	0	$-\frac{429}{2}$	0	$\frac{715}{6}$	0	$-\frac{91}{2}$	0	$\frac{35}{2}$	$-\frac{15}{2}$	1

COEFFICIENTS e_k OF THE EULER POLYNOMIALS $E_n(x)=\sum\limits_{k=0}^{n} e_k x^k$

$n\backslash k$	0	1	2	3	4	5	6	7	8	9	10	11	12	13	14	15
0	1															
1	$-\frac{1}{2}$	1														
2	0	-1	1													
3	$\frac{1}{4}$	0	$-\frac{3}{2}$	1												
4	0	1	0	-2	1											
5	$-\frac{1}{2}$	0	$\frac{5}{2}$	0	$-\frac{5}{2}$	1										
6	0	-3	0	5	0	-3	1									
7	$\frac{17}{8}$	0	$-\frac{21}{2}$	0	$\frac{35}{4}$	0	$-\frac{7}{2}$	1								
8	0	17	0	-28	0	14	0	-4	1							
9	$-\frac{31}{2}$	0	$\frac{153}{2}$	0	-63	0	21	0	$-\frac{9}{2}$	1						
10	0	-155	0	255	0	-126	0	30	0	-5	1					
11	$\frac{691}{4}$	0	$-\frac{1705}{2}$	0	$\frac{2805}{4}$	0	-231	0	$\frac{165}{4}$	0	$-\frac{11}{2}$	1				
12	0	2073	0	-3410	0	1683	0	-396	0	55	0	-6	1			
13	$-\frac{5461}{2}$	0	$\frac{26949}{2}$	0	$-\frac{22165}{2}$	0	$\frac{7293}{2}$	0	$-\frac{1287}{2}$	0	$\frac{143}{2}$	0	$-\frac{13}{2}$	1		
14	0	-38227	0	62881	0	-31031	0	7293	0	-1001	0	91	0	-7	1	
15	$\frac{929569}{16}$	0	$-\frac{573405}{2}$	0	$\frac{943215}{4}$	0	$-\frac{155155}{2}$	0	$\frac{109395}{8}$	0	$-\frac{3003}{2}$	0	$\frac{455}{4}$	0	$-\frac{15}{2}$	1

Table 23.2   BERNOULLI AND EULER NUMBERS

$$B_n = N/D$$

n	N	D	B_n
0	1	1	(0) 1.0000 00000
1	−1	2	(− 1)−5.0000 00000
2	1	6	(− 1) 1.6666 66667
4	−1	30	(− 2)−3.3333 33333
6	1	42	(− 2) 2.3809 52381
8	−1	30	(− 2)−3.3333 33333
10	5	66	(− 2) 7.5757 57576
12	−691	2730	(− 1)−2.5311 35531
14	7	6	(0) 1.1666 66667
16	−3617	510	(0)−7.0921 56863
18	43867	798	(1) 5.4971 17794
20	−1 74611	330	(2)−5.2912 42424
22	8 54513	138	(3) 6.1921 23188
24	−2363 64091	2730	(4)−8.6580 25311
26	85 53103	6	(6) 1.4255 17167
28	−2 37494 61029	870	(7)−2.7298 23107
30	861 58412 76005	14322	(8) 6.0158 08739
32	−770 93210 41217	510	(10)−1.5116 31577
34	257 76878 58367	6	(11) 4.2961 46431
36	−26315 27155 30534 77373	19 19190	(13)−1.3711 65521
38	2 92999 39138 41559	6	(14) 4.8833 23190
40	−2 61082 71849 64491 22051	13530	(16)−1.9296 57934
42	15 20097 64391 80708 02691	1806	(17) 8.4169 30476
44	−278 33269 57930 10242 35023	690	(19)−4.0338 07185
46	5964 51111 59391 21632 77961	282	(21) 2.1150 74864
48	−560 94033 68997 81768 62491 27547	46410	(23)−1.2086 62652
50	49 50572 05241 07964 82124 77525	66	(24) 7.5008 66746
52	−80116 57181 35489 95734 79249 91853	1590	(26)−5.0387 78101
54	29 14996 36348 84862 42141 81238 12691	798	(28) 3.6528 77648
56	−2479 39292 93132 26753 68541 57396 63229	870	(30)−2.8498 76930
58	84483 61334 88800 41862 04677 59940 36021	354	(32) 2.3865 42750
60	−121 52331 40483 75557 20403 04994 07982 02460 41491	567 86730	(34)−2.1399 94926

n	E_n
0	1
2	−1
4	5
6	− 61
8	1385
10	− 50521
12	27 02765
14	− 1993 60981
16	1 93915 12145
18	−240 48796 75441
20	37037 11882 37525
22	− 69 34887 43931 37901
24	15514 53416 35570 86905
26	−40 87072 50929 31238 92361
28	12522 59641 40362 98654 68285
30	− 44 15438 93249 02310 45536 82821
32	17751 93915 79539 28943 66647 89665
34	− 80 72329 92358 87898 06216 82474 53281
36	41222 06033 95177 02122 34707 96712 59045
38	− 234 89580 52704 31082 52017 82857 61989 47741
40	1 48511 50718 11498 00178 77156 78140 58266 84425
42	−1036 46227 33519 61211 93979 57304 74518 59763 10201
44	7 94757 94225 97592 70360 80405 10088 07061 95192 73805
46	−6667 53751 66855 44977 43502 84747 73748 19752 41076 84661
48	60 96278 64556 85421 58691 68574 28768 43153 97653 90444 35185
50	− 60532 85248 18862 18963 14383 78511 16490 88103 49822 51468 15121
52	650 61624 86684 60884 77158 70634 08082 29834 83644 23676 53855 76565
54	−7 54665 99390 08739 09806 14325 65889 73674 42122 40024 71169 98586 45581
56	9420 32189 64202 41204 20228 62376 90583 22720 93888 52599 64600 93949 05945
58	−126 22019 25180 62187 19903 40923 72874 89255 48234 10611 91825 59406 99649 20041
60	181089 11496 57923 04965 45807 74165 21586 88733 48734 92363 14106 00809 54542 31325

From H. T. Davis, Tables of the higher mathematical functions, vol. II. Principia Press, Bloomington, Ind., 1935 (with permission).

SUMS OF RECIPROCAL POWERS   Table 23.3

n	$\zeta(n)=\sum\limits_{k=1}^{\infty}k^{-n}$	$\eta(n)=\sum\limits_{k=1}^{\infty}(-1)^{k-1}k^{-n}$
1	∞	0. 69314 71805 59945 30942
2	1. 64493 40668 48226 43647	0. 82246 70334 24113 21824
3	1. 20205 69031 59594 28540	0. 90154 26773 69695 71405
4	1. 08232 32337 11138 19152	0. 94703 28294 97245 91758
5	1. 03692 77551 43369 92633	0. 97211 97704 46909 30594
6	1. 01734 30619 84449 13971	0. 98555 10912 97435 10410
7	1. 00834 92773 81922 82684	0. 99259 38199 22830 28267
8	1. 00407 73561 97944 33938	0. 99623 30018 52647 89923
9	1. 00200 83928 26082 21442	0. 99809 42975 41605 33077
10	1. 00099 45751 27818 08534	0. 99903 95075 98271 56564
11	1. 00049 41886 04119 46456	0. 99951 71434 98060 75414
12	1. 00024 60865 53308 04830	0. 99975 76851 43858 19085
13	1. 00012 27133 47578 48915	0. 99987 85427 63265 11549
14	1. 00006 12481 35058 70483	0. 99993 91703 45979 71817
15	1. 00003 05882 36307 02049	0. 99996 95512 13099 23808
16	1. 00001 52822 59408 65187	0. 99998 47642 14906 10644
17	1. 00000 76371 97637 89976	0. 99999 23782 92041 01198
18	1. 00000 38172 93264 99984	0. 99999 61878 69610 11348
19	1. 00000 19082 12716 55394	0. 99999 80935 08171 67511
20	1. 00000 09539 62033 87280	0. 99999 90466 11581 52212
21	1. 00000 04769 32986 78781	0. 99999 95232 58215 54282
22	1. 00000 02384 50502 72773	0. 99999 97616 13230 82255
23	1. 00000 01192 19925 96531	0. 99999 98808 01318 43950
24	1. 00000 00596 08189 05126	0. 99999 99403 98892 39463
25	1. 00000 00298 03503 51465	0. 99999 99701 98856 96283
26	1. 00000 00149 01554 82837	0. 99999 99850 99231 99657
27	1. 00000 00074 50711 78984	0. 99999 99925 49550 48496
28	1. 00000 00037 25334 02479	0. 99999 99962 74753 40011
29	1. 00000 00018 62659 72351	0. 99999 99981 37369 41811
30	1. 00000 00009 31327 43242	0. 99999 99990 68682 28145
31	1. 00000 00004 65662 90650	0. 99999 99995 34340 33145
32	1. 00000 00002 32831 18337	0. 99999 99997 67169 89595
33	1. 00000 00001 16415 50173	0. 99999 99998 83584 85805
34	1. 00000 00000 58207 72088	0. 99999 99999 41792 39905
35	1. 00000 00000 29103 85044	0. 99999 99999 70896 18953
36	1. 00000 00000 14551 92189	0. 99999 99999 85448 09143
37	1. 00000 00000 07275 95984	0. 99999 99999 92724 04461
38	1. 00000 00000 03637 97955	0. 99999 99999 96362 02193
39	1. 00000 00000 01818 98965	0. 99999 99999 98181 01084
40	1. 00000 00000 00909 49478	0. 99999 99999 99090 50538
41	1. 00000 00000 00454 74738	0. 99999 99999 99545 25268
42	1. 00000 00000 00227 37368	0. 99999 99999 99772 62633

For $n>42$, $\zeta(n+1)=\frac{1}{2}[1+\zeta(n)]$ $\eta(n+1)=\frac{1}{2}[1+\eta(n)]$

From H. T. Davis, Tables of the higher mathematical functions, vol. II. Principia Press, Bloomington, Ind., 1935 (with permission).

Table 23.3   **SUMS OF RECIPROCAL POWERS**

n	$\lambda(n)=\sum\limits_{k=0}^{\infty}(2k+1)^{-n}$	$\beta(n)=\sum\limits_{k=0}^{\infty}(-1)^k(2k+1)^{-n}$
1	∞	0. 78539 81633 97448 310
2	1. 23370 05501 36169 82735	0. 91596 55941 77219 015
3	1. 05179 97902 64644 99972	0. 96894 61462 59369 380
4	1. 01467 80316 04192 05455	0. 98894 45517 41105 336
5	1. 00452 37627 95139 61613	0. 99615 78280 77088 064
6	1. 00144 70766 40942 12191	0. 99868 52222 18438 135
7	1. 00047 15486 52376 55476	0. 99955 45078 90539 909
8	1. 00015 51790 25296 11930	0. 99984 99902 46829 657
9	1. 00005 13451 83843 77259	0. 99994 96841 87220 090
10	1. 00001 70413 63044 82549	0. 99998 31640 26196 877
11	1. 00000 56660 51090 10935	0. 99999 43749 73823 699
12	1. 00000 18858 48583 11958	0. 99999 81223 50587 882
13	1. 00000 06280 55421 80232	0. 99999 93735 83771 841
14	1. 00000 02092 40519 21150	0. 99999 97910 87248 735
15	1. 00000 00697 24703 12929	0. 99999 99303 40842 624
16	1. 00000 00232 37157 37916	0. 99999 99767 75950 903
17	1. 00000 00077 44839 45587	0. 99999 99922 57782 104
18	1. 00000 00025 81437 55666	0. 99999 99974 19086 745
19	1. 00000 00008 60444 11452	0. 99999 99991 39660 745
20	1. 00000 00002 86807 69746	0. 99999 99997 13213 274
21	1. 00000 00000 95601 16531	0. 99999 99999 04403 029
22	1. 00000 00000 31866 77514	0. 99999 99999 68134 064
23	1. 00000 00000 10622 20241	0. 99999 99999 89377 965
24	1. 00000 00000 03540 72294	0. 99999 99999 96459 311
25	1. 00000 00000 01180 23874	0. 99999 99999 98819 768
26	1. 00000 00000 00393 41247	0. 99999 99999 99606 589
27	1. 00000 00000 00131 13740	0. 99999 99999 99868 863
28	1. 00000 00000 00043 71245	0. 99999 99999 99956 288
29	1. 00000 00000 00014 57081	0. 99999 99999 99985 429
30	1. 00000 00000 00004 85694	0. 99999 99999 99995 143
31	1. 00000 00000 00001 61898	0. 99999 99999 99998 381
32	1. 00000 00000 00000 53966	0. 99999 99999 99999 460
33	1. 00000 00000 00000 17989	0. 99999 99999 99999 820
34	1. 00000 00000 00000 05996	0. 99999 99999 99999 940
35	1. 00000 00000 00000 01999	0. 99999 99999 99999 980
36	1. 00000 00000 00000 00666	0. 99999 99999 99999 993
37	1. 00000 00000 00000 00222	0. 99999 99999 99999 998
38	1. 00000 00000 00000 00074	0. 99999 99999 99999 999
39	1. 00000 00000 00000 00025	
40	1. 00000 00000 00000 00008	
41	1. 00000 00000 00000 00003	
42	1. 00000 00000 00000 00001	

SUMS OF POSITIVE POWERS $\sum\limits_{k=1}^{m} k^n$

Table 23.4

$m\backslash n$	1	2	3	4	5	6
1	1	1	1	1	1	1
2	3	5	9	17	33	65
3	6	14	36	98	276	794
4	10	30	100	354	1300	4890
5	15	55	225	979	4425	20515
6	21	91	441	2275	12201	67171
7	28	140	784	4676	29008	1 84820
8	36	204	1296	8772	61776	4 46964
9	45	285	2025	15333	1 20825	9 78405
10	55	385	3025	25333	2 20825	19 78405
11	66	506	4356	39974	3 81876	37 19966
12	78	650	6084	60710	6 30708	67 35950
13	91	819	8281	89271	10 02001	115 62759
14	105	1015	11025	1 27687	15 39825	190 92295
15	120	1240	14400	1 78312	22 99200	304 82920
16	136	1496	18496	2 43848	33 47776	472 60136
17	153	1785	23409	3 27369	47 67633	713 97705
18	171	2109	29241	4 32345	66 57201	1054 09929
19	190	2470	36100	5 62666	91 33300	1524 55810
20	210	2870	44100	7 22666	123 33300	2164 55810
21	231	3311	53361	9 17147	164 17401	3022 21931
22	253	3795	64009	11 51403	215 71033	4156 01835
23	276	4324	76176	14 31244	280 07376	5636 37724
24	300	4900	90000	17 63020	359 70000	7547 40700
25	325	5525	1 05625	21 53645	457 35625	9988 81325
26	351	6201	1 23201	26 10621	576 17001	13077 97101
27	378	6930	1 42884	31 42062	719 65908	16952 17590
28	406	7714	1 64836	37 56718	891 76276	21771 07894
29	435	8555	1 89225	44 63999	1096 87425	27719 31215
30	465	9455	2 16225	52 73999	1339 87425	35009 31215
31	496	10416	2 46016	61 97520	1626 16576	43884 34896
32	528	11440	2 78784	72 46096	1961 71008	54621 76720
33	561	12529	3 14721	84 32017	2353 06401	67536 44689
34	595	13685	3 54025	97 68353	2807 41825	82984 49105
35	630	14910	3 96900	112 68978	3332 63700	1 01367 14730
36	666	16206	4 43556	129 48594	3937 29876	1 23134 97066
37	703	17575	4 94209	148 22755	4630 73833	1 48792 23475
38	741	19019	5 49081	169 07891	5423 09001	1 78901 59859
39	780	20540	6 08400	192 21332	6325 33200	2 14089 03620
40	820	22140	6 72400	217 81332	7349 33200	2 55049 03620
41	861	23821	7 41321	246 07093	8507 89401	3 02550 07861
42	903	25585	8 15409	277 18789	9814 80633	3 57440 39605
43	946	27434	8 94916	311 37590	11284 89076	4 20654 02654
44	990	29370	9 80100	348 85686	12934 05300	4 93217 16510
45	1035	31395	10 71225	389 86311	14779 33425	5 76254 82135
46	1081	33511	11 68561	434 63767	16838 96401	6 70997 79031
47	1128	35720	12 72384	483 43448	19132 41408	7 78789 94360
48	1176	38024	13 82976	536 51864	21680 45376	9 01095 84824
49	1225	40425	15 00625	594 16665	24505 20625	10 39508 72025
50	1275	42925	16 25625	656 66665	27630 20625	11 95758 72025

From H. T. Davis, Tables of the higher mathematical functions, vol. II. Principia Press, Bloomington, Ind., 1935 (with permission).

Table 23.4   SUMS OF POSITIVE POWERS $\sum_{k=1}^{m} k^n$

$m \backslash n$	1	2	3	4	5	6
51	1326	45526	17 58276	724 31866	31080 45876	13 71721 59826
52	1378	48230	18 98884	797 43482	34882 49908	15 69427 69490
53	1431	51039	20 47761	876 33963	39064 45401	17 91071 30619
54	1485	53955	22 05225	961 37019	43656 10425	20 39020 41915
55	1540	56980	23 71600	1052 87644	48688 94800	23 15826 82540
56	1596	60116	25 47216	1151 22140	54196 26576	26 24236 61996
57	1653	63365	27 32409	1256 78141	60213 18633	29 67201 09245
58	1711	66729	29 27521	1369 94637	66776 75401	33 47888 01789
59	1770	70210	31 32900	1491 11998	73925 99700	37 69693 35430
60	1830	73810	33 48900	1620 71998	81701 99700	42 36253 35430
61	1891	77531	35 75881	1759 17839	90147 96001	47 51457 09791
62	1953	81375	38 14209	1906 94175	99309 28833	53 19459 45375
63	2016	85344	40 64256	2064 47136	1 09233 65376	59 44694 47584
64	2080	89440	43 26400	2232 24352	1 19971 07200	66 31889 24320
65	2145	93665	46 01025	2410 74977	1 31573 97825	73 86078 14945
66	2211	98021	48 88521	2600 49713	1 44097 30401	82 12617 64961
67	2278	1 02510	51 89284	2802 00834	1 57598 55508	91 17201 47130
68	2346	1 07134	55 03716	3015 82210	1 72137 89076	101 05876 29754
69	2415	1 11895	58 32225	3242 49331	1 87778 20425	111 85057 92835
70	2485	1 16795	61 75225	3482 59331	2 04585 20425	123 61547 92835
71	2556	1 21836	65 33136	3736 71012	2 22627 49776	136 42550 76756
72	2628	1 27020	69 06384	4005 44868	2 41976 67408	150 35691 46260
73	2701	1 32349	72 95401	4289 43109	2 62707 39001	165 49033 72549
74	2775	1 37825	77 00625	4589 29685	2 84897 45625	181 91098 62725
75	2850	1 43450	81 22500	4905 70310	3 08627 92500	199 70883 78350
76	2926	1 49226	85 61476	5239 32486	3 33983 17876	218 97883 06926
77	3003	1 55155	90 18009	5590 85527	3 61051 02033	239 82106 87015
78	3081	1 61239	94 92561	5961 00583	3 89922 76401	262 34102 87719
79	3160	1 67480	99 85600	6350 50664	4 20693 32800	286 64977 43240
80	3240	1 73880	104 97600	6760 10664	4 53461 32800	312 86417 43240
81	3321	1 80441	110 29041	7190 57385	4 88329 17201	341 10712 79721
82	3403	1 87165	115 80409	7642 69561	5 25403 15633	371 50779 51145
83	3486	1 94054	121 52196	8117 27882	5 64793 56276	404 20183 24514
84	3570	2 01110	127 44900	8615 15018	6 06614 75700	439 33163 56130
85	3655	2 08335	133 59025	9137 15643	6 50985 28825	477 04658 71755
86	3741	2 15731	139 95081	9684 16459	6 98027 99001	517 50331 06891
87	3828	2 23300	146 53584	10257 06220	7 47870 08208	560 86593 07900
88	3916	2 31044	153 35056	10856 75756	8 00643 27376	607 30633 94684
89	4005	2 38965	160 40025	11484 17997	8 56483 86825	657 00446 85645
90	4095	2 47065	167 69025	12140 27997	9 15532 86825	710 14856 85645
91	4186	2 55346	175 22596	12826 02958	9 77936 08276	766 93549 37686
92	4278	2 63810	183 01284	13542 42254	10 43844 23508	827 57099 39030
93	4371	2 72459	191 05641	14290 47455	11 13413 07201	892 27001 22479
94	4465	2 81295	199 36225	15071 22351	11 86803 47425	961 25699 03535
95	4560	2 90320	207 93600	15885 72976	12 64181 56800	1034 76617 94160
96	4656	2 99536	216 78336	16735 07632	13 45718 83776	1113 04195 83856
97	4753	3 08945	225 91009	17620 36913	14 31592 24033	1196 33915 88785
98	4851	3 18549	235 32201	18542 73729	15 21984 32001	1284 92339 69649
99	4950	3 28350	245 02500	19503 33330	16 17083 32500	1379 07141 19050
100	5050	3 38350	255 02500	20503 33330	17 17083 32500	1479 07141 19050

SUMS OF POSITIVE POWERS $\sum_{k=1}^{m} k^n$

Table 23.4

$m \backslash n$	7	8	9
1	1	1	1
2	129	257	513
3	2316	6818	20196
4	18700	72354	2 82340
5	96825	4 62979	22 35465
6	3 76761	21 42595	123 13161
7	12 00304	79 07396	526 66768
8	32 97456	246 84612	1868 84496
9	80 80425	677 31333	5743 04985
10	180 80425	1677 31333	15743 04985
11	375 67596	3820 90214	39322 52676
12	733 99404	8120 71910	90920 33028
13	1361 47921	16278 02631	1 96965 32401
14	2415 61425	31035 91687	4 03575 79185
15	4124 20800	56664 82312	7 88009 38560
16	6808 56256	99614 49608	14 75204 15296
17	10911 94929	1 69372 07049	26 61082 91793
18	17034 14961	2 79571 67625	46 44675 82161
19	25972 86700	4 49407 30666	78 71552 79940
20	38772 86700	7 05407 30666	129 91552 79940
21	56783 75241	10 83635 90027	209 34353 26521
22	81727 33129	16 32394 63563	330 07045 44313
23	1 15775 58576	24 15504 48844	510 18572 05776
24	1 61640 30000	35 16257 63020	774 36647 46000
25	2 22675 45625	50 42136 53645	1155 83620 11625
26	3 02993 55801	71 30407 18221	1698 78656 90601
27	4 07597 09004	99 54702 54702	2461 34631 75588
28	5 42526 37516	137 32722 53038	3519 19191 28996
29	7 15025 13825	187 35186 65999	4969 90651 04865
30	9 33725 13825	252 96186 65999	6938 20651 04865
31	12 08851 27936	338 25097 03440	9582 16872 65536
32	15 52448 66304	448 20213 31216	13100 60593 54368
33	19 78633 09281	588 84299 49457	17741 75437 56321
34	25 03866 59425	767 42238 54353	23813 45365 22785
35	31 47259 56300	992 60992 44978	31695 01751 94660
36	39 30901 20396	1274 72091 52434	41851 01318 63076
37	48 80219 97529	1625 96886 06355	54847 18716 58153
38	60 24375 80121	2060 74807 44851	71368 79729 21001
39	73 96685 86800	2595 94900 05332	92241 63340 79760
40	90 35085 86800	3251 30900 05332	1 18456 03340 79760
41	109 82628 60681	4049 80152 34453	1 51194 22684 73721
42	132 88021 93929	5018 06672 30869	1 91861 36523 23193
43	160 06208 05036	6186 88675 08470	2 42120 62642 60036
44	191 98986 14700	7591 70911 33686	3 03932 81037 69540
45	229 35680 67825	9273 22165 24311	3 79600 87463 47665
46	272 93857 25041	11277 98287 56247	4 71819 89090 16721
47	323 60088 45504	13659 11154 18008	5 83732 93821 19488
48	382 30771 87776	16477 03958 47064	7 18993 48427 14176
49	450 13002 60625	19800 33264 16665	8 81834 84406 24625
50	528 25502 60625	23706 58264 16665	10 77147 34406 24625

Table 23.4   SUMS OF POSITIVE POWERS $\sum_{k=1}^{m} k^n$

$m \backslash n$	7	8	9
51	617 99609 38476	28283 37709 87066	13 10563 86137 15076
52	720 80326 41004	33629 34995 18522	15 88554 44973 50788
53	838 27437 80841	39855 31899 29883	19 18530 80891 52921
54	972 16689 90825	47085 51512 69019	23 08961 40014 66265
55	1124 41042 25200	55458 90891 59644	27 69498 05854 50640
56	1297 11990 74736	65130 64007 33660	33 11115 00335 95536
57	1492 60965 67929	76273 55578 45661	39 46261 19889 79593
58	1713 40807 35481	89079 86395 63677	46 89027 07286 24521
59	1962 27322 20300	1 03762 90771 67998	55 55326 65472 79460
60	2242 20922 20300	1 20559 06771 67998	65 63096 25472 79460
61	2556 48350 56321	1 39729 79901 65279	77 32510 86401 13601
62	2908 64496 62529	1 61563 80957 50175	90 86219 51863 77153
63	3302 54303 01696	1 86379 38760 17696	106 49600 93432 30976
64	3742 34768 12800	2 14526 88527 28352	124 51040 78527 12960
65	4232 57047 03425	2 46391 36656 18977	145 22232 06906 03585
66	4778 08654 04481	2 82395 42718 88673	168 98500 07044 03521
67	5384 15770 09804	3 23002 19494 45314	196 19153 51006 98468
68	6056 45658 28236	3 68718 51890 98690	227 27863 53971 28036
69	6801 09190 80825	4 20098 35635 27331	262 73072 32327 04265
70	7624 63490 80825	4 77746 36635 27331	303 08433 02327 04265
71	8534 14692 39216	5 42321 71947 73092	348 93283 09511 53296
72	9537 20822 43504	6 14542 13310 81828	400 93152 87653 82288
73	10641 94807 62601	6 95188 14229 75909	459 80311 54736 50201
74	11857 07610 35625	7 85107 61631 79685	526 34352 62487 29625
75	13191 91497 07500	8 85220 53135 70310	601 42821 25280 26500
76	14656 43442 79276	9 96524 01010 25286	686 01885 63746 04676
77	16261 28675 46129	11 20097 63925 72967	781 17055 08237 76113
78	18017 84364 01041	12 57109 07632 56103	888 03947 17370 60721
79	19938 23453 87200	14 08819 95731 62664	1007 89106 77196 79040
80	22035 38653 87200	15 76592 11731 62664	1142 10879 57196 79040
81	24323 06578 42161	17 61894 13620 14505	1292 20343 10166 78161
82	26815 92048 98929	19 66308 22206 69481	1459 82298 14263 86193
83	29529 52558 88556	21 91537 44528 08522	1646 76323 66939 26596
84	32480 42905 44300	24 39413 33638 91018	1854 97898 52248 56260
85	35686 19993 72425	27 11903 86142 81643	2086 59593 15080 59385
86	39165 47815 94121	30 11121 78853 47499	2343 92334 88197 23001
87	42938 02610 81904	33 39333 46007 84620	2629 46750 30627 52528
88	47024 78207 18896	36 98967 98488 39916	2945 94588 48916 18576
89	51447 91556 14425	40 92626 86545 41997	3296 30228 85991 03785
90	56230 88456 14425	45 23094 07545 41997	3683 72277 75991 03785
91	61398 49475 50156	49 93346 60306 93518	4111 65257 77288 92196
92	66976 96076 73804	55 06565 47620 69134	4583 81394 10154 48868
93	72993 96947 34561	60 66147 28587 19535	5104 22502 40039 36161
94	79478 74541 53825	66 75716 22441 30351	5677 21982 62325 52865
95	86462 11837 63200	73 39136 65570 20976	6307 46923 59571 62240
96	93976 59315 74016	80 60526 23468 59312	7000 00323 17816 42496
97	1 02056 42160 52129	88 44269 59412 36273	7760 23429 04362 07713
98	1 10737 67693 76801	96 95032 61670 54129	8593 98205 25663 57601
99	1 20058 33041 67500	106 17777 31113 33330	9507 49930 00499 98500
100	1 30058 33041 67500	116 17777 31113 33330	10507 49930 00499 98500

SUMS OF POSITIVE POWERS $\sum_{k=1}^{m} k^n$ Table 23.4

$m\backslash n$	10		$m\backslash n$	10
1	1		51	613 38941 75112 62626
2	1025		52	757 94452 34603 19650
3	60074		53	932 83199 38258 32699
4	11 08650		54	1143 66451 30907 53275
5	108 74275		55	1396 95967 52098 93900
6	713 40451		56	1700 26516 43060 08076
7	3538 15700		57	2062 29849 57628 99325
8	14275 57524		58	2493 10270 26623 05149
9	49143 41925		59	3004 21945 59629 46550
10	1 49143 41925		60	3608 88121 59629 46550
11	4 08517 66526		61	4322 22412 76258 29151
12	10 27691 30750		62	5161 52349 34941 69375
13	24 06276 22599		63	6146 45378 53759 60224
14	52 98822 77575		64	7299 37528 99828 07200
15	110 65326 68200		65	8645 64962 44456 97825
16	220 60442 95976		66	10213 98650 53564 93601
17	422 20381 96425		67	12036 82430 99082 55050
18	779 25054 23049		68	14150 74713 00654 65674
19	1392 35716 80850		69	16596 94119 07202 25475
20	2416 35716 80850		70	19421 69368 07202 25475
21	4084 34526 59051		71	22676 93723 17301 06676
22	6740 33754 50475		72	26420 84347 43545 94100
23	10882 98866 64124		73	30718 46930 40581 51749
24	17223 32676 29500		74	35642 45970 14140 29125
25	26760 06992 70125		75	41273 81117 23612 94750
26	40876 77949 23501		76	47702 70010 47012 36126
27	61465 89270 18150		77	55029 38057 72874 36775
28	91085 56937 13574		78	63365 15640 85236 36199
29	1 33156 29270 13775		79	72833 43249 11504 83400
30	1 92205 29270 13775		80	83570 85073 11504 83400
31	2 74168 12139 94576		81	95728 51619 02074 12201
32	3 86758 11208 37200		82	1 09473 31932 38034 70825
33	5 39916 01061 01649		83	1 24989 36051 10093 24274
34	7 46353 78601 61425		84	1 42479 48338 76074 16050
35	10 22208 52136 77050		85	1 62166 92382 16796 81675
36	13 87824 36537 40026		86	1 84297 08171 04827 52651
37	18 68682 80261 57875		87	2 09139 42312 96263 21500
38	24 96503 98741 46099		88	2 36989 52073 05665 33724
39	33 10544 59593 37700		89	2 68171 24066 05327 17325
40	43 59120 59593 37700		90	3 03039 08467 05327 17325
41	57 01386 52694 90101		91	3 41980 69648 23434 62726
42	74 09406 33911 67925		92	3 85419 54190 47066 76550
43	95 70554 57044 52174		93	4 33817 77262 26359 94799
44	122 90290 66428 70350		94	4 87679 28403 21259 64975
45	156 95353 55588 85975		95	5 47552 97795 59638 55600
46	199 37428 30416 62551		96	6 14036 24155 51139 60176
47	251 97341 52774 92600		97	6 87778 65424 46067 86225
48	316 89847 73860 37624		98	7 69485 93493 33614 75249
49	396 69074 36836 49625		99	8 59924 14243 42419 24250
50	494 34699 36836 49625		100	9 59924 14243 42419 24250

Table 23.5 $x^n/n!$

$n\backslash x$	2	3	4	5
1	(0)2.0000 00000	(0)3.0000 00000	(0)4.0000 00000	(0)5.0000 00000
2	(0)2.0000 00000	(0)4.5000 00000	(0)8.0000 00000	(1)1.2500 00000
3	(0)1.3333 33333	(0)4.5000 00000	(1)1.0666 66667	(1)2.0833 33333
4	(− 1)6.6666 66667	(0)3.3750 00000	(1)1.0666 66667	(1)2.6041 66667
5	(− 1)2.6666 66667	(0)2.0250 00000	(0)8.5333 33333	(1)2.6041 66667
6	(− 2)8.8888 88889	(0)1.0125 00000	(0)5.6888 88889	(1)2.1701 38889
7	(− 2)2.5396 82540	(− 1)4.3392 85714	(0)3.2507 93651	(1)1.5500 99206
8	(− 3)6.3492 06349	(− 1)1.6272 32143	(0)1.6253 96825	(0)9.6881 20040
9	(− 3)1.4109 34744	(− 2)5.4241 07143	(− 1)7.2239 85891	(0)5.3822 88911
10	(− 4)2.8218 69489	(− 2)1.6272 32143	(− 1)2.8895 94356	(0)2.6911 44455
11	(− 5)5.1306 71797	(− 3)4.4379 05844	(− 1)1.0507 61584	(0)1.2232 47480
12	(− 6)8.5511 19662	(− 3)1.1094 76461	(− 2)3.5025 38614	(− 1)5.0968 64499
13	(− 6)1.3155 56871	(− 4)2.5603 30295	(− 2)1.0777 04189	(− 1)1.9603 32500
14	(− 7)1.8793 66959	(− 5)5.4864 22060	(− 3)3.0791 54825	(− 2)7.0011 87499
15	(− 8)2.5058 22612	(− 5)1.0972 84412	(− 4)8.2110 79534	(− 2)2.3337 29166
16	(− 9)3.1322 78265	(− 6)2.0574 08273	(− 4)2.0527 69883	(− 3)7.2929 03644
17	(−10)3.6850 33252	(− 7)3.6307 20481	(− 5)4.8300 46785	(− 3)2.1449 71660
18	(−11)4.0944 81392	(− 8)6.0512 00802	(− 5)1.0733 43730	(− 4)5.9582 54611
19	(−12)4.3099 80412	(− 9)9.5545 27581	(− 6)2.2596 71010	(− 4)1.5679 61740
20	(−13)4.3099 80412	(− 9)1.4331 79137	(− 7)4.5193 42021	(− 5)3.9199 04350
21	(−14)4.1047 43250	(−10)2.0473 98767	(− 8)8.6082 70516	(− 6)9.3331 05594
22	(−15)3.7315 84772	(−11)2.7919 07410	(− 8)1.5651 40094	(− 6)2.1211 60362
23	(−16)3.2448 56324	(−12)3.6416 18361	(− 9)2.7219 82772	(− 7)4.6112 18179
24	(−17)2.7040 46937	(−13)4.5520 22951	(−10)4.5366 37953	(− 8)9.6067 04540
25	(−18)2.1632 37549	(−14)5.4624 27542	(−11)7.2586 20725	(− 8)1.9213 40908
26	(−19)1.6640 28884	(−15)6.3028 01010	(−11)1.1167 10881	(− 9)3.6948 86361
27	(−20)1.2326 13988	(−16)7.0031 12233	(−12)1.6543 86490	(−10)6.8423 82151
28	(−22)8.8043 85630	(−17)7.5033 34535	(−13)2.3634 09271	(−10)1.2218 53955
29	(−23)6.0719 90090	(−18)7.7620 70209	(−14)3.2598 74857	(−11)2.1066 44751
30	(−24)4.0479 93393	(−19)7.7620 70209	(−15)4.3464 99809	(−12)3.5110 74585
31	(−25)2.6116 08641	(−20)7.5116 80847	(−16)5.6083 86851	(−13)5.6630 23524
32	(−26)1.6322 55400	(−21)7.0422 00794	(−17)7.0104 83563	(−14)8.8484 74256
33	(−28)9.8924 56972	(−22)6.4020 00722	(−18)8.4975 55835	(−14)1.3406 77918
34	(−29)5.8190 92337	(−23)5.6488 24167	(−19)9.9971 24511	(−15)1.9715 85173
35	(−30)3.3251 95621	(−24)4.8418 49286	(−19)1.1425 28516	(−16)2.8165 50247
36	(−31)1.8473 30901	(−25)4.0348 74405	(−20)1.2694 76128	(−17)3.9118 75343
37	(−33)9.9855 72435	(−26)3.2715 19788	(−21)1.3724 06625	(−18)5.2863 18031
38	(−34)5.2555 64440	(−27)2.5827 78780	(−22)1.4446 38553	(−19)6.9556 81620
39	(−35)2.6951 61251	(−28)1.9867 52907	(−23)1.4816 80567	(−20)8.9175 40538
40	(−36)1.3475 80626	(−29)1.4900 64681	(−24)1.4816 80567	(−20)1.1146 92567
41	(−38)6.5735 64027	(−30)1.0902 91230	(−25)1.4455 42017	(−21)1.3593 81180
42	(−39)3.1302 68584	(−32)7.7877 94498	(−26)1.3767 06683	(−22)1.6183 10928
43	(−40)1.4559 38876	(−33)5.4333 44998	(−27)1.2806 57379	(−23)1.8817 56893
44	(−42)6.6179 03984	(−34)3.7045 33408	(−28)1.1642 33981	(−24)2.1383 60106
45	(−43)2.9412 90659	(−35)2.4697 02272	(−29)1.0348 74650	(−25)2.3759 55673
46	(−44)1.2788 22026	(−36)1.6106 75395	(−31)8.9989 09998	(−26)2.5825 60514
47	(−46)5.4417 95855	(−37)1.0280 90678	(−32)7.6586 46807	(−27)2.7474 04802
48	(−47)2.2674 14939	(−39)6.4255 66735	(−33)6.3822 05673	(−28)2.8618 80003
49	(−49)9.2547 54855	(−40)3.9340 20450	(−34)5.2099 63814	(−29)2.9202 85717
50	(−50)3.7019 01942	(−41)2.3604 12270	(−35)4.1679 71052	(−30)2.9202 85717

For $x=1$, see **Table 6.3.**

$$x^n/n!$$

Table 23.5

$n\backslash x$	6	7	8	9
1	(0) 6.0000 00000	(0) 7.0000 00000	(0) 8.0000 00000	(0) 9.0000 00000
2	(1) 1.8000 00000	(1) 2.4500 00000	(1) 3.2000 00000	(1) 4.0500 00000
3	(1) 3.6000 00000	(1) 5.7166 66667	(1) 8.5333 33333	(2) 1.2150 00000
4	(1) 5.4000 00000	(2) 1.0004 16667	(2) 1.7066 66667	(2) 2.7337 50000
5	(1) 6.4800 00000	(2) 1.4005 83333	(2) 2.7306 66667	(2) 4.9207 50000
6	(1) 6.4800 00000	(2) 1.6340 13889	(2) 3.6408 88889	(2) 7.3811 25000
7	(1) 5.5542 85714	(2) 1.6340 13889	(2) 4.1610 15873	(2) 9.4900 17857
8	(1) 4.1657 14286	(2) 1.4297 62153	(2) 4.1610 15873	(3) 1.0676 27009
9	(1) 2.7771 42857	(2) 1.1120 37230	(2) 3.6986 80776	(3) 1.0676 27009
10	(1) 1.6662 85714	(1) 7.7842 60610	(2) 2.9589 44621	(2) 9.6086 43080
11	(0) 9.0888 31169	(1) 4.9536 20388	(2) 2.1519 59724	(2) 7.8616 17066
12	(0) 4.5444 15584	(1) 2.8896 11893	(2) 1.4346 39816	(2) 5.8962 12799
13	(0) 2.0974 22577	(1) 1.5559 44865	(1) 8.8285 52715	(2) 4.0819 93476
14	(− 1) 8.9889 53903	(0) 7.7797 24327	(1) 5.0448 87266	(2) 2.6241 38663
15	(− 1) 3.5955 81561	(0) 3.6305 38019	(1) 2.6906 06542	(2) 1.5744 83198
16	(− 1) 1.3483 43085	(0) 1.5883 60383	(1) 1.3453 03271	(1) 8.8564 67989
17	(− 2) 4.7588 57949	(− 1) 6.5403 07461	(0) 6.3308 38922	(1) 4.6887 18347
18	(− 2) 1.5862 85983	(− 1) 2.5434 52902	(0) 2.8137 06187	(1) 2.3443 59174
19	(− 3) 5.0093 24157	(− 2) 9.3706 15953	(0) 1.1847 18395	(1) 1.1104 85924
20	(− 3) 1.5027 97247	(− 2) 3.2797 15584	(− 1) 4.7388 73579	(0) 4.9971 86659
21	(− 4) 4.2937 06420	(− 2) 1.0932 38528	(− 1) 1.8052 85173	(0) 2.1416 51425
22	(− 4) 1.1710 10842	(− 3) 3.4784 86225	(− 2) 6.5646 73356	(− 1) 8.7613 01286
23	(− 5) 3.0548 10892	(− 3) 1.0586 69721	(− 2) 2.2833 64645	(− 1) 3.4283 35286
24	(− 6) 7.6370 27229	(− 4) 3.0877 86685	(− 3) 7.6112 15485	(− 1) 1.2856 25732
25	(− 6) 1.8328 86535	(− 5) 8.6458 02719	(− 3) 2.4355 88955	(− 2) 4.6282 52636
26	(− 7) 4.2297 38158	(− 5) 2.3277 16117	(− 4) 7.4941 19862	(− 2) 1.6020 87451
27	(− 8) 9.3994 18128	(− 6) 6.0348 19562	(− 4) 2.2204 79959	(− 3) 5.3402 91503
28	(− 8) 2.0141 61027	(− 6) 1.5087 04890	(− 5) 6.3442 28455	(− 3) 1.7165 22269
29	(− 9) 4.1672 29712	(− 7) 3.6417 01460	(− 5) 1.7501 31987	(− 4) 5.3271 38076
30	(−10) 8.3344 59424	(− 8) 8.4973 03406	(− 6) 4.6670 18633	(− 4) 1.5981 41423
31	(−10) 1.6131 21179	(− 8) 1.9187 45930	(− 6) 1.2043 91905	(− 5) 4.6397 65421
32	(−11) 3.0246 02210	(− 9) 4.1972 56723	(− 7) 3.0109 79763	(− 5) 1.3049 34025
33	(−12) 5.4992 76746	(−10) 8.9032 71836	(− 8) 7.2993 44881	(− 6) 3.5589 10976
34	(−13) 9.7046 06022	(−10) 1.8330 26554	(− 8) 1.7174 92913	(− 7) 9.4206 46701
35	(−13) 1.6636 46746	(−11) 3.6660 53109	(− 9) 3.9256 98087	(− 7) 2.4224 52009
36	(−14) 2.7727 44578	(−12) 7.1284 36600	(−10) 8.7237 73527	(− 8) 6.0561 30022
37	(−15) 4.4963 42559	(−12) 1.3486 23141	(−10) 1.8862 45303	(− 8) 1.4731 12708
38	(−16) 7.0994 88251	(−13) 2.4843 05785	(−11) 3.9709 92217	(− 9) 3.4889 51151
39	(−16) 1.0922 28962	(−14) 4.4590 10384	(−12) 8.1456 25061	(−10) 8.0514 25733
40	(−17) 1.6383 43442	(−15) 7.8032 68172	(−12) 1.6291 25012	(−10) 1.8115 70790
41	(−18) 2.3975 75769	(−15) 1.3322 65298	(−13) 3.1787 80512	(−11) 3.9766 18807
42	(−19) 3.4251 08242	(−16) 2.2204 42163	(−14) 6.0548 20022	(−12) 8.5213 26015
43	(−20) 4.7792 20803	(−17) 3.6146 73288	(−14) 1.1264 78144	(−12) 1.7835 33352
44	(−21) 6.5171 19277	(−18) 5.7506 16595	(−15) 2.0481 42079	(−13) 3.6481 36402
45	(−22) 8.6894 92369	(−19) 8.9454 03592	(−16) 3.6411 41474	(−14) 7.2962 72802
46	(−22) 1.1334 12048	(−19) 1.3612 57068	(−17) 6.3324 19956	(−14) 1.4275 31635
47	(−23) 1.4469 08998	(−20) 2.0274 04144	(−17) 1.0778 58716	(−15) 2.7335 71217
48	(−24) 1.8086 36247	(−21) 2.9566 31044	(−18) 1.7964 31193	(−16) 5.1254 46032
49	(−25) 2.2146 56629	(−22) 4.2237 58634	(−19) 2.9329 48887	(−17) 9.4140 84548
50	(−26) 2.6575 87955	(−23) 5.9132 62088	(−20) 4.6927 18219	(−17) 1.6945 35219

24. Combinatorial Analysis

K. GOLDBERG,[1] M. NEWMAN,[2] E. HAYNSWORTH [3]

Contents

[1, 2] National Bureau of Standards.
[3] National Bureau of Standards. (Presently, Auburn University.)

24. Combinatorial Analysis

Mathematical Properties

In each sub-section of this chapter we use a fixed format which emphasizes the use and methods of extending the accompanying tables. The format follows this form:

I. Definitions

   A. Combinatorial

   B. Generating functions

   C. Closed form

II. Relations

   A. Recurrences

   B. Checks in computing

   C. Basic use in numerical analysis

III. Asymptotic and Special Values

In general the notations used are standard. This includes the difference operator Δ defined on functions of x by $\Delta f(x) = f(x+1) - f(x)$, $\Delta^{n+1} f(x) = \Delta(\Delta^n f(x))$, the Kronecker delta δ_{ij}, the Riemann zeta function $\zeta(s)$ and the greatest common divisor symbol (m, n). The range of the summands for a summation sign without limits is explained to the right of the formula.

The notations which are not standard are those for the multinomials which are arbitrary shorthand for use in this chapter, and those for the Stirling numbers which have never been standardized. A short table of various notations for these numbers follows:

Notations for the Stirling Numbers

Reference		First Kind	Second Kind	
	This chapter	$S_n^{(m)}$	$\mathfrak{S}_n^{(m)}$	
[24.2]	Fort	$S_n^{(m)}$	$\mathscr{S}_n^{(m)}$	*
[24.7]	Jordan	S_n^m	$\mathfrak{S}_n^m$	*
[24.10]	Moser and Wyman	S_n^m	σ_n^m	
[24.9]	Milne-Thomson	$\binom{n-1}{m-1} B_{n-m}^{(n)}$	$\binom{n}{m} B_{n-m}^{(-m)}$	
[24.15]	Riordan	$s(n, m)$	$S(n, m)$	
[24.1]	Carlitz			
[24.3]	Gould	$(-1)^{n-m} S_1(n-1, n-m)$	$S_2(m, n-m)$	
	Miksa (Unpublished tables)	$S(n-m+1, n)$	$_m S_n$	
[24.17]	Gupta		$u(n, m)$	

We feel that a capital S is natural for Stirling numbers of the first kind; it is infrequently used for other notation in this context. But once it is used we have difficulty finding a suitable symbol for Stirling numbers of the second kind. The numbers are sufficiently important to warrant a special and easily recognizable symbol, and yet that symbol must be easy to write. We have settled on a script capital $\mathfrak{S}$ without any certainty that we have settled this question permanently.

We feel that the subscript-superscript notation emphasizes the generating functions (which are powers of mutually inverse functions) from which most of the important relations flow.

24.1. Basic Numbers

24.1.1 Binomial Coefficients

I. Definitions

A. $\binom{n}{m}$ is the number of ways of choosing m objects from a collection of n distinct objects without regard to order.

B. Generating functions

$$* \quad (1+x)^n = \sum_{m=0}^{n} \binom{n}{m} x^m \qquad n = 0, 1, \ldots$$

$$(1-x)^{-m-1} = \sum_{n=m}^{\infty} \binom{n}{m} x^{n-m} \qquad |x| < 1$$

C. Closed form

$$\binom{n}{m} = \frac{n!}{m!(n-m)!} = \binom{n}{n-m}$$

$$\qquad n \geq m$$

$$= \frac{n(n-1) \ldots (n-m+1)}{m!}$$

II. Relations

A. Recurrences

$$\binom{n+1}{m} = \binom{n}{m} + \binom{n}{m-1} \qquad n \geq m \geq 1$$

$$= \binom{n}{m} + \binom{n-1}{m-1} + \ldots + \binom{n-m}{0} \qquad n \geq m$$

B. Checks

$$\sum_{m=0}^{n} \binom{r}{m} \binom{s}{n-m} = \binom{r+s}{n} \qquad r+s \geq n$$

$$\sum_{m=0}^{n} (-1)^{n-m} \binom{r}{m} = \binom{r-1}{n} \qquad r \geq n+1$$

$$\binom{n}{m} \equiv \binom{n_0}{m_0} \binom{n_1}{m_1} \ldots \pmod{p} \qquad p \text{ a prime}$$

*See page II.

where

$$n=\sum_{k=0}^{\infty} n_k p^k, \qquad m=\sum_{k=0}^{\infty} m_k p^k \qquad p>m_k, \; n_k \geq 0$$

$$\sum_{m=0}^{s} (-1)^m \binom{n}{m} f(x-m)$$

$$=\sum_{k=0}^{s} (-1)^{s-k} \binom{n-k-1}{s-k} \Delta^k f(x-s) \qquad s<n$$

C. Numerical analysis

III. Special Values

$$\Delta^n f(x)=\sum_{m=0}^{n} (-1)^{n-m} \binom{n}{m} f(x+m)$$

$$\binom{n}{0}=\binom{n}{n}=1$$

$$=\sum_{k=0}^{r} \binom{r}{k} \Delta^{n+k} f(x-r)$$

$$\binom{2n}{n}=\frac{2^n(2n-1)(2n-3)\ldots 3\cdot 1}{n!}$$

24.1.2   Multinomial Coefficients

I. Definitions

A. $(n;\, n_1,\, n_2,\, \ldots,\, n_m)$ is the number of ways of putting $n=n_1+n_2+\ldots+n_m$ different objects into m different boxes with n_k in the k-th box, $k=1, 2, \ldots, m$.

$(n;\, a_1,\, a_2,\, \ldots,\, a_n)^*$ is the number of permutations of $n=a_1+2a_2+\ldots+na_n$ symbols composed of a_k cycles of length k for $k=1, 2, \ldots, n$.

$(n;\, a_1,\, a_2,\, \ldots,\, a_n)'$ is the number of ways of partitioning a set of $n=a_1+2a_2+\ldots+na_n$ different objects into a_k subsets containing k objects for $k=1, 2, \ldots, n$.

B. Generating functions

$$(x_1+x_2+\ldots+x_m)^n=\Sigma(n;\, n_1,\, n_2,\, \ldots,\, n_m)x_1^{n_1}x_2^{n_2}\ldots x_m^{n_m} \qquad \text{summed over } n_1+n_2+\ldots+n_m=n$$

$$\left(\sum_{k=1}^{\infty} \frac{x_k}{k}\, t^k\right)^m=m!\sum_{n=m}^{\infty} \frac{t^n}{n!}\, \Sigma(n;\, a_1, a_2, \ldots, a_n)^* x_1^{a_1}x_2^{a_2}\ldots x_n^{a_n}$$

summed over $a_1+2a_2+\ldots+na_n=n$

and $a_1+a_2+\ldots+a_n=m$

$$\left(\sum_{k=1}^{\infty} \frac{x_k}{k!}\, t^k\right)^m=m!\sum_{n=m}^{\infty} \frac{t^n}{n!}\, \Sigma(n;\, a_1, a_2, \ldots, a_n)' x_1^{a_1}x_2^{a_2}\ldots x_n^{a_n}$$

C. Closed forms

$$(n;\, n_1, n_2, \ldots, n_m)=n!/n_1!n_2!\ldots n_m! \qquad\qquad n_1+n_2+\ldots+n_m=n$$

$$(n;\, a_1, a_2, \ldots, a_n)^*=n!/1^{a_1}a_1!2^{a_2}a_2!\ldots n^{a_n}a_n! \qquad a_1+2a_2+\ldots+na_n=n$$

$$(n;\, a_1, a_2, \ldots, a_n)'=n!/(1!)^{a_1}a_1!(2!)^{a_2}a_2!\ldots (n!)^{a_n}a_n! \qquad a_1+2a_2+\ldots+na_n=n$$

II. Relations

A. Recurrence

$$(n+m;\, n_1+1, n_2+1, \ldots, n_m+1)=\sum_{k=1}^{m} (n+m-1;\, n_1+1, \ldots, n_{k-1}+1, n_k, n_{k+1}+1, \ldots, n_m+1)$$

B. Checks

$$\ast\quad \Sigma(n;\, n_1, n_2, \ldots, n_m)=\begin{cases} m^n & \text{all } n_i \geq 1 \\ m!\; \mathfrak{S}_n^{(m)} & \end{cases} \qquad \text{summed over } n_1+n_2+\ldots+n_m=n$$

$$\Sigma(n;\, a_1, a_2, \ldots, a_n)^*=(-1)^{n-m}S_n^{(m)} \qquad \text{summed over } a_1+2a_2+\ldots+na_n=n \text{ and } a_1+a_2+\ldots+a_n=m$$

$$\Sigma(n;\, a_1, a_2, \ldots, a_n)'=\mathfrak{S}_n^{(m)}$$

C. Numerical analysis (Faà di Bruno's formula)

$$\frac{d^n}{dx^n} f(g(x))=\sum_{m=0}^{n} f^{(m)}(g(x))\Sigma(n;\, a_1, a_2, \ldots, a_n)'\{g'(x)\}^{a_1}\{g''(x)\}^{a_2}\ldots\{g^{(n)}(x)\}^{a_n}$$

summed over $a_1+2a_2+\ldots+na_n=n$ and $a_1+a_2+\ldots+a_n=m$.

*See page II.

$$\begin{vmatrix} P_1 & 1 & 0 & \cdots & 0 \\ P_2 & P_1 & 2 & \cdots & \cdot \\ P_3 & P_2 & P_1 & \cdots & \cdot \\ \cdot & \cdot & \cdot & \cdots & 0 \\ \cdot & \cdot & \cdot & \cdots & n-1 \\ P_n & P_{n-1} & P_{n-2} & \cdots & P_1 \end{vmatrix} = \Sigma(-1)^{n-\Sigma a_i}(n; a_1, a_2, \ldots, a_n)*P_1^{a_1}P_2^{a_2}\ldots P_n^{a_n}$$

summed over $a_1+2a_2+\ldots+na_n=n$; e.g. if $P_k=\Sigma_{j=1}^r x_j^k$ for $k=1, 2, \ldots, n$ then the determinant and sum equal $n!\Sigma x_1 x_2 \ldots x_n$, the latter sum denoting the n-th elementary symmetric function of $x_1, x_2, \ldots, x_r$.

24.1.3 Stirling Numbers of the First Kind

I. Definitions

A. $(-1)^{n-m}S_n^{(m)}$ is the number of permutations of n symbols which have exactly m cycles.

B. Generating functions

$$x(x-1)\ldots(x-n+1)=\sum_{m=0}^{n} S_n^{(m)}x^m$$

$$\{\ln (1+x)\}^m=m!\sum_{n=m}^{\infty} S_n^{(m)}\frac{x^n}{n!} \qquad |x|<1$$

C. Closed form (see closed form for $\mathscr{S}_n^{(m)}$)

$$S_n^{(m)}=\sum_{k=0}^{n-m} (-1)^k \binom{n-1+k}{n-m+k}\binom{2n-m}{n-m-k} \mathscr{S}_{n-m+k}^{(k)}$$

II. Relations

A. Recurrences

$$S_{n+1}^{(m)}=S_n^{(m-1)}-nS_n^{(m)} \qquad n\geq m\geq 1$$

$$\binom{m}{r} S_n^{(m)}=\sum_{k=m-r}^{n-r} \binom{n}{k} S_{n-k}^{(r)}S_k^{(m-r)} \qquad n\geq m\geq r$$

B. Checks

$$\sum_{m=1}^{n} S_n^{(m)}=0 \qquad n>1$$

$$\sum_{m=0}^{n}(-1)^{n-m}S_n^{(m)}=n!$$

$$\sum_{k=m}^{n} S_{n+1}^{(k+1)}n^{k-m}=S_n^{(m)}$$

C. Numerical analysis

$$\frac{d^m}{dx^m}f(x)=m!\sum_{n=m}^{\infty} \frac{S_n^{(m)}}{n!} \Delta^n f(x)$$

if convergent.

III. Asymptotics and Special Values

$$|S_n^{(m)}|\sim(n-1)!(\gamma+\ln n)^{m-1}/(m-1)!$$
$$\text{for } m=o(\ln n)$$

$$\lim_{m\to\infty} \frac{S_{n+m}^{(m)}}{m^{2n}}=\frac{(-1)^n}{2^n n!}$$

$$\lim_{n\to\infty} \frac{S_{n+1}^{(m)}}{nS_n^{(m)}}=-1$$

$$S_n^{(0)}=\delta_{0n}$$

$$S_n^{(1)}=(-1)^{n-1}(n-1)!$$

$$S_n^{(n-1)}=-\binom{n}{2}$$

$$S_n^{(n)}=1$$

24.1.4 Stirling Numbers of the Second Kind

I. Definitions

A. $\mathscr{S}_n^{(m)}$ is the number of ways of partitioning a set of n elements into m non-empty subsets.

B. Generating functions

$$x^n=\sum_{m=0}^{n} \mathscr{S}_n^{(m)}x(x-1)\ldots(x-m+1)$$

$$(e^x-1)^m=m!\sum_{n=m}^{\infty} \mathscr{S}_n^{(m)}\frac{x^n}{n!}$$

$$(1-x)^{-1}(1-2x)^{-1}\ldots(1-mx)^{-1}=\sum_{n=m}^{\infty} \mathscr{S}_n^{(m)}x^{n-m}$$
$$|x|<m^{-1}$$

C. Closed form

$$\mathscr{S}_n^{(m)}=\frac{1}{m!}\sum_{k=0}^{m} (-1)^{m-k} \binom{m}{k} k^n$$

II. Relations

A. Recurrences

$$\mathfrak{S}_{n+1}^{(m)} = m\,\mathfrak{S}_n^{(m)} + \mathfrak{S}_n^{(m-1)} \qquad n\geq m\geq 1$$

$$\binom{m}{r}\mathfrak{S}_n^{(m)} = \sum_{k=m-r}^{n-r}\binom{n}{k}\mathfrak{S}_{n-k}^{(r)}\mathfrak{S}_k^{(m-r)} \qquad n\geq m\geq r$$

B. Checks

$$\sum_{m=0}^{n}(-1)^{n-m}m!\,\mathfrak{S}_n^{(m)}=1$$

$$\sum_{k=m}^{n}\mathfrak{S}_{k-1}^{(m-1)}m^{n-k}=\mathfrak{S}_n^{(m)}$$

$$\mathfrak{S}_n^{(m)}=\sum_{k=0}^{n-m}(-1)^k\binom{n-1+k}{n-m+k}\binom{2n-m}{n-m-k}S_{n-m+k}^{(k)}$$

$$\sum_{k=m}^{n}S_k^{(m)}\mathfrak{S}_n^{(k)}=\sum_{k=m}^{n}S_n^{(k)}\mathfrak{S}_k^{(m)}=\delta_{mn}$$

C. Numerical analysis

$$\Delta^m f(x)=m!\sum_{n=m}^{\infty}\frac{\mathfrak{S}_n^{(m)}}{n!}f^{(n)}(x) \qquad \text{if convergent}$$

$$\sum_{k=0}^{n}k^m=\sum_{k=0}^{m}k!\,\mathfrak{S}_m^{(k)}\binom{n+1}{k+1}$$

$$\sum_{k=0}^{n}k^m x^k=\sum_{j=0}^{m}\mathfrak{S}_m^{(j)}x^j\frac{d^j}{dx^j}\left\{\frac{1-x^{n+1}}{1-x}\right\}$$

III. Asymptotics and Special Values

$$\lim_{n\to\infty}m^{-n}\mathfrak{S}_n^{(m)}=m!^{-1}$$

$$\mathfrak{S}_{n+m}^{(m)}\sim\frac{m^{2n}}{2^n n!} \qquad \text{for } n=o(m^{\frac{1}{2}})$$

$$\lim_{n\to\infty}\frac{\mathfrak{S}_{n+1}^{(m)}}{\mathfrak{S}_n^{(m)}}=m$$

$$\mathfrak{S}_n^{(0)}=\delta_{0n}$$

$$\mathfrak{S}_n^{(1)}=\mathfrak{S}_n^{(n)}=1$$

$$\mathfrak{S}_n^{(n-1)}=\binom{n}{2}$$

24.2. Partitions

24.2.1 Unrestricted Partitions

I. Definitions

A. $p(n)$ is the number of decompositions of n into integer summands without regard to order. E.g., $5=1+4=2+3=1+1+3=1+2+2=1+1+1+2=1+1+1+1+1$ so that $p(5)=7$.

B. Generating function

$$\sum_{n=0}^{\infty}p(n)x^n=\prod_{n=1}^{\infty}(1-x^n)^{-1}=\left\{\sum_{n=-\infty}^{\infty}(-1)^n x^{\frac{3n^2+n}{2}}\right\}^{-1} \qquad |x|<1$$

C. Closed form

$$p(n)=\frac{1}{\pi\sqrt{2}}\sum_{k=1}^{\infty}\sqrt{k}A_k(n)\frac{d}{dn}\frac{\sinh\left\{\frac{\pi}{k}\sqrt{\frac{2}{3}}\sqrt{n-\frac{1}{24}}\right\}}{\sqrt{n-\frac{1}{24}}}$$

where

$$A_k(n)=\sum_{\substack{0<h\leq k\\(h,k)=1}}e^{\pi i s(h,k)}e^{-\frac{2\pi i h n}{k}}$$

$$s(h,k)=\sum_{j=1}^{k-1}\frac{j}{k}\left(\!\left(\frac{hj}{k}\right)\!\right)$$

$$((x))=x-[x]-\tfrac{1}{2} \text{ if } x \text{ is not an integer}$$
$$=0 \qquad\qquad \text{if } x \text{ is an integer}$$

II. Relations

A. Recurrence

$$p(n)=\sum_{1\leq\frac{3k^2\pm k}{2}\leq n}(-1)^{k-1}p\left(n-\frac{3k^2\pm k}{2}\right) \qquad p(0)=1$$

$$=\frac{1}{n}\sum_{k=1}^{n}\sigma_1(k)p(n-k)$$

B. Check

$$p(n)+\sum_{1\leq\frac{3k^2\pm k}{2}\leq n}(-1)^k\frac{3k^2\pm k}{2}p\left(n-\frac{3k^2\pm k}{2}\right)=\sigma_1(n)$$

III. Asymptotics

$$p(n)\sim\frac{1}{4n\sqrt{3}}e^{\pi\sqrt{2/3}\sqrt{n}}$$

24.2.2 Partitions Into Distinct Parts

I. Definitions

A. $q(n)$ is the number of decompositions of n into distinct integer summands without regard to order. E.g., $5=1+4=2+3$ so that $q(5)=3$.

B. Generating function

$$\sum_{n=0}^{\infty}q(n)x^n=\prod_{n=1}^{\infty}(1+x^n)=\prod_{n=1}^{\infty}(1-x^{2n-1})^{-1} \qquad |x|<1$$

C. Closed form

$$q(n)=\frac{1}{\sqrt{2}}\sum_{k=1}^{\infty}A_{2k-1}(n)\frac{d}{dn}J_0\left(\frac{\pi i}{2k-1}\sqrt{\frac{1}{3}}\sqrt{n+\frac{1}{24}}\right)$$

where $J_0(x)$ is the Bessel function of order 0 and $A_{2k-1}(n)$ was defined in part I.C. of the previous subsection.

II. Relations

A. Recurrences

$$\sum_{0 \le \frac{3k^2 \pm k}{2} \le n} (-1)^k q\left(n - \frac{3k^2 \pm k}{2}\right) = (-1)^r \text{ if } n = 3r^2 \pm r$$

$$q(0) = 1$$

$$= 0 \text{ otherwise}$$

$$q(n) = \frac{1}{n} \sum_{k=1}^{n} \left\{ \sigma_1(k) - 2\sigma_1\left(\frac{k}{2}\right) \right\} q(n-k)$$

B. Check

$$\sum_{0 \le 3k^2 \pm k \le n} (-1)^k q(n - (3k^2 \pm k)) = 1 \text{ if } n = \frac{r^2 - r}{2}$$

$$= 0 \text{ otherwise.}$$

III. Asymptotics

$$q(n) \sim \frac{1}{4 \cdot 3^{1/4} \cdot n^{3/4}} e^{\pi \sqrt{1/3} \sqrt{n}}$$

24.3. Number Theoretic Functions

24.3.1 The Möbius Function

I. Definitions

A. $\mu(n) = 1$ if $n = 1$

$= (-1)^k$ if n is the product of k distinct primes

$= 0$ if n is divisible by a square > 1.

B. Generating functions

$$\sum_{n=1}^{\infty} \mu(n) n^{-s} = 1/\zeta(s) \qquad \mathscr{R}s > 1$$

$$\sum_{n=1}^{\infty} \frac{\mu(n) x^n}{1 - x^n} = x \qquad |x| < 1$$

II. Relations

A. Recurrence

$$\mu(mn) = \mu(m)\mu(n) \text{ if } (m, n) = 1$$

$$= 0 \qquad \text{if } (m, n) > 1$$

B. Check

$$\sum_{d|n} \mu(d) = \delta_{n1}$$

C. Numerical analysis

$g(n) = \sum_{d|n} f(d)$ for all n if and only if

$$f(n) = \sum_{d|n} \mu(d) g(n/d) \text{ for all } n$$

$g(n) = \prod_{d|n} f(d)$ for all n if and only if

$$f(n) = \prod_{d|n} g(n/d)^{\mu(d)} \text{ for all } n$$

$g(x) = \sum_{n=1}^{[x]} f(x/n)$ for all $x > 0$ if and only if

$$f(x) = \sum_{n=1}^{[x]} \mu(n) g(x/n) \text{ for all } x > 0$$

$g(x) = \sum_{n=1}^{\infty} f(nx)$ for all $x > 0$ if and only if

$$f(x) = \sum_{n=1}^{\infty} \mu(n) g(nx) \text{ for all } x > 0$$

and if $\sum_{m=1}^{\infty} \sum_{n=1}^{\infty} |f(mnx)| = \sum_{n=1}^{\infty} \sigma_0(n) |f(nx)|$ converges.

The cyclotomic polynomial of order n is $\prod_{d|n} (x^d - 1)^{\mu(n/d)}$

III. Asymptotics

$$\sum_{n=1}^{\infty} \frac{\mu(n)}{n} = 0$$

$$\sum_{n=1}^{\infty} \frac{\mu(n)}{n} \ln n = -1$$

$$\sum_{n \le x} \mu(n) = 0(x e^{-c \sqrt{\ln x}})$$

24.3.2 The Euler Totient Function

I. Definitions

A. $\varphi(n)$ is the number of integers not exceeding and relatively prime to n.

B. Generating functions

$$\sum_{n=1}^{\infty} \varphi(n) n^{-s} = \frac{\zeta(s-1)}{\zeta(s)} \qquad \mathscr{R}s > 2$$

$$\sum_{n=1}^{\infty} \frac{\varphi(n) x^n}{1 - x^n} = \frac{x}{(1-x)^2} \qquad |x| < 1$$

C. Closed form

$$\varphi(n) = n \prod_{p|n} \left(1 - \frac{1}{p}\right)$$

over distinct primes p dividing n.

II. Relations

A. Recurrence

$$\varphi(mn) = \varphi(m)\varphi(n) \qquad (m, n) = 1$$

B. Checks

$$\sum_{d|n} \varphi(d) = n$$

$$\varphi(n) = \sum_{d|n} \mu\left(\frac{n}{d}\right) d$$

$$a^{\varphi(n)} \equiv 1 \pmod{n} \qquad (a, n) = 1$$

III. Asymptotics

$$\frac{1}{n^2} \sum_{k=1}^{n} \varphi(k) = \frac{3}{\pi^2} + O\left(\frac{\ln n}{n}\right)$$

24.3.3 Divisor Functions

I. Definitions

A. $\sigma_k(n)$ is the sum of the k-th powers of the divisors of n. Often $\sigma_0(n)$ is denoted by $d(n)$, and $\sigma_1(n)$ by $\sigma(n)$.

B. Generating functions

$$\sum_{n=1}^{\infty} \sigma_k(n) n^{-s} = \zeta(s)\zeta(s-k) \qquad \mathscr{R}s > k+1$$

$$\sum_{n=1}^{\infty} \sigma_k(n) x^n = \sum_{n=1}^{\infty} \frac{n^k x^n}{1-x^n} \qquad |x| < 1$$

C. Closed form

$$\sigma_k(n) = \sum_{d|n} d^k = \prod_{i=1}^{s} \frac{p_i^{k(a_i+1)}-1}{p_i^k-1} \qquad n = p_1^{a_1} p_2^{a_2} \cdots p_s^{a_s}$$

II. Relations

A. Recurrences

$$\sigma_k(mn) = \sigma_k(m)\sigma_k(n) \qquad (m, n) = 1$$

$$\sigma_k(np) = \sigma_k(n)\sigma_k(p) - p^k \sigma_k(n/p) \qquad p \text{ prime}$$

III. Asymptotics

$$\frac{1}{n}\sum_{m=1}^{n} \sigma_0(m) = \ln n + 2\gamma - 1 + O(n^{-\frac{1}{2}})$$

$$(\gamma = \text{Euler's constant})$$

$$\frac{1}{n^2}\sum_{m=1}^{n} \sigma_1(m) = \frac{\pi^2}{12} + O\left(\frac{\ln n}{n}\right)$$

24.3.4 Primitive Roots

I. Definitions

The integers not exceeding and relatively prime to a fixed integer n form a group; the group is cyclic if and only if $n=2$, 4 or n is of the form p^k or $2p^k$ where p is an odd prime. Then g is a primitive root of n if it generates that group; i.e., if g, g^2, . . ., $g^{\varphi(n)}$ are distinct modulo n. There are $\varphi(\varphi(n))$ primitive roots of n.

II. Relations

A. Recurrences. If g is a primitive root of a prime p and $g^{p-1} \not\equiv 1 \pmod{p^2}$ then g is a primitive root of p^k for all k. If $g^{p-1} \equiv 1 \pmod{p^2}$ then $g+p$ is a primitive root of p^k for all k.

If g is a primitive root of p^k then either g or $g+p^k$, whichever is odd, is a primitive root of $2p^k$.

B. Checks. If g is a primitive root of n then g^k is a primitive root of n if and only if $(k, \varphi(n)) = 1$, and each primitive root of n is of this form.

References

Texts

[24.1] L. Carlitz, Note on Nörlunds polynomial $B_n^{(z)}$, Proc. Amer. Math. Soc. 11, 452–455 (1960).

[24.2] T. Fort, Finite differences (Clarendon Press, Oxford, England, 1948).

[24.3] H. W. Gould, Stirling number representation problems, Proc. Amer. Math. Soc. 11, 447–451 (1960).

[24.4] G. H. Hardy, Ramanujan (Chelsea Publishing Co., New York, N.Y., 1959).

[24.5] G. H. Hardy and E. M. Wright, An introduction to the theory of numbers, 4th ed. (Clarendon Press, Oxford, England, 1960).

[24.6] L. K. Hua, On the number of partitions of a number into unequal parts, Trans. Amer. Math. Soc. 51, 194–201 (1942).

[24.7] C. Jordan, Calculus of finite differences, 2d ed. (Chelsea Publishing Co., New York, N.Y., 1960).

[24.8] K. Knopp, Theory and application of infinite series (Blackie and Son, Ltd., London, England, 1951).

[24.9] L. M. Milne-Thomson, The calculus of finite differences (Macmillan and Co., Ltd., London, England, 1951).

[24.10] L. Moser and M. Wyman, Stirling numbers of the second kind, Duke Math. J. 25, 29–43 (1958).

[24.11] L. Moser and M. Wyman, Asymptotic development of the Stirling numbers of the first kind, J. London Math. Soc. 33, 133–146 (1958).

[24.12] H. H. Ostmann, Additive Zahlentheorie, vol. I (Springer-Verlag, Berlin, Germany, 1956).

[24.13] H. Rademacher, On the partition function, Proc. London Math. Soc. 43, 241–254 (1937).

[24.14] H. Rademacher and A. Whiteman, Theorems on Dedekind sums, Amer. J. Math. 63, 377–407 (1941).

[24.15] J. Riordan, An introduction to combinatorial analysis (John Wiley & Sons, Inc., New York, N.Y., 1958).

[24.16] J. V. Uspensky and M. A. Heaslet, Elementary number theory (McGraw-Hill Book Co., Inc., New York, N.Y., 1939).

Tables

[24.17] British Association for the Advancement of Science, Mathematical Tables, vol. VIII, Number-divisor tables (Cambridge Univ. Press, Cambridge, England, 1940). $n \le 10^4$.

[24.18] H. Gupta, Tables of distributions, Res. Bull. East Panjab Univ. 13–44 (1950); 750 (1951).

[24.19] H. Gupta, A table of partitions, Proc. London Math. Soc. 39, 142–149 (1935) and II. 42, 546–549 (1937). $p(n)$, $n=1(1)300$; $p(n)$, $n=301(1)600$.

[24.20] G. Kaván, Factor tables (Macmillan and Co., Ltd., London, England, 1937). $n \le 256,000$.

[24.21] D. N. Lehmer, List of prime numbers from 1 to 10,006,721, Carnegie Institution of Washington, Publication No. 165, Washington, D.C. (1914).

[24.22] Royal Society Mathematical Tables, vol. 3, Table of binomial coefficients (Cambridge Univ. Press, Cambridge, England, 1954). $\binom{n}{r}$ for $r \le \frac{1}{2}n \le 100$.

[24.23] G. N. Watson, Two tables of partitions, Proc. London Math. Soc. 42, 550–556 (1937).

Table 24.1   **BINOMIAL COEFFICIENTS** $\binom{n}{m}$

$n\backslash m$	0	1	2	3	4	5	6	7	8
1	1	1							
2	1	2	1						
3	1	3	3	1					
4	1	4	6	4	1				
5	1	5	10	10	5	1			
6	1	6	15	20	15	6	1		
7	1	7	21	35	35	21	7	1	
8	1	8	28	56	70	56	28	8	1
9	1	9	36	84	126	126	84	36	9
10	1	10	45	120	210	252	210	120	45
11	1	11	55	165	330	462	462	330	165
12	1	12	66	220	495	792	924	792	495
13	1	13	78	286	715	1287	1716	1716	1287
14	1	14	91	364	1001	2002	3003	3432	3003
15	1	15	105	455	1365	3003	5005	6435	6435
16	1	16	120	560	1820	4368	8008	11440	12870
17	1	17	136	680	2380	6188	12376	19448	24310
18	1	18	153	816	3060	8568	18564	31824	43758
19	1	19	171	969	3876	11628	27132	50388	75582
20	1	20	190	1140	4845	15504	38760	77520	1 25970
21	1	21	210	1330	5985	20349	54264	1 16280	2 03490
22	1	22	231	1540	7315	26334	74613	1 70544	3 19770
23	1	23	253	1771	8855	33649	1 00947	2 45157	4 90314
24	1	24	276	2024	10626	42504	1 34596	3 46104	7 35471
25	1	25	300	2300	12650	53130	1 77100	4 80700	10 81575
26	1	26	325	2600	14950	65780	2 30230	6 57800	15 62275
27	1	27	351	2925	17550	80730	2 96010	8 88030	22 20075
28	1	28	378	3276	20475	98280	3 76740	11 84040	31 08105
29	1	29	406	3654	23751	1 18755	4 75020	15 60780	42 92145
30	1	30	435	4060	27405	1 42506	5 93775	20 35800	58 52925
31	1	31	465	4495	31465	1 69911	7 36281	26 29575	78 88725
32	1	32	496	4960	35960	2 01376	9 06192	33 65856	105 18300
33	1	33	528	5456	40920	2 37336	11 07568	42 72048	138 84156
34	1	34	561	5984	46376	2 78256	13 44904	53 79616	181 56204
35	1	35	595	6545	52360	3 24632	16 23160	67 24520	235 35820
36	1	36	630	7140	58905	3 76992	19 47792	83 47680	302 60340
37	1	37	666	7770	66045	4 35897	23 24784	102 95472	386 08020
38	1	38	703	8436	73815	5 01942	27 60681	126 20256	489 03492
39	1	39	741	9139	82251	5 75757	32 62623	153 80937	615 23748
40	1	40	780	9880	91390	6 58008	38 38380	186 43560	769 04685
41	1	41	820	10660	101270	7 49398	44 96388	224 81940	955 48245
42	1	42	861	11480	111930	8 50668	52 45786	269 78328	1180 30185
43	1	43	903	12341	123410	9 62598	60 96454	322 24114	1450 08513
44	1	44	946	13244	135751	10 86008	70 59052	383 20568	1772 32627
45	1	45	990	14190	148995	12 21759	81 45060	453 79620	2155 53195
46	1	46	1035	15180	163185	13 70754	93 66819	535 24680	2609 32815
47	1	47	1081	16215	178365	15 33939	107 37573	628 91499	3144 57495
48	1	48	1128	17296	194580	17 12304	122 71512	736 29072	3773 48994
49	1	49	1176	18424	211876	19 06884	139 83816	859 00584	4509 78066
50	1	50	1225	19600	230300	21 18760	158 90700	998 84400	5368 78650

From Royal Society Mathematical Tables, vol. 3, Table of binomial coefficients. Cambridge Univ. Press, Cambridge, England, 1954 (with permission).

BINOMIAL COEFFICIENTS $\binom{n}{m}$

Table 24.1

$n \backslash m$	9	10	11	12	13
9	1				
10	10	1			
11	55	11	1		
12	220	66	12	1	
13	715	286	78	13	1
14	2002	1001	364	91	14
15	5005	3003	1365	455	105
16	11440	8008	4368	1820	560
17	24310	19448	12376	6188	2380
18	48620	43758	31824	18564	8568
19	92378	92378	75582	50388	27132
20	1 67960	1 84756	1 67960	1 25970	77520
21	2 93930	3 52716	3 52716	2 93930	2 03490
22	4 97420	6 46646	7 05432	6 46646	4 97420
23	8 17190	11 44066	13 52078	13 52078	11 44066
24	13 07504	19 61256	24 96144	27 04156	24 96144
25	20 42975	32 68760	44 57400	52 00300	52 00300
26	31 24550	53 11735	77 26160	96 57700	104 00600
27	46 86825	84 36285	130 37895	173 83860	200 58300
28	69 06900	131 23110	214 74180	304 21755	374 42160
29	100 15005	200 30010	345 97290	518 95935	678 63915
30	143 07150	300 45015	546 27300	864 93225	1197 59850
31	201 60075	443 52165	846 72315	1411 20525	2062 53075
32	280 48800	645 12240	1290 24480	2257 92840	3473 73600
33	385 67100	925 61040	1935 36720	3548 17320	5731 66440
34	524 51256	1311 28140	2860 97760	5483 54040	9279 83760
35	706 07460	1835 79396	4172 25900	8344 51800	14763 37800
36	941 43280	2541 86856	6008 05296	12516 77700	23107 89600
37	1244 03620	3483 30136	8549 92152	18524 82996	35624 67300
38	1630 11640	4727 33756	12033 22288	27074 75148	54149 50296
39	2119 15132	6357 45396	16760 56044	39107 97436	81224 25444
40	2734 38880	8476 60528	23118 01440	55868 53480	1 20332 22880
41	3503 43565	11210 99408	31594 61968	78986 54920	1 76200 76360
42	4458 91810	14714 42973	42805 61376	1 10581 16888	2 55187 31280
43	5639 21995	19173 34783	57520 04349	1 53386 78264	3 65768 48168
44	7089 30508	24812 56778	76693 39132	2 10906 82613	5 19155 26432
45	8861 63135	31901 87286	1 01505 95910	2 87600 21745	7 30062 09045
46	11017 16330	40763 50421	1 33407 83196	3 89106 17655	10 17662 30790
47	13626 49145	51780 66751	1 74171 33617	5 22514 00851	14 06768 48445
48	16771 06640	65407 15896	2 25952 00368	6 96685 34468	19 29282 49296
49	20544 55634	82178 22536	2 91359 16264	9 22637 34836	26 25967 83764
50	25054 33700	1 02722 78170	3 73537 38800	12 13996 51100	35 48605 18600

Table 24.1   BINOMIAL COEFFICIENTS $\binom{n}{m}$

$n\backslash m$	14	15	16	17	18	19
14	1					
15	15	1				
16	120	16	1			
17	680	136	17	1		
18	3060	816	153	18	1	
19	11628	3876	969	171	19	1
20	38760	15504	4845	1140	190	20
21	1 16280	54264	20349	5985	1330	210
22	3 19770	1 70544	74613	26334	7315	1540
23	8 17190	4 90314	2 45157	1 00947	33649	8855
24	19 61256	13 07504	7 35471	3 46104	1 34596	42504
25	44 57400	32 68760	20 42975	10 81575	4 80700	1 77100
26	96 57700	77 26160	53 11735	31 24550	15 62275	6 57800
27	200 58300	173 83860	130 37895	84 36285	46 86825	22 20075
28	401 16600	374 42160	304 21755	214 74180	131 23110	69 06900
29	775 58760	775 58760	678 63915	518 95935	345 97290	200 30010
30	1454 22675	1551 17520	1454 22675	1197 59850	864 93225	546 27300
31	2651 82525	3005 40195	3005 40195	2651 82525	2062 53075	1411 20525
32	4714 35600	5657 22720	6010 80390	5657 22720	4714 35600	3473 73600
33	8188 09200	10371 58320	11668 03110	11668 03110	10371 58320	8188 09200
34	13919 75640	18559 67520	22039 61430	23336 06220	22039 61430	18559 67520
35	23199 59400	32479 43160	40599 28950	45375 67650	45375 67650	40599 28950
36	37962 97200	55679 02560	73078 72110	85974 96600	90751 35300	85974 96600
37	61070 86800	93641 99760	1 28757 74670	1 59053 68710	1 76726 31900	1 76726 31900
38	96695 54100	1 54712 86560	2 22399 74430	2 87811 43380	3 35780 00610	3 53452 63800
39	1 50845 04396	2 51408 40660	3 77112 60990	5 10211 17810	6 23591 43990	6 89232 64410
40	2 32069 29840	4 02253 45056	6 28521 01650	8 87323 78800	11 33802 61800	13 12824 08400
41	3 52401 52720	6 34322 74896	10 30774 46706	15 15844 80450	20 21126 40600	24 46626 70200
42	5 28602 29080	9 86724 27616	16 65097 21602	25 46619 27156	35 36971 21050	44 67753 10800
43	7 83789 60360	15 15326 56696	26 51821 49218	42 11716 48758	60 83590 48206	80 04724 31850
44	11 49558 08528	22 99116 17056	41 67148 05914	68 63537 97976	102 95306 96964	140 88314 80056
45	16 68713 34960	34 48674 25584	64 66264 22970	110 30686 03890	171 58844 94940	243 83621 77020
46	23 98775 44005	51 17387 60544	99 14938 48554	174 96950 26860	281 89530 98830	415 42466 71960
47	34 16437 74795	75 16163 04549	150 32326 09098	274 11888 75414	456 86481 25690	697 31997 70790
48	48 23206 23240	109 32600 79344	225 48489 13647	424 44214 84512	730 98370 01104	1154 18478 96480
49	67 52488 72536	157 55807 02584	334 81089 92991	649 92703 98159	1155 42584 85616	1885 16848 97584
50	93 78456 56300	225 08295 75120	492 36896 95575	984 73793 91150	1805 35288 83775	3040 59433 83200

$n\backslash m$	20	21	22	23	24	25
20	1					
21	21	1				
22	231	22	1			
23	1771	253	23	1		
24	10626	2024	276	24	1	
25	53130	12650	2300	300	25	1
26	2 30230	65780	14950	2600	325	26
27	8 88030	2 96010	80730	17550	2925	351
28	31 08105	11 84040	3 76740	98280	20475	3276
29	100 15005	42 92145	15 60780	4 75020	1 18755	23751
30	300 45015	143 07150	58 52925	20 35800	5 93775	1 42506
31	846 72315	443 52165	201 60075	78 88725	26 29575	7 36281
32	2257 92840	1290 24480	645 12240	280 48800	105 18300	33 65856
33	5731 66440	3548 17320	1935 36720	925 61040	385 67100	138 84156
34	13919 75640	9279 83760	5483 54040	2860 97760	1311 28140	524 51256
35	32479 43160	23199 59400	14763 37800	8344 51800	4172 25900	1835 79396
36	73078 72110	55679 02560	37962 97200	23107 89600	12516 77700	6008 05296
37	1 59053 68710	1 28757 74670	93641 99760	61070 86800	35624 67300	18524 82996
38	3 35780 00610	2 87811 43380	2 22399 74430	1 54712 86560	96695 54100	54149 50296
39	6 89232 64410	6 23591 43990	5 10211 17810	3 77112 60990	2 51408 40660	1 50845 04396
40	13 78465 28820	13 12824 08400	11 33802 61800	8 87323 78800	6 28521 01650	4 02253 45056
41	26 91289 37220	26 91289 37220	24 46626 70200	20 21126 40600	15 15844 80450	10 30774 46706
42	51 37916 07420	53 82578 74440	51 37916 07420	44 67753 10800	35 36971 21050	25 46619 27156
43	96 05669 18220	105 20494 81860	105 20494 81860	96 05669 18220	80 04724 31850	60 83590 48206
44	176 10393 50070	201 26164 00080	210 40989 63720	201 26164 00080	176 10393 50070	140 88314 80056
45	316 98708 30126	377 36557 50150	411 67153 63800	411 67153 63800	377 36557 50150	316 98708 30126
46	560 82330 07146	694 35265 80276	789 03711 13950	823 34307 27600	789 03711 13950	694 35265 80276
47	976 24796 79106	1255 17595 87422	1483 38976 94226	1612 38018 41550	1612 38018 41550	1483 38976 94226
48	1673 56794 49896	2231 42392 66528	2738 56572 81648	3095 76995 35776	3224 76036 83100	3095 76995 35776
49	2827 75273 46376	3904 99187 16424	4969 98965 48176	5834 33568 17424	6320 53032 18876	6320 53032 18876
50	4712 92122 43960	6732 74460 62800	8874 98152 64600	10804 32533 66600	12154 86600 36300	12641 06064 37752

Multinomials and Partitions

Table 24.2

$$\pi = 1^{a_1}, 2^{a_2}, \ldots, n^{a_n}, \quad n = a_1 + 2a_2 + \ldots + na_n, \quad m = a_1 + a_2 + \ldots + a_n$$

$$M_1 = (n; n_1, n_2, \ldots, n_m) = n!/(1!)^{a_1}(2!)^{a_2} \ldots (n!)^{a_n}$$

$$M_2 = (n; a_1, a_2, \ldots, a_n)^* = n!/1^{a_1}a_1!2^{a_2}a_2! \ldots n^{a_n}a_n!$$

$$M_3 = (n; a_1, a_2, \ldots, a_n)' = n!/(1!)^{a_1}a_1!(2!)^{a_2}a_2! \ldots (n!)^{a_n}a_n!$$

n	m	π	M_1	M_2	M_3		n	m	π	M_1	M_2	M_3
1	1	1	1	1	1		8	1	8	1	5040	1
								2	1, 7	8	5760	8
2	1	2	1	1	1				2, 6	28	3360	28
	2	1^2	2	1	1				3, 5	56	2688	56
									4^2	70	1260	35
3	1	3	1	2	1			3	$1^2, 6$	56	3360	28
	2	1, 2	3	3	3				1, 2, 5	168	4032	168
	3	1^3	6	1	1				1, 3, 4	280	3360	280
									$2^2, 4$	420	1260	210
4	1	4	1	6	1				$2, 3^2$	560	1120	280
	2	1, 3	4	8	4			4	$1^3, 5$	336	1344	56
		2^2	6	3	3				$1^2, 2, 4$	840	2520	420
	3	$1^2, 2$	12	6	6				$1^2, 3^2$	1120	1120	280
	4	1^4	24	1	1				$1, 2^2, 3$	1680	1680	840
									2^4	2520	105	105
5	1	5	1	24	1			5	$1^4, 4$	1680	420	70
	2	1, 4	5	30	5				$1^3, 2, 3$	3360	1120	560
		2, 3	10	20	10				$1^2, 2^3$	5040	420	420
	3	$1^2, 3$	20	20	10			6	$1^5, 3$	6720	112	56
		$1, 2^2$	30	15	15				$1^4, 2^2$	10080	210	210
	4	$1^3, 2$	60	10	10			7	$1^6, 2$	20160	28	28
	5	1^5	120	1	1			8	1^8	40320	1	1
							9	1	9	1	40320	1
6	1	6	1	120	1			2	1, 8	9	45360	9
	2	1, 5	6	144	6				2, 7	36	25920	36
		2, 4	15	90	15				3, 6	84	20160	84
		3^2	20	40	10				4, 5	126	18144	126
	3	$1^2, 4$	30	90	15			3	$1^2, 7$	72	25920	36
		1, 2, 3	60	120	60				1, 2, 6	252	30240	252
		2^3	90	15	15				1, 3, 5	504	24192	504
	4	$1^3, 3$	120	40	20				$1, 4^2$	630	11340	315
		$1^2, 2^2$	180	45	45				$2^2, 5$	756	9072	378
	5	$1^4, 2$	360	15	15				2, 3, 4	1260	15120	1260
	6	1^6	720	1	1				3^3	1680	2240	280
								4	$1^3, 6$	504	10080	84
7	1	7	1	720	1				$1^2, 2, 5$	1512	18144	756
	2	1, 6	7	840	7				$1^2, 3, 4$	2520	15120	1260
		2, 5	21	504	21				$1, 2^2, 4$	3780	11340	1890
		3, 4	35	420	35				$1, 2, 3^2$	5040	10080	2520
	3	$1^2, 5$	42	504	21				$2^3, 3$	7560	2520	1260
		1, 2, 4	105	630	105			5	$1^4, 5$	3024	3024	126
		$1, 3^2$	140	280	70				$1^3, 2, 4$	7560	7560	1260
		$2^2, 3$	210	210	105				$1^3, 3^2$	10080	3360	840
	4	$1^3, 4$	210	210	35				$1^2, 2^2, 3$	15120	7560	3780
		$1^2, 2, 3$	420	420	210				$1, 2^4$	22680	945	945
		$1, 2^3$	630	105	105			6	$1^5, 4$	15120	756	126
	5	$1^4, 3$	840	70	35				$1^4, 2, 3$	30240	2520	1260
		$1^3, 2^2$	1260	105	105				$1^3, 2^3$	45360	1260	1260
	6	$1^5, 2$	2520	21	21			7	$1^6, 3$	60480	168	84
	7	1^7	5040	1	1				$1^5, 2^2$	90720	378	378
								8	$1^7, 2$	181440	36	36
								9	1^9	362880	1	1

Table 24.2   **Multinomials and Partitions**

n	m	π	M_1	M_2	M_3	n	m	π	M_1	M_2	M_3
10	1	10	1	362880	1	10		$2^3, 4$	18900	18900	3150
	2	1, 9	10	403200	10			$2^2, 3^2$	25200	25200	6300
		2, 8	45	226800	45		5	$1^4, 6$	5040	25200	210
		3, 7	120	172800	120			$1^3, 2, 5$	15120	60480	2520
		4, 6	210	151200	210			$1^3, 3, 4$	25200	50400	4200
		5^2	252	72576	126			$1^2, 2^2, 4$	*37800	*56700	9450
	3	$1^2, 8$	90	226800	45			$1^2, 2, 3^2$	50400	50400	12600
		1, 2, 7	360	259200	360			$1, 2^3, 3$	75600	25200	12600
		1, 3, 6	840	201600	840			2^5	113400	945	945
		1, 4, 5	1260	181440	1260		6	$1^5, 5$	30240	6048	252
		$2^2, 6$	1260	75600	630			$1^4, 2, 4$	75600	18900	3150
		2, 3, 5	2520	*120960	2520			$1^4, 3^2$	100800	8400	2100
		$2, 4^2$	3150	56700	1575			$1^3, 2^2, 3$	151200	25200	12600
		$3^2, 4$	4200	50400	2100			$1^2, 2^4$	226800	4725	4725
	4	$1^3, 7$	720	86400	120		7	$1^6, 4$	151200	1260	210
		$1^2, 2, 6$	2520	151200	1260			$1^5, 2, 3$	302400	5040	2520
		$1^2, 3, 5$	5040	120960	2520			$1^4, 2^3$	453600	3150	3150
		$1^2, 4^2$	6300	56700	1575		8	$1^7, 3$	604800	240	120
		$1, 2^2, 5$	7560	90720	3780			$1^6, 2^2$	907200	630	630
		1, 2, 3, 4	12600	151200	12600		9	$1^8, 2$	1814400	45	45
		$1, 3^3$	16800	22400	2800		10	1^{10}	3628800	1	1

*See page II.

STIRLING NUMBERS OF THE FIRST KIND $S_n^{(m)}$

Table 24.3

$n\backslash m$	1	2	3
1	1		
2	−1	1	
3	2	−3	1
4	−6	11	−6
5	24	−50	35
6	−120	274	−225
7	720	−1764	1624
8	−5040	13068	− 13132
9	40320	−1 09584	1 18124
10	−3 62880	10 26576	−11 72700
11	36 28800	−106 28640	127 53576
12	−399 16800	1205 43840	−1509 17976
13	4790 01600	− 14864 42880	19315 59552
14	− 62270 20800	1 98027 59040	−2 65967 17056
15	8 71782 91200	−28 34656 47360	39 21567 97824
16	−130 76743 68000	433 91630 01600	−616 58176 14720
17	2092 27898 88000	−7073 42823 93600	10299 22448 37120
18	− 35568 74280 96000	1 22340 55905 79200	−1 82160 24446 24640
19	6 40237 37057 28000	−22 37698 80585 21600	34 01224 95938 22720
20	−121 64510 04088 32000	431 56514 68176 38400	−668 60973 03411 53280
21	2432 90200 81766 40000	−8752 94803 67616 00000	13803 75975 36407 04000
22	− 51090 94217 17094 40000	1 86244 81078 01702 40000	−2 98631 90286 32163 84000
23	11 24000 72777 76076 80000	−41 48476 77933 54547 20000	67 56146 67377 09306 88000
24	−258 52016 73888 49766 40000	965 38966 65249 30662 40000	−1595 39850 27606 68605 44000
25	6204 48401 73323 94393 60000	−23427 87216 39871 85664 00000	39254 95373 27809 77192 96000

$n\backslash m$	4	5	6
4	1		
5	−10	1	
6	85	−15	1
7	−735	175	−21
8	6769	−1960	322
9	− 67284	22449	−4536
10	7 23680	−2 69325	63273
11	−84 09500	34 16930	−9 02055
12	1052 58076	−459 95730	133 39535
13	− 14140 14888	6572 06836	−2060 70150
14	2 03137 53096	− 99577 03756	33361 18786
15	−31 09892 60400	15 97216 05680	−5 66633 66760
16	505 69957 03824	−270 68133 45600	100 96721 07080
17	−8707 77488 75904	4836 60092 33424	−1886 15670 58880
18	1 58331 39757 27488	− 90929 99058 44112	36901 26492 34384
19	−30 32125 40077 19424	17 95071 22809 21504	−7 55152 75920 63024
20	610 11607 57404 91776	−371 38478 73452 28000	161 42973 65301 18960
21	− 12870 93124 51509 88800	8037 81182 26450 51776	−3599 97951 79476 07200
22	2 84093 31590 18114 68800	−1 81664 97952 06970 76096	83637 38169 95448 02976
23	−65 48684 85270 30686 97600	42 80722 86535 71471 42912	−20 21687 37691 06827 41568
24	1573 75898 28594 15107 32800	−1050 05310 75591 74529 84576	507 79532 53430 28501 98976
25	−39365 61409 13866 31181 31200	26775 03356 42796 03823 62624	−13237 14091 57918 58577 60000

From unpublished tables of Francis L. Miksa, with permission.

Table 24.3

STIRLING NUMBERS OF THE FIRST KIND $S_n^{(m)}$

$n\backslash m$	7	8	9
7	1		
8	−28	1	
9	546	−36	1
10	−9450	870	−45
11	1 57773	− 18150	1320
12	−26 37558	3 57423	− 32670
13	449 90231	−69 26634	7 49463
14	−7909 43153	1350 36473	−166 69653
15	1 44093 22928	− 26814 53775	3684 11615
16	−27 28032 10680	5 46311 29553	− 82076 28000
17	537 45234 77960	−114 69012 83528	18 59531 77553
18	− 11022 84661 84200	2487 18452 97936	−430 81053 01929
19	2 35312 50405 49984	− 55792 16815 47048	10241 77407 32658
20	−52 26090 33625 12720	12 95363 69899 43896	−2 50385 87554 67550
21	1206 64780 37803 73360	−311 33364 31613 90640	63 03081 20992 94896
22	− 28939 58339 73354 47760	7744 65431 01695 76800	−1634 98069 72465 83456
23	7 20308 21644 09246 53696	−1 99321 97822 10661 37360	43714 22964 95944 12832
24	−185 88776 35505 19497 76576	53 04713 71552 54458 12976	−12 04749 26016 17376 32496
25	4969 10165 05554 96448 36800	−1459 01905 52766 26492 88000	342 18695 95940 71489 92880

$n\backslash m$	10	11	12
10	1		
11	−55	1	
12	1925	−66	1
13	− 55770	2717	−78
14	14 74473	− 91091	3731
15	−373 12275	27 49747	−1 43325
16	9280 95740	−785 58480	48 99622
17	−2 30571 59840	21850 31420	−1569 52432
18	57 79248 94833	−6 02026 93980	48532 22764
19	−1471 07534 08923	166 15733 86473	−14 75607 03732
20	38192 20555 02195	−4628 06477 51910	446 52267 57381
21	−10 14229 98655 11450	1 30753 50105 40395	− 13558 51828 99530
22	276 01910 92750 35346	−37 60053 50868 59745	4 15482 38514 30525
23	−7707 40110 12973 61068	1103 23088 11859 49736	−129 00665 98183 31295
24	2 20984 45497 94337 17396	− 33081 71136 85742 04996	4070 38405 70075 69521
25	−65 08376 17966 81468 50000	10 14945 52782 52146 37300	−1 30770 92873 67558 73500

$n\backslash m$	13	14	15	16
13	1			
14	−91	1		
15	5005	−105	1	
16	−2 18400	6580	−120	1
17	83 94022	−3 23680	8500	−136
18	−2996 50806	138 96582	−4 68180	10812
19	1 02469 37272	−5497 89282	223 23822	−6 62796
20	−34 22525 11900	2 06929 33630	−9739 41900	349 16946
21	1131 02769 95381	−75 61111 84500	4 01717 71630	− 16722 80820
22	− 37310 09998 02531	2718 86118 69881	−159 97183 88730	7 52896 68850
23	12 36304 58470 86207	− 97125 04609 39913	6238 24164 21941	−325 60911 03430
24	−413 35671 43013 14056	34 70180 64487 04206	−2 40604 60386 44556	13727 25118 00831
25	13990 94520 02391 06865	−1246 20006 90702 15000	92 44691 13761 73550	−5 70058 63218 64500

$n\backslash m$	17	18	19	20	21	22	23	24	25
17	1								
18	−153	1							
19	13566	−171	1						
20	−9 20550	16815	−190	1					
21	533 27946	−12 56850	20615	−210	1				
22	− 27921 67686	797 21796	−16 89765	25025	−231	1			
23	13 67173 57942	− 45460 47198	1168 96626	−22 40315	30107	−253	1		
24	−640 05903 36096	24 12764 43496	− 72346 69596	1684 23871	−29 32776	35926	−276	1	
25	29088 66798 67135	−1219 12249 80000	41 49085 13800	−1 12768 42500	2388 10495	−37 95000	42550	−300	1

STIRLING NUMBERS OF THE SECOND KIND $\mathfrak{S}_n^{(m)}$ Table 24.4

$n\backslash m$ 1	2	3	4	5	6
1 1					
2 1	1				
3 1	3	1			
4 1	7	6	1		
5 1	15	25	10	1	
6 1	31	90	65	15	1
7 1	63	301	350	140	21
8 1	127	966	1701	1050	266
9 1	255	3025	7770	6951	2646
10 1	511	9330	34105	42525	22827
11 1	1023	28501	1 45750	2 46730	1 79487
12 1	2047	86526	6 11501	13 79400	13 23652
13 1	4095	2 61625	25 32530	75 08501	93 21312
14 1	8191	7 88970	103 91745	400 75035	634 36373
15 1	16383	23 75101	423 55950	2107 66920	4206 93273
16 1	32767	71 41686	1717 98901	10961 90550	27349 26558
17 1	65535	214 57825	6943 37290	56527 51651	1 75057 49898
18 1	1 31071	644 39010	27988 06985	2 89580 95545	11 06872 51039
19 1	2 62143	1934 48101	1 12596 66950	14 75892 84710	69 30816 01779
20 1	5 24287	5806 06446	4 52321 15901	74 92060 90500	430 60788 95384
21 1	10 48575	17423 43625	18 15090 70050	379 12625 68401	2658 56794 62804
22 1	20 97151	52280 79450	72 77786 23825	1913 78219 12055	16330 53393 45225
23 1	41 94303	1 56863 35501	291 63425 74750	9641 68881 84100	99896 98579 83405
24 1	83 88607	4 70632 00806	1168 10566 34501	48500 07834 95250	6 09023 60360 84530
25 1	167 77215	14 11979 91025	4677 12897 38810	2 43668 49741 10751	37 02641 70000 02430

$n\backslash m$	7	8	9	10
7	1			
8	28	1		
9	462	36	1	
10	5880	750	45	1
11	63987	11880	1155	55
12	6 27396	1 59027	22275	1705
13	57 15424	18 99612	3 59502	39325
14	493 29280	209 12320	51 35130	7 52752
15	4087 41333	2166 27840	671 28490	126 62650
16	32818 82604	21417 64053	8207 84250	1937 54990
17	2 57081 04786	2 04159 95028	95288 22303	27583 34150
18	19 74624 83400	18 90360 65010	10 61753 95755	3 71121 63803
19	149 29246 34839	170 97510 03480	114 46146 26805	47 72970 33785
20	1114 35540 45652	1517 09326 62679	1201 12826 44725	591 75849 64655
21	8231 09572 14948	13251 10153 47084	12327 24764 65204	7118 71322 91275
22	60276 23799 67440	1 14239 90799 91620	1 24196 33035 33920	83514 37993 77954
23	4 38264 19991 17305	9 74195 50199 00400	12 32006 88117 96900	9 59340 12973 13460
24	31 67746 38518 04540	82 31828 21583 20505	120 62257 43260 72500	108 25408 17849 31500
25	227 83248 29987 16310	690 22372 11183 68580	1167 92146 10929 73005	1203 16339 21753 87500

$n\backslash m$	11	12	13	14
11	1			
12	66	1		
13	2431	78	1	
14	66066	3367	91	1
15	14 79478	1 06470	4550	105
16	289 36908	27 57118	1 65620	6020
17	5120 60978	620 22324	49 10178	2 49900
18	83910 04908	12563 28866	1258 54638	84 08778
19	12 94132 17791	2 34669 51300	28924 39160	2435 77530
20	190 08424 29486	41 10166 33391	6 10686 60380	63025 24580
21	2682 68516 89001	683 30420 30178	120 49092 18331	14 93040 04500
22	36628 25008 70286	10882 33560 51137	2249 68618 48481	329 51652 81331 *
23	4 86425 13089 51100	1 67216 27734 83930	40128 25603 41390	6862 91758 07115
24	63 10016 56957 75560	24 93020 45907 58260	6 88883 60579 22000	1 36209 10216 41000
25	802 35590 44384 62660	362 26262 07848 74680	114 48507 33437 44260	25 95811 03608 96000

$n\backslash m$	15	16	17	18	19
15	1				
16	120	1			
17	7820	136	1		
18	3 67200	9996	153	1	
19	139 16778	5 27136	12597	171	1
20	4523 29200	223 50954	7 41285	15675	190
21	1 30874 62580	8099 44464	349 52799	10 23435	19285
22	34 56159 43200	2 60465 74004	14041 42047	533 74629	13 89850
23	847 94044 29331	76 23611 27264	4 99169 88803	23648 85369	797 81779
24	19582 02422 47080	2067 71824 65555	161 09499 36915	9 24849 25445	38807 39170
25	4 29939 46553 47200	52665 51616 95960	4806 33313 93110	327 56785 94925	16 62189 69675

$n\backslash m$	20	21	22	23	24	25
20	1					
21	210	1				
22	23485	231	1			
23	18 59550	28336	253	1		
24	1169 72779	24 54606	33902	276	1	
25	62201 94750	1685 19505	32 00450	40250	300	1

From unpublished tables of Francis L. Miksa, with permission.

*See page II.

Table 24.5   **NUMBER OF PARTITIONS AND PARTITIONS INTO DISTINCT PARTS**

n	$p(n)$	$q(n)$	n	$p(n)$	$q(n)$	n	$p(n)$	$q(n)$	n	$p(n)$	$q(n)$
0		1	50	2 04226	3658	100	1905 69292	4 44793	150	4 08532 35313	194 06016
1	1	1	51	2 39943	4097	101	2144 81126	4 83330	151	4 50606 24582	207 92120
2	2	1	52	2 81589	4582	102	2412 65379	5 25016	152	4 96862 88421	222 72512
3	3	2	53	3 29931	5120	103	2712 48950	5 70078	153	5 47703 36324	238 53318
4	5	2	54	3 86155	5718	104	3048 01365	6 18784	154	6 03566 73280	255 40982
5	7	3	55	4 51276	6378	105	3423 25709	6 71418	155	6 64931 82097	273 42421
6	11	4	56	5 26823	7108	106	3842 76336	7 28260	156	7 32322 43759	292 64960
7	15	5	57	6 14154	7917	107	4311 49389	7 89640	157	8 06309 64769	313 16314
8	22	6	58	7 15220	8808	108	4835 02844	8 55906	158	8 87517 78802	335 04746
9	30	8	59	8 31820	9792	109	5419 46240	9 27406	159	9 76627 28555	358 39008
10	42	10	60	9 66467	10880	110	6071 63746	10 04544	160	10 74381 59466	383 28320
11	56	12	61	11 21505	12076	111	6799 03203	10 87744	161	11 81590 68427	409 82540
12	77	15	62	13 00156	13394	112	7610 02156	11 77438	162	12 99139 04637	438 12110
13	101	18	63	15 05499	14848	113	8513 76628	12 74118	163	14 27989 95930	468 28032
14	135	22	64	17 41630	16444	114	9520 50665	13 78304	164	15 69194 75295	500 42056
15	176	27	65	20 12558	18200	115	10641 44451	14 90528	165	17 23898 00255	534 66624
16	231	32	66	23 23520	20132	116	11889 08248	16 11388	166	18 93348 22579	571 14844
17	297	38	67	26 79689	22250	117	13277 10076	17 41521	167	20 78904 20102	610 00704
18	385	46	68	30 87735	24576	118	14820 74143	18 81578	168	22 82047 32751	651 39008
19	490	54	69	35 54345	27130	119	16536 68665	20 32290	169	25 04389 25115	695 45358
20	627	64	70	40 87968	29927	120	18443 49560	21 94432	170	27 47686 17130	742 36384
21	792	76	71	46 97205	32992	121	20561 48051	23 68800	171	30 13848 02048	792 29676
22	1002	89	72	53 92783	36352	122	22913 20912	25 56284	172	33 04954 99613	845 43782
23	1255	104	73	61 85689	40026	123	25523 38241	27 57826	173	36 23268 59895	901 98446
24	1575	122	74	70 89500	44046	124	28419 40500	29 74400	174	39 71250 74750	962 14550
25	1958	142	75	81 18264	48446	125	31631 27352	32 07086	175	43 51576 97830	1026 14114
26	2436	165	76	92 89091	53250	126	35192 22692	34 57027	176	47 67158 57290	1094 20549
27	3010	192	77	106 19863	58499	127	39138 64295	37 25410	177	52 21158 31195	1166 58616
28	3718	222	78	121 32164	64234	128	43510 78600	40 13544	178	57 17016 05655	1243 54422
29	4565	256	79	138 48650	70488	129	48352 71870	43 22816	179	62 58467 53120	1325 35702
30	5604	296	80	157 96476	77312	130	53713 15400	46 54670	180	68 49573 90936	1412 31780
31	6842	340	81	180 04327	84756	131	59645 39504	50 10688	181	74 94744 11781	1504 73568
32	8349	390	82	205 06255	92864	132	66208 30889	53 92550	182	81 98769 08323	1602 93888
33	10143	448	83	233 38469	101698	133	73466 29512	58 02008	183	89 66848 17527	1707 27424
34	12310	512	84	265 43660	111322	134	81490 40695	62 40974	184	98 04628 80430	1818 10744
35	14883	585	85	301 67357	121792	135	90358 36076	67 11480	185	107 18237 74337	1935 82642
36	17977	668	86	342 62962	133184	136	1 00155 81680	72 15644	186	117 14326 92373	2060 84096
37	21637	760	87	388 87673	145578	137	1 10976 45016	77 55776	187	128 00110 42268	2193 58315
38	26015	864	88	441 08109	159046	138	1 22923 41831	83 34326	188	139 83417 45571	2334 51098
39	31185	982	89	499 95925	173682	139	1 36109 49895	89 53856	189	152 72735 99625	2484 10816
40	37338	1113	90	566 34173	189586	140	1 50658 78135	96 17150	190	166 77274 04093	2642 88462
41	44583	1260	91	641 12359	206848	141	1 66706 89208	103 27156	191	182 07011 00652	2811 38048
42	53174	1426	92	725 33807	225585	142	1 84402 93320	110 86968	192	198 72768 56363	2990 16608
43	63261	1610	93	820 10177	245920	143	2 03909 82757	118 99934	193	216 86271 05469	3179 84256
44	75175	1816	94	926 69720	267968	144	2 25406 54445	127 69602	194	236 60227 41845	3381 04630
45	89134	2048	95	1046 51419	291874	145	2 49088 58009	136 99699	195	258 08402 12973	3594 44904
46	105558	2304	96	1181 14304	317788	146	2 75170 52599	146 94244	196	281 45709 87591	3820 75868
47	124754	2590	97	1332 30930	345856	147	3 03886 71978	157 57502	197	306 88298 78530	4060 72422
48	147273	2910	98	1501 98136	376256	148	3 35494 19497	168 93952	198	334 53659 83698	4315 13602
49	173525	3264	99	1692 29875	409174	149	3 70273 55200	181 08418	199	364 60724 32125	4584 82688
50	204226	3658	100	1905 69292	444793	150	4 08532 35313	194 06016	200	397 29990 29388	4870 67746

Values of $p(n)$ from H. Gupta, A table of partitions, Proc. London Math. Soc. **39**, 142–149, 1935 and II. **42**, 546–549, 1937 (with permission).

NUMBER OF PARTITIONS AND PARTITIONS INTO DISTINCT PARTS   Table 24.5

n	$p(n)$	$q(n)$	n	$p(n)$	$q(n)$
200	397 29990 29388	4870 67746	250	23079 35543 64681	85192 80128
201	432 83636 58647	5173 61670	251	24929 14511 68559	89949 26602
202	471 45668 86083	5494 62336	252	26923 27012 52579	94961 58208
203	513 42052 87973	5834 73184	253	29072 69579 16112	1 00243 00890
204	559 00883 17495	6195 03296	254	31389 19913 06665	1 05807 47264
205	608 52538 59260	6576 67584	255	33885 42642 48680	1 11669 59338
206	662 29877 08040	6980 87424	256	36574 95668 70782	1 17844 71548
207	720 68417 06490	7408 90786	257	39472 36766 55357	1 24348 95064
208	784 06562 26137	7862 12446	258	42593 30844 09356	1 31199 20928
209	852 85813 02375	8341 94700	259	45954 57504 48675	1 38413 23582
210	927 51025 75355	8849 87529	260	49574 19347 60846	1 46009 65705
211	1008 50658 85767	9387 48852	261	53471 50629 08609	1 54008 01856
212	1096 37072 05259	9956 45336	262	57667 26749 47168	1 62428 82560
213	1191 66812 36278	10558 52590	263	62183 74165 09615	1 71293 59744
214	1295 00959 25895	11195 55488	264	67044 81230 60170	1 80624 90974
215	1407 05456 99287	11869 49056	265	72276 09536 90372	1 90446 44146
216	1528 51512 48481	12582 38720	266	77905 06295 62167	2 00783 03620
217	1660 15981 07914	13336 40710	267	83961 17303 66814	2 11660 75136
218	1802 81825 16671	14133 83026	268	90476 01083 16360	2 23106 91192
219	1957 38561 61145	14977 05768	269	97483 43699 44625	2 35150 17984
220	2124 82790 09367	15868 61606	270	1 05019 74899 31117	2 47820 61070
221	2306 18711 73849	16811 16852	271	1 13123 85039 38606	2 61149 71540
222	2502 58737 60111	17807 51883	272	1 21837 43498 44333	2 75170 53882
223	2715 24089 25615	18860 61684	273	1 31205 18008 16215	2 89917 72486
224	2945 45499 41750	19973 57056	274	1 41274 95651 73450	3 05427 58738
225	3194 63906 96157	21149 65120	275	1 52098 04928 51175	3 21738 19904
226	3464 31263 22519	22392 29960	276	1 63729 39693 37171	3 38889 46600
227	3756 11335 82570	23705 13986	277	1 76227 84330 57269	3 56923 20960
228	4071 80636 27362	25091 98528	278	1 89656 41035 91584	3 75883 26642
229	4413 29348 84255	26556 84608	279	2 04082 58525 75075	3 95815 57440
230	4782 62397 45920	28103 94454	280	2 19578 63116 82516	4 16768 26624
231	5182 00518 38712	29737 72212	281	2 36221 91453 37711	4 38791 78240
232	5613 81486 70947	31462 84870	282	2 54095 25900 45698	4 61938 97032
233	6080 61354 38329	33284 23936	283	2 73287 31835 47535	4 86265 19094
234	6585 15859 70275	35207 06304	284	2 93892 97939 29555	5 11828 44672
235	7130 41855 14919	37236 75326	285	3 16013 78671 48997	5 38689 49522
236	7719 58926 63512	39379 02688	286	3 39758 40119 86773	5 66911 97084
237	8356 11039 25871	41639 89458	287	3 65243 08360 71053	5 96562 52987
238	9043 68396 68817	44025 67324	288	3 92592 21614 89422	6 27710 98024
239	9786 29337 03585	46543 00706	289	4 21938 85285 87095	6 60430 42088
240	10588 22467 22733	49198 87992	290	4 53425 31269 00886	6 94797 40554
241	11454 08845 53038	52000 62976	291	4 87203 80564 72084	7 30892 09120
242	12388 84430 77259	54955 97248	292	5 23437 10697 53672	7 68798 39744
243	13397 82593 44888	58073 01632	293	5 62299 26919 50605	8 08604 19136
244	14486 76924 96445	61360 27874	294	6 03976 38820 95515	8 50401 45750
245	15661 84125 27946	64826 71322	295	6 48667 41270 79088	8 94286 47940
246	16929 67223 91554	68481 72604	296	6 96585 01441 95831	9 40360 04868
247	18297 38898 54026	72335 19619	297	7 47956 50785 10584	9 88727 65938
248	19772 65166 81672	76397 50522	298	8 03024 83849 43040	10 39499 71456
249	21363 69198 20625	80679 55712	299	8 62049 62754 65025	10 92791 76298
250	23079 35543 64681	85192 80128	300	9 25308 29367 23602	11 48724 72064

Table 24.5   NUMBER OF PARTITIONS AND PARTITIONS INTO DISTINCT PARTS

n	p(n)	q(n)	n	p(n)	q(n)
300	9 25308 29367 23602	11 48724 72064	350	279 36332 84837 02152	126 91829 24648
301	9 93097 23924 03501	12 07425 10607	351	298 33006 30627 58076	132 93477 19190
302	10 65733 12325 48839	12 69025 30816	352	318 55597 37883 29084	139 22769 71520
303	11 43554 20778 22104	13 33663 83848	353	340 12281 00485 77428	145 80938 18816
304	12 26921 80192 29465	14 01485 59930	354	363 11751 20481 10005	152 69267 15868
305	13 16221 78950 57704	14 72642 18618	355	387 63253 29190 29223	159 89096 56578
306	14 11866 26652 80005	15 47292 17536	356	413 76618 09333 42362	167 41824 09148
307	15 14295 27388 57194	16 25601 42890	357	441 62298 19293 58437	175 28907 55072
308	16 23978 65358 29663	17 07743 43642	358	471 31406 42683 98780	183 51867 38752
309	17 41418 01331 47295	17 93899 64242	359	502 95756 65060 00020	192 12289 32216
310	18 67148 82996 00364	18 84259 79304	360	536 67907 03106 91121	201 11827 04478
311	20 01742 67625 76945	19 79022 32212	361	572 61205 88980 37559	210 52205 02772
312	21 45809 60373 52891	20 78394 72390	362	610 89840 37518 84101	220 35221 50410
313	23 00000 66554 87337	21 82593 94656	363	651 68887 99972 06959	230 62751 50210
314	24 65010 61508 30490	22 91846 82870	364	695 14371 34589 46040	241 36750 01278
315	26 41580 76335 66326	24 06390 52286	365	741 43315 98840 81684	252 59255 33946
316	28 30502 03409 96003	25 26472 94208	366	790 73811 96494 11319	264 32392 51488
317	30 32618 19898 42964	26 52353 25352	367	843 25078 85625 28427	276 58376 86784
318	32 48829 33514 66654	27 84302 35904	368	899 17534 83960 88349	289 39517 78822
319	34 80095 48694 40830	29 22603 40224	369	958 72869 79123 38045	302 78222 57408
320	37 27440 57767 48077	30 67552 32574	370	1022 14122 83673 45362	316 77000 44480
321	39 91956 55269 99991	32 19458 41664	371	1089 65764 44243 99782	331 38466 77248
322	42 74807 80359 54696	33 78644 88192	372	1161 53783 48499 62850	346 65347 41118
323	45 77235 85435 78028	35 45449 47722	373	1238 05779 41191 25085	362 60483 21048
324	49 00564 36352 37875	37 20225 12608	374	1319 51059 97274 73500	379 26834 76992
325	52 46204 42288 28641	39 03340 57172	375	1406 20744 65614 84054	396 67487 30794
326	56 15660 21128 74289	40 95181 08690	376	1498 47874 35905 81081	414 85655 73659
327	60 10534 98396 66544	42 96149 17632	377	1596 67527 44907 56791	433 84690 00206
328	64 32537 46091 14550	45 06665 31450	378	1701 16942 79758 13525	453 68080 55808
329	68 83488 59460 73850	47 27168 74732	379	1812 35649 97394 72950	474 39464 06976
330	73 65328 78618 50339	49 58118 28759	380	1930 65607 23504 65812	496 02629 40968
331	78 80125 53026 66615	51 99993 15040	381	2056 51347 53366 33805	518 61523 80864
332	84 30081 56362 25119	54 53293 85792	382	2190 40133 24237 65131	542 20259 26436
333	90 17543 49805 49623	57 18543 13990	383	2332 82119 85438 92336	566 83119 27092
334	96 45011 01922 02760	59 96286 87918	384	2484 30529 42654 18180	592 54565 72864
335	103 15146 63217 35325	62 87095 13216	385	2645 41834 06887 63701	619 39246 14094
336	110 30786 04252 92772	65 91563 14788	386	2816 75950 32179 42792	647 42001 16480
337	117 94949 15461 13972	69 10312 43770	387	2998 96444 77364 52194	676 67872 37064
338	126 10851 78337 96355	72 43991 92576	388	3192 70751 84335 32826	707 22110 32064
339	134 81918 06233 01520	75 93279 10200	389	3398 70404 13581 60275	739 10183 03854
340	144 11793 65278 73832	79 58881 23110	390	3617 71276 38676 04423	772 37784 71936
341	154 04359 73795 76030	83 41536 64940	391	3850 53843 46674 29186	807 10844 79444
342	164 63747 91657 61044	87 42016 06890	392	4098 03453 56265 94791	843 35537 42947
343	175 94355 98104 22753	91 61123 94270	393	4361 10617 07622 84114	881 18291 29614
344	188 00864 70522 92980	95 99699 92704	394	4640 71312 46996 23515	920 65799 74150
345	200 88255 62876 83159	100 58620 35461	395	4937 87309 67881 91655	961 85031 43424
346	214 61829 97432 86299	105 38799 77632	396	5253 66512 44169 75163	1004 83241 32444
347	229 27228 68712 17150	110 41192 60918	397	5589 23320 25954 04488	1049 67982 04736
348	244 90453 74553 82406	115 66794 79970	398	5945 79011 47078 74597	1096 47115 85280
349	261 57890 73511 44125	121 16645 56454	399	6324 62148 25042 94325	1145 28826 89344
350	279 36332 84837 02152	126 91829 24648	400	6727 09005 17410 41926	1196 21634 00706

NUMBER OF PARTITIONS AND PARTITIONS INTO DISTINCT PARTS

Table 24.5

n	$p(n)$	$q(n)$	n	$p(n)$	$q(n)$
400	6727 09005 17410 41926	1196 21634 00706	450	1 34508 18800 15729 23840	9893 14440 61528
401	7154 64022 26539 42321	1249 34404 08000	451	1 42573 13615 53474 04229	10307 93957 13070
402	7608 80284 33398 79269	1304 76365 81998	452	1 51112 26207 19173 13678	10739 65687 10144
403	8091 20027 64844 65581	1362 57124 07808	453	1 60152 90524 45537 15585	11188 96810 43072
404	8603 55175 93486 55060	1422 86674 81438	454	1 69723 95104 64580 40965	11656 57102 54336
405	9147 67906 88591 17602	1485 75420 52794	455	1 79855 91645 39582 67598	12143 19032 12544
406	9725 51251 37420 21729	1551 34186 29884	456	1 90581 04044 26519 31034	12649 57862 22432
407	10339 09726 71239 47241	1619 74236 54282	457	2 01933 37928 51146 88629	13176 51755 08648
408	10990 60006 37759 26994	1691 07292 29128	458	2 13948 90703 27330 69132	13724 81881 00782
409	11682 31627 71923 17780	1765 45549 15430	459	2 26665 62143 58313 45565	14295 32530 93376
410	12416 67740 31511 90382	1843 01696 07104	460	2 40123 65561 39251 92081	14888 91233 20640
411	13196 25896 69254 35702	1923 88934 65516	461	2 54365 39575 85741 99975	15506 48874 75476
412	14023 78888 35188 47344	2008 20999 30208	462	2 69435 60521 29549 94471	16148 99826 46592
413	14902 15629 03099 48968	2096 12178 16576	463	2 85381 55524 19619 86287	16817 42073 15550
414	15834 42088 44881 87770	2187 77334 80960	464	3 02253 16287 25766 36605	17512 77348 45952
415	16823 82278 71392 35544	2283 31930 70488	465	3 20103 13615 29932 90544	18236 11274 38194
416	17873 79296 96898 76004	2382 92048 69148	466	3 38987 12724 95254 32549	18988 53505 94524
417	18987 96426 73316 64557	2486 74417 20078	467	3 58963 89376 81628 76613	19771 17881 29024
418	20170 18301 88059 33659	2594 96435 42056	468	3 80095 46876 31205 98477	20585 22576 95744
419	21424 52136 02556 36320	2707 76199 52640	469	4 02447 33986 17114 75160	21431 90268 83034
420	22755 29021 65800 25259	2825 32529 77152	470	4 26088 63801 56524 13417	22312 48299 10884
421	24167 05302 14413 63961	2947 84998 62528	471	4 51092 33635 50960 99864	23228 28849 04960
422	25664 64021 38377 14846	3075 53960 09352	472	4 77535 45970 81641 15593	24180 69117 98586
423	27253 16454 62304 21739	3208 60580 00384	473	5 05499 30531 42046 29558	25171 11509 01902
424	28938 03725 70847 98150	3347 26867 45954	474	5 35069 67535 16072 62125	26201 03821 12696
425	30724 98514 70950 51099	3491 75707 60097	475	5 66337 12186 58055 99675	27271 99448 23232
426	32620 06861 74102 32189	3642 30895 45254	476	5 99397 20478 23018 52926	28385 57585 65430
427	34629 70071 39035 75934	3799 17171 07136	477	6 34350 76365 37870 28583	29543 43443 69603
428	36760 66724 18315 27309	3962 60256 14146	478	6 71304 20389 67318 07232	30747 28468 94368
429	39020 14800 02372 59665	4132 86891 79000	479	7 10369 79823 66282 38005	31998 90573 73738
430	41415 73920 71023 58378	4310 24877 85006	480	7 51666 00419 49931 25591	33300 14373 57056
431	43955 47717 05181 16534	4495 03113 72460	481	7 95317 79841 47582 32180	34652 91433 03468
432	46647 86328 42292 67991	4687 51640 62334	482	8 41457 02874 28236 49455	36059 20520 80640
433	49501 89040 94051 50715	4888 01685 40672	483	8 90222 78495 19280 88294	37521 07873 43946
434	52527 07072 91082 40605	5096 85706 20480	484	9 41761 78911 49976 98055	39040 67468 62530
435	55733 46514 46362 86656	5314 37439 57460	485	9 96228 80660 85734 11012	40620 21308 45496
436	59131 71430 91696 18645	5540 91949 44512	486	10 53787 07886 24553 46513	42261 99712 45764
437	62733 07137 60430 79215	5776 85678 02880	487	11 14608 77893 64264 84248	43968 41621 12802
438	66549 43656 69662 97367	6022 56498 45546	488	11 78875 49115 57358 02646	45741 94910 51264
439	70593 39364 65621 35510	6278 43769 39520	489	12 46778 71600 12729 19665	47585 16717 64998
440	74878 24841 94708 86233	6544 88391 85792	490	13 18520 40161 22702 33223	49500 73777 62304
441	79418 06934 64434 02240	6822 32867 92200	491	13 94313 50322 44478 16939	51491 42772 84172
442	84227 73040 77294 99781	7111 21361 67457	492	14 74382 57204 03639 53132	53560 10694 36938
443	89322 95632 13536 45667	7411 99762 56080	493	15 58964 37499 49778 06173	55709 75216 10170
444	94720 37025 78934 71820	7725 15750 89318	494	16 48308 54706 61724 38760	57943 45082 47040
445	1 00437 54417 17528 47604	8051 18865 81728	495	17 42678 27774 77609 81187	60264 40509 50309
446	1 06493 05190 52391 18581	8390 60575 94564	496	18 42351 03350 31598 91466	62675 93600 10788
447	1 12906 52519 91961 03354	8743 94352 40798	497	19 47619 31798 76580 64007	65181 48774 31176
448	1 19698 71278 27202 05954	9111 75744 62854	498	20 58791 47204 28849 01563	67784 63214 30326
449	1 26891 54269 09814 18000	9494 62459 05984	499	21 76192 51543 92874 61625	70489 07325 21792
450	1 34508 18800 15729 23840	9893 14440 61528	500	23 00165 03257 43239 95027	73298 65212 45024

Table 24.6 — ARITHMETIC FUNCTIONS

n	$\varphi(n)$	σ_0	σ_1	n	$\varphi(n)$	σ_0	σ_1	n	$\varphi(n)$	σ_0	σ_1	n	$\varphi(n)$	σ_0	σ_1	n	$\varphi(n)$	σ_0	σ_1	n	$\varphi(n)$	σ_0	σ_1
1	1	1	1	51	32	4	72	101	100	2	102	151	150	2	152	201	132	4	272				
2	1	2	3	52	24	6	98	102	32	8	216	152	72	8	300	202	100	4	306				
3	2	2	4	53	52	2	54	103	102	2	104	153	96	6	234	203	168	4	240				
4	2	3	7	54	18	8	120	104	48	8	210	154	60	8	288	204	64	12	504				
5	4	2	6	55	40	4	72	105	48	8	192	155	120	4	192	205	160	4	252				
6	2	4	12	56	24	8	120	106	52	4	162	156	48	12	392	206	102	4	312				
7	6	2	8	57	36	4	80	107	106	2	108	157	156	2	158	207	132	6	312				
8	4	4	15	58	28	4	90	108	36	12	280	158	78	4	240	208	96	10	434				
9	6	3	13	59	58	2	60	109	108	2	110	159	104	4	216	209	180	4	240				
10	4	4	18	60	16	12	168	110	40	8	216	160	64	12	378	210	48	16	576				
11	10	2	12	61	60	2	62	111	72	4	152	161	132	4	192	211	210	2	212				
12	4	6	28	62	30	4	96	112	48	10	248	162	54	10	363	212	104	6	378				
13	12	2	14	63	36	6	104	113	112	2	114	163	162	2	164	213	140	4	288				
14	6	4	24	64	32	7	127	114	36	8	240	164	80	6	294	214	106	4	324				
15	8	4	24	65	48	4	84	115	88	4	144	165	80	8	288	215	168	4	264				
16	8	5	31	66	20	8	144	116	56	6	210	166	82	4	252	216	72	16	600				
17	16	2	18	67	66	2	68	117	72	6	182	167	166	2	168	217	180	4	256				
18	6	6	39	68	32	6	126	118	58	4	180	168	48	16	480	218	108	4	330				
19	18	2	20	69	44	4	96	119	96	4	144	169	156	3	183	219	144	4	296				
20	8	6	42	70	24	8	144	120	32	16	360	170	64	8	324	220	80	12	504				
21	12	4	32	71	70	2	72	121	110	3	133	171	108	6	260	221	192	4	252				
22	10	4	36	72	24	12	195	122	60	4	186	172	84	6	308	222	72	8	456				
23	22	2	24	73	72	2	74	123	80	4	168	173	172	2	174	223	222	2	224				
24	8	8	60	74	36	4	114	124	60	6	224	174	56	8	360	224	96	12	504				
25	20	3	31	75	40	6	124	125	100	4	156	175	120	6	248	225	120	9	403				
26	12	4	42	76	36	6	140	126	36	12	312	176	80	10	372	226	112	4	342				
27	18	4	40	77	60	4	96	127	126	2	128	177	116	4	240	227	226	2	228				
28	12	6	56	78	24	8	168	128	64	8	255	178	88	4	270	228	72	12	560				
29	28	2	30	79	78	2	80	129	84	4	176	179	178	2	180	229	228	2	230				
30	8	8	72	80	32	10	186	130	48	8	252	180	48	18	546	230	88	8	432				
31	30	2	32	81	54	5	121	131	130	2	132	181	180	2	182	231	120	8	384				
32	16	6	63	82	40	4	126	132	40	12	336	182	72	8	336	232	112	8	450				
33	20	4	48	83	82	2	84	133	108	4	160	183	120	4	248	233	232	2	234				
34	16	4	54	84	24	12	224	134	66	4	204	184	88	8	360	234	72	12	546				
35	24	4	48	85	64	4	108	135	72	8	240	185	144	4	228	235	184	4	288				
36	12	9	91	86	42	4	132	136	64	8	270	186	60	8	384	236	116	6	420				
37	36	2	38	87	56	4	120	137	136	2	138	187	160	4	216	237	156	4	320				
38	18	4	60	88	40	8	180	138	44	8	288	188	92	6	336	238	96	8	432				
39	24	4	56	89	88	2	90	139	138	2	140	189	108	8	320	239	238	2	240				
40	16	8	90	90	24	12	234	140	48	12	336	190	72	8	360	240	64	20	744				
41	40	2	42	91	72	4	112	141	92	4	192	191	190	2	192	241	240	2	242				
42	12	8	96	92	44	6	168	142	70	4	216	192	64	14	508	242	110	6	399				
43	42	2	44	93	60	4	128	143	120	4	168	193	192	2	194	243	162	6	364				
44	20	6	84	94	46	4	144	144	48	15	403	194	96	4	294	244	120	6	434				
45	24	6	78	95	72	4	120	145	112	4	180	195	96	8	336	245	168	6	342				
46	22	4	72	96	32	12	252	146	72	4	222	196	84	9	399	246	80	8	504				
47	46	2	48	97	96	2	98	147	84	6	228	197	196	2	198	247	216	4	280				
48	16	10	124	98	42	6	171	148	72	6	266	198	60	12	468	248	120	8	480				
49	42	3	57	99	60	6	156	149	148	2	150	199	198	2	200	249	164	4	336				
50	20	6	93	100	40	9	217	150	40	12	372	200	80	12	465	250	100	8	468				

From British Association for the Advancement of Science, Mathematical Tables, vol. VIII, Number-divisor tables. Cambridge Univ. Press, Cambridge, England, 1940 (with permission).

ARITHMETIC FUNCTIONS

Table 24.6

n	$\varphi(n)$	σ_0	σ_1	n	$\varphi(n)$	σ_0	σ_1	n	$\varphi(n)$	σ_0	σ_1	n	$\varphi(n)$	σ_0	σ_1	n	$\varphi(n)$	σ_0	σ_1
251	250	2	252	301	252	4	352	351	216	8	560	401	400	2	402	451	400	4	504
252	72	18	728	302	150	4	456	352	160	12	756	402	132	8	816	452	224	6	798
253	220	4	288	303	200	4	408	353	352	2	354	403	360	4	448	453	300	4	608
254	126	4	384	304	144	10	620	354	116	8	720	404	200	6	714	454	226	4	684
255	128	8	432	305	240	4	372	355	280	4	432	405	216	10	726	455	288	8	672
256	128	9	511	306	96	12	702	356	176	6	630	406	168	8	720	456	144	16	1200
257	256	2	258	307	306	2	308	357	192	8	576	407	360	4	456	457	456	2	458
258	84	8	528	308	120	12	672	358	178	4	540	408	128	16	1080	458	228	4	690
259	216	4	304	309	204	4	416	359	358	2	360	409	408	2	410	459	288	8	720
260	96	12	588	310	120	8	576	360	96	24	1170	410	160	8	756	460	176	12	1008
261	168	6	390	311	310	2	312	361	342	3	381	411	272	4	552	461	460	2	462
262	130	4	396	312	96	16	840	362	180	4	546	412	204	6	728	462	120	16	1152
263	262	2	264	313	312	2	314	363	220	6	532	413	348	4	480	463	462	2	464
264	80	16	720	314	156	4	474	364	144	12	784	414	132	12	936	464	224	10	930
265	208	4	324	315	144	12	624	365	288	4	444	415	328	4	504	465	240	8	768
266	108	8	480	316	156	6	560	366	120	8	744	416	192	12	882	466	232	4	702
267	176	4	360	317	316	2	318	367	366	2	368	417	276	4	560	467	466	2	468
268	132	6	476	318	104	8	648	368	176	10	744	418	180	8	720	468	144	18	1274
269	268	2	270	319	280	4	360	369	240	6	546	419	418	2	420	469	396	4	544
270	72	16	720	320	128	14	762	370	144	8	684	420	96	24	1344	470	184	8	864
271	270	2	272	321	212	4	432	371	312	4	432	421	420	2	422	471	312	4	632
272	128	10	558	322	132	8	576	372	120	12	896	422	210	4	636	472	232	8	900
273	144	8	448	323	288	4	360	373	372	2	374	423	276	6	624	473	420	4	528
274	136	4	414	324	108	15	847	374	160	8	648	424	208	8	810	474	156	8	960
275	200	6	372	325	240	6	434	375	200	8	624	425	320	6	558	475	360	6	620
276	88	12	672	326	162	4	492	376	184	8	720	426	140	8	864	476	192	12	1008
277	276	2	278	327	216	4	440	377	336	4	420	427	360	4	496	477	312	6	702
278	138	4	420	328	160	8	630	378	108	16	960	428	212	6	756	478	238	4	720
279	180	6	416	329	276	4	384	379	378	2	380	429	240	8	672	479	478	2	480
280	96	16	720	330	80	16	864	380	144	12	840	430	168	8	792	480	128	24	1512
281	280	2	282	331	330	2	332	381	252	4	512	431	430	2	432	481	432	4	532
282	92	8	576	332	164	6	588	382	190	4	576	432	144	20	1240	482	240	4	726
283	282	2	284	333	216	6	494	383	382	2	384	433	432	2	434	483	264	8	768
284	140	6	504	334	166	4	504	384	128	16	1020	434	180	8	768	484	220	9	931
285	144	8	480	335	264	4	408	385	240	8	576	435	224	8	720	485	384	4	588
286	120	8	504	336	96	20	992	386	192	4	582	436	216	6	770	486	162	12	1092
287	240	4	336	337	336	2	338	387	252	6	572	437	396	4	480	487	486	2	488
288	96	18	819	338	156	6	549	388	192	6	686	438	144	8	888	488	240	8	930
289	272	3	307	339	224	4	456	389	388	2	390	439	438	2	440	489	324	4	656
290	112	8	540	340	128	12	756	390	96	16	1008	440	160	16	1080	490	168	12	1026
291	192	4	392	341	300	4	384	391	352	4	432	441	252	9	741	491	490	2	492
292	144	6	578	342	108	12	780	392	168	12	855	442	192	8	756	492	160	12	1176
293	292	2	294	343	294	4	400	393	260	4	528	443	442	2	444	493	448	4	540
294	84	12	684	344	168	8	660	394	196	4	594	444	144	12	1064	494	216	8	840
295	232	4	360	345	176	8	576	395	312	4	480	445	352	4	540	495	240	12	936
296	144	8	570	346	172	4	522	396	120	18	1092	446	222	4	672	496	240	10	992
297	180	8	480	347	346	2	348	397	396	2	398	447	296	4	600	497	420	4	576
298	148	4	450	348	112	12	840	398	198	4	600	448	192	14	1016	498	164	8	1008
299	264	4	336	349	348	2	350	399	216	8	640	449	448	2	450	499	498	2	500
300	80	18	868	350	120	12	744	400	160	15	961	450	120	18	1209	500	200	12	1092

Table 24.6 **ARITHMETIC FUNCTIONS**

n	$\varphi(n)$	σ_0	σ_1	n	$\varphi(n)$	σ_0	σ_1	n	$\varphi(n)$	σ_0	σ_1	n	$\varphi(n)$	σ_0	σ_1	n	$\varphi(n)$	σ_0	σ_1
501	332	4	672	551	504	4	600	601	600	2	602	651	360	8	1024	701	700	2	702
502	250	4	756	552	176	16	1440	602	252	8	1056	652	324	6	1148	702	216	16	1680
503	502	2	504	553	468	4	640	603	396	6	884	653	652	2	654	703	648	4	760
504	144	24	1560	554	276	4	834	604	300	6	1064	654	216	8	1320	704	320	14	1524
505	400	4	612	555	288	8	912	605	440	6	798	655	520	4	792	705	368	8	1152
506	220	8	864	556	276	6	980	606	200	8	1224	656	320	10	1302	706	352	4	1062
507	312	6	732	557	556	2	558	607	606	2	608	657	432	6	962	707	600	4	816
508	252	6	896	558	180	12	1248	608	288	12	1260	658	276	8	1152	708	232	12	1680
509	508	2	510	559	504	4	616	609	336	8	960	659	658	2	660	709	708	2	710
510	128	16	1296	560	192	20	1488	610	240	8	1116	660	160	24	2016	710	280	8	1296
511	432	4	592	561	320	8	864	611	552	4	672	661	660	2	662	711	468	6	1040
512	256	10	1023	562	280	4	846	612	192	18	1638	662	330	4	996	712	352	8	1350
513	324	8	800	563	562	2	564	613	612	2	614	663	384	8	1008	713	660	4	768
514	256	4	774	564	184	12	1344	614	306	4	924	664	328	8	1260	714	192	16	1728
515	408	4	624	565	448	4	684	615	320	8	1008	665	432	8	960	715	480	8	1008
516	168	12	1232	566	282	4	852	616	240	16	1440	666	216	12	1482	716	356	6	1260
517	460	4	576	567	324	10	968	617	616	2	618	667	616	4	720	717	476	4	960
518	216	8	912	568	280	8	1080	618	204	8	1248	668	332	6	1176	718	358	4	1080
519	344	4	696	569	568	2	570	619	618	2	620	669	444	4	896	719	718	2	720
520	192	16	1260	570	144	16	1440	620	240	12	1344	670	264	8	1224	720	192	30	2418
521	520	2	522	571	570	2	572	621	396	8	960	671	600	4	744	721	612	4	832
522	168	12	1170	572	240	12	1176	622	310	4	936	672	192	24	2016	722	342	6	1143
523	522	2	524	573	380	4	768	623	528	4	720	673	672	2	674	723	480	4	968
524	260	6	924	574	240	8	1008	624	192	20	1736	674	336	4	1014	724	360	6	1274
525	240	12	992	575	440	6	744	625	500	5	781	675	360	12	1240	725	560	6	930
526	262	4	792	576	192	21	1651	626	312	4	942	676	312	9	1281	726	220	12	1596
527	480	4	576	577	576	2	578	627	360	8	960	677	676	2	678	727	726	2	728
528	160	20	1488	578	272	6	921	628	312	6	1106	678	224	8	1368	728	288	16	1680
529	506	3	553	579	384	4	776	629	576	4	684	679	576	4	784	729	486	7	1093
530	208	8	972	580	224	12	1260	630	144	24	1872	680	256	16	1620	730	288	8	1332
531	348	6	780	581	492	4	672	631	630	2	632	681	452	4	912	731	672	4	792
532	216	12	1120	582	192	8	1176	632	312	8	1200	682	300	8	1152	732	240	12	1736
533	480	4	588	583	520	4	648	633	420	4	848	683	682	2	684	733	732	2	734
534	176	8	1080	584	288	8	1110	634	316	4	954	684	216	18	1820	734	366	4	1104
535	424	4	648	585	288	12	1092	635	504	4	768	685	544	4	828	735	336	12	1368
536	264	8	1020	586	292	4	882	636	208	12	1512	686	294	8	1200	736	352	12	1512
537	356	4	720	587	586	2	588	637	504	6	798	687	456	4	920	737	660	4	816
538	268	4	810	588	168	18	1596	638	280	8	1080	688	336	10	1364	738	240	12	1638
539	420	6	684	589	540	4	640	639	420	6	936	689	624	4	756	739	738	2	740
540	144	24	1680	590	232	8	1080	640	256	16	1530	690	176	16	1728	740	288	12	1596
541	540	2	542	591	392	4	792	641	640	2	642	691	690	2	692	741	432	8	1120
542	270	4	816	592	288	10	1178	642	212	8	1296	692	344	6	1218	742	312	8	1296
543	360	4	728	593	592	2	594	643	642	2	644	693	360	12	1248	743	742	2	744
544	256	12	1134	594	180	16	1440	644	264	12	1344	694	346	4	1044	744	240	16	1920
545	432	4	660	595	384	8	864	645	336	8	1056	695	552	4	840	745	592	4	900
546	144	16	1344	596	296	6	1050	646	288	8	1080	696	224	16	1800	746	372	4	1122
547	546	2	548	597	396	4	800	647	646	2	648	697	640	4	756	747	492	6	1092
548	272	6	966	598	264	8	1008	648	216	20	1815	698	348	4	1050	748	320	12	1512
549	360	6	806	599	598	2	600	649	580	4	720	699	464	4	936	749	636	4	864
550	200	12	1116	600	160	24	1860	650	240	12	1302	700	240	18	1736	750	200	16	1872

ARITHMETIC FUNCTIONS

Table 24.6

n	$\varphi(n)$	σ_0	σ_1	n	$\varphi(n)$	σ_0	σ_1	n	$\varphi(n)$	σ_0	σ_1	n	$\varphi(n)$	σ_0	σ_1	n	$\varphi(n)$	σ_0	σ_1
751	750	2	752	801	528	6	1170	851	792	4	912	901	832	4	972	951	632	4	1272
752	368	10	1488	802	400	4	1206	852	280	12	2016	902	400	8	1512	952	384	16	2160
753	500	4	1008	803	720	4	888	853	852	2	854	903	504	8	1408	953	952	2	954
754	336	8	1260	804	264	12	1904	854	360	8	1488	904	448	8	1710	954	312	12	2106
755	600	4	912	805	528	8	1152	855	432	12	1560	905	720	4	1092	955	760	4	1152
756	216	24	2240	806	360	8	1344	856	424	8	1620	906	300	8	1824	956	476	6	1680
757	756	2	758	807	536	4	1080	857	856	2	858	907	906	2	908	957	560	8	1440
758	378	4	1140	808	400	8	1530	858	240	16	2016	908	452	6	1596	958	478	4	1440
759	440	8	1152	809	808	2	810	859	858	2	860	909	600	6	1326	959	816	4	1104
760	288	16	1800	810	216	20	2178	860	336	12	1848	910	288	16	2016	960	256	28	3048
761	760	2	762	811	810	2	812	861	480	8	1344	911	910	2	912	961	930	3	993
762	252	8	1536	812	336	12	1680	862	430	4	1296	912	288	20	2480	962	432	8	1596
763	640	4	880	813	540	4	1088	863	862	2	864	913	820	4	1008	963	636	6	1404
764	380	6	1344	814	360	8	1368	864	288	24	2520	914	456	4	1374	964	480	6	1694
765	384	12	1404	815	648	4	984	865	688	4	1044	915	480	8	1488	965	768	4	1164
766	382	4	1152	816	256	20	2232	866	432	4	1302	916	456	6	1610	966	264	16	2304
767	696	4	840	817	756	4	880	867	544	6	1228	917	780	4	1056	967	966	2	968
768	256	18	2044	818	408	4	1230	868	360	12	1792	918	288	16	2160	968	440	12	1995
769	768	2	770	819	432	12	1456	869	780	4	960	919	918	2	920	969	576	8	1440
770	240	16	1728	820	320	12	1764	870	224	16	2160	920	352	16	2160	970	384	8	1764
771	512	4	1032	821	820	2	822	871	792	4	952	921	612	4	1232	971	970	2	972
772	384	6	1358	822	272	8	1656	872	432	8	1650	922	460	4	1386	972	324	18	2548
773	772	2	774	823	822	2	824	873	576	6	1274	923	840	4	1008	973	828	4	1120
774	252	12	1716	824	408	8	1560	874	396	8	1440	924	240	24	2688	974	486	4	1464
775	600	6	992	825	400	12	1488	875	600	8	1248	925	720	6	1178	975	480	12	1736
776	384	8	1470	826	348	8	1440	876	288	12	2072	926	462	4	1392	976	480	10	1922
777	432	8	1216	827	826	2	828	877	876	2	878	927	612	6	1352	977	976	2	978
778	388	4	1170	828	264	18	2184	878	438	4	1320	928	448	12	1890	978	324	8	1968
779	720	4	840	829	828	2	830	879	584	4	1176	929	928	2	930	979	880	4	1080
780	192	24	2352	830	328	8	1512	880	320	20	2232	930	240	16	2304	980	336	18	2394
781	700	4	864	831	552	4	1112	881	880	2	882	931	756	6	1140	981	648	6	1430
782	352	8	1296	832	384	14	1778	882	252	18	2223	932	464	6	1638	982	490	4	1476
783	504	8	1200	833	672	6	1026	883	882	2	884	933	620	4	1248	983	982	2	984
784	336	15	1767	834	276	8	1680	884	384	12	1764	934	466	4	1404	984	320	16	2520
785	624	4	948	835	664	4	1008	885	464	8	1440	935	640	8	1296	985	784	4	1188
786	260	8	1584	836	360	12	1680	886	442	4	1332	936	288	24	2730	986	448	8	1620
787	786	2	788	837	540	8	1280	887	886	2	888	937	936	2	938	987	552	8	1536
788	392	6	1386	838	418	4	1260	888	288	16	2280	938	396	8	1632	988	432	12	1960
789	524	4	1056	839	838	2	840	889	756	4	1024	939	624	4	1256	989	924	4	1056
790	312	8	1440	840	192	32	2880	890	352	8	1620	940	368	12	2016	990	240	24	2808
791	672	4	912	841	812	3	871	891	540	10	1452	941	940	2	942	991	990	2	992
792	240	24	2340	842	420	4	1266	892	444	6	1568	942	312	8	1896	992	480	12	2016
793	720	4	868	843	560	4	1128	893	828	4	960	943	880	4	1008	993	660	4	1328
794	396	4	1194	844	420	6	1484	894	296	8	1800	944	464	10	1860	994	420	8	1728
795	416	8	1296	845	624	6	1098	895	712	4	1080	945	432	16	1920	995	792	4	1200
796	396	6	1400	846	276	12	1872	896	384	16	2040	946	420	8	1584	996	328	12	2352
797	796	2	798	847	660	6	1064	897	528	8	1344	947	946	2	948	997	996	2	998
798	216	16	1920	848	416	10	1674	898	448	4	1350	948	312	12	2240	998	498	4	1500
799	736	4	864	849	564	4	1136	899	840	4	960	949	864	4	1036	999	648	8	1520
800	320	18	1953	850	320	12	1674	900	240	27	2821	950	360	12	1860	1000	400	16	2340

844
Table 24.7
000

COMBINATORIAL ANALYSIS
Factorizations

499

N	0	1	2	3	4	5	6	7	8	9
0	—	1	2	3	2^2	5	$2\cdot3$	7	2^3	3^2
1	$2\cdot5$	11	$2^2\cdot3$	13	$2\cdot7$	$3\cdot5$	2^4	17	$2\cdot3^2$	19
2	$2^2\cdot5$	$3\cdot7$	$2\cdot11$	23	$2^3\cdot3$	5^2	$2\cdot13$	3^3	$2^2\cdot7$	29
3	$2\cdot3\cdot5$	31	2^5	$3\cdot11$	$2\cdot17$	$5\cdot7$	$2^2\cdot3^2$	37	$2\cdot19$	$3\cdot13$
4	$2^3\cdot5$	41	$2\cdot3\cdot7$	43	$2^2\cdot11$	$3^2\cdot5$	$2\cdot23$	47	$2^4\cdot3$	7^2
5	$2\cdot5^2$	$3\cdot17$	$2^2\cdot13$	53	$2\cdot3^3$	$5\cdot11$	$2^3\cdot7$	$3\cdot19$	$2\cdot29$	59
6	$2^2\cdot3\cdot5$	61	$2\cdot31$	$3^2\cdot7$	2^6	$5\cdot13$	$2\cdot3\cdot11$	67	$2^2\cdot17$	$3\cdot23$
7	$2\cdot5\cdot7$	71	$2^3\cdot3^2$	73	$2\cdot37$	$3\cdot5^2$	$2^2\cdot19$	$7\cdot11$	$2\cdot3\cdot13$	79
8	$2^4\cdot5$	3^4	$2\cdot41$	83	$2^2\cdot3\cdot7$	$5\cdot17$	$2\cdot43$	$3\cdot29$	$2^3\cdot11$	89
9	$2\cdot3^2\cdot5$	$7\cdot13$	$2^2\cdot23$	$3\cdot31$	$2\cdot47$	$5\cdot19$	$2^5\cdot3$	97	$2\cdot7^2$	$3^2\cdot11$
10	$2^2\cdot5^2$	101	$2\cdot3\cdot17$	103	$2^3\cdot13$	$3\cdot5\cdot7$	$2\cdot53$	107	$2^2\cdot3^3$	109
11	$2\cdot5\cdot11$	$3\cdot37$	$2^4\cdot7$	113	$2\cdot3\cdot19$	$5\cdot23$	$2^2\cdot29$	$3^2\cdot13$	$2\cdot59$	$7\cdot17$
12	$2^3\cdot3\cdot5$	11^2	$2\cdot61$	$3\cdot41$	$2^2\cdot31$	5^3	$2\cdot3^2\cdot7$	127	2^7	$3\cdot43$
13	$2\cdot5\cdot13$	131	$2^2\cdot3\cdot11$	$7\cdot19$	$2\cdot67$	$3^3\cdot5$	$2^3\cdot17$	137	$2\cdot3\cdot23$	139
14	$2^2\cdot5\cdot7$	$3\cdot47$	$2\cdot71$	$11\cdot13$	$2^4\cdot3^2$	$5\cdot29$	$2\cdot73$	$3\cdot7^2$	$2^2\cdot37$	149
15	$2\cdot3\cdot5^2$	151	$2^3\cdot19$	$3^2\cdot17$	$2\cdot7\cdot11$	$5\cdot31$	$2^2\cdot3\cdot13$	157	$2\cdot79$	$3\cdot53$
16	$2^5\cdot5$	$7\cdot23$	$2\cdot3^4$	163	$2^2\cdot41$	$3\cdot5\cdot11$	$2\cdot83$	167	$2^3\cdot3\cdot7$	13^2
17	$2\cdot5\cdot17$	$3^2\cdot19$	$2^2\cdot43$	173	$2\cdot3\cdot29$	$5^2\cdot7$	$2^4\cdot11$	$3\cdot59$	$2\cdot89$	179
18	$2^2\cdot3^2\cdot5$	181	$2\cdot7\cdot13$	$3\cdot61$	$2^3\cdot23$	$5\cdot37$	$2\cdot3\cdot31$	$11\cdot17$	$2^2\cdot47$	$3^3\cdot7$
19	$2\cdot5\cdot19$	191	$2^6\cdot3$	193	$2\cdot97$	$3\cdot5\cdot13$	$2^2\cdot7^2$	197	$2\cdot3^2\cdot11$	199
20	$2^3\cdot5^2$	$3\cdot67$	$2\cdot101$	$7\cdot29$	$2^2\cdot3\cdot17$	$5\cdot41$	$2\cdot103$	$3^2\cdot23$	$2^4\cdot13$	$11\cdot19$
21	$2\cdot3\cdot5\cdot7$	211	$2^2\cdot53$	$3\cdot71$	$2\cdot107$	$5\cdot43$	$2^3\cdot3^3$	$7\cdot31$	$2\cdot109$	$3\cdot73$
22	$2^2\cdot5\cdot11$	$13\cdot17$	$2\cdot3\cdot37$	223	$2^5\cdot7$	$3^2\cdot5^2$	$2\cdot113$	227	$2^2\cdot3\cdot19$	229
23	$2\cdot5\cdot23$	$3\cdot7\cdot11$	$2^3\cdot29$	233	$2\cdot3^2\cdot13$	$5\cdot47$	$2^2\cdot59$	$3\cdot79$	$2\cdot7\cdot17$	239
24	$2^4\cdot3\cdot5$	241	$2\cdot11^2$	3^5	$2^2\cdot61$	$5\cdot7^2$	$2\cdot3\cdot41$	$13\cdot19$	$2^3\cdot31$	$3\cdot83$
25	$2\cdot5^3$	251	$2^2\cdot3^2\cdot7$	$11\cdot23$	$2\cdot127$	$3\cdot5\cdot17$	2^8	257	$2\cdot3\cdot43$	$7\cdot37$
26	$2^2\cdot5\cdot13$	$3^2\cdot29$	$2\cdot131$	263	$2^3\cdot3\cdot11$	$5\cdot53$	$2\cdot7\cdot19$	$3\cdot89$	$2^2\cdot67$	269
27	$2\cdot3^3\cdot5$	271	$2^4\cdot17$	$3\cdot7\cdot13$	$2\cdot137$	$5^2\cdot11$	$2^2\cdot3\cdot23$	277	$2\cdot139$	$3^2\cdot31$
28	$2^3\cdot5\cdot7$	281	$2\cdot3\cdot47$	283	$2^2\cdot71$	$3\cdot5\cdot19$	$2\cdot11\cdot13$	$7\cdot41$	$2^5\cdot3^2$	17^2
29	$2\cdot5\cdot29$	$3\cdot97$	$2^2\cdot73$	293	$2\cdot3\cdot7^2$	$5\cdot59$	$2^3\cdot37$	$3^3\cdot11$	$2\cdot149$	$13\cdot23$
30	$2^2\cdot3\cdot5^2$	$3\cdot101$	$2\cdot151$	$3\cdot101$	$2^4\cdot19$	$5\cdot61$	$2\cdot3^2\cdot17$	307	$2^2\cdot7\cdot11$	$3\cdot103$
31	$2\cdot5\cdot31$	311	$2^3\cdot3\cdot13$	313	$2\cdot157$	$3^2\cdot5\cdot7$	$2^2\cdot79$	317	$2\cdot3\cdot53$	$11\cdot29$
32	$2^6\cdot5$	$3\cdot107$	$2\cdot7\cdot23$	$17\cdot19$	$2^2\cdot3^4$	$5^2\cdot13$	$2\cdot163$	$3\cdot109$	$2^3\cdot41$	$7\cdot47$
33	$2\cdot3\cdot5\cdot11$	331	$2^2\cdot83$	$3^2\cdot37$	$2\cdot167$	$5\cdot67$	$2^4\cdot3\cdot7$	337	$2\cdot13^2$	$3\cdot113$
34	$2^2\cdot5\cdot17$	$11\cdot31$	$2\cdot3^2\cdot19$	7^3	$2^3\cdot43$	$3\cdot5\cdot23$	$2\cdot173$	347	$2^2\cdot3\cdot29$	349
35	$2\cdot5^2\cdot7$	$3^3\cdot13$	$2^5\cdot11$	353	$2\cdot3\cdot59$	$5\cdot71$	$2^2\cdot89$	$3\cdot7\cdot17$	$2\cdot179$	359
36	$2^3\cdot3^2\cdot5$	19^2	$2\cdot181$	$3\cdot11^2$	$2^2\cdot7\cdot13$	$5\cdot73$	$2\cdot3\cdot61$	367	$2^4\cdot23$	$3^2\cdot41$
37	$2\cdot5\cdot37$	$7\cdot53$	$2^2\cdot3\cdot31$	373	$2\cdot11\cdot17$	$3\cdot5^3$	$2^3\cdot47$	$13\cdot29$	$2\cdot3^3\cdot7$	379
38	$2^2\cdot5\cdot19$	$3\cdot127$	$2\cdot191$	383	$2^7\cdot3$	$5\cdot7\cdot11$	$2\cdot193$	$3^2\cdot43$	$2^2\cdot97$	389
39	$2\cdot3\cdot5\cdot13$	$17\cdot23$	$2^3\cdot7^2$	$3\cdot131$	$2\cdot197$	$5\cdot79$	$2^2\cdot3^2\cdot11$	397	$2\cdot199$	$3\cdot7\cdot19$
40	$2^4\cdot5^2$	401	$2\cdot3\cdot67$	$13\cdot31$	$2^2\cdot101$	$3^4\cdot5$	$2\cdot7\cdot29$	$11\cdot37$	$2^3\cdot3\cdot17$	409
41	$2\cdot5\cdot41$	$3\cdot137$	$2^2\cdot103$	$7\cdot59$	$2\cdot3^2\cdot23$	$5\cdot83$	$2^5\cdot13$	$3\cdot139$	$2\cdot11\cdot19$	419
42	$2^2\cdot3\cdot5\cdot7$	421	$2\cdot211$	$3^2\cdot47$	$2^3\cdot53$	$5^2\cdot17$	$2\cdot3\cdot71$	$7\cdot61$	$2^2\cdot107$	$3\cdot11\cdot13$
43	$2\cdot5\cdot43$	431	$2^4\cdot3^3$	433	$2\cdot7\cdot31$	$3\cdot5\cdot29$	$2^2\cdot109$	$19\cdot23$	$2\cdot3\cdot73$	439
44	$2^3\cdot5\cdot11$	$3^2\cdot7^2$	$2\cdot13\cdot17$	443	$2^2\cdot3\cdot37$	$5\cdot89$	$2\cdot223$	$3\cdot149$	$2^6\cdot7$	449
45	$2\cdot3^2\cdot5^2$	$11\cdot41$	$2^2\cdot113$	$3\cdot151$	$2\cdot227$	$5\cdot7\cdot13$	$2^3\cdot3\cdot19$	457	$2\cdot229$	$3^3\cdot17$
46	$2^2\cdot5\cdot23$	461	$2\cdot3\cdot7\cdot11$	463	$2^4\cdot29$	$3\cdot5\cdot31$	$2\cdot233$	467	$2^2\cdot3^2\cdot13$	$7\cdot67$
47	$2\cdot5\cdot47$	$3\cdot157$	$2^3\cdot59$	$11\cdot43$	$2\cdot3\cdot79$	$5^2\cdot19$	$2^2\cdot7\cdot17$	$3^2\cdot53$	$2\cdot239$	479
48	$2^5\cdot3\cdot5$	$13\cdot37$	$2\cdot241$	$3\cdot7\cdot23$	$2^2\cdot11^2$	$5\cdot97$	$2\cdot3^5$	487	$2^3\cdot61$	$3\cdot163$
49	$2\cdot5\cdot7^2$	491	$2^2\cdot3\cdot41$	$17\cdot29$	$2\cdot13\cdot19$	$3^2\cdot5\cdot11$	$2^4\cdot31$	$7\cdot71$	$2\cdot3\cdot83$	499

* See page II.

From G. Kaván, Factor tables. Macmillan and Co., Ltd., London, England, 1937 (with permission).

Factorizations

Table 24.7

Factorizations of N from 500 to 999. Rows give the first two digits of N; columns give the units digit (0–9), so $N = (\text{row})\times 10 + (\text{column})$. Primes are shown in **bold**.

	0	1	2	3	4	5	6	7	8	9
50	$2^2\cdot5^3$	$3\cdot167$	$2\cdot251$	**503**	$2^3\cdot3^2\cdot7$	$5\cdot101$	$2\cdot11\cdot23$	$3\cdot13^2$	$2^2\cdot127$	**509**
51	$2\cdot3\cdot5\cdot17$	$7\cdot73$	2^9	$3^3\cdot19$	$2\cdot257$	$5\cdot103$	$2^2\cdot3\cdot43$	$11\cdot47$	$2\cdot7\cdot37$	$3\cdot173$
52	$2^3\cdot5\cdot13$	**521**	$2\cdot3^2\cdot29$	**523**	$2^2\cdot131$	$3\cdot5^2\cdot7$	$2\cdot263$	$17\cdot31$	$2^4\cdot3\cdot11$	23^2
53	$2\cdot5\cdot53$	$3^2\cdot59$	$2^2\cdot7\cdot19$	$13\cdot41$	$2\cdot3\cdot89$	$5\cdot107$	$2^3\cdot67$	$3\cdot179$	$2\cdot269$	$7^2\cdot11$
54	$2^2\cdot3^3\cdot5$	**541**	$2\cdot271$	$3\cdot181$	$2^5\cdot17$	$5\cdot109$	$2\cdot3\cdot7\cdot13$	**547**	$2^2\cdot137$	$3^2\cdot61$
55	$2\cdot5^2\cdot11$	$19\cdot29$	$2^3\cdot3\cdot23$	$7\cdot79$	$2\cdot277$	$3\cdot5\cdot37$	$2^2\cdot139$	**557**	$2\cdot3^2\cdot31$	$13\cdot43$
56	$2^4\cdot5\cdot7$	$3\cdot11\cdot17$	$2\cdot281$	**563**	$2^2\cdot3\cdot47$	$5\cdot113$	$2\cdot283$	$3^4\cdot7$	$2^3\cdot71$	**569**
57	$2\cdot3\cdot5\cdot19$	**571**	$2^2\cdot11\cdot13$	$3\cdot191$	$2\cdot7\cdot41$	$5^2\cdot23$	$2^6\cdot3^2$	**577**	$2\cdot17^2$	$3\cdot193$
58	$2^2\cdot5\cdot29$	$7\cdot83$	$2\cdot3\cdot97$	$11\cdot53$	$2^3\cdot73$	$3^2\cdot5\cdot13$	$2\cdot293$	**587**	$2^2\cdot3\cdot7^2$	$19\cdot31$
59	$2\cdot5\cdot59$	$3\cdot197$	$2^4\cdot37$	**593**	$2\cdot3^3\cdot11$	$5\cdot7\cdot17$	$2^2\cdot149$	$3\cdot199$	$2\cdot13\cdot23$	**599**
60	$2^3\cdot3\cdot5^2$	**601**	$2\cdot7\cdot43$	$3^2\cdot67$	$2^2\cdot151$	$5\cdot11^2$	$2\cdot3\cdot101$	**607**	$2^5\cdot19$	$3\cdot7\cdot29$
61	$2\cdot5\cdot61$	$13\cdot47$	$2^2\cdot3^2\cdot17$	**613**	$2\cdot307$	$3\cdot5\cdot41$	$2^3\cdot7\cdot11$	**617**	$2\cdot3\cdot103$	**619**
62	$2^2\cdot5\cdot31$	$3^3\cdot23$	$2\cdot311$	$7\cdot89$	$2^4\cdot3\cdot13$	5^4	$2\cdot313$	$3\cdot11\cdot19$	$2^2\cdot157$	$17\cdot37$
63	$2\cdot3^2\cdot5\cdot7$	**631**	$2^3\cdot79$	$3\cdot211$	$2\cdot317$	$5\cdot127$	$2^2\cdot3\cdot53$	$7^2\cdot13$	$2\cdot11\cdot29$	$3^2\cdot71$
64	$2^7\cdot5$	**641**	$2\cdot3\cdot107$	**643**	$2^2\cdot7\cdot23$	$3\cdot5\cdot43$	$2\cdot17\cdot19$	**647**	$2^3\cdot3^4$	$11\cdot59$
65	$2\cdot5^2\cdot13$	$3\cdot7\cdot31$	$2^2\cdot163$	**653**	$2\cdot3\cdot109$	$5\cdot131$	$2^4\cdot41$	$3^2\cdot73$	$2\cdot7\cdot47$	**659**
66	$2^2\cdot3\cdot5\cdot11$	**661**	$2\cdot331$	$3\cdot13\cdot17$	$2^3\cdot83$	$5\cdot7\cdot19$	$2\cdot3^2\cdot37$	$23\cdot29$	$2^2\cdot167$	$3\cdot223$
67	$2\cdot5\cdot67$	$11\cdot61$	$2^5\cdot3\cdot7$	**673**	$2\cdot337$	$3^3\cdot5^2$	$2^2\cdot13^2$	**677**	$2\cdot3\cdot113$	$7\cdot97$
68	$2^3\cdot5\cdot17$	$3\cdot227$	$2\cdot11\cdot31$	**683**	$2^2\cdot3^2\cdot19$	$5\cdot137$	$2\cdot7^3$	$3\cdot229$	$2^4\cdot43$	$13\cdot53$
69	$2\cdot3\cdot5\cdot23$	**691**	$2^2\cdot173$	$3^2\cdot7\cdot11$	$2\cdot347$	$5\cdot139$	$2^3\cdot3\cdot29$	$17\cdot41$	$2\cdot349$	$3\cdot233$
70	$2^2\cdot5^2\cdot7$	**701**	$2\cdot3^3\cdot13$	$19\cdot37$	$2^6\cdot11$	$3\cdot5\cdot47$	$2\cdot353$	$7\cdot101$	$2^2\cdot3\cdot59$	**709**
71	$2\cdot5\cdot71$	$3^2\cdot79$	$2^3\cdot89$	$23\cdot31$	$2\cdot3\cdot7\cdot17$	$5\cdot11\cdot13$	$2^2\cdot179$	$3\cdot239$	$2\cdot359$	**719**
72	$2^4\cdot3^2\cdot5$	$7\cdot103$	$2\cdot19^2$	$3\cdot241$	$2^2\cdot181$	$5^2\cdot29$	$2\cdot3\cdot11^2$	**727**	$2^3\cdot7\cdot13$	3^6
73	$2\cdot5\cdot73$	$17\cdot43$	$2^2\cdot3\cdot61$	**733**	$2\cdot367$	$3\cdot5\cdot7^2$	$2^5\cdot23$	$11\cdot67$	$2\cdot3^2\cdot41$	**739**
74	$2^2\cdot5\cdot37$	$3\cdot13\cdot19$	$2\cdot7\cdot53$	**743**	$2^3\cdot3\cdot31$	$5\cdot149$	$2\cdot373$	$3^2\cdot83$	$2^2\cdot11\cdot17$	$7\cdot107$
75	$2\cdot3\cdot5^4$	**751**	$2^4\cdot47$	$3\cdot251$	$2\cdot13\cdot29$	$5\cdot151$	$2^2\cdot3^3\cdot7$	**757**	$2\cdot379$	$3\cdot11\cdot23$
76	$2^3\cdot5\cdot19$	**761**	$2\cdot3\cdot127$	$7\cdot109$	$2^2\cdot191$	$3^2\cdot5\cdot17$	$2\cdot383$	$13\cdot59$	$2^8\cdot3$	**769**
77	$2\cdot5\cdot7\cdot11$	$3\cdot257$	$2^2\cdot193$	**773**	$2\cdot3^2\cdot43$	$5^2\cdot31$	$2^3\cdot97$	$3\cdot7\cdot37$	$2\cdot389$	$19\cdot41$
78	$2^2\cdot3\cdot5\cdot13$	$11\cdot71$	$2\cdot17\cdot23$	$3^3\cdot29$	$2^4\cdot7^2$	$5\cdot157$	$2\cdot3\cdot131$	**787**	$2^2\cdot197$	$3\cdot263$
79	$2\cdot5\cdot79$	$7\cdot113$	$2^3\cdot3^2\cdot11$	$13\cdot61$	$2\cdot397$	$3\cdot5\cdot53$	$2^2\cdot199$	**797**	$2\cdot3\cdot7\cdot19$	$17\cdot47$
80	$2^5\cdot5^2$	$3^2\cdot89$	$2\cdot401$	$11\cdot73$	$2^2\cdot3\cdot67$	$5\cdot7\cdot23$	$2\cdot13\cdot31$	$3\cdot269$	$2^3\cdot101$	**809**
81	$2\cdot3^4\cdot5$	**811**	$2^2\cdot7\cdot29$	$3\cdot271$	$2\cdot11\cdot37$	$5\cdot163$	$2^4\cdot3\cdot17$	$19\cdot43$	$2\cdot409$	$3^2\cdot7\cdot13$
82	$2^2\cdot5\cdot41$	**821**	$2\cdot3\cdot137$	**823**	$2^3\cdot103$	$3\cdot5^2\cdot11$	$2\cdot7\cdot59$	**827**	$2^2\cdot3^2\cdot23$	**829**
83	$2\cdot5\cdot83$	$3\cdot277$	$2^6\cdot13$	$7^2\cdot17$	$2\cdot3\cdot139$	$5\cdot167$	$2^2\cdot11\cdot19$	$3^3\cdot31$	$2\cdot419$	**839**
84	$2^3\cdot3\cdot5\cdot7$	29^2	$2\cdot421$	$3\cdot281$	$2^2\cdot211$	$5\cdot13^2$	$2\cdot3^2\cdot47$	$7\cdot11^2$	$2^4\cdot53$	$3\cdot283$
85	$2\cdot5^2\cdot17$	$23\cdot37$	$2^2\cdot3\cdot71$	**853**	$2\cdot7\cdot61$	$3^2\cdot5\cdot19$	$2^3\cdot107$	**857**	$2\cdot3\cdot11\cdot13$	**859**
86	$2^2\cdot5\cdot43$	$3\cdot7\cdot41$	$2\cdot431$	**863**	$2^5\cdot3^3$	$5\cdot173$	$2\cdot433$	$3\cdot17^2$	$2^2\cdot7\cdot31$	$11\cdot79$
87	$2\cdot3\cdot5\cdot29$	$13\cdot67$	$2^3\cdot109$	$3^2\cdot97$	$2\cdot19\cdot23$	$5^3\cdot7$	$2^2\cdot3\cdot73$	**877**	$2\cdot439$	$3\cdot293$
88	$2^4\cdot5\cdot11$	**881**	$2\cdot3^2\cdot7^2$	**883**	$2^2\cdot13\cdot17$	$3\cdot5\cdot59$	$2\cdot443$	**887**	$2^3\cdot3\cdot37$	$7\cdot127$
89	$2\cdot5\cdot89$	$3^4\cdot11$	$2^2\cdot223$	$19\cdot47$	$2\cdot3\cdot149$	$5\cdot179$	$2^7\cdot7$	$3\cdot13\cdot23$	$2\cdot449$	$29\cdot31$
90	$2^2\cdot3^2\cdot5^2$	$17\cdot53$	$2\cdot11\cdot41$	$3\cdot7\cdot43$	$2^3\cdot113$	$5\cdot181$	$2\cdot3\cdot151$	**907**	$2^2\cdot227$	$3^2\cdot101$
91	$2\cdot5\cdot7\cdot13$	**911**	$2^4\cdot3\cdot19$	$11\cdot83$	$2\cdot457$	$3\cdot5\cdot61$	$2^2\cdot229$	$7\cdot131$	$2\cdot3^3\cdot17$	**919**
92	$2^3\cdot5\cdot23$	$3\cdot307$	$2\cdot461$	$13\cdot71$	$2^2\cdot3\cdot7\cdot11$	$5^2\cdot37$	$2\cdot463$	$3^2\cdot103$	$2^5\cdot29$	**929**
93	$2\cdot3\cdot5\cdot31$	$7^2\cdot19$	$2^2\cdot233$	$3\cdot311$	$2\cdot467$	$5\cdot11\cdot17$	$2^3\cdot3^2\cdot13$	**937**	$2\cdot7\cdot67$	$3\cdot313$
94	$2^2\cdot5\cdot47$	**941**	$2\cdot3\cdot157$	$23\cdot41$	$2^4\cdot59$	$3^3\cdot5\cdot7$	$2\cdot11\cdot43$	**947**	$2^2\cdot3\cdot79$	$13\cdot73$
95	$2\cdot5^2\cdot19$	$3\cdot317$	$2^3\cdot7\cdot17$	**953**	$2\cdot3^2\cdot53$	$5\cdot191$	$2^2\cdot239$	$3\cdot11\cdot29$	$2\cdot479$	$7\cdot137$
96	$2^6\cdot3\cdot5$	31^2	$2\cdot13\cdot37$	$3^2\cdot107$	$2^2\cdot241$	$5\cdot193$	$2\cdot3\cdot7\cdot23$	**967**	$2^3\cdot11^2$	$3\cdot17\cdot19$
97	$2\cdot5\cdot97$	**971**	$2^2\cdot3^5$	$7\cdot139$	$2\cdot487$	$3\cdot5^2\cdot13$	$2^4\cdot61$	**977**	$2\cdot3\cdot163$	$11\cdot89$
98	$2^2\cdot5\cdot7^2$	$3^2\cdot109$	$2\cdot491$	**983**	$2^3\cdot3\cdot41$	$5\cdot197$	$2\cdot17\cdot29$	$3\cdot7\cdot47$	$2^2\cdot13\cdot19$	$23\cdot43$
99	$2\cdot3^2\cdot5\cdot11$	**991**	$2^5\cdot31$	$3\cdot331$	$2\cdot7\cdot71$	$5\cdot199$	$2^2\cdot3\cdot83$	**997**	$2\cdot499$	$3^3\cdot37$

Table 24.7   **Factorizations**

N	9	8	7	6	5	4	3	2	1	0	N
100	**1009**	$2^4\cdot3^2\cdot7$	$19\cdot53$	$2\cdot503$	$3\cdot5\cdot67$	$2^2\cdot251$	$17\cdot59$	$2\cdot3\cdot167$	$7\cdot11\cdot13$	$2^3\cdot5^3$	100
101	**1019**	$2\cdot509$	$3^2\cdot113$	$2^3\cdot127$	$5\cdot7\cdot29$	$2\cdot3\cdot13^2$	**1013**	$2^2\cdot11\cdot23$	$3\cdot337$	$2\cdot5\cdot101$	101
102	$3\cdot7^3$	$2^2\cdot257$	$13\cdot79$	$2\cdot3^3\cdot19$	$5^2\cdot41$	2^{10}	$3\cdot11\cdot31$	$2\cdot7\cdot73$	**1021**	$2^2\cdot3\cdot5\cdot17$	102
103	**1039**	$2\cdot3\cdot173$	$17\cdot61$	$2^2\cdot7\cdot37$	$3^2\cdot5\cdot23$	$2\cdot11\cdot47$	**1033**	$2^3\cdot3\cdot43$	**1031**	$2\cdot5\cdot103$	103
104	**1049**	$2^3\cdot131$	$3\cdot349$	$2\cdot523$	$5\cdot11\cdot19$	$2^2\cdot3^2\cdot29$	$7\cdot149$	$2\cdot521$	$3\cdot347$	$2^4\cdot5\cdot13$	104
105	$3\cdot353$	$2\cdot23^2$	$7\cdot151$	$2^5\cdot3\cdot11$	$5\cdot211$	$2\cdot17\cdot31$	$3^4\cdot13$	$2^2\cdot263$	**1051**	$2\cdot3\cdot5^2\cdot7$	105
106	**1069**	$2^2\cdot3\cdot89$	$11\cdot97$	$2\cdot13\cdot41$	$3\cdot5\cdot71$	$2^3\cdot7\cdot19$	**1063**	$2\cdot3^2\cdot59$	**1061**	$2^2\cdot5\cdot53$	106
107	$13\cdot83$	$2\cdot7^2\cdot11$	$3\cdot359$	$2^2\cdot269$	$5^2\cdot43$	$2\cdot3\cdot179$	$29\cdot37$	$2^4\cdot67$	$3^2\cdot7\cdot17$	$2\cdot5\cdot107$	107
108	$3^2\cdot11^2$	$2^6\cdot17$	**1087**	$2\cdot3\cdot181$	$5\cdot7\cdot31$	$2^2\cdot271$	$3\cdot19^2$	$2\cdot541$	$23\cdot47$	$2^3\cdot3^3\cdot5$	108
109	$7\cdot157$	$2\cdot3^2\cdot61$	**1097**	$2^3\cdot137$	$3\cdot5\cdot73$	$2\cdot547$	**1093**	$2^2\cdot3\cdot7\cdot13$	**1091**	$2\cdot5\cdot109$	109
110	**1109**	$2^2\cdot277$	$3^3\cdot41$	$2\cdot7\cdot79$	$5\cdot13\cdot17$	$2^4\cdot3\cdot23$	**1103**	$2\cdot19\cdot29$	$3\cdot367$	$2^2\cdot5^2\cdot11$	110
111	$3\cdot373$	$2\cdot13\cdot43$	**1117**	$2^2\cdot3^2\cdot31$	$5\cdot223$	$2\cdot557$	$3\cdot7\cdot53$	$2^3\cdot139$	$11\cdot101$	$2\cdot3\cdot5\cdot37$	111
112	**1129**	$2^3\cdot3\cdot47$	$7^2\cdot23$	$2\cdot563$	$3^2\cdot5^3$	$2^2\cdot281$	**1123**	$2\cdot3\cdot11\cdot17$	$19\cdot59$	$2^5\cdot5\cdot7$	112
113	$17\cdot67$	$2\cdot569$	$3\cdot379$	$2^4\cdot71$	$5\cdot227$	$2\cdot3^4\cdot7$	$11\cdot103$	$2^2\cdot283$	$3\cdot13\cdot29$	$2\cdot5\cdot113$	113
114	$3\cdot383$	$2^2\cdot7\cdot41$	$31\cdot37$	$2\cdot3\cdot191$	$5\cdot229$	$2^3\cdot11\cdot13$	$3^2\cdot127$	$2\cdot571$	$7\cdot163$	$2^2\cdot3\cdot5\cdot19$	114
115	$19\cdot61$	$2\cdot3\cdot193$	$13\cdot89$	$2^2\cdot17^2$	$3\cdot5\cdot7\cdot11$	$2\cdot577$	**1153**	$2^7\cdot3^2$	**1151**	$2\cdot5^2\cdot23$	115
116	$7\cdot167$	$2^4\cdot73$	$3\cdot389$	$2\cdot11\cdot53$	$5\cdot233$	$2^2\cdot3\cdot97$	**1163**	$2\cdot7\cdot83$	$3^3\cdot43$	$2^3\cdot5\cdot29$	116
117	$3^2\cdot131$	$2\cdot19\cdot31$	$11\cdot107$	$2^3\cdot3\cdot7^2$	$5^2\cdot47$	$2\cdot587$	$3\cdot17\cdot23$	$2^2\cdot293$	**1171**	$2\cdot3^2\cdot5\cdot13$	117
118	$29\cdot41$	$2^2\cdot3^3\cdot11$	**1187**	$2\cdot593$	$3\cdot5\cdot79$	$2^5\cdot37$	$7\cdot13^2$	$2\cdot3\cdot197$	**1181**	$2^2\cdot5\cdot59$	118
119	$11\cdot109$	$2\cdot599$	$3^2\cdot7\cdot19$	$2^2\cdot13\cdot23$	$5\cdot239$	$2\cdot3\cdot199$	**1193**	$2^3\cdot149$	$3\cdot397$	$2\cdot5\cdot7\cdot17$	119
120	$3\cdot13\cdot31$	$2^3\cdot151$	$17\cdot71$	$2\cdot3^2\cdot67$	$5\cdot241$	$2^2\cdot7\cdot43$	$3\cdot401$	$2\cdot601$	**1201**	$2^4\cdot3\cdot5^2$	120
121	$23\cdot53$	$2\cdot3\cdot7\cdot29$	**1217**	$2^6\cdot19$	$3^5\cdot5$	$2\cdot607$	**1213**	$2^2\cdot3\cdot101$	$7\cdot173$	$2\cdot5\cdot11^2$	121
122	**1229**	$2^2\cdot307$	$3\cdot409$	$2\cdot613$	$5^2\cdot7^2$	$2^3\cdot3^2\cdot17$	**1223**	$2\cdot13\cdot47$	$3\cdot11\cdot37$	$2^2\cdot5\cdot61$	122
123	$3\cdot7\cdot59$	$2\cdot619$	**1237**	$2^2\cdot3\cdot103$	$5\cdot13\cdot19$	$2\cdot617$	$3^2\cdot137$	$2^4\cdot7\cdot11$	**1231**	$2\cdot3\cdot5\cdot41$	123
124	**1249**	$2^5\cdot3\cdot13$	$29\cdot43$	$2\cdot7\cdot89$	$3\cdot5\cdot83$	$2^2\cdot311$	$11\cdot113$	$2\cdot3^3\cdot23$	$17\cdot73$	$2^3\cdot5\cdot31$	124
125	**1259**	$2\cdot17\cdot37$	$3\cdot419$	$2^3\cdot157$	$5\cdot251$	$2\cdot3\cdot11\cdot19$	$7\cdot179$	$2^2\cdot313$	$3^2\cdot139$	$2\cdot5^4$	125
126	$3^3\cdot47$	$2^2\cdot317$	$7\cdot181$	$2\cdot3\cdot211$	$5\cdot11\cdot23$	$2^4\cdot79$	$3\cdot421$	$2\cdot631$	$13\cdot97$	$2^2\cdot3^2\cdot5\cdot7$	126
127	**1279**	$2\cdot3^2\cdot71$	**1277**	$2^2\cdot11\cdot29$	$3\cdot5^2\cdot17$	$2\cdot7^2\cdot13$	$19\cdot67$	$2^3\cdot3\cdot53$	$31\cdot41$	$2\cdot5\cdot127$	127
128	**1289**	$2^3\cdot7\cdot23$	$3^2\cdot11\cdot13$	$2\cdot643$	$5\cdot257$	$2^2\cdot3\cdot107$	**1283**	$2\cdot641$	$3\cdot7\cdot61$	$2^8\cdot5$	128
129	$3\cdot433$	$2\cdot11\cdot59$	**1297**	$2^4\cdot3^4$	$5\cdot7\cdot37$	$2\cdot647$	$3\cdot431$	$2^2\cdot17\cdot19$	**1291**	$2\cdot3\cdot5\cdot43$	129
130	$7\cdot11\cdot17$	$2^2\cdot3\cdot109$	**1307**	$2\cdot653$	$3^2\cdot5\cdot29$	$2^3\cdot163$	**1303**	$2\cdot3\cdot7\cdot31$	**1301**	$2^2\cdot5^2\cdot13$	130
131	**1319**	$2\cdot659$	$3\cdot439$	$2^2\cdot7\cdot47$	$5\cdot263$	$2\cdot3^2\cdot73$	$13\cdot101$	$2^5\cdot41$	$3\cdot19\cdot23$	$2\cdot5\cdot131$	131
132	$3\cdot443$	$2^4\cdot83$	**1327**	$2\cdot3\cdot13\cdot17$	$5^2\cdot53$	$2^2\cdot331$	$3^3\cdot7^2$	$2\cdot661$	**1321**	$2^3\cdot3\cdot5\cdot11$	132
133	$13\cdot103$	$2\cdot3\cdot223$	$7\cdot191$	$2^3\cdot167$	$3\cdot5\cdot89$	$2\cdot23\cdot29$	$31\cdot43$	$2^2\cdot3^2\cdot37$	11^3	$2\cdot5\cdot7\cdot19$	133
134	$19\cdot71$	$2^2\cdot337$	$3\cdot449$	$2\cdot673$	$5\cdot269$	$2^6\cdot3\cdot7$	$17\cdot79$	$2\cdot11\cdot61$	$3^2\cdot149$	$2^2\cdot5\cdot67$	134
135	$3^2\cdot151$	$2\cdot7\cdot97$	$23\cdot59$	$2^2\cdot3\cdot113$	$5\cdot271$	$2\cdot677$	$3\cdot11\cdot41$	$2^3\cdot13^2$	$7\cdot193$	$2\cdot3^3\cdot5^2$	135
136	37^2	$2^3\cdot3^2\cdot19$	**1367**	$2\cdot683$	$3\cdot5\cdot7\cdot13$	$2^2\cdot11\cdot31$	$29\cdot47$	$2\cdot3\cdot227$	**1361**	$2^4\cdot5\cdot17$	136
137	$7\cdot197$	$2\cdot13\cdot53$	$3^4\cdot17$	$2^5\cdot43$	$5^3\cdot11$	$2\cdot3\cdot229$	**1373**	$2^2\cdot7^3$	$3\cdot457$	$2\cdot5\cdot137$	137
138	$3\cdot463$	$2^2\cdot347$	$19\cdot73$	$2\cdot3^2\cdot7\cdot11$	$5\cdot277$	$2^3\cdot173$	$3\cdot461$	$2\cdot691$	**1381**	$2^2\cdot3\cdot5\cdot23$	138
139	**1399**	$2\cdot3\cdot233$	$11\cdot127$	$2^2\cdot349$	$3^2\cdot5\cdot31$	$2\cdot17\cdot41$	$7\cdot199$	$2^4\cdot3\cdot29$	$13\cdot107$	$2\cdot5\cdot139$	139
140	**1409**	$2^7\cdot11$	$3\cdot7\cdot67$	$2\cdot19\cdot37$	$5\cdot281$	$2^2\cdot3^3\cdot13$	$23\cdot61$	$2\cdot701$	$3\cdot467$	$2^3\cdot5^2\cdot7$	140
141	$3\cdot11\cdot43$	$2\cdot709$	$13\cdot109$	$2^3\cdot3\cdot59$	$5\cdot283$	$2\cdot7\cdot101$	$3^2\cdot157$	$2^2\cdot353$	$17\cdot83$	$2\cdot3\cdot5\cdot47$	141
142	**1429**	$2^2\cdot3\cdot7\cdot17$	**1427**	$2\cdot23\cdot31$	$3\cdot5^2\cdot19$	$2^4\cdot89$	**1423**	$2\cdot3^2\cdot79$	$7^2\cdot29$	$2^2\cdot5\cdot71$	142
143	**1439**	$2\cdot719$	$3\cdot479$	$2^2\cdot359$	$5\cdot7\cdot41$	$2\cdot3\cdot239$	**1433**	$2^3\cdot179$	$3^3\cdot53$	$2\cdot5\cdot11\cdot13$	143
144	$3^2\cdot7\cdot23$	$2^3\cdot181$	**1447**	$2\cdot3\cdot241$	$5\cdot17^2$	$2^2\cdot19^2$	$3\cdot13\cdot37$	$2\cdot7\cdot103$	$11\cdot131$	$2^5\cdot3^2\cdot5$	144
145	**1459**	$2\cdot3^6$	$31\cdot47$	$2^4\cdot7\cdot13$	$3\cdot5\cdot97$	$2\cdot727$	**1453**	$2^2\cdot3\cdot11^2$	**1451**	$2\cdot5^2\cdot29$	145
146	$13\cdot113$	$2^2\cdot367$	$3^2\cdot163$	$2\cdot733$	$5\cdot293$	$2^3\cdot3\cdot61$	$7\cdot11\cdot19$	$2\cdot17\cdot43$	$3\cdot487$	$2^2\cdot5\cdot73$	146
147	$3\cdot17\cdot29$	$2\cdot739$	$7\cdot211$	$2^2\cdot3^2\cdot41$	$5^2\cdot59$	$2\cdot11\cdot67$	$3\cdot491$	$2^6\cdot23$	**1471**	$2\cdot3\cdot5\cdot7^2$	147
148	**1489**	$2^4\cdot3\cdot31$	**1487**	$2\cdot743$	$3^3\cdot5\cdot11$	$2^2\cdot7\cdot53$	**1483**	$2\cdot3\cdot13\cdot19$	**1481**	$2^3\cdot5\cdot37$	148
149	**1499**	$2\cdot7\cdot107$	$3\cdot499$	$2^3\cdot11\cdot17$	$5\cdot13\cdot23$	$2\cdot3^2\cdot83$	**1493**	$2^2\cdot373$	$3\cdot7\cdot71$	$2\cdot5\cdot149$	149
N	9	8	7	6	5	4	3	2	1	0	N

COMBINATORIAL ANALYSIS

847

Factorizations

Table 24.7

1500

1999

Factorizations of n for $1500 \le n \le 1999$. Within each row the first cell gives $n \div 10$; the ten factorization cells give, from left to right, the last digit $9, 8, 7, 6, 5, 4, 3, 2, 1, 0$. Bold entries are primes.

n	9	8	7	6	5	4	3	2	1	0
150	$3\cdot503$	$2^2\cdot13\cdot29$	$11\cdot137$	$2\cdot3\cdot251$	$5\cdot7\cdot43$	$2^5\cdot47$	$3^2\cdot167$	$2\cdot751$	$19\cdot79$	$2^2\cdot3\cdot5^3$
151	$7^2\cdot31$	$2\cdot3\cdot11\cdot23$	$37\cdot41$	$2^2\cdot379$	$3\cdot5\cdot101$	$2\cdot757$	$17\cdot89$	$2^3\cdot3^3\cdot7$	**1511**	$2\cdot5\cdot151$
152	$11\cdot139$	$2^3\cdot191$	$3\cdot509$	$2\cdot7\cdot109$	$5^2\cdot61$	$2^2\cdot3\cdot127$	**1523**	$2\cdot761$	$3^2\cdot13^2$	$2^4\cdot5\cdot19$
153	$3^4\cdot19$	$2\cdot769$	$29\cdot53$	$2^9\cdot3$	$5\cdot307$	$2\cdot13\cdot59$	$3\cdot7\cdot73$	$2^2\cdot383$	**1531**	$2\cdot3^2\cdot5\cdot17$
154	**1549**	$2^2\cdot3^2\cdot43$	$7\cdot13\cdot17$	$2\cdot773$	$3\cdot5\cdot103$	$2^3\cdot193$	**1543**	$2\cdot3\cdot257$	$23\cdot67$	$2^2\cdot5\cdot7\cdot11$
155	**1559**	$2\cdot19\cdot41$	$3^2\cdot173$	$2^2\cdot389$	$5\cdot311$	$2\cdot3\cdot7\cdot37$	**1553**	$2^4\cdot97$	$3\cdot11\cdot47$	$2\cdot5^2\cdot31$
156	$3\cdot523$	$2^5\cdot7^2$	**1567**	$2\cdot3^3\cdot29$	$5\cdot313$	$2^2\cdot17\cdot23$	$3\cdot521$	$2\cdot11\cdot71$	$7\cdot223$	$2^3\cdot3\cdot5\cdot13$
157	**1579**	$2\cdot3\cdot263$	$19\cdot83$	$2^3\cdot197$	$3^2\cdot5^2\cdot7$	$2\cdot787$	$11^2\cdot13$	$2^2\cdot3\cdot131$	**1571**	$2\cdot5\cdot157$
158	$7\cdot227$	$2^2\cdot397$	$3\cdot23^2$	$2\cdot13\cdot61$	$5\cdot317$	$2^4\cdot3^2\cdot11$	**1583**	$2\cdot7\cdot113$	$3\cdot17\cdot31$	$2^2\cdot5\cdot79$
159	$3\cdot13\cdot41$	$2\cdot17\cdot47$	**1597**	$2^2\cdot3\cdot7\cdot19$	$5\cdot11\cdot29$	$2\cdot797$	$3^3\cdot59$	$2^3\cdot199$	$37\cdot43$	$2\cdot3\cdot5\cdot53$
160	**1609**	$2^3\cdot3\cdot67$	**1607**	$2\cdot11\cdot73$	$3\cdot5\cdot107$	$2^2\cdot401$	$7\cdot229$	$2\cdot3^2\cdot89$	**1601**	$2^6\cdot5^2$
161	**1619**	$2\cdot809$	$3\cdot7^2\cdot11$	$2^4\cdot101$	$5\cdot17\cdot19$	$2\cdot3\cdot269$	**1613**	$2^2\cdot13\cdot31$	$3^2\cdot179$	$2\cdot5\cdot7\cdot23$
162	$3^2\cdot181$	$2^2\cdot11\cdot37$	**1627**	$2\cdot3\cdot271$	$5^3\cdot13$	$2^3\cdot7\cdot29$	$3\cdot541$	$2\cdot811$	**1621**	$2^2\cdot3^4\cdot5$
163	$11\cdot149$	$2\cdot3^2\cdot7\cdot13$	**1637**	$2^2\cdot409$	$3\cdot5\cdot109$	$2\cdot19\cdot43$	$23\cdot71$	$2^6\cdot3\cdot17$	$7\cdot233$	$2\cdot5\cdot163$
164	$17\cdot97$	$2^4\cdot103$	$3^3\cdot61$	$2\cdot823$	$5\cdot7\cdot47$	$2^2\cdot3\cdot137$	$31\cdot53$	$2\cdot821$	$3\cdot547$	$2^3\cdot5\cdot41$
165	$3\cdot7\cdot79$	$2\cdot829$	**1657**	$2^3\cdot3^2\cdot23$	$5\cdot331$	$2\cdot827$	$3\cdot19\cdot29$	$2^2\cdot7\cdot59$	$13\cdot127$	$2\cdot3\cdot5^2\cdot11$
166	**1669**	$2^2\cdot3\cdot139$	**1667**	$2\cdot7^2\cdot17$	$3^2\cdot5\cdot37$	$2^7\cdot13$	**1663**	$2\cdot3\cdot277$	$11\cdot151$	$2^2\cdot5\cdot83$
167	$23\cdot73$	$2\cdot839$	$3\cdot13\cdot43$	$2^2\cdot419$	$5^2\cdot67$	$2\cdot3^4\cdot31$	$7\cdot239$	$2^3\cdot11\cdot19$	$3\cdot557$	$2\cdot5\cdot167$
168	$3\cdot563$	$2^3\cdot211$	$7\cdot241$	$2\cdot3\cdot281$	$5\cdot337$	$2^2\cdot421$	$3^2\cdot11\cdot17$	$2\cdot29^2$	41^2	$2^4\cdot3\cdot5\cdot7$
169	**1699**	$2\cdot3\cdot283$	**1697**	$2^5\cdot53$	$3\cdot5\cdot113$	$2\cdot7\cdot11^2$	**1693**	$2^2\cdot3^2\cdot47$	$19\cdot89$	$2\cdot5\cdot13^2$
170	**1709**	$2^2\cdot7\cdot61$	$3\cdot569$	$2\cdot853$	$5\cdot11\cdot31$	$2^3\cdot3\cdot71$	$13\cdot131$	$2\cdot23\cdot37$	$3^5\cdot7$	$2^2\cdot5^2\cdot17$
171	$3^2\cdot191$	$2\cdot859$	$17\cdot101$	$2^2\cdot3\cdot11\cdot13$	$5\cdot7^3$	$2\cdot857$	$3\cdot571$	$2^4\cdot107$	$29\cdot59$	$2\cdot3^2\cdot5\cdot19$
172	$7\cdot13\cdot19$	$2^6\cdot3^3$	$11\cdot157$	$2\cdot863$	$3\cdot5^2\cdot23$	$2^2\cdot431$	**1723**	$2\cdot3\cdot7\cdot41$	**1721**	$2^3\cdot5\cdot43$
173	$37\cdot47$	$2\cdot11\cdot79$	$3^2\cdot193$	$2^3\cdot7\cdot31$	$5\cdot347$	$2\cdot3\cdot17^2$	**1733**	$2^2\cdot433$	$3\cdot577$	$2\cdot5\cdot173$
174	$3\cdot11\cdot53$	$2^2\cdot19\cdot23$	**1747**	$2\cdot3^2\cdot97$	$5\cdot349$	$2^4\cdot109$	$3\cdot7\cdot83$	$2\cdot13\cdot67$	**1741**	$2^2\cdot3\cdot5\cdot29$
175	**1759**	$2\cdot3\cdot293$	$7\cdot251$	$2^2\cdot439$	$3^3\cdot5\cdot13$	$2\cdot877$	**1753**	$2^3\cdot3\cdot73$	$17\cdot103$	$2\cdot5^3\cdot7$
176	$29\cdot61$	$2^3\cdot13\cdot17$	$3\cdot19\cdot31$	$2\cdot883$	$5\cdot353$	$2^2\cdot3^2\cdot7^2$	$41\cdot43$	$2\cdot881$	$3\cdot587$	$2^5\cdot5\cdot11$
177	$3\cdot593$	$2\cdot7\cdot127$	**1777**	$2^4\cdot3\cdot37$	$5^2\cdot71$	$2\cdot887$	$3^2\cdot197$	$2^2\cdot443$	$7\cdot11\cdot23$	$2\cdot3\cdot5\cdot59$
178	**1789**	$2^2\cdot3\cdot149$	**1787**	$2\cdot19\cdot47$	$3\cdot5\cdot7\cdot17$	$2^3\cdot223$	**1783**	$2\cdot3^4\cdot11$	$13\cdot137$	$2^2\cdot5\cdot89$
179	$7\cdot257$	$2\cdot29\cdot31$	$3\cdot599$	$2^2\cdot449$	$5\cdot359$	$2\cdot3\cdot13\cdot23$	$11\cdot163$	$2^8\cdot7$	$3^2\cdot199$	$2\cdot5\cdot179$
180	$3^3\cdot67$	$2^4\cdot113$	$13\cdot139$	$2\cdot3\cdot7\cdot43$	$5\cdot19^2$	$2^2\cdot11\cdot41$	$3\cdot601$	$2\cdot17\cdot53$	**1801**	$2^3\cdot3^2\cdot5^2$
181	$17\cdot107$	$2\cdot3^2\cdot101$	$23\cdot79$	$2^3\cdot227$	$3\cdot5\cdot11^2$	$2\cdot907$	$7^2\cdot37$	$2^2\cdot3\cdot151$	**1811**	$2\cdot5\cdot181$
182	$31\cdot59$	$2^2\cdot457$	$3^2\cdot7\cdot29$	$2\cdot11\cdot83$	$5^2\cdot73$	$2^5\cdot3\cdot19$	**1823**	$2\cdot911$	$3\cdot607$	$2^2\cdot5\cdot7\cdot13$
183	$3\cdot613$	$2\cdot919$	$11\cdot167$	$2^2\cdot3^3\cdot17$	$5\cdot367$	$2\cdot7\cdot131$	$3\cdot13\cdot47$	$2^3\cdot229$	**1831**	$2\cdot3\cdot5\cdot61$
184	43^2	$2^3\cdot3\cdot7\cdot11$	**1847**	$2\cdot13\cdot71$	$3^2\cdot5\cdot41$	$2^2\cdot461$	$19\cdot97$	$2\cdot3\cdot307$	$7\cdot263$	$2^4\cdot5\cdot23$
185	$11\cdot13^2$	$2\cdot929$	$3\cdot619$	$2^6\cdot29$	$5\cdot7\cdot53$	$2\cdot3^2\cdot103$	$17\cdot109$	$2^2\cdot463$	$3\cdot617$	$2\cdot5^2\cdot37$
186	$3\cdot7\cdot89$	$2^2\cdot467$	**1867**	$2\cdot3\cdot311$	$5\cdot373$	$2^3\cdot233$	$3^4\cdot23$	$2\cdot7^2\cdot19$	**1861**	$2^2\cdot3\cdot5\cdot31$
187	**1879**	$2\cdot3\cdot313$	**1877**	$2^2\cdot7\cdot67$	$3\cdot5^4$	$2\cdot937$	**1873**	$2^4\cdot3^2\cdot13$	**1871**	$2\cdot5\cdot11\cdot17$
188	**1889**	$2^5\cdot59$	$3\cdot17\cdot37$	$2\cdot23\cdot41$	$5\cdot13\cdot29$	$2^2\cdot3\cdot157$	$7\cdot269$	$2\cdot941$	$3^2\cdot11\cdot19$	$2^3\cdot5\cdot47$
189	$3^2\cdot211$	$2\cdot13\cdot73$	$7\cdot271$	$2^3\cdot3\cdot79$	$5\cdot379$	$2\cdot947$	$3\cdot631$	$2^2\cdot11\cdot43$	$31\cdot61$	$2\cdot3^3\cdot5\cdot7$
190	$23\cdot83$	$2^2\cdot3^2\cdot53$	**1907**	$2\cdot953$	$3\cdot5\cdot127$	$2^4\cdot7\cdot17$	$11\cdot173$	$2\cdot3\cdot317$	**1901**	$2^2\cdot5^2\cdot19$
191	$19\cdot101$	$2\cdot7\cdot137$	$3^3\cdot71$	$2^2\cdot479$	$5\cdot383$	$2\cdot3\cdot11\cdot29$	**1913**	$2^3\cdot239$	$3\cdot7^2\cdot13$	$2\cdot5\cdot191$
192	$3\cdot643$	$2^3\cdot241$	$41\cdot47$	$2\cdot3^2\cdot107$	$5^2\cdot7\cdot11$	$2^2\cdot13\cdot37$	$3\cdot641$	$2\cdot31^2$	$17\cdot113$	$2^7\cdot3\cdot5$
193	$7\cdot277$	$2\cdot3\cdot17\cdot19$	$13\cdot149$	$2^4\cdot11^2$	$3^2\cdot5\cdot43$	$2\cdot967$	**1933**	$2^2\cdot3\cdot7\cdot23$	**1931**	$2\cdot5\cdot193$
194	**1949**	$2^2\cdot487$	$3\cdot11\cdot59$	$2\cdot7\cdot139$	$5\cdot389$	$2^3\cdot3^5$	$29\cdot67$	$2\cdot971$	$3\cdot647$	$2^2\cdot5\cdot97$
195	$3\cdot653$	$2\cdot11\cdot89$	$19\cdot103$	$2^2\cdot3\cdot163$	$5\cdot17\cdot23$	$2\cdot977$	$3^2\cdot7\cdot31$	$2^5\cdot61$	**1951**	$2\cdot3\cdot5^2\cdot13$
196	$11\cdot179$	$2^4\cdot3\cdot41$	$7\cdot281$	$2\cdot983$	$3\cdot5\cdot131$	$2^2\cdot491$	$13\cdot151$	$2\cdot3^2\cdot109$	$37\cdot53$	$2^3\cdot5\cdot7^2$
197	**1979**	$2\cdot23\cdot43$	$3\cdot659$	$2^3\cdot13\cdot19$	$5^2\cdot79$	$2\cdot3\cdot7\cdot47$	**1973**	$2^2\cdot17\cdot29$	$3^3\cdot73$	$2\cdot5\cdot197$
198	$3^2\cdot13\cdot17$	$2^2\cdot7\cdot71$	**1987**	$2\cdot3\cdot331$	$5\cdot397$	$2^6\cdot31$	$3\cdot661$	$2\cdot991$	$7\cdot283$	$2^2\cdot3^2\cdot5\cdot11$
199	**1999**	$2\cdot3^3\cdot37$	**1997**	$2^2\cdot499$	$3\cdot5\cdot7\cdot19$	$2\cdot997$	**1993**	$2^3\cdot3\cdot83$	$11\cdot181$	$2\cdot5\cdot199$

Table 24.7   **Factorizations**

N	9	8	7	6	5	4	3	2	1	0	N
200	$7^2\cdot41$	$2^3\cdot251$	$3^2\cdot223$	$2\cdot17\cdot59$	$5\cdot401$	$2^2\cdot3\cdot167$	**2003**	$2\cdot7\cdot11\cdot13$	$3\cdot23\cdot29$	$2^4\cdot5^3$	200
201	$3\cdot673$	$2\cdot1009$	**2017**	$2^5\cdot3^2\cdot7$	$5\cdot13\cdot31$	$2\cdot19\cdot53$	$3\cdot11\cdot61$	$2^2\cdot503$	**2011**	$2\cdot3\cdot5\cdot67$	201
202	**2029**	$2^2\cdot3\cdot13^2$	**2027**	$2\cdot1013$	$3^4\cdot5^2$	$2^3\cdot11\cdot23$	$7\cdot17^2$	$2\cdot3\cdot337$	$43\cdot47$	$2^2\cdot5\cdot101$	202
203	**2039**	$2\cdot1019$	$3\cdot7\cdot97$	$2^2\cdot509$	$5\cdot11\cdot37$	$2\cdot3^2\cdot113$	$19\cdot107$	$2^4\cdot127$	$3\cdot677$	$2\cdot5\cdot7\cdot29$	203
204	$3\cdot683$	2^{11}	$23\cdot89$	$2\cdot3\cdot11\cdot31$	$5\cdot409$	$2^2\cdot7\cdot73$	$3^2\cdot227$	$2\cdot1021$	$13\cdot157$	$2^3\cdot3\cdot5\cdot17$	204
205	$29\cdot71$	$2\cdot3\cdot7^3$	$11^2\cdot17$	$2^3\cdot257$	$3\cdot5\cdot137$	$2\cdot13\cdot79$	**2053**	$2^2\cdot3^3\cdot19$	$7\cdot293$	$2\cdot5^2\cdot41$	205
206	**2069**	$2^2\cdot11\cdot47$	$3\cdot13\cdot53$	$2\cdot1033$	$5\cdot7\cdot59$	$2^4\cdot3\cdot43$	**2063**	$2\cdot1031$	$3^2\cdot229$	$2^2\cdot5\cdot103$	206
207	$3^3\cdot7\cdot11$	$2\cdot1039$	$31\cdot67$	$2^2\cdot3\cdot173$	$5^2\cdot83$	$2\cdot17\cdot61$	$3\cdot691$	$2^3\cdot7\cdot37$	$19\cdot109$	$2\cdot3^2\cdot5\cdot23$	207
208	**2089**	$2^3\cdot3^2\cdot29$	**2087**	$2\cdot7\cdot149$	$3\cdot5\cdot139$	$2^2\cdot521$	**2083**	$2\cdot3\cdot347$	**2081**	$2^5\cdot5\cdot13$	208
209	**2099**	$2\cdot1049$	$3^2\cdot233$	$2^4\cdot131$	$5\cdot419$	$2\cdot3\cdot349$	$7\cdot13\cdot23$	$2^2\cdot523$	$3\cdot17\cdot41$	$2\cdot5\cdot11\cdot19$	209
210	$3\cdot19\cdot37$	$2^2\cdot17\cdot31$	$7^2\cdot43$	$2\cdot3^4\cdot13$	$5\cdot421$	$2^3\cdot263$	$3\cdot701$	$2\cdot1051$	$11\cdot191$	$2^2\cdot3\cdot5^2\cdot7$	210
211	$13\cdot163$	$2\cdot3\cdot353$	$29\cdot73$	$2^2\cdot23^2$	$3^2\cdot5\cdot47$	$2\cdot7\cdot151$	**2113**	$2^6\cdot3\cdot11$	**2111**	$2\cdot5\cdot211$	211
212	**2129**	$2^4\cdot7\cdot19$	$3\cdot709$	$2\cdot1063$	$5^3\cdot17$	$2^2\cdot3^2\cdot59$	$11\cdot193$	$2\cdot1061$	$3\cdot7\cdot101$	$2^3\cdot5\cdot53$	212
213	$3\cdot23\cdot31$	$2\cdot1069$	**2137**	$2^3\cdot3\cdot89$	$5\cdot7\cdot61$	$2\cdot11\cdot97$	$3^3\cdot79$	$2^2\cdot13\cdot41$	**2131**	$2\cdot3\cdot5\cdot71$	213
214	$7\cdot307$	$2^2\cdot3\cdot179$	$19\cdot113$	$2\cdot29\cdot37$	$3\cdot5\cdot11\cdot13$	$2^5\cdot67$	**2143**	$2\cdot3^2\cdot7\cdot17$	**2141**	$2^2\cdot5\cdot107$	214
215	$17\cdot127$	$2\cdot13\cdot83$	$3\cdot719$	$2^2\cdot7^2\cdot11$	$5\cdot431$	$2\cdot3\cdot359$	**2153**	$2^3\cdot269$	$3^2\cdot239$	$2\cdot5^2\cdot43$	215
216	$3^2\cdot241$	$2^3\cdot271$	$11\cdot197$	$2\cdot3\cdot19^2$	$5\cdot433$	$2^2\cdot541$	$3\cdot7\cdot103$	$2\cdot23\cdot47$	**2161**	$2^4\cdot3^3\cdot5$	216
217	**2179**	$2\cdot3^2\cdot11^2$	$7\cdot311$	$2^7\cdot17$	$3\cdot5^2\cdot29$	$2\cdot1087$	$41\cdot53$	$2^2\cdot3\cdot181$	$13\cdot167$	$2\cdot5\cdot7\cdot31$	217
218	$11\cdot199$	$2^2\cdot547$	3^7	$2\cdot1093$	$5\cdot19\cdot23$	$2^3\cdot3\cdot7\cdot13$	$37\cdot59$	$2\cdot1091$	$3\cdot727$	$2^2\cdot5\cdot109$	218
219	$3\cdot733$	$2\cdot7\cdot157$	13^3	$2^2\cdot3^2\cdot61$	$5\cdot439$	$2\cdot1097$	$3\cdot17\cdot43$	$2^4\cdot137$	$7\cdot313$	$2\cdot3\cdot5\cdot73$	219
220	47^2	$2^5\cdot3\cdot23$	**2207**	$2\cdot1103$	$3^2\cdot5\cdot7^2$	$2^2\cdot19\cdot29$	**2203**	$2\cdot3\cdot367$	$31\cdot71$	$2^3\cdot5^2\cdot11$	220
221	$7\cdot317$	$2\cdot1109$	$3\cdot739$	$2^3\cdot277$	$5\cdot443$	$2\cdot3^3\cdot41$	**2213**	$2^2\cdot7\cdot79$	$3\cdot11\cdot67$	$2\cdot5\cdot13\cdot17$	221
222	$3\cdot743$	$2^2\cdot557$	$17\cdot131$	$2\cdot3\cdot7\cdot53$	$5^2\cdot89$	$2^4\cdot139$	$3^2\cdot13\cdot19$	$2\cdot11\cdot101$	**2221**	$2^2\cdot3\cdot5\cdot37$	222
223	**2239**	$2\cdot3\cdot373$	**2237**	$2^2\cdot13\cdot43$	$3\cdot5\cdot149$	$2\cdot1117$	$7\cdot11\cdot29$	$2^3\cdot3^2\cdot31$	$23\cdot97$	$2\cdot5\cdot223$	223
224	$13\cdot173$	$2^3\cdot281$	$3\cdot7\cdot107$	$2\cdot1123$	$5\cdot449$	$2^2\cdot3\cdot11\cdot17$	**2243**	$2\cdot19\cdot59$	$3^3\cdot83$	$2^6\cdot5\cdot7$	224
225	$3^2\cdot251$	$2\cdot1129$	$37\cdot61$	$2^4\cdot3\cdot47$	$5\cdot11\cdot41$	$2\cdot7^2\cdot23$	$3\cdot751$	$2^2\cdot563$	**2251**	$2\cdot3^2\cdot5^3$	225
226	**2269**	$2^2\cdot3^4\cdot7$	**2267**	$2\cdot11\cdot103$	$3\cdot5\cdot151$	$2^3\cdot283$	$31\cdot73$	$2\cdot3\cdot13\cdot29$	$7\cdot17\cdot19$	$2^2\cdot5\cdot113$	226
227	$43\cdot53$	$2\cdot17\cdot67$	$3^2\cdot11\cdot23$	$2^2\cdot569$	$5^2\cdot7\cdot13$	$2\cdot3\cdot379$	**2273**	$2^5\cdot71$	$3\cdot757$	$2\cdot5\cdot227$	227
228	$3\cdot7\cdot109$	$2^4\cdot11\cdot13$	**2287**	$2\cdot3^2\cdot127$	$5\cdot457$	$2^2\cdot571$	$3\cdot761$	$2\cdot7\cdot163$	**2281**	$2^3\cdot3\cdot5\cdot19$	228
229	$11^2\cdot19$	$2\cdot3\cdot383$	**2297**	$2^3\cdot7\cdot41$	$3^3\cdot5\cdot17$	$2\cdot31\cdot37$	**2293**	$2^2\cdot3\cdot191$	$29\cdot79$	$2\cdot5\cdot229$	229
230	**2309**	$2^2\cdot577$	$3\cdot769$	$2\cdot1153$	$5\cdot461$	$2^8\cdot3^2$	$7^2\cdot47$	$2\cdot1151$	$3\cdot13\cdot59$	$2^2\cdot5^2\cdot23$	230
231	$3\cdot773$	$2\cdot19\cdot61$	$7\cdot331$	$2^2\cdot3\cdot193$	$5\cdot463$	$2\cdot13\cdot89$	$3^2\cdot257$	$2^3\cdot17^2$	**2311**	$2\cdot3\cdot5\cdot7\cdot11$	231
232	$17\cdot137$	$2^3\cdot3\cdot97$	$13\cdot179$	$2\cdot1163$	$3\cdot5^2\cdot31$	$2^2\cdot7\cdot83$	$23\cdot101$	$2\cdot3^3\cdot43$	$11\cdot211$	$2^4\cdot5\cdot29$	232
233	**2339**	$2\cdot7\cdot167$	$3\cdot19\cdot41$	$2^5\cdot73$	$5\cdot467$	$2\cdot3\cdot389$	**2333**	$2^2\cdot11\cdot53$	$3^2\cdot7\cdot37$	$2\cdot5\cdot233$	233
234	$3^4\cdot29$	$2^2\cdot587$	**2347**	$2\cdot3\cdot17\cdot23$	$5\cdot7\cdot67$	$2^3\cdot293$	$3\cdot11\cdot71$	$2\cdot1171$	**2341**	$2^2\cdot3^2\cdot5\cdot13$	234
235	$7\cdot337$	$2\cdot3^2\cdot131$	**2357**	$2^2\cdot19\cdot31$	$3\cdot5\cdot157$	$2\cdot11\cdot107$	$13\cdot181$	$2^4\cdot3\cdot7^2$	**2351**	$2\cdot5^2\cdot47$	235
236	$23\cdot103$	$2^6\cdot37$	$3^2\cdot263$	$2\cdot7\cdot13^2$	$5\cdot11\cdot43$	$2^2\cdot3\cdot197$	$17\cdot139$	$2\cdot1181$	$3\cdot787$	$2^3\cdot5\cdot59$	236
237	$3\cdot13\cdot61$	$2\cdot29\cdot41$	**2377**	$2^3\cdot3^3\cdot11$	$5^3\cdot19$	$2\cdot1187$	$3\cdot7\cdot113$	$2^2\cdot593$	**2371**	$2\cdot3\cdot5\cdot79$	237
238	**2389**	$2^2\cdot3\cdot199$	$7\cdot11\cdot31$	$2\cdot1193$	$3^2\cdot5\cdot53$	$2^4\cdot149$	**2383**	$2\cdot3\cdot397$	**2381**	$2^2\cdot5\cdot7\cdot17$	238
239	**2399**	$2\cdot11\cdot109$	$3\cdot17\cdot47$	$2^2\cdot599$	$5\cdot479$	$2\cdot3^2\cdot7\cdot19$	**2393**	$2^3\cdot13\cdot23$	$3\cdot797$	$2\cdot5\cdot239$	239
240	$3\cdot11\cdot73$	$2^3\cdot7\cdot43$	$29\cdot83$	$2\cdot3\cdot401$	$5\cdot13\cdot37$	$2^2\cdot601$	$3^3\cdot89$	$2\cdot1201$	7^4	$2^5\cdot3\cdot5^2$	240
241	$41\cdot59$	$2\cdot3\cdot13\cdot31$	**2417**	$2^4\cdot151$	$3\cdot5\cdot7\cdot23$	$2\cdot17\cdot71$	$19\cdot127$	$2^2\cdot3^2\cdot67$	**2411**	$2\cdot5\cdot241$	241
242	$7\cdot347$	$2^2\cdot607$	$3\cdot809$	$2\cdot1213$	$5^2\cdot97$	$2^3\cdot3\cdot101$	**2423**	$2\cdot7\cdot173$	$3^2\cdot269$	$2^2\cdot5\cdot11^2$	242
243	$3^2\cdot271$	$2\cdot23\cdot53$	**2437**	$2^2\cdot3\cdot7\cdot29$	$5\cdot487$	$2\cdot1217$	$3\cdot811$	$2^7\cdot19$	$11\cdot13\cdot17$	$2\cdot3^5\cdot5$	243
244	$31\cdot79$	$2^4\cdot3^2\cdot17$	**2447**	$2\cdot1223$	$3\cdot5\cdot163$	$2^2\cdot13\cdot47$	$7\cdot349$	$2\cdot3\cdot11\cdot37$	**2441**	$2^3\cdot5\cdot61$	244
245	**2459**	$2\cdot1229$	$3^3\cdot7\cdot13$	$2^3\cdot307$	$5\cdot491$	$2\cdot3\cdot409$	$11\cdot223$	$2^2\cdot613$	$3\cdot19\cdot43$	$2\cdot5^2\cdot7^2$	245
246	$3\cdot823$	$2^2\cdot617$	**2467**	$2\cdot3^2\cdot137$	$5\cdot17\cdot29$	$2^5\cdot7\cdot11$	$3\cdot821$	$2\cdot1231$	$23\cdot107$	$2^2\cdot3\cdot5\cdot41$	246
247	$37\cdot67$	$2\cdot3\cdot7\cdot59$	**2477**	$2^2\cdot619$	$3^2\cdot5^2\cdot11$	$2\cdot1237$	**2473**	$2^3\cdot3\cdot103$	$7\cdot353$	$2\cdot5\cdot13\cdot19$	247
248	$19\cdot131$	$2^3\cdot311$	$3\cdot829$	$2\cdot11\cdot113$	$5\cdot7\cdot71$	$2^2\cdot3^3\cdot23$	$13\cdot191$	$2\cdot17\cdot73$	$3\cdot827$	$2^4\cdot5\cdot31$	248
249	$3\cdot7^2\cdot17$	$2\cdot1249$	$11\cdot227$	$2^6\cdot3\cdot13$	$5\cdot499$	$2\cdot29\cdot43$	$3^2\cdot277$	$2^2\cdot7\cdot89$	$47\cdot53$	$2\cdot3\cdot5\cdot83$	249

Factorizations

Table 24.7

2500 – 2999

Factorizations of the integers 2500–2999 (row = first three digits; column = units digit). Primes in **bold**.

tens	0	1	2	3	4	5	6	7	8	9
250	$2^2 \cdot 5^4$	$41 \cdot 61$	$2 \cdot 3^2 \cdot 139$	**2503**	$2^3 \cdot 313$	$3 \cdot 5 \cdot 167$	$2 \cdot 7 \cdot 179$	$23 \cdot 109$	$2^2 \cdot 3 \cdot 11 \cdot 19$	$13 \cdot 193$
251	$2 \cdot 5 \cdot 251$	$3^4 \cdot 31$	$2^4 \cdot 157$	$7 \cdot 359$	$2 \cdot 3 \cdot 419$	$5 \cdot 503$	$2^2 \cdot 17 \cdot 37$	$3 \cdot 839$	$2 \cdot 1259$	$11 \cdot 229$
252	$2^3 \cdot 3^2 \cdot 5 \cdot 7$	**2521**	$2 \cdot 13 \cdot 97$	$3 \cdot 29^2$	$2^2 \cdot 631$	$5^2 \cdot 101$	$2 \cdot 3 \cdot 421$	$7 \cdot 19^2$	$2^5 \cdot 79$	$3^2 \cdot 281$
253	$2 \cdot 5 \cdot 11 \cdot 23$	**2531**	$2^2 \cdot 3 \cdot 211$	$17 \cdot 149$	$2 \cdot 7 \cdot 181$	$3 \cdot 5 \cdot 13^2$	$2^3 \cdot 317$	$43 \cdot 59$	$2 \cdot 3^3 \cdot 47$	**2539**
254	$2^2 \cdot 5 \cdot 127$	$3 \cdot 7 \cdot 11^2$	$2 \cdot 31 \cdot 41$	**2543**	$2^4 \cdot 3 \cdot 53$	$5 \cdot 509$	$2 \cdot 19 \cdot 67$	$3^2 \cdot 283$	$2^2 \cdot 7^2 \cdot 13$	**2549**
255	$2 \cdot 3 \cdot 5^2 \cdot 17$	**2551**	$2^3 \cdot 11 \cdot 29$	$3 \cdot 23 \cdot 37$	$2 \cdot 1277$	$5 \cdot 7 \cdot 73$	$2^2 \cdot 3^2 \cdot 71$	**2557**	$2 \cdot 1279$	$3 \cdot 853$
256	$2^9 \cdot 5$	$13 \cdot 197$	$2 \cdot 3 \cdot 7 \cdot 61$	$11 \cdot 233$	$2^2 \cdot 641$	$3^3 \cdot 5 \cdot 19$	$2 \cdot 1283$	$17 \cdot 151$	$2^3 \cdot 3 \cdot 107$	$7 \cdot 367$
257	$2 \cdot 5 \cdot 257$	$3 \cdot 857$	$2^2 \cdot 643$	$31 \cdot 83$	$2 \cdot 3^2 \cdot 11 \cdot 13$	$5^2 \cdot 103$	$2^4 \cdot 7 \cdot 23$	$3 \cdot 859$	$2 \cdot 1289$	**2579**
258	$2^2 \cdot 3 \cdot 5 \cdot 43$	$29 \cdot 89$	$2 \cdot 1291$	$3^2 \cdot 7 \cdot 41$	$2^3 \cdot 17 \cdot 19$	$5 \cdot 11 \cdot 47$	$2 \cdot 3 \cdot 431$	$13 \cdot 199$	$2^2 \cdot 647$	$3 \cdot 863$
259	$2 \cdot 5 \cdot 7 \cdot 37$	**2591**	$2^5 \cdot 3^4$	**2593**	$2 \cdot 1297$	$3 \cdot 5 \cdot 173$	$2^2 \cdot 11 \cdot 59$	$7^2 \cdot 53$	$2 \cdot 3 \cdot 433$	$23 \cdot 113$
260	$2^3 \cdot 5^2 \cdot 13$	$3^2 \cdot 17^2$	$2 \cdot 1301$	$19 \cdot 137$	$2^2 \cdot 3 \cdot 7 \cdot 31$	$5 \cdot 521$	$2 \cdot 1303$	$3 \cdot 11 \cdot 79$	$2^4 \cdot 163$	**2609**
261	$2 \cdot 3^2 \cdot 5 \cdot 29$	$7 \cdot 373$	$2^2 \cdot 653$	$3 \cdot 13 \cdot 67$	$2 \cdot 1307$	$5 \cdot 523$	$2^3 \cdot 3 \cdot 109$	**2617**	$2 \cdot 7 \cdot 11 \cdot 17$	$3^3 \cdot 97$
262	$2^2 \cdot 5 \cdot 131$	**2621**	$2 \cdot 3 \cdot 19 \cdot 23$	$43 \cdot 61$	$2^6 \cdot 41$	$3 \cdot 5^3 \cdot 7$	$2 \cdot 13 \cdot 101$	$37 \cdot 71$	$2^2 \cdot 3^2 \cdot 73$	$11 \cdot 239$
263	$2 \cdot 5 \cdot 263$	$3 \cdot 877$	$2^3 \cdot 7 \cdot 47$	**2633**	$2 \cdot 3 \cdot 439$	$5 \cdot 17 \cdot 31$	$2^2 \cdot 659$	$3^2 \cdot 293$	$2 \cdot 1319$	$7 \cdot 13 \cdot 29$
264	$2^4 \cdot 3 \cdot 5 \cdot 11$	$19 \cdot 139$	$2 \cdot 1321$	$3 \cdot 881$	$2^2 \cdot 661$	$5 \cdot 23^2$	$2 \cdot 3^3 \cdot 7^2$	**2647**	$2^3 \cdot 331$	$3 \cdot 883$
265	$2 \cdot 5^2 \cdot 53$	$11 \cdot 241$	$2^2 \cdot 3 \cdot 13 \cdot 17$	$7 \cdot 379$	$2 \cdot 1327$	$3^2 \cdot 5 \cdot 59$	$2^5 \cdot 83$	**2657**	$2 \cdot 3 \cdot 443$	**2659**
266	$2^2 \cdot 5 \cdot 7 \cdot 19$	$3 \cdot 887$	$2 \cdot 11^3$	**2663**	$2^3 \cdot 3^2 \cdot 37$	$5 \cdot 13 \cdot 41$	$2 \cdot 31 \cdot 43$	$3 \cdot 7 \cdot 127$	$2^2 \cdot 23 \cdot 29$	$17 \cdot 157$
267	$2 \cdot 3 \cdot 5 \cdot 89$	**2671**	$2^4 \cdot 167$	$3^5 \cdot 11$	$2 \cdot 7 \cdot 191$	$5^2 \cdot 107$	$2^2 \cdot 3 \cdot 223$	**2677**	$2 \cdot 13 \cdot 103$	$3 \cdot 19 \cdot 47$
268	$2^3 \cdot 5 \cdot 67$	$7 \cdot 383$	$2 \cdot 3^2 \cdot 149$	**2683**	$2^2 \cdot 11 \cdot 61$	$3 \cdot 5 \cdot 179$	$2 \cdot 17 \cdot 79$	**2687**	$2^7 \cdot 3 \cdot 7$	**2689**
269	$2 \cdot 5 \cdot 269$	$3^2 \cdot 13 \cdot 23$	$2^2 \cdot 673$	**2693**	$2 \cdot 3 \cdot 449$	$5 \cdot 7^2 \cdot 11$	$2^3 \cdot 337$	$3 \cdot 29 \cdot 31$	$2 \cdot 19 \cdot 71$	**2699**
270	$2^2 \cdot 3^3 \cdot 5^2$	$37 \cdot 73$	$2 \cdot 7 \cdot 193$	$3 \cdot 17 \cdot 53$	$2^4 \cdot 13^2$	$5 \cdot 541$	$2 \cdot 3 \cdot 11 \cdot 41$	**2707**	$2^2 \cdot 677$	$3^2 \cdot 7 \cdot 43$
271	$2 \cdot 5 \cdot 271$	**2711**	$2^3 \cdot 3 \cdot 113$	**2713**	$2 \cdot 23 \cdot 59$	$3 \cdot 5 \cdot 181$	$2^2 \cdot 7 \cdot 97$	$11 \cdot 13 \cdot 19$	$2 \cdot 3^2 \cdot 151$	**2719**
272	$2^5 \cdot 5 \cdot 17$	$3 \cdot 907$	$2 \cdot 1361$	$7 \cdot 389$	$2^2 \cdot 3 \cdot 227$	$5^2 \cdot 109$	$2 \cdot 29 \cdot 47$	$3^3 \cdot 101$	$2^3 \cdot 11 \cdot 31$	**2729**
273	$2 \cdot 3 \cdot 5 \cdot 7 \cdot 13$	**2731**	$2^2 \cdot 683$	$3 \cdot 911$	$2 \cdot 1367$	$5 \cdot 547$	$2^4 \cdot 3^2 \cdot 19$	$7 \cdot 17 \cdot 23$	$2 \cdot 37^2$	$3 \cdot 11 \cdot 83$
274	$2^2 \cdot 5 \cdot 137$	**2741**	$2 \cdot 3 \cdot 457$	$13 \cdot 211$	$2^3 \cdot 7^3$	$3^2 \cdot 5 \cdot 61$	$2 \cdot 1373$	$41 \cdot 67$	$2^2 \cdot 3 \cdot 229$	**2749**
275	$2 \cdot 5^3 \cdot 11$	$3 \cdot 7 \cdot 131$	$2^6 \cdot 43$	**2753**	$2 \cdot 3^4 \cdot 17$	$5 \cdot 19 \cdot 29$	$2^2 \cdot 13 \cdot 53$	$3 \cdot 919$	$2 \cdot 7 \cdot 197$	$31 \cdot 89$
276	$2^3 \cdot 3 \cdot 5 \cdot 23$	$11 \cdot 251$	$2 \cdot 1381$	$3^2 \cdot 307$	$2^2 \cdot 691$	$5 \cdot 7 \cdot 79$	$2 \cdot 3 \cdot 461$	**2767**	$2^4 \cdot 173$	$3 \cdot 13 \cdot 71$
277	$2 \cdot 5 \cdot 277$	$17 \cdot 163$	$2^2 \cdot 3^2 \cdot 7 \cdot 11$	$47 \cdot 59$	$2 \cdot 19 \cdot 73$	$3 \cdot 5^2 \cdot 37$	$2^3 \cdot 347$	**2777**	$2 \cdot 3 \cdot 463$	$7 \cdot 397$
278	$2^2 \cdot 5 \cdot 139$	$3^3 \cdot 103$	$2 \cdot 13 \cdot 107$	$11^2 \cdot 23$	$2^5 \cdot 3 \cdot 29$	$5 \cdot 557$	$2 \cdot 7 \cdot 199$	$3 \cdot 929$	$2^2 \cdot 17 \cdot 41$	**2789**
279	$2 \cdot 3^2 \cdot 5 \cdot 31$	**2791**	$2^3 \cdot 349$	$3 \cdot 7^2 \cdot 19$	$2 \cdot 11 \cdot 127$	$5 \cdot 13 \cdot 43$	$2^2 \cdot 3 \cdot 233$	**2797**	$2 \cdot 1399$	$3^2 \cdot 311$
280	$2^4 \cdot 5^2 \cdot 7$	**2801**	$2 \cdot 3 \cdot 467$	**2803**	$2^2 \cdot 701$	$3 \cdot 5 \cdot 11 \cdot 17$	$2 \cdot 23 \cdot 61$	$7 \cdot 401$	$2^3 \cdot 3^3 \cdot 13$	53^2
281	$2 \cdot 5 \cdot 281$	$3 \cdot 937$	$2^2 \cdot 19 \cdot 37$	$29 \cdot 97$	$2 \cdot 3 \cdot 7 \cdot 67$	$5 \cdot 563$	$2^8 \cdot 11$	$3^2 \cdot 313$	$2 \cdot 1409$	**2819**
282	$2^2 \cdot 3 \cdot 5 \cdot 47$	$7 \cdot 13 \cdot 31$	$2 \cdot 17 \cdot 83$	$3 \cdot 941$	$2^3 \cdot 353$	$5^2 \cdot 113$	$2 \cdot 3^2 \cdot 157$	$11 \cdot 257$	$2^2 \cdot 7 \cdot 101$	$3 \cdot 23 \cdot 41$
283	$2 \cdot 5 \cdot 283$	$19 \cdot 149$	$2^4 \cdot 3 \cdot 59$	**2833**	$2 \cdot 13 \cdot 109$	$3^4 \cdot 5 \cdot 7$	$2^2 \cdot 709$	**2837**	$2 \cdot 3 \cdot 11 \cdot 43$	$17 \cdot 167$
284	$2^3 \cdot 5 \cdot 71$	$3 \cdot 947$	$2 \cdot 7^2 \cdot 29$	**2843**	$2^2 \cdot 3^2 \cdot 79$	$5 \cdot 569$	$2 \cdot 1423$	$3 \cdot 13 \cdot 73$	$2^5 \cdot 89$	$7 \cdot 11 \cdot 37$
285	$2 \cdot 3 \cdot 5^2 \cdot 19$	**2851**	$2^2 \cdot 23 \cdot 31$	$3^2 \cdot 317$	$2 \cdot 1427$	$5 \cdot 571$	$2^3 \cdot 3 \cdot 7 \cdot 17$	**2857**	$2 \cdot 1429$	$3 \cdot 953$
286	$2^2 \cdot 5 \cdot 11 \cdot 13$	**2861**	$2 \cdot 3^3 \cdot 53$	$7 \cdot 409$	$2^4 \cdot 179$	$3 \cdot 5 \cdot 191$	$2 \cdot 1433$	$47 \cdot 61$	$2^2 \cdot 3 \cdot 239$	$19 \cdot 151$
287	$2 \cdot 5 \cdot 7 \cdot 41$	$3^2 \cdot 11 \cdot 29$	$2^3 \cdot 359$	$13^2 \cdot 17$	$2 \cdot 3 \cdot 479$	$5^3 \cdot 23$	$2^2 \cdot 719$	$3 \cdot 7 \cdot 137$	$2 \cdot 1439$	**2879**
288	$2^6 \cdot 3^2 \cdot 5$	$43 \cdot 67$	$2 \cdot 11 \cdot 131$	$3 \cdot 31^2$	$2^2 \cdot 7 \cdot 103$	$5 \cdot 577$	$2 \cdot 3 \cdot 13 \cdot 37$	**2887**	$2^3 \cdot 19^2$	$3^3 \cdot 107$
289	$2 \cdot 5 \cdot 17^2$	$7^2 \cdot 59$	$2^2 \cdot 3 \cdot 241$	$11 \cdot 263$	$2 \cdot 1447$	$3 \cdot 5 \cdot 193$	$2^4 \cdot 181$	**2897**	$2 \cdot 3^2 \cdot 7 \cdot 23$	$13 \cdot 223$
290	$2^2 \cdot 5^2 \cdot 29$	$3 \cdot 967$	$2 \cdot 1451$	**2903**	$2^3 \cdot 3 \cdot 11^2$	$5 \cdot 7 \cdot 83$	$2 \cdot 1453$	$3^2 \cdot 17 \cdot 19$	$2^2 \cdot 727$	**2909**
291	$2 \cdot 3 \cdot 5 \cdot 97$	$41 \cdot 71$	$2^5 \cdot 7 \cdot 13$	$3 \cdot 971$	$2 \cdot 31 \cdot 47$	$5 \cdot 11 \cdot 53$	$2^2 \cdot 3^6$	**2917**	$2 \cdot 1459$	$3 \cdot 7 \cdot 139$
292	$2^3 \cdot 5 \cdot 73$	$23 \cdot 127$	$2 \cdot 3 \cdot 487$	$37 \cdot 79$	$2^2 \cdot 17 \cdot 43$	$3^2 \cdot 5^2 \cdot 13$	$2 \cdot 7 \cdot 11 \cdot 19$	**2927**	$2^4 \cdot 3 \cdot 61$	$29 \cdot 101$
293	$2 \cdot 5 \cdot 293$	$3 \cdot 977$	$2^2 \cdot 733$	$7 \cdot 419$	$2 \cdot 3^2 \cdot 163$	$5 \cdot 587$	$2^3 \cdot 367$	$3 \cdot 11 \cdot 89$	$2 \cdot 13 \cdot 113$	**2939**
294	$2^2 \cdot 3 \cdot 5 \cdot 7^2$	$17 \cdot 173$	$2 \cdot 1471$	$3^3 \cdot 109$	$2^7 \cdot 23$	$5 \cdot 19 \cdot 31$	$2 \cdot 3 \cdot 491$	$7 \cdot 421$	$2^2 \cdot 11 \cdot 67$	$3 \cdot 983$
295	$2 \cdot 5^2 \cdot 59$	$13 \cdot 227$	$2^3 \cdot 3^2 \cdot 41$	**2953**	$2 \cdot 7 \cdot 211$	$3 \cdot 5 \cdot 197$	$2^2 \cdot 739$	**2957**	$2 \cdot 3 \cdot 17 \cdot 29$	$11 \cdot 269$
296	$2^4 \cdot 5 \cdot 37$	$3^2 \cdot 7 \cdot 47$	$2 \cdot 1481$	**2963**	$2^2 \cdot 3 \cdot 13 \cdot 19$	$5 \cdot 593$	$2 \cdot 1483$	$3 \cdot 23 \cdot 43$	$2^3 \cdot 7 \cdot 53$	**2969**
297	$2 \cdot 3^3 \cdot 5 \cdot 11$	**2971**	$2^2 \cdot 743$	$3 \cdot 991$	$2 \cdot 1487$	$5^2 \cdot 7 \cdot 17$	$2^5 \cdot 3 \cdot 31$	$13 \cdot 229$	$2 \cdot 1489$	$3^2 \cdot 331$
298	$2^2 \cdot 5 \cdot 149$	$11 \cdot 271$	$2 \cdot 3 \cdot 7 \cdot 71$	$19 \cdot 157$	$2^3 \cdot 373$	$3 \cdot 5 \cdot 199$	$2 \cdot 1493$	$29 \cdot 103$	$2^2 \cdot 3^2 \cdot 83$	$7^2 \cdot 61$
299	$2 \cdot 5 \cdot 13 \cdot 23$	$3 \cdot 997$	$2^4 \cdot 11 \cdot 17$	$41 \cdot 73$	$2 \cdot 3 \cdot 499$	$5 \cdot 599$	$2^2 \cdot 7 \cdot 107$	$3^4 \cdot 37$	$2 \cdot 1499$	**2999**

Table 24.7 **Factorizations**

3000 — 3499

N	9	8	7	6	5	4	3	2	1	0	N
300	$3\cdot17\cdot59$	$2^6\cdot47$	$31\cdot97$	$2\cdot3^2\cdot167$	$5\cdot601$	$2^2\cdot751$	$3\cdot7\cdot11\cdot13$	$2\cdot19\cdot79$	**3001**	$2^3\cdot3\cdot5^3$	300
301	**3019**	$2\cdot3\cdot503$	$7\cdot431$	$2^3\cdot13\cdot29$	$3^2\cdot5\cdot67$	$2\cdot11\cdot137$	$23\cdot131$	$2^2\cdot3\cdot251$	**3011**	$2\cdot5\cdot7\cdot43$	301
302	$13\cdot233$	$2^2\cdot757$	$3\cdot1009$	$2\cdot17\cdot89$	$5^2\cdot11^2$	$2^4\cdot3^3\cdot7$	**3023**	$2\cdot1511$	$3\cdot19\cdot53$	$2^2\cdot5\cdot151$	302
303	$3\cdot1013$	$2\cdot7^2\cdot31$	**3037**	$2^2\cdot3\cdot11\cdot23$	$5\cdot607$	$2\cdot37\cdot41$	$3^2\cdot337$	$2^3\cdot379$	$7\cdot433$	$2\cdot3\cdot5\cdot101$	303
304	**3049**	$2^3\cdot3\cdot127$	$11\cdot277$	$2\cdot1523$	$3\cdot5\cdot7\cdot29$	$2^2\cdot761$	$17\cdot179$	$2\cdot3^2\cdot13^2$	**3041**	$2^5\cdot5\cdot19$	304
305	$7\cdot19\cdot23$	$2\cdot11\cdot139$	$3\cdot1019$	$2^4\cdot191$	$5\cdot13\cdot47$	$2\cdot3\cdot509$	$43\cdot71$	$2^2\cdot7\cdot109$	$3^3\cdot113$	$2\cdot5^2\cdot61$	305
306	$3^2\cdot11\cdot31$	$2^2\cdot13\cdot59$	**3067**	$2\cdot3\cdot7\cdot73$	$5\cdot613$	$2^3\cdot383$	$3\cdot1021$	$2\cdot1531$	**3061**	$2^2\cdot3^2\cdot5\cdot17$	306
307	**3079**	$2\cdot3^4\cdot19$	$17\cdot181$	$2^2\cdot769$	$3\cdot5^2\cdot41$	$2\cdot29\cdot53$	$7\cdot439$	$2^{10}\cdot3$	$37\cdot83$	$2\cdot5\cdot307$	307
308	**3089**	$2^4\cdot193$	$3^2\cdot7^3$	$2\cdot1543$	$5\cdot617$	$2^2\cdot3\cdot257$	**3083**	$2\cdot23\cdot67$	$3\cdot13\cdot79$	$2^3\cdot5\cdot7\cdot11$	308
309	$3\cdot1033$	$2\cdot1549$	$19\cdot163$	$2^3\cdot3^2\cdot43$	$5\cdot619$	$2\cdot7\cdot13\cdot17$	$3\cdot1031$	$2^2\cdot773$	$11\cdot281$	$2\cdot3\cdot5\cdot103$	309
310	**3109**	$2^2\cdot3\cdot7\cdot37$	$13\cdot239$	$2\cdot1553$	$3^3\cdot5\cdot23$	$2^5\cdot97$	$29\cdot107$	$2\cdot3\cdot11\cdot47$	$7\cdot443$	$2^2\cdot5^2\cdot31$	310
311	**3119**	$2\cdot1559$	$3\cdot1039$	$2^2\cdot19\cdot41$	$5\cdot7\cdot89$	$2\cdot3^2\cdot173$	$11\cdot283$	$2^3\cdot389$	$3\cdot17\cdot61$	$2\cdot5\cdot311$	311
312	$3\cdot7\cdot149$	$2^3\cdot17\cdot23$	$53\cdot59$	$2\cdot3\cdot521$	5^5	$2^2\cdot11\cdot71$	$3^2\cdot347$	$2\cdot7\cdot223$	**3121**	$2^4\cdot3\cdot5\cdot13$	312
313	$43\cdot73$	$2\cdot3\cdot523$	**3137**	$2^6\cdot7^2$	$3\cdot5\cdot11\cdot19$	$2\cdot1567$	$13\cdot241$	$2^2\cdot3^3\cdot29$	$31\cdot101$	$2\cdot5\cdot313$	313
314	$47\cdot67$	$2^2\cdot787$	$3\cdot1049$	$2\cdot11^2\cdot13$	$5\cdot17\cdot37$	$2^3\cdot3\cdot131$	$7\cdot449$	$2\cdot1571$	$3^2\cdot349$	$2^2\cdot5\cdot157$	314
315	$3^5\cdot13$	$2\cdot1579$	$7\cdot11\cdot41$	$2^2\cdot3\cdot263$	$5\cdot631$	$2\cdot19\cdot83$	$3\cdot1051$	$2^4\cdot197$	$23\cdot137$	$2\cdot3^2\cdot5^2\cdot7$	315
316	**3169**	$2^5\cdot3^2\cdot11$	$3^2\cdot11^2$ 	$2\cdot1583$	$3\cdot5\cdot211$	$2^2\cdot7\cdot113$	**3163**	$2\cdot3\cdot17\cdot31$	$29\cdot109$	$2^3\cdot5\cdot79$	316
317	$11\cdot17^2$	$2\cdot7\cdot227$	$3^2\cdot353$	$2^3\cdot397$	$5^2\cdot127$	$2\cdot3\cdot23^2$	$19\cdot167$	$2^2\cdot13\cdot61$	$3\cdot7\cdot151$	$2\cdot5\cdot317$	317
318	$3\cdot1063$	$2^2\cdot797$	**3187**	$2\cdot3^3\cdot59$	$5\cdot7^2\cdot13$	$2^4\cdot199$	$3\cdot1061$	$2\cdot37\cdot43$	**3181**	$2^2\cdot3\cdot5\cdot53$	318
319	$7\cdot457$	$2\cdot3\cdot13\cdot41$	$23\cdot139$	$2^2\cdot17\cdot47$	$3^2\cdot5\cdot71$	$2\cdot1597$	$31\cdot103$	$2^3\cdot3\cdot7\cdot19$	**3191**	$2\cdot5\cdot11\cdot29$	319
320	**3209**	$2^3\cdot401$	$3\cdot1069$	$2\cdot7\cdot229$	$5\cdot641$	$2^2\cdot3^2\cdot89$	**3203**	$2\cdot1601$	$3\cdot11\cdot97$	$2^7\cdot5^2$	320
321	$3\cdot29\cdot37$	$2\cdot1609$	**3217**	$2^4\cdot3\cdot67$	$5\cdot643$	$2\cdot1607$	$3^3\cdot7\cdot17$	$2^2\cdot11\cdot73$	$13^2\cdot19$	$2\cdot3\cdot5\cdot107$	321
322	**3229**	$2^2\cdot3\cdot269$	$7\cdot461$	$2\cdot1613$	$3\cdot5^2\cdot43$	$2^3\cdot13\cdot31$	$11\cdot293$	$2\cdot3^2\cdot179$	**3221**	$2^2\cdot5\cdot7\cdot23$	322
323	$41\cdot79$	$2\cdot1619$	$3\cdot13\cdot83$	$2^2\cdot809$	$5\cdot647$	$2\cdot3\cdot7^2\cdot11$	$53\cdot61$	$2^5\cdot101$	$3^2\cdot359$	$2\cdot5\cdot17\cdot19$	323
324	$3^2\cdot19^2$	$2^4\cdot7\cdot29$	$17\cdot191$	$2\cdot3\cdot541$	$5\cdot11\cdot59$	$2^2\cdot811$	$3\cdot23\cdot47$	$2\cdot1621$	$7\cdot463$	$2^3\cdot3^4\cdot5$	324
325	**3259**	$2\cdot3^2\cdot181$	**3257**	$2^3\cdot11\cdot37$	$3\cdot5\cdot7\cdot31$	$2\cdot1627$	**3253**	$2^2\cdot3\cdot271$	**3251**	$2\cdot5^3\cdot13$	325
326	$7\cdot467$	$2^2\cdot19\cdot43$	$3^3\cdot11^2$	$2\cdot23\cdot71$	$5\cdot653$	$2^6\cdot3\cdot17$	$13\cdot251$	$2\cdot7\cdot233$	$3\cdot1087$	$2^2\cdot5\cdot163$	326
327	$3\cdot1093$	$2\cdot11\cdot149$	$29\cdot113$	$2^2\cdot3^2\cdot7\cdot13$	$5^2\cdot131$	$2\cdot1637$	$3\cdot1091$	$2^3\cdot409$	**3271**	$2\cdot3\cdot5\cdot109$	327
328	$11\cdot13\cdot23$	$2^3\cdot3\cdot137$	$19\cdot173$	$2\cdot31\cdot53$	$3^2\cdot5\cdot73$	$2^2\cdot821$	$7^2\cdot67$	$2\cdot3\cdot547$	$17\cdot193$	$2^4\cdot5\cdot41$	328
329	**3299**	$2\cdot17\cdot97$	$3\cdot7\cdot157$	$2^5\cdot103$	$5\cdot659$	$2\cdot3^3\cdot61$	$37\cdot89$	$2^2\cdot823$	$3\cdot1097$	$2\cdot5\cdot7\cdot47$	329
330	$3\cdot1103$	$2^2\cdot827$	**3307**	$2\cdot3\cdot19\cdot29$	$5\cdot661$	$2^3\cdot7\cdot59$	$3^2\cdot367$	$2\cdot13\cdot127$	**3301**	$2^2\cdot3\cdot5^2\cdot11$	330
331	**3319**	$2\cdot3\cdot7\cdot79$	$31\cdot107$	$2^2\cdot829$	$3\cdot5\cdot13\cdot17$	$2\cdot1657$	**3313**	$2^4\cdot3^2\cdot23$	$7\cdot11\cdot43$	$2\cdot5\cdot331$	331
332	**3329**	$2^8\cdot13$	$3\cdot1109$	$2\cdot1663$	$5^2\cdot7\cdot19$	$2^2\cdot3\cdot277$	**3323**	$2\cdot11\cdot151$	$3^4\cdot41$	$2^3\cdot5\cdot83$	332
333	$3^2\cdot7\cdot53$	$2\cdot1669$	$47\cdot71$	$2^3\cdot3\cdot139$	$5\cdot23\cdot29$	$2\cdot1667$	$3\cdot11\cdot101$	$2^2\cdot7^2\cdot17$	**3331**	$2\cdot3^2\cdot5\cdot37$	333
334	$17\cdot197$	$2^2\cdot3^3\cdot31$	**3347**	$2\cdot7\cdot239$	$3\cdot5\cdot223$	$2^4\cdot11\cdot19$	**3343**	$2\cdot3\cdot557$	$13\cdot257$	$2^2\cdot5\cdot167$	334
335	**3359**	$2\cdot23\cdot73$	$3^2\cdot373$	$2^2\cdot839$	$5\cdot11\cdot61$	$2\cdot3\cdot13\cdot43$	$7\cdot479$	$2^3\cdot419$	$3\cdot1117$	$2\cdot5^2\cdot67$	335
336	$3\cdot1123$	$2^3\cdot421$	$7\cdot13\cdot37$	$2\cdot3^2\cdot11\cdot17$	$5\cdot673$	$2^2\cdot29^2$	$3\cdot19\cdot59$	$2\cdot41^2$	**3361**	$2^5\cdot3\cdot5\cdot7$	336
337	$31\cdot109$	$2\cdot3\cdot563$	$11\cdot307$	$2^4\cdot211$	$3^3\cdot5^3$	$2\cdot7\cdot241$	**3373**	$2^2\cdot3\cdot281$	**3371**	$2\cdot5\cdot337$	337
338	**3389**	$2^2\cdot7\cdot11^2$	$3\cdot1129$	$2\cdot1693$	$5\cdot677$	$2^3\cdot3^2\cdot47$	$17\cdot199$	$2\cdot19\cdot89$	$3\cdot7^2\cdot23$	$2^2\cdot5\cdot13^2$	338
339	$3\cdot11\cdot103$	$2\cdot1699$	$43\cdot79$	$2^2\cdot3\cdot283$	$5\cdot7\cdot97$	$2\cdot1697$	$3^2\cdot13\cdot29$	$2^6\cdot53$	**3391**	$2\cdot3\cdot5\cdot113$	339
340	$7\cdot487$	$2^4\cdot3\cdot71$	**3407**	$2\cdot13\cdot131$	$3\cdot5\cdot227$	$2^2\cdot23\cdot37$	$41\cdot83$	$2\cdot3^5\cdot7$	$19\cdot179$	$2^3\cdot5^2\cdot17$	340
341	$13\cdot263$	$2\cdot1709$	$3\cdot17\cdot67$	$2^3\cdot7\cdot61$	$5\cdot683$	$2\cdot3\cdot569$	**3413**	$2^2\cdot853$	$3^2\cdot379$	$2\cdot5\cdot11\cdot31$	341
342	$3^3\cdot127$	$2^2\cdot857$	$23\cdot149$	$2\cdot3\cdot571$	$5^2\cdot137$	$2^5\cdot107$	$3\cdot7\cdot163$	$2\cdot29\cdot59$	$11\cdot311$	$2^2\cdot3^2\cdot5\cdot19$	342
343	$19\cdot181$	$2\cdot3^2\cdot191$	$7\cdot491$	$2^2\cdot859$	$3\cdot5\cdot229$	$2\cdot17\cdot101$	**3433**	$2^3\cdot3\cdot11\cdot13$	$47\cdot73$	$2\cdot5\cdot7^3$	343
344	**3449**	$2^3\cdot431$	$3^2\cdot383$	$2\cdot1723$	$5\cdot13\cdot53$	$2^2\cdot3\cdot7\cdot41$	$11\cdot313$	$2\cdot1721$	$3\cdot31\cdot37$	$2^4\cdot5\cdot43$	344
345	$3\cdot1153$	$2\cdot7\cdot13\cdot19$	**3457**	$2^7\cdot3^3$	$5\cdot691$	$2\cdot11\cdot157$	$3\cdot1151$	$2^2\cdot863$	$7\cdot17\cdot29$	$2\cdot3\cdot5^2\cdot23$	345
346	**3469**	$2^2\cdot3\cdot17^2$	**3467**	$2\cdot1733$	$3^2\cdot5\cdot7\cdot11$	$2^3\cdot433$	**3463**	$2\cdot3\cdot577$	**3461**	$2^2\cdot5\cdot173$	346
347	$7^2\cdot71$	$2\cdot37\cdot47$	$3\cdot19\cdot61$	$2^2\cdot11\cdot79$	$5^2\cdot139$	$2\cdot3^2\cdot193$	$23\cdot151$	$2^4\cdot7\cdot31$	$3\cdot13\cdot89$	$2\cdot5\cdot347$	347
348	$3\cdot1163$	$2^5\cdot109$	$11\cdot317$	$2\cdot3\cdot7\cdot83$	$5\cdot17\cdot41$	$2^2\cdot13\cdot67$	$3^4\cdot43$	$2\cdot1741$	59^2	$2^3\cdot3\cdot5\cdot29$	348
349	**3499**	$2\cdot3\cdot11\cdot53$	$13\cdot269$	$2^3\cdot19\cdot23$	$3\cdot5\cdot233$	$2\cdot1747$	$7\cdot499$	$2^2\cdot3^2\cdot97$	**3491**	$2\cdot5\cdot349$	349

Factorizations

Table 24.7

Numbers 3500–3549

units	3500–	3510–	3520–	3530–	3540–
9	11²·29	3²·17·23	**3529**	**3539**	3·7·13²
8	2²·877	2·1759	2³·3²·7²	2·29·61	2²·887
7	3·7·167	**3517**	**3527**	3³·131	**3547**
6	2·1753	2²·3·293	2·41·43	2⁴·13·17	2·3²·197
5	5·701	5·19·37	3·5²·47	5·7·101	5·709
4	2⁴·3·73	2·7·251	2²·881	2·3·19·31	2³·443
3	31·113	3·1171	13·271	**3533**	3·1181
2	2·17·103	2³·439	2·3·587	2²·883	2·7·11·23
1	3²·389	**3511**	7·503	3·11·107	**3541**
0	2²·5³·7	2·3³·5·13	2⁶·5·11	2·5·353	2²·3·5·59

Numbers 3550–3599

units	3550–	3560–	3570–	3580–	3590–
9	**3559**	43·83	3·1193	37·97	59·61
8	2·3·593	2⁴·223	2·1789	2²·3·13·23	2·7·257
7	**3557**	3·29·41	7²·73	17·211	3·11·109
6	2²·7·127	2·1783	2³·3·149	2·11·163	2²·29·31
5	3²·5·79	5·23·31	5²·11·13	3·5·239	5·719
4	2·1777	2²·3⁴·11	2·1787	2⁹·7	2·3·599
3	11·17·19	7·509	3²·397	**3583**	**3593**
2	2⁵·3·37	2·13·137	2²·19·47	2·3²·199	2³·449
1	53·67	3·1187	**3571**	**3581**	3³·7·19
0	2·5²·71	2³·5·89	2·3·5·7·17	2²·5·179	2·5·359

Numbers 3600–3649

units	3600–	3610–	3620–	3630–	3640–
9	3²·401	7·11·47	19·191	3·1213	41·89
8	2³·11·41	2·3³·67	2²·907	2·17·107	2⁶·3·19
7	**3607**	**3617**	3²·13·31	**3637**	7·521
6	2·3·601	2⁵·113	2·7²·37	2²·3²·101	2·1823
5	5·7·103	3·5·241	5³·29	5·727	3⁶·5
4	2²·17·53	2·13·139	2³·3·151	2·23·79	2²·911
3	3·1201	**3613**	**3623**	3·7·173	**3643**
2	2·1801	2²·3·7·43	2·1811	2⁴·227	2·3·607
1	13·277	23·157	3·17·71	**3631**	11·331
0	2⁴·3²·5²	2·5·19²	2²·5·181	2·3·5·11²	2³·5·7·13

Numbers 3650–3699

units	3650–	3660–	3670–	3680–	3690–
9	**3659**	3·1223	13·283	7·17·31	3³·137
8	2·31·59	2²·7·131	2·3·613	2³·461	2·43²
7	3·23·53	19·193	**3677**	3·1229	**3697**
6	2³·457	2·3·13·47	2²·919	2·19·97	2⁴·3·7·11
5	5·17·43	5·733	3·5²·7²	5·11·67	5·739
4	2·3²·7·29	2⁴·229	2·11·167	2²·3·307	2·1847
3	13·281	3²·11·37	**3673**	29·127	3·1231
2	2²·11·83	2·1831	2³·3³·17	2·7·263	2²·13·71
1	3·1217	7·523	**3671**	3²·409	**3691**
0	2·5²·73	2²·3·5·61	2·5·367	2⁵·5·23	2·3²·5·41

Numbers 3700–3749

units	3700–	3710–	3720–	3730–	3740–
9	**3709**	**3719**	3·11·113	**3739**	23·163
8	2²·3²·103	2·11·13²	2⁴·233	2·3·7·89	2²·937
7	11·337	3²·7·59	**3727**	37·101	3·1249
6	2·17·109	2²·929	2·3⁴·23	2³·467	2·1873
5	3·5·13·19	5·743	5²·149	3²·5·83	5·7·107
4	2³·463	2·3·619	2²·7²·19	2·1867	2⁵·3²·13
3	7·23²	47·79	3·17·73	**3733**	19·197
2	2·3·617	2⁷·29	2·1861	2²·3·311	2·1871
1	**3701**	3·1237	61²	7·13·41	3·29·43
0	2²·5²·37	2·5·7·53	2³·3·5·31	2·5·373	2²·5·11·17

Numbers 3750–3799

units	3750–	3760–	3770–	3780–	3790–
9	3·7·179	**3769**	**3779**	3²·421	29·131
8	2·1879	2³·3·157	2·1889	2²·947	2·3²·211
7	13·17²	**3767**	3·1259	7·541	**3797**
6	2²·3·313	2·7·269	2⁶·59	2·3·631	2²·13·73
5	5·751	3·5·251	5²·151	5·757	3·5·11·23
4	2·1877	2²·941	2·3·17·37	2³·11·43	2·7·271
3	3³·139	53·71	7³·11	3·13·97	**3793**
2	2³·7·67	2·3²·11·19	2²·23·41	2·31·61	2⁴·3·79
1	11²·31	**3761**	3²·419	19·199	17·223
0	2·3·5⁴	2⁴·5·47	2·5·13·29	2²·3³·5·7	2·5·379

Numbers 3800–3849

units	3800–	3810–	3820–	3830–	3840–
9	13·293	3·19·67	7·547	11·349	3·1283
8	2⁵·7·17	2·23·83	2²·3·11·29	2·19·101	2³·13·37
7	3⁴·47	11·347	43·89	3·1279	**3847**
6	2·11·173	2³·3²·53	2·1913	2²·7·137	2·3·641
5	5·761	5·7·109	3²·5²·17	5·13·59	5·769
4	2²·3·317	2·1907	2⁴·239	2·3³·71	2²·31²
3	**3803**	3·31·41	**3823**	**3833**	3²·7·61
2	2·1901	2²·953	2·3·7²·13	2³·479	2·17·113
1	3·7·181	37·103	**3821**	3·1277	23·167
0	2³·5²·19	2·3·5·127	2²·5·191	2·5·383	2⁸·3·5

Numbers 3850–3899

units	3850–	3860–	3870–	3880–	3890–
9	17·227	53·73	3²·431	**3889**	7·557
8	2·3·643	2²·967	2·7·277	2⁴·3⁵	2·1949
7	7·19·29	3·1289	**3877**	13²·23	3²·433
6	2⁴·241	2·1933	2²·3·17·19	2·29·67	2³·487
5	3·5·257	5·773	5³·31	3·5·7·37	5·19·41
4	2·41·47	2³·3·7·23	2·13·149	2²·971	2·3·11·59
3	**3853**	**3863**	3·1291	11·353	17·229
2	2²·3²·107	2·1931	2⁵·11²	2·3·647	2²·7·139
1	**3851**	3³·11·13	7²·79	**3881**	3·1297
0	2·5²·7·11	2²·5·193	2·3²·5·43	2³·5·97	2·5·389

Numbers 3900–3949

units	3900–	3910–	3920–	3930–	3940–
9	3·1303	**3919**	**3929**	3·13·101	11·359
8	2²·977	2·3·653	2³·491	2·11·179	2²·3·7·47
7	**3907**	**3917**	3·7·11·17	31·127	**3947**
6	2·3²·7·31	2²·11·89	2·13·151	2⁵·3·41	2·1973
5	5·11·71	3³·5·29	5²·157	5·787	3·5·263
4	2⁶·61	2·19·103	2²·3²·109	2·7·281	2³·17·29
3	3·1301	7·13·43	**3923**	3²·19·23	**3943**
2	2·1951	2³·3·163	2·37·53	2²·983	2·3³·73
1	47·83	**3911**	3·1307	**3931**	7·563
0	2²·3·5²·13	2·5·17·23	2⁴·5·7²	2·3·5·131	2²·5·197

Numbers 3950–3999

units	3950–	3960–	3970–	3980–	3990–
9	37·107	3⁴·7²	23·173	**3989**	3·31·43
8	2·1979	2⁷·31	2·3²·13·17	2²·997	2·1999
7	3·1319	**3967**	41·97	3²·443	7·571
6	2²·23·43	2·3·661	2³·7·71	2·1993	2²·3³·37
5	5·7·113	5·13·61	3·5²·53	5·797	5·17·47
4	2·3·659	2²·991	2·1987	2⁴·3·83	2·1997
3	59·67	3·1321	29·137	7·569	3·11³
2	2⁴·13·19	2·7·283	2²·3·331	2·11·181	2³·499
1	3²·439	17·233	11·19²	3·1327	13·307
0	2·5²·79	2³·3²·5·11	2·5·397	2²·5·199	2·3·5·7·19

Table 24.7   **Factorizations**

N	0	1	2	3	4	5	6	7	8	9
400	$2^5\cdot5^3$	$\mathbf{4001}$	$2\cdot3\cdot23\cdot29$	$\mathbf{4003}$	$2^2\cdot7\cdot11\cdot13$	$3^2\cdot5\cdot89$	$2\cdot2003$	$\mathbf{4007}$	$2^3\cdot3\cdot167$	$19\cdot211$
401	$2\cdot5\cdot401$	$3\cdot7\cdot191$	$2^2\cdot17\cdot59$	$\mathbf{4013}$	$2\cdot3^2\cdot223$	$5\cdot11\cdot73$	$2^4\cdot251$	$3\cdot13\cdot103$	$2\cdot7^2\cdot41$	$\mathbf{4019}$
402	$2^2\cdot3\cdot5\cdot67$	$\mathbf{4021}$	$2\cdot2011$	$3^3\cdot149$	$2^3\cdot503$	$5^2\cdot7\cdot23$	$2\cdot3\cdot11\cdot61$	$\mathbf{4027}$	$2^2\cdot19\cdot53$	$3\cdot17\cdot79$
403	$2\cdot5\cdot13\cdot31$	$29\cdot139$	$2^6\cdot3^2\cdot7$	$37\cdot109$	$2\cdot2017$	$3\cdot5\cdot269$	$2^2\cdot1009$	$11\cdot367$	$2\cdot3\cdot673$	$7\cdot577$
404	$2^3\cdot5\cdot101$	$3^2\cdot449$	$2\cdot43\cdot47$	$13\cdot311$	$2^2\cdot3\cdot337$	$5\cdot809$	$2\cdot7\cdot17^2$	$3\cdot19\cdot71$	$2^4\cdot11\cdot23$	$\mathbf{4049}$
405	$2\cdot3^4\cdot5^2$	$\mathbf{4051}$	$2^2\cdot1013$	$3\cdot7\cdot193$	$2\cdot2027$	$5\cdot811$	$2^3\cdot3\cdot13^2$	$\mathbf{4057}$	$2\cdot2029$	$3^2\cdot11\cdot41$
406	$2^2\cdot5\cdot7\cdot29$	$31\cdot131$	$2\cdot3\cdot677$	$17\cdot239$	$2^5\cdot127$	$3\cdot5\cdot271$	$2\cdot19\cdot107$	$7^2\cdot83$	$2^2\cdot3^2\cdot113$	$13\cdot313$
407	$2\cdot5\cdot11\cdot37$	$3\cdot23\cdot59$	$2^3\cdot509$	$\mathbf{4073}$	$2\cdot3\cdot7\cdot97$	$5^2\cdot163$	$2^2\cdot1019$	$3^3\cdot151$	$2\cdot2039$	$\mathbf{4079}$
408	$2^4\cdot3\cdot5\cdot17$	$7\cdot11\cdot53$	$2\cdot13\cdot157$	$3\cdot1361$	$2^2\cdot1021$	$5\cdot19\cdot43$	$2\cdot3^2\cdot227$	$61\cdot67$	$2^3\cdot7\cdot73$	$3\cdot29\cdot47$
409	$2\cdot5\cdot409$	$\mathbf{4091}$	$2^2\cdot3\cdot11\cdot31$	$\mathbf{4093}$	$2\cdot23\cdot89$	$3^2\cdot5\cdot7\cdot13$	2^{12}	$17\cdot241$	$2\cdot3\cdot683$	$\mathbf{4099}$
410	$2^2\cdot5^2\cdot41$	$3\cdot1367$	$2\cdot7\cdot293$	$11\cdot373$	$2^3\cdot3^3\cdot19$	$5\cdot821$	$2\cdot2053$	$3\cdot37^2$	$2^2\cdot13\cdot79$	$7\cdot587$
411	$2\cdot3\cdot5\cdot137$	$\mathbf{4111}$	$2^4\cdot257$	$3^2\cdot457$	$2\cdot11^2\cdot17$	$5\cdot823$	$2^2\cdot3\cdot7^3$	$23\cdot179$	$2\cdot29\cdot71$	$3\cdot1373$
412	$2^3\cdot5\cdot103$	$13\cdot317$	$2\cdot3^2\cdot229$	$7\cdot19\cdot31$	$2^2\cdot1031$	$3\cdot5^3\cdot11$	$2\cdot2063$	$\mathbf{4127}$	$2^5\cdot3\cdot43$	$\mathbf{4129}$
413	$2\cdot5\cdot7\cdot59$	$3^5\cdot17$	$2^2\cdot1033$	$\mathbf{4133}$	$2\cdot3\cdot13\cdot53$	$5\cdot827$	$2^3\cdot11\cdot47$	$3\cdot7\cdot197$	$2\cdot2069$	$\mathbf{4139}$
414	$2^2\cdot3^2\cdot5\cdot23$	$41\cdot101$	$2\cdot19\cdot109$	$3\cdot1381$	$2^4\cdot7\cdot37$	$5\cdot829$	$2\cdot3\cdot691$	$11\cdot13\cdot29$	$2^2\cdot17\cdot61$	$3^2\cdot461$
415	$2\cdot5^2\cdot83$	$7\cdot593$	$2^3\cdot3\cdot173$	$\mathbf{4153}$	$2\cdot31\cdot67$	$3\cdot5\cdot277$	$2^2\cdot1039$	$\mathbf{4157}$	$2\cdot3^3\cdot7\cdot11$	$\mathbf{4159}$
416	$2^6\cdot5\cdot13$	$3\cdot19\cdot73$	$2\cdot2081$	$23\cdot181$	$2^2\cdot3\cdot347$	$5\cdot7^2\cdot17$	$2\cdot2083$	$3^2\cdot463$	$2^3\cdot521$	$11\cdot379$
417	$2\cdot3\cdot5\cdot139$	$43\cdot97$	$2^2\cdot7\cdot149$	$3\cdot13\cdot107$	$2\cdot2087$	$5^2\cdot167$	$2^4\cdot3^2\cdot29$	$\mathbf{4177}$	$2\cdot2089$	$3\cdot7\cdot199$
418	$2^2\cdot5\cdot11\cdot19$	$37\cdot113$	$2\cdot3\cdot17\cdot41$	$47\cdot89$	$2^3\cdot523$	$3^3\cdot5\cdot31$	$2\cdot7\cdot13\cdot23$	$53\cdot79$	$2^2\cdot3\cdot349$	$59\cdot71$
419	$2\cdot5\cdot419$	$3\cdot11\cdot127$	$2^5\cdot131$	$7\cdot599$	$2\cdot3^2\cdot233$	$5\cdot839$	$2^2\cdot1049$	$3\cdot1399$	$2\cdot2099$	$13\cdot17\cdot19$
420	$2^3\cdot3\cdot5^2\cdot7$	$\mathbf{4201}$	$2\cdot11\cdot191$	$3^2\cdot467$	$2^2\cdot1051$	$5\cdot29^2$	$2\cdot3\cdot701$	$7\cdot601$	$2^4\cdot263$	$3\cdot23\cdot61$
421	$2\cdot5\cdot421$	$\mathbf{4211}$	$2^2\cdot3^4\cdot13$	$11\cdot383$	$2\cdot7^2\cdot43$	$3\cdot5\cdot281$	$2^3\cdot17\cdot31$	$\mathbf{4217}$	$2\cdot3\cdot19\cdot37$	$\mathbf{4219}$
422	$2^2\cdot5\cdot211$	$3^2\cdot7\cdot67$	$2\cdot2111$	$41\cdot103$	$2^7\cdot3\cdot11$	$5^2\cdot13^2$	$2\cdot2113$	$3\cdot1409$	$2^2\cdot7\cdot151$	$\mathbf{4229}$
423	$2\cdot3^2\cdot5\cdot47$	$\mathbf{4231}$	$2^3\cdot23^2$	$3\cdot17\cdot83$	$2\cdot29\cdot73$	$5\cdot7\cdot11^2$	$2^2\cdot3\cdot353$	$19\cdot223$	$2\cdot13\cdot163$	$3^3\cdot157$
424	$2^4\cdot5\cdot53$	$\mathbf{4241}$	$2\cdot3\cdot7\cdot101$	$\mathbf{4243}$	$2^2\cdot1061$	$3\cdot5\cdot283$	$2\cdot11\cdot193$	$31\cdot137$	$2^3\cdot3^2\cdot59$	$7\cdot607$
425	$2\cdot5^3\cdot17$	$3\cdot13\cdot109$	$2^2\cdot1063$	$\mathbf{4253}$	$2\cdot3\cdot709$	$5\cdot23\cdot37$	$2^5\cdot7\cdot19$	$3^2\cdot11\cdot43$	$2\cdot2129$	$\mathbf{4259}$
426	$2^2\cdot3\cdot5\cdot71$	$\mathbf{4261}$	$2\cdot2131$	$3\cdot7^2\cdot29$	$2^3\cdot13\cdot41$	$5\cdot853$	$2\cdot3^3\cdot79$	$11\cdot397$	$2^2\cdot11\cdot97$	$3\cdot1423$
427	$2\cdot5\cdot7\cdot61$	$\mathbf{4271}$	$2^4\cdot3\cdot89$	$\mathbf{4273}$	$2\cdot2137$	$3^2\cdot5^2\cdot19$	$2^2\cdot1069$	$7\cdot13\cdot47$	$2\cdot3\cdot23\cdot31$	$11\cdot389$
428	$2^3\cdot5\cdot107$	$3\cdot1427$	$2\cdot2141$	$\mathbf{4283}$	$2^2\cdot3^2\cdot7\cdot17$	$5\cdot857$	$2\cdot2143$	$3\cdot1429$	$2^6\cdot67$	$\mathbf{4289}$
429	$2\cdot3\cdot5\cdot11\cdot13$	$7\cdot613$	$2^2\cdot29\cdot37$	$3^4\cdot53$	$2\cdot19\cdot113$	$5\cdot859$	$2^3\cdot3\cdot179$	$\mathbf{4297}$	$2\cdot7\cdot307$	$3\cdot1433$
430	$2^2\cdot5^2\cdot43$	$11\cdot17\cdot23$	$2\cdot3^2\cdot239$	$13\cdot331$	$2^4\cdot269$	$3\cdot5\cdot7\cdot41$	$2\cdot2153$	$59\cdot73$	$2^2\cdot3\cdot359$	$31\cdot139$
431	$2\cdot5\cdot431$	$3^2\cdot479$	$2^3\cdot7^2\cdot11$	$19\cdot227$	$2\cdot3\cdot719$	$5\cdot863$	$2^2\cdot13\cdot83$	$3\cdot1439$	$2\cdot17\cdot127$	$7\cdot617$
432	$2^5\cdot3^3\cdot5$	$29\cdot149$	$2\cdot2161$	$3\cdot11\cdot131$	$2^2\cdot23\cdot47$	$5^2\cdot173$	$2\cdot3\cdot7\cdot103$	$\mathbf{4327}$	$2^3\cdot541$	$3^2\cdot13\cdot37$
433	$2\cdot5\cdot433$	$61\cdot71$	$2^2\cdot3\cdot19^2$	$7\cdot619$	$2\cdot11\cdot197$	$3\cdot5\cdot17^2$	$2^4\cdot271$	$\mathbf{4337}$	$2\cdot3^2\cdot241$	$\mathbf{4339}$
434	$2^2\cdot5\cdot7\cdot31$	$3\cdot1447$	$2\cdot13\cdot167$	$43\cdot101$	$2^3\cdot3\cdot181$	$5\cdot11\cdot79$	$2\cdot41\cdot53$	$3^3\cdot7\cdot23$	$2^2\cdot1087$	$\mathbf{4349}$
435	$2\cdot3\cdot5^2\cdot29$	$19\cdot229$	$2^8\cdot17$	$3\cdot1451$	$2\cdot7\cdot311$	$5\cdot13\cdot67$	$2^2\cdot3^2\cdot11^2$	$\mathbf{4357}$	$2\cdot2179$	$3\cdot1453$
436	$2^3\cdot5\cdot109$	$7^2\cdot89$	$2\cdot3\cdot727$	$\mathbf{4363}$	$2^2\cdot1091$	$3^2\cdot5\cdot97$	$2\cdot37\cdot59$	$11\cdot397$	$2^4\cdot3\cdot7\cdot13$	$17\cdot257$
437	$2\cdot5\cdot19\cdot23$	$3\cdot31\cdot47$	$2^2\cdot1093$	$\mathbf{4373}$	$2\cdot3^7$	$5^4\cdot7$	$2^3\cdot547$	$3\cdot1459$	$2\cdot11\cdot199$	$29\cdot151$
438	$2^2\cdot3\cdot5\cdot73$	$13\cdot337$	$2\cdot7\cdot313$	$3^2\cdot487$	$2^5\cdot137$	$5\cdot877$	$2\cdot3\cdot17\cdot43$	$41\cdot107$	$2^2\cdot1097$	$3\cdot7\cdot11\cdot19$
439	$2\cdot5\cdot439$	$\mathbf{4391}$	$2^3\cdot3^2\cdot61$	$23\cdot191$	$2\cdot13^3$	$3\cdot5\cdot293$	$2^2\cdot7\cdot157$	$\mathbf{4397}$	$2\cdot3\cdot733$	$53\cdot83$
440	$2^4\cdot5^2\cdot11$	$3^3\cdot163$	$2\cdot31\cdot71$	$7\cdot17\cdot37$	$2^2\cdot3\cdot367$	$5\cdot881$	$2\cdot2203$	$3\cdot13\cdot113$	$2^3\cdot19\cdot29$	$\mathbf{4409}$
441	$2\cdot3^2\cdot5\cdot7^2$	$11\cdot401$	$2^2\cdot1103$	$3\cdot1471$	$2\cdot2207$	$5\cdot883$	$2^6\cdot3\cdot23$	$7\cdot631$	$2\cdot47^2$	$3^2\cdot491$
442	$2^2\cdot5\cdot13\cdot17$	$\mathbf{4421}$	$2\cdot3\cdot11\cdot67$	$\mathbf{4423}$	$2^3\cdot7\cdot79$	$3\cdot5^2\cdot59$	$2\cdot2213$	$19\cdot233$	$2^2\cdot3^3\cdot41$	$43\cdot103$
443	$2\cdot5\cdot443$	$3\cdot7\cdot211$	$2^4\cdot277$	$11\cdot13\cdot31$	$2\cdot3\cdot739$	$5\cdot887$	$2^2\cdot1109$	$3^2\cdot17\cdot29$	$2\cdot7\cdot317$	$23\cdot193$
444	$2^3\cdot3\cdot5\cdot37$	$\mathbf{4441}$	$2\cdot2221$	$3\cdot1481$	$2^2\cdot11\cdot101$	$5\cdot7\cdot127$	$2\cdot3^2\cdot13\cdot19$	$\mathbf{4447}$	$2^5\cdot139$	$3\cdot1483$
445	$2\cdot5^2\cdot89$	$\mathbf{4451}$	$2^2\cdot3\cdot7\cdot53$	$61\cdot73$	$2\cdot17\cdot131$	$3^4\cdot5\cdot11$	$2^3\cdot557$	$\mathbf{4457}$	$2\cdot3\cdot743$	$7^3\cdot13$
446	$2^2\cdot5\cdot223$	$3\cdot1487$	$2\cdot23\cdot97$	$\mathbf{4463}$	$2^4\cdot3^2\cdot31$	$5\cdot19\cdot47$	$2\cdot7\cdot11\cdot29$	$3\cdot1489$	$2^2\cdot1117$	$41\cdot109$
447	$2\cdot3\cdot5\cdot149$	$17\cdot263$	$2^3\cdot13\cdot43$	$3^2\cdot7\cdot71$	$2\cdot2237$	$5^2\cdot179$	$2^2\cdot3\cdot373$	$11^2\cdot37$	$2\cdot2239$	$3\cdot1493$
448	$2^7\cdot5\cdot7$	$\mathbf{4481}$	$2\cdot3^3\cdot83$	$\mathbf{4483}$	$2^2\cdot19\cdot59$	$3\cdot5\cdot13\cdot23$	$2\cdot2243$	$7\cdot641$	$2^3\cdot3\cdot11\cdot17$	67^2
449	$2\cdot5\cdot449$	$3^2\cdot499$	$2^2\cdot1123$	$\mathbf{4493}$	$2\cdot3\cdot7\cdot107$	$5\cdot29\cdot31$	$2^4\cdot281$	$3\cdot1499$	$2\cdot13\cdot173$	$11\cdot409$

Factorizations

Table 24.7

N	0	1	2	3	4	5	6	7	8	9
4500	$2^2 \cdot 3^2 \cdot 5^3$	$7 \cdot 643$	$2 \cdot 2251$	$3 \cdot 19 \cdot 79$	$2^3 \cdot 563$	$5 \cdot 17 \cdot 53$	$2 \cdot 3 \cdot 751$	4507	$2^2 \cdot 7^2 \cdot 23$	$3^3 \cdot 167$
4510	$2 \cdot 5 \cdot 11 \cdot 41$	$13 \cdot 347$	$2^5 \cdot 3 \cdot 47$	4513	$2 \cdot 37 \cdot 61$	$3 \cdot 5 \cdot 7 \cdot 43$	$2^2 \cdot 1129$	4517	$2 \cdot 3^2 \cdot 251$	4519
4520	$2^3 \cdot 5 \cdot 113$	$3 \cdot 11 \cdot 137$	$2 \cdot 7 \cdot 17 \cdot 19$	4523	$2^2 \cdot 3 \cdot 13 \cdot 29$	$5^2 \cdot 181$	$2 \cdot 31 \cdot 73$	$3^2 \cdot 503$	$2^4 \cdot 283$	$7 \cdot 647$
4530	$2 \cdot 3 \cdot 5 \cdot 151$	$23 \cdot 197$	$2^2 \cdot 11 \cdot 103$	$3 \cdot 1511$	$2 \cdot 2267$	$5 \cdot 907$	$2^3 \cdot 3^4 \cdot 7$	$13 \cdot 349$	$2 \cdot 2269$	$3 \cdot 17 \cdot 89$
4540	$2^2 \cdot 5 \cdot 227$	$19 \cdot 239$	$2 \cdot 3 \cdot 757$	$7 \cdot 11 \cdot 59$	$2^6 \cdot 71$	$3^2 \cdot 5 \cdot 101$	$2 \cdot 2273$	4547	$2^2 \cdot 3 \cdot 379$	4549
4550	$2 \cdot 5^2 \cdot 7 \cdot 13$	$3 \cdot 37 \cdot 41$	$2^3 \cdot 569$	$29 \cdot 157$	$2 \cdot 3^2 \cdot 11 \cdot 23$	$5 \cdot 911$	$2^2 \cdot 17 \cdot 67$	$3 \cdot 7^2 \cdot 31$	$2 \cdot 43 \cdot 53$	$47 \cdot 97$
4560	$2^4 \cdot 3 \cdot 5 \cdot 19$	4561	$2 \cdot 2281$	$3^3 \cdot 13^2$	$2^2 \cdot 7 \cdot 163$	$5 \cdot 11 \cdot 83$	$2 \cdot 3 \cdot 761$	4567	$2^3 \cdot 571$	$3 \cdot 1523$
4570	$2 \cdot 5 \cdot 457$	$7 \cdot 653$	$2^2 \cdot 3^2 \cdot 127$	$17 \cdot 269$	$2 \cdot 2287$	$3 \cdot 5^2 \cdot 61$	$2^5 \cdot 11 \cdot 13$	$23 \cdot 199$	$2 \cdot 3 \cdot 7 \cdot 109$	$19 \cdot 241$
4580	$2^2 \cdot 5 \cdot 229$	$3^2 \cdot 509$	$2 \cdot 29 \cdot 79$	4583	$2^3 \cdot 3 \cdot 191$	$5 \cdot 7 \cdot 131$	$2 \cdot 2293$	$3 \cdot 11 \cdot 139$	$2^2 \cdot 31 \cdot 37$	$13 \cdot 353$
4590	$2 \cdot 3^3 \cdot 5 \cdot 17$	4591	$2^4 \cdot 7 \cdot 41$	$3 \cdot 1531$	$2 \cdot 2297$	$5 \cdot 919$	$2^2 \cdot 3 \cdot 383$	4597	$2 \cdot 11^2 \cdot 19$	$3^2 \cdot 7 \cdot 73$
4600	$2^3 \cdot 5^2 \cdot 23$	$43 \cdot 107$	$2 \cdot 3 \cdot 13 \cdot 59$	4603	$2^2 \cdot 1151$	$3 \cdot 5 \cdot 307$	$2 \cdot 7^2 \cdot 47$	$17 \cdot 271$	$2^9 \cdot 3^2$	$11 \cdot 419$
4610	$2 \cdot 5 \cdot 461$	$3 \cdot 29 \cdot 53$	$2^2 \cdot 1153$	$7 \cdot 659$	$2 \cdot 3 \cdot 769$	$5 \cdot 13 \cdot 71$	$2^3 \cdot 577$	$3^5 \cdot 19$	$2 \cdot 2309$	$31 \cdot 149$
4620	$2^2 \cdot 3 \cdot 5 \cdot 7 \cdot 11$	4621	$2 \cdot 2311$	$3 \cdot 23 \cdot 67$	$2^4 \cdot 17^2$	$5^3 \cdot 37$	$2 \cdot 3^2 \cdot 257$	$7 \cdot 661$	$2^2 \cdot 13 \cdot 89$	$3 \cdot 1543$
4630	$2 \cdot 5 \cdot 463$	$11 \cdot 421$	$2^3 \cdot 3 \cdot 193$	$41 \cdot 113$	$2 \cdot 7 \cdot 331$	$3^2 \cdot 5 \cdot 103$	$2^2 \cdot 19 \cdot 61$	4637	$2 \cdot 3 \cdot 773$	4639
4640	$2^5 \cdot 5 \cdot 29$	$3 \cdot 7 \cdot 13 \cdot 17$	$2 \cdot 11 \cdot 211$	4643	$2^2 \cdot 3^3 \cdot 43$	$5 \cdot 929$	$2 \cdot 23 \cdot 101$	$3 \cdot 1549$	$2^3 \cdot 7 \cdot 83$	4649
4650	$2 \cdot 3 \cdot 5^2 \cdot 31$	4651	$2^2 \cdot 1163$	$3^2 \cdot 11 \cdot 47$	$2 \cdot 13 \cdot 179$	$5 \cdot 7^2 \cdot 19$	$2^4 \cdot 3 \cdot 97$	4657	$2 \cdot 17 \cdot 137$	$3 \cdot 1553$
4660	$2^2 \cdot 5 \cdot 233$	$59 \cdot 79$	$2 \cdot 3^2 \cdot 7 \cdot 37$	4663	$2^3 \cdot 11 \cdot 53$	$3 \cdot 5 \cdot 311$	$2 \cdot 2333$	$13 \cdot 359$	$2^2 \cdot 3 \cdot 389$	$7 \cdot 23 \cdot 29$
4670	$2 \cdot 5 \cdot 467$	$3^3 \cdot 173$	$2^6 \cdot 73$	4673	$2 \cdot 3 \cdot 19 \cdot 41$	$5^2 \cdot 11 \cdot 17$	$2^2 \cdot 7 \cdot 167$	$3 \cdot 1559$	$2 \cdot 2339$	4679
4680	$2^3 \cdot 3^2 \cdot 5 \cdot 13$	$31 \cdot 151$	$2 \cdot 2341$	$3 \cdot 7 \cdot 223$	$2^2 \cdot 1171$	$5 \cdot 937$	$2 \cdot 3 \cdot 11 \cdot 71$	$43 \cdot 109$	$2^4 \cdot 293$	$3^2 \cdot 521$
4690	$2 \cdot 5 \cdot 7 \cdot 67$	4691	$2^2 \cdot 3 \cdot 17 \cdot 23$	$13 \cdot 19^2$	$2 \cdot 2347$	$3 \cdot 5 \cdot 313$	$2^3 \cdot 587$	$7 \cdot 11 \cdot 61$	$2 \cdot 3^4 \cdot 29$	$37 \cdot 127$
4700	$2^2 \cdot 5^2 \cdot 47$	$3 \cdot 1567$	$2 \cdot 2351$	4703	$2^5 \cdot 3 \cdot 7^2$	$5 \cdot 941$	$2 \cdot 13 \cdot 181$	$3^2 \cdot 523$	$2^2 \cdot 11 \cdot 107$	$17 \cdot 277$
4710	$2 \cdot 3 \cdot 5 \cdot 157$	$7 \cdot 673$	$2^3 \cdot 19 \cdot 31$	$3 \cdot 1571$	$2 \cdot 2357$	$5 \cdot 23 \cdot 41$	$2^2 \cdot 3^2 \cdot 131$	$53 \cdot 89$	$2 \cdot 7 \cdot 337$	$3 \cdot 11^2 \cdot 13$
4720	$2^4 \cdot 5 \cdot 59$	4721	$2 \cdot 3 \cdot 787$	4723	$2^2 \cdot 1181$	$3^3 \cdot 5^2 \cdot 7$	$2 \cdot 17 \cdot 139$	$29 \cdot 163$	$2^3 \cdot 3 \cdot 197$	4729
4730	$2 \cdot 5 \cdot 11 \cdot 43$	$3 \cdot 19 \cdot 83$	$2^2 \cdot 7 \cdot 13^2$	4733	$2 \cdot 3^2 \cdot 263$	$5 \cdot 947$	$2^7 \cdot 37$	$3 \cdot 1579$	$2 \cdot 23 \cdot 103$	$7 \cdot 677$
4740	$2^2 \cdot 3 \cdot 5 \cdot 79$	$11 \cdot 431$	$2 \cdot 2371$	$3^2 \cdot 17 \cdot 31$	$2^3 \cdot 593$	$5 \cdot 13 \cdot 73$	$2 \cdot 3 \cdot 7 \cdot 113$	$47 \cdot 101$	$2^2 \cdot 1187$	$3 \cdot 1583$
4750	$2 \cdot 5^3 \cdot 19$	4751	$2^4 \cdot 3^3 \cdot 11$	$7^2 \cdot 97$	$2 \cdot 2377$	$3 \cdot 5 \cdot 317$	$2^2 \cdot 29 \cdot 41$	$67 \cdot 71$	$2 \cdot 3 \cdot 13 \cdot 61$	4759
4760	$2^3 \cdot 5 \cdot 7 \cdot 17$	$3^2 \cdot 23^2$	$2 \cdot 2381$	$11 \cdot 433$	$2^2 \cdot 3 \cdot 397$	$5 \cdot 953$	$2 \cdot 2383$	$3 \cdot 7 \cdot 227$	$2^5 \cdot 149$	$19 \cdot 251$
4770	$2 \cdot 3^2 \cdot 5 \cdot 53$	$13 \cdot 367$	$2^2 \cdot 1193$	$3 \cdot 37 \cdot 43$	$2 \cdot 7 \cdot 11 \cdot 31$	$5^2 \cdot 191$	$2^3 \cdot 3 \cdot 199$	$17 \cdot 281$	$2 \cdot 2389$	$3^4 \cdot 59$
4780	$2^2 \cdot 5 \cdot 239$	$7 \cdot 683$	$2 \cdot 3 \cdot 797$	4783	$2^4 \cdot 13 \cdot 23$	$3 \cdot 5 \cdot 11 \cdot 29$	$2 \cdot 2393$	4787	$2^2 \cdot 3^2 \cdot 7 \cdot 19$	4789
4790	$2 \cdot 5 \cdot 479$	$3 \cdot 1597$	$2^3 \cdot 599$	4793	$2 \cdot 3 \cdot 17 \cdot 47$	$5 \cdot 7 \cdot 137$	$2^2 \cdot 11 \cdot 109$	$3^2 \cdot 13 \cdot 41$	$2 \cdot 2399$	4799
4800	$2^6 \cdot 3 \cdot 5^2$	4801	$2 \cdot 7^4$	$3 \cdot 1601$	$2^2 \cdot 1201$	$5 \cdot 31^2$	$2 \cdot 3^3 \cdot 89$	$11 \cdot 19 \cdot 23$	$2^3 \cdot 601$	$3 \cdot 7 \cdot 229$
4810	$2 \cdot 5 \cdot 13 \cdot 37$	$17 \cdot 283$	$2^2 \cdot 3 \cdot 401$	4813	$2 \cdot 29 \cdot 83$	$3^2 \cdot 5 \cdot 107$	$2^4 \cdot 7 \cdot 43$	4817	$2 \cdot 3 \cdot 11 \cdot 73$	$61 \cdot 79$
4820	$2^2 \cdot 5 \cdot 241$	$3 \cdot 1607$	$2 \cdot 2411$	$7 \cdot 13 \cdot 53$	$2^3 \cdot 3^2 \cdot 67$	$5^2 \cdot 193$	$2 \cdot 19 \cdot 127$	$3 \cdot 1609$	$2^2 \cdot 17 \cdot 71$	$11 \cdot 439$
4830	$2 \cdot 3 \cdot 5 \cdot 7 \cdot 23$	4831	$2^5 \cdot 151$	$3^3 \cdot 179$	$2 \cdot 2417$	$5 \cdot 967$	$2^2 \cdot 3 \cdot 13 \cdot 31$	$7 \cdot 691$	$2 \cdot 41 \cdot 59$	$3 \cdot 1613$
4840	$2^3 \cdot 5 \cdot 11^2$	$47 \cdot 103$	$2 \cdot 3^2 \cdot 269$	$29 \cdot 167$	$2^2 \cdot 7 \cdot 173$	$3 \cdot 5 \cdot 17 \cdot 19$	$2 \cdot 2423$	$37 \cdot 131$	$2^4 \cdot 3 \cdot 101$	$13 \cdot 373$
4850	$2 \cdot 5^2 \cdot 97$	$3^2 \cdot 7^2 \cdot 11$	$2^2 \cdot 1213$	$23 \cdot 211$	$2 \cdot 3 \cdot 809$	$5 \cdot 971$	$2^3 \cdot 607$	$3 \cdot 1619$	$2 \cdot 7 \cdot 347$	$43 \cdot 113$
4860	$2^2 \cdot 3^5 \cdot 5$	4861	$2 \cdot 11 \cdot 13 \cdot 17$	$3 \cdot 1621$	$2^8 \cdot 19$	$5 \cdot 7 \cdot 139$	$2 \cdot 3 \cdot 811$	$31 \cdot 157$	$2^2 \cdot 1217$	$3^2 \cdot 541$
4870	$2 \cdot 5 \cdot 487$	4871	$2^3 \cdot 3 \cdot 7 \cdot 29$	$11 \cdot 443$	$2 \cdot 2437$	$3 \cdot 5^3 \cdot 13$	$2^2 \cdot 23 \cdot 53$	4877	$2 \cdot 3^2 \cdot 271$	$7 \cdot 17 \cdot 41$
4880	$2^4 \cdot 5 \cdot 61$	$3 \cdot 1627$	$2 \cdot 2441$	$19 \cdot 257$	$2^2 \cdot 3 \cdot 11 \cdot 37$	$5 \cdot 977$	$2 \cdot 7 \cdot 349$	$3^3 \cdot 181$	$2^3 \cdot 13 \cdot 47$	4889
4890	$2 \cdot 3 \cdot 5 \cdot 163$	$67 \cdot 73$	$2^2 \cdot 1223$	$3 \cdot 7 \cdot 233$	$2 \cdot 2447$	$5 \cdot 11 \cdot 89$	$2^5 \cdot 3^2 \cdot 17$	$59 \cdot 83$	$2 \cdot 31 \cdot 79$	$3 \cdot 23 \cdot 71$
4900	$2^2 \cdot 5^2 \cdot 7^2$	$13^2 \cdot 29$	$2 \cdot 3 \cdot 19 \cdot 43$	4903	$2^3 \cdot 613$	$3^2 \cdot 5 \cdot 109$	$2 \cdot 11 \cdot 223$	$7 \cdot 701$	$2^2 \cdot 3 \cdot 409$	4909
4910	$2 \cdot 5 \cdot 491$	$3 \cdot 1637$	$2^4 \cdot 307$	17^3	$2 \cdot 3^3 \cdot 7 \cdot 13$	$5 \cdot 983$	$2^2 \cdot 1229$	$3 \cdot 11 \cdot 149$	$2 \cdot 2459$	4919
4920	$2^3 \cdot 3 \cdot 5 \cdot 41$	$7 \cdot 19 \cdot 37$	$2 \cdot 23 \cdot 107$	$3^2 \cdot 547$	$2^2 \cdot 1231$	$5^2 \cdot 197$	$2 \cdot 3 \cdot 821$	$13 \cdot 379$	$2^6 \cdot 7 \cdot 11$	$3 \cdot 31 \cdot 53$
4930	$2 \cdot 5 \cdot 17 \cdot 29$	4931	$2^2 \cdot 3^2 \cdot 137$	4933	$2 \cdot 2467$	$3 \cdot 5 \cdot 7 \cdot 47$	$2^3 \cdot 617$	4937	$2 \cdot 3 \cdot 823$	$11 \cdot 449$
4940	$2^2 \cdot 5 \cdot 13 \cdot 19$	$3^4 \cdot 61$	$2 \cdot 7 \cdot 353$	4943	$2^4 \cdot 3 \cdot 103$	$5 \cdot 23 \cdot 43$	$2 \cdot 2473$	$3 \cdot 17 \cdot 97$	$2^2 \cdot 1237$	$7^2 \cdot 101$
4950	$2 \cdot 3^2 \cdot 5^2 \cdot 11$	4951	$2^3 \cdot 619$	$3 \cdot 13 \cdot 127$	$2 \cdot 2477$	$5 \cdot 991$	$2^2 \cdot 3 \cdot 7 \cdot 59$	4957	$2 \cdot 37 \cdot 67$	$3^2 \cdot 19 \cdot 29$
4960	$2^5 \cdot 5 \cdot 31$	$11^2 \cdot 41$	$2 \cdot 3 \cdot 827$	$7 \cdot 709$	$2^2 \cdot 17 \cdot 73$	$3 \cdot 5 \cdot 331$	$2 \cdot 13 \cdot 191$	4967	$2^3 \cdot 3^3 \cdot 23$	4969
4970	$2 \cdot 5 \cdot 7 \cdot 71$	$3 \cdot 1657$	$2^2 \cdot 11 \cdot 113$	4973	$2 \cdot 3 \cdot 829$	$5^2 \cdot 199$	$2^4 \cdot 311$	$3^2 \cdot 7 \cdot 79$	$2 \cdot 19 \cdot 131$	$13 \cdot 383$
4980	$2^2 \cdot 3 \cdot 5 \cdot 83$	$17 \cdot 293$	$2 \cdot 47 \cdot 53$	$3 \cdot 11 \cdot 151$	$2^3 \cdot 7 \cdot 89$	$5 \cdot 997$	$2 \cdot 3^2 \cdot 277$	4987	$2^2 \cdot 29 \cdot 43$	$3 \cdot 1663$
4990	$2 \cdot 5 \cdot 499$	$7 \cdot 23 \cdot 31$	$2^7 \cdot 3 \cdot 13$	4993	$2 \cdot 11 \cdot 227$	$3^3 \cdot 5 \cdot 37$	$2^2 \cdot 1249$	$19 \cdot 263$	$2 \cdot 3 \cdot 7^2 \cdot 17$	4999

Table 24.7   **Factorizations**

N	9	8	7	6	5	4	3	2	1	0	N
500	**5009**	$2^4\cdot313$	$3\cdot1669$	$2\cdot2503$	$5\cdot7\cdot11\cdot13$	$2^2\cdot3\cdot139$	**5003**	$2\cdot41\cdot61$	$3\cdot1667$	$2^3\cdot5^4$	500
501	$3\cdot7\cdot239$	$2\cdot13\cdot193$	$29\cdot173$	$2^3\cdot3\cdot11\cdot19$	$5\cdot17\cdot59$	$2\cdot23\cdot109$	$3^2\cdot557$	$2^2\cdot7\cdot179$	**5011**	$2\cdot3\cdot5\cdot167$	501
502	$47\cdot107$	$2^2\cdot3\cdot419$	$11\cdot457$	$2\cdot7\cdot359$	$3\cdot5^2\cdot67$	$2^5\cdot157$	**5023**	$2\cdot3^4\cdot31$	**5021**	$2^2\cdot5\cdot251$	502
503	**5039**	$2\cdot11\cdot229$	$3\cdot23\cdot73$	$2^2\cdot1259$	$5\cdot19\cdot53$	$2\cdot3\cdot839$	$7\cdot719$	$2^3\cdot17\cdot37$	$3^2\cdot13\cdot43$	$2\cdot5\cdot503$	503
504	$3^3\cdot11\cdot17$	$2^3\cdot631$	$7^2\cdot103$	$2\cdot3\cdot29^2$	$5\cdot1009$	$2^2\cdot13\cdot97$	$3\cdot41^2$	$2\cdot2521$	71^2	$2^4\cdot3^2\cdot5\cdot7$	504
505	**5059**	$2\cdot3^2\cdot281$	$13\cdot389$	$2^6\cdot79$	$3\cdot5\cdot337$	$2\cdot7\cdot19^2$	$31\cdot163$	$2^2\cdot3\cdot421$	**5051**	$2\cdot5^2\cdot101$	505
506	$37\cdot137$	$2^2\cdot7\cdot181$	$3^2\cdot563$	$2\cdot17\cdot149$	$5\cdot1013$	$2^3\cdot3\cdot211$	$61\cdot83$	$2\cdot2531$	$3\cdot7\cdot241$	$2^2\cdot5\cdot11\cdot23$	506
507	$3\cdot1693$	$2\cdot2539$	**5077**	$2^2\cdot3^3\cdot47$	$5^2\cdot7\cdot29$	$2\cdot43\cdot59$	$3\cdot19\cdot89$	$2^4\cdot317$	$11\cdot461$	$2\cdot3\cdot5\cdot13^2$	507
508	$7\cdot727$	$2^5\cdot3\cdot53$	**5087**	$2\cdot2543$	$3^2\cdot5\cdot113$	$2^2\cdot31\cdot41$	$13\cdot17\cdot23$	$2\cdot3\cdot7\cdot11^2$	**5081**	$2^3\cdot5\cdot127$	508
509	**5099**	$2\cdot2549$	$3\cdot1699$	$2^3\cdot7^2\cdot13$	$5\cdot1019$	$2\cdot3^2\cdot283$	$11\cdot463$	$2^2\cdot19\cdot67$	$3\cdot1697$	$2\cdot5\cdot509$	509
510	$3\cdot13\cdot131$	$2^2\cdot1277$	**5107**	$2\cdot3\cdot23\cdot37$	$5\cdot1021$	$2^4\cdot11\cdot29$	$3^6\cdot7$	$2\cdot2551$	**5101**	$2^2\cdot3\cdot5^2\cdot17$	510
511	**5119**	$2\cdot3\cdot853$	$7\cdot17\cdot43$	$2^2\cdot1279$	$3\cdot5\cdot11\cdot31$	$2\cdot2557$	**5113**	$2^3\cdot3^2\cdot71$	$19\cdot269$	$2\cdot5\cdot7\cdot73$	511
512	$23\cdot223$	$2^3\cdot641$	$3\cdot1709$	$2\cdot11\cdot233$	$5^3\cdot41$	$2^2\cdot3\cdot7\cdot61$	$47\cdot109$	$2\cdot13\cdot197$	$3^2\cdot569$	$2^{10}\cdot5$	512
513	$3^2\cdot571$	$2\cdot7\cdot367$	$11\cdot467$	$2^4\cdot3\cdot107$	$5\cdot13\cdot79$	$2\cdot17\cdot151$	$3\cdot29\cdot59$	$2^2\cdot1283$	$7\cdot733$	$2\cdot3^3\cdot5\cdot19$	513
514	$19\cdot271$	$2^2\cdot3^2\cdot11\cdot13$	**5147**	$2\cdot31\cdot83$	$3\cdot5\cdot7^3$	$2^3\cdot643$	$37\cdot139$	$2\cdot3\cdot857$	$53\cdot97$	$2^2\cdot5\cdot257$	514
515	$7\cdot11\cdot67$	$2\cdot2579$	$3^3\cdot191$	$2^2\cdot1289$	$5\cdot1031$	$2\cdot3\cdot859$	**5153**	$2^5\cdot7\cdot23$	$3\cdot17\cdot101$	$2\cdot5^2\cdot103$	515
516	$3\cdot1723$	$2^4\cdot17\cdot19$	**5167**	$2\cdot3^2\cdot7\cdot41$	$5\cdot1033$	$2^2\cdot1291$	$3\cdot1721$	$2\cdot29\cdot89$	$13\cdot397$	$2^3\cdot3\cdot5\cdot43$	516
517	**5179**	$2\cdot3\cdot863$	$31\cdot167$	$2^3\cdot647$	$3^2\cdot5^2\cdot23$	$2\cdot13\cdot199$	$7\cdot739$	$2^2\cdot3\cdot431$	**5171**	$2\cdot5\cdot11\cdot47$	517
518	**5189**	$2^2\cdot1297$	$3\cdot7\cdot13\cdot19$	$2\cdot2593$	$5\cdot17\cdot61$	$2^6\cdot3^4$	$71\cdot73$	$2\cdot2591$	$3\cdot11\cdot157$	$2^2\cdot5\cdot7\cdot37$	518
519	$3\cdot1733$	$2\cdot23\cdot113$	**5197**	$2^2\cdot3\cdot433$	$5\cdot1039$	$2\cdot7^2\cdot53$	$3^2\cdot577$	$2^3\cdot11\cdot59$	$29\cdot179$	$2\cdot3\cdot5\cdot173$	519
520	**5209**	$2^3\cdot3\cdot7\cdot31$	$41\cdot127$	$2\cdot19\cdot137$	$3\cdot5\cdot347$	$2^2\cdot1301$	$11^2\cdot43$	$2\cdot3^2\cdot17^2$	$7\cdot743$	$2^4\cdot5^2\cdot13$	520
521	$17\cdot307$	$2\cdot2609$	$3\cdot37\cdot47$	$2^5\cdot163$	$5\cdot7\cdot149$	$2\cdot3\cdot11\cdot79$	$13\cdot401$	$2^2\cdot1303$	$3^3\cdot193$	$2\cdot5\cdot521$	521
522	$3^2\cdot7\cdot83$	$2^2\cdot1307$	**5227**	$2\cdot3\cdot13\cdot67$	$5^2\cdot11\cdot19$	$2^3\cdot653$	$3\cdot1741$	$2\cdot7\cdot373$	$23\cdot227$	$2^2\cdot3^2\cdot5\cdot29$	522
523	$13^2\cdot31$	$2\cdot3^3\cdot97$	**5237**	$2^2\cdot7\cdot11\cdot17$	$3\cdot5\cdot349$	$2\cdot2617$	**5233**	$2^4\cdot3\cdot109$	**5231**	$2\cdot5\cdot523$	523
524	$29\cdot181$	$2^7\cdot41$	$3^2\cdot11\cdot53$	$2\cdot43\cdot61$	$5\cdot1049$	$2^2\cdot3\cdot19\cdot23$	$7^2\cdot107$	$2\cdot2621$	$3\cdot1747$	$2^3\cdot5\cdot131$	524
525	$3\cdot1753$	$2\cdot11\cdot239$	$7\cdot751$	$2^3\cdot3^2\cdot73$	$5\cdot1051$	$2\cdot37\cdot71$	$3\cdot17\cdot103$	$2^2\cdot13\cdot101$	$59\cdot89$	$2\cdot3\cdot5^3\cdot7$	525
526	$11\cdot479$	$2^2\cdot3\cdot439$	$23\cdot229$	$2\cdot2633$	$3^4\cdot5\cdot13$	$2^4\cdot7\cdot47$	$19\cdot277$	$2\cdot3\cdot877$	**5261**	$2^2\cdot5\cdot263$	526
527	**5279**	$2\cdot7\cdot13\cdot29$	$3\cdot1759$	$2^2\cdot1319$	$5^2\cdot211$	$2\cdot3^2\cdot293$	**5273**	$2^3\cdot659$	$3\cdot7\cdot251$	$2\cdot5\cdot17\cdot31$	527
528	$3\cdot41\cdot43$	$2^3\cdot661$	$17\cdot311$	$2\cdot3\cdot881$	$5\cdot7\cdot151$	$2^2\cdot1321$	$3^2\cdot587$	$2\cdot19\cdot139$	**5281**	$2^5\cdot3\cdot5\cdot11$	528
529	$7\cdot757$	$2\cdot3\cdot883$	**5297**	$2^4\cdot331$	$3\cdot5\cdot353$	$2\cdot2647$	$67\cdot79$	$2^2\cdot3^3\cdot7^2$	$11\cdot13\cdot37$	$2\cdot5\cdot23^2$	529
530	**5309**	$2^2\cdot1327$	$3\cdot29\cdot61$	$2\cdot7\cdot379$	$5\cdot1061$	$2^3\cdot3\cdot13\cdot17$	**5303**	$2\cdot11\cdot241$	$3^2\cdot19\cdot31$	$2^2\cdot5^2\cdot53$	530
531	$3^3\cdot197$	$2\cdot2659$	$13\cdot409$	$2^2\cdot3\cdot443$	$5\cdot1063$	$2\cdot2657$	$3\cdot7\cdot11\cdot23$	$2^6\cdot83$	$47\cdot113$	$2\cdot3^2\cdot5\cdot59$	531
532	73^2	$2^4\cdot3^2\cdot37$	$7\cdot761$	$2\cdot2663$	$3\cdot5^2\cdot71$	$2^2\cdot11^3$	**5323**	$2\cdot3\cdot887$	$17\cdot313$	$2^3\cdot5\cdot7\cdot19$	532
533	$19\cdot281$	$2\cdot17\cdot157$	$3^2\cdot593$	$2^3\cdot23\cdot29$	$5\cdot11\cdot97$	$2\cdot3\cdot7\cdot127$	**5333**	$2^2\cdot31\cdot43$	$3\cdot1777$	$2\cdot5\cdot13\cdot41$	533
534	$3\cdot1783$	$2^2\cdot7\cdot191$	**5347**	$2\cdot3^5\cdot11$	$5\cdot1069$	$2^5\cdot167$	$3\cdot13\cdot137$	$2\cdot2671$	$7^2\cdot109$	$2^2\cdot3\cdot5\cdot89$	534
535	$23\cdot233$	$2\cdot3\cdot19\cdot47$	$11\cdot487$	$2^2\cdot13\cdot103$	$3^2\cdot5\cdot7\cdot17$	$2\cdot2677$	$53\cdot101$	$2^3\cdot3\cdot223$	**5351**	$2\cdot5^2\cdot107$	535
536	$7\cdot13\cdot59$	$2^3\cdot11\cdot61$	$3\cdot1789$	$2\cdot2683$	$5\cdot29\cdot37$	$2^2\cdot3^2\cdot149$	$31\cdot173$	$2\cdot7\cdot383$	$3\cdot1787$	$2^4\cdot5\cdot67$	536
537	$3\cdot11\cdot163$	$2\cdot2689$	$19\cdot283$	$2^8\cdot3\cdot7$	$5^3\cdot43$	$2\cdot2687$	$3^3\cdot199$	$2^2\cdot17\cdot79$	$41\cdot131$	$2\cdot3\cdot5\cdot179$	537
538	$17\cdot317$	$2^2\cdot3\cdot449$	**5387**	$2\cdot2693$	$3\cdot5\cdot359$	$2^3\cdot673$	$7\cdot769$	$2\cdot3^2\cdot13\cdot23$	**5381**	$2^2\cdot5\cdot269$	538
539	**5399**	$2\cdot2699$	$3\cdot7\cdot257$	$2^2\cdot19\cdot71$	$5\cdot13\cdot83$	$2\cdot3\cdot29\cdot31$	**5393**	$2^4\cdot337$	$3^2\cdot599$	$2\cdot5\cdot7^2\cdot11$	539
540	$3^2\cdot601$	$2^5\cdot13^2$	**5407**	$2\cdot3\cdot17\cdot53$	$5\cdot23\cdot47$	$2^2\cdot7\cdot193$	$3\cdot1801$	$2\cdot37\cdot73$	$11\cdot491$	$2^3\cdot3^3\cdot5^2$	540
541	**5419**	$2\cdot3^2\cdot7\cdot43$	**5417**	$2^3\cdot677$	$3\cdot5\cdot19^2$	$2\cdot2707$	**5413**	$2^2\cdot3\cdot11\cdot41$	$7\cdot773$	$2\cdot5\cdot541$	541
542	$61\cdot89$	$2^2\cdot23\cdot59$	$3^4\cdot67$	$2\cdot2713$	$5^2\cdot7\cdot31$	$2^4\cdot3\cdot113$	$11\cdot17\cdot29$	$2\cdot2711$	$3\cdot13\cdot139$	$2^2\cdot5\cdot271$	542
543	$3\cdot7^2\cdot37$	$2\cdot2719$	**5437**	$2^2\cdot3^2\cdot151$	$5\cdot1087$	$2\cdot11\cdot13\cdot19$	$3\cdot1811$	$2^3\cdot7\cdot97$	**5431**	$2\cdot3\cdot5\cdot181$	543
544	**5449**	$2^3\cdot3\cdot227$	$13\cdot419$	$2\cdot7\cdot389$	$3^2\cdot5\cdot11^2$	$2^2\cdot1361$	**5443**	$2\cdot3\cdot907$	**5441**	$2^6\cdot5\cdot17$	544
545	$53\cdot103$	$2\cdot2729$	$3\cdot17\cdot107$	$2^4\cdot11\cdot31$	$5\cdot1091$	$2\cdot3^3\cdot101$	$7\cdot19\cdot41$	$2^2\cdot29\cdot47$	$3\cdot23\cdot79$	$2\cdot5^2\cdot109$	545
546	$3\cdot1823$	$2^2\cdot1367$	$7\cdot11\cdot71$	$2\cdot3\cdot911$	$5\cdot1093$	$2^3\cdot683$	$3^2\cdot607$	$2\cdot2731$	$43\cdot127$	$2^2\cdot3\cdot5\cdot7\cdot13$	546
547	**5479**	$2\cdot3\cdot11\cdot83$	**5477**	$2^2\cdot37^2$	$3\cdot5^2\cdot73$	$2\cdot7\cdot17\cdot23$	$13\cdot421$	$2^5\cdot3^2\cdot19$	**5471**	$2\cdot5\cdot547$	547
548	$11\cdot499$	$2^4\cdot7^3$	$3\cdot31\cdot59$	$2\cdot13\cdot211$	$5\cdot1097$	$2^2\cdot3\cdot457$	**5483**	$2\cdot2741$	$3^3\cdot7\cdot29$	$2^3\cdot5\cdot137$	548
549	$3^2\cdot13\cdot47$	$2\cdot2749$	$23\cdot239$	$2^3\cdot3\cdot229$	$5\cdot7\cdot157$	$2\cdot41\cdot67$	$3\cdot1831$	$2^2\cdot1373$	$17^2\cdot19$	$2\cdot3^2\cdot5\cdot61$	549

Factorizations

Table 24.7

5500 — 5999

Each entry gives the number n and its prime factorization. Bold numbers are primes.

Numbers ending in 9

550–554	555–559	560–564	565–569	570–574	575–579	580–584	585–589	590–594	595–599
5509 $7\cdot787$	5559 $3\cdot17\cdot109$	5609 $71\cdot79$	5659 **5659**	5709 $3\cdot11\cdot173$	5759 $13\cdot443$	5809 $37\cdot157$	5859 $3^3\cdot7\cdot31$	5909 $19\cdot311$	5959 $59\cdot101$
5519 **5519**	5569 **5569**	5619 $3\cdot1873$	5669 **5669**	5719 $7\cdot19\cdot43$	5769 $3^2\cdot641$	5819 $11\cdot23^2$	5869 **5869**	5919 $3\cdot1973$	5969 $47\cdot127$
5529 $3\cdot19\cdot97$	5579 $7\cdot797$	5629 $13\cdot433$	5679 $3^2\cdot631$	5729 $17\cdot337$	5779 **5779**	5829 $3\cdot29\cdot67$	5879 **5879**	5929 $7^2\cdot11^2$	5979 $3\cdot1993$
5539 $29\cdot191$	5589 $3^5\cdot23$	5639 **5639**	5689 **5689**	5739 $3\cdot1913$	5789 **5789**	5839 **5839**	5889 $3\cdot13\cdot151$	5939 **5939**	5989 $53\cdot113$
5549 $31\cdot179$	5599 $11\cdot509$	5649 $3\cdot7\cdot269$	5699 $41\cdot139$	5749 **5749**	5799 $3\cdot1933$	5849 **5849**	5899 $17\cdot347$	5949 $3^2\cdot661$	5999 $7\cdot857$

Numbers ending in 8

550–554	555–559	560–564	565–569	570–574	575–579	580–584	585–589	590–594	595–599
5508 $2^2\cdot3^4\cdot17$	5558 $2\cdot7\cdot397$	5608 $2^3\cdot701$	5658 $2\cdot3\cdot23\cdot41$	5708 $2^2\cdot1427$	5758 $2\cdot2879$	5808 $2^4\cdot3\cdot11^2$	5858 $2\cdot29\cdot101$	5908 $2^2\cdot7\cdot211$	5958 $2\cdot3^2\cdot331$
5518 $2\cdot31\cdot89$	5568 $2^6\cdot3\cdot29$	5618 $2\cdot53^2$	5668 $2^2\cdot13\cdot109$	5718 $2\cdot3\cdot953$	5768 $2^3\cdot7\cdot103$	5818 $2\cdot2909$	5868 $2^2\cdot3^2\cdot163$	5918 $2\cdot11\cdot269$	5968 $2^4\cdot373$
5528 $2^3\cdot691$	5578 $2\cdot2789$	5628 $2^2\cdot3\cdot7\cdot67$	5678 $2\cdot17\cdot167$	5728 $2^5\cdot179$	5778 $2\cdot3^3\cdot107$	5828 $2^2\cdot31\cdot47$	5878 $2\cdot2939$	5928 $2^3\cdot3\cdot13\cdot19$	5978 $2\cdot7^2\cdot61$
5538 $2\cdot3\cdot13\cdot71$	5588 $2^2\cdot11\cdot127$	5638 $2\cdot2819$	5688 $2^3\cdot3^2\cdot79$	5738 $2\cdot19\cdot151$	5788 $2^2\cdot1447$	5838 $2\cdot3\cdot7\cdot139$	5888 $2^8\cdot23$	5938 $2\cdot2969$	5988 $2^2\cdot3\cdot499$
5548 $2^2\cdot19\cdot73$	5598 $2\cdot3^2\cdot311$	5648 $2^4\cdot353$	5698 $2\cdot7\cdot11\cdot37$	5748 $2^2\cdot3\cdot479$	5798 $2\cdot13\cdot223$	5848 $2^3\cdot17\cdot43$	5898 $2\cdot3\cdot983$	5948 $2^2\cdot1487$	5998 $2\cdot2999$

Numbers ending in 7

550–554	555–559	560–564	565–569	570–574	575–579	580–584	585–589	590–594	595–599
5507 **5507**	5557 **5557**	5607 $3^2\cdot7\cdot89$	5657 **5657**	5707 $13\cdot439$	5757 $3\cdot19\cdot101$	5807 **5807**	5857 **5857**	5907 $3\cdot11\cdot179$	5957 $7\cdot23\cdot37$
5517 $3^2\cdot613$	5567 $19\cdot293$	5617 $41\cdot137$	5667 $3\cdot1889$	5717 **5717**	5767 $73\cdot79$	5817 $3\cdot7\cdot277$	5867 **5867**	5917 $61\cdot97$	5967 $3^3\cdot13\cdot17$
5527 **5527**	5577 $3\cdot11\cdot13^2$	5627 $17\cdot331$	5677 $7\cdot811$	5727 $3\cdot23\cdot83$	5777 $53\cdot109$	5827 **5827**	5877 $3^2\cdot653$	5927 **5927**	5977 $43\cdot139$
5537 $7^2\cdot113$	5587 $37\cdot151$	5637 $3\cdot1879$	5687 $11^2\cdot47$	5737 **5737**	5787 $3^2\cdot643$	5837 $13\cdot449$	5887 $7\cdot29^2$	5937 $3\cdot1979$	5987 **5987**
5547 $3\cdot43^2$	5597 $29\cdot193$	5647 **5647**	5697 $3^3\cdot211$	5747 $7\cdot821$	5797 $11\cdot17\cdot31$	5847 $3\cdot1949$	5897 **5897**	5947 $19\cdot313$	5997 $3\cdot1999$

Numbers ending in 6

550–554	555–559	560–564	565–569	570–574	575–579	580–584	585–589	590–594	595–599
5506 $2\cdot2753$	5556 $2^2\cdot3\cdot463$	5606 $2\cdot2803$	5656 $2^3\cdot7\cdot101$	5706 $2\cdot3^2\cdot317$	5756 $2^2\cdot1439$	5806 $2\cdot2903$	5856 $2^5\cdot3\cdot61$	5906 $2\cdot2953$	5956 $2^2\cdot1489$
5516 $2^2\cdot7\cdot197$	5566 $2\cdot11^2\cdot23$	5616 $2^4\cdot3^3\cdot13$	5666 $2\cdot2833$	5716 $2^2\cdot1429$	5766 $2\cdot3\cdot31^2$	5816 $2^3\cdot727$	5866 $2\cdot7\cdot419$	5916 $2^2\cdot3\cdot17\cdot29$	5966 $2\cdot19\cdot157$
5526 $2\cdot3^2\cdot307$	5576 $2^3\cdot17\cdot41$	5626 $2\cdot29\cdot97$	5676 $2^2\cdot3\cdot11\cdot43$	5726 $2\cdot7\cdot409$	5776 $2^4\cdot19^2$	5826 $2\cdot3\cdot971$	5876 $2^2\cdot13\cdot113$	5926 $2\cdot2963$	5976 $2^3\cdot3^2\cdot83$
5536 $2^5\cdot173$	5586 $2\cdot3\cdot7^2\cdot19$	5636 $2^2\cdot1409$	5686 $2\cdot2843$	5736 $2^3\cdot3\cdot239$	5786 $2\cdot11\cdot263$	5836 $2^2\cdot1459$	5886 $2\cdot3^3\cdot109$	5936 $2^4\cdot7\cdot53$	5986 $2\cdot41\cdot73$
5546 $2\cdot47\cdot59$	5596 $2^2\cdot1399$	5646 $2\cdot3\cdot941$	5696 $2^6\cdot89$	5746 $2\cdot13^2\cdot17$	5796 $2^2\cdot3^2\cdot7\cdot23$	5846 $2\cdot37\cdot79$	5896 $2^3\cdot11\cdot67$	5946 $2\cdot3\cdot991$	5996 $2^2\cdot1499$

Numbers ending in 5

550–554	555–559	560–564	565–569	570–574	575–579	580–584	585–589	590–594	595–599
5505 $3\cdot5\cdot367$	5555 $5\cdot11\cdot101$	5605 $5\cdot19\cdot59$	5655 $3\cdot5\cdot13\cdot29$	5705 $5\cdot7\cdot163$	5755 $5\cdot1151$	5805 $3^3\cdot5\cdot43$	5855 $5\cdot1171$	5905 $5\cdot1181$	5955 $3\cdot5\cdot397$
5515 $5\cdot1103$	5565 $3\cdot5\cdot7\cdot53$	5615 $5\cdot1123$	5665 $5\cdot11\cdot103$	5715 $3^2\cdot5\cdot127$	5765 $5\cdot1153$	5815 $5\cdot1163$	5865 $3\cdot5\cdot17\cdot23$	5915 $5\cdot7\cdot13^2$	5965 $5\cdot1193$
5525 $5^2\cdot13\cdot17$	5575 $5^2\cdot223$	5625 $3^2\cdot5^4$	5675 $5^2\cdot227$	5725 $5^2\cdot229$	5775 $3\cdot5^2\cdot7\cdot11$	5825 $5^2\cdot233$	5875 $5^3\cdot47$	5925 $3\cdot5^2\cdot79$	5975 $5^2\cdot239$
5535 $3^3\cdot5\cdot41$	5585 $5\cdot1117$	5635 $5\cdot7^2\cdot23$	5685 $3\cdot5\cdot379$	5735 $5\cdot31\cdot37$	5785 $5\cdot13\cdot89$	5835 $3\cdot5\cdot389$	5885 $5\cdot11\cdot107$	5935 $5\cdot1187$	5985 $3^2\cdot5\cdot7\cdot19$
5545 $5\cdot1109$	5595 $3\cdot5\cdot373$	5645 $5\cdot1129$	5695 $5\cdot17\cdot67$	5745 $3\cdot5\cdot383$	5795 $5\cdot19\cdot61$	5845 $5\cdot7\cdot167$	5895 $3^2\cdot5\cdot131$	5945 $5\cdot29\cdot41$	5995 $5\cdot11\cdot109$

Numbers ending in 4

550–554	555–559	560–564	565–569	570–574	575–579	580–584	585–589	590–594	595–599
5504 $2^7\cdot43$	5554 $2\cdot2777$	5604 $2^2\cdot3\cdot467$	5654 $2\cdot11\cdot257$	5704 $2^3\cdot23\cdot31$	5754 $2\cdot3\cdot7\cdot137$	5804 $2^2\cdot1451$	5854 $2\cdot2927$	5904 $2^4\cdot3^2\cdot41$	5954 $2\cdot13\cdot229$
5514 $2\cdot3\cdot919$	5564 $2^2\cdot13\cdot107$	5614 $2\cdot7\cdot401$	5664 $2^5\cdot3\cdot59$	5714 $2\cdot2857$	5764 $2^2\cdot11\cdot131$	5814 $2\cdot3^2\cdot17\cdot19$	5864 $2^3\cdot733$	5914 $2\cdot2957$	5964 $2^2\cdot3\cdot7\cdot71$
5524 $2^2\cdot1381$	5574 $2\cdot3\cdot929$	5624 $2^3\cdot19\cdot37$	5674 $2\cdot2837$	5724 $2^2\cdot3^3\cdot53$	5774 $2\cdot2887$	5824 $2^6\cdot7\cdot13$	5874 $2\cdot3\cdot11\cdot89$	5924 $2^2\cdot1481$	5974 $2\cdot29\cdot103$
5534 $2\cdot2767$	5584 $2^4\cdot349$	5634 $2\cdot3^2\cdot313$	5684 $2^2\cdot7^2\cdot29$	5734 $2\cdot47\cdot61$	5784 $2^3\cdot3\cdot241$	5834 $2\cdot2917$	5884 $2^2\cdot1471$	5934 $2\cdot3\cdot23\cdot43$	5984 $2^5\cdot11\cdot17$
5544 $2^3\cdot3^2\cdot7\cdot11$	5594 $2\cdot2797$	5644 $2^2\cdot17\cdot83$	5694 $2\cdot3\cdot13\cdot73$	5744 $2^4\cdot359$	5794 $2\cdot2897$	5844 $2^2\cdot3\cdot487$	5894 $2\cdot7\cdot421$	5944 $2^3\cdot743$	5994 $2\cdot3^4\cdot37$

Numbers ending in 3

550–554	555–559	560–564	565–569	570–574	575–579	580–584	585–589	590–594	595–599
5503 **5503**	5553 $3^2\cdot617$	5603 $13\cdot431$	5653 **5653**	5703 $3\cdot1901$	5753 $11\cdot523$	5803 $7\cdot829$	5853 $3\cdot1951$	5903 **5903**	5953 **5953**
5513 $37\cdot149$	5563 **5563**	5613 $3\cdot1871$	5663 $7\cdot809$	5713 $29\cdot197$	5763 $3\cdot17\cdot113$	5813 **5813**	5863 $11\cdot13\cdot41$	5913 $3^4\cdot73$	5963 $67\cdot89$
5523 $3\cdot7\cdot263$	5573 **5573**	5623 **5623**	5673 $3\cdot31\cdot61$	5723 $59\cdot97$	5773 $23\cdot251$	5823 $3^2\cdot647$	5873 $7\cdot839$	5923 **5923**	5973 $3\cdot11\cdot181$
5533 $11\cdot503$	5583 $3\cdot1861$	5633 $43\cdot131$	5683 **5683**	5733 $3^2\cdot7^2\cdot13$	5783 **5783**	5833 $19\cdot307$	5883 $3\cdot37\cdot53$	5933 $17\cdot349$	5983 $31\cdot193$
5543 $23\cdot241$	5593 $7\cdot17\cdot47$	5643 $3^3\cdot11\cdot19$	5693 **5693**	5743 **5743**	5793 $3\cdot1931$	5843 **5843**	5893 $71\cdot83$	5943 $3\cdot7\cdot283$	5993 $13\cdot461$

Numbers ending in 2

550–554	555–559	560–564	565–569	570–574	575–579	580–584	585–589	590–594	595–599
5502 $2\cdot3\cdot7\cdot131$	5552 $2^4\cdot347$	5602 $2\cdot2801$	5652 $2^2\cdot3^2\cdot157$	5702 $2\cdot2851$	5752 $2^3\cdot719$	5802 $2\cdot3\cdot967$	5852 $2^2\cdot7\cdot11\cdot19$	5902 $2\cdot13\cdot227$	5952 $2^6\cdot3\cdot31$
5512 $2^3\cdot13\cdot53$	5562 $2\cdot3^3\cdot103$	5612 $2^2\cdot23\cdot61$	5662 $2\cdot19\cdot149$	5712 $2^4\cdot3\cdot7\cdot17$	5762 $2\cdot43\cdot67$	5812 $2^2\cdot1453$	5862 $2\cdot3\cdot977$	5912 $2^3\cdot739$	5962 $2\cdot11\cdot271$
5522 $2\cdot11\cdot251$	5572 $2^2\cdot7\cdot199$	5622 $2\cdot3\cdot937$	5672 $2^3\cdot709$	5722 $2\cdot2861$	5772 $2^2\cdot3\cdot13\cdot37$	5822 $2\cdot41\cdot71$	5872 $2^4\cdot367$	5922 $2\cdot3^2\cdot7\cdot47$	5972 $2^2\cdot1493$
5532 $2^2\cdot3\cdot461$	5582 $2\cdot2791$	5632 $2^9\cdot11$	5682 $2\cdot3\cdot947$	5732 $2^2\cdot1433$	5782 $2\cdot7^2\cdot59$	5832 $2^3\cdot3^6$	5882 $2\cdot17\cdot173$	5932 $2^2\cdot1483$	5982 $2\cdot3\cdot997$
5542 $2\cdot17\cdot163$	5592 $2^3\cdot3\cdot233$	5642 $2\cdot7\cdot13\cdot31$	5692 $2^2\cdot1423$	5742 $2\cdot3^2\cdot11\cdot29$	5792 $2^5\cdot181$	5842 $2\cdot23\cdot127$	5892 $2^2\cdot3\cdot491$	5942 $2\cdot2971$	5992 $2^3\cdot7\cdot107$

Numbers ending in 1

550–554	555–559	560–564	565–569	570–574	575–579	580–584	585–589	590–594	595–599
5501 **5501**	5551 $7\cdot13\cdot61$	5601 $3\cdot1867$	5651 **5651**	5701 **5701**	5751 $3^4\cdot71$	5801 **5801**	5851 **5851**	5901 $3\cdot7\cdot281$	5951 $11\cdot541$
5511 $3\cdot11\cdot167$	5561 $67\cdot83$	5611 $31\cdot181$	5661 $3^2\cdot17\cdot37$	5711 **5711**	5761 $7\cdot823$	5811 $3\cdot13\cdot149$	5861 **5861**	5911 $23\cdot257$	5961 $3\cdot1987$
5521 **5521**	5571 $3^2\cdot619$	5621 $7\cdot11\cdot73$	5671 $53\cdot107$	5721 $3\cdot1907$	5771 $29\cdot199$	5821 **5821**	5871 $3\cdot19\cdot103$	5921 $31\cdot191$	5971 $7\cdot853$
5531 **5531**	5581 **5581**	5631 $3\cdot1877$	5681 $13\cdot19\cdot23$	5731 $11\cdot521$	5781 $3\cdot41\cdot47$	5831 $7^3\cdot17$	5881 **5881**	5931 $3^2\cdot659$	5981 **5981**
5541 $3\cdot1847$	5591 **5591**	5641 **5641**	5691 $3\cdot7\cdot271$	5741 **5741**	5791 **5791**	5841 $3^2\cdot11\cdot59$	5891 $43\cdot137$	5941 $13\cdot457$	5991 $3\cdot1997$

Numbers ending in 0

550–554	555–559	560–564	565–569	570–574	575–579	580–584	585–589	590–594	595–599
5500 $2^2\cdot5^3\cdot11$	5550 $2\cdot3\cdot5^2\cdot37$	5600 $2^5\cdot5^2\cdot7$	5650 $2\cdot5^2\cdot113$	5700 $2^2\cdot3\cdot5^2\cdot19$	5750 $2\cdot5^3\cdot23$	5800 $2^3\cdot5^2\cdot29$	5850 $2\cdot3^2\cdot5^2\cdot13$	5900 $2^2\cdot5^2\cdot59$	5950 $2\cdot5^2\cdot7\cdot17$
5510 $2\cdot5\cdot19\cdot29$	5560 $2^3\cdot5\cdot139$	5610 $2\cdot3\cdot5\cdot11\cdot17$	5660 $2^2\cdot5\cdot283$	5710 $2\cdot5\cdot571$	5760 $2^7\cdot3^2\cdot5$	5810 $2\cdot5\cdot7\cdot83$	5860 $2^2\cdot5\cdot293$	5910 $2\cdot3\cdot5\cdot197$	5960 $2^3\cdot5\cdot149$
5520 $2^4\cdot3\cdot5\cdot23$	5570 $2\cdot5\cdot557$	5620 $2^2\cdot5\cdot281$	5670 $2\cdot3^4\cdot5\cdot7$	5720 $2^3\cdot5\cdot11\cdot13$	5770 $2\cdot5\cdot577$	5820 $2^2\cdot3\cdot5\cdot97$	5870 $2\cdot5\cdot587$	5920 $2^5\cdot5\cdot37$	5970 $2\cdot3\cdot5\cdot199$
5530 $2\cdot5\cdot7\cdot79$	5580 $2^2\cdot3^2\cdot5\cdot31$	5630 $2\cdot5\cdot563$	5680 $2^4\cdot5\cdot71$	5730 $2\cdot3\cdot5\cdot191$	5780 $2^2\cdot5\cdot17^2$	5830 $2\cdot5\cdot11\cdot53$	5880 $2^3\cdot3\cdot5\cdot7^2$	5930 $2\cdot5\cdot593$	5980 $2^2\cdot5\cdot13\cdot23$
5540 $2^2\cdot5\cdot277$	5590 $2\cdot5\cdot13\cdot43$	5640 $2^3\cdot3\cdot5\cdot47$	5690 $2\cdot5\cdot569$	5740 $2^2\cdot5\cdot7\cdot41$	5790 $2\cdot3\cdot5\cdot193$	5840 $2^4\cdot5\cdot73$	5890 $2\cdot5\cdot19\cdot31$	5940 $2^2\cdot3^3\cdot5\cdot11$	5990 $2\cdot5\cdot599$

Table 24.7   **Factorizations**

6000   6499

N	9	8	7	6	5	4	3	2	1	0	N
600	$3\cdot2003$	$2^3\cdot751$	**6007**	$2\cdot3\cdot7\cdot11\cdot13$	$5\cdot1201$	$2^2\cdot19\cdot79$	$3^2\cdot23\cdot29$	$2\cdot3001$	$17\cdot353$	$2^4\cdot3\cdot5^3$	600
601	$13\cdot463$	$2\cdot3\cdot17\cdot59$	$11\cdot547$	$2^7\cdot47$	$3\cdot5\cdot401$	$2\cdot31\cdot97$	$7\cdot859$	$2^2\cdot3^2\cdot167$	**6011**	$2\cdot5\cdot601$	601
602	**6029**	$2^2\cdot11\cdot137$	$3\cdot7^2\cdot41$	$2\cdot23\cdot131$	$5^2\cdot241$	$2^3\cdot3\cdot251$	$19\cdot317$	$2\cdot3011$	$3^3\cdot223$	$2^2\cdot5\cdot7\cdot43$	602
603	$3^2\cdot11\cdot61$	$2\cdot3019$	**6037**	$2^2\cdot3\cdot503$	$5\cdot17\cdot71$	$2\cdot7\cdot431$	$3\cdot2011$	$2^4\cdot13\cdot29$	$37\cdot163$	$2\cdot3^2\cdot5\cdot67$	603
604	$23\cdot263$	$2^5\cdot3^3\cdot7$	**6047**	$2\cdot3023$	$3\cdot5\cdot13\cdot31$	$2^2\cdot1511$	**6043**	$2\cdot3\cdot19\cdot53$	$7\cdot863$	$2^3\cdot5\cdot151$	604
605	$73\cdot83$	$2\cdot13\cdot233$	$3^2\cdot673$	$2^3\cdot757$	$5\cdot7\cdot173$	$2\cdot3\cdot1009$	**6053**	$2^2\cdot17\cdot89$	$3\cdot2017$	$2\cdot5^3\cdot11^2$	605
606	$3\cdot7\cdot17^2$	$2^2\cdot37\cdot41$	**6067**	$2\cdot3^2\cdot337$	$5\cdot1213$	$2^4\cdot379$	$3\cdot43\cdot47$	$2\cdot7\cdot433$	$11\cdot19\cdot29$	$2^2\cdot3\cdot5\cdot101$	606
607	**6079**	$2\cdot3\cdot1013$	$59\cdot103$	$2^2\cdot7^2\cdot31$	$3^5\cdot5^2$	$2\cdot3037$	**6073**	$2^3\cdot3\cdot11\cdot23$	$13\cdot467$	$2\cdot5\cdot607$	607
608	**6089**	$2^3\cdot761$	$3\cdot2029$	$2\cdot17\cdot179$	$5\cdot1217$	$2^2\cdot3^2\cdot13^2$	$7\cdot11\cdot79$	$2\cdot3041$	$3\cdot2027$	$2^6\cdot5\cdot19$	608
609	$3\cdot19\cdot107$	$2\cdot3049$	$7\cdot13\cdot67$	$2^4\cdot3\cdot127$	$5\cdot23\cdot53$	$2\cdot11\cdot277$	$3^2\cdot677$	$2^2\cdot1523$	**6091**	$2\cdot3\cdot5\cdot7\cdot29$	609
610	$41\cdot149$	$2^2\cdot3\cdot509$	$31\cdot197$	$2\cdot43\cdot71$	$3\cdot5\cdot11\cdot37$	$2^3\cdot7\cdot109$	$17\cdot359$	$2\cdot3^3\cdot113$	**6101**	$2^2\cdot5^2\cdot61$	610
611	$29\cdot211$	$2\cdot7\cdot19\cdot23$	$3\cdot2039$	$2^2\cdot11\cdot139$	$5\cdot1223$	$2\cdot3\cdot1019$	**6113**	$2^5\cdot191$	$3^2\cdot7\cdot97$	$2\cdot5\cdot13\cdot47$	611
612	$3^3\cdot227$	$2^4\cdot383$	$11\cdot557$	$2\cdot3\cdot1021$	$5^3\cdot7^2$	$2^2\cdot1531$	$3\cdot13\cdot157$	$2\cdot3061$	**6121**	$2^3\cdot3^2\cdot5\cdot17$	612
613	$7\cdot877$	$2\cdot3089$	$17\cdot19^2$	$2^3\cdot13\cdot59$	$3\cdot5\cdot409$	$2\cdot3067$	**6133**	$2^2\cdot3\cdot7\cdot73$	**6131**	$2\cdot5\cdot613$	613
614	$11\cdot13\cdot43$	$2^2\cdot29\cdot53$	$3^2\cdot683$	$2\cdot7\cdot439$	$5\cdot1229$	$2^{11}\cdot3$	**6143**	$2\cdot37\cdot83$	$3\cdot23\cdot89$	$2^2\cdot5\cdot307$	614
615	$3\cdot2053$	$2\cdot3079$	$47\cdot131$	$2^2\cdot3^4\cdot19$	$5\cdot1231$	$2\cdot17\cdot181$	$3\cdot7\cdot293$	$2^3\cdot769$	**6151**	$2\cdot3\cdot5^2\cdot41$	615
616	$31\cdot199$	$2^3\cdot3\cdot257$	$7\cdot881$	$2\cdot3083$	$3^2\cdot5\cdot137$	$2^2\cdot23\cdot67$	**6163**	$2\cdot3\cdot13\cdot79$	$61\cdot101$	$2^4\cdot5\cdot7\cdot11$	616
617	$37\cdot167$	$2\cdot3089$	$3\cdot29\cdot71$	$2^5\cdot193$	$5^2\cdot13\cdot19$	$2\cdot3^2\cdot7^3$	**6173**	$2^2\cdot1543$	$3\cdot11^2\cdot17$	$2\cdot5\cdot617$	617
618	$3\cdot2063$	$2^2\cdot7\cdot13\cdot17$	$23\cdot269$	$2\cdot3\cdot1031$	$5\cdot1237$	$2^3\cdot773$	$3^3\cdot229$	$2\cdot11\cdot281$	$7\cdot883$	$2^2\cdot3\cdot5\cdot103$	618
619	**6199**	$2\cdot3\cdot1033$	**6197**	$2^2\cdot1549$	$3\cdot5\cdot7\cdot59$	$2\cdot19\cdot163$	$11\cdot563$	$2^4\cdot3^2\cdot43$	$41\cdot151$	$2\cdot5\cdot619$	619
620	$7\cdot887$	$2^6\cdot97$	$3\cdot2069$	$2\cdot29\cdot107$	$5\cdot17\cdot73$	$2^2\cdot3\cdot11\cdot47$	**6203**	$2\cdot7\cdot443$	$3^2\cdot13\cdot53$	$2^3\cdot5^2\cdot31$	620
621	$3^2\cdot691$	$2\cdot3109$	**6217**	$2^3\cdot3\cdot7\cdot37$	$5\cdot11\cdot113$	$2\cdot13\cdot239$	$3\cdot19\cdot109$	$2^2\cdot1553$	**6211**	$2\cdot3^3\cdot5\cdot23$	621
622	**6229**	$2^2\cdot3^2\cdot173$	$13\cdot479$	$2\cdot11\cdot283$	$3\cdot5^2\cdot83$	$2^4\cdot389$	$7^2\cdot127$	$2\cdot3\cdot17\cdot61$	**6221**	$2^2\cdot5\cdot311$	622
623	$17\cdot367$	$2\cdot3119$	$3^4\cdot7\cdot11$	$2^2\cdot1559$	$5\cdot29\cdot43$	$2\cdot3\cdot1039$	$23\cdot271$	$2^3\cdot19\cdot41$	$3\cdot31\cdot67$	$2\cdot5\cdot7\cdot89$	623
624	$3\cdot2083$	$2^3\cdot11\cdot71$	**6247**	$2\cdot3^2\cdot347$	$5\cdot1249$	$2^2\cdot7\cdot223$	$3\cdot2081$	$2\cdot3121$	79^2	$2^5\cdot3\cdot5\cdot13$	624
625	$11\cdot569$	$2\cdot3\cdot7\cdot149$	**6257**	$2^4\cdot17\cdot23$	$3^2\cdot5\cdot139$	$2\cdot53\cdot59$	$13^2\cdot37$	$2^2\cdot3\cdot521$	$7\cdot19\cdot47$	$2\cdot5^5$	625
626	**6269**	$2^2\cdot1567$	$3\cdot2089$	$2\cdot13\cdot241$	$5\cdot7\cdot179$	$2^3\cdot3^3\cdot29$	**6263**	$2\cdot31\cdot101$	$3\cdot2087$	$2^2\cdot5\cdot313$	626
627	$3\cdot7\cdot13\cdot23$	$2\cdot43\cdot73$	**6277**	$2^2\cdot3\cdot523$	$5^2\cdot251$	$2\cdot3137$	$3^2\cdot17\cdot41$	$2^7\cdot7^2$	**6271**	$2\cdot3\cdot5\cdot11\cdot19$	627
628	$19\cdot331$	$2^4\cdot3\cdot131$	**6287**	$2\cdot7\cdot449$	$3\cdot5\cdot419$	$2^2\cdot1571$	$61\cdot103$	$2\cdot3^2\cdot349$	$11\cdot571$	$2^3\cdot5\cdot157$	628
629	**6299**	$2\cdot47\cdot67$	$3\cdot2099$	$2^3\cdot787$	$5\cdot1259$	$2\cdot3\cdot1049$	$7\cdot29\cdot31$	$2^2\cdot11^2\cdot13$	$3^3\cdot233$	$2\cdot5\cdot17\cdot37$	629
630	$3^2\cdot701$	$2^2\cdot19\cdot83$	$7\cdot17\cdot53$	$2\cdot3\cdot1051$	$5\cdot13\cdot97$	$2^5\cdot197$	$3\cdot11\cdot191$	$2\cdot23\cdot137$	**6301**	$2^2\cdot3^2\cdot5^2\cdot7$	630
631	$71\cdot89$	$2\cdot3^5\cdot13$	**6317**	$2^2\cdot1579$	$3\cdot5\cdot421$	$2\cdot7\cdot11\cdot41$	$59\cdot107$	$2^3\cdot3\cdot263$	**6311**	$2\cdot5\cdot631$	631
632	**6329**	$2^3\cdot7\cdot113$	$3^2\cdot19\cdot37$	$2\cdot3163$	$5^2\cdot11\cdot23$	$2^2\cdot3\cdot17\cdot31$	**6323**	$2\cdot29\cdot109$	$3\cdot7^2\cdot43$	$2^4\cdot5\cdot79$	632
633	$3\cdot2113$	$2\cdot3169$	**6337**	$2^6\cdot3^2\cdot11$	$5\cdot7\cdot181$	$2\cdot3167$	$3\cdot2111$	$2^2\cdot1583$	$13\cdot487$	$2\cdot3\cdot5\cdot211$	633
634	$7\cdot907$	$2^2\cdot3\cdot23^2$	$11\cdot577$	$2\cdot19\cdot167$	$3^3\cdot5\cdot47$	$2^3\cdot13\cdot61$	**6343**	$2\cdot3\cdot7\cdot151$	$17\cdot373$	$2^2\cdot5\cdot317$	634
635	**6359**	$2\cdot11\cdot17^2$	$3\cdot13\cdot163$	$2^2\cdot7\cdot227$	$5\cdot31\cdot41$	$2\cdot3^2\cdot353$	**6353**	$2^4\cdot397$	$3\cdot29\cdot73$	$2\cdot5^2\cdot127$	635
636	$3\cdot11\cdot193$	$2^5\cdot199$	**6367**	$2\cdot3\cdot1061$	$5\cdot19\cdot67$	$2^2\cdot37\cdot43$	$3^2\cdot7\cdot101$	$2\cdot3181$	**6361**	$2^3\cdot3\cdot5\cdot53$	636
637	**6379**	$2\cdot3\cdot1063$	$7\cdot911$	$2^3\cdot797$	$3\cdot5^3\cdot17$	$2\cdot3187$	**6373**	$2^2\cdot3^3\cdot59$	$23\cdot277$	$2\cdot5\cdot7^2\cdot13$	637
638	**6389**	$2^2\cdot1597$	$3\cdot2129$	$2\cdot31\cdot103$	$5\cdot1277$	$2^4\cdot3\cdot7\cdot19$	$13\cdot491$	$2\cdot3191$	$3^2\cdot709$	$2^2\cdot5\cdot11\cdot29$	638
639	$3^4\cdot79$	$2\cdot7\cdot457$	**6397**	$2^2\cdot3\cdot13\cdot41$	$5\cdot1279$	$2\cdot23\cdot139$	$3\cdot2131$	$2^3\cdot17\cdot47$	$7\cdot11\cdot83$	$2\cdot3^2\cdot5\cdot71$	639
640	$13\cdot17\cdot29$	$2^3\cdot3^2\cdot89$	$43\cdot149$	$2\cdot3203$	$3\cdot5\cdot7\cdot61$	$2^2\cdot1601$	$19\cdot337$	$2\cdot3\cdot11\cdot97$	$37\cdot173$	$2^8\cdot5^2$	640
641	$7^2\cdot131$	$2\cdot3209$	$3^2\cdot23\cdot31$	$2^4\cdot401$	$5\cdot1283$	$2\cdot3\cdot1069$	$11^2\cdot53$	$2^2\cdot7\cdot229$	$3\cdot2137$	$2\cdot5\cdot641$	641
642	$3\cdot2143$	$2^2\cdot1607$	**6427**	$2\cdot3^3\cdot7\cdot17$	$5^2\cdot257$	$2^3\cdot11\cdot73$	$3\cdot2141$	$2\cdot13^2\cdot19$	**6421**	$2^2\cdot3\cdot5\cdot107$	642
643	$47\cdot137$	$2\cdot3\cdot29\cdot37$	$41\cdot157$	$2^2\cdot1609$	$3^2\cdot5\cdot11\cdot13$	$2\cdot3217$	$7\cdot919$	$2^5\cdot3\cdot67$	$59\cdot109$	$2\cdot5\cdot643$	643
644	**6449**	$2^4\cdot13\cdot31$	$3\cdot7\cdot307$	$2\cdot11\cdot293$	$5\cdot1289$	$2^2\cdot3^2\cdot179$	$17\cdot379$	$2\cdot3221$	$3\cdot19\cdot113$	$2^3\cdot5\cdot7\cdot23$	644
645	$3\cdot2153$	$2\cdot3229$	$11\cdot587$	$2^3\cdot3\cdot269$	$5\cdot1291$	$2\cdot7\cdot461$	$3^3\cdot239$	$2^2\cdot1613$	**6451**	$2\cdot3\cdot5^2\cdot43$	645
646	**6469**	$2^2\cdot3\cdot7^2\cdot11$	$29\cdot223$	$2\cdot53\cdot61$	$3\cdot5\cdot431$	$2^6\cdot101$	$23\cdot281$	$2\cdot3^2\cdot359$	$7\cdot13\cdot71$	$2^2\cdot5\cdot17\cdot19$	646
647	$11\cdot19\cdot31$	$2\cdot41\cdot79$	$3\cdot17\cdot127$	$2^2\cdot1619$	$5^2\cdot7\cdot37$	$2\cdot3\cdot13\cdot83$	**6473**	$2^3\cdot809$	$3^2\cdot719$	$2\cdot5\cdot647$	647
648	$3^2\cdot7\cdot103$	$2^3\cdot811$	$13\cdot499$	$2\cdot3\cdot23\cdot47$	$5\cdot1297$	$2^2\cdot1621$	$3\cdot2161$	$2\cdot7\cdot463$	**6481**	$2^4\cdot3^4\cdot5$	648
649	$67\cdot97$	$2\cdot3^2\cdot19^2$	$73\cdot89$	$2^5\cdot7\cdot29$	$3\cdot5\cdot433$	$2\cdot17\cdot191$	$43\cdot151$	$2^2\cdot3\cdot541$	**6491**	$2\cdot5\cdot11\cdot59$	649

Factorizations

Table 24.7

	0	1	2	3	4	5	6	7	8	9
650	$2^2 \cdot 5^3 \cdot 13$	$3 \cdot 11 \cdot 197$	$2 \cdot 3251$	$7 \cdot 929$	$2^3 \cdot 3 \cdot 271$	$5 \cdot 1301$	$2 \cdot 3253$	$3^3 \cdot 241$	$2^2 \cdot 1627$	$23 \cdot 283$
651	$2 \cdot 3 \cdot 5 \cdot 7 \cdot 31$	$17 \cdot 383$	$2^4 \cdot 11 \cdot 37$	$3 \cdot 13 \cdot 167$	$2 \cdot 3257$	$5 \cdot 1303$	$2^2 \cdot 3^2 \cdot 181$	$7^3 \cdot 19$	$2 \cdot 3259$	$3 \cdot 41 \cdot 53$
652	$2^3 \cdot 5 \cdot 163$	**6521**	$2 \cdot 3 \cdot 1087$	$11 \cdot 593$	$2^2 \cdot 7 \cdot 233$	$3^2 \cdot 5^2 \cdot 29$	$2 \cdot 13 \cdot 251$	$61 \cdot 107$	$2^7 \cdot 3 \cdot 17$	**6529**
653	$2 \cdot 5 \cdot 653$	$3 \cdot 7 \cdot 311$	$2^2 \cdot 23 \cdot 71$	$47 \cdot 139$	$2 \cdot 3^3 \cdot 11^2$	$5 \cdot 1307$	$2^3 \cdot 19 \cdot 43$	$3 \cdot 2179$	$2 \cdot 7 \cdot 467$	$13 \cdot 503$
654	$2^2 \cdot 3 \cdot 5 \cdot 109$	$31 \cdot 211$	$2 \cdot 3271$	$3^2 \cdot 727$	$2^4 \cdot 409$	$5 \cdot 7 \cdot 11 \cdot 17$	$2 \cdot 3 \cdot 1091$	**6547**	$2^2 \cdot 1637$	$3 \cdot 37 \cdot 59$
655	$2 \cdot 5^2 \cdot 131$	**6551**	$2^3 \cdot 3^2 \cdot 7 \cdot 13$	**6553**	$2 \cdot 29 \cdot 113$	$3 \cdot 5 \cdot 19 \cdot 23$	$2^2 \cdot 11 \cdot 149$	$79 \cdot 83$	$2 \cdot 3 \cdot 1093$	$7 \cdot 937$
656	$2^5 \cdot 5 \cdot 41$	3^8	$2 \cdot 17 \cdot 193$	**6563**	$2^2 \cdot 3 \cdot 547$	$5 \cdot 13 \cdot 101$	$2 \cdot 7^2 \cdot 67$	$3 \cdot 11 \cdot 199$	$2^3 \cdot 821$	**6569**
657	$2 \cdot 3^2 \cdot 5 \cdot 73$	**6571**	$2^2 \cdot 31 \cdot 53$	$3 \cdot 7 \cdot 313$	$2 \cdot 19 \cdot 173$	$5^2 \cdot 263$	$2^4 \cdot 3 \cdot 137$	**6577**	$2 \cdot 11 \cdot 13 \cdot 23$	$3^2 \cdot 17 \cdot 43$
658	$2^2 \cdot 5 \cdot 7 \cdot 47$	**6581**	$2 \cdot 3 \cdot 1097$	$29 \cdot 227$	$2^3 \cdot 823$	$3 \cdot 5 \cdot 439$	$2 \cdot 37 \cdot 89$	$7 \cdot 941$	$2^2 \cdot 3^3 \cdot 61$	$11 \cdot 599$
659	$2 \cdot 5 \cdot 659$	$3 \cdot 13^3$	$2^6 \cdot 103$	$19 \cdot 347$	$2 \cdot 3 \cdot 7 \cdot 157$	$5 \cdot 1319$	$2^2 \cdot 17 \cdot 97$	$3^2 \cdot 733$	$2 \cdot 3299$	**6599**
660	$2^3 \cdot 3 \cdot 5^2 \cdot 11$	$7 \cdot 23 \cdot 41$	$2 \cdot 3301$	$3 \cdot 31 \cdot 71$	$2^2 \cdot 13 \cdot 127$	$5 \cdot 1321$	$2 \cdot 3^2 \cdot 367$	**6607**	$2^4 \cdot 7 \cdot 59$	$3 \cdot 2203$
661	$2 \cdot 5 \cdot 661$	$11 \cdot 601$	$2^2 \cdot 3 \cdot 19 \cdot 29$	$17 \cdot 389$	$2 \cdot 3307$	$3^3 \cdot 5 \cdot 7^2$	$2^3 \cdot 827$	$13 \cdot 509$	$2 \cdot 3 \cdot 1103$	**6619**
662	$2^2 \cdot 5 \cdot 331$	$3 \cdot 2207$	$2 \cdot 7 \cdot 11 \cdot 43$	$37 \cdot 179$	$2^5 \cdot 3^2 \cdot 23$	$5^3 \cdot 53$	$2 \cdot 3313$	$3 \cdot 47^2$	$2^2 \cdot 1657$	$7 \cdot 947$
663	$2 \cdot 3 \cdot 5 \cdot 13 \cdot 17$	$19 \cdot 349$	$2^3 \cdot 829$	$3^2 \cdot 11 \cdot 67$	$2 \cdot 31 \cdot 107$	$5 \cdot 1327$	$2^2 \cdot 3 \cdot 7 \cdot 79$	**6637**	$2 \cdot 3319$	$3 \cdot 2213$
664	$2^4 \cdot 5 \cdot 83$	$29 \cdot 229$	$2 \cdot 3^4 \cdot 41$	$7 \cdot 13 \cdot 73$	$2^2 \cdot 11 \cdot 151$	$3 \cdot 5 \cdot 443$	$2 \cdot 3323$	$17^2 \cdot 23$	$2^3 \cdot 3 \cdot 277$	$61 \cdot 109$
665	$2 \cdot 5^2 \cdot 7 \cdot 19$	$3^2 \cdot 739$	$2^2 \cdot 1663$	**6653**	$2 \cdot 3 \cdot 1109$	$5 \cdot 11^3$	$2^9 \cdot 13$	$3 \cdot 7 \cdot 317$	$2 \cdot 3329$	**6659**
666	$2^2 \cdot 3^2 \cdot 5 \cdot 37$	**6661**	$2 \cdot 3331$	$3 \cdot 2221$	$2^3 \cdot 7^2 \cdot 17$	$5 \cdot 31 \cdot 43$	$2 \cdot 3 \cdot 11 \cdot 101$	$59 \cdot 113$	$2^2 \cdot 1667$	$3^3 \cdot 13 \cdot 19$
667	$2 \cdot 5 \cdot 23 \cdot 29$	$7 \cdot 953$	$2^4 \cdot 3 \cdot 139$	**6673**	$2 \cdot 47 \cdot 71$	$3 \cdot 5^2 \cdot 89$	$2^2 \cdot 1669$	$11 \cdot 607$	$2 \cdot 3^2 \cdot 7 \cdot 53$	**6679**
668	$2^3 \cdot 5 \cdot 167$	$3 \cdot 17 \cdot 131$	$2 \cdot 13 \cdot 257$	$41 \cdot 163$	$2^2 \cdot 3 \cdot 557$	$5 \cdot 7 \cdot 191$	$2 \cdot 3343$	$3^2 \cdot 743$	$2^5 \cdot 11 \cdot 19$	**6689**
669	$2 \cdot 3 \cdot 5 \cdot 223$	**6691**	$2^2 \cdot 7 \cdot 239$	$3 \cdot 23 \cdot 97$	$2 \cdot 3347$	$5 \cdot 13 \cdot 103$	$2^3 \cdot 3^3 \cdot 31$	$37 \cdot 181$	$2 \cdot 17 \cdot 197$	$3 \cdot 7 \cdot 11 \cdot 29$
670	$2^2 \cdot 5^2 \cdot 67$	**6701**	$2 \cdot 3 \cdot 1117$	**6703**	$2^4 \cdot 419$	$3^2 \cdot 5 \cdot 149$	$2 \cdot 7 \cdot 479$	$19 \cdot 353$	$2^2 \cdot 3 \cdot 13 \cdot 43$	**6709**
671	$2 \cdot 5 \cdot 11 \cdot 61$	$3 \cdot 2237$	$2^3 \cdot 839$	$7^2 \cdot 137$	$2 \cdot 3^2 \cdot 373$	$5 \cdot 17 \cdot 79$	$2^2 \cdot 23 \cdot 73$	$3 \cdot 2239$	$2 \cdot 3359$	**6719**
672	$2^6 \cdot 3 \cdot 5 \cdot 7$	$11 \cdot 13 \cdot 47$	$2 \cdot 3361$	$3^4 \cdot 83$	$2^2 \cdot 41^2$	$5^2 \cdot 269$	$2 \cdot 3 \cdot 19 \cdot 59$	$7 \cdot 31^2$	$2^3 \cdot 29^2$	$3 \cdot 2243$
673	$2 \cdot 5 \cdot 673$	$53 \cdot 127$	$2^2 \cdot 3^2 \cdot 11 \cdot 17$	**6733**	$2 \cdot 7 \cdot 13 \cdot 37$	$3 \cdot 5 \cdot 449$	$2^4 \cdot 421$	**6737**	$2 \cdot 3 \cdot 1123$	$23 \cdot 293$
674	$2^2 \cdot 5 \cdot 337$	$3^2 \cdot 7 \cdot 107$	$2 \cdot 3371$	$11 \cdot 613$	$2^3 \cdot 3 \cdot 281$	$5 \cdot 19 \cdot 71$	$2 \cdot 3373$	$3 \cdot 13 \cdot 173$	$2^2 \cdot 7 \cdot 241$	$17 \cdot 397$
675	$2 \cdot 3^3 \cdot 5^3$	$43 \cdot 157$	$2^5 \cdot 211$	$3 \cdot 2251$	$2 \cdot 11 \cdot 307$	$5 \cdot 7 \cdot 193$	$2^2 \cdot 3 \cdot 563$	$29 \cdot 233$	$2 \cdot 31 \cdot 109$	$3^2 \cdot 751$
676	$2^3 \cdot 5 \cdot 13^2$	**6761**	$2 \cdot 3 \cdot 7^2 \cdot 23$	**6763**	$2^2 \cdot 19 \cdot 89$	$3 \cdot 5 \cdot 11 \cdot 41$	$2 \cdot 17 \cdot 199$	$67 \cdot 101$	$2^4 \cdot 3^2 \cdot 47$	$7 \cdot 967$
677	$2 \cdot 5 \cdot 677$	$3 \cdot 37 \cdot 61$	$2^2 \cdot 1693$	$13 \cdot 521$	$2 \cdot 3 \cdot 1129$	$5^2 \cdot 271$	$2^3 \cdot 7 \cdot 11^2$	$3^3 \cdot 251$	$2 \cdot 3389$	**6779**
678	$2^2 \cdot 3 \cdot 5 \cdot 113$	**6781**	$2 \cdot 3391$	$3 \cdot 7 \cdot 17 \cdot 19$	$2^7 \cdot 53$	$5 \cdot 23 \cdot 59$	$2 \cdot 3^2 \cdot 13 \cdot 29$	$11 \cdot 617$	$2^2 \cdot 1697$	$3 \cdot 31 \cdot 73$
679	$2 \cdot 5 \cdot 7 \cdot 97$	**6791**	$2^3 \cdot 3 \cdot 283$	**6793**	$2 \cdot 43 \cdot 79$	$3^2 \cdot 5 \cdot 151$	$2^2 \cdot 1699$	$7 \cdot 971$	$2 \cdot 3 \cdot 11 \cdot 103$	$13 \cdot 523$
680	$2^4 \cdot 5^2 \cdot 17$	$3 \cdot 2267$	$2 \cdot 19 \cdot 179$	**6803**	$2^2 \cdot 3^5 \cdot 7$	$5 \cdot 1361$	$2 \cdot 41 \cdot 83$	$3 \cdot 2269$	$2^3 \cdot 23 \cdot 37$	$11 \cdot 619$
681	$2 \cdot 3 \cdot 5 \cdot 227$	$7^2 \cdot 139$	$2^2 \cdot 13 \cdot 131$	$3^2 \cdot 757$	$2 \cdot 3407$	$5 \cdot 29 \cdot 47$	$2^5 \cdot 3 \cdot 71$	$17 \cdot 401$	$2 \cdot 7 \cdot 487$	$3 \cdot 2273$
682	$2^2 \cdot 5 \cdot 11 \cdot 31$	$19 \cdot 359$	$2 \cdot 3^2 \cdot 379$	**6823**	$2^3 \cdot 853$	$3 \cdot 5^2 \cdot 7 \cdot 13$	$2 \cdot 3413$	**6827**	$2^2 \cdot 3 \cdot 569$	**6829**
683	$2 \cdot 5 \cdot 683$	$3^3 \cdot 11 \cdot 23$	$2^4 \cdot 7 \cdot 61$	**6833**	$2 \cdot 3 \cdot 17 \cdot 67$	$5 \cdot 1367$	$2^2 \cdot 1709$	$3 \cdot 43 \cdot 53$	$2 \cdot 13 \cdot 263$	$7 \cdot 977$
684	$2^3 \cdot 3^2 \cdot 5 \cdot 19$	**6841**	$2 \cdot 11 \cdot 311$	$3 \cdot 2281$	$2^2 \cdot 29 \cdot 59$	$5 \cdot 37^2$	$2 \cdot 3 \cdot 7 \cdot 163$	$41 \cdot 167$	$2^6 \cdot 107$	$3^2 \cdot 761$
685	$2 \cdot 5^2 \cdot 137$	$13 \cdot 17 \cdot 31$	$2^2 \cdot 3 \cdot 571$	$7 \cdot 11 \cdot 89$	$2 \cdot 23 \cdot 149$	$3 \cdot 5 \cdot 457$	$2^3 \cdot 857$	**6857**	$2 \cdot 3^3 \cdot 127$	19^3
686	$2^2 \cdot 5 \cdot 7^3$	$3 \cdot 2287$	$2 \cdot 47 \cdot 73$	**6863**	$2^4 \cdot 3 \cdot 11 \cdot 13$	$5 \cdot 1373$	$2 \cdot 3433$	$3^2 \cdot 7 \cdot 109$	$2^2 \cdot 17 \cdot 101$	**6869**
687	$2 \cdot 3 \cdot 5 \cdot 229$	**6871**	$2^3 \cdot 859$	$3 \cdot 29 \cdot 79$	$2 \cdot 7 \cdot 491$	$5^4 \cdot 11$	$2^2 \cdot 3^2 \cdot 191$	$13 \cdot 23^2$	$2 \cdot 19 \cdot 181$	$3 \cdot 2293$
688	$2^5 \cdot 5 \cdot 43$	$7 \cdot 983$	$2 \cdot 3 \cdot 31 \cdot 37$	**6883**	$2^2 \cdot 1721$	$3^4 \cdot 5 \cdot 17$	$2 \cdot 11 \cdot 313$	$71 \cdot 97$	$2^3 \cdot 3 \cdot 7 \cdot 41$	83^2
689	$2 \cdot 5 \cdot 13 \cdot 53$	$3 \cdot 2297$	$2^2 \cdot 1723$	$61 \cdot 113$	$2 \cdot 3^2 \cdot 383$	$5 \cdot 7 \cdot 197$	$2^4 \cdot 431$	$3 \cdot 11^2 \cdot 19$	$2 \cdot 3449$	**6899**
690	$2^2 \cdot 3 \cdot 5^2 \cdot 23$	$67 \cdot 103$	$2 \cdot 7 \cdot 17 \cdot 29$	$3^2 \cdot 13 \cdot 59$	$2^3 \cdot 863$	$5 \cdot 1381$	$2 \cdot 3 \cdot 1151$	**6907**	$2^2 \cdot 11 \cdot 157$	$3 \cdot 7^2 \cdot 47$
691	$2 \cdot 5 \cdot 691$	**6911**	$2^8 \cdot 3^3$	$31 \cdot 223$	$2 \cdot 3457$	$3 \cdot 5 \cdot 461$	$2^2 \cdot 7 \cdot 13 \cdot 19$	**6917**	$2 \cdot 3 \cdot 1153$	$11 \cdot 17 \cdot 37$
692	$2^3 \cdot 5 \cdot 173$	$3^2 \cdot 769$	$2 \cdot 3461$	$7 \cdot 23 \cdot 43$	$2^2 \cdot 3 \cdot 577$	$5^2 \cdot 277$	$2 \cdot 3463$	$3 \cdot 2309$	$2^4 \cdot 433$	$13^2 \cdot 41$
693	$2 \cdot 3^2 \cdot 5 \cdot 7 \cdot 11$	$29 \cdot 239$	$2^2 \cdot 1733$	$3 \cdot 2311$	$2 \cdot 3467$	$5 \cdot 19 \cdot 73$	$2^3 \cdot 3 \cdot 17^2$	$7 \cdot 991$	$2 \cdot 3469$	$3^3 \cdot 257$
694	$2^2 \cdot 5 \cdot 347$	$11 \cdot 631$	$2 \cdot 3 \cdot 13 \cdot 89$	$53 \cdot 131$	$2^5 \cdot 7 \cdot 31$	$3 \cdot 5 \cdot 463$	$2 \cdot 23 \cdot 151$	**6947**	$2^2 \cdot 3^2 \cdot 193$	**6949**
695	$2 \cdot 5^2 \cdot 139$	$3 \cdot 7 \cdot 331$	$2^3 \cdot 11 \cdot 79$	$17 \cdot 409$	$2 \cdot 3 \cdot 19 \cdot 61$	$5 \cdot 13 \cdot 107$	$2^2 \cdot 37 \cdot 47$	$3^2 \cdot 773$	$2 \cdot 7^2 \cdot 71$	**6959**
696	$2^4 \cdot 3 \cdot 5 \cdot 29$	**6961**	$2 \cdot 59^2$	$3 \cdot 11 \cdot 211$	$2^2 \cdot 1741$	$5 \cdot 7 \cdot 199$	$2 \cdot 3^4 \cdot 43$	**6967**	$2^3 \cdot 13 \cdot 67$	$3 \cdot 23 \cdot 101$
697	$2 \cdot 5 \cdot 17 \cdot 41$	**6971**	$2^2 \cdot 3 \cdot 7 \cdot 83$	$19 \cdot 367$	$2 \cdot 11 \cdot 317$	$3^2 \cdot 5^2 \cdot 31$	$2^6 \cdot 109$	**6977**	$2 \cdot 3 \cdot 1163$	$7 \cdot 997$
698	$2^2 \cdot 5 \cdot 349$	$3 \cdot 13 \cdot 179$	$2 \cdot 3491$	**6983**	$2^3 \cdot 3^2 \cdot 97$	$5 \cdot 11 \cdot 127$	$2 \cdot 7 \cdot 499$	$3 \cdot 17 \cdot 137$	$2^2 \cdot 1747$	$29 \cdot 241$
699	$2 \cdot 3 \cdot 5 \cdot 233$	**6991**	$2^4 \cdot 19 \cdot 23$	$3^3 \cdot 7 \cdot 37$	$2 \cdot 13 \cdot 269$	$5 \cdot 1399$	$2^2 \cdot 3 \cdot 11 \cdot 53$	**6997**	$2 \cdot 3499$	$3 \cdot 2333$

Table 24.7   **Factorizations**

7000   7499

N	9	8	7	6	5	4	3	2	1	0
700	$43 \cdot 163$	$2^5 \cdot 3 \cdot 73$	$7^2 \cdot 11 \cdot 13$	$2 \cdot 31 \cdot 113$	$3 \cdot 5 \cdot 467$	$2^2 \cdot 17 \cdot 103$	$47 \cdot 149$	$2 \cdot 3^2 \cdot 389$	**7001**	$2^3 \cdot 5^3 \cdot 7$
701	**7019**	$2 \cdot 11^2 \cdot 29$	$3 \cdot 2339$	$2 \cdot 877$	$5 \cdot 23 \cdot 61$	$2 \cdot 3 \cdot 7 \cdot 167$	**7013**	$2^2 \cdot 1753$	$3^2 \cdot 19 \cdot 41$	$2 \cdot 5 \cdot 701$
702	$3^2 \cdot 11 \cdot 71$	$2^2 \cdot 7 \cdot 251$	**7027**	$2 \cdot 3 \cdot 1171$	$5^2 \cdot 281$	$2^4 \cdot 439$	$3 \cdot 2341$	$2 \cdot 3511$	$7 \cdot 17 \cdot 59$	$2^2 \cdot 3^3 \cdot 5 \cdot 13$
703	**7039**	$2 \cdot 3^2 \cdot 17 \cdot 23$	$31 \cdot 227$	$2 \cdot 1759$	$3 \cdot 5 \cdot 7 \cdot 67$	$2 \cdot 3517$	$13 \cdot 541$	$2^3 \cdot 3 \cdot 293$	$79 \cdot 89$	$2 \cdot 5 \cdot 19 \cdot 37$
704	$7 \cdot 19 \cdot 53$	$2^3 \cdot 881$	$3^5 \cdot 29$	$2 \cdot 13 \cdot 271$	$5 \cdot 1409$	$2^2 \cdot 3 \cdot 587$	**7043**	$2 \cdot 7 \cdot 503$	$3 \cdot 2347$	$2^7 \cdot 5 \cdot 11$
705	$3 \cdot 13 \cdot 181$	$2 \cdot 3529$	**7057**	$2^4 \cdot 3^2 \cdot 7^2$	$5 \cdot 17 \cdot 83$	$2 \cdot 3527$	$3 \cdot 2351$	$2^2 \cdot 41 \cdot 43$	$11 \cdot 641$	$2 \cdot 3 \cdot 5^2 \cdot 47$
706	**7069**	$2^2 \cdot 3 \cdot 19 \cdot 31$	$37 \cdot 191$	$2 \cdot 3533$	$3^2 \cdot 5 \cdot 157$	$2^3 \cdot 883$	$7 \cdot 1009$	$2 \cdot 3 \cdot 11 \cdot 107$	$23 \cdot 307$	$2^6 \cdot 5 \cdot 23$
707	**7079**	$2 \cdot 3539$	$3 \cdot 7 \cdot 337$	$2^2 \cdot 29 \cdot 61$	$5^2 \cdot 283$	$2 \cdot 3^3 \cdot 131$	$11 \cdot 643$	$2^5 \cdot 13 \cdot 17$	$3 \cdot 2357$	$2 \cdot 5 \cdot 7 \cdot 101$
708	$3 \cdot 17 \cdot 139$	$2^4 \cdot 443$	$19 \cdot 373$	$2 \cdot 3 \cdot 1181$	$5 \cdot 13 \cdot 109$	$2^2 \cdot 7 \cdot 11 \cdot 23$	$3^2 \cdot 787$	$2 \cdot 3541$	$73 \cdot 97$	$2^3 \cdot 3 \cdot 5 \cdot 59$
709	$31 \cdot 229$	$2 \cdot 3 \cdot 7 \cdot 13^2$	$47 \cdot 151$	$2^3 \cdot 887$	$3 \cdot 5 \cdot 11 \cdot 43$	$2 \cdot 3547$	$41 \cdot 173$	$2^2 \cdot 3^2 \cdot 197$	$7 \cdot 1013$	$2 \cdot 5 \cdot 709$
710	**7109**	$2^2 \cdot 1777$	$3 \cdot 23 \cdot 103$	$2 \cdot 11 \cdot 17 \cdot 19$	$5 \cdot 7^2 \cdot 29$	$2^6 \cdot 3 \cdot 37$	**7103**	$2 \cdot 5^3 \cdot 67$	$3^3 \cdot 263$	$2^2 \cdot 5^2 \cdot 71$
711	$3^2 \cdot 7 \cdot 113$	$2 \cdot 3559$	$11 \cdot 647$	$2^2 \cdot 3 \cdot 593$	$5 \cdot 1423$	$2 \cdot 3557$	$3 \cdot 2371$	$2^3 \cdot 7 \cdot 127$	$13 \cdot 547$	$2 \cdot 3^2 \cdot 5 \cdot 79$
712	**7129**	$2^3 \cdot 3^4 \cdot 11$	**7127**	$2 \cdot 7 \cdot 509$	$3 \cdot 5^3 \cdot 19$	$2^2 \cdot 13 \cdot 137$	$17 \cdot 419$	$2 \cdot 3 \cdot 1187$	**7121**	$2^4 \cdot 5 \cdot 89$
713	$11^2 \cdot 59$	$2 \cdot 43 \cdot 83$	$3^2 \cdot 13 \cdot 61$	$2^5 \cdot 223$	$5 \cdot 1427$	$2 \cdot 3 \cdot 29 \cdot 41$	$7 \cdot 1019$	$2^2 \cdot 1783$	$3 \cdot 2377$	$2 \cdot 5 \cdot 23 \cdot 31$
714	$3 \cdot 2383$	$2^2 \cdot 1787$	$7 \cdot 1021$	$2 \cdot 3^2 \cdot 397$	$5 \cdot 1429$	$2^3 \cdot 19 \cdot 47$	$3 \cdot 2381$	$2 \cdot 3571$	$37 \cdot 193$	$2^2 \cdot 3 \cdot 5 \cdot 7 \cdot 17$
715	**7159**	$2 \cdot 3 \cdot 1193$	$17 \cdot 421$	$2^2 \cdot 1789$	$3^3 \cdot 5 \cdot 53$	$2 \cdot 7^2 \cdot 73$	$23 \cdot 311$	$2^4 \cdot 3 \cdot 149$	**7151**	$2 \cdot 5^2 \cdot 11 \cdot 13$
716	$67 \cdot 107$	$2^{10} \cdot 7$	$3 \cdot 2389$	$2 \cdot 3583$	$5 \cdot 1433$	$2^2 \cdot 3^2 \cdot 199$	$13 \cdot 19 \cdot 29$	$2 \cdot 3581$	$3 \cdot 7 \cdot 11 \cdot 31$	$2^3 \cdot 5 \cdot 179$
717	$3 \cdot 2393$	$2 \cdot 37 \cdot 97$	**7177**	$2^3 \cdot 3 \cdot 13 \cdot 23$	$5^2 \cdot 7 \cdot 41$	$2 \cdot 17 \cdot 211$	$3^2 \cdot 797$	$2^2 \cdot 11 \cdot 163$	$71 \cdot 101$	$2 \cdot 3 \cdot 5 \cdot 239$
718	$7 \cdot 13 \cdot 79$	$2^2 \cdot 3 \cdot 599$	**7187**	$2 \cdot 3593$	$3 \cdot 5 \cdot 479$	$2^4 \cdot 449$	$11 \cdot 653$	$2 \cdot 3^3 \cdot 7 \cdot 19$	$43 \cdot 167$	$2^2 \cdot 5 \cdot 359$
719	$23 \cdot 313$	$2 \cdot 59 \cdot 61$	$3 \cdot 2399$	$2^2 \cdot 7 \cdot 257$	$5 \cdot 1439$	$2 \cdot 3 \cdot 11 \cdot 109$	**7193**	$2^3 \cdot 29 \cdot 31$	$3^2 \cdot 17 \cdot 47$	$2 \cdot 5 \cdot 719$
720	$3^4 \cdot 89$	$2^3 \cdot 17 \cdot 53$	**7207**	$2 \cdot 3 \cdot 1201$	$5 \cdot 11 \cdot 131$	$2^2 \cdot 1801$	$3 \cdot 7^4$	$2 \cdot 13 \cdot 277$	$19 \cdot 379$	$2^5 \cdot 3^2 \cdot 5^2$
721	**7219**	$2 \cdot 3^2 \cdot 401$	$7 \cdot 1031$	$2^4 \cdot 11 \cdot 41$	$3 \cdot 5 \cdot 13 \cdot 37$	$2 \cdot 3607$	**7213**	$2^2 \cdot 3 \cdot 601$	**7211**	$2 \cdot 5 \cdot 7 \cdot 103$
722	**7229**	$2^2 \cdot 13 \cdot 139$	$3^2 \cdot 11 \cdot 73$	$2 \cdot 3613$	$5^2 \cdot 17^2$	$2^3 \cdot 3 \cdot 7 \cdot 43$	$31 \cdot 233$	$2 \cdot 23 \cdot 157$	$3 \cdot 29 \cdot 83$	$2^2 \cdot 5 \cdot 19^2$
723	$3 \cdot 19 \cdot 127$	$2 \cdot 7 \cdot 11 \cdot 47$	**7237**	$2^2 \cdot 3^3 \cdot 67$	$5 \cdot 1447$	$2 \cdot 3617$	$3 \cdot 2411$	$2^6 \cdot 113$	$7 \cdot 1033$	$2 \cdot 3 \cdot 5 \cdot 241$
724	$11 \cdot 659$	$2^4 \cdot 3 \cdot 151$	**7247**	$2 \cdot 3623$	$3^2 \cdot 5 \cdot 7 \cdot 23$	$2^2 \cdot 1811$	**7243**	$2 \cdot 3 \cdot 17 \cdot 71$	$13 \cdot 557$	$2^3 \cdot 5 \cdot 181$
725	$7 \cdot 17 \cdot 61$	$2 \cdot 19 \cdot 191$	$3 \cdot 41 \cdot 59$	$2^3 \cdot 907$	$5 \cdot 1451$	$2 \cdot 3^2 \cdot 13 \cdot 31$	**7253**	$2^2 \cdot 7^2 \cdot 37$	$3 \cdot 2417$	$2 \cdot 5^3 \cdot 29$
726	$3 \cdot 2423$	$2^2 \cdot 23 \cdot 79$	$13^2 \cdot 43$	$2 \cdot 3 \cdot 7 \cdot 173$	$5 \cdot 1453$	$2^5 \cdot 227$	$3^3 \cdot 269$	$2 \cdot 3631$	$53 \cdot 137$	$2^2 \cdot 3 \cdot 5 \cdot 11^2$
727	$29 \cdot 251$	$2 \cdot 3 \cdot 1213$	$19 \cdot 383$	$2^2 \cdot 17 \cdot 107$	$3 \cdot 5^2 \cdot 97$	$2 \cdot 3637$	$7 \cdot 1039$	$2^3 \cdot 3^2 \cdot 101$	$11 \cdot 661$	$2 \cdot 5 \cdot 727$
728	$37 \cdot 197$	$2^3 \cdot 911$	$3 \cdot 7 \cdot 347$	$2 \cdot 3643$	$5 \cdot 31 \cdot 47$	$2^2 \cdot 3 \cdot 607$	**7283**	$2 \cdot 11 \cdot 331$	$3^2 \cdot 809$	$2^4 \cdot 5 \cdot 7 \cdot 13$
729	$3^2 \cdot 811$	$2 \cdot 41 \cdot 89$	**7297**	$2^7 \cdot 3 \cdot 19$	$5 \cdot 1459$	$2 \cdot 7 \cdot 521$	$3 \cdot 11 \cdot 13 \cdot 17$	$2^2 \cdot 1823$	$23 \cdot 317$	$2 \cdot 3^6 \cdot 5$
730	**7309**	$2^2 \cdot 3^2 \cdot 7 \cdot 29$	**7307**	$2 \cdot 13 \cdot 281$	$3 \cdot 5 \cdot 487$	$2^3 \cdot 11 \cdot 83$	$67 \cdot 109$	$2 \cdot 3 \cdot 1217$	$7^2 \cdot 149$	$2^2 \cdot 5^2 \cdot 73$
731	$13 \cdot 563$	$2 \cdot 3659$	$3^3 \cdot 271$	$2^2 \cdot 31 \cdot 59$	$5 \cdot 7 \cdot 11 \cdot 19$	$2 \cdot 3 \cdot 23 \cdot 53$	$71 \cdot 103$	$2^4 \cdot 457$	$3 \cdot 2437$	$2 \cdot 5 \cdot 17 \cdot 43$
732	$3 \cdot 7 \cdot 349$	$2^5 \cdot 229$	$17 \cdot 431$	$2 \cdot 3^2 \cdot 11 \cdot 37$	$5^2 \cdot 293$	$2^2 \cdot 1831$	$3 \cdot 2441$	$2 \cdot 7 \cdot 523$	**7321**	$2^3 \cdot 3 \cdot 5 \cdot 61$
733	$41 \cdot 179$	$2 \cdot 3 \cdot 1223$	$11 \cdot 23 \cdot 29$	$2^3 \cdot 7 \cdot 131$	$3^2 \cdot 5 \cdot 163$	$2 \cdot 19 \cdot 193$	**7333**	$2^2 \cdot 3 \cdot 13 \cdot 47$	**7331**	$2 \cdot 5 \cdot 733$
734	**7349**	$2^2 \cdot 11 \cdot 167$	$3 \cdot 31 \cdot 79$	$2 \cdot 3673$	$5 \cdot 13 \cdot 113$	$2^4 \cdot 3^3 \cdot 17$	$7 \cdot 1049$	$2 \cdot 3671$	$3 \cdot 2447$	$2^2 \cdot 5 \cdot 367$
735	$3 \cdot 11 \cdot 223$	$2 \cdot 13 \cdot 283$	$7 \cdot 1051$	$2^2 \cdot 3 \cdot 613$	$5 \cdot 1471$	$2 \cdot 3677$	$3^2 \cdot 19 \cdot 43$	$2^3 \cdot 919$	**7351**	$2 \cdot 3 \cdot 5^2 \cdot 7^2$
736	**7369**	$2^3 \cdot 3 \cdot 307$	$53 \cdot 139$	$2 \cdot 29 \cdot 127$	$3 \cdot 5 \cdot 491$	$2^2 \cdot 7 \cdot 263$	$37 \cdot 199$	$2 \cdot 3^2 \cdot 409$	$17 \cdot 433$	$2^6 \cdot 5 \cdot 23$
737	$47 \cdot 157$	$2 \cdot 7 \cdot 17 \cdot 31$	$73 \cdot 101$	$2^4 \cdot 461$	$5^3 \cdot 59$	$2 \cdot 3 \cdot 1229$	$3 \cdot 2459$	$2^2 \cdot 19 \cdot 97$	$3^4 \cdot 7 \cdot 13$	$2 \cdot 5 \cdot 11 \cdot 67$
738	$3^2 \cdot 821$	$2^2 \cdot 1847$	$83 \cdot 89$	$2 \cdot 3 \cdot 1231$	$5 \cdot 7 \cdot 211$	$2^3 \cdot 13 \cdot 71$	$3 \cdot 23 \cdot 107$	$2 \cdot 3691$	$11^2 \cdot 61$	$2^2 \cdot 3^2 \cdot 5 \cdot 41$
739	$7^2 \cdot 151$	$2 \cdot 3^3 \cdot 137$	$13 \cdot 569$	$2^2 \cdot 43^2$	$3 \cdot 5 \cdot 17 \cdot 29$	$2 \cdot 3697$	**7393**	$2^5 \cdot 3 \cdot 7 \cdot 11$	$19 \cdot 389$	$2 \cdot 5 \cdot 739$
740	$31 \cdot 239$	$2^4 \cdot 463$	$3^2 \cdot 823$	$2 \cdot 7 \cdot 23^2$	$5 \cdot 1481$	$2^2 \cdot 3 \cdot 617$	$11 \cdot 673$	$2 \cdot 3701$	$3 \cdot 2467$	$2^3 \cdot 5^2 \cdot 37$
741	$3 \cdot 2473$	$2 \cdot 3709$	**7417**	$2^3 \cdot 3^2 \cdot 103$	$5 \cdot 1483$	$2 \cdot 11 \cdot 337$	$3 \cdot 7 \cdot 353$	$2^2 \cdot 17 \cdot 109$	**7411**	$2 \cdot 3 \cdot 5 \cdot 13 \cdot 19$
742	$17 \cdot 19 \cdot 23$	$2^2 \cdot 3 \cdot 619$	$7 \cdot 1061$	$2 \cdot 47 \cdot 79$	$3^3 \cdot 5^2 \cdot 11$	$2^8 \cdot 29$	$13 \cdot 571$	$2 \cdot 3 \cdot 1237$	$41 \cdot 181$	$2^2 \cdot 5 \cdot 7 \cdot 53$
743	$43 \cdot 173$	$2 \cdot 3719$	$3 \cdot 37 \cdot 67$	$2^2 \cdot 11 \cdot 13^2$	$5 \cdot 1487$	$2 \cdot 3^2 \cdot 7 \cdot 59$	**7433**	$2^3 \cdot 929$	$3 \cdot 2477$	$2 \cdot 5 \cdot 743$
744	$3 \cdot 13 \cdot 191$	$2^3 \cdot 7^2 \cdot 19$	$11 \cdot 677$	$2 \cdot 3 \cdot 17 \cdot 73$	$5 \cdot 1489$	$2^2 \cdot 1861$	$3^2 \cdot 827$	$2 \cdot 61^2$	$7 \cdot 1063$	$2^4 \cdot 3 \cdot 5 \cdot 31$
745	**7459**	$2 \cdot 3 \cdot 11 \cdot 113$	**7457**	$2^5 \cdot 233$	$3 \cdot 5 \cdot 7 \cdot 71$	$2 \cdot 3727$	$29 \cdot 257$	$2^2 \cdot 3^4 \cdot 23$	**7451**	$2 \cdot 5^2 \cdot 149$
746	$7 \cdot 11 \cdot 97$	$2^2 \cdot 1867$	$3 \cdot 19 \cdot 131$	$2 \cdot 3733$	$5 \cdot 1493$	$2^3 \cdot 3 \cdot 311$	$17 \cdot 439$	$2 \cdot 7 \cdot 13 \cdot 41$	$3^2 \cdot 829$	$2^2 \cdot 5 \cdot 373$
747	$3^3 \cdot 277$	$2 \cdot 3739$	**7477**	$2^2 \cdot 3 \cdot 7 \cdot 89$	$5^2 \cdot 13 \cdot 23$	$2 \cdot 37 \cdot 101$	$3 \cdot 47 \cdot 53$	$2^4 \cdot 467$	$31 \cdot 241$	$2 \cdot 3^2 \cdot 5 \cdot 83$
748	**7489**	$2^6 \cdot 3^2 \cdot 13$	**7487**	$2 \cdot 19 \cdot 197$	$3 \cdot 5 \cdot 499$	$2^2 \cdot 1871$	$7 \cdot 1069$	$2 \cdot 3 \cdot 29 \cdot 43$	**7481**	$2^3 \cdot 5 \cdot 11 \cdot 17$
749	**7499**	$2 \cdot 23 \cdot 163$	$3^2 \cdot 7^2 \cdot 17$	$2^3 \cdot 937$	$5 \cdot 1499$	$2 \cdot 3 \cdot 1249$	$59 \cdot 127$	$2^2 \cdot 1873$	$3 \cdot 11 \cdot 227$	$2 \cdot 5 \cdot 7 \cdot 107$

Factorizations

Table 24.7

In each row the base number N (first column) is combined with the last digit at the column head to give the number $10N + d$; the cell gives its prime factorization. (Bold entries are primes.)

N	9	8	7	6	5	4	3	2	1	0
750	$3\cdot2503$	$2^2\cdot1877$	$\mathbf{7507}$	$2\cdot3^3\cdot139$	$5\cdot19\cdot79$	$2^4\cdot7\cdot67$	$3\cdot41\cdot61$	$2\cdot11^2\cdot31$	$13\cdot577$	$2^2\cdot3\cdot5^4$
751	$73\cdot103$	$2\cdot3\cdot7\cdot179$	$\mathbf{7517}$	$2^2\cdot1879$	$3^2\cdot5\cdot167$	$2\cdot13\cdot17^2$	$11\cdot683$	$2^3\cdot3\cdot313$	$7\cdot29\cdot37$	$2\cdot5\cdot751$
752	$\mathbf{7529}$	$2^3\cdot941$	$3\cdot13\cdot193$	$2\cdot53\cdot71$	$5^2\cdot7\cdot43$	$2^2\cdot3^2\cdot11\cdot19$	$\mathbf{7523}$	$2\cdot3761$	$3\cdot23\cdot109$	$2^5\cdot5\cdot47$
753	$3\cdot7\cdot359$	$2\cdot3769$	$\mathbf{7537}$	$2^4\cdot3\cdot157$	$5\cdot11\cdot137$	$2\cdot3767$	$3^5\cdot31$	$2^2\cdot7\cdot269$	$17\cdot443$	$2\cdot3\cdot5\cdot251$
754	$\mathbf{7549}$	$2^2\cdot3\cdot17\cdot37$	$\mathbf{7547}$	$2\cdot7^3\cdot11$	$3\cdot5\cdot503$	$2^3\cdot23\cdot41$	$19\cdot397$	$2\cdot3^2\cdot419$	$\mathbf{7541}$	$2^2\cdot5\cdot13\cdot29$
755	$\mathbf{7559}$	$2\cdot3779$	$3\cdot11\cdot229$	$2^2\cdot1889$	$5\cdot1511$	$2\cdot3\cdot1259$	$7\cdot13\cdot83$	$2^7\cdot59$	$3^2\cdot839$	$2\cdot5^2\cdot151$
756	$3^2\cdot29^2$	$2^4\cdot11\cdot43$	$7\cdot23\cdot47$	$2\cdot3\cdot13\cdot97$	$5\cdot17\cdot89$	$2^2\cdot31\cdot61$	$3\cdot2521$	$2\cdot19\cdot199$	$\mathbf{7561}$	$2^3\cdot3^3\cdot5\cdot7$
757	$11\cdot13\cdot53$	$2\cdot3^2\cdot421$	$\mathbf{7577}$	$2^3\cdot947$	$3\cdot5^2\cdot101$	$2\cdot7\cdot541$	$\mathbf{7573}$	$2^2\cdot3\cdot631$	$67\cdot113$	$2\cdot5\cdot757$
758	$\mathbf{7589}$	$2^2\cdot7\cdot271$	$3^3\cdot281$	$2\cdot3793$	$5\cdot37\cdot41$	$2^5\cdot3\cdot79$	$\mathbf{7583}$	$2\cdot17\cdot223$	$3\cdot7\cdot19^2$	$2^2\cdot5\cdot379$
759	$3\cdot17\cdot149$	$2\cdot29\cdot131$	$71\cdot107$	$2^2\cdot3^2\cdot211$	$5\cdot7^2\cdot31$	$2\cdot3797$	$3\cdot2531$	$2^3\cdot13\cdot73$	$\mathbf{7591}$	$2\cdot3\cdot5\cdot11\cdot23$
760	$7\cdot1087$	$2^3\cdot3\cdot317$	$\mathbf{7607}$	$2\cdot3803$	$3^2\cdot5\cdot13^2$	$2^2\cdot1901$	$\mathbf{7603}$	$2\cdot3\cdot7\cdot181$	$11\cdot691$	$2^4\cdot5^2\cdot19$
761	$19\cdot401$	$2\cdot13\cdot293$	$3\cdot2539$	$2^6\cdot7\cdot17$	$5\cdot1523$	$2\cdot3^4\cdot47$	$23\cdot331$	$2^2\cdot11\cdot173$	$3\cdot43\cdot59$	$2\cdot5\cdot761$
762	$3\cdot2543$	$2^2\cdot1907$	$29\cdot263$	$2\cdot3\cdot31\cdot41$	$5^3\cdot61$	$2^3\cdot953$	$3^2\cdot7\cdot11^2$	$2\cdot37\cdot103$	$\mathbf{7621}$	$2^2\cdot3\cdot5\cdot127$
763	$\mathbf{7639}$	$2\cdot3\cdot19\cdot67$	$7\cdot1091$	$2^2\cdot23\cdot83$	$3\cdot5\cdot509$	$2\cdot11\cdot347$	$17\cdot449$	$2^4\cdot3^2\cdot53$	$13\cdot587$	$2\cdot5\cdot7\cdot109$
764	$\mathbf{7649}$	$2^5\cdot239$	$3\cdot2549$	$2\cdot3823$	$5\cdot11\cdot139$	$2^2\cdot3\cdot7^2\cdot13$	$\mathbf{7643}$	$2\cdot3821$	$3^3\cdot283$	$2^3\cdot5\cdot191$
765	$3^2\cdot23\cdot37$	$2\cdot7\cdot547$	$13\cdot19\cdot31$	$2^3\cdot3\cdot11\cdot29$	$5\cdot1531$	$2\cdot43\cdot89$	$3\cdot2551$	$2^2\cdot1913$	$7\cdot1093$	$2\cdot3^2\cdot5^2\cdot17$
766	$\mathbf{7669}$	$2^2\cdot3^3\cdot71$	$11\cdot17\cdot41$	$2\cdot3833$	$3\cdot5\cdot7\cdot73$	$2^4\cdot479$	$79\cdot97$	$2\cdot3\cdot1277$	$47\cdot163$	$2^2\cdot5\cdot383$
767	$7\cdot1097$	$2\cdot11\cdot349$	$3^2\cdot853$	$2^2\cdot19\cdot101$	$5^2\cdot307$	$2\cdot3\cdot1279$	$\mathbf{7673}$	$2^3\cdot7\cdot137$	$3\cdot2557$	$2\cdot5\cdot13\cdot59$
768	$3\cdot11\cdot233$	$2^3\cdot31^2$	$\mathbf{7687}$	$2\cdot3^2\cdot7\cdot61$	$5\cdot29\cdot53$	$2^2\cdot17\cdot113$	$3\cdot13\cdot197$	$2\cdot23\cdot167$	$\mathbf{7681}$	$2^9\cdot3\cdot5$
769	$\mathbf{7699}$	$2\cdot3\cdot1283$	$43\cdot179$	$2^4\cdot13\cdot37$	$3^4\cdot5\cdot19$	$2\cdot3847$	$7^2\cdot157$	$2^2\cdot3\cdot641$	$\mathbf{7691}$	$2\cdot5\cdot769$
770	$13\cdot593$	$2^2\cdot41\cdot47$	$3\cdot7\cdot367$	$2\cdot3853$	$5\cdot23\cdot67$	$2^3\cdot3^2\cdot107$	$\mathbf{7703}$	$2\cdot3851$	$3\cdot17\cdot151$	$2^2\cdot5^2\cdot7\cdot11$
771	$3\cdot31\cdot83$	$2\cdot17\cdot227$	$\mathbf{7717}$	$2^2\cdot3\cdot643$	$5\cdot1543$	$2\cdot7\cdot19\cdot29$	$3^2\cdot857$	$2^5\cdot241$	$11\cdot701$	$2\cdot3\cdot5\cdot257$
772	$59\cdot131$	$2^4\cdot3\cdot7\cdot23$	$\mathbf{7727}$	$2\cdot3863$	$3\cdot5^2\cdot103$	$2^2\cdot1931$	$\mathbf{7723}$	$2\cdot3^3\cdot11\cdot13$	$7\cdot1103$	$2^3\cdot5\cdot193$
773	$71\cdot109$	$2\cdot53\cdot73$	$3\cdot2579$	$2^3\cdot967$	$5\cdot7\cdot13\cdot17$	$2\cdot3\cdot1289$	$11\cdot19\cdot37$	$2^2\cdot1933$	$3^2\cdot859$	$2\cdot5\cdot773$
774	$3^3\cdot7\cdot41$	$2^2\cdot13\cdot149$	$61\cdot127$	$2\cdot3\cdot1291$	$5\cdot1549$	$2^6\cdot11^2$	$3\cdot29\cdot89$	$2\cdot7^2\cdot79$	$\mathbf{7741}$	$2^2\cdot3^2\cdot5\cdot43$
775	$\mathbf{7759}$	$2\cdot3^2\cdot431$	$\mathbf{7757}$	$2^2\cdot7\cdot277$	$3\cdot5\cdot11\cdot47$	$2\cdot3877$	$\mathbf{7753}$	$2^3\cdot3\cdot17\cdot19$	$23\cdot337$	$2\cdot5^3\cdot31$
776	$17\cdot457$	$2^3\cdot971$	$3^2\cdot863$	$2\cdot11\cdot353$	$5\cdot1553$	$2^2\cdot3\cdot647$	$7\cdot1109$	$2\cdot3881$	$3\cdot13\cdot199$	$2^4\cdot5\cdot97$
777	$3\cdot2593$	$2\cdot3889$	$7\cdot11\cdot101$	$2^5\cdot3^5$	$5^2\cdot311$	$2\cdot13^2\cdot23$	$3\cdot2591$	$2^2\cdot29\cdot67$	$19\cdot409$	$2\cdot3\cdot5\cdot7\cdot37$
778	$\mathbf{7789}$	$2^2\cdot3\cdot11\cdot59$	$13\cdot599$	$2\cdot17\cdot229$	$3^2\cdot5\cdot173$	$2^3\cdot7\cdot139$	$43\cdot181$	$2\cdot3\cdot1297$	$31\cdot251$	$2^2\cdot5\cdot389$
779	$11\cdot709$	$2\cdot7\cdot557$	$3\cdot23\cdot113$	$2^2\cdot1949$	$5\cdot1559$	$2\cdot3^2\cdot433$	$\mathbf{7793}$	$2^4\cdot487$	$3\cdot7^2\cdot53$	$2\cdot5\cdot19\cdot41$
780	$3\cdot19\cdot137$	$2^7\cdot61$	$37\cdot211$	$2\cdot3\cdot1301$	$5\cdot7\cdot223$	$2^2\cdot1951$	$3^3\cdot17^2$	$2\cdot47\cdot83$	$29\cdot269$	$2^3\cdot3\cdot5^2\cdot13$
781	$7\cdot1117$	$2\cdot3\cdot1303$	$\mathbf{7817}$	$2^3\cdot977$	$3\cdot5\cdot521$	$2\cdot3907$	$13\cdot601$	$2^2\cdot3^2\cdot7\cdot31$	$73\cdot107$	$2\cdot5\cdot11\cdot71$
782	$\mathbf{7829}$	$2^2\cdot19\cdot103$	$3\cdot2609$	$2\cdot7\cdot13\cdot43$	$5^2\cdot313$	$2^4\cdot3\cdot163$	$\mathbf{7823}$	$2\cdot3911$	$3^2\cdot11\cdot79$	$2^2\cdot5\cdot17\cdot23$
783	$3^2\cdot13\cdot67$	$2\cdot3919$	$17\cdot461$	$2^2\cdot3\cdot653$	$5\cdot1567$	$2\cdot3917$	$3\cdot7\cdot373$	$2^3\cdot11\cdot89$	$41\cdot191$	$2\cdot3^3\cdot5\cdot29$
784	$47\cdot167$	$2^3\cdot3^2\cdot109$	$7\cdot19\cdot59$	$2\cdot3923$	$3\cdot5\cdot523$	$2^2\cdot37\cdot53$	$11\cdot23\cdot31$	$2\cdot3\cdot1307$	$\mathbf{7841}$	$2^5\cdot5\cdot7^2$
785	$29\cdot271$	$2\cdot3929$	$3^4\cdot97$	$2^4\cdot491$	$5\cdot1571$	$2\cdot3\cdot7\cdot11\cdot17$	$\mathbf{7853}$	$2^2\cdot13\cdot151$	$3\cdot2617$	$2\cdot5^2\cdot157$
786	$3\cdot43\cdot61$	$2^2\cdot7\cdot281$	$\mathbf{7867}$	$2\cdot3^2\cdot19\cdot23$	$5\cdot11^2\cdot13$	$2^3\cdot983$	$3\cdot2621$	$2\cdot3931$	$7\cdot1123$	$2^2\cdot3\cdot5\cdot131$
787	$\mathbf{7879}$	$2\cdot3\cdot13\cdot101$	$\mathbf{7877}$	$2^2\cdot11\cdot179$	$3^2\cdot5^3\cdot7$	$2\cdot31\cdot127$	$\mathbf{7873}$	$2^6\cdot3\cdot41$	$17\cdot463$	$2\cdot5\cdot787$
788	$7^3\cdot23$	$2^4\cdot17\cdot29$	$3\cdot11\cdot239$	$2\cdot3943$	$5\cdot19\cdot83$	$2^2\cdot3^3\cdot73$	$\mathbf{7883}$	$2\cdot7\cdot563$	$3\cdot37\cdot71$	$2^3\cdot5\cdot197$
789	$3\cdot2633$	$2\cdot11\cdot359$	$53\cdot149$	$2^3\cdot3\cdot7\cdot47$	$5\cdot1579$	$2\cdot3947$	$3^2\cdot877$	$2^2\cdot1973$	$13\cdot607$	$2\cdot3\cdot5\cdot263$
790	$11\cdot719$	$2^2\cdot3\cdot659$	$\mathbf{7907}$	$2\cdot59\cdot67$	$3\cdot5\cdot17\cdot31$	$2^5\cdot13\cdot19$	$7\cdot1129$	$2\cdot3^2\cdot439$	$\mathbf{7901}$	$2^2\cdot5^2\cdot79$
791	$\mathbf{7919}$	$2\cdot37\cdot107$	$3\cdot7\cdot13\cdot29$	$2^2\cdot1979$	$5\cdot1583$	$2\cdot3\cdot1319$	$41\cdot193$	$2^3\cdot23\cdot43$	$3^3\cdot293$	$2\cdot5\cdot7\cdot113$
792	$3^2\cdot881$	$2^3\cdot991$	$\mathbf{7927}$	$2\cdot3\cdot1321$	$5^2\cdot317$	$2^2\cdot7\cdot283$	$3\cdot19\cdot139$	$2\cdot17\cdot233$	89^2	$2^4\cdot3^2\cdot5\cdot11$
793	$17\cdot467$	$2\cdot3^4\cdot7^2$	$\mathbf{7937}$	$2^8\cdot31$	$3\cdot5\cdot23^2$	$2\cdot3967$	$\mathbf{7933}$	$2^2\cdot3\cdot661$	$7\cdot11\cdot103$	$2\cdot5\cdot13\cdot61$
794	$\mathbf{7949}$	$2^2\cdot1987$	$3^2\cdot883$	$2\cdot29\cdot137$	$5\cdot7\cdot227$	$2^3\cdot3\cdot331$	$13^2\cdot47$	$2\cdot11\cdot19^2$	$3\cdot2647$	$2^2\cdot5\cdot397$
795	$3\cdot7\cdot379$	$2\cdot23\cdot173$	$73\cdot109$	$2^2\cdot3^2\cdot13\cdot17$	$5\cdot37\cdot43$	$2\cdot41\cdot97$	$3\cdot11\cdot241$	$2^4\cdot7\cdot71$	$\mathbf{7951}$	$2\cdot3\cdot5^2\cdot53$
796	$13\cdot613$	$2^5\cdot3\cdot83$	$31\cdot257$	$2\cdot7\cdot569$	$3^3\cdot5\cdot59$	$2^2\cdot11\cdot181$	$\mathbf{7963}$	$2\cdot3\cdot1327$	$19\cdot419$	$2^3\cdot5\cdot199$
797	$79\cdot101$	$2\cdot3989$	$3\cdot2659$	$2^3\cdot997$	$5^2\cdot11\cdot29$	$2\cdot3^2\cdot443$	$7\cdot17\cdot67$	$2^2\cdot1993$	$3\cdot2657$	$2\cdot5\cdot797$
798	$3\cdot2663$	$2^2\cdot1997$	$7^2\cdot163$	$2\cdot3\cdot11^3$	$5\cdot1597$	$2^4\cdot499$	$3^2\cdot887$	$2\cdot13\cdot307$	$23\cdot347$	$2^2\cdot3\cdot5\cdot7\cdot19$
799	$19\cdot421$	$2\cdot3\cdot31\cdot43$	$11\cdot727$	$2^2\cdot1999$	$3\cdot5\cdot13\cdot41$	$2\cdot7\cdot571$	$\mathbf{7993}$	$2^3\cdot3^3\cdot37$	$61\cdot131$	$2\cdot5\cdot17\cdot47$

Table 24.7 **Factorizations**

N	9	8	7	6	5	4	3	2	1	0	N
800	**8009**	$2^3\cdot7\cdot11\cdot13$	$3\cdot17\cdot157$	$2\cdot4003$	$5\cdot1601$	$2^2\cdot3\cdot23\cdot29$	$53\cdot151$	$2\cdot4001$	$3^2\cdot7\cdot127$	$2^6\cdot5^3$	800
801	$3^6\cdot11$	$2\cdot19\cdot211$	**8017**	$2^4\cdot3\cdot167$	$5\cdot7\cdot229$	$2\cdot4007$	$3\cdot2671$	$2^2\cdot2003$	**8011**	$2\cdot3^2\cdot5\cdot89$	801
802	$7\cdot31\cdot37$	$2^2\cdot3^2\cdot223$	$23\cdot349$	$2\cdot4013$	$3\cdot5^2\cdot107$	$2^3\cdot17\cdot59$	$71\cdot113$	$2\cdot3\cdot7\cdot191$	$13\cdot617$	$2^2\cdot5\cdot401$	802
803	**8039**	$2\cdot4019$	$3^2\cdot19\cdot47$	$2^2\cdot7^2\cdot41$	$5\cdot1607$	$2\cdot3\cdot13\cdot103$	$29\cdot277$	$2^5\cdot251$	$3\cdot2677$	$2\cdot5\cdot11\cdot73$	803
804	$3\cdot2683$	$2^4\cdot503$	$13\cdot619$	$2\cdot3^3\cdot149$	$5\cdot1609$	$2^2\cdot2011$	$3\cdot7\cdot383$	$2\cdot4021$	$11\cdot17\cdot43$	$2^3\cdot3\cdot5\cdot67$	804
805	**8059**	$2\cdot3\cdot17\cdot79$	$7\cdot1151$	$2^3\cdot19\cdot53$	$3^2\cdot5\cdot179$	$2\cdot4027$	**8053**	$2^2\cdot3\cdot11\cdot61$	$83\cdot97$	$2\cdot5^2\cdot7\cdot23$	805
806	**8069**	$2^2\cdot2017$	$3\cdot2689$	$2\cdot37\cdot109$	$5\cdot1613$	$2^7\cdot3^2\cdot7$	$11\cdot733$	$2\cdot29\cdot139$	$3\cdot2687$	$2^2\cdot5\cdot13\cdot31$	806
807	$3\cdot2693$	$2\cdot7\cdot577$	$41\cdot197$	$2^2\cdot3\cdot673$	$5^2\cdot17\cdot19$	$2\cdot11\cdot367$	$3^3\cdot13\cdot23$	$2^3\cdot1009$	$7\cdot1153$	$2\cdot3\cdot5\cdot269$	807
808	**8089**	$2^3\cdot3\cdot337$	**8087**	$2\cdot13\cdot311$	$3\cdot5\cdot7^2\cdot11$	$2^2\cdot43\cdot47$	$59\cdot137$	$2\cdot3^2\cdot449$	**8081**	$2^4\cdot5\cdot101$	808
809	$7\cdot13\cdot89$	$2\cdot4049$	$3\cdot2699$	$2^5\cdot11\cdot23$	$5\cdot1619$	$2\cdot3\cdot19\cdot71$	**8093**	$2^2\cdot7\cdot17^2$	$3^2\cdot29\cdot31$	$2\cdot5\cdot809$	809
810	$3^2\cdot17\cdot53$	$2^2\cdot2027$	$11^2\cdot67$	$2\cdot3\cdot7\cdot193$	$5\cdot1621$	$2^3\cdot1013$	$3\cdot37\cdot73$	$2\cdot4051$	**8101**	$2^2\cdot3^4\cdot5^2$	810
811	$23\cdot353$	$2\cdot3^2\cdot11\cdot41$	**8117**	$2^2\cdot2029$	$3\cdot5\cdot541$	$2\cdot4057$	$7\cdot19\cdot61$	$2^4\cdot3\cdot13^2$	**8111**	$2\cdot5\cdot811$	811
812	$11\cdot739$	$2^6\cdot127$	$3^3\cdot7\cdot43$	$2\cdot17\cdot239$	$5^4\cdot13$	$2^2\cdot3\cdot677$	**8123**	$2\cdot31\cdot131$	$3\cdot2707$	$2^3\cdot5\cdot7\cdot29$	812
813	$3\cdot2713$	$2\cdot13\cdot313$	$79\cdot103$	$2^3\cdot3^2\cdot113$	$5\cdot1627$	$2\cdot7^2\cdot83$	$3\cdot2711$	$2^2\cdot19\cdot107$	$47\cdot173$	$2\cdot3\cdot5\cdot271$	813
814	$29\cdot281$	$2^2\cdot3\cdot7\cdot97$	**8147**	$2\cdot4073$	$3^2\cdot5\cdot181$	$2^4\cdot509$	$17\cdot479$	$2\cdot3\cdot23\cdot59$	$7\cdot1163$	$2^2\cdot5\cdot11\cdot37$	814
815	$41\cdot199$	$2\cdot4079$	$3\cdot2719$	$2^2\cdot2039$	$5\cdot7\cdot233$	$2\cdot3^3\cdot151$	$31\cdot263$	$2^3\cdot1019$	$3\cdot11\cdot13\cdot19$	$2\cdot5^2\cdot163$	815
816	$3\cdot7\cdot389$	$2^3\cdot1021$	**8167**	$2\cdot3\cdot1361$	$5\cdot23\cdot71$	$2^2\cdot13\cdot157$	$3^2\cdot907$	$2\cdot7\cdot11\cdot53$	**8161**	$2^5\cdot3\cdot5\cdot17$	816
817	**8179**	$2\cdot3\cdot29\cdot47$	$13\cdot17\cdot37$	$2^4\cdot7\cdot73$	$3\cdot5^2\cdot109$	$2\cdot61\cdot67$	$11\cdot743$	$2^2\cdot3^2\cdot227$	**8171**	$2\cdot5\cdot19\cdot43$	817
818	$19\cdot431$	$2^2\cdot23\cdot89$	$3\cdot2729$	$2\cdot4093$	$5\cdot1637$	$2^3\cdot3\cdot11\cdot31$	$7^2\cdot167$	$2\cdot4091$	$3^4\cdot101$	$2^2\cdot5\cdot409$	818
819	$3^2\cdot911$	$2\cdot4099$	$7\cdot1171$	$2^2\cdot3\cdot683$	$5\cdot11\cdot149$	$2\cdot17\cdot241$	$3\cdot2731$	2^{13}	**8191**	$2\cdot3^2\cdot5\cdot7\cdot13$	819
820	**8209**	$2^4\cdot3^3\cdot19$	$29\cdot283$	$2\cdot11\cdot373$	$3\cdot5\cdot547$	$2^2\cdot7\cdot293$	$13\cdot631$	$2\cdot3\cdot1367$	$59\cdot139$	$2^3\cdot5^2\cdot41$	820
821	**8219**	$2\cdot7\cdot587$	$3^2\cdot11\cdot83$	$2^3\cdot13\cdot79$	$5\cdot31\cdot53$	$2\cdot3\cdot37^2$	$43\cdot191$	$2^2\cdot2053$	$3\cdot7\cdot17\cdot23$	$2\cdot5\cdot821$	821
822	$3\cdot13\cdot211$	$2^2\cdot11^2\cdot17$	$19\cdot433$	$2\cdot3^2\cdot457$	$5^2\cdot7\cdot47$	$2^5\cdot257$	$3\cdot2741$	$2\cdot4111$	**8221**	$2^2\cdot3\cdot5\cdot137$	822
823	$7\cdot11\cdot107$	$2\cdot3\cdot1373$	**8237**	$2^2\cdot29\cdot71$	$3^3\cdot5\cdot61$	$2\cdot23\cdot179$	**8233**	$2^3\cdot3\cdot7^3$	**8231**	$2\cdot5\cdot823$	823
824	$73\cdot113$	$2^3\cdot1031$	$3\cdot2749$	$2\cdot7\cdot19\cdot31$	$5\cdot17\cdot97$	$2^2\cdot3^2\cdot229$	**8243**	$2\cdot13\cdot317$	$3\cdot41\cdot67$	$2^4\cdot5\cdot103$	824
825	$3\cdot2753$	$2\cdot4129$	$23\cdot359$	$2^6\cdot3\cdot43$	$5\cdot13\cdot127$	$2\cdot4127$	$3^2\cdot7\cdot131$	$2^2\cdot2063$	$37\cdot223$	$2\cdot3\cdot5^3\cdot11$	825
826	**8269**	$2^2\cdot3\cdot13\cdot53$	$7\cdot1181$	$2\cdot4133$	$3\cdot5\cdot19\cdot29$	$2^3\cdot1033$	**8263**	$2\cdot3^5\cdot17$	$11\cdot751$	$2^2\cdot5\cdot7\cdot59$	826
827	$17\cdot487$	$2\cdot4139$	$3\cdot31\cdot89$	$2^2\cdot2069$	$5^2\cdot331$	$2\cdot3\cdot7\cdot197$	**8273**	$2^4\cdot11\cdot47$	$3^2\cdot919$	$2\cdot5\cdot827$	827
828	$3^3\cdot307$	$2^5\cdot7\cdot37$	**8287**	$2\cdot3\cdot1381$	$5\cdot1657$	$2^2\cdot19\cdot109$	$3\cdot11\cdot251$	$2\cdot41\cdot101$	$7^2\cdot13^2$	$2^3\cdot3^2\cdot5\cdot23$	828
829	$43\cdot193$	$2\cdot3^2\cdot461$	**8297**	$2^3\cdot17\cdot61$	$3\cdot5\cdot7\cdot79$	$2\cdot11\cdot13\cdot29$	**8293**	$2^2\cdot3\cdot691$	**8291**	$2\cdot5\cdot829$	829
830	$7\cdot1187$	$2^2\cdot31\cdot67$	$3^2\cdot13\cdot71$	$2\cdot4153$	$5\cdot11\cdot151$	$2^4\cdot3\cdot173$	$19^2\cdot23$	$2\cdot7\cdot593$	$3\cdot2767$	$2^2\cdot5^2\cdot83$	830
831	$3\cdot47\cdot59$	$2\cdot4159$	**8317**	$2^2\cdot3^3\cdot7\cdot11$	$5\cdot1663$	$2\cdot4157$	$3\cdot17\cdot163$	$2^3\cdot1039$	**8311**	$2\cdot3\cdot5\cdot277$	831
832	**8329**	$2^3\cdot3\cdot347$	$11\cdot757$	$2\cdot23\cdot181$	$3^2\cdot5^2\cdot37$	$2^2\cdot2081$	$7\cdot29\cdot41$	$2\cdot3\cdot19\cdot73$	$53\cdot157$	$2^7\cdot5\cdot13$	832
833	$31\cdot269$	$2\cdot11\cdot379$	$3\cdot7\cdot397$	$2^4\cdot521$	$5\cdot1667$	$2\cdot3^2\cdot463$	$13\cdot641$	$2^2\cdot2083$	$3\cdot2777$	$2\cdot5\cdot7^2\cdot17$	833
834	$3\cdot11^2\cdot23$	$2^2\cdot2087$	$17\cdot491$	$2\cdot3\cdot13\cdot107$	$5\cdot1669$	$2^3\cdot7\cdot149$	$3^4\cdot103$	$2\cdot43\cdot97$	$19\cdot439$	$2^2\cdot3\cdot5\cdot139$	834
835	$13\cdot643$	$2\cdot3\cdot7\cdot199$	$61\cdot137$	$2^2\cdot2089$	$3\cdot5\cdot557$	$2\cdot4177$	**8353**	$2^5\cdot3^2\cdot29$	$7\cdot1193$	$2\cdot5^2\cdot167$	835
836	**8369**	$2^4\cdot523$	$3\cdot2789$	$2\cdot47\cdot89$	$5\cdot7\cdot239$	$2^2\cdot3\cdot17\cdot41$	**8363**	$2\cdot37\cdot113$	$3^2\cdot929$	$2^3\cdot5\cdot11\cdot19$	836
837	$3^2\cdot7^2\cdot19$	$2\cdot59\cdot71$	**8377**	$2^3\cdot3\cdot349$	$5^3\cdot67$	$2\cdot53\cdot79$	$3\cdot2791$	$2^2\cdot7\cdot13\cdot23$	$11\cdot761$	$2\cdot3^3\cdot5\cdot31$	837
838	**8389**	$2\cdot3^2\cdot233$	**8387**	$2\cdot7\cdot599$	$3\cdot5\cdot13\cdot43$	$2^6\cdot131$	$83\cdot101$	$2\cdot3\cdot11\cdot127$	$17^2\cdot29$	$2^2\cdot5\cdot419$	838
839	$37\cdot227$	$2\cdot13\cdot17\cdot19$	$3^3\cdot311$	$2^2\cdot2099$	$5\cdot23\cdot73$	$2\cdot3\cdot1399$	$7\cdot11\cdot109$	$2^3\cdot1049$	$3\cdot2797$	$2\cdot5\cdot839$	839
840	$3\cdot2803$	$2^3\cdot1051$	$7\cdot1201$	$2\cdot3^2\cdot467$	$5\cdot41^2$	$2^2\cdot11\cdot191$	$3\cdot2801$	$2\cdot4201$	$31\cdot271$	$2^4\cdot3\cdot5^2\cdot7$	840
841	**8419**	$2\cdot3\cdot23\cdot61$	$19\cdot443$	$2^5\cdot263$	$3^2\cdot5\cdot11\cdot17$	$2\cdot7\cdot601$	$47\cdot179$	$2^2\cdot3\cdot701$	$13\cdot647$	$2\cdot5\cdot29^2$	841
842	**8429**	$2^2\cdot7^2\cdot43$	$3\cdot53^2$	$2\cdot11\cdot383$	$5^2\cdot337$	$2^3\cdot3^4\cdot13$	**8423**	$2\cdot4211$	$3\cdot7\cdot401$	$2^2\cdot5\cdot421$	842
843	$3\cdot29\cdot97$	$2\cdot4219$	$11\cdot13\cdot59$	$2^2\cdot3\cdot19\cdot37$	$5\cdot7\cdot241$	$2\cdot4217$	$3^2\cdot937$	$2^4\cdot17\cdot31$	**8431**	$2\cdot3\cdot5\cdot281$	843
844	$7\cdot17\cdot71$	$2^8\cdot3\cdot11$	**8447**	$2\cdot41\cdot103$	$3\cdot5\cdot563$	$2^2\cdot2111$	**8443**	$2\cdot3^2\cdot7\cdot67$	$23\cdot367$	$2^3\cdot5\cdot211$	844
845	$11\cdot769$	$2\cdot4229$	$3\cdot2819$	$2^3\cdot7\cdot151$	$5\cdot19\cdot89$	$2\cdot3\cdot1409$	$79\cdot107$	$2^2\cdot2113$	$3^3\cdot313$	$2\cdot5^2\cdot13^2$	845
846	$3^2\cdot941$	$2^2\cdot29\cdot73$	**8467**	$2\cdot3\cdot17\cdot83$	$5\cdot1693$	$2^4\cdot23^2$	$3\cdot7\cdot13\cdot31$	$2\cdot4231$	**8461**	$2^2\cdot3^2\cdot5\cdot47$	846
847	$61\cdot139$	$2\cdot3^3\cdot157$	$7^2\cdot173$	$2^2\cdot13\cdot163$	$3\cdot5^2\cdot113$	$2\cdot19\cdot223$	$37\cdot229$	$2^3\cdot3\cdot353$	$43\cdot197$	$2\cdot5\cdot7\cdot11^2$	847
848	$13\cdot653$	$2^3\cdot1061$	$3^2\cdot23\cdot41$	$2\cdot4243$	$5\cdot1697$	$2^2\cdot3\cdot7\cdot101$	$17\cdot499$	$2\cdot4241$	$3\cdot11\cdot257$	$2^5\cdot5\cdot53$	848
849	$3\cdot2833$	$2\cdot7\cdot607$	$29\cdot293$	$2^4\cdot3^2\cdot59$	$5\cdot1699$	$2\cdot31\cdot137$	$3\cdot19\cdot149$	$2^2\cdot11\cdot193$	$7\cdot1213$	$2\cdot3\cdot5\cdot283$	849

Factorizations

Table 24.7

8500 – 8999

Each entry gives the factorization of $N = 10n + d$, where n is the row index (850–899) and d is the final-digit column (0–9). Bold entries are primes.

n	$+0$	$+1$	$+2$	$+3$	$+4$	$+5$	$+6$	$+7$	$+8$	$+9$
850	$2^2\cdot5^3\cdot17$	$\mathbf{8501}$	$2\cdot3\cdot13\cdot109$	$11\cdot773$	$2^3\cdot1063$	$3^5\cdot5\cdot7$	$2\cdot4253$	$47\cdot181$	$2^2\cdot3\cdot709$	$67\cdot127$
851	$2\cdot5\cdot23\cdot37$	$3\cdot2837$	$2^6\cdot7\cdot19$	$\mathbf{8513}$	$2\cdot3^2\cdot11\cdot43$	$5\cdot13\cdot131$	$2^2\cdot2129$	$3\cdot17\cdot167$	$2\cdot4259$	$7\cdot1217$
852	$2^3\cdot3\cdot5\cdot71$	$\mathbf{8521}$	$2\cdot4261$	$3^3\cdot947$	$2^2\cdot2131$	$5^2\cdot11\cdot31$	$2\cdot3\cdot7^2\cdot29$	$\mathbf{8527}$	$2^4\cdot13\cdot41$	$3\cdot2843$
853	$2\cdot5\cdot853$	$19\cdot449$	$2\cdot3^3\cdot79$	$7\cdot23\cdot53$	$2\cdot17\cdot251$	$3\cdot5\cdot569$	$2^3\cdot11\cdot97$	$\mathbf{8537}$	$2\cdot3\cdot1423$	$\mathbf{8539}$
854	$2^2\cdot5\cdot7\cdot61$	$3^2\cdot13\cdot73$	$2\cdot4271$	$\mathbf{8543}$	$2^5\cdot3\cdot89$	$5\cdot1709$	$2\cdot4273$	$3\cdot7\cdot11\cdot37$	$2^2\cdot2137$	$83\cdot103$
855	$2\cdot3^2\cdot5^2\cdot19$	$17\cdot503$	$2^3\cdot1069$	$3\cdot2851$	$2\cdot7\cdot13\cdot47$	$5\cdot29\cdot59$	$2^2\cdot3\cdot23\cdot31$	$43\cdot199$	$2\cdot11\cdot389$	$3^3\cdot317$
856	$2^4\cdot5\cdot107$	$7\cdot1223$	$2\cdot3\cdot1427$	$\mathbf{8563}$	$2^2\cdot2141$	$3\cdot5\cdot571$	$2\cdot4283$	$13\cdot659$	$2^3\cdot3^2\cdot7\cdot17$	$11\cdot19\cdot41$
857	$2\cdot5\cdot857$	$3\cdot2857$	$2^2\cdot2143$	$\mathbf{8573}$	$2\cdot3\cdot1429$	$5^2\cdot7^3$	$2^7\cdot67$	$3^2\cdot953$	$2^2\cdot4289$	$23\cdot373$
858	$2^2\cdot3\cdot5\cdot11\cdot13$	$\mathbf{8581}$	$2\cdot7\cdot613$	$3\cdot2861$	$2^3\cdot29\cdot37$	$5\cdot17\cdot101$	$2\cdot3^4\cdot53$	$31\cdot277$	$2\cdot19\cdot113$	$3\cdot7\cdot409$
859	$2\cdot5\cdot859$	$11^2\cdot71$	$2^4\cdot3\cdot179$	$13\cdot661$	$2\cdot4297$	$3^2\cdot5\cdot191$	$2^2\cdot7\cdot307$	$\mathbf{8597}$	$2\cdot3\cdot1433$	$\mathbf{8599}$
860	$2^3\cdot5^2\cdot43$	$3\cdot47\cdot61$	$2\cdot11\cdot17\cdot23$	$7\cdot1229$	$2^2\cdot3^2\cdot239$	$5\cdot1721$	$2\cdot13\cdot331$	$3\cdot19\cdot151$	$2^5\cdot269$	$\mathbf{8609}$
861	$2\cdot3\cdot5\cdot7\cdot41$	$79\cdot109$	$2^2\cdot2153$	$3^3\cdot11\cdot29$	$2\cdot59\cdot73$	$5\cdot1723$	$2^3\cdot3\cdot359$	$7\cdot1231$	$2\cdot31\cdot139$	$3\cdot13^2\cdot17$
862	$2^2\cdot5\cdot431$	$37\cdot233$	$2\cdot3^2\cdot479$	$\mathbf{8623}$	$2^4\cdot7^2\cdot11$	$3\cdot5^3\cdot23$	$2\cdot19\cdot227$	$\mathbf{8627}$	$2^3\cdot3\cdot719$	$\mathbf{8629}$
863	$2\cdot5\cdot863$	$3^2\cdot7\cdot137$	$2^3\cdot13\cdot83$	$89\cdot97$	$2\cdot3\cdot1439$	$5\cdot11\cdot157$	$2^2\cdot17\cdot127$	$3\cdot2879$	$2\cdot7\cdot617$	$53\cdot163$
864	$2^6\cdot3^3\cdot5$	$\mathbf{8641}$	$2\cdot29\cdot149$	$3\cdot43\cdot67$	$2^2\cdot2161$	$5\cdot7\cdot13\cdot19$	$2\cdot3\cdot11\cdot131$	$\mathbf{8647}$	$2^3\cdot23\cdot47$	$3^2\cdot31^2$
865	$2\cdot5^2\cdot173$	$41\cdot211$	$2^2\cdot3\cdot7\cdot103$	$17\cdot509$	$2\cdot4327$	$3\cdot5\cdot577$	$2^4\cdot541$	$11\cdot787$	$2\cdot3\cdot13\cdot37$	$7\cdot1237$
866	$2^2\cdot5\cdot433$	$3\cdot2887$	$2\cdot61\cdot71$	$\mathbf{8663}$	$2^3\cdot3\cdot19^2$	$5\cdot1733$	$2\cdot7\cdot619$	$3^4\cdot107$	$2^3\cdot11\cdot197$	$\mathbf{8669}$
867	$2\cdot3\cdot5\cdot17^2$	$13\cdot23\cdot29$	$2^5\cdot271$	$3\cdot7^2\cdot59$	$2\cdot4337$	$5^2\cdot347$	$2^2\cdot3^2\cdot241$	$\mathbf{8677}$	$2\cdot4339$	$3\cdot11\cdot263$
868	$2^3\cdot5\cdot7\cdot31$	$\mathbf{8681}$	$2\cdot3\cdot1447$	$19\cdot457$	$2^2\cdot13\cdot167$	$3^2\cdot5\cdot193$	$2\cdot43\cdot101$	$7\cdot17\cdot73$	$2^4\cdot3\cdot181$	$\mathbf{8689}$
869	$2\cdot5\cdot11\cdot79$	$3\cdot2897$	$2^2\cdot41\cdot53$	$\mathbf{8693}$	$2\cdot3^3\cdot7\cdot23$	$5\cdot37\cdot47$	$2^3\cdot1087$	$3\cdot13\cdot223$	$2\cdot4349$	$\mathbf{8699}$
870	$2^2\cdot3\cdot5^2\cdot29$	$7\cdot11\cdot113$	$2\cdot19\cdot229$	$3^2\cdot967$	$2^9\cdot17$	$5\cdot1741$	$2\cdot3\cdot1451$	$\mathbf{8707}$	$2^2\cdot7\cdot311$	$3\cdot2903$
871	$2\cdot5\cdot13\cdot67$	$31\cdot281$	$2^3\cdot3^2\cdot11^2$	$\mathbf{8713}$	$2\cdot4357$	$3\cdot5\cdot7\cdot83$	$2^2\cdot2179$	$23\cdot379$	$2\cdot3\cdot1453$	$\mathbf{8719}$
872	$2^4\cdot5\cdot109$	$3^3\cdot17\cdot19$	$2\cdot7^2\cdot89$	$11\cdot13\cdot61$	$2^2\cdot3\cdot727$	$5^2\cdot349$	$2\cdot4363$	$3\cdot2909$	$2^3\cdot1091$	$7\cdot29\cdot43$
873	$2\cdot3^2\cdot5\cdot97$	$\mathbf{8731}$	$2^2\cdot37\cdot59$	$3\cdot41\cdot71$	$2\cdot11\cdot397$	$5\cdot1747$	$2^5\cdot3\cdot7\cdot13$	$\mathbf{8737}$	$2\cdot17\cdot257$	$3^2\cdot971$
874	$2^2\cdot5\cdot19\cdot23$	$\mathbf{8741}$	$2\cdot3\cdot31\cdot47$	$7\cdot1249$	$2^3\cdot1093$	$3\cdot5\cdot11\cdot53$	$2\cdot4373$	$\mathbf{8747}$	$2^2\cdot3^7$	$13\cdot673$
875	$2\cdot5^4\cdot7$	$3\cdot2917$	$2^4\cdot547$	$\mathbf{8753}$	$2\cdot3\cdot1459$	$5\cdot17\cdot103$	$2^2\cdot11\cdot199$	$3^2\cdot7\cdot139$	$2\cdot29\cdot151$	$19\cdot461$
876	$2^3\cdot3\cdot5\cdot73$	$\mathbf{8761}$	$2\cdot13\cdot337$	$3\cdot23\cdot127$	$2^2\cdot7\cdot313$	$5\cdot1753$	$2\cdot3^2\cdot487$	$11\cdot797$	$2^5\cdot137$	$3\cdot37\cdot79$
877	$2\cdot5\cdot877$	$7^2\cdot179$	$2^2\cdot3\cdot17\cdot43$	$31\cdot283$	$2\cdot41\cdot107$	$3^3\cdot5^2\cdot13$	$2^3\cdot1097$	$67\cdot131$	$2\cdot3\cdot7\cdot11\cdot19$	$\mathbf{8779}$
878	$2^2\cdot5\cdot439$	$3\cdot2927$	$2\cdot4391$	$\mathbf{8783}$	$2^4\cdot3^2\cdot61$	$5\cdot7\cdot251$	$2\cdot23\cdot191$	$3\cdot29\cdot101$	$2^2\cdot13^3$	$11\cdot17\cdot47$
879	$2\cdot3\cdot5\cdot293$	$59\cdot149$	$2^3\cdot7\cdot157$	$3^2\cdot977$	$2\cdot4397$	$5\cdot1759$	$2^2\cdot3\cdot733$	$19\cdot463$	$2\cdot53\cdot83$	$3\cdot7\cdot419$
880	$2^5\cdot5^2\cdot11$	$13\cdot677$	$2\cdot3^3\cdot163$	$\mathbf{8803}$	$2^2\cdot31\cdot71$	$3\cdot5\cdot587$	$2\cdot7\cdot17\cdot37$	$\mathbf{8807}$	$2^3\cdot3\cdot367$	$23\cdot383$
881	$2\cdot5\cdot881$	$3^2\cdot11\cdot89$	$2^2\cdot2203$	$7\cdot1259$	$2\cdot3\cdot13\cdot113$	$5\cdot41\cdot43$	$2^4\cdot19\cdot29$	$3\cdot2939$	$2\cdot4409$	$\mathbf{8819}$
882	$2^2\cdot3^2\cdot5\cdot7^2$	$\mathbf{8821}$	$2\cdot11\cdot401$	$3\cdot17\cdot173$	$2^3\cdot1103$	$5^2\cdot353$	$2\cdot3\cdot1471$	$7\cdot13\cdot97$	$2^2\cdot2207$	$3^4\cdot109$
883	$2\cdot5\cdot883$	$\mathbf{8831}$	$2^7\cdot3\cdot23$	$11^2\cdot73$	$2\cdot7\cdot631$	$3\cdot5\cdot19\cdot31$	$2^2\cdot47^2$	$\mathbf{8837}$	$2\cdot3^2\cdot491$	$\mathbf{8839}$
884	$2^2\cdot5\cdot13\cdot17$	$3\cdot7\cdot421$	$2\cdot4421$	$37\cdot239$	$2^2\cdot3\cdot11\cdot67$	$5\cdot29\cdot61$	$2\cdot4423$	$3^2\cdot983$	$2^4\cdot7\cdot79$	$\mathbf{8849}$
885	$2\cdot3\cdot5^2\cdot59$	$53\cdot167$	$2^2\cdot2213$	$3\cdot13\cdot227$	$2\cdot19\cdot233$	$5\cdot7\cdot11\cdot23$	$2^3\cdot3^3\cdot41$	$17\cdot521$	$2\cdot43\cdot103$	$3\cdot2953$
886	$2^2\cdot5\cdot443$	$\mathbf{8861}$	$2\cdot3\cdot7\cdot211$	$\mathbf{8863}$	$2^5\cdot277$	$3^2\cdot5\cdot197$	$2\cdot11\cdot13\cdot31$	$\mathbf{8867}$	$2^2\cdot3\cdot739$	$7^2\cdot181$
887	$2\cdot5\cdot887$	$3\cdot2957$	$2^3\cdot1109$	$19\cdot467$	$2\cdot3^2\cdot17\cdot29$	$5^3\cdot71$	$2^2\cdot7\cdot317$	$3\cdot11\cdot269$	$2\cdot23\cdot193$	$13\cdot683$
888	$2^4\cdot3\cdot5\cdot37$	$83\cdot107$	$2\cdot4441$	$3^3\cdot7\cdot47$	$2^2\cdot2221$	$5\cdot1777$	$2\cdot3\cdot1481$	$\mathbf{8887}$	$2^3\cdot11\cdot101$	$3\cdot2963$
889	$2\cdot5\cdot7\cdot127$	$17\cdot523$	$2^2\cdot3^2\cdot13\cdot19$	$\mathbf{8893}$	$2\cdot4447$	$3\cdot5\cdot593$	$2^6\cdot139$	$7\cdot31\cdot41$	$2\cdot3\cdot1483$	$11\cdot809$
890	$2^2\cdot5^2\cdot89$	$3^2\cdot23\cdot43$	$2\cdot4451$	$29\cdot307$	$2^3\cdot3\cdot7\cdot53$	$5\cdot13\cdot137$	$2\cdot61\cdot73$	$3\cdot2969$	$2^2\cdot17\cdot131$	$59\cdot151$
891	$2\cdot3^4\cdot5\cdot11$	$7\cdot19\cdot67$	$2^4\cdot557$	$3\cdot2971$	$2\cdot4457$	$5\cdot1783$	$2^2\cdot3\cdot743$	$37\cdot241$	$2\cdot7^3\cdot13$	$3^2\cdot991$
892	$2^3\cdot5\cdot223$	$11\cdot811$	$2\cdot3\cdot1487$	$\mathbf{8923}$	$2^2\cdot23\cdot97$	$3\cdot5^2\cdot7\cdot17$	$2\cdot4463$	$79\cdot113$	$2^5\cdot3^2\cdot31$	$\mathbf{8929}$
893	$2\cdot5\cdot19\cdot47$	$3\cdot13\cdot229$	$2^2\cdot7\cdot11\cdot29$	$\mathbf{8933}$	$2\cdot3\cdot1489$	$5\cdot1787$	$2^3\cdot1117$	$3^3\cdot331$	$2\cdot41\cdot109$	$7\cdot1277$
894	$2^2\cdot3\cdot5\cdot149$	$\mathbf{8941}$	$2\cdot17\cdot263$	$3\cdot11\cdot271$	$2^4\cdot13\cdot43$	$5\cdot1789$	$2\cdot3^2\cdot7\cdot71$	$23\cdot389$	$2^2\cdot2237$	$3\cdot19\cdot157$
895	$2\cdot5^2\cdot179$	$\mathbf{8951}$	$2^3\cdot3\cdot373$	$7\cdot1279$	$2\cdot11^2\cdot37$	$3^2\cdot5\cdot199$	$2^2\cdot2239$	$13^2\cdot53$	$2\cdot3\cdot1493$	$17^2\cdot31$
896	$2^8\cdot5\cdot7$	$3\cdot29\cdot103$	$2\cdot4481$	$\mathbf{8963}$	$2^2\cdot3^3\cdot83$	$5\cdot11\cdot163$	$2\cdot4483$	$3\cdot7^2\cdot61$	$2^3\cdot19\cdot59$	$\mathbf{8969}$
897	$2\cdot3\cdot5\cdot13\cdot23$	$\mathbf{8971}$	$2^2\cdot2243$	$3^2\cdot997$	$2\cdot7\cdot641$	$5^2\cdot359$	$2^4\cdot3\cdot11\cdot17$	$47\cdot191$	$2\cdot67^2$	$3\cdot41\cdot73$
898	$2^2\cdot5\cdot449$	$7\cdot1283$	$2\cdot3^2\cdot499$	$13\cdot691$	$2^3\cdot1123$	$3\cdot5\cdot599$	$2\cdot4493$	$11\cdot19\cdot43$	$2^2\cdot3\cdot7\cdot107$	$89\cdot101$
899	$2\cdot5\cdot29\cdot31$	$3^5\cdot37$	$2^5\cdot281$	$17\cdot23^2$	$2\cdot3\cdot1499$	$5\cdot7\cdot257$	$2^2\cdot13\cdot173$	$3\cdot2999$	$2\cdot11\cdot409$	$\mathbf{8999}$

COMBINATORIAL ANALYSIS

Table 24.7
9000

Factorizations

N	9	8	7	6	5	4	3	2	1	0	N
900	$3^2 \cdot 11 \cdot 13$	$2^4 \cdot 563$	**9007**	$2 \cdot 3 \cdot 19 \cdot 79$	$5 \cdot 1801$	$2^2 \cdot 2251$	$3 \cdot 3001$	$2 \cdot 7 \cdot 643$	**9001**	$2^3 \cdot 3^2 \cdot 5^3$	900
901	$29 \cdot 311$	$2 \cdot 3^3 \cdot 167$	$71 \cdot 127$	$2^8 \cdot 7^2 \cdot 23$	$3 \cdot 5 \cdot 601$	$2 \cdot 4507$	**9013**	$2^2 \cdot 3 \cdot 751$	**9011**	$2 \cdot 5 \cdot 17 \cdot 53$	901
902	**9029**	$2^2 \cdot 37 \cdot 61$	$3^2 \cdot 17 \cdot 59$	$2 \cdot 4513$	$5^2 \cdot 19^2$	$2^6 \cdot 3 \cdot 47$	$7 \cdot 1289$	$2 \cdot 13 \cdot 347$	$3 \cdot 31 \cdot 97$	$2^2 \cdot 5 \cdot 11 \cdot 41$	902
903	$3 \cdot 23 \cdot 131$	$2 \cdot 4519$	$7 \cdot 1291$	$2^2 \cdot 3^2 \cdot 251$	$5 \cdot 13 \cdot 139$	$2 \cdot 4517$	$3 \cdot 3011$	$2^3 \cdot 1129$	$11 \cdot 821$	$2 \cdot 3 \cdot 5 \cdot 7 \cdot 43$	903
904	**9049**	$2^3 \cdot 3 \cdot 13 \cdot 29$	$83 \cdot 109$	$2 \cdot 4523$	$3^3 \cdot 5 \cdot 67$	$2^2 \cdot 7 \cdot 17 \cdot 19$	**9043**	$2 \cdot 3 \cdot 11 \cdot 137$	**9041**	$2^4 \cdot 5 \cdot 113$	904
905	**9059**	$2 \cdot 7 \cdot 647$	$3 \cdot 3019$	$2^5 \cdot 283$	$5 \cdot 1811$	$2 \cdot 3^2 \cdot 503$	$11 \cdot 823$	$2^2 \cdot 31 \cdot 73$	$3 \cdot 7 \cdot 431$	$2 \cdot 5^2 \cdot 181$	905
906	$3 \cdot 3023$	$2^2 \cdot 2267$	**9067**	$2 \cdot 3 \cdot 1511$	$5 \cdot 7^2 \cdot 37$	$2^3 \cdot 11 \cdot 103$	$3^2 \cdot 19 \cdot 53$	$2 \cdot 23 \cdot 197$	$13 \cdot 17 \cdot 41$	$2^2 \cdot 3 \cdot 5 \cdot 151$	906
907	$7 \cdot 1297$	$2 \cdot 3 \cdot 17 \cdot 89$	$29 \cdot 313$	$2^2 \cdot 2269$	$3 \cdot 5^2 \cdot 11^2$	$2 \cdot 13 \cdot 349$	$43 \cdot 211$	$2^4 \cdot 3^4 \cdot 7$	$47 \cdot 193$	$2 \cdot 5 \cdot 907$	907
908	$61 \cdot 149$	$2^7 \cdot 71$	$3 \cdot 13 \cdot 233$	$2 \cdot 7 \cdot 11 \cdot 59$	$5 \cdot 23 \cdot 79$	$2^2 \cdot 3 \cdot 757$	$31 \cdot 293$	$2 \cdot 19 \cdot 239$	$3^2 \cdot 1009$	$2^3 \cdot 5 \cdot 227$	908
909	$3^3 \cdot 337$	$2 \cdot 4549$	$11 \cdot 827$	$2^3 \cdot 3 \cdot 379$	$5 \cdot 17 \cdot 107$	$2 \cdot 4547$	$3 \cdot 7 \cdot 433$	$2^2 \cdot 2273$	**9091**	$2 \cdot 3^2 \cdot 5 \cdot 101$	909
910	**9109**	$2^2 \cdot 3^2 \cdot 11 \cdot 23$	$7 \cdot 1301$	$2 \cdot 29 \cdot 157$	$3 \cdot 5 \cdot 607$	$2^4 \cdot 569$	**9103**	$2 \cdot 3 \cdot 37 \cdot 41$	$19 \cdot 479$	$2^2 \cdot 5^2 \cdot 7 \cdot 13$	910
911	$11 \cdot 829$	$2 \cdot 47 \cdot 97$	$3^2 \cdot 1013$	$2^2 \cdot 43 \cdot 53$	$5 \cdot 1823$	$2 \cdot 3 \cdot 7^2 \cdot 31$	$13 \cdot 701$	$2^3 \cdot 17 \cdot 67$	$3 \cdot 3037$	$2 \cdot 5 \cdot 911$	911
912	$3 \cdot 17 \cdot 179$	$2^3 \cdot 7 \cdot 163$	**9127**	$2 \cdot 3^3 \cdot 13^2$	$5^3 \cdot 73$	$2^2 \cdot 2281$	$3 \cdot 3041$	$2 \cdot 4561$	$7 \cdot 1303$	$2^5 \cdot 3 \cdot 5 \cdot 19$	912
913	$13 \cdot 19 \cdot 37$	$2 \cdot 3 \cdot 1523$	**9137**	$2^4 \cdot 571$	$3^2 \cdot 5 \cdot 7 \cdot 29$	$2 \cdot 4567$	**9133**	$2^2 \cdot 3 \cdot 761$	$23 \cdot 397$	$2 \cdot 5 \cdot 11 \cdot 83$	913
914	$7 \cdot 1307$	$2^2 \cdot 2287$	$3 \cdot 3049$	$2 \cdot 17 \cdot 269$	$5 \cdot 31 \cdot 59$	$2^3 \cdot 3^2 \cdot 127$	$41 \cdot 223$	$2 \cdot 7 \cdot 653$	$3 \cdot 11 \cdot 277$	$2^2 \cdot 5 \cdot 457$	914
915	$3 \cdot 43 \cdot 71$	$2 \cdot 19 \cdot 241$	**9157**	$2^2 \cdot 3 \cdot 7 \cdot 109$	$5 \cdot 1831$	$2 \cdot 23 \cdot 199$	$3^4 \cdot 113$	$2^6 \cdot 11 \cdot 13$	**9151**	$2 \cdot 3 \cdot 5^2 \cdot 61$	915
916	$53 \cdot 173$	$2^4 \cdot 3 \cdot 191$	$89 \cdot 103$	$2 \cdot 4583$	$3 \cdot 5 \cdot 13 \cdot 47$	$2^2 \cdot 29 \cdot 79$	$7^2 \cdot 11 \cdot 17$	$2 \cdot 3^2 \cdot 509$	**9161**	$2^3 \cdot 5 \cdot 229$	916
917	$67 \cdot 137$	$2 \cdot 13 \cdot 353$	$3 \cdot 7 \cdot 19 \cdot 23$	$2^3 \cdot 31 \cdot 37$	$5^2 \cdot 367$	$2 \cdot 3 \cdot 11 \cdot 139$	**9173**	$2^2 \cdot 2293$	$3^2 \cdot 1019$	$2 \cdot 5 \cdot 7 \cdot 131$	917
918	$3^2 \cdot 1021$	$2^2 \cdot 2297$	**9187**	$2 \cdot 3 \cdot 1531$	$5 \cdot 11 \cdot 167$	$2^5 \cdot 7 \cdot 41$	$3 \cdot 3061$	$2 \cdot 4591$	**9181**	$2^2 \cdot 3^3 \cdot 5 \cdot 17$	918
919	**9199**	$2 \cdot 3^2 \cdot 7 \cdot 73$	$17 \cdot 541$	$2^2 \cdot 11^2 \cdot 19$	$3 \cdot 5 \cdot 613$	$2 \cdot 4597$	$29 \cdot 317$	$2^3 \cdot 3 \cdot 383$	$7 \cdot 13 \cdot 101$	$2 \cdot 5 \cdot 919$	919
920	**9209**	$2^3 \cdot 1151$	$3^3 \cdot 11 \cdot 31$	$2 \cdot 4603$	$5 \cdot 7 \cdot 263$	$2^2 \cdot 3 \cdot 13 \cdot 59$	**9203**	$2 \cdot 43 \cdot 107$	$3 \cdot 3067$	$2^4 \cdot 5^2 \cdot 23$	920
921	$3 \cdot 7 \cdot 439$	$2 \cdot 11 \cdot 419$	$13 \cdot 709$	$2^{10} \cdot 3^2$	$5 \cdot 19 \cdot 97$	$2 \cdot 17 \cdot 271$	$3 \cdot 37 \cdot 83$	$2^2 \cdot 7^2 \cdot 47$	$61 \cdot 151$	$2 \cdot 3 \cdot 5 \cdot 307$	921
922	$11 \cdot 839$	$2^2 \cdot 3 \cdot 769$	**9227**	$2 \cdot 7 \cdot 659$	$3^2 \cdot 5^2 \cdot 41$	$2^3 \cdot 1153$	$23 \cdot 401$	$2 \cdot 3 \cdot 29 \cdot 53$	**9221**	$2^2 \cdot 5 \cdot 461$	922
923	**9239**	$2 \cdot 31 \cdot 149$	$3 \cdot 3079$	$2^2 \cdot 2309$	$5 \cdot 1847$	$2 \cdot 3^5 \cdot 19$	$7 \cdot 1319$	$2^4 \cdot 577$	$3 \cdot 17 \cdot 181$	$2 \cdot 5 \cdot 13 \cdot 71$	923
924	$3 \cdot 3083$	$2^5 \cdot 17^2$	$7 \cdot 1321$	$2 \cdot 3 \cdot 23 \cdot 67$	$5 \cdot 43^2$	$2^2 \cdot 2311$	$3^2 \cdot 13 \cdot 79$	$2 \cdot 4621$	**9241**	$2^3 \cdot 3 \cdot 5 \cdot 7 \cdot 11$	924
925	$47 \cdot 197$	$2 \cdot 3 \cdot 1543$	**9257**	$2^3 \cdot 13 \cdot 89$	$3 \cdot 5 \cdot 617$	$2 \cdot 7 \cdot 661$	$19 \cdot 487$	$2^2 \cdot 3^2 \cdot 257$	$11 \cdot 29^2$	$2 \cdot 5^3 \cdot 37$	925
926	$13 \cdot 23 \cdot 31$	$2^2 \cdot 7 \cdot 331$	$3 \cdot 3089$	$2 \cdot 41 \cdot 113$	$5 \cdot 17 \cdot 109$	$2^4 \cdot 3 \cdot 193$	$59 \cdot 157$	$2 \cdot 11 \cdot 421$	$3^3 \cdot 7^3$	$2^2 \cdot 5 \cdot 463$	926
927	$3^2 \cdot 1031$	$2 \cdot 4639$	**9277**	$2^2 \cdot 3 \cdot 773$	$5^2 \cdot 7 \cdot 53$	$2 \cdot 4637$	$3 \cdot 11 \cdot 281$	$2^3 \cdot 19 \cdot 61$	$73 \cdot 127$	$2 \cdot 3^2 \cdot 5 \cdot 103$	927
928	$7 \cdot 1327$	$2^3 \cdot 3^3 \cdot 43$	$37 \cdot 251$	$2 \cdot 4643$	$3 \cdot 5 \cdot 619$	$2^2 \cdot 11 \cdot 211$	**9283**	$2 \cdot 3 \cdot 7 \cdot 13 \cdot 17$	**9281**	$2^6 \cdot 5 \cdot 29$	928
929	$17 \cdot 547$	$2 \cdot 4649$	$3^2 \cdot 1033$	$2^4 \cdot 7 \cdot 83$	$5 \cdot 11 \cdot 13^2$	$2 \cdot 3 \cdot 1549$	**9293**	$2^2 \cdot 23 \cdot 101$	$3 \cdot 19 \cdot 163$	$2 \cdot 5 \cdot 929$	929
930	$3 \cdot 29 \cdot 107$	$2^2 \cdot 13 \cdot 179$	$41 \cdot 227$	$2 \cdot 3^2 \cdot 11 \cdot 47$	$5 \cdot 1861$	$2^3 \cdot 1163$	$3 \cdot 7 \cdot 443$	$2 \cdot 4651$	$71 \cdot 131$	$2^2 \cdot 3 \cdot 5^2 \cdot 31$	930
931	**9319**	$2 \cdot 3 \cdot 1553$	$7 \cdot 11^3$	$2^2 \cdot 17 \cdot 137$	$3^4 \cdot 5 \cdot 23$	$2 \cdot 4657$	$67 \cdot 139$	$2^5 \cdot 3 \cdot 97$	**9311**	$2 \cdot 5 \cdot 7^2 \cdot 19$	931
932	$19 \cdot 491$	$2^4 \cdot 11 \cdot 53$	$3 \cdot 3109$	$2 \cdot 4663$	$5^2 \cdot 373$	$2^2 \cdot 3^2 \cdot 7 \cdot 37$	**9323**	$2 \cdot 59 \cdot 79$	$3 \cdot 13 \cdot 239$	$2^3 \cdot 5 \cdot 233$	932
933	$3 \cdot 11 \cdot 283$	$2 \cdot 7 \cdot 23 \cdot 29$	**9337**	$2^3 \cdot 3 \cdot 389$	$5 \cdot 1867$	$2 \cdot 13 \cdot 359$	$3^2 \cdot 17 \cdot 61$	$2^2 \cdot 2333$	$7 \cdot 31 \cdot 43$	$2 \cdot 3 \cdot 5 \cdot 311$	933
934	**9349**	$2^2 \cdot 3 \cdot 19 \cdot 41$	$13 \cdot 719$	$2 \cdot 4673$	$3 \cdot 5 \cdot 7 \cdot 89$	$2^7 \cdot 73$	**9343**	$2 \cdot 3^3 \cdot 173$	**9341**	$2^2 \cdot 5 \cdot 467$	934
935	$7^2 \cdot 191$	$2 \cdot 4679$	$3 \cdot 3119$	$2^2 \cdot 2339$	$5 \cdot 1871$	$2 \cdot 3 \cdot 1559$	$47 \cdot 199$	$2^3 \cdot 7 \cdot 167$	$3^2 \cdot 1039$	$2 \cdot 5^2 \cdot 11 \cdot 17$	935
936	$3^3 \cdot 347$	$2^3 \cdot 1171$	$17 \cdot 19 \cdot 29$	$2 \cdot 3 \cdot 7 \cdot 223$	$5 \cdot 1873$	$2^2 \cdot 2341$	$3 \cdot 3121$	$2 \cdot 31 \cdot 151$	$11 \cdot 23 \cdot 37$	$2^4 \cdot 3^2 \cdot 5 \cdot 13$	936
937	$83 \cdot 113$	$2 \cdot 3^2 \cdot 521$	**9377**	$2^5 \cdot 293$	$3 \cdot 5^5$	$2 \cdot 43 \cdot 109$	$7 \cdot 13 \cdot 103$	$2^2 \cdot 3 \cdot 11 \cdot 71$	**9371**	$2 \cdot 5 \cdot 937$	937
938	$41 \cdot 229$	$2^2 \cdot 2347$	$3^2 \cdot 7 \cdot 149$	$2 \cdot 13 \cdot 19^2$	$5 \cdot 1877$	$2^3 \cdot 3 \cdot 17 \cdot 23$	$11 \cdot 853$	$2 \cdot 4691$	$3 \cdot 53 \cdot 59$	$2^2 \cdot 5 \cdot 7 \cdot 67$	938
939	$3 \cdot 13 \cdot 241$	$2 \cdot 37 \cdot 127$	**9397**	$2^2 \cdot 3^4 \cdot 29$	$5 \cdot 1879$	$2 \cdot 7 \cdot 11 \cdot 61$	$3 \cdot 31 \cdot 101$	$2^4 \cdot 587$	**9391**	$2 \cdot 3 \cdot 5 \cdot 313$	939
940	97^2	$2^6 \cdot 3 \cdot 7^2$	$23 \cdot 409$	$2 \cdot 4703$	$3^2 \cdot 5 \cdot 11 \cdot 19$	$2^2 \cdot 2351$	**9403**	$2 \cdot 3 \cdot 1567$	$7 \cdot 17 \cdot 79$	$2^3 \cdot 5^2 \cdot 47$	940
941	**9419**	$2 \cdot 17 \cdot 277$	$3 \cdot 43 \cdot 73$	$2^3 \cdot 11 \cdot 107$	$5 \cdot 7 \cdot 269$	$2 \cdot 3^2 \cdot 523$	**9413**	$2^2 \cdot 13 \cdot 181$	$3 \cdot 3137$	$2 \cdot 5 \cdot 941$	941
942	$3 \cdot 7 \cdot 449$	$2^2 \cdot 2357$	$11 \cdot 857$	$2 \cdot 3 \cdot 1571$	$5^2 \cdot 13 \cdot 29$	$2^4 \cdot 19 \cdot 31$	$3^3 \cdot 349$	$2 \cdot 7 \cdot 673$	**9421**	$2^2 \cdot 3 \cdot 5 \cdot 157$	942
943	**9439**	$2 \cdot 3 \cdot 11^2 \cdot 13$	**9437**	$2^2 \cdot 7 \cdot 337$	$3 \cdot 5 \cdot 17 \cdot 37$	$2 \cdot 53 \cdot 89$	**9433**	$2^3 \cdot 3^2 \cdot 131$	**9431**	$2 \cdot 5 \cdot 23 \cdot 41$	943
944	$11 \cdot 859$	$2^3 \cdot 1181$	$3 \cdot 47 \cdot 67$	$2 \cdot 4723$	$5 \cdot 1889$	$2^2 \cdot 3 \cdot 787$	$7 \cdot 19 \cdot 71$	$2 \cdot 4721$	$3^2 \cdot 1049$	$2^5 \cdot 5 \cdot 59$	944
945	$3^2 \cdot 1051$	$2 \cdot 4729$	$7^2 \cdot 193$	$2^4 \cdot 3 \cdot 197$	$5 \cdot 31 \cdot 61$	$2 \cdot 29 \cdot 163$	$3 \cdot 23 \cdot 137$	$2^2 \cdot 17 \cdot 139$	$13 \cdot 727$	$2 \cdot 3^3 \cdot 5^2 \cdot 7$	945
946	$17 \cdot 557$	$2^2 \cdot 3^2 \cdot 263$	**9467**	$2 \cdot 4733$	$3 \cdot 5 \cdot 631$	$2^3 \cdot 7 \cdot 13^2$	**9463**	$2 \cdot 3 \cdot 19 \cdot 83$	**9461**	$2^2 \cdot 5 \cdot 11 \cdot 43$	946
947	**9479**	$2 \cdot 7 \cdot 677$	$3^6 \cdot 13$	$2^2 \cdot 23 \cdot 103$	$5^2 \cdot 379$	$2 \cdot 3 \cdot 1579$	**9473**	$2^8 \cdot 37$	$3 \cdot 7 \cdot 11 \cdot 41$	$2 \cdot 5 \cdot 947$	947
948	$3 \cdot 3163$	$2^4 \cdot 593$	$53 \cdot 179$	$2 \cdot 3^2 \cdot 17 \cdot 31$	$5 \cdot 7 \cdot 271$	$2^2 \cdot 2371$	$3 \cdot 29 \cdot 109$	$2 \cdot 11 \cdot 431$	$19 \cdot 499$	$2^3 \cdot 3 \cdot 5 \cdot 79$	948
949	$7 \cdot 23 \cdot 59$	$2 \cdot 3 \cdot 1583$	**9497**	$2^3 \cdot 1187$	$3^2 \cdot 5 \cdot 211$	$2 \cdot 47 \cdot 101$	$11 \cdot 863$	$2^2 \cdot 3 \cdot 7 \cdot 113$	**9491**	$2 \cdot 5 \cdot 13 \cdot 73$	949

Factorizations

Table 24.7

The entries below give the complete prime factorization of each integer from 9500 to 9999. Each row is labelled by the first three digits; the ten columns give units digits 0 through 9.

	0	1	2	3	4	5	6	7	8	9
950	$2^2 \cdot 5^3 \cdot 19$	$3 \cdot 3167$	$2 \cdot 4751$	$13 \cdot 17 \cdot 43$	$2^5 \cdot 3^3 \cdot 11$	$5 \cdot 1901$	$2 \cdot 7^2 \cdot 97$	$3 \cdot 3169$	$2^2 \cdot 2377$	$37 \cdot 257$
951	$2 \cdot 3 \cdot 5 \cdot 317$	**9511**	$2^3 \cdot 29 \cdot 41$	$3^2 \cdot 7 \cdot 151$	$2 \cdot 67 \cdot 71$	$5 \cdot 11 \cdot 173$	$2^2 \cdot 3 \cdot 13 \cdot 61$	$31 \cdot 307$	$2 \cdot 4759$	$3 \cdot 19 \cdot 167$
952	$2^4 \cdot 5 \cdot 7 \cdot 17$	**9521**	$2 \cdot 3^2 \cdot 23^2$	$89 \cdot 107$	$2^2 \cdot 2381$	$3 \cdot 5^2 \cdot 127$	$2 \cdot 11 \cdot 433$	$7 \cdot 1361$	$2^3 \cdot 3 \cdot 397$	$13 \cdot 733$
953	$2 \cdot 5 \cdot 953$	$3^3 \cdot 353$	$2^2 \cdot 2383$	**9533**	$2 \cdot 3 \cdot 7 \cdot 227$	$5 \cdot 1907$	$2^6 \cdot 149$	$3 \cdot 11 \cdot 17^2$	$2 \cdot 19 \cdot 251$	**9539**
954	$2^2 \cdot 3^2 \cdot 5 \cdot 53$	$7 \cdot 29 \cdot 47$	$2 \cdot 13 \cdot 367$	$3 \cdot 3181$	$2^3 \cdot 1193$	$5 \cdot 23 \cdot 83$	$2 \cdot 3 \cdot 37 \cdot 43$	**9547**	$2^2 \cdot 7 \cdot 11 \cdot 31$	$3^2 \cdot 1061$
955	$2 \cdot 5^2 \cdot 191$	**9551**	$2^4 \cdot 3 \cdot 199$	$41 \cdot 233$	$2 \cdot 17 \cdot 281$	$3 \cdot 5 \cdot 7^2 \cdot 13$	$2^2 \cdot 2389$	$19 \cdot 503$	$2 \cdot 3^4 \cdot 59$	$11^2 \cdot 79$
956	$2^3 \cdot 5 \cdot 239$	$3 \cdot 3187$	$2 \cdot 7 \cdot 683$	$73 \cdot 131$	$2^2 \cdot 3 \cdot 797$	$5 \cdot 1913$	$2 \cdot 4783$	$3^2 \cdot 1063$	$2^5 \cdot 13 \cdot 23$	$7 \cdot 1367$
957	$2 \cdot 3 \cdot 5 \cdot 11 \cdot 29$	$17 \cdot 563$	$2^2 \cdot 2393$	$3 \cdot 3191$	$2 \cdot 4787$	$5^2 \cdot 383$	$2^3 \cdot 3^2 \cdot 7 \cdot 19$	$61 \cdot 157$	$2 \cdot 4789$	$3 \cdot 31 \cdot 103$
958	$2^2 \cdot 5 \cdot 479$	$11 \cdot 13 \cdot 67$	$2 \cdot 3 \cdot 1597$	$7 \cdot 37^2$	$2^4 \cdot 599$	$3^3 \cdot 5 \cdot 71$	$2 \cdot 4793$	**9587**	$2^2 \cdot 3 \cdot 17 \cdot 47$	$43 \cdot 223$
959	$2 \cdot 5 \cdot 7 \cdot 137$	$3 \cdot 23 \cdot 139$	$2^3 \cdot 11 \cdot 109$	$53 \cdot 181$	$2 \cdot 3^2 \cdot 13 \cdot 41$	$5 \cdot 19 \cdot 101$	$2^2 \cdot 2399$	$3 \cdot 7 \cdot 457$	$2 \cdot 4799$	$29 \cdot 331$
960	$2^7 \cdot 3 \cdot 5^2$	**9601**	$2 \cdot 4801$	$3^2 \cdot 11 \cdot 97$	$2^2 \cdot 7^4$	$5 \cdot 17 \cdot 113$	$2 \cdot 3 \cdot 1601$	$13 \cdot 739$	$2^3 \cdot 1201$	$3 \cdot 3203$
961	$2 \cdot 5 \cdot 31^2$	$7 \cdot 1373$	$2^2 \cdot 3^3 \cdot 89$	**9613**	$2 \cdot 11 \cdot 19 \cdot 23$	$3 \cdot 5 \cdot 641$	$2^4 \cdot 601$	$59 \cdot 163$	$2 \cdot 3 \cdot 7 \cdot 229$	**9619**
962	$2^2 \cdot 5 \cdot 13 \cdot 37$	$3^2 \cdot 1069$	$2 \cdot 17 \cdot 283$	**9623**	$2^3 \cdot 3 \cdot 401$	$5^3 \cdot 7 \cdot 11$	$2 \cdot 4813$	$3 \cdot 3209$	$2^2 \cdot 29 \cdot 83$	**9629**
963	$2 \cdot 3^2 \cdot 5 \cdot 107$	**9631**	$2^5 \cdot 7 \cdot 43$	$3 \cdot 13^2 \cdot 19$	$2 \cdot 4817$	$5 \cdot 41 \cdot 47$	$2^2 \cdot 3 \cdot 11 \cdot 73$	$23 \cdot 419$	$2 \cdot 61 \cdot 79$	$3^4 \cdot 7 \cdot 17$
964	$2^3 \cdot 5 \cdot 241$	$31 \cdot 311$	$2 \cdot 3 \cdot 1607$	**9643**	$2^2 \cdot 2411$	$3 \cdot 5 \cdot 643$	$2 \cdot 7 \cdot 13 \cdot 53$	$11 \cdot 877$	$2^4 \cdot 3^2 \cdot 67$	**9649**
965	$2 \cdot 5^2 \cdot 193$	$3 \cdot 3217$	$2^2 \cdot 19 \cdot 127$	$7^2 \cdot 197$	$2 \cdot 3 \cdot 1609$	$5 \cdot 1931$	$2^3 \cdot 17 \cdot 71$	$3^2 \cdot 29 \cdot 37$	$2 \cdot 11 \cdot 439$	$13 \cdot 743$
966	$2^2 \cdot 3 \cdot 5 \cdot 7 \cdot 23$	**9661**	$2 \cdot 4831$	$3 \cdot 3221$	$2^6 \cdot 151$	$5 \cdot 1933$	$2 \cdot 3^3 \cdot 179$	$7 \cdot 1381$	$2^2 \cdot 2417$	$3 \cdot 11 \cdot 293$
967	$2 \cdot 5 \cdot 967$	$19 \cdot 509$	$2^3 \cdot 3 \cdot 13 \cdot 31$	$17 \cdot 569$	$2 \cdot 7 \cdot 691$	$3^2 \cdot 5^2 \cdot 43$	$2^2 \cdot 41 \cdot 59$	**9677**	$2 \cdot 3 \cdot 1613$	**9679**
968	$2^4 \cdot 5 \cdot 11^2$	$3 \cdot 7 \cdot 461$	$2 \cdot 47 \cdot 103$	$23 \cdot 421$	$2^2 \cdot 3^2 \cdot 269$	$5 \cdot 13 \cdot 149$	$2 \cdot 29 \cdot 167$	$3 \cdot 3229$	$2^3 \cdot 7 \cdot 173$	**9689**
969	$2 \cdot 3 \cdot 5 \cdot 17 \cdot 19$	$11 \cdot 881$	$2^2 \cdot 2423$	$3^3 \cdot 359$	$2 \cdot 37 \cdot 131$	$5 \cdot 7 \cdot 277$	$2^5 \cdot 3 \cdot 101$	**9697**	$2 \cdot 13 \cdot 373$	$3 \cdot 53 \cdot 61$
970	$2^2 \cdot 5^2 \cdot 97$	$89 \cdot 109$	$2 \cdot 3^2 \cdot 7^2 \cdot 11$	$31 \cdot 313$	$2^3 \cdot 1213$	$3 \cdot 5 \cdot 647$	$2 \cdot 23 \cdot 211$	$17 \cdot 571$	$2^2 \cdot 3 \cdot 809$	$7 \cdot 19 \cdot 73$
971	$2 \cdot 5 \cdot 971$	$3^2 \cdot 13 \cdot 83$	$2^4 \cdot 607$	$11 \cdot 883$	$2 \cdot 3 \cdot 1619$	$5 \cdot 29 \cdot 67$	$2^2 \cdot 7 \cdot 347$	$3 \cdot 41 \cdot 79$	$2 \cdot 43 \cdot 113$	**9719**
972	$2^3 \cdot 3^5 \cdot 5$	**9721**	$2 \cdot 4861$	$3 \cdot 7 \cdot 463$	$2^2 \cdot 11 \cdot 13 \cdot 17$	$5^2 \cdot 389$	$2 \cdot 3 \cdot 1621$	$71 \cdot 137$	$2^9 \cdot 19$	$3^2 \cdot 23 \cdot 47$
973	$2 \cdot 5 \cdot 7 \cdot 139$	$37 \cdot 263$	$2^2 \cdot 3 \cdot 811$	**9733**	$2 \cdot 31 \cdot 157$	$3 \cdot 5 \cdot 11 \cdot 59$	$2^3 \cdot 1217$	$7 \cdot 13 \cdot 107$	$2 \cdot 3^2 \cdot 541$	**9739**
974	$2^2 \cdot 5 \cdot 487$	$3 \cdot 17 \cdot 191$	$2 \cdot 4871$	**9743**	$2^4 \cdot 3 \cdot 7 \cdot 29$	$5 \cdot 1949$	$2 \cdot 11 \cdot 443$	$3^3 \cdot 19^2$	$2^2 \cdot 2437$	**9749**
975	$2 \cdot 3 \cdot 5^3 \cdot 13$	$7^2 \cdot 199$	$2^3 \cdot 23 \cdot 53$	$3 \cdot 3251$	$2 \cdot 4877$	$5 \cdot 1951$	$2^2 \cdot 3^2 \cdot 271$	$11 \cdot 887$	$2 \cdot 7 \cdot 17 \cdot 41$	$3 \cdot 3253$
976	$2^5 \cdot 5 \cdot 61$	$43 \cdot 227$	$2 \cdot 3 \cdot 1627$	$13 \cdot 751$	$2^2 \cdot 2441$	$3^2 \cdot 5 \cdot 7 \cdot 31$	$2 \cdot 19 \cdot 257$	**9767**	$2^3 \cdot 3 \cdot 11 \cdot 37$	**9769**
977	$2 \cdot 5 \cdot 977$	$3 \cdot 3257$	$2^2 \cdot 7 \cdot 349$	$29 \cdot 337$	$2 \cdot 3^3 \cdot 181$	$5^2 \cdot 17 \cdot 23$	$2^4 \cdot 13 \cdot 47$	$3 \cdot 3259$	$2 \cdot 4889$	$7 \cdot 11 \cdot 127$
978	$2^2 \cdot 3 \cdot 5 \cdot 163$	**9781**	$2 \cdot 67 \cdot 73$	$3^2 \cdot 1087$	$2^3 \cdot 1223$	$5 \cdot 19 \cdot 103$	$2 \cdot 3 \cdot 7 \cdot 233$	**9787**	$2^2 \cdot 2447$	$3 \cdot 13 \cdot 251$
979	$2 \cdot 5 \cdot 11 \cdot 89$	**9791**	$2^6 \cdot 3^2 \cdot 17$	$7 \cdot 1399$	$2 \cdot 59 \cdot 83$	$3 \cdot 5 \cdot 653$	$2^2 \cdot 31 \cdot 79$	$97 \cdot 101$	$2 \cdot 3 \cdot 23 \cdot 71$	$41 \cdot 239$
980	$2^3 \cdot 5^2 \cdot 7^2$	$3^4 \cdot 11^2$	$2 \cdot 13^2 \cdot 29$	**9803**	$2^2 \cdot 3 \cdot 19 \cdot 43$	$5 \cdot 37 \cdot 53$	$2 \cdot 4903$	$3 \cdot 7 \cdot 467$	$2^4 \cdot 613$	$17 \cdot 577$
981	$2 \cdot 3^2 \cdot 5 \cdot 109$	**9811**	$2^2 \cdot 11 \cdot 223$	$3 \cdot 3271$	$2 \cdot 7 \cdot 701$	$5 \cdot 13 \cdot 151$	$2^3 \cdot 3 \cdot 409$	**9817**	$2 \cdot 4909$	$3^2 \cdot 1091$
982	$2^2 \cdot 5 \cdot 491$	$7 \cdot 23 \cdot 61$	$2 \cdot 3 \cdot 1637$	$11 \cdot 19 \cdot 47$	$2^5 \cdot 307$	$3 \cdot 5^2 \cdot 131$	$2 \cdot 17^3$	$31 \cdot 317$	$2^2 \cdot 3^3 \cdot 7 \cdot 13$	**9829**
983	$2 \cdot 5 \cdot 983$	$3 \cdot 29 \cdot 113$	$2^3 \cdot 1229$	**9833**	$2 \cdot 3 \cdot 11 \cdot 149$	$5 \cdot 7 \cdot 281$	$2^2 \cdot 2459$	$3^2 \cdot 1093$	$2 \cdot 4919$	**9839**
984	$2^4 \cdot 3 \cdot 5 \cdot 41$	$13 \cdot 757$	$2 \cdot 7 \cdot 19 \cdot 37$	$3 \cdot 17 \cdot 193$	$2^2 \cdot 23 \cdot 107$	$5 \cdot 11 \cdot 179$	$2 \cdot 3^2 \cdot 547$	$43 \cdot 229$	$2^3 \cdot 1231$	$3 \cdot 7^2 \cdot 67$
985	$2 \cdot 5^2 \cdot 197$	**9851**	$2^2 \cdot 3 \cdot 821$	$59 \cdot 167$	$2 \cdot 13 \cdot 379$	$3^3 \cdot 5 \cdot 73$	$2^7 \cdot 7 \cdot 11$	**9857**	$2 \cdot 3 \cdot 31 \cdot 53$	**9859**
986	$2^2 \cdot 5 \cdot 17 \cdot 29$	$3 \cdot 19 \cdot 173$	$2 \cdot 4931$	$7 \cdot 1409$	$2^3 \cdot 3^2 \cdot 137$	$5 \cdot 1973$	$2 \cdot 4933$	$3 \cdot 11 \cdot 13 \cdot 23$	$2^2 \cdot 2467$	$71 \cdot 139$
987	$2 \cdot 3 \cdot 5 \cdot 7 \cdot 47$	**9871**	$2^4 \cdot 617$	$3^2 \cdot 1097$	$2 \cdot 4937$	$5^3 \cdot 79$	$2^2 \cdot 3 \cdot 823$	$7 \cdot 17 \cdot 83$	$2 \cdot 11 \cdot 449$	$3 \cdot 37 \cdot 89$
988	$2^3 \cdot 5 \cdot 13 \cdot 19$	$41 \cdot 241$	$2 \cdot 3^4 \cdot 61$	**9883**	$2^2 \cdot 7 \cdot 353$	$3 \cdot 5 \cdot 659$	$2 \cdot 4943$	**9887**	$2^5 \cdot 3 \cdot 103$	$11 \cdot 29 \cdot 31$
989	$2 \cdot 5 \cdot 23 \cdot 43$	$3^2 \cdot 7 \cdot 157$	$2^2 \cdot 2473$	$13 \cdot 761$	$2 \cdot 3 \cdot 17 \cdot 97$	$5 \cdot 1979$	$2^3 \cdot 1237$	$3 \cdot 3299$	$2 \cdot 7^2 \cdot 101$	$19 \cdot 521$
990	$2^2 \cdot 3^2 \cdot 5^2 \cdot 11$	**9901**	$2 \cdot 4951$	$3 \cdot 3301$	$2^4 \cdot 619$	$5 \cdot 7 \cdot 283$	$2 \cdot 3 \cdot 13 \cdot 127$	**9907**	$2^2 \cdot 2477$	$3^3 \cdot 367$
991	$2 \cdot 5 \cdot 991$	$11 \cdot 17 \cdot 53$	$2^3 \cdot 3 \cdot 7 \cdot 59$	$23 \cdot 431$	$2 \cdot 4957$	$3 \cdot 5 \cdot 661$	$2^2 \cdot 37 \cdot 67$	$47 \cdot 211$	$2 \cdot 3^2 \cdot 19 \cdot 29$	$7 \cdot 13 \cdot 109$
992	$2^6 \cdot 5 \cdot 31$	$3 \cdot 3307$	$2 \cdot 11^2 \cdot 41$	**9923**	$2^2 \cdot 3 \cdot 827$	$5^2 \cdot 397$	$2 \cdot 7 \cdot 709$	$3^2 \cdot 1103$	$2^3 \cdot 17 \cdot 73$	**9929**
993	$2 \cdot 3 \cdot 5 \cdot 331$	**9931**	$2^2 \cdot 13 \cdot 191$	$3 \cdot 7 \cdot 11 \cdot 43$	$2 \cdot 4967$	$5 \cdot 1987$	$2^4 \cdot 3^3 \cdot 23$	$19 \cdot 523$	$2 \cdot 4969$	$3 \cdot 3313$
994	$2^2 \cdot 5 \cdot 7 \cdot 71$	**9941**	$2 \cdot 3 \cdot 1657$	$61 \cdot 163$	$2^3 \cdot 11 \cdot 113$	$3^2 \cdot 5 \cdot 13 \cdot 17$	$2 \cdot 4973$	$7^3 \cdot 29$	$2^2 \cdot 3 \cdot 829$	**9949**
995	$2 \cdot 5^2 \cdot 199$	$3 \cdot 31 \cdot 107$	$2^5 \cdot 311$	$37 \cdot 269$	$2 \cdot 3^2 \cdot 7 \cdot 79$	$5 \cdot 11 \cdot 181$	$2^2 \cdot 19 \cdot 131$	$3 \cdot 3319$	$2 \cdot 13 \cdot 383$	$23 \cdot 433$
996	$2^3 \cdot 3 \cdot 5 \cdot 83$	$7 \cdot 1423$	$2 \cdot 17 \cdot 293$	$3^5 \cdot 41$	$2^2 \cdot 47 \cdot 53$	$5 \cdot 1993$	$2 \cdot 3 \cdot 11 \cdot 151$	**9967**	$2^4 \cdot 7 \cdot 89$	$3 \cdot 3323$
997	$2 \cdot 5 \cdot 997$	$13^2 \cdot 59$	$2^2 \cdot 3^2 \cdot 277$	**9973**	$2 \cdot 4987$	$3 \cdot 5^2 \cdot 7 \cdot 19$	$2^3 \cdot 29 \cdot 43$	$11 \cdot 907$	$2 \cdot 3 \cdot 1663$	$17 \cdot 587$
998	$2^2 \cdot 5 \cdot 499$	$3^2 \cdot 1109$	$2 \cdot 7 \cdot 23 \cdot 31$	$67 \cdot 149$	$2^8 \cdot 3 \cdot 13$	$5 \cdot 1997$	$2 \cdot 4993$	$3 \cdot 3329$	$2^2 \cdot 11 \cdot 227$	$7 \cdot 1427$
999	$2 \cdot 3^3 \cdot 5 \cdot 37$	$97 \cdot 103$	$2^3 \cdot 1249$	$3 \cdot 3331$	$2 \cdot 19 \cdot 263$	$5 \cdot 1999$	$2^2 \cdot 3 \cdot 7^2 \cdot 17$	$13 \cdot 769$	$2 \cdot 4999$	$3^2 \cdot 11 \cdot 101$

Table 24.8 — Primitive Roots, Factorization of $p-1$

g, G denote the least positive and least negative (respectively) primitive roots of p. ϵ denotes whether 10, −10 both or neither are primitive roots.

p	$p-1$	g	$-G$	ϵ	p	$p-1$	g	$-G$	ϵ	p	$p-1$	g	$-G$	ϵ
3	2	2	1	-10	359	$2\cdot179$	7	2	-10	821	$2^2\cdot5\cdot41$	2	2	±10
5	2^2	2	2	-----	367	$2\cdot3\cdot61$	6	2	10	823	$2\cdot3\cdot137$	3	2	10
7	$2\cdot3$	3	2	10	373	$2^2\cdot3\cdot31$	2	2	-----	827	$2\cdot7\cdot59$	2	3	-10
11	$2\cdot5$	2	3	-----	379	$2\cdot3^3\cdot7$	2	4	10	829	$2^2\cdot3^2\cdot23$	2	2	-----
13	$2^2\cdot3$	2	2	-----	383	$2\cdot191$	5	2	10	839	$2\cdot419$	11	2	-10
17	2^4	3	3	±10	389	$2^2\cdot97$	2	2	±10	853	$2^2\cdot3\cdot71$	2	2	-----
19	$2\cdot3^2$	2	4	10	397	$2^2\cdot3^2\cdot11$	5	5	-----	857	$2^3\cdot107$	3	5	±10
23	$2\cdot11$	5	2	10	401	$2^4\cdot5^2$	3	3	-----	859	$2\cdot3\cdot11\cdot13$	2	4	-----
29	$2^2\cdot7$	2	2	±10	409	$2^3\cdot3\cdot17$	21	21	-----	863	$2\cdot431$	5	2	10
31	$2\cdot3\cdot5$	3	7	-10	419	$2\cdot11\cdot19$	2	3	10	877	$2^2\cdot3\cdot73$	2	2	-----
37	$2^2\cdot3^2$	2	2	-----	421	$2^2\cdot3\cdot5\cdot7$	2	2	-----	881	$2^4\cdot5\cdot11$	3	3	-----
41	$2^3\cdot5$	6	6	-----	431	$2\cdot5\cdot43$	7	5	-10	883	$2\cdot3^2\cdot7^2$	2	4	-10
43	$2\cdot3\cdot7$	3	9	-10	433	$2^4\cdot3^3$	5	5	±10	887	$2\cdot443$	5	2	10
47	$2\cdot23$	5	2	10	439	$2\cdot3\cdot73$	15	5	-10	907	$2\cdot3\cdot151$	2	4	-----
53	$2^2\cdot13$	2	2	-----	443	$2\cdot13\cdot17$	2	3	-10	911	$2\cdot5\cdot7\cdot13$	17	3	-10
59	$2\cdot29$	2	3	10	449	$2^6\cdot7$	3	3	-----	919	$2\cdot3^3\cdot17$	7	5	-10
61	$2^2\cdot3\cdot5$	2	2	±10	457	$2^3\cdot3\cdot19$	13	13	-----	929	$2^5\cdot29$	3	3	-----
67	$2\cdot3\cdot11$	2	4	-10	461	$2^2\cdot5\cdot23$	2	2	±10	937	$2^3\cdot3^2\cdot13$	5	5	±10
71	$2\cdot5\cdot7$	7	2	-10	463	$2\cdot3\cdot7\cdot11$	3	2	-----	941	$2^2\cdot5\cdot47$	2	2	±10
73	$2^3\cdot3^2$	5	5	-----	467	$2\cdot233$	2	3	-10	947	$2\cdot11\cdot43$	2	3	-10
79	$2\cdot3\cdot13$	3	2	-----	479	$2\cdot239$	13	2	-10	953	$2^3\cdot7\cdot17$	3	3	±10
83	$2\cdot41$	2	3	-10	487	$2\cdot3^5$	3	2	10	967	$2\cdot3\cdot7\cdot23$	5	2	-----
89	$2^3\cdot11$	3	3	-----	491	$2\cdot5\cdot7^2$	2	4	10	971	$2\cdot5\cdot97$	6	3	10
97	$2^5\cdot3$	5	5	±10	499	$2\cdot3\cdot83$	7	5	10	977	$2^4\cdot61$	3	3	±10
101	$2^2\cdot5^2$	2	2	-----	503	$2\cdot251$	5	2	10	983	$2\cdot491$	5	2	10
103	$2\cdot3\cdot17$	5	2	-----	509	$2^2\cdot127$	2	2	±10	991	$2\cdot3^2\cdot5\cdot11$	6	2	-10
107	$2\cdot53$	2	3	-10	521	$2^3\cdot5\cdot13$	3	3	-----	997	$2^2\cdot3\cdot83$	7	7	-----
109	$2^2\cdot3^3$	6	6	±10	523	$2\cdot3^2\cdot29$	2	4	-10	1009	$2^4\cdot3^2\cdot7$	11	11	-----
113	$2^4\cdot7$	3	3	±10	541	$2^2\cdot3^3\cdot5$	2	2	±10	1013	$2^2\cdot11\cdot23$	3	3	-----
127	$2\cdot3^2\cdot7$	3	9	-----	547	$2\cdot3\cdot7\cdot13$	2	4	-----	1019	$2\cdot509$	2	3	10
131	$2\cdot5\cdot13$	2	3	10	557	$2^2\cdot139$	2	2	-----	1021	$2^2\cdot3\cdot5\cdot17$	10	10	±10
137	$2^3\cdot17$	3	3	-----	563	$2\cdot281$	2	3	-10	1031	$2\cdot5\cdot103$	14	2	-----
139	$2\cdot3\cdot23$	2	4	-----	569	$2^3\cdot71$	3	3	-----	1033	$2^3\cdot3\cdot43$	5	5	±10
149	$2^2\cdot37$	2	2	±10	571	$2\cdot3\cdot5\cdot19$	3	5	10	1039	$2\cdot3\cdot173$	3	2	-10
151	$2\cdot3\cdot5^2$	6	5	-10	577	$2^6\cdot3^2$	5	5	±10	1049	$2^3\cdot131$	3	3	-----
157	$2^2\cdot3\cdot13$	5	5	-----	587	$2\cdot293$	2	3	-10	1051	$2\cdot3\cdot5^2\cdot7$	7	5	10
163	$2\cdot3^4$	2	4	-10	593	$2^4\cdot37$	3	3	±10	1061	$2^2\cdot5\cdot53$	2	2	-----
167	$2\cdot83$	5	2	10	599	$2\cdot13\cdot23$	7	2	-10	1063	$2\cdot3^2\cdot59$	3	2	10
173	$2^2\cdot43$	2	2	-----	601	$2^3\cdot3\cdot5^2$	7	7	-----	1069	$2^2\cdot3\cdot89$	6	6	±10
179	$2\cdot89$	2	3	10	607	$2\cdot3\cdot101$	3	2	-----	1087	$2\cdot3\cdot181$	3	2	10
181	$2^2\cdot3^2\cdot5$	2	2	±10	613	$2^2\cdot3^2\cdot17$	2	2	-----	1091	$2\cdot5\cdot109$	2	4	10
191	$2\cdot5\cdot19$	19	2	-10	617	$2^3\cdot7\cdot11$	3	3	-----	1093	$2^2\cdot3\cdot7\cdot13$	5	5	-----
193	$2^6\cdot3$	5	5	±10	619	$2\cdot3\cdot103$	2	4	10	1097	$2^3\cdot137$	3	3	±10
197	$2^2\cdot7^2$	2	2	-----	631	$2\cdot3^2\cdot5\cdot7$	3	9	-10	1103	$2\cdot19\cdot29$	5	3	10
199	$2\cdot3^2\cdot11$	3	2	-10	641	$2^7\cdot5$	3	3	-----	1109	$2^2\cdot277$	2	2	±10
211	$2\cdot3\cdot5\cdot7$	2	4	-----	643	$2\cdot3\cdot107$	11	7	-----	1117	$2^2\cdot3^2\cdot31$	2	2	-----
223	$2\cdot3\cdot37$	3	9	10	647	$2\cdot17\cdot19$	5	2	10	1123	$2\cdot3\cdot11\cdot17$	2	4	-10
227	$2\cdot113$	2	3	-10	653	$2^2\cdot163$	2	2	-----	1129	$2^3\cdot3\cdot47$	11	11	-----
229	$2^2\cdot3\cdot19$	6	6	±10	659	$2\cdot7\cdot47$	2	3	10	1151	$2\cdot5^2\cdot23$	17	2	-10
233	$2^3\cdot29$	3	3	±10	661	$2^2\cdot3\cdot5\cdot11$	2	2	-----	1153	$2^7\cdot3^2$	5	5	±10
239	$2\cdot7\cdot17$	7	2	-----	673	$2^5\cdot3\cdot7$	5	5	-----	1163	$2\cdot7\cdot83$	5	3	-10
241	$2^4\cdot3\cdot5$	7	7	-----	677	$2^2\cdot13^2$	2	2	-----	1171	$2\cdot3^2\cdot5\cdot13$	2	4	10
251	$2\cdot5^3$	6	3	-----	683	$2\cdot11\cdot31$	5	10	-10	1181	$2^2\cdot5\cdot59$	7	7	±10
257	2^8	3	3	±10	691	$2\cdot3\cdot5\cdot23$	3	6	-----	1187	$2\cdot593$	2	3	-10
263	$2\cdot131$	5	2	10	701	$2^2\cdot5^2\cdot7$	2	2	±10	1193	$2^3\cdot149$	3	3	±10
269	$2^2\cdot67$	2	2	±10	709	$2^2\cdot3\cdot59$	2	2	±10	1201	$2^4\cdot3\cdot5^2$	11	11	-----
271	$2\cdot3^3\cdot5$	6	2	-----	719	$2\cdot359$	11	2	-10	1213	$2^2\cdot3\cdot101$	2	2	-----
277	$2^2\cdot3\cdot23$	5	5	-----	727	$2\cdot3\cdot11^2$	5	7	10	1217	$2^6\cdot19$	3	3	±10
281	$2^3\cdot5\cdot7$	3	3	-----	733	$2^2\cdot3\cdot61$	6	6	-----	1223	$2\cdot13\cdot47$	5	2	10
283	$2\cdot3\cdot47$	3	6	-10	739	$2\cdot3^2\cdot41$	3	6	-----	1229	$2^2\cdot307$	2	2	±10
293	$2^2\cdot73$	2	2	-----	743	$2\cdot7\cdot53$	5	2	10	1231	$2\cdot3\cdot5\cdot41$	3	2	-----
307	$2\cdot3^2\cdot17$	5	7	-10	751	$2\cdot3\cdot5^3$	3	2	-----	1237	$2^2\cdot3\cdot103$	2	2	-----
311	$2\cdot5\cdot31$	17	2	-10	757	$2^2\cdot3^3\cdot7$	2	2	-----	1249	$2^5\cdot3\cdot13$	7	7	-----
313	$2^3\cdot3\cdot13$	10	10	±10	761	$2^3\cdot5\cdot19$	6	6	-----	1259	$2\cdot17\cdot37$	2	3	10
317	$2^2\cdot79$	2	2	-----	769	$2^8\cdot3$	11	11	-----	1277	$2^2\cdot11\cdot29$	2	2	-----
331	$2\cdot3\cdot5\cdot11$	3	5	-----	773	$2^2\cdot193$	2	2	-----	1279	$2\cdot3^2\cdot71$	3	2	-10
337	$2^4\cdot3\cdot7$	10	10	±10	787	$2\cdot3\cdot131$	2	4	-10	1283	$2\cdot641$	2	3	-10
347	$2\cdot173$	2	3	-10	797	$2^2\cdot199$	2	2	-----	1289	$2^3\cdot7\cdot23$	6	6	-----
349	$2^2\cdot3\cdot29$	2	2	-----	809	$2^3\cdot101$	3	3	-----	1291	$2\cdot3\cdot5\cdot43$	2	4	10
353	$2^5\cdot11$	3	3	-----	811	$2\cdot3^4\cdot5$	3	5	10	1297	$2^4\cdot3^4$	10	10	±10

Primitive Roots, Factorization of $p-1$ — Table 24.8

g, G denote the least positive and least negative (respectively) primitive roots of p. ϵ denotes whether 10, -10 both or neither are primitive roots.

p	$p-1$	g	$-G$	ϵ	p	$p-1$	g	$-G$	ϵ	p	$p-1$	g	$-G$	ϵ
1301	$2^2\cdot5^2\cdot13$	2	2	±10	1831	$2\cdot3\cdot5\cdot61$	3	9	-----	2377	$2^3\cdot3^3\cdot11$	5	5	-----
1303	$2\cdot3\cdot7\cdot31$	6	2	10	1847	$2\cdot13\cdot71$	5	2	10	2381	$2^2\cdot5\cdot7\cdot17$	3	3	-----
1307	$2\cdot653$	2	3	-10	1861	$2^2\cdot3\cdot5\cdot31$	2	2	±10	2383	$2\cdot3\cdot397$	5	13	10
1319	$2\cdot659$	13	2	-10	1867	$2\cdot3\cdot311$	2	4	-10	2389	$2^2\cdot3\cdot199$	2	2	±10
1321	$2^3\cdot3\cdot5\cdot11$	13	13	-----	1871	$2\cdot5\cdot11\cdot17$	14	2	-10	2393	$2^3\cdot13\cdot23$	3	3	-----
1327	$2\cdot3\cdot13\cdot17$	3	9	10	1873	$2^4\cdot3^2\cdot13$	10	10	±10	2399	$2\cdot11\cdot109$	11	2	-10
1361	$2^4\cdot5\cdot17$	3	3	-----	1877	$2^2\cdot7\cdot67$	2	2	-----	2411	$2\cdot5\cdot241$	6	3	10
1367	$2\cdot683$	5	2	10	1879	$2\cdot3\cdot313$	6	2	-----	2417	$2^4\cdot151$	3	3	±10
1373	$2^2\cdot7^3$	2	2		1889	$2^5\cdot59$	3	3	-----	2423	$2\cdot7\cdot173$	5	2	10
1381	$2^2\cdot3\cdot5\cdot23$	2	2	±10	1901	$2^2\cdot5^2\cdot19$	2	2	-----	2437	$2^2\cdot3\cdot7\cdot29$	2	2	-----
1399	$2\cdot3\cdot233$	13	5	-10	1907	$2\cdot953$	2	3	-10	2441	$2^3\cdot5\cdot61$	6	6	-----
1409	$2^7\cdot11$	3	3	-----	1913	$2^3\cdot239$	3	3	±10	2447	$2\cdot1223$	5	2	10
1423	$2\cdot3^2\cdot79$	3	9		1931	$2\cdot5\cdot193$	3	3	-----	2459	$2\cdot1229$	2	3	10
1427	$2\cdot23\cdot31$	2	3	-10	1933	$2^2\cdot3\cdot7\cdot23$	5	5	-----	2467	$2\cdot3^2\cdot137$	2	4	
1429	$2^2\cdot3\cdot7\cdot17$	6	6	±10	1949	$2^2\cdot487$	2	2	±10	2473	$2^3\cdot3\cdot103$	5	5	±10
1433	$2^3\cdot179$	3	3	±10	1951	$2\cdot3\cdot5^2\cdot13$	3	2		2477	$2^2\cdot619$	2	2	-----
1439	$2\cdot719$	7	2	-10	1973	$2^2\cdot17\cdot29$	2	2	-----	2503	$2\cdot3^2\cdot139$	3	2	-----
1447	$2\cdot3\cdot241$	3	2	10	1979	$2\cdot23\cdot43$	2	3	10	2521	$2^3\cdot3^2\cdot5\cdot7$	17	17	-----
1451	$2\cdot5^2\cdot29$	2	3	-----	1987	$2\cdot3\cdot331$	2	4	-----	2531	$2\cdot5\cdot11\cdot23$	2	3	
1453	$2^2\cdot3\cdot11^2$	2	2	-----	1993	$2^3\cdot3\cdot83$	5	5	-----	2539	$2\cdot3^3\cdot47$	2	4	10
1459	$2\cdot3^6$	3	6	-----	1997	$2^2\cdot499$	2	2	-----	2543	$2\cdot31\cdot41$	5	2	10
1471	$2\cdot3\cdot5\cdot7^2$	6	5	-10	1999	$2\cdot3^3\cdot37$	3	5	-10	2549	$2^2\cdot7^2\cdot13$	2	2	±10
1481	$2^3\cdot5\cdot37$	3	3	-----	2003	$2\cdot7\cdot11\cdot13$	5	3	-10	2551	$2\cdot3\cdot5^2\cdot17$	6	2	-----
1483	$2\cdot3\cdot13\cdot19$	2	4	-----	2011	$2\cdot3\cdot5\cdot67$	3	5	-----	2557	$2^2\cdot3^2\cdot71$	2	2	-----
1487	$2\cdot743$	5	2	10	2017	$2^5\cdot3^2\cdot7$	5	5	±10	2579	$2\cdot1289$	2	3	10
1489	$2^4\cdot3\cdot31$	14	14	-----	2027	$2\cdot1013$	2	3	-10	2591	$2\cdot5\cdot7\cdot37$	7	2	-----
1493	$2^2\cdot373$	2	2	-----	2029	$2^2\cdot3\cdot13^2$	2	2	±10	2593	$2^5\cdot3^4$	7	7	±10
1499	$2\cdot7\cdot107$	2	3	-----	2039	$2\cdot1019$	7	2	-10	2609	$2^4\cdot163$	3	3	-----
1511	$2\cdot5\cdot151$	11	2	-10	2053	$2^2\cdot3^3\cdot19$	2	2	-----	2617	$2^3\cdot3\cdot109$	5	5	±10
1523	$2\cdot761$	2	3	-10	2063	$2\cdot1031$	5	2	10	2621	$2^2\cdot5\cdot131$	2	2	±10
1531	$2\cdot3^2\cdot5\cdot17$	2	4	10	2069	$2^2\cdot11\cdot47$	2	2	±10	2633	$2^3\cdot7\cdot47$	3	3	±10
1543	$2\cdot3\cdot257$	5	2	10	2081	$2^5\cdot5\cdot13$	3	3	-----	2647	$2\cdot3^3\cdot7^2$	3	2	-----
1549	$2^2\cdot3^2\cdot43$	2	2	±10	2083	$2\cdot3\cdot347$	2	4	-10	2657	$2^5\cdot83$	3	3	±10
1553	$2^4\cdot97$	3	3	±10	2087	$2\cdot7\cdot149$	5	2	-----	2659	$2\cdot3\cdot443$	2	4	-----
1559	$2\cdot19\cdot41$	19	2	-10	2089	$2^3\cdot3^2\cdot29$	7	7	-----	2663	$2\cdot11^3$	5	2	10
1567	$2\cdot3^3\cdot29$	3	2	10	2099	$2\cdot1049$	2	3	10	2671	$2\cdot3\cdot5\cdot89$	7	5	-10
1571	$2\cdot5\cdot157$	2	3	10	2111	$2\cdot5\cdot211$	7	2	-10	2677	$2^2\cdot3\cdot223$	2	2	-----
1579	$2\cdot3\cdot263$	3	5	10	2113	$2^6\cdot3\cdot11$	5	2	±10	2683	$2\cdot3^2\cdot149$	2	4	-----
1583	$2\cdot7\cdot113$	5	2	10	2129	$2^4\cdot7\cdot19$	3	3	-----	2687	$2\cdot17\cdot79$	5	3	10
1597	$2^2\cdot3\cdot7\cdot19$	11	11	-----	2131	$2\cdot3\cdot5\cdot71$	2	4	-----	2689	$2^7\cdot3\cdot7$	19	19	-----
1601	$2^6\cdot5^2$	3	3	-----	2137	$2^3\cdot3\cdot89$	10	10	±10	2693	$2^2\cdot673$	2	2	-----
1607	$2\cdot11\cdot73$	5	2	10	2141	$2^2\cdot5\cdot107$	2	2	±10	2699	$2\cdot19\cdot71$	2	3	10
1609	$2^3\cdot3\cdot67$	7	7	-----	2143	$2\cdot3^2\cdot7\cdot17$	3	9	10	2707	$2\cdot3\cdot11\cdot41$	2	4	-10
1613	$2^2\cdot13\cdot31$	3	3	-----	2153	$2^3\cdot269$	3	3	±10	2711	$2\cdot5\cdot271$	7	2	-10
1619	$2\cdot809$	2	3	10	2161	$2^4\cdot3^3\cdot5$	23	23	-----	2713	$2^3\cdot3\cdot113$	5	5	±10
1621	$2^2\cdot3^4\cdot5$	2	2	±10	2179	$2\cdot3^2\cdot11^2$	7	5	10	2719	$2\cdot3^2\cdot151$	3	2	-10
1627	$2\cdot3\cdot271$	3	6	-----	2203	$2\cdot3\cdot367$	5	7	-10	2729	$2^3\cdot11\cdot31$	3	3	-----
1637	$2^2\cdot409$	2	2	-----	2207	$2\cdot1103$	5	2	10	2731	$2\cdot3\cdot5\cdot7\cdot13$	3	5	10
1657	$2^3\cdot3^2\cdot23$	11	11	-----	2213	$2^2\cdot7\cdot79$	2	2	-----	2741	$2^2\cdot5\cdot137$	2	2	±10
1663	$2\cdot3\cdot277$	3	2	10	2221	$2^2\cdot3\cdot5\cdot37$	2	2	±10	2749	$2^2\cdot3\cdot229$	6	6	-----
1667	$2\cdot7^2\cdot17$	2	3	-10	2237	$2^2\cdot13\cdot43$	2	2	-----	2753	$2^6\cdot43$	3	3	±10
1669	$2^2\cdot3\cdot139$	2	2	-----	2239	$2\cdot3\cdot373$	3	2	-10	2767	$2\cdot3\cdot461$	3	9	10
1693	$2^2\cdot3^2\cdot47$	2	2	-----	2243	$2\cdot19\cdot59$	2	3	-10	2777	$2^3\cdot347$	3	3	±10
1697	$2^5\cdot53$	3	3	±10	2251	$2\cdot3^2\cdot5^3$	7	5	10	2789	$2^2\cdot17\cdot41$	2	2	±10
1699	$2\cdot3\cdot283$	3	6	-----	2267	$2\cdot11\cdot103$	2	3	-10	2791	$2\cdot3^2\cdot5\cdot31$	6	7	-----
1709	$2^2\cdot7\cdot61$	3	3	±10	2269	$2^2\cdot3^4\cdot7$	2	2	±10	2797	$2^2\cdot3\cdot233$	2	2	-----
1721	$2^3\cdot5\cdot43$	3	3	-----	2273	$2^5\cdot71$	3	3	±10	2801	$2^4\cdot5^2\cdot7$	3	3	-----
1723	$2\cdot3\cdot7\cdot41$	3	6	-----	2281	$2^3\cdot3\cdot5\cdot19$	7	7	-----	2803	$2\cdot3\cdot467$	2	4	-10
1733	$2^2\cdot433$	2	2	-----	2287	$2\cdot3^2\cdot127$	19	7	-----	2819	$2\cdot1409$	2	3	10
1741	$2^2\cdot3\cdot5\cdot29$	2	2	±10	2293	$2^2\cdot3\cdot191$	2	2	-----	2833	$2^4\cdot3\cdot59$	5	5	±10
1747	$2\cdot3^2\cdot97$	2	4	-----	2297	$2^3\cdot7\cdot41$	5	5	±10	2837	$2^2\cdot709$	2	2	-----
1753	$2^3\cdot3\cdot73$	7	7	-----	2309	$2^2\cdot577$	2	2	±10	2843	$2\cdot7^2\cdot29$	2	4	-10
1759	$2\cdot3\cdot293$	6	2	-10	2311	$2\cdot3\cdot5\cdot7\cdot11$	3	2	-----	2851	$2\cdot3\cdot5^2\cdot19$	2	4	10
1777	$2^4\cdot3\cdot37$	5	5	±10	2333	$2^2\cdot11\cdot53$	2	2	-----	2857	$2^3\cdot3\cdot7\cdot17$	11	11	-----
1783	$2\cdot3^4\cdot11$	10	2	10	2339	$2\cdot7\cdot167$	2	3	10	2861	$2^2\cdot5\cdot11\cdot13$	2	2	±10
1787	$2\cdot19\cdot47$	2	3	-10	2341	$2^2\cdot3^2\cdot5\cdot13$	7	7	±10	2879	$2\cdot1439$	7	2	-10
1789	$2^2\cdot3\cdot149$	6	6	±10	2347	$2\cdot3\cdot17\cdot23$	3	6	-10	2887	$2\cdot3\cdot13\cdot37$	5	2	10
1801	$2^3\cdot3^2\cdot5^2$	11	11	-----	2351	$2\cdot5^2\cdot47$	13	3	-10	2897	$2^4\cdot181$	3	3	±10
1811	$2\cdot5\cdot181$	6	3	10	2357	$2^2\cdot19\cdot31$	2	2	-----	2903	$2\cdot1451$	5	2	10
1823	$2\cdot911$	5	2	10	2371	$2\cdot3\cdot5\cdot79$	2	4	10	2909	$2^2\cdot727$	2	2	±10

Table 24.8   **Primitive Roots, Factorization of $p-1$**

g, G denote the least positive and least negative (respectively) primitive roots of p. ϵ denotes whether $10, -10$ both or neither are primitive roots.

p	$p-1$	g	$-G$	ϵ	p	$p-1$	g	$-G$	ϵ	p	$p-1$	g	$-G$	ϵ
2917	$2^2 \cdot 3^6$	5	5	-----	3527	$2 \cdot 41 \cdot 43$	5	2	10	4079	$2 \cdot 2039$	11	2	-10
2927	$2 \cdot 7 \cdot 11 \cdot 19$	5	2	10	3529	$2^3 \cdot 3^2 \cdot 7^2$	17	17	-----	4091	$2 \cdot 5 \cdot 409$	2	3	10
2939	$2 \cdot 13 \cdot 113$	2	3	10	3533	$2^2 \cdot 883$	2	2	-----	4093	$2^2 \cdot 3 \cdot 11 \cdot 31$	2	2	-----
2953	$2^3 \cdot 3^2 \cdot 41$	13	13	-----	3539	$2 \cdot 29 \cdot 61$	2	3	10	4099	$2 \cdot 3 \cdot 683$	2	4	10
2957	$2^2 \cdot 739$	2	2	-----	3541	$2^2 \cdot 3 \cdot 5 \cdot 59$	7	7	-----	4111	$2 \cdot 3 \cdot 5 \cdot 137$	12	2	-10
2963	$2 \cdot 1481$	2	3	-10	3547	$2 \cdot 3^2 \cdot 197$	2	4	-10	4127	$2 \cdot 2063$	5	2	10
2969	$2^3 \cdot 7 \cdot 53$	3	3	-----	3557	$2^2 \cdot 7 \cdot 127$	2	2	-----	4129	$2^5 \cdot 3 \cdot 43$	13	13	-----
2971	$2 \cdot 3^3 \cdot 5 \cdot 11$	10	5	10	3559	$2 \cdot 3 \cdot 593$	3	2	-10	4133	$2^2 \cdot 1033$	2	2	-----
2999	$2 \cdot 1499$	17	2	-10	3571	$2 \cdot 3 \cdot 5 \cdot 7 \cdot 17$	2	4	10	4139	$2 \cdot 2069$	2	3	10
3001	$2^3 \cdot 3 \cdot 5^3$	14	14	-----	3581	$2^2 \cdot 5 \cdot 179$	2	2	± 10	4153	$2^3 \cdot 3 \cdot 173$	5	5	± 10
3011	$2 \cdot 5 \cdot 7 \cdot 43$	2	3	10	3583	$2 \cdot 3^2 \cdot 199$	2	2	-----	4157	$2^2 \cdot 1039$	2	2	-----
3019	$2 \cdot 3 \cdot 503$	2	4	10	3593	$2^3 \cdot 449$	3	3	± 10	4159	$2 \cdot 3^3 \cdot 7 \cdot 11$	3	2	-----
3023	$2 \cdot 1511$	5	2	10	3607	$2 \cdot 3 \cdot 601$	5	11	10	4177	$2^4 \cdot 3^2 \cdot 29$	5	5	± 10
3037	$2^2 \cdot 3 \cdot 11 \cdot 23$	2	2	-----	3613	$2^2 \cdot 3 \cdot 7 \cdot 43$	2	2	-----	4201	$2^3 \cdot 3 \cdot 5^2 \cdot 7$	11	11	-----
3041	$2^5 \cdot 5 \cdot 19$	3	3	-----	3617	$2^5 \cdot 113$	3	3	± 10	4211	$2 \cdot 5 \cdot 421$	6	3	10
3049	$2^3 \cdot 3 \cdot 127$	11	11	-----	3623	$2 \cdot 1811$	5	2	10	4217	$2^3 \cdot 17 \cdot 31$	3	3	± 10
3061	$2^2 \cdot 3^2 \cdot 5 \cdot 17$	6	6	-----	3631	$2 \cdot 3 \cdot 5 \cdot 11^2$	15	10	-10	4219	$2 \cdot 3 \cdot 19 \cdot 37$	2	4	10
3067	$2 \cdot 3 \cdot 7 \cdot 73$	2	4	-10	3637	$2^2 \cdot 3^2 \cdot 101$	2	2	-----	4229	$2^2 \cdot 7 \cdot 151$	2	2	± 10
3079	$2 \cdot 3^4 \cdot 19$	6	2	-10	3643	$2 \cdot 3 \cdot 607$	2	4	-10	4231	$2 \cdot 3^2 \cdot 5 \cdot 47$	3	2	-10
3083	$2 \cdot 23 \cdot 67$	2	3	-10	3659	$2 \cdot 31 \cdot 59$	2	3	10	4241	$2^4 \cdot 5 \cdot 53$	3	3	-----
3089	$2^4 \cdot 193$	3	3	-----	3671	$2 \cdot 5 \cdot 367$	13	2	-----	4243	$2 \cdot 3 \cdot 7 \cdot 101$	2	4	-10
3109	$2^2 \cdot 3 \cdot 7 \cdot 37$	6	6	-----	3673	$2^3 \cdot 3^3 \cdot 17$	5	5	± 10	4253	$2^2 \cdot 1063$	2	2	-----
3119	$2 \cdot 1559$	7	2	-10	3677	$2^2 \cdot 919$	2	2	-----	4259	$2 \cdot 2129$	2	3	10
3121	$2^4 \cdot 3 \cdot 5 \cdot 13$	7	7	-----	3691	$2 \cdot 3^2 \cdot 5 \cdot 41$	2	4	-----	4261	$2^2 \cdot 3 \cdot 5 \cdot 71$	2	2	± 10
3137	$2^6 \cdot 7^2$	3	3	± 10	3697	$2 \cdot 43^2$	5	5	-----	4271	$2 \cdot 5 \cdot 7 \cdot 61$	7	3	-10
3163	$2 \cdot 3 \cdot 17 \cdot 31$	3	6	-10	3701	$2^2 \cdot 5^2 \cdot 37$	2	2	± 10	4273	$2^4 \cdot 3 \cdot 89$	5	5	-----
3167	$2 \cdot 1583$	5	2	10	3709	$2^2 \cdot 3^2 \cdot 103$	2	2	± 10	4283	$2 \cdot 2141$	2	3	-10
3169	$2^5 \cdot 3^2 \cdot 11$	7	7	-----	3719	$2 \cdot 11 \cdot 13^2$	7	2	-10	4289	$2^6 \cdot 67$	3	3	-----
3181	$2^2 \cdot 3 \cdot 5 \cdot 53$	7	7	-----	3727	$2 \cdot 3^4 \cdot 23$	3	2	10	4297	$2^3 \cdot 3 \cdot 179$	5	5	-----
3187	$2 \cdot 3^3 \cdot 59$	2	4	-----	3733	$2^2 \cdot 3 \cdot 311$	2	2	-----	4327	$2 \cdot 3 \cdot 7 \cdot 103$	3	2	10
3191	$2 \cdot 5 \cdot 11 \cdot 29$	11	5	-----	3739	$2 \cdot 3 \cdot 7 \cdot 89$	7	5	-----	4337	$2^4 \cdot 271$	3	3	± 10
3203	$2 \cdot 1601$	2	3	-10	3761	$2^4 \cdot 5 \cdot 47$	3	3	-----	4339	$2 \cdot 3^2 \cdot 241$	10	5	10
3209	$2^3 \cdot 401$	3	3	-----	3767	$2 \cdot 7 \cdot 269$	5	2	10	4349	$2^2 \cdot 1087$	2	2	± 10
3217	$2^4 \cdot 3 \cdot 67$	5	5	-----	3769	$2^3 \cdot 3 \cdot 157$	7	7	-----	4357	$2^2 \cdot 3^2 \cdot 11^2$	2	2	-----
3221	$2^2 \cdot 5 \cdot 7 \cdot 23$	10	10	± 10	3779	$2 \cdot 1889$	2	3	10	4363	$2 \cdot 3 \cdot 727$	2	4	-10
3229	$2^2 \cdot 3 \cdot 269$	6	6	-----	3793	$2^4 \cdot 3 \cdot 79$	5	5	-----	4373	$2^2 \cdot 1093$	2	2	-----
3251	$2 \cdot 5^3 \cdot 13$	6	3	10	3797	$2^2 \cdot 13 \cdot 73$	2	2	-----	4391	$2 \cdot 5 \cdot 439$	14	2	-10
3253	$2^2 \cdot 3 \cdot 271$	2	2	-----	3803	$2 \cdot 1901$	2	3	-10	4397	$2^2 \cdot 7 \cdot 157$	2	2	-----
3257	$2^3 \cdot 11 \cdot 37$	3	3	± 10	3821	$2^2 \cdot 5 \cdot 191$	3	3	± 10	4409	$2^3 \cdot 19 \cdot 29$	3	3	-----
3259	$2 \cdot 3^2 \cdot 181$	3	5	10	3823	$2 \cdot 3 \cdot 7^2 \cdot 13$	3	9	-----	4421	$2^2 \cdot 5 \cdot 13 \cdot 17$	3	3	± 10
3271	$2 \cdot 3 \cdot 5 \cdot 109$	3	5	-10	3833	$2^3 \cdot 479$	3	3	± 10	4423	$2 \cdot 3 \cdot 11 \cdot 67$	3	7	10
3299	$2 \cdot 17 \cdot 97$	2	3	10	3847	$2 \cdot 3 \cdot 641$	5	2	10	4441	$2^3 \cdot 3 \cdot 5 \cdot 37$	21	21	-----
3301	$2^2 \cdot 3 \cdot 5^2 \cdot 11$	6	6	± 10	3851	$2 \cdot 5^2 \cdot 7 \cdot 11$	2	4	-----	4447	$2 \cdot 3^2 \cdot 13 \cdot 19$	3	2	10
3307	$2 \cdot 3 \cdot 19 \cdot 29$	2	4	-10	3853	$2^2 \cdot 3^2 \cdot 107$	2	2	-----	4451	$2 \cdot 5^2 \cdot 89$	2	3	10
3313	$2^4 \cdot 3^2 \cdot 23$	10	10	± 10	3863	$2 \cdot 1931$	5	2	10	4457	$2^3 \cdot 557$	3	3	± 10
3319	$2 \cdot 3 \cdot 7 \cdot 79$	6	2	-----	3877	$2^2 \cdot 3 \cdot 17 \cdot 19$	2	2	-----	4463	$2 \cdot 23 \cdot 97$	5	2	10
3323	$2 \cdot 11 \cdot 151$	2	3	-10	3881	$2^3 \cdot 5 \cdot 97$	13	13	-----	4481	$2^7 \cdot 5 \cdot 7$	3	3	-----
3329	$2^8 \cdot 13$	3	3	-----	3889	$2^4 \cdot 3^5$	11	11	-----	4483	$2 \cdot 3^3 \cdot 83$	2	4	-----
3331	$2 \cdot 3^2 \cdot 5 \cdot 37$	3	5	10	3907	$2 \cdot 3^2 \cdot 7 \cdot 31$	2	4	-10	4493	$2^2 \cdot 1123$	2	2	-----
3343	$2 \cdot 3 \cdot 557$	5	11	10	3911	$2 \cdot 5 \cdot 17 \cdot 23$	13	2	-10	4507	$2 \cdot 3 \cdot 751$	2	4	-----
3347	$2 \cdot 7 \cdot 239$	2	3	-10	3917	$2^2 \cdot 11 \cdot 89$	2	2	-----	4513	$2^5 \cdot 3 \cdot 47$	7	7	-----
3359	$2 \cdot 23 \cdot 73$	11	2	-10	3919	$2 \cdot 3 \cdot 653$	3	2	-----	4517	$2^2 \cdot 1129$	2	2	-----
3361	$2^5 \cdot 3 \cdot 5 \cdot 7$	22	22	-----	3923	$2 \cdot 37 \cdot 53$	2	3	-10	4519	$2 \cdot 3^2 \cdot 251$	3	9	-----
3371	$2 \cdot 5 \cdot 337$	2	3	10	3929	$2^3 \cdot 491$	3	3	-----	4523	$2 \cdot 7 \cdot 17 \cdot 19$	5	3	-10
3373	$2^2 \cdot 3 \cdot 281$	5	5	-----	3931	$2 \cdot 3 \cdot 5 \cdot 131$	2	4	-----	4547	$2 \cdot 2273$	2	3	-10
3389	$2^2 \cdot 7 \cdot 11^2$	3	3	± 10	3943	$2 \cdot 3^3 \cdot 73$	3	9	10	4549	$2^2 \cdot 3 \cdot 379$	6	6	-----
3391	$2 \cdot 3 \cdot 5 \cdot 113$	3	5	-10	3947	$2 \cdot 1973$	2	3	-10	4561	$2^4 \cdot 3 \cdot 5 \cdot 19$	11	11	-----
3407	$2 \cdot 13 \cdot 131$	5	2	10	3967	$2 \cdot 3 \cdot 661$	6	2	10	4567	$2 \cdot 3 \cdot 761$	3	7	10
3413	$2^2 \cdot 853$	2	2	-----	3989	$2^2 \cdot 997$	2	2	± 10	4583	$2 \cdot 29 \cdot 79$	5	2	10
3433	$2^3 \cdot 3 \cdot 11 \cdot 13$	5	5	± 10	4001	$2^5 \cdot 5^3$	3	3	-----	4591	$2 \cdot 3^3 \cdot 5 \cdot 17$	11	2	-10
3449	$2^3 \cdot 431$	3	3	-----	4003	$2 \cdot 3 \cdot 23 \cdot 29$	2	4	-----	4597	$2^2 \cdot 3 \cdot 383$	5	5	-----
3457	$2^7 \cdot 3^3$	7	7	-----	4007	$2 \cdot 2003$	5	2	10	4603	$2 \cdot 3 \cdot 13 \cdot 59$	2	4	-10
3461	$2^2 \cdot 5 \cdot 173$	2	2	± 10	4013	$2^2 \cdot 17 \cdot 59$	2	2	-----	4621	$2^2 \cdot 3 \cdot 5 \cdot 7 \cdot 11$	2	2	-----
3463	$2 \cdot 3 \cdot 577$	3	9	10	4019	$2 \cdot 7^2 \cdot 41$	2	4	10	4637	$2^2 \cdot 19 \cdot 61$	2	2	-----
3467	$2 \cdot 1733$	2	3	-10	4021	$2^2 \cdot 3 \cdot 5 \cdot 67$	2	2	-----	4639	$2 \cdot 3 \cdot 773$	3	2	-10
3469	$2^2 \cdot 3 \cdot 17^2$	2	2	± 10	4027	$2 \cdot 3 \cdot 11 \cdot 61$	3	6	-10	4643	$2 \cdot 11 \cdot 211$	5	3	-10
3491	$2 \cdot 5 \cdot 349$	2	3	-----	4049	$2^4 \cdot 11 \cdot 23$	3	3	-----	4649	$2^3 \cdot 7 \cdot 83$	3	3	-----
3499	$2 \cdot 3 \cdot 11 \cdot 53$	2	4	-----	4051	$2 \cdot 3^4 \cdot 5^2$	10	5	10	4651	$2 \cdot 3 \cdot 5^2 \cdot 31$	3	5	10
3511	$2 \cdot 3^3 \cdot 5 \cdot 13$	7	2	-10	4057	$2^3 \cdot 3 \cdot 13^2$	5	5	± 10	4657	$2^4 \cdot 3 \cdot 97$	15	15	-----
3517	$2^2 \cdot 3 \cdot 293$	2	2	-----	4073	$2^3 \cdot 509$	3	3	± 10	4663	$2 \cdot 3^2 \cdot 7 \cdot 37$	3	9	-----

Primitive Roots, Factorization of $p-1$ — Table 24.8

g, G denote the least positive and least negative (respectively) primitive roots of p. ϵ denotes whether 10, -10 both or neither are primitive roots.

p	$p-1$	g	$-G$	ϵ	p	$p-1$	g	$-G$	ϵ	p	$p-1$	g	$-G$	ϵ
4673	$2^6 \cdot 73$	3	3	± 10	5297	$2^4 \cdot 331$	3	3	± 10	5867	$2 \cdot 7 \cdot 419$	5	3	-10
4679	$2 \cdot 2339$	11	2	-10	5303	$2 \cdot 11 \cdot 241$	5	2	10	5869	$2^2 \cdot 3^2 \cdot 163$	2	2	± 10
4691	$2 \cdot 5 \cdot 7 \cdot 67$	2	3	10	5309	$2^2 \cdot 1327$	2	2	± 10	5879	$2 \cdot 2939$	11	2	-10
4703	$2 \cdot 2351$	5	2	10	5323	$2 \cdot 3 \cdot 887$	5	10	-10	5881	$2^3 \cdot 3 \cdot 5 \cdot 7^2$	31	31	-----
4721	$2^4 \cdot 5 \cdot 59$	6	6	-----	5333	$2^2 \cdot 31 \cdot 43$	2	2	-----	5897	$2^3 \cdot 11 \cdot 67$	3	3	± 10
4723	$2 \cdot 3 \cdot 787$	2	4	-10	5347	$2 \cdot 3^5 \cdot 11$	3	6	-10	5903	$2 \cdot 3 \cdot 227$	5	2	10
4729	$2^3 \cdot 3 \cdot 197$	17	17	-----	5351	$2 \cdot 5^2 \cdot 107$	11	2	-10	5923	$2 \cdot 3^2 \cdot 7 \cdot 47$	2	4	-10
4733	$2^2 \cdot 7 \cdot 13^2$	5	5	-----	5381	$2^2 \cdot 5 \cdot 269$	3	3	± 10	5927	$2 \cdot 2963$	5	2	10
4751	$2 \cdot 5^3 \cdot 19$	19	3	-10	5387	$2 \cdot 2693$	2	3	-10	5939	$2 \cdot 2969$	2	3	10
4759	$2 \cdot 3 \cdot 13 \cdot 61$	3	5	-10	5393	$2^4 \cdot 337$	3	3	± 10	5953	$2^6 \cdot 3 \cdot 31$	7	7	-----
4783	$2 \cdot 3 \cdot 797$	6	2	10	5399	$2 \cdot 2699$	7	2	-10	5981	$2^2 \cdot 5 \cdot 13 \cdot 23$	3	3	± 10
4787	$2 \cdot 2393$	2	3	-10	5407	$2 \cdot 3 \cdot 17 \cdot 53$	3	2	-----	5987	$2 \cdot 41 \cdot 73$	2	3	-10
4789	$2^2 \cdot 3^2 \cdot 7 \cdot 19$	2	2	-----	5413	$2^2 \cdot 3 \cdot 11 \cdot 41$	5	5	-----	6007	$2 \cdot 3 \cdot 7 \cdot 11 \cdot 13$	3	9	-----
4793	$2^3 \cdot 599$	3	3	± 10	5417	$2^3 \cdot 677$	3	3	± 10	6011	$2 \cdot 5 \cdot 601$	2	4	10
4799	$2 \cdot 2399$	7	2	-10	5419	$2 \cdot 3^2 \cdot 7 \cdot 43$	3	5	10	6029	$2^2 \cdot 11 \cdot 137$	2	2	± 10
4801	$2^6 \cdot 3 \cdot 5^2$	7	7	-----	5431	$2 \cdot 3 \cdot 5 \cdot 181$	3	2	-10	6037	$2^2 \cdot 3 \cdot 503$	5	5	-----
4813	$2^2 \cdot 3 \cdot 401$	2	2	-----	5437	$2^2 \cdot 3^2 \cdot 151$	5	5	-----	6043	$2 \cdot 3 \cdot 19 \cdot 53$	5	6	-10
4817	$2^4 \cdot 7 \cdot 43$	3	3	± 10	5441	$2^6 \cdot 5 \cdot 17$	3	3	-----	6047	$2 \cdot 3023$	5	2	10
4831	$2 \cdot 3 \cdot 5 \cdot 7 \cdot 23$	3	2	-----	5443	$2 \cdot 3 \cdot 907$	2	4	-----	6053	$2^2 \cdot 17 \cdot 89$	2	2	-----
4861	$2^2 \cdot 3^5 \cdot 5$	11	11	-----	5449	$2^3 \cdot 3 \cdot 227$	7	7	-----	6067	$2 \cdot 3^2 \cdot 337$	2	4	-10
4871	$2 \cdot 5 \cdot 487$	11	3	-10	5471	$2 \cdot 5 \cdot 547$	7	3	-----	6073	$2^3 \cdot 3 \cdot 11 \cdot 23$	10	10	± 10
4877	$2^2 \cdot 23 \cdot 53$	2	2	-----	5477	$2^2 \cdot 37^2$	2	2	-----	6079	$2 \cdot 3 \cdot 1013$	17	7	-----
4889	$2^3 \cdot 13 \cdot 47$	3	3	-----	5479	$2 \cdot 3 \cdot 11 \cdot 83$	3	2	-10	6089	$2^3 \cdot 761$	3	3	-----
4903	$2 \cdot 3 \cdot 19 \cdot 43$	3	2	-----	5483	$2 \cdot 2741$	2	3	-10	6091	$2 \cdot 3 \cdot 5 \cdot 7 \cdot 29$	7	11	-----
4909	$2^2 \cdot 3 \cdot 409$	6	6	-----	5501	$2^2 \cdot 5^3 \cdot 11$	2	2	± 10	6101	$2^2 \cdot 5^2 \cdot 61$	2	2	-----
4919	$2 \cdot 2459$	13	2	-10	5503	$2 \cdot 3 \cdot 7 \cdot 131$	3	9	10	6113	$2^5 \cdot 191$	3	3	± 10
4931	$2 \cdot 5 \cdot 17 \cdot 29$	6	3	10	5507	$2 \cdot 2753$	2	3	-10	6121	$2^3 \cdot 3^2 \cdot 5 \cdot 17$	7	7	-----
4933	$2^2 \cdot 3^2 \cdot 137$	2	2	-----	5519	$2 \cdot 31 \cdot 89$	13	2	-10	6131	$2 \cdot 5 \cdot 613$	2	3	10
4937	$2^3 \cdot 617$	3	3	± 10	5521	$2^4 \cdot 3 \cdot 5 \cdot 23$	11	11	-----	6133	$2^2 \cdot 3 \cdot 7 \cdot 73$	5	5	-----
4943	$2 \cdot 7 \cdot 353$	7	2	10	5527	$2 \cdot 3^2 \cdot 307$	5	2	10	6143	$2 \cdot 37 \cdot 83$	5	2	10
4951	$2 \cdot 3^2 \cdot 5^2 \cdot 11$	6	2	-10	5531	$2 \cdot 5 \cdot 7 \cdot 79$	10	5	10	6151	$2 \cdot 3 \cdot 5^2 \cdot 41$	3	7	-----
4957	$2^2 \cdot 3 \cdot 7 \cdot 59$	2	2	-----	5557	$2^2 \cdot 3 \cdot 463$	2	2	-----	6163	$2 \cdot 3 \cdot 13 \cdot 79$	3	6	-----
4967	$2 \cdot 13 \cdot 191$	5	2	10	5563	$2 \cdot 3^3 \cdot 103$	2	4	-10	6173	$2^2 \cdot 1543$	2	2	-----
4969	$2^3 \cdot 3^3 \cdot 23$	11	11	-----	5569	$2^6 \cdot 3 \cdot 29$	13	13	-----	6197	$2^2 \cdot 1549$	2	2	-----
4973	$2^2 \cdot 11 \cdot 113$	2	2	-----	5573	$2^2 \cdot 7 \cdot 199$	2	2	-----	6199	$2 \cdot 3 \cdot 1033$	3	2	-10
4987	$2 \cdot 3^2 \cdot 277$	2	4	-10	5581	$2^2 \cdot 3^2 \cdot 5 \cdot 31$	6	6	± 10	6203	$2 \cdot 7 \cdot 443$	2	3	-----
4993	$2^7 \cdot 3 \cdot 13$	5	5	-----	5591	$2 \cdot 5 \cdot 13 \cdot 43$	11	2	-10	6211	$2 \cdot 3^3 \cdot 5 \cdot 23$	2	4	10
4999	$2 \cdot 3 \cdot 7^2 \cdot 17$	3	9	-----	5623	$2 \cdot 3 \cdot 937$	5	2	10	6217	$2^3 \cdot 3 \cdot 7 \cdot 37$	5	5	± 10
5003	$2 \cdot 41 \cdot 61$	2	3	-10	5639	$2 \cdot 2819$	7	2	-10	6221	$2^2 \cdot 5 \cdot 311$	3	3	± 10
5009	$2^4 \cdot 313$	3	3	-----	5641	$2^3 \cdot 3 \cdot 5 \cdot 47$	14	14	-----	6229	$2^2 \cdot 3^2 \cdot 173$	2	2	-----
5011	$2 \cdot 3 \cdot 5 \cdot 167$	2	4	-----	5647	$2 \cdot 3 \cdot 941$	3	2	-----	6247	$2 \cdot 3^2 \cdot 347$	5	2	10
5021	$2^2 \cdot 5 \cdot 251$	3	3	± 10	5651	$2 \cdot 5^2 \cdot 113$	2	3	10	6257	$2^4 \cdot 17 \cdot 23$	3	3	± 10
5023	$2 \cdot 3^4 \cdot 31$	3	2	-----	5653	$2^2 \cdot 3^2 \cdot 157$	5	5	-----	6263	$2 \cdot 31 \cdot 101$	5	2	10
5039	$2 \cdot 11 \cdot 229$	11	2	-10	5657	$2^3 \cdot 7 \cdot 101$	3	3	± 10	6269	$2^2 \cdot 1567$	2	2	± 10
5051	$2 \cdot 5^2 \cdot 101$	2	3	-----	5659	$2 \cdot 3 \cdot 23 \cdot 41$	2	4	10	6271	$2 \cdot 3 \cdot 5 \cdot 11 \cdot 19$	11	17	-----
5059	$2 \cdot 3^2 \cdot 281$	2	4	10	5669	$2^2 \cdot 13 \cdot 109$	3	3	± 10	6277	$2^2 \cdot 3 \cdot 523$	2	2	-----
5077	$2^2 \cdot 3^3 \cdot 47$	2	2	-----	5683	$2 \cdot 3 \cdot 947$	2	4	-10	6287	$2 \cdot 7 \cdot 449$	7	2	10
5081	$2^3 \cdot 5 \cdot 127$	3	3	-----	5689	$2^3 \cdot 3^2 \cdot 79$	11	11	-----	6299	$2 \cdot 47 \cdot 67$	2	3	-----
5087	$2 \cdot 2543$	5	2	10	5693	$2^2 \cdot 1423$	2	2	-----	6301	$2^2 \cdot 3^2 \cdot 5^2 \cdot 7$	10	10	± 10
5099	$2 \cdot 2549$	2	3	10	5701	$2^2 \cdot 3 \cdot 5^2 \cdot 19$	2	2	± 10	6311	$2 \cdot 5 \cdot 631$	7	2	-10
5101	$2^2 \cdot 3 \cdot 5^2 \cdot 17$	6	6	-----	5711	$2 \cdot 5 \cdot 571$	19	3	-----	6317	$2^2 \cdot 1579$	2	2	-----
5107	$2 \cdot 3 \cdot 23 \cdot 37$	2	4	-10	5717	$2^2 \cdot 1429$	2	2	-----	6323	$2 \cdot 29 \cdot 109$	2	3	-10
5113	$2^3 \cdot 3^2 \cdot 71$	19	19	-----	5737	$2^3 \cdot 3 \cdot 239$	5	5	± 10	6329	$2^3 \cdot 7 \cdot 113$	3	3	-----
5119	$2 \cdot 3 \cdot 853$	3	2	-----	5741	$2^2 \cdot 5 \cdot 7 \cdot 41$	2	2	± 10	6337	$2^6 \cdot 3^2 \cdot 11$	10	10	± 10
5147	$2 \cdot 31 \cdot 83$	2	3	-10	5743	$2 \cdot 3^2 \cdot 11 \cdot 29$	10	2	10	6343	$2 \cdot 3 \cdot 7 \cdot 151$	3	2	10
5153	$2^5 \cdot 7 \cdot 23$	5	5	± 10	5749	$2^2 \cdot 3 \cdot 479$	2	2	± 10	6353	$2^4 \cdot 397$	3	3	± 10
5167	$2 \cdot 3^2 \cdot 7 \cdot 41$	6	11	10	5779	$2 \cdot 3^3 \cdot 107$	2	4	10	6359	$2 \cdot 11 \cdot 17^2$	13	2	-10
5171	$2 \cdot 5 \cdot 11 \cdot 47$	2	4	-----	5783	$2 \cdot 7^2 \cdot 59$	7	2	10	6361	$2^3 \cdot 3 \cdot 5 \cdot 53$	19	19	-----
5179	$2 \cdot 3 \cdot 863$	2	4	10	5791	$2 \cdot 3 \cdot 5 \cdot 193$	6	2	-----	6367	$2 \cdot 3 \cdot 1061$	3	2	10
5189	$2^2 \cdot 1297$	2	2	± 10	5801	$2^3 \cdot 5^2 \cdot 29$	3	3	-----	6373	$2^2 \cdot 3^3 \cdot 59$	2	2	-----
5197	$2^2 \cdot 3 \cdot 433$	7	7	-----	5807	$2 \cdot 2903$	5	2	10	6379	$2 \cdot 3 \cdot 1063$	2	4	-----
5209	$2^3 \cdot 3 \cdot 7 \cdot 31$	17	17	-----	5813	$2^2 \cdot 1453$	2	2	-----	6389	$2^2 \cdot 1597$	2	2	± 10
5227	$2 \cdot 3 \cdot 13 \cdot 67$	2	4	-10	5821	$2^2 \cdot 3 \cdot 5 \cdot 97$	6	6	± 10	6397	$2^2 \cdot 3 \cdot 13 \cdot 41$	2	2	-----
5231	$2 \cdot 5 \cdot 523$	7	2	-10	5827	$2 \cdot 3 \cdot 971$	2	4	-10	6421	$2^2 \cdot 3 \cdot 5 \cdot 107$	6	6	-----
5233	$2^4 \cdot 3 \cdot 109$	10	10	± 10	5839	$2 \cdot 3 \cdot 7 \cdot 139$	6	2	-10	6427	$2 \cdot 3^3 \cdot 7 \cdot 17$	3	6	-----
5237	$2^2 \cdot 7 \cdot 11 \cdot 17$	3	3	-----	5843	$2 \cdot 23 \cdot 127$	2	4	-10	6449	$2^4 \cdot 13 \cdot 31$	3	3	-----
5261	$2^2 \cdot 5 \cdot 263$	2	2	-----	5849	$2^3 \cdot 17 \cdot 43$	3	3	-----	6451	$2 \cdot 3 \cdot 5^2 \cdot 43$	3	6	-----
5273	$2^3 \cdot 659$	3	3	± 10	5851	$2 \cdot 3^2 \cdot 5^2 \cdot 13$	2	4	-----	6469	$2^2 \cdot 3 \cdot 7^2 \cdot 11$	2	2	-----
5279	$2 \cdot 7 \cdot 13 \cdot 29$	7	3	-10	5857	$2^5 \cdot 3 \cdot 61$	7	7	± 10	6473	$2^3 \cdot 809$	3	3	± 10
5281	$2^5 \cdot 3 \cdot 5 \cdot 11$	7	7	-----	5861	$2^2 \cdot 5 \cdot 293$	3	3	± 10	6481	$2^4 \cdot 3^4 \cdot 5$	7	7	-----

Table 24.8   **Primitive Roots, Factorization of $p-1$**

g, G denote the least positive and least negative (respectively) primitive roots of p. ϵ denotes whether 10, -10 both or neither are primitive roots.

p	$p-1$	g	$-G$	ϵ	p	$p-1$	g	$-G$	ϵ	p	$p-1$	g	$-G$	ϵ
6491	$2\cdot5\cdot11\cdot59$	2	3	-----	7121	$2^4\cdot5\cdot89$	3	3	-----	7741	$2^2\cdot3^2\cdot5\cdot43$	7	7	-----
6521	$2^3\cdot5\cdot163$	6	6	-----	7127	$2\cdot7\cdot509$	5	2	-----	7753	$2^3\cdot3\cdot17\cdot19$	10	10	±10
6529	$2^7\cdot3\cdot17$	7	7	-----	7129	$2^3\cdot3^4\cdot11$	7	7	-----	7757	$2^2\cdot7\cdot277$	2	2	-----
6547	$2\cdot3\cdot1091$	2	4	-----	7151	$2\cdot5^2\cdot11\cdot13$	7	3	-----	7759	$2\cdot3^2\cdot431$	3	2	-10
6551	$2\cdot5^2\cdot131$	17	2	-10	7159	$2\cdot3\cdot1193$	3	2	-10	7789	$2^2\cdot3\cdot11\cdot59$	2	2	-----
6553	$2^3\cdot3^2\cdot7\cdot13$	10	10	±10	7177	$2^3\cdot3\cdot13\cdot23$	10	10	±10	7793	$2^4\cdot487$	3	3	±10
6563	$2\cdot17\cdot193$	5	10	-10	7187	$2\cdot3593$	2	3	-10	7817	$2^3\cdot977$	3	3	±10
6569	$2^3\cdot821$	3	3	-----	7193	$2^3\cdot29\cdot31$	3	3	±10	7823	$2\cdot3911$	5	2	10
6571	$2\cdot3^2\cdot5\cdot73$	3	7	10	7207	$2\cdot3\cdot1201$	3	2	10	7829	$2^2\cdot19\cdot103$	2	2	±10
6577	$2^4\cdot3\cdot137$	5	5	-----	7211	$2\cdot5\cdot7\cdot103$	2	3	-----	7841	$2^5\cdot5\cdot7^2$	12	12	-----
6581	$2^2\cdot5\cdot7\cdot47$	14	14	-----	7213	$2^2\cdot3\cdot601$	5	5	-----	7853	$2^2\cdot13\cdot151$	2	2	-----
6599	$2\cdot3299$	13	2	-10	7219	$2\cdot3^2\cdot401$	2	4	10	7867	$2\cdot3^2\cdot19\cdot23$	3	6	-10
6607	$2\cdot3^2\cdot367$	3	2	-----	7229	$2^2\cdot13\cdot139$	2	2	±10	7873	$2^6\cdot3\cdot41$	5	5	±10
6619	$2\cdot3\cdot1103$	2	4	10	7237	$2^2\cdot3^2\cdot67$	2	2	-----	7877	$2^2\cdot11\cdot179$	2	2	-----
6637	$2^2\cdot3\cdot7\cdot79$	2	2	-----	7243	$2\cdot3\cdot17\cdot71$	2	4	-10	7879	$2\cdot3\cdot13\cdot101$	3	2	-10
6653	$2^2\cdot1663$	2	2	-----	7247	$2\cdot3623$	5	2	10	7883	$2\cdot7\cdot563$	2	3	-10
6659	$2\cdot3329$	2	3	10	7253	$2^2\cdot7^2\cdot37$	2	2	-----	7901	$2^2\cdot5^2\cdot79$	2	2	±10
6661	$2^2\cdot3^2\cdot5\cdot37$	6	6	±10	7283	$2\cdot11\cdot331$	2	3	-10	7907	$2\cdot59\cdot67$	2	3	-10
6673	$2^4\cdot3\cdot139$	5	5	±10	7297	$2^7\cdot3\cdot19$	5	5	-----	7919	$2\cdot37\cdot107$	7	2	-10
6679	$2\cdot3^2\cdot7\cdot53$	7	5	-10	7307	$2\cdot13\cdot281$	2	3	-10	7927	$2\cdot3\cdot1321$	3	7	10
6689	$2^5\cdot11\cdot19$	3	3	-----	7309	$2^2\cdot3^2\cdot7\cdot29$	6	6	±10	7933	$2^2\cdot3\cdot661$	2	2	-----
6691	$2\cdot3\cdot5\cdot223$	2	4	10	7321	$2^3\cdot3\cdot5\cdot61$	7	7	-----	7937	$2^8\cdot31$	3	3	±10
6701	$2^2\cdot5^2\cdot67$	2	2	±10	7331	$2\cdot5\cdot733$	2	4	-----	7949	$2^2\cdot1987$	2	2	±10
6703	$2\cdot3\cdot1117$	5	2	10	7333	$2^2\cdot3\cdot13\cdot47$	6	6	-----	7951	$2\cdot3\cdot5^2\cdot53$	6	2	-10
6709	$2^2\cdot3\cdot13\cdot43$	2	2	±10	7349	$2^2\cdot11\cdot167$	2	2	±10	7963	$2\cdot3\cdot1327$	5	10	-10
6719	$2\cdot3359$	11	2	-10	7351	$2\cdot3\cdot5^2\cdot7^2$	6	5	-----	7993	$2^3\cdot3^2\cdot37$	5	5	-----
6733	$2^2\cdot3^2\cdot11\cdot17$	2	2	-----	7369	$2^3\cdot3\cdot307$	7	7	-----	8009	$2^3\cdot7\cdot11\cdot13$	3	3	-----
6737	$2^4\cdot421$	3	3	±10	7393	$2^5\cdot3\cdot7\cdot11$	5	5	±10	8011	$2\cdot3^2\cdot5\cdot89$	14	7	-----
6761	$2^3\cdot5\cdot13^2$	3	3	-----	7411	$2\cdot3\cdot5\cdot13\cdot19$	2	4	10	8017	$2^4\cdot3\cdot167$	5	5	±10
6763	$2\cdot3\cdot7^2\cdot23$	2	4	-----	7417	$2^3\cdot3^2\cdot103$	5	5	-----	8039	$2\cdot4019$	11	2	-10
6779	$2\cdot3389$	2	3	10	7433	$2^3\cdot929$	3	3	±10	8053	$2^2\cdot3\cdot11\cdot61$	2	2	-----
6781	$2^2\cdot3\cdot5\cdot113$	2	2	-----	7451	$2\cdot5^2\cdot149$	2	4	10	8059	$2\cdot3\cdot17\cdot79$	3	5	10
6791	$2\cdot5\cdot7\cdot97$	7	3	-----	7457	$2^5\cdot233$	3	3	±10	8069	$2^2\cdot2017$	2	2	±10
6793	$2^3\cdot3\cdot283$	10	10	±10	7459	$2\cdot3\cdot11\cdot113$	2	4	10	8081	$2^4\cdot5\cdot101$	3	3	-----
6803	$2\cdot19\cdot179$	2	3	-10	7477	$2^2\cdot3\cdot7\cdot89$	2	2	-----	8087	$2\cdot13\cdot311$	5	2	10
6823	$2\cdot3^2\cdot379$	3	2	10	7481	$2^3\cdot5\cdot11\cdot17$	6	6	-----	8089	$2^3\cdot3\cdot337$	17	17	-----
6827	$2\cdot3413$	2	3	-10	7487	$2\cdot19\cdot197$	5	3	10	8093	$2^2\cdot7\cdot17^2$	2	2	-----
6829	$2^2\cdot3\cdot569$	2	2	±10	7489	$2^6\cdot3^2\cdot13$	7	7	-----	8101	$2^2\cdot3^4\cdot5^2$	6	6	-----
6833	$2^4\cdot7\cdot61$	3	3	±10	7499	$2\cdot23\cdot163$	2	3	10	8111	$2\cdot5\cdot811$	11	2	-----
6841	$2^3\cdot3^2\cdot5\cdot19$	22	22	-----	7507	$2\cdot3^3\cdot139$	2	4	-10	8117	$2^2\cdot2029$	2	2	-----
6857	$2^3\cdot857$	3	3	±10	7517	$2^2\cdot1879$	2	2	-----	8123	$2\cdot31\cdot131$	2	3	-10
6863	$2\cdot47\cdot73$	5	2	10	7523	$2\cdot3761$	2	3	-10	8147	$2\cdot4073$	2	3	-10
6869	$2^2\cdot17\cdot101$	2	2	±10	7529	$2^3\cdot941$	3	3	-----	8161	$2^5\cdot3\cdot5\cdot17$	7	7	-----
6871	$2\cdot3\cdot5\cdot229$	3	9	-10	7537	$2^4\cdot3\cdot157$	7	7	-----	8167	$2\cdot3\cdot1361$	3	9	-----
6883	$2\cdot3\cdot31\cdot37$	2	4	-10	7541	$2^2\cdot5\cdot13\cdot29$	2	2	±10	8171	$2\cdot5\cdot19\cdot43$	2	3	10
6899	$2\cdot3449$	2	3	10	7547	$2\cdot7^3\cdot11$	2	3	-10	8179	$2\cdot3\cdot29\cdot47$	2	4	10
6907	$2\cdot3\cdot1151$	2	4	-----	7549	$2^2\cdot3\cdot17\cdot37$	2	2	-----	8191	$2\cdot3^2\cdot5\cdot7\cdot13$	17	11	-----
6911	$2\cdot5\cdot691$	7	2	-10	7559	$2\cdot3779$	13	2	-10	8209	$2^4\cdot3^3\cdot19$	7	7	-----
6917	$2^2\cdot7\cdot13\cdot19$	2	2	-----	7561	$2^3\cdot3^3\cdot5\cdot7$	13	13	-----	8219	$2\cdot7\cdot587$	2	3	10
6947	$2\cdot23\cdot151$	2	3	-10	7573	$2^2\cdot3\cdot631$	2	2	-----	8221	$2^2\cdot3\cdot5\cdot137$	2	2	-----
6949	$2^2\cdot3^2\cdot193$	2	2	±10	7577	$2^3\cdot947$	3	3	±10	8231	$2\cdot5\cdot823$	11	2	-10
6959	$2\cdot7^2\cdot71$	7	3	-10	7583	$2\cdot17\cdot223$	5	2	10	8233	$2^3\cdot3\cdot7^3$	10	10	±10
6961	$2^4\cdot3\cdot5\cdot29$	13	13	-----	7589	$2^2\cdot7\cdot271$	2	2	-----	8237	$2^2\cdot29\cdot71$	2	2	-----
6967	$2\cdot3^4\cdot43$	5	13	10	7591	$2\cdot3\cdot5\cdot11\cdot23$	6	2	-10	8243	$2\cdot13\cdot317$	2	3	-10
6971	$2\cdot5\cdot17\cdot41$	2	4	10	7603	$2\cdot3\cdot7\cdot181$	2	4	-----	8263	$2\cdot3^5\cdot17$	3	2	10
6977	$2^6\cdot109$	3	3	±10	7607	$2\cdot3803$	5	2	10	8269	$2^2\cdot3\cdot13\cdot53$	2	2	±10
6983	$2\cdot3491$	5	2	10	7621	$2^2\cdot3\cdot5\cdot127$	2	2	-----	8273	$2^4\cdot11\cdot47$	3	3	±10
6991	$2\cdot3\cdot5\cdot233$	6	2	-10	7639	$2\cdot3\cdot19\cdot67$	7	5	-10	8287	$2\cdot3\cdot1381$	3	7	10
6997	$2^2\cdot3\cdot11\cdot53$	5	5	-----	7643	$2\cdot3821$	2	3	-10	8291	$2\cdot5\cdot829$	2	3	10
7001	$2^3\cdot5^3\cdot7$	3	3	-----	7649	$2^5\cdot239$	3	3	-----	8293	$2^3\cdot1049$	3	3	-----
7013	$2^2\cdot1753$	2	2	-----	7669	$2^2\cdot3^3\cdot71$	2	2	-----	8297	$2^2\cdot2099$	3	3	±10
7019	$2\cdot11^2\cdot29$	2	3	10	7673	$2^3\cdot7\cdot137$	3	3	±10	8311	$2\cdot3\cdot5\cdot277$	3	2	-10
7027	$2\cdot3\cdot1171$	2	4	-----	7681	$2^9\cdot3\cdot5$	17	17	-----	8317	$2^2\cdot3^3\cdot7\cdot11$	6	6	-----
7039	$2\cdot3^2\cdot17\cdot23$	3	2	-----	7687	$2\cdot3^2\cdot7\cdot61$	6	2	10	8329	$2^3\cdot3\cdot347$	7	7	-----
7043	$2\cdot7\cdot503$	2	4	-----	7691	$2\cdot5\cdot769$	2	3	10	8353	$2^5\cdot3^2\cdot29$	5	5	±10
7057	$2^4\cdot3^2\cdot7^2$	5	5	±10	7699	$2\cdot3\cdot1283$	3	5	10	8363	$2\cdot37\cdot113$	2	3	-10
7069	$2^2\cdot3\cdot19\cdot31$	2	2	±10	7703	$2\cdot3851$	5	2	10	8369	$2^4\cdot523$	3	3	-----
7079	$2\cdot3539$	7	2	-10	7717	$2^2\cdot3\cdot643$	2	2	-----	8377	$2^3\cdot3\cdot349$	5	5	±10
7103	$2\cdot53\cdot67$	5	2	10	7723	$2\cdot3^3\cdot11\cdot13$	3	6	-----	8387	$2\cdot7\cdot599$	2	3	-----
7109	$2^2\cdot1777$	2	2	±10	7727	$2\cdot3863$	5	2	10	8389	$2^2\cdot3^2\cdot233$	6	6	±10

Primitive Roots, Factorization of $p-1$ — Table 24.8

g, G denote the least positive and least negative (respectively) primitive roots of p. ϵ denotes whether 10, -10 both or neither are primitive roots.

p	$p-1$	g	$-G$	ϵ	p	$p-1$	g	$-G$	ϵ	p	$p-1$	g	$-G$	ϵ
8419	$2 \cdot 3 \cdot 23 \cdot 61$	3	6	-----	8941	$2^2 \cdot 3 \cdot 5 \cdot 149$	6	6	-----	9463	$2 \cdot 3 \cdot 19 \cdot 83$	3	9	-----
8423	$2 \cdot 4211$	5	2	10	8951	$2 \cdot 5^2 \cdot 179$	13	2	-10	9467	$2 \cdot 4733$	2	3	-10
8429	$2^2 \cdot 7^2 \cdot 43$	2	2	± 10	8963	$2 \cdot 4481$	2	3	-10	9473	$2^8 \cdot 37$	3	3	± 10
8431	$2 \cdot 3 \cdot 5 \cdot 281$	3	2	-10	8969	$2^3 \cdot 19 \cdot 59$	3	3	-----	9479	$2 \cdot 7 \cdot 677$	7	2	-10
8443	$2 \cdot 3 \cdot 3^2 \cdot 7 \cdot 67$	2	4	-10	8971	$2 \cdot 3 \cdot 5 \cdot 13 \cdot 23$	2	4	10	9491	$2 \cdot 5 \cdot 13 \cdot 73$	2	3	10
8447	$2 \cdot 41 \cdot 103$	5	2	10	8999	$2 \cdot 11 \cdot 409$	7	2	-10	9497	$2^3 \cdot 1187$	3	3	± 10
8461	$2^2 \cdot 3^2 \cdot 5 \cdot 47$	6	6	-----	9001	$2^3 \cdot 3^2 \cdot 5^3$	7	7	-----	9511	$2 \cdot 3 \cdot 5 \cdot 317$	3	9	-----
8467	$2 \cdot 3 \cdot 17 \cdot 83$	2	4	-10	9007	$2 \cdot 3 \cdot 19 \cdot 79$	3	2	-----	9521	$2^4 \cdot 5 \cdot 7 \cdot 17$	3	3	-----
8501	$2^2 \cdot 5^3 \cdot 17$	7	7	± 10	9011	$2 \cdot 5 \cdot 17 \cdot 53$	2	4	10	9533	$2^2 \cdot 2383$	2	2	-----
8513	$2^6 \cdot 7 \cdot 19$	5	5	± 10	9013	$2^2 \cdot 3 \cdot 751$	5	5	-----	9539	$2 \cdot 19 \cdot 251$	2	3	10
8521	$2^3 \cdot 3 \cdot 5 \cdot 71$	13	13	-----	9029	$2^2 \cdot 37 \cdot 61$	2	2	± 10	9547	$2 \cdot 3 \cdot 37 \cdot 43$	2	4	-10
8527	$2 \cdot 3 \cdot 7^2 \cdot 29$	5	2		9041	$2^4 \cdot 5 \cdot 113$	3	3		9551	$2 \cdot 5^2 \cdot 191$	11	2	-----
8537	$2^3 \cdot 11 \cdot 97$	3	3	± 10	9043	$2 \cdot 3 \cdot 11 \cdot 137$	3	6	-10	9587	$2 \cdot 4793$	2	3	-10
8539	$2 \cdot 3 \cdot 1423$	2	4	-----	9049	$2^3 \cdot 3 \cdot 13 \cdot 29$	7	7	-----	9601	$2^7 \cdot 3 \cdot 5^2$	13	13	-----
8543	$2 \cdot 4271$	5	2	10	9059	$2 \cdot 7 \cdot 647$	2	4	10	9613	$2^2 \cdot 3^3 \cdot 89$	2	2	-----
8563	$2 \cdot 3 \cdot 1427$	2	4	-10	9067	$2 \cdot 3 \cdot 1511$	3	6	-10	9619	$2 \cdot 3 \cdot 7 \cdot 229$	2	4	-----
8573	$2^2 \cdot 2143$	2	2	-----	9091	$2 \cdot 3^2 \cdot 5 \cdot 101$	3	5	-----	9623	$2 \cdot 17 \cdot 283$	5	3	10
8581	$2^2 \cdot 3 \cdot 5 \cdot 11 \cdot 13$	6	6	-----	9103	$2 \cdot 3 \cdot 37 \cdot 41$	6	2	10	9629	$2^2 \cdot 29 \cdot 83$	2	2	± 10
8597	$2^2 \cdot 7 \cdot 307$	2	2	-----	9109	$2^2 \cdot 3^2 \cdot 11 \cdot 23$	10	10	± 10	9631	$2 \cdot 3^2 \cdot 5 \cdot 107$	3	9	-10
8599	$2 \cdot 3 \cdot 1433$	3	2	-----	9127	$2 \cdot 3^3 \cdot 13^2$	3	2	-----	9643	$2 \cdot 3 \cdot 1607$	2	4	-10
8609	$2^5 \cdot 269$	3	3	-----	9133	$2^2 \cdot 3 \cdot 761$	6	6	-----	9649	$2^4 \cdot 3^2 \cdot 67$	7	7	-----
8623	$2 \cdot 3^2 \cdot 479$	3	2	10	9137	$2^4 \cdot 571$	3	3	± 10	9661	$2^2 \cdot 3 \cdot 5 \cdot 7 \cdot 23$	2	2	-----
8627	$2 \cdot 19 \cdot 227$	2	3	-10	9151	$2 \cdot 3 \cdot 5^2 \cdot 61$	3	2	-----	9677	$2^2 \cdot 41 \cdot 59$	2	2	-----
8629	$2^2 \cdot 3 \cdot 719$	6	6	-----	9157	$2^2 \cdot 3 \cdot 7 \cdot 109$	6	6	-----	9679	$2 \cdot 3 \cdot 1613$	3	2	-----
8641	$2^6 \cdot 3^3 \cdot 5$	17	17	-----	9161	$2^3 \cdot 5 \cdot 229$	3	3	-----	9689	$2^3 \cdot 7 \cdot 173$	3	3	-----
8647	$2 \cdot 3 \cdot 11 \cdot 131$	3	2	10	9173	$2^2 \cdot 2293$	2	2	-----	9697	$2^5 \cdot 3 \cdot 101$	10	10	± 10
8663	$2 \cdot 61 \cdot 71$	5	2	10	9181	$2^2 \cdot 3 \cdot 5 \cdot 17$	2	2	-----	9719	$2 \cdot 43 \cdot 113$	17	3	-10
8669	$2^2 \cdot 11 \cdot 197$	2	2	± 10	9187	$2 \cdot 3 \cdot 1531$	3	6	-10	9721	$2^3 \cdot 3^5 \cdot 5$	7	7	-----
8677	$2^2 \cdot 3^2 \cdot 241$	2	2		9199	$2 \cdot 3^2 \cdot 7 \cdot 73$	3	2	-10	9733	$2^2 \cdot 3 \cdot 811$	2	2	-----
8681	$2^3 \cdot 5 \cdot 7 \cdot 31$	15	15	-----	9203	$2 \cdot 43 \cdot 107$	2	3	-10	9739	$2 \cdot 3^2 \cdot 541$	3	5	10
8689	$2^4 \cdot 3 \cdot 181$	13	13	-----	9209	$2^3 \cdot 1151$	3	3	-----	9743	$2 \cdot 4871$	5	2	10
8693	$2^2 \cdot 41 \cdot 53$	2	2	-----	9221	$2^2 \cdot 5 \cdot 461$	2	2	± 10	9749	$2^2 \cdot 2437$	2	2	± 10
8699	$2 \cdot 4349$	2	3	10	9227	$2 \cdot 7 \cdot 659$	2	3	-10	9767	$2 \cdot 19 \cdot 257$	5	2	10
8707	$2 \cdot 3 \cdot 1451$	5	7	-10	9239	$2 \cdot 31 \cdot 149$	19	2	-10	9769	$2^3 \cdot 3 \cdot 11 \cdot 37$	13	13	-----
8713	$2^3 \cdot 3^2 \cdot 11^2$	5	5	± 10	9241	$2^3 \cdot 3 \cdot 5 \cdot 7 \cdot 11$	13	13	-----	9781	$2^2 \cdot 3 \cdot 5 \cdot 163$	6	6	± 10
8719	$2 \cdot 3 \cdot 1453$	3	5	-10	9257	$2^3 \cdot 13 \cdot 89$	3	3	± 10	9787	$2 \cdot 3 \cdot 7 \cdot 233$	3	6	-10
8731	$2 \cdot 3^2 \cdot 5 \cdot 97$	2	4	10	9277	$2^2 \cdot 3 \cdot 773$	5	5	-----	9791	$2 \cdot 5 \cdot 11 \cdot 89$	11	2	-10
8737	$2^5 \cdot 3 \cdot 7 \cdot 13$	5	5	-----	9281	$2^6 \cdot 5 \cdot 29$	3	3	-----	9803	$2 \cdot 13^2 \cdot 29$	2	3	-10
8741	$2^2 \cdot 5 \cdot 19 \cdot 23$	2	2	± 10	9283	$2 \cdot 3 \cdot 7 \cdot 13 \cdot 17$	2	4	-----	9811	$2 \cdot 3^2 \cdot 5 \cdot 109$	3	5	10
8747	$2 \cdot 4373$	2	3	-10	9293	$2^2 \cdot 23 \cdot 101$	2	2	-----	9817	$2^3 \cdot 3 \cdot 409$	5	5	± 10
8753	$2^4 \cdot 547$	3	3	± 10	9311	$2 \cdot 5 \cdot 7^2 \cdot 19$	7	2	-10	9829	$2^2 \cdot 3^3 \cdot 7 \cdot 13$	10	10	± 10
8761	$2^3 \cdot 3 \cdot 5 \cdot 73$	23	23	-----	9319	$2 \cdot 3 \cdot 1553$	3	2	-10	9833	$2^3 \cdot 1229$	3	3	± 10
8779	$2 \cdot 3 \cdot 7 \cdot 11 \cdot 19$	11	22	-----	9323	$2 \cdot 59 \cdot 79$	2	3	-10	9839	$2 \cdot 4919$	7	2	-10
8783	$2 \cdot 4391$	5	2	10	9337	$2^3 \cdot 3 \cdot 389$	5	5	-----	9851	$2 \cdot 5^2 \cdot 197$	2	4	10
8803	$2 \cdot 3^3 \cdot 163$	2	4	-----	9341	$2^2 \cdot 5 \cdot 467$	2	2	± 10	9857	$2^7 \cdot 7 \cdot 11$	5	5	± 10
8807	$2 \cdot 7 \cdot 17 \cdot 37$	5	2	10	9343	$2 \cdot 3^3 \cdot 173$	5	2	10	9859	$2 \cdot 3 \cdot 31 \cdot 53$	2	4	-----
8819	$2 \cdot 4409$	2	3	10	9349	$2^2 \cdot 3 \cdot 19 \cdot 41$	2	2	-----	9871	$2 \cdot 3 \cdot 5 \cdot 7 \cdot 47$	3	2	-10
8821	$2^2 \cdot 3^2 \cdot 5 \cdot 7^2$	2	2	± 10	9371	$2 \cdot 5 \cdot 937$	2	3	10	9883	$2 \cdot 3^4 \cdot 61$	2	4	-10
8831	$2 \cdot 5 \cdot 883$	7	5	-10	9377	$2^5 \cdot 293$	3	3	± 10	9887	$2 \cdot 4943$	5	2	10
8837	$2^2 \cdot 47^2$	2	2	-----	9391	$2 \cdot 3 \cdot 5 \cdot 313$	3	2	-10	9901	$2^2 \cdot 3^2 \cdot 5^2 \cdot 11$	2	2	-----
8839	$2 \cdot 3^2 \cdot 491$	3	2	-10	9397	$2^2 \cdot 3^4 \cdot 29$	2	2	-----	9907	$2 \cdot 3 \cdot 13 \cdot 127$	2	4	-10
8849	$2^4 \cdot 7 \cdot 79$	3	3	-----	9403	$2 \cdot 3 \cdot 1567$	3	6	-----	9923	$2 \cdot 11^2 \cdot 41$	2	3	-10
8861	$2^2 \cdot 5 \cdot 443$	2	2	± 10	9413	$2^2 \cdot 13 \cdot 181$	3	3	-----	9929	$2^3 \cdot 17 \cdot 73$	3	3	-----
8863	$2 \cdot 3 \cdot 7 \cdot 211$	3	9	10	9419	$2 \cdot 17 \cdot 277$	2	3	-----	9931	$2 \cdot 3 \cdot 5 \cdot 331$	10	5	10
8867	$2 \cdot 11 \cdot 13 \cdot 31$	2	3	-10	9421	$2^2 \cdot 3 \cdot 5 \cdot 157$	2	2	± 10	9941	$2^2 \cdot 5 \cdot 7 \cdot 71$	2	2	-----
8887	$2 \cdot 3 \cdot 1481$	3	2	10	9431	$2 \cdot 5 \cdot 23 \cdot 41$	7	3	-10	9949	$2^2 \cdot 3 \cdot 829$	2	2	± 10
8893	$2^2 \cdot 3^2 \cdot 13 \cdot 19$	5	5	-----	9433	$2^3 \cdot 3^2 \cdot 131$	5	5	-----	9967	$2 \cdot 3 \cdot 11 \cdot 151$	3	2	10
8923	$2 \cdot 3 \cdot 1487$	2	4	-----	9437	$2^2 \cdot 7 \cdot 337$	2	2	-----	9973	$2^2 \cdot 3^2 \cdot 277$	11	11	-----
8929	$2^5 \cdot 3^2 \cdot 31$	11	11	-----	9439	$2 \cdot 3 \cdot 11^2 \cdot 13$	22	7	-----					
8933	$2^2 \cdot 7 \cdot 11 \cdot 29$	2	2	-----	9461	$2^2 \cdot 5 \cdot 11 \cdot 43$	3	3	± 10					

Table 24.9   **PRIMES**

n	0	1	2	3	4	5	6	7	8	9	10	11	12	13	14	15	16	17	18	19	20	21	22	23	24
1	2	547	1229	1993	2749	3581	4421	5281	6143	7001	7927	8837	9739	10663	11677	12569	13513	14533	15413	16411	17393	18329	19427	20359	21391
2	3	557	1231	1997	2753	3583	4423	5297	6151	7013	7933	8839	9743	10667	11681	12577	13523	14537	15427	16417	17401	18341	19429	20369	21397
3	5	563	1237	1999	2767	3593	4441	5303	6163	7019	7937	8849	9749	10687	11689	12583	13537	14543	15439	16421	17417	18353	19433	20389	21401
4	7	569	1249	2003	2777	3607	4447	5309	6173	7027	7949	8861	9767	10691	11699	12589	13553	14549	15443	16427	17419	18367	19441	20393	21407
5	11	571	1259	2011	2789	3613	4451	5323	6197	7039	7951	8863	9769	10709	11701	12601	13567	14551	15451	16433	17431	18371	19447	20399	21419
6	13	577	1277	2017	2791	3617	4457	5333	6199	7043	7963	8867	9781	10711	11717	12611	13577	14557	15461	16447	17443	18379	19457	20407	21433
7	17	587	1279	2027	2797	3623	4463	5347	6203	7057	7993	8887	9787	10723	11719	12613	13591	14561	15467	16451	17449	18397	19463	20411	21467
8	19	593	1283	2029	2801	3631	4481	5351	6211	7069	8009	8893	9791	10729	11731	12619	13597	14563	15473	16453	17467	18401	19469	20431	21481
9	23	599	1289	2039	2803	3637	4483	5381	6217	7079	8011	8923	9803	10733	11743	12637	13613	14591	15493	16477	17471	18413	19471	20441	21487
10	29	601	1291	2053	2819	3643	4493	5387	6221	7103	8017	8929	9811	10739	11777	12641	13619	14593	15497	16481	17477	18427	19477	20443	21491
11	31	607	1297	2063	2833	3659	4507	5393	6229	7109	8039	8933	9817	10753	11779	12647	13627	14621	15511	16487	17483	18433	19483	20477	21493
12	37	613	1301	2069	2837	3671	4513	5399	6247	7121	8053	8941	9829	10771	11783	12653	13633	14627	15527	16493	17489	18439	19489	20479	21499
13	41	617	1303	2081	2843	3673	4517	5407	6257	7127	8059	8951	9833	10781	11789	12659	13649	14629	15541	16519	17491	18443	19501	20483	21503
14	43	619	1307	2083	2851	3677	4519	5413	6263	7129	8069	8963	9839	10789	11801	12671	13669	14633	15551	16529	17497	18451	19507	20507	21517
15	47	631	1319	2087	2857	3691	4523	5417	6269	7151	8081	8969	9851	10799	11807	12689	13679	14639	15559	16547	17509	18457	19531	20509	21521
16	53	641	1321	2089	2861	3697	4547	5419	6271	7159	8087	8971	9857	10831	11813	12697	13681	14653	15569	16553	17519	18461	19541	20521	21523
17	59	643	1327	2099	2879	3701	4549	5431	6277	7177	8089	8999	9859	10837	11821	12703	13687	14657	15581	16561	17539	18481	19543	20533	21529
18	61	647	1361	2111	2887	3709	4561	5437	6287	7187	8093	9001	9871	10847	11827	12713	13691	14669	15583	16567	17551	18493	19553	20543	21557
19	67	653	1367	2113	2897	3719	4567	5441	6299	7193	8101	9007	9883	10853	11831	12721	13693	14683	15601	16573	17569	18503	19559	20549	21559
20	71	659	1373	2129	2903	3727	4583	5443	6301	7207	8111	9011	9887	10859	11833	12739	13697	14699	15607	16603	17573	18517	19571	20551	21563
21	73	661	1381	2131	2909	3733	4591	5449	6311	7211	8117	9013	9901	10861	11839	12743	13709	14713	15619	16607	17579	18521	19577	20563	21569
22	79	673	1399	2137	2917	3739	4597	5471	6317	7213	8123	9029	9907	10867	11863	12757	13711	14717	15629	16619	17581	18523	19583	20593	21577
23	83	677	1409	2141	2927	3761	4603	5477	6323	7219	8147	9041	9923	10883	11867	12763	13721	14723	15641	16631	17597	18539	19597	20599	21587
24	89	683	1423	2143	2939	3767	4621	5479	6329	7229	8161	9043	9929	10889	11887	12781	13723	14731	15643	16633	17599	18541	19603	20611	21589
25	97	691	1427	2153	2953	3769	4637	5483	6337	7237	8167	9049	9931	10891	11897	12791	13729	14737	15647	16649	17609	18553	19609	20627	21599
26	101	701	1429	2161	2957	3779	4639	5501	6343	7243	8171	9059	9941	10903	11903	12799	13751	14741	15649	16651	17623	18583	19661	20639	21601
27	103	709	1433	2179	2963	3793	4643	5503	6353	7247	8179	9067	9949	10909	11909	12809	13757	14747	15661	16657	17627	18587	19681	20641	21611
28	107	719	1439	2203	2969	3797	4649	5507	6359	7253	8191	9091	9967	10937	11923	12821	13759	14753	15667	16661	17657	18593	19687	20663	21613
29	109	727	1447	2207	2971	3803	4651	5519	6361	7283	8209	9103	9973	10939	11927	12823	13763	14759	15671	16673	17659	18617	19697	20681	21617
30	113	733	1451	2213	2999	3821	4657	5521	6367	7297	8219	9109	10007	10949	11933	12829	13781	14767	15679	16691	17669	18637	19699	20693	21647
31	127	739	1453	2221	3001	3823	4663	5527	6373	7307	8221	9127	10009	10957	11939	12841	13789	14771	15683	16693	17681	18661	19709	20707	21649
32	131	743	1459	2237	3011	3833	4673	5531	6379	7309	8231	9133	10037	10973	11941	12853	13799	14779	15727	16699	17683	18671	19717	20717	21661
33	137	751	1471	2239	3019	3847	4679	5557	6389	7321	8233	9137	10039	10979	11953	12889	13807	14783	15731	16703	17707	18679	19727	20719	21673
34	139	757	1481	2243	3023	3851	4691	5563	6397	7331	8237	9151	10061	10987	11959	12893	13829	14797	15733	16729	17713	18691	19739	20731	21683
35	149	761	1483	2251	3037	3853	4703	5569	6421	7333	8243	9157	10067	10993	11969	12899	13831	14813	15737	16741	17729	18701	19751	20743	21701
36	151	769	1487	2267	3041	3863	4721	5573	6427	7349	8263	9161	10069	11003	11971	12907	13841	14821	15739	16747	17737	18713	19753	20747	21713
37	157	773	1489	2269	3049	3877	4723	5581	6449	7351	8269	9173	10079	11027	11981	12911	13859	14827	15749	16759	17747	18719	19759	20749	21727
38	163	787	1493	2273	3061	3881	4729	5591	6451	7369	8273	9181	10091	11047	11987	12917	13873	14831	15761	16763	17749	18731	19763	20753	21737
39	167	797	1499	2281	3067	3889	4733	5623	6469	7393	8287	9187	10093	11057	12007	12919	13877	14843	15767	16787	17761	18743	19777	20759	21739
40	173	809	1511	2287	3079	3907	4751	5639	6473	7411	8291	9199	10099	11059	12011	12923	13879	14851	15773	16811	17783	18749	19793	20771	21751
41	179	811	1523	2293	3083	3911	4759	5641	6481	7417	8293	9203	10103	11069	12037	12941	13883	14867	15787	16823	17789	18757	19801	20773	21757
42	181	821	1531	2297	3089	3917	4783	5647	6491	7433	8297	9209	10111	11071	12041	12953	13901	14869	15791	16829	17791	18773	19813	20789	21767
43	191	823	1543	2309	3109	3919	4787	5651	6521	7451	8311	9221	10133	11083	12043	12959	13903	14879	15797	16831	17807	18787	19819	20807	21773
44	193	827	1549	2311	3119	3923	4789	5653	6529	7457	8317	9227	10139	11087	12049	12967	13907	14887	15803	16843	17827	18793	19841	20809	21787
45	197	829	1553	2333	3121	3929	4793	5657	6547	7459	8329	9239	10141	11093	12071	12973	13913	14891	15809	16871	17837	18797	19843	20849	21799
46	199	839	1559	2339	3137	3931	4799	5659	6551	7477	8353	9241	10151	11113	12073	12979	13921	14897	15817	16879	17839	18803	19853	20857	21803
47	211	853	1567	2341	3163	3943	4801	5669	6553	7481	8363	9257	10159	11117	12097	12983	13931	14923	15823	16883	17851	18839	19861	20873	21817
48	223	857	1571	2347	3167	3947	4813	5683	6563	7487	8369	9277	10163	11119	12101	13001	13933	14929	15859	16889	17863	18859	19867	20879	21821
49	227	859	1579	2351	3169	3967	4817	5689	6569	7489	8377	9281	10169	11131	12109	13003	13963	14939	15877	16901	17881	18869	19889	20887	21839
50	229	863	1583	2357	3181	3989	4831	5693	6571	7499	8387	9283	10177	11149	12113	13007	13967	14947	15881	16903	17891	18899	19891	20897	21841
51	233	877	1597	2371	3187	4001	4861	5701	6577	7507	8389	9293	10181	11159	12119	13009	13997	14951	15887	16921	17903	18911	19913	20899	21851
52	239	881	1601	2377	3191	4003	4871	5711	6581	7517	8419	9311	10193	11161	12143	13033	13999	14957	15889	16927	17909	18913	19919	20903	21859
53	241	883	1607	2381	3203	4007	4877	5717	6599	7523	8423	9319	10211	11171	12149	13037	14009	14969	15901	16931	17911	18917	19927	20921	21863
54	251	887	1609	2383	3209	4013	4889	5737	6607	7529	8429	9323	10223	11173	12157	13043	14011	14983	15907	16937	17921	18919	19937	20929	21871
55	257	907	1613	2389	3217	4019	4903	5741	6619	7537	8431	9337	10243	11177	12161	13049	14029	14989	15913	16943	17923	18947	19949	20939	21881
56	263	911	1619	2393	3221	4021	4909	5743	6637	7541	8443	9341	10247	11197	12163	13063	14033	15017	15919	16963	17929	18959	19961	20947	21893
57	269	919	1621	2399	3229	4027	4919	5749	6653	7547	8447	9343	10253	11213	12197	13093	14051	15031	15923	16979	17939	18973	19963	20959	21911
58	271	929	1627	2411	3251	4049	4931	5779	6659	7549	8461	9349	10259	11239	12203	13099	14057	15053	15937	16981	17957	18979	19973	20963	21929
59	277	937	1637	2417	3253	4051	4933	5783	6661	7559	8467	9371	10267	11243	12211	13103	14071	15061	15959	16987	17959	19001	19979	20981	21937
60	281	941	1657	2423	3257	4057	4937	5791	6673	7561	8501	9377	10271	11251	12227	13109	14081	15073	15971	16993	17971	19009	19991	20983	21943
61	283	947	1663	2437	3259	4073	4943	5801	6679	7573	8513	9391	10273	11257	12239	13121	14083	15077	15973	17011	17977	19013	19993	21001	21961
62	293	953	1667	2441	3271	4079	4951	5807	6689	7577	8521	9397	10289	11261	12241	13127	14087	15083	15991	17021	17981	19031	19997	21011	21977
63	307	967	1669	2447	3299	4091	4957	5813	6691	7583	8527	9403	10301	11273	12251	13147	14107	15091	16001	17027	17987	19037	20011	21013	21991
64	311	971	1693	2459	3301	4093	4967	5821	6701	7589	8537	9413	10303	11279	12253	13151	14143	15101	16007	17029	17989	19051	20021	21017	21997
65	313	977	1697	2467	3307	4099	4969	5827	6703	7591	8539	9419	10313	11287	12263	13159	14149	15107	16033	17033	18013	19069	20023	21019	22003
66	317	983	1699	2473	3313	4111	4973	5839	6709	7603	8543	9421	10321	11299	12269	13163	14153	15121	16057	17041	18041	19073	20029	21023	22013
67	331	991	1709	2477	3319	4127	4987	5843	6719	7607	8563	9431	10331	11311	12277	13171	14159	15131	16061	17047	18043	19079	20047	21031	22027
68	337	997	1721	2503	3323	4129	4993	5849	6733	7621	8573	9433	10333	11317	12281	13177	14173	15137	16063	17053	18047	19081	20051	21059	22031
69	347	1009	1723	2521	3329	4133	4999	5851	6737	7639	8581	9437	10337	11321	12289	13183	14177	15139	16067	17077	18049	19087	20063	21061	22037
70	349	1013	1733	2531	3331	4139	5003	5857	6761	7643	8597	9439	10343	11329	12301	13187	14197	15149	16069	17093	18059	19121	20071	21067	22039
71	353	1019	1741	2539	3343	4153	5009	5861	6763	7649	8599	9461	10357	11351	12323	13217	14207	15161	16073	17099	18061	19139	20089	21089	22051
72	359	1021	1747	2543	3347	4157	5011	5867	6779	7669	8609	9463	10369	11353	12329	13219	14221	15173	16087	17107	18077	19141	20101	21101	22063
73	367	1031	1753	2549	3359	4159	5021	5869	6781	7673	8623	9467	10391	11369	12343	13229	14243	15187	16091	17117	18089	19157	20107	21107	22067
74	373	1033	1759	2551	3361	4177	5023	5879	6791	7681	8627	9473	10399	11383	12347	13241	14249	15193	16097	17123	18097	19163	20113	21121	22073
75	379	1039	1777	2557	3371	4201	5039	5881	6793	7687	8629	9479	10427	11393	12373	13249	14251	15199	16103	17137	18119	19181	20117	21139	22079
76	383	1049	1783	2579	3373	4211	5051	5897	6803	7691	8641	9491	10429	11399	12377	13259	14281	15217	16111	17159	18121	19183	20123	21143	22091
77	389	1051	1787	2591	3389	4217	5059	5903	6823	7699	8647	9497	10433	11411	12379	13267	14293	15227	16127	17167	18127	19207	20129	21149	22093
78	397	1061	1789	2593	3391	4219	5077	5923	6827	7703	8663	9511	10453	11423	12391	13291	14303	15233	16139	17183	18131	19211	20143	21157	22109
79	401	1063	1801	2609	3407	4229	5081	5927	6829	7717	8669	9521	10457	11437	12401	13297	14321	15241	16141	17189	18133	19213	20147	21163	22111
80	409	1069	1811	2617	3413	4231	5087	5939	6833	7723	8677	9533	10459	11443	12413	13309	14323	15259	16183	17191	18143	19219	20149	21169	22123
81	419	1087	1823	2621	3433	4241	5099	5953	6841	7727	8681	9539	10463	11447	12421	13313	14327	15263	16187	17203	18149	19231	20161	21179	22129
82	421	1091	1831	2633	3449	4243	5101	5981	6857	7741	8689	9547	10477	11467	12433	13327	14341	15269	16189	17207	18169	19237	20173	21187	22133
83	431	1093	1847	2647	3457	4253	5107	5987	6863	7753	8693	9551	10487	11471	12437	13331	14347	15271	16193	17209	18181	19249	20177	21191	22147
84	433	1097	1861	2657	3461	4259	5113	6007	6869	7757	8699	9587	10499	11483	12451	13337	14369	15277	16217	17231	18191	19259	20183	21193	22153
85	439	1103	1867	2659	3463	4261	5119	6011	6871	7759	8707	9601	10501	11489	12457	13339	14387	15287	16223	17239	18199	19267	20201	21211	22157
86	443	1109	1871	2663	3467	4271	5147	6029	6883	7789	8713	9613	10513	11491	12473	13367	14389	15289	16229	17257	18211	19273	20219	21221	22159
87	449	1117	1873	2677	3469	4273	5153	6037	6899	7793	8719	9619	10529	11497	12479	13381	14401	15299	16231	17291	18217	19289	20231	21227	22171
88	457	1123	1877	2683	3491	4283	5167	6043	6907	7817	8731	9623	10531	11503	12487	13397	14407	15307	16249	17293	18223	19301	20233	21247	22189
89	461	1129	1879	2687	3499	4289	5171	6047	6911	7823	8737	9629	10559	11519	12491	13399	14411	15313	16253	17299	18229	19309	20249	21269	22193
90	463	1151	1889	2689	3511	4297	5179	6053	6917	7829	8741	9631	10567	11527	12497	13411	14419	15319	16261	17317	18233	19319	20261	21277	22229
91	467	1153	1901	2693	3517	4327	5189	6067	6947	7841	8747	9643	10589	11549	12503	13417	14423	15329	16273	17321	18251	19333	20269	21283	22247
92	479	1163	1907	2699	3527	4337	5197	6073	6949	7853	8753	9649	10597	11551	12511	13421	14431	15331	16301	17327	18253	19373	20287	21313	22259
93	487	1171	1913	2707	3529	4339	5209	6079	6959	7867	8761	9661	10601	11579	12517	13441	14437	15349	16319	17333	18257	19379	20297	21317	22271
94	491	1181	1931	2711	3533	4349	5227	6089	6961	7873	8779	9677	10607	11587	12527	13451	14447	15359	16333	17341	18269	19381	20323	21319	22273
95	499	1187	1933	2713	3539	4357	5231	6091	6967	7877	8783	9679	10613	11593	12539	13457	14449	15361	16339	17351	18287	19387	20327	21323	22277
96	503	1193	1949	2719	3541	4363	5233	6101	6971	7879	8803	9689	10627	11597	12541	13463	14461	15373	16349	17359	18289	19391	20333	21341	22279
97	509	1201	1951	2729	3547	4373	5237	6113	6977	7883	8807	9697	10631	11617	12547	13469	14479	15377	16361	17377	18301	19403	20341	21347	22283
98	521	1213	1973	2731	3557	4391	5261	6121	6983	7901	8819	9719	10639	11621	12553	13477	14489	15383	16363	17383	18307	19417	20347	21377	22291
99	523	1217	1979	2741	3559	4397	5273	6131	6991	7907	8821	9721	10651	11633	12569	13487	14503	15391	16369	17387	18311	19421	20353	21379	22297
100	541	1223	1987	2741	3571	4409	5279	6133	6997	7919	8831	9733	10657	11657	12553	13499	14519	15401	16381	17389	18313	19423	20357	21383	22307

From D. N. Lehmer, List of prime numbers from 1 to 10,006,721, Carnegie Institution of Washington, Publication No. 165, Washington, D.C., 1914 (with permission).

PRIMES

Table 24.9

	25	26	27	28	29	30	31	32	33	34	35	36	37	38	39	40	41	42	43	44	45	46	47	48	49
1	22343	23327	24317	25409	26407	27457	28513	29453	30577	31607	32611	33617	34651	35771	36787	37831	38923	39979	41113	42083	43063	44203	45317	46451	47533
2	22349	23333	24329	25411	26417	27479	28517	29473	30593	31627	32621	33619	34667	35797	36791	37847	38933	39983	41117	42089	43067	44207	45319	46457	47543
3	22367	23339	24337	25423	26423	27481	28537	29483	30631	31643	32633	33623	34673	35801	36793	37853	38953	39989	41131	42101	43093	44221	45329	46471	47563
4	22369	23357	24359	25439	26431	27487	28541	29501	30637	31649	32647	33629	34679	35803	36809	37861	38959	40009	41141	42131	43103	44249	45337	46477	47569
5	22381	23369	24371	25447	26437	27509	28547	29527	30643	31657	32653	33637	34687	35809	36821	37871	38971	40013	41143	42139	43117	44257	45341	46489	47581
6	22391	23371	24373	25453	26449	27527	28549	29531	30649	31663	32687	33641	34693	35831	36833	37879	38977	40031	41149	42157	43133	44263	45343	46499	47591
7	22397	23399	24379	25457	26459	27529	28559	29537	30661	31667	32693	33647	34703	35837	36847	37889	38993	40037	41161	42169	43151	44267	45361	46507	47599
8	22409	23417	24391	25463	26479	27539	28571	29567	30671	31687	32707	33679	34721	35839	36857	37897	39019	40039	41177	42179	43159	44269	45377	46511	47609
9	22433	23431	24407	25469	26489	27541	28573	29569	30677	31699	32713	33703	34729	35851	36871	37907	39023	40063	41179	42181	43177	44273	45389	46523	47623
10	22441	23447	24413	25471	26497	27551	28579	29573	30689	31721	32717	33713	34739	35863	36877	37951	39041	40087	41183	42187	43189	44279	45403	46549	47629
11	22447	23459	24419	25523	26501	27581	28591	29581	30697	31723	32719	33721	34747	35869	36887	37957	39043	40093	41189	42193	43201	44281	45413	46559	47639
12	22453	23473	24421	25537	26513	27583	28597	29587	30703	31727	32749	33739	34757	35879	36899	37963	39047	40099	41201	42197	43207	44293	45427	46567	47653
13	22469	23497	24439	25541	26539	27611	28603	29599	30707	31729	32771	33749	34759	35897	36901	37967	39079	40111	41203	42209	43223	44351	45433	46573	47657
14	22481	23509	24443	25561	26557	27617	28607	29611	30713	31741	32779	33751	34763	35899	36913	37987	39089	40123	41213	42221	43237	44357	45439	46589	47659
15	22483	23531	24469	25577	26561	27631	28619	29629	30727	31751	32783	33757	34781	35911	36919	37991	39097	40127	41221	42223	43261	44371	45481	46591	47681
16	22501	23537	24473	25579	26573	27647	28621	29633	30757	31769	32789	33767	34807	35923	36923	37993	39103	40129	41227	42227	43271	44381	45491	46601	47699
17	22511	23539	24481	25583	26591	27653	28627	29641	30763	31771	32797	33769	34819	35933	36929	37997	39107	40151	41231	42239	43283	44383	45497	46619	47701
18	22531	23549	24499	25589	26597	27673	28631	29663	30773	31793	32801	33773	34841	35951	36931	38011	39113	40153	41233	42257	43291	44389	45503	46633	47711
19	22541	23557	24509	25601	26627	27689	28643	29669	30781	31799	32803	33791	34843	35963	36943	38039	39119	40163	41243	42281	43313	44417	45523	46639	47713
20	22543	23561	24517	25603	26633	27691	28649	29671	30803	31817	32831	33797	34847	35969	36947	38047	39133	40169	41257	42283	43319	44449	45533	46643	47717
21	22549	23563	24527	25609	26641	27697	28657	29683	30809	31847	32833	33809	34849	35977	36973	38053	39139	40177	41263	42293	43321	44453	45541	46649	47737
22	22567	23567	24533	25621	26647	27701	28661	29717	30817	31849	32839	33811	34871	35983	36979	38069	39157	40189	41269	42299	43331	44483	45553	46663	47741
23	22571	23581	24547	25633	26669	27733	28663	29723	30829	31859	32843	33827	34877	35993	36997	38083	39161	40193	41281	42307	43391	44491	45557	46679	47743
24	22573	23593	24551	25639	26681	27737	28669	29741	30839	31873	32869	33829	34883	35999	37003	38113	39163	40213	41299	42323	43397	44497	45569	46681	47777
25	22613	23599	24571	25643	26683	27739	28687	29753	30841	31883	32887	33851	34897	36007	37013	38119	39181	40231	41333	42331	43399	44501	45587	46687	47779
26	22619	23603	24593	25657	26687	27743	28697	29759	30851	31891	32909	33857	34913	36011	37019	38149	39191	40237	41341	42337	43403	44507	45589	46691	47791
27	22621	23609	24611	25667	26693	27749	28703	29761	30853	31907	32911	33863	34919	36013	37021	38153	39199	40241	41351	42349	43411	44519	45599	46703	47797
28	22637	23623	24623	25673	26699	27751	28711	29789	30859	31957	32917	33871	34939	36017	37039	38167	39209	40253	41357	42359	43427	44531	45613	46723	47807
29	22639	23627	24631	25679	26701	27763	28723	29803	30869	31963	32933	33889	34949	36037	37049	38177	39217	40277	41381	42373	43441	44533	45631	46727	47809
30	22643	23629	24659	25693	26711	27767	28729	29819	30871	31973	32939	33893	34961	36061	37057	38183	39227	40283	41387	42379	43451	44537	45641	46747	47819
31	22651	23633	24671	25703	26713	27773	28751	29833	30881	31981	32941	33911	34963	36067	37061	38189	39229	40289	41389	42391	43457	44543	45659	46751	47837
32	22669	23663	24677	25717	26717	27779	28753	29837	30893	31991	32957	33923	34981	36073	37087	38197	39233	40343	41399	42397	43481	44549	45667	46757	47843
33	22679	23669	24683	25733	26723	27791	28759	29851	30911	32003	32969	33931	35023	36083	37097	38201	39239	40351	41411	42403	43487	44563	45673	46769	47857
34	22691	23671	24691	25741	26729	27793	28771	29863	30931	32009	32971	33937	35027	36097	37117	38219	39241	40357	41413	42407	43499	44579	45677	46771	47869
35	22697	23677	24697	25747	26731	27799	28789	29867	30937	32027	32983	33941	35051	36107	37123	38231	39251	40361	41443	42409	43517	44587	45691	46807	47881
36	22699	23687	24709	25759	26737	27803	28793	29873	30941	32029	32987	33961	35053	36109	37139	38237	39293	40387	41453	42433	43541	44617	45697	46811	47903
37	22709	23689	24733	25763	26759	27809	28807	29879	30949	32051	32993	33967	35059	36131	37159	38239	39301	40423	41467	42437	43543	44621	45707	46817	47911
38	22717	23719	24749	25771	26777	27817	28813	29881	30971	32057	32999	33997	35069	36137	37171	38261	39313	40427	41479	42443	43573	44623	45737	46819	47917
39	22721	23741	24763	25793	26783	27823	28817	29917	30977	32059	33013	34019	35081	36151	37181	38273	39317	40429	41491	42451	43577	44633	45751	46829	47933
40	22727	23743	24767	25799	26801	27827	28837	29921	30983	32063	33023	34031	35083	36161	37189	38281	39323	40433	41507	42457	43579	44641	45757	46831	47939
41	22739	23747	24781	25801	26813	27847	28843	29927	31013	32069	33029	34033	35089	36187	37199	38287	39341	40459	41513	42461	43591	44647	45763	46853	47947
42	22741	23753	24793	25819	26821	27851	28859	29947	31019	32077	33037	34039	35099	36191	37201	38299	39343	40471	41519	42463	43597	44651	45767	46861	47951
43	22751	23761	24799	25841	26833	27867	28867	29959	31033	32083	33049	34057	35107	36209	37217	38303	39359	40483	41521	42467	43607	44657	45779	46867	47963
44	22769	23773	24809	25847	26839	27893	28871	29983	31039	32089	33053	34061	35111	36217	37223	38317	39367	40487	41539	42473	43609	44683	45817	46877	47969
45	22777	23789	24821	25849	26849	27901	28879	29989	31051	32099	33071	34123	35117	36229	37253	38321	39371	40493	41543	42487	43613	44687	45821	46889	47977
46	22783	23801	24841	25867	26861	27917	28901	30011	31063	32117	33073	34127	35129	36241	37273	38327	39373	40499	41549	42491	43627	44699	45823	46901	47981
47	22787	23813	24847	25873	26863	27919	28909	30013	31069	32119	33083	34129	35141	36251	37277	38333	39383	40507	41579	42499	43633	44701	45827	46919	48017
48	22807	23819	24851	25889	26879	27941	28921	30029	31079	32141	33091	34141	35149	36263	37307	38351	39397	40519	41593	42509	43649	44711	45833	46933	48023
49	22811	23827	24859	25903	26881	27943	28927	30047	31081	32143	33107	34147	35153	36269	37313	38371	39409	40529	41597	42533	43651	44729	45841	46957	48029
50	22817	23831	24877	25913	26891	27947	28933	30059	31091	32159	33113	34157	35159	36277	37321	38373	39419	40531	41603	42557	43661	44741	45853	46993	48049
51	22853	23833	24889	25919	26893	27953	28949	30071	31121	32173	33119	34159	35171	36293	37337	38377	39439	40543	41609	42569	43669	44753	45863	46997	48073
52	22859	23857	24907	25931	26903	27961	28961	30089	31123	32183	33149	34171	35201	36299	37339	38431	39443	40559	41611	42571	43691	44771	45869	47017	48079
53	22861	23869	24917	25933	26921	27967	28979	30091	31139	32189	33151	34183	35221	36307	37357	38447	39451	40577	41617	42577	43711	44777	45887	47041	48091
54	22871	23873	24919	25939	26927	27983	29009	30097	31147	32191	33161	34211	35227	36313	37361	38449	39461	40583	41621	42589	43717	44789	45893	47051	48109
55	22877	23879	24923	25943	26947	27997	29017	30103	31151	32203	33179	34213	35251	36319	37363	38461	39499	40591	41627	42611	43721	44797	45943	47057	48119
56	22901	23887	24943	25951	26951	28001	29021	30109	31153	32213	33181	34217	35257	36341	37369	38453	39503	40597	41641	42641	43753	44809	45949	47059	48121
57	22907	23893	24953	25969	26953	28019	29023	30113	31159	32233	33191	34231	35267	36343	37379	38459	39509	40609	41647	42643	43759	44819	45953	47087	48131
58	22921	23899	24967	25981	26959	28027	29027	30119	31177	32237	33199	34253	35279	36353	37397	38501	39511	40627	41651	42649	43777	44839	45959	47093	48157
59	22937	23909	24971	25997	26981	28031	29033	30133	31181	32251	33203	34259	35281	36373	37409	38543	39521	40637	41659	42667	43781	44843	45971	47111	48163
60	22943	23911	24977	25999	26987	28051	29059	30137	31183	32257	33211	34261	35291	36383	37423	38557	39541	40639	41669	42677	43783	44857	45979	47119	48179
61	22961	23917	24979	26003	26993	28057	29063	30139	31189	32261	33223	34267	35311	36389	37441	38561	39551	40693	41681	42683	43787	44861	45989	47123	48187
62	22963	23929	24989	26017	27011	28069	29077	30161	31193	32297	33247	34273	35317	36433	37447	38567	39563	40697	41687	42689	43789	44867	46021	47129	48193
63	22973	23957	25013	26021	27017	28081	29101	30169	31219	32299	33287	34283	35323	36451	37463	38569	39569	40699	41719	42697	43793	44879	46027	47137	48197
64	22993	23971	25031	26029	27031	28087	29123	30181	31223	32303	33289	34297	35327	36457	37483	38581	39581	40709	41729	42701	43801	44887	46049	47143	48221
65	23003	23977	25033	26041	27043	28097	29129	30187	31231	32309	33301	34301	35339	36467	37489	38593	39607	40739	41737	42703	43853	44893	46051	47147	48239
66	23011	23981	25057	26053	27059	28099	29131	30197	31237	32321	33311	34303	35353	36469	37493	38603	39619	40751	41759	42709	43867	44909	46061	47149	48247
67	23017	23993	25057	26083	27061	28109	29137	30203	31247	32323	33317	34313	35363	36473	37501	38609	39623	40759	41761	42719	43889	44917	46073	47161	48259
68	23021	24001	25073	26099	27067	28111	29147	30211	31249	32327	33329	34319	35381	36479	37507	38611	39631	40763	41771	42727	43891	44927	46091	47189	48271
69	23027	24007	25087	26107	27073	28123	29153	30223	31253	32341	33331	34327	35393	36493	37511	38629	39659	40771	41777	42737	43913	44939	46093	47207	48281
70	23029	24019	25097	26111	27077	28151	29167	30241	31259	32353	33343	34337	35401	36497	37517	38639	39667	40787	41801	42743	43933	44953	46099	47221	48299
71	23039	24023	25111	26113	27091	28163	29173	30253	31267	32359	33347	34351	35419	36523	37529	38651	39671	40801	41809	42751	43943	44959	46103	47237	48311
72	23041	24029	25117	26119	27103	28181	29179	30259	31271	32363	33349	34361	35423	36527	37537	38653	39679	40813	41813	42767	43951	44963	46133	47251	48313
73	23053	24043	25121	26141	27107	28183	29191	30269	31277	32369	33353	34367	35437	36529	37547	38669	39703	40819	41843	42773	43961	44971	46141	47269	48337
74	23057	24049	25127	26153	27109	28201	29201	30271	31307	32371	33359	34369	35447	36541	37549	38671	39709	40823	41849	42787	43963	44983	46147	47279	48341
75	23059	24061	25147	26161	27127	28211	29207	30293	31319	32377	33377	34381	35449	36551	37561	38677	39719	40829	41851	42793	43969	44987	46153	47287	48353
76	23063	24071	25153	26171	27143	28219	29209	30307	31321	32381	33391	34403	35461	36559	37567	38693	39727	40841	41863	42797	43973	45007	46171	47293	48371
77	23071	24077	25163	26177	27179	28229	29221	30313	31327	32401	33403	34421	35491	36563	37571	38699	39733	40847	41879	42821	43987	45013	46181	47297	48383
78	23081	24083	25169	26183	27191	28277	29231	30319	31333	32411	33409	34429	35507	36571	37573	38707	39749	40849	41887	42829	43991	45053	46183	47303	48397
79	23087	24091	25171	26189	27197	28279	29243	30323	31337	32413	33413	34439	35509	36583	37579	38711	39761	40853	41893	42839	43997	45061	46187	47309	48407
80	23099	24097	25183	26203	27211	28283	29251	30341	31357	32423	33427	34457	35521	36587	37589	38713	39769	40867	41897	42841	44017	45077	46199	47317	48409
81	23117	24097	25189	26209	27239	28289	29269	30347	31379	32429	33457	34469	35527	36599	37591	38723	39779	40879	41903	42853	44021	45083	46219	47339	48413
82	23131	24103	25219	26227	27241	28297	29287	30367	31387	32441	33461	34471	35531	36607	37607	38729	39791	40883	41911	42859	44027	45119	46229	47351	48437
83	23143	24107	25229	26237	27253	28307	29297	30389	31391	32443	33469	34483	35533	36629	37619	38737	39799	40897	41927	42863	44029	45121	46237	47353	48449
84	23159	24109	25237	26249	27259	28309	29303	30391	31393	32467	33479	34487	35537	36637	37633	38747	39821	40903	41941	42899	44041	45127	46261	47363	48463
85	23167	24113	25243	26251	27271	28319	29311	30403	31397	32479	33487	34499	35543	36643	37643	38749	39827	40927	41947	42901	44053	45131	46271	47381	48473
86	23173	24121	25247	26261	27277	28349	29327	30427	31469	32491	33493	34501	35569	36653	37663	38767	39829	40933	41953	42923	44059	45137	46273	47387	48479
87	23189	24133	25253	26263	27281	28351	29333	30431	31477	32497	33503	34511	35573	36671	37691	38783	39839	40939	41957	42929	44071	45139	46279	47389	48481
88	23197	24137	25261	26267	27283	28387	29339	30449	31481	32503	33521	34513	35591	36677	37693	38791	39841	40949	41959	42937	44087	45161	46301	47407	48487
89	23201	24151	25301	26293	27299	28393	29347	30467	31489	32507	33529	34519	35593	36683	37699	38803	39847	40961	41969	42943	44089	45179	46307	47417	48491
90	23203	24169	25303	26297	27329	28403	29363	30469	31511	32531	33533	34537	35597	36691	37717	38821	39857	40973	41981	42953	44101	45181	46309	47419	48497
91	23209	24179	25307	26309	27337	28409	29383	30491	31513	32533	33547	34543	35603	36697	37747	38833	39863	40993	41983	42961	44111	45191	46327	47431	48523
92	23227	24181	25309	26317	27361	28411	29387	30493	31517	32537	33563	34549	35617	36709	37781	38839	39869	41011	41999	42967	44119	45197	46337	47441	48527
93	23251	24197	25321	26321	27367	28429	29389	30497	31531	32561	33569	34583	35671	36713	37783	38851	39877	41017	42013	42979	44123	45233	46349	47459	48533
94	23269	24203	25339	26339	27397	28433	29399	30509	31541	32563	33577	34589	35677	36721	37799	38867	39883	41023	42017	42989	44129	45247	46351	47491	48539
95	23279	24223	25343	26347	27407	28439	29401	30517	31543	32569	33581	34591	35677	36739	37747	38867	39887	41039	42019	43003	44131	45259	46381	47497	48541
96	23291	24229	25349	26357	27409	28447	29411	30529	31547	32573	33587	34603	35729	36749	37781	38873	39901	41047	42023	43013	44159	45263	46399	47501	48563
97	23293	24239	25357	26371	27427	28463	29423	30539	31567	32579	33589	34607	35731	36761	37783	38891	39929	41051	42043	43019	44171	45281	46411	47507	48571
98	23297	24247	25367	26387	27431	28477	29429	30553	31573	32587	33599	34613	35747	36767	37799	38903	39937	41057	42061	43037	44179	45289	46439	47513	48589
99	23311	24251	25373	26393	27437	28493	29437	30557	31583	32603	33601	34631	35753	36779	37811	38917	39953	41077	42071	43049	44189	45293	46441	47521	48593
100	23321	24281	25391	26399	27449	28499	29443	30559	31601	32609	33613	34649	35759	36781	37813	38921	39971	41081	42073	43051	44201	45307	46447	47527	48611

Table 24.9 **PRIMES**

	50	51	52	53	54	55	56	57	58	59	60	61	62	63	64	65	66	67	68	69	70	71	72	73	74
1	48619	49667	50767	51817	52937	54001	55109	56197	57193	58243	59369	60509	61637	62791	63823	65071	66107	67247	68389	69497	70663	71719	72859	73999	75083
2	48623	49669	50773	51827	52951	54011	55117	56207	57203	58271	59377	60521	61643	62801	63839	65089	66109	67261	68399	69499	70667	71741	72869	74017	75109
3	48647	49681	50777	51829	52957	54013	55127	56209	57221	58309	59387	60527	61651	62819	63841	65099	66137	67271	68437	69539	70687	71761	72871	74021	75133
4	48649	49697	50789	51839	52963	54037	55147	56237	57223	58313	59393	60539	61657	62827	63853	65101	66161	67273	68443	69557	70709	71777	72883	74027	75149
5	48661	49711	50821	51853	52967	54049	55163	56239	57241	58321	59399	60589	61667	62851	63857	65111	66169	67289	68447	69593	70717	71789	72889	74047	75161
6	48673	49727	50833	51859	52973	54059	55171	56249	57251	58337	59407	60601	61673	62861	63863	65119	66173	67307	68449	69623	70729	71807	72893	74051	75167
7	48677	49739	50839	51869	52981	54083	55201	56263	57259	58363	59417	60607	61681	62869	63901	65123	66179	67339	68473	69653	70753	71809	72901	74071	75169
8	48679	49741	50849	51871	52999	54091	55207	56267	57269	58367	59419	60611	61687	62873	63907	65129	66191	67343	68477	69661	70769	71821	72907	74077	75181
9	48731	49747	50857	51893	53003	54101	55213	56269	57271	58369	59441	60617	61703	62897	63913	65141	66221	67349	68483	69677	70783	71837	72911	74093	75193
10	48733	49757	50867	51899	53017	54121	55217	56299	57283	58379	59443	60623	61717	62903	63929	65147	66239	67369	68489	69691	70793	71843	72923	74099	75209
11	48751	49783	50873	51907	53047	54133	55219	56311	57287	58391	59447	60631	61723	62921	63949	65167	66271	67391	68491	69697	70823	71849	72931	74101	75211
12	48757	49787	50891	51913	53051	54139	55229	56333	57301	58393	59453	60637	61729	62927	63977	65171	66293	67399	68501	69709	70841	71861	72937	74131	75217
13	48761	49789	50893	51929	53069	54151	55243	56359	57329	58403	59467	60647	61751	62929	63997	65173	66301	67409	68507	69737	70843	71867	72949	74143	75227
14	48767	49801	50909	51941	53077	54163	55249	56369	57331	58411	59471	60649	61757	62939	64007	65179	66337	67411	68521	69739	70849	71879	72953	74149	75239
15	48779	49807	50923	51949	53087	54167	55259	56377	57347	58417	59473	60659	61781	62969	64013	65183	66343	67421	68531	69761	70853	71881	72959	74159	75253
16	48781	49811	50929	51971	53089	54181	55291	56383	57349	58427	59497	60661	61813	62971	64019	65203	66347	67427	68539	69763	70867	71887	72973	74161	75253
17	48787	49823	50951	51973	53093	54193	55313	56393	57367	58439	59509	60679	61819	62981	64033	65213	66359	67429	68543	69767	70877	71899	72977	74167	75269
18	48799	49831	50957	51977	53101	54217	55331	56401	57373	58441	59513	60689	61837	62983	64037	65239	66361	67433	68567	69779	70879	71909	72997	74177	75277
19	48809	49843	50969	51991	53113	54251	55333	56417	57383	58451	59539	60703	61843	62987	64063	65257	66373	67447	68581	69809	70891	71917	73009	74189	75289
20	48817	49853	50971	52009	53117	54269	55337	56431	57389	58453	59557	60719	61861	62989	64067	65267	66377	67453	68597	69821	70901	71933	73013	74197	75307
21	48821	49871	50989	52021	53129	54277	55339	56437	57397	58477	59561	60727	61871	63029	64081	65269	66383	67477	68611	69827	70913	71941	73019	74201	75323
22	48823	49877	50993	52027	53147	54287	55343	56443	57413	58481	59567	60733	61879	63031	64091	65287	66403	67481	68633	69829	70919	71947	73037	74203	75329
23	48847	49891	51001	52051	53149	54293	55351	56453	57427	58511	59581	60737	61909	63059	64109	65293	66413	67489	68639	69833	70921	71963	73039	74209	75337
24	48857	49919	51031	52057	53161	54311	55373	56467	57457	58537	59611	60757	61927	63067	64123	65309	66431	67493	68659	69847	70937	71971	73043	74219	75347
25	48859	49921	51043	52067	53171	54319	55381	56473	57467	58543	59617	60761	61933	63073	64151	65323	66449	67499	68669	69857	70949	71983	73061	74231	75353
26	48869	49927	51047	52069	53173	54323	55399	56477	57487	58549	59621	60763	61949	63079	64153	65327	66457	67511	68683	69859	70951	71987	73063	74257	75367
27	48871	49937	51059	52081	53189	54331	55411	56479	57493	58567	59627	60773	61961	63097	64157	65353	66463	67523	68687	69877	70957	71993	73079	74279	75377
28	48883	49939	51061	52103	53197	54347	55439	56489	57503	58573	59629	60779	61967	63103	64171	65357	66467	67531	68699	69899	70969	71999	73091	74287	75389
29	48889	49943	51071	52121	53201	54361	55441	56501	57527	58579	59651	60793	61979	63113	64187	65371	66491	67537	68711	69911	70979	72019	73121	74293	75391
30	48907	49957	51109	52127	53231	54367	55457	56503	57529	58601	59659	60811	61981	63127	64189	65381	66499	67547	68713	69929	70981	72031	73127	74297	75401
31	48947	49991	51131	52147	53233	54371	55469	56509	57557	58603	59663	60821	61987	63131	64217	65393	66509	67559	68729	69931	70991	72043	73133	74311	75403
32	48953	49993	51133	52153	53239	54377	55487	56519	57559	58613	59669	60859	61991	63149	64223	65407	66523	67577	68737	69941	70997	72047	73141	74317	75407
33	48973	49999	51137	52163	53267	54401	55501	56527	57571	58631	59671	60869	62003	63179	64231	65413	66529	67579	68743	69959	70999	72053	73181	74323	75431
34	48989	50021	51151	52177	53269	54403	55511	56531	57587	58657	59693	60887	62011	63197	64237	65419	66533	67589	68749	69991	71011	72073	73189	74353	75437
35	48991	50023	51157	52181	53279	54409	55529	56533	57593	58661	59699	60889	62017	63199	64271	65423	66541	67601	68767	69997	71023	72077	73237	74357	75479
36	49003	50033	51169	52183	53281	54413	55541	56543	57601	58679	59707	60899	62039	63211	64279	65437	66553	67607	68771	70001	71039	72089	73243	74363	75503
37	49009	50047	51193	52189	53299	54419	55547	56569	57637	58687	59723	60901	62047	63241	64283	65447	66569	67619	68777	70003	71059	72091	73259	74377	75511
38	49019	50051	51197	52201	53309	54421	55579	56591	57641	58693	59729	60913	62053	63247	64301	65449	66571	67631	68779	70009	71069	72101	73277	74381	75521
39	49031	50053	51199	52223	53323	54437	55589	56597	57649	58699	59743	60917	62057	63277	64303	65479	66587	67651	68813	70019	71081	72103	73291	74383	75527
40	49033	50069	51203	52237	53327	54443	55603	56599	57653	58711	59747	60919	62071	63281	64319	65497	66593	67679	68819	70039	71089	72109	73303	74411	75533
41	49037	50077	51217	52249	53353	54449	55609	56611	57667	58727	59753	60923	62081	63299	64333	65519	66601	67679	68821	70051	71119	72139	73309	74413	75539
42	49043	50087	51229	52253	53359	54469	55619	56629	57719	58733	59771	60937	62099	63311	64373	65521	66617	67699	68863	70061	71129	72161	73327	74419	75541
43	49057	50093	51239	52259	53377	54493	55621	56633	57689	58741	59779	60943	62119	63313	64381	65537	66629	67709	68879	70067	71143	72167	73331	74441	75553
44	49069	50101	51241	52267	53381	54497	55631	56659	57697	58757	59791	60953	62129	63317	64399	65539	66643	67723	68881	70079	71147	72169	73351	74449	75557
45	49081	50111	51257	52289	53401	54499	55633	56663	57709	58763	59797	60961	62131	63331	64399	65543	66653	67733	68891	70099	71153	72173	73361	74453	75571
46	49103	50119	51263	52291	53407	54503	55639	56671	57713	58771	59809	61001	62137	63337	64403	65551	66683	67741	68897	70111	71161	72211	73363	74471	75577
47	49109	50123	51283	52301	53411	54517	55661	56681	57719	58787	59833	61007	62141	63347	64433	65557	66697	67751	68899	70117	71167	72221	73369	74489	75583
48	49117	50129	51287	52313	53419	54521	55663	56687	57727	58789	59863	61027	62143	63353	64439	65563	66701	67757	68903	70121	71171	72223	73379	74507	75611
49	49121	50131	51307	52321	53437	54539	55667	56701	57731	58831	59879	61031	62171	63361	64451	65579	66713	67759	68909	70123	71191	72227	73387	74509	75617
50	49123	50147	51329	52361	53441	54541	55673	56711	57737	58889	59887	61043	62189	63367	64453	65581	66721	67763	68917	70139	71209	72229	73417	74521	75619
51	49139	50153	51341	52363	53453	54547	55681	56713	57751	58897	59921	61051	62191	63377	64483	65587	66733	67777	68927	70141	71233	72251	73421	74527	75629
52	49157	50159	51343	52369	53479	54559	55691	56731	57773	58901	59929	61057	62201	63389	64489	65599	66739	67783	68947	70157	71237	72253	73433	74531	75641
53	49169	50177	51347	52379	53503	54563	55697	56737	57781	58907	59951	61091	62207	63391	64499	65609	66749	67789	68963	70163	71249	72269	73453	74551	75653
54	49171	50207	51349	52387	53507	54577	55711	56747	57787	58909	59957	61099	62213	63397	64513	65617	66751	67801	68993	70177	71257	72271	73459	74561	75659
55	49177	50221	51361	52391	53527	54581	55717	56767	57791	58913	59971	61121	62219	63409	64553	65629	66763	67807	69001	70181	71261	72277	73471	74567	75679
56	49193	50227	51383	52433	53549	54583	55721	56773	57793	58921	59981	61129	62233	63419	64567	65633	66791	67819	69011	70183	71263	72287	73477	74573	75683
57	49199	50231	51407	52453	53551	54601	55733	56779	57803	58937	59999	61141	62273	63421	64577	65647	66797	67829	69019	70199	71287	72307	73483	74587	75689
58	49201	50261	51413	52457	53569	54617	55763	56783	57809	58943	60013	61151	62297	63439	64579	65651	66809	67843	69029	70201	71293	72313	73517	74597	75703
59	49207	50263	51419	52489	53591	54623	55787	56807	57829	58963	60017	61153	62299	63443	64591	65657	66821	67853	69031	70207	71317	72337	73523	74609	75707
60	49211	50273	51421	52501	53593	54629	55793	56809	57839	58967	60029	61169	62303	63463	64601	65677	66841	67867	69061	70223	71327	72341	73529	74611	75709
61	49223	50287	51427	52511	53597	54631	55799	56813	57847	58979	60037	61211	62311	63467	64609	65687	66851	67883	69067	70229	71329	72353	73547	74623	75721
62	49253	50291	51431	52517	53609	54647	55807	56821	57853	58991	60041	61223	62323	63473	64613	65699	66853	67891	69073	70237	71333	72367	73553	74653	75731
63	49261	50311	51437	52529	53611	54667	55813	56827	57859	58997	60077	61231	62327	63487	64621	65701	66863	67901	69109	70241	71339	72379	73561	74687	75743
64	49277	50321	51439	52541	53617	54673	55817	56843	57881	59009	60083	61253	62347	63493	64627	65707	66877	67927	69119	70249	71341	72383	73571	74699	75767
65	49279	50329	51449	52543	53623	54679	55819	56857	57901	59011	60089	61261	62351	63499	64633	65713	66883	67931	69127	70271	71347	72421	73583	74707	75773
66	49297	50333	51461	52553	53629	54709	55823	56873	57917	59021	60091	61283	62383	63521	64661	65717	66889	67933	69143	70289	71353	72431	73589	74713	75781
67	49307	50341	51473	52561	53633	54713	55829	56891	57923	59023	60101	61291	62401	63527	64663	65719	66919	67949	69149	70297	71359	72461	73597	74717	75787
68	49331	50359	51479	52567	53639	54721	55837	56893	57943	59029	60103	61297	62417	63533	64667	65729	66923	67957	69151	70309	71363	72467	73607	74719	75793
69	49333	50363	51481	52571	53653	54727	55843	56897	57947	59051	60107	61331	62423	63541	64679	65731	66931	67961	69163	70313	71387	72469	73609	74729	75797
70	49339	50377	51487	52579	53657	54751	55849	56909	57973	59053	60127	61333	62459	63559	64693	65761	66943	67967	69191	70321	71389	72481	73613	74731	75821
71	49363	50383	51503	52583	53681	54767	55871	56911	57977	59063	60133	61339	62467	63577	64709	65777	66947	67979	69193	70327	71399	72493	73637	74747	75833
72	49367	50387	51511	52609	53693	54773	55889	56921	57991	59069	60139	61343	62473	63587	64717	65789	66949	67987	69197	70351	71411	72497	73643	74759	75853
73	49369	50411	51517	52627	53699	54779	55897	56923	57997	59077	60149	61357	62477	63589	64747	65809	66959	67993	69203	70373	71413	72503	73651	74761	75869
74	49391	50417	51521	52631	53717	54787	55901	56929	58013	59083	60161	61363	62483	63599	64763	65827	66973	68023	69221	70379	71419	72533	73673	74771	75883
75	49393	50423	51539	52639	53719	54799	55903	56941	58031	59093	60167	61379	62497	63601	64781	65831	66977	68041	69233	70381	71429	72547	73679	74779	75913
76	49409	50441	51551	52667	53731	54829	55921	56951	58043	59107	60169	61381	62501	63607	64783	65837	67003	68053	69239	70393	71437	72551	73681	74797	75931
77	49411	50459	51563	52673	53759	54833	55927	56957	58049	59113	60209	61403	62507	63611	64793	65839	67021	68059	69247	70423	71443	72559	73693	74821	75937
78	49417	50461	51577	52691	53773	54851	55931	56963	58057	59119	60217	61409	62533	63617	64811	65843	67033	68071	69257	70429	71453	72577	73699	74827	75941
79	49429	50497	51581	52697	53777	54869	55933	56983	58061	59123	60223	61417	62539	63629	64817	65851	67043	68087	69259	70439	71471	72613	73709	74831	75967
80	49433	50503	51593	52709	53783	54877	55949	56989	58067	59141	60251	61441	62549	63647	64849	65867	67049	68099	69263	70451	71473	72617	73721	74843	75979
81	49451	50513	51599	52711	53791	54881	55967	56993	58073	59149	60257	61463	62563	63649	64853	65881	67057	68111	69313	70457	71479	72623	73727	74857	75983
82	49459	50527	51607	52721	53813	54907	55987	56999	58099	59159	60259	61469	62581	63659	64871	65899	67061	68141	69317	70459	71483	72643	73751	74861	75989
83	49463	50539	51613	52727	53819	54917	55997	57037	58109	59167	60271	61471	62591	63667	64877	65921	67073	68113	69337	70481	71503	72647	73757	74869	75991
84	49477	50543	51631	52733	53831	54919	56003	57041	58111	59183	60289	61483	62597	63671	64879	65927	67079	68147	69341	70487	71527	72649	73771	74873	75997
85	49481	50549	51637	52747	53849	54941	56009	57047	58129	59197	60293	61487	62603	63689	64891	65929	67103	68207	69371	70489	71537	72661	73783	74887	76001
86	49499	50551	51647	52757	53857	54949	56039	57059	58129	59207	60317	61493	62617	63691	64901	65951	67121	68161	69379	70501	71549	72671	73819	74891	76003
87	49523	50581	51659	52769	53861	54959	56041	57073	58147	59209	60331	61507	62627	63697	64919	65957	67129	68171	69383	70507	71551	72673	73823	74897	76031
88	49529	50587	51673	52783	53881	54973	56053	57077	58151	59219	60337	61511	62633	63703	64921	65963	67139	68207	69389	70529	71563	72679	73847	74903	76039
89	49531	50591	51679	52807	53887	54979	56081	57089	58153	59221	60343	61519	62639	63709	64927	65981	67141	68209	69401	70537	71569	72689	73849	74923	76079
90	49537	50593	51683	52813	53891	54983	56087	57097	58169	59233	60353	61543	62653	63719	64937	65983	67153	68213	69403	70549	71593	72701	73859	74929	76081
91	49547	50599	51691	52817	53897	55001	56093	57107	58171	59239	60373	61547	62659	63727	64951	65993	67157	68219	69427	70571	71597	72707	73867	74933	76091
92	49549	50627	51713	52837	53899	55009	56099	57119	58189	59243	60383	61553	62683	63737	64969	66029	67169	68227	69431	70573	71633	72719	73877	74959	76103
93	49559	50647	51719	52859	53917	55021	56101	57131	58193	59263	60397	61559	62687	63743	64997	66037	67181	68239	69439	70583	71647	72727	73883	74989	76123
94	49597	50651	51721	52861	53923	55049	56113	57139	58199	59273	60413	61561	62701	63761	65003	66041	67187	68261	69457	70589	71663	72733	73897	75011	76129
95	49603	50671	51749	52879	53927	55051	56123	57143	58207	59281	60427	61583	62723	63773	65011	66047	67189	68279	69463	70607	71671	72739	73907	75013	76147
96	49613	50683	51767	52883	53939	55057	56131	57149	58211	59333	60443	61603	62731	63781	65027	66067	67211	68281	69467	70619	71693	72763	73939	75017	76157
97	49627	50707	51769	52889	53951	55061	56149	57163	58217	59341	60449	61609	62743	63793	65029	66071	67213	68311	69473	70621	71699	72767	73943	75029	76159
98	49633	50723	51787	52901	53959	55073	56167	57173	58229	59351	60457	61613	62747	63799	65033	66083	67217	68329	69481	70627	71707	72797	73951	75037	76163
99	49639	50741	51797	52903	53987	55079	56171	57179	58231	59357	60493	61627	62761	63803	65053	66089	67219	68351	69491	70639	71711	72817	73961	75041	76207
100	49663	50753	51803	52919	53993	55103	56179	57191	58237	59359	60497	61631	62773	63809	65063	66103	67231	68371	69493	70657	71713	72823	73973	75079	76207

PRIMES

Table 24.9

	75	76	77	78	79	80	81	82	83	84	85	86	87	88	89	90	91	92	93	94	95
1	76213	77359	78487	79627	80737	81817	82903	84131	85243	86381	87557	88807	89867	90989	92177	93187	94351	95443	96587	97829	98953
2	76231	77369	78497	79631	80747	81839	82913	84137	85247	86389	87559	88811	89891	90997	92179	93199	94379	95461	96589	97841	98963
3	76243	77377	78509	79633	80749	81847	82939	84143	85259	86399	87583	88813	89897	91009	92189	93229	94397	95467	96601	97843	98981
4	76249	77383	78511	79657	80761	81853	82963	84163	85297	86413	87587	88817	89899	91019	92203	93239	94399	95471	96643	97847	98993
5	76253	77417	78517	79669	80777	81869	82981	84179	85303	86423	87589	88819	89909	91033	92219	93241	94421	95479	96661	97849	98999
6	76259	77419	78539	79687	80779	81883	82997	84181	85313	86441	87613	88843	89917	91079	92221	93251	94427	95483	96667	97859	99013
7	76261	77431	78541	79691	80783	81899	83003	84191	85331	86453	87623	88853	89923	91081	92227	93253	94433	95507	96671	97861	99017
8	76283	77447	78553	79693	80789	81901	83009	84199	85333	86461	87629	88861	89939	91097	92233	93257	94439	95527	96697	97871	99023
9	76289	77471	78569	79697	80803	81919	83023	84211	85361	86467	87631	88867	89959	91099	92237	93263	94441	95531	96703	97879	99041
10	76303	77477	78571	79699	80809	81929	83047	84221	85363	86477	87641	88873	89963	91121	92243	93281	94447	95539	96731	97883	99053
11	76333	77479	78577	79757	80819	81931	83059	84223	85369	86491	87643	88883	89977	91127	92251	93283	94463	95549	96737	97919	99079
12	76343	77489	78583	79769	80831	81937	83063	84229	85381	86501	87649	88897	89983	91129	92269	93287	94477	95561	96739	97927	99083
13	76367	77491	78593	79777	80833	81943	83071	84239	85411	86509	87671	88903	89989	91139	92297	93307	94483	95569	96749	97931	99089
14	76369	77509	78607	79801	80849	81953	83077	84247	85427	86531	87679	88919	90001	91141	92311	93319	94513	95581	96757	97943	99103
15	76379	77513	78623	79811	80863	81967	83089	84263	85429	86533	87683	88937	90007	91151	92317	93323	94529	95597	96763	97961	99109
16	76387	77521	78643	79813	80897	81971	83093	84299	85439	86539	87691	88951	90011	91153	92333	93329	94531	95603	96769	97967	99119
17	76403	77527	78649	79817	80909	81973	83101	84307	85447	86561	87697	88969	90017	91159	92347	93337	94541	95617	96779	97973	99131
18	76421	77543	78653	79823	80911	82003	83117	84313	85451	86573	87701	88993	90019	91163	92353	93371	94543	95621	96787	97987	99133
19	76423	77549	78691	79829	80917	82007	83137	84317	85453	86579	87719	88997	90023	91183	92357	93377	94547	95629	96797	98009	99137
20	76441	77551	78697	79841	80923	82009	83177	84319	85469	86587	87721	89003	90031	91193	92363	93383	94559	95633	96799	98011	99139
21	76463	77557	78707	79843	80929	82013	83203	84347	85487	86599	87739	89009	90053	91199	92369	93407	94561	95651	96821	98017	99149
22	76471	77563	78713	79847	80933	82021	83207	84349	85513	86627	87743	89017	90059	91229	92377	93419	94573	95701	96823	98041	99173
23	76481	77569	78721	79861	80953	82031	83219	84377	85517	86629	87751	89021	90067	91237	92381	93427	94583	95707	96827	98047	99181
24	76487	77573	78737	79867	80963	82037	83221	84389	85523	86677	87767	89041	90071	91243	92383	93463	94597	95713	96847	98057	99191
25	76493	77587	78779	79873	80989	82039	83227	84431	85531	86689	87793	89051	90073	91249	92387	93479	94603	95717	96851	98081	99223
26	76507	77591	78781	79889	81001	82051	83231	84401	85549	86693	87797	89057	90089	91253	92399	93481	94613	95723	96857	98101	99233
27	76511	77611	78787	79901	81013	82067	83233	84407	85571	86711	87803	89069	90107	91283	92401	93487	94621	95731	96893	98123	99241
28	76519	77617	78791	79903	81017	82073	83243	84421	85577	86719	87811	89071	90121	91291	92413	93491	94649	95737	96907	98129	99251
29	76537	77621	78797	79907	81019	82129	83257	84431	85597	86729	87833	89083	90127	91297	92419	93493	94651	95747	96911	98143	99257
30	76541	77641	78803	79939	81023	82139	83267	84437	85601	86743	87853	89087	90149	91303	92431	93497	94687	95773	96931	98179	99259
31	76543	77647	78809	79943	81031	82141	83269	84443	85607	86753	87869	89101	90163	91309	92459	93503	94693	95783	96953	98207	99277
32	76561	77659	78823	79967	81041	82153	83273	84449	85619	86767	87877	89107	90173	91331	92461	93523	94709	95789	96959	98213	99289
33	76579	77681	78839	79973	81043	82163	83299	84457	85621	86771	87881	89113	90187	91367	92467	93529	94723	95791	96973	98221	99317
34	76597	77687	78853	79979	81047	82171	83311	84463	85627	86783	87887	89119	90191	91369	92479	93553	94727	95801	96979	98227	99347
35	76603	77689	78857	79987	81049	82183	83339	84467	85639	86813	87911	89123	90197	91373	92489	93557	94747	95803	96989	98251	99349
36	76607	77699	78877	79997	81071	82189	83341	84481	85643	86837	87917	89137	90199	91381	92503	93559	94771	95813	96997	98257	99367
37	76631	77711	78887	79999	81077	82193	83357	84499	85661	86843	87931	89153	90203	91387	92507	93563	94777	95819	97001	98269	99371
38	76649	77713	78889	80021	81083	82207	83383	84503	85667	86851	87943	89189	90217	91393	92551	93581	94781	95857	97003	98297	99377
39	76651	77719	78893	80039	81097	82217	83389	84509	85669	86857	87959	89203	90227	91397	92567	93601	94789	95869	97007	98299	99391
40	76667	77723	78901	80051	81101	82219	83399	84521	85691	86861	87961	89209	90239	91411	92567	93607	94793	95873	97021	98317	99397
41	76673	77731	78919	80071	81119	82223	83401	84523	85703	86869	87973	89213	90247	91423	92569	93629	94811	95881	97039	98321	99401
42	76679	77743	78929	80077	81131	82231	83407	84533	85711	86923	87977	89227	90263	91433	92581	93637	94819	95891	97073	98323	99409
43	76697	77747	78941	80107	81157	82237	83417	84551	85717	86927	87991	89231	90271	91453	92593	93683	94823	95911	97081	98327	99431
44	76717	77761	78977	80111	81163	82241	83423	84559	85733	86929	88007	89237	90281	91457	92623	93701	94837	95917	97103	98347	99439
45	76733	77773	78979	80141	81173	82261	83431	84589	85751	86939	88003	89261	90289	91459	92627	93703	94841	95923	97117	98369	99469
46	76753	77783	78989	80147	81181	82267	83437	84629	85701	86951	88007	89269	90313	91463	92639	93719	94847	95929	97127	98377	99487
47	76757	77797	79031	80149	81197	82279	83443	84631	85793	86959	88019	89273	90353	91493	92641	93739	94849	95947	97151	98387	99497
48	76771	77801	79039	80153	81199	82301	83449	84649	85817	86969	88037	89293	90359	91499	92647	93761	94873	95957	97157	98389	99523
49	76777	77813	79043	80167	81203	82307	83459	84653	85819	86981	88069	89303	90371	91513	92657	93763	94889	95959	97159	98407	99527
50	76781	77839	79063	80173	81223	82339	83471	84659	85829	86993	88079	89317	90373	91529	92669	93787	94903	95971	97169	98411	99529
51	76801	77849	79087	80177	81233	82349	83477	84673	85831	87011	88093	89329	90379	91541	92671	93809	94907	95987	97171	98419	99551
52	76819	77863	79103	80191	81239	82351	83497	84691	85837	87013	88117	89363	90397	91571	92681	93811	94933	95989	97177	98429	99559
53	76829	77867	79111	80207	81281	82361	83537	84697	85843	87037	88129	89371	90401	91573	92683	93827	94949	96001	97187	98443	99563
54	76831	77893	79133	80209	81283	82373	83557	84701	85847	87041	88169	89381	90403	91577	92693	93851	94951	96013	97213	98453	99571
55	76837	77899	79139	80221	81293	82387	83561	84713	85853	87049	88177	89387	90407	91583	92699	93871	94961	96017	97231	98459	99577
56	76847	77929	79147	80231	81299	82393	83563	84719	85889	87071	88211	89393	90437	91591	92707	93887	94999	96053	97241	98467	99601
57	76871	77933	79151	80233	81307	82421	83579	84731	85903	87083	88223	89399	90439	91621	92717	93889	95003	96059	97259	98473	99607
58	76873	77951	79153	80239	81331	82457	83591	84737	85909	87103	88237	89413	90469	91631	92723	93893	95009	96079	97283	98479	99611
59	76883	77969	79159	80251	81343	82463	83597	84751	85931	87107	88241	89417	90473	91639	92737	93901	95021	96097	97301	98491	99623
60	76907	77977	79181	80263	81349	82469	83609	84761	85933	87119	88259	89431	90481	91673	92753	93911	95021	96137	97303	98507	99643
61	76913	77983	79187	80273	81353	82471	83617	84787	85991	87121	88261	89443	90499	91691	92761	93913	95027	96137	97327	98519	99661
62	76919	77999	79193	80279	81359	82483	83621	84793	85999	87133	88289	89449	90511	91703	92767	93923	95063	96149	97367	98533	99667
63	76943	78007	79201	80287	81371	82487	83639	84809	86011	87149	88301	89459	90523	91711	92779	93937	95071	96157	97369	98543	99679
64	76949	78017	79229	80309	81373	82493	83641	84811	86017	87151	88321	89477	90527	91733	92789	93941	95083	96167	97373	98561	99689
65	76961	78031	79231	80317	81401	82499	83663	84827	86027	87179	88327	89491	90529	91753	92791	93949	95087	96179	97379	98563	99707
66	76963	78041	79241	80329	81409	82507	83663	84857	86029	87181	88337	89501	90533	91757	92801	93967	95089	96181	97381	98573	99709
67	76991	78049	79259	80341	81421	82529	83689	84869	86069	87187	88339	89513	90547	91771	92809	93971	95093	96199	97387	98597	99713
68	77003	78059	79273	80347	81439	82531	83701	84877	86077	87211	88379	89519	90583	91781	92821	93979	95101	96211	97397	98621	99719
69	77017	78079	79279	80363	81457	82549	83717	84907	86083	87221	88397	89521	90599	91801	92831	93983	95107	96221	97423	98627	99721
70	77023	78101	79283	80369	81463	82559	83719	84913	86111	87223	88411	89527	90617	91807	92849	93997	95111	96223	97429	98639	99733
71	77029	78121	79301	80387	81509	82561	83737	84919	86113	87251	88423	89533	90619	91811	92857	94007	95131	96233	97441	98641	99761
72	77041	78137	79309	80407	81517	82567	83761	84947	86117	87253	88427	89561	90631	91813	92861	94009	95143	96259	97453	98663	99767
73	77047	78139	79319	80429	81527	82571	83773	84961	86131	87257	88463	89563	90641	91823	92863	94033	95153	96263	97459	98669	99787
74	77069	78157	79333	80447	81533	82591	83777	84967	86137	87277	88469	89567	90647	91837	92867	94049	95177	96269	97463	98689	99793
75	77081	78163	79337	80449	81547	82601	83791	84977	86143	87281	88471	89591	90659	91841	92893	94057	95189	96281	97499	98711	99809
76	77093	78167	79349	80471	81551	82609	83813	84979	86161	87293	88493	89597	90677	91867	92899	94063	95191	96289	97501	98713	99817
77	77101	78173	79357	80473	81553	82613	83833	84991	86171	87299	88499	89599	90679	91873	92921	94079	95203	96293	97511	98717	99823
78	77137	78179	79367	80489	81559	82619	83843	85009	86179	87313	88513	89603	90697	91909	92927	94099	95213	96323	97523	98729	99829
79	77141	78191	79379	80491	81563	82633	83857	85021	86183	87317	88523	89611	90703	91921	92941	94109	95219	96329	97547	98731	99839
80	77153	78193	79393	80513	81569	82651	83869	85027	86197	87323	88547	89627	90709	91939	92951	94111	95231	96331	97549	98737	99859
81	77167	78203	79397	80527	81611	82657	83873	85037	86201	87337	88589	89633	90731	91943	92957	94117	95233	96337	97553	98773	99859
82	77171	78229	79399	80537	81619	82699	83891	85049	86209	87359	88591	89653	90749	91951	92959	94121	95239	96353	97561	98779	99871
83	77191	78233	79411	80557	81629	82721	83903	85061	86239	87383	88607	89657	90787	91957	92987	94151	95257	96377	97571	98801	99877
84	77201	78241	79423	80567	81637	82723	83911	85081	86243	87403	88609	89659	90793	91961	92993	94153	95261	96401	97577	98807	99881
85	77213	78259	79427	80599	81647	82727	83921	85087	86249	87407	88643	89669	90803	91967	93001	94169	95267	96419	97579	98809	99901
86	77237	78277	79433	80603	81649	82729	83933	85091	86257	87421	88651	89671	90821	91969	93047	94201	95273	96431	97583	98837	99907
87	77239	78283	79451	80611	81667	82757	83939	85093	86263	87427	88657	89681	90823	91997	93053	94207	95279	96443	97607	98849	99923
88	77243	78301	79481	80621	81671	82759	83969	85103	86269	87433	88661	89689	90833	92003	93059	94219	95287	96451	97609	98867	99961
89	77249	78307	79493	80627	81677	82763	83983	85109	86287	87443	88663	89753	90841	92009	93077	94229	95311	96457	97613	98869	99971
90	77261	78311	79531	80629	81689	82781	83987	85121	86291	87473	88667	89759	90847	92033	93083	94253	95317	96461	97649	98873	99989
91	77263	78317	79537	80651	81701	82787	84011	85133	86293	87481	88681	89767	90863	92041	93089	94261	95327	96469	97651	98887	99989
92	77267	78341	79549	80657	81703	82793	84017	85147	86297	87491	88721	89779	90887	92051	93097	94273	95339	96479	97673	98893	99991
93	77269	78347	79559	80669	81707	82799	84047	85159	86311	87509	88729	89783	90901	92077	93103	94291	95369	96487	97687	98897	
94	77279	78367	79561	80671	81727	82811	84053	85193	86323	87511	88741	89797	90907	92083	93113	94307	95383	96493	97711	98899	
95	77291	78401	79579	80677	81737	82813	84059	85199	86341	87517	88747	89809	90931	92107	93131	94309	95393	96497	97729	98909	
96	77317	78427	79589	80681	81749	82837	84061	85201	86351	87523	88771	89819	90917	92111	93133	94321	95401	96517	97771	98911	
97	77323	78437	79601	80683	81761	82847	84067	85213	86353	87539	88789	89821	90931	92119	93139	94327	95413	96527	97777	98927	
98	77339	78439	79609	80687	81769	82883	84089	85223	86357	87541	88793	89833	90947	92143	93151	94331	95419	96553	97787	98929	
99	77347	78467	79613	80701	81773	82889	84121	85229	86369	87547	88799	89839	90971	92153	93169	94343	95429	96557	97789	98939	
100	77351	78479	79621	80713	81799	82891	84127	85237	86371	87553	88801	89849	90977	92173	93179	94349	95441	96581	97813	98947	

25. Numerical Interpolation, Differentiation, and Integration

Philip J. Davis[1] and Ivan Polonsky[2]

Contents

---

[1] National Bureau of Standards.

[2] National Bureau of Standards. (Presently, Bell Tel. Labs., Whippany, N.J.)

25. Numerical Interpolation, Differentiation, and Integration

Numerical analysts have a tendency to accumulate a multiplicity of tools each designed for highly specialized operations and each requiring special knowledge to use properly. From the vast stock of formulas available we have culled the present selection. We hope that it will be useful. As with all such compendia, the reader may miss his favorites and find others whose utility he thinks is marginal.

We would have liked to give examples to illuminate the formulas, but this has not been feasible. Numerical analysis is partially a science and partially an art, and short of writing a textbook on the subject it has been impossible to indicate where and under what circumstances the various formulas are useful or accurate, or to elucidate the numerical difficulties to which one might be led by uncritical use. The formulas are therefore issued together with a caveat against their blind application.

Formulas

Notation: Abscissas: $x_0 < x_1 < \ldots$; functions: $f, g, \ldots$; values: $f(x_i) = f_i$, $f'(x_i) = f'_i$. f', $f^{(2)}$, $\ldots$ indicate 1st, 2^{d}, $\ldots$ derivatives. If abscissas are equally spaced, $x_{i+1} - x_i = h$ and $f_p = f(x_0 + ph)$ (p not necessarily integral). R, R_n indicate remainders.

25.1. Differences

Forward Differences

25.1.1

$$\Delta(f_n) = \Delta_n = \Delta_n^1 = f_{n+1} - f_n$$

$$\Delta_n^2 = \Delta_{n+1}^1 - \Delta_n^1 = f_{n+2} - 2f_{n+1} + f_n$$

$$\Delta_n^3 = \Delta_{n+1}^2 - \Delta_n^2 = f_{n+3} - 3f_{n+2} + 3f_{n+1} - f_n$$

$$\Delta_n^k = \Delta_{n+1}^{k-1} - \Delta_n^{k-1} = \sum_{j=0}^{k} (-1)^j \binom{k}{j} f_{n+k-j}$$

Central Differences

25.1.2

$$\delta(f_{n+\frac{1}{2}}) = \delta_{n+\frac{1}{2}} = \delta_{n+\frac{1}{2}}^1 = f_{n+1} - f_n$$

$$\delta_n^2 = \delta_{n+\frac{1}{2}}^1 - \delta_{n-\frac{1}{2}}^1 = f_{n+1} - 2f_n + f_{n-1}$$

$$\delta_{n+\frac{1}{2}}^3 = \delta_{n+1}^2 - \delta_n^2 = f_{n+2} - 3f_{n+1} + 3f_n - f_{n-1}$$

$$\delta_n^{2k} = \sum_{j=0}^{2k} (-1)^j \binom{2k}{j} f_{n+k-j}$$

$$\delta_{n+\frac{1}{2}}^{2k+1} = \sum_{j=0}^{2k+1} (-1)^j \binom{2k+1}{j} f_{n+k+1-j}$$

$$\delta_{\frac{1}{2}n}^k = \Delta_{\frac{1}{2}(n-k)}^k \text{ if } n \text{ and } k \text{ are of same parity.}$$

Forward Differences

$$
\begin{array}{llll}
x_0 & f_0 & & \\
 & & \Delta_0 & \\
x_1 & f_1 & & \Delta_0^2 \\
 & & \Delta_1 & & \Delta_0^3 \\
x_2 & f_2 & & \Delta_1^2 \\
 & & \Delta_2 & \\
x_3 & f_3 & &
\end{array}
$$

Central Differences

$$
\begin{array}{llll}
x_{-1} & f_{-1} & & \\
 & & \delta_{-\frac{1}{2}} & \\
x_0 & f_0 & & \delta_0^2 \\
 & & \delta_{\frac{1}{2}} & & \delta_{\frac{1}{2}}^3 \\
x_1 & f_1 & & \delta_1^2 \\
 & & \delta_{3/2} & \\
x_2 & f_2 & &
\end{array}
$$

Mean Differences

25.1.3
$$\mu(f_n) = \tfrac{1}{2}(f_{n+\frac{1}{2}} + f_{n-\frac{1}{2}})$$

Divided Differences

25.1.4
$$[x_0, x_1] = \frac{f_0 - f_1}{x_0 - x_1} = [x_1, x_0]$$

$$[x_0, x_1, x_2] = \frac{[x_0, x_1] - [x_1, x_2]}{x_0 - x_2}$$

$$[x_0, x_1, \ldots, x_k] = \frac{[x_0, \ldots, x_{k-1}] - [x_1, \ldots, x_k]}{x_0 - x_k}$$

Divided Differences in Terms of Functional Values

25.1.5
$$[x_0, x_1, \ldots, x_n] = \sum_{k=0}^{n} \frac{f_k}{\pi'_n(x_k)}$$

25.1.6 where $\pi_n(x) = (x-x_0)\ (x-x_1)\ \ldots\ (x-x_n)$ and $\pi_n'(x)$ is its derivative:

25.1.7

$$\pi_n'(x_k) = (x_k-x_0)\ \ldots\ (x_k-x_{k-1})(x_k-x_{k+1})$$
$$\ldots\ (x_k-x_n)$$

Let D be a simply connected domain with a piecewise smooth boundary C and contain the points $z_0, \ldots, z_n$ in its interior. Let $f(z)$ be analytic in D and continuous in $D+C$. Then,

25.1.8 $[z_0, z_1, \ldots, z_n] = \dfrac{1}{2\pi i} \displaystyle\int_C \dfrac{f(z)}{\prod\limits_{k=0}^{n} (z-z_k)}\ dz$

25.1.9 $\quad \Delta_0^n = h^n f^{(n)}(\xi) \qquad (x_0 < \xi < x_n)$

25.1.10

$$[x_0, x_1, \ldots, x_n] = \frac{\Delta_0^n}{n!h^n} = \frac{f^{(n)}(\xi)}{n!} \qquad (x_0 < \xi < x_n)$$

25.1.11

$$[x_{-n}, x_{-n+1}, \ldots, x_0, \ldots, x_n] = \frac{\delta_0^{2n}}{h^{2n}(2n)!}$$

Reciprocal Differences

25.1.12

$$\rho(x_0, x_1) = \frac{x_0-x_1}{f_0-f_1}$$

$$\rho_2(x_0, x_1, x_2) = \frac{x_0-x_2}{\rho(x_0, x_1) - \rho(x_1, x_2)} + f_1$$

$$\rho_3(x_0, x_1, x_2, x_3) = \frac{x_0-x_3}{\rho_2(x_0, x_1, x_2) - \rho_2(x_1, x_2, x_3)} + \rho(x_1, x_2)$$

$$\vdots$$

$$\rho_n(x_0, x_1, \ldots, x_n) = \frac{x_0-x_n}{\rho_{n-1}(x_0, \ldots, x_{n-1}) - \rho_{n-1}(x_1, \ldots, x_n)} + \rho_{n-2}(x_1, \ldots, x_{n-1})$$

25.2. Interpolation

Lagrange Interpolation Formulas

25.2.1 $\quad f(x) = \displaystyle\sum_{i=0}^{n} l_i(x) f_i + R_n(x)$

25.2.2

$$l_i(x) = \frac{\pi_n(x)}{(x-x_i)\pi_n'(x_i)}$$
$$= \frac{(x-x_0)\ \ldots\ (x-x_{i-1})(x-x_{i+1})\ \ldots\ (x-x_n)}{(x_i-x_0)\ \ldots\ (x_i-x_{i-1})(x_i-x_{i+1})\ \ldots\ (x_i-x_n)}$$

Remainder in Lagrange Interpolation Formula

25.2.3

$$R_n(x) = \pi_n(x) \cdot [x_0, x_1, \ldots, x_n, x]$$
$$= \pi_n(x) \cdot \frac{f^{n+1}(\xi)}{(n+1)!} \qquad (x_0 < \xi < x_n)$$

25.2.4

$$|R_n(x)| \le \frac{(x_n-x_0)^{n+1}}{(n+1)!} \max_{x_0 \le x \le x_n} |f^{(n+1)}(x)|$$

25.2.5

$$R_n(z) = \frac{\pi_n(z)}{2\pi i} \int_C \frac{f(t)}{(t-z)(t-z_0)\ldots(t-z_n)}\ dt$$

The conditions of **25.1.8** are assumed here.

Lagrange Interpolation, Equally Spaced Abscissas

n Point Formula

25.2.6 $\quad f(x_0+ph) = \displaystyle\sum_k A_k^n(p) f_k + R_{n-1}$

For n even, $\quad \left(-\dfrac{1}{2}(n-2) \le k \le \dfrac{1}{2} n\right).$

For n odd, $\quad \left(-\dfrac{1}{2}(n-1) \le k \le \dfrac{1}{2}(n-1)\right).$

25.2.7

$$A_k^n(p) = \frac{(-1)^{\frac{1}{2}n+k}}{\left(\dfrac{n-2}{2}+k\right)!(\tfrac{1}{2}n-k)!(p-k)} \prod_{t=1}^{n} (p+\tfrac{1}{2}n-t)$$
$$n \text{ even.}$$

$$A_k^n(p) = \frac{(-1)^{\frac{1}{2}(n-1)+k}}{\left(\dfrac{n-1}{2}+k\right)!\left(\dfrac{n-1}{2}-k\right)!(p-k)}$$
$$\prod_{t=0}^{n-1} \left(p+\frac{n-1}{2}-t\right), \qquad n \text{ odd.}$$

25.2.8

$$R_{n-1} = \frac{1}{n!} \prod_k (p-k) h^n f^{(n)}(\xi)$$
$$\approx \frac{1}{n!} \prod_k (p-k)\Delta_0^n \qquad (x_0 < \xi < x_n)$$

k has the same range as in **25.2.6**.

Lagrange Two Point Interpolation Formula (Linear Interpolation)

25.2.9 $\quad f(x_0+ph) = (1-p)f_0 + pf_1 + R_1$

25.2.10 $\quad R_1(p) \approx .125h^2 f^{(2)}(\xi) \approx .125\Delta^2$

Lagrange Three Point Interpolation Formula

25.2.11

$$f(x_0+ph)=A_{-1}f_{-1}+A_0f_0+A_1f_1+R_2$$

$$\approx \frac{p(p-1)}{2}f_{-1}+(1-p^2)f_0+\frac{p(p+1)}{2}f_1$$

25.2.12

$$R_2(p)\approx .065h^3f^{(3)}(\xi)\approx .065\Delta^3 \qquad (|p|\leq 1)$$

Lagrange Four Point Interpolation Formula

25.2.13

$$f(x_0+ph)=A_{-1}f_{-1}+A_0f_0+A_1f_1+A_2f_2+R_3$$

$$\approx \frac{-p(p-1)(p-2)}{6}f_{-1}+\frac{(p^2-1)(p-2)}{2}f_0$$

$$-\frac{p(p+1)(p-2)}{2}f_1+\frac{p(p^2-1)}{6}f_2$$

25.2.14 $\qquad R_3(p)\approx$

$$.024h^4f^{(4)}(\xi)\approx .024\Delta^4 \qquad (0<p<1)$$

$$.042h^4f^{(4)}(\xi)\approx .042\Delta^4 \qquad (-1<p<0,\ 1<p<2)$$

$$(x_{-1}<\xi<x_2)$$

Lagrange Five Point Interpolation Formula

25.2.15

$$f(x_0+ph)=\sum_{i=-2}^{2} A_if_i+R_4$$

$$\approx \frac{(p^2-1)p(p-2)}{24}f_{-2}-\frac{(p-1)p(p^2-4)}{6}f_{-1}$$

$$+\frac{(p^2-1)(p^2-4)}{4}f_0-\frac{(p+1)p(p^2-4)}{6}f_1$$

$$+\frac{(p^2-1)p(p+2)}{24}f_2$$

25.2.16 $\qquad R_4(p)\approx$

$$.012h^5f^{(5)}(\xi)\approx .012\Delta^5 \qquad (|p|<1)$$

$$.031h^5f^{(5)}(\xi)\approx .031\Delta^5 \qquad (1<|p|<2) \qquad (x_{-2}<\xi<x_2)$$

Lagrange Six Point Interpolation Formula

25.2.17

$$f(x_0+ph)=\sum_{i=-2}^{3} A_if_i+R_5$$

$$\approx \frac{-p(p^2-1)(p-2)(p-3)}{120}f_{-2}$$

$$+\frac{p(p-1)(p^2-4)(p-3)}{24}f_{-1}$$

$$-\frac{(p^2-1)(p^2-4)(p-3)}{12}f_0$$

$$+\frac{p(p+1)(p^2-4)(p-3)}{12}f_1-\frac{p(p^2-1)(p+2)(p-3)}{24}f_2$$

$$+\frac{p(p^2-1)(p^2-4)}{120}f_3$$

25.2.18 $\qquad R_5(p)\approx$

$$.0049h^6f^{(6)}(\xi)\approx .0049\Delta^6 \qquad (0<p<1)$$

$$.0071h^6f^{(6)}(\xi)\approx .0071\Delta^6 \qquad (-1<p<0,\ 1<p<2)$$

$$.024h^6f^{(6)}(\xi)\approx .024\Delta^6 \qquad (-2<p<-1,\ 2<p<3)$$

$$(x_{-2}<\xi<x_3)$$

Lagrange Seven Point Interpolation Formula

25.2.19 $\qquad f(x_0+ph)=\sum_{i=-3}^{3} A_if_i+R_6$

25.2.20

$$R_6(p)\approx \begin{cases} .0025h^7f^{(7)}(\xi)\approx .0025\Delta^7 & (|p|<1) \\ .0046h^7f^{(7)}(\xi)\approx .0046\Delta^7 & (1<|p|<2) \\ .019h^7f^{(7)}(\xi)\approx .019\Delta^7 & (2<|p|<3) \end{cases}$$

$$(x_{-3}<\xi<x_3)$$

Lagrange Eight Point Interpolation Formula

25.2.21 $\qquad f(x_0+ph)=\sum_{i=-3}^{4} A_if_i+R_7$

25.2.22

$$R_7(p)\approx \begin{cases} .0011h^8f^{(8)}(\xi)\approx .0011\Delta^8 & (0<p<1) \\ .0014h^8f^{(8)}(\xi)\approx .0014\Delta^8 & \begin{array}{l}(-1<p<0)\\(1<p<2)\end{array} \\ .0033h^8f^{(8)}(\xi)\approx .0033\Delta^8 & \begin{array}{l}(-2<p<-1)\\(2<p<3)\end{array} \\ .016h^8f^{(8)}(\xi)\approx .016\Delta^8 & \begin{array}{l}(-3<p<-2)\\(3<p<4)\end{array} \end{cases}$$

$$(x_{-3}<\xi<x_4)$$

Aitken's Iteration Method

Let $f(x|x_0,x_1,\ldots,x_k)$ denote the unique polynomial of kth degree which coincides in value with $f(x)$ at $x_0,\ldots,x_k$.

25.2.23

$$f(x|x_0,x_1)=\frac{1}{x_1-x_0}\begin{vmatrix} f_0 & x_0-x \\ f_1 & x_1-x \end{vmatrix}$$

$$f(x|x_0,x_2)=\frac{1}{x_2-x_0}\begin{vmatrix} f_0 & x_0-x \\ f_2 & x_2-x \end{vmatrix}$$

$$f(x|x_0,x_1,x_2)=\frac{1}{x_2-x_1}\begin{vmatrix} f(x|x_0,x_1) & x_1-x \\ f(x|x_0,x_2) & x_2-x \end{vmatrix}$$

$$f(x|x_0,x_1,x_2,x_3)=\frac{1}{x_3-x_2}\begin{vmatrix} f(x|x_0,x_1,x_2) & x_2-x \\ f(x|x_0,x_1,x_3) & x_3-x \end{vmatrix}$$

Taylor Expansion

25.2.24

$$f(x) = f_0 + (x-x_0)f_0' + \frac{(x-x_0)^2}{2!}f_0^{(2)} + \cdots$$
$$+ \frac{(x-x_0)^n}{n!}f_0^{(n)} + R_n$$

25.2.25

$$R_n = \int_{x_0}^{x} f^{(n+1)}(t)\frac{(x-t)^n}{n!}\,dt$$
$$= \frac{(x-x_0)^{n+1}}{(n+1)!}f^{(n+1)}(\xi) \qquad (x_0 < \xi < x)$$

Newton's Divided Difference Interpolation Formula

25.2.26

$$f(x) = f_0 + \sum_{k=1}^{n} \pi_{k-1}(x)[x_0, x_1, \ldots, x_k] + R_n$$

x_0	f_0			
		$[\boldsymbol{x_0, x_1}]$		
x_1	f_1		$[\boldsymbol{x_0, x_1, x_2}]$	
		$[x_1, x_2]$		$[\boldsymbol{x_0, x_1, x_2, x_3}]$
x_2	f_2		$[x_1, x_2, x_3]$	
		$[x_2, x_3]$		
x_3	f_3			

25.2.27

$$R_n(x) = \pi_n(x)[x_0, \ldots, x_n, x] = \pi_n(x)\frac{f^{(n+1)}(\xi)}{(n+1)!}$$

$$(x_0 < \xi < x_n)$$

(For π_n see **25.1.6**.)

Newton's Forward Difference Formula

25.2.28

$$f(x_0+ph) = f_0 + p\Delta_0 + \binom{p}{2}\Delta_0^2 + \cdots + \binom{p}{n}\Delta_0^n + R_n$$

x_0	f_0			
		$\boldsymbol{\Delta_0}$		
x_1	f_1		$\boldsymbol{\Delta_0^2}$	
		Δ_1		$\boldsymbol{\Delta_0^3}$
x_2	f_2		Δ_1^2	
		Δ_2		
x_3	f_3			

25.2.29

$$R_n = h^{n+1}\binom{p}{n+1}f^{(n+1)}(\xi) \approx \binom{p}{n+1}\Delta_0^{n+1}$$

$$(x_0 < \xi < x_n)$$

Relation Between Newton and Lagrange Coefficients

25.2.30

$$\binom{p}{2} = A_{-1}^3(p) \qquad \binom{p}{3} = -A_{-1}^4(p) \qquad \binom{p}{4} = A_{-2}^5(1-p)$$

$$\binom{p}{5} = A_{-3}^6(2-p)$$

Everett's Formula

25.2.31

$$f(x_0+ph) = (1-p)f_0 + pf_1 - \frac{p(p-1)(p-2)}{3!}\delta_0^2$$
$$+ \frac{(p+1)p(p-1)}{3!}\delta_1^2 + \cdots - \binom{p+n-1}{2n+1}\delta_0^{2n}$$
$$+ \binom{p+n}{2n+1}\delta_1^{2n} + R_{2n}$$
$$= (1-p)f_0 + pf_1 + E_2\delta_0^2 + F_2\delta_1^2 + E_4\delta_0^4$$
$$+ F_4\delta_1^4 + \cdots + R_{2n}$$

x_0	f_0	δ_0^2	δ_0^4
	$\delta_{\frac{1}{2}}$	$\delta_{\frac{1}{2}}^3$	
x_1	f_1	δ_1^2	δ_1^4

25.2.32

$$R_{2n} = h^{2n+2}\binom{p+n}{2n+2}f^{(2n+2)}(\xi)$$
$$\approx \binom{p+n}{2n+2}\left[\frac{\Delta_{-n-1}^{2n+2} + \Delta_{-n}^{2n+2}}{2}\right] \qquad (x_{-n} < \xi < x_{n+1})$$

Relation Between Everett and Lagrange Coefficients

25.2.33

$$E_2 = A_{-1}^4 \qquad E_4 = A_{-2}^6 \qquad E_6 = A_{-3}^8$$

$$F_2 = A_2^4 \qquad F_4 = A_3^6 \qquad F_6 = A_4^8$$

Everett's Formula With Throwback (Modified Central Difference)

25.2.34

$$f(x_0+ph) = (1-p)f_0 + pf_1 + E_2\delta_{m,0}^2 + F_2\delta_{m,1}^2 + R$$

25.2.35

$$\delta_m^2 = \delta^2 - .184\delta^4$$

25.2.36

$$R \approx .00045|\mu\delta_{\frac{1}{2}}^4| + .00061|\delta_{\frac{1}{2}}^5|$$

25.2.37

$$f(x_0+ph) = (1-p)f_0 + pf_1 + E_2\delta_0^2 + F_2\delta_1^2$$
$$+ E_4\delta_{m,0}^4 + F_4\delta_{m,1}^4 + R$$

25.2.38

$$\delta_m^4 = \delta^4 - .207\delta^6 + \cdots$$

25.2.39

$$R \approx .000032|\mu\delta_{\frac{1}{2}}^6| + .000052|\delta_{\frac{1}{2}}^7|$$

25.2.40

$$f(x_0+ph) = (1-p)f_0 + pf_1 + E_2\delta_0^2 + F_2\delta_1^2$$
$$+ E_4\delta_0^4 + F_4\delta_1^4 + E_6\delta_{m,0}^6 + F_6\delta_{m,1}^6 + R$$

25.2.41

$$\delta_m^6 = \delta^6 - .218\delta^8 + .049\delta^{10} + \cdots$$

25.2.42

$$R \approx .0000037|\mu\delta_{\frac{1}{2}}^8| + \cdots$$

Simultaneous Throwback

25.2.43

$$f(x_0+ph)=(1-p)f_0+pf_1+E_2\delta^2_{m,0}+F_2\delta^2_{m,1}$$
$$+E_4\delta^4_{m,0}+F_4\delta^4_{m,1}+R$$

25.2.44 $\quad \delta^2_m=\delta^2-.01312\delta^6+.0043\delta^8-.001\delta^{10}$

25.2.45 $\quad \delta^4_m=\delta^4-.27827\delta^6+.0685\delta^8-.016\delta^{10}$

25.2.46 $\quad R\approx.00000083|\mu\delta^6_{\frac{1}{2}}|+.0000094\delta^7$

Bessel's Formula With Throwback

25.2.47

$$f(x_0+ph)=(1-p)f_0+pf_1+B_2(\delta^2_{m,0}+\delta^2_{m,1})$$
$$+B_3\delta^3_{\frac{1}{2}}+R,\ B_2=\frac{p(p-1)}{4},\ B_3=\frac{p(p-1)(p-\frac{1}{2})}{6}$$

25.2.48 $\qquad \delta^2_m=\delta^2-.184\delta^4$

25.2.49 $\qquad R\approx.00045|\mu\delta^4_{\frac{1}{2}}|+.00087|\delta^5_{\frac{1}{2}}|$

Thiele's Interpolation Formula

25.2.50

$$f(x)=f(x_1)+$$
$$\cfrac{x-x_1}{\rho(x_1,x_2)+\cfrac{x-x_2}{\rho_2(x_1,x_2,x_3)-f(x_1)+\cfrac{x-x_3}{\begin{pmatrix}\rho_3(x_1,x_2,x_3,x_4)\\-\rho(x_1,x_2)+\ldots\end{pmatrix}}}}$$

(For reciprocal differences, ρ, see **25.1.12**.)

Trigonometric Interpolation

Gauss' Formula

25.2.51 $\qquad f(x)\approx\sum_{k=0}^{2n}f_k\zeta_k(x)=t_n(x)$

25.2.52

$$\zeta_k(x)=\frac{\sin\frac{1}{2}(x-x_0)\ldots\sin\frac{1}{2}(x-x_{k-1})}{\sin\frac{1}{2}(x_k-x_0)\ldots\sin\frac{1}{2}(x_k-x_{k-1})}$$
$$\frac{\sin\frac{1}{2}(x-x_{k+1})\ldots\sin\frac{1}{2}(x-x_{2n})}{\sin\frac{1}{2}(x_k-x_{k+1})\ldots\sin\frac{1}{2}(x_k-x_{2n})}$$

$t_n(x)$ is a trigonometric polynomial of degree n such that $t_n(x_k)=f_k$ $\quad (k=0,1,\ldots,2n)$

Harmonic Analysis

Equally spaced abscissas

$$x_0=0,\qquad x_1,\ldots,x_{m-1},x_m=2\pi$$

25.2.53

$$f(x)\approx\frac{1}{2}a_0+\sum_{k=1}^{n}(a_k\cos kx+b_k\sin kx)$$

25.2.54 $\qquad m=2n+1$

$$a_k=\frac{2}{2n+1}\sum_{r=0}^{2n}f_r\cos kx_r;\qquad b_k=\frac{2}{2n+1}\sum_{r=0}^{2n}f_r\sin kx_r$$
$$(k=0,1,\ldots,n)$$

25.2.55 $\qquad m=2n$

$$a_k=\frac{1}{n}\sum_{r=0}^{2n-1}f_r\cos kx_r;\qquad b_k=\frac{1}{n}\sum_{r=0}^{2n-1}f_r\sin kx_r$$
$$(k=0,1,\ldots,n)\qquad (k=0,1,\ldots,n-1)$$

b_n is arbitrary.

Subtabulation

Let $f(x)$ be tabulated initially in intervals of width h. It is desired to subtabulate $f(x)$ in intervals of width h/m. Let Δ and $\overline{\Delta}$ designate differences with respect to the original and the final intervals respectively. Thus $\overline{\Delta}_0=f\left(x_0+\frac{h}{m}\right)-f(x_0)$. Assuming that the original 5th order differences are zero,

25.2.56

$$\overline{\Delta}_0=\frac{1}{m}\Delta_0+\frac{1-m}{2m^2}\Delta^2_0+\frac{(1-m)(1-2m)}{6m^3}\Delta^3_0$$
$$+\frac{(1-m)(1-2m)(1-3m)}{24m^4}\Delta^4_0$$

$$\overline{\Delta}^2_0=\frac{1}{m^2}\Delta^2_0+\frac{1-m}{m^3}\Delta^3_0+\frac{(1-m)(7-11m)}{12m^4}\Delta^4_0$$

$$\overline{\Delta}^3_0=\frac{1}{m^3}\Delta^3_0+\frac{3(1-m)}{2m^4}\Delta^4_0$$

$$\overline{\Delta}^4_0=\frac{1}{m^4}\Delta^4_0$$

From this information we may construct the final tabulation by addition. For $m=10$,

25.2.57

$$\overline{\Delta}_0=.1\Delta_0-.045\Delta^2_0+.0285\Delta^3_0-.02066\Delta^4_0$$
$$\overline{\Delta}^2_0=.01\Delta^2_0-.009\Delta^3_0+.007725\Delta^4_0$$
$$\overline{\Delta}^3_0=.001\Delta^3_0-.00135\Delta^4_0$$
$$\overline{\Delta}^4_0=.0001\Delta^4_0$$

Linear Inverse Interpolation

Find p, given $f_p(=f(x_0+ph))$.

Linear

25.2.58 $\qquad p\approx\dfrac{f_p-f_0}{f_1-f_0}$

Quadratic Inverse Interpolation

25.2.59

$$(f_1 - 2f_0 + f_{-1})p^2 + (f_1 - f_{-1})p + 2(f_0 - f_p) \approx 0$$

Inverse Interpolation by Reversion of Series

25.2.60   Given $f(x_0 + ph) = f_p = \sum_{k=0}^{\infty} a_k p^k$

25.2.61

$$p = \lambda + c_2 \lambda^2 + c_3 \lambda^3 + \ldots, \quad \lambda = (f_p - a_0)/a_1$$

25.2.62

$$c_2 = -a_2/a_1$$

$$c_3 = \frac{-a_3}{a_1} + 2\left(\frac{a_2}{a_1}\right)^2$$

$$c_4 = \frac{-a_4}{a_1} + \frac{5a_2 a_3}{a_1^2} - \frac{5a_2^3}{a_1^3}$$

$$c_5 = \frac{-a_5}{a_1} + \frac{6a_2 a_4}{a_1^2} + \frac{3a_3^2}{a_1^2} - \frac{21a_2^2 a_3}{a_1^3} + \frac{14a_2^4}{a_1^4}$$

Inversion of Newton's Forward Difference Formula

25.2.63

$$a_0 = f_0$$

$$a_1 = \Delta_0 - \frac{\Delta_0^2}{2} + \frac{\Delta_0^3}{3} - \frac{\Delta_0^4}{4} + \ldots$$

$$a_2 = \frac{\Delta_0^2}{2} - \frac{\Delta_0^3}{2} + \frac{11\Delta_0^4}{24} + \ldots$$

$$a_3 = \frac{\Delta_0^3}{6} - \frac{\Delta_0^4}{4} + \ldots$$

$$a_4 = \frac{\Delta_0^4}{24} + \ldots$$

(Used in conjunction with **25.2.62**.)

Inversion of Everett's Formula

25.2.64

$$a_0 = f_0$$

$$a_1 = \delta_{\frac{1}{2}} - \frac{\delta_0^2}{3} - \frac{\delta_1^2}{6} + \frac{\delta_0^4}{20} + \frac{\delta_1^4}{30} + \ldots$$

$$a_2 = \frac{\delta_0^2}{2} - \frac{\delta_0^4}{24} + \ldots$$

$$a_3 = \frac{-\delta_0^2 + \delta_1^2}{6} - \frac{\delta_0^4 + \delta_1^4}{24} + \ldots$$

$$a_4 = \frac{\delta_0^4}{24} + \ldots$$

$$a_5 = \frac{-\delta_0^4 + \delta_1^4}{120} + \ldots$$

(Used in conjunction with **25.2.62**.)

Bivariate Interpolation

Three Point Formula (Linear)

25.2.65

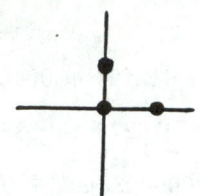

$$f(x_0 + ph, y_0 + qk) = (1 - p - q)f_{0,0}$$
$$+ pf_{1,0} + qf_{0,1} + O(h^2)$$

Four Point Formula

25.2.66

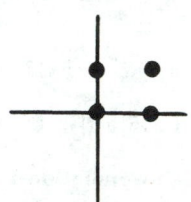

$$f(x_0 + ph, y_0 + qk) = (1-p)(1-q)f_{0,0} + p(1-q)f_{1,0}$$
$$+ q(1-p)f_{0,1} + pqf_{1,1} + O(h^2)$$

Six Point Formula

25.2.67

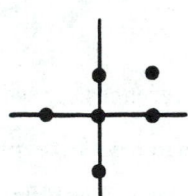

$$f(x_0 + ph, y_0 + qk) = \frac{q(q-1)}{2}f_{0,-1} + \frac{p(p-1)}{2}f_{-1,0}$$

$$+ (1 + pq - p^2 - q^2)f_{0,0}$$

$$+ \frac{p(p-2q+1)}{2}f_{1,0}$$

$$+ \frac{q(q-2p+1)}{2}f_{0,1} + pqf_{1,1} + O(h^3)$$

25.3. Differentiation

Lagrange's Formula

25.3.1 $$f'(x) = \sum_{k=0}^{n} l_k'(x)f_k + R_n'(x)$$

(See **25.2.1**.)

25.3.2 $$l_k'(x) = \sum_{\substack{j=0 \\ j \neq k}}^{n} \frac{\pi_n(x)}{(x - x_k)(x - x_j)\pi_n'(x_k)}$$

25.3.3

$$R'_n(x)=\frac{f^{(n+1)}}{(n+1)!}(\xi)\pi'_n(x)+\frac{\pi_n(x)}{(n+1)!}\frac{d}{dx}f^{(n+1)}(\xi)$$

$$\xi=\xi(x)\ (x_0<\xi<x_n)$$

Equally Spaced Abscissas

Three Points

25.3.4

$$f'_p=f'(x_0+ph)$$

$$=\frac{1}{h}\{(p-\tfrac{1}{2})f_{-1}-2pf_0+(p+\tfrac{1}{2})f_1\}+R'_2$$

Four Points

25.3.5

$$f'_p=f'(x_0+ph)=\frac{1}{h}\left\{-\frac{3p^2-6p+2}{6}f_{-1}\right.$$

$$+\frac{3p^2-4p-1}{2}f_0-\frac{3p^2-2p-2}{2}f_1$$

$$\left.+\frac{3p^2-1}{6}f_2\right\}+R'_3$$

Five Points

25.3.6

$$f'_p=f'(x_0+ph)=\frac{1}{h}\left\{\frac{2p^3-3p^2-p+1}{12}f_{-2}\right.$$

$$-\frac{4p^3-3p^2-8p+4}{6}f_{-1}+\frac{2p^3-5p}{2}f_0$$

$$-\frac{4p^3+3p^2-8p-4}{6}f_1$$

$$\left.+\frac{2p^3+3p^2-p-1}{12}f_2\right\}+R'_4$$

For numerical values of differentiation coefficients see **Table 25.2**.

Markoff's Formulas

(Newton's Forward Difference Formula Differentiated)

25.3.7

$$f'(a_0+ph)=\frac{1}{h}\left[\Delta_0+\frac{2p-1}{2}\Delta_0^2\right.$$

$$\left.+\frac{3p^2-6p+2}{6}\Delta_0^3+\ldots+\frac{d}{dp}\binom{p}{n}\Delta_0^n\right]+R'_n$$

25.3.8

$$R'_n=h^nf^{(n+1)}(\xi)\frac{d}{dp}\binom{p}{n+1}+h^{n+1}\binom{p}{n+1}\frac{d}{dx}f^{(n+1)}(\xi)$$

$$(a_0<\xi<a_n)$$

25.3.9 $\quad hf'_0=\Delta_0-\frac{1}{2}\Delta_0^2+\frac{1}{3}\Delta_0^3-\frac{1}{4}\Delta_0^4+\ldots$

25.3.10 $\quad h^2f_0^{(2)}=\Delta_0^2-\Delta_0^3+\frac{11}{12}\Delta_0^4-\frac{5}{6}\Delta_0^5+\ldots$

25.3.11

$$h^3f_0^{(3)}=\Delta_0^3-\frac{3}{2}\Delta_0^4+\frac{7}{4}\Delta_0^5-\frac{15}{8}\Delta_0^6+\ldots$$

25.3.12

$$h^4f_0^{(4)}=\Delta_0^4-2\Delta_0^5+\frac{17}{6}\Delta_0^6-\frac{7}{2}\Delta_0^7+\ldots$$

25.3.13

$$h^5f_0^{(5)}=\Delta_0^5-\frac{5}{2}\Delta_0^6+\frac{25}{0}\Delta_0^7-\frac{35}{0}\Delta_0^8+\ldots$$

Everett's Formula

25.3.14

$$hf'(x_0+ph)\approx-f_0+f_1-\frac{3p^2-6p+2}{6}\delta_0^2+\frac{3p^2-1}{6}\delta_1^2$$

$$-\frac{5p^4-20p^3+15p^2+10p-6}{120}\delta_0^4+\frac{5p^4-15p^2+4}{120}\delta_1^4$$

$$+\ldots-\left[\binom{p+n-1}{2n+1}\right]'\delta_0^{2n}+\left[\binom{p+n}{2n+1}\right]'\delta_1^{2n}$$

25.3.15

$$hf'_0\approx-f_0+f_1-\frac{1}{3}\delta_0^2-\frac{1}{6}\delta_1^2+\frac{1}{20}\delta_0^4+\frac{1}{30}\delta_1^4$$

Differences in Terms of Derivatives

25.3.16

$$\Delta_0\approx hf'_0+\frac{h^2}{2!}f_0^{(2)}+\frac{h^3}{3!}f_0^{(3)}+\frac{h^4}{4!}f_0^{(4)}+\frac{h^5}{5!}f_0^{(5)}$$

25.3.17

$$\Delta_0^2\approx h^2f_0^{(2)}+h^3f_0^{(3)}+\frac{7}{12}h^4f_0^{(4)}+\frac{1}{4}h^5f_0^{(5)}$$

25.3.18 $\quad \Delta_0^3\approx h^3f_0^{(3)}+\frac{3}{2}h^4f_0^{(4)}+\frac{5}{4}f_0^{(5)}$

25.3.19 $\quad \Delta_0^4\approx h^4f_0^{(4)}+2h^5f_0^{(5)}$

25.3.20 $\quad \Delta_0^5\approx h^5f_0^{(5)}$

Partial Derivatives

25.3.21

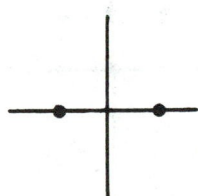

$$\frac{\partial f_{0,0}}{\partial x}=\frac{1}{2h}(f_{1,0}-f_{-1,0})+O(h^2)$$

25.3.22

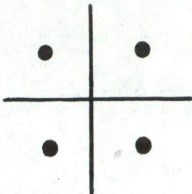

$$\frac{\partial f_{0,0}}{\partial x}=\frac{1}{4h}\ (f_{1,1}-f_{-1,1}+f_{1,-1}-f_{-1,-1})+O(h^2)$$

25.3.23

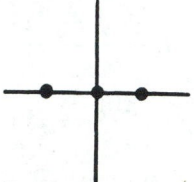

$$\frac{\partial^2 f_{0,0}}{\partial x^2}=\frac{1}{h^2}\ (f_{1,0}-2f_{0,0}+f_{-1,0})+O(h^2)$$

25.3.24

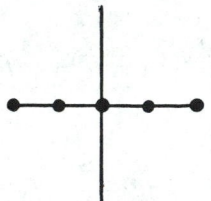

$$\frac{\partial^2 f_{0,0}}{\partial x^2}=\frac{1}{12h^2}\ (-f_{2,0}+16f_{1,0}-30f_{0,0}$$
$$+16f_{-1,0}-f_{-2,0})+O(h^4)$$

25.3.25

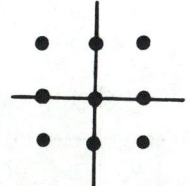

$$\frac{\partial^2 f_{0,0}}{\partial x^2}=\frac{1}{3h^2}\ (f_{1,1}-2f_{0,1}+f_{-1,1}+f_{1,0}-2f_{0,0}+f_{-1,0}$$
$$+f_{1,-1}-2f_{0,-1}+f_{-1,-1})+O(h^2)$$

25.3.26

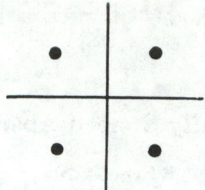

$$\frac{\partial^2 f_{0,0}}{\partial x\partial y}=\frac{1}{4h^2}\ (f_{1,1}-f_{1,-1}-f_{-1,1}+f_{-1,-1})+O(h^2)$$

25.3.27

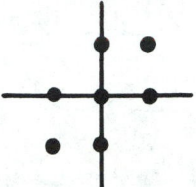

$$\frac{\partial^2 f_{0,0}}{\partial x\partial y}=\frac{-1}{2h^2}\ (f_{1,0}+f_{-1,0}+f_{0,1}+f_{0,-1}$$
$$-2f_{0,0}-f_{1,1}-f_{-1,-1})+O(h^2)$$

25.3.28

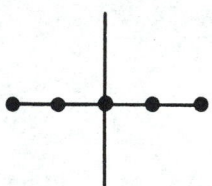

$$\frac{\partial^4 f_{0,0}}{\partial x^4}=\frac{1}{h^4}\ (f_{2,0}-4f_{1,0}+6f_{0,0}-4f_{-1,0}+f_{-2,0})+O(h^2)$$

25.3.29

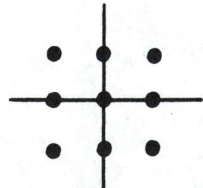

$$\frac{\partial^4 f_{0,0}}{\partial x^2\partial y^2}=\frac{1}{h^4}\ (f_{1,1}+f_{-1,1}+f_{1,-1}+f_{-1,-1}$$
$$-2f_{1,0}-2f_{-1,0}-2f_{0,1}-2f_{0,-1}+4f_{0,0})+O(h^2)$$

Laplacian

25.3.30

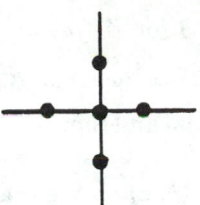

$$\nabla^2 u_{0,0} = \left(\frac{\partial^2 u}{\partial x^2} + \frac{\partial^2 u}{\partial y^2}\right)_{0,0}$$

$$= \frac{1}{h^2}(u_{1,0} + u_{0,1} + u_{-1,0} + u_{0,-1} - 4u_{0,0}) + O(h^2)$$

25.3.31

$$\nabla^2 u_{0,0} = \frac{1}{12h^2}[-60u_{0,0} + 16(u_{1,0} + u_{0,1} + u_{-1,0} + u_{0,-1})$$

$$-(u_{2,0} + u_{0,2} + u_{-2,0} + u_{0,-2})] + O(h^4)$$

Biharmonic Operator

25.3.32

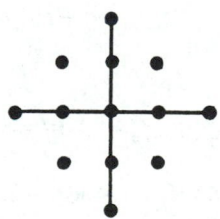

$$\nabla^4 u_{0,0} = \left(\frac{\partial^4 u}{\partial x^4} + 2\frac{\partial^4 u}{\partial x^2 \partial y^2} + \frac{\partial^4 u}{\partial y^4}\right)_{0,0}$$

$$= \frac{1}{h^4}[20u_{0,0} - 8(u_{1,0} + u_{0,1} + u_{-1,0} + u_{0,-1})$$

$$+ 2(u_{1,1} + u_{1,-1} + u_{-1,1} + u_{-1,-1})$$

$$+ (u_{0,2} + u_{2,0} + u_{-2,0} + u_{0,-2})] + O(h^2)$$

25.3.33

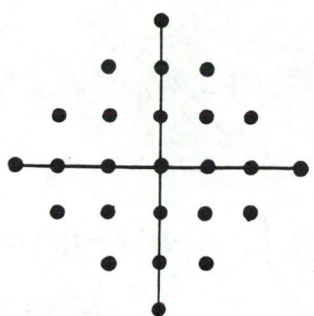

$$\nabla^4 u_{0,0} = \frac{1}{6h^4}[-(u_{0,3} + u_{0,-3} + u_{3,0} + u_{-3,0})$$

$$+ 14(u_{0,2} + u_{0,-2} + u_{2,0} + u_{-2,0})$$

$$- 77(u_{0,1} + u_{0,-1} + u_{1,0} + u_{-1,0})$$

$$+ 184u_{0,0} + 20(u_{1,1} + u_{1,-1} + u_{-1,1} + u_{-1,-1})$$

$$- (u_{1,2} + u_{2,1} + u_{1,-2} + u_{2,-1} + u_{-1,2} + u_{-2,1}$$

$$+ u_{-1,-2} + u_{-2,-1})] + O(h^4)$$

25.4. Integration

Trapezoidal Rule

25.4.1

$$\int_{x_0}^{x_1} f(x)\,dx = \frac{h}{2}(f_0 + f_1) - \frac{1}{2}\int_{x_0}^{x_1}(t - x_0)(x_1 - t)f''(t)\,dt$$

$$= \frac{h}{2}(f_0 + f_1) - \frac{h^3}{12}f''(\xi) \qquad (x_0 < \xi < x_1)$$

Extended Trapezoidal Rule

25.4.2

$$\int_{x_0}^{x_m} f(x)\,dx = h\left[\frac{f_0}{2} + f_1 + \ldots + f_{m-1} + \frac{f_m}{2}\right]$$

$$- \frac{mh^3}{12}f''(\xi)$$

Error Term in Trapezoidal Formula for Periodic Functions

If $f(x)$ is periodic and has a continuous k^{th} derivative, and if the integral is taken over a period, then

25.4.3
$$|\text{Error}| \leq \frac{\text{constant}}{m^k}$$

Modified Trapezoidal Rule

25.4.4

$$\int_{x_0}^{x_m} f(x)\,dx = h\left[\frac{f_0}{2} + f_1 + \ldots + f_{m-1} + \frac{f_m}{2}\right]$$

$$+ \frac{h}{24}[-f_{-1} + f_1 + f_{m-1} - f_{m+1}] + \frac{11m}{720}h^5 f^{(4)}(\xi)$$

Simpson's Rule

25.4.5

$$\int_{x_0}^{x_2} f(x)dx = \frac{h}{3}\,[f_0+4f_1+f_2]$$

$$+\frac{1}{6}\int_{x_0}^{x_1}(x_0-t)^2(x_1-t)f^{(3)}(t)dt$$

$$+\frac{1}{6}\int_{x_1}^{x_2}(x_2-t)^2(x_1-t)f^{(3)}(t)dt$$

$$=\frac{h}{3}\,[f_0+4f_1+f_2]-\frac{h^5}{90}f^{(4)}(\xi)$$

Extended Simpson's Rule

25.4.6

$$\int_{x_0}^{x_{2n}} f(x)dx = \frac{h}{3}\,[f_0+4(f_1+f_3+\ldots+f_{2n-1})$$

$$+2(f_2+f_4+\ldots+f_{2n-2})+f_{2n}]-\frac{nh^5}{90}f^{(4)}(\xi)$$

Euler-Maclaurin Summation Formula

25.4.7

$$\int_{x_0}^{x_n} f(x)dx = h\left[\frac{f_0}{2}+f_1+f_2+\ldots+f_{n-1}+\frac{f_n}{2}\right]$$

$$-\frac{B_2}{2!}\,h^2(f_n'-f_0')-\ldots-\frac{B_{2k}h^{2k}}{(2k)!}\,[f_n^{(2k-1)}-f_0^{(2k-1)}]+R_{2k}$$

$$R_{2k}=\frac{\theta n B_{2k+2}h^{2k+3}}{(2k+2)!}\max_{x_0\le x\le x_n}|f^{(2k+2)}(x)|,\qquad(-1\le\theta\le1)$$

(For B_{2k}, Bernoulli numbers, see chapter **23**.)

If $f^{(2k+2)}(x)$ and $f^{(2k+4)}(x)$ do not change sign for $x_0<x<x_n$ then $|R_{2k}|$ is less than the first neglected term. If $f^{(2k+2)}(x)$ does not change sign for $x_0<x<x_n$, $|R_{2k}|$ is less than twice the first neglected term.

Lagrange Formula

25.4.8

$$\int_a^b f(x)dx = \sum_{i=0}^{n}(L_i^{(n)}(b)-L_i^{(n)}(a))f_i+R_n$$

(See **25.2.1**.)

25.4.9

$$L_i^{(n)}(x)=\frac{1}{\pi_n'(x_i)}\int_{x_0}^{x}\frac{\pi_n(t)}{t-x_i}\,dt=\int_{x_0}^{x}l_i(t)dt$$

25.4.10 $\quad R_n=\dfrac{1}{(n+1)!}\displaystyle\int_a^b \pi_n(x)f^{(n+1)}(\xi(x))dx$

Equally Spaced Abscissas

25.4.11

$$\int_{x_0}^{x_k} f(x)dx = \frac{1}{h^n}\sum_{i=0}^{n}f_i\frac{(-1)^{n-i}}{i!(n-i)!}\int_{x_0}^{x_k}\frac{\pi_n(x)}{x-x_i}\,dx+R_n$$

25.4.12 $\quad\displaystyle\int_{x_m}^{x_{m+1}} f(x)dx = h\sum_{i=-\left[\frac{n-1}{2}\right]}^{\left[\frac{n}{2}\right]} A_i(m)f_i+R_n$ *

(See **Table 25.3** for $A_i(m)$.)

Newton-Cotes Formulas (Closed Type)

(For Trapezoidal and Simpson's Rules see **25.4.1–25.4.6**.)

25.4.13   (Simpson's $\frac{3}{8}$ rule)

$$\int_{x_0}^{x_3} f(x)dx = \frac{3h}{8}\,(f_0+3f_1+3f_2+f_3)-\frac{3f^{(4)}(\xi)h^5}{80}$$

25.4.14   (Bode's rule)

$$\int_{x_0}^{x_4} f(x)dx = \frac{2h}{45}\,(7f_0+32f_1+12f_2$$

$$+32f_3+7f_4)-\frac{8f^{(6)}(\xi)h^7}{945}$$

25.4.15

$$\int_{x_0}^{x_5} f(x)dx = \frac{5h}{288}\,(19f_0+75f_1+50f_2+50f_3$$

$$+75f_4+19f_5)-\frac{275f^{(6)}(\xi)h^7}{12096}$$

25.4.16

$$\int_{x_0}^{x_6} f(x)dx = \frac{h}{140}\,(41f_0+216f_1+27f_2+272f_3$$

$$+27f_4+216f_5+41f_6)-\frac{9f^{(8)}(\xi)h^9}{1400}$$

25.4.17

$$\int_{x_0}^{x_7} f(x)dx = \frac{7h}{17280}\,(751f_0+3577f_1+1323f_2$$

$$+2989f_3+2989f_4+1323f_5+3577f_6$$

$$+751f_7)-\frac{8183f^{(8)}(\xi)h^9}{518400}$$

25.4.18

$$\int_{x_0}^{x_8} f(x)dx = \frac{4h}{14175}\,(989f_0+5888f_1-928f_2$$

$$+10496f_3-4540f_4+10496f_5-928f_6+5888f_7$$

$$+989f_8)-\frac{2368}{467775}f^{(10)}(\xi)h^{11}$$

25.4.19

$$\int_{x_0}^{x_9} f(x)dx = \frac{9h}{89600}\,\{2857(f_0+f_9)$$

$$+15741(f_1+f_8)+1080(f_2+f_7)+19344(f_3+f_6)$$

$$+5778(f_4+f_5)\}-\frac{173}{14620}f^{(10)}(\xi)h^{11}$$

25.4.20

$$\int_{x_0}^{x_{10}} f(x)dx = \frac{5h}{299376} \{16067(f_0+f_{10})$$

$$+106300(f_1+f_9)-48525(f_2+f_8)+272400(f_3+f_7)$$

$$-260550(f_4+f_6)+427368f_5\}$$

$$-\frac{1346350}{326918592} f^{(12)}(\xi)h^{13}$$

Newton-Cotes Formulas (Open Type)

25.4.21

$$\int_{x_0}^{x_3} f(x)dx = \frac{3h}{2}(f_1+f_2) + \frac{f^{(2)}(\xi)h^3}{4}$$

25.4.22

$$\int_{x_0}^{x_4} f(x)dx = \frac{4h}{3}(2f_1-f_2+2f_3) + \frac{28f^{(4)}(\xi)h^5}{90}$$

25.4.23

$$\int_{x_0}^{x_5} f(x)dx = \frac{5h}{24}(11f_1+f_2+f_3+11f_4) + \frac{95f^{(4)}(\xi)h^5}{144}$$

25.4.24

$$\int_{x_0}^{x_6} f(x)dx = \frac{6h}{20}(11f_1-14f_2+26f_3-14f_4+11f_5)$$

$$+\frac{41f^{(6)}(\xi)h^7}{140}$$

25.4.25

$$\int_{x_0}^{x_7} f(x)dx = \frac{7h}{1440}(611f_1-453f_2+562f_3+562f_4$$

$$-453f_5+611f_6) + \frac{5257}{8640} f^{(6)}(\xi)h^7$$

25.4.26

$$\int_{x_0}^{x_8} f(x)dx = \frac{8h}{945}(460f_1-954f_2+2196f_3-2459f_4$$

$$+2196f_5-954f_6+460f_7) + \frac{3956}{14175} f^{(8)}(\xi)h^9$$

Five Point Rule for Analytic Functions

25.4.27

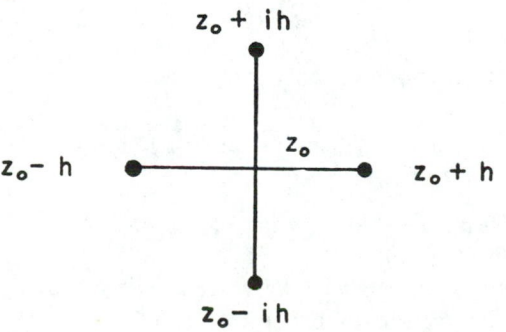

$$\int_{z_0-h}^{z_0+h} f(z)dz = \frac{h}{15}\{24f(z_0)+4[f(z_0+h)+f(z_0-h)]$$

$$-[f(z_0+ih)+f(z_0-ih)]\}+R$$

$|R| \leq \frac{|h|^7}{1890} \operatorname*{Max}_{z\epsilon S} |f^{(6)}(z)|$, S designates the square with vertices $z_0+i^kh(k=0,1,2,3)$; h can be complex.

Chebyshev's Equal Weight Integration Formula

25.4.28 $\qquad \int_{-1}^1 f(x)dx = \frac{2}{n}\sum_{i=1}^n f(x_i) + R_n$

Abscissas: x_i is the ith zero of the polynomial part of

$$x^n \exp\left[\frac{-n}{2\cdot 3x^2} - \frac{n}{4\cdot 5x^3} - \frac{n}{6\cdot 7x^4} - \cdots\right]$$

(See **Table 25.5** for x_i.)

For $n=8$ and $n\geq 10$ some of the zeros are complex.
Remainder:

$$R_n = \int_{-1}^{+1} \frac{x^{n+1}}{(n+1)!} f^{(n+1)}(\xi)dx$$

$$-\frac{2}{n(n+1)!}\sum_{i=1}^n x_i^{n+1}f^{(n+1)}(\xi_i)$$

where $\xi=\xi(x)$ satisfies $0\leq\xi\leq x$ and $0\leq\xi_i\leq x_i$

$$(i=1,\ldots,n)$$

Integration Formulas of Gaussian Type

(For Orthogonal Polynomials see chapter **22**)

Gauss' Formula

25.4.29 $\qquad \int_{-1}^1 f(x)dx = \sum_{i=1}^n w_i f(x_i) + R_n$

Related orthogonal polynomials: Legendre polynomials $P_n(x)$, $P_n(1)=1$

Abscissas: x_i is the ith zero of $P_n(x)$

Weights: $w_i = 2/(1-x_i^2)[P_n'(x_i)]^2$ *
(See **Table 25.4** for x_i and w_i.)

$$R_n = \frac{2^{2n+1}(n!)^4}{(2n+1)[(2n)!]^3} f^{(2n)}(\xi) \qquad (-1<\xi<1)$$

Gauss' Formula, Arbitrary Interval

25.4.30 $\qquad \int_a^b f(y)dy = \frac{b-a}{2}\sum_{i=1}^n w_i f(y_i) + R_n$

$$y_i = \left(\frac{b-a}{2}\right)x_i + \left(\frac{b+a}{2}\right)$$

Related orthogonal polynomials: $P_n(x)$, $P_n(1)=1$
Abscissas: x_i is the ith zero of $P_n(x)$

* Weights: $w_i = 2/(1-x_i^2) [P_n'(x_i)]^2$

$$R_n = \frac{(b-a)^{2n+1}(n!)^4}{(2n+1)[(2n)!]^3} f^{(2n)}(\xi)$$

Radau's Integration Formula

25.4.31

$$\int_{-1}^{1} f(x)dx = \frac{2}{n^2}f_{-1} + \sum_{i=1}^{n-1} w_i f(x_i) + R_n$$

Related polynomials:

$$\frac{P_{n-1}(x)+P_n(x)}{x+1}$$

Abscissas: x_i is the ith zero of

$$\frac{P_{n-1}(x)+P_n(x)}{x+1}$$

Weights:

$$w_i = \frac{1}{n^2}\frac{1-x_i}{[P_{n-1}(x_i)]^2} = \frac{1}{1-x_i}\frac{1}{[P_{n-1}'(x_i)]^2}$$

Remainder:

$$R_n = \frac{2^{2n-1}\cdot n}{[(2n-1)!]^3}[(n-1)!]^4 f^{(2n-1)}(\xi) \qquad (-1<\xi<1)$$

Lobatto's Integration Formula

25.4.32

$$\int_{-1}^{1} f(x)dx = \frac{2}{n(n-1)}[f(1)+f(-1)]$$
$$+ \sum_{i=2}^{n-1} w_i f(x_i) + R_n$$

Related polynomials: $P_{n-1}'(x)$

Abscissas: x_i is the $(i-1)^{\text{st}}$ zero of $P_{n-1}'(x)$

Weights:

$$w_i = \frac{2}{n(n-1)[P_{n-1}(x_i)]^2} \qquad (x_i \neq \pm 1)$$

(See **Table 25.6** for x_i and w_i.)

Remainder:

$$R_n = \frac{-n(n-1)^3 2^{2n-1}[(n-2)!]^4}{(2n-1)[(2n-2)!]^3} f^{(2n-2)}(\xi)$$
$$(-1<\xi<1)$$

25.4.33

$$\int_0^1 x^k f(x)dx = \sum_{i=1}^{n} w_i f(x_i) + R_n$$

Related orthogonal polynomials:

$$q_n(x) = \sqrt{k+2n+1}\,P_n^{(k,0)}(1-2x)$$

(For the Jacobi polynomials $P_n^{(k,0)}$ see chapter **22**.)

Abscissas:

$$x_i \text{ is the } i^{\text{th}} \text{ zero of } q_n(x)$$

Weights:

$$w_i = \left\{ \sum_{j=0}^{n-1}[q_j(x_i)]^2 \right\}^{-1} \qquad *$$

(See **Table 25.8** for x_i and w_i.)

Remainder:

$$R_n = \frac{f^{(2n)}(\xi)}{(k+2n+1)(2n)!}\left[\frac{n!(k+n)!}{(k+2n)!}\right]^2 \qquad (0<\xi<1)$$

25.4.34

$$\int_0^1 f(x)\sqrt{1-x}\,dx = \sum_{i=1}^{n} w_i f(x_i) + R_n$$

Related orthogonal polynomials:

$$\frac{1}{\sqrt{1-x}}P_{2n+1}(\sqrt{1-x}),\ P_{2n+1}(1)=1$$

Abscissas: $x_i = 1-\xi_i^2$ where ξ_i is the ith positive zero of $P_{2n+1}(x)$.
Weights: $w_i = 2\xi_i^2 w_i^{(2n+1)}$ where $w_i^{(2n+1)}$ are the Gaussian weights of order $2n+1$.
Remainder:

$$R_n = \frac{2^{4n+3}[(2n+1)!]^4}{(2n)!(4n+3)[(4n+2)!]^2} f^{(2n)}(\xi) \qquad (0<\xi<1)$$

25.4.35

$$\int_a^b f(y)\sqrt{b-y}\,dy = (b-a)^{3/2}\sum_{i=1}^{n} w_i f(y_i)$$
$$y_i = a+(b-a)x_i$$

Related orthogonal polynomials:

$$\frac{1}{\sqrt{1-x}}P_{2n+1}(\sqrt{1-x}),\ P_{2n+1}(1)=1$$

Abscissas: $x_i = 1-\xi_i^2$ where ξ_i is the ith positive zero of $P_{2n+1}(x)$.
Weights: $w_i = 2\xi_i^2 w_i^{(2n+1)}$ where $w_i^{(2n+1)}$ are the Gaussian weights of order $2n+1$.

25.4.36
$$\int_0^1 \frac{f(x)}{\sqrt{1-x}}\,dx=\sum_{i=1}^n w_i f(x_i)+R_n$$

Related orthogonal polynomials:

$$P_{2n}\left(\sqrt{1-x}\right),\ P_{2n}(1)=1$$

Abscissas: $x_i=1-\xi_i^2$ where ξ_i is the ith positive zero of $P_{2n}(x)$.
Weights: $w_i=2w_i^{(2n)}$, $w_i^{(2n)}$ are the Gaussian weights of order $2n$.
Remainder:

$$R_n=\frac{2^{4n+1}}{4n+1}\frac{[(2n)!]^3}{[(4n)!]^2}f^{(2n)}(\xi)\qquad(0<\xi<1)$$

25.4.37
$$\int_a^b \frac{f(y)}{\sqrt{b-y}}\,dy=\sqrt{b-a}\sum_{i=1}^n w_i f(y_i)+R_n$$
$$y_i=a+(b-a)x_i$$

Related orthogonal polynomials:

$$P_{2n}(\sqrt{1-x}),\ P_{2n}(1)=1$$

Abscissas:
$x_i=1-\xi_i^2$ where ξ_i is the ith positive zero of $P_{2n}(x)$.

Weights: $w_i=2w_i^{(2n)}$, $w_i^{(2n)}$ are the Gaussian weights of order $2n$.

25.4.38
$$\int_{-1}^{+1}\frac{f(x)}{\sqrt{1-x^2}}\,dx=\sum_{i=1}^n w_i f(x_i)+R_n$$

Related orthogonal polynomials: Chebyshev Polynomials of First Kind

$$T_n(x),\ T_n(1)=\frac{1}{2^{n-1}}$$

Abscissas:

$$x_i=\cos\frac{(2i-1)\pi}{2n}$$

Weights:

$$w_i=\frac{\pi}{n}$$

Remainder:

$$R_n=\frac{\pi}{(2n)!2^{2n-1}}f^{(2n)}(\xi)\quad(-1<\xi<1)$$

25.4.39
$$\int_a^b \frac{f(y)dy}{\sqrt{(y-a)(b-y)}}=\sum_{i=1}^n w_i f(y_i)+R_n$$
$$y_i=\frac{b+a}{2}+\frac{b-a}{2}x_i$$

Related orthogonal polynomials:

$$T_n(x),\ T_n(1)=\frac{1}{2^{n-1}}$$

Abscissas:

$$x_i=\cos\frac{(2i-1)\pi}{2n}$$

Weights:

$$w_i=\frac{\pi}{n}$$

25.4.40
$$\int_{-1}^{+1}f(x)\sqrt{1-x^2}\,dx=\sum_{i=1}^n w_i f(x_i)+R_n$$

Related orthogonal polynomials: Chebyshev Polynomials of Second Kind

$$U_n(x)=\frac{\sin\,[(n+1)\,\text{arccos}\,x]}{\sin\,(\text{arccos}\,x)} \qquad *$$

Abscissas:

$$x_i=\cos\frac{i}{n+1}\,\pi \qquad *$$

Weights:

$$w_i=\frac{\pi}{n+1}\sin^2\frac{i}{n+1}\,\pi \qquad *$$

Remainder:

$$R_n=\frac{\pi}{(2n)!2^{2n+1}}f^{(2n)}(\xi)\qquad(-1<\xi<1)$$

25.4.41
$$\int_a^b \sqrt{(y-a)(b-y)}f(y)dy=\left(\frac{b-a}{2}\right)^2\sum_{i=1}^n w_i f(y_i)+R_n$$
$$y_i=\frac{b+a}{2}+\frac{b-a}{2}\,x_i$$

Related orthogonal polynomials:

$$U_n(x)=\frac{\sin\,[(n+1)\,\text{arccos}\,x]}{\sin\,(\text{arccos}\,x)} \qquad *$$

Abscissas:

$$x_i=\cos\frac{i}{n+1}\,\pi \qquad *$$

Weights:

$$w_i=\frac{\pi}{n+1}\sin^2\frac{i}{n+1}\,\pi \qquad *$$

25.4.42
$$\int_0^1 f(x)\sqrt{\frac{x}{1-x}}\,dx=\sum_{i=1}^n w_i f(x_i)+R_n$$

Related orthogonal polynomials:

$$\frac{1}{\sqrt{x}}\,T_{2n+1}(\sqrt{x})$$

Abscissas:

$$x_i=\cos^2\frac{2i-1}{2n+1}\cdot\frac{\pi}{2}$$

Weights:

$$w_i=\frac{2\pi}{2n+1}\,x_i$$

Remainder:

$$R_n = \frac{\pi}{(2n)!2^{4n+1}} f^{(2n)}(\xi) \qquad (0 < \xi < 1)$$

25.4.43

$$\int_a^b f(x) \sqrt{\frac{x-a}{b-x}}\, dx = (b-a) \sum_{i=1}^n w_i f(y_i) + R_n$$
$$y_i = a + (b-a)x_i$$

Related orthogonal polynomials:

$$\frac{1}{\sqrt{x}}\, T_{2n+1}(\sqrt{x})$$

Abscissas:

$$x_i = \cos^2 \frac{2i-1}{2n+1} \cdot \frac{\pi}{2}$$

Weights:

$$w_i = \frac{2\pi}{2n+1}\, x_i$$

25.4.44 $\quad \int_0^1 \ln x\, f(x) dx = \sum_{i=1}^n w_i f(x_i) + R_n$

Related orthogonal polynomials: polynomials orthogonal with respect to the weight function $-\ln x$
Abscissas: See **Table 25.7**
Weights: See **Table 25.7**

25.4.45

$$\int_0^\infty e^{-x} f(x) dx = \sum_{i=1}^n w_i f(x_i) + R_n$$

Related orthogonal polynomials: Laguerre polynomials $L_n(x)$.
Abscissas: x_i is the ith zero of $L_n(x)$
Weights:

$$w_i = \frac{(n!)^2 x_i}{(n+1)^2 [L_{n+1}(x_i)]^2}$$

(See **Table 25.9** for x_i and w_i.)
Remainder:

$$R_n = \frac{(n!)^2}{(2n)!} f^{(2n)}(\xi) \qquad (0 < \xi < \infty)$$

25.4.46

$$\int_{-\infty}^\infty e^{-x^2} f(x) dx = \sum_{i=1}^n w_i f(x_i) + R_n$$

Related orthogonal polynomials: Hermite polynomials $H_n(x)$.
Abscissas: x_i is the ith zero of $H_n(x)$
Weights:

$$\frac{2^{n-1} n! \sqrt{\pi}}{n^2 [H_{n-1}(x_i)]^2}$$

(See **Table 25.10** for x_i and w_i.)

Remainder:

$$R_n = \frac{n! \sqrt{\pi}}{2^n (2n)!} f^{(2n)}(\xi) \qquad (-\infty < \xi < \infty)$$

Filon's Integration Formula [3]

25.4.47

$$\int_{x_0}^{x_{2n}} f(x) \cos tx\, dx = h\Big[\alpha(th)(f_{2n} \sin tx_{2n}$$
$$-f_0 \sin tx_0) + \beta(th)\cdot C_{2n} + \gamma(th)\cdot C_{2n-1}$$
$$+ \frac{2}{45}\, th^4 S'_{2n-1}\Big] - R_n$$

25.4.48

$$C_{2n} = \sum_{i=0}^n f_{2i} \cos (tx_{2i}) - \tfrac{1}{2}[f_{2n} \cos tx_{2n} + f_0 \cos tx_0]$$

25.4.49

$$C_{2n-1} = \sum_{i=1}^n f_{2i-1} \cos tx_{2i-1}$$

25.4.50

$$S'_{2n-1} = \sum_{i=1}^n f^{(3)}_{2i-1} \sin tx_{2i-1}$$

25.4.51

$$R_n = \frac{1}{90} nh^5 f^{(4)}(\xi) + O(th^7)$$

25.4.52

$$\alpha(\theta) = \frac{1}{\theta} + \frac{\sin 2\theta}{2\theta^2} - \frac{2 \sin^2 \theta}{\theta^3}$$
$$\beta(\theta) = 2\left(\frac{1 + \cos^2 \theta}{\theta^2} - \frac{\sin 2\theta}{\theta^3} \right)$$
$$\gamma(\theta) = 4\left(\frac{\sin \theta}{\theta^3} - \frac{\cos \theta}{\theta^2} \right)$$

For small θ we have

25.4.53

$$\alpha = \frac{2\theta^3}{45} - \frac{2\theta^5}{315} + \frac{2\theta^7}{4725} - \cdots$$
$$\beta = \frac{2}{3} + \frac{2\theta^2}{15} - \frac{4\theta^4}{105} + \frac{2\theta^6}{567} - \cdots$$
$$\gamma = \frac{4}{3} - \frac{2\theta^2}{15} + \frac{\theta^4}{210} - \frac{\theta^6}{11340} + \cdots$$

25.4.54

$$\int_{x_0}^{x_{2n}} f(x) \sin tx\, dx = h\Big[\alpha(th)(f_0 \cos tx_0 - f_{2n} \cos tx_{2n})$$
$$+ \beta S_{2n} + \gamma S_{2n-1} + \frac{2}{45}\, th^4 C'_{2n-1}\Big] - R_n$$

25.4.55

$$S_{2n} = \sum_{i=0}^n f_{2i} \sin (tx_{2i}) - \frac{1}{2}[f_{2n} \sin (tx_{2n}) + f_0 \sin (tx_0)]$$

---

[3] For certain difficulties associated with this formula, see the article by J. W. Tukey, p. 400, "On Numerical Approximation," Ed. R. E. Langer, Madison, 1959.

25.4.56 $\quad S_{2n-1}=\sum_{i=1}^{n} f_{2i-1}\sin\,(tx_{2i-1})$

25.4.57 $\quad C'_{2n-1}=\sum_{i=1}^{n} f^{(3)}_{2i-1}\cos\,(tx_{2i-1})$

(See **Table 25.11** for $\alpha,\,\beta,\,\gamma$.)

Iterated Integrals

25.4.58

$$\int_0^x dt_n \int_0^{t_n} dt_{n-1}\ldots\int_0^{t_3} dt_2 \int_0^{t_2} f(t_1)dt_1$$
$$=\frac{1}{(n-1)!}\int_0^x (x-t)^{n-1}f(t)dt$$

25.4.59

$$\int_a^x dt_n \int_a^{t_n} dt_{n-1}\ldots\int_a^{t_3} dt_2 \int_a^{t_2} f(t_1)dt_1$$
$$=\frac{(x-a)^n}{(n-1)!}\int_0^1 t^{n-1}f(x-(x-a)t)dt$$

Multidimensional Integration

Circumference of Circle Γ: $x^2+y^2=h^2$.

25.4.60

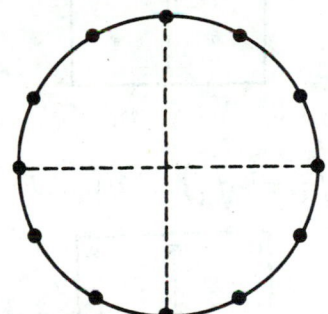

$$\frac{1}{2\pi h}\int_\Gamma f(x,y)ds=\frac{1}{2m}\sum_{n=1}^{2m} f\left(h\cos\frac{\pi n}{m},\,h\sin\frac{\pi n}{m}\right)$$
$$+O(h^{2m-2})$$

Circle C: $x^2+y^2\le h^2$.

25.4.61

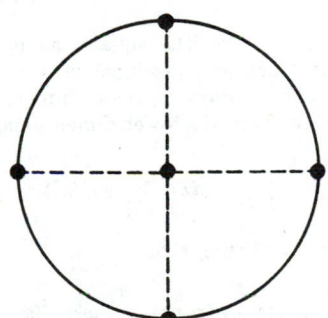

$$\frac{1}{\pi h^2}\iint_C f(x,y)dxdy=\sum_{i=1}^{n} w_i f(x_i,y_i)+R$$

(x_i,y_i)	w_i	
$(0,0)$	$1/2$	$R=O(h^4)$
$(\pm h,0),\,(0,\pm h)$	$1/8$	

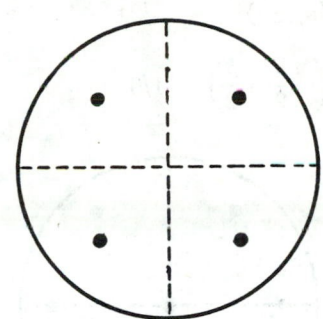

(x_i,y_i)	w_i	
$\left(\pm\dfrac{h}{2},\pm\dfrac{h}{2}\right)$	$1/4$	$R=O(h^4)$

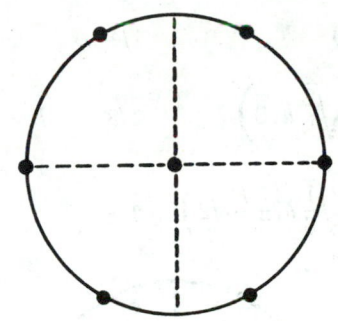

(x_i,y_i)	w_i	
$(0,0)$	$1/2$	
$(\pm h,0)$	$1/12$	$R=O(h^4)$
$\left(\pm\dfrac{h}{2},\pm\dfrac{h}{2}\sqrt{3}\right)$	$1/12$	

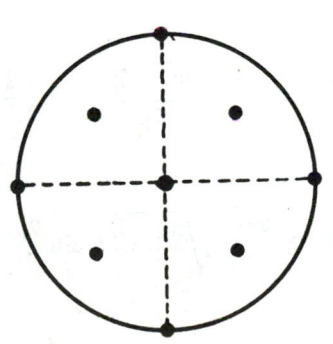

(x_i, y_i)	w_i	
$(0,0)$	$1/6$	
$(\pm h, 0)$	$1/24$	$R = O(h^6)$
$(0, \pm h)$	$1/24$	
$\left(\pm\dfrac{h}{2}, \pm\dfrac{h}{2}\right)$	$1/6$	

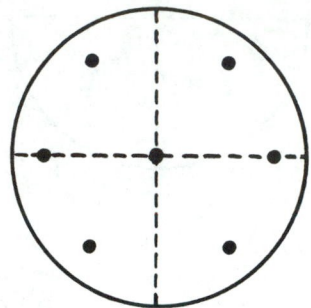

(x_i, y_i)	w_i	
$(0,0)$	$1/4$	
$\left(\pm\sqrt{\dfrac{2}{3}}\,h, 0\right)$	$1/8$	$R = O(h^6)$
$\left(\pm\sqrt{\dfrac{1}{6}}\,h, \pm\dfrac{h}{2}\sqrt{2}\right)$	$1/8$	

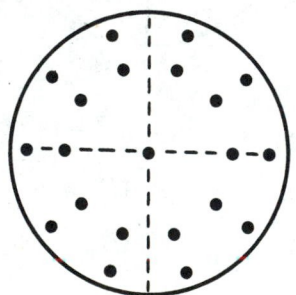

(x_i, y_i)	w_i
$(0,0)$	$1/9$
$\left(\sqrt{\dfrac{6-\sqrt{6}}{10}}\,h\,\cos\dfrac{2\pi k}{10},\ \sqrt{\dfrac{6-\sqrt{6}}{10}}\,h\,\sin\dfrac{2\pi k}{10}\right)$	$\dfrac{16+\sqrt{6}}{360}$
	$(k=1,\ldots,10)$
$\left(\sqrt{\dfrac{6+\sqrt{6}}{10}}\,h\,\cos\dfrac{2\pi k}{10},\ \sqrt{\dfrac{6+\sqrt{6}}{10}}\,h\,\sin\dfrac{2\pi k}{10}\right)$	$\dfrac{16-\sqrt{6}}{360}$
	$R = O(h^{10})$

Square[4] S: $|x| \le h, |y| \le h$

25.4.62

$$\frac{1}{4h^2}\iint_S f(x,y)\,dx\,dy = \sum_{i=1}^{n} w_i f(x_i, y_i) + R$$

(x_i, y_i)	w_i	
$(0,0)$	$4/9$	
$(\pm h, \pm h)$	$1/36$	$R = O(h^4)$
$(\pm h, 0)$	$1/9$	
$(0, \pm h)$	$1/9$	

(x_i, y_i)	w_i	
$\left(\pm h\sqrt{\dfrac{1}{3}},\ \pm h\sqrt{\dfrac{1}{3}}\right)$	$1/4$	$R = O(h^4)$

(x_i, y_i)	w_i
$(0,0)$	$16/81$

---

[4] For regions, such as the square, cube, cylinder, etc., which are the Cartesian products of lower dimensional regions, one may always develop integration rules by "multiplying together" the lower dimensional rules. Thus if

$$\int_0^1 f(x)\,dx \approx \sum_{i=1}^{n} w_i f(x_i)$$

is a one dimensional rule, then

$$\int_0^1\int_0^1 f(x,y)\,dx\,dy \approx \sum_{i,j=1}^{n} w_i w_j f(x_i, x_j)$$

becomes a two dimensional rule. Such rules are not necessarily the most "economical".

$$\left(\pm\sqrt{\tfrac{3}{5}}\,h,\pm\sqrt{\tfrac{3}{5}}\,h\right) \qquad 25/324$$

$$R=O(h^6)$$

$$\left(0,\pm\sqrt{\tfrac{3}{5}}\,h\right) \qquad 10/81$$

$$\left(\pm\sqrt{\tfrac{3}{5}}\,h,0\right) \qquad 10/81$$

Equilateral Triangle T

Radius of Circumscribed Circle $=h$

25.4.63

$$\frac{1}{\tfrac{3}{4}\sqrt{3}h^2}\iint_T f(x,y)\,dxdy=\sum_{i=1}^{n}w_i f(x_i,y_i)+R$$

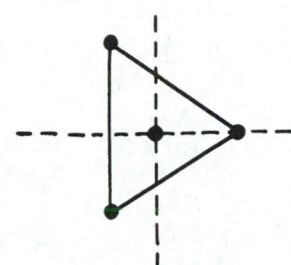

(x_i,y_i)	w_i
$(0,0)$	$3/4$
$(h,0)$	$1/12$
$\left(-\tfrac{h}{2},\pm\tfrac{h}{2}\sqrt{3}\right)$	$1/12$

$$R=O(h^3)$$

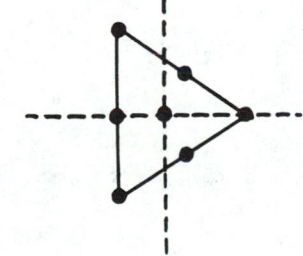

(x_i,y_i)	w_i
$(0,0)$	$27/60$
$(h,0)$	$3/60$
$\left(-\tfrac{h}{2},\pm\tfrac{h}{2}\sqrt{3}\right)$	$3/60$
$\left(-\tfrac{h}{2},0\right)$	$8/60$
$\left(\tfrac{h}{4},\pm\tfrac{h}{4}\sqrt{3}\right)$	$8/60$

$$R=O(h^4)$$

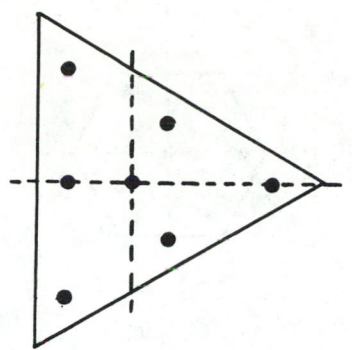

(x_i,y_i)	w_i
$(0,0)$	$270/1200$
$\left(\left(\tfrac{\sqrt{15}+1}{7}\right)h,0\right)$	
$\left(\left(\tfrac{-\sqrt{15}+1}{14}\right)h,\pm\left(\tfrac{\sqrt{15}+1}{14}\right)\sqrt{3}h\right)$	$\dfrac{155-\sqrt{15}}{1200}$
$\left(\left(-\tfrac{\sqrt{15}-1}{7}\right)h,0\right)$	
$\left(\left(\tfrac{\sqrt{15}-1}{14}\right)h,\pm\left(\tfrac{\sqrt{15}-1}{14}\right)\sqrt{3}h\right)$	$\dfrac{155+\sqrt{15}}{1200}$

$$R=O(h^6)$$

Regular Hexagon H

Radius of Circumscribed Circle $=h$

25.4.64

$$\frac{1}{\tfrac{3}{2}\sqrt{3}h^2}\iint_H f(x,y)\,dxdy=\sum_{i=1}^{n}w_i f(x_i,y_i)+R$$

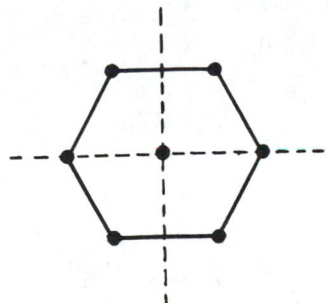

(x_i,y_i)	w_i
$(0,0)$	$21/36$
$\left(\pm\tfrac{h}{2},\pm\tfrac{h}{2}\sqrt{3}\right)$	$5/72$
$(\pm h,0)$	$5/72$

$$R=O(h^4)$$

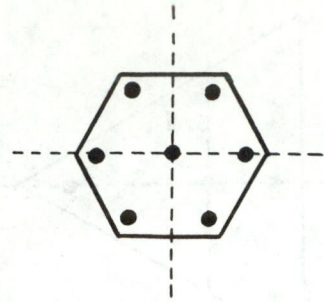

(x_i, y_i)	w_i	
$(0,0)$	$258/1008$	
$\left(\pm\dfrac{h}{10}\sqrt{14}, \pm\dfrac{h}{10}\sqrt{42}\right)$	$125/1008$	$R=O(h^6)$
$\left(\pm h\dfrac{\sqrt{14}}{5}, 0\right)$	$125/1008$	

Surface of Sphere Σ: $x^2+y^2+z^2=h^2$

25.4.65

$$\frac{1}{4\pi h^2}\int_\Sigma\!\!\int f(x,y,z)d\sigma = \sum_{i=1}^{n} w_i f(x_i,y_i,z_i)+R$$

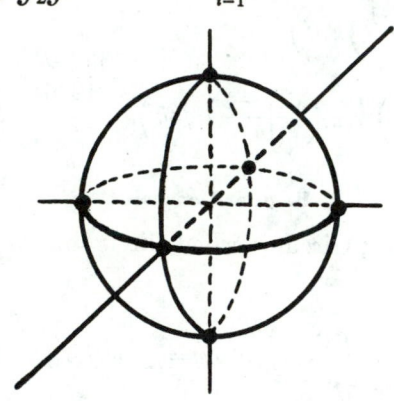

(x_i,y_i,z_i)	w_i	
$(\pm h,0,0)$	$1/6$	
$(0,\pm h,0)$	$1/6$	$R=O(h^4)$
$(0,0,\pm h)$	$1/6$	

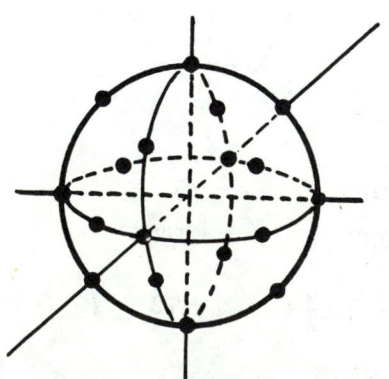

(x_i, y_i, z_i)	w_i	
$\left(\pm\sqrt{\dfrac{1}{2}}h, \pm\sqrt{\dfrac{1}{2}}h, 0\right)$		
$\left(\pm\sqrt{\dfrac{1}{2}}h, 0, \pm\sqrt{\dfrac{1}{2}}h\right)$	$1/15$	
$\left(0, \pm\sqrt{\dfrac{1}{2}}h, \pm\sqrt{\dfrac{1}{2}}h\right)$		$R=O(h^6)$
$(\pm h,0,0)$		
$(0,\pm h,0)$	$1/30$	
$(0,0,\pm h)$		

(x_i, y_i, z_i)	w_i	
$\left(\pm\sqrt{\dfrac{1}{3}}h, \pm\sqrt{\dfrac{1}{3}}h, \pm\sqrt{\dfrac{1}{3}}h\right)$	$27/840$	
$\left(\pm\sqrt{\dfrac{1}{2}}h, \pm\sqrt{\dfrac{1}{2}}h, 0\right)$		
$\left(\pm\sqrt{\dfrac{1}{2}}h, 0, \pm\sqrt{\dfrac{1}{2}}h\right)$	$32/840$	$R=O(h^8)$
$\left(0, \pm\sqrt{\dfrac{1}{2}}h, \pm\sqrt{\dfrac{1}{2}}h\right)$		
$(\pm h,0,0)$		
$(0,\pm h,0)$	$40/840$	
$(0,0,\pm h)$		

Sphere S: $x^2+y^2+z^2 \le h^2$

25.4.66

$$\frac{1}{\frac{4}{3}\pi h^3}\iiint_S f(x,y,z)dxdydz = \sum_{i=1}^{n} w_i f(x_i,y_i,z_i)+R$$

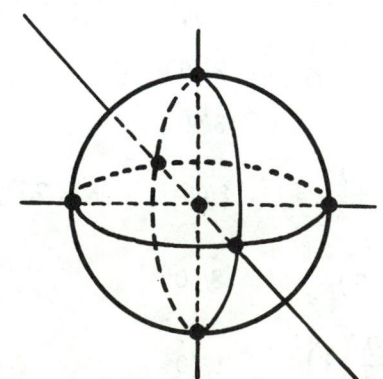

(x_i, y_i, z_i)	w_i
$(0,0,0)$	$2/5$
$(\pm h,0,0)$	$1/10$
$(0,\pm h,0)$	$1/10$
$(0,0,\pm h)$	$1/10$

$$R = O(h^4)$$

$$\text{Cube}^5 \; C: \; |x| \leq h$$
$$|y| \leq h$$
$$|z| \leq h$$

25.4.67

$$\frac{1}{8h^3} \iiint_C f(x,y,z)\,dxdydz = \sum_{i=1}^{n} w_i f(x_i, y_i, z_i) + R$$

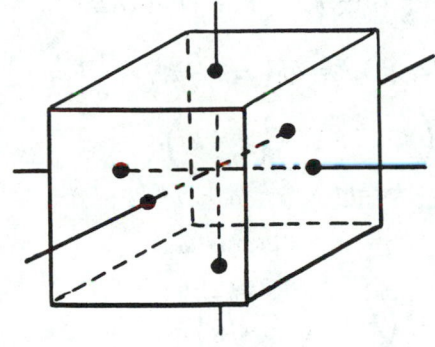

(x_i, y_i, z_i)	w_i
$(\pm h,0,0)$	$1/6$
$(0,\pm h,0)$	$1/6$
$(0,0,\pm h)$	$1/6$

$$R = O(h^4)$$

25.4.68

$$\frac{1}{8h^3} \iiint_C f(x,y,z)\,dxdydz$$

$$= \frac{1}{360}[-496 f_m + 128\sum f_r + 8\sum f_f + 5\sum f_v] + O(h^6)$$

25.4.69

$$= \frac{1}{450}[91\sum f_f - 40\sum f_e + 16\sum f_d] + O(h^6)$$

where $f_m = f(0,0,0)$.

⁵ See footnote to **25.4.62**.

$\sum f_r =$ sum of values of f at the 6 points midway from the center of C to the 6 faces.

$\sum f_f =$ sum of values of f at the 6 centers of the faces of C.

$\sum f_v =$ sum of values of f at the 8 vertices of C.

$\sum f_e =$ sum of values of f at the 12 midpoints of edges of C.

$\sum f_d =$ sum of values of f at the 4 points on the diagonals of each face at a distance of $\frac{1}{2}\sqrt{5}h$ from the center of the face.

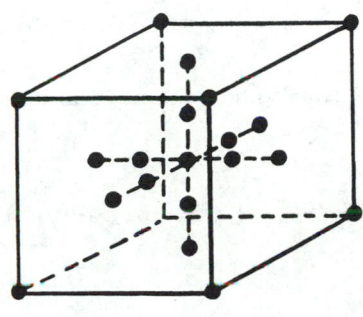

Tetrahedron: $\mathscr{T}$

25.4.70

$$\frac{1}{V}\iiint_{\mathscr{T}} f(x,y,z)\,dxdydz = \frac{1}{40}\sum f_v + \frac{9}{40}\sum f_f$$
$$+ \text{terms of 4}^{th}\text{ order}$$

$$= \frac{32}{60} f_m + \frac{1}{60}\sum f_v + \frac{4}{60}\sum f_e$$
$$+ \text{terms of 4}^{th}\text{ order}$$

where

V: Volume of $\mathscr{T}$

$\sum f_v$: Sum of values of the function at the vertices of $\mathscr{T}$.

$\sum f_e$: Sum of values of the function at midpoints of the edges of $\mathscr{T}$.

$\sum f_f$: Sum of values of the function at the center of gravity of the faces of $\mathscr{T}$.

f_m: Value of function at center of gravity of $\mathscr{T}$.

25.5. Ordinary Differential Equations[6]

First Order: $y'=f(x, y)$

Point Slope Formula

25.5.1 $$y_{n+1}=y_n+hy'_n+O(h^2)$$

25.5.2 $$y_{n+1}=y_{n-1}+2hy'_n+O(h^3)$$

Trapezoidal Formula

25.5.3 $$y_{n+1}=y_n+\frac{h}{2}(y'_{n+1}+y'_n)+O(h^3)$$

Adams' Extrapolation Formula

25.5.4
$$y_{n+1}=y_n+\frac{h}{24}(55y'_n-59y'_{n-1}+37y'_{n-2}-9y'_{n-3})+O(h^5)$$

Adams' Interpolation Formula

25.5.5
$$y_{n+1}=y_n+\frac{h}{24}(9y'_{n+1}+19y'_n-5y'_{n-1}+y'_{n-2})+O(h^5)$$

Runge-Kutta Methods

Second Order

25.5.6
$$y_{n+1}=y_n+\frac{1}{2}(k_1+k_2)+O(h^3)$$
$$k_1=hf(x_n,y_n),k_2=hf(x_n+h,y_n+k_1)$$

25.5.7
$$y_{n+1}=y_n+k_2+O(h^3)$$
$$k_1=hf(x_n,y_n),k_2=hf\left(x_n+\frac{1}{2}h,y_n+\frac{1}{2}k_1\right)$$

Third Order

25.5.8
$$y_{n+1}=y_n+\frac{1}{6}k_1+\frac{2}{3}k_2+\frac{1}{6}k_3+O(h^4)$$
$$k_1=hf(x_n,y_n),k_2=hf\left(x_n+\frac{1}{2}h,y_n+\frac{1}{2}k_1\right)$$
$$k_3=hf(x_n+h,y_n-k_1+2k_2)$$

---

[6] The reader is cautioned against possible instabilities especially in formulas **25.5.2** and **25.5.13**. See, e.g. [25.11], [25.12].

25.5.9
$$y_{n+1}=y_n+\frac{1}{4}k_1+\frac{3}{4}k_3+O(h^4)$$
$$k_1=hf(x_n,y_n),k_2=hf\left(x_n+\frac{1}{3}h,y_n+\frac{1}{3}k_1\right)$$
$$k_3=hf\left(x_n+\frac{2}{3}h,y_n+\frac{2}{3}k_2\right)$$

Fourth Order

25.5.10
$$y_{n+1}=y_n+\frac{1}{6}k_1+\frac{1}{3}k_2+\frac{1}{3}k_3+\frac{1}{6}k_4+O(h^5)$$
$$k_1=hf(x_n,y_n),k_2=hf\left(x_n+\frac{1}{2}h,y_n+\frac{1}{2}k_1\right)$$
$$k_3=hf\left(x_n+\frac{1}{2}h,y_n+\frac{1}{2}k_2\right),k_4=hf(x_n+h,y_n+k_3)$$

25.5.11
$$y_{n+1}=y_n+\frac{1}{8}k_1+\frac{3}{8}k_2+\frac{3}{8}k_3+\frac{1}{8}k_4+O(h^5)$$
$$k_1=hf(x_n,y_n),k_2=hf\left(x_n+\frac{1}{3}h,y_n+\frac{1}{3}k_1\right)$$
$$k_3=hf\left(x_n+\frac{2}{3}h,y_n-\frac{1}{3}k_1+k_2\right),$$
$$k_4=hf(x_n+h,y_n+k_1-k_2+k_3)$$

Gill's Method

25.5.12
$$y_{n+1}=y_n+\frac{1}{6}\left(k_1+2\left(1-\sqrt{\frac{1}{2}}\right)k_2\right.$$
$$\left.+2\left(1+\sqrt{\frac{1}{2}}\right)k_3+k_4\right)+O(h^5)$$
$$k_1=hf(x_n,y_n)$$
$$k_2=hf\left(x_n+\frac{1}{2}h,\ y_n+\frac{1}{2}k_1\right)$$
$$k_3=hf\left(x_n+\frac{1}{2}h,\ y_n+\left(-\frac{1}{2}+\sqrt{\frac{1}{2}}\right)k_1\right.$$
$$\left.+\left(1-\sqrt{\frac{1}{2}}\right)k_2\right)$$
$$k_4=hf\left(x_n+h,y_n-\sqrt{\frac{1}{2}}k_2+\left(1+\sqrt{\frac{1}{2}}\right)k_3\right)$$

Predictor-Corrector Methods

Milne's Methods

25.5.13

P: $$y_{n+1}=y_{n-3}+\frac{4h}{3}(2y'_n-y'_{n-1}+2y'_{n-2})+O(h^5)$$

C: $$y_{n+1}=y_{n-1}+\frac{h}{3}(y'_{n-1}+4y'_n+y'_{n+1})+O(h^5)$$

25.5.14

P: $\quad y_{n+1}=y_{n-5}+\dfrac{3h}{10}\,(11y_n'-14y_{n-1}'$
$$+26y_{n-2}'-14y_{n-3}'+11y_{n-4}')+O(h^7)$$

C: $\quad y_{n+1}=y_{n-3}+\dfrac{2h}{45}\,(7y_{n+1}'+32y_n'$
$$+12y_{n-1}'+32y_{n-2}'+7y_{n-3}')+O(h^7)$$

Formulas Using Higher Derivatives

25.5.15

P: $\quad y_{n+1}=y_{n-2}+3(y_n-y_{n-1})+h^2(y_n''-y_{n-1}'')+O(h^5)$

C: $\quad y_{n+1}=y_n+\dfrac{h}{2}\,(y_{n+1}'+y_n')-\dfrac{h^2}{12}\,(y_{n+1}''-y_n'')+O(h^5)$

25.5.16

P: $\quad y_{n+1}=y_{n-2}+3(y_n-y_{n-1})+\dfrac{h^3}{2}\,(y_n'''+y_{n-1}''')+O(h^7)$

C: $\quad y_{n+1}=y_n+\dfrac{h}{2}\,(y_{n+1}'+y_n')-\dfrac{h^2}{10}\,(y_{n+1}''-y_n'')$
$$+\dfrac{h^3}{120}\,(y_{n+1}'''+y_n''')+O(h^7)$$

Systems of Differential Equations

First Order: $y'=f(x,y,z),\ z'=g(x,y,z)$.

Second Order Runge-Kutta

25.5.17

$$y_{n+1}=y_n+\tfrac{1}{2}\,(k_1+k_2)+O(h^3),$$
$$z_{n+1}=z_n+\tfrac{1}{2}\,(l_1+l_2)+O(h^3)$$

$k_1=hf(x_n,y_n,z_n),\qquad l_1=hg(x_n,y_n,z_n)$

$k_2=hf(x_n+h,y_n+k_1,z_n+l_1),$
$$l_2=hg(x_n+h,y_n+k_1,z_n+l_1)$$

Fourth Order Runge-Kutta

25.5.18

$$y_{n+1}=y_n+\tfrac{1}{6}\,(k_1+2k_2+2k_3+k_4)+O(h^5),$$
$$z_{n+1}=z_n+\tfrac{1}{6}\,(l_1+2l_2+2l_3+l_4)+O(h^5)$$

$k_1=hf(x_n,y_n,z_n)\qquad l_1=hg(x_n,y_n,z_n)$

$k_2=hf\left(x_n+\tfrac{1}{2}\,h,y_n+\tfrac{1}{2}\,k_1,z_n+\tfrac{1}{2}\,l_1\right)$
$$l_2=hg\left(x_n+\tfrac{h}{2},y_n+\tfrac{k_1}{2},z_n+\tfrac{l_1}{2}\right)$$

$k_3=hf\left(x_n+\tfrac{1}{2}\,h,y_n+\tfrac{1}{2}\,k_2,z_n+\tfrac{1}{2}\,l_2\right)$
$$l_3=hg\left(x_n+\tfrac{h}{2},y_n+\tfrac{k_2}{2},z_n+\tfrac{l_2}{2}\right)$$

$k_4=hf(x_n+h,y_n+k_3,z_n+l_3)$
$$l_4=hg(x_n+h,y_n+k_3,z_n+l_3)$$

Second Order: $y''=f(x,y,y')$

Milne's Method

25.5.19

P: $\quad y_{n+1}'=y_{n-3}'+\dfrac{4h}{3}\,(2y_{n-2}''-y_{n-1}''+2y_n'')+O(h^5)$

C: $\quad y_{n+1}'=y_{n-1}'+\dfrac{h}{3}\,(y_{n-1}''+4y_n''+y_{n+1}'')+O(h^5)$

Runge-Kutta Method

25.5.20

$$y_{n+1}=y_n+h\left[y_n'+\tfrac{1}{6}\,(k_1+k_2+k_3)\right]+O(h^5)$$

$$y_{n+1}'=y_n'+\tfrac{1}{6}\,(k_1+2k_2+2k_3+k_4)$$

$k_1=hf(x_n,y_n,y_n')$

$k_2=hf\left(x_n+\tfrac{1}{2}\,h,y_n+\tfrac{h}{2}\,y_n'+\tfrac{h}{8}\,k_1,y_n'+\tfrac{k_1}{2}\right)$

$k_3=hf\left(x_n+\tfrac{1}{2}\,h,y_n+\tfrac{h}{2}\,y_n'+\tfrac{h}{8}\,k_1,y_n'+\tfrac{k_2}{2}\right)$ $\qquad$ *

$k_4=hf\left(x_n+h,y_n+hy_n'+\tfrac{h}{2}\,k_3,y_n'+k_3\right)$

Second Order: $y''=f(x,y)$

Milne's Method

25.5.21

P: $\quad y_{n+1}=y_n+y_{n-2}-y_{n-3}$
$$+\dfrac{h^2}{4}\,(5y_n''+2y_{n-1}''+5y_{n-2}'')+O(h^6)$$

C: $\quad y_n=2y_{n-1}-y_{n-2}+\dfrac{h^2}{12}\,(y_n''+10y_{n-1}''+y_{n-2}'')+O(h^6)$

Runge-Kutta Method

25.5.22 $\quad y_{n+1}=y_n+h\left(y_n'+\tfrac{1}{6}\,(k_1+2k_2)\right)+O(h^4)$

$$y_{n+1}'=y_n'+\tfrac{1}{6}\,k_1+\tfrac{2}{3}\,k_2+\tfrac{1}{6}\,k_3$$

$k_1=hf(x_n,y_n)$

$k_2=hf\left(x_n+\tfrac{h}{2},y_n+\tfrac{h}{2}\,y_n'+\tfrac{h}{8}\,k_1\right)$

$k_3=hf\left(x_n+h,y_n+hy_n'+\tfrac{h}{2}\,k_2\right).$

*See page II.

References

Texts

(For textbooks on numerical analysis, see texts in chapter 3)

[25.1] J. Balbrecht and L. Collatz, Zur numerischen Auswertung mehrdimensionaler Integrale, Z. Angew. Math. Mech. **38**, 1–15 (1958).

[25.2] Berthod-Zaborowski, Le calcul des intégrales de la forme: $\int_0^1 f(x) \log x \, dx$.
H. Mineur, Techniques de calcul numérique, pp. 555–556 (Librairie Polytechnique Ch. Béranger, Paris, France, 1952).

[25.3] W. G. Bickley, Formulae for numerical integration, Math. Gaz. **23**, 352 (1939).

[25.4] W. G. Bickley, Formulae for numerical differentiation, Math. Gaz. **25**, 19–27 (1941).

[25.5] W. G. Bickley, Finite difference formulae for the square lattice, Quart. J. Mech. Appl. Math., **1**, 35–42 (1948).

[25.6] G. Birkhoff and D. Young, Numerical quadrature of analytic and harmonic functions, J. Math. Phys. **29**, 217–221 (1950).

[25.7] L. Fox, The use and construction of mathematical tables, Mathematical Tables vol. I, National Physical Laboratory (Her Majesty's Stationery Office, London, England, 1956).

[25.8] S. Gill, Process for the step-by-step integration of differential equations in an automatic digital computing machine, Proc. Cambridge Philos. Soc. **47**, 96–108 (1951).

[25.9] P. C. Hammer and A. H. Stroud, Numerical evaluation of multiple integrals II, Math. Tables Aids Comp. **12**, 272–280 (1958).

[25.10] P. C. Hammer and A. W. Wymore, Numerical evaluation of multiple integrals I, Math. Tables Aids Comp. **11**, 59–67 (1957).

[25.11] P. Henrici, Discrete variable methods in ordinary differential equations (John Wiley & Sons, Inc., New York, N. Y., 1961).

[25.12] F. B. Hildebrand, Introduction to numerical analysis (McGraw-Hill Book Co., Inc., New York, N.Y., 1956).

[25.13] Z. Kopal, Numerical analysis (John Wiley & Sons, Inc., New York, N.Y., 1955).

[25.14] A. A. Markoff, Differenzenrechnung (B. G. Teubner, Leipzig, Germany, 1896).

[25.15] S. E. Mikeladze, Quadrature formulas for a regular function, Soobšč. Akad. Nauk Gruzin. SSR. **17**, 289–296 (1956).

[25.16] W. E. Milne, A note on the numerical integration of differential equations, J. Research NBS **43**, 537–542 (1949) RP2046.

[25.17] D. J. Panov, Formelsammlung zur numerischen Behandlung partieller Differentialgleichungen nach dem Differenzenverfahren (Akad. Verlag, Berlin, Germany, 1955).

[25.18] R. Radau, Études sur les formules d'approximation qui servent à calculer la valeur d'une intégrale définie, J. Math. Pures Appl. (3) **6**, 283–336 (1880).

[25.19] R. D. Richtmeyer, Difference methods for initial-value problems (Interscience Publishers, New York, N.Y., 1957).

[25.20] M. Sadowsky, A formula for approximate computation of a triple integral, Amer. Math. Monthly **47**, 539–543 (1940).

[25.21] H. E. Salzer, A new formula for inverse interpolation, Bull. Amer. Math. Soc. **50**, 513–516 (1944).

[25.22] H. E. Salzer, Formulas for complex Cartesian interpolation of higher degree, J. Math. Phys. **28**, 200–203 (1949).

[25.23] H. E. Salzer, Formulas for numerical integration of first and second order differential equations in the complex plane, J. Math. Phys. **29**, 207–216 (1950).

[25.24] H. E. Salzer, Formulas for numerical differentiation in the complex plane, J. Math. Phys. **31**, 155–169 (1952).

[25.25] A. Sard, Integral representations of remainders, Duke Math. J. **15**, 333–345 (1948).

[25.26] A. Sard, Remainders: functions of several variables, Acta Math. **84**, 319–346 (1951).

[25.27] G. Schulz, Formelsammlung zur praktischen Mathematik (DeGruyter and Co., Berlin, Germany, 1945).

[25.28] A. H. Stroud, A bibliography on approximate integration, Math. Comp. **15**, 52–80 (1961).

[25.29] G. J. Tranter, Integral transforms in mathematical physics (John Wiley & Sons, Inc., New York, N.Y., 1951).

[25.30] G. W. Tyler, Numerical integration with several variables, Canad. J. Math. **5**, 393–412 (1953).

Tables

[25.31] L. J. Comrie, Chambers' six-figure mathematical tables, vol. 2 (W. R. Chambers, Ltd., London, England, 1949).

[25.32] P. Davis and P. Rabinowitz, Abscissas and weights for Gaussian quadratures of high order, J. Research NBS **56**, 35–37 (1956) RP2645.

[25.33] P. Davis and P. Rabinowitz, Additional abscissas and weights for Gaussian quadratures of high order: Values for $n = 64$, 80, and 96, J. Research NBS **60**, 613–614 (1958) RP2875.

[25.34] E. W. Dijkstra and A. van Wijngaarden, Table of Everett's interpolation coefficients (Elcelsior's Photo-offset, The Hague, Holland, 1955).

[25.35] H. Fishman, Numerical integration constants, Math. Tables Aids Comp. **11**, 1–9 (1957).

[25.36] H. J. Gawlik, Zeros of Legendre polynomials of orders 2–64 and weight coefficients of Gauss quadrature formulae, A.R.D.E. Memo (B) 77/58, Fort Halstead, Kent, England (1958).

[25.37] Gt. Britain H.M. Nautical Almanac Office, Interpolation and allied tables (Her Majesty's Stationery Office, London, England, 1956).

[25.38] I. M. Longman, Tables for the rapid and accurate numerical evaluation of certain infinite integrals involving Bessel functions, Math. Tables Aids Comp. **11**, 166–180 (1957).

[25.39] A. N. Lowan, N. Davids, and A. Levenson, Table of the zeros of the Legendre polynomials of order 1–16 and the weight coefficients for Gauss' mechanical quadrature formula, Bull. Amer. Math. Soc.**48**, 739–743 (1942).

[25.40] National Bureau of Standards, Tables of Lagrangian interpolation coefficients (Columbia Univ. Press, New York, N.Y., 1944).

[25.41] National Bureau of Standards, Collected Short Tables of the Computation Laboratory, Tables of functions and of zeros of functions, Applied Math. Series 37 (U.S. Government Printing Office, Washington, D.C., 1954).

[25.42] P. Rabinowitz, Abscissas and weights for Lobatto quadrature of high order, Math. Tables Aids Comp. **69**, 47–52 (1960).

[25.43] P. Rabinowitz and G. Weiss, Tables of abscissas and weights for numerical evaluation of integrals of the form

$$\int_0^\infty e^{-x} x^n f(x)\,dx,$$

Math. Tables Aids Comp. **68**, 285–294 (1959).

[25.44] H. E. Salzer, Tables for facilitating the use of Chebyshev's quadrature formula, J. Math. Phys. **26**, 191–194 (1947).

[25.45] H. E. Salzer and R. Zucker, Table of the zeros and weight factors of the first fifteen Laguerre polynomials, Bull. Amer. Math. Soc. **55**, 1004–1012 (1949).

[25.46] H. E. Salzer, R. Zucker, and R. Capuano, Table of the zeros and weight factors of the first twenty Hermite polynomials, J. Research NBS **48**, 111–116 (1952) RP2294.

[25.47] H. E. Salzer, Table of coefficients for obtaining the first derivative without differences, NBS Applied Math. Series 2 (U.S. Government Printing Office, Washington, D.C., 1948).

[25.48] H. E. Salzer, Coefficients for facilitating trigonometric interpolation, J. Math. Phys. **27**, 274–278 (1949).

[25.49] H. E. Salzer and P. T. Roberson, Table of coefficients for obtaining the second derivative without differences, Convair-Astronautics, San Diego, Calif. (1957).

[25.50] H. E. Salzer, Tables of osculatory interpolation coefficients, NBS Applied Math. Series 56 (U.S. Government Printing Office, Washington, D.C., 1958).

Table 25.1   THREE-POINT LAGRANGIAN INTERPOLATION COEFFICIENTS

$$A_k^3(p) = (-1)^{k+1} \frac{p(p^2-1)}{(1+k)!(1-k)!(p-k)}$$

p	A_{-1}	A_0	A_1	p	A_{-1}	A_0	A_1
0.00	−0.00000	1.00000	0.00000	0.50	−0.12500	0.75000	0.37500
0.01	−0.00495	0.99990	0.00505	0.51	−0.12495	0.73990	0.38505
0.02	−0.00980	0.99960	0.01020	0.52	−0.12480	0.72960	0.39520
0.03	−0.01455	0.99910	0.01545	0.53	−0.12455	0.71910	0.40545
0.04	−0.01920	0.99840	0.02080	0.54	−0.12420	0.70840	0.41580
0.05	−0.02375	0.99750	0.02625	0.55	−0.12375	0.69750	0.42625
0.06	−0.02820	0.99640	0.03180	0.56	−0.12320	0.68640	0.43680
0.07	−0.03255	0.99510	0.03745	0.57	−0.12255	0.67510	0.44745
0.08	−0.03680	0.99360	0.04320	0.58	−0.12180	0.66360	0.45820
0.09	−0.04095	0.99190	0.04905	0.59	−0.12095	0.65190	0.46905
0.10	−0.04500	0.99000	0.05500	0.60	−0.12000	0.64000	0.48000
0.11	−0.04895	0.98790	0.06105	0.61	−0.11895	0.62790	0.49105
0.12	−0.05280	0.98560	0.06720	0.62	−0.11780	0.61560	0.50220
0.13	−0.05655	0.98310	0.07345	0.63	−0.11655	0.60310	0.51345
0.14	−0.06020	0.98040	0.07980	0.64	−0.11520	0.59040	0.52480
0.15	−0.06375	0.97750	0.08625	0.65	−0.11375	0.57750	0.53625
0.16	−0.06720	0.97440	0.09280	0.66	−0.11220	0.56440	0.54780
0.17	−0.07055	0.97110	0.09945	0.67	−0.11055	0.55110	0.55945
0.18	−0.07380	0.96760	0.10620	0.68	−0.10880	0.53760	0.57120
0.19	−0.07695	0.96390	0.11305	0.69	−0.10695	0.52390	0.58305
0.20	−0.08000	0.96000	0.12000	0.70	−0.10500	0.51000	0.59500
0.21	−0.08295	0.95590	0.12705	0.71	−0.10295	0.49590	0.60705
0.22	−0.08580	0.95160	0.13420	0.72	−0.10080	0.48160	0.61920
0.23	−0.08855	0.94710	0.14145	0.73	−0.09855	0.46710	0.63145
0.24	−0.09120	0.94240	0.14880	0.74	−0.09620	0.45240	0.64380
0.25	−0.09375	0.93750	0.15625	0.75	−0.09375	0.43750	0.65625
0.26	−0.09620	0.93240	0.16380	0.76	−0.09120	0.42240	0.66880
0.27	−0.09855	0.92710	0.17145	0.77	−0.08855	0.40710	0.68145
0.28	−0.10080	0.92160	0.17920	0.78	−0.08580	0.39160	0.69420
0.29	−0.10295	0.91590	0.18705	0.79	−0.08295	0.37590	0.70705
0.30	−0.10500	0.91000	0.19500	0.80	−0.08000	0.36000	0.72000
0.31	−0.10695	0.90390	0.20305	0.81	−0.07695	0.34390	0.73305
0.32	−0.10880	0.89760	0.21120	0.82	−0.07380	0.32760	0.74620
0.33	−0.11055	0.89110	0.21945	0.83	−0.07055	0.31110	0.75945
0.34	−0.11220	0.88440	0.22780	0.84	−0.06720	0.29440	0.77280
0.35	−0.11375	0.87750	0.23625	0.85	−0.06375	0.27750	0.78625
0.36	−0.11520	0.87040	0.24480	0.86	−0.06020	0.26040	0.79980
0.37	−0.11655	0.86310	0.25345	0.87	−0.05655	0.24310	0.81345
0.38	−0.11780	0.85560	0.26220	0.88	−0.05280	0.22560	0.82720
0.39	−0.11895	0.84790	0.27105	0.89	−0.04895	0.20790	0.84105
0.40	−0.12000	0.84000	0.28000	0.90	−0.04500	0.19000	0.85500
0.41	−0.12095	0.83190	0.28905	0.91	−0.04095	0.17190	0.86905
0.42	−0.12180	0.82360	0.29820	0.92	−0.03680	0.15360	0.88320
0.43	−0.12255	0.81510	0.30745	0.93	−0.03255	0.13510	0.89745
0.44	−0.12320	0.80640	0.31680	0.94	−0.02820	0.11640	0.91180
0.45	−0.12375	0.79750	0.32625	0.95	−0.02375	0.09750	0.92625
0.46	−0.12420	0.78840	0.33580	0.96	−0.01920	0.07840	0.94080
0.47	−0.12455	0.77910	0.34545	0.97	−0.01455	0.05910	0.95545
0.48	−0.12480	0.76960	0.35520	0.98	−0.00980	0.03960	0.97020
0.49	−0.12495	0.75990	0.36505	0.99	−0.00495	0.01990	0.98505
0.50	−0.12500	0.75000	0.37500	1.00	−0.00000	0.00000	1.00000
$-p$	A_1	A_0	A_{-1}	$-p$	A_1	A_0	A_{-1}

See **25.2.6.**

Compiled from National Bureau of Standards, Tables of Lagrangian interpolation coefficients.   Columbia Univ. Press, New York, N.Y., 1944 (with permission).

FOUR-POINT LAGRANGIAN INTERPOLATION COEFFICIENTS   Table 25.1

$$A_k^4(p) = (-1)^{k+2} \frac{p(p^2-1)(p-2)}{(1+k)!(2-k)!(p-k)}$$

p	A_{-1}	A_0	A_1	A_2	
0.00	0.00000 00	1.00000 00	0.00000 00	0.00000 00	1.00
0.01	−0.00328 35	0.99490 05	0.01004 95	−0.00166 65	0.99
0.02	−0.00646 80	0.98960 40	0.02019 60	−0.00333 20	0.98
0.03	−0.00955 45	0.98411 35	0.03043 65	−0.00499 55	0.97
0.04	−0.01254 40	0.97843 20	0.04076 80	−0.00665 60	0.96
0.05	−0.01543 75	0.97256 25	0.05118 75	−0.00831 25	0.95
0.06	−0.01823 60	0.96650 80	0.06169 20	−0.00996 40	0.94
0.07	−0.02094 05	0.96027 15	0.07227 85	−0.01160 95	0.93
0.08	−0.02355 20	0.95385 60	0.08294 40	−0.01324 80	0.92
0.09	−0.02607 15	0.94726 45	0.09368 55	−0.01487 85	0.91
0.10	−0.02850 00	0.94050 00	0.10450 00	−0.01650 00	0.90
0.11	−0.03083 85	0.93356 55	0.11538 45	−0.01811 15	0.89
0.12	−0.03308 80	0.92646 40	0.12633 60	−0.01971 20	0.88
0.13	−0.03524 95	0.91919 85	0.13735 15	−0.02130 05	0.87
0.14	−0.03732 40	0.91177 20	0.14842 80	−0.02287 60	0.86
0.15	−0.03931 25	0.90418 75	0.15956 25	−0.02443 75	0.85
0.16	−0.04121 60	0.89644 80	0.17075 20	−0.02598 40	0.84
0.17	−0.04303 55	0.88855 65	0.18199 35	−0.02751 45	0.83
0.18	−0.04477 20	0.88051 60	0.19328 40	−0.02902 80	0.82
0.19	−0.04642 65	0.87232 95	0.20462 05	−0.03052 35	0.81
0.20	−0.04800 00	0.86400 00	0.21600 00	−0.03200 00	0.80
0.21	−0.04949 35	0.85553 05	0.22741 95	−0.03345 65	0.79
0.22	−0.05090 80	0.84692 40	0.23887 60	−0.03489 20	0.78
0.23	−0.05224 45	0.83818 35	0.25036 65	−0.03630 55	0.77
0.24	−0.05350 40	0.82931 20	0.26188 80	−0.03769 60	0.76
0.25	−0.05468 75	0.82031 25	0.27343 75	−0.03906 25	0.75
0.26	−0.05579 60	0.81118 80	0.28501 20	−0.04040 40	0.74
0.27	−0.05683 05	0.80194 15	0.29660 85	−0.04171 95	0.73
0.28	−0.05779 20	0.79257 60	0.30822 40	−0.04300 80	0.72
0.29	−0.05868 15	0.78309 45	0.31985 55	−0.04426 85	0.71
0.30	−0.05950 00	0.77350 00	0.33150 00	−0.04550 00	0.70
0.31	−0.06024 85	0.76379 55	0.34315 45	−0.04670 15	0.69
0.32	−0.06092 80	0.75398 40	0.35481 60	−0.04787 20	0.68
0.33	−0.06153 95	0.74406 85	0.36648 15	−0.04901 05	0.67
0.34	−0.06208 40	0.73405 20	0.37814 80	−0.05011 60	0.66
0.35	−0.06256 25	0.72393 75	0.38981 25	−0.05118 75	0.65
0.36	−0.06297 60	0.71372 80	0.40147 20	−0.05222 40	0.64
0.37	−0.06332 55	0.70342 65	0.41312 35	−0.05322 45	0.63
0.38	−0.06361 20	0.69303 60	0.42476 40	−0.05418 80	0.62
0.39	−0.06383 65	0.68255 95	0.43639 05	−0.05511 35	0.61
0.40	−0.06400 00	0.67200 00	0.44800 00	−0.05600 00	0.60
0.41	−0.06410 35	0.66136 05	0.45958 95	−0.05684 65	0.59
0.42	−0.06414 80	0.65064 40	0.47115 60	−0.05765 20	0.58
0.43	−0.06413 45	0.63985 35	0.48269 65	−0.05841 55	0.57
0.44	−0.06406 40	0.62899 20	0.49420 80	−0.05913 60	0.56
0.45	−0.06393 75	0.61806 25	0.50568 75	−0.05981 25	0.55
0.46	−0.06375 60	0.60706 80	0.51713 20	−0.06044 40	0.54
0.47	−0.06352 05	0.59601 15	0.52853 85	−0.06102 95	0.53
0.48	−0.06323 20	0.58489 60	0.53990 40	−0.06156 80	0.52
0.49	−0.06289 15	0.57372 45	0.55122 55	−0.06205 85	0.51
0.50	−0.06250 00	0.56250 00	0.56250 00	−0.06250 00	0.50
	A_2	A_1	A_0	A_{-1}	p

Table 25.1   FOUR-POINT LAGRANGIAN INTERPOLATION COEFFICIENTS

$$A_k^4(p) = (-1)^{k+2} \frac{p(p^2-1)(p-2)}{(1+k)!(2-k)!(p-k)}$$

p	A_{-1}	A_0	A_1	A_2	
1.00	0.00000 00	0.00000 00	1.00000 00	0.00000 00	0.00
1.01	0.00166 65	−0.00994 95	1.00489 95	0.00338 35	0.01
1.02	0.00333 20	−0.01979 60	1.00959 60	0.00686 80	0.02
1.03	0.00499 55	−0.02953 65	1.01408 65	0.01045 45	0.03
1.04	0.00665 60	−0.03916 80	1.01836 80	0.01414 40	0.04
1.05	0.00831 25	−0.04868 75	1.02243 75	0.01793 75	0.05
1.06	0.00996 40	−0.05809 20	1.02629 20	0.02183 60	0.06
1.07	0.01160 95	−0.06737 85	1.02992 85	0.02584 05	0.07
1.08	0.01324 80	−0.07654 40	1.03334 40	0.02995 20	0.08
1.09	0.01487 85	−0.08558 55	1.03653 55	0.03417 15	0.09
1.10	0.01650 00	−0.09450 00	1.03950 00	0.03850 00	0.10
1.11	0.01811 15	−0.10328 45	1.04223 45	0.04293 85	0.11
1.12	0.01971 20	−0.11193 60	1.04473 60	0.04748 80	0.12
1.13	0.02130 05	−0.12045 15	1.04700 15	0.05214 95	0.13
1.14	0.02287 60	−0.12882 80	1.04902 80	0.05692 40	0.14
1.15	0.02443 75	−0.13706 25	1.05081 25	0.06181 25	0.15
1.16	0.02598 40	−0.14515 20	1.05235 20	0.06681 60	0.16
1.17	0.02751 45	−0.15309 35	1.05364 35	0.07193 55	0.17
1.18	0.02902 80	−0.16088 40	1.05468 40	0.07717 20	0.18
1.19	0.03052 35	−0.16852 05	1.05547 05	0.08252 65	0.19
1.20	0.03200 00	−0.17600 00	1.05600 00	0.08800 00	0.20
1.21	0.03345 65	−0.18331 95	1.05626 95	0.09359 35	0.21
1.22	0.03489 20	−0.19047 60	1.05627 60	0.09930 80	0.22
1.23	0.03630 55	−0.19746 65	1.05601 65	0.10514 45	0.23
1.24	0.03769 60	−0.20428 80	1.05548 80	0.11110 40	0.24
1.25	0.03906 25	−0.21093 75	1.05468 75	0.11718 75	0.25
1.26	0.04040 40	−0.21741 20	1.05361 20	0.12339 60	0.26
1.27	0.04171 95	−0.22370 85	1.05225 85	0.12973 05	0.27
1.28	0.04300 80	−0.22982 40	1.05062 40	0.13619 20	0.28
1.29	0.04426 85	−0.23575 55	1.04870 55	0.14278 15	0.29
1.30	0.04550 00	−0.24150 00	1.04650 00	0.14950 00	0.30
1.31	0.04670 15	−0.24705 45	1.04400 45	0.15634 85	0.31
1.32	0.04787 20	−0.25241 60	1.04121 60	0.16332 80	0.32
1.33	0.04901 05	−0.25758 15	1.03813 15	0.17043 95	0.33
1.34	0.05011 60	−0.26254 80	1.03474 80	0.17768 40	0.34
1.35	0.05118 75	−0.26731 25	1.03106 25	0.18506 25	0.35
1.36	0.05222 40	−0.27187 20	1.02707 20	0.19257 60	0.36
1.37	0.05322 45	−0.27622 35	1.02277 35	0.20022 55	0.37
1.38	0.05418 80	−0.28036 40	1.01816 40	0.20801 20	0.38
1.39	0.05511 35	−0.28429 05	1.01324 05	0.21593 65	0.39
1.40	0.05600 00	−0.28800 00	1.00800 00	0.22400 00	0.40
1.41	0.05684 65	−0.29148 95	1.00243 95	0.23220 35	0.41
1.42	0.05765 20	−0.29475 60	0.99655 60	0.24054 80	0.42
1.43	0.05841 55	−0.29779 65	0.99034 65	0.24903 45	0.43
1.44	0.05913 60	−0.30060 80	0.98380 80	0.25766 40	0.44
1.45	0.05981 25	−0.30318 75	0.97693 75	0.26643 75	0.45
1.46	0.06044 40	−0.30553 20	0.96973 20	0.27535 60	0.46
1.47	0.06102 95	−0.30763 85	0.96218 85	0.28442 05	0.47
1.48	0.06156 80	−0.30950 40	0.95430 40	0.29363 20	0.48
1.49	0.06205 85	−0.31112 55	0.94607 55	0.30299 15	0.49
1.50	0.06250 00	−0.31250 00	0.93750 00	0.31250 00	0.50
	A_2	A_1	A_0	A_{-1}	$-p$

FOUR-POINT LAGRANGIAN INTERPOLATION COEFFICIENTS　　Table 25.1

$$A_k^4(p) = (-1)^{k+2} \frac{p(p^2-1)(p-2)}{(1+k)!(2-k)!(p-k)}$$

p	A_{-1}	A_0	A_1	A_2	
1.50	0.06250 00	−0.31250 00	0.93750 00	0.31250 00	0.50
1.51	0.06289 15	−0.31362 45	0.92857 45	0.32215 85	0.51
1.52	0.06323 20	−0.31449 60	0.91929 60	0.33196 80	0.52
1.53	0.06352 05	−0.31511 15	0.90966 15	0.34192 95	0.53
1.54	0.06375 60	−0.31546 80	0.89966 80	0.35204 40	0.54
1.55	0.06393 75	−0.31556 25	0.88931 25	0.36231 25	0.55
1.56	0.06406 40	−0.31539 20	0.87859 20	0.37273 60	0.56
1.57	0.06413 45	−0.31495 35	0.86750 35	0.38331 55	0.57
1.58	0.06414 80	−0.31424 40	0.85604 40	0.39405 20	0.58
1.59	0.06410 35	−0.31326 05	0.84421 05	0.40494 65	0.59
1.60	0.06400 00	−0.31200 00	0.83200 00	0.41600 00	0.60
1.61	0.06383 65	−0.31045 95	0.81940 95	0.42721 35	0.61
1.62	0.06361 20	−0.30863 60	0.80643 60	0.43858 80	0.62
1.63	0.06332 55	−0.30652 65	0.79307 65	0.45012 45	0.63
1.64	0.06297 60	−0.30412 80	0.77932 80	0.46182 40	0.64
1.65	0.06256 25	−0.30143 75	0.76518 75	0.47368 75	0.65
1.66	0.06208 40	−0.29845 20	0.75065 20	0.48571 60	0.66
1.67	0.06153 95	−0.29516 85	0.73571 85	0.49791 05	0.67
1.68	0.06092 80	−0.29158 40	0.72038 40	0.51027 20	0.68
1.69	0.06024 85	−0.28769 55	0.70464 55	0.52280 15	0.69
1.70	0.05950 00	−0.28350 00	0.68850 00	0.53550 00	0.70
1.71	0.05868 15	−0.27899 45	0.67194 45	0.54836 85	0.71
1.72	0.05779 20	−0.27417 60	0.65497 60	0.56140 80	0.72
1.73	0.05683 05	−0.26904 15	0.63759 15	0.57461 95	0.73
1.74	0.05579 60	−0.26358 80	0.61978 80	0.58800 40	0.74
1.75	0.05468 75	−0.25781 25	0.60156 25	0.60156 25	0.75
1.76	0.05350 40	−0.25171 20	0.58291 20	0.61529 60	0.76
1.77	0.05224 45	−0.24528 35	0.56383 35	0.62920 55	0.77
1.78	0.05090 80	−0.23852 40	0.54432 40	0.64329 20	0.78
1.79	0.04949 35	−0.23143 05	0.52438 05	0.65755 65	0.79
1.80	0.04800 00	−0.22400 00	0.50400 00	0.67200 00	0.80
1.81	0.04642 65	−0.21622 95	0.48317 95	0.68662 35	0.81
1.82	0.04477 20	−0.20811 60	0.46191 60	0.70142 80	0.82
1.83	0.04303 55	−0.19965 65	0.44020 65	0.71641 45	0.83
1.84	0.04121 60	−0.19084 80	0.41804 80	0.73158 40	0.84
1.85	0.03931 25	−0.18168 75	0.39543 75	0.74693 75	0.85
1.86	0.03732 40	−0.17217 20	0.37237 20	0.76247 60	0.86
1.87	0.03524 95	−0.16229 85	0.34884 85	0.77820 05	0.87
1.88	0.03308 80	−0.15206 40	0.32486 40	0.79411 20	0.88
1.89	0.03083 85	−0.14146 55	0.30041 55	0.81021 15	0.89
1.90	0.02850 00	−0.13050 00	0.27550 00	0.82650 00	0.90
1.91	0.02607 15	−0.11916 45	0.25011 45	0.84297 85	0.91
1.92	0.02355 20	−0.10745 60	0.22425 60	0.85964 80	0.92
1.93	0.02094 05	−0.09537 15	0.19792 15	0.87650 95	0.93
1.94	0.01823 60	−0.08290 80	0.17110 80	0.89356 40	0.94
1.95	0.01543 75	−0.07006 25	0.14381 25	0.91081 25	0.95
1.96	0.01254 40	−0.05683 20	0.11603 20	0.92825 60	0.96
1.97	0.00955 45	−0.04321 35	0.08776 35	0.94589 55	0.97
1.98	0.00646 80	−0.02920 40	0.05900 40	0.96373 20	0.98
1.99	0.00328 35	−0.01480 05	0.02975 05	0.98176 65	0.99
2.00	0.00000 00	0.00000 00	0.00000 00	1.00000 00	1.00
	A_2	A_1	A_0	A_{-1}	$-p$

Table 25.1   **FIVE-POINT LAGRANGIAN INTERPOLATION COEFFICIENTS**

$$A_k^5(p)=(-1)^{k+2}\frac{p(p^2-1)(p^2-4)}{(2+k)!(2-k)!(p-k)}$$

p	A_{-2}	A_{-1}	A_0	A_1	A_2	
0.00	0.00000 00000	0.00000 00000	1.00000 00000	0.00000 00000	0.00000 00000	0.00
0.01	0.00082 90838	−0.00659 98350	0.99987 50025	0.00673 31650	−0.00083 74163	0.01
0.02	0.00164 93400	−0.01306 53600	0.99950 00400	0.01359 86400	−0.00168 26600	0.02
0.03	0.00246 02838	−0.01939 56350	0.99887 52025	0.02059 53650	−0.00253 52163	0.03
0.04	0.00326 14400	−0.02558 97600	0.99800 06400	0.02772 22400	−0.00339 45600	0.04
0.05	0.00405 23438	−0.03164 68750	0.99687 65625	0.03497 81250	−0.00426 01563	0.05
0.06	0.00483 25400	−0.03756 61600	0.99550 32400	0.04236 18400	−0.00513 14600	0.06
0.07	0.00560 15838	−0.04334 68350	0.99388 10025	0.04987 21650	−0.00600 79163	0.07
0.08	0.00635 90400	−0.04898 81600	0.99201 02400	0.05750 78400	−0.00688 89600	0.08
0.09	0.00710 44838	−0.05448 94350	0.98989 14025	0.06526 75650	−0.00777 40163	0.09
0.10	0.00783 75000	−0.05985 00000	0.98752 50000	0.07315 00000	−0.00866 25000	0.10
0.11	0.00855 76838	−0.06506 92350	0.98491 16025	0.08115 37650	−0.00955 38163	0.11
0.12	0.00926 46400	−0.07014 65600	0.98205 18400	0.08927 74400	−0.01044 73600	0.12
0.13	0.00995 79838	−0.07508 14350	0.97894 64025	0.09751 95650	−0.01134 25163	0.13
0.14	0.01063 73400	−0.07987 33600	0.97559 60400	0.10587 86400	−0.01223 86600	0.14
0.15	0.01130 23438	−0.08452 18750	0.97200 15625	0.11435 31250	−0.01313 51563	0.15
0.16	0.01195 26400	−0.08902 65600	0.96816 38400	0.12294 14400	−0.01403 13600	0.16
0.17	0.01258 78838	−0.09338 70350	0.96408 38025	0.13164 19650	−0.01492 66163	0.17
0.18	0.01320 77400	−0.09760 29600	0.95976 24400	0.14045 30400	−0.01582 02600	0.18
0.19	0.01381 18838	−0.10167 40350	0.95520 08025	0.14937 29650	−0.01671 16163	0.19
0.20	0.01440 00000	−0.10560 00000	0.95040 00000	0.15840 00000	−0.01760 00000	0.20
0.21	0.01497 17838	−0.10938 06350	0.94536 12025	0.16753 23650	−0.01848 47163	0.21
0.22	0.01552 69400	−0.11301 57600	0.94008 56400	0.17676 82400	−0.01936 50600	0.22
0.23	0.01606 51838	−0.11650 52350	0.93457 46025	0.18610 57650	−0.02024 03163	0.23
0.24	0.01658 62400	−0.11984 89600	0.92882 94400	0.19554 30400	−0.02110 97600	0.24
0.25	0.01708 98438	−0.12304 68750	0.92285 15625	0.20507 81250	−0.02197 26563	0.25
0.26	0.01757 57400	−0.12609 89600	0.91664 24400	0.21470 90400	−0.02282 82600	0.26
0.27	0.01804 36838	−0.12900 52350	0.91020 36025	0.22443 37650	−0.02367 58163	0.27
0.28	0.01849 34400	−0.13176 57600	0.90353 66400	0.23425 02400	−0.02451 45600	0.28
0.29	0.01892 47838	−0.13438 06350	0.89664 32025	0.24415 63650	−0.02534 37163	0.29
0.30	0.01933 75000	−0.13685 00000	0.88952 50000	0.25415 00000	−0.02616 25000	0.30
0.31	0.01973 13838	−0.13917 40350	0.88218 38025	0.26422 89650	−0.02697 01163	0.31
0.32	0.02010 62400	−0.14135 29600	0.87462 14400	0.27439 10400	−0.02776 57600	0.32
0.33	0.02046 18838	−0.14338 70350	0.86683 98025	0.28463 39650	−0.02854 86163	0.33
0.34	0.02079 81400	−0.14527 65600	0.85884 08400	0.29495 54400	−0.02931 78600	0.34
0.35	0.02111 48438	−0.14702 18750	0.85062 65625	0.30535 31250	−0.03007 26563	0.35
0.36	0.02141 18400	−0.14862 33600	0.84219 90400	0.31582 46400	−0.03081 21600	0.36
0.37	0.02168 89838	−0.15008 14350	0.83356 04025	0.32636 75650	−0.03153 55163	0.37
0.38	0.02194 61400	−0.15139 65600	0.82471 28400	0.33697 94400	−0.03224 18600	0.38
0.39	0.02218 31838	−0.15256 92350	0.81565 86025	0.34765 77650	−0.03293 03163	0.39
0.40	0.02240 00000	−0.15360 00000	0.80640 00000	0.35840 00000	−0.03360 00000	0.40
0.41	0.02259 64838	−0.15448 94350	0.79693 94025	0.36920 35650	−0.03425 00163	0.41
0.42	0.02277 25400	−0.15523 81600	0.78727 92400	0.38006 58400	−0.03487 94600	0.42
0.43	0.02292 80838	−0.15584 68350	0.77742 20025	0.39098 41650	−0.03548 74163	0.43
0.44	0.02306 30400	−0.15631 61600	0.76737 02400	0.40195 58400	−0.03607 29600	0.44
0.45	0.02317 73438	−0.15664 68750	0.75712 65625	0.41297 81250	−0.03663 51563	0.45
0.46	0.02327 09400	−0.15683 97600	0.74669 36400	0.42404 82400	−0.03717 30600	0.46
0.47	0.02334 37838	−0.15689 56350	0.73607 42025	0.43516 33650	−0.03768 57163	0.47
0.48	0.02339 58400	−0.15681 53600	0.72527 10400	0.44632 06400	−0.03817 21600	0.48
0.49	0.02342 70838	−0.15659 98350	0.71428 70025	0.45751 71650	−0.03863 14163	0.49
0.50	0.02343 75000	−0.15625 00000	0.70312 50000	0.46875 00000	−0.03906 25000	0.50
	A_2	A_1	A_0	A_{-1}	A_{-2}	$-p$

FIVE-POINT LAGRANGIAN INTERPOLATION COEFFICIENTS   Table 25.1

$$A_k^5(p) = (-1)^{k+2} \frac{p(p^2-1)(p^2-4)}{(2+k)!(2-k)!(p-k)}$$

p	A_{-2}	A_{-1}	A_0	A_1	A_2	p
0.50	0.02343 75000	−0.15625 00000	0.70312 50000	0.46875 00000	−0.03906 25000	0.50
0.51	0.02342 70838	−0.15576 68350	0.69178 80025	0.48001 61650	−0.03946 44163	0.51
0.52	0.02339 58400	−0.15515 13600	0.68027 90400	0.49131 26400	−0.03983 61600	0.52
0.53	0.02334 37838	−0.15440 46350	0.66860 12025	0.50263 63650	−0.04017 67163	0.53
0.54	0.02327 09400	−0.15352 77600	0.65675 76400	0.51398 42400	−0.04048 50600	0.54
0.55	0.02317 73438	−0.15252 18750	0.64475 15625	0.52535 31250	−0.04076 01563	0.55
0.56	0.02306 30400	−0.15138 81600	0.63258 62400	0.53673 98400	−0.04100 09600	0.56
0.57	0.02292 80838	−0.15012 78350	0.62026 50025	0.54814 11650	−0.04120 64163	0.57
0.58	0.02277 25400	−0.14874 21600	0.60779 12400	0.55955 38400	−0.04137 54600	0.58
0.59	0.02259 64838	−0.14723 24350	0.59516 84025	0.57097 45650	−0.04150 70163	0.59
0.60	0.02240 00000	−0.14560 00000	0.58240 00000	0.58240 00000	−0.04160 00000	0.60
0.61	0.02218 31838	−0.14384 62350	0.56948 96025	0.59382 67650	−0.04165 33163	0.61
0.62	0.02194 61400	−0.14197 25600	0.55644 08400	0.60525 14400	−0.04166 58600	0.62
0.63	0.02168 89838	−0.13998 04350	0.54325 74025	0.61667 05650	−0.04163 65163	0.63
0.64	0.02141 18400	−0.13787 13600	0.52994 30400	0.62808 06400	−0.04156 41600	0.64
0.65	0.02111 48438	−0.13564 68750	0.51650 15625	0.63947 81250	−0.04144 76563	0.65
0.66	0.02079 81400	−0.13330 85600	0.50293 68400	0.65085 94400	−0.04128 58600	0.66
0.67	0.02046 18838	−0.13085 80350	0.48925 28025	0.66222 09650	−0.04107 76163	0.67
0.68	0.02010 62400	−0.12829 69600	0.47545 34400	0.67355 90400	−0.04082 17600	0.68
0.69	0.01973 13838	−0.12562 70350	0.46154 28025	0.68486 99650	−0.04051 71163	0.69
0.70	0.01933 75000	−0.12285 00000	0.44752 50000	0.69615 00000	−0.04016 25000	0.70
0.71	0.01892 47838	−0.11996 76350	0.43340 42025	0.70739 53650	−0.03975 67163	0.71
0.72	0.01849 34400	−0.11698 17600	0.41918 46400	0.71860 22400	−0.03929 85600	0.72
0.73	0.01804 36838	−0.11389 42350	0.40487 06025	0.72976 67650	−0.03878 68163	0.73
0.74	0.01757 57400	−0.11070 69600	0.39046 64400	0.74088 50400	−0.03822 02600	0.74
0.75	0.01708 98438	−0.10742 18750	0.37597 65625	0.75195 31250	−0.03759 76563	0.75
0.76	0.01658 62400	−0.10404 09600	0.36140 54400	0.76296 70400	−0.03691 77600	0.76
0.77	0.01606 51838	−0.10056 62350	0.34675 76025	0.77392 27650	−0.03617 93163	0.77
0.78	0.01552 69400	−0.09699 97600	0.33203 76400	0.78481 62400	−0.03538 10600	0.78
0.79	0.01497 17838	−0.09334 36350	0.31725 02025	0.79564 33650	−0.03452 17163	0.79
0.80	0.01440 00000	−0.08960 00000	0.30240 00000	0.80640 00000	−0.03360 00000	0.80
0.81	0.01381 18838	−0.08577 10350	0.28749 18025	0.81708 19650	−0.03261 46163	0.81
0.82	0.01320 77400	−0.08185 89600	0.27253 04400	0.82768 50400	−0.03156 42600	0.82
0.83	0.01258 78838	−0.07786 60350	0.25752 08025	0.83820 49650	−0.03044 76163	0.83
0.84	0.01195 26400	−0.07379 45600	0.24246 78400	0.84863 74400	−0.02926 33600	0.84
0.85	0.01130 23438	−0.06964 68750	0.22737 65625	0.85897 81250	−0.02801 01563	0.85
0.86	0.01063 73400	−0.06542 53600	0.21225 20400	0.86922 26400	−0.02668 66600	0.86
0.87	0.00995 79838	−0.06113 24350	0.19709 94025	0.87936 65650	−0.02529 15163	0.87
0.88	0.00926 46400	−0.05677 05600	0.18192 38400	0.88940 54400	−0.02382 33600	0.88
0.89	0.00855 76838	−0.05234 22350	0.16673 06025	0.89933 47650	−0.02228 08163	0.89
0.90	0.00783 75000	−0.04785 00000	0.15152 50000	0.90915 00000	−0.02066 25000	0.90
0.91	0.00710 44838	−0.04329 64350	0.13631 24025	0.91884 65650	−0.01896 70163	0.91
0.92	0.00635 90400	−0.03868 41600	0.12109 82400	0.92841 98400	−0.01719 29600	0.92
0.93	0.00560 15838	−0.03401 58350	0.10588 80025	0.93786 51650	−0.01533 89163	0.93
0.94	0.00483 25400	−0.02929 41600	0.09068 72400	0.94717 78400	−0.01340 34600	0.94
0.95	0.00405 23438	−0.02452 18750	0.07550 15625	0.95635 31250	−0.01138·51563	0.95
0.96	0.00326 14400	−0.01970 17600	0.06033 66400	0.96538 62400	−0.00928 25600	0.96
0.97	0.00246 02838	−0.01483 66350	0.04519 82025	0.97427 23650	−0.00709 42163	0.97
0.98	0.00164 93400	−0.00992 93600	0.03009 20400	0.98300 66400	−0.00481 86600	0.98
0.99	0.00082 90838	−0.00498 28350	0.01502 40025	0.99158 41650	−0.00245 44163	0.99
1.00	0.00000 00000	0.00000 00000	0.00000 00000	1.00000 00000	0.00000 00000	1.00
	A_2	A_1	A_0	A_{-1}	A_{-2}	$-p$

Table 25.1   **FIVE-POINT LAGRANGIAN INTERPOLATION COEFFICIENTS**

$$A_k^5(p)=(-1)^{k+2}\frac{p(p^2-1)(p^2-4)}{(2+k)!(2-k)!(p-k)}$$

p	A_{-2}	A_{-1}	A_0	A_1	A_2	
1.00	0.00000 00000	0.00000 00000	0.00000 00000	1.00000 00000	0.00000 00000	1.00
1.01	-0.00083 74163	0.00501 61650	-0.01497 39975	1.00824 91650	0.00254 60838	1.01
1.02	-0.00168 26600	0.01006 26400	-0.02989 19600	1.01632 66400	0.00518 53400	1.02
1.03	-0.00253 52163	0.01513 63650	-0.04474 77975	1.02422 73650	0.00791 92838	1.03
1.04	-0.00339 45600	0.02023 42400	-0.05953 53600	1.03194 62400	0.01074 94400	1.04
1.05	-0.00426 01563	0.02535 31250	-0.07424 84375	1.03947 81250	0.01367 73438	1.05
1.06	-0.00513 14600	0.03048 98400	-0.08888 07600	1.04681 78400	0.01670 45400	1.06
1.07	-0.00600 79163	0.03564 11650	-0.10342 59975	1.05396 01650	0.01983 25838	1.07
1.08	-0.00688 89600	0.04080 38400	-0.11787 77600	1.06089 98400	0.02306 30400	1.08
1.09	-0.00777 40163	0.04597 45650	-0.13222 95975	1.06763 15650	0.02639 74838	1.09
1.10	-0.00866 25000	0.05115 00000	-0.14647 50000	1.07415 00000	0.02983 75000	1.10
1.11	-0.00955 38163	0.05632 67650	-0.16060 73975	1.08044 97650	0.03338 46838	1.11
1.12	-0.01044 73600	0.06150 14400	-0.17462 01600	1.08652 54400	0.03704 06400	1.12
1.13	-0.01134 25163	0.06667 05650	-0.18850 65975	1.09237 15650	0.04080 69838	1.13
1.14	-0.01223 86600	0.07183 06400	-0.20225 99600	1.09798 26400	0.04468 53400	1.14
1.15	-0.01313 51563	0.07697 81250	-0.21587 34375	1.10335 31250	0.04867 73438	1.15
1.16	-0.01403 13600	0.08210 94400	-0.22934 01600	1.10847 74400	0.05278 46400	1.16
1.17	-0.01492 66163	0.08722 09650	-0.24265 31975	1.11334 99650	0.05700 88838	1.17
1.18	-0.01582 02600	0.09230 90400	-0.25580 55600	1.11796 50400	0.06135 17400	1.18
1.19	-0.01671 16163	0.09736 99650	-0.26879 01975	1.12231 69650	0.06581 48838	1.19
1.20	-0.01760 00000	0.10240 00000	-0.28160 00000	1.12640 00000	0.07040 00000	1.20
1.21	-0.01848 47163	0.10739 53650	-0.29422 77975	1.13020 83650	0.07510 87838	1.21
1.22	-0.01936 50600	0.11235 22400	-0.30666 63600	1.13373 62400	0.07994 29400	1.22
1.23	-0.02024 03163	0.11726 67650	-0.31890 83975	1.13697 77650	0.08490 41838	1.23
1.24	-0.02110 97600	0.12213 50400	-0.33094 65600	1.13992 70400	0.08999 42400	1.24
1.25	-0.02197 26563	0.12695 31250	-0.34277 34375	1.14257 81250	0.09521 48438	1.25
1.26	-0.02282 82600	0.13171 70400	-0.35438 15600	1.14492 50400	0.10056 77400	1.26
1.27	-0.02367 58163	0.13642 27650	-0.36576 33975	1.14696 17650	0.10605 46838	1.27
1.28	-0.02451 45600	0.14106 62400	-0.37691 13600	1.14868 22400	0.11167 74400	1.28
1.29	-0.02534 37163	0.14564 33650	-0.38781 77975	1.15008 03650	0.11743 77838	1.29
1.30	-0.02616 25000	0.15015 00000	-0.39847 50000	1.15115 00000	0.12333 75000	1.30
1.31	-0.02697 01163	0.15458 19650	-0.40887 51975	1.15188 49650	0.12937 83838	1.31
1.32	-0.02776 57600	0.15893 50400	-0.41901 05600	1.15227 90400	0.13556 22400	1.32
1.33	-0.02854 86163	0.16320 49650	-0.42887 31975	1.15232 59650	0.14189 08838	1.33
1.34	-0.02931 78600	0.16738 74400	-0.43845 51600	1.15201 94400	0.14836 61400	1.34
1.35	-0.03007 26563	0.17147 81250	-0.44774 84375	1.15135 31250	0.15498 98438	1.35
1.36	-0.03081 21600	0.17547 26400	-0.45674 49600	1.15032 06400	0.16176 38400	1.36
1.37	-0.03153 55163	0.17936 65650	-0.46543 65975	1.14891 55650	0.16868 99838	1.37
1.38	-0.03224 18600	0.18315 54400	-0.47381 51600	1.14713 14400	0.17577 01400	1.38
1.39	-0.03293 03163	0.18683 47650	-0.48187 23975	1.14496 17650	0.18300 61838	1.39
1.40	-0.03360 00000	0.19040 00000	-0.48960 00000	1.14240 00000	0.19040 00000	1.40
1.41	-0.03425 00163	0.19384 65650	-0.49698 95975	1.13943 95650	0.19795 34838	1.41
1.42	-0.03487 94600	0.19716 98400	-0.50403 27600	1.13607 38400	0.20566 85400	1.42
1.43	-0.03548 74163	0.20036 51650	-0.51072 09975	1.13229 61650	0.21354 70838	1.43
1.44	-0.03607 29600	0.20342 78400	-0.51704 57600	1.12809 98400	0.22159 10400	1.44
1.45	-0.03663 51563	0.20635 31250	-0.52299 84375	1.12347 81250	0.22980 23438	1.45
1.46	-0.03717 30600	0.20913 62400	-0.52857 03600	1.11842 42400	0.23818 29400	1.46
1.47	-0.03768 57163	0.21177 23650	-0.53375 27975	1.11293 13650	0.24673 47838	1.47
1.48	-0.03817 21600	0.21425 66400	-0.53853 69600	1.10699 26400	0.25545 98400	1.48
1.49	-0.03863 14163	0.21658 41650	-0.54291 39975	1.10060 11650	0.26436 00838	1.49
1.50	-0.03906 25000	0.21875 00000	-0.54687 50000	1.09375 00000	0.27343 75000	1.50
	A_2	A_1	A_0	A_{-1}	A_{-2}	$-p$

FIVE-POINT LAGRANGIAN INTERPOLATION COEFFICIENTS   Table 25.1

$$A_k^5(p)=(-1)^{k+2}\frac{p(p^2-1)(p^2-4)}{(2+k)!(2-k)!(p-k)}$$

p	A_{-2}	A_{-1}	A_0	A_1	A_2	
1.50	−0.03906 25000	0.21875 00000	−0.54687 50000	1.09375 00000	0.27343 75000	1.50
1.51	−0.03946 44163	0.22074 91650	−0.55041 09975	1.08643 21650	0.28269 40838	1.51
1.52	−0.03983 61600	0.22257 66400	−0.55351 29600	1.07864 06400	0.29213 18400	1.52
1.53	−0.04017 67163	0.22422 73650	−0.55617 17975	1.07036 83650	0.30175 27838	1.53
1.54	−0.04048 50600	0.22569 62400	−0.55837 83600	1.06160 82400	0.31155 89400	1.54
1.55	−0.04076 01563	0.22697 81250	−0.56012 34375	1.05235 31250	0.32155 23438	1.55
1.56	−0.04100 09600	0.22806 78400	−0.56139 77600	1.04259 58400	0.33173 50400	1.56
1.57	−0.04120 64163	0.22896 01650	−0.56219 19975	1.03232 91650	0.34210 90838	1.57
1.58	−0.04137 54600	0.22964 98400	−0.56249 67600	1.02154 58400	0.35267 65400	1.58
1.59	−0.04150 70163	0.23013 15650	−0.56230 25975	1.01023 85650	0.36343 94838	1.59
1.60	−0.04160 00000	0.23040 00000	−0.56160 00000	0.99840 00000	0.37440 00000	1.60
1.61	−0.04165 33163	0.23044 97650	−0.56037 93975	0.98602 27650	0.38556 01838	1.61
1.62	−0.04166 58600	0.23027 54400	−0.55863 11600	0.97309 94400	0.39692 21400	1.62
1.63	−0.04163 65163	0.22987 15650	−0.55634 55975	0.95962 25650	0.40848 79838	1.63
1.64	−0.04156 41600	0.22923 26400	−0.55351 29600	0.94558 46400	0.42025 98400	1.64
1.65	−0.04144 76563	0.22835 31250	−0.55012 34375	0.93097 81250	0.43223 98438	1.65
1.66	−0.04128 58600	0.22722 74400	−0.54616 71600	0.91579 54400	0.44443 01400	1.66
1.67	−0.04107 76163	0.22584 99650	−0.54163 41975	0.90002 89650	0.45683 28838	1.67
1.68	−0.04082 17600	0.22421 50400	−0.53651 45600	0.88367 10400	0.46945 02400	1.68
1.69	−0.04051 71163	0.22231 69650	−0.53079 81975	0.86671 39650	0.48228 43838	1.69
1.70	−0.04016 25000	0.22015 00000	−0.52447 50000	0.84915 00000	0.49533 75000	1.70
1.71	−0.03975 67163	0.21770 83650	−0.51753 47975	0.83097 13650	0.50861 17838	1.71
1.72	−0.03929 85600	0.21498 62400	−0.50996 73600	0.81217 02400	0.52210 94400	1.72
1.73	−0.03878 68163	0.21197 77650	−0.50176 23975	0.79273 87650	0.53583 26838	1.73
1.74	−0.03822 02600	0.20867 70400	−0.49290 95600	0.77266 90400	0.54978 37400	1.74
1.75	−0.03759 76563	0.20507 81250	−0.48339 84375	0.75195 31250	0.56396 48438	1.75
1.76	−0.03691 77600	0.20117 50400	−0.47321 85600	0.73058 30400	0.57837 82400	1.76
1.77	−0.03617 93163	0.19696 17650	−0.46235 93975	0.70855 07650	0.59302 61838	1.77
1.78	−0.03538 10600	0.19243 22400	−0.45081 03600	0.68584 82400	0.60791 09400	1.78
1.79	−0.03452 17163	0.18758 03650	−0.43856 07975	0.66246 73650	0.62303 47838	1.79
1.80	−0.03360 00000	0.18240 00000	−0.42560 00000	0.63840 00000	0.63840 00000	1.80
1.81	−0.03261 46163	0.17688 49650	−0.41191 71975	0.61363 79650	0.65400 88838	1.81
1.82	−0.03156 42600	0.17102 90400	−0.39750 15600	0.58817 30400	0.66986 37400	1.82
1.83	−0.03044 76163	0.16482 59650	−0.38234 21975	0.56199 69650	0.68596 68838	1.83
1.84	−0.02926 33600	0.15826 94400	−0.36642 81600	0.53510 14400	0.70232 06400	1.84
1.85	−0.02801 01563	0.15135 31250	−0.34974 84375	0.50747 81250	0.71892 73438	1.85
1.86	−0.02668 66600	0.14407 06400	−0.33229 19600	0.47911 86400	0.73578 93400	1.86
1.87	−0.02529 15163	0.13641 55650	−0.31404 75975	0.45001 45650	0.75290 89838	1.87
1.88	−0.02382 33600	0.12838 14400	−0.29500 41600	0.42015 74400	0.77028 86400	1.88
1.89	−0.02228 08163	0.11996 17650	−0.27515 03975	0.38953 87650	0.78793 06838	1.89
1.90	−0.02066 25000	0.11115 00000	−0.25447 50000	0.35815 00000	0.80583 75000	1.90
1.91	−0.01896 70163	0.10193 95650	−0.23296 65975	0.32598 25650	0.82401 14838	1.91
1.92	−0.01719 29600	0.09232 38400	−0.21061 37600	0.29302 78400	0.84245 50400	1.92
1.93	−0.01533 89163	0.08229 61650	−0.18740 49975	0.25927 71650	0.86117 05838	1.93
1.94	−0.01340 34600	0.07184 98400	−0.16332 87600	0.22472 18400	0.88016 05400	1.94
1.95	−0.01138 51563	0.06097 81250	−0.13837 34375	0.18935 31250	0.89942 73438	1.95
1.96	−0.00928 25600	0.04967 42400	−0.11252 73600	0.15316 22400	0.91897 34400	1.96
1.97	−0.00709 42163	0.03793 13650	−0.08577 87975	0.11614 03650	0.93880 12838	1.97
1.98	−0.00481 86600	0.02574 26400	−0.05811 59600	0.07827 86400	0.95891 33400	1.98
1.99	−0.00245 44163	0.01310 11650	−0.02952 69975	0.03956 81650	0.97931 20838	1.99
2.00	0.00000 00000	0.00000 00000	0.00000 00000	0.00000 00000	1.00000 00000	2.00
	A_2	A_1	A_0	A_{-1}	A_{-2}	$-p$

Table 25.1   **SIX-POINT LAGRANGIAN INTERPOLATION COEFFICIENTS**

$$A_k^6(p)=(-1)^{k+3}\frac{p(p^2-1)(p^2-4)(p-3)}{(2+k)!(3-k)!(p-k)}$$

p	A_{-2}	A_{-1}	A_0	A_1	A_2	A_3	
0.00	0.00000 00000	0.00000 00000	1.00000 00000	0.00000 00000	0.00000 00000	0.00000 00000	1.00
0.01	0.00049 57921	−0.00493 33767	0.99654 20858	0.01006 60817	−0.00250 38746	0.00033 32917	0.99
0.02	0.00098 30066	−0.00973 36932	0.99283 67064	0.02026 19736	−0.00501 43268	0.00066 63334	0.98
0.03	0.00146 14085	−0.01440 12590	0.98888 64505	0.03058 41170	−0.00752 95922	0.00099 88752	0.97
0.04	0.00193 07725	−0.01893 64224	0.98469 39648	0.04102 89152	−0.01004 78976	0.00133 06675	0.96
0.05	0.00239 08828	−0.02333 95703	0.98026 19531	0.05159 27344	−0.01256 74609	0.00166 14609	0.95
0.06	0.00284 15335	−0.02761 11276	0.97559 31752	0.06227 19048	−0.01508 64924	0.00199 10065	0.94
0.07	0.00328 25281	−0.03175 15567	0.97069 04458	0.07306 27217	−0.01760 31946	0.00231 90557	0.93
0.08	0.00371 36794	−0.03576 13568	0.96555 66336	0.08396 14464	−0.02011 57632	0.00264 53606	0.92
0.09	0.00413 48096	−0.03964 10640	0.96019 46604	0.09496 43071	−0.02262 23873	0.00296 96742	0.91
0.10	0.00454 57500	−0.04339 12500	0.95460 75000	0.10606 75000	−0.02512 12500	0.00329 17500	0.90
0.11	0.00494 63412	−0.04701 25223	0.94879 81771	0.11726 71904	−0.02761 05290	0.00361 13426	0.89
0.12	0.00533 64326	−0.05050 55232	0.94276 97664	0.12855 95136	−0.03008 83968	0.00392 82074	0.88
0.13	0.00571 58827	−0.05387 09296	0.93652 53917	0.13994 05758	−0.03255 30217	0.00424 21011	0.87
0.14	0.00608 45585	−0.05710 94524	0.93006 82248	0.15140 64552	−0.03500 25676	0.00455 27815	0.86
0.15	0.00644 23359	−0.06022 18359	0.92340 14844	0.16295 32031	−0.03743 51953	0.00486 00078	0.85
0.16	0.00678 90995	−0.06320 88576	0.91652 84352	0.17457 68448	−0.03984 90624	0.00516 35405	0.84
0.17	0.00712 47422	−0.06607 13273	0.90945 23870	0.18627 33805	−0.04224 23240	0.00546 31416	0.83
0.18	0.00744 91654	−0.06881 00868	0.90217 66936	0.19803 87864	−0.04461 31332	0.00575 85746	0.82
0.19	0.00776 22787	−0.07142 60096	0.89470 47517	0.20986 90158	−0.04695 96417	0.00604 96051	0.81
0.20	0.00806 40000	−0.07392 00000	0.88704 00000	0.22176 00000	−0.04928 00000	0.00633 60000	0.80
0.21	0.00835 42553	−0.07629 29929	0.87918 59183	0.23370 76492	−0.05157 23583	0.00661 75284	0.79
0.22	0.00863 29786	−0.07854 59532	0.87114 60264	0.24570 78536	−0.05383 48668	0.00689 39614	0.78
0.23	0.00890 01118	−0.08067 98752	0.86292 38830	0.25775 64845	−0.05606 56760	0.00716 50719	0.77
0.24	0.00915 56045	−0.08269 57824	0.85452 30848	0.26984 93952	−0.05826 29376	0.00743 06355	0.76
0.25	0.00939 94141	−0.08459 47266	0.84594 72656	0.28198 24219	−0.06042 48047	0.00769 04297	0.75
0.26	0.00963 15055	−0.08637 77876	0.83720 00952	0.29415 13848	−0.06254 94324	0.00794 42345	0.74
0.27	0.00985 18513	−0.08804 60729	0.82828 52783	0.30635 20892	−0.06463 49783	0.00819 18324	0.73
0.28	0.01006 04314	−0.08960 07168	0.81920 65536	0.31858 03264	−0.06667 96032	0.00843 30086	0.72
0.29	0.01025 72328	−0.09104 28802	0.80996 76929	0.33083 18746	−0.06868 14711	0.00866 75510	0.71
0.30	0.01044 22500	−0.09237 37500	0.80057 25000	0.34310 25000	−0.07063 87500	0.00889 52500	0.70
0.31	0.01061 54844	−0.09359 45385	0.79102 48096	0.35538 79579	−0.07254 96127	0.00911 58993	0.69
0.32	0.01077 69446	−0.09470 64832	0.78132 84864	0.36768 39936	−0.07441 22368	0.00932 92954	0.68
0.33	0.01092 66459	−0.09571 08458	0.77148 74242	0.37998 63433	−0.07622 48054	0.00953 52378	0.67
0.34	0.01106 46105	−0.09660 89124	0.76150 55448	0.39229 07352	−0.07798 55076	0.00973 35295	0.66
0.35	0.01119 08672	−0.09740 19922	0.75138 67969	0.40459 28906	−0.07969 25391	0.00992 39766	0.65
0.36	0.01130 54515	−0.09809 14176	0.74113 51552	0.41688 85248	−0.08134 41024	0.01010 63885	0.64
0.37	0.01140 84054	−0.09867 85435	0.73075 46195	0.42917 33480	−0.08293 84077	0.01028 05783	0.63
0.38	0.01149 97774	−0.09916 47468	0.72024 92136	0.44144 30664	−0.08447 36732	0.01044 63626	0.62
0.39	0.01157 96219	−0.09955 14258	0.70962 29842	0.45369 33833	−0.08594 81254	0.01060 35618	0.61
0.40	0.01164 80000	−0.09984 00000	0.69888 00000	0.46592 00000	−0.08736 00000	0.01075 20000	0.60
0.41	0.01170 49786	−0.10003 19092	0.68802 43508	0.47811 86167	−0.08870 75421	0.01089 15052	0.59
0.42	0.01175 06306	−0.10012 86132	0.67706 01464	0.49028 49336	−0.08998 90068	0.01102 19094	0.58
0.43	0.01178 50351	−0.10013 15915	0.66599 15155	0.50241 46520	−0.09120 26598	0.01114 30487	0.57
0.44	0.01180 82765	−0.10004 23424	0.65482 26048	0.51450 34752	−0.09234 67776	0.01125 47635	0.56
0.45	0.01182 04453	−0.09986 23828	0.64355 75781	0.52654 71094	−0.09341 96484	0.01135 68984	0.55
0.46	0.01182 16375	−0.09959 32476	0.63220 06152	0.53854 12648	−0.09441 95724	0.01144 93025	0.54
0.47	0.01181 19546	−0.09923 64892	0.62075 59108	0.55048 16567	−0.09534 48621	0.01153 18292	0.53
0.48	0.01179 15034	−0.09879 36768	0.60922 76736	0.56236 40064	−0.09619 38432	0.01160 43366	0.52
0.49	0.01176 03961	−0.09826 63965	0.59762 01254	0.57418 40421	−0.09696 48548	0.01166 66877	0.51
0.50	0.01171 87500	−0.09765 62500	0.58593 75000	0.58593 75000	−0.09765 62500	0.01171 87500	0.50
	A_3	A_2	A_1	A_0	A_{-1}	A_{-2}	p

SIX-POINT LAGRANGIAN INTERPOLATION COEFFICIENTS

Table 25.1

$$A_k^6(p) = (-1)^{k+3} \frac{p(p^2-1)(p^2-4)(p-3)}{(2+k)!(3-k)!(p-k)}$$

p	A_{-2}	A_{-1}	A_0	A_1	A_2	A_3	
1.00	0.00000 00000	0.00000 00000	0.00000 00000	1.00000 00000	0.00000 00000	0.00000 00000	0.00
1.01	−0.00033 32917	0.00249 55421	−0.00993 27517	1.00320 79192	0.00506 67067	−0.00050 41246	0.01
1.02	−0.00066 63334	0.00498 10068	−0.01972 86936	1.00616 33736	0.01026 69732	−0.00101 63266	0.02
1.03	−0.00099 88752	0.00745 46597	−0.02938 43870	1.00886 39545	0.01560 09890	−0.00153 63410	0.03
1.04	−0.00133 06675	0.00991 47776	−0.03889 64352	1.01130 73152	0.02106 89024	−0.00206 38925	0.04
1.05	−0.00166 14609	0.01235 96484	−0.04826 14844	1.01349 11719	0.02667 08203	−0.00259 86953	0.05
1.06	−0.00199 10065	0.01478 75724	−0.05747 62248	1.01541 33048	0.03240 68076	−0.00314 04535	0.06
1.07	−0.00231 90556	0.01719 68621	−0.06653 73917	1.01707 15592	0.03827 68866	−0.00368 88606	0.07
1.08	−0.00264 53606	0.01958 58432	−0.07544 17664	1.01846 38464	0.04428 10368	−0.00424 35994	0.08
1.09	−0.00296 96742	0.02195 28547	−0.08418 61771	1.01958 81446	0.05041 91940	−0.00480 43420	0.09
1.10	−0.00329 17500	0.02429 62500	−0.09276 75000	1.02044 25000	0.05669 12500	−0.00537 07500	0.10
1.11	−0.00361 13426	0.02661 43965	−0.10118 26604	1.02102 50279	0.06309 70523	−0.00594 24737	0.11
1.12	−0.00392 82074	0.02890 56768	−0.10942 86336	1.02133 39136	0.06963 64032	−0.00651 91526	0.12
1.13	−0.00424 21011	0.03116 84892	−0.11750 24458	1.02136 74133	0.07630 90596	−0.00710 04152	0.13
1.14	−0.00455 27815	0.03340 12476	−0.12540 11752	1.02112 38552	0.08311 47324	−0.00768 58785	0.14
1.15	−0.00486 00078	0.03560 23828	−0.13312 19531	1.02060 16406	0.09005 30859	−0.00827 51484	0.15
1.16	−0.00516 35405	0.03777 03424	−0.14066 19648	1.01979 92448	0.09712 37376	−0.00886 78195	0.16
1.17	−0.00546 31415	0.03990 35915	−0.14801 84505	1.01871 52180	0.10432 62572	−0.00946 34747	0.17
1.18	−0.00575 85746	0.04200 06132	−0.15518 87064	1.01734 81864	0.11166 01668	−0.01006 16854	0.18
1.19	−0.00604 96051	0.04405 99092	−0.16217 00858	1.01569 68533	0.11912 49396	−0.01066 20112	0.19
1.20	−0.00633 60000	0.04608 00000	−0.16896 00000	1.01376 00000	0.12672 00000	−0.01126 40000	0.20
1.21	−0.00661 75284	0.04805 94258	−0.17555 59192	1.01153 64867	0.13444 47229	−0.01186 71878	0.21
1.22	−0.00689 39614	0.04999 67468	−0.18195 53736	1.00902 52536	0.14229 84332	−0.01247 10986	0.22
1.23	−0.00716 50719	0.05189 05435	−0.18815 99545	1.00622 53220	0.15028 04052	−0.01307 52443	0.23
1.24	−0.00743 06355	0.05373 94176	−0.19415 53152	1.00313 57952	0.15838 98624	−0.01367 91245	0.24
1.25	−0.00769 04297	0.05554 19922	−0.19995 11719	0.99975 58594	0.16662 59766	−0.01428 22266	0.25
1.26	−0.00794 42345	0.05729 69124	−0.20554 13048	0.99608 47848	0.17498 78676	−0.01488 40255	0.26
1.27	−0.00819 18324	0.05900 28458	−0.21092 35592	0.99212 19267	0.18347 46029	−0.01548 39838	0.27
1.28	−0.00843 30086	0.06065 84832	−0.21609 58464	0.98786 67264	0.19208 51968	−0.01608 15514	0.28
1.29	−0.00866 75509	0.06226 25385	−0.22105 61446	0.98331 87121	0.20081 86102	−0.01667 61653	0.29
1.30	−0.00889 52500	0.06381 37500	−0.22580 25000	0.97847 75000	0.20967 37500	−0.01726 72500	0.30
1.31	−0.00911 58993	0.06531 08802	−0.23033 30279	0.97334 27954	0.21864 94685	−0.01785 42169	0.31
1.32	−0.00932 92954	0.06675 27168	−0.23464 59136	0.96791 43936	0.22774 45632	−0.01843 64646	0.32
1.33	−0.00953 52378	0.06813 80729	−0.23873 94133	0.96219 21808	0.23695 77758	−0.01901 33784	0.33
1.34	−0.00973 35295	0.06946 57876	−0.24261 18552	0.95617 61352	0.24628 77924	−0.01958 43305	0.34
1.35	−0.00992 39766	0.07073 47266	−0.24626 16406	0.94986 63281	0.25573 32422	−0.02014 86797	0.35
1.36	−0.01010 63885	0.07194 37824	−0.24968 72448	0.94326 29248	0.26529 26976	−0.02070 57715	0.36
1.37	−0.01028 05783	0.07309 18752	−0.25288 72180	0.93636 61855	0.27496 46735	−0.02125 49379	0.37
1.38	−0.01044 63626	0.07417 79532	−0.25586 01864	0.92917 64664	0.28474 76268	−0.02179 54974	0.38
1.39	−0.01060 35618	0.07520 09929	−0.25860 48533	0.92169 42208	0.29463 99558	−0.02232 67544	0.39
1.40	−0.01075 20000	0.07616 00000	−0.26112 00000	0.91392 00000	0.30464 00000	−0.02284 80000	0.40
1.41	−0.01089 15052	0.07705 40096	−0.26340 44867	0.90585 44542	0.31474 60392	−0.02335 85111	0.41
1.42	−0.01102 19094	0.07788 20868	−0.26545 72536	0.89749 83336	0.32495 62932	−0.02385 75506	0.42
1.43	−0.01114 30487	0.07864 33273	−0.26727 73220	0.88885 24895	0.33526 89215	−0.02434 43676	0.43
1.44	−0.01125 47635	0.07933 68576	−0.26886 37952	0.87991 78752	0.34568 20224	−0.02481 81965	0.44
1.45	−0.01135 68984	0.07996 18359	−0.27021 58594	0.87069 55469	0.35619 36328	−0.02527 82578	0.45
1.46	−0.01144 93025	0.08051 74524	−0.27133 27848	0.86118 66648	0.36680 17276	−0.02572 37575	0.46
1.47	−0.01153 18292	0.08100 29296	−0.27221 39267	0.85139 24942	0.37750 42192	−0.02615 38871	0.47
1.48	−0.01160 43366	0.08141 75232	−0.27285 87264	0.84131 44064	0.38829 89568	−0.02656 78234	0.48
1.49	−0.01166 66877	0.08176 05223	−0.27326 67121	0.83095 38796	0.39918 37265	−0.02696 47286	0.49
1.50	−0.01171 87500	0.08203 12500	−0.27343 75000	0.82031 25000	0.41015 62500	−0.02734 37500	0.50
	A_3	A_2	A_1	A_0	A_{-1}	A_{-2}	$-p$

Table 25.1   **SIX-POINT LAGRANGIAN INTERPOLATION COEFFICIENTS**

$$A_k^6(p) = (-1)^{k+3} \frac{p(p^2-1)(p^2-4)(p-3)}{(2+k)!(3-k)!(p-k)}$$

p	A_{-2}	A_{-1}	A_0	A_1	A_2	A_3	
1.50	−0.01171 87500	0.08203 12500	−0.27343 75000	0.82031 25000	0.41015 62500	−0.02734 37500	0.50
1.51	−0.01176 03961	0.08222 90640	−0.27337 07954	0.80939 19629	0.42121 41848	−0.02770 40202	0.51
1.52	−0.01179 15034	0.08235 33568	−0.27306 63936	0.79819 40736	0.43235 51232	−0.02804 46566	0.52
1.53	−0.01181 19546	0.08240 35567	−0.27252 41808	0.78672 07483	0.44357 65921	−0.02836 47617	0.53
1.54	−0.01182 16375	0.08237 91276	−0.27174 41352	0.77497 40152	0.45487 60524	−0.02866 34225	0.54
1.55	−0.01182 04453	0.08227 95703	−0.27072 63281	0.76295 60156	0.46625 08984	−0.02893 97109	0.55
1.56	−0.01180 82765	0.08210 44224	−0.26947 09248	0.75066 90048	0.47769 84576	−0.02919 26835	0.56
1.57	−0.01178 50350	0.08185 32590	−0.26797 81855	0.73811 53530	0.48921 59897	−0.02942 13812	0.57
1.58	−0.01175 06306	0.08152 56932	−0.26624 84664	0.72529 75464	0.50080 06868	−0.02962 48294	0.58
1.59	−0.01170 49786	0.08112 13767	−0.26428 22208	0.71221 81883	0.51244 96721	−0.02980 20377	0.59
1.60	−0.01164 80000	0.08064 00000	−0.26208 00000	0.69888 00000	0.52416 00000	−0.02995 20000	0.60
1.61	−0.01157 96219	0.08008 12933	−0.25964 24542	0.68528 58217	0.53592 86554	−0.03007 36943	0.61
1.62	−0.01149 97774	0.07944 50268	−0.25697 03336	0.67143 86136	0.54775 25532	−0.03016 60826	0.62
1.63	−0.01140 84054	0.07873 10110	−0.25406 44895	0.65734 14570	0.55962 85377	−0.03022 81108	0.63
1.64	−0.01130 54515	0.07793 90976	−0.25092 58752	0.64299 75552	0.57155 33824	−0.03025 87085	0.64
1.65	−0.01119 08672	0.07706 91797	−0.24755 55469	0.62841 02344	0.58352 37891	−0.03025 67891	0.65
1.66	−0.01106 46105	0.07612 11924	−0.24395 46648	0.61358 29448	0.59553 63876	−0.03022 12495	0.66
1.67	−0.01092 66459	0.07509 51133	−0.24012 44942	0.59851 92617	0.60758 77354	−0.03015 09703	0.67
1.68	−0.01077 69446	0.07399 09632	−0.23606 64064	0.58322 28864	0.61967 43168	−0.03004 48154	0.68
1.69	−0.01061 54845	0.07280 88061	−0.23178 18796	0.56769 76471	0.63179 25427	−0.02990 16318	0.69
1.70	−0.01044 22500	0.07154 87500	−0.22727 25000	0.55194 75000	0.64393 87500	−0.02972 02500	0.70
1.71	−0.01025 72328	0.07021 09477	−0.22253 99629	0.53597 65304	0.65610 92010	−0.02949 94834	0.71
1.72	−0.01006 04314	0.06879 55968	−0.21758 60736	0.51978 89536	0.66830 00832	−0.02923 81286	0.72
1.73	−0.00985 18513	0.06730 29404	−0.21241 27483	0.50338 91158	0.68050 75083	−0.02893 49649	0.73
1.74	−0.00963 15055	0.06573 32676	−0.20702 20152	0.48678 14952	0.69272 75124	−0.02858 87545	0.74
1.75	−0.00939 94141	0.06408 69141	−0.20141 60156	0.46997 07031	0.70495 60547	−0.02819 82422	0.75
1.76	−0.00915 56045	0.06236 42624	−0.19559 70048	0.45296 14848	0.71718 90176	−0.02776 21555	0.76
1.77	−0.00890 01118	0.06056 57427	−0.18956 73530	0.43575 87205	0.72942 22061	−0.02727 92045	0.77
1.78	−0.00863 29786	0.05869 18332	−0.18332 95464	0.41836 74264	0.74165 13468	−0.02674 80814	0.78
1.79	−0.00835 42553	0.05674 30604	−0.17688 61883	0.40079 27558	0.75387 20883	−0.02616 74609	0.79
1.80	−0.00806 40000	0.05472 00000	−0.17024 00000	0.38304 00000	0.76608 00000	−0.02553 60000	0.80
1.81	−0.00776 22787	0.05262 32771	−0.16339 38217	0.36511 45892	0.77827 05717	−0.02485 23376	0.81
1.82	−0.00744 91654	0.05045 35668	−0.15635 06136	0.34702 20936	0.79043 92132	−0.02411 50946	0.82
1.83	−0.00712 47422	0.04821 15948	−0.14911 34570	0.32876 82245	0.80258 12540	−0.02332 28741	0.83
1.84	−0.00678 90995	0.04589 81376	−0.14168 55552	0.31035 88352	0.81469 19424	−0.02247 42605	0.84
1.85	−0.00644 23359	0.04351 40234	−0.13407 02344	0.29179 99219	0.82676 64453	−0.02156 78203	0.85
1.86	−0.00608 45585	0.04106 01324	−0.12627 09448	0.27309 76248	0.83879 98476	−0.02060 21015	0.86
1.87	−0.00571 58826	0.03853 73971	−0.11829 12617	0.25425 82292	0.85078 71516	−0.01957 56336	0.87
1.88	−0.00533 64326	0.03594 68032	−0.11013 48864	0.23528 81664	0.86272 32768	−0.01848 69274	0.88
1.89	−0.00494 63412	0.03328 93898	−0.10180 56471	0.21619 40145	0.87460 30590	−0.01733 44750	0.89
1.90	−0.00454 57500	0.03056 62500	−0.09330 75000	0.19698 25000	0.88642 12500	−0.01611 67500	0.90
1.91	−0.00413 48096	0.02777 85315	−0.08464 45304	0.17766 04979	0.89817 25173	−0.01483 22067	0.91
1.92	−0.00371 36794	0.02492 74368	−0.07582 09536	0.15823 50336	0.90985 14432	−0.01347 92806	0.92
1.93	−0.00328 25281	0.02201 42242	−0.06684 11158	0.13871 32833	0.92145 25246	−0.01205 63882	0.93
1.94	−0.00284 15335	0.01904 02076	−0.05770 94952	0.11910 25752	0.93297 01724	−0.01056 19265	0.94
1.95	−0.00239 08828	0.01600 67578	−0.04843 07031	0.09941 03906	0.94439 87109	−0.00899 42734	0.95
1.96	−0.00193 07725	0.01291 53024	−0.03900 94848	0.07964 43648	0.95573 23776	−0.00735 17875	0.96
1.97	−0.00146 14086	0.00976 73265	−0.02945 07205	0.05981 22880	0.96696 53223	−0.00563 28077	0.97
1.98	−0.00098 30066	0.00656 43732	−0.01975 94264	0.03992 21064	0.97809 16068	−0.00383 56534	0.98
1.99	−0.00049 57921	0.00330 80442	−0.00994 07558	0.01998 19233	0.98910 52046	−0.00195 86242	0.99
2.00	0.00000 00000	0.00000 00000	0.00000 00000	0.00000 00000	1.00000 00000	0.00000 00000	1.00
	A_3	A_2	A_1	A_0	A_{-1}	A_{-2}	$-p$

SIX-POINT LAGRANGIAN INTERPOLATION COEFFICIENTS

Table 25.1

$$A_k^6(p)=(-1)^{k+3}\frac{p(p^2-1)(p^2-4)(p-3)}{(2+k)!(3-k)!(p-k)}$$

p	A_{-2}	A_{-1}	A_0	A_1	A_2	A_3	
2.00	0.00000 00000	0.00000 00000	0.00000 00000	0.00000 00000	1.00000 00000	0.00000 00000	1.00
2.01	0.00050 41246	−0.00335 80392	0.01005 74108	−0.02001 52433	1.01076 97879	0.00204 19592	1.01
2.02	0.00101 63266	−0.00676 42932	0.02022 59064	−0.04005 52264	1.02140 82732	0.00416 90134	1.02
2.03	0.00153 63410	−0.01021 69214	0.03049 97755	−0.06011 12080	1.03190 90702	0.00638 29427	1.03
2.04	0.00206 38925	−0.01371 40224	0.04087 31648	−0.08017 42848	1.04226 57024	0.00868 55475	1.04
2.05	0.00259 86953	−0.01725 36328	0.05134 00781	−0.10023 53906	1.05247 16016	0.01107 86484	1.05
2.06	0.00314 04535	−0.02083 37276	0.06189 43752	−0.12028 52952	1.06252 01076	0.01356 40865	1.06
2.07	0.00368 88605	−0.02445 22191	0.07252 97708	−0.14031 46033	1.07240 44679	0.01614 37232	1.07
2.08	0.00424 35994	−0.02810 69568	0.08323 98336	−0.16031 37536	1.08211 78368	0.01881 94406	1.08
2.09	0.00480 43420	−0.03179 57264	0.09401 79854	−0.18027 30179	1.09165 32752	0.02159 31417	1.09
2.10	0.00537 07500	0.03551 62500	0.10485 75000	−0.20018 25000	1.10100 37500	0.02446 67500	1.10
2.11	0.00594 24737	−0.03926 61847	0.11575 15021	−0.22003 21346	1.11016 21335	0.02744 22100	1.11
2.12	0.00651 91526	−0.04304 31232	0.12669 29664	−0.23981 16864	1.11912 12032	0.03052 14874	1.12
2.13	0.00710 04151	−0.04684 45921	0.13767 47167	−0.25951 07492	1.12787 36409	0.03370 65686	1.13
2.14	0.00768 58785	−0.05066 80524	0.14868 94248	−0.27911 87448	1.13641 20324	0.03699 94615	1.14
2.15	0.00827 51484	−0.05451 08984	0.15972 96094	−0.29862 49219	1.14472 88672	0.04040 21953	1.15
2.16	0.00886 78195	−0.05837 04576	0.17078 76352	−0.31801 83552	1.15281 65376	0.04391 68205	1.16
2.17	0.00946 34747	−0.06224 39898	0.18185 57120	−0.33728 79445	1.16066 73385	0.04754 54091	1.17
2.18	0.01006 16854	−0.06612 86868	0.19292 58936	−0.35642 24136	1.16827 34668	0.05129 00546	1.18
2.19	0.01066 20112	−0.07002 16721	0.20399 00767	−0.37541 03092	1.17562 70208	0.05515 28726	1.19
2.20	0.01126 40000	−0.07392 00000	0.21504 00000	−0.39424 00000	1.18272 00000	0.05913 60000	1.20
2.21	0.01186 71878	−0.07782 06554	0.22606 72433	−0.41289 96758	1.18954 43042	0.06324 15959	1.21
2.22	0.01247 10986	−0.08172 05532	0.23706 32264	−0.43137 73464	1.19609 17332	0.06747 18414	1.22
2.23	0.01307 52443	−0.08561 65377	0.24801 92080	−0.44966 08405	1.20235 39865	0.07182 89394	1.23
2.24	0.01367 91245	−0.08950 53824	0.25892 62848	−0.46773 78048	1.20832 26624	0.07631 51155	1.24
2.25	0.01428 22266	−0.09338 37891	0.26977 53906	−0.48559 57031	1.21398 92578	0.08093 26172	1.25
2.26	0.01488 40255	−0.09724 83876	0.28055 72952	−0.50322 18152	1.21934 51676	0.08568 37145	1.26
2.27	0.01548 39838	−0.10109 57353	0.29126 26033	−0.52060 32358	1.22438 16841	0.09057 06999	1.27
2.28	0.01608 15514	−0.10492 23168	0.30188 17536	−0.53772 68736	1.22908 99968	0.09559 58886	1.28
2.29	0.01667 61653	−0.10872 45427	0.31240 50179	−0.55457 94504	1.23346 11915	0.10076 16184	1.29
2.30	0.01726 72500	−0.11249 87500	0.32282 25000	−0.57114 75000	1.23748 62500	0.10607 02500	1.30
2.31	0.01785 42169	−0.11624 12010	0.33312 41346	−0.58741 73671	1.24115 60498	0.11152 41668	1.31
2.32	0.01843 64646	−0.11994 80832	0.34329 96864	−0.60337 52064	1.24446 13632	0.11712 57754	1.32
2.33	0.01901 33784	−0.12361 55083	0.35333 87492	−0.61900 69817	1.24739 28571	0.12287 75053	1.33
2.34	0.01958 43305	−0.12723 95124	0.36323 07448	−0.63429 84648	1.24994 10924	0.12878 18095	1.34
2.35	0.02014 86797	−0.13081 60547	0.37296 49219	−0.64923 52344	1.25209 65234	0.13484 11641	1.35
2.36	0.02070 57715	−0.13434 10176	0.38253 03552	−0.66380 26752	1.25384 94976	0.14105 80685	1.36
2.37	0.02125 49379	−0.13781 02060	0.39191 59445	−0.67798 59770	1.25519 02548	0.14743 50458	1.37
2.38	0.02179 54974	−0.14121 93468	0.40111 04136	−0.69177 01336	1.25610 89268	0.15397 46426	1.38
2.39	0.02232 67544	−0.14456 40883	0.41010 23092	−0.70513 99417	1.25659 55371	0.16067 94293	1.39
2.40	0.02284 80000	−0.14784 00000	0.41888 00000	−0.71808 00000	1.25664 00000	0.16755 20000	1.40
2.41	0.02335 85111	−0.15104 25717	0.42743 16758	−0.73057 47083	1.25623 21204	0.17459 49727	1.41
2.42	0.02385 75506	−0.15416 72132	0.43574 53464	−0.74260 82664	1.25536 15932	0.18181 09894	1.42
2.43	0.02434 43676	−0.15720 92540	0.44380 88405	−0.75416 46730	1.25401 80027	0.18920 27162	1.43
2.44	0.02481 81965	−0.16016 39424	0.45160 98048	−0.76522 77248	1.25219 08224	0.19677 28435	1.44
2.45	0.02527 82578	−0.16302 64453	0.45913 57031	−0.77578 10156	1.24986 94141	0.20452 40859	1.45
2.46	0.02572 37575	−0.16579 18476	0.46637 38152	−0.78580 79352	1.24704 30276	0.21245 91825	1.46
2.47	0.02615 38870	−0.16845 51516	0.47331 12358	−0.79529 16683	1.24370 08004	0.22058 08967	1.47
2.48	0.02656 78234	−0.17101 12768	0.47993 48736	−0.80421 51936	1.23983 17568	0.22889 20166	1.48
2.49	0.02696 47286	−0.17345 50590	0.48623 14504	−0.81256 12829	1.23542 48077	0.23739 53552	1.49
2.50	0.02734 37500	−0.17578 12500	0.49218 75000	−0.82031 25000	1.23046 87500	0.24609 37500	1.50
	A_3	A_2	A_1	A_0	A_{-1}	A_{-2}	$-p$

Table 25.1   SIX-POINT LAGRANGIAN INTERPOLATION COEFFICIENTS

$$A_k^6(p)=(-1)^{k+3}\frac{p(p^2-1)(p^2-4)(p-3)}{(2+k)!(3-k)!(p-k)}$$

p	A_{-2}	A_{-1}	A_0	A_1	A_2	A_3	
2.50	0.02734 37500	−0.17578 12500	0.49218 75000	−0.82031 25000	1.23046 87500	0.24609 37500	1.50
2.51	0.02770 40203	−0.17798 45173	0.49778 93671	−0.82745 11996	1.22495 22660	0.25499 00635	1.51
2.52	0.02804 46566	−0.18005 94432	0.50302 32064	−0.83395 95264	1.21886 39232	0.26408 71834	1.52
2.53	0.02836 47616	−0.18200 05246	0.50787 49817	−0.83981 94142	1.21219 21734	0.27338 80221	1.53
2.54	0.02866 34225	−0.18380 21724	0.51233 04648	−0.84501 25848	1.20492 53524	0.28289 55175	1.54
2.55	0.02893 97109	−0.18545 87109	0.51637 52344	−0.84952 05469	1.19705 16797	0.29261 26328	1.55
2.56	0.02919 26835	−0.18696 43776	0.51999 46752	−0.85332 45952	1.18855 92576	0.30254 23565	1.56
2.57	0.02942 13812	−0.18831 33223	0.52317 39770	−0.85640 58095	1.17943 60710	0.31268 77026	1.57
2.58	0.02962 48294	−0.18949 96068	0.52589 81336	−0.85874 50536	1.16966 99868	0.32305 17106	1.58
2.59	0.02980 20377	−0.19051 72046	0.52815 19417	−0.86032 29742	1.15924 87533	0.33363 74461	1.59
2.60	0.02995 20000	−0.19136 00000	0.52992 00000	−0.86112 00000	1.14816 00000	0.34444 80000	1.60
2.61	0.03007 36943	−0.19202 17879	0.53118 67083	−0.86111 63408	1.13639 12367	0.35548 64894	1.61
2.62	0.03016 60826	−0.19249 62732	0.53193 62664	−0.86029 19864	1.12392 98532	0.36675 60574	1.62
2.63	0.03022 81107	−0.19277 70702	0.53215 26730	−0.85862 67055	1.11076 31190	0.37825 98730	1.63
2.64	0.03025 87085	−0.19285 77024	0.53181 97248	−0.85610 00448	1.09687 81824	0.39000 11315	1.64
2.65	0.03025 67891	−0.19273 16016	0.53092 10156	−0.85269 13281	1.08226 20703	0.40198 30547	1.65
2.66	0.03022 12495	−0.19239 21076	0.52943 99352	−0.84837 96552	1.06690 16876	0.41420 88905	1.66
2.67	0.03015 09704	−0.19183 24679	0.52735 96683	−0.84314 39008	1.05078 38166	0.42668 19134	1.67
2.68	0.03004 48154	−0.19104 58368	0.52466 31936	−0.83696 27136	1.03389 51168	0.43940 54246	1.68
2.69	0.02990 16317	−0.19002 52752	0.52133 32829	−0.82981 45154	1.01622 21240	0.45238 27520	1.69
2.70	0.02972 02500	−0.18876 37500	0.51735 25000	−0.82167 75000	0.99775 12500	0.46561 72500	1.70
2.71	0.02949 94834	−0.18725 41335	0.51270 31996	−0.81252 96321	0.97846 87823	0.47911 23003	1.71
2.72	0.02923 81286	−0.18548 92032	0.50736 75264	−0.80234 86464	0.95836 08832	0.49287 13114	1.72
2.73	0.02893 49650	−0.18346 16409	0.50132 74142	−0.79111 20467	0.93741 35896	0.50689 77188	1.73
2.74	0.02858 87545	−0.18116 40324	0.49456 45848	−0.77879 71048	0.91561 28124	0.52119 49855	1.74
2.75	0.02819 82422	−0.17858 88672	0.48706 05469	−0.76538 08594	0.89294 43359	0.53576 66016	1.75
2.76	0.02776 21555	−0.17572 85376	0.47879 65952	−0.75084 01152	0.86939 38176	0.55061 60845	1.76
2.77	0.02727 92044	−0.17257 53385	0.46975 38095	−0.73515 14420	0.84494 67873	0.56574 69793	1.77
2.78	0.02674 80814	−0.16912 14668	0.45991 30536	−0.71829 11736	0.81958 86468	0.58116 28586	1.78
2.79	0.02616 74609	−0.16535 90208	0.44925 49742	−0.70023 54067	0.79330 46696	0.59686 73228	1.79
2.80	0.02553 60000	−0.16128 00000	0.43776 00000	−0.68096 00000	0.76608 00000	0.61286 40000	1.80
2.81	0.02485 23376	−0.15687 63042	0.42540 83408	−0.66044 05733	0.73789 96529	0.62915 65462	1.81
2.82	0.02411 50946	−0.15213 97332	0.41217 99864	−0.63865 25064	0.70874 85132	0.64574 86454	1.82
2.83	0.02332 28741	−0.14706 19865	0.39805 47055	−0.61557 09380	0.67861 13352	0.66264 40097	1.83
2.84	0.02247 42605	−0.14163 46624	0.38301 20448	−0.59117 07648	0.64747 27424	0.67984 63795	1.84
2.85	0.02156 78203	−0.13584 92578	0.36703 13281	−0.56542 66406	0.61531 72266	0.69735 95234	1.85
2.86	0.02060 21015	−0.12969 71676	0.35009 16552	−0.53831 29752	0.58212 91476	0.71518 72385	1.86
2.87	0.01957 56335	−0.12316 96841	0.33217 19008	−0.50980 39333	0.54789 27329	0.73333 33502	1.87
2.88	0.01848 69274	−0.11625 79968	0.31325 07136	−0.47987 34336	0.51259 20768	0.75180 17126	1.88
2.89	0.01733 44751	−0.10895 31915	0.29330 65154	−0.44849 51479	0.47621 11402	0.77059 62087	1.89
2.90	0.01611 67500	−0.10124 62500	0.27231 75000	−0.41564 25000	0.43873 37500	0.78972 07500	1.90
2.91	0.01483 22068	−0.09312 80498	0.25026 16321	−0.38128 86646	0.40014 35985	0.80917 92770	1.91
2.92	0.01347 92806	−0.08458 93632	0.22711 66464	−0.34540 65664	0.36042 42432	0.82897 57594	1.92
2.93	0.01205 63881	−0.07562 08571	0.20286 00467	−0.30796 88792	0.31955 91059	0.84911 41956	1.93
2.94	0.01056 19265	−0.06621 30924	0.17746 91048	−0.26894 80248	0.27753 14724	0.86959 86135	1.94
2.95	0.00899 42734	−0.05635 65234	0.15092 08594	−0.22831 61719	0.23432 44922	0.89043 30703	1.95
2.96	0.00735 17875	−0.04604 14976	0.12319 21152	−0.18604 52352	0.18992 11776	0.91162 16525	1.96
2.97	0.00563 28077	−0.03525 82547	0.09425 94420	−0.14210 68745	0.14430 44035	0.93316 84760	1.97
2.98	0.00383 56534	−0.02399 69268	0.06409 91736	−0.09647 24936	0.09745 69068	0.95507 76866	1.98
2.99	0.00195 86242	−0.01224 75371	0.03268 74067	−0.04911 32392	0.04936 12858	0.97735 34596	1.99
3.00	0.00000 00000	0.00000 00000	0.00000 00000	0.00000 00000	0.00000 00000	1.00000 00000	2.00
	A_3	A_2	A_1	A_0	A_{-1}	A_{-2}	$-p$

SEVEN-POINT LAGRANGIAN INTERPOLATION COEFFICIENTS Table 25.1

$$A_k^7(p) = (-1)^{k+3}\frac{p(p^2-1)(p^2-4)(p^2-9)}{(3+k)!(3-k)!(p-k)}$$

p	A_{-3}	A_{-2}	A_{-1}	A_0	A_1	A_2	A_3	
0.0	0.00000 00000	0.00000 00000	0.00000 00000	1.00000 00000	0.00000 00000	0.00000 00000	0.00000 00000	0.0
0.1	-0.00159 10125	0.01409 18250	-0.06725 64375	0.98642 77500	0.08220 23125	-0.01557 51750	0.00170 07375	0.1
0.2	-0.00295 68000	0.02580 48000	-0.11827 20000	0.94617 60000	0.17740 80000	-0.03153 92000	0.00337 92000	0.2
0.3	-0.00400 28625	0.03445 94250	-0.15241 66875	0.88062 97500	0.28305 95625	-0.04662 15750	0.00489 23875	0.3
0.4	-0.00465 92000	0.03960 32000	-0.16972 80000	0.79206 40000	0.39603 20000	-0.05940 48000	0.00609 28000	0.4
0.5	-0.00488 28125	0.04101 56250	-0.17089 84375	0.68359 37500	0.51269 53125	-0.06835 93750	0.00683 59375	0.5
0.6	-0.00465 92000	0.03870 72000	-0.15724 80000	0.55910 40000	0.62899 20000	-0.07188 48000	0.00698 88000	0.6
0.7	-0.00400 28625	0.03291 24250	-0.13068 16875	0.42315 97500	0.74052 95625	-0.06835 65750	0.00643 93875	0.7
0.8	-0.00295 68000	0.02407 68000	-0.09363 20000	0.28089 60000	0.84268 80000	-0.05617 92000	0.00510 72000	0.8
0.9	-0.00159 10125	0.01283 78250	-0.04898 64375	0.13788 77500	0.93074 23125	-0.03384 51750	0.00295 47375	0.9
1.0	0.00000 00000	0.00000 00000	0.00000 00000	0.00000 00000	1.00000 00000	0.00000 00000	0.00000 00000	1.0
1.1	0.00170 07375	-0.01349 61750	0.04980 73125	-0.12678 22500	1.04595 35625	0.04648 68250	-0.00367 00125	1.1
1.2	0.00337 92000	-0.02661 12000	0.09676 80000	-0.23654 40000	1.06444 80000	0.10644 48000	-0.00788 48000	1.2
1.3	0.00489 23875	-0.03824 95750	0.13719 95625	-0.32365 02500	1.05186 33125	0.18031 94250	-0.01237 48625	1.3
1.4	0.00609 28000	-0.04730 88000	0.16755 20000	-0.38297 60000	1.00531 20000	0.26808 32000	-0.01675 52000	1.4
1.5	0.00683 59375	-0.05273 43750	0.18457 03125	-0.41015 62500	0.92285 15625	0.36914 06250	-0.02050 78125	1.5
1.6	0.00698 88000	-0.05358 08000	0.18547 20000	-0.40185 60000	0.80371 20000	0.48222 72000	-0.02296 32000	1.6
1.7	0.00643 93875	-0.04907 85750	0.16813 95625	-0.35606 02500	0.64853 83125	0.60530 24250	-0.02328 08625	1.7
1.8	0.00510 72000	-0.03870 72000	0.13132 80000	-0.27238 40000	0.45964 80000	0.73543 68000	-0.02042 88000	1.8
1.9	0.00295 47375	-0.02227 41750	0.07488 73125	-0.15240 22500	0.24130 35625	0.86869 28250	-0.01316 20125	1.9
2.0	0.00000 00000	0.00000 00000	0.00000 00000	0.00000 00000	0.00000 00000	1.00000 00000	0.00000 00000	2.0
2.1	-0.00367 00125	0.02739 08250	-0.09056 64375	0.17825 77500	-0.25523 26875	1.12302 38250	0.02079 67375	2.1
2.2	-0.00788 48000	0.05857 28000	-0.19219 20000	0.37273 60000	-0.51251 20000	1.23002 88000	0.05125 12000	2.2
2.3	-0.01237 48625	0.09151 64250	-0.29812 16875	0.57031 97500	-0.75677 04375	1.31173 54250	0.09369 53875	2.3
2.4	-0.01675 52000	0.12337 92000	-0.39916 80000	0.75398 40000	-0.96940 80000	1.35717 12000	0.15079 68000	2.4
2.5	-0.02050 78125	0.15039 06250	-0.48339 84375	0.90234 37500	-1.12792 96875	1.35351 56250	0.22558 59375	2.5
2.6	-0.02296 32000	0.16773 12000	-0.53580 80000	0.98918 40000	-1.20556 80000	1.28593 92000	0.32148 48000	2.6
2.7	-0.02328 08625	0.16940 54250	-0.53797 66875	0.98296 97500	-1.17089 04375	1.13743 64250	0.44233 63875	2.7
2.8	-0.02042 88000	0.14810 88000	-0.46771 20000	0.84633 60000	-0.98739 20000	0.88865 28000	0.59243 52000	2.8
2.9	-0.01316 20125	0.09508 88250	-0.29867 64375	0.53555 77500	-0.61307 26875	0.51770 58250	0.77655 87375	2.9
3.0	0.00000 00000	0.00000 00000	0.00000 00000	0.00000 00000	0.00000 00000	0.00000 00000	1.00000 00000	3.0
	A_3	A_2	A_1	A_0	A_{-1}	A_{-2}	A_{-3}	$-p$

EIGHT-POINT LAGRANGIAN INTERPOLATION COEFFICIENTS

$$A_k^8(p) = (-1)^{k+4}\frac{p(p^2-1)(p^2-4)(p^2-9)(p-4)}{(3+k)!(4-k)!(p-k)}$$

p	A_{-3}	A_{-2}	A_{-1}	A_0	A_1	A_2	A_3	A_4	
0.0	0.00000 00000	0.00000 00000	0.00000 00000	1.00000 00000	0.00000 00000	0.00000 00000	0.00000 00000	0.00000 00000	1.0
0.1	-0.00088 64213	0.00915 96863	-0.05246 00213	0.96176 70563	0.10686 30063	-0.03037 15913	0.00663 28763	-0.00070 45913	0.9
0.2	-0.00160 51200	0.01634 30400	-0.08988 67200	0.89886 72000	0.22471 68000	-0.05992 44800	0.01284 09600	-0.00135 16800	0.8
0.3	-0.00211 57988	0.02124 99787	-0.11278 83487	0.81458 25188	0.34910 67938	-0.08624 99137	0.01810 18337	-0.00188 70638	0.7
0.4	-0.00239 61600	0.02376 19200	-0.12220 41600	0.71285 76000	0.47523 84000	-0.10692 86400	0.02193 40800	-0.00226 30400	0.6
0.5	-0.00244 14063	0.02392 57812	-0.11962 89062	0.59814 45312	0.59814 45313	-0.11962 89062	0.02392 57812	-0.00244 14062	0.5
1.0	0.00000 00000	0.00000 00000	0.00000 00000	0.00000 00000	1.00000 00000	0.00000 00000	0.00000 00000	0.00000 00000	0.0
1.1	0.00070 45912	-0.00652 31512	0.02888 82412	-0.09191 71312	1.01108 84438	0.06740 58962	-0.01064 30362	0.00099 61462	-0.1
1.2	0.00135 16800	-0.01241 85600	0.05419 00800	-0.16558 08000	0.99348 48000	0.14902 27200	-0.02207 74400	0.00202 75200	-0.2
1.3	0.00188 70638	-0.01721 23088	0.07408 77638	-0.21846 39188	0.94667 69812	0.24343 12238	-0.03341 21288	0.00300 53238	-0.3
1.4	0.00226 30400	-0.02050 04800	0.08712 70400	-0.24893 44000	0.87127 04000	0.34850 81600	-0.04356 35200	0.00382 97600	-0.4
1.5	0.00244 14062	-0.02197 26562	0.09228 51562	-0.25634 76562	0.76904 29688	0.46142 57812	-0.05126 95312	0.00439 45312	-0.5
1.6	0.00239 61600	-0.02143 23200	0.08902 65600	-0.24111 36000	0.64296 96000	0.57867 26400	-0.05511 16800	0.00459 26400	-0.6
1.7	0.00211 57988	-0.01881 34538	0.07734 41988	-0.20473 46438	0.49721 27062	0.69609 77888	-0.05354 59838	0.00432 35888	-0.7
1.8	0.00160 51200	-0.01419 26400	0.05778 43200	-0.14981 12000	0.33707 52000	0.80898 04800	-0.04494 33600	0.00350 20800	-0.8
1.9	0.00088 64213	-0.00779 59613	0.03145 26712	-0.08001 11812	0.16891 24938	0.91212 74662	-0.02764 02263	0.00206 83163	-0.9
2.0	0.00000 00000	0.00000 00000	0.00000 00000	0.00000 00000	0.00000 00000	1.00000 00000	0.00000 00000	0.00000 00000	-1.0
2.1	-0.00099 61462	0.00867 37612	-0.03441 52462	0.08467 24312	-0.16164 73688	1.06687 26338	0.03951 38012	-0.00267 38662	-1.1
2.2	-0.00202 75200	0.01757 18400	-0.06918 91200	0.16773 12000	-0.30750 72000	1.10702 59200	0.09225 21600	-0.00585 72800	-1.2
2.3	-0.00300 53238	0.02592 96538	-0.10136 13738	0.24238 58938	-0.42883 65812	1.11497 51112	0.15928 21588	-0.00936 95388	-1.3
2.4	-0.00382 97600	0.03290 11200	-0.12773 37600	0.30159 36000	-0.51701 76000	1.08573 69600	0.24127 44800	-0.01292 54400	-1.4
2.5	-0.00439 45312	0.03759 76562	-0.14501 95312	0.33837 89062	-0.56396 48438	1.01513 67188	0.33837 89062	-0.01611 32812	-1.5
2.6	-0.00459 26400	0.03913 72800	-0.15002 62400	0.34621 44000	-0.56259 84000	0.90015 74400	0.45007 87200	-0.01837 05600	-1.6
2.7	-0.00432 35888	0.03670 45088	-0.13987 39388	0.31946 51688	-0.50738 58562	0.73933 36762	0.57503 73038	-0.01895 72738	-1.7
2.8	-0.00350 20800	0.02962 17600	-0.11225 08800	0.25390 08000	-0.39495 68000	0.53319 16800	0.71092 22400	-0.01692 67200	-1.8
2.9	-0.00206 83162	0.01743 29512	-0.06570 88162	0.14727 83812	-0.22479 33188	0.28473 82038	0.85421 46112	-0.01109 36962	-1.9
3.0	0.00000 00000	0.00000 00000	0.00000 00000	0.00000 00000	0.00000 00000	0.00000 00000	1.00000 00000	0.00000 00000	-2.0
3.1	0.00267 38662	-0.02238 70762	0.08354 20162	-0.18415 17562	0.27184 30688	-0.31138 38788	1.14174 08888	0.01812 28712	-2.1
3.2	0.00585 72800	-0.04888 57600	0.18157 56800	-0.39719 68000	0.57774 08000	-0.63551 48800	1.27102 97600	0.04539 39200	-2.2
3.3	0.00936 95388	-0.07796 16338	0.28827 67388	-0.62605 55438	0.89825 36062	-0.95353 07512	1.37732 21962	0.08432 58488	-2.3
3.4	0.01292 54400	-0.10723 32800	0.39481 34400	-0.85155 84000	1.20637 44000	-1.24084 22400	1.44764 92800	0.13787 13600	-2.4
3.5	0.01611 32812	-0.13330 07812	0.48876 95312	-1.04736 32812	1.46630 85938	-1.46630 85938	1.46630 85938	0.20947 26562	-2.5
3.6	0.01837 05600	-0.15155 71200	0.55351 29600	-1.17877 76000	1.63215 36000	-1.59134 97600	1.41453 31200	0.30311 42400	-2.6
3.7	0.01895 72738	-0.15598 17788	0.56750 81738	-1.20148 12688	1.66647 43312	-1.56899 31862	1.27013 73412	0.42337 91138	-2.7
3.8	0.01692 67200	-0.13891 58400	0.50356 99200	-1.06014 72000	1.43877 12000	-1.34285 31200	1.00713 98400	0.57550 84800	-2.8
3.9	0.01109 36962	-0.09081 78862	0.32805 64462	-0.68695 58062	0.92383 71188	-0.84604 03088	0.59536 16988	0.76546 50412	-2.9
4.0	0.00000 00000	0.00000 00000	0.00000 00000	0.00000 00000	0.00000 00000	0.00000 00000	0.00000 00000	1.00000 00000	-3.0
	A_4	A_3	A_2	A_1	A_0	A_{-1}	A_{-2}	A_{-3}	p

Table 25.2

COEFFICIENTS FOR DIFFERENTIATION

Differentiation Formula: $\dfrac{d^k f(x)}{dx^k}\bigg|_{x=x_j} \approx \dfrac{k!}{m!h^k}\sum_{i=0}^{m} A_i f(x_i)$

FIRST DERIVATIVE ($k=1$)

*	j	A_0	A_1	A_2	A_3	A_4	A_5	$\dfrac{h^k}{k!}$ Error
				Three Point ($m=2$)				
	0	−3	4	−1				1/3
	1	−1	0	1				−1/6 $h^3 f^{(3)}$
	2	1	−4	3				1/3
				Four Point ($m=3$)				
	0	−11	18	−9	2			−1/4
	1	−2	−3	6	−1			1/12
	2	1	−6	3	2			−1/12 $h^4 f^{(4)}$
	3	−2	9	−18	11			1/4
				Five Point ($m=4$)				
	0	−50	96	−72	32	−6		1/5
	1	−6	−20	36	−12	2		−1/20
	2	2	−16	0	16	−2		1/30 $h^5 f^{(5)}$
	3	−2	12	−36	20	6		−1/20
	4	6	−32	72	−96	50		1/5
				Six Point ($m=5$)				
	0	−274	600	−600	400	−150	24	−1/6
	1	−24	−130	240	−120	40	−6	1/30
	2	6	−60	−40	120	−30	4	−1/60 $h^6 f^{(6)}$
	3	−4	30	−120	40	60	−6	1/60
	4	6	−40	120	−240	130	24	−1/30
	5	−24	150	−400	600	−600	274	1/6

SECOND DERIVATIVE ($k=2$)

*	j	A_0	A_1	A_2	A_3	A_4	A_5	$\dfrac{h^k}{k!}$ Error
				Three Point ($m=2$)				
	0	1	−2	1				−1/2 $h^3 f^{(3)}$
	1	1	−2	1				−1/24 $h^4 f^{(4)}$
	2	1	−2	1				1/2 $h^3 f^{(3)}$
				Four Point ($m=3$)				
	0	6	−15	12	−3			11/24
	1	3	−6	3	0			−1/24
	2	0	3	−6	3			−1/24 $h^4 f^{(4)}$
	3	−3	12	−15	6			11/24
				Five Point ($m=4$)				
	0	35	−104	114	−56	11		−5/12 $h^5 f^{(5)}$
	1	11	−20	6	4	−1		1/24
	2	−1	16	−30	16	−1		1/180 $h^6 f^{(6)}$
	3	−1	4	6	−20	11		−1/24 $h^5 f^{(5)}$
	4	11	−56	114	−104	35		5/12
				Six Point ($m=5$)				
	0	225	−770	1070	−780	305	−50	137/360
	1	50	−75	−20	70	−30	5	−13/360
	2	−5	80	−150	80	−5	0	1/180 $h^6 f^{(6)}$
	3	0	−5	80	−150	80	−5	1/180
	4	5	−30	70	−20	−75	50	−13/360
	5	−50	305	−780	1070	−770	225	137/360

THIRD DERIVATIVE ($k=3$)

j	A_0	A_1	A_2	A_3	A_4	A_5	$\dfrac{h^k}{k!}$ Error	*
			Four Point ($m=3$)					
0	−1	3	−3	1			−1/4	
1	−1	3	−3	1			−1/12 $h^4 f^{(4)}$	
2	−1	3	−3	1			1/12	
3	−1	3	−3	1			1/4	
			Five Point ($m=4$)					
0	−10	36	−48	28	−6		7/24	
1	−6	20	−24	12	−2		1/24	
2	−2	4	0	−4	2		−1/24 $h^5 f^{(5)}$	
3	2	−12	24	−20	6		1/24	
4	6	−28	48	−36	10		7/24	
			Six Point ($m=5$)					
0	−85	355	−590	490	−205	35	−5/16	
1	−35	125	−170	110	−35	5	−1/48	
2	−5	−5	50	−70	35	−5	1/48 $h^6 f^{(6)}$	
3	5	−35	70	−50	5	5	−1/48	
4	−5	35	−110	170	−125	35	1/48	
5	−35	205	−490	590	−355	85	5/16	

FOURTH DERIVATIVE ($k=4$)

j	A_0	A_1	A_2	A_3	A_4	A_5	$\dfrac{h^k}{k!}$ Error	*
			Five Point ($m=4$)					
0	1	−4	6	−4	1		−1/12 $h^5 f^{(5)}$	
1	1	−4	6	−4	1		−1/24	
2	1	−4	6	−4	1		−1/144 $h^6 f^{(6)}$	
3	1	−4	6	−4	1		1/24 $h^5 f^{(5)}$	
4	1	−4	6	−4	1		1/12	
			Six Point ($m=5$)					
0	15	−70	130	−120	55	−10	17/144	
1	10	−45	80	−70	30	−5	5/144	
2	5	−20	30	−20	5	0	−1/144 $h^6 f^{(6)}$	
3	0	5	−20	30	−20	5	−1/144	
4	−5	30	−70	80	−45	10	5/144	
5	−10	55	−120	130	−70	15	17/144	

FIFTH DERIVATIVE ($k=5$)

j	A_0	A_1	A_2	A_3	A_4	A_5	$\dfrac{h^k}{k!}$ Error	*
			Six Point ($m=5$)					
0	−1	5	−10	10	−5	1	−1/48	
1	−1	5	−10	10	−5	1	−1/80	
2	−1	5	−10	10	−5	1	−1/240 $h^6 f^{(6)}$	
3	−1	5	−10	10	−5	1	1/240	
4	−1	5	−10	10	−5	1	1/80	
5	−1	5	−10	10	−5	1	1/48	

Compiled from W. G. Bickley, Formulae for numerical differentiation, Math. Gaz. **25**, 19–27, 1941 (with permission).

*See page II.

LAGRANGIAN INTEGRATION COEFFICIENTS

Table 25.3

$$\int_{x_m}^{x_{m+1}} f(x)\,dx \approx h \sum_k A_k(m) f(x_k) \qquad *$$

$$DA_k^n(m)$$

$n = \text{odd}$

n	$m\backslash k$	-4	-3	-2	-1	0	1	2	3	4		D
3	−1				5	8	−1				0	12
5	−2			251	646	−264	106	−19			1	720
	−1			−19	346	456	−74	11			0	
7	−3		19087	65112	−46461	37504	−20211	6312	−863		2	60480
	−2		−863	25128	46989	−16256	7299	−2088	271		1	
	−1		271	−2760	30819	37504	−6771	1608	−191		0	
9	−4	1070017	4467094	−4604594	5595358	−5033120	3146338	−1291214	312874	−33953	3	3628800
	−3	−33953	1375594	3244786	−1752542	1317280	−755042	294286	−68906	7297	2	
	−2	7297	−99626	1638286	2631838	−833120	397858	−142094	31594	−3233	1	
	−1	−3233	36394	−216014	1909858	2224480	−425762	126286	−25706	2497	0	
		4	3	2	1	0	−1	−2	−3	−4	$k\backslash m$	

$n = \text{even}$

n	$m\backslash k$	-4	-3	-2	-1	0	1	2	3	4	5		D
4	−1				9	19	−5	1				1	24
	0				−1	13	13	−1				0	
6	−2			475	1427	−798	482	−173	27			2	1440
	−1			−27	637	1022	−258	77	−11			1	
	0			11	−93	802	802	−93	11			0	
8	−3		36799	139849	−121797	123133	−88547	41499	−11351	1375		3	120960
	−2		−1375	47799	101349	−44797	26883	−11547	2999	−351		2	
	−1		351	−4183	57627	81693	−20227	7227	−1719	191		1	
	0		−191	1879	−9531	68323	68323	−9531	1879	−191		0	
10	−4	2082753	9449717	−11271304	16002320	−17283646	13510082	−7394032	2687864	−583435	57281	4	7257600
	−3	−57281	2655563	6872072	−4397584	3973310	−2848834	1481072	−520312	110219	−10625	3	
	−2	10625	−163531	3133688	5597072	−2166334	1295810	−617584	206072	−42187	3969	2	
	−1	−3969	50315	−342136	3609968	4763582	−1166146	462320	−141304	27467	−2497	1	
	0	2497	−28939	162680	−641776	4134338	4134338	−641776	162680	−28939	2497	0	
		5	4	3	2	1	0	−1	−2	−3	−4	$k\backslash m$	

Compiled from National Bureau of Standards, Tables of Lagrangian interpolation coefficients. Columbia Univ. Press, New York, N.Y., 1944 (with permission).

*See page II.

Table 25.4   ABSCISSAS AND WEIGHT FACTORS FOR GAUSSIAN INTEGRATION

$$\int_{-1}^{+1} f(x)dx \approx \sum_{i=1}^{n} w_i f(x_i)$$

Abscissas=$\pm x_i$ (Zeros of Legendre Polynomials)   Weight Factors=w_i

$\pm x_i$	w_i	$\pm x_i$	w_i
$n=2$		**$n=8$**	
0.57735 02691 89626	1.00000 00000 00000	0.18343 46424 95650	0.36268 37833 78362
$n=3$		0.52553 24099 16329	0.31370 66458 77887
0.00000 00000 00000	0.88888 88888 88889	0.79666 64774 13627	0.22238 10344 53374
0.77459 66692 41483	0.55555 55555 55556	0.96028 98564 97536	0.10122 85362 90376
$n=4$		**$n=9$**	
0.33998 10435 84856	0.65214 51548 62546	0.00000 00000 00000	0.33023 93550 01260
0.86113 63115 94053	0.34785 48451 37454	0.32425 34234 03809	0.31234 70770 40003
$n=5$		0.61337 14327 00590	0.26061 06964 02935
0.00000 00000 00000	0.56888 88888 88889	0.83603 11073 26636	0.18064 81606 94857
0.53846 93101 05683	0.47862 86704 99366	0.96816 02395 07626	0.08127 43883 61574
0.90617 98459 38664	0.23692 68850 56189	**$n=10$**	
$n=6$		0.14887 43389 81631	0.29552 42247 14753
0.23861 91860 83197	0.46791 39345 72691	0.43339 53941 29247	0.26926 67193 09996
0.66120 93864 66265	0.36076 15730 48139	0.67940 95682 99024	0.21908 63625 15982
0.93246 95142 03152	0.17132 44923 79170	0.86506 33666 88985	0.14945 13491 50581
$n=7$		0.97390 65285 17172	0.06667 13443 08688
0.00000 00000 00000	0.41795 91836 73469	**$n=12$**	
0.40584 51513 77397	0.38183 00505 05119	0.12523 34085 11469	0.24914 70458 13403
0.74153 11855 99394	0.27970 53914 89277	0.36783 14989 98180	0.23349 25365 38355
0.94910 79123 42759	0.12948 49661 68870	0.58731 79542 86617	0.20316 74267 23066
		0.76990 26741 94305	0.16007 83285 43346
		0.90411 72563 70475	0.10693 93259 95318
		0.98156 06342 46719	0.04717 53363 86512

$\pm x_i$	w_i
$n=16$	
0.09501 25098 37637 440185	0.18945 06104 55068 496285
0.28160 35507 79258 913230	0.18260 34150 44923 588867
0.45801 67776 57227 386342	0.16915 65193 95002 538189
0.61787 62444 02643 748447	0.14959 59888 16576 732081
0.75540 44083 55003 033895	0.12462 89712 55533 872052
0.86563 12023 87831 743880	0.09515 85116 82492 784810
0.94457 50230 73232 576078	0.06225 35239 38647 892863
0.98940 09349 91649 932596	0.02715 24594 11754 094852
$n=20$	
0.07652 65211 33497 333755	0.15275 33871 30725 850698
0.22778 58511 41645 078080	0.14917 29864 72603 746788
0.37370 60887 15419 560673	0.14209 61093 18382 051329
0.51086 70019 50827 098004	0.13168 86384 49176 626898
0.63605 36807 26515 025453	0.11819 45319 61518 417312
0.74633 19064 60150 792614	0.10193 01198 17240 435037
0.83911 69718 22218 823395	0.08327 67415 76704 748725
0.91223 44282 51325 905868	0.06267 20483 34109 063570
0.96397 19272 77913 791268	0.04060 14298 00386 941331
0.99312 85991 85094 924786	0.01761 40071 39152 118312
$n=24$	
0.06405 68928 62605 626085	0.12793 81953 46752 156974
0.19111 88674 73616 309159	0.12583 74563 46828 296121
0.31504 26796 96163 374387	0.12167 04729 27803 391204
0.43379 35076 26045 138487	0.11550 56680 53725 601353
0.54542 14713 88839 535658	0.10744 42701 15965 634783
0.64809 36519 36975 569252	0.09761 86521 04113 888270
0.74012 41915 78554 364244	0.08619 01615 31953 275917
0.82000 19859 73902 921954	0.07334 64814 11080 305734
0.88641 55270 04401 034213	0.05929 85849 15436 780746
0.93827 45520 02732 758524	0.04427 74388 17419 806169
0.97472 85559 71309 498198	0.02853 13886 28933 663181
0.99518 72199 97021 360180	0.01234 12297 99987 199547

Compiled from P. Davis and P. Rabinowitz, Abscissas and weights for Gaussian quadratures of high order, J. Research NBS **56**, 35–37, 1956, RP2645; P. Davis and P. Rabinowitz, Additional abscissas and weights for Gaussian quadratures of high order.   Values for n=64, 80, and 96, J. Research NBS **60**, 613–614, 1958, RP2875; and A. N. Lowan, N. Davids, and A. Levenson, Table of the zeros of the Legendre polynomials of order 1–16 and the weight coefficients for Gauss' mechanical quadrature formula, Bull. Amer. Math. Soc. **48**, 739–743, 1942 (with permission).

Table 25.4

ABSCISSAS AND WEIGHT FACTORS FOR GAUSSIAN INTEGRATION

$$\int_{-1}^{+1} f(x)\,dx \approx \sum_{i=1}^{n} w_i f(x_i)$$

Abscissas=$\pm x_i$ (Zeros of Legendre Polynomials)   Weight Factors=w_i

$\pm x_i$	w_i
n=32	
0.04830 76656 87738 316235	0.09654 00885 14727 800567
0.14447 19615 82796 493485	0.09563 87200 79274 859419
0.23928 73622 52137 074545	0.09384 43990 80804 565639
0.33186 86022 82127 649780	0.09117 38786 95763 884713
0.42135 12761 30635 345364	0.08765 20930 04403 811143
0.50689 99089 32229 390024	0.08331 19242 26946 755222
0.58771 57572 40762 329041	0.07819 38957 87070 306472
0.66304 42669 30215 200975	0.07234 57941 08848 506225
0.73218 21187 40289 680387	0.06582 22227 76361 846838
0.79448 37959 67942 406963	0.05868 40934 78535 547145
0.84936 76137 32569 970134	0.05099 80592 62376 176196
0.89632 11557 66052 123965	0.04283 58980 22226 680657
0.93490 60759 37739 689171	0.03427 38629 13021 433103
0.96476 22555 87506 430774	0.02539 20653 09262 059456
0.98561 15115 45268 335400	0.01627 43947 30905 670605
0.99726 38618 49481 563545	0.00701 86100 09470 096600
n=40	
0.03877 24175 06050 821933	0.07750 59479 78424 811264
0.11608 40706 75255 208483	0.07703 98181 64247 965588
0.19269 75807 01371 099716	0.07611 03619 00626 242372
0.26815 21850 07253 681141	0.07472 31690 57968 264200
0.34199 40908 25758 473007	0.07288 65823 95804 059061
0.41377 92043 71605 001525	0.07061 16473 91286 779695
0.48307 58016 86178 712909	0.06791 20458 15233 903826
0.54946 71250 95128 202076	0.06480 40134 56601 038075
0.61255 38896 67980 237953	0.06130 62424 92928 939167
0.67195 66846 14179 548379	0.05743 97690 99391 551367
0.72731 82551 89927 103281	0.05322 78469 83936 824355
0.77830 56514 26519 387695	0.04869 58076 35072 232061
0.82461 22308 33311 663196	0.04387 09081 85673 271992
0.86595 95032 12259 503821	0.03878 21679 74472 017640
0.90209 88069 68874 296728	0.03346 01952 82547 847393
0.93281 28082 78676 533361	0.02793 70069 80023 401098
0.95791 68192 13791 655805	0.02224 58491 94166 957262
0.97725 99499 83774 262663	0.01642 10583 81907 888713
0.99072 62386 99457 006453	0.01049 82845 31152 813615
0.99823 77097 10559 200350	0.00452 12770 98533 191258
n=48	
0.03238 01709 62869 362033	0.06473 76968 12683 922503
0.09700 46992 09462 698930	0.06446 61644 35950 082207
0.16122 23560 68891 718056	0.06392 42385 84648 186624
0.22476 37903 94689 061225	0.06311 41922 86254 025657
0.28736 24873 55455 576736	0.06203 94231 59892 663904
0.34875 58862 92160 738160	0.06070 44391 65893 880053
0.40868 64819 90716 729916	0.05911 48396 98395 635746
0.46690 29047 50958 404545	0.05727 72921 00403 215705
0.52316 09747 22233 033678	0.05519 95036 99984 162868
0.57722 47260 83972 703818	0.05289 01894 85193 667096
0.62886 73967 76513 623995	0.05035 90355 53854 474958
0.67787 23796 32663 905212	0.04761 66584 92490 474826
0.72403 41309 23814 654674	0.04467 45608 56694 280419
0.76715 90325 15740 339254	0.04154 50829 43464 749214
0.80706 62040 29442 627083	0.03824 13510 65830 706317
0.84358 82616 24393 530711	0.03477 72225 64770 438893
0.87657 20202 74247 885906	0.03116 72278 32798 088902
0.90587 91367 15569 672822	0.02742 65097 08356 948200
0.93138 66907 06554 333114	0.02357 07608 39324 379141
0.95298 77031 60430 860723	0.01961 61604 57355 527814
0.97059 15925 46247 250461	0.01557 93157 22943 848728
0.98412 45837 22826 857745	0.01147 72345 79234 539490
0.99353 01722 66350 757548	0.00732 75539 01276 262102
0.99877 10072 52426 118601	0.00315 33460 52305 838633

Table 25.4

ABSCISSAS AND WEIGHT FACTORS FOR GAUSSIAN INTEGRATION

$$\int_{-1}^{+1} f(x)dx \approx \sum_{i=1}^{n} w_i f(x_i)$$

Abscissas$=\pm x_i$ (Zeros of Legendre Polynomials)   Weight Factors$=w_i$

$\pm x_i$	w_i
$n=64$	
0.02435 02926 63424 432509	0.04869 09570 09139 720383
0.07299 31217 87799 039450	0.04857 54674 41503 426935
0.12146 28192 96120 554470	0.04834 47622 34802 957170
0.16964 44204 23992 818037	0.04799 93885 96458 307728
0.21742 36437 40007 084150	0.04754 01657 14830 308662
0.26468 71622 08767 416374	0.04696 81828 16210 017325
0.31132 28719 90210 956158	0.04628 47965 81314 417296
0.35722 01583 37668 115950	0.04549 16279 27418 144480
0.40227 01579 63991 603696	0.04459 05581 63756 563060
0.44636 60172 53464 087985	0.04358 37245 29323 453377
0.48940 31457 07052 957479	0.04247 35151 23653 589007
0.53127 94640 19894 545658	0.04126 25632 42623 528610
0.57189 56462 02634 034284	0.03995 37411 32720 341387
0.61115 53551 72393 250249	0.03855 01531 78615 629129
0.64896 54712 54657 339858	0.03705 51285 40240 046040
0.68523 63130 54233 242564	0.03547 22132 56882 383811
0.71988 18501 71610 826849	0.03380 51618 37141 609392
0.75281 99072 60531 896612	0.03205 79283 54851 553585
0.78397 23589 43341 407610	0.03023 46570 72402 478868
0.81326 53151 22797 559742	0.02833 96726 14259 483228
0.84062 92962 52580 362752	0.02637 74697 15054 658672
0.86599 93981 54092 819761	0.02435 27025 68710 873338
0.88931 54459 95114 105853	0.02227 01738 08383 254159
0.91052 21370 78502 805756	0.02013 48231 53530 209372
0.92956 91721 31939 575821	0.01795 17157 75697 343085
0.94641 13748 58402 816062	0.01572 60304 76024 719322
0.96100 87996 52053 718919	0.01346 30478 96718 642598
0.97332 68277 89910 963742	0.01116 81394 60131 128819
0.98333 62538 84625 956931	0.00884 67598 26363 947723
0.99101 33714 76744 320739	0.00650 44579 68978 362856
0.99634 01167 71955 279347	0.00414 70332 60562 467635
0.99930 50417 35772 139457	0.00178 32807 21696 432947
$n=80$	
0.01951 13832 56793 997654	0.03901 78136 56306 654811
0.05850 44371 52420 668629	0.03895 83959 62769 531199
0.09740 83984 41584 599063	0.03883 96510 59051 968932
0.13616 40228 09143 886559	0.03866 17597 74076 463327
0.17471 22918 32646 812559	0.03842 49930 06959 423185
0.21299 45028 57666 132572	0.03812 97113 14477 638344
0.25095 23583 92272 120493	0.03777 63643 62001 397490
0.28852 80548 84511 853109	0.03736 54902 38730 490027
0.32566 43707 47701 914619	0.03689 77146 38276 008839
0.36230 47534 99487 315619	0.03637 37499 05835 978044
0.39839 34058 81969 227024	0.03579 43939 53416 054603
0.43387 53708 31756 093062	0.03516 05290 44747 593496
0.46869 66151 70544 477036	0.03447 31204 51753 928794
0.50280 41118 88784 987594	0.03373 32149 84611 522817
0.53614 59208 97131 932020	0.03294 19393 97645 401383
0.56867 12681 22709 784725	0.03210 04986 73487 773148
0.60033 06228 29751 743155	0.03121 01741 88114 701642
0.63107 57730 46871 966248	0.03027 23217 59557 980661
0.66085 98989 86119 801736	0.02928 83695 83267 847693
0.68963 76443 42027 600771	0.02825 98160 57276 862397
0.71736 51853 62099 880254	0.02718 82275 00486 380674
0.74400 02975 83597 272317	0.02607 52357 67565 117903
0.76950 24201 35041 373866	0.02492 25357 64115 491105
0.79383 27175 04605 449949	0.02373 18828 65930 101293
0.81695 41386 81463 470371	0.02250 50902 46332 461926
0.83883 14735 80255 275617	0.02124 40261 15782 006389
0.85943 14066 63111 096977	0.01995 06108 78141 998929
0.87872 25676 78213 828704	0.01862 68142 08299 031429
0.89667 55794 38770 683194	0.01727 46520 56269 306359
0.91326 31025 71757 654165	0.01589 61835 83725 688045
0.92845 98771 72445 795953	0.01449 35080 40509 076117
0.94224 27613 09872 674752	0.01306 87615 92401 339294
0.95459 07663 43634 905493	0.01162 41141 20797 826916
0.96548 50890 43799 251452	0.01016 17660 41103 064521
0.97490 91405 85727 793386	0.00868 39452 69260 858426
0.98284 85727 38629 070418	0.00719 29047 68117 312753
0.98929 13024 99755 531027	0.00569 09224 51403 198649
0.99422 75409 65688 277892	0.00418 03131 24694 895237
0.99764 98643 98237 688900	0.00266 35335 89512 681669
0.99955 38226 51630 629880	0.00114 49500 03186 941534

Table 25.4

ABSCISSAS AND WEIGHT FACTORS FOR GAUSSIAN INTEGRATION

$$\int_{-1}^{+1} f(x)dx \approx \sum_{i=1}^{n} w_i f(x_i)$$

Abscissas= $\pm x_i$ (Zeros of Legendre Polynomials)   Weight Factors= w_i

$\pm x_i$	w_i
$n = 96$	

$\pm x_i$	w_i
0.01627 67448 49602 969579	0.03255 06144 92363 166242
0.04881 29851 36049 731112	0.03251 61187 13868 835987
0.08129 74954 64425 558994	0.03244 71637 14064 269364
0.11369 58501 10665 920911	0.03234 38225 68575 928429
0.14597 37146 54896 941989	0.03220 62047 94030 250669
0.17809 68823 67618 602759	0.03203 44562 31992 663218
0.21003 13104 60567 203603	0.03182 87588 94411 006535
0.24174 31561 63840 012328	0.03158 93307 70727 168558
0.27319 88125 91049 141487	0.03131 64255 96861 355813
0.30436 49443 54496 353024	0.03101 03325 86313 837423
0.33520 85228 92625 422616	0.03067 13761 23669 149014
0.36569 68614 72313 635031	0.03029 99154 20827 593794
0.39579 76498 28908 603285	0.02989 63441 36328 385984
0.42547 89884 07300 545365	0.02946 10899 58167 905970
0.45470 94221 67743 008636	0.02899 46141 50555 236543
0.48345 79739 20596 359768	0.02849 74110 65085 385646
0.51169 41771 54667 673586	0.02797 00076 16848 334440
0.53938 81083 24357 436227	0.02741 29627 26029 242823
0.56651 04185 61397 168404	0.02682 68667 25591 762198
0.59303 23647 77572 080684	0.02621 23407 35672 413913
0.61892 58401 25468 570386	0.02557 00360 05349 361499
0.64416 34037 84967 106798	0.02490 06332 22483 610288
0.66871 83100 43916 153953	0.02420 48417 92364 691282
0.69256 45366 42171 561344	0.02348 33990 85926 219842
0.71567 68123 48967 626225	0.02273 70696 58329 374001
0.73803 06437 44400 132851	0.02196 66444 38744 349195
0.75960 23411 76647 498703	0.02117 29398 92191 298988
0.78036 90438 67433 217604	0.02035 67971 54333 324595
0.80030 87441 39140 817229	0.01951 90811 40145 022410
0.81940 03107 37931 675539	0.01866 06796 27411 467385
0.83762 35112 28187 121494	0.01778 25023 16045 260838
0.85495 90334 34601 455463	0.01688 54798 64245 172450
0.87138 85059 09296 502874	0.01597 05629 02562 291381
0.88689 45174 02420 416057	0.01503 87210 26994 938006
0.90146 06353 15852 341319	0.01409 09417 72314 860916
0.91507 14231 20898 074206	0.01312 82295 66961 572637
0.92771 24567 22308 690965	0.01215 16046 71088 319635
0.93937 03397 52755 216932	0.01116 21020 99838 498591
0.95003 27177 84437 635756	0.01016 07705 35008 415758
0.95968 82914 48742 539300	0.00914 86712 30783 386633
0.96832 68284 63264 212174	0.00812 68769 25698 759217
0.97593 91745 85136 466453	0.00709 64707 91153 865269
0.98251 72635 63014 677447	0.00605 85455 04235 961683
0.98805 41263 29623 799481	0.00501 42027 42927 517693
0.99254 39003 23762 624572	0.00396 45543 38444 686674
0.99598 18429 87209 290650	0.00291 07318 17934 946408
0.99836 43758 63181 677724	0.00185 39607 88946 921732
0.99968 95038 83230 766828	0.00079 67920 65552 012429

Table 25.5   ABSCISSAS FOR EQUAL WEIGHT CHEBYSHEV INTEGRATION

$$\int_{-1}^{+1} f(x)dx \approx \frac{2}{n} \sum_{i=1}^{n} f(x_i)$$

Abscissas $= \pm x_i$

n	$\pm x_i$	n	$\pm x_i$	n	$\pm x_i$
2	0.57735 02692	5	0.83249 74870	7	0.88336 17008
			0.37454 14096		0.52965 67753
			0.00000 00000		0.32391 18105
					0.00000 00000
3	0.70710 67812				
	0.00000 00000			9	0.91158 93077
					0.60101 86554
		6	0.86624 68181		0.52876 17831
4	0.79465 44723		0.42251 86538		0.16790 61842
	0.18759 24741		0.26663 54015		0.00000 00000

Compiled from H. E. Salzer, Tables for facilitating the use of Chebyshev's quadrature formula, J. Math. Phys. **26**, 191–194, 1947 (with permission).

Table 25.6   ABSCISSAS AND WEIGHT FACTORS FOR LOBATTO INTEGRATION

$$\int_{-1}^{+1} f(x)dx \approx w_1 f(-1) + \sum_{i=2}^{n-1} w_i f(x_i) + w_n f(1)$$

Abscissas $= \pm x_i$ · · · · · Weight Factors $= w_i$

n	$\pm x_i$	w_i	n	$\pm x_i$	w_i
			7	1.00000 000	0.04761 904
				0.83022 390	0.27682 604
				0.46884 879	0.43174 538
				0.00000 000	0.48761 904
3	1.00000 000	0.33333 333			
	0.00000 000	1.33333 333	8	1.00000 000	0.03571 428
				0.87174 015	0.21070 422
				0.59170 018	0.34112 270
4	1.00000 000	0.16666 667		0.20929 922	0.41245 880
	0.44721 360	0.83333 333			
			9	1.00000 00000	0.02777 77778
				0.89975 79954	0.16549 53616
5	1.00000 000	0.10000 000		0.67718 62795	0.27453 87126
	0.65465 367	0.54444 444		0.36311 74638	0.34642 85110
	0.00000 000	0.71111 111		0.00000 00000	0.37151 92744
			10	1.00000 00000	0.02222 22222
				0.91953 39082	0.13330 59908
				0.73877 38651	0.22488 93420
6	1.00000 000	0.06666 667		0.47792 49498	0.29204 26836
	0.76505 532	0.37847 496		0.16527 89577	0.32753 97612
	0.28523 152	0.55485 838			

Compiled from Z. Kopal, Numerical analysis, John Wiley & Sons, Inc., New York, N.Y., 1955 (with permission).

Table 25.7   ABSCISSAS AND WEIGHT FACTORS FOR GAUSSIAN INTEGRATION FOR INTEGRANDS WITH A LOGARITHMIC SINGULARITY

$$\int_{0}^{1} f(x) \ln x \, dx = \sum_{i=1}^{n} w_i f(x_i) + \frac{f^{(2n)}(\xi)}{(2n)!} K_n$$

Abscissas $= x_i$ · · · · · Weight Factors $= w_i$

n	x_i	$-w_i$	K_n	n	x_i	$-w_i$	K_n	n	x_i	$-w_i$	K_n
2	0.112009	0.718539	0.00285	3	0.063891	0.513405	0.00017	4	0.041448	0.383464	0.00001
	0.602277	0.281461			0.368997	0.391980			0.245275	0.386875	
					0.766880	0.094615			0.556165	0.190435	
									0.848982	0.039225	

Compiled from Berthod-Zaborowski, Le calcul des intégrales de la forme $\int_{0}^{1} f(x) \log x \, dx$. H. Mineur, Techniques de calcul numérique, pp. 555–556. Librairie Polytechnique Ch. Béranger, Paris, France, 1952 (with permission).

*See page II.

ABSCISSAS AND WEIGHT FACTORS FOR GAUSSIAN INTEGRATION OF MOMENTS

Table 25.8

$$\int_0^1 x^k f(x)\,dx \approx \sum_{i=1}^n w_i f(x_i)$$

Abscissas $= x_i$ Weight Factors $= w_i$

n	$k=0$ x_i	w_i	$k=1$ x_i	w_i	$k=2$ x_i	w_i
1	0.50000 00000	1.00000 00000	0.66666 66667	0.50000 00000	0.75000 00000	0.33333 33333
2	0.21132 48654	0.50000 00000	0.35505 10257	0.18195 86183	0.45584 81560	0.10078 58821
	0.78867 51346	0.50000 00000	0.84494 89743	0.31804 13817	0.87748 51773	0.23254 74513
3	0.11270 16654	0.27777 77778	0.21234 05382	0.06982 69799	0.29499 77901	0.02995 07030
	0.50000 00000	0.44444 44444	0.59053 31356	0.22924 11064	0.65299 62340	0.14624 62693
	0.88729 83346	0.27777 77778	0.91141 20405	0.20093 19137	0.92700 59759	0.15713 63611
4	0.06943 18442	0.17392 74226	0.13975 98643	0.03118 09710	0.20414 85821	0.01035 22408
	0.33000 94782	0.32607 25774	0.41640 95676	0.12984 75476	0.48295 27049	0.06863 38872
	0.66999 05218	0.32607 25774	0.72315 69864	0.20346 45680	0.76139 92624	0.14345 87898
	0.93056 81558	0.17392 74226	0.94289 58039	0.13550 69134	0.95149 94506	0.11088 84156
5	0.04691 00770	0.11846 34425	0.09853 50858	0.01574 79145	0.14894 57871	0.00411 38252
	0.23076 53449	0.23931 43352	0.30453 57266	0.07390 88701	0.36566 65274	0.03205 56007
	0.50000 00000	0.28444 44444	0.56202 51898	0.14638 69871	0.61011 36129	0.08920 01612
	0.76923 46551	0.23931 43352	0.80198 65821	0.16717 46381	0.82651 96792	0.12619 89619
	0.95308 99230	0.11846 34425	0.96019 01429	0.09678 15902	0.96542 10601	0.08176 47843
6	0.03376 52429	0.08566 22462	0.07305 43287	0.00873 83018	0.11319 43838	0.00183 10758
	0.16939 53068	0.18038 07865	0.23076 61380	0.04395 51656	0.28431 88727	0.01572 02972
	0.38069 04070	0.23395 69673	0.44132 84812	0.09866 11509	0.49096 35868	0.05128 95711
	0.61930 95930	0.23395 69673	0.66301 53097	0.14079 25538	0.69756 30820	0.09457 71867
	0.83060 46932	0.18038 07865	0.85192 14003	0.13554 24972	0.86843 60583	0.10737 64997
	0.96623 47571	0.08566 22462	0.97068 35728	0.07231 03307	0.97409 54449	0.06253 87027
7	0.02544 60438	0.06474 24831	0.05626 25605	0.00521 43622	0.08881 68334	0.00089 26880
	0.12923 44072	0.13985 26957	0.18024 06917	0.02740 83567	0.22648 27534	0.00816 29256
	0.29707 74243	0.19091 50253	0.35262 47171	0.06638 46965	0.39997 84867	0.02942 22113
	0.50000 00000	0.20897 95918	0.54715 36263	0.10712 50657	0.58599 78554	0.06314 63787
	0.70292 25757	0.19091 50253	0.73421 01772	0.12739 08973	0.75944 58740	0.09173 38033
	0.87076 55928	0.13985 26957	0.88532 09468	0.11050 92582	0.89691 09709	0.09069 88246
	0.97455 39562	0.06474 24831	0.97752 06136	0.05596 73634	0.97986 72262	0.04927 65018
8	0.01985 50718	0.05061 42681	0.04463 39553	0.00329 51914	0.07149 10350	0.00046 85178
	0.10166 67613	0.11119 05172	0.14436 62570	0.01784 29027	0.18422 82964	0.00447 45217
	0.23723 37950	0.15685 33229	0.28682 47571	0.04543 93195	0.33044 77282	0.01724 68638
	0.40828 26788	0.18134 18917	0.45481 33152	0.07919 95995	0.49440 29218	0.04081 44264
	0.59171 73212	0.18134 18917	0.62806 78354	0.10604 73594	0.65834 80085	0.06844 71834
	0.76276 62050	0.15685 33229	0.78569 15206	0.11250 57995	0.80452 48315	0.08528 47692
	0.89833 32387	0.11119 05172	0.90867 63921	0.09111 90236	0.91709 93825	0.07681 80933
	0.98014 49282	0.05061 42681	0.98222 00849	0.04455 08044	0.98390 22404	0.03977 89578

Compiled from H. Fishman, Numerical integration constants, Math. Tables Aids Comp. **11**, 1–9, 1957 (with permission).

Table 25.8   ABSCISSAS AND WEIGHT FACTORS FOR GAUSSIAN INTEGRATION OF MOMENTS

$$\int_0^1 x^k f(x)\,dx \approx \sum_{i=1}^n w_i f(x_i)$$

Abscissas $=x_i$ Weight Factors $=w_i$

n	$k=3$ x_i	w_i	$k=4$ x_i	w_i	$k=5$ x_i	w_i
1	0.80000 00000	0.25000 00000	0.83333 33333	0.20000 00000	0.85714 28571	0.16666 66667
2	0.52985 79359	0.06690 52498	0.58633 65823	0.04908 24923	0.63079 15938	0.03833 75627
	0.89871 34927	0.18309 47502	0.91366 34177	0.15091 75077	0.92476 39617	0.12832 91039
3	0.36326 46302	0.01647 90593	0.42011 30593	0.01046 90422	0.46798 32355	0.00729 70036
	0.69881 12692	0.10459 98976	0.73388 93552	0.08027 66735	0.76162 39697	0.06459 66123
	0.93792 41006	0.12892 10432	0.94599 75855	0.10925 42844	0.95221 09767	0.09477 30507
4	0.26147 77888	0.00465 83671	0.31213 54928	0.00251 63516	0.35689 37290	0.00153 44797
	0.53584 64461	0.04254 17241	0.57891 56596	0.02916 93822	0.61466 93899	0.02142 84046
	0.79028 32300	0.10900 43689	0.81289 15166	0.08706 77121	0.83107 90039	0.07205 63642
	0.95784 70806	0.09379 55399	0.96272 39976	0.08124 65541	0.96658 86465	0.07164 74181
5	0.19621 20074	0.00152 06894	0.23979 20448	0.00069 69771	0.27969 31248	0.00036 97155
	0.41710 02118	0.01695 73249	0.46093 36745	0.01021 05417	0.49870 98270	0.00672 96904
	0.64857 00042	0.06044 49532	0.68005 92327	0.04402 44695	0.70633 38189	0.03376 77450
	0.84560 51500	0.10031 65045	0.86088 63437	0.08271 27131	0.87340 27279	0.07007 13397
	0.96943 57035	0.07076 05281	0.97261 44185	0.06235 52986	0.97519 38347	0.05572 81761
6	0.15227 31618	0.00056 17109	0.18946 95839	0.00021 94140	0.22446 89954	0.00010 13258
	0.33130 04570	0.00708 53159	0.37275 11560	0.00372 67844	0.40953 33505	0.00218 79257
	0.53241 15667	0.03052 61922	0.56757 23729	0.01995 62647	0.59778 90484	0.01396 96531
	0.72560 27783	0.06844 32818	0.74883 64975	0.05223 99543	0.76841 36046	0.04148 63470
	0.88161 66844	0.08830 09912	0.89238 51584	0.07464 91503	0.90135 07338	0.06445 88592
	0.97679 53517	0.05508 25080	0.97898 52313	0.04920 84323	0.98079 72084	0.04446 25560
7	0.12142 71288	0.00022 99041	0.15324 14389	0.00007 70737	0.18382 87683	0.00003 11046
	0.26836 34403	0.00314 75964	0.30632 65225	0.00144 70088	0.34080 75951	0.00075 53838
	0.44086 64606	0.01531 21671	0.47654 00930	0.00892 69676	0.50794 05240	0.00566 04137
	0.61860 40284	0.04099 51686	0.64638 93025	0.02854 78428	0.67036 34101	0.02095 92982
	0.78025 35520	0.06975 00981	0.79771 66898	0.05522 48742	0.81258 84660	0.04510 49816
	0.90636 25341	0.07655 65614	0.91421 99006	0.06602 18459	0.92085 64173	0.05790 76135
	0.98176 99145	0.04400 85043	0.98334 38305	0.03975 43870	0.98466 74508	0.03624 78712
8	0.09900 17577	0.00010 24601	0.12637 29744	0.00002 97092	0.15315 06616	0.00001 05316
	0.22124 35074	0.00148 56841	0.25552 90521	0.00059 89500	0.28726 44039	0.00027 83586
	0.36912 39000	0.00785 50738	0.40364 12989	0.00407 79241	0.43462 74067	0.00233 53415
	0.52854 54312	0.02363 15807	0.55831 66758	0.01490 99334	0.58451 85666	0.01004 46144
	0.68399 32484	0.04745 43798	0.70600 95429	0.03471 99507	0.72512 64097	0.02648 53011
	0.82028 39497	0.06736 18394	0.83367 15420	0.05491 00973	0.84518 94879	0.04588 56532
	0.92409 37129	0.06618 20353	0.92999 57161	0.05800 05653	0.93504 35075	0.05153 42238
	0.98529 34401	0.03592 69468	0.98646 31979	0.03275 28699	0.98746 05085	0.03009 26424

ABSCISSAS AND WEIGHT FACTORS FOR LAGUERRE INTEGRATION

Table 25.9

$$\int_0^\infty e^{-x} f(x)\,dx \approx \sum_{i=1}^n w_i f(x_i) \qquad\qquad \int_0^\infty g(x)\,dx \approx \sum_{i=1}^n w_i e^{x_i} g(x_i)$$

Abscissas $= x_i$ (Zeros of Laguerre Polynomials)   Weight Factors $= w_i$

x_i	w_i	$w_i e^{x_i}$	x_i	w_i	$w_i e^{x_i}$
$n=2$			**$n=9$**		
0.58578 64376 27	(−1)8.53553 390593	1.53332 603312	0.15232 22277 32	(− 1)3.36126 421798	0.39143 11243 16
3.41421 35623 73	(−1)1.46446 609407	4.45095 733505	0.80722 00227 42	(− 1)4.11213 980424	0.92180 50285 29
			2.00513 51556 19	(− 1)1.99287 525371	1.48012 790994
			3.78347 39733 31	(− 2)4.74605 627657	2.08677 080755
$n=3$			6.20495 67778 77	(− 3)5.59962 661079	2.77292 138971
0.41577 45567 83	(−1)7.11093 009929	1.07769 285927	9.37298 52516 88	(− 4)3.05249 767093	3.59162 606809
2.29428 03602 79	(−1)2.78517 733569	2.76214 296190	13.46623 69110 92	(− 6)6.59212 302608	4.64876 600214
6.28994 50829 37	(−2)1.03892 565016	5.60109 462543	18.83359 77889 92	(− 8)4.11076 933035	6.21227 541975
			26.37407 18909 27	(−11)3.29087 403035	9.36321 823771
$n=4$			**$n=10$**		
0.32254 76896 19	(−1)6.03154 104342	0.83273 91238 38	0.13779 34705 40	(− 1)3.08441 115765	0.35400 97386 07
1.74576 11011 58	(−1)3.57418 692438	2.04810 243845	0.72945 45495 03	(− 1)4.01119 929155	0.83190 23010 44
4.53662 02969 21	(−2)3.88879 085150	3.63114 630582	1.80834 29017 40	(− 1)2.18068 287612	1.33028 856175
9.39507 09123 01	(−4)5.39294 705561	6.48714 508441	3.40143 36978 55	(− 2)6.20874 560987	1.86306 390311
			5.55249 61400 64	(− 3)9.50151 697518	2.45025 555808
$n=5$			8.33015 27467 64	(− 4)7.53008 388588	3.12276 415514
0.26356 03197 18	(−1)5.21755 610583	0.67909 40422 08	11.84378 58379 00	(− 5)2.82592 334960	3.93415 269556
1.41340 30591 07	(−1)3.98666 811083	1.63848 787360	16.27925 78313 78	(− 7)4.24931 398496	4.99241 487219
3.59642 57710 41	(−2)7.59424 496817	2.76944 324237	21.99658 58119 81	(− 9)1.83956 482398	6.57220 248513
7.08581 00058 59	(−3)3.61175 867992	4.31565 690092	29.92069 70122 74	(−13)9.91182 721961	9.78469 584037
12.64080 08442 76	(−5)2.33699 723858	7.21918 635435			
			$n=12$		
$n=6$			0.11572 21173 58	(− 1)2.64731 371055	0.29720 96360 44
0.22284 66041 79	(−1)4.58964 673950	0.57353 55074 23	0.61175 74845 15	(− 1)3.77759 275873	0.69646 29804 31
1.18893 21016 73	(−1)4.17000 830772	1.36925 259071	1.51261 02697 76	(− 1)2.44082 011320	1.10778 139462
2.99273 63260 59	(−1)1.13373 382074	2.26068 459338	2.83375 13377 44	(− 2)9.04492 222117	1.53846 423904
5.77514 35691 05	(−2)1.03991 974531	3.35052 458236	4.59922 76394 18	(− 2)2.01023 811546	1.99832 760627
9.83746 74183 83	(−4)2.61017 202815	4.88682 680021	6.84452 54531 15	(− 3)2.66397 354187	2.50074 576910
15.98287 39806 02	(−7)8.98547 906430	7.84901 594560	9.62131 68424 57	(− 4)2.03231 592663	3.06532 151828
			13.00605 49933 06	(− 6)8.36505 585682	3.72328 911078
$n=7$			17.11685 51874 62	(− 7)1.66849 387654	4.52981 402998
0.19304 36765 60	(−1)4.09318 951701	0.49647 75975 40	22.15109 03793 97	(− 9)1.34239 103052	5.59725 846184
1.02666 48953 39	(−1)4.21831 277862	1.17764 306086	28.48796 72509 84	(−12)3.06160 163504	7.21299 546093
2.56787 67449 51	(−1)1.47126 348658	1.91824 978166	37.09912 10444 67	(−16)8.14807 746743	10.54383 74619
4.90035 30845 26	(−2)2.06335 144687	2.77184 863623			
8.18215 34445 63	(−3)1.07401 014328	3.84124 912249			
12.73418 02917 98	(−5)1.58654 643486	5.38067 820792			
19.39572 78622 63	(−8)3.17031 547900	8.40543 248683			
			$n=15$		
			0.09330 78120 17	(− 1)2.18234 885940	0.23957 81703 11
			0.49269 17403 02	(− 1)3.42210 177923	0.56010 08427 93
			1.21559 54120 71	(− 1)2.63027 577942	0.88700 82629 19
			2.26994 95262 04	(− 1)1.26425 818106	1.22366 440215
			3.66762 27217 51	(− 2)4.02068 649210	1.57444 872163
$n=8$			5.42533 66274 14	(− 3)8.56387 780361	1.94475 197653
0.17027 96323 05	(−1)3.69188 589342	0.43772 34104 93	7.56591 62266 13	(− 3)1.21243 614721	2.34150 205664
0.90370 17767 99	(−1)4.18786 780814	1.03386 934767	10.12022 85680 19	(− 4)1.11674 392344	2.77404 192683
2.25108 66298 66	(−1)1.75794 986637	1.66970 976566	13.13028 24821 76	(− 6)6.45992 676202	3.25564 334640
4.26670 01702 88	(−2)3.33434 922612	2.37692 470176	16.65440 77083 30	(− 7)2.22631 690710	3.80631 171423
7.04590 54023 93	(−3)2.79453 623523	3.20854 091335	20.77647 88994 49	(− 9)4.22743 038498	4.45847 775384
10.75851 60101 81	(−5)9.07650 877336	4.26857 551083	25.62389 42267 29	(−11)3.92189 726704	5.27001 778443
15.74067 86412 78	(−7)8.48574 671627	5.81808 336867	31.40751 91697 54	(−13)1.45651 526407	6.35956 346973
22.86313 17368 89	(−9)1.04800 117487	8.90622 621529	38.53068 33064 86	(−16)1.48302 705111	8.03178 763212
			48.02608 55726 86	(−20)1.60059 490621	11.52777 21009

Compiled from H. E. Salzer and R. Zucker, Table of the zeros and weight factors of the first fifteen Laguerre polynomials, Bull. Amer. Math. Soc. **55**, 1004–1012, 1949 (with permission).

Table 25.10 **ABSCISSAS AND WEIGHT FACTORS FOR HERMITE INTEGRATION**

$$\int_{-\infty}^{\infty} e^{-x^2} f(x)\,dx \approx \sum_{i=1}^{n} w_i f(x_i) \qquad\qquad \int_{-\infty}^{\infty} g(x)\,dx \approx \sum_{i=1}^{n} w_i e^{x_i^2} g(x_i)$$

Abscissas $=\pm x_i$ (Zeros of Hermite Polynomials) $\qquad$ Weight Factors $= w_i$

$\pm x_i$	w_i	$w_i e^{x_i^2}$
$n=2$		
0.70710 67811 86548	(−1)8.86226 92545 28	1.46114 11826 611
$n=3$		
0.00000 00000 00000	(0)1.18163 59006 04	1.18163 59006 037
1.22474 48713 91589	(−1)2.95408 97515 09	1.32393 11752 136
$n=4$		
0.52464 76232 75290	(−1)8.04914 09000 55	1.05996 44828 950
1.65068 01238 85785	(−2)8.13128 35447 25	1.24022 58176 958
$n=5$		
0.00000 00000 00000	(−1)9.45308 72048 29	0.94530 87204 829
0.95857 24646 13819	(−1)3.93619 32315 22	0.98658 09967 514
2.02018 28704 56086	(−2)1.99532 42059 05	1.18148 86255 360
$n=6$		
0.43607 74119 27617	(−1)7.24629 59522 44	0.87640 13344 362
1.33584 90740 13697	(−1)1.57067 32032 29	0.93558 05576 312
2.35060 49736 74492	(−3)4.53000 99055 09	1.13690 83326 745
$n=7$		
0.00000 00000 00000	(−1)8.10264 61755 68	0.81026 46175 568
0.81628 78828 58965	(−1)4.25607 25261 01	0.82868 73032 836
1.67355 16287 67471	(−2)5.45155 82819 13	0.89718 46002 252
2.65196 13568 35233	(−4)9.71781 24509 95	1.10133 07296 103
$n=8$		
0.38118 69902 07322	(−1)6.61147 01255 82	0.76454 41286 517
1.15719 37124 46780	(−1)2.07802 32581 49	0.79289 00483 864
1.98165 67566 95843	(−2)1.70779 83007 41	0.86675 26065 634
2.93063 74202 57244	(−4)1.99604 07221 14	1.07193 01442 480
$n=9$		
0.00000 00000 00000	(−1)7.20235 21560 61	0.72023 52156 061
0.72355 10187 52838	(−1)4.32651 55900 26	0.73030 24527 451
1.46855 32892 16668	(−2)8.84745 27394 38	0.76460 81250 946
2.26658 05845 31843	(−3)4.94362 42755 37	0.84175 27014 787
3.19099 32017 81528	(−5)3.96069 77263 26	1.04700 35809 767

$\pm x_i$	w_i	$w_i e^{x_i^2}$
$n=10$		
0.34290 13272 23705	(−1)6.10862 63373 53	0.68708 18539 513
1.03661 08297 89514	(−1)2.40138 61108 23	0.70329 63231 049
1.75668 36492 99882	(−2)3.38743 94455 48	0.74144 19319 436
2.53273 16742 32790	(−3)1.34364 57467 81	0.82066 61264 048
3.43615 91188 37738	(−6)7.64043 28552 33	1.02545 16913 657
$n=12$		
0.31424 03762 54359	(−1)5.70135 23626 25	0.62930 78743 695
0.94778 83912 40164	(−1)2.60492 31026 42	0.63962 12320 203
1.59768 26351 52605	(−2)5.16079 85615 88	0.66266 27732 669
2.27950 70805 01060	(−3)3.90539 05846 29	0.70522 03661 122
3.02063 70251 20890	(−5)8.57368 70435 88	0.78664 39394 633
3.88972 48978 69782	(−7)2.65855 16843 56	0.98969 90470 923
$n=16$		
0.27348 10461 3815	(−1)5.07929 47901 66	0.54737 52050 378
0.82295 14491 4466	(−1)2.80647 45852 85	0.55244 19573 675
1.38025 85391 9888	(−2)8.38100 41398 99	0.56321 78290 882
1.95178 79909 1625	(−2)1.28803 11535 51	0.58124 72754 009
2.54620 21578 4748	(−4)9.32284 00862 42	0.60973 69582 560
3.17699 91619 7996	(−5)2.71186 00925 38	0.65575 56728 761
3.86944 79048 6012	(−7)2.32098 08448 65	0.73824 56222 777
4.68873 89393 0582	(−10)2.65480 74740 11	0.93687 44928 841
$n=20$		
0.24534 07083 009	(−1)4.62243 66960 06	0.49092 15006 667
0.73747 37285 454	(−1)2.86675 50536 28	0.49384 33852 721
1.23407 62153 953	(−1)1.09017 20602 00	0.49992 08713 363
1.73853 77121 166	(−2)2.48105 20887 46	0.50967 90271 175
2.25497 40020 893	(−3)3.24377 33422 38	0.52408 03509 486
2.78880 60584 281	(−4)2.28338 63601 63	0.54485 17423 644
3.34785 45673 832	(−6)7.80255 64785 32	0.57526 24428 525
3.94476 40401 156	(−7)1.08606 93707 69	0.62227 86961 914
4.60368 24495 507	(−10)4.39934 09922 73	0.70433 29611 769
5.38748 08900 112	(−13)2.22939 36455 34	0.89859 19614 532

Compiled from H. E. Salzer, R. Zucker, and R. Capuano, Table of the zeros and weight factors of the first twenty Hermite polynomials, J. Research NBS **48**, 111–116, 1952, RP2294 (with permission).

Table 25.11 **COEFFICIENTS FOR FILON'S QUADRATURE FORMULA**

θ	α	β	γ
0.00	0.00000 000	0.66666 667	1.33333 333
0.01	0.00000 004	0.66668 000	1.33332 000
0.02	0.00000 036	0.66671 999	1.33328 000
0.03	0.00000 120	0.66678 664	1.33321 334
0.04	0.00000 284	0.66687 990	1.33312 001
0.05	0.00000 555	0.66699 976	1.33300 003
0.06	0.00000 961	0.66714 617	1.33285 340
0.07	0.00001 524	0.66731 909	1.33268 012
0.08	0.00002 274	0.66751 844	1.33248 020
0.09	0.00003 237	0.66774 417	1.33225 365
0.1	0.00004 438	0.66799 619	1.33200 048
0.2	0.00035 354	0.67193 927	1.32800 761
0.3	0.00118 467	0.67836 065	1.32137 184
0.4	0.00278 012	0.68703 909	1.31212 154
0.5	0.00536 042	0.69767 347	1.30029 624
0.6	0.00911 797	0.70989 111	1.28594 638
0.7	0.01421 151	0.72325 813	1.26913 302
0.8	0.02076 156	0.73729 136	1.24992 752
0.9	0.02884 683	0.75147 168	1.22841 118
1.0	0.03850 188	0.76525 831	1.20467 472

See **25.4.47**.

26. Probability Functions

Marvin Zelen[1] and Norman C. Severo[2]

Contents

---

[1] National Bureau of Standards. (Presently, National Institutes of Health.)
[2] National Bureau of Standards. (Presently, University of Buffalo.)

The authors gratefully acknowledge the assistance of David S. Liepman in the preparation and checking of the tables and graphs and the many helpful comments received from members of the Committee on Mathematical Tables of the Institute of Mathematical Statistics.

26. Probability Functions

Mathematical Properties[3]

26.1. Probability Functions: Definitions and Properties

Univariate Cumulative Distribution Functions

A real-valued function $F(x)$ is termed a (univariate) cumulative distribution function (c.d.f.) or simply distribution function if

  i) $F(x)$ is non-decreasing, i.e., $F(x_1) \leq F(x_2)$ for $x_1 \leq x_2$

  ii) $F(x)$ is everywhere continuous from the right, i.e., $F(x) = \lim\limits_{\epsilon \to 0+} F(x+\epsilon)$

  iii) $F(-\infty) = 0$, $F(\infty) = 1$.

The function $F(x)$ signifies the probability of the event "$X \leq x$" where X is a random variable, i.e., $Pr\{X \leq x\} = F(x)$, and thus describes the c.d.f. of X. The two principal types of distribution functions are termed *discrete* and *continuous*.

Discrete Distributions: Discrete distributions are characterized by the random variable X taking on an enumerable number of values . . ., x_{-1}, x_0, x_1, . . . with point probabilities

$$p_n = Pr\{X = x_n\} \geq 0$$

which need only be subject to the restriction

$$\sum_n p_n = 1.$$

The corresponding distribution function can then be written

26.1.1 $\qquad F(x) = Pr\{X \leq x\} = \sum\limits_{x_n \leq x} p_n$

---

[3] *Comment on notation and conventions.*

  a. We follow the customary convention of denoting a random variable by a capital letter, i.e., X, and using the corresponding lower case letter, i.e., x, for a particular value that the random variable assumes.

  b. For statistical applications it is often convenient to have tabulated the "upper tail area," $1-F(x)$, or the c.d.f. for $|X|$, $F(x)-F(-x)$, instead of simply the c.d.f. $F(x)$. We use the notation P to indicate the c.d.f. of X, $Q=1-P$ to indicate the "upper tail area" and $A = P-Q$ to denote the c.d.f. of $|X|$. In particular we use $P(x)$, $Q(x)$, and $A(x)$ to denote the corresponding functions for the normal or Gaussian probability function, see **26.2.2–26.2.4**. When these distributions depend on other parameters, say θ_1 and θ_2, we indicate this by writing $P(x|\theta_1, \theta_2)$, $Q(x|\theta_1, \theta_2)$, or $A(x|\theta_1, \theta_2)$. For example the chi-square distribution **26.4** depends on the parameter ν and the tabulated function is written $Q(\chi^2|\nu)$.

---

where the summation is over all values of x for which $x_n \leq x$. The set $\{x_n\}$ of values for which $p_n > 0$ is termed the domain of the random variable X. A discrete distribution of a random variable is called a *lattice distribution* if there exist numbers a and $b \neq 0$ such that every possible value of X can be represented in the form $a+bn$ where n takes on only integral values. A summary of some properties of certain discrete distributions is presented in **26.1.19–26.1.24**.

Continuous Distributions. Continuous distributions are characterized by $F(x)$ being absolutely continuous. Hence $F(x)$ possesses a derivative $F'(x) = f(x)$ and the c.d.f. can be written

26.1.2 $\qquad F(x) = Pr\{X \leq x\} = \int_{-\infty}^{x} f(t)dt.$

The derivative $f(x)$ is termed the *probability density function* (p.d.f.) or *frequency function*, and the values of x for which $f(x) > 0$ make up the domain of the random variable X. A summary of some properties of certain selected continuous distributions is presented in **26.1.25–26.1.34**.

Multivariate Probability Functions

The real-valued function $F(x_1, x_2, \ldots x_n)$ defines an n-variate cumulative distribution function if

  i) $F(x_1, x_2, \ldots x_n)$ is a non-decreasing function for each x_i

  ii) $F(x_1, x_2, \ldots x_n)$ is continuous from the right in each x_i; i.e., $F(x_1, x_2, \ldots x_n) = \lim\limits_{\epsilon \to 0+} F(x_1, \ldots, x_i+\epsilon, \ldots, x_n)$

  iii) $F(x_1, x_2, \ldots x_n) = 0$ when any $x_i = -\infty$; $F(\infty, \infty, \ldots, \infty) = 1$.   *

  iv) $F(x_1, x_2, \ldots, x_n)$ assigns nonnegative probability to the event $x_1 < X_1 \leq x_1+h_1$, $x_2 < X_2 \leq x_2+h_2$, . . ., $x_n < X_n \leq x_n+h_n$ for all $x_1, x_2, \ldots, x_n$ and all nonnegative $h_1, h_2, \ldots, h_n$, e.g., for $n=2$, $F(x_1+h_1, x_2+h_2) - F(x_1, x_2+h_2) - F(x_1+h_1, x_2) + F(x_1, x_2) \geq 0$ and in general for $x_i < X_i \leq x_i+h_i$ $(i=1, 2, \ldots, n)$, the kth order difference $\triangle_k F(x_1, x_2, \ldots, x_n) > 0$ for $k=1, 2, \ldots, n$.

The joint probability of the event $X_1 \leq x_1$, $X_2 \leq x_2$, . . ., $X_n \leq x_n$ is $F(x_1, x_2, \ldots x_n)$. Analogous to the one-dimensional case, *discrete* distributions assign all probability to an enumerable set of vectors $(x_1, x_2, \ldots, x_n)$ and *continuous* distributions are characterized by absolute continuity of $F(x_1, x_2, \ldots, x_n)$.

Characteristics of distribution functions: Moments, characteristic functions, cumulants

			Continuous distributions	Discrete distributions
26.1.3	nth moment about origin		$\mu_n' = \int_{-\infty}^{\infty} x^n f(x) dx$	$\mu_n' = \sum_s x_s^n p_s$
26.1.4	mean		$m = \mu_1' = \int_{-\infty}^{\infty} x f(x) dx$	$m = \mu_1' = \sum_s x_s p_s$
26.1.5	variance		$\sigma^2 = \mu_2' - m^2 = \int_{-\infty}^{\infty} (x-m)^2 f(x) dx$	$\sigma^2 = \mu_2' - m^2 = \sum_s (x_s-m)^2 p_s$
26.1.6	nth central moment		$\mu_n = \int_{-\infty}^{\infty} (x-m)^n f(x) dx$	$\mu_n = \sum_s (x_s-m)^n p_s$
26.1.7	expected value operator for the function $g(x)$		$E[g(X)] = \int_{-\infty}^{\infty} g(x) f(x) dx$	$E[g(X)] = \sum_s g(x_s) p_s$
26.1.8	characteristic function of X		$\phi(t) = E(e^{itX}) = \int_{-\infty}^{\infty} e^{itx} f(x) dx$	$\phi(t) = E(e^{itX}) = \sum_s e^{itx_s} p_s$
26.1.9	characteristic function of $g(X)$		$\phi_g(t) = E(e^{itg(X)}) = \int_{-\infty}^{\infty} e^{itg(x)} f(x) dx$	$\phi_g(t) = E(e^{itg(X)}) = \sum_s e^{itg(x_s)} p_s$
26.1.10	inversion formula		$f(x) = \frac{1}{2\pi} \int_{-\infty}^{\infty} e^{-itx} \phi(t) dt$	$p_n = \frac{b}{2\pi} \int_{-\pi/b}^{\pi/b} e^{-itx_n} \phi(t) dt$
				(lattice distributions only)

Relation of the Characteristic Function to Moments About the Origin

26.1.11
$$\phi^{(n)}(0) = \left[\frac{d^n}{dt^n} \phi(t) \right]_{t=0} = i^n \mu_n'$$

Cumulant Function

26.1.12
$$\ln \phi(t) = \sum_{n=0}^{\infty} \kappa_n \frac{(it)^n}{n!}$$

κ_n is called the nth cumulant.

26.1.13
$$\kappa_1 = m, \quad \kappa_2 = \sigma^2, \quad \kappa_3 = \mu_3, \quad \kappa_4 = \mu_4 - 3\mu_2^2$$

Relation of Central Moments to Moments About the Origin

26.1.14
$$\mu_n = \sum_{j=0}^{n} \binom{n}{j} (-1)^{n-j} \mu_j' m^{n-j}$$

Coefficients of Skewness and Excess

26.1.15
$$\gamma_1 = \frac{\kappa_3}{\kappa_2^{3/2}} = \frac{\mu_3}{\sigma^3} \qquad \text{(skewness)}$$

26.1.16
$$\gamma_2 = \frac{\kappa_4}{\kappa_2^2} = \frac{\mu_4}{\sigma^4} - 3 \qquad \text{(excess)}$$

Occasionally coefficients of skewness and excess (or kurtosis) are given by

26.1.17
$$\beta_1 = \gamma_1^2 = \left(\frac{\mu_3}{\sigma^3} \right)^2 \qquad \text{(skewness)}$$

26.1.18
$$\beta_2 = \gamma_2 + 3 = \frac{\mu_4}{\sigma^4}$$

(excess or kurtosis)

Name	Domain	Point Probabilities	Restrictions on Parameters	Mean	Variance	Skewness γ_1	Excess γ_2	Characteristic function	Cumulants
26.1.19 Single point or degenerate	$x=c$ (c a constant)	$p=1$	$-\infty < c < +\infty$	c	0	--------------	--------------	$e^{i\lambda_t}$	$\kappa_1 = \lambda$, $\kappa_r = 0$ for $r > 1$
26.1.20 Binomial	$x_s = s$, for $s = 0, 1, 2, \ldots, n$	$\binom{n}{s} p^s (1-p)^{n-s}$	$0 < p < 1$ $(q = 1-p)$	np	npq	$\dfrac{q-p}{\sqrt{npq}}$	$\dfrac{1-6pq}{npq}$	$(q+pe^{it})^n$	$\kappa_1 = np$ $\kappa_{r+1} = pq \dfrac{d\kappa_r}{dp}$ for $r \geq 1$
26.1.21 Hypergeometric	$x_s = s$, for $s = 0, 1, \ldots \min(n, N_1)$	$\dfrac{\binom{N_1}{s}\binom{N_2}{n-s}}{\binom{N_1+N_2}{n}}$	N_1 and N_2 integers, and $n \leq N_1 + N_2$ $(N = N_1 + N_2,$ $p = N_1/N$ and $q = 1-p = N_2/N)$	np	$npq \left(\dfrac{N-n}{N-1}\right)$	$\dfrac{q-p}{\sqrt{npq}} \left(\dfrac{N-1}{N-n}\right)^{\frac{1}{2}} \left(\dfrac{N-2n}{N-2}\right)$	Compli-cated	$\dfrac{\binom{N_2}{n}}{\binom{N}{n}} F(-n, -N_1; N_2-n+1; e^{it})$	Complicated
26.1.22 Poisson	$x_s = s$, for $s = 0, 1, 2, \ldots, \infty$	$\dfrac{e^{-m} m^s}{s!}$	$0 < m < \infty$	m	m	$m^{-\frac{1}{2}}$	m^{-1}	$e^{m(e^{it}-1)}$	$\kappa_r = m$ for $r = 1, 2, \ldots$
26.1.23 Negative binomial	$x_s = s$, for $s = 0, 1, 2, \ldots, \infty$	$\binom{n+s-1}{s} p^n (1-p)^s$	$n \geq 0$ and $0 < p < 1$ $(p = 1/Q,$ and $1-p = P/Q)$	nP	nPQ	$\dfrac{Q+P}{\sqrt{nPQ}}$	$\dfrac{1+6PQ}{nPQ}$	$(Q-Pe^{it})^{-n}$	$\kappa_1 = nP$ $\kappa_{r+1} = PQ \dfrac{d\kappa_r}{dQ}$ for $r \geq 1$
26.1.24 Geometric	$x_s = s$, for $s = 0, 1, 2, \ldots, \infty$	$p(1-p)^s$	$0 < p < 1$	$\dfrac{1-p}{p}$	$\dfrac{1-p}{p^2}$	$\dfrac{2-p}{\sqrt{1-p}}$	$6 + \dfrac{p^2}{1-p}$	$p[1-(1-p)e^{it}]^{-1}$	$\kappa_1 = \dfrac{1-p}{p}$, $\kappa_{r+1} = -(1-p) \dfrac{d\kappa_r}{dp}$, $r \geq 1$

Some one-dimensional continuous distribution functions

	Name	Domain	Probability Density Function $f(x)$	Restrictions on Parameters	Mean	Variance	Skewness γ_1	Excess γ_2	Characteristic function	Cumulants
26.1.25	Error function	$-\infty < x < \infty$	$\dfrac{h}{\sqrt{\pi}}\, e^{-h^2 x^2}$	$0 < h < \infty$	0	$\dfrac{1}{2h^2}$	0	0	$e^{\frac{-t^2}{4h^2}}$	$\kappa_1 = 0,\ \kappa_2 = \dfrac{1}{2h^2}$ $\kappa_n = 0$ for $n > 2$
26.1.26	Normal	$-\infty < x < \infty$	$\dfrac{1}{\sigma\sqrt{2\pi}}\, e^{-\frac{1}{2}\left(\frac{x-m}{\sigma}\right)^2}$	$-\infty < m < \infty$ $0 < \sigma < \infty$	m	σ^2	0	0	$e^{imt - \frac{\sigma^2 t^2}{2}}$	$\kappa_1 = m,\ \kappa_2 = \sigma^2,\ \kappa_n = 0$ for $n > 2$
26.1.27	Cauchy	$-\infty < x < \infty$	$\dfrac{1}{\pi\beta}\dfrac{1}{1+\left(\frac{x-\alpha}{\beta}\right)^2}$	$-\infty < \alpha < \infty$ $0 < \beta < \infty$	not defined	not defined	not defined	not defined	$e^{i\alpha t - \beta|t|}$	not defined
26.1.28	Exponential	$\alpha \le x < \infty$	$\dfrac{1}{\beta}\, e^{-\left(\frac{x-\alpha}{\beta}\right)}$	$-\infty < \alpha < \infty$ $0 < \beta < \infty$	$\alpha + \beta$	β^2	2	6	$e^{i\alpha t}(1 - i\beta t)^{-1}$	$\kappa_1 = \alpha + \beta,\ \kappa_n = \beta^n \Gamma(n)$ for $n > 1$
26.1.29	Laplace, or double exponential	$-\infty < x < \infty$	$\dfrac{1}{2\beta}\, e^{-\left|\frac{x-\alpha}{\beta}\right|}$	$-\infty < \alpha < \infty$ $0 < \beta < \infty$	α	$2\beta^2$	0	3	$e^{i\alpha t}(1 + \beta^2 t^2)^{-1}$	$\kappa_1 = \alpha,\ \kappa_2 = 2\beta^2$ $\kappa_{2n+1} = 0,\ \kappa_{2n} = \dfrac{(2n)!}{n}\beta^{2n}$ for $n = 1, 2, \ldots$
26.1.30	Extreme-Value,[4] (Fisher-Tippett Type I or doubly exponential)	$-\infty < x < \infty$	$\dfrac{1}{\beta}\exp\left(-y - e^{-y}\right)$ with $y = \dfrac{x-\alpha}{\beta}$	$-\infty < \alpha < \infty$ $0 < \beta < \infty$	$\alpha + \gamma\beta$	$\dfrac{(\pi\beta)^2}{6}$	1.3	2.4	$\Gamma(1 - i\beta t)e^{i\alpha t}$	$\kappa_1 = \gamma,\ \kappa_2 = \dfrac{(\pi\beta)^2}{6}$ $\kappa_n = \beta^n \Gamma(n)\sum\limits_{r=1}^{\infty}\dfrac{1}{r^n}$ for $n > 2$
26.1.31	Pearson Type III	$\alpha \le x < \infty$	$\dfrac{1}{\beta\Gamma(p)}\, y^{p-1}e^{-y}$ with $y = \dfrac{x-\alpha}{\beta}$	$-\infty < \alpha < \infty$ $0 < \beta < \infty$ $0 < p < \infty$	$\alpha + p\beta$	$p\beta^2$	$\dfrac{2}{\sqrt{p}}$	$6/p$	$e^{i\alpha t}(1 - i\beta t)^{-p}$	$\kappa_1 = \alpha + \beta p,\ \kappa_n = \beta^n p\, \Gamma(n)$ for $n > 1$
26.1.32	Gamma distribution	$0 \le x < \infty$	$\dfrac{1}{\Gamma(p)}\, x^{p-1}e^{-x}$	$0 < p < \infty$	p	p	$\dfrac{2}{\sqrt{p}}$	$6/p$	$(1 - it)^{-p}$	$\kappa_1 = p,\ \kappa_n = p\,\Gamma(n)$ for $n > 1$
26.1.33	Beta distribution	$0 \le x \le 1$	$\dfrac{1}{B(a,\,b)}\, x^{a-1}(1-x)^{b-1}$	$1 \le a < \infty$ $1 \le b < \infty$	$\dfrac{a}{a+b}$	$\dfrac{ab}{(a+b)^2(a+b+1)}$	$\dfrac{2(a-b)}{(a+b+2)}$	See footnote 5.	$M(a,\, a+b,\, it)$	
26.1.34	Rectangular, or uniform	$m - \dfrac{h}{2} \le x \le m + \dfrac{h}{2}$	$\dfrac{1}{h}$	$-\infty < m < \infty$ $0 < h < \infty$	m	$\dfrac{h^2}{12}$	0	−1.2	$\dfrac{2}{ht}\sin\left(\dfrac{ht}{2}\right)e^{imt}$	$\kappa_1 = m,\ \kappa_{2n+1} = 0$ $\kappa_{2n} = \dfrac{h^{2n}B_{2n}}{2n}$ B_{2n} (Bernoulli numbers), $B_2 = \dfrac{1}{6},\ B_4 = -\dfrac{1}{30},\ \ldots$

[4] γ (Euler's constant) $= .57721\ 56649\ \ldots$

[5] $\gamma_2 = \sqrt{\dfrac{a+b+1}{ab}}\left\{\dfrac{3(a+b+1)[2(a+b)^2 + ab(a+b-6)]}{ab(a+b+2)(a+b+3)} - 3\right\}$.

*See page II.

Inequalities for distribution functions

($F(x)$ denotes the c.d.f. of the random variable X and t denotes a positive constant; further m is always assumed to be finite and all expectations are assumed to exist.)

Inequality	Conditions
26.1.35 $\quad Pr\{g(X) \geq t\} \leq E[g(X)]/t$	(i) $g(X) \geq 0$
26.1.36 $\quad Pr\{X \geq t\} \leq m/t$ $F(t) \geq 1 - \dfrac{m}{t}$	(i) $Pr\{X < 0\} = 0$ (ii) $E(X) = m$
26.1.37 $\quad Pr\{\|X - m\| \geq t\sigma\} \leq 1/t^2$ $F(m+t\sigma) - F(m-t\sigma) \geq 1 - \dfrac{1}{t^2}$	(i) $E(X) = m$ (ii) $E(X-m)^2 = \sigma^2 \qquad *$
26.1.38 $Pr\{\|\overline{X} - \overline{m}\| > t\overline{\sigma}\} < \dfrac{1}{nt^2}$	(i) $E(X_i) = m_i$ (ii) $E(X_i - m_i)^2 = \sigma_i^2$ (iii) $E([X_i - m_i][X_j - m_j]) = 0 \ (i \neq j)$ (iv) $\overline{X} = \displaystyle\sum_{i=1}^n \frac{X_i}{n},$ $\overline{m} = \displaystyle\sum_{i=1}^n \frac{m_i}{n}; \ \overline{\sigma} = \left[\displaystyle\sum_{i=1}^n \frac{\sigma_i^2}{n}\right]^{\frac{1}{2}}$
26.1.39 $Pr\{\|X - m\| \geq t\sigma\} \leq \dfrac{4}{9}\left\{\dfrac{1 + \left(\dfrac{m - x_0}{\sigma}\right)^2}{\left(t - \left\|\dfrac{m - x_0}{\sigma}\right\|\right)^2}\right\}$ $F(m+t\sigma) - F(m-t\sigma) \geq 1 - \dfrac{4}{9}\left\{\dfrac{1 + \left(\dfrac{m - x_0}{\sigma}\right)^2}{\left(t - \left\|\dfrac{m - x_0}{\sigma}\right\|\right)^2}\right\}$	(i) $E(X-m)^2 = \sigma^2$ (ii) $F(x)$ is a continuous c.d.f. (iii) $F(x)$ is unimodal at x_0 [6]
26.1.40 $\quad Pr\{\|X - m\| \geq t\sigma\} \leq 4/9t^2$ $F(m+t\sigma) - F(m-t\sigma) \geq 1 - \dfrac{4}{9t^2}$	(i) $E(X-m)^2 = \sigma^2$ (ii) $F(x)$ is a continuous c.d.f. (iii) $F(x)$ is unimodal at x_0 [6] (iv) $m = x_0$
26.1.41 $\quad Pr\{\|X - m\| \geq t\sigma\} \leq \dfrac{\mu_4 - \sigma^4}{\mu_4 + t^4\sigma^4 - 2t^2\sigma^4}$ $F(m+t\sigma) - F(m-t\sigma) \geq 1 - \dfrac{\mu_4 - \sigma^4}{\mu_4 + t^4\sigma^4 - 2t^2\sigma^4}$	(i) $E(X-m)^2 = \sigma^2$ (ii) $E(X-m)^4 = \mu_4$

[6] x_0 is such that $F'(x_0) > F'(x)$ for $x \neq x_0$.

26.2. Normal or Gaussian Probability Function

26.2.1
$$Z(x) = \frac{1}{\sqrt{2\pi}} e^{-x^2/2}$$

26.2.2 $\quad P(x) = \dfrac{1}{\sqrt{2\pi}} \displaystyle\int_{-\infty}^{x} e^{-t^2/2} dt = \int_{-\infty}^{x} Z(t) dt$

26.2.3 $\quad Q(x) = \dfrac{1}{\sqrt{2\pi}} \displaystyle\int_{x}^{\infty} e^{-t^2/2} dt = \int_{x}^{\infty} Z(t) dt$

26.2.4 $\quad A(x) = \dfrac{1}{\sqrt{2\pi}} \displaystyle\int_{-x}^{x} e^{-t^2/2} dt = \int_{-x}^{x} Z(t) dt$

26.2.5 $\qquad P(x) + Q(x) = 1$

26.2.6 $\qquad P(-x) = Q(x)$

26.2.7 $\qquad A(x) = 2P(x) - 1$

Probability Integral with Mean m and Variance σ^2

A random variable X is said to be normally distributed with mean m and variance σ^2 if the probability that X is less than or equal to x is given by

26.2.8
$$Pr\{X \leq x\} = \frac{1}{\sigma\sqrt{2\pi}} \int_{-\infty}^{x} e^{-\frac{(t-m)^2}{2\sigma^2}} dt$$

$$= \frac{1}{\sqrt{2\pi}} \int_{-\infty}^{(x-m)/\sigma} e^{-t^2/2} dt = P\left(\frac{x-m}{\sigma}\right).$$

The corresponding probability density function is

26.2.9
$$\frac{\partial}{\partial x} P\left(\frac{x-m}{\sigma}\right) = \frac{1}{\sigma} Z\left(\frac{x-m}{\sigma}\right) = \frac{1}{\sigma\sqrt{2\pi}} e^{-\frac{(x-m)^2}{2\sigma^2}}$$

and is symmetric around m, i.e.

$$Z\left(\frac{m+x}{\sigma}\right) = Z\left(\frac{m-x}{\sigma}\right).$$

The inflexion points of the probability density function are at $m \pm \sigma$.

Power Series $(x \geq 0)$

26.2.10
$$P(x) = \frac{1}{2} + \frac{1}{\sqrt{2\pi}} \sum_{n=0}^{\infty} \frac{(-1)^n x^{2n+1}}{n!\, 2^n (2n+1)}$$

26.2.11
$$P(x) = \frac{1}{2} + Z(x) \sum_{n=0}^{\infty} \frac{x^{2n+1}}{1 \cdot 3 \cdot 5 \ldots (2n+1)}$$

Asymptotic Expansions $(x > 0)$

26.2.12
$$Q(x) = \frac{Z(x)}{x} \left\{ 1 - \frac{1}{x^2} + \frac{1 \cdot 3}{x^4} + \ldots \right.$$
$$\left. + \frac{(-1)^n 1 \cdot 3 \ldots (2n-1)}{x^{2n}} \right\} + R_n$$

where
$$R_n = (-1)^{n+1} 1 \cdot 3 \ldots (2n+1) \int_x^{\infty} \frac{Z(t)}{t^{2n+2}}\, dt$$

which is less in absolute value than the first neglected term.

26.2.13
$$Q(x) \sim \frac{Z(x)}{x} \left\{ 1 - \frac{a_1}{x^2+2} + \frac{a_2}{(x^2+2)(x^2+4)} \right.$$
$$\left. - \frac{a_3}{(x^2+2)(x^2+4)(x^2+6)} + \ldots \right\}$$

where $a_1 = 1$, $a_2 = 1$, $a_3 = 5$, $a_4 = 9$, $a_5 = 129$ and the general term is
$$a_n = c_0 1 \cdot 3 \ldots (2n-1) + 2c_1 1 \cdot 3 \ldots (2n-3)$$
$$+ 2^2 c_2 1 \cdot 3 \ldots (2n-5) + \ldots + 2^{n-1} c_{n-1}$$

and c_s is the coefficient of t^{n-s} in the expansion of $t(t-1) \ldots (t-n+1)$.

Continued Fraction Expansions

26.2.14
$$Q(x) = Z(x) \left\{ \frac{1}{x+} \frac{1}{x+} \frac{2}{x+} \frac{3}{x+} \frac{4}{x+} \cdots \right\} \quad (x > 0)$$

26.2.15
$$Q(x) = \frac{1}{2} - Z(x) \left\{ \frac{x}{1-} \frac{x^2}{3+} \frac{2x^2}{5-} \frac{3x^2}{7+} \frac{4x^2}{9-} \cdots \right\} \quad (x \geq 0)$$

Polynomial and Rational Approximations[7] for $P(x)$ and $Z(x)$

$$0 \leq x < \infty$$

26.2.16
$$P(x) = 1 - Z(x)(a_1 t + a_2 t^2 + a_3 t^3) + \epsilon(x), \qquad t = \frac{1}{1+px}$$
$$|\epsilon(x)| < 1 \times 10^{-5}$$
$$p = .33267 \qquad a_1 = .43618\ 36$$
$$a_2 = -.12016\ 76$$
$$a_3 = .93729\ 80$$

26.2.17
$$P(x) = 1 - Z(x)(b_1 t + b_2 t^2 + b_3 t^3 + b_4 t^4 + b_5 t^5) + \epsilon(x),$$
$$t = \frac{1}{1+px}$$
$$|\epsilon(x)| < 7.5 \times 10^{-8}$$
$$p = .23164\ 19$$
$$b_1 = .31938\ 1530 \qquad b_4 = -1.82125\ 5978$$
$$b_2 = -.35656\ 3782 \qquad b_5 = 1.33027\ 4429$$
$$b_3 = 1.78147\ 7937$$

26.2.18
$$P(x) = 1 - \frac{1}{2}(1 + c_1 x + c_2 x^2 + c_3 x^3 + c_4 x^4)^{-4} + \epsilon(x)$$
$$|\epsilon(x)| < 2.5 \times 10^{-4}$$
$$c_1 = .196854 \qquad c_3 = .000344$$
$$c_2 = .115194 \qquad c_4 = .019527$$

26.2.19
$$P(x) = 1 - \frac{1}{2}(1 + d_1 x + d_2 x^2 + d_3 x^3$$
$$+ d_4 x^4 + d_5 x^5 + d_6 x^6)^{-16} + \epsilon(x)$$
$$|\epsilon(x)| < 1.5 \times 10^{-7}$$
$$d_1 = .04986\ 73470 \qquad d_4 = .00003\ 80036$$
$$d_2 = .02114\ 10061 \qquad d_5 = .00004\ 88906$$
$$d_3 = .00327\ 76263 \qquad d_6 = .00000\ 53830$$

26.2.20
$$Z(x) = (a_0 + a_2 x^2 + a_4 x^4 + a_6 x^6)^{-1} + \epsilon(x)$$
$$|\epsilon(x)| < 2.7 \times 10^{-3}$$
$$a_0 = 2.490895 \qquad a_4 = -.024393$$
$$a_2 = 1.466003 \qquad a_6 = .178257$$

[7] Based on approximations in C. Hastings, Jr., Approximations for digital computers. Princeton Univ. Press, Princeton, N.J., 1955 (with permission).

26.2.21

$$Z(x) = (b_0 + b_2 x^2 + b_4 x^4 + b_6 x^6 + b_8 x^8 + b_{10} x^{10})^{-1} + \epsilon(x)$$

$$|\epsilon(x)| < 2.3 \times 10^{-4}$$

$b_0 = 2.50523\ 67$	$b_6 = \ \ \ .13064\ 69$
$b_2 = 1.28312\ 04$	$b_8 = -.02024\ 90$
$b_4 = \ \ \ .22647\ 18$	$b_{10} = \ \ \ .00391\ 32$

Rational Approximations [7] for x_p where $Q(x_p) = p$

$$0 < p \leq .5$$

26.2.22

$$x_p = t - \frac{a_0 + a_1 t}{1 + b_1 t + b_2 t^2} + \epsilon(p), \qquad t = \sqrt{\ln \frac{1}{p^2}}$$

$$|\epsilon(p)| < 3 \times 10^{-3}$$

$a_0 = 2.30753$	$b_1 = .99229$
$a_1 = \ \ .27061$	$b_2 = .04481$

26.2.23

$$x_p = t - \frac{c_0 + c_1 t + c_2 t^2}{1 + d_1 t + d_2 t^2 + d_3 t^3} + \epsilon(p), \qquad t = \sqrt{\ln \frac{1}{p^2}}$$

$$|\epsilon(p)| < 4.5 \times 10^{-4}$$

$c_0 = 2.515517$	$d_1 = 1.432788$
$c_1 = \ \ .802853$	$d_2 = \ \ .189269$
$c_2 = \ \ .010328$	$d_3 = \ \ .001308$

Bounds Useful as Approximations to the Normal Distribution Function

26.2.24

$$P(x) \leq \begin{cases} P_1(x) = \dfrac{1}{2} + \dfrac{1}{2}(1 - e^{-2x^2/\pi})^{\frac{1}{2}} & (x > 0) \\[2ex] P_2(x) = 1 - \dfrac{(4+x^2)^{\frac{1}{2}} - x}{2}(2\pi)^{-\frac{1}{2}} e^{-x^2/2} \end{cases}$$

$$(x > 1.4)$$

26.2.25

$$P(x) \geq \begin{cases} P_3(x) = \dfrac{1}{2} + \dfrac{1}{2}\left(1 - e^{-2x^2/\pi} - \dfrac{2(\pi-3)}{3\pi^2} x^4 e^{-x^2/2}\right)^{\frac{1}{2}} \\[2ex] \qquad\qquad\qquad\qquad\qquad\qquad\qquad (x > 0) \\[2ex] P_4(x) = 1 - \dfrac{1}{x}(2\pi)^{-\frac{1}{2}} e^{-x^2/2} \qquad (x > 2.2) \end{cases}$$

See **Figure 26.1** for error curves.

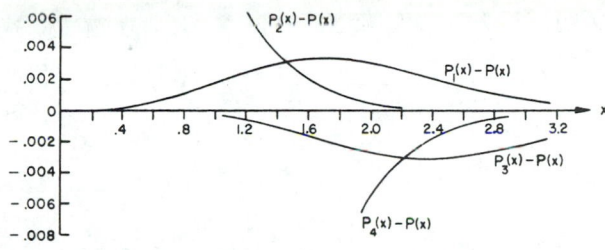

FIGURE 26.1. *Error curves for bounds on normal distribution.*

Derivatives of the Normal Probability Density Function

26.2.26

$$Z^{(m)}(x) = \frac{d^m}{dx^m} Z(x)$$

Differential Equation

26.2.27 $\quad Z^{(m+2)}(x) + x Z^{(m+1)}(x) + (m+1) Z^{(m)}(x) = 0$

Value at $x = 0$

26.2.28

$$Z^{(m)}(0) = \begin{cases} \dfrac{(-1)^{m/2} m!}{\sqrt{2\pi}\, 2^{m/2}\left(\dfrac{m}{2}\right)!} & \text{for } m = 2r,\, r = 0, 1, \ldots \\[3ex] 0 & \text{for odd } m > 0 \end{cases}$$

Relation of $P(x)$ and $Z^{(m)}(x)$ to Other Functions

Function	*Relation*

26.2.29 Error function

$$\operatorname{erf} x = 2P(x\sqrt{2}) - 1 \qquad (x \geq 0)$$

26.2.30 Incomplete gamma function (special case)

$$\frac{\gamma\left(\frac{1}{2}, x\right)}{\Gamma\left(\frac{1}{2}\right)} = [2P(\sqrt{2x}) - 1] \qquad (x \geq 0)$$

26.2.31 Hermite polynomial

$$He_n(x) = (-1)^n \frac{Z^{(n)}(x)}{Z(x)}$$

26.2.32   "

$$H_n(x) = (-1)^n 2^{n/2} \frac{Z^{(n)}(x\sqrt{2})}{Z(x\sqrt{2})}$$

26.2.33 Hh function

$$Hh_{-n}(x) = (-1)^{n-1}\sqrt{2\pi}\, Z^{(n-1)}(x) \qquad (n > 0)$$

26.2.34   "

$$Hh_n(x) = \frac{(-1)^n}{n!} Hh_{-1}(x) \frac{d^n}{dx^n}\left(\frac{Q(x)}{Z(x)}\right) \qquad (n > 0)$$

26.2.35 Tetrachoric function

$$\tau_n(x) = \frac{(-1)^{n-1}}{\sqrt{n!}} Z^{(n-1)}(x)$$

26.2.36 Confluent hypergeometric function (special case)

$$M\left(\frac{1}{2}, \frac{3}{2}, -\frac{x^2}{2}\right) = \frac{\sqrt{2\pi}}{x}\left\{P(x) - \frac{1}{2}\right\} \qquad (x > 0)$$

26.2.37   "

$$M\left(1, \frac{3}{2}, \frac{x^2}{2}\right) = \frac{1}{xZ(x)}\left\{P(x) - \frac{1}{2}\right\} \qquad (x > 0)$$

26.2.38   "

$$M\left(\frac{2m+1}{2}, \frac{1}{2}, -\frac{x^2}{2}\right) = \frac{Z^{(2m)}(x)}{Z^{(2m)}(0)} \qquad (x \geq 0)$$

26.2.39   "

$$M\left(\frac{2m+2}{2}, \frac{3}{2}, -\frac{x^2}{2}\right) = \frac{Z^{(2m-1)}(x)}{xZ^{(2m)}(0)} \qquad (x \geq 0)$$

26.2.40 Parabolic cylinder function

$$U\left(-n-\frac{1}{2}, x\right) = e^{-\frac{1}{4}x^2}(-1)^n \frac{Z^{(n)}(x)}{Z(x)} \qquad (n > 0)$$

Repeated Integrals of the Normal Probability Integral

26.2.41 $\quad I_n(x) = \int_x^\infty I_{n-1}(t)\,dt \qquad (n \geq 0)$

where $I_{-1}(x) = Z(x)$

26.2.42

$$I_{-n}(x) = \left(-\frac{d}{dx}\right)^{n-1} Z(x) = (-1)^{n-1} Z^{(n-1)}(x)$$
$$(n \geq -1)$$

26.2.43 $\quad \left(\frac{d^2}{dx^2} + x\frac{dx}{dn} - n\right) I_n(x) = 0$

26.2.44

$$(n+1)I_{n+1}(x) + xI_n(x) - I_{n-1}(x) = 0 \qquad (n > -1)$$

26.2.45

$$I_n(x) = \int_x^\infty \frac{(t-x)^n}{n!} Z(t)dt = e^{-x^2/2} \int_0^\infty \frac{t^n}{n!} Z(t)dt$$
$$(n > -1)$$

26.2.46
$$I_n(0) = I_{-n}(0) = \frac{1}{\left(\frac{n}{2}\right)! 2^{\frac{n+2}{2}}} \qquad (n \text{ even})$$

Asymptotic Expansions of an Arbitrary Probability Density Function and Distribution Function

Let Y_i ($i = 1, 2, \ldots, n$) be n independent random variables with mean m_i, variance σ_i^2, and higher cumulants $\kappa_{r,i}$. Then asymptotic expansions with respect to n for the probability density and cumulative distribution function of

$$X = \frac{\sum_{i=1}^m (Y_i - m_i)}{\left(\sum_{i=1}^m \sigma_i^2\right)^{\frac{1}{2}}} \text{ are}$$

26.2.47

$$f(x) \sim Z(x) - \left[\frac{\gamma_1}{6} Z^{(3)}(x)\right] + \left[\frac{\gamma_2}{24} Z^{(4)}(x) + \frac{\gamma_1^2}{72} Z^{(6)}(x)\right]$$

$$- \left[\frac{\gamma_3}{120} Z^{(5)}(x) + \frac{\gamma_1\gamma_2}{144} Z^{(7)}(x) + \frac{\gamma_1^3}{1296} Z^{(9)}(x)\right]$$

$$+ \left[\frac{\gamma_4}{720} Z^{(6)}(x) + \frac{\gamma_2^2}{1152} Z^{(8)}(x) + \frac{\gamma_1\gamma_3}{720} Z^{(8)}(x)\right.$$

$$\left. + \frac{\gamma_1^2\gamma_2}{1728} Z^{(10)}(x) + \frac{\gamma_1^4}{31104} Z^{(12)}(x)\right] + \ldots$$

26.2.48

$$F(x) \sim P(x) - \left[\frac{\gamma_1}{6} Z^{(2)}(x)\right] + \left[\frac{\gamma_2}{24} Z^{(3)}(x) + \frac{\gamma_1^2}{72} Z^{(5)}(x)\right]$$

$$- \left[\frac{\gamma_3}{120} Z^{(4)}(x) + \frac{\gamma_1\gamma_2}{144} Z^{(6)}(x) + \frac{\gamma_1^3}{1296} Z^{(8)}(x)\right]$$

$$+ \left[\frac{\gamma_4}{720} Z^{(5)}(x) + \frac{\gamma_2^2}{1152} Z^{(7)}(x) + \frac{\gamma_1\gamma_3}{720} Z^{(7)}(x)\right.$$

$$\left. + \frac{\gamma_1^2\gamma_2}{1728} Z^{(9)}(x) + \frac{\gamma_1^4}{31104} Z^{(11)}(x)\right] + \ldots$$

where

$$\gamma_{r-2} = \frac{1}{n^{\frac{r}{2}-1}} \frac{\left(\frac{1}{n}\sum_{i=1}^n \kappa_{r,i}\right)}{\left(\frac{1}{n}\sum_{i=1}^n \sigma_i^2\right)^{r/2}}$$

Terms in brackets are terms of the same order with respect to n. When the Y_i have the same distribution, then $m_i = m$, $\sigma_i^2 = \sigma^2$, $\kappa_{r,i} = \kappa_r$ and

$$\gamma_{r-2} = \frac{1}{n^{\frac{1}{2}r-1}} \left(\frac{\kappa_r}{\sigma^r}\right)$$

Asymptotic Expansion for the Inverse Function of an Arbitrary Distribution Function

Let the cumulative distribution function of $Y = \sum_{i=1}^n Y_i$ be denoted by $F(y)$. Then the (Cornish-Fisher) asymptotic expansion with respect to n for the value of y_p such that $F(y_p) = 1 - p$ is

26.2.49
$$y_p \sim m + \sigma w$$

where

$$w = x + [\gamma_1 h_1(x)]$$
$$+ [\gamma_2 h_2(x) + \gamma_1^2 h_{11}(x)]$$
$$+ [\gamma_3 h_3(x) + \gamma_1\gamma_2 h_{12}(x) + \gamma_1^3 h_{111}(x)]$$
$$+ [\gamma_4 h_4(x) + \gamma_2^2 h_{22}(x) + \gamma_1\gamma_3 h_{13}(x) + \gamma_1^2\gamma_2 h_{112}(x)$$
$$+ \gamma_1^4 h_{1111}(x)] + \ldots$$

and

$$Q(x) = p, \qquad \gamma_{r-2} = \frac{\kappa_r}{\kappa_2^{r/2}}, \qquad r = 3, 4, \ldots$$

26.2.50

$$h_1(x) = \frac{1}{6} He_2(x)$$

$$h_2(x) = \frac{1}{24} He_3(x)$$

$$h_{11}(x) = -\frac{1}{36} [2He_3(x) + He_1(x)]$$

$$h_3(x) = \frac{1}{120} [He_4(x)]$$

$$h_{12}(x) = -\frac{1}{24} [He_4(x) + He_2(x)]$$

$$h_{111}(x) = \frac{1}{324} [12He_4(x) + 19He_2(x)]$$

$$h_4(x) = \frac{1}{720} He_5(x)$$

$$h_{22}(x) = -\frac{1}{384} [3He_5(x) + 6He_3(x) + 2He_1(x)]$$

$$h_{13}(x) = -\frac{1}{180} [2He_5(x) + 3He_3(x)]$$

$$h_{112}(x) = \frac{1}{288} [14He_5(x) + 37He_3(x) + 8He_1(x)]$$

$$h_{1111}(x) = -\frac{1}{7776} [252He_5(x) + 832He_3(x)$$
$$+ 227He_1(x)]$$

Terms in brackets in **26.2.49** are terms of the same order with respect to n. The $He_n(x)$ are the Hermite polynomials. (See chapter **22**.)

26.2.51
$$He_n(x)=(-1)^n \frac{Z^{(n)}(x)}{Z(x)}=n! \sum_{m=0}^{\left[\frac{n}{2}\right]} \frac{(-1)^m}{2^m m!(n-2m)!} x^{n-2m}$$

In the following auxiliary table, the polynomial functions $h_1(x), h_2(x) \ldots h_{1111}(x)$ are tabulated for

$$p=.25, .1, .05, .025, .01, .005, .0025, .001, .0005.$$

Auxiliary coefficients[8] for use with Cornish-Fisher asymptotic expansion. **26.2.49**

	p								
	.25	.10	.05	.025	.01	.005	.0025	.001	.0005
x	.67449	1.28155	1.64485	1.95996	2.32635	2.57583	2.80703	3.09022	3.29053
$h_1(x)$	−.09084	.10706	.28426	.47358	.73532	.93915	1.14657	1.42491	1.63793
$h_2(x)$	−.07153	−.07249	−.02018	.06872	.23379	.39012	.57070	.84331	1.07320
$h_{11}(x)$	.07663	.06106	−.01878	−.14607	−.37634	−.59171	−.83890	−1.21025	−1.52234
$h_3(x)$	.00398	−.03464	−.04928	−.04410	−.00152	.06010	.14841	.30746	.46059
$h_{12}(x)$	.00282	.14644	.17532	.10210	−.17621	−.53531	−1.02868	−1.89355	−2.71243
$h_{111}(x)$	−.01428	−.11629	−.11900	−.02937	.25195	.59757	1.06301	1.86787	2.62337
$h_4(x)$	.00998	.00227	−.01082	−.02357	−.03176	−.02621	−.00666	.04591	.10950
$h_{22}(x)$	−.03285	.00776	.05985	.09659	.07888	−.01226	−.19116	−.59060	−1.03555
$h_{13}(x)$	−.05126	.01086	.09462	.16106	.16058	.05366	−.17498	−.70464	−1.30531
$h_{112}(x)$	.14764	−.10858	−.39517	−.55856	−.32621	.35696	1.60445	4.29304	7.23307
$h_{1111}(x)$	−.06898	.09585	.25623	.31624	.07286	−.46534	−1.39199	−3.32708	−5.40702

[8] From R. A. Fisher, Contributions to mathematical statistics, Paper 30 (with E. A. Cornish) Extrait de la Revue de l'Institute International de Statistique 4, 1–14 (1937) (with permission).

26.3. Bivariate Normal Probability Function

26.3.1
$$g(x,y,\rho)=[2\pi\sqrt{1-\rho^2}]^{-1}\exp-\frac{1}{2}\left(\frac{x^2-2\rho xy+y^2}{1-\rho^2}\right)$$

26.3.2 $\quad g(x,y,\rho)=(1-\rho^2)^{-\frac{1}{2}}Z(x)Z\left(\frac{y-\rho x}{\sqrt{1-\rho^2}}\right)$

26.3.3
$$L(h,k,\rho)=\int_h^\infty dx\int_k^\infty g(x,y,\rho)dy$$
$$=\int_h^\infty Z(x)dx\int_w^\infty Z(w)\,dw, \quad w=\left(\frac{k-\rho x}{\sqrt{1-\rho^2}}\right)$$

26.3.4 $\quad L(-h,-k,\rho)=\int_{-\infty}^h dx\int_{-\infty}^k g(x,y,\rho)dy$

26.3.5 $\quad L(-h,k,-\rho)=\int_{-\infty}^h dx\int_k^\infty g(x,y,\rho)dy$

26.3.6 $\quad L(h,-k,-\rho)=\int_h^\infty dx\int_{-\infty}^k g(x,y,\rho)dy$

26.3.7 $\quad\quad L(h,k,\rho)=L(k,h,\rho)$

26.3.8 $\quad L(-h,k,\rho)+L(h,k,-\rho)=Q(k)$

26.3.9 $\quad L(-h,-k,\rho)-L(h,k,\rho)=P(k)-Q(h)$

26.3.10
$$* \quad 2[L(h,k,\rho)+L(h,k,-\rho)+P(h)-Q(k)]-1$$
$$=\int_{-h}^h dx\int_{-k}^k g(x,y,\rho)dy$$

Probability Function With Means m_x, m_y, Variances σ_x^2, σ_y^2, and Correlation ρ

The random variables X, Y are said to be distributed as a bivariate Normal distribution with means and variances (m_x, m_y) and (σ_x^2, σ_y^2) and correlation ρ if the joint probability that X is less than or equal to h and Y less than or equal to k is given by

26.3.11
$$Pr\{X\leq h, Y\leq k\}=\frac{1}{\sigma_x\sigma_y}\int_{-\infty}^{\frac{h-m_x}{\sigma_x}}\int_{-\infty}^{\frac{k-m_y}{\sigma_y}}g(s,t,\rho)ds\,dt$$
$$=L\left(-\left(\frac{h-m_x}{\sigma_x}\right), -\left(\frac{k-m_y}{\sigma_y}\right), \rho\right)$$

The probability density function is

26.3.12
$$\frac{1}{2\pi\sigma_x\sigma_y\sqrt{1-\rho^2}}\exp\frac{-Q}{2(1-\rho^2)}=\frac{1}{\sigma_x\sigma_y}g\left(\frac{x-m_x}{\sigma_x},\frac{y-m_y}{\sigma_y},\rho\right)$$

where

$$Q=\frac{(x-m_x)^2}{\sigma_x^2}-\frac{2\rho(x-m_x)(y-m_y)}{\sigma_x\sigma_y}+\frac{(y-m_y)^2}{\sigma_y^2}$$

Circular Normal Probability Density Function

26.3.13
$$\frac{1}{\sigma^2}g\left(\frac{x-m_x}{\sigma},\frac{y-m_y}{\sigma},0\right)=$$
$$\frac{1}{2\pi\sigma^2}\exp-\frac{(x-m_x)^2+(y-m_y)^2}{2\sigma^2}$$

*See page II.

Special Values of $L(h, k, \rho)$

26.3.14
$$L(h, k, 0) = Q(h)\,Q(k)$$

26.3.15
$$L(h, k, -1) = 0 \qquad (h+k \geq 0)$$

26.3.16
$$L(h, k, -1) = P(h) - Q(k) \qquad (h+k \leq 0)$$

26.3.17
$$L(h, k, 1) = Q(h) \qquad (k \leq h)$$

26.3.18
$$L(h, k, 1) = Q(k) \qquad (k \geq h)$$

26.3.19
$$L(0, 0, \rho) = \frac{1}{4} + \frac{\arcsin \rho}{2\pi}$$

$L(h, k, \rho)$ as a Function of $L(h, 0, \rho)$

26.3.20
$$L(h, k, \rho) = L\left(h, 0, \frac{(\rho h - k)(\operatorname{sgn} h)}{\sqrt{h^2 - 2\rho h k + k^2}}\right)$$
$$+ L\left(k, 0, \frac{(\rho k - h)(\operatorname{sgn} k)}{\sqrt{h^2 - 2\rho h k + k^2}}\right)$$
$$- \begin{cases} 0 & \text{if } hk > 0 \text{ or } hk = 0 \\ & \text{and } h+k \geq 0 \\ \frac{1}{2} & \text{otherwise} \end{cases}$$

where $\operatorname{sgn} h = 1$ if $h \geq 0$ and $\operatorname{sgn} h = -1$ if $h < 0$.

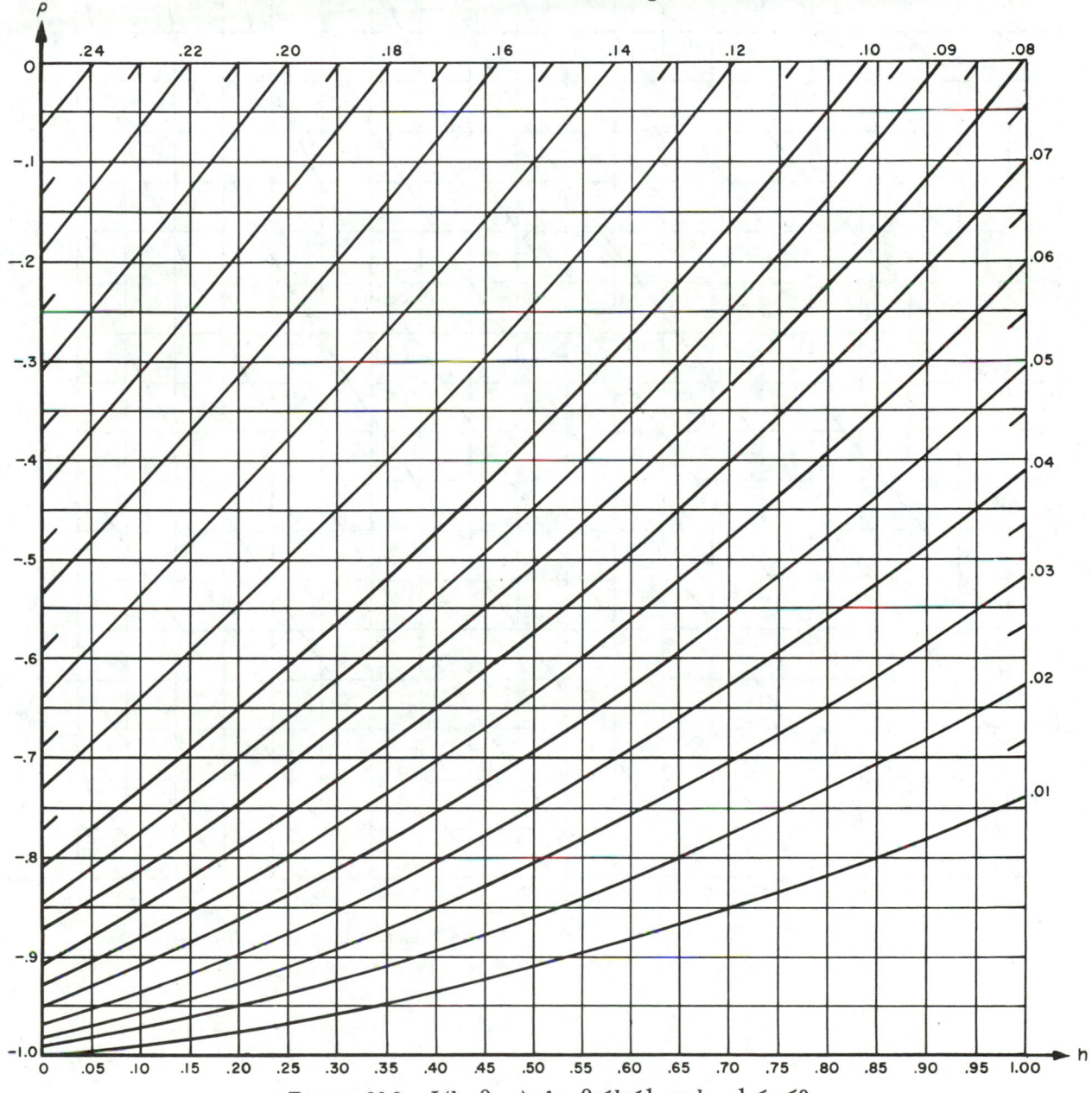

FIGURE 26.2. $L(h,\ 0,\ \rho)$ for $0 \leq h \leq 1$ and $-1 \leq \rho \leq 0$.

Values for $h<0$ can be obtained using $L(h, 0, -\rho) = \frac{1}{2} - L(-h, 0, \rho)$.

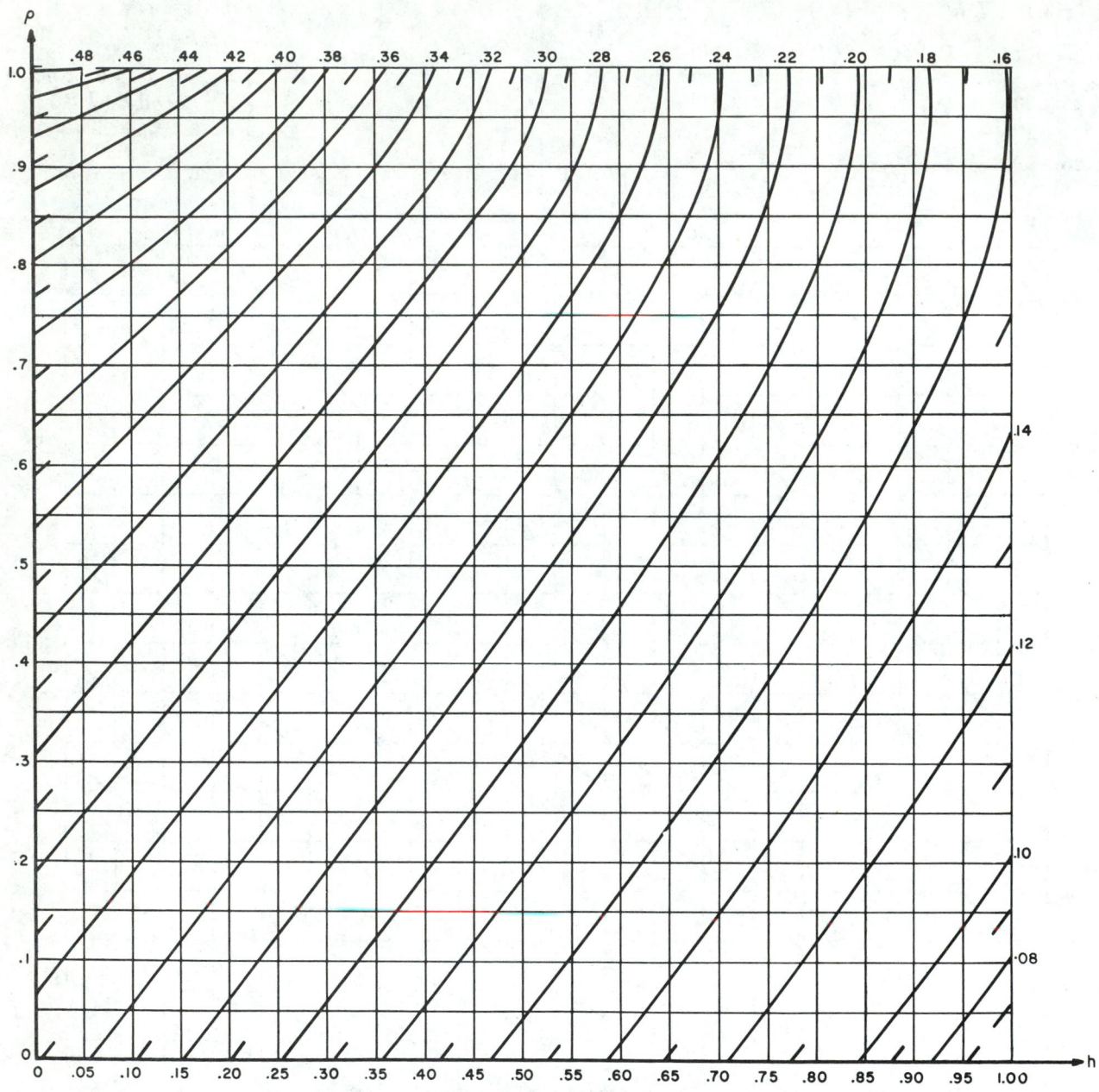

FIGURE 26.3.　$L(h,\ 0,\ \rho)$ for $0 \leq h \leq 1$ and $0 \leq \rho \leq 1$.

Values for $h<0$ can be obtained using $L(h,\ 0,\ -\rho)=\frac{1}{2}-L(-h,\ 0,\ \rho)$.

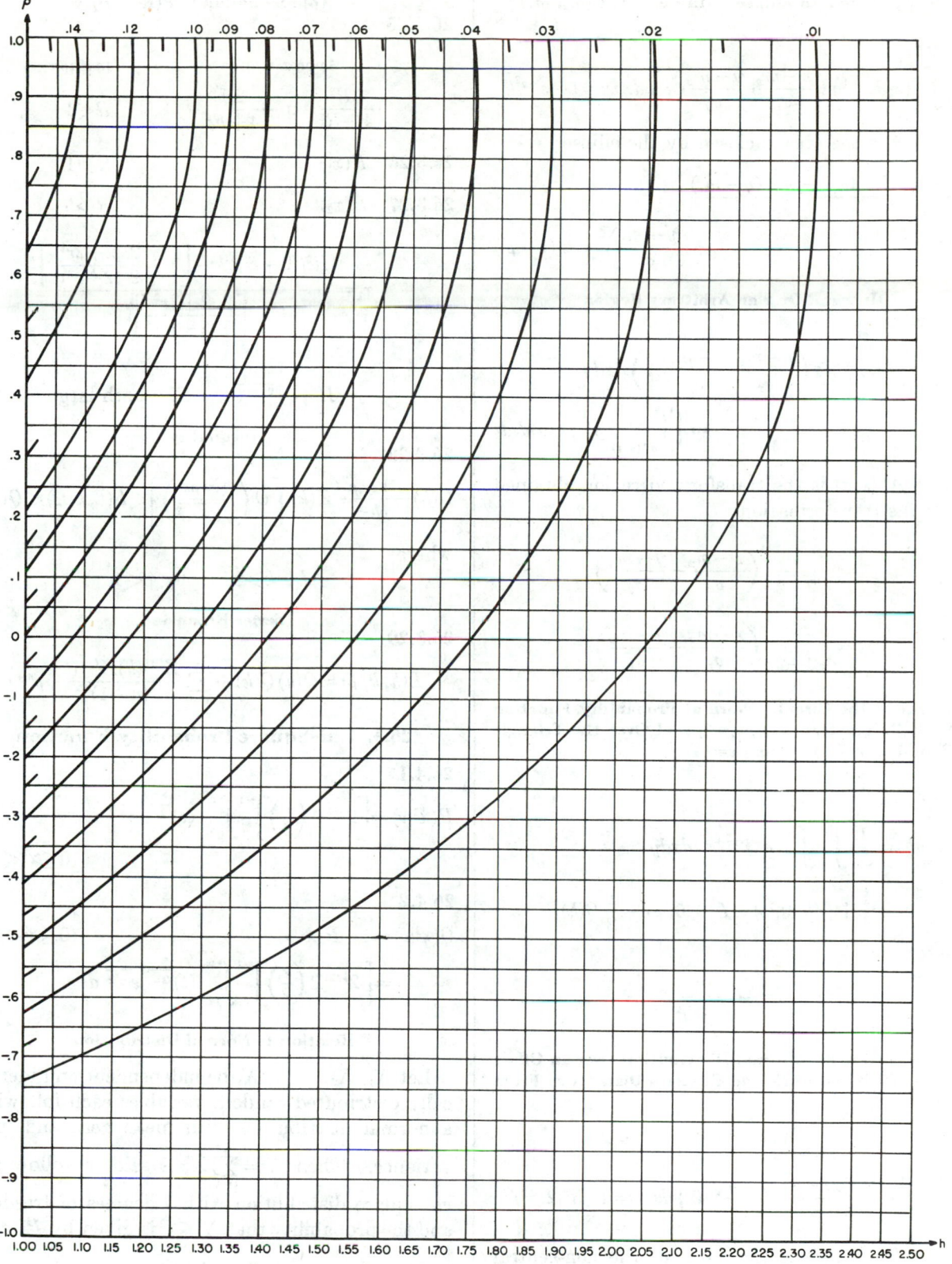

FIGURE 26.4. $L(h, 0, \rho)$ for $h \geq 1$ and $-1 \leq \rho \leq 1$.

Values for $h < 0$ can be obtained using $L(h, 0, -\rho) = \frac{1}{2} - L(-h, 0, \rho.)$

Integral Over an Ellipse With Center at (m_x, m_y)

26.3.21

$$\iint_A (\sigma_x\sigma_y)^{-1} g\left(\frac{x-m_x}{\sigma_x}, \frac{y-m_y}{\sigma_y}, \rho\right) dxdy = 1 - e^{-a^2/2}$$

where A is the area enclosed by the ellipse

$$\left(\frac{x-m_x}{\sigma_x}\right)^2 - \frac{2\rho(x-m_x)(y-m_y)}{\sigma_x\sigma_y}$$
$$+ \left(\frac{y-m_y}{\sigma_y}\right)^2 = a^2(1-\rho^2)$$

Integral Over an Arbitrary Region

26.3.22

$$\iint_{A(x,y)} (\sigma_x\sigma_y)^{-1} g\left(\frac{x-m_x}{\sigma_x}, \frac{y-m_y}{\sigma_y}, \rho\right) dxdy$$
$$= \iint_{A^*(s,t)} g(s, t, o)dsdt$$

where $A^*(s, t)$ is the transformed region obtained from the transformation

$$s = \frac{1}{\sqrt{2+2\rho}}\left(\frac{x-m_x}{\sigma_x} + \frac{y-m_y}{\sigma_y}\right)$$

$$t = \frac{-1}{\sqrt{2-2\rho}}\left(\frac{x-m_x}{\sigma_x} - \frac{y-m_y}{\sigma_y}\right)$$

Integral of the Circular Normal Probability Function With Parameters $m_x=m_y=0$, $\sigma=1$ Over the Triangle Bounded by $y=0$, $y=ax$, $x=h$

26.3.23

$$V(h, ah) = \frac{1}{2\pi}\int_0^h\int_0^{ax} e^{-\frac{1}{2}(x^2+v^2)}dxdy$$
$$= \frac{1}{4} + L(h, 0, \rho) - L(0, 0, \rho) - \frac{1}{2}Q(h)$$

where

$$\rho = -\frac{a}{\sqrt{1+a^2}}$$

Integral of Circular Normal Distribution Over an Offset Circle With Radius $R\sigma$ and Center a Distance $r\sigma$ From (m_x, m_y)

26.3.24

$$\int_A \int \sigma^{-2} g\left(\frac{x-m_x}{\sigma}, \frac{y-m_y}{\sigma}, 0\right) dxdy = P(R^2|2, r^2)$$

where $P(R^2|2, r^2)$ is the c.d.f. of the non-central χ^2 distribution (see **26.4.25**) with $\nu=2$ degrees of freedom and noncentrality parameter r^2.

Approximation to $P(R^2|2, r^2)$

26.3.25

Approximation	*Condition*
$\dfrac{2R^2}{4+R^2}\exp-\dfrac{2r^2}{4+R^2}$	$R<1$

26.3.26 $\quad P(x_1)$ $\qquad\qquad\qquad\qquad R>1$

26.3.27 $\quad P(x_2)$ $\qquad\qquad\qquad\qquad R>5$

$$x_1 = \frac{[R^2/(2+r^2)]^{1/3} - \left[1 - \dfrac{2}{9}\dfrac{2+2r^2}{(2+r^2)^2}\right]}{\left[\dfrac{2}{9}\dfrac{2+2r^2}{(2+r^2)^2}\right]^{\frac{1}{2}}}$$

$$x_2 = R - \sqrt{r^2-1} \qquad R, r \text{ both large} \qquad *$$

Inequality

26.3.28

$$Q(h) - \frac{1-\rho^2}{\rho h-k} Z(k)\left[Q\left(\frac{h-\rho k}{\sqrt{1-\rho^2}}\right)\right] < L(h, k, \rho) < Q(h)$$

where

$$\rho h - k > 0, \qquad 0 < \rho < 1.$$

Series Expansion

26.3.29

$$L(h, k, \rho) = Q(h)Q(k) + \sum_{n=0}^{\infty} \frac{Z^{(n)}(h)Z^{(n)}(k)}{(n+1)!}\rho^{n+1}$$

26.4. Chi-Square Probability Function

26.4.1

$$P(\chi^2|\nu) = \left[2^{\nu/2}\Gamma\left(\frac{\nu}{2}\right)\right]^{-1}\int_0^{\chi^2}(t)^{\frac{\nu}{2}-1}e^{-\frac{t}{2}}dt$$
$$(0 \le \chi^2 < \infty)$$

26.4.2

$$Q(\chi^2|\nu) = 1 - P(\chi^2|\nu) \qquad\qquad (0 \le \chi^2 < \infty)$$
$$= \left[2^{\nu/2}\Gamma\left(\frac{\nu}{2}\right)\right]^{-1}\int_{\chi^2}^{\infty}(t)^{\frac{\nu}{2}-1}e^{-\frac{t}{2}}dt$$

Relation to Normal Distribution

Let $X_1, X_2, \ldots, X_\nu$ be independent and identically distributed random variables each following a normal distribution with mean zero and unit variance. Then $X^2 = \sum_{i=1}^{\nu} X_i^2$ is said to follow the chi-square distribution with ν degrees of freedom and the probability that $X^2 \le \chi^2$ is given by $P(\chi^2|\nu)$.

Cumulants

26.4.3 $\qquad \kappa_{n+1} = 2^n n!\nu \qquad (n=0, 1, \ldots)$

Series Expansions

26.4.4

$$Q(\chi^2|\nu)=2Q(\chi)+2Z(\chi)\sum_{r=1}^{\frac{\nu-1}{2}}\frac{\chi^{2r-1}}{1\cdot 3\cdot 5\ldots(2r-1)}$$

$$(\nu\text{ odd})\text{ and }\chi=\sqrt{\chi^2}$$

26.4.5

$$Q(\chi^2|\nu)=\sqrt{2\pi}Z(\chi)\left\{1+\sum_{r=1}^{\frac{\nu-2}{2}}\frac{\chi^{2r}}{2\cdot 4\ldots(2r)}\right\}$$

$$(\nu\text{ even})$$

26.4.6

$$P(\chi^2|\nu)=\left(\frac{1}{2}\chi^2\right)^{\nu/2}\frac{e^{-\chi^2/2}}{\Gamma\left(\frac{\nu+2}{2}\right)}$$

$$*\quad\left\{1+\sum_{r=1}^{\infty}\frac{\chi^{2r}}{(\nu+2)(\nu+4)\cdots(\nu+2r)}\right\}$$

26.4.7 $\quad P(\chi^2|\nu)=\frac{1}{\Gamma\left(\frac{\nu}{2}\right)}\sum_{n=0}^{\infty}\frac{(-1)^n(\chi^2/2)^{\frac{\nu}{2}+n}}{n!\left(\frac{\nu}{2}+n\right)}$

Recurrence and Differential Relations

26.4.8 $\quad Q(\chi^2|\nu+2)=Q(\chi^2|\nu)+\frac{(\chi^2/2)^{\nu/2}e^{-\chi^2/2}}{\Gamma\left(\frac{\nu}{2}+1\right)}$

26.4.9 $\quad \frac{\partial^m Q(\chi^2|\nu)}{\partial(\chi^2)^m}=\frac{1}{2^m}\sum_{j=0}^{m}\binom{m}{j}(-1)^{m+j}Q(\chi^2|\nu-2j)$

Continued Fraction

26.4.10 $\quad *Q(\chi^2|\nu)=\frac{(\chi^2)^{\nu/2}e^{-\chi^2/2}}{2^{\nu/2}\Gamma(\nu/2)}$

$$\left\{\frac{1}{\chi^2/2+}\frac{1-\nu/2}{1+}\frac{1}{\chi^2/2+}\frac{2-\nu/2}{1+}\frac{2}{\chi^2/2+}\ldots\right\}$$

Asymptotic Distribution for Large ν

26.4.11 $\quad P(\chi^2|\nu)\sim P(x)\qquad$ where $x=\frac{\chi^2-\nu}{\sqrt{2\nu}}$

Asymptotic Expansions for Large χ^2

26.4.12

$$Q(\chi^2|\nu)\sim\frac{(\chi^2)^{\frac{\nu}{2}-1}e^{-\chi^2/2}}{2^{\nu/2}\Gamma(\nu/2)}\sum_{j=0}^{\infty}(-1)^j\frac{\Gamma\left(1-\frac{\nu}{2}+j\right)}{\Gamma\left(1-\frac{\nu}{2}\right)}\frac{2^{j+1}}{(\chi^2)^j}$$

*See page II.

Approximations to the Chi-Square Distribution for Large ν

26.4.13

	Approximation		*Condition*
$Q(\chi^2|\nu)\approx Q(x_1),$	$x_1=\sqrt{2\chi^2}-\sqrt{2\nu-1}$		$(\nu>100)$

26.4.14

$$Q(\chi^2|\nu)\approx Q(x_2),\qquad x_2=\frac{(\chi^2/\nu)^{1/3}-\left(1-\frac{2}{9\nu}\right)}{\sqrt{2/9\nu}}\qquad(\nu>30)$$

26.4.15

$$Q(\chi^2|\nu)\approx Q(x_2+h_\nu),\qquad h_\nu=\frac{60}{\nu}h_{60}\qquad(\nu>30)$$

Values of h_{60}

x	h_{60}	x	h_{60}	x	h_{60}
-3.5	$-.0118$	-1.0	$+.0006$	$+1.5$	$-.0005$
-3.0	$-.0067$	$-.5$	$-.0006$	2.0	$+.0002$
-2.5	$-.0033$	$.0$	$+.0002$	2.5	$.0017$
-2.0	$-.0010$	$+.5$	$-.0003$	3.0	$.0043$
-1.5	$+.0001$	1.0	$-.0006$	3.5	$.0082$

Approximations for the Inverse Function for Large ν

If $Q(\chi_p^2|\nu)=p$ and $Q(x_p)=1-P(x_p)=p$, then

	Approximation	*Condition*
26.4.16	$\chi_p^2\approx\frac{1}{2}\left\{x_p+\sqrt{2\nu-1}\right\}^2$	$(\nu>100)$
26.4.17	$\chi_p^2\approx\nu\left\{1-\frac{2}{9\nu}+x_p\sqrt{\frac{2}{9\nu}}\right\}^3$	$(\nu>30)$
26.4.18	$\chi_p^2\approx\nu\left\{1-\frac{2}{9\nu}+(x_p-h_\nu)\sqrt{\frac{2}{9\nu}}\right\}^3$	$(\nu>30)$

where h_ν is given by **26.4.15**.

Relation to Other Functions

26.4.19 Incomplete gamma function

$$\frac{\gamma(a,x)}{\Gamma(a)}=P(\chi^2|\nu),\qquad\nu=2a,\ \chi^2=2x$$

$$\frac{\Gamma(a,x)}{\Gamma(a)}=Q(\chi^2|\nu)$$

26.4.20 Pearson's incomplete gamma function

$$I(u,p)=\frac{1}{\Gamma(p+1)}\int_0^{u\sqrt{p+1}}t^pe^{-t}dt=P(\chi^2|\nu)$$

$$\nu=2(p+1),\ \chi^2=2u\sqrt{p+1}$$

26.4.21 Poisson distribution

$$Q(\chi^2|\nu)=\sum_{j=0}^{c-1}e^{-m}\frac{m^j}{j!},\qquad c=\frac{\nu}{2},\ m=\frac{\chi^2}{2},\ (\nu\text{ even})$$

$$Q(\chi^2|\nu)-Q(\chi^2|\nu-2)=e^{-m}\frac{m^{c-1}}{(c-1)!}$$

26.4.22 Pearson Type III

$$\left[\frac{ab}{e}\right]^{ab}\int_{-a}^{x}\left(1+\frac{t}{a}\right)^{ab}e^{-bt}dt = P(\chi^2|\nu)$$

$$\nu = 2ab+2,\ \chi^2 = 2b(x+a)$$

26.4.23 Incomplete moments of Normal distribution

$$\int_0^x t^n Z(t)dt = \begin{cases} (n-1)!!\,\dfrac{P(\chi^2|\nu)}{2} & (n\ \text{even}) \\[2ex] \dfrac{(n-1)!!}{\sqrt{2\pi}}\,P(\chi^2|\nu) & (n\ \text{odd}) \end{cases}$$

$$\chi^2 = x^2,\ \nu = n+1$$

26.4.24 Generalized Laguerre Polynomials

$$n!\,L_n^{(\alpha)}(x) = \frac{\sum_{j=0}^{n+1}(-1)^{n+j}\binom{n+1}{j}Q(\chi^2|\nu+2-2j)}{2^n[Q(\chi^2|\nu+2)-Q(\chi^2|\nu)]}$$

$$x = \chi^2/2,\ \alpha = \nu/2$$

Non-Central χ^2 Distribution Function

26.4.25

$$P(\chi'^2|\nu,\ \lambda) = \sum_{j=0}^{\infty} e^{-\lambda/2}\frac{(\lambda/2)^j}{j!}P(\chi'^2|\nu+2j)$$

where $\lambda \geq 0$ is termed the non-centrality parameter.

Relation of Non-Central χ^2 Distribution With $\nu=2$ to the Integral of Circular Normal Distribution ($\sigma^2=1$) Over an Offset Circle Having Radius R and Center a Distance $r=\sqrt{\lambda}$ From the Origin. (See 26.3.24–26.3.27.)

26.4.26

$$\iint_A g(x,\ y,\ 0)\,dxdy = P(\chi^2 = R^2|\nu=2,\ \lambda)$$

$$= 1 - \sum_{j=0}^{\infty}\frac{e^{-\lambda/2}\lambda^j}{2^j j!}Q(R^2|2+2j)$$

Approximations to the Non-Central χ^2 Distribution

$$a = \nu + \lambda \qquad b = \frac{\lambda}{\nu+\lambda}$$

Approximating Function   *Approximation*

26.4.27 χ^2 distribution

$$P(\chi'^2|\nu,\ \lambda) \approx P\left(\frac{\chi^2}{1+b}\Big|\nu^*\right),\qquad \nu^* = \frac{a}{1+b}$$

26.4.28 Normal distribution

$$P(\chi'^2|\nu,\ \lambda) \approx P(x),\qquad x = \frac{(\chi'^2/a)^{1/3}-\left[1-\dfrac{2}{9}\left(\dfrac{1+b}{a}\right)\right]}{\sqrt{\dfrac{2}{9}\left(\dfrac{1+b}{a}\right)}}$$

26.4.29 Normal distribution

$$P(\chi'^2|\nu,\ \lambda) \approx P(x),\qquad x = \left[\frac{2\chi'^2}{1+b}\right]^{\frac12} - \left[\frac{2a}{1+b}-1\right]^{\frac12}$$

Approximations to the Inverse Function of Non-Central χ^2 Distribution

If $Q(\chi_p'^2|\nu,\ \lambda)=p$, $Q(\chi_p^2|\nu^*)=p$, and $Q(x_p)=p$ then

Approximating Variable   *Approximation to the Inverse Function*

26.4.30 χ^2

$$\chi_p'^2 \approx (1+b)\chi_p^2$$

26.4.31 Normal

$$\chi_p'^2 \approx \frac{1+b}{2}\left[x_p + \sqrt{\frac{2a}{1+b}-1}\right]^2$$

26.4.32 Normal

$$\chi_p'^2 \approx a\left[x_p\sqrt{\frac{2}{9}\left(\frac{1+b}{a}\right)}+1-\frac{2}{9}\left(\frac{1+b}{a}\right)\right]^3$$

Properties of Chi-Square, Non-Central Chi-Square, and Related Quantities

$$a = \nu + \lambda \qquad b = \frac{\lambda}{\nu + \lambda}$$

$$\psi(z) = \frac{d}{dz} \ln \Gamma(z), \qquad \psi'(z) = \frac{d^2}{dz^2} \psi(z)$$

	Variable	Mean	Variance	Coefficient of skewness (γ_1)	Coefficient of excess (γ_2)
26.4.33	χ^2	ν	2ν	$\dfrac{2^{3/2}}{\sqrt{\nu}}$	$12\nu^{-1}$
26.4.34	$\sqrt{2\chi^2}$	$(2\nu-1)^{\frac{1}{2}}\{1+[16\nu(\nu-1)]^{-1}\}+O(\nu^{-7/2})$	$1-\dfrac{1}{4\nu}-\dfrac{1}{8\nu^2}+\dfrac{5}{64\nu^3}-O(\nu^{-4})$	$\dfrac{1}{\sqrt{2\nu}}\left[1+\dfrac{5}{8\nu}-\dfrac{1}{128\nu^2}\right]+O(\nu^{-7/2})$	$\dfrac{3}{2^2}\dfrac{1}{\nu^2}\left[1+\dfrac{3}{2\nu}\right]+O(\nu^{-4})$
26.4.35	$(\chi^2/\nu)^{1/3}$	$1-\dfrac{2}{3^2\nu}+\dfrac{80}{3^7\nu^2}+O(\nu^{-4})$	$\dfrac{2}{3^2\nu}-\dfrac{104}{3^7\nu^2}+O(\nu^{-4})$	$\dfrac{2^{7/2}}{3^5\nu^{3/2}}\left[1+\dfrac{8}{3^2\nu}\right]+O(\nu^{-7/2})$	$-\dfrac{4}{9\nu}\left[1+\dfrac{16}{9\nu}\right]+O(\nu^{-3})$
26.4.36	$\ln(\chi^2/\nu)$	$\psi\left(\dfrac{\nu}{2}\right)-\ln\left(\dfrac{\nu}{2}\right)=-\dfrac{1}{\nu}-\dfrac{1}{3\nu^2}+O(\nu^{-4})$	$\psi'\left(\dfrac{\nu}{2}\right)=\dfrac{2}{\nu-1}\left[1-\dfrac{1}{3(\nu-1)^2}\right]+O((\nu-1)^{-5})$	$\dfrac{\psi''\left(\frac{\nu}{2}\right)}{\psi'\left(\frac{\nu}{2}\right)^{3/2}}=-\sqrt{\dfrac{2}{\nu-1}}\left[1-\dfrac{1}{2(\nu-1)^2}\right]+O((\nu-1)^{-9/2})$	$\dfrac{\psi^{(3)}\left(\frac{\nu}{2}\right)}{\psi'\left(\frac{\nu}{2}\right)^2}=\dfrac{4}{\nu-1}\left[1+\dfrac{4}{3(\nu-1)^2}\right]+O((\nu-1)^{-5})$
26.4.37	χ'^2	a	$2a(1+b)$	$\left(\dfrac{2}{1+b}\right)^{3/2}(1+2b)a^{-\frac{1}{2}}$	$\dfrac{12}{a}\dfrac{(1+3b)}{(1+b)^2}$
26.4.38	$\sqrt{2\chi'^2}$	$[2a-(1+b)]^{\frac{1}{2}}+O(a^{-3/2})$	$(1+b)-\dfrac{a^{-1}}{4}[8b+(1+b)(1-7b)]+O(a^{-2})$	$\dfrac{a^{-\frac{1}{2}}(1-b)(1+3b)}{2^{\frac{1}{2}}(1+b)^{3/2}}+O(a^{-1})$	$\dfrac{3b(b+2)}{(1+b)^2 a}+O(a^{-2})$
26.4.39	$(\chi'^2/a)^{1/3}$	$1-\dfrac{2}{3^2}\dfrac{1+b}{a}-\dfrac{40}{3^4}\dfrac{b^2}{a^2}+O(a^{-3})$	$\dfrac{2}{9}a^{-1}(1+b)+\dfrac{16}{27}a^{-2}b^2+O(a^{-3})$	$\left(\dfrac{2}{1+b}\right)^{3/2}b^2a^{-\frac{1}{2}}+O(a^{-3/2})$	$-\dfrac{4}{3^2}\dfrac{(1+3b+12b^2-44b^3)}{a(1+b)^3}-O(a^{-2})$

26.5. Incomplete Beta Function

26.5.1

$$I_x(a,b)=\frac{1}{B(a,b)}\int_0^x t^{a-1}(1-t)^{b-1}dt \qquad (0\leq x\leq 1)$$

26.5.2

$$I_x(a,b)=1-I_{1-x}(b,a)$$

Relation to the Chi-Square Distribution

If X_1^2 and X_2^2 are independent random variables following chi-square distributions 26.4.1 with ν_1 and ν_2 degrees of freedom respectively, then $\frac{X_1^2}{X_1^2+X_2^2}$ is said to follow a beta distribution with ν_1 and ν_2 degrees of freedom and has the distribution function

26.5.3

$$P\left\{\frac{X_1^2}{X_1^2+X_2^2}\leq x\right\}=\frac{1}{B(a,b)}\int_0^x t^{a-1}(1-t)^{b-1}dt$$

$$=I_x(a,b) \qquad a=\frac{\nu_1}{2},\ b=\frac{\nu_2}{2}$$

Series Expansions ($0<x<1$)

26.5.4

$$* \quad I_x(a,b)=\frac{x^a(1-x)^b}{aB(a,b)}\left\{1+\sum_{n=0}^{\infty}\frac{B(a+1,n+1)}{B(a+b,n+1)}x^{n+1}\right\}$$

26.5.5

$$I_x(a,b)=\frac{x^a(1-x)^{b-1}}{aB(a,b)}$$

$$\left\{1+\sum_{n=0}^{\infty}\frac{B(a+1,n+1)}{B(b-n-1,n+1)}\left(\frac{x}{1-x}\right)^{n+1}\right\}$$

$$=\frac{x^a(1-x)^{b-1}}{aB(a,b)}$$

$$\left\{1+\sum_{n=0}^{s-2}\frac{B(a+1,n+1)}{B(b-n-1,n+1)}\left(\frac{x}{1-x}\right)^{n+1}\right\}$$

$$+I_x(a+s,b-s)$$

26.5.6

$$1-I_x(a,b)=I_{1-x}(b,a)$$

$$=\frac{(1-x)^b}{B(a,b)}\sum_{i=0}^{a-1}(-1)^i\binom{a-1}{i}\frac{(1-x)^i}{b+i}\ \text{(integer a)}$$

26.5.7

$$1-I_x(a,b)=I_{1-x}(b,a)$$

$$=(1-x)^{a+b-1}\sum_{i=0}^{a-1}\binom{a+b-1}{i}\left(\frac{x}{1-x}\right)^i\ \text{(integer a)}$$

Continued Fractions

26.5.8

$$I_x(a,b)=\frac{x^a(1-x)^b}{aB(a,b)}\left\{\frac{1}{1+}\frac{d_1}{1+}\frac{d_2}{1+}\cdots\right\} \quad *$$

$$d_{2m+1}=-\frac{(a+m)(a+b+m)}{(a+2m)(a+2m+1)}x$$

$$d_{2m}=\frac{m(b-m)}{(a+2m-1)(a+2m)}x$$

Best results are obtained when $x<\frac{a-1}{a+b-2}$. Also the $4m$ and $4m+1$ convergents are less than $I_x(a,b)$ and the $4m+2$, $4m+3$ convergents are greater than $I_x(a,b)$.

26.5.9

$$I_x(a,b)=\frac{x^a(1-x)^{b-1}}{aB(a,b)}\left[\frac{e_1}{1+}\frac{e_2}{1+}\frac{e_3}{1+}\cdots\right]$$

$$* \quad x<1 \qquad e_1=1$$

$$e_{2m}=-\frac{(a+m-1)(b-m)}{(a+2m-2)(a+2m-1)}\frac{x}{1-x}$$

$$e_{2m+1}=\frac{m(a+b-1+m)}{(a+2m-1)(a+2m)}\frac{x}{1-x}$$

Recurrence Relations

26.5.10

$$I_x(a,b)=xI_x(a-1,b)+(1-x)I_x(a,b-1)$$

26.5.11

$$I_x(a,b)=\frac{1}{x}\{I_x(a+1,b)-(1-x)I_x(a+1,b-1)\}$$

26.5.12

$$\left[I_x(a,b)=\right]\frac{1}{a(1-x)+b}\{bI_x(a,b+1)$$

$$+a(1-x)I_x(a+1,b-1)\} \quad *$$

26.5.13

$$I_x(a,b)=\frac{1}{a+b}\{aI_x(a+1,b)+bI_x(a,b+1)\}$$

26.5.14

$$I_x(a,a)=\frac{1}{2}I_{1-x'}\left(a,\frac{1}{2}\right), \qquad x'=4\left(x-\frac{1}{2}\right)^2\left[x\leq\frac{1}{2}\right]^*$$

26.5.15

$$I_x(a,b)=\frac{\Gamma(a+b)}{\Gamma(a+1)\Gamma(b)}x^a(1-x)^{b-1}+I_x(a+1,b-1)$$

26.5.16

$$I_x(a,b)=\frac{\Gamma(a+b)}{\Gamma(a+1)\Gamma(b)}x^a(1-x)^b+I_x(a+1,b)$$

Asymptotic Expansions

26.5.17

$$1-I_x(a,b)=I_{1-x}(b,a)\sim\frac{\Gamma(b,y)}{\Gamma(b)}$$

$$-\frac{1}{24N^2}\left\{\frac{y^b e^{-\nu}}{(b-2)!}(b+1+y)\right\}$$

$$+\frac{1}{5760N^4}\left\{\frac{y^b e^{-\nu}}{(b-2)!}[(b-3)(b-2)(5b+7)(b+1+y)\right.$$

$$\left.-(5b-7)(b+3+y)y^2]\right\}$$

$$y=-N\ln x,\qquad N=a+\frac{b}{2}-\frac{1}{2}$$

26.5.18

$$I_x(a,b)\sim\frac{\Gamma(a,w)}{\Gamma(a)}+\frac{e^{-w}w^a}{\Gamma(a)}\left\{\frac{(a-1-w)}{2b}\right.$$

$$+\frac{1}{(2b)^2}\left(\frac{a^3}{2}-\frac{5}{3}a^2+\frac{3}{2}a-\frac{1}{3}-w\left[\frac{3}{2}a^2-\frac{11}{6}a+\frac{1}{3}\right]\right.$$

$$\left.\left.+w^2\left(\frac{3}{2}a-\frac{1}{6}\right)-\frac{1}{2}w^3\right)\right\}$$

$$w=b\left(\frac{x}{1-x}\right)$$

26.5.19

$$I_x(a,b)\sim P(y)-Z(y)\left[a_1+\frac{a_2(y-a_1)}{1+a_2}\right.$$

$$\left.+\frac{a_3(1+y^2/2)}{1+a_2}+\cdots\right]$$

$$a_1=\frac{2}{3}(b-a)[(a+b-2)(a-1)(b-1)]^{-\frac{1}{2}}$$

$$a_2=\frac{1}{12}\left[\frac{1}{a-1}+\frac{1}{b-1}-\frac{13}{a+b-1}\right]$$

$$a_3=-\frac{8}{15}\left[a_1\left(a_2+\frac{3}{a+b-2}\right)\right]$$

$$y^2=2\left[(a+b-1)\ln\frac{a+b-1}{a+b-2}+(a-1)\ln\frac{a-1}{(a+b-1)x}\right.$$

$$\left.+(b-1)\ln\frac{b-1}{(a+b-1)(1-x)}\right]$$

and y is taken negative when $x<\dfrac{a-1}{a+b-2}$

Approximations

26.5.20 If $(a+b-1)(1-x)\le.8$

$$I_x(a,b)=Q(\chi^2|\nu)+\epsilon,$$

$$|\epsilon|<5\times10^{-3}\text{ if }a+b>6$$

$$\chi^2=(a+b-1)(1-x)(3-x)-(1-x)(b-1),$$

$$\nu=2b$$

26.5.21 If $(a+b-1)(1-x)\ge.8$

$$I_x(a,b)=P(y)+\epsilon,$$

$$|\epsilon|<5\times10^{-3}\text{ if }a+b>6$$

$$y=\frac{3\left[w_1\left(1-\frac{1}{9b}\right)-w_2\left(1-\frac{1}{9a}\right)\right]}{\left[\frac{w_1^2}{b}+\frac{w_2^2}{a}\right]^{\frac{1}{2}}},$$

$$w_1=(bx)^{1/3},\ w_2=[a(1-x)]^{1/3}$$

Approximation to the Inverse Function

26.5.22 If $I_{x_p}(a,b)=p$ and $Q(y_p)=p$ then

$$x_p\approx\frac{a}{a+be^{2w}}$$

$$w=\frac{y_p(h+\lambda)^{\frac{1}{2}}}{h}-\left(\frac{1}{2b-1}-\frac{1}{2a-1}\right)\left(\lambda+\frac{5}{6}-\frac{2}{3h}\right)$$

$$h=2\left(\frac{1}{2a-1}+\frac{1}{2b-1}\right)^{-1},\qquad\lambda=\frac{y_p^2-3}{6}$$

Relations to Other Functions and Distributions

Function	*Relation*
26.5.23 Hypergeometric function	$\dfrac{1}{B(a,b)}\dfrac{x^a}{a}F(a,1-b;a+1;x)=I_x(a,b)$
26.5.24 Binomial distribution	$\sum_{s=a}^{n}\binom{n}{s}p^s(1-p)^{n-s}=I_p(a,n-a+1)$
26.5.25 `"`	$\binom{n}{a}p^a(1-p)^{n-a}=I_p(a,n-a+1)-I_p(a+1,n-a)$ *
26.5.26 Negative binomial distribution	$\sum_{s=a}^{n}\binom{n+s-1}{s}p^n q^s=I_q(a,n)$
26.5.27 Student's distribution	$\frac{1}{2}[1-A(t|\nu)]=\frac{1}{2}I_x\left(\frac{\nu}{2},\frac{1}{2}\right),\qquad x=\dfrac{\nu}{\nu+t^2}$ *
26.5.28 F-(variance-ratio) distribution	$Q(F|\nu_1,\nu_2)=I_x\left(\frac{\nu_2}{2},\frac{\nu_1}{2}\right),\qquad x=\dfrac{\nu_2}{\nu_2+\nu_1 F}$

*See page II.

26.6. F-(Variance-Ratio) Distribution Function

26.6.1

$$P(F|\nu_1, \nu_2) = \frac{\nu_1^{\frac{1}{2}\nu_1} \nu_2^{\frac{1}{2}\nu_2}}{B\left(\frac{1}{2}\nu_1, \frac{1}{2}\nu_2\right)} \int_0^F t^{\frac{1}{2}(\nu_1-2)}(\nu_2+\nu_1 t)^{-\frac{1}{2}(\nu_1+\nu_2)}dt$$
$$(F \geq 0)$$

26.6.2

$$Q(F|\nu_1, \nu_2) = 1 - P(F|\nu_1, \nu_2) = I_x\left(\frac{\nu_2}{2}, \frac{\nu_1}{2}\right)$$

where

$$x = \frac{\nu_2}{\nu_2 + \nu_1 F}$$

Relation to the Chi-Square Distribution

If X_1^2 and X_2^2 are independent random variables following chi-square distributions **26.4.1** with ν_1 and ν_2 degrees of freedom respectively, then the distribution of $F = \frac{X_1^2/\nu_1}{X_2^2/\nu_2}$ is said to follow the variance ratio or F-distribution with ν_1 and ν_2 degrees of freedom. The corresponding distribution function is $P(F|\nu_1, \nu_2)$.

Statistical Properties

26.6.3

mean: $\quad m = \dfrac{\nu_2}{\nu_2 - 2} \qquad (\nu_2 > 2)$

variance: $\quad \sigma^2 = \dfrac{2\nu_2^2(\nu_1+\nu_2-2)}{\nu_1(\nu_2-2)^2(\nu_2-4)} \qquad (\nu_2 > 4)$

third central moment:

$$\mu_3 = \left(\frac{\nu_2}{\nu_1}\right)^3 \frac{8\nu_1(\nu_1+\nu_2-2)(2\nu_1+\nu_2-2)}{(\nu_2-2)^3(\nu_2-4)(\nu_2-6)} \qquad (\nu_2 > 6)$$

moments about the origin:

$$\mu_n' = \left(\frac{\nu_2}{\nu_1}\right)^n \frac{\Gamma\left(\frac{\nu_1+2n}{2}\right) \Gamma\left(\frac{\nu_1-2n}{2}\right)}{\Gamma\left(\frac{\nu_1}{2}\right) \Gamma\left(\frac{\nu_2}{2}\right)} \qquad (\nu_2 > 2n)$$

characteristic function:

$$\phi(t) = E(e^{iFt}) = M\left(\frac{\nu_1}{2}, -\frac{\nu_2}{2}, -\frac{\nu_2}{\nu_1} it\right)$$

Series Expansions

$$x = \frac{\nu_2}{\nu_2 + \nu_1 F}$$

26.6.4

$$* \quad Q(F|\nu_1, \nu_2) = x^{\nu_2/2}\left[1 + \frac{\nu_2}{2}(1-x) + \frac{\nu_2(\nu_2+2)}{2\cdot 4}(1-x)^2 + \ldots\right.$$
$$\left. + \frac{\nu_2(\nu_2+2)\ldots(\nu_2+\nu_1-4)}{2\cdot 4\ldots(\nu_1-2)}(1-x)^{\frac{\nu_1-2}{2}}\right] \quad (\nu_1 \text{ even})$$

26.6.5

$$Q(F|\nu_1, \nu_2) = 1 - (1-x)^{\nu_1/2}\left[1 + \frac{\nu_1}{2}x + \frac{\nu_1(\nu_1+2)}{2\cdot 4}x^2 + \ldots\right.$$
$$\left. + \frac{\nu_1(\nu_1+2)\ldots(\nu_2+\nu_1-4)}{2\cdot 4\ldots(\nu_2-2)}x^{\frac{\nu_2-2}{2}}\right] \quad (\nu_2 \text{ even})$$

26.6.6

$$Q(F|\nu_1, \nu_2) = x^{\frac{\nu_1+\nu_2-2}{2}}\left[1 + \frac{\nu_1+\nu_2-2}{2}\left(\frac{1-x}{x}\right)\right.$$
$$+ \frac{(\nu_1+\nu_2-2)(\nu_1+\nu_2-4)}{2\cdot 4}\left(\frac{1-x}{x}\right)^2 + \ldots$$
$$\left. + \frac{(\nu_1+\nu_2-2)\ldots(\nu_2+2)}{2\cdot 4\ldots(\nu_1-2)}\left(\frac{1-x}{x}\right)^{\frac{\nu_1-2}{2}}\right] \quad (\nu_1 \text{ even})$$

26.6.7

$$Q(F|\nu_1, \nu_2) = 1 - (1-x)^{\frac{\nu_1+\nu_2-2}{2}}\left[1 + \frac{\nu_1+\nu_2-2}{2}\left(\frac{x}{1-x}\right)\right.$$
$$\left. + \ldots + \frac{(\nu_1+\nu_2-2)\ldots(\nu_1+2)}{2\cdot 4\ldots(\nu_2-2)}\left(\frac{x}{1-x}\right)^{\frac{\nu_2-2}{2}}\right]$$
$$(\nu_2 \text{ even})$$

26.6.8

$$Q(F|\nu_1, \nu_2) = 1 - A(t|\nu_2) + \beta(\nu_1, \nu_2) \qquad (\nu_1, \nu_2 \text{ odd})$$

$$A(t|\nu_2)' = \begin{cases} \dfrac{2}{\pi}\cdot\left\{\theta + \sin\theta[\cos\theta + \dfrac{2}{3}\cos^3\theta + \ldots + \right. \\ \left. \dfrac{2\cdot 4\ldots(\nu_2-3)}{3\cdot 5\ldots(\nu_2-2)}\cos^{\nu_2-2}\theta]\right\} \text{ for } \nu_2 > 1 \\ \dfrac{2\theta}{\pi} \text{ for } \nu_2 = 1 \end{cases}$$

$$\beta(\nu_1, \nu_2) = \begin{cases} \dfrac{2}{\sqrt{\pi}}\dfrac{\left(\frac{\nu_2-1}{2}\right)!}{\left(\frac{\nu_2-2}{2}\right)!}\sin\theta\cos^{\nu_2}\theta\left\{1 + \right. \\ \dfrac{\nu_2+1}{3}\sin^2\theta + \ldots + \\ \left. \dfrac{(\nu_2+1)(\nu_2+3)\ldots(\nu_1+\nu_2-4)\sin^{\nu_1-3}\theta}{3\cdot 5\ldots(\nu_1-2)}\right\} \\ \qquad\qquad \text{for } \nu_2 > 1 \\ 0 \text{ for } \nu_1 = 1 \quad * \end{cases}$$

where

$$\theta = \arctan\sqrt{\frac{\nu_1}{\nu_2}F}$$

Reflexive Relation

If $F_p(\nu_1, \nu_2)$ and $F_{1-p}(\nu_2, \nu_1)$ satisfy

$$Q(F_p(\nu_1, \nu_2)|\nu_1, \nu_2) = p$$

$$Q(F_{1-p}(\nu_2, \nu_1)|\nu_2, \nu_1) = 1 - p$$

*See page II.

26.6.9 then

$$F_p(\nu_1, \nu_2) = \frac{1}{F_{1-p}(\nu_2, \nu_1)}$$

Relation to Student's t-Distribution Function (See 26.7)

26.6.10 $Q(F|\nu_1=1, \nu_2) = 1 - A(t|\nu_2)$ $\qquad t = \sqrt{F}$

Limiting Forms

26.6.11

$$\lim_{\nu_1 \to \infty} Q(F|\nu_1, \nu_2) = Q(\chi^2|\nu_1), \qquad \chi^2 = \nu_1 F$$

26.6.12

$$\lim_{\nu_1 \to \infty} Q(F|\nu_1, \nu_2) = P(\chi^2|\nu_2), \qquad \chi^2 = \frac{\nu_2}{F}$$

Approximations

26.6.13

$$Q(F|\nu_1, \nu_2) \approx Q(x),\atop (\nu_1 \text{ and } \nu_2 \text{ large})} \qquad x = \frac{F - \frac{\nu_2}{\nu_2 - 2}}{\frac{\nu_2}{\nu_2 - 2}\sqrt{\frac{2(\nu_1 + \nu_2 - 2)}{\nu_1(\nu_2 - 4)}}}$$

26.6.14

$$Q(F|\nu_1, \nu_2) \approx Q(x), \qquad x = \frac{\sqrt{(2\nu_2 - 1)\frac{\nu_1}{\nu_2}F} - \sqrt{2\nu_1 - 1}}{\sqrt{1 + \frac{\nu_1}{\nu_2}F}}$$

26.6.15

$$Q(F|\nu_1, \nu_2) \approx Q(x), \qquad x = \frac{F^{1/3}\left(1 - \frac{2}{9\nu_2}\right) - \left(1 - \frac{2}{9\nu_1}\right)}{\sqrt{\frac{2}{9\nu_1} + F^{2/3}\frac{2}{9\nu_2}}}$$

Approximation to the Inverse Function

26.6.16 If $Q(F_p|\nu_1, \nu_2) = p$, then

$$F_p \approx e^{2w} \text{ where } w \text{ is given by } \mathbf{26.5.22}, \text{ with}$$

$$\nu_1 = 2b, \ \nu_2 = 2a$$

Non-Central F-Distribution Function

26.6.17

$$P(F'|\nu_1, \nu_2, \lambda) = \int_0^{F'} p(t|\nu_1, \nu_2, \lambda)dt = 1 - Q(F'|\nu_1, \nu_2, \lambda)$$

where

$$p(t|\nu_1, \nu_2, \lambda) = \sum_{j=0}^{\infty} e^{-\lambda/2}\frac{(\lambda/2)^j}{j!}\frac{(\nu_1 + 2j)^{\frac{\nu_1 + 2j}{2}}\nu_2^{\nu_2/2}}{B\left(\frac{\nu_1 + 2j}{2}, \frac{\nu_2}{2}\right)}$$

$$\times t^{\frac{\nu_1 + 2j - 2}{2}}[\nu_2 + (\nu_1 + 2j)t]^{-(\nu_1 + 2j + \nu_2)/2}$$

and $\lambda \geq 0$ is termed the non-centrality parameter.

Relation of Non-Central F-Distribution Function to Other Functions

Function	*Relation*

26.6.18 F-distribution

$$P(F'|\nu_1, \nu_2, \lambda) = \sum_{j=0}^{\infty} e^{-\lambda/2}\frac{(\lambda/2)^j}{j!}P(F'|\nu_1 + 2j, \nu_2)$$

$$P(F'|\nu_1, \nu_2, \lambda = 0) = P(F'|\nu_1, \nu_2)$$

26.6.19 Non-central t-distribution

$$P(F'|\nu_1 = 1, \nu_2, \lambda) = P(t'|\nu, \delta), t' = \sqrt{F'}, \nu = \nu_2, \delta = \sqrt{\lambda}$$

26.6.20 Incomplete Beta function

$$P(F'|\nu_1, \nu_2) = \sum_{j=0}^{\infty} e^{-\lambda/2}\frac{(\lambda/2)^j}{j!}I_x\left(\frac{\nu_1}{2} + j, \frac{\nu_2}{2}\right),$$

$$x = \frac{\nu_1 F'}{\nu_1 F' + \nu_2} *$$

26.6.21 Confluent hypergeometric function

$$P(F'|\nu_1, \nu_2, \lambda) = \sum_{i=0}^{\frac{\nu_2}{2} - 1}\frac{2e^{-\lambda/2}}{(\nu_1 + \nu_2)B\left(\frac{\nu_1}{2} + i + 1, \frac{\nu_2}{2} - i\right)}\times$$

$$x^{\frac{\nu_1}{2} + 1}(1 - x)^{\frac{\nu_2}{2} - i - 1}M\left(\frac{\nu_1 + \nu_2}{2}, \frac{\nu_1}{2} + i + 1, \frac{\lambda x}{2}\right)$$

$$\left(\nu_2 \text{ even and } x = \frac{\nu_2}{\nu_1 F' + \nu_2}\right)$$

*See page II.

Series Expansion

26.6.22

$$P(F'|\nu_1,\nu_2,\lambda)=e^{-\frac{\lambda}{2}(1-x)}\,x^{\frac{1}{2}(\nu_1+\nu_2-2)}\sum_{i=0}^{\frac{\nu_2}{2}-1}T_i \quad (\nu_2 \text{ even})$$

where

$$T_0=1$$

$$T_1=\frac{1}{2}(\nu_1+\nu_2-2+\lambda x)\frac{1-x}{x}$$

$$T_i=\frac{1-x}{2i}[(\nu_1+\nu_2-2i+\lambda x)T_{i-1}+\lambda(1-x)T_{i-2}]$$

$$x=\frac{\nu_2}{\nu_1 F'+\nu_2}$$

Limiting Forms

26.6.23

$$\lim_{\nu_2\to\infty}P(F'|\nu_1,\nu_2,\lambda)=P(\chi'^2|\nu,\lambda), \qquad \chi'^2=\nu_1 F', \ \nu=\nu_1$$

26.6.24

$$\lim_{\nu_1\to\infty}P(F'|\nu_1,\nu_2,\lambda)=Q(\chi^2|\nu), \qquad \chi^2=\frac{\nu_2(1+c^2)}{F'}$$

where $\lambda/\nu_1\to c^2$ as $\nu_1\to\infty$.

Approximations to the Non-Central F-Distribution

26.6.25 $\quad P(F'|\nu_1,\nu_2,\lambda)\approx P(x_1), \qquad (\nu_1 \text{ and } \nu_2 \text{ large})$

where

$$x_1=\frac{F'-\dfrac{\nu_2(\nu_1+\lambda)}{\nu_1(\nu_2-2)}}{\dfrac{\nu_2}{\nu_1}\left[\dfrac{2}{(\nu_2-2)(\nu_2-4)}\left\{\dfrac{(\nu_1+\lambda)^2}{\nu_2-2}+\nu_1+2\lambda\right\}\right]^{\frac{1}{2}}}$$

26.6.26

$$P(F'|\nu_1,\nu_2,\lambda)\approx P(F|\nu_1^*,\nu_2),$$

$$F=\frac{\nu_1}{\nu_1+\lambda}F', \ \nu_1^*=\frac{(\nu_1+\lambda)^2}{\nu_1+2\lambda}$$

26.6.27

$$P(F'|\nu_1,\nu_2,\lambda)\approx P(x_2),$$

$$x_2=\frac{\left[\dfrac{\nu_1 F'}{(\nu_1+\lambda)}\right]^{1/3}\left[1-\dfrac{2}{9\nu_2}\right]-\left[1-\dfrac{2(\nu_1+2\lambda)}{9(\nu_1+\lambda)^2}\right]}{\left[\dfrac{2}{9}\dfrac{\nu_1+2\lambda}{(\nu_1+\lambda)^2}+\dfrac{2}{9\nu_2}\left(\dfrac{\nu_1}{\nu_1+\lambda}F'\right)^{2/3}\right]^{\frac{1}{2}}}$$

26.7. Student's t-Distribution

If X is a random variable following a normal distribution with mean zero and variance unity, and χ^2 is a random variable following an independent chi-square distribution with ν degrees of freedom, then the distribution of the ratio $\dfrac{X}{\sqrt{\chi^2/\nu}}$

is called Student's t-distribution with ν degrees of freedom. The probability that $\dfrac{X}{\sqrt{\chi^2/\nu}}$ will be less in absolute value than a fixed constant t is

26.7.1

$$A(t|\nu)=P_r\left\{\left|\frac{X}{\sqrt{\chi^2/\nu}}\right|\le t\right\}$$

$$=\left[\sqrt{\nu}B\left(\frac{1}{2},\frac{\nu}{2}\right)\right]^{-1}\int_{-t}^{t}\left(1+\frac{x^2}{\nu}\right)^{-\frac{\nu+1}{2}}dx$$

$$=1-I_x\left(\frac{\nu}{2},\frac{1}{2}\right), \qquad (0\le t<\infty) \quad *$$

where

$$x=\frac{\nu}{\nu+t^2}$$

Statistical Properties

26.7.2

mean: $\quad m=0$

variance: $\quad \sigma^2=\dfrac{\nu}{\nu-2} \hspace{3cm} (\nu>2)$

skewness: $\gamma_1=0$

excess: $\quad \gamma_2=\dfrac{6}{\nu-4} \hspace{3cm} (\nu>4)$

moments:

$$\mu_{2a}=\frac{1\cdot3\ldots(2n-1)\nu^n}{(\nu-2)(\nu-4)\ldots(\nu-2n)} \qquad (\nu>2n)$$

$$\mu_{2n+1}=0$$

characteristic function:

$$\phi(t)=E\left[\exp\left(it\frac{X}{\sqrt{\chi^2/\nu}}\right)\right]=\frac{\left(\dfrac{|t|}{2\sqrt{\nu}}\right)^{\nu/2}}{\pi\Gamma(\nu/2)}Y_{\frac{\nu}{2}}\left(\frac{|t|}{\sqrt{\nu}}\right)$$

Series Expansions

$$\left(\theta=\arctan\frac{t}{\sqrt{\nu}}\right)$$

26.7.3

$$A(t|\nu)=\begin{cases}\dfrac{2}{\pi}\left\{\theta+\sin\theta\left[\cos\theta+\dfrac{2}{3}\cos^3\theta+\ldots\right.\right.\\ \left.\left.\qquad+\dfrac{2\cdot4\ldots(\nu-3)}{1\cdot3\ldots(\nu-2)}\cos^{\nu-2}\theta\right]\right\} \quad * \\ \qquad\qquad (\nu>1 \text{ and odd}) \\[2mm] \dfrac{2}{\pi}\theta \qquad (\nu=1)\end{cases}$$

26.7.4

$$A(t|\nu)=\sin\theta\left\{1+\frac{1}{2}\cos^2\theta+\frac{1\cdot3}{2\cdot4}\cos^4\theta+\ldots\right.$$

$$\left.+\frac{1\cdot3\cdot5\ldots(\nu-3)}{2\cdot4\cdot6\ldots(\nu-2)}\cos^{\nu-2}\theta\right\} \qquad (\nu \text{ even}) \ *$$

*See page II.

Asymptotic Expansion for the Inverse Function

If $A(t_p|\nu)=1-2p$ and $Q(x_p)=p$, then

26.7.5

$$t_p \sim x_p + \frac{g_1(x_p)}{\nu} + \frac{g_2(x_p)}{\nu^2} + \frac{g_3(x_p)}{\nu^3} + \cdots$$

$$g_1(x)=\frac{1}{4}(x^3+x)$$

$$g_2(x)=\frac{1}{96}(5x^5+16x^3+3x)$$

$$g_3(x)=\frac{1}{384}(3x^7+19x^5+17x^3-15x)$$

$$g_4(x)=\frac{1}{92160}(79x^9+776x^7+1482x^5-1920x^3-945x)$$

Limiting Distribution

26.7.6

$$\lim_{\nu\to\infty} A(t|\nu)=\frac{1}{\sqrt{2\pi}}\int_{-t}^{t} e^{-x^2/2}dx=A(t)$$

Approximation for Large Values of t and $\nu \leq 5$

26.7.7
$$A(t|\nu) \approx 1-2\left\{\frac{a_\nu}{t^\nu}+\frac{b_\nu}{t^{\nu+1}}\right\}$$

ν	1	2	3	4	5
a_ν	.3183	.4991	1.1094	3.0941	9.948
b_ν	.0000	.0518	−.0460	−2.756	−14.05

Approximation for Large ν

26.7.8 $\quad A(t|\nu) \approx 2P(x)-1, \qquad x=\dfrac{t\left(1-\dfrac{1}{4\nu}\right)}{\sqrt{1+\dfrac{t^2}{2\nu}}}$

Non-Central t-Distribution

26.7.9

$$P(t'|\nu,\delta)=$$

$$\frac{1}{\sqrt{\nu}B\left(\frac{1}{2},\frac{\nu}{2}\right)}\int_{-\infty}^{t'}\left(\frac{\nu}{\nu+x^2}\right)^{\frac{\nu+1}{2}} e^{-\frac{1}{2}\frac{\nu\delta^2}{\nu+x^2}} Hh_\nu\left(\frac{-\delta x}{\sqrt{\nu+x^2}}\right)dx$$

$$=1-\sum_{j=0}^{\infty} e^{-\delta^2/2}\frac{(\delta^2/2)^j}{2j!} I_x\left(\frac{\nu}{2},\frac{1}{2}+j\right), \qquad x=\frac{\nu}{\nu+t'^2} \ *$$

where δ is termed the non-centrality parameter.

Approximation to the Non-Central t-Distribution

26.7.10

$$P(t'|\nu,\delta) \approx P(x) \qquad \text{where } x=\frac{t'\left(1-\dfrac{1}{4\nu}\right)-\delta}{\left(1+\dfrac{t'^2}{2\nu}\right)^{\frac{1}{2}}}$$

Numerical Methods

26.8. Methods of Generating Random Numbers and Their Applications [9]

Random digits are digits generated by repeated independent drawings from the population 0, 1, 2, . . ., 9 where the probability of selecting any digit is one-tenth. This is equivalent to putting 10 balls, numbered from 0 to 9, into an urn and drawing one ball at a time, replacing the ball after each drawing. The recorded set of numbers forms a collection of random digits. Any group of n successive random digits is known as a *random number*.

Several lengthy tables of random digits are available (see references). However, the use of random numbers in electronic computers has resulted in a need for random numbers to be generated in a completely deterministic way. The numbers so generated are termed pseudo-random numbers. The quality of pseudo-random numbers is determined by subjecting the numbers to several statistical tests, see [26.55], [26.56]. The purpose of these statistical tests is to detect any properties of the pseudo-random numbers which are different from the (conceptual) properties of random numbers.

[9] The authors wish to express their appreciation to Professor J. W. Tukey who made many penetrating and helpful suggestions in this section.

Experience has shown that the congruence method is the most preferable device for generating random numbers on a computer. Let the sequence of pseudo-random numbers be denoted by $\{X_n\}$, $n=0, 1, 2, \ldots$. Then the congruence method of generating pseudo-random numbers is

$$X_{n+1}=aX_n+b\,(\text{mod } T)$$

where b and T are relatively prime. The choice of T is determined by the capacity and base of the computer; a and b are chosen so that: (1) the resulting sequence $\{X_n\}$ possesses the desired statistical properties of random numbers, (2) the period of the sequence is as long as possible, and (3) the speed of generation is fast. A guide for choosing a and b is to make the correlation between the numbers be near zero, e.g., the correlation between X_n and X_{n+s} is

$$\rho_s=\frac{1-6\dfrac{b_s}{T}\left(1-\dfrac{b_s}{T}\right)}{a_s}+e$$

where

$$a_s=a^s\,(\text{mod } T)$$
$$b_s=(1+a+a^2+\ \cdots\ +a^{s-1})b\,(\text{mod } T)$$
$$|e|<a_s/T$$

*See page II.

which occur in

$$X_{n+s} = a_s X_n + b_s \ (\text{mod } T)$$

When a is chosen so that $a \approx T^{1/2}$, the correlation $\rho_1 \approx T^{-1/2}$.

The sequence defined by the multiplicative congruence method will have a full period of T numbers if

(i) b is relatively prime to T
(ii) $a = 1 \ (\text{mod } p)$ if p is a prime factor of T
(iii) $a = 1 \ (\text{mod } 4)$ if 4 is a factor of T.

Consequently if $T = 2^q$, b need only be odd, and $a = 1 \ (\text{mod } 4)$. When $T = 10^q$, b need only be not divisible by 2 or 5, and $a = 1 \ (\text{mod } 20)$. The most convenient choices for a are of the form $a = 2^s + 1$ (for binary computers) and $a = 10^s + 1$ (for decimal computers). This results in the fastest generation of random numbers as the operations only require a shift operation plus two additions. Also any number can serve as the starting point to generate a sequence of random digits. A good summary of generating pseudo-random numbers is [26.51].

Below are listed various congruence schemes and their properties.

Congruence methods for generating random numbers
$X_{n+1} = a X_n + b (\text{mod } T)$, T and b relatively prime

		a	b	T	Period	X_0	Special cases for which random numbers have passed statistical tests for randomness [10]
26.8.1		$1 + t^s$	odd	$T = t^q$	t^q	$0 \le X_0 < T$	$T = 2^{35}$, X_0 unknown; $a = 2^7 + 1$, $b = 1$; $T = 2^{47}$, $a = 2^9 + 1$, $b = 29741\ 09625\ 8473$, $X_0 = 76293\ 94531\ 25$.
26.8.2		$r2^s \pm 1$ (r odd, $s \ge 2$)	0	$T = t^q$	t^{q-s}	relatively prime to T	$T = 2^{40}$, 2^{42}, $X_0 = 1$; $a = 5^{17}(s = 2)$ $T = 2^{35}$, $X_0 = 1$; $T = 2^{39}$, $X_0 = 1 - 2^{-39}$, $.5478126193$; $a = 5^{13}(s = 2)$ $T = 2^{35}$, $X_0 = 1$; $a = 5^{15}(s = 2)$
26.8.3		$r2^s \pm 1$ (r odd, $s \ge 2$)	0	$T = t^q \pm 1$	(varies)	relatively prime to T	$T = 2^{35} + 1$, $X_0 = 10,987,654,321$; $a = 23$; period $\approx 10^6$ $T = 10^8 + 1$, $X_0 = 47,594,118$; $a = 23$; period $\approx 5.8 \times 10^6$
26.8.4		7^{4s+1}	0	$T = 10^q$	$5 \cdot 10^{q-3}$	relatively prime to T	$T = 10^{10}$, $X_0 = 1$; $a = 7$ $T = 10^{11}$, $X_0 = 1$; $a = 7^{13}$
26.8.5		3^{4s+1} ($s = 0, 2, 3, 4$)	0	$T = 10^q$	$5 \cdot 10^{q-2}$	relatively prime to T	

[10] X_0 given is the starting point for random numbers when statistical tests were made.

When the numbers are generated using a congruence scheme, the least significant digits have short periods. Hence the entire word length cannot be used. If one desired random numbers with as many digits as possible, one would have to modify the congruence schemes. One way is to generate the numbers mod $T \pm 1$. This unfortunately reduces the period.

Generation of Random Deviates

Let $\{X\}$ be a generated sequence of independent random numbers having the domain $(0, T)$. Then $\{U\} = \{T^{-1}X\}$ is a sequence of random deviates (numbers) from a uniform distribution on the interval $(0, 1)$. This is usually a necessary preliminary step in the generation of random deviates having a given cumulative distribution function $F(y)$ or probability density function $f(y)$. Below are summarized some general techniques for producing arbitrary random deviates. (In what follows $\{U\}$ will always denote a sequence of random deviates from a uniform distribution on the interval $(0, 1)$.)

1. Inverse Method

The solutions $\{y\}$ of the equations $\{u = F(y)\}$ form a sequence of independent random deviates with cumulative distribution function $F(y)$. (If $F(y)$ has a discontinuity at $y = y_0$, then whenever u is such that $F(y_0 - 0) < u < F(y_0)$, select y_0 as the corresponding deviate.) Generally the inverse method is not practical unless the inverse function $y = F^{-1}(u)$ can be obtained explicitly or can be conveniently approximated.

2. Generating a Discrete Random Variable

Let Y be a discrete random variable with point probabilities $p_i = Pr\{Y = y_i\}$ for $i = 1, 2, \ldots$.

*See page II.

The direct way to generate Y is to generate $\{U\}$ and put $Y = y_1$ if

$$p_1 + p_2 + \ldots + p_{i-1} < U < p_1 + p_2 + \ldots + p_i.$$

However, this method requires complicated machine programs that take too long.

An alternative way due to Marsaglia [26.53] is simple, fast, and seems to be well suited to high-speed computations. Let p_i for $i = 1, 2, \ldots, n$ be expressed by k decimal digits as $p_i = .\delta_{1i}\delta_{2i} \ldots \delta_{ki}$ where the δ's are the decimal digits. (If the domain of the random variable is infinite, it is necessary to truncate the probability distribution at p_n.) Define

$$P_0 = 0, \quad P_r = 10^{-r} \sum_{i=1}^{n} \delta_{ri} \text{ for } r = 1, 2, \ldots, k, \text{ and}$$

$$\Pi_s = \sum_{r=0}^{s} 10^r P_r, \quad s = 1, 2, \ldots, k.$$

Number the computer memory locations by 0, 1, 2, $\ldots$, $\Pi_k - 1$. The memory locations are divided into k mutually exclusive sets such that the sth set consists of memory locations Π_{s-1}, $\Pi_{s-1} + 1, \ldots, \Pi_s - 1$. The information stored in the memory locations of the sth set consists of y_1 in δ_{s1} locations, y_2 in δ_{s2} locations, $\ldots$, y_n in δ_{sn} locations.

Denote the decimal expansion of the uniform deviates generated by the computer by $u = \cdot d_1 d_2 d_3 \ldots$ and finally let $\sigma\{m\}$ be the contents of memory location m. Then if

$$\sum_{i=0}^{s-1} P_i \leq U < \sum_{i=0}^{s} P_i$$

put

$$y = a \left\{ d_1 d_2 \ldots d_s + \Pi_{s-1} - 10^s \sum_{i=1}^{s-1} P_i \right\}.$$

This method is perhaps the best all-around method for generating random deviates from a discrete distribution. In order to illustrate this method consider the problem of generating deviates from the binomial distribution with point probabilities

$$p_i = \binom{n}{i} p^i (1-p)^{n-i}$$

for $n = 5$ and $p = .20$. The point probabilities to 4 D are

Value of Random Variable	Point Probabilities
0	$p_0 = 0.3277$
1	$p_1 = .4096$
2	$p_2 = .2048$
3	$p_3 = .0512$
4	$p_4 = .0064$
5	$p_5 = .0003$

and thus $P_0 = 0$, $P_1 = .9$, $P_2 = .07$, $P_3 = .027$, $P_4 = .0030$ from which $\Pi_0 = 0$, $\Pi_1 = 9$, $\Pi_2 = 16$, $\Pi_3 = 43$, $\Pi_4 = 73$. The 73 memory locations are divided into 4 mutually exclusive sets such that

Set	Memory Locations
1	0, 1, $\ldots$, 8
2	9, 10, $\ldots$, 15
3	16, $\ldots$, 42
4	43, $\ldots$, 72

Among the nine memory locations of set 1, zero is stored $\delta_{10} = 3$ times, 1 is stored $\delta_{11} = 4$ times, 2 is stored $\delta_{12} = 2$ times; the seven locations of set 2 store 0 $\delta_{20} = 2$ times and 3 $\delta_{23} = 5$ times; etc. A summary of the memory locations is set out below:

	Value of Random Variable					
	0	1	2	3	4	5
Frequency (set 1)	3	4	2	0	0	0
Frequency (set 2)	2	0	0	5	0	0
Frequency (set 3)	7	9	4	1	6	0
Frequency (set 4)	7	6	8	2	4	3

Then to generate the random variables if

$0 \leq u < .9$	put	$y = a\{d_1\}$
$.9 \leq u < .97$		$y = a\{d_1 d_2 - 81\}$
$.97 \leq u < .997$		$y = a\{d_1 d_2 d_3 - 954\}$
$.997 \leq u < 1.000$		$y = a\{d_1 d_2 d_3 d_4 - 9927\}$

3. Generating a Continuous Random Variable

The method for generating deviates from a discrete distribution can be adapted to random variables having a continuous distribution. Let $F(y)$ be the cumulative distribution function and assume that the domain of the random variable is (a, b) where the interval is finite. (If the domain is infinite, it must be truncated at (say) the points a and b.) Divide the interval $(b-a)$ into n sub-intervals of length Δ $(n\Delta = b-a)$ such that the boundary of the ith interval is (y_{i-1}, y_i) where $y_i = a + i\Delta$ for $i = 0, 1, \ldots, n$. Now define a discrete distribution having domain

$$\left\{ z_i = \frac{y_i + y_{i-1}}{2} \right\}$$

with point probabilities $p_i = F(y_i) - F(y_{i-1})$. Finally, let W be a random variable having a uniform distribution on $\left(-\frac{\Delta}{2}, \frac{\Delta}{2}\right)$. This can be done by setting $W = \Delta\left(U - \frac{1}{2}\right)$. Then random deviates from the distribution function $F(y)$, can be generated (approximately) by setting $y = z + w = z + \Delta\left(u - \frac{1}{2}\right)$. This is simply an approximate decomposition of the continuous random variable into the sum of a discrete and continuous random variable. The discrete variable can be generated quickly by the method described previously. The smaller the value of Δ the better will be the approximation. Each number can be generated by using the leading digits of U to generate the discrete random variable Z and the remaining digits forming a uniformly distributed deviate having (0,1) domain.

4. Acceptance-Rejection Methods

In what follows the random variable Y will be assumed to have finite domain (a, b). If the domain is infinite, it must be truncated for computational purposes at (say) the points a and b. Then the resulting random deviates will only have this truncated domain.

a) Let f be the maximum of $f(y)$. Then the procedure for generating random deviates is: (1) generate a pair of uniform deviates U_1, U_2; (2) compute a point $y = a + (b-a)u_2$ in (a, b); (3) if $u_1 < f(y)/f$ accept y as the random deviate, otherwise reject the pair (u_1, u_2) and start again. The acceptance ratio of deviates actually produced is $[(b-a)f]^{-1}$. Hence the acceptance ratio decreases as the domain increases. One way to increase the acceptance ratio is to divide the interval (a, b) into mutually exclusive sub-intervals and then carry out the acceptance-rejection process. For this purpose let the interval (a, b) be divided into k sub-intervals such that the end points of the jth interval are (ξ_{j-1}, ξ_j) with $\xi_0 = a$, $\xi_k = b$ and $\int_{\xi_{j-1}}^{\xi_j} f(y)dy = p_j$; further let the maximum of $f(y)$ in the jth interval be f_j. Then to generate random deviates from $f(y)$, generate n pairs of deviates $(u_{1s}, u_{2s})s = 1, 2, \ldots, n$. Assign $[np_j]$ such pairs to the jth interval and compute $y_j = \xi_{j-1} + (\xi_j - \xi_{j-1})u_{2s}$. If $u_{1s} < f(y_j)/f_j$ accept y_j as a deviate. The acceptance ratio of this method is

$$\sum_{j=1}^{k} p_j [(\xi_j - \xi_{j-1})f_j]^{-1}$$

b) Let $F(y)$ be such that $f(y) = f_1(y)f_2(y)$ where the domain of y is (a, b). Let f_1 and f_2 be the maximum of $f_1(y)$ and $f_2(y)$ respectively. Then the procedure for generating random de-

viates having the probability density function $f(y)$ is: (1) generate U_1, U_2, U_3; (2) define $z = a + (b-a)u_3$; (3) if both $u_1 < \frac{f_1(z)}{f_1}$ and $u_2 < \frac{f_2(z)}{f_2}$, take z as the random deviate; otherwise take another sample of three uniform deviates. The acceptance ratio of this method is $[(b-a)f_1 f_2]^{-1}$ and can be increased by dividing (a, b) into sub-intervals as in the previous case.

c) Let the probability density function of Y be

$$f(y) = \int_\alpha^\beta g(y, t)dt, \quad (\alpha \le t \le \beta), \quad (a \le y \le b).$$

Let g be the maximum of $g(y, t)$. Then the procedure for generating random deviates having the probability density function $f(y)$ is: (1) generate U_1, U_2, U_3; (2) define $s = \alpha + (\beta - \alpha)u_2$; $z = a + (b-a)u_3$; (3) if $u_1 < \frac{g(z, s)}{g}$, take z as the random deviate; otherwise take another sample of three. The acceptance ratio for this method is $[(b-a)g]^{-1}$ and can be increased by dividing the domain of t and y into sub-domains.

5. Composition Method

Let $g_z(y)$ be a probability density function which depends on the parameter z; further let $H(z)$ be the cumulative distribution function for z. In order to generate random deviates Y having the frequency function

$$f(y) = \int_{-\infty}^\infty g_z(y)dH(z)$$

one draws a deviate having the cumulative distribution function $H(z)$; then draws a second sample having the probability density function $g_z(y)$.

6. Generation of Random Deviates From Well Known Distributions

a. Normal distribution

(1) *Inverse method:* The inverse method depends on having a convenient approximation to the inverse function $x = P^{-1}(u)$ where

$$u = (2\pi)^{-1/2} \int_{-\infty}^x e^{-t^2/2}dt.$$

Two ways of performing this operation are to (i) use **26.2.23** with $t = \left(\ln \frac{1}{u^2}\right)^{1/2}$ or (ii) approximate $x = P^{-1}(u)$ piecewise using Chebyshev polynomials, see [26.54].

(2) *Sum of uniform deviates:* Let $U_1, U_2, \ldots, U_n$ be a sequence of n uniform deviates. Then

$$X_n = \left(\sum_{i=1}^{n} U_i - \frac{n}{2} \right) \left(\frac{n}{12} \right)^{-1/2}$$

will be distributed asymptotically as a normal random deviate. When $n=12$, the maximum errors made in the normal deviate are 9×10^{-3} for $|X| < 2$, 9×10^{-1} for $2 < |X| < 3$. An improvement can be made by taking a polynomial function of X_n (say)

$$X_n^* = X_n \sum_{s=0}^{k} a_{2s} X_n^{2s}$$

as the normal deviate where a_{2s} are suitable coefficients. These coefficients may be calculated using (say) Chebyshev polynomials or simply by making the asymptotic random deviate agree with the correct normal deviate at certain specified points. When $n=12$, the maximum error in the normal deviate is 8×10^{-4} using the coefficients

> * $a_0 = 9.8746$ * $a_6 = (-7) - 5.102$
> * $a_2 = (-3) 3.9439$ * $a_8 = (-7) 1.141$
> * $a_4 = (-5) 7.474$

(3) *Direct method:* Generate a pair of uniform deviates (U_1, U_2). Then

$$X_1 = (-2 \ln U_1)^{1/2} \cos 2\pi U_2,$$

$X_2 = (-2 \ln U_1)^{1/2} \sin 2\pi U_2$ will be a pair of independent normal random deviates with mean zero and unit variance. This procedure can be modified by calculating $\cos 2\pi U$ and $\sin 2\pi U$ using an acceptance rejection method; e.g., (1) generate (U_1, U_2); (2) if $(2U_1 - 1)^2 + (2U_2 - 1)^2 \leq 1$ generate a third uniform deviate U_3, otherwise reject the pair and start over; (3) calculate

$$y_1 = (-\ln u_3)^{1/2} \frac{u_1^2 - u_2^2}{u_1^2 + u_2^2}, \quad y_2 = \pm 2 (-\ln u_3)^{1/2} \frac{u_1 u_2}{u_1^2 + u_2^2} (\pm$$

random). Both y_1 and y_2 are the desired random deviates.

(4) *Acceptance-rejection method:* 1) Generate a pair of uniform deviates (U_1, U_2); 2) compute
* $x = -\ln u_1$; 3) if $e^{-\frac{1}{2}(x-1)^2} \geq u_2$ (or equivalently $(x-1)^2 \leq -2 (\ln u_2)$ accept x, otherwise reject the

pair and start over. The quantity will be the required normal deviate with mean zero and unit variance.

b. Bivariate normal distribution

Let $\{X_1, X_2\}$ be a pair of independent normal deviates with mean zero and unit variance. Then $\{X_1, \rho X_1 + (1 - \rho^2)^{1/2} X_2\}$ represent a pair of deviates from a bivariate normal distribution with zero means, unit variances, and correlation coefficient ρ.

c. Exponential distribution

(1) *Inverse method:* Since $F(x) = e^{-x/\theta}$, $X = -\theta \ln U$ will be a deviate from the exponential distribution with parameter θ.

(2) *Acceptance-rejection method:* 1) Generate a pair of independent uniform deviates (U_0, U_1); 2) if $U_1 < U_0$ generate a third value U_2; 3) if $U_1 + U_2 < U_0$ generate a fourth value U_3, etc.; 4) continue generating uniform deviates until an n is obtained such that $U_1 + U_2 + \ldots + U_{n-1} < U_0 < U_1 + \ldots + U_n$; 5) if n is even reject the procedure and start a fresh trial with a new value of U_0, otherwise if n is odd take $X = \theta U_0$ as the desired deviate; 6) in general if t is the number of trials until an acceptable sequence is obtained $X = \theta(t + U_0)$. The random deviates produced in this way follow an exponential distribution with parameter θ. One can expect to generate approximately six uniform deviates for every exponential deviate.

(3) *Discrete Distribution Method:* Let Y and n be discrete random variables with point probabilities

> * $Pr\{Y = r\} = (e-1) e^{-(r+1)}$ $r = 0, 1, 2, \ldots$
> $Pr\{n = s\} = [s! (e-1)]^{-1}$ $s = 1, 2, 3, \ldots$

Then $X = Y + \min(U_1, U_2, \ldots, U_n)$ will follow an exponential distribution. The average value of n is 1.58 so that one needs, on the average, only 1.58 u's from which the minimum is selected.

26.9. Use and Extension of the Tables

Use of Probability Function Inequalities

Example 1. Let X be a random variable with finite mean and variance equal to m and σ^2, respectively. Use the inequalities for probability functions **26.1.37, 40, 41** to place lower bounds on

$$A(t) = F(t) - F(-t) = P\left\{ \frac{|X - m|}{\sigma} \leq t \right\}$$

for $t = 1(1)4$.

Lower bounds on $A(t) = F(t) - F(-t)$

$t=1$	2	3	4	Remarks
0	.7500	.8889	.9375	no knowledge of $F(t)$; **26.1.37**
.5556	.8889	.9506	.9722	$F(t)$ is unimodal and continuous; **26.1.40**
0	.8182	.9697	.9912	$F(t)$ is such that $\mu_4 = 3$; **26.1.41**

It is of interest to note that the standard normal distribution is unimodal, has mean zero, unit variance $\mu_4=3$, is continuous, and such that

$$A(t)=P(t)-P(-t)$$
$$=.6827, .9545, .9973, \text{ and } .9999$$

for $t=1, 2, 3$ and 4 respectively.

Interpolation for $P(x)$ in Table 26.1

Example 2. Compute $P(x)$ for $x=2.576$ to fifteen decimal places using a Taylor expansion.

Writing $x=x_0+\theta$ we have

$$P(x)=P(x_0)+Z(x_0)\theta+Z^{(1)}(x_0)\frac{\theta^2}{2!}$$
$$+Z^{(2)}(x_0)\frac{\theta^3}{3!}+Z^{(3)}(x_0)\frac{\theta^4}{4!}+\cdots$$

Taking $x_0=2.58$ and $\theta=-4\times10^{-3}$ we calculate the successive terms to 16D

```
+.99505  99842  42230
-     5  72204  35976  6
-         2952  57449  6
-            8  63097  8
-                1439  4
-                         9
─────────────────────────
 .99500  24676  84265  7
```

The result correct to 17D is

$$P(2.576)=.99500 \quad 24676 \quad 84264 \quad 98$$

Calculation for Arbitrary Mean and Variance

Example 3. Find the value to 5D of

$$P\{X\le.50\}=\frac{1}{2\sqrt{2\pi}}\int_{-\infty}^{.5}e^{-1/2\left(\frac{t-1}{2}\right)^2}dt$$

using **26.2.8** and **Table 26.1**.

This represents the probability of the random variable being less than or equal to .5 for a normal distribution with mean $m=1$ and variance $\sigma^2=4$. Using **26.2.8** we have

$$P\{X\le.5\}=P\left(\frac{.5-1}{2}\right)=P(-.25)$$

Since $P(-x)=1-P(x)$, we have

$$P(-.25)=1-P(.25)=1-.59871=.40129$$

where a two-term Taylor series was used for interpolation. Note that when interpolating for $P(x)$ for a value of x midway between the tabulated values we can write $x=x_0+.01$ and a two-term Taylor series is $P(x)=P(x_0)+Z(x_0)10^{-2}$. Thus one need only multiply $Z(x_0)$ by 10^{-2} and add the result to $P(x_0)$.

Calculation of $P(x)$ for x Approximate

Example 4. Using **Table 26.1**, find $P(x)$ for $x=1.96$, when there is a possible error in x of $\pm5\times10^{-3}$.

This is an example where the argument is only known approximately. The question arises as to how many decimal places one should retain in $P(x)$. If Δx and $\Delta P(x)$ denote the error in x and the resulting error in $P(x)$, respectively, then

$$\Delta P(x)\approx Z(x)\Delta x$$

Hence $\Delta P(1.960)=3\times10^{-4}$ which indicates that $P(1.960)$ need only be calculated to 4D. Therefore $P(1.960)=.9750$.

Inverse Interpolation for $P(x)$

Example 5. Find the value of x for which $P(x)=.97500 \ 00000 \ 00000$ using **Table 26.1** and determining as many decimal places as is consistent with the tabulated function.

For inverse interpolation the tabulated function $P(x)$ may be regarded as having a possible error of $.5\times10^{-15}$. Hence

$$\Delta x\approx\frac{\Delta P(x)}{Z(x)}=\frac{.5\times10^{-15}}{Z(x)}$$

Let $P(x_0)$ correspond to the closest tabulated value of $P(x)$. Then a convenient formula for inverse interpolation is

$$x=x_0+t+\frac{x_0t^2}{2}+\frac{2x_0^2+1}{6}t^3$$

where

$$t=\frac{P(x)-P(x_0)}{Z(x_0)}$$

If only the first two terms (i.e., $x=x_0+t$) are used, the error in x will be bounded by $\frac{x}{8}\times10^{-4}$ and the true value will always be greater than the value thus calculated.

With respect to this example, $\Delta x\approx10^{-14}$ and thus the interpolated value of x may be in error by one unit in the fourteenth place. The closest value to $P(x)=.97500 \ 00000 \ 00000$ is $P(x_0)=.97500 \ 21048 \ 51780$ with $x_0=1.96$. Hence using the preceding inverse interpolation formulas with

$$t = -.00003\ 60167\ 31129$$

and carrying fifteen decimals we have the successive terms

$$
\begin{aligned}
&+1.96000 \quad 00000 \quad 00000 \\
&-\ \ \ .00003 \quad 60167 \quad 31129 \\
&+\quad\quad\quad\quad\quad 12 \quad 71261 \\
&-\quad\quad\quad\quad\quad\quad\quad\quad\ 68 \\
&\quad\quad\quad\quad\quad\quad\quad\quad\quad 0 \\
\hline
&+1.95996 \quad 39845 \quad 40064
\end{aligned}
$$

Edgeworth Asymptotic Expansion

Example 6. Find the Edgeworth asymptotic expansion **26.2.49** for the c.d.f. of chi-square.

Method 1. Expansion for χ^2

Let

$$Q(\chi^2|\nu) = 1 - F(t)$$

where

$$t = \frac{\chi^2 - \nu}{(2\nu)^{\frac{1}{2}}}$$

Since the values of γ_1 and γ_2 **26.4.33** are

$$\gamma_1 = 2\sqrt{2}/\nu^{\frac{1}{2}}$$

$$\gamma_2 = 12/\nu,$$

we obtain, by using the first two bracketed terms of **26.2.49**

$$
F(t) \sim P(t) - \frac{1}{\nu^{\frac{1}{2}}}\left[\frac{\sqrt{2}}{3} Z^{(2)}(t)\right]
$$
$$
+ \frac{1}{\nu}\left[\frac{1}{2} Z^{(3)}(t) + \frac{1}{9} Z^{(5)}(t)\right]
$$

The Edgeworth expansion is an asymptotic expansion in terms of derivatives of the normal distribution function. It is often possible to transform a random variable so that the distribution of the transformed random variable more closely approximates the normal distribution function than does the distribution of the original random variable. Hence for the same number of terms, greater accuracy may be achieved by using the transformed variable in the expansion. Since the distribution of $\sqrt{2\chi^2}$ is more closely approximated by a normal distribution than χ^2 itself (as judged by a comparison of the values of γ_1 and γ_2), we would expect that the Edgeworth asymptotic expansion of $\sqrt{2\chi^2}$ would be superior to that of χ^2.

Method 2. Expansion for $\sqrt{2\chi^2}$. Let

$$
Q(\chi^2|\nu) = 1 - F(t) = 1 - F\left(\frac{\sqrt{2\chi^2} - (2\nu-1)^{\frac{1}{2}}}{\left(1 - \frac{1}{4\nu}\right)^{\frac{1}{2}}}\right)
$$

where $(2\nu-1)^{\frac{1}{2}}$ and $1 - \frac{1}{4\nu}$ are the mean and variance to terms of order ν^{-2} of $\sqrt{2\chi^2}$ (see **26.4.34**). The values of γ_1 and γ_2 for $\sqrt{2\chi^2}$ are

$$\gamma_1 \approx \frac{1}{\sqrt{2\nu}}\left[1 + \frac{5}{8\nu}\right] \qquad \gamma_2 \approx \frac{3}{4\nu^2}$$

Thus we obtain

$$
F(t) \sim P(t) - \frac{1}{\nu^{\frac{1}{2}}}\left[\frac{\sqrt{2}}{12}\left(1 + \frac{5}{8\nu}\right) Z^{(2)}(t)\right]
$$
$$
+ \frac{1}{\nu}\left[\frac{1}{32\nu} Z^{(3)}(t) + \frac{1}{144}\left(1 + \frac{5}{8\nu}\right)^2 Z^{(5)}(t)\right]
$$

For numerical examples using these expansions see **Example 12.**

Calculation of $L(h, k, \rho)$

Example 7. Find $L(.5, .4, .8)$. Using **26.3.20**

$$\sqrt{h^2 - 2\rho hk + k^2} = \sqrt{.09} = .3$$

$$L(.5, .4, .8) = L(.5, 0, 0) + L(.4, 0, -.6)$$

Reference to **Figure 26.2** yields

$$L(.5, 0, 0) + L(.4, 0, -.6) = .16 + .08 = .24$$

The answer to 3D is $L(.5, .4, .8) = .250$.

Calculation of the Bivariate Normal Probability Function

Example 8. Let X and Y follow a bivariate normal distribution with parameters $m_x = 3$, $m_y = 2$, $\sigma_x = 4$, $\sigma_y = 2$, and $\rho = -.125$. Find the value of $P_r\{X \geq 2,\ Y \geq 4\}$ using **26.3.20** and **Figures 26.2, 26.3**.

Since $P_r\{X \geq h, Y \geq k\} = L\left(\frac{h - m_x}{\sigma_x}, \frac{k - m_y}{\sigma_y}, \rho\right)$ we have $P\{X \geq 2,\ Y \geq 4\} = L(-.25, 1, -.125)$. Using **26.3.20**

$$L(-.25, 1, -.125) = L(-.25, 0, .969)$$
$$+ L(1, 0, .125) - \frac{1}{2}$$

Figure 26.2 only gives values for $h > 0$, however, using the relationship **26.3.8** with $k = 0$, $L(-h, 0, \rho) = \frac{1}{2} - L(h, 0, -\rho)$ and thus $L(-.25, 0, .969) = \frac{1}{2} - L(.25, 0, -.969)$. Therefore $L(-.25, 1, -.125) = -L(.25, 0, -.969) + L(1, 0, .125) = -.01 + .09 = .08$. The answer to 3D is $L(-.25, 1., -.125) = .080$.

Integral of a Bivariate Normal Distribution Over a Polygon

Example 9. Let the random variables X and Y have a bivariate normal distribution with parameters $m_x=5$, $\sigma_x=2$, $m_y=9$, $\sigma_y=4$, and $\rho=.5$. Find the probability that the point (X, Y) be inside the triangle whose vertices are $A=(7, 8)$, $B=(9, 13)$, and $C=(2, 9)$.

When obtaining the integral of a bivariate normal distribution over a polygon, it is first necessary to use **26.3.22** in order to transform the variates so that one deals with a circular normal distribution. The polygon in the region of the transformed variables is then divided into configurations such that the integral over any selected configuration can be easily obtained. Below are listed some of the most useful configurations.

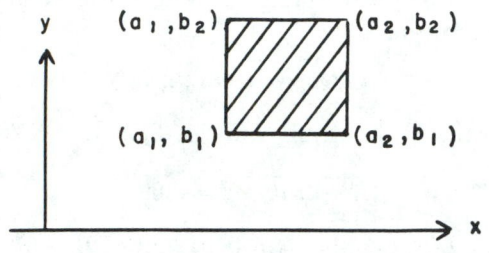

FIGURE 26.5

$$\int_{a_1}^{a_2}\int_{b_1}^{b_2} g(x,y,0)\,dxdy=[P(a_2)-P(a_1)][P(b_2)-P(b_1)]$$

FIGURE 26.6

$$\int_0^\infty \int_0^{ax} g(x,y,0)\,dxdy=\frac{\arctan a}{2\pi}$$

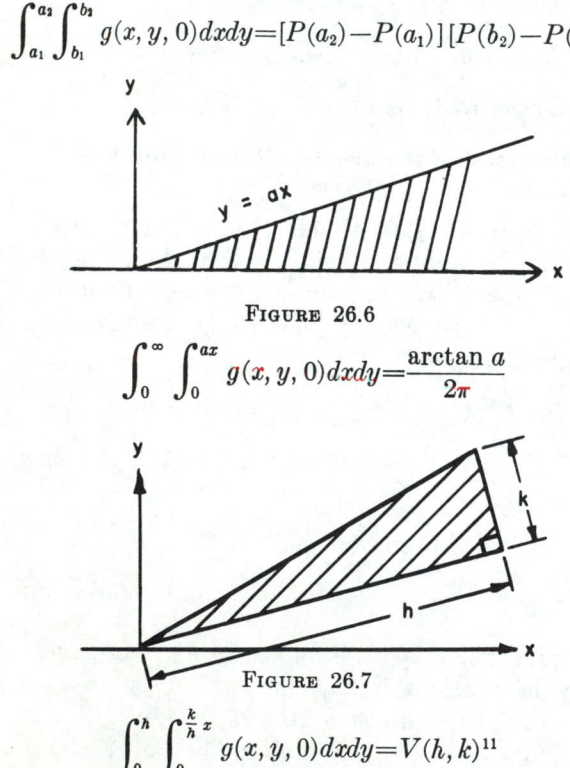

FIGURE 26.7

$$\int_0^h \int_0^{\frac{k}{h}x} g(x,y,0)\,dxdy=V(h,k)^{11}$$

---

[11] See **26.3.23** for definition of $V(h, k)$.

For the following two configurations we define

$$h=\frac{|t_2s_1-t_1s_2|}{[(s_2-s_1)^2+(t_2-t_1)^2]^{\frac{1}{2}}}$$

$$k_1=\frac{|s_1(s_2-s_1)+t_1(t_2-t_1)|}{[(s_2-s_1)^2+(t_2-t_1)^2]^{\frac{1}{2}}}$$

$$k_2=\frac{|s_2(s_2-s_1)+t_2(t_2-t_1)|}{[(s_2-s_1)^2+(t_2-t_1)^2]^{\frac{1}{2}}}$$

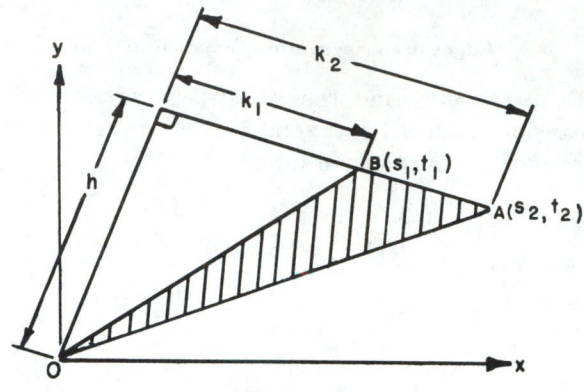

FIGURE 26.8

$$\iint_{\triangle AOB} g(x,y,0)\,dxdy=V(h,k_2)-V(h,k_1)$$

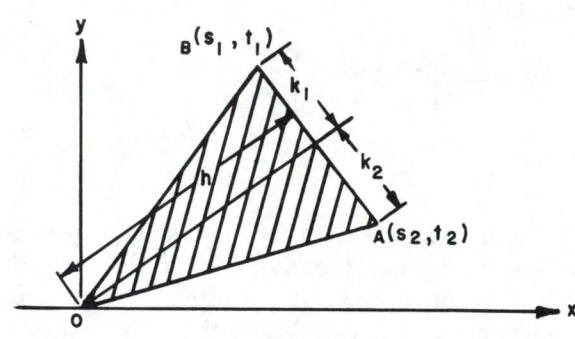

FIGURE 26.9

$$\iint_{\triangle AOB} g(x,y,0)\,dxdy=V(h,k_2)+V(h,k_1)$$

Using the circularizing transformation **26.3.22** for our example results in

$$s=\frac{1}{\sqrt{3}}\left(\frac{x-5}{2}+\frac{y-9}{4}\right)$$

$$t=-\frac{1}{1}\left(\frac{x-5}{2}-\frac{y-9}{4}\right)$$

The vertices of the triangle in the (s, t) coordinates become $A=(\sqrt{3}/4, \ -5/4)$, $B=(\sqrt{3}, \ -1)$ and $C=\left(-\frac{\sqrt{3}}{2}, \frac{3}{2}\right)$. These points are plotted below. From the figure it is seen that the desired probability is the sum of the probabilities that the point having the transformed variables as coordinates is inside the triangles AOB, AOC, and BOC.

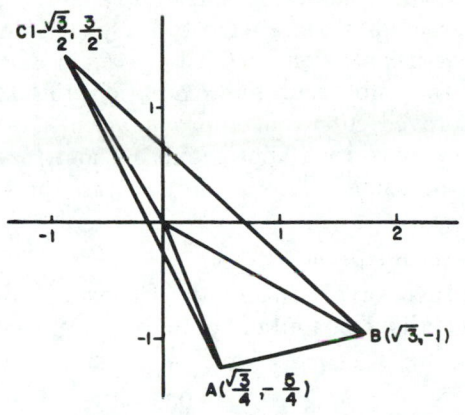

FIGURE 26.10

For these three triangles we have

	h	k_1	k_2
$\triangle AOB$	$\frac{2}{7}\sqrt{21}$	$\sqrt{7}/14$	$\frac{4}{7}\sqrt{7}$
$\triangle AOC$	$\frac{1}{74}\sqrt{111}$	$\frac{8}{37}\sqrt{37}$	$\frac{21}{74}\sqrt{37}$
$\triangle BOC$	$\frac{1}{13}\sqrt{39}$	$\frac{7}{13}\sqrt{13}$	$\frac{6}{13}\sqrt{13}$

From the graph it is seen that the probability over AOB may be found in the same manner as that over **Figure 26.8**, and over AOC and BOC the probabilities may be found as that over **Figure 26.9**.

Hence

$$\iint_{\triangle} g(x, y, .5)dxdy = \iint_{\triangle ABC} g(s, t, 0)dsdt$$

$$= \iint_{\triangle AOB} g(s, t, 0)dsdt + \iint_{\triangle AOC} g(s, t, 0)dsdt$$

$$+ \iint_{\triangle BOC} g(s, t, 0)dsdt$$

and consequently using **26.3.23** and **Figure 26.2**

$$\iint_{\triangle AOB} g(s, t, 0)dsdt = V\left(\frac{2}{7}\sqrt{21}, \frac{4\sqrt{7}}{7}\right) - V\left(\frac{2}{7}\sqrt{21}, \frac{\sqrt{7}}{14}\right)$$

$$= \left[\frac{1}{4} + L(1.31, 0, -.76) - L(0, 0, -.76) - \frac{1}{2}Q(1.31)\right]$$

$$- \left[\frac{1}{4} + L(1.31, 0, -.14) - L(0, 0, -.14) - \frac{1}{2}Q(1.31)\right]$$

$$= L(1.31, 0, -.76) - L(0, 0, -.76)$$

$$- L(1.31, 0, -.14) + L(0, 0, -.14)$$

$$= .00 - .11 - .04 + .23 = .08$$

$$\iint_{\triangle AOC} g(s, t, 0)dsdt = V\left(\frac{\sqrt{111}}{74}, \frac{8\sqrt{37}}{37}\right) + V\left(\frac{\sqrt{111}}{74}, \frac{21\sqrt{37}}{74}\right)$$

$$= \left[\frac{1}{4} + L(.14, 0, -.99) - L(0, 0, -.99) - \frac{1}{2}Q(.14)\right]$$

$$+ \left[\frac{1}{4} + L(.14, 0, -1) - L(0, 0, -1) - \frac{1}{2}Q(.14)\right]$$

$$= .01 + .02 = .03$$

$$\iint_{\triangle BOC} g(s, t, 0)dsdt = V\left(\frac{\sqrt{39}}{13}, \frac{7\sqrt{13}}{13}\right) + V\left(\frac{\sqrt{39}}{13}, \frac{6\sqrt{13}}{13}\right)$$

$$= \left[\frac{1}{4} + L(.48, 0, -.97) - L(0, 0, -.97) - \frac{1}{2}Q(.48)\right]$$

$$+ \left[\frac{1}{4} + L(.48, 0, -.96) - L(0, 0, -.96) - \frac{1}{2}Q(.48)\right]$$

$$= .05 + .04 = .09$$

Thus adding all parts, the probability that X and Y are in triangle ABC is $= .08 + .03 + .09 = .20$. The answer to 3D is .211.

Calculation of a Circular Normal Distribution Over an Offset Circle

Example 10. Let X and Y have a circular normal distribution with $\sigma = 1000$. Find the probability that the point (X, Y) falls within a circle having a radius equal to 540 whose center is displaced 1210 from the mean of the circular normal distribution.

In units of σ, the radius and displacement from the center are, respectively, $R = \frac{540}{1000} = .54$ and $r = \frac{1210}{1000} = 1.21$. The problem is thus reduced to finding the probability of X and Y falling in a circle of radius $R = .54$ displaced $r = 1.21$ from the center of the distribution where $\sigma = 1$.

Since $R<1$, the approximation **26.3.25** is used. This results in

$$P(R^2|2, r^2) = \frac{2(.54)^2}{4+(.54)^2} \exp \frac{-2(1.21)^2}{4+(.54)^2}$$

$$= (.1359)e^{-.6823} = .06869$$

The answer to 5D is .06870.

Interpolation for $Q(x^2|\nu)$

Example 11. Find $Q(25.298|20)$ using the interpolation formula given with **Table 26.7.**

Taking $x^2 = 25$, $\theta = .298$ and applying the interpolation formula results in

$$Q(25.298|20) = \frac{1}{8} \{ Q(25|16)\theta^2 + Q(25|18)(4\theta - 2\theta^2)$$
$$+ Q(25|20)(8-4\theta+\theta^2) \}$$
$$= \frac{1}{8} \{ (.06982)(.088804)$$
$$+ (.12492)(1.014392)$$
$$+ (.20143)(6.896804) \}$$
$$= .19027$$

A less accurate interpolate may be obtained by setting θ^2 equal to zero in the above formula. This results in the value .19003. The correct value to 6D is $Q(25.298|20) = .190259$.

On the other hand if $x^2 = 25.298$ is assumed to have an error of $\pm 5 \times 10^{-4}$, then how large an error arises in $Q(x^2|\nu)$? Denoting the error in x^2 by Δx^2 and the resulting error in $Q(x^2|\nu)$ by $\Delta Q(x^2|\nu)$, we then have the approximate relationship

$$\Delta Q(x^2|\nu) \approx \frac{\partial Q(x^2|\nu)}{\partial x^2} \Delta x^2$$

Using **26.4.8** we can write

$$\frac{\partial Q(x^2|\nu)}{\partial x^2} = \frac{1}{2} [Q(x^2|\nu-2) - Q(x^2|\nu)]$$

and

$$\Delta Q(x^2|\nu) \approx \frac{1}{2} [Q(x^2|\nu-2) - Q(x^2|\nu)]\Delta x^2$$

For practical purposes it is sufficient to evaluate the derivative to one or two significant figures. Consequently we can write

$$\frac{\partial Q(x^2|\nu)}{\partial x^2} \approx \frac{\partial Q(x_0^2|\nu)}{\partial x^2}$$

where x_0^2 is the closest value to x^2 for which Q is tabulated. Hence

$$\Delta Q(x^2|\nu) \approx \frac{1}{2} [Q(x_0^2|\nu-2) - Q(x_0^2|\nu)]\Delta x^2$$

For this example $\Delta x^2 = \pm 5 \times 10^{-4}$ and $x_0^2 = 25$. This results in

$$\Delta Q(x^2|\nu) = \frac{1}{2}(-.076)(\pm 5)10^{-4} = \pm 2 \times 10^{-5}$$

as the possible error in $Q(x^2|\nu)$.

Calculation of $Q(x^2|\nu)$ Outside the Range of Table 26.7

Example 12. Find the value of $Q(84|72)$.

Since this value is outside the range of **Table 26.7** we can approximate $Q(84|72)$ by (1) using the Edgeworth expansion for $Q(x^2|\nu)$ given in **Example 6,** (2) the cube root approximation **26.4.14,** (3) the improved cube root approximation **26.4.15** or (4) the square root approximation **26.4.13.** The results of using all four methods are presented below:

1. Edgeworth expansion

The successive terms of the Edgeworth expansion for the distribution of chi-square result in

$$1-Q(84|72) = .841345$$
$$.000000$$
$$.001120$$
$$\overline{\hspace{1.5cm}}$$
$$.842465$$

Hence $Q(84|72) = .15754$.

The successive terms of the Edgeworth expansion for the distribution of $\sqrt{2x^2}$ result in

$$1-Q(84|72) = .842544$$
$$-.000034$$
$$-.000138$$
$$\overline{\hspace{1.5cm}}$$
$$.842372$$

Hence $Q(84|72) = .15764$.

2. Cube root approximation 26.4.14

Using the cube root approximation we have

$$Q(84|72) = Q(x)$$

where

$$x = \frac{\left(\frac{84}{72}\right)^{1/3} \left[1 - \frac{2}{9(72)}\right]}{\left[\frac{2}{9(72)}\right]^{\frac{1}{2}}} = 1.0046$$

This results in $Q(84|72) = Q(1.0046) = 1 - P(1.0046) = .15754$.

3. Improved cube root approximation 26.4.15

The improved cube root approximation involves calculating a correction factor h_ν to x. Linearly interpolating for h_{60} (which appears below **26.4.15**) with $x = 1.0046$ results in $h_{60} = -.0006$ and hence

$$h_{72} = \frac{60}{72}(-.0006) = -.00049$$

Thus

$$Q(84|72) = Q(1.0046 - .0005) = Q(1.0041)$$
$$= 1 - P(1.0041) = .15766$$

4. Square root approximation 26.4.13

Using the square root approximation we have $Q(84|72) = Q(x)$ where

$$x = \sqrt{2(84)} - \sqrt{2(72) - 1} = 1.0032.$$

This results in

$$Q(84|72) = Q(1.0032) = 1 - P(1.0032) = .15788$$

The value correct to 6D is $Q(84|72) = .157653$. Generally the improved cube root approximation will be correct with a maximum error of a few units in the fifth decimal and is recommended for calculations which are outside the range of **Table 26.7.**

Calculation of x^2 for $Q(x^2|\nu)$ Outside the Range of Table 26.8

Example 13. Find the value of x^2 for which $Q(x^2|144) = .01$.

Since $\nu = 144$ is outside the range of **Table 26.8,** we can compute it by using (1) the Cornish-Fisher asymptotic expansion **26.2.50,** for x^2, (2) the cube approximation **26.4.17,** (3) the improved cube approximation **26.4.18,** or (4) the square approximation **26.4.16.** We shall compute the value by all four methods.

1. Cornish-Fisher asymptotic expansion 26.2.50

The Cornish-Fisher asymptotic expansion for x^2 with $\nu = 144$ can be written as

$$x^2 \sim 144 + 12\sqrt{2}x + 4h_1(x) + \frac{4\sqrt{2}}{12}[3h_2(x) + 2h_{11}(x)]$$

$$+ \frac{8}{12^2}[6h_3(x) + 3h_{12}(x) + 2h_{111}(x)] + \frac{16\sqrt{2}}{12^3}[30h_4(x)$$

$$+ 9h_{22}(x) + 12h_{13}(x) + 6h_{112}(x) + 4h_{1111}(x)]$$

Hence using the auxiliary table following **26.2.51** with $p = .01$ we have

$$
\begin{aligned}
&144.0000 \\
&39.4794 \\
&2.9413 \\
&-.0242 \\
&-.0019 \\
&+.0002 \\
\hline
x^2 = \ &186.395
\end{aligned}
$$

*See page II.

2. Cube approximation 26.4.17

Taking $x_{.01} = 2.32635$ we have

$$x^2 = 144 \left\{ \left[1 - \frac{2}{9(144)}\right] + (2.32635)\sqrt{\frac{2}{9(144)}} \right\}^3 = 186.405$$

3. Improved cube approximation 26.4.18

From the table for h_{60} we obtain using linear interpolation with $x = 2.33$ (approximately)

$$h_{60} = .0012 \text{ and thus } h_{144} = \frac{60}{144}(.0012) = .00049$$

Hence

$$x^2 = 144\left[1 - \frac{2}{9(144)} + (2.32635 - .00049)\sqrt{\frac{2}{9(144)}}\right]^3 = 186.394$$

4. Square approximation 26.4.16

$$x^2 = \frac{1}{2}[2.32635 + \sqrt{2(144) - 1}]^2 = 185.616$$

The correct answer to 3D is $x^2 = 186.394$. Generally the improved cube approximation will give results correct in the second or third decimal for $\nu > 30$.

Calculation of the Incomplete Gamma Function

Example 14. Find the value of

$$\gamma(2.5, .9) = \int_0^9 t^{1.5}e^{-t}dt$$

making use of **26.4.19** and **Table 26.7.** Using **26.4.19** we have

$$\gamma(2.5, .9) = \Gamma(2.5)P(1.8|5) = \Gamma(2.5)[1 - Q(1.8|5)]$$

$$\gamma(2.5, .9) = \frac{3}{4}\sqrt{\pi}[1 - .87607] = .16475$$

Poisson Distribution

Example 15. Find the value of m for which

$$\sum_{i=0}^{3} e^{-m}\frac{m^i}{i!} = .99$$

using **26.4.21** and **Table 26.8.**

From **Table 26.8** with $\nu = 2c = 8$ and $Q = .99$ we have $x^2 = 1.646482$. Hence $m = x^2/2 = .823241$.

Inverse of the Incomplete Beta Function

Example 16. Find the value of x for which $I_x(10, 6) = .10$ using **Table 26.9** and **26.5.28.** *
Using **26.5.28** we have

$$I_x(10,6)=Q(F|12,20)=.10 \text{ where } x=\frac{20}{20+12F}$$

From **Table 26.9** the upper 10 percent point of F with 12 and 20 degrees of freedom is $F=1.89$. Hence

$$x=\frac{20}{20+12(1.89)}=.469$$

The correct value to 4D is $x=.4683$.

Calculation of $I_x(a, b)$ for a or b Small Integers

Example 17.   Calculate $I_{.10}(3, 20)$.

Values of $I_x(a, b)$ for small integral a or b can conveniently be calculated using **26.5.6** or **26.5.7**. Using **26.5.6** we have

$$1-I_{.90}(20,3)=\frac{(.9)^{20}}{B(3,20)}\left\{\sum_{i=0}^{2}(-1)^i\binom{2}{i}\frac{.9^i}{20+i}\right\}$$

$$=\frac{.121576}{.216450\times10^{-3}}(.110390\times10^{-2})=.620040$$

Binomial Distribution

Example 18.   Find the value of p which satisfies

$$\sum_{s=0}^{20}\binom{50}{s}p^s q^{50-s}=.95, \qquad q=1-p$$

using **26.5.24** and **Table 26.9**.

*   Combining 26.5.24 and 26.5.28 we have

$$\sum_{s=a}^{n}\binom{n}{s}p^s q^{n-s}=Q(F|\nu_1, \nu_2)$$

where

$$\nu_1=2(n-a+1), \nu_2=2(a), \text{ and } p=\frac{a}{a+(n-a+1)F}$$

Hence

$$\sum_{s=0}^{20}\binom{50}{s}p^s q^{50-s}=1-\sum_{s=21}^{50}\binom{50}{s}p^s q^{50-s}$$

$$=1-Q(F|60,42)=.95$$

Harmonic interpolation on ν_2 in the table for which $Q(F|\nu_1, \nu_2)=.05$ results in $F=1.624$ for $\nu_1=60$, $\nu_2=42$, and thus $p=\frac{42}{42+60(1.624)}=.301$. The correct answer to 4D is $p=.3003$.

Approximating the Incomplete Beta Function

Example 19.   Find $I_{.60}(16, 10.5)$ using **26.5.21**.

Values of $I_x(a, b)$ can conveniently be calculated with good accuracy using the approximation given by **26.5.20** or **26.5.21**. For this example $(a+b-1)(1-x)=10.20$ which is greater than .8 and hence **26.5.21** will be used. Thus

$$w_1=[(10.5)(.60)]^{1/3}=1.8469, w_2=[16(.4)]^{1/3}=1.8566$$

$$y=\frac{3[(1.8469)(.98942)-(1.8566)(.99306)]}{\left[\frac{(1.8469)^2}{10.5}+\frac{(1.8566)^2}{16}\right]^{\frac{1}{2}}}=-.0668$$

and interpolating in **Table 26.1** gives

$$P(-.0668)=1-P(.0668)=.47336$$

The answer correct to 5D is $I_{.60}(16, 10.5)=.47332$.

Interpolation for F in Table 26.9

Example 20.   Find the value of F for which

$$Q(F|7, 20)=.05 \text{ using } \textbf{Table 26.9}.$$

Interpolation in **Table 26.9** is approximately linear when the reciprocals of the degrees of freedom (ν_1, ν_2) are used as the interpolating variable. For this example it is only necessary to interpolate with respect to $1/\nu_1$. Thus linear interpolation on $1/\nu_1$ results in $F=2.51$ which is the correct interpolate.

Calculation of F for $Q(F|\nu_1,\nu_2)>.50$

Example 21.   Find the value of F for which $Q(F|4,8)=.90$ using **26.6.9** and **Table 26.9**.

Table 26.9 only tabulates values of F for which $Q(F|\nu_1, \nu_2)=p$ where $p=.500, .250, .100, .050, .025, .010, .005, .001$. However making, use of **Table 26.9** we can find the values of F_p for which $p=.75, .9, .95, .975, .99, .995, .999$. For this example we have

$$F_{.90}(4,8)=\frac{1}{F_{.10}(8,4)}$$

and referring to the table for which $Q(F|\nu_1, \nu_2)=.10$ gives $F_{.10}(8,4)=3.95$ and thus $F_{.90}(4,8)=\frac{1}{3.95}=.253$.

Calculation of $Q(F|\nu_1,\nu_2)$ for Small Integral ν_1 or ν_2

Example 22.   Compute $Q(2.5|4, 15)$ using **26.6.4**.

Values of $Q(F|\nu_1, \nu_2)$ can be readily computed for small ν_1 or ν_2 using the expansions **26.6.4** to **26.6.8** inclusive. We have using **26.6.4**

$$x=\frac{15}{15+4(2.50)}=.60$$

and

$$Q(2.50|4,15)=(.6)^{7.5}\left[1+\frac{15}{2}(.4)\right]=.086\ 735$$

Approximating $Q(F|\nu_1, \nu_2)$

Example 23. Calculate $Q(1.714|10, 40)$ using **26.6.15**.

The approximation given by **26.6.15** will result in a maximum error of .0005. For this example we have

$$x = \frac{(1.714)^{1/3}\left(1 - \frac{2}{9(40)}\right) - \left(1 - \frac{2}{9(10)}\right)}{\left[\frac{2}{9(10)} + (1.714)^{2/3}\frac{2}{9(40)}\right]^{\frac{1}{2}}} = 1.2222$$

Interpolating in **Table 26.1** results in

$$Q(1.714|10, 40) \approx Q(1.2222) = 1 - P(1.2222) = .1108$$

The correct value to 5D is $Q(1.714|10, 40) = .11108$.

On the other hand the approximation given by **26.6.14** which is usually less accurate results in

$$x = \frac{\sqrt{[2(40) - 1]\left(\frac{10}{40}\right)(1.714)} - \sqrt{2(10) - 1}}{\sqrt{1 + \frac{10}{40}(1.714)}} = 1.2210$$

and interpolating in **Table 26.1** gives

$$Q(1.714|10, 40) \approx Q(1.2210) = 1 - P(1.2210) = .1112$$

Calculation of F Outside the Range of Table 26.9

Example 24. Find the value of F for which $Q(F|10, 20) \approx .0001$ using **26.6.16** and **26.5.22**.

For this problem we have $a = \frac{\nu_2}{2} = 10$, $b = \frac{\nu_1}{2} = 5$, $p = .0001$. The value of the normal deviate which cuts off .0001 in the tail of the distribution is

$y = 3.7190$ (i.e., $Q(3.7190) = .0001$). Hence substituting in **26.5.22** gives

$$h = 2\left[\frac{1}{19} + \frac{1}{9}\right]^{-1} = 12.2143$$

$$\lambda = \frac{3.7190^2 - 3}{6} = 1.8052$$

$$w = 3.7190 \frac{(12.2143 + 1.8052)^{\frac{1}{2}}}{12.2143}$$
$$- \left(\frac{1}{9} - \frac{1}{19}\right)\left[1.8052 + .8333 - \frac{2}{3(12.2143)}\right]$$

$$w = .9889$$

and thus $F \approx e^{2w} = 7.23$. The correct answer is $F = 7.180$.

Approximating the Non-Central F-Distribution

Example 25. Compute $P(3.71|3, 10, 4)$ using the approximation **26.6.27** to the non-central F-distribution.

Using **26.6.27** with $\nu_1 = 3$, $\nu_2 = 10$, $\lambda = 4$, $F' = 3.71$ we have

$$x = \frac{\left[\left(\frac{3}{3+4}\right)(3.71)\right]^{1/3}\left[1 - \frac{2}{9(10)}\right] - \left[1 - \frac{2}{9}\frac{(3+8)}{(3+4)^2}\right]}{\left[\frac{2}{9}\frac{3+8}{(3+4)^2} + \frac{2}{9(10)}\left[\left(\frac{3}{3+4}\right)(3.71)\right]^{2/3}\right]^{\frac{1}{2}}}$$
$$= .675$$

and interpolating in **Table 26.1** gives

$$P(3.71|3, 10, 4) \approx P(.675) = .750$$

The exact answer is $P(3.71|3, 10, 4) = .745$.

References

Texts

[26.1] H. Cramér, Mathematical methods of statistics (Princeton Univ. Press, Princeton, N.J., 1951).

[26.2] A. Erdélyi et al., Higher transcendental functions, vols. I, II, III. (McGraw-Hill Book Co., Inc., New York, N.Y., 1955).

[26.3] W. Feller, Probability theory and its applications, 2d ed. (John Wiley & Sons, Inc., New York, N.Y., 1957).

[26.4] R. A. Fisher, Contributions to mathematical statistics, Paper 30 (with E. A. Cornish), Moments and cumulants in the specification of distributions (John Wiley & Sons, Inc., New York, N.Y., 1950).

[26.5] C. Hastings, Jr., Approximations for digital computers (Princeton Univ. Press, Princeton, N.J., 1955).

[26.6] M. G. Kendall and A. Stuart, The advanced theory of statistics, vol. I, Distribution theory (Charles Griffin and Co. Ltd., London, England, 1958).

Tables

General Collections

[26.7] R. A. Fisher and F. Yates, Statistical tables for biological, agricultural and medical research (Oliver and Boyd, London, England, 1949).

[26.8] J. Arthur Greenwood and H. O. Hartley, Guide to tables in mathematical statistics (Princeton Univ. Press, Princeton, N.J., 1962). (Catalogues a large selection of tables used in mathematical statistics).

[26.9] A. Hald, Statistical tables and formulas (John Wiley & Sons, Inc., New York, N.Y., 1952).

[26.10] D. B. Owen, Handbook of statistical tables (Addison-Wesley Publishing Co., Inc., Reading, Mass., 1962).

[26.11] E. S. Pearson and H. O. Hartley (Editors). Biometrika tables for statisticians, vol. I (Cambridge Univ. Press, Cambridge, England, 1954).

[26.12] K. Pearson (Editor), Tables for statisticians and biometricians, parts I and II (Cambridge Univ. Press, Cambridge, England, 1914, 1931).

Normal Probability Integral and Derivatives

[26.13] J. R. Airey, Table of Hh functions, British Association for the Advancement of Science, Mathematical Tables I (Cambridge Univ. Press, Cambridge, England, 1931).

[26.14] Harvard University, Tables of the error function and of its first twenty derivatives (Harvard Univ. Press, Cambridge, Mass., 1952). $P(x) - \frac{1}{2}$, $Z(x)$, $Z^{(n)}(x)$, $n=1(1)4$ for $x=0(.004)$ 6.468, 6D; $Z^{(n)}(x)$, $n=5(1)10$ for $x=0(.004)$ 8.236, 6D; $Z^{(n)}(x)$, $n=11(1)15$ for $x=0(.002)$ 9.61, 7S; $Z^{(n)}(x)$, $n=16(1)20$ for $x=0(.002)$ 10.902, 7S or 6D.

[26.15] T. L. Kelley, The Kelley Statistical Tables (Harvard Univ. Press, Cambridge, Mass., 1948). x for $P(x)=.5(.0001).9999$ and corresponding values of $Z(x)$, 8D.

[26.16] National Bureau of Standards, A guide to tables of the normal probability integral, Applied Math. Series 21 (U.S. Government Printing Office, Washington, D.C., 1951).

[26.17] National Bureau of Standards, Tables of normal probability functions, Applied Math. Series 23 (U.S. Government Printing Office, Washington, D.C., 1953). $Z(x)$ and $A(x)$ for $x=0(.0001)$ 1(.001)7.8, 15D; $Z(x)$ and $2[1-P(x)]$ for $x=6(.01)10$, 7S.

[26.18] W. F. Sheppard, The probability integral, British Association for the Advancement of Science, Mathematical Tables VII (Cambridge Univ. Press, Cambridge, England, 1939). $A(x)/Z(x)$ for $x=0(.01)10$, 12D; $x=0(.1)10$, 24D.

Bivariate Normal Probability Integral

[26.19] Bell Aircraft Corporation, Table of circular normal probabilities, Report No. 02–949–106 (1956). Tabulates the integral of the circular normal distribution over an off-set circle having its center a distance r from the origin with radius R; $R=0(.01)4.59$, $r=0(.01)3$, 5D.

[26.20] National Bureau of Standards, Tables of the bivariate normal distribution function and related functions, Applied Math. Series 50 (U.S. Government Printing Office, Washington, D.C., 1959). $L(h, k, \rho)$ for h, $k=0(.1)4$, $\rho=0(.05).95$ (.01)1, 6D; $L(h, k, -\rho)$ for h, $k=0(.1)A$, $\rho=0(.05).95(.01)1$ where A is such that $L<.5\cdot10^{-7}$, 7D; $V(h, ah)$ for $h=0(.01)4(.02)4.6(.1)5.6$, ∞, 7D; $V(ah, h)$ for $a=.1(.1)1$, $h=0(.01)4(.02)5.6$, ∞, 7D.

[26.21] C. Nicholson, The probability integral for two variables, Biometrika **33**, 59–72 (1943). $V(h, ah)$ for $h=.1(.1)3$, $ah=.1(.1)3$, ∞, 6D.

[26.22] D. B. Owen, Tables for computing bivariate normal probabilities, Ann. Math. Statist. **27**, 1075–1090 (1956). $T(h, a)=\frac{1}{2\pi}$ arctan $a-V(h, ah)$ for $a=.25(.25)1$, $h=0(.01)2(.02)3$; $a=0(.01)1$, ∞, $h=0(.25)3$; $a=.1$, $.2(.05).5(.1).8$, 1, ∞, $h=3(.05)3.5(.1)4.7$, 6D.

[26.23] D. B. Owen, The bivariate normal probability function, Office of Technical Services, U.S. Department of Commerce (1957). $T(h, a)=\frac{1}{2\pi}$ arctan $a-V(h, ah)$ for $a=0(.025)1$, ∞; $h=0(.01)3.5(.05)4.75$, 6D.

[26.24] Tables VIII and IX, Part II of [26.12]. $L(h, k, \rho)$ for h, $k=0(.1)2.6$, $\rho=-1(.05)1$, 6D for $\rho>0$ and 7D for $\rho<0$.

Chi-Square, Non-Central Chi-Square, Probability Integral, Incomplete Gamma Function, Poisson Distribution

[26.25] G. A. Campbell, Probability curves showing Poisson's exponential summation, Bell System Technical Journal, 95–113 (1923). Tabulates values of $m=\frac{\chi^2}{2}$ for which $Q(\chi^2|\nu)=.000001$, 2D; .0001, .01, 3D; .1, .25, .5, .75, .9, 4D; .99, .9999, 3D; .999999, 2D for $c=\frac{\nu}{2}=1(1)101$.

[26.26] Table IV of [26.7]. Tabulates values of χ^2 for $Q(\chi^2|\nu)=.001$, .01, .02, .05, .1, .2, .3, .5, .7, .8, .9, .95, .98, .99 and $\nu=1(1)30$, 3D or 3S.

[26.27] E. Fix, Tables of noncentral χ^2, Univ. of California Publications in Statistics 1, 15–19 (1949). Tabulates λ for $P(\chi'^2|\nu, \lambda)=.1(.1).9$, $Q(\chi'^2|\nu)=.01$, .05; $\nu=1(1)20(2)40(5)60(10)100$, 3D or 3S.

[26.28] H. O. Hartley and E. S. Pearson, Tables of the χ^2 integral and of the cumulative Poisson distribution, Biometrika 37, 313–325 (1950). Also reproduced as Table 7 in [26.11]. $P(\chi^2|\nu)$ for $\nu=1(1)20(2)70$, $\chi^2=0(.001).01(.01).1(.1)2(.2)10$ (.5)20(1)40(2)134, 5D.

[26.29] T. Kitagawa, Tables of Poisson distribution (Baifukan, Tokyo, Japan, 1951). $e^{-m}m^s/s!$ for $m=.001(.001)1(.01)5$, 8D; $m=5(.01)10$, 7D.

[26.30] E. C. Molina, Poisson's exponential binomial limit (D. Van Nostrand Co., Inc., New York, N.Y., 1940). $e^{-m}m^s/s!$ and $P(\chi^2|\nu)=\sum_{j=c}^{\infty} e^{-m}m^j/j!$ for $m=\chi^2/2=0(.1)16(1)100$, 6D; $m=0(.001).01$ (.01)3, 7D.

[26.31] K. Pearson (Editor), Tables of the incomplete Γ-function, Biometrika Office, University College (Cambridge Univ. Press, Cambridge, England, 1934). $I(u,p)$ for $p=-1(.05)0(.1)5(.2)50$, $u=0(.1)$ $I(u,p)=1$ to 7D; $p=-1(.01)-.75$, $u=0(.1)6$, 5D; $\ln[I(u,p)|u^{p+1}]$, $p=-1(.05)0$ (.1)10, $u=0(.1)1.5$, 8D; $[x^{p+1}\Gamma(p+1)]^{-1}\gamma(p,x)$, $p=-1(.01)-.9, x=0(.01)3$, 7D.

[26.32] E. E. Sluckii, Tablitsy dlya vyčioleniya nepolnoi Γ-funktsii i funktsii veroyatnosti χ^2. (Izdat. Akad. Nauk SSSR, Moscow-Leningrad, U.S.S.R., 1950). $\Gamma(\chi^2,\nu)=\left(\frac{1}{2}\chi^2\right)^{-\nu/2} P(\chi^2|\nu)$, $\mathscr{P}(t,\nu)=$ $Q(\chi^2|\nu)$, $\Pi(t, x)=Q(\chi^2|\nu)$ where $t=(2\chi^2)^{\frac{1}{2}}-(2\nu)^{\frac{1}{2}}$, $x=(\nu/2)^{-\frac{1}{2}}$. $\Gamma(\chi^2,\nu)$, $\chi^2=0(.05)2(.1)10$, $\nu=0(.05)$ 2(.1)6; $Q(\chi^2|\nu)$, $\chi^2=0(.1)3.2$, $\nu=0(.05)2(.1)6$; $\chi^2=3.2(.2)7(.5)10(1)35$, $\nu=0(.1).4(.2)6$; $\mathscr{P}(t,\nu)$, $t=-4(.1)4.8$, $\nu=6(.5)11(1)32$; $\Pi(t,x)$: $t=-4.5$ (.1)4.8, $x=0(.02).22(.01).25$, 5D.

Incomplete Beta Function, Binomial Distribution

[26.33] Harvard University, Tables of the cumulative binomial probability distribution (Harvard Univ. Press, Cambridge, Mass., 1955).

$$\sum_{s=c}^{\infty} \binom{n}{s} p^s (1-p)^{n-s} \text{ for } p=.01(.01).5, \ 1/16, \ 1/12,$$

1/8, 1/6, 3/16, 5/16, 1/3, 3/8, 5/12, 7/16, $n=1(1)50(2)100(10)200(20)500(50)1000$, 5D.

[26.34] National Bureau of Standards, Tables of the binomial probability distribution, Applied Math. Series 6 (U.S. Government Printing Office, Washington, D.C., 1950). $\binom{n}{s} p^s (1-p)^{n-s}$ and

$$\sum_{s=c}^{n} \binom{n}{s} p^s (1-p)^{n-s} \text{ for } p=.01(.01).5, \ n=2(1)49,$$

7D.

[26.35] K. Pearson (Editor), Tables of the incomplete beta function, Biometrika Office, University College (Cambridge Univ. Press, Cambridge, England, 1948). $I_x(a,b)$ for $x=.01(.01)1$; $a,b=.5(.5)11(1)50$, $a \geq b$, 7D.

[26.36] W. H. Robertson, Tables of the binomial distribution function for small values of p, Office of Technical Services, U.S. Department of Commerce (1960).

$$\sum_{s=0}^{c} \binom{n}{s} p^s (1-p)^{n-s} \text{ for } p=.001(.001).02, \ n=2(1)$$

$100(2)200(10)500(20)1000$; $p=.021(.001).05$, $n=2(1)50(2)100(5)200(10)300(20)600(50)1000$, 5D.

[26.37] H. G. Romig, 50–100 Binomial tables (John Wiley & Sons, Inc., New York, N.Y., 1953).

$$\binom{n}{s} p^s (1-p)^{n-s} \text{ and } \sum_{s=0}^{c} \binom{n}{s} p^s (1-p)^{n-s} \text{ for }$$

$p=.01(.01).5$ and $n=50(5)100$, 6D.

[26.38] C. M. Thompson, Tables of percentage points of the incomplete beta function, Biometrika 32, 151–181 (1941). Also reproduced as Table 16 in [26.11]. Tabulates values of x for which
* $I_x(a, b)=.005, .01, .025, .05, .1, .25, .5$; $2a=1(1)30, 40, 60, 120, \infty$; $2b=1(1)10, 12, 15, 20, 24, 30, 40, 60, 120, \infty$, 5D.

[26.39] U.S. Ordnance Corps, Tables of the cumulative binomial probabilities, ORDP 20–1, Office of Technical Services, Washington, D.C. (1952).

$$\sum_{s=c}^{n} \binom{n}{s} p^s (1-p)^{n-s} \text{ for } p=.01(.01).5 \text{ and } n=1$$

$(1)150$, 7D.

F (Variance-Ratio) and Non-Central F Distribution

[26.40] Table V of [26.7]. Tabulates values of F and

$$Z=\frac{1}{2} \ln F \text{ for } Q(F|\nu_1, \nu_2)=.2, .1, .05, .01, .001;$$

$\nu_1=1(1)6, 8, 12, 24, \infty$; $\nu_2=1(1)30, 40, 60, 120, \infty$, 2D for F, 4D for Z.

[26.41] E. Lehmer, Inverse tables of probabilities of errors of the second kind, Ann. Math. Statist. 15, 388–398 (1944). $\phi=\sqrt{\lambda/(\nu_1+1)}$ for $\nu_1=1(1)10$, 12, 15, 20, 24, 30, 40, 60, 120, ∞; $\nu_2=2(2)20$, 24, 30, 40, 60, 80, 120, 240, ∞ and $P(F'|\nu_1, \nu_2, \phi)$ $=.2, .3$ where $Q(F'|\nu_1, \nu_2)=.01, .05$, 3D or 3S.

[26.42] M. Merrington and C. M. Thompson, Tables of percentage points of the inverted beta (F) distribution, Biometrika 33, 73–88 (1943). Tabulates values of F for which $Q(F|\nu_1, \nu_2)=.5, .25, .1, .05, .025, .01, .005$; $\nu_1=1(1)10, 12, 15, 20, 24, 30, 40, 60, 120, \infty$; $\nu_2=1(1)30, 40, 60, 120, \infty$.

[26.43] P. C. Tang, The power function of the analysis of variance tests with tables and illustrations of their use, Statistical Research Memoirs II, 126–149 and tables (1938). $P(F'|\nu_1, \nu_2, \phi)$ for $\nu_1=1(1)8, \nu_2=2(2)6(1)30, 60, \infty$ and $\phi=\sqrt{\lambda/\nu_1+1)}=1(.5)3(1)8$ where $Q(F'|\nu_1, \nu_2)=.01, .05$, 3D.

Student's t and Non-Central t-Distributions

[26.44] E. T. Federighi, Extended tables of the percentage points of Student's t-distribution, J. Amer. Statist. Assoc. 54, 683–688 (1959.) Values of

$$t \text{ for which } Q(t|\nu)=\frac{1}{2}\left[1-A(t|\nu)\right]=.25 \times 10^{-n},$$

$.1 \times 10^{-n}$, $n=0(1)6$, $.05 \times 10^{-n}$, $n=0(1)5$, $\nu=1(1)30(5)60(10)100, 200, 500, 1000, 2000, 10000, \infty$; 3D.

[26.45] Table III of [26.7]. Values of t for which $A(t|\nu)=.1(.1).9, .95, .98, .99, .999$ and $\nu=1(1)30, 40, 60, 120, \infty$; 3D.

[26.46] N. L. Johnson and B. L. Welch, Applications of the noncentral t-distribution, Biometrika 31, 362–389 (1939). Tabulates an auxiliary function which enables calculation of δ for given t' and p, or t' for given δ and p where $P(t'|\nu,\delta)=p=.005, .01, .025, .05, .1(.1).9, .95, .975, .99, .995$.

[26.47] J. Neyman and B. Tokarska, Errors of the second kind in testing Student's hypothesis, J. Amer. Statist. Assoc. 31, 318–326 (1936). Tabulates δ for $P(t'|\nu,\delta)=.01, .05, .1(.1).9$; $\nu=1(1)30, \infty$; $Q(t'|\nu)=.01, .05$.

[26.48] Table 9 of [26.11]. $P(t|\nu)=\frac{1}{2}[1+A(t|\nu)]$ for $t=0(.1)4(.2)8$; $\nu=1(1)20$, 5D; $t=0(.05)2(.1)4, 5$; $\nu=20(1)24, 30, 40, 60, 120, \infty$, 5D.

[26.49] G. S. Resnikoff and G. J. Lieberman, Tables of the noncentral t-distribution (Stanford Univ. Press, Stanford, Calif., 1957). $\partial P(t'|\nu,\delta)/\partial t'$ and $P(t'|\nu,\delta)$ for $\nu=2(1)24(5)49$, $\delta=\sqrt{\nu+1} \ x_p$ where $Q(x_p)=p=.25, .15, .1, .065, .04, .025, .01, .004, .0025, .001$ and $t'/\sqrt{\nu}$ covers the range of values such that throughout most of the table the entries lie between 0 and 1, 4D.

Random Numbers and Normal Deviates

[26.50] E. C. Fieller, T. Lewis and E. S. Pearson, Correlated random normal deviates, Tracts for Computers 26 (Cambridge Univ. Press, Cambridge, England, 1955).

[26.51] T. E. Hull and A. R. Dobell, Random number generators, Soc. Ind. App. Math. 4, 230–254 (1962).

[26.52] M. G. Kendall and B. Babington Smith, Random sampling numbers (Cambridge Univ. Press, Cambridge, England, 1939).

*See page II.

[26.53] G. Marsaglia, Random variables and computers, Proc. Third Prague Conference in Probability Theory 1962. (Also as Math. Note No. 260, Boeing Scientific Research Laboratories, 1962).

[26.54] M. E. Muller, An inverse method for the generation of random normal deviates on large scale computers, Math. Tables Aids Comp. **63.** 167–174 (1958).

[26.55] Rand Corporation, A million random digits with 100,000 normal deviates (The Free Press, Glencoe, Ill. 1955).

[26.56] H. Wold, Random normal deviates, Tracts for Computers 25 (Cambridge Univ. Press, Cambridge, England, 1948).

Table 26.1 NORMAL PROBABILITY FUNCTION AND DERIVATIVES

x	$P(x)$	$Z(x)$	$Z^{(1)}(x)$
0.00	0.50000 00000 00000	0.39894 22804 01433	0.00000 00000 00000
0.02	0.50797 83137 16902	0.39886 24999 23666	−0.00797 72499 98473
0.04	0.51595 34368 52831	0.39862 32542 04605	−0.01594 49301 68184
0.06	0.52392 21826 54107	0.39822 48301 95607	−0.02389 34898 11736
0.08	0.53188 13720 13988	0.39766 77055 11609	−0.03181 34164 40929
0.10	0.53982 78372 77029	0.39695 25474 77012	−0.03969 52547 47701
0.12	0.54775 84260 20584	0.39608 02117 93656	−0.04752 96254 15239
0.14	0.55567 00048 05907	0.39505 17408 34611	−0.05530 72437 16846
0.16	0.56355 94628 91433	0.39386 83615 68541	−0.06301 89378 50967
0.18	0.57142 37159 00901	0.39253 14831 20429	−0.07065 56669 61677
0.20	0.57925 97094 39103	0.39104 26939 75456	−0.07820 85387 95091
0.22	0.58706 44226 48215	0.38940 37588 33790	−0.08566 88269 43434
0.24	0.59483 48716 97796	0.38761 66151 25014	−0.09302 79876 30003
0.26	0.60256 81132 01761	0.38568 33691 91816	−0.10027 76759 89872
0.28	0.61026 12475 55797	0.38360 62921 53479	−0.10740 97618 02974
0.30	0.61791 14221 88953	0.38138 78154 60524	−0.11441 63446 38157
0.32	0.62551 58347 23320	0.37903 05261 52702	−0.12128 97683 68865
0.34	0.63307 17360 36028	0.37653 71618 33254	−0.12802 26350 23306
0.36	0.64057 64332 17991	0.37391 06053 73128	−0.13460 78179 34326
0.38	0.64802 72924 24163	0.37115 38793 59466	−0.14103 84741 56597
0.40	0.65542 17416 10324	0.36827 01403 03323	−0.14730 80561 21329
0.42	0.66275 72731 51751	0.36526 26726 22154	−0.15341 03225 01305
0.44	0.67003 14463 39407	0.36213 48824 13092	−0.15933 93482 61761
0.46	0.67724 18897 49653	0.35889 02910 33545	−0.16508 95338 75431
0.48	0.68438 63034 83778	0.35553 25285 05997	−0.17065 56136 82879
0.50	0.69146 24612 74013	0.35206 53267 64299	−0.17603 26633 82150
0.52	0.69846 82124 53034	0.34849 25127 58974	−0.18121 61066 34667
0.54	0.70540 14837 84302	0.34481 80014 39333	−0.18620 17207 77240
0.56	0.71226 02811 50973	0.34104 57886 30353	−0.19098 56416 32997
0.58	0.71904 26911 01436	0.33717 99438 22381	−0.19556 43674 16981
0.60	0.72574 68822 49927	0.33322 46028 91800	−0.19993 47617 35080
0.62	0.73237 11065 31017	0.32918 39607 70765	−0.20409 40556 77874
0.64	0.73891 37003 07139	0.32506 22640 84082	−0.20803 98490 13813
0.66	0.74537 30853 28664	0.32086 38037 71172	−0.21177 01104 88974
0.68	0.75174 77695 46430	0.31659 29077 10893	−0.21528 31772 43407
0.70	0.75803 63477 76927	0.31225 39333 66761	−0.21857 77533 56733
0.72	0.76423 75022 20749	0.30785 12604 69853	−0.22165 29075 38294
0.74	0.77035 00028 35210	0.30338 92837 56300	−0.22450 80699 79662
0.76	0.77637 27075 62401	0.29887 24057 75953	−0.22714 30283 89724
0.78	0.78230 45624 14267	0.29430 50297 88325	−0.22955 79232 34894
0.80	0.78814 46014 16604	0.28969 15527 61483	−0.23175 32422 09186
0.82	0.79389 19464 14187	0.28503 63584 89007	−0.23372 98139 60986
0.84	0.79954 58067 39551	0.28034 38108 39621	−0.23548 88011 05281
0.86	0.80510 54787 48192	0.27561 82471 53457	−0.23703 16925 51973
0.88	0.81057 03452 23288	0.27086 39717 98338	−0.23836 02951 82537
0.90	0.81593 98746 53241	0.26608 52498 98755	−0.23947 67249 08879
0.92	0.82121 36203 85629	0.26128 63012 49553	−0.24038 33971 49589
0.94	0.82639 12196 61376	0.25647 12944 25620	−0.24108 30167 60083
0.96	0.83147 23925 33162	0.25164 43410 98117	−0.24157 85674 54192
0.98	0.83645 69406 72308	0.24680 94905 67043	−0.24187 33007 55702
1.00	0.84134 47460 68543	0.24197 07245 19143	−0.24197 07245 19143
	$\left[\begin{smallmatrix}(-5)1\\10\end{smallmatrix}\right]$	$\left[\begin{smallmatrix}(-5)2\\10\end{smallmatrix}\right]$	$\left[\begin{smallmatrix}(-5)3\\10\end{smallmatrix}\right]$

$$Z(x)=\frac{1}{\sqrt{2\pi}}\,e^{-\frac{1}{2}x^2} \qquad P(x)=\int_{-\infty}^{x}Z(t)dt \qquad Z^{(n)}(x)=\frac{d^n}{dx^n}Z(x) \qquad He_n(x)=(-1)^n Z^{(n)}(x)/Z(x)$$

NORMAL PROBABILITY FUNCTION AND DERIVATIVES

Table 26.1

x	$Z^{(2)}(x)$	$Z^{(3)}(x)$	$Z^{(4)}(x)$	$Z^{(5)}(x)$	$Z^{(6)}(x)$
0.00	−0.39894 22804	0.00000 000	1.19682 684	0.00000 000	−5.98413 421
0.02	−0.39870 29549	0.02392 856	1.19563 029	−0.11962 684	−5.97575 893
0.04	−0.39798 54570	0.04780 928	1.19204 400	−0.23891 887	−5.95066 325
0.06	−0.39679 12208	0.07159 445	1.18607 800	−0.35754 249	−5.90893 742
0.08	−0.39512 26322	0.09523 664	1.17774 897	−0.47516 649	−5.85073 151
0.10	−0.39298 30220	0.11868 881	1.16708 019	−0.59146 327	−5.77625 460
0.12	−0.39037 66567	0.14190 445	1.15410 144	−0.70610 997	−5.68577 399
0.14	−0.38730 87267	0.16483 771	1.13884 890	−0.81878 968	−5.57961 395
0.16	−0.38378 53315	0.18744 353	1.12136 503	−0.92919 252	−5.45815 435
0.18	−0.37981 34631	0.20967 776	1.10169 839	−1.03701 674	−5.32182 895
0.20	−0.37540 09862	0.23149 727	1.07990 350	−1.14196 980	−5.17112 356
0.22	−0.37055 66169	0.25286 011	1.05604 063	−1.24376 938	−5.00657 387
0.24	−0.36528 98981	0.27372 555	1.03017 556	−1.34214 434	−4.82876 317
0.26	−0.35961 11734	0.29405 426	1.00237 941	−1.43683 568	−4.63831 979
0.28	−0.35353 15588	0.31380 836	0.97272 834	−1.52759 737	−4.43591 441
0.30	−0.34706 29121	0.33295 156	0.94130 327	−1.61419 723	−4.22225 716
0.32	−0.34021 78003	0.35144 923	0.90818 965	−1.69641 762	−3.99809 459
0.34	−0.33300 94659	0.36926 849	0.87347 711	−1.77405 617	−3.76420 646
0.36	−0.32545 17909	0.38637 828	0.83725 919	−1.84692 643	−3.52140 244
0.38	−0.31755 92592	0.40274 947	0.79963 298	−1.91485 840	−3.27051 871
0.40	−0.30934 69179	0.41835 488	0.76069 880	−1.97769 904	−3.01241 439
0.42	−0.30083 03372	0.43316 939	0.72055 987	−2.03531 269	−2.74796 802
0.44	−0.29202 55692	0.44716 995	0.67932 193	−2.08758 144	−2.47807 382
0.46	−0.28294 91055	0.46033 566	0.63709 291	−2.13440 537	−2.20363 810
0.48	−0.27361 78339	0.47264 779	0.59398 256	−2.17570 278	−1.92557 548
0.50	−0.26404 89951	0.48408 982	0.55010 207	−2.21141 033	−1.64480 520
0.52	−0.25426 01373	0.49464 748	0.50556 372	−2.24148 307	−1.36224 740
0.54	−0.24426 90722	0.50430 874	0.46048 050	−2.26589 443	−1.07881 949
0.56	−0.23409 38293	0.51306 383	0.41496 574	−2.28463 613	−0.79543 249
0.58	−0.22375 26107	0.52090 525	0.36913 279	−2.29771 801	−0.51298 749
0.60	−0.21326 37459	0.52782 777	0.32309 457	−2.30516 783	−0.23237 218
0.62	−0.20264 56463	0.53382 841	0.27696 332	−2.30703 091	+0.04554 255
0.64	−0.19191 67607	0.53890 643	0.23085 017	−2.30336 981	0.31990 583
0.66	−0.18109 55308	0.54306 327	0.18486 483	−2.29426 388	0.58988 999
0.68	−0.17020 03472	0.54630 259	0.13911 528	−2.27980 875	0.85469 355
0.70	−0.15924 95060	0.54863 016	0.09370 741	−2.26011 583	1.11354 405
0.72	−0.14826 11670	0.55005 386	0.04874 473	−2.23531 162	1.36570 074
0.74	−0.13725 33120	0.55058 359	+0.00432 808	−2.20553 714	1.61045 709
0.76	−0.12624 37042	0.55023 127	−0.03944 465	−2.17094 715	1.84714 311
0.78	−0.11524 98497	0.54901 073	−0.08247 882	−2.13170 944	2.07512 746
0.80	−0.10428 89590	0.54693 765	−0.12468 324	−2.08800 401	2.29381 943
0.82	−0.09337 79110	0.54402 952	−0.16597 047	−2.04002 228	2.50267 061
0.84	−0.08253 32179	0.54030 551	−0.20625 697	−1.98796 617	2.70117 643
0.86	−0.07177 09916	0.53578 644	−0.24546 336	−1.93204 726	2.88887 745
0.88	−0.06110 69120	0.53049 467	−0.28351 458	−1.87248 587	3.06536 044
0.90	−0.05055 61975	0.52445 403	−0.32034 003	−1.80951 008	3.23025 923
0.92	−0.04013 35759	0.51768 968	−0.35587 378	−1.74335 486	3.38325 538
0.94	−0.02985 32587	0.51022 310	−0.39005 463	−1.67426 103	3.52407 854
0.96	−0.01972 89163	0.50209 689	−0.42282 627	−1.60247 436	3.65250 673
0.98	−0.00977 36558	0.49332 478	−0.45413 732	−1.52824 456	3.76836 628
1.00	0.00000 00000	0.48394 145	−0.48394 145	−1.45182 435	3.87153 159
	$\begin{bmatrix} (-5)6 \\ 6 \end{bmatrix}$	$\begin{bmatrix} (-4)1 \\ 6 \end{bmatrix}$	$\begin{bmatrix} (-4)3 \\ 6 \end{bmatrix}$	$\begin{bmatrix} (-4)7 \\ 6 \end{bmatrix}$	$\begin{bmatrix} (-3)2 \\ 7 \end{bmatrix}$

$$P(-x) = 1 - P(x) \qquad Z(-x) = Z(x) \qquad Z^{(n)}(-x) = (-1)^n Z^{(n)}(x)$$

Table 26.1   **NORMAL PROBABILITY FUNCTION AND DERIVATIVES**

x	$P(x)$	$Z(x)$	$Z^{(1)}(x)$
1.00	0.84134 47460 68543	0.24197 07245 19143	−0.24197 07245 19143
1.02	0.84613 57696 27265	0.23713 19520 19380	−0.24187 45910 59767
1.04	0.85083 00496 69019	0.23229 70047 43366	−0.24158 88849 33101
1.06	0.85542 77003 36091	0.22746 96324 57386	−0.24111 78104 04829
1.08	0.85992 89099 11231	0.22265 34987 51761	−0.24046 57786 51902
1.10	0.86433 39390 53618	0.21785 21770 32551	−0.23963 73947 35806
1.12	0.86864 31189 57270	0.21306 91467 75718	−0.23863 74443 88804
1.14	0.87285 68494 37202	0.20830 77900 47108	−0.23747 08806 53704
1.16	0.87697 55969 48657	0.20357 13882 90759	−0.23614 28104 17281
1.18	0.88099 98925 44800	0.19886 31193 87276	−0.23465 84808 76986
1.20	0.88493 03297 78292	0.19418 60549 83213	−0.23302 32659 79856
1.22	0.88876 75625 52166	0.18954 31580 91640	−0.23124 26528 71801
1.24	0.89251 23029 25413	0.18493 72809 63305	−0.22932 22283 94499
1.26	0.89616 53188 78700	0.18037 11632 27080	−0.22726 76656 66121
1.28	0.89972 74320 45558	0.17584 74302 97662	−0.22508 47107 81008
1.30	0.90319 95154 14390	0.17136 85920 47807	−0.22277 91696 62150
1.32	0.90658 24910 06528	0.16693 70417 41714	−0.22035 68950 99062
1.34	0.90987 73275 35548	0.16255 50552 25534	−0.21782 37740 02216
1.36	0.91308 50380 52915	0.15822 47903 70383	−0.21518 57149 03721
1.38	0.91620 66775 84986	0.15394 82867 62634	−0.21244 86357 32434
1.40	0.91924 33407 66229	0.14972 74656 35745	−0.20961 84518 90043
1.42	0.92219 61594 73454	0.14556 41300 37348	−0.20670 10646 53034
1.44	0.92506 63004 65673	0.14145 99652 24839	−0.20370 23499 23768
1.46	0.92785 49630 34106	0.13741 65392 82282	−0.20062 81473 52131
1.48	0.93056 33766 66669	0.13343 53039 51002	−0.19748 42498 47483
1.50	0.93319 27987 31142	0.12951 75956 65892	−0.19427 63934 98838
1.52	0.93574 45121 81064	0.12566 46367 89088	−0.19101 02479 19414
1.54	0.93821 98232 88188	0.12187 75370 32402	−0.18769 14070 29899
1.56	0.94062 00594 05207	0.11815 72950 59582	−0.18432 53802 92948
1.58	0.94294 65667 62246	0.11450 48002 59292	−0.18091 75844 09682
1.60	0.94520 07083 00442	0.11092 08346 79456	−0.17747 33354 87129
1.62	0.94738 38615 45748	0.10740 60751 13484	−0.17399 78416 83844
1.64	0.94949 74165 25897	0.10396 10953 28764	−0.17049 61963 39173
1.66	0.95154 27737 33277	0.10058 63684 27691	−0.16697 33715 89966
1.68	0.95352 13421 36280	0.09728 22693 31467	−0.16343 42124 76865
1.70	0.95543 45372 41457	0.09404 90773 76887	−0.15988 34315 40708
1.72	0.95728 37792 08671	0.09088 69790 16283	−0.15632 56039 08007
1.74	0.95907 04910 21193	0.08779 60706 10906	−0.15276 51628 62976
1.76	0.96079 60967 12518	0.08477 63613 08022	−0.14920 63959 02119
1.78	0.96246 20196 51483	0.08182 77759 92143	−0.14565 34412 66014
1.80	0.96406 96808 87074	0.07895 01583 00894	−0.14211 02849 41609
1.82	0.96562 04975 54110	0.07614 32736 96207	−0.13858 07581 27097
1.84	0.96711 58813 40836	0.07340 68125 81657	−0.13506 85351 50249
1.86	0.96855 72370 19248	0.07074 03934 56983	−0.13157 71318 29989
1.88	0.96994 59610 38800	0.06814 35661 01045	−0.12810 99042 69964
1.90	0.97128 34401 83998	0.06561 58147 74677	−0.12467 00480 71886
1.92	0.97257 10502 96163	0.06315 65614 35199	−0.12126 05979 55581
1.94	0.97381 01550 59548	0.06076 51689 54565	−0.11788 44277 71856
1.96	0.97500 21048 51780	0.05844 09443 33451	−0.11454 42508 93565
1.98	0.97614 82356 58492	0.05618 31419 03868	−0.11124 26209 69659
2.00	0.97724 98680 51821	0.05399 09665 13188	−0.10798 19330 26376
	$\begin{bmatrix}(-5)1 \\ 10\end{bmatrix}$	$\begin{bmatrix}(-6)9 \\ 10\end{bmatrix}$	$\begin{bmatrix}(-5)2 \\ 10\end{bmatrix}$

$$Z(x) = \frac{1}{\sqrt{2\pi}} e^{-\frac{1}{2}x^2} \qquad P(x) = \int_{-\infty}^{x} Z(t)\,dt \qquad Z^{(n)}(x) = \frac{d^n}{dx^n} Z(x) \qquad He_n(x) = (-1)^n Z^{(n)}(x)/Z(x)$$

NORMAL PROBABILITY FUNCTION AND DERIVATIVES

Table 26.1

x	$Z^{(2)}(x)$	$Z^{(3)}(x)$	$Z^{(4)}(x)$	$Z^{(5)}(x)$	$Z^{(6)}(x)$
1.00	0.00000 00000	0.48394 145	−0.48394 145	−1.45182 435	3.87153 159
1.02	0.00958 01309	0.47397 745	−0.51219 739	−1.37346 846	3.96192 478
1.04	0.01895 54356	0.46346 412	−0.53886 899	−1.29343 272	4.03951 497
1.06	0.02811 52466	0.45243 346	−0.56392 521	−1.21197 312	4.10431 754
1.08	0.03704 95422	0.44091 805	−0.58734 012	−1.12934 487	4.15639 308
1.10	0.04574 89572	0.42895 094	−0.60909 290	−1.04580 155	4.19584 622
1.12	0.05420 47909	0.41656 552	−0.62916 776	−0.96159 420	4.22282 430
1.14	0.06240 90139	0.40379 549	−0.64755 390	−0.87697 050	4.23751 585
1.16	0.07035 42718	0.39067 467	−0.66424 543	−0.79217 397	4.24014 894
1.18	0.07803 38880	0.37723 697	−0.67924 129	−0.70744 317	4.23098 941
1.20	0.08544 18642	0.36351 629	−0.69254 515	−0.62301 100	4.21033 894
1.22	0.09257 28784	0.34954 639	−0.70416 524	−0.53910 399	4.17853 305
1.24	0.09942 22822	0.33536 083	−0.71411 427	−0.45594 161	4.13593 896
1.26	0.10598 60955	0.32099 205	−0.72240 928	−0.37373 571	4.08295 339
1.28	0.11226 09995	0.30647 534	−0.72907 143	−0.29268 993	4.02000 029
1.30	0.11824 43285	0.29184 071	−0.73412 591	−0.21299 916	3.94752 847
1.32	0.12393 40598	0.27712 083	−0.73760 168	−0.13484 911	3.86600 921
1.34	0.12932 88019	0.26234 695	−0.73953 132	−0.05841 584	3.77593 384
1.36	0.13442 77819	0.24754 965	−0.73995 087	+0.01613 459	3.67781 128
1.38	0.13923 08305	0.23275 873	−0.73889 953	0.08864 645	3.57216 556
1.40	0.14373 83670	0.21800 319	−0.73641 957	0.15897 463	3.45953 335
1.42	0.14795 13818	0.20331 117	−0.73255 600	0.22698 486	3.34046 152
1.44	0.15187 14187	0.18870 986	−0.72735 645	0.29255 386	3.21550 469
1.46	0.15550 05559	0.17422 548	−0.72087 087	0.35556 954	3.08522 283
1.48	0.15884 13858	0.15988 325	−0.71315 137	0.41593 103	2.95017 891
1.50	0.16189 69946	0.14570 730	−0.70425 193	0.47354 871	2.81093 657
1.52	0.16467 09400	0.13172 067	−0.69422 823	0.52834 425	2.66805 791
1.54	0.16716 72298	0.11794 528	−0.68313 742	0.58025 051	2.52210 132
1.56	0.16939 02982	0.10440 190	−0.67103 785	0.62921 147	2.37361 937
1.58	0.17134 49831	0.09111 010	−0.65798 890	0.67518 208	2.22315 681
1.60	0.17303 65021	0.07808 827	−0.64405 073	0.71812 810	2.07124 871
1.62	0.17447 04284	0.06535 359	−0.62928 410	0.75802 588	1.91841 857
1.64	0.17565 26667	0.05292 202	−0.61375 011	0.79486 211	1.76517 671
1.66	0.17658 94284	0.04080 829	−0.59751 005	0.82863 352	1.61201 862
1.68	0.17728 72076	0.02902 592	−0.58062 516	0.85934 661	1.45942 351
1.70	0.17775 27562	0.01758 718	−0.56315 647	0.88701 729	1.30785 296
1.72	0.17799 30597	+0.00650 315	−0.54516 459	0.91167 051	1.15774 966
1.74	0.17801 53128	−0.00421 632	−0.52670 954	0.93333 988	1.00953 633
1.76	0.17782 68955	−0.01456 254	−0.50785 061	0.95206 725	0.86361 469
1.78	0.17743 53495	−0.02452 804	−0.48864 614	0.96790 228	0.72036 463
1.80	0.17684 83546	−0.03410 647	−0.46915 342	0.98090 203	0.58014 345
1.82	0.17607 37061	−0.04329 263	−0.44942 853	0.99113 045	0.44328 526
1.84	0.17511 92921	−0.05208 243	−0.42952 621	0.99865 794	0.31010 045
1.86	0.17399 30717	−0.06047 285	−0.40949 971	1.00356 087	0.18087 536
1.88	0.17270 30539	−0.06846 193	−0.38940 073	1.00592 110	+0.05587 197
1.90	0.17125 72766	−0.07604 873	−0.36927 924	1.00582 548	−0.06467 219
1.92	0.16966 37866	−0.08323 327	−0.34918 347	1.00336 537	−0.18054 414
1.94	0.16793 06209	−0.09001 655	−0.32915 976	0.99863 613	−0.29155 530
1.96	0.16606 57874	−0.09640 044	−0.30925 250	0.99173 666	−0.39754 137
1.98	0.16407 72476	−0.10238 771	−0.28950 408	0.98276 891	−0.49836 204
2.00	0.16197 28995	−0.10798 193	−0.26995 483	0.97183 740	−0.59390 063
	$\begin{bmatrix}(-5)4\\6\end{bmatrix}$	$\begin{bmatrix}(-5)7\\6\end{bmatrix}$	$\begin{bmatrix}(-4)2\\6\end{bmatrix}$	$\begin{bmatrix}(-4)4\\6\end{bmatrix}$	$\begin{bmatrix}(-3)1\\7\end{bmatrix}$

$$P(-x) = 1 - P(x) \qquad Z(-x) = Z(x) \qquad Z^{(n)}(-x) = (-1)^n Z^{(n)}(x)$$

Table 26.1   NORMAL PROBABILITY FUNCTION AND DERIVATIVES

x	$P(x)$	$Z(x)$	$Z^{(1)}(x)$
2.00	0.97724 98680 51821	0.05399 09665 13188	−0.10798 19330 26376
2.02	0.97830 83062 32353	0.05186 35766 82821	−0.10476 44248 99298
2.04	0.97932 48371 33930	0.04980 00877 35071	−0.10159 21789 79544
2.06	0.98030 07295 90623	0.04779 95748 82077	−0.09846 71242 57079
2.08	0.98123 72335 65062	0.04586 10762 71055	−0.09539 10386 43794
2.10	0.98213 55794 37184	0.04398 35959 80427	−0.09236 55515 58897
2.12	0.98299 69773 52367	0.04216 61069 61770	−0.08939 21467 58953
2.14	0.98382 26166 27834	0.04040 75539 22860	−0.08647 21653 94921
2.16	0.98461 36652 16075	0.03870 68561 47456	−0.08360 68092 78504
2.18	0.98537 12692 24011	0.03706 29102 47806	−0.08079 71443 40218
2.20	0.98609 65524 86502	0.03547 45928 46231	−0.07804 41042 61709
2.22	0.98679 06161 92744	0.03394 07631 82449	−0.07534 84942 65037
2.24	0.98745 45385 64054	0.03246 02656 43697	−0.07271 09950 41882
2.26	0.98808 93745 81453	0.03103 19322 15008	−0.07013 21668 05919
2.28	0.98869 61557 61447	0.02965 45848 47341	−0.06761 24534 51938
2.30	0.98927 58899 78324	0.02832 70377 41601	−0.06515 21868 05683
2.32	0.98982 95613 31281	0.02704 80995 46882	−0.06275 15909 48766
2.34	0.99035 81300 54642	0.02581 65754 71588	−0.06041 07866 03515
2.36	0.99086 25324 69428	0.02463 12693 06382	−0.05812 97955 63063
2.38	0.99134 36809 74484	0.02349 09853 58201	−0.05590 85451 52519
2.40	0.99180 24640 75404	0.02239 45302 94843	−0.05374 68727 07623
2.42	0.99223 97464 49447	0.02134 07148 99923	−0.05164 45300 57813
2.44	0.99265 63690 44652	0.02032 83557 38226	−0.04960 11880 01271
2.46	0.99305 31492 11376	0.01935 62767 31737	−0.04761 64407 60073
2.48	0.99343 08808 64453	0.01842 33106 46862	−0.04568 98104 04218
2.50	0.99379 03346 74224	0.01752 83004 93569	−0.04382 07512 33921
2.52	0.99413 22582 84668	0.01667 01008 37381	−0.04200 86541 10200
2.54	0.99445 73765 56918	0.01584 75790 25361	−0.04025 28507 24416
2.56	0.99476 63918 36444	0.01505 96163 27377	−0.03855 26177 98086
2.58	0.99505 99842 42230	0.01430 51089 94150	−0.03690 71812 04906
2.60	0.99533 88119 76281	0.01358 29692 33686	−0.03531 57200 07583
2.62	0.99560 35116 51879	0.01289 21261 07895	−0.03377 73704 02686
2.64	0.99585 46986 38964	0.01223 15263 51278	−0.03229 12295 67374
2.66	0.99609 29674 25147	0.01160 01351 13703	−0.03085 63594 02449
2.68	0.99631 88919 90825	0.01099 69366 29406	−0.02947 17901 66807
2.70	0.99653 30261 96960	0.01042 09348 14423	−0.02813 65239 98941
2.72	0.99673 59041 84109	0.00987 11537 94751	−0.02684 95383 21723
2.74	0.99692 80407 81350	0.00934 66383 67612	−0.02560 97891 27258
2.76	0.99710 99319 23774	0.00884 64543 98237	−0.02441 62141 39135
2.78	0.99728 20550 77299	0.00836 96891 54653	−0.02326 77358 49935
2.80	0.99744 48696 69572	0.00791 54515 82980	−0.02216 32644 32344
2.82	0.99759 88175 25811	0.00748 28725 25781	−0.02110 17005 22701
2.84	0.99774 43233 08458	0.00707 11048 86019	−0.02008 19378 76295
2.86	0.99788 17949 59596	0.00667 93237 39203	−0.01910 28658 94119
2.88	0.99801 16241 45106	0.00630 67263 96266	−0.01816 33720 21246
2.90	0.99813 41866 99616	0.00595 25324 19776	−0.01726 23440 17350
2.92	0.99824 98430 71324	0.00561 59835 95991	−0.01639 86721 00294
2.94	0.99835 89387 65843	0.00529 63438 65311	−0.01557 12509 64014
2.96	0.99846 18047 88262	0.00499 28992 13612	−0.01477 89816 72293
2.98	0.99855 87580 82660	0.00470 49575 26934	−0.01402 07734 30263
3.00	0.99865 01019 68370	0.00443 18484 11938	−0.01329 55452 35814
	$\begin{bmatrix}(-6)5\\10\end{bmatrix}$	$\begin{bmatrix}(-6)8\\10\end{bmatrix}$	$\begin{bmatrix}(-6)7\\10\end{bmatrix}$

$$Z(x)=\frac{1}{\sqrt{2\pi}}\,e^{-\frac{1}{2}x^2} \qquad P(x)=\int_{-\infty}^{x}Z(t)dt \qquad Z^{(n)}(x)=\frac{d^n}{dx^n}Z(x) \qquad He_n(x)=(-1)^nZ^{(n)}(x)/Z(x)$$

NORMAL PROBABILITY FUNCTION AND DERIVATIVES

Table 26.1

x	$Z^{(2)}(x)$	$Z^{(3)}(x)$	$Z^{(4)}(x)$	$Z^{(5)}(x)$	$Z^{(6)}(x)$
2.00	0.16197 28995	-0.10798 193	-0.26995 483	0.97183 740	-0.59390 063
2.02	0.15976 05616	-0.11318 748	-0.25064 297	0.95904 873	-0.68406 360
2.04	0.15744 79574	-0.11800 948	-0.23160 454	0.94451 117	-0.76878 007
2.06	0.15504 27011	-0.12245 372	-0.21287 345	0.92833 417	-0.84800 114
2.08	0.15255 22841	-0.12652 667	-0.19448 137	0.91062 795	-0.92169 927
2.10	0.14998 40623	-0.13023 543	-0.17645 779	0.89150 307	-0.98986 750
2.12	0.14734 52442	-0.13358 762	-0.15882 997	0.87107 003	-1.05251 862
2.14	0.14464 28800	-0.13659 143	-0.14162 297	0.84943 890	-1.10968 436
2.16	0.14188 38519	-0.13925 550	-0.12485 967	0.82671 890	-1.16141 446
2.18	0.13907 48644	-0.14158 892	-0.10856 076	0.80301 811	-1.20777 570
2.20	0.13622 24365	-0.14360 115	-0.09274 478	0.77844 311	-1.24885 097
2.22	0.13333 28941	-0.14530 204	-0.07742 816	0.75309 866	-1.28473 023
2.24	0.13041 23633	-0.14670 170	-0.06262 527	0.72708 743	-1.31554 947
2.26	0.12746 67648	-0.14781 055	-0.04834 844	0.70050 969	-1.34140 971
2.28	0.12450 18090	-0.14863 922	-0.03460 801	0.67346 314	-1.36245 589
2.30	0.12152 29919	-0.14919 851	-0.02141 241	0.64604 257	-1.37883 587
2.32	0.11853 55915	-0.14949 939	-0.00876 819	0.61833 976	-1.39070 730
2.34	0.11554 46652	-0.14955 294	+0.00331 989	0.59044 323	-1.39823 661
2.36	0.11255 50482	-0.14937 032	0.01484 882	0.56243 808	-1.40159 796
2.38	0.10957 13521	-0.14896 273	0.02581 724	0.53440 589	-1.40097 220
2.40	0.10659 79642	-0.14834 137	0.03622 539	0.50642 453	-1.39654 584
2.42	0.10363 90478	-0.14751 744	0.04607 505	0.47856 812	-1.38851 010
2.44	0.10069 85430	-0.14650 207	0.05536 942	0.45090 689	-1.37705 991
2.46	0.09778 01675	-0.14530 633	0.06411 307	0.42350 717	-1.36239 299
2.48	0.09488 74192	-0.14394 118	0.07231 187	0.39643 129	-1.34470 892
2.50	0.09202 35776	-0.14241 744	0.07997 287	0.36973 759	-1.32420 833
2.52	0.08919 17075	-0.14074 579	0.08710 428	0.34348 039	-1.30109 199
2.54	0.08639 46618	-0.13893 674	0.09371 533	0.31771 001	-1.27556 010
2.56	0.08363 50852	-0.13700 058	0.09981 624	0.29247 277	-1.24781 146
2.58	0.08091 54185	-0.13494 742	0.10541 808	0.26781 102	-1.21804 284
2.60	0.07823 79028	-0.13278 711	0.11053 277	0.24376 323	-1.18644 824
2.62	0.07560 45843	-0.13052 927	0.11517 293	0.22036 399	-1.15321 833
2.64	0.07301 73197	-0.12818 326	0.11935 186	0.19764 415	-1.11853 985
2.66	0.07047 77809	-0.12575 818	0.12308 341	0.17563 084	-1.08259 509
2.68	0.06798 74610	-0.12326 282	0.12638 196	0.15434 760	-1.04556 139
2.70	0.06554 76800	-0.12070 569	0.12926 232	0.13381 449	-1.00761 072
2.72	0.06315 95904	-0.11809 501	0.13173 965	0.11404 817	-0.96890 932
2.74	0.06082 41838	-0.11543 869	0.13382 945	0.09506 206	-0.92961 727
2.76	0.05854 22966	-0.11274 431	0.13554 741	0.07686 640	-0.88988 829
2.78	0.05631 46165	-0.11001 916	0.13690 942	0.05946 846	-0.84986 942
2.80	0.05414 16888	-0.10727 020	0.13793 149	0.04287 262	-0.80970 080
2.82	0.05202 39229	-0.10450 406	0.13862 969	0.02708 053	-0.76951 553
2.84	0.04996 15987	-0.10172 706	0.13902 007	+0.01209 127	-0.72943 954
2.86	0.04795 48727	-0.09894 520	0.13911 867	-0.00209 857	-0.68959 143
2.88	0.04600 37850	-0.09616 416	0.13894 142	-0.01549 465	-0.65008 248
2.90	0.04410 82652	-0.09338 928	0.13850 412	-0.02810 482	-0.61101 661
2.92	0.04226 81389	-0.09062 562	0.13782 240	-0.03993 892	-0.57249 036
2.94	0.04048 31340	-0.08787 791	0.13691 166	-0.05100 863	-0.53459 292
2.96	0.03875 28865	-0.08515 058	0.13578 706	-0.06132 737	-0.49740 627
2.98	0.03707 69473	-0.08244 776	0.13446 347	-0.07091 012	-0.46100 520
3.00	0.03545 47873	-0.07977 327	0.13295 545	-0.07977 327	-0.42545 745
	$\begin{bmatrix}(-5)1\\6\end{bmatrix}$	$\begin{bmatrix}(-5)5\\6\end{bmatrix}$	$\begin{bmatrix}(-5)7\\6\end{bmatrix}$	$\begin{bmatrix}(-4)2\\6\end{bmatrix}$	$\begin{bmatrix}(-4)7\\6\end{bmatrix}$

$$P(-x) = 1 - P(x) \qquad Z(-x) = Z(x) \qquad Z^{(n)}(-x) = (-1)^n Z^{(n)}(x)$$

Table 26.1 NORMAL PROBABILITY FUNCTION AND DERIVATIVES

x	$P(x)$	$Z(x)$	$Z^{(1)}(x)$
3.00	0.99865 01020	(−3) 4.43184 8412	(−2) −1.32955 45
3.05	0.99885 57932	(−3) 3.80976 2098	(−2) −1.16197 74
3.10	0.99903 23968	(−3) 3.26681 9056	(−2) −1.01271 39
3.15	0.99918 36477	(−3) 2.79425 8415	(−3) −8.80191 40
3.20	0.99931 28621	(−3) 2.38408 8201	(−3) −7.62908 22
3.25	0.99942 29750	(−3) 2.02904 8057	(−3) −6.59440 62
3.30	0.99951 65759	(−3) 1.72256 8939	(−3) −5.68447 75
3.35	0.99959 59422	(−3) 1.45873 0805	(−3) −4.88674 82
3.40	0.99966 30707	(−3) 1.23221 9168	(−3) −4.18954 52
3.45	0.99971 97067	(−3) 1.03828 1296	(−3) −3.58207 05
3.50	0.99976 73709	(−4) 8.72682 6950	(−3) −3.05438 94
3.55	0.99980 73844	(−4) 7.31664 4628	(−3) −2.59740 88
3.60	0.99984 08914	(−4) 6.11901 9301	(−3) −2.20284 69
3.65	0.99986 88798	(−4) 5.10464 9743	(−3) −1.86319 72
3.70	0.99989 22003	(−4) 4.24780 2706	(−3) −1.57168 70
3.75	0.99991 15827	(−4) 3.52595 6824	(−3) −1.32223 38
3.80	0.99992 76520	(−4) 2.91946 9258	(−3) −1.10939 83
3.85	0.99994 09411	(−4) 2.41126 5802	(−4) −9.28337 33
3.90	0.99995 19037	(−4) 1.98655 4714	(−4) −7.74756 34
3.95	0.99996 09244	(−4) 1.63256 4088	(−4) −6.44862 81
4.00	0.99996 83288	(−4) 1.33830 2258	(−4) −5.35320 90
4.05	0.99997 43912	(−4) 1.09434 0434	(−4) −4.43207 88
4.10	0.99997 93425	(−5) 8.92616 5718	(−4) −3.65972 79
4.15	0.99998 33762	(−5) 7.26259 3030	(−4) −3.01397 61
4.20	0.99998 66543	(−5) 5.89430 6776	(−4) −2.47560 88
4.25	0.99998 93115	(−5) 4.77186 3654	(−4) −2.02804 21
4.30	0.99999 14601	(−5) 3.85351 9674	(−4) −1.65701 35
4.35	0.99999 31931	(−5) 3.10414 0706	(−4) −1.35030 12
4.40	0.99999 45875	(−5) 2.49424 7129	(−4) −1.09746 87
4.45	0.99999 57065	(−5) 1.99917 9671	(−5) −8.89634 95
4.50	0.99999 66023	(−5) 1.59837 4111	(−5) −7.19268 35
4.55	0.99999 73177	(−5) 1.27473 3238	(−5) −5.80003 62
4.60	0.99999 78875	(−5) 1.01408 5207	(−5) −4.66479 20
4.65	0.99999 83403	(−6) 8.04718 2456	(−5) −3.74193 98
4.70	0.99999 86992	(−6) 6.36982 5179	(−5) −2.99381 78
4.75	0.99999 89829	(−6) 5.02950 7289	(−5) −2.38901 60
4.80	0.99999 92067	(−6) 3.96129 9091	(−5) −1.90142 36
4.85	0.99999 93827	(−6) 3.11217 5579	(−5) −1.50940 52
4.90	0.99999 95208	(−6) 2.43896 0746	(−5) −1.19509 08
4.95	0.99999 96289	(−6) 1.90660 0903	(−6) −9.43767 45
5.00	0.99999 97133	(−6) 1.48671 9515	(−6) −7.43359 76
	$\begin{bmatrix} (-6)3 \\ 7 \end{bmatrix}$		

Table 26.2 NORMAL PROBABILITY FUNCTION FOR LARGE ARGUMENTS

x	$-\log Q(x)$	x	$-\log Q(x)$	x	$-\log Q(x)$
5	6.54265	15	50.43522	25	137.51475
6	9.00586	16	57.19458	26	148.60624
7	11.89285	17	64.38658	27	160.13139
8	15.20614	18	72.01140	28	172.09024
9	18.94746	19	80.06919	29	184.48283
10	23.11805	20	88.56010	30	197.30921
11	27.71882	21	97.48422	31	210.56940
12	32.75044	22	106.84167	32	224.26344
13	38.21345	23	116.63253	33	238.39135
14	44.10827	24	126.85686	34	252.95315
	$\begin{bmatrix} (-2)5 \\ 5 \end{bmatrix}$		$\begin{bmatrix} (-2)5 \\ 4 \end{bmatrix}$		$\begin{bmatrix} (-2)5 \\ 3 \end{bmatrix}$

From E. S. Pearson and H. O. Hartley (editors), Biometrika tables for statisticians, vol. I. Cambridge Univ. Press, Cambridge, England, 1954 (with permission). Known error has been corrected.

NORMAL PROBABILITY FUNCTION AND DERIVATIVES Table 26.1

x	$Z^{(2)}(x)$	$Z^{(3)}(x)$	$Z^{(4)}(x)$	$Z^{(5)}(x)$	$Z^{(6)}(x)$
3.00	(−2) 3.54547 87	(−2) −7.97732 71	(−1) 1.32955 45	(−2) −7.97732 71	(−1) −4.25457 45
3.05	(−2) 3.16305 50	(−2) −7.32336 28	(−1) 1.28470 92	(−2) −9.89017 82	(−1) −3.40704 15
3.10	(−2) 2.81273 12	(−2) −6.69403 89	(−1) 1.23133 27	(−1) −1.13951 58	(−1) −2.62416 45
3.15	(−2) 2.49317 71	(−2) −6.09312 50	(−1) 1.17138 12	(−1) −1.25260 09	(−1) −1.91121 33
3.20	(−2) 2.20289 75	(−2) −5.52345 55	(−1) 1.10663 65	(−1) −1.33185 47	(−1) −1.27124 77
3.25	(−2) 1.94027 72	(−2) −4.98701 97	(−1) 1.03869 82	(−1) −1.38096 14	(−2) −7.05366 66
3.30	(−2) 1.70362 07	(−2) −4.48505 27	(−2) 9.68981 20	(−1) −1.40361 69	(−2) −2.12970 34
3.35	(−2) 1.49118 76	(−2) −4.01812 87	(−2) 8.98716 85	(−1) −1.40345 00	(−2) +2.07973 11
3.40	(−2) 1.30122 34	(−2) −3.58625 07	(−2) 8.28958 19	(−1) −1.38395 76	(−2) 5.60664 85
3.45	(−2) 1.13198 62	(−2) −3.18893 82	(−2) 7.60587 84	(−1) −1.34845 27	(−2) 8.49222 78
3.50	(−3) 9.81768 03	(−2) −2.82531 02	(−2) 6.94328 17	(−1) −1.30002 45	(−1) 1.07844 49
3.55	(−3) 8.48913 69	(−2) −2.49416 18	(−2) 6.30753 35	(−1) −1.24150 96	(−1) 1.25359 25
3.60	(−3) 7.31834 71	(−2) −2.19403 56	(−2) 5.70302 39	(−1) −1.17547 44	(−1) 1.38019 58
3.65	(−3) 6.29020 46	(−2) −1.92328 53	(−2) 5.13292 98	(−1) −1.10420 53	(−1) 1.46388 44
3.70	(−3) 5.39046 16	(−2) −1.68013 34	(−2) 4.59935 51	(−1) −1.02970 80	(−1) 1.51024 21
3.75	(−3) 4.60578 11	(−2) −1.46272 12	(−2) 4.10347 00	(−2) −9.53712 78	(−1) 1.52468 79
3.80	(−3) 3.92376 67	(−2) −1.26915 17	(−2) 3.64564 64	(−2) −8.77684 95	(−1) 1.51237 96
3.85	(−3) 3.33297 22	(−2) −1.09752 68	(−2) 3.22558 66	(−2) −8.02840 11	(−1) 1.47814 11
3.90	(−3) 2.82289 42	(−3) −9.45977 49	(−2) 2.84244 39	(−2) −7.30162 14	(−1) 1.42641 04
3.95	(−3) 2.38395 17	(−3) −8.12688 36	(−2) 2.49493 35	(−2) −6.60423 39	(−1) 1.36120 56
4.00	(−3) 2.00745 34	(−3) −6.95917 17	(−2) 2.18143 27	(−2) −5.94206 20	(−1) 1.28610 85
4.05	(−3) 1.68555 79	(−3) −5.94009 36	(−2) 1.90007 05	(−2) −5.31924 82	(−1) 1.20426 03
4.10	(−3) 1.41122 68	(−3) −5.05408 43	(−2) 1.64880 65	(−2) −4.73847 30	(−1) 1.11837 07
4.15	(−3) 1.17817 42	(−3) −4.28662 75	(−2) 1.42549 82	(−2) −4.20116 64	(−1) 1.03073 50
4.20	(−4) 9.80812 65	(−3) −3.62429 14	(−2) 1.22795 86	(−2) −3.70770 95	(−2) 9.43258 69
4.25	(−4) 8.14199 24	(−3) −3.05473 83	(−2) 1.05400 40	(−2) −3.25762 18	(−2) 8.57487 24
4.30	(−4) 6.73980 59	(−3) −2.56671 38	(−3) 9.01492 78	(−2) −2.84973 34	(−2) 7.74638 98
4.35	(−4) 5.56339 62	(−3) −2.15001 71	(−3) 7.68355 55	(−2) −2.48233 98	(−2) 6.95640 04
4.40	(−4) 4.57943 77	(−3) −1.79545 89	(−3) 6.52618 76	(−2) −2.15333 90	(−2) 6.21159 79
4.45	(−4) 3.75895 76	(−3) −1.49480 91	(−3) 5.52421 34	(−2) −1.86035 13	(−2) 5.51645 66
4.50	(−4) 3.07687 02	(−3) −1.24073 79	(−3) 4.66025 95	(−2) −1.60082 16	(−2) 4.87356 75
4.55	(−4) 2.51154 32	(−3) −1.02675 14	(−3) 3.91825 60	(−2) −1.37210 59	(−2) 4.28395 39
4.60	(−4) 2.04439 58	(−4) −8.47126 22	(−3) 3.28346 19	(−2) −1.17154 20	(−2) 3.74736 21
4.65	(−4) 1.65953 02	(−4) −6.96842 75	(−3) 2.74245 97	(−3) −9.96506 67	(−2) 3.26252 61
4.70	(−4) 1.34339 61	(−4) −5.71519 82	(−3) 2.28312 43	(−3) −8.44460 51	(−2) 2.82740 22
4.75	(−4) 1.08448 75	(−4) −4.67351 25	(−3) 1.89457 22	(−3) −7.12981 28	(−2) 2.43937 50
4.80	(−5) 8.73070 32	(−4) −3.81045 28	(−3) 1.56709 63	(−3) −5.99788 09	(−2) 2.09543 47
4.85	(−5) 7.00939 74	(−4) −3.09767 67	(−3) 1.29209 13	(−3) −5.02757 21	(−2) 1.79232 68
4.90	(−5) 5.61204 87	(−4) −2.51088 57	(−3) 1.06197 25	(−3) −4.19931 11	(−2) 1.52667 62
4.95	(−5) 4.48098 88	(−4) −2.02933 60	(−4) 8.70091 63	(−3) −3.49521 92	(−2) 1.29508 77
5.00	(−5) 3.56812 68	(−4) −1.63539 15	(−4) 7.10651 93	(−3) −2.89910 31	(−2) 1.09422 56

NORMAL PROBABILITY FUNCTION FOR LARGE ARGUMENTS Table 26.2

x	$-\log Q(x)$	x	$-\log Q(x)$	x	$-\log Q(x)$
35	267.94888	45	441.77568	100	2173.87154
36	283.37855	46	461.54561	150	4888.38812
37	299.24218	47	481.74964	200	8688.58977
38	315.53979	48	502.38776	250	13574.49960
39	332.27139	49	523.45999	300	19546.12790
40	349.43701	50	544.96634	350	26603.48018
41	367.03664	60	783.90743	400	34746.55970
42	385.07032	70	1066.26576	450	43975.36860
43	403.53804	80	1392.04459	500	54289.90830
44	422.43983	90	1761.24604		
	$\begin{bmatrix}(-2)5\\3\end{bmatrix}$		$\begin{bmatrix}(0)5\\5\end{bmatrix}$		$\begin{bmatrix}(+2)1\\9\end{bmatrix}$

$$Q(x)=1-P(x)=\frac{1}{\sqrt{2\pi}}\int_x^\infty e^{-\frac{1}{2}t^2}dt \quad Z(x)=\frac{1}{\sqrt{2\pi}}e^{-\frac{1}{2}x^2} \quad P(x)=\int_{-\infty}^x Z(t)dt \quad Z^{(n)}(x)=\frac{d^n}{dx^n}Z(x)$$

$$He_n(x)=(-1)^n Z^{(n)}(x)/Z(x) \qquad P(-x)=1-P(x) \qquad Z(-x)=Z(x) \qquad Z^{(n)}(-x)=(-1)^n Z^{(n)}(x)$$

Table 26.3   **HIGHER DERIVATIVES OF THE NORMAL PROBABILITY FUNCTION**

x	$Z^{(7)}(x)$	$Z^{(8)}(x)$	$Z^{(9)}(x)$	$Z^{(10)}(x)$	$Z^{(11)}(x)$	$Z^{(12)}(x)$
0.0	0.00000 00	(1) 4.18889 39	0.00000 00	(2)−3.77000 46	0.00000 00	(3) 4.14700 50
0.1	(0) 4.12640 51	(1) 4.00211 42	(1)−3.70133 55	(2)−3.56488 94	(2) 4.05782 44	(3) 3.88080 01
0.2	(0) 7.88604 35	(1) 3.46206 56	(1)−7.00124 79	(2)−2.97583 41	(2) 7.59641 48	(3) 3.12148 92
0.3	(1) 1.09518 61	(1) 2.62702 42	(1)−9.54959 57	(2)−2.07783 39	(3) 1.01729 46	(3) 1.98042 89
0.4	(1) 1.30711 60	(1) 1.58584 37	(2)−1.10912 65	(1)−9.83608 69	(3) 1.14847 09	(2)+6.22581 20
0.5	(1) 1.40908 65	(0)+4.46820 41	(2)−1.14961 02	(1)+1.72666 73	(3) 1.14097 69	(2)−7.60421 83
0.6	(1) 1.39704 30	(0)−6.75565 29	(2)−1.07710 05	(2) 1.25426 91	(3) 1.00184 44	(3)−1.98080 26
0.7	(1) 1.27812 14	(1)−1.67416 58	(1)−9.05305 52	(2) 2.14046 31	(2) 7.55473 11	(3)−2.88334 06
0.8	(1) 1.06929 69	(1)−2.46111 11	(1)−6.58548 60	(2) 2.74183 89	(2) 4.39201 49	(3)−3.36738 39
0.9	(0) 7.94982 72	(1)−2.97666 59	(1)−3.68086 24	(2) 3.01027 69	(1)+9.71613 18	(3)−3.39874 98
1.0	(0) 4.83941 45	(1)−3.19401 36	(0)−6.77518 03	(2) 2.94236 40	(2)−2.26484 60	(3)−3.01011 58
1.1	(0)+1.65937 85	(1)−3.11962 40	(1)+2.10408 36	(2) 2.57621 24	(2)−4.93791 72	(3)−2.29066 27
1.2	(0)−1.31434 07	(1)−2.78951 64	(1) 4.39889 22	(2) 1.98269 77	(2)−6.77812 94	(3)−1.36759 19
1.3	(0)−3.85379 20	(1)−2.26227 70	(1) 6.02399 37	(2) 1.25293 01	(2)−7.65280 28	(3)−3.83358 74
1.4	(0)−5.79719 45	(1)−1.61006 61	(1) 6.89184 82	(1)+4.84200 76	(2)−7.56972 92	(2)+5.27141 25
1.5	(0)−7.05769 71	(0)−9.09001 03	(1) 7.00965 92	(1)−2.33347 96	(2)−6.65963 73	(3) 1.25562 83
1.6	(0)−7.62276 66	(0)−2.30231 44	(1) 6.46658 36	(1)−8.27445 07	(2)−5.14267 14	(3) 1.73301 70
1.7	(0)−7.54545 38	(0)+3.67230 07	(1) 5.41207 19	(2)−1.25055 93	(2)−3.28612 11	(3) 1.93425 58
1.8	(0)−6.92967 04	(0) 8.41240 26	(1) 4.02950 39	(2)−1.48242 69	(2)−1.36113 54	(3) 1.87567 40
1.9	(0)−5.91207 57	(1) 1.16856 49	(1) 2.50938 72	(2)−1.52849 20	(1)+3.94747 58	(3) 1.60633 92
2.0	(0)−4.64322 31	(1) 1.34437 51	(1)+1.02582 84	(2)−1.41510 32	(2) 1.80437 81	(3) 1.19573 79
2.1	(0)−3.27029 67	(1) 1.37966 95	(0)−2.81068 72	(2)−1.18267 82	(2) 2.76469 29	(2) 7.20360 48
2.2	(0)−1.92318 65	(1) 1.29729 67	(1)−1.31550 35	(1)−8.78156 27	(2) 3.24744 73	(2)+2.51533 48
2.3	(−1)−7.04932 91	(1) 1.12731 97	(1)−2.02888 89	(1)−5.47943 26	(2) 3.28915 84	(2)−1.53768 85
2.4	(−1)+3.13162 82	(0) 9.02423 01	(1)−2.41634 55	(1)−2.32257 79	(2) 2.97376 42	(2)−4.58219 83
2.5	(0) 1.09209 53	(0) 6.53922 01	(1)−2.50848 12	(0)+3.85905 05	(2) 2.41200 50	(2)−6.45450 80
2.6	(0) 1.62218 61	(0) 4.08745 39	(1)−2.36048 69	(1) 2.45855 73	(2) 1.72126 20	(2)−7.17969 42
2.7	(0) 1.91766 20	(0) 1.87558 77	(1)−2.04053 83	(1) 3.82142 44	(2) 1.00875 37	(2)−6.92720 18
2.8	(0) 2.00992 65	(−2)+4.01113 24	(1)−1.61917 24	(1) 4.49758 25	(1)+3.59849 29	(2)−5.95491 88
2.9	(0) 1.94057 71	(0)−1.35055 73	(1)−1.16080 01	(1) 4.58182 18	(1)−1.67928 25	(2)−4.55301 20
3.0	(0) 1.75501 20	(0)−2.28683 38	(0)−7.17959 44	(1) 4.21202 87	(1)−5.45649 18	(2)−2.99628 41
3.1	(0) 1.49720 05	(0)−2.80440 64	(0)−3.28394 42	(1) 3.54198 84	(1)−7.69621 99	(2)−1.51035 91
3.2	(0) 1.20591 21	(0)−2.96904 52	(−1)−1.46351 84	(1) 2.71897 33	(1)−8.55436 26	(1)−2.53474 56
3.3	(−1) 9.12450 33	(0)−2.86200 69	(0)+2.14502 00	(1) 1.86794 96	(1)−8.30925 36	(1)+6.87309 15
3.4	(−1) 6.39748 51	(0)−2.56761 03	(0) 3.61188 70	(1) 1.08280 77	(1)−7.29343 32	(2) 1.28867 88
3.5	(−1) 4.02558 98	(0)−2.16386 79	(0) 4.35306 57	(0)+4.23908 09	(1)−5.83674 40	(2) 1.57656 15
3.6	(−1) 2.08414 13	(0)−1.71642 80	(0) 4.51182 76	(−1)−7.94727 62	(1)−4.22572 56	(2) 1.60868 13
3.7	(−2)+5.90352 21	(0)−1.27559 98	(0) 4.24743 76	(0)−4.23512 06	(1)−2.68044 29	(2) 1.45762 72
3.8	(−2)−4.80932 87	(−1)−8.75911 24	(0) 3.71320 90	(0)−6.22699 31	(1)−1.34695 16	(2) 1.19681 09
3.9	(−1)−1.18202 76	(−1)−5.37496 49	(0) 3.04185 84	(0)−7.02577 94	(0)−3.01804 44	(1) 8.90539 46
4.0	(−1)−1.57919 67	(−1)−2.68597 26	(0) 2.33774 64	(0)−6.93361 02	(0)+4.35697 68	(1) 5.88418 05
4.1	(−1)−1.74223 60	(−2)−6.85427 28	(0) 1.67481 40	(0)−6.24985 27	(0) 8.87625 64	(1) 3.23557 28
4.2	(−1)−1.73706 08	(−2)+6.92844 60	(0) 1.09865 39	(0)−5.23790 66	(1) 1.10126 69	(1)+1.13637 65
4.3	(−1)−1.62110 76	(−1) 1.54828 96	(−1) 6.31121 50	(0)−4.10728 31	(1) 1.13501 02	(0)−3.62532 62
4.4	(−1)−1.44109 96	(−1) 1.99272 00	(−1) 2.76082 94	(0)−3.00821 29	(1) 1.04753 07	(1)−1.30010 10
4.5	(−1)−1.23261 24	(−1) 2.13525 86	(−2)+2.52235 61	(0)−2.03523 88	(0) 8.90633 89	(1)−1.76908 98
4.6	(−1)−1.02086 14	(−1) 2.07280 89	(−1)−1.36802 99	(0)−1.23623 43	(0) 7.05470 76	(1)−1.88530 78
4.7	(−2)−8.22202 74	(−1) 1.88517 13	(−1)−2.28268 33	(−1)−6.23793 04	(0) 5.21451 06	(1)−1.76464 76
4.8	(−2)−6.45935 81	(−1) 1.63368 76	(−1)−2.67421 39	(−1)−1.86696 14	(0) 3.57035 54	(1)−1.50840 48
4.9	(−2)−4.96112 66	(−1) 1.36227 87	(−1)−2.70626 44	(−1)+1.00018 72	(0) 2.21617 27	(1)−1.19594 52
5.0	(−2)−3.73166 60	(−1) 1.09987 51	(−1)−2.51404 27	(−1) 2.67133 76	(0) 1.17837 39	(0)−8.83034 08

$$Z(x)=\frac{1}{\sqrt{2\pi}}\,e^{-\frac{1}{2}x^2} \qquad Z^{(n)}(x)=\frac{d^n}{dx^n}Z(x) \qquad He_n(x)=(-1)^n Z^{(n)}(x)/Z(x) \qquad Z^{(n)}(-x)=(-1)^n Z^{(n)}(x)$$

NORMAL PROBABILITY FUNCTION—VALUES OF $Z(x)$ IN TERMS OF $P(x)$ AND $Q(x)$ Table 26.4

$Q(x)$	0.000	0.001	0.002	0.003	0.004	0.005	0.006	0.007	0.008	0.009	0.010	
0.00	0.00000	0.00337	0.00634	0.00915	0.01185	0.01446	0.01700	0.01949	0.02192	0.02431	0.02665	0.99
0.01	0.02665	0.02896	0.03123	0.03348	0.03569	0.03787	0.04003	0.04216	0.04427	0.04635	0.04842	0.98
0.02	0.04842	0.05046	0.05249	0.05449	0.05648	0.05845	0.06040	0.06233	0.06425	0.06615	0.06804	0.97
0.03	0.06804	0.06992	0.07177	0.07362	0.07545	0.07727	0.07908	0.08087	0.08265	0.08442	0.08617	0.96
0.04	0.08617	0.08792	0.08965	0.09137	0.09309	0.09479	0.09648	0.09816	0.09983	0.10149	0.10314	0.95
0.05	0.10314	0.10478	0.10641	0.10803	0.10964	0.11124	0.11284	0.11442	0.11600	0.11756	0.11912	0.94
0.06	0.11912	0.12067	0.12222	0.12375	0.12528	0.12679	0.12830	0.12981	0.13130	0.13279	0.13427	0.93
0.07	0.13427	0.13574	0.13720	0.13866	0.14011	0.14156	0.14299	0.14442	0.14584	0.14726	0.14867	0.92
0.08	0.14867	0.15007	0.15146	0.15285	0.15423	0.15561	0.15698	0.15834	0.15970	0.16105	0.16239	0.91
0.09	0.16239	0.16373	0.16506	0.16639	0.16770	0.16902	0.17033	0.17163	0.17292	0.17421	0.17550	0.90
0.10	0.17550	0.17678	0.17805	0.17932	0.18057	0.18184	0.18309	0.18433	0.18557	0.18681	0.18804	0.89
0.11	0.18804	0.18926	0.19048	0.19169	0.19290	0.19410	0.19530	0.19649	0.19768	0.19886	0.20004	0.88
0.12	0.20004	0.20121	0.20238	0.20354	0.20470	0.20585	0.20700	0.20814	0.20928	0.21042	0.21155	0.87
0.13	0.21155	0.21267	0.21379	0.21490	0.21601	0.21712	0.21822	0.21932	0.22041	0.22149	0.22258	0.86
0.14	0.22258	0.22365	0.22473	0.22580	0.22686	0.22792	0.22898	0.23003	0.23108	0.23212	0.23316	0.85
0.15	0.23316	0.23419	0.23522	0.23625	0.23727	0.23829	0.23930	0.24031	0.24131	0.24232	0.24331	0.84
0.16	0.24331	0.24430	0.24529	0.24628	0.24726	0.24823	0.24921	0.25017	0.25114	0.25210	0.25305	0.83
0.17	0.25305	0.25401	0.25495	0.25590	0.25684	0.25778	0.25871	0.25964	0.26056	0.26148	0.26240	0.82
0.18	0.26240	0.26331	0.26422	0.26513	0.26603	0.26693	0.26782	0.26871	0.26960	0.27049	0.27137	0.81
0.19	0.27137	0.27224	0.27311	0.27398	0.27485	0.27571	0.27657	0.27742	0.27827	0.27912	0.27996	0.80
0.20	0.27996	0.28080	0.28164	0.28247	0.28330	0.28413	0.28495	0.28577	0.28658	0.28739	0.28820	0.79
0.21	0.28820	0.28901	0.28981	0.29060	0.29140	0.29219	0.29298	0.29376	0.29454	0.29532	0.29609	0.78
0.22	0.29609	0.29686	0.29763	0.29840	0.29916	0.29991	0.30067	0.30142	0.30216	0.30291	0.30365	0.77
0.23	0.30365	0.30439	0.30512	0.30585	0.30658	0.30730	0.30802	0.30874	0.30945	0.31016	0.31087	0.76
0.24	0.31087	0.31158	0.31228	0.31298	0.31367	0.31436	0.31505	0.31574	0.31642	0.31710	0.31778	0.75
0.25	0.31778	0.31845	0.31912	0.31979	0.32045	0.32111	0.32177	0.32242	0.32307	0.32372	0.32437	0.74
0.26	0.32437	0.32501	0.32565	0.32628	0.32691	0.32754	0.32817	0.32879	0.32941	0.33003	0.33065	0.73
0.27	0.33065	0.33126	0.33187	0.33247	0.33307	0.33367	0.33427	0.33486	0.33545	0.33604	0.33662	0.72
0.28	0.33662	0.33720	0.33778	0.33836	0.33893	0.33950	0.34007	0.34063	0.34119	0.34175	0.34230	0.71
0.29	0.34230	0.34286	0.34341	0.34395	0.34449	0.34503	0.34557	0.34611	0.34664	0.34717	0.34769	0.70
0.30	0.34769	0.34822	0.34874	0.34925	0.34977	0.35028	0.35079	0.35129	0.35180	0.35230	0.35279	0.69
0.31	0.35279	0.35329	0.35378	0.35427	0.35475	0.35524	0.35572	0.35620	0.35667	0.35714	0.35761	0.68
0.32	0.35761	0.35808	0.35854	0.35900	0.35946	0.35991	0.36037	0.36082	0.36126	0.36171	0.36215	0.67
0.33	0.36215	0.36259	0.36302	0.36346	0.36389	0.36431	0.36474	0.36516	0.36558	0.36600	0.36641	0.66
0.34	0.36641	0.36682	0.36723	0.36764	0.36804	0.36844	0.36884	0.36923	0.36962	0.37001	0.37040	0.65
0.35	0.37040	0.37078	0.37116	0.37154	0.37192	0.37229	0.37266	0.37303	0.37340	0.37376	0.37412	0.64
0.36	0.37412	0.37447	0.37483	0.37518	0.37553	0.37588	0.37622	0.37656	0.37690	0.37724	0.37757	0.63
0.37	0.37757	0.37790	0.37823	0.37855	0.37888	0.37920	0.37951	0.37983	0.38014	0.38045	0.38076	0.62
0.38	0.38076	0.38106	0.38136	0.38166	0.38196	0.38225	0.38254	0.38283	0.38312	0.38340	0.38368	0.61
0.39	0.38368	0.38396	0.38423	0.38451	0.38478	0.38504	0.38531	0.38557	0.38583	0.38609	0.38634	0.60
0.40	0.38634	0.38659	0.38684	0.38709	0.38734	0.38758	0.38782	0.38805	0.38829	0.38852	0.38875	0.59
0.41	0.38875	0.38897	0.38920	0.38942	0.38964	0.38985	0.39007	0.39028	0.39049	0.39069	0.39089	0.58
0.42	0.39089	0.39109	0.39129	0.39149	0.39168	0.39187	0.39206	0.39224	0.39243	0.39261	0.39279	0.57
0.43	0.39279	0.39296	0.39313	0.39330	0.39347	0.39364	0.39380	0.39396	0.39411	0.39427	0.39442	0.56
0.44	0.39442	0.39457	0.39472	0.39486	0.39501	0.39514	0.39528	0.39542	0.39555	0.39568	0.39580	0.55
0.45	0.39580	0.39593	0.39605	0.39617	0.39629	0.39640	0.39651	0.39662	0.39673	0.39683	0.39694	0.54
0.46	0.39694	0.39703	0.39713	0.39723	0.39732	0.39741	0.39749	0.39758	0.39766	0.39774	0.39781	0.53
0.47	0.39781	0.39789	0.39796	0.39803	0.39809	0.39816	0.39822	0.39828	0.39834	0.39839	0.39844	0.52
0.48	0.39844	0.39849	0.39854	0.39858	0.39862	0.39866	0.39870	0.39873	0.39876	0.39879	0.39882	0.51
0.49	0.39882	0.39884	0.39886	0.39888	0.39890	0.39891	0.39892	0.39893	0.39894	0.39894	0.39894	0.50
	0.010	0.009	0.008	0.007	0.006	0.005	0.004	0.003	0.002	0.001	0.000	$P(x)$

Linear interpolation yields an error no greater than 5 units in the fifth decimal place.

$$Z(x)=\frac{1}{\sqrt{2\pi}}\,e^{-\frac{1}{2}x^2} \qquad P(x)=1-Q(x)=\int_{-\infty}^{x} Z(t)\,dt$$

Compiled from T. L. Kelley, The Kelley Statistical Tables. Harvard Univ. Press, Cambridge, Mass., 1948 (with permission).

Table 26.5   NORMAL PROBABILITY FUNCTION—VALUES OF x IN TERMS OF $P(x)$ AND $Q(x)$

$Q(x)$	0.000	0.001	0.002	0.003	0.004	0.005	0.006	0.007	0.008	0.009	0.010	
0.00	∞	3.09023	2.87816	2.74778	2.65207	2.57583	2.51214	2.45726	2.40892	2.36562	2.32635	0.99
0.01	2.32635	2.29037	2.25713	2.22621	2.19729	2.17009	2.14441	2.12007	2.09693	2.07485	2.05375	0.98
0.02	2.05375	2.03352	2.01409	1.99539	1.97737	1.95996	1.94313	1.92684	1.91104	1.89570	1.88079	0.97
0.03	1.88079	1.86630	1.85218	1.83842	1.82501	1.81191	1.79912	1.78661	1.77438	1.76241	1.75069	0.96
0.04	1.75069	1.73920	1.72793	1.71689	1.70604	1.69540	1.68494	1.67466	1.66456	1.65463	1.64485	0.95
0.05	1.64485	1.63523	1.62576	1.61644	1.60725	1.59819	1.58927	1.58047	1.57179	1.56322	1.55477	0.94
0.06	1.55477	1.54643	1.53820	1.53007	1.52204	1.51410	1.50626	1.49851	1.49085	1.48328	1.47579	0.93
0.07	1.47579	1.46838	1.46106	1.45381	1.44663	1.43953	1.43250	1.42554	1.41865	1.41183	1.40507	0.92
0.08	1.40507	1.39838	1.39174	1.38517	1.37866	1.37220	1.36581	1.35946	1.35317	1.34694	1.34076	0.91
0.09	1.34076	1.33462	1.32854	1.32251	1.31652	1.31058	1.30469	1.29884	1.29303	1.28727	1.28155	0.90
0.10	1.28155	1.27587	1.27024	1.26464	1.25908	1.25357	1.24808	1.24264	1.23723	1.23186	1.22653	0.89
0.11	1.22653	1.22123	1.21596	1.21072	1.20553	1.20036	1.19522	1.19012	1.18504	1.18000	1.17499	0.88
0.12	1.17499	1.17000	1.16505	1.16012	1.15522	1.15035	1.14551	1.14069	1.13590	1.13113	1.12639	0.87
0.13	1.12639	1.12168	1.11699	1.11232	1.10768	1.10306	1.09847	1.09390	1.08935	1.08482	1.08032	0.86
0.14	1.08032	1.07584	1.07138	1.06694	1.06252	1.05812	1.05374	1.04939	1.04505	1.04073	1.03643	0.85
0.15	1.03643	1.03215	1.02789	1.02365	1.01943	1.01522	1.01103	1.00686	1.00271	0.99858	0.99446	0.84
0.16	0.99446	0.99036	0.98627	0.98220	0.97815	0.97411	0.97009	0.96609	0.96210	0.95812	0.95416	0.83
0.17	0.95416	0.95022	0.94629	0.94238	0.93848	0.93458	0.93072	0.92686	0.92301	0.91918	0.91537	0.82
0.18	0.91537	0.91156	0.90777	0.90399	0.90023	0.89647	0.89273	0.88901	0.88529	0.88159	0.87790	0.81
0.19	0.87790	0.87422	0.87055	0.86689	0.86325	0.85962	0.85600	0.85239	0.84879	0.84520	0.84162	0.80
0.20	0.84162	0.83805	0.83450	0.83095	0.82742	0.82390	0.82038	0.81687	0.81338	0.80990	0.80642	0.79
0.21	0.80642	0.80296	0.79950	0.79606	0.79262	0.78919	0.78577	0.78237	0.77897	0.77557	0.77219	0.78
0.22	0.77219	0.76882	0.76546	0.76210	0.75875	0.75542	0.75208	0.74876	0.74545	0.74214	0.73885	0.77
0.23	0.73885	0.73556	0.73228	0.72900	0.72574	0.72248	0.71923	0.71599	0.71275	0.70952	0.70630	0.76
0.24	0.70630	0.70309	0.69988	0.69668	0.69349	0.69031	0.68713	0.68396	0.68080	0.67764	0.67449	0.75
0.25	0.67449	0.67135	0.66821	0.66508	0.66196	0.65884	0.65573	0.65262	0.64952	0.64643	0.64335	0.74
0.26	0.64335	0.64027	0.63719	0.63412	0.63106	0.62801	0.62496	0.62191	0.61887	0.61584	0.61281	0.73
0.27	0.61281	0.60979	0.60678	0.60376	0.60076	0.59776	0.59477	0.59178	0.58879	0.58581	0.58284	0.72
0.28	0.58284	0.57987	0.57691	0.57395	0.57100	0.56805	0.56511	0.56217	0.55924	0.55631	0.55338	0.71
0.29	0.55338	0.55047	0.54755	0.54464	0.54174	0.53884	0.53594	0.53305	0.53016	0.52728	0.52440	0.70
0.30	0.52440	0.52153	0.51866	0.51579	0.51293	0.51007	0.50722	0.50437	0.50153	0.49869	0.49585	0.69
0.31	0.49585	0.49302	0.49019	0.48736	0.48454	0.48173	0.47891	0.47610	0.47330	0.47050	0.46770	0.68
0.32	0.46770	0.46490	0.46211	0.45933	0.45654	0.45376	0.45099	0.44821	0.44544	0.44268	0.43991	0.67
0.33	0.43991	0.43715	0.43440	0.43164	0.42889	0.42615	0.42340	0.42066	0.41793	0.41519	0.41246	0.66
0.34	0.41246	0.40974	0.40701	0.40429	0.40157	0.39886	0.39614	0.39343	0.39073	0.38802	0.38532	0.65
0.35	0.38532	0.38262	0.37993	0.37723	0.37454	0.37186	0.36917	0.36649	0.36381	0.36113	0.35846	0.64
0.36	0.35846	0.35579	0.35312	0.35045	0.34779	0.34513	0.34247	0.33981	0.33716	0.33450	0.33185	0.63
0.37	0.33185	0.32921	0.32656	0.32392	0.32128	0.31864	0.31600	0.31337	0.31074	0.30811	0.30548	0.62
0.38	0.30548	0.30286	0.30023	0.29761	0.29499	0.29237	0.28976	0.28715	0.28454	0.28193	0.27932	0.61
0.39	0.27932	0.27671	0.27411	0.27151	0.26891	0.26631	0.26371	0.26112	0.25853	0.25594	0.25335	0.60
0.40	0.25335	0.25076	0.24817	0.24559	0.24301	0.24043	0.23785	0.23527	0.23269	0.23012	0.22754	0.59
0.41	0.22754	0.22497	0.22240	0.21983	0.21727	0.21470	0.21214	0.20957	0.20701	0.20445	0.20189	0.58
0.42	0.20189	0.19934	0.19678	0.19422	0.19167	0.18912	0.18657	0.18402	0.18147	0.17892	0.17637	0.57
0.43	0.17637	0.17383	0.17128	0.16874	0.16620	0.16366	0.16112	0.15858	0.15604	0.15351	0.15097	0.56
0.44	0.15097	0.14843	0.14590	0.14337	0.14084	0.13830	0.13577	0.13324	0.13072	0.12819	0.12566	0.55
0.45	0.12566	0.12314	0.12061	0.11809	0.11556	0.11304	0.11052	0.10799	0.10547	0.10295	0.10043	0.54
0.46	0.10043	0.09791	0.09540	0.09288	0.09036	0.08784	0.08533	0.08281	0.08030	0.07778	0.07527	0.53
0.47	0.07527	0.07276	0.07024	0.06773	0.06522	0.06271	0.06020	0.05768	0.05517	0.05266	0.05015	0.52
0.48	0.05015	0.04764	0.04513	0.04263	0.04012	0.03761	0.03510	0.03259	0.03008	0.02758	0.02507	0.51
0.49	0.02507	0.02256	0.02005	0.01755	0.01504	0.01253	0.01003	0.00752	0.00501	0.00251	0.00000	0.50
	0.010	0.009	0.008	0.007	0.006	0.005	0.004	0.003	0.002	0.001	0.000	$P(x)$

For $Q(x) > 0.007$, linear interpolation yields an error of one unit in the third decimal place; five-point interpolation is necessary to obtain full accuracy.

$$P(x) = 1 - Q(x) = \int_{-\infty}^{x} Z(t)\,dt$$

Compiled from T. L. Kelley, The Kelley Statistical Tables. Harvard Univ. Press, Cambridge, Mass., 1948 (with permission).

NORMAL PROBABILITY FUNCTION—VALUES OF x FOR EXTREME VALUES OF $P(x)$ AND $Q(x)$ Table 26.6

$Q(x)$	0.0000	0.0001	0.0002	0.0003	0.0004	0.0005	0.0006	0.0007	0.0008	0.0009	0.0010	
0.000	∞	3.71902	3.54008	3.43161	3.35279	3.29053	3.23888	3.19465	3.15591	3.12139	3.09023	0.999
0.001	3.09023	3.06181	3.03567	3.01145	2.98888	2.96774	2.94784	2.92905	2.91124	2.89430	2.87816	0.998
0.002	2.87816	2.86274	2.84796	2.83379	2.82016	2.80703	2.79438	2.78215	2.77033	2.75888	2.74778	0.997
0.003	2.74778	2.73701	2.72655	2.71638	2.70648	2.69684	2.68745	2.67829	2.66934	2.66061	2.65207	0.996
0.004	2.65207	2.64372	2.63555	2.62756	2.61973	2.61205	2.60453	2.59715	2.58991	2.58281	2.57583	0.995
0.005	2.57583	2.56897	2.56224	2.55562	2.54910	2.54270	2.53640	2.53019	2.52408	2.51807	2.51214	0.994
0.006	2.51214	2.50631	2.50055	2.49488	2.48929	2.48377	2.47833	2.47296	2.46765	2.46243	2.45726	0.993
0.007	2.45726	2.45216	2.44713	2.44215	2.43724	2.43238	2.42758	2.42283	2.41814	2.41350	2.40891	0.992
0.008	2.40891	2.40437	2.39989	2.39545	2.39106	2.38671	2.38240	2.37814	2.37392	2.36975	2.36562	0.991
0.009	2.36562	2.36152	2.35747	2.35345	2.34947	2.34553	2.34162	2.33775	2.33392	2.33012	2.32635	0.990
0.010	2.32635	2.32261	2.31891	2.31524	2.31160	2.30798	2.30440	2.30085	2.29733	2.29383	2.29037	0.989
0.011	2.29037	2.28693	2.28352	2.28013	2.27677	2.27343	2.27013	2.26684	2.26358	2.26034	2.25713	0.988
0.012	2.25713	2.25394	2.25077	2.24763	2.24450	2.24140	2.23832	2.23526	2.23223	2.22921	2.22621	0.987
0.013	2.22621	2.22323	2.22028	2.21734	2.21442	2.21152	2.20864	2.20577	2.20293	2.20010	2.19729	0.986
0.014	2.19729	2.19449	2.19172	2.18896	2.18621	2.18349	2.18078	2.17808	2.17540	2.17274	2.17009	0.985
0.015	2.17009	2.16746	2.16484	2.16224	2.15965	2.15707	2.15451	2.15197	2.14943	2.14692	2.14441	0.984
0.016	2.14441	2.14192	2.13944	2.13698	2.13452	2.13208	2.12966	2.12724	2.12484	2.12245	2.12007	0.983
0.017	2.12007	2.11771	2.11535	2.11301	2.11068	2.10836	2.10605	2.10375	2.10147	2.09919	2.09693	0.982
0.018	2.09693	2.09467	2.09243	2.09020	2.08798	2.08576	2.08356	2.08137	2.07919	2.07702	2.07485	0.981
0.019	2.07485	2.07270	2.07056	2.06843	2.06630	2.06419	2.06208	2.05998	2.05790	2.05582	2.05375	0.980
0.020	2.05375	2.05169	2.04964	2.04759	2.04556	2.04353	2.04151	2.03950	2.03750	2.03551	2.03352	0.979
0.021	2.03352	2.03154	2.02957	2.02761	2.02566	2.02371	2.02177	2.01984	2.01792	2.01600	2.01409	0.978
0.022	2.01409	2.01219	2.01029	2.00841	2.00653	2.00465	2.00279	2.00093	1.99908	1.99723	1.99539	0.977
0.023	1.99539	1.99356	1.99174	1.98992	1.98811	1.98631	1.98450	1.98271	1.98092	1.97914	1.97737	0.976
0.024	1.97737	1.97560	1.97384	1.97208	1.97033	1.96859	1.96685	1.96512	1.96340	1.96168	1.95996	0.975
	0.0010	0.0009	0.0008	0.0007	0.0006	0.0005	0.0004	0.0003	0.0002	0.0001	0.0000	$P(x)$

For $Q(x) > 0.0007$, linear interpolation yields an error of one unit in the third decimal place; five-point interpolation is necessary to obtain full accuracy.

$Q(x)$	x	$Q(x)$	x	$Q(x)$	x	$Q(x)$	x
(−4)1.0	3.71902	(−9)1.0	5.99781	(−14)1.0	7.65063	(−19)1.0	9.01327
(−5)1.0	4.26489	(−10)1.0	6.36134	(−15)1.0	7.94135	(−20)1.0	9.26234
(−6)1.0	4.75342	(−11)1.0	6.70602	(−16)1.0	8.22208	(−21)1.0	9.50502
(−7)1.0	5.19934	(−12)1.0	7.03448	(−17)1.0	8.49379	(−22)1.0	9.74179
(−8)1.0	5.61200	(−13)1.0	7.34880	(−18)1.0	8.75729	(−23)1.0	9.97305

$$P(x) = 1 - Q(x) = \int_{-\infty}^{x} Z(t)\,dt$$

Compiled from T. L. Kelley, The Kelley Statistical Tables. Harvard Univ. Press, Cambridge, Mass., 1948 (with permission) for $Q(x) > (−9)1$.

Table 26.7   PROBABILITY INTEGRAL OF χ^2-DISTRIBUTION, INCOMPLETE GAMMA FUNCTION CUMULATIVE SUMS OF THE POISSON DISTRIBUTION

ν	$\chi^2{=}0.001$	0.002	0.003	0.004	0.005	0.006	0.007	0.008	0.009	0.010
	$m{=}0.0005$	0.0010	0.0015	0.0020	0.0025	0.0030	0.0035	0.0040	0.0045	0.0050
1	0.97477	0.96433	0.95632	0.94957	0.94363	0.93826	0.93332	0.92873	0.92442	0.92034
2	0.99950	0.99900	0.99850	0.99800	0.99750	0.99700	0.99651	0.99601	0.99551	0.99501
3	0.99999	0.99998	0.99996	0.99993	0.99991	0.99988	0.99984	0.99981	0.99977	0.99973
4							0.99999	0.99999	0.99999	0.99999

ν	$\chi^2{=}0.01$	0.02	0.03	0.04	0.05	0.06	0.07	0.08	0.09	0.10
	$m{=}0.005$	0.010	0.015	0.020	0.025	0.030	0.035	0.040	0.045	0.050
1	0.92034	0.88754	0.86249	0.84148	0.82306	0.80650	0.79134	0.77730	0.76418	0.75183
2	0.99501	0.99005	0.98511	0.98020	0.97531	0.97045	0.96561	0.96079	0.95600	0.95123
3	0.99973	0.99925	0.99863	0.99790	0.99707	0.99616	0.99518	0.99412	0.99301	0.99184
4	0.99999	0.99995	0.99989	0.99980	0.99969	0.99956	0.99940	0.99922	0.99902	0.99879
5			0.99999	0.99998	0.99997	0.99995	0.99993	0.99991	0.99987	0.99984
6							0.99999	0.99999	0.99999	0.99998

ν	$\chi^2{=}0.1$	0.2	0.3	0.4	0.5	0.6	0.7	0.8	0.9	1.0
	$m{=}0.05$	0.10	0.15	0.20	0.25	0.30	0.35	0.40	0.45	0.50
1	0.75183	0.65472	0.58388	0.52709	0.47950	0.43858	0.40278	0.37109	0.34278	0.31731
2	0.95123	0.90484	0.86071	0.81873	0.77880	0.74082	0.70469	0.67032	0.63763	0.60653
3	0.99184	0.97759	0.96003	0.94024	0.91889	0.89643	0.87320	0.84947	0.82543	0.80125
4	0.99879	0.99532	0.98981	0.98248	0.97350	0.96306	0.95133	0.93845	0.92456	0.90980
5	0.99984	0.99911	0.99764	0.99533	0.99212	0.98800	0.98297	0.97703	0.97022	0.96257
6	0.99998	0.99985	0.99950	0.99885	0.99784	0.99640	0.99449	0.99207	0.98912	0.98561
7		0.99997	0.99990	0.99974	0.99945	0.99899	0.99834	0.99744	0.99628	0.99483
8			0.99998	0.99994	0.99987	0.99973	0.99953	0.99922	0.99880	0.99825
9				0.99999	0.99997	0.99993	0.99987	0.99978	0.99964	0.99944
10					0.99999	0.99998	0.99997	0.99994	0.99989	0.99983
11							0.99999	0.99998	0.99997	0.99995
12									0.99999	0.99999

ν	$\chi^2{=}1.1$	1.2	1.3	1.4	1.5	1.6	1.7	1.8	1.9	2.0
	$m{=}0.55$	0.60	0.65	0.70	0.75	0.80	0.85	0.90	0.95	1.00
1	0.29427	0.27332	0.25421	0.23672	0.22067	0.20590	0.19229	0.17971	0.16808	0.15730
2	0.57695	0.54881	0.52205	0.49659	0.47237	0.44933	0.42741	0.40657	0.38674	0.36788
3	0.77707	0.75300	0.72913	0.70553	0.68227	0.65939	0.63693	0.61493	0.59342	0.57241
4	0.89427	0.87810	0.86138	0.84420	0.82664	0.80879	0.79072	0.77248	0.75414	0.73576
5	0.95410	0.94488	0.93493	0.92431	0.91307	0.90125	0.88890	0.87607	0.86280	0.84915
6	0.98154	0.97689	0.97166	0.96586	0.95949	0.95258	0.94512	0.93714	0.92866	0.91970
7	0.99305	0.99093	0.98844	0.98557	0.98231	0.97864	0.97457	0.97008	0.96517	0.95984
8	0.99753	0.99664	0.99555	0.99425	0.99271	0.99092	0.98887	0.98654	0.98393	0.98101
9	0.99917	0.99882	0.99838	0.99782	0.99715	0.99633	0.99537	0.99425	0.99295	0.99147
10	0.99973	0.99961	0.99944	0.99921	0.99894	0.99859	0.99817	0.99766	0.99705	0.99634
11	0.99992	0.99987	0.99981	0.99973	0.99962	0.99948	0.99930	0.99908	0.99882	0.99850
12	0.99998	0.99996	0.99994	0.99991	0.99987	0.99982	0.99975	0.99966	0.99954	0.99941
13	0.99999	0.99999	0.99998	0.99997	0.99996	0.99994	0.99991	0.99988	0.99983	0.99977
14			0.99999	0.99999	0.99999	0.99998	0.99997	0.99996	0.99994	0.99992
15						0.99999	0.99999	0.99999	0.99998	0.99997
16									0.99999	0.99999

$$Q(\chi^2|\nu)=1-P(\chi^2|\nu)=\left[2^{\frac{\nu}{2}}\Gamma\left(\frac{\nu}{2}\right)\right]^{-1}\int_{\chi^2}^{\infty}e^{-\frac{t}{2}}t^{\frac{\nu}{2}-1}\,dt=\left[\Gamma\left(\frac{\nu}{2}\right)\right]^{-1}\int_{\frac{1}{2}\chi^2}^{\infty}e^{-t}t^{\frac{\nu}{2}-1}\,dt=\sum_{j=0}^{c-1}e^{-m}m^j/j!\quad(\nu\text{ even},\ c=\tfrac{1}{2}\nu,\ m=\tfrac{1}{2}\chi^2)$$

Compiled from E. S. Pearson and H. O. Hartley (editors), Biometrika tables for statisticians, vol. I. Cambridge Univ. Press, Cambridge, England, 1954 (with permission).

PROBABILITY INTEGRAL OF χ^2-DISTRIBUTION, INCOMPLETE GAMMA FUNCTION Table 26.7
CUMULATIVE SUMS OF THE POISSON DISTRIBUTION

ν	$\chi^2=2.2$ $m=1.1$	2.4 1.2	2.6 1.3	2.8 1.4	3.0 1.5	3.2 1.6	3.4 1.7	3.6 1.8	3.8 1.9	4.0 2.0
1	0.13801	0.12134	0.10686	0.09426	0.08327	0.07364	0.06520	0.05778	0.05125	0.04550
2	0.33287	0.30119	0.27253	0.24660	0.22313	0.20190	0.18268	0.16530	0.14957	0.13534
3	0.53195	0.49363	0.45749	0.42350	0.39163	0.36181	0.33397	0.30802	0.28389	0.26146
4	0.69903	0.66263	0.62682	0.59183	0.55783	0.52493	0.49325	0.46284	0.43375	0.40601
5	0.82084	0.79147	0.76137	0.73079	0.69999	0.66918	0.63857	0.60831	0.57856	0.54942
6	0.90042	0.87949	0.85711	0.83350	0.80885	0.78336	0.75722	0.73062	0.70372	0.67668
7	0.94795	0.93444	0.91938	0.90287	0.88500	0.86590	0.84570	0.82452	0.80250	0.77978
8	0.97426	0.96623	0.95691	0.94628	0.93436	0.92119	0.90681	0.89129	0.87470	0.85712
9	0.98790	0.98345	0.97807	0.97170	0.96430	0.95583	0.94631	0.93572	0.92408	0.91141
10	0.99457	0.99225	0.98934	0.98575	0.98142	0.97632	0.97039	0.96359	0.95592	0.94735
11	0.99766	0.99652	0.99503	0.99311	0.99073	0.98781	0.98431	0.98019	0.97541	0.96992
12	0.99903	0.99850	0.99777	0.99680	0.99554	0.99396	0.99200	0.98962	0.98678	0.98344
13	0.99961	0.99938	0.99903	0.99856	0.99793	0.99711	0.99606	0.99475	0.99314	0.99119
14	0.99985	0.99975	0.99960	0.99938	0.99907	0.99866	0.99813	0.99743	0.99655	0.99547
15	0.99994	0.99990	0.99984	0.99974	0.99960	0.99940	0.99913	0.99878	0.99832	0.99774
16	0.99998	0.99996	0.99994	0.99989	0.99983	0.99974	0.99961	0.99944	0.99921	0.99890
17	0.99999	0.99999	0.99998	0.99996	0.99993	0.99989	0.99983	0.99975	0.99964	0.99948
18			0.99999	0.99998	0.99997	0.99995	0.99993	0.99989	0.99984	0.99976
19				0.99999	0.99999	0.99998	0.99997	0.99995	0.99993	0.99989
20						0.99999	0.99999	0.99998	0.99997	0.99995
21								0.99999	0.99999	0.99998
22										0.99999

ν	$\chi^2=4.2$ $m=2.1$	4.4 2.2	4.6 2.3	4.8 2.4	5.0 2.5	5.2 2.6	5.4 2.7	5.6 2.8	5.8 2.9	6.0 3.0
1	0.04042	0.03594	0.03197	0.02846	0.02535	0.02259	0.02014	0.01796	0.01603	0.01431
2	0.12246	0.11080	0.10026	0.09072	0.08209	0.07427	0.06721	0.06081	0.05502	0.04979
3	0.24066	0.22139	0.20354	0.18704	0.17180	0.15772	0.14474	0.13278	0.12176	0.11161
4	0.37962	0.35457	0.33085	0.30844	0.28730	0.26739	0.24866	0.23108	0.21459	0.19915
5	0.52099	0.49337	0.46662	0.44077	0.41588	0.39196	0.36904	0.34711	0.32617	0.30622
6	0.64963	0.62271	0.59604	0.56971	0.54381	0.51843	0.49363	0.46945	0.44596	0.42319
7	0.75647	0.73272	0.70864	0.68435	0.65996	0.63557	0.61127	0.58715	0.56329	0.53975
8	0.83864	0.81935	0.79935	0.77872	0.75758	0.73600	0.71409	0.69194	0.66962	0.64723
9	0.89776	0.88317	0.86769	0.85138	0.83431	0.81654	0.79814	0.77919	0.75976	0.73992
10	0.93787	0.92750	0.91625	0.90413	0.89118	0.87742	0.86291	0.84768	0.83178	0.81526
11	0.96370	0.95672	0.94898	0.94046	0.93117	0.92109	0.91026	0.89868	0.88637	0.87337
12	0.97955	0.97509	0.97002	0.96433	0.95798	0.95096	0.94327	0.93489	0.92583	0.91608
13	0.98887	0.98614	0.98298	0.97934	0.97519	0.97052	0.96530	0.95951	0.95313	0.94615
14	0.99414	0.99254	0.99064	0.98841	0.98581	0.98283	0.97943	0.97559	0.97128	0.96649
15	0.99701	0.99610	0.99501	0.99369	0.99213	0.99029	0.98816	0.98571	0.98291	0.97975
16	0.99851	0.99802	0.99741	0.99666	0.99575	0.99467	0.99338	0.99187	0.99012	0.98810
17	0.99928	0.99902	0.99869	0.99828	0.99777	0.99715	0.99639	0.99550	0.99443	0.99319
18	0.99966	0.99953	0.99936	0.99914	0.99886	0.99851	0.99809	0.99757	0.99694	0.99620
19	0.99985	0.99978	0.99969	0.99958	0.99943	0.99924	0.99901	0.99872	0.99836	0.99793
20	0.99993	0.99990	0.99986	0.99980	0.99972	0.99962	0.99950	0.99934	0.99914	0.99890
21	0.99997	0.99995	0.99993	0.99991	0.99987	0.99982	0.99975	0.99967	0.99956	0.99943
22	0.99999	0.99998	0.99997	0.99996	0.99994	0.99991	0.99988	0.99984	0.99978	0.99971
23	0.99999	0.99999	0.99999	0.99998	0.99997	0.99996	0.99994	0.99992	0.99989	0.99986
24			0.99999	0.99999	0.99999	0.99998	0.99997	0.99996	0.99995	0.99993
25					0.99999	0.99999	0.99999	0.99998	0.99998	0.99997
26								0.99999	0.99999	0.99998
27									0.99999	0.99999

$$\phi=\tfrac{1}{2}\left(\chi^2-\chi_0^2\right) \qquad w=\nu-\nu_0>0$$

Interpolation on χ^2

$$Q(\chi^2|\nu)=Q\left(\chi_0^2\big|\nu_0-4\right)\left[\tfrac{1}{2}\phi^2\right]+Q\left(\chi_0^2\big|\nu_0-2\right)\left[\phi-\phi^2\right]+Q\left(\chi_0^2\big|\nu_0\right)\left[1-\phi+\tfrac{1}{2}\phi^2\right]$$

Double Entry Interpolation

$$Q\left(\chi^2\big|\nu\right)=Q\left(\chi_0^2\big|\nu_0-4\right)\left[\tfrac{1}{2}\phi^2\right]+Q\left(\chi_0^2\big|\nu_0-2\right)\left[\phi-\phi^2-w\phi\right]+Q\left(\chi_0^2\big|\nu_0-1\right)\left[\tfrac{1}{2}w^2-\tfrac{1}{2}w+w\phi\right]$$

$$+Q\left(\chi_0^2\big|\nu_0\right)\left[1-w^2-\phi+\tfrac{1}{2}\phi^2+w\phi\right]+Q\left(\chi_0^2\big|\nu_0+1\right)\left[\tfrac{1}{2}w^2+\tfrac{1}{2}w-w\phi\right]$$

Table 26.7 PROBABILITY INTEGRAL OF χ^2-DISTRIBUTION, INCOMPLETE GAMMA FUNCTION CUMULATIVE SUMS OF THE POISSON DISTRIBUTION

$\chi^2=6.2$	6.4	6.6	6.8	7.0	7.2	7.4	7.6	7.8	8.0
ν $m=3.1$	3.2	3.3	3.4	3.5	3.6	3.7	3.8	3.9	4.0
1 0.01278	0.01141	0.01020	0.00912	0.00815	0.00729	0.00652	0.00584	0.00522	0.00468
2 0.04505	0.04076	0.03688	0.03337	0.03020	0.02732	0.02472	0.02237	0.02024	0.01832
3 0.10228	0.09369	0.08580	0.07855	0.07190	0.06579	0.06018	0.05504	0.05033	0.04601
4 0.18470	0.17120	0.15860	0.14684	0.13589	0.12569	0.11620	0.10738	0.09919	0.09158
5 0.28724	0.26922	0.25213	0.23595	0.22064	0.20619	0.19255	0.17970	0.16761	0.15624
6 0.40116	0.37990	0.35943	0.33974	0.32085	0.30275	0.28543	0.26890	0.25313	0.23810
7 0.51660	0.49390	0.47168	0.45000	0.42888	0.40836	0.38845	0.36918	0.35056	0.33259
8 0.62484	0.60252	0.58034	0.55836	0.53663	0.51522	0.49415	0.47349	0.45325	0.43347
9 0.71975	0.69931	0.67869	0.65793	0.63712	0.61631	0.59555	0.57490	0.55442	0.53415
10 0.79819	0.78061	0.76259	0.74418	0.72544	0.70644	0.68722	0.66784	0.64837	0.62884
11 0.85969	0.84539	0.83049	0.81504	0.79908	0.78266	0.76583	0.74862	0.73110	0.71330
12 0.90567	0.89459	0.88288	0.87054	0.85761	0.84412	0.83009	0.81556	0.80056	0.78513
13 0.93857	0.93038	0.92157	0.91216	0.90215	0.89155	0.88038	0.86865	0.85638	0.84360
14 0.96120	0.95538	0.94903	0.94215	0.93471	0.92673	0.91819	0.90911	0.89948	0.88933
15 0.97619	0.97222	0.96782	0.96296	0.95765	0.95186	0.94559	0.93882	0.93155	0.92378
16 0.98579	0.98317	0.98022	0.97693	0.97326	0.96921	0.96476	0.95989	0.95460	0.94887
17 0.99174	0.99007	0.98816	0.98599	0.98355	0.98081	0.97775	0.97437	0.97064	0.96655
18 0.99532	0.99429	0.99309	0.99171	0.99013	0.98833	0.98630	0.98402	0.98147	0.97864
19 0.99741	0.99679	0.99606	0.99521	0.99421	0.99307	0.99176	0.99026	0.98857	0.98667
20 0.99860	0.99824	0.99781	0.99729	0.99669	0.99598	0.99515	0.99420	0.99311	0.99187
21 0.99926	0.99905	0.99880	0.99850	0.99814	0.99771	0.99721	0.99662	0.99594	0.99514
22 0.99962	0.99950	0.99936	0.99919	0.99898	0.99873	0.99843	0.99807	0.99765	0.99716
23 0.99981	0.99974	0.99967	0.99957	0.99945	0.99931	0.99913	0.99892	0.99867	0.99837
24 0.99990	0.99987	0.99983	0.99978	0.99971	0.99963	0.99953	0.99941	0.99926	0.99908
25 0.99995	0.99994	0.99991	0.99989	0.99985	0.99981	0.99975	0.99968	0.99960	0.99949
26 0.99998	0.99997	0.99996	0.99994	0.99992	0.99990	0.99987	0.99983	0.99978	0.99973
27 0.99999	0.99999	0.99998	0.99997	0.99996	0.99995	0.99993	0.99991	0.99989	0.99985
28	0.99999	0.99999	0.99999	0.99998	0.99998	0.99997	0.99996	0.99994	0.99992
29			0.99999	0.99999	0.99999	0.99999	0.99998	0.99998	0.99997	0.99996
30					0.99999	0.99999	0.99999	0.99999	0.99999	0.99998

$\chi^2=8.2$	8.4	8.6	8.8	9.0	9.2	9.4	9.6	9.8	10.0
ν $m=4.1$	4.2	4.3	4.4	4.5	4.6	4.7	4.8	4.9	5.0
1 0.00419	0.00375	0.00336	0.00301	0.00270	0.00242	0.00217	0.00195	0.00175	0.00157
2 0.01657	0.01500	0.01357	0.01228	0.01111	0.01005	0.00910	0.00823	0.00745	0.00674
3 0.04205	0.03843	0.03511	0.03207	0.02929	0.02675	0.02442	0.02229	0.02034	0.01857
4 0.08452	0.07798	0.07191	0.06630	0.06110	0.05629	0.05184	0.04773	0.04394	0.04043
5 0.14555	0.13553	0.12612	0.11731	0.10906	0.10135	0.09413	0.08740	0.08110	0.07524
6 0.22381	0.21024	0.19736	0.18514	0.17358	0.16264	0.15230	0.14254	0.13333	0.12465
7 0.31529	0.29865	0.28266	0.26734	0.25266	0.23861	0.22520	0.21240	0.20019	0.18857
8 0.41418	0.39540	0.37715	0.35945	0.34230	0.32571	0.30968	0.29423	0.27935	0.26503
9 0.51412	0.49439	0.47499	0.45594	0.43727	0.41902	0.40120	0.38383	0.36692	0.35049
10 0.60931	0.58983	0.57044	0.55118	0.53210	0.51323	0.49461	0.47626	0.45821	0.44049
11 0.69528	0.67709	0.65876	0.64035	0.62189	0.60344	0.58502	0.56669	0.54846	0.53039
12 0.76931	0.75314	0.73666	0.71991	0.70293	0.68576	0.66844	0.65101	0.63350	0.61596
13 0.83033	0.81660	0.80244	0.78788	0.77294	0.75768	0.74211	0.72627	0.71020	0.69393
14 0.87865	0.86746	0.85579	0.84365	0.83105	0.81803	0.80461	0.79081	0.77666	0.76218
15 0.91551	0.90675	0.89749	0.88774	0.87752	0.86683	0.85569	0.84412	0.83213	0.81974
16 0.94269	0.93606	0.92897	0.92142	0.91341	0.90495	0.89603	0.88667	0.87686	0.86663
17 0.96208	0.95723	0.95198	0.94633	0.94026	0.93378	0.92687	0.91954	0.91179	0.90361
18 0.97551	0.97207	0.96830	0.96420	0.95974	0.95493	0.94974	0.94418	0.93824	0.93191
19 0.98454	0.98217	0.97955	0.97666	0.97348	0.97001	0.96623	0.96213	0.95771	0.95295
20 0.99046	0.98887	0.98709	0.98511	0.98291	0.98047	0.97779	0.97486	0.97166	0.96817
21 0.99424	0.99320	0.99203	0.99070	0.98921	0.98755	0.98570	0.98365	0.98139	0.97891
22 0.99659	0.99593	0.99518	0.99431	0.99333	0.99222	0.99098	0.98958	0.98803	0.98630
23 0.99802	0.99761	0.99714	0.99659	0.99596	0.99524	0.99442	0.99349	0.99245	0.99128
24 0.99888	0.99863	0.99833	0.99799	0.99760	0.99714	0.99661	0.99601	0.99532	0.99455
25 0.99937	0.99922	0.99905	0.99884	0.99860	0.99831	0.99798	0.99760	0.99716	0.99665
26 0.99966	0.99957	0.99947	0.99934	0.99919	0.99902	0.99882	0.99858	0.99830	0.99798
27 0.99981	0.99977	0.99971	0.99963	0.99955	0.99944	0.99932	0.99917	0.99900	0.99880
28 0.99990	0.99987	0.99984	0.99980	0.99975	0.99969	0.99962	0.99953	0.99942	0.99930
29 0.99995	0.99993	0.99991	0.99989	0.99986	0.99983	0.99979	0.99973	0.99967	0.99960
30 0.99997	0.99997	0.99996	0.99994	0.99993	0.99991	0.99988	0.99985	0.99982	0.99977

$$Q(\chi^2|\nu)=1-P(\chi^2|\nu)=\left[2^{\frac{\nu}{2}}\Gamma\left(\frac{\nu}{2}\right)\right]^{-1}\int_{\chi^2}^{\infty}e^{-\frac{t}{2}}t^{\frac{\nu}{2}-1}\,dt=\left[\Gamma\left(\frac{\nu}{2}\right)\right]^{-1}\int_{\frac{1}{2}\chi^2}^{\infty}e^{-t}t^{\frac{\nu}{2}-1}\,dt=\sum_{j=0}^{c-1}e^{-m}m^j/j!\ (\nu\text{ even},\ c=\tfrac{1}{2}\nu,\ m=\tfrac{1}{2}\chi^2)$$

Table 26.7

PROBABILITY INTEGRAL OF χ^2-DISTRIBUTION, INCOMPLETE GAMMA FUNCTION
CUMULATIVE SUMS OF THE POISSON DISTRIBUTION

$\chi^2=$	10.5	11.0	11.5	12.0	12.5	13.0	13.5	14.0	14.5	15.0
ν $m=$	5.25	5.5	5.75	6.0	6.25	6.5	6.75	7.0	7.25	7.5
1	0.00119	0.00091	0.00070	0.00053	0.00041	0.00031	0.00024	0.00018	0.00014	0.00011
2	0.00525	0.00409	0.00318	0.00248	0.00193	0.00150	0.00117	0.00091	0.00071	0.00055
3	0.01476	0.01173	0.00931	0.00738	0.00585	0.00464	0.00367	0.00291	0.00230	0.00182
4	0.03280	0.02656	0.02148	0.01735	0.01400	0.01128	0.00907	0.00730	0.00586	0.00470
5	0.06225	0.05138	0.04232	0.03479	0.02854	0.02338	0.01912	0.01561	0.01273	0.01036
6	0.10511	0.08838	0.07410	0.06197	0.05170	0.04304	0.03575	0.02964	0.02452	0.02026
7	0.16196	0.13862	0.11825	0.10056	0.08527	0.07211	0.06082	0.05118	0.04297	0.03600
8	0.23167	0.20170	0.17495	0.15120	0.13025	0.11185	0.09577	0.08177	0.06963	0.05915
9	0.31154	0.27571	0.24299	0.21331	0.18657	0.16261	0.14126	0.12233	0.10562	0.09094
10	0.39777	0.35752	0.31991	0.28506	0.25299	0.22367	0.19704	0.17299	0.15138	0.13206
11	0.48605	0.44326	0.40237	0.36364	0.32726	0.29333	0.26190	0.23299	0.20655	0.18250
12	0.57218	0.52892	0.48662	0.44568	0.40640	0.36904	0.33377	0.30071	0.26992	0.24144
13	0.65263	0.61082	0.56901	0.52764	0.48713	0.44781	0.40997	0.37384	0.33960	0.30735
14	0.72479	0.68604	0.64639	0.60630	0.56622	0.52652	0.48759	0.44971	0.41316	0.37815
15	0.78717	0.75259	0.71641	0.67903	0.64086	0.60230	0.56374	0.52553	0.48800	0.45142
16	0.83925	0.80949	0.77762	0.74398	0.70890	0.67276	0.63591	0.59871	0.56152	0.52464
17	0.88135	0.85656	0.82942	0.80014	0.76896	0.73619	0.70212	0.66710	0.63145	0.59548
18	0.91436	0.89436	0.87195	0.84724	0.82038	0.79157	0.76106	0.72909	0.69596	0.66197
19	0.93952	0.92384	0.90587	0.88562	0.86316	0.83857	0.81202	0.78369	0.75380	0.72260
20	0.95817	0.94622	0.93221	0.91608	0.89779	0.87738	0.85492	0.83050	0.80427	0.77641
21	0.97166	0.96279	0.95214	0.93962	0.92513	0.90862	0.89010	0.86960	0.84718	0.82295
22	0.98118	0.97475	0.96686	0.95738	0.94618	0.93316	0.91827	0.90148	0.88279	0.86224
23	0.98773	0.98319	0.97748	0.97047	0.96201	0.95199	0.94030	0.92687	0.91165	0.89463
24	0.99216	0.98901	0.98498	0.97991	0.97367	0.96612	0.95715	0.94665	0.93454	0.92076
25	0.99507	0.99295	0.99015	0.98657	0.98206	0.97650	0.96976	0.96173	0.95230	0.94138
26	0.99696	0.99555	0.99366	0.99117	0.98798	0.98397	0.97902	0.97300	0.96581	0.95733
27	0.99815	0.99724	0.99598	0.99429	0.99208	0.98925	0.98567	0.98125	0.97588	0.96943
28	0.99890	0.99831	0.99749	0.99637	0.99487	0.99290	0.99037	0.98719	0.98324	0.97844
29	0.99935	0.99899	0.99846	0.99773	0.99672	0.99538	0.99363	0.99138	0.98854	0.98502
30	0.99963	0.99940	0.99907	0.99860	0.99794	0.99704	0.99585	0.99428	0.99227	0.98974

$\chi^2=$	15.5	16.0	16.5	17.0	17.5	18.0	18.5	19.0	19.5	20.0
ν $m=$	7.75	8.0	8.25	8.5	8.75	9.0	9.25	9.5	9.75	10.0
1	0.00008	0.00006	0.00005	0.00004	0.00003	0.00002	0.00002	0.00001	0.00001	0.00001
2	0.00043	0.00034	0.00026	0.00020	0.00016	0.00012	0.00010	0.00008	0.00006	0.00005
3	0.00144	0.00113	0.00090	0.00071	0.00056	0.00044	0.00035	0.00027	0.00022	0.00017
4	0.00377	0.00302	0.00242	0.00193	0.00154	0.00123	0.00099	0.00079	0.00063	0.00050
5	0.00843	0.00684	0.00555	0.00450	0.00364	0.00295	0.00238	0.00192	0.00155	0.00125
6	0.01670	0.01375	0.01131	0.00928	0.00761	0.00623	0.00510	0.00416	0.00340	0.00277
7	0.03010	0.02512	0.02092	0.01740	0.01444	0.01197	0.00991	0.00819	0.00676	0.00557
8	0.05012	0.04238	0.03576	0.03011	0.02530	0.02123	0.01777	0.01486	0.01240	0.01034
9	0.07809	0.06688	0.05715	0.04872	0.04144	0.03517	0.02980	0.02519	0.02126	0.01791
10	0.11487	0.09963	0.08619	0.07436	0.06401	0.05496	0.04709	0.04026	0.03435	0.02925
11	0.16073	0.14113	0.12356	0.10788	0.09393	0.08158	0.07068	0.06109	0.05269	0.04534
12	0.21522	0.19124	0.16939	0.14960	0.13174	0.11569	0.10133	0.08853	0.07716	0.06709
13	0.27719	0.24913	0.22318	0.19930	0.17744	0.15752	0.13944	0.12310	0.10840	0.09521
14	0.34485	0.31337	0.28380	0.25618	0.23051	0.20678	0.18495	0.16495	0.14671	0.13014
15	0.41604	0.38205	0.34962	0.31886	0.28986	0.26267	0.23729	0.21373	0.19196	0.17193
16	0.48837	0.45296	0.41864	0.38560	0.35398	0.32390	0.29544	0.26866	0.24359	0.22022
17	0.55951	0.52383	0.48871	0.45437	0.42102	0.38884	0.35797	0.32853	0.30060	0.27423
18	0.62740	0.59255	0.55770	0.52311	0.48902	0.45565	0.42320	0.39182	0.36166	0.33282
19	0.69033	0.65728	0.62370	0.58987	0.55603	0.52244	0.48931	0.45684	0.42521	0.39458
20	0.74712	0.71662	0.68516	0.65297	0.62031	0.58741	0.55451	0.52183	0.48957	0.45793
21	0.79705	0.76965	0.74093	0.71111	0.68039	0.64900	0.61718	0.58514	0.55310	0.52126
22	0.83990	0.81589	0.79032	0.76336	0.73519	0.70599	0.67597	0.64533	0.61428	0.58304
23	0.87582	0.85527	0.83304	0.80925	0.78402	0.75749	0.72983	0.70122	0.67185	0.64191
24	0.90527	0.88808	0.86919	0.84866	0.82657	0.80301	0.77810	0.75199	0.72483	0.69678
25	0.92891	0.91483	0.89912	0.88179	0.86287	0.84239	0.82044	0.79712	0.77254	0.74683
26	0.94749	0.93620	0.92341	0.90908	0.89320	0.87577	0.85683	0.83643	0.81464	0.79156
27	0.96182	0.95295	0.94274	0.93112	0.91806	0.90352	0.88750	0.87000	0.85107	0.83076
28	0.97266	0.96582	0.95782	0.94859	0.93805	0.92615	0.91285	0.89814	0.88200	0.86446
29	0.98071	0.97554	0.96939	0.96218	0.95383	0.94427	0.93344	0.92129	0.90779	0.89293
30	0.98659	0.98274	0.97810	0.97258	0.96608	0.95853	0.94986	0.94001	0.92891	0.91654

$$\phi=\tfrac{1}{2}\left(\chi^2-\chi_0^2\right) \qquad w=\nu-\nu_0>0$$

Interpolation on χ^2

$$Q(\chi^2|\nu)=Q\left(\chi_0^2|\nu_0-4\right)\left[\tfrac{1}{2}\phi^2\right]+Q\left(\chi_0^2|\nu_0-2\right)\left[\phi-\phi^2\right]+Q\left(\chi_0^2|\nu_0\right)\left[1-\phi+\tfrac{1}{2}\phi^2\right]$$

Double Entry Interpolation

$$Q(\chi^2|\nu)=Q\left(\chi_0^2|\nu_0-4\right)\left[\tfrac{1}{2}\phi^2\right]+Q\left(\chi_0^2|\nu_0-2\right)\left[\phi-\phi^2-w\phi\right]+Q\left(\chi_0^2|\nu_0-1\right)\left[\tfrac{1}{2}w^2-\tfrac{1}{2}w+w\phi\right]$$
$$+Q\left(\chi_0^2|\nu_0\right)\left[1-w^2-\phi+\tfrac{1}{2}\phi^2+w\phi\right]+Q\left(\chi_0^2|\nu_0+1\right)\left[\tfrac{1}{2}w^2+\tfrac{1}{2}w-w\phi\right]$$

Table 26.7 PROBABILITY INTEGRAL OF χ^2-DISTRIBUTION, INCOMPLETE GAMMA FUNCTION
CUMULATIVE SUMS OF THE POISSON DISTRIBUTION

ν	$\chi^2 = 21$ $m = 10.5$	22 11.0	23 11.5	24 12.0	25 12.5	26 13.0	27 13.5	28 14.0	29 14.5	30 15.0
1	0.00001									
2	0.00003	0.00002	0.00001	0.00001						
3	0.00011	0.00007	0.00004	0.00003	0.00002	0.00001	0.00001			
4	0.00032	0.00020	0.00013	0.00008	0.00005	0.00003	0.00002	0.00001	0.00001	0.00001
5	0.00081	0.00052	0.00034	0.00022	0.00014	0.00009	0.00006	0.00004	0.00002	0.00002
6	0.00184	0.00121	0.00080	0.00052	0.00034	0.00022	0.00015	0.00009	0.00006	0.00004
7	0.00377	0.00254	0.00171	0.00114	0.00076	0.00050	0.00033	0.00022	0.00015	0.00010
8	0.00715	0.00492	0.00336	0.00229	0.00155	0.00105	0.00071	0.00047	0.00032	0.00021
9	0.01265	0.00888	0.00620	0.00430	0.00297	0.00204	0.00140	0.00095	0.00065	0.00044
10	0.02109	0.01511	0.01075	0.00760	0.00535	0.00374	0.00260	0.00181	0.00125	0.00086
11	0.03337	0.02437	0.01768	0.01273	0.00912	0.00649	0.00460	0.00324	0.00227	0.00159
12	0.05038	0.03752	0.02773	0.02034	0.01482	0.01073	0.00773	0.00553	0.00394	0.00279
13	0.07293	0.05536	0.04168	0.03113	0.02308	0.01700	0.01244	0.00905	0.00655	0.00471
14	0.10163	0.07861	0.06027	0.04582	0.03457	0.02589	0.01925	0.01423	0.01045	0.00763
15	0.13683	0.10780	0.08414	0.06509	0.04994	0.03802	0.02874	0.02157	0.01609	0.01192
16	0.17851	0.14319	0.11374	0.08950	0.06982	0.05403	0.04148	0.03162	0.02394	0.01800
17	0.22629	0.18472	0.14925	0.11944	0.09471	0.07446	0.05807	0.04494	0.03453	0.02635
18	0.27941	0.23199	0.19059	0.15503	0.12492	0.09976	0.07900	0.06206	0.04838	0.03745
19	0.33680	0.28426	0.23734	0.19615	0.16054	0.13019	0.10465	0.08343	0.06599	0.05180
20	0.39713	0.34051	0.28880	0.24239	0.20143	0.16581	0.13526	0.10940	0.08776	0.06985
21	0.45894	0.39951	0.34398	0.29306	0.24716	0.20645	0.17085	0.14015	0.11400	0.09199
22	0.52074	0.45989	0.40173	0.34723	0.29707	0.25168	0.21123	0.17568	0.14486	0.11846
23	0.58109	0.52025	0.46077	0.40381	0.35029	0.30087	0.25597	0.21578	0.18031	0.14940
24	0.63873	0.57927	0.51980	0.46160	0.40576	0.35317	0.30445	0.26004	0.22013	0.18475
25	0.69261	0.63574	0.57756	0.51937	0.46237	0.40760	0.35588	0.30785	0.26392	0.22429
26	0.74196	0.68870	0.63295	0.57597	0.51898	0.46311	0.40933	0.35846	0.31108	0.26761
27	0.78629	0.73738	0.68501	0.63032	0.57446	0.51860	0.46379	0.41097	0.36090	0.31415
28	0.82535	0.78129	0.73304	0.68154	0.62784	0.57305	0.51825	0.46445	0.41253	0.36322
29	0.85915	0.82019	0.77654	0.72893	0.67825	0.62549	0.57171	0.51791	0.46507	0.41400
30	0.88789	0.85404	0.81526	0.77203	0.72503	0.67513	0.62327	0.57044	0.51760	0.46565

ν	$\chi^2 = 31$ $m = 15.5$	32 16.0	33 16.5	34 17.0	35 17.5	36 18.0	37 18.5	38 19.0	39 19.5	40 20.0
5	0.00001	0.00001								
6	0.00003	0.00002	0.00001	0.00001						
7	0.00006	0.00004	0.00003	0.00002	0.00001	0.00001				
8	0.00014	0.00009	0.00006	0.00004	0.00003	0.00002	0.00001	0.00001		
9	0.00030	0.00020	0.00013	0.00009	0.00006	0.00004	0.00003	0.00002	0.00001	0.00001
10	0.00059	0.00040	0.00027	0.00019	0.00012	0.00008	0.00006	0.00004	0.00003	0.00002
11	0.00110	0.00076	0.00053	0.00036	0.00025	0.00017	0.00012	0.00008	0.00005	0.00004
12	0.00197	0.00138	0.00097	0.00068	0.00047	0.00032	0.00022	0.00015	0.00011	0.00007
13	0.00337	0.00240	0.00170	0.00120	0.00085	0.00059	0.00041	0.00029	0.00020	0.00014
14	0.00554	0.00401	0.00288	0.00206	0.00147	0.00104	0.00074	0.00052	0.00036	0.00026
15	0.00878	0.00644	0.00469	0.00341	0.00246	0.00177	0.00127	0.00090	0.00064	0.00045
16	0.01346	0.01000	0.00739	0.00543	0.00397	0.00289	0.00210	0.00151	0.00109	0.00078
17	0.01997	0.01505	0.01127	0.00840	0.00622	0.00459	0.00337	0.00246	0.00179	0.00129
18	0.02879	0.02199	0.01669	0.01260	0.00945	0.00706	0.00524	0.00387	0.00285	0.00209
19	0.04037	0.03125	0.02404	0.01838	0.01397	0.01056	0.00793	0.00593	0.00442	0.00327
20	0.05519	0.04330	0.03374	0.02613	0.02010	0.01538	0.01170	0.00886	0.00667	0.00500
21	0.07366	0.05855	0.04622	0.03624	0.02824	0.02187	0.01683	0.01289	0.00981	0.00744
22	0.09612	0.07740	0.06187	0.04912	0.03875	0.03037	0.02366	0.01832	0.01411	0.01081
23	0.12279	0.10014	0.08107	0.06516	0.05202	0.04125	0.03251	0.02547	0.01984	0.01537
24	0.15378	0.12699	0.10407	0.08467	0.06840	0.05489	0.04376	0.03467	0.02731	0.02139
25	0.18902	0.15801	0.13107	0.10791	0.08820	0.07160	0.05774	0.04626	0.03684	0.02916
26	0.22827	0.19312	0.16210	0.13502	0.11165	0.09167	0.07475	0.06056	0.04875	0.03901
27	0.27114	0.23208	0.19707	0.16605	0.13887	0.11530	0.09507	0.07786	0.06336	0.05124
28	0.31708	0.27451	0.23574	0.20087	0.16987	0.14260	0.11886	0.09840	0.08092	0.06613
29	0.36542	0.31987	0.27774	0.23926	0.20454	0.17356	0.14622	0.12234	0.10166	0.08394
30	0.41541	0.36753	0.32254	0.28083	0.24264	0.20808	0.17714	0.14975	0.12573	0.10486

PROBABILITY INTEGRAL OF χ^2-DISTRIBUTION, INCOMPLETE GAMMA FUNCTION Table 26.7
CUMULATIVE SUMS OF THE POISSON DISTRIBUTION

ν	$\chi^2=42$ $m=21$	44 22	46 23	48 24	50 25	52 26	54 27	56 28	58 29	60 30
10	0.00001									
11	0.00002	0.00001								
12	0.00003	0.00002	0.00001							
13	0.00006	0.00003	0.00001	0.00001						
14	0.00012	0.00006	0.00003	0.00001	0.00001					
15	0.00023	0.00011	0.00005	0.00003	0.00001	0.00001				
16	0.00040	0.00020	0.00010	0.00005	0.00002	0.00001	0.00001			
17	0.00067	0.00034	0.00017	0.00009	0.00004	0.00002	0.00001	0.00001		
18	0.00111	0.00058	0.00030	0.00015	0.00008	0.00004	0.00002	0.00001		
19	0.00177	0.00094	0.00050	0.00026	0.00013	0.00007	0.00003	0.00002	0.00001	
20	0.00277	0.00151	0.00081	0.00043	0.00022	0.00011	0.00006	0.00003	0.00001	0.00001
21	0.00421	0.00234	0.00128	0.00069	0.00036	0.00019	0.00010	0.00005	0.00003	0.00001
22	0.00625	0.00355	0.00198	0.00109	0.00059	0.00031	0.00016	0.00009	0.00004	0.00002
23	0.00908	0.00526	0.00299	0.00167	0.00092	0.00050	0.00027	0.00014	0.00007	0.00004
24	0.01291	0.00763	0.00443	0.00252	0.00142	0.00078	0.00043	0.00023	0.00012	0.00006
25	0.01797	0.01085	0.00642	0.00373	0.00213	0.00120	0.00066	0.00036	0.00020	0.00011
26	0.02455	0.01512	0.00912	0.00540	0.00314	0.00180	0.00102	0.00056	0.00031	0.00017
27	0.03292	0.02068	0.01272	0.00768	0.00455	0.00265	0.00152	0.00086	0.00048	0.00026
28	0.04336	0.02779	0.01743	0.01072	0.00647	0.00384	0.00224	0.00129	0.00073	0.00041
29	0.05616	0.03670	0.02346	0.01470	0.00903	0.00545	0.00324	0.00189	0.00109	0.00062
30	0.07157	0.04769	0.03107	0.01983	0.01240	0.00762	0.00460	0.00273	0.00160	0.00092

ν	$\chi^2=62$ $m=31$	64 32	66 33	68 34	70 35	72 36	74 37	76 38
21	0.00001							
22	0.00001	0.00001						
23	0.00002	0.00001	0.00001					
24	0.00003	0.00002	0.00001					
25	0.00006	0.00003	0.00002	0.00001				
26	0.00009	0.00005	0.00003	0.00001	0.00001			
27	0.00014	0.00008	0.00004	0.00002	0.00001	0.00001		
28	0.00023	0.00012	0.00007	0.00004	0.00002	0.00001	0.00001	
29	0.00035	0.00019	0.00011	0.00006	0.00003	0.00002	0.00001	
30	0.00052	0.00029	0.00016	0.00009	0.00005	0.00003	0.00001	0.00001

$$Q(\chi^2|\nu)=1-P(\chi^2|\nu)=\left[2^{\frac{\nu}{2}}\Gamma\left(\frac{\nu}{2}\right)\right]^{-1}\int_{\chi^2}^{\infty}e^{-\frac{t}{2}}t^{\frac{\nu}{2}-1}dt=\left[\Gamma\left(\frac{\nu}{2}\right)\right]^{-1}\int_{\frac{1}{2}\chi^2}^{\infty}e^{-t}t^{\frac{\nu}{2}-1}dt=\sum_{j=0}^{c-1}e^{-m}m^j/j!\ (\nu\ \text{even},\ c=\tfrac{1}{2}\nu,\ m=\tfrac{1}{2}\chi^2)$$

$$\phi=\tfrac{1}{2}\left(\chi^2-\chi_0^2\right)\qquad w=\nu-\nu_0>0$$

Interpolation on χ^2

$$Q(\chi^2|\nu)=Q\left(\chi_0^2|\nu_0-4\right)\left[\tfrac{1}{2}\phi^2\right]+Q\left(\chi_0^2|\nu_0-2\right)\left[\phi-\phi^2\right]+Q\left(\chi_0^2|\nu_0\right)\left[1-\phi+\tfrac{1}{2}\phi^2\right]$$

Double Entry Interpolation

$$Q\left(\chi^2|\nu\right)=Q\left(\chi_0^2|\nu_0-4\right)\left[\tfrac{1}{2}\phi^2\right]+Q\left(\chi_0^2|\nu_0-2\right)\left[\phi-\phi^2-w\phi\right]+Q\left(\chi_0^2|\nu_0-1\right)\left[\tfrac{1}{2}w^2-\tfrac{1}{2}w+w\phi\right]$$

$$+Q\left(\chi_0^2|\nu_0\right)\left[1-w^2-\phi+\tfrac{1}{2}\phi^2+w\phi\right]+Q\left(\chi_0^2|\nu_0+1\right)\left[\tfrac{1}{2}w^2+\tfrac{1}{2}w-w\phi\right]$$

Table 26.8   **PERCENTAGE POINTS OF THE χ^2-DISTRIBUTION—VALUES OF χ^2 IN TERMS OF Q AND ν**

$\nu \backslash Q$	0.995	0.99	0.975	0.95	0.9	0.75	0.5	0.25
1	(−5)3.92704	(−4)1.57088	(−4)9.82069	(−3)3.93214	0.0157908	0.101531	0.454937	1.32330
2	(−2)1.00251	(−2)2.01007	(−2)5.06356	0.102587	0.210720	0.575364	1.38629	2.77259
3	(−2)7.17212	0.114832	0.215795	0.351846	0.584375	1.212534	2.36597	4.10835
4	0.206990	0.297110	0.484419	0.710721	1.063623	1.92255	3.35670	5.38527
5	0.411740	0.554300	0.831211	1.145476	1.61031	2.67460	4.35146	6.62568
6	0.675727	0.872085	1.237347	1.63539	2.20413	3.45460	5.34812	7.84080
7	0.989265	1.239043	1.68987	2.16735	2.83311	4.25485	6.34581	9.03715
8	1.344419	1.646482	2.17973	2.73264	3.48954	5.07064	7.34412	10.2188
9	1.734926	2.087912	2.70039	3.32511	4.16816	5.89883	8.34283	11.3887
10	2.15585	2.55821	3.24697	3.94030	4.86518	6.73720	9.34182	12.5489
11	2.60321	3.05347	3.81575	4.57481	5.57779	7.58412	10.3410	13.7007
12	3.07382	3.57056	4.40379	5.22603	6.30380	8.43842	11.3403	14.8454
13	3.56503	4.10691	5.00874	5.89186	7.04150	9.29906	12.3398	15.9839
14	4.07468	4.66043	5.62872	6.57063	7.78953	10.1653	13.3393	17.1170
15	4.60094	5.22935	6.26214	7.26094	8.54675	11.0365	14.3389	18.2451
16	5.14224	5.81221	6.90766	7.96164	9.31223	11.9122	15.3385	19.3688
17	5.69724	6.40776	7.56418	8.67176	10.0852	12.7919	16.3381	20.4887
18	6.26481	7.01491	8.23075	9.39046	10.8649	13.6753	17.3379	21.6049
19	6.84398	7.63273	8.90655	10.1170	11.6509	14.5620	18.3376	22.7178
20	7.43386	8.26040	9.59083	10.8508	12.4426	15.4518	19.3374	23.8277
21	8.03366	8.89720	10.28293	11.5913	13.2396	16.3444	20.3372	24.9348
22	8.64272	9.54249	10.9823	12.3380	14.0415	17.2396	21.3370	26.0393
23	9.26042	10.19567	11.6885	13.0905	14.8479	18.1373	22.3369	27.1413
24	9.88623	10.8564	12.4011	13.8484	15.6587	19.0372	23.3367	28.2412
25	10.5197	11.5240	13.1197	14.6114	16.4734	19.9393	24.3366	29.3389
26	11.1603	12.1981	13.8439	15.3791	17.2919	20.8434	25.3364	30.4345
27	11.8076	12.8786	14.5733	16.1513	18.1138	21.7494	26.3363	31.5284
28	12.4613	13.5648	15.3079	16.9279	18.9392	22.6572	27.3363	32.6205
29	13.1211	14.2565	16.0471	17.7083	19.7677	23.5666	28.3362	33.7109
30	13.7867	14.9535	16.7908	18.4926	20.5992	24.4776	29.3360	34.7998
40	20.7065	22.1643	24.4331	26.5093	29.0505	33.6603	39.3354	45.6160
50	27.9907	29.7067	32.3574	34.7642	37.6886	42.9421	49.3349	56.3336
60	35.5346	37.4848	40.4817	43.1879	46.4589	52.2938	59.3347	66.9814
70	43.2752	45.4418	48.7576	51.7393	55.3290	61.6983	69.3344	77.5766
80	51.1720	53.5400	57.1532	60.3915	64.2778	71.1445	79.3343	88.1303
90	59.1963	61.7541	65.6466	69.1260	73.2912	80.6247	89.3342	98.6499
100	67.3276	70.0648	74.2219	77.9295	82.3581	90.1332	99.3341	109.141
X	−2.5758	−2.3263	−1.9600	−1.6449	−1.2816	−0.6745	0.0000	0.6745

$$Q(\chi^2 | \nu) = \left[2^{\frac{\nu}{2}} \Gamma \left(\frac{\nu}{2} \right) \right]^{-1} \int_{\chi^2}^{\infty} e^{-\frac{t}{2}} t^{\frac{\nu}{2}-1} \, dt$$

From E. S. Pearson and H. O. Hartley (editors), Biometrika tables for statisticians, vol. I. Cambridge Univ. Press, Cambridge, England, 1954 (with permission) for $Q > 0.0005$.

PERCENTAGE POINTS OF THE χ^2-DISTRIBUTION—VALUES OF χ^2 IN TERMS OF Q AND ν

Table 26.8

$\nu \backslash Q$	0.1	0.05	0.025	0.01	0.005	0.001	0.0005	0.0001
1	2.70554	3.84146	5.02389	6.63490	7.87944	10.828	12.116	15.137
2	4.60517	5.99147	7.37776	9.21034	10.5966	13.816	15.202	18.421
3	6.25139	7.81473	9.34840	11.3449	12.8381	16.266	17.730	21.108
4	7.77944	9.48773	11.1433	13.2767	14.8602	18.467	19.997	23.513
5	9.23635	11.0705	12.8325	15.0863	16.7496	20.515	22.105	25.745
6	10.6446	12.5916	14.4494	16.8119	18.5476	22.458	24.103	27.856
7	12.0170	14.0671	16.0128	18.4753	20.2777	24.322	26.018	29.877
8	13.3616	15.5073	17.5346	20.0902	21.9550	26.125	27.868	31.828
9	14.6837	16.9190	19.0228	21.6660	23.5893	27.877	29.666	33.720
10	15.9871	18.3070	20.4831	23.2093	25.1882	29.588	31.420	35.564
11	17.2750	19.6751	21.9200	24.7250	26.7569	31.264	33.137	37.367
12	18.5494	21.0261	23.3367	26.2170	28.2995	32.909	34.821	39.134
13	19.8119	22.3621	24.7356	27.6883	29.8194	34.528	36.478	40.871
14	21.0642	23.6848	26.1190	29.1413	31.3193	36.123	38.109	42.579
15	22.3072	24.9958	27.4884	30.5779	32.8013	37.697	39.719	44.263
16	23.5418	26.2962	28.8454	31.9999	34.2672	39.252	41.308	45.925
17	24.7690	27.5871	30.1910	33.4087	35.7185	40.790	42.879	47.566
18	25.9894	28.8693	31.5264	34.8053	37.1564	42.312	44.434	49.189
19	27.2036	30.1435	32.8523	36.1908	38.5822	43.820	45.973	50.796
20	28.4120	31.4104	34.1696	37.5662	39.9968	45.315	47.498	52.386
21	29.6151	32.6705	35.4789	38.9321	41.4010	46.797	49.011	53.962
22	30.8133	33.9244	36.7807	40.2894	42.7956	48.268	50.511	55.525
23	32.0069	35.1725	38.0757	41.6384	44.1813	49.728	52.000	57.075
24	33.1963	36.4151	39.3641	42.9798	45.5585	51.179	53.479	58.613
25	34.3816	37.6525	40.6465	44.3141	46.9278	52.620	54.947	60.140
26	35.5631	38.8852	41.9232	45.6417	48.2899	54.052	56.407	61.657
27	36.7412	40.1133	43.1944	46.9630	49.6449	55.476	57.858	63.164
28	37.9159	41.3372	44.4607	48.2782	50.9933	56.892	59.300	64.662
29	39.0875	42.5569	45.7222	49.5879	52.3356	58.302	60.735	66.152
30	40.2560	43.7729	46.9792	50.8922	53.6720	59.703	62.162	67.633
40	51.8050	55.7585	59.3417	63.6907	66.7659	73.402	76.095	82.062
50	63.1671	67.5048	71.4202	76.1539	79.4900	86.661	89.560	95.969
60	74.3970	79.0819	83.2976	88.3794	91.9517	99.607	102.695	109.503
70	85.5271	90.5312	95.0231	100.425	104.215	112.317	115.578	122.755
80	96.5782	101.879	106.629	112.329	116.321	124.839	128.261	135.783
90	107.565	113.145	118.136	124.116	128.299	137.208	140.782	148.627
100	118.498	124.342	129.561	135.807	140.169	149.449	153.167	161.319
X	1.2816	1.6449	1.9600	2.3263	2.5758	3.0902	3.2905	3.7190

$$Q(\chi^2 \,|\, \nu) = \left[2^{\frac{\nu}{2}} \Gamma\left(\frac{\nu}{2}\right) \right]^{-1} \int_{\chi^2}^{\infty} e^{-\frac{t}{2}} t^{\frac{\nu}{2}-1} \, dt$$

Table 26.9   PERCENTAGE POINTS OF THE *F*-DISTRIBUTION—VALUES OF *F* IN TERMS OF Q, ν_1, ν_2

$$Q(F|\nu_1, \nu_2) = 0.5$$

$\nu_2 \backslash \nu_1$	1	2	3	4	5	6	8	12	15	20	30	60	∞
1	1.00	1.50	1.71	1.82	1.89	1.94	2.00	2.07	2.09	2.12	2.15	2.17	2.20
2	0.667	1.00	1.13	1.21	1.25	1.28	1.32	1.36	1.38	1.39	1.41	1.43	1.44
3	0.585	0.881	1.00	1.06	1.10	1.13	1.16	1.20	1.21	1.23	1.24	1.25	1.27
4	0.549	0.828	0.941	1.00	1.04	1.06	1.09	1.13	1.14	1.15	1.16	1.18	1.19
5	0.528	0.799	0.907	0.965	1.00	1.02	1.05	1.09	1.10	1.11	1.12	1.14	1.15
6	0.515	0.780	0.886	0.942	0.977	1.00	1.03	1.06	1.07	1.08	1.10	1.11	1.12
7	0.506	0.767	0.871	0.926	0.960	0.983	1.01	1.04	1.05	1.07	1.08	1.09	1.10
8	0.499	0.757	0.860	0.915	0.948	0.971	1.00	1.03	1.04	1.05	1.07	1.08	1.09
9	0.494	0.749	0.852	0.906	0.939	0.962	0.990	1.02	1.03	1.04	1.05	1.07	1.08
10	0.490	0.743	0.845	0.899	0.932	0.954	0.983	1.01	1.02	1.03	1.05	1.06	1.07
11	0.486	0.739	0.840	0.893	0.926	0.948	0.977	1.01	1.02	1.03	1.04	1.05	1.06
12	0.484	0.735	0.835	0.888	0.921	0.943	0.972	1.00	1.01	1.02	1.03	1.05	1.06
13	0.481	0.731	0.832	0.885	0.917	0.939	0.967	0.996	1.01	1.02	1.03	1.04	1.05
14	0.479	0.729	0.828	0.881	0.914	0.936	0.964	0.992	1.00	1.01	1.03	1.04	1.05
15	0.478	0.726	0.826	0.878	0.911	0.933	0.960	0.989	1.00	1.01	1.02	1.03	1.05
16	0.476	0.724	0.823	0.876	0.908	0.930	0.958	0.986	0.997	1.01	1.02	1.03	1.04
17	0.475	0.722	0.821	0.874	0.906	0.928	0.955	0.983	0.995	1.01	1.02	1.03	1.04
18	0.474	0.721	0.819	0.872	0.904	0.926	0.953	0.981	0.992	1.00	1.02	1.03	1.04
19	0.473	0.719	0.818	0.870	0.902	0.924	0.951	0.979	0.990	1.00	1.01	1.02	1.04
20	0.472	0.718	0.816	0.868	0.900	0.922	0.950	0.977	0.989	1.00	1.01	1.02	1.03
21	0.471	0.716	0.815	0.867	0.899	0.921	0.948	0.976	0.987	0.998	1.01	1.02	1.03
22	0.470	0.715	0.814	0.866	0.898	0.919	0.947	0.974	0.986	0.997	1.01	1.02	1.03
23	0.470	0.714	0.813	0.864	0.896	0.918	0.945	0.973	0.984	0.996	1.01	1.02	1.03
24	0.469	0.714	0.812	0.863	0.895	0.917	0.944	0.972	0.983	0.994	1.01	1.02	1.03
25	0.468	0.713	0.811	0.862	0.894	0.916	0.943	0.971	0.982	0.993	1.00	1.02	1.03
26	0.468	0.712	0.810	0.861	0.893	0.915	0.942	0.970	0.981	0.992	1.00	1.01	1.03
27	0.467	0.711	0.809	0.861	0.892	0.914	0.941	0.969	0.980	0.991	1.00	1.01	1.03
28	0.467	0.711	0.808	0.860	0.892	0.913	0.940	0.968	0.979	0.990	1.00	1.01	1.02
29	0.466	0.710	0.808	0.859	0.891	0.912	0.940	0.967	0.978	0.990	1.00	1.01	1.02
30	0.466	0.709	0.807	0.858	0.890	0.912	0.939	0.966	0.978	0.989	1.00	1.01	1.02
40	0.463	0.705	0.802	0.854	0.885	0.907	0.934	0.961	0.972	0.983	0.994	1.01	1.02
60	0.461	0.701	0.798	0.849	0.880	0.901	0.928	0.956	0.967	0.978	0.989	1.00	1.01
120	0.458	0.697	0.793	0.844	0.875	0.896	0.923	0.950	0.961	0.972	0.983	0.994	1.01
∞	0.455	0.693	0.789	0.839	0.870	0.891	0.918	0.945	0.956	0.967	0.978	0.989	1.00

$$Q(F|\nu_1, \nu_2) = 0.25$$

$\nu_2 \backslash \nu_1$	1	2	3	4	5	6	8	12	15	20	30	60	∞
1	5.83	7.50	8.20	8.58	8.82	8.98	9.19	9.41	9.49	9.58	9.67	9.76	9.85
2	2.57	3.00	3.15	3.23	3.28	3.31	3.35	3.39	3.41	3.43	3.44	3.46	3.48
3	2.02	2.28	2.36	2.39	2.41	2.42	2.44	2.45	2.46	2.46	2.47	2.47	2.47
4	1.81	2.00	2.05	2.06	2.07	2.08	2.08	2.08	2.08	2.08	2.08	2.08	2.08
5	1.69	1.85	1.88	1.89	1.89	1.89	1.89	1.89	1.89	1.88	1.88	1.87	1.87
6	1.62	1.76	1.78	1.79	1.79	1.78	1.78	1.77	1.76	1.76	1.75	1.74	1.74
7	1.57	1.70	1.72	1.72	1.71	1.71	1.70	1.68	1.68	1.67	1.66	1.65	1.65
8	1.54	1.66	1.67	1.66	1.66	1.65	1.64	1.62	1.62	1.61	1.60	1.59	1.58
9	1.51	1.62	1.63	1.63	1.62	1.61	1.60	1.58	1.57	1.56	1.55	1.54	1.53
10	1.49	1.60	1.60	1.59	1.59	1.58	1.56	1.54	1.53	1.52	1.51	1.50	1.48
11	1.47	1.58	1.58	1.57	1.56	1.55	1.53	1.51	1.50	1.49	1.48	1.47	1.45
12	1.46	1.56	1.56	1.55	1.54	1.53	1.51	1.49	1.48	1.47	1.45	1.44	1.42
13	1.45	1.55	1.55	1.53	1.52	1.51	1.49	1.47	1.46	1.45	1.43	1.42	1.40
14	1.44	1.53	1.53	1.52	1.51	1.50	1.48	1.45	1.44	1.43	1.41	1.40	1.38
15	1.43	1.52	1.52	1.51	1.49	1.48	1.46	1.44	1.43	1.41	1.40	1.38	1.36
16	1.42	1.51	1.51	1.50	1.48	1.47	1.45	1.43	1.41	1.40	1.38	1.36	1.34
17	1.42	1.51	1.50	1.49	1.47	1.46	1.44	1.41	1.40	1.39	1.37	1.35	1.33
18	1.41	1.50	1.49	1.48	1.46	1.45	1.43	1.40	1.39	1.38	1.36	1.34	1.32
19	1.41	1.49	1.49	1.47	1.46	1.44	1.42	1.40	1.38	1.37	1.35	1.33	1.30
20	1.40	1.49	1.48	1.47	1.45	1.44	1.42	1.39	1.37	1.36	1.34	1.32	1.29
21	1.40	1.48	1.48	1.46	1.44	1.43	1.41	1.38	1.37	1.35	1.33	1.31	1.28
22	1.40	1.48	1.47	1.45	1.44	1.42	1.40	1.37	1.36	1.34	1.32	1.30	1.28
23	1.39	1.47	1.47	1.45	1.43	1.42	1.40	1.37	1.35	1.34	1.32	1.30	1.27
24	1.39	1.47	1.46	1.44	1.43	1.41	1.39	1.36	1.35	1.33	1.31	1.29	1.26
25	1.39	1.47	1.46	1.44	1.42	1.41	1.39	1.36	1.34	1.33	1.31	1.28	1.25
26	1.38	1.46	1.45	1.44	1.42	1.41	1.38	1.35	1.34	1.32	1.30	1.28	1.25
27	1.38	1.46	1.45	1.43	1.42	1.40	1.38	1.35	1.33	1.32	1.30	1.27	1.24
28	1.38	1.46	1.45	1.43	1.41	1.40	1.38	1.34	1.33	1.31	1.29	1.27	1.24
29	1.38	1.45	1.45	1.43	1.41	1.40	1.37	1.34	1.32	1.31	1.29	1.26	1.23
30	1.38	1.45	1.44	1.42	1.41	1.39	1.37	1.34	1.32	1.30	1.28	1.26	1.23
40	1.36	1.44	1.42	1.40	1.39	1.37	1.35	1.31	1.30	1.28	1.25	1.22	1.19
60	1.35	1.42	1.41	1.38	1.37	1.35	1.32	1.29	1.27	1.25	1.22	1.19	1.15
120	1.34	1.40	1.39	1.37	1.35	1.33	1.30	1.26	1.24	1.22	1.19	1.16	1.10
∞	1.32	1.39	1.37	1.35	1.33	1.31	1.28	1.24	1.22	1.19	1.16	1.12	1.00

Compiled from E. S. Pearson and H. O. Hartley (editors), Biometrika tables for statisticians, vol. I. Cambridge Univ. Press, Cambridge, England, 1954 (with permission).

PERCENTAGE POINTS OF THE *F*-DISTRIBUTION—VALUES Table 26.9
OF *F* IN TERMS OF *Q*, ν_1, ν_2

$$Q(F|\nu_1,\nu_2)=0.1$$

$\nu_2\backslash\nu_1$	1	2	3	4	5	6	8	12	15	20	30	60	∞
1	39.86	49.50	53.59	55.83	57.24	58.20	59.44	60.71	61.22	61.74	62.26	62.79	63.33
2	8.53	9.00	9.16	9.24	9.29	9.33	9.37	9.41	9.42	9.44	9.46	9.47	9.49
3	5.54	5.46	5.39	5.34	5.31	5.28	5.25	5.22	5.20	5.18	5.17	5.15	5.13
4	4.54	4.32	4.19	4.11	4.05	4.01	3.95	3.90	3.87	3.84	3.82	3.79	3.76
5	4.06	3.78	3.62	3.52	3.45	3.40	3.34	3.27	3.24	3.21	3.17	3.14	3.10
6	3.78	3.46	3.29	3.18	3.11	3.05	2.98	2.90	2.87	2.84	2.80	2.76	2.72
7	3.59	3.26	3.07	2.96	2.88	2.83	2.75	2.67	2.63	2.59	2.56	2.51	2.47
8	3.46	3.11	2.92	2.81	2.73	2.67	2.59	2.50	2.46	2.42	2.38	2.34	2.29
9	3.36	3.01	2.81	2.69	2.61	2.55	2.47	2.38	2.34	2.30	2.25	2.21	2.16
10	3.29	2.92	2.73	2.61	2.52	2.46	2.38	2.28	2.24	2.20	2.16	2.11	2.06
11	3.23	2.86	2.66	2.54	2.45	2.39	2.30	2.21	2.17	2.12	2.08	2.03	1.97
12	3.18	2.81	2.61	2.48	2.39	2.33	2.24	2.15	2.10	2.06	2.01	1.96	1.90
13	3.14	2.76	2.56	2.43	2.35	2.28	2.20	2.10	2.05	2.01	1.96	1.90	1.85
14	3.10	2.73	2.52	2.39	2.31	2.24	2.15	2.05	2.01	1.96	1.91	1.86	1.80
15	3.07	2.70	2.49	2.36	2.27	2.21	2.12	2.02	1.97	1.92	1.87	1.82	1.76
16	3.05	2.67	2.46	2.33	2.24	2.18	2.09	1.99	1.94	1.89	1.84	1.78	1.72
17	3.03	2.64	2.44	2.31	2.22	2.15	2.06	1.96	1.91	1.86	1.81	1.75	1.69
18	3.01	2.62	2.42	2.29	2.20	2.13	2.04	1.93	1.89	1.84	1.78	1.72	1.66
19	2.99	2.61	2.40	2.27	2.18	2.11	2.02	1.91	1.86	1.81	1.76	1.70	1.63
20	2.97	2.59	2.38	2.25	2.16	2.09	2.00	1.89	1.84	1.79	1.74	1.68	1.61
21	2.96	2.57	2.36	2.23	2.14	2.08	1.98	1.87	1.83	1.78	1.72	1.66	1.59
22	2.95	2.56	2.35	2.22	2.13	2.06	1.97	1.86	1.81	1.76	1.70	1.64	1.57
23	2.94	2.55	2.34	2.21	2.11	2.05	1.95	1.84	1.80	1.74	1.69	1.62	1.55
24	2.93	2.54	2.33	2.19	2.10	2.04	1.94	1.83	1.78	1.73	1.67	1.61	1.53
25	2.92	2.53	2.32	2.18	2.09	2.02	1.93	1.82	1.77	1.72	1.66	1.59	1.52
26	2.91	2.52	2.31	2.17	2.08	2.01	1.92	1.81	1.76	1.71	1.65	1.58	1.50
27	2.90	2.51	2.30	2.17	2.07	2.00	1.91	1.80	1.75	1.70	1.64	1.57	1.49
28	2.89	2.50	2.29	2.16	2.06	2.00	1.90	1.79	1.74	1.69	1.63	1.56	1.48
29	2.89	2.50	2.28	2.15	2.06	1.99	1.89	1.78	1.73	1.68	1.62	1.55	1.47
30	2.88	2.49	2.28	2.14	2.05	1.98	1.88	1.77	1.72	1.67	1.61	1.54	1.46
40	2.84	2.44	2.23	2.09	2.00	1.93	1.83	1.71	1.66	1.61	1.54	1.47	1.38
60	2.79	2.39	2.18	2.04	1.95	1.87	1.77	1.66	1.60	1.54	1.48	1.40	1.29
120	2.75	2.35	2.13	1.99	1.90	1.82	1.72	1.60	1.55	1.48	1.41	1.32	1.19
∞	2.71	2.30	2.08	1.94	1.85	1.77	1.67	1.55	1.49	1.42	1.34	1.24	1.00

$$Q(F|\nu_1,\nu_2)=0.05$$

$\nu_2\backslash\nu_1$	1	2	3	4	5	6	8	12	15	20	30	60	∞
1	161.4	199.5	215.7	224.6	230.2	234.0	238.9	243.9	245.9	248.0	250.1	252.2	254.3
2	18.51	19.00	19.16	19.25	19.30	19.33	19.37	19.41	19.43	19.45	19.46	19.48	19.50
3	10.13	9.55	9.28	9.12	9.01	8.94	8.85	8.74	8.70	8.66	8.62	8.57	8.53
4	7.71	6.94	6.59	6.39	6.26	6.16	6.04	5.91	5.86	5.80	5.75	5.69	5.63
5	6.61	5.79	5.41	5.19	5.05	4.95	4.82	4.68	4.62	4.56	4.50	4.43	4.36
6	5.99	5.14	4.76	4.53	4.39	4.28	4.15	4.00	3.94	3.87	3.81	3.74	3.67
7	5.59	4.74	4.35	4.12	3.97	3.87	3.73	3.57	3.51	3.44	3.38	3.30	3.23
8	5.32	4.46	4.07	3.84	3.69	3.58	3.44	3.28	3.22	3.15	3.08	3.01	2.93
9	5.12	4.26	3.86	3.63	3.48	3.37	3.23	3.07	3.01	2.94	2.86	2.79	2.71
10	4.96	4.10	3.71	3.48	3.33	3.22	3.07	2.91	2.85	2.77	2.70	2.62	2.54
11	4.84	3.98	3.59	3.36	3.20	3.09	2.95	2.79	2.72	2.65	2.57	2.49	2.40
12	4.75	3.89	3.49	3.26	3.11	3.00	2.85	2.69	2.62	2.54	2.47	2.38	2.30
13	4.67	3.81	3.41	3.18	3.03	2.92	2.77	2.60	2.53	2.46	2.38	2.30	2.21
14	4.60	3.74	3.34	3.11	2.96	2.85	2.70	2.53	2.46	2.39	2.31	2.22	2.13
15	4.54	3.68	3.29	3.06	2.90	2.79	2.64	2.48	2.40	2.33	2.25	2.16	2.07
16	4.49	3.63	3.24	3.01	2.85	2.74	2.59	2.42	2.35	2.28	2.19	2.11	2.01
17	4.45	3.59	3.20	2.96	2.81	2.70	2.55	2.38	2.31	2.23	2.15	2.06	1.96
18	4.41	3.55	3.16	2.93	2.77	2.66	2.51	2.34	2.27	2.19	2.11	2.02	1.92
19	4.38	3.52	3.13	2.90	2.74	2.63	2.48	2.31	2.23	2.16	2.07	1.98	1.88
20	4.35	3.49	3.10	2.87	2.71	2.60	2.45	2.28	2.20	2.12	2.04	1.95	1.84
21	4.32	3.47	3.07	2.84	2.68	2.57	2.42	2.25	2.18	2.10	2.01	1.92	1.81
22	4.30	3.44	3.05	2.82	2.66	2.55	2.40	2.23	2.15	2.07	1.98	1.89	1.78
23	4.28	3.42	3.03	2.80	2.64	2.53	2.37	2.20	2.13	2.05	1.96	1.86	1.76
24	4.26	3.40	3.01	2.78	2.62	2.51	2.36	2.18	2.11	2.03	1.94	1.84	1.73
25	4.24	3.39	2.99	2.76	2.60	2.49	2.34	2.16	2.09	2.01	1.92	1.82	1.71
26	4.23	3.37	2.98	2.74	2.59	2.47	2.32	2.15	2.07	1.99	1.90	1.80	1.69
27	4.21	3.35	2.96	2.73	2.57	2.46	2.31	2.13	2.06	1.97	1.88	1.79	1.67
28	4.20	3.34	2.95	2.71	2.56	2.45	2.29	2.12	2.04	1.96	1.87	1.77	1.65
29	4.18	3.33	2.93	2.70	2.55	2.43	2.28	2.10	2.03	1.94	1.85	1.75	1.64
30	4.17	3.32	2.92	2.69	2.53	2.42	2.27	2.09	2.01	1.93	1.84	1.74	1.62
40	4.08	3.23	2.84	2.61	2.45	2.34	2.18	2.00	1.92	1.84	1.74	1.64	1.51
60	4.00	3.15	2.76	2.53	2.37	2.25	2.10	1.92	1.84	1.75	1.65	1.53	1.39
120	3.92	3.07	2.68	2.45	2.29	2.17	2.02	1.83	1.75	1.66	1.55	1.43	1.25
∞	3.84	3.00	2.60	2.37	2.21	2.10	1.94	1.75	1.67	1.57	1.46	1.32	1.00

Table 26.9   PERCENTAGE POINTS OF THE F-DISTRIBUTION—VALUES
OF F IN TERMS OF Q, ν_1, ν_2

$$Q(F|\nu_1,\nu_2)=0.025$$

$\nu_2\backslash\nu_1$	1	2	3	4	5	6	8	12	15	20	30	60	∞
1	647.8	799.5	864.2	899.6	921.8	937.1	956.7	976.7	984.9	993.1	1001	1010	1018
2	38.51	39.00	39.17	39.25	39.30	39.33	39.37	39.41	39.43	39.45	39.46	39.48	39.50
3	17.44	16.04	15.44	15.10	14.88	14.73	14.54	14.34	14.25	14.17	14.08	13.99	13.90
4	12.22	10.65	9.98	9.60	9.36	9.20	8.98	8.75	8.66	8.56	8.46	8.36	8.26
5	10.01	8.43	7.76	7.39	7.15	6.98	6.76	6.52	6.43	6.33	6.23	6.12	6.02
6	8.81	7.26	6.60	6.23	5.99	5.82	5.60	5.37	5.27	5.17	5.07	4.96	4.85
7	8.07	6.54	5.89	5.52	5.29	5.12	4.90	4.67	4.57	4.47	4.36	4.25	4.14
8	7.57	6.06	5.42	5.05	4.82	4.65	4.43	4.20	4.10	4.00	3.89	3.78	3.67
9	7.21	5.71	5.08	4.72	4.48	4.32	4.10	3.87	3.77	3.67	3.56	3.45	3.33
10	6.94	5.46	4.83	4.47	4.24	4.07	3.85	3.62	3.52	3.42	3.31	3.20	3.08
11	6.72	5.26	4.63	4.28	4.04	3.88	3.66	3.43	3.33	3.23	3.12	3.00	2.88
12	6.55	5.10	4.47	4.12	3.89	3.73	3.51	3.28	3.18	3.07	2.96	2.85	2.72
13	6.41	4.97	4.35	4.00	3.77	3.60	3.39	3.15	3.05	2.95	2.84	2.72	2.60
14	6.30	4.86	4.24	3.89	3.66	3.50	3.29	3.05	2.95	2.84	2.73	2.61	2.49
15	6.20	4.77	4.15	3.80	3.58	3.41	3.20	2.96	2.86	2.76	2.64	2.52	2.40
16	6.12	4.69	4.08	3.73	3.50	3.34	3.12	2.89	2.79	2.68	2.57	2.45	2.32
17	6.04	4.62	4.01	3.66	3.44	3.28	3.06	2.82	2.72	2.62	2.50	2.38	2.25
18	5.98	4.56	3.95	3.61	3.38	3.22	3.01	2.77	2.67	2.56	2.44	2.32	2.19
19	5.92	4.51	3.90	3.56	3.33	3.17	2.96	2.72	2.62	2.51	2.39	2.27	2.13
20	5.87	4.46	3.86	3.51	3.29	3.13	2.91	2.68	2.57	2.46	2.35	2.22	2.09
21	5.83	4.42	3.82	3.48	3.25	3.09	2.87	2.64	2.53	2.42	2.31	2.18	2.04
22	5.79	4.38	3.78	3.44	3.22	3.05	2.84	2.60	2.50	2.39	2.27	2.14	2.00
23	5.75	4.35	3.75	3.41	3.18	3.02	2.81	2.57	2.47	2.36	2.24	2.11	1.97
24	5.72	4.32	3.72	3.38	3.15	2.99	2.78	2.54	2.44	2.33	2.21	2.08	1.94
25	5.69	4.29	3.69	3.35	3.13	2.97	2.75	2.51	2.41	2.30	2.18	2.05	1.91
26	5.66	4.27	3.67	3.33	3.10	2.94	2.73	2.49	2.39	2.28	2.16	2.03	1.88
27	5.63	4.24	3.65	3.31	3.08	2.92	2.71	2.47	2.36	2.25	2.13	2.00	1.85
28	5.61	4.22	3.63	3.29	3.06	2.90	2.69	2.45	2.34	2.23	2.11	1.98	1.83
29	5.59	4.20	3.61	3.27	3.04	2.88	2.67	2.43	2.32	2.21	2.09	1.96	1.81
30	5.57	4.18	3.59	3.25	3.03	2.87	2.65	2.41	2.31	2.20	2.07	1.94	1.79
40	5.42	4.05	3.46	3.13	2.90	2.74	2.53	2.29	2.18	2.07	1.94	1.80	1.64
60	5.29	3.93	3.34	3.01	2.79	2.63	2.41	2.17	2.06	1.94	1.82	1.67	1.48
120	5.15	3.80	3.23	2.89	2.67	2.52	2.30	2.05	1.94	1.82	1.69	1.53	1.31
∞	5.02	3.69	3.12	2.79	2.57	2.41	2.19	1.94	1.83	1.71	1.57	1.39	1.00

$$Q(F|\nu_1,\nu_2)=0.01$$

$\nu_2\backslash\nu_1$	1	2	3	4	5	6	8	12	15	20	30	60	∞
1	4052	4999.5	5403	5625	5764	5859	5982	6106	6157	6209	6261	6313	6366
2	98.50	99.00	99.17	99.25	99.30	99.33	99.37	99.42	99.43	99.45	99.47	99.48	99.50
3	34.12	30.82	29.46	28.71	28.24	27.91	27.49	27.05	26.87	26.69	26.50	26.32	26.13
4	21.20	18.00	16.69	15.98	15.52	15.21	14.80	14.37	14.20	14.02	13.84	13.65	13.46
5	16.26	13.27	12.06	11.39	10.97	10.67	10.29	9.89	9.72	9.55	9.38	9.20	9.02
6	13.75	10.92	9.78	9.15	8.75	8.47	8.10	7.72	7.56	7.40	7.23	7.06	6.88
7	12.25	9.55	8.45	7.85	7.46	7.19	6.84	6.47	6.31	6.16	5.99	5.82	5.65
8	11.26	8.65	7.59	7.01	6.63	6.37	6.03	5.67	5.52	5.36	5.20	5.03	4.86
9	10.56	8.02	6.99	6.42	6.06	5.80	5.47	5.11	4.96	4.81	4.65	4.48	4.31
10	10.04	7.56	6.55	5.99	5.64	5.39	5.06	4.71	4.56	4.41	4.25	4.08	3.91
11	9.65	7.21	6.22	5.67	5.32	5.07	4.74	4.40	4.25	4.10	3.94	3.78	3.60
12	9.33	6.93	5.95	5.41	5.06	4.82	4.50	4.16	4.01	3.86	3.70	3.54	3.36
13	9.07	6.70	5.74	5.21	4.86	4.62	4.30	3.96	3.82	3.66	3.51	3.34	3.17
14	8.86	6.51	5.56	5.04	4.69	4.46	4.14	3.80	3.66	3.51	3.35	3.18	3.00
15	8.68	6.36	5.42	4.89	4.56	4.32	4.00	3.67	3.52	3.37	3.21	3.05	2.87
16	8.53	6.23	5.29	4.77	4.44	4.20	3.89	3.55	3.41	3.26	3.10	2.93	2.75
17	8.40	6.11	5.18	4.67	4.34	4.10	3.79	3.46	3.31	3.16	3.00	2.83	2.65
18	8.29	6.01	5.09	4.58	4.25	4.01	3.71	3.37	3.23	3.08	2.92	2.75	2.57
19	8.18	5.93	5.01	4.50	4.17	3.94	3.63	3.30	3.15	3.00	2.84	2.67	2.49
20	8.10	5.85	4.94	4.43	4.10	3.87	3.56	3.23	3.09	2.94	2.78	2.61	2.42
21	8.02	5.78	4.87	4.37	4.04	3.81	3.51	3.17	3.03	2.88	2.72	2.55	2.36
22	7.95	5.72	4.82	4.31	3.99	3.76	3.45	3.12	2.98	2.83	2.67	2.50	2.31
23	7.88	5.66	4.76	4.26	3.94	3.71	3.41	3.07	2.93	2.78	2.62	2.45	2.26
24	7.82	5.61	4.72	4.22	3.90	3.67	3.36	3.03	2.89	2.74	2.58	2.40	2.21
25	7.77	5.57	4.68	4.18	3.85	3.63	3.32	2.99	2.85	2.70	2.54	2.36	2.17
26	7.72	5.53	4.64	4.14	3.82	3.59	3.29	2.96	2.81	2.66	2.50	2.33	2.13
27	7.68	5.49	4.60	4.11	3.78	3.56	3.26	2.93	2.78	2.63	2.47	2.29	2.10
28	7.64	5.45	4.57	4.07	3.75	3.53	3.23	2.90	2.75	2.60	2.44	2.26	2.06
29	7.60	5.42	4.54	4.04	3.73	3.50	3.20	2.87	2.73	2.57	2.41	2.23	2.03
30	7.56	5.39	4.51	4.02	3.70	3.47	3.17	2.84	2.70	2.55	2.39	2.21	2.01
40	7.31	5.18	4.31	3.83	3.51	3.29	2.99	2.66	2.52	2.37	2.20	2.02	1.80
60	7.08	4.98	4.13	3.65	3.34	3.12	2.82	2.50	2.35	2.20	2.03	1.84	1.60
120	6.85	4.79	3.95	3.48	3.17	2.96	2.66	2.34	2.19	2.03	1.86	1.66	1.38
∞	6.63	4.61	3.78	3.32	3.02	2.80	2.51	2.18	2.04	1.88	1.70	1.47	1.00

PERCENTAGE POINTS OF THE *F*-DISTRIBUTION—VALUES
OF *F* IN TERMS OF *Q*, ν_1, ν_2
Table 26.9

$$Q(F|\nu_1,\nu_2)=0.005$$

$\nu_2 \backslash \nu_1$	1	2	3	4	5	6	8	12	15	20	30	60	∞
1	16211	20000	21615	22500	23056	23437	23925	24426	24630	24836	25044	25253	25465
2	198.5	199.0	199.2	199.2	199.3	199.3	199.4	199.4	199.4	199.4	199.5	199.5	199.5
3	55.55	49.80	47.47	46.19	45.39	44.84	44.13	43.39	43.08	42.78	42.47	42.15	41.83
4	31.33	26.28	24.26	23.15	22.46	21.97	21.35	20.70	20.44	20.17	19.89	19.61	19.32
5	22.78	18.31	16.53	15.56	14.94	14.51	13.96	13.38	13.15	12.90	12.66	12.40	12.14
6	18.63	14.54	12.92	12.03	11.46	11.07	10.57	10.03	9.81	9.59	9.36	9.12	8.88
7	16.24	12.40	10.88	10.05	9.52	9.16	8.68	8.18	7.97	7.75	7.53	7.31	7.08
8	14.69	11.04	9.60	8.81	8.30	7.95	7.50	7.01	6.81	6.61	6.40	6.18	5.95
9	13.61	10.11	8.72	7.96	7.47	7.13	6.69	6.23	6.03	5.83	5.62	5.41	5.19
10	12.83	9.43	8.08	7.34	6.87	6.54	6.12	5.66	5.47	5.27	5.07	4.86	4.64
11	12.23	8.91	7.60	6.88	6.42	6.10	5.68	5.24	5.05	4.86	4.65	4.44	4.23
12	11.75	8.51	7.23	6.52	6.07	5.76	5.35	4.91	4.72	4.53	4.33	4.12	3.90
13	11.37	8.19	6.93	6.23	5.79	5.48	5.08	4.64	4.46	4.27	4.07	3.87	3.65
14	11.06	7.92	6.68	6.00	5.56	5.26	4.86	4.43	4.25	4.06	3.86	3.66	3.44
15	10.80	7.70	6.48	5.80	5.37	5.07	4.67	4.25	4.07	3.88	3.69	3.48	3.26
16	10.58	7.51	6.30	5.64	5.21	4.91	4.52	4.10	3.92	3.73	3.54	3.33	3.11
17	10.38	7.35	6.16	5.50	5.07	4.78	4.39	3.97	3.79	3.61	3.41	3.21	2.98
18	10.22	7.21	6.03	5.37	4.96	4.66	4.28	3.86	3.68	3.50	3.30	3.10	2.87
19	10.07	7.09	5.92	5.27	4.85	4.56	4.18	3.76	3.59	3.40	3.21	3.00	2.78
20	9.94	6.99	5.82	5.17	4.76	4.47	4.09	3.68	3.50	3.32	3.12	2.92	2.69
21	9.83	6.89	5.73	5.09	4.68	4.39	4.01	3.60	3.43	3.24	3.05	2.84	2.61
22	9.73	6.81	5.65	5.02	4.61	4.32	3.94	3.54	3.36	3.18	2.98	2.77	2.55
23	9.63	6.73	5.58	4.95	4.54	4.26	3.88	3.47	3.30	3.12	2.92	2.71	2.48
24	9.55	6.66	5.52	4.89	4.49	4.20	3.83	3.42	3.25	3.06	2.87	2.66	2.43
25	9.48	6.60	5.46	4.84	4.43	4.15	3.78	3.37	3.20	3.01	2.82	2.61	2.38
26	9.41	6.54	5.41	4.79	4.38	4.10	3.73	3.33	3.15	2.97	2.77	2.56	2.33
27	9.34	6.49	5.36	4.74	4.34	4.06	3.69	3.28	3.11	2.93	2.73	2.52	2.29
28	9.28	6.44	5.32	4.70	4.30	4.02	3.65	3.25	3.07	2.89	2.69	2.48	2.25
29	9.23	6.40	5.28	4.66	4.26	3.98	3.61	3.21	3.04	2.86	2.66	2.45	2.21
30	9.18	6.35	5.24	4.62	4.23	3.95	3.58	3.18	3.01	2.82	2.63	2.42	2.18
40	8.83	6.07	4.98	4.37	3.99	3.71	3.35	2.95	2.78	2.60	2.40	2.18	1.93
60	8.49	5.79	4.73	4.14	3.76	3.49	3.13	2.74	2.57	2.39	2.19	1.96	1.69
120	8.18	5.54	4.50	3.92	3.55	3.28	2.93	2.54	2.37	2.19	1.98	1.75	1.43
∞	7.88	5.30	4.28	3.72	3.35	3.09	2.74	2.36	2.19	2.00	1.79	1.53	1.00

$$Q(F|\nu_1,\nu_2)=0.001$$

$\nu_2 \backslash \nu_1$	1	2	3	4	5	6	8	12	15	20	30	60	∞
1	(5)4.053	(5)5.000	(5)5.404	(5)5.625	(5)5.764	(5)5.859	(5)5.981	(5)6.107	(5)6.158	(5)6.209	(5)6.261	(5)6.313	(5)6.366
2	998.5	999.0	999.2	999.2	999.3	999.3	999.4	999.4	999.4	999.4	999.5	999.5	999.5
3	167.0	148.5	141.1	137.1	134.6	132.8	130.6	128.3	127.4	126.4	125.4	124.5	123.5
4	74.14	61.25	56.18	53.44	51.71	50.53	49.00	47.41	46.76	46.10	45.43	44.75	44.05
5	47.18	37.12	33.20	31.09	29.75	28.84	27.64	26.42	25.91	25.39	24.87	24.33	23.79
6	35.51	27.00	23.70	21.92	20.81	20.03	19.03	17.99	17.56	17.12	16.67	16.21	15.75
7	29.25	21.69	18.77	17.19	16.21	15.52	14.63	13.71	13.32	12.93	12.53	12.12	11.70
8	25.42	18.49	15.83	14.39	13.49	12.86	12.04	11.19	10.84	10.48	10.11	9.73	9.33
9	22.86	16.39	13.90	12.56	11.71	11.13	10.37	9.57	9.24	8.90	8.55	8.19	7.81
10	21.04	14.91	12.55	11.28	10.48	9.92	9.20	8.45	8.13	7.80	7.47	7.12	6.76
11	19.69	13.81	11.56	10.35	9.58	9.05	8.35	7.63	7.32	7.01	6.68	6.35	6.00
12	18.64	12.97	10.80	9.63	8.89	8.38	7.71	7.00	6.71	6.40	6.09	5.76	5.42
13	17.81	12.31	10.21	9.07	8.35	7.86	7.21	6.52	6.23	5.93	5.63	5.30	4.97
14	17.14	11.78	9.73	8.62	7.92	7.43	6.80	6.13	5.85	5.56	5.25	4.94	4.60
15	16.59	11.34	9.34	8.25	7.57	7.09	6.47	5.81	5.54	5.25	4.95	4.64	4.31
16	16.12	10.97	9.00	7.94	7.27	6.81	6.19	5.55	5.27	4.99	4.70	4.39	4.06
17	15.72	10.66	8.73	7.68	7.02	6.56	5.96	5.32	5.05	4.78	4.48	4.18	3.85
18	15.38	10.39	8.49	7.46	6.81	6.35	5.76	5.13	4.87	4.59	4.30	4.00	3.67
19	15.08	10.16	8.28	7.26	6.62	6.18	5.59	4.97	4.70	4.43	4.14	3.84	3.51
20	14.82	9.95	8.10	7.10	6.46	6.02	5.44	4.82	4.56	4.29	4.00	3.70	3.38
21	14.59	9.77	7.94	6.95	6.32	5.88	5.31	4.70	4.44	4.17	3.88	3.58	3.26
22	14.38	9.61	7.80	6.81	6.19	5.76	5.19	4.58	4.33	4.06	3.78	3.48	3.15
23	14.19	9.47	7.67	6.69	6.08	5.65	5.09	4.48	4.23	3.96	3.68	3.38	3.05
24	14.03	9.34	7.55	6.59	5.98	5.55	4.99	4.39	4.14	3.87	3.59	3.29	2.97
25	13.88	9.22	7.45	6.49	5.88	5.46	4.91	4.31	4.06	3.79	3.52	3.22	2.89
26	13.74	9.12	7.36	6.41	5.80	5.38	4.83	4.24	3.99	3.72	3.44	3.15	2.82
27	13.61	9.02	7.27	6.33	5.73	5.31	4.76	4.17	3.92	3.66	3.38	3.08	2.75
28	13.50	8.93	7.19	6.25	5.66	5.24	4.69	4.11	3.86	3.60	3.32	3.02	2.69
29	13.39	8.85	7.12	6.19	5.59	5.18	4.64	4.05	3.80	3.54	3.27	2.97	2.64
30	13.29	8.77	7.05	6.12	5.53	5.12	4.58	4.00	3.75	3.49	3.22	2.92	2.59
40	12.61	8.25	6.60	5.70	5.13	4.73	4.21	3.64	3.40	3.15	2.87	2.57	2.23
60	11.97	7.76	6.17	5.31	4.76	4.37	3.87	3.31	3.08	2.83	2.55	2.25	1.89
120	11.38	7.32	5.79	4.95	4.42	4.04	3.55	3.02	2.78	2.53	2.26	1.95	1.54
∞	10.83	6.91	5.42	4.62	4.10	3.74	3.27	2.74	2.51	2.27	1.99	1.66	1.00

*See page II.

Table 26.10
PERCENTAGE POINTS OF THE t-DISTRIBUTION— VALUES OF t IN TERMS OF A AND ν

ν \ A	0.2	0.5	0.8	0.9	0.95	0.98	0.99	0.995	0.998	0.999	0.9999	0.99999	0.999999
1	0.325	1.000	3.078	6.314	12.706	31.821	63.657	127.321	318.309	636.619	6366.198	63661.977	636619.772
2	0.289	0.816	1.886	2.920	4.303	6.965	9.925	14.089	22.327	31.598	99.992	316.225	999.999
3	0.277	0.765	1.638	2.353	3.182	4.541	5.841	7.453	10.214	12.924	28.000	60.397	130.155
4	0.271	0.741	1.533	2.132	2.776	3.747	4.604	5.598	7.173	8.610	15.544	27.771	49.459
5	0.267	0.727	1.476	2.015	2.571	3.365	4.032	4.773	5.893	6.869	11.178	17.897	28.477
6	0.265	0.718	1.440	1.943	2.447	3.143	3.707	4.317	5.208	5.959	9.082	13.555	20.047
7	0.263	0.711	1.415	1.895	2.365	2.998	3.499	4.029	4.785	5.408	7.885	11.215	15.764
8	0.262	0.706	1.397	1.860	2.306	2.896	3.355	3.833	4.501	5.041	7.120	9.782	13.257
9	0.261	0.703	1.383	1.833	2.262	2.821	3.250	3.690	4.297	4.781	6.594	8.827	11.637
10	0.260	0.700	1.372	1.812	2.228	2.764	3.169	3.581	4.144	4.587	6.211	8.150	10.516
11	0.260	0.697	1.363	1.796	2.201	2.718	3.106	3.497	4.025	4.437	5.921	7.648	9.702
12	0.259	0.695	1.356	1.782	2.179	2.681	3.055	3.428	3.930	4.318	5.694	7.261	9.085
13	0.259	0.694	1.350	1.771	2.160	2.650	3.012	3.372	3.852	4.221	5.513	6.955	8.604
14	0.258	0.692	1.345	1.761	2.145	2.624	2.977	3.326	3.787	4.140	5.363	6.706	8.218
15	0.258	0.691	1.341	1.753	2.131	2.602	2.947	3.286	3.733	4.073	5.239	6.502	7.903
16	0.258	0.690	1.337	1.746	2.120	2.583	2.921	3.252	3.686	4.015	5.134	6.330	7.642
17	0.257	0.689	1.333	1.740	2.110	2.567	2.898	3.223	3.646	3.965	5.044	6.184	7.421
18	0.257	0.688	1.330	1.734	2.101	2.552	2.878	3.197	3.610	3.922	4.966	6.059	7.232
19	0.257	0.688	1.328	1.729	2.093	2.539	2.861	3.174	3.579	3.883	4.897	5.949	7.069
20	0.257	0.687	1.325	1.725	2.086	2.528	2.845	3.153	3.552	3.850	4.837	5.854	6.927
21	0.257	0.686	1.323	1.721	2.080	2.518	2.831	3.135	3.527	3.819	4.784	5.769	6.802
22	0.256	0.686	1.321	1.717	2.074	2.508	2.819	3.119	3.505	3.792	4.736	5.694	6.692
23	0.256	0.685	1.319	1.714	2.069	2.500	2.807	3.104	3.485	3.768	4.693	5.627	6.593
24	0.256	0.685	1.318	1.711	2.064	2.492	2.797	3.090	3.467	3.745	4.654	5.566	6.504
25	0.256	0.684	1.316	1.708	2.060	2.485	2.787	3.078	3.450	3.725	4.619	5.511	6.424
26	0.256	0.684	1.315	1.706	2.056	2.479	2.779	3.067	3.435	3.707	4.587	5.461	6.352
27	0.256	0.684	1.314	1.703	2.052	2.473	2.771	3.057	3.421	3.690	4.558	5.415	6.286
28	0.256	0.683	1.313	1.701	2.048	2.467	2.763	3.047	3.408	3.674	4.530	5.373	6.225
29	0.256	0.683	1.311	1.699	2.045	2.462	2.756	3.038	3.396	3.659	4.506	5.335	6.170
30	0.256	0.683	1.310	1.697	2.042	2.457	2.750	3.030	3.385	3.646	4.482	5.299	6.119
40	0.255	0.681	1.303	1.684	2.021	2.423	2.704	2.971	3.307	3.551	4.321	5.053	5.768
60	0.254	0.679	1.296	1.671	2.000	2.390	2.660	2.915	3.232	3.460	4.169	4.825	5.449
120	0.254	0.677	1.289	1.658	1.980	2.358	2.617	2.860	3.160	3.373	* 4.025	* 4.613	* 5.158
∞	0.253	0.674	1.282	1.645	1.960	2.326	2.576	2.807	3.090	3.291	3.891	4.417	4.892

$$A = A(t|\nu) = \left[\sqrt{\nu} B\left(\frac{1}{2}, \frac{\nu}{2} \right) \right]^{-1} \int_{-t}^{t} \left(1 + \frac{x^2}{\nu} \right)^{-\left(\frac{\nu+1}{2} \right)} dx$$

From E. S. Pearson and H. O. Hartley (editors), Biometrika tables for statisticians, vol. I. Cambridge Univ. Press, Cambridge, England, 1954 for A 0.999, from E. T. Federighi, Extended tables of the percentage points of Student's t-distribution, J. Amer. Statist. Assoc. **54**, 683–688 (1959) for A 0.999 (with permission).

*See page II.

2500 FIVE DIGIT RANDOM NUMBERS — Table 26.11

53479	81115	98036	12217	59526	40238	40577	39351	43211	69255
97344	70328	58116	91964	26240	44643	83287	97391	92823	77578
66023	38277	74523	71118	84892	13956	98899	92315	65783	59640
99776	75723	03172	43112	83086	81982	14538	26162	24899	20551
30176	48979	92153	38416	42436	26636	83903	44722	69210	69117
81874	83339	14988	99937	13213	30177	47967	93793	86693	98854
19839	90630	71863	95053	55532	60908	84108	55342	48479	63799
09337	33435	53869	52769	18801	25820	96198	66518	78314	97013
31151	58295	40823	41330	21093	93882	49192	44876	47185	81425
67619	52515	03037	81699	17106	64982	60834	85319	47814	08075
61946	48790	11602	83043	22257	11832	04344	95541	20366	55937
04811	64892	96346	79065	26999	43967	63485	93572	80753	96582
05763	39601	56140	25513	86151	78657	02184	29715	04334	15678
73260	56877	40794	13948	96289	90185	47111	66807	61849	44686
54909	09976	76580	02645	35795	44537	64428	35441	28318	99001
42583	36335	60068	04044	29678	16342	48592	25547	63177	75225
27266	27403	97520	23334	36453	33699	23672	45884	41515	04756
49843	11442	66682	36055	32002	78600	36924	59962	68191	62580
29316	40460	27076	69232	51423	58515	49920	03901	26597	33068
30463	27856	67798	16837	74273	05793	02900	63498	00782	35097
28708	84088	65535	44258	33869	82530	98399	26387	02836	36838
13183	50652	94872	28257	78547	55286	33591	61965	51723	14211
60796	76639	30157	40295	99476	28334	15368	42481	60312	42770
13486	46918	64683	07411	77842	01908	47796	65796	44230	77230
34914	94502	39374	34185	57500	22514	04060	94511	44612	10485
28105	04814	85170	86490	35695	03483	57315	63174	71902	71182
59231	45028	01173	08848	81925	71494	95401	34049	04851	65914
87437	82758	71093	36833	53582	25986	46005	42840	81683	21459
29046	01301	55343	65732	78714	43644	46248	53205	94868	48711
62035	71886	94506	15263	61435	10369	42054	68257	14385	79436
38856	80048	59973	73368	52876	47673	41020	82295	26430	87377
40666	43328	87379	86418	95841	25590	54137	94182	42308	07361
40588	90087	37729	08667	37256	20317	53316	50982	32900	32097
78237	86556	50276	20431	00243	02303	71029	49932	23245	00862
98247	67474	71455	69540	01169	03320	67017	92543	97977	52728
69977	78558	65430	32627	28312	61815	14598	79728	55699	91348
39843	23074	40814	03713	21891	96353	96806	24595	26203	26009
62880	87277	99895	99965	34374	42556	11679	99605	98011	48867
56138	64927	29454	52967	86624	62422	30163	76181	95317	39264
90804	56026	48994	64569	67465	60180	12972	03848	62582	93855
09665	44672	74762	33357	67301	80546	97659	11348	78771	45011
34756	50403	76634	12767	32220	34545	18100	53513	14521	72120
12157	73327	74196	26668	78087	53636	52304	00007	05708	63538
69384	07734	94451	76428	16121	09300	67417	68587	87932	38840
93358	64565	43766	45041	44930	69970	16964	08277	67752	60292
38879	35544	99563	85404	04913	62547	78406	01017	86187	22072
58314	60298	72394	69668	12474	93059	02053	29807	63645	12792
83568	10227	99471	74729	22075	10233	21575	20325	21317	57124
28067	91152	40568	33705	64510	07067	64374	26336	79652	31140
05730	75557	93161	80921	55873	54103	34801	83157	04534	81368

Compiled from Rand Corporation, A million random digits with 100,000 normal deviates. The Free Press, Glencoe, Ill., 1955 (with permission).

Table 26.11   2500 FIVE DIGIT RANDOM NUMBERS

26687	74223	43546	45699	94469	82125	37370	23966	68926	37664
60675	75169	24510	15100	02011	14375	65187	10630	64421	66745
45418	98635	83123	98558	09953	60255	42071	40930	97992	93085
69872	48026	89755	28470	44130	59979	91063	28766	85962	77173
03765	86366	99539	44183	23886	89977	11964	51581	18033	56239
84686	57636	32326	19867	71345	42002	96997	84379	27991	21459
91512	49670	32556	85189	28023	88151	62896	95498	29423	38138
10737	49307	18307	22246	22461	10003	93157	66984	44919	30467
54870	19676	58367	20905	38324	00026	98440	37427	22896	37637
48967	49579	65369	74305	62085	39297	10309	23173	74212	32272
91430	79112	03685	05411	23027	54735	91550	06250	18705	18909
92564	29567	47476	62804	73428	04535	86395	12162	59647	97726
41734	12199	77441	92415	63542	42115	84972	12454	33133	48467
25251	78110	54178	78241	09226	87529	35376	90690	54178	08561
91657	11563	66036	28523	83705	09956	76610	88116	78351	50877
00149	84745	63222	50533	50159	60433	04822	49577	89049	16162
53250	73200	84066	59620	61009	38542	05758	06178	80193	26466
25587	17481	56716	49749	70733	32733	60365	14108	52573	39391
01176	12182	06882	27562	75456	54261	38564	89054	96911	88906
83531	15544	40834	20296	88576	47815	96540	79462	78666	25353
19902	98866	32805	61091	91587	30340	84909	64047	67750	87638
96516	78705	25556	35181	29064	49005	29843	68949	50506	45862
99417	56171	19848	24352	51844	03791	72127	57958	08366	43190
77699	57853	93213	27342	28906	31052	65815	21637	49385	75406
32245	83794	99528	05150	27246	48263	62156	62469	97048	16511
12874	72753	66469	13782	64330	00056	73324	03920	13193	19466
63899	41910	45484	55461	66518	82486	74694	07865	09724	76490
16255	43271	26540	41298	35095	32170	70625	66407	01050	44225
75553	30207	41814	74985	40223	91223	64238	73012	83100	92041
41772	18441	34685	13892	38843	69007	10362	84125	08814	66785
09270	01245	81765	06809	10561	10080	17482	05471	82273	06902
85058	17815	71551	36356	97519	54144	51132	83169	27373	68609
80222	87572	62758	14858	36350	23304	70453	21065	63812	29860
83901	88028	56743	25598	79349	47880	77912	52020	84305	02897
36303	57833	77622	02238	53285	77316	40106	38456	92214	54278
91543	63886	60539	96334	20804	72692	08944	02870	74892	22598
14415	33816	78231	87674	96473	44451	25098	29296	50679	07798
82465	07781	09938	66874	72128	99685	84329	14530	08410	45953
27306	39843	05634	96368	72022	01278	92830	40094	31776	41822
91960	82766	02331	08797	33858	21847	17391	53755	58079	48498
59284	96108	91610	07483	37943	96832	15444	12091	36690	58317
10428	96003	71223	21352	78685	55964	35510	94805	23422	04492
65527	41039	79574	05105	59588	02115	33446	56780	18402	36279
59688	43078	93275	31978	08768	84805	50661	18523	83235	50602
44452	10188	43565	46531	93023	07618	12910	60934	53403	18401
87275	82013	59804	78595	60553	14038	12096	95472	42736	08573
94155	93110	49964	27753	85090	77677	69303	66323	77811	22791
26488	76394	91282	03419	68758	89575	66469	97835	66681	03171
37073	34547	88296	68638	12976	50896	10023	27220	05785	77538
83835	89575	55956	93957	30361	47679	83001	35056	07103	63072

2500 FIVE DIGIT RANDOM NUMBERS   Table 26.11

55034	81217	90564	81943	11241	84512	12288	89862	00760	76159
25521	99536	43233	48786	49221	06960	31564	21458	88199	06312
85421	72744	97242	66383	00132	05661	96442	37388	57671	27916
61219	48390	47344	30413	39392	91365	56203	79204	05330	31196
20230	03147	58854	11650	28415	12821	58931	30508	65989	26675
95776	83206	56144	55953	89787	64426	08448	45707	80364	60262
07603	17344	01148	83300	96955	65027	31713	89013	79557	49755
00645	17459	78742	39005	36027	98807	72666	54484	68262	38827
62950	83162	61504	31557	80590	47893	72360	72720	08396	33674
79350	10276	81933	26347	08068	67816	06659	87917	74166	85519
48339	69834	59047	82175	92010	58446	69591	56205	95700	86211
05842	08439	79836	50957	32059	32910	15842	13918	41365	80115
25855	02209	07307	59942	71389	76159	11263	38787	61541	22606
25272	16152	82323	70718	98081	38631	91956	49909	76253	33970
73003	29058	17605	49298	47675	90445	68919	05676	23823	84892
81310	94430	22663	06584	38142	00146	17496	51115	61458	65790
10024	44713	59832	80721	63711	67882	25100	45345	55743	67618
84671	52806	89124	37691	20897	82339	22627	06142	05773	03547
29296	58162	21858	33732	94056	88806	54603	00384	66340	69232
51771	94074	70630	41286	90583	87680	13961	55627	23670	35109
42166	56251	60770	51672	36031	77273	85218	14812	90758	23677
78355	67041	22492	51522	31164	30450	27600	44428	96380	26772
09552	51347	33864	89018	73418	81538	77399	30448	97740	18158
15771	63127	34847	05660	06156	48970	55699	61818	91763	20821
13231	99058	93754	36730	44286	44326	15729	37500	47269	13333
50583	03570	38472	73236	67613	72780	78174	18718	99092	64114
99485	57330	10634	74905	90671	19643	69903	60950	17968	37217
54676	39524	73785	48864	69835	62798	65205	69187	05572	74741
99343	71549	10248	76036	31702	76868	88909	69574	27642	00336
35492	40231	34868	55356	12847	68093	52643	32732	67016	46784
98170	25384	03841	23920	47954	10359	70114	11177	63298	99903
02670	86155	56860	02592	01646	42200	79950	37764	82341	71952
36934	42879	81637	79952	07066	41625	96804	92388	88860	68580
56851	12778	24309	73660	84264	24668	16686	02239	66022	64133
05464	28892	14271	23778	88599	17081	33884	88783	39015	57118
15025	20237	63386	71122	06620	07415	94982	32324	79427	70387
95610	08030	81469	91066	88857	56583	01224	28097	19726	71465
09026	40378	05731	55128	74298	49196	31669	42605	30368	96424
81431	99955	52462	67667	97322	69808	21240	65921	12629	92896
21431	59335	58627	94822	65484	09641	41018	85100	16110	32077
95832	76145	11636	80284	17787	97934	12822	73890	66009	27521
99813	44631	43746	99790	86823	12114	31706	05024	28156	04202
77210	31148	50543	11603	50934	02498	09184	95875	85840	71954
13268	02609	79833	66058	80277	08533	28676	37532	70535	82356
44285	71735	26620	54691	14909	52132	81110	74548	78853	31996
70526	45953	79637	57374	05053	31965	33376	13232	85666	86615
88386	11222	25080	71462	09818	46001	19065	68981	18310	74178
83161	73994	17209	79441	64091	49790	11936	44864	86978	34538
50214	71721	33851	45144	05696	29935	12823	01594	08453	52825
97689	29341	67747	80643	13620	23943	49396	83686	37302	95350

PROBABILITY FUNCTIONS

Table 26.11   **2500 FIVE DIGIT RANDOM NUMBERS**

```
12367   23891   31506   90721   18710   89140   58595   99425   22840   08267
38890   30239   34237   22578   74420   22734   26930   40604   10782   80128
80788   55410   39770   93317   18270   21141   52085   78093   85638   81140
02395   77585   08854   23562   33544   45796   10976   44721   24781   09690
73720   70184   69112   71887   80140   72876   38984   23409   63957   44751

61383   17222   55234   18963   39006   93504   18273   49815   52802   69675
39161   44282   14975   97498   25973   33605   60141   30030   77677   49294
80907   74484   39884   19885   37311   04209   49675   39596   01052   43999
09052   65670   63660   34035   06578   87837   28125   48883   50482   55735
33425   24226   32043   60082   20418   85047   53570   32554   64099   52326

72651   69474   73648   71530   55454   19576   15552   20577   12124   50038
04142   32092   83586   61825   35482   32736   63403   91499   37196   02762
85226   14193   52213   60746   24414   57858   31884   51266   82293   73553
54888   03579   91674   59502   08619   33790   29011   85193   62262   28684
33258   51516   82032   45233   39351   33229   59464   65545   76809   16982

75973   15957   32405   82081   02214   57143   33526   47194   94526   73253
90638   75314   35381   34451   49246   11465   25102   71489   89883   99708
65061   15498   93348   33566   19427   66826   03044   97361   08159   47485
64420   07427   82233   97812   39572   07766   65844   29980   15533   90114
27175   17389   76963   75117   45580   99904   47160   55364   25666   25405

32215   30094   87276   56896   15625   32594   80663   08082   19422   80717
54209   58043   72350   89828   02706   16815   89985   37380   44032   59366
59286   66964   84843   71549   67553   33867   83011   66213   69372   23903
83872   58167   01221   95558   22196   65905   38785   01355   47489   28170
83310   57080   03366   80017   39601   40698   56434   64055   02495   50880

64545   29500   13351   78647   92628   19354   60479   57338   52133   07114
39269   00076   55489   01524   76568   22571   20328   84623   30188   43904
29763   05675   28193   65514   11954   78599   63902   21346   19219   90286
06310   02998   01463   27738   90288   17697   64511   39552   34694   03211
97541   47607   57655   59102   21851   44446   07976   54295   84671   78755

82968   85717   11619   97721   53513   53781   98941   38401   70939   11319
76878   34727   12524   90642   16921   13669   17420   84483   68309   85241
87394   78884   87237   92086   95633   66841   22906   64989   86952   54700
74040   12731   59616   33697   12592   44891   67982   72972   89795   10587
47896   41413   66431   70046   50793   45920   96564   67958   56369   44725

87778   71697   64148   54363   92114   34037   59061   62051   62049   33526
96977   63143   72219   80040   11990   47698   95621   72990   29047   85893
43820   13285   77811   81697   29937   70750   02029   32377   00556   86687
57203   83960   40096   39234   65953   59911   91411   55573   88427   45573
49065   72171   80939   06017   90323   63687   07932   99587   49014   26452

94250   84270   95798   13477   80139   26335   55169   73417   40766   45170
68148   81382   82383   18674   40453   92828   30042   37412   43423   45138
12208   97809   33619   28868   41646   16734   88860   32636   41985   84615
88317   89705   26119   12416   19438   65665   60989   59766   11418   18250
56728   80359   29613   63052   15251   44684   64681   42354   51029   77680

07138   12320   01073   19304   87042   58920   28454   81069   93978   66659
21188   64554   55618   36088   24331   84390   16022   12200   77559   75661
02154   12250   88738   43917   03655   21099   60805   63246   26842   35816
90953   85238   32771   07305   36181   47420   19681   33184   41386   03249
80103   91308   12858   41293   00325   15013   19579   91132   12720   92603
```

2500 FIVE DIGIT RANDOM NUMBERS Table 26.11

```
92630  78240  19267  95457  53497  23894  37708  79862  76471  66418
79445  78735  71549  44843  26104  67318  00701  34986  66751  99723
59654  71966  27386  50004  05358  94031  29281  18544  52429  06080
31524  49587  76612  39789  13537  48086  59483  60680  84675  53014
06348  76938  90379  51392  55887  71015  09209  79157  24440  30244

28703  51709  94456  48396  73780  06436  86641  69239  57662  80181
68108  89266  94730  95761  75023  48464  65544  96583  18911  16391
99938  90704  93621  66330  33393  95261  95349  51769  91616  33238
91543  73196  34449  63513  83834  99411  58826  40456  69268  48562
42103  02781  73920  56297  72678  12249  25270  36678  21313  75767

17138  27584  25296  28387  51350  61664  37893  05363  44143  42677
28297  14280  54524  21618  95320  38174  60579  08089  94999  78460
09331  56712  51333  06289  75345  08811  82711  57392  25252  30333
31295  04204  93712  51287  05754  79396  87399  51773  33075  97061
36146  15560  27592  42089  99281  59640  15221  96079  09961  05371

29553  18432  13630  05529  02791  81017  49027  79031  50912  09399
23501  22642  63081  08191  89420  67800  55137  54707  32945  64522
57888  85846  67967  07835  11314  01545  48535  17142  08552  67457
55336  71264  88472  04334  63919  36394  11196  92470  70543  29776
10087  10072  55980  64688  68239  20461  89381  93809  00796  95945

34101  81277  66090  88872  37818  72142  67140  50785  21380  16703
53362  44940  60430  22834  14130  96593  23298  56203  92671  15925
82975  66158  84731  19436  55790  69229  28661  13675  99318  76873
54827  84673  22898  08094  14326  87038  42892  21127  30712  48489
25464  59098  27436  89421  80754  89924  19097  67737  80368  08795

67609  60214  41475  84950  40133  02546  09570  45682  50165  15609
44921  70924  61295  51137  47596  86735  35561  76649  18217  63446
33170  30972  98130  95828  49786  13301  36081  80761  33985  68621
84687  85445  06208  17654  51333  02878  35010  67578  61574  20749
71886  56450  36567  09395  96951  35507  17555  35212  69106  01679

00475  02224  74722  14721  40215  21351  08596  45625  83981  63748
25993  38881  68361  59560  41274  69742  40703  37993  03435  18873
92882  53178  99195  93803  56985  53089  15305  50522  55900  43026
25138  26810  07093  15677  60688  04410  24505  37890  67186  62829
84631  71882  12991  83028  82484  90339  91950  74579  03539  90122

34003  92326  12793  61453  48121  74271  28363  66561  75220  35908
53775  45749  05734  86169  42762  70175  97310  73894  88606  19994
59316  97885  72807  54966  60859  11932  35265  71601  55577  67715
20479  66557  50705  26999  09854  52591  14063  30214  19890  19292
86180  84931  25455  26044  02227  52015  21820  50599  51671  65411

21451  68001  72710  40261  61281  13172  63819  48970  51732  54113
98062  68375  80089  24135  72355  95428  11808  29740  81644  86610
01788  64429  14430  94575  75153  94576  61393  96192  03227  32258
62465  04841  43272  68702  01274  05437  22953  18946  99053  41690
94324  31089  84159  92933  99989  89500  91586  02802  69471  68274

05797  43984  21575  09908  70221  19791  51578  36432  33494  79888
10395  14289  52185  09721  25789  38562  54794  04897  59012  89251
35177  56986  25549  59730  64718  52630  31100  62384  49483  11409
25633  89619  75882  98256  02126  72099  57183  55887  09320  73463
16464  48280  94254  45777  45150  68865  11382  11782  22695  41988
```

27. Miscellaneous Functions

IRENE A. STEGUN [1]

Contents

---

[1] National Bureau of Standards.

27. Miscellaneous Functions

27.1. Debye Functions

Series Representations

27.1.1

$$\int_0^x \frac{t^n dt}{e^t-1}=x^n[\frac{1}{n}-\frac{x}{2(n+1)}+\sum_{k=1}^{\infty}\frac{B_{2k}x^{2k}}{(2k+n)(2k)!}]$$
$$(|x|<2\pi, n\geq 1)$$

(For Bernoulli numbers B_{2k}, see chapter **23**.)

27.1.2

$$\int_x^{\infty}\frac{t^n dt}{e^t-1}=\sum_{k=1}^{\infty}e^{-kx}[\frac{x^n}{k}+\frac{nx^{n-1}}{k^2}+\frac{(n)(n-1)x^{n-2}}{k^3}$$
$$+\ldots+\frac{n!}{k^{n+1}}](x>0, n\geq 1)$$

Relation to Riemann Zeta Function (see chapter **23**)

27.1.3
$$\int_0^{\infty}\frac{t^n dt}{e^t-1}=n!\zeta(n+1).$$

[27.1] J. A. Beattie, Six-place tables of the Debye energy and specific heat functions, J. Math. Phys. **6**, 1–32 (1926).

$$\frac{3}{x^3}\int_0^x\frac{y^3 dy}{e^y-1},\ \frac{12}{x^3}[\int_0^x\frac{y^3 dy}{e^y-1}-\frac{3x}{e^x-1}],\ x=0(.01)24,\ \ 6S.$$

[27.2] E. Grüneisen, Die Abhängigkeit des elektrischen Widerstandes reiner Metalle von der Temperatur, Ann. Physik. (5) **16**, 530–540 (1933).

$$\frac{20}{x^4}\int_0^x\frac{t^4 dt}{e^t-1}-\frac{4x}{e^x-1},$$
$$x=0(.1)13(.2)18(1)20(2)52(4)80,\ \ 4S.$$

Table 27.1

Debye Functions

x	$\frac{1}{x}\int_0^x\frac{tdt}{e^t-1}$	$\frac{2}{x^2}\int_0^x\frac{t^2dt}{e^t-1}$	$\frac{3}{x^3}\int_0^x\frac{t^3dt}{e^t-1}$	$\frac{4}{x^4}\int_0^x\frac{t^4dt}{e^t-1}$
0.0	1.000000	1.000000	1.000000	1.000000
0.1	0.975278	0.967083	0.963000	0.960555
0.2	0.951111	0.934999	0.926999	0.922221
0.3	0.927498	0.903746	0.891995	0.884994
0.4	0.904437	0.873322	0.857985	0.848871
0.5	0.881927	0.843721	0.824963	0.813846
0.6	0.859964	0.814940	0.792924	0.779911
0.7	0.838545	0.786973	0.761859	0.747057
0.8	0.817665	0.759813	0.731759	0.715275
0.9	0.797320	0.733451	0.702615	0.684551
1.0	0.777505	0.707878	0.674416	0.654874
1.1	0.758213	0.683086	0.647148	0.626228
1.2	0.739438	0.659064	0.620798	0.598598
1.3	0.721173	0.635800	0.595351	0.571967
1.4	0.703412	0.613281	0.570793	0.546317
1.6	0.669366	0.570431	0.524275	0.497882
1.8	0.637235	0.530404	0.481103	0.453131
2.0	0.606947	0.493083	0.441129	0.411893
2.2	0.578427	0.458343	0.404194	0.373984
2.4	0.551596	0.426057	0.370137	0.339218
2.6	0.526375	0.396095	0.338793	0.307405
2.8	0.502682	0.368324	0.309995	0.278355
3.0	0.480435	0.342614	0.283580	0.251879
3.2	0.459555	0.318834	0.259385	0.227792
3.4	0.439962	0.296859	0.237252	0.205915
3.6	0.421580	0.276565	0.217030	0.186075
3.8	0.404332	0.257835	0.198571	0.168107
4.0	0.388148	0.240554	0.181737	0.151855
4.2	0.372958	0.224615	0.166396	0.137169
4.4	0.358696	0.209916	0.152424	0.123913
4.6	0.345301	0.196361	0.139704	0.111957
4.8	0.332713	0.183860	0.128129	0.101180
5.0	0.320876	0.172329	0.117597	0.091471
5.5	0.294240	0.147243	0.095241	0.071228
6.0	0.271260	0.126669	0.077581	0.055677
6.5	0.251331	0.109727	0.063604	0.043730
7.0	0.233948	0.095707	0.052506	0.034541
7.5	0.218698	0.084039	0.043655	0.027453
8.0	0.205239	0.074269	0.036560	0.021968
8.5	0.193294	0.066036	0.030840	0.017702
9.0	0.182633	0.059053	0.026200	0.014368
9.5	0.173068	0.053092	0.022411	0.011747
10.0	0.164443	0.047971	0.019296	0.009674
	$\begin{bmatrix}(-4)5\\5\end{bmatrix}$	$\begin{bmatrix}(-4)6\\5\end{bmatrix}$	$\begin{bmatrix}(-4)6\\5\end{bmatrix}$	$\begin{bmatrix}(-4)6\\5\end{bmatrix}$

Planck's Radiation Function

Table 27.2

$$f(x) = x^{-5}(e^{1/x}-1)^{-1}$$

x	$f(x)$	x	$f(x)$	x	$f(x)$	x	$f(x)$	x	$f(x)$
0.050	0.007	0.10	4.540	0.20	21.199	0.40	8.733	0.9	0.831
0.055	0.025	0.11	6.998	0.22	20.819	0.45	6.586	1.0	0.582
0.060	0.074	0.12	9.662	0.24	19.777	0.50	5.009	1.1	0.419
0.065	0.179	0.13	12.296	0.26	18.372	0.55	3.850	1.2	0.309
0.070	0.372	0.14	14.710	0.28	16.809	0.60	2.995	1.3	0.233
0.075	0.682	0.15	16.780	0.30	15.224	0.65	2.356	1.4	0.178
0.080	1.137	0.16	18.446	0.32	13.696	0.70	1.875	1.5	0.139
0.085	1.752	0.17	19.692	0.34	12.270	0.75	1.508	2.0	0.048
0.090	2.531	0.18	20.539	0.36	10.965	0.80	1.225	2.5	0.021
0.095	3.466	0.19	21.025	0.38	9.787	0.85	1.005	3.0	0.010
0.100	4.540	0.20	21.199	0.40	8.733	0.90	0.831	3.5	0.006

$$\begin{bmatrix} (-2)2 \\ 4 \end{bmatrix} \qquad \begin{bmatrix} (-2)5 \\ 5 \end{bmatrix} \qquad \begin{bmatrix} (-2)8 \\ 5 \end{bmatrix} \qquad \begin{bmatrix} (-2)7 \\ 5 \end{bmatrix} \qquad \begin{bmatrix} (-2)1 \\ 4 \end{bmatrix}$$

$x_{max} = .20140\ 52353 \qquad f(x_{max}) = 21.20143\ 58.$

[27.3] Miscellaneous Physical Tables, Planck's radiation functions and electronic functions, MT 17 (U.S. Government Printing Office, Washington, D.C., 1941).

$$R_\lambda = c_1 \lambda^{-5}(e^{c_2/\lambda T}-1)^{-1}, \quad R_{0-\lambda} = \int_0^\lambda R_\lambda d\lambda,$$

$$N_\lambda = 2\pi c \lambda^{-4}(e^{c_2/\lambda T}-1)^{-1}, \quad N_{0-\lambda} = \int_0^\lambda N_\lambda d\lambda$$

Table I: $\dfrac{R_\lambda}{R_{\lambda\ max}}$, $\dfrac{R_{0-\lambda}}{R_{0-\infty}}$, $\dfrac{N_\lambda}{N_{\lambda\ max}}$, $\dfrac{N_{0-\lambda}}{N_{0-\infty}}$ for $\lambda T = [.05(.001).1(.005).4(.01).6(.02)1(.05)2]$cm $K°$.

Table II: R_λ, $R_{0-\lambda}$, N_λ, $N_{0-\lambda}$ ($T = 1000°$ K) for $\lambda = [.5(.01)1(.05)4(.1)6(.2)10(.5)20]$ microns.

Table III: N_λ for $\lambda = [.25(.05)1.6(.2)3(1)10]$ microns, $T = [1000°(500°)3500°$ K and $6000°$ K].

Einstein Functions

Table 27.3

x	$\dfrac{x^2 e^x}{(e^x-1)^2}$	$\dfrac{x}{e^x-1}$	$\ln(1-e^{-x})$	$\dfrac{x}{e^x-1} - \ln(1-e^{-x})$
0.00	1.00000	1.00000	$-\infty$	∞
0.05	0.99979	0.97521	-3.02063	3.99584
0.10	0.99917	0.95083	-2.35217	3.30300
0.15	0.99813	0.92687	-1.97118	2.89806
0.20	0.99667	0.90333	-1.70777	2.61110
0.25	0.99481	0.88020	-1.50869	2.38888
0.30	0.99253	0.85749	-1.35023	2.20771
0.35	0.98985	0.83519	-1.21972	2.05491
0.40	0.98677	0.81330	-1.10963	1.92293
0.45	0.98329	0.79182	-1.01508	1.80690
0.50	0.97942	0.77075	-0.93275	1.70350
0.55	0.97517	0.75008	-0.86026	1.61035
0.60	0.97053	0.72982	-0.79587	1.52569
0.65	0.96552	0.70996	-0.73824	1.44820
0.70	0.96015	0.69050	-0.68634	1.37684
0.75	0.95441	0.67144	-0.63935	1.31079
0.80	0.94833	0.65277	-0.59662	1.24939
0.85	0.94191	0.63450	-0.55759	1.19209
0.90	0.93515	0.61661	-0.52184	1.13844
0.95	0.92807	0.59910	-0.48897	1.08809
1.00	0.92067	0.58198	-0.45868	1.04065
1.05	0.91298	0.56523	-0.43069	0.99592
1.10	0.90499	0.54886	-0.40477	0.95363
1.15	0.89671	0.53285	-0.38073	0.91358
1.20	0.88817	0.51722	-0.35838	0.87560
1.25	0.87937	0.50194	-0.33758	0.83952
1.30	0.87031	0.48702	-0.31818	0.80520
1.35	0.86102	0.47245	-0.30008	0.77253
1.40	0.85151	0.45824	-0.28315	0.74139
1.45	0.84178	0.44436	-0.26732	0.71168
1.50	0.83185	0.43083	-0.25248	0.68331

$$\begin{bmatrix} (-5)5 \\ 3 \end{bmatrix} \qquad \begin{bmatrix} (-5)5 \\ 3 \end{bmatrix}$$

Table 27.3   **Einstein Functions**

x	$\dfrac{x^2 e^x}{(e^x-1)^2}$	$\dfrac{x}{e^x-1}$	$\ln(1-e^{-x})$	$\dfrac{x}{e^x-1}$ $-\ln(1-e^{-x})$
1.6	0.81143	0.40475	-0.22552	0.63027
1.7	0.79035	0.37998	-0.20173	0.58171
1.8	0.76869	0.35646	-0.18068	0.53714
1.9	0.74657	0.33416	-0.16201	0.49617
2.0	0.72406	0.31304	-0.14541	0.45845
2.1	0.70127	0.29304	-0.13063	0.42367
2.2	0.67827	0.27414	-0.11744	0.39158
2.3	0.65515	0.25629	-0.10565	0.36194
2.4	0.63200	0.23945	-0.09510	0.33455
2.5	0.60889	0.22356	-0.08565	0.30921
2.6	0.58589	0.20861	-0.07718	0.28578
2.7	0.56307	0.19453	-0.06957	0.26410
2.8	0.54049	0.18129	-0.06274	0.24403
2.9	0.51820	0.16886	-0.05659	0.22545
3.0	0.49627	0.15719	-0.05107	0.20826
3.2	0.45363	0.13598	-0.04162	0.17760
3.4	0.41289	0.11739	-0.03394	0.15133
3.6	0.37429	0.10113	-0.02770	0.12883
3.8	0.33799	0.08695	-0.02262	0.10958
4.0	0.30409	0.07463	-0.01849	0.09311
4.2	0.27264	0.06394	-0.01511	0.07905
4.4	0.24363	0.05469	-0.01235	0.06705
4.6	0.21704	0.04671	-0.01010	0.05681
4.8	0.19277	0.03983	-0.00826	0.04809
5.0	0.17074	0.03392	-0.00676	0.04068
5.2	0.15083	0.02885	-0.00553	0.03438
5.4	0.13290	0.02450	-0.00453	0.02903
5.6	0.11683	0.02078	-0.00370	0.02449
5.8	0.10247	0.01761	-0.00303	0.02065
6.0	0.08968	0.01491	-0.00248	0.01739
	$\begin{bmatrix}(-4)3\\4\end{bmatrix}$	$\begin{bmatrix}(-4)3\\4\end{bmatrix}$	$\begin{bmatrix}(-4)4\\4\end{bmatrix}$	$\begin{bmatrix}(-4)6\\4\end{bmatrix}$

[27.4] H. L. Johnston, L. Savedoff and J. Belzer, Contributions to the thermodynamic functions by a Planck-Einstein oscillator in one degree of freedom, NAVEXOS p. 646, Office of Naval Research, Department of the Navy, Washington, D.C. (1949). Values of $x^2 e^x (e^x-1)^{-2}$, $x(e^x-1)^{-1}$, $-\ln(1-e^{-x})$ and $x(e^x-1)^{-1}-\ln(1-e^{-x})$ for $x=0(.001)3(.01)$ 14.99, 5D with first differences.

27.4. Sievert Integral

$$\int_0^\theta e^{-x \sec \phi} d\phi$$

Relation to the Error Function

27.4.1

$$\int_0^\theta e^{-x \sec \phi} d\phi \sim \sqrt{\frac{\pi}{2x}}\, e^{-x}\, \mathrm{erf}\left(\sqrt{\frac{x}{2}}\,\theta\right) \qquad (x \to \infty)$$

(For erf, see chapter 7.)

Representation in Terms of Exponential Integrals

27.4.2

$$\int_0^\theta e^{-x \sec \phi} d\phi = \int_0^{\frac{\pi}{2}} e^{-x \sec \phi} d\phi$$
$$-\sum_{k=0}^\infty \alpha_k (\cos \theta)^{2k+1} E_{2k+2}\left(\frac{x}{\cos \theta}\right)$$
$$\left(x \geq 0, 0 < \theta < \frac{\pi}{2}\right)$$

$$\alpha_0 = 1, \alpha_k = \frac{1 \cdot 3 \cdot 5 \ldots (2k-1)}{2 \cdot 4 \cdot 6 \ldots (2k)}$$

(For $E_{2k+2}(x)$, see chapter 5.)

Relation to the Integral of the Bessel Function $K_0(x)$

27.4.3

$$\int_0^{\frac{\pi}{2}} e^{-x \sec \phi} d\phi = \mathrm{Ki}_1(x) = \int_x^\infty K_0(t) dt \text{ where}$$

$$x^{\frac{1}{2}} e^x \mathrm{Ki}_1(x) \sim (\tfrac{1}{2}\pi)^{\frac{1}{2}}\left\{1 - \frac{5}{8x} + \frac{129}{128x^2}\right.$$
$$\left. - \frac{2655}{1024x^3} + \frac{301035}{32768x^4} - \cdots\right\}$$

(For $\mathrm{Ki}_1(x)$, see chapter 11.)

[27.5] National Bureau of Standards, Table of the Sievert integral, Applied Math. Series— (U.S. Government Printing Office, Washington, D.C. In press).

$x = 0(.01)2(.02)5(.05)10$, $\theta = 0°(1°)90°$, 9D.

[27.6] R. M. Sievert, Die v-Strahlungsintensität an der Oberfläche und in der nächsten Umgebung von Radiumnadeln, Acta Radiologica 11, 239–301 (1930).

$\int_0^\phi e^{-A \sec \phi} d\phi$, $\phi = 30°(1°)90°$, $A = 0(.01).5$, 3D.

Sievert Integral $\displaystyle\int_0^\theta e^{-x \sec \phi} d\phi$ **Table 27.4**

$x \backslash \theta$	10°	20°	30°	40°	50°	60°	75°	90°
0.0	0.174533	0.349066	0.523599	0.698132	0.872665	1.047198	1.308997	1.570796
0.1	0.157843	0.315187	0.471456	0.625886	0.777323	0.923778	1.123611	1.228632
0.2	0.142749	0.284598	0.424515	0.561159	0.692565	0.815477	0.968414	1.023680
0.3	0.129099	0.256978	0.382355	0.503165	0.617194	0.720366	0.837712	0.868832
0.4	0.116754	0.232040	0.344209	0.451198	0.550154	0.636769	0.727031	0.745203
0.5	0.105589	0.209522	0.309957	0.404629	0.490508	0.563236	0.632830	0.643694
0.6	0.095492	0.189191	0.279118	0.362893	0.437428	0.498504	0.552287	0.558890
0.7	0.086361	0.170833	0.251353	0.325486	0.390178	0.441478	0.483134	0.487198
0.8	0.078103	0.154256	0.226354	0.291957	0.348109	0.391204	0.423535	0.426062
0.9	0.070634	0.139289	0.203845	0.261901	0.310642	0.346851	0.371996	0.373579
1.0	0.063880	0.125775	0.183579	0.234956	0.277267	0.307694	0.327288	0.328286
1.2	0.052247	0.102553	0.148899	0.189138	0.221027	0.242523	0.254485	0.254889
1.4	0.042733	0.083620	0.120780	0.152298	0.176336	0.191533	0.198885	0.199051
1.6	0.034951	0.068183	0.097979	0.122667	0.140792	0.151541	0.156087	0.156156
1.8	0.028587	0.055597	0.079488	0.098829	0.112497	0.120105	0.122932	0.122961
2.0	0.023381	0.045335	0.064492	0.079644	0.089954	0.095342	0.097108	0.097121
2.2	0.019123	0.036967	0.052329	0.064201	0.071979	0.075797	0.076905	0.076911
2.4	0.015641	0.030145	0.042463	0.051766	0.057635	0.060342	0.061040	0.061043
2.6	0.012793	0.024582	0.034460	0.041750	0.046179	0.048100	0.048541	0.048542
2.8	0.010463	0.020045	0.027968	0.033680	0.037024	0.038387	0.038667	0.038668
3.0	0.008558	0.016347	0.022700	0.027177	0.029702	0.030670	0.030848	0.030848
3.5	0.005178	0.009817	0.013477	0.015912	0.017164	0.017576	0.017634	0.017634
4.0	0.003132	0.005896	0.008005	0.009330	0.009951	0.010128	0.010147	0.010147
4.5	0.001895	0.003542	0.004756	0.005478	0.005787	0.005862	0.005869	0.005869
5.0	0.001147	0.002127	0.002828	0.003221	0.003374	0.003407	0.003409	0.003409
5.5	0.000694	0.001278	0.001682	0.001896	0.001972	0.001986	0.001987	0.001987
6.0	0.000420	0.000768	0.001001	0.001117	0.001155	0.001162	0.001162	0.001162
6.5	0.000254	0.000461	0.000596	0.000659	0.000678	0.000681	0.000681	0.000681
7.0	0.000154	0.000277	0.000355	0.000389	0.000399	0.000400	0.000400	0.000400
7.5	0.000093	0.000167	0.000211	0.000230	0.000235	0.000235	0.000235	0.000235
8.0	0.000056	0.000100	0.000126	0.000136	0.000139	0.000139	0.000139	0.000139
8.5	0.000034	0.000060	0.000075	0.000081	0.000082	0.000082	0.000082	0.000082
9.0	0.000021	0.000036	0.000045	0.000048	0.000048	0.000048	0.000048	0.000048
9.5	0.000012	0.000022	0.000027	0.000028	0.000029	0.000029	0.000029	0.000029
10.0	0.000008	0.000013	0.000016	0.000017	0.000017	0.000017	0.000017	0.000017
	$\begin{bmatrix}(-3)2\\6\end{bmatrix}$	$\begin{bmatrix}(-4)5\\6\end{bmatrix}$	$\begin{bmatrix}(-4)8\\6\end{bmatrix}$	$\begin{bmatrix}(-3)1\\7\end{bmatrix}$	$\begin{bmatrix}(-3)1\\7\end{bmatrix}$	$\begin{bmatrix}(-3)2\\7\end{bmatrix}$	$\begin{bmatrix}(-3)4\\7\end{bmatrix}$	$\begin{bmatrix}(-2)2\\11\end{bmatrix}$

27.5. $f_m(x) = \displaystyle\int_0^\infty t^m e^{-t^2 - \frac{x}{t}} dt$ and

Related Integrals

$m = 0, 1, 2 \ldots$

Differential Equations

27.5.1 $\qquad x f_m''' - (m-1) f_m'' + 2 f_m = 0$

27.5.2 $\qquad f_m' = -f_{m-1} \qquad (m = 1, 2, \ldots)$

Recurrence Relation

27.5.3 $\qquad 2 f_m = (m-1) f_{m-2} + x f_{m-3} \qquad (m \geq 3)$

Power Series Representations

27.5.4 $\qquad 2 f_1(x) = \displaystyle\sum_{k=0}^\infty (a_k \ln x + b_k) x^k$

$a_k = \dfrac{-2 a_{k-2}}{k(k-1)(k-2)} \qquad b_k = \dfrac{-2 b_{k-2} - (3k^2 - 6k + 2) a_k}{k(k-1)(k-2)}$

$a_0 = a_1 = 0 \qquad a_2 = -b_0$

$b_0 = 1 \qquad b_1 = -\sqrt{\pi} \qquad b_2 = \dfrac{3}{2}(1 - \gamma)$

(For γ, see chapter 6.)

27.5.5

$$2f_1(x) = 1 - \sqrt{\pi}x + .6342x^2 + .5908x^3 - .1431x^4$$
$$- .01968x^5 + .00324x^6 + .000188x^7 \ldots$$
$$- x^2 \ln x(1 - .08333x^2 + .001389x^4 - .0000083x^6 + \ldots)$$

27.5.6

$$2f_2(x) = \frac{\sqrt{\pi}}{2} - x + \frac{\sqrt{\pi}}{2}x^2 - .3225x^3 - .1477x^4 + .03195x^5$$
$$+ .00328x^6 - .000491x^7 - .0000235x^8 \ldots$$
$$+ x^3 \ln x(\tfrac{1}{3} - .01667x^2 + .000198x^4 - \ldots)$$

27.5.7

$$2f_3(x) = 1 - \frac{\sqrt{\pi}}{2}x + \frac{x^2}{2} - .2954x^3 + .1014x^4 + .02954x^5$$
$$- .00578x^6 - .00047x^7 + .000064x^8 \ldots$$
$$- x^4 \ln x(.0833 - .00278x^2 + .000025x^4 - \ldots)$$

Asymptotic Representation

27.5.8

$$f_m(x) \sim \sqrt{\frac{\pi}{3}} \, 3^{-\frac{m}{2}} v^{\frac{m}{2}} e^{-v} \left(a_0 + \frac{a_1}{v} + \frac{a_2}{v^2} + \ldots + \frac{a_k}{v^k} + \ldots \right)$$
$$(x \to \infty)$$

$$v = 3\left(\frac{x}{2}\right)^{2/3}$$

$$a_0 = 1, \; a_1 = \frac{1}{12}(3m^2 + 3m - 1)$$

$$12(k+2)a_{k+2} = -(12k^2 + 36k - 3m^2 - 3m + 25)a_{k+1}$$
$$+ \tfrac{1}{2}(m - 2k)(2k + 3 - m)(2k + 3 + 2m)a_k$$
$$(k = 0, 1, 2 \ldots)$$

27.5.9 $$g_1(x) + ig_2(x) = \int_0^\infty t^3 e^{-t^2 + i\frac{x}{t}} dt$$

27.5.10 $$g_1(x) = \mathscr{R}f_3(ix) \qquad g_2(x) = -\mathscr{I}f_3(ix)$$

Asymptotic Representation

27.5.11

$$g_1(x) = \left(\frac{\pi}{3}\right)^{1/2} \frac{x}{2} \exp\left[-\frac{3}{2}\left(\frac{x}{2}\right)^{2/3}\right](A \sin\theta + B\cos\theta)$$

27.5.12

$$g_2(x) = -\left(\frac{\pi}{3}\right)^{1/2} \frac{x}{2} \exp\left[-\frac{3}{2}\left(\frac{x}{2}\right)^{2/3}\right](A\cos\theta - B\sin\theta)$$

$$\theta = \frac{3}{2}\sqrt{3}\left(\frac{x}{2}\right)^{2/3}$$

$$A \sim a_0 - a_3\left(\frac{2}{x}\right)^2 + \frac{1}{2}\left[a_1\left(\frac{2}{x}\right)^{2/3} - a_2\left(\frac{2}{x}\right)^{4/3}\right.$$
$$\left. - a_4\left(\frac{2}{x}\right)^{8/3} + a_5\left(\frac{2}{x}\right)^{10/3} - \ldots\right] \quad (x \to \infty)$$

$$B \sim \frac{\sqrt{3}}{2}\left[a_1\left(\frac{2}{x}\right)^{2/3} + a_2\left(\frac{2}{x}\right)^{4/3} - a_4\left(\frac{2}{x}\right)^{8/3}\right.$$
$$\left. - a_5\left(\frac{2}{x}\right)^{10/3} + \ldots\right] \quad (x \to \infty)$$

$$a_0 = 1 \qquad a_1 = .972222 \qquad a_2 = .148534$$
$$a_3 = -.017879 \qquad a_4 = .004594 \qquad a_5 = -.000762$$

[27.7] M. Abramowitz, Evaluation of the integral $\int_0^\infty e^{-u^2 - x/u} du$, J. Math. Phys. **32**, 188–192 (1953).

[27.8] H. Faxén, Expansion in series of the integral $\int_y^\infty \exp[-x(t \pm t^{-n})]t^2 dt$, Ark. Mat., Astr., Fys. **15**, 13, 1–57 (1921).

[27.9] J. E. Kilpatrick and M. F. Kilpatrick, Discrete energy levels associated with the Lennard-Jones potential, J. Chem. Phys. **19**, 7, 930–933 (1951).

[27.10] U. E. Kruse and N. F. Ramsey, The integral $\int_0^\infty y^3 \exp\left(-y^2 + i\frac{x}{y}\right) dy$, J. Math. Phys. **30**, 40 (1951).

[27.11] O. Laporte, Absorption coefficients for thermal neutrons, Phys. Rev. **52**, 72–74 (1937).

[27.12] H. C. Torrey, Notes on intensities of radio frequency spectra, Phys. Rev. **59**, 293 (1941).

[27.13] C. T. Zahn, Absorption coefficients for thermal neutrons, Phys. Rev. **52**, 67–71 (1937). $\int_0^\infty y^n e^{-y - x/\sqrt{y}} dy$ for $n = 0, \tfrac{1}{2}, 1$; $x = 0(.01).1(.1)1$.

$$f_m(x)=\int_0^\infty t^m e^{-t^2-\frac{x}{t}}\,dt$$

Table 27.5

x	$f_1(x)$	$f_2(x)$	$f_3(x)$	x	$f_1(x)$	$f_2(x)$	$f_3(x)$	x	$f_1(x)$	$f_2(x)$	$f_3(x)$
0.00	0.5000	0.4431	0.5000	0.1	0.4263	0.3970	0.4580	0.6	0.2255	0.2415	0.3025
0.01	0.4914	0.4382	0.4956	0.2	0.3697	0.3573	0.4204	0.7	0.2015	0.2202	0.2793
0.02	0.4832	0.4333	0.4912	0.3	0.3238	0.3227	0.3864	0.8	0.1807	0.2011	0.2584
0.03	0.4753	0.4285	0.4869	0.4	0.2855	0.2923	0.3557	0.9	0.1626	0.1839	0.2392
0.04	0.4676	0.4238	0.4826	0.5	0.2531	0.2654	0.3278	1.0	0.1466	0.1685	0.2215
0.05	0.4602	0.4191	0.4784								
	$\begin{bmatrix}(-5)5\\2\end{bmatrix}$	$\begin{bmatrix}(-5)5\\2\end{bmatrix}$	$\begin{bmatrix}(-5)5\\2\end{bmatrix}$		$\begin{bmatrix}(-3)1\\4\end{bmatrix}$	$\begin{bmatrix}(-4)7\\3\end{bmatrix}$	$\begin{bmatrix}(-4)5\\3\end{bmatrix}$		$\begin{bmatrix}(-4)6\\3\end{bmatrix}$	$\begin{bmatrix}(-4)4\\3\end{bmatrix}$	$\begin{bmatrix}(-4)4\\3\end{bmatrix}$

x	$\mathscr{R}f_3(ix)$	$-\mathscr{I}f_3(ix)$	x	$\mathscr{R}f_3(ix)$	$-\mathscr{I}f_3(ix)$	x	$\mathscr{R}f_3(ix)$	$-\mathscr{I}f_3(ix)$
0.0	0.50000	0.00000	4.0	−0.2626	0.0430	8.0	0.06078	−0.09808
0.2	0.49019	0.08754	4.2	−0.2552	+0.0094	8.5	0.07562	−0.07131
0.4	0.46229	0.16933	4.4	−0.2441	−0.0214	9.0	0.08221	−0.04496
0.6	0.41950	0.24139	4.6	−0.2299	−0.0490	9.5	0.08191	−0.02082
0.8	0.36543	0.30136	4.8	−0.2132	−0.0734	10.0	0.07626	−0.00010
1.0	0.30366	0.34805	5.0	−0.1945	−0.0944	10.5	0.06684	+0.01654
1.2	0.23746	0.38122	5.2	−0.1745	−0.1120	11.0	0.05507	0.02889
1.4	0.16972	0.40127	5.4	−0.1536	−0.1263	11.5	0.04224	0.03707
1.6	0.10288	0.40910	5.6	−0.1322	−0.1374	12.0	0.02937	0.04146
1.8	+0.03892	0.40592	5.8	−0.1108	−0.1455	12.5	0.01727	0.04259
2.0	−0.02062	0.39314	6.0	−0.0896	−0.1507	13.0	+0.00650	0.04109
2.2	−0.0746	0.3722	6.2	−0.0691	−0.1533	13.5	−0.00259	0.03758
2.4	−0.1221	0.3448	6.4	−0.0493	−0.1535	14.0	−0.00982	0.03268
2.6	−0.1629	0.3122	6.6	−0.0307	−0.1515	14.5	−0.01517	0.02696
2.8	−0.1966	0.2759	6.8	−0.0132	−0.1476	15.0	−0.01872	0.02089
3.0	−0.2233	0.2371	7.0	+0.00286	−0.14211	16.0	−0.02118	+0.00921
3.2	−0.2432	0.1971	7.2	0.01749	−0.13518	17.0	−0.01906	−0.00022
3.4	−0.2565	0.1569	7.4	0.03061	−0.12709	18.0	−0.01435	−0.00650
3.6	−0.2639	0.1173	7.6	0.04220	−0.11805	19.0	−0.00879	−0.00965
3.8	−0.2657	0.0792	7.8	0.05224	−0.10830	20.0	−0.00360	−0.01021
	$\begin{bmatrix}(-3)2\\6\end{bmatrix}$	$\begin{bmatrix}(-3)2\\5\end{bmatrix}$		$\begin{bmatrix}(-4)5\\3\end{bmatrix}$	$\begin{bmatrix}(-4)4\\4\end{bmatrix}$		$\begin{bmatrix}(-3)1\\5\end{bmatrix}$	$\begin{bmatrix}(-4)7\\5\end{bmatrix}$

Compiled from U. E. Kruse and N. F. Ramsey, The integral $\int_0^\infty y^3 \exp\left(-y^2+i\frac{x}{y}\right)dy$, J. Math. Phys. **30**, 40 (1951) (with permission).

27.6. $f(x)=\int_0^\infty \dfrac{e^{-t^2}}{t+x}\,dt$

Power Series Representation

27.6.1

$$f(x)=-e^{-x^2}\ln x + e^{-x^2}[\sqrt{\pi}\sum_{k=0}^\infty \frac{x^{2k+1}}{k!(2k+1)} - \sum_{k=1}^\infty \frac{x^{2k}}{k!\,2k}-\frac{\gamma}{2}]$$

27.6.2

$$=-e^{-x^2}\ln x + \frac{1}{2}\sum_{k=0}^\infty \frac{(-1)^k\psi(k+1)x^{2k}}{k!} + \sqrt{\pi}\sum_{k=0}^\infty \frac{(-2)^k x^{2k+1}}{1\cdot3\cdot5\ldots(2k+1)}$$

(For γ and the digamma function $\psi(x)$, see chapter **6**.)

Relation to the Exponential Integral

27.6.3 $f(x)=-\dfrac{1}{2}e^{-x^2}\,\mathrm{Ei}\,(x^2)+\sqrt{\pi}e^{-x^2}\int_0^x e^{t^2}dt$

(For Ei (x) see chapter **5**; $e^{-x^2}\int_0^x e^{t^2}\,dt$, see chapter **7**.

Asymptotic Representation

27.6.4

$$f(x)\sim\frac{\sqrt{\pi}}{2}[\frac{1}{x}+\frac{1}{2x^3}+\frac{1\cdot3}{4x^5}+\frac{1\cdot3\cdot5}{8x^7}+\ldots]$$

$$-\frac{1}{2}[\frac{1}{x^2}+\frac{1}{x^4}+\frac{2!}{x^6}+\frac{3!}{x^8}+\ldots]\qquad (x\to\infty)$$

[27.14] A. Erdélyi, Note on the paper "On a definite integral" by R. H. Ritchie, Math. Tables Aids Comp. **4**, 31, 179 (1950).

[27.15] E. T. Goodwin and J. Staton, Table of $\int_0^\infty \dfrac{e^{-u^2}}{u+x}\,du$, Quart. J. Mech. Appl. Math. **1**, 319 (1948). $x=0(.02)2(.05)3(.1)10$. Auxiliary function for $x=0(.01)1$.

[27.16] R. H. Ritchie, On a definite integral, Math. Tables Aids Comp. **4**, 30, 75 (1950).

Table 27.6
$$f(x)=\int_0^\infty \frac{e^{-t^2}}{t+x}\,dt$$

x	$f(x)+\ln x$	x	$f(x)+\ln x$	x	$f(x)$	x	$f(x)$	x	$f(x)$
0.00	-0.2886	0.50	0.2704	1.0	0.6051	2.0	0.3543	3.0	0.2519
0.05	-0.2081	0.55	0.3100	1.1	0.5644	2.1	0.3404	3.5	0.2203
0.10	-0.1375	0.60	0.3479	1.2	0.5291	2.2	0.3276	4.0	0.1958
0.15	-0.0735	0.65	0.3842	1.3	0.4980	2.3	0.3157	4.5	0.1762
0.20	-0.0146	0.70	0.4192	1.4	0.4705	2.4	0.3046	5.0	0.1602
0.25	$+0.0402$	0.75	0.4529	1.5	0.4460	2.5	0.2944	5.5	0.1468
0.30	0.0915	0.80	0.4854	1.6	0.4239	2.6	0.2848	6.0	0.1356
0.35	0.1398	0.85	0.5168	1.7	0.4040	2.7	0.2758	6.5	0.1259
0.40	0.1856	0.90	0.5472	1.8	0.3860	2.8	0.2673	7.0	0.1175
0.45	0.2290	0.95	0.5766	1.9	0.3695	2.9	0.2594	7.5	0.1102
0.50	0.2704	1.00	0.6051	2.0	0.3543	3.0	0.2519	8.0	0.1037

$$\begin{bmatrix} (-3)1 \\ 4 \end{bmatrix} \qquad \begin{bmatrix} (-4)2 \\ 3 \end{bmatrix} \qquad \begin{bmatrix} (-4)7 \\ 4 \end{bmatrix} \qquad \begin{bmatrix} (-4)1 \\ 3 \end{bmatrix} \qquad \begin{bmatrix} (-4)9 \\ 4 \end{bmatrix}$$

Compiled from E. T. Goodwin and J. Staton, Table of $\int_0^\infty \frac{e^{-u^2}}{u+x}\,du$, Quart. J. Mech. Appl. Math. **1**, 319 (1948) (with permission).

27.7. Dilogarithm

(Spence's Integral for $n=2$)

27.7.1
$$f(x)=-\int_1^x \frac{\ln t}{t-1}\,dt$$

Series Expansion

27.7.2 $\quad f(x)=\sum_{k=1}^\infty (-1)^k \frac{(x-1)^k}{k^2} \qquad (2\geq x\geq 0)$

Functional Relationships

27.7.3
$$f(x)+f(1-x)=-\ln x \ln (1-x)+\frac{\pi^2}{6} \qquad (1\geq x\geq 0)$$

27.7.4
$$f(1-x)+f(1+x)=\frac{1}{2}f(1-x^2) \qquad (1\geq x>0)$$

27.7.5 $\quad f(x)+f\left(\frac{1}{x}\right)=-\frac{1}{2}(\ln x)^2 \qquad (0\leq x\leq 1)$

27.7.6
$$f(x+1)-f(x)=-\ln x \ln (x+1)-\frac{\pi^2}{12}-\frac{1}{2}f(x^2)$$
$$(2\geq x\geq 0)$$

Relation to Debye Functions

27.7.7 $\quad f(e^{-t})=-f(e^t)-\frac{t^2}{2}=\int_0^t \frac{t\,dt}{e^t-1}$

[27.17] L. Lewin, Dilogarithms and associated functions (Macdonald, London, England, 1958).

[27.18] K. Mitchell, Tables of the function $\int_0^z \frac{-\log|1-y|}{y}\,dy$, with an account of some properties of this and related functions, Phil. Mag. **40**, 351–368 (1949). $x=-1(.01)1$; $x=0(.001).5$, 9D.

[27.19] E. O. Powell, An integral related to the radiation integrals, Phil. Mag. **7**, 34, 600–607 (1943). $\int_1^x \frac{\log y}{y-1}\,dy$, $x=0(.01)2(.02)6$, 7D.

[27.20] A. van Wijngaarden, Polylogarithms, by the Staff of the Computation Department, Report R24, Mathematisch Centrum, Amsterdam, Holland (1954). $F_n(z)=\sum_{h=1}^\infty h^{-n}z^h$ for $z=x=-1(.01)1$; $z=ix$, for $x=0(.01)1$; $z=e^{i\pi\alpha/2}$ for $\alpha=0(.01)2$, 10D.

Dilogarithm

Table 27.7

$$f(x) = -\int_1^x \frac{\ln t}{t-1}\, dt$$

x	$f(x)$	x	$f(x)$	x	$f(x)$	x	$f(x)$	x	$f(x)$
0.00	1.64493 4067	0.10	1.29971 4723	0.20	1.07479 4600	0.30	0.88937 7624	0.40	0.72758 6308
0.01	1.58862 5448	0.11	1.27452 9160	0.21	1.05485 9830	0.31	0.87229 1733	0.41	0.71239 5042
0.02	1.54579 9712	0.12	1.25008 7584	0.22	1.03527 7934	0.32	0.85542 7404	0.42	0.69736 1058
0.03	1.50789 9041	0.13	1.22632 0101	0.23	1.01603 0062	0.33	0.83877 6261	0.43	0.68247 9725
0.04	1.47312 5860	0.14	1.20316 7961	0.24	0.99709 9088	0.34	0.82233 0471	0.44	0.66774 6644
0.05	1.44063 3797	0.15	1.18058 1124	0.25	0.97846 9393	0.35	0.80608 2689	0.45	0.65315 7631
0.06	1.40992 8300	0.16	1.15851 6487	0.26	0.96012 6675	0.36	0.79002 6024	0.46	0.63870 8705
0.07	1.38068 5041	0.17	1.13693 6560	0.27	0.94205 7798	0.37	0.77415 3992	0.47	0.62439 6071
0.08	1.35267 5161	0.18	1.11580 8451	0.28	0.92425 0654	0.38	0.75846 0483	0.48	0.61021 6108
0.09	1.32572 8728	0.19	1.09510 3088	0.29	0.90669 4053	0.39	0.74293 9737	0.49	0.59616 5361
0.10	1.29971 4723	0.20	1.07479 4600	0.30	0.88937 7624	0.40	0.72758 6308	0.50	0.58224 0526

$$\begin{bmatrix} (-3)2 \end{bmatrix} \qquad \begin{bmatrix} (-4)1 \end{bmatrix} \qquad \begin{bmatrix} (-5)5 \\ 7 \end{bmatrix} \qquad \begin{bmatrix} (-5)3 \\ 6 \end{bmatrix} \qquad \begin{bmatrix} (-5)2 \\ 5 \end{bmatrix}$$

From K. Mitchell, Tables of the function $\int_0^z \frac{-\log|1-y|}{y}\, dy$, with an account of some properties of this and related functions, Phil. Mag. **40**, 351–368 (1949) (with permission).

27.8. Clausen's Integral and Related Summations

27.8.1

$$f(\theta) = -\int_0^\theta \ln\left(2\sin\frac{t}{2}\right) dt = \sum_{k=1}^\infty \frac{\sin k\theta}{k^2} \qquad (0 \le \theta \le \pi)$$

Series Representation

27.8.2

$$f(\theta) = -\theta \ln|\theta| + \theta + \sum_{k=1}^\infty \frac{(-1)^{k-1}}{(2k)!} B_{2k} \frac{\theta^{2k+1}}{2k(2k+1)}$$
$$\left(0 \le \theta < \frac{\pi}{2}\right)$$

27.8.3

$$f(\pi-\theta) = \theta \ln 2 - \sum_{k=1}^\infty \frac{(-1)^{k-1}}{(2k)!} B_{2k}(2^{2k}-1) \frac{\theta^{2k+1}}{2k(2k+1)}$$
$$(\pi/2 < \theta < \pi)$$

Functional Relationship

27.8.4 $\quad f(\pi-\theta) = f(\theta) - \frac{1}{2} f(2\theta) \qquad \left(0 \le \theta \le \frac{\pi}{2}\right)$

Relation to Spence's Integral

27.8.5

$$if(\theta) = g(e^{i\theta}) + \frac{\theta^2}{4} \text{ where } g(x) = \int_1^x \frac{dt}{t} \ln|1+t|$$

Summable Series

27.8.6

$$\sum_{n=1}^\infty \frac{\cos n\theta}{n} = -\ln\left(2\sin\frac{\theta}{2}\right) \qquad (0 < \theta < 2\pi)$$

$$\sum_{n=1}^\infty \frac{\cos n\theta}{n^2} = \frac{\pi^2}{6} - \frac{\pi\theta}{2} + \frac{\theta^2}{4} \qquad (0 \le \theta \le 2\pi)$$

$$\sum_{n=1}^\infty \frac{\cos n\theta}{n^4} = \frac{\pi^4}{90} - \frac{\pi^2\theta^2}{12} + \frac{\pi\theta^3}{12} - \frac{\theta^4}{48} \qquad (0 \le \theta \le 2\pi)$$

$$\sum_{n=1}^\infty \frac{\sin n\theta}{n} = \frac{1}{2}(\pi-\theta) \qquad (0 < \theta < 2\pi)$$

$$\sum_{n=1}^\infty \frac{\sin n\theta}{n^3} = \frac{\pi^2\theta}{6} - \frac{\pi\theta^2}{4} + \frac{\theta^3}{12} \qquad (0 \le \theta \le 2\pi)$$

$$\sum_{n=1}^\infty \frac{\sin n\theta}{n^5} = \frac{\pi^4\theta}{90} - \frac{\pi^2\theta^3}{36} + \frac{\pi\theta^4}{48} - \frac{\theta^5}{240} \qquad (0 \le \theta \le 2\pi)$$

[27.21] A. Ashour and A. Sabri, Tabulation of the function $\psi(\theta) = \sum_{n=1}^\infty \frac{\sin n\theta}{n^2}$, Math. Tables Aids Comp. **10**, 54, 57–65 (1956).

[27.22] T. Clausen, Über die Zerlegung reeller gebrochener Funktionen, J. Reine Angew. Math. 8, 298–300 (1832). $x = 0°(1°)180°$, 16D.

[27.23] L. B. W. Jolley, Summation of series (Chapman Publishing Co., London, England, 1925).

[27.24] A. D. Wheelon, A short table of summable series, Report No. SM–14642, Douglas Aircraft Co., Inc., Santa Monica, Calif. (1953).

Table 27.8   **Clausen's Integral**

$$f(\theta) = -\int_0^\theta \ln\left(2\sin\frac{t}{2}\right)dt$$

$\theta°$	$f(\theta)+\theta\ln\theta$	$\theta°$	$f(\theta)$	$\theta°$	$f(\theta)$	$\theta°$	$f(\theta)$	$\theta°$	$f(\theta)$
0	0.000000	15	0.612906	30	0.864379	60	1.014942	90	0.915966
1	0.017453	16	0.635781	32	0.886253	62	1.014421	95	0.883872
2	0.034908	17	0.657571	34	0.906001	64	1.012886	100	0.848287
3	0.052362	18	0.678341	36	0.923755	66	1.010376	105	0.809505
4	0.069818	19	0.698149	38	0.939633	68	1.006928	110	0.767800
5	0.087276	20	0.717047	40	0.953741	70	1.002576	115	0.723427
6	0.104735	21	0.735080	42	0.966174	72	0.997355	120	0.676628
7	0.122199	22	0.752292	44	0.977020	74	0.991294	125	0.627629
8	0.139664	23	0.768719	46	0.986357	76	0.984425	130	0.576647
9	0.157133	24	0.784398	48	0.994258	78	0.976776	135	0.523889
10	0.174607	25	0.799360	50	1.000791	80	0.968375	140	0.469554
11	0.192084	26	0.813635	52	1.006016	82	0.959247	145	0.413831
12	0.209567	27	0.827249	54	1.009992	84	0.949419	150	0.356908
13	0.227055	28	0.840230	56	1.012773	86	0.938914	160	0.240176
14	0.244549	29	0.852599	58	1.014407	88	0.927755	170	0.120755
15	0.262049	30	0.864379	60	1.014942	90	0.915966	180	0.000000

$$\begin{bmatrix}(-7)8\\3\end{bmatrix}\qquad\begin{bmatrix}(-4)1\\4\end{bmatrix}\qquad\begin{bmatrix}(-4)3\\4\end{bmatrix}\qquad\begin{bmatrix}(-4)1\\4\end{bmatrix}\qquad\begin{bmatrix}(-4)4\\6\end{bmatrix}$$

Compiled from A. Ashour and A. Sabri, Tabulation of the function $\psi(\theta)=\sum_{n=1}^{\infty}\frac{\sin n\theta}{n^2}$, Math. Tables Aids Comp. **10**, 54, 57–65 (1956) (with permission).

27.9. Vector-Addition Coefficients

(Wigner coefficients or Clebsch-Gordan coefficients)

Definition

27.9.1

$$(j_1 j_2 m_1 m_2 | j_1 j_2 jm) = \delta(m, m_1+m_2)\cdot\sqrt{\frac{(j_1+j_2-j)!(j+j_1-j_2)!(j+j_2-j_1)!(2j+1)}{(j+j_1+j_2+1)!}}$$

$$\cdot\sum_k\frac{(-1)^k\sqrt{(j_1+m_1)!(j_1-m_1)!(j_2+m_2)!(j_2-m_2)!(j+m)!(j-m)!}}{k!(j_1+j_2-j-k)!(j_1-m_1-k)!(j_2+m_2-k)!(j-j_2+m_1+k)!(j-j_1-m_2+k)!}$$

$$\delta(i,k)=\begin{cases}1, & i=k\\0, & i\neq k\end{cases}$$

Conditions

27.9.2 $\quad j_1, j_2, j = +n$ or $+\frac{n}{2}\quad$ (n=integer)

27.9.3 $\quad\quad j_1+j_2+j=n$

27.9.4 $\quad\quad j_1+j_2-j$

27.9.5 $\quad\quad j_1-j_2+j\quad\Big\}\geq 0$

27.9.6 $\quad\quad -j_1+j_2+j$

27.9.7 $\quad\quad m_1, m_2, m = \pm n$ or $\pm\frac{n}{2}$

27.9.8 $\quad |m_1|\leq j_1,\ |m_2|\leq j_2,\ |m|\leq j$

27.9.9 $\quad (j_1 j_2 m_1 m_2 | j_1 j_2 jm)=0\quad\quad m_1+m_2\neq m$

Special Values

27.9.10 $\quad (j_1 0 m_1 0 | j_1 0 jm) = \delta(j_1, j)\delta(m_1, m)$

27.9.11 $\quad (j_1 j_2 00 | j_1 j_2 j0)=0\quad\quad j_1+j_2+j=2n+1$

27.9.12 $\quad (j_1 j_1 m_1 m_1 | j_1 j_1 jm)=0\quad\quad 2j_1+j=2n+1$

Symmetry Relations

27.9.13

$$(j_1 j_2 m_1 m_2 | j_1 j_2 j m)$$

$$= (-1)^{j_1+j_2-j} (j_1 j_2 -m_1 -m_2 | j_1 j_2 j -m)$$

27.9.14 $$= (j_2 j_1 -m_2 -m_1 | j_2 j_1 j -m)$$

27.9.15 $$= (-1)^{j_1+j_2-j} (j_2 j_1 m_1 m_2 | j_2 j_1 j m)$$

27.9.16 $$= \sqrt{\frac{2j+1}{2j_1+1}} (-1)^{j_2+m_2} (j j_2 -m m_2 | j j_2 j_1 -m_1)$$

27.9.17
$$= \sqrt{\frac{2j+1}{2j_1+1}} (-1)^{j_1-m_1+j-m} (j j_2 m -m_2 | j j_2 j_1 m_1)$$

27.9.18
$$= \sqrt{\frac{2j+1}{2j_1+1}} (-1)^{j-m+j_1-m_1} (j_2 j m_2 -m | j_2 j j_1 -m_1)$$

27.9.19
$$= \sqrt{\frac{2j+1}{2j_2+1}} (-1)^{j_1-m_1} (j_1 j m_1 -m | j_1 j j_2 -m_2)$$

27.9.20
$$= \sqrt{\frac{2j+1}{2j_2+1}} (-1)^{j_1-m_1} (j j_1 m -m_1 | j j_1 j_2 m_2)$$

$(j_1 \tfrac{1}{2} m_1 m_2 | j_1 \tfrac{1}{2} j m)$ **Table 27.9.1**

$j=$	$m_2=\tfrac{1}{2}$	$m_2=-\tfrac{1}{2}$
$j_1+\tfrac{1}{2}$	$\sqrt{\dfrac{j_1+m+\tfrac{1}{2}}{2j_1+1}}$	$\sqrt{\dfrac{j_1-m+\tfrac{1}{2}}{2j_1+1}}$
$j_1-\tfrac{1}{2}$	$-\sqrt{\dfrac{j_1-m+\tfrac{1}{2}}{2j_1+1}}$	$\sqrt{\dfrac{j_1+m+\tfrac{1}{2}}{2j_1+1}}$

$(j_1 \, 1 \, m_1 \, m_2 | j_1 \, 1 \, j \, m)$ **Table 27.9.2**

$j=$	$m_2=1$	$m_2=0$	$m_2=-1$
j_1+1	$\sqrt{\dfrac{(j_1+m)(j_1+m+1)}{(2j_1+1)(2j_1+2)}}$	$\sqrt{\dfrac{(j_1-m+1)(j_1+m+1)}{(2j_1+1)(j_1+1)}}$	$\sqrt{\dfrac{(j_1-m)(j_1-m+1)}{(2j_1+1)(2j_1+2)}}$
j_1	$-\sqrt{\dfrac{(j_1+m)(j_1-m+1)}{2j_1(j_1+1)}}$	$\dfrac{m}{\sqrt{j_1(j_1+1)}}$	$\sqrt{\dfrac{(j_1-m)(j_1+m+1)}{2j_1(j_1+1)}}$
j_1-1	$\sqrt{\dfrac{(j_1-m)(j_1-m+1)}{2j_1(2j_1+1)}}$	$-\sqrt{\dfrac{(j_1-m)(j_1+m)}{j_1(2j_1+1)}}$	$\sqrt{\dfrac{(j_1+m+1)(j_1+m)}{2j_1(2j_1+1)}}$

Table 27.9.3 $\qquad\qquad (j_1\ \tfrac{3}{2}\ m_1\ m_2\ |\ j_1\ \tfrac{3}{2}\ j\ m)$

$j=$	$m_2=\tfrac{3}{2}$	$m_2=\tfrac{1}{2}$
$j_1+\tfrac{3}{2}$	$\sqrt{\dfrac{(j_1+m-\tfrac{1}{2})(j_1+m+\tfrac{1}{2})(j_1+m+\tfrac{3}{2})}{(2j_1+1)(2j_1+2)(2j_1+3)}}$	$\sqrt{\dfrac{3(j_1+m+\tfrac{1}{2})(j_1+m+\tfrac{3}{2})(j_1-m+\tfrac{3}{2})}{(2j_1+1)(2j_1+2)(2j_1+3)}}$
$j_1+\tfrac{1}{2}$	$-\sqrt{\dfrac{3(j_1+m-\tfrac{1}{2})(j_1+m+\tfrac{1}{2})(j_1-m+\tfrac{3}{2})}{2j_1(2j_1+1)(2j_1+3)}}$	$-(j_1-3m+\tfrac{3}{2})\sqrt{\dfrac{j_1+m+\tfrac{1}{2}}{2j_1(2j_1+1)(2j_1+3)}}$
$j_1-\tfrac{1}{2}$	$\sqrt{\dfrac{3(j_1+m-\tfrac{1}{2})(j_1-m+\tfrac{1}{2})(j_1-m+\tfrac{3}{2})}{(2j_1-1)(2j_1+1)(2j_1+2)}}$	$-(j_1+3m-\tfrac{1}{2})\sqrt{\dfrac{j_1-m+\tfrac{1}{2}}{(2j_1-1)(2j_1+1)(2j_1+2)}}$
$j_1-\tfrac{3}{2}$	$-\sqrt{\dfrac{(j_1-m-\tfrac{1}{2})(j_1-m+\tfrac{1}{2})(j_1-m+\tfrac{3}{2})}{2j_1(2j_1-1)(2j_1+1)}}$	$\sqrt{\dfrac{3(j_1+m-\tfrac{1}{2})(j_1-m-\tfrac{1}{2})(j_1-m+\tfrac{1}{2})}{2j_1(2j_1-1)(2j_1+1)}}$

$j=$	$m_2=-\tfrac{1}{2}$	$m_2=-\tfrac{3}{2}$
$j_1+\tfrac{3}{2}$	$\sqrt{\dfrac{3(j_1+m+\tfrac{3}{2})(j_1-m+\tfrac{1}{2})(j_1-m+\tfrac{3}{2})}{(2j_1+1)(2j_1+2)(2j_1+3)}}$	$\sqrt{\dfrac{(j_1-m-\tfrac{1}{2})(j_1-m+\tfrac{1}{2})(j_1-m+\tfrac{3}{2})}{(2j_1+1)(2j_1+2)(2j_1+3)}}$
$j_1+\tfrac{1}{2}$	$(j_1+3m+\tfrac{3}{2})\sqrt{\dfrac{j_1-m+\tfrac{1}{2}}{2j_1(2j_1+1)(2j_1+3)}}$	$\sqrt{\dfrac{3(j_1+m+\tfrac{3}{2})(j_1-m-\tfrac{1}{2})(j_1-m+\tfrac{1}{2})}{2j_1(2j_1+1)(2j_1+3)}}$
$j_1-\tfrac{1}{2}$	$-(j_1-3m-\tfrac{1}{2})\sqrt{\dfrac{j_1+m+\tfrac{1}{2}}{(2j_1-1)(2j_1+1)(2j_1+2)}}$	$\sqrt{\dfrac{3(j_1+m+\tfrac{1}{2})(j_1+m+\tfrac{3}{2})(j_1-m-\tfrac{1}{2})}{(2j_1-1)(2j_1+1)(2j_1+2)}}$
$j_1-\tfrac{3}{2}$	$-\sqrt{\dfrac{3(j_1+m-\tfrac{1}{2})(j_1+m+\tfrac{1}{2})(j_1-m-\tfrac{1}{2})}{2j_1(2j_1-1)(2j_1+1)}}$	$\sqrt{\dfrac{(j_1+m-\tfrac{1}{2})(j_1+m+\tfrac{1}{2})(j_1+m+\tfrac{3}{2})}{2j_1(2j_1-1)(2j_1+1)}}$

Table 27.9.4

$$(j_1\ 2\ m_1\ m_2 \mid j_1\ 2\ j\ m)$$

$j=$	$m_2=2$	$m_2=1$	$m_2=0$
j_1+2	$\sqrt{\dfrac{(j_1+m-1)(j_1+m)(j_1+m+1)(j_1+m+2)}{(2j_1+1)(2j_1+2)(2j_1+3)(2j_1+4)}}$	$\sqrt{\dfrac{(j_1-m+2)(j_1+m+2)(j_1+m+1)(j_1+m)}{(2j_1+1)(2j_1+2)(2j_1+3)(j_1+2)}}$	$\sqrt{\dfrac{3(j_1-m+2)(j_1-m+1)(j_1+m+2)(j_1+m+1)}{(2j_1+1)(2j_1+2)(2j_1+3)(j_1+2)}}$
j_1+1	$-\sqrt{\dfrac{(j_1+m-1)(j_1+m)(j_1+m+1)(j_1-m+2)}{2j_1(2j_1+1)(j_1+2)(2j_1+1)}}$	$-(j_1-2m+2)\sqrt{\dfrac{(j_1+m+1)(j_1+m)}{2j_1(2j_1+1)(j_1+1)(j_1+2)}}$	$m\sqrt{\dfrac{3(j_1-m+1)(j_1+m+1)}{j_1(2j_1+1)(j_1+1)(j_1+2)}}$
j_1	$\sqrt{\dfrac{3(j_1+m-1)(j_1+m)(j_1-m+1)(j_1-m+2)}{(2j_1-1)2j_1(2j_1+1)(2j_1+3)}}$	$(1-2m)\sqrt{\dfrac{3(j_1-m+1)(j_1+m)}{(2j_1-1)j_1(2j_1+2)(2j_1+3)}}$	$\dfrac{3m^2-j_1(j_1+1)}{\sqrt{(2j_1-1)j_1(j_1+1)(2j_1+3)}}$
j_1-1	$-\sqrt{\dfrac{(j_1+m+1)(j_1-m)(j_1-m+1)(j_1-m+2)}{2(2j_1-1)j_1(2j_1+1)(2j_1+1)}}$	$(j_1+2m-1)\sqrt{\dfrac{(j_1-m+1)(j_1-m)}{(j_1-1)j_1(2j_1+1)(2j_1+2)}}$	$-m\sqrt{\dfrac{3(j_1-m)(j_1+m)}{(j_1-1)j_1(2j_1+1)(j_1+1)}}$
j_1-2	$\sqrt{\dfrac{(j_1-m-1)(j_1-m)(j_1-m+1)(j_1-m+2)}{(2j_1-2)(2j_1-1)2j_1(2j_1+1)}}$	$-\sqrt{\dfrac{(j_1-m+1)(j_1-m)(j_1-m-1)(j_1+m-1)}{(2j_1-2)(2j_1-1)j_1(2j_1+1)}}$	$\sqrt{\dfrac{3(j_1-m)(j_1-m-1)(j_1+m)(j_1+m-1)}{(2j_1-2)(2j_1-1)j_1(2j_1+1)}}$

$j=$	$m_2=-1$	$m_2=-2$
j_1+2	$\sqrt{\dfrac{(j_1-m+2)(j_1-m)(j_1-m+1)(j_1+m+2)}{(2j_1+1)(2j_1+2)(2j_1+3)(j_1+2)}}$	$\sqrt{\dfrac{(j_1-m-1)(j_1-m)(j_1-m+1)(j_1-m+2)}{(2j_1+1)(2j_1+2)(2j_1+3)(2j_1+4)}}$
j_1+1	$(j_1+2m+2)\sqrt{\dfrac{(j_1-m+1)(j_1-m)}{2j_1(2j_1+1)(j_1+1)(j_1+2)}}$	$\sqrt{\dfrac{(j_1-m-1)(j_1-m)(j_1-m+1)(j_1+m+2)}{2j_1(2j_1+1)(j_1+2)(2j_1+1)}}$
j_1	$(1+2m)\sqrt{\dfrac{3(j_1+m+1)(j_1-m)}{(2j_1-1)j_1(2j_1+2)(2j_1+3)}}$	$\sqrt{\dfrac{3(j_1-m-1)(j_1-m)(j_1+m+1)(j_1+m+2)}{(2j_1-1)2j_1(2j_1+1)(2j_1+3)}}$
j_1-1	$(j_1-2m-1)\sqrt{\dfrac{(j_1+m+1)(j_1+m)}{(j_1-1)j_1(2j_1+1)(2j_1+2)}}$	$\sqrt{\dfrac{(j_1-m+1)(j_1+m)(j_1+m+1)(j_1+m+2)}{2(2j_1-1)j_1(2j_1+1)(2j_1+1)}}$
j_1-2	$-\sqrt{\dfrac{(j_1+m+1)(j_1+m)(j_1+m-1)(j_1-m-1)}{(2j_1-2)(2j_1-1)j_1(2j_1+1)}}$	$\sqrt{\dfrac{(j_1+m-1)(j_1+m)(j_1+m+1)(j_1+m+2)}{(2j_1-2)(2j_1-1)2j_1(2j_1+1)}}$

Table 27.9.5 [By use of symmetry relations, coefficients may be put in standard form $j_1 \leq j_2 \leq j$ and $m \geq 0$]

m_2	m	j_1	j	$(j_1j_2m_1m_2|j_1j_2jm)$	
				$j_2 = \frac{1}{2}$	
$-\frac{1}{2}$	0	$\frac{1}{2}$	1	$\sqrt{\frac{1}{2}}$	0.70711
$\frac{1}{2}$	0	$\frac{1}{2}$	1	$\sqrt{\frac{1}{2}}$	0.70711
$\frac{1}{2}$	1	$\frac{1}{2}$	1		1.00000
				$j_2 = 1$	
-1	0	1	1	$\sqrt{\frac{1}{2}}$	0.70711
0	0	1	1		0.00000
1	0	1	1	$-\sqrt{\frac{1}{2}}$	-0.70711
0	1	1	1	$\sqrt{\frac{1}{2}}$	0.70711
1	1	1	1	$-\sqrt{\frac{1}{2}}$	-0.70711
0	$\frac{1}{2}$	$\frac{1}{2}$	$\frac{3}{2}$	$\sqrt{\frac{2}{3}}$	0.81650
1	$\frac{1}{2}$	$\frac{1}{2}$	$\frac{3}{2}$	$\sqrt{\frac{1}{3}}$	0.57735
1	$\frac{3}{2}$	$\frac{1}{2}$	$\frac{3}{2}$		1.00000 *
-1	0	1	2	$\sqrt{\frac{1}{6}}$	0.40825
0	0	1	2	$\sqrt{\frac{2}{3}}$	0.81650
1	0	1	2	$\sqrt{\frac{1}{6}}$	0.40825
0	1	1	2	$\sqrt{\frac{1}{2}}$	0.70711
1	1	1	2	$\sqrt{\frac{1}{2}}$	0.70711
1	2	1	2		1.00000
				$j_2 = \frac{3}{2}$	
$-\frac{1}{2}$	$\frac{1}{2}$	1	$\frac{3}{2}$	$\sqrt{\frac{8}{15}}$	0.73030
$\frac{1}{2}$	$\frac{1}{2}$	1	$\frac{3}{2}$	$-\sqrt{\frac{1}{15}}$	-0.25820
$\frac{3}{2}$	$\frac{1}{2}$	1	$\frac{3}{2}$	$-\sqrt{\frac{2}{5}}$	-0.63246
$\frac{1}{2}$	$\frac{3}{2}$	1	$\frac{3}{2}$	$\sqrt{\frac{2}{5}}$	0.63246
$\frac{3}{2}$	$\frac{3}{2}$	1	$\frac{3}{2}$	$-\sqrt{\frac{3}{5}}$	-0.77460
$-\frac{1}{2}$	0	$\frac{1}{2}$	2	$\sqrt{\frac{1}{2}}$	0.70711
$\frac{1}{2}$	0	$\frac{1}{2}$	2	$\sqrt{\frac{1}{2}}$	0.70711
$\frac{1}{2}$	1	$\frac{1}{2}$	2	$\frac{1}{2}\sqrt{3}$	0.86603
$\frac{3}{2}$	1	$\frac{1}{2}$	2		0.50000
$\frac{3}{2}$	2	$\frac{1}{2}$	2		1.00000
$-\frac{3}{2}$	0	$\frac{3}{2}$	2		0.50000
$-\frac{1}{2}$	0	$\frac{3}{2}$	2		0.50000
$\frac{1}{2}$	0	$\frac{3}{2}$	2		-0.50000
$\frac{3}{2}$	0	$\frac{3}{2}$	2		-0.50000
$-\frac{1}{2}$	1	$\frac{3}{2}$	2	$\sqrt{\frac{1}{2}}$	0.70711
$\frac{1}{2}$	1	$\frac{3}{2}$	2		0.00000
$\frac{3}{2}$	1	$\frac{3}{2}$	2	$-\sqrt{\frac{1}{2}}$	-0.70711
$\frac{1}{2}$	2	$\frac{3}{2}$	2	$\sqrt{\frac{1}{2}}$	0.70711
$\frac{3}{2}$	2	$\frac{3}{2}$	2	$-\sqrt{\frac{1}{2}}$	-0.70711
$-\frac{1}{2}$	$\frac{1}{2}$	1	$\frac{5}{2}$	$\sqrt{\frac{3}{10}}$	0.54772
$\frac{1}{2}$	$\frac{1}{2}$	1	$\frac{5}{2}$	$\sqrt{\frac{3}{5}}$	0.77460
$\frac{3}{2}$	$\frac{1}{2}$	1	$\frac{5}{2}$	$\sqrt{\frac{1}{10}}$	0.31623
$\frac{1}{2}$	$\frac{3}{2}$	1	$\frac{5}{2}$	$\sqrt{\frac{3}{5}}$	0.77460
$\frac{3}{2}$	$\frac{3}{2}$	1	$\frac{5}{2}$	$\sqrt{\frac{2}{5}}$	0.63246
$\frac{3}{2}$	$\frac{5}{2}$	1	$\frac{5}{2}$		1.00000

Compiled from A. Simon, Numerical tables of the Clebsch-Gordan coefficients, Oak Ridge National Laboratory Report 1718, Oak Ridge, Tenn. (1954) (with permission).

[27.25] E. U. Condon and G. A. Shortley, Theory of atomic spectra (Cambridge Univ. Press, Cambridge, England, 1935).

[27.26] M. E. Rose, Elementary theory of angular momemtum (John Wiley & Sons, Inc., New York, N.Y., 1955).

[27.27] A. Simon, Numerical tables of the Clebsch-Gordan coefficients, Oak Ridge National Laboratory Report 1718, Oak Ridge, Tenn. (1954).
$C(j_1j_2j; m_1m_2m)$ for all angular moments $< \frac{9}{2}$, 10D.

*See page II.

28. Scales of Notation

S. Peavy,[1] A. Schopf[2]

Contents

The authors acknowledge the assistance of David S. Liepman in the preparation and checking of the tables.

---

[1] National Bureau of Standards.
[2] Guest worker, National Bureau of Standards, from The American University (deceased).

28. Scales of Notation

Representation of Numbers

Any positive real number x can be uniquely represented in the scale of some integer $b>1$ as

$$x=(A_m \ldots A_1 A_0 \cdot a_{-1} a_{-2} \ldots)_{(b)},$$

where every A_i and a_{-j} is one of the integers 0, $1, \ldots, b-1$, not all A_i, a_{-j} are zero, and $A_m>0$ if $x \geq 1$. There is a one-to-one correspondence between the number and the sequence

$$x=A_m b^m + \ldots + A_1 b + A_0 + \sum_1^\infty a_{-j} b^{-j}$$

where the infinite series converges. The integer b is called the base or radix of the scale.

The sequence for x in the scale of b may terminate, i.e., $a_{-n-1}=a_{-n-2}= \ldots =0$ for some $n \geq 1$ so that

$$x=(A_m \ldots A_1 A_0 \cdot a_{-1} a_{-2} \ldots a_{-n})_{(b)};$$

then x is said to be a finite b-adic number.

A sequence which does not terminate may have the property that the infinite sequence a_{-1}, a_{-2}, $\ldots$ becomes periodic from a certain digit $a_{-n}(n \geq 1)$ on; according as $n=1$ or $n>1$ the sequence is then said to be pure or mixed recurring.

A sequence which neither terminates nor recurs represents an irrational number.

Names of Scales

Base	Scale	Base	Scale
2	Binary	8	Octal
3	Ternary	9	Nonary
4	Quaternary	10	Decimal
5	Quinary	11	Undenary
6	Senary	12	Duodenary
7	Septenary	16	Hexadecimal

General Conversion Methods

Any number can be converted from the scale of b to the scale of some integer $\bar{b} \neq b$, $\bar{b}>1$, by using arithmetic operations in either the b-scale or the $\bar{b}$-scale. Accordingly, there are four methods of conversion, depending on whether the number to be converted is an integer or a proper fraction.

Integers $X=(A_m \ldots A_1 A_0)_{(b)}$

(I) b-scale arithmetic. Convert $\bar{b}$ to the b-scale and define

$$X/\bar{b}=X_1+\overline{A}_0'/\bar{b},$$
$$X_1/\bar{b}=X_2+\overline{A}_1'/\bar{b},$$
$$\vdots$$
$$X_{\overline{m}}/\bar{b}=0+\overline{A}_{\overline{m}}'/\bar{b},$$

where $\overline{A}_0'$, $\overline{A}_1'$, $\ldots$, $\overline{A}_{\overline{m}}'$ are the remainders and X_1, X_2, $\ldots$, $X_{\overline{m}}$ the quotients (in the b-scale) where X, X_1, $\ldots$, $X_{\overline{m}-1}$, respectively are divided by $\bar{b}$ in the b-scale. Then convert the remainders to the $\bar{b}$-scale,

$$(\overline{A}_0')_{(\bar{b})}=\overline{A}_0, \ (\overline{A}_1')_{(\bar{b})}=\overline{A}_1, \ \ldots, \ (\overline{A}_{\overline{m}}')_{(\bar{b})}=\overline{A}_{\overline{m}}$$

and obtain

$$X=(\overline{A}_{\overline{m}} \ldots \overline{A}_1 \overline{A}_0)_{(\bar{b})}.$$

(II) $\bar{b}$-scale arithmetic. Convert b and A_0, A_1, $\ldots$, A_m to the $\bar{b}$-scale and define, using arithmetic operations in the $\bar{b}$-scale,

$$X_{m-1}=A_m b+A_{m-1},$$
$$X_{m-2}=X_{m-1} b+A_{m-2},$$
$$X_1=X_2 b+A_1,$$

then

$$X=X_1 b+A_0.$$

Proper fractions $x=(0.a_{-1}a_{-2} \ldots)_{(b)}$

To convert a proper fraction x, given to n digits in the b-scale, to the scale of $\bar{b} \neq b$ such that inverse conversion from the $\bar{b}$-scale may yield the same n rounded digits in the b-scale, the representation of x in the $\bar{b}$-scale must be obtained to $\bar{n}$ rounded digits where n satisfies $\bar{b}^{\bar{n}}>b^n$.

(III) b-scale arithmetic. Convert $\bar{b}$ to the b-scale and define

$$x\bar{b}=x_1+\bar{a}_{-1}'$$
$$x_1\bar{b}=x_2+\bar{a}_{-2}'$$
$$x_{\overline{n}-1}\bar{b}=x_{\overline{n}}+\bar{a}_{-\overline{n}}'$$

where $\bar{a}_{-1}, \bar{a}'_{-2}, \ldots, \bar{a}_{-\bar{n}}$ are the integral parts and $x_1, x_2, \ldots, x_{\bar{n}}$ the fractional parts (in the b-scale) of the products $x\bar{b}, x_1\bar{b}, \ldots, x_{\bar{n}-1}\bar{b}$, respectively. Then convert the integral parts to the $\bar{b}$-scale,

$$(\bar{a}'_{-1})_{(\bar{b})} = \bar{a}_{-1}, \quad (\bar{a}'_{-2})_{(\bar{b})} = \bar{a}_{-2}, \quad \ldots, \quad (\bar{a}'_{-n})_{(\bar{b})} = \bar{a}_{-n},$$

and obtain

$$x = (0.\bar{a}_{-1}\bar{a}_{-2} \ldots \bar{a}_{-n})_{(\bar{b})}.$$

(IV) $\bar{b}$-scale arithmetic. Convert b and $a_{-1}, a_{-2}, \ldots, a_{-n}$ to the $\bar{b}$-scale and define, using arithmetic operations in the $\bar{b}$-scale,

$$x_{-n+1} = a_{-n}/b + a_{-n+1},$$
$$x_{-n+2} = x_{-n+1}/b + a_{-n+2},$$
$$x_{-1} = x_2/b + a_{-1};$$

then

$$x = x_{-1}/b.$$

Numerical Methods

The examples are restricted to the scales of 2, 8, 10 because of their importance to electronic computers.

Note that the octal scale is a power of the binary scale. In fact, an octal digit corresponds to a triplet of binary digits. Then, binary arithmetic may be used whenever a number either is to be converted to the octal scale or is given in the octal scale and is to be converted to some other scale.

Decimal	1	2	3	4	5	6	7	8	9	10
Octal	1	2	3	4	5	6	7	10	11	12
Binary	1	10	11	100	101	110	111	1 000	1 001	1 010

Example 1. Convert $X = (1369)_{(10)}$ to the octal scale. By (I) we have $b = 10$, $\bar{b} = 8_{(10)}$ and so, using decimal arithmetic,

$$1369/8 = 171 + 1/8,$$
$$171/8 = 21 + 3/8,$$
$$21/8 = 2 + 5/8,$$
$$2/8 = 0 + 2/8;$$

then

$$X = (2531)_{(8)}.$$

By (II) we have $b = (12)_{(8)}$ and $A_3 = 1_{(8)}$, $A_2 = 3_{(8)}$, $A_1 = 6_{(8)}$, $A_0 = (11)_{(8)}$. Hence, using octal arithmetic,

$$X_2 = 1 \cdot 12 + 3 = (15)_{(8)},$$
$$X_1 = 15 \cdot 12 + 6 = (210)_{(8)},$$
$$X = 210 \cdot 12 + 11 = (2531)_{(8)}.$$

Using binary arithmetic we have, by (II), $b = (1010)_{(2)}$ and $A_3 = 1_{(2)}$, $A_2 = (11)_{(2)}$, $A_1 = (110)_{(2)}$, $A_0 (1001)_{(2)}$. Thus

$$X_2 = 1 \cdot 1010 + 11 = (1101)_{(2)},$$
$$X_1 = 1101 \cdot 1010 + 110 = (10\ 001\ 000)_{(2)},$$
$$X = 10\ 001\ 000 \cdot 1010 + 1001 = (10\ 101\ 011\ 001)_{(2)},$$

whence, on converting to the octal scale,

$$X = (2531)_{(8)}.$$

Example 2. Convert $X = (2531)_{(8)}$ to the decimal scale. By (I) we have $\bar{b} = 10 = (12)_{(8)}$ and hence, using octal arithmetic,

$$2531/12 = 210 + 11/12$$
$$210/12 = 15 + 6/12$$
$$15/12 = 1 + 3/12$$
$$1/12 = 0 + 1/12$$

Thus, converting to the decimal scale,

$$\bar{A}_0 = (11)_{(8)} = 9, \quad \bar{A}_1 = 6_{(8)} = 6, \quad \bar{A}_2 = 3_{(8)} = 3, \quad \bar{A}_3 = 1,$$

and so

$$X = (1369)_{(10)}.$$

By (II) we have $\bar{b} = 10$, and the octal digits of X are unchanged in the decimal scale. Hence, using decimal arithmetic,

$$X_2 = 2 \cdot 8 + 5 = (21)_{(10)},$$
$$X_1 = 21 \cdot 8 + 3 = (171)_{(10)},$$
$$X = 171 \cdot 8 + 1 = (1369)_{(10)}.$$

Using binary arithmetic we have, by (II), $b = 8 = (1000)_{(2)}$ and $A_0 = 1$, $A_1 = (11)_{(2)}$, $A_2 = (101)_{(2)}$, $A_3 = (10)_{(2)}$. Then,

$$X_2 = 10 \cdot 1000 + 101 = (10\ 101)_{(2)},$$
$$X_1 = 10\ 101 \cdot 1000 + 11 = (10\ 101\ 011)_{(2)},$$
$$X = 10\ 101\ 011 \cdot 1000 + 1 = (10\ 101\ 011\ 001)_{(2)},$$

whence, on converting to the decimal scale,

$$X = (1369)_{(10)}.$$

Observe that in both examples above, octal arithmetic is used as an intermediate step to convert, according to (II), the given number to the binary scale. If, instead, the given number is first converted to the binary scale, then binary arithmetic may be applied directly to convert, according to (I), the given number from the binary scale to the scale desired.

For example, in converting $X=(2531)_{(8)}$ to the decimal scale, we find first $X=(10101011001)_{(2)}$ and then obtain, using (I) with $\bar{b}=10=(1010)_{(2)}$,

$$10\ 101\ 011\ 001/1010=10\ 001\ 000+1001/1010,$$
$$10\ 001\ 000/1010=1101+110/1010,$$
$$1101/1010=1+11/1010,$$
$$1/1010=0+1/1010.$$

Thus, on converting to the decimal scale,

$$A_0=(1001)_{(2)}=9,\quad A_1=(110)_{(2)}=6,$$
$$A_2=(11)_{(2)}=3,\quad A_3=1,$$

whence

$$X=(1369)_{(10)}.$$

Example 3. Convert $x=(0.355)_{(10)}$ to the binary scale.

We first convert to the octal scale, using decimal arithmetic. By (III), we find with $\bar{b}=8$

$$(0.355)\cdot8=2+0.840,\quad (0.080)\cdot8=0+0.640$$
$$(0.840)\cdot8=6+0.720,\quad (0.640)\cdot8=5+0.120$$
$$(0.720)\cdot8=5+0.760,\quad (0.120)\cdot8=0+0.960$$
$$(0.760)\cdot8=6+0.080,\quad (0.960)\cdot8=7+0.680$$

whence $x=(0.26560507\ldots)_{(8)}$. Thus, on converting to the binary scale,

$$x=(0.010\ 110\ 101\ 110\ 000\ 101\ 000\ 111\ \ldots)_{(2)}.$$

In order that inverse conversion of x from the binary to the decimal scale yield again x to the given number n of decimal digits, we must round x in the binary scale to at least $\bar{n}$ digits where $\bar{n}$ is chosen such that $2^{\bar{n}}>10^n$. As a working rule, we may take $\bar{n}\geq\frac{10}{3}n$. Hence, to obtain $x=(0.355)_{(10)}$ by inverse conversion, x must be rounded in the binary scale to $\bar{n}\geq\frac{10}{3}3=10$ digits. Thus,

$$x=(0.010\ 110\ 110\ 0)_{(2)}.$$

To carry out the inverse conversion we can first convert to the octal scale,

$$x=(0.266)_{(8)},$$

and then apply (IV) with $b=8$, using decimal arithmetic:

$$x_{-2}=6/8+6=6.75,$$
$$x_{-1}=6.75/8+2=2.84375,$$
$$x=2.84375/8=0.355\ 46875.$$

Alternatively, we can apply (III) with $\bar{b}=(1010)_{(2)}$, using binary arithmetic:

$$(0.010\ 110\ 11)\cdot1010=11+(0.100\ 011\ 1),$$
$$(0.100\ 011\ 1)\cdot1010=101+(0.100\ 011),$$
$$(0.100\ 011)\cdot1010=101+(0.011\ 11),$$
$$(0.011\ 11)\cdot1010=100+(0.101\ 1).$$

Converting the integral parts to the decimal scale, we find

$$\bar{a}_{-1}=(11)_{(2)}=3,\quad \bar{a}_{-2}=\bar{a}_{-3}=(101)_{(2)}=5,$$
$$\bar{a}_{-4}=(100)_{(2)}=4,$$

and thus

$$x=(0.3554)_{(10)}$$

Note that the fractional part in any step is the unconverted remainder. Thus, to round at any step, it is only necessary to ascertain whether the unconverted portion to be neglected is greater or less than $\frac{1}{2}$; i.e., whether, in the binary scale, the first neglected digit is 1 or 0.

Example 4. Convert $x=(3.141593)_{(10)}\cdot10^{-9}$ to the binary scale.

The desired representation is

$$x=(1.a_{-1}a_{-2}\ \ldots\ a_{-n})_{(2)}\cdot2^{-k}$$

where n and k are such that inverse conversion from the binary scale to the decimal scale will produce x to the same given 15 decimal digits. Accordingly, by the rule stated in **Example 3,** n and k are to be chosen so as to satisfy $n+k\geq\frac{10}{3}\cdot15=50$.

From **Table 28.1** we find

$$2^{-29}<(3.141593)_{(10)}\cdot10^{-9}<2^{-28}$$

Thus, we must take $k=29$ and, consequently, choose $n>21$. The conversion on a desk calculator thus proceeds as follows. First, we obtain by use of **Table 28.1**

$$2^{29}x=(1.686\ 629\ 899)_{(10)}$$

Then, for convenience's sake, we convert this number to the octal scale, using the method of **Example 3** and rounding as required, to at least 7 octal ($=21$ binary) digits. We find

$$2^{29}x=(1.537\ 4337)_{(8)}.$$

Hence

$$x=(1.537\ 433\ 7)_{(8)}\cdot2^{-29}$$

and, consequently,

$$x=(1.\ 101\ 011\ 111\ 100\ 011\ 011\ 111)_{(2)}\cdot2^{-29}.$$

To convert x back to the decimal scale we only need to obtain from **Table 28.1** the various powers of 2 which appear in the above representation and sum them. However, since $2^{-m}=2^{-m+1}-2^{-m}$ for any real constant m, it is more convenient to reduce first the binary representation of x to the form

$$x=2^{-28}-2^{-31}-2^{-33}-2^{-39}+2^{-42}-2^{-45}-2^{-50}$$

and then sum these powers of 2. (Note that the number of summands is thereby decreased from 16 to 7.) From **Table 28.1** we have

$$
\begin{aligned}
+2^{-28} &=+3.725 \; 290 \; 298 \; \cdot 10^{-9}\\
-2^{-31} &=- \;\;\; .465 \; 661 \; 287 \; \cdot 10^{-9}\\
-2^{-33} &=- \;\;\; .116 \; 415 \; 322 \; \cdot 10^{-9}\\
-2^{-39} &=- \;\;\; .001 \; 818 \; 989 \; \cdot 10^{-9}\\
+2^{-42} &=+ \;\;\; .000 \; 227 \; 374 \; \cdot 10^{-9}\\
-2^{-45} &=- \;\;\; .000 \; 028 \; 422 \; \cdot 10^{-9}\\
-2^{-50} &=- \;\;\; .000 \; 000 \; 888 \; \cdot 10^{-9}\\
\hline
x &= \;\;\;\; 3.141 \; 592 \; 764 \; \cdot 10^{-9}
\end{aligned}
$$

Nine decimal digits are used for sufficient accuracy reserve. Hence, rounding to seven significant figures, we find

$$x=(3.141593)_{(10)}\cdot 10^{-9}.$$

To convert a number such as

$$x=(\xi)_{(10)}\cdot 10^{k}$$

to the binary scale, where k is a positive integer so large that **Table 28.1** cannot be used, apply the following device: Compute

$$\log_2 x=\frac{\log_{10} x}{\log_{10} 2}=k+\frac{x_1}{\log_{10} 2}$$

where k is the quotient and x_1 the remainder, the division being carried out in the decimal scale. Then find $\eta=10^{x_1}$, i.e., $x_1=\log_{10}\eta$, so that ·

$$\log_2 x=k+\frac{\log_{10}\eta}{\log_{10} 2}=k+\log_2\eta$$

whence

$$x=(\eta)_{(10)}2^{k}.$$

Now convert $(\eta)_{(10)}$ to the binary scale by any of the methods described above.

A similar device may be used to convert to the decimal scale a binary number that is outside the range of **Table 28.1**.

Example 5. Convert $x=(2.773)_{(10)}\cdot 10^{83}$ to the binary scale.

We first compute, using **4.1.19** and **Table 4.1**,

$$\log_2 x=\frac{\log_{10} x}{\log_{10} 2}=\frac{83.44295}{.30103}=277+\frac{.05764}{.30103},$$

and find from **Table 4.1**, $.05764=\log_{10} 1.1419$. Hence

$$\log_2 x=277+\frac{\log_{10} 1.1419}{\log_{10} 2}=277+\log_2 1.1419$$

and so

$$x=(1.1419)_{(10)}\cdot 2^{277}.$$

Now we apply the methods of **Example 3** to obtain $(1.1419)_{(10)}=(1.110516)_{(8)}$ where octal notation is used for the sake of convenience.

To round such that inverse conversion will yield the same decimal digits of x, observe that the last non-zero decimal digit of x is $3\cdot 10^{80}$. **Table 28.4** shows that $2^{265}<10^{80}<2^{266}$. Hence, in the binary scale, x must be a binary integer times 2^{265}; i.e., $(1.110516)_{(8)}$ must be rounded to 4 octal ($=12$ binary) digits. As a result,

$$x=(1.1105)_{(8)}\cdot 2^{277}=(11105)_{(8)}\cdot 2^{265}$$
$$=(1 \; 001 \; 001 \; 000 \; 101)_{(2)} 2^{265}$$

Conversion back to the decimal scale proceeds as follows, we write

$$
\begin{aligned}
\log_{10} x &=\log_{10} 2 \; \log_2 x\\
&=\log_{10} 2\{265+\log_2 (11105)_{(8)}\}\\
&=\log_{10} 2\left\{265+\frac{\log_{10} (11105)_{(8)}}{\log_{10} 2}\right\}\\
&=265 \log_{10} 2+\log_{10} (11105)_{(8)}.
\end{aligned}
$$

Hence, converting $(11105)_{(8)}$ to the decimal scale by any of the methods of **Example 2,** we obtain

$$\log_{10} x=265 \log_{10} 2+\log_{10} 4677$$

which yields, using **Table 4.1**

$$\log_{10} x=83.44292$$

Thus, by **Table 4.1,** we find, rounded to four significant figures,

$$x=(2.773)_{(10)}\cdot 10^{83}.$$

References

[28.1] J. Malengreau, Étude des écritures binaires, Bibliothèque Sci. 32 Mathématique. Édition Griffon, Neuchâtel, Suisse (1958).

[28.2] D. D. McCracken, Digital computer programming (John Wiley & Sons, Inc., New York, N.Y., 1957).

[28.3] R. K. Richards, Arithmetic operation in digital computers (D. Van Nostrand Co., Inc., New York, N.Y., 1955).

Table 28.1　　　　　　　　　　　$2^{\pm n}$ **IN DECIMAL**

2^n	n	2^{-n}
1	0	1.0
2	1	0.5
4	2	0.25
8	3	0.125
16	4	0.0625
32	5	0.03125
64	6	0.01562 5
128	7	0.00781 25
256	8	0.00390 625
512	9	0.00195 3125
1024	10	0.00097 65625
2048	11	0.00048 82812 5
4096	12	0.00024 41406 25
8192	13	0.00012 20703 125
16384	14	0.00006 10351 5625
32768	15	0.00003 05175 78125
65536	16	0.00001 52587 89062 5
1 31072	17	0.00000 76293 94531 25
2 62144	18	0.00000 38146 97265 625
5 24288	19	0.00000 19073 48632 8125
10 48576	20	0.00000 09536 74316 40625
20 97152	21	0.00000 04768 37158 20312 5
41 94304	22	0.00000 02384 18579 10156 25
83 88608	23	0.00000 01192 09289 55078 125
167 77216	24	0.00000 00596 04644 77539 0625
335 54432	25	0.00000 00298 02322 38769 53125
671 08864	26	0.00000 00149 01161 19384 76562 5
1342 17728	27	0.00000 00074 50580 59692 38281 25
2684 35456	28	0.00000 00037 25290 29846 19140 625
5368 70912	29	0.00000 00018 62645 14923 09570 3125
10737 41824	30	0.00000 00009 31322 57461 54785 15625
21474 83648	31	0.00000 00004 65661 28730 77392 57812 5
42949 67296	32	0.00000 00002 32830 64365 38696 28906 25
85899 34592	33	0.00000 00001 16415 32182 69348 14453 125
1 71798 69184	34	0.00000 00000 58207 66091 34674 07226 5625
3 43597 38368	35	0.00000 00000 29103 83045 67337 03613 28125
6 87194 76736	36	0.00000 00000 14551 91522 83668 51806 64062 5
13 74389 53472	37	0.00000 00000 07275 95761 41834 25903 32031 25
27 48779 06944	38	0.00000 00000 03637 97880 70917 12951 66015 625
54 97558 13888	39	0.00000 00000 01818 98940 35458 56475 83007 8125
109 95116 27776	40	0.00000 00000 00909 49470 17729 28237 91503 90625
219 90232 55552	41	0.00000 00000 00454 74735 08864 64118 95751 95312 5
439 80465 11104	42	0.00000 00000 00227 37367 54432 32059 47875 97656 25
879 60930 22208	43	0.00000 00000 00113 68683 77216 16029 73937 98828 125
1759 21860 44416	44	0.00000 00000 00056 84341 88608 08014 86968 99414 0625
3518 43720 88832	45	0.00000 00000 00028 42170 94304 04007 43484 49707 03125
7036 87441 77664	46	0.00000 00000 00014 21085 47152 02003 71742 24853 51562 5
14073 74883 55328	47	0.00000 00000 00007 10542 73576 01001 85871 12426 75781 25
28147 49767 10656	48	0.00000 00000 00003 55271 36788 00500 92935 56213 37890 625
56294 99534 21312	49	0.00000 00000 00001 77635 68394 00250 46467 78106 68945 3125
112589 99068 42624	50	0.00000 00000 00000 88817 84197 00125 23233 89053 34472 65625

2^x IN DECIMAL

Table 28.2

x	2^x	x	2^x	x	2^x
0.001	1.00069 33874 62581	0.01	1.00695 55500 56719	0.1	1.07177 34625 36293
0.002	1.00138 72557 11335	0.02	1.01395 94797 90029	0.2	1.14869 83549 97035
0.003	1.00208 16050 79633	0.03	1.02101 21257 07193	0.3	1.23114 44133 44916
0.004	1.00277 64359 01078	0.04	1.02811 38266 56067	0.4	1.31950 79107 72894
0.005	1.00347 17485 09503	0.05	1.03526 49238 41377	0.5	1.41421 35623 73095
0.006	1.00416 75432 38973	0.06	1.04246 57608 41121	0.6	1.51571 65665 10398
0.007	1.00486 38204 23785	0.07	1.04971 66836 23067	0.7	1.62450 47927 12471
0.008	1.00556 05803 98468	0.08	1.05701 80405 61380	0.8	1.74110 11265 92248
0.009	1.00625 78234 97782	0.09	1.06437 01824 53360	0.9	1.86606 59830 73615

$10^{\pm n}$ IN OCTAL

Table 28.3

10^n	n	10^{-n}	10^n	n	10^{-n}
1	0	1.000 000 000 000 000 000 00	112 402 762 000	10	0.000 000 000 006 676 337 66
12	1	0.063 146 314 631 463 146 31	1 351 035 564 000	11	0.000 000 000 000 537 657 77
144	2	0.005 075 341 217 270 243 66	16 432 451 210 000	12	0.000 000 000 000 043 136 32
1 750	3	0.000 406 111 564 570 651 77	221 411 634 520 000	13	0.000 000 000 000 003 411 35
23 420	4	0.000 032 155 613 530 704 15	2 657 142 036 440 000	14	0.000 000 000 000 000 264 11
303 240	5	0.000 002 476 132 610 706 64	34 327 724 461 500 000	15	0.000 000 000 000 000 022 01
3 641 100	6	0.000 000 206 157 364 055 37	434 157 115 760 200 000	16	0.000 000 000 000 000 001 63
46 113 200	7	0.000 000 015 327 745 152 75	5 432 127 413 542 400 000	17	0.000 000 000 000 000 000 14
575 360 400	8	0.000 000 001 257 143 561 06	67 405 553 164 731 000 000	18	0.000 000 000 000 000 000 01
7 346 545 000	9	0.000 000 000 104 560 276 41			

$n \log_{10} 2$, $n \log_2 10$ IN DECIMAL

Table 28.4

n	$n \log_{10} 2$	$n \log_2 10$	n	$n \log_{10} 2$	$n \log_2 10$
1	0.30102 99957	3.32192 80949	6	1.80617 99740	19.93156 85693
2	0.60205 99913	6.64385 61898	7	2.10720 99696	23.25349 66642
3	0.90308 99870	9.96578 42847	8	2.40823 99653	26.57542 47591
4	1.20411 99827	13.28771 23795	9	2.70926 99610	29.89735 28540
5	1.50514 99783	16.60964 04744	10	3.01029 99566	33.21928 09489

ADDITION AND MULTIPLICATION TABLES

Table 28.5

Addition   Multiplication

Binary Scale

$$0 + 0 = 0$$
$$0 + 1 = 1 + 0 = 1$$
$$1 + 1 = 10$$

$$0 \times 0 = 0$$
$$0 \times 1 = 1 \times 0 = 0$$
$$1 \times 1 = 1$$

Octal Scale

0	01	02	03	04	05	06	07
1	02	03	04	05	06	07	10
2	03	04	05	06	07	10	11
3	04	05	06	07	10	11	12
4	05	06	07	10	11	12	13
5	06	07	10	11	12	13	14
6	07	10	11	12	13	14	15
7	10	11	12	13	14	15	16

1	02	03	04	05	06	07
2	04	06	10	12	14	16
3	06	11	14	17	22	25
4	10	14	20	24	30	34
5	12	17	24	31	36	43
6	14	22	30	36	44	52
7	16	25	34	43	52	61

MATHEMATICAL CONSTANTS IN OCTAL SCALE

Table 28.6

$\pi = (3.11037\ 552421)_{(8)}$

$\pi^{-1} = (0.24276\ 301556)_{(8)}$

$\sqrt{\pi} = (1.61337\ 611067)_{(8)}$

$\ln \pi = (1.11206\ 404435)_{(8)}$

$\log_2 \pi = (1.51544\ 163223)_{(8)}$

$\sqrt{10} = (3.12305\ 407267)_{(8)}$

$e = (2.55760\ 521305)_{(8)}$

$e^{-1} = (0.27426\ 530661)_{(8)}$

$\sqrt{e} = (1.51411\ 230704)_{(8)}$

$\log_{10} e = (0.33626\ 754251)_{(8)}$

$\log_2 e = (1.34252\ 166245)_{(8)}$

$\log_2 10 = (3.24464\ 741136)_{(8)}$

$\gamma = (0.44742\ 147707)_{(8)}$

$\ln \gamma = -(0.43127\ 233602)_{(8)}$

$\log_2 \gamma = -(0.62573\ 030645)_{(8)}$

$\sqrt{2} = (1.32404\ 746320)_{(8)}$

$\ln 2 = (0.54271\ 027760)_{(8)}$

$\ln 10 = (2.23273\ 067355)_{(8)}$

29. Laplace Transforms

Contents

29. Laplace Transforms

29.1. Definition of the Laplace Transform

One-dimensional Laplace Transform

29.1.1
$$f(s) = \mathscr{L}\{F(t)\} = \int_0^\infty e^{-st} F(t)\,dt$$

$F(t)$ is a function of the real variable t and s is a complex variable. $F(t)$ is called the original function and $f(s)$ is called the image function. If the integral in **29.1.1** converges for a real $s = s_0$, i.e.,

$$\lim_{\substack{A \to 0 \\ B \to \infty}} \int_A^B e^{-s_0 t} F(t)\,dt$$

exists, then it converges for all s with $\mathscr{R}s > s_0$, and the image function is a single valued analytic function of s in the half-plane $\mathscr{R}s > s_0$.

Two-dimensional Laplace Transform

29.1.2
$$f(u,v) = \mathscr{L}\{F(x,y)\} = \int_0^\infty \int_0^\infty e^{-ux-vy} F(x,y)\,dx\,dy$$

Definition of the Unit Step Function

29.1.3
$$u(t) = \begin{cases} 0 & (t<0) \\ \frac{1}{2} & (t=0) \\ 1 & (t>0) \end{cases}$$

In the following tables the factor $u(t)$ is to be understood as multiplying the original function $F(t)$.

29.2. Operations for the Laplace Transform[1]

	Original Function $F(t)$	Image Function $f(s)$
29.2.1	$F(t)$	$\int_0^\infty e^{-st} F(t)\,dt$
Inversion Formula		
29.2.2	$\dfrac{1}{2\pi i} \int_{c-i\infty}^{c+i\infty} e^{ts} f(s)\,ds$	$f(s)$
Linearity Property		
29.2.3	$AF(t) + BG(t)$	$Af(s) + Bg(s)$
Differentiation		
29.2.4	$F'(t)$	$sf(s) - F(+0)$
29.2.5	$F^{(n)}(t)$	$s^n f(s) - s^{n-1} F(+0) - s^{n-2} F'(+0) - \ldots - F^{(n-1)}(+0)$
Integration		
29.2.6	$\int_0^t F(\tau)\,d\tau$	$\dfrac{1}{s} f(s)$
29.2.7	$\int_0^t \int_0^\tau F(\lambda)\,d\lambda\,d\tau$	$\dfrac{1}{s^2} f(s)$
Convolution (Faltung) Theorem		
29.2.8	$\int_0^t F_1(t-\tau) F_2(\tau)\,d\tau = F_1 * F_2$	$f_1(s) f_2(s)$
29.2.9	$-tF(t)$	$f'(s)$ **Differentiation**
29.2.10	$(-1)^n t^n F(t)$	$f^{(n)}(s)$

---

[1] Adapted by permission from R. V. Churchill, Operational mathematics, 2d ed., McGraw-Hill Book Co., Inc., New York, N.Y., 1958.

	Original Function $F(t)$	Image Function $f(s)$

Integration

29.2.11 $\quad \dfrac{1}{t} F(t)$ $\qquad\qquad \displaystyle\int_s^\infty f(x)dx$

Linear Transformation

29.2.12 $\quad e^{at} F(t)$ $\qquad\qquad f(s-a)$

29.2.13 $\quad \dfrac{1}{c} F\left(\dfrac{t}{c}\right) \quad (c>0)$ $\qquad\qquad f(cs)$

29.2.14 $\quad \dfrac{1}{c} e^{(b/c)t} F\left(\dfrac{t}{c}\right) \quad (c>0)$ $\qquad\qquad f(cs-b)$

Translation

29.2.15 $\quad F(t-b)u(t-b) \quad (b>0)$ $\qquad\qquad e^{-bs}f(s)$

Periodic Functions

29.2.16 $\quad F(t+a)=F(t)$ $\qquad\qquad \dfrac{\displaystyle\int_0^a e^{-st}F(t)dt}{1-e^{-as}}$

29.2.17 $\quad F(t+a)=-F(t)$ $\qquad\qquad \dfrac{\displaystyle\int_0^a e^{-st}F(t)dt}{1+e^{-as}}$

Half-Wave Rectification of $F(t)$ in 29.2.17

29.2.18 $\quad F(t)\displaystyle\sum_{n=0}^\infty (-1)^n u(t-na)$ $\qquad\qquad \dfrac{f(s)}{1-e^{-as}}$

Full-Wave Rectification of $F(t)$ in 29.2.17

29.2.19 $\quad |F(t)|$ $\qquad\qquad f(s)\coth\dfrac{as}{2}$

Heaviside Expansion Theorem

29.2.20 $\quad \displaystyle\sum_{n=1}^m \dfrac{p(a_n)}{q'(a_n)}e^{a_n t}$ $\qquad\qquad \dfrac{p(s)}{q(s)}, \ q(s)=(s-a_1)(s-a_2)\ldots(s-a_m)$

$\qquad\qquad\qquad\qquad\qquad\qquad\qquad\qquad\quad p(s)$ a polynomial of degree$<m$

29.2.21 $\quad e^{at}\displaystyle\sum_{n=1}^r \dfrac{p^{(r-n)}(a)}{(r-n)!}\dfrac{t^{n-1}}{(n-1)!}$ $\qquad\qquad \dfrac{p(s)}{(s-a)^r}$

$\qquad\qquad\qquad\qquad\qquad\qquad\qquad\qquad\quad p(s)$ a polynomial of degree$<r$

29.3. Table of Laplace Transforms[2,3]

For a comprehensive table of Laplace and other integral transforms see [29.9]. For a table of two-dimensional Laplace transforms see [29.11].

	$f(s)$	$F(t)$
29.3.1	$\dfrac{1}{s}$	1
29.3.2	$\dfrac{1}{s^2}$	t

---

[2] The numbers in bold type in the $f(s)$ and $F(t)$ columns indicate the chapters in which the properties of the respective higher mathematical functions are given.

[3] Adapted by permission from R. V. Churchill, Operational mathematics, 2d. ed., McGraw-Hill Book Co., Inc., New York, N. Y., 1958.

	$f(s)$		$F(t)$

29.3.3 $\quad \dfrac{1}{s^n} \quad (n=1,2,3,\ldots)$

$\dfrac{t^{n-1}}{(n-1)!}$

29.3.4 $\quad \dfrac{1}{\sqrt{s}}$

$\dfrac{1}{\sqrt{\pi t}}$

29.3.5 $\quad s^{-3/2}$

$2\sqrt{t/\pi}$

29.3.6 $\quad s^{-(n+\frac{1}{2})} \quad (n=1,2,3,\ldots)$

$\dfrac{2^n t^{n-\frac{1}{2}}}{1\cdot 3\cdot 5\ldots(2n-1)\sqrt{\pi}}$

29.3.7 $\quad \dfrac{\Gamma(k)}{s^k} \quad (k>0)$

$\quad$ 6 $\quad t^{k-1}$

29.3.8 $\quad \dfrac{1}{s+a}$

e^{-at}

29.3.9 $\quad \dfrac{1}{(s+a)^2}$

te^{-at}

29.3.10 $\quad \dfrac{1}{(s+a)^n} \quad (n=1,2,3,\ldots)$

$\dfrac{t^{n-1}e^{-at}}{(n-1)!}$

29.3.11 $\quad \dfrac{\Gamma(k)}{(s+a)^k} \quad (k>0)$

$\quad$ 6 $\quad t^{k-1}e^{-at}$

29.3.12 $\quad \dfrac{1}{(s+a)(s+b)} \quad (a\neq b)$

$\dfrac{e^{-at}-e^{-bt}}{b-a}$

29.3.13 $\quad \dfrac{s}{(s+a)(s+b)} \quad (a\neq b)$

$\dfrac{ae^{-at}-be^{-bt}}{a-b}$

29.3.14 $\quad \dfrac{1}{(s+a)(s+b)(s+c)}$

$-\dfrac{(b-c)e^{-at}+(c-a)e^{-bt}+(a-b)e^{-ct}}{(a-b)(b-c)(c-a)}$

$(a,b,c \text{ distinct constants})$

29.3.15 $\quad \dfrac{1}{s^2+a^2}$

$\dfrac{1}{a}\sin at$

29.3.16 $\quad \dfrac{s}{s^2+a^2}$

$\cos at$

29.3.17 $\quad \dfrac{1}{s^2-a^2}$

$\dfrac{1}{a}\sinh at$

29.3.18 $\quad \dfrac{s}{s^2-a^2}$

$\cosh at$

29.3.19 $\quad \dfrac{1}{s(s^2+a^2)}$

$\dfrac{1}{a^2}(1-\cos at)$

29.3.20 $\quad \dfrac{1}{s^2(s^2+a^2)}$

$\dfrac{1}{a^3}(at-\sin at)$

29.3.21 $\quad \dfrac{1}{(s^2+a^2)^2}$

$\dfrac{1}{2a^3}(\sin at-at\cos at)$

	$f(s)$	$F(t)$	
29.3.22	$\dfrac{s}{(s^2+a^2)^2}$	$\dfrac{t}{2a}\sin at$	
29.3.23	$\dfrac{s^2}{(s^2+a^2)^2}$	$\dfrac{1}{2a}(\sin at+at\cos at)$	
29.3.24	$\dfrac{s^2-a^2}{(s^2+a^2)^2}$	$t\cos at$	
29.3.25	$\dfrac{s}{(s^2+a^2)(s^2+b^2)}\quad(a^2\neq b^2)$	$\dfrac{\cos at-\cos bt}{b^2-a^2}$	
29.3.26	$\dfrac{1}{(s+a)^2+b^2}$	$\dfrac{1}{b}\,e^{-at}\sin bt$	
29.3.27	$\dfrac{s+a}{(s+a)^2+b^2}$	$e^{-at}\cos bt$	
29.3.28	$\dfrac{3a^2}{s^3+a^3}$	$e^{-at}-e^{\frac{1}{2}at}\left(\cos\dfrac{at\sqrt{3}}{2}-\sqrt{3}\sin\dfrac{at\sqrt{3}}{2}\right)$	
29.3.29	$\dfrac{4a^3}{s^4+4a^4}$	$\sin at\cosh at-\cos at\sinh at$	
29.3.30	$\dfrac{s}{s^4+4a^4}$	$\dfrac{1}{2a^2}\sin at\sinh at$	
29.3.31	$\dfrac{1}{s^4-a^4}$	$\dfrac{1}{2a^3}(\sinh at-\sin at)$	
29.3.32	$\dfrac{s}{s^4-a^4}$	$\dfrac{1}{2a^2}(\cosh at-\cos at)$	
29.3.33	$\dfrac{8a^3s^2}{(s^2+a^2)^3}$	$(1+a^2t^2)\sin at-at\cos at$	
29.3.34	$\dfrac{1}{s}\left(\dfrac{s-1}{s}\right)^n$	$L_n(t)$	22
29.3.35	$\dfrac{s}{(s+a)^{\frac{3}{2}}}$	$\dfrac{1}{\sqrt{\pi t}}\,e^{-at}(1-2at)$	
29.3.36	$\sqrt{s+a}-\sqrt{s+b}$	$\dfrac{1}{2\sqrt{\pi t^3}}\,(e^{-bt}-e^{-at})$	
29.3.37	$\dfrac{1}{\sqrt{s}+a}$	$\dfrac{1}{\sqrt{\pi t}}-ae^{a^2t}\,\mathrm{erfc}\,a\sqrt{t}$	7
29.3.38	$\dfrac{\sqrt{s}}{s-a^2}$	$\dfrac{1}{\sqrt{\pi t}}+ae^{a^2t}\,\mathrm{erf}\,a\sqrt{t}$	7
29.3.39	$\dfrac{\sqrt{s}}{s+a^2}$	$\dfrac{1}{\sqrt{\pi t}}-\dfrac{2a}{\sqrt{\pi}}\,e^{-a^2t}\displaystyle\int_0^{a\sqrt{t}}e^{\lambda^2}d\lambda$	7
29.3.40	$\dfrac{1}{\sqrt{s}(s-a^2)}$	$\dfrac{1}{a}\,e^{a^2t}\,\mathrm{erf}\,a\sqrt{t}$	7

	$f(s)$	$F(t)$	
29.3.41	$\dfrac{1}{\sqrt{s}(s+a^2)}$	$\dfrac{2}{a\sqrt{\pi}}\,e^{-a^2t}\displaystyle\int_0^{a\sqrt{t}}e^{\lambda^2}d\lambda$	**7**
29.3.42	$\dfrac{b^2-a^2}{(s-a^2)(b+\sqrt{s})}$	$e^{a^2t}[b-a\ \mathrm{erf}\ a\sqrt{t}]-be^{b^2t}\ \mathrm{erfc}\ b\sqrt{t}$	**7**
29.3.43	$\dfrac{1}{\sqrt{s}(\sqrt{s}+a)}$	$e^{a^2t}\ \mathrm{erfc}\ a\sqrt{t}$	**7**
29.3.44	$\dfrac{1}{(s+a)\sqrt{s+b}}$	$\dfrac{1}{\sqrt{b-a}}\,e^{-at}\ \mathrm{erf}\ (\sqrt{b-a}\sqrt{t})$	**7**
29.3.45	$\dfrac{b^2-a^2}{\sqrt{s}(s-a^2)(\sqrt{s}+b)}$	$e^{a^2t}\left[\dfrac{b}{a}\ \mathrm{erf}\ (a\sqrt{t})-1\right]+e^{b^2t}\ \mathrm{erfc}\ b\sqrt{t}$	**7**
29.3.46	$\dfrac{(1-s)^n}{s^{n+\frac{1}{2}}}$	$\dfrac{n!}{(2n)!\sqrt{\pi t}}\,H_{2n}(\sqrt{t})$	**22**
29.3.47	$\dfrac{(1-s)^n}{s^{n+\frac{3}{2}}}$	$\dfrac{n!}{(2n+1)!\sqrt{\pi}}\,H_{2n+1}(\sqrt{t})$	**22**
29.3.48	$\dfrac{\sqrt{s+2a}}{\sqrt{s}}-1$	$ae^{-at}[I_1(at)+I_0(at)]$	**9**
29.3.49	$\dfrac{1}{\sqrt{s+a}\sqrt{s+b}}$	$e^{-\frac{1}{2}(a+b)t}I_0\left(\dfrac{a-b}{2}\,t\right)$	**9**
29.3.50	$\dfrac{\Gamma(k)}{(s+a)^k(s+b)^k}$ $(k>0)$ **6**	$\sqrt{\pi}\left(\dfrac{t}{a-b}\right)^{k-\frac{1}{2}}e^{-\frac{1}{2}(a+b)t}I_{k-\frac{1}{2}}\left(\dfrac{a-b}{2}\,t\right)$	**10**
29.3.51	$\dfrac{1}{(s+a)^{\frac{1}{2}}(s+b)^{\frac{3}{2}}}$	$te^{-\frac{1}{2}(a+b)t}\left[I_0\left(\dfrac{a-b}{2}\,t\right)+I_1\left(\dfrac{a-b}{2}\,t\right)\right]$	**9**
29.3.52	$\dfrac{\sqrt{s+2a}-\sqrt{s}}{\sqrt{s+2a}+\sqrt{s}}$	$\dfrac{1}{t}\,e^{-at}I_1(at)$	**9**
29.3.53	$\dfrac{(a-b)^k}{(\sqrt{s+a}+\sqrt{s+b})^{2k}}$ $(k>0)$	$\dfrac{k}{t}\,e^{-\frac{1}{2}(a+b)t}I_k\left(\dfrac{a-b}{2}\,t\right)$	**9**
29.3.54	$\dfrac{(\sqrt{s+a}+\sqrt{s})^{-2\nu}}{\sqrt{s}\sqrt{s+a}}$ $(\nu>-1)$	$\dfrac{1}{a^\nu}\,e^{-\frac{1}{2}at}I_\nu(\tfrac{1}{2}at)$	**9**
29.3.55	$\dfrac{1}{\sqrt{s^2+a^2}}$	$J_0(at)$	**9**
29.3.56	$\dfrac{(\sqrt{s^2+a^2}-s)^\nu}{\sqrt{s^2+a^2}}$ $(\nu>-1)$	$a^\nu J_\nu(at)$	**9**
29.3.57	$\dfrac{1}{(s^2+a^2)^k}$ $(k>0)$	$\dfrac{\sqrt{\pi}}{\Gamma(k)}\left(\dfrac{t}{2a}\right)^{k-\frac{1}{2}}J_{k-\frac{1}{2}}(at)$	**6, 10**

	$f(s)$	$F(t)$	
29.3.58	$(\sqrt{s^2+a^2}-s)^k \quad (k>0)$	$\dfrac{ka^k}{t}J_k(at)$	**9**
29.3.59	$\dfrac{(s-\sqrt{s^2-a^2})^\nu}{\sqrt{s^2-a^2}} \quad (\nu>-1)$	$a^\nu I_\nu(at)$	**9**
29.3.60	$\dfrac{1}{(s^2-a^2)^k} \quad (k>0)$	$\dfrac{\sqrt{\pi}}{\Gamma(k)}\left(\dfrac{t}{2a}\right)^{k-\frac{1}{2}}I_{k-\frac{1}{2}}(at)$	**6, 10**
29.3.61	$\dfrac{1}{s}e^{-ks}$	$u(t-k)$	
29.3.62	$\dfrac{1}{s^2}e^{-ks}$	$(t-k)u(t-k)$	
29.3.63	$\dfrac{1}{s^\mu}e^{-ks} \quad (\mu>0)$	$\dfrac{(t-k)^{\mu-1}}{\Gamma(\mu)}u(t-k)$	**6**
29.3.64	$\dfrac{1-e^{-ks}}{s}$	$u(t)-u(t-k)$	
29.3.65	$\dfrac{1}{s(1-e^{-ks})}=\dfrac{1+\coth\frac{1}{2}ks}{2s}$	$\displaystyle\sum_{n=0}^{\infty}u(t-nk)$	
29.3.66	$\dfrac{1}{s(e^{ks}-a)}$	$\displaystyle\sum_{n=1}^{\infty}a^{n-1}u(t-nk)$	
29.3.67	$\dfrac{1}{s}\tanh ks$	$u(t)+2\displaystyle\sum_{n=1}^{\infty}(-1)^n u(t-2nk)$	
29.3.68	$\dfrac{1}{s(1+e^{-ks})}$	$\displaystyle\sum_{n=0}^{\infty}(-1)^n u(t-nk)$	
29.3.69	$\dfrac{1}{s^2}\tanh ks$	$tu(t)+2\displaystyle\sum_{n=1}^{\infty}(-1)^n(t-2nk)u(t-2nk)$	
29.3.70	$\dfrac{1}{s\sinh ks}$	$2\displaystyle\sum_{n=0}^{\infty}u[t-(2n+1)k]$	
29.3.71	$\dfrac{1}{s\cosh ks}$	$2\displaystyle\sum_{n=0}^{\infty}(-1)^n u[t-(2n+1)k]$	

	$f(s)$	$F(t)$	

29.3.72 $\dfrac{1}{s}\coth ks$ $u(t)+2\sum\limits_{n=1}^{\infty} u(t-2nk)$

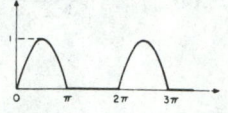

29.3.73 $\dfrac{k}{s^2+k^2}\coth\dfrac{\pi s}{2k}$ $|\sin kt|$

29.3.74 $\dfrac{1}{(s^2+1)(1-e^{-\pi s})}$ $\sum\limits_{n=0}^{\infty}(-1)^n\,u(t-n\pi)\sin t$ *

29.3.75 $\dfrac{1}{s}\,e^{-\frac{k}{s}}$ $J_0(2\sqrt{kt})$ **9**

29.3.76 $\dfrac{1}{\sqrt{s}}\,e^{-\frac{k}{s}}$ $\dfrac{1}{\sqrt{\pi t}}\cos 2\sqrt{kt}$

29.3.77 $\dfrac{1}{\sqrt{s}}\,e^{\frac{k}{s}}$ $\dfrac{1}{\sqrt{\pi t}}\cosh 2\sqrt{kt}$

29.3.78 $\dfrac{1}{s^{3/2}}\,e^{-\frac{k}{s}}$ $\dfrac{1}{\sqrt{\pi k}}\sin 2\sqrt{kt}$

29.3.79 $\dfrac{1}{s^{3/2}}\,e^{\frac{k}{s}}$ $\dfrac{1}{\sqrt{\pi k}}\sinh 2\sqrt{kt}$

29.3.80 $\dfrac{1}{s^{\mu}}\,e^{-\frac{k}{s}}\quad(\mu>0)$ $\left(\dfrac{t}{k}\right)^{\frac{\mu-1}{2}}J_{\mu-1}(2\sqrt{kt})$ **9**

29.3.81 $\dfrac{1}{s^{\mu}}\,e^{\frac{k}{s}}\quad(\mu>0)$ $\left(\dfrac{t}{k}\right)^{\frac{\mu-1}{2}}I_{\mu-1}(2\sqrt{kt})$ **9**

29.3.82 $e^{-k\sqrt{s}}\quad(k>0)$ $\dfrac{k}{2\sqrt{\pi t^3}}\exp\left(-\dfrac{k^2}{4t}\right)$

29.3.83 $\dfrac{1}{s}\,e^{-k\sqrt{s}}\quad(k\ge0)$ $\operatorname{erfc}\dfrac{k}{2\sqrt{t}}$ **7**

29.3.84 $\dfrac{1}{\sqrt{s}}\,e^{-k\sqrt{s}}\quad(k\ge0)$ $\dfrac{1}{\sqrt{\pi t}}\exp\left(-\dfrac{k^2}{4t}\right)$

29.3.85 $\dfrac{1}{s^{\frac{3}{2}}}\,e^{-k\sqrt{s}}\quad(k\ge0)$ $2\sqrt{\dfrac{t}{\pi}}\exp\left(-\dfrac{k^2}{4t}\right)-k\operatorname{erfc}\dfrac{k}{2\sqrt{t}}=2\sqrt{t}\ \mathrm{i}\operatorname{erfc}\dfrac{k}{2\sqrt{t}}$ **7**

29.3.86 $\dfrac{1}{s^{1+\frac{1}{2}n}}\,e^{-k\sqrt{s}}\quad(n=0,1,2,\ldots;\,k\ge0)$ $(4t)^{\frac{1}{2}n}\,\mathrm{i}^n\operatorname{erfc}\dfrac{k}{2\sqrt{t}}$ **7**

29.3.87 $s^{\frac{n-1}{2}}\,e^{-k\sqrt{s}}\quad(n=0,1,2,\ldots;\,k>0)$ $\dfrac{\exp\left(-\dfrac{k^2}{4t}\right)}{2^n\sqrt{\pi t^{n+1}}}H_n\left(\dfrac{k}{2\sqrt{t}}\right)$ **22**

29.3.88 $\dfrac{e^{-k\sqrt{s}}}{a+\sqrt{s}}\quad(k\ge0)$ $\dfrac{1}{\sqrt{\pi t}}\exp\left(-\dfrac{k^2}{4t}\right)-ae^{ak}e^{a^2t}\operatorname{erfc}\left(a\sqrt{t}+\dfrac{k}{2\sqrt{t}}\right)$ **7**

*See page II.

	$f(s)$		$F(t)$	
29.3.89	$\dfrac{ae^{-k\sqrt{s}}}{s(a+\sqrt{s})}$ $(k\geq 0)$		$-e^{ak}e^{a^2t}\,\mathrm{erfc}\left(a\sqrt{t}+\dfrac{k}{2\sqrt{t}}\right)+\mathrm{erfc}\,\dfrac{k}{2\sqrt{t}}$	7
29.3.90	$\dfrac{e^{-k\sqrt{s}}}{\sqrt{s}(a+\sqrt{s})}$ $(k\geq 0)$		$e^{ak}e^{a^2t}\,\mathrm{erfc}\left(a\sqrt{t}+\dfrac{k}{2\sqrt{t}}\right)$	7
29.3.91	$\dfrac{e^{-k\sqrt{s(s+a)}}}{\sqrt{s(s+a)}}$ $(k\geq 0)$		$e^{-\frac{1}{2}at}I_0(\tfrac{1}{2}a\sqrt{t^2-k^2})u(t-k)$	9
29.3.92	$\dfrac{e^{-k\sqrt{s^2+a^2}}}{\sqrt{s^2+a^2}}$ $(k\geq 0)$		$J_0(a\sqrt{t^2-k^2})u(t-k)$	9
29.3.93	$\dfrac{e^{-k\sqrt{s^2-a^2}}}{\sqrt{s^2-a^2}}$ $(k\geq 0)$		$I_0(a\sqrt{t^2-k^2})u(t-k)$	9
29.3.94	$\dfrac{e^{-k(\sqrt{s^2+a^2}-s)}}{\sqrt{s^2+a^2}}$ $(k\geq 0)$		$J_0(a\sqrt{t^2+2kt})$	9
29.3.95	$e^{-ks}-e^{-k\sqrt{s^2+a^2}}$ $(k>0)$		$\dfrac{ak}{\sqrt{t^2-k^2}}J_1(a\sqrt{t^2-k^2})u(t-k)$	9
29.3.96	$e^{-k\sqrt{s^2-a^2}}-e^{-ks}$ $(k>0)$		$\dfrac{ak}{\sqrt{t^2-k^2}}I_1(a\sqrt{t^2-k^2})u(t-k)$	9
29.3.97	$\dfrac{a^\nu e^{-k\sqrt{s^2+a^2}}}{\sqrt{s^2+a^2}(\sqrt{s^2+a^2}+s)^\nu}$ $(\nu>-1,k\geq 0)$		$\left(\dfrac{t-k}{t+k}\right)^{\frac{1}{2}\nu}J_\nu(a\sqrt{t^2-k^2})u(t-k)$	9
29.3.98	$\dfrac{1}{s}\ln s$		$-\gamma-\ln t\,(\gamma=.57721\,56649\ldots\text{Euler's constant})$	
29.3.99	$\dfrac{1}{s^k}\ln s$ $(k>0)$		$\dfrac{t^{k-1}}{\Gamma(k)}\,[\psi(k)-\ln t]$	6
29.3.100	$\dfrac{\ln s}{s-a}$ $(a>0)$		$e^{at}[\ln a+E_1(at)]$	5
29.3.101	$\dfrac{\ln s}{s^2+1}$		$\cos t\,\mathrm{Si}\,(t)-\sin t\,\mathrm{Ci}\,(t)$	5
29.3.102	$\dfrac{s\ln s}{s^2+1}$		$-\sin t\,\mathrm{Si}\,(t)-\cos t\,\mathrm{Ci}\,(t)$	5
29.3.103	$\dfrac{1}{s}\ln(1+ks)$ $(k>0)$		$E_1\left(\dfrac{t}{k}\right)$	5
29.3.104	$\ln\dfrac{s+a}{s+b}$		$\dfrac{1}{t}\,(e^{-bt}-e^{-at})$	
29.3.105	$\dfrac{1}{s}\ln(1+k^2s^2)$ $(k>0)$		$-2\,\mathrm{Ci}\left(\dfrac{t}{k}\right)$	5
29.3.106	$\dfrac{1}{s}\ln(s^2+a^2)$ $(a>0)$		$2\ln a-2\,\mathrm{Ci}\,(at)$	5

	$f(s)$		$F(t)$	
29.3.107	$\dfrac{1}{s^2}\ln(s^2+a^2) \quad (a>0)$		$\dfrac{2}{a}[at\ln a+\sin at-at\operatorname{Ci}(at)]$	**5**
29.3.108	$\ln\dfrac{s^2+a^2}{s^2}$		$\dfrac{2}{t}(1-\cos at)$	
29.3.109	$\ln\dfrac{s^2-a^2}{s^2}$		$\dfrac{2}{t}(1-\cosh at)$	
29.3.110	$\arctan\dfrac{k}{s}$		$\dfrac{1}{t}\sin kt$	
29.3.111	$\dfrac{1}{s}\arctan\dfrac{k}{s}$		$\operatorname{Si}(kt)$	**5**
29.3.112	$e^{k^2s^2}\operatorname{erfc} ks \quad (k>0)$	7	$\dfrac{1}{k\sqrt{\pi}}\exp\left(-\dfrac{t^2}{4k^2}\right)$	
29.3.113	$\dfrac{1}{s}e^{k^2s^2}\operatorname{erfc} ks \quad (k>0)$	7	$\operatorname{erf}\dfrac{t}{2k}$	**7**
29.3.114	$e^{ks}\operatorname{erfc}\sqrt{ks} \quad (k>0)$	7	$\dfrac{\sqrt{k}}{\pi\sqrt{t}(t+k)}$	
29.3.115	$\dfrac{1}{\sqrt{s}}\operatorname{erfc}\sqrt{ks} \quad (k\geq 0)$	7	$\dfrac{1}{\sqrt{\pi t}}u(t-k)$	
29.3.116	$\dfrac{1}{\sqrt{s}}e^{ks}\operatorname{erfc}\sqrt{ks} \quad (k\geq 0)$	7	$\dfrac{1}{\sqrt{\pi(t+k)}}$	
29.3.117	$\operatorname{erf}\dfrac{k}{\sqrt{s}}$	7	$\dfrac{1}{\pi t}\sin 2k\sqrt{t}$	
29.3.118	$\dfrac{1}{\sqrt{s}}e^{\frac{k^2}{s}}\operatorname{erfc}\dfrac{k}{\sqrt{s}}$	7	$\dfrac{1}{\sqrt{\pi t}}e^{-2k\sqrt{t}}$	
29.3.119	$K_0(ks) \quad (k>0)$	9	$\dfrac{1}{\sqrt{t^2-k^2}}u(t-k)$	
29.3.120	$K_0(k\sqrt{s}) \quad (k>0)$	9	$\dfrac{1}{2t}\exp\left(-\dfrac{k^2}{4t}\right)$	
29.3.121	$\dfrac{1}{s}e^{ks}K_1(ks) \quad (k>0)$	9	$\dfrac{1}{k}\sqrt{t(t+2k)}$	
29.3.122	$\dfrac{1}{\sqrt{s}}K_1(k\sqrt{s}) \quad (k>0)$	9	$\dfrac{1}{k}\exp\left(-\dfrac{k^2}{4t}\right)$	
29.3.123	$\dfrac{1}{\sqrt{s}}e^{\frac{k}{s}}K_0\left(\dfrac{k}{s}\right) \quad (k>0)$	9	$\dfrac{2}{\sqrt{\pi t}}K_0(2\sqrt{2kt})$	**9**
29.3.124	$\pi e^{-ks}I_0(ks) \quad (k>0)$	9	$\dfrac{1}{\sqrt{t(2k-t)}}[u(t)-u(t-2k)]$	
29.3.125	$e^{-ks}I_1(ks) \quad (k>0)$	9	$\dfrac{k-t}{\pi k\sqrt{t(2k-t)}}[u(t)-u(t-2k)]$	

	$f(s)$		$F(t)$
29.3.126	$e^{as}E_1(as)$ $(a>0)$	5	$\dfrac{1}{t+a}$
29.3.127	$\dfrac{1}{a}-se^{as}E_1(as)$ $(a>0)$	5	$\dfrac{1}{(t+a)^2}$
29.3.128	$a^{1-n}e^{as}E_n(as)$ $(a>0;n=0,1,2,\ldots)$	5	$\dfrac{1}{(t+a)^n}$
29.3.129	$\left[\dfrac{\pi}{2}-\mathrm{Si}\,(s)\right]\cos s+\mathrm{Ci}(s)\sin s$	5	$\dfrac{1}{t^2+1}$

29.4. Table of Laplace-Stieltjes Transforms [4]

	$\phi(s)$	$\Phi(t)$
29.4.1	$\displaystyle\int_0^\infty e^{-st}d\Phi(t)$	$\Phi(t)$
29.4.2	e^{-ks} $(k>0)$	$u(t-k)$
29.4.3	$\dfrac{1}{1-e^{-ks}}$ $(k>0)$	$\displaystyle\sum_{n=0}^\infty u(t-nk)$
29.4.4	$\dfrac{1}{1+e^{-ks}}$ $(k>0)$	$\displaystyle\sum_{n=0}^\infty (-1)^n u(t-nk)$
29.4.5	$\dfrac{1}{\sinh ks}$ $(k>0)$	$2\displaystyle\sum_{n=0}^\infty u[t-(2n+1)k]$
29.4.6	$\dfrac{1}{\cosh ks}$ $(k>0)$	$2\displaystyle\sum_{n=0}^\infty (-1)^n u[t-(2n+1)k]$
29.4.7	$\tanh ks$ $(k>0)$	$u(t)+2\displaystyle\sum_{n=1}^\infty (-1)^n u(t-2nk)$
29.4.8	$\dfrac{1}{\sinh (ks+a)}$ $(k>0)$	$2\displaystyle\sum_{n=0}^\infty e^{-(2n+1)a}u[t-(2n+1)k]$
29.4.9	$\dfrac{e^{-hs}}{\sinh (ks+a)}$ $(k>0,\,h>0)$	$2\displaystyle\sum_{n=0}^\infty e^{-(2n+1)a}u[t-h-(2n+1)k]$
29.4.10	$\dfrac{\sinh (hs+b)}{\sinh (ks+a)}$ $(0<h<k)$	$\displaystyle\sum_{n=0}^\infty e^{-(2n+1)a}\{e^b u[t+h-(2n+1)k]$ $-e^{-b}u[t-h-(2n+1)k]\}$
29.4.11	$\displaystyle\sum_{n=0}^\infty a_n e^{-k_n s}$ $(0<k_0<k_1<\ldots)$	$\displaystyle\sum_{n=0}^\infty a_n u(t-k_n)$

For the definition of the Laplace-Stieltjes transform see [29.7]. In practice, Laplace-Stieltjes transforms are often written as ordinary Laplace transforms involving Dirac's delta function $\delta(t)$. This "function" may formally be considered as the derivative of the unit step function, $du(t)=\delta(t)\,dt$, so that $\displaystyle\int_{-\infty}^x du(t)=\int_{-\infty}^x \delta(t)dt=\begin{cases}0 & (x<0)\\1 & (x>0).\end{cases}$ The correspondence **29.4.2**, for instance, then assumes the form $e^{-ks}=\displaystyle\int_0^\infty e^{-st}\delta(t-k)dt$.

---

[4] Adapted by permission from P. M. Morse and H. Feshbach, Methods of theoretical physics, vols. 1, 2, McGraw-Hill Book Co., Inc., New York, N.Y., 1953.

References

Texts

[29.1] H. S. Carslaw and J. C. Jaeger, Operational methods in applied mathematics, 2d ed. (Oxford Univ. Press, London, England, 1948).

[29.2] R. V. Churchill, Operational mathematics, 2d ed. (McGraw-Hill Book Co., Inc., New York, N.Y., Toronto, Canada, London, England, 1958).

[29.3] G. Doetsch, Handbuch der Laplace-Transformation, vols. I–III (Birkhäuser, Basel, Switzerland, 1950; Basel, Switzerland, Stuttgart, Germany, 1955, 1956).

[29.4] G. Doetsch, Einführung in Theorie und Anwendung der Laplace-Transformation (Birkhäuser, Basel, Switzerland, Stuttgart, Germany, 1958).

[29.5] P. M. Morse and H. Feshbach, Methods of theoretical physics, vols. I, II (McGraw-Hill Book Co., Inc., New York, N.Y., Toronto, Canada, London, England, 1953).

[29.6] B. van der Pol and H. Bremmer, Operational calculus, 2d. ed. (Cambridge Univ. Press, Cambridge, England, 1955).

[29.7] D. V. Widder, The Laplace transform (Princeton Univ. Press, Princeton, N.J., 1941).

Tables

[29.8] G. Doetsch, Guide to the applications of Laplace transforms (D. Van Nostrand, London, England; Toronto, Canada; New York, N.Y.; Princeton, N.J., 1961).

[29.9] A. Erdélyi et al., Tables of integral transforms, vols. I, II (McGraw-Hill Book Co., Inc., New York, N.Y., Toronto, Canada, London, England, 1954).

[29.10] W. Magnus and F. Oberhettinger, Formulas and theorems for the special functions of mathematical physics (Chelsea Publishing Co., New York, N.Y., 1949).

[29.11] D. Voelker and G. Doetsch, Die zweidimensionale Laplace-Transformation (Birkhäuser, Basel, Switzerland, 1950).

Subject Index

Index of Notations

Notation — Greek Letters

Miscellaneous Notations

A CATALOGUE OF SELECTED DOVER BOOKS
IN ALL FIELDS OF INTEREST

A CATALOGUE OF SELECTED DOVER
BOOKS IN ALL FIELDS OF INTEREST

CELESTIAL OBJECTS FOR COMMON TELESCOPES, T. W. Webb. The most used book in amateur astronomy: inestimable aid for locating and identifying nearly 4,000 celestial objects. Edited, updated by Margaret W. Mayall. 77 illustrations. Total of 645pp. 5⅜ x 8½.
20917-2, 20918-0 Pa., Two-vol. set $9.00

HISTORICAL STUDIES IN THE LANGUAGE OF CHEMISTRY, M. P. Crosland. The important part language has played in the development of chemistry from the symbolism of alchemy to the adoption of systematic nomenclature in 1892. ". . . wholeheartedly recommended,"—Science. 15 illustrations. 416pp. of text. 5⅝ x 8¼.　63702-6 Pa. $6.00

BURNHAM'S CELESTIAL HANDBOOK, Robert Burnham, Jr. Thorough, readable guide to the stars beyond our solar system. Exhaustive treatment, fully illustrated. Breakdown is alphabetical by constellation: Andromeda to Cetus in Vol. 1; Chamaeleon to Orion in Vol. 2; and Pavo to Vulpecula in Vol. 3. Hundreds of illustrations. Total of about 2000pp. 6⅛ x 9¼.
23567-X, 23568-8, 23673-0 Pa., Three-vol. set $27.85

THEORY OF WING SECTIONS: INCLUDING A SUMMARY OF AIR-FOIL DATA, Ira H. Abbott and A. E. von Doenhoff. Concise compilation of subatomic aerodynamic characteristics of modern NASA wing sections, plus description of theory. 350pp. of tables. 693pp. 5⅜ x 8½.
60586-8 Pa. $8.50

DE RE METALLICA, Georgius Agricola. Translated by Herbert C. Hoover and Lou H. Hoover. The famous Hoover translation of greatest treatise on technological chemistry, engineering, geology, mining of early modern times (1556). All 289 original woodcuts. 638pp. 6¾ x 11.
60006-8 Clothbd. $17.95

THE ORIGIN OF CONTINENTS AND OCEANS, Alfred Wegener. One of the most influential, most controversial books in science, the classic statement for continental drift. Full 1966 translation of Wegener's final (1929) version. 64 illustrations. 246pp. 5⅜ x 8½.　61708-4 Pa. $4.50

THE PRINCIPLES OF PSYCHOLOGY, William James. Famous long course complete, unabridged. Stream of thought, time perception, memory, experimental methods; great work decades ahead of its time. Still valid, useful; read in many classes. 94 figures. Total of 1391pp. 5⅜ x 8½.
20381-6, 20382-4 Pa., Two-vol. set $13.00

HISTORY OF BACTERIOLOGY, William Bulloch. The only comprehensive history of bacteriology from the beginnings through the 19th century. Special emphasis is given to biography-Leeuwenhoek, etc. Brief accounts of 350 bacteriologists form a separate section. No clearer, fuller study, suitable to scientists and general readers, has yet been written. 52 illustrations. 448pp. 5⅝ x 8¼. 23761-3 Pa. $6.50

THE COMPLETE NONSENSE OF EDWARD LEAR, Edward Lear. All nonsense limericks, zany alphabets, Owl and Pussycat, songs, nonsense botany, etc., illustrated by Lear. Total of 321pp. 5⅜ x 8½. (Available in U.S. only) 20167-8 Pa. $3.95

INGENIOUS MATHEMATICAL PROBLEMS AND METHODS, Louis A. Graham. Sophisticated material from Graham *Dial*, applied and pure; stresses solution methods. Logic, number theory, networks, inversions, etc. 237pp. 5⅜ x 8½. 20545-2 Pa. $4.50

BEST MATHEMATICAL PUZZLES OF SAM LOYD, edited by Martin Gardner. Bizarre, original, whimsical puzzles by America's greatest puzzler. From fabulously rare *Cyclopedia*, including famous 14-15 puzzles, the Horse of a Different Color, 115 more. Elementary math. 150 illustrations. 167pp. 5⅜ x 8½. 20498-7 Pa. $2.75

THE BASIS OF COMBINATION IN CHESS, J. du Mont. Easy-to-follow, instructive book on elements of combination play, with chapters on each piece and every powerful combination team—two knights, bishop and knight, rook and bishop, etc. 250 diagrams. 218pp. 5⅜ x 8½. (Available in U.S. only) 23644-7 Pa. $3.50

MODERN CHESS STRATEGY, Ludek Pachman. The use of the queen, the active king, exchanges, pawn play, the center, weak squares, etc. Section on rook alone worth price of the book. Stress on the moderns. Often considered the most important book on strategy. 314pp. 5⅜ x 8½. 20290-9 Pa. $4.50

LASKER'S MANUAL OF CHESS, Dr. Emanuel Lasker. Great world champion offers very thorough coverage of all aspects of chess. Combinations, position play, openings, end game, aesthetics of chess, philosophy of struggle, much more. Filled with analyzed games. 390pp. 5⅜ x 8½. 20640-8 Pa. $5.00

500 MASTER GAMES OF CHESS, S. Tartakower, J. du Mont. Vast collection of great chess games from 1798-1938, with much material nowhere else readily available. Fully annotated, arranged by opening for easier study. 664pp. 5⅜ x 8½. 23208-5 Pa. $7.50

A GUIDE TO CHESS ENDINGS, Dr. Max Euwe, David Hooper. One of the finest modern works on chess endings. Thorough analysis of the most frequently encountered endings by former world champion. 331 examples, each with diagram. 248pp. 5⅜ x 8½. 23332-4 Pa. $3.75

SECOND PIATIGORSKY CUP, edited by Isaac Kashdan. One of the greatest tournament books ever produced in the English language. All 90 games of the 1966 tournament, annotated by players, most annotated by both players. Features Petrosian, Spassky, Fischer, Larsen, six others. 228pp. 5⅜ x 8½. 23572-6 Pa. $3.50

ENCYCLOPEDIA OF CARD TRICKS, revised and edited by Jean Hugard. How to perform over 600 card tricks, devised by the world's greatest magicians: impromptus, spelling tricks, key cards, using special packs, much, much more. Additional chapter on card technique. 66 illustrations. 402pp. 5⅜ x 8½. (Available in U.S. only) 21252-1 Pa. $4.95

MAGIC: STAGE ILLUSIONS, SPECIAL EFFECTS AND TRICK PHOTOGRAPHY, Albert A. Hopkins, Henry R. Evans. One of the great classics; fullest, most authorative explanation of vanishing lady, levitations, scores of other great stage effects. Also small magic, automata, stunts. 446 illustrations. 556pp. 5⅜ x 8½. 23344-8 Pa. $6.95

THE SECRETS OF HOUDINI, J. C. Cannell. Classic study of Houdini's incredible magic, exposing closely-kept professional secrets and revealing, in general terms, the whole art of stage magic. 67 illustrations. 279pp. 5⅜ x 8½. 22913-0 Pa. $4.00

HOFFMANN'S MODERN MAGIC, Professor Hoffmann. One of the best, and best-known, magicians' manuals of the past century. Hundreds of tricks from card tricks and simple sleight of hand to elaborate illusions involving construction of complicated machinery. 332 illustrations. 563pp. 5⅜ x 8½. 23623-4 Pa. $6.00

MADAME PRUNIER'S FISH COOKERY BOOK, Mme. S. B. Prunier. More than 1000 recipes from world famous Prunier's of Paris and London, specially adapted here for American kitchen. Grilled tournedos with anchovy butter, Lobster a la Bordelaise, Prunier's prized desserts, more. Glossary. 340pp. 5⅜ x 8½. (Available in U.S. only) 22679-4 Pa. $3.00

FRENCH COUNTRY COOKING FOR AMERICANS, Louis Diat. 500 easy-to-make, authentic provincial recipes compiled by former head chef at New York's Fitz-Carlton Hotel: onion soup, lamb stew, potato pie, more. 309pp. 5⅜ x 8½. 23665-X Pa. $3.95

SAUCES, FRENCH AND FAMOUS, Louis Diat. Complete book gives over 200 specific recipes: bechamel, Bordelaise, hollandaise, Cumberland, apricot, etc. Author was one of this century's finest chefs, originator of vichyssoise and many other dishes. Index. 156pp. 5⅜ x 8.
23663-3 Pa. $2.75

TOLL HOUSE TRIED AND TRUE RECIPES, Ruth Graves Wakefield. Authentic recipes from the famous Mass. restaurant: popovers, veal and ham loaf, Toll House baked beans, chocolate cake crumb pudding, much more. Many helpful hints. Nearly 700 recipes. Index. 376pp. 5⅜ x 8½.
23560-2 Pa. $4.50

THE COMPLETE BOOK OF DOLL MAKING AND COLLECTING, Catherine Christopher. Instructions, patterns for dozens of dolls, from rag doll on up to elaborate, historically accurate figures. Mould faces, sew clothing, make doll houses, etc. Also collecting information. Many illustrations. 288pp. 6 x 9.   22066-4 Pa. $4.50

THE DAGUERREOTYPE IN AMERICA, Beaumont Newhall. Wonderful portraits, 1850's townscapes, landscapes; full text plus 104 photographs. The basic book. Enlarged 1976 edition. 272pp. 8¼ x 11¼.
23322-7 Pa. $7.95

CRAFTSMAN HOMES, Gustav Stickley. 296 architectural drawings, floor plans, and photographs illustrate 40 different kinds of "Mission-style" homes from *The Craftsman* (1901-16), voice of American style of simplicity and organic harmony. Thorough coverage of Craftsman idea in text and picture, now collector's item. 224pp. 8⅛ x 11.   23791-5 Pa. $6.00

PEWTER-WORKING: INSTRUCTIONS AND PROJECTS, Burl N. Osborn. & Gordon O. Wilber. Introduction to pewter-working for amateur craftsman. History and characteristics of pewter; tools, materials, step-by-step instructions. Photos, line drawings, diagrams. Total of 160pp. 7⅞ x 10¾.   23786-9 Pa. $3.50

THE GREAT CHICAGO FIRE, edited by David Lowe. 10 dramatic, eye-witness accounts of the 1871 disaster, including one of the aftermath and rebuilding, plus 70 contemporary photographs and illustrations of the ruins—courthouse, Palmer House, Great Central Depot, etc. Introduction by David Lowe. 87pp. 8¼ x 11.   23771-0 Pa. $4.00

SILHOUETTES: A PICTORIAL ARCHIVE OF VARIED ILLUSTRA-TIONS, edited by Carol Belanger Grafton. Over 600 silhouettes from the 18th to 20th centuries include profiles and full figures of men and women, children, birds and animals, groups and scenes, nature, ships, an alphabet. Dozens of uses for commercial artists and craftspeople. 144pp. 8⅜ x 11¼.
23781-8 Pa. $4.50

ANIMALS: 1,419 COPYRIGHT-FREE ILLUSTRATIONS OF MAM-MALS, BIRDS, FISH, INSECTS, ETC., edited by Jim Harter. Clear wood engravings present, in extremely lifelike poses, over 1,000 species of animals. One of the most extensive copyright-free pictorial sourcebooks of its kind. Captions. Index. 284pp. 9 x 12.   23766-4 Pa. $8.95

INDIAN DESIGNS FROM ANCIENT ECUADOR, Frederick W. Shaffer. 282 original designs by pre-Columbian Indians of Ecuador (500-1500 A.D.). Designs include people, mammals, birds, reptiles, fish, plants, heads, geometric designs. Use as is or alter for advertising, textiles, leathercraft, etc. Introduction. 95pp. 8¾ x 11¼.   23764-8 Pa. $3.50

SZIGETI ON THE VIOLIN, Joseph Szigeti. Genial, loosely structured tour by premier violinist, featuring a pleasant mixture of reminiscenes, insights into great music and musicians, innumerable tips for practicing violinists. 385 musical passages. 256pp. 5⅝ x 8¼.   23763-X Pa. $4.00

PRINCIPLES OF ORCHESTRATION, Nikolay Rimsky-Korsakov. Great classical orchestrator provides fundamentals of tonal resonance, progression of parts, voice and orchestra, tutti effects, much else in major document. 330pp. of musical excerpts. 489pp. 6½ x 9¼.   21266-1 Pa. **$7.50**

TRISTAN UND ISOLDE, Richard Wagner. Full orchestral score with complete instrumentation. Do not confuse with piano reduction. Commentary by Felix Mottl, great Wagnerian conductor and scholar. Study score. 655pp. 8⅛ x 11.   22915-7 Pa. **$13.95**

REQUIEM IN FULL SCORE, Giuseppe Verdi. Immensely popular with choral groups and music lovers. Republication of edition published by C. F. Peters, Leipzig, n. d. German frontmaker in English translation. Glossary. Text in Latin. Study score. 204pp. 9⅜ x 12¼.
23682-X Pa. **$6.00**

COMPLETE CHAMBER MUSIC FOR STRINGS, Felix Mendelssohn. All of Mendelssohn's chamber music: Octet, 2 Quintets, 6 Quartets, and Four Pieces for String Quartet. (Nothing with piano is included). Complete works edition (1874-7). Study score. 283 pp. 9⅜ x 12¼.
23679-X Pa. **$7.50**

POPULAR SONGS OF NINETEENTH-CENTURY AMERICA, edited by Richard Jackson. 64 most important songs: "Old Oaken Bucket," "Arkansas Traveler," "Yellow Rose of Texas," etc. Authentic original sheet music, full introduction and commentaries. 290pp. 9 x 12.   23270-0 Pa. **$7.95**

COLLECTED PIANO WORKS, Scott Joplin. Edited by Vera Brodsky Lawrence. Practically all of Joplin's piano works—rags, two-steps, marches, waltzes, etc., 51 works in all. Extensive introduction by Rudi Blesh. Total of 345pp. 9 x 12.   23106-2 Pa. **$14.95**

BASIC PRINCIPLES OF CLASSICAL BALLET, Agrippina Vaganova. Great Russian theoretician, teacher explains methods for teaching classical ballet; incorporates best from French, Italian, Russian schools. 118 illustrations. 175pp. 5⅜ x 8½.   22036-2 Pa. **$2.50**

CHINESE CHARACTERS, L. Wieger. Rich analysis of 2300 characters according to traditional systems into primitives. Historical-semantic analysis to phonetics (Classical Mandarin) and radicals. 820pp. 6⅛ x 9¼.
21321-8 Pa. **$10.00**

EGYPTIAN LANGUAGE: EASY LESSONS IN EGYPTIAN HIERO-GLYPHICS, E. A. Wallis Budge. Foremost Egyptologist offers Egyptian grammar, explanation of hieroglyphics, many reading texts, dictionary of symbols. 246pp. 5 x 7½. (Available in U.S. only)
21394-3 Clothbd. **$7.50**

AN ETYMOLOGICAL DICTIONARY OF MODERN ENGLISH, Ernest Weekley. Richest, fullest work, by foremost British lexicographer. Detailed word histories. Inexhaustible. Do not confuse this with *Concise Etymological Dictionary,* which is abridged. Total of 856pp. 6½ x 9¼.
21873-2, 21874-0 Pa., Two-vol. set **$12.00**

CATALOGUE OF DOVER BOOKS

GEOMETRY, RELATIVITY AND THE FOURTH DIMENSION, Rudolf Rucker. Exposition of fourth dimension, means of visualization, concepts of relativity as Flatland characters continue adventures. Popular, easily followed yet accurate, profound. 141 illustrations. 133pp. 5⅜ x 8½.
23400-2 Pa. $2.75

THE ORIGIN OF LIFE, A. I. Oparin. Modern classic in biochemistry, the first rigorous examination of possible evolution of life from nitrocarbon compounds. Non-technical, easily followed. Total of 295pp. 5⅜ x 8½.
60213-3 Pa. $4.00

PLANETS, STARS AND GALAXIES, A. E. Fanning. Comprehensive introductory survey: the sun, solar system, stars, galaxies, universe, cosmology; quasars, radio stars, etc. 24pp. of photographs. 189pp. 5⅜ x 8½. (Available in U.S. only)
21680-2 Pa. $3.75

THE THIRTEEN BOOKS OF EUCLID'S ELEMENTS, translated with introduction and commentary by Sir Thomas L. Heath. Definitive edition. Textual and linguistic notes, mathematical analysis, 2500 years of critical commentary. Do not confuse with abridged school editions. Total of 1414pp. 5⅜ x 8½.
60088-2, 60089-0, 60090-4 Pa., Three-vol. set $18.50

Prices subject to change without notice.

Available at your book dealer or write for free catalogue to Dept. GI, Dover Publications, Inc., 31 East Second Street, Mineola, N.Y. 11501. Dover publishes more than 175 books each year on science, elementary and advanced mathematics, biology, music, art, literary history, social sciences and other areas.